BEEF CATTLE SCIENCE

(Animal Agriculture Series)

by

M. E. Ensminger, B.S., M.A., Ph.D.

and

R. C. Perry, B.S., M.S., Ph.D.

SEVENTH EDITION

INTERSTATE PUBLISHERS, INC.
Danville, Illinois

Editions:

First	1951
Second	1955
Third	1960
Fourth	1968
Fifth	1976
Sixth	1987
Seventh	1997

Library of Congress Catalog Card No. 96-79327

ISBN 0-8134-3006-2

2 3
4 5 6
7 8 9

BEEF CATTLE SCIENCE

(Animal Agriculture Series)

All flesh is grass! Without photosynthesis and ruminants, there would be no oxygen, no plants, no food, no animals, and no people. This shows Hereford steers on bromegrass pasture in Nebraska. (Courtesy, CB&Q Railroad, Chicago, IL)

Other books by M. E. Ensminger
available from Interstate Publishers:

Animal Science
Animal Science Digest
Dairy Cattle Science
Feeds & Nutrition (with J. Oldfield & W. Heinemann)
Feeds & Nutrition Digest (with J. Oldfield & W. Heinemann)
Horses and Horsemanship
Poultry Science
Sheep & Goat Science
Swine Science (with R. Parker)
Stockman's Handbook, The
Stockman's Handbook Digest

Animal Science presents a perspective or panorama of the far-flung livestock industry; whereas each of the other books presents specialized material pertaining to the specific class of farm animals indicated by its title.

Feeds & Nutrition and *Feeds & Nutrition Digest* bring together both the art and the science of livestock feeding, narrow the gap between nutrition research and application, and assure more and better animals in the current era of biotechnology.

The Stockman's Handbook presents the "why" as well as the "how." It contains, under one cover, the pertinent things that a stockman needs to know in the daily operation of a farm or ranch. It covers the broad field of animal agriculture, concisely and completely, and wherever possible in tabular and outline form.

ABOUT THE SENIOR AUTHOR

DR. M. E. ENSMINGER

Marion Eugene Ensminger completed B.S. and M.A. degrees at the University of Missouri, and the Ph.D. at the University of Minnesota. Dr. Ensminger served, in order, as Manager of the Dixon Springs Agricultural Center (University of Illinois), Simpson, Illinois; and on the staffs of the University of Massachusetts, the University of Minnesota, and Washington State University. Dr. Ensminger also served as Consultant, General Electric Company, Nucleonics Department, and as the first President of the American Society of Agricultural Consultants. Since 1964, Dr. Ensminger has served as President of Agriservices Foundation, Clovis, California, a nonprofit foundation serving world agriculture in the area of World Food, Hunger, and Malnutrition.

Among Dr. Ensminger's many honors and awards are: Distinguished Teacher Award, American Society of Animal Science; the "Ensminger Beef Cattle Research Center" at Washington State University, Pullman, named after him in recognition of his contributions to the University; Faculty-Alumni Award of the University of Missouri; Outstanding Achievement Award of the University of Minnesota; Distinguished Service Award of the American Medical Association (with Mrs. Ensminger); Honorary Professor, Huazhong Agricultural College, Wuhan, China; Doctor of Laws (LL.D.) conferred by the National Agrarian University of Ukraine; and an oil portrait of him was placed in the 300-year-old gallery of the famed Saddle and Sirloin Club, Lexington, Kentucky.

In 1995, Cuba honored Dr. Ensminger by making him an Honorary Member of the Cuban Association of Animal Production; presenting him the 30th anniversary Gold Medal of the Institute of Animal Science, at Havana; making him an Honorary Guest Professor of the Agricultural University (ISCAH), at Havana; and making him an Honorary Guest Professor of the University of Camaguey, Camaguey, Cuba.

In 1995, Dr. Ensminger received the Distinguished Teacher Award, the highest honor of the National Association of Colleges and Teachers of Agriculture (NACTA).

In May 1996, Iowa State University awarded Dr. Ensminger the honorary degree, Doctor of Humane Letters for "extraordinary achievements in animal science, education, and international agriculture."

In July 1996, Dr. Ensminger was the recipient of the International Animal Agriculture Bouffault Award of the American Society of Animal Science.

Dr. Ensminger founded the International Ag-Tech Schools, which he directed for more than 50 years. He has directed schools, lectured, and/or conducted seminars in 70 countries. Dr. Ensminger is the author of more than 500 scientific articles, bulletins, and feature articles; and he is the author or co-author of 22 books, which are in several languages and used all over the world. He waives all royalties on the foreign editions of his books in order to help the people. The whole world is Dr. Ensminger's classroom.

ABOUT THE JUNIOR AUTHOR

DR. R. C. PERRY

Randy Charles Perry grew up on a family-owned purebred beef cattle operation in California, in which he has been a partner since 1974. He completed the B.S. degree at California Polytechnic State University, San Luis Obispo; and M.S. and Ph.D. degrees at Kansas State University, Manhattan. While at Kansas State University, he served as a Graduate Teaching Assistant and as Assistant Livestock Judging Team Coach; and he participated in a number of extension activities.

Presently, Dr. Perry is a Professor in the Department of Animal Science and Education at California State University, Fresno, where he teaches Advanced Beef Cattle Management, Beef Production, Animal Reproduction, Reproductive Physiology and Endocrinology, AI and Embryo Transfer, and General Animal Science. Dr. Perry has had research experience in the use of ultrasonography to monitor follicular development, uterine involution, and pregnancy, and to measure subcutaneous fat and rib eye in cattle; estrous synchronization and artificial insemination in cattle; pregnancy determination via rectal palpation in cattle; jugular cannulation and blood collection; weigh-suckle-weigh technique and oxytocin injection combined with milking machine to determine milk production in cattle; radioimmunoassay procedures for progesterone, estradiol, luteinizing hormone, and follicle stimulation hormone; and intake and digestibility of feedstuffs using internal markers.

Dr. Perry has received a number of honors and awards; among them, Allstar and National Winner in Beef National 4-H Program; Senior Achievement Award winner, California Cattlemen's Association; Outstanding Senior award, Boots and Spurs Club, California Polytechnic University, San Luis Obispo; Outstanding Animal Science Graduate Student, Kansas State University; Teacher Award of Merit, 1992 and 1993, National Association of Colleges and Teachers of Agriculture; and Nicolas Salgo Outstanding Teacher Award, 1993, California State University, Fresno.

Dr. Perry is the author or co-author of 19 publications and abstracts in referred journals, 14 field day reports, and 6 popular press articles.

Cover photo: Courtesy, American Hereford Association, Kansas City, MO

TO CATTLEMEN—artists and scientists. Not artists whose tools are the clay and marble of sculpture, but artists whose materials are the "green pastures and still waters" that have inspired musicians to capture their beauty in pastoral symphonies, and painters to reproduce their splendor in landscape designs; artists whose materials are the living flesh and blood of animals, molded to perfection through heredity and environment. Not scientists who look through a microscope or shake a test tube, but scientists who, from the remote day of domestication, have given attention to the breeding, feeding, care and management, and marketing of animals.

TO CATTLEMEN—who take pride in their brands, boots, hats, and canes. To them they are much more than a trademark of the profession; they are symbols of service, pledges of integrity of the men behind them, and marks of courage, character, and wisdom. They are indicative of the quality of the cattle raised, the class of bulls used, the condition of the pasture or range, and the kind of caretakers connected with the outfit.

SELECTED GENERAL REFERENCES ON BEEF CATTLE

Title of Publication	Author(s)	Publisher
Beef Cattle, Seventh Edition	A. L. Neumann	John Wiley & Sons, Inc., New York, NY, 1977
Beef Cattle Book	B. E. Fichte	The Progressive Farmer Company, Birmingham, AL, 1967
Beef Cattle in Florida	L. H. Lewis T. J. Cunha G. N. Rhodes	Department of Agriculture, State of Florida, Tallahassee, FL, 1962
Beef Cattle Production	J. F. Lasley	Prentice-Hall, Inc., Englewood Cliffs, NJ, 1981
Beef Cattle Production	K. A. Wagnon R. Albaugh G. H. Hart	The Macmillan Co., New York, NY, 1960
Beef Cattle Production	N. T. M. Yeates P. J. Schmidt	Butterworths Pty., Limited, Melbourne, Australia, 1974
Beef Cattle Science Handbook	Ed. by M. E. Ensminger	Agriservices Foundation, Clovis, CA, pub. Annually 1964–1981
Beef Production	R. V. Diggins C. E. Bundy	Prentice-Hall, Inc., Englewood Cliffs, NJ, 1958
Beef Production and Management Decisions	R. E. Taylor	Macmillan Publishing Company, New York, NY, 1994
Beef Production and Management, Second Edition	G. L. Minish D. G. Fox	Reston Publishing Company, Inc., Reston, VA, 1982
California Beef Production	H. R. Guilbert G. H. Hart	California Agricultural Extension Service, Berkeley, CA, 1946
Commercial Beef Cattle Production, Second Edition	Ed. by C. C. O'Mary I. A. Dyer	Lea & Febiger, Philadelphia, PA, 1978
Intensive Beef Production	T. R. Preston M. B. Willis	Pergamon Press, Oxford, England, 1974
Livestock Book, The	W. R. Thompson, *et al.*	Vulcan Service Co., Inc., Birmingham, AL, 1952
Practical Beef Production	J. Widmer	Charles Scribner's Sons, New York, NY, 1946
Problems and Practices of American Cattlemen, Wash. Ag. Exp. Sta. Bull. 562	M. E. Ensminger M. W. Galgan W. L. Slocum	Washington State University, Pullman, WA, 1955
World Cattle, Vols. I, II, and III	J. E. Rouse	University of Oklahoma Press, Norman, OK, Vols. I and II, 1970; Vol. III, 1973
World's Beef Business, The	J. R. Simpson D. E. Farris	Iowa State University Press, Ames, IA, 1982

PREFACE TO THE SEVENTH EDITION

The entire beef industry—cow-calf producers, stocker operators, cattle feeders, beef packers, and beef retailers—is at the crossroads! *The choices*: (1) meet the challenge to lessen the fat and improve the juiciness, tenderness, flavor, palatability, and consistency of beef; or (2) hear the victory crow of roosters and the victory grunt of pigs, louder and louder. It is the senior author's fond hope that the seventh edition of ***Beef Cattle Science*** will give the entire beef team a big assist in taking the high road. At the outset, however, it is important that the industry recognize that meeting the challenge and overcoming will be slow, costly, unpalatable, and paraphrasing Sir Winston Churchill, make for "blood, sweat, and tears."

Until the 1980's, beef was king. To professional members of the beef team and writers like me, beef was either "tender" or "less tender." It was never "tough." In the good old days, beef was well finished, carcasses were aged in the meat packing plant for three weeks or longer, and beef was consistently tender, uniform, and good—consumers could purchase the same kind of beef that they secured the previous week.

Then came a host of changes: (1) new beef breeds, crossbreeds, and composites sprang up like toadstools after a rain; (2) the grain-feeding binge and big commercial feedlots evolved: (3) the cholesterol scare was here and now; (4) beef that was no longer "tender" or "less tender"—it was "tough"; (5) broilers and turkeys that were attractively packaged—either whole or further processed, always tender, affordably priced, and consistently uniform; and (6) pork that was lean, attractively packaged—either cured or fresh, seldom tough and never bloody, and a better buy than beef. *Note:* In the meantime, the sheep and wool industry persisted in promoting and marketing 115 lb lambs, with the result that it became a minor industry.

The importation and U.S. development of numerous new breeds of beef cattle, along with the development of the composite breeds and increased crossbreeding, enhanced a number of production traits including weaning weight, mature size, milk production, adaptability to climate, age of puberty, scrotal circumference, and fertility. But the mushrooming of breeds, composites, and crossbreeding was a major causative factor in lack of consistency, or repeatability, of U.S. beef during the last third of the 20th century, causing beef eaters to bypass the beef counter. The spawning and impact of new breeds becomes evident in the following statistics in the editions of this book, ***Beef Cattle Science***. In 1968, the fourth edition of this book listed only 21 breeds of U.S. Beef cattle; in 1976, the fifth edition listed 53 breeds; in 1987, the sixth edition listed 59 breeds; in 1997, the seventh edition of ***Beef Cattle Science*** lists 83 breeds. Thus, from 1968 to 1997, a span of 29 years, 62 new breeds or composites of beef cattle evolved—a fourfold increase; and, simultaneously, crossbreeding proliferated.

The growth years of beef cattle feeding extended from 1947 to 1986. In 1947, only 6.9 million beef cattle were grain fed, representing 30.1% of the slaughter cattle that year, and per capita beef consumption (carcass basis) was 69.1 lb. In 1986, 26.2 million head of cattle were grain fed, representing 70% of the slaughter cattle that year, and per capita beef consumption (carcass basis) peaked at 107.6 lb.

Broilers and turkeys are on a roll, with production and per capita consumption going up and up. In 1934, 34 million broilers were produced in the United States. By 1960, the figure had increased to 1.8 billion, and in 1995, 7.0 billion broilers were produced in the united States. In 1910, 3.7 million turkeys were produced in the United States. In 1995, 289 million turkeys were marketed.

Currently, the entire beef industry is openly admitting to, and facing, its problems, as evidenced by the following reports, plus others:

1. *The War on Fat, 1990*, a report prepared by "The Value Based Task Force," assembled under the combined auspices of the National Cattlemen's Association (NCA) and the Beef Industry Council (BIC) of the National Live Stock and Meat Board.
2. The *National Beef Quality Audit, 1992*, published by the National Cattlemen's Association in cooperation with Colorado State University and Texas A&M University. This three-phase study listed 12 beef quality concerns, and gave an estimated monetary loss per steer/heifer to accrue from each.

3. *The Strategic Alliances Field Study, 1993*, managed by the National Cattlemen's Association in coordination with Colorado State University and Texas A&M University. This study showed that quality losses or non-conformities identified by the National Beef Quality Audit can be reduced, waste can be lessened, taste can be improved, management can be improved, and weight can be controlled.

4. The *National Beef Tenderness Conference Executive Summary, April 22-23, 1994*, in Denver, Colorado, sponsored by the National Cattlemen's Association (NCA). At the close of the conference, the 60 industry leaders/participants presented 11 recommendations to combat the industry's tenderness issue.

5. *The National Non-Fed Beef Quality Audit, December 1994*, managed by the National Cattlemen's Association, conducted by Colorado State University. This study showed a total quality loss of $69.90 for each non-fed animal marketed in 1994, of which $14.60 per head could have been saved had the cattle been managed properly, $27.65 per head could have been saved had the cattle been monitored correctly, and $27.65 per head could have been saved had the cattle been marketed properly.

6. *The National Beef Quality Audit, 1995*, conducted by Colorado State University, Oklahoma State University, and Texas A&M University, revealed that the trimming of excess fats on the outside of a cut satisfied the consumer's frugality, but that lessening beef quality by marketing leaner animals did nothing for America's taste buds. Consumers want juiciness, flavor, and tenderness in their beef cuts—qualities that are associated with a certain amount of fat marbled in beef. The 1995 audit also suggested 13 strategies for improving the quality, consistency, competitiveness, and market share of beef.

The entire beef industry was greatly concerned that a ten percentage point decline in beef's market share occurred between 1982 and 1992. Without doubt, the several programs now being launched by the industry will slow the tide. But the real challenge to the industry is how to get 909,130 (data from *Agricultural Statistics 1995-96*, p. VII-15, Table 389) independently-minded cattlemen to agree upon and follow any program on a voluntary basis. Additionally, who is going to tell several of the new and moneymaking breeders of the new exotic and composite breeds that they are the primary cause of the lack of consistency in today's beef; so, they should cease and desist in their respective-breeding programs? *Note well*: Some of the new exotics and composites are good; hence, they should be retained.

The most effective approach would be to make sufficient price differential at the market place between those animals that meet the new standards and those that do not. Some of this will come to pass. But people being people, it will be spotty and partial.

Another approach is to slaughter calves, hogs, and lambs at younger ages and lighter weights, which are lean and tender, with small cuts. At its best, this would involve a federally-graded Baby Beef Program consisting of marketing 500 to 700 lb heavy calves at weaning, from milk and grass, augmented by creep feeding if necessary. Even if only 10% of the calves went this route, it would help. At its best, this would call for slaughtering lambs at 65 lb live weight, similar to the Canterbury lambs of New Zealand which have long enjoyed a great demand abroad; and it would involve emulating Denmark which produces the highest quality pork in the world from hogs slaughter at 190 lb live weight. But U.S. packers favor heavier carcasses because of the lower per pound cost in processing.

In the meantime, broilers, turkey, and pork will continue to increase their share of the consumer's meat dollar. As consumers push their carts along the meat counter, they will search for attractively packaged meat that is consistently juicy, tender, and lean, and affordably priced; without allegiance to beef, pork, lamb, broilers, turkey, or fish, and with a long memory of meat that doesn't measure up.

From the above, it may be concluded: *As beef goes, the entire beef industry—cow-calf producers, stocker operators, cattle feeders, beef packers, and beef retailers—will go.*

National Live Stock & Meat Board, we shall miss thee.

Since its formation in 1922—75 years ago—the National Live Stock and Meat Board has imparted class, style, and elegance to the entire livestock industry. It has made a great difference. Its works will live on. Hopefully, the glory days of the National Live Stock and Meat Board will return soon.

I am Grateful to Many People.

The author gratefully acknowledges the help of the many people who contributed to the seventh edition of ***Beef Cattle Science.*** Special appreciation is expressed to the following who either served as critics of the entire book, or of a special chapter(s) or section(s) of their specialty, many of whom also provided pictures and data: Ron Baker, President, C & B Livestock, Hermiston, OR; Paul O. Brackelsberg, Ph.D., Department of Animal Science, Iowa State University, Ames, IA; Don Fender, Director, Field Sales Services, Excel Corporation, Wichita, KS; H. Kenneth Johnson, Vice President, and Dr. Terry Dockerty, National Live Stock and Meat Board, Chicago, IL; Martin Jorgensen, Jorgensen Farms, Ideal, SD; Wayman W. Watts, CPA, Fresno, CA; and Jimmy W. Wise, Ph.D., USDA/Livestock & Seed Division, Washington, DC.

Also, the author wishes to express his thanks to the following who provided special illustrations and/or data for this book: Garth W. Boyd, Ph.D., and Rhonda Campbell, Murphy Farms, Rose Hill, NC; Tom Brink and Shawn Walter, Cattle Fax, National Cattlemen's Beef Association, Englewood, CO; Randy Jones, Livestock Director, Houston Live Stock Show and Rodeo, Houston, TX; Stephen J. "Tio" Kleberg and Hal E. Hawkins, King Ranch, Kingsville, TX; Jim Robb, Program Director, USDA Livestock Marketing Information Center, Lakewood, CO; Charles W. Sylvester, General Manager, National Western Stock Show & Rodeo, Denver, CO; Wm. J. Waldrip, Spade Ranches, Lubbock, TX; and Harold Workman, President & CEO, Kentucky Fair & Exposition Center, Louisville, KY.

Also, I am grateful to my former student, Dr. Gary C. Smith, Distinguished Professor, Colorado State University, who kept me current on the frantic beef quality research studies and reports in the 1990's.

Special appreciation is expressed to the following persons who contributed richly to the revision of *Beef Cattle Science*: my co-author, Dr. Randy C. Perry, Professor, Department of Animal Sciences and Agricultural Education, California State University-Fresno, Fresno, California, who greatly pursued his assignment; Audrey H. Ensminger (Mrs. E), who shepherded it from my longhand revision to camera-ready and who designed the cover; Randall and Susan Rapp, who prepared the camera-ready copy and did the proofreading; Joan Wright, who handled the voluminous correspondence involved; and Lawrence A. Duewer, Ph.D., Agricultural Economist, USDA, ERS, who provided many of the statistical facts and figures, and to whom the Ensmingers will ever be grateful. Additionally, a host of individuals, associations, and companies provided pictures or made other notable contributions, which are gratefully acknowledged at appropriate places throughout the book. Most of all, I am grateful to the users of this book who cause me to keep on keeping on.

M. E. Ensminger

Clovis, California

CONTENTS

PART IV: BEHAVIOR & ENVIRONMENT/GLOSSARY

APPENDIX

PART I

General Beef Cattle

Broadly classified there are two phases of beef cattle production: (1) the cow-calf system, and (2) cattle finishing. Sometimes, both phases are conducted on a single farm or ranch as successive steps of a continuous process—they're integrated. For example, cow-calf operators may carry home-produced calves through the stocker stage, and they may even finish them out. More often, however, each phase is conducted to the exclusion of the other, not only on individual operations, but also in agricultural regions. Nevertheless, the fundamentals of beef cattle production are much the same in both cow-calf production and cattle finishing. That is, the general principles—such as business aspects, feeding, management, buildings and equipment, health, marketing, and meats—apply to both phases, regardless of whether they are conducted as an integrated operation or as separate phases. For this reason, organizationally, these are covered in Part I, General Beef Cattle—the first 16 chapters of this book; and they are not repeated. Instead, the practical application of these principles follows in Part II, Cow-Calf Systems and Stockers; and in Part III, Cattle Feedlots and Pasture Finishing. Part IV contains Chapter 36, Behavior/Environment, and Chapter 37, Glossary of Cattle Terms.

The way it used to be done! "The White Heifer That Traveled," painting by Thomas Weaver, 1811. This shows the noted heifer at seven years of age and a weight of 2,300 pounds. The caretaker has a basket of mangels, a root crop that was sliced and fed to cattle in that era. (Courtesy, Frank Harding, Harding & Harding, Geneva, IL)

HISTORY AND DEVELOPMENT OF THE BEEF CATTLE INDUSTRY

Contents *Page*

Cattle are the most important of all the animals domesticated, and, next to the dog, one of the most ancient. There are nearly 1.3 billion cattle in the world.[1]

The word *cattle* seems to have the same origin as *chattel*, which means "possession." This is a very natural meaning, for when Rome was in its glory, wealth was often computed in terms of cattle possessions, a practice which still persists among primitive people in Africa and Asia. That the ownership of cattle implied wealth is further attested by the fact that the earliest known coins bear an ox head; and the Roman word *pecunia* for money (preserved in our adjective *pecuniary*) was derived from the Latin word *pecus*, meaning "cattle." It is also noteworthy that the oldest known treatise on agriculture, written by the Greek poet Hesiod, referred to cattle. Apparently having had some disturbing experience with young oxen, Hesiod advised: "For draught and yoking together, nine-year-old oxen are best because, being past the mischievous and frolicsome age, they are not likely to break the pole and leave the plowing in the middle."

ORIGIN AND DOMESTICATION OF CATTLE

It seems probable that cattle were first domesticated in Europe and Asia during the New Stone Age. In the opinion of most authorities, today's cattle bear the blood of either or both of two ancient ancestors-namely, *Bos taurus* and *Bos indicus*. Other species or subspecies were frequently listed in early writings, but these are seldom referred to today. Perhaps most, if not all, of these supposedly ancestral species were also descendants of *Bos taurus* or *Bos indicus* or crosses between the two.

Fig. 1-1. Ancient drawing of a bison on a rock, made by Paleolithic (Old Stone Age) *Homo sapiens*. Even prior to their domestication, people revered animals, according them a conspicuous place in the art of the day. (Courtesy, The Bettmann Archive, Inc.)

BOS TAURUS

Bos taurus includes those domestic cattle common to the more temperate zones, and it, in turn, appears to be derived from a mixture of the descendants of the Aurochs (*Bos primigenius*) and the Celtic Shorthorn (*Bos longifrons*).

Most cattle, including the majority of the breeds found in the United States, are believed to have descended mainly from the massive Aurochs (also referred to as *Uri*, *Ur*, or *Urus*). This was the mighty wild ox that was hunted by our forefathers. It roamed the forests of central Europe down to historic times, finally becoming extinct about the year 1627. About the year 65 B.C. Caesar mentioned this ox in his writings, but it was domesticated long before (perhaps early in the Neolithic Age) and probably south of the Alps or in the Balkans or in Asia Minor. Caesar referred to these animals as *approaching the elephant in size, but presenting the figure of a bull.* Although this is somewhat of an exaggeration as to the size of the Aurochs, they were tremendous beasts, standing 6 to 7 ft high at the withers, as is proved by complete skeletons found in bogs.

In addition to the Aurochs, another progenitor of some of our modern breeds and the earliest known domestic race of cattle was the Celtic Shorthorn or Celtic Ox. These animals, which have never been found except in a state of domestication, were the only oxen in the British Isles until 500 A.D., when the Anglo-Saxons came, bringing with them animals derived from the Aurochs of Europe. The Celtic Shorthorn was of smaller size than the Aurochs and possessed a dished face. It may have had a still different wild ancestor, or

Fig. 1-2. Artist's conception of an Aurochs (*Bos primigenius*) based on historical information. This was the mighty wild ox that was hunted by our ancestors. Most cattle are believed to have descended mainly from the Aurochs. (Drawing by R. F. Johnson)

[1]*FAO Production Yearbook*, Food and Agriculture Organization of the United Nations, Rome, Italy, Vol. 48, 1994, p. 189, Table 89.

it may have been an independent domestication from the Aurochs.

BOS INDICUS

Bos indicus includes those humped cattle common to the tropical countries that belong to the Zebu (or Brahman) group. They are wholly domestic creatures, no wild ancestors having been found since historic times. It has been variously estimated that cattle of this type were first domesticated anywhere from 2100 to 4000 B.C. The Zebu is characterized by a hump of fleshy tissue over the withers (which sometimes weighs as much as 40 to 50 lb), a very large dewlap, large drooping ears, and a voice that is more of a grunt than a low. These animals seem to have more resistance to certain diseases and parasites and to heat than the descendants of *Bos taurus*. For this reason, they have been crossed with some of the cattle of Brazil and in the southern states of this country, especially in the region bordering the Gulf of Mexico.

Fig. 1-3. Zebu (*Bos indicus*). These wholly domestic animals were the ancestors of the humped cattle common to the tropical countries. (Drawing by R. F. Johnson)

POSITION OF OXEN IN THE ZOOLOGICAL CLASSIFICATION

Domesticated cattle belong to the family *Bovidae*, which includes ruminants with hollow horns. Members of this family possess one or more enlargements for food storage along the esophagus, and they chew their cuds. In addition to what are commonly called oxen or cattle, the family *Bovidae* (and the subfamily *Bovinae*) includes the true buffalo, the bison, musk-ox, banteng, gaur, gayal, yak, and Zebu.

The following outline shows the basic position of the domesticated cow in the zoological classification:

Kingdom *Animalia*: Animals collectively; the animal kingdom.

Phylum *Chordata*: One of approximately 21 phyla of the Animal Kingdom, in which there is either a backbone (in the vertebrates) or the rudiment of a backbone, the chorda.

Class *Mammalia*: Mammals or warm-blooded, hairy animals that produce their young alive and suckle them for a variable period on a secretion from the mammary glands.

Order *Artiodactyla*: Even-toed, hoofed mammals.

Family *Bovidae*: Ruminants having a polycotyledonary placenta; hollow, nondeciduous, up-branched horns; and nearly universal presence of a gall bladder.

Genus *Bos*: Ruminant quadrupeds, including wild and domestic cattle, distinguished by a stout body and hollow, curved horns standing out laterally from the skull.

Species *Bos taurus* and *Bos indicus*: *Bos taurus* includes the ancestors of the European cattle and of the majority of the cattle found in the United States; *Bos indicus* is represented by the humped cattle (Zebu) of India and Africa and by the Brahman breed of America.

USE OF CATTLE IN ANCIENT TIMES

Like other animals, cattle were first hunted and used as a source of food and other materials. As civilization advanced and people turned to tillage of the soil, it is probable that the domestication of cattle was first motivated because of their projected value for draft purposes. Large, well-muscled, powerful beasts were in demand; and any tendency to fatten excessively or to produce more milk than was needed for a calf was considered detrimental rather than desirable. Not all cattle were used for work purposes, however, in the era following their domestication. Instead of planting seeds, some races of people chose a pastoral existence—moving about with their herds as they required new pastures. These nomadic people lived mainly on the products of their herds and flocks.

As populations became more dense, feed became more abundant, and cattle became more plentiful, people became more interested in larger production of meat and milk. The pastoral people adopted a more settled life and began selecting those animals that possessed the desired qualities-including rapid growth, fat storage, and milk production. Following this transformation, Biblical and other literature referred to milk cows, the stall-fed ox, and the fatted calf.

In contrast with the very great importance of cattle in western Asia and Europe in both ancient and modern times, it is noteworthy that cattle were never very highly valued in China, Japan, or Korea. The people of these countries have never used much beef, milk, butter, and cheese. In India, on the other hand, cattle play as important a role as in our western civilization and still retain a great religious significance.

CATTLE IN MEDIEVAL FARMING

The best of medieval farms would excite the scorn or contempt of a modern farmer. Except for plowing and carting with oxen, all labor was done by hand. Although the fields were small, several oxen were often yoked to the plow. As few farmers owned many head, it frequently was necessary for an entire village to pool its oxen and plow the fields in common.

Cattle fared badly in these early days. Pastures were overgrazed and winter feed was scarce. In the fall of the year, it was the common practice to kill and salt the carcasses of all those animals not needed for draft or breeding purposes. Prior to slaughter, aged animals and worn-out oxen were grass fattened, after a fashion. Those that were wintered over were fed largely on straw and the forage they could glean from the fields. Often by spring they were so thin that they could hardly walk.

Very little cow's milk was available, most of it being produced during the grazing season. In fact, more goat's milk than cow's milk was consumed in liquid form. Even in the 13th century, when farming methods had improved, one writer indicated that three cows could be expected to produce only 3.5 lb of butter per week. Most cow's milk was used in cheese making.

LATE MIDDLE-AGES AGRICULTURE OF ENGLAND

During the Middle Ages (500–1500) in England, as elsewhere, rotation and improvement of crops and improved breeding methods were not a necessity because virgin soil was abundant and worn-out lands could be deserted for new. Increasing population and the establishment of settlements were later to make improved husbandry a dire necessity.

Examples of the open-field system could still be found in England up to the 18th century. However, shortly before 1500, feudalism in England practically ceased to exist, and with its passing the system of enclosures and individual ownership became more prevalent.

IMPROVEMENTS IN ENGLISH CATTLE

English agrarian conditions began to improve during the reign of Elizabeth (1558–1603). No well-directed efforts toward the improvement of cattle were made, however, even in England, until late in the 18th century. By 1700, from ⅓ to ½ of the arable land was still cultivated on the open field system. Individual owners could not attempt to improve their herds when all the cattle of the village grazed together on the same common.

Enclosing started about 1450, but progress in this direction was slow. Animals on the common were often half starved, and it was said that 5 acres of individually owned pasture was worth more than the pasture rights over 250 acres of common.

During the 18th century, agricultural progress in England quickened. With the coming of field cultivation of clover and seeded grasses, sometime after 1600, and the introduction and cultivation of the turnip somewhat later, a great impetus was given to agriculture and livestock breeding. Winter feed could now be had, more livestock kept, more manure produced, and better crops grown. Indeed, the progress in stock raising in the 18th century cannot be understood apart from the progress made at the same time in general agriculture.

In cattle, size was the main criterion in selection, though power at the yoke and milking quality were not overlooked. Perhaps the ultimate in cattle size was represented by the Lincolnshire ox, standing 19 hands high and measuring 4 yd from his face to his rump a worthy descendant of the Aurochs.

BAKEWELL'S IMPROVEMENT OF ENGLISH CATTLE

Robert Bakewell of Dishley (1726–1795)—an English farmer of remarkable sagacity and hard, common sense—was the first great improver of cattle in England. His objective was to breed cattle that would yield the greatest quantity of good beef rather than to obtain great size. Bakewell had the imagination to picture the future needs of a growing population in terms of meat

Fig. 1-4. Robert Bakewell of Dishley (1726–1795), noted agriculturalist and the first great improver of cattle in England. Bakewell also contributed greatly to the improvement of the Leicester breed of sheep and the Shire horse. (Courtesy, Picture Post Library, London, England)

and set about creating a low-set, blocky, quickmaturing type of beef cattle. He paid little or no attention to fancy points. Rather, he was intensely practical, and no meat animal met with his favor unless it had the ability to put meat on its back.

Bakewell's efforts with cattle were directed toward the perfection of the English Longhorn, a class of cattle common to the Tees River Area. He also contributed greatly to the improvement of the Leicester breed of sheep, and the Shire horse. Success crowned his patient skill and unwearied efforts. But success in breeding was no mere happenstance in Bakewell's program. Careful analysis of his methods reveals that three factors were paramount: (1) a definite goal as evidenced by the joints that he preserved in pickle and the skeletons of the more noted animals that adorned his halls; (2) a breeding system characterized by *breeding the best to the best* regardless of relationship, rather than crossing breeds as was the common practice of the time; and (3) a system of proving sires by leasing them at fancy prices to his neighbors, rather than selling them. Because of Bakewell's methods and success, he has often been referred to as the founder of animal breeding.

Bakewell's experiments were the top news of the day, and his successes the subject of much comment, both oral and written. The American poet Emerson, for example, said of the British farmer, "He created sheep, cows, and horses to order . . . the cow is sacrificed to her bag, the ox to his sirloin."

By the beginning of the Napoleonic Wars, Bakewell's methods were widely practiced in England, and sheep and cattle were raised more for their flesh than formerly. A new era in livestock improvement was born. As an indication of this change, it is interesting to observe the increase in weights of animals at the famous Smithfield market. In 1710, beeves had averaged 370 lb, calves 50 lb, sheep 28 lb, and lambs 18 lb; whereas in 1795 they had reached 800, 148, 80, and 50 lb, respectively. Although the effect of improved agriculture is not to be minimized, the main influence in this transformation can be attributed to Robert Bakewell, whose imagination, initiative, and courage put a firm foundation under improved methods of livestock breeding.

INTRODUCTION OF CATTLE TO AMERICA

Cattle are not native to the Western Hemisphere. They were first brought to the West Indies by Columbus on his second voyage in 1493. According to historians, these animals were intended as work oxen for the West Indies colonists. Cortéz took cattle from Spain to Mexico in 1519. Then, beginning about 1600, other Spanish cattle were brought over for work and milk purposes in connection with the chain of Christian missions which the Spaniards established among the Indians in the New World. These missions extended from the east coast of Mexico up the Rio Grande, thence across the mountains to the Pacific Coast. Here, in a land of abundant feed and water, these Longhorns multiplied at a prodigious rate. By 1833, the Spanish priests estimated that their missions owned a total of 424,000 head of cattle,[2] many of which were running in a semiwild state. The Longhorns, animals of Spanish extraction, were long on horns, hardiness, and longevity; and they had the ability to fight and fend for themselves when they reverted to the wild. In the movie *The Rare Breed*, Jimmy Stewart described them as meatless, milkless, and murderous.

The colonists first brought cattle from England in 1609. Other English importations followed, with Governor Edward Winslow bringing a notable importation to the Plymouth Colony in 1623. The latter shipment included three heifers and a bull. Three years later, at a public court, these animals and their progeny—and perhaps some subsequent importations—were appropriated among the Plymouth settlers on the basis of one cow to six persons. It is further reported that three ships carried cattle to the Massachusetts Bay Colony in 1625. Other colonists came to the shores of New England, bringing with them their oxen from the mother country. As would be expected, the settlers brought

[2] *Yearbook of Agriculture*, USDA, 1921, p. 233.

along the kind of cattle to which they had been accustomed in the mother country. This made for considerable differences in color, size, and shape of horns, but all of these colonial-imported cattle possessed ruggedness and the ability to perform work under the yoke.

Fig. 1-5. Texas Longhorn steer. (Courtesy, N. H. Rose Collection, San Antonio, TX)

For a number of years, there were very few cattle in the United States. Moreover, those animals that the colonists did possess went without winter feed and shelter, and the young suffered the depredations of the wolves. It was difficult enough for the settlers to build houses for themselves, and they could barely raise enough corn in their fields to sustain human life.

Conditions soon changed for the better. The cattle of earlier importations multiplied, new shipments were received, and feed supplies became more abundant. Cambridge, Massachusetts, enjoyed the double distinction of being the seat of Harvard College, the first institution of higher learning in what later came to be the United States, and the most prosperous cattle center in early New England. In order to provide ample grass and browse for the increased cattle population, it was necessary that the animals range some distance from the commons (the town pasture). Thus, the tale that the streets of Boston were laid out along former cowpaths is not legend but fact. Usually in their travels, the cattle were under the supervision of a paid "cowkeeper" whose chief duty consisted in safely escorting the cattle to and from pasture.

In the village economy, the bull was an animal of considerable importance. Usually the town fathers selected those animals that they considered most desirable to retain as sires, and those citizens who were so fortunate as to own animals of this caliber were paid an approved service fee on a per head basis.

DRAFT OXEN MORE PRIZED BY COLONISTS THAN BEEF

From the very beginning, the colonists valued cattle for their work, milk, butter, and hides, but little importance was attached to their value for meat. In fact, beef was considered as much a byproduct as hides are today. After all, wild game was plentiful, and the colonists had learned to preserve venison, fish, and other meats by salting, smoking, and drying. So necessary were cattle for draft purposes that, in some of the early-day town meetings, ordinances were passed making it a criminal offense to slaughter a work oxen before he had passed the useful work age of seven or more years. The work requirement led to the breeding of large, rugged cattle, with long legs, lean though muscular bodies, and heavy heads and necks. Patient oxen of this type were well adapted for clearing away the forest and turning the sod on the rugged New England hillsides, for hauling the harvested produce over the rough roads to the seaport markets, and for subsisting largely on forages.

Fig. 1–6. Oxen hauling logs to the saw mill. Draft oxen were more prized by the colonists than beef.

AMERICAN INTEREST IN BREEDING

Interest in obtaining well-bred cattle on this side of the Atlantic was a slow one. All through the colonial period, the American farmer let his animals shift for themselves, never providing shelter and rarely feeding them during the winter months. Eventually, the lot of the colonist improved, and with it came the desire to secure blooded stock. Fortunately for the United States, the producer could draw on the improved animals already developed in Britain and on the Continent. Thus, it is not surprising to find that the vast majority of the older breeds of U.S. beef cattle originated across the Atlantic.

In 1783, three Baltimore gentlemen—Messrs. Patton, Goff, and Ringold—sent to England for the best cattle obtainable. They could not have had any particular breed in mind, for at that time no distinct breed can be said to have existed in England, unless it was Bakewell's Longhorns. Hubback, the celebrated foundation Shorthorn bull, was only six years old in 1783, and the fame of the Devon and Hereford was purely local. Other importations followed. Gradually the native stock was improved.

In due time, animals of this improved breeding were to make their influence evident on the western range. The Texas Longhorns, which had thrived since the time of their importation by the Spaniards in the 16th century were decidedly lacking in early maturity and development in the regions of the high priced cuts. Thus, the infusion of blood of English ancestry cattle resulted in a marked improvement in the beef qualities of the range cattle, but it must be admitted that no admixture of breeding could have improved the Texas Longhorns in hardiness and in ability to fight and to fend for themselves.

In order to supply the increased range demand for high-class bulls, a considerable number of purebred herds were established, especially in the central states. In addition to selling range bulls, these breeders furnished foundation heifers and herd bulls for other purebred breeders.

EFFECT OF THE CIVIL WAR ON THE CATTLE INDUSTRY

In 1860, just prior to the Civil War, stock raising was on the threshold of becoming one of the nation's leading industries. At that time, the aggregate value of United States livestock was more than a billion dollars, representing an increase of more than 100% since 1850. Texas, which had been admitted to the Union in 1845, was the leading cattle-producing state, and Chicago was the foremost packing center.

With the outbreak of the Civil War, the cattle of the Southwest could no longer reach the normal markets to the north and east. Union gunboats patrolled the waterways and the Northern armies blocked land transportation. Not even the Confederate armies of the South could be used as an outlet. Prices slumped to where the best cattle in Texas could be purchased at $4 to $6 per head.

In sharp contrast to the conditions in the South and Southwest, the Civil War made for a very prosperous cattle industry in the North. The war-made industrial prosperity of the densely populated East and the food needs of the Union army produced an abnormal demand for beef. Inflated prices followed, with the result that choice steers were selling up to $100 per head at the close of hostilities. Many cattlemen amassed modest fortunes.

At the close of the Civil War, therefore, a wide difference existed between cattle values in the North and the Southwest. With the return of normal commerce between the states, this condition was soon rectified, only to receive another and more serious jolt with the outbreak of cattle tick fever (Texas fever) in 1868, which, a year later, was spread through cattle shipments northward to Illinois and eastward to the Atlantic Coast.

ABILENE, KANSAS, SHIPPING POINT

Until 1867, the only convenient shipping point for Texas cattle was at Sedalia, Missouri, on the Missouri Pacific Railroad. But distance was not the only hazard to early-day trailing. At that time, the Missouri Ozarks were the chief hideaway and point of operation of numerous bands of cattle thieves and robbers. Sometimes these outlaws operated under the guise of sheriffs or other local officials who pretended to be enforcing laws that prohibited the passage of Texas cattle. In any event, the end result of their handiwork was always the same: (1) stampeding the cattle and making away with a large number of them before the drovers could get them under control; (2) beating or otherwise torturing the drovers until they were glad to abandon the herd and flee for their very lives; or (3) killing those drovers who resisted. Because of these treacherous bands, it soon became necessary for the Texas drovers to travel farther to the west through eastern Kansas.

Finally, in a desperate effort to circumvent the outlaw hazards of the Missouri Ozarks, Mr. Joe G. McCoy, a prominent Illinois stockman, in 1867,

Fig. 1-7. *Trail Herd Flank Riders*, from an etching by Edward Borein. The six flank riders held the herd in line. The five riders at the tail of the herd were known as *drag* riders who kept watch on the crippled, sick, and exhausted animals. The chuck wagon is in the center background and the remuda (extra horses) in the left background. (Courtesy, Armour and Company)

conceived the idea of establishing a rail shipping point further to the west. To this end, Mr. McCoy personally inspected the Kansas Pacific Railroad route through eastern Kansas in an effort to select a site where cattle could be grazed while awaiting shipment. Abilene (now famous as the hometown of former President Eisenhower) was decided upon because (1) it was located on a railroad (the Kansas Pacific Railroad, which was then being extended to Denver), (2) the surrounding country was sparsely populated, and (3) there was an abundance of grass and water upon which cattle could be held pending shipment. At the time, Abilene merely consisted of 12 small log huts, most of them with dirt floors. In less than two months, Joe McCoy transformed Abilene into a thriving cattle town. He built stockyards, cattle pens, a livery stable, and an 80-room hotel, called the Drover's Cottage, for trail bosses and eastern buyers. Then the cattle came. (See Table 1–1.)

In 1871, Abilene received 600,000 head of cattle. Both shipping and grazing facilities had been seriously overtaxed. With the arrival of winter, thousands of head of cattle remained unsold, and insufficient feed supplies were available. The severe winter that followed brought heavy losses. With the coming of spring, it was estimated that 250,000 head of cattle had starved to death within sight of Abilene. This disaster, coupled with the opening up of more plentiful and convenient shipping points, marked the rapid decline of Abilene as a shipping center.

TABLE 1–1
CATTLE SHIPMENTS TO ABILENE KANSAS, 1867–1871

Year	Number of Cattle That Arrived at Abilene
1867	35,000
1868	75,000
1869	150,000
1870	300,000
1871	600,000

Fig. 1-8. Early Abilene, Kansas, shipping point. This end of the Texas cattle trail in eastern Kansas on the Kansas Pacific Railroad was established in 1867, for the purpose of providing safe transportation to the East, unmolested by the Ozark outlaws. (Courtesy, Abilene Chamber of Commerce)

FAMOUS CATTLE TRAILS

A number of famous cattle trails led from the Southwest to assembling and shipping points to the North, the best known of which were the Chisholm Trail, the Shawnee Trail, the Western Trail, the Goodnight-Loving Trail, and the Bozeman Trail (Fig. 1-9). It has been estimated that 10 million cattle were driven over these trails between 1866 and 1890.

The trails varied in length from 600 to 1,700 mi. Several months were required to cover the longer distances as the cattle could travel only 10 to 20 mi per day. But the tough, sinewy muscles made it possible for the Texas Longhorns to travel long distances to sparse water holes, and even longer distances to the railheads.

Around such drives center many of the most romantic incidents of American history. They were accompanied by an almost ceaseless battle with the elements, clashes with thieves and other drovers, and no small amount of bloodshed. But the historical records are all that remain today. The trails are gone, and the cattle of that era have given way to animals less adapted to travel on foot.

OUTBREAK OF CATTLE TICK FEVER (Texas Fever)

Following the Civil War, the Texas cattle trade had received real encouragement with the establishment of the Abilene, Kansas, shipping point in 1867, thereby alleviating the hazards of the Missouri outlaws. In 1868, however, the Texas cattle trade received a serious setback. That summer, a group of Chicago cattle producers shipped 40,000 head of cattle from Texas to Tolono, Illinois, in Champaign County, where they were sold to the local farmers (part of this shipment was taken over into neighboring Indiana) for grazing and wintering. Soon after the arrival of the Texas shipment, the native cattle, with which they were turned to pasture, became mysteriously sick and died in great numbers; whereas the southern cattle–although apparently responsible for spreading the disease–remained in perfect health. According to reports, in some infested areas nearly every native cow died. In one township only one milk cow survived. The cause

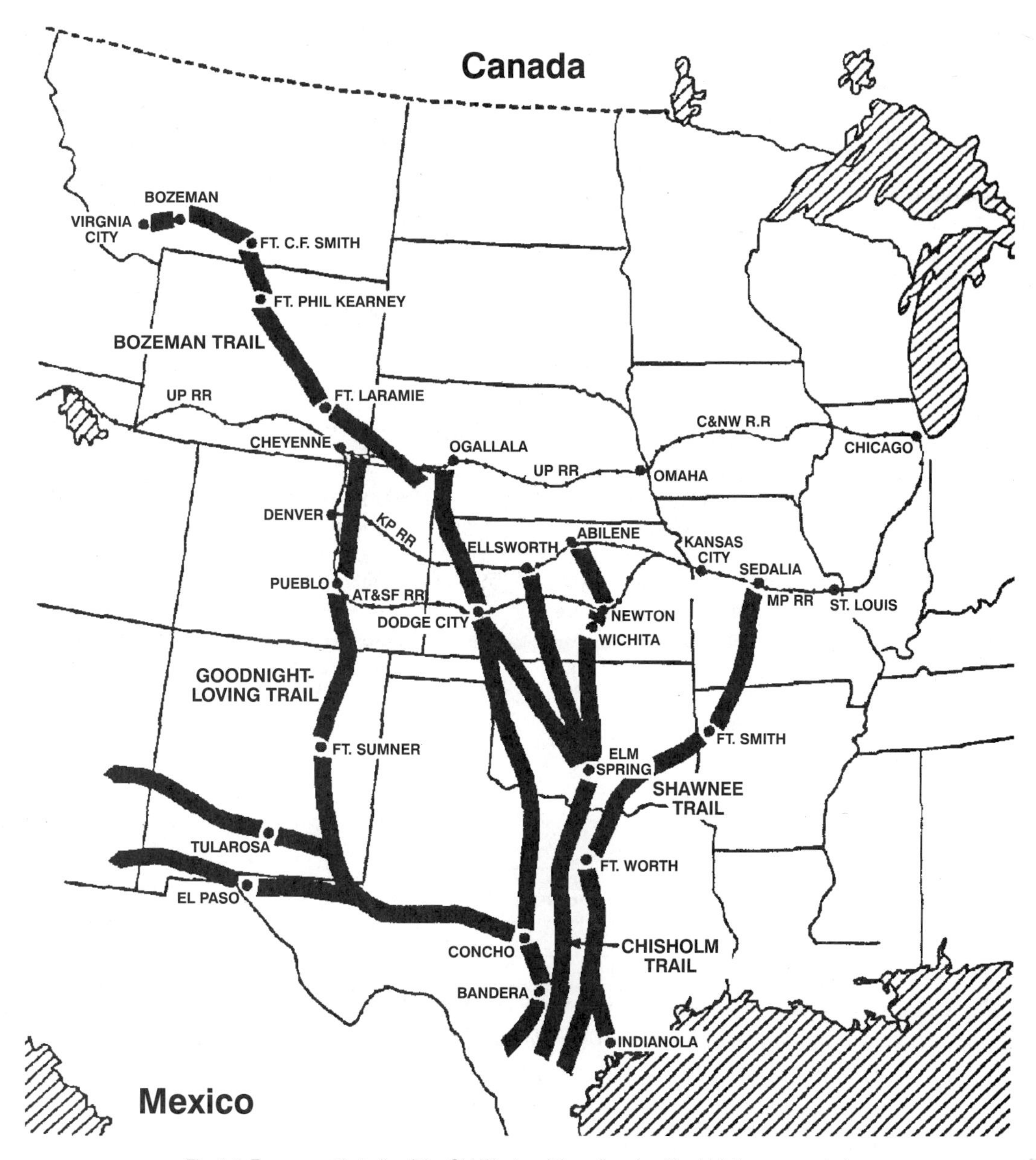

Fig. 1-9. Famous cattle trails of the Old West and the railroads with which they connected.

of the malady was unknown, and there was no cure. Farmers became panicky. In despair, many infected herds were shipped to eastern markets, thereby spreading the disease all the way to the Atlantic Coast.

Although the cause of the disease was unknown, the evidence pointed toward the Texas cattle as being the carriers. Strong prejudice against Texas cattle developed. Wild laws and regulations aimed at controlling the movement of Texas cattle and preventing the spread of the disease were enacted by several states. At this point, the price of Texas steers fell, and many cattle held at the Abilene, Kansas, shipping point could not be sold. The Texas cattle producers, however, were not to be outdone. Some of the more ingenious among them conceived a novel advertising campaign. This consisted of (1) shipping to St. Louis and Chicago a carload of buffaloes decorated with placards extolling the virtues of Texas cattle and beef; (2) pointing out that Texas cattle were more hardy than northern cattle, for none of them contracted the disease when shipped north; and (3) propagandizing the reputed superior carcasses and greater tenderness and palatability of beef from Texas cattle. Soon the pendulum swung back, with the result that Texas cattle became very popular, even commanding a premium over comparable animals native to the North or East.

FAR-WESTERN EXPANSION OF THE CATTLE INDUSTRY

From the very beginning, cattle raising on a large scale was primarily a frontier activity. As the population of the eastern United States became more dense, the stock raising industry moved farther inland. The great westward push came in the 19th century. By 1800, the center of the cow country was west of the Alleghenies, in Ohio and Kentucky; in 1860, it was in Illinois and Missouri; and by the 1880s, it was in the Great Plains. The ranches and cowboys of the Far West were the counterpart of the New England commons and cow-drivers of the 17th and 18th centuries.

The western range was recognized as one of the greatest cattle countries that the world had ever known. Plenty of water and unlimited grazing area were free to all comers, and the market appeared to be unlimited. Fantastic stories of the fabulous wealth to be made from cattle ranching caused a rush comparable to that of the gold diggers of 1849. All went well until the severe winter of 1886. It was the type of winter that is the bane of the cattle producer's existence. Then, but all too late, it was realized that too many cattle had been kept and too little attention had been given to storing up winter feed supplies. The inevitable happened. With the melting of the snow in the spring of 1887, thousands of cattle skeletons lay weathering on the western range, a grim reminder of overstocking and inadequate feed supplies. Many ranchers went broke, and the cattle industry of the West suffered a crippling blow that plagued it for the next two decades. Out of this disaster, however, the ranchers learned the never-to-be-forgotten lessons of avoiding over-expansion and too-close grazing, and the necessity of an adequate winter feed supply.

MILESTONES IN THE BEEF CATTLE INDUSTRY

The roots of the present lie deep in the past! Thus, historical events often help to interpret the present and to project into the future. It follows that, as the beef cattle industry races to the 21st century—and beyond, those engaged in it can best understand and chart their course if they have a knowledge of the past.

A brief chronology of some of the most important developments in the world cattle industry is presented in Table 1–2.

TABLE 1–2
CHRONOLOGY OF THE WORLD BEEF CATTLE INDUSTRY

Year	Event
5500 B.C.	Cattle were domesticated. Domestication of cattle was preceded by domestication of the dog about 8000 to 10000 B.C.; by domestication of sheep and goats about 8000 B.C.; and by domestication of the pig about 6000 B.C.
1493 A.D.	On his second voyage, Columbus took cattle to Santo Domingo.
1519	Cortéz took cattle from Spain to Mexico.
1609	U.S. colonists brought cattle from England.
1760	Robert Bakewell became the first great improver of cattle in England.
1783	Shorthorn cattle were imported from England.
1817	Hereford cattle were imported from England to Kentucky by Henry Clay.
1822	George Coates established the first herdbook for any breed of cattle—for Shorthorns in Great Britain and Ireland.
1836	First public livestock auction sale was held in Ohio by the Ohio Company, whose business was importing English cattle.
1849	Brahman cattle were imported to South Carolina.
1862	President Abraham Lincoln signed three acts into law which were to change all American agriculture forever, including the beef industry, namely: (1) the Act creating the U.S. Department of Agriculture; (2) the Morrill Land Grant College Act, which established a College of Agriculture in each state; and (3) the Homestead Act, which made frontier land available to those who committed themselves to living on, and developing, the land.
1866	Mendel, the Austrian monk, published the results of his experiments with garden peas which marked the founding of modern genetics.
1866	From 1866 to 1890, an estimated 10 million cattle were trailed from the U.S. Southwest to assembling and shipping points to the North, the most famous trails of which were the Chisholm Trail, the Shawnee Trail, the Western Trail, the Goodnight-Loving Trail, and the Bozeman Trail.

(Continued)

TABLE 1–2 (Continued)

Year	Event
1867	A rail cattle shipping point was established at Abilene, Kansas, for the purpose of providing safe transportation to the East, unmolested by the numerous bands of cattle thieves and robbers whose chief hideaway and point of operation was the Missouri Ozarks.
1871	First successful rail shipment of fresh meat by refrigeration was achieved.
1873	Angus cattle were imported from Scotland.
1886	In the severe winter of 1886, thousands of cattle perished on the western ranges, a grim reminder of overstocking and inadequate winter feed supplies.
1887	The Hatch Act was passed, creating Agricultural Experiment Stations as the research arms of the Agricultural Colleges.
1902	The University of Illinois published their recommended *Market Classes and Grades of Live Cattle*, based on a thorough study of livestock markets.
1906	The meat inspection service of the U.S. Department of Agriculture was inaugurated.
1911	Truck transportation of animals to market had its beginning.
1912	The Russians reported the first successful artificial insemination (AI) of cattle.
1914	The U.S. Congress established the Cooperative Extension Service, thereby providing for correlated U.S. agricultural teaching, research, and extension, which is the envy of, and without effective counterpart in, the world.
1923	Federal grading of beef was first started as a special service to U.S. steamship lines; and on February 10, 1925, the 68th Congress passed an act setting up a federal meat grading service.
1941	First performance bull testing station was established in Texas.
1949	British scientists reported that the addition of glycerine to diluters permitted them to freeze semen and still retain a high degree of fertility following thawing.
1951	First successful bovine embryo transfer.
1951	Snorter dwarfism in cattle was reported.
1953	Francis Crick and James Watson described the structure of the DNA molecule, the basic building block of life on earth, which ushered in the biotechnology era.
1955	First Beef Cattle Improvement Association was formed in Virginia.
1956	Performance Registry International was established in Texas.
1964	Trading in live cattle futures opened on the Chicago Mercantile Exchange.
1965	USDA adopted yield grading of beef.
1977	Gene splicing, also known as recombinant DNA technique, ushered in a new era of genetic engineering and biotechnology when scientists at the University of California-San Francisco, altered genes—making ordinary bacteria capable of producing insulin, a hormone essential to the survival of diabetics.
1982	First identical twins from embryo splitting produced in the U.S.
1987	EPDs gained industry-wide acceptance as a selection tool.
1989	USDA uncoupled yield grades and quality grades.
1996	National Cattlemen's Association and the Beef Industry Council of the National Live Stock and Meat Board merged to form the National Cattlemen's Beef Association (NCBA).

QUESTIONS FOR STUDY AND DISCUSSION

1. How do you account for the fact that most of the cattle in the U. S. are descendants of *Bos taurus* rather than *Bos indicus*?

2. What anatomical and physiological characteristics are common to members of the family *Bovidae*?

3. Throughout the ages, and in many sections of the world, cattle have been used for work purposes more than horses and mules. Why has this been so?

4. Compare Robert Bakewell's breeding methods with those used in modern production testing programs. What three factors contributed most to his success as an animal breeder?

5. Of what significance is each of the following in the history and development of the U.S. beef cattle industry?

a. Cortéz.
b. The Spanish missions.
c. Longhorn cattle.
d. The colonists.
e. The Civil War.
f. Abilene, Kansas.
g. Joe G. McCoy.
h. Famous cattle trails.
i. Cattle tick fever.
j. The winter of 1886.

6. List what you consider to be the six most significant milestones in the beef cattle industry. When did each of these events occur? Why do you consider them to be so significant?

SELECTED REFERENCES

Title of Publication	Author(s)	Publisher
American Cattle Trails 1540–1900	C. M. Brayer H. O. Brayer	Western Range Cattle Industry, Study and American Pioneer Trails Assn., Bayside, NY, 1952
Animals and Men	H. Dembeck	The American Museum of Natural History, The Natural History Press, Garden City, NY, 1965
Cattle and Men	C. W. Towne E. N. Wentworth	University of Oklahoma Press, Norman, OK, 1955
Cattlemen, The	M. Sandoz	Hastings House, New York, NY, 1958
Cowboys, The	Ed. by Time-Life Books, text by W. M. Forbis	Time-Life Books, New York, NY, 1973
Encyclopaedia Britannica		Encyclopaedia Britannica, Inc., Chicago, IL
History of Domesticated Animals, A	F. E. Zeuner	Harper & Row, New York, NY
History of Livestock Raising in the United States 1607–1860, Ag. History Series No. 5	J. W. Thompson	U.S. Department of Agriculture, Washington, D.C., Nov., 1942
Livestock Book, The	W. R. Thompson, *et al.*	Vulcan Service Co., Inc., Birmingham, AL, 1952
Our Friendly Animals and Whence They Came	K. P. Schmidt	M. A. Donohue & Co., Chicago, IL, 1938
Principles of Classification and a Classification of Mammals, The, Vol. 85	G. G. Simpson	American Museum of Natural History, New York, NY, 1945
Stock Raising in the North West 1884	H. O. Brayer	
G. Weis		
		The Branding Iron Press, Evanston, IL, 1951
Yearbook of Agriculture 1921		U.S. Department of Agriculture, Washington, D.C., 1921

Hunting the prehistoric European Bison, after a painting by Ernest Griset. (Courtesy, Smithsonian Institution)

WORLD AND U.S. CATTLE AND BEEF

Contents *Page*

Contents Page

World and U.S. cattle have a past to honor and a future to build in meeting the world food needs of the 21st century. Ponder the following:

- More than 700 million people in the world go to bed hungry each night.[1]
- More than 188 million children throughout the world suffer from malnutrition.[1]

The historical achievements of world agriculture and the current biotechnology era impart confidence that we can, and will, meet the world food needs in the decades to come.

People and cattle are not only inseparable in history—they were a major part of it. Without them, there would have been no history—no humankind.

World food problems, however, are more complex than just too many people and too little food. Around the world, food problems relate primarily to the unequal distribution of population, production, and wealth. With more people, there is need for more food production. Also, in countries where wealth increases, the diet also changes. People start to turn away from a starch-oriented diet to one based on animal protein, increasing the demand for beef and veal. In poor countries, with rapidly expanding populations, food production does not meet the demand and the countries lack the money to purchase food from the countries producing excesses.

Animal protein is costly. It is resource-expensive. Each calorie of animal-derived food requires an average energy input of about 27 calories. For example, 6 to 9 lb of feed (grain, roughage, and protein supplement) are necessary to produce 1 lb of beef. Additionally, energy is required for marketing, slaughtering, refrigeration, wholesale-retail, and preparation of the beef. Thus, most people in developed countries consume about 2,000 lb of cereals each year in the form of meat, milk, and eggs, whereas, if consumed directly, only 400 lb of cereals are sufficient to maintain an individual for 1 year.

Even though animal proteins are expensive, everywhere in the world where people have achieved higher income, the demand for animal protein has soared. In an effort to meet a world food crisis, however, traditional energy-intensive animal production (including beef) may need to change to make available a higher proportion of grains and seeds for humans. Nevertheless, cattle have a future in increasing food availability by converting roughage to high-quality protein.

WORLD HUMAN POPULATION

It took from the dawn of man until 1,600 years after the birth of Christ for the number of people in the world to reach ½ billion. The population of the world first topped 1 billion in 1830. And it took another 100 years—until 1930, for the population to reach 2 billion. In 1960, it was 3 billion; in 1975, it was 4 billion; and in 1994, it was 5.5 billion. Even more frightening in the people-numbers game are the forecasts of population experts relative to the years ahead. The United Nations (U.N.) reports that the world population is increasing at the rate of almost 100 million people per year, with 93% of the increase occurring in the developing countries. By the year 2020, the U.N. predicts that the world will have 8 billion more people to feed (see Fig. 2-1).

Actual numbers are, however, only part of the picture. As Table 2-1 shows, total population, growth rate of the population, and the wealth of various areas of the world differ.

[1]The Food and Agriculture Organization of the United Nations defines a diet containing fewer than 1,600 calories per day as inadequate. This low value is only about 20% above the basal metabolic rate. Additionally, in many areas of the world where calorie (energy) intake is low, protein intake is marginal and of poor quality (plant origin); thus, the protein eaten is not used for its essential functions. Estimates vary, but worldwide, 15 to 20% of the population is undernourished. Moreover, in populations subsisting primarily on millet and sorghum or roots and tubers, some degree of malnutrition is common.

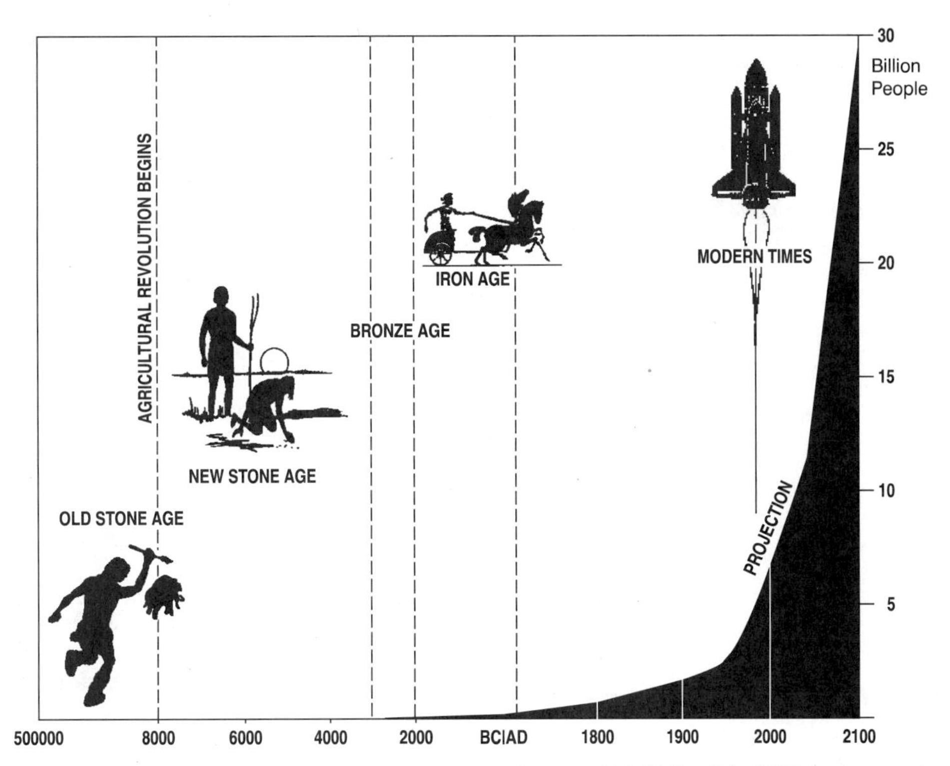

Fig. 2-1. A history of human population growth, with projections to 2100 A.D. *Note:* If the Old Stone Age were in scale, its base line would extend 8 ft to the left.

TABLE 2–1
POPULATION AND GNP PER CAPITA, BY RANK, OF 25 MOST HUMAN POPULOUS COUNTRIES[1]

Population Rank	Country	Human Population 1994	Projected Population 2000	Population Per Sq. Mile 1994	Gross National Product Per Capita 1991
		(1,000)	*(1,000)*		*($)*
1	China	1,190,431	1,260,154	331	1,327
2	India	919,903	1,018,105	801	303
3	United States	260,714	275,327	74	22,550
4	Indonesia	200,410	219,496	284	577
5	Brazil	158,739	169,543	49	2,601
6	Russia	149,609	151,460	23	8,639[2]
7	Pakistan	128,856	148,540	429	369
8	Bangladesh	125,149	143,548	2,421	195
9	Japan	125,107	127,554	821	27,300
10	Nigeria	98,091	118,620	279	242
11	Mexico	92,202	102,912	124	3,051
12	Germany	81,088	82,239	600	19,830
13	Vietnam	73,104	80,533	582	—
14	Philippines	69,809	77,747	606	694
15	Iran	65,615	78,350	104	1,689

(Continued)

TABLE 2-1 (Continued)

Population Rank	Country	Human Population 1994	Projected Population 2000	Population Per Sq. Mile 1994	Gross National Product Per Capita 1991
		(1,000)	*(1,000)*		*($)*
16	Turkey	62,154	69,624	209	1,790
17	Egypt	60,765	67,957	158	543
18	Thailand	59,510	63,620	301	1,618
19	Ethiopia	58,710	70,340	138	124
20	Italy	58,138	58,865	512	19,630
21	United Kingdom	58,135	58,951	623	17,400
22	France	57,840	59,354	275	20,900
23	Ukraine	51,847	51,931	222	—
24	South Korea	45,083	47,861	1,189	6,430
25	Burma	44,277	49,300	174	531
	World Total	5,643,290	6,165,079	112	

[1] *Statistical Abstract of the United States*, U.S. Department of Commerce, 1994, p. 850, Table 1351; and p. 862, Table 1366.

[2] Estimated.

East and south Asia have almost 60% of the world's population on about 22% of the world's land. Additionally, many countries in east and south Asia are characterized by a low gross national product and an annual growth rate greater than that for the world. Many other countries in other areas of the world are classified as less developed or developing countries. Their annual growth rate is above the average for the world and their gross national product is considerably lower than that of developed countries such as Sweden, the United Kingdom, France, Spain, Canada, Germany, the United States, and Japan.

Worldwide, the challenge is for production to meet the rapidly growing and diverse need for food—to prevent the fulfillment of the 1798 doomsday prophecy of the English clergyman Thomas Robert Malthus: "The power of population is infinitely greater than the power of the earth to provide subsistence for man." This will not be easy. In the future, the gap will widen between the developing and the developed countries. Through increased production, primarily in the developed countries, the supply of food could meet the worldwide demand—at least for a time, but the real question will be who is going to pay for the food and how will it be transported, stored, and distributed.

Farmers around the world have always demonstrated their willingness to respond to prices and profits. Farmers are people, and people do those things which are most profitable to them. They will expect to be paid to supply the food for the demands of a growing world population. For the beef producer, profits will need to be sufficient to pay for cattle, feed, shipping, interest, death losses, marketing, taxes, and other costs—all before cattle reach the packer.

WORLD CATTLE AND BEEF

Cattle producers and those who counsel with them must be well informed concerning worldwide beef production in order to know which countries are potential competitors. Like the price of all commodities in a free commerce, the price of beef is determined chiefly by supply and demand. A demand exists in those countries that do not produce enough to meet their domestic needs and the supply is that which can be spared by those nations producing a surplus.

WORLD CATTLE NUMBERS AND BEEF PRODUCTION

Fig. 2-2 shows the distribution of cattle by major world areas, while Table 2-2 shows the leading cattle- and buffalo-producing countries of the world. World cattle numbers average nearly 1.3 billion head, about 1 cow for every 4 people, or 27 head of cattle per square mile.

But cattle numbers alone do not tell the whole story. Large numbers of cattle are kept for work and milk in India, the Russian Federation, and China. Besides, cattle are sacred to the Hindus of India (about 80% of the population). The United States leads the world, by a wide margin, in the production of beef and

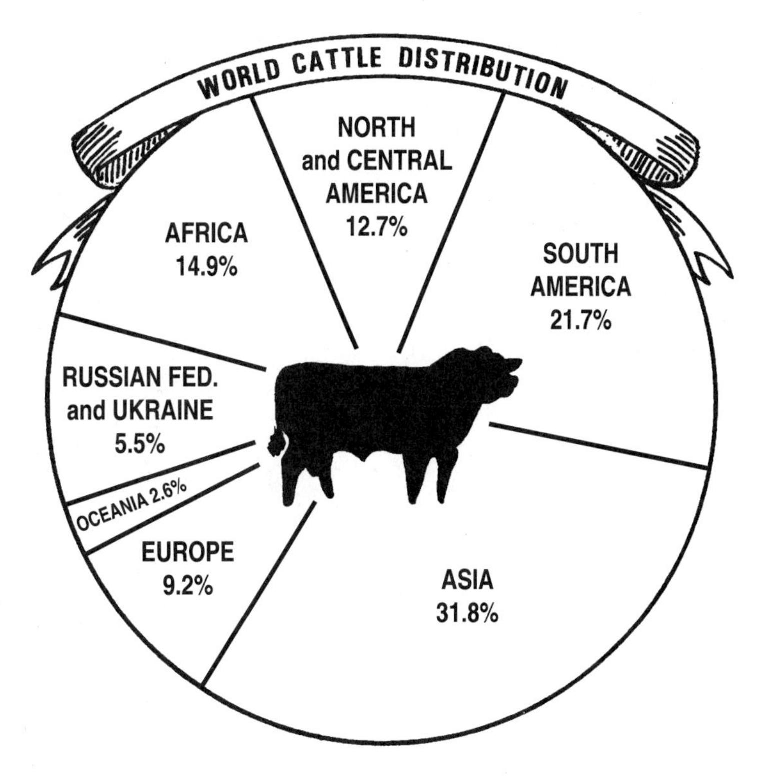

Fig. 2-2. Worldwide distribution of cattle, by major areas. Oceania includes primarily Australia and New Zealand, and North America includes Canada, the United States, and Mexico. (Based on data from *FAO Production Yearbook*, Food and Agriculture Organization of the United Nations, Rome, Italy, Vol. 48, 1994, p. 189, Table 89)

TABLE 2–2
LEADING CATTLE- AND BUFFALO-PRODUCING COUNTRIES OF THE WORLD

Country	Cattle and Buffalo 1993[1]	Human Population 1994[2]	Size of Country		Cattle Per Capita	Cattle Per Area	
	(1,000)	*(1,000)*	*(sq. mi)*	*(sq. km)*	*(no.)*	*(sq. mi)*	*(sq. km)*
India	272,655	919,903	1,269,338	*3,287,585*	0.30	214.8	*82.9*
Brazil	144,900	158,739	3,286,470	*8,511,957*	0.91	44.1	*17.0*
China	113,157	1,190,431	3,706,560	*9,599,851*	0.10	30.5	*11.8*
U.S.A.	100,988	260,714	3,615,102	*9,363,114*	0.39	27.9	*10.8*
Russian Federation	48,900	149,609	8,649,489	*22,402,176*	0.33	5.7	*2.2*
Argentina	54,875	33,913	1,068,296	*2,766,887*	1.62	51.4	*19.8*
Mexico	30,702	92,202	761,600	*1,972,544*	0.33	40.3	*15.6*
Australia	25,732	18,077	2,967,892	*7,686,840*	1.42	8.7	*3.3*
Ukraine	21,607	51,868	233,100	*603,729*	0.42	92.7	*35.8*
France	20,112	57,840	211,207	*547,026*	0.35	95.2	*36.8*
Colombia	16,614	35,578	439,735	*1,138,914*	0.47	37.8	*14.6*
Germany	15,897	81,088	137,838	*357,000*	0.20	115.3	*44.5*
World Total	1,436,922	5,643,290	52,444,043	*135,830,071*	0.25	27.4	*10.6*

[1] *Agricultural Statistics 1995–96*, p. VII-5, Table 374.

[2] *Statistical Abstract of the United States 1994*, p. 850, Table 1351.

TABLE 2–3
MAJOR BEEF- AND VEAL-PRODUCING COUNTRIES 1994[1]

Country	Beef & Veal Production	Cattle Population	Beef Production Per Inventory of Cattle Per Year	
	(1,000 MT)	*(1,000)*	*(lb/hd)*	*(kg/hd)*
United States	11,199	100,988	244	*111*
Russian Fed.	3,350[2]	48,900	151	*69*
Brazil	3,160	151,600	46	*21*
Argentina	2,590	50,000	114	*52*
China	2,253	90,906	55	*25*
Australia	1,825	24,732	162	*74*
France	1,667	20,112	182	*83*
Germany	1,500	15,891	208	*94*
Ukraine	1,421	21,607	145	*66*
Mexico	1,365	30,702	98	*44*
India	1,292[2]	192,980	15	*7*
Italy	1,180	7,683	338	*154*
World Total	50,509	1,288,124	86	*39*

[1]*FAO Production Yearbook 1994*, Vol. 48, p. 189, Table 89, and p. 197, Table 92.

[2]Estimated.

veal (Table 2-3), with the ranking of the other countries as shown.

Efficiency of production is indicated by the annual production of beef per head of inventory (lb/head, or kg/head), although this is not totally accurate (see Table 2-3).

BEEF PRODUCTION IN INDIA

India, land of sacred cows and native home of the U.S. Brahman breed, is the leading cattle country in the world in numbers, with 215 cattle per square mile vs 28 in the United States (see Table 2-2). But India's cattle are of very negligible importance from the standpoint of meat production, due to the large number that are either sacred or used for draft or milk purposes. The humped cattle of India pillage crops in rural areas and roam the streets of villages and cities—gentle and traffic-wise. Some are homeless, others are turned loose by owners who do not wish to pay for their keep, and still others are just loose. To the Hindus (approximately 80% of India's population), the cow is regarded as a mother and an object of reverence; and the eating of beef is taboo. Although India's cattle population puts a serious drain on the nation's resources, no politician dares twist a cow's tail, or even flick a hair.

Fig. 2-3. Cattle in a village in India. In India, Brahman-type cattle are usually herded by the young or the old. (Courtesy, Ford Foundation)

BEEF PRODUCTION IN AUSTRALIA AND NEW ZEALAND

Beef production in Australia and New Zealand increased sharply in the late 1960s and the early 1970s in response to high beef prices and low wool prices. Of all the beef-producing countries of the world, Australia and New Zealand have the best potential for increase.

Australia is a natural cattle country, and it is free from foot-and-mouth disease. Most of the cattle are grazed year-round on unfenced ranges, herded by musterers—the counterpart of the American cowboy. Slaughter animals consist of 2- to 4-year-old steers which are grass finished, although there is a growing trend to market younger animals and grain finish. The vast majority of the cattle operations in Australia are very large, ranging in size from 5,000 acres in the more developed southeastern part of Australia to over 3 million acres in the Northern Territory, and with 10,000 to 50,000 cattle per unit.

Shorthorn, Hereford, and Angus are the leading breeds. In the tropical areas of the North, Brahman and Santa Gertrudis have been introduced. Cross-breeding is practiced widely. Many stations (ranches) are inadequately fenced and watered, with much room for improvement in their nutrition and husbandry. For example, in Northwestern Queensland, Northern Territory, and parts of western Australia, only a 45 to 55% calf crop is raised to branding age. Also, on some properties it is standard practice to write off a 12 to 15% mortality each year.

The beef industry of Australia is subjected to recurrent droughts. Until recently, the chief obstacle to further expansion of the nation's beef production was the great distance to the consumer markets of Europe and the United States. However, improved technology in processing and transporting beef is gradually overcoming this handicap. Also, a new and relatively nearby market for beef in Japan has opened up. Thus, the improved market for beef, and government policies favorable toward the development of the cattle industry of the country, indicate a bright future for the expanding beef industry of Australia. The cost of beef production in Australia is much lower than in the United States.

Fig. 2-4. Australian cattle scene, showing Hereford cattle herded by musterers—the counterpart of the American cowboy. (Courtesy, Australian News and Information Bureau)

New Zealand is a small, picturesque country, about the size of Colorado. The climate is temperate, with plentiful sunshine, adequate rainfall, and no great extremes of heat or cold. Year-round grazing is available, on fenced holdings. Cattle and sheep share many areas, to the advantage of each other, with cattle utilizing the coarser vegetation and sheep the finer grasses and legumes. Very few New Zealand farmers devote themselves exclusively to beef production. In general, the raising and fattening of beef is carried on in conjunction with sheep farming. The dairy sector provides a large contribution from its cull cows and surplus calves.

BEEF PRODUCTION IN EUROPE, INCLUDING EC

Beef production throughout Europe is largely a by-product of milk production—mainly from dual-purpose cattle, animals bred to produce both milk and meat. This poses the problem of how to increase beef output from such herds without pushing up milk surpluses. This is being accomplished chiefly by fewer calf slaughtering and higher slaughter weights.

Fig. 2-5. Milking ability is a *must* in beef cows in the United Kingdom. These calves give evidence that their dams possess this trait in abundance. (Courtesy, Devon Cattle Breeders' Society of England)

There are substantial numbers of beef cows only in France, the United Kingdom, Italy, and Ireland.

In Belgium, France, Germany, Luxembourg, the Netherlands, and Italy, about 45% of the beef and veal production comes from cull dairy cows and milk-fed calves. But cull dairy cows and calves in the United Kingdom, Ireland, Denmark, and Greece account for only about 20% of the total beef and veal production.

By comparison, in the United States, dairy cows and calves merely account for about 15% of federally inspected beef and veal production.

In addition to obtaining beef from cull dairy cows, throughout Europe bull calves (not steers) are fattened out and slaughtered as yearlings.

Many European farmers would have difficulty making a transition from dairy to beef cattle because the farms are too small to produce grass-fed beef profitably, as is done in Argentina and Australia. Further, in the United Kingdom and Ireland, where there are larger farms and beef production from grass-fed animals is profitable, high EC grain prices may result in some pasture being diverted to grain or other crop production.

In much of Europe, dairy cattle are selected for their beef qualities. For example, stud bulls that are widely used in artificial insemination are often selected on the basis of rate and efficiency of gain, very much as beef bulls are selected on performance test in the United States. However, throughout much of Europe the drift is toward more specialized dairy types, with increasing neglect of suitability for beef.

The European Community (EC)—also called European Economic Community, Common Market, or European Union (EU)—which was formed to coordinate the economic programs of the member nations, has had several programs to pay farmers to convert from dairy to beef cattle, but payments were not large enough to equal current income from milk sales. Additionally, the EC has banned U.S. beef from their member nations because U.S. cattle may be treated with growth hormones. Most U.S. beef industry leaders consider this as a trade barrier camouflaged as a human health hazard.

Cattle numbers are not a limiting factor for beef production in the EC. For example, the original EC-6 (the original six member nations) has enough cattle to produce some 20% more beef if current calf slaughter were reduced to U. S. rates and the calves were fed in feedlots. However, to produce this much beef from grass and grain would require a longer turn-around period and larger cattle numbers. Also, this could result in increased milk production, which is in surplus already.

BEEF PRODUCTION IN CANADA

Canada is still a frontier type of country with almost unlimited opportunities for expansion of the beef cattle industry. In general, Canadian cattle are noted for their size, scale, and ruggedness, because in the great expanses of frontier agriculture, cattle production is on a cost-per-head rather than on a cost-per-pound basis. That is, to produce a sizable beast costs little more than to produce a small one. The main obstacles to increased beef production in Canada are: (1) the long, severe winters in much of the cattle country centered primarily in the eastern and western provinces where up to 7 months feeding is required; (2) the high duty and sometimes closed borders for exports to the United States, the most natural potential market; and (3) the need for a permanent outlet for stocker and feeder cattle.

Fig. 2-6. Cattle roundup in Canada. On their way to summer pasture, this herd fords the Milk River in southern Alberta. Canada had 12.3 million head of cattle in 1994. (Courtesy, The National Film Board of Canada)

The cattle producers of Canada appear to be optimistic about the future of the industry. More and more cattle will be finished on the small grains which are produced in great abundance.

BEEF PRODUCTION IN MEXICO

Mexico ranks seventh among the leading cattle countries of the world (Table 2-2).

Since January 1, 1955, Mexico has been free of foot-and-mouth disease, and the border has been open, subject to the usual quotas and duties.

Factors unfavorable to beef production in Mexico are: (1) the ravages of parasites, particularly screwworms and ticks; (2) the lack of improved breeding, which is made difficult because of the susceptibility of newly imported cattle to diseases and parasites; (3) frequent droughts; and (4) political uncertainties and government policies unfavorable to the development of cattle units of adequate size to permit practical and economic operation in the present era.

Despite all the difficulties now existing in Mexico, cattle are afforded a long grazing season and labor is cheap and abundant. Cattle can be produced very cheaply. Also, in recent years, the better cattle producers of Mexico have made marked progress in improving

Fig. 2-7. Part of a fine herd of Herefords on the ranch of Guillermo Finan, Hacienda Valle Colombia, Muzquiz, Coahuila, Mexico. This herd would be considered outstanding anywhere—in Mexico, in the United States, or in Canada. (Courtesy, Mr. and Mrs. Guillermo Finan)

both the quality of their cattle and the efficiency of their production.

Mexico has provided substantial numbers of feeder cattle for growing on the ranges of the Southwest or finishing in U.S. feedlots. However, Mexico has a growing domestic market. Already she is beginning to feel the drain of live cattle to the United States. As a result, Mexico will not be in a position to increase significantly beef or cattle exports in the near future. Rather, she will continue to control them through quotas established by the government.

BEEF PRODUCTION IN SOUTH AMERICA

Of the South American countries, Argentina, which ranks sixth in world cattle numbers, is recognized as the outstanding beef producer. In fact, taken as a whole, Argentine cattle probably possess better breeding and show more all-round beef excellence than do the cattle of any other country in the world. The excellence of the Argentine cattle can be attributed to two factors: (1) their superior breeding and (2) the lush pastures of the country. Beginning in 1850 and continuing to the present time, large numbers of purebred animals have been imported from England, Scotland, and the United States. No price has been considered too high for bulls of the right type. Often British and American breeders have been outbid by Argentine estancieros in the auction rings of Europe. These bulls and their progeny have been crossed on the native stock of Spanish extraction (Criollo cattle). Today, Herefords, Angus, and Shorthorns are the most numerous breeds of the country.

The finest cattle pastures of the Argentine are found along the La Plata River, in the region known as the Pampas, a vast, fertile plains area embracing about 250,000 sq mi, which slopes ever so gently toward the sea. It's a dreamland of cattle and grass, and the "beef basket" of South America. Much of this fertile area is seeded to alfalfa, providing year-round pasture. Instead of finishing cattle largely on grains, as in the United States, corn of the Pampas region, representing an acreage ½ as great as that devoted to alfalfa, is largely exported. Usually 2- and 3-year-old steers are finished by turning them into a lush alfalfa pasture for a period of 4 to 8 months prior to marketing. The surplus beef of Argentina is marketed as frozen or chilled beef to the European countries, especially to Great Britain. None of the frozen or chilled beef from Argentina is admitted into the United States because of the hazard of foot-and-mouth disease. It must be canned or fully cured (*e.g.*, corned beef).

Fig. 2-8. Well-bred cattle on lush pastures in Argentina. (Courtesy, Counselor Office Cultural Relations, Republic of Argentina, Washington, D.C.)

Other South American countries of importance in beef production are Brazil, Colombia, Uruguay, and Paraguay.

Generally speaking, Brazil, which is slightly larger than the United States, produces hardy cattle of rather low quality, predominately of Zebu breeding.

Colombia is handicapped by lack of improved breeding, poor transportation facilities, and limited refrigeration, although beef production is one of the nation's principal industries.

Uruguay, which is but little larger than the state of Missouri, is noted (1) as an ideal cattle country (because of its rich pastures, abundant water supply, and temperate climate), (2) for Hereford and Shorthorn cattle of good breeding, although they are not equal in quality to the cattle in Argentina, (3) as one of the most highly specialized beef cattle countries in the world, and (4) as a beef-exporting country, despite its small size (80% of the nation's exports consisting of animal products).

Paraguay, which is about 2½ times larger than Uruguay, produces cattle of similar breeding and quality to those in Brazil.

As in Argentina, year-round grazing constitutes the basis of the beef cattle industry of the other South

American countries. Virtually no grain is used in finishing animals, except for those being fitted for show. No attempt is made to finish steers until they are fully mature.

In general, the foremost obstacles, or unfavorable factors, affecting South American beef production are:

1. The ever-present foot-and-mouth disease, which, though seldom fatal, results in enormous economic losses through retarded growth and emaciation and which limits the foreign sale of both beef and cattle on foot.

2. Droughts are rather frequent in many of the cattle sections, and they are likely to be of rather long duration.

3. Parasites and certain diseases other than foot-and-mouth disease are rather prevalent in the warmer sections.

4. Prices are very much dependent upon the export trade, creating an uncertain market.

5. Local markets are often unsatisfactory; modern packing plants are not too plentiful; and refrigeration facilities are limited. Many of the cattle slaughtered in the more isolated areas of South America, especially in Brazil and Paraguay, are still made into jerked or salted beef.

6. Transportation facilities are few and far between.

7. Except for the cattle of Argentina and Uruguay, much improvement in breeding is needed; but the introduction of improved cattle is difficult because of the heavy infestation of diseases and parasites to which the native and Zebu cattle are more resistant.

BEEF PRODUCTION IN CENTRAL AMERICA AND PANAMA

Central America and Panama are the tropical land mass connecting North America and South America, consisting of Guatemala, Honduras, Belize (formerly British Honduras), Nicaragua, El Salvador, Costa Rica, and Panama. On an individual basis, none of these countries produces or exports sufficient beef to be much of a factor. However, as a group, in 1993 they had 19,784,000 cattle and they exported 119,000,000 lb of beef and veal to the United States. Moreover, it is estimated that their exports will increase in the future.

Compared with coffee, bananas, and cotton, grazing is a relatively old economy in Central America and Panama. It dates back to the early colonial period, and in terms of land utilization, quality of stock, and disposal of products, it has changed very little through the years. Primary emphasis has been on cattle of

Fig. 2-9. Part of a herd of 800 head of Criollo X Brahman cattle in a corral at Santa Clara Ranch, owned by Sr. Jorgé Cordero, Cordero Ranches, Guatemala City, Guatemala. (Photo by Audrey Ensminger)

unimproved breeding, low in both yields and quality, but high in resistance to ticks and other environmental handicaps of the region. Cattle are predominantly Criollo, although efforts have been made to improve their beef cattle in recent years by importing breeding stock from the United States, primarily Brahman and Santa Gertrudis.

Livestock are concentrated chiefly along the Pacific Coast, where they fatten on grass. Pasture is abundant, except during the dry season, when it is short for about 3 months. Some of the more progressive ranchers are (1) ensiling grass, corn, and/or sorghum, and/or (2) irrigating pastures to provide forage during the dry season.

Corn is high in price, because of the demand for human consumption. Whole cottonseed, cottonseed hulls, and cane molasses are relatively cheap. Rising land values are prompting interest in supplemental pasture finishing.

Factors **favorable** to beef production in Central America and Panama include: abundant grass, relatively cheap land and labor, freedom from foot-and-mouth disease, and tax deferment (up to 15 years).

Unfavorable factors include: scarce and high-priced cereal grains; heavy parasite infestation—especially flies and ticks; each country controlling its live cattle exports more carefully each year; little culling; and low percentage calf crop.

WORLD MEAT CONSUMPTION

The world's leading red meat and poultry eaters in 1994, in per capita consumption, ranked in descending order, were: United States, 238 lb; Canada, 219 lb; Australia, 209 lb; and Argentina, 192 lb (see Fig. 2-10).

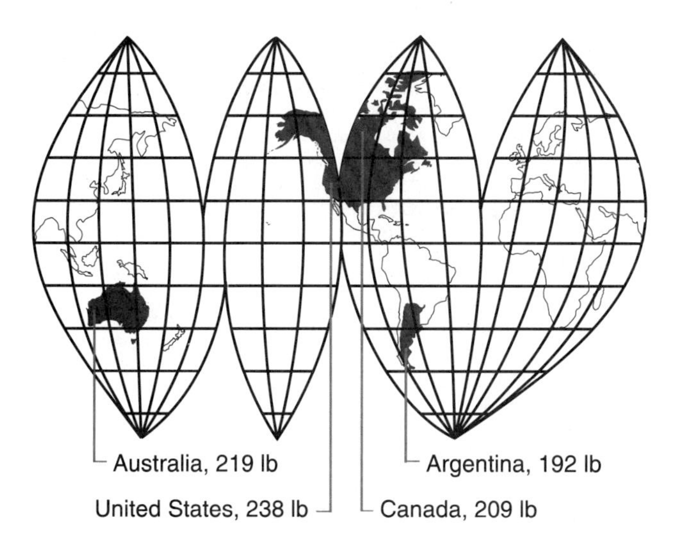

Fig. 2-10. Geography and per capita consumption of world's leading red meat and poultry eaters in 1994. One pound equals 0.45 kg. (Source: *Livestock and Poultry World Markets and Trade*, FAS-USDA, October, 1994, pp. 53-55 and 93)

MEAT CONSUMPTION IN MAJOR PRODUCING COUNTRIES

Pounds per Capita per Year

	U.S.	Australia	New Zealand	Canada	European Union 11 (Average)
Lamb, Mutton, and Goat	1.3	44.0	55.0		13.3
Poultry	71.5	53.2		58.0	31.7
Pork	67.5	39.8		73.0	91.5
Beef and Veal	97.5	81.6	63.6	78.0	44.8

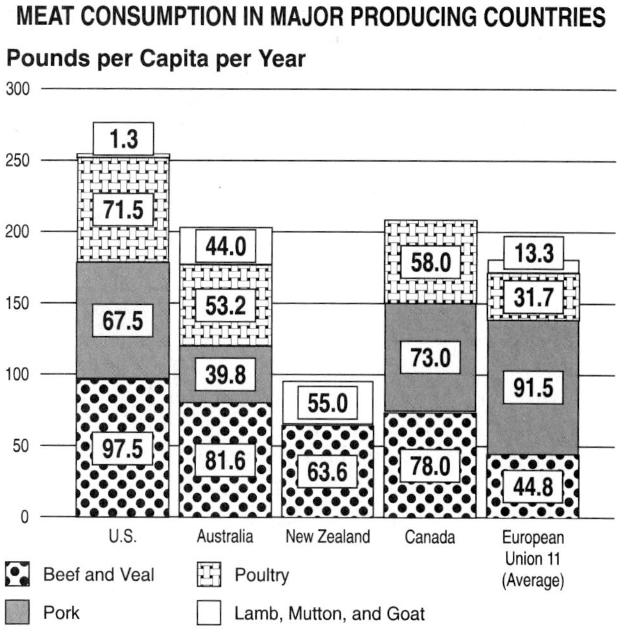

Fig. 2-11. Per capita meat consumption in carcass weight in major producing countries based on 1994 data. Poultry and pork consumption data not available for New Zealand; lamb, mutton, and goat data not available for Canada. One pound equals 0.45 kg. (Source: *Livestock and Poultry World Markets and Trade*, FAS-USDA, October, 1994, pp. 53-55 and 93)

In general, meat consumption is highest in countries that have extensive grasslands, temperate climates, well-developed livestock industries, and that are sparsely populated. In many of the older and more densely populated regions of the world, insufficient grain is produced to support the human population when consumed directly. This lessens the possibility of keeping animals, except for consuming forages and other inedible feeds. Certainly, when it is a choice between the luxury of meat and animal by-products or starvation, people will elect to accept a lower standard of living and go on a grain diet. In addition to the available meat supply, food habits and religious restrictions affect the kind and amount of meat consumed.

Fig. 2-11 presents the per capita consumption of meat, including poultry, in the major producing countries, by showing the proportion that the consumption of beef and veal contribute.

FACTORS AFFECTING WORLD BEEF CONSUMPTION

The major factors which influence the total demand for beef are (1) the increase in human population, and (2) the increase in the buying power of the population. The gross domestic product (GDP), which is highly related to domestic buying power, is expected to rise considerably in the years ahead, especially in the developed nations. The rate of growth of the national income in the developing nations is expected to be less than that in the developed nations.

People with low incomes consume relatively low priced carbohydrates. With higher incomes, higher quality foods, notably beef, are eaten. Thus, per capita beef consumption is a good barometer of GDP and standard of living of a country. Beef is a status symbol.

However, national demand for beef and veal is not a simple matter of merely multiplying the human population times the per capita gross domestic product. For example, meat prices are clearly a factor in determining variations in consumption levels. Also, such barriers to world trade as transport costs, import and export taxes, and tariffs prevent equalization of meat prices between countries.

Major meat exporters such as Uruguay, Argentina, New Zealand, and Australia have lower meat price levels and their people consume more meat in relation to income levels than the rest of the world. Also, in the South American countries, exports have been restricted to ensure adequate supplies, hold domestic prices down, and keep traditional high consumption levels intact.

Countries such as the United States, Canada, and the United Kingdom, which have internal grain prices at world levels and generally free access to their meat markets, can be considered to have meat consumption levels in undistorted relation to their income levels. Here, beef and pork prices are influenced by world grain prices and meat imports from other sources can compete freely.

Countries with sufficient protection in the grain and/or meat sector to put consumer meat prices above world levels—such as the European Community (EC) countries and Switzerland—have consumption levels below what disposable income would indicate.

Some countries maintain very tight import controls, often through quotas and/or high tariffs, and the resultant very high meat prices offset higher income levels, causing per capita meat consumption to lag. Japan and Sweden are two such countries.

In addition to price and income, traditional eating habits influence meat consumption levels.

WORLD BEEF PRICES

U.S. consumers concerned over rising food prices need not feel alone. Their views have been echoed around the world, from Tokyo to Bonn, and with increased fervor as inflation continues to mount (see Fig. 2-12). Governments have responded with stiffer price controls and, in some cases, freer import policies. But halting the food price spiral remains an elusive goal, complicated by such problems as soaring demand for meat and high-quality products desired by people as they become more affluent, crop failures, and protective trade policies of the European Union (EU), Japan, and other countries and regional groups.

Meat prices, beef and veal in particular, account for much of the increase in food prices. Such large price rises have not occurred in other meats; and the price of poultry meat has even declined. Of course, beef spiraled in price because people all over the world are eating more beef, and they are able and willing to pay for it.

Beef prices in many other countries are higher than in the United States. Because of differences in cuts and quality, prices are not strictly comparable, but Fig. 2-13 gives the retail beef prices in selected cities around the world.

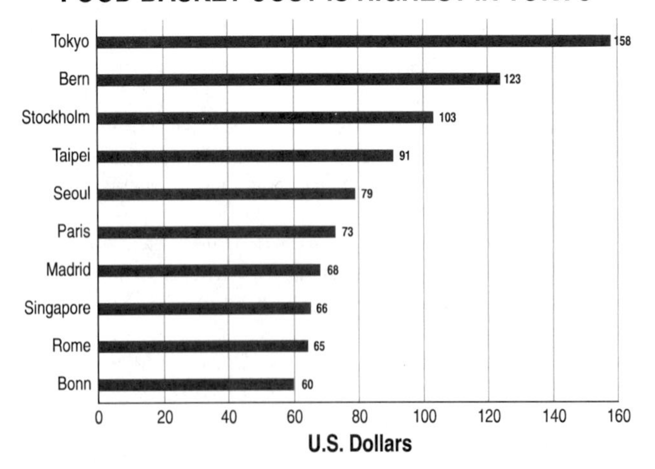

Fig. 2-12. The food basket contains 15 staple food items and tends towards foods found in a western diet. (From: *Trade Highlights*, January 1993, FAS-USDA, p. 13)

Fig. 2-13. Average retail price in U.S. dollars for a pound of sirloin steak in selected capital cities, 1993. (From: *Trade Highlights*, January 1993, FAS-USDA, p. 16)

When considering the comparative price of beef or other foods between countries, two points are pertinent:

1. **The cost of living in terms of working time required to purchase food.** For example, the working time to buy a sirloin steak varies from 45 to 48 minutes in the United States and Canada to 1.7 hours in France and 4.4 hours in Japan.

2. **The proportion of income spent for food.** As income rises, consumers spend a smaller proportion of their disposable income for food. U.S. consumers spend about 12% of their disposable income on food, compared with 53% in the Philippines (see Fig. 2-14).

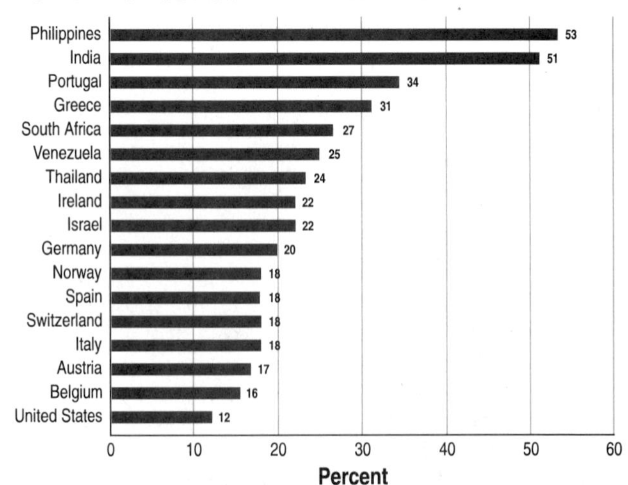

Fig. 2-14. Percentage of total expenditures (disposable income) spent on food in various countries. (From: *Statistical Abstract of the United States 1994*, U.S. Dept. of Commerce, p. 864, Table 1371; United States data from USDA)

WORLD BEEF CONSUMPTION

Fig. 2-15 shows the world's beef eaters, based on per capita consumption of beef and veal in 1993.

In terms of total consumption of beef and veal (Table 2-4), the major consuming nations are much the same as the major producing nations. Per capita beef consumption, however, shows a considerably different type of ranking of nations. Uruguay ranks first, with Argentina second, United States third, Australia fourth, and Canada fifth (Table 2-6).

The world production of beef and veal totaled 50,509,000 metric tons in 1994, of which the United States produced 11,199,000 metric tons, or about 22.2% of the world production.[2]

Tables 2-5 and 2-6 show the meat production and

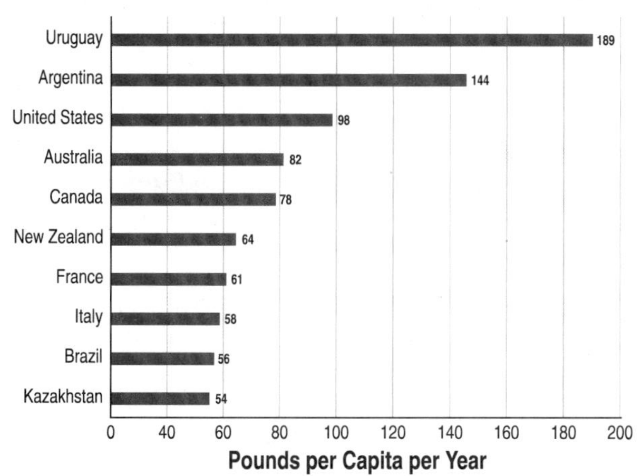

Fig. 2-15. Leading beef and veal consumers, in pounds per capita, 1994. One pound equals about 0.45 kg. (Source: *Statistical Abstract of the United States 1995*, p. 858, Table 1380)

TABLE 2-4
MAJOR BEEF- AND VEAL-CONSUMING NATIONS[1]

Country	Consumption 1993	Consumption Avg. of 1990–1992	Production 1993	Production Avg. of 1990–1992
	(1,000 metric tons)			
United States	11,019	11,090	10,584	10,537
European Union	7,517	7,506	7,798	8,461
Brazil	4,294	4,295	4,614	4,519
Russian Fed.	3,389	4,132	3,359	3,977
Argentina	2,273	2,244	2,550	2,607
China mainland	2,184	1,381	2,337	1,532
Mexico	1,805	1,777	1,710	1,677
Japan	1,302	1,135	593	572
Canada	954	991	883	909
India	825	827	945	924

[1]*Livestock and Poultry: World Markets and Trade*, October 1994, FAS-USDA, p. 56.

TABLE 2-5
BEEF AND VEAL PRODUCTION COMPARED TO
TOTAL MEAT PRODUCTION IN SPECIFIED COUNTRIES[1]

Country	Total Meat Production[2]	Beef and Veal Production[3]	Beef and Veal as a Percentage of All Meat Production
	(1,000 MT)	*(1,000 MT)*	*(%)*
China mainland	44,814	2,253	5.0
U.S.A.	32,965	11,199	34.0
Brazil	8,080	3,160	39.1
Russian Fed.	7,475	3,350[4]	44.8
France	6,140	1,667	27.1
Germany	5,789	1,500[4]	25.9
India	4,117[4]	1,292[4]	31.4
Italy	4,030	1,180	29.3
Spain	3,765	472	12.5
Australia	3,355	1,825	54.4
Japan	3,334	605	18.1
United Kingdom	3,285	877	26.7

[1]*FAO Production Yearbook 1994*, United Nations.

[2]*Ibid.* ,p. 209, Table 97 (includes buffalo, horse meat, and poultry in addition to beef, veal, pork, and lamb).

[3]*Ibid.*, p. 197, Table 92.

[4]Estimated.

TABLE 2-6
ANNUAL PER CAPITA BEEF AND VEAL CONSUMPTION
COMPARED TO PER CAPITA CONSUMPTION OF ALL MEATS
IN SELECTED COUNTRIES[1]

Country	1994 Per Capita Consumption: All Meat (lb)	All Meat (kg)	Beef and Veal (lb)	Beef and Veal (kg)	Beef and Veal as a Percentage of All Meat (%)
United States	238	*108*	97.5	*44.3*	41.0
Australia	219	*99*	81.6	*37.1*	37.3
Canada	209	*95*	78.1	*35.5*	37.4
Argentina	192	*87*	143.0	*65.1*	74.5
Uruguay	189	*86*	188.0	*85.8*	99.4
France	181	*82*	61.2	*27.8*	33.8
Germany	175	*79*	41.1	*18.7*	23.5
Italy	170	*77*	57.6	*26.2*	33.9
New Zealand	119	*54*	63.6	*28.9*	53.4
Brazil	112	*51*	55.7	*25.3*	49.7

[1]*Livestock and Poultry: World Markets and Trade*, FAS-USDA, October 1994, pp. 53, 54, 55, 93.

[2]Based on data from *FAO Production Yearbook 1994*, Vol. 48, p. 197, Table 92.

per capita consumption of meats in selected countries and the favored position of beef.

WORLD BEEF TRADE

Meat is produced principally for domestic markets; only about 5% of carcass meat enters world trade. In terms of value, agricultural products account for 25% of world trade, and meat and livestock about 3%. Nevertheless, meat has had a tremendous impact on international trade. Countries and people struggling to catch up, once they approach affluency, want meat. This has led to increased meat prices and trade. A long-term upward trend is noted for meat prices; much more than for any other class of agricultural product. World red meat exports rose from 3,175,000 metric tons to 12,072,000 metric tons from 1961 to 1994. Out of the 1994 total, about 58% of the world trade was for beef and veal.

Most movements of cattle are among European countries, though there are fairly large movements from Canada and Mexico into the United States. But carcass meat is the major component of world meat and livestock trade.

Basically, the trade in world meat reacts to the law of supply and demand, but with several country-to-country exceptions such as the following: protecting their own livestock industry by quotas and levies; agreements on nonagricultural products, which may hinder the normal flow of meat; differences in currency values; problems in transportation and storage; and even political decisions.

WORLD BEEF AND VEAL EXPORTS

Beef and veal are the most important meats in world trade. In 1994, bovine meat accounted for 58% of world exports of red meat (see Table 2-7).

The major beef- and veal-exporting nations are, with the exception of the EC, countries with extensive land areas and low labor costs. They are listed by rank in Table 2-7.

Australia alone accounts for 17% of the total beef and veal exports, while Australia, U.S.A., and New Zealand account for about 33% of all beef and veal exports.

Most of the beef exported by the EC is traded between member countries. Beef exports are along two major trade routes—Oceania (Australia and New Zealand) to North America and South America to the United Kingdom and the EC.

Beef exporters are divided into two foot-and-mouth disease groups—those that have it, and those that are free of the disease. Australia, New Zealand, and the North American countries north of the Panama Canal are the primary foot-and-mouth disease-free exporters. Argentina, Brazil, and Uruguay are the principal foot-and-mouth–infected exporting countries.

TABLE 2-7
MAJOR BEEF- AND VEAL-EXPORTING COUNTRIES[1, 2]

Country	Beef and Veal Exports		Beef and Veal as a Percentage of Red Meats
	Average 1990–1994	1994	
	- - - - - *(1,000 metric tons)* - - - - -		*(%)*
Australia	1,132	1,155	83.0
United States	572	685	76.4
New Zealand	425	464	48.0
Brazil	389	450	99.0
Argentina	341	290	97.0
Canada	163	245	33.8
China mainland	156	175	32.9
Uruguay	113	70	98.0
India	108	130	77.5
Austria	61	10	100.0
Total	6,863	7,002	58.0

[1] *Livestock and Poultry: World Markets and Trade*, FAS-USDA, October 1994, p. 56.

[2] 1992 from *Agricultural Statistics 1993*, USDA, p. 281, Table 442.

All of the beef- and veal-exporting nations are projected to increase their exports in the years ahead. Increased exports will be achieved primarily by higher calving percentages, lower mortality rates, marketing animals at younger ages, and finishing more cattle prior to slaughter, rather than greatly increasing cattle numbers. In an effort to keep up their exports in the early 1970s, Argentina, Uruguay, and Colombia had *meatless weeks and meatless days* to keep down domestic consumption.

WORLD BEEF AND VEAL IMPORTS

The major beef- and veal-consuming nations are also the major importing and the richer nations (see Table 2-8).

The EC, the United States, and Japan account for 77% of all beef imports. The United States is the largest importer of beef, importing 14% of all beef traded. As a unit, the EC is the largest importer of beef and veal, importing about 51% of all beef traded. Imports into the EC countries reflect both increased imports from overseas and the growth of intra-Community trade as the Common Agricultural Policy integrated the European market in beef and veal. The United States suspended meat import quotas in June 1972, as a means of encouraging more imports. Japan has been a growing market for beef and veal, particu-

TABLE 2–8
MAJOR BEEF- AND VEAL-IMPORTING COUNTRIES[1]

Country	Beef and Veal Imports		Beef and Veal as a Percentage of All Red Meats[2]
	Average 1990–1994	1994	
	- - - - - *(1,000 metric tons)* - - - - -		*(%)*
EC	2,083	2,124	40
United States	1,166	1,089	78
Japan	639	829	42
Canada	239	300	100
Korea	156	170	81
Egypt	129	153	90
Brazil	115	60	100
Mexico	103	110	65
Taiwan	55	58	100
Total	5,340	5,388	51

[1]*Livestock and Poultry: World Markets and Trade*, FAS-USDA, October 1994, p. 56.
[2]1992 from *Agricultural Statistics 1993*, USDA, p. 281, Table 442.

larly from Oceania (Australia and New Zealand). Since the mid-1960s, internal demand pressures in Japan have resulted in expanded import beef quotas, followed by sharply increased imports, although the nation's beef imports are still comparatively small.

Future U.S. beef imports will depend on the government's decision on the size of the quota or import restraint program.

BEEF TRADE PROBLEMS

The chief beef trade problems relate to the trade barriers of the main importing countries of North America and Western Europe and the quality and continuity of supplies from the exporting countries.

TRADE BARRIERS

It is particularly important to appreciate the reasons for trade barriers and the ways in which they operate. The two main purposes of trade regulations are: (1) to protect consumers from health risks; and (2) to protect the cattle industries of importing countries from both animal diseases and low-priced imports.

■ **Health and veterinary regulations**—In all the main meat-importing countries, imports are subject to licensing. To obtain a license, the slaughtering facilities of the exporting country must satisfy the sanitary regulations of the importing country. This also applies to processing plants in the case of processed meat. Thus, importing countries try to protect their consumers from the possibility of food-borne diseases and food poisoning. Veterinary regulations are designed to prevent bringing in animal diseases which could have serious economic consequences for the importing country.

■ **Tariffs and quotas**—In the developed countries, the income of farmers and ranchers tends to be lower than those of other professions. A prime objective of the agricultural policies is to raise farmers' incomes. Generally this is attempted by introducing measures designed to raise the prices which domestic farmers receive above those prevailing in world markets. In the United States, quotas and tariffs exist.

Quotas limit the number of pounds of meat and the number of animals brought into a country. For example, U.S. legislation of August 1964 established a basic limit on meat plus an added factor based on the nation's beef and veal production.

Tariffs are duties, or charges (on a per pound basis in the United States), imposed by a government on imported meat, designed to raise the prices which cattle producers receive above those ruling in world markets.

BEEF SUPPLIES FOR EXPORT

Some of the main beef-exporting countries in the Southern Hemisphere operate under two major handicaps:

1. Endemic cattle diseases, which are difficult and expensive to eradicate.
2. Serious periodic droughts, which not only prevent cattle fattening but deplete herds.

RECENT DEVELOPMENTS IN THE BEEF TRADE

Formerly, most beef sold on the world market was in the form of half or quarter carcasses. Processing into joints took place in the importing country. In recent years, more jointing has taken place before export. This development has been accelerated by importing countries as a result of (1) the spread of supermarkets selling prepackaged meats, and (2) the United Kingdom's ban, imposed in 1968, on imports of "bone in" beef from countries where foot-and-mouth disease is endemic (on the belief that the virus is most likely to be introduced in bone marrow, offal, and lymphatic glands).

In the main importing countries, the beef market is becoming increasingly split into two separate components: (1) high-quality fresh beef for direct sale to consumers; and (2) beef for processing, known as manufacturing grade beef, suitable for soups, quick-frozen foods, and hamburgers. Intensive feeding techniques, such as the grain-finished cattle of the United States, result in beef which is unsuitable for manufacturing—it is too fat. On the other hand, beef from the developing nations is generally grass fed, lean, and ideal for manufacturing. Also, countries with cattle disease problems can supply manufacturing beef in cooked or frozen form, which is sterile and will not spread disease. Thus, in the United States, and to a lesser but increasing extent in other beef-importing countries, most of the high-quality fresh beef comes from domestic sources, and more and more of the manufacturing grades come from imports.

EUROPEAN COMMUNITY (EC)

The European Community (EC), or European Economic Community (EEC), built from the ruins of World War II, is one of the great trade centers of the world. It all began in 1950 when Belgium, West Germany, Italy, Luxembourg, the Netherlands, and France pooled their coal and steel production under a common high authority, known as the European Coal and Steel Community (ECSC). This experiment was so successful that the six countries soon extended their cooperation to cover their economies as a whole, and the peaceful application of nuclear energy. They created the European Economic Community (EEC) or Common Market (commonly abbreviated Economic Community, or EC) and the European Atomic Energy Community (Euratom) in 1958.

Subsequently, four other European nations—the United Kingdom, Ireland, Denmark, and Greece—joined the original six member nations, bringing the total to 10—the EC-10. Subsequently, Spain and Sweden joined. It is now referred to as the European Union (EU-12).

The Common Agricultural Policy (CAP) has been described as the engine of the Common Market. The CAP seeks to ensure Community farmers a fair standard of living, stable markets, and improved methods. Its main features are (1) free trade in farm produce; (2) common price supports to raise and maintain farmers' incomes; (3) variable levies at the Community frontier, which adjust the price of imported farm products to the general level of Community prices as set by the Council of Ministers; and (4) a long-term, jointly financed plan to modernize farms and consolidate land-holdings that are too small to be efficient.

BEEF AND VEAL IN THE EC

The final step toward a common market in beef and veal was taken in August 1968, with the fixing of a single guide price for all member countries, and a common external rate of duty on imports. The cornerstone of the program as it relates to beef and veal is the regulation of quantities coming onto the market in such a way that market prices approximate a predetermined guide price. This is done principally by influencing imports through the variable levy system, which discourages imports when market prices are low relative to the guide price, and allows freer access when they are above the guide price. In addition, imports from most-favored nations (MFN) are subject to a duty of 1¢ per lb for live cattle and 2¢ per lb for beef and veal.

When, despite the discouragement of imports through the levy system, producer prices fall significantly below the guide price, intervention purchases are made by government-sponsored intervention agencies, thus effectively providing a floor below which market prices will not fall.

Levies on beef and beef cattle are fixed every week, comprising the difference between the price at which the consignments are imported (including duties) and the guide price, but the proportion of the levy payable by the importer depends on the state of the home market. This is assessed on representative market prices, usually known as reference prices, which are calculated in each member country and then brought together as a Community reference price, weighted for the size of the cattle population in each country. Thus, depressed markets in one country do not have a disproportionate effect on the Community reference price.

Today, the EC is the most important importer of beef in the world, importing more than 42% of all beef traded (see Table 2-8). Most of the beef exported by the EC is traded between member countries. For example, France uses hindquarters of domestic cattle for steaks and roasts and exports fores to Germany and Belgium (for sausage) and to the United Kingdom; then imports hinds from these countries to supplement domestic production, and, in addition, buys boneless beef cuts from South America.

CATTLE AND BEEF IN THE GLOBAL PERSPECTIVE

Rising incomes around the world are creating higher standards of living and increased demands for the good things of life, including animal protein—and to most people this means beef. This, along with more people to feed and constraints in expanding cattle production, is causing supply to lag behind demand.

WORLD BEEF DEMAND AHEAD

In recent years, per capita real incomes have risen in almost every nation on the globe. More calories—merely satisfying hunger—have first call on income. Above this point, a part of this income is spent for increased quantities of animal protein. For this reason, it is the developed countries that are demanding more beef.

■ **Japan**—The Japanese are turning away from their traditional fish diet, primarily because of greater affluency. It is expected that Japan will continue to be a growing market for beef and veal, particularly from Australia and New Zealand.

■ **European Community (EC)**—The EC is the most important importer of beef. Likely the real income in EC countries will increase each year and the demand for meat will rise each year. Domestic production can support a slight increase in demand for beef and veal. Increased beef production is being accomplished by shifting away from traditional calf and veal slaughter to older cattle, by feeding bulls, and by confinement finishing. However, the EC will always be a major beef importer, relying on South America, Australia, and New Zealand to fill the gap between its domestic supply and its demand.

■ **Eurasia**—With the disintegration of the former U.S.S.R., the economy of the Republics and the Baltics deteriorated. Until the economy improves, Eurasia will import precious little beef.

■ **United States**—The United States is the world's largest single nation beef importer. With a growing population, the demand for beef will continue to increase in the United States. Part of this demand will be met by increasing the imports of manufacturing-type beef suitable for hamburgers, hot dogs, and luncheon meats.

U.S. beef production cannot respond quickly to demand and price changes in the years ahead. The economics of cattle finishing and the consumer preference for lean beef do not favor grain feeding, calf slaughter has been greatly reduced, and the number of cull dairy cows slaughtered will stabilize below the level of the first half of the 1980s. Hence, most increased beef in the United States must come from more beef cows kept to produce more feeder calves. People do those things which are most profitable to them; and cattle producers can produce more beef, provided it makes money. Too many consumers fail to realize that a time lag exists in responding to prices. For example, a heifer cannot be bred until she is about 1½ years of age, the pregnancy period takes another 9 months, and, finally, the young are usually grown 6 to 12 months before being sold to cattle feeders, who finish them from 4 months to a year. Thus, under the most favorable conditions, this biological manufacturing process cannot be speeded up; it requires about 4 years in which to produce a new generation of market cattle.

MEETING DEMAND

As the world demand for beef increases, the major question is: How will this demand be met?

■ **Developing nations**—Right off, it would appear that there is great potential for increasing beef production in many of the developing nations—in Africa, Latin America, and the Far East. Most of them have a surplus of forages. Also, they desperately need more high-quality protein in the diet, and they need to improve their income. However, several bottlenecks restrict fast increases in these areas: (1) lack of infrastructure—railroads, roads, marketing facilities, slaughterhouses, and refrigeration; (2) periodic droughts; (3) diseases and parasites which take huge tolls of the animal population and/or restrict expansion; (4) lack of soil testing facilities, irrigation, and fertilizers; (5) inadequate agricultural extension service and coordination between teaching, research, and extension; (6) insufficient transportation, storage (including refrigeration), and markets; and (7) insufficient investment capital.

Rapid increases in beef production in the developing countries are difficult. Due to the extremely low fertility and late maturity of the cattle in many of these countries, they are forced to slaughter some of the heifers in order to maintain an adequate offtake rate to meet demand rather than being able to keep these heifers to build up their cattle population. In the developing countries, the main efforts to increase beef production need to be directed so as to accomplish the following: (1) to increase the fertility of the breeding animals: (2) to decrease death losses; (3) to decrease the age at slaughter; and (4) to increase the average carcass weight. This calls for improvement in nutrition (better pastures, especially during dry seasons, and/or supplementation), management, health, and selection. Such improvements require both capital and knowledgeable personnel, both of which are generally in short supply in the developing nations.

■ **Developed nations**—Quick mobilization for increased beef production can come about in the developed countries more readily than in the developing countries because they have the necessary infrastructure (transportation, marketing, slaughtering, and refrigeration), the capital for investment; adequate-sized breeding herds, and the management support and knowledge necessary for bettering production. An example of this mobilization is the dramatic growth in cattle feeding in the United States.

■ **Where will the beef come from?**—Perhaps, in the final analysis, increased beef will come from both the developing and the developed nations. Among the developing nations having high potential for increasing beef production are Argentina, Brazil, Uruguay, Paraguay, and Colombia. However, because of the presence of foot-and-mouth disease in each of these countries, the major importing countries do not wish to take their beef unless it is deboned, cooked, or pasteurized. Consequently, the countries which have an immediate advantage of increasing exports are Australia, New Zealand, the Central American countries, Mexico, Ireland, the United States, and Canada.

WORLD PROTEIN SITUATION

Diogenes, a Greek philosopher, when asked about the proper time to eat, said: "If a rich man, when you will; if a poor man, when you can." This statement is particularly applicable to protein foods. Well-fed nations are scrambling to eat better, and hungry nations to eat at all.

Protein malnutrition is the most serious and common cause of infant mortality and general debility in developing countries, and among the poor in developed countries. The diseases kwashiorkor (primarily protein deficiency) and marasmus (severe undernourishment) directly or indirectly account for 3 to 10 times greater infant mortality and as much as 20 to 50 times higher death rate among 1- to 4-year-old children in certain African countries than in the industrialized regions of the world. Children below 5 years of age may account for 40% of the total mortality. The low protein reserves of the body and generally poor nutritional status cannot sustain those children who develop fevers from respiratory and gastrointestinal infections, which markedly increase the body losses of protein.

Food and Agriculture Organization (FAO) figures show food supplies in the world are sufficient to furnish 68 g of protein per person per day, with 70% of it coming from plant foods and 30% from animal sources. This is enough to meet minimum needs if it were properly distributed according to individual needs of each person. But food supplies of different countries vary widely. Many of the same countries and regions that are short on animal protein and protein in general are also short on energy-rich foods, especially during the dry season, because of low production and limited food storage facilities.

Proteins from animal sources (meat, milk, and eggs) have a higher biological value than plant sources because they contain more of the essential amino acids (protein building blocks) needed for growth. They are rich in the amino acids lysine and methionine, in which vegetable sources are deficient. Also, they contain the essential vitamin B_{12}. Thus, the addition of even a small amount of animal protein will greatly improve the value of a diet rich in cereal grains, beans, or other plant foods. Soybean protein also has a high value, provided it is properly heat treated to improve its availability. But no soybean platter, no matter how

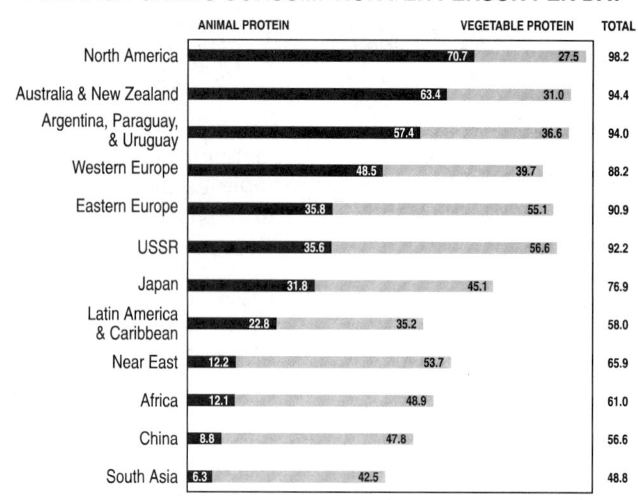

Fig. 2-16. Average grams protein consumption per person per day, with a breakdown into animal and vegetable protein, by geographic areas and countries. (From: *Ceres*, Food and Agriculture Organization of the United Nations, Vol. 8, No. 3)

well disguised, can ever inspire the toast or impart the status symbol of a roast of beef or a sizzling steak.

BEEF CATTLE AS A SOURCE OF PROTEIN

The primary reason for producing beef cattle is to obtain human food. Beef is good, and good for people. In addition to supplying the highest quality protein (along with milk and eggs), it is a rich source of the B vitamins. Some plant foods are deficient in these factors. Vitamin B_{12} does not occur in plant foods; only in animal sources and fermentation products. Beef is also a particularly rich source of iron, and the iron availability of beef is twice as high as in plants.

Some opinions to the contrary, there is as much protein in Prime beef as in the leanest piece. When finishing animals, fat merely replaces the moisture, without materially altering the protein content.

The efficiency of animals in converting feeds to food and their place in the economical production of human foods are subjects of much discussion. Some plant scientists, economists, and others, claim that food needs should be met by plant sources entirely—that animals should be eliminated. This might be desirable if animals subsisted entirely on cereal grains and other edible foods, or if animals and feeds were grown on land needed to produce foods for human consumption. However, much of the land of the world is unsuited to the production of food crops. In temperate climates, this nonarable land is used to grow cattle, sheep, and goats. These ruminants can eat pasture and coarse roughages that have no value except as animal feed. Thus, they contribute high-quality protein (meat), which is greatly prized for its nutritive qualities, plus its flavor and appetite appeal.

NEW PROTEIN SOURCES

New protein sources are being used, and others will be discovered. However, people like beef, so they will pay more for it than for plant food, or even other kinds of meats. There is an old saying in the southwestern part of the United States that "thin beef is better than fat beans." Nevertheless, no known nutritive essential exists in beef which cannot be provided from fish, vegetable sources, or by synthesis. Some promising developments include:

- **Fish protein**—The present world fish catch runs about 88 million metric tons per year, with an average protein content of about 15%. Fish fillets contain over 20% protein. They are one of the richest sources of protein. Each year, some 25 million metric tons of fish are processed into fish meal for animal feed. With improved handling and processing techniques, some of this fish could be converted into fish protein concentrate (FPC) suitable for human consumption. FPC has great potential as an inexpensive source of high-quality protein. It is produced from types of fish that are not popular in the usual channels of fresh fish trade. The fish are extracted to remove oil, dried, and ground to make a bland meal containing about 80% protein, 0.2% fat, and 13% mineral. Although nutritious, it is bland, tasteless, and odorless.

- **Plant protein concentrates**—Isolation of protein from oilseeds, nuts, and leaves provides another method of increasing the supply of quality protein. The proteins remaining following the extraction of oil from soybeans, cottonseeds, sunflower seeds, peanuts, and coconuts have been used as protein supplements in animal feeds for a number of years. For humans, the oil-free residue of soybeans holds great promise since it can be used to make soy grits, soy flour, and soy protein concentrates (SPCs). These SPCs are high in protein and can be used in a variety of products which include baked goods, dietary foods, meat products, and infant formulas. Moreover, soybean protein may be spun into fine filaments, flavored, and colored to resemble chicken, ham, or beef—a product referred to as texturized vegetable protein (TVP). Also, these proteins may be extruded to form ground meat extenders and meat analogs. Research is being conducted on the similar development of protein isolates from cottonseeds and peanuts.

  Isolating protein from plant leaves may provide a protein source which could be used as a complementary protein in countries where it is too rainy to dry seed crops. Green leaves are among the best sources of protein. By pressing the leaves, a protein-containing juice can be obtained which may be coagulated and dried forming a product that is 50% protein. This product is called leaf protein concentrate (LPC).

- **Genetic improvement of grains**—Scientists at Purdue University developed a special corn called Opaque-2 in which the incomplete protein, zein, is reduced to ½ the normal amount, while the complete protein, glutelin, is doubled. Moreover, lysine and tryptophan concentrations are increased about 50%, and leucine and isoleucine are in a better balance. Feeding trials on both animals and humans have demonstrated the increased value of Opaque-2. Breeding programs have also developed (1) high lysine sorghum and barley, (2) high protein rice and wheat, and (3) a cross between wheat and rye, called triticale. These genetic improvements bring plant protein quality closer to that of the animal proteins, and are very promising.

- **Single-cell protein (SCP)**—This refers to protein obtained from single-cell organisms or simple multicellular organisms such as yeast, bacteria, algae, and fungi. The most popular and most familiar of these are brewers' yeast and torula yeast, both of which are marketed. The potential for producing single-cell protein (SCP) is tremendous. Furthermore, many of the industrial by-products with little or no economic value may be used as a growing media. Some of these by-products include petroleum products, methane, alcohols, sulfite waste liquor, starch, molasses, cellulose, and animal waste. Still with all its potential, problems such as (1) safety, (2) acceptability, (3) palatability, (4) digestibility, (5) nutrient content, and (6) economics of production need to be solved.

- **Amino acid supplementation**—Since lysine is the limiting amino acid in wheat and other grains, the addition of limited amounts of lysine to cereal diets improves their protein quality. Indeed, studies in Peru and Guatemala demonstrated that growing children benefited by this addition. However, in most countries more can be gained by focusing on increasing the food supply in general.

UNITED STATES CATTLE AND BEEF

U.S. cattle producers and beef consumers must consider cattle and beef in the global perspective. To plan for the future, cattle producers need to know which countries are potential competitors, and they need to know historical and current trends. Beef consumers need to count their blessings!

Beef and veal demand, despite some ups and downs, will keep the beef industry in a favorable position, on a worldwide basis, for many years to come.

The role of the United States in worldwide beef and veal production and consumption is emphasized by the following salient points:[3]

1. The United States possesses only 5% of the world's human population (see Table 2-1).

2. The United States possesses only 7.0% of the world's cattle population (see Table 2-2).

3. The United States produces 22.1% of the world's beef and veal. (1994 figures from *FAO Production Yearbook 1994*, FAO, p. 197)

4. The United States consumes 25% of the world's beef and veal (see Table 2-4).

Think of it! Less than ⅒ of the world's cattle population producing more than 22% of the world's beef and veal. This shows the efficiency of U.S. cattle producers.

Fig. 2-17 shows that, from 1940 to 1975, the general trend (1) for U.S. cattle numbers on farms, and (2) for beef and veal production was upward, following which it leveled off; and that, for more than 50 years, U.S. cattle numbers and beef and veal production almost paralleled each other. Furthermore, the efficiency of producing beef has increased in the United States. For example, the pounds of liveweight of beef cattle marketed per breeding female amounted to 220 lb in 1925, 310 lb in 1950, 482 lb in 1975, and 524 lb in 1990.[4] Without doubt, much of this increased beef production on a per head basis has come about as a result of increased cattle feeding and decreased calf slaughter. The number of U.S. cattle on feed (full ration of grain or other concentrates) increased from 4,390,000 in 1950 to 14,433,000 in the peak year of 1973—they more than tripled during this 23-year period. Since 1973, the number of grain-fed cattle decreased until about 1982, following which it leveled off (Fig. 2-18).

But the United States is declining in its relative importance in human population, cattle numbers, and beef and veal production and consumption. Human population of other countries of the world is growing more rapidly. Also, their per capita disposable income is rising, and with it have come newly found affluent life-styles and demand for more beef. As a result, people all over the world are demanding more beef, and they are able and willing to pay for it. Not only are beef prices in many countries higher than in the United States, but they have gone up faster in most countries. These forces will continue to operate at an accelerated rate in the years ahead. Translated into reality, decreased supplies and increased prices are making for beef shortages in many countries. But cattle

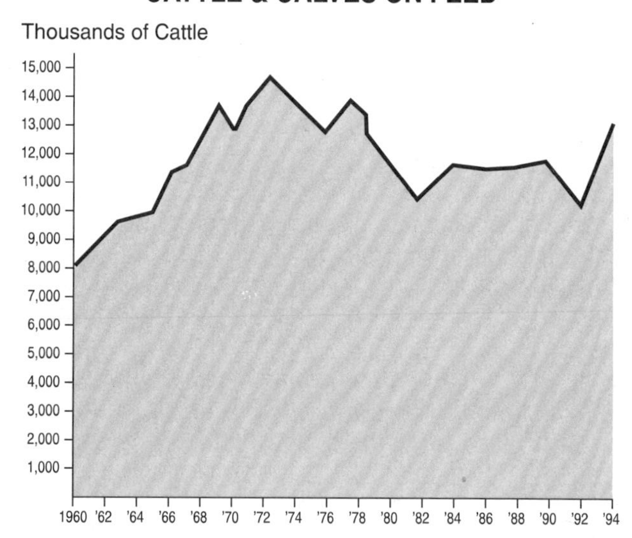

Fig. 2-18. Cattle and calves on feed in the United States, 1960-1994. (From: *Agricultural Statistics 1995–96*, USDA, p. VII-7, Table 377)

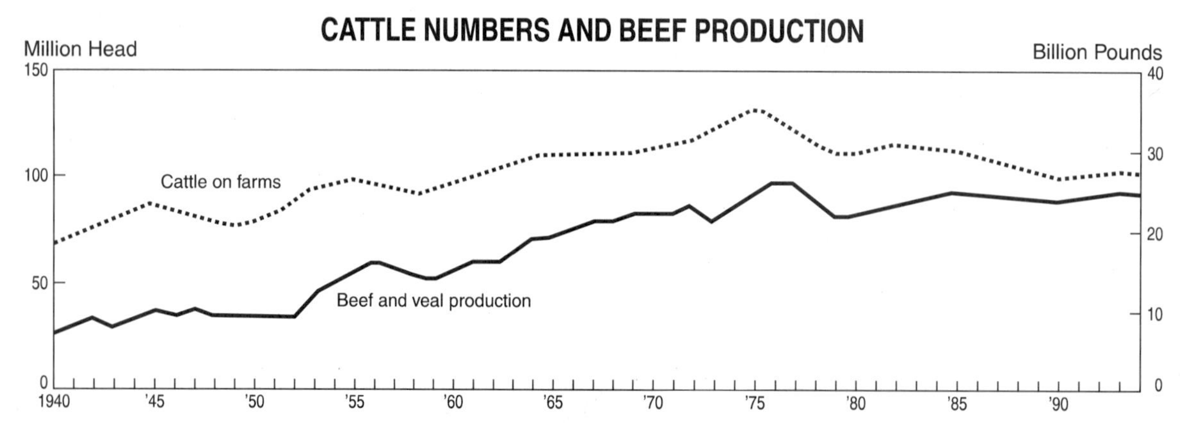

Fig. 2-17. Historical account of U.S. cattle numbers and beef and veal production. (From: *Agricultural Statistics 1995-96*, USDA, p. VII-2, Table 372, and p. VII-47, Table 436)

[3]Based on data from FAO, United Nations, and FAS-USDA.

[4]*Foods from Animals: Quantity, Quality and Safety*, Council for Agricultural Science and Technology (CAST), Rpt. No. 82, Mar. 1980, p. 13, Table 2; and USDA sources.

producers throughout the world are demonstrating their willingness to respond to prices. As never before, they are applying science and technology; in genetics, in management, and in the use of humanly inedible feeds and nonprotein nitrogen.

U.S. CATTLE PRODUCTION

The present and future importance of beef cattle in the agriculture of the United States rests chiefly upon their ability to convert coarse forage, grass, and by-product feeds, along with a minimum of grain, into a palatable and nutritious food for human consumption. As grain becomes scarcer and higher in price (due to increased human consumption), there will be less grain-fed beef and more grass-finished and short-fed cattle.

Data in Table 2-9 show the recent 20-year trend in U.S. cattle numbers, total value, and value per head. Peaks and valleys are evident as with many agricultural products. The value of cattle was low in 1975 when total numbers were high.

Fig. 2-19 further describes the U.S. cattle population. Since 1985, dairy cows that have calved have averaged about 10.4 million head; beef cows that have calved have averaged about 35 million head; and other beef and dairy cattle, including replacements, steers, heifers, and bulls, have averaged about 58 million head.

TABLE 2-9
TRENDS IN U.S. CATTLE NUMBERS AND VALUE[1]

Year	Total Number	Value	
		Total	Per Head
	(1,000s)	*($1,000)*	*($)*
1973	121,539	30,583,562	252
1975	132,028	20,999,808	159
1977	122,810	25,249,390	206
1979	110,864	44,697,773	403
1981	114,321	54,292,044	475
1983	115,001	46,708,350	406
1985	109,749	44,138,618	402
1987	102,000	41,482,765	407
1989	99,180	60,234,219	607
1991	98,896	64,661,865	654
1993	100,892	65,481,884	649
1995	102,755	63,156,538	615

[1]*Agricultural Statistics 1995–96*, USDA, p. VII-2, Table 372.

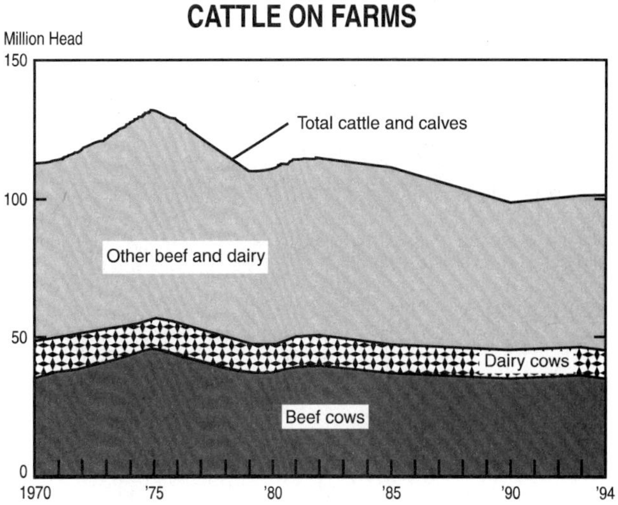

CATTLE ON FARMS

Fig. 2-19. Number by types of cattle on U.S. farms since 1970. (Courtesy, USDA)

The production of beef cattle differs from that of most other classes of livestock in that the operation is frequently a two-phase proposition: (1) the production of stockers and feeders; and (2) the finishing of cattle. In general, each of these phases is distinctive to certain areas.

COW-CALF PRODUCTION

Cow-calf production refers to the breeding of cows and the raising of calves. In this system, the calves run with their dams, usually on pasture, until they are weaned, and the cows are not milked.

Fig. 2-20 shows the geographic location of the nation's beef brood cows.

Approximately 53% of the U.S. beef cows are in the West North Central area (Missouri, Kansas, Nebraska, Iowa, South Dakota, North Dakota, and Minnesota) and the South and East (from Maryland, West Virginia, and Kentucky south and westward to and including Arkansas and Louisiana). The greatest expansion since 1950 has been in this area.

Some rather characteristic production practices are common to each cattle area:

- **West North Central area**—This seven-state area, accounting for about 27% of the nation's beef cows, embraces the four most westerly Corn Belt states (Missouri, Iowa, Kansas, and Nebraska) plus Minnesota and North and South Dakota. In many respects, it is a variable area. Iowa, north Missouri, eastern Kansas and South Dakota, and southern Minnesota are noted for fertile soil, medium-sized farms, high-priced land, and corn. Here hogs compete with cattle and other classes of livestock for the available feeds. The western portion of the area—commonly referred

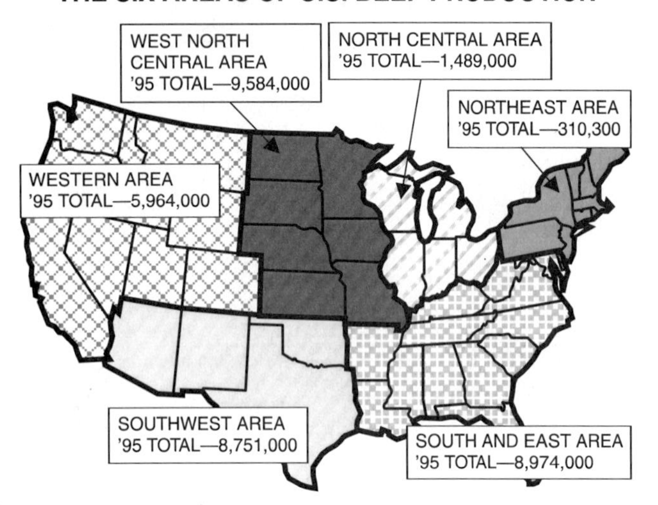

Fig. 2-20. The six areas of U.S. beef production, and the number of beef cows in 1995. In each area, the methods followed are determined largely by the climate, feed supply, and economic conditions. (Based on data from *Agricultural Statistics 1995–96*, USDA, p. V11-15, Table 389)

to as the Plains area—is predominantly range and wheat land. Beef cow numbers in the Northern Plains area (Nebraska, and North and South Dakota) more than doubled during the period 1950-70, with the greatest increase in the humid, eastern portion of the area. Included in the West North Central area are 2 famous pasture areas—the Sand Hills of Nebraska, and the Flint Hills of Kansas.

Much of the growth in the West North Central area came through (1) converting more marginal cropland to pasture, and (2) more effective utilization of relatively low-quality roughages, like cornstalks, to winter feed. The continuation of this shift will depend upon the relative profitability of grain crops vs beef cattle.

Many of the farmers and ranchers of the West North Central area maintain small commercial breeding herds, the offspring of which are sold as stockers or feeders, or finished out on homegrown feeds. Pastures furnish practically all the feed for the breeding herds from May to November, and cornstalks and other roughages are utilized as winter feeds.

In addition to keeping small breeding units, many West North Central farmers make a regular practice of buying feeder cattle from the western ranges. These cattle are usually purchased in the fall of the year and obtained at auctions, through public stockyards, through dealers, or direct from the range. Some of these are roughed through the first winter by utilizing stubble or stalk fields and then pastured the next season. Others are carried on a program of winter feeding, then sold in the spring to go into feedlots.

The West North Central area is also noted for its excellent purebred herds. Because of its proximity to the western ranges and the demands of the ranch owners for bulls, there has always been a good market for superior breeding stock.

■ **South and East area**—The South and East, with about 26% of U.S. beef cows, is the second most populous beef cattle area. Mild winters, adequate rainfall, and year-round grazing have encouraged many of the farmers of the area to turn to beef production. Other factors that have contributed greatly to the expansion of beef cattle in the area include: (1) the infusion of Brahman breeding, resulting in cattle better adapted to high summer temperatures and to resist the insects of the area; (2) improved pastures, suitable supplements for mineral deficiencies, and control of parasites; (3) adoption of better management practices; (4) increase in size of farms, with larger pasture acreage per farm; (5) increase in part-time farmers, who frequently choose beef cow enterprises because of the relatively small and flexible labor needs; and (6) more market outlets for feeder cattle to go into large finishing lots. All these improved conditions, together with year-round grazing, would indicate that the South and East area offers the greatest potential for future increase of beef cows of any area of the United States, and that the South and East will continue to expand in beef production on a sound basis. Some authorities estimate that this area could support 30 million beef cows—nearly three times present numbers, provided (1) current land was properly managed, and (2) idle and unproductive acres were put to work.

Factors that may have a restraining influence on beef cattle growth in the South and East are high cotton and soybean prices and shifts of land to nonfarm uses.

■ **Southwest area**—The Southwest has been a noted cow-calf area since the days of the Texas Longhorn. Grazing is the highest and best use for much of the land in this area. Except for some areas under irrigation, the units of operation are generally very large in size. In recent years, a noted cattle feeding area has developed in the High Plains area of Texas and Oklahoma. The future rate of expansion in beef cattle in the Southwest will be limited because it is already a well-stocked and highly specialized cattle area, with about 25% of the U.S. beef cattle. Modest increases in cow-calf numbers will come from improvement in pasture production, and some increase in cattle feeding may be expected.

■ **Western area**—The Western area, with about 17% of U.S. beef cattle, ranks fourth in cattle numbers. It is characterized by great diversity of topography, soil, rainfall, and temperature. Accordingly, the amount of vegetation and the resulting carrying capacity are variable factors. Combinations of private and public grazing land often prevail.

In general, the western ranges supply an abundance of cheap grass, but only a limited amount of grain. Under these conditions, the cow-calf system is the dominant type of enterprise.

Likely, beef cow numbers in the mountain region will expand at about the same rate as for the nation as a whole. Little change is expected in the arid portions of the West, because of current full utilization of existing forages. Likewise, little expansion of beef cattle numbers may be expected in the Pacific region because of competing land uses and rangeland limitations.

■ **North Central area**—The North Central region accounts for only 4.2% of the nation's beef cows, despite the inclusion of the three Corn Belt states of Illinois, Indiana, and Ohio. The area does have considerable production potential for grass-legume forages; and it is well adapted to beef production. However, so long as dairying remains profitable in the Lakes States, shifts to beef will be minimal. Likewise, the most fertile soils of Illinois, Indiana, and Ohio will continue to be devoted to crops which yield greater economic returns, primarily corn and soybeans, with only the marginal cropland left for grass and beef cattle. However, cornstalks will be more effectively utilized in the future.

■ **Northeast area**—The production of beef cattle was formerly an important and highly developed industry through the Northeast. However, the opening up of the western ranges; the rapid increase in population of the East, with the resulting industrialization; the division of the eastern farming lands into smaller units; and the adoption of more intensive systems of farming caused a decline of beef production in the eastern states. With these changes in economic conditions, beef production was to a large extent supplanted by dairying. With the existence of these favorable conditions for milk production, beef cattle cannot compete with dairy cattle, particularly on small farms. On the other hand, where the distance to market is too great, or where labor difficulties exist, beef cattle have a place. But the beef industry in the Northeast can never regain its former magnitude nor hold the place that it does in the rest of the United States.

Today, the Northeast area accounts for only about 0.9% of U.S. beef cattle. Although beef cow numbers in the area have increased slightly since 1950, the rise has not been sufficient to offset the decline in milk cow numbers. The beef herds of the Northeast are small and often operated as a supplementary enterprise by part-time farmers. Hence, the area is of minor importance in beef production.

LEADING STATES IN COW-CALF PRODUCTION

A ranking of the 10 leading states in beef cattle production is presented in Table 2-10.

TABLE 2-10
TEN LEADING STATES IN BEEF COW NUMBERS[1]

State	Cow Numbers
	(1,000s)
Texas	6,000
Missouri	2,105
Oklahoma	1,952
Nebraska	1,885
South Dakota	1,660
Montana	1,559
Kansas	1,509
Kentucky	1,165
Tenessee	1,130
Florida	1,130
U.S. Total	35,156

[1] *Agricultural Statistics 1995–96*, USDA, p.VII-15, Table 389.

Texas is by far the leading state. The large numbers of cattle in Texas may be attributed to its great range area, to the immense size of the state, and to increased cattle finishing.

CATTLE FEEDING AREAS

The center of cattle feeding has shifted from the Corn Belt to the West and Southwest (see Table 2-11). In 1995, about 80% of the fed cattle marketed in the 23 major producing states were fed in the 15 western states. The greatest expansion occurred in Texas, with the Lone Star State being the leading cattle feeding state in 1995. Two Corn Belt states rank second and third respectively; namely, Kansas and Nebraska. Colorado and Iowa rank fourth and fifth nationally, with California ranking sixth.

U.S. BEEF CONSUMPTION

If poultry consumption is separated into chicken and turkey, the annual per capita consumption of beef is more than that of any other meat; however, the consumption of beef has declined in recent years and the consumption of poultry has increased. Table 2-12 lists the annual per capita consumption of meat (fish and poultry included) since 1984, and compares the annual per capita consumption of beef, pork, poultry, fish, lamb and mutton, and veal. Figs. 2-21 and 2-22 further illustrate some of the shifts in U.S. food consumption since 1976.

From Table 2-12 and Figs. 2-21 and 2-22, the following consumer trends may be noted:

1. Beef has been the preferred red meat in the

TABLE 2-11
LEADING CATTLE FEEDING STATES, BY RANK[1]

State	Number on Feed January 1
Texas	2,380,000
Kansas	2,040,000
Nebraska	1,940,000
Colorado	990,000
Iowa	910,000
California	400,000
Oklahoma	380,000
South Dakota	340,000
Minnesota	310,000
Illinois	280,000
Idaho	270,000
Ohio	225,000
Arizona	210,000
Michigan	210,000
Indiana	200,000
Washington	156,000
New Mexico	155,000
Wisconsin	150,000
Wyoming	100,000
Oregon	100,000
Montana	100,000
North Dakota	100,000
Pennsylvania	80,000
Total (for 23 states)	12,026,000

[1]*Agricultural Statistics 1995–96*, USDA, p. VII-7, Table 376.

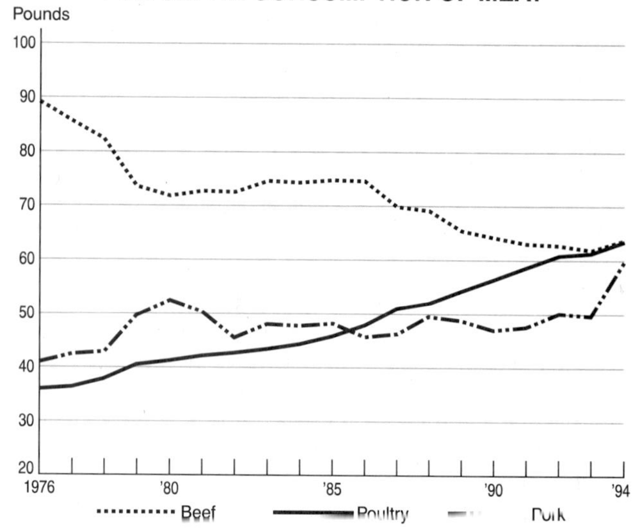

Fig. 2-21. Shifts in the per capita consumption of meat—beef, pork, and poultry—since 1976. In 1976, the per capita consumption of beef, pork, and poultry was 89 lb, 41 lb, and 36 lb, respectively. (Courtesy, USDA)

United States, having replaced pork in this position in the early 1950s.

2. Up to 1976, the trend in beef consumption was upward. Since 1977, beef consumption has trended downward.

3. Consumption of all meats (red meats, poultry and fish) has remained rather stable since 1984.

4. Consumers are eating more poultry, which picks up some of the losses in beef and pork consumption.

5. Consumers are eating more sugars and sweeteners, fruits, and vegetables.

TABLE 2-12
U.S. ANNUAL PER CAPITA MEAT CONSUMPTION[1]

Year	All Meats		Beef		Pork		Poultry		Fish		Lamb and Mutton		Veal	
	(lb)	*(kg)*	*(lb)*	*(kg)*	*(lb)*	*(kg)*	*(lb)*	*(kg)*	*(lb)*	*(kg)*	*(lb)*	*(kg)*	*(lb)*	*(kg)*
1984	182	*82*	73.9	*33.5*	47.2	*21.4*	43.7	*19.8*	14.1	*6.4*	1.1	*0.5*	1.5	*0.7*
1985	185	*84*	74.6	*33.8*	47.7	*21.6*	45.2	*20.5*	15.0	*6.8*	1.0	*0.5*	1.5	*0.7*
1986	185	*84*	74.4	*33.7*	45.2	*20.5*	47.1	*21.4*	15.4	*7.0*	1.0	*0.5*	1.6	*0.7*
1987	184	*84*	69.6	*31.6*	45.6	*20.7*	50.7	*23.0*	16.1	*7.3*	1.0	*0.5*	1.3	*0.6*
1988	186	*85*	68.6	*31.1*	48.8	*22.1*	51.7	*23.5*	15.1	*6.8*	1.0	*0.5*	1.1	*0.5*
1989	185	*84*	65.4	*29.7*	48.4	*22.0*	53.6	*24.3*	15.6	*7.1*	1.0	*0.5*	1.0	*0.5*
1990	183	*83*	64.0	*29.0*	46.4	*21.0*	56.0	*25.4*	15.0	*6.8*	1.0	*0.5*	0.9	*0.4*
1991	185	*84*	63.1	*28.6*	46.9	*21.3*	58.0	*26.3*	14.8	*6.7*	1.0	*0.5*	0.8	*0.4*
1992	189	*86*	62.8	*28.5*	49.5	*22.5*	60.0	*27.2*	14.7	*6.7*	1.0	*0.5*	0.8	*0.4*
1993	190	*86*	61.5	*27.7*	48.9	*22.0*	62.6	*28.2*	14.9	*6.7*	1.0	*0.5*	0.8	*0.4*

[1]*Agricultural Statistics 1985*, p. 497, and *1995-96*, p. XIII-5, Table 653.

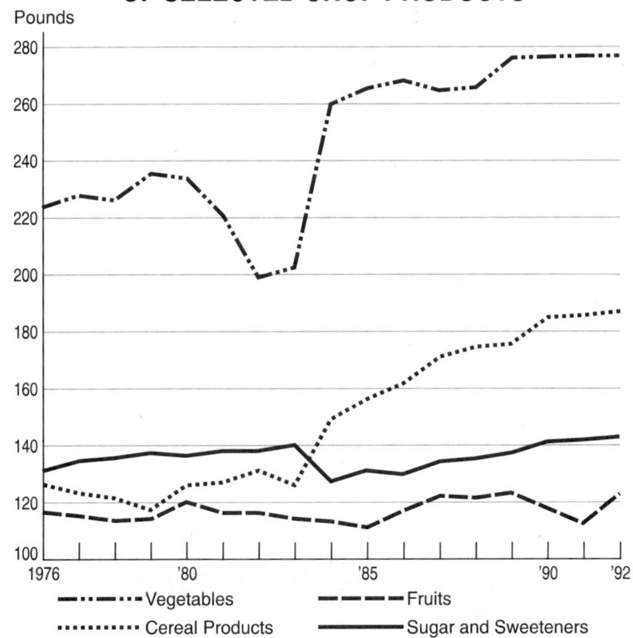

Fig. 2-22. Shifts in the per capita consumption of selected crop products since 1976. In 1976, the per capita consumption of sugar and sweeteners, fruits, vegetables, and cereal products was 131 lb, 116 lb, 224 lb, and 126 lb, respectively. (Courtesy, USDA)

Seemingly small changes in the per capita consumption of beef, or of any other product, are reflected in terms of production.

U.S. BEEF EXPORTS

Cattle producers are prone to ask why the United States buys cattle, and beef and veal, abroad. Conversely, consumers sometimes wonder why we export beef and veal. Occasionally, there is justification for such fears—on a temporary basis and in certain areas.

As shown in Table 2-13, we have imported far more beef than we exported.

The amount of beef exported from this country is dependent upon: (1) the volume of beef (and other meats) produced in the United states; (2) the volume of beef (and other meats) produced abroad, and (3) the price of and trade restrictions on beef abroad. With increased buying power abroad and higher prices for beef than in the United states, it is probable that more beef will be exported in the future.

U.S. exports of animals and animal products consist largely of lard and tallow, hides and skins, red meats, variety meats, live animals, and other livestock products such as wool and mohair (see Fig. 2-23).

TABLE 2-13
U.S. EXPORTS AND IMPORTS OF BEEF AND RED MEATS[1]

	Exports			Imports		
Year	Beef and Veal	All Red Meats	Beef and Veal as a Percentage of all Red Meats	Beef and Veal	All Red Meats	Beef and Veal as a Percentage of all Red Meats
	(1,000 metric tons)		*(%)*	*(1,000 metric tons)*		*(%)*
1983	93	417	22.3	642	912	70.4
1984	112	416	26.9	595	972	61.2
1985	110	431	25.5	677	1,130	59.9
1986	184	503	36.6	704	1,161	60.6
1987	211	535	39.4	745	1,240	60.1
1988	229	678	33.8	748	1,251	59.8
1989	382	813	47.0	709	1,091	65.0
1990	348	732	47.5	763	1,141	66.9
1991	406	870	46.7	783	1,129	69.4
1992	449	975	46.1	804	1,106	72.7
1993	425	996	42.6	794	1,143	69.5

[1]*Agricultural Statistics 1995–96*, USDA, p. VII-49, Table 438; and p. VII-50, Table 441.

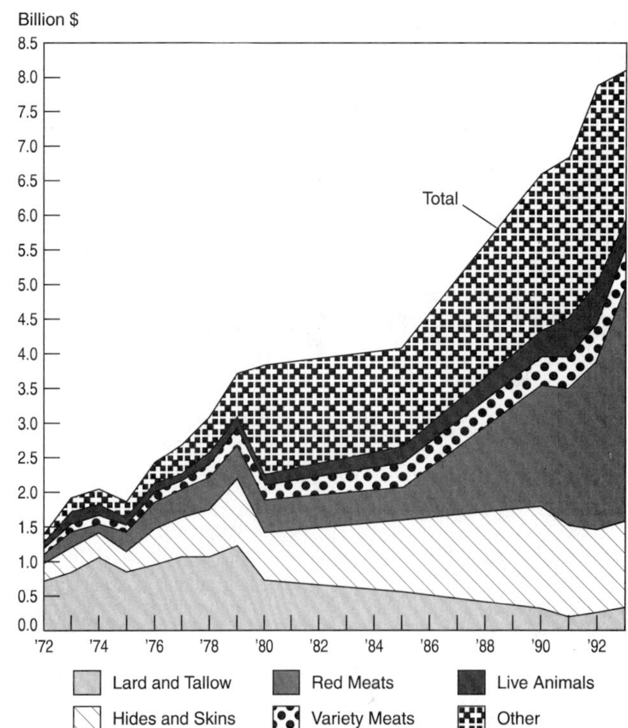

Fig. 2-23. Value of U.S. exports of livestock and livestock products since 1972. (Source: *Agricultural Statistics 1985*, p. 513, Table 704; *1988*, p. 508, Table 696; *1995–96*, p. XV-7, Table 671)

In dollar value, in 1993 animals and animal products accounted for 24.2% of all agricultural commodities imported and 18.5% of all agricultural commodities exported.[5] In 1993, live cattle, beef and veal, cattle hides, and tallow accounted for 45% of the animals and animal products exported; or almost $3.6 billion.[6]

Although not large in dollar sales when compared to other commodities, exports are extremely important to the cattle industry, even though most producers fail to recognize this fact.

The annual U.S. exports and imports of live cattle, given in dollar value, for the five year period 1989 to 1993, are given in Fig. 2-24. As noted, imports exceed exports. Most of the cattle exported are for breeding purposes, with the principal destinations being Canada and Mexico.

VALUE OF U.S. EXPORTS AND IMPORTS OF LIVE CATTLE

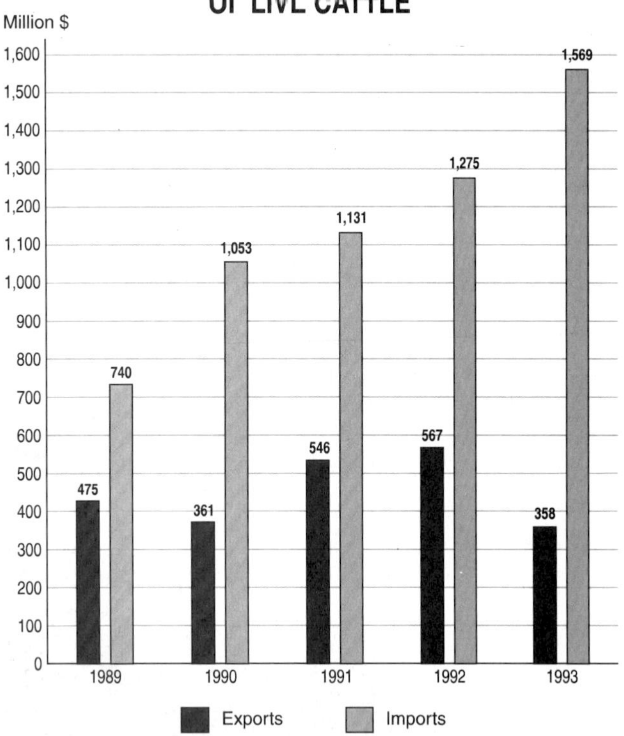

Fig. 2-24. The yearly exports and imports of live animals 1988-1993; given in dollar value. (Source: *Agricultural Statistics 1995–96*, USDA, pp. XV-7 and XV-8, Tables 671 & 672)

U.S. BEEF IMPORTS

Table 2-13 reveals that the United States imports more beef than it exports. Fig. 2-25 places beef imports and production in perspective. As shown, total beef imports have never exceeded 11% of our domestic production of beef.

BEEF AND VEAL IMPORTS AS A PERCENTAGE OF U.S. PRODUCTION

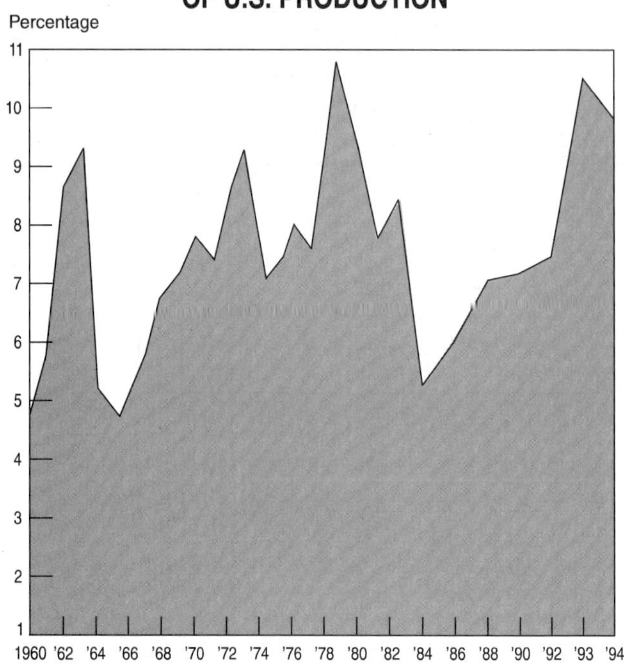

Fig. 2-25. Beef and veal imports as a percentage of U.S. production, 1960 to 1994. Basic production and import figures are expressed in carcass-weight equivalents. (Source: Annual *Agricultural Statistics*, USDA, reporting the years on which Fig. 2-25 is based)

The amount of beef imported from abroad depends to a substantial degree on (1) the level of U.S. beef (and meat) production, (2) consumer buying power, (3) cattle prices, (4) quotas and tariffs, and (5) need for manufacturing-type beef.

Fig. 2-26 shows that beef and veal dominate red meat imports. An average of 65% of all meats imported consists of beef and veal (see Table 2-13).

In dollars, U. S. agricultural imports represent a yearly average of about $22.4 billion, but only 4.7% of all commodities imported by the United States.[7] Meat and meat products, largely beef and veal, account for an average of 12.2% of all agricultural imports or almost $2.7 billion, while live cattle account for only 4.3% of agricultural imports or a yearly average of $966 million.[8]

---

[5]*Agricultural Statistics 1995–96*, USDA, p. XV-8, Table 672, and p. XV-7, Table 671.

[6]*Ibid.*, p, XV-7, Table 671.

[7]*Agricultural Statistics 1995–96*, USDA, p. XV-8, Table 672.

[8]*Ibid.*

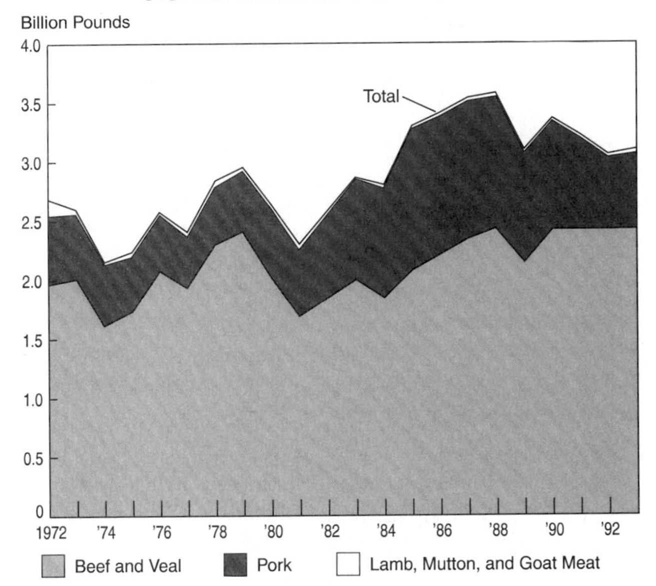

Fig. 2-26. U.S. imports of red meats since 1972, showing the dominant position of beef and veal. One billion pounds equals about 454,000 metric tons. (Source: *Agricultural Statistics 1993*, USDA, p. 284, Table 444; and *Agricultural Statistics 1995–96*, p. VII-53, Table 444)

Table 2-14 lists the amounts of the various kinds of beef and veal imported into the United States in recent years. Most of the fresh or frozen beef and veal (about 75%) comes from Australia and New Zealand, and most of the canned or other prepared or preserved meat (92%) comes from Brazil and Argentina.[9] Canada, Costa Rica, Denmark, Mexico, Uruguay, and Yugoslavia also contribute some to the U.S. imports of beef and veal.

TABLE 2-14
YEARLY U.S. IMPORTS OF BEEF AND VEAL BY TYPE OF PRODUCT[1]

Year	Product		
	Fresh or Frozen	Canned	Other Prepared or Preserved
	(metric tons)[2]		
1989	638,999	68,710	1,434
1990	694,163	67,054	1,520
1991	709,997	71,570	1,870
1992	728,922	72,631	2,112
1993	720,079	71,351	2,995
5-year avg.	698,432	70,263	1,986

[1]*Agricultural Statistics 1995–96*, USDA, p. VII-50, Table 441.

[2]One metric ton equals about 2,205 lb.

Because of restrictions designed to prevent the introduction of foot-and-mouth disease, neither fresh nor salted refrigerated beef can be imported to the United States from South America; beef importations from these countries must be canned, cooked, or fully cured (*e.g.*, corned beef).

But why import so much beef and so many cattle into the United States? This question has three basic answers:

1. **To provide supplies when domestic beef is in short supply and high.** Judiciously increasing imports of beef and cattle during times of scarcity and high prices, as an alternative to pricing beef out of the market, may be of some virtue. But, of course, beef imports can be overdone.

2. **To meet the demand for manufacturing meat.** Manufacturing-type beef is the kind that is boned and used in making hamburgers, franks, sausages, and bologna. The new generation of young people often prefers hamburgers to steaks. Thus, the demand for manufacturing meats began to exert itself in the early 1960s and was the primary factor in influencing the demand for imports. Hand in hand with this increase in demand, the domestic source of manufacturing meat—Utility and Canner dairy and beef cows, and older bulls—began to dwindle. Dairy cow numbers leveled off and fewer old bulls were needed in artificial insemination breeding. Per capita slaughter of cows tended downward from 40 to 50 lb per capita (liveweight basis) during the early 1950s to around 30 lb per capita in the late 1960s, but steer and heifer beef production increased steadily, with a slight tendency to vary cyclically.[10] In 1981, of the beef that was federally graded, 90% graded Choice. In 1994, only 65% graded Choice. These and other beef quality grades are shown in Fig. 2-27. *Note well:* Quality grades were revised in 1987, and quality and yield grades were uncoupled in 1989. So, the two pie diagrams are based on different standards and are not comparable.

To fill the gap for manufacturing beef, created by the increasing demand for hamburger (and other manufactured beef) and the decreasing domestic supply of Cutter and Canner cattle, the United States increased importations of frozen boneless beef, particularly from Australia and New Zealand.

3. **To provide more feeder cattle.** Beginning about 1950, the United States embarked upon an expansion program to produce more high-quality beef, primarily in the Choice and Select grades. The result

[9]*Ibid.*, p. VII-50, Table 440.

[10]Ehrich, R. L., and M. Usman, *Demand and Supply Functions for Beef Imports*, Ag. Exp. Sta. Bull. 604, Division of Agricultural Economics, University of Wyoming, Laramie, Jan. 1974, p. 5 Chart 2.

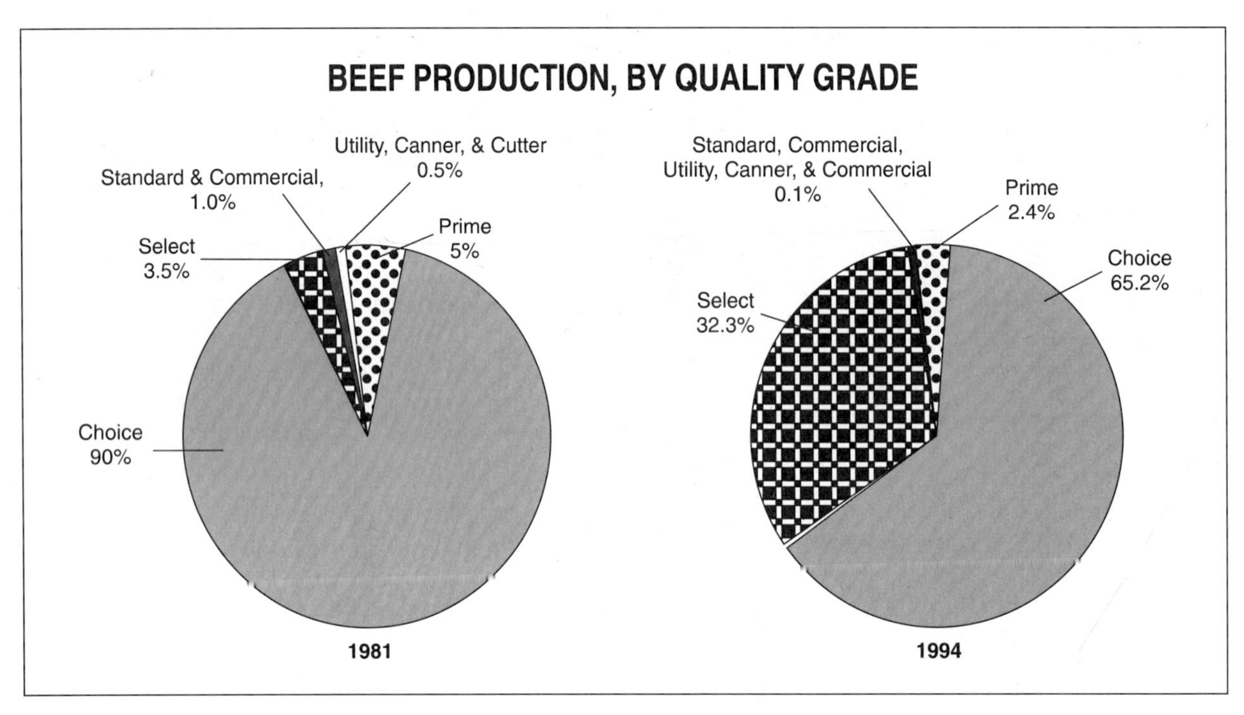

Fig. 2-27. U.S. beef production, by quality grade, 1981 vs 1994. (From: Agricultural Marketing Service, USDA)

was a spectacular rise in grain-fed cattle, increasing the demand for feeder calves faster than could be satisfied by domestic cow-calf producers. To meet this demand, increased importations of feeder cattle were made from Mexico and Canada. Most live cattle are imported for feeding and breeding purposes and relatively few are destined for immediate slaughter. The majority of feeders come from Mexico and Canada, but some breeding cattle enter the United States from a number of countries.

QUOTAS AND TARIFFS

U.S. beef quotas and tariffs are for the purpose of encouraging domestic cattle production through protecting it from competition from foreign sources.

In the early 1960s, cattle producers became deeply concerned over rising imports of beef. Instead of holding at the historical 1 to 5% level of U. S. beef production, imports spiraled to more than 10% of U.S. production. A number of factors caused the increased imports, the most significant of which was that the U.S. market was the most lucrative outlet in the world. Although the U.S. State Department favored voluntary agreements with major exporters, American cattle producers insisted on more stringent regulations. Finally, Congress enacted the Meat Import Law (Public Law 88-482) in August 1964, to become effective January 1, 1965. This bill provided for import quotas, based on a formula, for fresh, chilled, and frozen beef, veal, mutton (not lamb), and goat meat—including both carcass and boneless meat. It did not include lamb or canned meats. The law was for the purpose of limiting annual imports of the specified meats to a level comparable to the designated base period 1959– 1963, with an adjustment, or *growth factor*, based on changes in domestic production relative to the base period.

The base quota was established as the average annual quantity imported during the base period (1959–1963), which was 725,400,000 lb or 4.6% of domestic production during those years. Each year, the growth factor is determined by calculating the percentage by which the estimated U.S. commercial production of the specified meats in the current calendar year and the 2 years preceding (*i.e.*, a 3-year average) exceeds (or falls short of) the average annual U.S. production during the base period. Thus, to determine the growth factor at the beginning of a given year, it is necessary to estimate production for that year. Then, the calculated growth factor is multiplied times the base quantity to determine the amount of increase (or decrease) in the base. This increase (or decrease) added to the base gives an adjusted base quota. The Act allows a 10% leeway above the adjusted base quota before quotas are applied to individual countries. Thus, a quota trigger point is determined at 110% of the adjusted base quota.

The Secretary of Agriculture is required to estimate at the beginning of each quarter year the quantity of prospective imports. If the estimated quantity of prospective imports exceeds the trigger point, the president is required to invoke a quota on imports of these meats. In case quotas are imposed, the total import quota is allocated among the countries from whom the United States is importing on the basis of shares supplied by those countries during a representative period.

The law does contain provision under which the president is empowered to suspend or increase quotas when it is deemed in the best interest of the nation to do so because (1) of overriding economic or national security interest, (2) the supply of meat is inadequate to meet domestic demand at reasonable prices, or (3) international agreements have been entered into which will have the same effect as the Act.

In 1970, it appeared that the trigger point would be exceeded. Thereupon, the president, as required by the Act, issued a proclamation to place a limitation on imports of meats covered by the Act. At the same time, the president suspended the limitation, as authorized by law, after determining that this action was required by the overriding interest of the United States. Thus, quotas were triggered, but immediately rescinded.

Probably voluntary agreements will be the chief means of controlling future imports. But the existence of the law may have considerable psychological effect on negotiations.

The current quotas and tariffs of live animals and meats are given in Table 2-15.

TABLE 2-15
U.S. QUOTAS AND TARIFFS OF LIVE ANIMALS AND MEATS[1]

Import Item	Quotas (no. head/ year)[2]	Tariff (or duty)		
		1[3]	LDDC[4]	2[5]
Animals for breeding[6]	None	Free		Free
Cattle:				
Cattle weighing:				
under 200 lb	200,000	1¢/lb		2.5¢/lb
between 200 and 700 lb		1¢/lb		2.5¢/lb
Dairy cattle weighing:				
over 700 lb		Free		3¢/lb
Other cattle[7]	400,000	1¢/lb		3¢/lb
Beef and veal (fresh, chilled, or frozen)[7]	(see footnote 8)	2¢/lb		6¢/lb

[1] *Tariff Schedules of the United States Annotated (1987)*, USITC Publication 1906, United States International Trade Commission, Washington, D.C.

[2] Includes Canada, Mexico, and all other countries.

[3] Products of Canada and all other countries not designated LDDC or 2.

[4] Products of Least Developed Developing Countries.

[5] Products of communist countries.

[6] Must be purebreds of a recognized breed and registered in a recognized registry book.

[7] For not over 400,000 head entered in the 12-months period beginning Apr. 1, in any year, of which not over 120,000 shall be entered in any quarter beginning Apr. 1, Oct. 1, or Jan. 1.

[8] Legislation of Aug. 1964 establishes a basic limit of 725.4 million lb *(329.0 million kg)* plus an added factor based on U.S. production.

U.S. AGRICULTURAL TRADE AND BALANCE OF PAYMENTS

Trade balance, or balance of trade, is the difference in value over a period of time between exports and imports of commodities. When exports exceed imports, a favorable balance of trade is said to exist. Conversely, when imports exceed exports, there is an unfavorable balance of trade.

Historically, agriculture makes important contributions to the United States international balance of payments. From the time of the first shipment of 2,500 lb of tobacco from Jamestown, Virginia, to England in 1616 until 1916—300 years later—agricultural products accounted for the bulk of our exports. As late as 1900, agricultural products accounted for ⅔ of total merchandise exports.

U.S. agricultural exports dropped drastically in the 1920s and 1930s as European countries strived for self-sufficiency in agriculture. With the onset of World War II, our agricultural exports again rose significantly. They declined in the early 1950s, then moved upward quite steadily through the 1960s.

Beginning in 1973, the importance of U.S. agricultural products in world trade was hailed as the cure for the United States' nagging trade deficit which had persisted since 1970. Historically, the U.S. agricultural trade balance has been favorable. Further, its importance in recent years has been increasing significantly (see Table 2-16).

As indicated in Table 2-16, agricultural exports exceeded imports by $17.1 billion in 1994, while the nonagricultural trade deficit amounted to $162.7 billion that same year.

U.S. farm commodities are increasingly being exported; and increased exports are usually accompanied by higher prices. In 1992, production from 2 out of every 5 acres and 23% of farm marketings were exported.

Although high prices abroad for agricultural products are one reason for agriculture's contribution to a favorable balance of trade, other forces have also played a prominent role:

1. A very tight world grain supply situation, due to unfavorable growing conditions in many areas of the world.

TABLE 2-16
U.S. TRADE BALANCE[1]

Item	Year					
	1984	1986	1988	1990	1992	1994
	(billion $)					
Exports:						
Agricultural	32.2	23.5	33.8	35.1	40.1	43.5
Nonagricultural	187.7	199.8	286.5	353.6	399.2	469.0
Imports:						
Agricultural	21.9	24.4	24.9	26.7	27.8	26.4
Nonagricultural	310.5	344.0	422.3	470.9	507.7	631.7
Trade Balance:						
Agricultural	+ 10.3	− 0.9	+ 9.7	+ 8.5	+ 12.2	+ 17.1
Nonagricultural	−122.8	−144.2	−135.8	−117.3	−108.5	−162.7

[1] *Statistical Abstract of the United States 1994*, U.S. Dept. of Commerce, p. 804, Table 1306; and *Agricultural Statistics 1995–96*, USDA, p. XV-2, Table 664.

2. Continued rapid improvement in world economic conditions.

3. Continued efforts to upgrade diets, especially in regard to protein foods, are altering the eating habits—and demand patterns—of people all over the world.

4. Expansion of trade with the former U.S.S.R. and the People's Republic of China.

5. The devaluation of the dollar, which made U.S. commodities less expensive in terms of most other currencies.

6. The availability of U.S. supplies and the capability to move large quantities of grains and oilseeds into the international market.

UNIFICATION OF THE U.S. BEEF INDUSTRY

In 1993, a Task Force of beef industry leaders developed a long range plan which included a recommendation that the following four major organizations serving the U.S. beef industry establish a unified structure: (1) the Beef Industry Council of the National Live Stock and Meat Board (BIC), (2) the Cattlemen's Beef Promotion Research Board (CBB), (3) the National Cattlemen's Association (NCA), and (4) the U.S. Meat Export Federation (MEF). The stated goal: "To recapture beef's lost market share and keep beef competitive in the next century." Also, the Task Force noted that a united organization would lessen causes for friction and divisiveness within the industry, and make for considerable saving in administrative costs—dollars which could be spent on improving beef's competitive position.

In January of 1996, the National Cattlemen's Association and the Beef Industry Council of the National Live Stock and Meat Board officially merged to form the National Cattlemen's Beef Association (NCBA). The U.S. Meat Export Federation voted not to join the new organization because of its multi-species nature and the possibility that it would no longer be able to receive federal funds if merged with the other organizations. The Cattlemen's Beef Promotion Research Board was not allowed to join because of the rules of the beef checkoff, which did not allow the mixing of checkoff funds with dues funds and also did not allow checkoff funds to be used for lobbying efforts.

In the new consolidated structure, the NCA, which previopusly represented 230,000 professionals, will no longer exist. It is giving up nearly 100 years of organizational history and the highest name recognition in the cattle industry in order to implement the unification of the beef industry. Likewise, the National Live Stock and Meat Board has a long and proud history of yeoman services to the beef, pork, and lamb industries, dating to its formation in 1922.

WHAT'S AHEAD OF THE U.S. BEEF INDUSTRY?

The authors' crystal ball shows the following ahead of the U.S. beef industry.

- **Declining cattle prices due to increasing competition and changing consumer eating habits**—By properly positioning themselves, however, cow-calf operations can be profitable during such a period. Here's how: (1) by cutting production costs through becoming more efficient, and (2) by adding value to your product, such as by selling bred replacement heifers instead of heifer calves, and/or retaining ownership of steer calves through the feedlot to slaughter. *Note:* In 10 of the 13 years from 1980 to 1992, it was profitable to retain ownership through the feedlot.

- **Increased competition from poultry**—More poultry, primarily broilers, will be consumed in the future. With it, beef consumption will trend downward.

- **Cow-calf expansion in the (1) West North Central, (2) North Central, and (3) South and East areas will remain favorable**—This prediction is based on more effective utilization of low-quality roughages, such as cornstalks, in the first two areas, and more improved pastures and year-round grazing in the South and East.

- **Th e Southern Plains and the Central Plains will continue to dominate cattle feeding**—The major cattle feeding areas will continue to be (1) the Southern

Plains (Texas and Oklahoma), and (2) the Central Plains (Colorado, western Nebraska, and western Kansas). But future growth of cattle feeding in the Southern Plains will be slower than in the period of 1960–1980 because of limitations in irrigation water and feed grains.

■ **Beef imports no great threat to U.S. cattle feeders**—Limited grain feeding potential in other countries precludes any real foreign threat to U.S. cattle feeders from the standpoint of importations of high-quality beef. But the United States will continue to import considerable quantities of lean, frozen, grass-finished beef, known as manufacturing beef; and importation of any beef, regardless of quality, competes for the consumer's dollar.

■ **Beef shortages in many countries will provoke governments to apply all conceivable, and some inconceivable, means of increasing domestic supplies**—Beef shortages in some countries have provoked, and will continue to provoke, governments to invoke all the known methods, plus some new ones, in dealing with the situation, some of which will affect the U.S. beef industry. Governments, depending on their individual position in the world beef trade, have instituted consumer price ceilings and freezes—and higher producer price supports; consumption subsidies—and meatless days and weeks; lower tariffs—and higher export taxes; freer import quotas—and tighter export quotas; import subsidies—and export embargoes. All with the objective of increasing domestic supplies at lower consumer prices—and all counterproductive on a worldwide basis.

■ **Pressure for liberalizing or removing U.S. quotas**—American consumers will exert more and more pressure for liberalizing or removing quotas and tariffs. Also, in a time of world beef shortages, and when the United States is (1) eyeing exports for top grade beef, and (2) importing manufacturing beef suitable for hamburgers, hot dogs, and luncheon meats, there is less need for quotas and tariffs than in earlier years.

■ **Export of more high-quality beef**—The United States will continue to develop the high-quality portion of the Japanese beef market, because American grain-fed beef closely resembles the highly marbled Kobe beef which is so coveted in Japan.

With increased buying power in Japan and Europe, and with the price of high-quality beef in these countries higher than in the United States, it is inevitable that U.S. beef producers would like to export finished beef.

Finished beef, produced in U.S. feedlots, and fabricated, packaged, and frozen in U.S. packing plants, can be transported via refrigerated jet freight and marketed in Japan and Europe, at higher prices than can be secured at home.

All of the above could be realized on a free market. However, both Japan and Europe have trade barriers. Japan restricts U.S. beef by quotas; and the EC countries have a standard 20% duty on incoming beef in addition to a variable levy.

■ **Ever-lengthening delivery times**—A growing proportion of world beef trade will be in chilled form as modern technology—vacuum packing and temperature-controlled containers—make possible ever-lengthening delivery times.

■ **Trade balance from agricultural exports**—Agricultural exports will continue to be the hope for a favorable trade balance in the United States. Exporting a high value product, like grain-fed beef, gives a big assist.

■ **Increasingly difficult to satisfy people of beef-exporting countries**—Cattle expansion and development programs will continue around the world. But all people are demanding more beef. As a result, exporting countries will run into more and more difficulty satisfying their own people.

■ **Importation of more manufacturing beef**—The United States will import more and more manufacturing-type beef—suitable for hamburgers, hot dogs, and luncheon meats—to meet the growing demands of the younger generation and those with moderate incomes.

■ **Limitation of cattle and beef from Mexico and Canada to the United States**—Neighbors of the United States will restrict exports of both feeder cattle and beef in order to meet their ever-increasing domestic demands.

■ **Inefficient producers through trade barriers**—The more trade barriers (quotas, tariffs, subsidies, etc.)—the more protectionism—built around the beef industry (or any other agricultural economy), the more apt it is to become inefficient.

■ **High income and more beef eaters**—As personal incomes rise in countries around the world, and as people become more affluent, they will consume more meat—particularly beef. The two—beef consumption and affluency—will continue to rise together. Increasingly, beef will be a status symbol.

■ **Demand for more animal protein, especially beef**—Population growth and rising per capita disposable income will make for a more affluent life-style and a greater demand for animal protein—especially beef, followed by beef shortages and higher prices.

■ **Possible pricing of beef out of the market basket**—High prices for beef will do two things: (1) decrease demand; and (2) increase competition from poultry and pork.

■ **Mixed reactions to world beef shortages and high prices**—These will be expressed in (1) the form of consumer boycotts, pickets, and hoarding; (2) pressures from some for export controls, from others for import controls; and (3) pressures from some to import more beef, and from others to export more beef.

▪ **Competition from simulated meats**—The simulated meats (synthetic meats, or meat analogs) will likely become more competitive with beef in the future, as their price becomes relatively more favorable and their taste and texture are improved. To meet this type of competition, the cattle industry of the future will place increasing emphasis on the palatability and nutritive qualities of beef, or join forces by creating new products using beef and simulated meats.

▪ **Increasing health consciousness**—Nationwide concern about nutrition will continue, including warnings against consumption of foods high in fats and cholesterol. Health faddists are telling people to eat grains and vegetables instead of beef. This kind of thinking and publicity gives cause for cattle producers to keep a wary eye on long-time beef consumption trends.

▪ **Increased demand for beef due to inflation**—Continued inflation around the world will result in many countries importing more farm products to relieve the pressures of inflation.

▪ **Energy shortages affecting beef**—Energy shortages in many parts of the world will make for a decline in gross national product, which, in turn, will lessen buying power and demand for beef.

▪ **More pollution control**—Environmentalists will force more and more pollution control.

▪ **Production and profits together**—The American cattle producer can and will produce more beef, provided the business is profitable.

▪ **Increased human population**—The population of the United States continues to expand, even though it is at a slower rate. A reasonable assumption is that gradually less meat per capita will become available; and more and more grains will be consumed directly as human foods. This does not mean that the people of the United States are on the verge of going on an Asiatic grain diet. Rather, history often has an uncanny way of repeating itself—even though such changes come about ever so slowly. Certainly, these conditions would indicate the desirability of eliminating the less efficient animals.

▪ **Greater reliance on roughages**—Beef cattle will increasingly be expected to rely upon their ability to convert coarse forage, grass, and by-product feeds, along with a minimum of grain, into palatable and nutritious food for human consumption; thereby not competing for humanly edible grains.

When put into the feedlot as calves or short yearlings and long fed on a high-concentrate ration, their lifetime total feed (forage and grain combined) conversion is on the order of 10 to 1 liveweight.

Less grain-fed and more grass-fed beef are in the future.

▪ **Less emphasis on carcass quality; more emphasis on lean meat**—With emphasis on lean beef, the likely hazard is that carcass quality will be relegated to a position of minor importance. Should this happen to any appreciable degree, per capita beef consumption could eventually suffer.

▪ **Improved genetics, management, and feed**—Our best hope in meeting beef shortages and competition from other meats is the proper application of our knowledge of genetics and management and the maximum use of coarse forages, grasses, by-product feeds, and nonprotein nitrogen.

▪ **Increased productivity per animal unit must come**—Increased productivity per animal unit to offset higher production costs, plus the continued output of a more desirable kind of beef, will be necessary for the prosperity and survival of America's number one agricultural industry—beef production.

▪ **Worldwide cooperation needed**—Political boundaries between nations will always exist, but cattle producers the world over will continue to work together and serve humankind through beef production.

QUESTIONS FOR STUDY AND DISCUSSION

1. Discuss the impact on world food, hunger, and malnutrition of each of the following: human population, unequal distribution of human population, population growth rate, food production, wealth, and animal proteins.

2. Who was Thomas Robert Malthus? Discuss his prophecy.

3. Why should cattle producers be well informed relative to worldwide beef production?

4. Why do not cattle numbers alone reflect the beef and veal production of countries?

5. Discuss the beef production potential of Australia and New Zealand.

6. What is the Economic Community (EC)? What other names and abbreviations does it go by? Why cannot most of the EC countries make any marked transition from dairy to beef?

7. Discuss the increased beef production potential in each: Canada, Mexico, South America, and Central America and Panama.

8. How do you account for the fact that the following countries are the leading red meat and poultry eaters in per capita consumption, ranked in descending order: United States, Canada, Australia, and New Zealand?

9. Discuss the major factors which influence the total demand for beef.

10. Discuss the relationship of each of the following points to the price of beef and other foods: (a) the cost of living in terms of working time required to purchase food, and (b) the proportion of income spent for food.

11. List by rank the five leading beef- and veal-consuming nations, in pounds per capita. Why do the people of these countries eat so much beef?

12. List the major beef- and veal-exporting nations. What is the significance of extensive land and low labor costs, which characterize these countries, from the standpoint of beef exports?

13. How do you account for the fact that the United States imports more beef and veal than any other country in the world?

14. What are the objects of trade barriers? Are such barriers good or bad so far as beef is concerned? Justify your answer.

15. In the United States, most of our high-quality fresh beef comes from domestic sources; and more and more of the manufacturing grades come from imports. Why is this so?

16. What do you foresee in world beef demand ahead in Japan, the European Community (EC), Eurasia, and the United States?

17. Can beef cattle be justified as a source of protein, or should beef cattle be eliminated? Justify your answer.

18. List and discuss new protein sources.

19. The United States has 7.0% of the world's cattle population. Yet, it produces a fantastic 22% of the world's beef. What's the explanation for this situation?

20. Since 1973, the number of grain-fed cattle in the United States has decreased. What caused this decrease? Will this trend away from grain-fed beef continue? Justify your answer.

21. U.S. total cattle numbers have declined since the peak year of 1975. Will this trend downward continue? Justify your answer.

22. The greatest expansion in U.S. cow-calf operations since 1950 has been in the West North Central area, and in the South and East. What's the explanation of the expansion in these two areas?

23. What contributing factors cause Texas to be the leading cow-calf state?

24. What's the explanation of the shift in the center of cattle feeding from the Corn Belt to the West and Southwest?

25. Discuss consumer trends in each of the following: red meats, poultry, sugars and sweeteners, fruits, and vegetables.

26. Why does the United States buy live cattle, beef, and veal, from abroad?

27. What's the reason for U.S. beef quotas and tariffs?

28. Define *trade balance*, or *balance of trade*.

29. Has the unification of the U.S. beef industry been good or bad?

30. Will the United States export more high-quality beef in the future? Justify your answer.

31. Will the United States import more or import less manufacturing-type beef in the future—suitable for hamburgers, hot dogs, and luncheon meats? Justify your answer.

32. Can beef be priced out of the market basket? Justify your answer.

33. Should cattle producers be concerned about increased competition from simulated meats? Justify your answer.

34. Will more pollution control mitigate against beef production in the future? Justify your answer.

35. Will U.S. beef cattle of the future increasingly be roughage burners, and will there be less emphasis on carcass quality? Justify your answer.

36. How can U.S. cattle producers achieve increased productivity per cow unit?

SELECTED REFERENCES

Title of Publication	Author(s)	Publisher
Beef Cattle, Seventh Edition	A. L. Neumann	John Wiley & Sons, Inc., New York, NY, 1977
Beef Production and Distribution	H. Degraff	University of Oklahoma Press, Norman, OK, 1960
Cattlemen, The	Mari Sandoz	Hastings House Publishers, New York, NY, 1958
Chisholm Trail, The	Sam P. Ridings	Co-operative Publishing Co., Guthrie, OK, 1936
Criollo, The	John E. Rouse	University of Oklahoma Press, Norman, OK, 1977
World Cattle, Vols. I, II, and III	J. E. Rouse	University of Oklahoma Press, Norman, OK, Vols. I and II, 1970; Vol. III, 1973

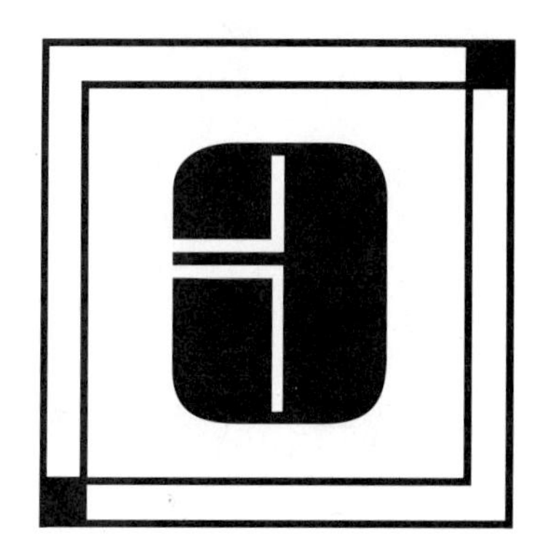

Lest we forget! Beef that the consumer enjoys and will buy, consistently available and competitively priced, is the ultimate objective of all cattle breeds and breeding. (Courtesy, American Murray Grey Assn., Kansas City, MO)

BREEDS OF CATTLE

Contents *Page*

Contents *Page*

A breed is a group of animals related by descent from common ancestors and visibly similar in most characters. A breed may come about as a result of planned matings, or, as has been more frequently the case, it may be pure happenstance. Once a breed has evolved, a breed association is usually organized.

One of the most frequently asked questions is: Who approves a breed? The answer: No person or department has authority to approve a breed. The only legal basis for recognizing a breed is contained in the U.S. Tariff Act of 1930, which provides for the duty-free admission of purebred breeding animals provided they are registered in the country of origin. But this applies to imported animals only. In this book, therefore, no official recognition of any breed is intended or implied. Rather, every effort has been made to present the factual story of the breeds. In particular, information about the new or less widely distributed breeds is needed and often difficult to obtain.

The importation and U.S. development of numerous new breeds of beef cattle, along with increased crossbreeding, enhanced a number of good production traits, including weaning weight, mature size, milk production, adaptability to climate, age of puberty, scrotal circumference, and fertility. But the mushrooming of breeds and crossbreeding was a major causative factor in lack of consistency, or repeatability, of U.S. beef in the last third of the 20th century, causing many meat eaters to bypass the beef counter. The impact of new breeds becomes evident in the following editions of this book, *Beef Cattle Science*. In 1968, the fourth edition of *Beef Cattle Science* listed only 21 breeds of U.S. beef cattle; in 1976, the fifth edition listed 53 breeds; in 1987, the sixth edition listed 59 breeds; in 1997, the seventh edition of *Beef Cattle Science* lists 83 breeds. Thus, from 1968 to 1997, a period of only 29 years, 62 new breeds or composites of beef cattle evolved—a four fold increase; and, simultaneously, crossbreeding proliferated.

HISTORY OF CATTLE BREEDS

The use of livestock pedigrees in the modern manner had its beginning in England late in the 18th century, and the general formation of breed registry societies began around the middle of the 19th century. The typical history of the formation of the various breeds of cattle may be summarized as follows:

1. A recognition of the existence of what was considered to be a more desirable and useful type.
2. The best animals of that type were gathered into one or a few herds which ceased to introduce outside blood.
3. Intense inbreeding to fix characters followed.
4. With greater numbers of animals, more herds were established.
5. When the breed became so numerous and the number of animal generations in the pedigrees increased until no person could remember all of the foundation animals far back in the pedigree, the necessity for a herd book arose. In addition to supplying a knowledge of foundation animals, the herd book was designed to prevent, insofar as possible, unscrupulous traders from exporting grades or common stock as purebreds.
6. The breed association or society that published the stud book was also organized for the purposes of improving the breed and promoting the general interests of the breeders.

Naturally, not all breed histories were identical. Circumstances often varied the pattern that molded the breed, but the end results and objectives were similar. The first herd book of any breed or class of animals, known *as An Introduction to the General Stud Book*, published in 1791, was for Thoroughbred horses. In it were included the pedigrees of horses winning the important races. Thus, it was really a record of performance. The Shorthorn herd book, first undertaken as the private venture of Mr. and Mrs. George

Coates, of Great Britain, followed in 1822. Other societies and herd books for the various breeds were originated in due time.

It is noteworthy that some of the early breeders objected to furnishing pedigrees of their sale animals, fearing that they would thus give away valuable trade secrets.

THE GOLDEN PUREBRED ERA

The improvement that Robert Bakewell (1726–1795) and his followers, of England, had made in their breeding stock came to be known in other lands. Agriculture was on the move, and the golden age of stock breeding was at hand. Animals possessing common characteristics were no longer to be confined to a small area and restricted to a few breeders.

Both in England and on the continent, the breeding of superior cattle engaged the interest of many persons of wealth and high position, even of royalty itself. Albert, Prince Consort of Queen Victoria, had a magnificent herd of 200 Shorthorns at the home farm near Windsor Castle, another herd of 90 Herefords at the Flemish Farm 2 mi distant, and 100 Devons on the Norfolk Farm. King William of Württemberg, as early as 1824, began to import and breed Shorthorns, as did his contemporaries, Nicholas II of Russia, Francis Joseph of Austria, and Louis Philippe of France. Napoleon III was also a heavy buyer of English blooded stock; and if he could not buy the bulls that he desired, he often leased them—particularly was this true of sires from the Booth Herd at Warlaby. The King of Sardinia and the King of Spain became interested in stock breeding, crossing English Shorthorns on the long-horned white Tuscan cattle and black Spanish cows.

Early in the 19th century, progressive U.S. cattle breeders, ever alert to their opportunity, proceeded to make large importations from England and Scotland. At this period in history, cattle were not valued solely for meat or milk. In 1817, Henry Clay of Kentucky, who made the first importation of Hereford cattle to the United States said of them: "My opinion is that the Herefords make better work cattle, are hardier, and will, upon being fattened, take themselves to market better than their rivals."

By 1850, *the battle of the breeds*, which for many years had divided English stockmen into rival camps, was transferred across the Atlantic, where it has raged ever since. Except for the occasional emphasis on fancy points to the detriment of utility values, perhaps this breed competition has been a good thing. Undoubtedly, the extensive importations of Shorthorns, Herefords, Ayrshires, and Jerseys in the decades before the Civil War materially improved the quality of American cattle, both for beef and milk production.

In the early 1900s, the trend among U.S. beef cattle producers was toward purebreds or high grades. Breeding was a matter of pride, a status symbol, and usually profitable, too. The choice of the breed was determined by individual preference or prejudice, and, for the most part, it was limited to Shorthorns, Herefords, and Angus.

CROSSBREEDING DATES TO THE LONGHORN

Crossbreeding is not new to American animal agriculture. An estimated 80% of the nation's hogs, sheep, and layers, and more than 95% of our broilers, are crossbreds.

When we trace the history of the cattle industry of the United States, we become aware that the commercial cattle producer has practiced crossbreeding and exploited the benefits of heterosis since the beginning of time. The commercial cattle industry of this country began with the Longhorn. Introduced by the Spaniards in the 16th century, they were long on horns, hardiness, and longevity; and they had the ability to fight and fend for themselves as they reverted to the wild. Also, their tough, sinewy muscles made it possible for them to travel long distances to the sparsely located water holes on the early-day western ranges, and the even longer distances—often 1,000 mi or more—to the railheads.

As the living standards improved and the desire for better steaks developed, the aristocratic Shorthorns from Scotland came on the American scene. They were admirably promoted, and their assets that would improve the deficient traits of the Longhorn were exploited. The Shorthorn lived up to its name—it shortened the murderous weapons on the head of the Longhorn. Additionally, the Shorthorn X Longhorn

Fig. 3-1. Longhorns, foundation of the American cattle industry. In the movie *The Rare Breed*, James Stewart described them as meatless, milkless, and murderous.

crossbreds showed marked improvement in mothering ability, milk production, early maturity, weaning weight, beef qualities, and disposition. Yet, the hybrids retained the hardiness of the Longhorns. It was a perfect example of one breed complementing another.

Gradually, the Longhorn became extinct and the commercial herds of this country were dominated by Shorthorn blood. At the beginning of the present century, the grand champion steers at the Chicago International weighed 2,500 lb and averaged 4 to 5 years of age. With the advent of USDA grading standards and the modern butcher, it became evident that carcass size of steers had to be reduced in keeping with consumer demand. Also, with the increase in Shorthorn blood and the decrease in heterosis, there was a decrease in the ability of the animals to withstand the rigors of the western range. At this point in history, the Hereford was lofted into prominence on the beef cattle scene. When crossed on the Shorthorn, the Hereford provided early maturity, smoothness of fleshing, hardiness, and ability to withstand severe winter conditions. The crossbreds retained good milk production from the Shorthorn, and, as a bonus, they got hybrid vigor.

Over the next 20 to 30 years, the cow herds of the beef-producing areas of the nation became predominantly Hereford, and the Longhorn and Shorthorn were relegated to minor importance. Later, the Angus entered, and the crossbreeding and accompanying heterosis resulted in the *black baldies.*

It is noteworthy that, for a considerable period of time, crossbreeding strengthened the position of purebred breeders and made for breed growth; first with the Shorthorn which was crossed on the Longhorn, second with the Hereford which was crossed on the Shorthorn, and third with the Angus which was crossed on the Hereford. Noteworthy, too, is the fact that, without a planned program to sustain heterosis, each breed brought on its own decline.

NEW BREEDS ARRIVE

From the time cattle of improved breeding began to make their influence evident in North America, the majority of the genetic material used was of British origin. During the early period, the most significant changes were in the proportions of the British breeds composing the national cow herd; first came the Shorthorns, then the Herefords, and finally the Angus.

The introduction of the Zebu type of cattle in the early part of the present century significantly improved the efficiency of beef production in subtropical and desert range areas of the southern United States. Also, the Brahman was used in crossbreeding and in developing the Santa Gertrudis, Beefmaster, Brangus, Charbray, and Braford breeds. But the gradual development and expansion of the Brahman and the new part-Brahman breeds were not considered a serious threat by supporters of the British breeds.

The beginning of the present exotic era must be credited to the Charolais, which found its way into the United States from Mexico in the late 1930s, thence spread north into Canada. In test stations, Charolais crosses demonstrated higher lean growth rate than straightbred or crossbred British breeds. Through active promotion based on performance facts, the enchantment of a new breed, and the momentum of a new registry, the Charolais was used widely enough to be regarded as a serious threat to the established breeds. Thus, the Charolais breed can rightfully be credited for accelerating genetic improvement programs in all other beef breeds, both old and new.

EXOTICS

Next came the exotics! According to Webster, the word *exotic* means, "from another country; not native to the place where found. Having the appeal of the unknown—mysterious, romantic, picturesque, glamorous. Strikingly unusual in color or design." Indeed, exotic cattle are all these things—and more.

Back of the rage for the exotics is the desire to use them in crossbreeding. Among commercial cattle producers, the tidal wave of exotics engendered enthusiasm and excitement such as had not been seen in recent times. But among some purebred breeders of the established breeds, it produced animosities reminiscent of the range wars in the days of intruding sheepmen and nesters. The *established breeders* were riled because crossbreeding was being extolled, and bulls of these stark newcomers were being used. Arguments waxed hot; many oldtime purebred breeders became emotional and explosive. This was surprising because, as pointed out earlier in this discussion, the practice of crossbreeding has been common in this country since the birth of the cattle business. The commercial cattle producer has always exploited the benefits of heterosis—first by using Shorthorns on Longhorns, thence followed by Herefords and Angus. Moreover, at one time or another, all breeds of cattle in North America, including the Longhorn and British breeds, were exotics—none of them were indigenous to this country. They are no more native than North American people, except the American Indians. In the present century, the Brahman was the first exotic; thence the Charolais. But there were two great differences about the recent tidal wave of exotics: (1) they came in more quickly; and (2) they came in greater numbers, because of the following circumstances:

1. **Canada's quarantine station.** Under the leadership of Canada's astute and progressive Minister of

Agriculture, Harry W. Hays, founder of the Hays Converter breed, a quarantine station was established on Grosse Isle in the St. Lawrence River, and opened in 1965. It was then possible to bring cattle from parts of Europe previously closed because of the disease situation. An influx soon followed.

2. **The time was ripe.** During much of the 1950s and 1960s, cow-calf operations were hurting financially. They recognized that profits could be increased by producing for less, as well as by selling for more. Crossbreeding with its demonstrated potential for producing for less through complementary genes and heterosis offered new hope—hope for survival. Many cattle producers had tried it with success in the '60s, and others were ready and anxious to do so.

3. **Artificial insemination (AI) facilitated it.** The use of AI in beef breeding herds increased substantially during the 1950s and 1960s, with many beef producers becoming competent in managing such programs. This lessened the number of bulls that had to be maintained in conducting a crossbreeding program and made it possible to ship semen great distances.

4. **The dollar value of the Okie.** Cattle feeders came to a realization that crossbred cattle common to the South—the Okies—often made more rapid and efficient gains (hence, more dollars profit) than straightbreds.

5. **The Charolais inferiority complex.** A common reaction was: "I missed out on the Charolais, but I'm not going to miss out on the other new breeds."

6. **Promoters help usher them in.** Although most of the exotics were, and still are, in the hands of bona-fide cattle producers intent on becoming constructive breeders, they attracted a considerable number of promoters whose main objective was to turn a big and quick profit. Their promotional know-how, along with money, fed the exotic boom.

The first Simmental was imported in 1967. Other breeds followed.

Indeed, the influx of exotics has disturbed the placid tranquillity of the pastoral scene. But a candid look at crossbreeding reveals the following:

1. The practice is as old as animal agriculture in North America—it has always been an economic necessity for commercial cattle producers.

2. Researchers haven't discovered anything new; rather, they have evolved with new tools for applying, and new methods of measuring the merits of, a very old practice.

3. It is not the solution to all the commercial producer's problems. It will give hybrid vigor, but—

a. The success of the resultant programs depends on the merits of the individual parents and the selection of breeds that are complementary.

b. The producer must sort out fact from fiction in the reports, both written and oral, extolling the virtues of both breeds and crossbreeding; and the producer's axiom, "well bought is half sold," must not be forgotten. This double-barreled caution is prompted because the exotic boom was ushered in by the greatest array of promoters, entrepreneurs, and *fast buck boys* ever to latch on to the cattle industry.

c. Without a longtime, planned breeding program, crossbreeding will almost inevitably end up with little to show for it, other than a motley collection of females and progeny varying in type and color.

d. The feeding and management of the exotics and crosses should satisfy their requirements for (1) increased growth rate, (2) greater milk production, (3) larger mature size, and/or (4) special needs because of differences in adaptability, winter hardiness, etc.

PUREBREDS CONTROL DESTINY

Despite the very considerable virtues of crossbreeding, there will always be purebreds, and they will control the destiny of cattle improvement. They are the only way to improve crossbreds and composites above their present level of performance. Commercial hybrids and composites must depend on selection in the purebreds to improve those traits that are high in heritability, such as rapid and efficient gains and carcass merit. Master purebred breeders are to improved crossbreds and composites what master designers are to better cars.

Yet, breed partisans cannot be complacent. Not all breeds—from among the traditional beef breeds, newly imported breeds, and newly synthesized breeds—are needed in a crossbreeding program, from the standpoint of complementary traits and adaptation. Thus, only those breeds which provide the most of the best in desired qualities as reflected in net returns to the commercial producer, and in meeting consumer demands, will survive over a period of time; the rest will be relegated to a position of minor importance, or fall by the wayside. Thus, purebred breeders of today who wish to be in business tomorrow must produce superior animals backed by meaningful production records. Also, they must engineer (through selection) breeds that not only will grow fast and calve annually with minimal assistance, but will produce beef that meets consumer demands.

TYPES OF CATTLE

Type may be defined as *an ideal or a standard of*

perfection combining all the characters that contribute to the animal's usefulness for a specific purpose. It should be noted that this definition of type does not embrace breed fancy points. These have certain value as breed trademarks and for promotional purposes, but in no sense can it be said that they contribute to an animal's utility value. There are four distinct types of cattle in the world: beef-type, dairy-type, dual-purpose–type, and draft-type.

Beef-type animals are characterized by meatiness. Their primary purpose is to convert feed efficiently into the maximum of high-quality meat for human consumption.

Dairy-type animals are characterized by a lean, angular form and a well-developed mammary system. Their type is especially adapted to convert feed efficiently into the maximum of high-quality milk.

Dual-purpose–type animals are intermediate between the beef-type and dairy-type in conformation, and also in the production of both meat and milk.

Although many breeders have the dual-purpose–type clearly in mind, and although many fine specimens of the respective breeds have been produced, there is less uniformity in dual-purpose cattle than in strictly beef- or dairy-type animals. This is as one would expect when two important qualifications, beef and milk, are combined.

Draft-type animals, when true to form, are characterized by great size and ruggedness with considerable length of leg. Although oxen are seldom seen

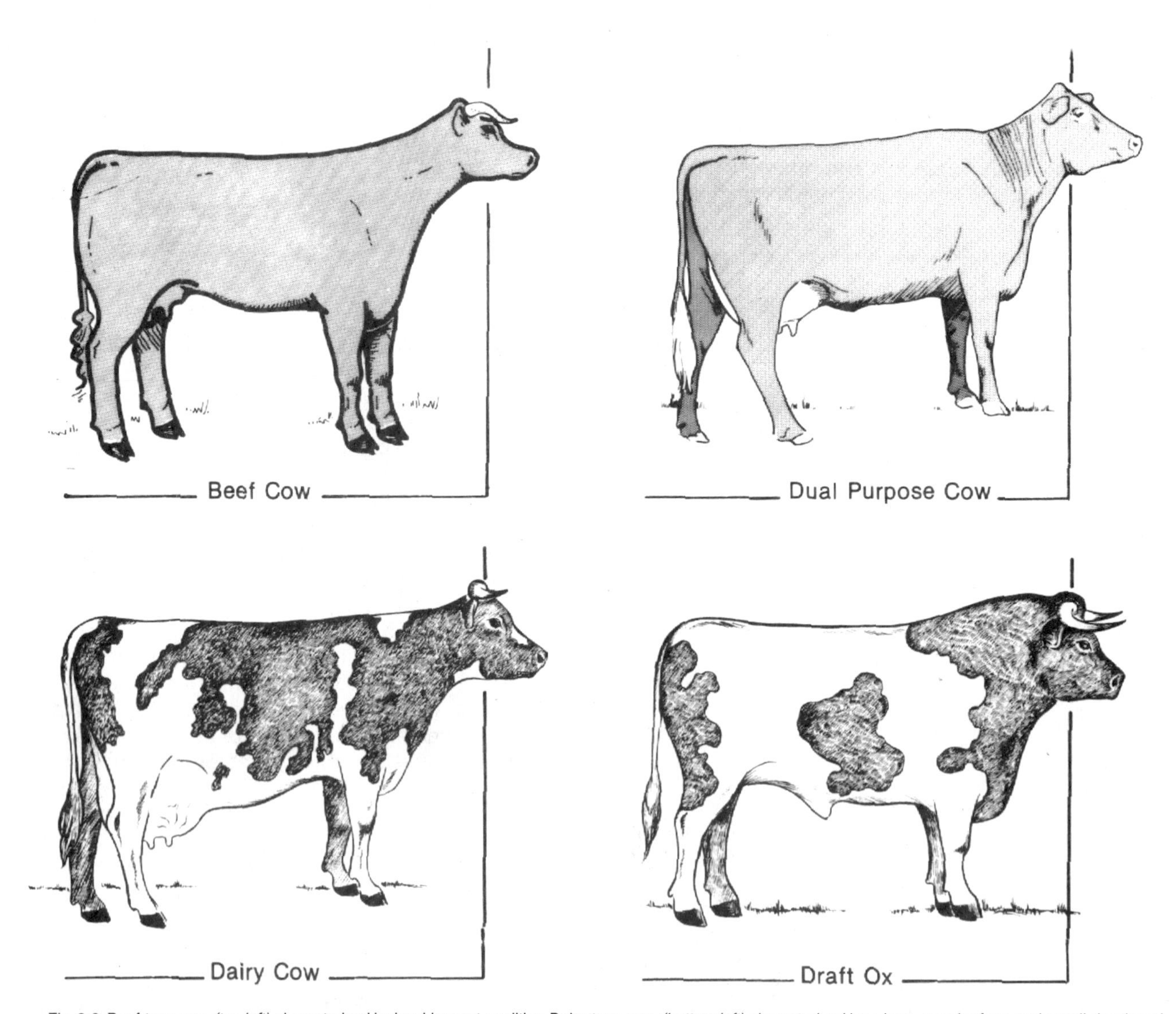

Fig. 3-2. Beef-type cow *(top left)*, characterized by bred-in meat qualities. Dairy-type cow *(bottom left)*, characterized by a lean, angular form and a well-developed mammary system. Dual-purpose–type cow *(top right)*, intermediate between the beef-type and the dairy-type in conformation. Draft-type ox *(bottom right)*, characterized by great size and ruggedness with considerable length of leg.

in the United States, it must be remembered that these patient, steady, plodding beasts are still the chief source of power in many parts of the world.

Several distinct breeds of cattle of each of the types have been developed in different parts of the world. Although each of these breeds possesses one or more characteristics peculiar to the group (breed characteristics), in general the type of cow that will produce a large flow of milk is the same the world over, despite acknowledged differences in size, color, shape of head and horns, or in any other distinctive breed characteristic. Likewise, there is a general similarity between all the beef-type breeds.

RELATIVE POPULARITY OF BREEDS OF CATTLE

Table 3-1 shows (1) the 1995 and (2) the total registration to date of the breeds of cattle. In these changing times, the annual figures are more meaningful than the all-time registrations, although it is recognized that only one year's data fail to show trends. Further, it is realized that some of the breeds are so new that they have not yet established trends.

Some breed associations either (1) did not respond to the authors' request, or (2) were not known to exist at the time the seventh edition of *Beef Cattle*

TABLE 3-1
ANNUAL AND TOTAL REGISTRATIONS OF CATTLE IN U.S. BREED ASSOCIATIONS[1]

Breed	Annual Registrations	Total Registrations	Breed	Annual Registrations	Total Registrations
Amerifax	NA	NA	Indu-Brazil (Zebu)	150	NA
Angus	224,700	12,570,230	Limousin	79,274	1,303,723
Ankole-Watusi	280	NA	Maine-Anjou	12,791	220,000
Barzona	1,200	NA	Marchigiana	95	5,606
Beefalo	649	NA	Murray Grey	915	24,821
Beef Friesian (1993)	378	8,396	Normande	300	NA
Beefmaster[2] (1993)	56,357	693,804	Piedmontese	687	8,136
Belted Galloway	1,215	NA	Pinzgauer	1,456	35,000
Blonde d'Aquitaine	981	19,500	Polled Hereford	45,500	5,802,539
Braford	5,182	81,853	Red Angus	34,812	511,000
Brahman	15,362	1,200,000	Red Brangus	3,000	NA
Brahmousin	427	12,618	Red Poll	1,500	NA
Brangus	28,599	648,793	Salers	20,000	NA
Braunvieh	1,990	12,000	Santa Gertrudis	13,000	733,550
British White	550	NA	Scotch Highland	1,215	27,921
Charbray	369	762,362	Shorthorn	20,000	5,005,458
Charolais	55,245	1,215,819	Simbrah	4,498	101,790
Chiangus	NA	NA	Simmental	64,568	1,726,720
Chianina	7,473	245,900	South Devon	1,400	NA
Devon	150	NA	Sussex	NA	NA
Dexter (1993)	648	6,200	Tarentaise	1,878	90,754
Galloway	413	NA	Texas Longhorn	12,000	239,500
Gelbvieh	33,768	525,000	Wagyu	800	3,900
Hays Converter	NA	NA	Welsh Black	350	NA
Hereford	63,000	19,600,538	White Park	350	NA

[1]Annual registrations in 1995, and total from date breed registry was formed through 1995, unless otherwise indicated.

[2]These are total figures for two Beefmaster associations. In 1993, Beefmaster Breeders Universal registered 45,750, with total registrations to date of 575,804; in 1993, the Foundation Beefmaster Assn. registered 10,607, with a total registration to date of 118,000.

Science was prepared; hence, they are not listed in Table 3-1.

MAJOR BREEDS

Brief, but pertinent, information pertaining to each of the breeds of cattle follows. Major U.S. breeds are covered, alphabetically, in narrative format while the newer and/or less populous breeds are listed, alphabetically, in Table 3-2. It should be noted that some of the exotic breeds such as Chianina, Limousin, Gelbvieh, Maine Anjou, and Simmental have purebred animals that are black in color and polled. These cattle were developed through breeding up programs in their respective breeds.

ANGUS

Sometimes Angus cattle are referred to as *doddies*, a Scotch term for polled or hornless.

ORIGIN AND NATIVE HOME

The breed originated in Scotland, in the northeastern counties of Aberdeen, Angus, Kincardine, and Forfar. The origin of Angus cattle is speculative. Some claim that they are a mutation from an earlier black, horned breed of Scotland. Others claim that they sprang from the polled cattle of Britain.

Fig. 3-3. George Grant Memorial, Victoria, Kansas. Mr. Grant imported the first Angus bulls into the United States from Scotland in 1873. (Courtesy, American Angus Assn., St. Joseph, MO)

EARLY AMERICAN IMPORTATIONS

In 1873, George Grant of Victoria, Kansas—a native of Banffshire, Scotland, and a retired London silk merchant—imported the first Angus bulls into the United States from Scotland, to use on his commercial range cattle. However, the first breeding herd—including animals of both sexes—to be imported into the New World was brought to Canada in 1876 by Professor Brown of Ontario Agricultural College.

Fig. 3-4. An excellent Angus cow, bred and owned by Jim Baldridge, Baldridge Angus, North Platte, Nebraska. (Courtesy, American Angus Assn., St. Joseph, MO)

CHARACTERISTICS

Angus cattle are black and polled, and they have a comparatively smooth hair coat and a somewhat cylindrical body.

Any of the following traits disqualifies an animal from registry: dwarfism, osteopetrosis, double muscling, syndactyly, heterochromia irides, white above underline or in front of the navel, or red hair.

BEEFMASTER

Beefmaster cattle are approximately ½ Brahman and ¼ each Shorthorn and Hereford.

ORIGIN AND NATIVE HOME

Beefmasters originated in the United States, on the Edward C. Lasater Ranch, in Fulfurrias, Texas, beginning in 1908. In 1931, Tom Lasater took over his father's herd and carried forward. In 1949, he moved the herd to its present location near Matheson, Colorado.

Tom Lasater patented the Beefmaster name in 1949. In 1954, the Beefmaster was recognized as an

Fig. 3-5. Beefmaster cow. (Courtesy, Beefmaster Breeders United, San Antonio, TX)

American breed by the U.S. Department of Agriculture. In 1996, the Beefmaster Breeders Universal and the Foundation Beefmaster Association merged to form the Beefmaster Breeders United.

CHARACTERISTICS

Red is the dominant color, but color is variable and is disregarded in selection. The majority are horned, although a few are naturally polled. During the entire period of development, selection has been practiced for six essentials: disposition, fertility, weight, conformation, hardiness, and milk production. Beefmaster cows are good milk producers under range conditions and wean off heavy calves.

In order that each Beefmaster may be permanently identified with the breeder thereof, breeders must use a prefix name, such as *Jones Beefmaster, Smith Beefmaster*, etc., to designate their cattle. Thus, in a unique way, the responsibility for the continued improvement of the breed is placed squarely upon the individual breeder.

Beefmaster Breeders United lists the following disqualifications: hernia, cryptorchid, wry nose, wry tail, double muscling, malformed genitalia, undershot and overshot jaw, dwarfism, and freemartin heifers.

BRAHMAN

Brahmans are the *sacred cattle of India*. The American Brahman originated in the United States from cattle imported from India. Brahman (Zebu) cattle are of the species *Bos indicus*.

ORIGIN AND NATIVE HOME

In India, 30 or more breeds or varieties of *indicus* cattle exist. No special effort was made to keep the different breeds or varieties separate in the United States, so the American Brahman is the result of the amalgamation of several Indian breeds or varieties, probably with a small infusion of European breeding.

EARLY AMERICAN IMPORTATIONS

There are conflicting reports on the introduction of Brahman cattle into the United States, but the following history appears to be authoritative: The first Indian cattle were imported in 1849 by Dr. James Bolton Davis of South Carolina, who became acquainted with *Bos indicus* cattle while serving as agricultural advisor to the Sultan of Turkey; but the identity of this first importation was lost during the Civil War. Subsequent importations followed. In 1854, Richard Barrow, cotton and sugar plantation owner of St. Francisville, Louisiana, received a gift of two Indian bulls from the British Crown, in recognition of his services for teaching cotton and sugar cane culture to British technicians, who, in turn, took these arts to India. In 1885, J. M. Frost and Albert Montgomery of Houston, Texas, imported two more Indian bulls. From time to time, Indian cattle were imported and used for circus purposes; among them, a red bull named *Prince*, which the Haggenbach Animal Show sold to A. M. McFaddin of Victoria, Texas, in 1904, and 12 head which Haggenbach sold to Dr. William States Jacobs of Houston. In 1905 and 1906, the Pierce Ranch of Pierce, Texas, assisted by Thomas M. O'Connor of Victoria, Texas, imported 30 bulls and 3 females of several Indian types. Subsequently, several *Bos*

Fig. 3-6. Brahman cow. (Courtesy, American Brahman Breeders Assn., Houston, TX)

indicus cattle were imported from Brazil and reached the United States by way of Mexico.

CHARACTERISTICS

Steel-gray is the most common color, usually with a tendency toward a darker color on both the fore and rear quarters of the bull. Solid red is also a popular color but some are brown, black, and even black and white or red and white and spotted.

Brahmans are easily identified by their drooping ears, long face, prominent hump over the shoulders, and an abundance of loose, pendulous skin under the throat and along the dewlap. They are moderately deep-bodied and well-muscled throughout. Further, Brahmans are comparatively trim in their middles, rather thin in their hides, and usually show a high dressing percentage. Bulls weigh 1,800 to 2,000 lb, and cows in ordinary condition weigh about 1,200 lb. The Brahman's voice resembles a grunt rather than a moo.

Disqualifications include any of the following: (1) brindle grulla (a smutty or blackish red) or albino color; (2) cryptorchidism; (3) freemartinism; (4) inherited lameness; or (5) dwarf or midget characteristics.

The Brahman breed is well adapted to areas characterized by hot climates, heavy insect infestations, and sparse vegetation.

BRANGUS

Beginning about 1932, the U.S. Department of Agriculture studied the crossing of Brahman and Angus cattle. The breed is now standardized so that the registered cattle carry 3/8 Brahman and 5/8 Angus breeding. However, the Brangus association enrolls registered Angus and Brahman cattle, and then certifies the resultant crosses of these two breeds as 3/4 bloods (3/4 Brahman and 1/4 Angus) or 1/4 bloods (1/4 Brahman and 3/4 Angus).

ORIGIN AND NATIVE HOME

Most of the development of the Brangus as a breed can be credited to Frank Buttram, owner of the Clear Creek Ranches at Welch, Oklahoma, and Granada, Mississippi, and to Raymond Pope of Vinita, Oklahoma; beginning in 1942.

CHARACTERISTICS

Brangus cattle are black and thin hided with a smooth, sleek coat of short hair. They are polled, wide muzzled, and strong headed. Their extra length of ear and looseness of hide indicate the Brahman influence. Also, Brangus possess a slight crest over the neck.

Fig. 3-7. Brangus cow and calf. (Courtesy, International Brangus Breeders Assn., Inc., San Antonio, TX)

Brangus are resistant to heat and high humidity, but in a cool and cold climate they seem to produce enough hair for adequate protection. The cows are good mothers, dropping medium-sized calves with ease.

Any one of several faults disqualifies Brangus cattle for registry, including: horns, any color other than black, white in front of the navel, small for age, extreme nervousness, too fine boned, too long ears and too loose hide, or thin hided and short hair coat.

All enrolled, certified, and/or registered Brangus cattle must be inspected before they are eligible for recordation.

CHAROLAIS

Charolais cattle are native to France; and like other cattle of continental Europe, they were used for draft, milk, and meat.

ORIGIN AND NATIVE HOME

The Charolais originated in west central to southeastern France in the old French provinces of Charolles and Nievre. Although the foundation stock is unknown, it must have been found in the area. One of the early herds that had considerable influence on the breed was that of Count Charles de Bouille, which was established in 1840, near the village of Magny-cours.

The breed society was founded in France in 1887.

EARLY AMERICAN IMPORTATIONS

The first Charolais brought into the United States

came from a herd established in Mexico in 1930, by Jean Pugibet, a Mexican industrialist. In World War I, Pugibet served as a volunteer in the French army and became acquainted with the Charolais. He imported Charolais to his hacienda in Mexico in 1930, 1931, and 1937. The King Ranch and other Texas ranches purchased Charolais from Pugibet's herd, introducing and developing the breed in the United States. An outbreak of hoof-and-mouth disease in the middle 1940s halted imports to the United States from Mexico and other countries with the disease.

In 1951, the American Charolais Breeders Association was organized; and in 1954, the International Charolais Association was organized. The two associations merged in 1957 under the existing name of American-International Charolais Association.

Fig. 3-8. Charolais cow. (Courtesy, American International Charolais Assn., Kansas City, MO)

CHARACTERISTICS

Charolais are white or creamy white, with pink skin and mucous membranes. They are noted for their large size, rapid gain, and bred-in red meat. Mature bulls weigh from 2,000 to over 2,500 lb and cows weigh 1,250 to over 2,000 lb. Most Charolais cattle are horned, although the polled characteristic is found in the breed and many breeders have selected for it.

The Association disqualifies any animal with (1) a dark nose, (2) spots, or (3) excessive dark skin pigmentation.

CHIANINA

The Chianina breed (pronounced *Key-a-nee-na*) is of very ancient origin, going back to the days of the Roman Empire, when animals of this breed were used for draft. Shortly after the birth of Christ, the Latin poet Columella described the cattle of Rome *as boves albos et vastos* ("white cattle of great size"). He went on to tell how they were used as sacrificial animals and for pulling the triumphal cart of the Emperor. Animal geneticists in Italy speculate that the breed is related to the *Bos indicus*, although it does not have the characteristic hump or excessive dewlap.

ORIGIN AND NATIVE HOME

The Chianina originated in central Italy, in the province of Tuscany, in the Chiana Valley (from which it takes its name). Most of the foundation animals imported to Canada were the largest of the breed, from the plains of Arezzo and Siena.

EARLY AMERICAN IMPORTATIONS

Like many of the exotic European breeds, the Chianina first came to the United States via Canada. For the most part, bulls and semen were imported from Italy to Canada, thence bulls and semen were imported from Canada to the United States. From semen imported from Canada, the first half-Chianina calf was born in the United States in 1972. Later, cattle were brought from Canada. The American Chianina Association began registering cattle in 1972.

CHARACTERISTICS

Chianinas have soft, short gray hair, a black switch, black hoofs and muzzle, short curved horns, and dark skin, which gives the breed resistance to heat—the white hair reflecting the sun's rays and the

Fig. 3-9. Chianina cow and calf. (Courtesy, American Chianina Assn., Platte City, MO)

dark pigment preventing sunburn. Unlike the Charolais, the white of the Chianina is recessive in inheritance. Crosses of Chianina X Holstein are black; and crosses of Chianina X Brown Swiss are also dark, with Swiss markings on the nose and ears.

Calves are born a tan color, which gradually turns to white at about 60 days of age.

The Chianina is the largest breed of cattle in the world. They are tall, long legged, long and round bodied, and heavy. Mature bulls may be 6 ft high at the withers and weigh up to 4,000 lb, and mature cows weigh up to 2,400 lb.

Other distinguishing characteristics of the breed are: trimness of middle, fineness of head, horn, and bone; and absence of excessive dewlap and brisket. However, cows often have poorly formed udders and are not known for their milk production. The growth rate and leanness of the breed give Chianina bulls an important role as a terminal cross in crossbreeding programs. Their greatest potential should accrue when Chianina bulls are crossed on cows that can provide enough milk to allow the calf to express its growth potential. Despite the large size of the breed, and large calves at birth (male calves average about 100 lb at birth, and females about 85 lb), calving difficulties are infrequent, perhaps due to the rather small heads and long, narrow bodies of the newborn.

The American Chianina Association does not attempt to establish weight or color specification. Registrable Chianinas need only be at least ¼ Chianina and sired by Chianina bulls registered with the Association.

GELBVIEH

The Gelbvieh is also known as the German Yellow.

ORIGIN AND NATIVE HOME

The Gelbvieh originated in Bavaria, in southern Germany. It is a descendant from the red-brown Keltic-German Landrace, on which Simmental and Shorthorn were crossed in the early 1800s. The Gelbvieh actually came into being when four breeds of German cattle—Franconian, Glan-Donnersberg, Lahn, and Limpurg—amalgamated around 1920. Since then, the breed has been selected for solid color, growth, and, later, for carcass quality. In 1952, a planned program was started for selection for milk production. The Gelbvieh claimed to be the best solid-colored, dual-purpose breed.

EARLY AMERICAN IMPORTATIONS

The Gelbvieh was introduced into the United States primarily by an artificial insemination program. Carnation Genetics started importing semen from Germany in July 1971. The first Gelbvieh animal was imported to North America in 1972.

In 1971, the American Gelbvieh Association was established. Most of the recordings of the breed have been of percentage cattle; ⅞ females and 15⁄16 bulls are considered purebreds.

Fig. 3-10. Gelbvieh cow and calf. (Courtesy, American Gelbvieh Assn., Westminster, CO)

CHARACTERISTICS

Gelbviehs are a solid golden red to rust color with strong pigmentation of the skin. They are horned, but polled cattle have evolved in the United States. Gelbviehs are large, long-bodied, well-muscled, fast-gaining cattle producing a high-quality carcass. Mature bulls weigh from 2,000 to 2,500 lb, and cows weigh from 1,200 to 1,500 lb. The breed is of dual-purpose origin; hence, milk production is good. Cows average about 7,800 lb of milk, with 4.07% fat.

Performance data are required for registration.

HEREFORD

Very early in their development, Hereford cattle were selected for maximum use of forage, preparing the breed for the prominent place it was to take on the ranges of the United States.

ORIGIN AND NATIVE HOME

Hereford cattle originated in Herefordshire, located on the western border and below the central part of England and bounded on the west by Wales and on the north by Shropshire. The earliest records of cattle in Herefordshire describe large, solid red cattle with widespread horns. However, the origin of the Hereford color pattern is uncertain, although it is one of the most characteristic and attractive traits of the breed.

Benjamin Tomkins began improvement of the Herefordshire cattle in 1742. He is often regarded as the founder of the breed. His son, Benjamin Tomkins, Jr., significantly improved the breed. The first volume of the herd book was published by T. C. Eyton of Shropshire in about 1846.

EARLY AMERICAN IMPORTATIONS

The earliest importation of Hereford cattle into the United States, of which there is authentic record, was made by Henry Clay of Ashland, Kentucky, in 1817, who imported one bull and two females. The first large importation was made by W. H. Sotham and Erastus Corning, who imported 21 cows and 1 bull in 1839 and 1840. They established a herd at Albany, New York. Over the next few years, the importation of Herefords gathered momentum and many people helped popularize the breed in the United States. In the period from 1846 to 1886 a total of 3,703 Herefords were imported and recorded in the *American Hereford Herd Book*. From 1886 to 1893 no Herefords were imported. But, another 1,100 head were brought from England between 1893 and 1916. Very few Herefords have been imported since 1916, because by this time American Herefords suited the requirements better than those in England.

CHARACTERISTICS

The color and the color pattern of the Hereford is a valuable and well-known trademark of the breed. Often Herefords are referred to as "white faces" because of the red body and white face with further white markings on the crest, brisket, underline, and switch. The red of the Hereford varies from a really dark red to a very light-yellowish cast. Also, Herefords have a thick hide and a thick coat of hair. Originally, the breed was horned, but a polled breed was developed in the United States.

Fig. 3-11. "The Young Bull," from one of the great paintings of the world, by Paul Potter. Painted in 1647. Note the white-faced, red-bodied cow—an individual resembling many plain-looking Herefords of the past. (Courtesy, The Netherlands Information Bureau)

Fig. 3-12. Hereford cow. (Courtesy, The American Hereford Assn., Kansas City, MO)

All characteristics that are associated with efficient utilization of feed and the production of desirable carcasses are important—a beefy animal with ample muscle and bone structure.

Herefords are highly adaptable. The breed is found in all 50 states and in 20 countries. In 1995, the American Polled Hereford Association and the American Hereford Association merged and retained the name of the latter.

LIMOUSIN

Cave paintings at Lascaux in the French countryside depict cattle remarkably similar to the Limousin cattle of today. The Limousin is one of the new European breeds raised primarily for meat production, rather than for dual-purpose—both meat and milk.

ORIGIN AND NATIVE HOME

The Limousin breed originated in southwestern France, in the 19th century. The Aquitaine region of France has had cattle that resemble present-day Limousin as long as there has been recorded history of the area. The breed takes its name from the Limousin Mountains. In 1857, the first carcass show held was at Limoges, France, for the purpose of demonstrating the superior quality of the Limousin cattle of the area.

U.S. soldiers returning from World War II duty in France told of the golden-red Limousin cattle, but nearly 25 years passed before the cattle came to America.

EARLY AMERICAN IMPORTATIONS

Prince Pompadour was the first full-blood Limousin animal imported by North American cattle producers in Canada in November 1969. His semen was distributed in Canada and the United States. Eventually, semen from other bulls and a few straight bred cattle were imported. In 1975, the island of St. Pierre et Miquelon was declared free of foot-and-mouth disease and 71 Limousins designated for export were allowed entry into the United States.

The North American Limousin Foundation was started in 1968. The herd book is still open, allowing breeders to record all heifer calves above 35% and bull calves above 50% if they are sired by a registered Limousin bull. All cattle are recorded at their actual percentage of Limousin blood enabling breeders to increase the percentage Limousin in their herd up to the purebred status (87% in females and 94% in bulls) or to maintain their herd at any given percentage of Limousin blood they desire.

Fig. 3-13. Limousin cow and calf. (Courtesy, North American Limousin Foundation, Englewood, CO)

CHARACTERISTICS

In France, Limousins are naturally horned and red, with the color varying from a deep red to a golden fawn. In the United States, however, many Limousins are black and/or polled, due to upgrading from black and/or polled animals. They carry a good, thick hair coat. Mature bulls weigh about 2,400 lb, and cows about 1,300 lb. The breed is noted for easy calving, rapid growth, good length, exceptional muscling and cutability, and high-quality carcasses.

MAINE-ANJOU

In France, Maine-Anjou cows average 6,600 lb of milk testing 3.7%. They were developed as a dual-purpose breed with an emphasis on beef.

ORIGIN AND NATIVE HOME

In the middle of the 19th century, Durham (Shorthorn) cattle from England were introduced into western France and crossed with the native Mancelle cows. Cattle from this cross were referred to as *Durham-Mancelle*. The *French Herd Book* for these cattle was established in 1908 in Chateau-Gontier, Mayenne. Then, in 1909, the name was changed to Maine-Anjou, from the names of two valleys in the Mancelle area.

EARLY AMERICAN IMPORTATIONS

The Maine-Anjous came to the United States about the same time and via the same route of other breeds. They first came to Canada, and shortly thereafter to the United States. In 1972, the first Maine-Anjous were registered in the United States. Bulls carrying $^{15}/_{16}$ Maine-Anjou blood and females carrying $^{7}/_{8}$ Maine-Anjou blood are registered as purebreds. Cattle carrying lesser amounts of Maine-Anjou breeding are recorded as percentage cattle.

CHARACTERISTICS

The color is usually a dark red with white on the underline, switch, and forehead, often with small, white patches on the body. Also, dark roans are found. Purebred black Maine-Anjous, some with white markings, were developed in America. For registration of Maine-Anjou cattle, no color restrictions exist.

Most Maine-Anjou heads are solid colored, with

Fig. 3-14. Maine-Anjou cow. (Courtesy, American Maine-Anjou Assn., Kansas City, MO)

the eyes always surrounded by color, and with a white marking on the forehead. The head is long, especially from the eyes to the nostrils.

The Maine-Anjou is the largest of the French breeds; mature bulls weighing 2,500 lb or more, and cows 1,900 lb or more.

Maine-Anjous are noted for rapid growth, early maturity, docility, and milking ability. They are long-bodied cattle, and very deep bodied. Maine-Anjous have been used in crossbreeding to improve maternal traits, including milking ability, fertility, docility, and growth. They have also been used in terminal cross-breeding systems to improve gainability and feed efficiency.

Fig. 3-15. Pinzgauer cow. (Courtesy, American Pinzgauer Assn., Jenera, OH)

PINZGAUER (Pinzgau)

This is a *beefy* breed—more so than most of the exotics, although it is classed as a dual-purpose breed in its native land. The breed is new in the United States.

ORIGIN AND NATIVE HOME

The breed originated in the Pinz Valley of Austria and in adjacent areas of Italy and Germany; in the Alpine region. Apparently the foundation stock were the native cattle of the area that had been in the Valley from very early times.

In Austria, all animals are subjected to and must pass a rigid conformation test before they can be registered. Additionally, performance of dams and daughters in milk production and butterfat content is a criterion in the selection of breed bulls.

EARLY AMERICAN IMPORTATIONS

Pinzgauer cattle in the United States are essentially crosses produced by using Pinzgauer semen on other breeds or on the produce of such crosses.

The American Pinzgauer Association was established in 1973.

CHARACTERISTICS

Pinzgauer cattle have brown sides, a white topline and underline, usually white feet, and deep orange pigment around the eyes and on the udder; and they are horned.

Mature bulls weigh 2,200 to 2,900 lb, and mature cows from 1,300 to 1,650 lb.

The breed is noted for hardiness, longevity (the oldest animals reach 17 to 18 years of age), fertility, and foraging ability.

POLLED HEREFORD

Before the 1960s, Polled Herefords were generally regarded as a segment of the Hereford. Today, they are recognized as having attained the status of a major beef breed. In 1995, the American Polled Hereford Association and the American Hereford Association merged and retained the name of the latter.

ORIGIN AND NATIVE HOME

Polled Herefords were developed in the United States, in Iowa, by Warren Gammon. A number of attempts were made to develop Polled Herefords, but not until 1898 was the first serious breeding program started with the idea of using registered Herefords exclusively. Warren Gammon's thinking was largely influenced by the writing of Charles Darwin, which prompted the idea of locating a few purebred Herefords that failed to develop horns. In 1901, Gammon contacted 2,500 members of The American Hereford Association to locate some naturally hornless purebred Herefords. Eventually he located 10 females and 1 bull which became the foundation for the Polled Herefords.

The bull included in the 11 head of cattle, which Gammon started with, established the breed more than any other animal. His name was Giant.

The American Polled Hereford Association began in 1907 with Warren Gammon as secretary, followed, in 1910, by his son, B. O. Gammon, who had been associated with his father in the breeding operations.

Grade cattle, resembling Herefords, but without horns, were originally registered as *single standard* and they were registered only in the forerunner of the American Polled Hereford Association—the American Polled Hereford Club. Herefords without horns, but eligible for registration in both The American Hereford

Fig. 3-16. Warren Gammon (1846–1923), who, in 1901 and 1902, assembled the 11 Hereford mutations from which the Polled Hereford breed developed. (Courtesy, B. O. Gammon)

Association and the American Polled Hereford Association, were recorded in both associations and called *double standard*. This distinction is no longer made, and present-day Polled Herefords trace their ancestral lines to the original English herd books of Hereford cattle, since single- and double-standard cattle were always kept in separate herd books.

CHARACTERISTICS

With the exception of polledness, Polled Herefords are very similar to Herefords. They are red with white markings—with a white face and white on the underline, the flank, the crest, the switch, the breast, and below the knees and hocks. White back of the crops, high on the flanks, or too high on the legs is objectionable. Likewise, dark or smutty noses are frowned upon.

Of course, horned animals are disqualified from registration. Also, no calf is eligible for registration unless its sire was at least 10 months of age at the time of conception, and its dam at least 20 months of age at the time of calving.

Fig. 3-17. The Polled Hereford bull Giant 101740 AHR 1APHR, the sire that Warren Gammon used most extensively, beginning in 1902. The occurrence of the polled characteristic within the horned Hereford breed is an example of a mutation of economic importance. Out of this gene change arose the Polled Hereford breed of cattle. (Courtesy, B. O. Gammon)

Fig. 3-18. Polled Hereford bull, Grand Champion at the National Western Stock Show, Denver, CO, 1994. (Courtesy, American Polled Hereford Assn., Kansas City, MO)

RED ANGUS

Red Angus cattle are not really a distinct breed. Rather, they evolved as a result of specific color in an established breed.

ORIGIN AND NATIVE HOME

That two Black Angus parents might produce a red calf was common knowledge among Angus breeders. Even in the breed's homeland of Scotland, Angus cattle with the red coat color are registered in the herd book along with the black cattle.

The red color results from the effects of a simple recessive gene. When the bull and the cow both pass a red color gene to the offspring, the coat color is red. Since red is a recessive gene, black animals can possess it and when mated to another black carrier of the red gene, they will, on the average, produce 25% Red Angus. A Red Angus bred to a Red Angus always produces a Red Angus.

EARLY AMERICAN IMPORTATIONS

In England and Scotland, both Red Angus and Black Angus are registered in the same association, without distinction. In the U.S., however, red-colored animals have been barred from registry in the American Angus Association since 1917. Then, in 1954, the Red Angus Association of America was formed. The Recordation Program of the Red Angus Association is designed for Red Angus or Black Angus with the red gene, and percentage Red or Black Angus cattle.

Fig. 3-19. Red Angus cow and calf. (Courtesy, Red Angus Assn. of America, Denton, TX)

CHARACTERISTICS

Except for the red color, Red Angus cattle are similar to Black Angus cattle.

Disqualifications include any of the following: horns, scurs, or any hornlike growth; white any place other than the underline; dwarfism, double muscling, cryptorchidism, syndactylism, hydrocephalus or other genetic defects; black skin pigmentation; and marble bone disease (osteopetrosis).

RED BRANGUS

Red Brangus are similar in breeding and type to Brangus cattle, except they are red instead of black.

ORIGIN AND NATIVE HOME

The Red Brangus breed originated in the United States from a Brahman X Angus cross, made in 1946. The breed registry was started in 1956.

Fig. 3-20. Red Brangus cow and calf. (Courtesy, American Red Brangus Assn., Austin, TX)

CHARACTERISTICS

As indicated by the name, the breed is red in color. Other distinguishing characteristics include: polled, smooth, sleek coat; broad head, with slightly curved forehead and straight profile; medium-sized and moderately drooping ears; and a crest immediately forward the shoulders of males.

Any of the following is a disqualification: hernia, one testicle, or malformed genitals; small infantile vulva; wry nose, wry tail, double muscling, undershot jaw, overshot jaw, dwarfism, mulefoot, marble bone disease; large, pendulous sheath, long and non retractable prepuce, lack of sheath; underdeveloped teats and/or udder, very large teats, meaty udder, loosely attached udder; extremely wild or nervous; extremely small cattle for their age; animals that are extremely compact, rangy, or light boned; hard horns; or excessive Brahman or Angus type.

SALERS

In France, Salers (pronounced *Sa-lair*) are used for meat, milk, and work; and they have given a good account of themselves in all three areas. Also, the severity of the environment in which they have lived has produced a breed that can endure hardships.

ORIGIN AND NATIVE HOME

The Salers breed acquired its name from the Salers district, a mountainous region in south-central France—one of the most rugged parts of France. The name *Salers* was first applied to the cattle of this area in 1880. The *Salers Herd Book* was started in 1906.

In its native land, the breed is noted for hardiness, adaptability, and rapid gains.

EARLY AMERICAN IMPORTATION

The Salers were brought to Canada before they were introduced into the United States. The first semen in the United States came from Canada in 1974, and the first bull, Jet, was brought to this country in 1975.

The American Salers Association was formed in 1974. Half bloods may be registered from all breeds except Holstein, Guernsey, Jersey, and Ayrshire. Cows that have had two outstanding performance calves may be used as the source for embryo transplants and the calves may be registered.

Fig. 3-21. Salers cow and calf. (Courtesy, American Salers Assn., Englewood, CO)

CHARACTERISTICS

Salers are deep cherry red, with white switch and sometimes white spots under the belly. The hair is medium to long, and often curly. The skin in exposed areas is rose-colored, and the hide is thick. The breed is horned, although there are polled strains. Mature bulls have an average weight of 2,530 lb, and mature cows average about 1,540 lb.

The breed registry makes the claim that no genetic defect and no double muscling have ever been reported in the Salers breed.

SANTA GERTRUDIS

The Santa Gertrudis breed was developed by the famed King Ranch of Texas and named after the Santa Gertrudis Land Grant, granted by the Crown of Spain, on which the breed evolved, now the headquarters division of the King Ranch.

Santa Gertrudis cattle carry approximately ⅝ Shorthorn and ⅜ Brahman breeding.

ORIGIN AND NATIVE HOME

The Santa Gertrudis was the first new beef breed created in North America. The development of the Santa Gertrudis breed dates back to the early 1900s when the King Ranch in Texas recognized the inability of Hereford and Shorthorn cattle to produce profitably under the adverse environmental conditions of South Texas. In 1910, experimental crossings of Shorthorns and Brahmans were initiated, and in 1920, the outstanding bull calf, Monkey (so named because of his playful antics), was born. In 1923, Monkey was used in a breeding herd of first-cross Brahman X Shorthorn red heifers. Selective line breeding, occasional inbreeding, and highly skilled mass selection were used to develop a uniform breeding herd which passed on Monkey's outstanding characteristics. At the time of his death in 1932, Monkey had produced over 150 valuable sons. By the mid-1930s, the King Ranch was selling a few bulls to other ranches, and in 1940 the U.S. Department of Agriculture recognized the Santa Gertrudis as America's first beef breed.

In 1951, Santa Gertrudis Breeders International (SGBI) was incorporated. A breeder must present his eligible animals to an SGBI field director, who classifies them according to the Standard of Excellence. Males and females with sufficient desirable breed characteristics to meet the minimum standard for the *certified purebred* category are branded with an *S*. Females with only three top crosses or four top-cross females that do not quite meet the purebred-type qualifications are recognized as *accredited* and branded with an S̲.

CHARACTERISTICS

Santa Gertrudis cattle are red or cherry red and generally horned, though some polled strains exist. Their hair should be short, straight, and slick. Their hide should be loose with surface area increased by

Fig. 3-22. Modern Santa Gertrudis bull. (Courtesy, Santa Gertrudis Breeders International, Kingsville, TX)

neck folds and sheath or navel flap, but neither should be excessive.

Brood cows are noted for ease of calving, milking ability, mothering instinct, and longevity.

Disqualifications include: hernia, cryptorchid or malformed genitalia; non retractable parietal prepuce, loose, 90° pendulous, close to the ground; excessive sheath development and large skinfold forward of orifice; wry nose, overshot or undershot jaw; wry tail; or double muscling. Also, an animal is disqualified for white spots out of the underline, white underline exceeding 50%; fawn or cream color, brindling or roan, or solid black skin.

Fig. 3-23. Sittyton, where Amos Cruickshank—the beloved herdsman of Aberdeenshire—developed the *Scotch* strain of Shorthorns. (Courtesy, Mr. Arnold Nicholson)

SHORTHORN

The Shorthorn was the first breed of cattle to have a herd book, the *Coates Herd Book*, founded in 1822 by George Coates. Also, the Shorthorn was one of the first beef breeds to be brought to America—the first importation came in 1783. The name *Shorthorn* is derived from the fact that, through selection and breeding, the early improvers of the breed shortened the horns of the original Longhorn cattle that were native to the district.

ORIGIN AND NATIVE HOME

Shorthorn cattle originated in England, in the northeastern counties of Durham, Northumberland, York, and Lincoln. The time of their introduction into England is not known, but northern England is said to have been their home for centuries.

The names of many illustrious English breeders loom large in the formation of the Shorthorn breed; among them, Robert Bakewell, the Colling brothers—Charles and Robert, the Booth family, and Thomas Bates.

Amos Cruickshank of Sittyton, born in 1808, became the greatest of all Scotch Shorthorn breeders. He spent considerable time and effort visiting herds of England and Scotland, and when he found the type of cattle he wished to raise at Sittyton he paid a good price. His brother, Anthony, served as his partner, keeping records and advertising the herd. After 1860, Cruickshank used the bull, Champion of England, calved in his herd, and began a system of inbreeding to concentrate his blood. Before his death in 1895, Amos Cruickshank had sold over 1,900 animals from his herd.

Polled Shorthorn cattle evolved in the United States, in the North Central area, chiefly in Minnesota. Until 1919, Polled Shorthorns were known as Polled Durhams. Today, Polled Shorthorns are registered in the *American Shorthorn Herd Book* along with horned Shorthorns. There is only one Shorthorn breed, which includes both horned and polled animals.

EARLY AMERICAN IMPORTATIONS

Records show that the first exportation of Shorthorn cattle to the United States left England in 1783, but no record of the number involved or their particular breeding exists. In 1791, a second shipment was sent to New York, but little is known of them. In 1812, a Mr. Cox imported a bull and two cows from England. Then, in 1817, Samuel M. Hopkins imported the first registered Shorthorn cattle to America, into the Genesee Valley, New York. Between 1820 and 1850, many Shorthorn cattle were imported into the United States, especially to the states of Kentucky and Ohio. From 1854 to 1856, Scotch Shorthorns of Cruickshank breeding were brought into Ohio by the Shaker Society.

Importations of Shorthorns increased during World War I, then declined with the recession in prices afterward. Another increase in importations began in 1939 and remained quite active until 1969.

The American Shorthorn Breeders' Association was established in 1882, and in 1889 the Polled Shorthorn Society was established as a branch of the Association. The Association is now the American Shorthorn Association, which registers both horned and polled Shorthorns.

CHARACTERISTICS

Shorthorn cattle are red, white, or any combination of red and white. Their horns are rather short, refined, and incurving. In form, Shorthorns are often more rectangular than other beef breeds. Further, they are deep in body, straight in their topline and considerably

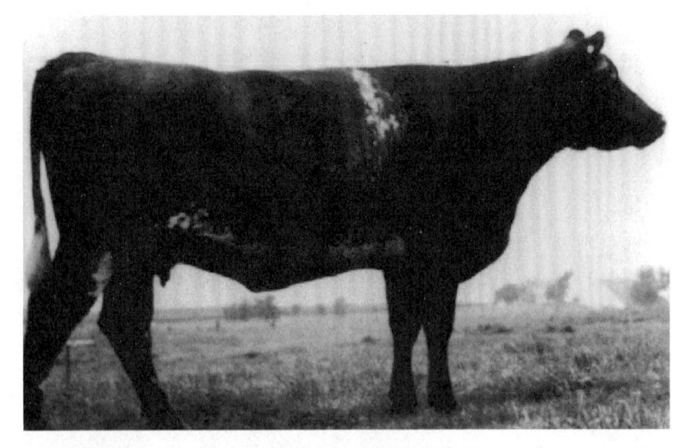

Fig. 3-24. Shorthorn female. (Courtesy, American Shorthorn Assn., Omaha, NE)

refined about the head, brisket, and dewlap. Bulls may reach a mature weight of 2,400 lb and cows a mature weight of 1,700 lb.

No calf is eligible for registration unless its sire and dam were each at least 18 months of age at the birth date of the calf. Other registration requirements include: (1) Animals must be under 18 months of age at the time of registration (otherwise, they must be approved by the Executive Committee); (2) animals cannot be polled; and (3) sire and dam must be registered in the herd book. A *smutty nose*, or *dark nose*, is objectionable.

SIMMENTAL

In Switzerland, their native home, Simmental cattle are considered triple-purpose cattle, noted for (1) their production of dairy products—milk, butter, and cheese; (2) their value as draft animals; and (3) their production of meat.

ORIGIN AND NATIVE HOME

This breed derives its name from the Simme Valley, in the canton of Berne, in Switzerland. Simmental cattle, or cattle similar to the breed, were in the valley in the 18th century, but, in 1806, the government of Berne set up a herd register, with a performance requirement for entry. As the breed spread throughout Europe, it became known by several different names: Pie Rouge in France, Austrovich in Austria, and Fleckvich in Germany.

EARLY AMERICAN IMPORTATIONS

Early importations of Simmentals occurred in 1890 and in 1914. However, it was not until 1967, when a group of Alberta cattle producers headed by Travers Smith of Cardston, Alberta, Canada, imported Simmentals, that the breed began to grow and become a major American breed. In 1968, semen from Canadian Simmental bulls became available in the United States. The first pure bred Simmental bull was imported into the United States in 1971.

The American Simmental Association was established in 1968. Percentage cattle are classified on the basic fractions $\frac{1}{2}$, $\frac{3}{4}$, $\frac{7}{8}$, and $\frac{15}{16}$. Third-cross heifers ($\frac{7}{8}$) and fourth-cross bulls ($\frac{15}{16}$) are considered purebreds. Performance records are required for registration.

CHARACTERISTICS

Simmental cattle are generally red-and-white spotted, although some are nearly solid in color. Usually the face, underline, and switch of the tail are white. There is often red around the eyes. Like the Hereford, the white face is dominant in inheritance. The red varies from dark to a more common dilute, almost yellow, shade. Also, black Simmental have emerged as a result of the upgrading process involving predominately Angus cattle. Breeders who prefer the black color will find increasing numbers of purebred Simmental available. Simmental cattle are horned, although polled strains exist in the United States.

Simmental cattle combine meat and milk to an unusually high degree, along with rapid growth rate. The breed milk production averages about 8,000 lb with a 4% butterfat test. Mature bulls in good condition may exceed 2,500 lb, and mature cows often exceed 1,600 lb. Size, weight gains, and performance records have contributed to their popularity.

Any genetic unsoundness disqualifies an animal from registry. Also, any animal whose sire cannot be confirmed through blood typing cannot be registered.

Fig. 3-25. Simmental cow and calf. (Courtesy, American Simmental Assn., Bozeman, MT)

TARENTAISE (Tarine)

The Tarentaise has been extolled chiefly for its maternal characteristics; the females calve with greater ease than many of the breeds and give more milk than most of the beef breeds. Tarentaise bulls are well suited for use on first-calf heifers, because the calves are small and vigorous at birth.

ORIGIN AND NATIVE HOME

The Tarentaise breed originated in France, in the Alps, where the cattle were known as Tarentaise beginning in about 1859. In 1888, a herd book was established for the breed.

EARLY AMERICAN IMPORTATIONS

Alpine, the first bull of the breed brought to North America, was imported to Canada in 1973. Relatively few purebred Tarentaise cattle have been brought to Canada and the United States. Most of the breed in America has evolved through a grading-up process, using semen.

Charter membership in the American Tarentaise Association was closed January 1, 1974.

CHARACTERISTICS

Tarentaise cattle are solid wheat-colored, ranging from light cherry to dark blond, with black pigmentation of the muzzle and around the eyes. Bulls tend to darken around the neck and shoulders with maturity, and frequently they have a dark dorsal stripe. The breed is polled.

The breed is noted for easy calving, due to adequate pelvic capacity and small calves; vigorous calves at birth; hardiness; rarity of cancer eye and sunburned udders, due to black hair around the eyes and pigmented udders and teats; good fertility; and superior milking ability, with cows averaging around 8,000 lb per lactation. The Tarentaise is smaller than most of the exotics; mature bulls average about 1,800 lb and cows about 1,150 lb. Tarentaise are not docile in disposition as compared to most breeds of beef cattle.

Fig. 3-26. Tarentaise bull. (Courtesy, American Tarentaise Assn., North Kansas City, MO)

In France, the breed registry advocates eliminating (although it does not disqualify) widespread patches of white hairs or badger gray coloring, bright red or mahogany overall color, a stripe on the back lighter than the general coloring; very dark or black parts of the coat (cheeks, dewlap, shoulders, etc.); total absence of black pigmentation on mucous membranes and extremities; and poor general conformation, particularly a crest shaped tail.

TEXAS LONGHORN

Although numerous in the mid to late 1800s, the Texas Longhorn was driven to near extinction by 1900, replaced by the European breeds—the Shorthorn, Hereford, and Angus. Then in 1927, the U.S. Congress appropriated $3,000 for the purpose of preserving the breed. Out of this a nucleus herd was established by the federal government on the Wichita Mountains Wildlife Refuge in Oklahoma. From the Wichita herd a second herd was started on the Fort Niobrara National Wildlife Refuge at Valentine, Nebraska. In the early 1930s, a small herd was donated to the State of Texas.

The Texas Longhorn Breeders Association of America was organized in 1964. At that time, surveys showed that approximately 1,500 head of genuine Texas Longhorn cattle were in existence—about one-third of which were in the two federal refuges, the State of Texas herd, and in zoos and parks. The rest were in private herds.

ORIGIN AND NATIVE HOME

Texas Longhorns originated in the United States, from cattle of Spanish extraction. On his second voyage in 1493, Columbus brought Spanish cattle to Santo Domingo. In the two decades following the Civil War, an estimated 10 million head of Texas Longhorns were trailed north, either for fattening on midwestern pastures or for slaughter for the eastern market.

For a time, Longhorns were kept for sentimental reasons, but recently interest in the breed has greatly increased. They are commanding record prices for breeding stock from serious cattle producers.

Fig. 3-27. Texas Longhorn bull, the 1994 TLBAA World Grand Champion bull, owned by Bob and Linda Moore, Gainesville, Texas; and John T. L. Jones, Jr., Lexington, Kentucky. (Courtesy, Texas Longhorn Breeders Assn. of America, Fort Worth, TX)

Animals on the Wichita Refuge represent only one of seven major families of Longhorn cattle. Several other foundation herds of Longhorns are maintained in privately owned herds and state parks.

CHARACTERISTICS

Texas Longhorns are characterized by a great array of colors, in all degrees of richness, and in all possible combinations and patterns. Their large, spreading horns curve upward; a pole measurement of 40 in. or more tip to tip is desirable on mature bulls and cows. The head on a Texas Longhorn should be of moderate width with pronounced length. Mature bulls average 1,300 to 1,900 lb, and mature cows average 850 to 1,200 lb.

Texas Longhorns were shaped by nature. They are noted for fertility, ease of calving, hardiness, resistance to many common diseases, rustling ability, good feet and legs, longevity, and adaptation to a wide range of environmental conditions.

Registry disqualifications include any of the following: (1) any hereditary deformities that occur in cattle, such as hernia, cryptorchid, or malformed genitalia; (2) extra large sheath or navel flap; (3) any evidence of a hump over the shoulder region; (4) extremely large, droopy ears; (5) wry nose, overshot or undershot jaw; (6) horns under 24 in. tip to tip at 4 to 5 years of age; (7) wry tail; or (8) double muscling.

COMPOSITE BREEDS (also called Synthetics or Hybrids)

A composite is a population made up of two or more component breeds, designed to retain heterosis (hybrid vigor) in future generations without crossbreeding. A composite is maintained like a pure breed. Most scientists agree that a 50-50 mix of British and Continental breeding is ideal in forming a composite for much of North America. In those regions where feed is limited, a higher percentage of British breeding should be used. In regions of abundant feed and/or where maximum lean is desired, more Continental breeding should be used. In subtropical regions, one-quarter to one-half *Bos indicus* breeding is recommended.

Properly done, developing a composite requires (1) a large population of females (500 to 750 cows) and 25 or more sires per generation, (2) three generations of *inter se* (within-herd matings or crosses) following the initial cross, and (3) liquidating the original parent stock.

Compared to traditional rotational crossbreeding systems, composites are attractive to commercial cattle producers because they (1) are less cumbersome to manage, especially in small herds, (2) are easier to manage under intensive, short-duration grazing systems, (3) avoid the wide swings in biological type (size, milk, carcass composition, etc.) that often occur from one generation to another in rotational systems, and (4) make it possible to maintain a relatively high percentage of heterosis as long as inbreeding is avoided.

The more breeds in a composite, the higher the heterosis that can be retained. It can range from 50% of maximum possible heterosis for a two-breed composite to 87.5% for an eight-breed composite.

Selection of the breeds that go into the composite is most important. Breed differences (complementary) should be fully exploited so as to match the composite with the environment in which it will be used, and with the market demands (carcass size, lean yield, marbling, etc.). Additionally, attention should be given to lowly heritable traits and to traits that are difficult to measure (*e.g.*, disposition, udder, sheath, etc.).

NEW AND/OR LESS POPULOUS BREEDS/COMPOSITES/BUFFALOS

In addition to the more widely distributed breeds of beef cattle covered in earlier sections of this chapter, the new and/or less populous breeds, composites, and buffalos of importance in different sections of the United States and Canada are presented in alphabetical order and in summary form in Table 3-2.

Cattle producers, cattle feeders, beef packers, and beef retailers are in agreement that the spawning of many U.S. breeds of beef cattle, including composites, along with increased crossbreeding, has made for lack of consistency in the end product—beef. This has turned consumers off. Rectifying this situation calls for (1) fewer and/or less diverse breeds and (2) value-based marketing. The beef "team" needs to identify precisely, and without fail, what beef consumers enjoy and will buy, then fill this need consistently and competitively. The future of beef is now!

TABLE 3-2
NEW AND/OR LESS POPULOUS BREEDS/COMPOSITES/BUFFALOS[1]

Breed	Place of Origin; First U.S. Importation	Color	Distinctive Head Characteristics	Other Distinguishing Characteristics	Disqualification; Comments
American Breed (1/2 Brahman, 1/4 Charolais, 1/8 buffalo, 1/16 Hereford, and 1/16 Shorthorn) The name, according to Art Jones, the breed founder, reflects the mixture of breeds from which the American Breed originated.	On Art Jones' Cactus Ranch, Portales, New Mexico.	Color characteristics of the breed defy description. Some of the cattle are white like their Charolais ancestors, whereas others are dark or marked with various color patterns, showing their buffalo, Hereford, and Shorthorn background.	The head is not distinctive.	The American Breed is moderate in size; mature bulls weigh from 1,800 to 2,400 lb *(818 to 1,091 kg)*, and mature cows from 1,100 to 1,500 lb *(500 to 682 kg)*. The American Breed was selected for (1) doing well on alkaline range grass, (2) ability to travel long distances for water, (3) high fertility, (4) easy calving, (5) high percentage calf crop of small calves with high weaning weights, (6) high natural immunity to most diseases and parasites, and (7) carcass cutability and quality.	Lack of performance is the only disqualification. Art Jones, founder of the breed, considers the infusion of 1/8 buffalo blood to be one of the most important ingredients in the mixture of breeds from which the American Breed originated.
Amerifax (5/8 Angus, 3/8 Beef Friesian)	U.S.; in Kansas, Nebraska, South Dakota, and Wyoming.	Solid black or red.	Polled.	Moderate mature cow weight of about 1,000 lb *(454 kg)*.	Name stands for American Friesian X Angus cross.
Ankole-Watusi	Ankole district of Uganda, also Rwanda, and Kenya, Africa. Imported to the U.S. in 1960 from Sweden to the Catskill Game Farm, New York. The Ankole-Watusi international Registry was formed in 1983.	Dark red, black, white, gray, brown, yellow dun, also spotted with small spots or large white splashes.	Ankole-Watusi cattle have the largest horns of any cattle in the world. They have large, uprising, outswept horns, 5 ft *(1.5 m)* long and 16 in. *(41 cm)* in diameter. Up to 6 ft *(1.8 m)* between horn tips. Head long and face straight or slightly dished.	Slight hump; moderate dewlap; good depth of body; strong topline; moderate to fine bone.	A variety of deleterious genetic recessives, such as arthrogryposis, double muscling, hydrocephalus, dwarfism, syndactylism, cryptorchidism; polled animals.
Barzona (1/4 each Africander, Hereford, Angus, and Santa Gertrudis)	U.S., on the Bard Ranches of Kirkland, Arizona; hence the name *Barzona* a contraction of *Bard* and *Arizona*). The foundation of the breed, which was laid in 1942, consisted of Africander, Hereford, Shorthorn, Brahman, and Angus. The Barzona Breeders Assn. of America was founded on March 28, 1968.	Red.	Horned, long head, straight profile.	Well adapted to arid and semi-arid ranges (the Bard Ranches, where the breed was developed, have an average rainfall of 12.76 in. *[80 cm]*).	The breeding program followed by Bard Ranches to create the Barzona breed consisted of (1) forming a large genetic pool by crossing breeds, then breeding within the herd, and (2) using records to eliminate undesirable genes and retain desirable genes.

(Continued)

TABLE 3-2 (Continued)

Breed	Place of Origin; First U.S. Importation	Color	Distinctive Head Characteristics	Other Distinguishing Characteristics	Disqualification; Comments
Barzona (continued)					In the formative stage of the breed, production records were maintained and rigid selectivity was carried out on fertility, rate of gain, and mothering characteristics. Disqualifications: Any heritable defects or deformities; solid black color; white on bull other than underline; white on cow other than head or underline.
Beefalo (3/8 buffalo [bison] and 5/8 domestic cattle)	U.S., by D. C. Basolo, Tracy, California. It is claimed that the first fertile Beefalo bull was produced in 1966.	Multicolor.	The head is not distinctive. Horned/polled.	The breed is promoted for its foraging ability, adaptability, hardiness, calving ease, low maintenance cost, and long life.	Earlier bison X domestic cattle crosses in Canada were not successful. Foundation sires are those developed by Basolo and those approved by the Association. The Association registers full-blooded American Beefalo and percentage offspring.
Beef Friesian	U.S.; based on dual-purpose Friesians brought from Europe, beginning with an importation from Ireland in 1971. In Europe, the Friesian has always been a dual-purpose animal, whereas the American descendant, the Holstein-Friesian, has been developed exclusively as a dairy breed. The Beef Friesian Society was organized in 1973.	Black and white, or black. Beef Friesians X Angus cattle are generally black.	Horned or polled. Broad muzzle, open nostrils, strong jaw, broad and moderately dished forehead, straight bridged nose.	Rate and efficiency of gains comparable to the exotics; little calving difficulty; good milking ability.	Beef Friesians are being developed by three approaches: (1) from purebred Beef Friesians, based on European stock; (2) through crossing Beef Friesian bulls on Holstein females, thence grading up; and (3) from Beef Friesian X Angus crosses.
Belgian Blue	Belgium	White, blue, or black and white.	Horned or polled.	Heavily muscled.	Belgian Blues were imported through Canada in 1986.
Belted Galloway	Scotland; in the southwestern district of Galloway. First imported into the U.S. in 1950. The Belted Galloway Society, Inc., was established in 1962.	Black with a brownish tinge, or dun; with a white belt completely encircling the body between the shoulders and the hooks.	Polled.	Striking white belt; heavy coat of hair.	Red color, incomplete belt, other white marks, or scurs.
Blonde d'Aquitane	Southwest France; in 1961, when three French strains of similar background—Garonne,	Yellow, brown, fawn, or wheat colored.	Horned.	The breed is long bodied, long rumped, and relatively fine boned. There is little calving	In France, Blonde d'Aquitanes are usually performance and progeny tested. Generally, the

(Continued)

TABLE 3-2 (Continued)

Breed	Place of Origin; First U.S. Importation	Color	Distinctive Head Characteristics	Other Distinguishing Characteristics	Disqualification; Comments
Blonde d'Aquitane (continued)	Quercy, and Pyrenee-nee—combined. Also, there were infusions of Shorthorn, Charolais, and Limousin blood. The American Blonde d'Aquitane Assn. was established in 1973.			difficulty, due to the width and shape of the pelvis. The breed is considered to be large.	top third of the bulls in a performance test are subsequently progeny tested.
Boran (Zebu type)	Ethiopia	White, grey, red.	Horned or polled.	Drought and heat resistance. Longevity.	
Braford (approx. 5/8 Hereford and 3/8 Brahman)	U.S.; on Adams Ranches, Fort Pierce, Florida, beginning about 1948. Breed registry formed in 1973. Evolved from crossing Brahmans and Herefords.	Red or brindle, with white markings on the head and pigmentation around the eyes.	Both horned and polled strains.	Short haired; heat tolerant; only a slight hump; fertile; good milk production. Mature bulls weigh 1,500 to 2,000 lb *(682 to 909 kg)* and cows 1,000 to 1,500 lb *(454 to 682 kg)*.	Offspring cannot be registered if they are from cows that have not calved annually, that required veterinary assistance at calving, or that have bad udders. For registration, the Association requires pedigree, performance records, and that the animal pass an inspection.
Brah-Maine (3/8 Brahman, 5/8 Maine-Anjou)	U.S.	Dark red to black and white.	Horned.		This is a composite breed.
Brahmental (min. 1/4 Brahman and min. 3/8 Simmental)	U.S.	Various colors.			This is a composite breed.
Brahmousin (5/8 Limousin and 3/8 Brahman)	Southern U.S.	Red, light red, or blonde.	Polled, or polled with scurs.	Optimum milk. Good mothering.	This is a composite breed.
Bralers (5/8 Salers, 3/8 Brahman)	In the U.S., on R-NOL-D Farms, Brenham, Texas.	Dark mahogany.	Horned. Long head, straight profile.	Efficient feed utilization, high fertility, and ease of calving.	Breed rules and disqualifications are the same as for Brangus.
Braunvieh	Switzerland. Imported to U.S. in 1983.	Various shades of brown.	Horned. Lighter colored around muzzle.	Lighter colored inside of legs. Small calves at birth.	
British White	Italy	White, with black ears, nose, and feet. Dark skin.	Polled.	Ease in calving. Abundant milk.	Horns.
Buffalo (*Bison bison*) Bison belong to the bovid family.	North America. In 1913, *Bison bison* was immortalized on the U.S. nickel.	Brownish-black front. Brown behind.	Long, coarse hair on head, with a beard. Horns.	Large head and neck. Humped shoulders.	Buffalo are not considered a breed. But two associations are registering them. In 1995, an estimated 200,000 buffalo roamed the western range. Some are raised for meat. Cattle producers fear buffalo will spread brucellosis.

(Continued)

TABLE 3-2 (Continued)

Breed	Place of Origin; First U.S. Importation	Color	Distinctive Head Characteristics	Other Distinguishing Characteristics	Disqualification; Comments
Cash	U.S.	Various colors.			Composite breed, combining Charolais, Angus, Swiss, and Hereford.
Charbray (3/4 Charolais X 1/4 Brahman to 7/8 Charolais X 1/8 Brahman)	U.S.; in the Rio Grande Valley of Texas, beginning in the late 1930s. From Charolais X Brahman crosses.	Light tan at birth, but usually change to a creamy white in a few weeks.	Horned or polled.	A slight hint of the Brahman dewlap remains. The Charbray has the growth thrust of the Charolais and the heat-insect tolerance of the Brahman.	Purebred Charbrays must be 3/8 Brahman and 5/8 Charolais.
Chargrey (Charolais X Murray Grey)	Australia	Various colors.			This is a composite breed.
Charswiss (Brahman X Brown Swiss)	U.S.	Various colors.			This is a composite breed.
Chiangus (3/4 Chianina X 1/4 Angus)	Clayton, Ohio	Black or dark brown.	Polled. Scurs acceptable.		A composite. Disqualified if any color other than black or dark brown, horned, or having genetic defects.
Chiford (Chianina X Hereford)	U.S.	Marked like a Hereford.	Horned or polled.		No more than 75% Chianina.
Chimaine	U.S.	Various colors.	Horned or polled.		Any parentage up to and including 1/4 Chianina.
Cracker Cattle (also called *Florida Scrub* or *Piney Woods*)	Florida	Various colors.			Not a breed. But a breeders association formed to preserve the descendants of the Spanish cattle brought to Florida.
Devon	England; in the counties of Devon and Somerset. Red cattle from the Devon area were brought to the Plymouth Colony in America in 1623. Vol. 1 of the *Devon Record* was published in 1881.	Red. A rich dark red is preferred; hence the name "Ruby Red."	Creamy white horns with black tips. Also, there are polled strains.	Switch varies from whitish red to nearly white at tip. Skin is orange-yellow with pigment especially noticeable around the eyes and muzzle.	Double muscling. Dwarfism. Excessive white color.
Dexter	Ireland; in the southern and southwestern parts. These animals were named after their founder, a Mr. Dexter.	Black or red.	Horned. Head is rather long.	Small size and short legs, with smallness accentuated by shortness of legs from knees and hocks down. Mature bulls should not exceed 1,000 lb *(454 kg)* and mature cows 800 lb *(364 kg)*. Some mature animals are less than 44 in. *(112 cm)* high.	Animals having white other than on the belly, switch, udder, or scrotum are disqualified for registry. Bulldog calves, a lethal condition, occurs in some animals of the breed; but it is a rarity.

(Continued)

TABLE 3-2 (Continued)

Breed	Place of Origin; First U.S. Importation	Color	Distinctive Head Characteristics	Other Distinguishing Characteristics	Disqualification; Comments
Galloway	Scotland; in the southwestern province of Galloway. In 1878, the Galloway Cattle Society of Great Britain was organized. The Galloway was first imported into Canada in 1853. The first importation of registered Galloways into the U.S. came via Canada in 1866. The American Galloway Breeders' Assn. was established in 1882.	Black, dun, belted, white, or red.	Polled.	Thick, wavy hair; hardiness and ability to rustle in cold weather. Easy calving.	Scurs or horns.
Gelbray (Gelbvieh X Brahman)	U.S.	Various colors.	Horned or polled.		Gelbvieh (max. 3/4, min. 5/8), Brahman (max. 3/8, min. 1/4).
Hays Converter	In Canada; by the former Minister of Agriculture, Harry Hays, beginning in 1957. Foundation breeds were Hereford, Brown Swiss, and Holstein. Claimed that the breed converts feed into profit; hence the name *Converter.*	Predominant color is black with a white face, white feet, and a white tail. About 30% are red with white faces. Color is not a factor in selection.		Traits upon which Sen. Hays built the breed are (1) growth; (2) fertility; (3) minimum calving problems; (4) well-attached udders; (5) abundant milk; (6) sound feet and legs; and (7) pigmentation. The Hays Converter is a beef breed. Mature bulls weigh about 2,200 lb *(999 kg)* and cows 1,400 lb *(636 kg).*	The system and steps used to produce the breed were: (1) selected from different breeds the important characteristics needed; (2) combined the genes into one large breeding population; (3) selected intensely for important characteristics and culled ruthlessly for several generations; and (4) measured the resulting animals after hybrid vigor was no longer important, to determine the transmissible genetic superiority. Only performance tested animals are used.
Hereans	Switzerland	Rusty brown or black, with some white on udder.	Horned.		Mountain cattle in Switzerland.
Indu-Brazil (Zebu)	Brazil	Light gray to silver gray; dun to red.	Metrical horns drawing upward and to the rear. Prominent forehead and long, drooping ears.	Prominent hump over the shoulders. An abundance of loose, pendulous skin under the throat and along the dewlap. A voice that resembles a grunt rather than a low.	Brindle color combinations. White markings on the nose or switch. Absence of loose, thick, mellow skin. Weak and improperly formed hump.
King Ranch Santa Cruz ("Cruz" is Spanish for cross) (1/2 Santa Gertrudis, 1/4 Red Angus, 1/4 Gelbvieh)	King Ranch, Kingsville, Texas. The development of the breed started in 1987. The breed was released in 1994, following seven years of research and development.	From light red or honey to Santa Gertrudis cherry red.	Polled or horned.	Rapid growth, early sexual maturity, fertility, ease in calving, moderate milk production, gentle, and superior carcasses.	This is a composite breed, which is adapted to south Texas' harsh climate. Mature cows weigh 1,100 to 1,200 lb *(500–545 kg).* Mature bulls weigh 1,800 to 2,00 lb *(818–909 kg).*

(Continued)

TABLE 3-2 (Continued)

Breed	Place of Origin; First U.S. Importation	Color	Distinctive Head Characteristics	Other Distinguishing Characteristics	Disqualification; Comments
Leachman Hybrids	U.S., in Montana.	Red.	Polled.		Combines Red Angus, Gelbvieh, Simmental, Salers, and South Devon breeds in several different composites.
Mandalong Special	Australia	Reddish tan.	Polled.		Composite of Brahman, Shorthorn, Charolais, Chianina, and British White breeds.
MARC I	Meat Animal Research Center, Clay Center, Nebraska.				Composite of Brown Swiss, Limousin, Charolais, Hereford, and Angus.
MARC II	Meat Animal Research Center, Clay Center, Nebraska.				Composite of Hereford, Angus, Simmental, and Gelbvieh.
MARC III	Meat Animal Research Center, Clay Center, Nebraska.				Composite of Red Poll, Hereford, Angus, and Pinzgauer.
Marchigiana (pronounced *Mar-key-jahna*)	Italy, in the Marche region, around Rome. With the fall of the Roman Empire in the 5th century, nomadic cattle were crossed with the two native Italian breeds of the time—the Chianina and the Romagnola. Out of these crosses evolved the basic foundation stock for the Marchigiana. The *Herd Book* was established in Italy in 1957. The American International Marchigiana Society was founded in 1973.	Grayish white, although bulls may be darker. Dark skin pigmentation, and dark muzzle, switch, and below or around the eyes. Calves are born tan but turn white at about 2 months of age.	Horns that appear small in proportion to the size of the cattle.	Ability to do well under adverse conditions. Mature bulls weigh 2,650 to 3,100 lb *(1,203 to 1,407 kg)*. Mature cows weigh from 1,400 to 1,800 lb *(636 to 817 kg)*.	In Italy, Marchigianas have been very popular in crossbreeding programs with dairy cattle. All 100% Marchigianas must be blood typed to verify parentage.
Murray Grey	Australia, along the Murray River, from a mating of a very light roan (almost white) Shorthorn cow and an Angus bull, first made by the Sutherlands on *Thologolong*, in Murray Valley, near Wodonga, Victoria, Australia, in 1905. Because of the use of Angus bulls following the first cross, the Murray Grey is predominantly Angus. The Murray Grey Beef Cattle Society, of Australia, was formed in 1962; and the American Murray Grey Assn., Inc. was organized in 1970.	Silver-gray color, which adapts them to sunny areas, as well as colder areas.	Polled.	Ease of calving, because of small calves at birth; dark skin pigmentation, which lessens cancer eye; superior carcass; good dispositions. Bulls weigh around 2,000 lb *(909 kg)* at maturity, and females from 1,100 to 1,300 lb *(500 to 590 kg)*.	In the American Murray Grey Assn., Inc., females with 7/8 Murray Grey blood can be registered. Bulls are eligible for registry with 15/16 Murray Grey blood. In addition, Recordation Certificates can be obtained on any crosses of 1/2 or more Murray Grey breeding. A ranch prefix (the owner's last name, the ranch name, or whatnot) is required of each breeder.

(Continued)

TABLE 3-2 (Continued)

Breed	Place of Origin; First U.S. Importation	Color	Distinctive Head Characteristics	Other Distinguishing Characteristics	Disqualification; Comments
Normande	France; in the area of Normandy, Brittany, and Maine. The breed registry was established in 1883. The American Normande Assn. was formed in 1974.	Primarily dark red and white. Colored patches around the eyes give them a "bespectacled" appearance and resistance to cancer eye and pinkeye; and dark pigmentation on the udder prevents sunburn.	Bespectacled eyes, due to dark coloring.	The Normande is known as a dual-purpose breed in France. Mature bulls in good condition average about 2,425 lb *(1,101 kg)* although weights up to 2,850 lb *(1,294 kg)* have been reported. Cows weigh 1,550 to 1,750 lb *(704 to 795 kg)* and produce an average of about 8,800 lb *(3,895 kg)* of milk per year.	For purebred registration, females cannot have less than 7/8 Normande blood and bulls not less than 15/16.
Norwegian Red	Norway, where they are known as Norwegian red-and-Whites. The first importation of the breed into the U.S. was made by the Southern Cattle Corporation, Memphis, Tennessee, in 1973.	Red; red and white.	Horned.	Abundant milk production; excellent feed conversion; and good carcasses. Mature bulls weigh from 2,200 to 2,640 lb *(999 to 1,199 kg)* and mature cows from 1,210 to 1,430 lb *(549 to 649 kg)*.	In Norway, the Norwegian Red-and-White is a dual-purpose breed, kept for both milk and beef. Today, the Norwegian Red-and-White is the dominant breed of Norway; there are only 200 cattle of other breeds. It is noteworthy, too, that 58% of the cattle of Norway are registered purebreds.
Piedmontese (Piedmont)	Italy, where they are the most popular breed.	White or pale gray with black points.		About 80% of the bulls of the Piedmontese breed are double muscled to some degree. In Italy, Italian breeders report 9% higher dressing percentage and twice the steaks from double muscled cattle over normal cattle. Piedmontese cattle command a very considerable premium on the Italian market. Originally, the Piedmontese was considered a dual-purpose breed in Italy, but today it is selected and bred for beef qualities.	In Italy, under their system of intensive care, breeding and calving problems of double-muscled Piedmontese cattle do not appear to be serious. Double-muscled Piedmontese cattle mean to the cattle industry of Italy what broad-breasted turkeys and Cornish cross broilers mean to the U.S. poultry industry; all are meat producers par excellence.
Ranger	U.S.; beginning in 1950, on the following three ranches: Barnes Livestock Company, Riverton, Wyoming; W. W. Ritchie and Family, Buffalo, Wyoming; and Watson Cattle Company, Cedarville, California. The following breeds were used in developing the Ranger: Hereford, Milking Shorthorn, Red Angus, Shorthorn, Beef-	They run the gamut of cattle colors, including both solid and broken colors.		Medium size; hardy; fertile—animals have been selected to calve at an early age, at yearly intervals or less, and without assistance; adequate and persistent milk production; heavy weaning weight; good carcass quality.	The developers of the breed refer to it as a "a 'cow' breed, for the cowman who must have a profitable commercial operation."

(Continued)

TABLE 3-2 (Continued)

Breed	Place of Origin; First U.S. Importation	Color	Distinctive Head Characteristics	Other Distinguishing Characteristics	Disqualification; Comments
Ranger (continued)	master, Scotch Highland, and Brahman. The name *Ranger* was was selected because the breed was developed on, and is adapted to, the range areas of the West.				
Red Poll	England. Imported to U.S. in 1873.	Red, from light to dark.	Polled.	A dual-purpose breed.	Disqualifications: Bulls with one testicle, scurs.
Romagnola	Italy. Imported to U.S. in 1974.	Off-white to light gray.	Horned. Black muzzle.	Heavily muscled.	
RX_3	U.S.	Red.	Polled.		Composite of 1/4 Hereford, 1/2 Red Angus, and 1/4 Red and White Holstein.
Salorn	U.S.	Various colors (but predominantly red).	Horned.		Composite of 5/8 Salers, 3/8 Longhorn.
Scotch Highland (or Highland)	Scotland. The first importation into the U.S. was made in 1883. The American Scotch Highland Breeders Assn. was established in 1948.	Red, yellow, white, dun, black, or brindle.	Short head; long, widespread horns; and heavy foretop.	Long, shaggy hair; short legs; hardiness and ability to rustle in cold weather.	Polled and spotted animals are disqualified. They should be solid color except for an occasional white tip on switch, or white on the underline or udder. Curly hair is a fault.
Senepol	St. Croix, Virgin Islands, by crossing Red Poll and N'Dawa breeds.	Cherry red.	Predominantly polled.	Heat and insect resistant; docile; fertile; good mothers; and good rustlers.	
Simbrah (5/8 Simmental and 3/8 Brahman)	The southern part of the United States. Cattle producers began making the first crosses of Simmental and Brahman in 1968-1969.	Red, straw colored, or gray with white intermixed.	Simbrah may be polled, scurred, or horned.	The breed exhibits some characteristics of both Brahman and Simmental, but less of the Brahman extremes for sheath, hump, and looseness of skin folds. Simbrahs are fertile, hardy, adaptable, disease and parasite resistant, and fast-gaining.	Both percentage and purebred Simbrah are recorded by the registry association.
South Devon	England; originated in southern Devonshire, through infusion of Guernsey blood into the Devon breed. The South Devon has had its own *Herd Book* in England since 1891. Henry Wallace, former U.S. Vice President,	Medium light red color.	Horned.	The South Devon is a dual-purpose breed. Mature bulls weigh 2,000 to 2,800 lb *(909 to 1,271 kg)* and mature cows weigh 1,200 to 1,700 lb *(545 to 772 kg)*. Cows are heavy milkers; they average about 6,550 lb *(2,974 kg)* of	The South Devon is the only breed in England that both (1) receives a Milk Marketing Board premium for rich milk, and (2) qualifies for the British Beef Subsidy. The breed is extolled as a *dam breed*, superior in maternal traits.

(Continued)

TABLE 3-2 (Continued)

Breed	Place of Origin; First U.S. Importation	Color	Distinctive Head Characteristics	Other Distinguishing Characteristics	Disqualification; Comments
South Devon (continued)	brought the first shipment of South Devons to the U.S. in 1936. The North American South Devon Assn. was formed in 1968.			milk per lactation, with 4.2% fat.	During 1969 and 1970, Big Beef Hybrids, Stillwater, Minnesota, imported nearly 200 purebred South Devons.
Sussex	England; descended from the indigenous red cattle that inhabited Sussex and Kent Counties at the time of the Norman Conquest (1066). First registered in England in 1840. First imported into the U.S. in 1883.	Deep mahogany red, white switch, and ivory horns with dark tips.	A high percentage of Sussex cattle are polled. The head is moderately long, and the nose is pale pink.	In England, the breed has earned the reputation as the "butcher's beast," because of the evenness of fleshing, predominance of lean meat, and high dressing percentage.	In the U.S., the Sussex cattle Assn. of America registers cattle in the *English Herd Book*. American entries in the *English Herd Book* commenced in 1967. Sussex cattle were immortalized by Rudyard Kipling in the poem "Alnascher and the Oxen."
Texon (Texas Longhorn X Devon)	Southern U.S. The major breed developers: Dr. Robert M. Simpson, Duncan, Oklahoma; Dr. Gerald A. Engh, Alexandria, Virginia; and cooperating breeders in Texas and Alabama.	Red.	Polled and horned.	Docile; medium-sized bone of good quality; strong muscle pattern; trimness in the middle; females with pronounced femininity; bulls with pronounced masculinity; and medium size (mature cows weigh 900 to 1,100 lb).	The Texon is a genetically engineered, synthetic breed, combining the "made by U.S. nature" of the Texas Longhorn and the grazing ability of the Devon.
Tuli	Africa	Red predominates.	Horned or polled.	Early maturing, heat and tick resistant (adapted to a tropical environment).	
Wagyu	Japan	Dull black to light brown.	Horned or polled.	Highly marbled beef.	Native cattle of Japan are called Wagyu of which there are 3 breeds: Japanese black, Japanese brown, and Japanese polled.
Water Buffalo (there are several types and breeds)	Asia, probably India.	Black, gray, and dun are most common.	The horns sweep out and back.	They are adapted to wet, muddy areas.	In Asia, water buffalo are are used for power, milk, meat, and leather. In the U.S., a limited number of water buffalo are raised for meat.
Watusi	U.S. A breed registry was formed in 1971.	Variable.	Large horns, but horns are not as large as the horns of Ankole-Watusi.		Originated from Ankole-Watusi and Texas Longhorn.
Welsh Black	Wales; where they have long been bred as dual-purpose cattle. The Canadian Welsh Black Cattle Society was formed in 1970. The U.S. Welsh Black Cattle Assn. was formed in 1974.	Black.	Horned, although there is a polled strain in Wales.	High fertility; little calving difficulty; good milk production; adapted to harsh conditions of climate and forage; longevity; relative freedom from sunburned udders and cancer eye.	The major impact of the breed is expected to be on brood cows—as maternal sires in a crossbreeding program. In Wales, white except on udder or scrotal area is a disqualification.

(Continued)

TABLE 3-2 (Continued)

Breed	Place of Origin; First U.S. Importation	Color	Distinctive Head Characteristics	Other Distinguishing Characteristics	Disqualification; Comments
Welsh Black (continued)				Mature bulls weigh 1,800 to 2,000 lb *(817 to 909 kg)*, and cows from 1,000 to 1,300 lb *(454 to 590 kg)*. Cows give 6,000 to 7,700 lb *(2,724 to 3,495 kg)* of milk per lactation.	
Western Red (Hereford X Red Brahman X Red Angus X Gelbvieh)	IL Ranch, Nevada. Owned by Agri Beef Company	Red.	Horned or polled.	Early sexual maturity, ease in calving, excellent mothering ability, hardy and superior carcasses.	This is a composite breed, which is adapted to high desert conditions.
White Park	England. The ancestors of the breed were brought to England by the Romans, in 55 B.C. They were domesticated in the 18th century. During World War II, 5 females and 1 bull of the White Park breed were shipped to the U.S., to insure preservation of the seed stock. The White Park Cattle Assn. of America was formed in 1975.	White, with black or red pigmentation and markings around the eyes, ears, nose, feet, legs, teats, and anal area.	Predominantly polled. White head with black ears, eyes, and nose.	Comparable in size to Angus, Hereford, and Shorthorn cattle. Strong maternal instinct; ease of calving; excellent milking ability; docile and easy to handle.	Overmarking and undermarking are undesirable. The name *White Park* came from large game preservelike enclosures in which these cattle were kept after being domesticated.
Zebu *(Bos indicus)*	India	Variable colors.	Horns.	A large hump on the shoulders.	Used for work in Asia. There are approximately 60 strains (breeds) of Zebu.

[1]The Names and address of the major breed registry associations are given in the Appendix, Table VI-1.

COMPARATIVE BREED RATING CHART[1]

Table 3-3, in which 59 breeds are listed, points up (1) the number of economic traits that should be considered in evaluating breeds, and (2) breed differences in economic traits. As shown, there are wide differences between breeds. *But no attempt is made,*

[1]Documentation and further information on breed comparisons are contained in the following reports:

■ Gregory, K. E., L. V. Cundiff, and R. M. Koch, *Composite breeds to use heterosis and breed differences to improve efficiency of beef production*, USDA-ARS, Roman L. Hruska U.S. Meat Animal Research Center, Clay Center, NE, August, 1992.

■ Cundiff, Larry V., Ferenc Szabo, Keith E. Gregory, R. M. Koch, M. E. Dikeman, and J. D. Crouse, *Breed comparisons in the germplasm evaluation program at MARC*, USDA-ARS, Roman L. Hruska U.S. Meat Animal Research Center, Clay Center, NE, May, 1993.

■ *Beef Research Progress Report No. 4*, USDA-ARS-71, Roman L. Hruska U.S. Meat Animal Research Center, Clay Center, NE, May, 1993.

■ Barkhouse, Kristin L., L. D. Van Vleck, and L. V. Cundiff, *Breed comparisons for growth and maternal traits adjusted to a 1992 base*, USDA-ARS, Roman L. Hruska U.S. Meat Animal Research Center, Clay Center, NE, June, 1994.

■ *Germ Plasm Evaluation Program Progress Report No. 13*, USDA-ARS, Roman L. Hruska U.S. Meat Animal Research Center, Clay Center, NE, June, 1994.

■ Gregory, K. E., L. V. Cundiff, R. M. Koch, M. E. Dikeman, and M. Koohmaraie, *Breed effect, retained heterosis, and estimates of genetic and phenotypic parameters for carcass and meat traits of beef cattle*, USDA-ARS, Roman L. Hruska U.S. Meat Animal Research Center, Clay Center, NE, J. Animal Science, 1994, 72:1174-1183.

or should be made, to rank the breeds by assuming that the traits are of equal value (which is not true) and averaging them. It is important that straightbred or purebred breeders—the breeders of seed stock—recognize the strong points and the weak points of the breed that they choose. Armed with this information, they are in a better position to bring about breed improvement. Also, they will know what the well-informed crossbreeding producers will be looking for in a particular breed. To commercial producers, knowledge of such information as is presented in Table 3-3 makes it possible for them (1) to select more intelligently the breeds to use in a crossbreeding program which will complement each other, and (2) to identify those traits that should receive major attention when selecting individual breeding animals from a particular breed.

In using Table 3-3, it should be recognized that the characterization of breeds for the economically important traits that are controlled by many pairs of genes—such as weight, carcass quality, maternal ability, etc.—is more difficult than a classification based on characters controlled by only a few genes. This is so because wide variation exists within breeds. Nevertheless, breed differences do exist, and knowledge of these differences is important in deciding (1) which breed to raise in different areas of the country, and (2) which breeds to use in crossbreeding programs.

Some producers may be confused because there are "too many breeds" from which to choose. However, a cattle breeding program is slow at best, and major mistakes are costly, simply because it takes so long to replace a cow herd by its own reproduction. It is far better, therefore, to consider all available alternatives at the beginning. Also, it is comforting to know that there is a large assortment of the world's most useful cattle germ plasm from which to choose for the divergent climates, objectives, and economic conditions found in North America.

In addition to being acquainted with the breeds as such, commercial cattle producers who are following a crossbreeding program should select breeds that are complementary; that is, they should select those breeds that possess the favorable expression of traits which they desire in the crossbred cattle that they intend to produce.

The Table 3-3 values are not static. They will change as breeds change and as more experimental work becomes available. Moreover, new breeds are constantly being imported, and several different composite (synthetic or hybrid) breeds are currently being developed. In the meantime, Table 3-3 is presented for the guidance of producers and students who are interested in studying the breeds from the standpoint of crossbreeding programs. Also, the table may serve as a challenge to purebred breeders regarding the traits that need improvement in their respective breeds.

There is no halo around any breed. In the future, the competition among breeds will be intense, based on their performance as both straightbreds and crossbreds. The exotics and composites must prove that their establishment is justified through (1) the value of broadening the germ plasm base, and (2) higher performance compared to the established breeds. Hopefully, many breeds will meet the tests for total performance. Without doubt, some will fail and pass into oblivion. Still others may serve as part of the foundation material for developing entirely new breeds provided they possess one or two desirable traits that can be extracted for use through crossing and selection.

CHOOSING THE BREED

No one breed of cattle can be said to excel all others in all points of beef production for all conditions. Hence, some choices must be made.

The purebred breeder can no longer choose a breed on the basis of breed preference based on (1) fancy points, (2) imparted status symbol, and/or (3) numbers—because it is either very populous or very scarce. Today's cattle industry is too sophisticated and too profit-oriented to permit such luxury. Enlightened breed choice calls for anticipating the future needs of the beef cattle industry and choosing a breed for which there will be great demand (and profit). Breed choice calls for recognizing that few purebred breeders make money from selling cattle to each other; rather, that the vast majority of purebreds, especially bulls, must be sold to commercial cattle producers. Breed choice calls for choosing a breed that will be widely used by the commercial producers of the area under consideration.

Fewer and fewer commercial cattle of the future will be straightbreds. Rather, they will be crossbreds in order to take advantage of complementary traits and heterosis. Once a decision to crossbreed has been reached, the most critical part of the plan is choosing the right parent breeds to fit the producer's objective and environment; then using the chosen breeds in a designed breeding system that will maximize the expression of their desirable traits and minimize the influence of their undesirable traits.

In addition to taking advantage of hybrid vigor or heterosis, crossbreeding should result in a desired combination of traits not available in any one breed. You cannot, for example, breed a bull from a cow family in which poor milkers are commonplace to a poor milking cow and get a heifer that is a superior milker. Neither can you cross breeds that are low gainers and get fast-gaining offspring, although, due to hybrid vigor, the gaining ability of such offspring will be higher than the average of the parents. Those who are crossbreeding cattle also need to know which breeds to

TABLE
COMPARATIVE RATING (for Beef Purposes)

Breed	Mature Size	Cow Traits							Calf Traits	
		Age of Puberty	Conception Rate	Gestation Period	Milk Production	Mothering Ability	Adaptation to Beef Management	Efficiency Under Minimal Management	Calf Birth Weight	Hardiness
American	A	3	2	A	3	2	1	1	1	1
Amerifax	L	2	2	A	2	2	3	3	2	3
Angus	A	2	2	S	3	2	2	2	2	2
Ankina	A	2	2	A	3	3	3	3	3	3
Ankole-Watusi	A	3	3	A	3	2	2	2	2	2
Ayrshire	A	1	3	A	1	3	4	5	2	3
Barzona	A	3	3	A	3	3	1	1	2	1
Beefalo	L	3	3	A	3	3	3	2	2	1
Beef Friesian	L	2	2	A	1	3	4	4	4	3
Beefmaster	A	3	2	A	3	1	1	1	2	1
Belted Galloway . . .	A	2	2	A	2	2	2	2	2	2
Blonde d'Aquitane .	L	4	4	L	3	4	3	3	3	3
Braford	A	4	2	A	3	2	1	1	2	1
Brahman	A	5	5	L	2	1	2	2	2	2
Bralers	A	3	3	A	3	2	2	2	3	2
Brangus	A	3	2	A	3	1	1	1	2	1
Brown Swiss	L	3	4	A	1	2	3	4	4	3
Charbray	L	4	4	L	3	3	3	3	3	2
Charolais	L	4	4	L	3	4	4	4	4	3
Chianina	L	4	3	L	3	4	4	4	5	3
Devon	A	2	2	A	3	2	2	2	2	2
Dexter	S	1	2	S	3	3	4	5	1	3
Galloway	A	2	2	A	3	1	2	1	2	2
Gelbvieh	A	2	3	A	1	3	3	3	3	3
Guernsey	S	1	3	S	1	3	4	4	2	4
Hays Converter . . .	L	2	2	A	2	2	2	4	3	3
Hereford	A	3	2	A	4	3	2	2	3	2
Holstein-Friesian . .	L	2	2	A	1	3	4	5	4	3
Indu-Brazil (Zebu) . .	A	5	5	L	3	1	2	2	2	2
Jersey	S	1	2	S	2	3	3	4	1	3
Limousin	A	4	3	L	3	3	2	3	3	3
Lincoln Red	A	3	2	A	2	2	2	3	1	2
Maine-Anjou	L	3	3	A	2	3	2	3	5	3
Marchigiana	L	4	3	L	3	3	3	3	3	3
Milking Shorthorn . .	A	2	3	A	2	3	3	4	2	2
Murray Grey	A	2	3	A	2	3	2	2	2	3
Normande	A	3	3	A	2	1	2	2	3	2
Norwegian Red . . .	A	2	2	A	2	3	4	4	2	3
Piedmontese	S	2	2	A	2	3	4	5	2	4
Pinzgauer	A	2	3	A	2	2	2	3	3	2

Plate 1. American Breed heifer, shown at 6 months of age and weighing 550 lb (*250 kg*). (Courtesy, Art Jones, Cactus Ranch, Portales, NM)

Plate 2. Amerifax bull.
(Courtesy, Amerifax Cattle Assn., Hastings, NE)

Plate 3. Angus bull, owned by Van Dyke and Sydenstricker Angus Farm, Mexico, MO. (Courtesy, American Angus Assn., St. Joseph, MO)

Plate 4. Ankole-Watusi bull.
(Courtesy, Ankole-Watusi International Registry, Hebron, ND)

Plate 5. Barzona bull.
(Courtesy, Barzona Breeders Assn. of America, Prescott, AZ)

Plate 6. Beefalo bull.
(Courtesy, American Beefalo World Registry, North Kansas City, MO)

Plate 7. Beef Friesian bull. Polled.
(Courtesy, Beef Friesian Society, Johnstown, CO)

Plate 8. Beefmaster bull.
(Courtesy, Foundation Beefmaster Assn., Denver, CO)

Plate 9. Belted Galloway bull, owned by R.M. Rolan Cattle Co., New London, WI.
(Courtesy, Belted Galloway Society, Leeds, AL)

Plate 10. Blonde d'Aquitane bull.
(Courtesy, American Blonde d'Aquitane Assn., North Kansas City, MO)

Plate 11. Braford bull.
(Courtesy, Adams Ranch, Fort Pierce, FL)

Plate 12. Brahman bull.
(Courtesy, American Brahman Breeders Assn., Houston, TX)

Plate 13. Brahmousin cow.
(Courtesy, American Brahmousin Council, Midlothian, TX)

Plate 14. Brangus bull.
(Courtesy, International Brangus Breeders Assn., Inc. San Antonio, TX)

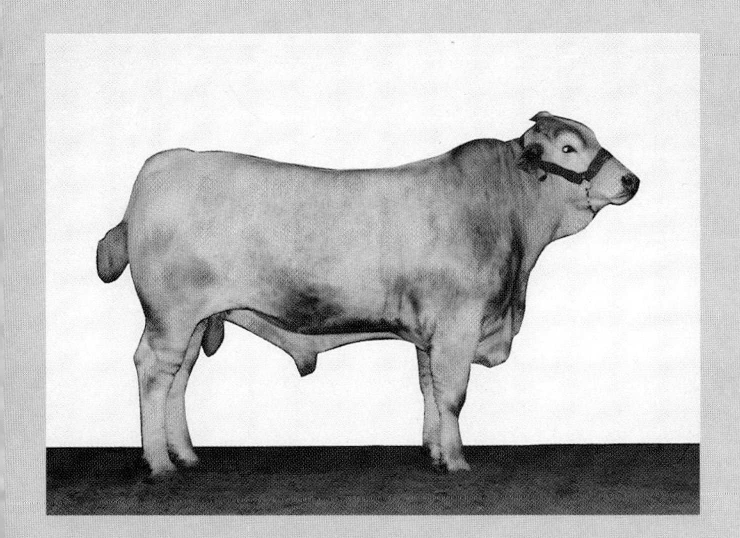

Plate 15. Charbray bull, *TPR Grande 2045.*
(Coutesy, American-International Charolais Assn., Kansas City, MO)

Plate 16. Charolais bull.
(Courtesy, American-International Charolais Assn., Kansas City, MO)

Plate 17. Chiangus cow and calf.
(Courtesy, American Chianina Assn., Platte City, MO)

Plate 18. Chianina bull. Full-blooded. Italian breeding.
(Courtesy, American Chianina Assn., Platte City, MO)

Plate 19. Devon bull, a polled bull owned by Effingham Plantation, Nokesville, VA.
(Courtesy, Dr. S.H. Fowler, Pensacola, FL)

Plate 20. Dexter bull.
(Courtesy, Wes Patton, Orland, CA)

Plate 21. Galloway female.
(Courtesy, Robert G. Mullendore, Greenough Cattle Co., Greenough, MT)

Plate 22. Gelbvieh bull.
(Courtesy, American Gelbvieh Assn., Westminster, CO)

Plate 23. Hays Converter bull, *Harry 20601 F.*
(Courtesy, The Canadian Hays Converter Assn., Calgary, Alberta, Canada)

Plate 24. Hereford bull.
(Courtesy, The American Hereford Assn., Kansas City, MO)

Plate 25. Indu-Brazil (Zebu) bull.
(Courtesy, *American Zebu Journal*, Mexia, TX)

Plate 26. Limousin bull.
(Courtesy, North American Limousin Foundation, Englewood, CO)

Plate 27. Maine-Anjou bull.
(Courtesy, American Maine-Anjou Assn., Kansas City, MO)

Plate 28. A mature 100% Marchigiana cow.
(Courtesy, Marky Cattle Assn., Walton, KS)

Plate 29. Murray Grey bull.
(Courtesy, American Murray Grey Assn., North Kansas City, MO)

Plate 30. Normande cow.
(Courtesy, American Normande Assn., Kearney, MO)

Plate 31. Piedmontese bull, Mesquite Istinto, imported from Italy.
(Courtesy, Ovan Muelker, Muelker Farms, Cost, TX.

Plate 32. Pinzgauer cow and calf.
(Courtesy, Canadian Pinzgauer Assn., Calgary, Alberta, Canada)

Plate 33. Polled Hereford cow and her bull calf.
(Courtesy, American Polled Hereford Assn., Kansas., City, MO)

Plate 34. Red Angus bull.
(Courtesy, Red Angus Assn. of America, Denton, TX)

Plate 35. Red Brangus bull, at 2 years of age.
(Courtesy, American Red Brangus Assn., Austin, TX)

Plate 36. Red Poll cow and calf.
(Courtesy, American Red Poll Assn., Louisville, KY)

Plate 37. Salers bull.
(Courtesy, American Salers Assn., Englewood, CO)

Plate 38. King Ranch Santa Cruz bull.
(Courtesy, King Ranch, Kingsville, TX)

Plate 39. Santa Gertrudis bull, with cow and calves.
(Courtesy, King Ranch, Kingsville, TX)

Plate 40. Scotch Highland female, owned by Glenn & Carol Bluhm, Reed City, MT.
(Courtesy, American Scotch Highland Breeders Assn., Denver, CO)

Plate 41. Shothorn bull, Millbrook Marc IV, syndicated for $100,000.
(Courtesy, American Shorthorn Assn., Omaha, NE)

Plate 42. Simbrah bull.
(Courtesy, *Simbrah World*)

Plate 43. Simmental bull, C&B Junior, foundation sire. C&B Livestock, Inc., Hermiston, OR. (Courtesy, Ron Baker, Pres., C&B Livestock)

Plate 44. South Devon cow and calf.
(Courtesy, North American South Devon Assn., Lynville, IA)

Plate 45. Tarentaise female.
(Courtesy, American Tarentaise Assn., North Kansas City, MO)

Plate 46. Texas Longhorn bull, owned by Robert Simpson, M.D., and S.H. Fowler, Ph.D. (Courtesy, Dr. Fowler)

Plate 47, Welsh Black cow.
(Courtesy, S.H. Fowler, Pensacola, FL)

Plate 48. White Park cow and calf.
(Courtesy, White Park Cattle Assn. of America, Madrid, IA)

3-3
OF ECONOMIC TRAITS OF 59 BREEDS OF CATTLE[1]

Calf Traits		Carcass			Bull Traits			Breed's Place in Crossbreeding		
Growth Rate	Optimum Slaughter Wt., Lb[2]	Cutability (Muscle to Fat Ratio)	Marbling	Tenderness	Fertility	Freedom from Genital Defects	Calving Ease (Size Effect)	Maternal	Rotational	Terminal
3	1,000	3	3	3	2	2	1	x	x	
2	1,050	2	2	2	2	2	2	x	x	
3	1,150	4	1	2	2	2	2	x	x	
2	1,250	2	3	2	2	2	2	x	x	
3	1,000	4	4	3	3	3	2	x	x	
3	925	3	3	2	2	2	2	x		
3	1,050	3	3	3	2	2	2	x	x	
3	1,150	3	3	3	4	2	2	x	x	x
1	1,300	2	3	2	2	1	3	x	x	
2	1,100	3	3	4	2	2	1	x	x	
4	1,000	4	2	2	2	1	2	x	x	
1	1,225	1	4	2	3	2	3		x	x
3	1,100	3	3	4	3	2	1	x	x	
3	1,150	2	4	5	4	4	1	x	x	
3	1,100	2	2	4	3	2	1	x	x	
3	1,075	3	2	4	3	2	2	x	x	
1	1,300	2	3	2	2	2	3	x	x	
2	1,250	2	4	4	3	3	4		x	x
1	1,300	1	3	2	3	2	5		x	x
1	1,350	1	4	3	3	2	2			x
3	1,100	3	2	2	2	1	1	x	x	
5	700	3	3	2	3	2	5			
3	1,100	2	2	2	2	1	4	x	x	x
3	1,250	2	3	3	2	1	2	x	x	x
4	900	2	2	1	2	3	1	x		
2	1,150	2	3	2	2	1	3	x	x	
3	1,150	4	2	2	2	1	2	x	x	
1	1,300	2	3	2	2	1	3	x	x	
3	1,150	2	4	5	4	4	1	x	x	
5	850	4	1	1	1	3	1	x		
2	1,150	1	3	3	3	2	3		x	x
3	1,050	3	2	2	2	1	1	x	x	
1	1,350	1	2	2	3	1	4	x	x	x
3	1,150	3	3	3	2	1	3			x
3	1,000	4	2	2	3	1	2	x	x	
3	1,025	3	2	2	3	2	2	x	x	
3	1,100	3	3	3	3	3	3	x	x	
3	950	3	3	2	2	2	2	x	x	
3	1,050	1	3	2	2	2	3			x
3	1,250	2	2	2	2	2	3	x	x	

(Continued)

TABLE 3-3

Breed	Mature Size	Cow Traits							Calf Traits	
		Age of Puberty	Conception Rate	Gestation Period	Milk Production	Mothering Ability	Adaptation to Beef Management	Efficiency Under Minimal Management	Calf Birth Weight	Hardiness
Polled Hereford . . .	A	3	2	A	4	3	2	2	3	2
Polled Shorthorn . .	A	2	2	A	3	3	2	3	3	3
Ranger	A	3	3	A	3	3	2	3	3	3
Red Angus	A	2	2	S	3	2	2	2	2	3
Red Brangus	A	3	2	A	3	1	1	1	2	1
Red Danish	A	2	2	A	2	3	4	4	2	3
Red Poll	A	2	2	A	2	3	2	3	2	3
Salers	A	2	2	A	2	2	2	2	3	2
Santa Gertrudis . . .	L	3	4	A	3	1	2	3	3	2
Scotch Highland . .	S	2	3	A	4	1	1	1	1	1
Shorthorn	A	2	2	A	3	3	2	3	3	2
Simbrah	L	4	3	L	2	3	2	2	3	2
Simmental	L	2	2	A	1	3	2	3	4	3
South Devon	L	2	3	A	2	3	2	3	3	3
Sussex	A	2	3	A	3	3	3	2	3	3
Tarentaise	A	2	1	A	2	2	3	3	2	1
Texas Longhorn . . .	S	3	1	L	4	1	1	1	1	1
Welsh Black	A	3	2	A	2	2	2	2	2	1
White Park	A	2	2	A	3	2	2	2	2	2

[1]This chart was prepared for the seventh edition of *Beef Cattle Science* by Larry V. Cundiff, Ph.D., Research Leader Genetics and Breeding, USDA, Roman L. Hruska U.S. Meat Animal Research Center, Clay Center, Nebraska.

In the columns above, the following terms and values are used:

- In the "Mature Size" column, A = average, L = large, and S = small.
- In the columns carrying numerical grades, grade 1 is the highest or most desirable, and grade 5 is the lowest or least desirable. For example, in the column headed "Age of Puberty," the number 1 indicates very early puberty, number 5 indicates very late puberty, and number 3 is intermediate.

cross for a certain size animal, for improved carcass quality, and for every conceivable trait. Additionally, they need to consider all the points that follow:

- **Plan and follow a sound system of crossbreeding**—For maximum continuous high expression of heterosis, and maximum beef output per cow unit, it is necessary that producers plan and follow a sound system of crossbreeding. The systems of crossbreeding are fully covered in Chapter 5 of this book, Some Principles of Cattle Genetics; hence, repetition at this point is unnecessary.

- **Start with breed of present cow herd, then complement it**—Practical considerations favor utilizing the breed of the present cow herd (if the producer is already in business), with the additional breed choice made (or with additional breeds chosen) to complement the present breed so as to meet the objectives of the owner. Some producers sell weaner calves, others market yearlings, and still others retain ownership through the feedlot (either their own lot or by custom feeding).

From the above, it may be concluded that the *best breed* is determined by the intended purpose. Smaller cattle have an advantage in efficiency for producing weaning calves, whereas larger cattle have the advantage if they are taken to heavier weights. Certain breeds are noted for superior carcasses. The cattle for use as a cow line in a hybrid operation are not the best for a sire line. Thus, different cattle are required for different markets, production systems, and breeding plans.

- **Cattle tend to grow and develop from conception to maturity in proportion to their mature size**—This means that—

1. As mature size increases, birth weight, weaning weight, yearling weight, and 18-month weight tend to increase; that is, at any given age, weight tends to increase.

(Continued)

Calf Traits		Carcass			Bull Traits			Breed's Place in Crossbreeding		
Growth Rate	Optimum Slaughter Wt., Lb[2]	Cutability (Muscle to Fat Ratio)	Marbling	Tenderness	Fertility	Freedom from Genital Defects	Calving Ease (Size Effect)	Maternal	Rotational	Terminal
3	1,150	4	2	2	2	2	2	x	x	
3	1,150	4	1	2	3	2	2	x	x	
3	1,150	2	3	2	3	2	2	x	x	
3	1,150	4	1	2	2	2	2	x	x	
3	1,150	3	2	4	3	2	2	x	x	
3	950	3	3	2	2	2	2	x	x	
3	1,050	4	2	2	3	2	2	x	x	
2	1,100	2	3	3	2	1	2	x	x	
2	1,150	3	2	4	4	3	2	x	x	x
4	900	3	2	2	3	1	1	x	x	
3	1,150	4	1	2	3	1	2	x	x	
1	1,300	2	3	4	3	2	3	x	x	x
1	1,350	2	3	3	2	1	5	x	x	x
2	1,200	3	2	2	3	2	3	x	x	
3	1,100	3	3	3	4	3	3	x	x	
4	1,100	2	3	4	2	2	1	x	x	
4	1,000	3	3	3	2	1	1	x	x	
3	1,000	3	3	2	2	2	2	x	x	
3	1,050	3	3	2	2	2	2	x	x	

- In the "Gestation Period" column, A = average, L = long, and S = short.
- The check marks in the last three columns are the authors' opinion as to the best use of the breed in selected systems of crossbreeding—as maternal foundation, rotational cross, or terminal cross.

[2]To convert to kg, divide by 2.2.

2. As mature size increases, the age of puberty and size at any given level of finish increases; that is, at any given age, degree of maturity tends to decrease.

As a result of the above relationships, cattle of genetic potential for large size tend to gain faster and more efficiently (require less feed per pound of gain at any given weight). But they have to be carried to older ages and heavier weights in order to obtain market finish. Cows of the larger breeds tend to require more total feed for maintenance (nutrient requirements for maintenance are probably closely proportional to weight to the ¾ power).

- **Use high-performing parent stock**—Both the selection of breeds and a planned program are important to the success of crossbreeding. However, commercial producers must go further—they must use high-performing parent stock.

- **Match or complement the breeds in crossbreeding**—In crossbreeding, the breeder can choose matching breeds. This is important because one breed simply cannot be all things. For example, a steer with a very high rate of gain simply cannot be expected from small cattle; likewise, a cow cannot be expected to have low feed intake for maintenance (efficient maintenance per cow unit) if she is very large. Characteristics which are especially desirable in the cow are: adaptation to area; small size and low maintenance requirement; high fertility; early sexual maturity; ease of calving; good milking ability; and longevity. Bulls should be selected for: large size and high rate of gain; high fertility, with libido; siring calves that make for ease in calving; lean and muscular, with high cutout; and tender palatable beef. Both cows and bulls should be sound and docile.

- **Traits low in heritability affected most by heterosis; highly heritable traits affected most by selection**—In using Table 3-3 it should be understood that the greatest amount of heterosis is expressed in those traits that are rather low in heritability, such as cow fertility, calf survivability, and weaning weight. Highly heritable traits—like rate and efficiency of gain, and carcass quality—are affected most by selection and reflect the average performance of the parent breeds.

■ **Select less related breeds for maximum heterosis**—Crossing less closely related breeds will yield maximum heterosis. An example of choosing breeds to maximize heterosis among the British breeds can be found in the Hereford, Angus, and Shorthorn crossbred data. Since Shorthorns resemble Angus more than Herefords in several economic traits, the Hereford X Angus cross and the Hereford X Shorthorn cross calves have excelled Shorthorn X Angus calves in growth traits.

■ **Adaptation is important**—The ranking of each breed on traits of economic importance (in Table 3-3) is based upon its performance as a straightbred in a region where it is adapted and most frequently used. Hence, it would not be expected that Scotch Highland cattle would perform well in the South, or that Brahmans would perform so well in the North.

Different breeds or breed combinations for straightbreeding or crossbreeding will be desirable for different areas of the country and management systems. In the western range country, a medium-sized cow that is hardy, calves easily, has good mothering ability, and can utilize the maximum of range grass is desired. In the Corn Belt, where feed is more abundant and available with less walking, and cattle can be watched more closely, the larger, less active, and heavier milking breeds may be used. In three-breed fixed or static crossbreeding programs, maternal breeds may be selected for producing crossbred females and a sire of a larger breed with more growth potential and high carcass cutability can be used as a terminal cross for market animals.

■ **Advantag es of the exotics**—In general, the exotics differ from the standard British breeds in one or more of the following ways: they grow faster, produce more milk; and/or have leaner carcasses.

■ **Disadvantages of the exotics**—A major disadvantage shared by many of the exotics is the heavier calves at birth, which can lead to greater calving difficulty. Also, market discrimination, both as feeders and as slaughter animals, frequently occurs with any new breed or cross; but this generally disappears rather fast after satisfaction in the product is evident.

■ **Recommended ways of using each breed in a crossbreeding program**—It is important that the choice of breeds be made with a designed breeding system in mind—one that will maximize the expression of their desirable traits and minimize the influence of their undesirable traits; thus, the last three columns of Table 3-3 indicate recommended ways of using each breed in a crossbreeding system. As noted, most of the breeds are recommended for use in both rotational crossbreeding and as a maternal component towards specific three-breed crossing. The latter are the breeds which have the genetic capacity for exceptional post-weaning muscle growth rate.

FEEDING AND MANAGING THE NEW BREEDS AND CROSSES

The feeding and managing of the new breeds and crosses should satisfy their requirements for (1) more rapid growth rate, (2) larger mature size, (3) increased milk production, and (4) special requirements because of difference in adaptability, winter hardiness, etc.

Bigger cattle produce faster growing calves that finish out at heavier weights. In order to meet current market requirements for finished cattle weighing 1,000 to 1,250 lb, big, rapidly growing calves must receive more milk and be fed more liberally (with consideration given to creep feeding). Also, animals with more red meat and less fat likely require relatively more protein and less energy than the traditional British breeds.

A larger cow has larger maintenance requirements. If her minimum nutritional requirements are not met, poor conception and a lower calf crop will result. Such nutritional inadequacies are likely to become especially serious when rebreeding heifers that calved as 2-year-olds.

The use of a dairy or dual-purpose breed in a crossbreeding program also calls for a special feeding and management program. Heavy-milking cows should be in good condition at calving; restricted somewhat immediately after calving, to lessen the milk flow (so as to avoid hand-milking and spoiled udders, and lessen calf scouring); flushed preceding breeding, so as to insure good conception; and fed rather liberally the latter part of the lactation period.

BREED REGISTRY ASSOCIATIONS

A breed registry association consists of a group of breeders banded together for the purposes of: (1) recording the lineage of their animals; (2) protecting the purity of the breed; (3) encouraging further improvement of the breed; and (4) promoting the interest of the breed. A list of the cattle breed registry associations is given in Table VII-1 in the Appendix.

REGISTERING ANIMALS PRODUCED THROUGH ARTIFICIAL INSEMINATION

The regulations pertaining to the registration of calves resulting from artificial insemination vary considerably among the different breed registry associations. This variation reflects the different policies and attitudes of the management personnel and members of the respective associations.

Table 3-4 is a summary of the artificial insemination regulations of the various breed registry associations. Generally speaking, the information contained therein pertains specifically to the registration of the progeny from out-of-herd artificial insemination; out-of-herd artificial insemination being the use of artificial insemination by a breeder (owner of record of the dam at the time of breeding) who is not the owner or co-owner of the recorded bull being used.

Table 3-4 was compiled by Certified Semen Services, Inc., a subsidiary of the National Association of Animal Breeders, P.O. Box 1033, Columbia, MO 65205. For convenience, the information in Table 3-4 is arranged in alphabetical order by breed name.

TABLE 3-4
ARTIFICIAL INSEMINATION REGULATIONS OF BEEF BREED ASSOCIATIONS

Association	Blood Typing Required		Affiliated Laboratory	Breeding Receipt Required	Non-Owner Certificate Required	Restrictions on Certificates	Disposition of Semen after Death of Sire	General Comments
	Sire	Dam						
Amerifax								
Amerifax Cattle Assn. P.O. Box 149 Hastings, NE 68901	Yes	No	Ohio State Univ.	No	No	None	None	Registration requirements same as natural service.
Angus								
American Angus Assn. 3201 Frederick Blvd. St. Joseph, MO 64506	Yes	Yes	Ohio State Univ.	No	Yes	None	May be used provided the assn. is made aware of the date of death of bull.	
Ankole-Watusi								
Ankole-Watusi International Registry 22484 W. 239th St. Spring Hill, KS 66083	Yes	Yes	Univ. of Cal.-Davis	No	No	None	None	
Barzona								
Barzona Breeders Assn. of America, Inc. P.O. Box 631 Prescott, AZ 86302	Yes	Yes	Imm Gen, Inc.	Yes	Yes	Rights reserved by Board of Directors.	No limit if owner of sire notifies assn. in writing; date of bulls death and inventory of semen.	
Beefalo								
American Beefalo World Registry P.O. Box 12315 North Kansas City, MO 64116	Yes	No	Stormont Laboratories	No	Yes	None	None	

(Continued)

TABLE 3-4 (Continued)

Association	Blood Typing Required		Affiliated Laboratory	Breeding Receipt Required	Non-Owner Certificate Required	Restrictions on Certificates	Disposition of Semen after Death of Sire	General Comments
	Sire	Dam						
Beefmaster								
Beefmaster Breeders United 6800 Park Ten Blvd., Suite 290 West San Antonio, TX 78213	Yes	No	Imm Gen, Inc.	No	Yes	None	None	
Belgian Blue								
American Belgian Blue Assn. Box 307 Sulphur Springs, TX 75482-0307	Yes	Yes	Imm Gen, Inc.	Yes	Yes	None	Would like to be contacted about amount of semen available.	Strongly wants purity in breeds.
Belgian Blue Assn. of the Americas P.O. Box 6111 Sarasota, FL 34278-6111	Yes	No	Imm Gen, Inc.	No	No	None	None	Registration requirements same as natural service.
Belted Galloway								
Belted Galloway Society, Inc. 7118 Elliott Lane Leeds, AL 35094	No	No		Yes	No	None	None	
Blonde d'Aquitane								
American Blonde d'Aquitane Assn. P.O. Box 12341 Kansas City, MO 64116	Yes	Yes	Imm Gen, Inc.	No	No	None	None	Bull's blood type is required to be on file with assn. lab for AI sired calves.
Braford								
United Braford Breeders 422 E. Main, Suite 218 Nacagdoches, TX 75961	Yes	No	Imm Gen, Inc.	No	Yes	None	Report inventory within 60 days.	Currently reviewing AI regulations. Physical certificate may be replaced by simple signing fee.
Brahman								
American Brahman Breeders Assn. 1313 LaConcha Lane Houston, TX 77054	Yes	Yes out of herd AI or ET	Imm Gen, Inc.	Yes	No	None	Report within 90 days in writing with inventory of semen and location(s) where stored.	
Brahmousin								
American Brahmousin Council Box 12363 Kansas City, MO 64116	Yes	Yes for ET	Stormont Laboratories	No	No	None	None	

(Continued)

TABLE 3-4 (Continued)

Association	Blood Typing Required		Affiliated Laboratory	Breeding Receipt Required	Non-Owner Certificate Required	Restrictions on Certificates	Disposition of Semen after Death of Sire	General Comments
	Sire	Dam						
Bralers								
American Bralers Assn. HC 61, Box 41 Ganado, TX 77962	Yes out of herd	Yes out of herd	Imm Gen, Inc.	No	Yes	None	Has to be reported to assn.	
Brangus								
International Brangus Breeders P.O. Box 696020 San Antonio, TX 78269-6020	Yes	Yes	Imm Gen, Inc.	Yes	Yes	Only owners who have interest in bull can order AI certificates.	None	
Braunvieh								
Braunvieh Assn. of America P.O. Box 6396 Lincoln, NE 68506	Yes	Yes ET donor	Imm Gen, Inc.	No	No	None	Contact assn. for current requirements.	Contact assn. for further information.
British White								
British White Cattle Assn. of America P.O. Box 12702 North Kansas City, MO 64116	Yes	No	No specific laboratory	No	No	None	None	
Charolais								
American International Charolais Assn. P.O. Box 20247 Kansas City, MO 64195	Yes	No	Imm Gen, Inc.	No	Yes	Require sire owner signature when calf is registered for semen transfer from sire owner.	None	AI sires born after January 1, 1993 need to be parentage verified. (Both sire and dam must be blood typed.)
Chianina								
American Chianina Assn. P.O. Box 890 Platte City, MO 64079	Yes	Yes	Ohio State Univ.	No	No	AI certificate required on bulls used that have been enrolled in voluntary AI program.	None	All bulls used in AI, donor cows, and ET calves submitted for registration are required to be blood typed.
Devon								
Devon Cattle Assn., Inc. P.O. Box 61 The Plains, VA 22171	Yes	No	No specific laboratory	No	Yes	None	None	
Dexter								
American Dexter Cattle Assn. Route 1, Box 378 Concordia, MO 64020	Yes	No	Ohio State Univ.	Yes	No	None	None	

(Continued)

TABLE 3-4 (Continued)

Association	Blood Typing Required		Affiliated Laboratory	Breeding Receipt Required	Non-Owner Certificate Required	Restrictions on Certificates	Disposition of Semen after Death of Sire	General Comments
	Sire	Dam						
Gelbray								
Gelbray International, Inc. P.O. Box 2177 Ardmore, OK 73402	Yes	No	Imm Gen, Inc.	No	No	None	None	Registration requirements same as natural service. Contact assn. for current requirements.
Gelbvieh								
American Gelbvieh Assn. 10900 Dover Street Westminster, CO 80021	Yes	Yes	Imm Gen, Inc.	No	No	Has a voluntary AI certificate program.	None	
Hereford								
American Hereford Assn. P.O. Box 014059 Kansas City, MO 64101	Yes	Yes out of herd AI	Imm Gen, Inc.	Yes	Yes	No limitations on number.	No special requirements.	
Limousin								
North American Limousin Foundation 7383 S. Alton Way P.O. Box 4467 Englewood, CO 80155	Yes bulls born after 1/1/91 must be parent verified	Yes	Stormont Laboratories	No	No	None	None	
Maine-Anjou								
American Maine-Anjou Assn. 528 Livestock Exchange Building Kansas City, MO 64102	Yes	No	Univ. of Cal.-Davis	No	No	None	None	
Marchigiana								
American International Marchigiana Society Marky Cattle Assn. Box 198 Walton, KS 67151-0198	Yes	Yes	Imm Gen, Inc.	No	No	All full blooded Marchigianas must be blood typed.	No policies or requirements.	
Murray Grey								
American Murray Grey Assn. P.O. Box 34590 North Kansas City, MO 64116	Yes	No	No specific laboratory	No	No	None	None	Blood type submitted for embryo transplants, sire, dam, and calf.
Piedmontese								
Piedmontese Assn. of the U.S. 108 Livestock Exchange Building Denver, CO 80216	Yes	Yes	Stormont Laboratories	No	No	None	None	

(Continued)

TABLE 3-4 (Continued)

Association	Blood Typing Required		Affiliated Laboratory	Breeding Receipt Required	Non-Owner Certificate Required	Restrictions on Certificates	Disposition of Semen after Death of Sire	General Comments
	Sire	Dam						
Pinzgauer								
American Pinzgauer Assn. 21555 St., Rt. 698 Jenera, OH 45841	Yes	No	Imm Gen, Inc.	No	No	Certificate required for bulls enrolled in optional certificate program.	None	
Polled Hereford								
American Hereford Assn. P.O. Box 014059 Kansas City, MO 64101	Yes	Yes	Imm Gen, Inc.	No	Yes	None	No special requirements.	A parentage qualification is required via blood typing of subject animal and both parents prior to issuing AI permits.
Red Angus								
Red Angus Assn. of America, Inc. P.O. Box 776 4201 I-35 North Denton, TX 76207	Yes	No	Ohio State Univ.	No	Yes	Must be ordered by the recorded owner of the bull.	None	
Red Brangus								
American Red Brangus P.O. Box 1326 Austin, TX 78767	Yes	No	Imm Gen, Inc.	No	Yes	None	Report to assn. within 60 days; number of units stored and location of units.	Contact breed assn. for current information on registration.
Red Poll								
American Red Poll Assn. P.O. Box 35519 Louisville, KY 40232	Yes	No	Imm Gen, Inc.	Yes	Yes	None	No time limit on use providing bull is blood typed, notice is given to assn. within 60 days, inventory of semen.	
Romagnola								
American Romagnola Assn. Box 450 Navasota, TX 77868	No	No	Ohio State Univ.	No	Yes	None	Contact assn. immediately.	
Salers								
American Salers Assn. 5600 South Quebec, Suite 220A Englewood, CO 80111-2207	Yes	Yes	Stormont Laboratories	Yes	Yes	None	Notify assn. within 30 days; date of death and semen inventory. No limit on AI certficiates as long as semen available.	

(Continued)

TABLE 3-4 (Continued)

Association	Blood Typing Required		Affiliated Laboratory	Breeding Receipt Required	Non-Owner Certificate Required	Restrictions on Certificates	Disposition of Semen after Death of Sire	General Comments
	Sire	Dam						
Santa Gertrudis								
Santa Gertrudis Breeders International P.O. Box 1257 Kingsville, TX 78363	Yes	Yes	Imm Gen, Inc.	Yes	Yes	AI Certificates are only made out to record owner of bull or semen.	Has no bearing on sale of semen.	
Scotch Highland								
American Highland Breeders Assn. 200 Livestock Exchange Building 4701 Marion Street Denver, CO 80216	Yes	Yes	No specific laboratory	Yes	Yes	None	None	Inseminator must be registered with assn. Current health certificate for bull must be on file at the assn. Calves sired by AI must be registered before one year of age.
Senepol								
North American International Senepol Assn. 11020 NW Ambassador Drive Kansas City, MO 64153-1149	Yes	No	Gen Mark (blood profiling)	No	No	None	None	
Shorthorn								
American Shorthorn Assn. 8288 Hascall Street Omaha, NE 68124	Yes	Yes	Ohio State Univ.	No	Yes	None	Report to assn. within 60 days; date of death and amount of semen on inventory.	All bulls used for AI and all ET donor females (born on or after 1/1/92) must be parentage verified as of 1/1/94.
Simmental								
American Simmental Assn. 1 Simmental Way Bozeman, MT 59715	Yes	Yes	Univ. of Cal.-Davis	No	No	Sire must be registered in assn. and be blood typed—confirmed to parents.	None	
South Devon								
North American South Devon Assn. Box 68 Lynnville, IA 50153	Yes	No	Imm Gen, Inc.	No	No	None	No time restriction on use of semen.	Registration requirements same as natural service.

(Continued)

TABLE 3-4 (Continued)

Association	Blood Typing Required		Affiliated Laboratory	Breeding Receipt Required	Non-Owner Certificate Required	Restrictions on Certificates	Disposition of Semen after Death of Sire	General Comments
	Sire	Dam						
Tarentaise								
American Tarentaise Assn. P.O. Box 34705 North Kansas City, MO 64116	Yes	No	Imm Gen, Inc.	No	No	None	None	
Texas Longhorn								
Texas Longhorn Breeders Assn. of America 2315 N. Main Street, Suite 402 Fort Worth, TX 76106	Yes	No	No specific laboratory	No	No	None	Notify assn. of death of sire.	
Wagyu								
American Wagyu Assn. P.O. Box 12626 Shawnee Mission, KS 66282	Yes	Yes	Gen Mark (blood profiling) Only through assn. requests and forms.	No	Yes	Sold by assn. to owner of sire only. Unlimited number available.	None	DNA profile must be established. Assn. permit required for "out of herd" use of any bull.
Welsh Black								
Welsh Black Cattle Assn., U.S. RR 1, Box 768 Shelburn, IN 47879	Yes	Yes	No specific laboratory	Yes	Yes	None	Death of sire must be reported to assn. within 60 days.	
White Park								
White Park Registry HC87, Box 2214 Big Timber, MO 59011	Yes	No	Imm Gen, Inc.	No	No	Record of semen transfer to applicant from owner of bull or semen must be on record with WPR. Owner of AI sire must be member of WPR. Requirements do not apply for within herd use of applicant's own bull.	None	

BLOOD TYPING/PROFILING[2]

Most breed registry associations require that a record of the blood type of each bull used in artificial insemination be on file with the breed registry. It is recommended that each bull be blood typed at the time that semen is first collected. Some associations are now requiring blood profiling using DNA technology.

Stormont Laboratories claim 99% accuracy in their cattle and horse blood-typing tests in solving paternity cases.

When having a bull blood typed/profiled for purposes of recording his blood type/profile with the breed association, the blood should be sent to the laboratory with which the particular breed association is affiliated; information which can be secured from each breed registry. Before collecting blood samples for blood typing/profiling, one should obtain the necessary instructions, report forms, and blood tubes containing anticoagulant solution or other required materials from the particular breed registry association.

[2]This section was authoritatively reviewed by Dr. Clyde Stormont, Chairman and Director of Laboratory Services, Stormont Laboratories, Inc., 1237 E. Beamer St., Suite D, Woodland, CA 95776.

A list of U.S. affiliated laboratories follows:

Cattle Blood Typing Laboratory
Ohio State University
Columbus, OH 43210
(614) 292-6659
Fax (614) 292-7116

Gen Mark
421 Wakara Way, Suite 201
Salt Lake City, UT 84108
(801) 582-2600
Fax (801) 582-2637

Imm Gen, Inc.
P.O. Box 10135
College Station, TX 77842
(409) 696-5382
Fax (409) 764-0996

Serology Laboratory
University of California
Davis, CA 95616
(916) 752-2211
Fax (916) 752-3556

Stormont Laboratories
1237 E. Beamer Street
Woodland, CA 95776
(916) 661-3078
Fax (916) 661-0391

QUESTIONS FOR STUDY AND DISCUSSION

1. Define *breed* as applied to animals. Must a new breed of cattle be approved by someone, or can anyone start a new breed?

2. Give the step by step typical history of the formation of the various breeds of cattle.

3. Trace the crossbreeding of U.S. beef cattle, beginning with the Longhorn. Why didn't cattle producers continue to upgrade Longhorns with Shorthorns only; why did they turn to Herefords and eventually to Angus?

4. Why did supporters of the British breeds (Shorthorns, Herefords, and Angus) not consider the introduction of the Brahman a serious threat, whereas the introduction of the Charolais disturbed them greatly?

5. What breeds would you classify as among the exotics? What prompted the bringing in of the exotics? How do you account for the fact that the exotics came quickly and in large numbers?

6. Are purebreds important in an era of crossbreeding? If so, why?

7. Why did we bring to North America so many dual-purpose–type and draft-type cattle?

8. Why have beef-type and dairy-type (a) live animals, and (b) carcasses become more alike in recent years?

9. In outline form, list the (a) distinguishing characteristics, and (b) disqualifications of six breeds of beef cattle; then discuss the importance of these listings.

10. With what breed(s) is each of the following associated?

a. George Grant.
b. Tom Lasater.
c. Clear Creek Ranch.
d. Robert Bakewell.
e. Henry Clay.
f. Warren Gammon.
g. Polled Durhams.
h. King Ranch.
i. Monkey.
j. Amos Cruickshank.
k. Wichita Mountains Wildlife Refuge.

11. Justify any preference that you may have for one particular breed of beef cattle.

12. Obtain breed registry association literature and a sample copy of a magazine of your favorite breed of cattle. (See Appendix Section VII for addresses.) Evaluate the soundness and value of the material that you receive.

13. What factors should be considered in selecting the breeds for a crossbreeding program?

14. What is a composite breed? Under what circumstances would you recommend (a) a composite breed, and (b) crossbreeding?

15. Knowledgeable members of the beef team from producers to retailers report that consumers complain that beef of the 1990s lacks consistency. Further, the members of the meat team attribute this problem to too many breeds, including composites and too much crossbreeding. How serious is this problem? If it is serious, how may it be rectified?

16. Of what value is Table 3-3, Comparative Rating (for Beef Purposes) of 59 Breeds of Cattle?

17. What factors should be considered in choosing a purebred breed? What factors should be considered in choosing breeds for crossbreeding?

18. How should the feeding of the exotics differ from the feeding of Shorthorns, Herefords, and Angus?

19. Are the rules of cattle registry associations relative to registering young produced artificially becoming less rigid? If so, why?

20. Give the names and addresses of three different laboratories where a cattle producer may send a blood sample for typing.

SELECTED REFERENCES

Title of Publication	Author(s)	Publisher
Aberdeen-Angus Breed: A History, The	J. R. Barclay A. Keith	The Aberdeen-Angus Cattle Society, Aberdeen, Scotland, UK, 1958
Birth of a Breed—The History of Polled Herefords	O. K. Sweet	American Polled Hereford Assn., Kansas City, MO, 1975
Breeds of Cattle	H. R. Purdy	Chanticleer Press, Inc., New York, NY
Breeds of Livestock, The	C. W. Gay	The Macmillan Co., New York, NY, 1918
Breeds of Livestock in America	H. W. Vaughan	R. G. Adams and Co., Columbus, OH, 1937
Dairy Cattle Breeds	R. B. Becker	University of Florida Press, Gainesville, FL, 1973
Hereford in America, The	D. R. Ornduff	The author, Kansas City, MO, 1957
Hereford Heritage	B. R. Taylor	The author, The University of Arizona, Tucson, AZ, 1953
History of Linebred Anxiety 4th Herefords, A	J. M. Hazelton	Associated Breeders of Anxiety 4th Herefords, Graphic Arts Bldg., Kansas City, MO, 1939
Lasater Philosophy of Cattle Raising, The	L. M. Lasater	Texas Western Press, The University of Texas, El Paso, TX, 1972
Modern Breeds of Livestock, Fourth Edition	H. M. Briggs D. M. Briggs	The Macmillan Co., New York, NY, 1980
Santa Gertrudis Breed, The	A. O. Rhoad	Inter-American Institute of Ag. Sciences, Turrialba, Costa Rica, 1949
Santa Gertrudis Breeders International Recorded Herds	R. J. Kleberg, Jr.	Santa Gertrudis Breeders International, Kingsville, TX, 1953
Shorthorn Cattle	A. H. Sanders	Sanders Publishing Co., Chicago, IL, 1918

Title	Author(s)	Publisher
Stockman's Handbook, The, Seventh Edition	M. E. Ensminger	Interstate Publishers, Inc., Danville, IL, 1992
Study of Breeds in America, The	T. Shaw	Orange Judd Co., New York, NY, 1900
Story of the Herefords, The	A. H. Sanders	*Breeders Gazette*, Chicago, IL, 1914
Types and Breeds of African Cattle	N. R. Joshi E. A. McLaughlin R. W. Phillips	Food and Agriculture Organization of the United Nations, Rome, Italy, 1957
Types and Breeds of Farm Animals	C. S. Plumb	Ginn and Company, Boston, MA, 1920
World Cattle, Vols. I, II, and III	J. E. Rouse	University of Oklahoma Press, Norman, OK, Vols. I and II, 1970; Vol. III, 1973
World Dictionary of Livestock Breeds, Third Edition	I. L. Mason	Commonwealth Agricultural Bureaux, Slough, Bucks, England, 1951
Zebu Cattle of India and Pakistan	N. R. Joshi R. W. Phillips	Food and Agriculture Organization of the United Nations, Rome, Italy, 1988

Also, breed literature pertaining to each breed may be secured by writing to the respective breed registry associations. (See Appendix Section VII for the name and address of each association.)

Match cow size and milk production to the feed/environment, and match the bull to the market. (Courtesy, American Hereford Assn., Kansas City, MO)

SELECTING BEEF CATTLE[1]

Contents *Page*

[1]The authors are most grateful to Professor Emeritus Richard F. Johnson, Dept. of Animal Science, California Polytechnic State University, San Luis Obispo, California, a fine artist and able animal scientist, who did all the drawings for this chapter.

Contents *Page*

Virgil, great Roman poet who was born on a farm in Italy in 70 B.C., in his *Georgics*—Book III, a poem written in 37 to 30 B.C., dealing with the raising of herds and flocks, had the following to say about the selection of cattle:[2]

> Distinguish all betimes with branding fire,
> To note the tribe, the lineage and the sire;
> Whom to reserve for husband of the herd;
> Or who shall be to sacrifice preferred;
> Or whom thou shalt to turn thy glebe allow,
> To smooth the furrows, and sustain the plough:
> The rest, for whom no lot is yet decreed,
> May run in pastures, and at pleasure feed.

The foregoing verse gives conclusive evidence that the power of selection in cattle breeding was known and practiced long ago.

Whether establishing or maintaining a herd, cattle producers must constantly appraise or evaluate animals; they must buy, sell, retain, and cull. Where the beef cattle herd is being neither increased nor decreased in size, each year about 46% of the heifers, on the average, are retained in order to replace about 20% of the old cows.[3] In addition, bulls must be selected and culled, and steers and other surplus animals must be marketed. Thus, in normal operations, producers are constantly called upon to cull out animals, to select replacements, and to market surpluses. Each of these decisions calls for an evaluation or appraisal.

Producers are ever aware of market demands as influenced by consumer preferences. Also, the great livestock shows throughout the land have exerted a powerful influence in molding cattle types.

However, only a comparatively few animals on the farms and ranches are subjected annually to the scrutiny of market specialists or experienced show-ring judges. Rather, the vast majority of purebred animals and practically all commercial herds are evaluated by practical operators who select their own foundation or replacement stock and conduct their own culling operations. They have no interest in the so-called breed fancy points. These practical operators may not be able to express fluently their reasons for selecting certain animals while culling others, but usually they become quite deft in their evaluations. Whether young animals are being raised for market or for breeding stock, successful livestock operators are generally good judges of livestock.

[2]From *Georgics*—Book III, written by Virgil (Publius Vergilius Maro, 70-19 B.C.), Dryden's translation. Of *Georgics*, the poem of the land, it has been said, "The stateliest measure ever molded by the lips of man."

[3]A herd of 100 cows will produce about 44 heifers each year where there is an 88% calf crop. Hence, 46% of the heifers will be needed to replace 20% of the cows ($44 \times 0.46 = 20$).

ESTABLISHING THE HERD

Except for the comparatively few persons who keep animals merely as a hobby, farmers and ranchers raise cattle because, over a period of years, they have been profitable, provided the production and marketing phases were conducted in an enlightened and intelligent manner. Therefore, after it has been ascertained that the feeds and available labor are adapted to cattle production, and that suitable potential markets exist, the next assignment is that of establishing a herd that is efficient from the standpoint of production and that meets market demands. This involves a number of considerations.

PUREBRED OR COMMERCIAL CATTLE

Broadly classified, cow-calf producers are either (1) purebred breeders, or (2) commercial producers. Purebred breeders are a small, but select, group. The vast majority of cattle operations are commercial. An estimated 96% of the cattle of America are nonpurebreds.

Purebred breeders produce *seedstock* for other purebred breeders, and both bulls and purebred females for F_1 heifer programs of commercial producers.

Purebred breeders need to be more than good cattle producers. They should be knowledgeable relative to breeding systems, pedigrees and registration, production testing, advertising, sales, and other special marketing methods, and perhaps fitting and showing. Also, they should be thoroughly knowledgeable relative to breeding commercial cattle the modern way in order that the needs of the commercial producer will be reflected in their breeding programs. Indeed, both types—purebred and commercial—are interdependent.

For the person with experience and adequate capital, the breeding of purebreds offers unlimited opportunities. It has been well said that honor, fame, and fortune are all within the realm of possible realization in the purebred business, but it should also be added that only a few achieve this high calling.

The goal of most commercial beef operations is to convert the production of the land—grass and crops—into dollars through the traditional cow-calf operation. Usually the product is marketed at the weaning stage, although some commercial cattle producers carry them to the yearling stage, or even finish them for market. More and more commercial herds will be crossbreds, simply because the economics favor crossbreeding accompanied by complementary genes and heterosis, and because the crossbreds can be used to produce beef according to specification.

As a group, commercial cattle producers are intensely practical. No animal meets with their favor unless it produces meat over the block at a profit. The commercial cattle business requires less outlay of cash than the purebred business on a per animal basis; and less knowledge relative to the many facets of the purebred business—pedigrees, promotion, etc.

SELECTION OF THE BREED OR CROSS

Since no one breed of cattle or no breed cross excels all others in all points of beef production under all conditions, where no strong preference exists smart cattle producers select the breed or cross on the following bases:

Match the cow size and milk production to the available feed/environment; and match the sire breed to the market.

For the purebred breeder, the selection of a particular breed is most often a matter of personal preference, and usually the preferred breed will make for the greatest success. Where no definite preference exists, however, it is well to choose the breed that is most popular in the community—if any one breed predominates. If this procedure is followed, it is often possible to arrange for an exchange of animals, especially bulls. Moreover, if a given community is noted for producing good cattle of a particular breed, there are many advantages from the standpoints of advertising and sales.

Germ plasm choice for the commercial cattle producer is becoming increasingly difficult because of the large number of breeds and breed cross combinations now available. With only 3 main breeds (Shorthorns, Herefords, and Angus), there are 3 single-cross combinations and 3 three-way cross combinations from which to choose. However, with 10 breeds there are 45 single-cross combinations and 360 possible three-way combinations from which to select. Of course, there are more than 10 breeds; 81 major and minor breeds are described in this book (see Chapter 3, Breeds of Cattle).

MILKING ABILITY

Cattle producers are admonished to—

> Match the level of milk production to the environment, of which available feed is important.

Weaning weight is the most important trait affecting net income in a cow-calf operation; and weaning weight of beef calves is influenced more by the dam's milk production than by any other single factor. For this reason, the pressure is on to increase the milk production of beef cows. For this reason, also, this subject "Milking Ability"—is fully covered herein.

Performance testing programs which emphasize weaning weight automatically result in selection for higher milk production. However, in order to more rapidly increase beef production, dual-purpose and

Fig. 4-1. "Lotta grass, lotta milk, lotta calf." A Charolais X Brangus heifer calf, sired by a Charolais bull and out of a Brangus cow. (Courtesy, American-International Charolais Assn., Kansas City, MO)

dairy breeding are being infused into many commercial beef herds.

Research has shown a strong correlation between the level of milk production of cows and the weaning weight of their calves. Also, conversion of milk to beef is rather efficient—on the order of 10 lb of milk to 1 additional lb of weaned calf, although conversion may not be quite as efficient at higher levels of milk production.

■ **Will more milk produce more beef?**—Beef cattle exhibitors have long known that milk is an important ingredient in a calf's diet. As a result, they use milk replacers when fitting calves. Now there is unmistakable experimental evidence that more milk does, indeed, produce more beef.

Research at the Arkansas, Texas, Georgia, Oklahoma, Beltsville (USDA) and Alberta (Canada) stations has shown that 50 to 80% of the variability in the weaning weight of calves is due to the milk production of their dams—that weaning weight is influenced by milk production more than all other effects combined.

1. **What the Arkansas experiment showed.** In a study involving Hereford cows, the Arkansas Station reported the following results:[4]

Average Daily Milk Production/Cow		Weaning Weight of Calves at 8 Months	
(lb)	*(kg)*	*(lb)*	*(kg)*
5.0	*2.3*	354	*161*
15.6	*7.1*	475	*216*

Thus, a 10-lb per day difference in milk production made for 121 lb greater weaning weight. With 60¢ calves, that's $72.60 per head greater returns.

2. **What the Georgia Station found.** Using Hereford cows, the Georgia Station found the following:[5]

Average Daily Milk Production/Cow		Weaning Weight of Calves at 205 Days	
(lb)	*(kg)*	*(lb)*	*(kg)*
4.9	*2.2*	350	*159*
12.2	*5.5*	475	*216*

Hence, 7.3 lb more milk per day during the suckling period produced 125 lb more calf at weaning time.

■ **How much milk will a beef cow give?**—There's more difference within than between breeds when it comes to milk production. Also, it's difficult, if not impossible from a practical standpoint, to secure a truly representative sample of a widely scattered, populous breed of beef cattle.

Studies reveal that beef cattle range from less than 1 lb of milk per day up to 25 lb, but it appears that most of them average around 11 lb per day. By contrast, it's noteworthy that the nation's 11 million dairy cows average about 43 lb per day. That's a big difference!

Table 4-1 shows what some experiment stations have found relative to the milk production of beef cows.

■ **How many pounds of milk required to produce a pound of gain?**—The pounds of milk required for a pound of gain vary. Breed, age of calf, quantity of milk, and many other factors enter in. For example, the Texas Station reported that Hereford calves required fewer pounds of milk per pound of gain than Angus or Charolais.[6] But, based on a number of studies, it appears that, on the average, each additional 10 lb of milk will produce 1 lb of weaned calf. Hence, 1 gal (8.6 lb) more milk per day may be expected to produce about a 150-lb heavier calf at weaning age.

■ **How can you get more milk?**—Of course, if cows are not fed, either enough or properly, they will not produce. But, from a breeding standpoint, two main approaches may be used to get more milk: (1) selection; and (2) infusion of dairy breeding.

1. **Selection.** In dairy cattle, we know that milk yield is 25% heritable. Of course, few studies on milk production in beef cattle have been made. But recent work indicates that in beef cattle milk yield averages 32% heritable. So, for purposes of this discussion, let's use the 32% figure. This means that, if from a beef herd that averages 1,000 lb of milk in 205 days, you select top milk-producing beef cows that average 2,000 lb, then mate them to a beef bull which you select in similar manner for 2,000-lb production, you may expect 32% apparent superiority of the parents to be expressed in the offspring. Since the selected parents in this case averaged 1,000 lb of milk higher than the herd, 32% of 1,000 lb = 320 lb. Thus, the offspring could be expected to average 1,320 lb of milk. This shows that you can soon improve the milking ability of a beef herd by giving proper attention to milk production when selecting the herd bull. The best method is to select a bull whose sire and dam have produced good milking daughters. In beef cattle, where milk production is not actually measured, selecting individuals with superior weaning weights indirectly selects for higher milk production.

---

[4]Gifford, Warren, *Record of Performance Tests for Beef Cattle in Breeding Herds*, Ag. Exp. Sta. Bull. 531, University of Arkansas, College of Agriculture, Fayetteville, p. 28.

[5]Neville, W. E., Jr., "To Spread Beef Gains You Need Milk," University of Georgia, reported in *Livestock Breeder Journal*, Sept. 1971, pp. 40, 42.

[6]Melton, A. A., *et al.*, "Milk Production, Composition and Calf Gains of Angus, Charolais and Hereford Cows, *Journal of Animal Science*, Vol. 26, No. 4, p. 804.

TABLE 4-1
MILK PRODUCTION OF BEEF COWS

Station	Breed	Average Daily Milk Production/Cow		Comments
		(lb)	*(kg)*	
Arkansas[1]	Angus	8.5	*3.9*	The 8.5 figure was for a lactation period of 8 months. During the first 3 months, the Angus cows averaged 9.7 lb *(4.4 kg)* of milk per day.
Texas[2]	1. Hereford	7.3	*3.3*	Crossbred dams yielded more milk and had faster growing calves than did Hereford or Brahman dams.
	2. Brahman	9.6	*4.4*	
	3. Brahman X Hereford	13.6	*6.2*	
USDA, Beltsville[3]	Shorthorn	17.5	*7.9*	Lactation period of 252 days. Milk production estimated by weighing calves before and after nursing.
In coop. with Kansas State[3]	Shorthorn	13.0	*5.9*	Hand milked; complete lactation of 365 days or less.
Louisiana[4]	1. Angus	10.5	*4.8*	Brahmans gave the most milk; Herefords gave the least, and Angus and Charolais were intermediate producers.
	2. Brahman	14.0	*6.4*	
	3. Charolais	11.7	*5.3*	
	4. Hereford	8.1	*3.4*	
	Average	11.1	*5.0*	

[1]Clifford, Warren, *Record of Performance Tests for Beef Cattle in Breeding Herds*, Ag. Exp. Sta. Bull. 531, University of Arkansas College of Agriculture, Fayetteville, Arkansas.

[2]Todd, J. C., H. A. Fitzhugh, Jr., and J. K. Riggs, "Effect of Breed and Age of Dam on Milk Yield and Progeny Growth," *Beef Cattle Research in Texas*, Texas A&M University, College Station, Texas.

[3]Dawson, W. M., A. C. Cook, and Bradford Knapp, Jr., "Milk Production of Beef Shorthorn Cows," *Journal of American Science*, Vol. 19, No. 2.

[4]*Louisiana Agriculture*, Vol. 26, No. 4, Summer 1983.

2. **Infuse dairy breeding, or higher milking strains of beef cattle.** Of course, it's possible rapidly to increase milk production to a much higher level by infusing dairy breeding than by selecting from within most existing beef breeds. The following results with 4-year-old Hereford, Hereford X Holstein, and Holstein cows, reported by the Oklahoma Station, show the effect of infusing Holstein breeding:[7]

Breed	Av. Milk/Day; 205 Days		Av. Wt. of Calves; 240 Days	
	(lb)	*(kg)*	*(lb)*	*(kg)*
Hereford	13.5	*6.1*	575	*261*
Hereford X Holstein	20.3	*9.2*	642	*292*
Holstein	27.2	*12.4*	708	*322*

As noted, the Holstein calves had a weight advantage of 133 lb over the straightbred Herefords, whereas the Hereford X Holsteins enjoyed a weight advantage of 67 lb over the Herefords. Clearly, milk and beef go hand in hand, and it doesn't make much difference whether the cow is red, magpie, black, or polka dot.

[7]Lusby, K. S., *et al.*, "Performance of Four-Year-Old Hereford, Hereford X Holstein and Holstein Females As Influenced by Level of Winter Supplementation Under Range Conditions," *Animal Sciences and Industry Research Report*, Apr. 1974, p. 56.

■ **What else do the experiments and experiences tell us?**—The experiments and experiences also tell us the following about milk and beef:

1. The highest relationship of milk to calf weight gains is during the first 60 days of the calf's life. This is because the calf has need for a highly concentrated source of energy at a time when its consumption capacity is limited.

2. Conversion of milk to beef may not be as efficient at high levels of milk production. Thus, according to the Georgia Station, instead of going beyond 3,000 lb of milk, in beef cows it may be more efficient to creep feed the calves.

3. Maximum milk production of beef cows occurs during the first two months, then declines.

4. A good rule of thumb is that a beef calf will consume 1 lb of milk daily for each 10 lb body weight. Hence, a 300-lb calf should get 30 lb of milk daily, or approximately 3½ gal—more than twice as much as most beef cows give. In this connection, it is note-

worthy that the California Station reported that calves under three months of age consumed up to 50 lb of milk per day. If a cow produces more milk than her calf will take, the pressure will build up in the udder and the drying-up process will usually take care of the situation; and there won't be as many spoiled udders as most cattle producers fear. So, unless there is an udder problem (like a sunburned udder or a big teat), it's not necessary to milk her out. Nevertheless, an excessive flow of milk in commercial cattle is undesirable, for no one likes to milk a wild cow.

5. In selecting for milk, some valuable beef characteristics may be lessened because the higher the milk production the greater the dairy temperament, or angularity; they're built to convert a larger proportion of their feed to milk. But, of course, with the emphasis on red meat in recent years, beef type and dairy type have moved closer together.

6. There is no indication that postweaning gains are affected by milk production of the dam up to 3,500 lb. Beyond this may be another story, because it is generally recognized that creep feeding results in slower and costlier gains during the feeding period following weaning.

7. Bull calves suckle more frequently than steer or heifer calves. Consequently, cows with bull calves tend to give more milk.

8. Cows nursing crossbred calves give more milk than cows suckling straightbred calves, perhaps due to the fact that the crossbreds are more vigorous nursers.

9. Replacement heifers that become too fat as calves or yearlings will produce less milk during their first lactation, due to fat deposits in the mammary gland area. The solution: select replacement heifers when they're three to four months of age, then limit their preweaning and postweaning gains to 1.5 lb per head per day.

10. Cows which tend to produce large amounts of milk may become very thin on poor pasture and may have a longer interval between calving and rebreeding than poorer milking cows on similar pasture.

■ **Summary relative to milking ability**—There may well be several answers to the two-pronged question, "How much milk should a beef cow produce, and what's the best way to increase milk production in a beef herd?" On a poor range where feed is sparse, a relatively low level of milk production may be necessary to allow good reproduction, while on improved pastures a very high level of milk production may be desirable. Also, more study needs to be given to selecting the high-producing strains of beef cattle and to the use of *dam breeds.* Reasonable goals for the more successful beef herds are: a cow averaging 20 lb of milk per day (about double the present level); and weaning off a 600-lb calf. Cattle producers can be sure of one thing: "Little milk—little calf; 'lotta' milk—'lotta' calf."

SIZE OF THE HERD

No minimum or maximum figures can be given as to the best size for the herd. Rather, each case is one for individual consideration. It is to be pointed out, however, that labor costs differ very little whether the herd numbers 100 or 300. The cost of purchasing and maintaining a herd bull also comes rather high when too few females are kept. Other efficiencies can be achieved through size, provided the operation is under competent management. For this reason, bigness in every kind of business, including the cattle business, is a sign of the times.

The extent and carrying capacity of the pasture, the amount of hay and other roughage produced, and the facilities for wintering stock are factors that should be considered in determining the size of herd for a particular farm unit. The system of disposing of the young stock will also be an influencing factor. For example, if the calves are disposed of at weaning time or finished as baby beef, practically no cattle other than the breeding herd are maintained. On the other hand, if the calves are carried over as stockers and feeders or are finished at an older age, more feed, pasture, and shelter are required.

Then, too, whether the beef herd is to be a major or minor enterprise will have to be decided upon. Here again, each case is one for individual consideration. In most instances, replacements should be made from heifers raised on the farm.

UNIFORMITY

Uniformity in a herd has reference to the animals looking alike—"like peas in a pod," as a cattle producer is prone to remark; particularly from the standpoints of size, type, and color.

Uniformity in color is still important in purebred herds. For the most part, however, the desire for uniformity in color of commercial herds went out as crossbreeding came in. So long as a quality product is produced efficiently, consumers and most cattle feeders have no interest in color of hair. It should be added for the benefit of the color conscious, however, that it is still possible to obtain uniformity of color even in crossbreds through making certain crosses; for example, uniform-colored animals with black bodies and white faces can be produced by crossing Herefords and Angus.

Size and type in any given lot of cattle are still important to cattle feeders and packer buyers. Cattle

Fig. 4-2. Hereford X Angus crossbreds; uniform-colored animals with black bodies and white faces—*black baldies.* (Courtesy, University of Missouri)

of uniform size and type feed better in a lot; and packers must provide their retail outlets with carcasses and cuts that meet their exacting specifications and grades. But buyer appeal, for both the feedlot and packer buyer, can be imparted amazingly well by sorting and grading prior to offering cattle for sale. Properly done, this practice can make for more uniformity faster in a few minutes than can be achieved through years of selective breeding.

HEALTH

All animals selected should be in a thrifty, vigorous condition and free from diseases and parasites. They should give every evidence of a life of usefulness ahead of them. The cows should appear capable of producing good calves, and the bull should be able to withstand a normal breeding season. Tests should be made to make certain of freedom from both tuberculosis and contagious abortion, and perhaps certain other diseases in some areas. In fact, all purchases should be made subject to the animals being free from contagious diseases. With costly purebred animals, a health certificate should be furnished by a licensed veterinarian. Newly acquired animals should be isolated for several days before being turned with the rest of the herd.

(Also see Chapter 12, Beef Cattle Health, Disease Prevention, and Parasite Control.)

CONDITION

Although an extremely thin and emaciated condition, which may lower reproduction, is to be avoided, it must be remembered that an overfat condition may be equally harmful from the standpoint of reproduction.

It takes a unique ability to project the end result of feeding a few hundred pounds of grain or hay to a thin animal, and fortunate indeed is the producer who possesses this quality. This applies alike to both the purebred and the commercial producer. In fact, it is probably of greater importance with the commercial producer, for replacement females and stocker and feeder steers are usually in very average condition.

AGE AND LONGEVITY

In establishing the herd, it is usually advisable to purchase a large proportion of mature cows (cows 4 to 5 years of age) that have a record of producing uniformly high-quality calves. Perhaps it can be said that not over ½ of the newly founded herd should consist of untried heifers. Aside from the fact that some of the heifers may prove to be nonbreeders, they require more assistance during calving time than do older cows. Perhaps the best buy of all, when they are available, consists of buying cows with promising calves at side and rebred to a good bull—a 3-in-1 proposition.

Once the herd has been established, replacement females should come from the top heifers raised on the farm or ranch. Old cows, irregular breeders, and poor milkers sell to best advantage before they become thin and "shelly."

A sound practice in buying a bull is to seek one of serviceable age that is known to have sired desirable calves—a proved sire. However, with limited capital, it may be necessary to consider the purchase of a younger bull. Usually a wider selection is afforded with the latter procedure, and, also, such an individual has a longer life of usefulness ahead. Naturally, the time and number of services demanded of the bull will have considerable bearing on the age of the animal selected.

Since most beef females do not reproduce until they are about two years of age, their regular and prolonged reproduction thereafter has an important bearing upon the overhead cost of developing breeding stock in relation to the number of calves produced. The longer the good, proved, producing cows can be kept without sacrifice of the calf crop or too much decrease in salvage value, the less the percentage replacement required. Moreover, the proportion of younger animals that can be marketed is correspondingly increased. Selection and improvement in longevity are possible in all breeds and should receive more attention.

In a nationwide survey made by Washington State University, it was found that old cows are culled or removed from the beef breeding herd at an average

age of 9.6 years, and bulls at 6.3 years.[8] It is recognized that the severity of culling will vary somewhat from year to year, primarily on the basis of whether cattle numbers are expanding or declining; and that purebred cattle are usually retained longer than commercial cattle.

ADAPTATION

As has already been indicated—except in those localities where a certain breed predominates, thus making possible the exchange of breeding stock and joint benefits in selling surplus stock—one will usually do best to select that breed for which the producer may have a decided preference. On the other hand, there are certain areas and conditions wherein the adaptation of the breed or class of animals should be given consideration. For example, in the South, Brahman cattle and certain breeds with Brahman blood are able to thrive despite the extreme heat, heavy insect infestation, and less abundant vegetation common to the area. Because of this, Brahman blood has been added to many herds of the South and Southwest, and new strains of beef cattle have evolved.

The Missouri Station conducted some classical studies designed to show breed differences between Shorthorn, Santa Gertrudis (5/8 Shorthorn and 3/8 Brahman), and Brahman cattle.[9] The animals were housed in "climatic chambers," in which the temperature, humidity, and air movements were regulated as desired. The ability of representatives of the different breeds to withstand different temperatures was then determined by studying the respiration rate and body temperature, the feed consumption, and the productivity in growth, milk, beef, etc. Dr. Brody reported the following pertinent points:

1. The most comfortable temperature for the Shorthorns was in the range of 30° to 60°F while for the Brahmans it was 50° to 80°F, and for the Santa Gertrudis it was intermediate between the ideal temperatures for the Shorthorn and Brahman.

2. The Brahman cattle could tolerate more heat—they could withstand higher temperature better than Shorthorns, whereas the Santa Gertrudis approached the Brahman in heat tolerance.

3. The Shorthorn cattle could tolerate more cold—they could withstand a lower temperature better than the other two breeds, while the Santa Gertrudis were more cold-tolerant than Brahman cattle.

Dr. Brody attributed the higher heat tolerance of Brahman cattle to their lower heat production, greater surface area (their loose skin) per unit weight, shorter hair, and "other body-temperature regulating mechanisms not visually apparent."

Translated into practicality, the Missouri experiment proved what is generally suspected; namely, (1) that Brahman and Santa Gertrudis cattle are better equipped to withstand tropical and subtropical temperatures than the European breeds; (2) that mature breeding animals of acclimatized European breeds do not need expensive, warm barns—they merely need protection from wind, snow, and rain; and (3) that more attention needs to be given to providing summer shades and other devices to assure warm weather comfort for cattle.

The cattle producer must always breed for a strong constitution—the power to live and thrive under the adverse conditions to which most animals are subjected sometime during their lifetime. Under natural conditions, selection occurs for this characteristic by the elimination of the unfit. In domestic herds, however, the constitution of foundation or replacement animals should receive primary consideration.

PRICE

With a commercial herd, it is seldom necessary to pay much in excess of market prices for the cows. However, additional money paid for a superior bull, as compared to a mediocre sire, is always a good investment. In fact, a poor bull is high at any price.

With the purebred breeder, the matter of price for foundation stock is one of considerable importance. Though higher prices can be justified in the purebred business, sound judgment should always prevail.

BASES OF SELECTION

In simple terms, selection in cattle breeding is an attempt to secure or retain the best of those animals in the current generation as parents of the next generation. Obviously, the skill with which selections are made is all important in determining the future of the herd. It becomes perfectly clear, therefore, that the destiny of herd improvement is dependent upon the selection for breeding purposes of those animals which are genetically superior. Making the wrong selections and using genetically inferior animals for breeding purposes has ruined many a herd. Under the latter circumstances, the producer would be better off to let the cattle decide on the breeding program by random sampling.

The ultimate objective of beef production is selling

---

[8]Ensminger, M. E., M. W. Galgan, and W. L. Slocum, *Problems and Practices of American Cattlemen*, Wash. Ag. Exp. Sta. Bull. 562.

[9]Brody, Samuel, *Climate Physiology of Cattle*, Jour. Series No. 1607, Mo. Ag. Exp. Sta.

beef over the block. Thus, fads or fancies in beef cattle selection that stray too far from this objective will, sooner or later, bring discredit and a penalty.

Strictly from the standpoint of the consumer, a beef animal should produce a carcass which has a high proportion of lean meat, no excess fat, and a minimum of bone—plus *eating quality*, which includes tenderness, flavor, and juiciness. Additionally, for efficiency of production under practical farm or ranch conditions, the producer must have animals that reproduce regularly throughout a long life, and that utilize feed efficiently. From this it may be deducted that the profitability of any one animal, or of a herd, is determined by the following two factors:

1. **Individuality.** Which is based upon the ability of the animal to produce beef for a discriminating market.

2. **Performance or efficiency of production.** Which means the ability to reproduce regularly and utilize feed efficiently.

Regardless of the method of selection, the following points must be observed if maximum genetic progress is to be made:

■ **Selection should be from among animals kept under an environment similar to that which you expect them and their offspring to perform**—This requisite applies to animals brought in from another herd, either foundation or replacement animals. For example, animals that are going into a range herd should be selected from among animals handled under range conditions, rather than from among stall-fed animals. This recommendation is based on the results of a long-time experiment conducted by the senior author and his colleagues at Washington State University.[10]

■ **Selection should be for heritable traits of economic importance**—The traits upon which selection is based should be both highly heritable and of economic importance. (See Table 5-5, Economically Important Traits in Beef Cattle, and Their Heritability.) It stands to reason that the more highly heritable characters should receive higher priority in selection than those which are less heritable, for more progress can be made thereby.

By economic importance of traits is meant their dollars and cents value. Thus, those characteristics which have the greatest effect on profits should receive the most attention.

■ **Selection must be accompanied by an orderly and accurate method of scoring or evaluating**—In order to determine whether progress is being made in a breeding program—in order to ascertain whether, through selection, each generation is actually better than the preceding one—a measure or yardstick must be applied to each trait to be evaluated. Moreover, individual measurements must be accurate. For example, weighing conditions must be alike; you cannot weigh some cattle *full* and others *shrunk*. Likewise, if cutability is determined, it should be defined so that it can be repeated.

Four methods of selection are at the disposal of the cattle producer: (1) selection based on individuality or appearance; (2) selection based on pedigree; (3) selection based on show-ring winnings; and (4) selection based on Expected Progeny Difference (EPD). Since each method of selection has its place, a producer, especially a purebred breeder, may make judicious use of more than one of them.

SELECTION BASED ON INDIVIDUALITY OR APPEARANCE

In starting a new herd, certain matters pertinent thereto must be decided—like the breed or cross, whether to start with open heifers or bred cows, etc. Of equal importance to the success of the operation is the selection of the individuals—the choice of the cows and bulls that constitute the foundation herd, and the selection of replacement heifers, usually from within the herd.

Visual appearance has been, and still is, the basis of both feeder cattle and slaughter cattle trade. Likewise, individuality or appearance is the usual method followed by both commercial and purebred breeders. For the most part, it has been responsible for the transformation of the Texas Longhorn to the present-day bullock.

In making selections based on individuality, it must be borne in mind that the characteristics found in the parents are likely to be reflected in the offspring, for here, as in any breeding program, a fundamental principle is that "like tends to produce like." From a practical standpoint, this points up two things:

1. Only those animals which are at least average, or preferably better than average, should be used for breeding purposes.

2. A cow's inheritance will influence only one calf each year, whereas the herd bull may influence as many as 25 to 50 animals in a given season. Hence, in any selection based on individuality, the selection of the herd bull merits maximum attention.

Selection based on individuality or appearance involves selecting animals for the following six traits:

1. Reproductive efficiency
2. Muscling

---

[10]Wash. Ag. Exp. Sta. Bull. 34.

3. Size
4. Freedom from waste
5. Structural soundness
6. Breed type if purebreds

REPRODUCTIVE EFFICIENCY

Reproductive efficiency calls for females producing and weaning a calf each year, beginning at two years of age. A nonexistent or dead calf cannot be scored or evaluated. Thus, being born and born alive, or reproductive efficiency, is the most important trait affecting profit in the cattle business. Sometimes cattle producers lose sight of this simple fact for two reasons: (1) it is overshadowed by emphasis on such traits as rate and efficiency of gain; and (2) fertility in cattle of low heritability—only about 10%. Most cattle producers lie awake nights trying to figure ways in which to improve weight gains, feed efficiency, and carcass quality; and they extol production records involving these traits. But similar concern over breeding efficiency is seldom evidenced, despite the fact that, when it comes to improving the dollar return on investment in beef cattle, no trait is so important as conception and getting a live calf.

Without a calf being born, and being born alive, the other economic traits are of academic interest only—they cannot be scored or evaluated because there is no calf. For example, size is economically important, but no matter how big they are, sterile cows and bulls are not producers. A "mating of the gods," involving the greatest genes in the world, is of no value unless those genes result in (1) the successful joining of the sperm and egg, and (2) the birth of a live calf.

Because reproductive efficiency is essential to successful beef production, it is important that heifers be selected from a herd with a consistent high calving percentage (certainly above 90%, and preferably above 95%). Also, heifers should be selected from a herd with a short calving season (calves coming in less than a 45-day period), and not to exceed a 12-month calving interval.

For the United States as a whole, out of each 100 cows bred, only about 88% drop calves annually.[11] The other 12% are nonproducers, either temporarily or permanently. In some herds, the calf crop percentage runs as low as 50%. With a 50% calf crop, it simply means that two cows are being maintained an entire year to produce one calf. As reproductive ability is fundamental to economical beef production, it can be readily understood that reproductive failure constitutes a major annual loss in the cattle business. In fact, cattle producers acknowledge that the calf crop percentage is the biggest single factor affecting profit in beef cattle production. Improper feeding and disease are the two most common causes of low percentage calf crops in cattle.

Overfeeding accompanied by extremely high condition, or underfeeding accompanied by an emaciated and run-down condition, usually results in temporary sterility that may persist until the condition is corrected. Lack of exercise, inflammation and infection of the reproductive tract resulting from retained afterbirth or other difficulties encountered at calving, and infections of various other kinds may also result in temporary sterility. The most common causes of permanent sterility in cattle are: old age; diseased reproductive organs, such as cystic ovaries; diseased Fallopian tubes; and heredity. The reproductive ability of an individual or an entire herd may also be greatly affected, either temporarily or permanently, by the presence of brucellosis and certain other diseases.

Sterility may be present in either sex. Occasionally bulls are sterile even though sexually active. Differences in the fertility of bulls are especially revealed by the records kept and the semen studies made.

In addition to the above factors affecting reproduction, the percentage calf crop may be affected by the proportion of bulls to cows and their distribution on the range, by the season of breeding, and by diseases.

Of course, the most obvious indication of reproductive efficiency of a cow is a calf at side; and of a bull, it's a large number of calves from a season's service. However, breeding records are not always available and reliable; and they are nonexistent in animals that are not of breeding age. Hence, animals should be scored for reproductive efficiency, although it is recognized that fertility is a *zero* or *one* proposition so far as any one cow is concerned.

Fig. 4-3 illustrates and describes the difference in appearance between a highly fertile female and a lowly fertile female. Females should show femininity at all stages of development. Feminine females are trim in the jaw, throat, and dewlap, and they are smooth shouldered. Avoid coarse, heavy-fronted females that are excessively muscular and give an impression of masculinity rather than femininity.

Fig. 4-4 illustrates and describes the difference in appearance between a highly fertile bull and a lowly fertile bull. A mature bull should look masculine in front. But do not discriminate against a bull calf that lacks masculinity. Bull calves that show extreme masculinity at an early age are apt to mature too early and quit growing too soon. When selecting a bull, observe his

---

[11]Methods of computing calf crop percentages vary, but the three most common methods are: (a) number of calves born alive in comparison with the number of cows bred; (b) number of calves marketed or branded in relation to the cows bred: and (c) number of calves that reach weaning age as compared with number of cows bred. Perhaps the first method is the proper and most scientific one, but, for convenience reasons, some ranchers use the other two.

testicles for size and scrotal configuration. Make certain that the testicles are of normal size, and that they are well defined in the scrotum, rather than surrounded by excess fat.

Although masculinity in a yearling or older bull is an indication of fertility and is desirable, extreme masculinity in a bull under one year of age is undesirable. Development of the masculine sex characteristic at a

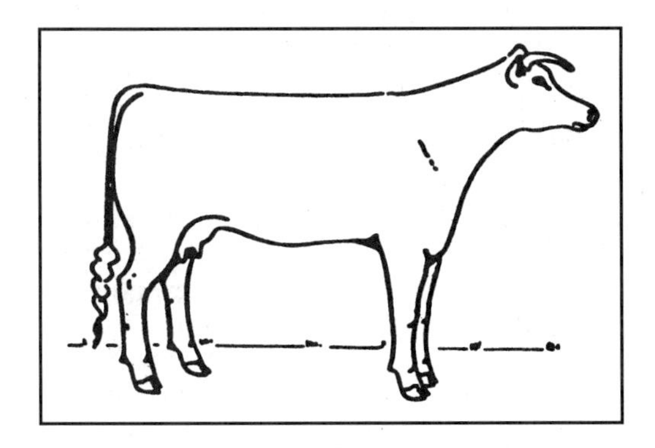

A. Feminine female—long body, lean, smooth muscled; refined, feminine head; lean cheek, jaw, neck, brisket, shoulder, and hindquarters; and a good functional udder (or promise of udder development in a heifer). Length of head in keeping with size of body.

B. Steery female—coarse, heavy front, masculine rather than feminine; protruding brisket; bristly hair on neck and top of shoulders; rounded hindquarters; and fat deposits on the face, brisket, shoulders, hips, rump, pins, below the vulva, and in front of the udder.

Fig. 4-3. Fertility makes the difference! Highly fertile female **(A)** vs lowly fertile female **(B)**.

A. Bull with a masculine front and a sound pair of testicles behind; alert—he's *on the look*, with head up and ears cocked; well-developed crest; muscles well developed and clearly defined, especially in the regions of the neck, loin, and thigh; and well-developed external genitalia, with testicles of equal size and well defined, and a proper neck to the scrotum. Length of head in keeping with size of body.

B. Bull lacking in masculinity; sleepy and droopy; ears not alert; undeveloped crest; muscles lacking development and not clearly defined; testicles may be small, unbalanced, or have one carried high; and the scrotum may be twisted or filled with fat.

Fig. 4-4. Fertility makes the difference! Highly fertile bull **(A)** vs lowly fertile bull **(B)**.

young age is indicative of early maturity, which may result in reduced size and increased fat disposition.

SCROTAL MEASUREMENT

Testicular size is influenced by age, breed, and condition. But, within a given breed and at the same age and condition, testicular size and semen production are highly correlated.

The recommended method for taking scrotal circumference measurements is shown in Fig. 4-5. Note that the tape measurement is made at the largest circumference. The recommended minimum scrotal circumferences by the Society for Theriogenology (American Veterinary Society for the Study of Breeding Soundness) for bulls of different ages in good condition are as follows:

Age	Scrotal Circumference	
(months)	*(in.)*	*(cm)*
12-15	12.0	*30*
15-18	12.4	*31*
18-21	12.8	*32*
21-24	13.2	*33*
24	13.6	*34*

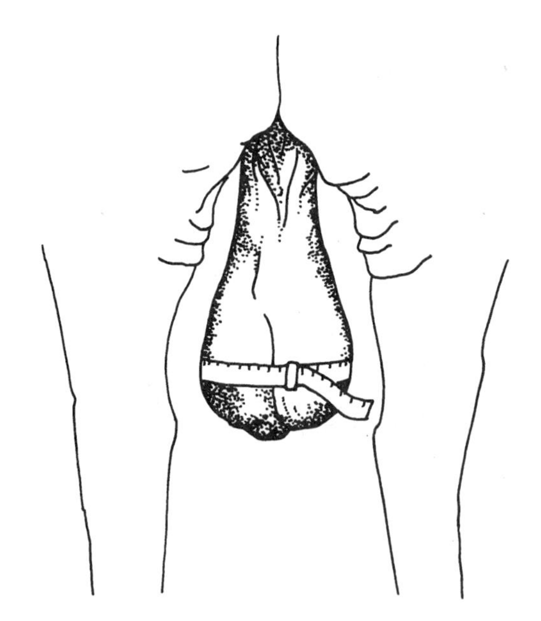

Fig. 4-5. Method of determining scrotal circumference; using a tape and marking the reading at the largest circumference.

MUSCLING

Consumers are demanding, and getting, beef with a maximum amount of lean and a minimum of fat. Fortunately, cattle can be bred for heavy muscling, even to the point that it can be overdone, resulting in double muscling.

Old-time draft horsemen always selected horses for heavy muscling by looking at those parts of the body where there was muscle alone, without fat—the arm, forearm, stifle, and gaskin (between the hock and the stifle). Shrewd cattle producers have done the same thing. But it remained for Butterfield of Australia to confirm the logic of this procedure experimentally; he showed a correlation of 0.93 to 0.99 between the weight of a single muscle or group of muscles and the total muscle in a carcass.[12] Thus, cattle that are heavily muscled in the regions of the arm, forearm, stifle, and gaskin are heavily muscled throughout the body. Butterfield also exploded a myth: he showed that the proportion of muscle in the high-priced cuts in all breeds remains constant at 56%—that it is not possible to increase the proportion of muscle in the regions of the high priced cuts through selection.

A cattle producer can increase total muscle per animal by selection, and thereby increase value based on cutout.

When evaluating animals for muscling, it should be recognized that condition or fatness can create a false impression of muscling. For example, a very thin animal seldom has the appearance of muscularity, with the result that it is apt to be scored rather low in muscling. When the same animal is fattened, it is likely to score higher on muscling. Yet, when finish is overdone, the extra condition may cause the animal to lose its muscular appearance.

In addition to selecting cattle for heavy muscling, producers have at their disposal many ways in which they can tailor-make their product so as to have a higher proportion of lean (muscling) to fat at slaughter time; among them:

- **Use of different breeds and crosses**—Breeds differ in the stage at which they begin to lay down fat. For the most part, the British breeds are early fatteners, whereas the exotics are late fatteners. Thus, the optimum slaughter weight for the British breeds is lighter than for the exotics. Also, the feedlot finishing should differ; the British breeds may be grown out (on a growing ration or handled as stockers) prior to fattening, whereas, the large, late-maturing exotics should be placed in the feedlot immediately following weaning and put on a high energy ration as soon as possible.
- **Delayed fattening**—Heavier slaughter weights without excess fat may be secured in types of cattle inclined to be early fatteners (1) by handling them as

[12]Butterfield, R. M. "The Relationship of Carcass Measurements and Dissection Data to Beef Carcass Composition," *Research in Veterinary Science*, Vol. 6, No. 1, Jan. 1965, p. 26, Table III.

stockers, or backgrounding, before putting them on a "hot" finishing ration, (2) by feeding a low energy ration throughout the finishing period, or (3) by limiting feed intake.

- **Short feed**—Short-fed cattle (on feed less than 100 days) have less finish than long-fed cattle.

- **Lighter slaughter weight**—Marketing feedlot cattle at lighter slaughter weights will cut down on the number of overfinished, low cutability animals. Although this stage and weight will vary with the size of cattle and the ration fed, cattle should always be slaughtered when they have reached the optimum level of fatness for the market requirements.

- **Alter fat by energy of feed**—High energy rations can produce more fat at an early age. But *no amount or kind of feed will alter the amount of muscle.*

- **Heifers vs steers**—The muscles of heifers and steers grow much the same, but females start to lay down fat a little earlier in life than males. Hence, heifers should be slaughtered at lighter weights than steers. It is noteworthy that heifers finished following first calving produce very acceptable carcasses at carcass weights approaching those achieved in steers. Hence, if heavier heifer carcass weights are desired, a one-calf system followed by finishing may be considered.

- **Bulls vs steers**—In comparison with steers, bulls grow faster, are more efficient, and produce leaner carcasses of equal and acceptable quality. Also, they can be carried to heavier slaughter weights than steers without becoming excessively fat. Because of these several advantages, the feeding of young bulls is an accepted practice in many countries; and it will increase in North America.

- **Double muscling**—The proportion of muscle to fat and bone is greater in double muscled cattle, but there are many disadvantages of double muscling (see Chapter 5 section on "Double Muscling"). Moreover, selecting too strongly for heavy muscle development in any breed will, sooner or later, result in more double muscled cattle being produced, provided genes for double muscling are present in the breed.

Muscling is evaluated by looking at the points on the skeleton when there is little other than muscle—the arm, forearm, stifle, and gaskin. Look for the bulge in the muscle in these regions. When the animal walks, look for muscle movement and bulging in the shoulder and stifle regions. Fat just hangs and shakes on an animal, like it does on a fat person. Really muscular cattle are not smooth; rather, they show some creases and indentations between muscles. Also, they are slightly narrower through the heart girth and loin than through the shoulder and round.

Long, smooth muscling is preferred in calves, because such animals will usually grow for a longer period of time and get thicker with age.

Since muscling is a masculine trait, it is more important in bulls and steers than in heifers.

Breeding cattle with coarse shoulders (very heavy muscled shoulders) should be avoided since this condition is frequently associated with calving problems.

Fig. 4-6 shows four bulls that are the same in all traits except muscling. Note that E and F are front and rear views, respectively, of bulls A and D.[13]

ULTRASOUND SCANNING

Ultrasound scanning, a machine that sends high-frequency sound waves into the animal and records these waves as they bounce off the tissues, may be used to record the fat thickness, rib-eye area, and to estimate marbling of breeding animals.

Since the accuracy of ultrasound scanning is dependent upon the experience and technique of the evaluator, The Beef Improvement Federation started training and certifying ultrasound scanners in 1994. Those taking the training and passing the test are awarded *Beef Improvement Certification*, which is good for two years, following which it must be renewed by retesting and certification. The use of the ultrasound scanner on live animals should be augmented by the operator evaluating and recording the predisposition of fat of the animal in the brisket and flank areas.

SIZE

The question of size of beef cattle has been a point of considerable controversy among producers. Generally speaking, arguments are advanced more in defense of a certain size (strain or breed) to which the producer is partial than on the basis of experimental results. Also, it is recognized that available feeds and markets are so variable as to preclude any standard size. Thus, the best size cattle for a given farm or ranch may not be the best size for another, even for a neighbor. Also, cow size must be considered as it relates to the production efficiency of the cow herd as well as the gain and finishing qualities of the progeny.

The following points are pertinent to the question of size in cattle:

- **Size has changed**—Fashions in beef cattle size have changed rather radically during the present century, moving from the big, rugged, beefy—but often-

[13]Four animals are pictured side view in this section, and under each subsequent trait, because this is the traditional number used in judging contests.

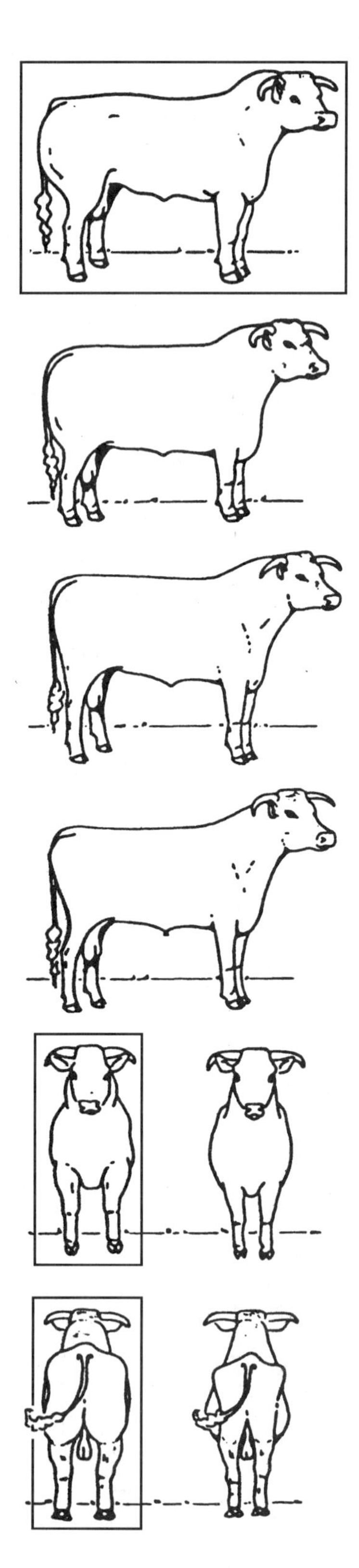

A. Very heavily muscled bull. Note bulging arm, forearm, stifle, and gaskin; rounded loin and round—all good reference points for predicting the total amount of muscle. Note, too, the crease in the thigh. There's a well-defined groove down the topline as a result of the loin eye bulging on each side of the backbone. You can see the muscle move as the animal walks; unlike muscle, fat is inanimated and flabby.

B. Above average muscled bull of the kind that progressive purebred breeders and commercial cattle producers are using. This animal shows the same evidence of heavy muscling as animal **A** above, but to a lesser degree.

C. Below average muscled bull, but not as deficient as **D**.

D. Very light muscled bull. Note narrow, straight, and flat forearm; absence of bulge over the shoulder, over the gaskin, and in the stifle; flat loin; and concave round.

E. Front view of heavily muscled bull (*left;* bull **A** above) and poorly muscled bull (*right;* bull **D** above). Note that the heavily muscled bull stands wide and has a prominent shoulder and bulging forearm.

F. Hind view of heavily muscled bull (*left;* bull **A** above) and poorly muscled bull (*right;* bull **D** above). Note that the heavily muscled bull is wide through the stifle region and has wide stance.

Fig. 4-6. Muscling makes the difference! Four bulls showing variations in muscling. Note that **E** and **F** are front and rear views of bulls **A** and **D**.

Fig. 4-7. Cattle fashions have changed! "Firly," a prize ox of Britain in 1835, shown at 4 years and 8 months of age and weighing 3,000 lb. The near animal is the current concept of the *ideal steer.*

times rough *farmer or rancher type* that our grandfathers produced up through the first third of the present century—to the smaller, earlier-maturing, blockier, smoother types in vogue from about 1935 to 1950. (The extremes in this era were known as *comprests* in Herefords and *compacts* in Shorthorns.) Then, in the 1960s, the pendulum swung back to greater size. This time, the transition was not limited to selection within the existing breeds. It was speeded up through crossbreeding and the introduction of the exotics.

The reason for these shifts in cattle size were many and varied. The smaller types evolved principally because of the demand on the part of the consumer for smaller cuts of meat. Likewise, the show-ring was an important factor, and show-ring fashions for both the finished product and breeding animals tended to follow consumer preferences.

Unfortunately, the tremendously important utility factor or economy of production was largely overlooked in this shift, and little information was available in regard to whether smaller cattle could be produced as economically as larger cattle. Many producers felt that bigger cattle could be produced more economically, especially under conditions where operating costs are on a per head rather than a per pound basis (it requires no more labor to take care of a big cow than a little one), and where profit in a cattle enterprise depends primarily upon the ability of the animals to utilize efficiently large amounts of roughage. The opinion was also prevalent that show-ring fashions toward the low-set, blocky, earlier-maturing, pony-type cattle went further than consumer demand justified. Additionally, there was the desire to get away from dwarfism, and to produce more meat during a period of world beef shortages and good prices. As a result, the pendulum swung toward larger cattle; and currently, medium- or medium-to-large type cattle are favored by most breeders.

Yet, the right size cow to maintain in the beef breeding herd remains unresolved. To answer this question, cow size must be looked at in the total context of beef production; *i.e.*, the production efficiency of the cow as well as the gain and finishing qualities of the progeny. On this basis, small cows appear to be more efficient producers than large cows.

■ **What the experiments show**—Fortunately, some experimental work, designed to answer these and other questions, has been conducted. In a study involving large-, intermediate-, and small-type Hereford cattle, Stonaker of the Colorado Station reported that wide variations in mature size were not antagonistic with the market demands of the 1950s, efficiency of feed use, or carcass cutout values and grades.[14] However, the larger cattle produced a higher percentage calf crop and required some less fixed cost expenditure per pound of beef produced. In the Colorado experiments, all groups of steers were fed to about the same degree of finish. This is important because fat is high in energy; and the fatter an animal becomes the less efficiently it utilizes its feed. Thus it stands to reason that smaller-type animals, when fed to the same weight as large-type animals, will be fatter and, therefore, will require more feed per 100 lb of gain.

In a study conducted cooperatively by the Kansas, Oklahoma, and Ohio Agricultural Experiment Stations, involving steers sired by small-, medium-, and large-size bulls, medium-size steers were favored. It was found that medium-size cattle tend to combine the gaining ability of large cattle and the finishing ability of small cattle without sacrifice of efficiency of gain.[15]

The New Mexico Station compared the gains of, and carcasses produced from, compact, medium, and rangy steers.[16] They found that the rangy steers weighed more when put on feed, gained more, and yielded a higher dressing percentage than the compact steers; and that the medium type was intermediate in each case. There was no indication of any differences in economy of gain, however.

Brungardt, at the Wisconsin Station, made a study of efficiency and profit differences of Angus, Charolais, and Hereford cattle varying in size and growth. He concluded that (1) feed efficiency does not differ greatly among cattle of different sizes when they are fed to the same grade or degree of finish, and (2) at the same weight (but not at the same degree of finish)

---

[14]Stonaker, H. H., Colorado State Univ. Bull. 501-S.

[15]Weber, A. D., A. E. Darlow, and Paul Gerlaugh, *American Hereford Journal*, Vol. 41, No. 22, p. 20.

[16]Knox, J. H., and Marvin Koger, *Journal of Animal Science*, Vol. 5. No. 4, p. 331.

faster-gaining cattle are more efficient than smaller and slower-gaining cattle.[17]

Brown and Brown calculated maintenance costs for small, early-maturing cows and large, late-maturing cows. They concluded that the greater salvage value of larger cows may offset the greater yearly maintenance costs.[18]

Dr. Berg, of Canada, made calculations of 1,100-lb vs 900-lb cows.[19] Based on University of Alberta data, he projected that a cow with a 100-lb weight advantage would produce a 7-lb heavier calf at weaning. At the feed and calf prices that existed at the time he made the calculations, the Canadian investigator concluded: "The heavier cow hardly pays her way in extra calf produced at weaning. At lower feed prices and high calf prices, extra cow size would be more profitable." Then, Dr. Berg very wisely extended his projections to market weight, at which time his studies showed an advantage of 40 lb for the calf from the larger cow. His final conclusion relative to the bigger cow at the prices of feed and cattle then prevailing, *definitely profitable*.

An Oklahoma Station experiment, involving 14 calf crops for a total of 3,298 calves from 863 Hereford and Angus cows, showed that for each 100-lb increase in the yearling weight of cows, a producer can expect 14-lb increase in the weaning weight of his calves.[20]

▪ **Factors of importance in cattle size**—In the final analysis, therefore, the most practical size cattle will vary according to conditions. Among the factors that should be considered when choosing between large medium, and small cattle are:

1. **Plane of nutrition and fertility.** Unless a cow's maintenance requirements are met, fertility (reproduction) will suffer. Thus, the genetic capabilities of the cow should match the available feed. Translated into practicality, this means that the better the pasture or range (or supplemental feeding), the bigger the cow; or conversely, the sparser the range, the smaller the cow.
2. **Plane of nutrition and milk production.** Milking ability is positively correlated with weaning weight of calves. Hence, suckling calves should receive adequate milk. But cows milk to their genetic potential only after their nutritional needs for body maintenance are met. This means that the size cow should match the available feed on the pasture or range. Small cows will milk better on poorer ranges than big cows.
3. **Rapid gains of progeny.** Large cows pass greater growth potential, or more rapid gains, to their progeny than small cows. This is a major reason why the exotics have become so popular.
4. **Labor costs.** Under most conditions, the labor costs for handling larger cattle with their greater pounds are no greater than for the smaller ones.
5. **Relative price of feed and cattle.** Relatively low feed costs and high cattle prices tend to favor big cattle, whereas the reverse conditions favor small cattle.
6. **Weaning vs market weight of progeny.** Big cows and little cows differ very little on the basis of calf produced at weaning. But on the basis of progeny carried to market weight (steers 1,000 lb or more), big cows are definitely more profitable than small cows.
7. **Market weight.** Heavy market weights (steers weighing around 1,200 lb) favor big cattle. On the other hand, where the consumers desire high-quality but smaller cuts of meat, the situation may favor the production of smaller cattle.
8. **Salvage value.** The greater salvage value of big cows than of small cows may offset their greater maintenance cost.
9. **Estimating production efficiency of cows of various sizes.** The two most important factors determining what size beef cow to keep are (a) the weight of the calf weaned, and (b) the total feed required to produce a weaner calf. Of course, the price of each of these is variable.

Also, bear in mind that light calves generally sell at a higher price per pound than heavy calves. This, too, favors small cows.

More research on different size cows is needed, embracing (a) the production efficiency of the cows, and (b) the efficiency of their progeny when fed to a constant finish for slaughter.

Certainly animals can be too big. Huge size is usually accompanied by coarseness, poor fleshing qualities, loose rather than compact conformation, and slow maturity and finishing. Perhaps, under most conditions and in the final analysis, medium-type cattle are best from the standpoint of widest adaptability, general vigor, reproductive efficiency, milk production, longevity, and marketability. Yet, it is reasonable to conjecture that there are environmental and market conditions under which one or the other extremes in beef type (the large type or the small type) may be more profitable.

Size refers to height to top of hip and length from top of shoulders to the tailhead. Young breeding animals and steers should be long, tall, and not excessively fat—indications that they will continue to grow.

---

[17]Brungardt, V. H., "Efficiency and Profit Differences of Angus, Charolais, and Hereford Cattle Varying in Size and Growth," *Research Report*, R 2400, University of Wisconsin-Madison, p. 2.

[18]Brown, C. J., and J. E. Brown, *Arkansas Research*, University of Arkansas, Vol. 20, p. 3.

[19]Berg, R. T., "How to Feed and Manage the 'New Look' Beef Animal," *Beef Cattle Science Handbook*, Vol. 11 edited by M. E. Ensminger and published by Agriservices Foundation.

[20]"Selecting Replacement Females by Growth Boosts and Weaning Performance of Calf Crop," *Better Beef Business*, Vol. 14, No. 9, pp. 16-17.

Bulls should not be too masculine too early in life because the more masculine the animal, the more male hormones secreted. Male hormones, in addition to their reproductive function, inhibit long bone growth (and body growth). They cause a narrowing of the epiphyseal cartilage and a more rapid fusion or ossification of the epiphysis (break-joint). It follows that bulls that show signs of early sexual maturity are not likely to make continued rapid growth and reach large mature size.

Fig. 4-8 shows four bulls that are the same in all major characteristics except size as determined by height and length.

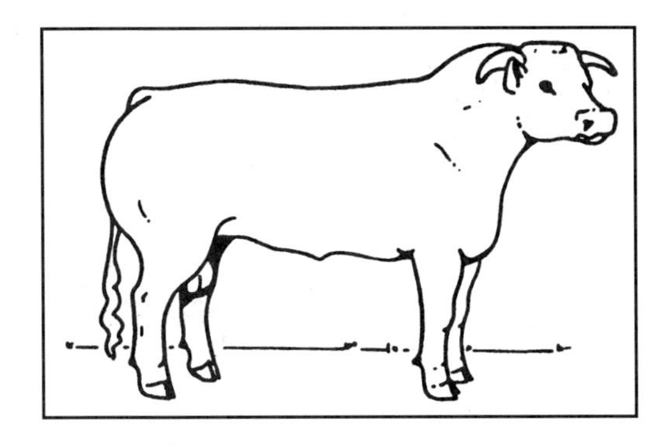

A. A very large bull—tall and long-bodied, the kind for which both purebred and commercial breeders were striving in the 1970s and 1980s. At this stage in life, this bull lacks sexual maturity, which is not a good sign—it is indicative of continued rapid growth and possibly too large of a mature size.

B. Above average-sized bull. Progressive purebred breeders and commercial cattle producers are using bulls of this size.

C. Below desired-sized bull, too short bodied and not enough height.

D. Small bull—lacking in height and length, the kind that was in vogue from 1935 to 1950. This bull is the most masculine appearing and has the most crest of the four bulls. But too much masculinity too early in life is undesirable, because the production of male sex hormones stops long bone growth and makes for small mature size. This bull appears to be deeper than the first bull, but as may be seen by applying a ruler, this is an illusion due to the shortness of the bull.

Fig. 4-8. Size makes the difference! Four bulls of different sizes.

MEASURING CATTLE

Beginning in the 1960s, measurements became a descriptive supplement to many herd testing programs. Adjusted weights and weight ratios accompanied by linear measurements have added another dimension to evaluating the fat-lean ratio of an individual animal in a performance measure program.

Linear measurements are objective. They serve as supplemental information for comprehensive performance testing. How much emphasis breeders should place on linear measurement information should depend on their goals relative to shape and growth patterns, the extent to which certain shape relationships may be important to them, and any advantage these shape relationships give them in marketing beef cattle.

A linear measurement should never be interpreted as a replacement for the weight of an animal at a given age. Instead, linear measurements should be used with growth information as a supplement to selection. No one frame size for an animal will be best for all feed resources, breeding systems, and feed costs. Reproductive efficiency and market weight will determine the optimum frame size range within a given set of feed resources, breeding systems, and production costs.

The most common measurement is over the hip halfway between the hook and the pin. Some producers also measure height at the shoulders (the highest point over the shoulders), and body length from the shoulders (highest point) to the tailhead. Fig. 4-9 shows these points, which may be measured by tape or calipers. When making height measurements, the animal should be stood squarely on a level area.

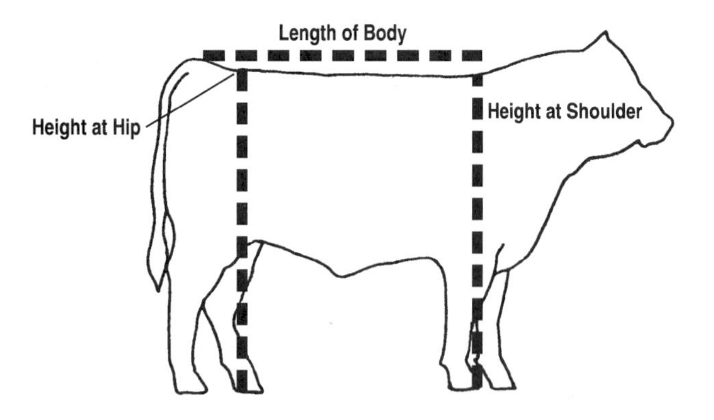

Fig. 4-9. Measurement points pertinent to cattle.

FREEDOM FROM WASTE

From the days of the Texas Longhorn through the 1960s, most producers and scientists described their ideal beast about as follows: "The ideal beef animal should be low set and compact. The body should have great width and depth throughout, with good lines, and with all parts smoothly blended together."

How wrong we were! We now know that the smoother the loin, the squarer the rump, and the straighter the underline, the wastier the animal. We also know that high dressing percentage is *not* an indication of high carcass merit. Rather high dressing percentage is largely due to increase in carcass fat. Worse yet, as the fat increases there is a tendency for the greater proportion of it to be deposited on the cheaper cuts of meat (in the flanks, brisket, and along the underline) or in lumps where it will constitute fat trim. Thus, smoothness of form and straightness of lines should be penalized because they are indicative of fat deposition.

It is generally recognized that there is less fat deposition at early ages in late-maturing animals. This has been a factor in the rise in popularity of late-maturing Holsteins and exotics. This also indicates that when scoring weaner calves the too early-maturing kind—those that look old for their age—should be scored down on "Freedom from Waste."

Also, it is noteworthy that excess fat has the greatest influence on cutability. Thus, estimating cutability involves estimating fatness. Basically, the federal grading of beef in both the United states and Canada relies on measuring fatness on the carcass to estimate cutability.

Freedom from waste, or trimness, is important in both breeding and slaughter cattle. Excessively fat breeding cattle usually have lowered reproduction. Excessively fat slaughter cattle have reduced carcass value.

Look for fat over the following parts of the body: point of shoulder, back ribs, and along the top line directly above the backbone. No muscle should ever be found at these places, so if you feel something, it can be just one thing—fat.

Also, look for fat at those points where fat is deposited at a faster-than-average rate. Look for fat in the rear flank; and look for predisposition to waste as indicated by loose hide on the throat, dewlap, brisket, fore flank, navel or sheath, cod or udder, and twist.

Freedom from waste can be scored on the basis of (1) trimness of middle, and (2) freedom from fat or loose hide in the brisket and fore flank, for no muscle is ever found at these two points.

Of course, freedom from waste in young cattle does not necessarily measure how fat they are at the time of scoring. Rather, it's a score of their estimated, or projected, tendency to get wasty (1) following a normal feeding period, or (2) if fed to normal slaughter weights. Generally, calves that have large briskets, and that are deep in their flanks and twist will have a tendency to get overly fat as they mature.

Fig. 4-10 shows four bulls that are the same for all traits except waste. These bulls are the same length and the same height (as can be determined by use of a ruler). The only difference is in the progressive increase in depth of body from bull A to bull D. However, this is not due to a more capacious chest and middle, which is desirable in all cattle, and which is obtained through long, well-sprung ribs. Rather, it is

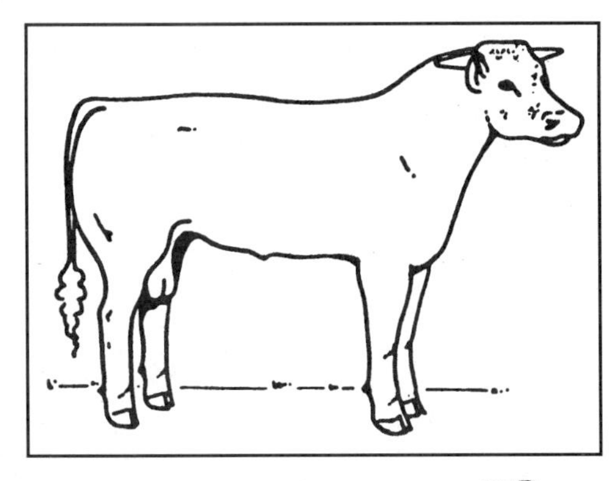

A. A very trim bull, free from waste fat. Note that this bull appears to be longer and taller, and to have a longer neck, than the other three bulls. However, this is only an illusion, as can be proven by use of a ruler. All four bulls are the same in height and body length. They differ only in depth of body and freedom from waste.

B. A trim bull of the kind that most progressive purebred breeders and commercial cattle producers are using.

C. A slightly wasty bull, but not as wasty as **D**.

D. A very wasty bull. Note that the rear flank is loose, hangs low, and is likely filled with fat—a definite fault. Note too, that this bull is smoother over the loin, squarer in the rump, straighter in the underline, and deeper in the body than the other three bulls. But the added depth of body is not due to a more capacious chest and middle; rather, it is due to a loose hide that is serving, or will serve, no more useful purpose than as a pouch for unwanted fat.

Also, note that bull **D** appears to be shorter necked than the rest. However, this is only an illusion created by the loose skin in the dewlap and the added fat in the shoulder vein. As a measurement with a ruler will reveal, the necks of all the bulls are the same length.

Fig. 4-10. Wastiness makes the difference! Four bulls varying in freedom from waste.

due to wastiness—to a loose hide that is filled, or will fill, with fat.

STRUCTURAL SOUNDNESS

Correct skeletal structure is fundamental to a long and productive life in beef cattle, and to efficient beef production. Beef cattle must be able to travel freely and long distances over pasture or range, in order to harvest their feed—chiefly grass; and bulls must be able to follow and breed cows. This means that skeletal defects—particularly in the legs, feet, and joints—can shorten the productive life of any cow or bull. This calls for structural soundness, especially in the underpinning. The legs should be straight, true, and squarely set. The feet should be large, wide and deep at the heel, and have toes of equal size and shape that point straight ahead. The joints, particularly the hock and knee joints which are subject to great wear, should be large and correctly set, and should be clean—without tendency toward puffiness or swelling. Sickle-hocked, post-legged, back at the knees (calf-kneed), or over at the knees (buck-kneed) are faults, and such animals should be scored down.

Fig. 4-11 illustrates structural soundness, which is very important in breeding cattle, but only moderately so in slaughter steers.

Fig. 4-10a. BB Washington, Senior Champion Hereford bull at the National Western Stock Show, Denver, 1995, bred by Bill Bennett, BB Cattle Co., Connell, Washington. This bull weighed 2,880 lb. Half interest in him was sold to Ankony Angus, Minatare, Nebraska, for $25,000. (*Note:* Bill Bennett is one of my former students of whom I am very proud. Dr. E, senior author of *Beef Cattle Science*.)

A. Correct skeletal structure. Note that the legs are squarely set, and that there is sufficient set or angle at the hocks to provide spring and flexion.

B. Sickle-hocked, which means that the hindfeet are placed too far forward beneath the body; and back at the knees (calf-kneed). Sickle-hocked cattle frequently develop hock weaknesses. Cattle that are back at the knees stand and go uphill all the time.

C. Post-legged (too straight hock) and **buck-kneed** (over at the knees). Post-legged cattle have a short, stilted stride; and they're likely to be puffy in the hocks and predisposed to stifle injury. Buck-kneed cattle are less stable on their front legs.

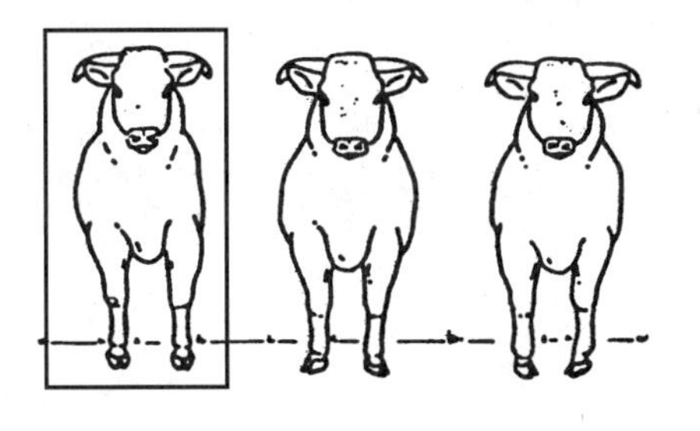

D. Left to right: (1) Correct front feet; (2) toes out; (3) toes in. Cattle that toe out swing the moving foot inward toward the supporting leg, whereas animals that toe in swing their feet outward when moving.

E. Left to right: (1) Correct hindfeet; (2) cow-hocked; (3) wide at the hocks. Cow-hocks and wide at the hocks predispose to a twisting, rotating action when the animal is moving.

Fig. 4-11. Soundness makes the difference!

BREED TYPE

Breed type may be defined as the distinctive characteristics which distinguish one breed from another. Usually, it involves color and markings, shape of head, presence or absence of horns (and shape of horns if present), set of ears, body shape, and size. The combination of these characteristics constitutes the breed *trademark*. Generally they are highly valued by owners of purebred herds and used for promotional purposes.

So long as breed characteristics are not associated with undesirable traits, and so long as their selection does not lower the utility value of the animal, there is nothing wrong with them. Moreover, the choice of a breed is purely voluntary. Hence, when a breed is selected, purebred breeders are morally obligated to comply with the rules of the registry association. Otherwise, they should either select another breed more compatible with their thinking or start another registry association.

First and foremost, emphasis in selection should be on those traits which fit a beef animal for the functions that it is expected to perform—reproductive efficiency, size, muscling, freedom from waste, and structural soundness. Additionally, a registered animal (or a purebred which it is intended shall be registered) should be scored on breed type. Commercial operators may disregard the latter trait. Likewise, it is unimportant in steers.

Fig. 4-12. Hereford breed type vividly portrayed in "Hereford Madonna," a painting by the noted artist, Tom Phillips. Also, the artist captured the behavior of a mother cow and her calf. (Courtesy, Tom Phillips, 915 Fulton Street, San Francisco, CA 94117)

THE SCORECARD

A scorecard is a listing of the different parts of an animal, with a numerical value assigned to each part according to its relative importance. It is a standard of excellence. The use of the scorecard involves studying each part—head, neck, back, loin, rump, round, fore and rear flanks, etc.—and assigning a rank or score to each.

A scorecard is a valuable teaching aid for beginners. It systematizes judging and avoids any part of the animal being overlooked.

SELECTION BASED ON PEDIGREE

Selection based on pedigree refers to the selection of animals to be the parents of the next generation, based upon their ancestors. This method is used in most purebred herds, usually in combination with one or more of the other bases of selection: individuality, production test, or show-ring winnings.

Pedigree selection is of special importance (1) when progeny performance data are not available, or (2) when the animals are either so thin or so young that their individual merit cannot be ascertained with any degree of certainty. Also, when selection is being made between animals of comparable individual merit, the pedigree may be the determining factor.

In making use of pedigree selections, it must be remembered that the ancestors close up in the pedigree are much more important than those many generations removed. Also, pedigree fads as such should be avoided, especially if there has not also been rigid culling and selection based on utility value; and in all instances, poor individuals should be culled, regardless of the excellence of the relatives. Likewise, one should not be misled by or overestimate the value of such pedigree information as the following:

1. **Family names or favorite animals many generations removed.** The value of family names is generally overrated. Obviously, if the foundation animal (the one giving the family name) is very many generations removed, the genetic superiority of this family head is halved so many times by subsequent matings that there is little reason to think that one family is superior to another. Worse yet, some breeders extol a family name in a pedigree on the basis that there are few members of the elite family, little realizing that, in some cases at least, there may be unfortunate reasons for the scarcity in numbers.

2. **Percentage of blood.** With the need to build up populations of a new breed by upgrading, questionable pedigree schemes are sometimes devised by new breed associations. Among them:

a. **The use of *blood* fractions instead of percentages.** Instead of the use of such *blood* fractions as ½ blood, ¾ blood, ⅞ blood, etc., actual percentage should be calculated. To calculate the percentage of certain breeding in the offspring, add that of the sire and dam, then divide by 2. For example, if a 72% bull is mated to a 48% cow, the offspring will be 60% (72 + 48 ÷ 2 = 60).

b. **Setting an arbitrary fraction for purebreds.** Some associations set arbitrary fractions like ⅞ or 15/16, and call such animals as reach this point purebred. Of course, there is nothing magical about reaching such a point, nor is any aura of superiority imparted. It would appear, therefore, that breeders and buyers should be allowed to make their own decisions on how much importance to place on different percentages. Simply list them, and let them decide.

■ **Performance pedigree**—For maximum genetic progress, the pedigree should be more than a mere listing of birth dates and names of ancestors. A pedigree should combine genealogy and performance. Such a pedigree should contain a complete listing of an animal's performance record and its ancestors' performance and progeny records.

The Beef Improvement Federation (BIF) recommends that a performance pedigree contain the following basic information, with the format of the pedigree left to each recording organization:[21]

1. Birth weight, birth weight ratio, EPD, and accuracy.
2. Adjusted 205-day weight ratio, EPD, and accuracy.
3. 365-, 452-, or 550-day yearling weight ratio, EPD, and accuracy.
4. Number of contemporaries at weaning and as yearlings.

Individual weights should not be reported on performance pedigrees due to large environmental variations.

EPDs and progeny ratios may be added for additional traits that are considered important and may be included on the performance pedigree.

Note: See Chapter 20, "Bulls," section headed "Pedigree" for an American Hereford Association "Performance Pedigree."

[21] *Guidelines for Uniform Beef Improvement Programs*, Beef Improvement Federation Recommendations, Seventh Edition, 1996.

SELECTION BASED ON SHOW-RING WINNINGS

For years, many cattle producers (purebred breeders and commercial producers alike) looked favorably upon and used show-ring winnings as a basis of selection. Purebred breeders were quick to recognize this appeal and to extol their champions through advertising. In most instances, the selection of foundation and replacement cattle, and herd bulls, on the basis of show-ring winnings and standards was for the good. On some occasions, however, purebred and commercial breeders alike came to regret selection based on show-ring winnings. A case at point was the period from 1935 to 1946, when the smaller, earlier-maturing, blockier, and smoother types of cattle were winning. Among many, this debacle brought disrepute, from which livestock shows may never fully recover. This would indicate that some scrutiny should be exercised relative to the type of animals winning in the show, especially to ascertain whether they are the kind that are efficient from the standpoint of the producer, and whether, over a period of years, they will command a premium on a discriminating market.

Perhaps the principal value of selections based on show-ring winnings lies in the fact that shows direct the attention of new breeders to those types and strains of cattle that are meeting with the approval of the better breeders and judges.

(Also, see Chapter 14, Fitting, Showing, and Judging Beef Cattle.)

SELECTION BASED ON EXPECTED PROGENY DIFFERENCE (EPD)

Note: Further information relative to production testing and traits of importance in beef cattle is presented in Chapter 5.

The modern era in beef cattle breeding was ushered in with the weighing of animals and the keeping of written production records. This led to performance testing programs, most of which were aimed at improvement of growth rate and feed efficiency. The ease of measuring these two characteristics made performance testing acceptable to producers.

Simply stated, performance testing is a record-keeping system for the purpose of collecting data to be used in selection. It has been an important selection tool in the hands of producers. It brought an awareness that some animals are more efficient than others, and that such characteristics as rate of gain and feed efficiency are at least partially under genetic control and can be passed on to offspring.

However, a performance testing program based only on rate of gain and feed efficiency is not adequate. To fill this need, Expected Progeny Differences (EPDs) evolved.

Expected Progeny Differences (EPDs) are the differences in performance expected from the offspring of one individual compared to the offspring of another individual within the same breed.

The first use of EPDs came through *National Sire Evaluations (NSE)* conducted by some breed associations. These programs involved a *sire summary* comparing sires across an entire breed. Comparisons were made possible only if sires had progeny in more than one contemporary group where other sires were represented. Widespread use of AI, especially in new breeds available in the United States beginning in the late 1960s, made these comparisons possible. The first National Sire Summary, published by one of these registry associations in 1972, compared 13 sires. However, only sires with adequate progeny in more than one comparison group were included in these programs. Also, some incorrect assumptions reduced their accuracy.

Fig. 4-13. Selection based on EPD is the modern way. This shows a purebred Maine-Anjou herd on pasture. (Courtesy, American Maine-Anjou Assn., Kansas City, MO)

Finally, refined mathematical techniques and additional data resulted in the *National Cattle Evaluation (NCE)* where all animals, or even planned matings, in a breed can be more accurately compared than with progeny NSE. All major breed associations now have such programs and print sire summaries, many of which are available through the Internet.

Until now, information in a performance pedigree has been of primary use for within-herd comparisons. But new techniques and new analyses have been implemented so that values in a performance pedigree are now comparable to those in the National Sire Summary; that is, these values are comparable across the entire breed. In fact, estimates can be made of Expected Progeny Difference of individuals that would result from a possible mating, *i.e.*, an EPD can be estimated for an individual before a mating is even made.

EPD TRAITS

All breed programs evaluate certain traits. Presently, those traits common to all breeds are as follows:

- **Birth weight**—This is the expected birth weight deviation of calves from this individual, excluding maternal influence. In other words, this is the relative birth weight with equal uterine conditions assumed. This, then, is the genetics passed on directly to progeny by this individual only for weight at birth.

- **Weaning weight**—This is the weight at 205 days of calves sired by, or out of, this individual excluding maternal influence, *i.e.*, as if produced by and reared by the exact same cow.

- **Yearling weight**—This is the weight at 365 days with maternal influence excluded.

- **Maternal influence**—This is generally called *maternal milk*; actually, it is any maternal influence. This is expressed in pounds of weaning weight (not pounds of milk) produced by daughters of this individual due to maternal influence, excluding genetics for growth. Total Maternal EPD is also reported by most breeds, which is actually milk EPD + ½ weaning weight EPD. Again, this is what is expected from daughters of this individual.

Other traits calculated by some breeds are:

- **Direct calving ease**—This is an estimate of calving ease of calves sired by or out of this individual; basically, it is determined by the size, shape, etc., of the calf produced by this individual. For bulls, this is the expected calving ease (or ease of birth) compared to other bulls when mated to equal cows. For females, this is the expected calving ease of a calf she might produce excluding her own maternal influence (estimated as maternal calving ease below). In other words, this is the relative ease of being born of calves conceived by the female with maternal calving factors equalized. Another way of looking at this is the calving ease expected from embryos implanted into exactly equal recipients.

- **Maternal calving ease**—This is the relative ease of calving experienced by daughters of this individual; basically, it is determined by the size, internal structure, uterine environment, etc., of the calving female. This is an estimate of the ease with which daughters of this individual would give birth compared to daughters of other individuals, as if all daughters were mated to the same sire and managed equally.

- **Gestation length**—In days.

- **Yearling height**—In inches.

- **Scrotal circumference**—In centimeters. This is an estimate of age at puberty in males (and their female relatives) and estimated volume of sperm production.

- **Carcass weight**—In pounds.

- **Marbling**—In USDA marbling degrees.

- **Ribeye area**—In square inches.

- **Fat thickness**—In inches.

EPD VALUES

EPDs are generally reported in plus or minus values of the actual units of measurement. Weights are reported in pounds, physical measurements in inches or centimeters, and time periods in days. Only in the subjective estimate of calving ease is a ratio generally used, with 100 being the base average. One association now calculates calving ease figures using both birth weight and subjective calving ease data, where EPDs are expressed as deviations in percent unassisted births.

Some examples of EPDs from a breed sire summary follow:

Example No. 1: Assume that one bull has a birth weight EPD of +3.0 with an accuracy of 0.67 (accuracies range from 0 to 1); and that another bull has an EPD of –0.5 with an accuracy of 0.70. From this information we are able to estimate that, if these bulls were used on equal sets of cows under equal conditions, the first bull would be expected to sire, on the average, calves weighing 3.5 lb heavier at birth than the other bull. The accuracy of the bulls is similar and is in what might be called a moderate range (0.5 to 0.75) of accuracy. Accuracy is influenced by number of records, source of records, and number of contemporary comparison groups.

Example No. 2: Assume that one bull has an EPD for yearling weight of +42 and another bull has an EPD for yearling weight of +17. In comparing the two bulls, we would estimate a difference of 25 lb in weight of their calves at yearling age.

EPD CAUTIONS

When using EPDs, the following cautions should be noted:

1. Each breed association conducts its own analysis, and the information is not comparable *between* or *across* breeds.
2. Since the base or zero value for each breed analysis varies, and breeds may change over time in

trait performance, the average EPD of *recently born* individuals in a breed may not be zero or even near zero. For example, the most recent average EPD for yearling weight in one breed is +1.4 lb but in another breed it is +37.6 lb. Again, this does not mean the second breed is 36.2 lb (37.6 minus 1.4) heavier at yearling.

3. An EPD cannot be used to predict absolute performance. The use of a sire +15 in yearling weight does not mean that yearling weights in your herd will be increased by an average of 15 lb, over what they have been. It only means that this sire would be predicted to produce progeny that average 15 lb heavier than another sire in the same breed with a yearling weight EPD of 0. Perhaps more importantly, birth weight is often misused. If a bull has a birth weight EPD of +3, this does not mean that he would increase your birth weight 3 lb. Nor would a bull with a –2 EPD decrease your birth weights by 2 lb. Both of these example bulls might still result in calving difficulty in your herd, especially if the bulls are from breeds of large physical size and your cows (or heifers) are small. What EPD tells us in this example is that the two bulls would be expected to sire calves 5 lb different in birth weight (the difference in +3 and –2). But the actual birth weights might be 65 and 60 lb, or 95 and 90 lb, or any other possible values differing by 5 lb. In short, EPD can simply *not* be used to predict actual performance level but, rather, to predict comparative differences in performance.

4. A breeder may not wish to select continuously for maximum genetic expression. For example, continued emphasis on heavier yearling weight and higher milk production, without regard for other factors, would not necessarily be advisable under marginal nutritional conditions.

5. Note that the word *expected* appears in Expected Progeny Difference. Perhaps *average* should have been used instead. Lower accuracies mean that greater deviation from the EPD value is likely. Higher accuracy equals higher predictability, or less deviation, from the EPD value.

In conclusion, modern EPD values are the most useful and most accurate of all performance records. These are the only records which can be directly compared for all animals in an entire breed.

EPDs OF YOUNG BULLS

Because of new analysis procedures used by breed associations, data on young (non-parent) bulls is now available which is superior to individual performance records alone. This new data is in the form of *Expected Progeny Differences (EPDs)* for such traits as birth weight, weaning weight, yearling weight, maternal weaning weight, and maternal milk. In addition, some breed associations present EPDs for height, scrotal circumference, gestation length, calving ease direct and maternal calving ease. An EPD is a prediction of how future progeny of a sire are expected to perform for a particular trait as compared to a fixed breed average. "Difference" is the key to understanding EPDs. Each EPD has an accuracy figure associated with it which is a reliability measure and on non-parent bulls will be relatively low. The same kind of data is available for progeny-proven bulls in breed sire summaries and the accuracies on their data may be much higher. Comparisons of EPDs should be made within breed only.

Sometimes EPDs may not be available on yearling bulls. However, if we assume an individual receives half of his genetic component from his sire and half from his dam, we can estimate his EPD if we have access to pedigree performance for the various traits as follows:

$$\text{Estimated EPD}_{\text{Young Bull}} = \frac{\text{EPD}_{\text{Sire}} + \text{EPD}_{\text{Dam}}}{2}$$

Thus, to make the fewest mistakes and the greatest progress, progeny proven bulls with relatively high accurate EPDs should be used AI when possible.

In selecting young sires for natural service, the best avenue is to select sons of superior, progeny-proven sires which are superior based upon non-parent EPDs and individual performance records. Groups of young sires selected in this manner are more apt to produce as expected and will cause real genetic improvement in a commercial herd.

HERD IMPROVEMENT THROUGH SELECTION

Once the herd has been established, the primary objective should be to improve it so as to obtain the maximum production of quality offspring. In order to accomplish this, there must be constant culling and careful selection of replacements. The breeders who have been most constructive in such a breeding program have usually used great breeding bulls and they have obtained their replacements by selecting some of the outstanding, early-maturing heifers from the more prolific families.

Improvements through selection are really twofold: (1) the immediate gain in increased calf production from the better animals that are retained; and (2) the genetic gain in the next generation. The first is important in all herds, whereas the second is of special importance in purebred herds and in all herds where replacement females are raised. Most of the immediate gain is attained in selecting the cows, which are more numerous than the bulls, whereas the majority of the genetic gain comes from the careful selection of bulls. The genetic gain is small, but it is permanent and can be considered a capital investment.

Many good cattle breeders consider it a sound practice to make about a 20% replacement each year. Under such a system of management, 46% of the heifer calves are retained each year.[22]

[22]A herd of 100 cows will produce about 44 heifers each year (with an 88% calf crop—the national average). Hence, 46% of the heifers will be needed to replace 20% of the cows ($44 \times 0.46 = 20$).

QUESTIONS FOR STUDY AND DISCUSSION

1. Discuss the importance of matching cow size and milk production to the feed/environment, and matching the bull to the market.

2. What is the object of selection (a) in breeding cattle, (b) in feeder cattle, and (c) in slaughter cattle? Wherein does selection for each of these three purposes differ from the other two?

3. Select a certain farm or ranch (either your home farm or ranch, or one with which you are familiar). Assume that there are no beef cattle on this establishment at the present time. Outline, step by step, how you would go about (a) establishing a herd, and (b) selecting the individuals. Justify your decisions.

4. Should the selection of individuals for a purebred herd and for a commercial herd differ? If so, how?

5. How would you go about getting more milking ability (a) in a purebred herd, and (b) in a commercial herd?

6. Is uniformity of color in a herd important in a commercial herd where a crossbreeding program is being followed? If so, how could you obtain uniformity? Has increased crossbreeding of beef cattle been a major cause for the current lack of consistency in beef?

7. Discuss the economics of longevity in a cow herd. Use as the examples 2 cows dropping their first calves as 2-year-olds, each producing a calf each year thereafter—but one producing through age 8, and the other producing through age 10.

8. Discuss the practical importance of the Missouri Experiment Station study by Dr. Samuel Brody in which cattle were housed in "climatic chambers."

9. Why should selection be from among animals kept under an environment similar to that under which the cattle producer expects them and their offspring to perform?

10. Do all methods of selection, including production testing, rely in part at least on visual appearance or "eyeballing"?

11. Why is the selection of the bull of greater importance than the selection of the female?

12. Under selection based on individuality or appearance, six traits are listed. Rank these traits in order of economic importance. Justify your ranking.

13. Why is being born and born alive the most important trait affecting profit in the cattle business?

14. Why are testicular size and scrotal configuration important in the selection of bulls?

15. Is it possible through selection to increase (a) the overall, total muscling of cattle, and (b) the proportion of muscle in the high-priced cuts? Justify your answer.

16. In addition to selecting cattle that are heavy muscled genetically, cattle producers have at their disposal other ways through which they can tailor-make this product so as to have a higher proportion of lean to fat at slaughter time. List and discuss each of these.

17. Why does Europe feed bulls whereas America feeds steers?

18. With emphasis on selecting cattle for heavy muscling, will there be more and more double muscling? Is double muscling good or bad? Justify your answer.

19. What size beef cattle do you feel is best for most farms and ranches of America—large, medium, or small? Justify your answer.

20. Should breeders measure cattle? If so, how should they be measured?

21. If fat is unwanted, and largely waste, why are cattle finished out in a feedlot?

22. Should breed type be considered in a selection program?

23. How could pedigrees be made more useful from a selection standpoint?

24. Cite examples as proof of the fact that showring standards have not always been practical.

25. What are EPDs?

26. What EPD traits are common to all breeds?

27. When using EPDs, what cautions should be noted?

28. Once the herd is improved, how may it be further improved through selection?

SELECTED REFERENCES

Title of Publication	Author(s)	Publisher
Animal Growth and Nutrition	Ed. by E. S. E. Hafez I. A. Dyer	Lea & Febiger, Philadelphia, PA, 1969
Beef Cattle, Seventh Edition	A. L. Neumann	John Wiley & Sons, Inc., New York, NY, 1977
Beef Cattle Production	J. F. Lasley	Prentice-Hall, Inc., Englewood Cliffs, NJ, 1981
Beef Production and the Beef Industry	R. E. Taylor	Burgess Publishing Company, Minneapolis, MN, 1984
Beef Production in the South, Modified Edition	S. H. Fowler	The Interstate Printers & Publishers, Inc., Danville, IL, 1979
Bonsma Lectures	Jan C. Bonsma	Agriservices Foundation, 648 W. Sierra Ave., Clovis, CA 93612, 1979
Breeding and Improving Farm Animals, Seventh Edition	E. J. Warwick J. E. Legates	McGraw-Hill Book Co., New York, NY, 1979
Commercial Beef Cattle Production, Second Edition	Ed. by C. C. O'Mary I. A. Dyer	Lea & Febiger, Philadelphia, PA, 1978
Genetics of Livestock Improvement, Third Edition	J. F. Lasley	Prentice-Hall, Inc., Englewood Cliffs, NJ, 1978
Guidelines for Uniform Beef Improvement Programs, Seventh Edition	Beef Improvement Federation	Beef Improvement Federation, Colby, KS, 1996
Improvement of Livestock	R. Bogart	The Macmillan Co., New York, NY, 1959
Improving Reproductive Efficiency in Beef Cattle	J. R. Beverly, *et al.*	Glidwell Printers, Bryan, TX, 1972
Lasater Philosophy of Cattle Raising, The	L. M. Lasater	Texas Western Press, The University of Texas, El Paso, TX, 1972
Man Must Measure	J. Bonsma	Agri Books, Cody, WY, 1983
Proceedings—Second World Conference on Animal Production	R. E. Hodgson, *et al.*	American Dairy Science Association, Urbana, IL, 1969
Stockman's Handbook, The, Seventh Edition	M. E. Ensminger	Interstate Publishers, Inc., Danville, IL, 1992

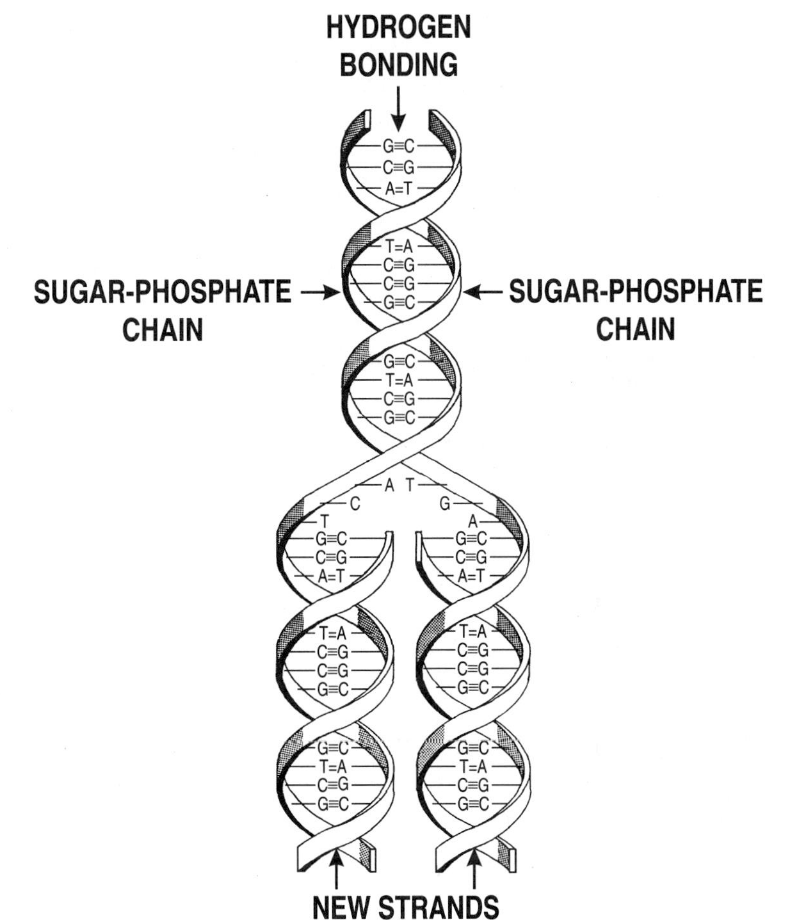

The spiral structure of deoxyribonucleic acid, or DNA—the basic building block of life on earth. It is a double helix (a double spiral structure), with the sugar (deoxyribos)—phosphate (phosphoric acid) *backbone* represented by two spiral ribbons. Connecting the *backbone* are four nitrogenous bases (a base is the nonacid part of a salt): adenine (A) paired with thymine (T), and guanine (G) paired with cytosine (C); with the parallel spiral ribbons held together by hydrogen bonding between these base pairs. Adenine and guanine are purines, while thymine and cytosine are pyrimidines.

SOME PRINCIPLES OF CATTLE GENETICS

Contents *Page*

Contents Page

Biotechnology is having an exciting and dramatic impact on cattle breeding. It is being applied right alongside traditional methods.

During the decades ahead, biotechnology will dominate and pace the entire beef cattle industry. The final goal will be the realization of more efficiency, more disease resistance, more milk, more growth, more high-quality carcasses, and more and better everything.

The fruition of many facets of biotechnology will take time and money. But the end results will justify what it takes to get there, and be worth waiting for.

DEFINITION AND HISTORY OF BIOTECHNOLOGY

Biotechnology refers to methods of using plants, animals, and microbes to produce useful substances or improve existing species.

Although the term is relatively new, biotechnology has served humankind for thousands of years, dating to the use of yeasts, molds, and bacteria to make fermented foods like beer, wine, and bread; and to preserve foods such as turning milk into cheese and yogurt.

Biotechnology was scientifically founded by Gregor Johann Mendel, an Austrian monk.

But modern biotechnology as such was ushered in with the discovery of deoxyribonucleic acid (DNA). The history of DNA began with a young Swiss chemist, Friedrich Miescher, who in 1868 isolated nucleic acid from pus cells from a discarded bandage. Despite considerable subsequent research work on DNA, and although it was known that human cells had 46 chromosomes and that genes were aligned on these, the biochemical mechanism of heredity remained obscure for the next 85 years. Finally, in 1953, scientists James Watson, an American geneticist, and Francis Crick, an English physicist, working at the University of Cambridge, proposed a double spiral structure for DNA. For their landmark work, Watson and Crick shared the 1962 Nobel Prize for Physiology and Medicine.

MENDEL'S CONTRIBUTION TO GENETICS

Genetics was really founded by Gregor Johann Mendel, an Austrian monk, who smoked long, black cigars and gardened because of his obesity. He conducted breeding experiments with garden peas from 1857 to 1865, during the time of the Civil War in the United States. In his monastery at Brünn, Mendel applied a powerful curiosity and a clear mind to reveal some of the basic principles of hereditary transmission. In 1866, he published in the proceedings of a local scientific society a report covering 8 years of his studies, but for 34 years his findings went unheralded and ignored. Finally, in 1900, 16 years after Mendel's death, 3 European biologists working independently, duplicated his findings. This led to the dusting off of the original paper published by the monk 34 years earlier.

Fig. 5-1. Gregor Johann Mendel (1822–1884), a cigar-smoking Austrian monk, whose breeding experiments with garden peas founded genetics. (Courtesy, The Bettmann Archives, Inc.)

The essence of Mendelism is that inheritance is by particles or units (called genes), that these genes are present in pairs—one member of each pair having come from each parent—and that each gene maintains its identity generation after generation. Thus, Mendel's work with peas laid the basis for two of the general laws of inheritance: (1) the law of segregation, and (2) the independent assortment of genes. Later genetic principles have been added; yet all the phenomena of inheritance, based upon the reactions of genes, are generally known under the collective term *Mendelism*.

Thus, genetics is really unique in that it was founded by an amateur who was not trained as a geneticist and who did his work merely as a hobby. During the years since the rediscovery of Mendel's principles (in 1900), many additional genetic principles have been added, but the fundamentals as set forth by Mendel have been proved correct in every detail. It can be said, therefore, that inheritance in both plants and animals follows the biological laws discovered by Mendel.

SOME FUNDAMENTALS OF HEREDITY IN CATTLE

In the sections which follow, no attempt will be made to cover all of the diverse field of genetics. Rather, the authors will present a condensation of a few of the known facts in regard to the field and briefly summarize their application to beef cattle.

CELLS

The bodies of all animals are made up of tiny cells, microscopic in size. It has been estimated that the human body is made up of approximately 250 trillion cells; so, animals, which are larger than humans probably have more cells.

Most cells are made up of three major parts: the *cell membrane*, the *nucleus*, and the *cytoplasm*.

The cell membrane, which surrounds the cell (1) serves as the framework and maintains the shape of the cell, and (2) controls the passage of substances into and out of the cell.

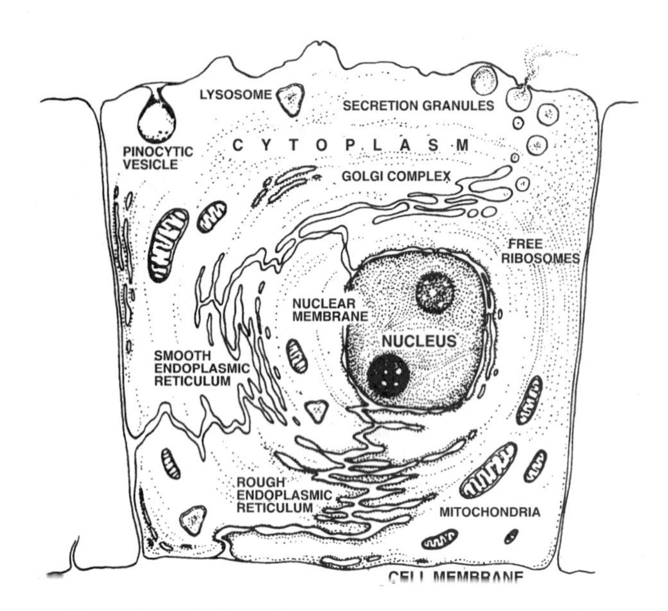

Fig. 5-2. A typical cell showing some of the structural components that most cells have in common, although cells of the body are variable in structure and function.

The nucleus is the spherical body near the center of the cell which contains the chromosomes.

The cytoplasm is the material between the nucleus and the cell membrane. It contains many secretion products and many highly specialized structures.

CHROMOSOMES

The nucleus contains threadlike bodies called *chromosomes.*

In the body cells of an animal, each of the chromosomes is duplicated (geneticists call these pairs *homologous chromosomes*), whereas in the formation of the sex cells (gametes), the egg and the sperm, a reduction division occurs and only one chromosome and one gene of each pair goes into a sex cell. This means that only half the number of chromosomes and genes present in the body cells of the animal go into each egg and sperm, but each sperm or egg cell has genes for every characteristic of its species. When mating and fertilization occur, the single chromosomes from the germ cell of each parent unite to form new pairs, and the genes are again present in duplicate in the body cells of the embryo.

The central inner portion of each chromosome contains a long double helical structure called a deoxyribonucleic acid molecule (DNA). (See the lead illustration at the beginning of Chapter 5.) This molecule resembles a ladder that has been twisted in opposite ways at the ends. The DNA molecule varies in length, depending upon the specific chromosome in which it is found.

KARYOTYPE

Chromosomes vary in shape, size, and location of centromeres (the point at which a dividing chromosome is still attached before complete division occurs), both within and between species. The chromosome pairs, or homologous chromosomes, are alike. Much work has been done to characterize a group of chromosomes according to their morphology. This grouping is called *karyotype.*

Recent research in cytogenics has made it possible to identify each pair of human, cattle, sheep, and swine chromosomes by a specific and reproducible banding pattern when stained by means of a number of different techniques. The banding pattern for each species is different. The banding patterns for each of the chromosome pairs in female cattle are shown in Fig. 5-3.

Fig. 5-3. Photomicrograph of the 30 pairs of chromosomes of a heifer arranged in karyotype, with the chromosomes in pairs. Note the XX chromosomes (the female sex chromosomes), lower right. (Courtesy, Dr. Dale Vogt, University of Missouri–Columbia)

CHROMOSOME NUMBER

Each species of animals has a characteristic number of chromosomes, although in some cases 2 species may have the same number of chromosomes (for example, cattle and horses). The nucleus of each body cell of cattle contains 30 pairs of chromosomes, or a total of 60.[1] The karyotype of European cattle *(Bos taurus)*, Zebu *(Bos indicus)*, and the American bison *(Bison bison)* are very similar in numbers and appearance. However, crosses between *Bos taurus* and *Bison bison* usually result in sterile F_1 males, because the testicles are carried too close to the body and spermatogenesis cannot take place at this elevated temperature.

GENES; DNA

The study of genes and DNA actually began in 1857 with the Austrian monk Gregor Mendel experimenting with garden peas. He showed that heredity could be understood in terms of simple mathematical ratios. Although Mendel was not able to explain how hereditary characteristics were transmitted to offspring, he was observing genes and DNA in action.

The gene is a portion of deoxyribonucleic acid (DNA) molecule; and hundreds or even thousands of genes are carried on a single chromosome. DNA in animal cells is found in the nucleus and extends the length of the chromosome, near its center. The DNA molecule is a long, helical (spiral) structure, resembling a long, twisted ladder with two sides or strands joined together by rungs, or steps. The fact that chromosomes occur in pairs also causes genes to occur in pairs. Members of the homologous pairs of chromosomes have a loci, or the same position where similar genes (alleles) are carried. So, *alleles are genes that occupy the same location on homologous chromosomes.*

More than one alternative form of gene may exist, with each one affecting the same trait but in a different manner. These forms are called alleomorphos, or alleles.

Each of the paired autosomes in the male and female of a species normally carries the same loci. However, the sex chromosomes do not always carry the same locus, because in mammals the X chromosome is much larger than the Y chromosome. For this reason the X chromosome carries loci not carried on the Y chromosome. Since female mammals are XX and males are XY, the male possesses less genetic material than the female.

The DNA of the chromosomes in the nuclei of cells carries the coded master plans for all of the inherited characteristics—size, shape, and orderly development from conception to birth to death. DNA is different for each species, and even for each individual within a species. These differences consist of minor rearrangements of sequences among the nitrogenous bases, which constitute a code containing all the information on the heritable characteristics of cells, tissues, organs, and individuals.

The messages carried by DNA are put into action in the cells by the other nucleic acid, RNA. To do this, DNA serves as a template (as the pattern or guide) for the formation of RNA. The genetic message is coded by the sequence of purine and pyrimidine bases attached to the *backbone* of the DNA structure—a long chain of the sugar deoxyribose and phosphoric acid. Purine bases in DNA include adenine and guanine, while pyrimidine bases include cytosine and thymine. One molecule of DNA may contain 500 million bases. The *backbone* of RNA is also a sugar, the sugar ribose, plus phosphoric acid. However, in RNA the pyrimidine base thymine is replaced by uracil, another pyrimidine. RNA molecules are considerably smaller than DNA, containing from less than a hundred to hundreds of bases—not millions.

GENETIC ENGINEERING

Changing the characteristics of an animal or plant by altering or rearranging its DNA is known as genetic engineering.

Long before the genetic composition of domestic animals was known, farmers bred them to enhance their usefulness to humans. Since their domestication, cattle have been bred for their power at the plow, for their fighting spirit, and for milk or beef.

Gene splicing involves taking specific genes from the cells of one organism and transferring them to another. The development of gene splicing (also known as recombinant DNA) ushered in genetic engineering, the essential steps of which include: (1) identifying a particular gene that encodes for a desired trait; (2) isolating the gene; (3) studying the gene's function and regulation; (4) modifying the gene by altering the nucleotide sequence; and (5) introducing the hybrid DNA (known as recombinant DNA) into a selected cell for reproduction and synthesis. An animal that integrates recombinant DNA into its genome (the total number of genes in a species) is called *transgenic.*

Microinjection is the predominant method used to transfer genes in farm animals. Current research is devoted primarily to improvement of production traits, enhancement of animal health, and production of biomedically useful human health products.

Genetic engineering provides tools beyond those of conventional genetics for use in livestock improvement. Its advantages include: (1) easier manipulation of the genome, (2) circumvention of incompatibility between species, and (3) more rapid genetic change. Its limitations include: (1) lack of useful cloned genes

---

[1]Horses also have 60 chromosomes; sheep have 54; swine have 40; and people have 46.

for productivity traits and disease resistance, (2) lack of gene maps (genetic markers) of the overall makeup of animals, (3) insufficient knowledge of mechanisms involved in regulation of transgenes, and (4) the cost and time required for research and development.

Currently, animal scientists are focusing on gene mapping (genetic markers) of the overall genetic makeup of cattle. Once the specific genes that control certain traits have been pinpointed, this information can be used selectively to breed animals with those genes. With this information, cattle of the desired type can be produced—animals tailored for rapid growth, ease of calving, resistance to disease, consistently tender, lean beef, and any other desired traits.

A few years ago, genetic engineering was science fiction. Today, it is reality.

On May 23, 1977, scientists at the University of California-San Francisco reported a major breakthrough ao a rooult of altoring gonoo turning ordinary bacteria into factories capable of producing insulin, a valuable hormone previously extracted at slaughter from pigs, sheep, and cattle, so essential to the survival of 1.8 million diabetics. The feat opened the door to further genetic engineering or splicing. Already, this genetic wizardry has been used in transplanting into bacteria (and recently into yeast cells) genes responsible for many critical biochemicals in addition to insulin; among them, endorphin, somatotrophin, interferon, and vaccines.

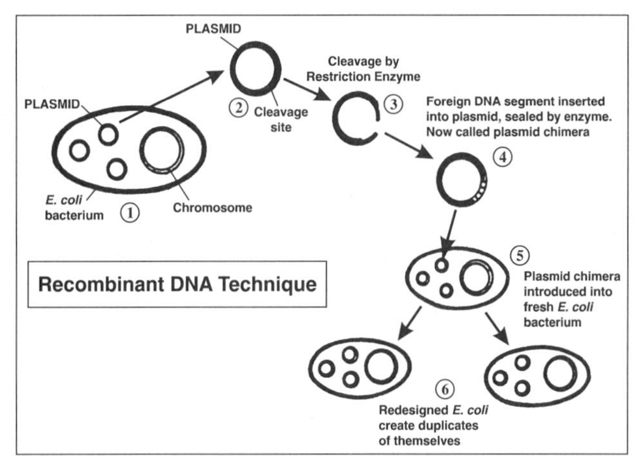

Fig. 5-4. Redesigning *E. coli*, common bacteria of animal and human intestines. The steps:

1. The scientist places the bacterium in a test tube with a detergent. This dissolves the microbe's outer membrane, causing its DNA strand to spill out.
2. The plasmids (the closed loops), which are genetic particles found in bacteria, are separated from the chromosomal DNA in a centrifuge.
3. The plasmids are placed in a solution with a chemical catalyst called a restriction enzyme, which cuts through the plasmids' DNA strips at specific points.
4. The opened plasmid loops are then mixed in a solution with genes—also removed by the use of restriction enzymes—from the DNA of a plant, animal, bacterium, or virus. In the solution is another enzyme called a DNA ligase, which cements the foreign gene into place in the opening of the plasmids. These new loops of DNA are called plasmid chimeras because, like the chimera—the mythical lion-goat-serpent after which they are named—they contain the components of more than one organism.
5. The chimeras are placed in a cold solution of calcium chloride containing normal *E. coli* bacteria. Then the solution is suddenly heated, at which time the membranes of the *E. coli* become permeable, allowing the plasmid chimeras to pass through and become a part of the microbe's new genetic structure.
6. When the redesigned *E. coli* reproduce, they create duplicates of themselves, new plasmids—and DNA sequences—and all.

Genetic manipulations to create new forms of life make biologists custodians of a great power. Despite different schools of thought, scare headlines, and political hearings, molecular biologists will continue recombinant DNA studies, with reasonable restraints, and work ceaselessly away at making the world a better place in which to live. (See Fig. 5-4.)

In summary, it may be said that scientists are coming closer and closer to understanding the very essence of life—DNA.

GENETIC RESISTANCE TO DISEASE

Genetics as a tool for eliminating or controlling certain diseases holds promise. In this area, plant breeders have led the way. In 1905, it was discovered that certain varieties of wheat were more resistant to mycotic stem rust than others, thereby laying the foundation for important advances in the knowledge of genetic resistance to disease. Subsequently, scientists have evolved with many varieties of plants showing genetic resistance to disease. Evidence that similar genetic resistance to disease holds for animals has been demonstrated by experiences and experiments. For example, Brahman cattle are more resistant to certain parasites, notably Texas fever, than the British breeds.

In the future, scientists may be able to genetically engineer animals resistant to some of the most costly diseases. The goal is to improve the overall health of livestock without compromising desirable production traits like reproduction, growth, or meat quality. But for this to be a reality, the genes responsible for disease resistance, or susceptibility, must be identified and understood.

Already scientists have engineered transgenic chickens with great resistance to leukosis.

The application of genetics to disease control in animals presents greater problems than in plants; it is more expensive and time consuming. Also, to be of greatest practical value, it would be necessary to develop strains or breeds of animals that are genetically resistant to several diseases. Nevertheless, the stakes are high and this approach is worthy of greater attention than it has received in the past.

PATENTED ANIMALS

In April 1987, the U.S. Patent and Trademark Office (PTO) ruled that patents could be issued on genetically engineered animals. Subsequently, the PTO (1) put the new animals on a par with mechanical inventions by decreeing that livestock producers must pay fees to those who patent genetically altered animals, and (2) ruled that livestock producers must pay royalties to the patent holder on each generation of patented animals for the life of the patent—which may be as long as 17 years.

Stud fees and one-time payments for animals have always been a part of livestock farming, but farmers have the rights to, and complete control over, subsequent generations without any additional payments.

In 1988, the U.S. Patent Office granted a patent to Harvard University for a genetically engineered mouse, involving developing a way to add cancer genes to the embryo of mice, thus making them more likely to get cancer, thereby facilitating cancer experiments.

Farmers and ranchers face a dilemma: Generally speaking, they want to reap the benefits of genetically improved livestock, but they do not want to pay royalties on each succeeding generation.

CLONING

Cloning has been around a long time. With most plants, the process is fairly easy—just take a cutting and let it root in damp soil.

Cloning of an animal is the production of an exact genetic copy. In a technical sense, identical twins are clones; they are derived from a single cell, as a result of the embryo splitting early in development to yield what is essentially two carbon copies.

Through cloning, it will be possible for all dairy animals in an entire herd to look alike, be genetically alike, have the same nutritive requirements, and produce the same quantity of milk of the same composition.

The exciting and much sought technological breakthrough in the cloning of mammals involves the manipulation of embryos.

The dream of cloning is based on the following two pieces of scientific evidence:

1. With few exceptions, all cells in the body of an animal appear to contain the same genetic information. This information is contained in the DNA, a molecule that is located in a sac inside cells called the nucleus. Thus, within an animal, the DNA sequence in the nucleus of a liver cell is identical to that in a skin cell. These cells differ in appearance and function because they make use of different parts of the genetic information, not because the total amount of information differs. Further, all of these cells have the genetic information that was present in the one-cell embryo that developed in the animal. Therefore, if the nucleus of any of these cells was used to replace the genetic information in any one-cell embryo, an exact genetic copy of the animal whose cells donated the nucleus would develop. With such an approach, thousands of cloned copies could be made.

2. The second piece of scientific evidence is that nuclear transplantation experiments have been done successfully with several species of animals, especially frogs and fish, which have the big advantage of their eggs being thousands of times larger than mammalian eggs.

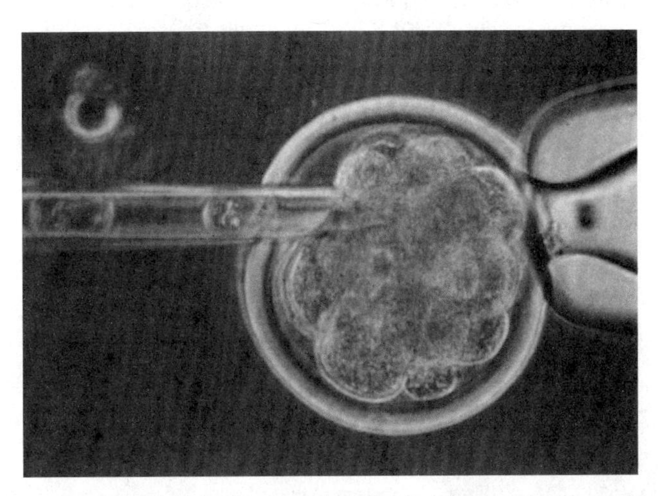

Fig. 5-5. This shows the removal of cells of a valuable embryo to use in cloning, to produce 8 to 20 cloned embryos from this embryo by transfer of each of these cells to an enucleated oocyte. (Courtesy, Dr. Robert Walton, American Breeders Service, a division fo W. R. Grace & Co., De Forest, WI)

Historically, research and development in cloning has passed through the following stages, in order and period of time:

1. Identical twin calves were produced by microsurgically splitting embryos, then transferring half embryos to recipients.

2. A bull calf was born as a result of using the laboratory culturing technique *(in vitro)* of maturing an egg, fertilizing the egg *in vitro*, then transferring the fertilized egg to a surrogate mother.

Nuclei have been taken from 16-cell bovine embryos and placed in one-cell bovine eggs whose nuclei had already been removed. The new one-cell embryos were matured and transferred into recipient cows, which subsequently gave birth to the cloned female calves.

During the latter part of the 20th century, the most

advanced cloning procedure consisted of flushing the embryo out of the donor cow at day five of its development (at the 32-cell stage); followed by putting the embryo under a microscope and manually removing one of the cells, then freezing the remaining 31-cell embryo (much like semen is frozen) and putting it away until an order is received. Next, the technician takes an unfertilized egg that has been flushed from a low-grade donor cow, removes the nucleus from it, inserts the borrowed cell into the egg, patches up the hole with a short zap of electricity, and the cloned embryo divides day after day and develops into a genetically identical duplicate of the heifer that results from the 31-cell embryo.

The goal ahead is (1) to let embryos grow to the 32-cell stage, then split them all into 32 more embryos each, resulting in 1,024 (32 × 32) genetically identical copies of the same pedigree—*the exact same*; (2) to let 31 of the embryos grow up, split apart the 32nd, and keep carrying out the chain for as long as desired; and the first animal born would be identical to the last one—*identical.* The goal ahead is illustrated in Fig. 5-6.

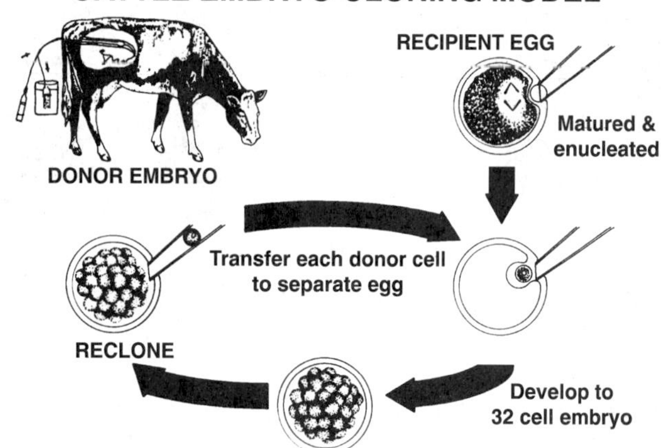

Fig. 5-6. Cattle embryo cloning model. (Prepared by Dr. Robert Walton, American Breeders Service, a division of W. R. Grace & Co., De Forest, WI)

MILESTONES IN ANIMAL BIOTECHNOLOGY

A chronology of the history of biotechnology, beginning with Mendel, is given in Table 5-1.

TABLE 5-1
ANIMAL BIOTECHNOLOGY MILESTONES

Year	Milestone
1866	Scientific biotechnology was founded by Gregor Johan Mendel, an Austrian monk, who published a report covering eight years (1857-1865) of his studies with garden peas.
1868	Friedrich Miescher, a young Swiss chemist, isolated nucleic acid from pus cells from a discarded bandage.
1900	Sixteen years after Mendel's death, three European biologists—Hugo de Vries of the Netherlands, Erich Von Tschermak of Austria, and Karl Correns of Germany—working independently, rediscovered Mendel's work; and Mendelism was reborn.
1906	The term *genetics* was coined.
1909	The term *gene* was first used to replace Mendel's term *factor.*
1909	The Russians began artificially inseminating cattle and mares. In 1928, they inseminated 1.2 million cattle and 15 million sheep.
1950	Jim Rowson of Cambridge University, in England, successfully transferred sheep embryos.
1953	Scientists James Watson, an American geneticist, and Francis Crick, an English physicist, working at the University of Cambridge, discovered the double-helix structure of DNA. For their landmark work, Watson and Crick shared the 1962 Nobel Prize for Physiology and Medicine.
1973	Dr. Stanley N. Cohen of Stanford and Herbert W. Boyer of the University of California at San Francisco inserted recombinant DNA into host bacteria that reproduced, or cloned. Genetic engineering began.
1976	Genentech, Inc., South San Francisco, California, a commercial company specializing in DNA technology, was formed. In 1980, its initial public offering set a Wall Street record for price per share increase ($35 to $89 in 20 minutes).
1977	Scientists at the University of California at San Francisco reported turning ordinary bacteria into factories capable of producing insulin. This feat opened the door to further genetic engineering or splicing—the transferring of a gene from one individual to another.
1980	A broad patent covering a gene-splicing and cloning method developed by Dr. Stanley N. Cohen at Stanford University and Dr. Herbert W. Boyer at the University of California was issued to the developing universities by the U.S. Patent Office.
1982	Mighty mice, double the normal mature size, were produced by inserting genes for either a rat or human growth hormone. Transgenic animals, by gene transfer, were here and now.
1988	The U.S. Patent and Trademark Office awarded Harvard University the world's first patent for an animal, a mouse whose cells had been genetically engineered to carry a cancer-promoting gene for facilitating cancer research.
1993	The U.S. Food and Drug Administration (FDA) approved the first commercial bovine somatotropin (BST) product, Posilac, developed by Monsanto. Sales of this product were initiated in 1994.

APPLYING CATTLE GENETICS IN MODERN CATTLE BREEDING

It remains for the modern cattle breeder to put it all together—cells, chromosomes, genes, and DNA.

The cattle breeder knows that the bodies of all cattle are made up of trillions of tiny cells, microscopic in size; that each cell contains a nucleus in which there are a number of pairs of bundles, called chromosomes; that, in turn, the chromosomes carry pairs of minute particles, called genes, which are the basic hereditary material; that the nucleus of each body cell of cattle contains 30 pairs of chromosomes, or a total of 60, whereas there are thousands of pairs of genes which determine all the hereditary characteristics of cattle—-from the body type to the color of the hair; and that inheritance goes by units, rather than by the blending of two fluids as our grandparents thought.

The modern cattle breeder also knows that the job of transmitting qualities from one generation to the next is performed by germ cells—a sperm from the male and an ovum or egg from the female. All animals, therefore, are the result of the union of two such tiny cells, one from each of its parents. These two germ cells contain all the anatomical, physiological, and psychological characters that the offspring will inherit. They determine whether a calf shall be polled instead of horned, black instead of white, a bull instead of a heifer, etc.

Only half the number of chromosomes and genes present in the body cells of the animal go into each egg and sperm (thus, each reproductive cell of cattle has 30 chromosomes), but each sperm or egg cell has genes for every characteristic of its species. As will be explained later, the particular half that any one germ cell gets is determined by chance. When mating and fertilization occur, the single chromosomes from the germ cell of each parent unite to form new pairs, and the genes are again present in duplicate in the body cells of the embryo.

With all possible combinations in 30 pairs of chromosomes (the species number in cattle) and the genes that they bear, any bull or cow can transmit over 1 billion different samples of its inheritance; and the combination from both parents makes possible 1 billion times 1 billion genetically different offspring. It is not strange, therefore, that no two animals within a given breed (except identical twins from a single egg split after fertilization) are exactly alike. Rather, we can marvel that the members of a given breed bear as much resemblance to each other as they do.

Even between such closely related individuals as full sisters, it is possible that there will be quite wide differences in size, growth rate, temperament, conformation, and in almost every conceivable character. Admitting that many of these differences may be due to undetected differences in environment, it is still true that in such animals much of the variation is due to hereditary differences. A bull, for example, will sometimes transmit to one offspring much better inheritance than he does to most of his get, simply as the result of chance differences in the genes that go to different sperm at the time of the reduction division. Such differences in inheritance in offspring have been called both the hope and the despair of the livestock breeder.

If an animal gets similar determiners or genes from each parent, it will produce uniform germ cells; because any half of its inheritance is just like any other half. For example, regardless of what combination of chromosomes goes into a particular germ cell, it will be just like any other egg or sperm from the same individual. Such animals are referred to as being homozygous. Few, if any, of our animals are in this hereditary state at the present time. Instead of being homozygous, they are heterozygous. This explains why there may be such wide variation within the offspring of any given sire or dam. The wise and progressive breeder recognizes this fact, and insists on the production records of all get rather than that of just a few meritorious individuals.

Variation between the offspring of animals that are not pure or homozygous, to use the technical term, is not to be marveled at, but is rather to be expected. No one would expect to draw exactly 20 sound apples and 10 rotten ones every time a random sample of 30 is taken from a barrel containing 40 sound ones and 20 rotten ones, although, on the average—if enough samples were drawn—about that proportion of each would be expected. Individual drawings would, of course, vary rather widely. Exactly the same situation applies to the relative numbers of *good* and *bad* genes that may be present in different germ cells from the same animal. Because of this situation, the mating of a cow with a fine show record to a bull that on the average transmits relatively good offspring will not always produce calves of merit equal to that of their parents. The calves could be markedly poorer than the parents or, happily, they could in some cases be better than either parent.

Selection and close breeding are the tools through which the cattle producer can obtain bulls and cows whose chromosomes and genes contain similar hereditary determiners—animals that are genetically more homozygous.

Actually, a completely homozygous state would be undesirable and unfortunate. This is so because economic and environmental changes are apt to dictate animal changes from time to time. Were complete homozygosity to exist, such shifts would be impossible except through the slow and uncertain process of mutations; thus, making it extremely difficult to effect a change in phenotype. Fortunately, enough heterozygosity exists in our improved breeds so that they are flexible in the hands of breeders; they can be molded in any desired direction. (In fact, enough heterozygosity exists that it is nigh impossible to fix complete homozygosity.) It may be said, therefore, that variation in the biological material permits the animal breeder to mold, change, and improve animals.

MUTATIONS

A gene mutation may be defined as a change in the code transmitted by the DNA molecule (gene) on a chromosome. New mutations may be carried on the autosomes (chromosomes other than the sex chromosomes), or in those carried on the sex chromosomes. They may also occur in body cells or in cells of the germinal epithelium of the testicles and ovaries. Reverse mutations may also occur; *i.e.*, genes may mutate from dominant to recessive, then back again from recessive to dominant.

Not only are mutations rare, but they are prevailingly harmful, and most of them are recessive. Further, one cannot induce a particular kind of mutation. For all practical purposes, therefore, the genes can be thought of as unchanged from one generation to the next. The observed differences between animals are usually due to different combinations of genes being present rather than to mutations. Each gene probably changes only about once in each 100,000 to 1,000,000 animals produced.

Once in a great while a mutation occurs in a farm animal, and it produces a visible effect in the animal carrying it. Such mutations are occasionally of practical value. The occurrence of the polled characteristic within the horned Hereford and Shorthorn breeds of cattle is an example of a mutation of economic importance.[2] Out of this has arisen the Polled Hereford breed and Polled Shorthorn cattle.

Gene changes can be accelerated by exposure to X-rays, radium, mustard gas, and ultraviolet light rays. Such changes may eventually be observed in the offspring of both people and animals of Japan who were exposed to the atomic bombs unleashed in World War II.

Although induced mutations have been used successfully in developing commercial varieties of plants, the technique does not appear very promising for the improvement of animals. This is so because (1) an enormous number of mutations would have to be induced in order to have much chance of getting one which had commercial value, and (2) its frequency would have to be increased by selection and as many of the concurrent undesirable mutations as possible eliminated from the stock.

[2]The horned gene mutates to the polled gene at a fairly high frequency; apparently at the rate of about 1 in 20,000.

SIMPLE GENE INHERITANCE (Qualitative Traits)

In the simplest type of inheritance, only one pair of genes is involved. Thus, a pair of genes is responsible for the color of hair in Shorthorn cattle. This situation is illustrated by Fig. 5-7.

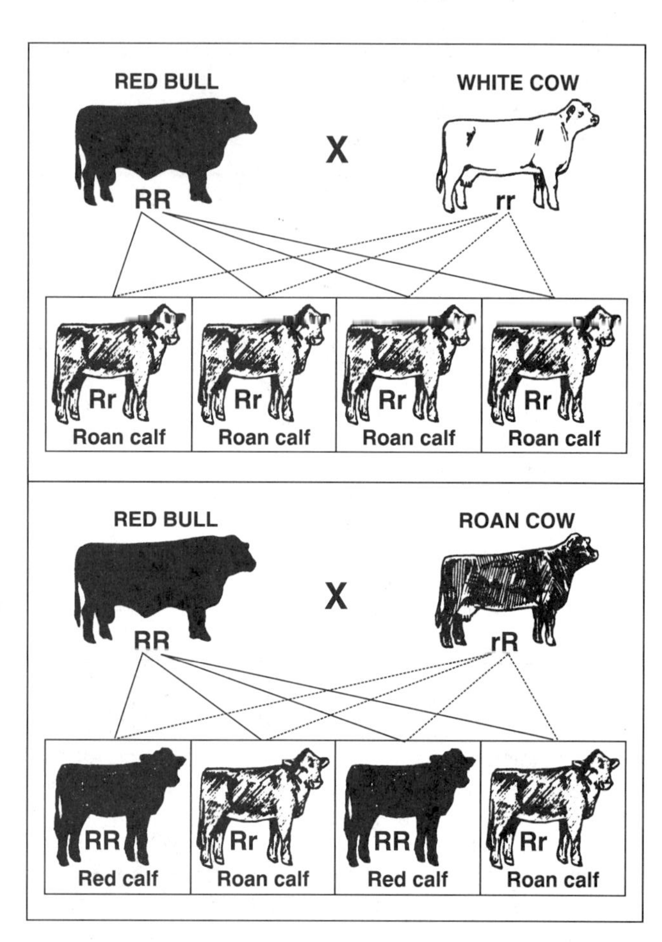

Fig. 5-7. Diagrammatic illustration of the inheritance of color in Shorthorn cattle. Red X white matings *(upper)* in Shorthorn cattle usually produce roan offspring, whereas red X roan matings *(lower)* produce ½ red offspring and ½ roan offspring.

An animal having 2 genes for red (RR) is actually red in color, whereas an animal having 2 genes for white (rr) is white in color. On the other hand, a Shorthorn which has 1 gene for red (R) and 1 for white (r) is neither red nor white but roan (Rr), which is a mixture of red and white. Thus, red X white matings in Shorthorn cattle usually produce roan offspring. Likewise, white X white matings generally produce white offspring; but it must be remembered that white in Shorthorns is seldom pure, for the face bristles, eyelashes, and ears usually carry red hairs. Roans, having 1 gene for red and 1 for white on the paired chromosomes will never breed true and, if mated together, will produce calves in the proportion of 1 red, 2 roans, and 1 white. If one wishes to produce roans, the most certain way is to mate red cows with a white bull or vice versa, for then all the calves will be roan. If a roan animal is bred to a red one, ½ the offspring will be red, whereas the other ½ will be roan. Likewise when a roan animal is bred to a white one, approximately an equal number of roan and white calves will be produced.

This example illustrates the most important principles of inheritance; namely, (1) genes occur in animals in pairs because one member of each pair comes from each parent, and (2) each reproductive cell contains a sample half of the genes of that particular animal.

It should be borne in mind that there is no way to sort out the numerous genes so as to get the most desirable ones into the same reproductive cell except as it occurs by chance during the formation of eggs and sperm. Thus, it follows that the various gene combinations, such as referred to above, occur at random and that the various colors will appear in the offspring in the proportions indicated only when relatively large numbers are concerned. The possible gene combinations, therefore, are governed by the laws of chance, operating in much the same manner as the results obtained from flipping coins. For example, if a penny is flipped often enough, the number of heads and tails will come out about even. However, with the laws of chance in operation, it is possible that out of any 4 tosses one might get all heads, all tails, or even 3 to 1. In exactly the same manner, a Shorthorn breeder may be highly elated in obtaining 4 red calves from roan X roan matings only to be greatly depressed when the next 4 calves, from the same matings, are white in color.

In addition to color of hair, other examples of simple gene inheritance in animals (sometimes referred to as qualitative traits) include color of eyes, presence or absence of horns, type of blood, and lethals.

DOMINANT AND RECESSIVE FACTORS

In the example of Shorthorn colors, each gene of the pair (R and r) produced a visible effect, whether paired as identical genes (2 reds or 2 whites) or as 2 different genes (red and white).

This is not true of all genes; some of them have the ability to prevent or mask the expression of others, with the result that the genetic makeup of such animals cannot be recognized with perfect accuracy. This ability to cover up or mask the presence of one member of a set of genes is called dominance. The gene which masks the one is the dominant gene; the one which is masked is the recessive gene.

In cattle, the polled character is dominant to the horned character. Thus, if a *pure polled* bull is used on horned cows (or vice versa), the resulting progeny are not midway between two parents but are of polled character.[3] It must be remembered, however, that not all hornless animals are pure for the polled character; many of them carry a factor for horns in the hidden or recessive condition. In genetic terminology, animals that are pure for a certain character—for example the polled characteristic—are termed *homozygous*, whereas those that have one dominant and one recessive factor are termed *heterozygous*. A simple breeding test can be used in order to determine whether a polled bull is homozygous or heterozygous, but it is impossible to determine such purity or impurity through inspection. The breeding test consists of mating the polled sire with a number of horned females. If the bull is pure or homozygous for the polled character, all of the calves will be polled; whereas if he is impure or heterozygous, only half of the resulting offspring will, on the average, be polled and half will have horns like the horned parents. Many breeders of polled cattle of various breeds test their herd sires in this manner, mating the prospective sire to several horned animals.

It is clear, therefore, that a dominant character will cover up a recessive. Hence, an animal's breeding performance cannot be recognized by its phenotype (how it looks), a fact which is of great significance in practical breeding.

Another example of dominance is that of the white face of Hereford cattle—the white face being dominant over the type of coloration in which the head and body are of the same color. Undoubtedly, this condition of dominance, which constitutes a trademark of the

[3]It is noteworthy, however, that when a homozygous polled animal is crossed with a homozygous horned animal, some *scurs* or small loosely attached horns usually appear. There are conflicting reports and opinions concerning the inheritance of scurs, with the following theories prevailing:

a. That the gene for scurs is recessive and independent of the major genes for horns. According to this theory, scurs appear only in individuals homozygous for the scurred gene (sc sc).

b. That scurs are a sex-influenced character. According to this theory, scurs will occur in males either homozygous (Sc Sc) or heterozygous (Sc sc) for the character, but only in females homozygous (Sc Sc) for the character; in other words it acts as a dominant in polled males and a recessive in polled females.

c. That the major gene (P) for polled condition is only partially dominant, with heterozygous individuals (Pp) tending to be scurred, especially in bulls.

Horns prevent the expression of any genes an animal may have for scurs, and thus complicate studies designed to determine the exact mode of inheritance of scurs.

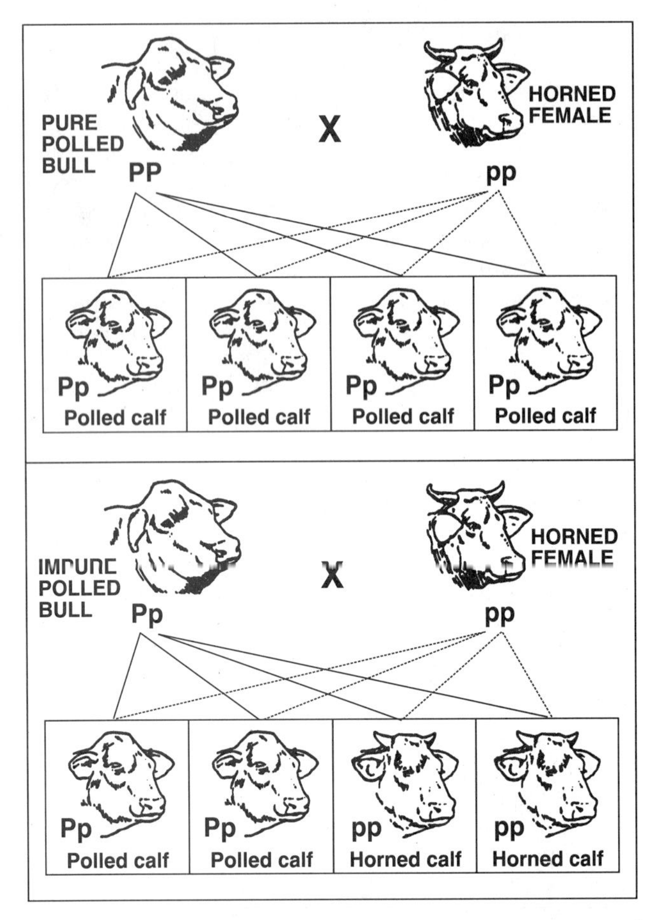

Fig. 5-8. Diagrammatic illustration of the inheritance of horns in cattle of European extraction. Although there may be a very occasional exception, if a bull that is considered pure or homozygous for the polled character is mated with a number of horned females, all of the calves will be polled *(upper)*; whereas if a bull that is impure or heterozygous for the polled character is mated with a number of horned females, only half of the calves will, on the average, be polled *(lower)*.

breed, has been of importance from a promotional standpoint.

As can be readily understood, dominance often makes the task of identifying and discarding all animals carrying an undesirable recessive factor a difficult one. Recessive genes can be passed on from generation to generation, appearing only when 2 animals, both of which carry the recessive factor, happen to mate. Even then, only 1 out of 4 offspring produced will, on the average, be homozygous for the recessive factor and show it.

In Angus cattle, the red color is an example of such an undesirable recessive factor. Black polled cattle have been known in Scotland since 1523; and since the days of Hugh Watson, William McCombie, and George McPherson Grant, black has been the accepted color of the breed. Yet, down through the years, a recessive factor for red coat color has persisted in the breed. For this reason, a red calf occasionally and unexpectedly shows up in a purebred Angus herd (about 1 red calf appears in every 200 to 500 Angus calves dropped[4]). This occasional appearance of a red calf in the Angus breed does not signify any impurity of breeding but merely the outcropping of a long hidden recessive gene. When a red calf does appear, one can be very certain that both the sire and dam contributed equally to the condition and that each of them carried the recessive gene for red color. This fact should be given consideration in the culling program.

As the factor for red is recessive, the red animals are pure for color. The mating of two red animals, therefore, will always produce red calves.[5]

Other examples of recessives are red color in Holstein cattle and dwarfism in cattle.

Assuming that a hereditary defect or abnormality has occurred in a herd and that it is recessive in nature, the breeding program to be followed to prevent or minimize the possibility of its future occurrence will depend somewhat on the type of herd involved—especially on whether it is a commercial or purebred herd. In a commercial herd, the breeder can usually guard against further reappearance of the undesirable recessive simply by using an outcross (unrelated) sire within the same breed or by crossbreeding with a sire from another breed. With this system, the breeder is fully aware of the recessive being present, but he has taken action to keep it from showing up.

On the other hand, if such an undesirable recessive appears in a purebred herd, the action should be more drastic. Reputable purebred breeders have an obligation not only to themselves but to their customers among both the purebred and commercial herds. Purebred animals must be purged of undesirable genes and lethals. This can be done by:

1. Eliminating those sires and dams that are known to have transmitted the undesirable recessive character.

2. Eliminating both the abnormal and normal offspring produced by these sires and dams (approximately half of the normal animals will carry the undesirable character in the recessive condition).

---

[4]In order to obtain 1 red calf out of 200, 1 parent out of every 7 must be a carrier of the red gene. Actually, to get $\frac{1}{196}$ red calves, there must be $\frac{1}{14}$ b (red gene) reproductive cells in both males and females because $\frac{1}{14} \times \frac{1}{14} = \frac{1}{196}$. Thus, $\frac{1}{7}$ of the parents must be Bb (black in color, but carrying the red gene) while $\frac{6}{7}$ are BB (pure for black).

[5]A separate U.S. breed registry association for these Red Angus cattle was organized in 1954 (see the Appendix, Table VII-1). To these breed enthusiasts, the recessive gene for red is desirable.

3. Breeding a prospective herd sire to a number of females known to carry the factor for the undesirable recessive, thus making sure that the new sire is free from the recessive.

Such action in a purebred herd is expensive, and it calls for considerable courage. Yet it is the only way in which the purebred livestock of the country can be freed from such undesirable genes.

INCOMPLETE OR PARTIAL DOMINANCE

The results of crossing polled with horned cattle are clear-cut because the polled character is completely dominant over its allele (horned). If, however, a cross is made between a red and a white Shorthorn, the result is a roan (mixture of red and white hairs) color pattern. In the latter cross, the action of a gene is such that it does not cover the allele, which is known as incomplete dominance; or, stated differently, the roan color is the result of the action of a pair of genes (joint action) neither of which is dominant. This explains the futility of efforts to develop Shorthorns pure for roan.

The above discussion also indicates that there are varying degrees of dominance—from complete dominance to an entire lack of dominance. In the vast majority of cases, however, dominance is neither complete nor absent, but incomplete or partial. Also, it is now known that dominance is not the result of single-factor pairs but that the degree of dominance depends upon the animal's whole genetic make-up together with the environment to which it is exposed, and the various interactions between the genetic complex (genotype) and the environment.

MULTIPLE GENE INHERITANCE (Quantitative Traits)

Relatively few characters of economic importance in farm animals are inherited in as simple a manner as the coat color or polled conditions described. Important characters—such as meat production and milk production—are due to many genes; thus, they are called multiple-factor characters or multiple-gene characters. Because such characters show all manner of gradation—from high to low performance, for example—they are sometimes referred to as quantitative traits.

In quantitative inheritance, the extremes (both good and bad) tend to swing back to the average. Thus, the offspring of a grand champion bull and a grand champion cow are not apt to be as good as either parent. Likewise, and happily so, the progeny of two very mediocre parents will likely be superior to either parent.

Estimates of the number of pairs of genes affecting each economically important characteristic vary greatly, but the majority of geneticists agree that for most such characters 10 or more pairs of genes are involved. Growth rate in cattle, therefore, is affected by the following: (1) the animal's appetite; (2) feed consumption; (3) feed utilization—that is, the proportion of the feed eaten that is absorbed into the blood stream; (4) feed assimilation—the use to which the nutrients are put after absorption; and (5) feed conversion—whether used for muscle, fat, or bone formation. This should indicate clearly enough that such a characteristic as growth rate is controlled by many genes and that it is difficult to determine the mode of inheritance of such characters.

INHERITANCE OF SOME CHARACTERS IN CATTLE

Mendelian characters are inherited in alternative pairs (or series). These alternative forms of a gene, which are located at the same point on each one of a pair of chromosomes, are called *alleles*—for example, horns (recessive) and polled (dominant).

An individual that is heterozygous with respect to 1 pair of allelic genes is a *monohybrid*; one that is heterozygous with respect to 2 pairs of allelic genes is a *dihybrid*; and one that is heterozygous with respect to 3 pairs of allelic genes is a *trihybrid*. (See Fig. 5-9 for a trihybrid cross.)

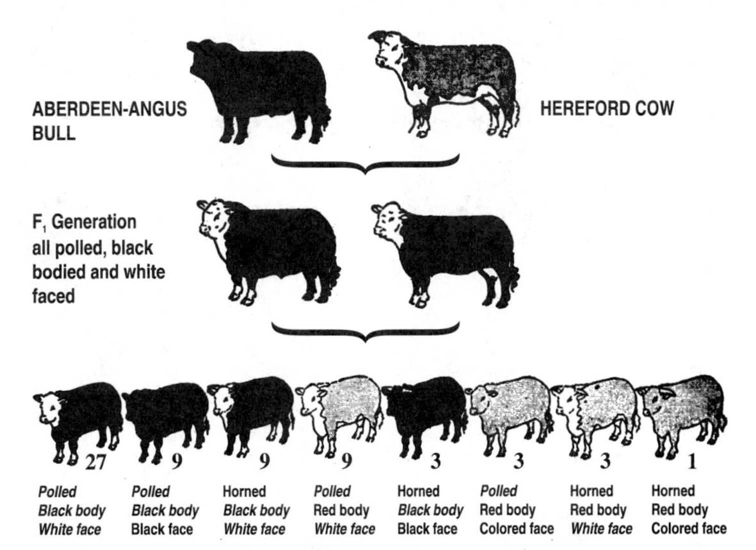

Fig. 5-9. Diagram showing a trihybrid cross; the inheritance of polled, white face, and black body characteristics in an Angus X Hereford cross. Note that all first cross (F_1) animals are polled, black bodied, and white faced, whereas on the average, the F_1 X F_1 cross results in the 27:9:9:9:3:3:3:1 ratio shown. (Drawing by R. F. Johnson)

MULTIPLE BIRTHS

Multiple births among cattle have been observed since their domestication.

A review of the literature reveals that, on the average, such multiple births occur at the frequencies shown in Table 5-2.

TABLE 5-2
FREQUENCY OF TWINS IN CATTLE

Breed	Total Number of Births	Percent of Twin Births
Several breeds		0.44
Angus	1,111	0.81
Grade Angus	586	1.71
Brown Swiss	14,111	2.70
Holstein	18,736	3.08
Jersey	87,926	1.02
Simmental	12,625	4.61

Fig. 5-10. Quadruplet purebred Angus heifers, at two weeks of age. Bred and owned by O. H. Delchamps, Point Clear, Alabama. (Courtesy, Mr. Delchamps)

Normally, the breeds of beef cattle of British origin produce twins from about 1% of the pregnancies.

In the experimental herds at USDA's Meat Animal Research Center (MARC), Clay Center, Nebraska, twinning has been increased by about two percentage points per year. In 1996, the MARC experimental herd had an overall twinning rate of about 37%. Ultimately, the USDA researchers expect to increase the MARC herd's twinning rate to more than 40%.

In 1994, under a cooperative marketing agreement with USDA, American Breeders Service, De Forest, Wisconsin, started selling semen and embryos carrying predicted breeding values for twinning of at least 40% for bulls and 30% for cows.

The most likely beneficiaries of twinning are those cattle breeders who can provide their herds feeds of high nutritive value, along with extra management and labor. These breeders could reduce their cost to produce a pound of beef by as much as 30%.

Twins may be produced in any of the following five ways:

1. By two eggs being produced at the same heat period, with both fertilized and carried to term.
2. By two eggs being shed at the same heat period, but the cow being bred to two different bulls with a sperm from each of the bulls uniting with an egg.
3. By a cow coming in heat and being bred, then three weeks later coming in heat again and being rebred; with both matings resulting in viable offspring.
4. By a single fertilized ovum splitting during the early stage of development.
5. By the use of hormones to induce superovulation.

Twins may be either fraternal (dizygotic) or identical (monozygotic). Fraternal twins are produced from 2 separate ova that were fertilized by 2 different sperm. Identical twins result when a single fertilized egg divides very early in its embryology, into 2 separate individuals.

In humans, nearly half of the like-sexed twins are identical, whereas in cattle only 5 to 12% of such births are identical. Such twins are always of the same sex, a pair of males or a pair of females, and alike genetically—their chromosomes and genes are alike; they are 100% related. When identical twins are not entirely separate, they are known as Siamese twins.

Genetically, fraternal twins are no more alike than full brothers and sisters born at different times; they are only 50% related. They usually resemble each other more, however, because they were subjected to the same intrauterine environment before birth and generally they are reared under much the same environment. Also, fraternal twins may be of different sexes.

Distinguishing between identical and fraternal twin calves is not easy, but the following characteristics of identical twins will be helpful:

1. Identical twins are usually born in rapid succession, and frequently there is only one placenta.
2. The calves are necessarily of the same sex.
3. The coat colors are identical; i.e., if there is a broken color, there must be a strong degree of resemblance in this respect.

4. There is little variation in birth weights, general conformation and, more particularly, the shape of the head, position of the horns and occurrence of skin pigmentation, rudimentary teats, etc.

5. Muzzle prints show a degree of resemblance.

6. The shape, twisting, and position of the horns and behavior of the twins can be observed at a later stage. Identical twins are inclined to keep together when grazing, walking, lying down, or ruminating.

7. Identical twins have the same blood group.

Most cattle producers prefer single births to twins, for the following reasons:

1. The high incidence of stillbirths in twins. Herefords on the range show 3.6% stillbirths among singles vs 15.7% stillbirths among twins. Despite this fact, twinning would result in more live calves per 100 cows calving; 96 live calves from singles ($100 \times 3.6\% = 3.6$; then, $100 - 3.6 = 96$) vs 168 from twins ($200 \times 15.7\% = 31.4$; then, $200 - 31.4 = 168$).

2. About 85% of all heifers born twin with a bull are apt to be freemartins (sterile heifers).

3. Twin calves average 20 to 30% lighter weights at birth than singles.

4. The tendency of cows that have produced twins to have a lowered conception rate following twinning.

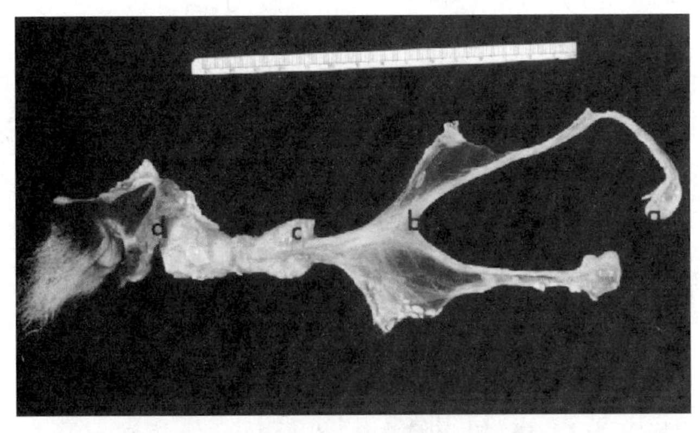

Fig. 5-11. Freemartin calf born co-twin to a male partner. Gonads **(a)**, cordlike Mullerian ducts **(b)**, seminal vesicles **(c)**, and short vagina **(d)**. (Courtesy, Professor H. W. Leipold, Kansas State University)

HOW SEX IS DETERMINED

The possibility of sex determination and control has fascinated humankind since time immemorial. For example, in a book published in 1662,[6] studmasters were admonished as follows: "For a male colt you must bind back with a cord, or pull back his left stone, and for the female, bind back the right stone, and thus you may do unto all other kinds of cattle." The same book revealed that "if the rams be put with the ewes when the wind is in the north, the ewes will bring males and if the wind be in the south when the ewes are covered, they will be females." These are but two of the hundreds of superstitions that have evolved concerning the control of sex.

On the average, and when considering a large population, approximately equal numbers of males and females are born in all common species of animals. To be sure, many notable exceptions can be found in individual herds. The history of the Washington State University Angus herd, for example, reads like a story book. The entire herd was built up from one foundation female purchased in 1910. She produced 7 daughters. In turn, her first daughter produced 6 females. Most remarkable yet and extremely fortunate from the standpoint of building up the cow herd, it was 4 years before any bull calves were dropped in the herd.[7]

Such unusual examples often erroneously lead breeders to think that something peculiar to the treatment or management of a particular herd resulted in a preponderance of males or females, as the case may be. In brief, through such examples, the breeder may get the impression that variation in sex ratio is not random—that it is under the control of some unknown and mysterious influence. Under such conditions, it can be readily understood why the field of sex control is a fertile one. Certainly, any "foolproof" method of controlling sex would have tremendous commercial possibilities. For example, producers wishing to build up a herd could then secure a high percentage of heifer calves. On the other hand, commercial producers would then elect to produce only enough heifers for replacement purposes; from an economical standpoint, they would want a preponderance of bull calves for the reason that commercial steers sell for a higher price than do commercial heifers.

Sex is determined by the chromosomal makeup of the individual. One particular pair of the chromosomes is called the sex chromosomes. In farm animals, the female has a pair of similar chromosomes (usually called X chromosomes), whereas the male has a pair of unlike sex chromosomes (usually called X and Y chromosomes). In the bird, this condition is reversed, the females having the unlike pair and the male having the like pair.

---

[6]Mascal, Leonard, *The Government of Cattel,* London printed for John Stafford and William Gilbertson, ". . . and are to be sold at the George-Yard near Fleet-bridge; and at the Bible without New Gate . . ." 1662.

[7]The probability of getting 7 heifer calves in a row can be figured by multiplying ½ by itself 7 times. This is 1 out of 128, meaning that all 7 calves in a group of 7 will be heifers in less than 1% of the cases.

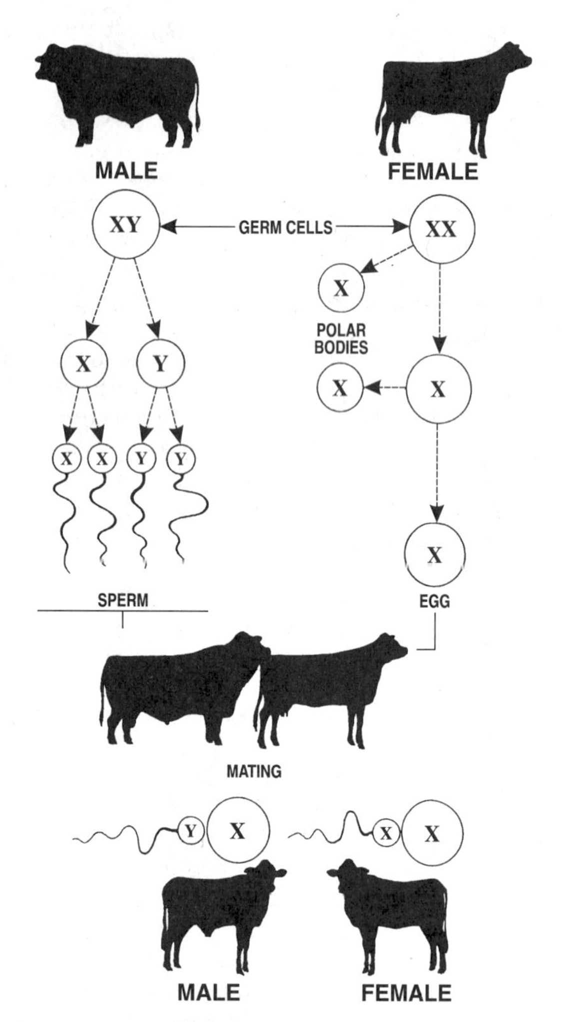

Fig. 5-12. Diagrammatic illustration of the mechanism of sex determination in cattle showing how sex is determined by the chromosomal makeup of the individual. The cow has a pair of like sex chromosomes, whereas the bull has a pair of unlike sex chromosomes. Thus, if an egg and a sperm of like sex chromosomal makeup unite, the offspring will be female; whereas if an egg and a sperm of unlike sex chromosomal makeup unite; the offspring will be male.

The pairs of sex chromosomes separate out when the germ cells are formed. Thus, the ovum or egg produced by the cow contains the X chromosome; whereas the sperm of the bull are of 2 types, ½ containing the X chromosome and the other ½ the Y chromosome. Since, on the average, the egg and sperm unite at random, it can be understood that ½ of the progeny will contain the chromosomal makeup XX (females) with the other ½ XY (males).[8]

[8]The scientists' symbols for the male and female, respectively, are: ♂ (the sacred shield and spear of Mars, the Roman God of War), and ♀ (the looking glass of Venus, the Roman Goddess of Love and Beauty).

SEX CONTROL

Research workers have employed many techniques designed to change the sex ratio. Most efforts have been focused upon separating the sperm cells containing X chromosomes from those containing Y chromosomes prior to artificial insemination through the following approaches: (1) sedimentation, in which it is hoped that the heavy X sperm will settle to the bottom and the Y sperm float to the top, then draw off the *male* and/or *female* fraction of the semen; (2) electrophoresis, in which an anode and a cathode are placed in the semen to draw either X or Y sperm into its field of influence; (3) treatment of sperm with a substance that is deadly to the chromosomes that determine maleness; (4) treatment of semen with solutions causing male and female sperm to give off distinct and different fluorescence, producing one-sex sperm up to 95% accuracy, but reducing potency; (5) monoclonal antibody treatment; (6) laminar flow (streamline flow) technique; and (7) determining the DNA—sperm bearing X chromosomes (which lead to females) have about 4% more DNA than bull sperm carrying Y chromosomes (which lead to males). Currently (1996), there is no commercially available method of sexing semen.

It would also be advantageous to transfer embryos of the desired sex. Fortunately, embryos can be sexed, although it is not easy.

Predetermination of the sex of 6- to 12-day-old embryos with 90% accuracy has been achieved. This does not alter the initial ratios at fertilization; it simply allows the selection of the particular sex.

Among the theoretical approaches to controlling sex are the following, although no calves have been produced by these methods to date:

1. Cloning by cultivating a particular cell type and producing a new individual, which would result in the control of both sex and the whole genetic makeup of the individual.
2. Oocyte fusion, involving treating two oocytes with an enzyme and causing them to fuse together. Because female pronuclei contain no male sex hormones, oocytes fertilized in this manner will develop into an embryo which is always female. It follows that combining oocyte fusion with embryo transplant offers a way in which to guarantee female offspring.

It appears that nature is about to yield to sex control. Predetermination of the sex of 6- to 12-day-old embryos is a reality, and progress is being made in the separation of sperm cells containing X chromosomes from those containing Y chromosomes. Procedures for sexing sperm may not be far off.

ABNORMAL DEVELOPMENT OF SEX IN CATTLE

Sex abnormalities occasionally occur in cattle; freemartins, intersexes, and hermaphrodites are the most common ones. Each of these is discussed in Chapter 7.

LETHALS AND OTHER ABNORMALITIES IN CATTLE

Many abnormal animals are born on the nation's farms and ranches each year. Unfortunately, purebred breeders, whose chief business is that of selling breeding stock, are likely to "keep mum" about the appearance of any defective animals in their herds because of the justifiable fear that it might hurt sales. With commercial producers, however, the appearance of such abnormalities is simply so much economic loss, with the result that they generally, openly and without embarrassment, admit the presence of such defects and seek correction.

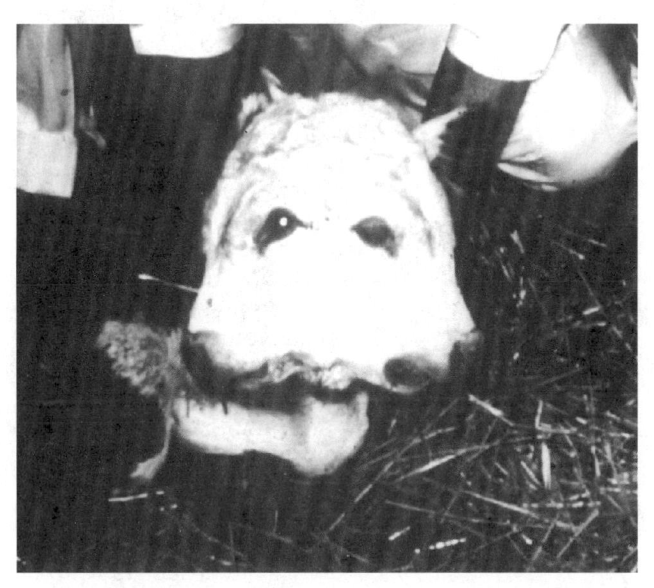

Fig. 5-13. Two-headed calf (diprosopus). (Courtesy, Dr. Clair M. Hibbs, College of Agriculture, North Platte Station, The University of Nebraska)

The embryological development—the development of the young from the time that the egg and the sperm unite until the animal is born—is very complicated. Thus, the oddity probably is that so many of the offspring develop normally rather than that a few develop abnormally.

Many such abnormalities (commonly known as monstrosities or freaks) are hereditary, being caused by certain "bad" genes. Most lethals are recessive and may, therefore, remain hidden for many generations. The prevention of such genetic abnormalities requires that the germ plasm be purged of the "bad" genes. This means that, where recessive lethals are involved, the breeder must be aware of the fact that both parents carry the gene. For the total removal of the lethals, test matings[9] and rigid selection must be practiced.

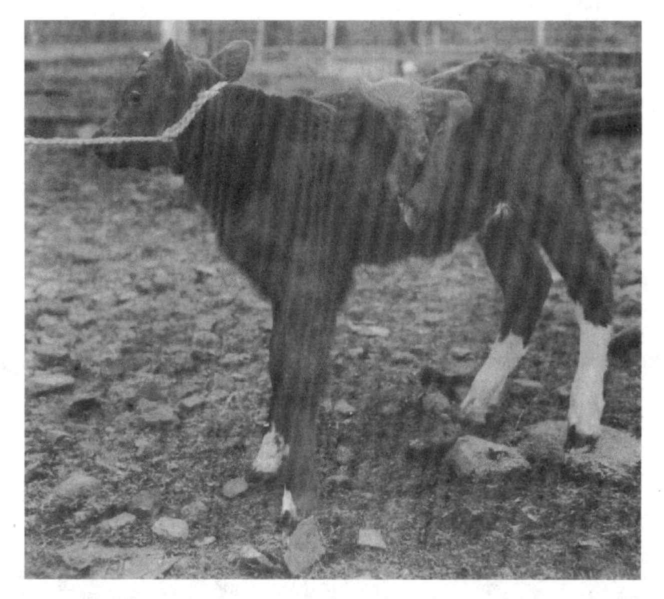

Fig. 5-14. Freak calf at 2 weeks of age. An accident of development, with three extra legs attached to the back. (Courtesy of the owner, C. M. Barker, White Salmon, WA)

In addition to hereditary abnormalities, there are certain abnormalities that may be due to nutritional deficiencies, or to accidents of development—the latter including those which seem to occur sporadically and for which there is no well-defined reason. When only a few defective individuals occur within a particular herd, it is often impossible to determine whether their occurrence is due to: (1) defective heredity; (2) defective nutrition; or (3) merely accidents of development. If the same abnormality occurs in any appreciable number of animals, however, it is probably either hereditary or nutritional. In any event, the diagnosis of the condition is not always a simple matter.

The following conditions would tend to indicate a hereditary defect:

---

[9]To test a bull at the 0.05 level of significance for heterozygosity for a single autosomal recessive gene, the following numbers of progeny are required: (a) 5 from homozygous females; (b) 11 from known heterozygous females; or (c) 23 from matings on daughters.

1. If the defect had previously been reported as hereditary in the same breed of livestock.

2. If it occurred more frequently within certain families or when there had been inbreeding.

3. If it occurred in more than one season and when different rations had been fed.

The following conditions might be accepted as indications that the abnormality was due to a nutritional deficiency:

1. If previously it had been reliably reported to be due to a nutritional deficiency.

2. If it appeared to be restricted to a certain area.

3. If it occurred when the ration of the mother was known to be deficient.

4. If it disappeared when an improved ration was fed.

If there is suspicion that the ration is defective, it should be improved, not only from the standpoint of preventing such deformities, but from the standpoint of good and efficient management.

If there is good and sufficient evidence that the abnormal condition is hereditary, the steps to be followed in purging the herd of the undesirable gene are identical to those for ridding the herd of any other undesirable recessive factor. An inbreeding program, of course, is the most effective way in which to expose hereditary lethals in order that purging may follow.

Table 5-3 includes most of the well-authenticated abnormal conditions that have been reported in cattle. As noted, to date many of these defects are limited to specific breeds and certain countries. For convenience, these abnormalities are classified into five categories as follows:

1. *Lethals are genetic factors which cause death of the calves carrying them prior to or shortly after birth.* (See Table 5-3, Part I.) Except for Achondroplasia I, which is incompletely dominant, all lethals listed in Part I are recessive, either autosomal or sex-linked.

2. *Semilethals (or sublethals) are genetic factors which cause the death of the young after birth or sometime later in life if environmental situations aggravate the conditions.* (See Table 5-3, Part II.) Semilethals are probably far more serious than lethals because they are more likely to go unnoticed; yet they may be causing considerable damage.

3. *Delayed lethals are gene changes which are expressed later in life.* (See Table 5-3, Part III.) Delayed lethals interfere with normal reproduction of the cow or the bull.

4. *Harmful and defective genes cause a host of minor abnormalities which may afflict cattle, some of which are probably of hereditary nature.* (See Table 5-3, Part IV.) These genetic factors are not lethals, but they interfere with the usefulness of the animals. Genetic factors reducing body size (dwarfism) are representative cases. Other developmental defects such as notched ears and screw tail are hereditary, but they do not have marked effects on production and fertility.

5. *Congenital defects refer to qualities that an animal shows evidence of having at birth—natal conditions.* (See Table 5-3, Part IV.) When a calf is born with an abnormality, the abnormality may be due to heredity; or it may be due to some freak of nature during prenatal development that had nothing whatsoever to do with chromosomes and genes. Congenital abnormalities may occur in any external or internal organ.

Fig. 5-15. Any observed defect may be hereditary, nutritional, or a freak of nature (as a result of faulty development in the embryological life). Six types of defects sometimes observed in cattle are shown: (1) bulldog; (2) hairless streaks; (3) muscle contraction; (4) Siamese twins; (5) short spine; (6) dwarfism. (Drawing by Dr. E. S. E. Hafez)

TABLE 5-3
LETHALS AND OTHER ABNORMALITIES IN CATTLE
PART I. HEREDITARY LETHALS
(Except for Achondroplasia I, which is incompletely dominant, all lethals listed in Part I are recessive, either autosomal or sex-linked)

Lethal	Symptoms of Abnormality	Breed Reported In	Pertinent References
1. **Achondroplasia I**	Short legs and head; often have hernia and cleft palate. Most of them are absorbed from the fourth to eighth month of pregnancy. Delivery when close to term is usually difficult because of the extremely large head. Completely dominant.	Dexter, British Friesian, Hereford, Jersey	Crew, 1923, *Proc. Royal Soc.* (London), 95B, 228. Punnet, 1936, *J. Genetics*, 32, 65. Mead, *et al.*, 1946, *J. Hered.*, 37, 183. Berger & Innes, 1948, *Vet. Record*, 60, 57.
2. **Achondroplasia II**	This type is similar to Achondroplasia I but less extreme. Calves are usually carried to term but die soon after.	Ayrshire, Guernsey, Jersey, Telemark	Mohr & Wriedt, 1925, *Z. fur Zuchtung*, B. 3, 223. Mohr, 1930, *Naturens Verden*, 14, 1. Brandt, 1941, *J. Hered.*, 32, 183. Gregory, *et al.*, 1942, *J. Hered.*, 33, 317.
3. **Achondroplasia III**	Less extreme than Achondroplasia II. Calves may live several hours.	Jersey, Swedish Red-and-White	Gregory, *et al.*, 1942, *J. Hered.*, 33, 317. Johansson, 1953, *Hereditas*, 39, 75.
4. ***Acroteriasis congenita*** (amputated)	Appendages short or absent. The calves are stillborn or die shortly after birth.	Swedish Holstein	Rieck & Buhr, 1967, *Deutsch, Tierazll, Wschs.* Wriedt & Mohr, 1928, *J. Genetics*, 20, 187.
5. **Agnathia**	Very short lower jaw. (See Fig. 5-17.)	Angus, Jersey	Annett, 1939, *J. Genetics*, 37, 301. Ely, *et al.*, 1939, *J. Hered.*, 30, 105. Grant, 1956, *J. Hered.*, 47, 165.
6. **Amputated** (See *Acroteriasis congenita*)			
7. **Ankylosis**	Ossification of joints. Legs rigid due to joints fusing; some cases of lower jaw fusions.	German, Norwegian Lyngdal	Mohr, 1930, *Naturens Verden*, 14, 1. Eaton, 1937, *J. Hered.*, 28, 320. Stang, 1940, *Z fur Zuchtung*, 36, 280. Tuff, 1948, *Skand. Vet. Tid.*, 38, 379.
8. **Arthrogryposis** (crooked calf)	Front legs, hind legs, all four legs rigid. Cleft palate. Simple autosomal recessive. (See Figs. 5-20 and 5-21.)	Charolais, Simmental, Maine-Anjou, Holstein, Hereford	Leipold, *et al.*, 1970, *J. Am. Vet. Res.*, 31, 367. Greene, *et al.*, 1973, *J. Am. Vet. Res.*, 34, 887.
9. **Bulldog head** (prognathism)	Short, broad skull; eye sockets large; upper jaw short; impaired vision. (See Fig. 5-15.)	Jersey	Becker & Arnold, 1928, *J. Hered.*, 10, 281.

(Continued)

TABLE 5-3, PART I (Continued)

Lethal	Symptoms of Abnormality	Breed Reported In	Pertinent References
10. **Congenital dropsy**	Water in tissues and cavities. The calves are born 1 or 2 months previous to term or at term.	Ayrshire, Swedish Lowland Black-and-White	Johansson, 1941, *Proc. 7th Int. Cong. Genet.* (Edinburgh), 169. Donald, *et al.*, 1952, *Brit. Vet. J.*, 108, 227. Herrick & Eldridge, 1955, *J. Dairy Sci.*, 38, 440.
11. **Congenital ichthyosis**	Scaly, cracked skin.	Norwegian Red Poll	Tuff & Gladitsch, 1949, *Nordisk Vet. Med.*, I, 619.
12. **Crooked calf** (See Arthrogryposis)			
13. **Digital abnormality**	One toe shorter; toes spread; animal lame. Dominant.		Mead, *et al.*, 1949, *J. Hered.*, 40, 151–155.
14. **Doddler cattle**	Extreme muscle spasm, convulsions, nystagmation, and dilation of the eyes. A lethal recessive.	Hereford	High, 1958, *J. Hered.*, 49, 250.
15. **Edema**	Watery enlargement of legs, muzzle, and belly.	Ayrshire	Eldridge & Atkinson, 1953, *J. Dairy Sci.*, 35, 598.
16. **Epithelia defects**	Defective formation of skin below knees, one or more claws undeveloped, deformed integument of muzzle and the mucous membranes of nostrils, tongue, palate, and cheek.	Holstein	Hadley & Cole, 1928, Wisc. Ag. Exp. Sta. Bull. No. 86. Hutt & Frost, 1934, *J. Hered.* 25, 41.
17. **Epitheliogenesis imperfecta**	Imperfect skin and partly hairless below knees; one or more claws or dewclaws undeveloped. The calves are usually born at term but die as a result of bacterial invasion. (See Fig. 5-23.)	Holstein, Jersey	Hadley & Cole, 1928, Wisc. Ag. Exp. Sta. Bull. No. 86. Regan, *et al.*, 1935, *J. Hered.*, 26, 357. Hutt & Frost, 1948, *J. Hered.*, 39, 131. Leipold, *et al.*, 1973, *Can. Vet. J.*, 14, 114.
18. **Fetal resorption**	Various stages from decomposed masses to only bones or dried mummies have been reported.	Dairy and beef breeds	
19. **Hairless** (See *Hypotrichosis congenita*)			
20. ***Hernia cerebri***	Opening in skull because of failure of frontal bones to fuse. Affected calves are stillborn or die soon after birth.	Holstein-Friesian	Shaw, 1983, *J. Hered.*, 29, 319.
21. ***Hypotrichosis congenita***	A little hair is found on the muzzle, eyelids, ears, pasterns, and end of the tail. Most afflicted calves die shortly after birth; some live but grow slowly and never have normal hair. Recessive.		Mohr & Wriedt, 1928, *J. Genetics*, 19, 314. Regan, *et al.*, 1935, *J. Hered.*, 26, 357. Kidwell & Guilbert, 1950, *J. Hered.*, 41, 190. Surrarrer, 1943, *J. Hered.*, 24, 175. Hutt & Saunders, *J. Hered.*, 44, 97.

(Continued)

TABLE 5-3, PART I (Continued)

Lethal	Symptoms of Abnormality	Breed Reported In	Pertinent References
22. **Hydrocephalus** (water brain)	Affected animals have bulging forehead and enlargement of the cranial vault; excess fluid in portions of the brain. Simple autosomal recessive. (See Fig. 5-24.)	Several breeds	Houck, 1930, *Anat. Record*, 45, 83. Innes, *et al.*, 1940, *J. Path. & Bact.*, 50, 456. Cole & Moore, 1942, *J. Ag. Res.*, 65, 483. Godgluck, 1942, *Monat, Fur Vet.*, 7, 250. Giannotti, 1952, *Mem. Soc. Tosc. Sci. Nat.*, B. 59, 32. Kobozieff, *et al.*, 1955, *Bull. Biol.*, 89, 189–210. Gilman, 1956, *Cornell Vet.*, 45, 487. Blackwell & Knox, 1959, *J. Hered.*, 50, 143–148. Baker, *et al.*, 1961, *J. Hered.*, 52, 135–138. Leipold, *et al.*, 1971, *J. Am. Vet. Res.*, 32, 1019. Greene, *et al.*, 1973, *Irish Vet. J.*, 27, 37.
23. **Impacted molars**	Short lower jaw, impacted teeth; die within week.	Milking Shorthorn	Heizer & Harvey, 1937, *J. Hered.*, 28, 123.
24. **Lameness in hind limbs**	Calves unable to stand.	Red Danish	Christensen & Christensen, 1952, *Norsk. Vet. Tid.*, 4, 861.
25. **Limber leg**	The calf cannot stand, due to the incompletely formed muscles and joints. Many limber leg calves are born dead. Those born alive appear healthy and may be kept for some time. Inherited as a simple recessive.	Jersey	Lamb, 1971, *J. Dairy Sci.*, 54, No. 4, 544–546.
26. **Mummification**	They have a short neck, stiff legs, and prominent joints. Fetuses die at about 8 months' gestation but carried to term.	Red Danish	Loje, 1930, *Tidssk. for Landok.*, 10, 517.
27. **Muscle contracture**	Head and legs drawn; joints stiff. Calves born at full term. Recessive. (See Fig. 5-15.)	Holstein, Norwegian breeds	Hutt, 1934, *J. Hered.*, 25, 41; Nes, 1953, *Nordisk Vet. Med.*, 5, 869.
28. **Night blindness**	Animals see poorly in the twilight and at night, but have good daytime vision.		Craft, 1927, *J. Hered.*, 15, 255.
29. **Osteopetrosis** (marble-bone disease)	Stillborn 1 to 4 weeks premature; solid bones; no bone marrow cavities; short lower jaw; protruding tongue; and impacted molar teeth. Osteopetrosis is diagnosed by splitting a long bone lengthwise, and demonstrating a bone marrow cavity filled with solid "bony" tissue. Simple autosomal recessive. (See Figs. 5-25 and 5-26.)	Angus, Hereford	Leipold, *et al.*, 1971, *An. Sel. Genet. Anim.*, 3, 245. Greene, *et al.*, 1974, *J.A.Y.M.A.*
30. **Paralysis**	Posterior paralysis—may have muscular tumors and blindness, die shortly after birth.		Tufl, 1948, *Vet.-Tid.*, 38, 379. Cranek & Ralson, 1953, *J. An. Sci.*, 12, 892.
31. **Paralyzed hindquarters**	Calves appear normal except that they cannot stand on hindquarters.	Norwegian Red Poll, Red Danish	Loje, 1930, *Tidssk. for Landok.*, 10, 517.

(Continued)

TABLE 5-3, PART I (Continued)

Lethal	Symptoms of Abnormality	Breed Reported In	Pertinent References
32. **Prolonged pregnancy I**	The gestation period is from 311 to 403 days. Fetus grows in uterus almost as a calf born normally, with the result that parturition is difficult or impossible.	Japanese breeds	Mead, *et al.*, 1949, *JDS*, 32, 705. Gregory, *et al.*, 1951, *Portigualie Acta Biol.*, A., 861. Jasper, 1951, *Cornell Vet.*, 40, 165.
33. **Prolonged pregnancy II**	Calves are carried up to 500 days. Calves show immature bone development, various degrees of hairlessness and lack of thyroid development; and they do not have pituitary gland.		Hallgren, 1951, *Nord. Vet. Med.*, 3, 1043.
34. **Short limbs**	Limbs short; hoofs undeveloped.	Russian breeds, Swiss breeds, Shorthorn	Leipold, *et al.*, 1970, *Can. Vet. J.*, 11, 258.
35. **Short spine**	Ribs and vertebrae fused; back bent down. Calves are stillborn or die soon after birth. Recessive. (See Fig. 5-15.)	Norwegian, Mountain Angus	Mohr & Wriedt, 1930, *J. Genetics*, 22, 279. Leopold & Dennis, 1972, *Cornell Vet.*, 62, 507.
36. **Skinless** (See Epitheliogenesis imperfecta)			
37. **Spasms**	Calves first appear normal, but soon develop spasmodic muscular contractions. Affected animals die within a few weeks after birth. A lethal recessive.	Jersey	Shrode & Lush, 1947, *Advances in Genetics*, Academic Press, Inc., NY. Gregory, *et al.*, 1944, *J. Hered.*, 35, 195. Sauders, *et al.*, 1952, *Cornell Vet.*, 42, 559.
38. **Streaked hairlessness**	Gene carried only in females. Carrier females exhibit streaked hairlessness and produce a sex ratio of 2 females to 1 male. (See Fig. 5-15.)	Holstein	Eldridge & Atkeson, 1953, *J. Hered.*, 44, 265.
39. **Syndactyly** (mulefoot)	Right front foot, both front feet, or 3 feet with a single toe. Rarely 4 feet have a single toe. Affected animals cannot withstand stress. (See Fig. 5-27.)	Holstein, Simmental, Chianina	Leipold, *et al.*, 1973, *Vet. Bull.*
40. **Tendon contracture**	Tendons pulled rigidly, calves either are born dead or die after birth.	Milking Shorthorn	Dale & Moxley, 1952, *Can. J. Comp. Med.*, 16, 399.

PART II. HEREDITARY SEMILETHALS

Semilethal	Symptoms of Abnormality	Probable Mode of Interitance	Breed Reported In	Pertinent References
1. **Albinism**	Complete absence of pigmentation in the skin, muzzle, and hoofs, and in the walls of the rumen. However, the eyes are not pink, due to a slight grayish coloration of the cortex. (See Fig. 5-18.)	Probably recessive	Hereford, Brown Swiss	Hafez, *et al.*, 1958, *J. Hered.*, 49, 111. Greene, *et al.*, 1973, *J. Hered.*
2. **Albinism** (partial)	White animals; spots in skin; blue eyes.	Dominant		Leipold, *et al.*, 1968, *J. Hered.*, 59, 2.
3. ***Atresia ani***	Closed anus. Calves do not survive corrective surgery.			Kuppuswami, 1937, *IND. J. Vet. Sci. & Ani. Husb.*, 7, 305. Lerner, 1944, *J. Hered.*, 35, 21.9.

(Continued)

TABLE 5-3, PART II (Continued)

Semilethal	Symptoms of Abnormality	Probable Mode of Interitance	Breed Reported In	Pertinent References
4. **Epilepsy**	Cattle subject to epileptic type of attacks.	Dominant	Brown Swiss	Atkeson, *et al.*, 1944, *J. Hered.*, 35, 45.
5. **Imperforate anus** (See *Atresia ani*)				
6. **Muscular dystrophy**	Genetic abnormality, develops over 2 years.	Dominant	Brown Swiss	Leipold, *et al.*, 1968, *Vet. Med.*, 68, 645.

PART III. HEREDITARY DELAYED LETHALS

Delayed Lethal	Symptoms of Abnormality	Probable Mode of Interitance	Breed Reported In	Pertinent References
1. **Atrophy of testicles**	a. Gross microscopic changes to testicles, including atrophy, calcification, degeneration of the seminiferous tubules, and varying degrees of fibriosis. b. Epithelium layers of seminal ducts are underdeveloped; in either one or both testicles.	? ?	Swedish Highland	Rollinson, 1955, *Ani. Breed. Abstr.*, 23, 215.
2. **Cystic ovaries**	Sterility; nymphomania.	?	Swedish Highland	Rollinson, 1955, *Ani. Breed. Abstr.*, 23, 215.
3. **Female sterility**	Cows fail to settle.	Simple sex-limited		Gregory, *et al.*, 1945, *Genetics,* 30, 506. Gregory, *et al.*, 1951, *J. Dairy Sci.*, 34, 1047.
4. **Gonad hypoplasis**	Atrophy of 1 or 2 ovaries.	Autosomal		
5. **Gonadless**	Absence of ovaries.	?		
6. **Hypoplasia of the ovary**	An underdeveloped condition of the gonads in both sexes. When both gonads are involved, the animal is sterile. When one is involved, the animal is less fertile than normal.	Recessive		Erickson, 1943, *Biological and Genetic Investigations* (I. Lund: Hakan Ohlssons, "Boktryckeri").
7. ***Impotentia coeundia***	Bull does not possess ability to copulate due to failure of sigmoid curve of penis to straighten during coitus.	Autosomal recessive		Rollinson, 1955, *Ani. Breed. Abstr.*, 23, 215.
8. **Knobbed spermatozoa**	Abnormal formation of sperm. It may be related to some unknown changes in nucleic acid metabolism of the sperm head. Absence of any chromosomal aberration, quantitative or structural.	Autosomal sex-limited		Rollinson, 1955, *Ani. Breed. Abstr.*, 23, 215.
9. ***Mannosidosis***	Affected calves stop growing, develop incoordination, don't follow their mothers, and become aggressive when crowded. The animals actually go crazy. Usually calves are born alive and function normally at birth. Most affected calves die within a year.	Recessive	Angus, Murray-Grey	Jolly & Leipold, 1973, *New Zealand Vet. J.* Barr, 1981, *Angus J.*, 23.

(Continued)

TABLE 5-3, PART III (Continued)

Delayed Lethal	Symptoms of Abnormality	Probable Mode of Interitance	Breed Reported In	Pertinent References
10. **Marble bone disease** (See Osteopetrosis, Part I)				
11. **Multiple lipomatosis**	A large growth of adipose tissue in the peritoneal area appears at about 3.5 years of age and becomes progressively larger. In some cases, the fat deposition invades the udder and prevents it from functioning. It occurs in both males and females.	Dominant		Albright, 1960, *J. Hered.*, 51, 231.
12. **Turned sperm tails**	Sperm tails turned back past the head.	?		
13. **Umbilical hernia**	It appears to be limited to males.	Dominant	Holstein	Warren & Atkeson, 1931, *J. Hered.*, 22, 345.
14. **White heifer disease**	Persistent hymen or incomplete cervix; horns of uterus become distended with fluid. Most commonly found in white heifers of the Shorthorn breed, but it has been reported in roan and red Shorthorns and in colored animals of other breeds.	Sex-limited recessive	Shorthorn, other breeds	Gilmore, 1949, *J. Dairy Sci.*, 32, 71. Spriggs, *et al.*, 1946, *Vet. Record*, 58, 405. Bennet & Olds, 1971, Ky. An. Sci. Res. Rpt., 196.

PART IV. HARMFUL AND DEFECTIVE (NONLETHALS) GENES; CONGENITAL DEFECTS

Nonlethal	Symptoms of Abnormality	Probable Mode of Interitance	Breed Reported In	Pertinent References
1. **Achondroplasia** (stumpy)	Short legs and head, curly coat, and thin.	Recessive	Shorthorn	Mead, et al., 1946, *J. Hered.*, 37, 183. Baker, et al., 1950, *J. Hered.*, 41, 243.
2. **Cancer eye**	Herefords without pigmented eyelids more susceptible than other cattle. Usually occurs in older animals.	?	Hereford	Frank, 1943, *J.A.V.M.A.*, 102, 200. Guilbert, *et al.*, 1948, *J. Ani. Sci.*, 7, 426. Woodward & Knapp, 1950, *J. Ani. Sci.*, 9, 580. Anderson, *et al.*, 1957, *J. Ani. Sci.*, 16, 739.
3. **Cataract**	Opaque conditon of the lens of the eye.	Recessive	Jersey	Detlefson & Yapp, 1920, *Am. Naturalist*, 54, 277. Gregory, 1943, *J. Hered.*, 34, 125.
4. **Cerebellar hypoplasia** (cerebellar cortical atrophy)	Poor coordination, Cerebellum rudimentary, excessive fluid. Some walk like ballet dancers.	Probably recessive	Holstein-Friesian, Jersey	Anderson & Davis, 1950, *J.A.V.M.A.*, 17, 460. Sunders, *et al.*, 1952, *Cornell Vet.*, 42, 559.
5. **Comprest**	The animal is extremely bowlegged, often being unable to stand. The vertebrae do not have the spines which are pronounced in the "snorter dwarf."	Incomplete dominancy	Hereford	Lucas, 1950, thesis (M.S.) Colo. State Univ. (Fort Collins). Chambers, 1954, Okla. AESMP No. MP-34. Stonaker, 1958, Colo. Exp. Sta. Bull. 501-S. Hafez, 1959, unpublished data.

(Continued)

TABLE 5-3, PART IV (Continued)

Nonlethal	Symptoms of Abnormality	Probable Mode of Interitance	Breed Reported In	Pertinent References
6. **Crooked legs**	Front legs crooked. Certain lines of breeding show this affliction whereas others do not, even in areas where it is prevalent. Those blood lines manifesting the condition in an area where it is prevalent are not afflicted in the areas where it is not found.	Nutritional-genetic interaction	Hereford	Stonaker, 1958, Colo. Exp. Sta. Bull. 501-2. Bogart, 1959, personal communication. Hafez, 1959, unpublished data.
7. **Cross eyes** (See Strabismus)				
8. **Cryptorchidism**	Failure of one or both testicles to descend to their normal positions in the scrotum; they remain hidden inside the abdominal cavity or inguinal canal. Affected testicle(s) usually sterile.	Simple autosomal recessive, with sex limitation to bull. Other opinions postulate cryptorchidism to be a multiple gene trait.	Hereford	Wheat, 1961, *J. Hered.*, 52, 5.
9. **Curly hair**	Hair in tight curls, viable.	Dominant	Ayrshire	Eldridge, *et al.*, 1949, *J. Hered.*, 40, 205.
10. **Curved limbs**	Hind limbs grossly deformed, with the hocks held close to the body and scarcely flexed forward.	Probably a recessive	Guernsey	Freeman, 1958, *J. Hered.*, 5, 229.
11. **Double ear**	A thin, flat piece of cartilage lies parallel to the long axis of the ear, extends beyond the ear tip.	Dominant	Brahman	Lush, 1924, *J. Hered.*, 15, 93.
12. **Double muscled**	Characterized by abnormally wide thighs, with this extreme width extending forward to include the loin. Deep grooves between the muscles are conspicuous externally. Little fat covering is present. (The "Dopple-lander" calf is a mutation of doubtful value; the valuable muscles in the back and loin are doubled, but the animals are sterile [Hammond, 1935, *Emp. J. Exp. Ag.*, 3m Bi, 9].)	Single recessive gene, masked by dominant gene in heterozygous carriers.	Angus, Charolais, Hereford, Piedmont, Shorthorn	Weber & Ibsen, 1934, *Proc. Am. Soc. Ani. Prod.*, 228. Kidwell, *et al.*, 1952, *J. Hered.* 43, 62. Oliver & Cartwright, 1969, Texas Ag. Exp. Sta. Bull. 12.
13. **Duck legged**	Legs shorter than normal.	Dominant		Lush, 1927, *J. Hered.*, 21, 85.
14. **Dwarfism**	Animals small in size. Several kinds have been identified: snorter dwarfs, long-headed dwarfs, and compress dwarfs are caused by 3 different genes. (See Fig. 5-22.)	Recessive for some; partially dominant for others (comprest)	Angus, Hereford, Shorthorn	Mead, *et al.*, 1942, *J. Hered.*, 33, 411. Johnson, *et al.*, 1950, *J. Hered.*, 41, 177. Lindley, 1951, *J. Hered.*, 42, 273. Baker, *et al.*, 1951, *J. Hered.*, 42, 141. Gregory, *et al.*, 1953, *Hilgardia*, 22, 407. Pahinsh, *et al.*, 1955, *J. Ani. Sci.*, 14, 200, 1025.
15. **Extra toes** (See Polydactylism)				

(Continued)

TABLE 5-3, PART IV (Continued)

Nonlethal	Symptoms of Abnormality	Probable Mode of Interitance	Breed Reported In	Pertinent References
16. **Flexed pasterns**	Toes turned under.	Recessive	Jersey	Hable, 1948, *J. Vet. Res.*, 9, 131. Atkeson, *et al.*, 1943, *J. Hered.*, 34, 25. Mead, *et al.*, 1943, *J. Hered.*, 35, 367.
17. **Fused teats**	The front and rear teats on the same side are fused.	Recessive	Hereford	Johnson, 1945, *J. Hered.*, 36, 317. Hiezer, 1932, *J. Hered.*, 23, 111.
18. **Harelip**	Unilaterally harelipped, and the dental pad on that side is missing. Calves have difficulty nursing.	Epistasis may be involved	Shorthorn	Wheat, 1960, *J. Hered.*, 51, 99.
19. **Missing teat**	One teat on left side.	Recessive	Holstein-Friesian	Rollinson, 1955, *Ani. Breed. Abstr.*, 23, 215.
20. **Muscular hypertrophy**	First described in a crossbred Afrikander X Angus, but present in other breeds. Also, the thighs are extremely thick and full. Animals often assume an unusual stance with the forelegs extended foreward and the back legs extended backward.	Probably recessive	Several breeds	Kidwell & Vernon, 1952, *J. Hered.*, 43, 62.
21. **Notched ears**	Ears imperfect in shape.	Dominant		
22. **Pink teeth** (See Porphyrinuria)				
23. **Polydactylism I** (Weavers polydactyly)	Affected animals have an extra toe on each front foot, with accompanying tenderness and lameness.	Dominant autosomal	Hereford (males), Brown Swiss	Roberts, 1921, *J. Hered.*, 12, 84. Morrill, 1945, *J. Hered.*, 36, 81. Shrode & Lush, 1947, *Advances in Genetics*, Academic Press, NY.
24. **Polydactylism II**	A three-toed condition.	Dominant autosomal	Holstein-Friesian	
25. **Porphyrinuria**	a. Cattle are photosensitive and develop lesions in unpigmented areas of skin. b. Excessive coproporphyrin and uroporphyrin.	?	Shorthorn, Holstein-Friesian	Fourie, 1939, Onderst. *J. Vet. Sci.*, Ani. Indus., 13, 383. Jorenson & With, 1955, *Nature* (London), 176, 156. Clare, 1955, *Advan. Vet. Sci.*, 2, 191.
26. **Protoporphyria**	Animals sensitive to sunlight. Calves that have this malady avoid sunlight and seek shade. The skin on ears and nose shows lesions or weeping sores, then scabs.	Recessive	Limousin	Martin, 1978, *Inter. Limousin J.*, 115.
27. **Rectal and vaginal constriction (RVC)**	Inseminators unable to insert forearm into the rectum; vaginal palpation is impossible; vaginal constriction usually makes first calving impossible without Caesarean section or episiotomy (incision of the vulva).	Recessive	Jersey	Leipold, *et al.*, 1975, *The Am. Vet. Med. Assn. Mag.*

(Continued)

TABLE 5-3, PART IV (Continued)

Nonlethal	Symptoms of Abnormality	Probable Mode of Interitance	Breed Reported In	Pertinent References
28. **Screw tail**	Tail appears to be broken due to fusion of two or more vertebrae.	Recessive	Red Poll	Knapp, 1936, *J. Hered.*, 27, 269.
29. **Semihairlessness**	Absence of hair from margin of the ears, along the underline, on the inside of the legs, the side of the neck, shoulder vein, sides, and thighs.	Recessive	Hereford, Polled Hereford	Cole, 1919, *J. Hered.*, 22, 345. Craft & Blizzard, 1934, *J. Hered.*, 25, 385.
30. **Spread hoofs**	Hoofs spread greatly. Painful; animals often walk on knees.	Recessive	Jersey	
31. **Strabismus**	Cattle have a cross-eyed condition. Not evident at birth but identified by 12 months of age.	Recessive		Regan, *et al.*, 1944, *J. Hered.*, 35, 233.
32. **Stumpy** (See Achondroplasia)				
33. **Syndactylism** (mulefoot)	One or two toes affected; seldom three toes. The uncloven foot resembles a mule's foot. (See Fig. 5-27.)	Simple autosomal recessive	Holstein-Friesian, Angus	Eldridge, *et al.*, 1951, *J. Hered.*, 42, 241.
34. **Tailless**	Calves born without tails. (See Fig. 5-28.)	Probably recessive	Common	Gilmore & Fechheimer, 1957, *Ohio Farm and Home Res.*, 32–33.
35. **Umbilical hernia**	Hernia involving the umbilical area. Covered by skin. Varies from an inch to several inches in diameter.	Incompletely dominant; or recessive	Common	Leipold & Dickenson, 1978, *Inter. Limousin J.*, 114–123.
36. **Weaver**	A degeneration of the spinal cord causes the animal to weave when it walks, and the hindquarters of the animal to collapse in uncontrolled disarray when it tries to run.	Probably recessive	Brown Swiss	Leipold, 1982, *The Advanced Animal Breeder*, 7.
37. **Wrytail**	Distortion of the tail head, with the base of the tail set at an angle to the backbone instead of in line with it.	Recessive	Several breeds	Atkeson, 1944, *J. Hered.*, 35, 11.

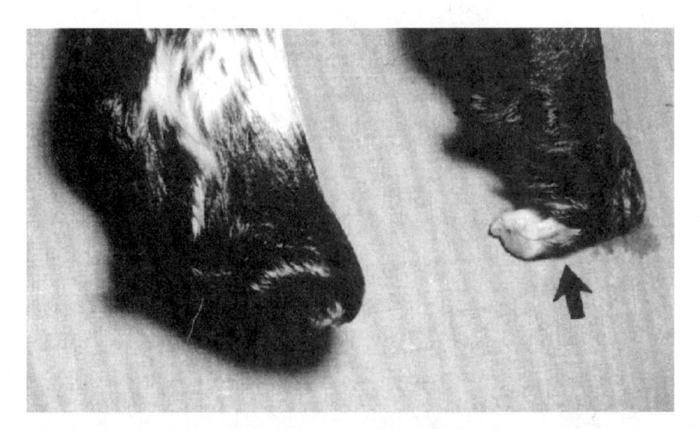

Fig. 5-16. Adactylia (absence of digital rays) in a Shorthorn calf. Note small uncloven hoof at distal end of leg (arrow). (Courtesy, Professor H. W. Leipold, Kansas State University)

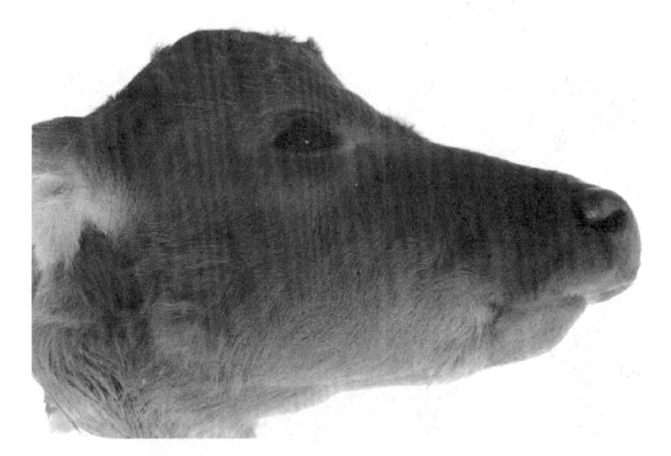

Fig. 5-17. Agnathia (short lower jaw) in a calf. (Courtesy, Professor H. W. Leipold, Kansas State University)

Fig. 5-18. Albino Hereford cow. Note lack of pigment. (Courtesy, Professor H. W. Leipold, Kansas State University)

Fig. 5-19. Anophthalmia in a Hereford calf. Note lack of eye development (arrow). (Courtesy, Professor H. W. Leipold, Kansas State University)

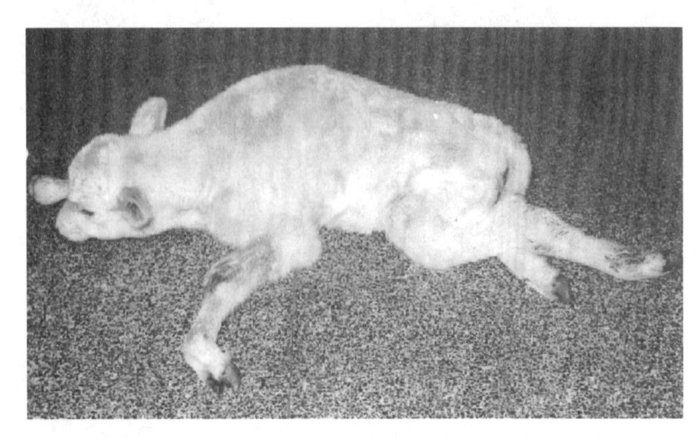

Fig. 5-20. Arthrogryposis, showing rigid legs, in a Charolais calf. (Courtesy, Professor H. W. Leipold, Kansas State University)

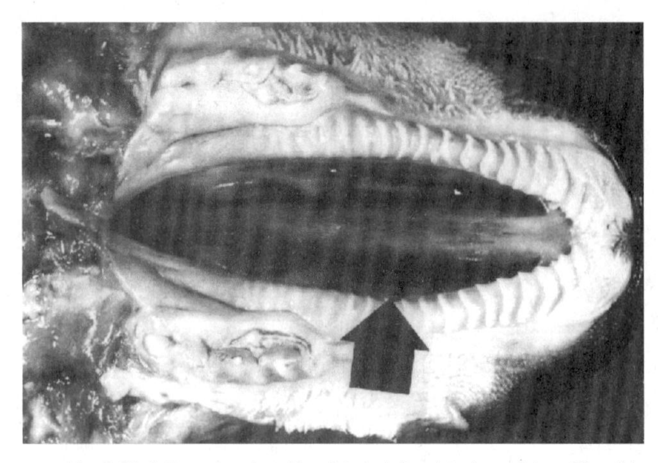

Fig. 5-21. Arthrogryposis, with related cleft palate (arrow) in a Charolais calf. (Courtesy, Professor H. W. Leipold, Kansas State University)

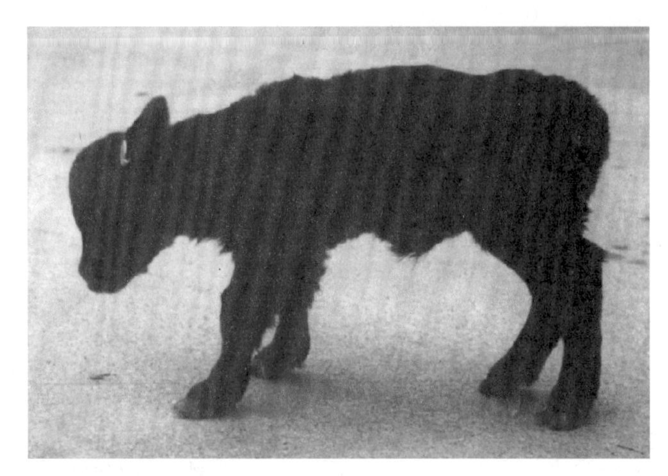

Fig. 5-22. Dwarf Angus calf. Note the stocky conformation, the short legs, the short head and the dished face, and the protruding lower jaw. (Courtesy, University of Minnesota, Agricultural Extension Service, St. Paul, MN)

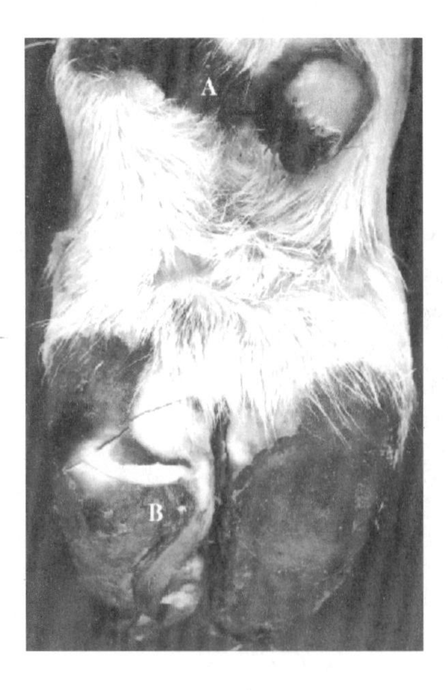

Fig. 5-23. Epitheliogenesis imperfecta in a Holstein-Friesian calf. Note imperfectly developed dew claw **(a)** and defective claw **(b)**. Note, too, imperfect skin below knees. (Courtesy, Professor H. W. Leipold, Kansas State University)

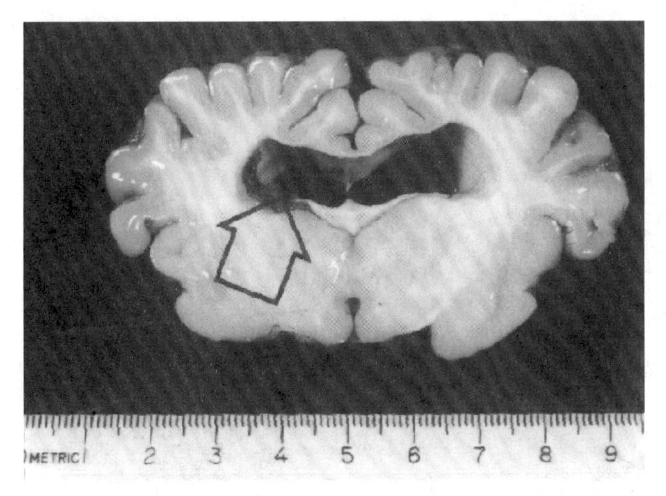

Fig. 5-24. Hydrocephalus (internal) in a Hereford calf. Note dilation of internal compartments of brain (arrow). (Courtesy, Professor H. W. Leipold, Kansas State University)

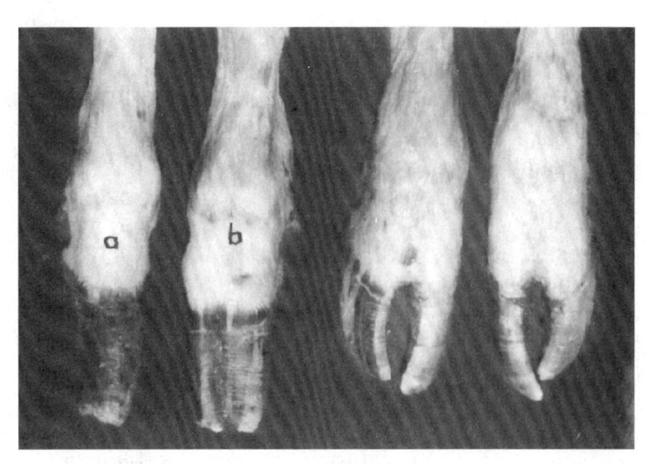

Fig. 5-27. Syndactyly (mulefoot) in cattle. Note complete mulefoot of right front foot **(a)** and partial mulefoot of left front foot **(b)** compared to normal hind feet. (Courtesy, Professor H. W. Leipold, Kansas State University)

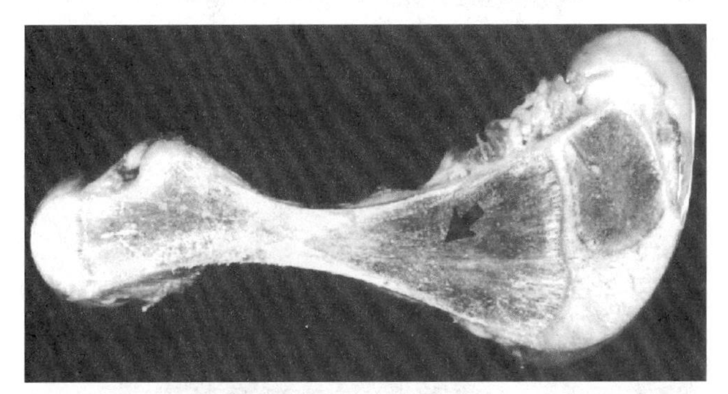

Fig. 5-25. Osteopetrosis. Bisectioned femur of an Angus affected with osteopetrosis (marble-bone disease). Note dense bone without formation of bone marrow cavity (arrow). (Courtesy, Professor H. W. Leipold, Kansas State University)

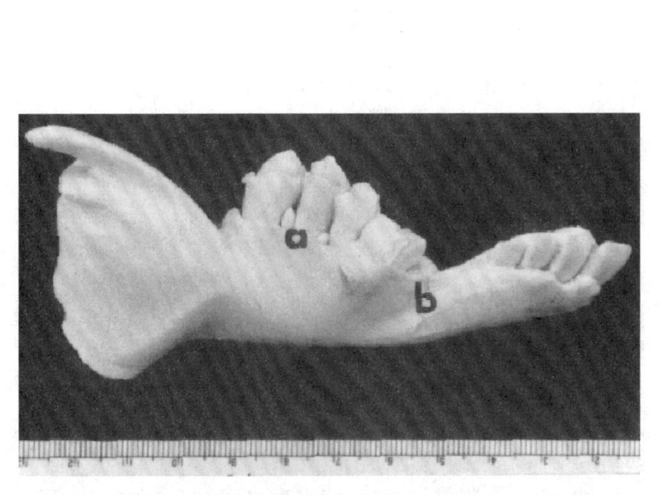

Fig. 5-26. Osteopetrosis, with related short lower jaw of a calf affected with osteopetrosis. Note impaction of molar teeth **(a)** and lack of the foramen mentale **(b)**. (Courtesy, Professor H. W. Leipold, Kansas State University)

Fig. 5-28. Tailless cow and her normal-tailed calf sired by a tailless bull. (Courtesy, Professor H. W. Leipold, Kansas State University)

DWARFISM IN CATTLE

There are several different types of dwarfs, of which the short-headed, short-legged, pot-bellied dwarf—commonly referred to as the snorter dwarf—is the most frequent. The discussion that follows applies specifically to snorter dwarfism.

Though very small (usually weighing about half as much as normal calves), dwarf calves are exceedingly stocky and well-built. The eyes protrude, giving a characteristic pop-eyed appearance. Some dwarfs are weak and unsteady in gait at birth. Others appear to be strong enough, but soon develop a large stomach, heavy shoulders, crooked hind legs, and sometimes labored breathing. Survival is somewhat lower than with normal calves, although most purebred breeders make no attempt to raise them.

There is complete agreement among scientists (1) that the dwarf condition is of genetic origin, and (2) that it is inherited as a simple autosomal recessive (the word *autosomal* merely means that it is not carried on the sex chromosomes) and conditioned by at least two pairs of modifying genes. Thus, the birth of a dwarf calf identifies both the sire and the dam as carriers of the dwarf gene.

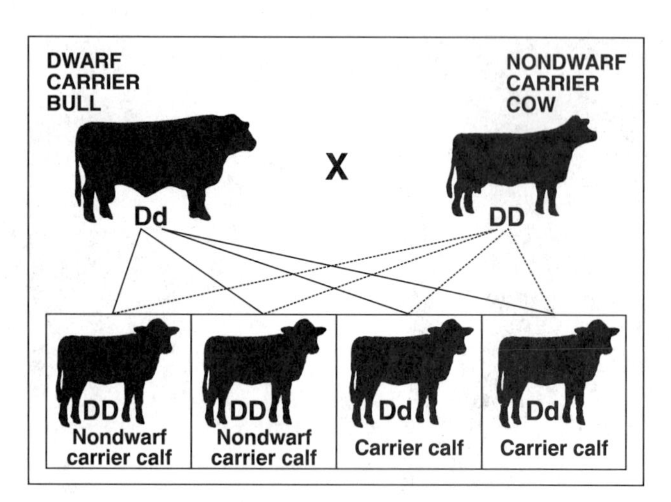

Fig. 5-29. Diagrammatic illustration of the inheritance of the most common type of dwarfism, showing what to expect when a carrier (heterozygous) bull(s) is mated to a noncarrier (homozygous normal) cow(s); or the sexes may be reversed. As shown, the carrier X noncarrier matings will, *on the average*, produce calves of which (a) 50% are carriers, although not dwarfs, and (b) 50% are noncarriers and nondwarfs. Unfortunately, the two groups look alike and cannot be detected by sight.

CONDITIONS PREVAILING IN DWARF-AFFLICTED HERDS

One of the conditions (or perhaps both conditions) shown in Figs. 5-29 and 5-30 prevails in any herd of cattle in which dwarf-carrying animals are being used.[10]

From Fig. 5-29, it may be seen that 100 offspring from matings of carrier bulls X noncarrier cows will, on the average, possess the following genetic picture from the standpoint of dwarfism:

50 carriers, although not dwarfs
50 noncarriers and nondwarfs
100 total

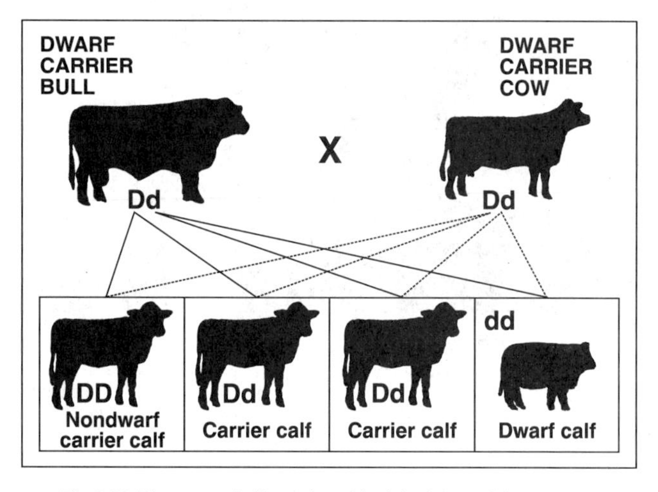

Fig. 5-30. Diagrammatic illustration of the inheritance of the most common type of dwarfism, showing what to expect when a carrier (heterozygous) bull(s) is mated to a carrier (heterozygous) cow(s); or the sexes may be reversed. As shown, carrier X carrier mating will, *on the average*, produce calves of which (a) 25% are dwarfs, (b) 50% are carriers, although not dwarfs, and (c) 25% are noncarriers and nondwarfs. Unfortunately, only the dwarfs can be detected by sight; the two nondwarf groups look alike and cannot be distinguished by sight.

From Fig. 5-30, it may be seen that 100 offspring from matings of carrier bulls X carrier cows will, on the average, possess the following genetic picture from the standpoint of dwarfism:

25 dwarfs
50 carriers, although not dwarfs
25 noncarriers and nondwarfs
100 total

On the basis of these facts, it may be concluded that the following dwarfism genetic picture applies to any given calf having (1) one carrier parent, or (2) both carrier parents:

1. A calf out of parents one of which is a known carrier and the other a noncarrier has a 50% chance of being free of the dwarf factor.
2. A calf both of whose parents are carriers has only one chance in four of being free of the dwarf factor.

[10]All ratios are averages based on large numbers; thus, they may not apply to any given herd.

It is recognized that the percentage of carrier females in any given herd will vary. Obviously, where dwarf calves have appeared, there are both carrier bulls and carrier cows. Some breeders may remove carrier animals, especially cows, once they have dropped a dwarf calf, thus selecting away from the trait. Others may unwittingly select for, rather than against, animals of the carrier type, if such a carrier type exists and if it is associated with some much sought characteristic, such as a markedly dished face.

From Figs. 5-29 and 5-30 the following deductions of value to practical operators may be made: (1) where a carrier bull is mated to noncarrier cows, no dwarfs will be produced, but, on the average, one-half of the calves will be carriers, and (2) the carrier heifers from this first cross can, and likely will, produce one-fourth dwarfs if they are mated back to a carrier bull. In other words, although the use of a carrier bull in a clean herd will not produce any dwarf calves, the seed for dwarfism is sown and it will crop out providing a second carrier bull is used in the herd.

Figs. 5-29 and 5-30 also indicate the futility of continuing the use of carrier bulls or females. Also, it should be recognized that any animal producing a dwarf is a carrier, regardless of the number of dwarfs produced (one or several).

PURGING THE HERD OF DWARFS

The breeding program followed to remove or minimize the dwarf condition will depend somewhat on the type of herd involved—especially on whether it is a commercial or purebred herd.

In a commercial herd, the breeder may lessen the chances of obtaining dwarfs by using an outcross (unrelated) sire within the same breed or by crossbreeding with a sire from another breed. With this system, the dwarf-carrying cows will remain, but—because of the recessive condition of the dwarf factor—it will be covered up.

In a purebred herd, the action taken in handling the dwarf situation should be more drastic. Reputable purebred breeders have an obligation, not only to themselves, but to their customers among both the purebred and the commercial herds. Purebred herds should be purged of the undesirable dwarf genes. This can be done through pursuing any one of the following three breeding systems.[11]

1. **Using sires of families free of the dwarf factor; pedigree-clean animals.** Within each breed where dwarfs have appeared, families exist that are free of the dwarf factor. Securing and using bulls from such families will "cover up" the dwarf situation.

Pedigree information is especially useful for early screening of prospective breeding animals and for small breeders who cannot afford the expense of progeny testing.

2. **Testing bulls of present breeding in the present herd.** Continue to select bulls from within the herd or from herds from which purchases have been made previously, but select bulls that appear to be free from the dwarf factor as judged by pedigree and family background. Next, test these bulls by mating each of them to cows in the present herd that have produced dwarf calves. If each bull tested is mated to 15 known carrier cows and all the progeny are normal, there would be only 1.3 chances out of 100 (1.3%), or 1 chance in 75, that the bull is a dwarf carrier and yet passed the test undetected.[12] A test of about the same validity can be secured by breeding a sire to 30 of his daughters. Accordingly, such a tested bull could be used in any herd with reasonable certainty that he is free of the dwarf-producing factor.

If rigidly adhered to, this system will eventually produce the desired results, but it has the following limitations: (a) It necessitates retaining carrier cows in the existing herd for bull testing purposes, thus giving doubting fellow breeders an opportunity to question the entire herd; and (b) there is always the temptation to retain outstanding calves although they are likely carriers.

3. **Using a commercial herd of test cows for the purpose of proving bulls.** If carefully followed, this is the most desirable of all the methods herein

[11]It is recognized that it is practically impossible to eliminate completely the dwarf factor from a herd or breed of cattle once it has appeared. With a proper breeding program, however, the incidence of dwarfism will become so small as to be unimportant—much as the occurrence of the red color (caused by another recessive factor) has been minimized in the Angus and Holstein breeds of cattle.

[12]A lesser number than 15 might be used but the breeder could not be so sure of the results. For example, with 10 or 5 such matings the chances of failing to detect a carrier bull are increased from 1.3% to 5.6% and 23.7%, respectively. Thus, with 10 such matings and all normal calves, there is only 1 chance in 18 that the bull is a dwarf carrier and yet passed the test undetected.

This is computed as follows: the chance of obtaining a dwarf calf from a dwarf-carrier bull bred to known carrier cows is 1 in 4, or $\frac{1}{4}$. The chance of a dwarf not occurring from this type of mating is $1 - \frac{1}{4}$, or $\frac{3}{4}$. Thus, if 5 such matings are made, $(\frac{3}{4})^5 = 23.7$, or the chances of failing to detect a carrier bull with 5 matings is 23.7%; with 10 matings it is $(\frac{3}{4})^{10}$, or 5.6%; and with 15 matings $(\frac{3}{4})^{15}$, or 1.3%.

If only a limited number of carrier cows are available, it may be desirable to breed each prospective herd sire to 4 to 6 carrier cows initially, followed by more thorough testing of those passing the initial screening.

If the factor becomes so rare that it is impossible to secure sufficient carrier cows for bull testing, the incidence of dwarfism will be so small as to be unimportant.

proposed, and over a period of years the one which will pay the most handsome dividends. At first glance, this method will appear drastic and expensive, but, in the end, the approach is the soundest of the three proposals. Under this system, it is suggested that the breeder assemble a herd of "tester cows" (from either purebreds or grades) each known to have dropped at least one dwarf calf, *with these animals operated strictly as a commercial herd.*

Prior to using any bull in the purebred herd, bulls that are otherwise desirable would be tested by mating each of them to approximately 15 of the dwarf-factor–carrying cows in this commercial herd. Then the top bulls from among those whose get are free of dwarfs could be used in the purebred herd with reasonable certainty.[13]

Carrier cows mated to bulls as indicated would, on the average, not produce in excess of 25% dwarfs if all the bulls were dwarf carriers (see section relative to "Conditions Prevailing in Dwarf-afflicted Herds"). Of course, fewer dwarfs would be produced if some of the bulls were dwarf-free, as expected. Thus there would be considerable remuneration from the sale of calves in the operation of such a commercial herd. Further, and most important, the merits of young sires, from the standpoint of type and efficiency of production, could be determined in the commercial herd prior to using them in the purebred herd—thus making it possible to select sires by modern record of performance methods.

Under any of the three systems herein proposed, it would be wise to eliminate those bulls and cows that are known to have produced dwarf calves as soon as desirable replacement animals proved to be free of the factor are available.

Providing one does not select a larger than normal percentage of carrier heifers as replacements for the herd, each generation of calves sired by dwarf-free bulls would halve (lessen by 50%) the number of carriers in the herd. True enough, there would always be some of the dwarf carriers present, for the incidence of carriers is halved with each generation of such matings, but not eliminated. Yet, after two generations of such matings, the incidence of dwarf carriers would be small. Also, it is noteworthy that the use of proved dwarf-factor–free bulls would give assurance that no more dwarf calves would be produced in the herd.

It is perfectly obvious that the elimination of the dwarf-producing factor is both slow and costly. Yet it is the only way in which cattle can be freed from dwarfs. Also it may require real courage to recognize openly the situation and discard outstanding animals that are known carriers.

Because dwarfs represent an almost complete economic loss, the problem deserves careful attention.

DOUBLE MUSCLING (Muscular Hypertrophy)

Double muscling refers to cattle characterized by bulging muscles of the shoulder and thigh, a very rounded rear end (as viewed from the side), a wide but shallow body throughout, appearance of intermuscular grooves, and fine bones.

Since beef cattle are produced primarily for their muscle, it's logical that selection should be centered around muscularity. It follows that cattle with *double muscles*, or these tendencies, have appeal and have increased in frequency in the United States during recent years. But there are disadvantages as well as advantages to double muscling. Hence, producers should be familiar with the characteristics and genetics of double muscling, and its side effects.

Double muscling is really a misnomer. Likewise, the scientific name, muscular hypertrophy, is incorrect because it implies increased size of fibers in each muscle, which is not the case. Rather, it has been shown that double muscled cattle have more fibers, not larger fibers.

In Germany, the trait is known as *doppellender* (double rump); in Italy, it's *doppia* (horse rump); and in France, it's *culard.*

Double muscling is a genetically controlled character. It appears to be caused by a single recessive gene, which tends to be *masked* by the dominant gene in the heterozygous carriers. Other examples of a character controlled by one pair of genes are: polledness and hornedness, and dwarfism. The genetics, therefore, are relatively simple. Since each animal has two genes for such characters, all cattle can be classified as follows:

DM DM—Homozygous normal; two dominant normal genes—a normal animal.

DM dm—Heterozygous; one dominant normal gene (DM) which tends to cover up the one recessive gene (dm)—these are called carriers. This cover up is not complete; hence, there is a tendency toward double muscling.

dm dm—Homozygous recessive; two recessive double muscle genes—a double muscled animal.

The progeny from a sire and dam of all these three genotypes are predictable, on the average, but not necessarily for any one offspring. The possible matings and progeny are:

---

[13]The chances of avoiding a dwarf-factor–carrying bull through such testing on 15 carrier cows is covered in the section immediately preceding.

Sire	Dam	Progeny
DM DM	DM DM	DM DM, ALL normal.
DM DM or DM dm	DM dm or DM DM	½ DM DM, ½ DM dm ALL normal, but ½ carriers.
DM dm	DM dm	¼ DM DM, ½ DM dm, ¼ dm dm Of the ¾ normal, 2 out of 3 are carriers; the remaining ¼ are double muscled.
DM DM or dm dm	dm dm or DM DM	DM dm, ALL carriers.
DM dm or dm dm	dm dm or DM dm	½ DM dm, ½ dm dm ½ carriers, ½ double muscled.
dm dm	dm dm	dm dm, ALL double muscled.

APPEARANCE OF HOMOZYGOUS DOUBLE MUSCLED (dm dm) CATTLE

Obviously, the problem is to determine if an animal is DM dm (a carrier), rather than DM DM (a normal animal). There are two ways to do this: (1) appearance; and (2) breeding tests. Detection by appearance is not 100% sure, but the experienced observer does not make many mistakes.

The appearance of homozygous double muscled (dm dm) cattle follows, in summary form. But remember that the double muscle character is really a syndrome of many characteristics. Remember, too, that all of these traits may not be present to the same degree in any one animal.

Body Part: Trait	Appearance in Double Muscled Cattle
Rump and round	Protruding and rounding; definite grooves, or creases, between the thigh muscles.
Tail	Short. Attached far forward. Prominent tailhead.
Middle; heart girth; flank	Shallow bodied, light heart girth, tucked up flank; animal appears leggy and cylindrical.
Head	Small, long, carried lower than top of shoulders.
Shoulder	Large and bulging; grooves, or creases, evident in arm and forearm.
Cannon bone	Short and fine.
Stance	Animal stands camped out; forelegs extended to front and hindlegs stretched.
Vulva	The vulva of females is small, high, and far forward.
Testicles	Small and carried close to the stomach.
Lying down	Double muscled cattle spend a lot of time lying down; they're *muscle laden*, with a high proportion of flesh to bone.
Age	Double muscling is most conspicuous in young animals. It becomes less apparent with advancing age.
Sex	General lack of masculinity, other than muscularity, in bulls; and a lack of a femininity in heifers and cows. At breeding age, double muscling is more marked in males than females.
Birth; early growth	Heavy birth weight. Calves often have enlarged tongues and crooked legs, and are weak. Good early growth. But growth markedly slower by one year of age; and small mature size.
Environment	Double muscling is more marked with superior environment; it shows up more in well-fed animals.

APPEARANCE OF CARRIER (DM dm) CATTLE

Generally speaking, homozygous double muscled animals can be identified. But it isn't easy to pick out the heterozygotes—the carriers of the double muscle gene, due to the wide variation in expression. Some of the carriers look quite normal, others look like homozygous double muscled animals, and still others are intermediate between these two extremes. Also, identity is further complicated because few, if any, double muscled animals show all of the characteristics. Nevertheless, carriers are characterized by general overall trim appearance, thicker quarter with bulging, thicker round, and a higher tailhead setting than normal animals.

DOUBLE MUSCLED CATTLE—GOOD OR BAD?

Pose the above question in Italy, and the answer is "good." Pose the same question in the United States, and the answer is "bad." Why the difference? Can both be right?

Here is what the senior author found in Italy:

Double muscled Piedmontese (Piedmont) cattle mean to the Italian cattle industry what broad-breasted turkeys and Cornish cross broilers mean to the U.S. poultry industry. All are meat producers par excellence.

Italian authorities estimate that 80% of the bulls of the Piedmontese breed of cattle are double muscled to some degree. Moreover, producers select for the trait; and both feeders and slaughterers vie for double muscled bulls. (In Italy, they feed bulls; not steers.) The reason: All of them make more money from double muscled cattle than from normal cattle. Piedmontese cattle are the most popular breed in Italy. The fact that 80% of the Piedmontese bulls are double muscled to some degree indicates that the character responds to selection. Knowledgeable Piedmontese cattle breeders in Italy told the senior author that they expect the following results in their Piedmontese breeding programs (see Fig. 5-31):

1. Phenotypically normal heifers mated to double muscled bulls will produce 80% double muscled bull calves.

2. By culling out the first calf heifers whose bull calves from the above mating were not double muscled, then mating only proved heterozygotes (or carriers) to double muscled bulls, 95% double muscled bull progeny will be produced.

If double muscling is caused by a single recessive gene, as seems to be the case in the British breeds (Angus, Hereford, and Shorthorn), (1) mating phenotypically normal (noncarrier) cows to a homozygous double muscled bull would produce 100% heterozygotes (carriers), none of which would be double muscled, and (2) mating known carrier cows (heterozygotes) to a homozygous double muscled bull would produce 50% homozygous double muscled calves and 50% heterozygous carriers. However, the Italians report breeding results from their Piedmontese breed which suggest that modifier genes common in that breed tend to endow the double muscling gene with dominance (partial dominance). As a result of these modifier genes, the 100% heterozygotes (carriers) referred to could, in the Piedmontese breed, easily be classified as 80% double muscled and 20% normal, thereby explaining Fig. 5-31, alternate breeding program no. 1. Likewise, in the Piedmontese breed, mating known carrier cows to double muscled bulls could result in 95% of the offspring being classed as double muscled (Fig. 5-31, alternate breeding program no. 2), as a result of the action of modifier genes causing a large part of the 50% heterozygotes to be classed as double muscled.

Also, it would seem reasonable to suspect that, if 80% of the bulls of the Piedmontese breed are double muscled to some degree, a large part of the heifers are, also. Hence, many of the heifers classed as *phenotypically normal* in Fig. 5-31, alternate breeding program no. 1, would actually be carriers of the double muscled gene. However, due to the presence or absence of various modifier genes, perhaps they may appear completely normal or only moderately double muscled, with the result that they are classed as "normals."

In Italy, all members of the beef team are making money from double muscled cattle. Here are the facts:

1. When double muscled bull calves are 5 days of age, producers contract them to feeders at 2½ times the going price for calves that are not double muscled, for owner pickup at 1 month to 6 weeks of age.

2. The cattle feeder has a ready market for finished double muscled bulls. Double muscled Piedmontese animals weighing 1,100 lb sell for about $29/cwt more than can be obtained from Charolais cattle, or any other breed, of comparable finish and quality. Good cattle feeders net about $240 more per double muscled animal than for cattle that are not double muscled.

Fig. 5-31. Double muscling predictability in Piedmontese cattle, in Italy.

Alternate breeding programs:

No. 1:	Piedmontese cow phenotypically normal	➡	X (bred to)	➡	Piedmontese bull double muscled homozygous	➡ (will produce)	80% double muscled bull calves
No. 2:	Piedmontese cow a proved heterozygote (or carrier)	➡	X (bred to)	➡	Piedmontese bull double muscled homozygous	➡ (will produce)	95% double muscled bull calves

The Piedmontese breed differs from other breeds in having a very high frequency of double muscled cattle.

Fig. 5-32. Double muscled bulls of the Piedmontese (Piedmont) breed fed in confinement by Ernest Gerbi, Asti, Italy. In Italy, cattle are never steered. During the senior author's visit, Mr. Gerbi sold this lot of bulls to a *country buyer* at $29/cwt more than Charolais cattle were selling at on the same market. Net profit to Mr. Gerbi, U.S. $242/head. Mr. Gerbi feeds about 1,200 double muscled bulls each year. Double muscling—good or bad?

3. The slaughterer and the retailer are happy, too. Double muscled cattle dress 72% (vs 63% for normal). Moreover, with this method of cutting, the retailer gets 80% steaks from double muscled carcasses vs 40% for normal carcasses. Steaks can literally be cut from end to end on a double muscled carcass.

ADVANTAGES OF DOUBLE MUSCLED CATTLE

Most producers, in both Italy and in the United States, are in business to make money—that's as it should be. So, based on the figures already presented, it's obvious that all members of the beef team in Italy are profiting from double muscled cattle. But, they are quick to point out that there are other advantages, too. Here are some of the pluses that they list in favor of double muscled over normal cattle:

1. The calves grow more rapidly up to one year of age.

2. They convert feed more efficiently; it requires fewer pounds of feed to produce 1 lb of beef.

3. They (a) have a higher dressing percentage, and (b) yield a higher proportion of the more desirable cuts—more steaks. In comparison with normal cattle, double muscled cattle have a larger ribeye; produce less brisket, plate, flank; and produce less kidney and pelvic fat.

In summary: Double muscled cattle are superior to normal cattle in (1) rate and efficiency of gain to 1 year of age, and (2) general carcass desirability.

DISADVANTAGES OF DOUBLE MUSCLED CATTLE

It is recognized that there may be breed differences when it comes to the advantages and disadvantages of double muscled cattle. Nevertheless, here are some of the disadvantages to double muscling that have been reported in different countries:

1. The conception rate is lower, due to (a) the infantile reproductive tracts or slow sexual maturity of some animals, and (b) the flat vulva, which makes copulation difficult.

2. The gestation period is about 10 days longer.

3. There is more calving difficulty (caesarian section; pulling calves; stillborn), due to heavier calves at birth (Piedmontese double muscled calves average 108 lb at birth vs 99 lb for normal calves), along with the enlarged rump and round regions.

4. Double muscled calves are more difficult to raise; due to such things as (a) enlarged tongues (macroglossia), and (b) greater susceptibility to disease.

5. Double muscled cows are poor milkers; they produce 30 to 50% less milk than normal cows.

6. Double muscled cattle must be fed a higher proportion of concentrate to roughage, simply because they cannot utilize roughage effectively.

7. There is less marbling.

It's unlikely that all of the above disadvantages will occur in any one herd at any one time. Moreover, the degree to which they occur among breeds, and within breeds, will vary according to the "background" genes or modifying genes. Nevertheless, cattle producers should be apprised of the possibilities.

WHY ARE WE GETTING MORE DOUBLE MUSCLING?

Why are more double muscled cattle cropping up in recent years—in all breeds? The answer: In selecting breeding animals with more bred-in meat type, producers are, unconsciously, selecting more carrier animals; simply because the carriers, or heterozygotes, are the heavier muscled ones. (They're the ones with the big ribeyes.)

Of course, double muscling has been around for a very long time—at least 200 years, in Europe, Australia, and the United States. Also, the condition has been reported in almost every breed. Hence, when selecting the heavier muscled animals for breeding purposes, more and more carriers are being used; and more and more double muscled cattle are showing up, and will continue to show up.

SUMMARY RELATIVE TO DOUBLE MUSCLING

Double muscled cattle—good or bad? Obviously, in Italy it's good—very good. If it were not so, no breed that produced 80% double muscled bull calves could survive. The main reason that double muscling in Italy is so good is that their slaughterers pay a premium of $29/cwt on foot for double muscled cattle at market time. With a 1,100-lb animal, that's a premium of $319/head. The reason for the premium: the reputed 9% higher dressing percentage (72 vs 63%) and twice the steaks (80 vs 40%) of double muscled cattle over normal cattle. Also, the Italians give the impression (without scientific proof) that many of the disadvantages that U.S. cattle producers attribute to double muscling have been minimized—they're less nettlesome—in the Piedmontese breed and to Italian producers. For example, they don't seem to complain too much about breeding or calving problems. Maybe they have overcome this through selection. Obviously, the expression of the trait in many approved heterozygote (carrier) Piedmontese cows is nil (see Fig. 5-31). Moreover, their system of handling early-weaned calves—nurse-cowing (2 to 5 calves/cow) plus a starter ration, until 4 to 6 weeks of age—makes for a good start in life; even if double muscled calves are less vigorous at birth.

Indeed, Italian producers have something going for them. If the senior author were in Italy, he would breed and feed double muscled Piedmontese cattle. In the United States, it's another story. Until, and unless, a premium is paid for double muscled beef over the butcher's block, there's no incentive; there's insufficient reason to risk the disadvantages of double muscled cattle—even if they could be minimized through selection. Yet, there are some aspects of double muscling in cattle that might be used in improving efficiency of beef production. Also, producers may well emulate poultry breeders—the broad-breasted turkey and Cornish chicken counterpart, by producing double muscled cattle (perhaps the heterozygotes) in a well-planned breeding program designed to minimize their production weaknesses and maximize their higher dressing percentage, more red meat, and more steaks; then promote and sell them at a premium. Experimental work conducted by Rollins, *et al.*, of the California Station, indicates this possibility. They found that, in comparison with normal cattle, calves from double muscled X normal parents (calves heterozygous for double muscling) had a 10% advantage in terms of pounds of trimmed retail cuts per day of age at marketing, with no undesirable side effects in either production or performance, and little or no reduction in carcass quality grade at marketing.[14] Broad-breasted turkeys and chickens; double muscled cattle! Why not?

THE RELATIVE IMPORTANCE OF SIRE AND DAM

As a sire can have so many more offspring during a given season or a lifetime than a dam, he is from a hereditary standpoint a more important individual than any one female so far as the whole herd is concerned, although both the sire and the dam are of equal importance so far as concerns any one offspring. Because of their wider use, therefore, sires are usually culled more rigidly than females, and the breeder can well afford to pay more for an outstanding sire than for an equally outstanding female.

Experienced producers have long felt that sires often more closely resemble their daughters than their sons, whereas dams resemble their sons. Some sires and dams, therefore, enjoy a reputation based almost exclusively on the merit of their sons, whereas others owe their prestige to their daughters. Although this situation is likely to be exaggerated, any such phenomenon as may exist is due to sex-linked inheritance which may be explained as follows: the genes that determine sex are carried on one of the chromosomes. The other genes that are located on the same chromosome will be linked or associated with sex and will be transmitted to the next generation in combination with sex. Thus, because of sex linkage, there are more color-blind men than color-blind women. In poultry breeding, the sex-linked factor is used in a practical way for the purpose of distinguishing the pullets from the cockerels early in life, through the process known as "sexing" the chicks. When a black cock is crossed with barred hens, all the cocks come barred and all the hens come black. It should be emphasized, however, that under most conditions it appears that the influence of the sire and dam on any one offspring is about equal. Most breeders, therefore, will do well to seek excellence in both sexes of breeding animals.

PREPOTENCY

Prepotency refers to the ability of the animal, either male or female, to stamp its own characteristics

---

[14]Rollins, W. C., R. B. Thiessen, and Moira Tanaka, "Usefulness of Market Calves Heterozygous for Double Muscling Gene," *California Agriculture*, Vol. 28, No. 3, Mar. 1974, p. 8.

on its offspring. The offspring of a prepotent bull, for example, resemble both their sire and each other more closely than usual. The only conclusive and final test of prepotency consists of the inspection of the get.

From a genetic standpoint, there are two requisites that an animal must possess in order to be prepotent: (1) dominance, and (2) homozygosity. Every offspring that receives a dominant gene or genes will show the effect of that gene or genes in the particular character or characters which result therefrom. Moreover, a perfectly homozygous animal would transmit the same kind of genes to all of its offspring. Although entirely homozygous animals probably never exist, it is realized that a system of inbreeding is the only way to produce animals that are as nearly homozygous as possible.

Popular beliefs to the contrary, there is no evidence that prepotency can be predicted by the appearance of an animal. To be more specific, there is no reason why a vigorous, masculine-appearing sire will be any more prepotent than one less desirable in these respects.

It should also be emphasized that it is impossible to determine just how important prepotency may be in animal breeding, although many sires of the past have enjoyed a reputation of being extremely prepotent. Perhaps these animals were prepotent, but there is also the possibility that their reputation for producing outstanding animals may have rested upon the fact that they were mated to some of the best females of the breed.

In summary, it may be said that if a given sire or dam possesses a great number of genes that are completely dominant for desirable type and performance and if the animal is relatively homozygous, the offspring will closely resemble the parent and resemble each other, or be uniform. Fortunate, indeed, is the breeder who possesses such an animal.

NICKING

If the offspring of certain matings are especially outstanding and in general better than their parents, breeders are prone to say that the animals "nicked" well. For example, a cow may produce outstanding calves to the service of a certain bull, but when mated to another bull of apparent equal merit as a sire, the offspring may be disappointing. Or sometimes the mating of a rather average bull to an equally average cow will result in the production of a most outstanding individual both from the standpoint of type and performance.

So-called successful nicking is due, genetically speaking, to the fact that the right combination of genes for good characters is contributed by each parent, although each of the parents within itself may be lacking in certain genes necessary for excellence. In other words, the animals nicked well because their respective combinations of good genes were such as to complement each other.

The history of animal breeding includes records of several supposedly favorable nicks, one of the most famous of which was the favorable result secured from crossing sons of Anxiety 4th with daughters of North Pole in the Gudgell and Simpson herd of Hereford cattle. At this late date, it is impossible to determine whether these Anxiety 4th X North Pole matings were successful because of nicking or whether the good results should be more rightfully attributed to the fact that the sons of Anxiety 4th were great breeding bulls and that they merely happened to be mated, for the most part, with daughters of North Pole because the available females in the Gudgell and Simpson herd were of this particular breeding.

Fig. 5-33. Anxiety 4th 9904, whose sons were alleged to "nick" exceedingly well on the daughters of North Pole in the Gudgell and Simpson herd of Hereford cattle. (Courtesy, *The American Hereford Journal*)

Because of the very nature of successful nicks, outstanding animals arising therefrom must be carefully scrutinized from a breeding standpoint, because, with their heterozygous origin, it is quite unlikely that they will breed true.

FAMILY NAMES

In cattle, depending upon the breed, family names are traced through either the males or females. In Angus and Shorthorn cattle, the family names had their origin with certain great foundation females, whereas in Herefords the family names trace through the sires.

Unfortunately, the value of family names is generally grossly exaggerated. Obviously, if the foundation male or female, as the case may be, is very many generations removed, the genetic superiority of this head of a family is halved so many times by subsequent matings that there is little reason to think that one family is superior to another. For example, if a present-day Queen Mother (an old and well-known Angus family) is 18 generations removed from the founder, she would carry the following relationship to the head of the family: $(\frac{1}{2})^{18}$ or $\frac{1}{262{,}144}$ or 0.0004%. Obviously, this Queen Mother may not have inherited a single gene from the foundation cow, and merely being a Queen Mother does not differentiate her much from other families which make up the breed.

The situation relative to family names is often further distorted by breeders placing a premium on family names of which there are few members, little realizing that, in at least some cases, there may be unfortunate reasons for the scarcity in numbers.

Such family names have about as much significance as human family names. Who would be so foolish as to think that the Joneses as a group are alike and different from the Smiths? Perhaps, if the truth were known, there have been many individuals with each of these family names who have been of no particular credit to the clan, and the same applies to all other family names.

Family names lend themselves readily to speculation. Because of this, the history of livestock breeding has often been blighted by instances of unwise pedigree selection on the basis of not too meaningful family names. The most classical example of a situation of this type occurred with the Duchess family of Shorthorn cattle, founded by the noted pioneer English Shorthorn breeder, Thomas Bates. Bates, and more especially those later breeders who emulated him, followed preferences in bloodlines within increasingly narrow limits, until ultimately they were breeding cattle solely according to fashionable pedigrees, using good, bad, and indifferent animals. Fad and fancy in pedigrees dominated the situation, and the fundamental importance of good individuality as the basis of selecting animals for breeding purposes was for the time largely ignored. The sole desire of these breeders was to concentrate the Duchess blood. The climax of the Duchess boom (or Bates boom) came in September 1873, when the New York Mills herd was sold at auction with English and American breeders competing for the offering. At this memorable event, 109 head of Duchess-bred cattle averaged $3,504 per head, with the 7-year-old 8th Duchess of Geneva selling at the world's record price of $40,600.

As with most booms, the New York Mills sale was followed by a rather critical reaction, and eventually the bottom dropped out of values. Even more tragic, the owners of Duchess Shorthorns suddenly came to a realization that indiscriminate inbreeding and a lack of selection had put the family name in disrepute. As a result, the strain became virtually extinct a few years later.

On the other hand, certain linebred families—linebred to a foundation sire or dam so that the family is kept highly related to it—do have genetic significance. Moreover, if the programs involved have been accompanied by rigid culling, many good individuals may have evolved, and the family name may be in good repute. The Anxiety 4th family of Hereford cattle is probably the best known family of this kind in meat animals. Even so, there is real danger in assuming that an *airtight* or *straightbred* Anxiety 4th pedigree is within itself meritorious and that this family is superior to that of any other family in Hereford cattle.

HEREDITY AND ENVIRONMENT

A massive purebred bull, standing belly deep in straw and with a manger full of feed before him, is undeniably the result of two forces—heredity and environment (with the latter including training). If turned out on the range, an identical twin to the placid bull would present an entirely different appearance. By the same token, optimum environment could never make a champion out of a bull with scrub ancestry, but it might well be added that "fat and hair will cover up a multitude of sins."

Artificial treatments sometimes conceal undesirable traits to the extent that improvements in heredity are impeded. For example, the practice of cutting ties on cattle may be of immediate benefit in improving the appearance of treated animals, but the trait continues to be transmitted. Likewise, the use of nurse cows for developing young show animals in beef herds tends to favor the perpetuation of the genes for poor milkers.

Fig. 5-34. Galloway females on Greenough Ranch winter range in Montana, at 3,800 to 5,500 ft elevation. With its thick, wavy hair and hardiness, this breed is well adapted to cold weather. (Courtesy, Robert G. Mullendore, Greenough Cattle Company, Greenough, MT)

These are extreme examples, but they do emphasize the fact that any particular animal is the product of heredity and environment. Stated differently, heredity may be thought of as the foundation, and environment as the structure. Heredity has already made its contribution at the time of fertilization, but environment works ceaselessly away until death.

Experimental work has long shown conclusively enough that the vigor and size of animals at birth is dependent upon the environment of the embryo from the minute the ovum or egg is fertilized by the sperm, and now we have evidence to indicate that newborn animals are affected by the environment of the egg and sperm long before fertilization has been accomplished. In other words, perhaps due to storage of nutritive factors, the kind and quality of the ration fed to young, growing females may later affect the quality of their progeny. Generally speaking, then, environment may inhibit the full expression of potentialities from a time preceding fertilization until physiological maturity has been attained.

Admittedly, after looking over an animal or studying its production record, a breeder cannot with certainty know whether it is genetically a high or a low producer. There can be no denying the fact that environment—including feeding, management, and disease—plays a tremendous part in determining the extent to which hereditary differences that are present will be expressed in animals. In general, however, the results of a longtime experiment conducted at Washington State University support the contention that selection of breeding animals should be carried on under the same environmental conditions as those under which commercial animals are produced.[15]

Within the pure breeds of livestock—managed under average or better than average conditions—it has been found that, in general, only 30 to 45% of the observed variation in a characteristic is actually brought about by hereditary variations. (See Table 5-4.) To be sure, if we contrast animals that differ very greatly in heredity—for example, a champion bull and a scrub—90% or more of the apparent differences in type may be due to heredity. The point is, however, that extreme cases such as the one just mentioned are not involved in the advancement within improved breeds of livestock. Here the comparisons are between animals of average or better than average quality, and the observed differences are often very minor.

The problem of the progressive breeder is that of selecting the very best animals available genetically—these to be parents of the next generation of offspring in the herd. The fact that only 30 to 45% of the observed variation is due to differences in inheritance and that environmental differences can produce misleading variations makes mistakes in the selection of breeding animals inevitable. However, if the purebred breeder has clearly in mind a well-defined ideal and adheres rigidly to it in selecting breeding stock, very definite progress can be made, especially if mild inbreeding is judiciously used as a tool through which to fix the hereditary material.

[15]Fowler, Stewart H., and M. E. Ensminger, Wash. Ag. Exp. Sta. Bull. 34, Jan. 1961.

HYBRID VIGOR OR HETEROSIS

Hybrid vigor or heterosis is a name given to the biological phenomenon which causes crossbreds to outproduce the average of their parents. For numerous traits, the performance of the cross is superior to the average of the parental breeds. This phenomenon has been well known for years, and has been used in many breeding programs. The production of hybrid seed corn by developing inbred lines and then crossing them is probably the most important attempt by agriculturalists to take advantage of hybrid vigor. Also, heterosis is being used extensively in commercial swine, sheep, layer, and broiler production today; an estimated 80% of market hogs, market lambs, and layers are crossbreds, and 95% of broilers are crosses.

The genetic explanation for the hybrid's extra vigor is basically the same, whether it be cattle, hogs, sheep, layers, broilers, hybrid corn, hybrid sorghum, or whatnot. Heterosis is produced by the fact that the dominant gene of a parent is usually more favorable than its recessive partner. When the genetic groups differ in the frequency of genes they have and dominance exists, then heterosis will be produced.

Heterosis is measured by the amount the crossbred offspring exceeds the average of the two parent breeds or inbred lines for a particular trait, using the following formula for any one trait:

Fig. 5-35. First cross Texon heifer (Texas Longhorn X Devon) at Wild Plum Ranch, Duncan, OK. (Photo by Dr. Robert M. Simpson, M.D., owner)

$$\frac{\text{Crossbred average} - \text{Purebred average}}{\text{Purebred average}} \times 100 = \text{Percent hybrid vigor}$$

Thus, if the average of the two parent populations for weaning weight of calves at 205 days of age is 400 lb and the average of their crossbred offspring is 420 lb, application of the above formula shows that the amount of heterosis is 20 lb, or 5%.

Traits high in heritability—like tenderness of rib eye—respond consistently to selection, but show little response in hybrid vigor. Traits low in heritability—like mothering ability, calving interval, and conception rate—usually show good response in hybrid vigor.

The level of hybrid vigor for all traits depends on the breeds crossed. The greater the genetic difference between two breeds, the greater the hybrid vigor expected. The genetic difference between a British breed and a breed of Indian origin is greater than the difference between one British breed and another British breed.

It is most important to have hybrid vigor in the cow herd where it results in increased fertility, survivability of the calves, milk production, growth rate of calves, and longevity of the cow—all factors that mean more profit to the producer.

It is noteworthy that purebreds must be constantly tapped to renew the vigor of crossbreds; otherwise, the vigor is dissipated.

COMPLEMENTARY COMBINATIONS

Complementary combinations refer to the advantage of a cross over another cross or a purebred resulting from the manner in which two or more characters combine or complement each other. They are a matching of breeds so that they compensate each other; the objective being to get the desirable traits of each. Thus, in a crossbreeding program, breeds that complement each other should be selected, thereby maximizing the desirable traits and minimizing the undesirable traits. Since breeds which are selected because they tend to express a maximum of some trait (*e.g.*, high daily gain) will have some undesirable traits (*e.g.*, large mature cow size and high maintenance cost), different breeds must be selected for different purposes. A well-known example of breed complementation for improving overall carcass desirability in the market animal is the Angus X Charolais cross, combining the higher carcass grade of the Angus with the higher cutability of the Charolais.

TRAITS OF BEEF CATTLE; THEIR (1) ECONOMIC VALUE, (2) HERITABILITY, AND (3) HETEROSIS

In Table 5-4, the economically important traits of beef cattle are grouped into three broad classes—reproduction, production, and product; and the heritability and heterosis of each of these groups is given.

TABLE 5-4
ECONOMIC VALUE, SELECTION VARIATION, AND HETEROSIS INCREASE OF THREE CLASSES OF TRAITS OF BEEF CATTLE[1]

Class of Traits	Relative Economic Value	Selection Variation	Heterosis Increases
		(%)	(%)
Reproduction (calf crop, etc.)	20	10	10
Production (avg. daily gain, etc.)	2	40	5
Product (retail meat yield, etc.)	1	50	0

[1] *Beef Cattle Science Handbook*, Vol. 10, 1973, p. 194. From a paper on "Beef Breeding Programs," by Richard L. Willham, Professor of Animal Science, Iowa State University, Ames, Iowa.

BREEDING PROGRAMS

A breeding program is a complete system of management designed to bring about genetic change in a group of animals. Modern cattle breeding programs have the dual objectives of (1) breeding better beef cattle, and (2) better beef cattle breeding. Two types of breeding programs exist—commercial and purebred. Generally speaking, commercial cattle producers crossbreed—they mate animals of different breeds—although a gradually decreasing number still have straightbreds (males and females of the same genetic background, or breed). The seed stock producer is almost always a purebred breeder (straightbreds). All of them are more specialized and precise than formerly.

Most commercial breeding programs utilize crossbreeding because of hybrid vigor—which results in better phenotypic performance than straight breeding. But they rely on purebred breeders for seed stock (especially bulls), because the only way to improve crossbreds above their present level of performance is through the development of better straightbreds. Commercial producers usually market their end product (calves) on a per pound basis.

Purebred breeding programs are those which produce breeding stock. The cattle are merchandised on their breeding value rather than their actual performance. How the animals perform is not at issue; only as performance predicts breeding value at the commercial level is it of importance. This type of breeding program is restrictive in the kind of selection that can be practiced once the genetic groups have been chosen. Only traits having moderate to high heritability (production and product) can be improved appreciably.

The general form of beef breeding programs, both commercial and purebred, involves the production and measurement of a calf crop followed by the selection of the parents for the next generation. These matters are detailed in the section headed "Production Testing Beef Cattle", which follows later in this chapter.

In the future, most breeds will be either (1) propagated as straightbreds or composites, or (2) used for crossbreeding. The vast majority of the straightbreds will be purebreds and production tested. Fewer and fewer grade herds will be maintained for commercial purposes, simply because the economics favor crossbreeding accompanied by complementary genes and heterosis. The straightbreds will be used primarily to produce (1) bulls and semen, and (2) purebred females for F_1 heifer programs. The crossbreds will be used to produce beef according to specification.

Since purebred breeding programs have as their function the supplying of breeding stock for commercial programs, the latter type will be presented first in order that its needs will be reflected in the purebred programs. However, both types are interdependent.

COMMERCIAL BREEDING PROGRAMS

The goal of most commercial beef operations is to convert the production of the land—grass and/or crops—into dollars through the traditional cow-calf operation. Usually, the product is marketed at the weaning stage, although some commercial producers carry them to the yearling stage, or even finish them out for market. In any event, the goal of increased net returns must involve a critical analysis of current and future economics and the integration of this with the production potential available from the land. These are business and management aspects, rather than genetic; nevertheless, they must be analyzed and decisions must be reached. The level of genetic potential should be compatible with the resources available. Table 5-4 groups the classes of traits—reproduction, production, and product—and gives their relative economic values as they involve the goal of increased net returns. Improvement in reproductive output—a higher percent calf crop, born alive—is the key to increased net returns in most commercial cattle operations. Next in economic value comes improvement in pounds of beef produced per cow. Last is improvement in yield

Fig. 5-36. Commercial Charolais on a Brown County, Texas, ranch. (Courtesy, *Livestock Weekly*, San Angelo, TX)

in retail beef of acceptable quality. In the long run, the latter may be more important economically than indicated by Table 5-4, based on what people are willing to pay for, simply because the future of beef production is only partially dependent upon economical protein production; it is more dependent upon quality of the product and the utilization of roughage by the ruminant.

Choosing and keeping performance records in a commercial cattle operation is not easy. Yet, records are necessary. Specification of the product offered for sale (calves, in the case of the commercial cattle producer) is becoming the rule. Thus, complete records, adequately analyzed and utilized, will help any commercial operation. But records cost money. A large commercial cow-calf operation (of 300 cows or more) may not be able to justify more than a simple feeder calf program involving the sampling of the product that it is offering for sale. On an every-other-year basis, this might involve a random sampling of calves which are fed out and slaughtered. The gain and carcass data are then used by the producer (1) in the development of performance reputation (the production of reputation feeder cattle), and (2) in the selection of herd sires to improve performance. Small commercial operations (with 50 to 300 cows) can well afford to keep more detailed records on their cowherds since this is a means by which they can more effectively compete. Their produce of dam records can be an aid in developing a high producing cow herd, and in adjusting the management to optimize production. Most state Beef Cattle Improvement Associations have such programs available at nominal cost. The use of such programs, even with multi-sire pastures, allows producers at least to evaluate groups of sires purchased. In this manner, they can study the sources of breeding stock supply and be more critical in their future selection.

Germ plasm choice is becoming increasingly difficult because of the large number of breeds and breed cross combinations now available. With only 3 main breeds (as was once the case with the Shorthorn, Hereford, and Angus), there are 3 single-cross combinations and 3 three-way cross combinations from which to select. However, with 10 breeds there are 45 single-cross combinations and 360 possible three-way combinations from which to select. Think for a moment of the combinations involved with the 83 breeds listed in Chapter 3. Remember, too, that more breeds will come. Thus, commercial producers need critically to evaluate the best information available in order to determine which breeds and breed crosses can be used in their particular programs. Much money and effort is being, and will continue to be, put into promotion of breeds and their cross combinations. This promotional information must be carefully considered and evaluated by the commercial operator, especially if the producer has a longstanding good herd of a traditional breed.

PUREBRED BREEDING PROGRAMS

Purebred breeding programs are the key to continued genetic improvement of cattle. Thus, seed stock producers should establish and follow a modern, sophisticated program. Also, commercial producers should be aware of exactly what constitutes a sound breeding program at the purebred level; otherwise, their choices among herds will likely be both limited and unwise.

The goal of the purebred producer is to provide stock with superior breeding values. Obviously, such a goal involves being paid according to the superiority of the breeding animals being offered for sale, along with the cost involved in promoting these animals. The goal of the purebred breeder must involve the improvement of the production and product traits (Table 5-4), along with at least maintaining the reproductive performance. Today, the emphasis is on improving production, especially yearling weight and efficiency of feed utilization, and carcass merit.

Fig. 5-37. A purebred Braunvieh herd on arid New Mexico range. (Courtesy, Wm. J. Waldrip, Spade Ranches, Lubbock, TX)

In the past, breeding stock production could be accomplished in smaller units than commercial production. During the first half of the present century, it was not uncommon to find that many of the top purebred herds, in Europe as well as in America, only had 20 to 40 cows. Today, in order to take advantage of numbers in the development of selection schemes, purebred breeding is becoming centralized into larger units, including corporate structures. Since performance is fast becoming the rule in evaluating breeding stock, the use of animal breeding technology will be a necessary requisite to remaining in the purebred business. Fortunately, many of the breed registry associations are presently conducting cooperative sire evaluation programs within the breeds, thereby permitting the 100- to 200-head purebred breeder to participate in such a program and effectively compete with larger breeding units.

The various performance programs of the breeds and other organizations which are a part of the Beef Improvement Federation dictate a rather standard timing of records and a specification of the test. The important things in breeding stock production are that (1) they be as uniform from one calf crop to the next as is possible so that the complete set of records developed can be used in making selections, and (2) they conform as nearly as possible to the conditions under which the sale animals are to be used. This means that artificial conditions, other than those necessary to produce uniformity from year to year, should be avoided.

The recordkeeping system and the tests conducted should be as complete as possible. Current breed registry association programs, state programs, and national programs are relatively complete in terms of providing a means of recording important measures from breeding to slaughter. However, progressive breeders will measure additional traits that need improvement in herds. Also, they will recognize that just keeping records is of little importance if they are not in a form in which they can be used in a creative selection program—and if they are not used.

Today, following the importation of numerous new European breeds, a wide choice in breeds is possible. Economics, timing, and promotion must be seriously considered before embarking on a program to breed one of the newly introduced breeds. Grading up to a new breed by the use of sires only is a long and costly process. It is rapid at first, but it becomes slower with each succeeding generation since purity is halved each time. Because of the long generation interval of cattle, the development of a seed stock herd through grading up will take approximately 20 years. Coupled with this, the heterosis produced in the initial single cross will be halved with each succeeding generation. thereby making it difficult to maintain the initial level of performance through selection alone. Such grading up to a purebred program will require perseverance, time, and money; and the reward may or may not be commensurate with it.

The choice of the selection program in the purebred herd is of great importance, since it determines the success of the program in producing genetic improvement. With heritability of 40 to 50% for production and product traits, selection based on own performance, coupled with sequential progeny testing among the selected individuals, provides near maximum selection advance. Using additional information on relatives, such as sibs, increases the accuracy of selection somewhat, provided the generation interval is not lengthened. Progeny testing is worthwhile only when used in combination with own performance selection. The accuracy of progeny testing can more than compensate for the extra two years added to the generation interval of the bulls. The importance of thoroughly testing bulls is pointed up by the fact that over 90% of the selection advance will come from sire selection, with the remainder coming from heifer selection. This results from the sex ratio being 50-50 while only 1 male to 20 females is required even under natural service. Of course, under AI, the ratio can be 1 to 1,000 or greater.

Cooperative sire evaluation programs can be of distinct benefit to the small purebred breeder—the producer with 100 to 300 cows. Such a program might consist of performance testing the bull crop, then selecting from these the top 1 or 2 bulls to go into a reference sire progeny test conducted by the breed.

If possible, the heifers in a purebred herd should also be tested—especially those that may become the dams of herd bulls. Such a program could make the difference between ordinary and superior. Testing heifer replacement for production traits (by saving more for breeding than is necessary) would enhance a breeding program. Also, a young cow herd—the product of the last superior sires—will increase the selection response per year.

In conclusion, it may be said that performance will be the rule in evaluating breeding stock of the future, and the use of animal breeding technology will be a necessary requisite to remaining in the business. The production and selling of breeding stock in the beef industry will become a specification business. But promotion will still be necessary. Central bull tests, correctly conducted, provide an excellent means by which to compete and promote. Also, breed sire evaluation programs will be a good promotional tool if the progeny of bulls compete with equal opportunity. Finally it may be said that testing programs, whatever the kind, are no better than the accuracy of the records kept, the use made of the records in a selection program, and the honesty and integrity of the people back of the test.

SYSTEMS OF BREEDING

The many diverse types and breeds among each class of farm animals in existence today originated from only a few wild types within each species. These early domesticated animals possessed the pool of genes, which, through controlled matings and selection, proved flexible in the hands of breeders. In cattle, for example, through various systems of breeding, there evolved animals especially adapted to draft purposes, beef production, milk production, and dual-purpose needs.

Successful breeders follow a breeding system, the purpose of which is to give greater control of heredity than if selection alone is used. Thus, breeders need to know about the different breeding systems; what they can do well, and what they can do poorly or not at all.

Perhaps at the outset it should be stated that there is no one best system of breeding or secret of success for any and all conditions. Each breeding program is an individual case, requiring careful study. The choice of the system of breeding should be determined primarily by size and quality of herd, finances and skill of the operator, and the ultimate goal ahead.

PUREBREEDING

A purebred animal may be defined as a member of a breed, the animals of which possess a common ancestry and distinctive characteristics; and it is either registered or eligible for registry in the herd book of that breed.

It must be emphasized that pure breeding and homozygosity may bear very different connotations. The term *purebred* refers to animals whose entire lineage, regardless of the number of generations removed, traces back to the foundation animals accepted by the breed or to any animals which have been subsequently approved for infusion. On the other hand, *homozygosity* refers to the likeness of the genes.

Yet there is some interrelationship between purebreds and homozygosity. Because most breeds had a relatively small number of foundation animals, the unavoidable inbreeding and linebreeding during the formative stage resulted in a certain amount of homozygosity. Moreover, through the normal sequence of events, it is estimated that purebreds become more homozygous by from ¼ to ½% per animal generation. It should be emphasized that the word *purebred* does not necessarily guarantee superior type or high productivity. That is to say, the word *purebred* is not, within itself, magic, nor is it sacred. Many producers have found to their sorrow that there are such things as *purebred scrubs*. Yet, on the average, purebred animals are superior to nonpurebreds.

For the person with experience and adequate capital, the breeding of purebreds may offer unlimited opportunities. It has been well said that honor, fame, and fortune are all within the realm of possible realization of the purebred breeder; but it should also be added that only a few achieve this high calling.

Purebred breeding is a highly specialized type of production. Generally speaking, only the experienced breeder should undertake the production of purebreds with the intention of furnishing foundation or replacement stock to other purebred breeders or purebred bulls to the producer of grades. Although we have had many constructive cattle breeders and great progress has been made, it must be remembered that only a few achieve sufficient success to classify as master breeders.

INBREEDING

Most scientists divide inbreeding into various categories, according to the closeness of the relationship of the animals mated and the purpose of the matings. There is considerable disagreement, however, as to both the terms used and the meanings that it is intended they should convey. For purposes of this book and the discussion which follows, the following definitions will be used.

Inbreeding *is the mating of animals more closely related than the average of the population from which they came.*

Closebreeding *is the mating of closely related animals; such as sire to daughter, son to dam, and brother to sister.*

Linebreeding *is the mating of animals more distantly related than in closebreeding, and in which the matings are usually directed toward keeping the offspring closely related to some highly admired ancestor; such as half-brother and half-sister, female to grandsire, and cousins.*

CLOSEBREEDING

Closebreeding is rarely practiced among present-day cattle producers, although it was common in the foundation animals of most of the breeds. For example, it is interesting to note that Comet (155), an illustrious sire and noted as the first Shorthorn to sell for $5,000, came from the mating of Favorite and Young Phoenix, a heifer that had been produced from the union of Favorite with his own dam. Such was the program of the Collings Brothers and many other early-day beef cattle breeders, including those in all breeds.

Closebreeding is that system of breeding in which closely related animals are mated. In it there is a minimum number of different ancestors. In the repeated mating of a brother with his full sister, for example, there are only 2 grandparents instead of 4, only 2 great grandparents instead of 8, and only 2 different ancestors in each generation farther back—instead of the theoretically possible 16, 32, 64, 128, etc. The most intensive form of in-breeding is self-fertilization. It occurs in some plants, such as wheat and garden peas, and in some of the lower animals; but domestic animals are not self-fertilized.

The reasons for practicing closebreeding are:

1. It increases the degree of homozygosity within animals, making the resulting offspring pure or homozygous in a larger proportion of their gene pairs than in the case of linebred or outcross animals. In so doing, the less desirable recessive genes are brought to light so that they can be more readily culled. Thus, closebreeding together with rigid culling, affords the surest and quickest method of fixing and perpetuating a desirable character or group of characters.
2. If carried on for a period of time, it tends to create lines or strains of animals that are uniform in type and other characteristics.
3. It keeps the relationship to a desirable ancestor highest.
4. Because of the greater homozygosity, it makes for greater prepotency. That is, selected closebred animals are more homozygous for desirable genes (genes which are often dominant), and they, therefore, transmit these genes with greater uniformity.
5. Through the production of inbred lines or families by closebreeding and the subsequent crossing of certain of these lines, it affords a modern approach to livestock improvement. Moreover, the best of the inbred animals are likely to give superior results in outcrosses.
6. Where a breeder is in the unique position of having a herd so far advanced that to go on the outside for seed stock would merely be a step backward, it offers the only sound alternative for maintaining existing quality or making further improvement.

The precautions in closebreeding may be summarized as follows:

1. As closebreeding greatly enhances the chances that recessives will appear during the early generations in obtaining homozygosity, it is almost certain to increase the proportion of worthless breeding stock produced. This may include such things as reduction in size, fertility, and general vigor. Also, lethals and other genetic abnormalities often appear with increased frequency in closebred animals.

2. Because of the rigid culling necessary to avoid the *fixing* of undesirable characters, especially in the first generations of a closebreeding program, it is almost imperative that this system of breeding be confined to a relatively large herd and to instances when the owner has sufficient finances to stand the rigid culling that must accompany such a program.

3. It requires skill in making planned matings and rigid selection, thus being most successful when applied by *master breeders*.

4. It is not adapted for use by the breeder with average or below average stock because the very fact that the animals are average means that a goodly share of undesirable genes are present. Closebreeding would merely make the animals more homozygous for undesirable genes, and, therefore, worse.

Judging from outward manifestations alone, it might appear that closebreeding is predominantly harmful in its effects—often leading to the production of defective animals lacking in the vitality necessary for successful and profitable production. But this is by no means the whole story. Although closebreeding often leads to the production of animals of low value, the resulting superior animals can confidently be expected to be homozygous for a greater than average number of good genes and thus more valuable for breeding purposes. Figuratively speaking, therefore, closebreeding may be referred to as *trial by fire*, and the breeder who practices it can expect to obtain many animals that fail to measure up and have to be culled. On the other hand, if closebreeding is properly handled, the breeder can also expect to secure animals of exceptional value.

Although closebreeding has been practiced less during the past century than in the formative period of the different pure breeds of livestock, it has real merit when its principles and limitations are fully understood. Perhaps closebreeding had best be confined to use by the skilled master breeder who is in a sufficiently sound financial position to endure rigid and intelligent culling and delayed returns and whose herd is both large and above average in quality.

LINEBREEDING

From a biological standpoint, closebreeding and linebreeding are the same thing, differing merely in intensity. In general, closebreeding has been frowned upon by producers, but linebreeding (the less intensive form) has been looked upon with favor in many quarters.

In a linebreeding program, the degree of relationship is not closer than half-brother and half-sister or matings more distantly related; cousin matings, grandparent to grand offspring, etc.

Linebreeding is usually practiced in order to conserve and perpetuate the good traits of a certain outstanding sire or dam. Because such descendants are of similar lineage, they have the same general type of germ plasm and therefore exhibit a high degree of uniformity in type and performance. During the past 5 decades, for example, a great many Hereford herds have been linebred to Prince Domino, that immortal Gudgell and Simpson bred bull who, in the hands of Otto Fulscher and the Wyoming Hereford Ranch, contributed so much to the improvement of the Hereford breed.

In a more limited way, a linebreeding program has the same advantages and disadvantages of a closebreeding program. Stated differently, linebreeding offers fewer possibilities both for good and harm than closebreeding. It is a more conservative and safer type of program, offering less probability either to *hit the jackpot* or *sink the ship*. It is a middle-of-the-road program that the vast majority of average and small breeders can safely follow to their advantage. Through it, reasonable progress can be made without taking any great risk. A degree of homozygosity of certain desirable genes can be secured without running too great a risk of intensifying undesirable ones.

Fig. 5-38. Prince Domino 499611, calved September 13, 1914, and died April 4, 1930. Many great Herefords have been produced from linebreeding to this immortal sire. Prince Domino's final resting place at Wyoming Hereford Ranch, Cheyenne, Wyoming, is marked with the following epitaph: "He lived and died and won a lasting name." This is a rare tribute, indeed, to any beast—or human being. (Courtesy, *The American Hereford Journal*)

Usually a linebreeding program is best accomplished through breeding to an outstanding sire rather than to an outstanding dam because of the greater number of offspring of the former. If a breeder is in possession of a great bull—proved great by the performance records of a large number of his get—a linebreeding program might be initiated in the following way: select two of the best sons of the noted bull and mate them to their half-sisters, balancing all possible defects in the subsequent matings. The next generation matings might well consist of breeding the daughters of one of the bulls to the son of the other, etc. If, in such a program, it seems wise to secure some outside blood (genes) to correct a common defect or defects in the herd, this may be done through selecting one or more outstanding proved cows from the outside—animals whose get are strong where the herd may be deficient—and then mating this female(s) to one of the linebred bulls with the hope of producing a son that may be used in the herd.

The owner of a small purebred herd with limited numbers can often follow a linebreeding program by buying all of the sires from a large breeder who follows such a program—thus in effect following the linebreeding program of the larger breeder.

Naturally, a linebreeding program may be achieved in other ways. Regardless of the actual matings used, the main objective in such a system of breeding is that of rendering the animals homozygous—in desired type and performance—to some great and highly regarded ancestor, while at the same time weeding out homozygous undesirable characteristics. The success of the program, therefore, is dependent upon having desirable genes with which to start and an intelligent intensification of these good genes.

It should be emphasized that there are some types of herds in which one should almost never closebreed or linebreed. These include grade or commercial herds and purebred herds of only average quality.

The owners of grade or commercial herds run the risk of undesirable results, and, even if successful, as commercial breeders, they cannot sell their stock at increased prices for breeding purposes.

With purebred herds of only average quality, more rapid progress can usually be made by introducing superior outcross sires. Moreover, if the animals are of only average quality they must have a preponderance of "bad" genes that would only be intensified through a closebreeding or linebreeding program.

OUTCROSSING

Outcrossing is the mating of animals that are members of the same breed but which show no relationship close up in the pedigree (for at least the first 4 to 6 generations).

Most of our purebred animals of all classes of livestock are the result of outcrossing. It is a relatively safe system of breeding, for it is unlikely that two such unrelated animals will carry the same "undesirable" genes and pass them on to their offspring.

Perhaps it might well be added that the majority of purebred breeders with average or below average herds had best follow an outcrossing program, because, in such herds, the problem is that of retaining a heterozygous type of germ plasm with the hope that genes for undesirable characters will be counteracted by genes for desirable characters. With such average or below average herds, an inbreeding program would merely make the animals homozygous for the less desirable characters, the presence of which already makes for their mediocrity. In general, continued outcrossing offers neither the hope for improvement nor the hazard of retrogression of linebreeding or closebreeding programs.

Judicious and occasional outcrossing may well be an integral part of linebreeding or closebreeding programs. As closely inbred animals become increasingly homozygous with germ plasm for good characters, they may likewise become homozygous for certain undesirable characters even though their general overall type and performance remains well above the breed average. Such defects may best be remedied by introducing an outcross through an animal or animals known to be especially strong in the character or characters needing strengthening. This having been accomplished, the wise breeder will return to the original closebreeding or linebreeding program, realizing full well the limitations of an outcrossing program.

GRADING UP

Grading up is that system of breeding in which purebred sires of a given pure breed are mated to native or grade females. Its purpose is to develop uniformity and quality and to increase performance in the offspring.

Many breeders will continue to produce purebred stock. However, the vast majority of animals in the United states—probably more than 97%—are not eligible for registry. In general, however, because of the obvious merit of using well-bred sires, farm animals are sired by purebreds. In comparison with the breeding of purebreds, such a system requires less outlay of cash, and less experience on the part of the producer.

Naturally, the greatest single step toward improved quality and performance occurs in the first cross. The first generation of such a program results in offspring carrying 50% of the hereditary material of the purebred parent (or 50% of the *blood* of the purebred parent, as many breeders speak of it). The next generation gives offspring carrying 75% of the *blood* of the purebred breed, and in subsequent generations the proportion of inheritance remaining from the original scrub females is halved with each cross. Later crosses usually increase quality and performance still more, though in less marked degree. After the third or fourth cross, the offspring compare very favorably with purebred stock in conformation, and only exceptionally good sires can bring about further improvement. This is especially so if the males used in grading up successive generations are derived from the same strain within a breed. High grade animals that are the offspring of several generations of outstanding purebred sires can be and often are superior to average or inferior purebreds.

CROSSBREEDING

Crossbreeding is the mating of animals of different breeds. In a broad sense, crossbreeding also includes the mating of purebred sires of one breed with high grade females of another breed.

Today, there is great interest in crossbreeding cattle, and increased research is under way on the subject. Crossbreeding is being used by producers to (1) increase productivity over straightbreds, because of the resulting hybrid vigor or heterosis, just as is being done by commercial corn and poultry producers; (2) produce commercial cattle with a desired combination of traits not available in any one breed; and (3) produce composite or new breeds.

The motivating forces back of increased crossbreeding in cattle are: (1) more artificial insemination in beef cattle, thereby simplifying the rotation of bulls of different breeds; and (2) the necessity for producers to become more efficient in order to meet their competition, from both inside and outside the cattle industry.

Crossbreeding will play an increasing role in the production of market cattle in the future, because it offers the following advantages:

1. **Hybrid vigor in reproduction, survival, early growth rate, and maternal ability.** Crossbreeding results in hybrid vigor because the desirable genes from both breeds are combined and the undesirable genes from each tend to be overshadowed as recessives. That is to say, there has been an inevitable, though small, amount of inbreeding in all purebreds during the period of the last 100 to 150 years. This has been partly intentional and partly due to geographical limitations upon the free exchange of breeding stock from one part of the country to another. As a result of this slight degree of inbreeding, there has been a slow but rather constant increase in homozygosis within each of the pure breeds of livestock. Most of the factors fixed in the homozygous state are desirable; but inevitably some undesirable genes have probably been fixed, resulting in lowered vigor, slower growth rate, decreased livability, etc.

Theoretically, then, crossbreeding should be an aid in relegating these undesirable genes to a recessive position and in allowing more dominant genes to express themselves. Practical observation and experiments indicate that this does occur in crossbreeding.

An example of choosing a breed to maximize heterosis among the British breeds is found in the Hereford, Angus, and Shorthorn crossbred data. Since the Shorthorn resembles the Angus more than the Hereford in several traits, the Hereford X Angus and the Hereford X Shorthorn calves have excelled Shorthorn X Angus calves in growth traits.

2. **Opportunity to introduce new desired genes and incorporate them in the cow herd and market animal at a faster rate than by selection within a breed.** An example of introducing new genes at a fast rate for milk production would be crossing a dairy breed with a beef breed, then selecting females from within the crossbred foundation for the future cow herd. Some of the American-created breeds were formed by using similar techniques; no doubt other new breeds will be developed in this manner in the future.

3. **Breed complementation or combining the desirable traits of two or more breeds to achieve a more desirable combination in the cow herd or the market animals than may be available in one breed.** The Angus X Charolais represents such a cross; the offspring combine the high carcass grade of the Angus and the high cutability of the Charolais.

4. **Opportunity to get hybrid vigor expressed in the cow.** Except for a two-breed cross, crossbreeding offers an opportunity to have hybrid vigor expressed by the cow. This enhances the cow's ability to *rough it*, conceive, give birth, and nurse her calf well.

Many other examples of all factors could be cited. It should be noted, however, that the magnitude of the advantage of all these factors—achieving the 15 to 25% potential immediate increase in beef yield per cow unit through continuous crossbreeding compared to continuous straight breeding—depends upon the following:

1. **Making wide crosses.** The wider the cross, the greater the heterosis.

2. **Selecting breeds that are complementary.** A crossbreeding program should involve breeds that possess the favorable expression of traits desired in the crossbred offspring that will be produced.

3. **Using high performing stock.** Once a crossbreeding program is initiated, further genetic improve-

ment is primarily dependent upon the use of high-indexing production-tested bulls.

4. **Following a sound crossbreeding system.** For a continuous high expression of heterosis and maximum beef output per cow, a sound system of crossbreeding must be followed. This should include the use of crossbred cows, for research clearly indicates that over one-half the higher profits from a crossbreeding program results therefrom.

CROSSBREEDING SYSTEMS

Without a planned breeding program, crossbreeding will almost inevitably end up with (1) a motley collection of females and progeny varying in type and color, and (2) minimum benefits from hybrid vigor or heterosis.

"Where do I go from here?" This is the question that many producers frequently ask, almost frantically, after having heifers of breeding age sired by exotic bulls. Others get worried when they notice that calves out of their crossbred cows aren't doing as well as the first-cross calves. Of course, what these producers really want to know is how they can maintain satisfactory hybrid vigor (heterosis) in animals when a herd is on a continuous crossbreeding program. They want to know how they can maintain 15 to 25% greater total efficiency in the crossbreds than the average of their parents; in production rate, calf livability, growth rate, and feed conversion.

Several different systems of crossbreeding may be used. Among them are the following:

1. **Two-breed cross.** This consists of mating purebred bulls to purebred or high-grade cows of another breed. An example would be using Angus bulls on Hereford cows, to give crossbred Angus X Hereford offspring—black baldies. This system of crossing has been used with success by cattle producers for many years.

In the two-breed cross, only the calves are crossbred—the breeding of the sires and dams remains the same. Hence, the two-breed cross imparts hybrid vigor only in the calf. On the average, it gives about an 8 to 10% increase in pounds of calf weaned per cow bred, plus another 2 to 3% advantage in rate of gain in the feedlot. In order to follow the two-breed cross indefinitely, the purebred females must be replaced with other purebreds sooner or later. They may either be purchased from another breeder or the breeder may want to produce purebred heifers within the herd.

The two-breed cross is relatively simple. However, it has one major deficiency; it does not make use of the crossbred cow.

2. **Two-breed backcross or crisscross.** This system involves the use of bulls of breed A on cows of breed B then backcrossing the progeny to bulls of

Fig. 5-39. Two-breed cross; Angus bull X Hereford cow, to produce black baldies.

either breed A or B. The rotation is accomplished by using bulls of the breed least related to the particular set of cows. For example, if Charolais bulls are mated to Hereford cows the crossbred Charolais X Hereford heifers could he retained and bred to either a Charolais or a Hereford bull. If Hereford bulls were used, the calves produced would be ¼ Charolais and ¾ Hereford. Later, if the heifers of this breeding are saved, they should be bred to a Charolais bull. The two-breed backcross results in about 67% of the maximum heterosis being attained in the crossbred calves. But since crossbred cows are used, overall performance should be a little better in pounds of calf weaned per cow bred than in the two-breed cross.

3. **Three-breed rotation cross.** This system calls for the selection of three breeds (*e.g.*, breeds A, B, and C, which might represent Herefords, Brahmans, and Charolais), possessing the combination of maternal, growth, and carcass traits desired in the crossbred cows and the slaughter cattle produced. Crossbred females selected for growth rate, are retained for breeding and bred to a purebred bull of one of the three breeds. Each new generation of crossbred females is retained for breeding and mated to a purebred bull until bulls of all three breeds have been used in rotation. Thus, such a system would operate as follows: Mate the existing B cow herd continuously to bulls of breed A; select crossbred heifers for growth rate and mate them continuously to bulls of breed C; mate the selected C (AB) females to bulls of breed B. After the rotation of bulls from the three breeds is completed, the rotation of purebred sires begins all over again. Thus, mate the selected B X (ABC) females to bulls of breed A.

Continue the same system indefinitely, always selecting the best performing crossbred females to be mated to the breed of sire in the program to which they are least related.

In addition to the genetic advantages of this system, commercial cattle producers select their own replacements; hence, the only outside cattle purchases are production tested bulls. The major disadvantage is that after the first four years it is necessary to maintain bulls of all three breeds simultaneously (unless AI is used).

A three-way rotation system results in about 87% of the maximum heterosis being attained.

4. **Three-breed fixed or static cross (terminal cross).** In this system, crossbred cows from a two-breed cross (F_1s) are used as females and are mated to a bull of a third breed. All offspring from this cross are sold. When replacement females are needed, they

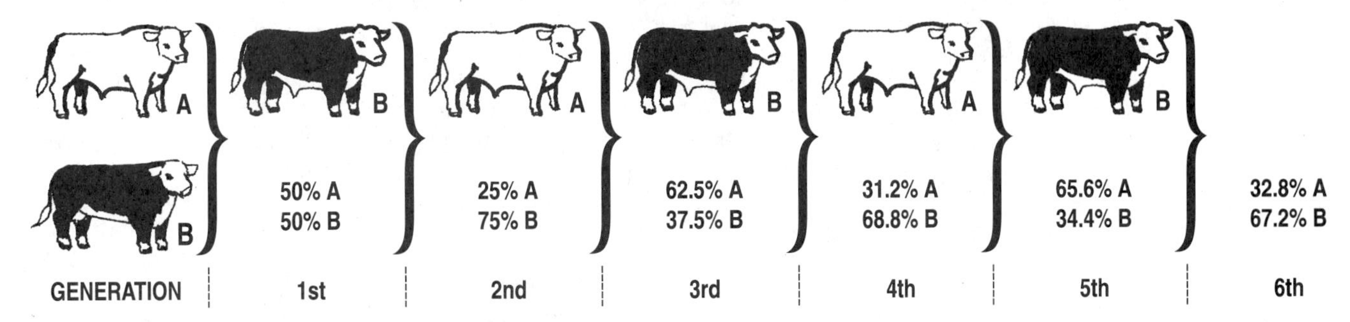

Fig. 5-40. Two-breed backcross or crisscross; Charolais bull X Hereford cow, thence female offspring bred to Hereford bull, thence female offspring to Charolais bull.

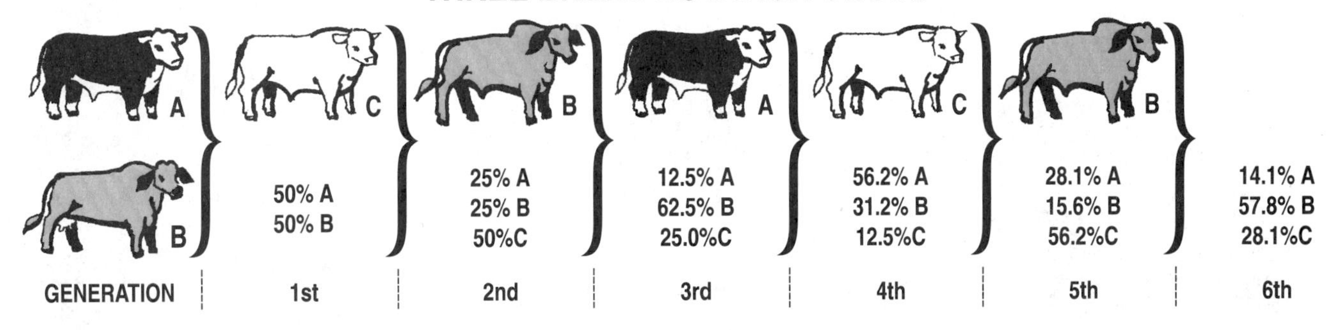

Fig. 5-41. Three-breed rotation cross; Hereford bull X Brahman female, thence female offspring to Charolais bull, thence female offspring bred to Brahman bull.

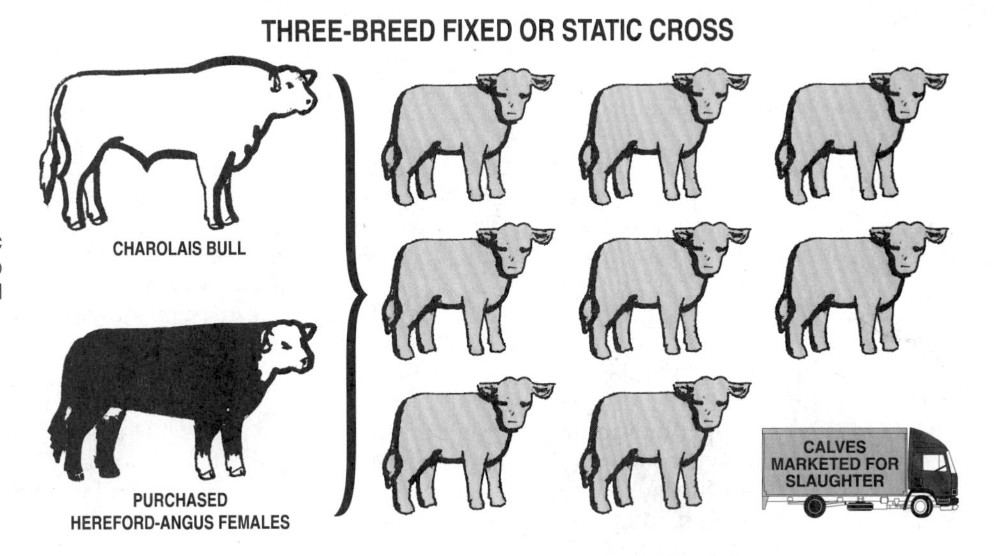

Fig. 5-42. Three-breed fixed or static cross (terminal cross); Charolais bull bred to crossbred Hereford-Angus female, with all the offspring marketed for slaughter.

are purchased. Thus, crossbred cows are used and crossbred calves with a fixed percentage of inheritance from three breeds are always produced.

In addition to realizing 100% of the maximum heterosis in each calf crop, this system allows the selection of maternal breeds to go into the production of the crossbred female and the selection of growthy breeds having desirable carcasses for the terminal cross sire breed. It allows the breeds to be used for their strong points without regard to some of their weaker points. A breeder can tailor-make the crossbred market animal, putting together in one animal desirable traits of several breeds. Such specification is not possible in the rotational system because all breeds contribute to maternal performance and calf performance.

The mechanics of this system consist in selecting three breeds for crossbreeding—two breeds (A and B) that will produce crossbred cows with outstanding maternal characteristics for fertility, milking ability, mothering ability, and adaptation. Select a third breed (breed C) with rapid, efficient, postweaning muscle growth rate. Breed C would be considered a "terminal" sire breed. All crossbred progeny of bull C are marketed for slaughter.

The problem with this system is the acquisition of production tested, crossbred (F_1) heifers for replacements in such a program, since all the three-way crosses are marketed. The system is perpetuated by having specialized multipliers produce crossbred (F_1) replacement females. Small operators (those with under 100 cows) might well use such a system where heifers are purchased along with the bulls. Large operators might well produce their own F_1 heifers in a specialized portion of their herd. Purebred breeders (seed stock breeders) would supply production selected terminal sires which the commercial producer would purchase for such a program.

Four, or more breeds may be used in a rotation crossbreeding system if the commercial producer so desires. However, the maximum hybrid vigor is usually realized with the three-breed cross.

Also, it is noteworthy that all of these crossbreeding systems rely upon the use of purebred bulls. Additionally, the two-breed cross relies on the use of purebred females, and the three-breed fixed or static cross relies on purebred females to produce the F_1 heifers necessary for the program.

Before going into a long-range crossbreeding program, the owner should know what is involved and what to expect. Plans should be developed before committing all available cattle and resources to a crossbreeding program. Consideration should be given to size of herd, markets, number of pastures, natural vs AI breeding, availability of breeding stock, etc. Sound management and sound selection of breeding stock based on performance, potential carcass characteristics, and overall productivity are just as important in crossbreeding as in any other breeding program.

Crossbreeding is no magic or *cure-all*, but it will give a powerful assist to the pocketbook if properly used. Also—and this point bears emphasis—sound management and sound selection of breeding stock based on performance, potential carcass characteristics, and overall productivity are just as important in crossbreeding as in any other breeding program.

All crossbreeding programs involve some animal identification system so that (1) growth rate of heifers may be determined, with selection of replacements made on this basis; and (2) where more than 1 breed is involved, the cow herd can be sorted for assignment to specific sire breeds for mating. Unless AI is used, it is necessary to maintain bulls of whatever breeds are involved, along with separate breeding pastures for each sire breed. Also, it should be recognized that where bulls of two or three different breeds are used, the crossbred slaughter progeny will vary considerably in performance and carcass traits, because they will be sired by bulls of different breeds and be produced by cows of divergent breed backgrounds.

BULL SELECTION FOR CROSSBREEDING

Once the system of crossbreeding has been chosen, the most important recurring genetic decision is that of selecting bulls. It is just as important in commercial production to select superior performance bulls as it is in purebred breeding herds. This is so because the traits of production and product (see Table 5-4) being highly heritable can be transmitted directly from parent to commercial offspring. In order intelligently to buy bulls, the commercial producers must at least know where herds stand in terms of average performance for the production and product traits. Then they can select bulls strong in the weaker points of their herds and get bulls which, on the average, will improve their herds. The performance test of bulls to a year of age, plus possibly a sib carcass test, is an adequate estimate of breeding value for commercial producers. On the average, the performance test will predict breeding value for the group, but for a particular individual it is not so good. Thus, the average performance of 10 yearling bulls will predict the average performance of their calves very well, but just which of the 10 is really the best bull is not well estimated. Of course, commercial producers using AI can afford to use progeny tested bulls available to them.

The selection of breeding herds from which to select commercial bulls is made difficult since only a fraction of herd differences are genetic—that is, the genetic differences between herds is probably less than 40%. On the other hand, research evidence suggests that for the production and product traits, the heritability of within herd differences among animals treated alike is between 40 and 50%. Thus, so long as the better herds are being patronized, selection of the best performing bulls within a herd is a safer bet than spending too much time deciding on which herd from which to buy bulls.

CROSSBRED BULLS

Cattle producers are sometimes tempted to use hybrid bulls as herd sires. This inclination is understandable since the performance of hybrid bulls, as individuals, is often superior to either parent breed. But, before putting a hybrid bull with cows, some cautions are in order.

Unlike improved production resulting from selection, advantages from hybrid vigor are not transmitted from parent to progeny. Hence, to the extent to which superior performance is due to hybrid vigor, a hybrid bull will not breed true. If the hybrid bull is out of purebreds which have been selected for superior performance, this portion of his inherited superiority may be transmitted to his progeny. However, individual performance of hybrids is a less accurate indicator of their breeding value than is the performance of purebreds.

Work at the Experiment Station, Miles City, Montana, indicates that there may be some gain from the use of crossbred bulls over straightbreds, in increased fertility, vigor, and livability of the calves. But the disadvantages of using a crossbred bull outweigh the advantages. Among the disadvantages are the following:

1. **Hybrid vigor may mask the true breeding worth of a hybrid bull.** The sire's effect on profitability is basically indirect through the performance of his progeny. Thus, it is important that sires be accurately selected for the characters desired in their progeny. Hybrid vigor is not transmitted; hence, it may mask the true breeding worth of a hybrid bull. In other words, selection of purebreds is expected to be more effective.

2. **The progeny of hybrid sires tend to be more variable.** Since hybrid sires are less prepotent, their progeny will tend to be more variable in all measures of performance. Also, variation in color and conformation will tend to be more evident. As a result of this lack of uniformity, the market price of their progeny will be lower, especially when they are sold as feeders.

3. **Crossbred bulls have less effect on performance than crossbred cows.** Cows affect offspring through milk production and mothering ability; hence, they affect the performance of offspring more than the sire.

4. **The likelihood that crossbred bulls will be produced by breeders who haven't the best cattle.** A major problem in considering crossbred bulls is that there is a strong likelihood that they will be produced by breeders who haven't the best cattle from which to produce bulls.

Despite the above, there are special situations in which crossbred bulls may be considered. Two of these circumstances are:

1. **For the creation of a new breed.** The use of crossbred bulls is necessary in the creation of new breeds especially adapted to certain conditions. For example, the Santa Gertrudis breed of cattle, a breed derived from ⅝ Shorthorn and ⅜ Brahman, was developed to meet a need in the hot, dry, insect-infested area of the Southwest. Experienced producers of the area will vouch for the fact that this is a practical example of a planned system of crossbreeding which has high utility value under the environmental conditions common to the country. Also, crossbred Charolais bulls were used extensively and successfully for many years in the United States, during the formation of the breed by grading up. Still other examples of crossbreeding in the creation of breeds may be cited, including the breeding up to one of the new exotics in which purebred sires may be too expensive or scarce.

2. **For coping with harsh environmental conditions.** Under certain conditions, it may be desirable to incorporate hybrid vigor and adaptability in the bull, as well as in the brood cow, in order to cope with harsh environmental conditions. For this reason, Gulf Coast operators often prefer bulls that have ⅛ to ¼ Brahman breeding.

From the above, it may be concluded that, except for special circumstances, crossbred bulls should not be used. They cannot be counted on to be herd improvers like comparable purebreds.

NOT ALL HYBRIDS EXCEL PUREBREDS

Cattle producers can learn from poultry producers when it comes to breeding, for the breeding of chickens has passed through the total presently known systems. In fact, each method has been, and still is, being used successfully.

The vast majority of chickens in America today are hybrids of one form or another—they're either strain crosses, breed crosses, or crosses between inbred lines. But they're exceptions!

Despite the fact that hybrids are widely used as commercial layers, it is noteworthy that egg-laying tests show that purebred Single Comb White Leghorns compete on even terms with hybrids under test conditions. Certainly, the hybrids are equal to the purebreds, but the point is that they do not excel them. The same principle applies to purebred vs crossbred cattle, *provided* the purebred cattle reach the pinnacle enjoyed by White Leghorns.

In broiler production, the main objective is the improvement of growth rate to eight weeks of age, although improvement in other economic factors is sought. Generally, growth rate and hybrid vigor are obtained by systematic matings that may involve crossing different breeds, different strains of the same breed, or the crossing of inbred lines. Most of the strains used as sires trace their ancestry to the broad-breasted Cornish breed. But there is some question whether heterosis, as obtained through hybrid breeding, contributes substantially to broiler weight. It is noteworthy, for example, that the best purebred New Hampshire strains generally equal the most rapid gaining crosses. The latter point is of great significance to purebred cattle breeders.

Body conformation is especially important in turkeys, because they are marketed at heavier weights than broilers and their carcass is usually left whole rather than cut up. Since conformation, size, and color of turkeys are highly heritable, they have responded well to simple methods of breeding and selection. As a result, most turkeys are bred as purebreds, rather than crossbreds. Also, it should be of more than passing interest to cattle producers to know that in turkey breeding programs (1) selections are largely based on physical appearances (phenotype), and (2) mass-matings (in which a number of males are allowed to run with the entire flock of hens) are the common practice.

BUFFALO X CATTLE HYBRIDS[16]

From time to time, American buffalo *(Bison bison)* and domestic beef cattle *(Bos taurus)* have been crossed, in Canada and the United States. Out of such crosses have evolved Cattalo (cattle of less than ½ bison parentage), Beefalo (⅜ buffalo, ⅝ domestic cattle), and the American Breed (⅛ buffalo, ½ Brahman, ¼ Charolais, 1⁄16 Shorthorn, and 1⁄16 Hereford). These breeds are extolled on the basis of their adaptability to cold, snowy climates; ability to thrive on poor vegetation which domestic cattle pass up; small birth weights (straight buffalo calves weigh only about 25 lb at birth); and leaner and more flavorful meat.

Pertinent information relative to the reproductive ability of the American buffalo *(Bison bison)* X domestic cattle *(Bos taurus)* hybrids follows:

1. Bison and domestic cattle interbreed.

[16]Buffalo breeders are banded together in the National Buffalo Association, Box 995, Pierre, SD 57501.

Fig. 5-43. Cattalo (¼ buffalo, ¾ domestic cattle) cow. The initial Cattalo breeding experiment was started by the Dominion Experimental Station, Scott, Saskatchewan, Canada, in 1915. The foundation herd consisted of 16 female and 4 male hybrids. (Courtesy, Research Station, Canada Department of Agriculture, Lethbridge, Alberta, Canada)

2. Fewer maternal calving losses occur when domestic bulls are used on bison cows, although the reciprocal mating may be made.
3. Half-buffalo bull calves (F_1 hybrids) show normal sexual behavior, but they are always sterile. The scrotum is held close to the body cavity, as in the bison.
4. The half-buffalo heifers (F_1 hybrids) are fertile.
5. A few backcross bull hybrids have produced semen containing some sperm.
6. Reproductive ability improves in both sexes of further generations as the percentage of domestic blood increases.

More scientific research on buffalo X cattle hybrids is needed.

COMPOSITE BREEDS (also called Synthetics or Hybrids)

Composite breeds are breeds that have been formed by crossing two or more breeds. Composites combine breeds and heterosis, thereby alleviating many of the frustrations associated with conventional crossbreeding; for example, composites simplify the breeding program by reducing the number of breeding pastures. Also, a composite can be produced for a given environment; and heterosis can be maintained year after year provided there is no inbreeding. It is noteworthy that several existing breeds (*e.g.*, Brahman crosses) are composites or hybrids. In a broad sense, most breeds may be considered as composites.

The goal of composite breeders is to blend the

Fig. 5-44. Salorn heifer, a composite (5/8 Salers X 3/8 Texas Longhorn), a lean beef breed adapted to a subtropical and tropical environment. (Courtesy, International Salorn Assn., Ardmore, OK)

desirable traits from two or more breeds into one improved model, and to obtain the added boost of hybrid vigor from crossing unlike breeds.

With composites, it is essential to avoid inbreeding. A composite breed that becomes inbred is just another pure breed without hybrid vigor. Inbreeding is easily circumvented by using a new bull every two years, thereby avoiding breeding heifers back to their sire.

Composites and crossbreeding have enhanced a number of good traits, including weaning weight, mature size, milk production, adaptability to climate, age of puberty, scrotal circumference, and fertility. But they have not been an unmixed blessing! They have made for a lack of consistency, or repeatability, in beef. Since consumers want beef like they purchased last time, this deficiency is of great concern to the entire beef team—producers, feeders, packers, and retailers. It is hoped that increased branded beef and individual animal identification tracked from producer to consumer will result in increased consistency of beef. Eventually, two factors—profitability and consumer demand—will determine which breeds, composites, and crossbreds survive, and which fade away.

(Also see Chapter 3, section headed "Composite Breeds.")

PRODUCTION TESTING BEEF CATTLE[17]

Nearly every state now has an approved Beef Cattle Improvement Association (BCIA) program. Also most breed registry associations have established programs. The Beef Improvement Federation (BIF) is a composite organization, responsible for standardization and uniformity in program systems.

Production testing embraces both (1) performance testing, and (2) progeny testing. The distinction between and the relationship of these terms is set forth in the following definitions:

1. ***Performance testing*** *is the practice of evaluating and selecting animals on the basis of their individual merit.*

2. ***Progeny testing*** *is the practice of selecting animals on the basis of the merit of their progeny.* It is usually costly and can be justified only for bulls of outstanding merit. Emphasis on progeny testing should be on traits not measurable in the bull himself—such as carcass traits and maternal ability of offspring. Generally speaking, the cost of progeny testing can be justified only for selecting bulls to be used extensively in artificial insemination or in very top seedstock herds.

3. ***Production testing*** *is a more inclusive term, including performance testing and/or progeny testing.*

Production testing involves the taking of accurate records rather than casual observation. Also, in order to be most effective, the accompanying selection must be based on characteristics of economic importance and high heritability (see Table 5-5), and an objective measure or *yardstick*, such as pounds, should be placed upon each of the traits to be measured. Finally, those breeding animals that fail to meet the high standards set forth must be removed from the herd promptly and unflinchingly.

In comparison with chickens, or even swine, production testing of beef cattle is slow, and, like most investigational work with large animals, it is likely to be expensive. Even so, in realization that such testing is absolutely necessary if maximum improvement is to be made, progressive purebred beef cattle breeders will have their herds on production test.

ECONOMICALLY IMPORTANT TRAITS IN BEEF CATTLE AND THEIR HERITABILITY

Table 5-5 lists the economically important traits in beef cattle; those that contribute to both productive efficiency and desirability of product. Regularity of production, rapid growth, efficient use of feed, and carcass qualities preferred by consumers are economic traits of major importance. Performance testing offers beef cattle breeders a way of measuring differ-

[17]In the preparation of this section, the authors drew heavily from *Guidelines for Uniform Beef Improvement Programs*, Beef Improvement Federation Recommendations, 1996.

TABLE 5–5
ECONOMICALLY IMPORTANT TRAITS IN BEEF CATTLE, AND THEIR HERITABILITY[1]

Economically Important Traits	Approximate Heritability of Traits	Comments
	(%)	
Mature cow weight	50	Mature cow weight is important because it relates to cost of herd maintenance.
Calving interval (fertility)	10	Fertility is economically the most important trait in beef cattle. Without a calf being born, and born alive, cattle are self-eliminating.
Calving ease	10	Calving losses at birth are the second most important reason for lower percent calf crops. Calving difficulty (dystocia) accounts for most calf deaths within the first 24 hours after calving, and most calving difficulty occurs in 2-year-old heifers.
Calf survival	10	Expressed as percent surviving to weaning.
Birth weight	30	Birth weight is associated with calf survival. Also, it has a positive correlation of 0.39 with growth rate. Selecting for increased birth weight is generally avoided because of likely increased calving difficulty.
Weaning weight	25	Heavy weaning weight is important because: 1. It is indicative of the milking ability of the cow. 2. Gains made before weaning are cheaper than those made after weaning. 3. Those who sell calves at weaning usually make more profit due to the heavier weight available to sell.
Post-weaning gain	30	Daily rate of gain is important because: 1. It is highly correlated with efficiency of gain. 2. It makes for a shorter time in reaching market weight and condition, thereby effecting a saving in labor and making for a more rapid turnover in capital.
Efficiency of gain	35	Efficiency of feed conversion is expressed as pounds of feed intake per 100 lb *(45 kg)* of gain. It is seldom measured in performance and progeny tests, because a positive relationship exists between rate and efficiency of gain. Hence, selection for rate of gain automatically selects for efficiency of gain.
Yearling weight (feedlot)	40	Final feedlot weight is usually referred to as *weight per day of age*. It is generally computed at 1 year of age or at the end of the performance test. It is probably the most important measurement of the estimated value of a beef bull. It is composed of birth weight, weaning weight, and postweaning weight.
Yearling weight (pasture)	35	Most beef animals spend a good part of their lives on grass; hence, pasture gain is important.
Carcass weight	25	Quantity of product is one factor determining total carcass value.
Dressing percent	40	Dressing percentage reflects the percentage yield of hot carcass in relation to the weight of the animal on foot. A high carcass yield without excess fat is desirable because the carcass is more valuable than the by-products.
Carcass grade	45	High carcass grade is important because it determines selling price and eating quality.
Ribeye area	40	The ribeye (the large muscle which lies in the angle of the rib and vertebra) is indicative of the bred-in muscling of the entire carcass. Thus, a large area of ribeye is much sought.
Thickness of outside carcass fat	45	Fat thickness taken over the twelfth rib should not exceed 1/4 inch. The less trimmable fat, the more the carcass is worth.
Marbling	40	Higher marbled beef is generally more tender, juicy, and flavorful.
Lean percent	55	Consumers prefer lean beef, without reducing marbling.
Lean/bone ratio	60	Consumers prefer a maximum amount of lean and a minimum of bone. Lean-to-bone ratio varies with age; from 2:1 (lb muscle to lb bone) in a young calf to 4:1 in an 1,100 lb slaughter steer.
Tenderness	30	Consumers rate tenderness as the most important palatability trait of beef. Warner-Bratzler shear test and taste panel test are recommended as methods of measuring tenderness.

[1]The heritability estimates in Table 5–5 were provided by Dr. Koots, *et al.*, Department of Animal and Poultry Science, University of Guelph, Guelph, Ontario, Canada. These values are heritability estimates obtained from a comprehensive review of the literature (Koots, *et al.*, 1994, *Animal Breeding Abstracts*, 62:309–338).

ences among animals in heritable characters. Performance levels for these characters are related to ability to transmit desired traits to offspring.

Differences among animals in traits of economic value are, to a considerable extent, inherited differences. Thus, systematic measurement of these differences, the recording of the measurements, and the use of records in selection will increase the rate of genetic improvement in individual herds, and eventually in the breed and in the total cattle population.

Research has shown that when cattle are kept under nearly like conditions, and records are kept of traits of economic value and adjusted for known sources of variation—such as age, age of dam, and sex—genetically superior animals can be identified.

The rate of improvement in a herd, breed, and population is dependent on (1) the percentage of observed differences between animals that is due to heredity (heritability), (2) the difference between selected individuals and the average of the herd or group from which they come (selection differential), (3) the genetic association among traits upon which selection is based (genetic correlations), and (4) the average age of parents when the offspring are born (generation interval).

The essentials of effective record of performance programs are:

1. All animals of a given sex and age are given equal opportunity through uniform feeding and management.
2. Systematic written records are kept of important traits of economic value of all animals.
3. Records are adjusted for known sources of variation, such as age of dam, age of calf, and sex.
4. Records are used in selecting replacements (bulls and heifers) and in culling poor producers.
5. Nutritional program and management practices are practical and uniform for the entire herd and are similar to those where progeny of the herd are expected to perform.

RATIOS

Ratios—*weight ratio*, *gain ratio*, and *conformation score ratio*—are used to refer to the performance of an individual relative to the average of all animals of the same group. Ratio is calculated as follows:

$$\frac{\text{Individual record}}{\text{Average of animals in group}} \times 100$$

It is a record of individual deviation from the group average expressed in terms of percentage. Thus, if an average test station gain was 3.00 lb per day, the gain ratio of a bull gaining 3.30 lb per day would be 110.

EVALUATION OF GROWTH RATE AND EFFICIENCY OF GAIN

Growth rate and efficiency of gain for beef cattle are of primary economic importance to the beef industry. Growth rate has a direct effect on net return and is positively correlated with efficiency of gain, weight, and value of retail product. Efficiency of gain has a direct effect on cost of production and net return. However, realized heritability for measures of preweaning and postweaning growth depends on how they are handled with respect to sex of the animal and age of the dam and in relation to contemporary animals.

PREWEANING PHASE

It is recommended that the preweaning phase incorporate the following: measurement of birth weight, measurement of weaning weight (205 days), weaning weight ratio, product-of-dam summary, and most probable producing ability (MPPA). For a detailed discussion of each of these, cattle producers and those who counsel with them should secure a copy of *Guidelines for Uniform Beef Improvement Programs*, from Dr. Ron Bolze, BIF Executive Director, Northwest Research Extension Center, 105 Experiment Farm Road, Colby, KS 67701; or contact the Beef Cattle Improvement Association of their state.

POSTWEANING PHASE—ON THE FARM AND RANCH

It is recommended that the postweaning phase on the farm and ranch incorporate the following: measurement of yearling weight (365 days) or long yearling weight (452 or 550 days), and weight ratios. Detailed information on each of these may be secured from the same sources as listed for the preweaning phase.

EFFICIENCY OF GAIN

Efficiency of gain in beef production is the ratio of nutrient input to beef output or the ratio of beef output to nutrient input. On a life cycle basis for either herds or individual cows, about 70% of the energy required for beef production is consumed by cows and bulls in the breeding herd and by their calves prior to weaning. About 30% of the energy required for beef production from weaning to slaughter is consumed by steers and heifers during the postweaning period. Depending on the cost of feed resources for the cow herd in relation to those for the feedlot, 45 to 55% of the total feed costs are incurred in the period from weaning to slaughter. Thus, efficiency of growth is usually measured by (1) cow efficiency, and (2) postweaning efficiency

- **Cow efficiency**—This measures the ratio of beef output per unit of nutrient input. The beef output includes the beef from cull cows as well as progeny slaughtered. Beef output is relatively easy to assess. However, it has not been possible to measure nutrient input of reproducing animals under practical economical management systems.

- **Postweaning feed efficiency**—Choice of interval of evaluation or end points has an important influence on ranking of animals and breeding groups for feed efficiency. The interval chosen depends on whether the objective is to maximize production efficiency of lean meat (weight or time end points) or meat of a constant fat-to-lean ratio (fatness end point). For either weight-constant (for example, 1,000 lb) or time-constant (for example, 365 days) end points, faster gaining cattle have greater feed efficiency. These cattle are characterized by larger mature size, larger carcasses, higher percentage of retail product, a lower percentage of fat trim, and lower marbling levels. Variation in both—feed efficiency and the relationship between feed efficiency and growth rate—is maximized in weight-constant intervals (for example, 500 to 1,000 lb). Cattle with more rapid growth rates require less feed per unit of gain as measured in weight-constant intervals primarily because they require fewer days and less feed for maintenance in a weight-constant interval.

The ranking of animals for feed efficiency in a time-constant interval is correlated positively with that for feed efficiency in a weight-constant interval. Faster gaining cattle tend to be more efficient in time-constant evaluations (such as the 140-day postweaning test), but the variation in feed efficiency and the association with growth rate is reduced because larger faster gaining animals that maintain heavier weights throughout the time-constant test require more feed for maintenance.

Feed efficiency can also be evaluated to fatness end points (such as a small degree of marbling corresponding to USDA Choice, or to a fat thickness of 0.3). Animals with the greatest propensity to fatten are the most efficient when evaluated to a fatness end point because they require fewer days and less feed for maintenance to reach that point. It appears that feed efficiency to fatness end points is not strongly associated with growth rate or size.

BEEF CARCASS EVALUATION

Edible beef products are the goal of all beef cattle improvement programs and activities. Quality of product and quantity of edible portion are the basic factors used to judge carcass merit. However, the relative values of quality and quantity are subject to change as market demands change.

Carcass evaluation is the technique by which the components of quality (overall palatability of the edible portion) and the components of quantity (amount of salable meat) are measured.

Not all beef producers will need complete carcass data. Careful thought should be given to the data a producer can really use. Increasing the amount of data on large numbers of cattle adds to the time required, costs, and likelihood of errors, and reduces packers' interest in cooperating.

If quality grade, yield grade, and warm carcass weight are sufficient, do not request more. Evaluating and recording accurate data on all specific factors relating to quality and yield grades requires considerably more time.

Ultrasound can be used to estimate carcass characteristics in live animals; however, the accuracy of the measurements is highly dependent upon the experience and expertise of the operator.

CENTRAL TEST STATIONS

Central test stations are locations where animals are assembled from several herds to evaluate differences in certain performance traits under uniform conditions. Uses of central test stations include: (1) comparing individual performance of potential seedstock herd sires to similar animals from other herds; (2) comparing bulls being readied for sale to commercial producers; (3) finishing steers or heifers scheduled for slaughter as part of progeny test programs for growth and carcass traits; (4) acquainting breeders with record of performance; and (5) estimating genetic differences

Fig. 5-45. A central test station at Western Illinois University Beef Evaluation Station, Macomb, IL.

between herds or between sire progenies in gaining ability, feed conversion, and carcass characteristics.

The following procedures and policies are recommended for central testing of bulls:

1. Age of calves at time of delivery to test stations should be at least 180 days (6 months) and not more than 305) days (10 months).

2. Herds from which bulls are consigned should be on a herd testing program for preweaning and postweaning performance. Calves should have completed the weaning phase of the performance records program, and the following information should be submitted to the test station: sire, birth date, age of dam, actual weaning weight and date, adjusted 205-day weight, within-herd weaning weight ratio (based on average of all bull calves in same weaning season and management group), and the number of calves making up this average.

3. There should be an adjustment, or a pretest, period of 21 days or more at the test station immediately prior to the test period.

4. The length of the feeding test may be influenced by feeding conditions and breed goals and should be 140 days or more.

5. The initial and final test weights may be either full or shrunk weights. If full weights are taken, initial and final test weights should be an average of two weights taken on successive days to minimize fill effect. If shrunk weights are taken, a single weight after a shrink of 24 hours will be adequate.

6. All bulls sold in a test sale should be examined by competent personnel for reproductive and structural soundness.

7. Test rations will vary according to locally available feeds and test objectives. Feeding should be free choice—keeping animals supplied with all the feed they want. Rations of between 60 and 70% total digestible nutrients (TDN) should be adequate for the expression of genetic differences in growth.

8. Sire group testing of bulls is more desirable than individual testing because it provides more information to both breeders and prospective buyers.

■ **Test station reports**—Test station reports may include the following:

1. **140-day average daily gain and gain ratio.** These are the most important figures in test station results, because they measure growth during the period when the bulls are together under test conditions. Selection for 140-day gains should improve weaning weights and feedlot performance because some of the genes which affect feedlot growth rate also affect preweaning growth rate. The gain ratio is obtained by dividing the individual animal's gain by test group average and multiplying by 100. A ratio of 100 means the bull is exactly average in his group, 115 means he is 15% above the average, and 90 means he is 10% below the average. This ratio makes comparisons of animals easier, and it is as informative as the actual measurement.

2. **Weaning weights and within-herd weaning weight ratios.** These provide good comparisons of bulls which come from the same herd but are not particularly useful for comparing bulls from different herds. Weaning weight is the best available measurement of the dam's milk production. Thus, it is desirable for the calf to have a weaning weight above the average of the herd in which the calf was produced (that is, within-herd weaning weight ratio above 100).

3. **The 365-day adjusted weight and 365-day weight ratio.** This combines adjusted weaning weight and postweaning gain into one composite measurement. The 365-day weight ratio is the best measure for comparing growth of calves from the same herd. This trait is highly heritable (approximately 40%). However, among bulls from different herds in a central test, use care with this measurement because the weaning weight portion was not made under comparable conditions.

4. **Efficiency of feed conversion.** This is expressed as pounds of feed per pound of gain. It is difficult to measure. Most tests do not attempt to get individual feed conversion data because it requires individual feeding or expensive electronic equipment. Where sire progeny groups are fed in separate pens, a good measure of the sire's ability to sire *efficient-gaining* sons may be obtained. This procedure also provides some information on the individual half-brothers in the pen.

5. **Conformation score or grade.** This is an optimal measure for test stations. This measurement should be based strictly on skeletal soundness and indications of carcass desirability (including carcass weight and cutability).

6. **Ration.** The composition of the ration should be stated in station reports. Rations vary considerably among test stations, particularly in level of energy they contain. This variation from one station to another causes some differences in average daily gains that are not heritable.

NATIONAL SIRE EVALUATION PROGRAMS

Formerly, progeny testing was considered valid only when done on a within-herd basis. However, artificial insemination and some refined data analysis allow comparison of bulls between herds. As a result, several breed associations now have National Sire

Evaluation Programs which follow the guidelines established by the Beef Improvement Federation. These guidelines are based on comparing the progeny of bulls being tested against the progeny of bulls designated as *reference sires*. This allows widespread use of reference sires in many different herds and in many different environments. The purpose of the program is to provide breeders with an aid to the selection of sires based on the performance of their progeny.

Reference sires are selected on the basis of a large number of progeny evaluated for each trait, semen availability, and willingness on the part of owners to cooperate. Reference sires are just benchmarks; they provide a common link between all herds and breeder-owned sires regardless of the breed, year, or season breeder-owned calves are born.

Several types of National Sire Evaluation Programs have evolved. The element common to all of them is the use of reference sires as a basis of comparing sires as shown in Fig. 5-46.

▪ **The progeny test**—The basics of a sound progeny test are:

1. **Comparable cows.** All bulls to be compared must be mated to comparable cows, in order to eliminate differences in cows from the differences between averages for sire progeny.

2. **Equal treatment of progeny.** Progeny from all bulls must be given equal treatment, in order to eliminate environmental differences from the differences between averages for sire progeny.

▪ **Expected progeny difference (EPD)**—The expected progeny difference (EPD) is an estimate of how future progeny of the sire are expected to perform relative to the progeny performance of the reference sires, when both are mated to comparable cows and the resulting progeny are treated alike. For example, the EPD for yearling weight on a bull might be +40 lb. This bull may be expected to sire calves approximately 40 lb heavier at a year of age than a sire with a yearling weight EPD of 0. Accuracy figures are provided to show how much confidence can be placed in the EPD.

▪ **Traits**—The selection of the particular traits to be evaluated in a National sire Evaluation Program is the prerogative of the organization conducting the program for the breed. The Beef Improvement Federation (BIF) suggests the following traits:

1. **Reproduction.** An adequate measure of calving ease would be beneficial to some breeds. It would be useful to include provisions to evaluate the maternal performance of daughters as to their overall reproductive potential.

2. **Production.** BIF recommends several measures of growth during the relevant commercial period, such as weaning weight and several measures of yearling weight (365-day, 452-day, or 550-day). Again, provisions to include the weaning weights of daughters are desirable.

3. **Product.** The amount (yield grade) and quality (quality grade) of the product produced is not measurable directly for the sires. Information on carcass evaluation adds new information in a sequential selection scheme. Such carcass progeny tests can be used effectively as sib tests on the sons from the tested sires.

▪ **Undesirab le genes**—Bulls may be progeny tested

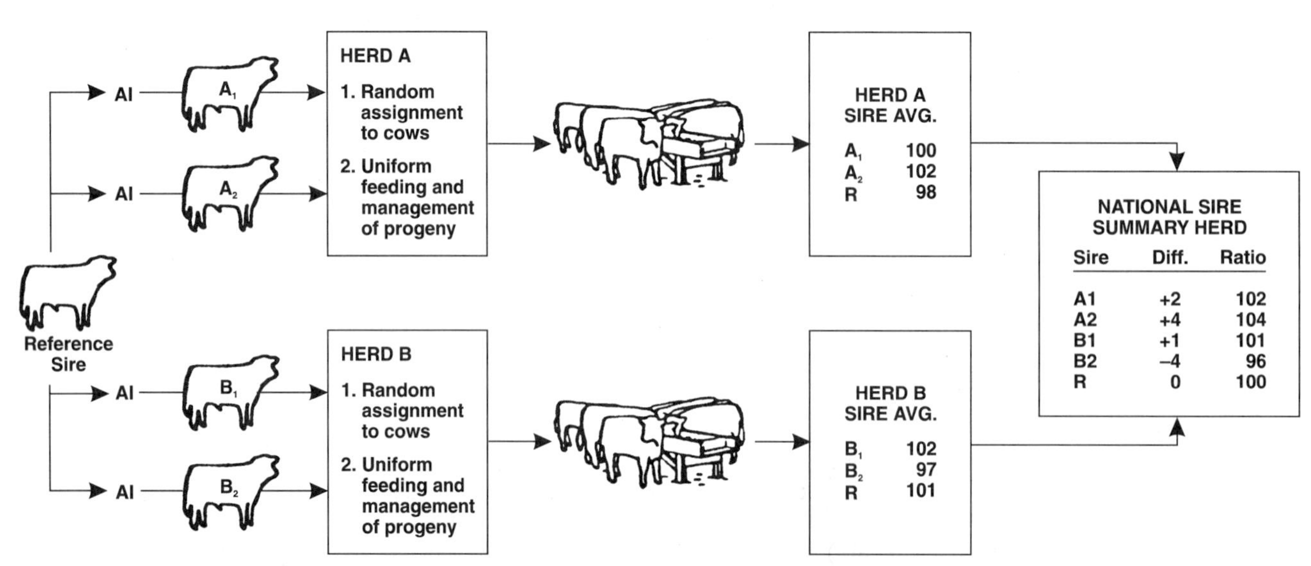

Fig. 5-46. National Sire Evaluation is possible by comparison to reference sires used through artificial insemination in different herds. (Adapted from *Beef Cattle Breeding*, USDA Bull. 20-286, p. 18)

for all undesirable recessive genes by either of the following two methods:

1. By breeding a bull to a large cross-section of cows. The probability of detection is a function of the existing gene frequency:

$$\text{Probability of detection} = 1 - (1 - \tfrac{1}{2}q)^n$$

where (q) is the gene frequency in the cow, and (n) is the number of progeny. This procedure allows a short generation interval, yet keeps undesirable, recessive genes at a low frequency.

2. By breeding a bull to his daughters. The probability of detection uses the same formula as above, with (q) equal to ¼ The production of normal offspring from 22 daughters gives a probability of 19 in 20 that the sire does not contain a specific recessive gene. From 35 daughters, the probability is 99 in 100.

- **Publication of the National Sire Summary**—The results of bulls tested are usually reported by the different breed associations in a National Sire Summary. The purpose of such a sire summary is to describe the germ plasm available for the traits considered of major economic importance to the breed. Selection of sires from among those described is the prerogative of the breeder. Many breed sire summaries are available through the Internet.

RECORD FORMS

A prerequisite for any production data is that each animal be positively identified—by means of ear notches, ear tags, or tattoos. For purebred breeders, who must use a system of animal identification anyway, this does not constitute an additional detail. But the taking of weights and grades does require additional time and labor—an expenditure which is highly worthwhile, however.

In order not to be burdensome, the record forms should be relatively simple. Also, they should be in a form that will permit easy summarization—for example, the record of one cow should be on one sheet if possible. Suggested record forms are shown in Figs. 5-47, 5-48a, and 5-48b.

Information on the productivity of *close relatives* (the sire and the dam and the brothers and sisters) can supplement that on the animal itself and thus be a distinct aid in selection. The production records of more distant relatives are of little significance, because, individually, due to the sampling nature of inheritance, they contribute only a few genes to an animal many generations removed.

RECORDS BY COMPUTER

Beef production is becoming more sophisticated. Today, much of the pencil work normally involved in production testing is being eliminated by computer. A number of good computer record programs are available on a modest charge basis, including some sponsored by breed registry associations. In many areas, commercial producers have available to them computerized record systems through their state's Beef Cattle Improvement Association.

HOW TO USE HERD RECORDS IN SELECTION

Herd records have little value unless they are intelligently used in culling operations and in deciding upon herd replacements. Also, most cattle producers can and should use production records for purposes of estimating the rate of progress and for determining the relative emphasis to place on each trait.

APPRAISING PERCENT OF CHANGE IN CHARACTERS DUE TO (1) HEREDITY, AND (2) ENVIRONMENT

Cattle producers are well aware that there are differences in birth weight, in weaning weight, in daily rate of gain, in body type, etc. If those animals which excel in the desired traits would, in turn, transmit without loss these same improved qualities to their offspring, progress would be simple and rapid. Unfortunately, this is not the case. Such economically important traits are greatly affected by environment (by feeding, care, management, type of birth [singles vs twins] age, diseases, etc.). Thus, only part of the apparent improvement in certain animals is hereditary, and can be transmitted on to the next generation.

As would be expected, improvements due to environment are not inherited. This means that if most of the improvement in an economically important trait is due to an improved environment, the heritability of that trait will be low and little progress can be made through selection. On the other hand, if the trait is highly heritable, marked progress can be made through selection. Thus, body color in cattle—*e.g.*, red, roan, and white—is a highly heritable trait, for environment appears to have little or no part in determining the difference between animals that are red and those that are roan or white. On the other hand, such a trait as condition or degree of finish is of low heritability because, for the most part, it is affected by environment (feed, care, management, etc.).

GET OF SIRE RECORD

Calf Crop for Year of ______________________ Sire's Name ______________ Reg. No. ______________

Sex of Get[1] ______________________ Date of Birth ______________________

Owner and Address ______________________

		Calf Data							Yearling Data					Dam Data				
Herd No. of Calf	Date of Birth	Wean-ing Date	Wean-ing Age in Days	Wt. in Lb	Daily Gain from Birth Wt., Lb	Adj. 205-Day Wean-ing Wt., Lb	Wean-ing Wt. Ratio[2]	Conf. Score	Date Weighed	Wt., Lb	Wt. Adj. to Days	Yr.-Wt. Ratio[2]	Conf. Score	Herd No.	Age This Yr.	Mature Wt., Lb	Conf. Score	Remarks
Totals																		
Averages																		

[1]One sheet should be used to record all the bull calves and another sheet to record all the heifer calves by the same sire.

[2]Ratio calculated as follows: $\frac{\text{individual record}}{\text{avg. of all calves on same farm and same season}} \times 100$

Fig. 5-47. Get of Sire Record.

INDIVIDUAL COW RECORD

Tattoo____________________

Name __________ Reg. No. __________ Bred by __________ Birth Date __________

Purchased from __________ Birth Wt., Lb __________

Address __________ Weaning Wt., Lb _____ Age _____ Conf. Score _____

Sire __________

Purchase date _____ Price, $_____ Yearling Wt., Lb _____ Age _____ Conf. Score _____

Disposition _____ Price, $_____ Two-Year Wt., Lb _____ Age _____ Conf. Score _____

Avg. Daily Gain Weaning to 1 Yr., Lb __________

Reason for Disposal __________ Feed Efficiency __________ Lb Feed/100 Lb Gain

Dam __________

Date _____ Temperament __________

Faults & Abnormalities __________

PRODUCE OF DAM RECORD

Calf Data													Yearling Data					Production Testing				
							Weaning Wt., Lb							Yr., Wt., Lb								
Birth Date	Sex	Tattoo	Sire	Birth Wt., Lb	Vigor at Birth[1]	Wean-ing Age, in Days	Act.	205 Day Adj.	Wean-ing Wt. Ratio[2]	Wean-ing Cond.	Conf. Score	Date	Adj.	Days	Year-ling Wt. Ratio[2]	Conf. Score[1]	Days on Feed	Avg. Daily Gain, Lb	Gain Ratio[2]	Lb Feed/ 100 Lb Gain	Disposition; Price Remarks

[1]0 = dead at birth; 1 = definitely undersized at birth; 2 = unthrifty, define indications of disorders; 3 = moderately thrifty, slight indications of disorders; 4 = thrifty, no signs of disorders, dry hair coat; 5 = thrifty, no signs of disorders, sleek hair coat; 6 = very large, healthy, and vigorous.

[2]Ratio calculated as follows: $\frac{\text{individual record}}{\text{avg. of all calves on same farm and same season}} \times 100$

Horn Brand or Neck Chain No. __________

Fig. 5-48a. Individual Cow Record (see Fig. 5-48b for reverse side of record form).

IMMUNIZATION AND TEST RECORD

Immunizations					Health Tests							Remarks
Date[1]	Blklg.	M. Edema	Bangs	Misc.	TB-Bangs	Johnes	Lepto.	Anaplas.	Vib.	Trich.	Misc.	

[1]Indicate vaccinations by check in appropriate column opposite date given; indicate test results by P (positive), N (negative), or S (suspect) opposite date of test.

GENERAL INFORMATION

Record all facts pertinent to the history of this cow, viz.: veterinary treatment (except immunizations), udder condition, mothering instinct, calving peculiarities, etc.

Date	Remarks

Fig. 5-48b. Individual Cow Record. This is the reverse side of the record form shown in Fig. 5-48a.

There is need, therefore, to know the approximate amount or percentage of change in each economically important trait which is due to heredity and the amount which is due to environment. Table 5-5 gives this information for beef cattle in terms of the approximate percentage heritability of each of the economically important traits. The heritability figures given therein are averages based on large numbers; thus, some variations from these may be expected in individual herds. Even though the heritability of many of the economically important traits listed in Table 5-5 is disappointingly small, it is gratifying to know that much of it is cumulative and permanent.

ESTIMATING RATE OF PROGRESS

For purposes of illustrating the way in which the heritability figures in Table 5-5 may be used in practical breeding operations, the following example is given:

In a certain beef cattle herd, the calf crop in a given year averages a weaning weight of 500 lb, with a range of 400 to 700 lb. There are available sufficient of the heavier weaning calves weighing 600 lb from which to select replacement breeding stock. What amount of this heavier weaning weight (100 lb above the average) is likely to be transmitted to the offspring of these heavier weaning calves?

Step by step, the answer to this question is secured as follows:

1. 600 − 500 = 100 lb, the amount by which the selected calves exceed the average from which they arose.

2. By referring to Table 5-5, it is found that weaning weight is 25% heritable. This means that 25% of the 100 lb can be expected to be due to the superior heredity of the stock saved as breeders, and that the other 75% is due to environment (feed, care, management, etc.).

3. 100 × 25% = 25 lb; which means that for weaning weight the stock saved for the breeding herd is 25 lb superior, genetically, to the stock from which it was selected.

4. 500 + 25 = 525 lb weaning weight; which is the expected performance of the next generation.

It is to be emphasized that the 525 lb weaning weight is merely the expected performance. The actual outcome may be altered by environment (feed, care, management, etc.) and by chance. Also, it should be recognized that where the heritability of a trait is lower less progress can be made. The latter point explains why the degree to which a character is heritable has a very definite influence in the effectiveness of mass selection.

APPRAISING FACTORS INFLUENCING RATE OF PROGRESS

Cattle breeders need to be informed relative to the factors which influence the rate of progress that can be made through selection. They are:

1. **The heritability of the character.** When heritability is high, much of that which is selected for will appear in the next generation, and marked improvement will be evident.

2. **The number of characters selected for at the same time.** The greater the number of characters selected for at the same time, the slower the progress in each. In other words, greater progress can be attained in 1 character if selection is made for it alone. For example, if selection of equal intensity is practiced for 4 independent traits, the progress in any 1 will be only ½ of that which would occur if only 1 trait were considered; whereas selection for 9 traits will reduce the progress in any 1 to ⅓. This emphasizes the importance of limiting the traits in selection to those which have greatest importance as determined by economic value and heritability. At the same time, it is recognized that it is rarely possible to select for 1 trait only, and that income is usually dependent upon several traits.

3. **The genotypic and phenotypic correlation between traits.** The effectiveness of selection is lessened by (a) negative correlation between two desirable traits, or (b) positive correlation of desirable with undesirable traits.

4. **The amount of heritable variation measured in such specific units as pounds, inches, numbers, etc.** If the amount of heritable variation—measured in such specific units as pounds, inches, or numbers—is small, the animals selected cannot vary much above the average of the entire herd, and progress will be slow. For example, there is much less spread, in pounds, in the birth weights of calves than in weaning weights. Therefore, more marked progress in selection can be made in the older weights than in birth weights, when measurements at each stage are in pounds.

5. **The accuracy of records and adherence to an ideal.** It is a well-established fact that a breeder who maintains accurate records and consistently selects toward a certain ideal or goal can make more rapid progress than one whose records are inaccurate and whose ideals change with fads and fancies.

6. **The number of available animals.** The greater the number of animals available from which to select, the greater the progress that can be made. In other

words, for maximum progress, enough animals must be born and raised to permit rigid culling. For this reason, more rapid progress can be made with swine than with animals that have only one offspring, and more rapid progress can be made when a herd is either being maintained at the same numbers or being reduced than when it is being increased in size.

7. **The age at which selection is made.** Progress is more rapid if selection is practiced at an early age. This is so because more of the productive life is ahead of the animal, and the opportunity for gain is then greatest.

8. **The generation interval.** Generation interval refers to the period of time required for parents to be succeeded by their offspring, from the standpoint of reproduction. The minimum generation interval of farm animals is about as follows: horses, 4 years; cattle, 3 years; sheep, 2 years; and swine, 1 year. In actual practice, the generation intervals are somewhat longer. By way of comparison, it is noteworthy that the average length of a human generation is 33 years.

Shorter generation intervals will result in greater progress per year, provided the same proportion of animals is retained after selection.

Usually it is possible to reduce the generation interval of sires, but it is not considered practical to reduce materially the generation interval of females. Thus, if progress is being made, the best young males should be superior to their sires. Then the advantage of this superiority can be gained by changing to new generations as quickly as possible. To this end, it is recommended that the breeder change to younger sires whenever their records equal or excel those of the older sires. In considering this procedure, it should be recognized, however, that it is very difficult to compare records made in different years or at different ages.

9. **The caliber of the sires.** Since a much smaller proportion of males than of females is normally saved for replacements, it follows that selection among the males can be more rigorous and that most of the genetic progress in a herd will be made from selection of males. Thus, if 2% of the males and 50% of the females in a given herd become parents, then about 75% of the hereditary gain from selection will result from the selection of males and 25% from the selection of females, provided their generation lengths are equal. If the generation lengths of males are shorter than the generation lengths of females, the proportion of hereditary gain due to the selection of males will be even greater.

DETERMINING RELATIVE EMPHASIS TO PLACE ON EACH CHARACTER

A replacement animal seldom excels in all of the economically important traits. The cattle producer must decide, therefore, how much importance will be given to each factor. Thus, the producer will have to decide how much emphasis shall be placed on birth weight, how much on weaning weight, how much on daily rate of gain, how much on efficiency of feed utilization, and how much on body type and carcass evaluation.

Perhaps the relative emphasis to place on each trait should vary according to the circumstances. Under certain conditions, some characters may even be ignored. Among the factors determining the emphasis to place on each trait are the following:

1. **The economic importance of the trait to the producer.** Table 5-5 lists the economically important traits in cattle, and summarizes (see comments column) their importance to the producer.

By economic importance is meant their dollars and cents value. Thus, those traits which have the greatest effect on profits should receive the most attention.

2. **The heritability of the trait.** It stands to reason that the more highly heritable traits should receive higher priority than those which are less heritable, for more progress can be made thereby.

3. **The genetic correlation between traits.** One trait may be so strongly correlated with another that selection for one automatically selects for the other. For example, rate of gain and economy of gain in beef cattle are correlated to the extent that selection for rate of gain tends to select for the most economical gains as well; thus, economy of gain may be largely disregarded if rate of gain is given strong consideration. Conversely, one trait may be negatively correlated with another so that selection for one automatically selects against the other.

4. **The amount of variation in each trait.** Obviously, if all animals were exactly alike in a given trait, there could be no selection for that trait. Likewise, if the amount of variation in a given trait is small, the selected animals cannot be very much above the average of the entire herd, and progress will be slow.

5. **The level of performance already attained.** If a herd has reached a satisfactory level of performance for a certain trait, there is not much need for further selection for that trait.

It should be recognized, however, that sufficient selection pressure should be exerted to maintain the desired excellence of a given trait; for once selection for many of the economic quantitative traits is relaxed, there is a tendency for the trait to regress rather rapidly toward the average of the breed. For simple quantitative traits (controlled by a single pair of genes), it may be possible to rid the herd of the undesired gene, following which selection against the trait could be dropped, except when adding outside animals to the herd.

EXPECTED PROGENY DIFFERENCE (EPD)

EPDs are the difference in performance expected from the offspring of one sire compared to the offspring of another sire in the same breed.

Sire summaries are easily obtained from the various breed associations.

Characteristics that can be selected from EPDs may include: birth weight, weaning weight, milk production, yearling weight, frame score, scrotal circumference, and carcass quality. Each EPD is assigned an accuracy value which can help to determine its reliability. The more data available for the calculation of an EPD, the greater the accuracy. Each EPD has a different accuracy value which may range from 0.00 to 1.00, with values closest to 1.00 indicating greatest reliability.

(Also see Chapter 4, section headed "Selection Based on Expected Progeny Difference [EPD].")

TYPES OF SELECTION

Cattle producers need to use a type of selection which will result in maximum total progress over a period of several years or animal generations. The three common types are:

1. **Tandem selection.** *This refers to that type of selection in which there is selection for only one trait at a time until the desired improvement in that particular trait is reached,* following which selection is made for another trait, etc. This system makes it possible to make rapid improvement in the trait for which selection is being practiced, but it has two major disadvantages: (a) usually it is not possible to select for one trait only; and (b) generally income is dependent on several traits.

Tandem selection is recommended only in those rare herds where one trait only is primarily in need of improvement.

2. **Establishing minimum standards for each trait, and selecting simultaneously, but independently, for each trait.** This system, in which several of the most important traits are selected for simultaneously, is without doubt the most common system of selection. It involves establishing minimum standards for each trait and culling animals which fall below these standards. For example, it might be decided to cull all calves weighing less than 55 lb at birth, or weighing less than 375 lb at weaning, or gaining less than 1¼ lb daily, or requiring more than 900 lb of feed per 100-lb gain, or grading 12 or less. Of course, the minimum standards may have to vary from year to year if environmental factors change markedly (for example, if calves average light at weaning time due to a severe drought and poor pasture).

The chief weakness of this system is that an individual may be culled because of being faulty in one character only, even though he is well nigh ideal otherwise.

3. **Selection index.** *A selection index combines all important traits into one overall value or index.* Theoretically, a selection index provides a more desirable way in which to select for several traits than either (a) the tandem type, or (b) the method of establishing minimum standards for each character and selecting simultaneously, but independently, for each character.

Selection indexes are designed to accomplish the following:

a. To give emphasis to the different traits in keeping with their relative importance.

b. To balance the strong points against the weak points of each animal.

c. To obtain an overall total score for each animal, following which all animals can be ranked from best to poorest.

d. To assure a constant and objective degree of emphasis on each trait being considered, without any shifting of ideals from year to year.

e. To provide a convenient way in which to correct for environmental effects, such as age of dam, etc.

Despite their acknowledged virtues, selection indexes are not perfect. Among their weaknesses are the following:

a. Their use may result in covering up or masking certain bad faults or defects.

b. They do not allow for year to year differences.

c. Their accuracy is dependent upon (1) the correct evaluation of the net worth of the economic traits considered, (2) the correctness of the estimate of heritability of the traits, and (3) the genetic correlation between the traits; and these estimates are often difficult to make.

In practice, the selection index is best used as a partial guide or tool in the selection program. For example, it may be used to select twice as many animals as are needed for herd or flock replacements, and this number may then be reduced through rigid culling on the basis of a thorough visual inspection for those traits that are not in the index, which may include such things as quality, freedom from defects, and market type.

BEEF IMPROVEMENT FEDERATION (BIF) GUIDELINES[18]

Performance testing offers those engaged in beef production a way of measuring heritable differences among animals so producers can select those individuals which are expected to transmit superior performance to their offspring. But, for maximum value, there is need for uniformity in measuring, recording, and evaluating beef cattle performance data. The latter role has been filled by the Beef Improvement Federation, and is detailed in *BIF Guidelines for Uniform Beef Improvement Programs*, seventh edition, 1996. Some pertinent points from this valuable, but voluminous, report are briefed in the sections that follow.

Fig. 5-49. A linebred Angus bull from a performance program in the 36th year. (Courtesy, M. Jorgensen, Jorgensen Ranches, Ideal, SD)

■ **Features of performance records**—The principal features of effective records or performance programs are as follows:

1. All animals of a given sex and age are given equal opportunity to perform through uniform feeding and management.
2. Records of economically important traits on all animals are systematically maintained.
3. Records are adjusted for known sources of variation, such as age of dam, age of calf, and sex.
4. Records are used in selecting replacements (bulls and heifers) and in eliminating poor producers.
5. The nutritional regime and management practices are practical and comparable to those where the progeny of the herd are expected to perform.

■ **Traits**—The Beef Improvement Federation (BIF) suggests that the following traits be measured:

1. Reproduction.
2. Postweaning feed efficiency.
3. Beef carcass evaluation.
4. Live animal.

[18]The Beef Improvement Federation Executive Director is Ron Bolze of Kansas State University.

REPRODUCTION

Reproduction or fertility is the most important economic trait in beef cattle. Breeders should record reproductive performance in both the female and the male.

■ **Female**—Breeders can use specific measures of reproductive performance in the female to monitor overall reproductive performance, identify genetic and environmental areas in which to concentrate improvement efforts, and make routine selection and culling decisions. Data to record or calculate on the female are as follows:

1. Breeding dates.
2. Pregnancy status.
3. Calving date.
4. Calving difficulty or ease.
5. Birth weight of calves.
6. Calf history.
7. Cow reproduction or disposal history.
8. Gestation length.
9. Age at first calving.
10. Calving interval.
11. Yearly prolificacy.

■ **Male**—An examination of bulls for breeding soundness before the breeding season can detect the majority of bulls which have obvious potential fertility problems. This examination should be performed by a veterinarian or other experienced, competent personnel. Techniques presently available do not allow for accurate predictions for degrees of fertility. Results from an actual breeding season remain the only test of a bull's breeding ability. Guidelines for examination of bulls follow:

1. **Physical examination.**
   a. Palpate scrotum and testes.
   b. Rectally palpate internal glands.
   c. Examine extended penis and prepuce.
2. **Scrotal circumference.**
3. **Semen evaluation.**

■ **Pelvic measurements**. Many producers today are interested in using pelvic measurements as a management tool to assist in reducing the incidence and severity of calving difficulty. Many factors are associated with calving difficulty, including small first-calf heifers, large calves, male calves, small pelvic size of heifer, long gestation, condition score of cow and abnormal presentation. Research indicates that a disproportion between the calf size (birth weight) and female birth canal (pelvic area) can be a big contributor to calving difficulty.

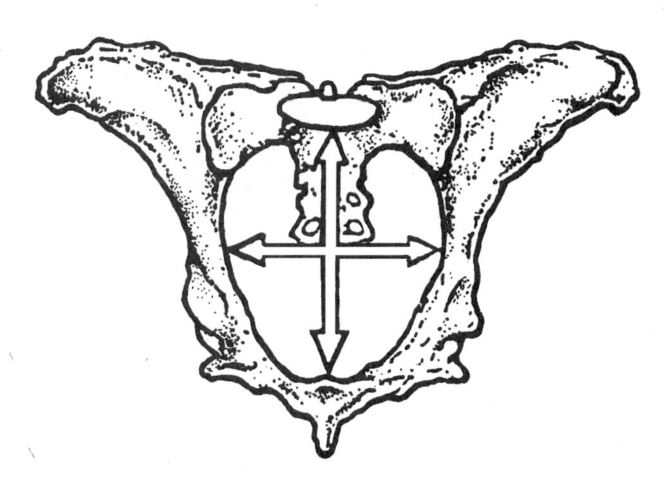

Fig. 5-50. Vertical and horizontal measurements are obtained to determine pelvic area.

1. **Measurement of heifers.** Since a large majority of the calving difficulty occurs in first calf heifers (calving at 22 to 28 months of age), many producers take pelvic measurements in their heifers at 12 months of age.

2. **Measurement of bulls.** To increase the pelvic area of their heifers, many producers are selecting bulls with a larger pelvic area. Pelvic size has been shown to be 60% heritable, indicating that selection for large pelvic size in bulls should result in increasing pelvic size in females.

GROWTH RATE AND EFFICIENCY OF GAIN

Growth rate and efficiency of gain for beef cattle are of primary economic importance to the beef industry. Growth rate has a direct effect on net return and is positively correlated with efficiency of gain, weight, and value of retail product. Efficiency of gain has a direct effect on cost of production and net return.

■ **Growth rate**—The following measurements of growth rate are recommended:

1. **Preweaning phase.** The preweaning phase is evaluated by—
   a. Measurement of birth weight.
   b. Measurement of weaning weight (205 days).
2. **Postweaning phase—on the farm and ranch.** Measurement of yearling weight (365 days) or long yearling weight (452 or 550 days).

■ **Postweaning feed efficiency**—Efficiency of gain in beef production is the ratio of nutrient input to beef output or the ratio of beef output to nutrient input. On a life cycle basis for either herds or individual cows, about 70% of the energy required for beef production is consumed by cows and bulls in the breeding herd and by their calves prior to weaning. About 30% of the energy required for beef production from weaning to slaughter is consumed by steers and heifers during the postweaning period. Depending on the cost of feed resources for the cowherd in relation to those for the feedlot, 45 to 55% of the total feed costs are incurred in the period from weaning to slaughter.

BEEF CARCASS EVALUATION

Edible beef products are the goal of all beef cattle improvement programs and activities. Quality of product and quantity of edible portion are the basic factors used to judge carcass merit. However, the relative values of quality and quantity are subject to change as market demands change.

LIVE ANIMAL

Evaluation of live animals takes into consideration any measurements or subjective evaluations that help describe an animal. For example, evaluation involves physical examination of bulls to include penis, rectal examination, and scrotum (including scrotal circumference).

Some other common measurements of cattle include: backfat, pelvic size, height at the shoulder, height at the hip, and length of body.

In recent years, measurements for height have become a descriptive supplement to many herd testing programs. Adjusted weights and weight ratios accompanied by linear measurements for height have added another dimension to evaluating the fat-lean ratio of an individual animal in a performance program.

The recommended point for linear measurement for height is to a point directly over the hooks (Fig. 5-51). This measurement is adjusted to relatively logical production end points at 205 days and 365 days (with the BIF ranges currently used for adjusted weights).

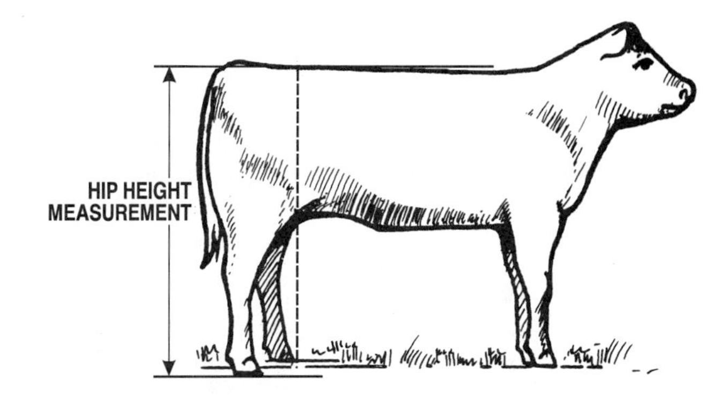

Fig. 5-51. Height measurement.

GUIDELINES FOR SEEDSTOCK (PUREBRED) PRODUCERS

The primary responsibility for maintaining and improving the traits associated with efficient production lies in the hands of the purebred or seedstock producers. Only 3% of the national cowherd is owned by registered cattle breeders. Genetic change in the commercial beef cattle population is controlled by the genetic merit of bulls produced by these breeders. It is their responsibility to know the specific needs of their customers and to produce registered cattle that can help to fulfill these needs.

There are two aspects of performance programs for seedstock producers: (1) programs for individual herds which are planned and controlled by individual breeders, and (2) the programs of seedstock organizations which are planned and controlled by groups of breeders.

GUIDELINES FOR COMMERCIAL PERFORMANCE PROGRAMS

Commercial beef producers need performance record programs that are both workable and affordable.

A performance program for a purebred breeder differs from that of a commercial producer in that the commercial producer sells *pounds* whereas the purebred breeder sells *breeding value*. Breeding value is the value of an individual as a parent. Commercial producers sell pounds and need to buy breeding value as well as combine breeds in logical combinations to obtain the crossbred advantage, especially for the traits in the reproductive complex. Thus, both commercial and seedstock breeders can benefit from understanding what breeding value is and from using performance records in their herds.

■ **Expected Progeny Differences (EPDs) of young bulls**—Because of new analysis procedures used by breed associations, data on young (non-parent) bulls is now available which is superior to individual performance records alone. This new data is in the form of *Expected Progeny Differences (EPDs)* for such traits as birth weight, weaning weight, yearling weight, maternal weaning weight, and maternal milk. In addition, some breed associations present EPDs for height, scrotal circumference, gestation length, calving ease direct, maternal calving ease, stayability, and carcass characteristics. An EPD is a prediction of how future progeny of a sire are expected to perform for a particular trait as compared to other individuals in a breed. "Difference" is the key to understanding EPDs. Each EPD has an accuracy figure associated with it which is a reliability measure and on non-parent bulls will be relatively low. The same kind of data is available for progeny-proven bulls in breed sire summaries and the accuracies on their data may be much higher. Comparisons of EPDs should be made within breed only.

Sometimes EPDs may not be available on yearling bulls. However, if we assume an individual receives half of his genetic component from his sire and half from his dam, we can estimate his EPD if we have access to pedigree performance for the various traits as follows:

$$\text{Estimated } EPD_{\text{Young Bull}} = \frac{EPD_{\text{Sire}} + EPD_{\text{Dam}}}{2}$$

Thus, to make the fewest mistakes and the greatest progress, progeny proven bulls with relatively high accurate EPDs should be used by AI when possible.

In selecting young sires for natural service, the best avenue is to select sons of superior, progeny-proven sires which are superior based upon non-parent EPDs and individual performance records. Groups of young sires selected in this manner are more apt to produce as expected and will cause real genetic improvement in a commercial herd.

CENTRAL TEST STATIONS

Central test stations are locations where animals from several herds are assembled to evaluate differences in certain performance traits under uniform conditions. Uses of central test stations include: (1) comparing individual performance of seedstock herd sire prospects and commercial bulls, (2) finishing steers and/or heifers in progeny-testing programs, and (3) educating purebred breeders and commercial producers on the value and utilization of performance records.

The following general procedures and policies are recommended for all types of central bull tests:

1. Calves should be 180–270 days old at time of delivery to the test.
2. Test groups should have a maximum 90-day age spread.
3. Ratios should be calculated only in each breed within a test group.
4. Consignors' herds should be enrolled on their respective breed association performance program. State beef cattle improvement programs are acceptable for herds whose breed associations do not have performance record programs.
5. Calves should have completed the weaning phase of the performance record program with their contemporary group.
6. Initial and final test weights may be either full or shrunk weights. If full weights are taken, initial and final weights should be an average of two weights taken on consecutive days to minimize fill effects. If shrunk weights are taken, a single weight after a shrink of 12 hours would be adequate.

7. All bulls sold in a central test station sale should be evaluated by competent personnel for structural soundness.

8. All bulls sold in a central test station sale should be required to pass a complete Breeding Soundness Exam conducted by a qualified veterinarian or reproductive physiologist.

9. Nutritional programs should provide adequate levels of protein and energy for the expression of genetic differences in growth between bulls. Test ration composition and analysis should be stated in all test station sale catalogs.

10. Sire group testing of bulls is more desirable than individual testing because it provides more information to both breeders and prospective buyers.

11. Expected Progeny Differences (EPDs) for birth weight, weaning weight, yearling weight, maternal milk, and direct maternal calving ease (if available), with their corresponding accuracies, should be provided in all test station sale catalogs.

12. EPD distributions (percentile rankings) by trait within breed for bulls of a given birth year should be presented in all test station sale catalogs to allow entire breed population comparisons.

NATIONAL CATTLE EVALUATION PROGRAM

The current national cattle evaluation program employs expected progeny differences (EPDs). So, it is important that cattle producers know what EPDs are and how to use them.

An expected progeny difference (EPD) is a prediction, based on available data, of one-half the breeding value of an animal. It is a prediction of what the animal is expected to transmit to its future offspring, expressed as a difference from the average in the units of measure.

Data from breed performance programs are used to predict EPDs for all animals of the breed on traits that are routinely measured.

Designed sire evaluations for specific traits, especially measures of carcass value, is being done.

The magnitude of the numbers of EPDs generated in national animal evaluation programs dictates that computers be used to make selections. Coupled with this are published sire summaries that include young bulls because of the importance of sire selection. Summaries of breeding herd EPDs, including interim predictions, need to be sent in timely fashion.

Major genes that enhance performance discovered by breeders can now be exploited using the tools of biotechnology. The problem of undesirable genes is always present. The use of sires on a large cross-section of cows and the elimination of carriers of undesirable genes and their offspring are satisfactory methods of control. Also, sire-daughter matings provide another procedure when the sire is suspected of being a carrier, yet is valuable in terms of his EPD values.

EMBRYO TRANSFER

Recipient dam selection and post-birth environment of embryo transfer (ET) calves have created problems and raised important questions concerning the collection, analysis, and interpretation of accurate, meaningful performance records for ET-born calves.

Currently, each performance recording organization has its own method of dealing with performance records for ET calves with little uniformity in handling the records between organizations. Until such time when sufficient data for ET-born calves have been researched and analyzed where better recommendations can be made for handling ET performance data, the Beef Improvement Federation has detailed recommended guidelines.

THE INTEGRATED SYSTEMS CONCEPT FOR CATTLE PRODUCTION AND IMPROVEMENT

The systems concept of beef production incorporates an awareness that there is more to consider in a beef cattle enterprise than simply the level of production. What is most important is the overall efficiency of the enterprise—in other words, net return. While level of production is an important factor affecting profitability, costs of production are equally important.

The "systems" part of the concept implies that a beef operation is really a system of many components, all of which play a part in determining net return. These components might be categorized in the following way: natural environment; costs, prices, and market requirements; cattle type; crossbreeding system (examples: rotational crossbreeding, use of large terminal sires to produce market calves only); and management practices (examples: supplementation, backgrounding, retained ownership through slaughter).

A beef production system is highly complex, both because of the large number of factors affecting the system, and because of the high degree of interaction of these factors. For example, the management practice of creep feeding might be advisable for one type of cattle in one environment given certain ranges of costs for creep feed and prices for feeder cattle. Change the cattle, the environment or the economics, however, and creep feeding may no longer pay.

To use the systems concept is to see the beef cattle operation in its entirety—to see how its component parts interact with one another to ultimately affect profitability. Good cattle producers have been doing this for years.

Determining exactly what is the *best* animal for a specific situation is difficult because there are so many traits of importance in beef cattle and so many trade-offs among these traits. For example, increased size and milk production contribute to heavier weaning weights, but may create stresses which sometimes can depress fertility. Cattle which are more productive in the sense that they produce larger, leaner, faster growing calves can be more risky. For this reason, a major element of the systems concept as it applies to cattle type is the avoidance of extremes in production traits. The largest, leanest, heaviest milking cattle are not always the most profitable. For these traits, intermediate levels of performance may be optimal.

The systems concept of beef production presents challenges to both the commercial producer and seedstock producer. For the commercial operator, the challenge is to combine cattle and management alternatives in a way which maximizes net return. For the seedstock producer, the challenge is to breed the kind of cattle which best fit the commercial cow-calf production system. This implies breeding cattle for specific purposes. One breeder may be producing cattle for the Corn Belt, another for the Arizona desert. One may specialize in bulls for virgin heifers, another in terminal sires, and another in general purpose cattle. All, however, can be breeders of systems cattle.

QUESTIONS FOR STUDY AND DISCUSSION

1. Define biotechnology. List and discuss two examples of its application in the beef cattle industry.

2. Who were James Watson and Francis Crick? What did they do to be awarded the Nobel Prize in Physiology and Medicine in 1962?

3. What unique circumstances surrounded the founding of genetics by Mendel?

4. Why did Mendel's great research work remain unheralded for 34 years?

5. Name the three main parts of cells, and describe each part.

6. What is meant by *karyotype*?

7. How many chromosomes are there in each of the following species: cattle, horses, sheep, swine, humans?

8. Discuss the relationship of genes and DNA, and explain how they function.

9. What is *genetic engineering*? Is it good or bad? Justify your answer.

10. Do you approve or disapprove of animals being patented? Defend your stand.

11. Define *cloning*. What advantages would accrue from having a herd of 500 cloned beef cows?

12. Under what conditions might a theoretically completely homozygous state in cattle be undesirable and unfortunate?

13. What is a mutation? Give an example of a helpful cattle mutation.

14. Diagram the inheritance of color in Shorthorn cattle of a red X roan mating.

15. How can you determine whether a polled bull is pure for the polled character?

16. Give examples of monohybrid, dihybrid, and trihybrid crosses in cattle; and describe each: (a) the first cross animals, and (b) the second cross animals, and give the ratio of the second crosses.

17. Why do not fraternal (dizygotic) twins look more alike?

18. Why do most cattle producers prefer single births to twins?

19. Explain how sex is determined.

20. When abnormal animals are born, what conditions tend to indicate each: (a) a hereditary defect, and (b) a nutritional deficiency?

21. Give the expected genetic picture of dwarfism of 100 offspring from matings of (a) carrier bulls X noncarrier cows, and (b) carrier bulls X carrier cows. What steps can be taken to get rid of dwarfism?

22. Is double muscling good or bad? Justify your answer.

23. The "sire is half the herd"! Is this an understatement or an overstatement? Justify your answer.

24. Define and evaluate the importance of each of the following in cattle breeding: *prepotency*, *nicking*, and *family names*.

25. In order to make intelligent selections and breed progress, is it necessary to fit, stall-feed, or place animals in show condition; or may they be selected in their "work clothes" off the farm or ranch?

26. Define *heterosis*. What is the genetic explanation for hybrid vigor?

27. Define the term *complementarity*, and give an example.

28. Challenge the following statements: In the future most breeds of cattle either will be (a) propagated as straightbreds or composites, or (b) used for crossbreeding. The vast majority of straightbreds will be purebreds and production tested.

29. Discuss each of the following systems of breeding: purebreeding inbreeding, outcrossing grading up, crossbreeding, and composites. What system of breeding do you consider to be best adapted to your herd, or to a herd with which you are familiar? Justify your choice.

30. Crossbreeding and composites will continue to play an important role in the production of market cattle in the future because of their several advantages. List these advantages.

31. Under what circumstances would you recommend each of the following: (a) a two-breed cross (b) a two-breed backcross or crisscross, (c) a three-breed rotation cross, and (d) a three-breed fixed or static cross?

32. List the disadvantages of using a crossbred bull.

33. Cite examples in poultry that show that not all hybrids excel purebreds.

34. Present pertinent information relative to the reproductive ability of the American buffalo.

35. It is generally agreed that crossbreeding and composites have made for a lack of consistency in beef, and that this is a serious matter. How can this situation be rectified?

36. How would you go about performance testing a herd of beef cattle? List and discuss each step.

37. List the factors which should be considered in determining the relative emphasis to place on each trait in a beef cattle selection program.

38. List the characteristics for which selection can be made in EPDs.

39. Evaluate the cattle record forms that are presented in Chapter 5.

40. What type of selection—(a) tandem, (b) establishing minimum culling levels, or (c) selection index—would you recommend and why?

41. What is the Beef Improvement Federation? Of what importance are its guidelines?

42. Discuss the systems concept for cattle production and improvement.

SELECTED REFERENCES

Title of Publication	Author(s)	Publisher
Agricultural Biotechnology	B. R. Baumgardt M. A. Martin	Purdue University, West Lafayette, IN, 1991
Animal Breeding	A. L. Hagedoorn	Crosby Lockwood & Son, Ltd., London, England, 1950
Animal Breeding	L. M. Winters	John Wiley & Sons, Inc., New York, NY, 1948
Animal Breeding Plans	J. L. Lush	Collegiate Press, Inc., Ames, IA, 1965
Artificial Insemination and Embryo Transfer of Dairy and Beef Cattle	H. A. Herman J. R. Mitchell G. A. Doak	Interstate Publishers, Inc., Danville, IL, 1994
Beef Production and Management Decisions	R. E. Taylor	Macmillan Publishing Co., New York, NY, 1994
Breeding Better Livestock	V. A. Rice F. N. Andrews E. J. Warwick	McGraw-Hill Book Co. Inc., New York, NY, 1953
Breeding and Improving Farm Animals, Seventh Edition	E. J. Warwick J. E. Legates	McGraw-Hill Book Co., Inc., New York, NY, 1979

Title	Author(s)	Publisher
Breeding Livestock Adapted to Unfavorable Environments	R. W. Phillips	Food and Agriculture Organization of the United Nations, Rome, Italy, 1949
Crossbreeding Beef Cattle, Series 2	M. Koger T. J. Cunha A. B. Warnick	University of Florida Press, Gainesville, FL, 1973
Elements of Genetics, The	C. D. Carlington K. Mather	The Macmillan Co. , New York, NY, 1950
Farm Animals	J. Hammond	Edward Arnold & Co., London, England, 1952
Genetic Resistance to Disease in Domestic Animals	F. B. Hutt	Comstock Publishing Assn., Cornell University Press, Ithaca, NY, 1958
Genetics, Second Edition	M. W. Strickberger	The Macmillan Co., New York, NY, 1976
Genetics and Animal Breeding	I. Johansson J. Rendel	Oliver and Boyd, Ltd., Edinburgh, Scotland, 1968
Genetics Is Easy, Fourth Revised Edition	P. Goldstein	Lantern Press, Inc., New York, NY, 1967
Genetics of Livestock Improvement, Third Edition	J. F. Lasley	Prentice Hall, Inc., Englewood Cliffs, NJ, 1978
Guidelines for Uniform Beef Improvement, Seventh Edition	Beef Improvement Federation	Beef Improvement Federation, Colby, KS, 1996
How Life Begins	J. Power	Simon and Schuster, Inc., New York, NY, 1965
Improvement of Livestock	R. Bogart	The Macmillan Co., New York, NY, 1959
Journal of Animal Science, Supplement 3, Vol. 71, 1993	Ed. by H. D. Hafs	American Society of Animal Science, Champaign-Urbana, IL, 1993
Lasater Philosophy of Cattle Raising, The	L. M. Lasater	Texas Western Press, The University of Texas, El Paso, TX, 1972
Livestock Improvement, Fourth Edition	J. E. Nichols	Oliver and Boyd, Ltd., Edinburgh, Scotland, 1957
Modern Developments in Animal Breeding	I. M. Lerner H. P. Donald	Academic Press, Inc., New York, NY, 1966
Principles of Genetics, Second Edition	I. H. Herskowitz	The Macmillan Co., New York, NY, 1977
Problems and Practices of American Cattlemen, Wash. Ag. Exp. Sta. Bull. 562	M. E. Ensminger M. W. Galgan W. L. Slocum	Washington State University, Pullman, WA, 1955
Reproduction in Farm Animals, Fourth Edition	Ed. by E. S. E. Hafez	Lea & Febiger, Philadelphia, PA, 1980
Robert Bakewell—Pioneer Livestock Breeder	H. C. Pawson	Crosby Lockwood & Son Ltd., London, England, 1957

Progeny of one superior female achieved by superovulation, followed by embryo transfer. (Courtesy, Granada Land and Livestock, Co., Wheelock, TX)

PHYSIOLOGY OF REPRODUCTION IN CATTLE

Contents *Page*

Contents *Page*

Contents Page

Contents Page

Fig. 6-1. Cow with newborn calf. (Courtesy, *The American Hereford Journal*, Kansas City, MO)

Reproduction is the first and most important requisite of cattle breeding. Without a calf being born and born alive, the other economic traits are of academic interest only. Yet, 12% of the nation's cows never calve, and there is an appalling calf loss of 6% at birth.

Many outstanding individuals, and even whole families, are disappointments because they either are sterile or reproduce poorly. The subject of physiology of reproduction is, therefore, of great importance.

THE REPRODUCTIVE ORGANS OF THE BULL

The bull's functions in reproduction are: (1) to produce the male reproductive cells, the *sperm* or *spermatozoa*; and (2) to introduce sperm into the female reproductive tract at the proper time. In order that these functions may be fulfilled, operators should have a clear understanding of the anatomy of the reproductive system of the bull and of the functions of each of its parts. Fig. 6-3 shows the reproductive organs of the bull. A description of each part follows:

1. **Scrotum.** This is a diverticulum of the abdomen, which encloses the testicles. Its chief function is thermoregulatory; to maintain the testicles at temperatures several degrees lower than that of the body proper.

2. **Testicles.** The testicles of the mature bull measure 4 to 5 in. in length and 2 to 3 in. in width. Their primary functions are (a) the production of sperm, and (b) the production of the male hormone testosterone.

Once the animal reaches sexual maturity, sperm production in the seminiferous tubules—the glandular portion of the testicles, in which are situated the spermatogonia (sperm-producing cells)—is a continuous process. Around and between the seminiferous tubules are the interstitial cells which produce testosterone or androgen.

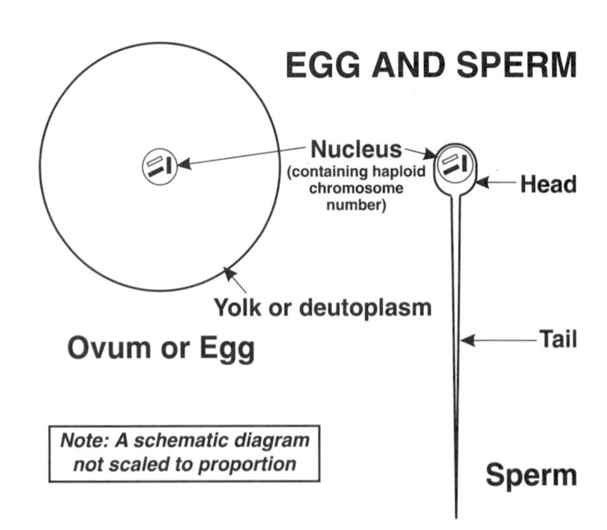

Fig. 6-2. Egg and sperm. The parent germ cells, the egg from the female and the sperm from the male, unite and transmit to the offspring all the characteristics that it will inherit.

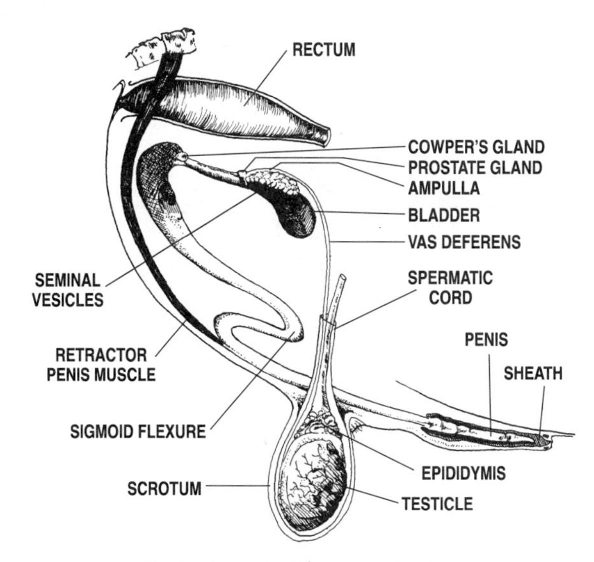

Fig. 6-3. Reproductive organs of the bull.

A sperm is a small (less than 1/500 in. in length), tadpole-shaped living entity, in which the head contains the unit of inheritance and the tail provides the means of locomotion.

Testosterone is essential for the development and function of male reproductive organs, male characteristics, and sexual drive.

Cryptorchids *are males one or both of whose testicles have not descended to the scrotum.* The undescended testicle(s) is usually sterile because of the high temperature in the abdomen.

The testicles communicate through the inguinal canal with the pelvic cavity, where accessory organs and glands are located. A weakness of the inguinal canal sometimes allows part of the viscera to pass out into the scrotum—a condition called *scrotal hernia.*

3. **Epididymis.** The efferent ducts of each testicle unite and form the epididymis. This long and greatly coiled tube consists of three parts:

a. **The head.** Consisting of several tubules which are grouped into lobules.

b. **The body.** The part of the epididymis which passes down along the sides of the testicle.

c. **The tail.** The part located at the bottom of the testicle.

The epididymis has four functions; namely, (1) as a passage way for sperm from the seminiferous tubules; (2) the storage of sperm; (3) the secretion of a fluid which probably nourishes the sperm; and (4) the maturation or ripening of the sperm.

4. **Vas deferens (ductus deferens).** This slender tube, which is lined with ciliated cells, leads from the tail of the epididymis to the pelvic part of the urethra. Its primary function is to move sperm into the urethra at the time of ejaculation.

The cutting or closing off of the vas deferens, known as *vasectomy,* is the most usual operation performed to produce sterility, where sterility without castration is desired.

5. **Spermatic cord.** The vas deferens—together with the longitudinal strands of smooth muscle, blood vessels, and nerves; all encased in a fibrous sheath—makes up the spermatic cord (two of them) which pass up through an opening in the abdominal wall, the inguinal canal, into the pelvic cavity.

6. **Seminal vesicles (or vesicula seminalis).** These compact glandular organs with a lobulated surface flank the vas deferens near its point of termination. They are the largest of the accessory glands of reproduction in the male. In the mature bull they measure 4 to 5 in. in length and 2 in. in width at their largest part and are located in the pelvic cavity.

7. **Prostate gland.** This gland is located at the neck of the bladder, surrounding or nearly surrounding the urethra and ventral to the rectum. The secretion of the prostate gland is thick and rich in proteins and salts. It is alkaline, and it has a characteristic odor.

It cleanses the urethra prior to and during ejaculation, and provides bulk and a suitable medium for the transport of sperm.

8. **Cowper's glands (bulbo-urethral glands).** These two glands, which are deeply imbedded in muscular tissue in the bull, are located on either side of the urethra in the pelvic region. They communicate with the urethra by means of a number of small ducts.

It is thought that these glands produce an alkaline secretion for the purpose of neutralizing or cleansing the urethra prior to the passage of semen.

9. **Urethra.** This is a long tube which extends from the bladder to the glans penis. The vas deferens and seminal vesicle open to the urethra close to its point of origin.

The urethra serves for the passage of both urine and semen.

10. **Penis.** This is the bull's organ of copulation. Also, it conveys urine to the exterior. It is composed of a small amount of erectile tissue, which, at the time of erection, becomes gorged with blood. Just behind the scrotum it forms an S-shaped curve, known as the sigmoid flexure, which allows for extension of the penis during erection. In the mature bull, the erected penis is about 3 ft long.

In total, the reproductive organs of the bull are designed to produce semen and to convey it to the female at the time of mating. The semen consists of two parts; namely (1) the sperm which are produced by the testicles, and (2) the liquid portion, or seminal plasma, which is secreted by the accessory sex glands: the seminal vesicles, the prostate, the ampulla, and the Cowper's glands. Actually, the sperm make up only a small portion of the ejaculate. On the average, at the time of each service, a bull ejaculates 4 to 7 cc of semen, containing about 6 to 10 billion sperm. The sperm concentration is about 1 to 1½ billion per cc.

THE REPRODUCTIVE ORGANS OF THE COW

The cow's functions in reproduction are: (1) to produce the female reproductive cells, the *eggs* or *ova*; (2) to develop the new individual, the *embryo*, in the uterus; (3) to expel the fully developed young at time of *birth* or *parturition*; and (4) to produce milk for the nourishment of the young. Actually, the part played by the cow in the generative process is much more complicated than that of the bull. It is imperative, therefore, that the modern cattle producer have a full understanding of the anatomy of the reproductive organs of the cow and the functions of each part. Figs. 6-4 and 6-5 show the reproductive organs of the cow, and a description of each part follows:

1. **Ovaries.** The two irregular-shaped ovaries of the cow are supported by a structure called the broad ligament, and lie rather loosely in the abdominal cavity 16 to 18 in. from the vulvar orifice. They average about 1½ in. in length, 1 in. in width, and ½ in. in thickness.

The ovaries have three functions: (a) to produce the female reproductive cells, the *eggs* or *ova*; (b) to secrete the female sex hormones, *estrogen* and *progesterone* (the latter is the hormone of the corpus luteum); and (c) to form the *corpora lutea*. The ovaries may alternate somewhat irregularly in the performance of these functions.

The ovaries differ from the testicles in that generally one egg is produced at intervals, toward the end

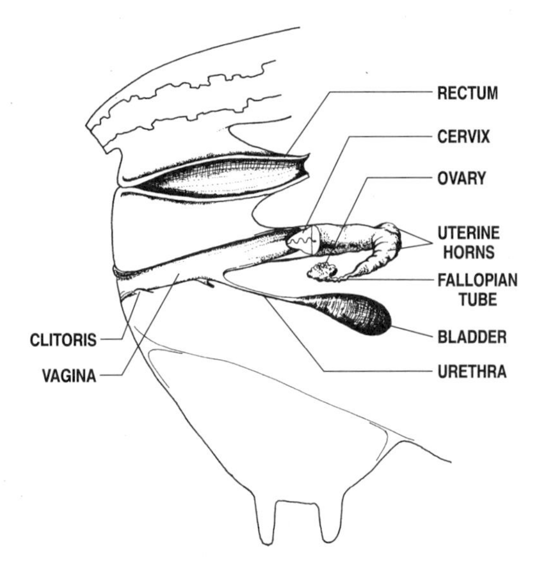

Fig. 6-4. Reproductive organs of the cow.

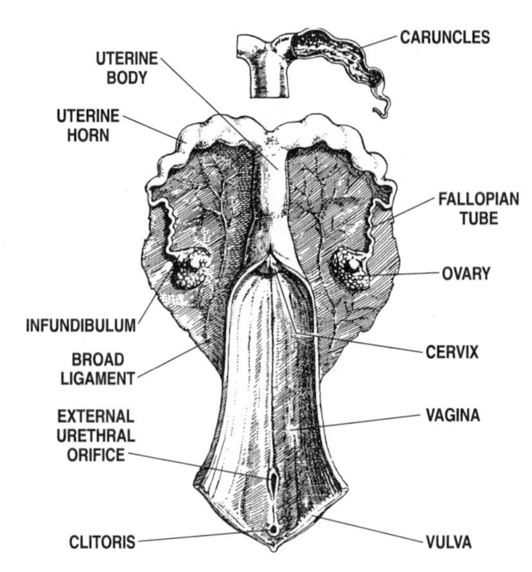

Fig. 6-5. Reproductive tract of the cow.

of the heat period or after heat symptoms have passed. Each miniature egg is contained in a sac, called *Graafian follicle*, a large number of which are scattered throughout the ovary. (It has been estimated that there are more than 75,000 follicles in the ovaries of a heifer calf at birth.) Generally, the follicles remain in an unchanged state until the advent of puberty, at which time some of them begin to enlarge through an increase in the follicular liquid within. Toward the end of heat, a follicle ruptures and discharges an egg, which process is known as *ovulation*. The corpus luteum develops in the ovulation site after ovulation. This corpus luteum secretes a hormone called progesterone, which (a) acts on the uterus so that it implants and nourishes the embryo, (b) prevents other eggs from maturing and keeps the animal from coming in heat during pregnancy, (c) maintains the animal in a pregnant condition, and (d) assists the female hormone in the development of the mammary glands. If the egg is not fertilized however, the corpus luteum atrophies and allows a new follicle to ripen and a new heat to appear. Occasionally the corpus luteum fails to atrophy at the normal time, thus inducing temporary sterility. This persistent corpus luteum can be treated with hormone therapy or manually ruptured.

The egg-containing follicles also secrete into the blood the female sex hormone estrogen. Estrogen is necessary for the development of the female reproductive system, for the mating behavior or heat of the female, for the development of the mammary glands, and for the development of the secondary sex characteristics, or femininity, in the cow.

From the standpoint of the practical cattle breeder, the ripening of the first Graafian follicle in a heifer generally coincides with puberty, and this marks the beginning of reproduction.

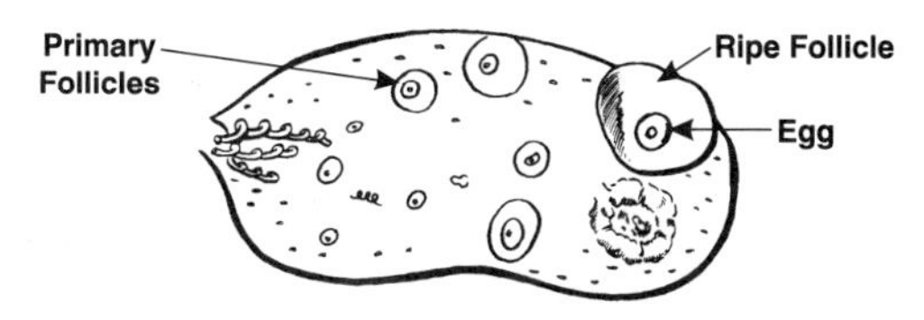

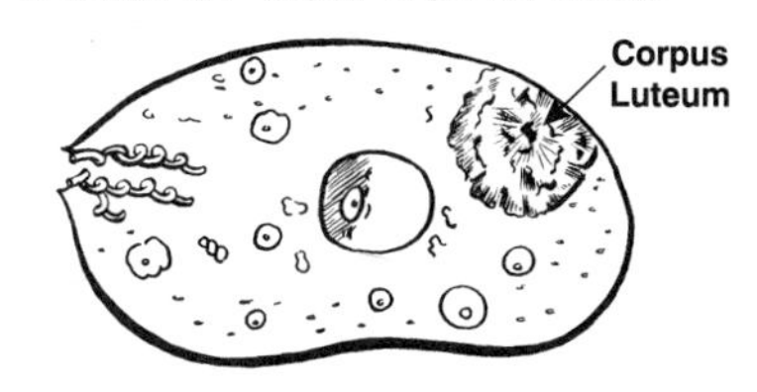

Fig. 6-6. Diagrammatic illustration of the ovary of the cow. *(Top)* Ovary of cow in heat. The ripe follicle secretes the estrogenic hormone responsible for the symptoms of heat. *(Bottom)* Ovary of cow not in heat. The corpus luteum is a glandular structure and secretes the hormone progesterone which maintains pregnancy in the pregnant cow.

2. **Fallopian tubes (or oviducts).** These small, cilia-lined tubes or ducts lead from the ovaries to the horns of the uterus. They are about 5 to 6 in. long in the cow, and the end of each tube nearest the ovary, called *infundibulum*, flares out like a funnel. They are not attached to the ovaries but lie so close to them that they seldom fail to catch the released eggs.

At ovulation, the egg passes into the infundibulum where, within a few minutes, the ciliary movement within the tube, assisted by the muscular movements of the tube itself, carries it down into the oviduct. If mating has taken place, the union of the sperm and egg usually takes place in the upper third of the Fallopian tube. Thence, the fertilized egg moves into the uterine horn. All this movement from ovary to the uterine horn takes place in approximately 5 days.

3. **Uterus.** The uterus is the muscular sac, connecting the Fallopian tubes and the vagina, in which the fertilized egg attaches itself and develops until expelled from the body of the cow at the time of parturition. The uterus consists of the 2 horns (cornua), the body and the neck (or cervix) of the womb. In the cow, the horns are about 15 in. long, the body about 1½ in. long, and the cervix about 4 in. long. In the mature cow, the uterus lies almost entirely within the abdominal cavity.

In the cow, the fetal membranes or placenta that surround the developing embryo are in contact with the lining of the uterus through placentomes, which consist of cotyledons (fetal side) and caruncles (maternal side).

The thick, muscular, fold-containing portion of the uterus, known as the cervix, forms an effective seal at the posterior end of the uterus. Cells within the cervix secrete copious amounts of mucus, forming the cervical plug. This gelatin-like material is discharged just prior to the occurrence of a normal heat period.

4. **Vagina.** The vagina admits the penis of the bull at the time of service and receives the semen. At the time of birth, it expands and serves as the final passageway for the fetus. In the nonpregnant cow, the vagina is 10 to 12 in. in length, but it is somewhat longer in the pregnant animal.

5. **Clitoris.** The clitoris is the erectile and sensory organ of the female, which is homologous to the penis in the male. It is situated just inside the portion of the vulva farthest removed from the anus.

6. **Urethra.** The urine makes its exit through the opening of the urethra.

7. **Vulva (urogenital sinus).** The vulva is the external opening of both the urinary and genital tracts.

The reproductive system of the cow is regulated by a complex endocrine system. The functions of the reproductive organs and the occurrence of estrus, conception, pregnancy, parturition and lactation are all regulated and coordinated by the hormones of the hypothalamus, the pituitary, the ovarian follicle, the corpus luteum, the uterus, and the placenta. The cyclic nature of these phenomena is shown in graphic form in Fig. 6-7.

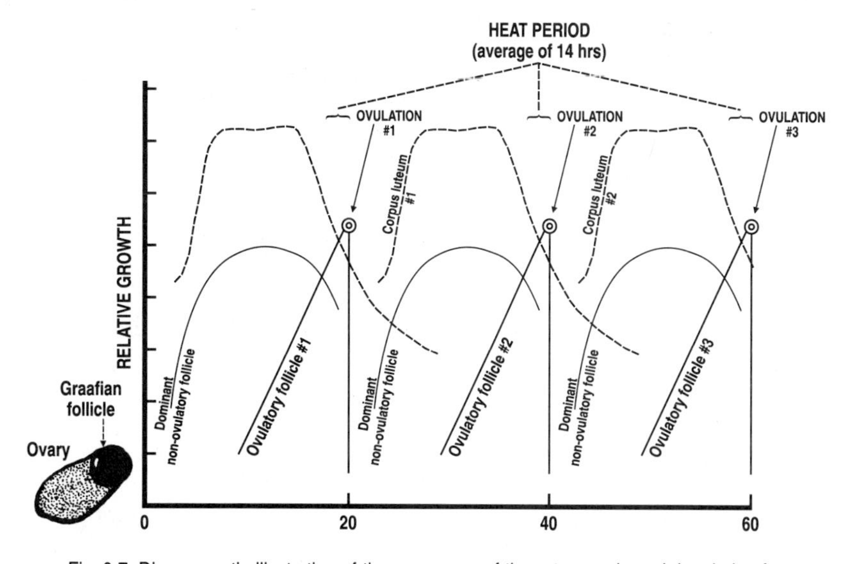

Fig. 6-7. Diagrammatic illustration of the recurrence of the estrus cycle and the chain of events which take place in the ovary of the nonpregnant cow. Continuous line shows the relative growth of the Graafian follicle. When maximum growth of the follicle is attained, rupture takes place and the ova is released (see double circles on top of diagram). At rupture, the size of the follicle reaches zero and the follicle is replaced by the corpus luteum (dotted line) which increases in size, reaches a maximum, then declines before the subsequent follicle starts to develop. Note that the rupture of the follicles coincides with the end of heat symptoms. The interval between two ovulations is 21 days on an average.

FERTILIZATION

Fertilization is the union of the male and female germ cells, the sperm and ovum. The sperm are deposited in the vagina at the time of service and from there ascend the female reproductive tract. Under favorable conditions, they meet the egg and one of them fertilizes it in the upper part of the oviduct near the ovary.

In cows, fertilization is an all or none phenomenon, since only one ovum is ordinarily involved. Thus, the breeder's problem is to synchronize ovulation and insemination; to ensure that large numbers of vigorous, fresh sperm will be present in the Fallopian tubes at the time of ovulation. This is very difficult, because (1) there is no reliable way of predicting the length of heat or the time of ovulation (it is known that ovulation generally takes place toward the end of or following the heat period; however, it may occur during the heat period or after it); (2) like all biological phenomena, there is considerable individual variation; (3) the sperm cells of the bull live only 24 to 48 hours in the reproductive tract of the female; (4) an unfertilized egg will not live over about 12 hours; and (5) it may require less than 1 minute for sperm cells to ascend the female reproductive tract of a cow.

From the above, it is perfectly clear that a series of delicate time relationships must be met; that breeding must take place at the right time. For the maximum rate of conception, therefore, it is recommended that breeding be done the latter part of the heat period; but, since the duration of heat in cattle is very short (seldom exceeding 20 hours), to delay too long may result in the cow being out of heat when mating is attempted. Cows that are detected in heat in the morning should be bred during the afternoon or evening of the same day, and those detected in the afternoon or evening should be bred early the next morning.

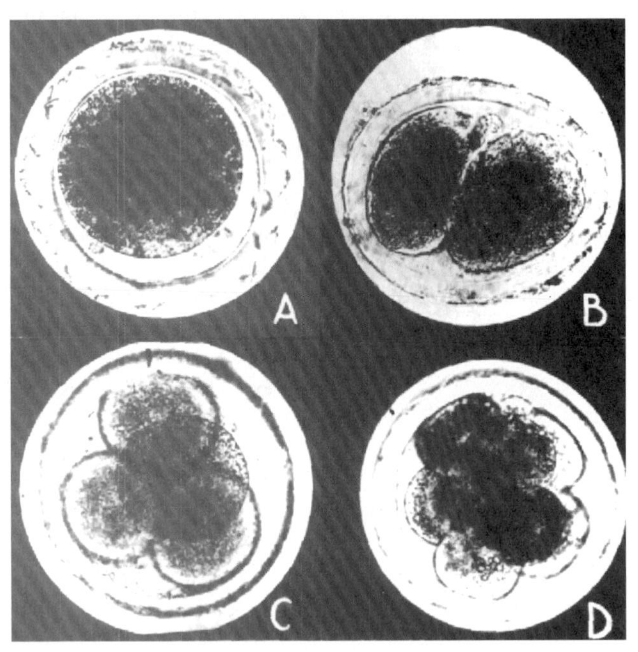

Fig. 6-8. Developmental changes in the egg at and after fertilization (these 4 stages all occur in the Fallopian tube).

A. A healthy egg in the 1-cell stage. Several spermatozoa attempt to fertilize the egg, but only 1 sperm will be able to do so. A few may succeed but die in the zona pellucida (outer circle) where the tails of the dead sperms may be seen.

B. Fertilization is followed by the multiplication of cells. An egg in the 2-cell stage; this stage in the cow takes place 24 to 30 hours after fertilization, *i.e.,* some 46 hours after the last symptoms of estrus.

C. The egg cleaves further (an increase in number of cells) without any increase in cellular mass. An egg in the 4-cell stage; this stage in the cow takes place 10 hours following stage **B**.

D. An egg in the 8-cell stage; this stage in the cow takes place 16 hours after stage **C**. After this stage the developing ovum migrates to the uterus where it will increase in cellular and protoplasmic mass.

(Courtesy, Dr. J. A. Laig, University of Bristol, Bristol, England)

NORMAL BREEDING HABITS OF COWS

In general, cattle that are bred when out on pasture or range are mated under environmental conditions approaching those which existed in nature prior to domestication. Less breeding trouble is generally encountered among such animals than among beef or dairy animals that are kept in confined conditions and under forced production.

AGE OF PUBERTY

The age of puberty in beef cattle ranges from 10 to 20 months. It varies according to (1) breed—British breeds, 10 to 12 months; Continental breeds, 12 to 16 months; Zebu breeds, 15 to 20 months; and (2) nutritional and environmental factors—occurring when beef heifers reach approximately 55 to 60% of mature weight.

HEAT PERIODS

The period of duration of heat—that is, the time during which the cow will take the bull—is very short, usually not over 16 to 20 hours, although it may vary from 6 to 30 hours. Cows tend to have a characteristic pattern of estrus behavior; for example, they come in heat during the morning hours, go out of heat in the evening or early part of night, and then ovulate approximately 28 hours after the beginning of heat (14 hours after the end of heat).

Females of all species bred near the end of the heat period are much more likely to conceive than if bred at any other time. The heat period recurs approximately at 21 day intervals, but it may vary from 19 to 23 days. In most cases, cows do not show signs of estrus until some 6 to 8 weeks after parturition, or in some instances even longer. Occasionally, an abnormal condition develops in cows that makes them remain in heat constantly. Such animals are known as *nymphomaniacs*.

SIGNS OF ESTRUS

It has been well said that "The cow that stands is the cow in heat," for this is the best single indicator of the heat period. Also, cows in heat usually exhibit one or more of the following characteristic symptoms: (1) rough hair on tailhead; (2) mud marks on side when the ground is wet; (3) nervousness; (4) bawling; (5) frequent urination; (6) mucus on rump and tail; and (7) a moist and swollen vulva. Dry cows and heifers usually show a noticeable swelling or enlargement of the udder during estrus, whereas in lactating cows a rather sharp decrease in milk production is often noted. A day or two following estrus, a bloody discharge is sometimes present.

The subject of "Heat Detection Methods and Devices" is presented in Chapter 10 of this book.

GESTATION PERIOD

The average gestation period of cows is 283 days, or roughly about 9½ months. Though there may be considerable breed and individual variation in the length of the gestation period, it is estimated that two-thirds of all cows will calve between 278 and 288 days after breeding.

INDUCED CALVING (Shortened Gestation)

Instead of letting "nature take its course," scientists are now artificially shortening gestation. The objectives of induced early calving are: (1) lowering birth weight of calves, thereby lessening parturition difficulty; (2) predicting calving dates in order to pool labor and concentrate watching; and (3) gaining a longer period from calving until rebreeding.

Females that have passed the 269th day of pregnancy will calve within 24 to 72 hours if injected intramuscularly with an adrenal steroid. Experimental work indicates that such induced calving will result in (1) 5 to 8 days earlier than normal calving, and (2) 6 to 8 lb lighter birth weight than calves carried to term. However, a higher incidence of retained placentas and lowered milk production accompany early calving. Antibiotics are recommended when the fetal membrane remains attached to the uterus longer than normal. Failure to expel the membranes after induced calving appears to have little effect on fertility as cows suffering this problem usually have no trouble breeding back.

As a result of a study involving 73 first calf heifers and 33 second calf cows, bred to the same Hereford sire, Bellows, *et al.*, USDA, Miles City, Montana, concluded that, when used correctly, creating brief labor by early obstetrical assistance improves subsequent pregnancy rate, and that induced parturition and early obstetrical assistance can be used separately or combined without detrimental effects on either the dam or the calf.[1]

FERTILITY IN BEEF CATTLE

Fertility refers to the ability of the male or female to produce viable germ cells capable of uniting with the germ cells of the opposite sex and of producing vigorous, living offspring. Fertility is lacking in very young animals, manifests itself first at puberty, increases for a time, then levels out, and finally recedes with the onset of senility. In cattle, as with other classes of farm animals, fertility is determined by heredity and environment.

In the wild state, when a female was served several times during her heat period, an annual conception rate of 90 to 100% was common rather than the exception. Aside from frequency of service, the outdoor exercise, vigor, good nutrition, and regular breeding habits and lack of contamination were conducive to conception. On the other hand, when handled under unnatural conditions in confinement, when the female is generally bred as soon as she starts to show

Fig. 6-9. The goal of every cattle producer is to have a 100% calf crop. This shows British White cows and calves on pasture. (Courtesy, Walter and Nancy Boharty, Bellwood, NE)

[1]Bellows, R. A., *et al.*, "Effects of Induced Parturition and Early Obstetrical Assistance in Beef Cattle," *Journal of Animal Science*, 1988, 66:1073–1080.

signs of heat, it is not surprising that the conception rate of cattle is rarely higher than 88% and is frequently much less.

Of course, the final test for fertility is whether young are produced, but unfortunately this test is both slow and expensive. Through evaluation of the quality of semen, it is possible to make a fairly satisfactory appraisal of the male's fertility; but no comparable measure of the female's relative fertility has yet been devised, although a pregnancy test may be made.

METHODS OF MATING

The two methods of mating beef cattle by natural service are: (1) hand mating; and (2) pasture mating. It has been found that 98% of the commercial cattle producers and 68% of the purebred breeders use pasture mating; the rest use hand mating.[2]

HAND MATING

In hand mating, the bull is kept separate from the cows at all times, except when an individual cow is to be bred and is turned in with him for this purpose. As a rule, in hand mating, only a single service is allowed, the cow being removed immediately after service. In the breeding of purebred cattle, when breeding records are so important, this method is frequently followed. Hand mating allows for a more accurate check on whether the bull is settling the cows. It also permits a larger number of cows to be served by a bull, an especially important consideration with a proved sire.

PASTURE MATING

In this system the bull is turned in with the herd, either throughout the entire year or during the breeding season. Even with pasture breeding, when it is desired to have the calves all come within a few weeks of each other thereby assuring more uniformity in size and offspring, the herd bull should be separated from the cows except during the breeding season. Uniformity in size is very important from the standpoint of marketing the calves advantageously. Also, by having the calves come as nearly as possible at one time, closer observation may be given the herd at the time of parturition.

Pasture breeding is most often followed with a commercial herd. As a rule, this system requires less labor, and there is less danger of missing cows when they are in heat. However, the convenience of pasture mating should not result in neglect to check whether the cows are being settled during the breeding season.

[2]Ensminger, M. E., M. W. Galgan, and W. L. Slocum, *Problems and Practices of American Cattlemen*, Wash. Ag. Exp. Sta. Bull. 562.

PREGNANCY TEST

Pregnancy tests can cut the wintering bill and make for increased profits. It is expensive business to overwinter a cow that will not produce a calf; necessitating feed, interest, labor, and other costs.

Barren cows can usually be marketed satisfactorily following testing; most feeders and packers will actually pay a premium for cows known to be open. Where valuable purebred animals are involved, bred cows can be sold with a more certain guarantee of being safely in calf, and barren cows may be accorded special care or hormone treatment. Additionally, and most important, early pregnancy detection gives early warning of breeding trouble, such as infertility in males and problem breeders in females.

Absence of heat is not always a sign of pregnancy, but a positive diagnosis can be made. By about 45 days of gestation, the uterus becomes enlarged, especially in the pregnant horn. An experienced technician can ascertain this sign of pregnancy by *feeling with the hand through the rectum wall.* Application of this method depends upon the recognition of changes in tone, size, and location of the uterine horns and changes in the uterine arteries. This is the most common test of pregnancy. It is popular because it affords early diagnosis, and there is little hazard when per-

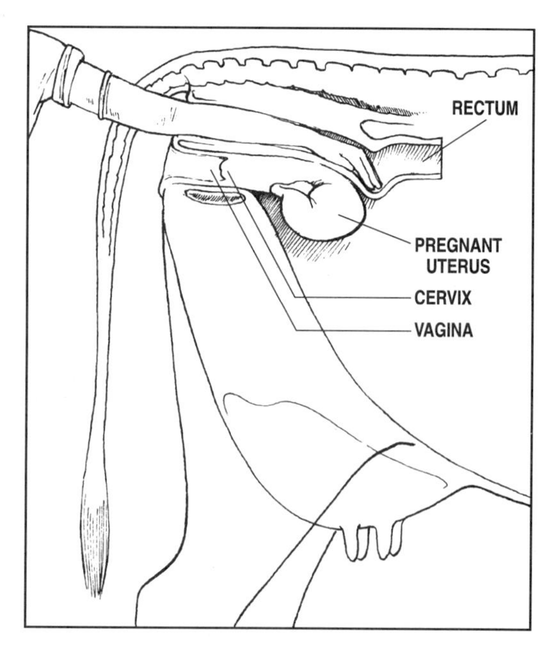

Fig. 6-10. Rectal method for determining pregnancy in the cow.

formed by experienced operators. It is recommended that cows be pregnancy tested, by this method, about two months after the bulls have been removed. Palpation takes only a few seconds. The speed with which pregnancy is determined depends largely on management of the cows as they come through the chutes, stage of pregnancy, and the experience of the tester. As many as 800 head of cows, or more, can be palpated in a normal working day under ideal conditions. When performed by an experienced person, rectal palpation is about 98% accurate in diagnosing pregnancy.

Other tests for pregnancy in cows are:

1. **Abdominal ballottement (bumping).** This may be used from the fifth to the seventh months of pregnancy. This consists in feeling the fetus by the following techniques: (a) place the hand or fist against the abdomen in the lower right flank region; and (b) execute a short, vigorous, inward-upward thrust in this region and retain the hand in place. The hard fetus may be felt. Because of the amniotic fluid, the technique described above will make the fetus recede, but it will fall back in place almost immediately.

2. **The formation of a uterine seal, or mucous plug, in the cervix.** The uterine seal, or mucous plug, which can be observed with the aid of a speculum and light, and which can also be determined by rectal palpation, is of some value in diagnosing pregnancy; but it is not always reliable.

3. **The fetal heart beat.** Sometimes this can be detected after the sixth month of pregnancy, though this method is not as certain in the cow as in other classes of farm animals. Use of a stethoscope is preferred, though good results are sometimes secured by merely placing the ear against the right lower abdominal region and listening. The fetal heart beat can be distinguished from that of the mother because of its greater frequency and lighter and higher pitch.

4. **Fetal movements.** Sometimes these can be observed through the abdominal wall during the latter half of pregnancy. This method of detecting pregnancy requires much patience. The observer simply must wait until voluntary movement of the fetus on the right side of the cow is observed. The practice of trying to induce movement of the fetus by allowing a very thirsty cow to take on a fill of cold water is cruel and is to be condemned.

5. **Progesterone in the milk.** Right off, it should be recognized that this pregnancy test involves taking a sample of milk; hence, it is best suited for use in dairy cattle.

In the early 1970s, English and American workers established that progesterone could be found, in milk, and that its levels therein reflected variations in the estrus cycle and the stage of pregnancy.

New York investigators reported that (a) cows diagnosed as pregnant averaged 7.12 nanograms of progesterone per milliliter (ng/ml); (b) nonpregnant cows averaged 2.36 ng/ml; and (c) at the time of estrus, cows averaged about 1.49 ng/ml.[3] (The term *ng/ml* refers to nanograms and indicates parts per billion in metric terms.)

No doubt, palpating the reproductive tract will continue to be the method of choice for determining pregnancy in beef cows. However, the milk progesterone test will be used increasingly in dairy herds.

6. **Electronic devices.** Several instruments are on the market for detecting pregnancy in animals. When used by an experienced operator, ultrasound is 95 to 98% accurate in determining pregnancy and accurate diagnosis can be made as early as 22 days after conception.

CARE OF THE PREGNANT COW

The nutritive requirements of the pregnant cow are less rigorous than those during lactation. In general, pregnant cows should be provided as nearly year-round pasture as possible. During times of inclement weather or when deep snows or droughts make supplemental feeding necessary, dry roughages and silage are the common feeds. If produced on fertile soils, such forage will usually provide all the needed nutrients for reproduction. Further discussion of the nutritive needs of pregnant cows is contained in Chapters 8 and 19.

No shelter is necessary except during periods of inclement weather. Normally, the cows will prefer to run outdoors. This desire is to be encouraged—in order to provide exercise, fresh air, and sunshine. Where and when shelter is necessary, it should be neither elaborate nor expensive. An open shed facing away from the direction of prevailing winds is quite as satisfactory for the protection of cows as a warm barn with individual box stalls—and it is far less expensive. The chief requirements are that the shelter be tight overhead, that it be sufficiently deep to afford protection from inclement weather and remain dry (depths of 34 to 36 ft are preferred), that it is well drained, and that it is of sufficient size to allow the animals to move about and lie down in comfort.

CARE OF THE COW AT CALVING TIME

The careful and observant caretaker will be ever

---

[3]Herman, H. A., and F. W. Madden, *The Artificial Insemination and Embryo Transfer of Dairy and Beef Cattle*, seventh edition, The Interstate Printers & Publishers, Inc., Danville, IL, 1987

alert and make definite preparations for calving in ample time. It is especially important that first-calf heifers be watched at calving time, for frequently they will need some assistance. Older cows that habitually have trouble in parturition may well be culled from the herd.

SIGNS OF APPROACHING PARTURITION

Perhaps the first sign of approaching parturition is a distended udder, which may be observed some weeks before calving time. Near the end of the gestation period, the content of the udder changes from a watery secretion to a thick, milky colostrum. As parturition approaches, there generally will be a marked shrinkage or falling away of the muscular parts in the region of the tail head and pin bones, together with a noticeable enlargement and swelling of the vulva.

The immediate indications that the cow is about to calve are extreme nervousness and uneasiness, separation from the rest of the herd, and muscular exertion and distress.

PREPARATION FOR CALVING

At the time the signs of approaching parturition seem to indicate that the calf may be expected within a short time, arrangements for the place of calving should be completed.

During the seasons of the year when the weather is warm, the most natural and ideal place for calving is a clean, open pasture away from other livestock. Hogs should not be allowed in the same place with the cow, for they are likely to injure or kill the young calf. They have even been known to injure the cow.

Under pasture conditions, there is decidedly less danger of either infection or mechanical injury to the cow and calf. In commercial range operations, it is common practice to ride the range more frequently at calving time. A better procedure consists of having a smaller pasture adjoining headquarters into which heavy springing cows are placed a few days before calving.

With the added convenience of such an arrangement, the animals can be given more careful attention.

During inclement weather, the cow should be placed in a roomy (10 or 12 sq ft), well-lighted, well-ventilated, comfortable box stall or maternity pen which should first be carefully cleaned, disinfected, and bedded for the occasion.

NORMAL PRESENTATION

Labor pains in a mild form usually start some hours before actual parturition. After a time, the water bag appears on the outside, usually increasing in size until it ruptures from the weight of its own contents. This is closely followed by the appearance of the amniotic bladder (the second water bag), with the fetus. With the rupture of the second water bag, the straining becomes more violent, and presentation soon follows. Normal presentation for the fetus is the anterior presentation, with the head lying on the forelegs and with the calf's back pointed toward the cow's back. The posterior presentation with the rear legs extended out through the vagina and the calf's back pointed toward the cow's back is not considered normal in cattle (although it is considered normal in other species of livestock). Any other position is considered abnormal and will probably cause dystocia (calving difficulty).

Malpresentation of the fetus is the second most common cause of dystocia, being exceeded only by a large fetus or a small maternal pelvis. Bellows reported that 67% of the calves were presented in the normal position at calving, but 20% of the dead calves were presented backward or breech.[4]

With posterior presentation (hind feet first), there is likely to be difficulty in calving. Moreover, there is considerably more danger of having the calf suffocate through rupture of the umbilical cord and strangulation.

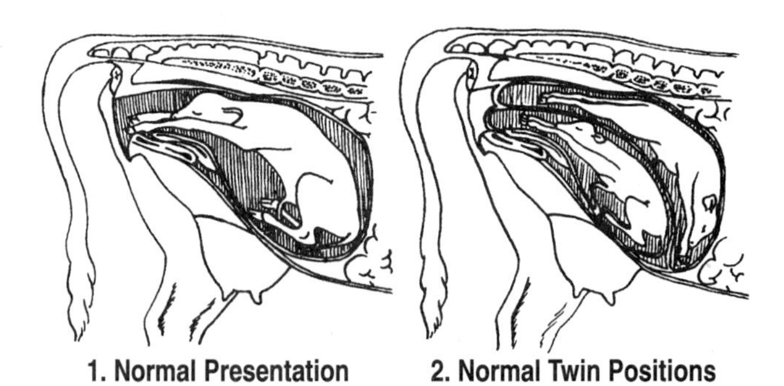

Fig. 6-11. Normal single presentation and normal twin positions.

1. Normal single presentation; the back of the fetus is directly toward that of the mother, the forelegs are extended toward the vulva, and the head rests between the forelegs. If it is necessary to render assistance, apply ropes above the angle joints and pull alternatively downward on each leg as the cow strains.

2. Normal twin positions. If delivery does not proceed normally, this is a case for a veterinarian.

---

[4]Bellows, R. A., "Calf Losses: What We Can Do About Them," *Beef Cattle Science Handbook*, Vol. 19, edited by Frank H. Baker, Winrock International, published by Westview Press, Inc., Boulder, CO, 1983, p. 385. Table 5.

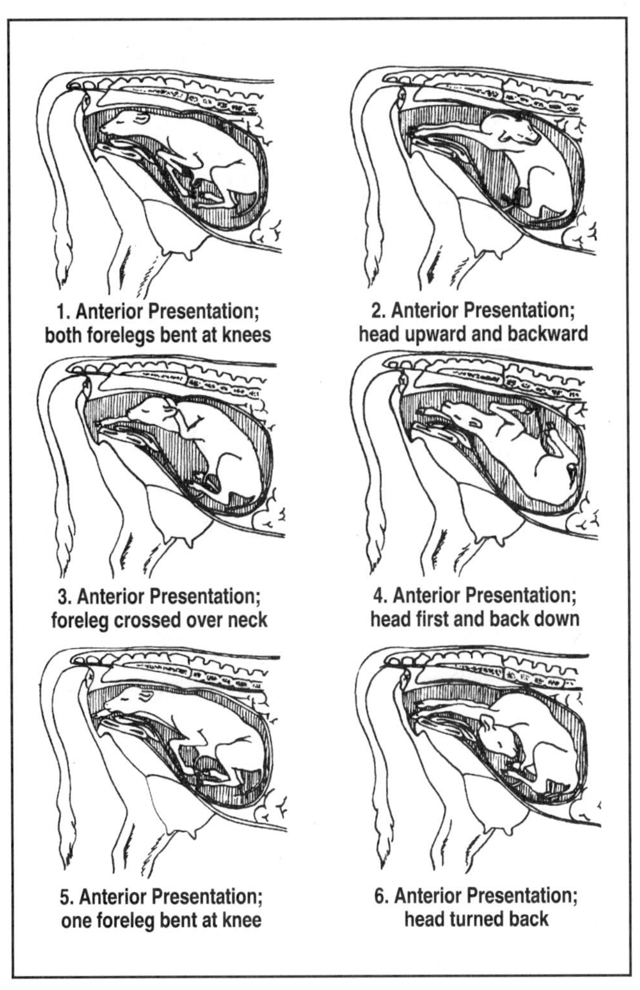

Fig. 6-12. Some abnormal presentations with suggestions for correction:

1. Extend the legs so that delivery can be accomplished.
2. Push back the fetus, which will often bring the head into its normal position.
3. Grasp the crossed leg a little above the ankle, raise it, draw it to the proper side, and extend it in the genital canal.
4. Rotate the fetus, extend the forelegs, and deliver by traction.
5. Lift the head and draw up and rope the leg so that it does not slip back again.
6. Rope the forelegs and then push them forward; place the head in normal position.

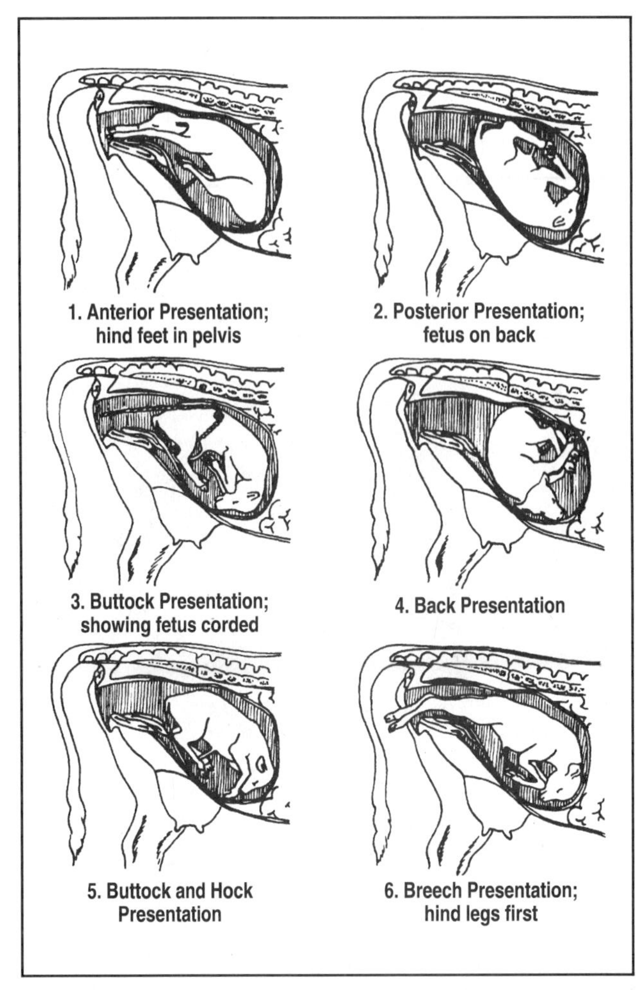

Fig. 6-13. Some abnormal presentations with suggestions for correction:

1. Force back the hind feet. This is a very serious malpresentation, in which it is generally impossible to save the fetus if delivery is far advanced.
2. Rotate the fetus, extend the rear legs, and deliver by traction.
3. Push the fetus forward and bring the legs properly into the genital passage.
4. Turn the fetus so that either the head and forelegs or the rear legs can be started through the pelvis.
5. Push the fetus forward and bring the legs properly into the genital passage.
6. Usually delivery is normal but traction may facilitate; beware of prolonged labor because calf may suffocate due to rupture of the navel cord.

RENDERING ASSISTANCE

A good rule for the attendant is to be near but not in sight. After the calf and membranes are in the birth canal, the cervix dilated, and abdominal straining has started, the calf should be born within an hour.

If the cow has labored for some time with little progress or is laboring rather infrequently, it is usually time to give assistance. Such aid will usually consist of fastening small ropes around the pasterns and pulling the young outward and downward as the cow strains. This should be done by an experienced caretaker or a competent veterinarian. It is always well to be reminded that rough, careless, or unsanitary methods at such a time may do more harm than good.

THE NEWBORN CALF

If parturition has been normal, the cow can usually take care of the newborn calf, and it is best not to interfere. However, in unusual cases, it may be necessary to wipe the mucus from the nostrils to permit breathing; or, more rarely yet, artificial respiration methods may have to be applied to some calves. This may be done by blowing into the mouth, working the ribs, rubbing the body rather vigorously, and permitting the calf to fall gently. The cow should be permitted to lick the calf dry.

With calves born in sanitary quarters or out on clean pastures, there is little likelihood of navel infection. To lessen the danger of such infection, the navel cord of the newborn calf should be treated at once with a 2% solution of tincture of iodine.

Fig. 6-14. A good start in life. When the weather is warm, the most natural and ideal place for calving is a clean, open pasture, away from other livestock. Under pasture conditions, there is less danger of either infection or mechanical injury to the cow and calf.

A vigorous calf will attempt to rise in about 15 minutes and usually will be nursing in half an hour to an hour. The weaker the calf, the longer the time before it will be able to be up and nursing. Sometimes it may even become necessary to assist the calf by holding it up to the cow's udder.

The colostrum (the milk yielded by the mother for a short period following the birth of the young) is most important for the well-being of the newborn calf. Experiments have shown that it is almost impossible to raise a calf that has not received any colostrum. Aside from the difference in chemical composition, compared with later milk, the colostrum seems to have the following functions:

1. It contains antibodies which temporarily protect the calf against certain infections, especially those of the digestive tract.

2. It serves as a natural purgative, removing fecal matter which has accumulated in the digestive tract.

3. It contains a very high content of vitamin A from 10 to 100 times that of ordinary milk. This provides the young calf, which is born with little body storage of this vitamin, with as much vitamin A on the first day as it would secure in some weeks from normal milk.

Usually it is best to keep the cow and calf in a small pasture for a few days. After this, they may be turned back with the main herd. Nothing is better for the cow at calving time than plenty of grass, and both the cow and calf will be helped by an abundance of fresh air and sunshine. The cow may deliberately hide the calf for the first few days, and the job may be so thoroughly done as to require considerable cleverness on the part of the caretaker to find it.

THE AFTERBIRTH

Under normal conditions, the fetal membranes (placenta or afterbirth) are expelled from 3 to 6 hours after parturition. Should they remain as long as 24 hours after calving, competent assistance should be given by an experienced caretaker or a licensed veterinarian. The operation of removing a retained afterbirth requires skill and experience; and, if improperly done, the cow may be made a nonbreeder. Before attempting to do so, the fingernails should be trimmed closely, and the hands and arms should be thoroughly washed with soap and warm water, disinfected, and then lubricated with Vaseline or linseed oil. In no case should a weight be tied to the placenta in an attempt to force removal.

As soon as the afterbirth is ejected, it should be removed and burned or buried in lime, thereby preventing the development of bacteria and foul odors. This step is less necessary on the open range, where animals travel over a wide area.

CARE AND MANAGEMENT OF THE BULL

Outdoor exercise throughout the year is one of the first essentials in keeping the bull virile and in a thrifty, natural condition. The finest and easiest method of providing such exercise is to arrange for a well-fenced, grassy paddock (about 2 acres is a good size for 1 bull). Many valuable sires have been ruined through close confinement in small stalls—or more likely yet—through being kept knee deep in mud within small, filthy enclosures. In addition to the valuable exercise obtained in the grassy paddock, the animal gets succulent pasture, an ideal feed for the herd bull.

A satisfactory and inexpensive shelter should be provided for the bull. The most convenient arrangement is to have this within or adjacent to the paddock, so that the bull may run in and out at will. Sufficient storage space for feed and conveniences for caring for the bull should be provided in this building. Normally, purebred bulls are kept in separate stalls and enclosures, though some successful purebred breeders regularly run several valuable bulls in one enclosure. Bulls used in commercial herds are usually run together, both on the range and when separated from the cows. Because of their scuffling and fighting, there is more injury hazard when bulls are handled in a group.

Under range conditions, it is rather difficult to give the bulls much attention during the breeding season. Usually the proper number of bulls is simply turned with the cow herd. During the balance of the year, however, the bulls are usually kept separate. Thus, if the producer desires calves that are dropped from February 1 to June 1, the bulls are turned with the cows about May 1 and are removed September 1.

The feeding of the herd bull is fully covered in Chapters 8 and 20. In brief, it may be said that the feeding program should be such as to keep the bull in a thrifty, vigorous condition at all times.

Fig. 6-15. Yearling Hereford bulls on the range. (Courtesy, *Livestock Weekly*, San Angelo, TX)

AGE AND SERVICE OF THE BULL

The number and quality of calves that a bull sires in a given season is more important than the total number of services. The number of services allowed will vary with the age, development, temperament, health, breeding condition, distribution of services, and system of mating (pasture or hand mating). With pasture mating, size of area, carrying capacity of the range, and the size of the herd are important factors. Therefore, no definite best number of services can be recommended for any and all conditions, and yet the practices followed by good operators do not differ greatly. For best results, a bull should be at least 15 months old and well grown for his age before being put into service.

Table 6-1 gives pertinent information relative to the use of the bull, including consideration that should be given to age and method of mating.

TABLE 6-1
HANDY BULL MATING GUIDE

Age	No. of Cows/Year		Comments
	Hand Mating	Pasture Mating	
Yearling	20–25	15–20	Most western ranchers use 1 bull to about 25 cows.
Two-year-old	25–30	20–25	A bull should remain a vigorous and reliable breeder up to 10 years or older; up to 6 to 7 years under range conditions.
Three-year-old or over	40–50	25–40	

In a survey conducted by Washington State University, it was found that 1 bull was used for every 21.5 cows and heifers bred.[5]

Should the bull prove to be an uncertain breeder, he should be given rest from service, forced to take plenty of exercise, and then placed in proper condition—neither fat nor thin. Sometimes a bull that is being let down in condition following showing will be temporarily sterile during the reducing process. Even though this lack of fertility may last for a year, usually such animals bounce back.

NORMAL BREEDING SEASON AND TIME OF CALVING

The season at which the cows are bred depends primarily on the facilities at hand, taking into consideration the feed supply, pasture, equipment, labor, and weather conditions; and whether the cattle are being produced for commercial or for purebred purposes.

The purebred breeder who exhibits cattle should plan the breeding program so that maximum advantage will be taken of various age groups. In most livestock shows throughout the country, the classifications are based upon the dates of January 1, March 1, May 1, and September 1. Further information relative to show classifications is presented in Chapter 14.

In commercial herds of beef cattle, two systems of breeding are commonly practiced in regard to the season of the year. In one system, the bulls are allowed to run with the cows throughout the year so that calving is on a year-round basis. This system results in greater

[5]Ensminger, M. E., *et al.*, Wash. Ag. Exp. Sta. Bull. 562.

use of the bull, and there is less delay in the first breeding of the heifers as soon as they are sufficiently mature. On the other hand, often the calves arrive at undesired and poorly adapted times; the breeding system is without order and regularity; and the calves usually lack uniformity. This system is sometimes followed in the southern states.

The other system of breeding followed in commercial herds, and the most widely used system today, is that of having all of the breeding done within a restricted season (of about three months) so that the calves arrive within a short spread of time—usually in the spring. Having the calves born about the same time, whether it be fall or spring, results in greater uniformity. Thus, it is easier to care for (brand, dehorn, castrate, vaccinate, etc.) and market such animals. Each farm has its individual problems, and the decision must be made accordingly. A rule of thumb of some range management specialists is: The optimum time to calve beef cows in order to match their nutritional requirements with the feed resources is approximately 45 to 60 days before the start of the good natural forage supply.

There is a marked area difference in the breeding season followed by commercial cattle producers, due to weather conditions. Thus, in the South about a third of the operators leave the bulls with the cow herd the year around; where this is not done, the breeding season is prolonged and the peak of the breeding season is in January. In the West, the breeding season is later and more restricted; most cattle producers do all of the breeding within a season of 3 to 4 months.

ARTIFICIAL INSEMINATION

Artificial insemination is, by definition, *the deposition of spermatozoa in the female genitalia by artificial rather than by natural means.*

Legend has it that artificial insemination had its origin in 1322, at which time an Arab chieftain used artificial methods to impregnate a prized mare with semen stealthily collected by night from the sheath of a stallion belonging to an enemy tribe. There is no substantial evidence, however, to indicate that the Arabs practiced artificial insemination to any appreciable degree.

The first scientific research in artificial insemination of domestic animals was conducted with dogs by the Italian physiologist, Lazarro Spallanzani, in 1780. A century later, American veterinarians employed artificial means to get mares in foal that persistently had failed to settle to natural service. They noticed that, because of obstructions, the semen was often found in the vagina and not in the uterus following natural service. By collecting the semen into a syringe from the floor of the vagina and injecting it into the uterus, they were able to impregnate mares with these anatomical difficulties.

The Russian physiologist, Ivanoff, began a study of artificial insemination of farm animals, particularly horses, in 1899; and in 1922, he was called upon by the Russian government to apply his findings in an effort to reestablish the livestock industry following its depletion during World War I. Crude as his methods were, his work with horses must be considered the foundation upon which the success of the more recent work is based.

The shifting of the large-scale use of artificial insemination to cattle and sheep, two decades after it was first introduced for horses, was not caused by the fading importance of the horse and the increased demand for cattle and sheep. Rather, it was found that progress was quicker and more easily achieved with these animals, because the exact time of ovulation in relation to signs of heat is more easily detected in the cow and ewe than in the mare. It was also discovered that the sperm of bulls and rams survive better in storage than stallion sperm.

Following World War II, British scientists were called upon to make wide use of artificial insemination in reestablishing the livestock industry of England. They looked upon it as: (1) a way in which to increase more rapidly the efficiency and utility value of their animals through making wider use of outstanding sires; (2) a means of controlling certain diseases; and (3) the best way in which to increase breeding efficiency.

Today, artificial insemination is more extensively practiced with dairy cattle than with any other class of four-footed farm animals—more than 80% of all dairy cows are bred artificially.

In 1938, when the program first began in America, only 7,359 cows and 646 herds were bred by this means in organized groups in the United States. In

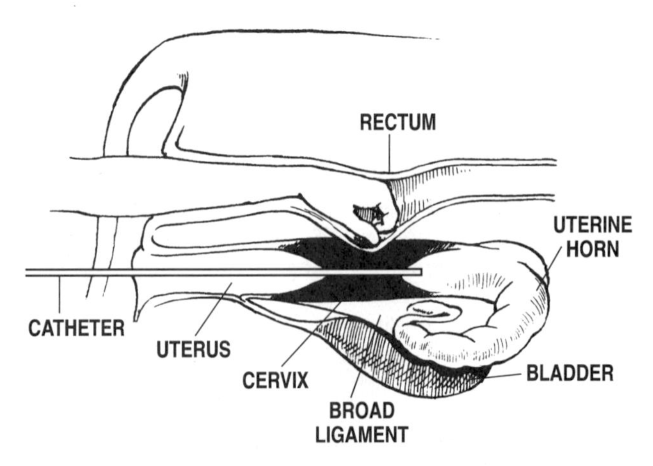

Fig. 6-16. Deep uterine insemination of the cow. The cervix is grasped *per rectum*, and the inseminating tube is carefully worked into and through the cervical canal.

1993, a total of 12,957,368 dairy units of semen, and a total of 1,117,798 beef units of semen, were marketed in the United States. *Note:* Currently, most reporting is in units of semen, not in number of cows bred. It takes one to two units of semen per cow bred.

Also, it is noteworthy that the number of cows bred per bull averaged only 228 in 1939—when the first artificial breeding association was organized in this country. Today, proved bulls in AI sire 5,000 to 10,000 calves in a single year; and some bulls have sired over 100,000 progeny in a lifetime.

It is recognized that—due primarily to their more frequent handling (in milking) and greater accessibility—artificial insemination is vastly easier to apply to dairy cattle than to beef cattle; and, therefore, more common in dairy cattle enterprises.

Subsequent to a Consent Decree between the American Angus Association and the U.S. Government, which was signed on July 13, 1970, there have been major changes liberalizing the rules for registration of offspring from AI by most purebred beef associations. The current attitudes and rulings of the cattle registry associations relative to artificial insemination are summarized in Chapter 3 of this book.

Fig. 6-17. A semen tank—a thermos container built to withstand liquid nitrogen temperatures of –320°F.

FROZEN SEMEN

The freezing of semen, particularly slow freezing such as might occur from exposure to ordinary subfreezing weather conditions or from placing a vial of semen in the freezing compartment of a refrigerator, will kill all spermatozoa immediately. Also, dry ice at –110°F isn't cold enough.

The technique of adding glycerine to bull semen, followed by freezing, was discovered by Dr. Christopher Polge at Cambridge University, in England, in the early 1950s. Today, spermatozoa can be preserved for several years provided glycerine is first added and it is frozen at a certain rate for storage in liquid nitrogen at –320°F. Liquid nitrogen is the universally used refrigerant in the United States, because uniform temperatures may be maintained for long periods of time and the method is more convenient in shipping and storing frozen semen.

Frozen bull semen has been stored as long as 30 years and conception obtained. There is a small reduction in fertility of semen stored long periods, but it is usual to find semen stored 1 year or longer in routine use.

Frozen semen, refrigerated with liquid nitrogen, can be shipped to all parts of the world. The thawing of semen is accomplished by placing the vial in a container of water containing thawing ice (34–38°F) immediately prior to insemination.

Frozen semen is potentially the most valuable breeding technique yet known. Through it, the following may be achieved:

1. The usefulness of outstanding bulls can be extended far beyond their lifetime; also, it insures the proven sire should he die.
2. Outstanding bulls can be used, nationwide and worldwide.
3. A multiherd progeny test can be completed at a much earlier age.
4. A stock of semen can be built up while waiting for a progeny record assessment.
5. Long-term storage of semen lessens semen wastage, and facilitates long-distance transport.
6. Semen from valuable sires may be fully utilized.
7. A herd owner can usually obtain the sire of his or her choice at any time.

ADVANTAGES OF ARTIFICIAL INSEMINATION

Some of the advantages of artificial insemination are:

1. **It increases the use of outstanding sires.** Through artificial insemination, many breeders can avail themselves of the use of an outstanding sire,

whereas the services of such an animal were formerly limited to one owner, or, at the most, to a partnership.

2. **It alleviates the danger and bother of keeping a sire.** Some hazard and bother is usually involved in keeping a bull. The ready availability of semen rules out this necessity.

3. **It makes it possible to overcome certain physical handicaps to mating.** Artificial insemination is of value (a) in mating animals of greatly different sizes; for example, in using heavy mature bulls on yearling heifers, and (b) in using stifled or otherwise crippled sires that are unable to perform natural service.

4. **It makes it possible to use a bull that is not alive at the time.** Since frozen semen can be stored for many years, it is possible to use a bull far beyond his lifetime.

5. **It lessens sire costs.** In smaller herds (in herds with fewer than 50 cows), artificial insemination is less expensive than the ownership of a worthwhile sire together with the accompanying building, feed, and labor costs.

6. **It reduces the likelihood of costly delays through using sterile sires.** Because the breeding efficiency of sires used artificially is constantly checked, it reduces the likelihood of breeding females to a sterile sire for an extended period of time.

7. **It makes it feasible to prove more sires.** Because of the small size of the herds in which they are used in natural service, many sires are never proved. Still others are destroyed before their true breeding worth is known. Through artificial insemination, it is possible to determine the genetic worth of a sire at an earlier age and with more certainty than in natural service.

8. **It creates large families of animals.** The use of artificial insemination makes possible the development of large numbers of animals within a superior family, thus providing uniformity and giving a better basis for a constructive breeding program.

9. **It increases pride of ownership.** The ownership of progeny of outstanding sires inevitably makes for pride of ownership, with accompanying improved feeding and management.

10. **It helps control diseases.** Artificial insemination is a valuable tool in preventing and controlling the spread of certain types of cattle diseases, especially those associated with the organs of reproduction, such as vibriosis and trichomoniasis, *provided it is properly done*. For disease control it is essential (a) that all males be carefully examined for symptoms of transmissible diseases, (b) that bacterial contamination be avoided during the collection and storage of semen, and (c) that clean, sterile equipment be used in the insemination. Artificial insemination organizations that are members of the National Association of Animal Breeders, Inc., are expected to follow the rigid Sire Health Code approved by the American Veterinary Medical Association and adopted by the National Association of Animal Breeders.

11. **It increases profits.** The offspring of outstanding sires are usually higher and more efficient producers, and thus more profitable. Artificial insemination provides a means of using such sires more widely.

LIMITATIONS OF ARTIFICIAL INSEMINATION

Like many other wonderful techniques, artificial insemination is not without its limitations. A full understanding of such limitations, however, will merely accentuate and extend its usefulness. Some of the limitations of artificial insemination are:

1. **It must conform to physiological principles.** One would naturally expect that the practice of artificial insemination must conform to certain physiological principles. Unfortunately, some false information concerning the usefulness of artificial insemination has been encountered—for example, the belief that females will conceive if artificially inseminated at any time during the estrus cycle. Others have even accepted exaggerated claims that the quality of semen may be improved through such handling, only to be disappointed.

2. **It requires skilled technicians.** In order to be successful, artificial insemination must be carried out by skilled technicians who have had considerable training and experience.

3. **It necessitates considerable capital to initiate and operate an artificial insemination organization.** Considerable money is necessary to initiate an artificial insemination enterprise, and still more is needed to expand and develop it properly.

4. **It may accentuate the damage of a poor sire.** It must be realized that when a male sires the wrong type of offspring, his damage is accentuated because of the increased number of progeny possible. For this reason, untried or untested bulls are seldom used extensively in a stud.

5. **It may restrict the sire market.** The fact that the market demand for poor or average sires will decrease if artificial insemination is widely adopted should be considered an attribute rather than a limitation. Also, it is noteworthy that a sizable number of the nation's beef cattle are still bred to scrub and nondescript bulls.

6. **It may increase the spread of disease.** As previously indicated, the careful and intelligent use of artificial insemination will lessen the spread of disease. To date, no outbreaks of disease traceable to the use of artificial insemination have been reported in the

United States. However, it must be recognized that carelessness and ignorance may result in the rapid spread of disease. Thus, semen should always be obtained from a source known to observe recommended health and sanitation procedures.

7. **It may be subject to certain abuses.** If semen is transported from farm to farm, the character of the technician must be above reproach. Trained workers can detect differences in the spermatozoa of the bull, ram, boar, stallion, or cock; but even the most skilled scientist is unable to differentiate between the semen of a Hereford and a Shorthorn, to say nothing of the difference between two bulls of the same breed. However, it appears that such abuse is more suspicioned than real. In a blood type study[6] of cattle, Rendel found 4.2% family records in error out of 615 animals by natural service, compared to 4% family records in error out of 199 sired by artificial insemination.

Of course, with skilled workers performing the techniques required in artificial insemination, there usually is more check on the operations and perhaps less likelihood of dishonesty than when only the owner is involved, such as is usually the situation with natural service. Also, bulls used in artificial insemination are blood typed in order to make possible investigation of suspected errors.

SOME PRACTICAL CONSIDERATIONS

Based on present knowledge, gained through research and practical observation, it may be concluded that artificial insemination can be made more successful through the following:

Fig. 6-18. Adequate nutrition is needed for proper cycling. (Courtesy, The Upjohn Company, Kalamazoo, MI)

[6]Rendel, J., "Studies of Cattle Blood Groups, II. Parentage Tests," *Acta. Ag. Scand.*, Vol. 8, No. 131, p. 140.

1. Give the female a reasonable rest following parturition and before rebreeding; in cows this should be about 60 days.
2. Keep record of heat periods and note irregularities.
3. Watch carefully for heat signs, especially at the approximate time.
4. Notify the insemination technician promptly when an animal comes in heat.
5. Avoid breeding diseased females or females showing pus in their mucus. The latter condition indicates an infection somewhere in the reproductive tract.
6. Have the veterinarian examine females that have been bred three times without conception or that show other reproductive abnormalities.
7. Have a pregnancy diagnosis made.

HORMONAL CONTROL OF ESTRUS IN COWS

Planned parenthood is not new. It has long been practiced among females of all species, women included. For several years, a product has been available and used by veterinarians to defer heat in dogs. Back in the 1930s, much progress was made in the isolation and identification of progesterone from the corpus luteum. In the 1940s, researchers at the University of Wisconsin injected progesterone into various species of farm animals and successfully synchronized estrus. They found that, during the treatment, the corpus luteum regressed; and, upon discontinuing it, heat followed. In the 1950s, certain pharmaceutical houses chemically synthesized progestogens (progesterone like compounds)—and the race was on. Many drugs have been administered in attempts to control the estrus cycle of cows.

SLY CONTROLS OF HEAT HAVE BEEN USED A LONG TIME

Animal breeders have long "tampered with" the breeding and parturition season that was common in the wild state. Prior to domestication, animals brought forth their young in the fields and glens, inhibited only by age and feed, and influenced somewhat by seasons. But caretakers changed all this—even without the use of hormones. Sly controls have been exercised over breeding for a very long time. For example, farm flock owners controlled reproduction in chickens by the simple act of putting eggs under an old setting hen—unless she hid out. Today, modern poultry producers regulate chick hatchings by controlling when, and how many, eggs go into the incubator. It's more difficult to accomplish the same thing in four-footed animals.

The motivating reasons back of people-made changes vary somewhat by species. Horse owners, especially those who race or show, want their mares to foal as soon after January 1 as possible, because a horse's age is computed on a January 1 basis, regardless of how late in the year it may have been born. Sheep and hog producers strive for two crops of offspring per year, and for multiple births. Purebred cattle breeders who show, plan their breeding programs to take maximum advantage of show classifications; commercial cattle producers are concerned with weather and feed supply; and milk producers want the largest flow of milk at a time when the product is likely to bring the highest price. Also, cattle producers recognize that controlled estrus would greatly facilitate both artificial insemination and ova transplantation.

The keepers of herds and flocks have altered nature's way in farm animals (1) by confining the male at certain times, or hand mating, (2) by emulating spring conditions—through providing better feed, shelter, and/or blankets when breeding at other times of the year, (3) by flushing—through feeding females more liberally 2 to 3 weeks ahead of the breeding season, and (4) by artificially controlling the hours of light per day—through use of ordinary electric lights, which activate hormone production. Each of these methods has been used with varying degrees of success. All have fallen short of achieving the hoped-for goal—that of bringing females in heat at will, followed by a high conception rate. Hormonal control appears to be the answer.

ADVANTAGES TO ACCRUE FROM BRINGING COWS IN HEAT AT WILL

Many obvious advantages would accrue from a sure method of controlling the breeding dates of cows, and, consequently, the birth dates of calves; among them, the following:

1. More cows would calve early in the breeding season and wean off heavier calves.
2. Cattle producers could have their calves come within a restricted period of time. They could even swap help with the neighbors, according to a predetermined schedule.
3. Calves of uniform age would also be more uniform in size, thereby making them easier to handle, feed, and sell.
4. It would greatly facilitate such management practices as pregnancy testing, dehorning, castrating, and vaccinating; all of which could be done at one time, rather than piecemeal.

Fig. 6-19. Synchronized breeding results in older, more uniform, heavier calves at weaning. (Courtesy, American Polled Hereford Assn., Kansas City, MO)

5. Cattle producers could plan their breeding programs so that calves would be ready for market when seasonal prices are highest. They could contract a certain number of calves of the same age on a specified date.
6. Artificial insemination would be simplified if the time of ovulation could be controlled to the extent that the hour and date of insemination of a breeding herd could be precisely determined, thereby alleviating the problem of heat detection and lowering the cost of semen distribution. This would greatly facilitate breeding range cattle by AI.
7. Ova transfer would be facilitated. One of the major hurdles to overcome in ova transfer is the synchronization of the estrus cycles of both the donor and the recipient females. Both must be in similar stages in their reproductive cycles to allow a successful transfer to take place.

HORMONES FOR ESTRUS CONTROL

Many different drugs have been administered (either orally, by injection, or by implantation) in attempts to control the estrus cycle of cattle, with prostaglandins, Syncro-Mate-B, and MGA heading the list.

■ **Prostaglandins (Lutalyse)**—Prostaglandin is a naturally occurring compound produced in the uterus that causes regression of the corpus luteum (CL) and subsequently results in the onset of heat or estrus. Synthetic prostaglandin products are also effective for inducing estrus in mid-cycle cows.

In 1979, the Upjohn Company, Kalamazoo, Michigan, received FDA approval on the use of Lutalyse, a prostaglandin product, for synchronization of estrus in beef cows and heifers and nonlactating dairy heifers.

Lutalyse contains the naturally occurring prostaglandin F_2 alpha, a fatty acid composed of carbon, hydrogen, and oxygen. Its generic name is *dinoprost tromethamine*. Subsequently, several other prostaglandin products have been approved for use by the Food and Drug Administration. All of them must be administered by injection, and all of them are available only on prescription from a veterinarian.

Prostaglandin is ineffective in heifers that have not reached puberty and in noncycling mature cows. It is effective only in heifers and cows that are in days 5 to 18 of their estrous cycle.

Prostaglandin is given in either a one-injection, or two-injection system.

In the one injection system, heat is detected for the first five days and the in-heat females are bred. If 20–25% of the animals show heat during this period of time, it can be assumed that the herd is cycling normally; and on the fifth day each of the remaining animals is given one injection of prostaglandin. Animals are then bred as they come in heat, with almost all cycling animals bred within 10 days.

In the two-injection system, all animals are injected with prostaglandin. On days 11, 12, or 13, all animals are again injected with prostaglandin. Within 5 days following the second injection, all cycling animals should come into heat, and be heat detected and inseminated.

▪ **Syncro-Mate-B (SMB)**—This is a trade name estrus synchronization product, manufactured by Rhone Merieux, Inc., and approved by the FDA in 1982, containing Norgestomet, a patented, potent synthetic progestin, and estradiol valerate, a synthetic estrogen. It is a nonprescription product.

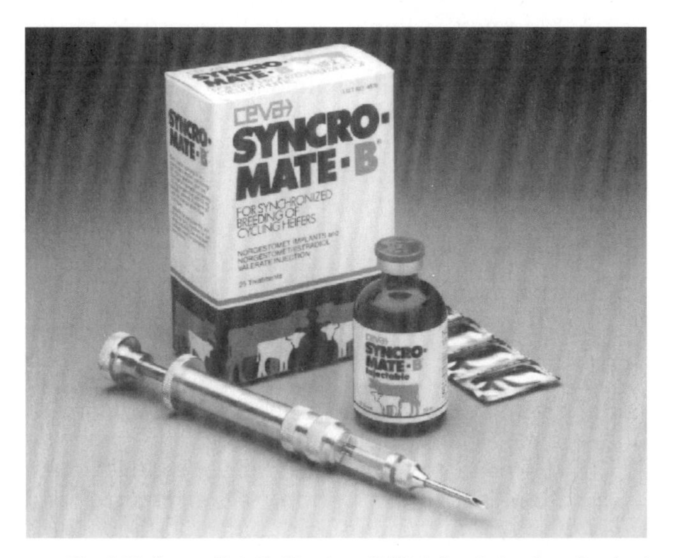

Fig. 6-20. Syncro-Mate-B. (Courtesy, CEVA Laboratories, Inc., Overland Park, KS)

SMB is designed to cause cows or heifers to ovulate in a predictable period of time.

In the SMB treatment, an implant containing 6 mg of Norgestomet is placed subcutaneously in the back portion of the ear for 9 days. Simultaneously, heifers also receive an intramuscular injection of 3 mg of Norgestomet and 5 mg of estradiol valerate and cows receive 3 mg of Norgestomet and 6 mg of estradiol valerate. The effect is to halt temporarily the estrus cycle. After 9 days, the implant is removed; and 48 to 54 hours later the cows should be bred. No heat checking is necessary, thereby resulting in a saving of time and labor. The manufacturer of SMB makes the claim that "you can get as many animals pregnant on that one time breeding, without heat detection, as you can over a 25-day period breeding on observed estrus." Test data show an impressive 70% pregnancy rate at the 25-day mark.

SMB effectively synchronizes the estrus cycle in cattle. It eliminates heat checking; works anytime in the cycle, and is not an abortive agent.

▪ **MGA and Prostaglandin**—Melengestrol acetate (MGA) is a synthetic progesterone. It is used in combination with prostaglandin. In this system, 0.5 mg per head, per day, of MGA is fed as a topdressing or mixed in with the feed for 14 days, then removed; and prostaglandin is injected 16 to 18 days later, following which the majority of the females will come in heat within 5 days. In this system, cattle have to be handled only once. But success is dependent upon all females consuming a constant quantity of MGA during the 14-day feeding period. A disadvantage is that the producer must begin feeding MGA 32 to 33 days prior to the breeding period.

WHAT ESTRUS SYNCHRONIZATION WILL AND WILL NOT DO

With any new program that offers improvement, hope springs eternal. Progressive cattle producers are always on the alert for something that will increase efficiency still further. On the other hand, submarginal producers are usually hunting for a panacea for all their troubles—for a "crutch," or an easy way out. A heat-grouping program is not likely to do all that is expected of it, but it will do more for the good producer than for the poor one. Here are some of the things that it will and will not do:

It will:

1. Add a management tool that will better the opportunity for improved fertility.
2. Reduce or eliminate heat detection, thereby decreasing labor requirements, inaccurate heat detection, and stress on people and animals.

3. Shorten the breeding and calving seasons.
4. Make for greater ease in maintaining a 12-month calving interval.
5. Give as good, and likely better, conception rate than can be obtained without hormone synchronization.
6. Make for earlier breeding.
7. Improve calf survival.
8. Increase weaning weight of calves.
9. Help improve the management program.
10. Make for a more uniform calf crop.
11. Facilitate AI and embryo transfer.

It will not:

1. Start a cow or heifer cycling.
2. Improve fertility.
3. Make it possible to breed sooner than 60 days following calving
4. Lower feed requirements.
5. Overcome sterility or delayed breeding due to (a) genital infections and disease, (b) poor management and feeding, (c) inherited abnormalities, (d) anatomical defects and injuries, and (e) shy or difficult breeders.
6. Assure good weather at a time when all of the calves are coming, for it pretty much "puts all the eggs in one basket."

RESULTS OF HORMONE-INDUCED ESTRUS

The major criteria for measuring the success of hormone-induced estrus are: (1) the percentage of the cows that come in heat; and (2) the percentage of conception.

Admittedly, the effectiveness of hormone-induced heat varies widely from herd to herd. This is expected, for the results are affected by reproductive abnormalities and management. For any such product to be most effective, the cows must be having normal estrus prior to treatment; the caretaker must be capable of detecting cows in heat in order to breed them; and the plane and kind of nutrition must be satisfactory. In fact, the better the level of the herd fertility without hormone treatment, the better the results with treatment. It follows, therefore, that hormone-planned parenthood in the cow business is not for the poor manager.

In view of the above statement, prior to undertaking any estrus synchronization program, the following questions should be answered for each herd, honestly and unflinchingly:

1. Are the females having normal heat periods, and are they reproducing normally?
2. Is the management program good?
3. Is the nutrition satisfactory?
4. Is the herd health good?
5. Has at least 60 days elapsed since calving?
6. Are heifers of proper age and size?
7. Will it be possible to administer the treatment according to directions?
8. Is the caretaker capable of detecting the cows in heat?
9. Are the facilities adequate to allow breeding (natural or artificial insemination) on the same day that cows are in heat?
10. Is the semen of good quality; and, if artificial insemination is used, is the inseminator experienced and properly trained?

If the answer to each of the above questions is affirmative, you are ready to proceed and the results will be good. If the answer to one or more of these questions is negative or in doubt, the results will be just as questionable, or even disappointing.

SUMMARY RELATIVE TO HORMONE-CONTROLLED ESTRUS

Researchers in both colleges and industries are in general agreement that hormone-controlled estrus synchronization will work, and that it offers promise of good returns when properly used in a well-managed cow herd. Scientists also realize that we don't know all the answers; that further research work is necessary. Among other things:

1. **We need to know which hormone(s) to use, and how to give it;** whether to feed, inject, or implant it; what dosage to give; and when to give it.
2. **We need to know how to lower costs exclusive of the drugs**—primarily, added labor and perhaps feed, when administering hormones.
3. **We need to know how to obtain a higher conception rate following hormone-induced heat**—although it is recognized that conception is as good, or better, in hormone-treated cows than in untreated cows.
4. **We need to know the effect of the stage of the estrus cycle on treatment**—For example, when treatment is started at or near estrus, some animals show estrus and/or ovulate during treatment.
5. **We need to know more about the effect of lactation and suckling,** both on sexual behavior and on the endocrine control of the ovary.
6. **We need to know more about the influence of location, season, nutrition, social factors, and time of day;** on suppression, synchronization, and fertility.
7. **We need to lower costs or increase the**

gross—for regardless of product, it is net returns that count.

It appears that planned parenthood in the cow business is here to stay; that its wide use only awaits getting the technique perfected and lowering costs, both of which will come. In the meantime, cattle producers are admonished to keep abreast of developments and to rely on well-informed advisors.

SUPEROVULATION

The bull is capable of producing from several thousand to millions of sperm daily whereas the cow normally produces 1 ovum (occasionally 2 ova) every 18 to 21 days. Yet, a heifer calf is born with a lifetime supply of about 200,000 oocytes in each of her 2 ovaries. Now it is possible, through the administration of hormones, to obtain 10 to 30 oocytes from a cow at 1 estrus cycle. It is also feasible to obtain a large number of eggs from very young calves, by injection of hormones.

Superovulation begins with the selection of healthy heifers or cows which are cycling normally. The animal is ready to be injected with hormone, starting about 16 days after the last heat; although this time may be varied by using prostaglandins or other compounds which cause degeneration of the corpus luteum, or yellow body. (See section "Hormones for Estrus Control.")

Eggs which are shed from the ovaries are stored in large follicles. The basic principle of superovulation is to stimulate extensive follicular development through the use of a hormone preparation, given intramuscularly or subcutaneously, with follicle-stimulating hormone (FSH) activity. The most common sources of such a hormone are pregnant mares' serum (PMSG) and FSH extracts from pituitaries of slaughtered animals. Many animals so treated will come into estrus about 5 days after initiation of treatment and ovulate, through release of their own luteinizing hormone (LH). However, to help assure that multiple ovulations occur, the ovulating LH from pituitaries or human chorionic gonadotropin (HCG) is injected. The multiple ovulations occur at about the same time the cow would have normally ovulated one egg (21 days after the previous ovulation).

Early studies have confirmed that FSH should be administered twice daily over a period of about 5 days. PMSG has a longer biological life and a single subcutaneous injection is normally used. Five or six days after the original FSH or PMSG "shot," LH or HCG is given intravenously.

The heifers should ovulate by the seventh day after starting hormone treatment.

An example of a superovulated ovary is shown in Fig. 6-21.

Since ovulation occurs over a period of time, not all the eggs are fertilized unless the donor is inseminated repeatedly. A yield of eight good fertilized eggs per donor is about average.

Of course, the real economic value of superovulation lies in the successful transfer of excess eggs from more valuable donor cows to less valuable recipient cows. As a result of this technique, litter-bearing cows are now a reality.

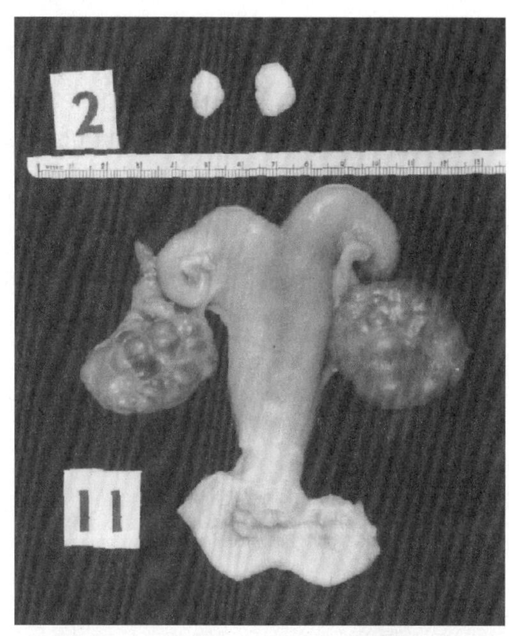

Fig. 6-21. Superovulation of the calf. **Calf No. 2**—Ovaries of a control 4-month-old calf. The ovaries contain thousands of dormant Graafian follicles. **Calf No. 11**—Ovaries and genital organs of a superovulated 4-month-old calf. The calf was not showing any signs of heat, since it had not reached sexual maturity. However, it was injected with 10 rat units of the gonadotropic hormone Vetrophin on each of 3 successive days and slaughtered 5 days after the last injection. The 2 ovaries contained 97 ripe follicles. Note the size of the superovulated ovaries in relation to the immature uterus. (Courtesy, Washington State University)

EMBRYO TRANSFER

Artificial insemination has given a means for the widespread distribution of desirable genes via the sperm. Similar genetic selection through high-quality females has, however, been limited since, normally, one cow will produce one calf per year and the average number of offspring per female will seldom exceed five in a lifetime. Out of the latter arose the idea that a marked increase in the production of offspring from desirable cows might be effected by superovulation, followed by transfer of the fertilized embryo to less desirable cows, with the latter serving as host-mothers or foster-mothers to the developing embryo.

Embryo transfer in cattle developed as a result of research done at Cambridge, England, in the late 1800s. This research, and that which followed during the early 1900s, involved using rabbits. In the 1940s, embryo transfer was performed successfully in sheep and goats, and, in 1951, Dr. Willet and his co-workers performed the first successful embryo transfer in cattle.

The first commercial embryo transfers in the United States were done in the early 1970s. Initially embryos were recovered from valuable donors and transferred into recipient animals, using surgical procedures. During the mid-1970s, nonsurgical methods for the recovery of embryos were developed. In the late 1970s, nonsurgical transfers grew in popularity.

A major factor responsible for the growth of embryo transfer is the success that has been achieved in producing offspring from valuable donor cows. Many donors have produced more than 20 calves, and a few more than 50 calves, within a year. On the average, 8 good embryos are recovered from normal donors. However, about ⅓ of all recoveries result in fewer than 3 embryos, and about ⅓ of them produce 8 or more embryos. It is noteworthy, too, that commercial embryo transfer services report up to 65% embryo transfer conception rates.

Embryo transfer is an eight-step process as follows:

Fig. 6-22. Fertilized ovum in dividing stage, ready to be flushed from the donor cow's oviduct and transferred to recipient cow. (Courtesy, A. H. J. Rajamannan, International Cryo-Biological Services, Inc., St. Paul, MN)

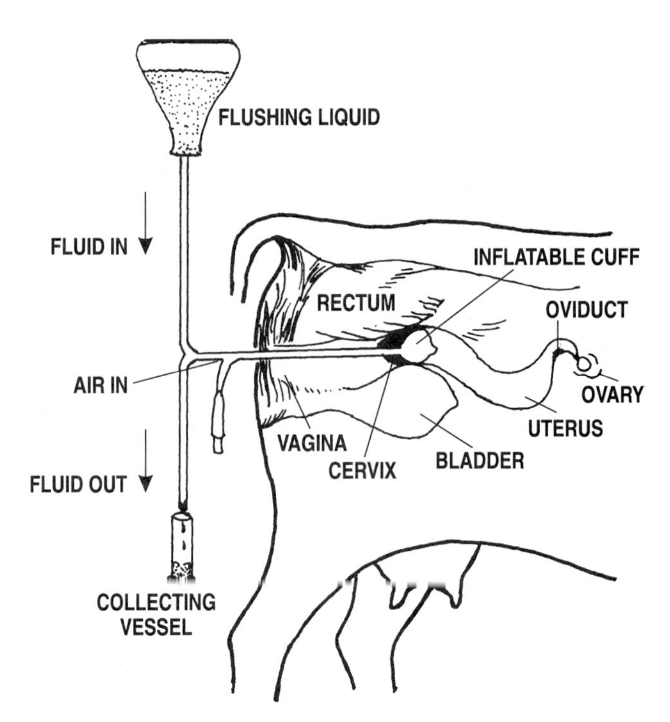

Fig. 6-23. Diagram showing the technique of nonsurgical embryo recovery. The collection procedure involves filling the uterus with fluid through a series of flexible tubes inserted in the uterus through the cervix. (Source: Wright, R. W., "The State of the Art of Cattle Embryo Transfer," *Beef Cattle Science Handbook*, Vol. 18, edited by M. E. Ensminger and published by Agriservices Foundation, 1981, p. 478, Fig. 1)

1. **Select the donor cow.** Needless to say a donor cow should be outstanding.
2. **Inject hormones to cause superovulation of the donor cow.** This should result in the release of from 3 to 30 eggs, instead of the normal release of only 1 egg without superovulation.
3. **Synchronize estrus of donor and recipient cows.** Recipients should be in estrus at the same time as the donor, plus or minus 12 hours.
4. **Breed donor cow.** Breed (usually by AI) donor cow with adequate semen to fertilize all eggs released (usually 4 units of semen).
5. **Collect embryos.** Make nonsurgical recovery of embryos from 7 to 8 days after breeding of donor.
6. **Examine embryos.** Immediately take the embryos into the laboratory and examine them microscopically for normal appearance and fertilization.
7. **Transfer eggs to recipients.** Make surgical or nonsurgical transfer of embryos into synchronized recipients.
8. **Freeze excess embryos.** Excess embryos should be frozen and stored.

Pregnancy in the recipients can be diagnosed in about 30 days. Full-term pregnancies result in full sibs (brothers and sisters) with the genetic traits of the donor cow and the bull to which she was bred. Recipi-

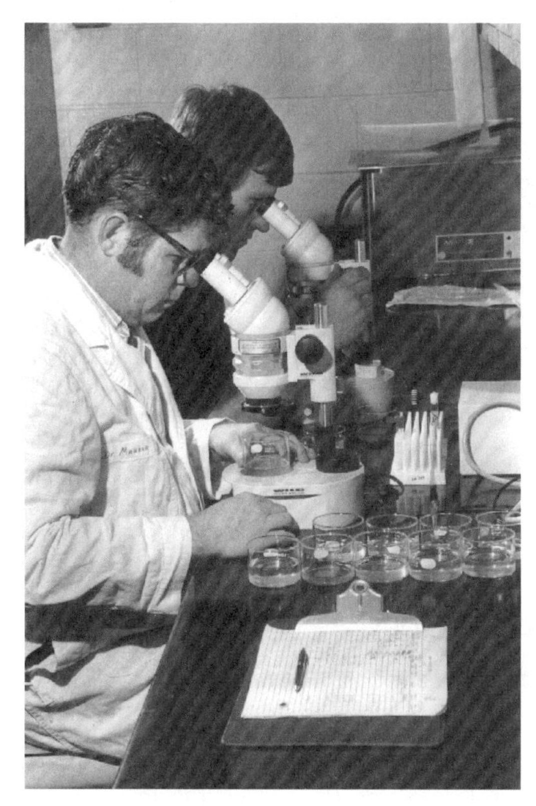

Fig. 6-24. Examining embryos prior to transfer. (Courtesy, USDA, Meat Animal Research Center, Clay Center, NE)

Fig. 6-25. Eight purebred calves (6 bulls and 2 heifers), all from the Maine-Anjou cow, Cetlela; the result of ova transplantation to 7 Jersey and Holstein cows (1 cow had twins). All embryos transferred in this operation resulted in live calves. (Courtesy, the owner—Harold Biensch, Nilburg, Saskatchewan, Canada)

ents have no genetic influence on the calves they carry—they merely serve as "incubators."

The following advantages accrue from embryo transfer:

1. Up to 50 calves may be obtained from a valuable cow during a year's time.

2. The rate of progress in genetic improvement is speeded up, because of the increased number of progeny from valuable cows.

3. Valuable cows that produce normal ova but fail to conceive due to some hormonal or anatomic defects need not be culled because of sterility; such animals may be used as donors for supplying ova for transplantation.

4. Embryo transfer offers a practical method through which to import and export cattle genetics, with three primary advantages: (a) reduced transportation costs, (b) reduced disease hazard, and (c) easy adaptation to the climate and environment in which the animal is born.

5. Heifers may be effectively progeny tested at an early age. If large numbers of fertilized eggs could be procured from calves and transplanted to sexually mature recipients, the generation time of cattle could be reduced by one year or more.

6. It is possible to produce calves of the beef breed of preference from dairy cows.

The following four advancements in bovine embryo transfer technology are in various stages and use:

1. **The one-step freezing process.** This is a new method of freezing and thawing embryos in a single

Fig. 6-26. Dr. T. J. Williams of the Embryo Transfer Laboratory, Colorado State University, with identical twin calves. These calves were produced by (a) fertilizing one ovum; (b) collecting the ovum at approximately the 60-cell stage, when it was about the size of the sharp end of a pen; (c) splitting the embryo, by using a high-powered microscope and micromanipulator equipped with a special tiny knife blade, into identical twin half-embryos, leaving one-half in the original shell, and placing the other half in a shell from which an unfertilized egg has been removed; and (d) transferring each half-embryo nonsurgically to a different female. (Courtesy, International Embryo Transfer Society, La Porte, CO)

straw container. This process enables embryos to be implanted in a recipient cow in much the same manner as breeding a cow by artificial insemination.

2. **The embryo splitting or cloning procedures.** If cloning is defined as asexual reproduction accomplished by replicating genetic material, then embryo splitting is a form of cloning. Embryo splitting is what the designation implies. After recovery from donor cows at approximately six days following insemination, the embryos are actually cut in half and each half is implanted into a recipient cow, or both halves can be implanted in the same cow to produce twins from one recipient. Splitting embryos doubles the number of potential offspring from embryos recovered from donor females. Also, one-half of a split embryo can be held in reserve by freezing, then brought out later if progeny from the first half-embryo possess superior characteristics. In 1984, the Louisiana Agricultural Experiment Station reported on the first ever calves produced from an embryo split four ways.

3. **Embryo sexing.** Experienced technicians report 90% accuracy in embryo sexing, thereby permitting embryos of known sex to be transferred.

4. ***In vitro* fertilization (IVF).** *The term in vitro refers to outside the live body. In vitro* fertilization is accomplished by placing eggs (oocytes) and sperm in culture dishes or test tubes so that, under suitable conditions, fertilization can take place. This technique has long been used to circumvent certain human infertility problems. As techniques improve and experience is gained, greater use will be made of *in vitro* fertilization.

An estimated 60,000 embryo transfer calves were born in the United States in 1993. Transfers cost up to $2,000 when the commercial transfer operator provided recipient cows; and as low as $400 to $500 per pregnancy when the breeder supplied the recipient.

In summary, it may be said that embryo transfer can be done very successfully by skilled teams. But the high cost of present techniques limits the application to the most elite stock. With more research, techniques will become more efficient, simple, and economical, and embryo transfer will be more widely used.

REGISTRATION OF EMBRYO TRANSFER OFFSPRING

Breed associations are in the process of reformulating regulations governing the registration of embryo transfer offspring. Policies are still in a state of flux; so, each association should be contacted before assigning a donor to a superovulation regimen since, in some cases, there are requirements which must be satisfied before treatment is begun.

The chief concern of breed associations in registering embryo transfer offspring is assuring parentage. For this reason, the most common requirement is the identification of the donor, sire, recipient and calf by blood type to eliminate the possibility that the recipient is the genetic mother. Unfortunately, the recipient cannot always be excluded on the basis of blood type. In this event, many of the associations rule on the registration. The regulations also specify the laboratory which must do the analysis, official analysis report forms, and the time when the blood sample is to be collected. Usually it is required that the donor be blood typed before embryos are recovered and that the recipient be blood typed either before transfer or when pregnancy is diagnosed.

In addition to positive identification of the genetic parents, some breed associations require certification by the embryo transfer unit of embryo removal, embryo transfer, or breeding of the donor. The registration numbers of identification of the donor, sire and recipient, date of transfer and affirmation that the transfer was made must be recorded on the certification forms.

Breed associations are also concerned about the impact of embryo transfer on management practices and breed quality, as they were about artificial insemination. Many of the purebred dairy cattle associations have ruled that advance permission must be obtained to propagate a given dam and sire by embryo transfer.

The fee for registering embryo transfer offspring in most cases is the same as for regular registration. Some associations, however, assess an extra fee, usually around $100, to compensate for the additional administrative burden.

While a few breed associations feel that embryo transfer will not affect them, most are discussing the impact which this technique will have on their industry and are establishing controls for its proper application.

INTERNATIONAL EMBRYO TRANSFER SOCIETY (IETS)

The address, along with phone and FAX numbers, of this society follows:

International Embryo Transfer Society
309 W. Clark Street
Champaign, IL 61820
Phone: 217-356-3182
FAX: 217-398-4119

Most cattle registry associations have embryo transfer regulations to insure the accuracy of parentage and the proper identification of the resulting offspring.

The International Embryo Transfer Society has prepared standardized forms for embryo transfer. The IETS encourages all breed associations throughout the world to use the standardized forms, which include:

Certificate of Embryo Recovery
Certificate of Embryo Transfer
Certificate of Freezing
Application for Embryo Transport

When transferable embryos are recovered from a donor, the "Certificate of Embryo Recovery" should be completed and signed by the practitioner recovering the embryos. When fresh or frozen embryos are transferred to a recipient, the "Certificate of Embryo Transfer" should be completed and signed by the practitioner transferring the embryos. When embryos are frozen, the "Certificate of Freezing" should be completed and signed by the person responsible for the freezing and labeling or identification of the frozen embryos. The "Application for Embryo Export" is a modified Certificate of Recovery which includes space for the signature of the seller, the name and address of the buyer, and the identification of the exporter. This certificate should be submitted to the appropriate breed registry of the exporting country. The breed registry will use the certificate to verify the identity of the embryo; and to provide information, such as blood type of the donor, for the herd book or recording agency of the receiving country.

The subsets of the above four forms accommodate nuclear transfer, in vitro fertilization, and oocyte aspiration.

Although IETS forms and requirements are used as a guide for many breed registry associations, there are many variations among breed registries. Each breeder of registered cattle who decides to use embryo transfer should contact the breed registry of the breed involved and follow the requirements set forth for the recording of an embryo transfer animal.

A major ongoing contribution of the IETS to the livestock industry is that of developing procedures for international movement of genetics in and out of countries where movement of semen is difficult and the movement of animals is prohibited.

The IETS procedures are outlined in Chapter 5, Section IV of the *Manual* of the International Embryo Transfer Society. Also, the IETS prints the four different standardized forms which they recommend.

BLOOD TYPING CATTLE

Cattle blood typing was developed at the University of Wisconsin during the decade 1940–1950. It involves a study of the components of the blood, which are inherited according to strict genetic rules that have been established in the research laboratory. By determining the genetic "markers" in each sample and then applying the rules of inheritance, parentage can be determined. To qualify as the offspring of a given cow and bull, an animal must not possess any genetic markers not present in its alleged parents. If it does, it constitutes grounds for illegitimacy.

Blood typing is used for the following purposes:

- **To verify parentage**—The test is used in instances where the offspring may bear some unusual color or markings or carry some undesirable recessive characteristic. It may also be used to verify a registration certificate. When one considers that about 5% of all registered cattle in the United States are illegitimate, there is need to use blood typing much more extensively as a bulwark of breed integrity. Through blood typing, parentage can be verified with 99% accuracy.[7] Although this means that 1% of the cases can't be settled, it's not possible to do any better than that in human blood typing.

- **To identify identical twins**—This may be important when identical twins are needed for experimental work.

- **To determine which of two bulls**—When a cow has been served by two or more bulls during one breeding season, blood typing can identify the sire.

- **To provide a permanent blood type record for identification purposes**—Two samples of blood are required for each animal to be studied; and the samples must be taken in tubes and in keeping with detailed instructions provided by the laboratory. In parentage cases, this calls for blood samples from the calf and both parents; in paternity cases, samples must be taken from the calf, the cow, and all the bulls.

- **To detect fertile heifers born co-twin with bulls**—About 15% of all heifers born twin with a bull are potentially fertile; the other 85% are sterile, or freemartins. Cattle producers need not wait until such heifers reach breeding age in order to ascertain their breeding potentialities. Instead, they can submit blood samples from each of the twins (the bull and the heifer) to a service-typing laboratory and request a diagnosis. If the bull and heifer have *like* blood types (except possible differences in the J system), the heifer is diagnosed as a freemartin and nonbreeder. If the bull and heifer have *unlike* blood types (except possible differences in the J system alone), the heifer is diagnosed as potentially fertile.

The basis for this remarkable method of diagnosing the breeding potentialities of a heifer born twin with a bull goes back to the early events in the embryology of cattle twins. In about 85% of the cattle

---

[7]In a personal communication to the senior author, Dr. Clyde Stormont, Professor of Immunogenetics, Department of Reproduction, School of Veterinary Medicine, University of California-Davis, reported that in the California Laboratory they have been able to solve cattle and horse parentage with 99% accuracy.

embryos, some of the chorionic blood vessels become anastomosed, or joined together. This results in a communal blood vascular system. Hence, the twins come to share each other's blood forming tissues. As a result they have like blood types.

- **To substitute for fingerprinting**—Much attention is now being given to the idea of utilizing blood typing as a positive means of identification of stolen animals, through proving their parentage.

For a list of blood typing laboratories, see Chapter 3 section headed "Blood Typing/Profiling."

DNA TYPING CATTLE

DNA typing is similar to traditional blood typing in that the goal of both is to identify and track genetic variation in an animal from one generation to the next. But they differ in that traditional blood typing is based on variation in proteins found on the surface of the red blood cells, whereas DNA typing analyzes the most fundamental difference that can exist among individuals—the difference in the individual's DNA. The DNA is the genetic code that determines all genetic characteristics of an animal, from hair coat to size. Since an individual receives one-half of its DNA from each parent, it is possible to verify an individual's parentage through DNA typing.

The power of DNA typing is being documented daily in the field of human forensics. It is being used in criminal cases throughout the world to identify individuals committing violent crimes.

In addition to routine parent verification for registration purposes, DNA typing can be used to identify the sire of each calf produced in multiple sire pastures.

FACTORS AFFECTING REPRODUCTION

Two major problems plague the cow-calf producer: (1) the long calving season; and (2) a poor calf crop. Both can be lessened very materially.

Nature ordained that the gestation period of a beef cow be about 283 days. This means that a cow must be pregnant again within 80 days of calving if she is going to produce a calf each year (283 + 80 = 363). This is not easy, primarily because the interval from calving to first estrus is long. The average interval from calving to first estrus is 61 days. But, as shown in Table 6-2, it is longer in heifers than in older cows.

TABLE 6-2
PROPORTION OF COWS IN HEAT AT VARIOUS TIMES AFTER CALVING: FIRST-CALF HEIFERS VS MATURE COWS[1]

No. of Days After Calving	Percent of Cows in Heat at the Time	
	3-Year-Old Heifers	Cows 5 Years Old or Older
40	15	30
50	24	53
60	47	72
70	62	82
80	68	89
90	79	94

[1]Wiltbank, J. N., "Reproductive Losses in the Beef Cow," *Beef Cattle Science Handbook*, Vol. 17, edited by M. E. Ensminger and published by Agriservices Foundation, 1980, p. 383, Table 9.

Table 6-2 reveals that, 50 days after calving, 53% of the mature cows had shown heat, in comparison with 24% of the heifers; hence, more than twice as many of the old cows were in heat. By 70 days after calving, 82% of the old cows had shown heat, in comparison with 62% of the first-calf heifers. Clearly, this shows that the interval from calving to first heat is too long, particularly in first-calf heifers.

This long interval from calving to first heat makes for reproductive problems such as (1) a high proportion of 2- and 3-year-old cows which are open or calving late, and (2) calving intervals of 13 to 14 months.

Since the pattern for a heifer's entire reproductive life is established with her first calf, it is important to get her off to the right start. To do so, 50% more replacement heifers should be selected than are needed as brood cows in order to get those which will produce early calves. Next, heifers should be kept separate from the brood cows, because younger animals don't compete well with older animals. From this stage on, getting heifers to calve early is a matter of condition, nutrition, and planning.

A poor calf crop is the other major problem affecting profits. Calving losses at birth have been estimated to run more than 6% of the potential calf crop, with these losses ranking second in magnitude only to cows failing to conceive. Fortunately, over 50% of these losses could be prevented by improved management.

Dystocia, or calving difficulty, is defined as calving that is prolonged or that requires artificial aid to deliver the calf. Since dystocia is so important to cattle producers, the Miles City Station made extensive studies of its causes. As a result of a 15-year study, involving 13,296 calvings and 893 calf losses (for a loss of 6.7%), the station researchers identified the major causes of dystocia (see Fig. 6-27).

Fig. 6-27 shows that large calf birth weight is the most important cause of dystocia, followed, in order,

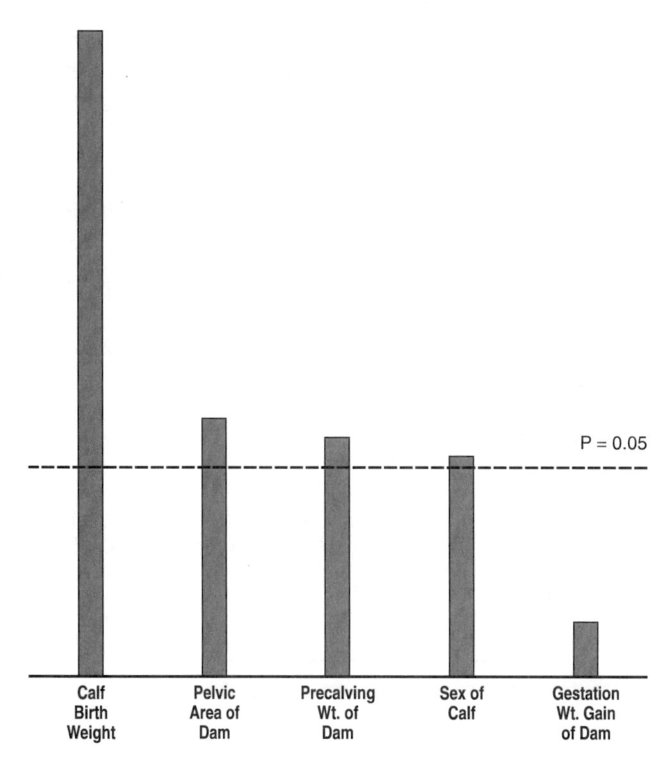

Fig. 6-27. Major causes of calving difficulty on the Miles City Station, Montana. Bars extending above the broken line are statistically significant. (Source: Bellows, R. A., "Causes of Dystocia," *Beef Cattle Science Handbook*, Vol. 18, edited by M. E. Ensminger and published by Agriservices Foundation, 1981, p. 404, Fig. 1)

by pelvic area of dam, precalving weight of dam, sex of calf, and gestation weight gain of dam.

Because of the economic importance of reproduction, a review of the experiments dealing therewith follows.

HEIFERS VS OLDER COWS

As every producer knows, calving difficulty and losses, known technically as dystocia, are higher in 2-year-old heifers than in older cows. Surveys indicate that approximately 50% of calving 2-year-olds require help at calving and 15% of the calves and 5% of the heifers are lost at parturition.[8]

At the U.S. Range Livestock Experiment Station, Miles City, Montana, where there is better than average care, data covering 14 Years showed (1) that 10.8% of the calves born to first-calf heifers were lost; but (2) that when these same dams had their second and third calves, the losses dropped to 4.1 and 4.8%, respectively (Table 6-3).

[8]Moore, D. G., D. Chambers, J. A. Whatley, Jr., and W. D. Campbell, "Some Factors Affecting Difficulty at Parturition of Two-Year-Old Hereford Heifers," *Journal of Animal Science*, Vol. 15, No. 4, p. 1225.

TABLE 6-3
EFFECT OF AGE OF DAM ON CALF LOSSES[1]

Age of Dam at Calving	Number Calving	Number of Calves Lost	Percent of Loss
			(%)
2-year-old, first calf	2,257	245	10.8
3-year-old, second calf	1,461	60	4.1
4-year-old, third calf	1,032	50	4.8

[1]Bellows, R. A., "Calf Losses: What We Can Do About Them," *Beef Cattle Science Handbook*, Vol. 19, edited by Frank H. Baker, Winrock International, and published by Westview Press, Boulder, CO, 1983, p. 382, Table 1.

Fig. 6-28. Two-year-old heifers have a much higher incidence of calving difficulty than older cows. (Courtesy, American Breeders Service, De Forest, WI)

BREED, WEIGHT, AND AGE

The three factors affecting the number of heifers showing heat early in the breeding season are: breed, weight, and age. These were confirmed in a study conducted at the Texas A&M Experiment Station, Beeville; the proportion of Hereford, Angus, and crossbred heifers that showed heat at different weights and ages is given in Table 6-4.

Table 6-4 shows that breed, weight, and age affected time of puberty. Angus heifers came in heat at an earlier age and lighter weight than Herefords. Most of the two breeds, along with the crossbreds, showed heat at 14 to 15 months of age *provided* they had sufficient weight; with the weight needed to reach puberty varying according to the breed of the heifer.

TABLE 6-4
PROPORTION OF HEIFERS IN HEAT
AT VARIOUS WEIGHTS AND AGES[1]

Breed	Weight		Age in Months			
			12	13	14	15
	(lb)	*(kg)*	*(% in heat)*	*(% in heat)*	*(% in heat)*	*(% in heat)*
Hereford	500	*227*	0	0	0	0
	600	*273*	15	20	30	37
	700	*318*	—	65	90	90
Angus	500	*227*	0	33	57	77
	600	*273*	40	65	70	80
	700	*318*	—	80	100	100
Angus X Hereford crossbreds	500	*227*	27	36	73	91
	600	*273*	40	75	82	96
	700	*318*	—	78	96	100

[1]Wiltbank, J. N., S. Roberts, and K. J. Nix, "Improving Pounds of Calf Weaned by Heifer Selection," *Beef Cattle Science Handbook*, Vol. 17, edited by M. E. Ensminger and published by Agriservices Foundation, 1980, p. 394, Table 3.

NUTRITIONAL FACTORS AFFECTING REPRODUCTION

Improper feeding—either (1) uncommonly high or low feed intake, or (2) a deficiency of specific nutrients—can affect reproduction. Experiments pertaining thereto are summarized in the sections that follow.

(Also see Chapter 19, section entitled, "Nutritional Reproductive Failures in Cows.")

LIBERAL FEEDING MAKES FOR EARLY SEXUAL MATURITY

The onset of heat and ovulation in young heifers is definitely and positively correlated with the level of nutrient intake. Thus, if feed intake is too low for normal rate of growth and development, the onset of reproductive function is delayed.

■ **A long-term experiment to determine the effect of plane of nutrition on sexual maturity**—Cornell University raised three groups of Holstein heifers from birth to first calving on three different nutritive levels—62% (low), 100% (medium), and 146% (high)—of the standard amount of total digestible nutrients (TDN).[9] An extremely important finding in the Cornell study was that regardless of plane of nutrition, the heifers (Holsteins) came into heat at about 600 lb body weight; showing that size and weight, not age, determine the time of sexual maturity. Thus, most researchers and knowledgeable cattle producers recommend breeding according to size and not age.

■ **The effect of three winter feeding levels on early breeding of heifers**—The U. S. Range Experiment Station, Miles City, Montana, studied the effects of three winter feeding levels—*low*, average daily gain 0.6 lb; *medium*, average daily gain 1.0 lb; and *high*, average daily gain 1.5 lb. The effect on the number of heifers bred and pregnant early in the breeding season is shown in Table 6-5.

TABLE 6-5
EFFECT OF THREE WINTERING LEVELS OF ANGUS X HEREFORD CROSSBRED HEIFERS ON NUMBER BRED AND PREGNANT[1]

	(1) Low Level		(2) Medium Level		(3) High Level	
Number of heifers	30		29		30	
	(lb)	*(kg)*	*(lb)*	*(kg)*	*(lb)*	*(kg)*
Winter gain, daily	0.6	*0.27*	1.0	*0.45*	1.5	*0.68*
Summer gain, daily	1.3	*0.58*	1.2	*0.54*	0.9	*0.41*
(June 15)	458	*208*	527	*239*	584	*265*
Age at puberty	434 days		412 days		388 days	
Percent in heat:			*(%)*			
Prior to breeding season (June 15)	7		31		83	
During breeding season	73		66		17	
After breeding season	20		3		0	
Percent bred and conceived:						
First 20 days	30		62		60	
Second 20 days	10		21		20	
Third 20 days	10		3		7	
Not bred	20		3		0	

[1]Bellows, R. A., "Prevention of Dystocia," *Beef Cattle Science Handbook*, Vol. 18, edited by M. E. Ensminger and published by Agriservices Foundation, 1981, p. 406, Table 1.

In the Miles City study, the heifers on the low level winter ration averaged 458 lb at the beginning of breeding. Of the low-level heifers, only 7% of them were cycling prior to the breeding season, only 30% of them conceived during the first 20 days of breeding, and only 80% came in heat during the breeding season. The failure of 20% of the low-level heifers to come in heat automatically reduced the maximum attainable calf crop to 80%; compared to a 97% and 100% maximum for the medium-level and high-level groups, respectively. There was no significant difference be-

[9]Reid, J. T. *et al.*, *Causes and Prevention of Reproductive Failures in Dairy Cattle, IV. Effect of Plane of Nutrition During Early Life on Growth, Reproduction, Production, Health, and Longevity of Holstein Cows, 1. Birth to Fifth Calving*, Cornell Univ. Ag. Exp. Sta. Bull. 987.

tween the medium-level and high-level groups in the percent of heifers that conceived.

Both the Cornell and Miles City experiments showed that well-fed heifers developed faster sexually than those raised on a lower plane of nutrition, and that size and weight, not age, determine the time of sexual maturity.

FLUSHING

Flushing refers to the practice of feeding thin cows and heifers to gain approximately 1.5 lb per head daily beginning 20 days before the start of and continuing throughout the breeding season. It may be accomplished either by providing more lush pasture or range or by feeding grain.

By 30-day grain (corn) flushing during the spring breeding season, the Louisiana Station increased the calf crop by 11%.[10]

Wiltbank, *et al.*, increased pregnancy by 10% over flushing alone by employing both flushing and 48-hour calf removal (Table 6-6). As shown, neither practice alone was as beneficial as a combination of the two.

TABLE 6-6
EFFECT OF FLUSHING AND CALF REMOVAL ON PREGNANCY[1]

	Control	Flushed[2]	Calf Removal[3]	Flushing and Calf Removal
No. of cows	18	21	21	21
Pregnancy	(%)			
21 days	28	14	38	57
24 days	56	52	62	72
63 days	72	76	62	86

[1]Wiltbank, J. N., R. Anderson, and H. L. Fillmore, "Beef Cattle Reproduction and Management," *Beef Cattle Science Handbook*, Vol. 20, edited by Frank H. Baker and Mason E. Miller, Winrock International, and published by Westview Press, Boulder, CO, 1984, p. 410, Table 7.

[2]Flushed by feeding 10 lb *(4.5 kg)* of corn/day for 2 weeks before breeding and first 3 weeks of breeding.

[3]Calf removed for 48 hours at start of breeding.

Although it is not likely that all of the benefits ascribed to flushing will be fully realized under all conditions, the general feeling persists that the practice will cause more cows to come into heat, to breed early in the season, and to conceive at first service. Certainly, following calving a cow should maintain her weight and/or make small gains.

[10]Loyacano, A F., "Grain Flushing Ups Calfing Percentage," *Better Beef Business*.

ENERGY (FEED) LEVEL

Restricted rations often occur during periods of drought, when pastures or ranges are overstocked, or when winter rations are skimpy. When such feed shortages are extreme, there may be lowered reproductive efficiency. Likewise, too liberal feeding and high condition may cause sterility. Summaries of several experiments pertaining to the energy (feed) level of cows follow.

ENERGY (FEED) LEVEL DURING GESTATION

The best calf crop is produced by cows that are kept in vigorous breeding condition—that are neither overfat nor thin and run-down.

■ **Effect of feed (energy) level of heifers during gestation on calf birth weight, calving difficulty, and when the dam comes back in heat after calving**—The U.S. Range Livestock Experiment Station, Miles City, Montana, conducted studies designed to answer this question. The results are given in Table 6-7. Crossbred Angus X Hereford heifers were bred to Charolais, Hereford, or Angus bulls. Ninety days prior to the predicted calving date, all heifers were placed in a feedlot on either a low or a high feed level until calving. During this period, the low feed-level group received 7.0 to 8.0 lb TDN; and the high feed-level group received 13.8 to 15 lb of TDN. Thus, the high feed-level animals received nearly twice as many pounds of TDN daily as the low feed-level animals.

All heifers were observed 24 hours per day during calving. Each calving was given a numerical score,

TABLE 6-7
EFFECTS OF GESTATION FEED LEVEL OF HEIFERS ON CALF BIRTH WEIGHT, CALVING DIFFICULTY, AND WHEN HEIFERS COME BACK IN HEAT[1]

Sire Breeds[2]	Gestation Feed Level[3]	Post-calving Body Weight		Calf Birth Weight		Dystocia[4]		Pregnancy Rate
		(lb)	*(kg)*	*(lb)*	*(kg)*	*(%)*	*(score)*	*(%)*
Charolais	Low	694	*315*	68	*31*	71	2.3	65
Hereford or Angus	High	794	*361*	71	*32*	66	2.2	83
	Difference	100	*46*	3	*1*	5	0.1	18

[1]Bellows, R. A. "Prevention of Dystocia," *Beef Cattle Science Handbook*, Vol. 18, edited by M. E. Ensminger and Published by Agriservices Foundation, 1981, p. 408, Table 4.

[2]Results of three studies combined; 133 head of animals.

[3]Low = 7.0 to 8.0 lb *(3.2 to 3.6 kg)* TDN; high = 13.8 to 15.0 lb *(6.3 to 6.8 kg)* TDN, last 90 days of gestation.

[4]Score range: 1 = no difficulty to 4 = extreme difficulty.

ranging from "1" where there was no difficulty to "4" where there was extreme difficulty, including Caesarean delivery. Calf birth weights were taken immediately after calving.

It is noteworthy that the feed level during the last 90 days of gestation made quite a difference in post-calving weights. The heifers on the high feed level weighed 100 lb more than their counterparts on the low level.

These studies showed that cutting the feed level prior to calving had little effect on either calf birth weight or calving difficulty, but it did have a marked depressing effect on subsequent reproduction. Pregnancy rates for the heifers from the high feed levels during gestation averaged 18% higher than that for heifers on the low feed levels. The latter finding is very important. If cows are not cycling before the breeding season starts, they tend to breed late or not at all.

A word of caution relative to the above experiment appears to be in order. If the gestation feed level is very high and heifers become excessively fat, the effects on calving difficulty will be drastically different. Studies conducted by the U.S. Department of Agriculture at Beltsville, Maryland, showed that extreme calving difficulty and high calf and dam losses resulted when heifers were too fat at calving. Also, excessively fat heifers do not milk as well as heifers in medium flesh, and the depressing effect on milk production of a high plane of nutrition becomes more evident with each lactation.

■ **Effect of two levels from 120 days prior to calving until calving as 2-year-olds**—In a study conducted at the Fort Robinson Beef Cattle Research Station, with the U. S. Department of Agriculture and the University of Nebraska cooperating, the effect of two levels of energy—*moderate* and *low*—on calving difficulty and calf losses of 2-year-old heifers was determined. Beginning 120 days prior to calving, the moderate-level energy group received 8 lb of TDN and gained about 1 lb per day, while the low-level group received 4.5 lb of TDN and barely held their weight the first year and lost weight the second year. The results of this study are summarized in Table 6-8.

At calving time, the heifers on the moderate level of feed weighed approximately 145 lb more than the ones on the low level of feed. At calving, the moderate plane heifers lost 125 lb weight in comparison with 99 lb loss for the low-level group. Calf birth weight was 7 lb heavier for the heifers on the moderate-level feed than those on the low level of feed. There was no significant difference between the 2 groups in calves alive at birth and 2 weeks after birth.

Although the Fort Robinson Station reduced birth weight of calves and calving difficulty by feeding low levels of energy, calving difficulties were not eliminated. Also, and most important, the heifers receiving the low energy diet did not return to heat and conceive as readily as heifers on the moderate level of feed.

As a result of this study, the investigators (1) concluded that, "you cannot starve calving losses out of a group of heifers"; and (2) recommended that heifers be fed so as to gain approximately 1 lb per head per day for the last 100 to 120 days before calving.

■ **Effect of energy level before calving on interval from calving to first estrus**—The Fort Robinson Beef Research Station, Crawford, Nebraska, found that energy level prior to calving had a marked effect on the interval from calving to first estrus. (See Table 6-9.)

Table 6-9 shows that more heifers receiving 8.0 lb of total digestible nutrients (TDN) prior to calving were in heat by 40, 60, 80, or 100 days after calving than heifers receiving 4.3 lb of TDN.

TABLE 6-8
EFFECT OF LEVEL OF ENERGY LAST 120 DAYS OF GESTATION ON CALVING DIFFICULTY AND CALF LOSSES OF 2-YEAR-OLD ANGUS AND HEREFORD HEIFERS[1]

	First Year		Second Year	
	Moderate (8 lb *[3.6 kg]* TDN)	Low (4.5 lb *[2.0 kg]* TDN)	Moderate (8 lb *[3.6 kg]* TDN)	Low (4.5 lb *[2.0 kg]* TDN)
Number of cows	140	94	123	111
120-day weight gain[2] . . . (lb)	125 *(57 kg)*	2 *(0.9 kg)*	108 *(49 kg)*	–31 *(–14 kg)*
Cow weight change at calving[3] . . (lb)	–119 *(–54 kg)*	–102 *(–46 kg)*	–130 *(–59 kg)*	–96 *(–44 kg)*
Birth weight of calf . . . (lb)	73 *(33 kg)*	67 *(30 kg)*	71 *(32 kg)*	63 *(29 kg)*
Difficult births . . . (%)	36	34	33	16

[1]Wiltbank, J. N., "Influence of Nutrition on Birth Weight, Pelvic Area, and Calving Difficulty," *Beef Cattle Science Handbook*, Vol. 18, edited by M. E. Ensminger and published by Agriservices Foundation, 1981, pp. 389–390, Tables 1, 2, and 3.

[2]120 days prior to calving.

[3]Weight lost between 7 days before calving and after calving.

TABLE 6-9
EFFECT OF ENERGY LEVEL BEFORE CALVING ON POSTPARTUM INTERVAL TO FIRST ESTRUS IN 2-YEAR-OLD HEIFERS[1]

No. of Days After Calving	Percent 2-Year-Old Heifers Showing Estrus at the Time	
	4.3 lb *(1.9 kg)* TDN Daily	8.0 lb *(3.6 kg)* TDN Daily
40	7	22
60	49	81
80	73	92
100	88	97

[1]Dunn, T. G., J. E. Ingalls, D. R. Zimmerman, and J. N. Wiltbank, "Reproductive Performance of 2-Year-Old Hereford and Angus Heifers As Influenced by Pre- and Post-calving Energy Intake," *Journal of American Science*, Vol. 29, p. 719.

CONDITION OF HEIFERS AT CALVING

Excessively fat heifers have calving problems, with heavy calf losses at and soon after birth. However, the losses at birth are not the result of increased birth weight. Rather, they are due to the pelvic area being filled with fat and the calves being presented backwards.

Fig. 6-29. Very fat heifers often have calving difficulty, because of excess fat decreasing the size of the birth canal. (Courtesy, American Breeders Service, De Forest, WI)

■ **Effect of three levels of energy on calf losses of heifers**—Table 6-10 summarizes an experiment in which three levels of energy were fed from weaning until calving as 2-year-olds.

As shown in Table 6-10, within 24 hours after birth, 12 calves were dead in the high energy, overly fat, heifer group compared to 3 calves in each of the other groups; and the fat heifers weaned a 48% calf crop. This means that 2 heifers were kept to produce 1 weaner calf. This does not indicate that a caretaker should attempt to alleviate calving difficulties by starving heifers. Rather, they should be in medium condition, neither too fat nor too thin.

TABLE 6-10
EFFECT OF LEVEL OF ENERGY ON CALVING OF 2-YEAR-OLD HEIFERS[1]

	Level of Energy					
	Low		Medium		High	
Number of heifers calving	30		34		33	
Gestation length	279		279		277	
Condition at calving	thin		good flesh		extremely fat	
Calves living:						
When born	27		31		27	
24 hours after birth	27		31		21	
2 weeks after birth	27		31		18	
At weaning time	27		29		16	
	(lb)	*(kg)*	*(lb)*	*(kg)*	*(lb)*	*(kg)*
Average birth weight	45.1	*20*	59.5	*27*	56.2	*26*
Heifer weight before calving	698	*317*	921	*419*	1,104	*502*
Heifer weight after calving	618	*281*	831	*378*	1,012	*460*

[1]Wiltbank, J. N., *et al.*, *Influence of Total Feed and Protein Intake on Reproductive Performance of the Beef Female Through Second Calving*, USDA Tech. Bull. 1314, pp. 20–21.

ENERGY (FEED) LEVEL AFTER CALVING

Cows with calves at side should be fed for the production of milk, for which requirements are more rigorous than those during pregnancy. The energy requirements of a cow nursing a calf are about 50% higher than those of a dry pregnant cow.

■ **Effect of energy level following calving on estrus and pregnancy**—Energy level after calving affects the proportion of cows showing heat and becoming pregnant (see Table 6-11).

TABLE 6-11
EFFECT OF WEIGHT CHANGE AFTER CALVING ON HEAT AND PREGNANCY RATE[1]

Calving Time to Breeding	Cows Not Showing Heat	Pregnant		
		From First Service	After Breeding	
			20 Days	90 Days
	(%)	*(%)*	*(%)*	*(%)*
Losing weight	14	43	29	72
Gaining weight	0	60	57	82
Difference	14	17	28	10

[1]Wiltbank, J. N., "Reproductive Losses in the Beef Cow," *Beef Cattle Science Handbook*, Vol. 17, edited by M. E. Ensminger and published by Agriservices Foundation, 1980, p. 385, Table 14.

Table 6-11 shows that 14% of the cows losing weight following calving failed to show heat during a 90-day breeding season. Also, Cows losing weight after calving have a lower conception rate than cows gaining weight. Only 43% of the cows losing weight conceived on first service compared to 60% of the cows gaining weight. After 20 days of breeding, there was a 28% difference in pregnancy rate, and after 90 days of breeding there was a 10% difference in pregnancy rate.

■ **Effect of level of energy following calving on early pregnancy**—Table 6-12 summarizes the results of a series of experiments designed to determine the effect of energy following calving on reproductive performance.

Table 6-12 shows that the low level of energy after calving had the following marked effects on both older cows and heifers: (1) a much greater number of females did not show heat during the breeding season; and (2) a sharp decrease in the number of females pregnant at the end of 20 days of breeding, from first service, and at the end of the breeding season.

In experiment no. 1, of the older cows which received 8 lb of TDN (16 to 18 lb of hay) following calving, 34% were pregnant after 20 days of breeding compared to 60% of cows receiving 16 lb of TDN (32 to 35 lb of hay). Similar results were obtained in 2-year-old heifers (experiments 2 and 3), although pregnancy rates were lower than in the older cows. The latter situation is understandable because the older cows were bred for 90 days, whereas the heifers were bred for 60 days only. Also, breeding on both groups started 60 days after calving, which disadvantaged the heifers.

Table 6-12 also reveals that there are two reasons for the poor reproductive performance in cows which are on low levels of feed and losing weight following calving: (1) some cows do not show heat during the breeding; and (2) the conception rate at first service is low.

PROTEIN

Cattle producers are prone to skimp on protein, especially (1) when cows are grazed on dried grass cured on the stalk in times of drought or during the winter months, or (2) when protein supplements are high in price. Sometimes both producers and those who counsel with them suspicion that such practice may cause cows to abort, have calving difficulty, drop dead or weak calves, fail to rebreed, and/or not milk well. Fortunately, experiments designed to answer these questions have been conducted.

TABLE 6-12
EFFECT OF TWO LEVELS OF ENERGY FOLLOWING CALVING ON EARLY PREGNANCY[1]

Exp. No.	Level of Feed: Before Calving TDN		Level of Feed: After Calving TDN		No. Cows	Did Not Show Heat During Breeding Season	Pregnant: At End of 20 Days of Breeding	Pregnant: From First Services	Pregnant: At End of Breeding Season
	(lb)	*(kg)*	*(lb)*	*(kg)*		*(%)*	*(%)*	*(%)*	*(%)*
					Cows 5 years or older				
1	9	*4.1*	16	*7.3*	21	0	60	67	95
	9	*4.1*	8	*3.6*	22	14	34	42	77
					Cows 2 years of age				
2	8	*3.6*	13	*5.9*	37	3	54	63	71
	8	*3.6*	7	*3.2*	42	19	33	53	64
					Cows 2 years of age				
3	8	*3.6*	13	*5.9*	24	0	54	50	79
	8	*3.6*	7	*3.2*	13	8	23	37	76

[1]Wiltbank, J. N., W. W. Rowden, J. E. Ingalls, K. E. Gregory, and R. M. Koch, "Effect of Energy Level on Reproductive Phenomena of Mature Hereford Cows," *Journal of Animal Science*, Vol. 21, p. 219.

Wiltbank, J. N., W. W. Rowden, J. E. Ingalls, and D. R. Zimmerman, "Influence of Post-partum Energy Intake on Reproductive Performance of Hereford Cows Restricted in Energy Intake Prior to Calving," *Journal of Animal Science*, Vol. 21, p. 658.

Dunn, T. G., J. E. Ingalls, D. R. Zimmerman, and J. N. Wiltbank, "Reproductive Performance of 2-Year-Old Hereford and Angus Heifers As Influenced by Pre- and Post-calving Energy Intake," *Journal of Animal Science*, Vol. 29, p. 719.

■ **Effect of protein level on number of calves born, calf birth weight, and calf losses**—Three levels of protein were fed to cows from conception through calving: *low*, 0.26 lb of digestible protein; *medium*, 0.89 lb of digestible protein; and *high*, 1.38 lb of digestible protein. The results are summarized in Table 6-13.

Except for the lower birth weight of the calves born to cows on a low level of protein, there were no adverse effects from a low protein ration on the calf crop—in number of calves born and calf survival.

TABLE 6-13
EFFECT OF PROTEIN LEVEL ON NUMBER OF CALVES BORN, CALF BIRTH WEIGHT, AND CALF LOSSES[1]

	Level of Digestible Protein Consumed per Head Daily, Conception Through Weaning					
	Low (0.26 lb) *(0.13 kg)*		Medium (0.89 lb) *(0.40 kg)*		High (1.38 lb) *(0.63 kg)*	
Number of heifers calving	35		32		30	
Gestation length	277 days		278 days		280 days	
Condition at calving	thin		good flesh		good flesh	
Calves living:						
When born	29		30		26	
24 hours after birth	29		26		24	
2 weeks after birth	28		26		22	
At weaning time	25		26		21	
	(lb)	*(kg)*	*(lb)*	*(kg)*	*(lb)*	*(kg)*
Average birth weight	49.4	*22*	55	*25*	56.4	*26*
Heifer weight before calving	840	*382*	948	*431*	935	*425*
Heifer weight after calving	758	*344*	856	*389*	848	*385*

[1]Wiltbank, J. N., *et al., Influence of Total Feed and Protein Intake on Reproductive Performance of the Beef Female Through Second Calving*, USDA Tech. Bull. 1314, pp. 22 and 23.

■ **Effect of protein level on rebreeding and on calf gains of first-calf heifers**—Among cattle producers the feeling persists that a shortage of protein following calving will have an adverse effect on rebreeding and on calf gains. This question was pursued by the same investigators who conducted the experiment summarized in Table 6-13.

Cows suckling their first calves received three levels of digestible protein: *low*, 0.4 lb; *medium*, 1.3 lb; and *high*, 2.0 lb. The results are summarized in Table 6-14.

The onset of heat was delayed in the cows on the low level of protein but conception was not adversely affected. The weight gains on calves were lowered by the low level of protein. The medium level of protein was as satisfactory as the high level; hence, there is no need to go to the added cost of providing more protein than is needed.

TABLE 6-14
EFFECT OF PROTEIN LEVEL ON REBREEDING, AND CALF GAINS OF COWS SUCKLING THEIR FIRST CALVES[1]

	Level of Digestible Protein Consumed per Head Daily During Lactation					
	Low (0.4 lb) *(0.18 kg)*		Medium (1.3 lb) *(0.59 kg)*		High (2.0 lb) *(0.9 kg)*	
Number of cows	21		21		21	
Proportion in heat:						
60 days after calving . (%)	36		48		71	
90 days after calving . (%)	54		90		81	
Pregnant . . . . (%)	91		90		100	
Average daily gains of calves:	*(lb)*	*(kg)*	*(lb)*	*(kg)*	*(lb)*	*(kg)*
First 60 days of life	1.1	*0.5*	1.7	*0.77*	1.7	*0.77*
First 150 days of life	1.3	*0.6*	1.7	*0.77*	1.8	*0.8*

[1]Adapted from: Wiltbank, J. N., *et al., Influence of Total Feed and Protein Intake on Reproductive Performance of the Beef Female Through Second Calving*, USDA Tech. Bull. 1314, pp. 24–27. Low energy data not included.

PHOSPHORUS

Generally speaking, range forages are low in phosphorus. This is especially true of dried grass cured on the stalk, such as exists during droughts and in the winter.

■ **Effect of phosphorus on reproduction**—Black and coworkers showed an increase in reproductive efficiency from phosphorus supplementation of cows in a five-year study on King Ranch in Texas, where the range is deficient in phosphorus (Table 6-15).

In the King Ranch study, phosphorus was supplied by (1) bone meal, self-fed, (2) dicalcium phosphate in the drinking water, or (3) fertilized pasture. As shown, phosphorus supplementation markedly increased the percentage of calves dropped (by 26%) and weaned (by 28%), the weaning weight of calves (54 lb), and the pounds calf weaned per cow (179 lb).

Other experiments have also shown that beef cows generally respond favorably to phosphorus supplementation; among them, studies conducted by the New Mexico Station and the Oklahoma Station. Also, there is general agreement that phosphorus should be available to beef cows throughout the year.

TABLE 6-15
EFFECT OF PHOSPHORUS ON REPRODUCTION AT KING RANCH IN TEXAS[1]

Group	No. of Cows	Cows Calving	Calves Weaned	Calf Weaning Weight		Calf Weaned/ Cow	
		(%)	*(%)*	*(lb)*	*(kg)*	*(lb)*	*(kg)*
No phosphorus supplement	42	69	64	489	*222*	319	*145*
Three phosphorus supplemented groups:	141	95	92	543	*247*	498	*226*
Bone meal, self-fed	42	92	88	535	*243*	468	*213*
Disodium phosphate in drinking water	42	96	92	542	*246*	500	*227*
Fertilized pasture	57	98	96	551	*250*	527	*239*

[1]Black, W. H., *et al., Comparison of Methods of Supplying Phosphorus to Range Cattle*, USDA Tech. Bull. 981.

CALVING DIFFICULTY

Difficult births may cause pathological conditions resulting in uterine tissue damage. Such conditions may or may not affect milk production, but they very likely will result in greater difficulty in getting affected cows settled.

■ **Effect of calving difficulty on rebreeding fertility**—In a study of 1,889 Angus and Hereford cows and heifers at the Meat Animal Research Center (MARC), Clay Center, Nebraska, it was found that cows experiencing difficult calving had significantly lower rebreeding fertility (Table 6-16).

Each of the measures of fertility shown in Table 6-16 indicates that calving difficulty increases the percentage of cows that do not resume their normal estrus cycle before the end of the breeding season.

TABLE 6-16
EFFECT OF CALVING DIFFICULTY ON REBREEDING[1]

	Cows Experiencing	
	Calving Difficulty	No Calving Difficulty
Number of cows	466	1,423
Detected in heat (45 days AI) (%)	60	74
Producing an AI calf (%)	54	69
Producing a calf (70-day breeding season) (%)	69	85

[1]Bellows, R. A., "Dystocia and Fertility," *Beef Cattle Science Handbook*, Vol. 18, edited by M. E. Ensminger and published by Agriservices Foundation, 1981, p. 413, Table 1.

SUCKLING STIMULUS

The length of interval from calving to first heat has been shown to be 20 to 42 days longer among cows suckling calves than among cows being milked (Table 6-17). Methods that may be used to shorten the interval from calving to first heat are: (1) early weaning, or (2) decreasing the frequency of suckling.

TABLE 6-17
EFFECT OF SUCKLING ON THE INTERVAL FROM CALVING TO FIRST HEAT[1]

Type of Cow	Suckled	Not Suckled	Difference
	(days)	*(days)*	*(days)*
Holstein	58	38	20
Milking Shorthorn	94	64	30
Beef	73	31	42

[1]Wiltbank, J. N., R. Anderson, and H. L. Fillmore, "Beef Cattle Production and Management," *Beef Cattle Science Handbook*, Vol. 20, edited by Frank H. Baker and Mason E. Miller, Winrock International, and published by Westview Press, Boulder, CO, 1984, p. 410, Table 6.

SIZE OF PELVIC OPENING

The calf must come through the birth canal. Consequently, when the calf is too big or the pelvic opening is too small, calving difficulties result. It would appear, therefore, that measurement of the pelvic area should provide a helpful index to use in decreasing calving difficulty.

The pelvic opening measurement technique consists of rectally inserting the hand containing the instrument (either a pelvimeter or a pair of sliding calipers) and determining the height and width of the birth canal. These results are multiplied together to determine the area.

■ **The size of the pelvic opening in 2-year-old heifers is small compared to that in older cows**—Studies conducted at the U.S. Range Livestock Experiment Station, Miles City, Montana, show that the pelvic opening of first-calf heifers is small (Table 6-18).

TABLE 6-18
PELVIC OPENING OF COWS OF DIFFERENT AGES[1]

Age	No.	Pelvic Area		Body Weight	
(yr.)		*(sq in.)*	*(cm²)*	*(lb)*	*(kg)*
2[2]	198	38.75	*250*	813	*369*
3[2]	251	45.26	*292*	1,023	*465*
4 and 5	75	51.46	*332*	1,045	*475*

[1]Short, R. E., and R. A. Bellows, "U.S. Range Livestock Experiment Station, Miles City, Montana, Factors Affecting Calf Losses and Calving Difficulty," *Beef Cattle Science Handbook*, Vol. 11, edited by M. E. Ensminger and published by Agriservices Foundation, p. 259.

[2]First-calf heifers.

Table 6-18 shows that the precalving pelvic area of 2-year-old heifers studied was 6.51 sq in. smaller than the pelvic area of the first-calf 3-year-olds, and 12.71 sq in. smaller than the pelvic area of the 4- and 5-year-old cows. The Miles City workers point out that (1) numbers were small in the group recorded as 4- and 5-year-old cows, and (2) the measurements of the older cows were not obtained from the same animals represented in the 2- and 3-year-old age groups. Nevertheless, there is strong indication that the size of the birth canal increases with age to a certain stage of maturity; hence, it suggests a reason for less calving difficulty in older animals.

Experiments show that the percentage of heifers experiencing calving difficulty decreases steadily as the pelvic area increases, particularly where the calf is presented normally. The only very difficult births encountered are in those heifers having a pelvic area less than 220 sq cm. This would indicate that it would be possible to cull heifers with pelvic openings of less than 220 sq cm and decrease calving difficulty markedly. However, in some herds this would result in culling half, or more, of the heifers, which would be impossible. So, two approaches are suggested: (1) breed heifers with pelvic areas smaller than 220 sq cm to a bull that will sire calves with low birth weights; and (2) include in the replacement heifer selection program data on pelvic area, then cull those with small pelvic openings.

SIRE

Most cattle producers assume that the bull is a major factor contributing to calving difficulty, and many of them will tell you that the way to reduce calving difficulty in heifers calving as 2-year-olds is to use Angus bulls, because, so they say, they sire small calves at birth. The first assumption is correct, but the second is not necessarily so.

■ **Effect of sire breed on calving difficulty**—Many experiments have shown that the sire breed used has a marked effect on birth weight and calving difficulty. The results of one such study, conducted at the U.S. Range Livestock Experiment Station, Miles City, Montana, are reported in Table 6-19.

TABLE 6-19
EFFECTS OF SIRE BREED ON BIRTH WEIGHT AND CALF LOSSES[1]

Sire Breed	Birth Weight: Straightbred Matings		Crossbred Matings		Brown Swiss Matings		Calf Losses: Heifers[2]	Cows
	(lb)	*(kg)*	*(lb)*	*(kg)*	*(lb)*	*(kg)*	*(%)*	*(%)*
Hereford	77	*35*	78	*35*	87	*39*	3.6	1.1
Angus	70	*32*	78	*35*	90	*41*	3.1	3.5
Charolais	90	*41*	89	*40*	109	*49*	6.8	3.9

[1]Short, R. E., and R. A. Bellows, "U.S. Range Livestock Experiment Station, Miles City, Montana, Factors Affecting Calf Losses and Calving Difficulty," *Beef Cattle Science Handbook*, Vol. 11, edited by M. E. Ensminger and published by Agriservices Foundation, p. 262.

[2]First-calf 3-year-olds.

The Miles City data (Table 6-19) show that the birth weights of calves sired by Charolais bulls were higher than calves sired by Hereford or Angus bulls. Likewise, the Charolais gave the highest calf losses.

■ **Effect of breed of sire on birth weight and calving difficulty**—Data from the U.S. Meat Animal Research Center, Clay Center, Nebraska, show the effect of breed of sire on calving difficulty in 2-year-old Hereford and Angus heifers (Table 6-20).

TABLE 6-20
EFFECT OF SIRE BREED ON BIRTH WEIGHT AND CALVING DIFFICULTY IN 2-YEAR-OLD HEREFORD AND ANGUS HEIFERS[1]

Sire Breed	Average Birth Weight		Difficult Births	Dead at Birth
	(lb)	*(kg)*	*(%)*	*(%)*
Jersey	58	*26*	20	4
Angus	66	*30*	41	8
Hereford	69	*31*	46	5
South Devon	72	*33*	55	10
Limousin	74	*34*	75	9
Simmental	76	*34*	74	13
Charolais	77	*35*	77	14

[1]Thedford, T. R., and M. R. Putnam, "Stages of the Birth Process and Causes of Dystocia in Cattle," *Beef Cattle Science Handbook*, Vol. 20, edited by Frank H. Baker and Mason E. Miller, Winrock International, and published by Westview Press, Boulder, CO, 1984, p. 387, Table 1.

The Clay Center data (Table 6-20) show that the larger breeds with higher birth weights had a greater percentage of difficult births and of deaths at birth. This would indicate that, in some cases, a Jersey sire should be seriously considered for use on heifers.

Clearly, therefore, choosing sires of breeds with lower birth weights will reduce the percentage of difficult births.

■ **Angus bulls do not necessarily sire small calves**—In an experiment in which Angus bulls were bred to Angus or Hereford heifers, the investigators found wide differences in calving difficulty and birth weight within the Angus breed (Table 6-21).

TABLE 6-21
CALVING DIFFICULTY IN CALVES SIRED BY ANGUS BULLS BRED TO ANGUS AND HEREFORD HEIFERS[1]

Angus Bulls	No. Calves	Calving Difficulty	Birth Weight		Calves Alive at 2 Weeks
		(%)	*(lb)*	*(kg)*	*(%)*
1	23	39	68	*31*	96
2	16	38	68	*31*	95
3	30	37	70	*32*	87
4	28	33	66	*30*	97
5	22	32	71	*32*	96
6	25	30	70	*32*	88
7	29	24	62	*28*	90
8	30	16	60	*27*	96

[1]Wiltbank, J. N., "Using Sire Selection to Alter Calving Difficulty," *Beef Cattle Science Handbook*, Vol. 18, edited by M. E. Ensminger and published by Agriservices Foundation, 1981, p. 397, Table 3.

As shown in Table 6-21, calving difficulty in calves sired by Angus bulls (1) varied from 16 to 39%, and (2) appeared to be somewhat related to birth weight. Calves sired by bull no. 8 had the least calving difficulty (16%) and the lightest birth weight (60 lb). Based on this study, it may be concluded (1) that Angus bulls causing the greatest calving difficulty tended to sire calves which had the highest birth weights, (2) that Angus bulls do not necessarily sire small calves, and (3) that Angus bulls can be selected that will decrease the level of calving difficulty.

■ **Within breed differences of bulls in weight of calves sired**—Even though Table 6-20 shows wide differences in birth weights and calving difficulty by breed of sire, the authors wish to emphasize that sires should not be selected by breed alone; rather, they should be selected on the basis of individual records. Although there are breed differences in weight of calves sired, much variation exists between sires within a breed. This situation is pointed up in Tables 6-21 and 6-22.

Table 6-22 shows that birth weights range from 61 to 72 lb of calves sired by eight different Hereford bulls. Other breeds show similar differences. Thus, to lessen calving difficulty, the important thing is to select a bull known to throw light calves, or in case of an untried sire, select one with EPDs that indicate he will produce calves with light birth weights.

TABLE 6-22
EFFECT OF DIFFERENT HEREFORD BULLS ON WEIGHT OF CALVES[1]

Hereford Bulls	No. Calves	Calving Difficulty	Birth Weight		Calves Alive at 2 Weeks
		(%)	*(lb)*	*(kg)*	*(%)*
9	23	54	72	*33*	90
10	34	41	67	*30*	100
11	12	33	71	*32*	92
12	29	32	70	*32*	100
13	29	29	66	*30*	97
14	17	24	70	*32*	98
15	35	20	69	*31*	88
16	23	15	61	*28*	100

[1]Wiltbank, J. N., "Using Sire Selection to Alter Calving Difficulty," *Beef Cattle Science Handbook*, Vol. 18, edited by M. E. Ensminger and published by Agriservices Foundation, 1981, p. 397, Table 3.

EXERCISE

Most cattle producers feel that adequate exercise in the pregnant female is of great importance in preventing calving difficulty. Their reasoning: Without sufficient exercise, muscle tone is not maintained, with the result that the muscles involved in the abdominal press exerted during calving do not have sufficient force to expel the calf.

■ **Effect of exercise on gestation length, birth weight, and calving difficulty**—The U.S. Range Livestock Experiment Station, Miles City, Montana, conducted a study of the effects of exercise on calving difficulty.

Angus-Hereford crossbred cows and heifers were used, and they were bred to the same Charolais sire. Ninety days prior to calving, pregnant dams were divided into two groups. One group was confined to a small feedlot, where activity was restricted, and fed at the rate of 10 lb of TDN daily. The second group was maintained under range conditions and fed the same type of feed as the drylot group in amounts necessary to obtain the same weight gains as their counterparts in the drylot. The pasture animals were fed 1 mi from the only water source, thereby forcing them to walk at least 2 mi per day.

All animals were calved in corrals. The results are summarized in Table 6-23.

TABLE 6-23
EFFECT OF EXERCISE ON CALVING DIFFICULTY[1]

Activity	No. of Animals	Gestation Length	Birth Weight		Calving Difficulty	
		(days)	*(lb)*	*(kg)*	*(%)*	*(score)*
Forced[2]	57	276	70	*31.8*	22.4	1.4
Restricted[3]	54	276	71	*32.5*	26.7	1.5

[1]Bellows, R. A., *et al.*, "Exercise and Induced-Parturition Effects on Dystocia and Breeding in Beef Cattle," *Journal of Animal Science*, 1994, 72:1667–1674.

[2]Walk of 2 mi *(3.2 km)* daily last 90 days of gestation.

[3]Confined to feedlot last 90 days of gestation.

Forced or restricted activity during gestation had little or no effect on gestation length, calf birth weight, incidence of calving difficulty, or calving difficulty score. But exercise did have a positive carryover effect on subsequent pregnancy rate. The researchers postulated that the increased subsequent pregnancy may have resulted from a feed-endocrine effect related to body weight gain, body composition-metabolic changes, or subsequent higher feed intake.

It is noteworthy that, in a Utah study, dairy heifers benefited from exercise (Table 6-24).

As shown in Table 6-24, exercise of dairy heifers resulted in a lower calving ease score and a decrease in the time for the placenta to be shed. There was also a decrease in the average number of days open, but that difference was not significant.

In summary, exercise was not effective with beef heifers at the Miles City Station, but it was effective with dairy heifers at Logan, Utah. The conjectured reason for the difference: The beef heifers in the Miles City study were in good physical condition when they went into the study, whereas the dairy heifers had been kept under confined conditions prior to initiation of the study, and their muscle tone was probably minimal.

TABLE 6-24
EFFECT OF EXERCISE DURING GESTATION ON CALVING AND REPRODUCTION IN DAIRY HEIFERS[1]

Group	No. of Animals	Calving Ease Score	Placenta Release Time	Days Open
Control	14	2.1	4.2	159
Exercise[2]	26	1.4	2.5	111

[1]Study conducted by Dr. Lamb, USDA–SEA at Logan, Utah. Reported by Bellows, R. A., "Prevention of Dystocia," *Beef Cattle Science Handbook*, Vol. 18, edited by M. E. Ensminger and published by Agriservices Foundation, 1981, p. 410, Table 7.

[2]Walk of 1 mi *(1.6 km)* daily at 3.5 mph for 4 weeks prior to calving.

LATE CALVING HEIFERS USUALLY REMAIN LATE CALVERS

Late calving heifers have little chance to calve early the next year because early in the breeding season in subsequent years (1) too few of them show heat, and (2) too few of them conceive at first service. Also, it is recognized that the time for young cows to show heat after calving is longer than that required by older cows.

Nature ordained that the uterus must return to normal for good conception rates. Available data indicate that 50 to 60 days after calving are required for conception rates to reach their maximum level.

If a 90 to 95% calf crop is expected each year, with 75 to 80% of the calves born early in the season, the management system must allow each cow sufficient time for the uterus to return to normal after calving and some time to come back in heat. The only feasible way to provide this time is to shorten the calving season. And since it takes a longer time for heifers to show heat after calving than older cows, the only feasible way of providing this added time is to breed heifers earlier than the rest of the herd, so that they will drop their calves earlier and be ready to rebreed on schedule each year thereafter.

The solution to the above problems appears to be as follows: Start with about 50% more heifers than needed, breed them 21 days earlier than the old cows, and shorten their breeding season to 45 days.

A MANAGEMENT PROGRAM FOR IMPROVED REPRODUCTION

Any management program for improved reproduction must begin with heifers, for the pattern of a cow's entire productive life is established with her first calf. Based on experiments and experiences, the following management system is recommended for replacement heifers:

- Select 50% more heifers than are actually needed as replacements, thereby permitting culling the late breeders—those that fail to conceive in a 45-day breeding season.

- Separate heifers from older cows.

- Feed heifers so that they weigh 650 to 750 lb at breeding time. (Remember that there are breed differences. For example, Angus heifers reach puberty at about 650 lb, and Herefords at about 750 lb.) Thus, gauge the feeding program accordingly. If a Hereford heifer weighs 500 lb at weaning (7 months), and it is planned to breed her at 14 months of age, she needs to gain 250 lb—that's 1.2 lb per day (250 ÷ 210= 1.2).

■ Breed heifers 21 days earlier than older cows and limit their first breeding season to 45 days. Cull those that do not conceive.

■ Feed pregnant heifers to gain about 1 lb per day during the 120 days prior to calving. That way they will retain their normal weight after calving. This is important because a cow's nutritional requirements double after calving, making it difficult to put weight on a thin first-calf heifer.

■ A management system like the above has been designed and tested experimentally by the Colorado Station in a 5-year study, the results of which are summarized in Tables 6-25, 6-26, 6-27, 6-28, and 6-29.[11]

■ **The experimental design**—The Colorado workers studied the effect of early calving of heifers on their future reproductive performance and pounds of calf weaned. One hundred and forty registered yearling heifers were divided into a new management (NM) group designed to ensure that the heifers would calve early, and a control group (C) in which there was no attempt to get the heifers to calve early. On the NM group, 170% more heifers than needed as replacements were bred; breeding lasted only 45 days; breeding started 21 days earlier than the cow herd; and estrus synchronization was used. In the C group approximately the same number of heifers as needed for replacements were bred; breeding lasted 90 days and started the same day as the cow herd (Table 6-25). The replacement heifers came from their respective groups and were handled the same as their dams had been the first year. The criteria for selection of NM replacements was early pregnancy while C replacements were selected on the basis of 205-day adjusted weights and conformation. All heifers and cows were bred by artificial insemination to one bull each year.

TABLE 6-25
EXPERIMENTAL DESIGN

Group	New Management	Control
No. pregnant replacements needed	50	50
No. exposed for breeding	85	54
Breeding season started:		
As heifers	4–22	5–12
3-year-olds and greater	5–12	5–12
Length of breeding, days	45	90
Estrus synchronization	yes	no
No. cycling 4/22	54 (64%)	35 (63%)
Selection criteria	Early pregnancy	Pregnancy, adjusted weaning weight, and conformation score

■ **More 2-year-old heifers calved early**—The NM program caused more heifers to drop their first calf early in the calving season. This can be seen in Table 6-26.

TABLE 6-26
CALVING TIME WHEN HEIFERS WERE 2 YEARS OLD

Calving Time	New Management Group	Control Group
	(no.)	*(no.)*
Feb. 9 or before	31	0
Feb. 10 to Mar. 6	19	23
Mar. 7 to Mar. 26	0	7
Mar. 27 to Apr. 16	0	7
Apr. 17 and after	0	3
Total no. calved	50	40

Heifers calving for the first time in the short concentrated period were compared to those calving over a long period. The question answered was the effect of this short calving period on future reproductive performance. The methods of measuring reproductive performance were cows pregnant and cows in heat. The following years, replacement heifers which were retained were handled as in Table 6-25 and were not switched between herds.

■ **More cows became pregnant early in the breeding season**—In subsequent years, more cows in the NM group became pregnant early in the breeding season than in the C group (Table 6-27).

As shown in Table 6-27, an average of 70% of the NM cows were pregnant after 21 days of breeding compared to 46% in the controls; a difference of 24%.

A 12% difference in cows pregnant after 45 days of breeding was noted between NM cows and control cows.

At the end of the breeding season, the difference in pregnancy rate between the NM and control groups was 3%.

[11]Wiltbank, J. N., S. Roberts, and K. J. Nix, "Improving Pounds of Calf Weaned by Heifer Selection," *Beef Cattle Science Handbook*, Vol. 17, edited by M. E. Ensminger and published by Agriservices Foundation, 1980, pp. 398–399, Tables 11–15.

TABLE 6-27
WHEN COWS BECAME PREGNANT

Pregnant	New Management Group	Control Group	Difference
	(%)	(%)	(%)
After 21 days of breeding	70	46	24
After 45 days of breeding	87	75	12
At end of breeding season[1]	87	90	3

[1]Forty-five days in NM group and 90 days in C group.

■ **More cows showed heat early in the breeding season**—The reason more cows became pregnant early in the breeding season in the NM group was because more cows showed heat early in the breeding season (Table 6-28). This was accomplished (1) by selecting early calving cows so that they had time for the uterus to clean up and time for the cows to come back in heat during the first 21 days of breeding; and (2) by proper nutrition.

TABLE 6-28
COWS SHOWING HEAT AT VARIOUS TIMES

Showing Heat	New Management Group	Control Group	Difference
	(%)	(%)	(%)
After 21 days of breeding	95	77	18
After 45 days of breeding	100	96	4

An average of 95% of the NM cows showing heat the first 21 days of the breeding season compared to an average of 77% in the controls. Between the two groups, there was an average difference of 18%. The differences after 45 days of breeding were small as all the cows in the NM group, and nearly all in the control group, had been detected in heat by this time.

■ **Increase in reproductive performance led to an increase in weaning weight**—The increase in reproductive performance led to an increase in the weaning weight of the calves in the NM group (Table 6-29).

Calves in the NM group averaged 433 lb at weaning compared to 396 lb in the C group, or a difference of 37 lb. This shows that the pounds of calf in the cow herd can be increased by using simple, inexpensive techniques. It also shows that good reproductive performance can be achieved in short breeding periods.

TABLE 6-29
WEANING WEIGHT OF CALVES

New Management Group		Control Group		Difference	
(lb)	(kg)	(lb)	(kg)	(lb)	(kg)
433	197	396	180	37	17

■ **Conclusion**—The two most important factors affecting pounds of calf weaned in a cow herd are (1) early calving, and (2) number of cows weaning a calf. More replacement heifers than needed should be bred, and those calving early in the calving season should be selected as replacements. This will help increase (1) calf crop, and (2) the number of early calving cows. To make certain that this happens, heifers should be fed to reach specific target weights for each breed.

It is emphasized that a 45-day breeding season may be disastrous unless it is accompanied by the proper nutrition level so that cows will show estrus and have a high conception rate at first service. However, experiments and experiences indicate that under optimum conditions, a 45-day breeding season will not decrease the number of cows calving.

SUMMARY OF FACTORS AFFECTING REPRODUCTION

Two major problems plague the cow-calf producer: (1) the long calving season; and (2) a poor calf crop.

Any factor which increases the size of the calf, and/or decreases the size of the birth canal of the mother, causes an increase in the percentage of difficult births.

Breeds differ widely in the incidence of calving difficulty. Also, there is considerable variation between sires within a breed.

The following conclusions appear to be justified based on experiments and experiences:

1. **Magnitude of calf losses.** Calf losses at birth have been estimated to run more than 6% of the potential calf crop, with these losses ranking second in magnitude only to cows failing to conceive. Fortunately, over 50% of these losses could be prevented by improved management.
2. **Causes of dystocia (difficult births).** Large calf birth weight is the most important cause of dystocia, followed, in order, by pelvic area of dam, precalving weight of dam, sex of calf (bull calves average 5 to 7 lb heavier at birth than heifers), and gestation weight gain of dam.
3. **Gestation length.** Length of gestation is positively correlated (0.30) with birth weight; hence, it is a cause of larger calves at birth. However, short gesta-

tions do not automatically mean small calves at birth, as evidenced by the Holstein breed, which has a short gestation period (average of 277 days), but relatively heavy birth weights.

4. **Age of cow.** Difficult births and calf losses from heifers are higher than in older cows; 10% of the calves from first-calf heifers may be lost, in comparison with 6% for cows of all ages in the United States.

5. **Liberal feeding of growing heifers.** Liberal feeding makes for early sexual maturity. This is important because size and weight, not age, determine the time of sexual maturity.

6. **Flushing.** Flushing by feeding cows and heifers to gain approximately 1.5 lb per head daily beginning 20 days before the start of and continuing through the breeding season will result in an increase in the calf crop and in the cows breeding both earlier and more nearly at the same time.

7. **Low levels of energy during gestation.** Low levels of energy the last 90 to 120 days of gestation may reduce calf birth weight and calving difficulty. But they will have a marked adverse effect on rebreeding—fewer heifers will be in heat at the beginning of the breeding season, and fewer will become pregnant—unless they are fed high energy diets immediately after calving and until breeding.

8. **Energy level after calving.** Both older cows and heifers that receive a low level of energy following calving show a marked decrease in the number pregnant after 20 days of breeding and at the end of the breeding season. There are two reasons for the poor reproductive performance in cows on low levels of feed and losing weight following calving: (a) Some of them do not show heat during the breeding season, and (b) the conception rate at first service is low.

9. **Low level of protein.** A low level of protein during gestation results in lighter birth weight of calves, but it has no adverse effect on number of calves born and calf survival.

A low level of protein following calving delays heat and results in lowered calf weight gains.

No beneficial effect is derived from feeding a high level of protein over a medium level; hence, there is no need to go to the added expense of providing more protein than is needed.

10. **Low phosphorus.** Low phosphorus will markedly decrease the percentage of calves dropped and weaned, the weaning weight of calves, and the pounds calf weaned per cow.

It is recommended that phosphorus supplementation be provided free-choice throughout the year, and that supplementation of other minerals be provided if local conditions and feeds indicate a deficiency.

11. **Effect of calving difficulty on rebreeding fertility.** Difficult births will likely result in greater difficulty in getting affected cows settled.

12. **Suckling stimulus.** The interval from calving to first heat is 20 to 42 days longer among cows suckling calves than among cows being milked. This interval may be shortened by (a) early weaning, and (b) decreasing the frequency of suckling.

13. **Over-fed and over-fat.** Excessively fat heifers often have difficult calvings, with heavy calf losses at and soon after birth, apparently as a result of excess fat decreasing the size of the birth canal.

14. **Pelvic opening.** The pelvic opening in 2-year-old heifers is small compared to that in older cows. Heifers with pelvic areas smaller than 220 sq cm should be bred to a bull that sires calves with small birth weights. Also, data on pelvic area should be included in the replacement heifer selection program.

15. **Difference between sire breeds.** Choosing sire breeds with lower birth weights will reduce the percentage of difficult births.

16. **Differences within the same sire breed.** Choosing among sires of the same breed can affect calving difficulty, but to a far lesser extent than choosing among breeds.

17. **Exercise during gestation.** Exercise during pregnancy was not effective with beef heifers at the Miles City Station, in Montana, but it was effective with dairy heifers at Logan, Utah.

18. **Late calving heifers usually remain late throughout life.** Cows that calve late have low conception rates at first service when they are bred early in the breeding season; hence, late calving cows usually remain late calving throughout life.

19. **Management system for first-calf heifers.** A management system, including proper nutrition, in which 50% more replacement heifers than needed are bred 21 days earlier than the cows, and in which the first breeding season is limited to 45 days, will result in early heat and conception throughout life and more total pounds of calf weaned.

QUESTIONS FOR STUDY AND DISCUSSION

1. Twelve percent of the nation's cows never calve, and there is an appalling calf loss of 6% between birth and weaning. Discuss the causes and economics of this situation.

2. Diagram and label the reproductive organs of the bull.

3. Diagram and label the reproductive organs of the cow.

4. How do you account for the difference in the age of puberty of the (a) British breeds, (b) Continental breeds, and (c) Zebu breeds?

5. In order to synchronize ovulation and insemination, when should cows be bred with relation to the heat period?

6. What are the objectives of induced calving (shortened gestation)? What are the hazards? What did the study conducted by the USDA, Miles City Station, show relative to early obstetrical assistance and induced parturition?

7. Discuss the economic aspects of pregnancy testing cows. Describe the most common test of pregnancy of beef cows.

8. Describe the normal presentation of a calf at birth.

9. For your home farm or ranch (or one with which you are familiar), what do you consider to be the most desirable breeding season and time of calving? Justify your answer.

10. List the advantages of artificial insemination. List the limitations of artificial insemination.

11. List the advantages to accrue from bringing cows in heat at will. List and discuss the main drugs used in estrus control.

12. What advantages accrue from the practical and extensive use of (a) superovulation, and (b) embryo transfer in beef cattle?

13. For what purposes may blood typing be used? Compare traditional blood typing and DNA typing.

14. List in order the major causes of dystocia (calving difficulty) as reported by the Miles City Station, in Montana.

15. Why are more calving difficulties encountered in 2-year-old heifers than in older cows?

16. Discuss the significance of the Texas A&M experiment which showed that breed, weight, and age affect time of puberty.

17. Discuss the effect of each of the following nutritional factors on reproduction: (a) liberal feeding, (b) flushing, (c) energy level, (d) protein, (e) phosphorus, and (f) calving difficulty.

18. Discuss the effect of each of the following factors on reproduction: (a) suckling stimulus, (b) size of pelvic opening, (c) sire, and (d) exercise.

19. Experiments clearly show that late calving heifers usually remain late calvers throughout life. Outline a program to lessen late calvers.

20. Summarize the factors affecting reproduction of beef cattle.

SELECTED REFERENCES

Title of Publication	Author(s)	Publisher
Anatomy of the Domestic Animals, The, Fourth Edition	S. Sisson J. D. Grossman	W. B. Saunders Company, Philadelphia, PA, 1953
Anatomy and Physiology of Farm Animals, Third Edition	R. D. Frandson	Lea & Febiger, Philadelphia, PA , 1981
Animal Breeding	A. L. Hagedoorn	Crosby Lockwood & Son, Ltd., London, England, 1950
Animal Breeding Plans	J. L. Lush	Collegiate Press, Inc., Ames, IA, 1965
Artificial Insemination and Embryo Transfer of Dairy and Beef Cattle, The, Eighth Edition	H. A. Herman F. W. Madden	Interstate Publishers, Inc., Danville, IL, 1994
Artificial Insemination of Farm Animals, The	Ed. by E. J. Perry	State Mutual Book, New York, NY, 1981

Title	Author	Publisher
Breeding Beef Cattle for Unfavorable Environments	Ed. by A. O. Rhoad	University of Texas Press, Austin, TX, 1955
Breeding Better Livestock	V. A. Rice F. N. Andrews E. J. Warwick	McGraw-Hill Book Co., Inc., New York, NY, 1953
Breeding and Improving Farm Animals, Seventh Edition	E. J. Warwick J. E. Legates	McGraw-Hill Book Co., Inc., New York, NY, 1979
Developmental Anatomy, Fourth Edition	L. B. Arey	W. B. Saunders Company, Philadelphia, PA, 1940
Dukes' Physiology of Domestic Animals, Eighth Edition	Ed. by M. J. Swenson	Cornell University Press, Ithaca, NY, 1970
Factors Affecting Calf Crop	Ed. by T. Cunha A. C. Warnick M. Koger	University of Florida Press, Gainesville, FL, 1967
Farm Animals	J. Hammond	Edward Arnold & Co., London, England, 1952
Hammond's Farm Animals, Fourth Edition	J. Hammond, Jr. I. L. Mason T. J. Robinson	Edward Arnold (Publishers) Ltd., London, England, 1971
How Life Begins	J. Power	Simon and Schuster, Inc., New York, NY, 1965
Improvement of Livestock	R. Bogart	The Macmillan Co., New York, NY, 1959
Improving Cattle by the Millions	H. A. Herman	University of Missouri Press, Columbia, MO, 1981
Livestock Improvement	J. E. Nichols	Oliver and Boyd, Ltd., London, England, 1957
Modern Developments in Animal Breeding	I. M. Lerner H. P. Donald	Academic Press, Inc., New York, NY, 1966
Prenatal and Postnatal Mortality in Cattle	Subcommittee on Prenatal and Postnatal Mortality in Bovines, Committee on Animal Health, National Research Council	National Academy of Sciences, Washington, DC, 1968
Problems and Practices of American Cattlemen, Wash. Ag. Exp. Sta. Bull. 562	M. E. Ensminger M. W. Galgan W. L. Slocum	Washington State University, Pullman, WA, 1955
Progress in the Physiology of Farm Animals, Vols. I, II, and III	Ed. by J. Hammond	Butterworths Scientific Publications, Ltd., London, England, 1954–1957
Reproduction in Farm Animals, Fourth Edition	Ed. by E. S. E. Hafez	Lea & Febiger, Philadelphia, PA, 1980
Reproductive Physiology	A. V. Nalbandov	W. H. Freeman & Co., San Francisco, CA, 1958
Veterinary Endocrinology and Reproduction, Third Edition	Ed. by L. E. McDonald	Lea & Febiger, Philadelphia, PA, 1980

A chronic *buller* or nymphomaniac cow, showing the characteristic sagging of the loin region and elevation of the tailhead. (Source: *Physiology of Reproduction*, by Marshall, p. 667; courtesy, The Royal Society of Edinburgh)

STERILITY AND DELAYED BREEDING IN BEEF CATTLE[1]

Contents — *Page*

[1]In the preparation of this chapter, the authors had the authoritative help of Dr. Robert F. Behlow, Professor and Extension Veterinarian, North Carolina State University, Raleigh, NC.

Contents Page

Sterility (infertility or barrenness) may be defined as temporary or permanent reproductive failure; resulting from anestrus (lack of heat), failure to conceive, or abortion. Animals are not simply fertile or sterile; rather, all degrees of fertility exist in both sexes.

In practical operations the breeding efficiency of most beef cattle is expressed in terms of the annual calf crop[2]—the most fertile herds being those in which the highest percentage of all cows conceive on a schedule which results in the spacing of calves each 12 months. Based on number of calves born in comparison with number of cows bred, the U.S. calf crop for all cattle (beef and dairy) is 88%. This means that the other 12% are nonproducers, either temporarily or permanently. Since reproductive ability is fundamental to economical beef production, it can be readily understood that sterility constitutes a major annual loss in the cattle business. In fact, most operators acknowledge that the calf crop percentage is the biggest single factor affecting profit in beef cattle production.

It is recognized that the problem of sterility is difficult to study, especially under range conditions where the majority of beef cattle are found. For this reason, most of the experimental studies on this problem have been conducted with dairy cattle. It is believed, however, that most principles apply to all breeds of *Bos taurus* and *Bos indicus*, regardless of type.

The incidence of sterility varies greatly from herd to herd, and within the same herd from year to year. Despite this fact, cattle producers should establish arbitrary standards by which breeding performance may be gauged. To this end, the following reasonable averages are proposed: not more than 10% breeding difficulty in the cows at any one time;[3] not more than an average of 1.85 service per conception,[4] and not lower than a 94% calf crop. Of course, the better managed and the more fortunate herds will do better. If this standard is accepted, however, there should be reason for concern if performance falls below these averages—it should then be assumed that something is wrong and that investigation is needed.

Fig. 7-1. Every cattle producer is concerned with getting a live, healthy calf from each cow every 12 months. (Courtesy, *Livestock Weekly*, San Angelo, TX)

Fortunately, comparatively few barren cows and sterile bulls are totally and permanently infertile. Those that are should be sold for slaughter without further delay or expense. Most of the others will regain their breeding abilities with good care and management and appropriate treatment. Mating at the proper stage of estrus and the correct training and use of bulls will do

[2]Methods of computing calf crop percentages vary. But the four most common methods are (a) number of calves born in comparison with the number of cows bred; (b) the number of calves marketed or branded in relation to the cows bred; (c) the number of calves that reach weaning age as compared with the number of cows bred; and (d) the number of calves at weaning divided by the number of cows kept the previous winter. The latter method reflects over-wintering costs, which are usually high. Perhaps the first method is the proper and most scientific one, but, for convenience reasons, some ranchers use the other three.

[3]These were the average figures obtained in a survey of dairy cattle in New York, as reported in the Northeastern States Regional Bulletin 32 (Cornell Univ. Ag. Exp. Sta. Bull. 924), p. 5.

[4]*Ibid.*

much to maintain a high conception rate. Care in the selection of disease-free breeding stock, isolation of newly purchased animals, and periodic health examinations are effective preventive measures. When breeding irregularities are noted or disease strikes, however, treatment should be prompt. In general, diagnosis and treatment should be left to veterinarians who possess training, experience, and skill in handling reproductive failures; and since infertility constitutes one of the major problems with which veterinarians must deal, they will wish to be well informed.

Cattle producers should also be well informed relative to reproductive failures, because the enlightened producer will (1) encounter less trouble as a result of the application of preventive measures, (2) more readily recognize serious trouble when it is encountered, and (3) be more competent in carrying out the treatment prescribed by the veterinarian.

Cattle reproductive diseases cause an estimated annual loss of $2.6 billion.[5]

STERILITY IN THE COW

Sterility is a primary reason for culling cows. In a study involving 13,818 cows in 155 Maryland Holstein herds, University of Maryland researchers found that sterility accounted for 30.7%, nearly one-third, of all cow removals.

Usually the failure of a female to have a heat period before 18 months of age, or to come in heat within 3 months after calving, or to conceive after 3 matings should constitute sufficient basis for assuming that an abnormal condition exists and that the services of a veterinarian should be obtained for diagnosis and possible treatment. Occasionally such conditions will correct themselves without treatment; in other cases, they subside for a time only to recur later—they become irregular breeders.

Repeat breeders—cows which exhibit regular or irregular heat periods, but fail to conceive—are most perplexing. The condition may be due to failure of fertilization or to early embryonic death.

When a cow fails to come in heat, she should first be checked for pregnancy (see section entitled "Pregnancy Test," Chapter 6); approximately 1 cow in 20 thought to be sterile will be found safely in calf.

For convenience, the common causes of sterility and delayed breeding in beef cows are herein classified as (1) genital infections and diseases, (2) poor management and feeding, (3) physiological and endocrine disturbances, (4) inherited (genetic) abnormalities, (5) anatomical defects and injuries, and (6) miscellaneous and unknown causes. Of course, at the outset it is recognized that no definite demarcation exists between these classifications; that many of the causes of sterility may be, and are by some authorities, listed under other classifications than those given herein; and that there may be interaction between two or more forces. Also, whatever the cause of sterility, there are no *cure-alls*; rather, each individual case requires careful diagnosis and specific treatment for whatever is wrong.

[5]Report prepared by the associate deans and directors of the veterinary research programs for the Council of Deans, Association of American Veterinary Medical Colleges, based on information available on disease losses as of Feb. 1, 1981.

GENITAL INFECTIONS AND DISEASES

Genital diseases account for about one-third of all sterility in cattle. In addition, nonspecific infections of the cervix and/or uterus are common causes of sterility.

NONSPECIFIC INFECTIONS OF THE GENITAL TRACT

Studies reveal that the invasion of the cervix and uterus by a variety of microorganisms normally follows parturition, but that such infections usually clear up, without treatment, within 40 to 50 days after calving. The return of estrus cycles of normal length and duration is, therefore, a reasonably good indication that the reproductive tract is again normal.

Because bacteria are likely to be present in the reproductive tract immediately following calving, and since breeding at this time may interfere with the normal breeding process and extend the period of infertility, it is recommended that cows not be rebred within 60 days after calving. This will give the cow sufficient time to recover. If the placenta was retained or other calving difficulties were encountered, the cow should be allowed to go through at least one normal heat period before rebreeding.

The veterinarian-prescribed treatment for nonspecific genital infections will depend upon their nature and the extent to which they have invaded the reproductive tract; antiseptics, antibiotics, and other drugs may be indicated for local or systemic use, and/or sexual rest may be recommended.

SPECIFIC GENITAL DISEASES

Bovine virus diarrhea (BVD, mucosal-disease complex), brucellosis, epizootic bovine abortion (EBA, foothill abortion), infectious bovine rhinotracheitis (IBR), leptospirosis, metritis, trichomoniasis, vaginitis, and vibriosis are the most troublesome specific genital diseases of cattle.

■ **Bovine virus diarrhea (BVD, mucosal-disease complex)**—This is an infectious disease of cattle caused by a myxovirus, characterized by diarrhea and dehydration. Also, the BVD virus may cause abortions in pregnant cows, generally in the first 3 months of pregnancy. This disease can be effectively prevented by immunization with a modified live virus vaccine. Immunization is commonly performed when cattle are 8 to 12 months of age. Booster vaccinations are often given on an annual basis.

■ **Brucellosis**—This is a serious genital disease in cattle caused primarily by the *Brucella abortus* bacteria, although the suis and melitensis types are also seen in cattle. It is characterized by (1) abortions at any stage of pregnancy, but most commonly between the fifth and eighth months, (2) above normal incidence of retained placenta, and (3) lowered conception rate. Brucellosis can be readily and accurately detected by both blood (the serum agglutination test) and milk tests (the milk ring test), or by the new rapid card test (the use of a disposable card on which blood serum plasma is mixed with a buffered whole-cell suspension of *Br. abortus* [antigen] which reacts [agglutinates] with antibodies in the blood serum of animals infected with brucellosis). Prevention consists of (1) calfhood vaccination of heifer calves between 4 and 12 months of age; (2) use of artificial insemination; and (3) the purchase only of vaccinated or tested (and found clean) animals.

■ **Epizootic bovine abortion (EBA, foothill abortion)**—This is an infectious disease of cattle, epizootic to California, where it is known as *foothill abortion* because of the high incidence in cows which are pastured on foothill terrain. It is caused by *Chlamydia psittaci*. Infected cows usually abort between the fifth and seventh months of gestation. In epizootic areas, only first-calf heifers and new cattle introduced from areas free of the disease are affected. The abortion rate varies from 25 to 75%. Aborted animals appear to be immune and should be retained in a herd. Under experimental conditions, it has been shown that feeding 2 g per head per day of chlortetracycline will prevent pregnant cows from aborting, but field application of this preventive measure has been limited because of cost and the problem of maintaining adequate dosage levels.

■ **Infectious bovine rhinotracheitis (IBR)**—This is an acute contagious viral infection characterized by inflammation of the upper respiratory tract. It was first diagnosed in the United States in 1950. IBR is most prevalent where there are large concentrations of cattle under confinement. The virus may invade the placenta and fetus via the maternal bloodstream, causing abortion or stillbirth from 2 to 3 months subsequent to the respiratory infection. The best method of control of IBR is the routine immunization of all potential replacement animals when they are 6 months to 1 year old; after calves have lost their maternally conferred immunity at about 6 months of age, and before breeding age. Booster inoculations may be given.

■ **Leptospirosis**—This disease is caused by several species of corkscrew-shaped organisms of the spirochete group. Among other symptoms, it is apt to produce a large number of abortions anywhere from the sixth month of pregnancy to term. In newly infected herds, abortions may approach 30%. The disease can be diagnosed by a blood test. For control, cattle owners rely on annual vaccinations.

■ **Metritis**—This disease is caused by various types of bacteria. Lacerations at the time of calving, wounds inflicted by well-meaning but inexperienced operators, and/or retention of afterbirth are the principal predisposing factors. Metritis usually develops soon after parturition. It is characterized by a foul-smelling discharge from the vulva that may be brownish or blood-stained and finally becomes thick and yellow. An acute infection may develop into the chronic form, producing sterility. Treatment should be left to the veterinarian. Most cases are treated by the introduction of an antibiotic or a sulfa into the uterus.

■ **Trichomoniasis**—This is a genital-tract infection caused by the protozoan organism *Trichomonas foetus*. The disease is characterized by (1) irregular sexual cycles, (2) early abortions, usually between 60 and 120 days, (3) a whitish vaginal discharge, and (4) resorption of the fetus, while the uterus becomes filled with a thin, grayish fluid. When these symptoms are observed in a herd known to be free of brucellosis, trichomonad infection should be suspected. The diagnosis can be confirmed microscopically by finding the organism. Trichomoniasis may be eliminated from the herd by adoption of a hygienic breeding program. Such a program depends upon using semen from bulls free of trichomoniasis, which can be accomplished by artificial insemination. In cows, the disease appears to be self-limiting; that is, cows appear to acquire an immunity after about three months' sexual rest. If natural service must be used, infected bulls should be replaced with clean animals. Trichomonad-infected bulls can be successfully treated, provided the time-consuming prescribed treatment is followed. The preferred treatment consists of the local application of bovoflavin ointment. A satisfactory treatment for bulls consists of the combined use of sodium iodide, acriflavin, and bovoflavin; administered as an ointment. Dimetridazole is an effective systemic treatment for bulls when administered orally at a dose of 50 mg/kg daily for 5 days.

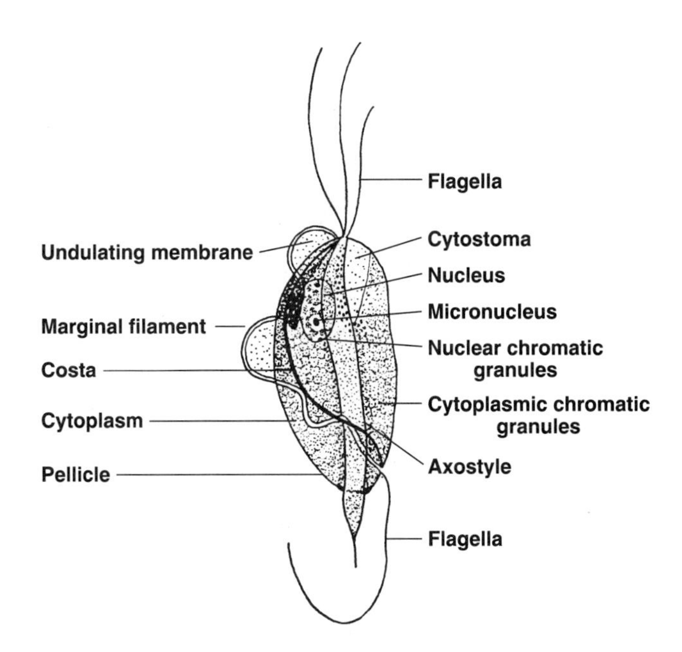

Fig. 7-2. Protozoan *Trichomonas foetus.* (Source: Herman, H. A., and F. W. Madden, *The Artificial Insemination and Embryo Transfer of Dairy and Beef Cattle,* The Interstate Printers & Publishers, Inc., Danville, IL, 1987)

■ **Vaginitis**—This is an infection of the vagina and vulva which causes an inflammation of varying intensity and results in difficult breeders. It may be caused by bruising or laceration of the vagina and vulva at parturition, service from a large and vigorous bull, or prolapse of the vagina. The inflamed vagina is painful, swollen, and often there is an offensive-smelling discharge, indicating infection. The veterinarian may place antibiotics in the uterus and vagina, or it may be treated by injecting antibiotics or sulfa drugs.

Granular venereal disease (granular vaginitis) is characterized by small spherical nodules on the vulva mucosa of cattle. A similar condition may occur in the lymphatic follicles of the bull's penis. The condition is a response of the lymphatic tissue in the affected area to an irritant or antigen; it is not a disease in the classic sense. Losses in females are in terms of lower percent calf crop and decreased milk production. Bulls may refuse to breed.

Treatment of females is not recommended. It will clear up spontaneously, although it may take several weeks. The condition in the bull is more persistent and should be treated. The prolapsed penis and sheath should be massaged with a suitable antibiotic ointment, repeated sufficiently often to assure elimination of any existing infection.

■ **Vibriosis**—This is caused by the microorganism *Vibrio fetus*, which is transmitted at the time of breeding. The disease is characterized by (1) several (4 or more) services per conception, (2) cows exhibiting irregular heat periods, but finally settling without much difficulty and carrying calves to normal term, and (3) 3 to 5% abortions, usually between the fifth and seventh months of pregnancy. For positive diagnosis, laboratory methods must be used. Infected cows may be treated by injecting drugs into the uterus and/or by allowing sexual rest. Artificial insemination with semen (1) from known noninfected bulls, or (2) which has been antibiotic-treated, is a rapid and practical method of stopping the transmission of infection from cow to cow. The disease tends to be self-limiting in the cow; a cow seems to be free of the disease once she has had a calf following an infection. Vaccination of females is effective in controlling the disease, especially in beef herds kept under range conditions. Usually vaccination is done about two months before breeding and is repeated annually. Bulls may be vaccinated but the value is unknown.

(Also see Chapter 12, Beef Cattle Health, Disease Prevention, and Parasite Control.)

POOR MANAGEMENT AND FEEDING

The term *management* is somewhat elusive and all inclusive. As used in the discussion that follows, reference is made only to those beef cattle management practices pertinent to breeding efficiency. It is noteworthy, however, that management is very important in determining breeding efficiency, and that breeders largely determine their own destiny in this regard.

OVULATION AND BREEDING NOT PROPERLY SYNCHRONIZED

If the sperm are introduced in the female reproductive tract much in advance of the egg's release, the chances of fertilization are greatly reduced. It has been shown that the best conception rates are obtained when cows are bred during the final 10 hours of standing heat, or during the first 10 hours after the end of standing heat. To meet these timing relationships, cows that are detected in heat in the morning should be bred during the afternoon or evening of the same day, and those detected in the afternoon should be bred late that evening or early the next morning.

Where hand mating is practiced, this means that failure to detect accurately when a cow is in heat may account for some breeding failures. Experience and careful observation are the answers to this problem. In addition to recognizing the signs of estrus (see Chapter 6, section on "Signs of Estrus"), where hand mating is to be followed, the caretaker should keep a

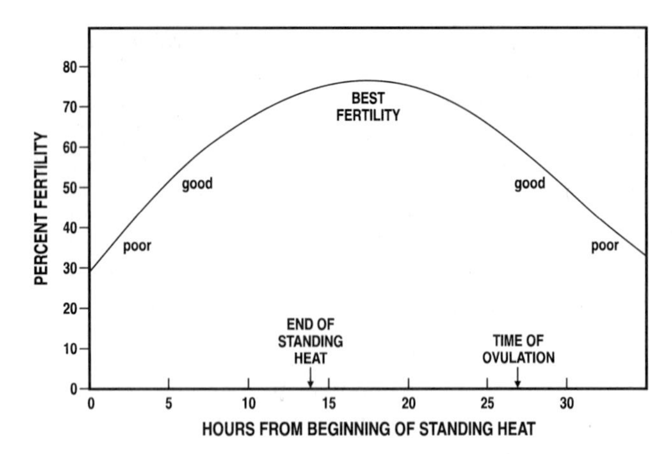

Fig. 7-3. For best conception, breed near the end of standing heat.

record of breeding dates and observed heat periods for each cow.

IMPROPER FEEDING

Improper feeding may imply (1) uncommonly high or low feed intake, or (2) a deficiency of specific nutrients.

Restricted rations often occur during periods of drought, when pastures or ranges are overstocked, or when winter rations are skimpy. When such deprivations are extreme, there may be lowered reproductive efficiency on a temporary basis. It is noteworthy, however, that experimental results to date fail to show that the fertility of the germ cells is seriously impaired by uncommonly high or low feed and nutrient intakes.[6] In contrast to these findings, it is equally clear that the onset of heat and ovulation in young heifers is definitely and positively correlated with the level of nutrient intake; thus, if feed or nutrient intake is too low for normal rates of growth and development, the onset of reproductive function is delayed.[7]

Research shows that cows losing body weight are much more difficult to get pregnant than those gaining weight. In one study, it was reported that 67% of the cows bred when gaining weight conceived, and required only 1.5 services per conception, compared to 44% conceived and 2.3 services per conception in cows bred during a period when losing weight. Since most body weight is lost in early lactation, a good early lactation feeding program will help cows to reestablish a positive energy balance, as well as to increase milk production.

Short, *et al.*, concluded that insufficient energy intake will delay puberty, lengthen postpartum anestrus, and cause anestrus in cows or heifers that are having estrous cycles.[8]

Also, too liberal feeding and high condition may cause sterility. Sometimes the presence of an excessive amount of fat in the pelvic region leads to a partial protrusion of the vagina and eventual inflammation. Then, too, where a female remains in an excessively high condition for an extended period of time, degeneration of the ovarian follicles may occur, thereby producing a prolonged state of sterility. (See Chapter 6, section on "Nutritional Factors Affecting Reproduction," along with the subsections thereunder, for supporting experimental work.)

Under natural conditions, a deficiency of vitamin A is the only vitamin likely to be lacking in cattle rations; and cattle obtain plenty of the precursor (carotene) when they are on green pasture or receive reasonable amounts of green hay not over one year old. A severe deficiency of vitamin A may result in a low conception rate, a small calf crop—with many calves weak or stillborn and with some calves born blind or without eyeballs, but estrus may remain normal.

Under practical conditions, it has been observed that when cows are on a phosphorus-deficient ration there is a marked inhibition of estrus and a tendency to reproduce every other year. Also, selenium is involved in fertility in general, and in sperm mobility; and numerous research studies have shown a reduction in retained placenta when selenium is added to the ration to meet the requirements. Deficiencies of iodine, zinc, cobalt, copper, and manganese have also been implicated in reproductive disturbances. However, experimental results have failed to show that the fertility of cows is seriously impaired by low trace-mineral content rations.[9]

Although an adequate supply of minerals and vitamins is essential for normal growth and health, adding an excess of these nutrients to a well-balanced beef cattle ration fed according to recommended practices has no known value in curing breeding troubles.

(Also see Chapter 19, section on "Nutritional Reproductive Failure in Cows.")

EXERCISE

Although heat periods are more easily detected when cows are out in the open, exercise is not essential for normal reproduction, provided cows are in good physical condition and have good muscle tone prior to

---

[6]Northeastern States Regional Bulletin 32 (Cornell Univ. Ag. Exp. Sta. Bull. 924), p. 30.

[7]*Ibid*, p. 29.

[8]Short, R. E., *et al.*, "Nutritional and Hormonal Interrelationships in Beef Cattle Reproduction," *Canadian Journal of Animal Science*, 68:29–39 (March, 1988).

[9]Northeastern States Regional Bulletin 32 (Cornell Univ. Ag. Exp. Sta. Bull. 924), p. 26.

confinement. It is recognized, however, that exercise is necessary for the normal well-being of the individual. (See Chapter 6, section on "Exercise" for supporting experimental work.)

SEXUAL REST

For maximum reproductive efficiency, cows should not be rebred too soon after calving. Although the reproductive tract usually returns to normal within about 6 weeks, barring infection or other abnormalities, it is inadvisable to rebreed a cow earlier than 60 days following parturition. This will still allow time for a second service if needed without exceeding the 12-month calving interval. A New York study revealed that the highest conception rate was obtained when breeding was between 70 and 90 days after calving.[10] In case of retained placenta or other calving difficulties, the cow should be allowed to go through at least one normal heat period before rebreeding.

SEASON AND LIGHT

Seasonal variations in conception rate have been observed in artificial insemination associations; poorest conception is obtained in the winter and maximum in the spring. From this it may be concluded that the amount of daylight (and perhaps the temperature) has an influence on fertility—as the amount of daylight increases, breeding efficiency increases proportionately until temperatures become too hot. In the South, the poorest conception rates occur in the hot summer months.

PHYSIOLOGICAL AND ENDOCRINE DISTURBANCES

The development of the reproductive organs, the production of ova, sexual behavior, the attachment and development of the fetus, parturition, and lactation are primarily regulated by hormones. Many cases of reproductive failure, particularly of conception and early development, may be due to hormone imbalance.

If neither infection of the genital tract nor any unusual condition is observed, the administration of hormones may be indicated. A wide variety of both natural and synthetic hormones is available, but none should be used except under the direction of a well-informed practitioner. The temptation is always strong to administer a mixture of hormones in the hope that one of them will correct the trouble. A wiser plan is to prescribe the specific hormone which it is believed will produce the desired results; however, this is often difficult or impossible, with the result that "trial and error" methods may be the only alternative.

[10]*Ibid.*, p. 5.

ANESTRUS (Failure to Come in Heat)

Anestrus is the prolonged period of sexual inactivity with no manifestations of estrus. The term is used to describe a cow which is not showing any external signs of heat, which may or may not be associated with ovarian inactivity. Although anestrus is observed during certain physiologic states, *e.g.*, before puberty and during pregnancy and lactation, it is most often a sign of temporary or permanent depression of ovarian activity caused by seasonal changes in the environment, nutritional deficiencies, lactation, or aging. Certain pathologic conditions of the ovaries or of the uterus also suppress estrus. This condition is normal in cows immediately following calving—during lactation. It may also occur in cows in the late winter and early spring when nutritional levels are low. Estrus can be readily induced in such animals by the administration of one of several natural or synthetic estrogens. Although it is unlikely that ovulation will accompany the induced heat, normal cycles are often reestablished thereby and conception may occur at the next heat period.

Short, *et al.*, USDA, Miles City, Montana, concluded that the primary cause of anestrus probably is different for different stages of anestrus. Further, they list the following management options to decrease the impact of anestrus and infertility: (1) restrict the breeding season to 45 days; (2) manage nutrition so that the body condition score is 5 to 7 before calving; (3) minimize effects of dystocia, and stimulate estrous activity with a sterile bull and estrous synchronization; and (4) make judicious use of complete, partial, or short-term weaning.[11]

DISTURBED ESTRUS CYCLES

Numerous variations of the normal estrus cycle occur, all of which are explainable in terms of improper gonadotropin-estrogen-progesterone relationships. Among such conditions are the following:

1. Ovulation without estrus; silent heats; or estrus of such low intensity that recognition is difficult. It has been estimated that 20% of ovulations in cows are not accompanied by external signs of heat.
2. Animals showing estrus cycles, but with a delay in ovulation.
3. Long or short heat periods.

[11]Short, R. E., *et al.*, "Physiological Mechanisms Controlling Anestrus and Infertility in Postpartum Beef Cattle," *Journal of Animal Science*, 1990, 68:799–816.

4. Abnormal intervals between heat periods.
5. Estrus without ovulation, known as an anovulatory cycle.

If it is definitely determined that cows with disturbed estrus cycles are not pregnant, and if infections have been ruled out, the judicious and careful use of hormones may be appropriate. Such treatment may restore the endocrine balance, which, in turn, will condition the reproductive system for estrus, conception, and pregnancy.

SEXUAL INFANTILISM

In this condition, the entire reproductive tract remains small. Ovulation may not occur; if it does, affected heifers may exhibit silent heat periods or be irregular in their sexual cycles. Sometimes sexual development is delayed but reproduction is normal after puberty. Heifers with this condition may become excessively fat and resemble spayed heifers or steers.

Sexual infantilism appears to be due to lack of gonadotropic hormone secretion by the anterior pituitary gland, but, unfortunately, treatment with gonadotropins is not often successful. If malnutrition does not appear to account for the condition, the possibility of a heritable factor should be suspected.

RETAINED CORPUS LUTEUM (Retained Yellow Body)

After the ovarian follicle ruptures and the egg is released, the cells within the follicular cavity change in character and function, forming a corpus luteum or yellow body in the cavity of the ruptured follicle. If the corpus luteum persists, subsequent heat periods do not usually occur. The corpus luteum produces a hormone (progesterone) which suppresses the pituitary output of the follicle-stimulating hormone (FSH). Thus, future follicular development and ripening is inhibited and the estrogens which would induce heat periods are not produced.

A retained corpus luteum may be suspected if a cow is not seen in heat within 60 days after calving. A few years ago, these were removed manually, but this procedure usually results in some bleeding. Injections of prostaglandin have replaced the old method.

CYSTIC OVARIES

Cystic ovaries may result when the ovarian follicle fails to rupture. The follicle persists, increases in size, and forms a cyst. It is believed that this condition is due to a derangement of gonadotropic-hormone secretion. An excessive secretion of FSH without adequate luteinizing hormone (LH) to produce ovulation causes continued follicular development and estrogen production.

When the condition is allowed to persist in cows, they frequently become chronic "bullers" or nymphomaniacs. Such individuals may show pronounced anatomic and psychologic changes; the pelvic ligaments relax so that there is a sagging of the loin region and elevation of the tailhead. Affected cows may acquire such male characteristics as thickened forequarters and may bellow and behave like a bull. As long as this condition exists, it is impossible for the cow to conceive because no ovum is shed to be fertilized.

Recommended treatment for cystic ovaries consists of the administration of gonadotropin releasing hormone. The use of progesterone for 14 consecutive days will clear up this condition. In some cases, a veterinarian may physically rupture the cyst and the cow may return to normal heat periods.

Generally, manually expressing cysts as a treatment is not as effective as the administration of gonadotropin.

There is evidence that the tendency toward cystic ovaries is inherited in cattle. Also, there is a higher incidence in the dairy breeds than in the beef breeds.

RETAINED PLACENTA (Retained Afterbirth)

Normally, the placenta is expelled within 3 to 6 hours after parturition. If it is retained as long as 12 hours after calving, competent assistance should be rendered.

Retained placenta occurs in about 5% of the parturitions in beef cattle and 10% in dairy cattle. It is more common following abnormally short or abnormally long pregnancies, among older cows, and following twinning. Experimentally, it has been found that a high incidence of retained afterbirth occurs when premature calving is induced by the administration of glucocorticoid drugs.

While infections such as brucellosis, vibriosis, and others have been associated with abortion and retained afterbirth, these are by no means the only causes. Nutritionally, deficiencies of carotene or vitamin A have been incriminated. Also, it appears that fewer cases of retained placenta occur (1) when calves stay with their dams and nurse for 12 to 24 hours, and (2) when cows are kept on pasture the year around. Among cows which have previously retained the placenta, 20% are likely to do so again.

Calves born when the placenta is retained are likely to be weak. A retained placenta may cause pathological conditions resulting in uterine tissue destruction. This condition may or may not affect milk production, but it very likely will result in 5 to 10% lower fertility than for normal cows.

When a retained placenta is encountered, appropriate treatment should be administered by the veterinarian, who will likely use either antibiotics or sulfonamides, either by direct infusion into the uterus or by other routes (or both).

It is seldom advisable to attempt removal of a retained placenta. If the membranes are dragging on the ground, they should be cut off at the hocks. But never, never tie bricks or other objects to them. In most instances, the membranes will fall out by themselves in 1 to 2 weeks.

It is desirable to have all cows which have had retained afterbirth examined at about 30 days after calving. If pus is present, they may be treated with estrogenic hormones to induce heat and then the uterus can be infused with an antibiotic solution or perhaps with a dilute Lugol's (iodine) solution. Such examination and treatment may save considerable time with regard to the onset of normal cycles and may result in a higher conception rate.

INTERSEXES AND HERMAPHRODITES

A brief description of two types of abnormal reproductive developments follow:

- **Intersex**—*An individual showing both maleness and femaleness; in which the sex differences are not confined to clearly demarcated parts of the body, but blend more or less with one another.*

- **Hermaphrodite**—*An individual whose genital organs have, in greater or lesser degree, the characters of both male and female.*

These conditions which occur only rarely in cattle are a result of (1) imbalance in the maternal fetal endocrine system, or (2) genetic factors.

INHERITED (GENETIC) ABNORMALITIES

The development of breeds or families within breeds which differ in prolificacy is good evidence that fertility may have a genetic basis. Thus, it is common knowledge that some once-popular families have become extinct because of the high incidence of infertility. A classical example of a situation of this type occurred in the Duchess family of Shorthorn cattle founded by the noted pioneer English Shorthorn breeder, Thomas Bates. Bates, and more especially those later breeders who emulated him, followed preferences in bloodlines until, ultimately, they were selecting cattle solely on the basis of fashionable pedigrees, without regard to fertility. Ironically, during their heyday, the scarcity of this strain of cattle contributed to their value. Eventually, but all too late, the owners of Duchess Shorthorns suddenly came to a realization that indiscriminate inbreeding and lack of selection had increased sterility to the point that the family name was in disrepute; a high incidence of sterility had actually contributed to their scarcity. As a result, Duchess Shorthorns became virtually extinct a few years later.

Of course, reproductive disorders of a heritable nature should not knowingly be perpetuated.

LETHAL GENES

Lethal genes, and other recessive genes causing sterility, belong in the group of highly heritable characters. Among such hereditary lethals are mummification, cystic ovaries, and gonad hypoplasia or gonadless. (See Chapter 5, Table 5-3, for a summary relative to lethal characters.) Generally these abnormalities are easily recognized by their actions, and their mode of transmission can often be analyzed and determined. Although lethals are of interest to the geneticist, they are generally no great problem to the cattle producer because their distribution remains under control. Usually recessive, their gene frequency is kept low by the self-destruction of the double recessive: thus they are self-selective.

WHITE HEIFER DISEASE

This name is a misnomer, for, although the condition is most commonly found in white heifers of the Shorthorn breed, it has been reported in roan and red Shorthorns and in colored animals of other breeds. It appears to be due to faulty development of the Mullerian ducts. Some of the more common characteristics are: closed hymen or hymen persisting in

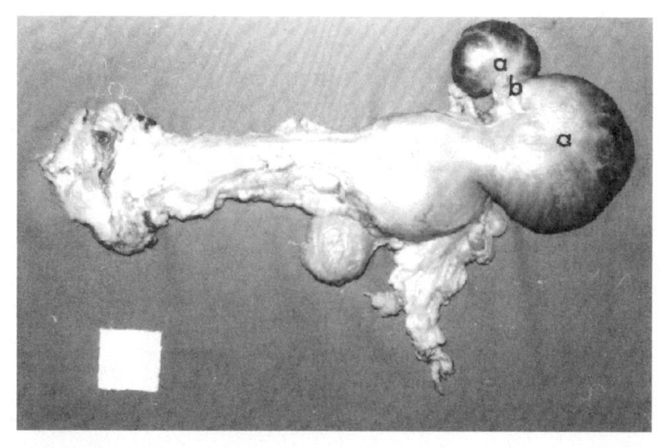

Fig. 7-4. White heifer disease. Note **(a)** balloonlike distention of uterine horns, and **(b)** segmental blockage. (Courtesy, Professor H. W. Leipold, Kansas State University)

varying degrees, distention of one or both uterine horns, and uterine body present in rudimentary form, complete absence of cervix, and anterior vagina.

White heifer disease appears to be caused by two pairs of autosomal recessive genes, one pair affecting the left uterine horn and the other the right uterine horn.

ANATOMICAL DEFECTS AND INJURIES

A long list of anatomical defects and injuries to the genital organs has been reported. Some of these are so severe as to cause sterility; others affect the degree of fertility. A Pennsylvania study of repeat breeders—cows which had failed to conceive after four services—revealed that 13% were anatomically abnormal.[12] A brief account of some of the more general anatomical defects follows.

FREEMARTIN HEIFERS

A sterile heifer that is born twin with a bull is known as a freemartin.

Freemartins occur in about 85 out of 100 twin births when a calf of each sex is involved. The fetal circulations fuse, and the male hormones get into the circulation of the unborn female where they interfere with the normal development of sex and modify the female fetus in the direction of the male. However, it has become increasingly clear that the hormonal theory alone is inadequate to explain the mode of origin of freemartins, and that cellular mechanisms may also be involved. In approximately 15% of twin births of unlike sexes, fusion of the circulation does not occur, and the animal is normal and fertile.

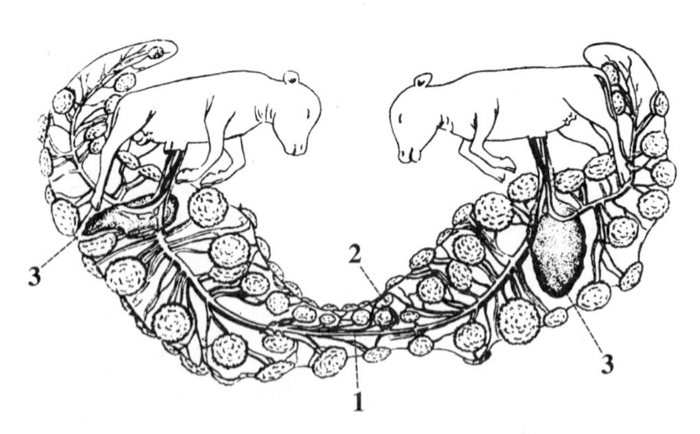

Fig. 7-5. Diagram showing fused fetal circulation of twin calves of opposite sex. Note **(1)** the fetal circulation of the male fused with that of the female, **(2)** fetal cotyledon free yolk sac, and **(3)** normal bull on the left and freemartin heifer on the right. (Source: *Physiology of Reproduction*, by Marshall; courtesy, the publisher, Longmans, Green and Co., Ltd., London, England)

[12]Northeastern States Regional Bulletin 32 (Cornell Univ. Ag. Exp. Sta. Bull. 924), p. 21.

Since only about 15% of such heifers are fertile, it is usually best to assume that they are sterile and market them, unless (1) an experienced person determined at the time of birth that their circulatory systems were not fused, (2) an examination of the vagina reveals that the animal is normal (in freemartin heifers, the vagina is usually about one-third normal length), or (3) skin-grafting or blood-typing, or chromosomal analysis, techniques show that they are not freemartins and that they may, therefore, be regarded as reproductively normal.

MECHANICAL INJURIES TO THE GENITAL ORGANS

Mechanical injuries may occur in the female at service and at parturition. A large, vigorous male may inflict injury when breeding; and complicated parturition may result in the loss of the offspring, permanent damage to the reproductive organs, or even death of the cow herself. Perforation of the uterine or vaginal walls, laceration of the cervix, eversion of the vagina, cervix, and uterus or of the rectum may all be sequelae of complicated birth. In some cases, adhesions or secondary infections of the genitalia cause tissue damage which prevents further reproduction.

MISCELLANEOUS AND UNKNOWN CAUSES

Unfortunately, many of the causes of sterility are unrecognized and unknown. A discussion of one of these follows.

EMBRYONIC MORTALITY (Fetal Death or Prenatal Mortality)

It is well known that early embryonic mortality occurs normally in the pig and the rabbit, where there is a surplus production of female gametes. Although it is not so common in the cow, it appears that 20 to 30% of the ova fertilized may meet embryonic death in 2 to 6 weeks. In such cases, the embryo may be absorbed or be expelled unobserved from the female reproductive tract. The cow may assume normal sexual cycles with the conclusion the fertilization did not take place.

Fetal death, followed by resorption or abortion, may occur at any stage of pregnancy. Such fetuses may range from decomposed masses to bones or dried mummies.

The cause or causes of prenatal death are not known.

STERILITY IN THE BULL

Any bull of breeding age that is purchased should be a guaranteed breeder; in fact, this is usually understood among reputable cattle producers.

The most reliable and obvious indication of fertility in a bull is a large number of healthy calves from a season's service. However, a good evaluation of a bull's fertility may be obtained through a microscopic examination of the semen made by an experienced person. It is recommended that all bulls be semen tested prior to the breeding season; and, where valuable purebred bulls are involved, periodic tests during the breeding season are desirable. Such procedure may alleviate much loss in time, feed, and labor, and avoid delayed and small calf crops.

For purposes of convenience, the common causes of sterility and delayed breeding in bulls are herein classified as (1) poor semen, (2) physical defects and injuries, (3) psychological, (4) genital infections and disease, (5) poor management and feeding, (6) physiological and endocrine disturbances, and (7) inherited (genetic) abnormalities.

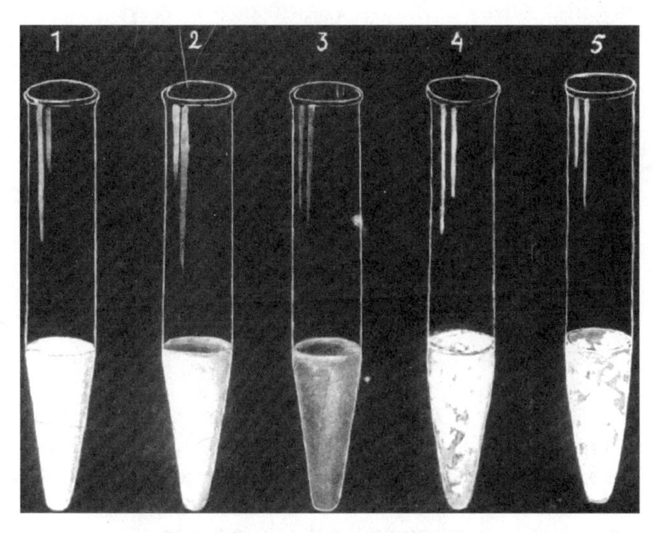

Fig. 7-6. Semen samples of different kinds.

1. Semen with normal appearance, about 1,000,000 spermatozoa per cubic mm.

2. & 3. Semen from a bull with hypoplastic testicles. Sample 2 contains about 200,000 spermatozoa per cubic mm. Sample 3, which is almost transparent contains about 25,000 spermatozoa per cubic mm.

4. & 5. Semen from a bull with inflammation in the seminal vesicles. In the semen, which is almost transparent, there are big, purulent flocci.

(Courtesy, Department of Obstetrics and Gynecology, Royal Veterinary College, Stockholm, Sweden)

POOR SEMEN

It is always well to obtain a sample of semen and to make a laboratory examination of the number and condition of the sperm. The four main criteria of semen quality are (1) volume, (2) sperm count, (3) progressive movement, and (4) morphology (shape). Although this technique is not infallible, an experienced person can predict with reasonable accuracy the relative fertility of bulls so examined.

PHYSICAL DEFECTS AND INJURIES

The most common defects involving the reproductive system are degenerating testicles, abscessed testicles, fibromas of the penis, broken penis, hematoma, adhesions within the sheath, and paralysis of the retractor muscle. Sometimes certain of these conditions can be corrected by skilled surgery. Infrequent, temporary, or permanent sterility may result from bruising, inflammation, and lacerations of the scrotum and

NORMAL AND ABNORMAL TYPES OF BOVINE SPERMATOZOA

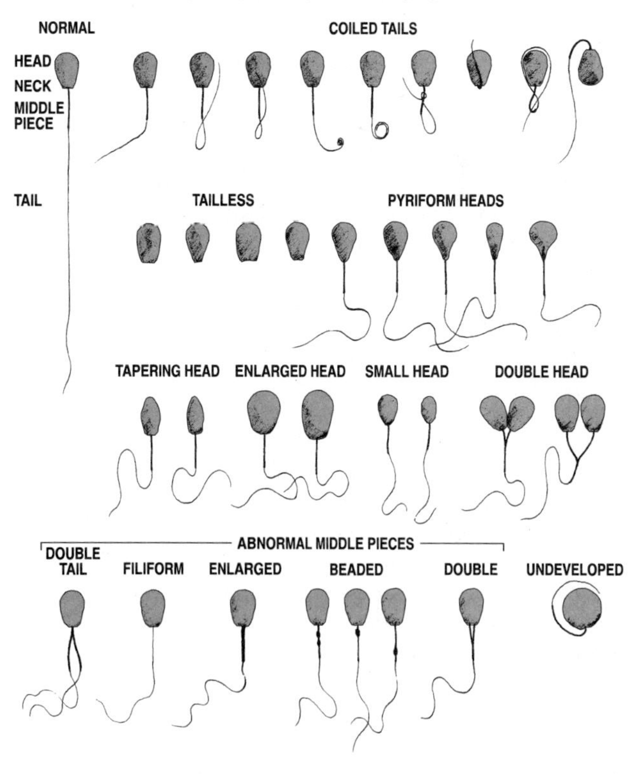

Fig. 7-7. Normal and morphologically abnormal spermatozoa of the bull. (Source, Herman, H. A., and F. W. Madden, *The Artificial Insemination and Embryo Transfer of Dairy and Beef Cattle*, seventh edition, The Interstate Printers & Publishers, Inc., Danville, IL, 1987)

testicles. Also, the penis may become bruised or lacerated during service.

Natural service is sometimes interfered with by unsound limbs and feet. This includes such conditions as broken limbs, bad sickle hocks, sore, overgrown or malformed feet, and arthritic or rheumatic joints. The latter condition is more prevalent in old bulls. Defects of the hind limbs are especially troublesome, because they make the bull unstable or cause severe pain when he shifts much of his weight onto them when mounting.

Keratitis (pinkeye) and blindness, especially if in both eyes, will interfere with mating.

Frequently, very short-legged, compact bulls that are in high condition and are somewhat awkward and clumsy are unable to serve cows. Also, an extremely paunchy bull may be unable to mate, because the "potbelly" acts as a mechanical obstruction and causes the penis to be directed too low. Usually such inability is temporary and may be corrected by reducing the ration and increasing the exercise.

Improvements can be effected in some of these physical defects and injuries, while others are irreparable.

PSYCHOLOGICAL

Sterile bulls are frequently victims of psychological sterility. Usually, this condition expresses itself in either (1) absence of or lowered sex drive, or (2) faulty reflex behavior during mating or ejaculation.

There is a wide individual difference in the reaction of bulls to their environment; and, likewise, these factors are, in part, hereditary. Thus, there are four major types of temperament in bulls: the nervous, the sulky, the placid, and the treacherous. The treacherous types should always be culled at an early age. Then, in selecting from among the other three types, consideration should be given to the herd requirements—whether a commercial or purebred herd is involved, whether pasture or hand mating is to be used, etc. Also, it is generally recognized that bulls lacking in masculinity or secondary sexual characteristics (bulls that are very docile, very fine boned, and lacking development of the crest, etc.) are likely to possess poor psycho-physiological sexual activity.

Young, inexperienced bulls mating for the first time are usually awkward to handle. They approach the cow hesitantly, spend a long time exploring the genitalia, mount hesitantly without erection, descend and try to mount again. Extreme patience and careful handling should be exercised during this critical first service; otherwise, difficult breeding habits may be established.

CAUSES OF PSYCHOLOGICAL STERILITY

Some common causes of psychological sterility are:

1. **Excitement.** Shouting, noises, distractions during mounting, and the presence of strangers may cause low sex drive. When restrained with a dog, a highly fertile bull may become sexually impotent. Bulls show evidence of such excitement when they urinate more frequently, as they do following the visit of strangers or the introduction of new animals.

   Also, it is important to keep sires as quiet as possible in the nonbreeding season. Undue excitement causes sexual impulses and results in the flow of semen into the ampullae.

2. **Transportation.** Psychological sterility is not uncommon in bulls that have been transported long distances by truck or rail.

3. **Animal management.** Young bulls that are isolated from all cows for long periods of time manifest homosexual reflexes and may become impotent.

   Also, inadequate sex drive may be due to an attempt to use a young bull on a cow that is too large for him to mount successfully. As a result of his failure to copulate, he develops a mild sense of frustration. Frequently, such a bull can be restored to a high state of breeding efficiency by giving him assistance and trying him on a heifer selected for her small size and willingness to stand quietly; or the same effect may be obtained by standing a large cow in a pit.

   Where hand mating is practiced, the bull should not be used immediately following feeding.

4. **Wrong technique during semen collection.** Certain inhibitory reactions may develop from improper use of the artificial vagina, including wrong timing in applying it, too hot or too cold water, or holding it at a wrong angle.

GENITAL INFECTIONS AND DISEASES

The presence of bacteria in the semen of bulls will not only tend to decrease the viability of the sperm cells but may very easily infect the cows to which they are mated, thereby preventing their conception even when bred to other bulls. If this condition is due to inflammation of the seminal vesicles or prostate, the systematic use of the appropriate antibiotic or sulfa drug may be recommended by the veterinarian. Occasionally, rectal massage of an infected prostate or seminal vesicle will aid the passage of the purulent exudate into the urethra and thus hasten recovery.

Brucellosis, trichomoniasis, granular venereal disease (granular vaginitis) and vibriosis are the specific genital diseases of most concern in bulls. Each of these diseases is fully covered in Chapter 12; and, likewise, a pertinent discussion of each disease from the standpoint of the cow appears earlier in this chapter. Thus, the ensuing comments will be limited to effects on the bull.

The *Brucella* organism may localize in the testicles, seminal vesicles, or vas deferens, and bulls may spread the disease by copulation. Sex drive is generally reduced if testicular involvement occurs.

In trichomoniasis, the bull is the source of infection. Positive diagnosis of infection in the bull may be made by means of (1) a microscopic examination of smears taken from the prepuce, or (2) mating him with a virgin heifer and checking her cervical and uterine smears for the organism as she approaches the next heat period.

It is believed that granular venereal disease is commonly transmitted by the bull at the time of service, but this is not the only means of transmission since virgin heifers may be infected. Infected bulls should be either treated or sold for slaughter. Treatment consists in massaging the prolapsed penis and sheath with a suitable antibiotic ointment, repeated sufficiently often to assure elimination of infection.

In vibriosis, no clinical lesions are observed in bulls. However, *V. fetus* appears to persist indefinitely in the genital tract of carrier bulls and the bull may transmit the infection. For diagnosis of vibriosis in individual bulls, one should (1) culture the semen, and, where possible, (2) breed to one or more virgin heifers and collect vaginal mucus for culture 10 to 20 days later. Effective control can be obtained by (1) adding antibiotics to semen from infected bulls and breeding artificially, or (2) establishing a new herd of sexually immature animals.

POOR MANAGEMENT AND FEEDING

In all too many cases, little thought is given to the management and feeding of the bull, other than during the breeding season. Instead, the program throughout the entire year should be such as to keep the bull in a vigorous, thrifty condition at all times. Also, lack of fertility in the bull may often be traced back to his early care and feeding. He may have been small and weak at birth, he may have been improperly fed during the first year of his life, or he may have been prey to infection.

IMPROPER FEEDING

Severe undernutrition (a poor, thin, run-down condition) and vitamin A deficiency are the two most common causes of lowered fertility in the bull, with young bulls more affected than mature bulls. Also, overfat, heavy bulls should be regarded with suspicion, for they may be uncertain breeders.

Vitamin A deficiency inhibits spermatogenesis and adversely affects semen quality and fertility in a manner similar to undernutrition. The common deficiency symptoms are degeneration of the germinal epithelium, reduced diameter of the seminiferous tubules, and testicular atrophy, usually accompanied with cystic changes in the pituitary. Also, there have been responses in sperm production, semen quality, and fertility in cattle to the following trace elements: copper, cobalt, zinc, manganese, and iodine.

Fitting bulls for show and sale results in a variable effect upon their semen-producing ability; in some it has no detectable detrimental effect, while others are severely affected by such practices. Many highly fitted bulls show a complete lack of sperm cells; others produce semen comparable in quality to a bull exhibiting testicular degeneration. Many, perhaps most, such fat bulls eventually reach normal breeding efficiency if they are properly let down—primarily by reducing the grain and by increasing the exercise.

EXERCISE

Exercise is necessary for the normal well-being of the bull, although it is difficult to show precisely to what extent lack of exercise induces low fertility. For example, in one study involving dairy bulls used in artificial insemination, eight bulls which were force exercised were compared with a like number which were kept in box or tie stalls, without forced exercise. The exercised group showed a nonreturn rate of 63.8%, whereas the bulls that were not exercised showed a nonreturn rate of 65%. Hence, the bulls without exercise were actually a little more fertile than the exercised ones.

Generally speaking, beef bulls should be exercised by allowing them the run of the pasture or large corral. However, it may be necessary to lead fat show bulls from 2 to 4 mi daily following the show season.

RETARDED SEXUAL MATURITY

As stated earlier in this chapter, the normal age of puberty in cattle is 12 months. But there is wide variation in the age of sexual maturity of bulls as measured by semen quality. Some yearling bulls are fully equipped to produce a healthy percentage of calves; others are not. At the present time, it is not known whether such retarded sexual development is due to hereditary, nutritional, and/or management factors. There is need, therefore, for additional research on this subject.

THE OVERWORKED BULL

Low fertility of the bull is frequently caused by overservice. Table 6-1 of Chapter 6 may be used as a bull mating guide for different age animals, under both hand-mating and pasture-mating practices.

In pasture breeding, the bull may copulate 4 or 5 times in succession, thereby correspondingly reducing his powers and lessening the size of the herd on which he should be used. However, it is recognized that this situation is compensated for, in part at least, by the fact that bulls on pasture have sexual vigor not possessed by stall-fed bulls.

PHYSIOLOGICAL AND ENDOCRINE DISTURBANCES

The development of the reproductive organs, the production of sperm, and sexual behavior are primarily regulated by hormones; thus, the possibilities of endocrine disturbances are endless. Examples of some of the more common abnormalities follow.

SEXUAL INFANTILISM

In this condition, the entire reproductive tract remains small and the testicles are visibly reduced in size. Affected animals lack sex drive. If a low plane of nutrition does not appear to account for the condition, the possibility of a genetic factor should be suspected.

SEX DRIVE (Libido)

Lack of sexual drive (lack of libido) may be due to nutrition, endocrine imbalance, environmental factors, or heredity.

Sex drive, or libido, is important, for if the bull is unwilling to perform and in-heat cows are not bred, there will be no reproduction.

Libido is affected before spermatogenesis. In bulls, both undernutrition and longtime vitamin A deficiency have led to depressed libido. In Europe, the sexual activity of bulls on trace-element deficient soils has been improved by supplementation with iodine, cobalt, copper, and zinc.

Also, overfeeding and obesity in bulls can lead to loss of libido, especially in hot weather.

INHERITED (GENETIC) ABNORMALITIES

Lack of fertility in the bull may often be traced right back to his own sire and dam. He may have been sired by a bull of low vigor and fertility and out of a cow of equally low fertility.

Reproductive disorders of a heritable nature should not knowingly be perpetuated. In addition to those which follow, Table 5-3 of Chapter 5 contains a summary of some fertility-affecting hereditary abnormalities in bulls, including impotentia, atrophied testicles, knobbed spermatozoa, and turned sperm tails.

CRYPTORCHIDISM

When one or both of the testicles of a bull have not descended to the scrotum, the animal is known as a cryptorchid. The undescended testicle(s) is usually sterile because of the high temperature in the abdomen. Since this condition may be heritable, it is recommended that animals so affected not be retained for breeding purposes.

SCROTAL HERNIA

When a weakness of the inguinal canal allows part of the viscera to pass out into the scrotum, the condition is called scrotal hernia. This abnormality may interfere with the circulation in the testicles and result in their atrophy.

TESTICULAR HYPOPLASIA

Hypoplasia of the testicles is an inherited defect in which the potential for development of the spermatogenic epithelium is lacking. It is best known in Swedish Highland cattle, although it occurs in other breeds. It is caused by a recessive autosomal gene with incomplete (about 50%) penetrance.

Testicular hypoplasia is suspected only at puberty, or later, because of reduced fertility or sterility. One or both testicles may be hypoplastic. A hypoplastic testicle is reduced in size, and the semen is watery and contains few or no spermatozoa.

UMBILICAL HERNIA

This condition, which may interfere with breeding efficiency, has been reported as due to (1) a sex-limited dominant gene, or (2) one or more pairs of autosomal recessive factors.

BULL BREEDING EVALUATION

Evaluation of breeding bulls is very important. The Society of Theriogenology, P.O. Box 2118, Hastings, Nebraska, has evolved with an evaluation form which lists guidelines for this purpose. A form provides assurance that all important points are evaluated. Some, but not all, of the points listed on this form follow:

Physical Examination

Body condition score:

Thin ☐ Moderate ☐ Good ☐ Obese ☐

Pelvic Height _____ Width _____ Area _____

Feet/legs ______________________________

Eyes ______________________________

Penis/prepuce ______________________________

Scrotum (shape) ______________________________

Scrotal circumference (CM) ______________________________

Semen Evaluation

Collection Method: EE ☐ AV ☐ Massage ☐

Response: Erection ☐ Protrusion ☐ Ejaculation ☐

Semen characteristics

Motility:

Gross ______________________________

Individual ______________________________

% Normal sperm ______________________________

% Primary abnormalities ______________________________

% Secondary abnormalities ______________________________

Note: The above evaluation is for physical soundness and quality of semen only. It does not include any tests for libido, mating ability, or infectious diseases.

Based on an interpretation of the data resulting from executing the above form, the bull is classified as—

☐ Satisfactory potential breeder

☐ Unsatisfactory potential breeder

☐ Classification deferred

The physical examination should give assurance of the bull's ability to follow the cows. The feet should be structurally sound; there should be two good eyes; and the scrotum should be of adequate size.

Semen should be collected and evaluated for motility and morphology.

Additionally, all bulls should be evaluated for sex drive (or libido), for if the bull is unwilling to perform and the in-heat cows are not bred, there will be no reproduction.

(Also see Chapter 4, section headed "Scrotal Measurement.")

In addition to evaluating breeding soundness when buying bulls, or immediately prior to the breeding season, constant vigilance is necessary during the breeding season. Foot, eye, and other injury problems take their toll on the herd bull battery.

A PROGRAM OF IMPROVED FERTILITY AND BREEDING EFFICIENCY

A program designed to give improved fertility and breeding efficiency in beef cattle follows, in summary form.[13]

1. Keep complete breeding records. Maintain complete fertility records on each animal, examine them periodically, and cull low producers. Where hand-mating is followed, keep a record of dates bred and observed heat periods of each cow; calculate the expected estrus cycle by adding 21 days to the date of last estrus.

2. Before the breeding season, check the bull for physical defects and quality of semen.

3. Breed only healthy cows to healthy bulls.

4. Avoid either an overfat or a thin, emaciated condition in all breeding animals; and feed balanced rations.

5. Provide plenty of exercise for bulls and pregnant cows, preferably by allowing them to graze in well-fenced pastures in which plenty of shade and water are available.

6. Do not breed young animals until they are sufficiently mature.

7. Provide an ample rest period between pregnancies; do not rebreed within 60 days after calving. Where the placenta was retained or other calving difficulties encountered, allow the cow to go through at least one normal heat period before rebreeding.

8. Do not overwork the bull.

9. Observe breeding females carefully during the breeding season; otherwise, heat periods of short duration may be missed. Also, keep a close watch for shy breeders; expose them to the bull often.

10. Diagnose cows for pregnancy.

11. Handle the newborn so that its health shall be assured, and in order that it may have uninterrupted development.

12. Retain as future replacements only those animals which are the progeny of healthy parents, that were carried in utero for a normal gestation period of from 279 to 288 days, and that were born without difficult calving, retained afterbirth, or metritis.

---

[13]In this section special emphasis is placed on increased fertility and breeding efficiency; thus, there is some repetition and there are some additions to the section entitled, "A Program of Beef Cattle Health, Disease Prevention, and Parasite Control," as given in Chapter 12 of this book.

13. Isolate newly acquired animals for a minimum of 3 weeks, during which time they should be tested for brucellosis, leptospirosis, trichomoniasis, and vibriosis. However, first make every reasonable effort to ascertain that they came from herds which are known to be free from these and other diseases.

14. When possible, purchase virgin heifers and bulls. Isolate nonvirgin bulls for a period of 3 weeks, and then turn them with a limited number of virgin heifers; observe these heifers for 30 to 60 days after breeding as an aid in preventing the introduction of breeding diseases.

15. When sterility is encountered, promptly call upon the veterinarian for treatment; do not delay action until the condition is of long standing.

Fig. 7-8. Reproduction is the first and most important requisite of beef cattle breeding. Without a calf being born—and born alive—the other economic traits are of academic interest only. (Courtesy, USDA)

QUESTIONS FOR STUDY AND DISCUSSION

1. Compute the following for your beef cattle herd, or for a herd with which you are familiar:

   a. How much is being lost in annual gross sales of calves at weaning time, when considering the current calf crop percentage vs a 100% calf crop?

   b. How much does it cost to maintain all of the barren cows for a year?

2. If a certain cattle producer is experiencing (a) 15% breeding difficulty in the cow herd at a given time, (b) 2.5 services per conception, and (c) a 75% calf crop, outline, step by step, your recommendations for determining the difficulties and improving the situation.

3. What precautions should a cattle producer take to avoid the introduction into the herd of genital infections and diseases?

4. What symptoms characterize each of the following specific genital diseases in cattle: bovine virus diarrhea, brucellosis, epizootic bovine abortion, infectious bovine rhinotracheitis, leptospirosis, metritis, trichomoniasis, vaginitis, and vibriosis? What positive diagnosis, if any, can be made of each? What control program should be initiated when the presence of each disease is known?

5. How do practical cattle producers synchronize ovulation and breeding?

6. May sterility be caused by (a) lack of feed as sometimes occurs during droughts, or (b) high condition as when fitted for show?

7. Is exercise necessary for normal reproduction? Justify your answer.

8. Why should the cattle producer call upon a well-informed practitioner if hormone injections are to be given to cattle?

9. On the basis of experimental evidence, should the corpus luteum be removed by an experienced technician by pressure applied through the rectal wall (rectal palpation) if the cow does not come in heat within 60 days after calving?

10. How should a cattle producer handle a case of retained placenta?

11. Define and describe the following: (a) nymphomania; (b) intersex; (c) hermaphrodite; and (d) white heifer disease.

12. How can you tell whether a heifer is a freemartin?

13. List and discuss the seven common causes of sterility and delayed breeding in bulls.

14. How would you go about testing a bull for breeding soundness?

15. Write out a *program of improved fertility and breeding efficiency* for your herd or for a herd with which you are familiar.

SELECTED REFERENCES

Title of Publication	Author(s)	Publisher
Artificial Insemination and Embryo Transfer of Dairy and Beef Cattle, The, Eighth Edition	H. A. Herman J. R. Mitchell G. A. Doak	Interstate Publishers, Inc., Danville, IL, 1994
Breeding Difficulties in Dairy Cattle, Cornell Univ. Ag. Exp. Sta. Bull. 924	S. A. Asdell	Cornell University, Ithaca, NY, 1957
Breeding Difficulties of Cattle	C. Staff	General Mills, Larro Feeds, Chicago, IL
Cattle Fertility and Sterility	S. A. Asdell	Little, Brown and Co., Boston, MA, 1968
Dairy Cattle Sterility	H. D. Hays L. J. Boyd	*Hoard's Dairyman*, W. E. Hoard and Sons, Company, Fort Atkinson, WI
Factors Affecting Reproductive Efficiency in Dairy Cattle, Kentucky Ag. Exp. Sta. Bull.	D. Olds D. M. Seath	University of Kentucky, Lexington, KY, 1954
Farm Animals	J. Hammond	Edward Arnold & Co., London, England, 1952

Physiology of Reproduction and Artificial Insemination of Cattle, Second Edition	G. W. Salisbury N. L. VanDemark J. R. Lodge	W. H. Freeman & Co., San Francisco, CA, 1978
Problems and Practices of American Cattlemen, Wash. Ag. Exp. Sta. Bull. 562	M. E. Ensminger M. W. Galgan W. L. Slocum	Washington State University, Pullman, WA, 1955
Reproduction in Dairy Cattle, Ext. Bull. 115	C. H. Boynton	University of New Hampshire, Durham, NH
Reproduction in Farm Animals, Fourth Edition	Ed. by E. S. E. Hafez	Lea & Febiger, Philadelphia, PA, 1980
Reproduction and Infertility	Centennial Symposium	Agricultural Experiment Station, Michigan State University, MI, 1955
Reproduction and Infertility III Symposium	Ed. by F. X. Gassner	Pergamon Press, New York, NY
Reproductive Physiology	A. V. Nalbandov	W. H. Freeman & Co., San Francisco, CA, 1958
Veterinary Endocrinology and Reproduction, Third Edition	Ed. by L. E. McDonald	Lea & Febiger, Philadelphia, PA, 1980

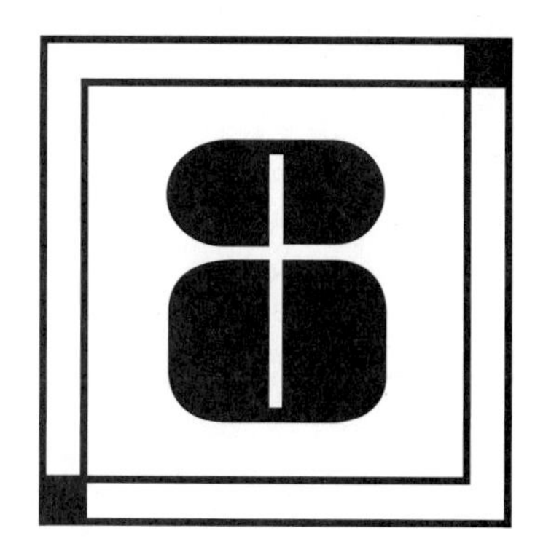

Feed is the most important influence in the environment of beef cattle. This shows a Braford cow and calf on good pasture. (Courtesy, United Braford Breeders, Nacogdoches, TX)

FEEDING BEEF CATTLE

Contents *Page*

Contents	Page

Although Webster defines the noun *ration* as *"the amount of food supplied to an animal for a definite period, usually for a day,"* to most cattle producers the word implies the feeds fed to an animal or animals without limitation to the time in which they are consumed. In this and other chapters of *Beef Cattle Science*, the authors accede to the common usage of the word rather than to dictionary correctness. Likewise, the authors accede to the common usage of the word *ration* instead of the word *diet*.

Cattle inherit certain genetic potentialities, but how well they develop depends upon the environment to which they are subjected; and the most important factor in the environment is the feed. The feeding of cattle also constitutes the greatest single cost item of their production. It is important, therefore, that the feeding practices be as satisfactory and economical as possible.

Fig. 8-1. Under the old system of unforced production and marketing at 3 to 5 years of age, reasonably good pasture and hay sufficed. Not so today! Meeting consumer demand for smaller cuts, more lean, less fat, and greater tenderness, calls for forced production, growing and finishing simultaneously, and, frequently, confinement. Consequently, the nutritive requirements are more critical than formerly, especially from the standpoint of proteins, minerals, and vitamins. (Courtesy, Pennsylvania Millers and Feed Dealers)

Pastures and other roughages, preferably with a maximum of the former, are the very foundation of successful beef cattle production. In fact, it may be said that the principal function of beef cattle is to harvest vast acreages of forages, and, with or without supplementation, to convert these feeds into more nutritious and palatable products for human consumption. It is estimated (1) that 84.5% of the total feed of beef cattle is derived from roughages (see Table 8-1), and (2) that 29.2% of the land area of the United States (50 states) is grassland, much of which is utilized by beef cattle. If produced on well-fertilized soils, green grass and well-cured, green, leafy hay can supply all of the nutrient requirements of beef cattle, except the need for common salt and whatever energy-rich feeds may be necessary for additional conditioning or drylot finishing.

TABLE 8-1
PERCENTAGE OF FEED FOR DIFFERENT CLASSES OF LIVESTOCK DERIVED FROM (1) CONCENTRATES, AND (2) ROUGHAGES, INCLUDING PASTURES[1]

Class of Livestock	Concentrates	Roughages
	(%)	*(%)*
Beef cattle	15.5	84.5
Dairy cattle	41.3	58.7
Sheep and goats	6.2	93.8
Swine	95.7	4.3
Horses and mules	27.0	73.0
Poultry	100.0	0.0
All livestock	38.3	61.7

[1]Senior author's estimates for 1996.

DIGESTIVE SYSTEM

An understanding of the principal parts and functions of the digestive system is essential to intelligent feeding of cattle. Fig. 8-2 shows the digestive tract of a cow. Table 8-2 gives capacity figures.

To be useful to animals, nutrients must enter the bloodstream for transport to various parts of the body. *The process whereby the animal releases feed nutrients from feed is termed digestion.* As commonly used, the process also includes absorption of food from the digestive tract into the bloodstream. Most of the unused portion of the feed is eliminated in the feces, although a considerable proportion is also given off as gas through the mouth and nose.

Cattle belong to the ruminant or cud-chewing group of animals (which includes cattle and sheep), whereas swine and chickens are monogastric animals (those having only one stomach).

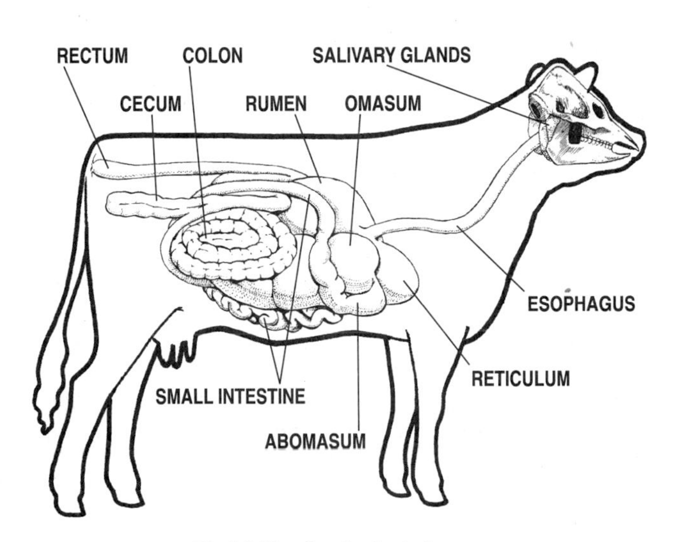

Fig. 8-2. The digestive tract of a cow.

The primary nutritional difference between cows, or ruminants, and simple-stomached animals comes about through the normal functioning of the rumen and its billions of microorganisms—the bacteria and protozoa. These microorganisms live on the food material supplied in the cow's diet. The rumen provides a highly desirable environment in which they can grow and reproduce themselves as rapidly as every half hour. In turn, they release to the cow during their life and from their bodies as protoplasm upon death, many of the cow's required nutrients. This is how the cow gets many nutrients needed for production that are not supplied in the diet.

PARTS AND FUNCTIONS OF THE RUMINANT STOMACH

The feed taken in by cows is mixed with a heavy flow of saliva, which is needed to help in chewing and swallowing of dry materials. Saliva of ruminants, unlike that of nonruminants, does not contain enzymes to aid in the digestion of starches. However, the saliva of cows, estimated to be about 120 lb per day in a mature animal, does have enough buffers (sodium bicarbonate) to neutralize the fatty acids produced in the rumen and maintain the rumen contents at approximately a neutral pH (7.0).

The ruminant stomach consists of four distinct compartments: (1) the rumen, or *paunch*, as it is commonly called; (2) the reticulum, or *honeycomb*; (3) the omasum, or *manyplies* (so called because of the plies or folds); and (4) the abomasum, or *true stomach*. When feeds are ingested, the normal pathway they follow is in the order just listed, with portions being returned to the mouth for chewing before they enter the omasum. These four compartments do not lie in a straight arrangement; rather, they are bunched and joined together to form a compact structure. A

TABLE 8-2
PARTS AND CAPACITIES OF DIGESTIVE TRACTS

Parts	Cow		Horse		Pig	
	(qt)	*(l)*	*(qt)*	*(l)*	*(qt)*	*(l)*
Stomach:	200	*190.0*	8–16	*7.6–15.2*	6–8	*5.7–7.6*
Rumen (paunch)	160	*151.0*				
Reticulum (honeycomb)	10	*9.5*				
Omasum (manyplies)	15	*14.3*				
Abomasum (true stomach)	15	*14.3*				
Small intestine	62	*58.9*	48	*45.6*	9	*8.5*
Cecum			28–32	*26.6–30.4*		
Large intestine	40	*38.0*	80	*76.0*	10	*9.5*

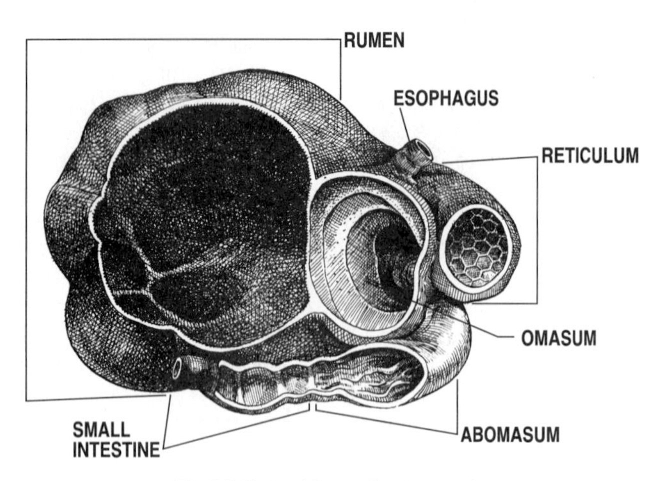

Fig. 8-3. Parts of the ruminant stomach.

discussion of each of the parts of the ruminant's stomach follows:

1. **Rumen.** In large, mature cows, the rumen has a capacity of about 40 gal, or as much as 300 lb of material. The functions of the rumen are: (a) to act as a storage place (it holds the feed which the cow gathers during her feeding period; later, when she rests, she regurgitates the larger particles back to the mouth, to grind them more completely, so that they pass from the rumen more quickly and make room for more feed); and (b) to provide a place for fermentation—there is a continual flow of feed material into and out of the rumen, constant introduction of saliva which controls the pH, absorption of the end products of the microbial action, and a population of microorganisms—bacteria and yeasts, both of which are plants, and one-celled animals called protozoa—which develop in accordance with the amount and type of feed consumed. These organisms in the paunch (a) help digest the crude fiber, (b) form essential amino acids and build up complete proteins, and (c) manufacture the B-complex vitamins.

For satisfactory rumen function in the digestive tract and for synthesis of B vitamins and proteins, rumen bacteria and protozoa require adequate nutrients from beef cattle rations at all times, including (a) energy, involving small amounts of available energy, such as sugars and starches; (b) ammonia-bearing ingredients, such as proteins, urea, and ammonium salts; (c) major minerals, especially sodium, potassium, and phosphorus; (d) cobalt and possibly other trace minerals; and (e) miscellaneous growth factors.

The fermentation process produces large amounts of carbon dioxide, methane, and ammonia; smaller amounts of hydrogen, hydrogen sulfide, and carbon monoxide; and probably trace amounts of other gases. Normally, these gases are passed off by the reflex action of belching. However, sometimes a cow is unable to get rid of this gas, and bloating results.

2. **Reticulum.** The reticulum lies directly in front of the rumen. Actually, the rumen and the reticulum are not completely separated; hence, food particles pass freely from one to the other. The interior of the reticulum is lined like a honeycomb; hence, the popular name—*the honeycomb.*

The main function of the reticulum is its action as a screening device. Heavy objects, such as nails or wire eaten with the feed, have a tendency to settle in this compartment. Therefore, it is sometimes referred to as the *hardware stomach.*

3. **Omasum.** The third compartment is lined with plies or folds of tissue; hence, it is sometimes called the *manyplies.* Less is known about the functions of the omasum than of any of the other compartments of the ruminant stomach. It is generally believed, however, that the primary functions of this compartment are (a) to reduce the water content of the feedstuffs, and (b) to exert a regrinding and squeezing action on the material.

4. **Abomasum.** The abomasum is often referred to as the *true stomach* because its action is similar to the stomach action in monogastric animals. As in the monogastric stomach, digestive juices are added and the moisture content of the feed material is increased. The digestive juices contain enzymes, resulting in protein digestion in the abomasum. Little or no digestion of fat, cellulose, or starch occurs in this organ.

The feed material leaving the abomasum is highly fluid in nature. It is then passed out of the stomach through the small intestine, where additional digestion occurs, and the unabsorbed material is then excreted by way of the large intestine.

RUMINATION

The process known as rumination, or chewing the cud, occupies about eight hours of the cow's time each day. When the cow regurgitates, a soft mass of coarse feed particles, called a *bolus*, passes from the rumen through the esophagus in a fraction of a second. She chews each bolus for about one minute, then swallows the entire mass again. A placid cow lying under a tree slowly chewing her cud conveys a special sense of contentment, symbolic of the tranquillity of the countryside. But this activity is most significant to the cow. Originally, it was thought that the regrinding which occurred during rechewing helped the digestion by exposing a greater surface area to fiber-digesting microflora. But recent experiments indicate that rechewing does not improve digestibility. Instead, rumination has an important effect on the amount of feed the animal can utilize. Feed particle size must be reduced to allow passage of the material from the rumen. It follows that high-quality forages require much less rechewing and pass out of the rumen at a faster rate; hence, they allow a cow to eat more. This concept is very important to the production of beef and milk because a cow will eat only as much coarse material as she can grind up by ruminating not more than 9 hours per day.

CALF'S STOMACH

When the calf is born, the rumen is small and the fourth stomach is by far the largest of the compartments. Thus, digestion in the young calf is more like that of a single-stomached animal than that of a ruminant. The milk which the calf normally consumes by-passes the first two compartments and goes almost directly to the fourth stomach in which the rennin and other compounds for the digestion of milk are produced. If the calf gulps too rapidly, or gorges itself, the milk may go into the rumen where it is not digested properly and may cause upset of the calf's digestive system. As the calf nibbles at hay, small amounts of material get into the rumen. When certain bacteria become established, the rumen develops and the calf gradually becomes a full-fledged ruminant.

FEED CONSUMPTION AND RATE OF GAIN

If they don't eat it, they won't gain. But feed consumption and rate of gain are affected by many things.

The daily consumption of dry matter (feed) by cattle is primarily dependent upon the following:

1. **Size.** Breeds and individuals maturing at heavier weights consume more feed, animal for animal, than small cattle. But they may or may not be more efficient than small cattle when consideration is given to (a) the production efficiency of their dams, and (b) carrying them to the same degree of finish.
2. **Age and condition.** Older and more fleshy feedlot cattle consume less feed per unit of liveweight than do younger, leaner animals. Mature animals in good condition may be expected to consume amounts of dry matter equal to 2% or more of their liveweight, whereas thin animals eating high-quality roughage may eat amounts equal to 3% of their liveweight per day.
3. **Digestible nutrient content (energy density).** As digestible nutrient content (energy density) increases, consumption of feed dry matter is usually reduced. It follows that feed efficiency is improved in high- and all-concentrate rations, due to their high energy. In the final analysis, however, the comparative price of concentrates and roughages—the economics of the situation—will be a major determining factor.
4. **Environmental stress.** Cattle producers have long known that environmental stress caused by high and low temperatures, mud, and other adverse environmental factors can affect the voluntary consumption of feed. For example, feedlot cattle consume less during very hot weather than in cold weather.

The rate of gain of feedlot cattle is influenced by the following factors:

1. **Sex.** Under feedlot conditions, at comparable weight and finish, bulls can be expected to make about 10% greater gains than steers, and steers can be expected to make about 10% greater gains than heifers.
2. **Implants and growth stimulants.** The use of certain implants and growth stimulants in finishing steers and heifers usually increase feed intake by approximately 6% and gains by 8 to 12%.

Elucidation and application of the subject "Feed Consumption and Rate of Gain" as it pertains to each of the respective classes of cattle is presented in the following chapters of this book: Chapter 19, Feeding and Managing Brood Cows; Chapter 20, Bulls; Chapter 21, Feeding and Handling Calves; Chapter 22, Replacement Heifers; and Chapter 30, Feeding Finishing (Fattening) Cattle.

NUTRITIVE NEEDS OF BEEF CATTLE

The nutritive requirements of beef cattle have become more critical with the shift in beef production practices. Steers were formerly permitted to make their growth primarily on roughages—pastures in the summertime and hay and other forages in the winter. After making moderate and unforced growth for 2 to 4 years, usually the animals were either turned into the feedlot or placed on more lush pastures for a reasonable degree of finishing. With this system, the growth and finishing requirements of cattle came largely at two separate periods in the life of the animal.

Under the old system of moderate growth rate, reasonably good pastures and good-quality hay fully met the protein requirements, as well as the mineral and vitamin needs. As the feeding period was not so long with these older cattle, and the stress was not so great, in comparison with the period required in the finishing of calves or yearlings in a drylot, there also was less tendency for vitamin deficiencies to develop in the feedlot; and the protein requirements were less important during the finishing period.

The preference of the consumer for smaller cuts

of beef—meats that are more tender and have less fat—has caused a shift in management and marketing. Today, increasing numbers of cattle are finished at younger ages, in a shorter period than formerly, in bigger feedlots, and marketed as choice beef. Such animals are in forced production. Their bodies are simultaneously laying on fat and growing rapidly in protein tissues and skeleton. Consequently, the nutritive requirements are more critical than those of older cattle, especially from the standpoint of proteins, minerals, and vitamins.

In recent years, the introduction of crossbreeding and the exotic breeds has produced faster-gaining calves, late-maturing cattle, and heavier-milking cows. Also, more and more heifers are being bred to calve as 2-year-olds. All of these factors influence the nutrient requirements of beef cattle.

In this book, nutrient requirements are presented using two different systems or editions of the *Nutrient Requirements of Beef Cattle* published by the National Research Council (NRC) of the National Academy of Sciences. The previous edition of this book used the 1984 NRC Nutrient Requirements of Beef Cattle as a reference for nutrient requirements. These requirements are presented in Appendix, Section I, Table I-1, and they are referred to, and used in the the section concerning balancing rations in this chapter.

The Seventh Revised Edition of the *Nutrient Requirements of Beef Cattle* was published in 1996. The original intent of the senior author was to use this publication as a reference for nutrient requirements in this book. However, significant changes have been made concerning how requirements are generated and presented, which, in the opinion of the authors, may make the information more difficult for producers to understand and more importantly, more difficult for them to use. Therefore, the requirements from the previous edition of the NRC publication are presented and used along with the more current information.

In the 1996 NRC publication, nutrient requirements are not listed in tables by class and size of animal as they had been in previous publications. Rather, a computer program that is provided as part of the publication must be used to generate requirements for different production situations. The computer model uses many more factors such as mature weight (actual or estimated), body condition score, breed type, implant strategy, and numerous environmental influences in generating requirements. Therefore, it is not practical to publish tables of nutrient requirements because with the new system they are potentially different for each production situation.

Tables listing nutrient requirements for animals for four different production situations are presented in Appendix, Section I, Table I-2. These tables are presented only as examples, because, as previously discussed, the computer program uses imputed information to generate nutrient requirements for each production situation. It should be noted that in determining nutrient requirements for growing and finishing cattle and growing bulls, requirements for maintenance need to be added to requirements for the desired level of performance to determine the total or actual requirement for each nutrient characteristic.

If producers do not have access to, or know how to use a computer, then they have no way of estimating or determining nutrient requirements of an animal or group of animals without using the tables in the 1984 NRC publication. Because the computer program also has the ability to evaluate rations, actual requirements may not be as important with this edition of the NRC publication. However, if producers are balancing rations by hand, or using a computer program in which requirements need to be imputed, then having access to actual nutrient requirement tables is still a necessity.

The 1996 NRC publication does provide tables of diet nutrient densities for nine different production situations. These tables are presented in Appendix, Section I, Table I-3. For growing and finishing cattle, the tables are categorized according to estimated finishing or slaughter weight. For replacement heifers, the tables are categorized according to estimated mature weight of the heifers. Body weight and desired average daily gain are then used in the appropriate table to determine the needed concentration of energy (either TDN or net energy), crude protein, calcium, and phosphorus for the desired level of performance with a particular class of animals.

Tables for pregnant replacement heifers are categorized according to estimated mature weight and months since conception. Tables for beef cows are categorized according to mature weight, level of milk production, and months since calving.

The other major change in the 1996 NRC publication involves how protein requirements are expressed. In the diet nutrient density tables discussed in the previous two paragraphs, needed concentrations of protein are expressed as percent of crude protein, similar to previous NRC publications. However, animal nutrient requirements for protein are expressed as units of metabolizable protein. Metabolizable protein is defined as the true protein absorbed by the intestine, supplied by microbial protein and undegraded intake (bypass) protein. This system accounts for rument degradation of protein and separates nutrient requirements into needs of the microorganisms and needs of the animal.

When evaluating rations, the computer program uses the percent crude protein in a feedstuff along with the percent of protein that is degraded versus undegraded in the rumen. Although this system offers advantages in terms of characterizing and evaluating protein metabolism in the animal, the problem with this system is that curde protein is the only information that is available on many feedstuffs. Information concerning the percent of protein degraded versus undegraded (bypass) in the rumen is not available on many feedstuffs. The authors of the 1996 NRC publication realized and admitted to these shortcomings, however, they expressed their belief that the new system using metabolizable versus crude protein is required if we are to move forward in the area of rumen nutrition.

In this book, crude protein percent of numerous feedstuffs is presented in Appendix, Section II, Table II-1 and percent undegraded protein (bypass protein) in some common feeds is presented in Appendix, Section II, Table II-3.

As feeds represent by far the greatest cost item in beef production, it is important that there be a basic understanding of the nutritive requirements. For convenience, these needs will be discussed under the following groups: (1) energy, (2) protein, (3) minerals, (4) vitamins, and (5) water.

ENERGY

Carbohydrates, which constitute about 75% of all the dry matter of plants, are the chief sources of energy of cattle feeds. In the usual chemical analysis of feeds, carbohydrates are divided into nitrogen-free extract (NFE) and fiber. The NFE, which is the more soluble part, includes sugars, starches, organic acids such as lactic and acetic acid (which are present in silage), and other more complex carbohydrates. Fiber includes the relatively insoluble carbohydrates, such as cellulose.

Next to carbohydrates, fats are important as energy sources. Because of their larger proportion of carbon and hydrogen than the carbohydrates, they liberate more energy; furnishing approximately 2.25 times as much heat or energy per pound as do the carbohydrates.

In addition to supplying nitrogen, natural plant protein compounds also supply a certain amount of energy.

A relatively large portion of the feeds consumed by beef cattle is used in meeting the energy needs, regardless of whether the animals are merely being maintained (as in wintering) or fed for growth, finishing, or reproduction.

The first and most important function of feeds is that of meeting the maintenance needs. *The maintenance requirement for energy can be defined as the amount of feed energy that will result in no loss or gain in body energy.* For some beef animals near their mature size (adult bulls, for example) maintenance may be the practical feeding goal. There are variations in maintenance requirements based on sex, breed, frame size, age, and environmental conditions. In general, breeds and individuals maturing at heavier weights may require the most energy, and *Bos indicus* breeds and crosses may require the least energy for maintenance. If there is not sufficient feed, as is frequently true during periods of drought or when winter rations are skimpy, the energy needs of the body are met by the breakdown of tissue. This results in loss of condition and body weight.

After the energy needs for body maintenance have been met, any surplus energy may be used for growth, finishing, reproduction, or lactation. With the present practice of finishing cattle at early ages, growth and finishing are in most instances simultaneous, and, therefore, not easily separated. The net energy for gain (NE_g) of finishing animals is the amount of energy deposited as nonfat organic matter (mostly protein) plus that deposited as fat.

In the finishing process, the percentage of protein, ash, and water steadily decreases as the animal matures and fattens, whereas the percentage of fat increases. Thus the body of a calf at birth may contain about 70% water and 4% fat; whereas the body of a fat 2-year-old steer may contain only 45 to 50% water but from 30 to 35% fat. This storage of fat requires a liberal allowance of energy feeds.

Through bacterial action in the rumen, cattle are able to utilize a considerable portion of roughages as sources of energy. Yet it must be realized that with extremely bulky rations the animal cannot consume sufficient quantities to produce the maximum amount of fat. For this reason, finishing rations contain a considerable proportion of concentrated feeds, mostly cereal grains. On the other hand, when the energy requirements are primarily for maintenance, roughages are usually the most economical sources of energy for beef cattle.

At times, fats may be cheap enough to merit consideration as partial substitutes for standard energy feeds. Also, it is probable that very small amounts of fatty acids are essential for beef cattle, as is true in certain other species, but no requirements have thus far been established.

ENERGY DEFICIENCY (Underfeeding)

Many cattle throughout the world are underfed all or some part of the year. In fact, lack of sufficient total feed is probably the most common deficiency suffered by beef cattle, although it is recognized that underfeeding is frequently complicated by concomitant shortage of protein and other nutrients. Restricted rations often occur during periods of drought, when pastures or ranges are overstocked, or when winter rations are skimpy. Also, many range operators regularly plan that cows in good flesh should lose some condition during the winter months; they feel that it is uneconomical to feed sufficient to retain the fleshy condition. Fortunately, during such times of restricted feed intake, animals have nutritive reserves upon which they can draw. Although they may survive for a considerable period of time under these conditions, there is an inevitable loss in body weight and condition; and, varying with the degree of underfeeding, there may be a slowing or cessation of growth (including skeletal growth), failure to conceive, and increased mortality. Low feed intake also commonly results in increased deaths from toxic plants and from lowered resistance to parasites and diseases.

During times of energy shortage when animals are withdrawing stored energy—mostly from body fat—the mobilized fat may not be completely metabolized, with the result that ketosis (partially metabolized fatty acids) develops. Mild cases of ketosis may not affect production or health. But animals with severe ketosis go off feed, which further aggravates the malady by lowering energy intake still more.

Research workers of the U.S. Department of Agriculture conducted an experiment to determine some of the economic effects and possible harm to animals of limited rations.[1] Identical twin calves were used and the following planes of nutrition were studied: (1) full feed—gains of more than 1.5 lb daily, (2) 75% of full feed—gains of 1.0 lb per day, (3) 62% of full feed—gains of 0.5 lb a day, and (4) a maintenance ration of about 50% of full feed—they neither gained nor lost in weight.

All animals—including those on the low energy rations—received ample protein, vitamins, minerals, and other nutrients. At the end of the period of retarded feeding, the steers were fed liberally until they reached a slaughter weight of 1,000 lb.

Although the low-plane-of-nutrition animals reached slaughter weight from 10 to 20 weeks later than did their twins, the former attained their weight on approximately the same total feed intake as the latter; which means that, after limited feeding ended, the retarded animals made more economical gains than did their twins. Carcass quality, amount of lean meat, and grade were not affected.

This experiment showed that, under conditions of feed scarcity, beef cattle between the ages of 6 and 12 months can be carried on a maintenance ration—so they will neither gain nor lose in weight—provided the nutrient needs other than energy are supplied—without subsequent loss in feed efficiency, carcass quality, or quantity of lean meat. Also, it shows that compensatory gains occur following a low plane of nutrition; it shows why feedlot finishers prefer feeder cattle that have not been backgrounded at a high rate of gain.

[1]Winchester, C. F., and Paul E. Howe, USDA Tech. Bull. No. 1108.

METHODS OF MEASURING ENERGY

Broadly speaking, two methods of measuring energy are employed in this country—the total digestible nutrient system (TDN), and the calorie or net energy system. Each system has its advantages and advocates. But, more and more feedstuffs are being evaluated in calories.

TOTAL DIGESTIBLE NUTRIENTS (TDN) SYSTEM

Total digestible nutrients (TDN) is the sum of the digestible protein, fiber, nitrogen-free extract, and fat × 2.25. It has been the most extensively used measure for energy in the United States.

Back of TDN values are the following steps:

1. **Digestibility.** The digestibility of a particular feed for a specific species is determined by a digestion trial.

2. **Computation of digestible nutrients.** Digestible nutrients are computed by multiplying the percentage of each nutrient in the feed (protein, fiber, nitrogen-free extract [NFE], and fat) by its digestion coefficient. The result is expressed as digestible protein, digestible fiber, digestible NFE, and digestible fat. For example, if No. 2 corn contains 8.9% protein of which 77% is digestible, the percent of digestible protein is 6.9.

3. **Computation of total digestible nutrients** (TDN). The TDN is computed by use of the following formula:

$$\%\,TDN = \%\,DCP + \%\,DCF + \%\,DNFE + (\%\,DEE \times 2.25)$$

where DCP = digestible crude protein; DCF = digestible crude fiber; DNFE = digestible nitrogen-free extract; and DEE = digestible ether extract.

TDN is ordinarily expressed as a percent of the ration or in units of weight (Lb or kg).

The main **advantage** of the TDN system is that it has been used for a very long time and many people are familiar with it.

The main **disadvantages** of the TDN system are:

1. It is really a misnomer, because TDN is not an actual total of the digestible nutrients in a feed. It does not include the digestible mineral matter (such as salt, limestone, and defluorinated phosphate—all of which are digestible); and the digestible fat is multiplied by the factor 2.25 before being included in the TDN figure, because its energy value is higher than carbohydrates and protein. As a result of multiplying fat by the factor 2.25, feeds high in fat will sometimes exceed 100 in percentage TDN (a pure fat with a coefficient of digestibility of 100% would have a theoretical value of 225%—100% × 2.25).

2. It is an empirical formula based upon chemical determinations that are not related to actual metabolism by the animal.

3. It is expressed as a percent or in weight (lb or kg), whereas energy is expressed in calories.

4. It takes into consideration only digestive losses; it does not take into account other important losses, such as losses in the urine, gases, and increased heat production (heat increment).

5. It over evaluates roughages in relation to concentrates when fed for high rates of production, due to the higher heat loss per pound of TDN in high-fiber feeds.

6. It does not recognize that zero net energy is equal to 25 to 30% TDN. For example, a feed such as peanut hulls with 30% TDN actually has a negative net energy value and requires energy for its digestion.

Because of these several limitations, in the United States the TDN system is gradually being replaced by other energy evaluation systems, particularly net energy. However, due to the voluminous TDN data on many feeds and long-standing tradition, it will continue to be used by many people for a long time to come.

CALORIE SYSTEM

Calories are used to express the energy value of feedstuffs. Nutritionists now standardize their combustion (bomb) calorimeters using specifically purified benzoic acid, the caloric value of which has been determined in electrical units and computed in terms of joules/g mole. *One calorie (always written with a small c) is the amount of heat required to raise the temperature of 1 g of water 1°C (precisely from 16.5°C to 17.5°C) at a pressure of 1 atmosphere.*

To measure this heat, an instrument known as the bomb calorimeter is used, in which the feed (or other substance) tested is placed and burned in the presence of oxygen.

Through various digestive and metabolic processes, much of the energy in feed is dissipated as it passes through the animal's digestive system. About 60% of the total combustible energy in grain and about 80% of the total combustible energy in roughage is lost as feces, urine, gases, and heat. These losses are illustrated in Figs. 8-4 and 8-5.

As shown in Figs. 8-4 and 8-5, energy losses occur in the digestion and metabolism of feed. Measures that are used to express animal requirements and the energy content of feeds differ primarily in the digestive and metabolic losses that are included in their determination. Thus, the following terms are used to express the energy value of feeds:

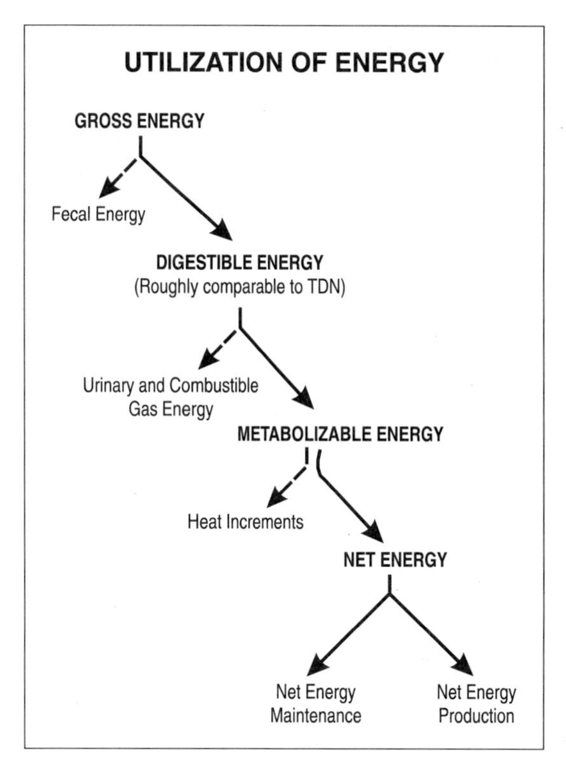

Fig 8-4. Utilization of energy.

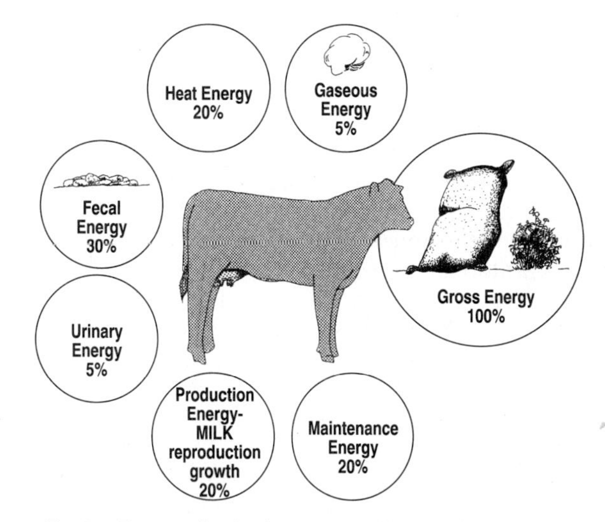

Fig. 8-5. Energy utilization by a cow suckling a calf, showing average partition of feed energy by the animal.

■ **Gross energy (GE)**—Gross energy represents the total combustible energy in a feedstuff. It does not differ greatly between feeds, except for those high in fat. For example, 1 lb of corncobs contains about the same amount of GE as 1 lb of shelled corn. Gross energy does not provide information regarding the availability of the energy to the animal.

■ **Digestible energy (DE)**—Digestible energy is that portion of the GE in a feed that is not excreted in the feces. The major weakness of DE as a basis for feeding systems is that it overestimates the available energy of high-fiber feedstuffs (hays, straws) relative to low-fiber, highly digestible products, such as grains.

■ **Metabolizable energy (ME)**—Metabolizable energy represents that portion of the GE that is not lost in the feces, urine, and gas. Although ME more accurately describes the useful energy in the feed than does GE or DE, it does not take into account the energy lost as heat.

■ **Net energy (NE)**—Net energy represents the energy fraction in a feed that is left after the fecal, urinary, gas, and heat losses are deducted from the GE. The net energy, because of its greater accuracy, is being used increasingly in ration formulations, especially in computerized formulation for large operations.

Two net energy values are assigned to each feedstuff. Likewise, animal requirements for energy are similarly subdivided. Net energy available or required for maintenance is termed NE_m, and net energy available or required for growth is termed NE_g. Two major advantages of the net energy system are: (1) animal requirements stated as net energy are independent of the diet; *i.e.*, they do not have to be adjusted for different roughage-concentrate ratios; and (2) feed requirements for maintenance are estimated separately from feed needed for productive functions.

PROTEIN

Proteins are complex organic compounds made up chiefly of amino acids, which are present in characteristic proportions for each specific protein. This nutrient always contains carbon, hydrogen, oxygen, and nitrogen; and, in addition, it usually contains sulfur and frequently phosphorus. Proteins are essential in all plant and animal life as components of the active protoplasm of each living cell.

At least 22 amino acids have been found in proteins; and they may occur in different combinations to form an almost limitless number of proteins.

The protein allowance for beef cattle, regardless of age or system of production, should be ample to replace the daily breakdown of the tissues of the body including the growth of hair, horns, and hoofs. In general, the protein needs are greatest for the growth of the young calf and for the gestating-lactating cow.

As discussed previously in this chapter, protein requirements are listed in this book using two different systems. The first is the system used in the 1984 NRC Nutrient Requirements of Beef Cattle which lists requirements for crude protein. These requirement tables are listed in Appendix, Section I, Table I-1. Crude protein refers to all the nitrogenous compounds in a feed. It is determined by finding the nitrogen content and multiplying the result by 6.25. The nitrogen content of protein averages about 16% ($100 \div 16 = 6.25$).

The second system is the system used in the 1996 NRC Nutrient Requirements of Beef Cattle which lists requirements for metabolizable protein (Appendix, Section I, Table I-2). Metabolizable protein is defined as the true protein absorbed by the intestine, supplied by microbial protein and undegraded intake (bypass) protein. This system accounts for rumen degradation of protein and separates requirements into needs of microorganisms and needs of the animal.

In order to use this system in balancing rations, one must have not only the percent crude protein of a feedstuff, but also the percent of the protein that is degraded and undegraded (bybass) in the rumen. Information concerning the latter two characteristics is not available for many feedstuffs.

Methods of feeding, feed preparation, and various feed additives do not appear to alter protein requirements. Feed consumption is reduced when high-concentrate rations are fed; as consumption declines, the percentage of protein in such rations should be increased proportionally.

Protein supplements are regularly in shorter supply and higher priced than cereal grains and other high-energy feeds used in livestock feeding. Normally, the United States produces about 40 million tons of protein supplements, exclusive of urea (Fig. 8-6). But it is estimated that an additional 2 to 5 million tons of these products could be used advantageously if all animals were supplied an adequate amount of protein.

As protein supplements ordinarily cost more per ton than grains, normally beef cattle should not be fed larger quantities of these supplements than actually needed to balance the ration.

With stocker cattle, or in the maintenance of the beef breeding herd, it usually does not pay to add a protein supplement when a legume hay is fed. With feedlot cattle on high-concentrate rations, or when the breeding herd is being wintered on a nonlegume roughage, sufficient protein supplement—usually 1 to 2 lb daily—should be added to the ration.

PROTEIN DEFICIENCIES AND TOXICITIES

Depressed appetite is the primary symptom of protein deficiency in beef cattle rations. Depressed appetite may, in turn, lead to an inadequate intake of energy; hence, protein deficiency and energy deficiency often occur together. Other symptoms of protein deficiency are loss of weight, poor growth, irregular or delayed estrus, and reduced milk production.

Rations containing up to 40% protein have been fed to steers. Feed intake was reduced for several days when protein was added, but no signs of ammonia toxicity were evident.[2] However, excesses of

[2]Fenderson, C L, and W. G. Bergen, "Effect of Excess Dietary Protein on Feed Intake and Nitrogen Metabolism in Steers," *Journal of Animal Science*, Vol. 42, p. 1323.

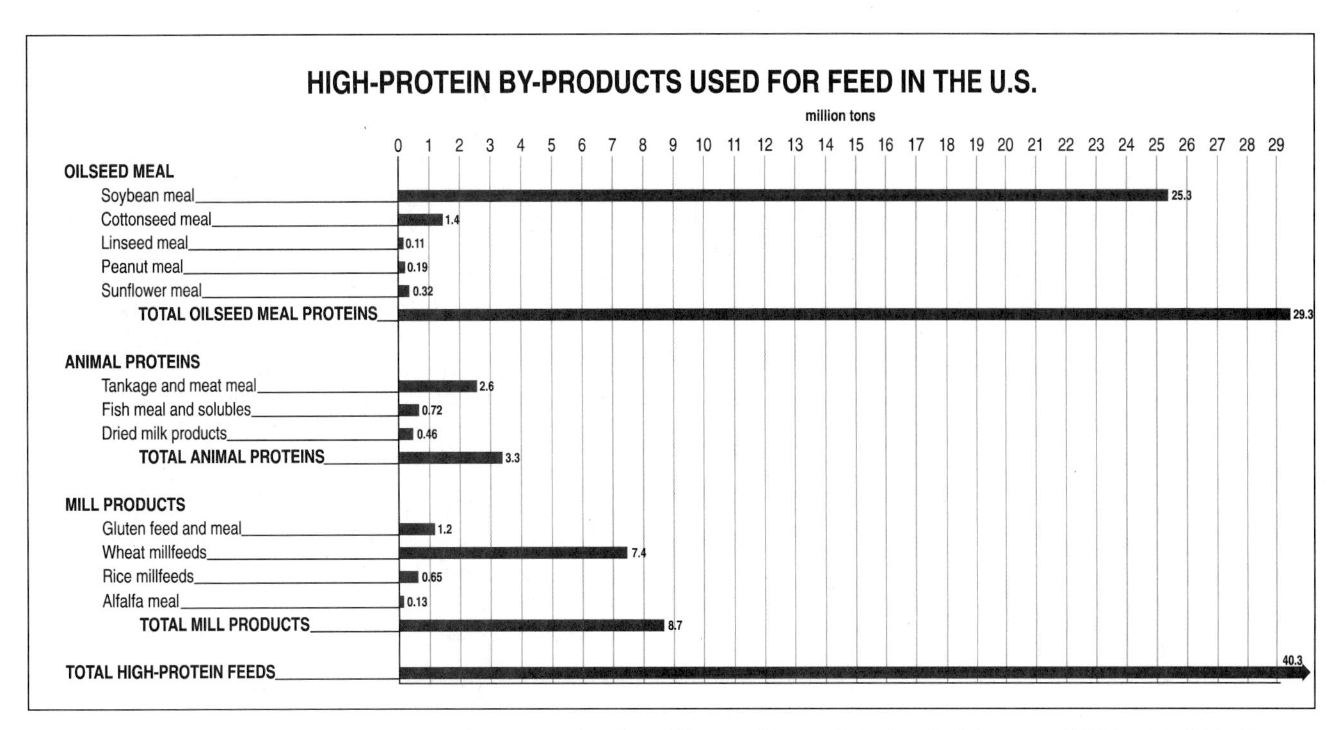

Fig. 8-6. Tonnage of different high-protein by-products used for feed in the U.S., 1993. (Source: *Agricultural Statistics 1995-96*, USDA, p. I-45, Table 69)

nonprotein nitrogen or soluble protein may precipitate ammonia toxicity.

PRERUMINANT CALF PROTEIN

The preruminant calf has little rumen function; therefore, its protein nutrition parallels that of the nonruminant. The amino acid requirements of the calf can be met through milk or milk replacers. Supplementing with certain amino acids may prove beneficial, but the usefulness of urea is limited. Milk replacers should meet protein requirements (Appendix, Section I, Table I-1 and I-2) while creep rations should be formulated to complement the energy and protein supplied by milk and/or pasture. After the rumen becomes functional, at 6 to 8 weeks of age—depending on the ration, the crude protein needs for two systems must be met—(1) the need for nitrogen for microbial fermentation in the reticulo-rumen, and (2) the need for postruminal amino acids for tissues of the host ruminant.

MICROBIAL SYNTHESIS AND REQUIREMENTS

Because of rumen synthesis of essential amino acids by microorganisms, the quality of proteins (or balance of essential amino acids) is of less importance in the feeding of beef cattle than in feeding some other classes of stock. Proteins from plant sources, therefore, are quite satisfactory. Also, these microorganisms—which are a low form of plant life and are able to use inorganic compounds such as ammonia just as plants utilize chemical fertilizers—build body proteins of high quality in their cells from sources of inorganic nitrogen that nonruminants cannot use.

About 75% of the carbohydrates digested by ruminants is fermented by microbes in the rumen. During fermentation, volatile fatty acids, ammonia, methane, and CO_2 are released, and energy is liberated for microbe growth and multiplication. A wide variety of microbe types including many species of anaerobic bacteria, protozoa, and even fungi, thrive in the rumen.

Swept out of the rumen to the abomasum and small intestine with fluid and particles, ruminants digest these microbes and obtain about half of their protein (amino acid) therefrom.

Ruminant bacteria can use various sources of nitrogen (primarily ammonia and some amino acids and peptides), energy (derived from fermentation), and minerals for growth. Any of these factors can limit bacterial growth because requirements are interrelated. The supply of ammonia can be inadequate when either the intake of protein or the ruminal degradation of protein is low. Ammonia deficiency in the rumen reduces the efficiency of bacterial growth and may reduce the rate and extent of digestion of organic matter in the rumen, which, in turn, may reduce feed intake.

Ammonia is derived from degradation of protein or nonprotein nitrogen (NPN) in the rumen. Although most bacterial species in the rumen can survive when using ammonia as the sole source of nitrogen, added protein may stimulate bacterial growth by providing amino acids, essential branch-chained fatty acids, or unidentified factors for ruminal bacteria to incorporate or use. Generally the least costly dietary source of ruminal ammonia is some form of NPN.

In ruminant nutrition, therefore, even such nonprotein sources of nitrogen as urea and ammonia have a protein replacement value. An exception is the very young ruminant in which the rumen and its ability to synthesize are not yet well developed. For such an animal, high-quality proteins in the diet are requisite to normal development.

The microbial synthesis in ruminants is in marked contrast to the situation with humans, dogs, swine, and poultry where considerable care must be exercised to provide all the essential amino acids in proper proportion for efficient and economical production.

NONPROTEIN NITROGEN

Certain nonprotein nitrogen (NPN) sources may be substituted for all or much of the supplemental protein required in most beef cattle rations, provided such rations are adequate in minerals and readily available carbohydrates. Among such products are urea, ammoniated molasses, ammoniated beet pulp, ammoniated cottonseed meal, ammoniated citrus pulp, and ammoniated rice hulls.

UREA

The nonprotein nitrogen (NPN) source most commonly fed to ruminants is urea. Urea is a white, crystalline, odorless, nonprotein nitrogen compound of the formula N_2H_4CO. It is manufactured in chemical plants that produce anhydrous ammonia by fixing some of the nitrogen of the air. Some of the ammonia gas is combined with gaseous carbon dioxide to produce the white, crystalline solid urea which is quite stable. Also, it is noteworthy that urea occurs as the principal end product of nitrogen metabolism in nearly all mammals; it is found in the urine of humans and farm animals.

Approximately 265,000 tons of urea are fed annually in the United States, as a source of protein for cattle, sheep, and goats. Initially, the protein equivalent value of feed grade urea was 42 (% nitrogen) times 6.25 (common protein factor), or 262% protein. Today, more concentrated 45% nitrogen ($45 \times 6.25 \div 281$) urea has replaced most of the 42% grade, at a lower unit cost.

Urea is rapidly hydrolyzed to ammonia in the rumen; as a result, excessive amounts of absorbed ammonia can prove toxic to ruminant animals. Proper management procedures are necessary when NPN is fed in order to prevent ammonia toxicity and to avoid reduction in feed intake. Single doses of urea at 0.14 to 0.36 g/lb of body weight have toxic effects. Toxicity can be avoided by thoroughly mixing urea in the ration and limiting the maximum concentration to 1% of the ration dry matter or one-third of the total protein in the ration. In typical rations for beef cattle, this concentration usually exceeds the amount that is needed. Slowly degraded sources of NPN also help avoid ammonia intoxication. High-concentrate diets provide more energy for synthesis of bacterial protein from ammonia and increase the amount of ammonia retained in ruminal fluid due to a lower ruminal pH. Ammonia absorption from the rumen and the likelihood of toxicity decrease as ruminal pH declines.

When NPN is substituted for protein in a ration, special care in mineral supplementation is necessary, because most sources of protein provide substantial amounts of sulfur, potassium, and phosphorus, which are absent in NPN sources.

NPN addition to a ration is useful only when the ruminal concentration of ammonia is inadequate for optimal bacterial action to (1) digest organic matter or (2) supply ammonia for microbial synthesis of protein.

Additional pertinent facts about urea follow.

■ **Feeding value of urea**—Attempts have been made to equate urea to oil meals by various thumb rules. One such thumb rule is that 1 lb of urea plus 6 lb of corn equals 7 lb of soybean meal. This combination of corn and urea supplies as much nitrogen as does soybean meal and, thus, could be considered equal to it in crude protein content. This is true if the rumen microorganisms can convert the urea nitrogen to protein. But this doesn't tell the whole story! Table 8-3 shows the inequalities in mineral content of these two feeds.

As noted in Table 8-3, the corn-urea combination is a little low in TDN, or energy. Additionally, the combination supplies only 13.5% as much calcium and 38.1% as much phosphorus as does soybean meal; and only in the case of cobalt does the combination supply as high a mineral level as does the soybean meal. Unless these deficiencies are met, poor utilization of the urea-containing ration can be expected. Of course, minerals can be added, but all ingredients cost money.

■ **Urea is best utilized in well-balanced high-energy rations**—Urea is not well utilized for supplementing low-quality roughages. The explanation is that the carbohydrates in grasses and hays appear to be so slowly available that the bacteria have difficulty in using the energy from roughages to make use of urea in preparing bacterial protein. It is generally held that some preformed protein should be present in the feed, also. Part of this will be provided by the grains, and

TABLE 8-3
COMPARISON OF A CORN:UREA (6:1) MIXTURE WITH SOYBEAN MEAL

Nutrient	Corn:Urea (6:1) as a % of Soybean Meal
TDN	88.9
N	100.0
Ca	13.5
P	38.1
Mg	36.0
K	14.7
S	9.1
Mn	14.6
Co	95.2
Cu	20.0
Fe	13.1
Zn	63.9

frequently some oil meals are used in preparing the formula feeds.

Other components of a balanced feed include calcium, phosphorus, iron, copper, cobalt, manganese, iodine, and perhaps zinc, sulfur, and magnesium. The need for these minerals as well as for vitamin A will depend upon local conditions with respect to the types of roughages produced and the influence of weather upon the quality of such roughages.

Another factor influencing the formulation includes the purpose of the feed, *e.g.*, for the breeding herd, creep-feeding, or for use in the finishing or feedlot.

■ **Factors essential for optimum use of urea**—Urea can be successfully and effectively used, or it can be abused. Observance of the following pointers will assure optimum use of urea:

1. Mix the urea thoroughly.
2. Feed urea only to cattle, sheep, or other ruminants. Never feed it to swine, poultry, or horses.[3]
3. Limit the intake of urea to recommended maximum levels. (See Table 8-4.)
4. Provide a readily available energy source, such as molasses or grain.
5. Supply adequate and balanced levels of minerals, including calcium, phosphorus, and trace minerals (especially cobalt and zinc).
6. Achieve a nitrogen-sulfur ratio not wider than 15:1.
7. Incorporate alfalfa meal as a source of unidentified factors to stimulate the microbial synthesis of protein.

[3]Most state laws restrict the use of urea to ruminant rations, hence, it is illegal to add it to swine, poultry, or horse rations. Urea may be toxic to foals.

TABLE 8-4
GUIDELINES FOR THE USE OF UREA FOR CATTLE[1]

	For Finishing Cattle	For Grower (Stocker) Cattle	For Wintering Pregnant and Lactating Cows
Percent of total protein in ration from urea (%)	33⅓	25.0	25.0
Percent of urea, by weight of total air-dry feed consumed (%)	1.0	1.0	1.0
Percent of urea, by weight, of concentrate mix (grain plus protein supplement)[2] (%)	2.0–3.0	3.0	3.0
Percent of urea, by weight, of the protein supplement (%)	20–30[3]	10.0[4]	10.0
Percent of supplemental nitrogen in high protein supplement from urea[5] (%)	60–90[6]	30.0	30.0
Pounds of urea added/ton of corn silage at ensiling time[7] (lb)	10.0 *(4.5 kg)*	10.0 *(4.5 kg)*	10.0 *(4.5 kg)*

[1]Based on a consensus obtained by the authors.

[2]Feed intake may be depressed if over 1% is used. Yet, many beef producers are successfully using 2%.

[3]This means that as much as 60–90% of the protein value of the supplement may come from nonprotein sources. However, since such a supplement will constitute only 2–5% of the total ration fed, the first rule of thumb given in Table 8–4 still applies; namely, only ⅓ of the total protein in the ration will be supplied from nonprotein sources.

[4]A protein supplement containing 10% urea provides 28.1% of protein equivalent (281% × 0.10) from nonprotein nitrogen.

[5]High urea supplements are best fed in complete mixed rations, which are *thoroughly* mixed. *Supplements containing 20–30% urea require extreme caution when being hand fed.*

[6]In a feedlot ration, this may be equivalent to 25–40% of the total nitrogen from all sources.

[7]On a dry matter basis, corn silage ensiled at the well-dented stage runs about 8% protein. The addition of 10 lb of urea per ton *(4.5 kg/1,000 kg)* of silage increases the protein content to 13%. However, there is loss of flexibility in feeding such a ration, and the rate of gain will be less than can be secured from higher, more dense rations. Also, it is extremely important that the urea be well mixed in the silage, otherwise there is hazard of toxicity.

8. Include adequate salt for palatability; 0.5% in complete rations and 3.5% in protein supplements.

9. Provide the proper level of vitamin A, and of such other vitamins, hormones, antibiotics, and additives as desired.

10. Use a free-flowing urea, and mix it thoroughly; avoid sifting or sorting of ingredients. Never use high urea-containing supplements as a topdressing.

11. Accustom animals gradually to urea-containing feeds (over a period of 5 to 7 days), exercise caution in feeding very hungry animals, and feed at frequent intervals.

12. Never use raw soybeans or beans of any kind, lespedeza seed, alfalfa seed, or wild mustard seed in a grain mixture containing urea for the reason that an enzyme (urease) in the beans will break down urea into ammonia and carbon dioxide. The liberated ammonia may be strong enough to be objectionable to cattle. Eventually, the animals will eat the feed, but the protein level will be reduced.

■ **Quantity of urea that may be fed**—Urea may constitute up to 33⅓% of the total protein of growing-finishing rations and 25% of the total for pregnant and lactating cows, provided additional energy is added in the form of molasses or grain to compensate for the lack of energy in the urea,[4] in order to feed the rumen bacteria properly. By total protein is meant the protein intake of the entire ration—including forage, grain, and protein supplements. Urea should not constitute more than 1.0% of a total mixed ration.

Guidelines relative to the use of urea are given in Table 8-4.

Less urea is recommended in range cubes or pellets because of (a) the more limited grain and the poor-quality roughage usually fed, and (b) the uncertainty of feeds being consumed regularly under adverse weather conditions. Thus, it is recommended that urea be limited to 5%, by weight, of range cubes or pellets used primarily to supplement dried range grass cured on the stalk.[5] Also, when feeding on the range, it is important that the supplement be spread out evenly and in such manner that the gluttonous animals do not get more than their share and the weak ones, that need help the most, are denied the benefits of the supplementary feed.

■ **Toxicity**—The symptoms of animals reacting from high-urea intake include uneasiness, muscular incoordination, bloat, prostration, convulsions, and even death.

The veterinarian should be called to treat cases of urea toxicity. As an emergency measure, 1 gal of vinegar may be administered to cattle as a drench. The acetic acid furnished by the vinegar lowers rumen pH and neutralizes ammonia, thus preventing further absorption of ammonia into the bloodstream.

■ **Palatability**—Although various opinions exist relative to the palatability of urea and urea-containing feeds, most feeders feel that urea is not palatable and, therefore, that feed consumption may be lowered in comparison with rations in which oil meal protein supplements are used entirely. For this reason, care should be exercised in selecting an appetizing urea-supplemented mixture.

In contrast to the above opinion, it should be noted that, occasionally, cattle will consume straight fertilizer urea or ammonium nitrate in sufficient amounts to poison themselves.

Sometimes cattle will consume a urea-containing feed for a few days or weeks and then refuse it. This has occurred in drought areas where farmers have tried to extend their roughage supplies by feeding straw and other mineral-poor, low-quality roughages. Appropriately increasing phosphorus and trace minerals has corrected the latter problem.

■ **Dry vs liquid high-urea supplements**—Liquid urea supplements are available in most areas; and some large feedlots are mixing them for their own use. Such supplements normally contain molasses, urea, phosphoric acid, vitamins, and trace minerals. Also, some of them contain alcohol, and other ingredients.

Liquid supplements are usually fed as a topdressing. Sometimes, they are self-fed.

Feedlot tests with beef steers have shown rather conclusively that there is no significant difference in the nutritional value or cattle response to high-urea dry or liquid supplements provided the supplement and/or ration contains the same essential nutrients in proper balance. In other words, cattle do not distinguish between the same nutrients fed in dry or liquid form. There is no nutritional advantage in liquid supplements or liquid feeds; it is just another way to balance the ration for growing or finishing cattle.

Several experiments have shown that for optimum performance with cattle, urea supplements must contain some source of unidentified factors; hence, it is recommended that high-urea dry supplements contain a ratio of 2 parts of either dehydrated alfalfa meal or distillers' dried grain solubles to 1 part of urea; and that liquid supplements contain distillers, solubles at a level of 2.5% on a dry matter basis.

In summary, there is no difference in the nutritional value of liquid and dry supplements built around urea if the supplements contain the same basic nu-

[4]For every pound of urea added to the ration of pregnant and lactating cows, 5 to 6 lb of a cereal grain or molasses should be added in order to replace the energy lost.

[5]The balance of the ingredients in range cubes or pellets usually consists of ground grain, molasses, oil meal proteins, and, under certain conditions, minerals (including trace minerals) and vitamin A supplements. Generally, the urea and molasses are first mixed, and then added to the rest of the concentrate.

trients. Thus, it is a matter of personal choice, convenience, and ingredient costs as to which is used by cattle feeders.

■ **How to compute how much urea is in a feed**—The level of urea in a feed may be noted in the following ways:

1. **Percent of urea in the feed.** When the percent of urea is given, one can calculate the amount of protein furnished by urea by multiplying the percent urea by 281 (the protein equivalent of urea). For example, if a 40% supplement contains 5% urea, then 14% protein is furnished by urea (281 × 5% = 14%). To determine the percent of the total protein furnished by urea, divide the percent of protein as urea by the percent of protein in the supplement (14 ÷ 40% = 35%). In this case, slightly more than one-third of the protein in the supplement is furnished by urea.
2. **Percent protein as urea.** When the urea in the supplement is expressed in percent protein as urea, one can determine the amount of urea by dividing this value by 281%. For example, if a 36% protein supplement has 12% protein as urea, it contains 4.3% urea (12 ÷ 281% = 4.3%). One-third of the protein in the supplement is furnished by urea (12 ÷ 36% = 33.33%).

POSTRUMINAL PROTEIN SUPPLY AND REQUIREMENTS

The supply of protein to the small intestine is the total of the fed protein that escapes the bypass ruminal destruction and the microbial protein synthesized within the rumen. The efficiency of microbial growth varies with specific culture conditions, such as pH, dilution rate, and limiting nutrients. Microbial protein synthesis is usually correlated with the amount of organic matter digested in the reticulo-rumen. Higher values are characteristic of high-roughage rations and faster bacterial growth rates.

The efficiency of converting feed protein or nitrogen to nonammonia nitrogen leaving the rumen is variable. Ruminal output as a percentage of protein intake may exceed 100%. Nitrogen recycled to the rumen can be used for protein synthesis by ruminal microbes, resulting in ruminal output of protein being greater than protein intake, especially when feeding low-protein rations or rations containing large amounts of protein which is not degraded in the rumen. On the other hand, when ruminal degradation of protein is high, or the amount of ammonia available exceeds the bacterial need, ruminal protein output will be less than the amount of protein consumed. In most experiments, protein flow out of the rumen has been between 75 and 120% of protein fed, and generally near 100% with typical diets containing 11 to 12% crude protein.

PROTEIN SOLUBILITY, DEGRADATION, AND BYPASS

The value of a protein for ruminant animals is related to how soluble it is in the rumen and how much of it will be degraded by rumen microorganisms. Some high-quality protein sources, such as fish meal, are not degraded to any great extent in the rumen. If such proteins pass into the intestinal tract and are digested rather completely there, they are able to supply the animal with a greater abundance of amino acids. There must, of course, be an adequate supply of soluble and degradable protein to nourish the rumen microorganisms. But, over and above meeting the latter requirement, for animals with a higher than normal need for some of the limiting amino acids, it would be desirable that any excess dietary protein be digested and absorbed in the small intestine. Proteins which are digestible, yet not degraded in the rumen, are commonly called *bypass* or *escape* proteins. The amount of protein which escapes destruction in the rumen and passes to the omasum and abomasum is affected by feed, bacterial, animal, and time conditions.

Current information suggests that protein from various feedstuffs may be classified into three bypass or escape categories as follows: (1) *low bypass* (under 40%)—soybean meal, peanut meal; (2) *medium bypass* (40 to 60%)—cottonseed meal, dehydrated alfalfa meal, corn, brewers' dried grains; and (3) *high bypass* (over 60%)—meat meal, corn gluten meal, blood meal, feather meal, fish meal, distillers' dried grains. The percent bypass protein in some common feeds are presented in Appendix II, Table II-3.

The solubility of proteins in feedstuffs may be slowed or reduced by (1) heat treatment; (2) formic acid, formaldehyde, or tannin treatments; and (3) utilizing nonprotein nitrogen ingredients along with a carbohydrate, such as (a) simple sugars (molasses), (b) starch (grain), or (c) cellulose (forage).

Increased bypass or escape does not ensure increased animal production, however, because (1) by-passed protein may be poorly digested in the small intestine, (2) the balance of amino acids in postruminal protein may be poor, or (3) the energy supply of nutrients other than amino acids may be limiting animal production.

Although solubility and rumen degradability of proteins appears to be an important factor in efficient nitrogen utilization by ruminants, further research is required to obtain better utilization in most practical situations.

MINERALS

Beef cattle are susceptible to the usual inefficiencies and ailments when exposed to (1) prolonged and severe mineral deficiencies, or (2) excesses of fluorine, selenium, or molybdenum (see Chapter 5, Nutritional Disorders/Toxins).

Needed minerals may be incorporated in beef cattle rations or in the water. In addition, it is recommended that all classes and ages of cattle be allowed free access to a two-compartment mineral box, with (1) salt (iodized salt in iodine-deficient areas) in one side, and (2) a suitable mineral mixture in the other side. Free-choice feeding is in the nature of cheap insurance, with the animals consuming the minerals if they are needed.

BEEF CATTLE MINERAL CHART

Table 8-5, Beef Cattle Mineral Chart, presents in summary form pertinent information relative to the mineral needs of beef cattle.

MAJOR OR MACROMINERALS

A discussion of the major or macromineral needs of beef cattle follows.

SALT (NaCl)

Salt should be available at all times. It may be fed in the form of granulated, half ground, or block salt, but because of weathering losses, flake salt is usually not satisfactory for feeding in the open. If block salt is used, the softer types should be selected.

Most ranchers compute the yearly salt requirements on the basis of about 25 lb for each cow. Mature animals will consume 3 to 5 lb of salt per month when pastures are lush and succulent, and 1 to 1½ lb per month during the balance of the season.

The careful location of the salt supply is recognized as an important adjunct in proper range management. Through judicious scattering of the salt supply and the moving of it at proper intervals, the animals can be distributed more properly; and overgrazing of certain areas can be minimized.

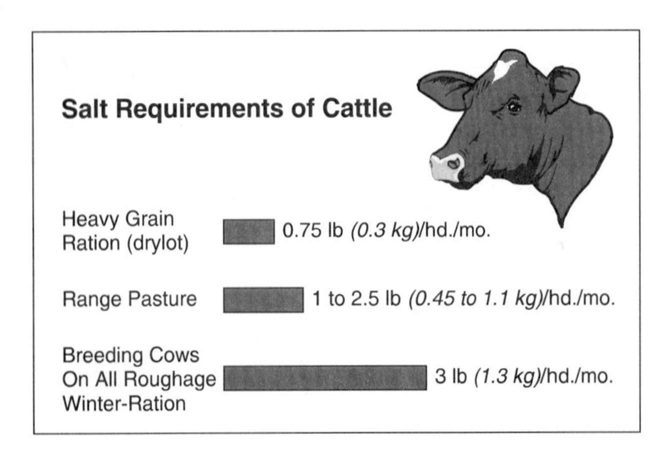

Fig. 8-7. The average salt requirements of cattle.

CALCIUM (Ca)

In contrast to phosphorus deficiency, calcium deficiency in beef cattle is relatively rare. In general, when the forage consists of at least one-third legume (legume hay, pasture, or silage), ample calcium will be provided. But even nonlegume forages contain more calcium than cereal grains. This indicates that a mineral source of calcium is less necessary when large quantities of roughage are being consumed. Also, plants grown on calcium-rich soils may contain a higher content of this element.

As finishing cattle consume a high proportion of grains to roughages—and the grains are low in calcium—they have a greater need for a calcium supplement than do beef cattle that are being fed largely on roughages. This is especially true of cattle of the younger ages and when a long feeding period is involved.

When the ration of beef cattle is suspected of being low in calcium, the animals should be given free access to a calcium supplement, with salt provided

Fig. 8-8. This heifer developed rickets early in life due to a deficiency of calcium and phosphorus. Note the bowed front legs and enlarged joints. (Courtesy, USDA)

separately; or a calcium supplement may be added to the ration in keeping with nutrient requirements.

Appendix, Section II, Table II-2 gives the composition of mineral supplements for beef cattle.

PHOSPHORUS (P)

Phosphorus deficiencies in cattle are widespread. In some sections of the United States and other countries, the soils are so deficient in phosphorus that the feeds produced thereon do not provide enough of this mineral for cattle or other classes of stock. As a result, the cattle produced in these areas may have depraved appetites, may fail to breed regularly, and may produce markedly less milk. Growth and development are slow, and the animals become emaciated and fail to reach normal adult size. Death losses are abnormally high.

In range areas where the soils are either known or suspected to be deficient in phosphorus, cattle should always be given free access to a suitable phosphorus supplement.

To be on the safe side, the general recommendation for beef cattle on both the range and in the finishing lot is to allow free choice of a suitable phosphorus supplement in a mineral box, or to add a phosphorus supplement to the ration in keeping with nutrient requirements.

When phosphorus is added to the water, either of the following methods may be employed:

1. If added by hand, add ¼ oz of monosodium phosphate per 8 gal of water, or ¼ oz per head daily.
2. If added by automatic dispenser, a stock solution of 2½ lb of monosodium phosphate per gal of water (or 100 lb to 40 gal of water) is recommended. The machine automatically proportions the stock solution to the water.

Fig. 8-9. Bone chewing by cattle is a common sign of phosphorus deficiency. (From Texas Sta. Bull. 344, courtesy of The Fertilizer Institute, Washington, DC)

Appendix Section II, Table II-2 gives the composition of mineral supplements for beef cattle.

MAGNESIUM (Mg)

Certain pastures in early spring are inadequate in magnesium, with the result that grass tetany may occur in cattle grazing on them. Lactating cows are most commonly affected. In problem areas, as high as 0.7 oz of supplemental magnesium per head daily may be required to prevent this malady.

POTASSIUM (K)

Potassium deficiencies are rare, but they occasionally occur in drylot finishing cattle fed high-concentrate rations. Forages are extremely good sources of potassium. For this reason, potassium is not generally added to feeds for cattle.

SULFUR (S)

Sulfur is a component of protein, some vitamins, and several important hormones. The common sulfur-containing amino acids are methionine, cysteine, and cystine. Also, the following amino acid derivatives contain sulfur: cystathionine, taurine, and cysteic acid. Methionine is a key amino acid, because all other sulfur compounds, except the B-vitamins thiamin and biotin, can be synthesized from methionine.

All feeds contain some sulfur, but the amount usually depends on the protein content of the feed—generally speaking, the higher the protein count, the higher the sulfur content. Availability of the sulfur in the feed to microbial reduction in the rumen may be of as much concern as the actual amount that is present.

TRACE OR MICROMINERALS

A discussion of the trace or micromineral needs of beef cattle follows.

COBALT (Co)

Deficiencies of cobalt in cattle are costly, for the affected animals become weak and emaciated and eventually die. Florida is without doubt the most serious cobalt-deficient area in the United States, but similar deficiencies of a lesser order have been observed in Michigan, Wisconsin, Massachusetts, New Hampshire, Pennsylvania, and New York. Cattle in these affected areas should have access to a cobaltized mineral mixture, made by mixing 0.2 oz of cobalt chloride, cobalt sulfate, cobalt oxide, or cobalt car-

TABLE
BEEF CATTLE

Mineral; Absorption; Excretion	Conditions Usually Prevailing Where Deficiencies Are Reported	Functions of Mineral	Deficiency Symptoms; Toxicity*
Major or Macrominerals:			
SALT (NaCl, sodium and chloride)—The requirements for sodium and chlorine are commonly expressed as salt requirements because salt is an effective, economical way of supplementing rations with these elements. **Absorption**—Sodium and chlorine are mainly absorbed from the proximal portion of the small intestine, but they may also be absorbed from the distal section of the small intestine and from the large intestine. Also, some absorption of sodium and chlorine may occur from the rumen. **Excretion**—Excess salt is excreted in the urine.	Negligence; for salt is inexpensive. Deficiencies of sodium and chlorine may occur because plants have low sodium contents, because sodium losses caused by perspiration may occur in animals maintained in warm environments or used for hard work, and because sodium needs increase during lactation and during periods of rapid growth.	Sodium (Na) functions in maintaining osmotic pressure, acid-base balance, and body-fluid balance; is involved in nerve transmission and active transport of amino acids; is required for cellular uptake of glucose through activation of the glucose carrier protein; and is a major cation of extracellular fluid and provides a majority of the alkaline reserve in plasma. Chlorine (Cl) is necessary for the activation of amylase; is essential for the formation of gastric hydrochloric acid; and is involved in respiration and regulation of blood pH, through the chloride shift.	**Deficiency symptoms**—Intensive craving of salt, manifested by the animals chewing and licking various objects, and by muscle cramps. Prolonged deficiency results in lack of appetite, unthrifty appearance, and decreased production. High-producing milk cows may collapse and die when salt deficiency has been of long duration. It is noteworthy that when salt is omitted, sodium expresses its deficiency first. **Toxicity**—*The NRC gives the maximum tolerable level of salt (NaCl) as 9% of ration dry matter. as much as 3 lb *(1.4 kg)* can be consumed per cow daily without harm provided animals have access to plenty of water.
CALCIUM (Ca)—Calcium is the most abundant mineral in the body. Most of the calcium in the body is found in the bones and teeth. It constitutes 2% of the body weight. In blood, calcium is found mostly in the plasma, with a controlled concentration of 10 mg/100 ml. **Absorption**—Calcium is absorbed actively from both the duodenum and the jejunum; but most calcium is absorbed in the proximal portion of the duodenum. **Excretion**—Calcium is excreted mainly in feces with only small quantities appearing in urine.	A calcium deficiency may occur when finishing cattle are fed heavily on concentrate and limited quantities of nonlegume roughage, especially young cattle on a long feed. Adding calcium to such a ration increases the rate of gain, improves feed utilization, results in heavier, stronger bones, and enhances market grades. Also, a calcium deficiency may occur when the ration consists chiefly of dried mature grasses or cereal straws, and when cows are in heavy lactation. Osteomalacia may occur when there are high metabolic demands on calcium and phosphorus stores, such as occur during pregnancy and lactation.	Essential for bone formation, development of teeth, production of milk, transmission of nerve impulses, maintenance of normal muscle excitability (along with sodium and potassium), regulation of heart beat, movement of muscles, blood clotting (conversion of prothrombin to thrombin), and activation and stabilization of enzymes (*i.e.*, pancreatic amylase).	**Deficiency symptoms**—A deficiency of calcium results in rickets in young animals and osteomalacia in older animals. Rickets may be caused by a deficiency of calcium, phosphorus, or vitamin D. It is characterized by improper calcification of the organic matrix of bones of young, growing animals. Thus, the bones are weak, soft, and lack density. Signs include swollen tender joints; enlargement of the ends of bones; and arched back; stiffness of the legs; and development of beads on the ribs. If the cause is not corrected, calves develop bowed and deformed legs. Also, rachitic bones are highly susceptible to fracture. Osteomalacia is the result of demineralization of the bones of adult animals. This condition is characterized by weak, brittle bones that may break when stressed. **Toxicity**—Ruminants tolerate high levels of calcium. **The NRC gives the maximum calcium level as 2% of ration dry matter. However, when high levels of calcium are fed, there may be reduced feed consumption and daily gains; reduced protein and energy digestibilities; reduced absorption of tetracyclines, manganese, and zinc; and stimulation of the production of calcitonin of the thyroids. Calcitonin inhibits bone resorption, with the result that the bones may thicken (osteopetrosis) because of continued deposition but limited resorption.

8-5
MINERAL CHART

Nutrient Requirements[1]				
Daily Nutrients/ Animal	**Percentage of Ration**	**Recommended Allowances[1]**	**Practical Sources of the Mineral**	**Comments**
For young, growing animals: 2–3 g of sodium, and less than 5 g of chlorine. For lactating cows: 11 g of sodium, and 15 g of chlorine.	*Sodium concentrations of 0.06–0.10% of ration dry matter for nonlactating yearlings and calves, and no more than 0.1% dry matter for lactating beef cows.	Cows on pasture or fed high-roughage winter rations will consume from 1–3 lb *(0.45–1.36 kg)* salt per head per month; finishing steers fed heavy grain rations in drylot will consume 1–3.5 lb *(0.45–1.59 kg)* per head per month; a wide range due to differences in age, rations, form of salt (rock, coarse ground, or block). Most ranchers compute the yearly salt requirements on the basis of 25 lb *(10 kg)* per cow. The careful location of the salt supply is an important adjunct in range management.	Salt should be available at all times. It should be both (1) self-fed, free-choice, and (2) mixed with other ration ingredients. Free access to salt in the form of loose rock, coarse ground, or block salt. Cattle prefer loose salt to block salt, because it can be eaten more rapidly and with less effort. However, experiments with lactating cows have shown fully as good results with block salt as with loose salt even though smaller quantities were consumed. This means that the additional intake of loose salt over block salt does not appear to benefit cattle. Commercial mineral mixes (in block, or loose form) may contain ⅓ or more salt.	The salt requirements of cattle differ (1) between individuals, (2) according to whether milk is produced (being higher for lactating cows than for dry cows, because of the salt in the milk), (3) from season to season, (4) between block and loose salt (animals often consuming twice as much easy-to-get loose salt as block salt), and (5) according to the salt content of the soil, feed, and water (being higher when vegetable proteins are fed, higher on predominantly forage rations than on predominantly concentrate rations, and higher on lush early pasture than on more mature grasses). These are some of the reasons why free-choice feeding of salt is advocated.
*Variable, according to age, weight, and type and level of production of cattle. Because true digestibilities of calcium in feedstuffs vary, the dietary calcium requirements shown in the tables may in some instances need to be adjusted.	*Variable, according to age, weight, and type and level of production of cattle. Because true digestibilities of calcium in feedstuffs vary, the dietary calcium requirements shown in the tables may in some instances need to be adjusted.	Free access to a calcium supplement, or a calcium supplement incorporated in the ration.	Legumes are high in calcium. Also, several of the oilseed meals are good sources of calcium. Sources of supplemental calcium include calcium carbonate, ground limestone, bone meal, dicalcium phosphate, defluorinated phosphate, monocalcium phosphate, and calcium sulfate. Where both calcium and phosphorus need to be supplemented, they should be provided in a readily available and palatable form such as dicalcium phosphate, defluorinated phosphate, or bone meal.	In addition to an adequate supply of calcium, proper utilization is dependent upon (1) a highly available source of the mineral, (2) a suitable ration between calcium and phosphorus (somewhere between 1 and 2 parts of calcium to 1 part of phosphorus). Calcium-phosphorus ratios of 2:1 have been shown to be beneficial in reducing urinary calculi. When calculi problems are encountered, even higher levels of calcium may be advisable. Ratios between calcium and phosphorus of 7:1 have been reported to be satisfactory for cattle. Generally, when cattle receive a ration with at least ⅓ of a legume forage, ample calcium will be provided. But even nonlegume forages contain more calcium than cereal grains. Plants grown on calcium-rich soils are high in calcium. Calcium availability of 70% is generally assumed for all feedstuffs.

(Continued)

TABLE 8-5

Mineral; Absorption; Excretion	Conditions Usually Prevailing Where Deficiencies Are Reported	Functions of Mineral	Deficiency Symptoms; Toxicity*
PHOSPHORUS (P)—Phosphorus has varied, but extremely important, biochemical and physiological roles. **Absorption**—Phosphorus absorption is dependent on source, intestinal pH, age of animal, and ration levels of sodium, calcium, iron, aluminum, manganese, potassium, magnesium, and fat. **Excretion**—Excess phosphorus is excreted primarily in the feces.	Semiarid regions are commonly associated with soils deficient in phosphorus. The phosphorus content of plants generally decreases markedly with maturity, with the result that deficiencies often occur in cattle subsisting for long periods on mature dried forage. High iron levels result in the formation of insoluble iron phosphate. Also, aluminum forms insoluble, unavailable phosphates.	Phosphorus is deposited in bones. It is also found in high concentrations in brain, muscle, liver, spleen, and kidneys. Phosphorus, as a component of phospholipids, influences cell permeability and is a component of myelin sheathing of nerves. Also, many energy transfers in cells involve the high-energy phosphate bonds in ATP. Phosphorus plays an important role in blood buffer systems. Activation of several B-vitamins (thiamin, niacin, pyridoxine, riboflavin, biotin, and pantothenic acid) to form coenzymes requires their initial phosphorylation. Phosphorus is also a part of the genetic materials DNA and RNA.	**Deficiency symptoms**—Phosphorus deficiencies in cattle are widespread. A deficiency of phosphorus results in decreased growth rates, in inefficient feed utilization, and in a depraved appetite (chewing of wood, soil, and bones—called pica); anestrus, low conception rate, and reduced milk production; low plasma phosphorus levels, and weak, fragile bones and stiffness of joints. **Toxicity**—*The NRC gives the maximum phosphorus level as 1% of ration dry matter. High phosphorus intakes may cause bone resorption, elevated plasma phosphorus levels, and urinary calculi.
MAGNESIUM (Mg)—Magnesium is the fourth most abundant cation in the body. **Absorption**—Absorption of magnesium occurs prior to the intestines, from the small intestine, and some from the large intestine. **Excretion**—Excretion of endogenous magnesium is primarily via feces. However, excess magnesium is disposed of primarily via urine.	When milk feeding of calves is prolonged without grain or hay. (Milk is rather low in magnesium.) When there is grass tetany, which is most likely to occur when beef cows in early lactation graze early spring pastures containing less than 0.2% magnesium.	Magnesium is required for skeletal development as a constituent of bone; plays an important role in neuromuscular transmission and activity; is required to activate many enzyme systems, including those involving ATP; and is required as a cofactor in decarboxylation and an activator of many peptidases. Approximately 65% of total body magnesium is contained in bone, the other 35% is distributed among various tissues and organs.	**Deficiency symptoms**—Grass tetany or grass staggers, characterized by anorexia, hyperemia, hyperirritability, convulsions, and death. **Magnesium**-deficient cattle exhibit loss of appetite and reduced dry matter digestibilities. Deficiencies in young cattle may result in defective bones and teeth. **Toxicity**—Normal rations will not cause toxicity. The maximum tolerable level of magnesium is considered to be 0.4% of the ration. Toxicity is characterized by loss of appetite, reduced performance, and occasional diarrhea. Also, cattle experiencing toxicity may exhibit lack of reflexes and respiration depression.
POTASSIUM (K)—Potassium is the third most abundant mineral element in the body. **Absorption**—Potassium is primarily absorbed in the small intestine. **Excretion**—Excretion is mainly via the kidneys.	When drylot finishing cattle receive high- or all-concentrate rations.	Essential for proper enzyme, muscle, and nerve function, rumen microorganism activity, and appetite.	**Deficiency symptoms**—Poor appetite and feed conversion, slow growth, stiffness, and emaciation. **Toxicity**—*The NRC gives the maximum tolerable level of potassium as 3% of ration dry matter. Toxicity from excessive intake is unlikely except (1) when water intake is restricted or water is saline, or (2) when the kidneys are not functioning properly.
SULFUR (S)—Sulfur is a component of protein, some vitamins, and several important hormones.	Cattle fed high-grain rations supplemented with nonprotein nitrogen.	Body functions that involve sulfur include protein synthesis and metabolism, fat and carbohydrate metabolism, blood clotting, endocrine function, and intra- and extracellular fluid acid-base balance. Sulfur has both structural and metabolic functions; it is found in virtually every tissue and organ of the body. Muscle has a fairly constant nitrogen to sulfur ratio of 15.3:1. The total body content of sulfur is approximately 0.15%.	**Deficiency symptoms**—Depressed appetite, loss of weight, weakness, excessive salivation, watery eyes, dullness, emaciation, and death. A lack of sulfur also results in a microbial population that does not utilize lactate. **Toxicity**—*The NRC gives the toxic level of sulfur as 0.40% of the ration dry matter. Sulfur toxicity is characterized by restlessness, diarrhea, muscular twitching, dyspnea, and in prolonged cases of inactivity followed by death.

(Continued)

Nutrient Requirements[1]				
Daily Nutrients/ Animal	**Percentage of Ration**	**Recommended Allowances[1]**	**Practical Sources of the Mineral**	**Comments**
*Variable, according to age, weight, and type and level of production.	*Variable, according to age, weight, and type and level of production.	Free access to a phosphorus supplement, or a phosphorus supplement added to the daily ration in keeping with the nutrient requirements. Where phosphorus is added to water, either of the following methods may be employed: 1. Added by hand at rate of ¼ [illegible] oz [illegible] *(7 g)* of monosodium phosphate/8 gal *(30 liter)* water, or ¼ oz/head/ day. 2. Added by dispenser, using stock solution of 2½ lb *(1.13 kg)* of monosodium phosphate/gal *(3.8 liter)* water (or 100 lb/40 gal *[45 kg/151 liter]* water).	Common sources of phosphorus are: dicalcium phosphate, defluorinated phosphate, bone meal, soft phosphate, sodium phosphate, ammonium polyphosphate, orthophosphates, metaphosphates, pyrophosphates, and tripolyphosphate. Oilseed meals and animal and fish products contain large amounts of phosphorus. Phytate phosphorus is not well utilized by nonruminants, but ruminants appear to use considerable quantities of this form of phosphorus.	Grains, grain by-products and high-protein supplements are fairly high in phosphorus; hence, rations high in such ingredients require little or no phosphorus supplementation. Calcium-phosphorus ratios of 2:1 are beneficial in reducing urinary calculi; and even higher levels of calcium may be necessary when urinary calculi is encountered. Ratios between calcium and phosphorus of 7:1 have been reported to be satisfactory for cattle.
*Young calves and growing-finishing cattle, 12–30 mg/kg body weight. *Beef cows, 7–9 g/ day during gestation and 18–21 g/day during lactation.	*Growing and finishing cattle, 0.10% of dry matter; gestating cows, 0.12% of dry matter; and lactating cows, 0.20% of dry matter.		Commonly used feedstuffs vary widely in magnesium content and availability. Magnesium carbonate, oxide, and sulfate are good sources of supplemental magnesium.	Supplemental feeding of magnesium (20 g/day) reduces the incidence of grass tetany in many outbreaks. Magnesium requirements are increased by feeding high levels of aluminum, potassium, phosphorus, or calcium; by younger cattle and magnesium-deficient cattle; and by high levels of milk production.
	*0.60% of the total ration dry matter. The needs for potassium vary with amounts of protein, phosphorus, calcium, and sodium consumed.	*The NRC suggested maximum level of magnesium in the ration is 0.40%.	Roughages usually contain ample potassium. Potassium chloride is the supplement of choice.	Grains often contain less than 0.5% potassium. Excessive levels of potassium have been found to interfere with magnesium absorption. Also, excessive levels of potassium, along with high levels of phosphorus, increase the incidence of phosphatic urinary calculi.
	*0.15% of ration dry matter.	*The NRC suggested maximum level of sulfur in the ration is 0.40%.	Feeds high in protein are usually high in sulfur. The microbial population of the rumen has the ability to convert inorganic sulfur into organic sulfur compounds that can be used by the animal. So, either organic or inorganic sulfur can be utilized by cattle. Most feedstuffs provided to beef cattle contain sufficient sulfur to meet their needs.	Copper requirements are increased by both sulfur and molybdenum. Selenium can replace sulfur in some organic compounds.

(Continued)

TABLE 8-5

Mineral; Absorption; Excretion	Conditions Usually Prevailing Where Deficiencies Are Reported	Functions of Mineral	Deficiency Symptoms; Toxicity*
Trace or Microminerals:			
COBALT (Co)—The cobalt requirement of cattle is actually a cobalt requirement of rumen microorganisms. The microbes incorporate cobalt into vitamin B-12, which is utilized by both microorganisms and animal tissues. **Absorption**—Vitamin B-12 is absorbed in the lower part of the small intestine. **Excretion**—Cobalt and vitamin B-12 are mainly excreted in the feces, although variable amounts are excreted in urine.	In cobalt-deficient soils where this element is not provided. Cobalt-deficient soils occur in many parts of the world, with large deficient areas in Australia, New Zealand, and along the southeast Atlantic Coast of the U.S.	The main function of cobalt is to serve as an integral part of vitamin B-12 (cobalamin). Vitamin B-12 is of importance in the metabolism of propionic acid, needed for the activity of the enzyme methylamalonyl-CoA isomerase. Vitamin B-12 is also a part of the enzyme that catalyzes the recycling of methionine from homocysteine after the loss of its labile methyl group. Vitamin B-12 is also needed for normal liver folate metabolism.	**Deficiency symptoms**—Loss of appetite and body weight, muscular wasting, severe anemia, followed by death. In severe deficiency, the mucous membranes become blanched, the skin turns pale, a fatty liver develops, and the body becomes almost totally devoid of fat. **Toxicity**—Cobalt toxicity is rare because toxic levels are about 300 times requirement levels. *The NRC gives the maximum tolerable level of cobalt as 10 ppm of the ration dry matter.
COPPER (Cu) **Absorption**—Copper is absorbed from the upper portion of the duodenum. Zinc and silver are antagonistic to copper absorption. **Excretion**—Copper is released into bile, thence into feces. Trace amounts of copper are excreted in urine, perspiration, and milk.	In copper-deficient areas (soils), as in Florida and the Coastal Plain region. On peat and muck soils, or where soil molybdenum levels are high. Deficiencies have occurred in calves kept on an exclusive milk diet for long periods.	Copper is necessary in hemoglobin formation, iron absorption from the small intestine, iron mobilization from tissue stores, and for the oxidation of iron, permitting it to bind with the iron transport—transferrin. Copper is essential in enzyme systems, hair development and pigmentation, bone development, reproduction, and lactation.	**Deficiency symptoms**—Emaciation, depigmentation (cattle turn yellowish) and loss of hair, stunted growth, anemia, and brittle and malformed bones. Also, heat periods are suppressed, and there may be depraved appetite and diarrhea. Young calves may have straight pasterns and stand forward on their toes. Low copper intake reduces the synthesis and activity of the copper-containing enzyme, tyrosinase, which is required for pigmentation of hair, wool, and feathers. **Toxicity**—*Maximum tolerable levels for cattle are 100 ppm of dry matter. Acute toxicity may cause nausea, vomiting, salivation, abdominal pain, convulsions, paralysis, collapse, and death. Also, high copper levels may predispose animals to anemia, muscular dystrophy, decreased growth, and impaired reproduction.
IODINE(I) **Absorption**—In ruminants, the rumen is the primary absorption site. **Excretion**—Two-thirds of ingested inorganic iodine is excreted by the kidneys.	In iodine-deficient areas (soils) where iodized salt is not fed (in northwestern U.S. and in the Great Lakes Region). Where feeds come from iodine-deficient areas. Substances that interfere with iodine metabolism. Rapeseed meal, soybean meal, and cottonseed meal have goitrogenic effects.	Inorganic iodine is taken up by the thyroid gland for the synthesis of thyroid hormones. Thyroid hormones have an active role in thermoregulation, intermediary metabolism, reproduction, growth and development, circulation, and muscle function.	**Deficiency symptoms**—Goiter, hairlessness in the young; retarded growth and maturity, lowered metabolic rate, and increased water retention. Occasional borderline cases may survive; in these, the moderate thyroid enlargement disappears in a few weeks. **Toxicity**—*50 ppm is the maximum tolerable level for calves. Symptoms of iodine toxicity include loss of appetite, coma, and death.
IRON (Fe) **Absorption**—Iron may be absorbed from all sections of the small intestine, but the principal site of absorption is the duodenum. Ferrous iron is absorbed to a much greater extent than ferric iron. **Excretion**—Excretion of iron occurs in urine, feces, sweat, dermis, and blood.	Calves on an exclusive milk ration (milk contains less than 10 ppm iron). Animals with excessive blood loss.	Iron has important biochemical functions in animals since it is a component of hemoglobin, myoglobin, cytochrome, and the enzymes catylase and peroxidase. Iron in these materials exists in porphyrin rings. Iron is involved in the transport of oxygen to cells and in cellular respiration.	**Deficiency symptoms**—Signs of lack of iron include anemia, reduced saturation of transferrin, listlessness, pale mucous membrane, reduced appetite and weight gain, and atrophy of the papillae of the tongue. **Toxicity**—*An iron level of 1,000 ppm is considered as the maximum tolerable level for cattle. Iron toxicity is characterized by reduced feed intake, reduced daily gain, diarrhea, hypothermia, and metabolic acidosis.

(Continued)

Nutrient Requirements[1]				
Daily Nutrients/ Animal	**Percentage of Ration**	**Recommended Allowances[1]**	**Practical Sources of the Mineral**	**Comments**
	*0.10 ppm of ration dry matter.	Free access to a cobaltized mineral mixture in cobalt-deficient areas; or administering a cobalt pellet.	A cobaltized mineral mixture may be prepared by adding cobalt at the rate of 0.2 oz/100 lb *(1.25 mg/kg)* of salt as cobalt chloride or cobalt sulfate, cobalt carbonate, cobalt oxide, or a good commercial mineral mixture or salt product may be used. Also, cobalt sulfate and cobalt oxide are effective as a drench; and a cobalt pellet (composed of cobalt oxide and finely divided iron) that lodges in the reticulum is an effective preventive.	Several good commercial cobalt-containing minerals are on the market. A vitamin B-12 injection will relieve a cobalt deficiency.
	*10 ppm of ration dry matter. For presence of high levels of molybdenum and inorganic sulfate, increase the copper requirements.	*Copper deficiency can be prevented by adding 0.25–0.5% copper sulfate to salt fed free-choice. *Copper (Cu) added to total feed (dry basis) 4 ppm. Copper may also be injected as glycinate to meet the nutritional needs for the mineral.	*Salt containing 0.25–0.5% copper sulfate.	Copper deficient cattle can be returned to normal by feeding 3 g of copper sulfate or blue vitriol every 10 days. An interesting interrelation exists between copper and molybdenum. An excess of molybdenum (in the presence of sulfate) causes a condition which can be cured only by administering copper. Excess copper is toxic; it accumulates in the liver, and death may result.
*1 mg/day for a 1,100-lb *(500-kg)* cow.	*0.5 ppm of ration dry matter.	Free access to stabilized iodized salt containing 0.01% potassium iodide (0.0076% iodine).	Stabilized iodized salt containing 0.01% potassium iodide. Feed additives that supply iodine are: ethylenediamine dihydroiodide (EDDI), calcium iodate, cuprous iodide, potassium iodate, sodium iodate, potassium iodide, sodium iodide, and pentacalcium periodate.	The enlargement of the thyroid gland (goiter) is nature's way of trying to make enough thyroxin, when there is insufficient iodine in the feed. Eighty percent of hormonal iodine stored in the thyroid is thyroxin. The amount of iodine in milk is influenced by iodine intake, season, level of milk production, and use of iodine disinfectants.
	*50 ppm of ration dry matter.		Levels of iron in common feed believed to be ample. Sources of supplemental iron in decreasing order of availability are: ferrous sulfate, ferrous carbonate, ferric chloride, and ferric oxide.	After calves are past 20 weeks of age, iron does not seem to be beneficial. About 30% of all calves are affected by prenatal iron deficiency. In cattle, a majority of body iron is in the form of hemoglobin, with lesser amounts existing as protein-bound stored iron, myoglobin, and cytochrome.

(Continued)

TABLE 8-5

Mineral; Absorption; Excretion	Conditions Usually Prevailing Where Deficiencies Are Reported	Functions of Mineral	Deficiency Symptoms; Toxicity*
MANGANESE (Mn) **Absorption**—Ruminants regulate manganese levels in blood and tissue via intestinal absorption. **Excretion**—Manganese is excreted via feces, with little in the urine.	In northwestern U.S. All-concentrate rations based on corn supplemented with nonprotein nitrogen.	Manganese is essential for normal reproduction in both males and females, for bone formation, and for the functioning of the central nervous system. Also, manganese is a preferred metal cofactor for many enzymes involved in carbohydrate metabolism and in mucopolysaccharide synthesis.	**Deficiency symptoms**—*In males:* impaired spermatogenesis, testicular and epididymal degeneration, sex hormone inadequacy, and sterility. *In females:* irregular and absent estrus, delayed conception, abortion, and deformed young at birth—crooked calves. **Toxicity**—For ruminants, manganese is among the least toxic of required minerals. *With balanced rations, about 1,000 ppm is the maximum tolerable level on a short-term basis for cattle.
MOLYBDENUM (Mo)—Molybdenum is found in nearly all body cells and fluids. **Absorption**—Molybdenum is well absorbed by cattle, chiefly from the small intestine. **Excretion**—Excretion of molybdenum is primarily via urine, with small amounts excreted in bile and milk.	Molybdenum toxicity occurs only occasionally in cattle and appears to be an area problem.	Molybdenum is a constituent of the enzymes xathine oxidase, aldehydo oxidase, and sulfide oxidase, enzymes involved in the oxidation of purines and reduction of cytochrome C.	**Deficiency symptoms**—Molybdemum deficiencies have not been demonstrated in cattle. **Toxicity**—*The NRC gives the maximum tolerable level as 5 ppm. Clinical signs of molybdenum toxicity in cattle are diarrhea, loss of appetite, anemia, ataxia, and bone malformation.
SELENIUM (Se)—Initially, interest in selenium was confined to the problem of toxicity in animals. **Absorption**—Most selenium is absorbed in the duodenum. **Excretion**—Selenium is excreted in the feces and urine; fecal excretion is greater than urinary excretion in ruminants.	Low selenium forage and low vitamin E. It is an area problem, but it occurs in many parts of the U.S.	Selenium functions (1) as a component of glutathione peroxidase, an enzyme that destroys peroxides in tissues, and (2) intertwined with vitamin E in a mutual sparing effect.	**Deficiency symptoms**—White muscle disease; characterized by white muscle, heart failure, and paralysis evidenced by lameness or inability to stand. Depression of glutathione peroxidase in tissues of selenium-deficient animals may account for many of the manifestations of selenium deficiency. **Toxicity**—The NRC suggests that 2 ppm of ration dry matter is the maximum tolerable level for all species. Signs of toxicity include loss of appetite, loss of tail hair, sloughing of hoofs, and eventual death. Two types of selenium poisoning have been observed: (1) acute, blind staggers; and (2) chronic, alkali disease. Selenium toxicity can be counteracted by feeding some forms of sulfur. Toxic levels have been reported in South Dakota, North Dakota, Montana, Wyoming, Utah, Nebraska, Kansas, and Colorado.
ZINC (Zn) **Absorption**—Absorption of zinc occurs primarily from the abomasum and lower small intestine. **Excretion**—The primary route of excretion is the feces.	Zinc deficiencies have been reported in ruminants grazing forages low in zinc or high in compounds interfering with zinc utilization.	Zinc functions as both an activator and a constituent of several dehydrogenases, peptidases, and phosphates that are involved in nucleic acid metabolism, protein synthesis, and carbohydrate metabolism.	**Deficiency symptoms**—Deficiencies are characterized by decreased performance and listlessness, followed by development of swollen feet and a dermatitis that is most severe on the neck, head, and legs. Deficiencies may also result in vision impairment, excessive salivation, decreased rumen volatile fatty acid production, failure of wounds to heal normally, and impaired reproductive performance in both bulls and cows. **Toxicity**—The maximum zinc tolerance level is dependent on the ration, particularly concentrations of minerals that affect zinc absorption and utilization. *The NRC lists the maximum tolerable level of zinc as 500 ppm.

[1]As used herein, the distinction between *nutrient requirements* and *recommended allowances* is as follows: In nutrient requirements, no margins of safety are included intentionally; whereas in recommended allowances, margins of safety are provided to compensate for variations in feed composition, environment, and possible losses during storage or processing.

*Where preceded by an asterisk, the toxicity levels, nutrient requirements, and recommended allowances listed herein were taken from *Nutrient Requirements of Beef Cattle*, seventh revised edition, National Research Council-National Academy of Sciences, Washington, DC, 1996.

**Where preceded by two asterisks, the toxicity levels, nutrient requirements, and recommended allowances listed herein were taken from *Nutrient Requirements of Beef Cattle*, sixth revised edition, National Research Council-National Academy of Sciences, Washington, DC, 1984.

(Continued)

Nutrient Requirements[1]				
Daily Nutrients/ Animal	**Percentage of Ration**	**Recommended Allowances[1]**	**Practical Sources of the Mineral**	**Comments**
	*40 ppm of ration dry matter for mature cows and bulls and 20 ppm of ration dry matter for growing-finishing cattle. *Note well:* Requirements for manganese are increased by elevated dietary levels of calcium and phosphorus.	An intake of 40 ppm for mature breeding cattle and 20 ppm for growing-finishing cattle.	Most forages contain high levels of manganese. Manganous oxide, sulfate, and carbonate are good sources of supplemental manganese.	The manganese levels in pastures, grains, and forages are variable because of variations in plant species, soil types, soil pH, and fertilization practices. A deficiency of manganese exists in northwestern U.S., where it has been shown to cause *crooked calves.*
Requirements for molybdenum are not established. Because copper and sulfate alter molybdenum metabolism, arriving at the molybdenum requirement is impossible.	Requirements for molybdenum are not established. Because copper and sulfate alter molybdenum metabolism, arriving at the molybdenum requirement is impossible.	As a feed additive, molybdenum is not cleared by the Food and Drug Administration.	Many feeds contain 6.8–13.6 mg/lb of ration dry matter.	Excess molybdenum may cause a copper deficiency. Sulfur, in the absence of molybdenum, also may cause a copper deficiency. Increasing copper level in ration to 1 g/head daily is effective in overcoming molybdenum toxicity in beef cattle.
	The selenium requirement of beef cattle depends on the amount of vitamin E in the ration, but ranges are suggested as follows by the NRC: *0.10 ppm of ration dry matter.	*An intake of 0.10 ppm of ration dry matter.	In 1979, the FDA approved the addition of selenium as either sodium selenite or sodium selenate at the rate of 0.1 ppm complete feed for beef cattle, dairy cattle, and sheep. In 1987, FDA increased the allowance of selenium in complete feeds for cattle (beef and dairy), sheep, swine, chickens, turkeys, and ducks from 0.1 ppm to 0.3 ppm. In 1989, FDA approved the use of the selenium bolus for beef and dairy cattle at a level of 3 mg of selenium per day.	Selenium toxicity may occur when cattle consume feeds containing 10–30 ppm of selenium on a dry matter basis for an extended period. In Israel, in a series of experiments extending over 3 years, low doses of selenium injected intramuscularly reduced the incidence of retained placenta to half that of the controls. (Eger, S., *et al.*, "Effect of Selenium and Vitamin E on the Incidence of Retained Placenta," *Journal of Dairy Science*, Vol. 68, No. 8, Aug. 1985, p. 219.)
	*30 ppm of ration dry matter. Beef cows with high levels of milk production have higher requirements, because milk contains 300–500 mg (300–500 ppm) of zinc per liter.	*An intake of 30 ppm of ration dry matter.	Feedstuffs vary widely in zinc concentrations, with legumes usually having higher concentrations than grasses, and with protein supplements of animal origin being higher than other protein supplements.	Mild zinc deficiency in feedlot cattle results in lowered weight gains without the development of a specific syndrome. Requirements vary according to age and growth rate, since zinc absorption decreases with age and as growth rate decreases. Requirements may be altered by dietary levels of cadmium, calcium, iron, magnesium, manganese, molybdenum, and selenium, since these minerals affect zinc absorption and/or utilization.

Mineral recommendations for all classes and ages of cattle: Provide free access to a two-compartment mineral box, with (1) salt (iodized salt in iodine-deficient areas) in one side and (2) dicalcium phosphate, defluorinated phosphate, or a mixture of ⅓ salt (salt added for purposes of palatability) and ⅔ steamed bone meal in the other side. Also, the mineral requirements may be met by using a good commercial mineral, in either block or loose form. If desired, the mineral supplement may be incorporated in the ration in keeping with the recommended allowances given in this table.

bonate per 100 lb of either (1) salt, or (2) mineral mix.

In other areas of the world, cobalt deficiency is known as *Denmark disease, coastal disease, enzootic marasmus, bush sickness, salt sickness, nakuritis,* and *pining disease.*

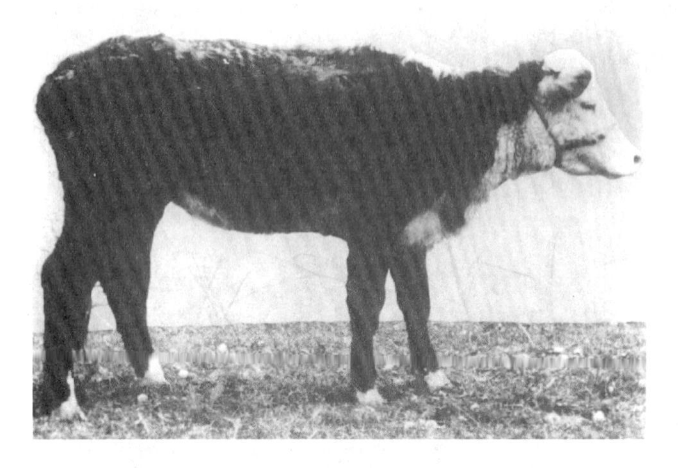

Fig. 8-10. Cobalt deficiency.

The upper picture shows a heifer suffering from cobalt deficiency. Anemia, loss of appetite, and roughness of hair coat characterize the malady.

The lower picture illustrates the remarkable recovery in the same animal brought about by the administration of cobalt.

(Courtesy, Michigan State University)

COPPER (Cu)

Copper is sometimes deficient in the soils of certain areas, notably in the state of Florida. In such areas, 0.25 to 0.5% of copper sulfate or copper oxide should be incorporated in the salt or mineral mixture. In addition to being an area disease, copper deficiencies have occurred in beef calves kept on nurse cows for periods extending beyond normal weaning age. Also, high levels of molybdenum in forage may interfere with metabolism of copper and cause copper deficiency.

Fig. 8-11. Copper deficiency in calf. Note rough coat and bleaching of hair. (Courtesy, University of Florida, Gainesville)

IODINE (I)

Iodized salt should always be fed to cattle in iodine-deficient areas (such as the northwestern United States and the Great Lakes region). This can be easily and cheaply accomplished by providing stabilized iodized salt containing 0.01% potassium iodide (0.0076% iodine). Under some conditions, organic iodine appears to be an effective aid in the prevention and treatment of foot rot and lumpy jaw (soft tissues) in cattle.

IRON (Fe)

Iron has important functions in the transport of oxygen to cells and in cellular respiration. With the exception of milk, the level of iron in common feeds is believed to be ample. Milk contains less than 10 ppm of iron, with the result that calves fed an exclusive milk diet are apt to show deficiencies. Signs of lack of iron include anemia, loss of appetite, reduced weight gains, listlessness, pale mucous membranes, and atrophy of the papillae of the tongue. Several supplemental sources of iron are available, including ferrous sulfate, ferrous carbonate, and ferric chloride.

MANGANESE (Mn)

A deficiency of manganese exists in some areas of the northwestern United States, where it has been shown to be one cause of *crooked calves*—calves born with enlarged joints, stiffness, twisted legs, overknuckling, and weak and shortened bones.

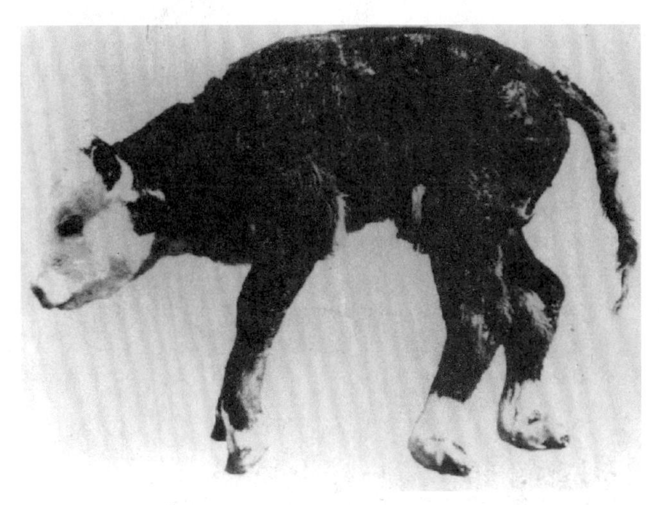

Fig. 8-12. Manganese deficiency in a newborn calf. Note weak legs and over-knuckling. (Courtesy, Washington State University)

MOLYBDENUM (Mo)

Molybdenum deficiencies have not been demonstrated in mammals. The greatest concern about molybdenum is its toxicity, which has been observed in areas where pastures are grown on high-molybdenum soils—known as *teart* pastures in England, Canada, and the United States. As noted earlier, excess molybdenum interferes with copper metabolism.

SELENIUM (Se)

Cows grazing low-selenium pastures may be affected as follows, in comparison with cows grazing similar pastures supplemented with selenium: (1) have a higher incidence of retained placenta, (2) have higher calf death losses, (3) produce more calves with nutritional muscular dystrophy (white muscle disease), and (4) wean off lighter weight calves. Also, it has been shown that the performance of feedlot cattle fed a selenium-deficient ration is improved by providing selenium.

Fig. 8-13. Selenium toxicity in cow grazing on forage produced on alkali soil containing excessive selenium. Note emaciated condition, curvature of back, and deformed hoofs. (Courtesy, Wyoming Ag. Exp. Sta., Laramie, WY)

ZINC (Zn)

Added zinc intake has been shown to increase the rate and efficiency of gains by feedlot cattle in certain areas. This may be due to correcting a deficiency of zinc, or it may be due to the relationship between (1) phytic acid and zinc, and (2) calcium and zinc, improper ratios of which may create a need for supplemental zinc.

COMMERCIAL MINERAL MIXTURES

Commercial mineral mixtures are just what the name implies—minerals mixed by manufacturers who specialize in the commercial mineral business, either handling minerals alone or a combination feed and mineral business. Most commercial minerals are very good.

The commercial mineral manufacturer has the distinct advantages over farm- or ranch-mixing of (1) purchase of minerals in quantity lots, thereby obtaining price advantages, (2) economical and controlled mixing, (3) the hiring of scientifically trained personnel for use in determining the formulations, and (4) quality control. Additionally, most farmers and ranchers do not have the equipment with which to mix minerals properly. Besides, mineral mixes have become more complicated with recognition of the increasing importance of trace elements and interrelationships. For these reasons, commercial minerals are finding a place of increasing importance in all livestock feeding.

Good mineral mixtures supply only the specific minerals that are deficient, and in the quantities necessary. Excesses and mineral imbalances are avoided. Thus, the value of any mineral mixture can easily be determined by how well it meets the needs.

HOW TO SELECT AND BUY COMMERCIAL MINERAL MIXES

Informed cattle producers will know what constitutes the best commercial mineral mix for their needs, and how to determine the best buy. Here are the factors to consider when buying a commercial mineral:

1. **The reputation of the manufacturer.** This can be determined by (a) checking on who is back of it, (b) conferring with other cattle producers who have used the particular product, and (c) checking on whether the product under consideration has consis-

tently met its guarantees. The latter can be determined by reading the bulletins or reports published by the respective state departments in charge of enforcing feed laws.

2. **Determining your needs.** The mineral requirements of cattle are much the same everywhere, although it is recognized that age, pregnancy, and lactation make for differences in mineral needs within a given herd. Additionally, there are some area differences. For example, the northern Great Plains and the Southwest are generally recognized as phosphorus-deficient areas—their grasses and hays are usually low in phosphorus. Accordingly, a high phosphorus mineral is needed for such areas—one containing 10 to 15% phosphorus. Also, unless there is a concomitant deficiency of calcium (which is not likely), the calcium-phosphorus ratio of a mineral should not be wider than 2:1.

The minimum daily phosphorus need for 533-kg Angus cows four months after calving, established by the National Research Council, is 21 grams/day or about 0.75 oz per head. In phosphorus-deficient areas, these minimum recommendations should be exceeded; thus, the daily recommended phosphorus allowances will approximate 23 grams/day, or 0.8 oz per head. However, under most conditions it is not necessary to supply more than one-third of the daily phosphorus need in supplemental form because cattle will get the rest of it from available feeds.

3. **Choose method of supplying minerals.** The daily phosphorus allowance recommended above, 0.8 oz/head/day, can be met in any of the following ways:

a. **Self-feeding a 10% phosphorus supplement.** If the monthly consumption of such a supplement is 3 to 4 lb per head, the average daily intake of phosphorus will equal one-third of the requirement. Actually, in phosphorus-deficient areas, the use of a 10 to 15% phosphorus mineral for cows is in the nature of good insurance; thereby affording protection during droughts and other periods of low feed consumption.

b. **Feeding a high phosphorus range supplement.** A daily consumption of 1 to 2 lb of a supplement containing 1 to 1.5% phosphorus will meet the needs.

c. **Adding phosphorus to the drinking water.** If cattle drink an average of 10 gal of phosphorus-treated water per head daily, adding ½ lb of monosodium phosphate (22% phosphorus) per 100 gal of water will meet the requirements.

4. **What's on the tag?** Cattle producers should study and be able to interpret what's on the tag. Does it contain what you need?

5. **Determine the best buy.** When buying a mineral, the cattle producers should check price against value received. For example, let's assume that the main need is for phosphorus and that we wish to compare two minerals, which we shall call brands "X" and "Y." Brand X contains 12% phosphorus and sells at $340.00 per ton or $17.00/cwt; whereas brand Y contains 10% phosphorus and sells at $320.00 per ton or $16.00/cwt. Which is the better buy?

COMPARATIVE VALUE OF BRANDS "X" AND "Y"
(Based on Phosphorus Content Alone)

Brand	Phosphorus	Price/Cwt	Cost/Lb Phosphorus
	(%)	*($)*	*($)*
X	12	17.00	1.41
Y	10	16.00	1.60

Hence, brand X is the better buy, even though it costs $1.00 more per hundred, or $20.00 more per ton.

One other thing is important. As a usual thing, the more scientifically formulated mineral mixes will have plus values in terms of (a) trace mineral (needs and balance), and (b) palatability. (Cattle will eat just the right amount of a good mineral, but they won't overdo it—due to appetizers, rather than needs.)

Commercial mineral mixtures costing $1.40 to $1.60 per lb of phosphorus are not excessively priced. If the average consumption per head per month of a mineral mix costing $17.00/cwt is 3 lb, the monthly per head cost will be about $0.51, or less than $0.02 per cow per day.

HOME-MIXED MINERALS FOR CATTLE

When buying and home-mixing minerals, as when buying commercial mineral mixes, operators should first determine their needs, based on (1) available feeds, (2) area (for example, the Northern Great Plains and the Southwest are phosphorus-deficient areas), and (3) the age and reproduction status (pregnancy and lactation make a difference) of the animals for which the mineral mix is intended. Of course, the available feeds, and the age and reproduction status of animals, on a given farm or ranch vary from time to time; and usually not all animals on a given establishment for which the mineral is to be used are of the same age and reproduction status. For the ultimate in exactness, therefore, there would have to be many changes and many different mineral mixes. Fortunately, a reasonable range in the allowances of the different mineral elements is permissible. As a result, the selection of the mineral mix, or mixes, for a particular farm or ranch usually involves a compromise, reached primarily on the basis of what will meet the needs of most of the animals, particularly during most critical stages, for the specific farm or ranch under consideration.

One of the mineral mixes given in Table 8-6 will

usually suffice for free-choice feeding most cattle. Moreover, where special circumstances necessitate some other formulation, Table 8-6 will serve as a useful guide. As noted, provision is made for cattle receiving rations containing different roughages and different levels of cereal grains. For example, a ration consisting of 15 lb of shelled corn and 5 lb of alfalfa-brome hay contains 75% cereal grain; hence, the mineral mix should contain 20% ground limestone and 80% trace mineralized salt. If the 5 lb of alfalfa-brome hay were replaced with 5 lb of grass hay (brome or timothy, for example), the mineral mix should be 50% ground limestone and 50% trace mineralized salt. If the 5 lb of alfalfa-brome were replaced with 15 lb of silage (wet basis), the mineral mix should be 60% ground limestone and 40% trace mineralized salt. As noted, the Table 8-6 mineral mixes for silage are on a wet basis; hence, a ration of 15 lb of wet silage and 15 lb of corn is considered as having 50% cereal grain.

TABLE 8-6
MINERAL MIXTURES FOR CATTLE FOR FREE-CHOICE FEEDING WITH DIFFERENT RATIONS

Mineral Supplement	Type of Roughage		
	Alfalfa, or Alfalfa-Grass Hay	Grass Hay	Corn Silage
	Rations containing up to 20% cereal grains[1]		
Dicalcium phosphate or bone meal[2]	—	40%	30%
Ground limestone	—	—	—
Trace mineralized salt	100%	60%	70%
	Rations containing 20% to 40% cereal grains[1]		
Dicalcium phosphate or bone meal[2]	—	25%	—
Ground limestone	—	25%	50%
Trace mineralized salt	100%	50%	50%
	Rations containing 40% to 60% cereal grains[1]		
Dicalcium phosphate or bone meal[2]	—	20%	—
Ground limestone	—	40%	60%
Trace mineralized salt	100%	40%	40%
	Rations containing 60% to 80% cereal grains[1]		
Dicalcium phosphate or bone meal[2]	—	—	—
Ground limestone	20%	50%	80%
Trace mineralized salt	80%	50%	20%
	Rations containing 80% to 100% cereal grains[1]		
Dicalcium phosphate or bone meal[2]	—	—	—
Ground limestone	50%	70%	80%
Trace mineralized salt	50%	30%	20%

[1]Percent on an *as-fed* basis. Examples: Ration composed of 15 lb *(6.8 kg)* shelled corn and 5 lb *(2.3 kg)* hay contains 75% cereal grains. Ration composed of 30 lb *(13.6 kg)* corn silage and 10 lb *(4.5 kg)* shelled corn contains 25% cereal grain.

[2]Defluorinated rock phosphate may also be used to replace bone meal and dicalcium phosphate.

Cattle grazing on native grass should be offered a free-choice mineral mix consisting of 40% dicalcium phosphate or bone meal and 60% trace mineralized salt. Minerals that are self-fed on pastures or in corrals should be in boxes protected from the weather.

Where a high proportion of urea is used as the protein source, it is necessary to add minerals to the ration unless the remainder of the ration contains adequate minerals to make up for the complete lack of minerals in the urea.

Where there is need to increase, or assure, the palatability of a mineral mix, usually molasses is added; and where there may be wind losses, a bland oil is usually added.

Salt should always be available on a free-choice basis in addition to whatever mineral mix is provided.

VITAMINS

The absence of one or more vitamins in the ration may lead to a failure in growth or reproduction, or to characteristic disorders known as vitamin deficiency diseases. In severe cases, death itself may follow. Although the occasional deficiency symptoms are the most striking result of vitamin deficiencies, it must be emphasized that in practice, mild deficiencies probably cause higher total economic losses than do severe deficiencies. It is relatively uncommon for a ration, or diet, to contain so little of a vitamin that obvious symptoms of a deficiency occur. When one such case does appear, it is reasonable to suppose that there must be several cases that are too mild to produce characteristic symptoms but which are sufficiently severe to lower the state of health and the efficiency of production.

Cattle have physiological requirements for most vitamins needed by other mammals. Synthesis by microorganisms in the rumen, supplies in natural feedstuffs, and synthesis in tissues meet most of the usual requirements. Although colostrum is rich in vitamins, providing immediate protection to the newborn calf, calves have minimal stores of vitamins at birth. The ability of the calf to synthesize B vitamins and vitamin K in the rumen develops rapidly when solid feed is introduced into the ration. Vitamin D is synthesized by animals exposed to direct sunlight and is found in large amounts in sun-cured forages. High-quality forages contain large amounts of vitamin A precursors and vitamin E.

Table 8-7 lists the vitamin requirements of beef cattle.

TABLE 8-7
VITAMIN REQUIREMENTS OF BEEF CATTLE
(in Percentage or Amount per Kilogram of Dry Ration)[1]

Nutrient	Growing and Finishing Steers and Heifers	Dry Pregnant Cows	Breeding Bulls and Lactating Cows
	- - - - - - - (% *per kg dry ration)* - - - - - - -		
Vitamin A activity . . . . . (IU)[2]	2,200	2,800	3,900
Vitamin D . . . . (IU)	275	275	275
Vitamin E . . . . (IU)	15–60	—	—

[1]From *Nutrient Requirements of Beef Cattle*, 7th rev. ed., National Research Council–National Academy of Sciences, Washington, DC, 1996.

[2]May be vitamin A or provitamin A equivalent.

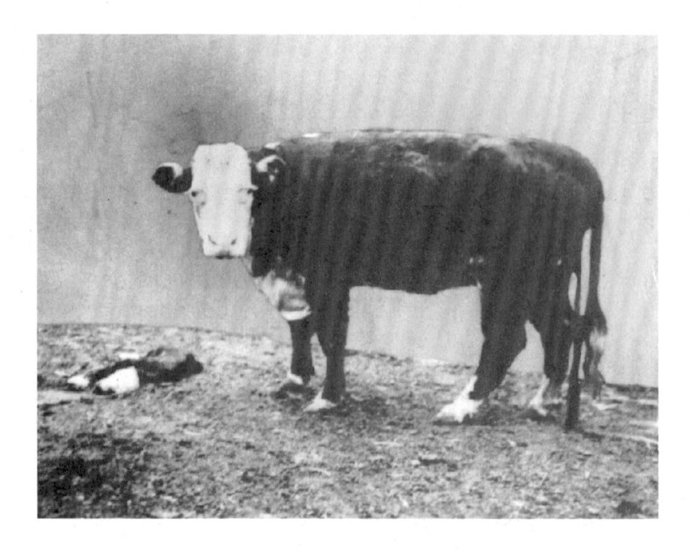

Fig. 8-14. Effect of vitamin A deficiency on reproduction. The heifer in the upper picture received a ration deficient in vitamin A, but otherwise complete. She became night blind and aborted during the last month of pregnancy; also, note the retained placenta. The heifer in the lower picture received the same ration, but during the latter part of the gestation period, a daily supplement of 1 lb of dehydrated alfalfa meal containing 50 mg of carotene was added. She produced a normal, vigorous calf. (Courtesy, California Ag. Exp. Sta., Davis, CA)

BEEF CATTLE VITAMIN CHART

Table 8-8, Beef Cattle Vitamin Chart, presents in summary form pertinent information relative to the vitamin needs of beef cattle.

FAT-SOLUBLE VITAMINS

A discussion of the fat-soluble vitamin needs of beef cattle follows.

VITAMIN A

The vitamin most likely to be deficient in beef cattle rations is vitamin A. True vitamin A is a chemically formed compound, which does not occur in plants. It is furnished in most beef cattle rations in the form of its precursor, carotene. However, plants are a variable, and sometimes undependable, source of carotene due to oxidation. Also, cattle are relatively inefficient converters of carotene to vitamin A. The latter fact was taken into consideration in the development of international standards for vitamin A, which are based on the rate at which the rat converts beta-carotene to vitamin A. The conversion rate for the rat is 1 mg of beta-carotene to 1,667 IU of vitamin A, whereas it is estimated that 1 mg of beta-carotene is equal to only 400 IU of vitamin A in cattle. Moreover, the conversion rate for cattle varies under different conditions; it is influenced by type of carotenoid, breed, individual differences in animals, and level of carotene intake. Stress conditions— such as high temperature and elevated nitrogen intake have also been suggested as causes for reduced conversion.

Under practical feeding conditions, cattle producers should consider (1) previous feeding as it influences body stores of vitamin A; (2) vitamin A destruction during feed processing or when mixed with oxidizing materials; and (3) carotene destruction in feeds during storage.

The possibility that beta-carotene may have a role in reproduction independent of its role as a vitamin A precursor has received considerable recent attention. However, further studies and evaluation are necessary.

VITAMIN D

When exposed to enough direct sunlight, beef cattle normally acquire their vitamin D needs, for the ultraviolet rays in sunlight penetrate the skin and produce vitamin D from traces of sterols in the tissues.

Also, cattle obtain vitamin D from sun-cured roughages. However, the addition of vitamin D to the ration is important when cattle, especially calves, are kept in a barn most of the day, when there is limited sunshine, when the calcium:phosphorus ratio is not correct, and/or when little or no sun-cured hay is fed. Vitamin D helps build strong bones and sturdy frames.

VITAMIN E

Added vitamin E may be necessary under certain conditions because of its relationship to vitamin A utilization and the prevention of white muscle disease.

VITAMIN K

Under normal conditions, adequate vitamin K is synthesized in the rumen of cattle. However, symptoms of inadequacy (a bleeding syndrome known as *sweet clover disease*) occur when moldy sweet clover hay, high in dicoumarol content, is fed, since dicoumarol is a metabolic antagonist that interferes with the normal action of vitamin K.

WATER-SOLUBLE VITAMINS

A discussion of the water soluble vitamin needs of beef cattle follows. Note that this discussion is limited to the B vitamins, which are needed by the young calf, and possibly by some feedlot cattle.

B VITAMINS

Dietary requirements for the B vitamins (thiamin, biotin, niacin, pyridoxine, pantothenic acid, riboflavin, and vitamin B-12) have been demonstrated experimentally for the young calf during the first 8 weeks of life, prior to the development of the functioning rumen. At this stage in life, these requirements are usually met by the milk of the dam. Later, the B vitamins appear to be synthesized in sufficient quantities by rumen bacterial fermentation, with the possible exception of feedlot cattle in which thiamin deficiences have been reported. However, inadequacy of protein or other nutrients in the ration may impair rumen fermentation, with the result that sufficient quantities of the B vitamins will not be synthesized.

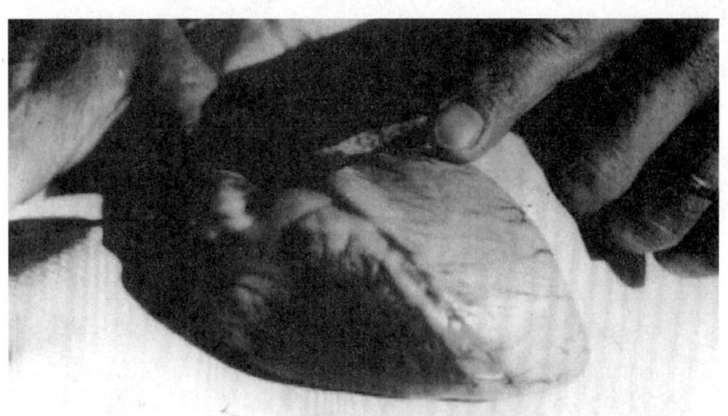

Fig. 8-15. White muscle disease in a calf.

Upper picture shows the generalized weakness of muscles, lameness, and difficulty in locomotion of an afflicted calf. Calf is about 3 months old.

Lower picture shows abnormal white areas in the heart muscles of a 6-week-old calf afflicted with white muscle disease.

(Courtesy, Oregon Ag. Exp. Sta., Corvallis, OR)

TABLE
BEEF CATTLE

Vitamins Which May Be Deficient Under Normal Conditions	Conditions Usually Prevailing Where Deficiencies Are Reported	Functions of Vitamin	Deficiency Symptoms
Fat-Soluble Vitamins:			
A—Vitamin A is found only in animals; plants contain the precursor—carotene. Vitamin A is the vitamin most likely to be of practical importance in feeding cattle.	Vitamin A deficiency is most likely to occur when cattle are fed (1) high-concentrate rations; (2) bleached pasture or hay grown under drought conditions; (3) feeds that have had excess exposure to sunlight, air, and high temperature; (4) feeds that have been heavily processed or mixed with oxidizing materials such as minerals; and (5) feeds that have been stored for long periods of time. Cattle particularly susceptible to vitamin A deficiency are: newborn calves deprived of colostrum; cattle that have been prevented from establishing or maintaining good liver stores through exposure to drought; cattle wintered without high-quality forage, and cattle exposed to stresses such as high temperatures or elevated nitrate intake.	Vitamin A functions as a component of the visual purple required for dim-light vision, and is essential for normal growth, reproduction, and maintenance of healthy epithelial tissue.	Signs of vitamin A deficiency include reduced feed intake, rough hair coat, edema of the joints and brisket, lacrimation, xeropthalmia, night blindness, slow growth, diarrhea, convulsive seizures, improper bone growth, blindness, low conception rates, abortion, stillbirths, blind calves, abnormal semen, reduced libido, and susceptibility to respiratory and other infections. Of these symptoms, only night blindness has proved unique to vitamin A deficiency. Clinical verification may include ophthalmoscopic examination, liver biopsy and assay, blood assay, testing spinal fluid pressure, conjunctival smears, and response to vitamin A therapy.
D	Young calves kept indoors, especially in the wintertime. Finishing cattle in northern U.S. on high silage and grain rations and a minimum of suncured hay.	Vitamin D is required for calcium and phosphorus absorption, normal mineralization of bone, and mobilization of calcium from bone.	Rickets in young calves, the symptoms of which are: decreased appetite, lowered growth rate, digestive disturbances, stiffness in gait, labored breathing, irritability, weakness, and occasionally, tetany and convulsions. Later, enlargement of the joints, slight arching of the back, bowing of the legs, and the erosion of the joint surfaces cause difficulty in locomotion. Posterior paralysis may follow fracture of vertebrae. In older animals with vitamin D deficiency, bones become weak and easily fractured, and posterior paralysis may accompany vertebral fractures. Vitamin D deficiency in the pregnant animal may result in dead, weak, or deformed calves at birth.
E	Where soils are very low in selenium. When unsatured fats are fed.	Vitamin E is an antioxidant. It has been widely used to protect and to facilitate the uptake and storage of vitamin A. In metabolism, it is linked closely with selenium. Some deficiency signs, particularly in white muscle disease, may respond to either selenium or vitamin E, or may require both.	Muscular dystrophy (commonly called white muscle disease) in calves 2 to 12 weeks of age; characterized by heart failure and paralysis varying in severity from slight lameness to inability to stand. Also, a dystrophic tongue is often seen in affected animals. A deficiency of vitamin E may be precipitated or accentuated by feeding unsaturated fats.
K	When moldy sweet clover hay high in dicoumarol content is fed. Vitamin K deficiency results from the antagonistic action of dicoumarol that is formed in moldy sweet clover hay.		Sweet clover disease, characterized by prolonged blood clotting. Mild cases can be treated effectively with vitamin K.

8-8
VITAMIN CHART

Nutrient Requirements[1]				
Daily Nutrients/ Animal (or Injection)	**Amount/Lb (or /kg) of Feed**	**Recommended Allowances[1]**	**Practical Sources of the Vitamin**	**Comments**
*Variable according to class, age, and weight of cattle. Injection of 1 million IU of vitamin A intramuscularly will prevent deficiency symptoms for 2–4 months in growing or breeding cattle.	*Variable according to class, age, and weight of cattle. On a dry ration basis, the vitamin A requirements are about as follows: *1. Growing-finishing steers and heifers, 1,000 IU/lb *(2,200 IU/kg)*. *2. Pregnant heifers and cows, 1,270 IU/lb *(2,800 IU/kg)*. *3. Lactating cows and breeding bulls, 1,770 IU/lb *(3,900 IU/kg)*.	Inject newborn calves (at birth) with 250,000–1,000,000 IU of vitamin A (use the higher level under confinement production or when scours may be a problem).	Stabilized vitamin A. Green pasture. Grass or legume silages. Yellow corn. Green hay not over 1 year old. The average carotene content of some common feeds is as follows: **mg Carotene per** *(lb)* *(kg)* Legume hays (including alfalfa) avg. quality ..... 9–14 20–31 Nonlegume hays, avg. quality .... 4–8 9–18 Dehydrated alfalfa meal, avg. quality .... 50–70 110–154 Yellow corn............. 0.8–1.0 1.8–2.2 Silages, corn, or sorghum . 2–10 4–22	Carotene is rapidly destroyed by exposure to sunlight and air, especially at high temperatures. Hay over 1 year old, regardless of green color, is usually not an adequate source of carotene or vitamin A activity. Ensiling effectively preserves carotene, but the availability of carotene from corn silage may be low. The younger the animal, the quicker vitamin A deficiencies will occur. Mature animals may store sufficient vitamin A to last 6 months. When deficiency symptoms appear, they can be corrected (1) by increasing carotene intake through the introduction of high-quality forage, or (2) by supplying vitamin A in the feed or by injection.
	*125 IU/lb *(275 IU/kg)* of dry ration.	Normally, beef cattle receive sufficient vitamin D from exposure to direct sunlight or from sun-cured hay.	Exposure to direct sunlight. Sun-cured hay. Irradiated yeast.	Sun-cured alfalfa hay contains 300–1,000 IU/lb *(661–2,204 IU/kg)*.
	*dl-alpha-tocopherol acetate added to dry ration at level of 6.8 to 27.3 IU/lb *(15 to 60 IU/kg)*.	Generally natural feeds supply adequate quantities of alpha-tocopherol for mature cattle, although muscular dystrophy in calves occurs in certain areas.	Alpha-tocopherol, added to the ration or injected intramuscularly. Commercial vitamin E supplements. Grains containing 6–15 mg vitamin E/lb *(13–33 mg/kg)*.	The incidence of white muscle disease appears to be lower when the cows receive 2–3 lb (0.91–1.36 kg) of grain during last 60 days of pregnancy. Where supplemental vitamin E is needed, it may be added to the ration or injected intramuscularly.
			Vitamin K_1 is abundant in pasture and green roughage. Vitamin K_2 is synthesized in large amounts in the rumen. Either K_1 or K_2 effectively fulfill the vitamin K role in blood clotting mechanism.	Except when the dicoumoural content of hay is excessively high (as in moldy sweet clover hay), sufficient vitamin K is synthesized in the rumen of cattle.

(Continued)

TABLE 8-8

Vitamins Which May Be Deficient Under Normal Conditions	Conditions Usually Prevailing Where Deficiencies Are Reported	Functions of Vitamin	Deficiency Symptoms
Water-Soluble Vitamins:			
B Vitamins	When an antagonist is present. When ruminal synthesis is limited by lack of precursors or other problems.	Most of the established metabolic functions of B vitamins are important to cattle, as well as to other animals. Consequently, a physiological need for most B vitamins can be assumed for cattle of all ages. Vitamin B-12 is of special interest because of its role in propionate metabolism, and the practical incidence of vitamin B-12 deficiency as a secondary result of cobalt deficiency. Niacin has been reported to enhance protein synthesis by ruminal microorganisms.	Deficiency signs in young calves have been clearly demonstrated for thiamin, riboflavin, pyridoxine, pantothenic acid, biotin, nicotinic acid, vitamin B-12, and choline. Polioencephalomalacia in grain-fed cattle has been linked to thiaminase activity or production of a thiamin antimetabolite in the rumen. Affected animals have responded to intravenous administration of thiamin (2.2 mg/kg body weight).

[1]As used herein, the distinction between *nutrient requirements* and *recommended allowances* is as follows: in nutrient requirements, no margins of safety are included intentionally; whereas in recommended allowances, margins of safety are provided in order to compensate for variations in feed composition, environment, and possible losses during storage or processing.

(See Appendix Section II, Table II-1 for the vitamin content of feeds commonly used in beef cattle rations.)

WATER

Water is the most vital of all nutrients. It is needed for all the essential processes of the body, such as the digestion and absorption of food nutrients, the removal of waste, and in regulating body temperature. Animals can survive for a longer period without feed than they can without water. Yet, under ordinary conditions, it can be readily provided in abundance and at little cost. Beef cattle should have an abundant supply of water before them at all times.

The water requirement is influenced by several factors, including weight, environmental temperature, rate and composition of gain, pregnancy, lactation, activity, type of ration, and feed intake. Because feeds contain some water and the oxidation of certain nutrients in feeds produces water, not all water needs must be provided by drinking. Such feeds as silages, green chop, or lush pastures are high in moisture content, while grains, hays, and dormant pasture forage are low. High-energy feeds produce much metabolic water, while low-energy feeds produce little metabolic water. These are among the complications in the matter of assessing water requirements.

Saline water containing 1% soluble salts may be toxic. Excessive nitrates or alkalinity may make water unsatisfactory for cattle.

Fig. 8-16. Windmill, an energy-conserving watering facility. (Courtesy, *Livestock Weekly*, San Angelo, TX)

In the northern latitudes, heaters must be provided to make the water available, but they are not needed to warm the water further.

Table 8-9 may be used as a guide to the water requirements of beef cattle.

(Continued)

Nutrient Requirements[1]		Recommended Allowances[1]	Practical Sources of the Vitamin	Comments
Daily Nutrients/ Animal (or Injection)	Amount/Lb (or /kg) of Feed			
		Usually, no dietary B vitamins need be supplied to cattle.	B-vitamins are abundant in milk and many other feeds, and synthesis of B vitamins by ruminal microorganisms is extensive. Calves begin microbial synthesis of B vitamins very soon after the introduction of dry feed in the ration.	

*Where preceded by an asterisk, the nutrient requirements listed herein were taken from *Nutrient Requirements of Beef Cattle*, seventh revised edition, National Research Council–National Academy of Sciences, Washington, DC, 1996.

TABLE 8-9
DAILY WATER INTAKE OF BEEF CATTLE[1]

Weight		Temperature in °F (°C)[2]											
		40 *(4.4)*		50 *(10.0)*		60 *(14.4)*		70 *(21.1)*		80 *(26.6)*		90 *(32.2)*	
(lb)	*(kg)*	*(gal)*	*(liter)*	*(gal)*	*(liter)*	*(gal)*	*(liter)*	*(gal)*	*(liter)*	*(gal)*	*(liter)*	*(gal)*	*(liter)*
Growing heifers, steers, and bulls													
400	*182*	4.0	*15.1*	4.3	*16.3*	5.0	*18.9*	5.8	*22.0*	6.7	*25.4*	9.5	*36.0*
600	*273*	5.3	*20.1*	5.8	*22.0*	6.6	*25.0*	7.8	*29.5*	8.9	*33.7*	12.7	*48.1*
800	*364*	6.3	*23.0*	6.8	*25.7*	7.9	*29.9*	9.2	*34.8*	10.6	*40.1*	15.0	*56.8*
Finishing cattle													
600	*273*	6.0	*22.7*	6.5	*24.6*	7.4	*28.0*	8.7	*32.9*	10.0	*37.9*	14.3	*54.1*
800	*364*	7.3	*27.6*	7.9	*29.9*	9.1	*34.4*	10.7	*40.5*	12.3	*46.6*	17.4	*65.9*
1,000	*454*	8.7	*32.9*	9.4	*35.6*	10.8	*40.9*	12.6	*47.7*	14.5	*54.9*	20.6	*78.0*
Wintering pregnant cows[3]													
900	*409*	6.7	*25.4*	7.2	*27.3*	8.3	*31.4*	9.7	*36.7*	—	—	—	—
1,100	*500*	6.0	*22.7*	6.5	*24.6*	7.4	*28.0*	8.7	*32.9*	—	—	—	—
Lactating cows													
900	*409*	11.4	*43.1*	12.6	*47.7*	14.5	*54.9*	16.9	*64.0*	17.9	*67.8*	16.2	*61.3*
Mature bulls													
1,400	*636*	8.0	*30.3*	8.6	*32.6*	9.9	*37.5*	11.7	*44.3*	13.4	*50.7*	19.0	*71.9*
1,600	*727*	8.7	*32.9*	9.4	*35.6*	10.8	*40.9*	12.6	*47.7*	14.5	*54.9*	20.6	*78.0*

[1]From *Nutrient Requirements of Beef Cattle*, seventh revised edition, National Research Council–National Academy of Sciences, 1996.

[2]Water intake of a given class of cattle in a specific management regime is a function of dry matter intake and ambient temperature. Water intake is quite constant up to 40°F *(4.4°C)*.

[3]Dry matter intake has a major influence on water intake. Heavier cows are assumed to be higher in body condition and to require less dry matter and, thus, less water intake.

WEATHER INFLUENCES ON NUTRITIVE NEEDS OF BEEF CATTLE

The nutrient requirements of beef cattle are influenced by weather. Details follow.

1. **Feed consumption.** Feed intake can be significantly affected by environment, particularly by temperatures outside the comfort zone 59°F to 77°F. With high temperature and humidity, intake may be lowered by up to 30%. High-temperature stress is reduced if accompanied by low humidity and/or relief through more comfortable nighttime conditions. Drop in feed consumption caused by elevated temperature and humidity is more severe on high-roughage rations than on low-roughage rations.

With low temperature, feed intake may be increased by up to 30%, **provided** cattle remain relatively dry; the extent of this increase is negatively associated with temperature. But extensive precipitation and muddy conditions can depress feed intake up to 30%. Forage intake of beef cows on winter range may be depressed up to 50% following a storm period that produces cold temperatures and snow cover.

2. **Availability of feedstuffs.** The availability of nutrients from feedstuffs can be altered by environmental temperature; with temperature and digestibility positively related. This is particularly true of roughages; effects of temperature on digestibility of concentrates are small.

3. **Maintenance energy requirement.** The maintenance energy requirements of beef cattle increase as temperature, humidity, and air movement depart from the comfort zone. Likewise, the heat loss from animals is affected by these three items.

For acute heat stress, maintenance energy requirements should be adjusted according to severity. Severity may be determined by the respiration of the animal. For rapid shallow breathing, the maintenance energy requirement should be increased by 7%; for deep open-mouth panting, the requirement should be increased from 11 to 25%.

For acute cold stress, maintenance energy expenditures should be increased according to exposure to temperature below their lower critical temperature (LCT). The LCT for any given animal depends on the animal's insulation, heat production, and age. Wind speed influences insulation and is an important environmental factor.

FEEDS FOR BEEF CATTLE

Beef cattle feeding practices vary according to the relative availability of grasses, dry roughages, and grains. Where roughages are abundant and grain is limited, as in the western range states, cattle are primarily grown out or finished on roughages. On the other hand. where grain is relatively more abundant, as in the Corn Belt and in the High Plains area of Texas and Oklahoma, finishing with more concentrates is common.

PASTURES

Good pasture is the cornerstone of successful beef cattle production. In fact, there has never been a great beef cattle country or area which did not produce good grass. It has been said that good farmers or ranchers can be recognized by the character of their pastures and that good cattle graze good pastures. Thus, the three go hand in hand—good farmers, good pastures, and good cattle. The relationship and importance of cattle and pastures has been further extolled in an old Flemish proverb which says, "No grass, no cattle; no cattle, no manure; no manure, no crops."

A total of 882 million acres in the United States is grassland. Much of this area, especially in the Far West, can be utilized only by beef cattle or sheep. Although the term *pasture* usually suggests growing plants, it is correct to speak of pasturing stalk and stubble fields. In fact, in the broad sense, pastures include all crops that are harvested directly by animals.

The type of pasture, as well as its carrying capacity and seasonable use, varies according to topography, soil, and climate. Because of the hundreds of species of grasses and legumes that are used as beef cattle pastures, each with its own best adaptation, no attempt is made to discuss the respective virtues of each variety. Instead, it is recommended that the farmer or rancher seek the advice of the local county agricultural agent, or write to the state agricultural college.

No method of harvesting has yet been devised

Fig. 8-17. King Ranch Santa Gertrudis cows and calves on pasture. (Courtesy, King Ranch, Kingsville, TX)

Fig. 8-18. Cows grazing on tall fescue fall regrowth and field-stored round bales, in southeastern Ohio. (Courtesy, Ohio Agricultural Research and Development Center, Wooster, OH)

that is as cheap as that which can be accomplished through grazing by animals. Accordingly, successful beef cattle management necessitates as nearly year-round grazing as possible. In the northern latitudes of the United States, the grazing season is usually of about 6 months' duration, whereas in the deep South, yearlong grazing is approached. In many range areas of the West, the breeding herds obtain practically all their forage the year-round from the range, being given supplemental roughage only if the grass or browse is buried deep in snow.

During the winter months, and in periods of drought, the pasture utilized by beef cattle may consist of dried grass cured on the stalk. On a dry basis, the crude protein content of mature, weathered grasses may be 3% or less. To supplement such feed, cattle producers commonly feed cake or cubes. The use of cake or cubes instead of meal reduces losses from wind, an especially important factor on the range.

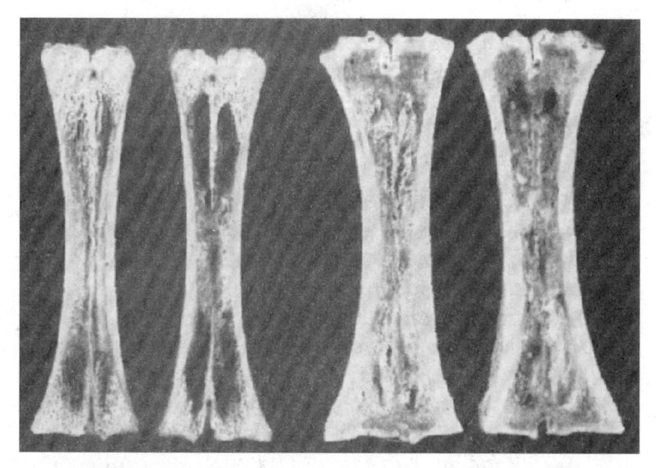

Fig. 8-19. Shin bones from two calves of similar breeding and age. Small, fragile, pitted bones *(left pair)* obtained from calf pastured on "belly deep" grasses grown on highly weathered but untreated soil. Heavy bones *(right pair)* obtained from calf pastured on grasses grown on moderately weathered but fertilized soil, including trace elements. (Courtesy, University of Missouri-Columbia)

In some instances, cattle on pasture fail to make the proper growth or gain in condition because the soil is seriously deficient in fertility or the pasture has not been well managed. In such instances, striking improvement will result from proper fertilization and management.

HAY AND OTHER DRY ROUGHAGES

Hay is the most important harvested roughage fed to beef cattle, although many other dry roughages can be and are utilized.

Fig. 8-20. Haying has gone modern! This shows a tightly packed, round-topped, virtually weatherproof, movable haystack. It lessens labor and makes it possible to deposit the stacks in the field, fence them, and feed cattle right from the stack. (Courtesy, Deere & Company, Moline, IL)

The dry roughages are all high in fiber and, therefore, lower than concentrates in total digestible nutrients. Hay averages about 28% fiber and straw approximately 38%, whereas such concentrates as corn and wheat contain only 2 to 3% fiber. Fortunately, cattle are equipped to handle large quantities of roughages. In the first place, the paunch of a mature cow has a capacity of 3 to 5 bu, thus providing ample storage for large quantities of less concentrated feeds. Secondly, the billions of microorganisms in the rumen attack the cellulose and pentosans of the fibrous roughages, such as hay, breaking them down into available and useful nutrients. In addition to providing nutrients at low cost, the roughages add needed bulk to cattle rations.

Roughages, like concentrates, may be classified as carbonaceous or nitrogenous, depending on their protein content. The principal dry carbonaceous roughages used by cattle include hay from the grasses, the straws and hays from cereal grains, corncobs, and the stalks and leaves of corn and the grain sorghums. Cured nitrogenous roughages include the various legume hays such as alfalfa, the clover hays, peanut hay, soybean hay, cowpea hay, and velvet bean hay.

Although leguminous roughages are preferable, weather conditions and soils often make it more practical to produce the nonlegumes. Also, in many areas, such feeds as dry grass cured on the stalk, cereal straws, corncobs, and cottonseed hulls are abundantly available and cheap. Under such circumstances, these feeds had best be used as part of the ration for wintering beef cows, for wintering stockers that are more than one year of age, or for finishing beef cattle.

In comparison with good-quality legume hays, the carbonaceous roughages are lower in protein content and in quality of proteins, lower in calcium, and generally deficient in carotene (provitamin A). Thus, where nonlegume roughages are used for extended periods, these nutritive deficiencies should be corrected; this is especially true with the gestating-lactating cow or the young, growing calf. To the end that the feeding value of some of the common nonlegumes may be enhanced for beef cattle, the following facts are pertinent:

1. The feeding value of nonlegume hays can be increased by cutting them at an early stage of maturity and curing so as to retain as much of the carotene content as possible.
2. Where dry and bleached pastures are grazed for an extended period of time, or where there is an unusually long winter, it is important that at least part of the roughage be a legume, either silage or hay, or that vitamin A be either added to the ration or injected intramuscularly; and that suitable energy, protein, and mineral supplements be provided.
3. Potentially, corncobs—which were formerly considered a waste product and of little worth—have a feeding value approaching that of hay. However, their energy cannot be utilized unless they are fortified with certain nutrients which help the bacteria and other organisms of the rumen break them down into a form which can be digested. Also, corncobs are low in palatability.
4. Cereal straws and cottonseed hulls may be incorporated in the wintering ration of pregnant cows or in the ration of finishing cattle provided their fundamental characteristics and nutritional limitations are recognized and corrected.

Fig. 8-21. Upright silos used for storing feed for beef cattle. (Courtesy, A. O. Smith, Harvestore Products, Inc., Arlington Heights, IL)

SILAGES AND ROOTS

Silage is an important adjunct to pastures in beef cattle production, it being possible to use a combination of the two forages in furnishing green, succulent feeds on a year-round basis. Extensive use of silage for beef cattle dates back only to about 1910. Prior to that time, it was generally thought of as a feed for dairy cows. Even today, only a relatively small percentage of the beef cattle of the United States are fed silage.

Where silage has been used, it has proved very popular. Some of the more important reasons advanced in favor of silage are as follows:

1. On most beef cattle–producing farms, silage is the cheapest form in which a good, succulent winter feed can be provided.
2. It is the cheapest form in which the whole stalk of 1 acre of corn or sorghum can be processed and stored.
3. Good silage can be made during times of rainy weather when it would be impossible to cure properly hay or fodder.
4. It helps to control weeds, which are often spread through hay or fodder.
5. Grass silage is a better source of vitamins, especially carotene and perhaps some of the unknown factors, than dried forages.
6. There is no danger of fire loss to silage.
7. Silage is a very palatable feed and a mild laxative.
8. Converting the crop into silage clears the land earlier than would otherwise be possible.
9. Silage makes for less waste, the entire plant being eaten with relish.
10. The removal of cornstalks, as is required in making silage, is one of the best methods of controlling the European corn borer.

11. Silage increases the number of animals that can be fed from the produce of a given area of land.

Corn was the first and still remains the principal crop used in the making of silage, but many other crops are ensiled in various sections of the country. The sorghums are the leading ensilage crop in the Southwest, and grasses and legumes are the leading ensilage crops in the Northeast. Also, in different sections of the country to which they are adapted, the following feeds are ensiled: cereal grains, field peas, cowpeas, soybeans, potatoes, and numerous fruit and vegetable refuse products. A rule of thumb is that crops that are palatable and nutritious to animals as pasture, as freshly harvested feed, or as dry forage, also make palatable and nutritious silage. Conversely, crops that are unpalatable and nonnutritious as pasture, as green feed, or as dry forage, also make unpalatable and nonnutritious silage.

Grass silage can be produced in those areas where the climate is too cool and the growing season too short for corn or sorghum silage. It is generally higher in protein and carotene, but lower in total digestible nutrients and vitamin D than corn or sorghum silage. Generally grass silage contains about 90% as much total digestible nutrients (TDN) as corn silage, but it is equal in TDN when 150 lb of grain per ton have been added as a preservative. Thus, grass silage generally requires the addition to the ration of less protein supplement but more total concentrates than corn or sorghum silage. This would indicate that corn or sorghum silage would be slightly preferable to grass silage in high-roughage finishing rations for beef cattle, whereas grass silage would be preferable in high-roughage rations for young, growing beef cattle.

When silage is fed to cattle, it must be remembered that, because of its high moisture content, about 3 lb of silage are generally considered equivalent to 1 lb of dry roughage of comparable quality. A ration of 55 to 60 lb of corn silage plus ½ to ¾ lb of a protein concentrate daily will carry a dry cow through the winter. The ration may be improved, however, by replacing ⅓ to ½ of the silage with an equivalent amount of a dry roughage, adding 1 lb of dry roughage for each 3 lb of silage replaced.

Silage may be successfully used for finishing steers. Long yearling steers will eat 25 to 35 lb a day at the beginning of the feeding period, the larger amounts being consumed when no hay is fed with it. Better results are obtained, however, if hay is included in the ration. The amount of silage is gradually decreased as the concentrates are increased. At the end of the feeding period, the cattle should be getting around 4 to 6 lb of silage and 1 or 2 lb of hay. Because of the more limited digestive capacity, the allowance of silage fed to calves should be correspondingly less.

Usually, silage provides a much cheaper succulent feed for beef cattle than roots. For this reason, the use of roots for beef cattle is very limited, being confined almost entirely to the northern areas.

CONCENTRATES

The concentrates include those feeds which are low in fiber and high in energy. For purposes of convenience, concentrates are often further classified as (1) carbonaceous feeds, and (2) nitrogenous feeds.

In general, the use of concentrates for beef cattle is limited to (1) the finishing of cattle, (2) the development of young stock, and (3) supplements in the winter ration. Over most of the United States, the cereal grains are the chief concentrates fed to beef cattle—these grains being combined, if necessary, with protein supplements to balance the ration.

The chief carbonaceous concentrates used for beef cattle are the cereal grains and such processed feeds as hominy feed, beet pulp, and molasses. The choice of the particular feeds is usually determined primarily by price and availability.

For best results, the feeder should correct the nutritive deficiencies of the cereal grains. All of them are low in protein, low in calcium, and lacking in vitamin D. All except yellow corn are also deficient in carotene. Regardless of whether the cereal grains are fed to growing, breeding, or finishing animals, their nutritive deficiencies can be corrected in a very effective and practical way by adding either (1) a good-quality legume hay to the ration, or (2) a protein concentrate plus suitable minerals and vitamins.

POULTRY WASTE

Approximately 100 million tons of poultry waste (from layers, broilers, and turkeys) are produced annually in the United States, most of which is potential beef cattle feed.

Because poultry production is highly intensive, with many birds in a small area, waste disposal is a major problem. Most cage-layer operations produce manure free of litter as the primary form of waste. Generally, broiler operations produce litter.

On a moisture-free basis, cage-layer manure generally contains 25 to 35% crude protein and minimal fiber, while broiler litter contains somewhat less protein—about 18 to 30% and substantially more fiber due to the presence of absorbent materials.

Poultry litter is the most collectable and the most nutritious of all animal wastes. Also, on a dry-matter basis, it generally makes for least cost rations and highest net returns.

PROCESSING POULTRY WASTE FOR FEED

Several methods have been used to process poultry waste for feed; among them, dehydration, ensiling, pelleting, preparation for liquid feeding, oxidation-ditch aerobic processing, commercial (patented) systems, and the use of wastes as substrates for single-cell protein production.

■ **Dehydrating poultry waste**—When voided, layer waste contains about 75% water. Reduction of the moisture content from 75% to 15% in dehydrators (698 to 1,292°F) requires removal of 1,284 lb of water per ton of dry solids, at an energy cost of $25 to $50/ton of dehydrated material.

The main **advantages** to artificial dehydration are: (1) reducing pathogens to low levels or eliminating them entirely, (2) lessening or removing odors, and (3) facilitating storing and handling.

The chief **disadvantages** of artificial drying are: (1) the high energy cost, and (2) the considerable loss of nitrogen and certain other components due to heating.

It may be concluded that artificial dehydration of animal waste results in excellent products, but the process may not be economically feasible due to high energy costs.

■ **Ensiling poultry waste**—Ensiling of animal waste is a controlled anaerobic fermentation process during which the carbohydrates in the mixture are converted to lactic and other acids. Once sufficient acids are produced, bacterial action ceases and the ensilage is stable. Heat is generated in the process. Processing animal wastes by ensiling has the **advantages** of (1) being economical, (2) diminishing the hazards from certain potentially pathogenic organisms, (3) rendering the waste mixture more palatable, and (4) producing a product with a pleasant aroma.

Because of the considerable expense and energy required for drying animal wastes, the trend is toward ensiling. Except for the cost of the silo, ensiling of wastes involves few expenses. Poultry wastes can be ensiled in bunker-type silos as well as oxygen-limiting tower silos.

South Carolina Experiment Station researchers conducted extensive studies on ensiling poultry wastes. They found that manure mixed with forage or litter takes about 6 weeks to ensile adequately. In an experiment designed to test the proper moisture level for ensiling a manure-forage combination, they found that hay and manure ensiled at 44% moisture produced the most desirable combination from a pH standpoint. Based on their studies, the South Carolina Agricultural Experiment Station workers recommended the following practices for ensiling poultry litter in an upright silo:

1. The litter should be ensiled at about 37% moisture. Although maximum fermentation takes place at higher moisture levels, it is difficult to blow wet litter into a tall silo because it clogs the blower pipe. At 37% moisture, there is adequate moisture to promote good fermentation; and blower difficulties are minimized. Bunker or trench silos do not pose any moisture problem and can, therefore, be used to ensile litter of higher moisture content.
2. In order to remove metal objects which commonly get into the litter, a magnet should be included in the ensiling and the feeding systems.
3. The easiest place to add enough water to obtain the desired moisture is in the poultry house. A portable moisture tester can be used to check the moisture content. However, a preliminary check on the moisture content of litter may be obtained by squeezing it; litter first begins to stick together at 35% moisture.
4. A front-end loader can be used to clean out the poultry trucks. This clean-out process facilitates mixing of the litter, thereby evenly distributing the moisture.

■ **Other methods of processing poultry waste**—Other processes that have been used follow:

1. **Pelleting.** Pelleting animal wastes prevents ingredient sorting by animals. However, the waste must be dried before pelleting; hence, pelleting is costly.
2. **Commercial patented systems.** Several commercial (patented) systems have been developed for processing animal wastes for feeding, but details relative to these are proprietary to the companies involved.
3. **Substrates for protein production.** The use of wastes as substrates for the production of protein supplements for livestock feeds is feasible, with systems using algae, yeasts, bacteria, and fungi all showing promise.

SUMMARY

The feeding value of poultry waste varies widely. But based on experiments and experiences, the following recommendations and evaluations appear to be justified:

1. Beef producers should feed litter that has at least 18% crude protein (with less than 25% of that protein in insoluble or bound form), and not more than 28% ash.
2. Poultry wastes are best suited to feeding gestating cows or stocker cattle.
3. To minimize risks from drug residues in the tissues of market cattle, feeding of poultry wastes should be discontinued 15 days before animals are marketed for slaughter.

PROCESSING FEEDSTUFFS

In recent years, many sophisticated techniques for processing beef cattle feeds have evolved. Most of the grains are processed for the purpose of improving the nutritive qualities, whereas most of the roughages are processed to facilitate handling and storage and to lessen wastage.

ROUGHAGE PROCESSING

Roughages may be chopped, shredded, or ground; pelleted; cubed (wafered); ensiled; or treated with alkali.

■ **Chopping, shredding, or grinding roughage**—Hay is frequently chopped because (1) it is easier to handle, (2) it can be stored in a smaller area at less cost, (3) it is fed with less waste, and (4) it may be handled mechanically. Also, cutting or shredding such coarse forages as corn fodder and stover makes them easier to handle and store, and results in less waste. Additional points pertinent to roughage preparation follow:

1. Chopping forages for cattle is more common in areas where forages are relatively more abundant and cheaper than the grains, with the result that a higher proportion of them are fed. Also, it follows that there is apt to be greater waste of forage under liberal feeding, unless precautions are taken to alleviate it.
2. For cattle, roughages should be coarsely chopped, not less than 2 in. in length.
3. In preparing forages, avoid (a) processing those with high moisture, which may heat and produce spontaneous combustion, and (b) processing those in which there are foreign objects (wire and other hardware), which the animals may not be able to select out and which may ignite a fire when being processed.

Fig. 8-22. Field-chopped hay stored in long, rectangular stacks at C & B Livestock, Inc., a modern cattle feedyard, at Hermiston, Oregon. (Courtesy, Ron Baker, President, C & B Livestock, Inc., Hermiston, OR)

4. Coarse chopping of long roughage, with or without cubing, does not affect nutritive value substantially and is recommended primarily to improve handling.

■ **Hay cubes**—This refers to the practice of compressing long or coarsely cut hay into wafers or cubes, which are larger and coarser than pellets. Most cubes are about 1¼ in. square and 2 in. long, with a bulk density of 30 to 32 lb per cu ft. Cubing costs about $5 per ton more than baling.

This method of haymaking is increasing, because it offers most of the advantages of pelleted forages, with few of the disadvantages. Cubed forage (1) simplifies haymaking, (2) alleviates fine grinding, (3) facilitates automation in both haymaking and feeding, (4) lessens transportation costs and storage space—cubed roughages require about one-third as much space as when the forage is baled and stacked, (5) reduces labor, (6) decreases nutrient losses, and (7) eliminates dust.

■ **Pelleting**—Fine grinding of long roughage is usually followed by pelleting.

Roughages alone, concentrates alone, or a combination of roughages and concentrates may be, and are, pelleted. Many cattle producers prefer to feed pellets or cubes on the range, primarily for reasons of convenience and reducing losses from wind blowing. The practice of pelleting will increase for the following reasons:

1. Pelleted feeds are less bulky and easier to store and handle, thus lessening storage and labor costs.
2. Pelleting prevents animals from selectively wasting ingredients likely to be high in certain dietary essentials; each bite is a balanced feed.
3. Pelleting alleviates wastage of relatively unpalatable feeds, such as rye and ground alfalfa.
4. Pelleting of roughages increases intake by 50% or more. Larger responses in intake are associated with poor-quality roughage, high-roughage rations, and younger cattle.
5. Pelleting of roughages depresses digestibility by up to 5%, with digestibility depressed most when intake of the processed material is high and when the roughage processed is grass.
6. Pelleting of roughages improves utilization of the digestible energy, partially because processing causes a higher percentage of the roughage to be digested postruminally.

The biggest deterrent to pelleting forages is that of being able to process chopped forage which is coarse enough to allow for optimum cellulose digestion in the rumen, and which will not increase the incidence of bloat.

Of course, the increased value of pellets should be appraised against the increased cost of pelleting.

■ **All-pelleted rations (grain and forage combined)**—Among the virtues ascribed to an all-pelleted (grain and forage combined) ration are: (1) it prevents selective eating—if properly formulated, each mouthful is a balanced diet; (2) it alleviates waste; (3) it eliminates dust; (4) it lessens labor and equipment—just fill self-feeders; (5) it lessens storage; and (6) it lends itself to automation. In all-pelleted rations, the ratio of roughage to concentrate should be higher than where long hay is fed.

With cubing or pelleting, the spread between high- and low-quality roughage is narrowed; that is, the poorer the quality of the roughage, the greater the advantage from pelleting or cubing. This is so because such preparation assures complete consumption of the roughage.

■ **Ensiling**—Ensiling is a fermentation process, which takes place when certain feeds with sufficient moisture are stored in a silo in the absence of air. Its greatest use is in preserving forages. Pertinent facts relative to the feeding value of ensiled roughages follow:

1. Ensiling may depress feed intake relative to other methods of preservation, such as drying.
2. Ensiling has little effect on digestibility and utilization of digested energy.
3. Ensiling may heat and depress protein digestibility if the ensiled material is dry and poorly packed.
4. Processing of forages prior to ensiling may influence feeding value; for example, the intake of grass silage chopped to 0.8 in. may be greater than the intake of unchopped material, and corn silage chopped to 0.5 in. or less is more digestible than corn silage chopped at longer lengths.

■ **Alkali treatment**—High-priced feeds and more stringent burning regulations have spurred research to find a practical method of improving the feeding value of crop residues and wastes. Of the various chemicals and treatments studied, alkali appears to be the most effective and practical. Results vary substantially, but based on presently available information, the following deductions can be made:

1. The effectiveness of alkali treatment depends on the residue or waste being treated and the technique employed. On the basis of efficacy of the treatment, the cereal straws rank as follows: wheat straw, barley straw, and oat straw.
2. Alkali treatment can (a) increase the rate of passage of indigestible material through the digestive tract, and (b) improve the intake of low-quality roughage by up to 50%.
3. Alkali treatment increases potential digestion of cell walls.
4. Alkali treatment increases digestibility of dry matter or energy up to 10%; and in a high straw ration it may be even greater.
5. Improvements in intake and digestibility may be small when treated straws constitute 50% or less of the ration.
6. The alkali treatment, through its heating effect, depresses the nitrogen digestibility by ruminants.
7. Because energy availability in the rumen is enhanced by alkali treatment, supplementation of treated roughages with more extensively degraded protein sources is usually beneficial.

GRAIN PROCESSING

The physical preparation of cereal grains for cattle by soaking and cooking has been practiced by cattle exhibitors for a very long time. In recent years, many sophisticated techniques for the processing of grains have been developed, especially for feedlot cattle.

Many factors other than processing have a bearing on the nutritive value of grains and may modify the effects of processing; among them, the following:

1. **Intake.** High intakes of grain reduce digestibility.
2. **Varieties of grain.** The variety of grain may have a major impact on digestibility; for example, the net energy of bird-resistant sorghum is considerably less than regular sorghum grain.
3. **Within varieties.** Within varieties of the same grain, the total digestible nutrients (TDN) may vary by 7%.

Fig. 8-23. This advanced electronic feed mill, which houses the computerized central control facility and utilizes a continuous flow mixing process, is capable of producing more than a ton of mixed feed per minute for an 80,000-head-per-year cattle feedlot near Greeley, Colorado. The continuous flow system, incorporating the use of augers and weight belts, ensures ration accuracy within 1% on each ingredient. (Courtesy, Farr Feeders, Inc., Greeley, CO)

4. **Available energy and feed intake.** The effect of processing may be less pronounced when the available energy in the unprocessed grain is relatively high or when feed intake is relatively low.

5. **Age of cattle.** Younger cattle digest unprocessed grain better than older cattle; hence, processing of grain usually increases the nutritive value of rations for yearling steers more than for calves less than a year old.

6. **Level of roughage.** The response to processing grain depends on the level of roughage in the ration.

Details relative to processing individual grains follows:

1. **Corn.** Steam processing and flaking (a) increase the net energy of an all-corn ration; (b) increase the net energy by 10% or more of rations containing at least 80% dry corn grain; and (c) increase the net energy by approximately 5% of rations containing intermediate levels of corn (65 to 80%).

In low-roughage rations, feeding corn in the unprocessed form (whole corn grain) maximizes intake and facilitates cattle management.

Steam processing and flaking improve energy retention in cattle from 6 to 10% more than from cracked corn, when the grain is incorporated into finishing rations composed of 70 to 80% corn.

Coarse grinding of corn is more desirable than fine grinding from the standpoints of favorable digestion of the corn and maintaining nutrient value of the roughage.

The digestible dry matter and energy of rations containing either high-moisture corn or dry corn reconstituted with moisture and stored for a short period of time before feeding are at least equal to, and may be as much as 5% higher than, the same ration containing dry corn. However, high-moisture corn may result in reduced feed consumption, and most, if not all, of its vitamin E may be lost during storage.

2. **Sorghum.** In low-roughage rations (less than 20%), steam processing and flaking improve energy (DE and TDN) by 5 to 10% and starch digestibility by 3 to 5%. Despite favorable results in the feeding value of steam processed, flaked sorghum grain, such processing does not improve the net energy of sorghum grain over dry grinding. Properly steam processed, flaked sorghum should have a density of about 25 lb/bu, or slightly less than one-half that of the original grain.

Reconstitution of whole grain sorghum with water, followed by an incubation period, then grinding, can improve energy and starch digestibility to the same extent as steam processing and flaking. However, steam processing and flaking may promote higher intake.

Energy values (NE_m and NE_g) of sorghum are increased by 8% through fine grinding as opposed to coarse rolling; thus, fine grinding may improve the value of sorghum to an extent comparable to steam processing and flaking.

In intermediate- and high-roughage rations, dry-rolled sorghum is better utilized than in low-roughage rations; thus, it is unlikely that steam processing and flaking will have beneficial effects in these circumstances.

The nutritive value of dry sorghum grain in low-roughage rations may be improved by heat treatments other than steam processing and flaking, including popping, micronizing, exploding, and roasting. However, these treatments may not be as successful as steam processing and flaking for promotion of high feed intake of low-roughage rations.

3. **Barley.** Grinding or rolling improves the digestibility and utilization of barley. To maximize intake and minimize digestive disturbances such as bloat, it is recommended that in high- and all-concentrate rations, barley be medium grind (to avoid fines).

No benefit for steam processing and flaking of barley vs dry rolling has been demonstrated.

High-moisture barley has a nutritive value for cattle comparable to dry ground or rolled barley. It is recommended that high-moisture barley be medium rather than fine grind to reduce the proportion of fines.

4. **Wheat**. The nutritive value of wheat is improved by processing. Steam processed and flaked wheat is well utilized by cattle provided the flakes are not thin. Fine grinding of wheat reduces feed intake and will likely cause acidosis, although acidosis depends on the variety of wheat fed. Heat treatment of wheat improves its digestibility, but does not significantly improve its net energy. From the above, it may be concluded that the nutritive value of wheat is optimized by dry rolling, coarse grinding, or steam processing to produce a thick flake.

Fig. 8-24. Flaked wheat. (Courtesy, Benedict Feeding Co., Casa Grande, AZ)

5. **Oats.** Although whole oats are better utilized by cattle than whole barley, rolling or grinding improves their utilization, also.

FEED SUBSTITUTION TABLES

Successful cattle producers are keen students of values. They recognize that feeds of similar nutritive properties can and should be interchanged in the ration as price relationships warrant, thus making it possible at all times to obtain a balanced ration at the lowest cost.

Table 8-10, Feed Substitution Table for Beef Cattle, As-fed Basis, is a summary of the comparative values of the most common U. S. feeds. In arriving at these values, two primary factors besides chemical composition and feeding value have been considered—namely, palatability and carcass quality.

In using this feed substitution table, the following facts should be recognized:

1. That, for best results, different ages and groups of animals within classes should be fed differently.

2. That individual feeds differ widely in feeding value. Barley and oats, for example, vary widely in feeding value according to the hull content and the test weight per bushel, and forages vary widely according to the stage of maturity at which they are cut and how well they are cured and stored.

3. That nonlegume forages may have a higher relative value to legumes than herein indicated provided the chief need of the animal is for additional energy rather than for supplemented protein. Thus, the nonlegume forages of low value can be used to better advantage for wintering mature, dry beef cows than for young calves.

On the other hand, legumes may actually have higher value relative to nonlegumes than herein indicated provided the chief need is for additional protein rather than for added energy. Thus, no protein supplement is necessary for breeding beef cows provided a good-quality legume forage is fed.

4. That, based primarily on available supply and price, certain feeds—especially those of medium protein content, such as brewers' dried grains, corn gluten feed (gluten feed), distillers' dried grains, distillers' dried solubles, peanuts, and peas (dried)—may be used interchangeably as (a) grains and by-product feeds, and/or (b) protein supplements.

5. That the feeding value of certain feeds is ma-

TABLE 8-10
FEED SUBSTITUTION TABLE FOR BEEF CATTLE (AS-FED BASIS)

Feedstuff	Relative Feeding Value (lb for lb) In Comparison With the Designated (underlined) Base Feed Which = 100	Maximum Percentage of Base Feed (or comparable feed or feeds) Which It Can Replace For Best Results	Remarks
GRAINS, BY-PRODUCT FEEDS, ROOTS, AND TUBERS:[1] (Low and Medium Protein Feeds)			
Corn, No. 2	*100*	*100*	The most important concentrate for cattle in the U.S. Grind coarsely or flake.
Almond hulls, dried, no shells	70–75	15–30	
Almond hulls and shell meal	35	15–20	
Apple pomace, air-dry	78	33⅓	Values given are for apple pomace with paper or rice hulls as press aids.
Bakery products, dried	110	15–30	
Bakery waste, not dried (30% water)	75	15–30	
Barley	90	25–100	The heavier the barley and the smaller the proportion of hulls, the higher the feeding value. Grind coarsely or roll for cattle. In Canada, where considerable barley is fed, it is often used as the only basal feed in the ration once animals are accustomed to it.
Beans (cull)	80	10	Best when cooked, but can also be fed raw. Beans should be ground. When cooked, 3–4 lb *(1.4–1.8 kg)*/head daily; when raw, 1–2 lb *(0.45–0.91 kg)*. Scouring may occur if they constitute more than 15% of total ration.

(Continued)

TABLE 8-10 (Continued)

Feedstuff	Relative Feeding Value (lb for lb) In Comparison With the Designated (underlined) Base Feed Which = 100	Maximum Percentage of Base Feed (or comparable feed or feeds) Which It Can Replace For Best Results	Remarks
GRAINS, BY-PRODUCT FEEDS, ROOTS, AND TUBERS:[1] (Continued)			
Beet pulp, dried	90	50	
Beet pulp, molasses, dried	90–95	50	
Beet pulp, wet	25	40	50% the value of corn silage. May compose 40% of ration on dry matter basis.
Brewers' dried grains	80	33⅓	Not very palatable. Fed chiefly to dairy cattle. Too bulky and usually too costly to be used in finishing rations.
Brewers' grains (wet)	13–15	33⅓	Grains usually come from barley. Best to haul and feed directly. Can be stored in silo if salt is added at rate of 25 lb *(11.4 kg)* per ton of grains.
Buckwheat	55–75	33⅓	Should be ground and mixed with other grains.
Carrots (cull)	10–15	20–25	Store 3–4 weeks before using; fresh carrots cause scouring. Feed whole or sliced.
Citrus pulp, dried	80–88	25–50	
Corn-and-cob meal	85–90	100	
Corn gluten feed (gluten feed)	85–90	50	
Distillers' dried grains	73–90	33⅓	Rye distillers' dried grains are of lower value than similar products made from corn or wheat. Distillers' dried grains are used chiefly for dairy cattle.
Distillers' dried solubles	73–90	33⅓	The chief difference between distillers' dried grains and distillers' dried solubles is the higher B vitamin content of the latter. Normally this is not important for cattle.
Fat (animal or vegetable)	225	5	Fat has 203 megacalories energy/100 lb *(45.4 kg)* for maintenance and 127 megacalories for weight gain, as compared to 92 and 60, respectively, for corn.
Hominy feed	100	50	
Manure, cattle, without bedding	75	50	Approximately 80% of the total nutrients of feeds is excreted as animal manure. However, the feeding value of manure will vary according to (1) the nutritive value of the feeds initially fed, (2) the class, age, and individuality of the animal to which the feeds were initially fed, and (3) the handling and processing of the manure.
Manure, poultry (see poultry house litter)			
Molasses, beet	75	10–40	Value is highest when used as an appetizer. May be laxative if fed at levels above 6 lb *(2.7 kg)* daily.
Molasses, cane	75	10–40	Value is highest when used as an appetizer.
Molasses, citrus	65–75	10–40	
Molasses, wood	26–30	10–20	Unpalatable.
Oats	70–90	10–100	Valuable for young stock, for breeding stock and for getting animals on feed. Oats have lowest value for finishing cattle and should be limited to ⅓ of such rations. Also, the feeding value of oats varies according to the test weight per bushel. Grind or roll for cattle.
Paunch, dried (also see "paunch-blood" under Protein Supplements of this table)	90	5–10	Dried paunch is not palatable, with the result that it depresses appetite. Rate of gain is not affected, but feed efficiency is slightly lowered.

(Continued)

TABLE 8-10 (Continued)

Feedstuff	Relative Feeding Value (lb for lb) In Comparison With the Designated (underlined) Base Feed Which = 100	Maximum Percentage of Base Feed (or comparable feed or feeds) Which It Can Replace For Best Results	Remarks
GRAINS, BY-PRODUCT FEEDS, ROOTS, AND TUBERS:[1] (Continued)			
Peas (cull), dried	88	40	Because of lack of palatability, peas will lower feed intake if they constitute more than 20% of the total ration. Also, there is bloat hazard if they exceed 40% of the ration.
Pear waste, air-dry	75	40	
Potatoes (Irish), wet	20–25	85	When fed with alfalfa hay, they are worth about 80% as much per ton as corn silage. Do not feed frozen. Sunburned, decomposed, or sprouted potatoes should not make up more than 10% of potatoes fed. Keep steers' heads down while eating to prevent choking.
Potatoes (Irish), dehydrated	88	50	Excellent source of energy, but deficient in protein, minerals, and vitamins.
Potatoes (sweet)	25	85	
Potatoes (sweet), dehydrated	95–100	50	Dehydrated sweet potatoes are more palatable than dehydrated Irish potatoes.
Poultry house litter	10–40	15–25	Poultry house litter may also be used as a protein source (see Protein Supplements, this table).
Prunes	62	15	Because of the laxative quality of prunes, they should be limited to 7% of the total ration.
Raisins (cull)	70	33⅓	
Raisin pulp	53	25	
Rice (rough rice)	80	100	
Rice bran	66⅔–75	33⅓	
Rice polishings	88	25	
Rye	96	33⅓	Not palatable when fed in large amounts.
Screenings, refuse	62–70	25–35	Should be finely ground in order to kill noxious weed seeds. Quality varies; good-quality screenings are equal to oats whereas poor-quality screenings resemble straw.
Sorghum (milo, kafir), grain	90–95	100	Varieties vary in protein content. Grind or roll for cattle.
Spelt and emmer	70–90	30–100	Similar to oats.
Wheat	100–105	50	Grind coarsely, or roll.
Wheat bran	70–90	25–33⅓	Because of its bulk and fiber, bran is not desirable for finishing rations. Bran is valuable for young animals, for breeding animals, and for starting animals on feed.
Wheat-mixed feed (mill run)	95	33⅓	Sometimes fed to the breeding herd, to young calves, and to finishing cattle being started on feed.
Wheat screenings	85	50	
Wood (cooked)	75–80	70	Wood products, which are largely cellulose and lignin, must be cooked before animals can digest them.
PROTEIN SUPPLEMENTS:			
Soybean meal (41%)	100	100	Slightly laxative effect.
Alfalfa or clover screenings	70–75	50	Grind finely to destroy weed seeds.

(Continued)

TABLE 8-10 (Continued)

Feedstuff	Relative Feeding Value (lb for lb) In Comparison With the Designated (underlined) Base Feed Which = 100	Maximum Percentage of Base Feed (or comparable feed or feeds) Which It Can Replace For Best Results	Remarks
PROTEIN SUPPLEMENTS: (Continued)			
Brewers' dried grains	55–65	50	Not very palatable. Fed chiefly to dairy cattle.
Copra meal (coconut oil meal), 21% . . .	90–100	50	
Corn gluten feed (gluten feed)	65–75	50–100	
Corn gluten meal (gluten meal)	90–100	50	Somewhat unpalatable.
Cottonseed meal (41%)	100	100	
Distillers' dried grains	65–70	100	Rye distillers' grains are about 10% lower in protein than similar products made from corn or wheat. Low in palatability.
Distillers' dried solubles	70	100	
Feather meal (hydrolyzed; 84% protein)	175	50	Feather meal is unpalatable; hence, cattle must be accustomed to it gradually and it must be limited in quantity. It is best used for wintering brood cows and stocker cattle.
Legume screenings	75	75	Satisfactory, but less palatable than soybean or cottonseed meal.
Linseed meal (35%)	95	100	Linseed meal has laxative effect. Some cattle will not tolerate more than 5–8% linseed meal in the ration.
Paunch-blood feed (also see "paunch, dried" under Grains section of this table)	100	100	At slaughter, each bovine yields about 20 lb *(9.1 kg)* of paunch and 20 lb *(9.1 kg)* of blood. Dried paunch runs around 10% protein, dried blood around 80%, and 50–50 mixture of two products, around 45%.
Peanut meal (45%)	100	100	Peanut meal may become rancid if stored too long, especially in warm, moist climates.
Peas (cull), dried	65–75	50	
Poultry house litter	50–55	25	Poultry house litter may also be used as an energy source (see Grains section of this table).
Rapeseed meal (37%)	88	75	Rapeseed meal should be limited to not more than 2 lb *(0.91 kg)* per cow.
Safflower meal, well hulled (42%)	92	100	
Safflower meal, with hulls (20%)	40–45	100	Safflower meal with hulls is unpalatable. Thus, it should be mixed with more palatable feeds.
Sesame meal	90–95	25	
Soybeans, whole	95–100	95	Not satisfactory for finishing calves. Soybean allowance should be limited to amount necessary to balance the ration. Larger amounts may be unduly laxative and cause cattle to go off feed.
Sunflower meal (39%)	95–100	100	If poorly hulled and lower protein content than 39%, feeding value will be lowered accordingly. It is well liked by cattle and keeps well in storage.
DRY FORAGES AND SILAGES:[2]			All the dry nonlegume forages listed herein are satisfactory when needed minerals and either a limited amount of legume hay or a protein supplement are supplied to balance the ration.
Alfalfa hay, all analyses	***100***	***100***	Does away with or lessens protein supplement requirements.
Alfalfa silage	33⅓–50	50–85	When alfalfa silage replaces corn silage, more energy feed must be provided but less protein.
Alfalfa straw	37	50	Feed with good hay.

(Continued)

TABLE 8-10 (Continued)

Feedstuff	Relative Feeding Value (lb for lb) In Comparison With the Designated (underlined) Base Feed Which = 100	Maximum Percentage of Base Feed (or comparable feed or feeds) Which It Can Replace For Best Results	Remarks
DRY FORAGES AND SILAGES:[2] (Continued)			
Apple pomace silage	17–25	50–85	Usually fed as a substitute for corn or grass silage. 50% the value of corn silage. Sometimes fed out of a stack or trench silo.
Apples	17–25	50–85	Do not feed more than 25 lb *(11.4 kg)*/mature bovine. Not recommended for finishing cattle. Danger of choking when fed whole. Relatively high handling cost.
Bagasse, dried; sugarcane or sorghum	10–20	5–10	
Barley hay	70	100	Avoid bearded varieties.
Barley silage	25–40	50–80	In silage, there is no problem with bearded varieties, which usually outyield beardless.
Barley straw	40	70	Of the cereal straws, barley ranks next to oat straw in feeding value. Feed to dry pregnant cows. Supplement with 5–6 lb *(2.3–2.7 kg)* alfalfa hay or 1–2 lb *(0.45–0.91 kg)* of 30–40% protein supplement.
Bean straw	34	50	Feed with good hay.
Beet tops, fresh	20	33⅓–50	In the West, large acreages of fresh beet tops are grazed by cattle and sheep. Bloat may be a problem when tops are frozen. Tops are laxative. Add 2½ lb *(1.1 kg)* of ground limestone/ton of feed.
Beet top silage, sugar	17–25	33⅓–50	Feed 2 oz *(56.7 g)* of finely ground limestone or chalk with each 100 lb *(45.4 kg)* of tops, as calcium changes the oxalic acid to insoluble calcium oxalate.
Clover hay, crimson	90–100	100	Crimson clover hay has a considerably lower value if not cut at an early stage.
Clover hay, red	90–100	100	If the rest of the ration is adequate in protein, clover hay will be equal to alfalfa in feeding value; otherwise, it will be lower.
Clover straw	37	50	Feed with good hay.
Clover-timothy hay	80–90	100	Value of clover-timothy mixed hay depends on the proportion of clover present and the stage of maturity at which it is cut.
Corncobs, ground	70	90	Ground corncobs can be used as the only roughage for beef cattle if properly supplemented with proteins, minerals, and vitamins.
Corn fodder	75	80–90	
Corn husklage (shucklage)	50	80–90	Highest and best use is for dry pregnant cows. It is slightly higher in energy and more palatable than corn stover.
Corn silage	33⅓–50	50–85	
Corn (sweet) silage, cannery waste	26–40	50–85	
Corn stover	45	70–90	Corn stover will meet the energy needs of dry pregnant cows, but is deficient in protein and low in phosphorus and vitamin A. Two acres of cornstalks will carry a cow 100–120 days.
Corn (sweet) stover	50	80–90	
Cottonseed hulls	66⅔	75	Use for dry pregnant cows. Supplement daily with 4–6 lb *(1.8–2.7 kg)* of good legume hay or 1–2 lb *(0.45–0.91 kg)* of a 30–40% protein supplement.
Cowpea hay	90–100	100	
Gin trash, cotton	75	75	

(Continued)

TABLE 8-10 (Continued)

Feedstuff	Relative Feeding Value (lb for lb) In Comparison With the Designated (underlined) Base Feed Which = 100	Maximum Percentage of Base Feed (or comparable feed or feeds) Which It Can Replace For Best Results	Remarks
DRY FORAGES AND SILAGES:[2] (Continued)			
Grape pomace or meal	5–15	10–15	Pomace including stems is of little value as a feed.
Grass-legume mixed hay	80–90	100	Value depends on the proportion of legume present and the stage of maturity at which it is cut.
Grass-legume silage	32–47	50–85	Unless grain is added as a preservative, grass silage requires more energy feed, but less protein supplement than corn silage when fed to finishing cattle.
Grass silage	30–45	50–85	For finishing cattle, grass silage must be supplemented with additional energy feeds, such as cereal grain or molasses, to be of the same value as corn silage.
Hop vine silage	20	50–75	It should be chopped when placed in the silo.
Hops, spent, dehydrated	80	50–65	Devoid of carotene; feed with legume hay.
Johnsongrass hay	70	100	
Lespedeza hay	80–100	100	Feeding value of lespedeza hay varies considerably with stage of maturity at which it is cut.
Mint hay	70–80	75	Cattle tire of mint hay when it is fed as the only roughage for extended periods.
Oat hay	75	100	
Oat silage	32–47	50–85	Must be chopped finely to exclude air from silo.
Oat straw	50	75	Oat straw is the best of the cereal straws. Use for dry pregnant cows. Supplement daily with 4–6 lb *(1.8–2.7 kg)* of good legume hay or 1–2 lb *(0.45–0.91 kg)* of 30–40% protein supplement.
Paper (newspaper; waste paper)	66⅔	50	Paper varies in feeding value in proportion to the cellulose (most paper is 60–90% cellulose) and lignin content. Magazine and bookstock papers are higher in cellulose and lower in lignin than newspapers; hence, of higher feeding value. Pelleting or cubing may increase the value of paper. *Caution:* Some newspapers contain heavy metals (boron, lead, barium, and antimony), sometimes used as a dye carrier in printer's ink, which may be toxic to animals. This is especially true of "funny" papers because of the quantity of heavy metals carried on the colored ink of the comics.
Pea straw	45–75	60–75	
Pea-vine hay	100–110	75–90	Can constitute the only roughage for finishing cattle.
Pea-vine silage	33⅓–50	50–85	Unless grain is added as a preservative, pea-vine silage requires more energy feed, but less protein supplement than corn silage when fed to finishing cattle.
Potato silage	25–30	50–75	About 75% the value of corn silage.
Prairie hay	65–70	100	
Reed canarygrass hay	70	100	
Rice straw	47	70	High levels of rice straw can be used for wintering cattle if the straw is properly fortified.
Sawdust	75–80	70	Feeding value varies among species of trees. Digestibility is increased by cooking and other treatments.
Sorghum fodder	70	100	
Sorghum silage (grain varieties)	32–47	50–85	For finishing cattle, 85–90% as valuable as corn silage and must be supplemented in the same manner as corn silage.
Sorghum silage (sweet varieties)	25–30	50–85	Nearly equal to grain varieties in value per acre because of greater yield.

(Continued)

TABLE 8-10 (Continued)

Feedstuff	Relative Feeding Value (lb for lb) In Comparison With the Designated (underlined) Base Feed Which = 100	Maximum Percentage of Base Feed (or comparable feed or feeds) Which It Can Replace For Best Results	Remarks
DRY FORAGES AND SILAGES:[2] (Continued)			
Sorghum (milo) stover	35	70–90	Can be grazed or harvested and stored either as dry feed or silage. About 2% higher in protein, but less palatable, than corn stover.
Soybean hay	85–90	50–75	Lower value than alfalfa hay, largely due to greater wastage in feeding. It may cause scouring when fed alone.
Sudangrass hay	70	100	
Sunflower silage	25–35	50–85	65–75% value of corn silage. Somewhat unpalatable and may cause constipation. Harvest for silage when ½–⅔ of heads are in bloom.
Sweet clover hay	100	100	Value of sweet clover hay varies widely. Moldy or spoiled sweet clover hay may cause sweet clover disease.
Timothy hay	70	100	
Vetch-oat hay	80–90	100	The higher the proportion of vetch, the higher the value.
Wheat hay	70	100	
Wheat straw	35	65	Of the cereal straws, wheat ranks third in nutritive value, behind oat straw and barley straw. Highest and best use is for dry pregnant cows. Supplement daily with 6 lb *(2.7 kg)* of alfalfa or 2 lb *(0.91 kg)* of a 30–40% protein supplement.

[1]Roots and tubers are of lower value than the grain and by-product feeds due to their higher moisture content.

[2]Silages are of lower value than dry forages due to their higher moisture content.

terially affected by preparation. Thus, wheat must be coarsely ground or rolled for cattle. The values herein reported are based on proper feed preparation in each case.

For the reasons noted above, the comparative values of feeds shown in the feed substitution table are not absolute. Rather, they are reasonably accurate approximations based on average-quality feeds, together with experiences and experiments.

HOME-MIXED VS COMMERCIAL FEEDS

The value of farm-grown grains—plus the cost of ingredients which need to be purchased to balance the ration, and the cost of grinding and mixing—as compared to the cost of commercial ready-mixed feeds laid down on the farm, should determine whether it is best to mix feeds at home or depend on ready-mixed feeds.

Although there is nothing about the mixing of feeds which is beyond the capacity of the intelligent farmer or rancher, under many conditions a commercial mixed feed supplied by a reputable dealer may be the most economical and the least irksome. The commercial dealer has the distinct advantages of (1) purchase of feeds in quantity lots, making possible price advantages, (2) economical and controlled mixing, and (3) the hiring of scientifically trained personnel for use in determining the rations. Because of these advantages, commercial feeds are finding a place of increasing importance in American agriculture.

Also, it is to the everlasting credit of reputable feed dealers that they have been good teachers, often getting livestock producers started in the feeding of balanced rations, a habit which is likely to remain with them whether or not they continue to buy commercial feeds.

HOW TO SELECT COMMERCIAL FEEDS

There is a difference in commercial feeds! That is, there is a difference from the standpoint of what operators can purchase with their feed dollars. Smart operators will know how to determine what constitutes the best in commercial feeds for their specific needs. They will not rely solely on how the feed looks and smells or on the feed salesperson. The most important factors to consider or look for in buying a commercial feed follow.

1. **The reputation of the manufacturer.** This should be determined by (a) conferring with other livestock producers who have used the particular products, and (b) checking on whether or not the commercial feed under consideration has consistently met its guarantees. The latter can be determined by reading the bulletins or reports published by the respective state departments in charge of enforcing feed laws.

2. **The specific needs.** Feed needs vary according to (a) the class, age, and productivity of the animals, and (b) whether the animals are fed primarily for maintenance, growth, finishing (or show-ring fitting), reproduction, lactation, or work. The wise operator will buy different formula feeds for different needs.

3. **The feed tag.** Most states require that mixed feeds carry a tag that guarantees the ingredients and the chemical makeup of the feed. Feeds with more protein and fat are better, and feeds with less fiber are better.

In general, if the fiber content is less that 8%, the feed may be considered as top quality; if the fiber is more than 8 but less than 12% the feed may be considered as medium quality; while feeds containing more than 12% fiber should be considered carefully. Of course, many feeds are high in fiber simply because they contain generous quantities of alfalfa; yet they may be perfectly good feeds for the purpose intended. On the other hand, if oat hulls and similar types of high fiber ingredients are responsible for the high fiber content of the feed, the quality should be questioned. The latter type of fiber is poorly digested and does not provide the nutrients required to stimulate the digestion of the fiber in roughages.

4. **Flexible formulas.** Feeds with flexible formulas are usually the best buy. This is because the price of feed ingredients in different source feeds varies considerably from time to time. Thus, a good feed manufacturer will shift formulas as prices change, so as to give the operator the most for the money. This is as it should be, for (a) there is no one best ration, and (b) if substitutions are made wisely, the price of the feed can be kept down and the feeder will continue to get equally good results.

BALANCED RATIONS FOR BEEF CATTLE

A balanced ration is one which provides an animal the proper proportions and amounts of all the required nutrients for a period of 24 hours.

Several suggested rations for different classes of cattle are listed in Table 8-13 of this chapter. Generally these rations will suffice, but it is recognized that rations should vary with conditions, and that many times they should be formulated to meet the conditions of a specific farm or ranch, or to meet the practices common to an area. Thus, where cattle are on pasture, or are receiving forage in the drylot, the added feed (generally grains, by-product feeds, and/or protein supplements), if any, should be formulated so as to meet the nutritive requirements not already provided by the forage.

Rations may be formulated by the methods which follow, but first the following pointers are noteworthy:

1. In computing rations, more than simple arithmetic should be considered, for no set of figures can substitute for experience. Compounding rations is both an art and a science—the art comes from cattle know-how and experience, and keen observation; the science is largely founded on chemistry, physiology, and bacteriology. Both are essential for success.

Also, a good producer should know how to balance a ration. Then, if the occasion demands, it can be done. Perhaps of even greater importance, the producer will then be able to select and buy rations with informed appraisal; to check on how well the manufacturer, dealer, or consultant is contributing to the business; and to evaluate the results.

2. Before attempting to balance a ration for cattle, the following major points should be considered:

a. **Availability and cost of the different feed ingredients.** Preferably, cost of ingredients should be based on delivery to the mill and after processing—because delivery and processing costs are quite variable. A simple method of evaluating feeds is presented in the section headed, "How to Determine the Best Buy in Feeds."

b. **Moisture content.** When considering costs and balancing rations, feeds should be placed on a comparable moisture basis; usually an air-dry basis, or 10% moisture content, is used. This is especially important in the case of silage. Here's how silage may be converted to an air-dry (10% moisture) basis:

If Silage Has a Moisture Content of—	Divide by
(%)	
75	3.6
70	3.0
65	2.6
60	2.25

c. **Composition of the feeds under consideration.** Feed composition tables *(book values)*, or average analysis, should be considered only as guides, because of wide variations in the composition of feeds. For example, the protein and moisture contents of milo, hay, and silages are quite variable. Whenever possible, especially with large operations, it is best to take a representative sample of each major feed ingredient and have a chemical analysis made of it for the more common constituents—protein, fat, fiber, nitrogen-free extract, and moisture; and often calcium, phosphorus, and carotene. Such ingredients as oil meals and prepared supplements, which must meet specific standards, need not be analyzed so often, except as quality control measures.

Despite the recognized value of a chemical analysis, it is not the total answer. It does not provide information on the availability of nutrients to the animal; it varies from sample to sample, because feeds vary and a representative sample is not always easily obtained, and it does not tell anything about the associated effects of feedstuffs—for example, the apparent way in which beet pulp enhances the value of ground milo. Nor does a chemical analysis tell anything about taste, palatability, texture, or undesirable physiological effects such as bloat and laxativeness.

However, a chemical analysis does give a solid foundation on which to start the evaluation of feeds. Also, with chemical analysis at hand, and bearing in mind that it's the composition of the total feed (the finished ration) that counts, the person formulating the ration can more intelligently determine the quantity of protein to buy, and the kind and amounts of minerals and vitamins to add.

d. **Soil analysis.** If the origin of a given feed ingredient is known, a soil analysis or knowledge of the soils of the area can be very helpful; for example, (1) the phosphorus content of soils affects plant composition, (2) soils high in molybdenum and selenium affect the composition of the feeds produced, (3) iodine- and cobalt-deficient areas are important in animal nutrition, and (4) other similar soil-plant-animal relationships exist.

e. **The nutrient allowances.** This should be known for the particular class of cattle for which a ration is to be formulated; and, preferably, it should be based on controlled feeding experiments. Also, it must be recognized that nutrient requirements and allowances must be changed from time to time, as a result of new experimental findings.

3. In addition to providing a proper quantity of feed and to meeting the protein and energy requirements, a well-balanced and satisfactory ration should be:

a. Palatable and digestible.

b. Economical. Generally speaking, this calls for the maximum use of feeds available in the area, especially forages.

c. Adequate in protein content, but not higher than is actually needed. Generally speaking, medium and high protein feeds are in scarcer supply and higher in price than high energy feeds. In this connection, it is noteworthy that the newer findings in nutrition indicate (1) that much of the value formerly attributed to proteins, as such, was probably due to the vitamins and minerals which they furnished, and (2) that lower protein content rations may be used successfully provided they are fortified properly with the needed vitamins and minerals.

d. Well fortified with the needed minerals, or free access to suitable minerals should be provided; but mineral imbalances should be avoided.

e. Well fortified with the needed vitamins.

f. So formulated as to nourish the billions of bacteria in the paunch of ruminants that there will be satisfactory (1) digestion of roughages, (2) utilization of lower-quality and cheaper proteins and other nitrogenous products (thus, it is possible to use urea to constitute up to one-third of the total protein of the ration of ruminants, provided care is taken to supply enough carbohydrates and other nutrients to assure adequate nutrition for rumen bacteria), and (3) synthesis of B vitamins.

This means that rumen microorganisms must be supplied adequate (1) energy, including small amounts of readily available energy such as sugars or starches; (2) ammonia-bearing ingredients such as proteins, urea, and ammonium salts; (3) major minerals, especially sodium, potassium, and phosphorus; (4) cobalt and possibly other trace minerals; and (5) unidentified factors found in certain natural feeds rich in protein or nonprotein nitrogenous constituents.

g. One that will enhance, rather than impair, the quality of meat produced.

4. In addition to considering changes in availability of feeds and feed prices, ration formulation should be altered at stages to correspond to changes in weight and productivity of animals.

The above points are pertinent to the balancing of rations, regardless of the mechanics of computation used. In the sections that follow, four different methods of ration formulation are presented: (1) the square method; (2) the trial and error method; (3) the computer method, and (4) prediction equations for estimating nutrient requirements and feed intake. Despite the sometimes confusing mechanics of each system, if done properly, the end result of all four methods is the same—a ration that provides the desired allowances of nutrients in correct proportions economically (or at least cost), but, more important, so as to achieve the greatest net returns—for it's net profit rather than cost per ton that counts. Because feed represents by far the greatest cost item in beef production (about 80% of the cost of finishing feedlot cattle, exclusive of the purchase price of the animals) the importance of balanced rations is evident.

To compute balanced rations by whatever method, it is first necessary to have available both

TABLE 8-11
BEEF CATTLE FEEDING RECOMMENDATIONS

		In Drylot, with Following Types of Forages:			On Pasture of the Following Grades:		
Description of Animals (1)	Recommendations[1] (2)	Legume and/or Legume-Nonlegume Mixed Forages of High Quality; Consisting of Dry Forages and/or Silage (High Protein Forages) (3)	Legume and Nonlegume Forages Mixed; Consisting of Dry Forages and/or Silage (Medium Protein Forages) (4)	Nonlegume Forages; Consisting of Dry Forges and/or Silage (Low Protein Forages) (5)	Excellent (6)	Fair to Good (7)	Poor, Including Winter Pasture Consisting of Dry Grass Cured on the Stalk[2] (8)
Mature pregnant beef breeding cows (avg. wt. 1,100 lb *[500 kg]*). Medium and low protein forages may be used for pregnant cows.	Forage per head daily, in lb.	18–20 *(8.2–9.1 kg)*	18–20 *(8.2–9.1 kg)*	18–20 (8.2–9.1 kg)			
	Concentrate: (1) Supplement allowance of soybean meal (or equivalent 41–45% crude protein) per head daily, in lb.[3]			0.5–1.5 *(0.23–0.68 kg)*			0.5–1.5 *(0.23–0.68 kg)*
Mature lactating beef breeding cows (avg. wt. 1,100 lb *[500 kg]*). When possible, use high quality, high protein forage for nursing cows.	Forage per head daily, in lb.	26 *(11.8 kg)*	24 *(10.9 kg)*	22 *(10 kg)*			
	Concentrate: (1) Total concentrate allowance per head daily, including protein supplement, in lb.		2.5 *(1.1 kg)*	5 *(2.3 kg)*		2.5 *(1.1 kg)*	5 *(2.3 kg)*
	(2) Supplement, allowance of soybean meal (or equivalent 41–45% crude protein) per head daily, in lb.[3, 4]		1.5 *(0.68 kg)*	3 *(1.4 kg)*		1.5 *(0.68 kg)*	3 *(1.4 kg)*
	(3) Crude protein composition of total concentrate, in %.	10–14	14–18	18–20	10–14	14–18	18–20
Replacement heifers (weighing 400–500 lb *[181–227 kg]*); to be bred to calve as 2-year-olds. Heifers bred to calve as 3-year-olds can be wintered at a lower level.	Forage per head daily, in lb.	12–18 *(5.4–8.2 kg)*	12–18 *(5.4–8.2 kg)*	12–18 *(5.4–8.2 kg)*			
	Concentrate: (1) Total concentrate allowance per head daily, including protein supplement, in lb.	2–4 *(0.91–1.8 kg)*	2.5–4 *(1.1–1.8 kg)*	2.5–4.5 *(1.1–2.0 kg)*			2.5–4.5 *(1.1–2.0 kg)*
	(2) Supplement, allowance of soybean meal (or equivalent 41–45% crude protein) per head daily, in lb.[3, 4]		0.5–1 *(0.23–0.45 kg)*	1.25–1.5 *(0.57–0.68 kg)*			1.25–1.5 *(0.57–0.68 kg)*
	(3) Crude protein composition of total concentrate, in %.	9–13 (Cereal grains only will suffice.)	14–18	17–22			17–22

(Continued)

feeding standards and feed composition tables. Several feeding standards can be and are used, and there is practically no limit to the number of nutrients that can be listed in feed composition tables.

For purposes of simplification, the authors have prepared Table 8-11, Beef Cattle Feeding Recommendations. Then, the crude protein content of most common feeds can be obtained from the Appendix, Sec-

TABLE 8-11 (Continued)

		In Drylot, with Following Types of Forages:			On Pasture of the Following Grades:		
Description of Animals **(1)**	**Recommendations[1]** **(2)**	**Legume and/or Legume-Nonlegume Mixed Forages of High Quality; Consisting of Dry Forages and/or Silage (High Protein Forages)** **(3)**	**Legume and Nonlegume Forages Mixed; Consisting of Dry Forages and/or Silage (Medium Protein Forages)** **(4)**	**Nonlegume Forages; Consisting of Dry Forges and/or Silage (Low Protein Forages)** **(5)**	**Excellent** **(6)**	**Fair to Good** **(7)**	**Poor, Including Winter Pasture Consisting of Dry Grass Cured on the Stalk[2]** **(8)**
Stocker calves: roughed through the winter and generally grazed the following summer. Fed for winter gains of 0.75–1 lb *(0.34–0.45 kg)* per head daily (weighing 400–500 lb *[181–227 kg]*, start of period).	Forage per head daily, in lb.	12–18 *(5.4–8.2 kg)*	12–18 *(5.4–8.2 kg)*	12–18 *(5.4–8.2 kg)*			
	Concentrate: (1) Supplement allowance of soybean meal (or equivalent 41–45% crude protein) per head daily, in lb.[4]		0.25–1 *(0.1–0.45 kg)*	1.25–1.5 *(0.57–0.68 kg)*			1.25–1.5 *(0.57–0.68 kg)*
Finishing calves (weighing 400–500 lb *[181–227 kg]*, start of feeding, and 750–850 lb *[340–386 kg]*, at marketing).	Forage per head daily, in lb.	2–6 *(0.9–2.7 kg)*	2–6 *(0.9–2.7 kg)*	2–5 (0.9–2.3 kg)			
	Concentrate: (1) Total concentrate allowance per head daily, including protein supplement, in lb.	12–15 *(5.4–6.8 kg)*	12–15 *(5.4–6.8 kg)*	12–15 (5.4–6.8 kg)	10–12 *(4.5–5.4 kg)*	11–13 *(5–5.9 kg)*	12–14 *(5.4–6.4 kg)*
	(2) Supplement, allowance of soybean meal (or equivalent 41–45% crude protein) per head daily, in lb.[3, 4]	1–1.5 *(0.45–0.68 kg)*	1.5–1.75 *(0.68–0.8 kg)*	1.75–2.25 (0.8–1.0 kg)		1.5–1.75 *(0.68–0.8 kg)*	1.75–2.25 *(0.8–1.0 kg)*
	(3) Crude protein composition of total concentrate, in %.	9–11 (Cereal grains only will suffice.)	12–13	13–15	9–11 (Cereal grains only will suffice.)	12–13	13–15
Yearlings: roughed through the winter, and pasture finished the following summer. Fed for winter gains of 1–1.25 lb *(0.43–0.57 kg)* per head daily (weighing about 600 lb *[272 kg]*, start of wintering).	Forage per head daily, in lb.	16–24 *(7.3–10.9 kg)*	16–24 *(7.3–10.9 kg)*	16–24 *(7.3–10.9 kg)*			
	Concentrate: (1) Supplement allowance of soybean meal (or equivalent 41–45% crude protein) per head daily, in lb.[3]		1–1.5 *(0.45–0.68 kg)*	1.5–1.75 *(0.68–0.8 kg)*			1.5–1.75 *(0.68–0.8 kg)*

TABLE 8-11 (Continued)

		In Drylot, with Following Types of Forages:			On Pasture of the Following Grades:		
Description of Animals (1)	Recommendations[1] (2)	Legume and/or Legume-Nonlegume Mixed Forages of High Quality; Consisting of Dry Forages and/or Silage (High Protein Forages) (3)	Legume and Nonlegume Forages Mixed; Consisting of Dry Forages and/or Silage (Medium Protein Forages) (4)	Nonlegume Forages; Consisting of Dry Forges and/or Silage (Low Protein Forages) (5)	Excellent (6)	Fair to Good (7)	Poor, Including Winter Pasture Consisting of Dry Grass Cured on the Stalk[2] (8)
Finishing yearlings (weighing about 600 lb *[272 kg]*, start of feeding, and 900–1,100 lb *[409–500 kg]*, at marketing).	Forage per head daily, in lb.	2–8 *(0.9–3.6 kg)*	2–8 *(0.9–3.6 kg)*	2–8 *(0.9–3.6 kg)*			
	Concentrate: (1) Total concentrate allowance per head daily, including protein supplement, in lb.	15–19.5 *(6.8–8.9 kg)*	15–19.75 *(6.8–9.0 kg)*	15–20 *(6.8–9.1 kg)*	12–18 *(5.4–8.2 kg)*	13–19 *(5.9–8.6 kg)*	14–20 *(8.4–9.1 kg)*
	(2) Supplement, allowance of soybean meal (or equivalent 41–45% crude protein) per head daily, in lb.[3, 4]	1–1.5 *(0.45–0.68 kg)*	1.25–1.75 *(0.57–0.79 kg)*	1.5–2.5 *(0.68–1.1 kg)*		1.25–1.75 *(0.57–0.79 kg)*	1.5–2.5 *(0.68–1.1 kg)*
	(3) Crude protein composition of total concentrate, in %.	8–10 (Cereal grains only will suffice.)	11–12	12–13	8–10 (Cereal grains only will suffice.)	11–12	12–13
Finishing long-yearling steers (weighing about 800 lb *[363 kg]*, start of feeding and 1,100–1,200 lb *[500–545 kg]*, at marketing).	Forage per head daily, in lb.	2–12 *(0.9–5.4 kg)*	2–12 *(0.9–5.4 kg)*	2–12 *(0.9–5.4 kg)*			
	Concentrate: (1) Total concentrate allowance per head daily, including protein supplement, in lb.	16–22 *(7.3–10 kg)*	16–22 *(7.3–10 kg)*	16.5–22.75 *(7.5–10.3 kg)*	13–19 *(5.9–8.6 kg)*	14–20 *(6.4–9.1 kg)*	15–21 *(6.8–9.5 kg)*
	(2) Supplement, allowance of soybean meal (or equivalent 41–45% crude protein) per head daily, in lb.[3, 4]		0.5–0.75 *(0.23–0.34 kg)*	1.5–1.75 *(0.68–0.79 kg)*		0.5–0.75 *(0.23–0.3 kg)*	1.5–1.75 *(0.68–0.79 kg)*
	(3) Crude protein composition of total concentrate, in %.	9–12 (Cereal grains only will suffice.)	10–11	11–12	9–10 (Cereal grains only will suffice.)	10–11	11–12

[1]The daily forage recommendations given herein are based on dry forage. When silage is included in the ration, figure 3 lb (1.3 kg) of silage equivalent to 1 lb *(0.45 kg)* of dry forage, due to the higher moisture content of silage. Many cattle producers do not winter feed as liberally as herein recommended. In general, these operators feel that it is more profitable (1) to let cattle "hold their own" or even lose in condition during the winter months (so long as they remain healthy), to keep winter feed and labor costs at a minimum, and (2) to make all or most of the gains on grass.

[2]On a dry basis, the crude protein content of mature, weathered grasses may be 3% or less. The upper limit of the concentrate allowance recommended in column 8 should be fed on winter range when (1) the grass is less abundant, and/or (2) the grass is relatively low in protein.

[3]Soybean meal, which usually ranges from 41 to 45% protein content, is herein used as a standard merely because it is the leading U.S. protein supplement. It is to be emphasized, however, (1) that other protein supplements, including numerous commercial products, may be used, (2) that, in general, those supplements should be purchased which provide a unit of protein at the lowest cost, and those feeds which are highest in protein content are usually the most economical, and (3) that where other protein feeds are substituted for the soybean meal recommended herein (41–45% protein), an equivalent amount of crude protein should be provided—for example, approximately 2 lb *(0.9 kg)* of a 20% crude protein supplement should be provided to replace each 1 lb *(0.45 kg)* of soybean meal (although it is recognized that 2 lb *[0.9 kg]* of a 20% protein feed will generally provide more energy, and may supply more of certain other important nutrients, than 1 lb *[0.45 kg]* of soybean meal).

[4]The recommended supplement allowance is based on the assumptions (1) that cereal grains, averaging 9–13% crude protein content, comprise the major part of the concentrate mix, and (2) that the forage is not comprised entirely or predominantly of nonlegume silage. Naturally, less protein supplement will need to be added where feeds of higher protein content than the cereal grains predominate. Also, less protein supplement is required to balance a ration consisting predominantly of barley (of 12.7% crude protein content) than one consisting mostly of corn (of 8.7% crude protein content). Likewise, the upper limit of protein supplement recommended herein (or even a higher figure) is required to balance a ration where the forage is comprised entirely or largely of very low protein forages such as those that are mature and weathered.

tion II, Table II-1, Composition of Some Beef Cattle Feeds. These two tables are adequate for balancing most rations by the square method.

SQUARE METHOD OF BALANCING RATIONS

The so-called *square method* (or the Pearson Square Method) is one of several methods that may be employed to balance rations.

The square method is simple, direct, and easy. Also, it permits quick substitution of feed ingredients in keeping with market fluctuations, without disturbing the protein content.

In balancing rations by the square method, it is recognized that protein content alone receives major consideration. Correctly speaking, therefore, it is a method of balancing the protein requirement, with only incidental consideration given to the vitamin, mineral, and other nutritive requirements.

With the instructions given herein, the square method may be employed to balance rations.

In using Table 8-11 and Appendix, Section II, Table II-1, the following points should be noted:[6]

1. Under "Description of Animals"—column 1 of Table 8-11—are sufficient groups to cover the vast majority of cattle found on the nation's farms and ranches.

2. Columns 2 to 8 give pertinent recommendations relative to both forages and concentrates. These recommendations are in keeping with those advocated by scientists, and with the actual practices followed by successful operators.

In particular, it should be noted that all protein recommendations are in terms of *crude protein* content,[7] rather than digestible protein.

3. It is recognized that most farmers and ranchers generally grow their own forages, and purchase part or all of the concentrates. Thus, they generally wish to know what crude protein content of concentrate alone (including grains, by-product feeds, and/or protein supplements) they need to feed to balance the forage which is available. Likewise, feed manufacturers have need for this information in compounding mixes. For these reasons, harvested forages in Table 8-11 are classified as (a) high protein forages, (b) medium protein forages, and (c) low protein forages; and specific recommendations are made for each. Similar classifications and recommendations are made for (a) excellent, (b) fair to good, and (c) poor pastures.

4. It is often hazardous to formulate rations for excellent pastures that are different from those for poor pastures, because (a) cattle producers may be in error in appraising the quality of their pastures, and (b) pastures are generally excellent in the early spring, but become progressively poorer as the season advances unless they are irrigated and fertilized.

For purposes of illustration, let us refer to Table 8-11. Under column 5, it is noted that a mature beef breeding cow (avg. wt. 1,100 lb) that is being fed a daily ration of somewhere between 18 and 20 lb of grass hay or other nonlegume dry roughage should receive, in addition, ½ to 1½ lb daily of a protein supplement of soybean meal (or some other protein supplement which will provide an amount equivalent to 41 to 45% crude protein). To be sure, it is entirely proper to meet this recommended crude protein content of concentrate by feeding double the allowance of some protein supplement with approximately 20% crude protein content. Many times the latter may be more economical, and even advisable—for example when the forage is of poor quality and added energy feed is needed. In general, however, those feeds should be purchased which furnish a unit of protein at the lowest cost, and those feeds which supply the protein in the most concentrated form are usually the most economical.

Under column 2 of Table 8-11, additional information, of value to both feed manufacturers and cattle producers who mix their own rations, is given. For example, in Table 8-11, under "Finishing long-yearling steers . . . ," recommendations are given relative to the following:

"(3) Crude protein composition of total concentrate, in %."

The application of the square method will be illustrated by solving a practical problem.

Problem:

A cattle producer wishes to compute a balanced ration for 800-lb finishing yearling steers in drylot. Grass hay is on hand, and corn (No. 2 grade; 10.1% protein) and soybean meal (solvent process; 49.9% protein) are the cheapest concentrate feeds available. The producer wishes to know (1) the pounds each of forage and of concentrate to feed daily, and (2) the proportions of corn and of soybean meal to put in the concentrate mixture.

Step by step, the answers may be calculated as follows:

1. Table 8-11, Beef Cattle Feeding Recommenda-

---

[6]In addition, see pertinent footnotes which accompany Table 8-11.

[7]Also, it is recognized (a) that beef cattle consume a large proportion of forage, and (b) that the percentage digestibility of protein of forages differs tremendously—for example, the percent digestibility of protein of wheat straw is 11 whereas for alfalfa hay it is 71. On the other hand, the grains do not differ greatly in percent digestibility of protein.

tions (column 5, Nonlegume Forage), gives the following recommendations for 800-lb finishing steers in drylot:

a. Forage per head per day = 2 to 8 lb.

b. Concentrate per head per day = 15 to 20 lb.

c. Crude protein content of the concentrate alone where a grass hay is fed = 12 to 13%.

2. Thus, when on full feed the steers should receive daily feed allowances of somewhere between 2 and 8 lb of the grass hay, and between 15 and 20 lb of the concentrate mixture. A range is given, because (a) individual animals and different lots of cattle differ in feed capacity, (b) feeds differ in composition and feeding value, and (c) the proportion of forage should decrease whereas the proportion of concentrate should increase as the finishing period advances.

3. The proportions of corn and of soybean meal to put in the concentrate mixture may be obtained by the square method as follows:

a. Place in the center of the square the percentage of crude protein needed in the mixture; in this case 13% (using the upper limit).

b. Place at the upper left-hand corner of the square the percentage of crude protein in the soybean meal; in this case 49.9%.

c. Place at the lower left-hand corner of the square the percentage of protein in the corn (maize); in this case 10.1%.

d. Connect the diagonal corners of the square with lines, and subtract, diagonally across the square, the smaller 49.9 – 13 = 36.9 and 13 – 10.1 = 2.9) figure from the larger. Place the answers at the opposite corners. This gives the

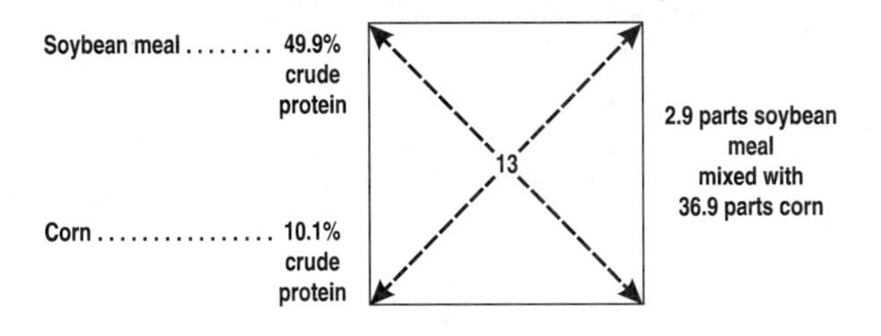

following:

e. Thus, a mixture of 2.9 parts of soybean meal and 36.9 parts of corn (39.8 parts total) will provide a concentrate mix with the desired 13% crude protein content.

f. The proportions of soybean meal and corn can be converted to 100 lb of mixture as follows (or to a ton basis by using 2,000 instead of 100):

$$\frac{2.9 \text{ parts soybean meal}}{39.8 \text{ (total of soybean meal and corn)}} \times 100 = \text{7.3 lb soybean meal}$$

$$\frac{36.9 \text{ parts corn}}{39.8 \text{ (total of soybean meal and corn)}} \times 100 = 92.7 \text{ lb corn}$$

Total . . . . 100.00 lb mix

g. Thus, to make a 13% crude protein concentrate mixture from corn and soybean meal, the cattle producer will need to mix 7.3 lb of soybean meal and 92.7 lb of corn for each 100 lb of mix.

TRIAL AND ERROR METHOD OF BALANCING

As stated, balancing rations by the square method is simple, direct, and easy, but protein alone receives major consideration. Balancing by the trial and error method goes further. In it, consideration is given to meeting whatever allowances are decided upon of each of the nutrients that one cares to list and consider. A problem-solving example follows.

The 1996 NRC Nutrient Requirements of Beef Cattle could be used to determine nutrient requirements in the following example; however, the computer program would need to be used to determine actual requirements. For ease of example, information from the 1984 NRC Nutrient Requirements of Beef Cattle is used to determine requirements. The same steps would be followed in balancing the ration, regardless of which publication was used to determine nutrient requirements for the specific animal.

Problem:

A cattle feeder desires to finish 700-lb medium-frame steers in drylot, gaining 3.0 lb per head per day. The following feeds are available: corn No. 2, alfalfa hay (midbloom), cottonseed hulls, molasses (sugarcane), soybean meal (49%), and salt. It is planned that the ration consist of 85% concentrate and 15% roughage.

The step by step application of the trial and error method to balancing a ration to meet the above needs follows.

1. **Set down the allowances.** Refer to Appendix, Section I, Table I-1, 1984 NRC Nutrient Requirements of Beef Cattle, but bear in mind that it lists requirements, not allowances. Requirements do not provide for margins of safety to compensate for variations in feed composition, and possible losses during storage and processing. Nor do they recognize that the needs of cattle do not necessarily remain the same from day to day or from period to period; that the age and size of animal, the stage of gestation and lactation, the kind and degree of activity, and the amount of stress; the system of management; the health, condition, and

temperament of the animal; and the kind, quality, and amount of feed are all exerting a powerful influence in determining nutritive needs. How well the feeder understands, anticipates, interprets, and meets these variable requirements usually determines the success or failure of the ration and the results obtained. Because the effects of each of these factors on nutritive needs vary in degree, and so little is known about many of them, the authors suggest that each cattle producer or feed company arrive at the nutritive allowances for a specific class of cattle and set of conditions

For purposes of this illustration, let's assume that the following allowances are needed for finishing 700-lb medium-frame yearling steers in drylot, gaining 3.0 lb/head/day:

Daily Feed		Proportion Concentrate and Roughage[8]	Crude Protein	TDN	Calcium	Phosphorus
(lb)	*(kg)*		*(%)*	*(%)*	*(%)*	*(%)*
16.9	*7.7*	85% concentrate, 15% roughage	10.5	76.5	0.44	0.23

2. **Apply the trial and error method.** Considering (a) available feeds, and (b) common feeding practices, the next step is arbitrarily to set down a ration and see how well it measures up to the desired allowances. The approximate composition of the available feeds may be arrived at from feed composition tables (Appendix, Section II, Table II-1, Composition of Some Beef Cattle Feeds) if an actual chemical analysis is not available. Where commercial supplements are used, the guarantee on the feed tag may be used.

Let's try the following ration:[9]

Ingredients	Percent	Per Short Ton	Per Metric Ton
	(%)	*(lb)*	*(kg)*
Corn (no. 2)	69.7	1,394	*697*
Alfalfa hay (midbloom)	5.0	100	*50*
Cottonseed hulls	10.0	200	*100*
Molasses (sugarcane)	10.0	200	*100*
Soybean meal (49%)	5.0	100	*50*
Salt	0.3	6	*3*
Total	100.0	2,000	*1,000*

Here's a listing of the desired allowances, followed by the composition of the proposed ration:

	Daily Feed		Concentrate to Roughage	Crude Protein	TDN	Calcium	Phosphorus
	(lb)	*(kg)*		*(%)*	*(%)*	*(%)*	*(%)*
Desired allowances	16.9	*7.7*	85% conc., 15% rough.	10.5	76.5	0.44	0.23
Approx. analysis of proposed ration	16.9	*7.7*	85% conc., 15% rough.	10.9	73.1	0.21	0.29

Thus, the proposed ration is slightly high in protein (0.4% over), low in TDN (3.4% under), low in calcium (0.23% under), and slightly high in phosphorus (0.06%). To correct these deficiencies, let's increase the concentrate, and decrease the roughage; increase the corn and alfalfa, reduce the cottonseed hulls, delete the molasses and add fat and calcium carbonate. Thus, our second trial ration is:

Ingredients	Percent	Per Short Ton	Per Metric Ton
	(%)	*(lb)*	*(kg)*
Corn (no. 2)	82.2	1,644	822.0
Alfalfa hay (midbloom)	10.0	200	100.0
Soybean meal (49%)	3.4	67	34.0
Cottonseed hulls	2.45	49	24.5
Fat	0.9	18	9.0
Calcium carbonate	0.75	16	7.5
Salt	0.3	6	3.0
Total	100.0	2,000	1,000.0

	Daily Feed		Concentrate to Roughage	Crude Protein	TDN	Calcium	Phosphorus
	(lb)	*(kg)*		*(%)*	*(%)*	*(%)*	*(%)*
Desired allowances	16.9	*7.7*	87.55% conc., 12.45 rough.	10.5	76.5	0.44	0.23
Approx. comp. of 2nd proposed allowances	16.9	*7.7*	87.55% conc., 12.45% rough.	10.5	76.5	0.44	0.26

This ration approximates the desired allowances and may be considered satisfactory.

NET ENERGY METHOD OF BALANCING

In order to apply the net energy method to the feeding of livestock, the following net energy values must be available:

[8]The proportion of roughage to concentrate is an arbitrary decision which producers must make.

[9]All calculations are on an as-fed basis.

1. A table showing the net energy requirements of the particular class of animal. Appendix, Section I, Table I-1, 1984 NRC Nutrient Requirements of Beef Cattle, shows the net energy requirements for growing-finishing beef cattle. Again, the 1996 NRC Nutrient Requirements of Beef Cattle could be used to determine nutrient requirements in the following example; however, the computer program would need to be used to determine actual requirements. For ease of example, information from the 1984 NRC Nutrient Requirements of Beef Cattle is used to determine requirements. The same steps would be followed in balancing the ration, regardless of which publication was used to determine nutrient requirements for the specific animal.

2. A table showing the nutrient composition of feeds, with the net energy of each feed partitioned into energy used for body maintenance and for gain; thus, the net energy values in megacalories (Mcal) per unit (kg) are needed for each feed for maintenance (NE_m) and for gain (NE_g) (see Appendix, Section II, Table II-1, Composition of Some Common Beef Cattle Feeds).

The two examples that follow will show how to apply the net energy method. In the first example, net energy values of feeds are used to calculate the number of kilograms of a given ration that a steer would need to consume to make a specified daily gain. In the second example, net energy is used to predict average daily gain based on consuming a certain number of kilograms of a specified ration. Bear in mind that the ration in both cases (in these examples, the ration in Table 8-12) must be balanced for protein, minerals, and vitamins, in order for these net energy values to have validity for calculating daily consumption and predicting average daily gain.

EXAMPLE 1. Using net energy values to calculate the number of pounds of the ration that must be consumed to produce a specific gain. How many calories would a medium-frame 660-lb yearling steer need to consume to gain 2.6 lb daily?

Step 1. Calculate the net energy for maintenance (NE_m) and gain (NE_g) values for a kilogram of the ration shown in Table 8-12.

By referring to Appendix, Section II, Table II-1, Composition of Some Beef Cattle Feeds, it is determined that 1 lb of the Table 8-12 ration supplies 0.70 Mcal of net energy for maintenance (Mcal NE_m) and 0.45 Mcal of net energy for gain (Mcal NE_g).

Step 2. From Appendix, Section I, Table I-1, 1984 NRC Nutrient Requirements of Beef Cattle, we find that the requirement for a 660-lb medium-frame steer calf to gain 2.61 lb daily are as follows:

	Mcal/Day
Ne_m	5.55
Ne_g	4.90

Step 3. Pounds of feed to meet the daily maintenance requirement:

5.55 Mcal ÷ 0.70 Mcal = 7.9 lb

Step 4. Pounds of feed to meet the requirement for 2.61 lb gain:

4.90 Mcal ÷ 0.45 Mcal = 10.9 lb

TABLE 8-12
RATION FOR FINISHING CATTLE

Ration Ingredient		Composition of Ingredients (as-fed) NE_m[1]	Ration Supplies NE_m[1]	Composition of Ingredients (as-fed) NE_g[2]	Ration Supplies NE_g[2]
	(lb)	*(Mcal/lb)*[3]	*(Mcal)*[3]	*(Mcal/lb)*[3]	*(Mcal)*[3]
Corn, grain	68.6	0.86	59.4[4]	0.59	40.5[5]
Soybean meal, solvent, 49% protein	4.0	0.72	2.9	0.47	1.9
Alfalfa hay (midbloom)	27.0	0.52	14.0	0.28	7.6
Salt	0.4	—	—	—	—
Total	100.0	2.10 ÷ 3 = 0.70	75.9	1.34 ÷ 3 = 0.45	50.0

[1]NE_m = net energy for maintenance.

[2]NE_g = net energy for gain.

[3]Mcal stands for megacalorie.

[4]68.6 × 0.86 = 59.0.

[5]68.6 × 0.59 = 40.5.

Step 5. Total pounds of feed steer must eat daily to gain 2.6 lb:

7.9 + 10.9 = 18.8 lb

EXAMPLE 2. Using net energy to predict the average daily gain of a 660-lb steer that is consuming a certain number of kilograms of a specified ration. Let's assume that we have a 660-lb steer that is consuming 20 lb of the ration shown in Table 8-12. What daily gain should be expected?

Step 1. Pounds of feed to meet the daily maintenance requirement = 7.9 lb. (See prior example)

Step 2. Pounds of feed left for gain:

20.0 − 7.9 = 12.1 lb

Step 3. Mcal of NE_g supplied by remaining feed:

12.1 lb × 0.45 Mcal = 5.4 Mcal

Step 4. Daily gain expected from 4.90 Mcal of NE_g = 2.6 lb. (See previous example.) Therefore, 5.4 Mcal will produce 2.9 lb daily gain (4.9:2.6::5.4:X. So, X = 2.9).

COMPUTER METHOD OF BALANCING

Until the late 1970s, only those beef producers with access to a large mainframe computer could formulate a ration using the computer. Usually, this was limited to those associated with a university or subscribing to a time-sharing system. Then came the microcomputer! By the mid-1980s, most sizeable beef producers owned or had access to a microcomputer. Today, practically all large cattle feedlots and cow-calf operations use computers for ration formulations.

Despite their sophistication, there is nothing magical or mysterious about balancing rations by computer. Although they can alleviate many human errors in calculations, the data which come out of a machine are no better than those which go into it. The people back of the computer—the producer and the nutritionist who prepare the data that go into it, and who evaluate and apply the results that come out of it—become more important than ever. This is so because an electronic computer doesn't know anything about (1) feed palatability; (2) bloat prevention; (3) limitations that must be imposed on certain feeds to obtain maximum utilization; (4) the goals in the feeding program—such as growing or finishing; (5) homegrown feeds for which there may not be a suitable market; (6) feed processing and storage facilities; (7) the health, environment, and stress of the animals; and (8) those responsible for actual feed preparation and feeding. Additionally, it must be recognized that a computer may even reflect, without challenge, the prejudices and whims of those who prepare the data for it.

Hand in hand with the use of computers in balancing rations, the term *least-cost ration formulation* evolved. In some respects this designation was unfortunate, for the use of least-cost rations does not necessarily assure the highest net returns—and net profit is more important than cost per ton. For example the least-cost ration may not produce the desired daily gain or carcass quality.

An electronic computer can do little more than a good mathematician, but it can do it much faster, and it can check all possible combinations. It alleviates the endless calculations and many hours of time required for hand calculations. For example, it is estimated that there may be as many as 500 practical solutions when 6 quality specifications and 10 feedstuffs are considered for a ration.

Two basic approaches to ration formulation are practiced with computers: (1) trial-and-error formulation, and (2) linear programming.

It does not take specialized computer software to use the trial-and-error method. With the advent of the microcomputer and electronic spreadsheets, this time-consuming chore can be easily and quickly accomplished.

Linear programming is a mathematical technique in which a large number of simultaneous equations are solved in such a way as to meet the minimum and maximum levels of nutrients and levels of feedstuffs specified by the user at the lowest possible cost.

COMPUTER SOFTWARE AND HARDWARE

Computer type and size of memory and disk drive must meet the criteria of the software developer; otherwise, the software may not be usable. So, the selection of the software should precede the selection of the hardware.

Numerous companies market computer software for ration formulation. The software varies from the very simple and straightforward to the very complex packages intended for large feed manufacturers. University personnel and nutrition consultants are good sources of information on software and hardware.

A computer program is provided with, and is one of the important compenents of, the 1996 NRC Requirements of Beef Cattle. It is used to generate animal nutrient requirements and to evaluate rations for beef cattle in different production situations. Two of the well known university-developed computer ration programs for beef cattle are:

1. The *Taurus Ration Program* developed by the University of California, at Davis.

2. The *Autonrca Ration Program* (for formulating cattle rations on an "as is" moisture basis), and the *Autonrcd* (for formulating beef cattle rations on a dry matter basis) developed by Oklahoma State University, Stillwater, Oklahoma.

The computer program that is detailed in the section that follows was developed by John Carbonniere, Carnier Research, an agricultural consultant. Mr. Carbonniere now resides in Carson, California. It combines beef cattle ration balancing, least cost formulation, and feed buying and selling assistance.

SETUP OF THE COMPUTER FORMULATION

A program is a precise series of directives given to a computer which enables it to solve problems. In least-cost formulation, all the computer does is solve a series of simultaneous equations through a sophisticated system of matrix algebra.

Nutrient requirements and certain restrictions are listed in what are called *rows*. The ingredients of the individual feeds are listed under the term *columns*. The values called for in the solution of the formulation are listed under what is called the *right-hand side*.

For example, the various components of the ration that are to be looked at in the ration, such as cost, energy, protein, and minerals, would constitute the rows. The various feeds to be reviewed for the ration would have their respective cost, energy, protein, and mineral values listed under columns. If we wanted the ration to equal 1 ton, with 400 lb of protein, 1,400 kcal of NE_g/kg of feed, and certain mineral specifications, we would list the specifications under the section called right-hand side.

In addition to setting up the rows, columns, and right-hand side, any limitations concerning minimum, maximum, or fixed amounts of any particular feedstuff must be included in the program. For example, the producer may want fish meal in the ration at a level of at least 5% but not to exceed 10%. These restrictions would have to be included in the information fed into the computer. (See Fig. 8-25.)

PROCEDURE FOR COMPUTER FORMULATION

Generally speaking, electronic feed formulation (1) effects a greater saving when first applied to a ration than in subsequent applications, and (2) is of most use where a wide selection of feed ingredients is available and/or prices shift rather rapidly.

The information needed and the procedure followed in formulating rations by computer are exactly the same as in the hand method of ration formulation; namely, (1) the nutritive requirements for the particular class and kind of animal, (2) nutritive content of the feeds, and (3) ingredient costs. Sometimes this simple

LEAST-COST FORMULATION WORKSHEET

Specifications (Rows)	Ingredient A	Ingredient B	Ingredient C	Restrictions (Right-hand side)
Cost				Minimize
Total weight				1,000 lb *(454 kg)*
Crude protein				133 lb *(60 kg)*
Digestible protein				100 lb *(45 kg)*
Ether extract				25 to 80 lb *(11 to 36 kg)*
Net energy lactation				900 Mcal
Calcium				5 to 10 lb *(2.2 to 4.5 kg)*
Phosphorus				7 lb *(3.1 kg)*
Vitamin A equivalent				35,000
Vitamin D				60,000
Limits on ingredients				
Minimum				
Maximum				

Fig. 8-25. Sample worksheet for a least-cost formulation. The column labeled "Specifications" corresponds to what is termed *rows* in least-cost formulation. The various feedstuffs to be considered are listed in the succeeding columns with their respective costs and nutritive values. The last column lists the restrictions desired on the final formulation. This column corresponds to the *right-hand side*.

fact is overlooked because of the awesomeness of the computer, and the jargon used by computer experts.

Step by step, the procedure in formulating rations by computer is:

1. **List available feed ingredients, and the cost of each.** It is necessary that all of the available feeds be listed along with the unit cost (usually per ton) of each; preferably, ingredient cost should be based on market price plus delivery, storage, and processing cost.

2. **Record nutrient composition of feeds.** The more that is known about the quality of feed, the better. This is so because of the wide variation in composition and feeding value within ingredients; for example, between two samples of alfalfa hay.

Whenever possible, an actual chemical analysis of a representative sample of each ingredient under consideration should be available and used. However, the imperfections of a chemical analysis of a feedstuff should be recognized; chiefly, (a) it does not provide information on the availability of nutrients to animals, and (b) there are variations between samples.

3. **Establish ration specifications.** Set down the ration specifications—the nutrients and the levels of each that are to be met. This is exactly the same procedure as is followed in the hand method. For example; in arriving at ration specifications for feedlot cattle, the nutritionist considers (a) age, weight, and grade of cattle; (b) length of feeding period; (c) the probable market; (d) season of year; (e) background and stress of animals; and (f) other pertinent factors.

4. **Give restrictions.** Usually it is necessary to establish certain limitations on the use of ingredients; for example, with feedlot cattle the restrictions would likely show (a) the maximum amount of roughage; (b) the maximum amount of urea; (c) the minimum and maximum amounts of fat; (d) the proportion of cottonseed hulls to alfalfa hay; (e) the proportion of one grain to another—such as 60% barley and 40% milo; (f) an upper limit of some ingredients—such as 20% rye; (g) the exact amount of the premix; and (h) the lower and upper limits of molasses, as between 5 and 10%.

It must be recognized that the narrower the limitations imposed on the computer, the less the choice it will have in ration formulation and the higher the cost.

5. **Stipulate feed additives.** Generally speaking, nutritionists make rigid stipulations as to amounts of these ingredients, much as they do with added vitamins and minerals. All of them cost money, and many of them must be used in compliance with the Food and Drug Administration regulations.

6. **Obtain program.** Take data, ration specifications, and restrictions to an experienced computer programmer or systems analyst, who will either (a) *tailor-make* a program for a given situation, or (b) suggest one of the *canned* linear programs available from the larger computer companies. The canned programs are, by necessity, general in nature, because they are written for a wide variety of applications; but they cost less than a tailor-made program.

7. **Entering data.** With the advent of the personal computer, feed ingredient data plus nutrient specifications and restrictions changes can be updated or entered into a computer miles away. Many programs now exist that allow a personal computer to communicate with a main frame computer miles away with the capability to update files or enter new data.

8. **Run the program.** When the program is run, it treats the data as one gigantic algebra problem and arrives at the ration formulation very quickly. Based on available feeds, analysis, and price, the computer evolves with the mix that will meet the desired nutritive allowances at the least possible cost.

9. **Formulate as necessary.** All rations should be reviewed at frequent intervals and reformulated when there are shifts in (a) availability of ingredients (certain ingredients may no longer be available, but new ones may have evolved), (b) price, and/or (c) chemical composition.

10. **Validate the restrictions.** That is, test or confirm them.

USE OF THE COMPUTER AS A FEED BUYING AND SELLING AID

When some computer programs formulate a ration, they also give a complete set of *shadow prices*, which may be used as follows:

1. If a certain ingredient does not enter the formula due to its chemical analysis as related to price, the shadow price will indicate how much the market price of this ingredient must go down in order for it to enter the formula. For example, if soybean meal is selling for $175 per ton and the shadow price is $143, soybean meal will not be included in the formula until the price goes down to, or falls below, $143; unless, of course, a minimum amount of soybean meal is specified.

2. If an ingredient is homegrown and on hand, and the feeder desires to use it, despite the fact that its market value is out of line, the shadow price will indicate the penalty that will be paid for using it. Sometimes, it may become obvious that it is good business to sell a certain homegrown product and buy something else to replace it.

3. The shadow price provides a technique for determining the value of each ingredient based on its chemical analysis, thereby making it possible to determine which ingredient is the best buy. As a result,

it is an excellent management tool for buying and selling feed ingredients.

4. The shadow price is used in price mapping (the range of prices over which an ingredient will stay in the formula). By considering one ingredient at a time, this is an excellent buying guide.

5. The shadow price is used in determining the cost of restrictions; that is, the decrease in price that will occur if a restriction is released.

EXAMPLE OF A LEAST-COST RATION

Fig. 8-26 shows a sample printout of a least-cost ration for finishing beef cattle. In the column of "Ingredient Name," all feeds considered for this formulation are listed. Some of the ingredients are repeated as A, B, and C, where the A value is used for starting rations, B for intermediate rations, and C for final finishing.

The breakdown of the format of this printout is as follows:

Column 1. This column lists the amount of each ingredient to be included in the ration.

Column 2. This column gives the cost per pound of each feed being considered in the program. It should be kept up to date and reflect the cost of the ingredient processed into the mixed feed.

Column 3. This column evaluates the total nutritive economic value of each feed. For example, hominy feed is selling for 0.061 per pound but its relative value is only 0.0485 per pound. Hence, hominy would not be included in the ration until the price falls below the 0.04849 per pound level. The price of milo, however, was favorable and was, therefore, incorporated into the ration. Beet pulp posed a different problem. In this example, beet pulp was available and had to be incorporated at a 10% level despite the fact that it was not worth the price. The cost per pound of beet pulp was 0.0495, but by using it at a 10% level, it was only worth 0.04828 per pound. Thus, beet pulp at this level is costing the feeding enterprise money.

Columns 4 and 5. These columns list the minimum and maximum restrictions of utilization for each feed to be considered.

Column 6. All of the nutrient factors to be considered in the program are listed in this column. The programmer must supply values of each nutrient factor for each feed ingredient to be considered. These values must be as current and accurate as possible. In this column, *ENE* refers to estimated net energy, *CRENME* refers to Carnier (the name of the formulator) estimated net energy for maintenance, and *CRENPE* refers to Carnier estimated net energy for gain.

Columns 7 and 8. These columns list the mini-

JOHN DOE FEEDLOT   C A R N I E R   R E S E A R C H

90% FINISHER

FEED TO FINISH   COST $ 4.7612

INGREDIENT NAME	(1) AMOUNT USED	(2) COST PER LB	(3) REL. VALUE	(4) MINIMUM SPEC.	(5) MAXIMUM SPEC.
BARLEY ROLLED A	.000	.05400	.04872	.000	.000
BARLEY ROLLED B	.000	.05400	.04872	.000	.000
BARLEY ROLLED C	.000	.05400	.04871	.000	.000
HOMINY FEED	.000	.06100	.04849	***	.000
MILO, ROLLED A	.000	.04825	.04825	***	.000
MILO, ROLLED B	.000	.04825	.04825	***	.000
MILO, ROLLED C	69.710	.04825	.04825	***	100.000
WHEAT, ROLLED A	.000	.02500	.04922	***	.000
WHEAT, ROLLED B	.000	.02500	.04922	***	.000
WHEAT, ROLLED C	.000	.02500	.04922	***	.000
BEET PULP W/MOLASSES DRY	9.990	.04950	.04828	9.990	10.000
FAT, ANIMAL	3.000	.14050	.04728	3.000	3.000
ALFALFA HAY A, CHOPPED	.000	.03650	.03650	.000	.000
ALFALFA HAY B, CHOPPED	.000	.03650	.03650	.000	.000
ALFALFA HAY C, CHOPPED	4.200	.03650	.03650	.000	10.000
BARLEY STRAW A	.000	.01500	.01500	***	.000
BARLEY STRAW B	.000	.01500	.01500	***	.000
BARLEY STRAW C	2.800	.01500	.01500	***	100.000
COTTONSEED HULLS	3.000	.02000	.04728	3.000	3.000
SALT PLAIN	.800	.01100	.01100	***	***
TRACE MINERALS	.010	.40000	.40000	***	***
LIMESTONE, GROUND	1.490	.00900	.00900	***	***
VITAMIN A + D	.009	.49000	.49000	***	***
UREA, 45(N.	1.177	.07200	.07200	***	***
PHOSPHORIC ACID, 23.7% P	.168	.09500	.09500	***	***
MOLASSES, CANE	3.645	.03150	.03150	.000	5.000
TOTAL WEIGHT	100.000				

(6) NUTRIENT FACTORS	(7) MINIMUM SPEC.	(8) MAXIMUM SPEC.	(9) NUTR. CONTENT	(10) SHADOW VALUE
CRUDE PROTEIN (	.000	13.000	12.054	.00000
CR.PROT.EQUIV.NPN(	.030	3.500	3.308	.00000
DIGEST. PROTEIN (	7.800	9.200	7.800	.01696
CRUDE FAT (	1.500	8.000	5.608	.00000
CRUDE FIBER (	.000	13.000	6.238	.00000
TOTAL ASH (	3.500	***	4.792	.00000
CALCIUM -CA- (	.450	1.000	.759	.00000
PHOSPHORUS -P- (	.280	***	.315	.00000
INORGANIC PHOS. (	.040	***	.040	.26681
ENE --MEGCAL/CWT	64.000	***	68.510	.00000
CRENME-MEGCAL/CWT	.000	***	62.568	.00000
CRENPE-MEGCAL/CWT	.000	***	54.451	.00000
SUGAR (	.000	3.500	.000	.00000
ROUGHAGE,AS FED (	10.000	10.000	10.000	.00000
ROUGHAGE, ADM (	7.000	7.000	7.000	.02041
MOISTURE, AS FED (	.000	100.000	12.209	.00000
DRY MATTER (	85.000	90.000	87.789	.00000
VITAMIN A 1.0E6	.120	***	.150	.00000
VITAMIN D 1.0E6	.015	***	.015	.28656
CARBONATE BUFFER	1.000	1.500	1.490	.00000
D E S GRAMS	.000	***	.000	.00000
DRY PREMIX	1.500	1.500	1.500	.03828
BARLEY MILO RATIO	.000	.000	.000	.00000
TRACE MINERALS (	.010	***	.010	.39100
SALT, MIXING	.500	.800	.800	.03628
BARLEY MILO RATIO	.000	.000	.000	.00000
AB RATIO	.000	.000	.000	.00000
AB RATIO	.000	.000	.000	.00000
AB RATIO	.000	.000	.000	.00000
BARLEY MILO RATIO	.000	.000	.000	.00000
MAGNESIUM, ADDED (	.000	***	.000	.00000
MAGNESIUM, TOTAL (	.000	***	.169	.00000
SODIUM (	.000	***	.371	.00000
POTASSIUM (	.000	***	.301	.00000
PREMIX PORTION (	5.000	5.000	5.000	.01578
TOTAL WEIGHT	100.000	100.000	100.000	.04728

THIS FORMULATION COST AT WHICH ANY ONE INGREDIENT COULD BE USED IN	PER LB.	PER TON
BIOFOS	.09287	185.73

Fig. 8-26. Sample printout of a least-cost ration. (Courtesy, John G. Carbonniere, Agricultural Consultant, Carson, California)

mum and maximum restrictions of nutrient content to be used in the formulated feed.

Column 9. This column lists the nutrient content of the ration that was formulated.

Column 10. Shadow value refers to the value of what it cost to get the last unit of the corresponding nutrient factor. For example, digestible protein had minimum and maximum requirements of 7.8% and 9.2%, respectively. The shadow value states that the last unit of digestible protein cost 0.01696¢. The final digestible protein content fulfilled the minimum restriction and yet was not limited by the maximum restriction—hence, a positive shadow value. If the nutrient content had exceeded the minimum restriction and was subsequently stopped by the maximum restriction, the shadow value would have been negative. The reason for the maximum restriction is that excessive levels of the nutrient factor might incur detrimental effects physiologically. The relative value is the sum of the shadow value and nutrient content of each feed ingredient.

In the bottom left-hand corner of the printout is a section for what are termed opportunity prices. These are ingredients that can be included in the ration provided their costs are low enough. When the cost of the ingredient reaches the cost listed in this section, the ingredient can be incorporated economically into the ration, assuming other ingredient costs remain the same.

PRECAUTIONS AND LIMITATIONS OF COMPUTER FORMULATION

When utilizing the computer in the formulation of least-cost rations, the producer needs to consult an experienced nutritionist. The nutritionist can then interpret the printout from the computer and make any adjustments necessary to make the ration more realistic. It must be reiterated that the computer formulates rations objectively from the information that is fed into it. What comes out of the computer may be the best solution to the mathematical problem, but it may not be practical or realistic.

One of the major costs involved in computer formulations is the continual review and revision of the information that must go into the program. The user must be constantly updating costs in order to maximize the use of the computer.

The producer should be aware that radical changes in ration composition cannot be made without causing digestive disorders—especially in ruminant animals. Livestock need time to adapt to changes in rations, a fact which the computer does not consider.

When the computer formulates rations, it is using average values for the nutrient composition of the various feeds. We know that feeds can often vary in their nutrient composition so there is a good possibility that the chemical analysis of the formulated feed will not be the same as the formulated analysis. However, in most cases, this difference is not of sufficient magnitude to create problems.

Finally, it must be remembered that the results obtained from the computer are only as good as the person who feeds the information into the machine. If the data given to the computer are outdated or wrong, the ration that is formulated will be of little value.

FEED ALLOWANCE AND SOME SUGGESTED RATIONS[10]

Some general rules of feeding may be given, but it must be remembered that *the eye of the master fattens his cattle*. Nevertheless, the beginner may well profit from the experience of successful feeders. It is with this hope that the suggested rations are herewith presented.

Table 8-13 is a beef cattle feeding guide for different classes and ages of cattle.[11] All of these are merely intended as general guides. Variations can and should be made in the rations used. The feeder should give consideration to (1) the supply of homegrown feeds, (2) the availability and price of purchased feeds, (3) the class and age of cattle, (4) the health and condition of the animals, and (5) the length of the grazing season.

In using Table 8-13 as a guide, it is to be recognized that feeds of similar nutritive properties can and should be interchanged as price relationships warrant. Thus, (1) the cereal grains may consist of corn, barley, wheat, oats, and/or sorghum; (2) the protein supplement may consist of soybean, cottonseed, peanut, sunflower, and/or linseed meal; (3) the roughage may include many varieties of hays and silages; and (4) a vast array of byproduct feeds may be utilized.

FEED TO FOOD EFFICIENCY

Table 8-14 points up the continued need for greater efficiency of animal production, the urgency of which becomes altogether too apparent in light of population increases and impending world food shortages. Even the broiler, the most efficient of farm animal protein converters, is only (1) 52% efficient in converting the protein of feed to food, and (2) 8% efficient in

---

[10]Insofar as possible, these rations were computed from the requirements as reported by the National Research Council and applied by the authors.

[11]Recommendations relative to feeding show animals are included in Chapter 14 of this book.

converting energy of feed to food. However, there is a wide gap between broiler and beef steers, with the former having a 43% higher protein efficiency than the latter. Table 8-14 also reveals that the beef steer is the least efficient of all feed converters (requiring 9 lb of feed to produce 1 lb of gain), and that only sheep are lower on the totem pole from the standpoint of protein conversion. Granted, feed is only one factor responsible for the productivity rating given in Table 8-14; other environmental factors and heredity play a part. Nevertheless, feed is a costly and important item, and, historically, hungry people throughout the world have been forced to eliminate some animals and to consume grains directly to avoid famine.

In recent years, science and technology have teamed up and made for great strides in improving the productive efficiency of poultry. No doubt, further progress with all animals lies ahead. Also, as increasing quantities of cereal grains are needed for human consumption, ruminants will utilize higher proportions of roughages to concentrates.

STATE COMMERCIAL FEED LAWS

Nearly all the states have laws regulating the sale of commercial feeds. These benefit both cattle producers and reputable feed manufacturers. In most states the laws require that every brand of commercial feed sold in the state be licensed, and that the chemical composition be guaranteed.

Samples of each commercial feed are taken each year, and analyzed chemically in the state's laboratory to determine if manufacturers live up to their guarantee. Additionally, skilled microscopists examine the sample to ascertain that the ingredients present are the same as those guaranteed. Flagrant violations on the latter point may be prosecuted.

Results of these examinations are generally published, annually, by the state department in charge of such regulatory work. Usually, the publication of the guarantee alongside any *short-changing* is sufficient to cause the manufacturer promptly to rectify the situation, for such public information soon becomes known to both users and competitors.

The sections that follow brief some of the pertinent information required on the label of commercial feeds by most states.

■ **Medicated feed tags and labels**—Medicated feeds (those which contain drug ingredients intended or represented for the cure, mitigation, treatment, or prevention of diseases of animals) must also carry the following information in their labeling: (1) the purpose of the medication; (2) directions for the use of the feed; (3) the names and amounts of all active drug ingredients; (4) a warning or caution statement for a withdrawal period prior to marketing when required for a particular drug; and (5) warnings against misuse.

■ **Vitamin product labels**—When a product is marketed as a vitamin supplement *per se*, the quantitative guarantees (unit/lb) of vitamins A and D are expressed in USP units; of E in IU; and of other vitamins in mg/lb.

■ **Mineral product labels**—Some states require that all minerals except salt (NaCl) be quantitatively guaranteed in terms of percentage of the element(s); others require mg/lb.

■ **Other rules and regulations**—Generally, the following rules and regulations also apply in the different states:

1. The brand or product name must not be misleading.
2. The sliding scale or range (for example, 15 to 18% crude protein) method of expressing guarantees is prohibited.
3. Ingredient names are those adopted by the Association of American Feed Control Officials.
4. The term *dehydrated* may precede the name of any product that has been artificially dried.
5. Urea and other nonprotein nitrogen products are acceptable ingredients for ruminant animals only.

■ **Terms used in analyses and guarantees**—Knowledge of the following terms is requisite to understanding analyses and guarantees:

Dry matter is found by determining the percentage of water and subtracting the water content from 100%.

Crude protein is used to designate the nitrogenous constituents of a feed. The percentage is obtained by multiplying the percentage of total nitrogen by the factor 6.25. The nitrogen is derived chiefly from complex chemical compounds called amino acids.

Crude fat is the material that is extracted from moisture-free feeds by ether. It consists largely of fats and oils with small amounts of waxes, resins, and coloring matter. In calculating the heat and energy value of the feed, the fat is considered 2.25 times that of either nitrogen-free extract or protein.

Crude fiber is the relatively insoluble carbohydrate portion of a feed consisting chiefly of cellulose. It is determined by its insolubility in dilute acids and alkalies.

Ash is the mineral matter of a feed. It is the residue remaining after complete burning of the organic matter.

Nitrogen-free extract consists principally of sugars, starches, pentoses and nonnitrogenous organic acids. The percentage is determined by subtracting the sum of the percentages of moisture, crude protein, crude fat, crude fiber, and ash from 100.

TABLE
BEEF CATTLE

Suggested Rations With all rations and for all classes and ages of cattle, provide free access in separate containers to (1) salt (iodized salt in iodine-deficient areas), and (2) a suitable mineral mixture	Wintering mature pregnant beef breeding cows (avg. wt. 1,100 lb *[499 kg]*)		Wintering mature lactating beef breeding cows (avg. wt. 1,100 lb *[499 kg]*)		Wintering replacement heifers (weighing 400 to 500 lb *[181 to 227 kg]* start of wintering)	
	Per Day		Per Day		Per Day	
	(lb)	*(kg)*	*(lb)*	*(kg)*	*(lb)*	*(kg)*
1. Legume hay or grass-legume mixed hay, good-quality	18–20	*8.2–9.1*	30	*13.6*	13–15[3]	*5.9–6.8*[3]
Grain					2–3	*0.91–1.36*
Protein supplement						
2. Grass hay or other nonlegume dry roughage	18–20	*8.2–9.1*	24–26	*10.9–11.8*	12–18[3]	*5.4–8.2*[3]
Grain			2	*0.91*	2.5–4.5	*1.13–2.04*
Protein supplement	0.5–1	*0.23–0.45*	3	*1.36*	1.25–1.5	*0.57–0.68*
3. Legume hay or grass-legume mixed hay, good-quality	7–11	*3.2–5.0*	26–28	*11.8–12.7*	8–12[3]	*3.6–5.4*[3]
Grass hay or other nonlegume dry roughage	9–11	*4.1–5.0*			4–6	*1.8–2.7*
Grain			1	*0.45*	2.5–4	*1.13–1.81*
Protein supplement			1	*0.45*	0.5–1	*0.23–0.54*
4. Corn or sorghum silage	50–55	*22.7–25*	55	*25*	25–40	*11.3–18.1*
Grain			2	*0.91*		
Protein	0–0.5	*0–0.22*	3	*1.36*	1.5–1.75	*0.68–0.79*
5. Grass silage, half or more legume	50	*22.7*	50	*22.7*	25–40	*11.4–18.2*
Grain			4	*1.81*	3–4	*1.36–1.81*
Protein supplement					0.5	*0.23*
6. Silage (corn or sorghum silage fed with legume hay or legume silage fed with grass hay)	35	*15.9*	40	*18.1*	15–30	*6.8–13.6*
Hay	5–6	*2.3–2.7*	10	*4.5*	3–4	*1.4–1.8*
Grain					1–2	*0.45–0.91*
Protein supplement	0–0.5	*0.23*			0.5–1	*0.22–0.45*

[1]If stocker calves are late or the roughage is fair to poor in quality, it may be desirable to add 2–4 lb *(0.91–1.81 kg)* of grain per head daily. If farm scales are available, monthly weights may be used as the criterion for grain feeding. Keep in mind that the calves should gain 0.75–1 lb *(0.34–0.45 kg)* daily.

[2]In general, the experienced feeder plans that cattle on full feed shall consume (1) feeds in amounts (daily: air-dry basis) equal to about 2.5–3.0% of their liveweight, (2) 70–90% concentrates, and (3) a minimum of 2.0–4 lb *(0.9–1.8 kg)* roughage for each 100 lb *(45 kg)* liveweight. In areas where roughage is more abundant and comparatively cheaper than grain, the proportions of roughage to grain should be somewhat higher than indicated. In computing roughage consumption, 3 lb *(1.36 kg)* of silage are considered equivalent to 1 lb *(0.45 kg)* of hay.

8-13
FEEDING GUIDE

Wintering stocker calves roughed through winter and grazed the following summer. Fed for winter gain of 0.75 to 1 lb *(0.34 to 0.45 kg)* per head daily (weighing 400 to 500 lb *[181 to 227 kg]* start of wintering)[1]		Finishing calves in drylot, generally in winter (weighing 400 to 500 lb *[181 to 227 kg]* start of feeding and 750 to 850 lb *[340 to 386 kg]* at marketing)[2]		Wintering yearlings; roughed through the winter, and generally pasture finished the following summer. Fed for winter gains of 1 to 1.25 lb *[0.45 to 0.57 kg]* per head daily (weighing about 600 lb *[272 kg]* start of wintering)		Finishing yearlings in drylot, generally in winter (weighing about 600 lb *[272 kg]* start of feeding, and 900 to 1,100 lb *[409 to 500 kg]* at marketing)[2]		Finishing long-yearling steers in drylot generally in winter (weighing about 850 lb *[386 kg]* start of feeding and 1,100 to 1,200 lb *[500 to 545 kg]* at marketing)[2]	
Per Day		Per Day		Per Day		Per Day		Per Day	
(lb)	*(kg)*	*(lb)*	*(kg)*	*(lb)*	*(kg)*	*(lb)*	*(kg)*	*(lb)*	*(kg)*
12–18[3]	*5.4–8.2*	4–6	*1.8–2.7*	16–24	*7.2–10.9*	4–8	*1.8–3.6*	6–12	*2.7–5.4*
		12–15	*5.4–6.8*			15–19.5	*6.8–8.8*	16–22	*7.2–10.0*
		1–1.5	*0.45–0.68*			1–1.5	*0.45–0.68*		
12–18[3]	*5.4–8.2*	4–5	*1.8–2.3*	16–24	*7.2–10.9*	4–8	*1.8–3.6*	6–12	*2.7–5.4*
		12–15	*5.4–6.8*			15–20	*6.8–9.1*	16.5–22.25	*7.5–10.3*
1.25–1.5	*0.57–0.68*	1.75–2	*0.79–0.91*	1.5–1.75	*0.68–0.79*	1.5–2.5	*0.68–1.1*	1.5–1.75	*0.68–0.79*
8–12[3]	*5.4–8.2*	2–3	*0.91–1.36*	6–8	*2.7–3.6*	2–4	*0.91–1.81*	3–6	*1.4–2.7*
4–6	*1.8–2.7*	2–3	*0.91–1.36*	10–16	*4.5–7.2*	2–4	*0.91–1.81*	3–6	*1.4–2.7*
		12–15	*5.4–6.8*			15–19.75	*6.8–9.0*	16–22	*7.2–10.0*
0.25–1	*0.11–0.45*	1.5–1.75	*0.68–0.79*	1–1.5	*0.45–0.68*	1.25–1.75	*0.57–0.79*	0.5–0.75	*0.23–0.34*
25–40	*11.3–18.2*	6–16	*2.7–7.2*	40–55	*18.2–24.9*	6–25	*2.7–11.4*	6–35	*2.7–5.9*
		8–12	*3.6–5.4*			11–16	*5.0–7.2*	15–21	*6.8–9.5*
1–1.25	*0.45–0.57*	2	*0.91*	1.25–1.5	*0.57–0.68*	2	*0.91*	1.25–1.5	*0.57–0.68*
25–40	*11.3–18.2*	6–16	*2.7–7.2*	40–55	*18.1–24.9*	6–25	*2.7–11.3*	6–35	*2.7–15.9*
2–3	*0.90–1.36*	8–12	*3.6–5.4*	4–5	*1.8–2.3*	11–16	*5.0–7.3*	15–21	*6.8–9.5*
0.5	*0.23*	1–2	*0.45–0.91*	0.5	*0.23*	1–1.5	*0.45–0.68*	1	*0.45*
15–30	*6.8–13.6*	3–8	*1.4–3.6*	20–35	*9.1–15.9*	3–15	*1.4–6.8*	3–15	*1.4–6.8*
3–4	*1.4–1.8*	1–3	*0.45–1.4*	7	*3.2*	1–4	*0.45–1.8*	1–7	*0.45–3.2*
1–2	*0.45–0.91*	8–12	*2.6–5.4*			11–16	*5.0–7.2*	15–21	*6.8–9.5*
0.5	*0.23*	1–2	*0.45–0.91*	0.5–0.75	*0.23–0.34*	1–1.75	*0.45–0.79*	1–1.25	*0.45–0.57*

[3]With calves (both replacement heifers and stockers) an extra 2 lb *(0.91 kg)* of hay daily, over and above requirements, are herewith indicated to allow for wastage. Practical operators generally feed stemmy or other hay left over by calves to the cow herd.

TABLE
FEED TO FOOD EFFICIENCY RATING BY SPECIES OF ANIMALS,
(Based on Energy as TDN or DE and Crude Protein in Feed Eaten by Various Kinds of

		Feed Required to Produce One Production Unit				Dressing Yield	
Species	**Unit of Production (on foot)**	**Pounds**	**TDN[1]**	**DE[2]**	**Protein**	**Percent**	**Net Left**
		(lb)	(lb)	(kcal)	(lb)	(%)	(lb)
Broiler	1 lb chicken	2.1[7]	1.7[8]	3,400	0.21[8]	72[13]	0.72
Fish	1 lb fish	1.6[9]	0.98	1,960	0.57	65[10]	0.65
Dairy cow	1 lb milk	1.11[7]	0.9[8]	1,800	0.1[8]	100	1.0
Turkey	1 lb turkey	5.2[7]	4.21[8]	8,420	0.46[8]	79.7[13]	0.797
Layer	1 lb eggs (8 eggs)	4.6[7]	3.73[8]	7,460	0.41[8]	100	1.0
Hog (birth to market weight)	1 lb pork	4.0[18]	3.2	6,400	0.36	70[16]	0.7
Rabbit	1 lb fryer	3.0[19]	2.2	4,400	0.48	55[19]	0.55
Beef steer (yearling finishing period in feedlot)	1 lb beef	9.0[18]	5.85	11,700	0.90	58[16]	0.58
Lamb (finishing period in feedlot)	1 lb lamb	8.0[18]	4.96	9,920	0.86	47[16]	0.47

[1]TDN pounds computed by multiplying pounds feed (column to left) times percent TDN in normal rations. Normal ration percent TDN taken from M. E. Ensminger's books and rations, except for the following: dairy cow, layer, broiler, and turkey from *Agricultural Statistics 1974*, p. 358, Table 518. Fish based on averages recommended by Michigan and Minnesota Stations and U.S. Fish and Wildlife Service.

[2]Digestible Energy (DE) in this column given in kcal, which is 1 Calorie (written with a capital C), or 1,000 calories (written with a small c). Kilocalories computed from TDN values in column to immediate left as follows: 1 lb TDN = 2,000 kcal.

[3]From *Lessons on Meat*, National Live Stock and Meat Board, 1965.

[4]Feed efficiency as used herein is based on pounds of feed required to produce 1 lb of product. Given in both percent and ratio.

[5]Kilocalories in ready-to-eat food = kilocalories in feed consumed, converted to percentage. Loss = kcal in feed ÷ kcal in product.

[6]Protein in ready-to-eat food = protein in feed consumed, converted to percentage. Loss = pounds protein in feed ÷ pounds protein in product.

[7]*Agricultural Statistics 1974*, p. 358, Table 518. Pounds feed per unit of production is expressed in equivalent feeding value of corn.

[8]Pounds feed (column No. 2) per unit of production (column No. 1) is expressed in equivalent feeding value of corn. Therefore, the values for corn were used in arriving at these computations. No. 2 corn values are TDN, 81%; protein, 8.9%. Hence, for the dairy cow 81% × 1.11 = 0.9 lb TDN; and 8.9% × 1.11 = 0.1 lb protein.

[9]Data from report by Dr. Philip J. Schaible, Michigan State University, *Feedstuffs*, April 15, 1967.

[10]*Industrial Fishery Technology*, edited by Maurice E. Stansby, Reinhold Pub. Corp., 1963, Ch. 26, Table 26-1.

8-14
RANKED BY PROTEIN CONVERSION EFFICIENCY
Animals Converted into Calories and Protein Content of Ready-to-Eat Human Food)

Ready-to-Eat; Yield of Edible Product (meat & fish deboned & after cooking)				Feed Efficiency[4]		Efficiency Rating			
As % of Raw Product (carcass)	Amount Remaining from One Unit of Production	Calorie[3]	Protein[3]	(lb feed to produce one lb product)		Calorie Efficiency[5]		Protein Efficiency[6]	
(%)	*(lb)*	*(kcal)*	*(lb)*	*(%)*	*(ratio)*	*(%)*	*(ratio)*	*(%)*	*(ratio)*
54[14]	0.39	274	0.11	47.6	2.1:1	8.1	12.4:1	52.4	1.9:1
57[11]	0.37	285	0.27	62.5	1.6:1	14.5	6.9:1	47.6	2.1:1
100	1.0	309	0.037	90.0	1.11:1	17.2	5.8:1	37.0	2.7:1
57[15]	0.45	446	0.146	19.2	5.2:1	5.3	18.9:1	31.7	3.2:1
100[12]	1.0[12]	616	0.106	21.8	4.6:1	8.3	12.1:1	25.9	3.9:1
44[17]	0.31	341	0.088	0.25	4.0:1	5.3	18.8:1	24.4	4.1:1
79[19]	0.43	301	0.08	35.7	2.8:1	6.8	14.6:1	16.7	6.0:1
49[17]	0.28	342	0.085	11.1	9.0:1	2.9	34.2:1	9.4	10.6:1
40[17]	0.19	225	0.052	12.5	8.0:1	2.3	44.1:1	6.0	16.5:1

[11]*Ibid.* Reports that "Dressed fish averages about 73% flesh, 21% bone, and 6% skin." In limited experiments conducted by A. Ensminger, it was found that there was a 22% cooking loss on filet of sole. Hence, these values—73% flesh from dressed fish, minus 22% cooking losses—give 57% yield of edible fish after cooking, as a percent of the raw, dressed product.

[12]Calories and protein computed basis per egg; hence, the values herein are 100% and 1.0 lb, respectively.

[13]*Marketing Poultry Products*, 5th Ed., by E. W. Benjamin, *et al.*, John Wiley & Sons, 1960, p. 147.

[14]*Factors Affecting Poultry Meat Yields*, University of Minnesota Sta. Bull. 476, 1964, p. 29, Table 11 (fricassee).

[15]*Ibid.* Page 28, Table 10.

[16]Ensminger, M. E., *The Stockman's Handbook*, 6th Ed., Sec. XII.

[17]Allowance made for both cutting and cooking losses following dressing. Thus, values are on a cooked, ready-to-eat basis of lean and marbled meat, exclusive of bone, gristle, and fat. Values provided by National Live Stock and Meat Board (personal communication to the senior author of June 5, 1967, from Dr. Wm. C. Sherman, Director, Nutrition Research), and based on the data from *The Nutritive Value of Cooked Meat*, by Ruth M. Leverton and George V. Odell, Misc. Pub. MP-49, Appendix C, March 1958.

[18]Estimates by the senior author.

[19]Based on information in *Commercial Rabbit Raising*, Ag. Hdbk. No. 309, USDA, 1966, and *A Handbook on Rabbit Raising*, by H. M. Butterfield, Washington State University Ext. Bull. No. 411.

Carbohydrates represent the sum of the crude fiber and nitrogen-free extract.

Calcium and phosphorus are essential mineral elements that are present in feeds in varying quantities. Mineral feeds are usually high in source materials of these elements.

TDN—The digestible nutrients of any ingredient are obtained by multiplying the percentage of each nutrient by the digestion coefficient. For example, dent corn contains 10% protein of which 77% is digestible. Therefore, the digestible protein is 7.7%.

The TDN is the sum of all the digestible organic nutrients—protein, fiber, nitrogen-free extract, and fat (the latter multiplied by 2.25).

HOW TO DETERMINE THE BEST BUY IN FEEDS

Feed prices vary widely. For profitable production, therefore, feeds with similar nutritive properties should be interchanged as price relationships warrant.

In buying feeds, the cattle producer should check prices against values received. This may be done by computing the cost per pound of protein and TDN. The use of this method can best be illustrated by the examples that follow:

If 44% protein (crude) soybean meal is selling at $8.00 per 100 lb whereas 35% protein (crude) linseed meal sells for $6.00 per 100 lb, which is the better buy? Divide $8.00 by 44 to get 18.2¢ per lb of crude protein for the soybean meal. Then divide $6.00 by 35 to get 17.1¢ per lb for the linseed meal. Thus, at these prices linseed meal is the better buy—by 1.1¢ per lb of crude protein.

When buying energy feed, one can compare the cost per lb of total digestible nutrients (TDN). For example, if corn is priced at $6.00 per 100 lb and has a TDN of 90%, divide $6.00 by 90 and the result is 6.67¢ per lb of TDN.

If barley with 84% TDN sells for $6.50 per 100 lb, divide $6.50 by 84 and the price is 7.74¢ per lb of TDN. Thus, corn would be the better buy by 1.24¢ per lb of TDN.

Of course, it is recognized that many other factors

TABLE
NUTRITIONAL DISEASES

Disease	Cause	Symptoms (and Age or Group Most Affected)	Distribution and Losses Caused by
Acetonemia (See Ketosis)			
Alkali disease (See Selenium poisoning)			
Anemia, nutritional	Commonly an iron deficiency, but it may be caused by a deficiency of copper, cobalt, and/or certain vitamins.	Loss of appetite, progressive emaciation, and death. Most prevalent in suckling young.	Worldwide. Losses consist of slow and inefficient gains, and deaths.
Aphosphorosis	Low phosphorus in feed.	Decreased growth rate; inefficient feed utilization; depraved appetite—chewing bones, wood, hair, rags, etc.; stiff joints and fragile bones; and breeding problems.	Worldwide. In S.W. United States.
Bloat—feedlot	High-concentrate rations increase numbers of slime-producing bacteria in rumen. Slime traps fermentation gas and produces bloat.	Symptoms same as pasture bloat. Occurs when cattle have been fed high-concentrate, low-roughage rations for approximately 60 days or longer.	Survey of Kansas feedlots showed 0.1% died of bloat, 0.2% bloated severely, and 0.6% mildly to moderately. Mild bloat probably affects animal performance.

affect the actual feeding value of each feed, such as (1) palatability, (2) grade of feed, (3) preparation of feed, (4) ingredients with which each feed is combined, and (5) quantities of each feed fed.

NUTRITIONAL DISEASES AND AILMENTS

More animals (and people) throughout the world suffer from hunger—from just plain lack of sufficient feed—than from the lack of one or more specific nutrients. Therefore, it is recognized that nutritional deficiencies may be brought about by either (1) too little feed, or (2) rations that are too low in one or more nutrients.

Also, forced production (such as finishing animals at early ages) and the feeding of forages and grains which are often produced on leached or depleted soils have created many problems in nutrition. This condition has been further aggravated through the increased confinement of stock, many animals being confined to lots or buildings all or a large part of the year. Under these unnatural conditions, nutritional diseases and ailments have become increasingly common.

Although the cause, prevention, and treatment of most of these nutritional diseases and ailments are known, they continue to reduce profits in the livestock industry simply because the available knowledge is not put into practice. Moreover, those widespread nutritional deficiencies which are not of sufficient proportions to produce clear-cut deficiency symptoms cause even greater economic losses because they go unnoticed and unrectified. Nutritional and metabolic diseases of cattle cause estimated annual losses of $200 million.[12]

Table 8-15 contains a summary of the important nutritional diseases and ailments affecting cattle.

[12]Estimate by George Lambert, Associate Director, USDA, Agricultural Research Service, National Animal Disease Center, Ames, Iowa.

8-15
AND AILMENTS OF CATTLE

Treatment	Control and Eradication	Prevention	Remarks
Provide dietary sources of the nutrient or nutrients, the deficiency of which is known to cause the condition.	When nutritional anemia is encountered, it can usually be brought under control by supplying dietary sources of the nutrient or nutrients, the deficiency of which is known to cause the condition.	Supply dietary sources of iron, copper, cobalt, and certain vitamins. Levels of iron in feed believed to be ample, since feeds contain 40 to 400 mg/lb.	Anemia is a condition in which the blood is deficient either in quality or in quantity. (A deficient quality refers to a deficiency in hemoglobin and/or red cells.)
Provide phosphorus supplement.	Controlled by feeding phosphorus, either free-choice or added to the ration.	Feed phosphorus in feed and/or as mineral supplement (free-choice). Keep the Ca:P ratio within the range 2:1 or 1:1.	Generally caused by lack of phosphorus in the pasture. Phosphorus fertilizing may help.
Drench with 1–2 oz *(0.028–0.056 kg)* poloxalene, and then relieve free gas with stomach tube 10 minutes after treatment.	If feasible, increase proportion of roughage in ration. However, good-quality legume hay may increase incidence of feedlot bloat. In this instance poloxalene is an effective preventive. Use according to manufacturer's directions.	No effective preventive drug available.	Feedlot bloat may occur during any month of the year; however, more common during hot, humid weather.

(Continued)

TABLE 8-15

Disease	Cause	Symptoms (and Age or Group Most Affected)	Distribution and Losses Caused by
Bloat—pasture	Most common on lush legume pastures. Incidence on wheat pasture has been increasing in recent years. Pasture bloat is a frothy bloat caused by interaction of several factors—plant, animal, and microbial. Soluble plant proteins play a prominent role in permitting stable froth formation.	First observed as distention of paunch on left side in front of hip bone. This is followed by distention of right side, protrusion of anus, respiratory distress, cyanosis of tongue, struggling, and death if not treated.	Widespread, although some areas appear to have more bloat than others. Often results in death. 36% of all mortality due to nutritional diseases and ailments is attributed to bloat.
Crooked calves	Manganese deficiency.	Calves born with crooked necks and legs.	Northwestern U.S.
Fluorine poisoning (fluorosis)	Ingesting excessive quantities of fluorine through either feed or water.	Abnormal teeth (especially mottled enamel) and bones (bones become thickened and softened), stiffness of joints, loss of appetite, emaciation, reduction in milk flow, diarrhea, and salt hunger.	The water in parts of Arkansas, California, South Carolina, and Texas has been reported to contain excess fluorine. Occasionally throughout the U.S., high fluorine phosphates are used in mineral mixtures.
Founder	Overeating, overdrinking, or inflammation of the uterus following parturition. Also intestinal inflammation. Too rapid a change in the ration.	Extreme pain, fever (103°–106°F *[39.4°–41.4°C]*) and reluctance to move. If neglected, chronic laminitis will develop, resulting in a dropping of the hoof soles and a turning up of the toe walls.	Worldwide. Actual death losses from founder are not very great.
Goiter (See Iodine deficiency)			
Grass tetany (grass staggers)	Magnesium deficiency. When beef cows in early lactation graze spring pasture containing less than 0.2% magnesium.	Generally occurs during first 2 weeks of pasture season. Nervousness, twitching of muscles (usually of head and neck), head held high, accelerated respiration, high temperature, gnashing of teeth, and abundant salivation. Slight stimulus may precipitate a crash to the ground, and finally death.	Reported in California, Nebraska, Kentucky, Missouri, Iowa, Washington, and other states. Also found in New Zealand, England, and the Netherlands. Highly fatal if not treated quickly.
Iodine deficiency (goiter)	A failure of the body to obtain sufficient iodine from which the thyroid gland can form thyroxin (an iodine-containing compound). Rapeseed meal, soybean meal, and cottonseed meal have goitrogenic effects.	Goiter (big neck) is the most characteristic symptom in humans, calves, lambs, and kids. Also, there may be reproductive failures and weak offspring that fail to survive. Pigs may be born hairless, and foals may be weak.	Northwestern U.S. and the Great Lakes region.

(Continued)

Treatment	Control and Eradication	Prevention	Remarks
Time permitting, severe cases of bloat should be treated by a veterinarian. Puncturing of the paunch should be a last resort. Mild cases may be home-treated by (1) keeping the animal on its feet and moving, and (2) drenching with (a) ½–1 qt *(450–900 ml)* mineral oil, (b) poloxalene, or (c) Laureth-23. Follow the manufacturer's directions when administering poloxalene or Laureth-23.	When there is high incidence of bloat, it may be desirable to change the feed. Where legume bloat is encountered, use poloxalene, oxytetracycline, or polyoxyethylene (23) lauryl ether (Laureth-23/Enproal Bloat Blox) according to the respective manufacturer's directions.	The incidence is lessened by (1) avoiding straight legume pastures, (2) feeding dry forage along with pasture, (3) avoiding a rapid fill from an empty start, (4) keeping animals continuously on pasture after they are once turned out, (5) keeping salt and water conveniently accessible at all times, and (6) avoiding frosted pastures. Use poloxalene (a nonionic surfactant), oxytetracycline (an antibiotic), or Laureth-23 (a detergent) according to the respective manufacturer's directions for the control of legume bloat.	Legume pastures, alfalfa hay, and barley appear to be associated with a higher incidence of bloat than many other feeds. Legume pastures are particularly hazardous when moist, after a light rain or dew. Molasses blocks containing poloxalene or Laureth-23 are now on the market.
		Feed manganese, 40 ppm of total feed for mature cows and bulls, and 20 ppm for growing-finishing cattle.	The Utah station has also produced crooked calves by feeding lupine.
Any damage may be permanent, but animals which have not developed severe symptoms may be helped to some extent, if the sources of excess fluorine are eliminated.	Discontinue the use of feeds, water, or mineral supplements containing excessive fluorine.	Avoid the use of feeds, water, or mineral supplements containing excessive fluorine. 100 ppm (0.01%) fluorine of the total dry ration is the borderline in toxicity for cattle. At levels of 25–100 ppm, some mottling of the teeth may occur over periods of 3–5 years. In breeding animals, therefore, the permissible level is 40 ppm of the total dry ration. Not more than 65–100 ppm fluorine should be present in dry matter of rations when rock phosphate is fed.	Fluorine is a cumulative poison. Undefluorinated rock phosphate often contains 3.5–4.0% fluorine.
Pending arrival of the veterinarian, the attendant should stand the animal's feet in a cold water bath. Antihistamines and anti-inflammatory agents may speed recovery and alleviate serious aftereffects.	Alleviate the causes; namely (1) overeating, (2) overdrinking, and/or (3) inflammation of the uterus following parturition.	Avoid animals overeating and overdrinking (especially when hot).	Unless foundered animals are quite valuable, it is usually desirable to dispose of them following a case of severe founder.
Intravenous injection of a solution of calcium and/or magnesium salt by a veterinarian. Keep animal quiet after treatment.	(See Prevention.)	Grass tetany can be prevented by not turning animals to pasture, but this is not practical. Feeding hay at night during the first 2 weeks of the pasture season is helpful. A salt lick of 10 parts each of magnesium sulfate and calcium diphosphate with 80 parts of salt will aid in prevention. Also, a mixture of 2 parts of magnesium oxide to 1 part of salt as the only source of salt is effective.	Affected animals show low blood magnesium, often low serum calcium. Treated cattle may be aggressive on arising, so watch out!
Occasionally borderline cases may survive; in these the moderate thyroid enlargement disappears in a few weeks.	At the first signs of iodine deficiency, iodized salt should be fed to all farm animals.	In iodine-deficient areas, feed iodized salt to all farm animals throughout the year. Stabilized iodized salt containing 0.01% potassium iodide is recommended.	The enlarged thyroid gland (goiter) is nature's way of attempting to make sufficient thyroxin under conditions where a deficiency exists.

(Continued)

TABLE 8-15

Disease	Cause	Symptoms (and Age or Group Most Affected)	Distribution and Losses Caused by
Ketosis (acetonemia)	A metabolic disorder, thought to be a disturbance in the carbohydrate metabolism.	In cows, ketosis or acetonemia is usually observed within first 1–6 weeks after calving. Affected animals show loss in appetite and condition, a marked decline in milk production, and the production of a peculiar, sweetish chloroform-like odor of acetone that may be present in the milk and pervade the barn.	Worldwide. Ketosis or acetonemia affects dairy cattle throughout the U.S.
Milk fever	Low blood calcium concentration. Too much calcium in the ration can cause this condition. In milking cows, the Ca:P ratio should not exceed 2:1.	Commonly occurs soon after calving in high-producing cows. Rarely occurs at first calving. First symptoms are loss of appetite, constipation, and general depression. This is followed by nervousness and finally collapse and complete loss of consciousness. The head is usually turned back.	A common, widespread disease of dairy cows. But, good milking beef cows may be affected. Losses are not great, although untreated animals are likely to die.
Molybdenum toxicity (commonly called teartness)	As little as 10–20 ppm in forages result in toxic symptoms.	Toxic levels of molybdenum interfere with copper metabolism, thereby increasing the copper requirement and producing typical copper deficiency symptoms. The physical symptoms are anemia and extreme diarrhea, with consequent loss in weight and milk yield.	England, and in Florida, California, and Manitoba, Canada.
Nitrate poisoning (oat hay poisoning, corn stalk poisoning)	1. The forages (seeds do not appear to accumulate nitrate nitrogen) of most grain crops (oats, wheat, barley, rye, corn, and sorghum), Sudangrass, and numerous weeds; especially (a) under stress—drought, insufficient sunlight, following spraying by weed killers, or after frost, or (b) following the application of high soil nitrate nitrogen (through nitrogen fertilizer, green manure crops, or barnyard manure) which may boost the nitrate nitrogen of plants to dangerous levels. Also, sometimes nitrate appears to be formed after forage is stacked. 2. Inorganic salts of nitrate or nitrite (including fertilizers) carelessly applied to fields or left where animals have access to them. Sometimes these chemicals are also mistakenly used in place of common salt. 3. Pond or shallow well water into which heavy rains have washed a high concentration of nitrate from (a) fertilizer from heavily fertilized fields, or (b) feedlot drainage (ammonium nitrate).	Accelerated respiration and pulse rate; diarrhea; frequent urination; loss of appetite; general weakness, trembling, and a staggering gait; frothing from the mouth; lowered milk production; abortion; blue color of the mucous membrane, muzzle, and udder due to lack of oxygen; and death in 4½–9 hrs. after eating lethal doses of nitrate. A rapid and accurate diagnosis of nitrate poisoning may be made by drawing and examining a venous (jugular) blood sample. Normal blood is red and becomes brighter on standing. Brown-colored blood, due to the formation of methemoglobin, is characteristic of animals suffering from nitrate poisoning; chemically, the nitrate oxidizes the ferrous hemoglobin (oxyhemoglobin) to ferric hemoglobin (methemoglobin) which cannot transport oxygen, with the result that death due to nitrate poisoning may be compared to asphyxiation or strangulation. Death occurs when about ¾ of the oxyhemoglobin (the oxygen carrier in the blood) has been converted to methemoglobin.	Excessive nitrate content of feeds is an increasingly important cause of poisoning in farm animals, due primarily to more and more high nitrogen fertilization. But nitrate toxicity is not new, having been reported as early as 1850, and having occurred in semiarid regions of this and other countries for years.

(Continued)

Treatment	Control and Eradication	Prevention	Remarks
Give affected animal ½–1 lb *(0.2–0.45 kg)* of either propylene glycol or sodium propionate per day, with the dose divided into 2 administrations per day. Treat for 5–10 days. Put treatment on grain if cow is eating; otherwise, give as drench. Intravenous injection of glucose (500 ml of a 50% glucose solution) is a rapid way of getting outside supply of sugar in blood.	Maintain relatively high energy intake before calving; increase energy intake substantially after calving.	The incidence of ketosis can be lessened by avoiding excessively fat cows at calving; (2) increasing the level of concentrates rapidly after calving; (3) feeding good-quality roughage after calving, and avoiding abrupt changes in roughage; (4) feeding adequate proteins, minerals, and vitamins; and (5) providing comfort, exercise, and ventilation. In problem herds, feeding ¼ lb *(7 g)* daily of propylene glycol or sodium propionate may be helpful.	The clinical findings are similar in the case of affected cattle and sheep, but it usually strikes ewes just before lambing, whereas cows are usually affected within the first 1–6 weeks after calving.
Treatment consists of having the veterinarian give an intravenous injection of a calcium salt.	(See Prevention.)	Each of the following measures will lessen the incidence of milk fever: 1. *Calcium-phosphorus, ratio and amounts*—Approximately a 2.3:1 Ca:P ratio. Feed a ration that contains 0.5–0.7% Ca and 0.3–0.4% P. 2. *Calcium shock treatment*—10–14 days before calving, feed a Ca deficient ration with a Ca:P ratio of 1:2. This activates the cow's calcium-mobilizing mechanism for drawing calcium from the bones, with the result that it is functioning before calving and, milk fever is avoided. 3. *High vitamin D*—This consists in feeding 20 million units of vitamin D/cow/day starting about 5 days before calving and continuing through the first day postpartum, with a maximum dosage period of 7 days.	The name milk fever is a misnomer, because the disease is not accompanied by fever; the temperature really being below normal.
1 g of copper sulfate per head daily will cure symptoms of molybdenum toxicity.		1 g of copper sulfate per head daily will prevent molybdenum toxicity.	When feeds are high in sulfate, toxic symptoms will be produced on lower levels of molybdenum and, conversely, higher levels of molybdenum can be tolerated with low levels of sulfate.
A 4% solution of methylene blue (in a 5% glucose or a 1.8% sodium sulfate solution) administered by a veterinarian intravenously at the rate of ¼ cup/1,000 lb *(100 cc/454 kg)* liveweight.	(See Prevention.)	Regard any amount of nitrate nitrogen over 0.5% of the total ration (moisture-free basis) as a potential source of trouble. When in doubt, have the feed analyzed (first make a rapid, qualitative field test, using a commercial test kit according to direction; then, if high-nitrate samples are spotted, follow with a quantitative laboratory chemical test). Nitrate poisoning may be lowered by (1) feeding high levels of carbohydrates or energy feeds (grain or molasses) and vitamin A, (2) feeding limited amounts of high-nitrate forage, (3) alternating or mixing high- and low-nitrate forages, and (4) ensiling forages high in nitrates, since fermentation reduces some of the nitrates to gas (but beware of nitric oxide and nitrogen dioxide gas, which is released as yellow-red fumes in the early stages of fermentation and may cause silo gas poisoning to both humans and animals). After 3–4 weeks the silage has usually lost most of its nitrates and is safe to feed.	Nitrate-form nitrogen does not appear to cause the actual toxicity. During digestion, the nitrate is reduced to nitrite, a far more toxic form (10–15 times more toxic than nitrates). In cows and sheep, this conversion takes place in the rumen (paunch). Lethal dose varies with (1) nutritional state, size, and type of animal; and (2) the consumption of feed other than nitrate-containing material. Nitrate over 5% of total ration is a potential source of trouble; 0.75% content nitrate forages must be fed with caution, and milk production will be lowered; and at 1.5% death will likely occur. Where nitrate troubles are suspected, consult the local veterinarian or county agent.

(Continued)

TABLE 8-15

Disease	Cause	Symptoms (and Age or Group Most Affected)	Distribution and Losses Caused by
Oat hay poisoning (See Nitrate poisoning)			
Osteomalacia	Inadequate phosphorus (sometimes inadequate calcium). Lack of vitamin D in confined cattle. Incorrect ratio of calcium to phosphorus.	Phosphorus deficiency symptoms are: depraved appetite (gnawing on bones, wood, or other objects, or eating dirt); lack of appetite, stiffness of joints, failure to breed regularly, decreased milk production, and an emaciated appearance. Calcium deficiency symptoms are: fragile bones, reproductive failures, and lowered lactations. Mature animals most affected. Most of the acute cases occur during pregnancy and lactation.	Southwestern U.S. is classed as a phosphorus-deficient area, whereas calcium-deficient areas have been reported in parts of Florida, Louisiana, Nebraska, and West Virginia.
Pine needle abortion	Needle of *Pinus ponderosa*, commonly called ponderosa pine, or western yellow pine. In 1994, USDA researchers at the ARS Poisonous Plant Research Laboratory, Logan, Utah, reported that isocupressic acid, a yellow, oily substance in ponderosa pine needles, is the toxin that triggers abortions in pregnant cows.	Pregnant cows, free of brucellosis, abort. At calving, excessive hemorrhage; retained placenta; sceptic metritis, often followed by peritonitis. If cow is affected near parturition, calf may be born normal or weak.	British Columbia, Canada, and in the states of Washington, Idaho, and Oregon.
Rickets	Lack of calcium, phosphorus, or vitamin D; or an incorrect ratio of the two minerals.	Enlargement of the knee and hock joints, and the animal may exhibit great pain when moving about. Irregular bulges (beaded ribs) at juncture of ribs with breastbone, and bowed legs. Rickets is a disease of young animals—calves, foals, pigs, lambs, kids, pups, and chicks.	Worldwide. It is seldom fatal.
Salt deficiency (sodium chloride)	Lack of salt (sodium chloride).	Intense craving for salt, loss of appetite, retarded growth, loss of weight, a rough coat, lowered production of milk, and muscle cramps.	Worldwide.
Salt sick (cobalt deficiency)	Cobalt deficiency, associated with copper and perhaps iron deficiencies.	Loss of appetite, depraved appetite, scaliness of skin, listlessness, and lack of thrift.	Australia, Western Canada, Florida, New Hampshire, Michigan, Wisconsin, New York, and North Carolina, especially on sandy soils.
Selenium poisoning (alkali disease)	Consumption of plants grown on soils containing selenium.	Loss of hair from the tail in cattle. In severe cases, the hoofs slough off, lameness occurs, feed consumption decreases, and death may occur by starvation.	In certain regions of western U.S.—especially certain areas in South Dakota, Montana, Wyoming, Nebraska, Kansas, and perhaps areas in other states in the Great Plains and Rocky Mountains. Also in Canada.
Sweet clover disease	Usually produced only by moldy or spoiled sweet clover hay or silage. Caused by presence of dicumarol which interferes with vitamin K in blood clotting.	Loss of clotting power of the blood. As a result, blood forms soft swellings beneath skin on different parts of body. Serious or fatal bleeding may occur at time of dehorning, castration, parturition, or following injury. All ages affected. A newborn animal may also have the condition at birth.	Wherever sweet clover is grown.

(Continued)

Treatment	Control and Eradication	Prevention	Remarks
Increase the calcium and phosphorus content of feed through fertilizing the soils. Select natural feeds that contain sufficient quantities of calcium and phosphorus. Feed a special mineral supplement or supplements. If this disease is advanced, treatment will not be successful.	(See Treatment.)	Feed balanced rations, and allow animals free access to a suitable phosphorus and calcium supplement.	Calcium deficiencies are much more rare than phosphorus deficiencies in cattle.
No treatment known.		Keep pregnant cows away from yellow pine trees.	Lodgepole pine *(Pinus contorta)*—commonly called black pine, jack pine, western jack pine, white pine *(Pinus monticola)* or cypress *(Cypressus arizonica)*—does not appear to cause abortion in cattle. Pregnant cows will consume quantities of pine needles even though fed an adequate ration.
If the disease has not advanced too far, treatment may be successful by supplying adequate amounts of vitamin D, calcium, and phosphorus, and/or adjusting the ratio of calcium to phosphorus.	(See Prevention.)	Provide (1) sufficient calcium, phosphorus, and vitamin D, and (2) a correct ratio of the two minerals.	Rickets is characterized by a failure of growing bone to ossify, or harden, properly.
Salt-starved animals should be gradually accustomed to salt, slowly increasing the hand-fed allowance until the animals may be safely allowed free access to it.	(See Treatment and Prevention.)	Provide plenty of salt at all times, preferably by free-choice feeding.	Common salt is one of the most essential minerals for grass-eating animals, and one of the easiest and cheapest to provide. Excessive salt intake can result in toxicity if animals are deprived of water.
A vitamin B-12 injection will relieve a cobalt deficiency.	Provide 0.2–0.5 oz *(6–14 g)* cobalt salt/100 lb *(45 kg)* of salt—or feed a suitable trace mineral supplement.	Mix 0.2 oz *(6 g)* of cobalt chloride, cobalt sulfate, or cobalt carbonate/100 lb *(45 kg)* of either (1) salt, or (2) the mineral other than salt.	Cobalt is needed for rumen microbial synthesis of vitamin B-12.
The use of salt containing 37.5 ppm of arsenic may reduce the incidence of chronic selenium poisoning in cattle on seleniferous range. Pasture rotation and use of supplemental feeds from nonseleniferous areas are practical solutions to the problem. There is no known treatment for acute selenium poisoning.	(Control measures based on Prevention.)	Abandon areas where soils contain excess selenium, because crops produced on such soils constitute a menace to both animals and people.	The NRC suggests that 2 mg/kg dry weight ration is the maximum tolerable level for all species. Chronic cases of selenium poisoning occur when cattle consume feeds containing 8.5 ppm of selenium over an extended period; acute cases occur on 500–1,000 ppm.
Remove the offending materials and administer menadione (vitamin K_3). The veterinarian usually gives the affected animal an injection of plasma or whole blood from a normal animal that was not fed on the same feed.	When a case of sweet clover disease is observed in the herd, either (1) discontinue feeding the damaged product or (2) alternate it with a better quality hay, especially alfalfa.	Properly cure an sweet clover hay or ensilage.	The disease has also been produced from feeding moldy lespedeza hay and from sweet clover pasture.

(Continued)

TABLE 8-15

Disease	Cause	Symptoms (and Age or Group Most Affected)	Distribution and Losses Caused by
Urinary calculi (gravel, stones, water belly)	Unknown, but it does seem to be nutritional. Experiments have shown a higher incidence of urinary calculi when there is (1) a high potassium or phosphorus intake, (2) an incorrect Ca:P ratio, or (3) a high proportion of beet pulp or grain sorghum in the ration.	Frequent attempts to urinate, dribbling or stoppage of the urine, pain, and renal colic. Usually only males affected, the females being able to pass the concretions. Bladder may rupture, with death following. Otherwise, uremic poisoning may set in.	Worldwide. Affected animals seldom recover completely.
Vitamin A deficiency (night blindness and xerophthalmia)	Vitamin A deficiency. High levels of nitrate in hay or silage.	Night blindness, the first symptom of vitamin A deficiency, is characterized by faulty vision, especially noticeable when the affected animal is forced to move about in twilight in strange surroundings. Xerophthalmia develops in the advanced stages of vitamin A deficiency. The eyes become severely affected, and blindness may follow. Severe diarrhea in young calves and intermittent diarrhea in advanced stages in adults. In finishing cattle, generalized edema or anasarca with lameness in hock and knee joints and swelling in the brisket area.	Worldwide. Especially prevalent in western U.S. where one of the following conditions frequently prevails: (1) extended drought, and (2) winter feeding on bleached grass cured on the stalk or on bleached hay.
White muscle disease (muscular dystrophy)	A deficiency of selenium and/or vitamin E. Where unsaturated fats are fed.	Symptoms range from mild "founderlike" stiffness to sudden death. Calves continue to nurse as long as they can reach the cow's teats. Many calves stand or lie with protruded tongue, fighting for breath against a severe pulmonary edema. It seems that more calves than lambs develop fatal heart damage. Affected calves show pathological lesions similar to those of "stiff lambs" (white muscle disease in lambs), namely, whitish areas or streaks in the heart and other muscles. Affects calves from birth to 3 months of age.	Throughout the U.S., but the incidence appears to be highest in the intermountain area, between the Rocky and Cascade mountains.

TABLE 8-16
MINERAL REQUIREMENTS AND MAXIMUM TOLERABLE LEVELS FOR BEEF CATTLE[1]

Mineral	Requirements			Maximum Tolerable Level	Mineral	Requirements			Maximum Tolerable Level
	Growing and Finishing Cattle	Cows				Growing and Finishing Cattle	Cows		
		Gestating	Lactating				Gestating	Lactating	
Calcium (%)	See Appendix, Tables I-1 and I-2				Phosphorus (%)	See Appendix, Tables I-1 and I-2			
Cobalt (ppm)	0.10	0.10	0.10	10.00	Potassium (%)	0.60	0.60	0.70	3.00
Copper (ppm)	10.00	10.00	10.00	100.00	Selenium (ppm)	0.10	0.10	0.10	2.00
Iodine (ppm)	0.50	0.50	0.50	50.00	Sodium (%)	0.06–0.08	0.06–0.08	0.10	—
Iron (ppm)	50.00	50.00	50.00	1,000.00	Chlorine (%)	—	—	—	—
Magnesium (%)	0.10	0.12	0.20	0.40	Sulfur (%)	0.15	0.15	0.15	0.40
Manganese (ppm)	20.00	40.00	40.00	1,000.00	Zinc (ppm)	30.00	30.00	30.00	500.00
Molybdenum (ppm)	—	—	—	5.00					

[1]Adapted from *Nutrient Requirements of Beef Cattle*, seventh revised edition, National Research Council-National Academy of Sciences, 1996, p. 54, Table 5-1.

(Continued)

Treatment	Control and Eradication	Prevention	Remarks
Once calculi develop, dietary treatment appears to be of little value. Smooth muscle relaxants may allow passage of calculi if used before rupture of bladder. Surgery may save the animal, but such treatment will result in bulls becoming nonbreeders. When the condition strikes in feedlot cattle, increase the salt content of the diet if the animals are not ready to market.	If severe outbreaks of urinary calculi occur in finishing steers, it is usually well to dispose of them if they are carrying acceptable finish. When an outbreak occurs in cattle, one of the following treatments is recommended: 1. Add ammonium chloride at the rate of 1¼–1½ oz *(35–42 g)*/head/day. 2. Increase the salt content of the total ration to a level of 1½%. Too high levels will lower feed consumption. 3. Incorporate 20% alfalfa in the ration.	Good feed and management appear to lessen the incidence. Delayed castration (castration of bull calves at 4–5 mo. of age) and high salt rations of feedlot cattle (1–3% salt in the grain ration, using the upper limits in the winter months) in order to induce more water consumption are effective preventive measures. Avoid high phosphorus and low calcium. Provide adequate vitamin A, salt and water.	Calculi are stonelike concretions in the urinary tract which almost always originate in the kidneys. These stones block the passage of urine. Ammonium chloride (see Control and Eradication) appears to be the product of choice. However, ammonium sulfate may be used at the rate of 1.7–2.0 oz *(48–57 g)*/head/day. Add it to the ration when an outbreak occurs.
Treatment consists of correcting the dietary deficiencies and (1) adding vitamin A to the ration, or (2) injecting intramuscularly or intraruminal 500,000–1,000,000 IU of vitamin A.	(See Prevention and Treatment.)	Provide good sources of carotene (vitamin A) through green, leafy hays; silage; lush, green pastures; yellow corn; or add stabilized vitamin A to the ration.	High levels of nitrates interfere with the conversion of carotene to vitamin A.
Confine affected animals to a stall, and give plenty of rest. The veterinarian may administer alpha-tocopherol.	(See Prevention and Treatment.)	Add 0.1 ppm selenium to the complete feed, or use a mineral mix containing 20–30 ppm selenium.	White muscle disease is often overlooked in calves.

TABLE 8-17
MAXIMUM TOLERABLE LEVELS OF CERTAIN TOXIC ELEMENTS[1, 2]

Element	Maximum Tolerable Level
	(ppm)
Aluminum	1,000.0
Arsenic	50.0
Bromine	200.0
Cadmium	0.5
Fluorine	40.0–100.0
Lead	30.0
Mercury	2.0
Strontium	2,000.0

[1]Adapted from *Nutrient Requirements of Beef Cattle*, seventh revised edition, National Research Council-National Academy of Sciences, 1996, p. 55, Table 5-2.

[2]National Research Council (1980), Table 1, Mineral Tolerance of Domestic Animals.

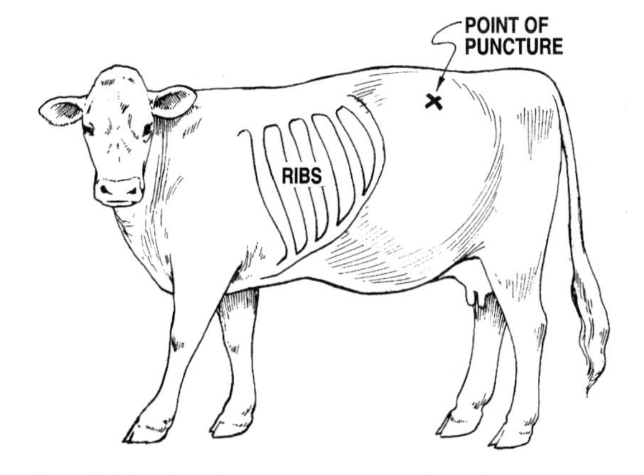

Fig. 8-27. When bloat is encountered, puncturing of the paunch should be a last resort. However, if such treatment must be administered to save the life of an animal, the cattle producer should know where to make the puncture. As noted in the above drawing, it should be made on the left side at the location shown.

Fig. 8-28. Grazing cattle approaching a molasses block containing poloxalene, a bloat-preventive. (Courtesy, Smith Kline Animal Health Products, Philadelphia, PA)

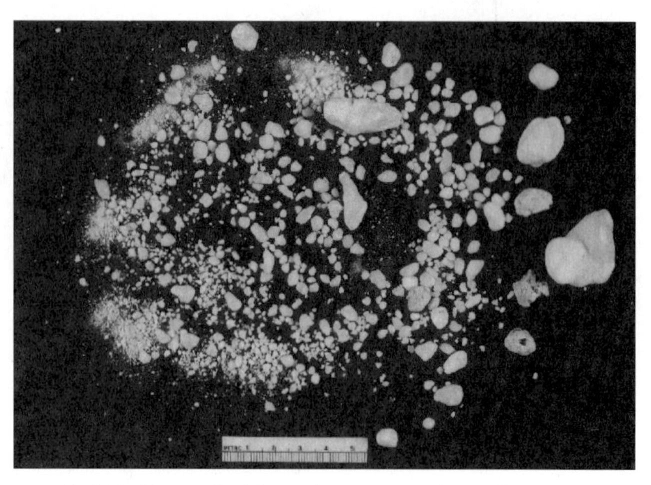

Fig. 8-30. Urinary calculi. Concretions or stones obtained from an imported bull that died from urinary calculi. (Courtesy, Washington State University)

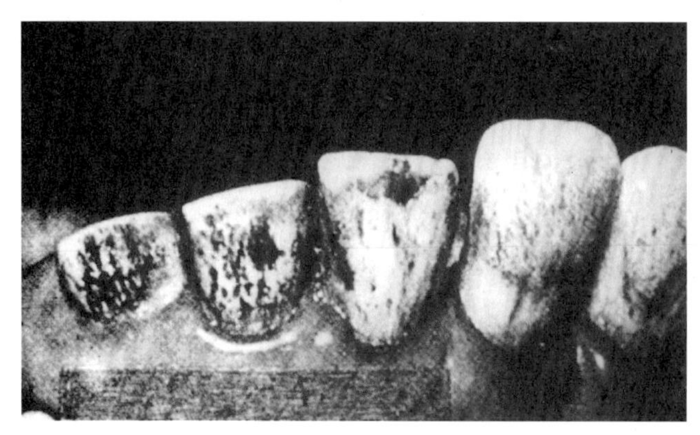

Fig. 8-29. Fluorosis in cattle. Incisors of 5-year-old cow fed ration averaging 37 ppm fluorine content. Note mottled enamel. (Courtesy, American Institute of Nutrition, Bethesda, MD)

NUTRIENT REQUIREMENTS OF BEEF CATTLE

Efficient beef production cannot be achieved unless nutrient requirements are met.

Although the basic biology of all beef cattle remains the same, differences in rate of maturity and in mature size have a marked influence on the application of the basic nutrition principles to the wide range of environmental and management conditions to which beef cattle are subjected. In recent years, types of beef cattle have changed in response to economic pressures and consumer demand for leaner

Fig. 8-31. Sweet clover disease. Note the collection of blood at the point of the left shoulder. (Courtesy, Dept. of Veterinary Pathology and Hygiene, College of Veterinary Medicine, University of Illinois)

cuts—larger and faster gaining cattle have evolved. For this reason, frame size should be considered in calculating the nutritive requirements of beef cattle.

The tables in the Appendix, Section I, Table I-1, were adapted by the authors from the National Research Council, *Nutrient Requirements of Beef Cattle*, Sixth Revised Edition, 1984. The Tables in the Appendix, Section I, Table I-2 and Table I-3, were adapted by the authors from the National Research Council, Nutrient Requirements of Beef Cattle, Seventh Revised Edition, 1996.

FEED COMPOSITION

Nutrient compositions of feedstuffs are necessary for intelligent ration preparation, animal health, and feed efficiency. Appendix, Section II, Table II-1, contains the composition of the most commonly used feeds for beef cattle; Appendix, Section II, Table II-2 gives the composition of common mineral supplements for beef cattle; and Appendix, Section II, Table II-3 lists the amount of bypass or undegraded protein in some common feeds.

QUESTIONS FOR STUDY AND DISCUSSION

1. Why is knowledge of beef cattle feeding so important?

2. Compare the nutritive needs of beef cattle with those of simple-stomached animals.

3. Sketch the ruminant's stomach and show the route of digestion followed by most feeds.

4. How does the calf's stomach differ from the ruminant's stomach in both structure and digestion?

5. List and discuss the factors upon which the daily consumption of dry matter (feed) is primarily dependent.

6. In recent years, crossbreeding and the exotic breeds have produced faster-gaining calves later-maturing cattle, and heavier-milking cows. How has this changed nutritive needs?

7. Describe the symptoms of (a) energy deficiency, and (b) protein deficiency in cattle.

8. Discuss the advantages and disadvantages of each: (a) the TDN system, and (b) the calorie system.

9. List the leading oilseed meals, and explain why soybean meal occupies such a dominant position as a protein supplement.

10. Discuss preruminant calf nutrition.

11. Discuss microbial synthesis and requirements.

12. Explain the place and importance of urea in feeding programs. What factors are essential for optimum use of urea?

13. Discuss postruminal protein supply and requirements.

14. Discuss protein solubility, degradation, and bypass.

15. Explain how you would go about determining mineral needs and buying a mineral mix.

16. Give the functions and deficiency symptoms of each of the following minerals for beef cattle: salt, calcium, phosphorus, cobalt, iodine, and selenium.

17. List the vitamins most apt to be deficient for beef cattle; then (a) list some of the deficiency symptoms, and (b) give practical sources of each vitamin for use on a farm or ranch.

18. Water is said to be the most vital of all nutrients. Why is this so?

19. Discuss the role of each of the following feeds for beef cattle: pastures, hay and other dry roughages, silages and roots, concentrates, and poultry waste.

20. Why should a cattle producer be knowledgeable relative to feed substitutions?

21. How would you go about selecting a commercial feed for beef cattle?

22. Which of the four methods listed for balancing rations do you prefer? Justify your choice.

23. Table 8-14 shows that beef cattle are near the bottom of the *animal totem pole* in *feed to food efficiency*. In light of this fact, will beef cattle decline in numbers throughout the world?

24. Give the cause, symptoms, treatment, and prevention of each of the following cattle nutritional diseases and ailments: bloat, fluorine poisoning, grass tetany, molybdenum toxicity, nitrate poisoning, rickets, selenium poisoning, sweet clover disease, vitamin A deficiency, and white muscle disease.

25. In order to feed beef cattle efficiently and economically, one must thoroughly understand the nutrients furnished by the available feeds, the extent to which cattle can utilize each feed, and the actual feeding value of these feeds. This can be accomplished only through careful and thorough study of the different feeds. So, the following exercises are designed better to acquaint the student with the feeds commonly fed to beef cattle:

a. **A study of available roughages.** Most of the feed of beef cattle is derived from roughages. Also, it is well known that a relatively large portion of the feed consumed by beef cattle is used in meeting the energy needs. Thus, a convenient and reasonably accurate way of determining which roughages are most economical under the conditions existing in a particular area at any given time is to compute the cost at which each of the available roughages furnishes 100 lb of total digestible nutrients. This is a measure of the economy with which the various feeds furnish fuel or energy. Refer to Table II-1 in the Appendix of this book for analyses, and obtain prices of available roughages locally. Then fill out the following table:

AVAILABLE ROUGHAGES

Feed	Farm Price per Ton	TDN per 100 lb *(45 kg)*	Cost per 100 lb (45 kg) TDN	Crude Protein per 100 lb *(45 kg)*	Calcium Content (%)	Phosphorus Content (%)	Carotene Vitamin A (1,000 IU/kg)

The student should also become familiar with the protein, calcium, phosphorus, and carotene content of the different roughages. The above table is so designed.

b. **A study of available grains and by-products.** At least one of the cereal grains is grown in almost every section of the country, and all of them are used quite widely as beef cattle feeds. As a group, the cereals and their by-products are high in energy. However, they possess certain nutritive deficiencies which may prove to be quite limiting if they are not properly used. Refer to Table II-1 in the Appendix of this book for analysis, and obtain prices of available grain and byproduct feeds locally. Then fill out the following table:

AVAILABLE GRAIN AND BY-PRODUCT FEEDS

Feed	Retail Price per Cwt	TDN per 100 lb *(45 kg)*	Cost per 100 lb *(45 kg)* TDN	Crude Protein per 100 lb *(45 kg)*	Cost per 100 lb *(45 kg)* Crude Protein	Calcium Content (%)	Phosphorus Content (%)	Carotene Vitamin A (1,000 IU/kg)

In addition to considering the cost per 100 lb of each TDN and digestible protein give consideration to calcium, phosphorus, and carotene content of each of these feeds. Also, in studying the byproduct feeds, be sure to understand the source of the feed and just what part of the original grain or seed goes into the byproduct.

26. Select a specific class of beef cattle and prepare a balanced ration, using those feeds that are available at the lowest cost.

SELECTED REFERENCES

Title of Publication	Author(s)	Publisher
AFMA Liquid Feed Symposium Proceedings	American Feed Manufacturers Association	AFMA, Chicago, IL, 1971
Alternative Sources of Protein for Animal Production	National Research Council	National Academy of Sciences, Washington, DC, 1973
Animal Feeds	M . H. Gutcho	Noves Data Corporation, Park Ridge, NJ, 1970
Animal Growth and Nutrition	Ed. by E. S. E. Hafez I. A. Dyer	Lea & Febiger, Philadelphia, PA, 1969
Animal Nutrition	L. A. Maynard, *et al.*	McGraw-Hill Book Co., New York, NY, 1979
Animal Nutrition	P. McDonald R. A. Edwards J. F. D. Greenhalgh	Oliver and Boyd, Ltd., Edinburgh, Scotland, 1972

Antibiotics in Nutrition	Thomas H. Jukes	Medical Encyclopedia, Inc., New York, NY, 1955
Applied Animal Feeding and Nutrition	M. H. Jurgens	Kendall/Hunt Publishing Company, Dubuque, IA, 1972
Applied Animal Nutrition, Second Edition	E . W. Crampton L. E. Harris	W. H. Freeman and Co,. San Francisco, CA, 1969
Association of American Feed Control Officials Incorporated, Official Publication	Association of American Feed Control Officials, Inc.	Association of American Feed Control Officials Inc., annual
Atlas of Nutritional Data on United States and Canadian Feeds	National Research Council U.S.A.; Committee on Feed Composition, Research Branch, Canada Dept. of Agriculture	National Academy of Sciences, Washington, DC, 1971
Beef Cattle Feeding and Nutrition, Second Edition	T. W. Perry M. J. Cecava	Academic Press, Inc., San Diego, CA 1995
Body Composition in Animals and Man, Pub. 1598	National Research Council	National Academy of Sciences, Washington, DC, 1968
Cereal Processing and Digestion	U.S. Feed Grains Council, USDA Foreign Agricultural Service	U.S. Feed Grains Council, London, England, 1972
Composition of Cereal Grains and Forages, Pub. 585	National Research Council	National Academy of Sciences, Washington, DC
Composition of Concentrate By-product Feeding Stuffs, Pub. 449	National Research Council	National Academy of Sciences, Washington, DC, 1956
Digestive Physiology and Nutrition of the Ruminant	Ed. by D. Lewis	Butterworth & Co., Ltd., London, England, 1961
Digestive Physiology and Nutrition of Ruminants, Vols. 1 and 3	D. C. Church	D. C. Church, Dept. of Animal Science, Oregon State University, Corvallis, OR, 1979
Digest of Research on Urea and Ruminant Nutrition	E. I. du Pont de Nemours & Company (Inc.), Polychemicals Department	E. I. du Pont de Nemours & Company, Wilmington, DE, 1958
Effect of Processing on the Nutritional Value of Feeds	National Research Council	National Academy of Sciences, Washington, DC, 1973
Energy Metabolism of Ruminants	K. L. Blaxter	Hutchinson & Co., Ltd., London, England, 1962
Evaluation of Feeds Through Digestibility Experiments	B. H. Schneider W. P. Flatt	The University of Georgia Press, Athens, GA, 1975
1974 Feed Additive Compendium	Ed. by D. Natz, The Animal Health Institute, Washington, D.C.	The Miller Publishing Company, Minneapolis, MN, annual
Feed Composition, Tables of, Pub. 1232	National Research Council	National Academy of Sciences, Washington, DC, 1964

Title	Author(s)	Publisher
Feed Flavor and Animal Nutrition	T. B. Tribble	Agriaids, Inc., Chicago, IL, 1962
Feed Formulations, Third Edition	T. W. Perry	The Interstate Printers & Publishers, Inc., Danville IL, 1982
Feeds and Feeding, Abridged	F. B. Morrison	The Morrison Publishing Company, Ithaca, NY, 1958
Feeds and Feeding, 22nd Edition	F. B. Morrison	The Morrison Publishing Company, Ithaca, NY, 1956
Feeds & Nutrition, Second Edition	M. E. Ensminger J. E. Oldfield W. W. Heinemann	The Ensminger Publishing Co., Clovis, CA, 1990
Feeds & Nutrition Digest	M. E. Ensminger J. E. Oldfield W. W. Heinemann	The Ensminger Publishing Co., Clovis, CA, 1990
Feeds for Livestock, Poultry and Pets	M . H. Gutcho	Noyes Data Corporation, Park Ridge, NJ, 1973
Feeds of the World	B. H. Schneider	Agricultural Experiment Station, West Virginia University, Morgantown, WV, 1947
Fundamentals of Nutrition, Second Edition	L. E. Lloyd B. E. McDonald E. W. Crampton	W. H. Freeman and Co., San Francisco, CA, 1978
Handbook of Feedstuffs, The	R. Seiden W. H. Pfander	Springer Publishing Company, Inc., New York, NY, 1957
International Conference on the Use of Antibiotics in Agriculture, Proceedings	American Cyanamid Company, Scientific Sessions	American Cyanamid Company, Wayne, NJ
International Feed Nomenclature and Methods for Summarizing and Using Feed Data to Calculate Diets, An, Bull. 479	L. E. Harris J. M. Asplund E. W. Crampton	Agricultural Experiment Station, Utah State University, Logan, UT, 1968
Livestock Feeds and Feeding	D. C. Church, *et al.*	O & B Books, Inc., Corvallis, OR, 1984
Lysine in Animal Nutrition, Annotated Bibliography	Merck Sharp & Dohme Research Laboratories	Merck & Co., Inc., Rahway, NJ, 1960
Manual of Clinical Nutrition	R. S. Goodhart M. G. Wohl	Lea & Febiger, Philadelphia PA, 1964
Manual for the Computer Formulation of Livestock Feed Mixtures	W. K. McPherson	M. L. McPherson, Gainesville, FL, 1971
Mineral Metabolism, An Advanced Treatise, Vol. I, Part A; Vol. II, Part A; and Vol. III	Ed. by C. L. Comar Felix Bronner	Academic Press, Inc., New York, NY, Vol. I, 1960; Vol. II, 1964; and Vol. III, 1969
Mineral Nutrition of Livestock, The	E. J. Underwood	Food and Agriculture Organization of the United Nations, Commonwealth Agricultural Bureaux, Rome, Italy, 1966
Mineral Nutrition of Plants and Animals	F. A. Gilbert	University of Oklahoma Press, Norman, OK, 1948

Title	Author	Publisher
Nonprotein Nitrogen in the Nutrition of Ruminants	J. K. Loosli I. W. McDonald	Food and Agriculture Organization of the United Nations, Rome, Italy, 1968
Nutrient Requirements of Beef Cattle, Sixth Revised Edition	National Research Council	National Academy of Sciences, Washington, DC, 1984
Nutrient Requirements of Beef Cattle, Seventh Revised Edition	National Research Council	National Academy of Sciences, Washington, DC, 1996
Nutrient Requirements of Farm Livestock, The, No. 3, Pigs	Agricultural Research Council	Agricultural Research Council, London, England, 1967
Nutritional Data	H. A. Wooster, Jr. F. C. Blanck	H. J. Heinz Company, Pittsburgh, PA, 1950
Nutritional Deficiencies in Livestock	R. T. Allman T. S. Hamilton	Food and Agriculture Organization of the United Nations, Rome, Italy, 1952
Nutrition of Animals of Agricultural Importance, Parts 1 and 2	Ed. by D. Cuthbertson	Pergamon Press, London, England, 1969
Nutrition of Plants, Animals, Man	College of Agriculture, Michigan State University	Michigan State University, East Lansing, MI, 1955
Nutrition Research Techniques for Domestic and Wild Animals, Vol. 1	L. E. Harris	The author, Logan, UT, 1970
Physiology of Digestion in the Ruminant	Dougherty, *et al.*	Butterworth, Inc., Washington, DC, 1964
Proceedings of the Fifth International Congress on Nutrition	National Research Council	Waverly Press, Inc., Baltimore, MD, 1961
Processed Plant Protein Foodstuffs	Ed. by A. M. Altschul	Academic Press, Inc., New York, NY, 1958
Processing and Utilization of Animal By-Products	I. Mann	Food and Agriculture Organization of the United Nations, Rome, Italy, 1962
Proteins—Their Chemistry and Politics	A. M. Altschul	Basic Books, Inc., New York, NY, 1965
Rations for Livestock, Bull. No. 48	R. E. Evans	Her Majesty's Stationery Office, London, England, 1960
Rumen and Its Microbes, The	R. E. Hungate	Academic Press, Inc., New York, NY, 1966
Selenium in Nutrition	National Research Council	National Academy of Sciences, Washington, DC, 1971
Single-Cell Protein	Ed. by R. I. Mateles S. R. Tannenbaum	The M.I.T. Press, Cambridge, MA, 1968
Stockman's Handbook, The, Seventh Edition	M. E. Ensminger	Interstate Publishers, Inc., Danville IL, 1992
Trace Elements in Agriculture	V. Sauchelli	Van Nostrand Reinhold Company, New York, NY, 1969
Urea and Non-protein Nitrogen in Ruminant Nutrition, Second Edition	Ed. by H. J. Stangel	Nitrogen Division, Allied Chemical Corporation, Morristown, NJ, 1963
Urea as a Protein Supplement	Ed. by M. H. Briggs	Pergamon Press, New York, NY, 1967

Use of Drugs in Animal Feeds, The, Pub. 1679	National Academy of Sciences	National Academy of Sciences, Washington, DC, 1969
Vitamin B_{12} in Animal Nutrition	Merck & Co., Inc., Chemical Division	Merck & Co., Inc., Rahway, NJ, 1957
Vitamin B_{12}, Selected Annotated Bibliography	Merck & Co., Inc., Chemical Division	Merck & Co., Inc., Rahway, NJ, 1954
Vitamins, The—Chemistry Physiology, Pathology, Methods, Second Edition, Vols. I-III	Ed. by W. H. Sebrell, Jr. R. S. Harris	Academic Press, Inc., New York, NY, 1967, 1968, 1971
Vitamins in Feeds for Livestock	F. C. Aitken R. G. Hankin	Commonwealth Agricultural Bureaux, Farnham Royal, Bucks, England, 1970
Vitamins and Hormones, Vol. XV	Ed. by R. S. Harris G. F. Marrian K. V. Thimann	Academic Press, Inc., New York, NY, 1957

All flesh is grass! Hereford steers on bromegrass pasture in Nebraska. (Courtesy, C. B. & Q. Railroad)

PASTURES AND RANGES

Contents — Page

Contents — Page

Contents	Page

Fig. 9-1. On pasture! Polled Hereford heifers on seeded pasture in South Carolina. (Courtesy, American Polled Hereford Association)

Fig. 9-2. On the range! Beefmaster cows and calves on the Lasater Ranch, Matheson, Colorado. (Photo by Wyatt M. Casey, Jr.; Courtesy, The Lasater Ranch)

The great cattle-producing areas of the world are characterized by good pastures. Good cattle producers, good cattle, and good pastures go hand in hand. Indeed, pasture is the cornerstone of successful beef cattle production. The economic importance of pastures is further attested by the following facts:

1. A total of 29.2% of the total land area of the United States (50 states) is used solely for grassland.
2. A total of 84.5% of the feed supply of all U.S. beef cattle is derived from forage; in season this means pasture.
3. Good pasture alone will produce 200 to 400 lb of beef per acre annually (in weight of calves weaned, or in added weight of older cattle); superior pastures will do much better.
4. No method of harvesting has yet been devised which is as cheap as that which can be accomplished by animals, although nutrient yields are generally less when highly productive forage crops are pastured rather than harvested as hay or silage. The difference in yield is generally more than offset by the added expense of harvesting, storing, and feeding.
5. Pasture gains are generally cheaper than drylot gains because (a) less labor is required, (b) grass is the cheapest of all roughages, (c) less expensive protein supplement is required, (d) the animals scatter their own droppings, thus alleviating hauling manure, and (e) fewer buildings and less equipment are necessary.

But grass—the nation's largest crop—should not be taken for granted. Again and again scientists and practical farmers and ranchers have demonstrated that the following desired goals in pasture production are well within the realm of possibility:

- To produce higher yields of palatable and nutritious forage.

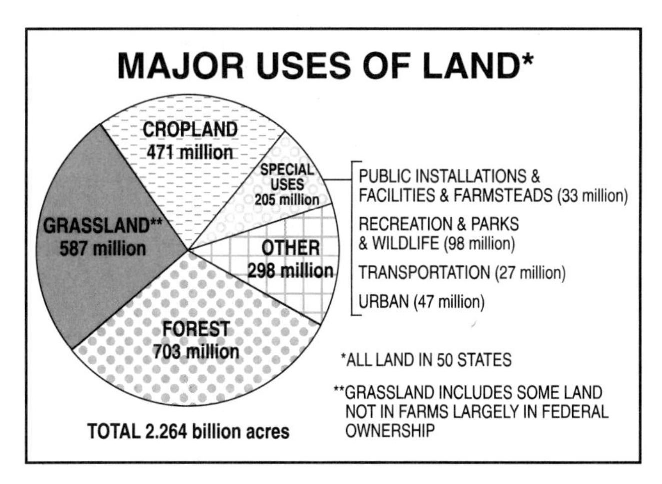

Fig. 9-3. About 587 million acres of pasture and range area in the U.S. are used solely as grazing land. (Courtesy, USDA)

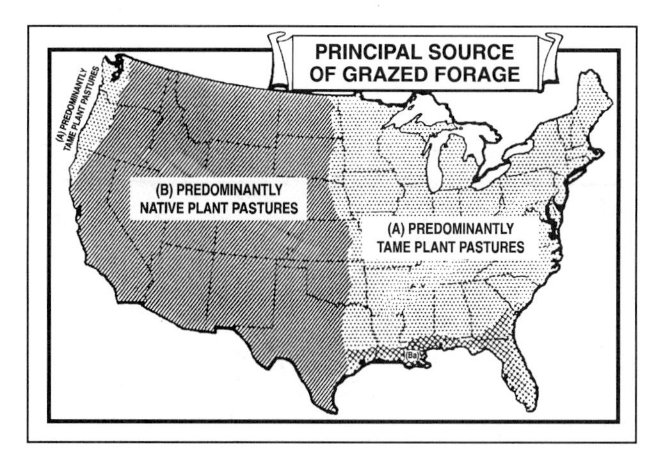

Fig. 9-4. The two major U.S. pasture areas—**(A)** tame (seeded), and **(B)** native (range)—about equally divide the 48 contiguous states into eastern and western halves. (Courtesy, USDA)

- To extend the grazing season from as early in the spring to as late in the fall as possible.
- To provide a fairly uniform supply of feed throughout the entire season.

At the outset, it should be recognized that no one plant embodies all the desirable characteristics necessary to meet the above goals. None of them will grow year-round, or during extremely cold or dry weather. Each of them has a period of peak growth which must be conserved for periods of little growth. Consequently, progressive cattle producers will find it desirable (1) to grow more than one species, and (2) to plan pastures for each season of the year. In general, a combination of permanent, rotation, and temporary pastures—accompanied by scientific management—will best achieve these ends.

Broadly speaking, all U.S. pastures may be classified as either (1) tame (seeded) pastures, or (2) native pastures (see Fig. 9-4). Although no sharp line of demarcation exists between the two groups, tame pastures include those which either receive more than approximately 20 in. of rainfall annually or are irrigated, whereas the latter group includes those range pastures which receive less than 20 in. of rainfall annually. The general principles and the objectives sought are the same, but, as will be discussed later, there are considerable differences in the recommended seeding and management practices for the two groups.

Pastures may be further classified as follows:

1. **Permanent pastures.** Those which, with proper care, last for many years. They are most commonly found on land that cannot be used profitably for cultivated crops, mainly because of topography, moisture, or fertility. The vast majority of the farms of the United States have one or more permanent pastures, and most range areas come under this classification.
2. **Rotation pastures.** Those that are used as a part of the established crop rotation. They are generally used for 2 to 7 years before plowing.
3. **Temporary and supplemental pastures.** Those that are used for a short period, usually annuals such as Sudangrass, millet, rye, wheat, oats, rape, or soybeans. They are seeded for the purpose of providing supplemental grazing during the season when the regular permanent or rotation pastures are relatively unproductive.

PART I
TAME (SEEDED) PASTURES

Tame pastures—either those which receive over 20 in. of rainfall annually or those which are irri-

Fig. 9-5. Cows and calves grazing a lush millet supplemental pasture in northern Florida. (Courtesy, *The Florida Cattleman*)

Fig. 9-6. Tame (seeded) irrigated pasture in the Gallatin Valley in southwestern Montana near Bozeman. (Courtesy, USDA)

gated—include the seeded (cultivated) pastures of the Corn Belt, the South, the East, and the irrigated areas, and smaller and scattered moderate- to high-rainfall areas throughout the West.

ADAPTED VARIETIES AND SUITABLE MIXTURES

The specific grass or grass-legume mixture will vary from area to area, according to differences in soil, temperature, and rainfall. A complete listing of all adapted and recommended grasses and legumes for cattle pastures would be too lengthy for this book. However, Fig. 9-7 shows the 10 generally recognized U.S. pasture areas and Table 9-1 lists the adapted grasses and legumes for each area. In using Table 9-1, bear in mind that many species of forages have wide geographic adaptation, but subspecies or varieties often have rather specific adaptation. Alfalfa, for example, is represented by many varieties which give this species adaptation to nearly all states. Variety then, within species, makes many forages adapted to the widely varying climate and geographic areas. The county agricultural agent and the state agricultural college can furnish recommendations for the areas that they serve.

Most cattle pastures can be improved by fertilizing and management. Also, cattle producers need to give attention to balancing pastures nutritionally. Early-in-the-season grasses are of high water content and lack energy. Mature weathered grass is almost always deficient in protein (being as low as 3% or less) and low in carotene. But these deficiencies can be corrected by proper supplemental feeding.

ESTABLISHING A NEW PASTURE

The following practices are usually adhered to in successfully establishing a new pasture in the tame pasture area:

1. **Adapted varieties and suitable mixtures are selected.** The first requisite of successful pastures is that adapted varieties of grasses and/or legumes shall

LEGUMES AND GRASSES ADAPTED TO 10 AREAS OF THE 48 CONTIGUOUS STATES

1. Northern Humid Area
2. Central Humid Area
3. Southern Humid Area
4. Eastern Coastal Area
5. Northern Great Plains Area
6. Southern Great Plains Area
7. Northwest Intermountain Area
8. Southwest Area
9. Northwest Coastal Area
10. California Coastal Area

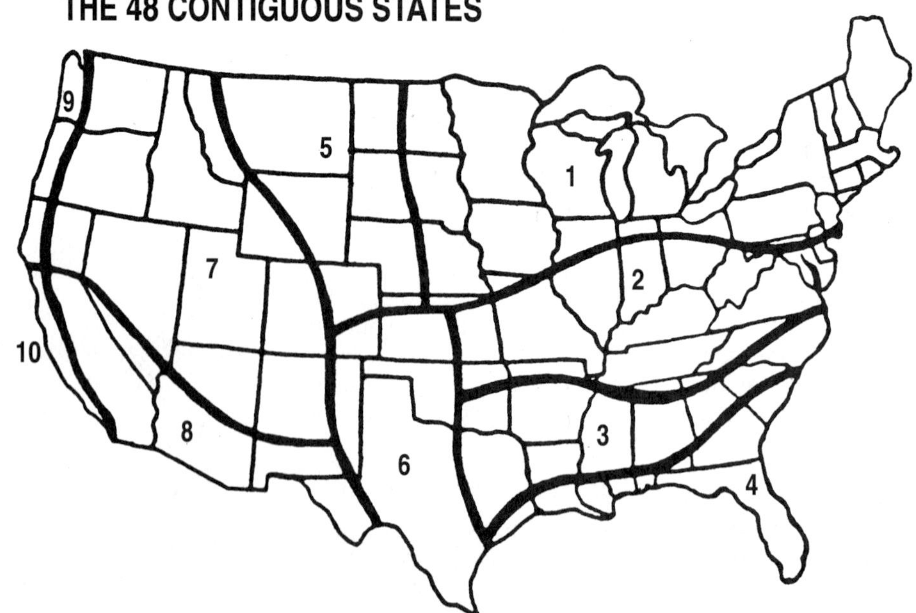

Fig. 9-7. The 10 generally recognized U.S. pasture areas.

TABLE 9-1

ADAPTED GRASSES AND LEGUMES (INCLUDING BROWSE AND FORBS) FOR CATTLE PASTURES, BY 10 GEOGRAPHICAL AREAS OF THE UNITED STATES (SEE FIG. 9-7 FOR GEOGRAPHICAL AREAS)

	Areas of the United States									
	1	2	3	4	5	6	7	8	9	10
Grasses, shrubs, forbs:										
Bahiagrass			x	x						
Bermudagrass		x	x	x		x		x		x
Bluegrass, big							x		x	
Bluegrass, Kentucky	x	x	x		x		x		x	
Bluestem, big	x	x	x	x	x	x				
Bluestem, Caucasian		x	x			x				
Bluestem, little	x	x	x	x	x	x				
Bluestem, sand	x	x			x	x				
Bristlegrass, plains						x		x		
Bromegrass, meadow					x		x	x	x	
Bromegrass, smooth	x	x			x		x	x	x	x
Buckwheat (wild)								x		
Buffalograss					x	x				
Buffelgrass						x				
Canarygrass, reed	x	x					x		x	
Cottontop, Arizona								x		
Curly mesquite						x		x		
Dallisgrass			x	x						
Digitgrass, pangola			x	x						
Dropseed, sand						x		x		
Fescue, tall	x	x	x				x	x	x	x
Foxtail, creeping							x		x	
Galleta						x		x		
Gamagrass, eastern	x	x	x	x	x	x				
Grama, black								x		
Grama, blue					x	x	x	x		
Grama, sideoats	x	x	x	x	x	x	x	x		
Hardinggrass						x			x	x
Indiangrass	x	x			x	x				
Indianwheat								x		
Johnsongrass			x	x		x				
Kleingrass						x				
Koleagrass, Perla									x	x
Limpograss				x						
Lovegrass, Lehmann							x	x		
Lovegrass, sand	x	x			x	x				
Lovegrass, weeping			x			x				
Maidencane				x						
Millet	x	x	x	x		x				
Muhly, spike							x	x		
Needle-and-thread	x				x					
Needlegrass, green	x				x					
Oatgrass, tall									x	
Oats	x	x	x	x	x	x		x	x	x
Orchardgrass	x	x	x	x	x		x	x	x	x
Paragrass				x						
Pearlmillet		x	x	x		x				
Redtop	x						x		x	
Rescuegrass			x	x					x	
Rhodesgrass			x	x						
Ricegrass, Indian							x	x		
Rye	x	x	x	x	x	x		x	x	x
Ryegrass, annual		x	x	x		x			x	
Ryegrass, perennial	x	x	x						x	
Sacaton, alkali						x	x	x		
Sage, pitchers	x	x	x		x	x				
Saltbrush, fourwing					x	x	x	x		
Sorghum-Sudan hybrids	x	x	x	x	x	x	x			
Stargrass			x							
Sudangrass	x	x	x	x	x	x	x	x	x	x
Sunflower, Maximilian	x	x	x		x	x				
Switchgrass	x	x	x		x	x				
Three-awn				x	x	x	x	x		
Timothy	x	x					x		x	
Tobosa grass						x				
Wheat	x	x	x	x	x	x	x	x	x	x
Wheatgrass, bluebunch					x		x		x	
Wheatgrass, crested					x		x			
Wheatgrass, intermediate	x				x		x		x	x
Wheatgrass, pubescent					x	x	x			x
Wheatgrass, tall	x				x	x	x	x	x	
Wheatgrass, western	x	x			x	x	x		x	
Wild-rye, basin					x		x			
Wild-rye, Canada	x	x	x		x	x	x			
Wild-rye, Russian					x		x			
Winterfat (white sage)								x		
Wintergrass, Texas						x				
Legumes:										
Alfalfa (lucerne)	x	x	x	x	x	x	x	x	x	x
Alyceclover			x	x						
Black medic (yellow trefoil)			x			x		x		
Bur-clover		x						x		x
Clover, alsike	x	x			x		x	x	x	
Clover, arrowleaf		x	x							
Clover, crimson		x	x							
Clover, Hubam (white sweet clover)	x	x						x	x	
Clover, Kura	x	x			x		x			x
Clover, Ladino	x	x	x	x			x	x	x	x
Clover, prairie						x		x		
Clover, red	x	x	x	x			x	x	x	
Clover, strawberry					x		x	x		x
Clover, subterranean			x	x				x	x	x
Clover, white	x	x	x	x			x	x	x	x
Cowpeas			x	x						
Crown vetch	x	x								
Flat pea			x	x			x			
Hairy indigo				x						
Lespedeza (annual)		x	x	x						
Lespedeza (perenial, sericea)		x	x	x						
Milk vetch, cicer	x				x		x			
Peas, field									x	
Pea shrub								x		
Prairie clover, purple	x	x	x		x	x	x			
Ratany								x		
Soybeans	x	x	x	x			x			
Sweet clover, white	x	x						x	x	
Sweet clover, yellow	x	x			x	x	x	x	x	
Trefoil, birdsfoot	x	x	x				x		x	x
Velvet bean			x	x						
Vetch		x	x	x	x	x		x	x	

be selected for the area and for the purposes intended. Table 9-1 gives the general recommendations. For more specific recommendations for a particular farm or ranch, the cattle producer should consult a local authority such as the county agent, vocational agriculture instructor, or successful neighbors.

Where grass-legume mixtures are to be grown, a 50-50 mixture is satisfactory for most purposes and conditions.

2. **The soil is tested and fertilized.** The soil is tested and fertilized (and limed if necessary) according to needs. The three elements required by all grasses and legumes in greatest abundance are nitrogen, phosphorus, and potassium. In addition, where legumes are grown, acid soils need lime. The pH of the soil should be about 6.5. It is best to work lime into the soil considerably in advance of seeding, but commercial fertilizers should be applied at seeding time.

A thin, uniform mulch of barnyard manure is especially valuable in establishing a new seeding.

3. **High-quality seed is purchased.** The seed should be of good quality, of high germination and purity as indicated on the tag, and free of noxious weeds. Also, proof of origin is of prime importance when an imported variety is secured. Certified seed carries a little more assurance of being high quality than noncertified seed, and gives proof of its origin much as a registration certificate does on a purebred animal.

4. **Scarified legume seed is used.** In the purchase of certain legume seed, it is important that it be scarified, which breaks the seed coat and allows faster moisture penetration—thus assuring quicker and more uniform germination and a better stand the first year.

5. **Legume seed is inoculated.** Since legumes can use nitrogen from the air provided they are inoculated with the proper bacteria, it is important that legume seed be inoculated.

Inoculant comes in several different forms, usually in a can with directions given thereon. It is important that the seed not be treated more than a few hours before seeding because these nitrogen-fixing bacteria are easily killed by drying, heat, sunlight, or by chemical seed treatment.

6. **A good seed bed is prepared.** A good seed bed is free from weeds, fine-textured, firm, and moist.

Weeds are usually destroyed by growing row crops or a small grain the year preceding seeding to pasture and by cultivating frequently following the harvesting of this crop.

There are many different ways in which to prepare a good seed bed. Perhaps as good a method as any consists in (a) plowing as far in advance of seeding as possible, (b) disking, (c) harrowing one or more times to level up the field and smooth down the surface, and (d) cultipacking or rolling. A properly prepared seed bed should be so firm that one barely leaves a footprint when walking across it; the firmer the better from the standpoint of moisture conservation of small seeds.

7. **The seeding operation is timed and carried out properly.** The seeding time will vary, being determined primarily by the area and by the species or mixture used.

The actual seeding operation may be (a) by broadcasting, with a whirlwind seeder or by hand, or (b) by drilling, with any one of several types of conventional seeders. Drilling is the preferred method, for it insures more uniform placement of seed in both depth and amount of seed per acre and results in a more uniform stand.

Since most grass and legume seeds are very small, they should not be covered deeply. A good rule of thumb is that they should not be covered more than 4 or 5 times the width of the seed; usually this means not more than ¼ in.

8. **A companion or nurse crop may or may not be included.** The value of planting a "companion" or nurse crop—usually consisting of annuals—with new seed crops is controversial.

The advantages are: (a) It furnishes a crop of value while the new seeding is being established; (b) it lessens erosion; and (c) it reduces the weed population.

The disadvantages are: (a) It may retard the growth of the seedlings for whose protection it is grown; and (b) it may rob the new seeding of so much moisture that it kills them during dry spells unless the companion crop is harvested early as pasture, hay, or silage.

IMPROVING OR RENOVATING AN OLD PASTURE

In altogether too many cases old permanent pastures are merely gymnasiums for livestock. Generally this condition exists because the least productive areas are used for pastures and because little attention is given to fertility and pasture management.

Permanent pastures in the tame pasture area that are run-down may be brought back into production by either of the following methods:

1. **By reseeding without growing a crop in the interim.** Poor, run-down permanent pastures are frequently renovated by reseeding without growing a crop in the interim; in other words, pasture follows pasture. This kind of renovation is designed to increase pasture yields without subjecting the soil to excessive erosion and without keeping the area out of pasture production any longer than necessary. The actual operations involved in renovating will vary from area to area, and

from field to field. In general, it involves (a) cultivating (preferably by plowing, but by disking or other methods in unplowable areas) so as to destroy all existing vegetation, (b) fertilizing and liming, and (c) preparing the seed bed and seeding with an adapted high-yielding pasture mixture.

2. **By fertilizing, overseeding, and managing.** Where a fair but unproductive permanent pasture stand exists, pasture improvement or renovation may consist of (a) fertilizing (and liming where needed), (b) seeding (overseeding) with desirable and adapted varieties, and (c) managing in accordance with the outline which follows. Usually the fertilizer and the seed are worked into the soil with a disk and springtooth harrow, but a minimum of the existing sod is destroyed.

MANAGEMENT OF PASTURES

Many good pastures have been established only to be lost through careless management. Good pasture management in the tame pasture area involves the following practices.

CONTROLLED GRAZING

Nothing contributes more to good pasture management than controlled grazing. At its best, it embraces the following:

1. **Protection of first-year seedings.** First-year seedings should be grazed lightly or not at all in order that they may get a good start in life. Where practical, instead of grazing, it is preferable to mow a new first-year seeding about 3 in. above the ground and to utilize it as hay, provided there is sufficient growth to justify this procedure.

2. **Rotation or alternate grazing.** Rotation or alternate grazing is accomplished by dividing a pasture into fields (usually two to four) of approximately equal size, so that one field can be grazed while the others are allowed to make new growth. This results in increased pasture yields, more uniform grazing, and higher quality forage.

Generally speaking, rotation or alternate grazing is (a) more practical and profitable on rotation and supplemental pastures than on permanent pastures, and (b) more beneficial where parasite infestations are heavy than where little or no parasitic problems are involved.

3. **Shifting the location of salt, shade, and water.** Where portable salt containers are used, more uniform grazing and scattering of the droppings may be obtained simply by the practice of shifting the location of the salt to the less grazed areas of the pasture. Where possible and practical, the shade and the water should be shifted, also.

4. **Deferred spring grazing.** Allow 6 to 8 in. of growth before turning cattle out to pasture in the spring, thus giving grass a needed start. Besides, the early spring growth of pastures is washy and high in moisture.

5. **Avoiding close late fall grazing.** Pastures that are grazed closely late in the fall start late in the spring. With most pastures, 3 to 5 in. of growth should be left for winter cover.

6. **Avoiding overgrazing.** Never graze more closely than 2 to 3 in. during the pasture season. Continued close grazing reduces the yield, weakens the plants, allows weeds to invade, and increases soil erosion. The use of temporary and supplemental pastures, may *spell off* regular pastures through seasons of drought and other pasture shortages and thus alleviate overgrazing.

7. **Avoiding undergrazing.** Undergrazing seeded pastures should also be avoided, because (a) mature forage is unpalatable and of low nutritive value, (b) tall-growing grasses may drive out such low-growing plants due to shading, and (c) weeds, brush, and coarse grasses are more apt to gain a foothold when the pasture is grazed insufficiently. It is a good rule, therefore, to graze the pasture fairly close at least once each year.

CLIPPING PASTURES AND CONTROLLING WEEDS

Pastures should be clipped at such intervals as necessary to control weeds (and brush) and to get rid of uneaten clumps and other unpalatable coarse growth left after incomplete grazing. Pastures that are grazed continuously may be clipped at or just preceding the usual haymaking time; rotated pastures may be clipped at the close of the grazing period. Weeds and brush may also be controlled by chemicals, by burning, etc.

TOPDRESSING

Like animals, for best results grasses and legumes must be fed properly throughout their lifetime. It is not sufficient that they be fertilized (and limed if necessary) at or prior to seeding time. In addition, in most areas it is desirable and profitable to topdress pastures with fertilizer annually, and, at less frequent intervals, with lime (lime to maintain a pH of about 6.5). Such treatments should be based on soil tests, and are usually applied in the spring or fall.

SCATTERING DROPPINGS

The droppings should be scattered three or four times each year and at the end of each grazing season in order to prevent animals from leaving ungrazed clumps and to help them fertilize a larger area. This can best be done by the use of a brush harrow or chain harrow.

GRAZING BY MORE THAN ONE CLASS OF ANIMALS

Grazing by two or more classes of animals makes for more uniform pasture utilization and fewer weeds and parasites, provided the area is not overstocked. Different kinds of livestock have different habits of grazing; they show preference for different plants and graze to different heights.

IRRIGATING WHERE PRACTICAL AND FEASIBLE

Where irrigation is practical and feasible, it alleviates the necessity of depending on the weather.

EXTENDING THE GRAZING SEASON

In the South and in Hawaii, year-round grazing is a reality on many successful cattle establishments. By careful planning and by selecting the proper combination of crops, other areas can approach this desired goal.

In addition to lengthening the grazing season through the selection of species, earlier spring pastures can be secured by avoiding grazing too late in the fall and by the application of a nitrogen fertilizer in the fall or early spring. Nitrogen fertilizers will often stimulate the growth of grass so that it will be ready for grazing 10 days to 2 weeks earlier than unfertilized areas.

PART II THE WESTERN RANGE

There are approximately 622 million acres of rangeland in the United States, excluding Alaska. About 214 million acres, or 34% is federal land. The bulk of rangelands occurs in the intermountain portions of the 11 western states. Nearly 408 million acres of rangeland, or 66%, is nonfederal lands, mostly privately owned.

Various geographical divisions are assumed in referring to the range area. Sometimes reference is made to the *17 range states*, embracing a land area of approximately 1.16 billion acres. At other times this larger division is broken down chiefly on the basis of topography into (1) the *11 western states* (Arizona, California, Colorado, Idaho, Montana, Nevada, New Mexico, Oregon, Utah, Washington, and Wyoming); and (2) the *Great Plains States* (the six states of Kansas, Nebraska, North Dakota, Oklahoma, South Dakota, and Texas). Specific references to the Great Plains Area includes, in addition to the six Great Plains States, that portion of Colorado, Montana, New Mexico, and Wyoming lying east of the Rocky Mountains.

Since the major portion of rangeland is located in the west, it is often referred to as the *Western Range*. Actually, rangelands occur in 25 of the 50 states. All states west of the Mississippi River, with the exception of Iowa, contain rangelands. They also occur in the east, extending through the Gulf Coast states from Louisiana to Florida. These eastern rangelands, although minor in extent, have high production potentials and are important to local economies.

Moreover, variation in climate, soils, and topography in the range country is accompanied by a great diversity in the kind and production of vegetation, and in the use made of it.

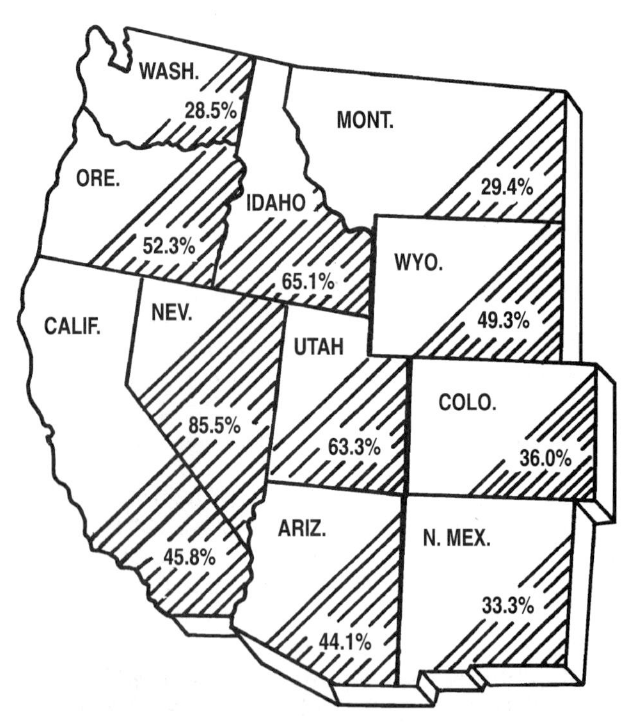

Fig. 9-8. A map showing the 11 western states and the proportion of land in each of these states that is owned by the U.S. Government. (Source: *Public Land Statistics*, Vol. 169, p. 10)

Most rangeland areas are suitable for year-round livestock operations, although supplements in forage or nutrients may be required at different seasons. In parts of the intermountain area, rangelands are grazed at different times of the year and herds and flocks migrate with the season, moving to higher elevations in summer and returning to lower elevations in winter.

Although some operations rely solely on range for forage, in most locations rangelands are used in combination with cultivated pastures and cropland forages in varying proportions.

There are striking differences in the nature and use of rangelands when compared to seeded pastures and in the principles and techniques required for management. From the standpoint of vegetation and utilization by livestock, ranges differ significantly from cultivated pastures as follows:

1. **They are less productive.** Generally, their productive capacity is lower. This is as one would expect, for they are largely made up of the residue remaining after the usable agricultural lands have been taken up. Also, plant growth on range lands frequently is limited by low and undependable rainfall (even drought), short growing seasons, shallow or rocky soil, alkali or salt accumulations, steep topography, etc. Under such conditions, forage plants are usually less resistant to grazing damage than those growing under a more favorable environment.
2. **They are more likely to progress to less palatable plants.** Range vegetation consists of a mixture of native and introduced plants, varying greatly in palatability, nutritive value, and productive ability. Grazing animals select the most palatable plants first. Thus, unless careful management is practiced, the best plants are crowded out through a combination of grazing injury and competition from the ungrazed, low-value plants. Continued poor management can result in good forage plants being almost completely replaced by low-value annual, weedy, or shrubby vegetation, or left denuded and subject to severe erosion.
3. **They are more difficult to restore when depleted.** Once a range becomes depleted, it is a slow process to rebuild it. Plowing and drilling are impractical on most rangelands, thus, very often the only feasible way of restoring a range to good condition is to stock it conservatively and manage it well.
4. **They often serve multiple uses.** Rangelands often have other uses in addition to grazing values. Among such uses are: water production, timber production, mineral production, wildlife production, and recreation (camping, hiking, picnicking, etc.)

Thus, many people, in addition to the livestock producer, have an interest in the grazing management practiced on ranges. This is part of the justification given for federal government ownership of large tracts of rangeland.

WILD ANIMALS OF THE RANGE

While cattle producers were expanding to the West, they had to compete for forage with great herds of wild grazing animals, the most numerous of which were buffalo, wild horses, and elk. Contrary to many present-day opinions, these animals were so numerous as to be a major factor in the utilization of the available forage. For example, it has been estimated, by those competent to judge, that there were 15 million buffalo in the West in 1864 and that the combined number of buffalo, wild horses, and elk in 1873 amounted to not less than 100 million. It is reasonable to assume, therefore, that the forage consumed annually by wild animals exceeded the amount utilized by the cattle and sheep grazed on these same ranges today. Except for the heavy and destructive grazing which occurred in the vicinity of strategic watering places and near salt licks, these wild animals alone were not particularly destructive to the virgin vegetation, primarily because their normal seasonal migrations permitted recovery of the range. But when cattle and sheep were added, overgrazing and range deterioration occurred for a time.

Fig. 9-9. Herd of elk at feed. It has been estimated by those competent to judge that the combined number of buffalo, wild horses, and elk on the western ranges in 1873 amounted to not less than 100 million. (Courtesy, U.S. Forest Service)

HISTORY OF THE RANGE CATTLE INDUSTRY

Animals of Spanish extraction served as sturdy foundation stock for the great cattle herds which were eventually to populate the western range. Although cattle have grazed intermittently on the southern plains of the United States since 1540, at which time they were introduced by Coronado, the period of continuous grazing accompanied the establishment of several Jesuit missions in Arizona, New Mexico, and Texas in the period from 1670 to 1690.

The growth of the industrial East and the subsequent development and extension of the railroads provided the necessary stimulus for further expansion of the range cattle industry. The grass supply of the vast ranges seemed unlimited, and the region was regarded as a permanent paradise for cattle. About 1880, the lure of the grass bonanza fired the imagination of investors, big and little. Cowboys, lawyers, farmers, merchants, laborers, and bankers—many of them English and Scotch investors in great companies—rushed in to seek their fortunes. The number of cattle increased rapidly, and soon the range was overstocked. Regulations were few, and the guiding philosophy was "to get what you can while the getting is good, and let the devil take the hindmost."

Then, suddenly, it became apparent that greed had taken its toll. The supply of tall grass was exhausted. Even more tragic, the winter of 1886–87 was unusually severe, and few owners had made provisions for winter feed. Cattle perished by the thousands. In some herds, 85% of the animals starved. And this was not all! The prolonged drought of 1886–95 brought further losses to the cattle companies, and the inroads of the homesteaders on the range and the growth of the range sheep industry contributed other difficulties. These circumstances marked the beginning of the end of the large cattle companies, the gradual growth of smaller operators, and increased attention to management.

TYPES OF RANGE VEGETATION

Cattle producers and students alike—whether they reside in the East, West, North, or South—should be well informed concerning range grasslands, the very foundation of the range livestock industry. This is so because this vast area, comprising 37% of the total land area and 84% of the total pasture and grazing lands of the United States (exclusive of Alaska and Hawaii), is one of the greatest cattle countries in the world and a potential competitor of every American cattle producer. Since cattle and sheep compete successfully with each other in utilizing most range forages, both classes of animals necessarily will be mentioned in the discussion which follows relative to types and uses of range vegetation.

Chiefly because of climate, topography, and soil, the character and composition of native range vegetation is quite variable. Ten broad types of vegetation native to the western ranges of the United States are discussed in this chapter.

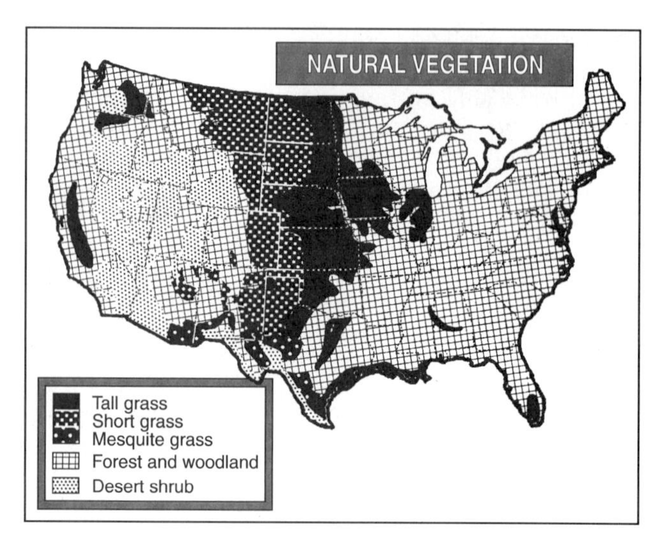

Fig. 9-10. Five major types of natural vegetation of importance to U.S. beef cattle production. (Courtesy, USDA)

TALL-GRASS TYPE

It is estimated that about 20 million acres of tall-grass range remain, most of which is in the eastern Great Plains region, in a rainfall area varying from 20 to 40 in. Although differing with the soil, topography, and rainfall, the dominant native tall-grass species include the bluestem (big and little), Indiangrass, switchgrass, side oats grama, and slough grass. Such famous grazing areas as the Flint Hills of Kansas (which encompasses 500 million acres), the Osage Pastures of Oklahoma, and the Sand Hills of Nebraska belong to the tall-grass type of vegetation. For the most part, this type of range is utilized by cattle, although sheep graze some of it. At one time, each fall, thousands of fat cattle were marketed after being finished on tall-grass vegetation without a grain supplement. The carrying capacity of these ranges is very high.

Fig. 9-11. Commercial Herefords on the famous Osage Pastures near Pawhuska, Oklahoma. (Courtesy, *Livestock Weekly*, San Angelo, TX)

SHORT-GRASS TYPE

The short-grass range, which is the largest and most important grassland type in the United States, embraces an area of approximately 280 million acres. It extends from the Texas Panhandle to the Canadian border and from the foothills of the Rocky Mountains eastward midway into the Dakotas. The common grasses of this area include the grama grasses, buffalograss, and western wheatgrass, all of which are well adapted to making their growth during the time of favorable moisture conditions in the late spring and early summer. Although they become bleached and cured on the stalk, because of the small amount of leaching in the fall and winter months, these plants retain sufficient nutrients to furnish valuable winter grazing. Because the forages in the short-grass area are dry during much of the year and droughts are rather frequent, cow-calf operations predominate in the cattle industry of the area; and most of the calves and older steers are finished in feedlots prior to slaughter. The smaller fine-wool breeds of sheep are also most numerous, and most of the lambs go the feeder route.

Fig. 9-13. Santa Gertrudis cattle on semidesert-grass type range on Chaparrosa Ranch, La Pryer, Texas. (Courtesy, *Livestock Weekly*, San Angelo, TX)

Fig. 9-12. Short-grass type range in Montana. Cow-calf operations predominate throughout this area, which represents the largest and most important grassland type in the United States. (Courtesy, U.S. Forest Service)

SEMIDESERT-GRASS TYPE

The semidesert-grass type—which predominates in an area characterized by low rainfall, frequent droughts, and mild winters—embraces about 93 million acres of grasslands in central and southwestern Texas, Arizona, and New Mexico. It provides year-round grazing. Because of great differences in climate and soil, the vegetation is quite variable. The most common grasses are grama, curly-mesquite, and black grama. Scattered among the more or less sparse grasses are many scraggly shrubs, dwarf trees, yuccas, and cacti. Some of these—especially saltbush, mesquite, ratany, and scrub oak—are rather palatable and are browsed effectively by goats. For the most part, the semidesert-grass area is utilized by commercial cattle as a cow-calf proposition. But bands of breeding sheep are found throughout the area. Both sheep and goats are common in southwestern Texas. The sheep of this area are kept primarily for wool production, and production of feeder lambs is secondary.

PACIFIC BUNCHGRASS TYPE

The Pacific bunchgrass area embraces about 60 million acres in western Montana, eastern Washington and Oregon, northern and southwestern Idaho, and central California. Much of the original bunchgrass area, including the famous Palouse area of eastern

Fig. 9-14. Hereford cows on Pacific bunchgrass on Tejon Ranch, Lebec, California. (Courtesy, *California Cattleman*, photo by Phil Raynard)

Washington and northern Idaho, is now devoted to the production of wheat and peas. Though well adapted to the dry summers and moist winters of the area, the native tall bunch or tuft-growing grasses of this area—bluebunch wheatgrass, Idaho fescue, Sandberg bluegrass, and California needlegrass—did not withstand overgrazing and have largely been replaced by such annuals as alfileria, bur-clover, cheatgrass, and wild oatgrasses. These ranges furnish excellent grazing in the spring and fall months but are too dry for summer use. The Pacific bunchgrass area is best adapted for spring, fall, and winter grazing by cattle and sheep.

SAGEBRUSH-GRASS TYPE

The sagebrush area, which is the third largest of all range types, embraces between 90 and 100 million acres extending from northern New Mexico and Arizona northwestward into Montana and to the east slope of the Cascades in the Pacific Northwest. This type of vegetation is characteristic of low rainfall areas where most of the meager precipitation occurs during the winter and spring seasons. Interspersed among the ever-present sagebrush, of which there are several kinds, are many species of native grasses among which are bluebunch and western wheatgrasses, needle-and-thread, Indian ricegrass, Sandberg bluegrass, and numerous species of weeds. The sagebrush, which varies from 2 to 7 ft in height, provides little forage except when winter snows blanket the grasses. For the most part, the sagebrush type of vegetation is used for early spring and late fall grazing for cattle and sheep. It furnishes interim pasture until more distant summer and winter grazing areas may be used. Studies have shown that the carrying capacity of sagebrush areas may be increased by destruction of the sagebrush, which encourages greater growth of the grasses.

Fig. 9-15. Sagebrush-grass type range in Idaho, showing big sagebrush with excellent understory of bluebunch wheatgrass, bluegrass, and palatable perennial weeds. Generally, this type of range is used for early spring and late fall grazing for cattle and sheep. (Courtesy, U.S. Forest Service)

SALT-DESERT SHRUB TYPE

About 40 million acres in central Nevada, Utah, southwestern Wyoming, western Colorado, and southern Idaho are covered with a mixture of low shrubs and scattered grasses. The common browse species of the area are shadscale, saltbush, black sagebrush, winterfat, rabbit-brush, greasewood, spiny hopsage, and horsebrush; and the rather sparse grass species include blue grama, sand dropseed, galleta, and Indian ricegrass. Because there is not any dependable source of water and because of the high temperature and dryness during the summer months, the use of much of this area by cattle and sheep is restricted to the winter months when there is snow. Other areas cannot be grazed because of the high alkali content of the soil.

Fig. 9-16. Salt-desert shrub range vegetation. Note cattle eating saltbush in preference to grass. (Courtesy, Bureau of Land Management, U.S. Department of the Interior)

SOUTHERN-DESERT SHRUB TYPE

Approximately 50 million acres, located chiefly in southeastern California, southern Nevada, and southwestern Arizona, are classed as southern-desert shrub vegetation. The common shrubs are the creosote bush and different kinds of cacti. Normally, the scant rainfall, extremely high temperatures, and sparse vegetation of this area make it rather poor grazing for cattle or sheep. However, when moisture conditions are favorable, there is growth of such annuals as alfileria, Indian wheat, bur-clover, black grama, tobosa, dropseed, and 6 weeks' fescue. When forage and water are available, nearby ranchers make use of the southern-desert shrub area, primarily for winter grazing, although it is used for spring and fall grazing and in a few cases throughout the year.

PINYON-JUNIPER TYPE

The pinyon-juniper type of vegetation forms the transition zone from the shrub and grass areas of the lower elevations to the forests of the mountains. The 76 million acres in this general type area extend all the way from southwestern Texas to south central Oregon. As the name would indicate, pinyon and juniper trees are common to the area. These scattered trees range in height from 20 to 40 ft, and interspersed among them are such low-growing shrubs as sagebrush, bitterbrush, mountain-mahogany and cliff-rose, and grasses like the gramas, bluebunch and bluestem wheatgrass, and galleta. For the most part, this area is used for spring and fall grazing by cattle and sheep; but in the Southwest, where the forage cures on the ground and retains much of its nutritive value through the winter, yearlong grazing is prevalent.

Fig. 9-17. Pinyon-juniper type range in Arizona, showing young juniper trees encroaching on the area. For the most part, this type of range is used for spring and fall grazing by cattle and sheep, although in the Southwest yearlong grazing is prevalent. (Courtesy, U.S. Forest Service)

WOODLAND-CHAPARRAL TYPE

This type of vegetation is characteristic of parts of California and Arizona. It varies all the way from an open forest of parklike oak and other hardwood trees with an undergrowth of herbaceous plants and shrubs to dense *chaparral* thickets of no value to animals. Alfileria, slender oatgrass, and bur-clover have been introduced in the more open areas. Though somewhat restricted, woodland-chaparral is used for fall, winter, and spring grazing by cattle and sheep.

OPEN-FOREST TYPE

Fig. 9-18. Open-forest type range in ponderosa pine timber of Oregon. (Courtesy, U.S. Forest Service)

The 130 million acres of open forests, found scattered in practically all the mountain ranges, constitute the second largest range-type vegetation. This is the summer range of the West, which provides grazing for large numbers of cattle, sheep, and big game. Grass-fat cattle and lambs are sent to market directly off these cool, lush, high-altitude ranges. For the most part, the tree growth common to the area consists of pine, fir, and spruce; and the grasses include blue grama, fescues, bluestem, wheatgrasses, timothy, bluegrasses, sedges, and many others. More than half of these mountain ranges are federally owned as national forestlands. In addition to serving as valuable grazing areas, the open forests are important for lumbering and recreational purposes.

GRAZING PUBLICLY OWNED LANDS

The ownership of U.S. land is summarized in Table 9-2.

About one-third of U.S. public lands are in Alaska. Because of its remoteness and northern location, land development has been slow in this state. As a result, the Federal Government still owns almost 67% of all the lands in Alaska.

The other two-thirds of the public lands are located in the 48 contiguous states, but are not evenly distributed across the country. About 93% of these

TABLE 9-2
OWNERSHIP OF U.S. LAND (50 STATES)[1]

Ownership	Area		Percentage of Total
	(million acres)	*(million ha)*	*(%)*
Private ownership	1,329	*538.0*	58.7
Indian land	51	*20.6*	2.2
Public ownership	885	*358.3*	39.1
Federal	730	*295.5*	32.2
State and local governments	155	*62.7*	6.8

[1]*Statistical Abstract of the United States*, 1987, p. 182, Table 318.

federal lands outside Alaska are in the 11 western states.

Today, in the 11 western public land states, the federal government owns and administers approximately 320 million acres on which grazing is allowed. At one time or another during the year, domestic cattle and sheep graze on about half of these public lands. More of the public lands are used for this purpose than for any other economic activity. In 1988, lands in the 11 western states administered by the Bureau of Land Management and the U.S. Forest Service provided grazing all or part of the year for an estimated 5,655,845 head of all classes of livestock. The number of livestock grazed on the publicly owned rangeland annually accounts for approximately 4 to 5% of the nation's total cattle numbers and slightly over 50% of sheep numbers.

AGENCIES ADMINISTERING PUBLIC LANDS

Because much of the grazing land that some ranchers rely upon to maintain their cattle and sheep enterprises is built up into operating units by leasing or by obtaining use permits from several federal and state agencies, private corporations, and individuals, it is imperative that the owner have a working knowledge of the most important of these agencies. Some range operators are placed in the position of using range rented from as many as six landlords—either private, state, and/or federal.

The bulk of federal land is administered by the following six agencies: the Bureau of Land Management, the U.S. Forest Service, the Bureau of Indian Affairs, the Department of Defense, the National Park Service, and the Bureau of Reclamation. The largest land area from the standpoint of grazing permits and utilization of grazing areas by animals is administered by the first two of these agencies; hence, each is discussed at this point, followed by pertinent information relative to Indian lands and state and local government-owned lands, and railroad-owned lands.

1. **Bureau of Land Management.** The Bureau of Land Management of the U.S. Department of the Interior administers more than 40% of all federal lands. More than one-third of the land it manages is in Alaska. The remainder is almost entirely in the 11 western states.

From the standpoint of the livestock producers, the most important function of the Bureau of Land Management is its administration of the grazing district established under the Taylor Grazing Act of 1934 and the unreserved public land situation outside of these districts which are subject to grazing lease under Section 15 of the Act. This federal act and its amendments authorize the withdrawal[1] of public domain from homestead entry and its organization into grazing districts administered by the Department of the Interior. Also, this legislation, as amended, allows the Bureau of Land Management to administer state and privately owned lands under a cooperative agreement.

Fig. 9-19. Hereford cattle on the ranch of a grazing permittee of the Bureau of Land Management. (Courtesy, Bureau of Land Management, U.S. Department of the Interior)

In 1988, the Bureau of Land Management had 54 grazing districts, operating in the 11 western states and totaling 146 million acres of public lands. In these districts, 11,853 operators were granted privileges to graze 3,787,332 head of livestock for an average of about 5 months each year. These operators paid the United States, as grazing fees for this range use, a total of $12,416,598. In addition to this livestock use,

[1]On May 28, 1954, a bill was signed by President Eisenhower lifting the 142 million acre limitation on public domain lands that can be included in Taylor Grazing Act districts.

in 1988, public lands supported an estimated 1.9 million big game animals, of which approximately 1.1 million were deer.

In addition to, and outside of, the grazing districts, in 1988 the Bureau of Land Management supervised 16 million acres of public domain in the western states, most of which was leased to 7,016 livestock producers for 604,758 head of livestock for about 5 months. These operators paid rentals in the amount of $2,039,033 for the use of these lands.

Each district is administered by a District Manager, who is a technically trained employee of the Bureau of Land Management. The District Manager is responsible to the state bureau office for the proper use, management, and welfare of the public land resources of the district. In turn, the state office is responsible to the Director's office in Washington, DC.

Grazing privileges are allocated to individual operators, associations, and corporations on the basis of (1) priority of use; (2) ownership or control of base property dependent on grazing district land for forage during certain seasons of the year, or control of permanent water needed to graze district land; (3) proximity of base property to public lands outside home ranch to the grazing district; and (4) adequate property to supply the feed needed along with grazing privileges, to maintain throughout the year the livestock permitted on public range.

All of these lands are subject to classification and disposal under Sections 7 and 14 of the Taylor Grazing Act, for any higher use or other appropriate purposes. Grazing privileges may, therefore, be cancelled whenever such lands are determined to be more suitable for other purposes.

A fee is charged for grazing privileges. In 1996, the basic fee was equivalent to $1.35 per animal unit month (AUM). An AUM is the equivalent of the grazing of a mature cow, 5 sheep, or 1 horse, for 1 month.

The Taylor Grazing Act has been responsible for many changes, not all of which have been popular. Some livestock producers complain about the loss of their ranges; others tell of increased costs; and there are those who resent government controls, and, above all, the confusions which results from dealing with several agencies. Without doubt, many of these criticisms are justified, and some errors in administration should be rectified; but those who would be fair are agreed that the ranges as a whole have improved under the supervision of the Bureau of Land Management and that further improvements are in the offing.

2. **USDA Forest Service.** Almost one-fourth of the federal lands are administered by the Forest Service. Over 100 million acres of the national forests and national grasslands are used for grazing under a system of permits issued to local farmers and ranchers by the Forest Service of the U.S. Department of Agriculture. In 1988, lands in the 11 western states

Fig. 9-20. Cows and calves on the Santa Rita Range Reserve, in Arizona. (Courtesy, U.S. Forest Service)

administered by the USDA Forest Service provided grazing for all or part of the year for 1,868,513 head of all classes of livestock.

The Forest Service issues term grazing permits and annual permits. Among other things, the permit prescribes the boundaries of the range which they may use, the maximum number of animals allowed, the season in which grazing is permitted, and the expiration date of term permits.

Temporary permits may be waived back to the government when the permittees sell livestock or base property. Then, the purchaser of the permitted livestock or base property may apply for and be issued a permit if qualified.

The requisites in order to qualify for a term permit are:

a. U.S. citizenship.

b. Ownership. The ownership of both the livestock and commensurate ranch property.

A term grazing permit is not a property right. Rather, it is approved for the exclusive use and benefit of the person to whom it is issued. Permits may be revoked in whole or in part for a clearly established violation of the terms of the permit, the regulations upon which it is based, or the instructions of forest officers issued thereunder.

A ranger administers the grazing use on each National Forest Ranger District. Several districts (usually 3 to 6 or more) comprise a national forest. A forest supervisor administers the national forest. Several national forests, under the direction of a regional forester and staff, comprise a Forest Service region. The chief administers the Forest Service from Washington, DC, under the supervision of the Secretary of Agriculture.

Local farmers and ranchers act in an advisory capacity in reviewing allotment management plans and the use of range betterment funds.

Forest Service grazing fees are based on a formula which takes into account livestock prices over the past 10 years, the quality of forage on the allotment, and the cost of range operation. In 1996, average charges were $1.35 per animal unit month (AUM); or $1.35 for a mature cow or horse, or for 5 sheep, for a month.

Although shortcomings exist in the management of the national forests, it is generally agreed that these ranges have been vastly improved under the administration of the Forest Service. Some of them now approach the quality that existed in their virgin state. Perhaps the most heated arguments between livestock producers and the Forest Service arise over the relative importance attached to the multiple use of big game and other wildlife, recreation, etc.

3. **Bureau of Indian Affairs.** Most Indian lands, comprising 51 million acres, are really not public lands. Rather, these lands are held in trust for the benefit or use of the Indians and are merely administered by the Bureau of Indian Affairs of the Department of the Interior. Because over 80% of Indian lands are in the range area of the West, they are suited primarily to livestock. Thus, it is noteworthy that the sale of livestock and animal by-products regularly accounts for two-thirds of the total Indian agricultural income. Although the Indians themselves own most of the stock that graze these lands, animals owned by non-Indians utilize one-fourth of the Indian lands devoted to grazing. Provision for such use is handled under lease agreement jointly approved by the Indian owners and the Bureau of Indian Affairs.

Many of the Indian lands have suffered serious vegetative depletion, but a concerted effort is now being made to decrease livestock numbers in keeping with available feed supplies and to improve the quality of animals produced. However, overstocking continues to be a difficult problem on the Navajo, Hopi, and Papago Reservations.

4. **State and local government-owned lands.** A total of 134 million acres are owned by state and local governments. For the most part, the management of these areas is diverse and confused, each state and local government having established different regulations relative to the lands under its ownership. In general, however, such lands are operated on a stipulated lease arrangement. On many such areas, range depletion has been severe.

5. **Railroad-owned lands.** Recognizing that the main deterrent to rapid settlement and development of the West was lack of adequate transportation facilities, the Federal Government very early encouraged the construction and westward extension of the railroads by means of large grants of land. It was intended that the railroads should sell or otherwise utilize these lands in financing their costs of construction. These initial grants, totaling 94,355,739 acres, consisted of alternate sections extending in a checkerboard fashion for a distance of from 10 to 40 miles on each side of the right-of-way. Today, less than 20 million acres of these lands are held by railroads. Many of these holdings are leased to livestock producers; but because of the inconvenience, past abuses, or other reasons, some of these lands are considered worthless grazing. In general, railroad lease agreements do not restrict the number of stock to be grazed or the season during which the land may be so used.

Fig. 9-21. Mescalero Apache tribe commercial Herefords, Mescalero, New Mexico. (Courtesy, *Livestock Weekly*, San Angelo, TX)

ENVIRONMENTAL EFFECTS OF GRAZING PUBLIC LANDS

Little pollution potential exists from pasture systems with low animal densities or numbers, or where pastures are rotated. So, except for high density pasture systems involving a number of animals, pollution is no problem. Nevertheless, some environmentalists have centered their attack on the grazing of public lands.

Grazing influences the environment on federal lands. Under poor range management, the environment is affected adversely; under good range management, such as exists on most ranges today, grazing actually improves the environment.

Eating of plant materials by animals is a natural process in earthly and aquatic systems. Thus, the coming of the white man to what is now the United States, along with the introduction of domestic animals, did not constitute an entirely new component in the environment. Rather, domestic animals replaced, or added to, the wild animals that were already there.

Mistakes in grazing practices have occurred in the United States in the past, the most significant of which was the exploitative grazing practices between

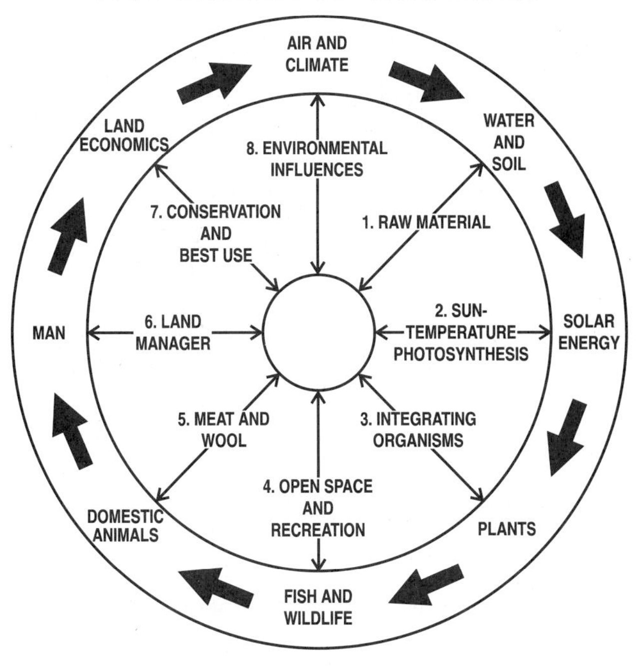

Fig. 9-22. The wheel of ecology. Cattle producers share today's increasing national concern for the quality of our environment, and through scientific range management, they are doing much to improve it.

1865 and the 1930s. The effects were almost catastrophic. Nevertheless they were not the result of grazing ranges that had never been grazed before. Rather, they resulted from several decades of grazing the western ranges with too many animals for too long, and often at the wrong season of the year. Most range livestock operators of that period were not aware of the benefits that could accrue to them from improved range management.

Scientific management of rangeland began at the turn of the century. Range managers and livestock operators found that controlling grazing improved both range conditions and livestock production. Development of this new concept marked the beginning of the end of the exploitative period of grazing and the introduction of managed grazing on the western ranges.

The environmental effects of grazing depend upon the kind of range, the intensity of grazing, and the kind of management employed to control livestock on the range. It is generally recognized that unregulated heavy grazing results in loss of desirable forage plants, increased runoff and erosion, and other indications of range deterioration. On the other hand, planned seasonal grazing and controlled animal distribution foster rapid vegetational growth. Most grazing experiments have shown that ranges may be improved more rapidly under proper grazing management than with no grazing at all.

There is no evidence that well-managed grazing of domestic livestock is incompatible with a high-quality environment. But there is ample evidence that managed grazing by livestock enhances certain uses and that poor management detracts from them. Properly managed grazing is a reasonable and beneficial use of the range.

Ecologists tell us that good range management will support more wildlife than the wilderness. This explains why big game numbers on federal lands have increased during recent years, and why wildlife production is an increasingly important use of rangelands.

Indeed, ranges actually improve while being properly utilized by domestic livestock. The benefits which accrue to the range include increased vegetation cover, improved plant species composition, improved soil fertility and soil structure, and greater yield of high quality water. When cattle and sheep go, rank underbrush takes over, and fire becomes a real hazard.

Both upland game birds and big game animals are benefited by grazing that promotes good cover for mating sites and enhances food supply and other habitat requirements.

On ranges with mixed types of vegetation, herbaceous species increase and browse species decline when grazed only by game. The converse is true when cattle graze the land. The combined grazing by two groups of animals maintains a better balance of browse species, preferred by game animals, and of herbaceous species, preferred by cattle.

Heavy livestock grazing is beneficial to irrigated pastures used by geese and other migratory waterfowl. Unless the vegetation is closely cropped, these areas are unattractive to the birds.

Thus, livestock grazing of the public lands is contributing to improved wildlife habitat conditions and increased numbers of game animals. Range development programs, particularly livestock water developments, have made more public land usable by game

Fig. 9-23. Aerial topdressing. Scientific range management benefits the environment.

animals and is partly responsible for the vast increase in game numbers over the years.

On many grass-shrub ranges, livestock grazing reduces the danger of fire by preventing a buildup of dry grass, which is highly inflammable.

Grazing systems and manipulation of vegetation can create contrast in vegetation color and pattern, thereby improving the aesthetical value of the landscape. Also, the livestock industry is traditional to the West; hence, a well-managed range with its cattle herd and roundup, or with its sheep camp, has recreational values. Indeed, cattle and sheep on the landscape are pleasing to tourists who come to view the *Old West*.

Ranges properly grazed by hoofed animals produce safe water. Counts of fecal coliform organisms, as indicators of water pollution by warm-blooded animals, relate more closely to the quantity of fecal material than to the kind of animal. Investigations have shown that the count of harmful bacteria in streams is no greater in areas grazed by livestock than in areas grazed by wild animals alone, and that modern livestock grazing has little effect upon the chemical and physical quality of the water.

It is noteworthy, too, that few western ranges are ever in a stable, natural condition, whether or not they are grazed by domestic animals. Rather, most of them are in a stage of vegetational development following disturbances by such phenomena as drought, flood, avalanche, frost, or fire. Also, cyclic phenomena, such as large numbers of deer, rodent epidemics, or insect plagues, temporarily change the natural ecosystems. Thus, an absolutely stable rangeland is seldom attained or maintained.

Significantly, the greatest diversity of animal and plant species and the highest rates of reproduction occur when the landscape supports many stages of ecosystem development. Fire, grazing, and drought stimulate plants and animals to new growth. Each stage of vegetational development is more productive of certain animal species than of others.

Finally, in an era of world food shortages, the contribution of properly managed federal lands in terms of food and fiber production needs to be recognized. More and more grains will be used for direct human consumption. As a result, there will be an increased reliance on ranges for meat and wool production. It just makes sense to preserve all the natural food and fiber that we can. Remember that petroleum is not needed to make wool. Remember, too, that cattle and sheep are completely recyclable. It takes thousands of years to create coal, oil, and natural gas; and when they're gone, they're gone forever. But animals produce a new crop each year and perpetuate themselves through their offspring.

Today, in the 11 western public land states, the federal government owns and administers approximately 320 million acres on which grazing is allowed.

Both cattle producers and environmentalists need to recognize (1) that forage is a renewable natural resource, which regrows each year and is wasted unless it is utilized annually; (2) that grazing on federal rangelands helps to keep the natural environmental systems active and productive; (3) that we cannot allow overgrazing by domestic livestock, bison, deer, or wild horses; and (4) that grazing must be scientifically controlled and responsive to the needs of all users.

Indeed, it may be said that humanity's influence on and use of the environment will determine how well we live—and how long we live.

SEASONAL USE OF THE RANGES

A prime requisite of successful range management for both cattle and sheep is that there shall be as nearly year-round grazing as possible and that both animals and the range shall thrive. In some areas, especially in the southwestern Great Plains region, these conditions are met without necessitating extensive migration of animals. The winter climate is mild, and the native forages cure well on the stalk, thus providing nutritious dry feed at times when green vegetation is not available. Generally speaking, however, most of the cattle and sheep from such areas are marketed via the feeder route rather than as grass-finished slaughter animals.

In general, the most desirable management, both from the standpoint of the animals and the vegetation, consists of the proper seasonal use of the range. Although there is wide variety in the customs and requirements for seasonal use of the range—because of the spread in climate, topography, and vegetative types included in the vast expanse of range country—seasonal-use ranges are usually placed in four major classes: (1) spring-fall, (2) winter, (3) spring-fall-winter, and (4) summer.

Because a range band of sheep can be moved and herded on unenclosed areas with greater ease than a herd of cattle and because investigations in range livestock management have been conducted more extensively with sheep, greater seasonal use of ranges is made with sheep. On the other hand, the more progressive cattle producers are finding ways and means of adopting many of the same methods.

Despite the values of yearlong grazing, it is recognized that the prevalence of severe winters in some parts of the West precludes winter grazing except to a limited degree, and stock must be fed during at least a part of the winter season. Where these conditions prevail, cattle and sheep are usually wintered

SEASONAL USE OF WESTERN RANGE

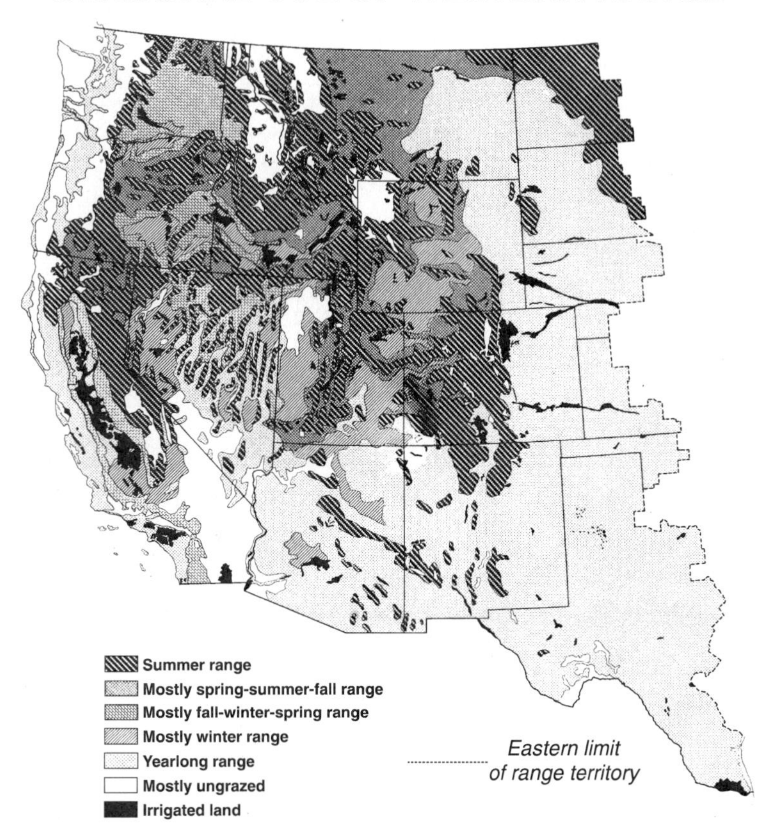

Fig. 9-24. Seasonal use of the western range. In general, the most desirable management, both from the standpoint of animals and vegetation, consists of the proper seasonal use of the range. (Courtesy, Bureau of Agricultural Economics)

Fig. 9-25. Trailing cattle from a winter range in Utah. Calves are from one day to several weeks of age. (Courtesy, U.S. Forest Service)

in the irrigated valleys, close to the feed supply, especially a supply of alfalfa or meadow hay.

Some pertinent points in determining the proper season of use of the range follow:

1. **Elevation.** Generally speaking, vegetative development is delayed 10 or 15 days by each 1,000-ft increase in elevation. Also. severe storms occur later in the spring and earlier in the fall at higher altitudes than at low, desert locations.

2. **Availability of water.** Certain desert areas are so poorly watered that only the occurrence of winter snows makes their use practical.

3. **Early forage *washy*.** Early spring forage is extremely *washy*, and may be incapable of supporting stock. Spring grazing should be delayed until the plants are developed enough to meet the nutritive needs of animals.

4. **Soil tramping.** Soil tramping may be serious in early spring. In order to avoid plant damage and soil compaction, grazing should be delayed until the soil is firm.

5. **Poisonous plants grow early.** Most poisonous plants are very early growers and cause their greatest damage when animals are turned out too early. Larkspur, which affects cattle, and death camas, which affects sheep, are two examples. Poisoning losses from these two plants are usually negligible if stock are detained until the best forage plants have made suitable growth.

6. **Winter range should be saved.** If stock are allowed to remain on winter ranges too long after spring growth begins, the next winter's feed will be reduced, because the forage produced on these ranges grows mainly during spring and early summer.

SPRING-FALL RANGE

The foothills and plateaus, lying between the plains or desert sections and mountains, furnish desirable spring and fall grazing for both cattle and sheep. Areas of this type are, therefore, commonly referred to as spring and fall range. The sagebrush-grass type is the leading range vegetation used for spring and fall range. Compared with the vegetation of the plains and desert regions, that growing on the foothills and plateaus in season is more luxuriant and succulent and better suited to milk production and finishing. Although spring and fall ranges are accessible most of the year—because such areas are not too extensive—usually they are not grazed for a very extended period. Sometimes fall grazing for the herd or flock is provided through using owned or leased stubble fields and meadow aftermath.

Fig. 9-26. Cattle on rich fall range in Oregon. (Courtesy, Bureau of Land Management, U.S. Department of the Interior)

WINTER RANGE

Areas adapted to yearlong grazing possess desirable climatic and vegetative conditions for winter range. Many such areas are used only during the winter season because of their proximity to more desirable spring-fall and summer ranges. Desert areas, such as the salt-desert shrub type, which would otherwise be suited to yearlong grazing, must be restricted to use as winter range. This is due to the absence of any reliable water supply except that

Fig. 9-27. Cattle on winter range. (Courtesy, *Livestock Weekly*, San Angelo, TX)

secured from the snow or rain during the winter months. In some sections, winter wheat is used for late fall and winter grazing.

SPRING-FALL-WINTER RANGE

In general, spring-fall-winter–type ranges are adapted to year-round grazing. But because of their proximity to more lush mountain ranges, they are not used during the summer months. On typical spring-fall-winter ranges, the annual grass species dry up during the summer months.

SUMMER RANGE

The most desirable summer ranges are located in the cooler and higher altitudes of the mountains where the vegetation is lush, palatable, and nutritious during the 3 to 6 summer months. There is usually a bountiful supply of water. Much of this range is of the open-forest type on national forest lands, where the average grazing period is 5.6 months for cattle and 3.1 months for sheep. Although summer range is usually very rugged, the conditions are generally ideal for finishing, so that year after year it is the most consistent type of range area for the production of grass-finished cattle and lambs. Sheep are sometimes trailed or shipped a distance of 200 mi in order to have access to a desirable summer range, but cattle are usually not moved so far. These summer ranges supplement the lower spring, fall, and winter ranges (either publicly or privately owned). Thus, they hold a key position in the yearlong operations of many livestock producers. Sometimes seeded grass-legumes

Fig. 9-28. A summer range for cattle in the Sawtooth National Forest of Idaho. (Courtesy, U.S. Forest Service)

Fig. 9-29. A range herd in New Mexico. (Courtesy, *Livestock Weekly*, San Angelo, TX)

or irrigated pastures are used for summer grazing instead of mountain ranges.

THE RANGE CATTLE HERD

The range herd may be owned by an individual, a cattle company, or a corporation. Usually the ranch headquarters consists of a ranch-type house, bunk house, sheds, corrals, and a water supply and there is ownership of sufficient adjacent land for the production of winter forage and a limited amount of grain. Additional and more distant rangeland is either owned or leased.

SIZE OF HERD AND CARRYING CAPACITY

Range herds vary in size from about a hundred head up to several thousands. In general, the most important single criterion in determining the size of herd is the number of animals that the unit will support each season over a period of years without injury to the range. When the range is stocked more heavily than its true grazing capacity, usually three conditions become evident: (1) the animals fail to thrive as they should; (2) the vegetation gets thinner and of less desirable species; and (3) erosion and soil losses occur. Practical observations and controlled experiments would indicate that somewhere between 20 and 30% of the palatable growth of the more important forage species should be left ungrazed each year. Additional considerations in arriving at the size of herd and carrying capacity are: (1) available reserve feed supply for drought and winter feeding, (2) the long-range economic conditions, and (3) the topography, water supply, and poisonous plants of the area.

STOCKING RATE

The key to successful long-term operation of rangeland lies in a good estimate of the grazing capacity of the individual units. Stocking too lightly wastes forage, while stocking too heavily results in reduced plant vigor and less forage produced per plant, as well as a change of forage plants to an abundance of worthless plants.

Of course, the stocking rate for any given unit may vary widely from year to year, depending on the forage production as affected by weather and other factors. For this reason, stocking either should be adjusted to forage yield each year, or should be set at a constant rate that will assure a sustained yield of the most valuable forage plants. (Constant stocking at about 25% below average capacity will usually achieve the latter.)

Recognition must also be given to the fact that animals do not graze uniformly over a range unit—that certain areas are more attractive to them. Consequently, some areas produce most of the grazed forage, while others may go practically unused. Cattle tend to congregate on fairly level creek bottoms, ridge tops, and around water and shade; whereas sheep, if herded, can be moved more uniformly over a unit. But even sheep graze some areas more heavily than others if not herded properly. For the purpose of determining grazing capacity, the key areas—those rather extensive parts of the range which are most heavily grazed—must be given greatest consideration.

If preferred or key areas are maintained in good condition, the whole unit will generally remain in good condition. Conversely, if key areas are allowed to deteriorate, the grazing capacity of the whole unit will be endangered.

Grazing capacity determinations are relatively complex and require careful study over a period of several years. They are arrived at most simply and accurately by observing soil conditions and changes in plant cover. If the best plants are being destroyed and soil movement is observed, numbers of animals should be reduced or season of use changed; conversely, if excessive forage remains at the end of the grazing season, numbers should be slowly increased until a balance is struck.

The following rule of thumb, applied to the more heavily grazed key areas, may be used in arriving at the proper stocking rate: "Use half and save half, and the half you save will grow bigger and bigger." The rule refers to half the weight, which is concentrated at the bottom of the plant, and not to half the height. Thus, when the 50% rule of thumb is applied to bluebunch wheatgrass, a common range plant, it means that approximately 75% of the bunches have been grazed to an average stubble height of about 4 in. and the remaining 25% of the plants left relatively ungrazed.

In arriving at grazing capacity determinations, it is generally wise to seek assistance from county agents, soil conservation service technicians, or other trained specialists. These specialists need to know-

1. The potential of the particular range.
2. The present state of the vegetation as it relates to potential on each site.
3. The alternative methods of changing present conditions to meet management objectives, including such things as flexible stocking, seeding, brush control, fences, watering, and trails.

The commonly used terms for describing range condition are: (1) excellent, (2) good, (3) fair, and (4) poor. If the range is covered with 75% or more high value forage plants, it is classified as being in excellent condition. If the best plants constitute less than 25% of the total cover, it is classified as being in poor condition. Good and fair classifications are intermediate. The trend in condition is also important; if the range condition is improving, the trend is upward, and vice versa. Actually, the range condition reflects the kind of management practiced in the past.

WATER SUPPLY

Cattle should always be provided with a dependable and adequate supply of good water. On some ranges, this problem presents a difficult, costly, or even impossible situation. The source(s) of water varies between ranges and areas, with reservoirs or ponds, streams, deep wells, and springs being utilized.

Fig. 9-30. Commercial Santa Gertrudis cattle at water tank on the Bill Maltsberger Ranch, Cotulla, Texas. (Courtesy, *Livestock Weekly*, San Angelo, TX)

Fig. 9-31. A cattle water hole in New Mexico. (Courtesy, Bureau of Land Management, U.S. Department of the Interior)

Some cattle producers haul water to the range. Although there is considerable variation in the equipment, the better outfits use the following: (1) a main storage tank that will hold at least 1 week's supply; (2) a 1,000-gal tank truck; and (3) galvanized rectangular troughs, approximately 8 ft long and 2 ft wide, with sloping sides so as to permit nesting and easy transportation. On unwatered ranges, hauling can make the difference between grazing and no grazing. Also, hauling helps obtain more uniform use of forage, reduces trailing damage, permits grazing at the most appropriate time, and makes for more animal weight gains.

For cattle, the ideal arrangement consists of having sufficient watering places distributed over the range so that the animals never have to travel over ½ mi, and not to exceed 2 mi at the most. Sheep can travel about twice as far to water as cattle. If cattle must travel great distances to water, the amount of walking will use up much of the energy which would otherwise be available for the production of meat or milk. They also will water infrequently and overgraze and tramp out the grass near the water supply.

RANGE MANAGEMENT

Range management consists chiefly of determining the proper stocking rate and time of range use that will allow the desirable species to remain vigorous and persistent. Many studies indicate the tendency of animals on range to graze selectively—grazing one plant, avoiding the adjacent ones, then moving to another, always selecting the most palatable. This results in uneven grazing and leads to a thinning out of the more preferred plants and a predominance of those less desirable.

Fig. 9-32. Savory Grazing Method, showing the controlled-type layout, with the water and handling facilities located in the central hub, and the grazing cell fence lines radiating out to the perimeter. (Courtesy, Snell Systems, Inc., San Antonio, TX)

It is noteworthy that the short duration grazing system, popularized by Allan Savory under the name Savory Grazing Method, has been around for a very long time. It is fully covered in the book *Grass Productivity* by André Voisin, a Frenchman, published in 1959. In this rather remarkable volume, M. Voisin quotes, in turn, as follows, from *The Agronomist*, published in France in 1760:

> Grass that is too mature becomes hard and loses much of feeding substances. Grass that is not mature does not possess enough of these substances as beasts always go to the most tender herbage, it is essential when managing grassland, so that all the grass will be grazed at maturity and regrow, that the pastures be divided up into sections, the size of which is in proportion to the number of beasts they are to carry; the aim being that each section contains sufficient keep for three or four days, after which the stock are put on to another section so that the first can bear fruit.

It is noteworthy, too, that the *Essays Relating to Agriculture and Rural Affairs* by James Anderson, a Scot, also published in the 18th century, described short duration grazing as follows:

> To obtain this constant supply of fresh grass, let us suppose that a farmer who has any extent of pasture ground should have it divided into 15 or 20 divisions, nearly of equal value; and that, instead of allowing his beasts to roam indiscriminately through the whole area at once, he collects the whole number of beasts that he intends to feed into one flock, and turns them all at once into one of these divisions; which, being quite fresh, and of a sufficient length for a full bite, would please their palates so much as to induce them to eat it greedily, and fill their bellies before they thought of roaming about, and thus destroying it with their feet. And if the number of beasts were so great as to consume the best part of the grass of one of these inclosures in one day, they might be allowed to remain there no longer—giving them a fresh park every morning, so as that the same delicious repast might again be repeated. And if there were just so many parks as the required days to make the grass of these fields advance to a proper length after being eaten bare down, the first field would be ready to receive them by the time they had gone over all the other; so that they might thus be carried round in a constant rotation. . . .

In the mid-1890s, H. L. Bentley, a special agent with the U.S. Department of Agriculture, initiated the first range experiments concerning correct stocking rates and range seeding. Soon after the turn of the century, most state and federal agencies in the West had range research programs. Between 1910 and 1915, Arthur Sampson conducted the first grazing system experiments in Oregon. He reasoned that deferment of grazing until seed maturity would allow for seedling establishment and replenishment of carbohydrate reserves.

Government intervention into grazing problems on the western range began in 1898, when the Department of the Interior granted grazing permits to limit the number of livestock on federal lands. In 1905, the Forest Service was set up in the Department of Agriculture and the process of forage allotment was initiated. Between 1910 and 1920, grazing laws were put into effect on National Forest lands. Because the price of cattle was high during World War I, a period of very severe overgrazing took place between 1915 and 1920. National Forest ranges began to improve again after World War I because scientific range management practices and controlled grazing were once again implemented. However, other rangelands in the West continued to deteriorate.

The discipline of range management flowered and developed in the 1920s. By 1925, approximately 15 colleges were offering courses in range management.

The biggest event of the 1930s in range management was the passage of the Taylor Grazing Act of 1934, which placed the administration of remaining public lands under the Grazing Service which later became the Bureau of Land Management. In 1933, the Soil Erosion Service was established in the Department of the Interior. It was transferred to the Department of Agriculture in 1935 and renamed the Soil Conservation Service. This organization was formed because of alarm over the drought in the Great Plains in the mid 1930s.

In 1948, the Society for Range Management was formed.

During the 1960s, the multiple use concept of range management on federal lands developed. Wildlife, water, and recreation became recognized as important range products as well as red meat. Previously, range research and management had been geared towards producing forage for livestock.

Toward the end of the 1970s and into the present, concern over the world population explosion generated renewed interest in using public rangeland for livestock production, primarily because lower energy inputs are required to produce red meat from rangeland than from cropland. In addition, only grazing animals can convert range forage into products usable by people. Range improvement on private lands accelerated during the 1970s because of improved information and education programs by state and federal agencies. Another factor of considerable importance was that the 1970s' ranchers were much better educated than those from previous periods.

Fig. 9-33. Bruce Corman of the Matador Cattle Co. watches cattle "mother up" after being moved to fresh feed in a deferred pasture, under rest-rotation management, on the Long Creek range allotment. (Courtesy, U.S. Forest Service)

RANGE GRAZING SYSTEMS

Ranges may be grazed continuously throughout the entire grazing season without rest, or the area may be subdivided and the pasture grazed rotationally with alternating periods of grazing and rest.

Theoretically, based on how plants grow and how animals graze, rotation grazing systems should improve the vigor and productivity of desirable plants, prevent the invasion of undesirable plants, and increase the carrying capacity and rate of gain of animals. In practice, however, all the benefits ascribed to rotation grazing may not accrue; perhaps due to human failure in management, or because there is not as much difference between the results of continuous and rotation grazing as had been thought. However, rotational systems usually make for increased plant vigor and carrying capacity, whereas individual animal gains has usually been in favor of continuous grazing.

Range grazing systems vary somewhat from unit to unit, depending on kind and class of livestock, kind and type of forage, mixture of range sites, time and amount of rainfall, pasture and corral layout, available water supply, the condition of the range, the long-time goals for improvement, the prevailing economics, and the time necessary and available to supervise and conduct the operation.

Some of the basic range grazing systems, of which there are many variations and adaptations from ranch to ranch, follow:

Continuous Grazing

Rotation Grazing

- Deferred rotation grazing systems
  1. Two pastures—one herd system
  2. Three pastures—one herd system
  3. Four pastures—three herds
- Short duration grazing systems
  1. Conventional (rectangular) grazing system
  2. Savory (or cell) grazing system

Intensive Early-Season Stocking

CONTINUOUS GRAZING

Continuous grazing is the simplest and most common grazing system of the western ranges; and varying the number of animals allowed to graze pasture is the most commonly used means for grazing management. In comparison with rotation systems, continuous grazing requires less fence, water, development, less labor in moving animals and fixing fences, and less knowledge of livestock and range management. Also, continuous grazing may be more suitable and practical when used in conjunction with a seasonal range, than a complicated rotation system.

The major **disadvantages** of continuous grazing are: poor flexibility; lower stocking rate; less animal gains per acre; poorer livestock distribution on the range caused by animals concentrating around water, bedding grounds, and feed grounds, and overgrazing such areas; and less opportunity to use such improvement practices as burning, brush control, and livestock management.

ROTATION GRAZING

Rotation grazing is a system in which pastures are grazed and rested in a planned sequence. It gives the more desirable plants a chance to regrow, compete, and multiply, thus gradually increasing the number and production of high quality plants.

The objectives of any rotation grazing system are to favor the growth and survival of desired plants; to obtain greater use of the less palatable plants; and to improve range conditions. The improved range increases livestock production, improves the habitat of wildlife, reduces erosion, and conserves water.

The two main types of rotation grazing systems are deferred rotation grazing and short duration grazing.

DEFERRED ROTATION GRAZING SYSTEMS

In deferred rotation grazing systems, the range is usually divided into two to four units. There are different ways in which to apply deferred rotation grazing, with the following three basic systems, or some variations therefrom, most common:

1. **Two pastures—one herd system (switchback system).** With this system one herd of livestock is rotated between two pastures. Each pasture is grazed or rested at a different time during the two-year period required to complete the grazing cycle.

2. **Three pasture—one herd system.** This system is similar to the two pastures-one herd system, except the herd is moved through three pastures instead of two. In any one year, one pasture may be used during the growing season; the second pasture may be used at a later stage of vegetative maturity, such as seed ripe; and the third pasture may be rested and not grazed by livestock. The length of each grazing period may be as short as 30 days or as long as 90 days. This sequence is rotated among years. By treating a unit in this manner, the entire area receives the equivalent of a year-long rest.

3. **Four pastures—three herd system (the Merrill system).** In Texas, and in much of the Southwest, ranges may be grazed throughout the year. Under these circumstances, a different type of rotation grazing system should be considered than where grazing is not year-round.

Where 4 pastures are available, or can be arranged, a 3-herds system is popular, with each pasture grazed 12 months and rested 4 months. This system is summarized in chart form (see Table 9-3).

Here is how the system outlined in Table 9-3 works:

1. Divide the grazing area into 4 units.
2. Rest pasture 1 for 4 months. Place all livestock on the remaining 3 pastures (2, 3, and 4), with little or no attempt to force animals to graze undesirable plants. Plants are not set back while being grazed, with the result that they make maximum improvement when they are rested.
3. At the end of 4 months, move livestock from pasture 2 (which has been grazed 4 months) to pasture 1 (which has been rested 4 months). For the next 4 months, rest pasture 2 and graze pastures 1, 3 and 4)
4. The next 4 months, graze pastures 2, 4, and 1, and rest pasture 3.
5. The next 4 months, graze pastures 3, 1, and 2, and rest pasture 4.

Note that the rest period of each pasture will

TABLE 9-3
FOUR PASTURES—THREE HERDS SYSTEM

Year and Season				Pastures			
				1	2	3	4
1996:							
Mar.	Apr.	May	June	Rest	Graze	Graze	Graze
July	Aug.	Sept.	Oct.	Graze	Rest	Graze	Graze
1997:							
Nov.	Dec.	Jan.	Feb.	Graze	Graze	Rest	Graze
Mar.	Apr.	May	June	Graze	Graze	Graze	Rest
July	Aug.	Sept.	Oct.	Rest	Graze	Graze	Graze
1998:							
Nov.	Dec.	Jan.	Feb.	Graze	Rest	Graze	Graze
Mar.	Apr.	May	June	Graze	Graze	Rest	Graze
July	Aug.	Sept.	Oct.	Graze	Graze	Graze	Rest
1999:							
Nov.	Dec.	Jan.	Feb.	Rest	Graze	Graze	Graze
Mar.	Apr.	May	June	Graze	Rest	Graze	Graze
July	Aug.	Sept.	Oct.	Graze	Graze	Rest	Graze
2000:							
Nov.	Dec.	Jan.	Feb.	Graze	Graze	Graze	Rest

begin 4 months later each succeeding year until in 4 years all pastures will have been rested 1 complete year.

The Texas station reports that, in comparison with conventional yearlong grazing on the same area, the "4 pasture/3 herds system" results in greater livestock gains and 25% increase in carrying capacity.

SHORT DURATION GRAZING SYSTEMS

Short duration grazing, as the name implies, employs frequent movement of animals, with the speed of the rotation adjusted according to the growth rate of the plants. During the peak of the growing season, animals are moved at shorter intervals, with longer intervals during the remainder of the year when plant growth slows. In practice, a pasture may be grazed 3 days and rested 20 days in one part of the year and grazed 20 days and rested 60 days in another season. The short duration technique uses rest periods within the growing season in order to restore plant vigor. This system will usually give more rapid range improvement than deferred rotation grazing.

1. **Conventional (rectangular) grazing system.** This system involves the use of conventional, rectangular pastures, along with a 16 to 18 ft alley for cattle to get water and minerals and move from one pasture to another. The principle and practices (grazing time, resting time, central watering, and easy movement of cattle between pastures) for both the rectangular and cell systems were first described by André Voison of France in the early 1950s. The only difference is the layout and design; the cell system uses the wagon wheel design, whereas the conventional (rectangular) system uses the design identified by the name—rectangular pastures.

2. **Savory (cell) grazing system.** This system is named after Allan Savory, who originated and popularized it in the United States. Ideally, grazing is from 1 to 5 days, and resting is from 30 to 60 days. It usually involves 12 or more pastures, and generally, although not always, the pastures are arranged as a grazing cell, with pastures formed by fence lines fanning out from the hublike spokes on a wagon wheel, and with the water, minerals, and handling facilities at the hub. When the animals come to the center, or hub, for water and minerals, they can be moved, between pastures by opening and closing gates. Producers using the Savory system generally use electric fences in order to reduce costs. The Savory system is designed to lessen movement stress. However, it can add to nutritional stress if (1) stock are held too long in pastures, or (2) if livestock numbers and forage production are out of balance.

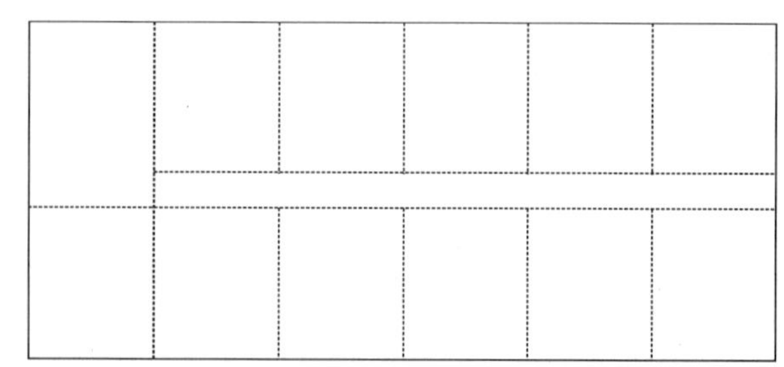

A–CONVENTIONAL (RECTANGULAR) SYSTEM

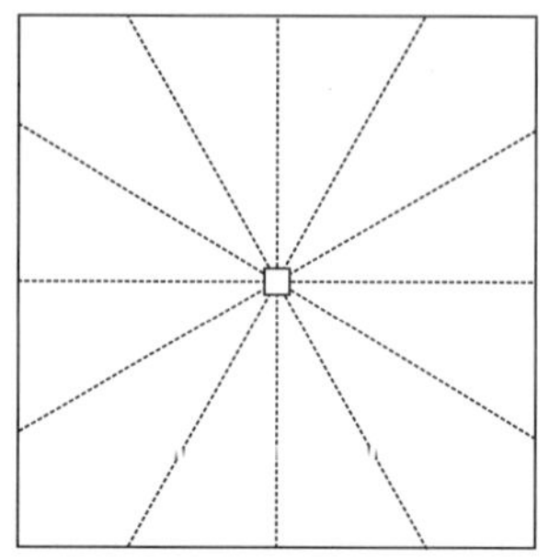

B–SAVORY (CELL) SYSTEM

Fig. 9-34. Two basic layouts (designs) for short duration grazing: **A**, the conventional (rectangular) system; and **B**, the Savory (cell) system.

INTENSIVE EARLY SEASON STOCKING

This grazing system was developed in the Flint Hills of Kansas where large numbers of stocker steers are custom grazed annually. In this area, nearly three-fourths of the total annual production of the bluestem grasses is produced during the 3-month period from mid-April to mid-July. Traditionally, livestock gains are highest during this same period and then taper off the remainder of the season.

This system was designed to take advantage of the high rates in daily gains by doubling the normal

Fig. 9-35. Stocker cattle on bluestem pasture in the famed Flint Hills of Kansas, the area in which intensive early season stocking was developed. (Courtesy, A. Solomon, Humboldt, KS)

stocking for three months. The pastures are emptied in mid-July by selling the cattle or moving to feedlots. Pasture grasses then have the remainder of the summer to regrow and regain vigor for the next year's cycle.

In this tall grass area greater steer gains are also obtained when pastures are burned. This has averaged nearly 30 lb per animal seasonally over unburned pastures. Prescribed burning, as a rule, is incorporated into this system.

The **advantages** of this system are (1) more pounds of beef per acre, (2) lower interest charges, because of owning the cattle for a shorter period of time, and (3) higher net returns. The main **limitation** is the lack of flexibility relative to removal of the herd; they must either be sold or moved into the feedlot as scheduled—in midsummer—or overgrazing and damage to the grasses will result.

DISTRIBUTION OF ANIMALS ON THE RANGE

Next to the proper rate of stocking and proper seasonal use, distribution of the animals on the range is the most important feature in range management. Proper distribution of animals is reflected in more even utilization of the forage. This assignment is more difficult with cattle than with sheep, especially on rough or mountainous land. Cattle have more of a tendency to utilize the flatter areas and to congregate around watering places. Better distribution of cattle on the range may be accomplished through: (1) fencing or riding the range (or herding); (2) providing water at short intervals (under ideal conditions, the distance between water in rough country should not exceed ½ mi for cattle, in level country to 1½ mi); (3) systematically locating salt grounds away from watering areas and salting at the proper intervals and in the right quantities, (4) building trails into inaccessible parts of the range, and (5) controlling livestock pests such as grubs and flies, which cause animals to congregate and seek protection.

Fig. 9-36. Showing how systematically locating salt grounds away from watering areas may be used as a means of obtaining better distribution of cattle on the range. (Courtesy, U.S. Forest Service)

RANGE RIDING

On a cattle spread, the range rider is the counterpart of the sheep herder. Riders prevent straying, force better distribution of animals on the range and more even utilization of forage, provide salt, service watering facilities, give minor repairs to fences, herd the animals away from areas infested with poisonous plants, see that *bogged down* animals are removed and cared for, dispose of dead carcasses, administer such assistance as may be required when animals are injured and at calving time, and warn the owner of any unusual parasite or disease problems. The number of cattle cared for by one rider with saddle and pack animals will vary from 150 to 1,200 head. The number will be determined chiefly by the character of the country and the carrying capacity of the range.

FENCING ON THE RANGE

Some fences are essential to the improvement of both cattle and ranges. Cattle fences are less costly than sheep fences. Moreover, cattle are less well adapted to herding methods than sheep. For these reasons, cattle ranges are more frequently fenced than sheep ranges.

The cattle rancher commonly refers to two types of fences: (1) drift fences, and (2) division fences. Drift fences are those that are not intended as enclosures, but which serve as barriers to retain animals within a certain area and to prevent drifting into an area where it is not desired that the animals shall travel. Usually drift fences extend between such natural barriers as steep ridges, ravines, etc. Frequently drift fences are useful for such things as preventing animals from following the snow line back in the spring, avoiding poison infested areas, confining animals to the area owned or leased by the operator, and holding stock during the roundup.

Division fences are those which enclose the boundaries of the range or field. They are usually used in keeping animals on the area owned or leased, in segregating different age and sex groups, and in conducting deferred or alternate systems of grazing.

Fig. 9-37. A drift fence on the Uncompahgre National Forest of Colorado's western slope. Drift fences are not intended as enclosures, but serve as barriers to retain animals within a certain area and to prevent drifting into an area where it is not desired that the animals travel. (Courtesy, *The Record Stockman*, Denver, CO)

BREEDING SEASON AND CALVING TIME

The breeding season on the range is usually timed so that the calves will be dropped in the spring of the year, with the coming of mild weather and green forage. Quite naturally, therefore, the calendar date of calving will vary in different sections of the United States. It is earlier on the ranges of the Southwest (February to April) and later in the Northwest (April to June).

Some range producers let the bulls run with the cows throughout the year, reasoning that a calf born out of season is better than no calf at all. The vast majority of the better producers, however, prefer to have the calves born in the shortest possible time. They make every effort to get calving over within 2 to 3 months.

WINTER FEEDING CATTLE ON THE RANGE

Cattle must be given rather close supervision when on the winter range, and usually some supplemental feeding is desirable. Unless they are on fenced range or herded, they are inclined to drift great distances during blizzards and in cold weather. Unlike

Fig. 9-38. Winter feeding cattle on the range. (Courtesy, Oklahoma State University, Stillwater, OK)

horses, they seldom learn to paw snow off the ground in search of forage.

In the early days of the range cattle industry, the animals were usually moved to the lower winter ranges and turned loose to get their feed as best they could. There was precious little feeding of supplemental forage or concentrates. If the winter happened to be mild and if a reasonable amount of grass had cured on the stalk, the herd came through in pretty good shape. During an exceedingly cold winter, however, the losses were severe and often disastrous. Today, the practical and successful rancher winter feeds, generally providing 1 to 2 lb daily of a protein supplement and an adequate supply of fodder or hay for those periods when the range is covered with snow. Young animals and cows with calves are more liberally fed than mature animals.

The progressive rancher is also equipped to meet emergency feeding periods, of which droughts are the most common in the West. Concentrates and roughages should be available for such emergencies.

THE ROUNDUP

Because of the large territory over which range cattle graze, it is common practice to gather them together at least twice each year, in the spring and again in the fall, for the purpose of carrying out certain routine assignments.

The spring roundup takes place between April and June, the exact time depending upon the earliness of the forage and the time of calving. The objects of this roundup are: (1) to get an accurate count on the stock; (2) to castrate, brand, and vaccinate calves, and (3) to separate the breeding animals from the steers and heifers.

The fall roundup, which usually takes place in

Fig. 9-39. Cattle roundup on a Wyoming ranch. Note animals being worked in the distance. (Courtesy, Bureau of Land Management, U.S. Department of the Interior)

September and October, is for the purpose of: (1) castrating, branding, and vaccinating calves born since the spring roundup; (2) culling out and marketing barren, old, or otherwise undesirable breeding stock which it is not desired to winter; and (3) weaning the calves.

QUESTIONS FOR STUDY AND DISCUSSION

1. Give facts and figures pointing up the economic importance of pastures in the United States.

2. What are the primary differences between (a) tame (seeded) pastures, and (b) native pastures?

3. What are the primary differences between (a) permanent pastures, (b) rotation pastures, and (c) temporary and supplemental pastures?

4. How would you go about determining what grass and/or legume to seed on a particular farm or ranch?

5. Outline, step by step, for tame pastures, the procedure (a) for establishing a new pasture and (b) for managing a new pasture.

6. Discuss methods by which run-down permanent pastures may be renovated—brought back into production.

7. Discuss the economic importance of the western range for cattle production.

8. Why is so much of the range area of the West publicly owned and unenclosed? Is it good or bad to have so much public domain? Justify your answer.

9. Do you concur in the policy which permits a sizable number of wild animals to feed on privately owned land? Justify your answer.

10. If you live in the West, classify the range grass of your ranch or area as to (a) type (which of the 10 broad types), and (b) dominant native species.

11. What similarities and differences characterize the various agencies administering public lands?

12. How do the grazing fees charged by the Bureau of Land Management and the U.S. Forest Service compare with charges for pasture rental on privately owned land?

13. Some environmentalists are agitating for a ban on grazing rights of public lands. What are the pros and cons for such action. and what is your recommendation?

14. Discuss the seasonal use of western ranges.

15. How can you detect if a given range has been overstocked? If overstocking is apparent, how would you rectify the situation?

16. What is *range management*? Discuss the history of range management in the United States.

17. List the five major U.S. range grazing systems, and tell how each of them works and where and how it is used/adapted.

18. Would you change the grazing system followed on your ranch, or in your area? Justify your answer.

19. Wherein does a modern cattle roundup differ from the roundup portrayed in most western movies?

20. How has the handling of cattle on the western range been mechanized? What further automation and mechanization of the western range can you suggest?

SELECTED REFERENCES

Title of Publication	Author(s)	Publisher
Commercial Beef Cattle Production, Second Edition	Ed. by C. C. O'Mary I. A. Dyer	Lea & Febiger, Philadelphia, PA, 1978
Crop Production, Fifth Edition	R. J. Delorit H. L. Ahlgren	Prentice-Hall, Inc., Englewood Cliffs, NJ, 1984
Ecological Implications of Livestock Herbivory in the West	M. Vavra W. A. Laycock R. D. Pieper	Society of Range Management, Denver, CO 1994
Forage and Pasture Crops	W. A. Wheeler	D. Van Nostrand Company, Inc., New York, NY, 1950
Forages, The Science of Grassland Agriculture, Fourth Edition	M. E. Heath R. F. Barnes D. S. Metcalfe	The Iowa State University Press, Ames, IA, 1980
Grass, The Yearbook of Agriculture, 1948	U.S. Department of Agriculture	U.S. Government Printing Office, Washington, DC, 1948
Holistic Resource Management	Allan Savory	Island Press, Covelo, CA, 1988
Intensive Grazing Management: Forage, Animals, Men, Profits	B. Smith P. Leung G. Love	The Graziers, Hui Kamuela, HI, 1986
Livestock Husbandry on Range and Pasture	A. W. Sampson	John Wiley & Sons, Inc., New York, NY, 1928
Manual of the Grasses of the United States, Second Edition	A. S. Hitchcock Rev. by A. Chase	U.S. Government Printing Office, Washington, DC, 1950
One Third of the Nation's Land	Public Land Law Review Commission	U.S. Government Printing Office, Washington, DC, 1970
Pasture and Range Plants	Phillips Petroleum Company	Phillips Petroleum Company, Bartlesville, OK, 1963
Pasture Book, The	W. R. Thompson	The author, State College, MI, 1950
Practical Grassland Management	B. W. Allred Ed. by H. M. Phillips	*Sheep and Goat Raiser Magazine*, San Angelo, TX, 1950
Problems and Practices of American Cattlemen, Wash. Ag. Exp. Sta. Bull. 562	M. E. Ensminger M. W. Galgan W. L. Slocum	Washington State University, Pullman, WA, 1955
Rangeland Management	H. F. Heady	McGraw-Hill Book Co., New York, NY, 1952
Range Management, Principles and Practices	A. W. Sampson	John Wiley & Sons, Inc., New York, NY, 1952
Stockman's Handbook, The, Seventh Edition	M. E. Ensminger	Interstate Publishers, Inc., Danville, IL, 1992
Veld and Pasture Management in South Africa	N. M. Tainton	Shuter & Shooter, Pietermaritzburg, South Africa, 1984
Western Range Livestock Industry, The	M. Clawson Ed. by S. Bruchey	Ayer Co., Salem, NH, 1979

Where cattle are worked with horses, no one expects the time-honored tradition of hot-iron branding to be obsoleted any time soon. This shows cowboy and cutting-horse sorting Santa Gertrudis cattle on world-famed King Ranch of Texas. (Courtesy, King Ranch, Kingsville, TX)

BEEF CATTLE MANAGEMENT

Contents — *Page*

Contents	Page

According to Webster, *management is the act, or art, of managing, handling, controlling or directing.*

Three major ingredients are essential to success in the cattle business: (1) good cattle; (2) a sound feeding program; and (3) good and aggressive management. The senior author called upon selected cattle producers and beef cattle specialists to rank these three factors in order of importance. Over 80% of the respondents put management at the top of the list. Further, their consensus was that a poor cow-calf manager can half the calf crop and lower weaning weights by a third, and a poor cattle feedlot manager can lower average daily gains by as much as ¾ of a pound and increase costs as much as 9¢ per pound. Indeed, a manager can make or break a cattle outfit. Unfortunately, this fact is often overlooked in the present era, primarily because the accent is on scientific findings, automation, and new products.

Management gives point and purpose to everything else. The skill of the manager materially affects how well cattle are bought and sold, the health of the animals, the results of the ration, the stress of the cattle, the percent calf crop and the weaning weights of the calves, the rate of gain and feed efficiency, the performance of labor, the public relations of the outfit, and even the expression of the genetic potential of the cattle. Indeed, a cattle manager must wear many hats—and wear each of them well.

Fig. 10-1. Cattle management has gone modern! On Tequesquite Ranch, Albert, New Mexico, this helicopter is used for observation, transporting crews, and rounding up cattle in large pastures. (Courtesy, Tequesquite Ranch, Albert, NM)

The bigger and the more complicated the cattle operation, the more competent the management required. This point merits emphasis because, currently, (1) bigness is a sign of the times, and (2) the most common method of attempting to *bail out* of an unprofitable cattle venture is to increase its size. Although it's easier to achieve efficiency of equipment, labor, purchases, and marketing in big operations, bigness alone will not make for greater efficiency as some owners have discovered to their sorrow, and others will experience. Management is still the key to success. When in financial trouble, owners should have no illusions on this point.

In manufacturing and commerce, the importance and scarcity of top managers are generally recognized and reflected in the salaries paid to persons in such positions. Unfortunately, agriculture as a whole has lagged; and altogether too many owners still subscribe to the philosophy that the way to make money out of the cattle business is to hire a manager cheap, with the result that they usually get what they pay for—a *cheap* manager.

Without attempting to cover all management practices, facts relative to—and methods of accomplishing—some simple beef cattle management practices follow.

HEAT DETECTION METHODS AND DEVICES

The problem of heat detection becomes more

important as artificial insemination replaces natural service, herds get larger, good hired help is more difficult to come by, cows produce more milk, and animal value increases.

Under ordinary farm conditions, operators miss an estimated 25 to 50% of the heat periods. On the average, a missed heat period prolongs the calving interval by 30 to 40 days and reduces profits. Some owners pay their employees a bonus for catching a cow in heat. For these reasons, cattle producers are interested in heat detection methods. Among them are the following:

1. **Chin-Ball Marker.** This device was developed in New Zealand. It is similar to a ball-point pen attached to a halter under the chin of a surgically modified, teaser bull, often called a *Gomer.* (One of the first ranches in North America to use the Chin-Ball Marker gave this name to the bull on which it was used.) During preservice sex play, it is usual for a bull to place his head over the shoulders, back, and rump of the cow. This causes a smearing of the colored ink from the ballpoint onto the cow.

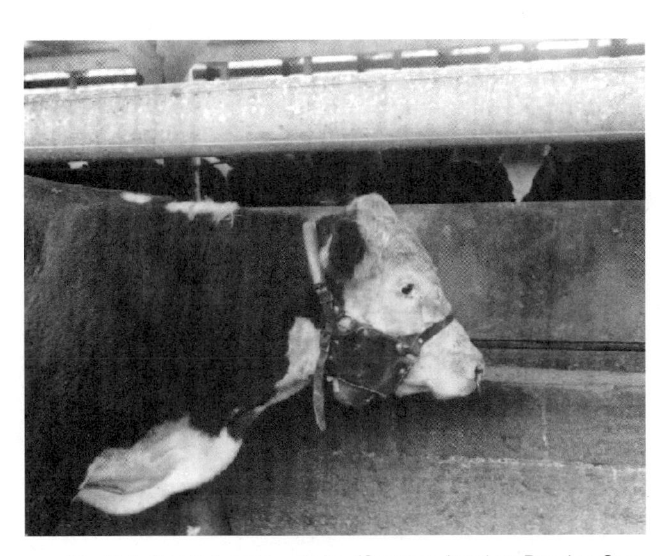

Fig. 10-2. Chin-Ball marking device. (Courtesy, American Breeders Service, De Forest, WI)

One filling of the stainless steel container is sufficient to mark 15 to 25 cows. Experience indicates that one Gomer bull can work approximately 80 cows. In large pastures and in larger sized herds, it is best to have two bulls.

This method of heat detection is a most dependable management tool.

2. **Heat patch.** The heat patch is glued to the top of the back between the tailhead and the hocks (Fig. 10-3).

The heat patch detector is a 2-in. × 4-in. fabric base to which is attached a white plastic capsule.

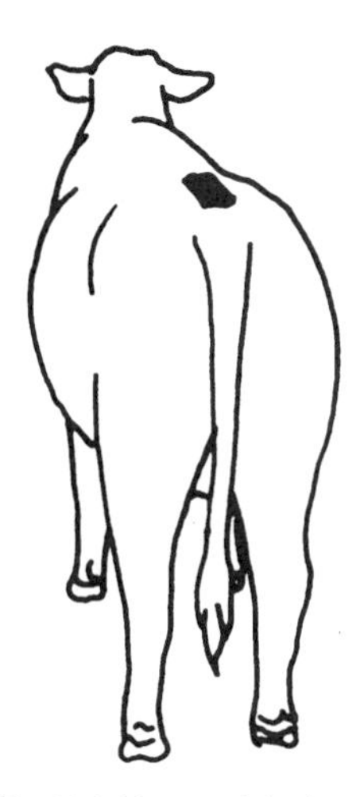

Fig. 10-3. Heat patch in place.

Inside the capsule is a small plastic tube containing red dye. The tube is constructed so the dye is released slowly by moderate pressure. When enough dye is released from the tube (after about 4 to 5 seconds of pressure), it spreads over the inner lining of the capsule, causing it to turn red (see Fig. 10-4).

The detector relies on the natural bovine instinct

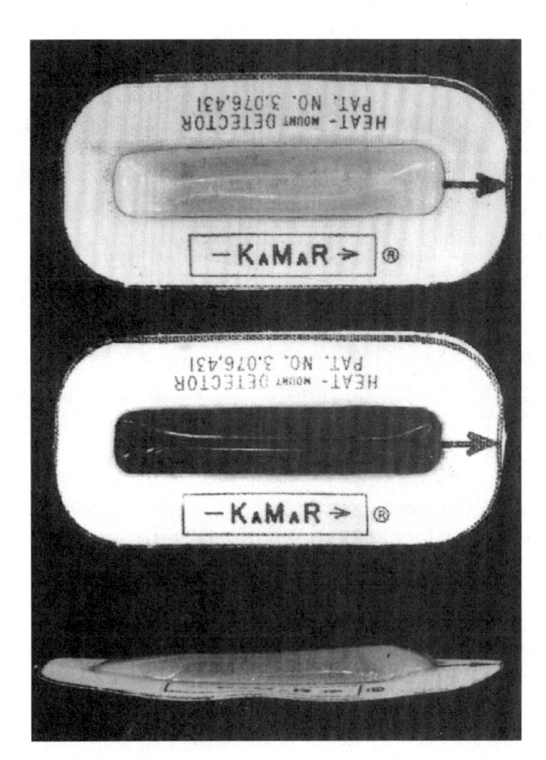

Fig. 10-4. The heat patch, a device for heat detection, is an aid in the artificial insemination of beef and dairy cows. At the top, the KAMAR Heat-Mount Detector is shown before activation. Center shows detector bright red after activation, indicating that cow is in heat. Lower view shows side or profile view of the device, which is applied to cow by an adhesive. (Courtesy, KAMAR, Inc., Steamboat Springs, CO)

of *bulling* or mounting during estrus. The pressure from the brisket of a mounting animal causes the dye to be released and the detector to turn red. If the cow does not stand for the mounting animal, there will not be enough pressure to release the dye and turn the detector red. This device has resulted in catching 95% of the heat periods.

3. **Pen-O-Block.** The Pen-O-Block is a plastic tube placed within the bull's sheath and held in place with a stainless steel pin. The bull can detect cows in heat and mount them in a normal way, but the device mechanically prevents him from making contact with the cow.

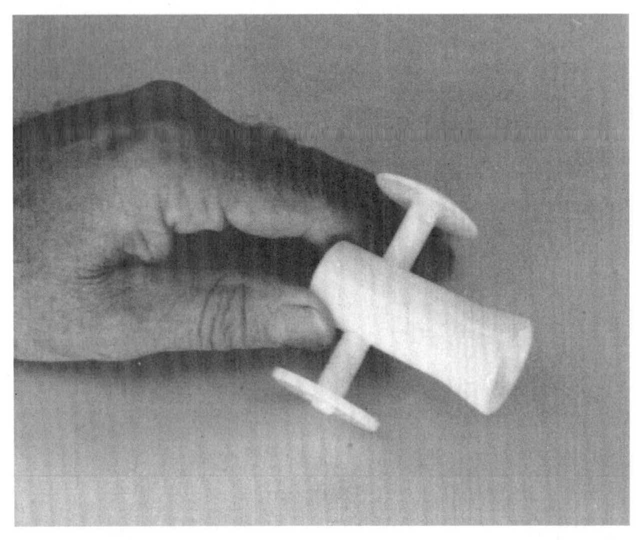

Fig. 10-5. Pen-O-Block marking device. (Courtesy, American Breeders Service, De Forest, WI)

The Pen-O-Block consists of a white plastic tube, the pin or cannula, two washers, and a cotter pin. The device is inserted within the bull's sheath and held in place by the cannula. The procedure is best carried out by a veterinarian, as it requires skill.

The **advantages** of the Pen-O-Block are (a) that venereal disease cannot be transmitted; and (b) that the Pen-O-Block can be removed anytime, following which the bull can be used in natural service. The **disadvantage** is that the bull may lose his sex drive.

4. **Vasectomized bull; deviated penis and sheath.** The vasectomized bull has been used as a heat detector for many years. The vasectomy operation, which should he performed by a veterinarian, involves removing part of the vas deferens from the spermatic cord. The **advantages** of the vasectomized bull are: the blood and nerve supply of the spermatic cord are left intact; the testicles and penis function normally, but transport of spermatozoa to the urethra is blocked; and the sexual activity of the bull remains unaltered. The **disadvantage** is that the bull continues to copulate; hence, he can transmit venereal diseases from one cow to another.

The deviated penis and sheath results from a surgical technique that transplants the penis and sheath from their normal position to the folds of the flank. The **advantage** to this method is that it permits normal erection, but does not allow copulation. The **disadvantage** is that the bull may become frustrated and lose his sex drive.

5. **Other heat detection methods and devices.** Other methods include (a) a surgically altered bull which serves as a teaser animal, but is incapable of serving a female; (b) hormone-treated steers or heifers; and (c) electronic monitoring.

Properly used, these aids will improve heat detection. They are by no means replacements for visual heat detection; nor will they solve all the problems in breeding a beef herd artificially. Other factors that need attention are:

1. **Nutrition.** Cows must have adequate nutrition to cycle at a satisfactory rate for successful breeding.

2. **Rest interval.** This is very important, as cows must have calved at least 60 days prior to breeding for satisfactory performance.

3. **AI facilities.** Facilities should be adequate for handling and breeding the cow herd. Locate them where the cows tend to gather, such as the watering hole.

4. **Personnel.** Trained personnel are needed to do heat detection, gather the in-heat cows, and inseminate the herd.

PREGNANCY TESTING

(See Chapter 6, Physiology of Reproduction in Cattle, section on "Pregnancy Test.")

CALVING SEASON

Most cattle producers favor spring calves over fall calves because they involve less intensive and expensive management practices than at any other season. Spring calving is more *in tune* with mother nature, because forage production is at its best and cows are generally more fertile.

Cows can usually drop spring calves outdoors, away from buildings, which helps reduce health problems in newborn calves. Also, pregnant cows can be wintered more economically than lactating cows with fall calves. Then, too, spring-born calves are ready for sale in the fall when demand for feeder calves is greatest.

The advantages and disadvantages of spring and fall calves follow:

■ **Advantages of spring calves**—The production of spring calves has the following advantages:

1. The cows are bred during the most natural breeding season—at a time when they are on pasture, gaining in flesh, and most likely to conceive. As a result, the calving percentage is usually higher with a system of spring calving.
2. The calves are old enough to use the cow's abundant milk supply when spring pastures are lush.
3. Weather conditions are usually favorable.
4. There are fewer calf diseases than among fall calves.
5. The calves will be in shape to sell directly from the cows in the fall, at which time there is a good demand for feeder calves.
6. Spring calves usually have higher 205-day adjusted weaning weights than fall calves.
7. If the calves are to be sold as yearlings, one wintering is saved; or if they are to be sold at weaning time, no wintering is required.
8. Because of greater utilization of cheap roughage, dry cows may be wintered more cheaply.
9. Spring calves require little or no supplemental feeding and utilize the maximum amount of pasture and roughage if marketed at weaning.
10. Less labor and attention is required in caring for the calves the first winter.

■ **Disadvantages of spring calves**—Labor is needed for calving when it is least available, due to spring work.

■ **Advantages of fall calves**—The production of fall calves has the following advantages:

1. The cows are in better condition at calving time.
2. Labor is more readily available for late fall calving than for spring calving.
3. The cows give more milk for a longer period.
4. The calves make better use of the grass during their first summer.
5. The calves escape flies, screwworms, and heat while they are small. (This is especially important in the South.)
6. Upon being weaned the following spring, the calves can be placed directly on pasture instead of in a drylot; or, if it is desired to sell, they usually find a ready market ahead of the influx of fall feeder calves from the range area.
7. When it is intended to sell market milk from dual-purpose cows, fall calves are usually best. The greater flow of milk is obtained during the period of highest prices.

■ **Disadvantages of fall calves**—The two main disadvantages of fall calves are:

1. Supplemental feed for the lactating cow must be provided in the drylot.
2. Although calves are heavier the next fall than spring calves sold at weaning, the price per hundredweight is often lower.

DAYTIME CALVING

The following benefits would accrue if the majority of cows would calve during the daylight hours:

1. Improved calf survival due to birth during warmer daylight hours and readily available assistance.
2. Reduced nighttime labor from fewer cows calving.

Limited research supports the theory of altering calving time by late feeding (feeding 5 p.m. to 9 p.m., starting about 2 weeks before calving), with the bulk of calves born during daylight. Scientists at the Fort Keogh Livestock and Range Research Laboratory, Miles City, Montana, report that heifers fed between 8 and 9 p.m. had 17% more daytime births than heifers fed between 8 and 9 a.m.

Iowa State University conducted a survey of 15 producers who fed either before noon or after 5 p.m. The producers who fed during the evening reported 85.1% of their cows gave birth during daytime. The early feeding group reported their calving times about equally divided between day and night. Other researchers report a 10 to 15% increase in daytime calving from nighttime feeding. The reason late feeding causes daytime calving has not been established.

Daytime calving has many advocates in the cow-calf industry. As a management practice, it does not require capital outlay, and, most important, it can make for increased profits from more live calves. But further experiments and experiences on the effect of feeding time on calving time are needed. In the meantime, producers may try nighttime feeding, but they shouldn't eliminate their night calving crews, because calving at night has not been totally eliminated.

MILKING A BEEF COW

Occasionally one of the following situations necessitates that a beef cow be milked: (1) when the cow produces more milk than the calf can consume; (2) when a calf is orphaned at birth and requires colostrum from another cow; and (3) when injury occurs to the cow or calf, thereby preventing the calf from nursing.

Milking a beef cow is accomplished by hand milking with the cow restrained in some manner.

DEHORNING

Although the presence of well-trained and properly polished horns may add to the attractiveness of the horned breeds in the show-ring or of the purebred herd, horns are objectionable on animals in the commercial herd and should always he removed.

■ **The chief reasons for dehorning are—**

1. Less shed and feeding spaces are required for dehorned cattle.
2. Dehorned cattle are less likely to inflict injury upon other cattle or upon the attendant.
3. Dehorned cattle are quieter and easier to handle.
4. Feeders prefer to buy dehorned cattle.
5. Dehorned cattle suffer fewer bruised carcasses and damaged hides in shipment to market, thus commanding a premium of 50 to 75¢ per hundredweight more than horned cattle of similar market class and grade.

Although the advantages in dehorning commercial cattle far outweigh the disadvantages, and progressive farmers and ranchers regularly dehorn their calf crop, there are certain unfavorable aspects. The **disadvantages** are:

1. Dehorning gives animals a setback, especially when the operation is performed on older animals. With yearling steers, the South Dakota Station[1] found (a) that about 2 weeks were needed for dehorned steers to equal their initial weight, and (b) that dehorned steers failed to catch up with horned animals; that, due to shrink at dehorning time, dehorned steers weighed slightly less at marketing time than horned steers.
2. Labor and equipment are required in dehorning.
3. There are some death losses which result from excessive bleeding, screwworms, or infection.
4. Scurs may result if the operation is not carefully done.
5. Diseases may be spread unless equipment (except the hot iron) is disinfected between animals.

AGE AND SEASON TO DEHORN

Cattle should be dehorned early in life, preferably before they are 2 months old, so as to minimize shock. Young calves are easier to handle, lose less blood, and suffer less setback. The danger of screwworm trouble and infection is also reduced when calves are dehorned while young.

In a study[2] which included more than a half million cattle in a 24-state area, it was found that, on the average, calves are dehorned at 5.2 months of age, which is considerably later in life than is desirable.

Less insect trouble is encountered when dehorning is done in the early spring or late fall—an especially important consideration in dehorning animals in an area where screwworms exist.

METHODS OF DEHORNING

There are several methods of dehorning, the most important of which are: breeding polled cattle, chemicals, saws and clippers, the hot iron, and the dehorning spoon and tube. Age and the presence of screwworms are important factors, and often determine the method used. Most cattle producers routinely do their own dehorning. Perhaps the most important thing is that dehorning be done at the proper time.

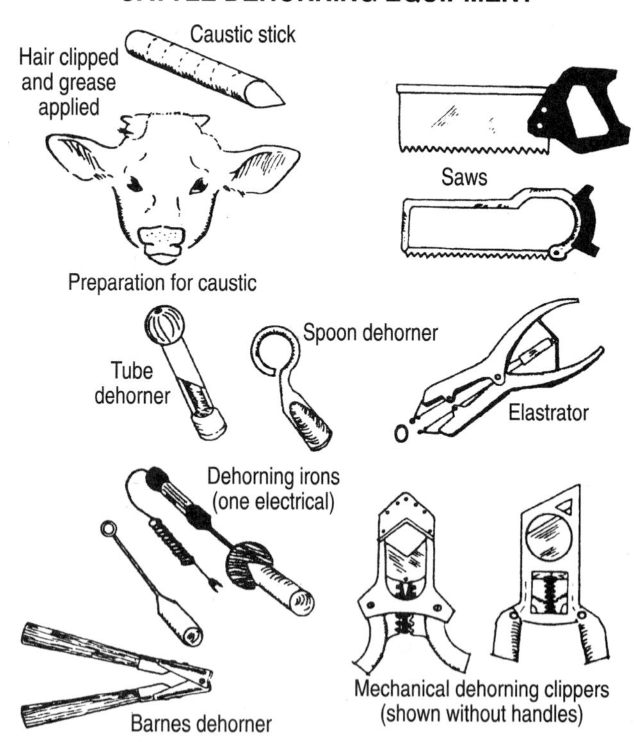

Fig. 10-6. Common instruments used for dehorning cattle. (Drawing by R. F. Johnson)

[1] Luther, Richard M., *South Dakota Farm and Home Research*, Vol. IX, No. 2, pp. 16-19.

[2] Ensminger, M. E., M. W. Galgan, and W. L. Slocum, *Problems and Practices of American Cattlemen*, Wash. Ag. Exp. Sta. Bull. 562.

BREEDING POLLED CATTLE

The use of polled bulls is the most humane method of securing cattle without horns. Also, it is especially popular in the South, because of the screwworm problem. If a polled bull is "pure" polled (carrying in his blood no tendency to produce horns), practically all of his calves will be polled, even though their dams have horns.[3] This system of securing hornless calves saves labor and avoids pain and possible setback to the calves. Without doubt, the breeding of polled calves will increase in popularity.

CHEMICALS

With small herds kept under close supervision, newborn calves may be dehorned satisfactorily by the use of caustic potash (potassium hydroxide) or caustic soda (sodium hydroxide). These chemicals can be purchased at almost any drug store and come in the form of a stick, paste, or a lacquer base (the accompanying use of petroleum jelly [Vaseline] is not necessary with the lacquer base). This method really prevents horn growth and does not actually remove the horns. The treatment should be applied when the calf is from 3 to 10 days old and when only small buttons are present. After the hair around the buttons has been clipped or sheared closely, smear a ring of heavy grease or petroleum jelly around the clipped area to keep the caustic from running into the calf's eyes. Then rub the caustic material over the button or little horn until the blood appears. This should be done carefully, otherwise some of the horn cells may not be destroyed and a scur may develop. The caustic should be wrapped in a cloth or paper to protect the operator's hands from serious burns. Within a week or 10 days, the thick scab that appears over the horn buttons will drop off, and the calf will suffer little inconvenience. Calves treated with caustic should be protected from rain for a day following the application, for the caustic may wash down and injure the side of the face. Also, it is best not to turn the calves back with the dams for a few hours following the application of the caustic.

[3]This applies to cattle of European extraction. See section in Chapter 5 of this book relative to "Dominant and Recessive Factors" and the footnote therewith, for a more complete treatment of this subject.

SAWS; CLIPPERS

(Barnes-Type Dehorner)

Saws or various forms of shears and clippers (including the Barnes-type dehorner) are used almost exclusively for dehorning in the range country, and for dehorning calves over 4 months old. Even on the general livestock farms of the central and eastern states, mechanical methods are more generally used than chemicals. Whatever the instrument used (saws or clippers), it is necessary to remove the horn with about ¼ to ½ in. of the skin around its base to make certain that the horn-forming cells are destroyed. The skin should then be allowed to grow over the wound. Dehorning of young animals can be done with greater ease, and there is less shock to the animals. However, some attention must be given to the season.

Ordinarily, clippers are satisfactory for removing the horns of younger cattle; but the hard, brittle horns of mature cattle can best be removed with a saw. With older animals, clippers are likely to sliver or crack the bone that forms the horn core. Moreover, the saw results in less loss of blood, for the action of the saw blade produces a lacerating of the blood vessels rather than a clean-cut cross section. On the other hand, the

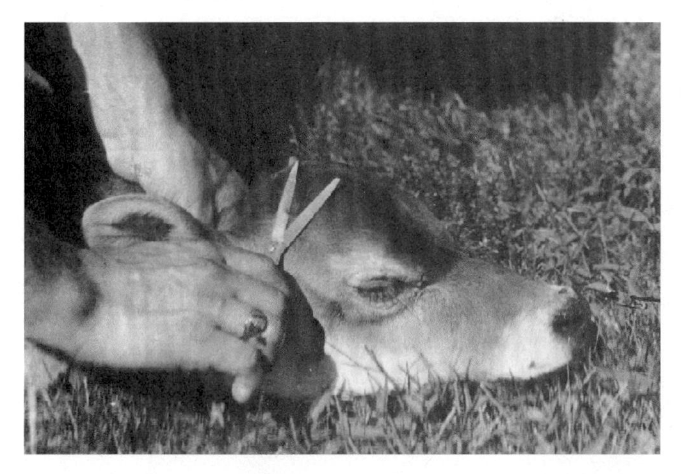

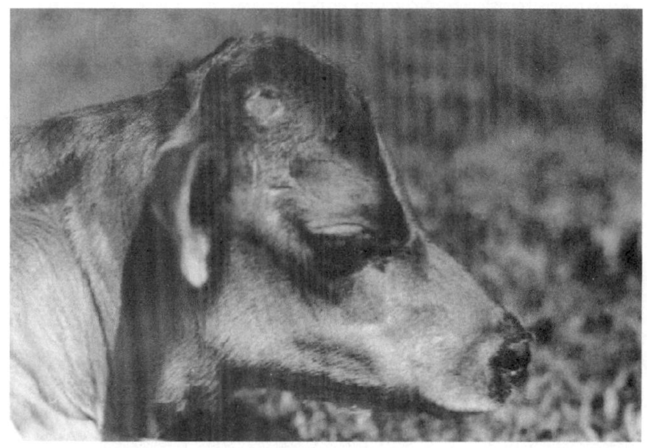

Fig. 10-7. Use of chemicals in dehorning. First the hair around the *button* is clipped *(left)*, following which caustic material should be rubbed over the little horn until blood appears *(right)*. (Courtesy, USDA)

ragged wound made by a saw heals more slowly, and the operation is less rapid.

While the dehorning operation with older cattle is being performed, it is necessary to have some device for confining or restraining animals. For this purpose, various types and arrangements of dehorning chutes, pinch gates, squeeze pens, and cattle stocks have been devised. Calves may be handled by throwing or snubbing them to a fence post and tying one side of the body against a strong fence or solid wall. Such methods are more difficult, however, for both people and beasts.

HOT IRON (Fire Irons; Electrically Heated Irons)

The hot-iron method of dehorning consists of the application of a specially designed hot iron to the horn of young calves. The iron is fashioned with one end cupped out (bell-shaped) so that it fits over the small horn button. Some ranchers first cut out the horn with a sharp knife before applying the hot iron; whereas others use a larger iron and fit it over the small horn. This system of dehorning is bloodless and may be used any time of the year, but it can be used on young calves only, with horn buttons less than ¾ in. in length.

Where electricity is available, the electric hot iron may be used. It keeps an even temperature, without getting too hot or too cold.

DEHORNING SPOON AND DEHORNING TUBE

The dehorning spoon (or gouge) is a small instrument with which the horns of young calves can he gouged out. In the hands of an experienced operator, it is both fast and effective. The use of the spoon leaves the head slightly rounded, and very seldom do scurs occur. In comparison with the spoon, the tube is easier, faster, and less tiresome to use, and more certain to avoid regrowth.

Dehorning tubes come in four sizes, varying in diameter from ¾ in. for the smallest to 1⅛ in. for the largest. All four sizes should be available.

The steps and directions for using the dehorning tube are as follows:

1. Restrain the calf.
2. Select a sharp tube of proper size to fit over the base of the horn and include about ⅛ in. of skin all the way around.
3. Place the cutting edge straight down over the horn and then push and twist, first one way and then the other, until the skin has been cut through. A cut from ⅛ to ⅜ in. deep is required, the greater depth being necessary with calves about 3 months of age. Going deeper than necessary will cause excessive bleeding.
4. Turn the tube to about a 45° angle and rapidly shove and turn the cutting edge until the button comes off.

Most ranchers who use the dehorning spoon or tube do so at the time of branding, thus avoiding extra handling. Either instrument can be used on calves up to 60 days of age, with horns less than 1½ in. long.

TREATMENT AFTER DEHORNING

If dehorning is done in cool weather (spring or fall), when there are no flies, no treatment is required. On the other hand, if the operation is performed when flies are present, it is important that a good repellent be applied to the wound. As a rule, there will be no danger from excess bleeding. The danger of infection will be materially reduced if cleanliness is practiced and the instruments (except hot irons) are disinfected at intervals.

CASTRATING

Castrating is the unsexing of a male animal. The practice of castrating males in animal species used for food production purposes is universally practiced and is one of the oldest surgical operations known.

Most male calves not intended for breeding purposes are castrated. It makes animals quieter and easier to handle.

The California Station measured the effects on healthy calves of castration by removal of one-third of the scrotum and pulling the testicle with the cord. The study extended 28 days following castration. In comparison with the noncastrated animals, the castrated calves gained 18 lb less per head, consumed 12% less feed, and required 22% more feed to produce a pound of body weight.[4] This study, along with other similar studies, indicates that, if market discrimination is removed in the future, more bulls will be fed.

AGE AND SEASON FOR CASTRATING

Bull calves are generally castrated at about 2½ months of age, although they can be desexed at any age. The older the animal at the time of the operation, the greater the shock and risk. Moreover, if a bull calf

[4]Addis, D. G., *et al.*, *Research Reports*, University of California.

is not castrated before he is 10 to 12 months old, he may become *staggy*—a very objectionable characteristic in the feeder or finished steer.

Currently, there is a trend toward castrating at slightly older ages than formerly, primarily (1) to take advantage of the higher gains and greater efficiency of bulls, and (2) to lessen the hazard of urinary calculi.

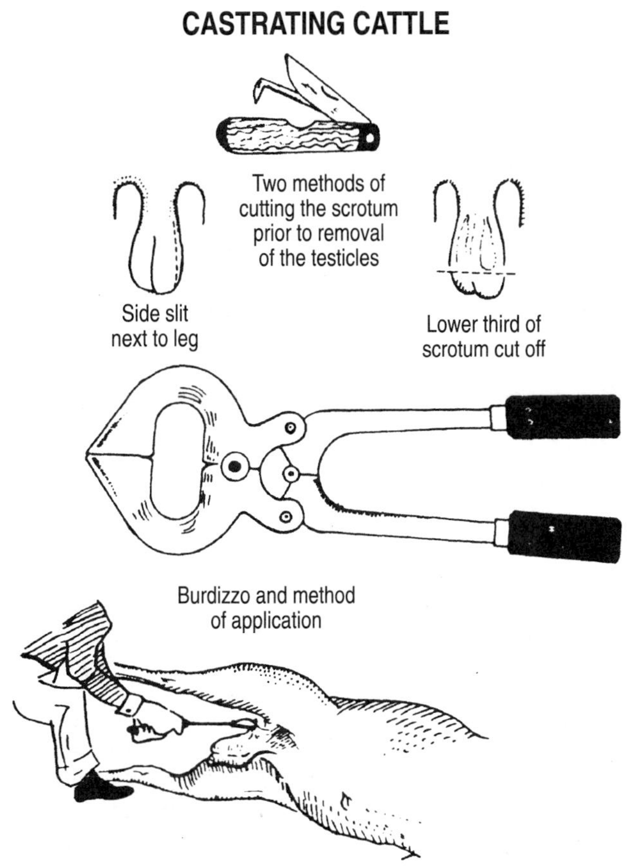

CASTRATING CATTLE

Two methods of cutting the scrotum prior to removal of the testicles

Side slit next to leg

Lower third of scrotum cut off

Burdizzo and method of application

Fig. 10-8. Common castrating equipment (the knife and the Burdizzo), and the method(s) of using each. (Drawing by R. F. Johnson)

As in dehorning, it is best to perform this operation in the early spring or late fall so as to avoid infestation from flies. Moreover, it is unwise to castrate during periods of inclement weather. If castration is performed in an area where screwworms exist, either (1) a fly repellent should be applied to the wound and the animal should be kept under close observation until the wound is healed over; or (2) a bloodless method of castration should be used.

METHODS OF CASTRATING

As is true in dehorning, most cattle producers routinely castrate their calves; others call upon the veterinarian. Perhaps the most important thing is that it be done at the proper time. Each of the common methods of castrating is discussed in the sections that follow.

REMOVAL OF LOWER END OF SCROTUM

In this method, approximately the lower one-third of the scrotum is removed (exposing the testicles from below); the membrane covering each testicle (the tunica vaginalis) is slit (if desired, the membrane need not be slit; simply remove it along with the testicle), and the testicles are removed by pulling them out. In older cattle, excessive bleeding may be prevented through severing the partially withdrawn cord by scraping with a knife or clamping with an emasculator. Although the removal of the end of the scrotum allows for excellent drainage, this method is not recommended for calves intended for show purposes because the cod will not be so large and shapely. Most ranchers castrate calves by this method.

SLITTING SCROTUM DOWN THE SIDES

In this method, one testicle is pulled down at a time and is held firmly to the outside so that the skin of the scrotum is tight over the testicle. With a sharp knife (or Newberry knife), an incision is then made on the outside of the scrotum next to the leg. It is important that the incision extend well down to the end of the scrotum to allow for proper drainage and that it extend both through the scrotum and membrane (if desired, the membrane, or tunica vaginalis, need not be slit; simply remove it along with the testicle). The testicle may be removed as previously indicated.

BURDIZZO PINCERS

Burdizzo pincers (named after their inventor, Dr. Burdizzo, and manufactured in Italy) are sometimes used in making a *bloodless castration*. In this method, the cords and associated blood vessels are crushed or severed so completely that the testicles waste away from want of circulation. After the animal is thrown, the cord is worked to the side of the scrotum, and the Burdizzo is clamped on about 1½ to 2 in. above the testicle, where it is held for a few seconds. Then, this operation is repeated on the same cord at a location about ¼ in. removed from the first one. The same procedure is then followed on the other testicle. In using the Burdizzo, it is important that the cord not slip out, that only one cord be clamped at a time, and that there be no interference with the circulation of the blood through the central portion of the scrotum.

This method is a satisfactory means of castration if done properly and by an experienced operator. But if the operation is not performed correctly, the cord may be incompletely crushed and the animal may develop stagginess later. This is especially disturbing when discovered in experimental or show animals or steers that are nearly ready for market or show. Because there is no break in the skin with this type of castration, there is no external bleeding. Nor can there be any trouble from screwworms—an important consideration in the South. Furthermore, steers so castrated usually develop very large and shapely cods, a characteristic that is very desirable in well-finished steers.

ELASTRATOR

The elastrator (developed in New Zealand) may be used in stretching a specially made rubber ring over the scrotum to castrate young calves. It works best on calves under 2 months of age. The directions are as follows: (1) hold the calf in either a sitting or a lying position; and (2) press both testicles through the ring and to the lower end of the scrotum, then release the rubber ring.

The **advantage** of the elastrator is that it is bloodless. The **disadvantage** is that tetanus and infection may be problems as the scrotum atrophies and sloughs off. When this method is used, tetanus antitoxin should be administered simultaneously.

SHORT SCROTUM BULLS

This method of rendering the intact male infertile was developed by the New Mexico Station. The animal is really a pseudocryptorchid in that the scrotum is shortened by bringing it through a distended rubber band with an elastrator when the calf is 1 to 3 months of age. Before the band is released, the testicles are moved near the abdominal wall. The scrotum below the rubber band sloughs off after 3 or 4 weeks.

Shortening the scrotum requires considerably less time than castration. It can be done in 15 to 30 seconds. Another advantage over castration is that there is no weight loss. As a result of the shortened scrotum, the testicles lie close to the abdominal wall and the animal is sterile. The testicles then develop to about half the weight of those from fertile bulls of the same age and weight. This technique eliminates indiscriminate breeding while retaining and even enhancing the testosterone level in the pseudocryptorchid male. The short scrotum treatment does not change either the temperament or the urge of the animal; hence, there will be riding if sexes are not kept separate.

Experimental work to date indicates that the rate and efficiency of gains of short scrotum bulls and intact bulls are about the same; and that the carcasses of short scrotum bulls are leaner than the carcasses of steers.

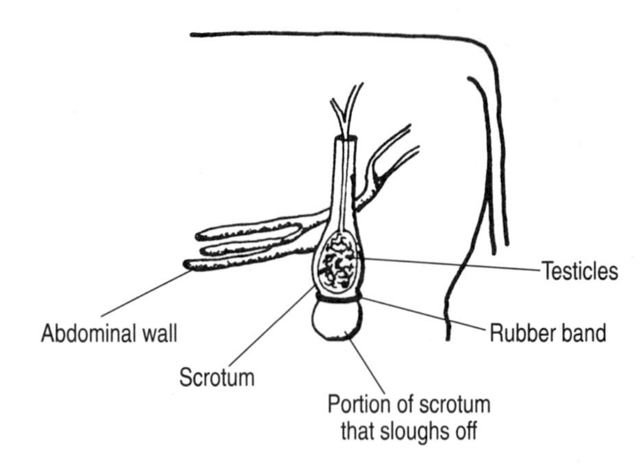

Fig. 10-9. Short scrotum method of castrating.

INJECTABLE CHEMICAL CASTRATION (CHEM-CAST)

A patented, injectable chemical solution, sold under the brand name *Chem-Cast*, was approved by FDA in 1983 and is now available through veterinarians. When used according to label directions, it painlessly destroys the testicles and spermatic cords of bull calves weighing up to 150 lb. There is no weight loss, going off-feed, high risk infection, hemorrhaging, or pain; and it is nearly 100% effective.

The procedure consists of injecting the prescribed dosage of Chem-Cast (for calves weighing up to 100 lb, the dosage is 1 ml per testicle; for calves weighing 101 to 150 lb, 1.5 ml per testicle) from the top into the middle third of each testicle via a small hypodermic needle (see Fig. 10-10).

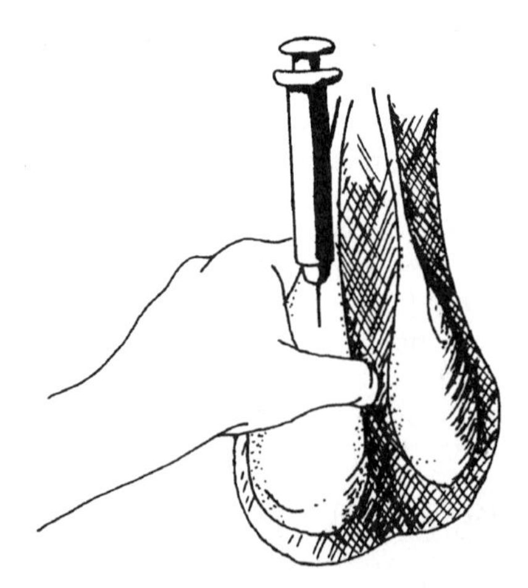

Fig. 10-10. Diagram showing syringe injection of Chem-Cast. Note that it is injected from the top into the middle one-third of each testicle.

About 24 hours following the injection, there is a swelling of the testicle, testicular vessels, and the spermatic cords. The swelling disappears within 2 weeks and the testicles begin to be reabsorbed and to be reduced in size. At the end of 60 days, only small, hard nodules are left in the cod; and, in time, these disappear, also.

Any cattle producer can use Chem-Cast after receiving a minimum of instruction and training.

POSITION AND TREATMENT IN CASTRATING

Young animals are usually thrown to be castrated, whereas animals eight months of age or older may be more easily operated on in a standing position.

Before making any incision, the hands of the operator and the knife should be thoroughly cleaned and washed in a good disinfectant. With these precautions, there is usually little danger of infection in castrating under range conditions; for normally the calves are turned back on the range where there is plenty of sunshine, fresh air, and no contamination. If the operation is performed in fly season and in an area where screwworms are prevalent, a good repellent should be applied.

SPAYING

In females, the operation corresponding to castration is known as spaying. Under most conditions, desexing heifers by the traditional surgical method is not recommended because (1) the operation is complicated and difficult, requiring a very experienced technician; (2) it is attended with more danger than castration; (3) it lowers both rate and efficiency of gains; (4) it eliminates the heifers for possible replacement purposes or sale as breeding stock; and (5) experiments and practical operations with spayed heifers have generally shown that the selling price obtained is not sufficiently higher to compensate for the lower and less efficient gains plus the attendant risk of the operation.

A summary of 11 experiments in which spayed heifers and intact open heifers were compared revealed that spaying made for 9.9% slower rate of gain and increased the feed required per 100 lb gain by 8.5%.[5]

However, spaying by one of the new methods may be justified because of the following advantages of spayed heifers: (1) They may move freely across state lines without being tested for brucellosis when destined for a feedlot, (2) they are easier to manage in the presence of bulls or steers, (3) it avoids the losses generally associated with pregnancy, and (4) they generally bring a premium price.

Spaying by an experienced technician, using one of the new techniques, followed by implanting the spays with a growth promotant(s) provides a sure way of preventing in-heat heifers and pregnancy, and assures top gains.

MARKING OR IDENTIFYING

It is of historical interest to note that one of the first uses of branding in many countries of the world was to identify human criminals permanently. Also, of interest is the fact that the noun *Maverick*, meaning "unbranded cattle," and now a recognized part of the American lexicon, originated with lawyer-cattleman Samuel A. Maverick (1803-1870). Maverick, who accepted 600 head of cattle as an attorney's fee, failed to brand his young stock. As a result, year after year, his unbranded yearlings fell into the hands of other cattlemen who promptly placed their brands on them. After 10 discouraging years of such operation, Maverick sold his depleted herd for the amount of the original fee.

Ranchers take pride in their brands, for to them it is much more than a sign of ownership; it is a symbol of service—a pledge of integrity of the person behind it and a mark of courage, character, and wisdom. It is indicative of the quality of stock that the owner raises, the class of bulls used, the con-

Fig. 10-11. Branding is a very old management practice. This picture made from an Egyptian tomb shows Egyptian drivers branding their cattle. (Courtesy, The Bettmann Archive, Inc.)

[5]*Montana Farmer-Stockman*, summary prepared by R. A. Bellows, U.S. Range Livestock Experiment Station, Miles City, MT.

dition of the range, and the kind of cowhands connected with the outfit.

On the western range, marking or branding is primarily a method of establishing ownership and/or age. In modern beef production, it is an important factor in managing herd health, artificial insemination, and performance testing. In the purebred herd, it is a means of ascertaining ancestry or pedigree. The method of marking employed will depend primarily upon the objective sought and the area.

The three leading methods of identification, by rank, are: hide brands, plastic ear tags, and tattoos. However, there are some differences between areas and between purebred and commercial herds. Hide brands and plastic ear tags are popular in the West and in commercial herds; and tattooing is a requisite to registration in most beef cattle registry associations.

HIDE BRANDS

When properly applied, hide brands are permanent. Throughout the range country, the hide brand is recognized as the cattle producer's trademark. Most of the western states require that each brand be recorded as to both type and location in order to avoid duplication. When stock are run close to a state boundary, the same brand may be recorded in two states.

In addition to the regular brand, many ranchers identify the age of the females by adding the last number of the year in which they were born (usually at a different location). Thus, heifer calves born in 1986 might be identified by adding the number 6 to the regular brand. At the end of 10 years, the numbers are used over again, for there is seldom any difficulty in determining ages where there is a 10-year spread. In those states where brands are recorded, these added numbers or brands must also be approved by and recorded with the Registrar of Brands.

Fig. 10-12. Branding time at Tequesquite Ranch, Albert, New Mexico. With this size crew, 700 calves are branded in one day. (Photo by H. D. Dolcater, Amarillo, TX. (Courtesy, Tequesquite Ranch, Albert, NM)

Hide brands have the disadvantage of being unsightly, and hot iron brands lower the market value of the hide. For these reasons, they are not recommended except when necessary for identification purposes. Even then, it is desirable that their size be as small as possible, consistent with serving the primary objective of the brand. The pertinent facts relative to branding are:

1. **Time.** In the range country, the usual practice is to brand calves at the same time they are castrated and vaccinated against blackleg.
2. **Location on animal.** The brand is located on a body area where it may be easily seen and where it will do the least possible damage.[6] Hips and thighs are favorite body areas for brands.
3. **Preparation.** Usually calves are thrown for branding—roped by the hind legs and dragged to the place of branding. Older cattle, however, are restrained in a chute. Some ranchers now prefer to use the specially designed branding chutes for calves.
4. **Four methods of applying brands are:**

   a. **Hot iron.** To date, this has been the preferred method. Ranchers heat the irons to a temperature that will burn sufficiently deep to make the scab peel, but which will not leave deep scar tissue. The proper temperature of the hot iron is indicated by a yellowish color. Branding is accomplished by placing the heated branding iron firmly against the body area which it is desired to mark and by not allowing it to slip for the few seconds when the hide is burned. The branding iron should be kept free from dirt and adhering hair at all times.

   The California Station measured the effect of hot iron branding (in one group they branded the rib; in the other they branded the hip). They found that branding had no effect on 28-day weight gains, pounds of feed consumed daily, or pounds of feed required to produce a pound of gain.

   Other methods of identification do not provide the excitement of roping and branding cattle with a hot iron. Hot branding, like the TV western, is a symbol of pride and tradition of cattle ranchers.

   Where electricity is available, the electric iron may be used; it keeps an even temperature, and if properly used, makes a clear, uniform brand.

   b. **Branding fluids.** Branding fluids, which are less widely used in making hide brands, consist of caustic material applied by means of a cold

[6] In arriving at both the kind and location of the brand, the owner should first check with the brand inspector or the local county agent to determine if any part of the animal is reserved for state or federal disease control programs; for example, the cheek of cattle is used for brucellosis reactor identification.

iron. Best results are secured if the area is first clipped. The chemical method of producing hide brands is slower; the results are generally less satisfactory, particularly if the operator is inexperienced with the method; and the resulting brand is less permanent.

c. **Freeze branding.** This method, developed by the U. S. Department of Agriculture, at Washington State University, makes use of a superchilled (by dry ice or liquid nitrogen) copper branding *iron* which is applied to the closely clipped surface for about 45 seconds, thereby depigmenting the hair follicles, following which the hair grows out white. When properly done, this method is painless, permanent, and there is no hide damage.

Fig. 10-13. Freeze-branded replacement heifers. (Courtesy, Martin Jorgensen, Jr., Ideal, SD)

On white cattle, deliberate overbranding (60 seconds or more) will produce a bald brand suitable for identification after clipping.

d. **Laser branding,** which is permanent, but which needs further development and experimental study.

5. **Characteristics of a good brand.** A good brand is one that is easily read, that is of simple design and yet cannot be easily changed or tampered with, that has no welds or thick points in the iron, and that interferes with the circulation as little as possible. Thick points mean deeper burning and slower healing; whereas small enclosed areas, such as a small *O*, will slough out entirely.

HOT-IRON BRANDS—NEED FOR, AND LOCATION OF

The National Beef Audit—launched in 1991, checkoff-funded, and spearheaded by the National Cattlemen's Association—showed that cow-calf producers should rethink the need for, and the location of, hot-iron brands. Their audit showed that the U.S. beef industry could recapture $39.08 million in lost value simply by switching from a rib brand to a hip brand. This saving would increase dramatically if no brands were applied, as rib-branded and butt-branded hides were discounted $14.69 and $9.46 per hide, respectively. These figures indicate that the beef cattle industry should vigorously seek acceptable alternatives to hot iron branding of cattle. In the meantime, cattle producers should give consideration to the following steps to reduce the incidence of hide defects due to hot-iron branding:

1. Eliminate rib branded hides.
2. Limit hot iron brands to replacement heifers.

PLASTIC EAR TAGS

Plastic ear tags, which can he applied at any age, are a popular method of animal identification. Their **advantages:** they are economical; they can be read at a distance; they are flexible—for example, the individual animal number can be placed on the bottom of

Fig. 10-14. Plastic tags. (Courtesy, American Angus Assn., St. Joseph, MO)

the tag, and the sire/or dam numbers can be placed on the top. The major **disadvantages** are: plastics tend to become hard and brittle in cold weather; loss of tags; prenumbered tags with block-type numbers are difficult to read if tags get soiled; therefore, most herd owners prefer to purchase blank tags and then number them with the largest possible numbers.

Because of the frequency of loss, plastic ear tags should be duplicated by having a tag in each ear, or used in combination with a permanent means of identification.

Plastic ear tags come in a variety of styles, sizes, and colors. Basically, there are 3 styles: 1-piece plastic, 2-piece plastic, and over-top-of-ear or on-top-of-ear tags. The latter have the advantage of being located in the toughest part of the ear; also, the on-top-of-ear tags may be numbered on both sides for ease of reading. The larger plastic tags are most popular because they are economical, easy to install, easy to read from a distance, stay pliable in cold weather, and stay in the ear longer than metal tags. Yellow tags with black numerals are the most readable; red and white tags tend to fade or discolor.

Plastic tags can be purchased prenumbered or blank.

METAL EAR TAGS AND BUTTONS

Metal ear tags are difficult to read at a distance, and, due to their sharpness, cut through the ears rather easily. Metal buttons stay put, but they cannot be read at a distance. Both metal tags and metal buttons frequently rub and scratch the skin, thereby making openings for screwworm infestation. Like plastic tags, metal tags can be purchased numbered or blank.

TATTOOS

Tattooing is a permanent method of identification, required by most purebred beef cattle registry associations. It consists of marking by piercing the skin with instruments equipped with needle points which form letters or numbers. This operation is followed by rubbing indelible ink or paste into the freshly pierced area. After healing, the tattoo is permanent. It is well to disinfect the tattooing instrument carefully between each operation in order to alleviate the hazard of spreading warts to the pierced area, for warts make it impossible to read the tattoo.

Animals may be tattooed at any age, but it is most convenient to tattoo baby calves.

The **advantages** of a tattoo are: it is permanent, and it does not disfigure an animal.

The major **disadvantage**: cattle must be confined in order to read tattoo numbers. Even then, tattoos are difficult to decipher on dark-skinned animals. For this reason, most producers apply a brand or ear tag, in addition to the tattoo, so that the animal can be easily identified from a distance.

EARMARKS

Earmarks are permanent and easily recognized, but they're unsightly. They may be administered with either a sharp knife or a regular ear notcher. Sometimes polled animals are individually identified through ear notches. In such instances a definite value is assigned to each area location. When earmarks are used in commercial operations, however, they are uniform and recorded for any given ranch. Some of the more common earmarks are *crops, swallow forks, bobs, over-bits, under-bits*, and *splits*.

Fig. 10-15. A show of numbers! Like mother, like calf. Also, ear tags of calves may be color-coded by sires. Hence, a calf and its mother can be readily paired, and the calf's sire identified by color of tag.

NECK CHAINS OR STRAPS

Prior to the advent of plastic ear tags, neck chains or straps were the most frequently used means of identifying polled cattle. Occasionally, chains or straps may be lost, but this is not particularly serious if the caretaker is on the alert and immediately replaces each one that is lost, without allowing several losses to accumulate before taking action. In rare instances, an animal will hang itself by the chain.

Neck chains or straps must be adjusted, for young animals grow, or animals change in condition.

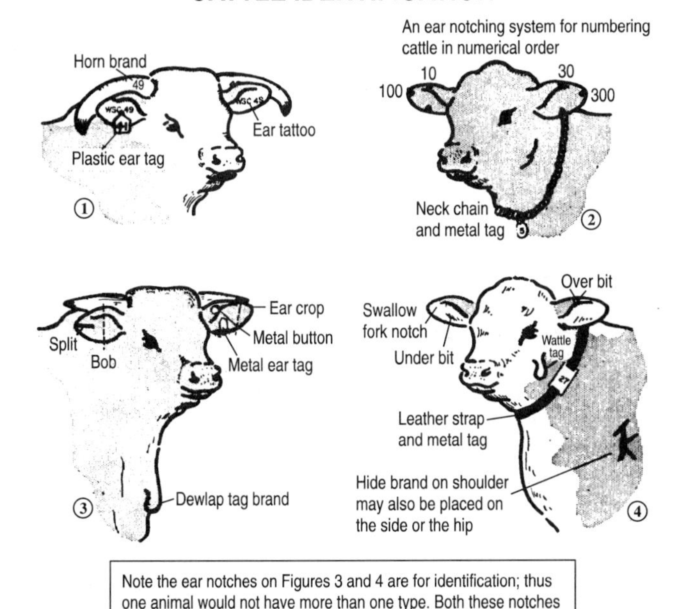

Fig. 10-16. Composite drawings showing a number of methods of cattle identification. It is unlikely that any individual animal will carry more than one or two of these methods of identification. (Drawing by R. F. Johnson)

HORN BRANDS

Horn branding for individual identification is commonly used among breeding or sale animals of the horned breeds. Usually horn brands are made by heating small copper numbers with a blow torch or charcoal burner. On mature animals, this method of branding works fairly well, but it cannot be used on young animals while the horns are still growing, unless it is repeated at intervals.

OTHER IDENTIFICATIONS

Other identification marks used on the range include: (1) *buds* formed by making a strip incision through the nose; (2) *wattles* made by cutting down a strip of skin on the jaw bone; and (3) *dewlaps* formed by cutting down a strip of skin on the brisket.

The U.S. Department of Agriculture requires that most cattle two years of age or older be backtagged or eartagged to identify the animals to their herd of origin before they are shipped across state lines.

South African officials are encouraging cattle raisers to keep records of the noseprints of cattle. Like human fingerprints, tiny ridges on cattle noses always form differing patterns. Of course, it is difficult to get a calf to hold still long enough to get a reliable print of its nose.

Various electronic devices are in different stages of research and development; among them-

1. **Radio transmitter in the second stomach.** The animal swallows a small radio transmitter enclosed in a ¾-in. × 2½-in. plastic capsule, which lodges in the second stomach. From there, it transmits a coded number when signaled by a receiving unit to do so. The transmitter can be retrieved at slaughter and reused.

2. **Implant under the skin or eartag.** Basically, the transponder is a silicone-coated mini-circuit which is implanted under the animal's skin (for example, behind the poll or it can be placed in an eartag) and powered by a microwave beam from a portable receiver unit. In addition to the animal's identification number, this device may include such information as a birth date, original owner, state of origin, year of implant, and temperature of animal. This method holds great promise as a means of combating cattle theft.

3. **Futuristic electronic device.** The consensus of the livestock industry is that current animal identifications, including electronic devices, are not adequate. There seems to be rather general agreement that the futuristic electronic devices should meet the following needs:

   a. Individual animal identification.
   b. Species adaptation, because different species have different ID needs and problems.
   c. Uniform standardization.
   d. Ease of use and reasonable cost.
   e. Identification of the animal from the farm or ranch to slaughter.
   f. Removable at slaughter.
   g. Improve food safety and the quality of products for the consumer.
   h. Driven largely by economics.

No one expects electronic devices to replace a form of individual identification where cattle are worked with horses.

MARKING PUREBRED CATTLE

Table 10-1, Marking or Identifying Guide for Registered Beef and Dual-Purpose Cattle, summarizes the pertinent regulations of the beef and dual-purpose cattle registry associations relative to marking or identifying.

Breeds not listed in Table 10-1 is indicative that the breed registry associations either did not respond to the authors' request or were not known to exist at the time the authors prepared the seventh edition of *Beef Cattle Science*.

TABLE 10-1
MARKING OR IDENTIFYING GUIDE FOR REGISTERED BEEF AND DUAL-PURPOSE CATTLE

Breed	Association Rules Relative to Marking
Angus	Each animal, for which application for registry is submitted, must be tattooed alike in both ears. Each breeder may devise a special tattooing system, using a series of numbers or letters, or a combination of numbers and letters. Tattoo marks are limited to 4 units in each ear, and only standard numerals or letters are acceptable. Each animal of the same sex to be registered by any one breeder must be tattooed differently.
Ankole-Watusi	All registered animals must be branded by fire or acid indicating the holding brand and private herd number. Brands must be a minimum of 3 in. *(7.6 cm)* on adult animals. No two people can use the same brand. A private herd number must be branded on the animal, with the location given prior to registration. The year of birth must be incorporated in the private herd number; for example, calves born in 1996 must be numbered starting with *6*. When branding of purebred cattle intended for exhibition purposes is not desirable or acceptable, a permit to tattoo must be applied for; and the Association will assign tattoos. Owners must brand all percentage cattle. Two colored pictures taken from one side of the animal are required.
Barzona	The holding brand and private herd number must be fire or freeze branded on the animal before it may be registered as a Barzona. Before a calf is weaned, it must have an ear tattoo if the private herd number brand has not been applied by that time.
Beefalo	Animal must be tattooed in both ears (except where prohibited by state law) with not more than four (4) standard numbers and/or letters.
Beefmaster	The animal must be branded and tattooed at the time of application for registry is made. Application must indicate type and location of identification; and method of ID, with any one of the following being acceptable: fire brand, freeze brand, and/or tattoo, preferably in left ear. Also, it should be indicated what the ranch brand looks like, preferably by submitting a drawing or sketch.
Belted Galloway	A tattoo in either or both ears is required, which must include the year code. Also, breeders are encouraged to use individual herd identifications such as ear tags or chains.
Blonde d'Aquitaine	All animals must be tattooed in the left ear, with their herd ID and individual number.
Braford	Must be permanently identified with brand or tattoo, with location and type of ID indicated on registration certificate.
Brahman	A holding brand (a symbol, a letter, a combination of letters and/or symbols, numerals, a replica of some object, etc., to denote ownership or breeder) and private herd number (both branded by fire) are required on a calf before it may be registered.
Brahmousin	Individual hot brand, or tattoo in either ear.
Brangus	Each animal must be tattooed with an approved prefix or suffix, or branded with a holding brand identifying the breeder and/or the first owner. The Association recommends the International System of identifying. This includes using the alphabet for year born instead of numbers, thereby eliminating the duplication of years every 10 years. The application for registration or enrollment must show where this brand is located on the body. The Association suggests starting with the number 1 on the private herd number and numbering consecutively. The holding brand is any mark, initial, or number, or combination of all three which the breeder chooses to use, and which is approved by the breed registry (IBBA)

(Continued)

TABLE 10-1 (Continued)

Breed	Association Rules Relative to Marking
Braunvieh	All registered animals must be permanently identified, with the location of the tattoo noted. Identification must include Breeder Herd Code Letters (2 or 3 letters assigned by the Association) and letter of year ("E" in 1995).
British Whites	Animals must be tattooed in both ears.
Charolais	Private herd number (tattoo or fire brand) and holding brand (tattoo or fire brand) required. The breeder may devise a special system, using a series of numbers or letters, or a combination of numbers and letters.
Chianina	All animals must be individually tattooed, and the tattoo must have the first owner's prefix and the year letter as the last digit.
Devon	Each breeder is assigned a herd tattoo code of 3 letters, which must be applied to the right ear. The individual herd number plus the letter code for the year of birth must be tattooed in the left ear.
Dexter	Tattoo in either or both ears.
Galloway	Must have ear tattoo.
Gelbvieh	Each calf must be permanently identified with ear tattoo showing breeder's 3-letter herd prefix, a 3- or 4-digit number, and the international year code letter suffix. Example: *RFR 3136C*, where *RFR* stands for *Rocky Ford Ranch*, the 3136 is the number within the individual herd; and the *C* the 1993 year code.
Hays Converter	Animals must be identified by tattoo markings in the ear specified by the Canadian Live Stock Records, with the registered identification letters of the owner of the animal at birth and a serial number followed by the designated year letter to signify year of birth. All calves must be identified by eartag or equivalent method as soon as practical after birth and by tattoo markings before they are 9 months of age and before applying for registration. No two animals may be tattooed alike.
Hereford	Each animal, for which application for registry is submitted, must be tattooed in one ear. Tattooing in both ears is recommended. Many breeders use the year number as the first digit; for example, the 25th calf in 1994 would be 425.
Limousin	All animals must have the first owner's herd prefix (4 letters). Each individual tattoo number must also have a letter at the end indicating the year of birth. The herd prefix may be in one ear and the calf's herd number and year code in the opposite ear.
Maine-Anjou	The permanent identification, brand or tattoo, of an animal must consist of 3 parts: (1) herd prefix—a set of 3 or 4 letters to identify the first owner; (2) animal number—which is an individual identification; and (3) year letter—which represents the year the animal was born.
Marchigiana	All registered animals must have either an ear tattoo or a brand. The Association prefers that breeders also use herd letters, which they will reserve on request. The Association uses the tattoo as part of the animal's registered name; for example, *XYZ Miss Letargo 338*.
Murray Grey	Each animal must be tattooed in the left ear. If member so desires, a corresponding tattoo may also be placed in the right ear. The international year and letter designations are required.
Normande	Animals must be tattooed.
Pinzgauer	Any 5 numbers, or letters and 1 additional letter tattoo character must be inserted in either ear of each animal prior to registration. The last character of the tattoo must be the international letter denoting the year of birth. Tattoo should be unique within the herd.

(Continued)

TABLE 10-1 (Continued)

Breed	Association Rules Relative to Marking
Polled Hereford	Each animal, for which application for registry is submitted, must be identified by a permanent tattoo. The Association recommends 2 tattoos; a herd code in one ear, and numerals indicating the calf's number and a letter indicating the year of birth in the other. Each animal of the same sex registered by any one breeder should be tattooed differently.
Polled Shorthorn	Each animal must have a tattoo number in its ear. No two animals of the same sex in a herd may be tattooed identically.
Red Angus	Calves must be tattooed, not exceeding 4 digits in one ear as follows: Right ear to have owner's assigned (by Sec.-Treas.) letters and last digit of year of animal's date of birth. Left ear to have herd identification number of owner's own system, but the animal must be definitely identified and without duplication in the herd. Special symbols, diagonals, brands, bars, joined letters, etc., cannot be recorded.
Red Brangus	Permanently identified by fire brand, including the owner's or breeder's brand and the animal's private herd number.
Red Poll	Breeders and first owners select their own tattoo marking system. However, identifying numbers must be tattooed in both ears, but either the same or different marks may be used in each ear (if different marks are used, they must be so specified on application for registry). Animals of the same sex and near the same age must be tattooed with a different number, but animals of different sexes and widely different ages may have the same number.
Salers	Calf must have individual tattoo number prior to registration. Only alphabetical letters and/or numbers may be used. Tags may be used to augment tattoos, but they cannot be used to replace them.
Santa Gertrudis	Prior to registration, each animal must be numbered by fire brand, freeze brand, or ear tattoo in both ears so that it can be individually identified, showing the animal's individual number and the year in which it was born. Two or more animals of the same sex may not bear the same number in a given herd for a minimum period of 10 years.
Scotch Highland (or Highland)	Owners must submit herd designation to the registry association, which must have prior approval before use. The herd designation (herd letters) must be tattooed in the ear along with the year of birth and the individual animal number.
Shorthorn	Each animal, for which application for registry is submitted, must be tattooed with a number in one ear. A letter or an initial may or may not precede the number. The application for registry must show whether the tattoo appears in the calf's right or left ear. Duplication of numbers in the same sex and herd is not permissible.
Simbrah	Each animal to have a private and permanent herd number (brand or tattoo), which shall include the international year/letter designation.
Simmental	Each animal to have a private and permanent herd number (brand or tattoo), which shall include the international year/letter designation.
South Devon	Each animal must be tattooed or branded.
Sussex	A designation mark comprised of 3 letters will be allocated by the Association for the exclusive use of each breeder in tattooing calves. A "year letter" denotes the year of birth. Calving season is also indicated by numbers.
Tarentaise	Each animal must be tattooed or branded with the breeder's member/nonmember number or herd brand which has been reported to the Association registry office (Example: *004* or brand *PDR* could be used as the breeder's identification), a year letter designate (Example: *N* for 1981, *P* for 1982, etc.), and a herd number (Example: *1, 2, 3,* etc.). The animal's private herd number (tattoo) can contain a total of 5 digits, plus the international year/letter designation to make a total of 6 characters available for the private herd number (tattoo). An example of a properly identified animal then would be (1) animal with *004* tattooed in right ear and *233N* in left ear, or an (2) animal with *PDR* brand on hip or side and tattoo *233N* in either or both ears.

(Continued)

TABLE 10-1 (Continued)

Breed	Association Rules Relative to Marking
Texas Longhorn	Fire, freeze, or acid brands, showing ownership and private herd number, and location of brand, are required.
Wagyu	All animals must be tattooed with a letter prefix followed by the private herd number. All $^{15}/_{16}$ and greater percentage must be parent verified by DNA to both sire and dam.
Welsh Black	All purebred animals must be ear tattooed. Tattoo shall be in the ear (left or right) as designated by the Association secretary and shall include the following: herd letters, assigned by the secretary; and the herd numbers of the animal, along with the designated year letter to indicate the year of birth. In grading up animals, a brand and an identification number may be used or the animal must be ear tattooed with both the herd letters and an identification number as in the purebred book.
White Park	Animals submitted for registry must be identified in one of two ways: (1) tattoo, with the breeder tattoo always carried in the right ear, and with identical tattoo in the left ear optional; or (2) brand (hot iron), including a holding brand and a herd number.

CLIPPING

Cattle are usually clipped to aid in identification or for show or sale purposes. It is best done with electric clippers.

The most common areas clipped on commercial cattle are the long hairs inside the ears so that ear tags can be seen, the area over a hot brand so that it can be read, and the area over light colored cattle that have been "overbranded" to produce a bald brand (freeze brands on dark cattle do not need clipping). These areas generally need clipping during the winter or just before calving.

Clipping show and sale cattle requires expertise, as well as a mental picture of how the ideal animal should look. Since this subject is fully covered in Chapter 14, Fitting, Showing, and Judging Cattle, in the section headed "Clipping," the reader is referred thereto.

CREEP FEEDING

(See Chapter 21, Feeding and Handling Calves, section on "Creep Feeding.")

WEANING

Early spring calves should be weaned in the fall, preferably before the forage becomes dry or just before moving the breeding herd to the winter range or into winter quarters. Calves are generally weaned when they are 6 to 8 months of age. Usually a cow will wean her calf by the time it reaches 10 to 11 months of age, even if the 2 animals are not separated.

Since the subject of weaning is fully covered in Chapter 21, Feeding and Handling Calves, section on "Weaning," the reader is referred thereto.

WEIGHING

In modern cattle operations, cattle are weighed many times during their lives. They may be weighed at birth; at weaning; at 205 days of age; at 365 days of age; before, during, and after the growing period as stockers; before, during, and after the feedlot finishing period; and/or at marketing time.

Animal weights are important in cow-calf, stocker, and cattle finishing operations. They are a must in performance testing; they can give a big assist in ration changes and herd health programs; they can be used to monitor rate of gain and feed efficiency of stocker and cattle finishing operations; and they can be used in calculating percentage shrink when transporting cattle. Many different kinds of scales may be used to meet these diverse needs.

HOOF TRIMMING

Hoof troubles make animals reticent to walk for feed and water. Hence, regular inspection and care of the feet of all beef cattle is important. It will reduce the incidence of foot rot, lameness, and other foot troubles.

Cattle in confinement tend to grow long toes and build up excessive tissue on the soles of the feet. As a result, more weight is carried by the heels and the hocks, and the pasterns are subjected to extra stress. If these conditions are not corrected by proper hoof trimming, permanent damage may result in the form of crooked legs and weak pasterns, and the productive life of the animal may be lessened.

Young animals should not be neglected. Corrective trimming can be accomplished when animals are young and growing rapidly; for example, the hoofs can be trimmed to correct excessive growth and to rectify pigeon-toed and splay-footed (toe-in or toe-out) conditions.

Herd bulls should receive special attention. The hoofs should be checked and trimmed 1 to 2 months before the breeding season. A well-planned foot trimming program can result in improved reproductive performance.

If the hoofs are trimmed properly, the animal stands squarely and walks properly, with each leg directly under the weight it supports.

The reason for and the technique of trimming hoofs are the same whether an animal is in a commercial or purebred herd, or being fitted for show. Since this subject is fully covered in Chapter 14, Fitting, Showing, and Judging Cattle, the reader is referred thereto.

LOADING AND TRANSPORTING

Modern cattle operations require much movement of cattle from one location to another; animals are moved to and from (1) pasture, (2) a farm or ranch, (3) a livestock show, and/or (4) a terminal market or auction.

Transportation of cattle is expensive; and usually the producer pays the bill, directly or indirectly. In addition to actual transportation costs, there may be hidden costs from shrinkage, bruising, crippling, disease (especially shipping fever), and death. Much of this loss can be alleviated or minimized if proper skills are used in preparing, handling, working, and transporting cattle.

Proper cattle loading and transporting techniques are the same whether animals are being moved to and from pasture, a farm or ranch, a livestock show, or a terminal market or auction. Since this subject is fully covered in Chapter 15, Marketing and Slaughtering Cattle and Calves, the reader is referred thereto.

HARDWARE DISEASE PREVENTION

The term *hardware disease* (traumatic gastritis) is used to describe the condition that results from swallowing foreign materials, usually metal (nails, wire, screws, pins, etc.). Cattle are involved more than other classes of animals. In most cases, the metal is found only in the reticulum (second stomach).

About 7,000 cattle are condemned each year by the Federal Meat Inspection Service as unfit for food because of hardware disease. Clinical reports indicate that the problem is increasing due to the use of more chopped feeds and more contamination. Sharp objects will injure the lining of the stomach and cause infection and inflammation, a condition known as traumatic gastritis.

Hardware disease is a problem in cattle because of their eating habits and stomach arrangement. The usual source of metals is the feed. The animals eat rapidly and are not able to sort foreign objects from their feed.

- **Preventio n**—Avoid foreign objects getting into the feed through good management. Also, install strong magnets, in keeping with the manufacturer's directions, (1) at the outlets of mechanical silo unloaders, and (2) in feed processing equipment.

- **Symptoms**—The most common symptoms are: loss of appetite and digestive disturbance, slow and stiff movement and arched back; elbows that bow outward; decreased rumen movement and chewing; possible diarrhea; tendency to stand with the front feet elevated so as to lessen the pressure of the viscera on the inflamed area; rise in body temperature; and swellings under the jaw, at the brisket, and at the hock joints. Bulls may be reluctant to mate.

- **Treatment**—Powerful magnets may be permanently placed in the cow's second stomach, for the purpose of holding objects that have not penetrated the stomach wall. However, the only sure cure for traumatic gastritis is veterinary surgery. Surgery will be successful only if performed before the condition has progressed to the point that damage has been done to the heart or other organs.

GENTLING CATTLE

Easy does it! Tom Lasater, famed cattleman and founder of the Beefmaster breed, uses the following method to gentle cattle; and it works.

He places up to 12 animals in a relatively small pen where he can walk among and stay fairly close to them. Then, he (1) carries a small bucket half-full of pellets, which he shakes to attract the animals' attention; (2) holds 1 pellet between his thumb and forefinger, and feeds each animal that will take the pellet from him, one by one; (3) repeats this procedure daily, or more often, until the animals get sufficiently gentle that they can be hand-fed and back-rubbed; and (4) *never, never* touches the animals on the forehead or horns, as this will teach them to butt.

The Lasater gentling procedure substitutes the application of animal behavior and human patience for rough handling.

BEDDING CATTLE

Bedding or litter is used primarily for the purposes of keeping animals clean and comfortable. But bedding has the following added values from the standpoint of the manure:

1. It soaks up the urine which contains about ½ the total plant food of manure.

2. It makes manure easier to handle.

3. It absorbs plant nutrients, fixing both ammonia and potash in relatively insoluble forms that protects them against losses by leaching. This characteristic of bedding is especially important in peat moss, but of little significance with sawdust and shavings.

KIND AND AMOUNT OF BEDDING

The kind of bedding material selected should be determined primarily by (1) availability and price, (2) absorptive capacity, and (3) plant nutrient content. In addition, a desirable bedding should not be dusty, should not be excessively coarse, and should remain well in place and not be too readily kicked aside. Table 10-2 summarizes the characteristics of some common bedding materials.

In addition to the bedding materials listed in Table 10-2, many other products can be and are successfully used for this purpose, including leaves of many kinds, tobacco stalks, buckwheat hulls, processed manure (made by separating solid fibers from the liquid and water soluble material in animal wastes), and shredded paper.

Naturally the availability and price per ton of various bedding materials vary from area to area, and from year to year. Thus, in the New England states shavings and sawdust are available, whereas other forms of bedding are scarce, and straws are more plentiful in the central and western states.

Table 10-2 shows that bedding materials differ considerably in their relative capacities to absorb liquid. Also, it is noteworthy that chopped straw will absorb more liquid than long straw. But there are disadvantages to chopping; chopped straws do not stay in place, and they may be dusty.

The suspicion that sawdust or shavings will hurt the land is rather widespread, but unfounded. It is true that these products decompose slowly. But this process can be expedited by the addition of nitrogen fertilizers.

The minimum desirable amount of bedding to use is the amount necessary to absorb completely the liquids in manure. Some helpful guides to the end that this may be accomplished follow:

TABLE 10-2
WATER ABSORPTION CAPACITY OF BEDDING MATERIALS

Material	Lb *(Kg)* of Water per Lb *(Kg)* of Bedding
Barley straw	2.10
Cocoa shells	2.70
Corn stover (shredded)	2.50
Corncobs (crushed or ground)	2.10
Cottonseed hulls	2.50
Flax straw	2.60
Hay (mature, chopped)	3.00
Leaves (broadleaf)	2.00
(pine needles)	1.00
Oat hulls	2.00
Oat straw (long)	2.80
(chopped)	3.75
Peanut hulls	2.50
Peat moss	10.00
Rye straw	2.10
Sand	0.25
Sugarcane bagasse	2.20
Vermiculite[1]	3.50
Wheat straw (long)	2.20
(chopped)	2.95
Wood	
Dry fine bark	2.50
Tanning bark	4.00
Pine chips	3.00
Sawdust	2.50
Shavings	2.00
Needles	1.00
Hardwood chips	1.50
Shavings	1.50
Sawdust	1.50

[1]This is a mica-like mineral mined chiefly in South Carolina and Montana.

1. With 24-hour stabling, the minimum daily bedding requirements, based on uncut wheat or oats straw, of cattle is as follows: cows, 9 lb; steers, 7 to 10 lb. With other bedding materials, these quantities will vary according to their respective absorptive capacities (see Table 10-2). Also, more than these minimum quantities of bedding may be desirable where cleanliness and comfort of the animal are important. Comfortable animals lie down more and utilize a higher proportion of the energy of the feed for productive purposes. (Cattle require 9% less energy when lying down than when standing.)

2. Under average conditions, about 500 lb of bedding is used for each ton of excrement.

3. Where the liquid excrement is collected separately in a cistern or tank—as is the common practice in Denmark, Germany, and France, and on some dairy farms in the United States—less bedding is required than where the liquid and solid excrement are kept together.

Many of today's improved crops are *shorty* varieties, with the result that smaller quantities of stalks and straws are available for bedding. Hand in hand with this transition, confinement animal production has increased. As a result, bedding materials have become scarce and high in price. This situation is prompting interest in slatted (slotted) floors and other means of lessening or alleviating bedding.

CATTLE MANURE

The term *manure* refers to *a mixture of animal excrements (consisting of undigested feeds plus certain body wastes) and bedding.* Increased confinement production has made for manure disposal problems. This has been particularly acute in most cattle feedlots. As a result, in many areas manure is unwanted. Worse yet, it may be looked upon as a foul-smelling, fly-breeding, dusty, unattractive centerpiece in a feedlot. In due time, however, science and technology will evolve with new methods of using and handling manure; and we shall learn to live with it. From the standpoint of soils and crops, barnyard manure contains the following valuable ingredients:

1. **Organic matter.** It supplies valuable organic matter which cannot be secured in commercial fertilizers. Organic matter—which constitutes 3 to 6%, by weight, of most soils—improves soil tilth, increases water-holding capacity, lessens water and wind erosion, improves aeration, and has a beneficial effect on soil microorganisms. It is the *lifeblood* of the land.

Fig. 10-17. Colorado feedlot showing heavy manure production. (Courtesy, The Great Western Sugar Co.)

2. **Plant food.** It supplies plant food or fertility—especially nitrogen, phosphorus, and potassium. In addition to these 3 nutrients, manure contains organic matter, calcium, and trace elements such as boron, manganese, copper, and zinc. A ton of well-preserved manure, including bedding, contains plant food nutrients equal to about 100 lb of 10-5-10 fertilizer. Thus, when manure is spread at the rate of 8 tons per acre, it is like applying 800 lb of 10-5-10 commercial fertilizer.

Modern beef cattle facilities and equipment should be designed to handle the manure produced by the animals which they serve; and this should be done efficiently, with a minimum of labor and pollution, so as to retrieve the maximum value of the manure, and make for maximum animal sanitation and comfort.

DAILY MANURE PRODUCTION AND STORAGE

Table 10-3 shows the approximate daily manure production of each class of animals.

TABLE 10-3
APPROXIMATE DAILY MANURE PRODUCTION[1]

Animal	Solids and Liquids/Day[2]		Liquids/Day[3]	
	(cu ft)	*(m^3)*	*(gal)*	*(l)*
1,000-lb *(454-kg)* cow	1.50	*0.042*	11.0	*46.3*
1,000-lb *(454-kg)* steer	1.00	*0.028*	7.5	*28.4*
1,000-lb *(454-kg)* horse	0.75	*0.021*	5.5	*20.8*
10 head of hogs:				
50 lb *(22.7 kg)*	0.67	*0.018*	5.0	*19.0*
100 lb *(45.4 kg)*	1.33	*0.037*	10.0	*37.9*
150 lb *(68 kg)*	2.25	*0.063*	17.0	*64.4*
200 lb *(90.9 kg)*	2.75	*0.077*	20.5	*77.7*
250 lb *(113.6 kg)*	3.50	*0.098*	26.0	*98.5*
10 head of sheep	0.50	*0.014*	4.0	*15.2*
1,000 5-lb *(2.2-kg)* layers	3.00	*0.084*	22.5	*85.3*

[1]Adapted by the authors from Michigan State University Circ. Bull. 231.

[2]There are about 34 cu ft in a ton of manure.

[3]One cu ft = 7.5 gal.

AMOUNT, COMPOSITION, AND VALUE OF MANURE

The quantity, composition, and value of manure produced vary according to species, weight, kind and amount of feed, and kind and amount of bedding. The authors' computations are on a fresh manure (exclusive of bedding) basis. Table 10-4 presents data by species per 1,000 lb liveweight, whereas Table 10-5 gives yearly tonnage and value.

The data in Table 10-4 and Fig. 10-18 are based on animals confined to stalls the year around. Actually, the manure recovered and available to spread where desired is considerably less than indicated because (1) animals are kept on pasture and along roads and lanes much of the year, where the manure is dropped, and (2) losses in weight often run as high as 60% when manure is exposed to the weather for a considerable time.

As shown in Fig. 10-19, about 75% of the nitrogen, 80% of the phosphorus, and 85% of the potassium contained in animal feeds are returned as manure. In addition, about 40% of the organic matter in feeds is excreted as manure. As a rule of thumb, it is commonly estimated that 80% of the total nutrients in feeds are excreted by animals as manure.

Naturally, it follows that the manure from well-fed animals is higher in nutrients and worth more than that from poorly fed ones. For example, the manure produced from steers liberally fed on nutritious concentrates is more valuable than that produced from cattle wintered on hay.

Although varying with (1) the kind of feed, (2) the class of animal, (3) the age, condition, and individuality of the animal, and (4) the kind and amount of bedding

TABLE 10-4
QUANTITY, COMPOSITION, AND VALUE OF FRESH MANURE (FREE OF BEDDING) EXCRETED BY 1,000 LB *(454 KG)* LIVEWEIGHT OF VARIOUS KINDS OF FARM ANIMALS

		Composition and Value of Manure on a Tonnage Basis[2]										
(1)	(2)	(3)	(4)		(5)	(6)		(7)		(8)		(9)
Animal	Tons Excreted/ Year/1,000 Lb *(454 Kg)* Liveweight[1]	Excrement	Lb or *Kg*/Ton[3]		Water	N		P_2O_5[4]		K_2O[4]		Value/ Ton[5]
	(tons)		*(lb)*	*(kg)*	*(%)*	*(lb)*	*(kg)*	*(lb)*	*(kg)*	*(lb)*	*(kg)*	*($)*
Cow (beef or dairy)	12.0	Liquid	600	*300*	79	11.2	*5.1*	4.6	*2.1*	12.0	*5.5*	5.99
		Solid	1,400	*700*								
		Total	2,000	*1,000*								
Steer (finishing cattle)	8.5	Liquid	600	*300*	80	14.0	*6.4*	9.2	*4.2*	10.8	*4.9*	7.64
		Solid	1,400	*700*								
		Total	2,000	*1,000*								
Sheep	6.0	Liquid	660	*352*	65	28.0	*12.7*	9.6	*4.4*	24.0	*10.9*	13.48
		Solid	1,340	*648*								
		Total	2,000	*1,000*								
Swine	16.0	Liquid	800	*400*	75	10.0	*4.5*	6.4	*2.9*	9.1	*4.2*	5.65
		Solid	1,200	*600*								
		Total	2,000	*1,000*								
Horse	8.0	Liquid	400	*200*	60	13.8	*6.3*	4.6	*2.1*	14.4	*6.5*	7.05
		Solid	1,600	*800*								
		Total	2,000	*1,000*								
Poultry	4.5	Total	2,000	*1,000*	54	31.2	*14.2*	18.4	*8.4*	8.4	*3.8*	13.83

[1]*Manure is Worth Money—It Deserves Good Care*, University of Illinois Circ. 595, 1953, p. 4.

[2]Columns 5, 6, 7, and 8 from *Farm Manures*, University of Kentucky Circ. 593, 1964, p. 5, Table 2.

[3]From *Reference Material for 1951 Saddle and Sirloin Essay Contest*, compiled by M. E. Ensminger, p. 43: data from *Fertilizers and Crop Production*, by Van Slyke, published by Orange Judd Publishing Co.

[4]P_2O_5 can be converted to phosphorus (P) by dividing the figure given above by 2.29, and K_2O can be converted to potassium (K) by dividing by 1.2.

[5]Calculated on the assumption that nitrogen (N) retails at 25¢, P_2O_5 at 25¢, and K_2O at 17¢ per pound in commercial fertilizers.

TABLE 10-5
TONNAGE AND VALUE OF MANURE (EXCLUSIVE OF BEDDING) EXCRETED BY U.S. LIVESTOCK[1]

Class of Livestock	Number of Animals on Farms[2]	Average Liveweight		Tons Manure Excreted/ Year/1,000 Lb *(454 Kg)* Liveweight[3]	Total Manure Production	Total Value of Manure[4]
	(no.)	*(lb)*	*(kg)*	*(tons)*	*(tons)*	*($)*
Cattle (beef and dairy; including steers) . . .	100,892,000	900	*409*	11	998,830,800	5,982,996,492
Sheep	10,191,000	100	*45*	6	6,114,600	82,424,808
Swine	54,477,000	200	*90*	16	174,326,400	984,944,160
Chickens—layers	281,644,000	4.5	*2*	4.5	281,644	3,895,137
—broilers	6,388,990,000	3.5	*1.6*	4.5	100,626,593	1,391,665,781
Turkeys	288,980,000	22	*10*	4.5	28,609,020	395,662,747
Horses	10,840,000	1,000	*454*	8	86,720,000	611,376,000
Total					1,307,509,057	9,452,965,125

[1]In these computations, no provision was made for animals that died or were slaughtered during the year. Rather, it was assumed that their places were taken by younger animals, and that the population of each species was stable throughout the year. The total value of manure for all cattle (beef, dairy, and steers) was computed on the basis of $5.99 per ton.

[2]*Agricultural Statistics 1993*, USDA, pp. 229, 247, 327, 333, 338.

[3]*Manure is Worth Money—It Deserves Good Care*, University of Illinois Circ. 595, 1953, p. 4.

[4]Computed on the basis of the value per ton given in the right-hand column of Table 10-4.

Fig. 10-18. Manure is a big *crop*! On the average, each class of stall-confined animals produces per year per 1,000-lb weight the tonnages shown above.

1,000 BUSHELS OF CORN CONTAIN:	ANIMALS RETAIN:	RETURNED IN MANURE:
1,000 LB N	250 LB N	750 LB N
170 LB P	34 LB P	136 LB P
190 LB K	19 LB K	171 LB K

Fig. 10-19. Animals retain about 20% of the nutrients in feed; the rest is excreted in manure.

used, a ton of *fresh* barnyard manure has approximately the composition shown in Fig. 10-20.

Table 10-4 gives the nutrients, by classes of animals in 1 ton of fresh manure. As shown, in terms of nutrients contained, chicken manure is the most valuable. Sheep manure ranks second, and steer manure third. Cow manure ranks lower than steer manure, because cows are fed largely on roughages, rather than concentrates.

The urine makes up 20% of the total weight of the excrement of horses, and 40% of that of hogs; these figures represent the two extremes in farm animals. Yet the urine, or liquid manure, contains nearly 50% of the nitrogen, 6% of the phosphorus, and 60% of the potassium of average manure; roughly one-half the total plant food of manure (see Fig. 10-21). Also, it is noteworthy that the nutrients in liquid manure are more readily available to plants than the nutrients in the solid excrement. These are the reasons why it is important to conserve the urine.

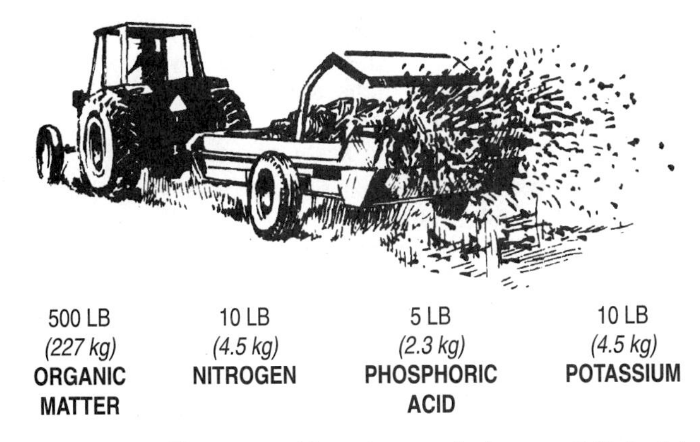

Fig. 10-20. The contents of 1 ton of average *fresh* manure. Cattle feedlot manure is much more nutritious; on a per ton basis, it contains about 27 lb of nitrogen, 24 lb of P_2O_5, and 36 lb of K_2O.

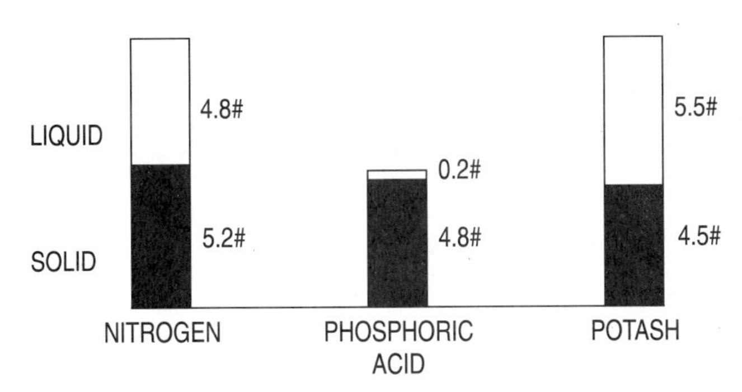

Fig. 10-21. Distribution of plant nutrients between liquid and solid portions of a ton of average farm manure. As noted, the urine contains about half the fertility value of manure.

MANURE GASES

When stored inside a building, gases from liquid wastes create a hazard and undesirable odors. Most (95% or more) of the gas produced by manure decomposition is methane, ammonia, hydrogen sulfide, and carbon dioxide. Several have undesirable odors or possible animal toxicity, and some promote corrosion of equipment. Table 10-6 gives some properties of the more abundant gases.

Animals and people can be killed (asphyxiated) because methane and carbon dioxide displace oxygen.

Most gas problems occur when manure is agitated or when ventilation fans fail.

No one should enter a storage tank, unless (1) the space over the wastes is first ventilated with a fan, (2) another person is standing by to give assistance if needed, and (3) they are wearing self-contained breathing equipment—the kind used for fire fighting or scuba diving.

It is important that maximum building ventilation be provided when agitating or pumping wastes from a pit. Also, an alarm system (loud bell) to warn of power failures in tightly enclosed buildings is important, because there can be a rapid buildup of gases when forced ventilation ceases.

TABLE 10-6
PROPERTIES OF THE MORE ABUNDANT MANURE GASES[1]

Gas	Weight Air = 1	Physiologic Effect	Other Properties
CH_4 Methane	½	Anesthetic	Odorless, explosive
NH_3 Ammonia	⅔	Irritant	Strong odor, corrosive
H_2S Hydrogen sulfide	1+	Poison	Rotten-egg odor, corrosive
CO_2 Carbon dioxide	1⅓	Asphyxiant	Odorless, mildly corrosive

[1]*Beef Housing and Equipment Handbook*, MWPS-6, Midwest Plan Service, Ames, Iowa, p. 10.

VALUE OF MANURE AS A FERTILIZER

The actual monetary value of manure can and should be based on (1) equivalent cost of a like amount of commercial fertilizer, and (2) increased crop yields. Tables 10-4 and 10-5 give the equivalent cost of a like amount of commercial fertilizer. Numerous experiments and practical observations have shown the measurable monetary value of manure in increased crop yields.

When the senior author was a boy on a Missouri farm, we fed livestock to produce manure, to grow more crops to feed more livestock, to produce more manure. But things changed! The use of chemical fertilizer expanded many fold; labor costs rose to the point where it was costly to conserve and spread manure on the land; more animals were raised in confinement; and a predominantly urban population didn't appreciate what they referred to as "foul-smelling, fly-breeding stuff." As a result, what to do with manure became a major problem on many livestock establishments.

Currently, we are producing manure (exclusive of bedding) at the rate of 1.3 billion tons annually (see Table 10-5). That's sufficient manure to add two-thirds of a ton each year to every acre of the total land area (1.9 billion acres) of the 48 contiguous states of the United States.

Based on equivalent fertilizer prices (see Table 10-4, right-hand column), the yearly manure crop is worth \$9.4 billion (Table 10-5). That's a potential annual income of \$4,571 for each of the nation's 2.1 million farms.

The value of manure varies according to (1) the

class of animals, (2) the kind of feed consumed and the kind of bedding used, (3) the method of handling, (4) the rate and method of application, and (5) the kind of soil and crops on which it is used.

Of course, the value of manure cannot be measured alone in terms of increased crop yields and equivalent cost of a like amount of commercial fertilizer. It has additional value for the organic matter which it contains, which almost all soils need, and which farmers and ranchers cannot buy in a sack or tank.

Also, it is noteworthy that, due to the slower availability of its nitrogen and to its contribution to the soil humus, manure produces rather lasting benefits which may continue for many years. Approximately one-half of the plant nutrients in manure are available to and effective upon the crops in the immediate cycle of the rotation to which the application is made. Of the unused remainder, about one-half in turn, is taken up by the crops in the second cycle of the rotation, one-half of the remainder in the third cycle, etc. Likewise, the continuous use of manure through several rounds of a rotation builds up a backlog which brings additional benefits, and a measurable climb in yield levels.

Cattle producers sometimes fail to recognize the value of this barnyard crop because (1) it is produced whether or not it is wanted, and (2) it is available without cost. Most of all, no one is selling it. Who ever heard of a traveling manure salesperson?

MODERN WAYS OF HANDLING MANURE

Modern handling of manure involves maximum automation and a minimum loss of nutrients. Among the methods being used, with varying degrees of success, are: slotted floors emptying or pumping into irrigation systems; storage vats; spreaders (including those designed to handle liquids alone or liquids and solids together); dehydration; power loaders; conveyors; industrial-type vacuums; lagoons; and oxidation ditches. Actually, there is no one best manure management system for all situations; rather, it is a matter of designing and using that system which will be most practical for a particular set of conditions.

PRECAUTIONS WHEN USING MANURE AS FERTILIZER

The following precautions should be observed when returning manure to the land:

1. Avoid applying waste closer than 100 ft to waterways, streams, lakes, wells, springs, or ponds.

2. Do not apply where downward movement of water is not good, or where irrigation water is very salty or inadequate to move salts down.

3. Incorporate (preferably by plowing or disking) manure into the soil as quickly as possible after application. This will maximize nutrient conservation, reduce odors, and minimize runoff pollution.

4. Distribute the waste as uniformly as possible on the area to be covered.

5. Irrigate thoroughly to leach excess salts below the root zone.

6. Allow about a month after irrigation before planting, to enable soil microorganisms to begin decomposition of manure.

7. Minimize odor problems by—

   a. Spreading raw manure frequently, especially during the summer.

   b. Spreading early in the day as the air is warming up, rather than late in the day when the air is cooling.

   c. Not spreading on days when the wind is blowing toward populated areas.

HOW MUCH MANURE CAN BE APPLIED TO THE LAND?

With today's heavy animal concentration in one location, the question is being asked: How can high rates of manure be applied to the land without depressing crop yields, making for salt problems in the soil, making for nitrate problems in the feed, contributing excess nitrate to groundwater or surface streams, or violating state regulations?

Based on earlier studies in the Midwest, before the rise of commercial fertilizers, it would appear that one can apply from 5 to 20 tons of manure/acre, year after year, with benefit.

Heavier applications can be made, but probably should not be repeated every year. With higher rates than 20 tons per year, there may be excess salt and nitrate buildup. Excess nitrate from manure can pollute streams and groundwater and result in toxic levels of nitrate in crops. Without doubt the maximum rate at which manure can be applied to the land will vary widely according to soil type, rainfall, and temperature.

State regulations differ in limiting the rate of manure application. Missouri draws the line at 30 tons/acre on pasture, and 40 tons/acre on cropland. Indiana limits manure application according to the amount of nitrogen applied, with the maximum limit set at 225 lb/acre per year. Nebraska requires only ½ acre of land for liquid manure disposal per acre of feedlot, which appears to be the least acreage for manure disposal required by any state.

One of the big problems in applying animal waste

to the land as a fertilizer is knowing the plant nutrient content of the material. If this is known, the amount of manure necessary to supply the needed nutrients can be added. So, representative samples of the manure should be analyzed for nitrogen, phosphorus, potash, and moisture content. Then the application rates should be based on soil tests, crop requirements, and composition of the manure sample. This relatively inexpensive procedure will avoid errors in application rate.

The amount of manure to be applied can usually be geared to the amount of nitrogen that the crop needs. Thus, if 150 lb of available nitrogen would be adequate for maximum crop production, the manure containing 300 lb of total nitrogen should be applied for the first year of use (twice the amount of nitrogen needed the first year is applied, because only half the nitrogen is available the first year).

When farmers have sufficient land, they should use rates of manure which supply only the nutrients needed by the crop rather than the maximum possible amounts suggested for pollution control.

OBJECTIONABLE FEATURES OF MANURE AS A FERTILIZER

Despite the recognized virtues of manure as a fertilizer, it does possess the following objectionable features:

1. **It is costly to haul and apply.** In the past, a main deterrent to the use of manure as a fertilizer has been the high cost of hauling and applying. But high priced chemical fertilizers have made manure more attractive.

2. **It can create salt (sodium chloride) problems.** Because manure contains appreciable quantities of salt, excessive applications of animal waste can result in salt accumulations in the soil, which will harm crops.

3. **It can create nitrate problems.** Excessive nitrogen in manure can result in excessive nitrates in the groundwater, which can be hazardous for animal or human consumption. Normally, the nitrate content of manure is low, but microorganisms rapidly convert ammonium to nitrate when conditions are favorable for crop growth. Nitrate is the mobile form of nitrogen-the form usually associated with pollution.

4. **It can create phosphorus problems.** Excessive phosphorus can interfere with zinc nutrition of crops and pollute streams.

5. **It can create potassium problems.** The high potassium content of manure tends to reduce the absorption of magnesium and calcium from the soil by plants and results in a greater possibility of grass tetany because of the lowered magnesium content of the forage.

6. **It may propagate insects.** Unless precautions are taken, often manure is the preferred breeding place for flies and other insects. It is noteworthy, however, that comparatively few houseflies are reared in cow manure.

7. **It may spread diseases and parasites.** Where animals are allowed to come in contact with their own excrement there is always danger of infections from diseases and parasites.

8. **It may produce undesirable odors and dust.** Where manure is stored improperly, there may be a nuisance from odors and dust.

9. **It may scatter weed seeds.** Even when fermented, manure usually contains a certain quantity of viable weed seeds which may be scattered over the land.

MANURE USES OTHER THAN FERTILIZER

Today, some manure is being recycled as a livestock feed. Various processing methods are being employed; some are even feeding manure without processing. Further experimentation and Food and Drug Administration approval will be required before the use of manure-feeds becomes widespread; but some researchers predict that eventually wastes may supply up to 20% of the nation's livestock feed, thereby freeing an equivalent amount of grain for human consumption.

Manure is also being recycled for bedding material; the solids are separated from the liquids, then it is dried. Several large dairies are processing manure for this purpose.

Manure may also serve as a source of energy, which, of course, is not new. The pioneers burned dried bison dung, which they dubbed "buffalo chips," to heat their sod shanties. In this century, methane from manure has been used for power in European farm hamlets when natural gas was hard to get. While the costs of constructing plants to produce energy from manure on a large-scale basis may be high, some energy specialists feel that a prolonged fuel shortage will make such plants economical. India now has about 10,000 anaerobic digestion plants in operation. There is nothing new or mysterious about this process. Sanitary engineers have long known that a family of bacteria produces methane when they ferment organic matter under strictly anaerobic conditions. (The senior author's granddad called it swamp gas; his city cousin called it sewer gas.) However, it should be added that both capital and technical resources are needed for the production of methane from manure. If all animal manure were converted to energy, it has been estimated that it would produce energy equal to 10% of the petroleum requirements or 12½% of our natural gas requirements.

Researchers have come up with ways in which to combine manure with broken glass to produce bricks, decorative and roofing tiles, wall core material, paint, and garden stones.

INTEGRATED RESOURCE MANAGEMENT (IRM)

Modern beef cattle production is big and important business. Therefore, it should be conducted in a businesslike manner. The National Cattlemen's Beef Association (NCBA) recognizes (1) that the cattle industry is a competitive business; (2) that the key to good business decisions is to have complete and accurate records of actual production, as well as the costs associated with that production; and (3) that the records can then be used for enterprise analysis, and to evaluate profit and loss areas of the current year, as well as monitor progress over longer periods of time.

Out of the above thinking, the NCBA evolved with the following detailed reports from which all cattle producers may benefit:

- IRM Desk Record for the Cow-Calf Enterprise
- Cow-Calf Standardized Performance Analysis (SPA)
- Seedstock Beef Cattle (SPA)
- Stocker-Feeder Program (SPA)

The analysis of the financial condition and the performance of the cattle enterprise is dependent upon keeping good records and preparing accurate financial statements. The NCBA records are designed for this purpose. These reports/forms may be secured from the NCBA at the following address and phone number:

National Cattlemen's Beef Association
5420 South Quebec Street
P.O. Box 3469
Englewood, CO 80155
Phone: (303) 694-0305
FAX: (303) 694-2851

Fig. 10-22. Good management calls for matching cow size and milk production to the feed and environment, and for matching the bull to the market. (Courtesy, American Hereford Assn., Kansas City, MO)

QUESTIONS FOR STUDY AND DISCUSSION

1. Under ordinary conditions, operators fail to detect 25 to 50% of the heat periods. What are the consequences of this failure, and how would you recommend that it be rectified?

2. List and discuss each of the heat detection methods.

3. What are the advantages and disadvantages of each (a) spring calving, (b) fall calving, and (c) daytime calving?

4. Under what circumstances may it be necessary that a beef cow be milked?

5. By what method and at what age and season would you dehorn cattle in your area? Give reasons for your answers.

6. By what method and at what age and season would you castrate cattle in your area? Give reasons for your answers.

7. What did the California Station find relative to the effects of castrating healthy calves, on the basis of a 28-day study?

8. Do you feel that injected chemical castration will be widely used in the future? Justify your answer.

9. Are there any circumstances under which you would recommend that heifers be spayed? Justify your answer.

10. Discuss the National Beef Audit findings and recommendations relative to hot-iron branding. Where cattle are worked with horses, will electronic branding replace traditional hot-iron branding?

11. Should all beef and dual-purpose breed registry associations use the same standardized rules relative to marking?

12. At what times during their lives are cattle commonly weighed?

13. Why is foot trimming important?

14. In addition to actual transportation costs as such, what hidden costs may be involved in transporting cattle?

15. Why is there more and more hardware disease? What can be done to lessen it?

16. Discuss the Lasater method of gentling cattle.

17. What type of cattle bedding is commonly used on your farm or ranch (or on a farm or ranch with which you are familiar)? Would some other type of bedding be more practical? If so, why?

18. For your farm or ranch (or one with which you are familiar) is it preferable and practical to apply manure to the land or should commercial fertilizers be used instead?

19. How would you recommend that manure be handled?

20. What precautions should be taken when using manure as a fertilizer?

21. What tonnage per acre of manure can be applied to the land?

22. What nonfertilizer use for manure offers the most profit potential?

23. India has 10,000 anaerobic digestion plants in operation to produce methane gas, whereas there are few such facilities in the United States. Why the difference?

24. What is Integrated Resource Management (IRM)? How may a cattle producer benefit from IRM?

SELECTED REFERENCES

Title of Publication	Author(s)	Publisher
Beef Cattle, Seventh Edition	A. L. Neumann	John Wiley & Sons, Inc., New York, NY, 1977
Beef Cattle Production	K. A. Wagnon R. Albaugh G. H. Hart	The Macmillan Company, New York, NY, 1960
Beef Cattle Science Handbook	Ed. by M. E. Ensminger	Agriservices Foundation, Clovis, CA, pub. annually 1964–1981
Beef Production and Management Decisions, Second Edition	R. E. Taylor	Macmillan Publishing Co., New York, NY, 1994
Beef Production and the Beef Industry	R. E. Taylor	Burgess Publishing Company, Minneapolis, MN, 1984
Beef Production in the South, Modified Edition	S. H. Fowler	The Interstate Printers & Publishers, Inc., Danville, IL, 1979
Commercial Beef Cattle Production, Second Edition	Ed. by C. C. O'Mary I. A. Dyer	Lea & Febiger, Philadelphia, PA, 1978
Handbook of Livestock Management Techniques	R. A. Battaglia V. B. Mayrose	Burgess Publishing Company, Minneapolis, MN, 1981
Problems and Practices of American Cattlemen, Wash. Ag. Exp. Sta. Bull. 562	M. E. Ensminger M. W. Galgan W. L. Slocum	Washington State University, Pullman, WA, 1955
Stockman's Handbook, The, Seventh Edition	M. E. Ensminger	Interstate Publishers, Inc., Danville, IL, 1992

Up, up and away! A heifer lunges from the squeeze chute after treatment by a veterinarian. (Courtesy, USDA)

BUILDINGS AND EQUIPMENT FOR BEEF CATTLE

Contents *Page*

Contents Page

The economical production of beef cattle in most sections of the United States depends largely upon the investment in practical, durable, and convenient buildings and equipment, as well as upon the care, feeding, and management of the herd. As would be expected in a country so large and diverse as the United States, there are wide differences in the system of beef production. In a broad general way, a major difference in management exists between the farm herd method and the range cattle method. In addition, further management differences exist within each area according to whether the enterprise is commercial or purebred, whether it is a cow-calf proposition or devoted to one of the many methods of growing stockers and feeders or cattle for finishing, or whether it is a combination of two or more of these systems of beef production. Climatic differences also vary, all the way from nearly year-round grazing in the deep South to a long winter-feeding period in the northern part of the United States. Then, too, the size of the herd may vary all the way from a few animals up to an operation involving many thousands of head. Finally, there is the matter of availability of materials and labor and individual preferences.

Fig. 11-1. Home on the range. Attractive ranch headquarters near Lowry, South Dakota, showing beef cattle grazing western wheatgrass in the foreground. (Courtesy, USDA; photo by Erwin Cole)

Except for the classic experiment conducted by the University of Missouri, little experimental work has been done on the basic building requirements of beef cattle.

Brody, of the Missouri Station,[1] placed cattle in *climatic chambers*, in which the temperature, humidity, and air movements were regulated as desired. The ability of the animals to withstand different temperatures was then determined by studying the respiration rate and body temperature, the feed consumption, and the productivity in growth, milk, beef, etc. Out of these experiments came the following pertinent findings: The most comfortable temperature for Shorthorns was within the range of 30 to 60°F; for the Brahmans it was 50 to 80°F; and for the Santa Gertrudis (5/8 Shorthorn X 3/8 Brahman) it was intermediate between the ideal temperatures given above. Translated into practicality, the Missouri experiment proved what is generally suspected; namely:

1. That there are breed differences; that Brahman and Santa Gertrudis cattle can tolerate more heat than the European breeds—thus, they are better equipped to withstand tropical and subtropical temperatures.
2. That acclimated European cattle do not need expensive, warm barns—they merely need protection from wind, snow, and rain.
3. That more attention needs to be given to providing summer shades and other devices to assure warm weather comfort for cattle.

The basic building and equipment requirements for beef cattle include: (1) some type of protection from weather; (2) adequate water supply and watering equipment; (3) feed storage, and possibly some equipment for feed processing and conveying; (4) feeding equipment; and (5) waste handling.

---

[1]Brody, Samuel, *Climate Physiology of Cattle*, Jour. Series No. 1607, Mo. Ag. Exp. Sta.

No standard set of buildings and equipment can be expected to be adapted to all the diverse conditions and systems of beef production. In presenting the discussion and illustrations that follow, it is intended, therefore, that they be considered as guides only. Detailed plans and specifications for buildings and equipment can usually be obtained through the local county agricultural agent, FFA instructor, or lumber dealer, or through writing the college of agriculture in the state.

The right kind of beef cattle buildings and equipment can materially lower the work required to do the job. Although beef cattle normally require a relatively small amount of labor, a great deal can be done to lower costs and shorten the hours of work.

WEATHER INFLUENCE ON BEEF CATTLE PERFORMANCE

Extreme weather can cause wide fluctuations in cattle performance. The research data clearly show that winter shelters and summer shades almost always improve gains and feed efficiency. However, the additional costs incurred by shelters have frequently exceeded the benefits gained by improved performance. For this reason, except for extreme weather stress of long duration, it becomes a question of whether to shelter cattle at all.

The effects of winter shelter have been widely tested in the Midwest. Table 11-1 shows (1) the reductions in performance, and (2) the feed cost increases resulting from lack of shelter in 13 separate tests at various research stations.

For the 13 trials summarized in Table 11-1, no shelter decreased the average daily gain by 12% and increased feed cost by 14%. The spreads of 4 to 22% decrease in daily gain and 4 to 28% increase in feed cost are typical of fluctuations that occur from one year to the next. Of course, the dollars and cents benefits to accrue from cattle shelters should be based on the added value of the cattle gains and the decreased feed cost.

Summer shade has resulted in less improvement in performance of cattle than winter shelter. Table 11-2 shows the results of eight trials.

For the Plains and the Midwest, it is doubtful that summer shade will pay.

Studies in the desert Southwest demonstrate a greater advantage to shade than the results presented in Table 11-2. In that section of the country, it would appear that shade is not only a paying proposition but a necessity.

TABLE 11-1
EFFECT OF NO SHELTER ON WINTER CATTLE PERFORMANCE[1]

Station	Trial No.	Decrease in Average Daily Gain	Increase in Feed Cost
		(%)	*(%)*
Iowa	1	18	21
Iowa	2	15	17
Iowa	3	15	15
Iowa	4	11	8
Iowa	5	13	14
Iowa	6	9	7
Ohio	7	2	8
Michigan	8	14	23
Michigan	9	10	14
Kansas	10	22	28
South Dakota	11	4	4
Connecticut	12	7	12
Saskatchewan	13	15	14
Average of 13 trials		12	14

[1]*Doane's Agricultural Report*, Vol. 33, No. 45-5.

TABLE 11-2
EFFECT OF NO SUMMER SHADE ON CATTLE PERFORMANCE[1]

Station	Trial No.	Decrease in Average Daily Gain	Increase in Feed Cost
		(%)	*(%)*
Iowa	1	6	4
Iowa	2	7	7
Iowa	3	7	5
Iowa	4	4	1
Iowa	5	3	3
Kansas	6	7	8
South Dakota	7	2	5
Ohio	8	7	2
Average of 8 trials		5	4

[1]*Doane's Agricultural Report*, Vol. 33, No. 45-5.

ENVIRONMENTAL CONTROL OF BEEF CATTLE

Environment may be defined as all the conditions, circumstances, and influences surrounding and affecting the growth, development, and production of a living thing. In beef cattle, this includes the air temperature, relative humidity, air velocity, wet bedding, dust, light, ammonia buildup, odors, and space requirements. Control or modification of these factors offers possibilities for improving cattle performance. There is still much to be learned about environmental control, but the gap between awareness and application is becoming smaller.

Cattle producers were little concerned with the effect of environment on beef cattle so long as they grazed on pastures or ranges. Space requirements, wet bedding, ammonia buildup, odors, and manure disposal were no problem. But the concentration of cattle into smaller spaces changed all this. With the shift to confinement structures and high density production operations, building design became more critical.

In addition to improved performance, the primary reasons cited for increased confinement housing of cattle are: (1) saving in land cost, and (2) saving in labor.

Before an environmental system can be designed for beef cattle, it is important to know (1) their heat production, (2) vapor production, and (3) space requirements. This information is as pertinent to designing beef cattle buildings as nutrient requirements are to balancing rations.

The tables accompanying the three sections that follow cover all classes of livestock (1) for comparative purposes, and (2) because most cattle producers have more than one species.

(Also see Chapter 36, Beef Cattle Behavior and Environment.)

HEAT PRODUCTION OF ANIMALS

The heat production of cattle and calves is given in Table 11-3, along with the heat production of other classes of animals for comparative purposes. Table

TABLE 11-3
HEAT PRODUCTION OF ANIMALS[1]

Heat Source	Unit		Heat Production			Heat Production		
			Temperature	Total	Sensible	Temperature	Total	Sensible
	(lb)	*(kg)*	*(°F)*	*(Btu/hr)*	*(Btu/hr)*	*(°C)*	*(kcal/hr)*	*(kcal/hr)*
Cow	1,000	*453.6*	40	3,600	2,640	*4*	*907.2*	*665.3*
			70	3,000	1,550	*21*	*756.0*	*390.6*
Calves (6–10 months)	—	—	60	780	660	*16*	*196.6*	*166.3*
			80	720	420	*27*	*181.4*	*105.8*
Hog:								
Sow and litter (3 weeks after farrowing)	400	*181.4*	—	2,000	1,000	—	*504.0*	*252.0*
Finishing	200	*90.7*	35	860	740	*2*	*216.7*	*186.5*
			70	610	435	*21*	*153.7*	*109.6*
Layer hen	4.5	*2.04*	50	40	28	*10*	*10.1*	*7.1*
Sheep	100	*45.4*		(0.039-in. fleece length)			*(0.1-cm fleece length)*	
			45	560	500	*7*	*141.1*	*126.0*
			70	320	245	*21*	*80.6*	*61.7*
				(3.937-in. fleece length)			*(10.0-cm fleece length)*	
			45	245	185	*7*	*61.7*	*46.6*
			70	260	125	*21*	*65.5*	*31.5*
Horse	1,000	*454.0*	70	1,800–2,500[2]	—	*21*	*453.6–630*	—

[1]Adapted by the authors from *Agricultural Engineers Yearbook*, St. Joseph, MI, ASAE Data Sheet D-249.2, p. 424, except for horse. Data for horse from *Farm Buildings*, by John C. Wooley, McGraw-Hill Book Company, Inc., Table 24, p. 140.

[2]Armsby and Kriss, in a paper entitled, "Some Fundamentals of Stable Ventilation," published in the *Journal of Agricultural Research*, Vol. 21, June 1921, p. 343, list the total heat output as follows: a 1,000-lb horse, 1,500 Btu per hour; a 1,500-lb horse, 2,450 Btu per hour.

11-3 may be used as a guide, but in doing so, consideration should be given to the fact that heat production varies with age, body weight, ration, breed, activity, house temperature, and humidity at high temperatures. As noted, Table 11-3 gives both total heat production and sensible heat production. Total heat production includes both sensible heat and latent heat combined. Latent heat refers to the energy involved in a change of state and cannot be measured with a thermometer; evaporation of water or respired moisture from the lungs are examples. Sensible heat is that portion of the total heat, measurable with a thermometer, that can be used for warming air, compensating for building losses, etc. Heat is measured in British thermal units (Btu). One Btu is the amount of heat required to raise the temperature of 1 lb of water 1°F.

VAPOR PRODUCTION OF ANIMALS

Cattle give off moisture during normal respiration; the higher the temperature the greater the moisture. This moisture should be removed from buildings through the ventilation system. Most building designers govern the amount of winter ventilation by the need for moisture removal. Also, cognizance is taken of the fact that moisture removal in the winter is lower than in the summer; hence, less air is needed. However, lack of heat makes moisture removal more difficult in the winter time. Table 11-4 gives the information necessary for determining the approximate amount of moisture to be removed.

Since ventilation also involves a transfer of heat, it is important to conserve heat in the building to maintain desired temperatures and reduce the need for supplemental heat. In a well-insulated building, mature animals may produce sufficient heat to provide a desirable balance between heat and moisture; but young animals will usually require supplemental heat. The major requirement of summer ventilation is temperature control, which requires moving more air than in the winter.

TABLE 11-4
VAPOR PRODUCTION OF ANIMALS[1]

Vapor Source	Unit		Temperature		Vapor Production		Vapor Production	
	(lb)	*(kg)*	*(°F)*	*(°C)*	*(lb/hr)*	*(Btu/hr)*	*(kg/hr)*	*(kcal/hr)*
Cow	1,000	*453.6*	40	*4*	0.92	960	*0.42*	*241.9*
			70	*21*	1.38	1,450	*0.63*	*365.4*
Calves (6–10 months)	—	—	60	*16*	0.11	120	*0.05*	*30.2*
			80	*27*	0.29	300	*0.13*	*75.6*
Hog:								
Sow and litter (3 weeks after farrowing)	400	*181.4*	—	—	0.97	1,020	*0.44*	*257.0*
Finishing	200	*90.7*	35	*2*	0.11	120	*0.05*	*30.2*
			70	*21*	0.16	175	*0.07*	*44.1*
Layer hen	4.5	*2.04*	50	*10*	0.012	12	*0.005*	*3.0*
Sheep	100	*45.4*			(0.039-in. fleece length)		*(0.1-cm fleece length)*	
			45	*7*	0.06	60	*0.03*	*15.1*
			70	*21*	0.07	75	*0.03*	*18.9*
					(3.937-in. fleece length)		*(10.0-cm fleece length)*	
			45	*7*	0.06	60	*0.03*	*15.1*
			70	*21*	0.13	135	*0.06*	*34.0*
Horse	1,000	*454.0*	70	*21*	0.729	—	*0.33*	—

[1]Adapted by the authors from *Agricultural Engineers Yearbook*, St. Joseph, MI, ASAE Data Sheet D-249.2, p. 424, except for horse. Data for horse from *Farm Buildings*, by John C. Wooley, McGraw-Hill Book Company, Inc., Table 25, p. 141.

RECOMMENDED ENVIRONMENTAL CONDITIONS FOR ANIMALS

The comfort of animals (or people) is a function of temperature, humidity, and air movement. Likewise, the heat loss from animals is a function of these three items.

The prime function of the winter ventilation system is to control moisture, whereas the summer ventilation system is primarily for temperature control. If air in beef cattle barns is supplied at a rate sufficient to control moisture—that is, to keep the inside relative humidity in winter below 75%—then this will usually provide the needed fresh air, help suppress odors, and prevent an ammonia buildup.

Some typical temperature, humidity, and ventilation recommendations for different classes of livestock are given in Table 11-5. This table will be helpful in obtaining a satisfactory environment in confinement livestock buildings, which require careful planning and design.

SPACE REQUIREMENTS OF BUILDINGS AND EQUIPMENT FOR BEEF CATTLE

One of the first and frequently one of the most difficult problems confronting the farmer or rancher who wishes to construct a building or piece of equipment for beef cattle is that of arriving at the proper size or dimensions. Table 11-6 contains some conservative average figures which, it is hoped, will prove helpful. In general, less space than indicated may jeopardize the health and well-being of the animals,

TABLE
RECOMMENDED ENVIRONMENTAL

Class of Animal	Temperature				Acceptable Humidity
	Comfort Zone		Optimum		
	(°F)	*(°C)*	*(°F)*	*(°C)*	*(%)*
Beef cow	40–70	*5–21*	50–60	*10–15*	50–75
Steer (enclosed bldg. on slotted floor)	40–70	*5–21*	50–60	*10–15*	50–75
Dairy cow	40–70	*5–21*	50–60	*10–15*	50–75
Dairy calves	50–75	*10–24*	65	*17*	
Sheep:					
Ewe	45–75	*7–24*	55	*13*	50–75
Feeder lamb	40–70	*5–21*	50–60	*10–15*	50–75
Newborn lamb	75–80	*24–27*			
Swine:					
Sow (farrowing house)	60–70	*15–20*	65	*17*	60–85
Newborn pigs (brooder area)	80–90	*27–32*	85	*29*	60–85
Growing-finishing hogs	60–65	*15–17*	60	*15*	60–85
Horse	45–75	*7–24*	55	*13*	50–75
Newborn foal	75–80	*24–27*			
Poultry:					
Layers	50–75	*10–24*	55–70	*13–20*	50–75
Broilers	85–95	*21–27*	70	*24*	50–75
Turkeys	95–100 (beginning poults)	*35–38*			

[1]Generally two different ventilating systems are provided; one for winter, and an additional one for summer. Hence, as shown in Table 11-5, the winter ventilating system in a beef cow barn should be designed to provide 100 cfm (cubic feet/minute) for each 1,000-lb *(454-kg)* cow. Then, the summer system should be designed to provide an added 100 cfm, thereby providing a total of 200 cfm for summer ventilation.

whereas more space may make the buildings and equipment more expensive than necessary.

BEEF CATTLE BUILDINGS

Beef cattle are not as sensitive to extremes in temperature—heat and cold—as are dairy cattle or swine. In fact, mature beef animals will withstand extremely cold weather if kept dry.

It is especially noteworthy that finishing steers, whose bodies generate considerable heat from the digestion and assimilation of their rations, do not need a warm shelter even during the cold winter months. Their chief need is for a dry bed and protection against cold winds, rains, and snow. About the same thing can be said about the sheltering of dry cows and stockers and feeders. Young stock require more protection. A

Fig. 11-2. Beef cattle can withstand extremely cold weather if kept dry. Cows grazing cornstalks covered with a foot of snow. (Courtesy, The University of Nebraska)

11-5
CONDITIONS FOR ANIMALS

Commonly Used Ventilation Rates[1]					Drinking Water			
Basis	Winter[2]		Summer		Winter		Summer	
	(cfm)	*(m³/min.)*	*(cfm)*	*(m³/min.)*	*(°F)*	*(°C)*	*(°F)*	*(°C)*
1,000 lb *(454 kg)*	100	*2.8*	200	*5.7*	50	*10*	60–75	*15–24*
1,000 lb *(454 kg)*	100	*2.1–2.3*	200	*4.2*	50	*10*	60–75	*15–24*
1,000 lb *(454 kg)*	100	*2.8*	200	*5.7*	50	*10*	60–75	*15–24*
per 100 lb *(45 kg)*	10		25					
	20–25	*0.6–0.7*	40–50	*1.1–1.4*	40–45	*5–8*	60–75	*15–24*
	15	*0.3*	30	*0.65*	40–45	*5–8*	60–75	*15–24*
Sow and litter	80	*1.4*	210	*2.8*	50	*10*	60–75	*15–24*
125 lb *(57 kg)*	15	*0.7*	75	*2.1*	50	*10*	60–75	*15–24*
1,000 lb *(454 kg)*	60	*1.7*	160	*4.5*	40–45	*5–8*	60–75	*15–24*
per bird	2		5		50	*10*	60–75	*15–24*
per lb body weight	½		1		50	*10*	60–75	*15–24*
per lb body weight	½		1		50	*10*	60–75	*15–24*

In practice, in many buildings added summer ventilation is provided by opening (1) barn doors, and (2) high-up hinged walls.

[2]Provide approximately one-fourth the winter rate continuously for moisture removal.

TABLE
SPACE REQUIREMENTS OF BUILDINGS

Class, Age, and Size of Animal	Barn or Shed		Shades		Feedlots[1]	
	Floor Area per Animal	Height of Ceiling[3]	Shade per Animal	Shade Height	Area If Ordinary Dirt Lot	Area If Paved Lot
	(sq ft)	*(ft)*	*(sq ft)*	*(ft)*	*(sq ft)*	*(sq ft)*
Cows, 2 years or over	40–50	$8\frac{1}{2}$–10	30–40	10–12	300[5]	50–100
Yearling finishing cattle . . .	Solid floor: 30–40 Slotted floor: 20–25	"	25–35	"	125–200[5]	30–50
Calves, 350 to 500 lb	20–30	"	15–25	"	130–175[5]	20–50
Cows in maternity stall . . .	100–120	"	35–40	"	1–2 acre pasture paddock	—
Herd bulls	100–150	"	35–45	"		—

[1]Allow slope of ⅝ in./ft in paved lots, and ½ in. or more in dirt lots (depending on soil and climate conditions).

[2]Feed bunks should be about 8 in. deep for calves and 12 in. for older cattle.

[3]Minimum ceiling height of 9 ft necessary where a power-operated manure loader is to be used.

[4]With liberal grain or other concentrate feeding, half the recommended space given herein. With bunker or self-feeder silos, allow 6 in./animal.

[5]More space is desirable under some soil and climatic conditions.

shelter also permits the caretakers to do their work in greater comfort. In the deep South and in the Southwest, barns are not necessary for cattle, except on rare occasions.

Beef cattle shelters are of two kinds, natural and artificial. The former includes hills and valleys, timber, and other natural windbreaks. The artificial shelters include those structures (solid fences, stacks, barns, and sheds) designed to protect cattle against the elements—heat, cold, wind, rains, and snows.

It is with beef cattle barns and sheds that this discussion will deal.

Fig. 11-3. Cattle barn on an Iowa farmer-feedlot. With finishing cattle, whose bodies generate considerable heat from the digestion and assimilation of their rations, an open barn may suffice. (Courtesy, *Beef* magazine)

In addition to protecting the animals during severe cold and stormy weather and at winter calving time, these structures should (1) provide a reasonably dry bed for the animals, (2) simplify feeding and management, (3) provide storage for feed and bedding when necessary, and (4) protect young calves.

Although the discussion describes beef cattle barns and sheds, it must be recognized that on small farms, which have a limited number of beef cattle, the animals are usually housed in a general-purpose barn or shed or in extensions to other barns rather than in separate and specially designed beef cattle structures.

BEEF CATTLE BARNS

Barns are more substantial structures than sheds and provide more complete protection for stock in the colder areas. In addition to housing the animals, such structures usually provide adequate facilities for all of the roughage and bedding needed during the winter season and for a considerable proportion of the concentrates. Stalls, pens, and storerooms may also be included—additions which are especially important where a breeding herd is to be served. In general, beef cattle barns effect a saving of labor and time in feeding, and save feed.

11-6
AND EQUIPMENT FOR BEEF CATTLE

Hay or Silage Manger, or Rack					Feed Bunk or Trough for Hand-Feeding Grain[2]				Self-feeder	Water	
Length per Animal[4]	Width If Feeds from 1 Side	Width If Feeds from 2 Sides	Width If Attached to Side of Barn	Height at Throat	Length per Animal	Width If Feeds from 1 Side	Width If Feeds from 2 Sides	Height at Throat	Trough Length If Feeder Is Kept Filled	Water per Animal per Day	Water Trough
(in.)	*(in.)*	*(in.)*	*(in.)*	*(in.)*	*(in.)*	*(in.)*	*(in.)*	*(in.)*	*(in.)*	*(gal)*	
24–30	30	48–60	30	24	24–30	18–30	48	24	6–12 per animal	12	Allow 1 linear ft of open water tank space for each 10 cattle; or 1 automatic watering bowl for each 25 cattle. A satisfactory water temperature range in winter is 40–45°F; in summer, 60–80°F.
20	"	"	"	20	18–24	18	"	22	6–9	10	
18	"	"	"	20	18	"	"	18	6–8	8	
30	—	—	—	26	30	—	—	30	9–12	15	
"	30	36–40	30	"	"	30	36–40	"	"	"	

Remarks:

Animals with horns require about 1 linear ft more manger or trough space per animal than the figures given in this table. Movable hayracks or feed bunks are usually 12 to 16 ft in length.

Provide a paved area of at least 10 ft around waterers, feed bunks, and roughage racks.

For specifications on slotted floors see Chapter 28, section entitled, "Design Requirements for Cattle Confinement Buildings and Slotted Floors."

Re: Water per animal per day. A minimum of 20 gal/day is needed for continuous flow to keep the water clean, and to keep it from freezing in the winter months.

To make metric conversions, see the Appendix, "Section IV—Weights and Measures."

KINDS OF BEEF CATTLE BARNS

The type of barn is determined by the kind of stock and the method of handling and management. In general, the following two types of beef cattle barns, or modifications thereof, are in common use:

1. **The pen, stall, or general utility barn.** These barns range up to 60 ft in width and may be either 1- or 2-stories, the only essential difference between the two being the overhead loft for hay and bedding storage in the latter. If feeding is done inside, mangers are placed along the alley and hayracks along the side walls; or the location of the racks and bunks may be reversed. Frequently, especially in a feeder cattle operation, hay is fed inside the barn and grain and silage outside. If this type of barn is to be used for breeding stock, the plan usually provides for grain bins and stalls and pens for calves, cows, and bulls. The barn may be an oblong structure; or it may have wings or extensions to form an L, T, or U shape.

2. **The central storage type with attached sheds or livestock sections.** This is a popular and economical type of cattle barn. These barns vary in width, but perhaps the average structure is about 60 ft wide and consists of a 24-ft center and two 18-ft wings. Sometimes the wings may extend around three sides of the building, thus providing more shelter space in relation to storage. The general floor plan for this type of barn consists of ground-to-roof storage for hay and bedding in the center and cattle sheds on each side. Racks for hay are adjacent to the central storage area. Troughs for grain and silage may be along the outside wall of the barn or outside in the lot.

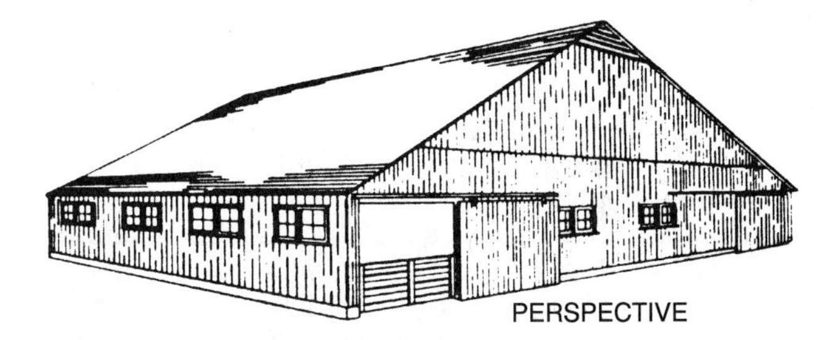

Fig. 11-4. Enclosed beef barn, pole-type construction. (From: *Handbook of Building Plans*, p. 137. Courtesy, Midwest Plan Service, Iowa State University)

BEEF CATTLE SHEDS

Sheds are the most versatile and widely used beef cattle shelters throughout the United States. They are used for cattle in the feedlot, as a range shelter for dry cows with calves, and for housing young stock. They usually open to the south or east, preferably opposite to the direction of the prevailing winds and

Fig. 11-5. An open beef cattle shed. Sheds are the most versatile and widely used beef cattle shelters throughout the United States. (Courtesy, Oklahoma State University)

toward the sun. They are enclosed on the ends and sides. Sometimes the front is partially closed, and in severe weather dropdoors may be used. The latter arrangement is especially desirable when the ceiling height is sufficient to accommodate a power manure loader.

So that the bedding can be kept reasonably dry, it is important that sheds be located on high, well-drained ground; that eave troughs and down spouts drain into suitable tile lines, or surface drains; and that the structures have sufficient width to prevent rain and snow from blowing to the back end. Sheds should be a minimum of 24 ft in depth, front to back, with depths up to 36 ft preferable. As a height of 8½ ft is necessary to accommodate some power-operated manure loaders, when this type of equipment is to be used in the shed, a minimum ceiling height of 9 ft is recommended. The extra 6 in. allow for the accumulation of manure. Lower ceiling heights are satisfactory when it is intended to use a blade or a pitchfork in cleaning the building.

The length of the shed can be varied according to needed capacity. Likewise, the shape may be either a single long shed or in the form of an L or a T. The long arrangement permits more corral space. When an open shed is contemplated, thought should be given to feed storage and feeding problems.

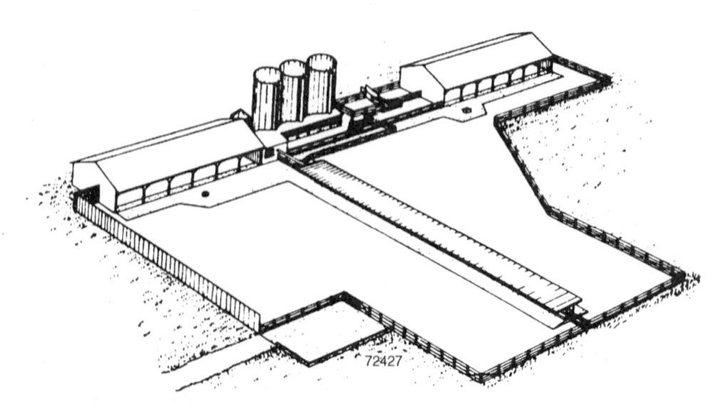

Fig. 11-6. Beef cattle layout, with opensheds and outside feeding. (From: *Handbook of Building Plans*, p. 133. Courtesy, Midwest Plan Service, Iowa State University)

Sometimes hayracks are built along the back wall of sheds, or next to an alley, if the shed is very wide or if there is some hay storage overhead. Most generally, however, hayracks, feed bunks, and watering troughs are placed outside the structure.

A MODERN MULTIPLE-USE BARN

The two-story red barn, long a traditional American trademark, is fast giving way to cheaper one-story structures of more flexible design and lower operating costs.

Figs. 11-7 and 11-8 show a modern low-cost, labor-saving, multiple-use barn. It may be used for beef cattle, for dairy cattle (as a loafing barn), for sheep, for swine, for horses, and/or for storage of feed, seed, fertilizer, and machinery. This barn is flexible and versatile.

Fig. 11-7. A modern multiple use barn. It is flexible and versatile.

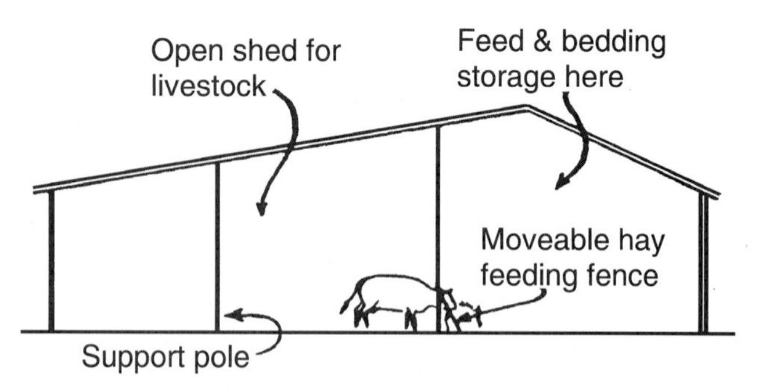

Fig. 11-8. Cross section of the modern multiple-use barn shown in Fig. 11-7. Note the movable hay feeding fence which makes it possible (a) to decrease the feed and bedding storage area and to increase the animal area as winter advances, and (b) to keep the feed and bedding in close proximity to the supply, thereby lessening labor and drudgery.

IT HAS DESIRABLE FEATURES

If properly designed and constructed, this modern

multiple-use barn should possess the following desirable features:

1. **Meet needed animal space and feed storage requirements.** It should meet the specific animal space and feed storage requirements of the farm or ranch.

2. **Face in the right direction.** Except in the deep South, it should open to the south or east, preferably opposite to the direction of the prevailing winds and toward the sun.

3. **Be relatively dry.** It should be constructed to assist in providing a reasonably dry bed for animals.

4. **Be of proper height.** A minimum ceiling height of 9 ft is recommended in order to accommodate a power-operated manure loader.

5. **Possess movable equipment.** Wherever possible, barn equipment should be as movable as the furniture in a home. Thus, where the entire floor of the barn is on the same ground level, movable feed facilities make it possible (a) to decrease the feed and bedding storage area and to increase the animal area as winter advances, and (b) to keep the feed and bedding in close proximity to the supply, thereby lessening labor and drudgery.

6. **Promote animal health.** This is most important, for healthy animals are the profitable and efficient ones. Today, it is recognized (a) that open sheds provide the cheapest type of ventilation, and usually the best, and (b) that, except for newborn or sick animals, the inside temperature of the barn should be as close to the outside temperature as possible. The modern multiple-use type of barn shown in Fig. 11-7 meets these requisites.

7. **Be flexible.** The need for flexibility is best illustrated by referring to the great number of obsolete draft horse and mule barns throughout the country, usually two-story structures with built-in stalls and permanent feed mangers. With the passing of the draft animal, many of these old barns either have remained unused or have been modernized at high cost. But this is not the end of such changes: Who can predict, with certainty for example, what the future holds relative to methods of harvesting, curing, and storing hay? Thus, it is important that buildings be of such flexible design that they can be easily and inexpensively modernized to meet changes in a changing world.

This desired flexibility is best obtained by constructing a single-story building with movable equipment—features of the modern barn shown in Fig. 11-8.

8. **Possess such added rooms and stalls as are needed.** Maternity stalls, feed, seed, or fertilizer rooms, a milking room, and/or a tack room may be incorporated in this type of barn if they are needed.

9. **Possess adjacent corrals and a loading chute.** Suitable corrals and a loading chute should be provided adjacent to the building.

IT IS REASONABLE IN COST

On many livestock establishments, the cost of the improvements is about equal to the value of the land alone. Thus, it is important that every consideration should be given to effecting savings in the construction of buildings and equipment. Among the most important reasons why the modern-type barn shown in Fig. 11-7 is reasonable in cost are the following:

1. **It is of simple pole-frame construction.** It is supported by poles chemically treated under pressure and set like fence posts 4 to 5 ft in the ground and spaced at least 12 ft apart. This barn (a) eliminates scaffolding, for even the highest points can be reached from ladders, trucks, or wagons, (b) lessens bracing; and (c) saves labor, for this simple type of construction can be built with any farm labor capable of building a movable hog house.

2. **It uses the ground to support feed and bedding.** This method alleviates the necessity of heavy construction to support this weight overhead.

3. **It is built low to the ground.** This makes for less wind pressure to resist, and requires less bracing.

4. **It leaves one side open.** This saves on material, and desirable ventilation is obtained without cost.

IT SAVES LABOR

The importance of designing farm buildings for efficiency of operation becomes apparent when it is realized that more than half of the average operator's working hours are spent in and around the farm buildings.

The type of barn shown in Fig. 11-7 is labor-saving in comparison with conventional two-story barns for the following reasons:

1. **Feed is stored on the ground level.** The truck, or other vehicle, may be driven directly into the ground-level storage area and unloaded without hoisting or elevating.

2. **Feed and bedding are stored where used.** A movable hay and feed bunk can be so designed that it can be moved back as the animals eat the feed and use the bedding. This eliminates the necessity of climbing into mows, poking feed and bedding down a chute, and carrying it some distance to where it is to be used. Instead, the feed may be tossed directly and easily into the rack or bunk and the bedding into the area where it is used.

3. **Overflow feed and bedding storage is convenient.** If the farm production of feed and bedding is higher than normal, the added tonnage may be temporarily stored in the animal area. Then as it is used more and more of the animal space becomes available

as winter advances. With a conventional two-story barn, such flexibility is not easily obtained.

4. **Manure may be removed by power loader.** The ceiling height and open shed arrangement make for ease in operating a manure loader and in getting in and out with a spreader. No hand labor is required.

BEEF CATTLE EQUIPMENT

It is not proposed that all of the numerous types of beef cattle equipment will be described herein; rather, only those articles that are most common. Suitable equipment saves feed and labor, conserves manure, and makes for increased production.

HAYRACKS

Various sizes and designs of racks are used in feeding hay and other forages. In general, these structures are of two types: overhead racks, and low mangers. Overhead racks are easily moved, but they are difficult to clean; and often there is considerable wastage, especially when cows have horns and do considerable fighting. Low mangers result in less wastage of the leaves and other fine particles than overhead racks because the cattle must work down from the top. Low mangers may be satisfactorily used in the feeding of kafir and other coarse forages, also.

It is preferable that racks be of sufficient size so that one filling will last several days, thus lessening the labor requirements. It is desirable also that the rack be mounted on runners or wheels, thus making it convenient to move it from place to place.

Fig. 11-9. A portable, circular steel hayrack. Note that the cattle must work down, thereby making for less wastage. (Courtesy, American Polled Hereford Assn., Kansas City, MO)

FEED TROUGHS OR BUNKS

Feed troughs or bunks are used for feeding both grain and silage. Bunks are usually built with legs of the desired height, and they're portable. They should be well braced, free from sharp corners, and made so as to prevent cattle from throwing feed out. They may be made stationary by extending the legs into the ground or by setting the legs on concrete foundations. Stationary troughs should be located on a well-drained site, or, preferably, placed on a concrete or other hard-surfaced platform. The dimensions of the feed bunk or trough should be in keeping with those given in Table 11-6. Bunks of a height for mature cattle may be used for calves by digging holes for the legs or runners. In many large cattle feeding yards, especially in the West, a "manger-type" trough forms a part of the enclosure and is filled from outside the corral along a service lane or road. Fig. 11-10 shows a concrete fenceline bunk.

Fig. 11-10. A concrete fenceline feed bunk. (Courtesy, *Livestock Weekly*, San Angelo, TX)

Stationary troughs in barns should be provided with adjustments for height, thus making it possible to adjust for cattle of different ages and for the accumulation of manure.

SELF-FEEDERS

Most cattle are hand-fed, but the use of self-feeders for finishing cattle is increasing. A self-feeder does a better job of feeding than a careless operator. Self-feeders may be either stationary or portable. The latter are usually equipped with runners or are mounted on low wheels. The most desirable cattle self-feeders have feed troughs along both sides, have a capacity ade-

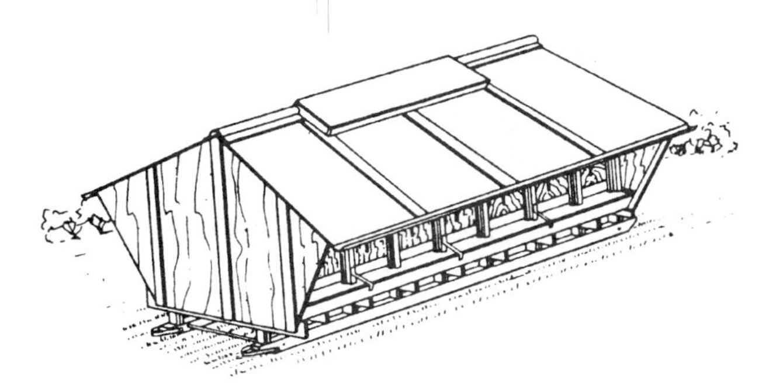

Fig. 11-11. A self-feeder for cattle. Note the door in the roof for filling and the runners for moving. (From: *Beef Housing and Equipment Handbook*, p. 58. Courtesy, Midwest Plan Service, Iowa State University)

quate to hold a feed supply that will last 10 days to 2 weeks, and are easily filled.

WATERING FACILITIES

Water is more essential than feed, for animals will subsist longer without feed than without water. As indicated in Table 11-6, on the average, mature cattle consume approximately 12 gal of water per head daily, with variations according to size of animal, season, and type of feed. Cattle are frequently watered from reservoirs, springs, lakes, and streams; but if surface water is not available, labor can be saved by having reliable power for pumping and by piping water under pressure to tanks or troughs where it will be available at all times. When wells are used as a source of supply, windmills are the most commonly used power unit, although electric motors and gas engines are sometimes used. In cold areas, outside tanks are generally provided with tank heaters or covers during the winter months.

Fig. 11-12. A good concrete water tank for cattle. Note boards over float. (Courtesy, *Livestock Weekly*, San Angelo, TX)

Regardless of the source of the water supply—wells, springs, streams, or surface-storage supplies—it is important that it be abundantly available at all times. It should also be fresh and well protected. When watering tanks and troughs are used, they should be of adequate size, and there should be provision for keeping animals out of the tank. In areas in which mud is a problem, it is desirable that tanks be surrounded by a pavement at least 6 ft in width, and provision should be made to pipe the overflow away from the vicinity of the tank.

SHADES

Cattle should be provided with suitable shade during the hot summer months. An unshaded cow standing in an air temperature of 100°F has to dispose of enough heat in a 10-hour period to bring 9 gal of ice water to the boiling point. At the Imperial Valley Field Station, El Centro, California, it was found (1) that the difference in heat load in an animal under a shade 10 or 12 ft high at 100°F, and one in the sun, was 1,334 Btu per hour, and (2) that to make 100 lb gain during midsummer required 200 to 300 lb more feed without shade than with shade.

Fig. 11-13. Cattle shades. (Courtesy, *California Cattleman*, photo by Phil Raynard)

The most satisfactory cattle shades are (1) oriented with a north-south placement, as such shades are drier underneath than those with east-west orientation, because the sun can get underneath to dry out the manure and urine; (2) at least 10 to 12 ft in height (in addition to being cooler, high shades allow a truck to be driven under and cattle to be worked by a person on a horse); and (3) open all around.

CORRALS

No equipment adds more to the ease and pleasure

Fig. 11-14. Corral system with a curved chute, curved alley, and diagonal sorting pens. Cattle are gathered in the large, round gathering pens and all handling functions are carried out in the curved alley and diagonal pens. Corners where cattle can bunch up are almost eliminated. (Courtesy, Temple Grandin, Grandin Livestock handling Systems, Inc., Urbana, IL.)

of handling cattle than a convenient system of well-constructed corrals. In addition, such a system saves money by reducing the shrinkage resulting from sorting and handling cattle. On the western range, this type of equipment is considered a virtual necessity. To be sure, it is not presumed that each operator will have need for the same size, number, and arrangement of lots. These will vary according to the size of the herd, management practices, and individual preferences. There are, however, certain salient features that should be observed in planning a system of corrals: (1) They should be large enough to accommodate easily the number of cattle involved; (2) they should be conveniently located with respect to the pasture or range and a nearby water supply; and (3) they should include at least one large pen for holding the herd, a chute or an alley to be used for separating or crowding work, and two smaller pens in which to put the separated cattle. Many modern cattle handling facilities are designed with round pens and curved alleys to take advantage of cattle's natural tendency to circle.

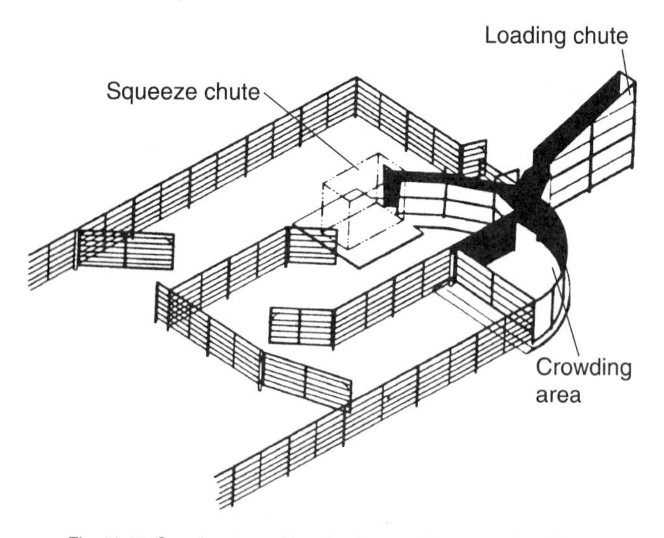

Fig. 11-15. Corral system with a circular crowding pen and working chute. A good arrangement for loading and sorting. This layout can be a hospital or receiving lot. (From: *Beef Housing and Equipment Handbook*, p. 34. Courtesy, Midwest Plan Service, Iowa State University)

LOADING CHUTE

The extensive use of trucks makes it desirable that the stock farm and ranch be equipped with a chute for loading and unloading stock. Such equipment may be either portable or stationary. In the latter case, it is usually desirable to attach the chute to the corrals or feedlot. A loading chute of sufficiently durable construction for cattle is equally satisfactory for sheep and swine. The main essentials are: (1) That the chute have proper height for the truck commonly served (or preferably have an adjustable height arrangement); (2) that it have adequate width to accommodate animals; (3) that it have sufficient slope and cleating to the platform approach to prevent slipping. Most chutes are about 28 to 30 in. in width and 46 in. high.

Fig. 11-16. Curved concrete loading chute. (Courtesy, *Feedlot Management*, Minneapolis, MN)

SQUEEZE

A squeeze for handling cattle can be profitably used on any farm or ranch for dehorning, branding, castrating, testing for tuberculosis, vaccinating, or in performing minor surgical operations. Lack of such equipment usually entails a great deal of labor in catching and throwing animals and is hard on both the caretaker and the animal. Numerous designs of homemade and commercial cattle squeezes are available, but the essential features of all of them are: (1) durability, (2) thorough restraint of the animal, and (3)

Fig. 11-17. Cattle squeeze. (Courtesy, Ken Gill, Thompson & Gill, Inc., Madera, CA)

convenience for the operator. When cattle corrals are constructed, the cattle squeeze is normally a part of the pen arrangement.

STOCKS

Cattle stocks are primarily an item of equipment for the purebred herd and the feedlot. Usually they are so located that animals must be led into them rather than driven, thus limiting their use to cattle that may be rather easily handled. Cattle stocks may be used for trimming and treating hoofs, dehorning, horn branding, drenching, ringing bulls, swinging injured animals, and restraining animals during surgical operations. The essential features of cattle stocks are: (1) durability; (2) thorough restraint of the animals; (3) convenience for the operator; (4) a canvas sling to place under the animal to prevent it from lying down while in the stocks (the swing may be wound up on side rollers by means of turning rods); and (5) wooden sills that extend along either side at a height of 15 in. from the floor and on which the feet may be rested and tied while being trimmed or treated.

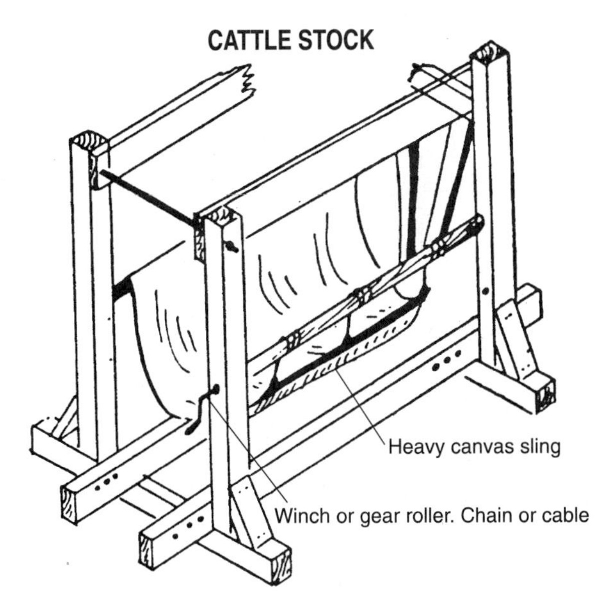

Fig. 11-18. Cattle stock. Detailed plans and specifications for such equipment can usually be obtained through the local lumber dealer, county agricultural agent or FFA instructor, or through writing the college of agriculture in the state. (Drawing by R. F. Johnson)

BREEDING RACK

Breeding racks are sometimes used by purebred operators who desire to breed young heifers to mature, heavy bulls. Fig. 11-19 shows a very satisfactory type of breeding rack.

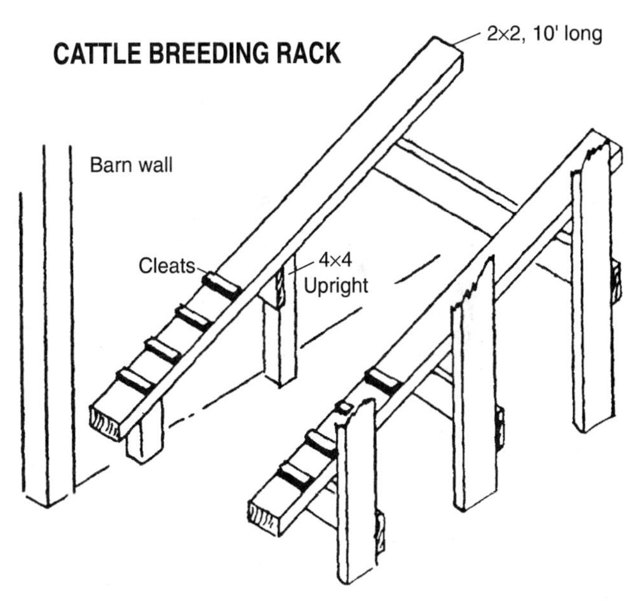

Fig. 11-19. Cattle breeding rack. Detailed plans and specifications for such equipment can usually be obtained through the local lumber dealer, county agricultural agent or FFA instructor, or through writing the college of agriculture in the state. (Drawing by R. F. Johnson)

DIPPING VAT AND SPRAYING EQUIPMENT

Dipping vats have long been used successfully and rather extensively on the western range in treating cattle, sheep, and sometimes horses, for external parasites. The vat is usually built at one side of the corral system. The chief virtue of the dipping vat lies in the fact that animals so treated are thoroughly covered. On the other hand, the vats are costly to construct and lack mobility and flexibility; and there is always considerable leftover dip at the finish of the operation. Frequently, dipping vats are built as a cooperative enterprise by a group of producers rather than by an individual operator. With the development of modern insecticides and improved spraying equipment, it appears probable that in the future spraying operations will increase and that fewer expensive dipping vats will be constructed.

OTHER BEEF CATTLE EQUIPMENT

There is hardly any limit to the number of different articles of beef cattle equipment, and the design of each. In addition to those already listed, Figs. 11-20 and 11-21, respectively, show a calf creep and a salt-mineral feeder.

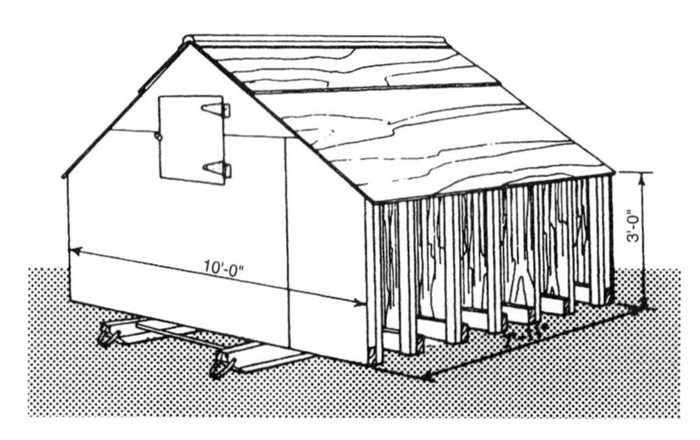

Fig. 11-20. Calf creep feeder, 42-bu capacity. Note the entrances for the calves. (From: *Beef Housing and Equipment Handbook*, p. 54. Courtesy, Midwest Plan Service, Iowa State University)

Fig. 11-21. A salt-mineral feeder on skids. Note that it is a three-compartment arrangement. (Courtesy, Kindred P. Coskey, Bar K Ranch, Weslaco, TX)

SILOS

The general kinds of silos are: tower silos, pit silos, trench silos, self-feeder or bunker silos, above-ground temporary silos, and plastic silos. The kind of silo decided upon and the choice of construction material should be determined primarily by the cost, the silage storage losses, and the suitability to the particular needs of the farm or ranch.

Fig. 11-22. Tower silos. (Courtesy, A. O. Smith Harvestore Products, Inc., Arlington Heights, IL)

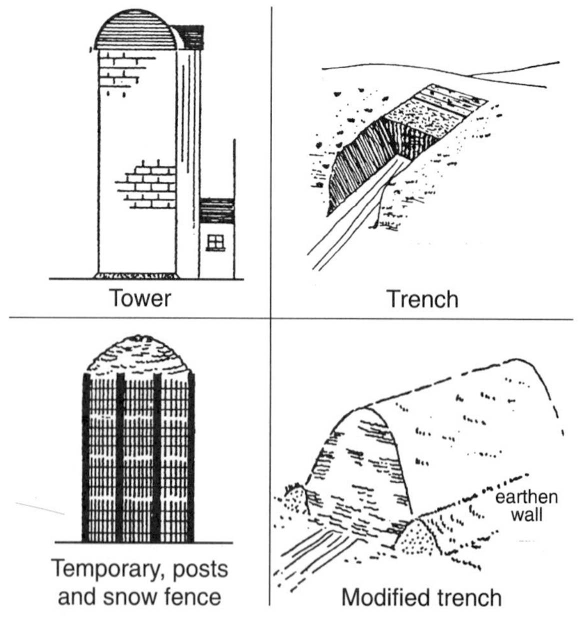

Fig. 11-23. Four kinds of silos: (a) tower silo *(upper left)*; (b) trench silo *(upper right)*; (c) enclosed stack silo *(lower left)*; and (d) modified trench-stack silo *(lower right)*. The latter two are both above-ground temporary silos. (Drawing by R. F. Johnson)

RECOMMENDED MINIMUM WIDTH OF SERVICE PASSAGES

In general, the requirements for service passages are similar, regardless of the kind of animals. Accordingly, the suggestions contained in Table 11-7 are equally applicable to cattle, sheep, swine, and horse barns.

TABLE 11-7
RECOMMENDED MINIMUM WIDTHS OF SERVICE PASSAGES

Kind of Passage	Use	Minimum Width	
		(ft)	*(m)*
Feed alley	For feed cart	4	*1.2*
Driveway	For wagon, spreader, or truck	9	*2.8*
Doors and gate	Drive-through	9	*2.8*
Doors and gate	To small pens	4	*1.2*

STORAGE SPACE REQUIREMENTS FOR FEED AND BEDDING

The space requirements for feed storage for the livestock enterprise—whether it be for cattle, sheep, hogs, or horses, or as is more frequently the case, a combination of these—vary so widely that it is difficult to provide a suggested method of calculating space requirements applicable to such diverse conditions. The amount of feed to be stored depends primarily upon (1) length of pasture season, (2) method of feeding and management, (3) kind of feed, (4) climate, and (5) the proportion of feeds produced on the farm or ranch in comparison with those purchased. Normally, the storage capacity should be sufficient to handle all feed grain and silage grown on the farm and to hold purchased supplies. Forage and bedding may or may not be stored under cover. In those areas where weather conditions permit, hay and straw are frequently stacked in the fields or near the barns in loose, baled, or chopped form. Sometimes poled framed sheds or a cheap cover of waterproof paper or wild grass is used for protection. Other forms of low-cost storage include temporary upright silos, trench silos, and temporary grain bins.

Table 11-8 gives the storage space requirements for feed and bedding. This information may be helpful to the individual operator who desires to compute the barn space required for a specific livestock enterprise. This table also provides a convenient means of estimating the amount of feed or bedding in storage.

Fig. 11-24. Feeding fence and hay storage. (From: *Beef Housing and Equipment Handbook*, p. 44. Courtesy, Midwest Plan Service, Iowa State University)

HOW TO DETERMINE THE SIZE BARN TO BUILD

The length and depth of the barn may be varied according to needs. The size barn to build for any given farm or ranch may be determined as follows:

1. Estimate the number and kind of animals to be quartered and compute their total animal space requirements. (See Table 11-6.)
2. Compute the yearly feed requirements of the animals to be fed and quartered by referring to Table 8-13, giving consideration to the length of the pasture season and the quantity and quality of the grass.
3. Estimate the farm production of feeds and bedding to be stored in the barn. In most operations this should coincide reasonably close to the total animal requirements (point 2), but there may be circumstances where the feed and bedding storage requirements are more or less than the animal feed requirements.
4. Estimate the total tonnage of feed and bedding to be stored by correlating the animal feed needs and the farm or ranch production. (Correlate the results of points 2 and 3.) Then determine the total storage space requirements for feed and bedding from Table 11-8.
5. Determine the size of barn to build from the total animal space requirements and the total yearly feed and bedding storage requirements (points 1 and 4).

In general, modern multiple-use barns of the type shown in Fig. 11-7 are 52 ft deep, with 26 ft devoted to feed and bedding storage and 26 ft to animals; although both the depth and the length can be varied to meet specific needs.

TABLE 11-8
STORAGE SPACE REQUIREMENTS FOR FEED AND BEDDING[1]

Kind of Feed or Bedding	Pounds per Cubic Foot	Kilograms per Cubic Meter	Cubic Feet per Ton	Cubic Meters per Ton	Pounds per Bushel of Grain	Kilograms per Hectoliter of Grain
				(avg.)		
Hay-Straw:						
1. Loose						
Alfalfa	4.4–4.0	*0.056*	450–500	*14.6*		
Nonlegume	4.4–3.3	*0.056*	450–600	*16.2*		
Straw	3.0–2.0	*0.028*	670–1,000	*25.7*		
2. Baled						
Alfalfa	10.0–6.0	*0.100*	200–330	*8.2*		
Nonlegume	8.0–6.0	*0.100*	250–330	*8.5*		
Straw	5.0–4.0	*0.056*	400–500	*13.9*		
3. Chopped						
Alfalfa	7.0–5.5	*0.078*	285–360	*9.7*		
Nonlegume	6.7–5.0	*0.074*	300–400	*10.8*		
Straw	8.0–5.7	*0.100*	250–350	*9.2*		
Corn:						
15.5% moisture						
Shelled	44.8	*0.570*			56.0	*70*
Ear	28.0	*0.356*			70.0	*90*
Shelled, ground	38.0	*0.484*			48.0	*62*
Ear, ground	36.0	*0.459*			45.0	*56*
30% moisture						
Shelled	54.0	*0.686*			67.5	*87*
Ear, ground	35.8	*0.456*			89.6	*115*
Barley, 15% moisture	38.4	*0.490*			48.0	*62*
ground	28.0	*0.356*			37.0	*48*
Flax, 11% moisture	44.8	*0.570*			56.0	*70*
Oats, 16% moisture	25.6	*0.325*			32.0	*42*
ground	18.0	*0.227*			23.0	*28*
Rye, 16% moisture	44.8	*0.570*			56.0	*70*
ground	38.0	*0.484*			48.0	*62*
Sorghum, grain 15% moisture	44.8	*0.570*			56.0	*70*
Soybeans, 14% moisture	48.0	*0.610*			60.0	*76*
Wheat, 14% moisture	48.0	*0.610*			60.0	*76*
ground	43.0	*0.546*			50.0	*64*

[1]*Beef Housing and Equipment Handbook*, MWPS-6, Midwest Plan Service, Iowa State University.

HOW TO DETERMINE THE SIZE SILO TO BUILD

The size of silo to build should be determined by needs. With tower type and pit silos, this means (1) that the diameter should be determined by quantity of silage to be fed daily, and (2) that the height (depth in a pit silo) should be determined by the length of the silage feeding period. Similar consideration should be accorded with trench silos.

SIZE OF TOWER SILO

If the diameter is too great, the silage will be exposed too long before it is fed; and unless a quantity is thrown away each day, spoiled silage will be fed.

The minimum recommended rate of removal of silage varies with the temperature. In most sections of the United States, it is desirable that a minimum of 1½ in. of silage be removed from tower silos daily during the winter feeding period, with the quantity increased to a minimum of 3 in. when summer feeding is practiced. Of course, the total daily silage consumption on any given farm or ranch will be determined by (1) the class and size of animals, (2) the number of animals, and (3) the rate of silage feeding. Some suggestions on how much silage to feed cattle are found in Table 8-13.

Silo height should be determined primarily by the length of the intended feeding period. In general, however, the height should not be less than twice, nor more than 3½ times the diameter. The greater the depth, the greater the unit capacity. Extreme height is to be avoided because of (1) the excessive power required to elevate the cut silage material, and (2) the heavier construction material required. Also, it is noteworthy that, with silos of the larger diameters, more labor is required in carrying the silage to the silo door for removal.

Table 11-9 may be used as a guide in computing the proper diameter of tower silo for any given farm or ranch.

Fig. 11-25 shows capacities of tower silos of different heights and diameters. It is based on well-eared corn silage harvested in the early dent stage, cut in ¼-in. lengths, well tramped when filled, and with the silo refilled once after settling for a day.

Fig. 11-25 can be adapted for corn silage of different stages of maturity and grain content, and for other kinds of silage, by applying the following rules of thumb:

Kind of Silage:	Changes to Be Made in the Number of Tons Shown in Fig. 11-25
1. For corn silage ensiled when less mature than usual . .	Add 5 to 10%
2. For corn ensiled when dry or overripe	Deduct 5 to 10%
3. For corn very rich in grain	Add 5 to 10%
4. For corn with very little grain	Deduct 5 to 10%
5. For sorghum silage	Use the same weights as used for corn silage of comparable grain and maturity
6. For sunflower silage	Add 5 to 10%
7. For grass silage	Add 10 to 15%[1]

[1]For this reason, a stronger structure is necessary where grass silage is stored.

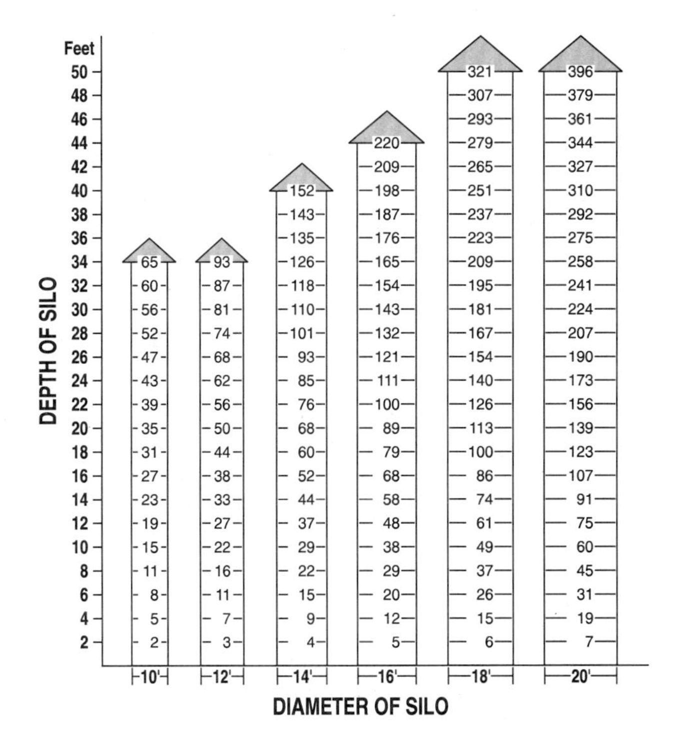

Fig. 11-25. Capacity in tons of settled corn silage in tower silos of varying sizes. (Based on data reported in USDA Circ. 603; drawing by R. F. Johnson)

TABLE 11-9
MINIMUM DIAMETER OF TOWER SILO TO BUILD IF SILAGE IS TO BE KEPT FRESH

Inches of Silage Removed Daily	Total Silage Removed Daily with an Inside Silo Diamter of—											
	10 ft *(3.1 m)*		12 ft *(3.7 m)*		14 ft *(4.3 m)*		16 ft *(4.9 m)*		18 ft *(5.5 m)*		20 ft *(6.1 m)*	
	(lb)	*(kg)*	*(lb)*	*(kg)*	*(lb)*	*(kg)*	*(lb)*	*(kg)*	*(lb)*	*(kg)*	*(lb)*	*(kg)*
Summer: 3 in. *(7.6 cm)* daily will remove[1]	786	*357*	1,312	*596*	1,539	*700*	2,010	*914*	2,545	*1,157*	3,142	*1,428*
Winter: 1.5 in. *(3.8 cm)* daily will remove[1]	393	*179*	656	*298*	770	*350*	1,005	*457*	1,272	*578*	1,571	*714*

[1]The pounds listed in each of the columns to the right are approximations based on an average constant weight of 40 lb of silage per cu ft *(18 kg per 0.5 m^3)*.

The following example will serve to illustrate how to determine the size tower silo to build:

Over a period of years, a farmer plans to winter 34 head of 425-lb stocker calves on a ration of corn silage and protein supplement. There is a 240-day wintering period. No increase in the herd is planned. What size tower silo should be built?

The answer is obtained as follows.

1. First, here are the silage requirements:

a. Table 8-13, Beef Cattle Feeding Guide, indicates that 425-lb stocker calves on a ration of corn silage and protein supplement should receive about 30 lb of silage per head per day.

b. 34 × 30 = 1,020 lb of silage required daily for the 34 calves.

c. 1,020 × 240 = 244,800 lb or 122.4 tons, of silage required for the 240-day wintering period for the 34 calves.

2. Next, here is the size silo to build:

a. Table 11-9 shows that in order to remove 1,005 lb of silage daily (which is only slightly less than the 1,020 lb needed daily), with 1½ in. removed from the top of the silo each day, diameter of the silo should not be greater than 16 ft.

b. Fig. 11-25 can now be used as a guide in determining both the proper height (or depth) and diameter of the silo. Fig. 11-25 shows that a silo 16 ft in diameter and 27 ft high will hold 127 tons of silage, which would allow for 4.6 tons spoilage in excess of the required 122.4 tons. However, the height of a silo should not be less than twice the diameter. It appears best, therefore, to plan on a 14-ft diameter silo. As noted in Fig. 11-25, 34 ft of settled silage in a 14-ft diameter silo will provide 126 tons of silage, which would allow for 3.6 tons spoilage in excess of the required 122.4 tons. To allow for settling, an additional 4 to 6 ft should be added to the height, thus making a 38- to 40-ft height.

c. The size silo to build to meet the needs outlined in this example, therefore, is one that is 14 ft in diameter and 38 to 40 ft high.

SIZE OF TRENCH SILO

As in an upright silo, the cross sectional area of a trench silo should be determined by the quantity of silage to be fed daily. The length is determined by the number of days of the silage feeding period. The only difference is that generally greater allowance for spoilage is made in the case of trench silos, though this factor varies rather widely.

Under most conditions, it is recommended that a minimum 4-in. slice be fed daily from the face (from the top to the bottom of the trench) of a trench silo during the winter months, with a somewhat thicker slice preferable during the summer months.

The dimensions, areas, and capacities given in Table 11-10 are based on the assumption that the silage weighs 35[2] lb per cu ft, which is an average figure for corn or sorghum silage. Thus, a trench silo 8 ft deep, 6 ft wide at the bottom, and 10 ft at the top has a cross sectional area of 64 ft. This size silo will hold 747 lb of silage for each 4-in. slice, or 2,240 lb of silage for each 1-ft slice, or 112 tons in a trench 100 ft long.

For illustrative purposes, let us use the same example and silage requirements as given in the section on "Size of Tower Silo," but this time determine the size trench silo to build rather than the size tower silo. Briefly, the requirements are for 1,020 lb of silage daily for a 240-day wintering period. As noted in Table 11-10, 1 day's feed or 1,020 lb of silage (1,062 lb to be exact) can be obtained in each 4-in. slice of a trench silo 8 ft wide at the bottom, 14 ft, 8 in. wide at the top, and 8 ft deep; or a 91 sq ft cross sectional area. The cross sectional area should not be larger than this if a 4-in. slice is to be removed daily in order to alleviate spoilage.

In order to obtain a 240-day feed supply, the filled trench must be 80 ft long (⅓ of 240; the ⅓ represents ⅓ ft or 4 in.).

The size trench silo to build to meet the specified needs, therefore, is one that is 8 ft wide at the bottom, 14 ft, 8 in. wide at the top, 8 ft deep, and 80 ft long. In order to take care of spoilage and to provide a measure of safety, it is recommended that the actual length be from 85 to 90 ft.

About 8 ft for a trench silo is the most economical depth from the standpoint of cost and feeding. Of course, in filling it is desirable to pile silage 3 ft higher over the center of the trench and round it off. This provides for settlement.

[2]Because the silage in trench silos is generally not so deep and well packed as the silage in tower silos, an average figure of 35 lb/cu ft is used herein for trench silos and 40 lb for upright silos. With all types of silos—including above-ground and below-ground types—the weight of a cubic foot of silage varies with the kind and maturity of the material, moisture content, length of cut, rate of filling, and depth of the silo. Corn silage harvested when about 74% of the grain has passed the milk stage and containing approximately 70% moisture is considered average silage. Volume for volume, sorghum silage weighs about the same as corn silage. Grass or grass-legume silage is 10 to 15% heavier than corn silage.

TABLE 11-10
DIMENSIONS, CROSS SECTIONAL AREA OF TRENCH SILO, AND WEIGHT OF SILAGE IN 4-IN. *(10.2 CM)* SLICE AND PER LINEAL FOOT *(30.5 CM)*[1]

Side Slope per Foot of Depth		Depth		Bottom Width		Top Width		Cross Sectional Area		Weight of Silage 4-in. *(10.2-cm)* Slice		Weight of Silage 1-ft *(30.5-cm)* Slice	
(in.)	*(cm)*	*(ft)*	*(m)*	*(ft)*	*(m)*	*(ft)*	*(m)*	*(sq ft)*	*(m²)*	*(lb)*	*(kg)*	*(lb)*	*(kg)*
3	*7.6*	4	*1.2*	5	*1.5*	7.0	*4.4*	24	*2.2*	280	*127*	840	*382*
4	*10.2*	4	*1.2*	6	*1.8*	8.6	*2.6*	29	*2.7*	338	*154*	1,015	*461*
5	*12.7*	4	*1.2*	7	*2.1*	10.3	*3.1*	33	*3.1*	385	*175*	1,155	*525*
3	*7.6*	6	*1.8*	6	*1.5*	9.0	*2.7*	45	*4.2*	525	*239*	1,575	*716*
4	*10.2*	6	*1.8*	7	*1.8*	11.0	*3.4*	54	*5.0*	630	*286*	1,890	*859*
5	*12.7*	6	*1.8*	8	*2.1*	13.0	*4.0*	63	*5.9*	735	*334*	2,205	*1,002*
3	*7.6*	8	*2.4*	6	*1.5*	10.0	*3.1*	64	*6.0*	747	*340*	2,240	*1,018*
4	*10.2*	8	*2.4*	7	*1.8*	12.3	*3.8*	77	*7.2*	898	*408*	2,695	*1,225*
5	*12.7*	8	*2.4*	8	*2.1*	14.6	*4.5*	91	*8.5*	1,062	*483*	3,185	*1,448*
3	*7.6*	10	*3.1*	6	*1.5*	11.0	*3.4*	85	*7.6*	992	*451*	2,975	*1,352*
4	*10.2*	10	*3.1*	8	*2.1*	14.6	*4.5*	113	*10.5*	1,318	*599*	3,955	*1,798*
5	*12.7*	10	*3.1*	10	*3.1*	18.3	*5.6*	142	*13.2*	1,657	*753*	4,970	*2,259*

[1] *Silos, Types and Construction*, USDA, Farmers Bull. No. 1820, p. 55.

FENCES FOR CATTLE

Good fences (1) maintain farm boundaries, (2) make livestock operations possible, (3) reduce losses to both animals and crops, (4) increase land values, (5) promote better relationships between neighbors, (6) lessen accidents from animals getting on roads, and (7) add to the attractiveness and distinctiveness of the premises.

The discussion which follows will be limited primarily to wire fencing, although it is recognized that such materials as rails, poles, boards, stone, and hedge have a place and are used under certain circumstances. Also, where there is a heavy concentration of animals, such as in corrals and feed yards, there is need for a more rigid type of fencing material than wire. Moreover, certain fencing materials have more artistic appeal than others; and this is an especially important consideration on the purebred establishment.

The kind of wire to purchase should be determined primarily by the class of animals to be confined. Tables 11-11 and 11-12 are suggested guides.

The following additional points are pertinent in the selection of wire:

1. **Styles of woven wire.** The standard styles of woven wire fences are designated by numbers as 958, 1155, 849, 1047, 741, 939, 832, and 726.

The first one or two digits represent the number of line (horizontal) wires; the last two, the height in inches; *i.e.*, 1155 has 11 horizontal wires and is 55 in.

Fig. 11-26. Woven wire cattle fence. (Courtesy, Keystone Steel and Wire, Peoria, IL)

in height. Each style can be obtained in either (a) 12-in. spacing of stays (or mesh), or (b) 6-in. spacing of stays. Also, a special 2-in. mesh is available for horses.

2. **Mesh.** Generally, a close-spaced fence with stay or vertical wires 6 in. apart (6-in. mesh) will give better service than a wide-spaced (12-in. mesh) fence. However some fence manufacturers believe that 12-in.

TABLE 11-11
WOVEN WIRE FENCE

Kind of Stock	Recommended Woven Wire Height	Recommended Weight of Stay Wire	Recommended Mesh or Spacing Between Stays	Recommended Number of Strands of Barbed Wire to Add to Woven Wire[1]	Comments
	(in.)	(gauge)	(in.)		
Cattle	47, 48, or 55	9 or 11	12	1 strand 2 in. to 3 in. above top of woven wire, with points 4 in. or 5 in. apart, to prevent animals from breaking down woven wire.	Also satisfactory for all farm animals, except young pigs. Fences for cattle feedlots should be constructed of wood, cable, pipe, or other strong material, and should be 60 in. high.
Sheep	32	11 or 12½	12	2 strands on top.	Sheep fences should total 39 in. in height. Twelve-in. mesh is best for sheep as they will not get their heads caught if they attempt to reach through. With a heavy concentration of feeder lambs, use wooden fence 39 in. high.
Swine	[illegible]	9 or 11	6	1 strand on bottom	Barbed wire on bottom prevents rooting under.
Horses	55 or 58	9 or 11	12	1 strand on top; with points 4 in. or 5 in. apart.	Also satisfactory for all farm animals except young pigs. Cyclone, wood pole, or other durable and attractive materials are usually used around the headquarters.
All farm animals	26 or	9 or 11	6	3 strands on top; 1 strand on bottom.	
	32	9 or 11	6	2 strands on top; 1 strand on bottom.	

[1]The American Society of Agricultural Engineers' standard for barbed wire calls for 4 in. spacing with 2-point wire and 5 in. spacing with 4-point wire.

TABLE 11-12
BARBED WIRE FENCE CHART[1]

Kind of Stock	Recommended Number of Points	Recommended Spacing Between Points	Recommended Weight of Strands	Recommended No. of Lines of Barbed Wire to Install	Comments
		(in.)	(gauge)		
Cattle or horses; in farm pastures	2 or 4	4 or 5	12½	4-5	Two point barbs are 4 in. apart; 4-point are 5 in. apart.
Cattle or horses; on the range	2 or 4	4 or 5	12½	2 or 4	Not all animals will be restrained by 2 or 3 strands.
Sheep	Barbed wire is not considered suitable for sheep because it tears the fleece.				
Swine	2 or 4	4 or 5	12½	6	A 6-strand barbed wire fence for swine may cost more to build and maintain than a woven wire fence.

[1]The American Society of Agricultural Engineers' standard for barbed wire calls for 4 in. spacing with 2-point wire and 5 in. spacing with 4-point wire.

spacing with No. 9 wire is superior to a 6-in. spacing with No. 11 filler wire (about the same amount of material is involved in each case).

3. **Weight of wire.** A fence made of heavier weight wires will usually last longer and prove cheaper than one made of light wires. Heavier or larger size wire is designated by a smaller gauge number. Thus, No. 9 gauge wire is heavier and larger than No. 11 gauge. Woven wire fencing comes in Nos. 9, 11, 12½ and 16 gauges—which refers to the gauge of the wires other than the top and bottom wires. Heavy barbed wire is 12½ gauge. But there is a lighter, high-tensile barbed wire which comes in 15½ to 16½ gauge.

Heavier or larger wire than normal should be used in those areas subject to (a) salty air from the ocean,

(b) smoke from industries of close proximity, which give off chemical fumes into the atmosphere, (c) rapid temperature changes, or (d) overflow or flood.

Likewise, heavier wire than normal should be used in fencing (a) small areas, (b) where a dense concentration of animals is involved, and (c) where animals have already learned to get out.

4. **Styles of barbed wire.** Styles of barbed wire differ in the shape and number of the points of the barb, and the spacing of the barbs on the line wires. The two-point barbs are commonly spaced 4 in. apart while four-point barbs are generally spaced 5 in. apart. Since any style is satisfactory, selection is a matter of personal preference.

5. **Standard size rolls or spools.** Woven wire comes in 20 and 40 rod rolls; barbed wire in 80 rod spools.

6. **Wire coating.** The kind and amount of coating on wire definitely affects its lasting qualities. Galvanized coating is most commonly used to protect wire from corrosion. Coatings are specified as Class I, Class II, and Class III. The higher the class number, the greater the coating thickness and performance.

SELECTING POSTS

Three kinds of material are commonly used for fence posts: wood, metal, and concrete. The selection of the particular kind of posts should be determined by (1) the availability and cost of each, (2) the length of service desired (posts should last as long as the fencing material attached to it, or the maintenance cost may be too high), (3) the kind and amount of livestock to be confined, and (4) the cost of installation.

■ **Wood posts**—Osage orange, black locust, chestnut, red cedar, black walnut, mulberry, and catalpa—each with an average life of 15 to 30 years without treatment—are the most durable wood posts, but they are not available in all sections. Untreated posts of the other and less durable woods will last 3 to 8 years only, but they are satisfactory if properly butt treated (to 6 to 8 in. above the ground line) with a good wood preservative.

The proper size of wood posts varies considerably with the strength and durability of the species used. In general, however, large posts last longer than small ones. Satisfactory line posts of osage orange or of other woods that have been pressure treated may be as small as 2½ in. in diameter; whereas line posts of other woods should be 4 to 8 in. in diameter at the smaller end. Split posts should be a minimum of 5 in. in diameter. Line posts are generally 7 to 8 ft in length, depending on the height of the fence to be constructed.

Wood corner, end, and gate posts should be substantial, usually not less than 8 in. in diameter. Also they should be long enough so that they can be set in the ground to a depth of at least 36 in.

■ **Metal posts**—Metal posts (made of steel or wrought iron) last longer, require less storage space when not in use, require less labor in setting than wood posts, and are fire resistant. Also, they may give protection against lightning by grounding the current. However, such protection is questionable in dry weather or in areas with a low water table. Metal posts are usually higher in price than wood posts.

Metal line posts are made in different styles and cross sections. Heavier studded "T" or "Y" section posts are most popular for livestock, although lighter channel posts ("U" posts) may be used for temporary and movable fences. Line posts are available in lengths of 5 to 8 ft in increments of 6 in. Metal corner, end, and gate posts are commonly made from angle sections, and come in 7 to 9 ft lengths.

■ **Concrete posts**—When properly made, concrete posts give excellent service over many years. In general, however, they are expensive.

TREATING POSTS

The less durable types of fence posts will last about five times longer when treated than when untreated. This affects yearling savings in two ways: (1) in the cost of posts, and (2) in the labor involved in fence construction.

Although the relative durability of posts does not materially affect initial fencing costs, the length of life of the posts is the greatest single factor in determining the cost of a fence on an annual basis. The length of life of posts can be increased by the proper use of an approved post preservative used in keeping with the manufacturer's directions.

Pressure treating is preferable because it forces the preservative to the center of the post—leaving none of the wood untreated.

ELECTRIC FENCES

Where a temporary enclosure is desired or where existing fences need bolstering from roguish or breachy animals, it may be desirable to install an electric fence, which can be done at minimum cost.

The following points are pertinent in the construction of an electric fence:

1. **Safety.** If an electric fence is to be installed and used, (a) necessary safety precautions against accidents to both persons and animals should be

taken, and (b) farmers or ranchers should first check into the regulations of their own state relative to the installation and use of electric fences. *Remember that an electric fence can be dangerous.* Fence controllers should be purchased from a reliable manufacturer; homemade controllers may be dangerous.

Fig. 11-27. High-tensile smooth wire electric fence attached to fiberglass "T" posts, effective in managing livestock and controlling predators. (Courtesy, Snell Systems, Inc., San Antonio, TX)

2. **Charger.** The charger should be safe and effective. (Purchase one made by a reputable manufacturer.)There are four types of chargers: (a) **the battery charger,** which uses different types of batteries; (b) **the inductive discharge system,** in which the current is fed to an interrupter device called a circuit breaker or chopper which energizes a current limiting transformer; (c) **the capacitor discharge system,** in which the power line is rectified to direct current and the current is stored in the capacitor; and (d) **the continuous current type,** in which a transformer regulates the flow of current from the power line to the fence.

3. **Wire height.** As a rule of thumb, the correct wire height for an electric fence is about three-fourths the height of the animal; with two wires provided for sheep and swine. Following are average fence heights above the ground for cattle and calves:

Cattle: 30 to 40 in. *(76 to 102 cm)*
Calves: 12 to 18 in. *(30.5 to 46 cm)*
Mixed livestock: 3 wires—8, 12, and 32 in. *(20, 30, and 81 cm)*

4. **Posts.** Either wood or steel posts may be used for electric fencing. Corner posts should be as firmly set and well braced as required for any nonelectric fence so as to stand the pull necessary to stretch the wire tight. Line posts (a) need only be heavy enough to support the wire and withstand the elements, and (b) may be spaced 40 to 50 ft apart for cattle.

5. **Wire.** The most popular galvanized and aluminized smooth wires range from 12½-gauge high-tensil wire for permanent fences to conventional 16 gauge wire for fences used in rotational grazing systems. In those states where barbed wire is legal for electric fences, 12½-gauge hog wire is preferred. Never use rusty wire, because rust is an insulator.

6. **Insulators.** Wire should be fastened to the posts by insulators and should not come into direct contact with posts, weeds, or the ground. Inexpensive solid glass, porcelain, or plastic insulators should be used, rather than old rubber or necks of bottles.

Fig. 11-28. Recommended height for electric fence for: **(A)**, cattle, and **(B)** calves. Generally, two or more wires are recommended. (Drawing by R. F. Johnson)

7. **Grounding.** Inadequate grounding is the root of power-surge problems. Always check with the manufacturer or distributor to determine proper grounding methods. *An electric fence should never be grounded to a water pipe, because it could carry lightning directly to connecting buildings.* A lightning arrester should be installed on the ground wire. If electric fences are installed in dry areas with little soil moisture, ground wires need to be installed on the fence along with hot wires to insure that animals will receive an electrical shock when touching the hot wire.

GUARDS OR BUMPER GATES

Cattle guards or bumper gates set in a fence permit convenient passage of automobiles and trucks but deter cattle, hogs, sheep, and most horses and mules. Where cattle guards are installed, gates should be constructed nearby in order to allow for the movement of animals.

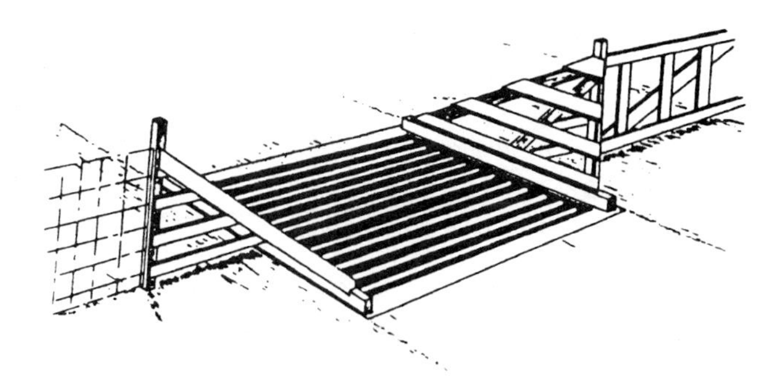

Fig. 11-29. Cattle guards.

QUESTIONS FOR STUDY AND DISCUSSION

1. Brody, of the Missouri Station, studied the basic building requirements of beef cattle. What was the nature of this classic study, what was found, and what is the practical application of this experiment?

2. Except for extremely cold or hot weather, does it normally pay to provide shelter for cattle? Justify your answer.

3. Define *environment.* In beef cattle, what environmental factors are involved?

4. Define the following terms: *total heat, sensible heat, latent heat, Btu,* and *vapor.*

5. The major requirement of winter ventilation is moisture removal, whereas the major requirement of summer ventilation is temperature control. Why the difference?

6. One of the first and frequently one of the most difficult problems confronting the cattle producer who wishes to construct a building or piece of equipment is that of arriving at the proper size or dimensions. In planning to construct new buildings and equipment for beef cattle, what factors and measurements for buildings and equipment should be considered?

7. Which is more important in beef cattle buildings and equipment—initial cost or labor saving in operation? Justify your answer.

8. Make a critical evaluation of your own beef cattle barn(s), or one with which you are very familiar. Determine its (a) desirable and (b) undesirable features.

9. List and discuss the factors determining the type and size of beef cattle buildings.

10. List and discuss the needed and desirable features of a modern multiple-use barn.

11. Make a critical evaluation of your own beef cattle equipment, or of the equipment on a beef cattle establishment with which you are familiar. Determine the (a) desirable and (b) undesirable features of each item.

12. Discuss the need for and the design of the following types of beef cattle equipment: hayracks, feed troughs or bunks, self-feeders, watering facilities, shades, corrals, loading chute, squeeze, stocks, breeding rack, and dipping vat and spraying equipment.

13. List and discuss the steps in determining the size beef cattle barn to build on any given farm or ranch.

14. List and discuss the steps in determining the size silo to build on any given farm or ranch.

15. In the selection of woven wire fence, what is meant by the number *1155*? What other factors should be considered in the selection of woven wire? What specifications may be used in ordering barbed wire?

16. Under what circumstances would you use electric fences for cattle?

SELECTED REFERENCES

Title of Publication	Author(s)	Publisher
Agricultural Engineers Yearbook	Ed. by R. H. Hahn, Jr.	American Society of Agricultural Engineers, St. Joseph, MI, annual
Beef Housing and Equipment Handbook, Fourth Edition	Midwest Plan Service	Midwest Plan Service, Iowa State University, Ames, IA, 1987

Bibliography of Livestock Waste Management	J. R. Miner D. Bundy G. Christenbury	Office of Research and Monitoring, U.S. Environmental Protection Agency, Washington, DC, 1972
Facts & Figures for Farmers	Doane Agricultural Service, Inc.	Doane Agricultural Service, Inc., St. Louis, MO, 1972
Farm Builder's Handbook, Second Edition	R. J. Lytle	Structures Publishing Company, Farmington, MI, 1973
Farm Building Design	L. W. Neubauer H. B. Walker	Prentice-Hall, Inc., Englewood Cliffs, NJ, 1961
Farm Buildings	R. E. Phillips	Doane-Western, Inc., St. Louis, MO, 1981
Farm Buildings, Second Edition	J. C. Wooley	McGraw-Hill Book Company, Inc., New York, NY, 1946
Farm Buildings, Third Edition	D. G. Carter W. A. Foster	John Wiley & Sons, Inc., New York, NY, 1947
Farm Service Buildings	H. E. Gray	McGraw-Hill Book Company, Inc., New York, NY, 1955
Farmstead Engineering	Proceedings of the Farmstead Engineering Conference, Dec. 1-2, 1980	American Society of Agricultural Engineers, St. Joseph, MI, 1980
Farm Structures	H. J. Barre L. L. Sammet	John Wiley & Sons, Inc., New York, NY, 1950
Handbook of Livestock Equipment, Second Edition	E. M. Juergenson	The Interstate Printers & Publishers, Inc., Danville, IL, 1979
Housing of Animals	A. Maton J. Daclemans J. Lambrecht	Elsevier, Amsterdam, The Netherlands, 1985
Latest Developments in Livestock Housing	Seminar Report	American Society of Agricultural Engineers, St. Joseph, MI, 1987
Livestock Environment IV	Ed. by E. Collins C. Boon	American Society of Agricultural Engineers, St. Joseph, MI, 1993
Livestock Waste Management and Pollution Abatement	Proceedings of the International Symposium of Livestock Wastes	American Society of Agricultural Engineers, St. Joseph, MI, 1971
Livestock Waste Management System Design Conference for Consulting and SCS Engineers		University of Nebraska Cooperative Extension Service, Lincoln, NE, 1973
Machines for Power Farming, Second Edition	A. A. Stone H. E. Gulvin	John Wiley & Sons, Inc., New York, NY, 1967
Practical Farm Buildings, Second Edition	J. S. Boyd	The Interstate Printers & Publishers, Inc., Danville, IL, 1979
Principles of Animal Environment	M. L. Esmay	The Avi Publishing Company, Inc., Westport, CT, 1978
The Stockman's Handbook, Seventh Edition	M. E. Ensminger	Interstate Publishers, Inc., Danville, IL, 1992
Structures and Environment Handbook, Eleventh Edition	Midwest Plan Service	Midwest Plan Service, Iowa State University, Ames, IA, 1983

Plans and specifications for beef cattle buildings and equipment can also be obtained from the local county agricultural agent, your state college of agriculture, and materials and equipment manufacturers and dealers.

The search goes on! Today, electronic microscopes help scientists see disease-causing viruses in complex with natural antibodies. (Courtesy, California State University, Fresno)

BEEF CATTLE HEALTH, DISEASE PREVENTION, AND PARASITE CONTROL[1]

by

Robert F. Behlow, D.V.M.[2]
Professor and Extension Veterinarian
North Carolina State University-Raleigh

and

M. E. Ensminger, Ph.D.
President, Agriservices Foundation

[1]The material presented in this chapter is based on factual information. However, when the instructions and precautions given herein are in disagreement with those of competent local authorities and/or reputable manufacturers, always follow the latter.

[2]Dr. Robert F. Behlow, D.V.M., was not involved in *Beef Cattle Science*, seventh edition.

Contents *Page*

Fig. 12-1. The cow doctor. From a woodcut, 1875. (Courtesy, The Bettmann Archive, Inc., New York, NY)

Without doubt, one of the most serious menaces threatening the livestock industry is animal ill-health, of which the largest loss is a result of the diseases that are due to a common factor transmitted from animal to animal. Today, with modern rapid transportation facilities and the dense livestock population centers, the opportunities for animals to become infected are greatly increased compared with a generation ago.

Each year, cattle producers suffer staggering losses from diseases and parasites—internal and external. Death takes a tremendous toll. Even greater economic losses—hidden losses—result from failure to reproduce living young, and from losses due to retarded growth and poor feed efficiency, carcass condemnations and decreases in meat quality, and labor and drug costs. Also, considerable cost is involved in keeping out diseases that do not exist in the United States; and quarantine of a diseased area may cause depreciation of land values or even restrict whole agricultural programs.

The authors estimated (1) that in 1992 the annual U.S. losses from diseases, parasites, and pests of livestock and poultry aggregated $12.95 billion; and (2) that in 1992 the annual U.S. losses from diseases, parasites, and pests of cattle and calves aggregated $5.7 billion.[3] Thus, the potential for increasing animal profits through disease prevention and parasite control is great.

Cattle producers should also be well informed relative to the relationship of cattle diseases and parasites to other classes of animals and to people, because many of them are transmissible between species. For example, approximately 200 different types of infectious and parasitic diseases can be transmitted from animals to human beings.[4] Of most concern in the latter respect are such animal diseases as brucellosis (undulant fever), leptospirosis, anthrax, rabies, trichinosis, and tuberculosis. Thus, rigid meat and milk inspection is necessary for the protection of human health. This is an added expense which the producer, processor, and consumer must share.

Fortunately for the producer, beef animals are out in the open much of the time, with a minimum of confinement in shelters and small enclosures and close contact with each other. On the other hand, the very fact that beef herds are usually inspected infrequently may result in serious and widespread losses before the ravages of diseases or parasites are observed and control measures instituted.

---

[3]These figures were estimated by the authors, and were derived as follows:

Two authoritative earlier studies were made of the annual U.S. losses from diseases, parasites, and pests of livestock and poultry. Here is what they showed:

a. *Losses in Agriculture, Agriculture Handbook No. 291, ARS, USDA, 1965.* This study showed estimated annual losses from diseases, parasites, and pests of livestock and poultry of $2.8 billion for the period 1951–60. During this same period, (1951–60), annual cash receipts from farm marketings of total livestock and products averaged $17.76 billion (*Agricultural Statistics*). So, a $2.8 billion disease/parasite loss was equivalent to 15.8% of the average annual farm cash receipts for the 10-year period 1951–60.

b. *Report of the Council of Deans, Association of American Veterinary Medical Colleges, 1981.* This study showed 1980 losses from diseases, parasites and pests of livestock and poultry of $10 billion. During the same year (1980), cash receipts from farm marketings of total livestock and products amounted to $67.991 billion (*Agricultural Statistics*). So, a $10 billion disease/parasite loss was equivalent to 14.7% of the annual farm cash receipts from marketing livestock and livestock products that year.

Note: The first study showed animal disease/parasite losses equivalent to 15.8% of the annual farm cash receipts. The second study showed animal disease/parasite losses equivalent to 14.7% of the annual farm cash receipts. So, based on the two studies, the authors concluded that annual livestock disease and parasite losses average about 15% of the cash farm receipts for any given year.

In 1992, the cash farm receipts from marketing U.S. livestock and livestock products totaled $86,358 billion (*Agricultural Statistics 1993,* p. 379, Table 559); hence, the authors estimated that the losses from animal diseases and parasites totaled $12.95 billion.

In 1992, the cash farm receipts from marketing cattle and cattle products totaled $37,882 billion (*Agricultural Statistics 1993*, p. 379, Table 559); hence, the authors estimated that the losses from cattle diseases and parasites totaled $5.7 billion.

[4]Hull, Thomas G., editor, *Diseases Transmitted from Animals to Man*, sixth edition, Charles C Thomas, Publisher Springfield, IL, 1975.

NORMAL TEMPERATURE, PULSE RATE, AND RESPIRATION RATE OF FARM ANIMALS

Table 12-1 gives the normal temperature, pulse rate, and breathing rate of farm animals. In general, any marked and persistent deviations from these normals may be looked upon as a sign of animal ill health.

Cattle producers should provide themselves with an animal thermometer, which is heavier and more rugged than the ordinary human thermometer. The temperature is measured by inserting the thermometer full length in the rectum, where it should be left a minimum of three minutes. Prior to inserting the thermometer, a long string should be tied to the end. Where a large number of animals is involved, a more rapid electronic thermometer may be used.

In general, infectious diseases are ushered in with a rise in body temperature, but it must be remembered that body temperature is affected by stable or outside temperature, exercise, excitement, age, feed, etc. It is lower in cold weather, in older animals, and at night.

The pulse rate indicates the rapidity of the heart action. The pulse of cattle is taken either on the outside of the jaw just above its lower border, on the soft area immediately above the inner dewclaw, or just above the hock joint. It should be pointed out that the younger, the smaller, and the more nervous the animal, the higher the pulse rate. Also, the pulse rate increases with exercise, excitement, digestion, and high outside temperature.

The respiration rate can be determined by placing the hand on the flank, by observing the rise and fall of the flanks, or, in the winter, by watching the breath condensate when coming from the nostrils. Rapid breathing due to recent exercise, excitement, hot weather, or poorly ventilated buildings should not be confused with disease. Respiration is accelerated in pain and in febrile conditions.

A PROGRAM OF BEEF CATTLE HEALTH, DISEASE PREVENTION, AND PARASITE CONTROL

Although the exact program will and should vary according to the specific conditions existing on each individual farm or ranch, the basic principles will remain the same. With this thought in mind, the following programs of beef cattle health, disease prevention, and parasite control is presented with the hope that beef cattle producers will use it (1) as a yardstick with which to compare their existing programs, and (2) as a guidepost so that they and their local veterinarians, and other advisors, may develop a similar and specific program for their own enterprises:

- **General beef cattle programs—**

1. Breed only healthy cows to healthy bulls.
2. Avoid either an overfat or a thin, emaciated condition in all breeding animals.
3. Flush cows by providing more lush pasture or by feeding grain so that they gain approximately 1.5 lb per head daily beginning 20 days before the start and continuing throughout the breeding season. This practice will cause more cows to come into heat, to breed early in the season, and to conceive at first service.
4. Provide plenty of exercise for bulls and pregnant cows, preferably by allowing them to graze in well-fenced pastures in which plenty of shade and water are available.
5. Keep lots and corrals well drained and as dry as practical to prevent breeding places for foot rot, other diseases, and parasites. Fence cattle out of pasture mudholes for the same reason.
6. If possible, divert drainage from adjacent premises and avoid across-the-fence contact with the neighbors' cattle unless they are definitely disease free. Do not visit farms where infectious diseases exist,

TABLE 12-1
NORMAL TEMPERATURE, PULSE RATE, AND RESPIRATION RATE OF FARM ANIMALS

Animal	Normal Rectal Temperature[1]				Normal Pulse Rate	Normal Respiration Rate
	Average		Range			
	(°F)	(°C)	(°F)	(°C)	(rate/minute)	(rate/minute)
Cattle	101.5	38.6	100.4–102.8	38.0–39.3	60–70	10–30
Sheep	102.3	39.1	100.9–103.8	38.3–39.9	70–80	12–20
Goats	103.8	39.9	101.7–105.3	38.7–40.7	70–80	12–20
Swine	102.6	39.2	102.0–103.6	38.9–39.8	60–80	8–13
Horses	100.5	38.1	99.0–100.8	37.2–38.2	32–44	8–16
Poultry	106.0	39.4	105.0–107.0	40.5–41.7	200–400	15–36

[1]To change Fahrenheit to degrees Centigrade, subtract 32, then multiply by 5/9.

as the germs may be brought home on shoes, clothing, or vehicles. For the same reason, feeds should not be brought from such farms, and one should beware of used feed bags.

7. If rented pastures must be used, avoid areas on which cattle have overwintered; and, preferably, use only those rented pastures that have not had cattle on them for one year or that have been plowed in the interim.

8. Eliminate the breeding ground of parasites as far as practical and use the proper insecticide or anthelmintic for their control.

9. Keep commercial cattle—such as stocker and feeder cattle, and finishing cattle—in isolated areas away from breeding animals.

10. Have all cows checked for pregnancy 45 to 60 days after the breeding season is over. When problems are encountered, immediately consult a veterinarian.

11. When disease troubles are encountered, isolate infected animals and follow the instructions and prescribed treatment of a veterinarian.

■ **New stock—**

1. When cattle are being brought in from out-of-state, comply with the specific health requirements of the state which they are entering.

2. Isolate newly acquired animals for a minimum of three weeks, during which time they should be cared for by a separate caretaker.

3. While in isolation, test all newly acquired breeding animals for anaplasmosis, brucellosis, Johne's disease, leptospirosis, tuberculosis, and vibriosis; first, however, make every reasonable effort to ascertain that they come from herds which are known to be free from these and other diseases.

4. Spray newly acquired animals for lice control; and check them for internal parasites, and treat where indicated.

5. When possible, it is preferable to purchase virgin heifers and bulls, from a disease control standpoint. Isolate *tried* (nonvirgin) bulls for a period of three weeks, and then turn them with a limited number of virgin heifers; observe these heifers for 30 to 60 days after breeding, as an aid in preventing the introduction of breeding diseases.

6. Thoroughly clean and disinfect the isolation stall after each animal(s) is removed and before a new animal(s) is placed therein.

■ **Calving time—**

1. When weather conditions permit, allow parturient cows to calve in a clean, uncontaminated, open pasture. During inclement weather, place the cows in isolated, roomy, light, well-ventilated maternity stalls—which should first be carefully cleaned, thoroughly disinfected, and provided with clean bedding for the occasion. After calving, all wet, stained, or soiled bedding should be removed and the floor sprinkled with lime; the afterbirth should be burned or buried deep in lime; and, if there has been trouble, the cows should be kept isolated until all discharges have ceased.

2. Unless the calves are born on a clean pasture away from possible infection, treat the navel cord of each newborn animal with tincture of iodine.

3. See that the newborn calf gets colostrum milk as soon as possible. But bear in mind that the antibodies of colostrum depend upon the dam's disease history, either directly or through vaccinations.

■ **Suckling calves—**

1. If the baby calves are confined to stalls, scrub stalls thoroughly twice each week with warm soap solution and disinfect the walls and feed bunks and/or mangers.

2. Vaccinate calves with blackleg and malignant edema bacterin at 2 to 3 months of age in areas that are endemic for these diseases.

CALF HEALTH-TREATMENT PROGRAM (PRECONDITIONING)

This is the process of preparing the calf to withstand the stress and rigors of leaving its mother, learning to eat new kinds of feed, and shipping from the farm or ranch to a feedlot or another pasture. A good program alerts buyers to the predictable performance of the calves, and avoids the duplication of certain vaccinations. Such a program is also known as *preconditioning.*

A good Calf Health-Treatment Record provides the following information:

1. The name, address, and telephone number of the producer of the calves.

2. Information about the calves that are for sale, including number in the group, tag numbers, sex and birth dates of calves, brand and location of brand, castration method, dehorning method, and breed description of the calves.

3. Vaccination information, including treatment date, treatment given, product used, lot number of product, company selling the product used, dosage, route of administration, and the processor's identification. Also, line drawings of the left and right sides of a calf provide a place on which to show the location of sites of injections.

4. Implant information, including product used, date given, and right/left ear.

5. Weaning/pre-conditioning information, including date weaned, bunkbroke, water-tank trained, ration fed—and number of days it was fed.

6. Signatures, including the owner's signature and date verifying the accuracy of the information, and the signature and telephone number of the producer's veterinarian.

The National Cattlemen's Beef Association's (NCBA) excellent one-page Calf Health-Treatment Record form is herewith reproduced as Fig. 12-2. This form, which is available without charge, may be obtained from the NCBA, P.O. Box 3469, Englewood, CO 90155, telephone (303) 694-0305. Many of the state cattle associations have a similar Calf Health-Treatment Record.

FATIGUE, STRESS, SHRINK, AND DISEASE RESISTANCE

Changed environment, excitement of sorting, loading, and shipping; long periods without feed and/or water; movement through one or more assembly points; change of feed; and exposure to disease—all add up to *fatigue, stress, shrink*, and *lowered disease resistance*.

Calf health-treatment is the answer. The steps used may, and should, vary somewhat among areas, farms, and ranches. The important thing is that the program be written down, adhered to rigidly, then certified to by both the owner and the veterinarian.

■ **Where does stress take place?**—Stress takes place when a calf is weaned, at which time the social structure is disrupted. Also, there is stress when an animal is placed in a different environment, whether it be in a pen on the home ranch or on a loaded truck on the way to market. If stress is to be lessened, therefore, it is obvious that measures must be taken before any of these steps happen.

■ **Losses attributed to prefeedlot stress**—On the average, 25% of the feedlot calves sicken, and 5% of the calves die. Losses from sickness are even greater—they run 2 to 5 times as great as the actual death loss. Sickness losses accrue from the expense and treatment plus the resulting inefficiency. Thus, the combined losses—death losses, and losses due to sickness—add approximately $55 per head onto the cost of every feedlot-finished animal.

■ **What are health-treated calves?**—Opinions differ rather widely as to what constitutes properly health-treated calves. However, the following pertinent provisions of such a program are presented with the hope that the beef producer will use it (1) as a yardstick with which to compare the existing program, or (2) as a guide so that the animal health team (the producer, the veterinarian, and other advisers) may develop a similar and specific program for their own enterprise.

1. **Handle quietly.** Calves should be handled quietly, with a minimum of excitement.

2. **Dehorn and castrate.** All calves that will eventually go into feedlots should be dehorned (although tipping of horns is acceptable), and they should be castrated unless they are to be fed out as bulls. There is far less stress if calves are dehorned and castrated well ahead of weaning—about two months of age is best.

3. **Wean.** Calves should be weaned 30 days ahead of shipment.

4. **Start on feed.** Adjust to feed bunks and water troughs and start on a ration similar to that which they will get in the feedlot. For the first three days following weaning, calves should have access to loose grass hay. Additionally, they should be started on a ration of about the following composition:

Crude protein, minimum %	12.0
Calcium, %	0.5
Phosphorus, %	0.3
Vitamin A, IU/lb	5,000
Net Energy for production, (NE_p), Mcal	38
Roughage-concentrate ratio, approx.	40:60

If weaning is totally impractical, calves should be started on a creep feed similar to the above ration.

This type of ration will be very similar to the starting ration that calves will receive when they arrive in the feedlot.

Use medicated feed only on the recommendation of your veterinarian.

5. **Vaccinate.** Vaccinate either two weeks before or after weaning. If calves were vaccinated for blackleg, malignant edema, and leptospirosis before three months of age, revaccinate. Simultaneously, vaccinate for *red nose* (infectious bovine rhinotracheitis, or IBR), bovine virus diarrhea (BVD), and bovine respiratory disease complex. In some instances, clostridial toxoids for types C and D are needed. Follow your veterinarian's advice for vaccination procedures. If a direct sale to a feedlot is involved, the calves should be vaccinated in keeping with the regular program of the feedlot.

Injection-site lesions are a serious problem in the beef industry. Therefore, it is important that proper needle size and length, route of administration, and sanitation are selected and followed in performing vaccinations.

6. **Treat for parasites.** At the time of weaning, and prior to shipment, calves should be checked for both internal and external parasites, and treated as necessary. Usually this involves (a) treating for grubs, through either spray, pour-on, or feed: (b) spraying for lice; and (c) checking for worm eggs, and worming if necessary.

7. **Reduce time from farm or ranch to feedlot.** Every effort should be made to reduce the total time

NCA

Calf Health-Treatment Record

Beef Quality Assurance National Cattlemen's Association

Owner ______ Ranch name ______

Address ______ City ______ State ______ Zip ______

Phone (Area Code ______) ______ Best time to call ______

Information on Cattle

Number of cattle in group ______ Tag numbers ______ Color of tags ______ Sex of cattle ______

Birthdate of calves ______ Castration method ______

Brand (draw picture) ______ Location of brand ______

Polled/dehorned/horned ______ Date of dehorning ______ Dehorning method ______

Breed/description of cattle ______

Location of Vaccination

(Preferred location indicated)

Instructions: Write "1" where "Treatment 1" given; write "2" where "Treatment 2" given; etc.

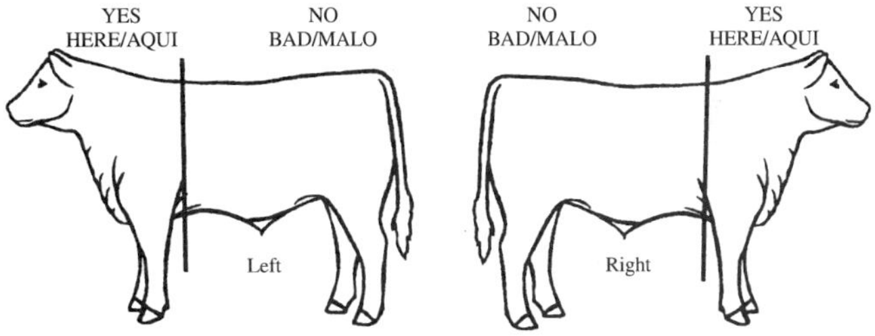

YES HERE/AQUI | NO BAD/MALO | Left | NO BAD/MALO | YES HERE/AQUI | Right

Vaccination Information

	Treatment Date	Treatment	Product	Lot No.	Company	Dosage	Route of Admin.	Processor's Initials
1		7-Way						
2		Brucell. Calfhd. Vac.						
3		Deworm						
4		Pour-on						
5		IBR						
6		PI3						
7		BRSV						
8		BVD						
9		H. Somnus						

Tag numbers of individuals requiring additional treatment & description of treatment given ______

Implant Information

Product used ______ Date given ______ Right/left ear ______

Weaning, Pre-Conditioning Information

Date weaned ______ Bunkbroke (yes/no) ______ Water-tank trained (yes/no) ______

Ration ______ No. of days ______

Signed ______ Date ______

Veterinarian ______ Phone ______

Fig. 12-2. Calf Health-Treatment Record.

between the moment calves leave the farm or ranch and when they arrive at the feedlot.

Where either truck or rail shipments are longer than 36 hours (the 28-hour law governing rail shipments may be extended to 36 hours upon written request of the owner), unload en route for the purpose of giving feed, water, and rest for a period of at least 5 consecutive hours before resuming transportation.

8. **Reduce stress and exposure to infection.** The stress and exposure to infection during the marketing and transportation periods should be reduced to a minimum.

9. **Provide Calf Health-Treatment Certificate.** It is extremely important that records be kept of all husbandry, nutritional, and medical histories, and that the seller of the calves provide the person receiving them with a written record of all of them. This will help the next owner fit the calves to the existing program and minimize costly and unnecessary procedures.

▪ **Cost of calf health-treatment**—The cost of calf health-treatment will run about $15 to $18 per head, the amount depending primarily upon the number and kind of vaccinations administered and the cost of feed. Some or all of this cost will be recovered by gains made during the treatment period.

▪ **Marketing agencies can lessen stress and exposure**—The above discussion pertains to calf health-treatment at the farm or ranch level. It is recognized, however, that not all cattle are moved directly from the cow-calf producer to the feedlot. Many of them pass through auction markets, terminal markets, and other similar intermediate locations. Hence, the marketing agencies handling feeder cattle can do much to cut down on stress and exposure to disease. Recommendations to this end are:

1. Refuse to accept sick animals, or at least isolate them. This is important because one sick animal may cause a disease outbreak in hundreds of other cattle.
2. Isolate cattle from individual producers whenever possible. Here again, this is recommended in order to cut down the spread of infectious diseases.
3. Keep dust to a minimum.
4. Feed animals well-balanced rations similar to the rations to which they have been accustomed.
5. Move cattle to their final destination as expeditiously as possible.
6. Prevent bruises.
7. Improve sanitation in all handling facilities.
8. Reduce weather stress.
9. Keep records which will help pinpoint problem areas and unscrupulous individuals.

▪ **Calf health-treatment should increase**—On an industrywide basis, health-treatment of calves could save millions of dollars now lost in sickness, shrink, and death.

Feeder buyers will determine how quickly health-treated calves become generally available. They will speed their availability when they ask the seller:

1. When were they weaned (if calves are involved?)
2. Are they accustomed to bunk feeding and trough watering?
3. Have they been treated for grubs?
4. Have they been examined for internal and external parasites, and treated if necessary?
5. Have they been vaccinated for IBR, BVD, shipping fever, blackleg?
6. Is certification for the above procedures available?

Buyers will increasingly favor those producers who follow such a program. Health-treatment, along with improved breeding based on production testing, will be the trademark of the producer of reputation feeder calves in the future.

A CATTLE FEEDLOT DISEASE AND PARASITE CONTROL PROGRAM

Cattle feedlot health is fully covered in Chapter 33 of this book: hence, the reader is referred thereto.

DISEASES OF BEEF CATTLE

Disease may be defined as an illness, a sickness, or any deviation from a state of health.

Seventy-five percent of the losses from beef cattle diseases are due to common, well-known diseases.

Fig. 12-3. When disease strikes, correct diagnosis and prompt treatment will save many animals. This shows a steer receiving a sustained release sulfa bolus through a balling gun. (Courtesy, Smith Kline Animal Health Products, Philadelphia, PA)

Modern science makes it possible effectively to prevent many of these diseases through a combination of good management, proper nutrition, vaccination, and other disease prevention practices.

It is not intended that this book shall serve as a source of home remedies. Rather, the enlightened cattle producer will institute a program designed to assure herd health, disease prevention, and parasite control. When animal disease troubles are encountered, the caretaker will not attempt to diagnose or treat but will call upon the local veterinarian in exactly the same manner as the family doctor is called when human ill health is encountered.

This chapter is limited to nonnutritional diseases and ailments; the nutritional diseases and ailments of cattle are covered in Table 8-15, Chapter 8.

ANAPLASMOSIS

Note: Although anaplasmosis is a parasitic disease of the blood, it is usually listed as a disease, not as a parasite.

Anaplasmosis is an infectious disease whose etiology and symptomatology are similar to cattle tick fever, except that more carriers are involved.

The disease is widely distributed in warm climates throughout the world. In the United States, it has been prevalent throughout the southern states, but it is slowly spreading to the northern states.

The mortality rate may vary from 2 to 5% to 50 to 60%. The most severe losses are found in older animals and in hot weather.

SYMPTOMS AND SIGNS

The symptoms may be those of a mild, acute, or chronic condition. Calves usually have the mild type of infection, simply becoming "dumpy" for a few days and then apparently recovering, though their blood remains the permanent abode of the parasite.

Fig. 12-4. Cow exhibiting typical symptoms of acute anaplasmosis. (Courtesy, USDA)

The more characteristic symptoms in mature animals include rapid, pounding heart action, labored and difficult breathing, rise in temperature (up to 107°F), dry muzzle, marked depression, tremors of the muscles, loss of appetite, and a great reduction in the milk flow. Animals usually show yellowing of the eye and other mucous membranes and of the skin, as in jaundice. Depraved appetite, evidenced by the eating of bones or dirt, is not uncommon. Sick animals may also show brain symptoms and an inclination to fight. Unlike cattle tick fever, bloody urine is not common in anaplasmosis. In severe acute cases, death may follow in one to a few days. Recovery is usually very slow, and although no clinical symptoms remain, such animals continue as permanent carriers of the parasite.

CAUSE, PREVENTION, AND TREATMENT

It is caused by a minute parasite, *Anaplasma marginale*, which invades the red blood cells. The parasite is transmitted from infected to healthy animals by ticks, horseflies, stable flies, mosquitoes, deer flies, and probably by other biting insects.

In infected animals, the causative parasite, *Anaplasma marginale*, lives in the red blood cells. The parasite and, consequently, the disease may be transmitted from animal to animal by means of biting insects and by such mechanical agencies as needles, dehorning instruments, etc. Any animal that has once contracted the disease permanently retains the parasite in the blood, though no signs of ill health may be evident. Such animals are *carriers*, and are potential sources of danger to others.

In addition to carrier animals, there is another reservoir of anaplasmosis infection in the western range states—the wood tick. This insect is a biological vector, since the disease will overwinter in its body.

Importing infected carrier cattle into clean areas and clean herds is the most common method of spreading the disease. Once an infected animal has been introduced, the disease is spread within the herd by insect vectors, primarily the wood tick and biting flies, or by mechanical means such as castrating, ear tagging, dehorning, and vaccination. Infection can be transferred any time fresh blood is transferred from an infected to a noninfected animal.

Once an animal becomes a carrier, it is immune to subsequent exposure to the disease agent. However, carrier animals that have been treated and freed of infection become susceptible again.

Prevention in lightly infected areas consists of (1) killing or repelling vectors on the host with chemical dusts and sprays; and (2) testing the herd and finding infected animals, and then either removing the infected animals by culling them for slaughter or feeding an antibiotic.

The following control measures will reduce losses due to anaplasmosis:

1. **Anaplasmosis vaccination.** A vaccine containing killed *Anaplasma*, marketed under the trade name of Anaplaz, is used in the United States. The initial vaccination (first year) consists of 2 doses given 4 weeks apart. Protection from clinical anaplasmosis is provided 2 weeks after the second injection. The following year, at least 2 weeks before the vector season, a single booster injection should be administered. Following the first booster, animals should receive subsequent single booster injections every 2 years to maintain practical herd protection. *Note well:* (1) a vaccinated animal is still capable of becoming infected with anaplasmosis and subsequently can become a carrier; and (2) the vaccine does not prevent infection, but aids in the prevention of the clinical symptoms of anaplasmosis.

2. **Test and segregation.** The rapid card agglutination (CT) test can be used to identify the carrier animals in a herd, following which they can be segregated from the "clean" animals and treated to cure the carrier state.

3. **Continuous use of oral tetracycline medication in the feed.** The tetracyclines (chlortetracycline, oxytetracycline, and tetracycline) in feed, mineral blocks, or salt-mineral mixes, at an oral dosage of 1.1 mg/kg body weight/day of tetracycline, will (a) prevent transmission of anaplasmosis, and (b) effectively halt the infection if given within the first week of the incubation period. Medicated feed should be provided free-choice beginning l month prior to the vector season and continuing until l month after the vectors disappear.

The tetracyclines (chlortetracycline, oxytetracycline, and tetracycline) are the only effective drugs for treatment of bovine anaplasmosis that are approved for use in food animals by the U.S. Food and Drug Administration.

ANTHRAX (Charbon, or Splenic Fever)

Anthrax, also referred to as charbon or splenic fever, is an acute, infectious disease affecting all warm-blooded animals and humans; but mature cattle are most susceptible. It usually occurs as scattered outbreaks or cases, but hundreds of animals may be involved. Certain sections are known as anthrax districts because of the repeated appearance of the disease. Grazing animals are particularly subject to anthrax, especially when pasturing closely following a drought or on land that has been recently flooded. In the United States, most human infections of anthrax result from handling diseased or dead animals on the farm or from handling hides, hair, and wool in factories.

Historically, anthrax is of great importance. It was one of the first scourges to be described in ancient and Biblical literature; it marked the beginning of modern bacteriology, being described by Koch in 1876; and it was the first disease in which immunization was effected by means of an attenuated culture, Pasteur having immunized animals against anthrax in 1881.

SYMPTOMS AND SIGNS[5]

The mortality is usually quite high. It runs a very short course and is characterized by a blood poisoning (septicemia). The first indication of the disease may be the presence of severe symptoms of colic accompanied by high temperature, loss of appetite, muscular weakness, depression, and the passage of blood-stained feces. Swellings may be observed over the body, especially around the neck region. Milk secretion may turn bloody or cease entirely; and there may be a bloody discharge from all body openings.

CAUSE, PREVENTION, AND TREATMENT

The disease is identified by a microscopic examination of the blood in which will be found the typical large, rod-shaped organisms *(Bacillus anthracus)* causing anthrax. The bacillus can survive for years in a spore stage, resisting most destructive agents. As a result, it may remain in the soil for extremely long periods.

This disease is one that can largely be prevented by immunization. In the so-called anthrax regions, vaccination should he performed annually, usually in the spring, and well in advance of the time when the disease normally makes its appearance. The nonencapsulated Sterne-strain vaccine is used almost universally for immunization.

In infected areas, adequate fly control should be obtained by spraying animals during the insect season.

Herds that are infected should be quarantined, and all milk and other products should be withheld from the market until the danger of disease transmission is past. The farmer or rancher should never open the carcass of a dead animal suspected of having died from anthrax, because the organism is infectious to

[5]Currently, many veterinarians prefer the word *signs* rather than *symptoms*, but throughout this chapter the authors accede to the more commonly accepted terminology among cattle producers and include the word *symptoms*.

humans; instead, the veterinarian should be summoned at the first sign of an outbreak.

When the presence of anthrax is suspected or proved, all carcasses and contaminated material should be completely burned or deeply buried and covered with quicklime, preferably on the spot. This precaution is important because the disease can be spread by dogs, coyotes, buzzards, and other flesh eaters, and by flies and other insects.

When an outbreak of anthrax is discovered, all sick animals should be isolated promptly and treated. All exposed healthy animals should be vaccinated, pastures should be rotated, and a rigid program of sanitation should be initiated. Anthrax is a reportable disease, requiring quarantine. Hence, control measures will be carried out under the supervision of state or federal regulatory officials.

Early treatment of affected animals, with massive doses of penicillin or the tetracyclines may be effective if given soon enough.

BACILLARY HEMOGLOBINURIA (or Red Water Disease)

Bacillary hemoglobinuria, which is an acute, infectious disease, is often confused with other cattle diseases in which blood-colored urine is seen. The disease usually occurs in cattle that are pastured on meadows or irrigated lands where drainage is poor, and during the summer and early fall months. Sheep are also affected, but to a lesser degree than cattle. A mortality rate up to 100% occurs in untreated cases.

SYMPTOMS AND SIGNS

All ages are affected, but most losses occur in cows over one year of age.

The course of the disease is usually two days or less. Appetite, rumination, and milk flow suddenly cease. The animal hesitates to move and stands apart from the herd. The eyes are sunken, bloodshot, and may appear yellow. Breathing is rapid, and the temperature is high. The urine and feces are usually both blood tinged from destroyed red cells. It must be understood, however, that bloody urine (red water) may also be one of the symptoms in such conditions and diseases as lack of phosphorus, leptospirosis, Texas fever, plant poisoning, and anthrax. As red water disease can easily be confused with other conditions producing bloody urine, laboratory assistance is usually indicated in the event of unknown hemoglobinuria (bloody urine).

CAUSE, PREVENTION, AND TREATMENT

An anaerobic bacterium called *Clostridium hemolyticum* is the primary causative agent, and its toxin causes the blood breakdown. It is found in moist alkaline soils, especially in the low lying valley land of the Sierra Nevada, and the Pacific Coast and Cascade ranges in Nevada, California, and Oregon; and in Washington, Louisiana, Florida, Montana, Idaho, and Texas.

Inoculations with a bacterin (inactivated bacteria) or toxoid (inactivated toxin) to stimulate immunity are valuable in communities where annual losses occur. This vaccination should occur two weeks prior to the time of the previous annual outbreak. Unless unavoidable, cattle should not be pastured on areas of known infection. Pools of stagnant water should be drained as such areas provide a favorable environment for the growth of the causative agent.

Treatment with large doses of penicillin or broad-spectrum antibiotics repeated at 12-hour intervals may be effective if started early.

Clostridium hemolyticum bacterin will confer immunity for six months. Where the disease occurs throughout the year, semiannual immunization is necessary.

BLACKLEG (Black Quarter, Emphysematous Gangrene, Quarter-Ill, or Symptomatic Anthrax)

This is a very infectious, highly fatal disease of cattle, and less frequently of sheep and goats. The disease is widespread, especially in the western range states. It occurs at almost any season, predominating in the spring and fall months among pastured cattle; but it may occur in the winter in stabled cattle. Once prevalent in a community, the disease remains there as a permanent hazard, the infected territory being referred to as a "hot area." It is seen most frequently in cattle ranging in age from 3 months to 2 years, but it may occur in older animals.

SYMPTOMS AND SIGNS

The incubation period is from 1 to 5 days, and its course is from 1 to 3 days. The first symptom noted is lameness, usually accompanied by or followed by swellings of gas under the skin in the areas of the neck, shoulder, flanks, thighs, and breast, which crackle under pressure. High fever, loss of appetite, and severe depression accompany the symptoms. Although there are a few recoveries, death is the usual termination, occurring within 3 days of the onset of symptoms.

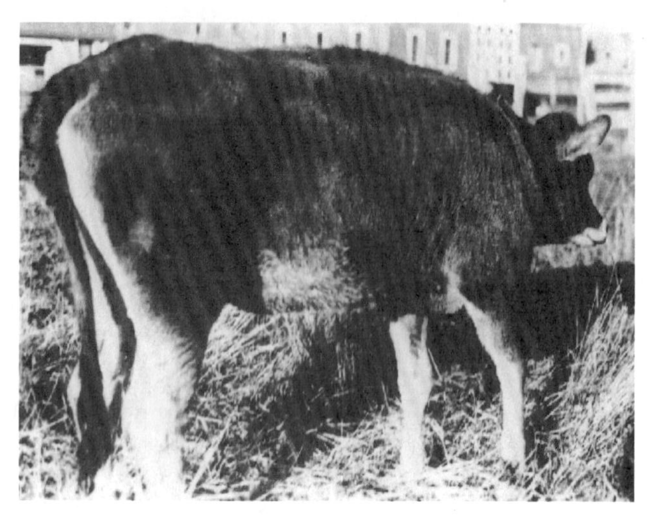

Fig. 12-5. Heifer with blackleg, six hours before death. Note the lameness and swelling over the neck and shoulder. (Courtesy, Veterinary Research Laboratory, Montana State University)

CAUSE, PREVENTION, AND TREATMENT

This disease is caused by an anaerobic bacterium, called *Clostridium chauvoei*, although it is often accompanied by other of the *Clostridia* genus. Infection is usually the result of wound contamination or ingestion of the organisms.

Prevention consists of vaccination (currently, most vaccines contain both blackleg and malignant edema) of all animals at approximately 3 to 4 months of age, followed by a second, or booster, vaccination at about 12 months of age, using one of the approved vaccines. In endemic areas, the first vaccination should be given at 1 month of age. A natural resistance tends to develop when the animal is about 2 years of age.

Animals that die from blackleg should not be cut open unless under the direction of a qualified veterinarian. The carcasses should be burned or deeply buried and the contaminated area disinfected. Eradication of blackleg from pastures is difficult if not impossible.

In the early stages of the disease, massive doses of penicillin or tetracyclines will sometimes save an animal. But a good immunization program is the key to preventing losses due to blackleg.

BLUETONGUE

Cattle are sometimes mildly affected with bluetongue. But they are an important reservoir for sheep and other susceptible ruminants.

SYMPTOMS AND SIGNS

If cows become infected during gestation, they may abort or give birth to abnormal calves. Some cattle may develop mild clinical signs similar to those in infected sheep, characterized by shortness of breath and panting, a blue tongue, high temperature (104–107°F), loss of appetite and weight, reddened mucous membrane of the mouth which turns purplish or blue in color, and lameness.

CAUSE, PREVENTION, AND TREATMENT

Bluetongue is caused by a virus transmitted by biting insects of the *Culicoides* sp.

Vaccines are available for sheep.

Treatment of cattle is not necessary because of the mildness of the disease in cattle.

BOVINE PULMONARY EMPHYSEMA

This disease is also known as cow asthma, panters, lungers, fog fever, skyline fever, summer pneumonia, green grass poisoning, and grunts.

Bovine pulmonary emphysema usually occurs when cattle are abruptly changed from dry, mature feed to green, immature pasture. Typically, cases develop when cattle are moved from dry range to green mountain or irrigated pastures in the fall or late summer. It may also occur when cattle are first turned to lush pasture in the spring.

SYMPTOMS AND SIGNS

The disease is characterized by rapid and labored breathing (the animal forces air from the lungs, and many grunt with each breath). Affected animals may breathe through, and froth at, the mouth. The temperature remains normal or only slightly elevated, and the appetite is good.

CAUSE, PREVENTION, AND TREATMENT

The cause is unknown. It is noteworthy, however, that workers at Washington State University have produced the condition experimentally by giving cattle the essential amino acid tryptophan, leading to speculation that rapidly growing pastures may be high in tryptophan.

Prevention consists of avoiding sudden changes from dry or poor pasture to immature, green feed. Some dry hay should be fed while making the transition.

No drug has proved effective in controlled trials, but aminophylline, corticosteroids, and epinephrine are widely used. Even severely affected animals may recover if removed from the offending pasture and handled quietly.

BOVINE RESPIRATORY DISEASE COMPLEX (BRDC, Shipping Fever, or Hemorrhagic Septicemia)

The term *shipping fever* is losing favor because it is misleading and the term *hemorrhagic septicemia* is best reserved for the septicemic *Pasteurella* infections seen in cattle.

Bovine respiratory disease complex (BRDC) is seldom the result of a single factor. It is usually caused by a combination of stress, virus infection, and invasion of the lungs by certain bacteria. The disease is most common in calves and following shipment. It occurs widely throughout the world, especially among thin and poorly nourished young animals that are subjected to shipment by truck or rail during periods of inclement weather, though it may occur in animals in good condition. The disease is a serious problem to both shippers and receivers of cattle.

SYMPTOMS AND SIGNS

The first sign of the disease (which may appear within 2 to 21 days after moving cattle) is a tired appearance and reduced appetite. The affected animal may show signs of depression, watery to slimelike nasal discharge, increased body temperature (rising to 105–107°F), occasional soft or hacking cough, rapid breathing, loss of appetite, followed by loss of body weight and drop in milk production. In very acute forms, animals may die showing no symptoms. Death losses may be high in untreated cases.

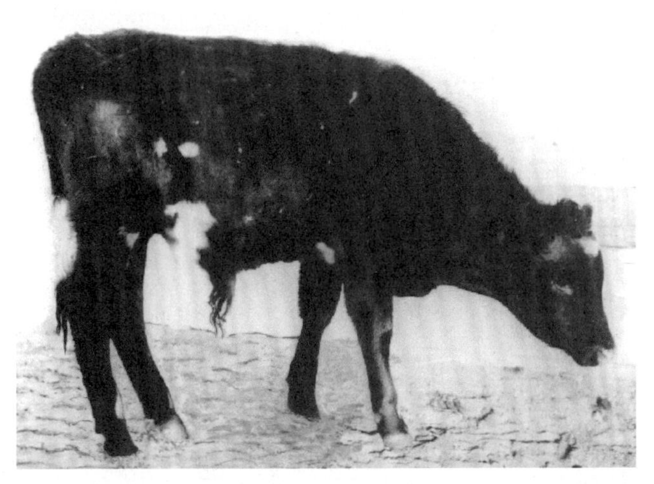

Fig. 12-6. An animal with bovine respiratory disease complex. The disease is most frequently associated with animals whose resistance has been lowered due to travel; hence, the name *shipping fever.* (Courtesy, USDA)

Calves are more susceptible than older animals, but cattle of all ages are affected.

CAUSE, PREVENTION, AND TREATMENT

Bovine respiratory disease complex is caused by multiple infection due to the interaction of viruses and bacteria, accentuated by environmental conditions creating physical tension or stress. Changes in weather and feed, overcrowding, hard driving, lack of rest, and improper shelter all help usher in the disease.

The five viruses which cause most bovine respiratory infections are: IBR (infectious bovine rhinotracheitis), BVD (bovine virus diarrhea), PI3 (parainfluenza 3), BRSV (bovine respiratory syncytial virus), and BVDV (bovine viral diarrhea virus, or Type 2 BVD).

The BVDV variant appeared in Canada and the United States in the early 1990s. It differs from the classical BVD syndrome in its rapid morbidity and mortality. Detection of the new BVDV strain from the classical strain can only be made by molecular techniques following isolation of the virus from tissues.

Other viruses which may cause respiratory problems include adenovirus, rhinovirus, and rotavirus. Unfortunately, two or more of these organisms may infect a herd at the same time. But viruses aren't the only agents of respiratory infection; bacteria can cause problems, too, especially in cattle already weakened by infections. For example, infection by *Pasteurella multocida* and *Pasteurella haemolytica* is thought to be a major cause of shipping fever (hemorrhagic septicemia). Other bacteria which may infect weakened cattle include *Haemophilus somnus* and species of *Salmonella*, *Pseudomonas*, and *Leptospira*.

As a preventive measure, one should eliminate as many as possible of the predisposing factors that lower the animal's vitality. Also, newly purchased animals should be isolated for 2 to 3 weeks before being placed in the herd.

Immunity can be achieved by administration of the proper vaccine according to label directions.

Bacterial infections can be controlled by bacterins (vaccines) which are composed of killed organisms. Two injections 14 or more days apart will provide adequate protection.

The cattle producer who plans to ship young stock should first confer with a veterinarian relative to the choice of and the time to administer the vaccine(s).

Where calves have been subjected to great stress—weaning, long shipment, extensive handling, and/or exposure to severe weather conditions—it is recommended that they be handled as follows: given long grass or oat hay and a calf starter ration, plus access to plenty of clean, fresh water at all times. Also, the incidence of BRDC can be reduced by feeding a combination of an antibiotic (chlortetracycline) and a sulfonamide (sulfamethazine) for 30 days; at the end of the 30 days, the medicated feed should be slowly withdrawn so as to prevent bloat. Newly arrived cattle should also receive 50,000 IU of vitamin A per head daily and have free access to a good mineral mixture.

Antibiotics (*e.g.*, oxytetracycline) and sulfa drugs (*e.g.*, sulfamethazine) are effective treatments if given early in the course of the disease. Treatment after BRDC develops is often disappointing.

BOVINE SPONGIFORM ENCEPHALOPATHY (BSE/Mad Cow Disease)

BSE is a fatal degenerative neurological disease of adult cattle which belongs to a family of diseases known as transmissible spongiform encephalopathies (TSEs) that includes scrapie of sheep and goats. It was first diagnosed in Great Britain in 1986. By December 1993, it had affected more than 112,000 cattle in over 24,000 herds in Britain. By September 2, 1994, a staggering total of 134,202 cases of BSE in 31,269 herds had been confirmed in Great Britain (England, Scotland, and Wales); and these confirmed cases affected 52.1% of the dairy herds and 13.7% of the beef herds of Britain.

On March 20, 1996, the Spongiform Encephalopathy Advisory Committee in Great Britain expressed concern to the British government in regards to 10 cases of Creutzfeldt-Jacob disease (CJD) that had affected youth in the United Kingdom. Although the committee concluded that there was no direct scientific evidence linking BSE and CJD based on current data, the most likely explanation in their opinion was that the humans were exposed to BSE. CJD is a slow, degenerative disease that affects the central nervous system in humans, causing dysfunction, progressive dementia, vacuolar degeneration of the brain, and eventually death.

This announcement caused panic in Great Britain concerning consumption of beef and beef products, and the news quickly spread around the world. On March 28, 1996, the European Union imposed an export ban on British beef, which caused cattle prices to plummet in Great Britain. Great Britain currently has a program in place in which all cattle older than 30 months of age eventually will be slaughtered and destroyed.

In addition to Great Britain, BSE has been reported in Canada (one case only), Portugal, Ireland, Switzerland, France, Germany, Oman, Denmark, and Falkland Islands.

BSE has not been discovered or reported in the United States. In 1989, the United States government imposed a ban on the imports of British cattle and processed beef. In addition, we have a testing and surveillance program for BSE in this country.

In the period between 1981 and 1989, 499 head of cattle were imported into the United States from Great Britain. These cattle are being monitored on a regular basis.

A number of national livestock and professional health organizations in the United States have recommended a voluntary ban on using ruminant-derived protein in rations for beef and dairy cattle. Also, for a number of years, rendering plants in this country have not accepted sheep carcasses because of concern regarding scrapie in sheep.

SYMPTOMS AND SIGNS

BSE-affected cattle show a progressive central nervous system degeneration and exhibit changes in temperament, including nervousness or aggression, abnormal posture, diminished coordination and difficulty in rising, decreased milk production, loss of body weight despite continued appetite, and eventually death. Diagnosis can be confirmed by a post-mortem examination of brain tissue.

CAUSE, PREVENTION, AND TREATMENT

The cause of BSE and other TSEs in other species is not fully understood. Three main theories have been proposed: (1) a prion, an abnormal protein capable of causing a cell to produce abnormal protein, (2) a very small unconventional virus, or (3) an incomplete virus "virino," lacking a protein coat. Most scientists working in the field feel the cause is a prion or abnormal protein.

Prevention consists of notification of suspected clinical cases to regulatory authorities, destroying affected animals, and not feeding meat meal made from rendered sheep infected with scrapie or from cattle infected with BSE. It is noteworthy that to date (1994) all confirmed cases of BSE in the U.K. were traced to meat meal made from rendered sheep carcasses infected with scrapie. Obviously, sheep scrapie crossed the species boundary and infected cattle that ate the infectious agent. Since 1989, the British government has imposed a ban on the feeding of ruminant-derived proteins in cattle, which has decreased the incidence of BSE.

Treatment is ineffective; however, a simple test for BSE was developed in 1996.

BOVINE VIRUS DIARRHEA (BVD, or Mucosal Disease)

Bovine virus diarrhea is not new, having first been described in 1946.

The disease is widespread in the United States. The greatest losses are in weight, condition, and feed. Mortality is low, rarely exceeding 5%.

Bovine virus diarrhea (BVD) may also be involved in bovine respiratory disease complex (BRDC); so, see also the latter.

SYMPTOMS AND SIGNS

The incubation period is 7 to 9 days following exposure to the virus. The disease is characterized by high temperature (104 to 107°F) for 2 to 5 days, nasal discharge, rapid breathing, depression, and loss of appetite. Some animals make a prompt recovery. In other cases, signs persist, including nasal discharge and diarrhea. Sometimes blood flecks occur in the feces. Coughing, eye lesions, and lameness may affect 10% of the herd. In pregnant cows, abortions may appear 3 to 6 weeks after infection; and, in lactating cows, a marked loss in milk production occurs.

Bovine virus diarrhea is not a good name because not all animals exhibit diarrhea.

CAUSE, PREVENTION, AND TREATMENT

As indicated by the name, the disease is caused by a virus.

The most effective preventive measures consist in avoiding contact with affected animals and in keeping away from contaminated feed and water. Also, all incoming animals should be isolated for at least 30 days. Once the disease makes its appearance, sick animals should be isolated and rigid sanitary measures should be initiated.

Where virus diarrhea is a constant problem, cows and feedlot cattle should be vaccinated. Immunity against BVD can be achieved by the intramuscular administration of modified live or inactivated vaccines. But two **don'ts** should be observed: (1) **Don't** use the vaccine on pregnant cows because of possible abortions and birth defects; and (2) **don't** vaccinate calves under 6 months of age because it may be ineffective due to the temporary immunity from colostrum of immune dams.

Antibiotics or sulfonamides effectively combat the secondary bacterial invaders that accompany the disease. Administration of balanced electrolytes and fluid is indicated to rehydrate animals with diarrhea.

BRUCELLOSIS (Bang's Disease)

Brucellosis, which occurs throughout the world, is an insidious (hidden) disease in which the lesions frequently are not evident. Although the medical term *brucellosis* is used in a collective way to designate the disease caused by the three different but closely related *Brucella* organisms, the specie names further differentiate the organisms as: (1) *Br. abortus*; (2) *Br. suis*; and (3) *Br. melitensis*.

The disease is known as brucellosis, Bang's disease (after Professor Bang, noted Danish research worker, who, in 1896, first discovered the organism responsible for bovine brucellosis), or contagious abortion in cattle, caused by *Brucella abortus*. In swine, it's Traum's disease, or infectious abortion, caused by *Brucella suis*. In goats, it's Malta fever, or abortion, caused by *Brucella melitensis*. In humans, it's Mediterranean fever, or undulant fever. The causative organism is also associated with fistuous withers and poll-evil in horses.

In the United States, the incidence of brucellosis-reactor cattle has been reduced from 11.5% in 1935 to 0.015% in 1994. Control and eradication of the disease are important for two reasons: (1) the danger of human infection, it being one of the most important U.S. animal-human diseases; and (2) the economic cattle loss in the form of fewer live calves, more retained placentas, more breeding trouble, more arthritis, more mastitis, and lowered milk production.

The following tests are used for diagnosis of the disease in cattle:

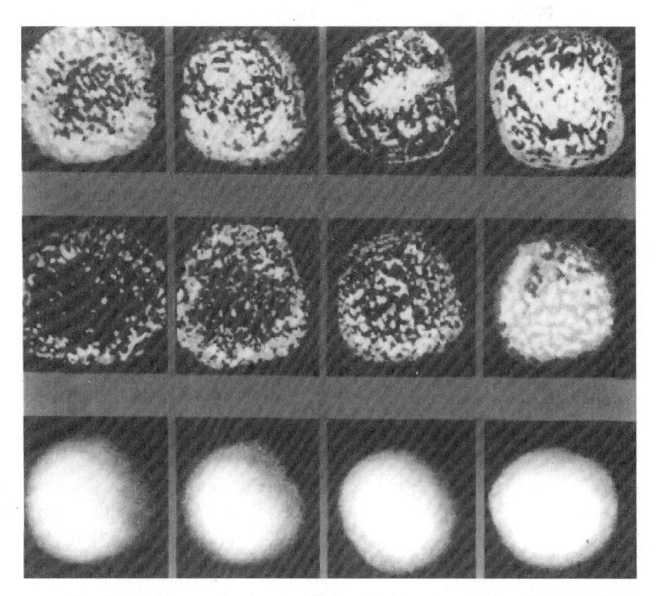

Fig. 12-7. Microscopic picture showing the blood serum (agglutination) test for diagnosis of brucellosis. Top row shows clumping (agglutination), indicating brucellosis. Center row shows complete clumping in the first three dilutions and partial clumping in the 1:200 dilution. Bottom row shows a negative test, indicating brucellosis free. (Courtesy, Lederle Laboratories)

1. *Agglutinatioin test*, of which there are two common methods:

   a. *The tube*, or *"slow" method*—in which a blood sample is taken from the jugular vein; the blood is allowed to clot and the serum to separate; and the serum is mixed in small test tubes with a suspension of specially selected strain of *B. abortus*. Complete agglutination in dilutions of 1:100 and higher are positive.

   b. *The plate* or *rapid test*—This is a rapid agglutination test which is done on a glass slide or plate. The antigen consists of specially selected strains of *B. abortus* stained with gentian violet and brilliant green.

2. *Milk ring test*—This is a modification of the agglutination test which is done with milk. The test involves mixing the antigen with fresh milk. The test depends on the fact that clumps of agglutinated organisms are carried to the surface by rising fat globules. A positive test is indicated by a purple cream layer with white milk below. The milk ring test is a highly efficient and accurate screening test for locating infected dairy herds.

3. *Card test*—This test involves the use of a disposable card on which blood serum or plasma is mixed with buffered whole-cell suspension of *B. abortus* (antigen), which reacts (agglutinates) with antibodies in the blood serum of animals infected with brucellosis.

SYMPTOMS AND SIGNS

Unfortunately, the symptoms of brucellosis are often rather indefinite. While abortion is the most readily observed symptom in cows, it should be borne in mind that not all animals that abort are affected with brucellosis and that not all animals affected with brucellosis will necessarily abort. On the other hand, every case of abortion should be regarded with suspicion until proved noninfectious.

The infected animal may prematurely give birth to a dead fetus, usually during the last third of pregnancy. On the other hand, the birth may be entirely normal; but the calf may be weak, or there may be retention of the afterbirth, inflammation of the uterus, and/or difficulty in future conception. The milk production is usually reduced. There may be abscess formation in the testicles of the male and swelling of the joints (arthritis). The observed symptoms in humans include weakness, joint pains, undulating (varying) fever, and occasionally orchitis (inflammation of the testicles).

CAUSE, PREVENTION, AND TREATMENT

The disease is caused by a bacteria called *Brucella abortus* in cattle, *Brucella suis* in swine, and

Fig. 12-8. Cow with aborted fetus. Every case of abortion should be regarded with suspicion until proved noninfectious. (Courtesy, USDA)

Brucella melitensis in goats. The suis and melitensis types are seen in cattle, but the incidence is rare; swine are infected with both the suis and melitensis types; and horses may become infected with all three types.

People are susceptible to all three species of brucellosis. In most areas, the vast majority of undulant fever cases in humans are due to *Brucella suis*. The swine organism causes a more severe disease in people than the cattle organism, although not so severe as that induced by the goat type (*Brucella melitensis*). Fortunately, far fewer people are exposed to the latter simply because of the limited number of goats and the rarity of the disease in goats in the United

SOURCES OF INFECTION

Dotted lines indicate sometimes a source.

Fig. 12-9. Sources of brucellosis infection (Drawing by R. F. Johnson)

States. Livestock producers are aware of the possibility that human beings may contract undulant fever from handling affected animals, especially at the time of parturition; from slaughtering operations or handling raw meats from affected animals; or from consuming raw milk or other raw by-products from cows or goats, and eating uncooked meats infected with brucellosis organisms. The simple precautions of pasteurizing milk and cooking meat, however, make these foods safe for human consumption.

The *Brucella* organism is quite resistant to drying but is killed by the common disinfectants and by pasteurization. It is found in immense numbers in the various tissues of the aborted young and in the discharges and membranes from the aborted animal. It is harbored indefinitely in the udder and may also be found in the sex glands, spleen, liver, kidneys, bloodstream, joints, and lymph nodes.

Brucellosis appears to be commonly acquired through the mouth in feed and water contaminated with the bacteria, or by licking infected animals, contaminated feeders, or other objects to which the bacteria may adhere. Venereal transmission by infected bulls to susceptible cows through natural service may occur, but it is rare.

Freedom from disease should be the goal of all control programs. An annual blood test (and more frequently if the disease is encountered), the removal of infected animals, strict sanitation, proper and liberal use of disinfectants, isolation at the time of parturition and the control of animals, feed, and water brought into the premises are the key to the successful control or eradication of brucellosis.

Sound management practices, which include either buying replacement animals that are free of the disease or raising all females, are a necessary adjunct in prevention. Drainage from infected areas should be diverted or fenced off, and visitors (human and animal) should be kept away from animal barns and feed lots. Feeds should not be bought from farms that have infected animals, and one should beware of used feed bags. Animals taken to livestock shows and fairs should be isolated on their return and tested 30 days later.

In 1934, a brucellosis control and eradication program was initiated in connection with the cattle-reduction program necessitated by the drought of that year. This program has continued to operate, with provision for the slaughter of animals that react positively to the test. Under the slaughter provision, several states pay partial indemnity to farmers whose animals are condemned under the program.

Strain 19 was the only approved vaccine in the United States for *B. abortus* infections prior to 1996.

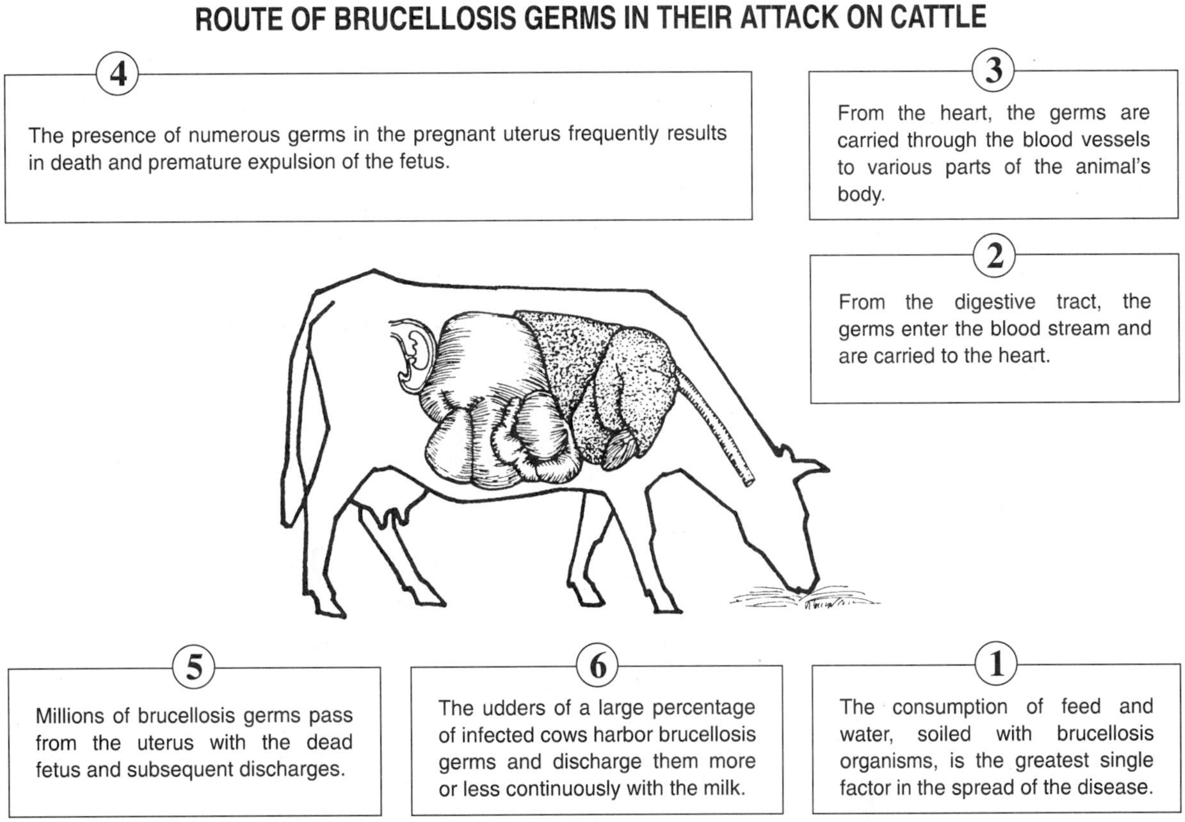

ROUTE OF BRUCELLOSIS GERMS IN THEIR ATTACK ON CATTLE

(1) The consumption of feed and water, soiled with brucellosis organisms, is the greatest single factor in the spread of the disease.

(2) From the digestive tract, the germs enter the blood stream and are carried to the heart.

(3) From the heart, the germs are carried through the blood vessels to various parts of the animal's body.

(4) The presence of numerous germs in the pregnant uterus frequently results in death and premature expulsion of the fetus.

(5) Millions of brucellosis germs pass from the uterus with the dead fetus and subsequent discharges.

(6) The udders of a large percentage of infected cows harbor brucellosis germs and discharge them more or less continuously with the milk.

Fig. 12-10. Diagram showing how cattle become infected with the *Brucella* organism and its route of attack.

In 1996, RB51 brucellosis vaccine was approved for use. It provides similar protection; however, it does not induce antibodies that react in the standard serological tests, one of the major problems with Strain 19 vaccine. An attenuated strain of *B. melitensis*, known as Rev 1, is used in other countries and is very effective.

Vaccination is very important in brucellosis control. Vaccine should be given to all replacement heifers between 4 and 12 months of age. *Note well:* Twenty-six states require calfhood vaccination for all breeding females. Also, many states now have restrictions on incoming heifers or cows that are not officially vaccinated; hence, brucellosis vaccination is a virtual necessity for interstate movement of breeding heifers or cows.

Slaughter or quarantine of infected cattle, together with rigid sanitation, must be a part of any successful eradication program once the disease has made its appearance.

The Uniform Methods and Regulations necessary to carry out a successful bovine brucellosis and eradication program are published by the Veterinary Service, APHIS. These methods and regulations are revised annually by the Brucellosis Committee of the U.S. Animal Health Association. The Committee recommends changes in the Uniform Methods and Regulations to cope with changes in the status of bovine brucellosis within the country. Probably the most important regulations are the ones that control the interstate and intrastate movement of reactor cattle or cattle of unknown status.

To date, there is no known medicinal agent that is completely effective in the treatment of brucellosis in any class of farm animals. Thus, the farmer and rancher should not waste valuable time and money on so-called cures that are advocated by fraudulent operators.

Great progress has been made in the control of brucellosis. In 1994, fewer than 200 U.S. cattle herds, involving 10,000 head of beef cattle and 5,000 head of dairy cattle, were under quarantine for brucellosis; and 32 states were brucellosis-free. In 1996, the number of herds under quarantine had decreased to 46.

On October 17, 1994, indemnity payments on contaminated cattle were increased to $750 per animal (average slaughter of $475, plus $250 indemnity), with the announced goal of eradicating brucellosis in the U.S. by 1998.

CALF DIPHTHERIA

Calf diphtheria is an acute, infectious disease of housed suckling calves and young feedlot cattle. The disease sometimes attacks these young animals as early as the third or fourth day after birth. If untreated, the mortality rate is very high. The disease is not to be confused with the sore mouth virus of sheep or the diphtheria of humans, with which it has no relationship.

SYMPTOMS AND SIGNS

The affected animal shows difficulty in breathing, eating, and drinking. Drooling and swallowing movements may also be noted. Inspection of the mouth reveals yellowish crumbling masses and patches of dead tissue (diphtheritic membranes) on the borders of the tongue, adjacent to the molar teeth, and in the throat. Once established and unchecked, it will spread rapidly, eventually causing the death of the animal.

CAUSE, PREVENTION, AND TREATMENT

The cause of this malady is the soil organism *Spherophorus necrophorus*, the same organism that is often found in foot rot. The organism gains entrance to the tissues through wounds or eruptions in the mouth, and within five days after entrance the animal will develop the symptoms noted.

Prevention consists of segregating the sick animals from the healthy ones and cleaning and disinfecting the quarters, not only after infection breaks out but before the calf is born. After outbreaks, all well animals should be checked daily.

Treatment consists of using sulfa drugs or antibiotics (penicillin or tetracyclines). The local application of a proteolytic enzyme for removal of the dead tissue is indicated.

CALF SCOURS

Calf scours is not a single disease; it is a clinical sign associated with several diseases characterized by diarrhea. Regardless of the cause, diarrhea prevents the absorption of fluids from the intestines, and body fluids pass from the scouring calf's body into the intestines. The scouring calf, which is approximately 70% water at birth, becomes dehydrated and suffers from electrolyte (sodium and potassium) loss and acidosis. Infectious agents cause the primary damage to the intestine, but death from scours usually results from dehydration, acidosis, and loss of electrolytes.

Most affected calves are less than two weeks of age. Outbreaks of calf scours are most common during fall, winter, and early spring.

Scours is the cause of more calf deaths than all other diseases combined. It is estimated that 10% of all calves in the United States are affected by the disease, and that 8% of beef calves and 18% of dairy calves so affected die.

SYMPTOMS AND SIGNS

Calf scours can vary from a mild to a severe disease. In the mild form, the main symptom is softer than normal feces. The severely affected calf initially appears depressed and has a lack of appetite. Then begins a severe diarrhea which consists of yellowish, foul smelling, watery or foamy feces. These calves can have a rough hair coat, sunken eyes, appear emaciated, and suffer from hypothermia (body temperature lower than normal).

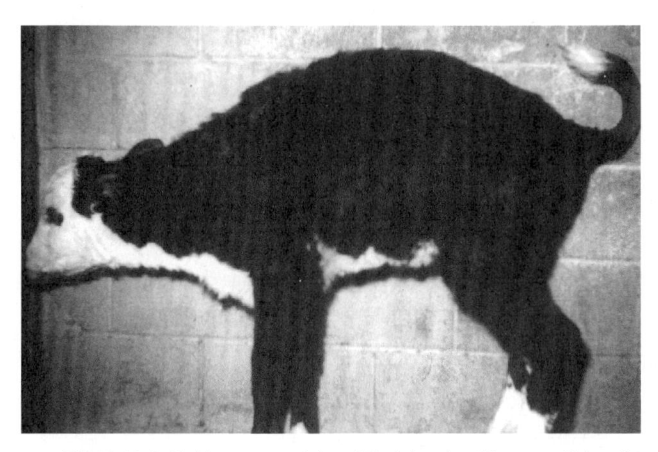

Fig. 12-11. Calf with scours straining while defecating. (Courtesy, University of Minnesota, Agricultural Extension Service, St. Paul, MN)

Clinical signs of dehydration first occur when the fluid loss reaches 5 to 6% of the body weight. Fluid loss of 8% results in depression, sunken eyes, dry skin, and the calf being unable to stand. A 12% loss of fluids usually results in death. Oral fluids used when scouring is first observed are quite successful. Consult your veterinarian relative to the choice of electrolytes. In very acute cases, death can occur before diarrhea is observed; however, death usually occurs 2 to 3 days after the onset of diarrhea. Some degree of associated pneumonia occurs more frequently in stabled dairy calves than in beef calves on the range.

It is sometimes difficult to distinguish the infectious disease from diarrhea caused by noninfectious factors—such as overfeeding, irregular feeding, use of unclean utensils, too rapid changes in feed, or exposure to drafts and cold or damp floors. With the infectious type of scours, however, several calves are usually affected; and some animals may die quickly.

CAUSE, PREVENTION, AND TREATMENT

The causes of scours may be grouped into two categories: (1) noninfectious causes, and (2) infectious causes.

The **noninfectious** causes are often referred to as predisposing or contributing factors. Nevertheless, there is a dramatic interaction between noninfectious causes and infectious causes. Any effort to prevent infectious causes is usually fruitless unless serious control of contributing (noninfectious) factors is part of the overall program.

The most commonly encountered noninfectious causes of scours are:

1. Inadequate nutrition of the pregnant dam, especially during the last third of gestation.
2. Inadequate environment stressing the newborn calf, including muddy lots, crowding, contaminated lots, wintering and calving in the same area, storms, heavy rain or snowfall, etc.
3. Insufficient attention to the newborn calf, particularly during difficult birth or adverse weather conditions. Also, the caretaker should see that the calf receives colostrum before it is 4 hours old; colostrum given to calves 24 to 36 hours old is practically useless—antibodies are seldom absorbed this late in life.

The **infectious** causes of calf scours may be grouped as follows:

1. **Bacterial causes:**
   Escherichia coli
   Salmonella spp
   Clostridium perfringens
   Other bacteria
2. **Viral causes:**
   Coronavirus
   Rotavirus

A brief discussion of each of the most common bacterial causes of calf scours follows:

- **Escherichia coli**—*E. coli*, of which there are numerous serotypes (kinds), is the single most important cause of bacterial scours in calves. Most newborn calves have a chance to pick up *E. coli* scours infection from the environment, particularly when the sanitation is marginal.

- ***Salmonellae***—*Salmonellae* produce a potent toxin or an endotoxin (poison) within their own cells.

- ***Clostridium perfringens***—These infections, commonly known as enterotoxemia, caused by various types of *C. perfringens* (types B, C, and D have been reported), are fatal.

Little was known about the viral causes of calf scours until 1968, when researchers at the University of Nebraska published their findings. A brief discussion of each of the main viral causes of calf scours follows:

- **Coronavirus**—Both coronavirus and rotavirus possess the ability to disrupt the cells which line the small intestine, with resulting diarrhea and dehydration. Coronavirus also damages the cells in the intestinal crypts and slows down the healing process in the intestinal lining.

■ **Rotavirus**—Rotavirus was originally known as reovirus. The damage caused by either coronavirus or rotavirus is often compounded by bacterial infections; the risk of fatal diarrhea is increased by such mixed infections.

Other viral causes of calf scours are: bovine virus diarrhea (BVD) and infectious bovine rhinotracheitis (IBR).

Also, there are two protozoan causes of calf scours: *Cryptosporidium* and *Coccidia*.

To keep the disease away from the herd, one must prevent primary infection of the newborn. This rests on strict sanitary measures and isolation. The disease can be introduced by adding calves or adult animals from another herd. Calf diarrhea frequently occurs when a newly assembled herd begins to calf.

Weather conditions permitting, birth should preferably take place in the open on an uncontaminated, sun-exposed pasture. Otherwise, a clean, disinfected maternity stall should be provided, and the navel cord of the newborn calf should be treated with tincture of iodine. The newborn animal should be segregated from other animals and the contaminated quarters thoroughly cleaned and disinfected. Prevention should also include proper feeding of pregnant cows and giving colostrum to calves that are subsequently to be raised on a milk replacer. When calf scours appears, infected animals should be segregated and the premises and feed containers should be thoroughly cleaned and disinfected.

■ **Vaccination programs**—A good vaccination program is an effective tool in preventing calf scours, providing (1) the management aspects are good, and (2) the calf nurses sufficient colostrum early in life.

The resistance of the newborn calf can be achieved by vaccinating the dam 2 to 6 weeks before calving to stimulate antibodies which are then passed on to the newborn through the colostrum. The vaccination program should be based on a good diagnostic knowledge of diseases present in the herd, provided by qualified veterinary assistance.

An *E. coli* antibody is now available for oral administration to calves immediately after birth.

Treatment of severely affected diarrheic calves should include: (1) isolating them from the rest of the herd, (2) changing the diet, (3) replacing fluids and electrolytes, (4) administering antimicrobial drugs and immunoglobulin, and possibly administering intestinal protectants.

Fig. 12-12. Aided by bacterium and viral vaccinations, this Montana calf weathered a wet snowstorm. (Courtesy, The Upjohn Company)

CIRCLING DISEASE (Encephalitis, or Listeriosis)

Circling disease, also called listerellosis, listeriosis, or encephalitis, is an infectious disease affecting mainly cattle, sheep, and goats; but it has been reported in swine, foals, and other animals, and in humans. Cattle of all ages are susceptible. One to seven percent of the herd may be infected, and the mortality rate of affected animals is extremely high.

SYMPTOMS AND SIGNS

This disease affects the nervous system. Depression, staggering, circling, and strange awkward movements are noted. One eye and one ear may be paralyzed. The animal may be seen holding a mouthful of hay for hours. There may be inflammation around the

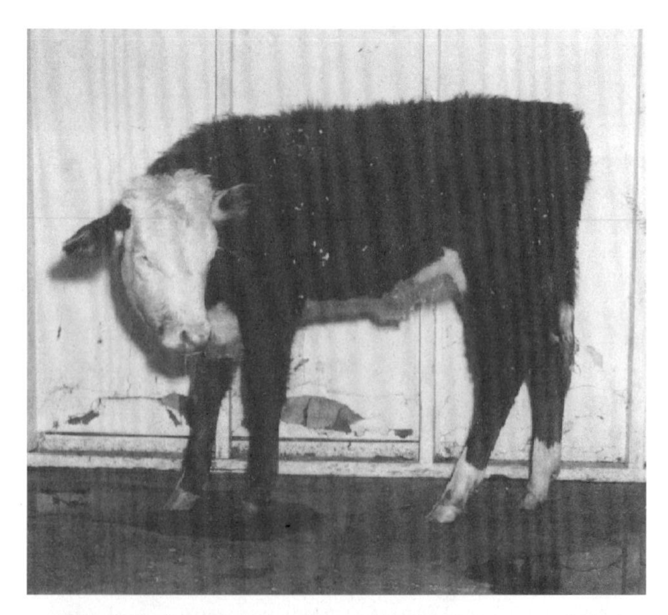

Fig. 12-13. Calf with circling disease. Note the drooping ear, slobbering, and tendency to turn to the left. (Courtesy, Dept. of Veterinary Pathology and Hygiene, College of Veterinary Medicine, University of Illinois)

eye, and abortion may occur. The course of the disease is very short, with paralysis and death the usual termination. Positive diagnosis can be made only by laboratory examination of the brain.

CAUSE, PREVENTION, AND TREATMENT

Circling disease results from the invasion of the central nervous system by bacteria called *Listeria monocytogenes*. The method of transmission is unknown. In an outbreak, affected animals should be segregated. If silage is being fed, discontinue that particular silage on a trial basis. Spoiled silage should be avoided, routinely.

Various antibiotics have shown beneficial results if given early. Penicillin is the drug of choice.

FESCUE FOOT (Fescue Toxicity)

Fescue is a valuable pasture grass that grows best in cool and cold weather. Under some conditions which are not completely understood, a fescue pasture may become toxic. Cattle that graze on such pastures do not perform well. In cold climates, they occasionally develop a crippling disease, known as fescue foot or fescue toxicity. Both beef and dairy cattle are susceptible; and this disease has been reported in sheep in Australia.

Most cases of fescue toxicity occur among cattle that graze pure stands of fescue during late fall and winter; and most toxic stands of fescue pasture are several years old. Fescue toxicity is more prevalent in animals suffering from malnutrition or parasitism.

SYMPTOMS AND SIGNS

There are variations in the severity of symptoms in cattle on toxic pastures. Some animals show no apparent lameness, whereas others show varying degrees of sloughing (necrosis) on the ends of their tails.

During the summer, cattle grazing toxic pastures show a poor growth rate, increased temperatures, and increased pulse and respiratory rates. The only complaint that cattle producers make is the fact that the cattle are not doing as well as in previous years. In some herds, the weaning weights decline for 2 or 3 years before cattle producers realize that they have a problem.

CAUSE, PREVENTION, AND TREATMENT

Fescue foot is associated with a fungus, which lives in the leaves, stems, and seed of the tall fescue plant and is not visible externally. The toxin is a vaso constrictor that affects the blood vessels. Cold weather is also a constrictor of blood vessels, which explains the extreme lameness found in the winter months. Occasionally, the circulation is closed to a degree that causes the entire foot to slough off. Such cattle walk on stumps of bone.

The seeding of fungus-free fescue is the best way to prevent fescue foot. Toxic pastures should be renovated and some legume should be seeded with the fescue. It requires good pasture management, along with fertilization, to maintain a good fescue-legume pasture.

No medication is effective for cattle with fescue foot. In severe cases where sloughing has occurred, the animal should be destroyed for humane reasons.

Cattle usually recover completely if they are removed from fescue pasture or fescue hay and are given other feed or pasture as soon as the first signs of the disease appears.

FOOT-AND-MOUTH DISEASE

This is a highly contagious disease of cloven-footed animals (mainly cattle, sheep, and swine) characterized by the appearance of watery blisters in the mouth (and in the snout in the case of hogs), on the skin between and around the claws of the hoof, and on the teats and udder. Fever, diminished rumination and reduced appetite are other signs of the disease.

People are mildly susceptible but very rarely infected, whereas the horse is immune.

Unfortunately, one attack does not render the animal permanently immune, but the disease has a tendency to recur, perhaps because there are several strains of the causative virus. The disease is not present in the United States, but there were at least 9 outbreaks (some authorities claim 10) in this country between 1870 and 1929, each of which was stamped out by the prompt slaughter of every affected and

Fig. 12-14. Cow with foot-and-mouth disease. The animal is reluctant to stand because of sore feet. The characteristic profuse flow of saliva is caused by blisters in the mouth. (Courtesy, USDA)

exposed animal. No U.S. outbreak has occurred since 1929, but the disease is greatly feared. Drastic measures are exercised in preventing the introduction of the disease into the United States, or, in the case of actual outbreak, in eradicating it.

In December, 1946, an outbreak of foot-and-mouth disease was confirmed in Mexico, and the border was closed from that date to September 1, 1952. Then the border remained open from September 1, 1952, until May 23, 1953, at which time another outbreak occurred and it was again closed. The U.S. Secretary of Agriculture again opened the border on January 1, 1955.

On February 25, 1952, an outbreak of foot-and-mouth disease was diagnosed in Saskatchewan, Canada. This resulted in the U.S.-Canada border being closed from this date until March, 1953.

Foot-and-mouth disease is constantly present in Europe, Asia, Africa, and South America. It has not been reported in New Zealand or Australia.

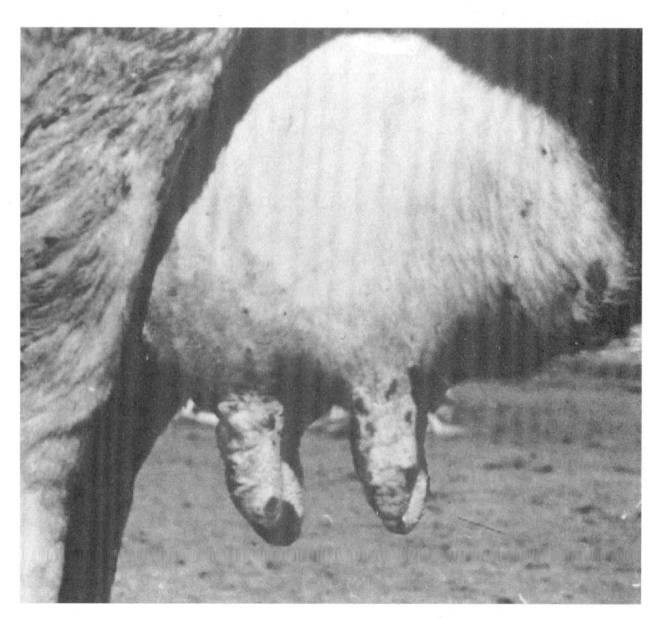

Fig. 12-15. Blisters (vesicles) on the teats of a cow with foot-and-mouth disease. (Courtesy, USDA)

SYMPTOMS AND SIGNS

The disease is characterized by the formation of blisters (vesicles) and a moderate fever 3 to 6 days following exposure. These blisters are found on the mucous membranes of the tongue, lips, palate, cheeks, and on the skin around the claws of the feet, and on the teats and udder.

Presence of these vesicles, especially in the mouth of cattle, stimulates a profuse flow of saliva that hangs from the lips in strings. Complicating or secondary factors are infected feet, caked udder, abortion, and great loss of weight. The mortality of adult animals is not ordinarily high, but the usefulness and productivity of affected animals is likely to be greatly damaged, thus causing great economic loss.

CAUSE, PREVENTION, AND TREATMENT

The infective agent of this disease is one of the smallest of the filterable viruses. In fact, it now appears that there are at least seven strains of the virus. Infection with one strain does not protect against the other strains.

The virus is present in the fluid and coverings of the blisters, in the blood, meat, milk, saliva, urine, and other secretions of the infected animal. The virus may be excreted in the urine for over 200 days following experimental inoculation. The virus can also be spread through infected biological products, such as smallpox vaccine and hog cholera virus and serum, and by the cattle fever tick.

Except for the nine outbreaks mentioned, the disease has been kept out of the United States by extreme precautions, such as quarantine at ports of entry and assistance with eradication in neighboring lands when introduction appears imminent. Neither live cloven-hoofed animals nor their fresh, frozen, or chilled meats can be imported from any country in which it has been determined that foot-and-mouth disease exists (meat imports from these countries must be canned or fully cured).[6]

Two methods have been applied in control: the slaughter method, and the quarantine procedure. Then, if the existence of the disease is confirmed by diagnosis, the area is immediately placed under strict quarantine; infected and exposed animals are slaughtered and buried, with owners being paid indemnities based on their appraised value. Everything is cleaned and thoroughly disinfected.

In countries where the disease is endemic, drastic measures of eradication are not always economically feasible. Control is based on a modified system of vaccination and quarantine, using vaccines specific for the type and subtype of virus involved. Vaccines have not been used in the outbreaks in the United States because they have not been regarded as favorable to rapid, complete eradication of the infection.

Fortunately, the foot-and-mouth disease virus is quickly destroyed by a solution of the cheap and

[6]Effective April 11, 1974, this rule was altered to permit dependent territories or possessions to be determined free of foot-and-mouth disease and rinderpest, regardless of the disease status of the mother country. Thus dependent territories or possessions that are geographically separated from their mother country, such as colonies or former colonies, can be judged as to their livestock disease status by the same criteria previously applied to politically separate countries.

common chemical sodium hydroxide (lye). Because quick control action is necessary, state or federal authorities must be notified the very moment the presence of the disease is suspected.

No effective treatment is known.

FOOT ROT (or Foul Foot)

This disease is an inflammation of the hoofs of cattle, sheep, and goats; but cross-infections of foot rot between cattle and sheep do not occur. It is a potential hazard wherever animals of these species are kept; especially in wet, muddy areas.

SYMPTOMS AND SIGNS

A shrewd observer will first notice a reddening and swelling of the skin just above the hoof, between the toes, or in the bulb of the heel. As the infection progresses, lameness will be noted. If not arrested, the infection will invade the soft tissue and cause a discharge of pus from the infected breaks in the skin. At this stage, a characteristic foul odor is present. Later, the joint cavities may be involved, and the animal may show fever and depression characteristic of a general infection. Affected animals lose weight, and, if lactating, produce less milk; and they may die if unattended.

CAUSE, PREVENTION, AND TREATMENT

Because the feet of animals are continually being exposed to all types of filth containing millions of microorganisms, it is difficult to incriminate the causative agent or agents. The soil organism *Spherophorus necrophorus* is most frequently recovered from cases of foot rot in cattle; but pus-forming bacteria, colon bacilli, and others may lend support to the destructive process. Similar types of infection may have different causative agents depending on area, soil, and other factors.

The prevention of this disease is much more effective than the treatment, because once established it is difficult to control the spread. Draining of muddy corrals and the segregation of infected and new animals is recommended. If the disease appears, a good cleaning is in order, and unaffected animals should be moved to clean quarters and pastures if possible. Also, effective prevention may be obtained through subjecting animals to (1) a foot bath of 2 to 5% bluestone (copper sulfate), or (2) a walk way of air-slaked lime. In some cases, the inclusion of an organic iodide compound (ethylene diamine dihydriodide; EDDI) in the feed or salt has markedly reduced the incidence of foot rot in problem herds. For prevention, use 50 mg of EDDI per head daily on a continuous basis (or mix 1 part of iodine and 9 parts of salt, fed free-choice continuously); for treatment, once cattle have foot rot, use 500 mg per head daily for 2 to 3 weeks.

The success of treatment seems to depend on the stage of infection and perhaps on the causative agent. The usual treatment consists of the following:

1. Place in a clean, dry place.
2. If necessary, trim away the affected part of the foot. Also, check for, and if necessary eliminate, foreign bodies in the hoof, or wire that might be wrapped around just above the hoof.

If these steps fail, astringent and antiseptic packs may be applied; they should be changed after 48 hours because it is important that the area be kept dry.

Systemic and local treatment with antibiotics and sulfonamides is recommended. Zinc methionine has also been used as a treatment.

Repeated paring and treatment are usually necessary because healing may require several weeks. It is important that an affected animal be kept in a dry stall. In advanced stages, best results are obtained by surgical amputation of the affected claw. Animals so treated soon walk as before on the one remaining healthy claw.

INFECTIOUS BOVINE RHINOTRACHEITIS (IBR, or Red Nose)

Infectious bovine rhinotracheitis (IBR) was first found in a Colorado feedlot in 1950. Since then, it has occurred throughout the United States. The main economic losses from the disease are in time, weight, milk production, and drugs.

IBR may also be involved in bovine respiratory disease complex (BRDC); so, see also the latter.

SYMPTOMS AND SIGNS

Affected animals go off feed and lose weight; generally cough; may show pain in swallowing; usually slobber and show a nasal discharge; breathe rapidly, with difficulty, and in severe cases through the mouth; show severe inflammation of the nostrils and trachea; have a high fever, 104 to 107°F, and may remain sick for as long as a week. When the disease breaks out, 25 to 100% of the animals are affected. Death loss rarely exceeds 5%.

Although IBR is usually thought of as a respiratory disease, it may cause inflammation of the eyes and/or vagina. Also, it may cause abortion.

Fig. 12-16. Cattle with infectious bovine rhinotracheitis (IBR) have elevated temperatures, rapid respiration, and eye and nasal discharges. (Courtesy, University of Minnesota, Agricultural Extension Service, St. Paul, MN)

CAUSE, PREVENTION, AND TREATMENT

The disease is caused by a herpes virus.

Immunity against IBR can be achieved by the intramuscular or intranasal administration of modified live or inactivated vaccines.

Feedlot cattle should be vaccinated as part of a preconditioning program since IBR may occur in unvaccinated cattle. Only healthy cattle should be vaccinated.

There is no known treatment, but sulfonamides and antibiotics effectively combat the secondary bacterial invaders that accompany the disease.

INFECTIOUS EMBOLIC MENINGOENCEPHALITIS (Haemophilosis, Thromboembolic Meningoencephalitis, TEME, or Sleeper Syndrome)

Infectious embolic meningoencephalitis is an acute, febrile disease of feedlot cattle, characterized by incoordination and coma. Only 1 or 2 cases develop in a lot at a time, but 10% of the cattle may be affected before the disease runs its course.

Twelve outbreaks of the disease were reported in feedlot cattle in Colorado from 1949 to 1956. Subsequently, it has been reported throughout the West.

SYMPTOMS AND SIGNS

The disease is most common in feedlot cattle, but it may occur in pastured animals. It is most prevalent in the fall and winter months. It affects both sexes, but there is a higher incidence of the disease in heifers than in steers. It is characterized by incoordination, coma, sometimes blindness, and always fever (near 107°F). Death usually follows in 2 to 4 days. Positive diagnosis can be made upon autopsy by the inflamed areas of infection observed in the brain.

The disease should not be confused with *polioencephalomalacia*, which also affects feedlot cattle and causes incoordination, but with which fever is rarely associated.

CAUSE, PREVENTION, AND TREATMENT

It is caused by a microaerophilic gram-negative bacterium, *Hemophilus somnus*.

There is a higher probability of TEME in young calves, in recently transported calves, and in herds with a history of recent respiratory disease outbreaks. Also, cold weather stress seems to result in a higher incidence of TEME.

Affected animals should be isolated and treated immediately with penicillin and streptomycin, or oxytetracycline. High levels of tetracyclines in the feed or drinking water are recommended if many animals are afflicted. Feedlot pens should be checked frequently to detect newly affected animals and to provide prompt treatment. Most outbreaks will run their course in 2 to 3 weeks. Bacterins may reduce morbidity and mortality and decrease the number of animals requiring treatment.

JOHNE'S DISEASE (Chronic Bacterial Dysentery, or Paratuberculosis)

This is a chronic, incurable, infectious disease seen chiefly in cattle; also found in sheep and goats, and more rarely in swine and horses. It resembles tuberculosis in many respects. The disease is very widespread, having been observed in practically every country where cattle are raised on a large scale. It is one of the most difficult diseases to eradicate from a herd.

In affected herds of cattle, from 2 to 10% of the adult animals die each year.

SYMPTOMS AND SIGNS

The disease seems to involve exposure with no evidence of infection for 6 to 18 months. At the end of this time, the animal loses flesh and displays intermittent diarrhea and constipation, the former becoming more prevalent. Affected animals may retain a good appetite and normal temperature. The feces are watery

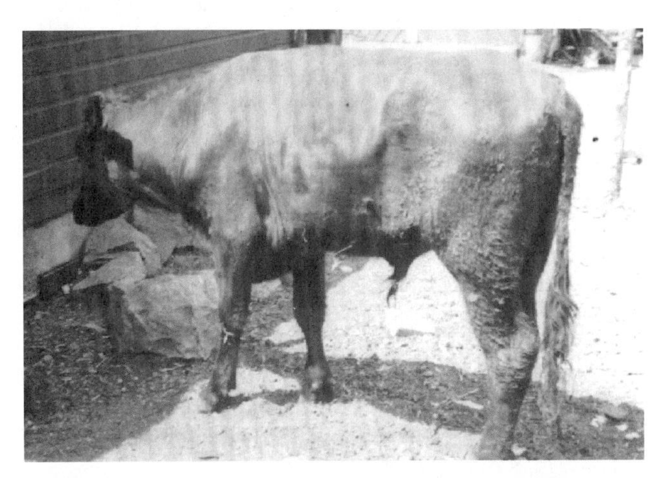

Fig. 12-17. Steer with Johne's disease, characterized by progressive emaciation and intermittent diarrhea and constipation. (Courtesy, University of Minnesota, Agricultural Extension Service, St. Paul, MN)

but contain no blood and have a normal odor. The disease is almost always fatal, but with the animal living from a month to two years.

Upon autopsy, the thickening of the infected part of the intestines, covered by a slimy discharge, is all that is evident. This thickening prevents the proper digestion and absorption of feed and explains the emaciation.

CAUSE, PREVENTION, AND TREATMENT

The disease is caused by the ingestion of a bacterium, *Mycobacterium paratuberculosis*. Inasmuch as this organism is acid-fast (that is, it retains certain dyes during a staining procedure), it resembles tuberculosis.

Effective prevention is accomplished by keeping the herd away from infected animals. If it is necessary to introduce new animals into a herd, they should be purchased from reputable breeders; and the owner should be questioned regarding the history of the herd.

It must be borne in mind that apparently healthy animals can spread the disease. Testing at regular intervals of 3 to 6 months with "Johnin," removing reactors, disinfecting quarters, and removing newborn animals from their dams immediately after birth, without allowing them to nurse, and raising them away from mature animals should be practiced in infected herds. In using the Johnin test, however, it should be realized that it is not entirely accurate as a diagnostic agent. Some affected animals fail to react to the test.

A vaccine for Johne's has been approved by the USDA, but individual approval of each state is required for its use.

No satisfactory treatment for Johne's disease has been found.

LEPTOSPIROSIS

Leptospirosis was first observed in people in 1915–16, in dogs in 1931, and in cattle in 1934. It has also been reported in hogs, horses, and sheep.

It was first reported in cattle in the United States in 1944, although it had been found in dogs in the United States since 1939. Bovine leptospirosis has been reported in Europe, Australia, and the United States.

Human infections may be contracted through skin abrasions when handling or slaughtering infected animals, by swimming in contaminated water, through consuming raw beef or other uncooked foods that are contaminated, or through drinking unpasteurized milk.

SYMPTOMS AND SIGNS

In most herds, leptospirosis is a mild disease. However, the symptoms may vary from herd to herd, or even within a herd. In general, the symptoms noted in cattle are: (1) high fever; (2) poor appetite; (3) abortion at any time; (4) bloody urine; (5) anemia; and (6) ropy milk.

All ages of cattle, and both sexes (including steers) are affected.

CAUSE, PREVENTION, AND TREATMENT

The disease is caused by several species of corkscrew-shaped organisms of the spirochete group; primarily *Leptospira pomona* in cattle, although six pathogenic serotypes have been isolated from cattle in the United States.

The following preventive measures are recommended:

1. Blood test animals prior to purchase, isolate for 30 days and then retest prior to adding them to the herd.
2. Do not allow animals to consume contaminated feed or water, or to breathe contaminated urinal mist.
3. Keep premises clean, and avoid use of stagnant water.
4. Vaccinate susceptible animals with a bacterin of an appropriate serotype if the disease is present in the area. Because there is often more than one serotype in a bovine population or in local wildlife carriers, polyvalent bacterins are recommended for effective control. Control through vaccination should include all cattle in the herd that are 6 months or older. In open herds, vaccination should be repeated every 6 months; in closed herds, annual vaccination is adequate.

Where a herd is infected, the following control measures should be initiated:

1. Blood test at least 10% of the cattle in the herd annually to provide continuing information about the status of the herd.

The same blood sample used in a brucellosis test may also be used for a leptospirosis test, by simply dividing the serum.

2. Spread the cattle over a large area; avoid congestion in a corral or barn.

3. Do not let animals drink from ponds, swamps, or slow-running streams, and avoid contaminated feed.

4. Clean and disinfect the premises; exterminate the rodents.

It should be recognized that carrier animals—that have had leptospirosis and survived—may spread the infection by shedding the organism in the urine. The infected urine may then either (1) be breathed as a mist in cow barns, or (2) contaminate feed and/or water and thus spread the infection. It is known that such recovered animals may remain carriers for 2 to 3 months or longer after getting over the marked symptoms. Fortunately, the organisms seldom survive for more than 30 days outside the animal. However, stagnant water and mild temperatures favor their survival.

Antibiotics have been used extensively in the treatment of leptospirosis, but the response has been inconsistent, perhaps due to the stage of the disease at the time of administration. To be effective, the antibiotic must be administered early in the course of the disease, and prior to toxic changes associated with loss of kidney function.

LUMPY JAW AND WOODEN TONGUE

These two infections are chronic diseases affecting mainly the head of cattle—hence, the name *big head.* They occur most frequently in young cattle during the period of changing teeth. At one time, both of the conditions were referred to as actinomycosis, but now this term is used only for lumpy jaw. Actinobacillosis is the synonym for wooden tongue and soft tissue lesions.

SYMPTOMS AND SIGNS

Because of the area involved in these diseases there is usually emaciation resulting from the difficulty encountered in chewing and swallowing.

Lumpy jaw only rarely attacks the soft tissue. It is usually confined to the bones of the lower jaw, although the upper jaw and nasal bones may be involved. The affected bone becomes enlarged and spongy and filled

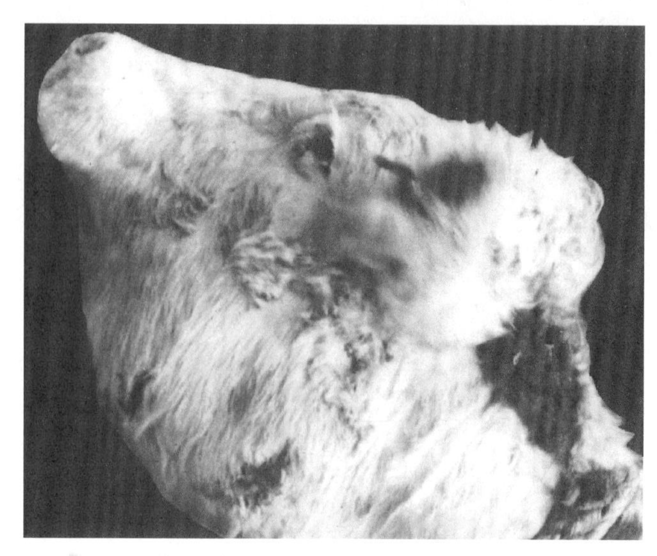

Fig. 12-18. Head of steer showing a bad case of lumpy jaw. It is usually confined to the bones of the lower jaw, although the upper jaw and nasal bones may be involved. (Courtesy, USDA)

with creamy pus. As the disease progresses, inflamed cauliflower masses of tissue spread out and may appear on the surface, discharging pus of foul odor. The surrounding flesh will also show inflammation, and the teeth may become loosened.

The same organism that causes lumpy jaw occasionally attacks the udder of sows, where it is characterized by many small abscesses filled with calcified granules. There may be fistulous tracts to the outside of the udder, discharging pus. On rare occasions, the organism causing the disease has also been found in fistulous withers of the horse, in conjunction with *Brucella* organisms.

Wooden tongue attacks chiefly the tissue in the throat area of cattle, but is also often seen in the tongue, stomach, lungs, and lymph glands. The first lesion usually observed is a movable, tumorlike swelling about the size of a small egg under the skin in the infected area. The enlargements usually break open and discharge a light colored and very sticky pus. An involved tongue may or may not be ulcerated but will show an increase in size and hardness. The tongue may become quite immobile and may protrude from the mouth.

With this wooden tongue condition, there will be constant drooling, and the animal will lose weight and condition through inability to take feed. Although any chronic swelling in the region of the head should lead one to suspect the presence of this infection, a positive diagnosis depends upon a microscopic examination of the yellowish, granular pus material that will eventually discharge from the swelling.

The same organism that causes wooden tongue occasionally attacks the lips and face of sheep.

CAUSE, PREVENTION, AND TREATMENT

The organism causing actinomycosis (lumpy jaw) is called *Actinomyces bovis*, and the organism causing actinobacillosis (wooden tongue) is called *Actinobacillus lignceresei*. In each case, they may be assisted by secondary invaders. Both organisms lack invasive power, often being found in a normal oral cavity. They are thought to enter the tissue only by wound infections—for example, they may be carried in by the sharp awns of fox tail, barley, rye, bearded wheat, or oats.

Prevention consists of segregation and proper treatment or elimination of infected animals and the restricted feeding of material having sharp awns that might injure the animal's mouth. The latter precaution is important as the organism is a normal inhabitant of the mucous membranes of the mouth and nasal cavity of animals and humans.

Lumpy jaw (actinomycosis) and wooden tongue (actinobacillosis) of soft tissue have responded well to surgery and iodine therapy; and antibiotics have been recommended to supplement other treatments. However, present treatments for lumpy jaw of bone and adjacent tissues are not entirely satisfactory.

MALIGNANT EDEMA (Gas Gangrene)

This is an acute infectious, but noncontagious, disease characterized by gangrene and emphysema around a wound. The incidence in a single herd may be high following castration, dehorning, or accidental wounds.

SYMPTOMS AND SIGNS

The affected animal goes off feed, breathes rapidly, and is profoundly depressed. A swelling forms around the wound. A gaseous and malodorous fluid exudes from the wound. In advanced stages of the disease, the animal is prostrated and often disoriented. There may or may not be a rise in temperature. Death occurs after a course of 12 to 48 hours. The mortality rate is high.

CAUSE, PREVENTION, AND TREATMENT

Malignant edema is caused by *Clostridium septicum*, often accompanied by other organisms.

Since malignant edema is associated with contamination of wounds, the disease can be partially prevented by minimizing wounds and by castrating and dehorning under hygienic conditions.

Vaccination of young cattle with a mixed bacterin containing *Cl. septicum* (for malignant edema) along with *Cl. chauvoei* (for blackleg) at the time of the blackleg vaccination(s) will give some protection against malignant edema. Where the disease is known to exist, calves and newly acquired animals should be vaccinated. Calves should be vaccinated at weaning time, when castrating and dehorning are commonly done.

In the early stages of the disease, treatment with massive doses of penicillin or broad-spectrum antibiotics may be effective.

METRITIS

Metritis is an inflammation of the uterus, which affects cattle, horses, sheep, and swine.

SYMPTOMS AND SIGNS

Metritis usually develops soon after the animal has given birth. It is characterized by a foul smelling discharge from the vulva that becomes thick and yellow or white, and finally brownish or blood-stained. Also there is chilling, high temperature, rapid breathing, marked thirst, loss of appetite, and lowered milk production. Pressure on the right flank may produce pain. The animal may lie down and refuse to get up. Affected animals may die in 1 or 2 days; or the acute infection may develop into a chronic form, producing sterility.

CAUSE, PREVENTION, AND TREATMENT

Metritis is caused by various types of bacteria. Laceration at the time of calving, wounds caused by inexperienced operators and/or retention of the afterbirth are the principal predisposing causes.

Preventive measures consist of alleviating as many of the predisposing factors as possible, including bruises and tears while giving birth, exposure to wet and cold, and the actual introduction of disease-causing bacteria during delivery or the manual removal of the afterbirth. Clean maternity stalls should be provided. If assistance at calving time becomes necessary, caretakers should first disinfect their hands and arms as well as the animal's external genitals.

Difficult parturition should be left to the veterinarian. Nothing is so distressing to the veterinarian as a history of long labor and well-meaning but ill-guided attempts to remove a calf. Most cases are treated by introducing (in solution or tablets) an antibiotic or a sulfa into the uterus.

NAVEL INFECTION (Joint-Ill, or Navel-Ill)

Navel infection is an infectious disease of newborn calves, foals, and lambs. It occurs less frequently in calves and lambs than in foals.

SYMPTOMS AND SIGNS

Navel infection is characterized by loss of appetite, by swelling, soreness and stiffness in the joints, by umbilical swelling and discharge, and by general listlessness.

CAUSE, PREVENTION, AND TREATMENT

Navel infection is caused by several kinds of bacteria.

The recommended preventive measures are: sanitation and hygiene at mating and parturition, painting the navel cord of the newborn animal with iodine, and the administration of bacterins.

For treatment, veterinarians may give a blood transfusion, or they may administer a combination of penicillin and sulfonamide.

PINKEYE (Infectious Bovine Keratoconjunctivitis)

This is the common name for an infectious disease that affects the eyes of cattle. *Moraxella bovis* is the primary cause of the disease. Other bacteria have been isolated from the conjunctiva of cattle, but their importance as causal factors is not known. Also, a group of viruses has been isolated from animals with pinkeye, but their role as causes is purely secondary.

Pinkeye attacks animals of any age, but it is not common in young animals. It seems to become more virulent in certain years and in certain communities. The disease is widespread throughout the United States, especially among range and feedlot cattle. Pinkeye is encountered in nearly half of the U.S. beef cattle herds and affects about 3% of all beef cattle.

SYMPTOMS AND SIGNS

The first thing one may notice in bacterial pinkeye is the liberal flow of tears and the tendency to keep the eyes closed. There will be redness and swelling of the lining membrane of the eyelids and sometimes of the visible part of the eye. There may also be a discharge of pus. Ulcers may form on the cornea. If unchecked, they may cause blindness and even loss of the eye. The attack may also be marked by slight fever, reduction in milk flow, and slight digestive upset.

Fig. 12-19. Cow with pinkeye. Note eye discharge and the cloudiness or milkiness of the cornea or covering of the eyeball. (Courtesy, Dept. of Veterinary Pathology and Hygiene, College of Veterinary Medicine, University of Illinois)

About one-half of the animals in a herd become infected regardless of the treatment or control measures employed.

Affected steers on pasture gain an average of 50 lb less during the grazing season than those not affected.

In viral pinkeye, the eyeball itself is only slightly affected. Infectious bovine rhinotracheitis (IBR), a viral infection of the eyes of cattle, mainly affects the eyelids and the tissues surrounding the eyes. It causes a severe swelling of the lining of the lids.

CAUSE, PREVENTION, AND TREATMENT

■ **Bacterial pinkeye**—The most prevalent bacterial form of the disease is caused by *Moraxella bovis.* This organism produces a toxin which irritates and erodes the covering of the eye. Bacterial pinkeye occurs mainly during warm weather. Bright sunlight, wind, and dust may contribute to the cause of the disease. Cattle with white faces or lack of pigment around the eyes are rather susceptible. Transmission is mainly by flies and other insects that feed on eye discharges of infected animals and then carry the infection to susceptible animals. Also, the disease is spread by direct contact, from animal to animal.

Prevention of bacterial pinkeye consists of the following: controlling face flies and other insects that feed around the eyes; good nutrition, including adequate vitamin A; vaccination; and isolation of affected animals.

The most common treatment for bacterial pinkeye is the application of antibiotics or sulfonamides to the affected eye as ointments, powders, or sprays; prefer-

Fig. 12-20. An eye patch may be used in pinkeye treatment to protect the inflamed eye from sunlight. Periodic treatment should be used with the patch. (Courtesy, University of Illinois)

ably, with treatment made twice daily. Foreign protein therapy, which is the subcutaneous or intramuscular injection of such things as sterile milk, has been used to treat pinkeye for years, but its value is difficult to establish. Cortisone is sometimes combined with antibiotics and injected under the covering (at the outer edge) of the eyeball. The cortisone aids in reducing inflammation and pain and lessens the tears. Recovery is speeded up by keeping the infected animals in a dark barn or by placing a patch over the infected eye. Protecting the cornea by suturing the eyelids together, by using adhesive patches, or by using an ocular insert (resembling a soft contact lens; developed by researchers at the University of Illinois) is indicated if there is severe corneal damage.

- **Virus pinkeye**—The best known virus infection of the eyes of cattle is caused by the *red nose* or infectious bovine rhinotracheitis (IBR) virus. It is much less common than bacterial pinkeye. When this organism infects the eyes of cattle, there may or may not be other signs of disease, such as respiratory infection, vaginitis, or abortion commonly associated with infectious bovine rhinotracheitis (IBR). IBR conjunctivitis occurs most frequently in the winter, but it may be seen at other times of the year. The disease is highly contagious by direct and indirect contact of infected animals with susceptible animals.

IBR conjunctivitis may be prevented by proper vaccination of animals prior to onset of the disease. The herd should not be vaccinated once the disease appears; nor should pregnant cows be vaccinated. Affected animals should be isolated. (See section relative to IBR.)

Treatment of IBR conjunctivitis is seldom of value, although antibiotics sometimes help reduce the secondary bacterial infection.

PNEUMONIA (Calf Pneumonia)

Pneumonia is an inflammation of the lungs in which the alveoli (air sacs) fill up with an inflammatory exudate or discharge. The disease is often secondary to many other conditions. It is difficult to describe and classify for the lung is subject to more forms of inflammation than any other organ in the body. It affects all animals. In cattle, it is seen most commonly as calf pneumonia. If untreated, 50 to 75% of affected animals die. Pneumonia causes one-fifth of all nonnutritional mortality in U.S. beef cattle.

SYMPTOMS AND SIGNS

The disease is ushered in by a chill, followed by elevated temperature. There is quick, shallow respiration, with discharge from the nostrils and perhaps from the eyes. A cough may be present. The animal appears distressed, stands with legs wide apart, drops in milk production, shows no appetite, and is constipated. There may be crackling noises with breathing, and gasping for breath may be noted. If the disease terminates favorably, the cough loosens and the appetite picks up.

Fig. 12-21. Calf with pneumonia. Note characteristic spread of front legs in an effort to ease breathing. (Courtesy, Dept. of Veterinary Pathology and Hygiene, College of Veterinary Medicine, University of Illinois)

CAUSE, PREVENTION, AND TREATMENT

The causes are numerous. Many microorganisms found in other acute and chronic diseases, such as mastitis and metritis, have been incriminated; and pneumonia can be caused by a number of different viruses, with bacteria as secondary invaders. One common cause that should be stressed is the inhalation of water or medicines that well-meaning but untrained persons give to animals in drenches. Also, it is generally recognized that changeable weather during the spring and fall, as well as damp barns, is conducive to pneumonia.

Prevention includes providing good, hygienic surroundings and practicing good, sound husbandry.

Sick animals should be segregated and placed in quiet, clean quarters away from drafts. At present there are no useful vaccines.

Calves can be treated with a broad spectrum antibiotic, such as oxytetracycline, for 4 or 5 days. Secondary bacterial pneumonia may also be treated with sulfonamides or antibiotics.

POLIOENCEPHALOMALACIA (Cerebrocortical Necrosis, or Forage Poisoning)

This is a noninfectious disease of pasture and feedlot cattle, affecting animals between 3 months and 2 years of age.

SYMPTOMS AND SIGNS

Some sudden deaths occur in the feedlots. Sick animals are excitable, uncoordinated, and have impaired vision. On driving, these animals go down into convulsions.

Fig. 12-22. Calf with polioencephalomalacia, showing blindness, anorexia, incoordination, and depression. (Courtesy, University of Minnesota, Agricultural Extension Service, St. Paul, MN)

CAUSE, PREVENTION, AND TREATMENT

This disease is believed to be due to a thiamin (B-1) deficiency. It is possible that enzymes or toxins of fungi or other microbes in the rumen may destroy the vitamin before absorption can take place.

Intramuscular or intravenous thiamin injections should be administered to sick animals. Supplementary fluids should be given by a stomach tube.

RABIES (Hydrophobia, or Madness)

Rabies (hydrophobia, or *madness*) is an acute infectious disease of all warm-blooded animals and people. It is characterized by deranged consciousness and paralysis, and terminates fatally. This disease is one that is far too prevalent, and, if present knowledge were applied, it could be controlled and even eradicated.

When a human being is bitten by a dog that is suspected of being rabid, the first impulse is to kill the dog immediately. This is a mistake. Instead, it is important to confine the animal under the observation of a veterinarian until the disease, if it is present, has a chance to develop and run its course. If no recognizable symptoms appear in the animal within a period of two weeks after it inflicted the bite, it is safe to assume that there was no rabies at the time. Death occurs within a few days after the symptoms appear, and the dog's brain can be examined for specific evidence of rabies.

With this procedure, unless the bite is in the region of the neck or head, there will usually be ample time in which to administer the Pasteur treatment to exposed human beings. As the virus has been found in the saliva of a dog at least five days before the appearance of the clinically recognizable symptoms, the bite of a dog should always be considered potentially dangerous until proved otherwise. In any event, when a human being is bitten by a dog, it is recommended that a physician be consulted immediately. Each year about 30,000 persons in the United States undergo the Pasteur treatment.

But not all animals that have bitten humans should be held for observation. Wild animals (skunks, raccoons, foxes, etc.) should be killed immediately and examined for evidence of rabies infection, because the signs of rabies in wild animals are variable and the duration of the virus excretion before clinical rabies develops may be longer than in dogs.

The fact that people are susceptible to the disease

Fig. 12-23. Cow with rabies. Note the violent butting with the head; a characteristic of the furious form which is seen most often in cattle. At this stage, the animal is insane and is very dangerous, for it may attack and bite itself, other animals, or people. (Courtesy, Pitman-Moore, Inc., Indianapolis, IN)

makes it more important than the economic losses would indicate.

SYMPTOMS AND SIGNS

Less than 10% of the rabies cases appear in cattle, horses, swine, and sheep. The disease usually manifests itself in two forms: the furious, irritable, or violent form, or the dumb or paralytic form. It is often difficult to distinguish between the two forms, however. The furious type usually merges into the dumb form because paralysis always occurs just before death.

The furious form is seen most often in cattle. In its early stages, the disease is marked by loss of appetite, cessation in milk secretion, anxiety, restlessness, and a change in disposition. This initial phase is followed by a stage of madness and extreme excitation indicated by a loud bellowing marked by a change in the voice, pawing of the ground, inability to swallow, and violent butting with the head. In all respects, the animal is insane and is very dangerous, for it may attack and bite itself or other animals and even people. On the fourth or fifth day, the animal becomes quieter and unsteady. This indicates approach of posterior paralysis. Loss of flesh is already very evident. On the sixth day, the animal may go into a coma and die.

CAUSE, PREVENTION, AND TREATMENT

Rabies is caused by a filterable virus which is usually carried into a bite wound by the infested saliva of a rabid animal. The malady is generally transmitted to farm animals by dogs and certain wild animals, such as the fox and skunk.

Rabies can best be prevented by attacking it at its chief source, the dog. With the advent of an improved anti-rabies vaccine for the dog, it should be a requirement that all dogs be immunized. This should be supplemented by regulations governing the licensing, quarantine, and transportation of dogs. For understandable reasons, the control of rabies in wild animals and bats is extremely difficult. In areas where rabies is present, all cattle should be vaccinated.

Rabies vaccines are available for cattle. Seek the advice of a veterinarian for more information.

After the disease is fully developed in cattle, there is no known treatment.

Persons bitten by a rabid animal should immediately report to the family doctor who usually administers Semple type vaccine, although irradiated vaccines are used to some extent. With severe bites, especially those around the head, antiserum is particularly indicated.

TETANUS (Lockjaw)

Tetanus is chiefly a wound infection disease that attacks the nervous system of horses (and other equines) and humans, although it does occur in cattle, swine, sheep and goats. In the Southwest, it is quite common in sheep after shearing, docking, and castrating. In the Central States, tetanus frequently affects calves, lambs, and pigs, following castration or other wounds. It is generally referred to as lockjaw.

In the United States, the disease occurs most frequently in the South, where precautions against tetanus are an essential part of the routine treatment of wounds. The disease is worldwide in distribution.

SYMPTOMS AND SIGNS

The incubation period of tetanus varies from 1 to 2 weeks, but may be from 1 day to many months. It is usually associated with a wound but may not directly follow an injury. The first noticeable sign of the disease is a stiffness first observed about the head. The animal often chews slowly and weakly and swallows awkwardly. The third eyelid is seen protruding over the forward surface of the eyeball (called "haws"). Violent spasms or contractions of groups of muscles may be brought on by the slightest movement or noise. The animal usually attempts to remain standing throughout the course of the disease until close to death. If recovery occurs, it will take a month or more. In over 80% of the cases, however, death ensues, usually because of sheer exhaustion or paralysis of vital organs.

CAUSE, PREVENTION, AND TREATMENT

The disease is caused by an exceedingly powerful

toxin (more than 100 times as toxic as strychnine) liberated by the tetanus organism *Clostridium tetani*. This organism is an anaerobe (lives in absence of oxygen) which forms the most hardy spores known. It may be found in certain soils, horse dung, and sometimes in human excreta. The organism usually causes trouble when it gets into a wound that rapidly heals or closes over it. In the absence of oxygen, it then grows and liberates the toxin which follows up nerve trunks. Upon reaching the spinal cord, the toxin excites the symptoms noted above.

The disease can largely be prevented by reducing the probability of wounds, by general cleanliness, by proper wound treatment, and by vaccination with tetanus toxoid in the so-called hot areas. When an animal has received a wound from which tetanus may result, short-term immunity can be conferred immediately by use of tetanus antitoxin, but the antitoxin is of little or no value after the symptoms have developed. All valuable animals should be protected with tetanus toxoid, given annually.

All perceptible wounds should be properly treated, and the animal should be placed in a quiet, dark stall and given tranquilizing agents, sedatives, immobilizing agents, and antitoxin. Supportive treatment is of great importance and will contribute towards a favorable course. This may entail artificial feeding. The animal should be placed under the care of a veterinarian.

TUBERCULOSIS

Tuberculosis is a chronic infectious disease of people and animals, which occurs worldwide. It is characterized by the development of nodules (tubercules) that may calcify or turn into abscesses. The disease spreads very slowly, and affects mainly the lymph nodes. There are three kinds of tuberculosis bacilli—the human, the bovine, and the avian (bird) types. Practically every species of animal is subject to one or more of the three kinds, as shown in Table 12-2.

In general, the incidence of tuberculosis is steadily declining in the United States, both in animals and humans.

SYMPTOMS AND SIGNS

Tuberculosis may take one or more of several forms. Human beings get tuberculosis of the skin (lupus), of the lymph nodes (scrofula), of the bones and joints, of the lining of the brain (tuberculous meningitis), and of the lungs. For the most part, tuberculosis in animals involves the lungs and lymph nodes. In cows, the udder becomes infected in chronic cases.

Fig. 12-24. Cow in the last stages of tuberculosis. Cattle are susceptible to all three kinds of tuberculosis. (Courtesy, USDA)

TABLE 12-2
RELATIVE SUSCEPTIBILITY OF PEOPLE AND ANIMALS TO THREE DIFFERENT KINDS OF TUBERCULOSIS BACILLI

Species	Susceptibility to Three Kinds of Tuberculosis Bacilli			Comments
	Humantype	Bovinetype	Aviantype (Bird)	
Humans	Susceptible	Moderately susceptible	Questionable	Pathogenicity of avian type for humans is practically nil.
Cattle	Slightly susceptible	Susceptible	Slightly susceptible	
Swine	Moderately susceptible	Susceptible	Susceptible	90% of all swine cases are due to the avian type.
Chickens	Resistant	Resistant	Very susceptible	Chickens only have the avian type.
Horses and mules	Relatively resistant	Moderately susceptible	Relatively resistant	Rarely seen in these animals in the U.S.
Sheep	Fairly resistant	Susceptible	Susceptible	Rarely seen in these animals.
Goats	Marked resistance	Highly susceptible	Susceptible	Rarely seen in these animals in the U.S.
Dogs	Susceptible	Susceptible	Resistant	Highly resistant.
Cats	Quite resistant	Susceptible	Quite resistant	Usually obtained from milk of tubercular cows.

Many times an infected animal will show no outward physical signs of the disease. There may be a gradual loss of weight and condition and swelling of joints, especially older animals. If the respiratory system is affected, there may be a chronic cough and labored breathing. Next to the lungs and lymph nodes, the udder is most frequently affected, showing increased size and swelling of the supra mammary lymph gland. Other seats of infection are genitals, central nervous system, and the digestive system.

CAUSE, PREVENTION, AND TREATMENT

The causative agent is a rod-shaped organism belonging to the acid-fast group known as *Mycobacterium tuberculosis*. The disease is usually contracted by eating feed or drinking fluids contaminated by the discharges of infected animals. Hogs may also contract the disease by eating part of a tubercular chicken.

With cattle, periodic testing and removal of the reactors is the only effective method of control. Also, avoid housing or pasturing cattle with chickens. It is well to abide by the old adage, *once a reactor, always a reactor.*

The test consists of the introduction of tuberculin—a standardized solution of the products of the tubercle bacillus—into an approved location on the animal.

There are three principal methods of tuberculin testing—the intradermic, subcutaneous, and ophthalmic. The first of these is the method now principally used. It consists of the injection of tuberculin into the dermis (the true skin).

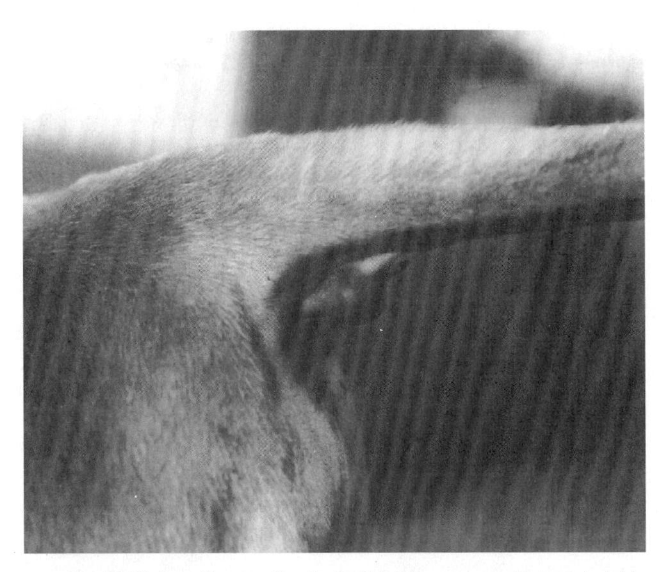

Fig. 12-25. A positive reaction (indicating the presence of tuberculosis) to the intradermic (into the true skin) tuberculin test in a cow. Reactors show a noticeable swelling, varying from the size of a pea to the size of a walnut, at the point of injection. The reading is made approximately 72 hours after injection. (Courtesy, Dept. of Veterinary Pathology and Hygiene, College of Veterinary Medicine, University of Illinois)

Upon injection into an infected animal, tuberculin will produce a reaction characterized by a swelling at the site of injection. In human beings the X-ray is usually used for purposes of detecting the presence of the disease.

As a part of the federal-state tuberculosis eradication campaign of 1917, provision was made for indemnity payments on animals slaughtered.

Preventive treatment for both humans and animals consists of pasteurization of milk and creamery byproducts and the removal and supervised slaughter of reactor animals. Various types of vaccines have been advocated for cattle, but none has provided effective immunity. Studies indicate that B.C.G. vaccine (Bacille' Calmette-Guerin, after two French scientists who first prepared the vaccine) reduces the severity of the initial disease in cattle, but does not completely prevent infection. Besides, vaccination induces hypersensitivity to tuberculin, and, thus, interferes with the diagnostic test. Countries that have attempted to use B.C.G. vaccination as the basis of a control program have ultimately abandoned the procedure in favor of the *test and slaughter* method.

In human beings, tuberculosis can be arrested by hospitalization and complete rest, along with drug therapy. But in animals this method of treatment is neither effective nor practical. Infected animals should be sent to slaughter.

All 50 states will accept for entry cattle meeting either of the following T.B. tests: (1) accredited herds tested within 12 months, or (2) individual negative test within 30 days. Title 9 of the Code of Federal Regulations, part 77, tuberculosis of cattle, requires the branding of reactors and tuberculin testing cattle which originate in areas which are not modified accredited.

Cattle to be exported must be tested for tuberculosis and found free of the disease within 90 days of shipment.

(Also see later section of this chapter entitled, "Federal and State Regulations Relative to Disease Control.")

VAGINITIS (Granular Vaginitis, or Granular Venereal Disease)

This is an infectious disease which localizes in the cow's vulva and on the penis and prepuce of the bull, causing an inflammation of varying intensity. It occurs throughout the United States.

SYMPTOMS AND SIGNS

The tissue of the vagina is reddish, roughened,

and granular in appearance. Infected animals are usually difficult breeders. Economic losses are in terms of lower percentage calf crop and decreased milk production.

CAUSE, PREVENTION, AND TREATMENT

The cause of vaginitis is unknown, but indications are that it is an infectious agent. There is substantial evidence that *Mycoplasma genitalium* isolated from the vaginal mucus of affected cattle has a casual role.

Prevention consists in purchasing clean animals from clean herds and avoiding the use of bulls that have been exposed to the infection. Artificial insemination can be effectively employed in a control program. Also, in problem herds, vaccination 30 to 60 days prior to the breeding season should be considered.

Because the lesions are aggravated by sexual activity and they tend to heal in time, sexual rest is recommended, particularly for affected bulls.

Treatment of females is not indicated. The condition will clear up by itself in several weeks. Vaginitis in females is not directly related to fertility. However, the predisposing agents (viruses, bacteria, fungi) may affect fertility.

The condition in bulls tends to be more persistent; and affected bulls may refuse to breed. They should be treated by massage of the anesthetized prolapsed penis and sheath with a suitable antibiotic ointment.

VIBRIONIC ABORTION (Vibrio Fetus, or Vibriosis)

This is an infectious venereal disease of cattle, which causes infertility and abortion. For diagnosis, laboratory methods must be used.

SYMPTOMS AND SIGNS

Infected herds are characterized by (1) abortions in the middle third of pregnancy, (2) several services per conception, and (3) irregular heat periods.

CAUSE, PREVENTION, AND TREATMENT

The disease is caused by the microorganism *Vibrio fetus*, which is transmitted at the time of coitus.

The best method of bringing the disease under control is the use of artificial insemination. Semen for this purpose should either be from known uninfected bulls, or be treated with penicillin or streptomycin.

Where natural service must be continued, as in large beef herds, vaccination is important in controlling the disease. The most important animals to vaccinate are herd bulls and replacement heifers. Vaccination is both curative and preventive in bulls. They should be given two subcutaneous injections initially and an annual booster injection thereafter. Vaccination of females is done at least six weeks before breeding and may need to be repeated annually.

Aborting cows should be isolated, and aborted fetuses and membranes should be burned or buried. Contaminated quarters should be thoroughly cleaned and disinfected.

Dihydrostreptomycin injected subcutaneously at 11 mg/lb body weight and 5 g of 50% solution applied locally to the penis and prepuce will eliminate the infection from bulls.

WARTS (Papillomatosis)

Warts, which are small tumors, are an infectious disease of cattle and other animals and humans. Young animals, under two years of age, are most often affected.

SYMPTOMS AND SIGNS

Warts are protruding growths on the skin, varying from very small to quite large, pendulous growths weighing several pounds. They may appear anywhere on the body, but are especially common on the teats and/or around the head.

Although warts are a nuisance, their presence does not normally interfere with the animal's health. However, they damage the hide, making the leather derived therefrom weak in the affected area.

Fig. 12-26. An extreme case of warts. (Courtesy, Fort Dodge Laboratories, Inc., Fort Dodge, IA)

CAUSE, PREVENTION, AND TREATMENT

Warts are caused by a virus. It appears that each animal species is attacked by a specific virus.

The following preventive measures are recommended:

1. Segregate *warty* cattle.
2. Clean and disinfect all exposed pens, stables, chutes, and rubbing posts.
3. Administer wart vaccine.

The most common treatment among cattle producers consists of softening the wart with oil for several days, and then tying off the growth with thread or a rubber band or snipping it off with sterile scissors. The stump should then be treated with tincture of iodine. Wart vaccines help in some cases, but, generally speaking, they are more effective in prevention than in treatment. The veterinarian may resort to surgical removal of extremely large warts.

WINTER DYSENTERY (Winter Scours)

This is an acute infectious disease of stabled cattle, both dairy and beef, most frequently occurring between the months of November and March.

SYMPTOMS AND SIGNS

It causes few death losses, but afflicted animals lose in condition; and, in lactating animals, there is a sharp reduction in milk flow.

The period of incubation is extremely short, varying from 3 to 5 days. A profuse watery diarrhea is the main symptom. Often the feces are dark brown in color, and tend to become darker when intestinal hemorrhages occur. Usually the temperature remains normal, and the appetite is unchanged. Calves and young animals are least susceptible, but animals of all ages are affected. The seasonal incidence of the disease, the age and number of animals affected, together with the suddenness of the onset, are helpful in arriving at a correct diagnosis.

CAUSE, PREVENTION, AND TREATMENT

Until recently, *Vibrio jejuni* had been accepted as the cause. It is now suspected that a virus is the cause of the disease or that several infections exist concurrently.

Prevention consists in isolating new or replacement animals. Also, any animal suffering from an acute attack of dysentery should be separated from the herd. Where the disease is encountered, rigid sanitation should be practiced.

Many mild cases of winter dysentery require no treatment, and the course of the disease does not appear to be altered since return to normal occurs in 2 to 3 days with or without treatment. Treatment of severely affected animals seems to hasten recovery and may prevent fatal outcome. Intestinal astringents and antiseptics, along with fluid and electrolyte therapy, are usually employed.

PARASITES OF BEEF CATTLE[7]

Beef cattle are attacked by a wide variety of internal and external parasites, with the losses about equally divided between internal and external parasites.

The prevention and control of parasites is one of the quickest, cheapest, and most dependable methods of increasing beef and milk production with no extra cattle, no additional feed, and little more labor. This is

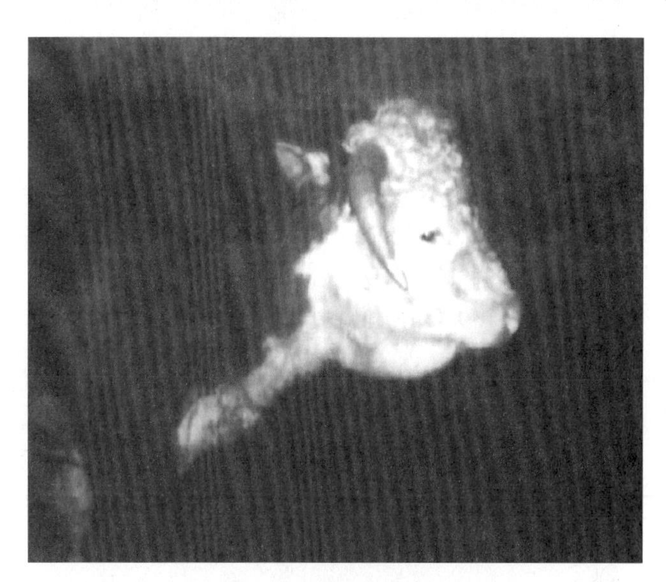

Fig. 12-27. Hereford bull with bottle jaw, indicative of a heavy infestation of several species of internal parasites. (Courtesy, University of Minnesota, Agricultural Extension Service, St. Paul, MN)

[7]From time to time, new insecticides and vermifuges are approved and old ones are banned or dropped. When parasitism is encountered, therefore, it is suggested that the cattle producer obtain from local authorities the current recommendation relative to the choice and concentration of the insecticide and vermifuge to use.

The use of trade names of wormers and insecticides in this section does not imply endorsement, nor is any criticism implied of similar products not named; rather, it is recognition of the fact that cattle producers, and those who counsel with them, are generally more familiar with the trade names than the generic names.

important, for, after all, the farmer or rancher bears the brunt of this reduced meat and milk production, wasted feed, and damaged hides. It is hoped that the discussion that follows may be helpful in (1) preventing the propagation of parasites, and (2) causing the destruction of parasites through the use of the most effective anthelmintic or insecticide.

For guidance in the selection of an anthelmintic or insecticide, the user should seek the counsel of the veterinarian, county agent, extension entomologist, or agricultural consultant; and for instruction on the use of an anthelmintic or insecticide, the user should follow the directions on the label.

BLOWFLY

The flies of the blowfly group include a number of species that find their principal breeding ground in dead and putrifying flesh, although they sometimes infest wounds or unhealthy tissues of live animals and fresh or cooked meat. Black blowfly larvae frequently infest dehorning wounds during winter months and occasionally the navel of newborn animals. All the important species of blowflies except the flesh flies, which are grayish and have three dark stripes on their backs, have a more or less metallic luster.

DISTRIBUTION AND LOSSES CAUSED BY BLOWFLIES

Although blowflies are widespread, they present the greatest problem in the Pacific Northwest and in the South and southwestern states. Death losses from blowflies are not excessive, but they cause much discomfort to affected animals and lower production.

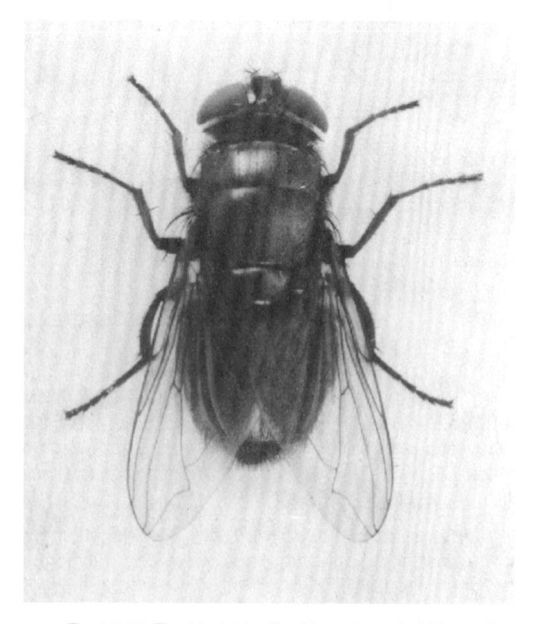

Fig. 12-28. The black blowfly *(phormia regina).* These flies are characterized by a metallic luster. (Courtesy, USDA)

LIFE HISTORY AND HABITS

With the exception of the group known as gray flesh flies, which deposit tiny living maggots instead of eggs, the blowflies have a similar life cycle to the screwworm, except that the cycle is completed in about one-half the time.

DAMAGE INFLICTED; SYMPTOMS AND SIGNS OF AFFECTED ANIMALS

The blowfly causes its greatest damage by infesting wounds and the soiled hair or fleece of living animals. Such damage, which is largely limited to the black blowfly, is similar to that caused by screwworms. Sheep are especially susceptible to attacks of blowflies, because their wool frequently becomes soiled or moistened by rain and accumulation of feces and urine. The maggots spread over the body, feeding on the skin surface, where they produce a severe irritation and destroy the ability of the skin to function. Infested animals rapidly become weak and fevered; and, although they recover, they may remain in an unthrifty condition for a long period.

Because blowflies infest fresh or cooked meat, they are often a problem of major importance around packing houses or farm homes.

PREVENTION, CONTROL, AND TREATMENT

Prevention of blowfly damage consists of eliminating the pest and decreasing the susceptibility of animals to infestation.

As blowflies breed principally in dead carcasses, the most effective control is effected by promptly destroying all dead animals by burning or deep burial. The use of traps, poisoned baits, and electrified screens is also helpful in reducing trouble from blowflies. Suitable repellents, such as pine tar oil, help prevent the fly from depositing its eggs.

When animals become infested with blowfly maggots, their wounds should be treated with a preparation such as ronnel (Korlan) applied according to directions on the label.

BOVINE TRICHOMONIASIS

This is a protozoan venereal disease of cattle characterized by early abortions (usually between the second and fourth months of pregnancy) and temporary sterility. The protozoa that cause the disease, known as *Trichomonas foetus*, are one-celled, microscopic in size, and capable of movement. They are found in aborted fetuses, fetal membranes and fluids, vaginal secretions of infected animals, and the sheaths of infected bulls. Diagnosis can be confirmed microscopically. The infected bull is the source of the infection. On the other hand, the disease appears to be self-limiting in the cow.

DISTRIBUTION AND LOSSES CAUSED BY BOVINE TRICHOMONIASIS

This disease is being reported with increasing frequency throughout the United States and has become a serious problem in many herds. The economic loss is primarily due to the low percentage of calf crops in infected herds.

LIFE HISTORY AND HABITS

The protozoa that cause the disease are one-celled microscopic organisms with three threadlike whips (flagella) at the front and one at the rear. The evidence indicates that the disease is spread from the infected to the clean cow by an infected bull at the time of service and that other types of contact infection do not occur. Following one or perhaps two abortions, cows appear to be immune to reinfection. Further than these facts, little is known of the life history and habits of *Trichomonas foetus*.

DAMAGE INFLICTED; SYMPTOMS AND SIGNS OF AFFECTED ANIMALS

There is no systemic disturbance manifested by the infected bull. There may be some mucus discharged from the sheath, and the latter may be slightly inflamed. The only clinical evidence of infection is the transmission of the disease to the cattle serviced.

Infected cows frequently show a whitish vaginal discharge, and the following characteristic conditions usually exist when a herd is infected: (1) abortions in the first third of pregnancy; (2) uterine infections; (3) irregular heat periods; and (4) several services per conception.

Early abortions or erratic heat periods in individuals or herds that are known to be free of Bang's disease should lead one to suspect the presence of the trichomonad infection. Definite diagnosis of infection in the bull is made by means of a microscopic examination of smears taken from the prepuce of the bull or the vagina of the cow.

Fig. 12-29. Bull with trichomoniasis. This animal appeared normal, but spread the disease during the breeding act. (Courtesy, Dept. of Veterinary Pathology and Hygiene, College of Veterinary Medicine, University of Illinois)

PREVENTION, CONTROL, AND TREATMENT

Prevention lies in the use of clean bulls or artificial insemination. If practical, infected animals should be sold for slaughter or allowed 90 days of sexual rest. Otherwise, treatment should be attempted.

Effective control consists of exercising great precaution in introducing new animals into the herd, in breeding outside cows, and in taking cows outside the herd for breeding purposes.

Trichomoniasis is usually self-limiting in cows; suspension of breeding is usually sufficient.

Most trichomonad-infected bulls should be marketed for slaughter. Only bulls of exceptional value should be treated, but treatment and testing are long and laborious. Drugs of known efficacy that are used elsewhere in the world are either not registered (available commercially) or not cleared for use in cattle in this country. Currently, there are no drugs approved by the Food and Drug Administration available for the treatment of trichomoniasis.

CATTLE TICK FEVER (Splenetic Fever, Texas Fever, etc.)

This is an infectious protozoan disease of adult cattle caused by *Babesia bigemina*, which depends upon the tick, chiefly *Boophilus annulatus*, for its survival and transmission.

DISTRIBUTION AND LOSSES CAUSED BY CATTLE TICK FEVER

Prior to 1906, at which time a concerted effort was initiated to eradicate the cattle fever tick, this infectious disease of cattle and the parasite which transmits it were the most serious obstacles faced by the cattle industry in the 15 southern and southwestern states, representing a combined area of nearly one-

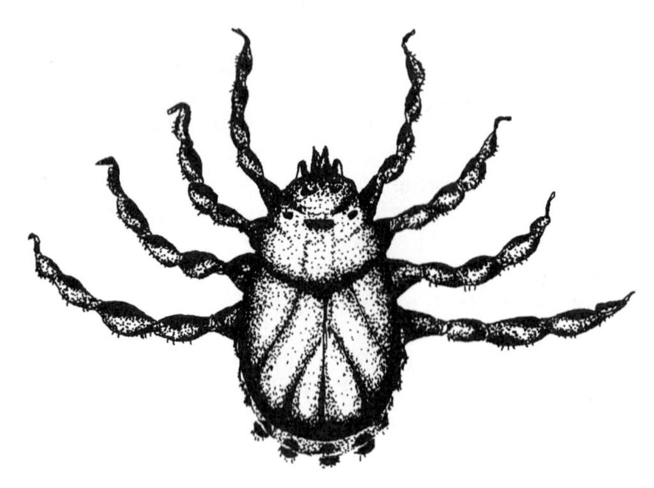

Fig. 12-30. An adult cattle tick *(Boophilus annulatus)*, chief transmitter of cattle tick fever.

fourth of the United States. At that time, conservative estimates placed the yearly losses at $40 million. Today, 99% of the formerly infested area has been freed (for the most part, the tick is confined to the Texas-Mexico border), and the once appalling losses have been practically eliminated. Cattle tick fever is not uncommon in Central and South America.

In addition to the serious death losses encountered in infected herds, the loss of blood—the only food of the cattle fever tick—results in serious damage. Infected young animals are stunted, mature animals are emaciated, and the milk flow of infected dairy animals is greatly reduced. Death occurs in about 10% of the chronic and 90% of the acute causes.

LIFE HISTORY AND HABITS

In 1889 and 1890, investigators of the USDA, Bureau of Animal Industry, established (1) that intracellular one-celled parasites, or protozoa, known as *Babesia* are the direct causative agents of the disease, and (2) that cattle tick infestation is necessary in the transmission of the disease. Thus, for the first time in either human or veterinary medicine, the discovery was made that an intermediate biological carrier may transmit a disease. It is noteworthy that this pioneer work opened up an entirely new field in medical science, pointing the way for studies that later solved the problems of the spread of such dreaded diseases as malaria, yellow fever, Rocky Mountain spotted fever, typhus, and others.

The life history and habits of the protozoa which causes cattle tick fever are as follows: infected ticks, which have sucked blood from an infected cow, pass along the protozoa *Babesia (Piroplasma) bigemina* to their eggs. The female tick falls to the ground and deposits from 2,000 to 4,000 eggs. In 2 or 3 weeks, these eggs hatch into young ticks or larvae. The larvae climb on nearby vegetation to await the passing of cattle to which they attach themselves, biting and sucking blood from the host. In the latter process, the protozoa *(Piroplasma)* are passed into the blood of cattle—the protozoa of infected ticks having been passed into the eggs of the tick and through all stages of its growth.

DAMAGE INFLICTED; SYMPTOMS AND SIGNS OF AFFECTED ANIMALS

The incubation period of cattle tick fever is about 10 days. The disease is characterized by high temperature, rapid breathing, enlarged spleen, engorged liver, pale and yellow membranes, and red to black urine. Sometimes the symptoms subside only to reoccur at another time. Although immune, the recovered animals are permanent carriers of the disease. In infected areas, native cattle are either immune or only slightly affected.

PREVENTION, CONTROL, AND TREATMENT

Prevention of the disease consists of avoiding contact with the cattle fever tick, the only natural agent by which cattle tick fever is transmitted from animal to animal.

The most effective control measures are directed at the eradication of the fever ticks, either by killing them on the pastures or on the cattle. Pastures may be rendered tick-free by excluding all the host animals—cattle, horses, and mules—until all the ticks have died of starvation in 8 to 10 months.

Dipping or spraying with an approved insecticide is the most effective method of control.

Fig. 12-31. Cattle with cattle tick fever. The disease is characterized by high temperature, rapid breathing, enlarged spleen, engorged liver, pale and yellow membranes, and red to black urine. (Courtesy, USDA)

Successful treatment of sick animals depends upon early recognition of the disease and prompt treatment.

Although immune, recovered animals are permanent carriers of the disease.

Title 9 of the Code of Federal Regulations, part 72, Texas fever in cattle, indicates areas under quarantine and requires dipping in a permitted dip prior to movement from an infected area.

(Also see later section in this chapter entitled, "Federal and State Regulations Relative to Disease Control.")

COCCIDIOSIS

Coccidiosis—a parasitic disease affecting cattle, sheep, goats, swine, pet stock, and poultry—is caused by microscopic protozoan organisms known as coccidia, which live in the cells of the intestinal lining. Each class of domestic livestock harbors its own species of coccidia, thus there is no cross-infection between animals; and the coccidia of animals cannot be transmitted to people.

Cattle in North America are known to be affected by 12 species of coccidia. But only *Eimeria bovis* and *E. zuerni* are important in the United States, with *E. zuerni* tending to cause the most serious infections.

DISTRIBUTION AND LOSSES CAUSED BY COCCIDIOSIS

The distribution of the disease is worldwide. Except in very severe infections, or where a secondary bacterial invasion develops, infested farm animals usually recover. The chief economic loss is in lowered gain and production. It is most severe in feeder cattle and young dairy calves.

LIFE HISTORY AND HABITS

Infected animals may eliminate daily with their droppings thousands of coccidia organisms (in the resistant öocyst stage). Under favorable conditions of temperature and moisture, coccidia sporulate to maturity in 3 to 5 days, and each öocyst contains 8 infective sporozoites. The öocyst then gains entrance into an animal by being swallowed with contaminated feed or water. In the host's intestine, the outer membrane of the öocyst, acted on by the digestive juices, ruptures and liberates the 8 sporozoites within. Each sporozoite then attacks and penetrates an epithelial cell, ultimately destroying it. While destroying the cell, however, the parasite undergoes sexual multiplication and fertilization with the formation of new öocysts. The parasite (öocyst) is then expelled with the feces and is again in a position to reinfest a new host.

The coccidia parasite abounds in wet, filthy surroundings; resists freezing and ordinary disinfectants; and can be carried long distances in streams.

DAMAGE INFLICTED; SYMPTOMS AND SIGNS OF AFFECTED ANIMALS

A severe infection with coccidia produces diarrhea, and the feces may be bloody. The bloody discharge is due to the destruction of the epithelial cells lining the intestines. Ensuing exposure and rupture of the blood vessels then produces hemorrhage into the intestinal lumen.

In addition to a bloody diarrhea, affected animals usually show pronounced unthriftiness and weakness.

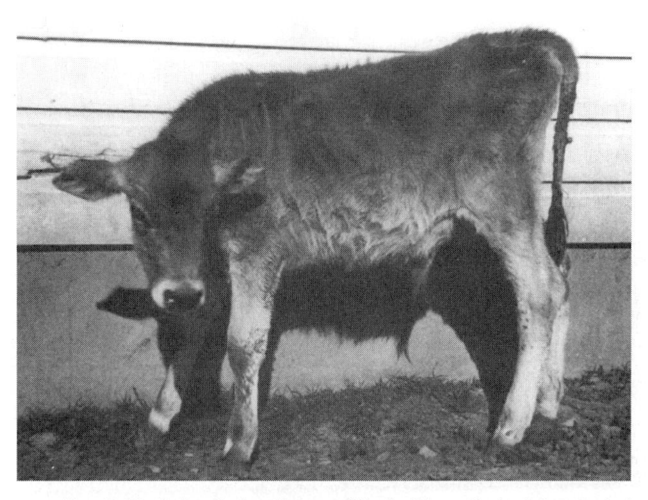

Fig. 12-32. Calf suffering from coccidiosis. The soiled tail is a typical symptom. (Courtesy, USDA, Regional Laboratory An. Dis. Res., Auburn, AL)

PREVENTION, CONTROL, AND TREATMENT

Coccidiosis can be prevented by protecting animals from feed or water that is contaminated with the protozoa that causes the disease. Prompt segregation of affected animals is important and should be done if practical. Manure and contaminated bedding should be removed daily. Low, wet areas should be drained. If possible, segregation and isolation of animals by age should be used in controlling the disease. All precautions should be undertaken to keep droppings from contaminating the feed. Although the öocysts resist freezing and certain disinfectants and may remain viable outside the body for 1 or 2 years, they are readily destroyed by direct sunlight and complete drying.

Animals suffering from coccidiosis should be treated with an approved anticoccidial according to directions on the label.

In addition to specific anticoccidial therapy, supportive fluids and good care of sick calves are helpful. Antibiotics and sulfonamides given orally may help prevent secondary bacterial infections of the denuded intestinal mucosa.

FACE FLY

The face fly, *Musca autumnalis*, was first found in this country in New York in 1953. It is a close relative of and similar in appearance to the housefly.

The face fly is primarily a pest of cattle—particularly animals in open fields and away from shade, although it will attack horses and sheep.

DISTRIBUTION AND LOSSES CAUSED BY FACE FLIES

By 1970, face flies had been reported in most of the states. Only a few of the states in the southern United States have remained free of face flies.

DAMAGE INFLICTED; SYMPTOMS AND SIGNS OF AFFECTED ANIMALS

The face fly does not bite. It feeds on moist areas around the eyes and muzzle and on blood that oozes from cuts and wounds. It is extremely annoying to animals; interfering with vision and breathing, and preventing normal grazing. Large populations force animals to leave pastures and seek relief in wooded areas and shelter.

When cattle enter a barn or darkened area, the fly leaves the animal's face and rests on fence posts, gates, sides of barns, etc. Adult flies hibernate in attics and other protected places during the winter.

The life cycle of the face fly closely resembles that of the horn fly in that fresh manure is required for the development of each; both lay their eggs in fresh manure, where the larvae develop.

PREVENTION, CONTROL, AND TREATMENT

Prevention consists of scattering or removing fresh cow manure.

The control of face flies is less effective than the control of other flies because face flies do not follow cattle into barns. Spraying with approved insecticides and using dust bags, mists, and face rubbers have given variable results. Halters, ear tags, collars, and other devices impregnated with insecticides give face fly reduction, but not control; they are less effective on face flies than on horn flies. (See Horn Fly, section on "Prevention, Control, and Treatment.")

GASTROINTESTINAL NEMATODE WORMS

More than 20 species of harmful parasites inhabit the fourth stomach or abomasum, the small intestine, the large intestine, the lungs, and the liver of cattle.

When their presence is suspected, their diagnosis should be confirmed by microscopic examination by the veterinarian of fresh manure samples. Such examination of the samples will show if parasite eggs of öocysts are present, and in what numbers.

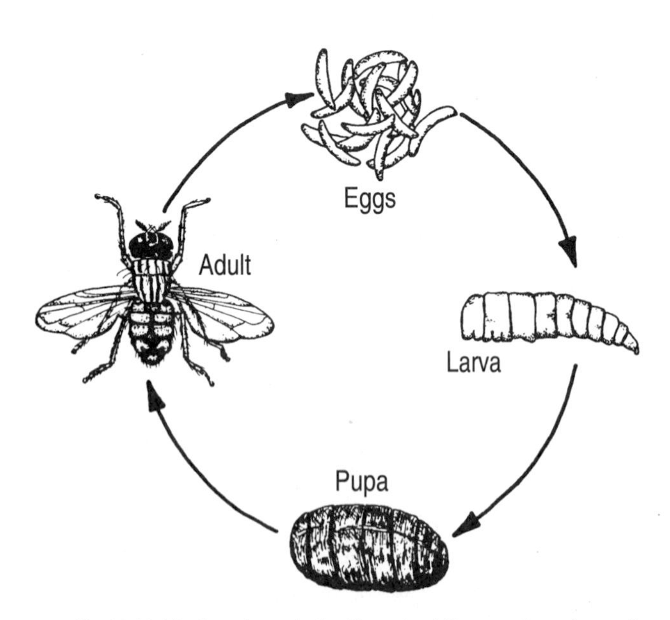

Fig. 12-33. The four stages in the life cycle of flies: egg, larva (maggot), pupa, and adult. Under optimum conditions during warm weather, the various species complete their life cycles in the following number of days: housefly, 7; stable fly, 21; face fly, 8; black blowfly, 11; green blowfly, 8; screwworm fly, 14.

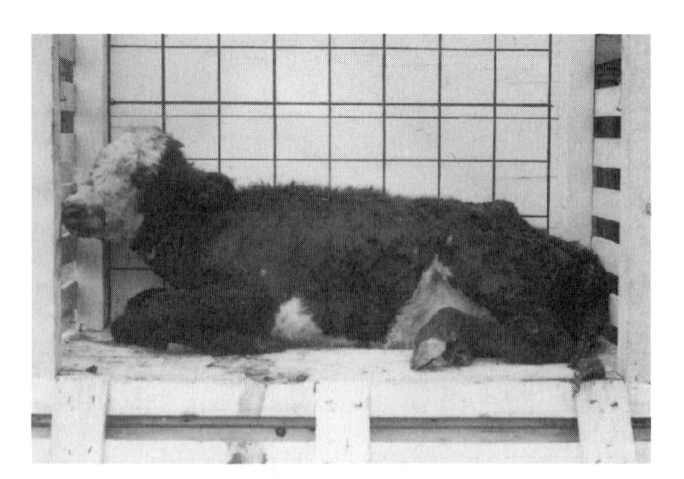

Fig. 12-34. Calf with stomach worms. (Courtesy, Dept. of Veterinary Pathology and Hygiene, College of Veterinary Medicine, University of Illinois)

DISTRIBUTION AND LOSSES CAUSED BY GASTROINTESTINAL NEMATODE WORMS

One or more species of internal parasites of cattle are found throughout the United States, but they may be especially severe in the South and on irrigated and permanent pastures.

The losses are in terms of lowered feed efficiency caused by disturbed digestion, lowered meat and milk production, and some death losses. Young animals are more severely affected than mature cattle.

LIFE HISTORY AND HABITS

Most of the gastrointestinal nematode worms have similar life cycles, of which the common stomach worm is typical. Infected cattle carry the mature worms in the fourth stomach. Eggs from these worms are expelled with manure and develop on the pasture into infective larvae. Then cattle become infected by eating grass infected with the larvae. The latter develop into mature worms in the stomach and these again produce eggs that recontaminate the pasture.

DAMAGE INFLICTED; SYMPTOMS AND SIGNS OF AFFECTED ANIMALS

An animal with a light infection rarely shows any outward symptoms, and the symptoms are not specific. But infected animals generally show loss of weight, retarded growth, anemia, diarrhea, and/or lowered resistance to other diseases. With a heavy infection there may be a swelling under the jaw (bottle jaw); and the parasites may even cause the death of the animals.

PREVENTION, CONTROL, AND TREATMENT

Preventive and control measures include (1) rotating pastures, (2) segregating calves from mature animals, (3) cross-grazing with cattle and horses, (4) avoiding overstocking or overgrazing pastures since most of the infective larvae are on the bottom inch of grass, and (5) keeping feeders and waterers sanitary.

Treatment consists of therapeutic doses in the spring and again in the fall (or at such other times as necessary) of one of the approved compounds (generic drugs) according to the manufacturers' directions.

From an approved list of drugs, the cattle producer and the veterinarian can (1) rotate drugs to prevent parasitic resistance and (2) select the method of administration easiest to follow (although varying according to drug, they may be given as a drench, a bolus, a feed or mineral mix, a paste, or an injection).

GRUBS (Heel Flies, or Warbles)

Cattle grubs are the maggot stage of insects known as heel flies, warble flies, or gadflies. Two species of cattle grubs are present in the United States. The northern cattle grub *(Hypoderma bovis)* occurs mainly in the north though it is found as far south as southern California, northern Arizona, Oklahoma, Tennessee, South Carolina, and Hawaii. The common cattle grub *(Hypoderma lineatum)* occurs throughout the 48 contiguous states and in Hawaii and Alaska. The cattle grub or heel fly is probably the most destructive insect attacking beef and dairy animals.

DISTRIBUTION AND LOSSES CAUSED BY CATTLE GRUBS

The species *Hypoderma lineatum* is widely distributed throughout the United States, whereas *Hypoderma bovis* is chiefly confined to the northern states.

The damage inflicted by cattle grubs affects cattle producers, packers, tanners, and, finally, consumers. The kinds of losses include the following:

1. **Decreased gains or milk production, mechanical injury, or even death.** Though the fly does not bite or sting, when it lays its eggs on the lower leg, it usually terrifies the animal, causing it to run with tail hoisted, seeking relief. It may run through fences, over cliffs, or become hopelessly bogged down in a mudhole or swamp. Beef animals suffer weight losses,

Fig. 12-35. Adult female heel fly *(Hypoderma lineatum)* whose maggot stage is responsible for the common cattle grub. (Courtesy, USDA)

Fig. 12-36. Heifers running away from heel flies. Though the fly does not bite or sting, when it lays its eggs on the lower leg, it usually terrifies the animal, causing it to run with tail hoisted, seeking relief. (Courtesy, Livestock Conservation, Inc.)

and milk production from dairy cows may be reduced from 10 to 25% during the period heel flies are laying their eggs. Livestock Conservation, Inc., is the authority for the statement that grub treatment can mean the following to cattle producers: 20 to 25 lb more per head in the feedlot, 30 to 40 lb heavier weaning weights, and 50 lb heavier yearlings.

2. **Carcass damage.** According to meat packers, about 35% of all beef carcasses are damaged by grubs. The yellowish, watery patches caused by the migration of the larvae under the skin are referred to by butchers as *pilled* or *licked beef.* Two to three pounds of *jellied* beef must be trimmed from the loins and ribs of each *grubby* animal, and the damaged cut of meat is devalued 2¢ per pound because of the ragged and unattractive appearance.

3. **Injury of hides.** Approximately one-third of all cattle hides produced in the United States are damaged by grubs. This loss is caused by the migration of the grub through the back, which leaves a scar in the most valuable part of the hide. According to trade custom, if a hide has as many as five grub holes, it is classed as Grade No. 2 and is subject to a discount of 1¢ per pound. Commonly as many as 40, and occasionally 100 or more, grub holes are found in a single hide. Hides of the latter quality are not considered worth tanning and are sold for by-products.

4. **Shock to animal.** In certain older animals that have been previously sensitized, the breaking of a grub under their skin may cause a terrific reaction (anaphylaxis or allergic reaction). The area may be greatly swollen and form an abscess, and there may be such a general reaction that the animal may die from shock. At the first signs of shock, the local veterinarian should be summoned quickly, to administer appropriate stimulants.

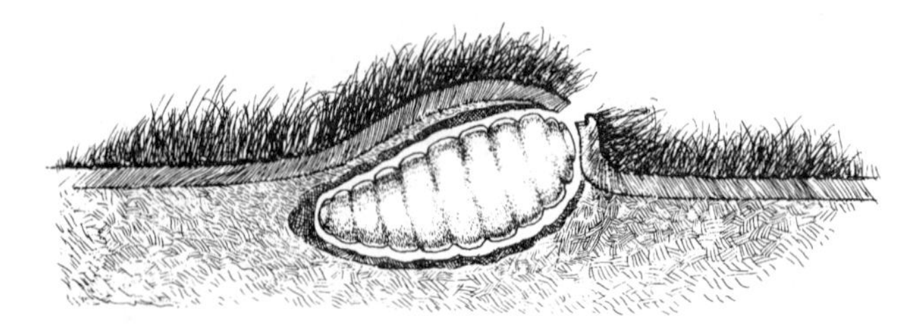

Fig. 12-37. Hole in the hide on the back of an animal, with a grub inside. Note hide opening through which the grub obtains air and finally escapes. (Courtesy, USDA)

LIFE HISTORY AND HABITS

Basically the two species have a similar life cycle. The female flies, called heel flies, attach their eggs to the hairs of the legs and bodies of cattle. The eggs hatch into larvae after about 3 days and enter the animals at the bases of the hairs. Once inside, the common cattle grub migrates from the point of entry to the gullet; the northern cattle grub migrates to the spinal column (both migrations take 2 to 4 months). After some additional months in the gullet or spinal column, grubs of both species migrate to the animals' backs, cut breathing holes in the hide, and remain there for about 6 weeks while they increase greatly in size. The resultant swellings are often called wolves or warbles. Fully grown grubs leave the hide through

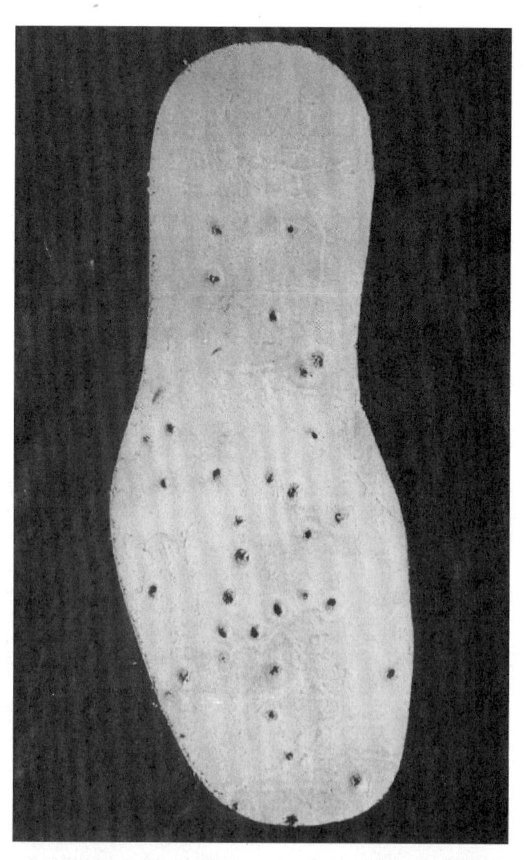

Fig. 12-38. Shoe sole damaged by cattle grubs. About one-third of all cattle hides produced in the U.S. are damaged by grubs. (Courtesy, Livestock Conservation, Inc.)

the breathing holes and drop to the ground where they pupate. Then, in a few weeks they transform to non-feeding adult flies that emerge and mate. On bright sunny days the female then seeks cattle for egg laying, which causes gadding. The entire life cycle takes about 1 year, and the same stages are usually found at about the same time each year in any given area. (See Fig. 12-39 for diagram of the life history and habits of cattle grubs.)

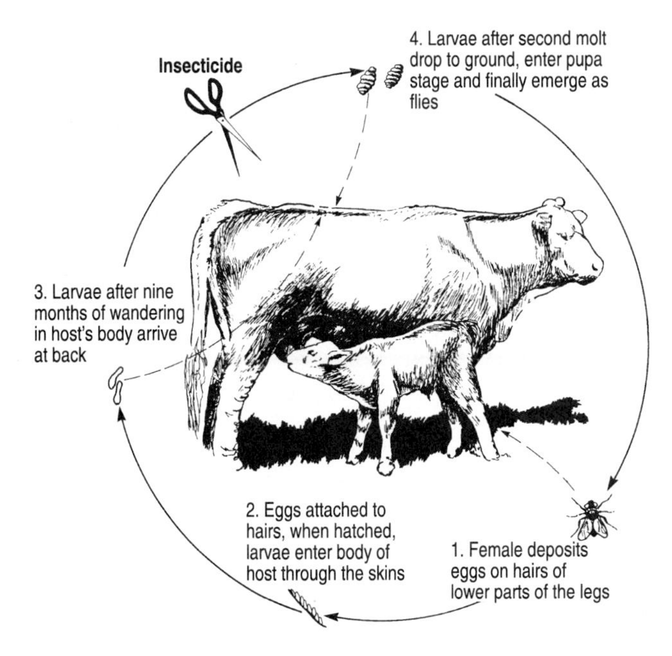

Fig. 12-39. Diagram showing the life history and habits of cattle grubs. As noted (see scissors) effective control and treatment (cutting the cycle of the parasite) may be obtained by the application of a suitable insecticide (spray, dip, or dust). The first application should be made 25 to 30 days after grubs first appear in the back, with subsequent treatments, at 30-day intervals thereafter as long as grubs are present in the back. Control may also be secured by the use of a systemic insecticide.

DAMAGE INFLICTED; SYMPTOMS AND SIGNS OF AFFECTED ANIMALS

The attack of the heel fly is unmistakable when, in the spring or early summer, cattle are seen madly running with their tails hoisted high over their backs in an attempt to escape. The presence of the grub (larva) in the back, usually from December to May, causes a characteristic swelling (and an opening in the skin, from which pus is discharged), which usually becomes conspicuous, so that a grubby back has a lumpy appearance.

PREVENTION, CONTROL, AND TREATMENT

Complete prevention of cattle grub damage within any given herd cannot be obtained unless all cattle grubs throughout the country are exterminated. This means a nationwide campaign in which all cooperate to eradicate the menace, farm by farm, ranch by ranch, county by county, and state by state.

The use of an approved compound, applied according to the directions on the label, is the best method of control and treatment of cattle grubs.

HORN FLY (Cattle Fly, Cow Fly, or Stock Fly)

This fly is one of the most numerous and worst annoyances of cattle. It is often found resting at the base of the horn; hence, the name *horn fly*. But by no means does it confine itself to this location on the animal. Horn flies may congregate by the hundreds or even thousands on the backs, shoulders, and bellies of cattle.

DISTRIBUTION AND LOSSES CAUSED BY THE HORN FLY

The horn fly is widely distributed throughout the United States. It is one of the most numerous of the biting flies affecting cattle in North America. An average of 4,000 to 5,000 flies per animal is not uncommon, and individual animals may support as many as 10,000 to 20,000 flies. Horn flies are more of a problem on grazing cattle than of cattle held in lots. The presence of this insect produces irritation and worry, loss of blood, reduced vitality, and, in the South, sores that may become infested with screwworms. From an economic standpoint, beef cattle gains are sharply reduced, and dairy cattle suffer lowered milk production. Studies indicate that the weight gains of beef animals may be reduced by as much as 50 lb per animal during

Fig. 12-40. Cattle heavily infested with horn flies at the base of the horn; hence, the name *horn fly*. (Courtesy, USDA)

fly season when high populations of horn flies go uncontrolled.

LIFE HISTORY AND HABITS

This fly, *Haematobia irritans*, is about one-half the size of an ordinary housefly or the stablefly and possesses a piercing beak. Unlike the housefly and stablefly, it remains on cattle throughout the day and night. Horn flies usually feed twice a day, sometimes more frequently.

The life cycle and habits of the horn fly are as follows:

1. The adult fly leaves the cow only for a brief 5-to 10-minute period to lay eggs. The eggs are laid beneath the droppings, where they are protected from the sun and rain and hatch out into tiny white maggots in about 16 hours. Each female fly is capable of laying 375 to 400 eggs during a lifetime.
2. The maggots crawl into the droppings where they feed and grow. They become fully grown in another four days; then they crawl down into the lower part of the droppings or into the soil to pupate and later emerge as a fly.
3. About an hour after the fly emerges from its pupal case, it seeks the nearest cow where it settles and starts feeding. The fly may mate as early as the second day after emergence and may deposit eggs on the third day.

The entire life cycle of the horn fly averages about 9 to 12 days during the summer months, and the adult fly lives about 7 weeks. Thus, it is small wonder that such hordes of this insect exist. (See Fig. 12-42 for diagram of the life history and habits of the horn fly.)

Fig. 12-41. Enlargement of an adult female horn fly. It is about one-half the size of an ordinary housefly. (Courtesy, USDA)

DAMAGE INFLICTED; SYMPTOMS AND SIGNS OF AFFECTED ANIMALS

The tormented cattle often refuse to graze during the day and seek protection by hiding in dark buildings, brush, or tall grass. Heavily infested cattle may also have a rough, sore skin, and they suffer an inevitable loss in condition.

PREVENTION, CONTROL, AND TREATMENT

Prevention and control rest chiefly in disturbing the main breeding ground of the horn fly. This is best accomplished by spreading fresh droppings with a spring tooth harrow in order to hasten their drying. Running pigs with cattle will accomplish the same purpose. These methods of control are not practical on extensive grazing areas, but they may be used in small pastures.

Horn flies are not difficult to control because they live on their hosts during most of their adult life. Insecticides are applied to the animal by the following methods: (1) direct application—sprays, dust, dips, and

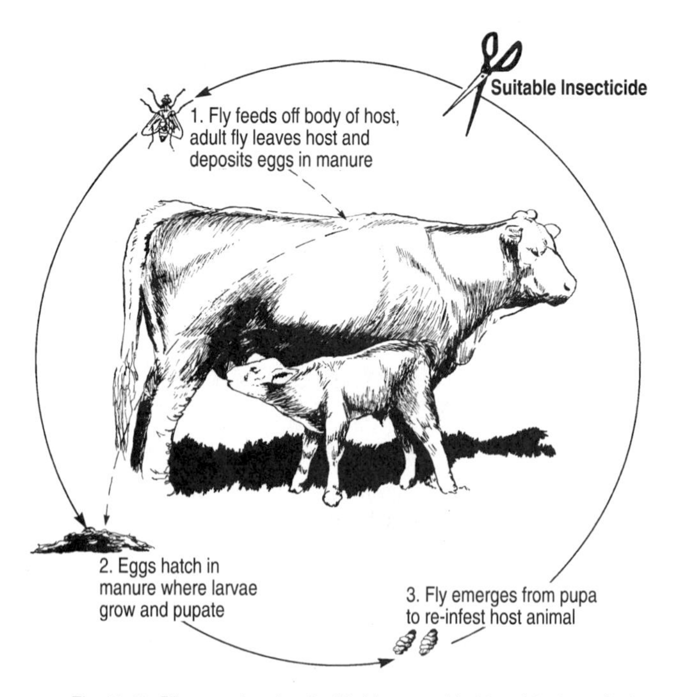

Fig. 12-42. Diagram showing the life history and habits of the horn fly. As noted (see scissors) effective control (cutting the life cycle of the parasite) consists of thorough spraying of animals with a suitable insecticide throughout the fly season. First and foremost, however, it is important that fresh droppings be spread with a spring-tooth harrow to hasten their drying and that accumulations of cattle manure around barns be hauled out at frequent intervals and spread thinly on the land.

pour-ons; (2) free-choice and self-application devices—backrubbers, treadle sprayers, and dust bags; (3) low-volume applications from aerial or ground units; (4) feeding to prevent the development of the larval stages in the manure; (5) slow-release products impregnated in plastic ear tags.

When properly used, insecticide-impregnated ear tags will give effective horn fly control and face fly reduction in cattle. Experiments and trials have shown that one tag per ear in either the cows or the calves, but not both, will give 95 to 100% horn fly control and some face fly reduction. Where fly resistance to pyrethroid tags develops, it is recommended that cattle producers use an organophosphate-type tag on every other animal or every third animal. Although the organophosphate tag will not last the full season, it will control flies that are not killed by the synthetic pyrethroid chemical. Another recommended practice is to continue to use backrubbers, dust bags, etc., with organophosphate insecticides to kill any flies not killed by pyrethroid tags. Under no circumstances are two tags per ear recommended for horn fly control.

Cautions: (1) Only federal and state approved chemicals should be used; (2) the manufacturers' directions should be carefully followed; and (3) the withdrawal periods should be heeded. Specific controls should be used only after checking with local regulatory officials.

HORSE FLIES AND DEER FLIES

An annoying group of biting flies that attack cattle, the Tabanidae, has two particularly troublesome genera, *Tabanus* (horse flies) and *Chrysops* (deer flies).

DISTRIBUTION AND LOSSES CAUSED BY HORSE FLIES AND DEER FLIES

Tabanids are found in all parts of the United States, and large numbers may be expected wherever there are extended areas of permanently wet, undeveloped land and a mild climate. Generally, horseflies are more of a problem to livestock than deer flies, but deer flies are often extremely annoying in the coastal areas of the South and the mountain areas of the West.

LIFE HISTORY AND HABITS

Females of both genera lay masses of eggs (as many as 1,000 eggs/mass) on foliage or on other objects that project over water or moist ground. The eggs hatch in 5 to 7 days, and the larvae drop into the water or upon the moist soil. Then they burrow into soil and feed on organic matter that may include the juices of other insect larvae, other tabanid larvae, and earthworms. Larvae can ordinarily be found in the top 2 or 3 in. of soil in swamps and around lakes, ponds, and permanent streams; however, recent studies indicate that they may also be widely distributed in drier soils such as forest floors and decaying wood. Generally, tabanids remain in the larval stage during the summer, fall, and winter months; then in the spring, they move to drier soil where they pupate. The pupal stage lasts 2 to 3 weeks. Most species complete one generation a year.

DAMAGE INFLICTED; SYMPTOMS AND SIGNS OF AFFECTED ANIMALS

Male horse flies and deer flies feed on vegetable sap, and some may suck juices of soft-bodied insects, but they do not attack warmblooded animals. Female horse flies and deer flies require a blood meal soon after emergence. They feed primarily during the daytime, but some feed at dusk and dawn, and a few species will attack animals in complete darkness. The bite from the slashing mouthparts of these insects is very painful, and animals generally try to dislodge the fly with their tail or tongue or by stamping their feet. Heavily attacked animals stop grazing and tend to bunch together or seek shelter. Severe outbreaks can seriously affect weight. Moreover, tabanids are also implicated in disease transmission because their habit of feeding on one animal and immediately attacking another can result in the direct mechanical transfer of pathogenic organisms that live in blood.

PREVENTION, CONTROL, AND TREATMENT

Horse flies and deer flies are among the most difficult to control of any of the blood-sucking flies. The majority of the new organic insecticides will kill horse flies and deer flies even at low dosages, but they will not do so fast enough to prevent biting. Presently, repellents are the only answer; and the synergized pyrethrins have given the best results. Check with local regulatory officials relative to specific controls.

If possible, avoid pasturing cattle near swampy wooded areas when these flies are numerous. Also, sheltering animals is often beneficial since tabanids do not ordinarily enter enclosures.

HOUSEFLIES

Houseflies *(Musca domestica)* are nonbiting flies that are common around barns and lots.

DISTRIBUTION AND LOSSES CAUSED BY HOUSEFLIES

Houseflies become numerous both inside and outside barns and farm buildings. Perhaps they are the most abundant insect pest of feedlots. Houseflies are annoying to cattle and people, and they can spread human and animal diseases.

LIFE HISTORY AND HABITS

Houseflies breed in manure, garbage, and decaying vegetable matter. The eggs hatch after an incubation period of 12 to 36 hours. The larvae feed on the organic medium and grow to full size in 6 to 11 days.

DAMAGE INFLICTED; SYMPTOMS AND SIGNS OF AFFECTED ANIMALS

Although houseflies are nonbiting, they cause serious economic losses through annoyance of cattle and by disease transmission. Also, they create public health problems.

PREVENTION, CONTROL, AND TREATMENT

Insecticides alone will not control houseflies. Adequate sanitary measures, including proper disposition or handling of manure, are necessary to eliminate fly breeding areas. Spread manure thinly in fields so fly eggs and larvae will be killed by drying and heat.

Several insecticides in spray or bait forms may be used to control adult flies in barns.

LICE

The louse is a small, flattened, wingless insect parasite of which there are several species.

Cattle are attacked by four species of blood-sucking lice—the short-nosed cattle louse *(Haematopinus eurysternus)*, the cattle tail louse *(Haematopinus quadripertusus)*, the long-nosed cattle louse *(Linognathus vituli)*, and the little blue louse *(Solenopotes capillatus)*, and by one species of biting louse, the cattle biting louse *(Bovicola bovis)*. Cattle lice will not remain on other farm animals, nor will lice from other animals infest cattle. Lice are always more abundant on weak, unthrifty animals and are more troublesome during the winter months than during the rest of the year.

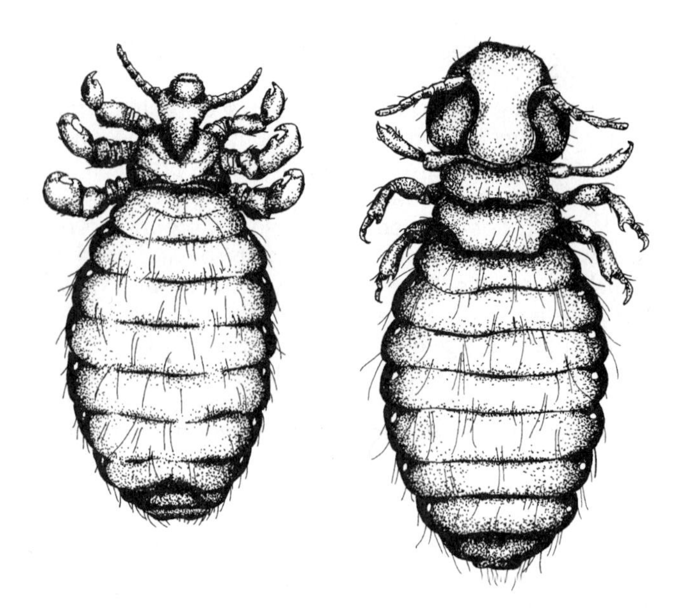

Fig. 12-43. Two species of cattle lice: (a) *(left)* hairy cattle louse—a blood-sucking louse; and (b) *(right)* chewing, or biting, cattle louse.

DISTRIBUTION AND LOSSES CAUSED BY LICE

The presence of lice upon animals is almost universal, but the degree of infestation depends largely upon the state of animal nutrition and the extent to which the producer will tolerate parasites. The irritation caused by the presence of lice retards growth, gains, and/or production of milk.

LIFE HISTORY AND HABITS

Lice spend their entire life cycle on the host's body. They attach their eggs or nits to the hair near the skin where they hatch in about two weeks. Two weeks later the young females begin laying eggs, and after reproduction they die on the host. Lice do not survive more than a week when separated from the host, but, under favorable conditions, eggs clinging to detached hairs may continue to hatch for 2 or 3 weeks.

DAMAGE INFLICTED; SYMPTOMS AND SIGNS OF AFFECTED ANIMALS

Infestation shows up most commonly in winter in ill-nourished and neglected animals. There is intense irritation, restlessness, and loss of condition. As many lice are blood suckers, they devitalize their host. There may be severe itching and the animal may be seen scratching, rubbing, and gnawing at the skin. The hair may be rough, thin, and lack luster; and scabs may be evident. In cattle, favorite locations for lice are the root of the tail, on the inside of the thighs over the

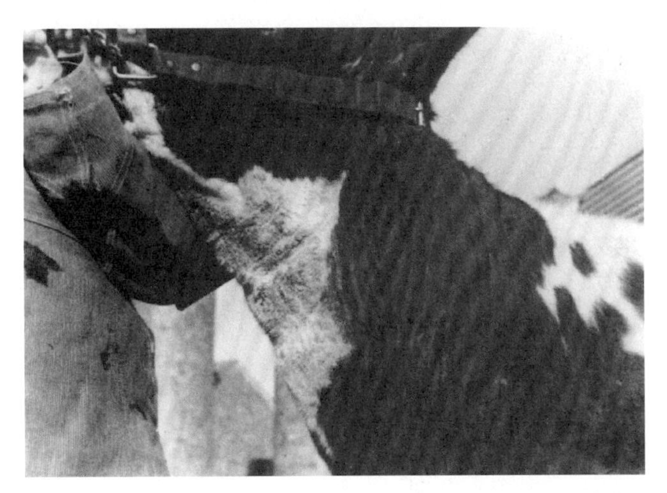

Fig. 12-44. Cow's neck heavily infested with short-nosed sucking lice. These pests seek the sheltered parts of the body on which to feed. (Courtesy, Dept. of Pathology and Hygiene, College of Veterinary Medicine, University of Illinois)

fetlock region, and along the neck and shoulders. In some cases, the symptoms may resemble that of mange and it must be kept in mind that the two may occur simultaneously.

With the coming of spring, when the hair sheds and the animals go to pasture, the problem of lice is greatly diminished.

PREVENTION, CONTROL, AND TREATMENT

Because of the close contact of domesticated animals, especially during the winter months, it is practically impossible to prevent entire herds from becoming slightly infested with the pests. Nevertheless, lice can be kept under control.

For effective control, all members of the herd must be treated simultaneously at intervals, and this is especially necessary during the fall months about the time they are placed in winter quarters. Cattle should be inspected for lice periodically throughout the winter and spring and retreated when necessary. Space will not permit listing all the available insecticides for the control of lice. Suffice it to say that all of them appear to be satisfactory when used in keeping with the directions on the label. Insecticides applied by spraying or dipping are the most effective against lice, but during cold weather, dusts and pour-ons are useful. Because approved products change from time to time, always read the directions on the label before use.

LIVER FLUKE

The liver fluke, *Fasciola hepatica*, is a flattened, leaf-like, brown worm, usually about 1 in. long. It affects cattle, sheep, goats, and other animals.

DISTRIBUTION AND LOSSES CAUSED BY THE LIVER FLUKE

The liver fluke is distributed throughout the world, wherever there are low-lying wet areas and suitable snails. In the United States, it is most common in some of the areas of the western and southwestern range country and in Florida.

Lowered gains and milk production and feed inefficiency are the chief losses. In addition, vast quantities of liver are condemned each year at the time of slaughter—an estimated 1,200 tons annually. In packing houses, such livers are referred to as *fluky livers* or *rotten livers*.

LIFE HISTORY AND HABITS

Flukes reproduce by means of eggs which, after passing from the host, hatch into embryos equipped with cilia that enable them to move about. Upon encountering certain kinds of snails, they penetrate into the body of the intermediate host and develop into cercariae (flukes in the larval stage), which leave the snails and become encysted on the nearby vegetation. The encysted cercariae are then ingested by animals during grazing. The fluke is liberated from the cyst, penetrates the intestinal wall, migrates about the ab-

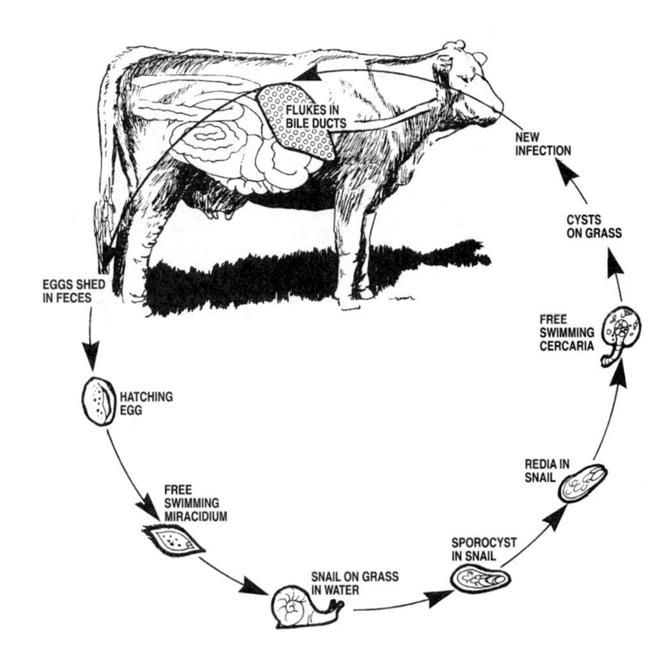

Fig. 12-45. Diagram showing the life history and habits of the liver fluke, *Fasciola hepatica.* When the eggs of the liver fluke hatch, they liberate larvae called miracidium. Each miracidium is ciliated and swims about in the water until it comes in contact with a suitable snail, into which it bores. Then it changes a number of times, finally becoming what is known as a redia. Large numbers of minute stages, shaped like tadpoles and known as cercariae, are produced in the redia. These eventually escape from the snail, swim about, and become encysted on grass or other vegetation.

dominal cavity, and finally reaches the liver where maturity is attained 2 or 3 months after infestation.

DAMAGE INFLICTED; SYMPTOMS AND SIGNS OF AFFECTED ANIMALS

Infested cattle show anemia, as indicated by pale mucous membranes, digestive disturbances, loss of weight and general weakness. As with most parasites, positive diagnosis consists of finding eggs in the feces by microscopic examination.

PREVENTION, CONTROL, AND TREATMENT

The following measures are recommended for the prevention and control of the liver fluke:

1. Drainage or avoidance of wet pastures.
2. Where relatively small snail-infested areas are involved, it may be practical to destroy the snail (carrier of liver fluke), preferably in the spring season, through—
   a. Applying 3 to 6 lb of copper sulfate (bluestone or blue vitrol) per acre of grassland, mixing and applying the small quantity of copper sulfate with a suitable carrier (such as a mixture of 1 part of the copper sulfate to 4 to 8 parts of either sand or lime), and
   b. Treating ponds or sloughs with 1 part of copper sulfate to 500,000 parts of water.

Fig. 12-46. Brahman bull dying from fluke infestation. (Courtesy, Pitman-Moore, Inc., Indianapolis, IN)

When copper sulfate is used in the dilutions indicated, it is not injurious to grasses and will not poison farm animals, but it may kill fish.

Snail-infested pastures should not be used for making hay.

The recommended treatment should be under the direction of a veterinarian or other local regulatory official, and the directions on the label of the product should be followed.

LUNGWORM

The lungworm, *Dictyocaulus viviparus*, is a white, threadlike worm 1½ to 3 in. long, found in the trachea and bronchi of cattle—especially calves.

DISTRIBUTION AND LOSSES CAUSED BY LUNGWORMS

Lungworms are distributed throughout the United States, especially on wet pastures.

The losses are chiefly in lowered feed efficiency, milk, and meat production. With a heavy infestation, there may be death losses.

LIFE HISTORY AND HABITS

In the bronchial tubes of the lungs, female lungworms produce large numbers of eggs. Usually, these hatch in the air passages and liberate larvae that are coughed up, swallowed, and eliminated in the feces. Sometimes the coughed-up eggs hatch in the stomach or intestines, but they may pass unhatched from the host, particularly when there is severe diarrhea.

Under favorable conditions, the larvae eliminated with the feces develop into the infective stage in about a week. Then they crawl upon blades of grass, where they are ingested by grazing cattle. Thence, they penetrate the intestinal wall and reach the lymph glands, from which they are eventually carried to the lungs.

Cattle lungworms mature in 3 to 4 weeks, at which time larvae appear in the feces. The worms live from 2 to 4 months in the host.

DAMAGE INFLICTED; SYMPTOMS AND SIGNS OF AFFECTED ANIMALS

Typical symptoms include coughing, labored breathing, loss of appetite, unthriftiness, and intermittent diarrhea. Death may follow, probably from suffocation or pneumonia.

PREVENTION, CONTROL, AND TREATMENT

Where lungworms are found, the following prevention and control measures are recommended:

1. Practice rigid sanitation.
2. Where practical, segregate calves from older animals.
3. Keep calves on a good ration.
4. Do not spread infested manure on pastures.
5. Utilize dry pasture if possible.

Several drugs are highly effective in lungworm

control. Secure the recommendation of local regulatory officials, and use the product of choice in keeping with the directions on the label.

MEASLES (Cysticercosis, or Measly Beef)

A parasitic disease of cattle, cysticercosis is an invasion of the musculature and viscera by larvae, *Cysticercus bovis*, *Taenia saginata*, the beef tapeworm of humans. Cattle are the intermediate host and people the definite host. The name is a misnomer, for it has no relationship, and little resemblance, to human measles. In humans, the disease is caused by a virus; in cattle the disease is caused by a tapeworm cyst which lodges and grows in the muscle tissue.

DISTRIBUTION AND LOSSES CAUSED BY MEASLES

Beef measles is worldwide. However, the incidence is highest in Africa, the Middle East, Asia, and South America. In the United States, the measles problem is largely confined to the Southwest.

At the time of slaughter, losses result from extensive trimming and prolonged storing of mildly infected carcasses and from condemning heavily infected carcasses.

Fig. 12-47. The cycle by which the beef tapeworm passes between cattle and humans.

LIFE HISTORY AND HABITS

Beef measles is transmitted to cattle by contamination of their feed and water with viable tapeworm eggs. The eggs are ingested by cattle. Within the bovine intestine, the embryo hatches, penetrates the intestinal wall, enters the bloodstream, wends its way to muscle tissue where it develops into *measles* or cysticerci. Each cysticercus appears like a small (when fully developed, it is about ¼ in. in diameter) white balloon filled with fluid. Within the cyst is the head of a new tapeworm. People become infected by eating rare beef.

Humans are the sole host for the adult tapeworm, and cattle are the only intermediate host. No other animals are involved in the life cycle. Thus, control consists of disposing of human excrement in such manner that it cannot come in contact with cattle.

DAMAGE INFLICTED; SYMPTOMS AND SIGNS OF AFFECTED ANIMALS

Most cases of beef measles produce few signs in live cattle. In the carcass, the cysticerci (cysts) are readily discernible. Carcasses that are excessively infested are unsatisfactory for food and should be condemned. If only a few cysticerci are found, the entire carcass is frozen sufficiently long to insure that the cysticerci are killed.

PREVENTION, CONTROL, AND TREATMENT

In endemic areas, workers employed in and around feedlots, cattle pastures, and dairies should be medically examined for beef tapeworm parasitism, and infested individuals should be treated for removal of the worms. Sanitary latrines should be provided for caretakers, and people should be forbidden to defecate in feedlots or pastures where cattle feed. At slaughter, cysticercus-infested meat should be disposed of in a manner which avoids the inclusion of viable cysticerci in human food. Meats should be thoroughly cooked to destroy viable cysticerci.

No effective treatment for bovine measles is known.

MITES (Mange, or Scabies)

Scabies in cattle, also known as scab, mange, or itch, is caused by mites living on or in the skin. Three of the more important species are the psoroptic or common scab mite *(Psoroptes equi* var. *bovis)*, the sarcoptic mange mite *(Sarcoptes scabei* var. *bovis)*, and the chorioptic or symbiotic scab mite *(Chorioptes bovis)*.

DISTRIBUTION AND LOSSES CAUSED BY MITES

Injury from mites is caused by blood sucking and the formation of scabs and other skin affections. In a severe attack, the skins may be much less valuable for leather. Growth is retarded, and production of meat and milk is lowered.

LIFE HISTORY AND HABITS

The mites that attack cattle breed exclusively on the bodies of their hosts, and will live for only 2 or 3 weeks when removed therefrom. The female mite which produces sarcoptic mange—the most severe form of scabies—lays from 10 to 25 eggs during the egg-laying period, which lasts about 2 weeks. At the end of another 2 weeks, the eggs have hatched and the mites have reached maturity. A new generation of mites may be produced every 15 days.

Mites are more prevalent during the winter months, when animals are confined and in close contact with each other.

DAMAGE INFLICTED; SYMPTOMS AND SIGNS OF AFFECTED ANIMALS

When the mite pierces the skin to feed on cells and lymph, there is marked irritation, itching, and scratching. Exudate forms on the surface, and this coagulates, crusting over the surface. The crusting is often accompanied or followed by the formation of a thick, tough, wrinkled skin. Often there are secondary skin infections. The only certain method of diagnosis is to demonstrate the presence of the mites.

PREVENTION, CONTROL, AND TREATMENT

Prevention consists of avoiding contact with diseased animals or infested premises. Mange is a reportable disease; hence, infestations should be reported to the proper livestock inspection authorities. The presence of certain species of cattle scab mites will result in the herd being quarantined. This prohibits their movement until they are inspected and found free of scabies.

Fig. 12-48. A severe case of mange on a cow, caused by sarcoptic mites. Note the rough and wrinkled condition of the skin, with crusting over the surface. (J. W. McManigal, Agricultural Photographer, Horton, KS)

Mites can be controlled by treating infested animals with an approved insecticide according to directions.

Hot lime-sulfur, 2%, has been used for many years and is usually used on lactating dairy cattle because it does not cause any measurable residue in the milk.

Title 9 of the Code of Federal Regulations, part 73, scabies in cattle, requires that affected cattle be treated prior to interstate movement.

(Also see later section in this chapter entitled, "Regulations Relative to Disease Control.")

MOSQUITOES

Mosquitoes, particularly species of the genera *Aedes, Psorophora*, and *Culex*, are a severe nuisance to cattle in many areas, especially in swampy regions that have permanent pools of water or that are exposed to frequent flooding.

DISTRIBUTION AND LOSSES CAUSED BY MOSQUITOES

Mosquitoes are rather widely distributed, but they are most numerous in southeastern United States. Sometimes they kill cattle, although this is rare.

LIFE HISTORY AND HABITS

Almost all female mosquitoes must take a blood meal before they can lay eggs. (The males do not suck blood, but feed on nectar and other plant juices.) Eggs are laid singly or in rafts on the surface of the water or on the ground in depressions that are flooded by

Fig. 12-49. Dipping cattle. The chief virtue of dipping vats lies in the fact that animals so treated are thoroughly covered. (Courtesy, Washington State University)

tidal waters, seepage, overflow, or rain water. The larvae and pupae are aquatic.

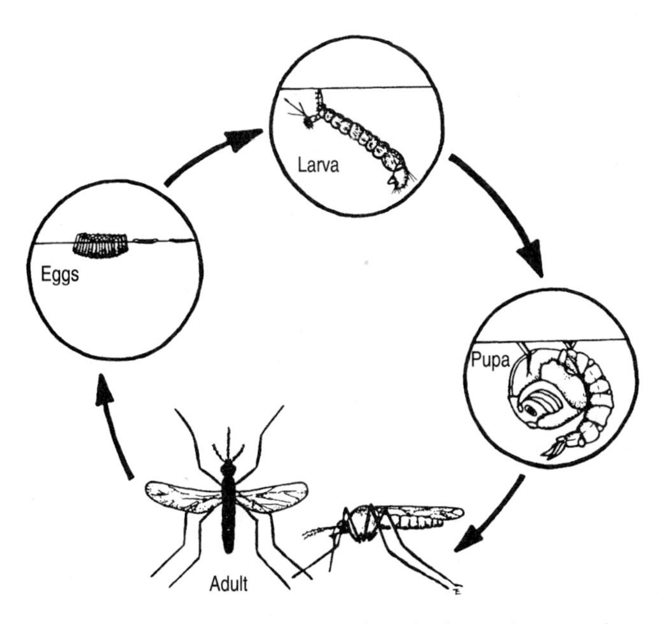

Fig. 12-50. The four stages in the life cycle of mosquitoes: egg, larva (wiggler), pupa (tumbler), and adult.

DAMAGE INFLICTED; SYMPTOMS AND SIGNS OF AFFECTED ANIMALS

Mosquitoes may occur in such abundance that cattle refuse to graze. Instead, they bunch together or stand neck deep in water to protect themselves from attack. Moreover, mosquitoes will annoy cattle day and night, so they can cause serious losses in meat production—or even death in extreme cases. Also, they may be disease carriers.

PREVENTION, CONTROL, AND TREATMENT

Mosquitoes can be controlled in several ways: (1) by elimination of breeding places, through providing fills, ditches, impoundments, improved irrigation methods, and other means of water manipulation; (2) by chemical destruction of larvae, by treating the relatively restricted breeding areas with proper larvicides; and (3) by chemical destruction of adults. Elimination of breeding sites is by far the most satisfactory and effective method of control. However, either this method or chemical destruction of larvae may not be economically practical if the breeding area is extensive. When the latter is the case, control can best be accomplished through group action, such as mosquito abatement districts. The cattle producer can achieve some control by daily spraying of cattle with an approved insecticide, used in keeping with the directions on the label.

RINGWORM

Ringworm, or barn itch, is a contagious disease of the outer layers of skin. It is caused by certain microscopic molds or fungi *Trichophyton, Achorion*, or *Microsporon*. All animals and people are susceptible.

DISTRIBUTION AND LOSSES CAUSED BY RINGWORM

Ringworm is widespread throughout the United States. It is a contagious disease of the outer layer of skin caused by certain microscopic molds or fungi, which affect all animals and humans. Though it may appear among animals on pasture, it is far more prevalent as a stable disease. It is unsightly, and affected animals may experience considerable discomfort; but the actual economic losses attributed to the disease are not too great.

LIFE HISTORY AND HABITS

The period of incubation for this disease is about

1 week. The fungi form seeds or spores that may live 18 months or longer in barns or elsewhere.

Ringworm is usually a winter disease, with recovery the following summer when the animals are on pasture.

DAMAGE INFLICTED; SYMPTOMS AND SIGNS OF AFFECTED ANIMALS

Round, scaly areas almost devoid of hair appear mainly in the vicinity of the eyes, ears, side of the neck, or the root of the tail. Crusts may form, and the skin may have a gray, powdery, asbestos-like appearance. The infected patches, if not checked, gradually increase in size. Mild itching usually accompanies the disease.

Fig. 12-51. Heifer with ringworm. The fungi causing these raised circular areas may be transmitted to humans. (Courtesy, Dept. of Pathology and Hygiene, College of Veterinary Medicine, University of Illinois)

PREVENTION, CONTROL, AND TREATMENT

The organisms are spread from animal to animal or through the medium of contaminated fence posts, curry combs, and brushes. Thus prevention and control consists of disinfecting everything that has been in contact with infected animals. The infected animals should also be isolated. Strict sanitation is essential in the control of ringworm.

For treatment, an approved topical medication should be applied according to the label directions.

In preparation for local treatment, the hair should be clipped, the scabs removed, and the area brushed and washed with soap. The diseased parts should then be painted with tincture of iodine and glycerin, lugol's solution, a 20% solution of sodium caprylate, or 75% thiabendazole. Certain proprietary remedies available only from veterinarians have proved very effective in treatment.

SCREWWORM

The screwworm *(Cochliomyia hominivorax)* was eradicated from the southeastern United States in the late 1950s. Now, many millions of sterile flies are being released to eradicate completely this pest from Mexico and adjoining states of the United States. Since the screwworm fly may infest an animal through wounds and also through lesions caused by ticks, horse flies, or even horn flies, there is an added impetus to control these parasites. They are not found in cold-blooded animals such as turtles, snakes, and lizards.

Unnatural wounds resulting from branding, castrating, and dehorning afford a breeding ground for this parasite. Add to this the wounds from some types of vegetation, from fighting, and from blood-sucking insects and ample places for propagation are provided.

DISTRIBUTION AND LOSSES CAUSED BY SCREWWORMS

Screwworm infestations occur occasionally in certain areas of Arizona, New Mexico, and Texas, and occur often in many areas of Mexico and other Latin American countries to the south.

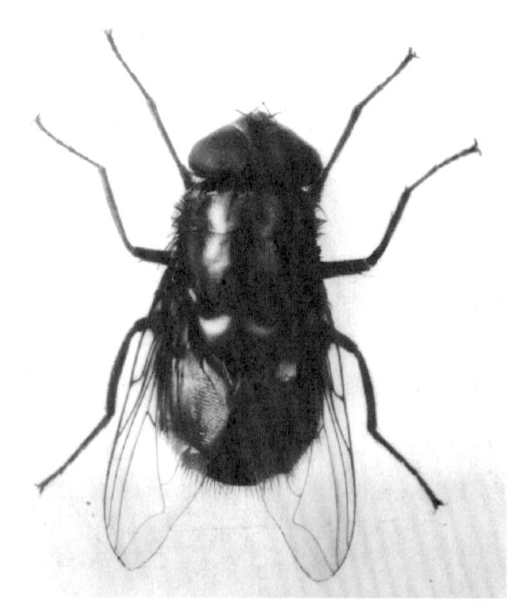

Fig. 12-52. The screwworm fly *(Cochliomyia americana)*. This fly is bluish green in color, with three dark stripes on its back and reddish or orange color below the eyes. (Courtesy, USDA)

LIFE HISTORY AND HABITS

The primary screwworm fly is bluish green in color, with three dark stripes on its back and reddish or orange color below the eyes. The fly generally deposits its eggs in shinglelike masses on the edges or the dry portion of wounds. From 50 to 300 eggs are laid at one time, with a single female being capable of laying about 3,000 eggs in a lifetime. Hatching of the eggs occurs in 11 hours, and the young, whitish worms (larvae or maggots) immediately burrow into the living flesh. There they feed and grow for a period of 4 to 7 days, shedding their skin twice during this period.

When these worms have reached their full growth, they assume a pinkish color, leave the wound, and drop to the ground, where they dig beneath the surface of the soil and undergo a transformation to the hard-skinned, dark-brown, motionless pupa. It is during the pupa stage that the maggot changes to the adult fly.

After the pupa has been in the soil from 7 to 60 days, the fly emerges from it, works its way to the surface of the ground, and crawls up on some nearby object (bush, weed, etc.) to allow its wings to unfold and otherwise to mature. Under favorable conditions, the newly emerged female fly becomes sexually mature and will lay eggs 5 days later. During warm weather, the entire life cycle is usually completed in 21 days, but under cold, unfavorable conditions the cycle may take as many as 80 days or longer.

DAMAGE INFLICTED; SYMPTOMS AND SIGNS OF AFFECTED ANIMALS

The injury caused by this parasite is inflicted chiefly by the maggots. The early symptoms in affected animals are loss of appetite and condition, and listlessness. Unless proper treatment is administered, the great destruction of many tissues kills the host in a few days.

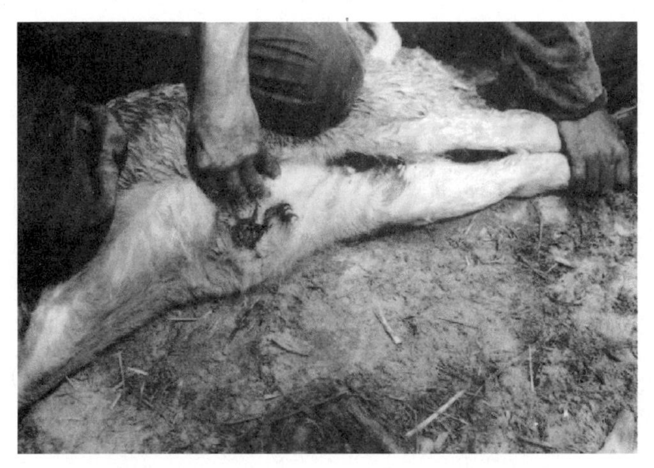

Fig. 12-53. Screwworm in navel of calf. (Courtesy, USDA)

PREVENTION, CONTROL, AND TREATMENT

Prevention in infested areas consists mainly of keeping animal wounds to a minimum and of protecting those that do materialize.

In 1958, the U.S. Department of Agriculture initiated an eradication program. Screwworm larvae were reared on artificial media. Two days before fly emergence, the pupae were exposed to gamma irradiation at a dosage which caused sexual sterility but no other deleterious effects. Sterile flies were distributed over the entire screwworm-infested region in sufficient quantity to outnumber the native flies, at an average rate of 400 males per square mile per week. The female mates only once and, therefore, when mated with a sterile male does not reproduce. There was a decline in the native population each generation until the native males were so outnumbered by sterile males that no fertile matings occurred and the native flies were eliminated. This program has virtually eliminated all the losses caused by screwworms in the United States. Unfortunately, the states bordering on Mexico are periodically reinfested by mated female flies from Mexico. For this reason, permanent elimination of screwworms from the United States by the sterile-male technique cannot be hoped for until they are also eradicated in Mexico.

When maggots (larvae) are found in an animal, they should be removed and sent to the proper authorities for identification, and the wound should be treated with an approved smear (wound dressing) applied in keeping with the directions on the label.

Title 9 of the Code of Federal Regulations, part 83, screwworms, gives the screwworm control zone, the areas of recurring infestation, and the inspection and treatment requirements for movement from these areas.

(Also see later section in this chapter entitled, "Regulations Relative to Disease Control.")

STABLEFLY (Biting Housefly, Dog Fly, or Stock Fly)

The stablefly, *Stomoxys calcitrans*, which is about the size of a housefly, is usually found in the vicinity of animals and derives its nourishment from sucking blood. It attacks people and all classes of farm animals.

DISTRIBUTION AND LOSSES CAUSED BY THE STABLEFLY

Stableflies are found in all parts of the United States and in many other countries. They are especially numerous in the central and southeastern states and some coastal states. They are not usually a problem on the open range, but large numbers can occur around barns where livestock congregate and where there are accumulations of decaying plant material. The chief economic loss from the stablefly is in terms of decreased gains in beef cattle and lowered milk production in dairy cattle, amounting to as much as 50% in seasons when the numbers of flies become large. In addition to causing such lowered production losses, the stablefly has been incriminated in the mechanical transmission of anthrax, swamp fever, surra, and even infantile paralysis of humans. It is also known to be an intermediate host for the peritoneal roundworm of cattle and the small-mouth stomach worm of horses. Along certain beaches, the annoyance to people has been so great that resort interests have been affected and real estate values have been lowered.

LIFE HISTORY AND HABITS

Like other true flies, the stablefly has four stages—egg, larva, pupa, and adult fly.

Although cattle feces are a suitable habitat for the development of stablefly larvae, they do not seem to be attractive to adult flies for egg laying unless they are mixed with straw, feed, or similar materials. Therefore, the eggs are commonly deposited in soggy hay or grain in the bottoms of and underneath feed racks, in piles of moist fermenting weeds and grass cuttings, in piles of moist fermenting peanut litter, in deposits of grass along beaches (especially in northwestern Florida), in straw in calf pens, and in rotting hay at the edge of haystacks. The eggs hatch in 2 days, and larvae complete development in about 10 days. The pupal stage lasts 4 to 5 days. The entire life cycle takes about 3 weeks so several generations can develop during the summer.

In northern climates, stableflies overwinter in either the larval or pupal stage. In the warmer southern regions, the length of each developmental period may be increased in the winter months, but flies may be present during warmer days, and breeding is generally continuous throughout the year. (See Fig. 12-54 for diagram of the life history and habits of the stablefly.)

DAMAGE INFLICTED; SYMPTOMS AND SIGNS OF AFFECTED ANIMALS

Stableflies commonly rest on the vertical surfaces of fences, buildings, trees, or other structures near cattle. They visit the animals only long enough to obtain a blood meal that they generally prefer to take from the legs and lower parts of the animal's body. The fly usually takes 5 or 10 minutes to engorge on blood, but it may puncture the skin several

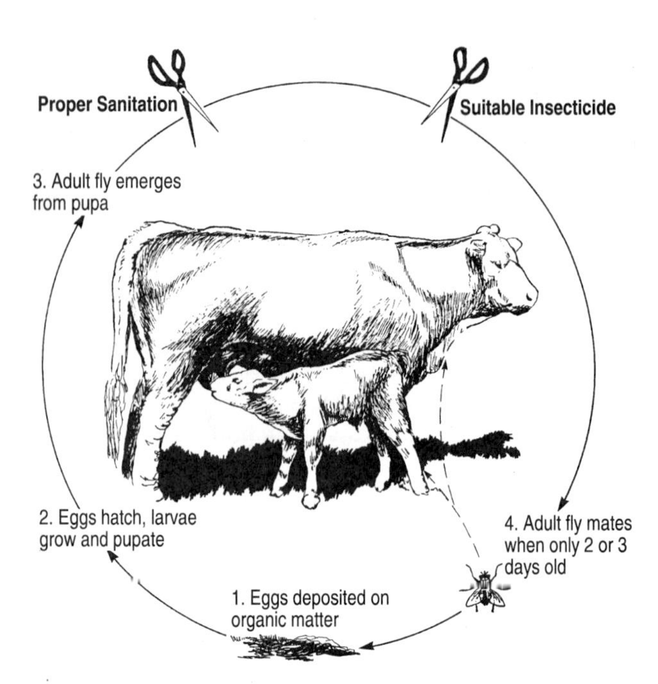

Fig. 12-54. Diagram showing the life history and habits of the stablefly. As noted (see scissors) effective control (cutting the life cycle of the parasite) consists of thorough spraying—especially the legs and lower part of the body—with a suitable insecticide throughout the fly season. First and foremost, however, it is important that there be proper disposal of manure and elimination of all fermenting or decaying organic material.

times with its proboscis in the process. The severe pain caused by this piercing and probing action causes cattle to fight the flies constantly by licking themselves and stomping their feet. As the fly feeds, its abdomen swells to almost twice the original size and turns a reddish color. The fully engorged fly then leaves the animal and finds a suitable resting place to digest the blood meal. Both sexes take one or more such feeds daily.

PREVENTION, CONTROL, AND TREATMENT

Control of stableflies by direct application of insecticides to cattle is usually not satisfactory. They are best controlled by sanitation and by application of insecticides to the resting surfaces. Sanitation is undoubtedly the most effective method of controlling stableflies in such areas as feedlots and barnyards because it breaks the life cycle by removing the breeding sites. Barnyards and feedlots should be well drained; manure and decaying organic matter should be removed from inside and outside buildings and disposed of weekly or more often, if possible, by spreading it out to dry (this kills developing larvae). If manure cannot be spread, it should be placed in compact piles where the surface will dry quickly and become unattractive to females.

With stableflies, approved insecticides should be

used as a supplement to good sanitation, rather than as the principal method of control, because alone they may not do a satisfactory job.

Although the animals may be annoyed by only five or more flies each, control measures should be initiated when even these few flies are present because the number of stableflies biting an animal at any one time may represent only 2 to 3% of the total number of stableflies in the area. As a blood-feeding fly, the stablefly is implicated in carrying disease, and high populations cause reduced weight gains in cattle.

TICKS

The Lone Star tick *(Amblyomma americanum)*, the Gulf Coast tick *(Amblyomma maculatum)*, the Rocky Mountain wood tick *(Dermacentor andersoni)*, the Pacific Coast tick *(Dermacentor occidentalis)*, and the American dog tick *(Dermacentor variabilis)* are 3-host species that attack cattle during the summer months. The black-legged tick *(Ixodes scapularis)* is a 3-host tick that is common in late winter and early spring. The winter tick *(Dermacentor aibipictus)* is a 1-host species found on cattle and horses in the fall and winter. In addition, larvae and nymphs of the so-called "spinose" ear tick *(Octobius megnini)*, a 1-host species, attach deep in the ears of cattle and feed there for several months.

DISTRIBUTION AND LOSSES CAUSED BY TICKS

Ticks are widely distributed, especially throughout the southern part of the United States. But they are usually seasonal in their activities.

Ticks suck blood. They cause economic losses by transmitting diseases; by restlessness, anemia, and inefficient feed utilization; and by necessitating expensive treatments. Among the diseases transmitted to or produced in cattle by ticks are anaplasmosis, piroplasmosis, Q fever, Texas fever, and tick paralysis.

In addition to animal losses from ticks, additional, but unmeasured, losses result from incapacitation of people affected with diseases transmitted by ticks of cattle.

LIFE HISTORY AND HABITS

Generally, all species of ticks have similar stages of development. The females lay eggs that hatch into 6-legged larvae (seed ticks). The larvae attach to a host, engorge on blood, and molt to 8-legged nymphs. The nymphs attach to a host, engorge on blood, and molt to 8-legged adults. Mating usually occurs on the host. The female then engorges fully, drops off the host, lays several thousand eggs, and dies.

DAMAGE INFLICTED; SYMPTOMS AND SIGNS OF AFFECTED ANIMALS

Generally speaking, injury to cattle from tick parasitism varies directly with numbers of parasites. Ticks feed exclusively on blood. Thus, when several hundred ticks feed, the host becomes anemic, unthrifty, and loses weight. In addition, some female ticks generate a paralyzing toxin. The spinose ear tick, commonly called the "ear tick," takes up residence along the inner surfaces of the ears and in the external ear canals, where they are extremely annoying. Cattle heavily parasitized by spinose ear ticks droop their heads, rub and shake their ears, and turn their heads to one side.

PREVENTION, CONTROL, AND TREATMENT

Because most species of ticks, except the ear tick, attach to the external surfaces of cattle, dipping and spraying are the most effective methods of control; however, dusts may be used. An approved chemical should be selected and used according to directions.

To treat animals for ear ticks, the chemical should be applied into the ears of the cattle.

CONTROL OF EXTERNAL PARASITES OF BEEF CATTLE

Losses to the beef cattle industry due to insect pests can be considerably reduced by treating animals with insecticides that give economical control.

The cost of controlling external parasites on beef cattle with insecticides is usually very small compared with losses incurred when the infestations go uncontrolled. A good program for controlling external parasites of cattle should (1) be initiated during the early stages of infestation, (2) include use of good sanitation practices and manipulation of standing water in addition to the application of insecticides (not all external parasites can be effectively controlled with insecticides), and (3) provide for the use of approved insecticides in complete accordance with the labels and instructions.

Generally speaking, suggestions concerning the specific insecticides that should be used are not presented in this book because of (1) the diversity of environments and management practices represented by this group, (2) the varying restrictions on the use of insecticides from area to area, and (3) the fact that registered use of insecticides change from time to time. Information about what insecticide is available and registered for use in a specific area can be obtained from the veterinarian, county agent, extension entomologist, or agricultural consultant.

FORMS OF INSECTICIDES

Insecticides for use on animals may be purchased in several forms. The most common are emulsifiable concentrates, dusts, wettable powders, oil solutions, injectables, feed additives, and impregnated ear tags. When treating animals, be sure to use only insecticide formulations that are prepared specifically for livestock.

- **Emulsifiable concentrates (EC)**—Emulsifiable concentrates, which are probably the most common type of formulation, are solutions of insecticides in petroleum oils or other solvents. An emulsifier has been added so that the solution will mix well with water. On occasion, usually after extended storage, an EC may separate into its various parts; in that case, it should be discarded. An emulsion may also separate if it is allowed to stand after the concentrate has been added to the water; periodic agitation will help prevent this.

- **Dusts**—Dusts are applied directly to animals or in self-application devices in the dry form and cannot be used as sprays.

- **Wettable powders**—Wettable powders are also dry, but the addition of a dispersing and wetting agent allows them to be suspended in water for application to cattle. Continuous agitation of the mixture is important when treating with wettable powders.

- **Oil solutions**—Oil solutions are insecticides dissolved in oil; no emulsifier is added. These materials are usually ready for use and should not be added to water.

APPLICATION OF INSECTICIDES

The availability of an insecticide and the type of application(s) for which it was formulated are of prime importance, but the treatment of a herd or flock is dictated pretty much by the animal species, number of animals, available handling facilities, time or season, the target pest, management practices, and cost. The common methods of insecticide application are (1) spraying, (2) dipping, (3) back rubbers, (4) pour-on, (5) feed additives, (6) injectables, (7) insecticide impregnated ear tags, and (8) boluses.

USE INSECTICIDES SAFELY

Certain basic precautions must be observed when insecticides are to be used because, used improperly, they can be injurious to people, domestic animals, wildlife, and beneficial insects. Follow the directions and heed all the precautions on the labels.

- **Selecting insecticides**—Always select the formulation and insecticide labeled for the purpose for which it is to be used.

- **Storing insecticides**—Always store insecticides in original containers. Never transfer them to unlabeled containers or to food or beverage containers. Store insecticides in a dry place out of reach of children, animals, or unauthorized persons.

- **Disposing of empty containers and unused insecticides**—The Environmental Protection Agency (EPA) has ruled that they consider triple rinsed containers as ordinary trash, not to be classified as a pesticide container. So, if a container is rinsed three times, the empty container can be disposed of as ordinary solid waste and sent to a landfill. Always observe local regulations in the disposal of unused insecticides.

- **Mixing and handling**—Mix and prepare insecticides in the open or in a well-ventilated place. Wear rubber gloves and clean dry clothing (respirator device may be necessary with some products). If any insecticide is spilled on you or your clothing, wash with soap and water immediately and change clothing. Avoid prolonged inhalation. Do not smoke, eat, or drink when mixing and handling insecticides.

- **Applying**—Use only amounts recommended. Apply at the correct time to avoid unlawful residues in meat. Avoid treating animals younger than specified on the label. Avoid retreating more often than label restrictions. Avoid drift on nearby crops, pastures, livestock, or other nontarget areas. Avoid prolonged contact with all insecticides. Do not eat, drink, or smoke until all operations have ceased and hands and face are thoroughly washed. Change and launder clothing after each day's work.

- **In case of an emergency**—If you accidentally swallow an insecticide, induce vomiting by taking one tablespoonful of salt in a glass of water. Repeat if necessary. Call a doctor.

- **Withdrawal**—After treating animals with pesticides, observe the prescribed number of days interval between the last treatment and slaughter. Refer to the product labels for this information.

DISINFECTANTS

Disinfectants are bactericidal or microbicidal agents that are free from infection (usually a chemical agent which destroys disease germs or other micro-

organisms, or inactive viruses). Related terms include the following:

1. *Antiseptics, which are chemicals which kill or prevent the growth of microorganisms.* They are formulated to minimize tissue irritation and damage.
2. *Germicides, which refers to the ability of an antiseptic or disinfectant to destroy pathogens.*
3. *Sanitizers, which refers to products used to clean.* Sanitizers may or may not have germicidal properties.

The high concentration of animals and continuous use of modern livestock buildings often results in a condition referred to as disease buildup. As disease-producing organisms—viruses, bacteria, fungi, and parasite eggs—accumulate in the environment, disease problems can become more severe and be transmitted to each succeeding group of animals raised on the same premises. Under these circumstances, cleaning and disinfection become extremely important in breaking the life cycle. Also, in the case of a disease outbreak, the premises must be disinfected.

Under ordinary conditions, proper cleaning of barns removes most of the microorganisms, along with the filth, thus eliminating the necessity of disinfection.

Effective disinfection depends on five things:

1. Thorough cleaning before application.
2. The phenol coefficient of the disinfectant, which indicates the killing strength of a disinfectant as compared to phenol (carbolic acid). It is determined by a standard laboratory test in which the typhoid fever germ often is used as the test organism.
3. The dilution at which the disinfectant is used.
4. The temperature; most disinfectants are much more effective if applied hot.
5. Thoroughness of application, and time of exposure.

In all cases, disinfection must be preceded by a very thorough cleaning, for organic matter serves to protect disease germs and otherwise interferes with the activity of the disinfecting agent.

Sunlight possesses disinfecting properties, but is variable and superficial in its action. Heat and some of the chemical disinfectants are more effective.

The application of heat by steam, by hot water, by burning, or by boiling is an effective method of disinfection. In many cases, however, it may not be practical to use heat.

A good disinfectant should (1) have the power to kill disease-producing organisms, (2) remain stable in the presence of organic matter (manure, hair, soil), (3) dissolve readily in water and remain in solution, (4) be nontoxic to animals and humans, (5) penetrate organic matter rapidly, (6) remove dirt and grease, and (7) be economical to use.

The number of available disinfectants is large because the ideal universally applicable disinfectant does not exist. Table 12-3 gives a summary of the limitations, usefulness, and strength of some of the common disinfectants.

When using a disinfectant, *always read and follow the manufacturer's directions.*

TABLE 12-3
DISINFECTANT GUIDE[1]

Kind of Disinfectant	Usefulness	Strength	Limitations and Comments
Alcohol (ethyl-ethanol, isopropyl, methanol)	Primarily as a skin disinfectant and for emergency purposes on instruments.	70% alcohol—the content usually found in rubbing alcohol.	They are too costly for general disinfection. They are ineffective against bacterial spores.
Boric acid[2]	As a wash for eyes, and other sensitive parts of the body.	1 oz in 1 pt water (about 6% solution).	It is a weak antiseptic. It may cause harm to the nervous system if absorbed into the body in large amounts. For this and other reasons, antibiotic solutions and saline solutions are replacing it.
Chlorines (sodium hypochlorate, chloramine-T)	They will kill all kinds of bacteria, fungi, and viruses, providing the concentration is sufficiently high. Used as a deodorant.	Generally used at about 200 ppm.	They are corrosive to metals and neutralized by organic materials. Not effective against TB organisms and spores.
Cresols (many commercial products available)	A generally reliable class of disinfectant. Effective against brucellosis, shipping fever, swine erysipelas, and tuberculosis. Cresols give good results in foot baths.	Cresol is usually used as a 2 to 4% solution (1 cup to 2 gal. of water makes a 4% solution).	Effective on organic material. Cannot be used where odor may be absorbed.

(Continued)

TABLE 12-3 (Continued)

Kind of Disinfectant	Usefulness	Strength	Limitations and Comments
Formaldehyde	Formaldehyde will kill anthrax spores, TB organisms, and animal viruses in a 1 to 2% solution. It is often used to disinfect buildings following a disease outbreak.	As a liquid disinfectant, it is usually used as a 1 to 2% solution. As a gaseous disinfectant (fumigant), use 1½ lb of potassium permanganate plus 3 pt of formaldehyde. Also, gas may be released by heating paraformaldehyde.	It has a disagreeable odor, destroys living tissue, and can be extremely poisonous. The bactericidal effectiveness of the gas is dependent upon having the proper relative humidity (above 75%) and temperature (above 86°F and preferably near 140°F, above 30°C and preferably near 60°C).
Heat (by steam, hot water, burning, or boiling)	In the burning of rubbish or articles of little value, and in disposing of infected body discharges. The steam "Jenny" is effective for disinfection *if properly employed*, particularly if used in conjunction with a phenolic germicide.	10 minutes' exposure to boiling water is usually sufficient.	Exposure to boiling water will destroy all ordinary disease germs, but sometimes fails to kill the spores of such diseases as anthrax and tetanus. Moist heat is preferred to dry heat, and steam under pressure is the most effective. Heat may be impractical or too expensive.
Iodine[2] (tincture)	Extensively used as skin disinfectant, for minor cuts and bruises.	Generally used as tincture of iodine, either 2% or 7%.	Never cover with a bandage. Clean skin before applying iodine. It is corrosive to metals.
Iodophors (tamed iodine)	Effective against all bacteria (both gram-negative and gram-positive), fungi, and most viruses.	Usually used as disinfectants at concentrations of 50 to 75 ppm titratable iodine, and as sanitizers at levels of 12.5 to 25 ppm. At 12.5 ppm titratable iodine, they can be used as an antiseptic in drinking water.	Iodophors are a combination of iodine and detergents. They are inhibited in their activity by organic matter. They are quite expensive. They should not be used near heat.
Lime (quicklime, burnt lime, calcium oxide)	As a deodorant when sprinkled on manure and animal discharges; or as a disinfectant when sprinkled on the floor or used as a newly made "milk of lime" or as a whitewash.	Use as a dust; as "milk of lime"; or as a whitewash, but *use fresh.*	Not effective against anthrax or tetanus spores. Wear goggles when adding water to quicklime.
Lye (sodium hydroxide, caustic soda)	On concrete floors; against microorganisms of brucellosis and the viruses of foot-and-mouth disease, hog cholera, and vesicular exanthema. In strong solution (5%), effective against anthrax.	Lye is usually used as either a 2% or a 5% solution. To prepare a 2% solution, add 1 can of lye to 5 gal of water. To prepare a 5% solution, add 1 can of lye to 2 gal of water. A 2% solution will destroy the organisms causing foot-and-mouth disease, but a 5% solution is necessary to destroy the spores of anthrax.	Damages fabrics, aluminum, and painted surfaces. Be careful, for it will burn the hands and face. Not effective against organisms of TB or Johne's disease. Lye solutions are most effective when used hot. It is relatively cheap. *Diluted vinegar can be used to neutralize lye.*
Lysol (the brand name of a product of cresol plus soap)	For disinfecting surgical instruments and instruments used in castrating and tattooing. Useful as a skin disinfectant before surgery, and for use on the hands before castrating.	0.5 to 2.0%	Has a disagreeable odor. Does not mix well with hard water. Less costly than phenol.
Phenols (carbolic acids): 1. Phenolics—coal tar derivatives. 2. Synthetic phenols.	They are ideal general-purpose disinfectants. Effective and inexpensive. They are very resistant to the inhibiting effects of organic residue; hence, they are suitable for barn disinfection, and foot and wheel dip-baths.	Both phenolics (coal tar) and synthetic phenols vary widely in efficacy from one compound to another. So, note and follow manufacturers' directions. Generally used in a 5% solution.	They are corrosive, and they are toxic to animals and humans. Ineffective on fungi and viruses. Effective against all bacteria including TB organisms.

(Continued)

TABLE 12-3 (Continued)

Kind of Disinfectant	Usefulness	Strength	Limitations and Comments
Quarternary ammonium compounds (QAC)	Very water soluble, ultra-rapid kill rate, effective deodorizing properties, and moderately priced. Good detergent characteristics and harmless to skin.	Follow manufacturers' directions.	They can corrode metal. Not very potent in combating viruses. Adversely affected by organic matter; hence, they are of limited use for disinfecting livestock facilities. Not effective against TB organisms and spores. Not effective against anthrax and tetanus.
Sal soda	It may be used in place of lye against foot-and-mouth disease and vesicular exanthema.	10½% solution (13½ oz to 1 gal water).	
Sal soda and soda ash (or sodium carbonate)	They may be used in place of lye against foot-and-mouth disease and vesicular exanthema.	4% solution (1 lb to 3 gal water). Most effective in hot solution.	Commonly used as cleansing agents, but have disinfectant properties, especially when used as a hot solution.
Soap	Its power to kill germs is very limited. Greatest usefulness is in cleansing and dissolving coatings from various surfaces, including the skin, prior to application of a good disinfectant.	As commercially prepared.	Although indispensable to sanitizing surfaces, soaps should not be used as disinfectants. They are not regularly effective; staphylococci and organisms which cause diarrheal disease are resistant.

[1]For metric conversions, see Glossary, "Weights and Measures."

[2]Sometimes loosely classed as a disinfectant but actually an antiseptic and useful only on living tissue.

POISONOUS PLANTS

Poisonous plants have been known to people since time immemorial. Biblical literature alludes to the poisonous properties of certain plants, and history records that hemlock (a poison made from the plant from which it takes its name) was administered by the Greeks to Socrates and other state prisoners.

No section of the United States is entirely free of poisonous plants, for there are hundreds of them. But the heaviest livestock losses from them occur on the western ranges because (1) there has been less cultivation and destruction of poisonous plants in range areas, and (2) the frequent overgrazing on some of the western ranges has resulted in the elimination of some of the more nutritious and desirable plants, and these have been replaced by increased numbers of the less desirable and poisonous species. It is estimated that poisonous plants account for 8 to 10% of all range animal losses each year; and even more in some areas.

Fig. 12-55. Cow with white snakeroot poisoning. Marked weakness results in the "tremble" characteristic of this condition. (Courtesy, Dept. of Pathology and Hygiene, College of Veterinary Medicine, University of Illinois)

DIAGNOSIS OF PLANT POISONING

The diagnosis of plant poisoning in animals is not an easy or precise procedure. Any case of sudden illness or death with no apparent cause is commonly considered to be a poisoning. This may not always be correct. When large numbers of animals are suddenly affected, however, a suspicion of poisoning is justified until it has been proven otherwise.

Symptoms or signs induced by eating poisonous plants may include (1) sudden death; (2) transitory illness; (3) general body weakness; (4) disturbance of the central nervous, vascular, and endocrine systems; (5) photosensitization; (6) frequent urination; (7) diarrhea; (8) bloating; (9) chronic debilitation and death; (10) embryonic death; (11) fetal death; (12) abortion; (13) extensive liver necrosis and/or cirrhosis; (14) edema and/or abdominal dropsy; (15) tumor growths in tissues; (16) congenital deformities; (17) metabolic deficiencies; and (18) physical injury.

No general set of symptoms and signs per se irrefutably provides all the information necessary to make a diagnosis of plant poisonings. Nevertheless, a careful description of the toxic signs coupled with information pertaining to available plants provides a meaningful basis for a tentative diagnosis. Additional information essential to a poisonous plant diagnosis includes (1) type of feed, site grazed, and availability of water; (2) identification and relative abundance of all poisonous plants available to animals; (3) amount and stage of growth of the various poisonous plants being grazed; (4) the toxicity and palatability of the plants in relation to their stage of growth; (5) time from eating the plants until onset of toxic signs; (6) species, age, and sex of animals affected; (7) clinical signs of toxic reactions; (8) chemical analysis of plants; and (9) a careful evaluation of all the information relative to the etiology of the disease.

WHY ANIMALS EAT POISONOUS PLANTS

A frequently asked question is: Why do animals eat poisonous plants? The answer is not simple, but among the reasons are the following: (1) total lack of sufficient palatable forage—the animals are hungry; (2) decrease in palatability and nutrients of mature, weathered range grasses, with the result that poisonous plants become more appealing, comparatively speaking; (3) insufficient spring grass; (4) rain, melting snow, and heavy dew may enhance the palatability of some poisonous plants; and (5) going without water too long, which results in a reduction in feed intake, then, after watering, they develop a ravenous appetite and eat anything in sight—including less palatable poisonous plants.

Poisonous plants vary in palatability—between species, and within species, and at different stages of growth. For example, poisonous hemlock is never palatable and is eaten only as a last resort—when palatable forage is not available. Locoweed and black nightshade are eaten at any stage of growth or when mixed with hay. Others, such as lupines, horsebrush, and death camas may be eaten only at certain stages of growth. Still others, such as milk vetch, larkspur, and halogeton, are highly palatable to livestock at any and all times, with the result that if they're present, animals will seek them out and there will be losses. Then, too, certain plants are poisonous to cattle but not to sheep (and vice versa), as shown in Table 12-4.

PREVENTING LOSSES FROM POISONOUS PLANTS

With poisonous plants, the emphasis should be on prevention of losses rather than on treatment, no matter how successful the latter. The following are effective preventative measures:

1. **Follow good pasture or range management in order to improve the quality of the pasture or range.** Plant poisoning is nature's sign of a ""sick" pasture or range, usually resulting from misuse. When a sufficient supply of desirable forage is available, poisonous plants may not be eaten, for they are usually less palatable. On the other hand, when overgrazing reduces the available supply of the more palatable and safe vegetation, animals may, through sheer hunger, consume toxic plants.

2. **Know the poisonous plants common to the area.** This can usually be accomplished through (a) studying drawings, photographs, and/or descriptions; (b) checking with local authorities; or (c) sending two or three fresh whole plants (if possible, include the roots, stems, leaves, and flowers) to the state agricultural college—first wrapping the plants in several thicknesses of moist paper.

TABLE 12-4
TYPE OF RANGE ANIMAL SUSCEPTIBLE TO POISONOUS PLANTS AT DEFINITE SEASONS

Poisonous to Cattle	Time of Year	Poisonous to Sheep	Time of Year	Poisonous to Cattle & Sheep	Time of Year
Low larkspur	Spring	Death camas	Spring	Broomweed	Spring and summer
Oak	Spring	Greasewood	Fall	Chokecherry	Spring
Tall larkspur	Early summer & early fall	Horsebrush	Spring	Copperweed	Summer
Timber milk vetch	Spring	Rubberweed	Summer	Desert parsley	Spring
Water hemlock	Spring	Sneezeweed	Summer	Halogeton	All year
				Loco	Spring
				Lupine	Summer and fall
				Milkweeds	Summer
				Veratrum	Summer

By knowing the poisonous plants common to the area, it will be possible—

a. To avoid areas heavily infested with poisonous plants which, due to animal concentration and overgrazing, usually include waterholes, salt grounds, bed grounds, and trails.

b. To control and eradicate the poisonous plants effectively, by mechanical or chemical means (as recommended by local authorities) or by fencing off.

c. To recognize more surely and readily the particular kind of plant poisoning when it strikes, for time is important.

d. To know what first aid, if any, to apply, especially when death is imminent or where a veterinarian is not readily available.

e. To graze with a class of livestock not harmed by the particular plant or plants, where this is possible. Many plants seriously poisonous to one kind of livestock are not poisonous to another, at least under practical conditions.

f. To shift the grazing season to a time when the plant is not dangerous, where this is possible. That is, some plants are poisonous at certain seasons of the year, but comparatively harmless at other seasons.

g. To avoid cutting poison-infested meadows for hay when it is known that the dried cured plant is poisonous. Some plants are poisonous in either green or dry form, whereas others are harmless when dry. When poisonous plants (or seeds) become mixed with hay (or grain), it is difficult for animals to separate the safe from the toxic material.

3. **Know the symptoms that generally indicate plant poisoning,** thus making for early action.

4. **Avoid turning to pasture in early spring.** Nature has ordained most poisonous plants as early growers—earlier than the desirable forage. For this reason, as well as from the standpoint of desirable pasture management, animals should not be turned to pasture in the early spring before the usual forage has become plentiful.

5. **Provide supplemental feed during droughts, after plants become mature, and after early frost.** Otherwise, hungry animals may eat poisonous plants in an effort to survive.

6. **Avoid turning very hungry animals where there are poisonous plants,** especially those that have been in corrals for branding, etc.; that have been recently shipped or trailed long distances; or that have been wintered on dry forage. First feed the animals to satisfy their hunger or allow a fill on an area known to be free from poisonous plants.

7. **Avoid driving animals too fast when trailing.** On long drives, either allow them to graze along the way or stop frequently and provide supplemental feed.

8. **Remove promptly all animals from infested areas when plant poisoning strikes.** Hopefully, this will check further losses.

9. **Treat promptly,** preferably by a veterinarian.

TREATMENT OF PLANT-POISONED ANIMALS

Unfortunately, plant-poisoned animals are not generally discovered in sufficient time to prevent loss. Thus, prevention is decidedly superior to treatment.

When trouble is encountered, the owner or caretaker should *promptly* call a veterinarian. In the meantime, the animal should be (1) placed where adequate care and treatment can be given, (2) protected from excessive heat and cold, and (3) allowed to eat only feeds known to be safe.

The veterinarian may determine the kind of poisonous plant involved (1) by observing the symptoms, and/or (2) by finding out exactly what poisonous plant was eaten through looking over the pasture and/or hay and identifying leaves or other plant parts found in the animal's digestive tract at the time of autopsy.

It is to be emphasized, however, that many poisoned animals that would have recovered had they been left undisturbed, have been killed by attempts to administer home remedies by well-meaning but untrained persons.

REGULATIONS RELATIVE TO DISEASE CONTROL

Certain animal diseases are so devastating that no individual farmer or rancher could long protect privately owned herds and flocks against their invasion. Moreover, where human health is involved, the problem is much too important to be entrusted to individual action. In the United States, therefore, certain regulatory activities in animal disease control are under the supervision of various federal and state organizations. Federally, this responsibility is entrusted to the following agency:

Veterinary Service
Animal and Plant Health Inspection Service
U.S. Department of Agriculture
Federal Center Building
Hyattsville, MD 20782

FOOD AND DRUG ADMINISTRATION (FDA)

In 1906, the U.S. Congress enacted the Pure Food

and Drug Law. Concurrently, the Federal Meat Inspection Act was passed. Both laws became effective in 1907. The U.S. Food and Drug Administration (FDA) was established as a separate unit of the U.S. Department of Agriculture in 1927. Then, in 1940, the FDA was transferred from the USDA to the Federal Security Agency, presently, the U.S. Department of Health and Human Services.

The FDA is charged with the responsibility of safeguarding American consumers against injury, unsanitary food, and fraud. It inspects and analyzes samples and conducts independent research on such things as toxicity (using laboratory animals), disappearance curves for pesticides, and long-range effects of drugs.

U.S. DEPARTMENT OF AGRICULTURE (USDA)

The following four divisions of the U.S. Department of Agriculture have primary responsibilities in the area of animal and human health.

1. **The Animal and Plant Health Inspection Service.** This division is charged with maintaining the wholesomeness and safety of meats processed in packing plants that ship meat and meat products, including poultry and poultry products, interstate. Veterinarians and other trained personnel make the inspections. Its purpose it to protect consumers against infected meats and fraudulent and unsanitary preparation of meat products. The inspection first consists of an examination of the live animals so that any unfit beast may be removed and disposed of properly. Secondly, the carcasses and internal organs are inspected for any abnormalities of animals carrying infectious diseases. Centers of infection sources may be located, thus assisting the livestock owners in the vicinity. The records of meat inspection also serve a useful purpose to the research scientist.

2. **The Labeling and Registration Section.** This section in the USDA has responsibility for the proper labeling and safe use of pesticides. Manufacturers of pesticides must present new products with their proposed labels for approval before they are authorized to sell them. The label must indicate, as a minimum, the following: the name of the product; the active and inactive ingredients, together with percentages of each, in the formulation; the pest(s) controlled; directions for use—including the method and rate of application; any restrictions to be observed in application and handling; and an antidote—if known.

It is the responsibility of FDA, however, to set legal tolerances for pesticides on or in raw agricultural products. Also, it sets the safe interval between last application of the insecticide and the time of harvest of the crop or the slaughter of the animal.

Thus, through cooperative supervision of the USDA and the FDA, both the pesticide user and the consumer of the product are safeguarded.

3. **The Veterinary Services Division (VSD).** This division of the USDA is responsible for programs to control and eradicate (if possible), certain diseases of livestock, e.g., brucellosis, tuberculosis, scabies, and hog cholera. It does the following things: conducts nationwide federal-state cooperative programs for the control and eradication of animal diseases; suppresses spread of disease through control of interstate and international movements of livestock; keeps informed of the overall disease situation nationally and internationally; administers laws to ensure human treatment of livestock and certain laboratory animals; collects and disseminates information on morbidity and mortality; and provides training for USDA employees and others in related government agencies.

4. **Stockyard Inspection.** With the advent of large public markets, public stockyards inspection was initiated. This is an addition to the regular inspection performed on animals by meat inspectors prior to slaughter. Among the principal diseases for which inspections are made are: anthrax, scabies of cattle and sheep, tick or splenetic fever, hog cholera, and erysipelas of swine.

Not only are the incoming shipments of livestock inspected, but a reinspection is made of outgoing shipments. Tests for tuberculosis and brucellosis are accomplished, and dipping for scabies is performed before shipments are allowed to return to farms and ranches.

U.S. PUBLIC HEALTH SERVICES (USPHS)

This section of the Department of Health and Human Services is concerned with the prevention and treatment of disease. It works in the areas of vector control, pollution control, and control of communicable diseases of people. A part of this important complex is the National Institute of Health (NIH), which was formed in 1930, and which is composed of the following nine sister institutes: the National Cancer Institute, the National Heart Institute, the National Institute of Allergy and Infectious Diseases, the National Institute of Arthritis and Metabolic Diseases, the National Institute of Dental Research, the National Institute of Mental Health, the National Institute of Neurological Diseases and Blindness (including multiple sclerosis, epilepsy, cerebral palsy, and blindness), the National Institute of Child Health and Human Development, and the National Institute of General Medical Science. In addition to its own research program, the USPHS provides grants for health-related research at many universities and research institutes in the United States.

STATE VETERINARIANS, SANITARY COMMISSIONS, AND BOARDS

Most states have state veterinarians, or comparable officials, who direct the livestock sanitary and regulatory programs within their respective states. Livestock producers may secure the regulations applicable to the state in which they reside by writing their state department of agriculture.

QUARANTINE

Many highly infectious diseases are prevented by quarantine from (1) gaining a foothold in this country, or (2) spreading. By quarantine is meant (1) segregation and confinement of one or more animals in the smallest possible area to prevent any direct or indirect contact with animals not so restrained; or (2) regulating movement of animals at points of entry.

When an infectious disease outbreak occurs, drastic quarantine must be imposed to restrict movement out of an area or within areas. The type of quarantine varies from one involving a mere physical examination and movement under proper certification to the complete prohibition against the movement of animals, produce, vehicles, and even human beings.

FOREIGN DISEASE PROTECTION

Distance no longer provides a buffer against the invasion of foreign diseases. More than 90% of animals imported into the United States arrive by air. An airplane can outpace the development of clinical signs of diseases in an animal that has been exposed to infection just prior to shipment. This prompts great concern for epizootic diseases capable of crippling or destroying entire livestock populations. Such diseases still exist in Europe, Asia, Africa, and Latin America; among them, are such dreaded diseases as rinderpest, contagious bovine pleuropneumonia, foot-and-mouth disease, hog cholera, Africa horse sickness, Africa swine fever, exotic Newcastle disease, African trypanosomiasis, East Coast fever, and piroplasmosis.

Until 1875, the importation of livestock into the United States was free and easy. But, that year the United States prohibited the importation of cattle and hides from Spain, where foot-and-mouth disease was rampant. By 1880, European countries were refusing to buy cattle or beef from the United States for fear of getting contagious pleuropneumonia. Then, in 1884, Congress established the Bureau of Animal Industry in the U.S. Department of Agriculture (USDA) and gave the Secretary of Agriculture authority to enforce quarantine laws.

Today, there are stations at several entry points, where inspectors of the USDA's Animal and Plant Health Inspection Service (APHIS) inspect all animals and poultry to be imported to the United States. If no communicable diseases are found, the animals may be quarantined for a period of time, during which time they are treated for external parasites and subjected to various tests—*e.g.*, horses are tested for glanders and equine infectious anemia, and cattle are tested for brucellosis. At the end of the quarantine period, if no communicable diseases are found, they are released to the purchaser. APHIS is also charged with the responsibility of safeguarding against diseases introduced by the importation of zoo animals into this country. Wild animals brought into this country must undergo an extensive quarantine period abroad, followed by a further quarantine period at the animal quarantine station at Newburg, New York. Moreover, they are allowed to go only to certain approved zoos, where the zoo animals are isolated from domestic livestock and where proper measures are taken to dispose of waste to prevent the spread of diseases.

FEDERAL QUARANTINE CENTER

A Federal Quarantine Center was authorized in Public Law 91-239, signed by the President on May 6, 1970; and a 16.1-acre site for the Center was selected at Fleming Key, near Key West, Florida.

The quarantine center is designed to hold some 400 head of cattle, or other species in equivalent numbers, at one time, for a five-month quarantine period. This maximum security station enables American livestock producers to import breeding animals from all parts of the world, while at the same time safeguarding our domestic herds and flocks from such diseases as foot-and-mouth disease, rinderpest, piroplasmosis, and others.

INDEMNITY PAYMENTS

Where certain animal diseases are involved, the livestock producer can obtain financial assistance in eradication programs through Federal and State sources.

Note well: Both Federal and State indemnity payments are subject to change. So, for current regulations, the livestock producer should contact the local veterinarian or State Department of Agriculture.

FEDERAL INDEMNITY PAYMENTS

Information relative to indemnities paid to owners by the federal government for animals disposed of as a result of outbreaks of certain diseases is given in Chapter 1, Subchapter B, Title 9 of the Code of Federal Regulations, a summary of which follows:

■ **Brucellosis and tuberculosis**—The indemnity payments to owners by the federal government where brucellosis and tuberculosis are involved change from time to time. But the pertinent regulations that existed when this book was written follow:

1. **Brucellosis.**

a. **Affected cattle.** Effective October 17, 1994, owners of cattle or bison destroyed which are affected with brucellosis may be paid an indemnity by the USDA of $750 per animal (average slaughter value of $475 plus $250 indemnity).

2. **Tuberculosis.**

a. **Affected cattle.** The Department may pay owners an indemnity for cattle and bison affected with tuberculosis not to exceed $750 for each animal, but any joint state-federal indemnity payment, plus salvage, must not exceed the appraised value of each animal.

Also, the following regulations apply specifically to tuberculosis: The deputy administrator may authorize the payment of indemnity to owners of cattle which are destroyed because of tuberculosis not to exceed $450 for any animal which is a part of a known infected herd when it has been determined by the deputy administrator that the destruction of all the exposed cattle and bison in the herd will contribute to the tuberculosis eradication program; but the joint state-federal indemnity payments, plus salvage, must not exceed the appraised value of each animal.

b. **Appraisals of cattle destroyed because of tuberculosis.** Cattle to be destroyed because of tuberculosis shall be appraised by an independent, professional appraiser at veterinary services' expense, except that the veterinarian in charge may waive the requirement for an independent professional appraiser for reasons which are considered satisfactory. Due consideration shall be given to their breeding value as well as to their dairy or meat value. Where purebreds are involved, the owner shall either (1) see that the animals are accompanied by their registration papers at time of appraisal; or (2) be granted a reasonable time, by the veterinarian in charge, in which to present papers. Veterinary services may decline to accept any appraisal that appears to be unreasonable or out of proportion to the market value of cattle of like quality.

3. **Marking (or identifying) and slaughtering brucellosis and tuberculosis reactors.** Prior to marketing, the cattle must be marked and identified as follows:

a. Brucellosis reactor cattle must be branded with a 2- to 3-in. high letter "B" on the left jaw, and tagged with a metal federal or state reactor tag in the left ear; provided, however, that in lieu of branding and tagging, reactors and exposed cattle and bison in herds scheduled for herd depopulation may be identified by USDA-approved backtags and either accompanied to slaughter by a veterinary service or state representative, or moved directly to slaughter in vehicles closed with official seals.

b. Tuberculosis reactor cattle must be branded with a 2- to 3-in. high letter "T" on the left jaw, and tagged with a metal federal or state reactor tag in the left ear.

c. The cattle on which indemnity payments are made must be slaughtered within 15 days after the appraisal is made, unless an extension of time is granted by APHIS.

■ **Foot-and-mouth disease, pleuropneumonia, rinderpest, and other contagious or infectious animal diseases which constitute an emergency and threaten the livestock industry of the country.** Under Title 9, Part 53, of the U.S. Code of Federal Regulations, the Secretary of Agriculture of the U.S. Department of Agriculture may declare a national emergency due to the existence of foot-and-mouth disease, rinderpest, contagious pleuropneumonia, or any other communicable disease of livestock and poultry which threatens the livestock industry of the country.

Information relative to indemnities paid to the owners by the federal government for animals disposed of as a result of outbreaks of certain diseases can be obtained from the U.S. Department of Agriculture.

Information pertaining to indemnities paid by each state can be secured by writing to the respective State Department of Agriculture.

Animals affected by or exposed to disease shall be destroyed promptly after appraisal and disposed of by burial or burning, unless otherwise specifically authorized by veterinary services of the Animal and Plant Health Inspection Service of the U.S. Department of Agriculture.

In order to reduce the cost of eradicating emergency disease to the livestock producer and to the state and federal governments, it is essential that suspicious cases be promptly reported. If such a disease is suspected, a report should promptly be made to your practicing veterinarian and to state and federal animal health officials.

QUESTIONS FOR STUDY AND DISCUSSION

1. Why are good beef cattle health records of value to beef cattle producers?

2. What is the normal temperature, pulse rate, and respiration rate of beef cattle, and how would you determine each?

3. Select a specific farm or ranch (either your own or one with which you are familiar) and outline (in 1, 2, 3, order) a program of beef cattle health, disease prevention, and parasite control.

4. What is meant by a calf health-treatment program (preconditioning)? List the steps that are generally involved in such a program.

5. In the U.S., brucellosis has not been eradicated, although the incidence of reactor cattle has been reduced from 11.5% in 1935 to 0.015% in 1994. Has the program been a success or a failure? Justify your stand.

6. Outline a program for the prevention and treatment of calf scours.

7. Some critics of the U.S. foot-and-mouth disease control program accuse U.S. cattle producers of favoring closing our borders to countries in which foot-and-mouth disease is present as a means of lessening beef importations; hence, a means of lessening competition. Is there basis for this accusation? Justify your stand.

8. Give the importance, symptoms, cause, prevention, and treatment of each of the following diseases: bovine respiratory disease complex, brucellosis, Johne's disease, leptospirosis, pinkeye, and rabies.

9. Give the symptoms, control, and treatment of each of the following parasites: anaplasmosis coccidiosis, grubs, lice, and liver fluke.

10. When mange is confirmed in a herd, why is that herd quarantined? Who does the quarantining?

11. Explain (a) how screwworm flies are sterilized, and (b) how the screwworm control program works.

12. Assume that you have, during a period of a year, encountered cattle death losses from three different diseases (you name them). What kind of disinfectant would you use in each case?

13. List the five things upon which effective disinfection depends. Name three disinfectants, and discuss the usefulness, strength, and limitations of each.

14. Assume that you have encountered death losses from a certain poisoning plant (you name it). What steps would you take to meet the situation (list in 1, 2, 3, order; be specific)?

15. In the U.S., why are certain regulatory activities in animal disease control under the supervision of various federal and state organizations?

16. Of what value is the Federal Quarantine Center?

17. When people are ill, they call the family doctor. Isn't it just as logical that a veterinarian be called when cattle are sick? Justify your stand.

SELECTED REFERENCES

Title of Publication	Author(s)	Publisher
Abortion Diseases of Livestock	Ed. by L. C. Faulkner	Charles C Thomas Publisher, Springfield, IL, 1968
Animal Agents and Vectors of Human Disease	E. C. Faust	Lea & Febiger, Philadelphia, PA, 1956
Animal Diseases	Ed. by A. Stefferud	U.S. Department of Agriculture, Washington, DC, 1956
Animal Health: A Layperson's Guide to Disease Control, Second Edition	W. J. Greer J. K. Baker	Interstate Publishers, Inc., Danville, IL, 1992
Animal Parasitism	C. P. Read	Prentice-Hall Inc., Englewood Cliffs, NJ, 1972
Bovine Medicine and Surgery, Vols. I and II	Ed. by H. E. Amstutz	American Veterinary Publications, Inc., Santa Barbara, CA, 1980
Bovine Medicine & Surgery and Herd Health Management	W. J. Gibbons E. J. Catcott J. F. Smithcors	American Veterinary Publications, Inc., Wheaton, IL, 1970
Brucellosis	National Institute of Health of the Public Health Service Federal Security Agency U.S. Department of Agriculture National Research Council	American Association for the Advancement of Science, Washington, DC, 1950
Contagious Bovine Pleuropneumonia	J. R. Hudson	Food and Agriculture Organization of the United Nations, Rome, Italy, 1971
Control of Ticks on Livestock, The	S. F. Barnett	Food and Agriculture Organization of the United Nations, Rome, Italy, 1968
Current Veterinary Therapy: Food Animal Practice	Ed. by J. L. Howard	W. B. Saunders Company, Philadelphia, PA, 1981
Diseases of Cattle, Revised Second Edition	Ed. by W. J. Gibbons R. Jensen	American Veterinary Publications, Inc., Wheaton, IL, 1963
Diseases of Feedlot Cattle, Third Edition	D. R. Mackey	Lea & Febiger, Philadelphia, PA, 1979
Diseases of Livestock, Sixth Edition	T. G. Hungerford	Angus and Robertson Ltd., Sydney, Australia, 1967
Diseases Transmitted from Animals to Man, Sixth Edition	Ed. by T. G. Hull, *et al.*	Charles C Thomas, Publisher, Springfield, IL, 1975
Disinfection, Sterilization, and Preservation, Third Edition	Ed. by C. A. Lawrence S. S. Block	Lea & Febiger, Philadelphia, PA, 1983
Farm Animal Health and Disease Control, Second Edition	J. K. Winkler	Lea & Febiger, Philadelphia, PA, 1982
Farmer's Veterinary Handbook	J. J. Haberman	Prentice-Hall, Inc., New York, NY, 1953

Title of Book	Author(s)	Publisher
Foreign Animal Diseases, Their Prevention, Diagnosis and Control, Official Report of the U.S. Livestock Sanitary Association		U.S. Livestock Sanitary Association, Trenton, NJ, 1954
Hagan's Infectious Diseases of Domestic Animals, Sixth Edition	D. W. Bruner J. H. Gillespie	Comstock Publishing Associates, Ithaca, NY, 1973
Home Veterinarian's Handbook, The	E. T. Baker	The Macmillan Company, New York, NY, 1949
Immunity to Animal Parasites	Ed. by E. J. L. Soulsby	Academic Press, Inc., New York, NY, 1972
Keeping Livestock Healthy, Yearbook of Agriculture, 1942		U.S. Department of Agriculture, Washington, DC, 1942
Lecture Outline of Preventive Veterinary Medicine for Animal Science Students, Third Edition	I. A. Schipper	Burgess Publishing Company, Minneapolis, MN, 1962
Livestock Health Encyclopedia, Third Edition	Ed. by R. Seiden	Springer Publishing Co., Inc., New York, NY, 1968
Losses in Agriculture, Ag. Hdbk. No. 291		U.S. Department of Agriculture, Washington, DC, 1965
Merck Veterinary Manual, The, Seventh Edition	Ed. by C. M. Fraser	Merck & Co., Inc., Rahway, NJ, 1991
Nationwide System for Animal Health Surveillance, A	National Research Council	National Academy of Sciences, Washington, DC, 1974
New Zealand Farmer's Veterinary Guide, The	D. G. Edgar, *et al.*	The New Zealand Dairy Exporter, Wellington, New Zealand, 1962
Pathology of Domestic Animals, Third Edition, Vols. 1 and 2	K. V. F. Jubb, *et al.*	Academic Press, Inc., New York, NY, 1984
Practical Parasitology: General Laboratory Techniques and Parasitic Protozoa	C. J. Price J. E. Reed	United Nations Development Programme and Food and Agriculture Organization of the United Nations, Rome, Italy, 1970
Principles of Veterinary Science, Fourth Edition	F. B. Hadley	W. B. Saunders Company, Philadelphia, PA, 1949
Some Diseases of Animals Communicable to Man in Britain	Ed. by O'Graham Jones	Pergamon Press, Ltd., London, England, 1968
Some Important Animal Diseases in Europe, FAO Agricultural Studies No. 10	Ed. by K. V. L. Kesteven	Food and Agriculture Organization of the United Nations, Rome, Italy, 1948
Special Report on Diseases of Cattle	V. T. Atkinson, *et al.*	U.S. Department of Agriculture, Washington, DC, 1942
Stockman's Handbook, The, Seventh Edition	M. E. Ensminger	Interstate Publishers, Inc., Danville, IL, 1992
Veterinary Guide for Farmers	G. W. Stamm Ed. by D. S. Burch	Windsor Press, Chicago, IL, 1950
Veterinary Handbook for Cattlemen, Second Edition	J. W. Bailey	Springer Publishing Co., Inc., New York, NY, 1958

Veterinary Medicine	D. C. Blood J. A. Henderson	The Williams and Wilkins Company, Baltimore, MD, 1960
Veterinary Parasitology, Second Revised Edition	G. Lapage	Charles C Thomas, Publisher, Springfleld, IL, 1968

In addition to the above selected references, valuable publications on different subjects pertaining to beef cattle diseases, parasites, disinfectants, and poisonous plants can be obtained from the following sources:

1. Division of Publications
   Office of Information
   U.S. Department of Agriculture
   Washington, D.C. 20250
2. Your state agricultural college.
3. Several biological, pharmaceutical, and chemical companies.

Computers in the cattle business. Big and complex cattle businesses have outgrown hand record keeping. (Courtesy, California State University, Fresno)

BUSINESS ASPECTS OF BEEF PRODUCTION

Contents *Page*

Contents Page

Fig. 13-1. Beef cattle business.

The cattle producers of the future will be good business people as well as knowledgeable cattle producers. They also will need operations that are large enough to provide their families an adequate standard of living and generate enough capital to keep expanding. Since profit margins will likely decline further, there will be greater stress on business and financial management skill.

Generating both equity and debt capital, or risk and borrowed capital, will be one of the main concerns of future producers. The large investment, plus the need to keep competitive by utilizing new and usually expensive technological advances, will cause capital to be very important.

To obtain capital several things will be necessary. The owner or manager will have to prepare (1) profit and loss statements to show that the operation is profitable, (2) financial statements to show that progress is being made, and (3) cash-flow projections to show loan repayability. Then, and then only, will the producer be ready to go looking for funds.

Skill in capital budgeting and analyzing alternative investment opportunities will be needed to see that the limited capital is invested where payoff will be the greatest. The producer will also have to exercise budget and cost controls of the business. Skill in building sound credit will be needed.

The greatest payoffs in the future are likely to come from the efforts devoted to improving producer's skills in business and financial management.

TYPES OF BUSINESS ORGANIZATION

The success of today's cattle business is very dependent on the type of business organization. No one type of organization is superior under all circumstances; rather, each situation must be considered individually. The size of the operation, the family situation, the enterprises, the objectives—all these, and more, are important in determining the best way in which to organize the cattle business.

Three major types of business organizations are commonly found among cattle enterprises: (1) **individuals**; (2) **partnerships**; and (3) **corporations**. Additionally, there are agency services. As shown in Fig. 13-2, the breakdown in types of business organization of U.S. farms is as follows: individual or family owned, 86.7%; partnership, 9.6%; and corporations, 3.2%. Among the factors which should be considered when deciding which business form best fits a given set of circumstances are the following:

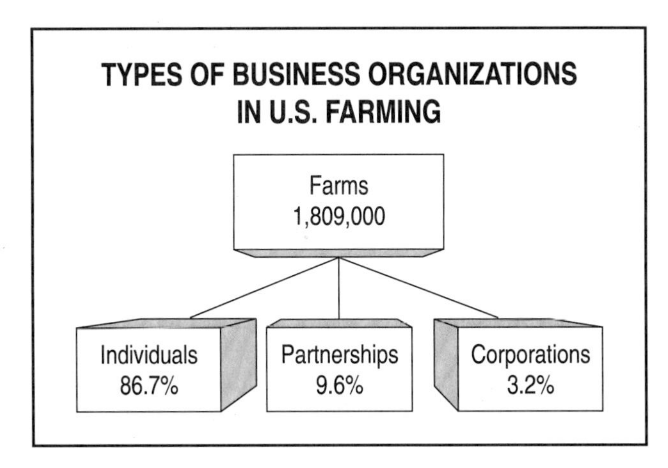

Fig. 13-2. Types of business organizations in U.S. farming. (Source: U.S. Bureau of the Census, *Census of Agriculture*, Vol. 1)

1. Which type of organization is most likely to be looked upon favorably from the standpoint of more credit and capital?
2. How much capital will be required of each individual involved?
3. Are there tax advantages to be gained from the business organization?
4. Is expansion of the business feasible and facilitated?
5. Which type of organization reduces risks and liability most?
6. Which type of organization can be terminated most easily and readily?
7. Which type of ownership provides for the most continuity and ease of transfer?
8. What costs for legal and accounting fees are involved, in setting up the organization and in the preparation of the annual reports required by law?
9. Who will manage the business?

Most cow-calf enterprises are operated as sole proprietorships; not necessarily because this is the best type of organization, but with no effort to form some other type of organization it naturally results. Both the partnership and the corporation, which require special planning and effort to bring about, are well suited to the operation of large commercial cattle feedlots.

PROPRIETORSHIP (Individual)

This is the most common type of business organization in U. S. farming—86.7% of the nation's farms are individually or family owned. (See Fig. 13-2.) Under the sole proprietorship, or individual (or family) ownership, one person controls the business. The owner may not provide all the capital used in the business; in fact, the owner usually does not. However, the owner has sole management and control of the operation, although this may be modified and delegated somewhat through lease agreements, contracts, etc. Basically, the sole proprietor gets all the profits of the business. Likewise, the sole proprietor must absorb all the losses.

In comparison with other forms of organization, the sole proprietorship has two major limitations: (1) It may be more difficult to acquire new capital for expansion; and (2) not much can be done to provide for continuity and to keep the present business going as a unit, with the result that it usually goes out of existence with the passing of the owner.

PARTNERSHIP (General Partnership)

A partnership is an association of two or more persons who as co-owners, operate the business. About 9.6% of U.S. farms are partnerships. (See Fig. 13-2.)

The basic idea of two or more persons joining together to carry out a business venture can be traced back to the syndicates that were used in major trading centers in Western Europe in the Middle Ages. Many of the early efforts to colonize the New World were also partnerships, or "companies" which provided venture capital, ships, provisions, and trade goods to induce settlement of large land grants.

Most farm partnerships involve family members who have pooled land, machinery, working capital, and often their labor and management to operate a larger business than would be possible if each member limited the operation to the resources of one person. It is a good way in which to bring a son or daughter, who is usually short on capital, into the business, yet keep the parent(s) in active participation. Although there are financial risks to each member of such a partnership, and potential conflicts in management decisions, the existence of family ties tends to minimize such problems.

In order for a partnership to be successful, the enterprise must be sufficiently large to utilize the abilities and skills of the partners and to compensate each adequately in keeping with his/her contribution to the business.

A partnership has the following **advantages**:

1. **Combining resources.** A partnership often increases returns from the operation due to combining resources. For example, one partner may contribute labor and management skills, whereas another may provide the capital. Under such an arrangement, it is very important that the partners agree on the value of each person's contribution to the business, and that this be clearly spelled out in the partnership agreement.

2. **Equitable management.** Unless otherwise agreed upon, all partners have equal rights, regardless of financial interest. Any limitations, such as voting rights proportionate to investments, should be a written part of the agreement.

3. **Tax savings.** A partnership does not pay any tax on its income, but it must file an informational return. The tax is paid as part of the individual tax returns of the respective partners, usually at lower tax rates.

4. **Flexibility.** Usually, the partnership does not need outside approval to change its structure or operation—the vote of the partners suffices.

Partnerships may have the following **disadvantages**:

1. **Liability for debts and obligations of the partnership.** In a partnership, each partner is liable for all the debts and obligations of the partnership.

2. **Uncertainty of length of agreement.** A partnership ceases with the death or withdrawal of any partner, unless the agreement provides for continuation by the remaining partners.

3. **Difficulty of determining value of partner's interest.** Since a partner owns a share of every individual item involved in the partnership, it is often very difficult to judge value. This tends to make transfer of a partnership difficult. This disadvantage may be lessened by determining market values regularly.

The above is what is known as a partnership or general partnership. It is characterized by (1) management of the business being shared by the partners, and (2) each partner being responsible for the activities and liabilities of all of the partners, in addition to their own activities within the partnership.

LIMITED PARTNERSHIP

A limited partnership is an arrangement in which two or more parties supply the capital, but only one partner is involved in the management. This is a special type of partnership with one or more *general* partners and one or more *limited* partners.

The limited partnership avoids many of the problems inherent in a partnership (general partnership) and has become the chief legal device for attracting outside investor capital into farm and ranch ventures. Although this device has been widely used in the soil and gas industries, and for acquiring income-producing urban real estate for a number of years, its application to agricultural ventures on a national scale is quite new. As the term implies, the financial liability of each partner is limited to their original investment, and the partnership does not require, and in fact prohibits, direct involvement of the limited partners in management. In many ways, a limited partner is in a similar position to a stockholder in a corporation.

A limited partnership must have at least one general partner who is responsible for managing the business and who is fully liable for all obligations.

Like most other business organizations, the limited partnership has both advantages and disadvantages.

The **advantages** of a limited partnership are:

1. **It facilitates bringing in outside capital.** A limited partnership provides a way in which to bring outside capital into an agricultural operation, without giving up any control of the operation.

2. **It need not dissolve with the loss of a partner.** Unlike a regular partnership, a limited partnership is not necessarily dissolved upon the death, bankruptcy, or withdrawal of a limited partner.

3. **Interests may be sold or transferred.** A limited partner may sell or transfer his/her interests without disrupting the partnership, so long as it is done in keeping with the agreement that is drawn up at the time of organizing. Usually, it is specified that a limited partner who wishes to sell his/her interest in the business must give the other partners first refusal before offering it to outsiders.

4. **The business is taxed as a partnership.** This allows profits and expenses to be passed on to the investor on a pro rata share based on investment.

5. **Liability is limited.** The investor's liability is limited to the amount of the investment.

6. **It may be used as a tax shelter.** It makes possible the deferral of taxable income to the next year through expending such as the normal feeding expenses in the year paid, along with the purchase of grain and other storable commodities in the current tax year for use in the next year. These expenses can be used to offset other income which in many cases includes profits from cattle fed during the prior year.

The **disadvantages** of a limited partnership are:

1. **The general partner has unlimited liability.** Although this fact should be recognized, it need not be reason for concern *provided* the operation is sound and well managed.

2. **The limited partners have no voice in management.** This is not an unmixed blessing, for it alleviates the hazard of compromise and weakness, which can, and often does, mitigate against a business.

CORPORATIONS

A corporation is a device for carrying out a farming or ranching enterprise as an entity entirely distinct from the persons who are interested in and control it. Each state authorizes the existence of corporations. As long as the corporation complies with the provisions of the law, it continues to exist—irrespective of changes in its membership.

Until about 1960, few farms and ranches were operated as corporations. In recent years, however, there has been increased interest in the use of corporations for the conduct of farm and ranch business. In 1987, the U. S Department of Agriculture reported the following relative to corporate farms: number of corporate farms, 67,000; percent of all farms that are corporate farms, 3.2%; total acreage in corporate farms, 119 million.

From an operational standpoint, a corporation possesses many of the privileges and responsibilities of a real person. It can own property; it can hire labor; it can sue and be sued; and it pays taxes.

Separation of ownership and management is a unique feature of corporations. The owners' interest in a corporation is represented by shares of stock. The shareholders elect the board of directors who, in turn, elect the officers. The officers are responsible for the day to day operation of the business. Of course, in a close family corporation, shareholders, directors, and officers can be the same persons.

The major **advantages** of a corporate structure are:

1. **Continuity.** It provides for continuity of the business despite the death of a stockholder.
2. **It facilitates transfer of ownership.** Since stock (rather than physical property) is sold, exchanged, or given, transfer of stock is easy.
3. **The liability of shareholders is limited to the value of their stock.** In an incorporated business, the shareholders are liable for the debts of the corporation and for any liability caused by negligent employees who injure others in the course of their employment, but only to the extent of their investment in the corporation.
4. **There may be some savings in income taxes.** For example, salaries paid to owners who also are employees of the corporation, along with costs of insurance, health, and retirement plans, can be deducted as business expenses, thereby reducing the amount of taxable income.

The **disadvantages** of a corporation are:

1. **Restricted to charter.** Corporations are restricted to doing only what is specified in their charter.
2. **Must register in each state.** They must register in each state in which they do business and abide by the regulations of each state.
3. **Must comply with regulations.** They must comply with specific regulations regarding meetings, records, and reorganizations; and they are subject to certain legal fees, payroll taxes, accumulated earnings taxes on profits not distributed to stockholders, and other fees. Thus, a corporation involves considerable paper work and expense.
4. **The hazard of higher taxes.** A corporation is always a ripe target for higher income taxes, particularly federal income taxes.
5. **Control can be lost.** Any person or group owning 51%, or more, of the stock can elect the Board of Directors and gain control.

FAMILY OWNED (Privately Owned) CORPORATION

Still another type of corporation is family owned (privately owned). It enjoys most of the advantages of its generally larger outside-investor counterpart, with few of the disadvantages. The chief **advantages** of the family owned corporation over a partnership arrangement are:

1. **It alleviates unlimited liability.** For this reason, a lawsuit cannot destroy the entire business and all the individual partners with it.
2. **It facilitates estate planning and ownership transfer.** It makes it possible to handle the estate and keep the business in the family and going if one of the partners should die. Each of the heirs can be given shares of stock—which are easy to sell or transfer and can be used as collateral to borrow money—while leaving the management of the enterprise to those heirs interested in operating it, or even outsiders.

TAX-OPTION CORPORATION (Subchapter S Corporation)

Instead of paying a corporate tax, a corporation with no more than 35 stockholders may elect to be taxed as a partnership, with the income or losses passed directly to the shareholders each of whom pays taxes on his/her share of the profits. This special type of corporation is variously referred to as a *tax-option* corporation, *subchapter S* corporation, pseudo-corporation, or elective corporation.

For income tax purposes, the owners of a tax-option corporation are taxed as if they were a partnership. That is, income earned by the corporation passes through the corporation to the personal income tax returns of the individual shareholders. Thus, the corporation does not pay any income tax. Instead, each shareholder pays tax on his/her share of corporate income at his/her individual tax rate, and each shareholder reports his/her share of long-term capital gains and receives his/her deductions therefore. Although each shareholder's portion of any corporate losses from current operations is deducted from his/her personal return, capital losses incurred by the corporation cannot be passed through to the shareholders.

Thus, there are some very real advantages to be gained from a *subchapter S* or *tax-option* corporation. However, in order to qualify as a subchapter S corporation, the following requisites must be met:

1. There cannot be more than 35 stockholders.
2. All stockholders must agree to be taxed as partnership.
3. Nonresident aliens cannot own stock.
4. There can be only one class of stock.
5. Not more than 20% of the gross receipts of the corporation can be from royalties, rents, dividends, interest, or annuities plus gains from sale or exchange of stock and securities; and not more than 80% of gross receipts can be from sources outside the United States.

ADVANTAGES OF LIMITED PARTNERSHIPS AND CORPORATIONS

In addition to the advantages peculiar to (1) limited partnerships, and (2) corporations, and covered under each, limited partnerships and corporations have the following advantages over individual ownership in the acquisition of capital:

1. They make it possible for several producers to pool their resources and develop an economic-sized operation, which might be too large for any one of them to finance individually.
2. They make it possible for persons outside agriculture to invest through purchase of shares or stock in the business.
3. They can generally borrow money easier because the strength of the loan is not dependent on the financial and managerial capability of one person.
4. They give assurance that the business will continue, even if one of the owners should die or decide to sell their interest.
5. They provide built-in management, with continuity; and, generally speaking, they attract very able management.

Thus, cattle producers—both cow-calf producers and cattle feeders—can and do use either of these two business organizations—a limited partnership or a corporation—to develop and maintain an economically sound operation. Actually, no one type of business organization is best suited for all purposes. Rather, each case must be analyzed, with the assistance of qualified specialists, to determine whether there is an advantage to using one of these types of organizations, and, if so, which organization is best suited to the proposed business.

AGENCY SERVICES

As an alternative to entering into the limited partnership arrangement or owning stock in a corporation, nonfarm investors wishing to engage in cattle feeding can utilize the services of several firms which specialize in purchasing feeders, contracting with feedlots, and selling market cattle. Also, similar services are available for acquiring and managing commercial breeding herds. Under the agency arrangement, the investor establishes a drawing account for the agent, arranges the financing, and can withdraw any profits realized. Because the investor obtains legal title to specific lots of cattle, the cattle may be used as collateral for loans.

Firms offering such services charge a flat fee per head, or a percentage of gross sales. They usually do not have financial interests in the feedlots or ranches with whom they contract.

CAPITAL

In 1994, U.S. farm assets—investments in land, improvements, machinery, equipment, animals, feed, and supplies—totaled $920 billion, which (1) ranked it as America's biggest single industry, and (2) is equivalent to about 3/5 the value of all the stocks of all corporations represented on the New York Stock Exchange. Another noteworthy statistic is that it takes about $33 in farm assets to produce $1 of net farm income.

There is no available breakdown of total capital

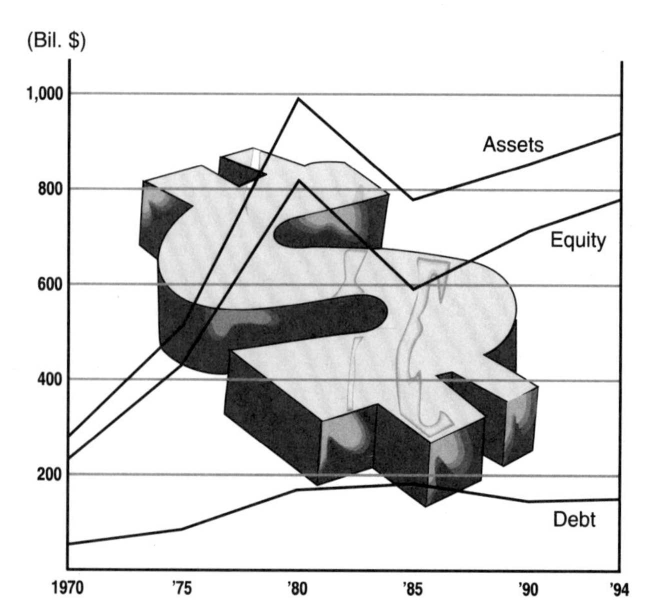

Fig. 13-3. Balance sheet of U.S. farming, showing (1) assets, (2) debts, and (3) equities. (Source: *Farm & Food Facts 1994*, from the Editors of Kiplinger Agriculture Letter, p. 23)

investment (including land, buildings and equipment, as well as animals) by classes of livestock, but it is reasonable to assume that beef cattle head the list.

In comparison with most other businesses, cow-calf operations are characterized by slow turnover. Nature ordained that heifers not reach puberty until 8 to 12 months of age, that they not be bred before about 15 months of age, and that there be a pregnancy period of 9 months. This means that the minimum generation interval of cattle is about 3 years. The producer can do very little to speed it up.

A faster capital turnover exists in cattle finishing than in cow-calf operations, but there is also higher risk.

SOURCES OF CAPITAL (Where Farmers Borrow)

Traditionally, agriculture has been financed by two kinds of capital, known as equity capital and credit (debt) capital. Formerly, equity capital came only from farmers—those who operated farms and ranches; debt capital came from a variety of sources. This was one of a number of characteristics of farm businesses that differentiated this sector from the rest of the economy.

Table 13-1 shows where farmers borrow, the amount of loans from each source, and the percent of the total held by each type of lender. Fig. 13-4 shows the percent of the total loan held by each type of lender.

TABLE 13-1
WHERE FARMERS BORROW[1]

Type and Source of Loan	Amount of Loan	Percent of Total
	(million $)	*(%)*
Real estate mortgage loans:		
Farm Credit Administration	24,770	33.01
Commercial banks	18,407	24.53
Individuals and others	16,004	21.33
Insurance companies	9,467	12.62
Farmers Home Administration	6,378	8.50
Total	75,028	100.0
Nonreal estate loans:		
Commercial banks	33,724	49.14
Individuals and others	13,230	19.28
Farm Credit Administration	10,464	15.25
Farmers Home Administration	7,216	10.51
Commodity Credit Corp.	4,000	5.83
Total	68,634	100.0

Total loans	143,663,000,000
Percent real estate	52.2
Percent nonreal estate	47.8

[1]1992 data from *A Primer of U.S. Farm Debt and the Market Share of the Farm Credit System*, Farm Credit Administration, September 28, 1993.

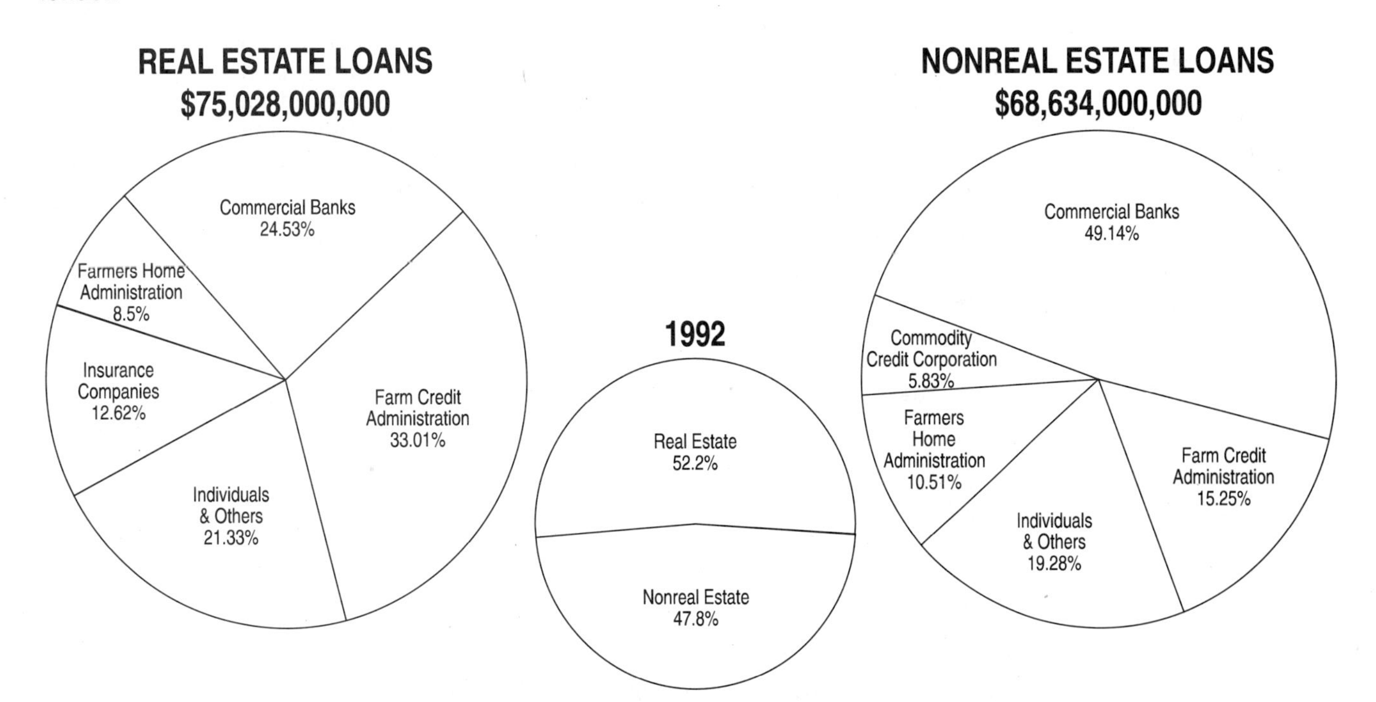

Fig. 13-4. Where farmers borrow. (Based on 1992 data provided by the Farm Credit Administration, Washington, DC)

But, agricultural financing is changing, and it will continue to change even more in the years ahead. Today, cattle producers are tapping the vast supply of farm equity or risk capital that is constantly seeking investment opportunities—nonfarm equity capital is being used in the cattle business.

Sometime or other most producers find it necessary to borrow money to buy land; to construct buildings and other improvements; to purchase equipment, seed, and cattle; and/or to pay for seasonal labor. They should know something, therefore, about the lending organizations available to them in order that they may determine which one will best serve their needs. The leading sources of farm credit are:

1. **Commercial banks.** Commercial bank funds for loans are provided by bank deposit. Since these fluctuate rather widely, the number of long-term loans banks can make is limited.

Progressive bankers are willing to lend to people who are beginning or expanding their cow herds. Most commonly, they will write a note for 6 to 12 months; however, some of them will make intermediate-term loans on breeding animals. A 6- to 12-month note is reviewed annually. Terms of the loan usually call for payment of interest due and some payment on the principal. In most cases, the note is renewed for the smaller principal amount.

Notes written for a year or less enable both the banker and the borrower to evaluate the progress being made by the borrower and permit the lender to update the interest rate being charged. A financial statement is required by bank examiners. This statement enables both the borrower and the lender to visualize how financing the cow loan may affect the financial status of a total operation.

Loans may be more easily obtained by the producer who already has a cash income from land, cash grain, or other livestock. Progressive commercial bankers also are more willing to lend to producers who are using management techniques that will increase their level of efficiency. Good record keeping on cow herds, use of performance tested bulls, proper range and/or pasture management that could increase carrying capacity are but a few of the management techniques presently viewed as profit boosters.

Some commercial banks have special agricultural representatives who are qualified to assist the borrower in many ways.

2. **Farm Credit System (FCS).** The structure of the Farm Credit System was changed substantially as a result of legislation during the period 1985–87. Among the changes, the Federal Land Banks and the Federal Intermediate Credit Banks were merged out of existence, replaced by the Farm Credit Banks (FCB). Thus, the following four types of associations exist currently:

a. **Federal Land Bank Associations.** About 72 of this traditional type of association remain, primarily in the Wichita and Texas districts. They act as agents for the district bank, which makes mortgage loans.

b. **Federal Land Credit Associations (FLCA).** About 31 associations continue to deal only with farm mortgage loans, but have been granted authority to be the lender, rather than the agent for the bank. This became possible with 1987 legislation.

c. **Agricultural Credit Associations.** About 66 associations have both the direct mortgage lending authority of an FLCA and the direct lending for other purposes of the PCA. Their authorities thus match the lending authorities of the FCB and the farm part of the ACB.

d. **Production Credit Associations (PCA).** About 69 of these traditional type associations remain in the western two-thirds of the United States.

The Farm Credit System obtains its loan funds primarily through the sale of its bonds in the nation's private money market.

3. **Farmers Home Administration (FmHA).** The Farmers Home Administration has now been split into two parts. The part that deals with farmers is now part of the Consolidated Farm Service Agency of the USDA. In recent years, their program has shifted from direct lending to guaranteeing loans made by other lenders.

The Farmers Home Administration makes short-, intermediate-, and long-term loans to farmers and ranchers who are unable to obtain adequate credit from other sources at reasonable rates and terms; and provides supervision for its loans. Applicants who are veterans and have farm experience receive preference. The FmHA helps applicants determine their credit needs, work out debt repayment schedules, and solve other financial problems even though they may not be eligible for a loan.

Farmers Home Administration loans are made to farmers who carry on farming operations on a scale large enough to support their families and to farmers on small farms who obtain income from off-farm employment. Each farm-ownership loan is based on a plan that when followed will provide enough income from the farm and other sources to enable the family to have a reasonable standard of living and make payments on its debts when due.

The county supervisors of the Farmers Home Administration help borrowers prepare the plan and provide on-the-farm assistance with management problems. Farm-ownership loans are scheduled for repayment according to the borrowers' ability to repay, over a period not exceeding 40 years.

In addition to farm-ownership loans, the Farmers

Home Administration makes loans to purchase cows, operating loans, housing, soil and water conservation, and emergency. Generally, cow loans are drawn up when the cows are purchased. The number of cows purchased is based on the grass and feed available on the farm or ranch; and the loan is set up on a 7-year note. The legal maximums are: $200,000 direct operating, $400,000 guaranteed operating loans, $200,000 direct real estate, $300,000 guaranteed real estate loans.

Where a natural disaster has occurred, under the Emergency Loan Program a borrower may borrow up to 80% of the loss, but not to exceed $500,000.

To obtain loans or detailed information about loans, apply to the Farmers Home Administration. The address of this local office can be obtained from the county agricultural agent or the Farmers Home Administration, Washington, DC 20250.

4. **Individuals.** Individuals with money, including retired farmers, landlords, and relatives, are a very important source of real estate credit.

Today, a number of sellers loan the amount of equity they have in land which they sell to a friend, an associate, or a member of the family. A retiring farmer, for example, may sell the farm to a young man who is capable of running it properly, but who can't make a down payment or obtain a large enough mortgage. The seller may be willing to take a second mortgage on part of the purchase price at a reasonable rate of interest. Frequently, the selling farmer will do this under an *installment land contract*, in which the down payment may vary from nothing to 29% of the purchase price. The seller then retains title until a certain agreed-to part of the price has been paid. By staying at 30% or under on the down payments, the sales qualify as installment sales and the seller(s) doesn't have to pay the capital gains tax except as proration when the money is received.

Under a land contract, the selling farmer usually gets a favorable price, along with a tax break on the profits of the sale. The buying farmer obtains the advantage of using the land which could not be financed through any other credit source.

After a land contract is run for some years and the new farm operator has an improved financial condition, the new operator will often refinance the land contract with an institutional lender.

One disadvantage of a loan from an individual is that the arrangement may be complicated by the lender's death unless adequate provision has been made for this eventuality. Individuals also make some nonreal estate loans to farmers. Here again, the terms and repayment provisions vary widely depending upon the agreement reached by the individual parties.

5. **Life insurance companies.** Many life insurance companies invest a part of their policyholders' reserve funds in loans secured by first mortgages on farm real estate. Terms usually range between 5 and 40 years and amortized repayment plans are now common. They usually offer a competitive rate of interest and amortized repayment plans. Generally insurance companies will make loans up to 60% of the appraised value of the farm or up to 50% of the sale value.

Life insurance activity in farm and ranch loans varies from region to region. In some states, life insurance companies hold close to one-half of the farm and ranch mortgage debt. In others, they have less than ½ of 1%. Their contacts are through branch offices or other local agents. Realtors can almost always provide information concerning insurance agencies actively making loans in the area.

6. **Merchants and dealers.** Many cattle producers buy machinery, feed, and supplies on time. This is usually the easiest type of credit to get, but it may be the most costly. If farmers and ranchers use this type of credit, they should try to find out how much it costs and compare it with costs of other sources of credit.

It must be realized that dealers and merchants extend credit to farmers and ranchers primarily for the purpose of promoting the sale of products and services. To a machinery dealer, the interest return from a loan is less important than the sale of a tractor. This is so because profits come from both sales and interest, and the dealer is selling *hard goods* that can be repossessed; hence, relatively high risk loans can be made. Installments over the expected productive life of the equipment, often several years, are the general practice.

The sellers of *soft goods*, like feed and fertilizer, must look at credit differently. Their products, once used, cannot be repossessed. Thus, the debt for feed and fertilizer customarily must be paid from the income of a single year.

PUBLIC OFFERINGS

Public offerings are the way the Farm Credit System raises its funds, and one way that both commercial banks and merchants/dealers obtain funds which they lend directly to farmers. They may also be a vehicle for large farming enterprises to raise funds directly in the market—as in the case of bull syndications, cattle feedlots, etc. In the case of the FCS, Securities and Exchange Commission (SEC) registration is not required, because the institutions are government sponsored enterprises. However, the Farm Credit Administration has created disclosure requirements that are near-equivalent to those created by the SEC.

Another form of market offering is the securities based on pools of loans, called mortgage backed or asset backed securities. Although these are best known in the area of housing mortgages, automobile leases, trade receivables, and the like, the Federal Agricultural Mortgage Corporation (Farmer Mac) was created in 1988 to provide such possibilities for farm mortgages. To date, only a small market has developed.

Three different kinds of public offerings are commonly used: (1) SEC registered offerings, (2) regulation A offerings, and (3) intrastate offerings.

SEC REGISTERED OFFERING

Where a public offering of limited partner interests (such as participation in a cattle feeding fund) in excess of $500,000 is to be marketed, a registration statement providing full and fair disclosure of the character of the securities must be filed with the SEC, in keeping with the Securities Act of 1933. The SEC considers selling interests in an agricultural enterprise, such as in a cattle feeding fund, similar to selling stock in a corporation. Thus, a prospectus, which reveals all pertinent facts of the securities offered, must be printed. The latter is used as informational matter to explain partnership operations to potential investors. Also, the services of a lawyer knowledgeable in the area of public securities offerings is necessary in preparing the prospectus, filing the registration statement, and negotiating with the underwriter. Generally speaking, an investment banker, or underwriter, is needed to market the offering—as a middleman—to bring buyers and sellers together, for which a commission is charged. Although there is nothing to keep the person or persons offering securities from marketing them privately, they usually do not have the necessary time, staff, or expertise. In selling limited partnership interests, underwriters generally work on a best effort basis. This means that they are not obliged to market all the securities. The limited partnership interests are sold to the public at a price previously established by the agricultural company. The time involved and the cost of registering a fund are very considerable. Normally, registration will require 4 to 8 months; and the total cost for legal, accounting, and printing, will run anywhere from $50,000 to $150,000, depending upon the size of the offering, legal fees, and underwriter fees. Other costs of organizing a limited partnership include registering with the state in which business will be conducted. While these costs may be passed on to the purchasers of the limited partners interests, such intention must be stated in the prospectus. Moreover, there is no assurance that the offering will sell, in which case the general partner must stand all costs.

The question of whether or not a registration fee must be filed with the SEC should be answered by an attorney as there are a number of exemptions for which provision is made in the SEC regulations. It is possible that the sale of the securities may be exempt from the registration provisions of the SEC if the offering does not involve a *public offering*, if the aggregate amount of the offering to the public does not exceed $500,000, or if the issue is to be sold on an intrastate basis.

REGULATION A OFFERING

Regulation A, issued by the Commission, provides for the exemption of certain classes of domestic and Canadian securities where the aggregate offering to the public does not exceed $500,000. While a registration statement need not be filed with the SEC under Regulation A, notification and reports are required. Also, the regulation requires that offering circulars containing information prescribed by the Commission must be furnished to buyers. Although a filing under Regulation A is not as difficult, time-consuming, or expensive as a registration, it still involves expense and labor.

INTRASTATE OFFERING

Some persons or companies may wish to sell their stock within the confines of a state since costs, fees, and time can be saved. Quite often the savings in filing fees, printing, etc., may bring the cost down to 25% of the expenditures that are incurred in a SEC registration. Certain states are in a position to qualify an issue within a week if it is presented properly, as compared to the 6 weeks or longer that are normally required in filing with the SEC.

It is necessary to determine whether an issue is exempt from registration with the SEC under the provisions of Section 3A (11) of the Securities Exchange Act of 1933 which provides:

> Any security which is a part of an issue offered and sold only to persons resident within a single State or Territory, where the issuer of such security is a person resident and doing business within, or, if a corporation, incorporated by and doing business within such State and Territory.

State laws differ as to registration of intrastate issues. Thus, if the offering can be sold within the confines of a state, it would be well to consider with an attorney the advantages and disadvantages of this type of registration.

CENTRAL MONEY MARKETS

This refers to financing operations, such as cattle feeding, by using central money markets through the medium of commercial paper in competition with other lending institutions. The use of Central Money Markets was prompted because the public agencies—Federal Intermediate Credit Banks, Production Credit Associations, and other federal and quasi-federal agencies—were not always able to provide the necessary financing for cattle feeders. Also, commercial banks were somewhat stymied by the limit of their deposits. The answer, according to this school of thought, was to put agricultural financing on a par with commercial financing—into the Central Money Markets with commercial paper as the basis for securing money.

Those arranging Central Money Market financing sell commercial paper in the open market at rates competitive to the market, and use the proceeds to make loans generally of one year or less maturity to agricultural producers. This is a new concept in agricultural financing. It may well prove to be the most desirable method of financing large commercial cattle feedlots, and other similar operations, in the future.

CREDIT IN THE CATTLE BUSINESS

Credit, upon which the whole vast structure of our commercial world rests, has been around for a very long time. According to the Greek philosopher Plato, the last words of Socrates were: "I owe a cock to Asclepius. Will you remember to pay the debt?" But it was the Romans who coined the word *credit*, which comes from the Latin *credere*, meaning "to believe" or trust.

Without credit, great businesses, including the cattle business, wouldn't make it, for few could supply all the capital needed.

Total farm assets in 1994, were estimated at $920 billion.

Total farm debt at the same time was estimated at $143 billion. This means that, in the aggregate, farmers had an 84.5% equity in their business, and 15.5% borrowed capital. Perhaps they have been too conservative, for it is estimated that ¼ to ⅓ of American farmers could profit from the use of more credit in their operations.

Credit is an integral part of today's cattle business. Wise use of it can be profitable, but unwise use of it can be disastrous. Accordingly, producers should know more about it. They need to know something about the lending agencies available to them, the types of credit, and how to go about obtaining a loan.

TYPES OF CREDIT OR LOANS

Getting the needed credit through the right kind of loan is an important part of sound financial farm management. The following three general types of agricultural credit are available, based on length of life and type of collateral needed:

- **Short-term loans**—This type of loan is made for operating expenses and is usually for one year or less. It is used for the purchase of feeders, feed, seed, fertilizer gasoline, and family living expenses. Security such as a chattel mortgage on the feeders or crop may be required by the lender; and the loan is repaid when the animals or crop are sold.

- **Intermediate-term loans**—These loans are used to buy equipment and breeding stock, for making land improvements, and for remodeling existing buildings. They are paid back in 1 to 7 years. Generally, they are secured by a chattel mortgage on livestock and machinery.

- **Long-term loans**—These loans are secured by mortgage on real estate and are used to buy land or make major improvements to farmland and buildings or to finance construction of new buildings. They may be for as long as 40 years. Usually they are paid off in regular annual or semiannual payments. The best sources of long-term loans are: an insurance company, the Federal Land Bank, the Farmers Home Administration, or an individual.

CATEGORIES OF CATTLE LOANS

There are three categories of cattle loans: (1) cow-calf loans; (2) cattle feeder loans (the financing of stocker cattle and finishing cattle is much the same); and (3) cattle on pasture loans.

COW-CALF LOANS

Many cow-calf operators use considerable borrowed capital. Their loans are for three purposes:

1. **Purchase of farm or ranch.** This is a long-term loan, with repayments made over several years.
2. **Production loans.** This refers primarily to initial loans for purchase of the breeding herd, or to loans made for the production of a crop. It also includes loans made against the breeding herd to refinance other debts, to purchase machinery and equipment, and to make minor ranch improvements. Usually, these loans are set up on an intermediate-term basis, mostly for 1 to 5 years, although they may range from 1 to 7 years. This means that the cash flow should be such that the operator can reduce the intermediate credit loan ⅕ or ⅐ each year, depending on the term.
3. **Operating budget loans.** These loans, which usually do not exceed a 12-month period, are for the

purpose of financing the recurring expenses during the year's operation. Barring disaster (drought, insects), the borrower should be able to pay back operating loans from current income.

CATTLE FEEDER LOANS

Adequate financing to carry on a cattle feeding program is extremely important. Thus, it behooves the feeder to have credit well established before jumping into cattle feeding. Consideration should be given to (1) the cash requirements for plant and equipment, for purchase of feeder cattle and feed, and for operating; and (2) the program to be followed—including kind and source of feeder cattle and feed, and market outlets. Both the financial needs and the program should be carefully conceived and put in writing. Then they should be reviewed by and discussed with a lender who has an understanding of cattle feeding and in whom the feeder has confidence. Once agreed upon, both the feeder and the lender should adhere to the program.

Many cattle feeders make use of long-term and intermediate credit for the purchase of land, the construction of the feedlot and mill, and the purchase of equipment. Additionally, cattle feeder loans fall into the following four categories:

1. **Loans to farmer-feeders.** These feeders generally grow most of their feeds. Frequently, they also produce part, or all, of their calves. They generally buy protein supplements, minerals, and feed additives. Usually, loans to farmer-feeders are modest in size and on a short-term basis.
2. **Revolving loans to year-round feeders.** These feeders maintain cattle in the feedlot the year round. As finished cattle are sold, feeders are brought in. Revolving loans (also referred to as line of credit) are needed for this type of operation.
3. **Loans to short-term feeders.** These feeders buy a certain number of cattle, feed them out, and then pay their loans in full. Short-term loans serve this need.
4. **Loans to grower-feeders.** These are generally calf-type programs, in which the cattle are grown to 600 to 700 lb weight, on relatively high-roughage rations, then sold to finishers. These feeders either produce or buy roughages, then buy calves to utilize the feed. Some grower-feeders operate on a year-round basis and require revolving loans.

Buying feeder calves involves less risk than finishing yearlings, simply because less money is tied up in cattle and more in feed. When the purchase price of animals is included, feed costs make up about 40% of the total cost of calves and about 35% of the total cost of yearlings.

The following pointers are pertinent to each of the above categories of cattle feeder loans:

- Many feedyard operators make the mistake of overcapitalizing in feedlot facilities and equipment, with the result that they have insufficient working capital to operate. Generally speaking, financial institutions require a margin of 20 to 25% of the feeder cattle and feed.

  Sound operators will generally maintain a safe margin of 30 to 40% so that they can keep going after a market reversal, which could cost more than $100 per head. They will be careful not to dip into the safe margin to finance long-term fixed assets; rather, they will finance such fixed assets through either (1) current margins in excess of their safe margins, or (2) intermediate or long-term loans.

- Lenders normally lend up to 100% of the purchase price of cattle if the borrower can cover feed and operating expenses.

- Lenders look upon the following as highly important factors when considering a feedlot loan: the ability, experience, and past record of the feeder; feeder's knowledge of cattle markets, and of buying and selling cattle; adequacy of the feedlot and facilities; availability and source of cattle and feed; and adequacy of margin, and the liquid assets of the operation.

- Feedlot expansion and improvement should be tied directly to the amount of the margin in the livestock and to the net worth of the feeder. In some cases, cattle equities are used for new homes, cars, and travel (for what is known as consumption purposes), with the result that insufficient money is available to operate on or to meet loans, particularly when some adversity strikes. As a result, the lender has no alternative but to stop financing the feeder.

CATTLE ON PASTURE LOANS

Pasture loans are very similar to cattle feeder loans. They are usually made in the spring and paid off in the fall.

Lenders will usually advance 75 to 80% of the cost of stockers for going on pasture provided the borrower (1) can put up the rest, and (2) has ample pasture and capital for operating expenses.

CREDIT FACTORS CONSIDERED AND EVALUATED BY LENDERS

Potential money borrowers sometimes make their first big mistake by going in *cold* to see a lender, without adequate facts and figures, with the result that they already have two strikes against them.

Fig. 13-5. Cattle producers can usually borrow up to 75 to 80% of the cost of stockers for pasture finishing. (Courtesy, USDA)

When considering and reviewing cattle loan requests, the lender tries to arrive at the repayment ability of the potential borrower. Likewise, the borrower has no reason to obtain money unless it will make money.

Lenders need certain basic information in order to evaluate the soundness of a loan request. To this end, the following information should be submitted:

1. **Analysis and feasibility study.** Lenders are impressed with borrowers who have a written-down program; showing where they are now, where they are going, and how they expect to get there. In addition to spelling out the goals, this should give assurance of the necessary management skills to achieve them. Such an analysis of the present and a projection into the future is imperative in big operations.

2. **About the applicant.** The applicant should furnish the following information:

   a. Name of applicant and spouse; age of applicant

   b. Number of children (minors, legal age)

   c. Partners in business, if any

   d. Years in area

   e. Experience—practical and educational

   f. Estate planning (will, trust, etc.)

   g. References

3. **About the present farm or ranch, if any.** If a loan is being obtained for the operation or expansion of the present ranch, the following information should be provided:

   a. Owner or tenant

   b. Location; legal description and county, and direction and distance from nearest town

   c. Type of ranch enterprise; cow-calf, feedlot, etc.

4. **Financial statement.** This document indicates the borrower's financial record and current financial position, potential ahead, and liability to others. The net worth statement records the financial status of a business at a particular point in time, whereas the profit and loss statement (P&L) measures a flow-through time. Borrowers should always have sufficient solvency to absorb reasonable losses due to such unforeseen happenstances as storms, droughts, diseases, and poor markets; thereby permitting lenders to stay with them in adversity and to give them a chance to recoup their losses in the future.

5. **Profit and Loss (P&L) statement.** *The profit and loss statement (P&L) is a measure of the income generated, and the cost incurred, during a specific period of time—usually one year.* It is completed at the end of the farm business year to arrive at actual returns for the year. To prepare an estimated profit and loss statement for the coming year, you need to estimate your farm sales, operating expenses, depreciation, and net inventory change. It is much better to know what is expected and change plans if necessary, than to wait until the end of the year and discover that your farm business did not make a satisfactory profit.

   The P&L statement serves as a valuable guide to the potential ahead; hence, it is important to the lender. Preferably, it should cover the previous 3 years. Also, most lenders prefer that the P&L statement be on an accrual basis (even if the producer is on a cash basis in reporting to the Internal Revenue Service).

6. **The potential borrower.** Most lenders agree that the potential borrower is the most important part of the loan.

   Lenders consider the borrower's—

   a. Character

   b. Honesty and integrity

   c. Experience and ability

   d. Moral and credit rating

   e. Age and health

   f. Family cooperation

   g. Continuity, or line of succession

   Lenders are quick to sense the *high livers*—those who live beyond their means; the poor manager—the kind who would have made it except for hard luck, and to whom the hard luck happened many times; and the dishonest, lazy, and incompetent.

   In recognition of the importance of the person back of the loan, *key person* insurance on the owner or manager should be considered by both the lender and the borrower.

7. **Production records.** This refers to a good set of records showing efficiency of production. Such records should show weight and price of products sold, calf-crop percentage and weaning weight, efficiency of feed utilization and rate of gain on feedlot cattle, age of livestock, heifer replacement program, depreciation schedule, average crop yield, and other pertinent information. Lenders will increasingly insist on good records.

8. **Progress with previous loans.** Has the borrower paid back previous loans plus interest; has the borrower reduced the amount of the loan, thereby giving evidence of progress?

9. **Physical plant:**
   a. Is it an economic unit?
   b. Does it have adequate water, and is it well balanced in feed and livestock?
   c. Is there adequate diversification?
   d. Is the right kind of livestock being produced?
   e. Are the right crops and varieties grown; and are approved methods of tillage and fertilizer practices being followed?
   f. Is the farmstead neat and well kept?

10. **Collateral (or security):**
    a. Adequate to cover loan, with margin
    b. Quality of security:
       (1) Grade and age of livestock
       (2) Type and condition of machinery
       (3) If grain storage is involved, adequacy to protect from moisture and rodents
       (4) Government participation
    c. Identification of security:
       (1) Brands, ear tags, tattoo marks of livestock
       (2) Serial numbers on machinery

THE LOAN REQUEST

Farmers and ranchers are in competition for money from urban business people. Hence, it is important that their request for a loan be well presented and supported. The potential borrower should tell the purpose of the loan; how much money is needed, when it's needed, and what it's needed for; the soundness of the venture; and the repayment schedule.

Here is a hypothetical example of the kind of loan request that a lender appreciates:

> In 1995, the Bar-None Ranch consisted of 2,000 acres and a 200-head cow herd, all unencumbered. Beginning in 1996, it was desired to expand the cow herd to 500 head, and to establish a 1,000-head cattle finishing operation, with a master plan for the feedlot to go to 2,000 head should the economics of the operation so warrant. Except for replacement heifers, it was proposed that the annual calf crop would be wintered over as stockers, grazed and grained on irrigated pastures from May 15 through September, then put in the dry lot October 1 and finished for market before Christmas. Further, it was proposed that the purchased stockers would be handled similar to the stockers raised—that is, purchased in October and marketed the next December. How sound is the proposal, and how much capital will be needed, for what will it be needed, and when will it be needed?

Step by step, here is the answer:

1. **Soundness.** This requires an analysis of: (a) the cow-calf enterprise (see Table 13-2), and (b) the feedlot operation (see Table 13-3). They are analyzed separately because each enterprise should stand on its own.

2. **Milling considerations.** Before feed milling and storage facilities are purchased, it is necessary to arrive at the quantity of feed needed and the method of processing it (see Table 13-4).

3. **Feedlot, mill, and equipment costs.** Next, the feedlot, mill, and equipment costs should be determined (see Table 13-5).

4. **Annual cash expense.** This should be pre-

TABLE 13-2
COW-CALF OPERATIONS ON BAR-NONE RANCH

(per cow basis; total for year)

Basis: 500-head cow herd	
Investment/acre/animal unit (one mature cow) in land and improvement (real estate)	$ 2,400.00
Cost to produce a 400-lb *(181-kg)* calf to weaning age	$ 200.00
One calf, 400 lb *(181 kg)*, sold @ 70¢/lb	$ 280.00
Management income/calf	$ 80.00
A total of 450 calves @ $80.00	
Total management income	$36,000.00

TABLE 13-3
FEEDLOT OPERATION ON BAR-NONE RANCH

	On Calves Raised	On Calves Bought
(per head basis; total for year)		
Expenses:		
Initial cost	$ 200.00	$ 260.00[1]
Feed and pasture as stocker	80.00	80.00
Feed for finishing	180.00	180.00
Nonfeed costs	35.00	35.00
	$ 495.00	$ 555.00
Miscellaneous costs	10.00	10.00
Total cost to produce	$ 505.00	$ 565.00
Income/steer[2]	$ 682.50	$ 682.50
For management	177.50	117.50
Total for management:		
360 head @ $117.50	$63,900.00	
640 head @ $117.50		$75,200.00

[1]400-lb *(181-kg)* calves @ 65¢/lb *(29.5¢/kg)*.

[2]1,050 lb *(477 kg)* @ 65¢/lb *(29.5¢/kg)*.

TABLE 13-4
MILLING CONSIDERATIONS ON BAR-NONE RANCH

For stockers:	
Two lb *(0.9 kg)* daily, 7 mo., 500 lb *(227 kg)*/animal for 1,000 animals	250 tons
For finishing:	
80% grain and 20% hay + protein supplement, or 21 lb *(10 kg)* grain/head/day, 75 days, 1,575 lb *(716 kg)* (rounded 3/4 ton)/animal for 1,000 animals	750 tons
Total	1,000 tons
If proper steam roller mill increases feeding value $10.00/ton, that's	$10,000/year
Daily capacity:	
For 1,000 stockers and 1,000 finishers	12 tons/day
For 2,000 stockers and 2,000 finishers	24 tons/day
Storage for 50 tons protein supplement/year (but storage for 2 months will be adequate)	10 tons

sented by the borrower in tabular form, using a table similar to Table 13-6.

5. **Annual cash income budget.** This should be presented by the borrower in tabular form, using a table similar to Table 13-7.

6. **Annual cash expense and income budget (cash flow).** This should be presented by the borrower in tabular form, using a table similar to Table 13-8. This

TABLE 13-5
FEEDLOT, MILL, AND EQUIPMENT COSTS ON BAR-NONE RANCH

Mill	$ 30,000
Feedlot (corrals, bunks, apron, scales, and waterers)	20,000
Storage	18,000
Truck and box	18,000
Tractor and loader	18,000
Total	$104,000

TABLE 13-6
ANNUAL CASH EXPENSE BUDGET[1]

________________________ for 19_____
(name of farm or ranch)

Item	Total	Jan.	Feb.	Mar.	Apr.	May	June	July	Aug.	Sept.	Oct.	Nov.	Dec.
Labor hired													
Feed purchased													
Gas, fuel, grease													
Taxes													
Insurance													
Interest													
Utilities													
etc.													
Total													

[1]The Annual Cash Expense Budget should show the monthly breakdown of various recurring items—everything except the initial loan and capital improvements. It includes labor, feed, supplies, fertilizer, taxes, interest, utilities, etc.

TABLE 13-7
ANNUAL CASH INCOME BUDGET[1]

________________________ for 19_____
(name of farm or ranch)

Item	Total	Jan.	Feb.	Mar.	Apr.	May	June	July	Aug.	Sept.	Oct.	Nov.	Dec.
500 steers													
430 bu oats													
etc.													
Total													

[1]The Annual Cash Income Budget is just what the name implies—an estimated cash income by months.

TABLE 13-8
ANNUAL CASH EXPENSE AND INCOME BUDGET (Cash Flow)[1]

______________________________ for 19_____

(name of farm or ranch)

Item	Total	Jan.	Feb.	Mar.	Apr.	May	June	July	Aug.	Sept.	Oct.	Nov.	Dec.
Gross income													
Gross expense													
Difference													
Surplus (+) or Deficit (–)	+					+	–						

[1]The Annual Cash Expense and Incme Budget is a cash flow budget, obtained from the first two forms. It's a money "flow" summary by months. From this can be ascertained when, and how much, money will need to be borrowed, the length of the loan and a suitable repayment schedule. It makes it possible to avoid tying up capital unnecessarily, and to avoid unnecessary interest.

shows, by months, the estimated expenses and income, and the particular months when money will need to be borrowed—and how much.

CREDIT FACTORS CONSIDERED BY BORROWERS

Credit is a two-way street; it must be good for both the borrower and the lender. If a borrower is the right kind of person and on a sound basis, more than one lender will want the business. Thus, it is usually well that borrowers shop around a bit; that they be familiar with several sources of credit and see what they have to offer. There are basic differences in length and type of loan, repayment schedules, services provided with the loan, interest rate, and the ability and willingness of lenders to stick by the borrower in emergencies and times of adversity. Thus, interest rates and willingness to loan are only two of the several factors to consider. If at all possible, all borrowing should be done from one source; a one-source lender will know more about the borrower's operations and be in a better position to help. Lenders strongly discourage *split borrowings*, and, in most cases, will not lend under such conditions.

HELPFUL HINTS FOR

BUILDING AND MAINTAINING

A GOOD CREDIT RATING

Cattle producers who wish to build up and maintain good credit are admonished to do the following:

1. **Keep credit in one place, or in a few places.** Generally, lenders frown upon "split financing." Borrowers should shop around for creditors (a) who are able, willing, and interested in extending the kind and amount of credit needed, and (b) who will lend at a reasonable rate of interest; then stay with them.
2. **Get the right kind of credit.** Don't obtain short-term credit to finance long-term improvements or other capital investments. Also, use the credit for the purpose intended.
3. **Be frank with the lender.** Be completely open and aboveboard. Mutual confidence and esteem should prevail between borrower and lender.
4. **Keep complete and accurate records.** Complete and accurate records should be kept by enterprises. By knowing the cost of doing business, decision-making can be on a sound basis.
5. **Keep annual inventory.** Take an annual inventory for the purpose of showing progress made during the year.
6. **Repay loans when due.** Borrowers should work out a repayment schedule on each loan, then meet payments when due. Sale proceeds should be promptly applied on loans.
7. **Plan ahead.** Analyze the next year's operation and project ahead.

BORROW MONEY TO MAKE MONEY

Cattle producers should never borrow money unless they are reasonably certain that it will make or save money. With this in mind, borrowers should ask, "How much should I borrow?" rather than, "How much will you lend me?"

CALCULATING INTEREST

The charge for the use of money is called interest. The basic charge varies from year to year and is strongly influenced by the following:

1. The *basic cost* of money in the money market.

2. The *servicing costs* of making, handling, collecting, and keeping necessary records on loans.
3. The *risk* of loans.

Interest rates vary among lenders and can be quoted and applied in several different ways. The quoted rate is not always the basis for proper comparison and analysis of credit costs. Even though several lenders may quote the same interest rate, the effective or simple annual rate of interest may vary widely. The more common procedures for determining the actual annual interest rate, or the equivalent of simple interest on the unpaid balance, follow.

1. **Simple or true annual interest on the unpaid balance.** A $1,200 note payable at maturity (12 months) with 12% interest:

Interest paid . . . 0.12 × $1,200 = $144

Average use of the money $1,200 for the entire year

Actual rate of interest . . . . . . $\frac{\$144 \text{ (interest)}}{\$1{,}200 \text{ (used for 1 year)}} = 12\%$

2. **Installment loan (with interest on unpaid balance).**[1] A $1,200 note payable in 12 monthly installments with 12% interest on the unpaid balance:

Interest paid ranges from:

First month . . . . $\frac{0.12 \times \$1{,}200}{12} = \$12$

to

Twelfth month . . $\frac{0.12 \times \$100}{12} = \$1$

Total for 12 months is $78

Average use of the money ranges from $1,200 for the first month down to $100 for the twelfth month, an average of $650 for 12 months.

Effective rate of interest . . . $\frac{\$78}{\$650} = 12\%$

3. **Add-on installment loan (with interest on face amount).** A $1,200 note payable in 12 monthly installments with 12% interest on face amount of loan:

Interest paid . . . 0.12 × $1,200 = $144

Average use of the money ranges from $1,200 for the first month down to $100 for the twelfth month, an average of $650 for 12 months.

Effective rate of interest . . . $\frac{\$144}{\$650} = 22.15\%$

4. **Points and interest.** Some lenders now charge *points*. A point is 1% of the face value of the loan. Thus, if 4 points are being charged on a $1,200 loan, $48 will be deducted and the borrower will receive only $1,152. But the borrower will have to repay the full $1,200. Obviously, this means that the actual interest rate will be more than the stated rate. But how much more.

Assume that a $1,200 loan is for 1 year and the annual rate of interest is 12%. Then the payment by the borrower of 4 points would make the actual interest rate as follows:

Interest . . . . . 0.12 × $1,200 = $144

Average use of the money . . . . $1,152 for one year

Effective rate of interest . . . . . $\frac{\$144 \text{ (interest)}}{\$1{,}152 \text{ (used for 1 year)}} = 12.5\%$

5. **If interest is not stated, use this formula to determine the effective annual interest rate:**

$$\text{Effective rate of interest} = \frac{\text{Number of payment periods in 1 year}^{2} \times 2 \times \text{Finance charges}^{3}}{\text{Balance owed}^{4} \times \text{Number of payments in contract plus one}}$$

For example, a store advertises a refrigerator for $500. It can be purchased on the installment plan for $80 down and monthly payments of $35 for 12 months. What is the actual rate of interest if you buy on the time payment plan?

$$\text{Effective rate of interest} = \frac{2 \times 12 \times \$35}{\$420 \times (12 \text{ plus } 1)} = \frac{\$840}{\$5{,}460} = 15.4\%$$

HIDDEN CHARGES

Before using credit, the borrower should thoroughly understand *all* the costs of borrowing. Many of

[1]This method is used for amortized loans.

[2]Regardless of the total number of payments to be made, use 12 if the payments are monthly, use 6 if payments are every other month, or use 2 if payments are semiannual.

[3]Use either the time payment price less the cash price, or the amount you pay the lender less the amount you receive if negotiating for a loan.

[4]Use cash price less down payment or, if negotiating for a loan, the amount you receive.

these costs are hidden; hence, it is important to investigate, ask questions, and consider every expense connected with the use of credit. Some lenders, for example, require the purchase of stock based on a percentage of the loan; others require minimum balances; still others pay commissions or *finder's fees* for loans. All these things, and more, point up the importance of comparing actual dollar charges when shopping for credit.

MANAGEMENT

Fortunes have been made and lost in the beef cattle industry. Although it is not possible to arrive at any overall, certain formula for success, in general those operators who have made money have paid close attention to the details of management—they have been good managers.

The almost innumerable beef cattle management practices vary widely between both areas and individual farmers and ranchers. In a general sort of way, however, the principles of good management are much alike. The main differences arise from the sheer size of a big cattle enterprise, which means that things must be done in a big way.

TRAITS OF A GOOD CATTLE MANAGER

There are established bases for evaluating many articles of trade, including cattle, hay, and grain. They are graded according to well-defined standards. Additionally, we chemically analyze feeds and conduct feeding trials. But no such standard or system of evaluation has evolved for cattle managers, despite their acknowledged importance.

The senior author has prepared the Cattle Manager Checklist, given in Table 13-9, which (1) employers may find useful when selecting or evaluating a manager, and (2) managers may apply to themselves for self-improvement purposes. No attempt has been made to assign a percentage score to each trait, because this will vary among cattle establishments.

TABLE 13-9
CATTLE MANAGER CHECKLIST

☐ **CHARACTER—**
Absolute sincerity, honesty, integrity, and loyalty; ethical.

☐ **INDUSTRY—**
Work, work, work; enthusiasm, initiative, and aggressiveness.

☐ **ABILITY—**
Cattle know-how and experience, business acumen—including ability systematically to arrive at the financial aspects and convert this information into sound and timely management decisions, knowledge of how to automate and cut costs, common sense, organized, growth potential.

☐ **PLANS—**
Sets goals, prepares organization chart and job description, plans work, and works plans.

☐ **ANALYZES—**
Identifies the problem, determines pros and cons, then comes to a decision.

☐ **COURAGE—**
To accept responsibility, to innovate, and to keep on keeping on.

☐ **PROMPTNESS AND DEPENDABILITY—**
A self-starter, has "T.N.T."; does it "today, not tomorrow."

☐ **LEADERSHIP—**
Stimulates subordinates, and delegates responsibility.

☐ **PERSONALITY—**
Cheerful, not a complainer.

JOB DESCRIPTIONS on Bar-None Ranch

Owner	Manager	Cow-Calf Foreman	Cattle Feedlot Foreman
Responsible for:	Responsible for:	Responsible for:	Responsible for:
1. Selecting management. 2. Making policy decisions. 3. Borrowing capital. 4. (List others.)	1. Supervising all staff. 2. Preparing proposed longtime plan. 3. Budgets. 4. (List others.)	1. Directing cow-calf staff. 2. Selecting and culling. 3. Breeding cows, including AI work. 4. Feeding the herd. 5. Calving. 6. Branding (marking), dehorning, castrating, vaccinating. 7. Herd health. 8. Preconditioning. 9. Marketing calf crop. 10. (List others.)	1. Directing feedlot staff. 2. Buying and selling cattle. 3. Processing incoming cattle. 4. Animal health. 5. Feedlot rations. 6. Feeding. 7. (List others.)

Fig. 13-6. Job descriptions.

Rather, it is hoped that this checklist will serve as a useful guide (1) to the traits of a good manager, and (2) to what the boss wants.

ORGANIZATION CHART AND JOB DESCRIPTION

It is important that all workers know to whom they are responsible and for what they are responsible; and the bigger and the more complex the operation, the more important this becomes. This should be written down in an organization chart and a job description.

AN INCENTIVE BASIS FOR THE HELP

Big farms and ranches must rely on hired labor, all or in part. Good help—the kind that everyone wants—is hard to come by; it's scarce, in strong demand, and difficult to keep. And the farm labor situation is going to become more difficult in the years ahead. There is need, therefore, for some system that will (1) give a big assist in getting and holding top-flight help, and (2) cut costs and boost profits. An incentive basis that makes hired help partners in profit is the answer.

Many manufacturers have long had an incentive basis. Executives are frequently accorded stock option privileges, through which they prosper as the business prospers. Common laborers may receive bonuses based on piecework or quotas (number of units, pounds produced). Also, most factory workers get overtime pay and have group insurance and a retirement plan. A few industries have a true profit-sharing arrangement based on net profits as such, a specified percentage of which is divided among employees. No two systems are alike. Yet, each is designed to pay more for labor, provided labor improves production and efficiency. In this way, both owners and laborers benefit from better performance.

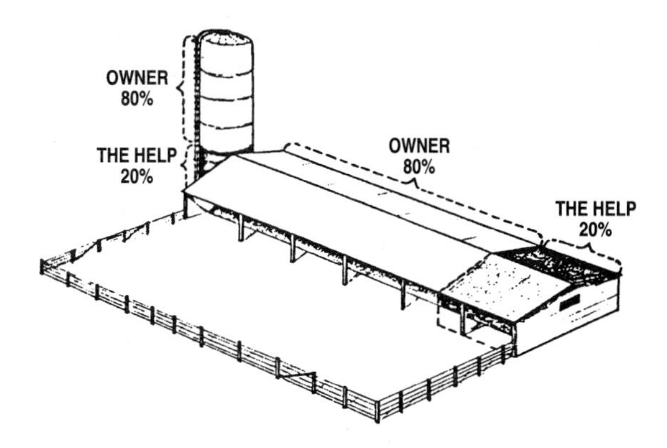

Fig. 13-7. The senior author favors an 80/20% split of the net between the owner and the help, provided several helpers are involved.

Family-owned and family-operated farms have a built-in incentive basis; there is pride of ownership, and all members of the family are fully cognizant that they prosper as the business prospers.

Many different incentive plans can be, and are, used. There is no best one for all operations. The various plans given in Table 13-10 are intended as guides only.

TABLE 13-10
INCENTIVE PLANS FOR BEEF CATTLE ESTABLISHMENTS

Types of Incentives	Pertinent Provisions of Some Known Incentive Systems in Use	Advantages	Disadvantages	Comments
1. Bonuses	A flat, arbitrary bonus; at Christmas time, year-end, or quarterly or other intervals. A tenure bonus such as (1) 5 to 10% of the base wage or 2 to 4 weeks additional salary paid at Christmas time or year-end, (2) 2 to 4 weeks vacation with pay, depending on length and quality of service, or (3) $10 to $20/week set aside and to be paid if employee stays on the job a specified time.	It's simple and direct.	Not very effective in increasing production and profits.	
2. Equity-building plan	Employee is allowed to own a certain number of animals. In cow-calf operations, these are usually fed without charge.	It imparts pride of ownership to the employee.	The hazard that the owner may feel that employees accord their animals preferential treatment; suspicioned if not proved. It is difficult to control animal numbers, feed, etc.	

(Continued)

TABLE 13-10 (Continued)

Types of Incentives	Pertinent Provisions of Some Known Incentive Systems in Use	Advantages	Disadvantages	Comments
3. Production sharing	$2 to $6/calf weaned $1/cwt of gain on feeder cattle. 50¢ to $2/head on fed cattle marketed.	It is an effective way to achieve higher production.	Net returns may suffer. For example, a higher rate of gain than is economical may be achieved by feeding stockers more concentrated and expensive feeds than are practical. This can be alleviated by (1) specifying the ration, and (2) setting an upper limit on the gains to which the incentive will apply. If a high performance level already exists, further gains or improvements may be hard to come by.	Incentive payments for production above certain levels—for example, above 450 lb *(205 kg)* calf weaned/cow bred—are more effective than paying for all units produced.
4. Profit sharing				
a. Percent of gross income	1 to 2% of the gross.			
b. Percent of net income	10 to 20% of the net after deducting all costs.	Net income sharing works better for managers, supervisors, and foremen than for common laborers, because fewer hazards are involved in opening up the books to them. It is an effective way to get hired help to cut costs. It is a good plan for a hustler.	Percent of gross does not impart cost of production consciousness. Both (1) percent of gross income, and (2) percent of net income expose the books and accounts to workers, who may not understand accounting principles. This can lead to suspicion and distrust. Controversy may arise (1) over accounting procedure—for example, from the standpoint of the owner a fast tax write-off may be desirable on new equipment, but this reduces the net shared with the worker; and (2) because some owners are prone to overbuild and overequip, thereby decreasing net. It is difficult to define what constitutes net income.	There must be prior written agreement on what constitutes gross or net receipts, as the case may be, and how it is figured.
5. Production sharing and prevailing price.	See Chapter 24, section entitled, "An Incentive Basis for the Help." See Chapter 27, section entitled, "An Incentive Basis for the Help."	It embraces the best features of both production sharing and profit sharing, without the major disadvantages of each. It (1) encourages high productivity and likely profits, (2) is tied in with prevailing prices, (3) does not necessitate opening the books, and (4) is flexible—it can be split between owner and employee on any basis desired, and the production part can be adapted to a sliding scale or escalator arrangement—for example, the incentive basis can be higher for the ¼ lb *(1/10 kg)* of feedlot gain made in excess of 2¾ lb *(1.3 kg)* than for a ¼ lb *(1/10 kg)* gain in excess of 2¼ lb *(1.0 kg)*.	It is a bit more complicated than some other plans, and it requires more complete records.	When properly done, and all factors considered, this is the most satisfactory incentive basis for a cattle enterprise, for both cow-calf and feedlot.

The incentive basis chosen should be tailored to fit the specific operation; with consideration given to kind and size of operation, extent of owner's supervision, present and projected productivity levels, mechanization, and other factors.

For most cow-calf and cattle feedlot operations, the senior author favors a "production sharing and prevailing price" type of incentive.

HOW MUCH INCENTIVE PAY?

After (1) reaching a decision to go on an incentive basis, and (2) deciding on the kind of incentive, it is necessary to arrive at how much to pay. Here are some guidelines that may be helpful in determining this:

1. Pay the going base, or guaranteed, salary; then add the incentive pay above this.
2. Determine the total stipend (the base salary plus incentive) to which you are willing to go.
3. Before making any offers, always check the plan on paper to see (a) how it would have worked out in past years based on your records, and (b) how it will work out as you achieve the future projected production.

Let's take the following example:

> A foreman of a 500-cow herd is now producing an average of 400 lb of calf weaned per cow bred. He is receiving a base salary of $1,000/month plus house, garden, and 600 lb of dressed beef/year. The owner prefers a *production sharing and prevailing price* type of incentive.

Step by step, here is the procedure for arriving at an incentive arrangement based on increased production:

1. By checking with local sources, it is determined that the present salary of $1,000/month plus *extras* is the going wage; and, of course, the foreman receives this regardless of what the year's calf production or price turns out to be—it's guaranteed.
2. A study of the cow-calf records reveals that with a little extra care on the part of the foreman—particularly in pregnancy testing, at calving time, and in rotating pastures—the average weaning weight of calves per cow bred can be boosted enough to permit paying $1,300/month, or $300/month more than is now being paid. That's $3,600 more per year. This can be fitted into the incentive plan.
3. An average increase of 60 lb of calf weaned per cow bred at 60¢/lb would mean $36.00/cow, or $15,000 on a 500-cow herd. With an 80:20 split between owner and manager, the foreman would get $3,000 or $250/month.

REQUISITES OF AN INCENTIVE BASIS

Owners who have not previously had experience with an incentive basis are admonished not to start with any plan until they are sure of both their plan and their help. Also, it is well to start with a conservative plan; then a change can be made to a more inclusive and sophisticated plan after experience is acquired.

Regardless of the incentive plan adopted for a specific operation, it should encompass the following essential features:

1. Good owner (or manager) and good workers. No incentive basis can overcome a poor manager. A good manager must be a good supervisor and fair to the help. Also, on big establishments, there must be a written-down organization chart and job description so the help know (a) to whom they are responsible, and (b) for what they are responsible. Likewise, no incentive basis can spur employees who are not able, interested, and/or willing. This necessitates that employees must be selected with special care where they will be on an incentive basis. Hence, the three—good owner (manager), good employees, and good incentive—go hand in hand.
2. It must be fair to both employer and employees.
3. It must be based on and make for mutual trust and esteem.
4. It must compensate for extra performance, rather than substitute for a reasonable base salary and other considerations (house, utilities, and certain provision).
5. It must be as simple, direct, and easily understood as possible.
6. It should compensate all members of the team; from cowboys to manager on a cow-calf outfit, and from feeders and feed processors to manager in a cattle feedlot.
7. It must be put in writing, so that there will be no misunderstanding. If some production-sharing plan is used in a cattle feedlot, it should stipulate the ration (or who is responsible for ration formulation), the maximum gain of stocker cattle, and the grade to which finishing cattle are to be carried. On a cow-calf outfit, it should stipulate the ration, the culling of cows, and other pertinent factors.
8. It is preferable, although not essential, that workers receive incentive payments (a) at rather frequent intervals, rather than annually, and (b) immediately after accomplishing the extra performance.
9. It should give the hired help a certain amount of responsibility, from the wise exercise of which they will benefit through the incentive arrangements.
10. It must be backed up by good records; other-

wise, there is nothing on which to base incentive payments.

11. It should be a two-way street. If employees are compensated for superior performance, they should be penalized (or, under most circumstances, fired) for poor performance. It serves no useful purpose to reward the unwilling, the incompetent, and the stupid. For example, no overtime pay should be given to employees who must work longer because of slowness or correcting mistakes of their own making. Likewise, if the reasonable break-even point on a cow-calf operation is an average of a 400-lb calf weaned per cow bred, and this production level is not reached because of obvious neglect (for example, not being on the job at calving time), the employee(s) should be penalized (or fired).

INDIRECT INCENTIVES

Normally, we think of incentives as monetary in nature—as direct payments or bonuses for extra production or efficiency. However, there are other ways of encouraging employees to do a better job. The latter are known as indirect incentives. Among them are: (1) good wages; (2) good labor relations; (3) adequate house plus such privileges as the use of the farm truck or car, payment of electric bill, use of a swimming pool, hunting and fishing, use of horse, and furnishing meat, milk, and eggs; (4) good buildings and equipment; (5) vacation time with pay, time off, and sick leave; (6) group health; (7) security; (8) the opportunity for self-improvement that can accrue from working for a top individual; (9) the right to invest in the business; (10) an all-expense–paid trip to a short course, show, or convention; and (11) year-end bonus for staying all year. These indirect incentives will be accorded to the help of more and more establishments, especially the big ones.

ANNUAL INVENTORY

The annual inventory is the most valuable record that a cattle producer can keep. It gives the most complete statement of financial conditions. A complete farm inventory is usually necessary before any other kind of records or accounts can be kept and analyzed; it should be considered the first and most important record to be assembled and used by all producers. The inventory should include a list and value of real estate, livestock, equipment, feed, supplies, and all other property, including cash on hand, notes, bills receivable and growing crops. Also, it should include a list of mortgages, notes, and bills payable. It shows farmers what they own and what they owe; whether they are getting ahead or falling behind. The following pointers may be helpful relative to the annual inventory.

1. **Time to take inventory.** The inventory should be taken at the beginning of the account year; usually this means Dec. 31 or Jan. 1.

2. **Proper and complete listing.** It is important that each item be properly and separately listed.

3. **Method of arriving at inventory values.** It is difficult to set up any hard and fast rule to follow in estimating values when taking inventories. Perhaps the following guides are as good as any:

   a. **Real estate.** Estimating the value of farm real estate is, without doubt, the most difficult of all. It is suggested that the farmer use either (1) the cost of the farm, (2) the present sale value of the farm, or (3) the capitalized rent value according to its productive ability with an average operator.

   b. **Buildings.** Buildings are generally inventoried on the basis of cost less observed depreciation and obsolescence. Once the original value of a building is arrived at, it is usually best to take depreciation on a straight line basis by dividing the original value by the estimated life in terms of years. Usually 4% or more depreciation is charged off each year for income tax purposes (which means that buildings are normally depreciated in 25 years).

   c. **Livestock.** Animals are usually not too difficult to inventory because there are generally sufficient current sales to serve as a reliable estimate of value.

   d. **Machinery.** The inventory value of machinery is usually arrived at by either of two methods: (1) the original cost less a reasonable allowance for depreciation each year; or (2) the probable price that it would bring at a well-attended auction.

   Under conditions of ordinary wear and reasonable care, it can be assumed that the general run of farm machinery (except trucks and autos) will last about 10 years. Thus, with new machinery, the annual depreciation will be the original cost divided by 10.

   e. **Feed and supplies.** The value of feed and supplies can be based on market price.

Two further points are important. Whatever method is used in arriving at inventory value (1) should be followed at both the beginning and the end of the year, and (2) should reflect the operator's opinion of the value of the property involved.

BUDGETS IN THE CATTLE BUSINESS

A budget is a projection of records and accounts and a plan for organizing and operating ahead for a specific period of time. A short-time budget is usually for one year, whereas a long-time budget is for a period of years. The principal value of a farm budget is that it provides a working plan through which the operation can be coordinated. Changes in prices, droughts, and other factors make adjustments necessary. But these adjustments are more simply and wisely made if there is a written budget to use as a reference.

HOW TO SET UP A BUDGET

It's unimportant whether a printed form (of which there are many good ones) is used or a form made up on an ordinary ruled 8½" × 11" sheet placed sidewise. The important things are: (1) that a budget is kept; (2) that it be on a monthly basis; and (3) that the operator be *comfortable* with whatever forms or system is used.

No budget is perfect. But it should be as good an estimate as can be made—despite the fact that it will be affected by such things as droughts, diseases, markets, and many other unpredictables.

A simple, easily kept, and adequate budget can be prepared by using the following three types of budget forms, all of which are shown earlier in this chapter in the section entitled, "The Loan Request":

1. Annual cash expense budget (see Table 13-6).
2. Annual cash income budget (see Table 13-7).
3. Annual cash expense and income budget—cash flow (see Table 13-8).

The annual cash expense budget should show the monthly breakdown of various recurring items—everything except the initial loan and capital improvements. It includes labor, feed, supplies, fertilizer, taxes, interest, utilities, etc.

The annual cash income budget is just what the name implies—an estimated cash income by months.

The annual cash expense and income budget is a cash flow budget, obtained from the first two forms. It's a money *flow* summary by months. From this, it can be ascertained when, and how much, money will need to be borrowed, and the length of the loan along with a repayment schedule. It makes it possible to avoid tying up capital unnecessarily, and to avoid unnecessary interest.

HOW TO FIGURE NET INCOME

Table 13-8 shows a gross income statement. There are other expenses that must be taken care of before net profit is determined; namely—

1. **Depreciation on buildings and equipment.** It is suggested that the "useful life" of buildings and equipment be as follows:

Buildings	15 years
Machinery and equipment	5 years

Sometimes a higher depreciation, or amortization, is desirable because it produces tax savings, and for protection against obsolescence due to scientific and technological developments.

2. **Interest on owner's money invested in farm and equipment.** This should be computed at the going rate in the area, say 12%.

Here's an example of how the above works:

> Let's assume that on a given farm there was a gross income of $200,000 and a gross expense of $125,000, or a surplus of $75,000 Let's also assume that there are $60,000 worth of machinery, $60,000 worth of buildings, and $200,000 of the owner's money invested in farm and equipment. Let's further assume that buildings are being depreciated in 15 years and machinery in 5 years.

Here is the result:

Gross profit					$75,000
Depreciation—					
Machinery:	$ 60,000	@	20% =	$12,000	
Buildings:	$ 60,000	@	6.67% =	4,002	
				$16,002	
Interest:	$200,000	@	12% =	$24,000	
					40,002
Return to labor and management					$34,998

Some people prefer to measure management in terms of return on invested capital, and not wages. This approach may be accomplished by paying management wages first, then figuring return on investment.

ENTERPRISE ACCOUNTS

Where a cattle enterprise is diversified—for example, a farm or ranch having a cow-calf operation, a feedlot, and crops—enterprise accounts should be kept, in this case three different accounts for three different enterprises. The reasons for keeping enterprise accounts are:

1. It makes it possible to determine which enterprises have been most profitable, and which least profitable.
2. It makes it possible to compare a given enterprise with competing enterprises of like kind, from the standpoint of ascertaining comparative performance.
3. It makes it possible to determine the profitability of an enterprise at the margin (the last unit of

production). This will give an indication as to whether to increase the size of a certain enterprise at the expense of an alternative existing enterprise when both enterprises are profitable in total.

COMPUTERS IN THE CATTLE BUSINESS

Accurate and up-to-the-minute records and controls have taken on increasing importance in all agriculture, including the cattle business, as the investment required to engage in farming and ranching has risen and profit margins have narrowed. Today's successful farmers and ranchers must have, and use, as complete records as any other business. Also, records must be kept current; it no longer suffices merely to know the bank balance at the end of the year.

Fig. 13-8. A computer suitable for a farm or ranch. (Courtesy, California State University, Fresno)

Big and complex enterprises have outgrown hand record keeping. It is too time consuming, with the result that it does not allow management enough time for planning and decision making. Additionally, it does not permit an all-at-once consideration of the complex interrelationships which affect the economic success of the business. This prompted linear programming.

Linear programming is similar to budgeting, in that it compares several plans simultaneously and chooses from among them the one likely to yield the greatest overall profit. It is a way in which to analyze a great mass of data and consider many alternatives. It is not a managerial genie; nor will it replace decision-making managers. However, it is a modern and effective tool in the present age, when just a few dollars per head can spell the difference between profit and loss.

There is hardly any limit to what computers can do if fed the proper information. Among the difficult questions that they can answer for a specific farm or ranch are:

1. **How is the entire operation doing so far?** It is preferable to obtain quarterly or monthly progress reports; often making it possible to spot trouble before it's too late.

2. **What farm enterprises are making money; which ones are freeloading or losing?** By keeping records by enterprises—cow-calf, cattle feedlot, wheat, corn—it is possible to determine strengths and weaknesses; then either rectify the situation or shift labor and capital to a more profitable operation. Through *enterprise analysis* some operators have discovered that one part of the farm business may earn $10, or more, per hour for labor and management, whereas another may earn only $2 per hour, and still another may lose money.

3. **Is each enterprise yielding maximum returns?** By having profit or performance indicators in each enterprise (see Chapter 24, Tables 24-1, -2, -3, and -4 for size and weight of calf crop weaned as profit indicators), it is possible to compare these (a) with the historical average of the same farm or ranch, or (b) with the same indicators of other farms or ranches.

4. **How does this ranch stack up with its competition?** Without revealing names, the computing center (local, state, area, or national) can determine how a given ranch compares with others—either the average, or the top (say 5%).

5. **How can you plan ahead?** By using projected prices and costs, computers can show what moves to make for the future—they can be a powerful planning tool. They can be used in determining when to plant, when to schedule farm machine use, etc.

6. **How can income taxes be cut to the legal minimum?** By keeping an accurate record of expenses and figuring depreciations accurately, computers make for a saving in income taxes on most farms and ranches.

7. **What are the *least cost* and *highest net return* rations?** Instructions on how to balance a ration by computer are given in Chapter 8 of this book; hence, the reader is referred thereto.

For providing answers to the above questions, and many more, computer accounting costs an average of about 1% of the gross farm income. By comparison, it is noteworthy that many city businesses pay double this amount.

There are three requisites for linear programming a farm or ranch; namely—

Fig. 13-9. Using computer printouts to analyze the farm/ranch business. When fed the proper information, computers can answer the seven questions posed in this section—and many more. (Courtesy, USDA)

1. Access to a computer.
2. Computer know-how, so as to set the program up properly and be able to analyze and interpret the results.
3. Good records; preferably, covering a period of years.

The pioneering computer services available to farmers and ranchers were operated by universities, trade associations, and government—most of them were on an experimental basis. Subsequently, others have entered the field, including commercial data processing firms, banks, machinery companies, feed and fertilizer companies, farm suppliers, breed registry associations, consultants, and computer specialists. Some of these are available as a *service sell* along with products, others are available on a charge basis.

A partial list of computer programs known to the authors follows: (*Note:* Without doubt, other computer programs not known to the authors exist, and new ones will evolve.)

Commercial Cow-Calf—Record Keeping

BeefUp
St. Benedict's Farm
P.O. Box 366
Waelder, TX 78959
(210) 540-4814

Chaps II
North Dakota State University Extension Office
Fargo, ND 58105
(701) 237-7397

CowBoss
Department of Animal Science
University of California
Davis, CA 95616
(916) 752-5650

PC-Cowcard
University of Nebraska Cooperative Extension
Lincoln, NE 68583
(308) 632-1245

RanchMaster-2
Agri-Soft, Inc.
P.O. Box 11231
Spring, TX 77391-1231
(800) 299-COWS

Commercial Cow-Calf—Financial

Beefpro
Extension Animal Science
Kansas State University
Manhattan, KS 66506-0201
(913) 532-7059

Commercial Cow-Calf SPA (Standardized Performance Analysis)
Texas Agricultural Extension Service
Department of Agricultural Economics
Texas A&M University
College Station, TX 77843-2124
(409) 845-7967

Commercial or Purebred Cow-Calf—Nutritional

Beef Cow Ration Balancer
Extension Animal Science
Kansas State University
Manhattan, KS 66506-0201
(913) 532-7059

Taurus
Department of Animal Science
University of California
Davis, CA 95616
(916) 752-5650

Purebred/Seedstock—Record Keeping

Cow-Calf Record Keeping System
Animal Science Department
Oklahoma State University
Stillwater, OK 74078-0425
(405) 744-6070

The Cattle Manager
Vertical Solutions, Inc.
203 11th Ave. SW
Minot, ND 58701
(800) 396-3279

Managing Partner
Robert C. de Baca
P.O. Box 400
Huxley, IA 50124
(515) 597-2727

RanchMaster-3
Agri-Soft, Inc.
P.O. Box 11231
Spring, TX 77391-1231
(800) 299-COWS

Many breed specific programs available from purebred cattle associations.

Purebred/Seedstock—Financial

Seedstock SPA (Standardized Performance Analysis)
Texas Agricultural Extension Service
Department of Agricultural Economics
Texas A&M University
College Station, TX 77843-2124
(409) 845-7967

Stocker/Grower

Pasture
Animal Science Department
Oklahoma State University
Stillwater, OK 74078-0425
(405) 744-6070

Feeder Cattle Breakeven Analysis
Department of Animal Science
University of Illinois
Urbana, IL 61801
(217) 333-2660

Stocker/Feeder SPA (Standardized Performance Analysis)
Texas Agricultural Extension Service
Department of Agricultural Economics
Texas A&M University
College Station, TX 77843-2124
(409) 845-7967

Feedlot—Ration Formulation

Mixit-2, Mixit-2+, and Mixit-4
Agricultural Software Consultants, Inc.
P.O. Box 32
Kingsville, TX 78364
(512) 595-1937

Master Ration Calculator
Animal Science Department
Oklahoma State University
Stillwater, OK 74078-0425
(405) 744-6070

Professional Nutritionist
University of Minnesota Extension Service
St. Paul, MN 55108
(612) 624-2703

Taurus
Department of Animal Science
University of California
Davis, CA 95616
(916) 752-5650

ANALYZING A CATTLE BUSINESS: IS IT PROFITABLE?

Most people are in business to make money—and cattle producers are people. In some areas, particularly near cities and where the population is dense, land values may appreciate so as to be a very considerable profit factor. Also, a tax angle may be important. But neither of these should be counted upon. The cattle operation should make a reasonable return on the investment; otherwise, the owner should not be in the cattle business.

A cattle owner or manager needs to analyze the business—to determine how well it is doing. With big operations, it's no longer possible to base such an analysis on the bank balance at the end of the year. In the first place, once a year is not frequent enough, for it is possible to go broke, without really knowing it, in that period of time. Secondly, a balance statement gives no basis for analyzing an operation—for ferreting out its strengths and weaknesses. In large cattle feedlots, it is strongly recommended that progress be charted by means of monthly or quarterly closings of financial records.

Also, not only must cattle producers compete with other producers down the road, but they must compete with themselves—with their records last year and the year before. They must work ceaselessly at making progress, improving the end product, and lowering costs of production.

To analyze a cattle business, two things are essential: (1) good records, and (2) yardsticks, or profit indicators, with which to measure an operation.

A profit indicator is a gauge for measuring the primary factors contributing to profit. In order for producers to determine how well they're doing, they must be able to compare their own operations with something else; for example, (1) their own historical 5-year average, (2) the average for the United States or for their particular area, or (3) the top 5%. The senior author favors the latter, for high goals have a tendency to spur superior achievements.

Like most profit indicators, the ones presented in this book (see Chapter 24, Table 24-4 for size and weight of calf crop weaned as profit indicators) are not perfect. But they will serve as useful guides. Also, on some establishments, there may be reason for adding or deleting some of the indicators; and this can be done. The important thing is that each cattle operation have adequate profit indicators, and that these be applied as frequently as possible; in a cattle feedlot, this may be done monthly with some indicators.

GUIDELINES RELATIVE TO FACILITY AND EQUIPMENT COSTS

Overinvestment is a rather common mistake. Cow-calf producers are prone to invest more in land and improvements than reasonably can be expected to make a satisfactory return; and cattle feedlot operators frequently invest too much in feed mills and equipment. Sometimes small cattle feeders fail to recognize that it may cost half as much to mechanize to feed 500 head as it costs to mechanize to feed 2,000 head.

In order to lessen over investment by the uninformed, guidelines are useful. Here are some:

1. **Guideline No. 1.** The break-even point on how much you can afford to invest in equipment to replace hired labor can be arrived at by the following formula:

$$\frac{\text{Annual saving in hired labor from new equipment}}{0.15} = \text{Amount you can afford to invest}$$

Example:

If saving in hired labor costs is $10,000 per year, this becomes—

$$\frac{\$10{,}000}{0.15} = \$66{,}667\text{, break-even point on new equipment}$$

Since labor costs are going up faster than machinery and equipment costs, it may be good business to exceed this limitation under some circumstances. Nevertheless, the break-even point, $66,667 in this case, is probably the maximum expenditure that can be economically justified at the time.

2. **Guideline No. 2.** The break-even point on new facility-equipment costs is five times the annual salary of each person replaced.

Assuming an annual cost plus operation of power machinery and equipment equal to 20% of new cost, the break-even point to justify replacement of one hired person is as follows:

If Annual Cost of One Hired Person Is—	The Break-even Point on New Investment is—
$ 8,000 (20%) × 5	$40,000
10,000 (20%) × 5	50,000
12,000 (20%) × 5	60,000

In the above figures, it is assumed that the productivity of laborers at different salaries is the same, which may or may not be the case.

Example:

Assume that the new cost of added equipment comes to $10,000, that the annual cost is 20% of this amount, and that the new equipment would save 2 hours of labor per day for 6 months of the year. Here's how to figure the value of labor to justify an expenditure of $10,000 for this item:

$10,000 (new cost) × 20% = $2,000, which is the annual ownership use cost

$2,000 ÷ 360 hours (labor saved) = $5.56/hour

So, if labor costs less than $5.56/hour, you probably shouldn't buy the new item.

CATTLE INSURANCE[5]

The ownership of cattle (or any kind of animal) constitutes a risk and a chance of financial loss. Unless the owner is in such strong financial position that he/she can assume this financial risk, the animal should be insured.

Several good companies write livestock insurance. In general, their policies and rates do not differ greatly.

The beef and dairy cattle rates of American Live Stock Insurance Company, Geneva, Illinois, which is rated A+ (superior) by A. M. Best Company, the independent rating service, follow:

[5]This section was authoritatively reviewed by Frank Harding and Duncan Alexander, American Live Stock Insurance Company, 200 South Fourth St., P.O. Box 520, Geneva, IL 60134-0520.

Conditions	Rate
Age limits, 3 mo—7 yr	
15-day term	$1.50/$100
(A 15-day policy may be endorsed "15-day cover to have its effective starting date at time of actual shipment.")	
1-mo term	$2.50/$100
2-mo term	$3.00/$100
3-mo term	$3.50/$100
6-mo term	$4.00/$100
1-yr term	$6.00/$100
Age exceptions—note added premium	
Calves, 2–7 wk old	$4.00/$100
Calves, 7 wk–3 mo old	$2.00/$100
(This additional premium to be added to period coverage and considered earned in entirety when written.)	

Bulls past 6, up to eighth birthday, eligible for insurance after amount of cover has been confirmed by company. One dollar per hundred additional premium charged for each year or part over seventh birthday. *At ninth birthday, insurance not available.*

Cows past 7 may be covered in the same manner. *At tenth birthday annual insurance not available.*

Generally, special stipulations and rates apply to (1) group (herd) insurance for six or more head of cattle, and (2) 4-H and FFA calves. For information relative to these, or other special types of coverage, the owner should make inquiry of a livestock insurance agent.

In order to obtain insurance, the following information is generally required: Name, registry number, ear tag or tattoo number, breed, sex, date of birth, amount to be insured for and period of insurance required, and a statement of health examination (made not more than 5 days prior to insuring) by an approved federal or state veterinarian to the effect "that the animal(s) (referring to it by name) is at the time of applying for insurance in a state of good physical health and condition."

LIABILITY[6]

Most farmers are in such financial position that they are vulnerable to damage suits. Moreover, the number of damage suits arising each year is increasing at an almost alarming rate, and astronomical damages are being claimed. Studies reveal that about 95% of the court cases involving injury result in damages being awarded.

Comprehensive personal liability insurance protects a farm operator who is sued for alleged damages suffered from an accident involving his/her property or family. The kinds of situations from which a claim might arise are quite broad, including suits for personal injuries caused by animals, equipment, or personal acts.

Both workers' compensation insurance and employer's liability insurance protect farmers against claims or court awards resulting from injury to hired help. Workers' compensation usually costs slightly more than straight employer's liability insurance, but it carries more benefits to the worker. An injured employee must prove negligence by the employer before the company will pay a claim under employer's liability insurance, whereas workers' compensation benefits are established by state law, and settlements are made by the insurance company without regard to who was negligent in causing the injury. Conditions governing participation in workers' compensation insurance vary among the states.

WORKERS' COMPENSATION

Workers' compensation laws, now in full force in every one of the 50 states, cover on-the-job injuries and protect disabled workers regardless of whether their disabilities are temporary or permanent. Although broad differences exist among the individual states in their workers' compensation laws, principally in their benefit provisions, all statutes follow a definite pattern as to employment covered, benefits, insurance and the like.

Workers' compensation is a program designed to provide employees with assured payment for medical expenses or lost income due to injury on the job. Whenever an employment-related injury results in death, compensation benefits are generally paid to the worker's surviving dependents.

Generally all employment is covered by workers' compensation, although a few states provide exemptions for farm labor, or exempt farm employers of fewer than 10 full-time employees, for example. Farm employers in these states, however, may elect workers' compensation protection. Livestock producers in these states may wish to consider coverage as a financial protection strategy because under workers' compensation, the upper limits for settlement of lawsuits are set by state law.

This government-required employee benefit is costly for livestock producers. Costs vary among insurance companies due to dividends paid, surcharges and minimum premiums, and competitive pricing. Some companies, as a matter of policy, will not write workers' compensation in agricultural industries. Some states have a quasi-government provider of workers' compensation to assure availability of coverage for small businesses and high-risk industries.

[6]The sections on Liability, Workers' Compensation, Social Security Law, Tax Management and Reporting, and Estate Planning were authoritatively reviewed by Waymon E. Watts, CPA, Fresno, California.

For information, contact your area extension farm management or personnel management advisor and an insurance agent experienced in marketing workers' compensation and liability insurance.

SOCIAL SECURITY LAW

The Social Security Law covers stipulated agricultural workers, including workers on farms and ranches. Thus, owners and managers of beef cattle operations should be familiar with, and follow pertinent provisions of, the Social Security Law.

The number on his/her social security card is very important to the farm operator as well as to the hired farm worker. It identifies the individual's social security record and is key to future benefit payments. It is important, therefore, that a person's social security number is on the social security reports for both the self-employed farmer and the agricultural worker.

Those who expect to draw social security payments later should check with the Social Security Administration every three years, especially if they change jobs frequently, to make sure that their records are in order and that their correct earnings are credited to their individual social security records.

For a social security card—either a new card or a duplicate of one that has been lost—or for more information about retirement, survivors, and disability insurance, Medicare health insurance, or Supplemental Security Income, get in touch with the nearest social security office or call Social Security's toll-free number: 1-800-772-1213.

TAX MANAGEMENT AND REPORTING

Good tax management and reporting consists in complying with the law, but in paying no more tax than is required. It is the duty of revenue agents to see that taxpayers pay the correct amount, and it is the business of taxpayers to make sure that they do not pay more than is required. From both standpoints, it is important that farmers and ranchers should familiarize themselves with as many of the tax laws and regulations as possible.

The cardinal principles of good tax management are: (1) maintenance of adequate records, and (2) conduct of business affairs to the end that the tax required is no greater than necessary. Good tax management and good farm management do not necessarily go hand in hand, and may sometimes be in conflict. When the latter condition prevails, the advantages of one must be balanced against the disadvantages of the other to the end that there shall be the greatest net return.

It is recognized that tax matters constitute a highly specialized and complex field, and each farm or ranch will need separate considerations in appropriate planning. The recent rounds of federal tax legislation have made significant changes in the procedures livestock producers must use in accounting, as well as in their approaches to financial and estate planning. More than ever, it is important that they consult competent professionals before embarking upon any business operation involving livestock. It is noteworthy that, if a livestock producer's return is to be audited, under the recently enacted Taxpayer Bill of Rights, the taxpayer is entitled to be represented at the audit by a representative. Though the IRS can require the taxpayer's attendance with a special summons, this is not likely to be used at the initial meeting.

Increasingly, as local governments must make up for decreased federal support, local tax matters become more important in planning; this also makes consultation with a specialist knowledgeable in state and local tax law, crucial for effective management.

Some tax pointers of particular interest to stock producers follow:

1. File an estimate or file your current return.
2. Keep adequate and accurate records and accounts.
3. Separate the farm home from the farm business.
4. Keep year-to-year income as steady as possible.
5. Select the best method of accounting—cash or accrual.

CASH BASIS

Under this system, farm income includes all cash or value of merchandise or other property received during the tax year. It includes all receipts from the sale of items produced on the farm and profits from the sales of items that have been sold. It does not include proceeds from sales if the proceeds were not actually available during the tax year.

Allowable deductions include those business expenses incurred that were actually paid during the year, and depreciation on depreciable items.

ACCRUAL BASIS

This system requires the keeping of complete annual inventories. Tax is paid on all income earned during the taxable year, regardless of whether payment was actually received, and on increases of inventory values of livestock, crops, feed, produce, etc., at the end of the year as compared with the beginning of the year. All expenses incurred during the year's business are deducted from gross income regardless of whether payment is actually made, and deductions are made for any decrease in inventory values of livestock, etc., during the year.

Four methods of inventorying are available to the accrual basis farmer or rancher.

1. **Cost.** Inventory items are valued at the actual cost of producing or purchasing them.
2. **The lower of cost or market value.** The comparison is made separately for each item in the inventory, not for the entire inventory. The entire stock should *not* be valued at cost and then at market, with the lower selected.
3. **Farm price.** Each item, raised or purchased, is valued at its market price less estimated direct cost of disposition. This method must be used for the entire inventory, except that livestock may be inventoried by the next method.
4. **Unit livestock price.** Animals are classified according to kind and age, and a standard unit price is used for each animal within a class. All raised livestock must be included in inventory under this method. Unit prices must reflect any costs required to be capitalized under the uniform capitalization rules. This method is usually chosen by many large operations. Producers using the unit-livestock method are permitted to elect a simplified production method for determining costs required to be capitalized.

DISTINGUISH CAPITAL GAINS FROM ORDINARY INCOME

There is a difference in the tax rates applied to ordinary income and capital gains. Income reported as capital gains will be taxed at a lower rate. Thus, livestock held for sale in inventory, and livestock held for breeding purposes, may produce different tax effects when sold, even if the sale prices are the same. But there continues to be developments in this area.

SET UP DEPRECIATION SCHEDULES PROPERLY

Depreciation is estimated operating expense covering wear, tear, exhaustion, and obsolescence of property used in a farm business.

Depreciation may be taken on all farm buildings (except the livestock producer's personal residence), and on everything from grain elevators to horse clippers, including tile drains, water systems, fences, machinery and equipment.

Those who file returns on a cash basis may also take depreciation on dairy cattle, breeding and work stock which were purchased, but they cannot take depreciation on livestock they raised because all costs of raising are deducted as operating expenses. On the accrual basis, depreciation may be taken on purchased animals that are not included in inventory.

Taxpayers should list each building, and each piece of machinery on which depreciation is to be computed on the depreciation schedule. Such items as cows and small implements may be grouped together, but such groupings should be derived from totaling of a detailed individual list kept current in a permanent farm record book.

Depreciation is not available for inventory, which would include livestock held for sale to customers. After 1986, depreciable property is placed in specific classes. Because the period over which property is amortized affects the overall tax revenues. Congress has tended to lengthen recovery periods as a means of increasing tax collections without *raising* taxes.

1. **Three-year property.** This includes horses that are more than 12 years old when placed in service (racehorses more than 2 years old when placed in service).
2. **Five-year property.** This includes automobiles and light-general purpose trucks, certain technological equipment and research and experimentation property.
3. **Seven-year property.** This includes breeding and work horses, 12 years or younger, and any horse not in any other category.
4. **Ten-year property.** Horticultural or single-purpose agricultural structures were originally recovered over 7 years, but after 1988 have a 10-year recovery period. A companion requirement limits recovery on such items to the 150% declining balance method (discussed below). Orchards, groves and vineyards placed in service after 1988 are depreciated on a straight-line basis over 10 years.
5. **Fifteen-year property.** This includes equipment used for two-way exchange of voice and data communications.
6. **27.5-year property.** This covers residential rental property.
7. **31.5-year property.** This covers nonresidential real property. This will include most farm buildings.

For property in the 3-, 5-, 7-, and 10-year classes, depreciation was, prior to 1989, calculated on the double declining balance method, switching to the straight-line method at the time where depreciation is maximized. For property in the 15- and 20-year classes, the 150% declining balance method is used. For the 27.5- and 31.5-year classes, the straight-line method is used. However, for personal property (i.e., nonreal property) placed in service in a farming business after 1988, the 150% declining balance must be used regardless of the recovery period.

A horse owner can elect to depreciate a 2-year-old race horse under the straight-line method provided the election is made for all property in the same class. Once the election is made it is irrevocable. Special provisions apply to property which is not placed in service at the beginning of the year. If property depreciated under certain methods is sold, the gain will be characterized as ordinary income, a factor which may become relevant if differential capital gains rates are reintroduced.

For purchased animals, the price paid will generally determine the amount which can be depreciated. Inherited or gift animals can be depreciated. However, their value may have to be established by a qualified appraiser, if the IRS contests the taxpayer's valuation.

DO NOT OVERLOOK ADDITIONAL DEDUCTIONS

- **Annual expensing**—The annual expensing limitation is $10,000 for property placed in service after 1986. However, this election is not available for taxpayers whose aggregate cost of qualifying property exceeds $210,000 (reduced dollar-for-dollar over $200,000). The amount which can be expensed is limited to taxable income derived from the trade or business. The repeal of the Investment Tax Credit and the longer recovery periods for most classes of property increases the value of this provision for the livestock producer.

- **Soil and water conservation**—Farmers and ranchers can deduct soil and water conservation expenditures only if the expenditures are consistent with a conservation plan approved by the USDA or a comparable state agency. Such expenditures include treatment or movement of earth, such as leveling, terracing or restoration of fertility, construction and protection of diversion channels, drainage ditches and earthen dams, planting of windbreaks, etc. Though land clearing expenses are no longer deductible, ordinary maintenance, including brush clearing, remains deductible. Costs of fertilizing and other conditioning of land remain deductible. The amount deducted under this election can't exceed 25% of the taxpayer's gross income from farming for the year. Part of the amount deducted may be recovered if the land is sold within ten years of the deduction.

- **Education expenses**—Educational expenses, such as the cost of short courses, are deductible if they are taken to maintain or improve the skills of the person in conducting the operation, or, if the person is employed by a farming operation and they are taken as a requirement of continuing that employment. However, if taken to allow the person to enter another trade or business, such expenses will not be deductible. For instance, a physician who owns a few horses, who takes a course of study with the idea of eventually managing a horse operation, will not likely be able to deduct the education expenses.

- **Pay children for farm work**—The farmer/rancher must be able to show that a true employer-employee relationship exists. To do so, children should, as much as possible, be treated as are other employees. They should be assigned definite jobs at agreed-upon wages, and paid regularly.

TREAT LOSSES APPROPRIATELY

On the cash basis, no death deduction can be made for an animal that was born and raised on the farm, because the cost of raising the animal has been deducted already with operating expenses. On the accrual basis, when the value of an animal appears in the beginning-of-year inventory but not in the end-of-year inventory, the loss is automatically accounted for in the change in inventory value. Any money received from insurance or indemnity is entered as other farm income. Other death losses are listed on line 34, Part II of Form 1040 F as "other deductions," with an explanation.

- **Losses from destruction, theft, and condemnation**—Special treatment is available for certain gains or losses that are netted. The gains and losses can arise from the sale or exchange of property used in the trade or business, involuntary conversion or condemnation. If gains exceed losses, the net gain is treated as long-term capital gain. If losses exceed gains, the loss is ordinary. While this has limited significance as long as there is no tax differential between capital gain and ordinary income, the likelihood of a reintroduced capital gain preference makes the matter important to keep in mind. Gains and losses from these causes include those involving (1) cattle and horses, regardless of age, which are held for draft, breeding, dairy or sporting purposes, and held for at least 24 months from the date of acquisition, and (2) other livestock, regardless of age, held for draft, breeding, dairy or sporting purposes, and held at least 12 months from the date of acquisition. The fact the livestock is included in inventory doesn't prevent this treatment if the animal is held for the required purposes and for the specified time.

- **Passive activity losses**—Perhaps the most complicated addition to tax law in recent years was the passive activity loss concept, a development which will take tax lawyers years to decipher, with untold questions yet to be answered. Under this concept, all income and losses are divided between passive and nonpassive activities. A passive activity is one which involves the trade or business in which the livestock producer does *not* materially participate. Losses and credits from passive trade or business activities are disallowed to the extent they exceed aggregate passive income. Passive income does not include portfolio income (interest, dividends or royalties). However, rental activities are (if within the definition provided in the Internal Revenue Code) always passive.

■ **Material participation**—The IRS has provided seven exclusive tests for meeting the material participation requirement as to a particular activity:

1. **The 500 hours test.** The livestock owner participates more than 500 hours in the operation during the year. Obviously, full-time livestock producers will not have significant difficulties in meeting this requirement.
2. **Substantially all test.** The producer's participation constitutes substantially all participation in the activity. Given the requirements for the care of animals, it is unlikely that this test is even necessary for producers, as they would then satisfy the first test in any case.
3. **The 100 hours test.** The individual participates for more than 100 hours and no other person participates for a greater number of hours. Again, this will not generally be relevant to livestock producers. Nevertheless, a physician who owns animals and has a full-time employee to take care of them will often fail to be an active participant under this test.
4. **The related activities test.** The livestock producer participates in a group of activities for more than 500 hours, more than 100 hours in each. This may apply where a producer has a number of operations, but only limited involvement in each.
5. **The 5 of 10 years test.** This allows a livestock producer who has materially participated in the particular activity in the past to qualify as materially participating presently, even if his/her direct involvement has fallen off somewhat.
6. **Personal service activities.** This would apply to consultants involved in the livestock industry, but not to livestock producers running their own operations.
7. **Facts and circumstances test.** This test is, according to many experts, essentially similar to the 100 hours test.

Also, certain retired livestock producers will qualify in the event of death during the year. Though these requirements will have no effect on the full-time producer, they are important factors in terms of investment planning for anyone who is considering investments in rental real estate activities and other ventures.

■ **At-risk rules**—Another provision in the Internal Revenue Code limits losses to the extent that a taxpayer is at risk with respect to a particular activity. This means generally that a taxpayer is limited to the amount of his/her personal investment and the amount as to which he/she is personally liable. This provision specifically applies to farming, which includes livestock activities. The provision was designed principally to preclude losses from tax shelters and other leveraged investments where there may be no real chance that the taxpayer will have to cover the losses. Thus, it will seldom affect livestock producers whose credit is generally limited to the amount of collateral they can provide.

AVOID OPERATING THE BUSINESS AS A HOBBY

If an activity is not engaged in for profit, deductions are generally not available for the conduct of the activity except to the extent of income from it. This requirement has often been applied when the IRS determines that a livestock operation is actually a hobby. Though the problem will generally not apply to full-time livestock producers, others who devote a smaller amount of their time to an operation may find their activity is classified by the IRS as a hobby.

The general presumption for activities is that if an activity is profitable for 3 of the 5 consecutive years before the year being audited, it will be presumed to be engaged in for profit. Recognizing that horse operations often depend on the success of a rare horse, in such an operation Congress has allowed the activity to be presumed to be engaged in for profit if only 2 of 7 years are profitable. The IRS has indicated that an activity cannot be considered as engaged in for profit until there is a profit year.

Horse owners can delay the determination of whether a horse operation is engaged in for profit until the seventh taxable year of the activity. This election also keeps open the statute of limitations for those years.

In determining whether a livestock operation is a business or a hobby, the IRS will examine the following factors:

1. **The manner in which the operator carries on the activity.** The more businesslike the conduct of the activity, the more likely it is to be recognized as a business. This includes the keeping of accurate records of income and expenses. If the operation is conducted in a manner similar to other profit-making livestock operations, it is more likely to be recognized

as a business. If operating methods and procedures are changed because of losses, the impression is enhanced that the operation is a business. If the operation is typical of the other operations in the vicinity, it may indicate an attempt to fit into the livestock industry.

2. **The expertise of the operator and the employees.** A study of the industry and of other successful operations indicates a profit-making approach. If operators tend to ignore advice, they may have to establish that their expertise is even greater than that of their advisors.

3. **Time and effort spent in carrying out the operation.** The more time the owner devotes to the activity as a business and not as a recreational pursuit, the more likely the Service will find that the operation is a business. If the owner hires a full-time assistant to run day-to-day operations, he/she will be in a stronger position to argue that an attempt is being made to turn a profit. If the assistant is an inexperienced family member, the owner's position may, on the other hand, be weakened. Proper and rigid culling of herds will enhance the evidence for business conduct.

4. **The expectation that assets used in the activity will appreciate in value.** Even if current operations do not produce much income, the investment in land and buildings may support an argument that the owner has taken other businesslike factors into consideration. If the primary focus of the operation is breeding, it may take considerable time to get the necessary stock.

5. **Prior successes of the livestock producer.** The more experienced the producer and the more successful his/her prior livestock operations, the more he/she is likely to be seen as a serious business person. It may be important that the producer comes from a family of successful livestock producers.

6. **The operation's history of income and losses.** If losses are due to unforeseen circumstances (drought, disease, fire, theft, weather damages or other involuntary conversions, or from depressed markets), it may be possible to argue that there was nevertheless a profit motive in the operation.

7. **Occasional profits.** An occasional profit may indicate a profit motive if the investment or the losses of other years are comparatively small. The more speculative the venture, the more the livestock producer may be able to show that the losses were not due to a lack of profit intent.

8. **Financial status of the livestock producer.** The more the producer relies on the livestock operation, the more likely the producer is able to justify it as a business. If there are substantial profits from other sources, it may appear that the operation is nothing more than a private tax shelter. If this is the case, the producer may also have to worry about the effect of the limits on passive activity losses.

9. **Elements of recreation or pleasure.** Though having fun does not mean an operation is a hobby, the more the recreational element dominates the livestock producer's involvement, the more likely the livestock producer will have difficulty convincing the IRS that he/she is trying to make a profit. The presence of fishing holes, tennis courts and guest houses may indicate that the producer has a country club (a different sort of business, but not a livestock operation).

ESTATE PLANNING

Human nature being what it is, most livestock producers shy away from suggestions that someone help plan the disposition of their property and other assets after they are gone. Also, they have a longstanding distrust of lawyers, legal terms, and trusts; and to them the subject of taxes on death seldom makes for pleasant conversation.

If a farmer has prepared a valid will, or placed the property in joint tenancy, the estate will be distributed as intended. If not, it goes to the heirs, according to

Fig. 13-10. Estate planning is a way in which to preserve the farm or ranch for use by your chosen successors.

the laws governing intestate (without a will) succession. The heirs are those persons whom the law appoints to succeed to the property in the event of intestacy, and are not necessarily the persons to whom the farmer would want to leave the property. These laws vary somewhat from state to state.

If no plans are made, estate taxes and settlement costs often run considerably higher than if proper estate planning is done. Today, livestock business is big business; many have well over $1 million invested in land, animals and equipment. Thus, it is not a satisfying thought to one who has worked hard to build and maintain a good livestock establishment during their lifetime to feel that the heirs will have to sell the facilities and animals to raise enough cash to pay estate and inheritance taxes. Therefore, livestock producers should go to an estate planning specialist—a lawyer or company specializing in this work, or the trust department of a commercial bank. A limited discussion of some of the major considerations follows:

- **Valuation can be based on farming use.** Owners of farms and small businesses have been granted an estate planning advantage by means of what is called *special use valuation.* Under this concept, a farm or ranch can escape valuation for estate tax purposes at the highest and best use. Thus, a farm located in an area undergoing development may be considerably more valuable to developers than it is as a farm. Nevertheless, if the family is willing to continue the farming or ranching use for ten years, the farm can be included in the estate at its value as a farm. The aggregate reduction in fair market value cannot exceed $750,000.

In order to qualify for special use valuation, the decedent must have been a U.S. citizen or resident and the farm must be located in the U.S. The farm must have been used by the decedent or a family member at the date of the decedent's death. A lease to a nonfamily member, if not dependent on production, will not satisfy this requirement. At least 50% of the value of the decedent's estate must consist of the farm and more than 25% of the estate must consist of the farm and real property. It may be possible to split up a farm and take the special valuation for only part of it, but this part must involve real property worth at least 25% of the estate.

The property must be passed to a qualified heir, including ancestors of the decedent, the spouse and lineal descendants, lineal descendants of the spouse or parents, and the spouse of any lineal descendant. Aunts, uncles and first cousins are excluded. Legally adopted children are included.

The property must have been owned by the decedent or a family member for five of the eight years preceding the decedent's death and used as a farm in that period. The decedent or a family member must have participated in the farming operation for such a period prior to the decedent's death or disability.

- **Electing special use valuation.** Though the procedures are clear as to how special use valuation is elected, the frequency with which mistakes are made indicates the importance of having a competent tax attorney or CPA firm prepare the estate tax return. A procedural failure denying the estate the considerable savings that can be gained by the election may give sufficient grounds for a malpractice suit against the return preparer.

- **Recapture tax.** If the farm ceases to be operated by the heir or a family member within ten years, an additional estate tax will be imposed and the advantage of the election will be substantially lost. Partition among qualified heirs will not bring about recapture. A recent change allows the surviving spouse of the decedent to lease a farm on a net cash basis to a family member without being subject to the recapture tax.

- **Longer time to pay estate taxes.** Estates eligible for special use valuation may often be able to defer payment of estate taxes. Where more than 35% of an estate of a U.S. citizen or resident consists of a farm, the estate tax liability may be paid in up to ten annual installments beginning as late as 5 years from when the tax might otherwise be due. Thus, a portion of the estate taxes is deferred as much as 15 years. For purposes of the 35% requirement, the residential buildings and improvements on them which are on the farm are considered to be part of the farming operation.

If more than 50% of the decedent's interest in the farm is disposed of in the deferral period, then the entire unpaid portion of the estate tax liability is accelerated. The transfer of the decedent's interest in a closely held business on the death of the original heir will not cause an acceleration if the transferee is a family member of the transferor.

- **Use the gift tax exclusion for lifetime transfers.** The nontaxable gift tax exclusion remains at $10,000 per donee per year. A husband and wife who elect gift-splitting may jointly give $20,000 per recipient per year. These gifts may be in the form of interests in the farming operation.

- **Plan with the unlimited marital deduction.** An unlimited deduction is permitted for the value of all property included in the gross estate that passes to the decedent's surviving spouse in the specified manner. Certain *terminable* interests do not qualify for such a deduction—that is, interests as to which of the surviving spouse's interests will terminate on the happening of some event. Surviving spouses may be given *qualified terminable interests.* The most common arrangement involves the surviving spouse receiving a lifetime interest in the farm, with the remainder passing on the spouse's death to others, perhaps the children of the decedent. No marital deduction is allowed if the surviving spouse is not a U.S. citizen, unless a specific trust arrangement is used.

- **Consult a professional.** The preparation of wills, trusts, redemption agreements (if the farm is incorporated), partnership agreements, etc., requires consideration of the effects of federal and state tax law, as well as state law governing the various potential arrangements. Consequently, it is strongly advised that competent professionals be consulted in order to achieve an effective and cost-saving estate plan.

WILLS

A will is a set of instructions drawn up by or for an individual which details how the individual wishes the estate to be handled after death.

Despite the importance of a will in distributing property in keeping with the individual's wishes, about 50% of farmers and ranchers pass away without having written a will. This means that state law determines property distribution in such cases.

Every farmer/rancher should have a will. By so doing, (1) the property will be distributed in keeping with their wishes, (2) they can name the executor of the estate, and (3) sizable tax savings can be made by the way in which the property is distributed. Because technical and legal rules govern the preparation, validity, and execution of a will, it should be drawn up by an attorney. Wills can and should be changed and updated from time to time. This can be done either by (1) a properly drawn-up codicil (formal amendment to a will), or (2) a completely new will which revokes the old one.

The same attorney should prepare both the husband's and wife's wills so that a common disaster clause can be incorporated and the estate planning of each can be coordinated.

TRUSTS

A trust is a written agreement by which an owner of property (the trustor) transfers title to a trustee for the benefit of persons called beneficiaries. Both real and personal property may be placed in trust.

The trustee may be an individual(s), bank, or corporation, or a combination of two or three of these. Management skill should be considered carefully in choosing a trustee.

A trust can continue for any period of time set by the owner—for a lifetime, until the youngest child reaches age 21, etc. If the trust extends beyond a lifetime, there are limitations which should be explained by an attorney.

KINDS OF TRUSTS

Basically, there are two kinds of trusts, the *living* and the *testamentary*. The living or *inter vivos* trust is in essence an agreement between the trustor and the trustee and may be revocable or irrevocable.

The *revocable trust* can be terminated or altered; under it the trustor is concerned about the here and now, rather than only the hereafter. The trustor continues to make decisions, and can call off the whole arrangement (it's revocable) if it doesn't work out as expected. The revocable trust offers no special estate tax advantage; the assets of a revocable trust are included in the estate of the deceased creating the trust. However, it can be written in such a manner as to reduce substantially the estate taxes of the beneficiaries. Also, the revocable trust will eliminate the cost of probate—costs which may include executor's fees, attorney's fees, court costs, and appraisal fees.

The *irrevocable trust* cannot be amended, altered, revoked, or terminated. Under an irrevocable trust, the trustor must be willing to part with the trust property forever (irrevocably) and have nothing further to do with it and its administration. However, the irrevocable trust has many favorable aspects in estate planning; it will reduce estate taxes in both the estate of the trustor and the estate(s) of the life beneficiaries, and it avoids probate.

The *testamentary trust* is so-called because it is established under the provisions of the trustor's last will and testament. The testamentary trust does not become effective until after death of the trustor, followed by probate. There is no tax saving in the trustor's estate. However, the trust may be drafted to save estate taxes in the estates of the beneficiaries. A testamentary trust is useful when the heirs are minors or inexperienced in money matters.

PREDICTING WHAT'S AHEAD

Most cattle producers are doing a good job when it comes to feeding and management. But they do little or nothing about scientifically projecting what's ahead. Worse yet, they are prone to flock with other producers, in Judas fashion, simply because they have no watchdog or barometer. As a result, they buy and sell when everyone else does. Altogether too often they buy when cattle are up and sell when they're down.

The trouble with most existing cattle forecasts is that they come too late. Information on numbers and weights of cattle on feed, for example, only confirm what has already happened. Additionally, even the best prediction may be upset by wars, droughts, and politics.

Ideally, cattle producers need to spot a zig or zag in the graphs that may mark the beginning of a trend. They need signs that indicate trouble in advance of an actual downturn. Then they can shift their program to take advantage of the situation. The authors feel that this can be achieved by applying a new sophistication to certain indicators. By using computers, the time lag for most information gathering can be reduced. To be sure of what the machine-produced information means requires experience in analysis and interpretation. Also, the relative importance of the different factors must be weighed and evaluated—there is need for an index figure.

The senior author predicts that, eventually, progressive individual producers, partnerships, and corporations will do it by hiring economists or professional analysts. They will utilize such public information figures as are available, but they will supplement them as necessary and evolve with their own barometer. Personnel for such an assignment will cost money, for it requires great competence; and there will be added charges for computers. But it will pay handsome dividends. For example, a saving of 7/10¢ per lb on a 14,000-capacity feedlot, handling 30,000 cattle annually and adding an average of 425-lb gain per animal, would mean $89,250 annually. That's a sizable sum! Yet, increased profits of this amount, or more, are well within the realm of possibility through the system herein proposed.

Here are some of the harbingers—the indicators of what's ahead in the cattle business, that should be considered:

1. **Cattle numbers.** Total cattle numbers, and number and weight of cattle on feed, are very important indicators. When cattle production begins to outrun beef consumption, it's time to slow down.

2. **Beef imports.** Beef imports are a factor in determining domestic cattle and beef prices, although opinions differ as to their importance. According to one U.S. Department of Agriculture study, when imports equal about 10% of total domestic beef production, a further increase of 10% in imports would cause about a 1% drop in Choice steer prices. But there are those who feel that the effect of imports is more marked. Certainly, psychological reactions are difficult to measure.

3. **Beef Quality and Price.** Consideration must be given to how beef compares with broilers, pork, lamb, and fish in being consistently good and competitively priced. This is so because when shoppers walk along the meat counter, they determine what they shall purchase. If they feel that beef is not consistently good and/or too high, they simply do not buy it. Instead, they may move down the counter a few feet and select some broilers, or perhaps pork, lamb, or fish. This simply means that beef must be consistently good and competitively priced. Thus, the producer's barometer needs to ferret out how the quality and price of beef compares with its competition. As is generally known, the relative importance of competitive meats changes from time to time. In particular, the broiler industry has enjoyed a tremendous growth, largely due to technological developments.

4. **Employment and wages.** This indicator is of particular importance to the cattle feeders. This is so because in periods of prosperity—when employment is high and wages are good—consumers place a premium on choice cuts and top grades of beef.

5. **Feed grain; production, prices, and federal programs.** Since about 80% of the cost of finishing cattle (exclusive of the initial purchase price of animals) is for feed, it is obvious that feed prices are a major item in determining profits and losses. The producer's barometer needs to keep tab on the factors which influence feed prices—acreages and yields, droughts, federal programs, and carry-over stocks.

6. **The feeder's margin** (also see Chapter 27, section on "Computing Break-even Prices for Cattle"). The wise feeder projects the necessary break-even margin before laying in cattle; keeping in mind that (a) the better the grade and the younger the cattle, the smaller the necessary margin; (b) the higher the cost of feed, the greater the necessary margin; (c) the heavier the initial weight of the feeder, the smaller the necessary margin; and (d) the longer the feeding period and the fatter the cattle, the greater the margin. Generally speaking, those feeders who ignore these facts and simply play for a rising slaughter cattle market don't stay in business very long.

7. **The buying power of the dollar.** This, of course, affects the price of all commodities. Consideration of it is a must in predicting what's ahead in the cattle business.

8. **The beef futures market, including both live cattle and carcass beef.** This brings into focus both hedgers and speculators. Since their money is involved, it reflects their thinking relative to what lies ahead. As such, it merits careful study.

9. **High- and low-cost producers and cattle cycles.** There will always be high- and low-cost cow-calf producers and cattle cycles.

Tom Brink, Director of Market Research for National Cattlemen's Beef Association Cattle Fax, reported that, in 1993, 25% of the lowest-cost cow-calf producers had an average debt load of $137 per cow, had an average calf break-even price of $59 per cwt, and had an average net income of $146 per cow. By contrast, the 25% with the highest cost had an average debt load of $514 per cow, had an average break-even of $146 per cow, and had a average net income of zero dollars per cow. This shows the dramatic difference in debt and break-even between low- and high-cost producers. When cattle numbers are up and prices are down, low-cost producers can buckle up and endure a rough ride.

10. **Contract growers.** Many economists and "so-called" industry experts have suggested that in order to survive the economic pressures of the future, many beef producers will have to form contractual arrangements with some of the large companies involved in the beef industry and basically become contract growers or producers for these companies, similar to many poultry producers in the southeastern United States. Although these arrangements may offer advantages in terms of financing and security, they result in loss of control of the management of an operation and many times lead to poverty and not wealth for individual producers.

Despite all the above, no one can consistently outguess the market. Thus, cattle producers are admonished to adopt a strategy that will minimize high losses.

QUESTIONS FOR STUDY AND DISCUSSION

1. Why have the business aspects of beef production become so important in recent years?

2. Discuss the advantages and the disadvantages of each of the three major types of business organizations common to cattle enterprises.

3. It takes about $33 in farm assets to generate $1 of net farm income. How does this ratio compare with other businesses? (Check with the managers of two successful businesses not in agriculture.)

4. List and discuss each of the major sources from which farmers borrow money.

5. List and discuss the three kinds of public offerings which are commonly used.

6. List and discuss the three general types of agriculture credit or loans available, based on length of life and type of collateral needed.

7. Compare the types and sources of credit for each (a) cow-calf loans, (b) cattle feeder loans, and (c) cattle on pasture loans.

8. Assume a certain kind and size of cattle operation—either cow-calf, or cattle feedlot, or a combination of the two—then prepare a detailed request to a lender for a loan.

9. Using Tables 13-6, 13-7, and 13-8 as guides, develop a budget for the year ahead for your own farm or ranch, or for one with which you are familiar.

10. What helpful hints would you give to a cattle producer who wishes to build and maintain a good credit rating?

11. Some lenders now charge points. If 4 points are being charged, and you borrow $20,000, what will this mean?

12. Use Table 13-9 to rate yourself as a cattle manager. How do you stack up?

13. Prepare an organization chart and a job description for your farm or ranch, or for a farm or ranch with which you are familiar.

14. Take your own farm or ranch, or one with which you are familiar, and develop a workable incentive basis for the help. Does the incentive plan that you developed meet the *requisites of an incentive basis*?

15. What should a farm inventory include? Give pertinent pointers relative to annual inventory.

16. Why should enterprise accounts be kept?

17. How may computers be used on a practical basis in the cattle business?

18. Why does a cattle producer need profit indicators?

19. Cite one guideline relative to facility and equipment costs. Give an example to show how it works.

20. Assume that you have paid $25,000 for a 6-year-old bull and that you wish to insure him for a year. How much will it cost?

21. Describe each of the following systems of accounting: (a) cash basis, and (b) accrual basis.

22. What is a depreciation schedule? What schedule applies to most farm buildings?

23. Why is estate planning important?

24. Why are wills important?

25. List and discuss the two basic kinds of trusts.

26. List and discuss the harbingers (indicators) of what's ahead in the cattle business, which cattle producers should consider.

SELECTED REFERENCES

Title of Publication	Author(s)	Publisher
Agricultural Law, Vols. I and II	J. C. Juergensmeyer J. B. Wadley	Little, Brown and Company, Boston, MA, 1982
Anatomy of an American Agricultural Credit Crisis	K. N. Peoples, *et al.*	Roman & Littlefield Pub., Inc., Lanham, MD, 1992
Business & Planning	N. E. Harl	Century Communications, Inc., Miles, IL, 1994
Business Management for Farmers	J. W. Looney	Doane Publishing, St. Louis, MO, 1983
Contract Farming and Economic Integration, Second Edition	E. P. Roy	The Interstate Printers & Publishers, Inc., Danville, IL, 1972
Cooperatives: Development, Principles and Management, Fourth Edition	E. P. Roy	The Interstate Printers & Publishers, Inc., Danville, IL, 1981
Cowboy Arithmetic: "Cattle as an Investment," Fourth Edition	H. L. Oppenheimer	The Interstate Printers & Publishers, Inc., Danville, IL, 1985
Cowboy Economics: "Rural Land as an Investment," Third Edition	H. L. Oppenheimer	The Interstate Printers & Publishers, Inc., Danville, IL, 1976
Cowboy Litigation: "Cattle and the Income Tax," Second Edition	H L. Oppenheimer J. D. Keast	The Interstate Printers & Publishers, Inc., Danville, IL, 1972
Doane's Tax Guide for Farmers	J. C. O'Byrne	Doane Agricultural Service, Inc., St. Louis, MO, 1977
Exploring Agribusiness, Third Edition	E. P. Roy	The Interstate Printers & Publishers, Inc., Danville, IL, 1980
Farm Management Handbook, Third Edition	Queensland Department of Primary Industries	Queensland Department of Primary Industries, Brisbane, Queensland, Australia, 1970
Farm Records and Accounting	J. A. Hopkins E. O. Heady	Iowa State University Press, Ames, IA, 1962
Going Public	Corplan Associates	Corplan Associates, IIT Research Institute, Chicago, IL
How to Do a Private Offering-Using Venture Capital	A. A. Sommer, Jr.	Practicing Law Institute, New York, NY, 1970
Introduction to Agribusiness Management, An, Second Edition	W. J. Wills	The Interstate Printers & Publishers, Inc., Danville, IL, 1979
Land Speculation: "An Evaluation and Analysis"	H. L. Oppenheimer	The Interstate Printers & Publishers, Inc., Danville, IL, 1972
Life Insurance Company Farm-Mortgage Loans, ERS-439		Economic Research Service, U.S. Department of Agriculture, Washington, DC, 1970
Microcomputing in Agriculture	J. Legacy T. Stilt F. Reneau	Reston Publishing Co., Reston, VA, 1984
Stockman's Handbook, The, Seventh Edition	M. E. Ensminger	Interstate Publishers, Inc., Danville, IL, 1992
Tax Loss Farming, ERS-546		Economic Research Service, U.S. Department of Agriculture, Washington, DC, 1974
1971 U.S. Master Tax Guide		Commerce Clearing House, Inc., Chicago, IL, 1970

Grand Champion steer of the 1995 Houston Livestock Show and Rodeo, shown by 4-H Club member, Morgan Moylan, Eastland, Texas. This Chianina steer sold for the record price of $500,000, of which the exhibitor got to keep $60,000 and $400,000 went to the educational scholarships for which the Houston Show is famous. (Photo by Barker Livestock Photography, Chappell Hill, TX. Courtesy, Houston Show)

FITTING, SHOWING, AND JUDGING BEEF CATTLE

Contents — *Page*

Contents *Page*

Fig. 14-1. Grand Champion steer in the 1994 National Western Stock Show, Denver, Colorado; exhibited by Sara Lewis, Magnolia, Texas; purchased by Lombardi Bros. Meat Packers, for $50,000.

The show-ring has long been a major force in shaping beef cattle type. The first American livestock show was held at Pittsfield, Massachusetts, in 1810, but livestock exhibitions had been initiated in Europe many years earlier.

The mere mention of a livestock show causes the senior author to become nostalgic. He grew up in an era when the show-ring was as traditional as Mom's apple pie. It was the way—the only way—to establish livestock standards.

Then came the cattle producers' mid-century wrangle relative to the merits or demerits of livestock shows—eyeball vs science, which has raged ever since. Those who would do away with livestock shows argue that we know how they should look. What we need to know, they add, is how they will perform on the pasture or range, in the feedlot, and, finally, on the grill or in the roaster. Production records, bolstered by computers and readings made by other sophisticated equipment, are the answer, according to this school of thought. Then they clinch their argument by citing the debacle which brought disrepute to cattle shows from which they may never fully recover. In the period 1935 to 1946, small, early-maturing, blocky, smooth cattle were sweeping shows from one end of the country to the other, with this beauty contest (it wasn't a beef contest) reaching the ultimate in the *compests* in Herefords and the *compacts* in Shorthorns, and culminating in dwarfism.

It's not all that bad, counter the traditionalists. Individual selection, or eyeballing, is the usual method followed by both commercial and purebred producers when selecting foundation animals, when adding new blood, and when selecting replacements from within the herd. Also, it's the basis of evaluating the vast majority of feeder and slaughter cattle. Then they add

Fig. 14-2. Ringside view of the North American International Livestock Exposition, Louisville, Kentucky. (Courtesy, American Polled Hereford Assn., Kansas City, MO)

this unassailable fact: eyeballing and livestock shows have been largely responsible for the vast improvements to date.

Both schools of thought have a point. Livestock shows need to change; progress in the beef industry bypassed them. To something old, they need to add something new. The show-ring has been, and can continue to be, an important vehicle for getting cattle and people together in one place and at one time to compare, design, and engineer the most economically productive and desirable beef model.

In recent years, shows have aided in the genetic improvement of beef cattle. The cattle winning in the major shows have been the growthy, meaty, productive, useful kind that fit the mold and standards of today's producers and consumers. Research has clearly shown that well-trained personnel can estimate lean to fat ratio (cutability) very well. With accurate performance and carcass information available, shows will be a major force in identifying cattle that have the genetic potential to improve the pounds of edible Choice grade lean beef produced per day of age, at a minimum cost.

ADVANTAGES AND DISADVANTAGES OF SHOWING

Livestock shows have both advantages and disadvantages.

Human nature being what it is, not all exhibitors share equally in the many advantages that may accrue from showing. In general, however, cattle shows offer the following **advantages**:

1. They afford the best medium yet discovered for molding breed type. For this reason, it behooves the breed registry associations and the purebred breeders alike to accept their rightful responsibility in seeing that the animals winning top honors are those which most nearly meet the efficiency of production requirements of the producer and the demands of the consumer.
2. They provide an incentive to breed better cattle, for breeders can determine how well they are keeping pace with their competitors only after securing an impartial appraisal of their entries in comparison with others.
3. They offer an opportunity to study the progress being made within other breeds and classes of livestock.
4. They serve as one of the very best advertising or promotional mediums for both the breed and the breeder.
5. They give breeders an opportunity to exchange ideas, thus serving as an educational event.
6. They offer an opportunity to sell a limited number of breeding animals.
7. They set sale values for the animals back home, for such values are based on the sale of show animals.
8. They direct the attention of new breeders to those types and strains of cattle that are meeting with the approval of the better breeders and judges.
9. They are an effective means of distributing new breeds to new geographic locations, throughout the United States and the world.
10. They provide instruction, training, and encouragement for young people to join the expanding and dynamic beef industry.
11. They attract new producers and their dollars, both of which are needed if the industry is to grow in numbers and economically.

Like many good things in life, livestock shows also have some **disadvantages**, among which are the following:

1. Breed fancy points—involving such things as color and markings, shape of head, shape of horns (if present), and set of ears—may overshadow utility value.
2. They may mislead, as occurred during the period 1935 to 1946—the era of the compresst and compacts. And, the emphasis on tall animals (hip height) during the 1980s has haunted us in the 1990s.
3. The desire to win sometimes causes exhibitors to resort to *surgical means* and *filling* in order to correct defects. Admittedly, such artificial corrections are not hereditary, and their effects are often not too durable—as is belatedly discovered by some innocent purchaser.
4. Valuable animals are sometimes kept out of reproduction in order to enhance their likelihood of winning in the show-ring.
5. Their educational value has not been exploited.

Fig. 14-3. A beautifully fitted and perfectly groomed Hereford bull. This is ABC L1 Domino 0246 "Super II", Reserve Champion Hereford Bull at the 1983 National Western Stock Show, Denver, owned by Adams Bros. & Co., Kilgore, Nebraska. (Courtesy, The American Hereford Assn., Kansas City, MO)

Many people who attend a cattle show for the first time are bewildered by the procedure and the many breeds and classes that they see. Others have never gone to a cattle show because they feel that their lack of knowledge of showing and judging cattle would prevent them from enjoying it.

RECENT IMPROVEMENTS IN LIVESTOCK SHOWS

Recently, there has been a concerted effort to improve livestock shows; to maximize their advantages and to minimize their disadvantages. To this end, there has been a trend toward the following changes:

1. To provide breeding classes with a maximum of a one month age span per class.
2. To divide the cattle into evenly sized classes of 15 or less after they arrive at the show. For example, the show committee might decide to have two April heifer classes of 15 animals each instead of one class of 30.
3. To line up by age, from oldest to youngest.
4. To eliminate the older heifer classes (beyond early junior yearling), because of the feeling that senior yearling and two-year-old females should be left home to raise their first calves.
5. To weigh all cattle after they arrive at the show.
6. To provide the judge and the spectators with the following information pertaining to each animal:
    a. Birth date.
    b. Weight per day of age.
    c. Fat thickness and ribeye measurements.
    d. Height measurements, now generally taken at the hip.
7. To combine live animal and carcass evaluation for steers, by requiring that the top 3 to 5 head of every class be slaughtered with carcass information obtained on these cattle.
8. To specify minimum carcass weight in carcass show classes, and to calculate and consider pounds of carcass per day of age.
9. To terminal steer shows, in which steers are evaluated both on the hoof and on the rail prior to making a final placing. These shows consider lean and fat per day of age, and quality grade, in addition to visual appraisal.
10. To include more performance information, such as EPDs, and objective measurements, such as scrotal circumference in bulls and pelvic area in heifers, in the evaluation criteria of breeding cattle shows.

SELECTING SHOW CATTLE

Producers who are planning to show have the difficult job of selecting animals which, following fitting, will meet the show-ring ideal. Requiring as it does a projection into the future, no judging assignment is quite so difficult. But the success or failure attained in the show-ring depends to a very great extent on the animals selected. Inasmuch as several months are required to bring an animal into the peak of show condition, the selection of animals for the show should be made as early in the season as possible, and from among animals that are healthy and vigorous and have good dispositions.

Fig. 14-4. No judging assignment is quite so difficult as that of selecting show animals for further development, for it requires a projection into the future.

When selecting steers for fitting and showing, look for growthiness, skeletal size, muscling, trimness, and correct structure. The primary objective is to identify steers that will efficiently produce the greatest quantity of Choice lean beef per day of age. Purchasing calves from purebred or commercial herds that emphasize growth rate and use fast-gaining, performance-tested bulls will give added assurance of rapid and efficient gains.

The ideal show steer should meet the following specifications:

1. Be structurally sound.
2. Have adequate body capacity, width of chest, and good middle for extra feed intake.
3. Have adequate frame size.
4. Weigh at least 550 lb at weaning (205 days).
5. Make an average daily gain of 3.0 lb from weaning to slaughter.
6. Have feed efficiency of less than 6.0 lb of feed per pound of gain.
7. Weigh over 950 lb at 12 months.
8. Grade Choice or above when at slaughter weight of 1,000 to 1,350 lb.

9. Have outside fat thickness of 0.3 to 0.5 in.
10. Yield grade less than 2.9.

The selection of show bulls and females differs from the selection of show steers primarily because sex character, along with greater emphasis on structural soundness, is important in breeding animals. Like steers, they should possess growthiness, skeletal size, muscling, and trimness. Additionally, when selecting bulls, add masculinity and structural soundness to the list; and when selecting heifers, add femininity and structural soundness.

The essential factors to be considered when selecting individual show prospects are: (1) type; (2) breeding and performance; and (3) show classification.

(Also, see Chapter 4, Selecting Beef Cattle.)

TYPE

When they make their placings, livestock judges endeavor to select cattle that are most efficient at producing quality beef. Steers that are growthy, muscular, and lack excessive finish (external fat) are usually placed at the top. Judges look for the same things in breeding classes, but, in addition, they look for structural soundness and sex character—traits that indicate that they will be efficient producers of offspring in the future.

Both breeding cattle and steers should be growthy. They should be long, tall, and not excessively fat if they are to continue to grow. Above all, avoid selecting a young animal that looks mature (that looks old for its age) and is overfat, because such characteristics are indicative of an animal that will not grow enough to be competitive in the show-ring.

Muscling, being a masculine trait, is more important in bulls and steers than in heifers. Long, smooth muscling is preferred in young animals because they will usually grow for a longer period of time and get thicker as they get older. Truly muscular cattle are not smooth all over. They show some creases and indentations between the muscles; they are prominent in the forearm and stifle; and they are slightly narrower through the heart girth and loin than through the shoulder and round. Breeding cattle with coarse shoulders (very heavy muscled shoulders) should be avoided since this condition is frequently associated with calving problems.

Trimness, or freedom from predisposition to waste, is important in both breeding and slaughter cattle. Excessively fat breeding cattle will usually have poorer reproduction than trimmer ones; and very fat slaughter cattle will have reduced carcass value. Usually, prospective calves are not fat to begin with; hence, the exhibitor must estimate whether they are the kind that are likely to get overfat. Generally, calves that have large briskets and that are deep in the flanks and twist will have a tendency to be overfat when fitted to mature weight, and should be avoided.

Structural soundness is especially important in breeding cattle, but only moderately so in show steers. Nevertheless, the skeletal structure of steers should be sufficiently correct that they will not develop any serious unsoundness as fitting progresses. Breeding cattle must be structurally sound so that they will calve easily and be able to travel well on the pasture or range. Proper structural soundness involves a big foot, a deep heel, toes that are the same size and point straight ahead, clean joints, and correctly set legs. Avoid cattle that are sickle-hocked or post-legged. Breeding cattle should be wide through the pins.

Signs of fertility and reproductive efficiency are extremely important in breeding cattle. Bulls should have two testicles that are of normal size, and that are well defined in the scrotum, rather than surrounded by excess fat. A mature bull should be masculine in front. But bull calves should not show extreme masculinity at a young age, because such animals are apt to mature too early and stop growing.

When selecting females for show, it is well to look for femininity at all ages. Feminine females are trim in the jaw, throat, and dewlap, and smooth shouldered. Avoid masculine females—the kind that are coarse, heavy fronted, and excessively muscular.

In summary, when selecting prospective show steers, emphasize growthiness, skeletal size, muscling, and trimness. Add masculinity and soundness to the list when selecting bulls; and add femininity and structural soundness when selecting females.

(Also see Chapter 4, Selecting Beef Cattle, section entitled, "The Functional Scoring System.")

BREEDING AND PERFORMANCE

Animals selected for show should always be from good ancestry, for this is added assurance of satisfactory future development.

Projecting the growthiness of cattle is more certain if they have been performance tested. If performance records are available, the exhibitor should select cattle that have weight ratios of 100 or above and above average EPDs, which indicates above average growth. However, it should be recognized that cattle with weight ratios below 100 may be very growthy if the herd has been selected for superior performing cattle for a number of years.

(Also see Chapter 5, Some Principles of Cattle Genetics.)

SHOW CLASSIFICATIONS

Distinct and separate show classifications are provided for breeding animals and for steers. Also, there is a breakdown of several classes within each category. These follow:

- **Breeding beef cattle**—Beef cattle show classes vary according to breed. Most breed associations make available to livestock shows, and to breeders, a "Standard Show Classification" setting forth their recommended classes and ages. A common beef cattle show classification is given in Table 14-1. Some breed differences exist; hence, the exhibitor should secure from the breed registry and/or show(s) the classification for the breed being exhibited.

TABLE 14-1
TYPICAL SHOW CLASSIFICATION FOR BREEDING BEEF CATTLE

Class	Age of Bulls	Age of Females
Junior calves	Born after January 1 of the current show year. In shows between January 1 and May 1, this class for bulls is divided into spring calves, and those born after April 1.	Same as bulls.
Winter calves	Born between November 1 and December 31 of the previous year.	Same as bulls.
Senior calves	Born between September 1 and October 31 of the previous year.	Same as bulls.
Summer yearlings	Born between May 1 and August 31 of the previous year.	Same as bulls.
Spring yearlings	Born between March 1 and April 30 of the previous year.	Same as bulls.
Junior yearlings	Born between January 1 and February 28 of the previous year.	No class.
Senior yearlings	Born between September 1 and December 31, 2 years prior to the show.	No class.
Two-year-olds	Born between March 1 and August 31, 2 years prior to the show.	No class.

Table 14-1 shows that there are no classes for breeding females older than spring yearlings. Older female classes have been eliminated from the classification because it is believed that the place for females of such age should be on the farm or ranch where they can start production rather than continue under fitted show condition with the possible impairment in future productivity and possible loss to the breed.

It should be noted that there is a trend in the major cattle shows throughout the country to have breeding classes with a maximum of 2- to 3-month age span per class.

In addition to providing for individual classifications for each sex as shown in Table 14-1, the major shows also make provision for championships and for various group classes. Since these differ somewhat between both shows and breeds, no attempt will be made to list them herein. Instead, the exhibitor is admonished to study the premium list of the show or shows in which it is planned to exhibit. Entries must be made for both individual and group classifications, but no entries for championship classes are required.

- **Steers**—The steer classification varies considerably; with some shows following age divisions, others following weight divisions, others following hip height divisions, and still others following combinations of age, weight, and/or height.

The National Western Stock Show of Denver, Colorado, provides the following **Junior Market Steer** rules and classifications:

- **This is a terminal show for market beef animals.**
- **To be eligible to exhibit in the Market Beef Show, the youth must be 8 years of age but not 20 years of age, and formally enrolled in 4-H or FFA.**
- **Owner of any market beef that is condemned for any reason by the slaughter house will not receive payment for their animal.**
- **Exhibitors will be permitted to enter a maximum of three market beef.**
- **No substitutions will be allowed.**
- **Market beef may have permanent central incisors up, but not in wear.**
- **Market steers will be shown by hip height. Market beef will be weighed and measured and divided into classes of approximately 20 head per class at the discretion of management. Steer weight minimum of 1,000 lb.**
- **At measuring time, each entry must be accompanied by original brand inspection certificate or bill of sale.**
- **Any violation of the rules will automatically result in disqualification with no recourse.**
- **National Western Stock Show awards ribbons to winners in each class through 8th place.**
- **NATIONAL WESTERN JUNIOR MARKET BEEF CHAMPIONS:**

The Champion and Reserve Champion of each division of the Show will compete for Grand Champion and Reserve Grand Champion of the Junior Market Beef Show.

GRAND CHAMPION — Banner by the National Western Stock Show

RESERVE GRAND CHAMPION — Banner by the National Western Stock Show

MANAGEMENT

Following the selection of show animals, they should receive the best possible care in order to protect the very considerable investment.

BUSINESS RECORDS

Experienced exhibitors consider showing a part of the cost of doing business—as advertising and promotion. It follows that business records should be kept.

4-H and FFA members should also keep good business records. They should—

1. **Record expenses.** They should record the weight and purchase price of the calf; and they should record all other expenses such as feed, trucking, insurance, veterinary fees, supplies, and miscellaneous items.
2. **Keep performance records.** This calls for keeping an accurate account of how much feed the calf consumes in relation to the pounds of gain. From this, the cost of feed per pound of gain can be calculated. Performance records directly affect profit or loss.
3. **Record sale price.** When the animal is sold, 4-H and FFA members should record selling weight, price per pound, and total price.
4. **Determine profit or loss.** The sale price (No. 3 above) less the expenses (No. 1 above) will give the profit or loss.

SHELTER AND CARE

The shelter and care accorded to show animals may be as important a factor as the animals.

Experienced exhibitors generally have a show barn, or other special facilities.

4-H and FFA members need to emulate older exhibitors, but on a smaller scale, and usually in a less expensive way. They should—

1. **Provide suitable quarters.** The quarters should be comfortable, clean, and dry; and they should allow approximately 25 sq ft of space per calf. A clean, sanitary place will minimize disease and parasite problems.
2. **Prevent diseases.** It is much easier and cheaper to prevent diseases than to cure them. So, the caretaker should seek and follow the counsel and advice of the local veterinarian in the health program.
3. **Control parasites.** Lice, flies, and grubs are the most common parasites. Lice may be controlled by applying a suitable insecticide in late winter or early spring. Flies are controlled by sanitation, and spraying the animal's quarters with a suitable insecticide once a week during warm weather. Grubs are controlled by spraying or dusting the animal in late fall.

   4-H and FFA members should consult their county agent, vo-ag teacher, or veterinarian, or an experienced cattle producer, relative to the choice of chemicals and products for specific parasite control. Also, they should always follow the label directions.
4. **Dehorn and castrate.** If the animal has not been dehorned and/or castrated, this should be done. Unless the 4-H or FFA member is experienced, it is best to have an experienced person do the dehorning and castrating.
5. **Vitamin A; growth promotants.** If the animal has been on poor-quality forage, it may be advisable to inject it with 1 to 2 million units of Vitamin A upon arrival. Also, approved growth promoting implants or feed additives may be used.

FEEDING FOR THE SHOW

The subject of nutrition and feeding is fully covered in Chapter 8 of this book, so needless repetition will be avoided at this point; only the application of feeding for the show will be discussed.

All animals intended for show purposes should be placed in the proper state of condition—they should be neither too fat nor too thin. Steers should be fed to a degree of finish that will help ensure their ability to grade at least Low Choice on the rail. Requirements differ slightly between breeds and lines of cattle. For example, the larger, leaner continental European breeds do not mature as quickly nor fatten as readily as the British breeds. The essentials in feeding cattle for the show might be described as similar to those in feeding steers for market, except that more attention must be given to the smallest details. A suitable ration must be selected and the animal or animals must be fed with care over a sufficiently long period.

At the beginning of show preparation, check for both internal and external parasites. If any are present, apply the recommended treatment. (See Chapter 12, Beef Cattle Health, Disease Prevention, and Parasite Control.)

RULES OF FEEDING SHOW CATTLE

Some general rules of feeding may be given, but it must be remembered that *the eye of the master conditions cattle.* The most successful cattle fitters have worked out systems of their own through years of practical experience and close observation—they do not follow set rules. Nevertheless, the beginner may well profit by the experience of successful fitters, and it is with this hope that the following general rules of feeding show cattle are presented:

1. **Practice economy, but avoid false economy.** Although the ration should be as economical as possible, it must be remembered that rapid growth and ideal condition are primary objectives, even at somewhat additional expense. It is beneficial to start feeding prospective show calves for a short time prior to weaning so that they are accustomed to eating grain when they are eventually weaned from their dams. A suitable creep ration is given in Chapter 21.

2. **Use care in getting animals on feed.** In starting animals on feed, use extreme caution to see that they do not get digestive disturbances. Animals that are not accustomed to grain or other concentrates must be started on feed gradually, or digestive trouble may result. Until the animal gets on full feed, at no time should it be given more grain than can be cleaned up in about 30 to 60 minutes' time. In starting, a safe plan is to feed not more than *1 lb of grain at the first feed, or 2 lb for the day. This may be increased approximately by ¼ to ½ lb daily until the animal is on full feed about 3 weeks later.* From the beginning, it is safe to full-feed grass hay or the hay to which the animal is accustomed. Oats are the best concentrate for the beginning ration. As the grain feed is increased according to the directions given, gradually (1) replace the oats with the mixed ration selected, and (2) decrease the hay. Whenever the calf does not seem to have a good appetite and does not clean up the grain within an hour's time, the allowance should be reduced for the next few days and then gradually worked up to a full feed again. With careful observation and good judgment, it is possible to have an animal on full feed in 3 weeks' time. After full feeding is reached, the animal may be fed according to either of two plans: (a) by governing the allowance according to the amount of feed that the animal will clean up in about 30 to 60 minutes' time; or (b) by more or less self-feeding, with some feed being kept before the animal most of the time. Perhaps, in the final analysis, the method of feeding decided upon should vary according to the individual feeder.

Ordinarily it is best to start the animal on feed with grass hays and then gradually change to legumes, if legumes are to be included in the ration. By this method, the animal may be allowed a full feed of hay at the beginning of the feeding period with no danger of scouring. When forced feeding is being sought, it is usually preferable to use a grass hay or a mixed hay with a limited amount of legume in it because of (a) the laxative effect of a straight legume under heavy feeding, and (b) the possible bloat hazard of legumes.

3. **Provide a variety of feeds.** A good variety of feeds increases the palatability of the ration and makes it easier to supply the proper balance of nutrients. Furthermore, a finishing ration consisting of only 1 or 2 feeds may lose its palatability during a long feeding period.

4. **Feed a balanced ration.** A balanced ration will be more economical and will result in better gains. That is to say, the ration should contain the proper balance of energy, protein, minerals, and vitamins. It must also be remembered that growing animals require a higher proportion of grain than mature ones. Then, too, because of its high cost, it is not economical under most conditions to feed more protein than is required.

5. **Do not overfeed.** Feed plenty but do not overfeed. Overfeeding is usually caused by the desire of the inexperienced caretaker to push the animal too rapidly. After the animal has reached full feed, it may be given all the grain that it will clean up, provided the appetite and well-being seem to so warrant and the droppings are of the proper consistency.

6. **Keep the feed box clean.** Never leave uneaten feed in the box or trough. It may become sour and cause the animal to go off feed.

7. **Do not underfeed.** It never pays to starve an animal. Gains and profits result from feeds consumed in excess of the maintenance requirement. A common expression among cattle feeders is: "Get every bit of feed under their hides that you can." When animals are consuming too little grain to grow-finish, the feeder may look for several causes, such as the consumption of too much roughage, unpalatability of the ration, or discontentment on the part of the animal—that is, if the caretaker is not deliberately withholding grain (false economy).

8. **Full feed for economical finishing.** When on full feed, the average animal will eat from 2 to 2½ lb of grain for each 100 lb of liveweight. The exact amount will depend primarily upon the age, size, and individuality of the animal; the bulkiness and palatability of the ration; and the amount of fat that the animal is carrying. A full feed of grain is the amount that an animal will clean up nicely in ½ to 1 hour's time at each feeding period.

9. **Supply palatable feeds.** In order to consume the maximum amount, the animal must relish the feed. Unpalatable feeds may be fed in limited quantities, provided that they are mixed with more palatable ingredients. Blackstrap molasses is relished by animals and is excellent for increasing the palatability of the ration (although blackstrap molasses is preferable, beet molasses is satisfactory). Usually the molasses is added by diluting with water (warm water in winter) and mixing it with the grain ration just before feeding. One-half to one pint of molasses diluted with an equal volume of water will be entirely satisfactory for this purpose. Most commercial feeds contain some molas-

ses in the mixture. Cooking certain feeds, especially barley, also makes for increased palatability.

10. **Provide succulent feeds.** Succulence is provided in such feeds as silage, root crops, and grasses. These feeds have a beneficial effect in the ration. They increase the palatability and produce a laxative effect on the animal's digestive system.

11. **The ration must provide the correct amount of bulk.** The beef animal is a ruminant, and, therefore, requires some bulk in order to distend the digestive tract. Mature animals can handle more roughage than calves. Furthermore, more bulk may be fed at the beginning of the finishing period than at the end. As the grain ration is increased, the animal will consume less roughage. When on full feed, 3 to 6 lb of hay daily are ample. Consumption of too much bulk will cause the animal to become paunchy.

12. **Do not feed damaged feeds.** Moldy, musty, or spoiled feeds may cause digestive disturbances and should not be fed to animals being fitted for show or sale.

13. **Prepare grains.** The grain ration of cattle intended for show purposes is almost always coarsely ground or rolled. Most caretakers prefer steamed flaked grains. The preparation of hay is neither necessary nor advisable.

14. **Feed regularly.** Animals intended for show should be fed with exacting regularity. In the earlier part of the feeding period, two feedings per day may be adequate. Later, the animals may be fed three times a day, particularly if they are rather thin and rapid improvement in condition is desired.

15. **Avoid sudden changes.** Sudden changes in either the kind or the amount of feed are apt to cause digestive disturbances. Any necessary changes should be gradual.

16. **Provide minerals.** All animals should be given free access to salt at all times. For feeding where the salt is not exposed to the weather, loose salt, rock salt, and block salt are all satisfactory. Loose salt, however, is preferred if kept under shelter. Stabilized iodized salt should be used in iodine-deficient areas; and trace mineralized salt should be used in areas where other trace mineral deficiencies exist.

When a nonlegume roughage is provided, it is especially likely that there will be a deficiency of calcium and phosphorus. Under these conditions and even when a legume roughage is used, the addition of calcium and phosphorus gives protection at very little cost. Probably as satisfactory and inexpensive a mineral as can be provided for cattle is steamed bone meal or dicalcium phosphate. The mineral may be placed in a box to which the animals have free access. The best arrangement is to provide a double-compartment mineral box with salt in one side and a mixture of ⅓ salt and either steamed bone meal or dicalcium phosphate, or a commercial mineral mix, in the other.

17. **Keep the animal quiet and contented.** Quiet and contentment are necessary for profit in feeding. The restless animal rarely makes good gains, whereas the quiet animal that will "eat and lie down" will show superior gains. This is not due to differences in digestive or assimilative powers, but rather to the fact that the quiet animal is putting on weight while the wild, nervous animal is using surplus energy for nonproductive purposes. Uncomfortable quarters, isolation from other animals, annoyance by parasites, sudden changes in quarters or feeds, improper handling, and unnecessary noise are the most common causes of discontentment.

18. **Provide exercise.** A certain amount of exercise is necessary in order to promote good circulation and to increase the thrift and vigor of the animal. Exercise also tends to stimulate the appetite and makes for greater feed consumption. Animals can usually be kept in condition by turning them in a paddock at night.

19. **Avoid scouring.** If the droppings are too thin or there is scouring, (a) decrease the grain allowance, and (b) clean up the quarters. If trouble still persists, decrease the legume roughage and the protein supplement (especially linseed meal).

20. **Avoid sudden water changes.** Frequently, show cattle fail to drink enough water while at the show. As a result, they become gaunt and show at a disadvantage. Usually this situation is caused by (a) the sudden change in drinking from a trough or tank at home to drinking out of a bucket at the show, and (b) the different taste of the water, primarily due to chlorine or mineral content. This problem can be alleviated by (a) getting the animal accustomed to drinking from the same bucket that will be used at the show, beginning 7 to 12 days before leaving home, and (b) adding a tiny bit of molasses to each bucket of water, from the time bucket watering is started until the show is over, thus avoiding any flavor or taste change in the water.

SOME SUGGESTED RATIONS

Variations can and should be made in the rations, depending upon the individual animal, the relative prices of feeds, and the supply of homegrown feeds. To secure the correct state of condition, a suitable ration must be selected and the animal or animals must be fed with care over a sufficiently long period. The rations that follow are ones that have been used, and are being used, by successful fitters. In general, when show animals are being forced fed on any one of these concentrate mixtures, experienced caretakers prefer to feed a grass hay or a grass-legume mixed hay to a straight legume, because of the laxative effect

Grower (Starter) Rations:

Ration No. 1	*(lb)*	*(kg)*
Rolled barley	50	*23.0*
Rolled oats	20	*9.1*
Wheat bran	20	*9.1*
Protein supplement[1]	10	*4.5*

Ration No. 2	*(lb)*	*(kg)*
Rolled corn	20	*9.1*
Rolled barley	30	*13.6*
Rolled oats	20	*9.1*
Wheat bran	20	*9.1*
Protein supplement[1]	10	*4.5*

Ration No. 3	*(lb)*	*(kg)*
Rolled oats	30	*13.6*
Rolled barley	30	*13.6*
Rolled corn	10	*4.5*
Wheat bran	20	*9.1*
Protein supplement[1]	10	*4.5*

Ration No. 4	*(lb)*	*(kg)*
Rolled oats	30	*13.6*
Rolled corn	60	*27.3*
Protein supplement[1]	10	*4.5*

Ration No. 5	*(lb)*	*(kg)*
Rolled corn	40	*18.2*
Rolled oats	30	*13.6*
Wheat bran	20	*9.1*
Protein supplement[1]	10	*4.5*

Ration No. 6	*(lb)*	*(kg)*
Crimped oats	70	*31.8*
Cracked corn	10	*4.5*
Wheat bran	10	*4.5*
Protein supplement[1]	10	*4.5*

Finisher Rations:

Ration No. 7	*(lb)*	*(kg)*
Rolled corn or sorghum chop	50	*23.0*
Rolled barley	40	*18.2*
Protein supplement[1]	10	*4.5*

Ration No. 8	*(lb)*	*(kg)*
Rolled corn	60	*27.3*
Rolled oats	20	*9.1*
Dry beet pulp	10	*4.5*
Protein supplement[1]	10	*4.5*

Ration No. 9	*(lb)*	*(kg)*
Rolled corn	40	*18.2*
Rolled barley	20	*9.1*
Rolled oats	10	*4.5*
Dry beet pulp	10	*4.5*
Wheat bran	10	*4.5*
Protein supplement[1]	10	*4.5*

Ration No. 10	*(lb)*	*(kg)*
Rolled oats	25	*11.4*
Rolled barley	20	*9.1*
Rolled wheat	20	*9.1*
Rolled corn	20	*9.1*
Wheat bran	10	*4.5*
Protein supplement[1]	5	*2.3*

Ration No. 11	*(lb)*	*(kg)*
Rolled barley	35	*15.9*
Rolled oats	20	*9.1*
Rolled wheat	20	*9.1*
Dry beet pulp	15	*6.8*
Protein supplement[1]	10	*4.5*

Ration No. 12	*(lb)*	*(kg)*
Crimped oats	40	*18.2*
Cracked corn	25	*11.4*
Wheat bran	25	*11.4*
Protein supplement[1]	10	*4.5*

Ration No. 13	*(lb)*	*(kg)*
Rolled barley	20	*9.1*
Rolled corn	20	*9.1*
Rolled oats	20	*9.1*
Whole barley (dry wt. basis, but cooked before feeding)	13	*5.9*
Beet pulp, dried molasses	4	*1.8*
Wheat bran	6	*2.7*
Commercial supplement[1]	8	*3.6*
Linseed oil meal (pellets)	8	*3.6*
Salt	1	*0.5*

[1]The protein supplement may consist of linseed, cottonseed, peanut, or soybean meal, or a commercial supplement may be used. With most fitters, linseed meal is the preferred protein supplement. It gives the animal a sleek hair coat and a pliable hide. However, it is a laxative feed. Caution should be exercised in feeding it.

Although it is true that an animal getting good clover or alfalfa hay needs less protein supplement than does one eating nonleguminous roughage, it is not possible to supply all the needed protein with hay and still get enough grain into young animals to finish them quickly.

and possible bloat hazard of the latter. Also, salt and a suitable mineral should be offered free-choice.

Rations 1 to 6 are bulky. They are recommended for use in starting prospective show animals on feed. Rations 7 to 13 are less bulky and more fattening. They are recommended for use during the latter part of the fitting period.

Ration 13 is the one which the senior author uses in fitting show steers. The cooked barley is prepared by (1) adding water in the proportion of 2 to 2½ gal to each gal of dry barley, and (2) cooking until the kernels are thoroughly swelled and can be easily crushed between the thumb and forefinger. Each young steer also receives 4 lb daily of a supplement high in milk by-products. As the animal approaches show finish, the ration is changed by decreasing the rolled barley by 7 lb and increasing the rolled oats by 5 lb and the wheat bran by 2 lb.

EQUIPMENT FOR FITTING AND SHOWING CATTLE

Every exhibitor should have a durable and attractive box in which to keep the necessary equipment. The attractiveness of the exhibit can be enhanced by the presence of a nicely painted box on which the name of the exhibitor is neatly printed.

The equipment for the show may include the following:

Basic Equipment
- Paper towels
- Feed pan
- Water bucket
- Broom
- Pitchfork
- Blocking chute
- Tie-out neck rope/band
- Extra rope halter
- Sprayer
- Blower
- Extension cord
- 4-way socket type outlet
- First-aid kit
- Hose and nozzle
- Leather hole punch

Equipment for Grooming
- Neckchain or nylon halter
- Clippers (3-types):
  straight edge
  sheep heads
  small clippers
- Soft bursh
- Multi toothed brush
- Hoof brush
- Scrub brush
- Riceroot brush
- Scotch comb
- Hand rubber comb
- Curry comb
- Tail comb
- Spray bottle for coat dressing
- Wool rags for coat dressing

Supplies for Grooming
- Saddle soap
- Cold water soap
- Insecticide
- Disinfectant
- Body adhesive
- Tail adhesive
- Leg adhesive
- Light oil mix
- Purple oil
- Streaks N' Tips
- Paint
- Show foam
- Hair spray
- Lanolin spray
- Alcohol
- Vinegar dip
- Rag oil
- Tail ties
- Final mist

Equipment for Showing
- Show halter
- Show stick

Also, include clothes and other articles for the personal use of the caretaker. The size of the box and the amount of equipment one carries will depend upon the number of cattle being shown and the length of the show circuit.

A ROPE HALTER AND HOW TO MAKE IT

Attractive leather halters with lead straps and chains are ideal for showing cattle, but they are rather expensive. Rope halters are much more practical for everyday use.

MATERIALS NEEDED

The following materials are needed for making a rope halter:

1. **Rope.** Thirteen feet (15 ft when making halters to use in breaking cattle to lead, thereby having a longer lead) of 3-strand manila or nylon rope (nylon will not draw up when it gets wet). Use ½-in. rope for cattle over 6 months of age and ⅜-in. rope for calves under 6 months.
2. **Fid (marlinspike).** This is needed for opening up the strands of rope. It may be made by taking a piece of ¾-in. round, hardwood stick 6 in. long, and tapering it to a point on one end; or a small, pointed piece of iron may be used.
3. **Measure and pencil.** A rule or tape and a pencil.

DIRECTIONS FOR MAKING

Here are the directions for making an eye-loop rope halter (see Fig. 14-5), which is adjustable in every respect except for the nose band. This type of halter keeps its adjustments. By contrast, the double-loop rope halter is objectionable for use in tying because it adjusts too easily with the result that many tied calves free themselves.

The steps in making an eye-loop rope halter are:

1. **Whip-slice one end.** The end which is whip-sliced, to prevent it from unraveling, is known as the short end. For this purpose, use a waxed cord, fish line, or strong piece of string about 40 in. long.

Double the whipping cord from one end to form a loop approximately 6 in. long. Lay this loop on top of the rope (Fig. 14-6, Step A), which is held in the left hand, with the end of the loop about 2 in. from the end of the rope. Now with the thumb holding the cord about ½ in. from the end of the rope (Fig. 14-6, Step B), make the first turn from front to back or clockwise around the rope, and be sure that this first turn locks the first wind in place.

Continue to wind tightly and neatly for a length of 1 in. Then run the last wind up through the loop (Fig.

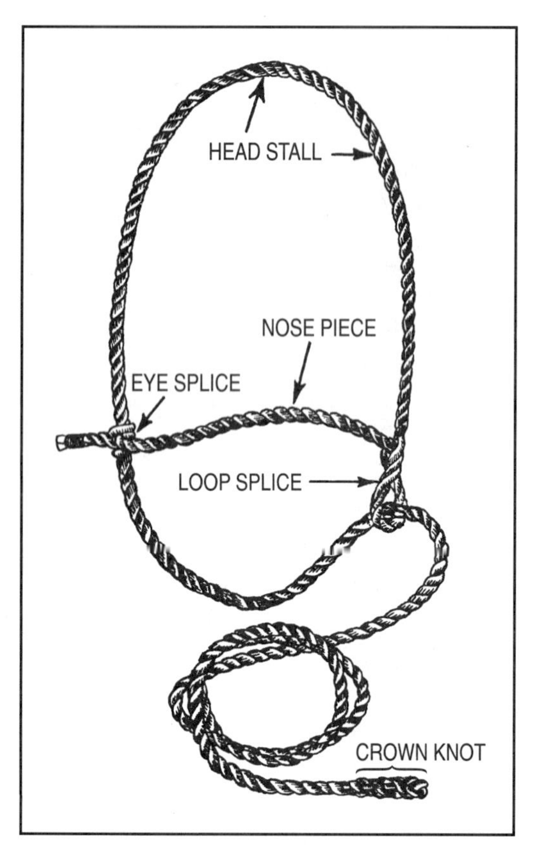

Fig. 14-5. Completed eye-loop rope halter.

14-6, Step C) and draw up tightly against the last turn and hold in place. Next grasp the cord out over the end of the rope and pull the loop under the windings for about ½ in. or half the length of the windings. These will now stay in place. With a knife, cut the remaining cord off even with the first and last windings. Soaking the whipping in waterproof cement makes the job more secure.

2. **Make the loop splice.** The loop splice is made as follows (see Fig. 14-7): The short end of the rope will form the nosepiece of the halter and will be its only permanent dimension. For average size cattle, use 15 to 16 in. for the short end (the future nosepiece); for large bulls use 18 to 20 in. Measure off this amount (the 15 to 16 in., or up to 18 to 20 in. as decided upon) from the whipping on the short end that you have just completed whip-splicing, and pencil-mark the 2 strands to be raised. Lay the rope in front of you, short end to the right. Bend the long end up or away from you clockwise. With the aid of the fid, raise the 2 strands on the top of the short end at the 15- to 16-in. mark and pull the long end down toward you and under the strands. Draw this long end through until the loop is about 1½ to 2 in. (make a 2-in. loop where the halter may get wet) inside diameter (see Fig. 14-7, Step A). A rule of thumb is that the inside diameter of the loop should be at least twice the thickness of the rope, a loop that is too small will close too tightly when the halter becomes wet and shrinks.

Next, take the short end and, with the help of the fid, pass it under a top strand on the long rope as close to the loop as possible (see Fig. 14-7, Step B).

Pull the lead end snug, and the loop splice is complete as shown in Fig. 14-7, Step C. This makes for a loop with an equal number of strands on each side of the splice and leaves the inside of the splice fairly smooth where it bears against the jaw.

3. **Make the eye splice.** Grasp the whip-spliced end in your left hand with the left thumb a couple of inches from the end and on top of the rope. Grasp the nosepiece (short end) with your right-hand thumb on

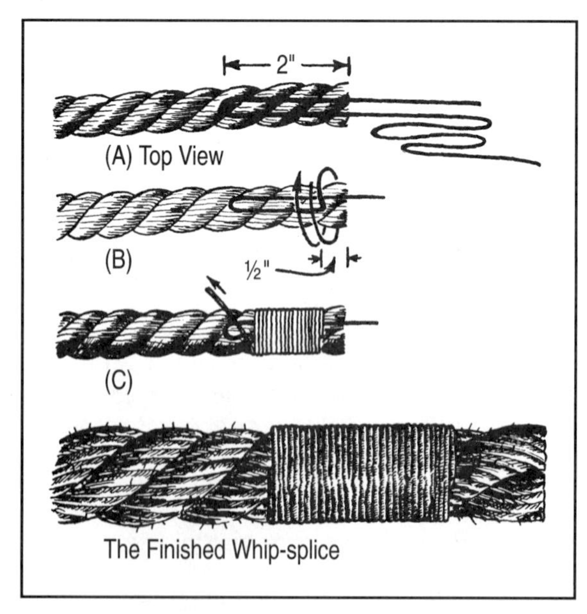

Fig. 14-6. Proper method of whip-splicing.

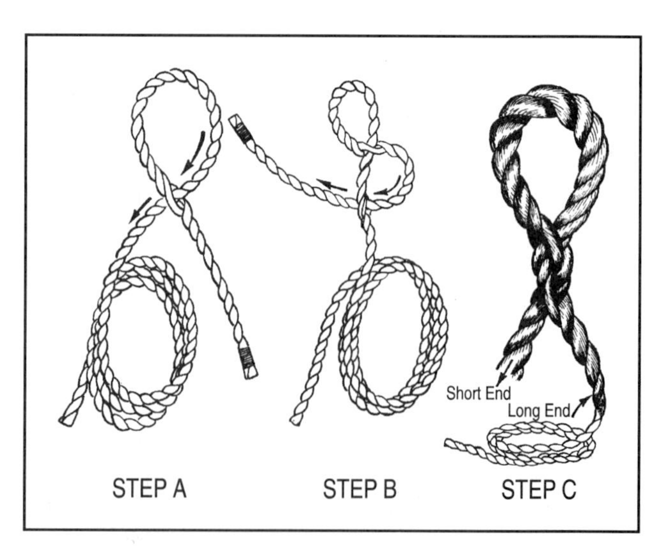

Fig. 14-7. Steps in making the loop splice.

the bottom of the rope, a couple of inches from the left hand. Twist your right hand away from you (Fig. 14-8, Step A), pushing while you twist, so that the strands separate and finally kink (Fig. 14-8, Step B). Take each kink on the same side of the rope and in line. Place the fid through the 3 kinks (as shown in Fig. 14-8, Step C, or 14-8, Step D—either of which is correct), and then follow with the long rope through all 3 (Fig. 14-8, Step E). Draw the long rope through the eye splice until it is free of kinks and then adjust it back until it is approximately the size of the headpiece. Arrange the kinks neatly on the rope.

Pass the long rope through the loop splice (Fig. 14-8, Step F) and the halter is complete except for preparing the end of the lead rope. The halter is placed on the animal so that the loop splice is at the left of the jaw. Needed adjustments may be readily made after fitting it on the animal. (When properly fitted on the animal, the nosepiece should fall about two-thirds the distance from the muzzle to the eyes.)

4. **Prepare the end of the rope.** The end of the lead rope may be prepared in one of three ways: (a) by whipping; (b) by making a crown knot; or (c) by making a wall knot.

Whipping leaves the end of the lead sufficiently small that it can be passed easily through tie rings. On the other hand, both the crown knot and the wall knot make it easier to hold on to an animal that is trying to get away. The procedure for preparing the end of the rope by each of these methods follows:

a. **Whipping the end of the rope.** If the lead end of the rope is to be whipped, this step is generally done at the very beginning as already directed for the short end of the rope (see Fig. 14-6).

b. **Make the crown knot.** Make the crown knot as follows (see Fig. 14-9):

(1) Unlay the rope for 5 to 7 turns, depending upon the length of finish desired (see Fig. 14-9, Step A) and throw up a bight in z between x and y and lock with the forefinger (see Fig. 14-9, Step A).

(2) Pull x over z and lock with the thumb (see Fig. 14-9, Step B).

(3) Pass y over x and linewards through bight z (see Fig. 14-9, Step C).

(4) Hold the formation in the tips of the fingers and set by pulling on x, y, and z in turn until they are uniformly snug (see Fig. 14-9, Step D).

(5) Tuck by passing each strand over 1 strand and under 1 strand, straight down the line, removing 1 or more yarns of the strand at each tuck after the first, according to the taper desired (see Fig. 14-9).

(6) Moisten (barely) the completed crown knot and roll the splice under the foot.

c. Make the wall knot. If desired, a wall knot may be made as follows (see Fig. 14-10):

(1) Unlay the rope from 5 to 7 turns (see Fig. 14-10, Step A) and throw down a bight in z and lock with the thumb (see Fig. 14-10, Step B).

(2) Throw a bight in y around z and lock under the first finger (see Fig. 14-10, Step B).

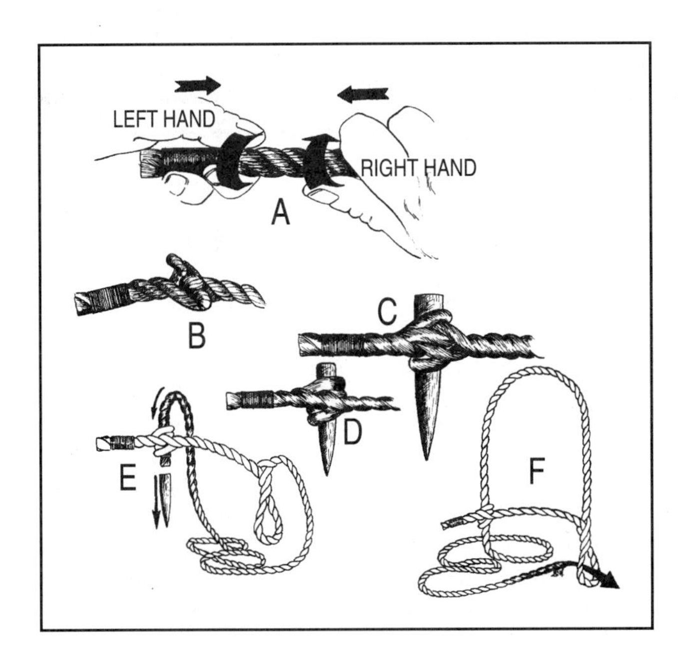

Fig. 14-8. Steps in making the eye splice.

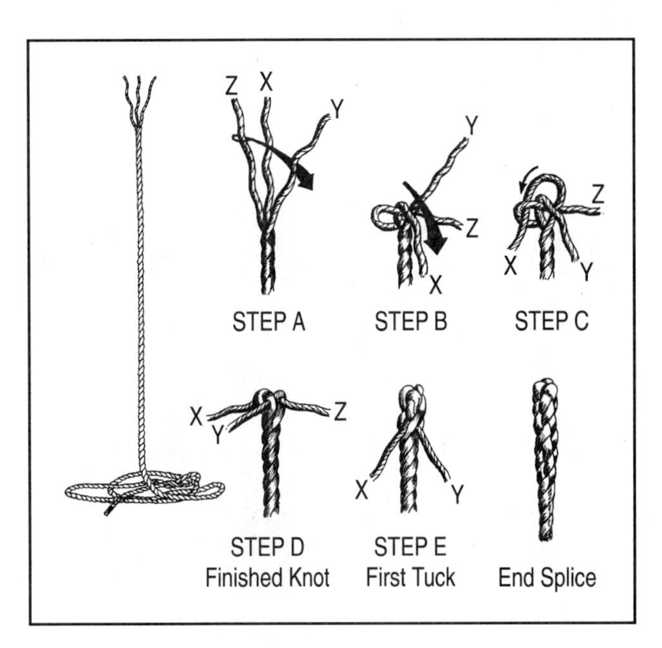

Fig. 14-9. Steps in making the crown knot.

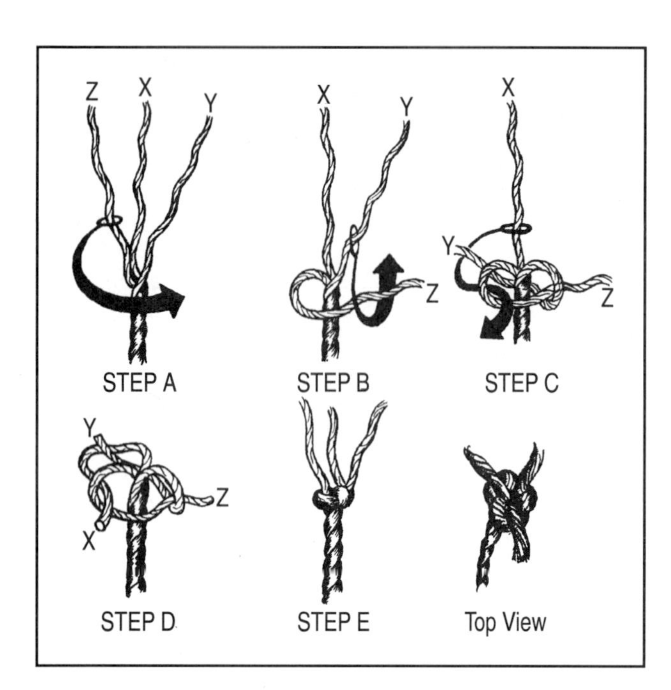

Fig. 14-10. Steps in making the wall knot.

(3) Throw a bight in x around y, extending the end through the bight in z (see Fig. 14-10, Steps C and D).

(4) Hold in the tips of the fingers and set by pulling alternately on each strand until it is uniformly snug (see Fig. 14-10, Step E).

TRAINING, GROOMING, AND SHOWING

Assuming that the animal has been carefully selected and properly fed, there yet remains the assignment of parading before the judge. In order to present a pleasing appearance in the show-ring, the animal must be well trained, thoroughly groomed, and properly shown. Competition is keen, and often the winner will be selected by a very narrow margin. Close attention to details may, therefore, be a determining factor in the decision.

Fig. 14-11. Junior exhibitors grooming their heifers at the Junior National in Wisconsin. (Courtesy, American Polled Hereford Assn., Kansas City, MO)

GENTLING, HALTER BREAKING, AND POSING

Proper training of the animal requires time, patience, firmness, and persistence. Such schooling makes it possible for the judge to see the animal at its best.

First, the animal should be gentled. Easy does it! Here is how: Place the animal(s) in a relatively small pen where you can walk fairly close to it. Then, (1) carry a small bucket containing a few pellets of feed, which you should shake to attract the animal's attention; (2) hold one pellet between your thumb and forefinger, and feed the animal when it will take the pellet from you; and (3) repeat this procedure daily, or more often, until the animal gets sufficiently gentle that it can be hand fed and backrubbed. Never, never touch the animal on the forehead or horns, as this will teach it to butt.

Following the gentling procedure outlined above, the animal may be haltered. Halter breaking can most easily be accomplished at an early age, rather than waiting until the animal is stronger and has *a mind of its own.* Begin by tying the calf in the stall or corral with a rope halter. Leave it tied for a period of 1 to 2 hours, brushing and talking to it occasionally, so that it realizes that you aren't going to hurt it. When you tie the animal, be sure to leave just the right amount of slack, so that it is able to lie down without getting its legs caught in the rope. After the animal has learned to stand tied, which may take 3 to 4 days, or much longer, start leading it around in the stall or corral. Later, lead the animal outside the stall or corral.

Caution: Never let an animal break away from you while you are training it; if you do, it will discover that it is master and you will have a problem animal.

Leading should be correctly done from the left side and with the halter strap or rope in the right hand. Beef exhibitors follow the custom of walking forward, glancing back over the right shoulder at frequent intervals. The rope halter is preferable when starting the training program, but it is very important that the animal become accustomed to being led with the leather show halter well in advance of the show. The latter precaution is important because the animal reacts differently when led with the show halter than when led with the rope halter.

The next step is that of teaching the animal to stand or *pose* properly, so that the judge may have an opportunity to examine it carefully. For correct posing the animal must stand squarely on all 4 feet (preferably with the forefeet on slightly higher ground than the rear feet). The back should be held perfectly straight and the head held on a level with the top of the back. At first, this position may be quite strained and unnatural for the animal. For this reason, it should not be required to hold this position too long. Later, it should be possible to *pose* the animal for 15 to 20 minutes at a time. In *posing*, the exhibitor should hold the strap in the left hand and face toward the animal. A show stick is usually used in placing the hindfeet but the exhibitor's foot is best used in obtaining correct placement of the front feet.

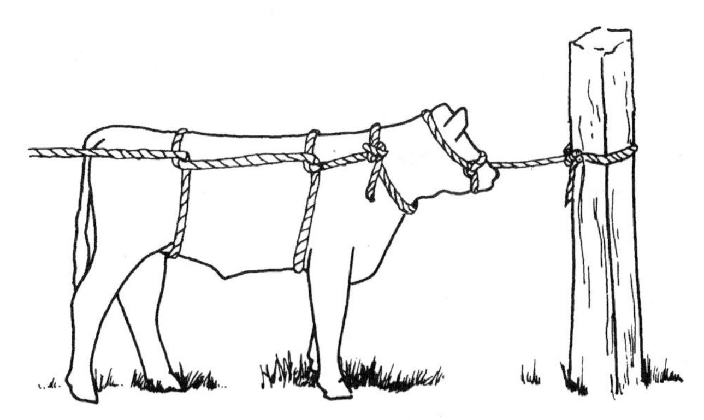

Fig. 14-13. A simple method of throwing an animal is illustrated here using a rope halter and a rope about 40 ft long. With the animal haltered, tie the halter shank or lead to a stout post, tree, etc. Then, with one end of the rope around the neck, tie a bowline (nonslip) knot. Next, circle the animal's body just behind the shoulder and a half hitch at the withers; continue the rope back to the loin and make a second half hitch and circle the rope around the body at the flanks. Make sure the second half hitch is just in front of the hooks. A strong backward pull on the rope will cause the animal to sink, and a shift in the pull to the side on which the animal is to fall will result in an easy, soft fall to the ground. Maintaining the strong pull on the rope will keep the animal lying on the ground, making it possible to do minor *doctoring*, foot trimming, etc.

TRIMMING AND CARING FOR THE FEET

So that the animal will stand squarely and walk properly, the feet should be trimmed regularly. Long toes or unevenly worn hoofs are unsightly in appearance. Trimming can best be done with the animal either in a set of stocks or on a trimming table with the animal turned on its side. With this method, it is possible to square up the sole and the sides of the feet as well as to cut back the ends of the toes. The practice of merely shortening the toes by standing the animal on a hard surface and cutting off the ends with a hammer and chisel gives only temporary relief, very often not really correcting the difficulty. If a set of stocks or a table cannot be made available, it may be advisable to throw the animal as shown in Fig. 14-13, thereby making it possible to work on the bottoms of the feet.

Fig. 14-12. Trimming the feet. (Drawing by R. F. Johnson)

The feet of some animals should be trimmed regularly as often as every two months. Too much trimming at any one time, however, may result in lameness. For this reason, it is not advisable to work on the feet immediately before the show.

Among the tools that may be used for trimming are electric sander, chisel, nippers, farrier's knife, and rasp. However, not all these need be available.

■ **Pointers on general trimming follow—**

1. Trim the inside toe and heel before trimming the outside toe and heel, because it generally grows faster and longer. Trim the toe before the heel to ensure that the animal will walk up on its toes.

2. Remove the outgrowth or rim of the sole around the edge of the toes and along the side of the foot with the nippers. Be careful to keep the foot level while trimming.

3. When the bottom of the foot is springy to the touch, the next cut will probably draw blood—and you have gone too far.

4. Shape the foot and all rough edges with a rasp. An electric sander often generates too much heat and may seal the pores of the foot, thus inhibiting proper growth of the foot. However, a sander may be used for steers, because they are slaughtered at an early age.

5. The bottom of the foot, between the toes, should be hollowed out to allow mud, etc., to ooze up through the toes. This functions as a self-cleaning mechanism.

6. Make the side of the toes relatively straight on the inside by rasping between them.

7. Apply a disinfectant to any cracks or cuts in the foot, especially between the toes and along the hoof head, to aid healing.

■ **Common leg problems and how to correct them follow—**

1. Bow-legged behind.
   a. Trim inside heel down.
   b. Trim rim off the inside claw out to the toe, but leave the toe long.
   c. Build up the outside heel.
   d. Trim the outside toe short.
2. Toe-out in front. Trim both the inside toe and heel short.
3. Pigeon-toed in front. Same as trimming for the bow-legged condition (see No. 1).
4. Cow-hocked behind. Same as trimming for the toe-out condition (see No. 2).

Note well: To correct leg problems, trimming should be performed regularly every 30 days.

■ **Treating dry, brittle hoofs**—Quite often, when cattle are kept constantly in stables, the feet may become dry and brittle. This condition can usually be corrected by turning the animals out in a pasture paddock at night when there is dew on the grass. Packing the hoofs with wet clay, or applying neat's-foot oil will also be helpful in such cases. If the animal gets sore feet from standing in a filthy stable, the soreness should first be corrected. Following this, the feet should be washed and disinfected.

DEHORNING/HORNS

Hereford heifers (and heifers of most other horned breeds) are usually shown dehorned. Hair is left on the face, but clipped shorter where the horns were, and around the sides of the cheeks—then blended into the neck.

Hereford bulls (and bulls of most other horned breeds) are shown with shaped horns. It must also be remembered that there is a difference in the desired shape with different breeds. The horns of the Hereford should curve downward, whereas the horns of the Shorthorn should curve slightly forward and inward.

As soon as the horns are long enough (3 to 4 in. long) and sufficiently strong to bear the weight, it is time to begin training. For this purpose ½-lb weights (the correct size can best be determined by study and experience) are usually used. Care should be taken to prevent making a sharp turn in the horn by using a weight that is too heavy or by allowing the weights to remain on for too long a time. If the horns yield too readily, it is best to remove the weights and give the horns a rest of from 10 days to a month, the length of time depending upon their condition. Then replace the weights until the desired effect is obtained. Weights should be removed when the horns become level with the top of the head, or not more than 1 in. below this level. Leaving the weights on longer tends to cause the horns to curve inward too much, thus causing a problem later. If the screw type of fastener is used, one should be careful not to force the screw into the horn so deeply that the depression cannot be removed. Horn weight losses may be reduced by tying a strong cord around the screws on the two weights; then if one weight is knocked off, it will not be lost. Horns may be pulled forward when they are 3 to 4 in. long by using a suitable spring or strap device for the purpose.

Fig. 14-14. Horn weights may be used in order to obtain symmetrical, properly curved and attractive horns. As soon as the horns are long enough and sufficiently strong, it is time to begin with a light weight. There is a variation in the degree of training that is desired in the different breeds. (Courtesy, USDA)

Horns may also be shaped by cutting, which is easier and less time consuming than the use of weights. Cutting is done at about one year of age. The horns are cut at a 45° angle at the natural break of the horn, followed by burning the ends to reduce bleeding.

Extremely long horns may appear out of proportion and unsightly. In such cases, they can often be cut back as much as 2 to 3 in., provided not more than ½ in. is removed at any one time and at no more frequent intervals than a month or 6 weeks. As a rule, most of the black tip can be removed without harming the sensitive part.

After the horns have been properly shaped, the next job is that of trimming and polishing. Usually it is best to smooth them down a week or two before the show. The rough surface may be smoothed with a sharp knife, a rasp, or a steel scraper; always scrape from the base toward the top. The final smoothing or finishing touches may then be given by using sandpaper, fine emery cloth, steel wool, or a flannel cloth and emery dust.

Horns are usually polished just before the show. An excellent polish that will not collect dust can be obtained as follows: Apply a paste which is made by mixing olive oil or sweet oil with pumice stone or tripoli. Polish by rubbing briskly with a flannel cloth. A quick and more simple polish can be obtained by the use of glycerine, linseed oil, or mineral oil. However, a polish obtained in this manner is rather temporary, and the oil will collect dust quickly.

WASHING

Frequent washing in the months preceding the show keeps the animal clean; stimulates a heavy growth of loose, fluffy hair; and keeps the skin smooth and mellow. However, most experienced fitters discontinue the use of soap 1 to 3 days before the show so that the hair will be more manageable on show day. During this period, they keep the animal as clean as possible, spot wash or rinse with clean water when necessary, and use a blower to remove dust and dandruff.

In preparation for washing, place a chain about the neck; never use a rope about the animal's neck when washing, for a wet rope cannot be easily loosened should the animal fall or otherwise get into trouble. Immediately prior to washing, remove all possible dirt and manure before wetting the animal. The latter may be accomplished with a curry comb and brush.

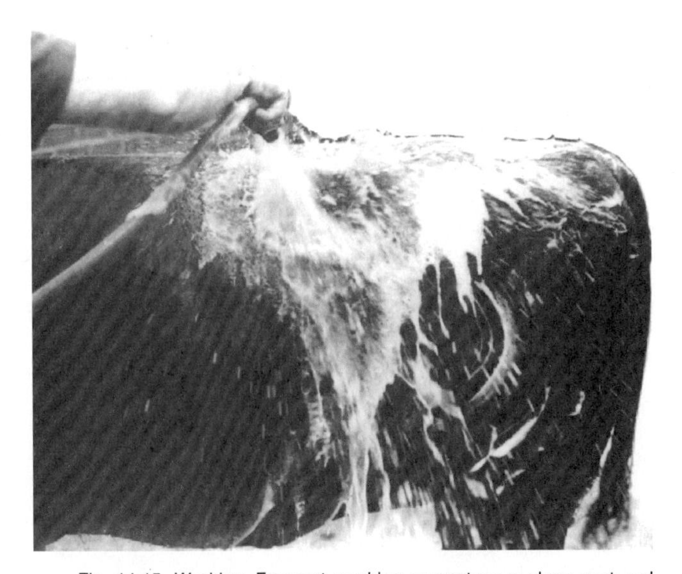

Fig. 14-15. Washing. Frequent washing guarantees a clean coat and stimulates hair growth. Show animals should be washed at least once a month until 6 weeks before the show and at least once a week thereafter until 1 to 3 days before show day. Use a mild, cold water soap followed by a thorough lathering, rinsing, and brushing to remove all excess soap, as the latter can cause dandruff. (Courtesy, American Angus Assn., St. Joseph, MO)

Most exhibitors use a high sudsing soap, such as Orvas or Castile, or a special livestock soap. However, if desired, an excellent preparation may be made as follows: mix 1 to 2 cups of good concentrated liquid coconut oil shampoo in 1½ gal of lukewarm water. With a bristle wash brush, thoroughly wet the dry animal with the soap solution. Then, with the hands and brush work the soap into a good lather, making sure that all parts of the body are well scrubbed and clean. Parts of the animal that are frequently neglected in washing are the head, tail, legs, brisket, and belly. Unless the animal is free of dandruff and the hide is thoroughly clean, double soaping is recommended.

Always wash the animal from head-to-tail and top-to-bottom. In washing the head, avoid getting soap or water in the eyes, ears, nostrils, and mouth. Cattle do not like to have their heads washed. A precaution commonly used in washing the head is to wash one side at a time while firmly holding the ear on the side being washed. Death of the animal may result from getting water into the lungs through the nostrils or mouth.

Following washing, the animal should be rinsed off very thoroughly in order to remove all traces of soap from the hair and skin, because soap left on the animal causes dandruff. For animals with light parts, a little bluing added to the last rinse water will improve the results. At this stage some exhibitors rinse the animal with a vinegar and water solution to lessen dandruff, as the vinegar helps to remove any soap residue. (Use ½ cup of vinegar per 1 gallon of water.) Excess rinse water may be removed by using the back of a Scotch comb, a water scraper, or by brushing downward. If you have an electric groomer or blower, use it to blow out the hair and dry the animal.

GROWING AND CARING FOR HAIR

A good coat of hair can, in the hands of an experienced exhibitor, cover up a multitude of defects.

Proper blocking and hair care necessitates that there first be sufficient hair.

Growing hair, especially in hot weather, involves the application of the following practices, beginning several months in advance of the show:

1. Keeping the animal as cool as possible.
2. Stimulating hair follicle growth with plenty of brushing.

Hair growth can be stimulated by wetting (*not* washing with soap) an animal down every morning and evening, during the coolest part of the day. After completely soaking the animal, the bar of the Scotch comb should be used to remove excess water before combing and brushing. Dampening the hair with a dip

Fig. 14-16. Grooming with Scotch comb and brush. (Courtesy, American Polled Hereford Assn., Kansas City, MO)

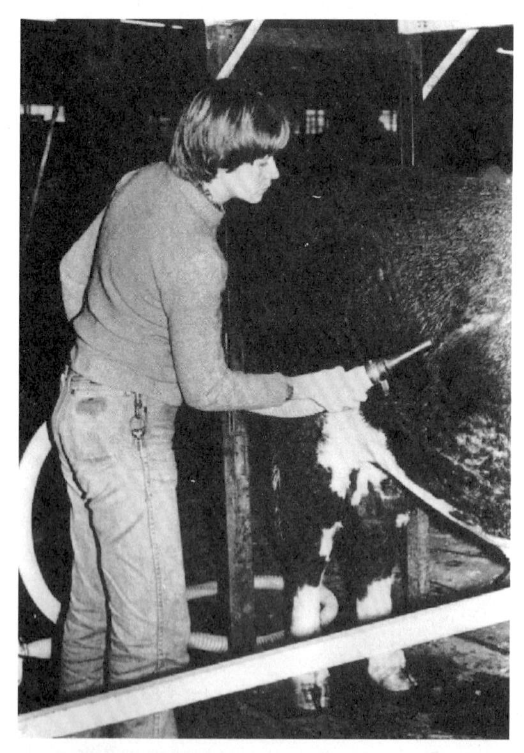

Fig. 14-17. Blowing the hair. This is done (1) to dry the hair, (2) to train the hair, or (3) to remove dust at the time of final grooming for the show. (Courtesy, American Polled Hereford Assn., Kansas City, MO)

solution, followed by combing, before turning the animal out at night will greatly help to train the hair.

While the hair is still damp, spend at least 20 minutes brushing and 10 minutes Scotch-combing. Never brush or comb down on long-haired animals; the object is to get the hair to stand up and out so as to add extra thickness and dimension. If you have a blower, alternate the brush and comb routine with some blowing in the direction you're brushing. When the hair is dry, apply a mist of water or hairset mixture; then brush, Scotch comb, and blow some more. For the first 2 weeks, work the hair forward and slightly down; during the second 2 weeks, work the hair straight forward; during the third 2 weeks, work the hair forward and upward to a 45° angle.

Remember that the key to a well-trained haircoat is the three Bs—brush, brush, brush. You cannot overbrush. Brushing stimulates the hair follicles and causes extra hair growth.

After working the hair, tie the animal in the shade where there is some air circulation. Better yet, tie it in an open barn or shed with a fan blowing on it to increase air circulation.

If the caretaker does not have sufficient time to grow hair and block, as outlined above, a popular alternative is to show *slick*. This consists of brushing all the hair on the body straight down, except for the legs and tailhead; saddle soap or adhesive is applied to the legs and tailhead and the hair is pulled up with a Scotch comb (see Fig. 14-18).

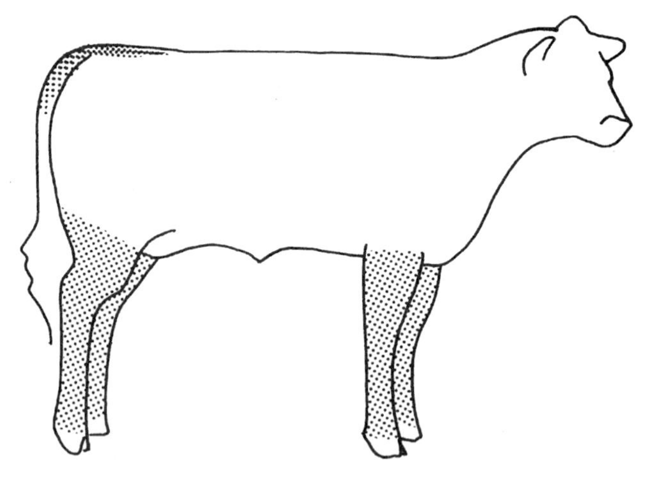

Fig. 14-18. For cattle with short hair, train and pull the hair up as shown in the shaded areas.

CLIPPING AND BLOCKING

Professional show people agree that no other assignment in fitting and showing cattle requires so much skill, patience, and time as does the art of proper clipping and blocking. The best way in which to master this art is to watch and emulate the experts.

Final clipping is best done about a week before the show, so that the clipped hair will lose its stubby, *fresh haircut* appearance.

Clipping and blocking are designed to accentuate the animal's most desirable qualities and to minimize its weaknesses. To this end, successful fitters study each animal; they observe it when walking, when standing relaxed, and when posed with the show stick. The fitter forms a picture of the animal as nature made it, and visualizes how it can be changed to approach the ideal. Following this preliminary study, the careful fitter plies the art, step by step, to emulate the ideal counterpart. Each bit of clipping and blocking is designed to assist in molding the final form. Each animal is treated as an individual, and the job of clipping and blocking is tailor made for that individual.

In preparation for clipping and blocking, the animal may be tied to a stout fence. However, if many animals are involved, most show people build or buy a simple metal pipe blocking chute with a headgate.

CLIPPING

Standard straight-edge clippers are used to clip the head, brisket, underline, and tail. Clipping these areas makes the animal look neat and trim, and not wasty.

Custom decrees certain breed differences in clipping and grooming (in *haircuts* and *hairdos*), and differences between steers and breeding animals. These will be pointed up in the discussion that follows:

1. **Head.** Head clipping styles vary, depending upon breed and sex. For cattle on which heads are normally clipped (show steers, and Angus), use an electric clipper and clip in front of a line that starts directly back of the ear; however, the long hairs on the poll should be left so as (a) to give the poll more prominence, and (b) to give the head a longer appearance (see Fig. 14-19). On horned Herefords and Shorthorns, do not clip the head. Other breeds vary; hence, professional advice should be sought relative to clipping them. Do not clip hair from the ears, the eyelashes, and around the nose.

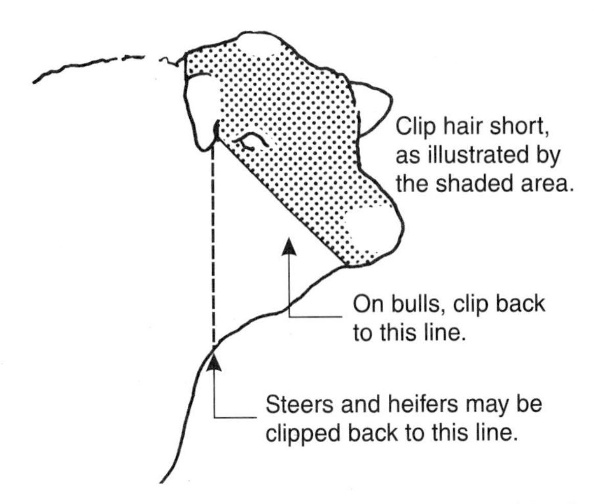

Fig. 14-19. Clip the head and face as shown (shaded areas), but do not clip the ears, the eyelashes, and around the nose.

2. **Brisket and underline.** Most beef animals need to have their briskets and underlines clipped, so that they appear trimmer. Clipping the belly also imparts the impression that the animal has more altitude. It is especially important that steers look trim, and not wasty. Also, the belly, brisket, shoulders, and front legs of breeding heifers may be clipped to emphasize femininity. If the heifer is shallow bodied, just clip the long hairs, and do not clip too close on the belly.

The brisket and underline should be clipped 1 to 2 weeks in advance of the show. Also, such clipped areas should gradually blend into the longer hair of the middle, neck, and shoulders, thereby avoiding any unartistic clipper line. This requires a steady hand. Also, the blending can be facilitated by use of a clipper guard (a plastic comb-like attachment) or by use of thinning shears.

3. **Tail.** One of the main objects in clipping the tail is to show the fullness of the twist and the thickness or beefiness of the hindquarters. In order to do this to best advantage with each individual, good judgment should be exercised in determining the extreme points of clipping that will show these characteristics to advantage. A general guide is to clip the tail from the high point of the twist to the tailhead (see Fig. 14-20). At the tailhead, the hair should be gradually tapered off near

Fig. 14-20. Clipping the hair from the tail of an Angus. One of the main objects in clipping the tail is to show the thickness or beefiness of the hindquarters. Clipping should begin above the switch of the tail opposite the point where the twist is fullest; and extend to the tailhead where it gradually tapers off, giving a blended effect with the rump. (Courtesy, American Angus Assn., St. Joseph, MO)

the body so that the tail blends nicely with the rump. To avoid leaving ridges of long hair extending down the center of the tail, the clipper should be run across the tail after it has been run upward to the tailhead.

BLOCKING

After the head, brisket, underline, and tail have been clipped, the animal should be blocked. Blocking is done to emphasize the animal's strong points—to make the animal appear at its best. Usually, a different type of clipper other than the standard straight-edge type is used for this job; the preferred clipper for use in blocking is known as a sheep-head clipper. However, when putting the finishing touches on an animal, a smaller, quieter set of animal clippers may be preferred.

Blocking the animal consists of clipping over the top and down the sides, and in blending in the points where hair has been closely clipped.

If the animal is to be blocked all over, care must be taken to insure that the hair is in the proper position before clipping. Sweeping the hair forward—from the rear quarters, along the sides, and to the shoulders—makes an animal look fatter.

Start at the animal's underside, and blend the underline into the hair on the side by gradually tapering the length of the hair until there is no longer a definite line. Shorten the hair through the middle, fore and rear flanks, and through the neck and shoulders to make the animal appear trimmer. Leave the hair on the rear quarters a little longer in order to make the animal appear very heavily muscled. In order to achieve the desired smooth look, professionals work both horizontally and vertically with the clippers, until they get to the topline of the animal.

Fig. 14-21. Trimming the underline and blocking the animal along the sides and back is practiced by most exhibitors to help complement the good points and diminish the bad points of an animal. (Courtesy, American Angus Assn., St. Joseph, MO)

Work the topline hair forward with a Scotch comb, getting it to stand up. Clip the topline hair so that the topline is as straight and level as possible as viewed from the side, and is rounding and muscular as viewed from the rear. Also, the topline hair should be blended into the tailhead, shoulder, and crest.

Professionals also block the following:

1. **Front legs.** The hair from above the knees should be clipped so as to blend smoothly into the forearm. Likewise, the hair should taper from the knees down; and long and bunchy hair should be lightly trimmed.
2. **Hind legs.** Pull the hair on the back legs up with a hard rubber brush or Scotch comb; and lightly trim the long and bunchy hair. Also, the legs should be trimmed to give the appearance of proper set and heavy-bone. Trim the area between the hock and twist, and also above the hock on the outside of the quarter so as to give it more bulge and expression of muscle. As viewed from behind, clip the top to give a round muscular appearance rather than a flat, square one.

Caution: Go slowly, and don't take off too much hair. Remember that you can always go back and take off more hair, but you cannot put back what you have already cut off. At intervals, step back and take a critical look at your work. Remember, too, that a good blocking job is an art, and that it takes plenty of time and patience.

Note well: Always check and comply with the rules of the livestock show. For example, the Houston Livestock Show and Rodeo, the world's largest livestock exhibition, has the following rule relative to the clipping of market steers:

> RULE 21: All steers will be clipped to have no more than ¼" of hair on any part of the animal's body upon arrival to show grounds. The only exception will be the tail switch. The switch may be ratted and balled or may be bobbed off. Steers will be checked at the time of weigh-in for conformity of the ¼" or less of hair. *NO CLIPPING* on steers (electrical or manual) will be allowed once steers are on the grounds at The Houston Livestock Show and Rodeo. All steers found in violation of this rule will be eliminated from competition and removed from the grounds. The exhibitor will also be eliminated from competition for the current show in the steer division.

MAKING SHOW ENTRIES

Well in advance of the show, the exhibitor should request that the show manager or secretary provide a premium list and entry blanks. All rules and regulations of the show should be studied carefully and followed to the letter—including requirements relative to entrance, registration certificates, vaccinations, health certificates, stall fees, exhibitor's and helper's tickets, and other matters pertaining to the show.

Generally, entries must be filed with the show about 30 days in advance of the opening date. Most shows specify that entries be made out on printed forms and in accordance with instructions thereon. The class, age, breed, registry number, and usually the name and registry number of the sire and dam must be given. Entries must be made in all individual and group classes in which it is intended to show, but no entries are made in the championship classes, the first place winners being eligible for the latter.

PROVIDING HEALTH CERTIFICATES

Health certificates, signed by an accredited veterinarian of the state of origin, are usually required for show animals. For cattle, most shows specify that this certificate indicate the following:

1. **Brucellosis.** That cattle over 6 months of age were tested and found negative to brucellosis within 30 days of entry, or that they were brucellosis vaccinated at 4 to 12 months of age, or that they meet other specified standards indicating that they are brucellosis free. Steers and spayed heifers are excepted.
2. **Tuberculosis.** That they are free of tuberculosis.
3. **Scabies.** That they are free of scabies, or that they have been dipped within 10 days prior to entry into the state. This provides reasonable assurance that diseases are not being spread. In addition, some states require that a special permit issued by the state veterinarian must accompany cattle on their trip home from the show.

SHIPPING TO THE SHOW

Show animals are usually shipped via truck or trailer. It is important that the following details receive consideration:

1. Schedule the transportation so that the cattle will arrive within the limitations imposed by the show and at least 2 to 3 days in advance of the date that they vie for awards.
2. Before using, thoroughly clean and disinfect any public conveyances.
3. Use long, clean, bright straw for bedding in order not to soil the hair or introduce foreign matter into it. It is also a good plan to sand the floor so that cattle will not slip.
4. In transporting by truck, cattle are generally stood crosswise of the truck, with the largest animal near the cab and tied facing to one side. The direction of facing the remaining animals is alternated; the second animal is faced in the opposite direction from the first, and so on. Some prefer to tie the animals so that all of them face the same direction. It takes more space, but they stay cleaner that way. In either case, it is best to tie animals fairly short and near enough together so that they will not lie down.
5. If space is at a premium, place the feed supply, bedding, and show equipment on a deck or platform in the truck, preferably at least 5½ ft above the floor. Allow for air circulation and tying of smaller animals under the deck.
6. When mixed feeds are used, as is usually the case in fitting rations, a supply adequate for the entire trip should be taken along in the truck. This will reduce the hazard of animals going off feed because of feed changes.
7. Limit show cattle to half feed at the last feeding before loading out and while in transit.
8. In transit, the animals should be handled quietly and should not be allowed to become hot nor to be in a draft.

STALL SPACE, FEEDING, AND MANAGEMENT AT THE SHOW

As soon as the show is reached, the animals should be unloaded and placed in clean stalls that are freshly bedded with clean straw. The cattle should be arranged in order of size so as to make the exhibit as attractive as possible.

While at the show, it is preferable that the cattle receive the same ration to which they were accustomed at home. Usually only a half ration is allowed for the first 24 hours after arrival at the show, and a normal ration is provided thereafter. So that the animals will maintain their appetite, however, it is necessary that they receive exercise while at the show. It is usually best to exercise the animals one-half hour or more in the cool of the evening and morning, when the animals are being led to and from their nightly tieouts. This also is a convenient time to clean out the stalls.

It is customary for exhibitors to identify their exhibits by means of neat and attractive signs, the size of which must be within the limitations imposed by the show. This sign usually gives the name of the breed of cattle and the name and address of the exhibitor.

SHOW DAY

On show day, the exhibitor should be well rested and get up early. The show animal(s) should be rinsed (*not* soaped) early and dried (a blower-dryer hastens the drying). Feed and water at the regular time. If the animal has a heavy middle, limit the feed and water, or do not feed and water. If the animal is tight middled, feed and water well. After feeding, let the animal rest.

Keep in mind that the natural look is in vogue now. *Note well:* In the past, shaving cream and butch wax served as grooming aids. Today, these products are considered outdated.

About 45 minutes before entering the ring, put the show halter on the animal; make sure that the nose strap is adjusted correctly. Blow or brush all dust and straw off the animal. Then proceed as follows:

1. **Brush the body.** The method of brushing the hair on the body depends on the length of the hair.

A long-haired animal should be groomed (brushed and Scotch-combed) forward and up (at about a 45° angle), following which the hair should be sprayed with a hair-setting product. (Several of these products are available through livestock supply companies.) If the hair is unruly, spray lightly with adhesive. But work it quickly with a Scotch comb, because adhesive sets fast.

If the animal has short hair, the body hair should be brushed down and smooth. Brush the hair up on the lower round; brush the hair on the shoulder and neck of heifers down to emphasize femininity; comb the hair on top of the tailhead up so that the tailhead looks square.

2. **Topknot the poll.** Pull the poll topknot up with your hand, and spray with adhesive.

3. **Prepare the tailhead.** The tailhead needs special treatment, also. Brush up the tailhead hair; spray with adhesive; bring up again with your hand and a Scotch comb; and trim with the clippers and scissors so as to level out the topline. The objective is a natural look.

4. **Bone the legs.** Many judges still try to select heavy boned cattle in breeding classes, despite the fact that studies reveal (a) that bone is of little importance in a selection program, (b) that bone varies little between animals of the same age, and (c) that bone cannot be determined by *eyeballing*, because what appears to be bone is really a combination of hair, hide, connective tissue, tendon, and bone (see Chapter 4, under the section entitled, "The Functional Scoring System"). Nevertheless, the exhibitor is exhibiting before the judge; not educating. Hence, the exhibitor should, to the extent possible, give what the judge wants. Since many judges insist on evaluating bone in breeding classes, and since no judge will place an animal down for appearing to have heavy bone (even though the judge is well aware of the facts stated above), it is good business to accentuate the bone. This is done by *boning the legs*.

The legs are boned for two reasons: (a) to impart an illusion of bigger bone, and (b) to make the legs appear more correctly set. First, spray on one area of a leg at a time a conservative amount of adhesive. (On normal days, use Easy Comb or Prime Time. On humid or rainy days, use Heat Wave or Tail Adhesive.) Next, pull the hair on the legs upward, to the forearms in front and to the stifle behind; using a Scotch comb. *Note well:* With heifers, only bone the hind legs. Brush the hair on the front legs straight down to give a feminine appearance (see Fig. 14-22).

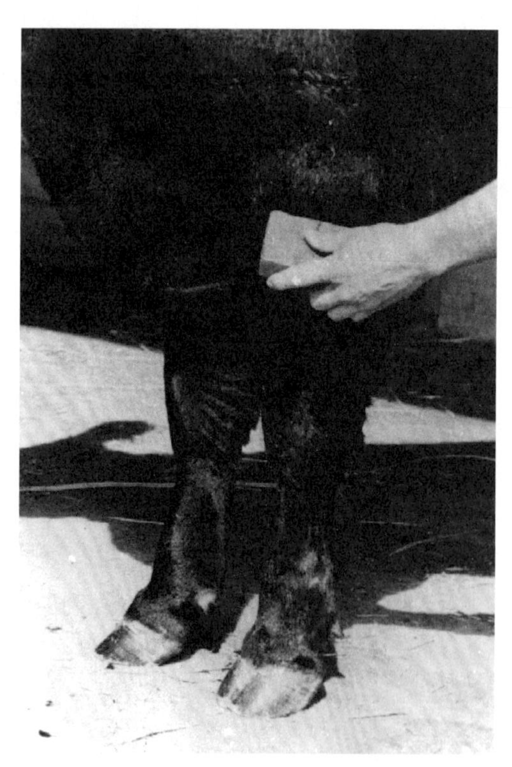

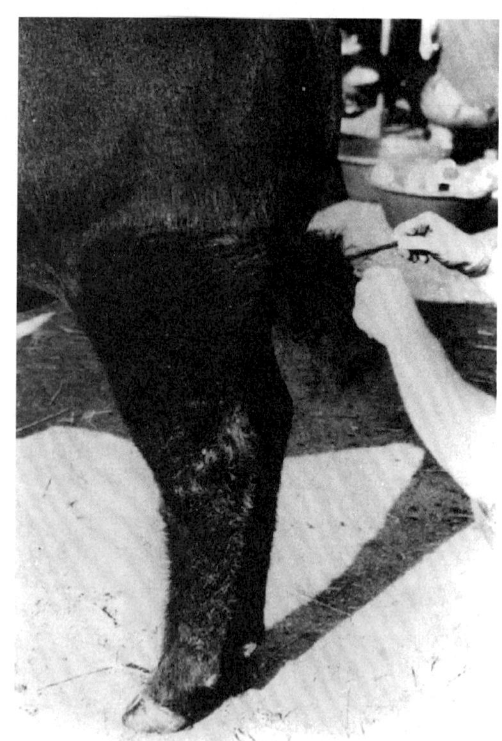

Fig. 14-22. Boning. By using a bar of glycerine saddle soap and a Scotch comb, an animal can be made to appear to be heavy boned. The bar is rubbed on the animal from the hoof to a point about 10 in. into the forearm *(left)*, followed by the use of a Scotch comb to pull the hair on the legs. This practice is also applied to the hind legs to show more bone and quarter in the rear of the animal *(right)*. (Courtesy, American Angus Assn., St. Joseph, MO)

5. **Ball or bob the tail.** *To ball:* Rat the tail into a tight ball by starting at the top of the tail and ratting everything high and to the center. Trim off excess long strands and spray with adhesive, and bag with a plastic bag. The ball should be done up so that it falls just below the quarter when the tail hangs naturally. Also, the ball may be shaped according to what will look best on the individual. On cattle that are long and flat quartered, the long, narrow ball is best. On cattle that are short quartered, the fuller ball will give more depth to the quarter when viewed from the side.

To bob: The switch is cut off at the hocks or slightly lower. Then, the hair is teased and tail adhesive is applied liberally.

Fig. 14-23. Balled tail of the Grand Champion Polled Hereford bull at the Kentucky Fairy & Exposition Center, Louisville, Kentucky. (Courtesy, Harold Workman, President and CEO, Kentucky Fair & Exposition Center)

Fig. 14-24. Bobbed tail of the grand Champion Limousin bull at the Kentucky Fair & Exposition Center, Louisville, Kentucky. (Courtesy, Harold Workman, President & CEO, Kentucky Fair & Exposition Center)

Breeding animals (males and females) are shown with the tail either balled or bobbed. However, professional fitters seem to favor bobbed tail steers.

6. **Oil the coat lightly.** Apply a light coat of oil either by (a) using a rag, oiled with a spray bottle; or (b) spraying oil on the animal, then brushing with a soft bristled brush. Do not show an animal with too much oil. This will give the coat an extra sheen so that the animal will stand out in the show-ring. **Caution:** Do not get oil on any area with adhesive on it, because the oil will take the adhesive out.

7. **Paint the hoofs.** The hoofs may be spray painted; using black lacquer on black cattle and clear lacquer on the rest.

8. **Make a final inspection.** Before entering the ring, give the animal a final check. Use the clipper and scissors to remove unruly hairs. Are the ears brushed and cleaned out? Did you remember to oil the face? Is the tail up and in shape to stay? *Remember, a natural-looking animal is desired.*

SHOWING BEEF CATTLE

Expert showmanship cannot be achieved through reading any set of instructions. Each show and each ring will be found to present unusual circumstances. However, there are certain guiding principles which are always adhered to by the most successful cattle exhibitors. Some of these are:

1. Train the animal long before entering the ring.
2. Have the animal carefully groomed and ready for the parade before the judge.
3. Dress neatly for the occasion.
4. Enter the ring promptly and in clockwise direction when the class is called.
5. Lead the animal from the left side (walking near the left shoulder), with the halter strap in the right hand.
6. When asked to line up, go quickly but not brashly.
7. When stopped, pose the animal correctly, and so as to minimize faults. Take the strap in the left hand and set the animal up with a leg under each corner. Generally, it is best to set the hind feet before setting the front feet. Keep the animal's head up and the back straight. A firm pressure near the navel, applied with the show stick, will help keep the weak-backed animal straighter. Animals with high loins can be pinched down with your fingers to straighten their tops. Cow-hocked animals can be made to look straight by pulling on the hocks with your hands.
8. Stroke the animal under the belly while posing, calming it.
9. When the judge handles your animal, react properly. If you feel that your animal may be slightly

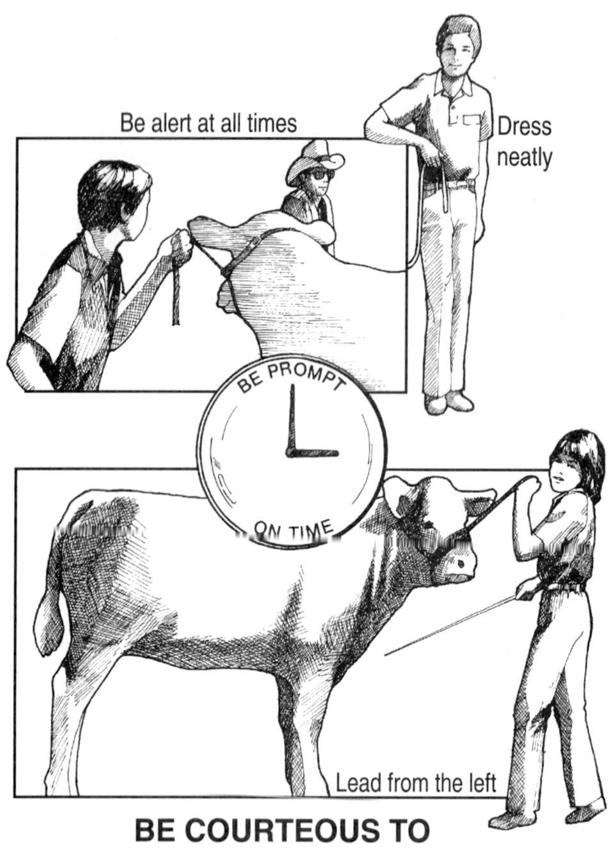

Fig. 14-25. Some of the guiding principles observed by the most successful exhibitors.

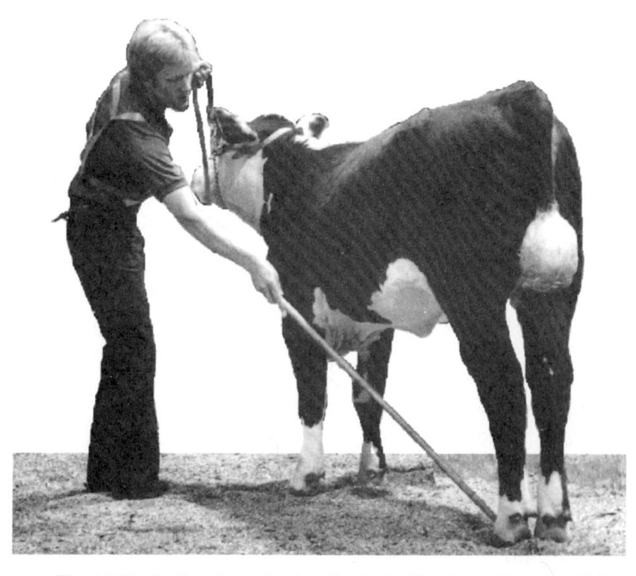

Fig. 14-26. Posing the animal at its best. (Courtesy, American Polled Hereford Assn., Kansas City, MO)

Fig. 14-27. Showing Charolais heifers. (Courtesy, National Western Stock Show, Denver, CO)

overdone, or too soft, turn its head away from the judge—thereby imparting firmness to the touch. If you think your animal is too bare, turn its head toward the judge—thereby imparting softness. After the judge handles the animal, comb the hair up where it was handled.

10. Keep one eye on the judge and the other on the animal. Center your attention entirely on showing the animal. The animal may be under the observation of the judge at a time when you least suspect it.

11. Let the animal stand *at ease* if you are in a big class and the judge is working at the other end of the ring. Calm the animal by scratching it with the show stick.

12. Never stand so that you block the judge's view; the judge is interested in seeing your animal—not you.

13. If you find that you are hemmed in and that the judge cannot see your animal, move to another location of vantage, unless, of course, the judge has asked you to hold your position.

14. Keep calm and collected. Remember that the nervous exhibitor creates an unfavorable impression.

15. Work in close partnership with the animal.

16. Be courteous and respect the rights of other exhibitors.

17. Do not enter into conversation with the judge. Speak to the judge only when you are asked a question; and never question the judge's placings.

18. Be a good sport. *Win without bragging and lose without squealing.*

19. Be ethical in how you prepare and show the animal, and practice sound animal husbandry in raising and caring for the animal.

BEEF CATTLE SHOWMANSHIP SCORECARD

Beef cattle showmanship contests are commonly conducted in 4-H, FFA, and colleges, for training purposes.

A scorecard listing what constitutes good showmanship, with a numerical value assigned to each point, (1) is a valuable teaching aid for beginners; and (2) systemizes judging such contests and avoids any point being overlooked. The Beef Cattle Showmanship Scorecard which follows was adapted from North Central Regional Extension Publication 156.

BEEF CATTLE SHOWMANSHIP SCORECARD

A. APPEARANCE OF ANIMAL — **Perfect Score = 40 points**

1. **Cleanliness** 10
   a. Hair clean and free of stains. Areas such as the switch, legs, belly, and head should be given special attention.
   b. Hide free from dirt and dandruff.
   c. Legs and feet clean and hoofs scraped.
   d. Leather or plastic halter clean.
2. **Grooming** 20
   a. Hair handled in a manner best suited to the individual animal. Long hair may be blocked and fluffed. Short, stiff hair may be shown smooth. In some cases, some areas may be brushed smooth, while other areas such as the rounds may be fluffed or pulled up to emphasize thickness or perhaps minimize some deficiency.
   b. Hoofs trimmed and shaped so the animal will stand straight; trimming done early enough so the animal's feet are not sore and so the animal has had time to adjust. Hoofs should be oiled.
   c. Horns should be curved, shaped, scraped, and polished. (All steers should have been dehorned.)
   d. Switch ratted into a tight ball or brushed out as fluffy as possible.
   e. Halter properly adjusted, nose strap midway between eyes and nose.
3. **Clipping** 10
   a. Except for minor touch-up, clipping should be done 5 to 10 days before the show.
   b. Heads clipped on market animals (no horns should be left on market animals).
   c. Heads of horned breeding heifers not clipped; heads of polled breeding heifers clipped unless breed custom dictates otherwise.
   d. If clipping is appropriate, the hair is clipped in front of a line starting about 1 in. back of the halter at the top of the head and extending almost straight down to the bottom of the throat. This includes the poll.
   e. Ears never clipped.
   f. Tail clipped from point half way between the twist upward to the tailhead. Tailhead not clipped short, but long hairs trimmed. Clipped portion should be blended into unclipped areas.

B. SHOWING THE ANIMAL — **Perfect Score = 40 points**

1. **Parading and Changing Positions** 10
   a. Animal led from left side; lead strap held in the right hand from 1 to 2 ft from the head and at height of animal's head. Extra lengths of lead strap held loosely or looped in hand to fit exhibitor's needs. Strap should not touch ground or interfere with exhibitor's showing. Exhibitor never leads animal while walking backward.
   b. Backward pressure applied with the lead strap to back animal out of line or move animal promptly to new location.
   c. Animal led in clockwise direction when necessary to parade it or move it to a different position.
2. **Posing in Ring** 15
   a. Animal kept from contact with a competitor or encroachment on space rightfully in possession of another.
   b. Exhibitor faces animal and holds lead strap in left hand while showing animal.
   c. Animal stands alert with head up, back level, and legs placed squarely under the body.
   d. Show stick used to place animal's feet. Show stick not overused. Show stick has bolt or dulled nail at end of stick to aid in feet placement.
   e. Exhibitor keeps whereabouts of judge in mind and has animal in position when judge is looking.
   f. It is proper to let animal relax, and to brush or comb animal when judge is at other end of ring. Grooming comb stored in handy pocket for use in ring.
   g. Animal shown until the entire class has been placed and the class is dismissed.
3. **Cooperation with Judge** 15
   a. Awareness of position of judge maintained but not made obvious.
   b. Body not permitted to obstruct view of judge.
   c. Animal maneuvered into improved position for benefit of judge's inspection prior to but not during inspection.
   d. Exhibitor steps aside if judge desires front view inspection.

C. APPEARANCE AND MERITS OF EXHIBITOR — **Perfect Score = 20 points**

1. **Appearance** 10
   a. Exhibitor well groomed with clothes neat and clean.
   b. All colors except white are appropriate for trousers. Extremes in color and fit should be avoided.
   c. White blouse appropriate for girls; white shirt or T-shirt appropriate for boys. Other colors satisfactory.
   d. Shoes hard sole.
2. **Merits** 10
   a. Brings animal into ring promptly.
   b. Recognizes and quickly corrects faults of animal.
   c. Alert and responsive to judge's and ringmaster's requests.
   d. Not distracted by persons and things outside ring.
   e. Shows animal, not self. Does not assume extreme, unnatural posture or gestures.
   f. Displays a courteous and sportsmanlike attitude.
   g. Prepared to give prompt answers to any questions pertaining to the animal.
   h. Does not leave ring until released by a ring official.

ETHICAL PROBLEMS AT LIVESTOCK SHOWS

The 4-H and FFA junior livestock show programs were developed to teach young people about agriculture and good animal husbandry practices. These programs also provide an arena for young people to be involved in good, clean competition. During the early 1990s, the reputation of these programs was tarnished as champion market animals at some of the major junior livestock shows in the country were found to have been administered illegal substances or tampered with in other ways prior to and during competition.

Testing of carcasses after slaughter revealed that some of the following substances were being used in show animals: (1) muscle-enhancing drugs or substances such as clenbuterol, administered to increase muscle growth (clenbuterol is a steroid-like drug that is banned in this country which poses a serious health risk to consumers if residues get into the food chain); (2) vegetable oil injected under the hide and into the muscle to give the appearance of increased muscularity; or (3) diuretics administered to cause rapid and severe weight loss. These and other problems led the federal government to issue warnings to meat packers concerning the slaughter of show animals, and this caused some packers to refuse to slaughter show animals.

These and other ethical abuses have resulted in exhibitors having to forfeit premiums and awards, and in some instances, exhibitors have been banned from shows for the rest of their competitive careers. Most of these unethical practices typically are being performed either by, or in association with, an adult (parent, leader, advisor, or professional fitter). In a few of the more serious cases, adults have been convicted of illegal activities and have received prison sentences.

It is a sad state of affairs when something as sacred as "mom and apple pie" is tarnished in this way. However, it should be stressed that these ethical abuses are being committed by a very small percent of the people involved in junior livestock show programs. The vast majority of exhibitors and adults are not being exposed to, or involved in committing, these kinds of abuses.

Many people involved in junior livestock show programs throughout the country have become extremely active and very aggressive in addressing these issues. In 1995, the Livestock Conservation Institute, with the help of Dr. Jeff Goodwin of the Texas Agricultural Extension Service, hosted a national symposium on this topic. Dr. Goodwin is the person who has led the charge in terms of addressing the problems in this area. Many fairs and expositions throughout the country have initiated testing and alternative show programs to insure that their junior livestock exhibitors are practicing good animal husbandry in the management and care of their livestock projects.

JUDGING BEEF CATTLE

Livestock judging is the art of visual appraisal, or the making of a subjective evaluation, of an animal. Cattle producers use it every day to select herd sires and replacement females for their breeding programs and to determine when animals are ready for market.

In addition to individual merit, the word *judging* implies the comparative appraisal or placing of several animals. In most judging contests, four animals are used in each class; and they are numbered 1, 2, 3, and 4, left to right as viewed from the rear. In livestock shows, a great number of animals may be ranked, or placed, in each class.

Judging beef cattle is an art, the rudiments of which must be obtained through patient study and long practice. The master breeders throughout the years have been competent livestock judges.

THE JUDGE

The judge(s) is the person chosen by the show management to determine the relative merits of the animals entered in the show. Judging is hard work and a great responsibility. Not only does the judge pick the winners, but the judge leads or misleads many people. For better or worse, the judge may be the cause of changing breeding programs and affecting the traits of the entire breed.

The essential qualifications that a good judge of beef cattle must possess, and the recommended procedure to follow in selecting or judging are as follows:

1. **Knowledge of the parts of cattle and the relative importance of each.** This consists of mastering the language that describes and locates the different parts of cattle (see Fig. 14-28). In addition, it is necessary to know which of these parts are of major importance in terms of form relating to function. In a slaughter animal, the latter necessitates knowledge of cutability and quality grades. In breeding animals, the Functional Scoring System (see Chapter 4 for pictures and description) may serve as a useful guide.

2. **A clearly defined ideal or standard of perfection.** The successful cattle judge must know for what to look; that is, the judge must have in mind an ideal or a standard of perfection based on a combination of (a) the efficient performance of the animal from the standpoint of the producer, and (b) the desirable

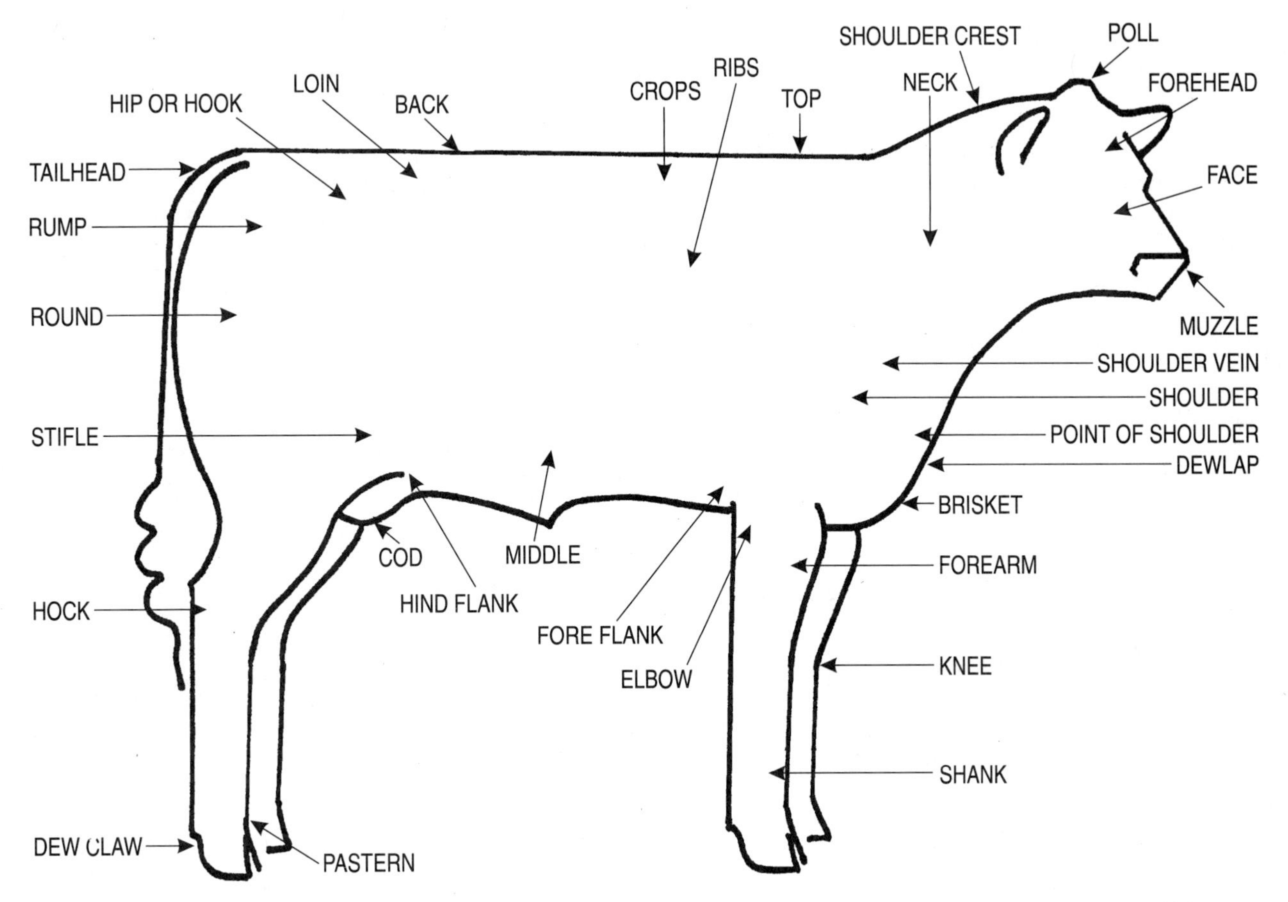

Fig. 14-28. Parts of a steer. The first step in preparation for judging beef cattle consists of mastering the language that describes and locates the different parts of the animal.

carcass characteristics of market animals as determined by the consumer. The major factors to consider are:

a. **Beef type.** This refers to the sum total of those economic traits included in the makeup of a commonly accepted ideal bull, female, or steer.

b. **Conformation.** This refers to the parts which make up the animal structure. Each part should be in proper form and proportion to lend balance in body conformation. Faulty conformation affects animal structure, form, and utility.

c. **Substance.** This refers to muscle and bone. Beef cattle need muscling to yield high cutability carcasses, and they need well-placed bone for skeletal soundness.

d. **Constitution.** This refers to capacity of chest and middle. Beef cattle should have ample capacity to house the digestive, respiratory, and reproductive systems.

e. **Size.** This refers to the skeletal frame and weight for age. The frame should be large enough to make possible rapid growth and efficiency in production and reproduction.

f. **Quality.** This is indicated by hide, hair, bone, and general bloom of the animal. Quality is not of significant importance, but it does add to animal appeal and attractiveness.

g. **Breed character.** This relates to the features of an animal which identify it with a particular breed.

h. **Sex character.** This refers to masculinity and femininity in breeding animals. Bulls should be strong, burly, and well developed in male sex features, with much importance placed on testicle development and position. Females should be feminine, indicating good mothering and milking ability.

i. **Disposition and temperament.** This is important in handling cattle on the range or in the feedlot. Good disposition is of economic value in beef production.

All of the above factors are important. The trained judge can evaluate them in a matter of seconds, using an objectively developed mental index.

It must be recognized, however, that the perfect specimen has never been produced. Each animal possesses one or several faults. In appraising an individual animal, therefore, its good points and its faults must be recognized, weighed, and evaluated in terms of an ideal. In comparative judging—that is, in judging a class of animals—the good points and the faults of each animal must be compared with the good points and the faults of every other animal in the class. In no other manner can they be ranked.

In addition to recognizing the strong and weak points in an animal, it is necessary that the successful judge recognize the degree to which the given points are good or bad. A sound evaluation of this kind requires patient study and long experience.

3. **Keen observation and sound judgment.** The good judge possesses the ability to observe both good conformation and defects, and to weigh and evaluate the relative importance of the various good and bad features.

4. **Honesty and courage.** The good judge of any class of livestock must possess honesty and courage, whether it be in making a show-ring placing or conducting a breeding and marketing program. For example, it often requires considerable courage to place a class of cattle without regard to: (a) placings in previous shows; (b) ownership; and (c) public applause. It may take even greater courage and honesty with oneself to discard from the herd a costly animal whose progeny has failed to measure up.

5. **Tact.** In discussing either (a) a show-ring class, or (b) cattle on a farm or ranch, it is important that the judge be tactful. The owner is likely to resent any remarks that imply that the animal(s) is inferior.

Having acquired the above referred to knowledge long hours must be spent in patient study and practice in comparing animals. Even this will not make expert and proficient judges in all instances, for there may be a grain of truth in the statement that "the best judges are born and not made." Nevertheless, training in judging and selecting cattle is effective when directed by a competent instructor or experienced cattle producer.

In the major shows, usually there is a separate judge(s) for each breed; in small shows, one judge may place all breeds. Each show must select its own judge or judges. The judge is usually selected because of expertise. An able judge knows modern beef type, and, in breeding classes, is familiar with the breed type, or character, of the particular breed being judged. Breed registry associations usually conduct judging clinics to train judges and provide literature setting

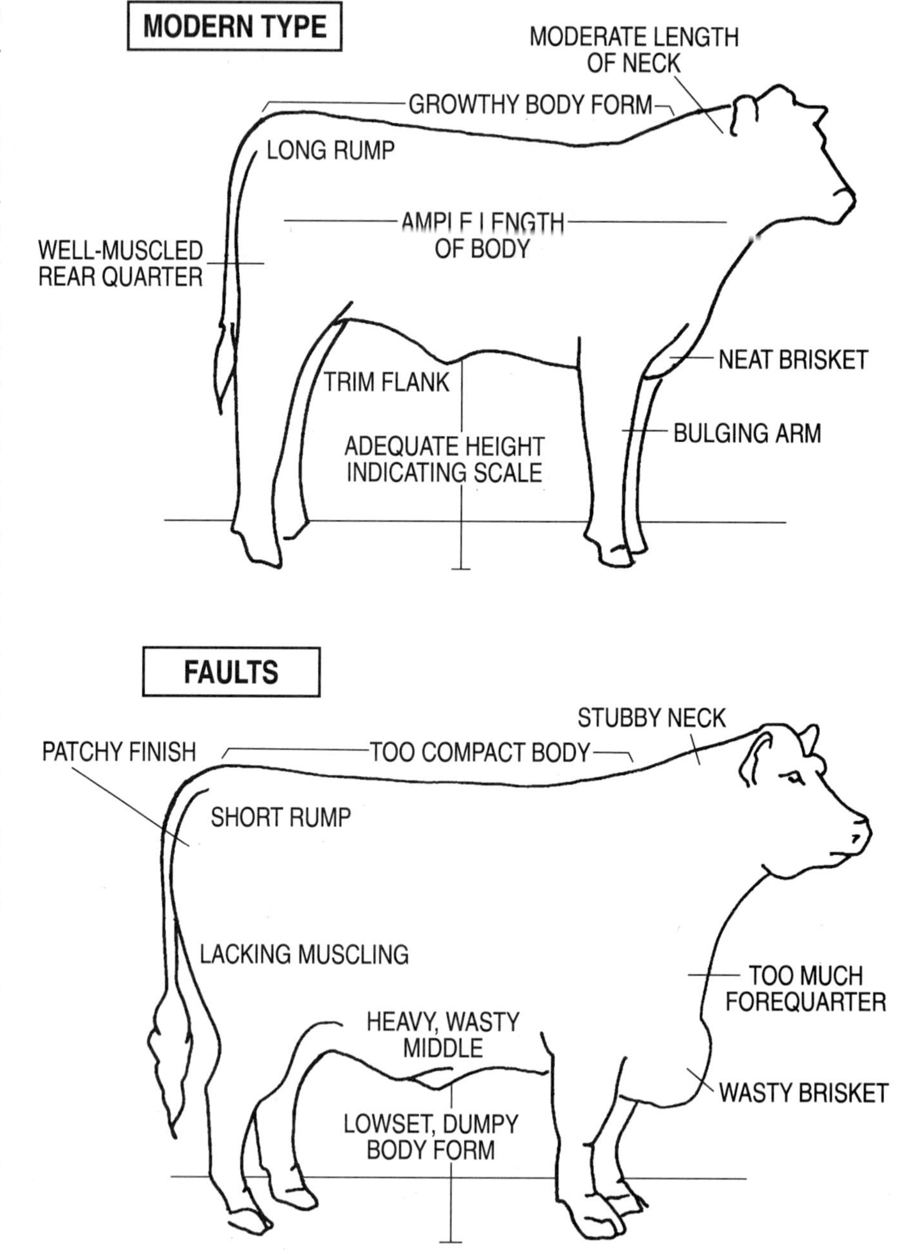

Fig. 14-29. Ideal beef type vs common faults. After mastering the language that describes and locates the different parts of an animal, the next requisite in judging is to have clearly in mind a standard or an ideal. In brief, the successful beef cattle judge must know what to look for, and be able to recognize and appraise the common faults. (Drawing by R. F. Johnson)

forth their breed standards, generally detailing the relative importance of each trait and listing faults and disqualifications. Some breed registry associations compile a certified list of judges with proven ability and make their names available to cattle show officials, as a recommended list from which to choose a judge(s).

Shows vary in the compensation of judges; generally speaking, the bigger the show the better the compensation. Most shows pay a professional fee plus expenses.

PLACING THE CLASS

When placing a class, the judge picks for the top animal the one which most nearly approaches the ideal, all traits considered. Then, the rest of the class is ranked on the same basis, placing as many animals as there are premium monies or ribbons.

Of course, opinions differ. As a result, two equally competent judges may come up with different placings on the same animals exhibited in successive shows. The more faults animals have, the more judges' opinions will differ. This is because they evaluate faults differently.

When officiating, most judges follow a sequence. Actually, it makes little difference as to the order of the views in inspecting cattle; what is important is that the same procedure be followed each time. With some slight variations, most judges proceed about as follows:

1. They start working a class after the ringmaster has lined them up, side by side, and advised that it is ready.
2. The animals in a class are examined as follows:
   a. Observed at a distance, in order to obtain a panoramic view. Also, the animals are viewed from all directions—(1) rear view, (2) front view, and (3) side view, thereby avoiding overlooking anything and making it easier to retain the observations that are made.
   b. Handled (with the hands), in order to determine the degree and quality of finish.
   c. Moved, in order to study their action—to see how they walk. The judge usually stands in the center of the ring, and the ringmaster asks that the exhibitors circle clockwise of the ring, thereby avoiding the necessity of the exhibitors who are leading the animals (from the left side) from being between the animal and the judge.
3. The judge evolves with the ranking of the class, based on the nearness with which each approaches the breed standard of perfection, along with each animal's merits and demerits.
4. The judge signals the animals to line up, in the order of the placings.
5. The judge may walk the class *Indian file* according to the placings, to double check the decision and let the crowd see the ranking.
6. The judge may give oral reasons on the class, defending the placings.
7. The ringmaster hands out the ribbons.
8. The ringmaster announces the next class. Then, class by class, this same procedure is followed until the class winners in each of the male and female classes are judged, followed by the selection of champions.
9. Usually after the show is completed, the judge goes over to the judge's table and signs the judge's books, in which the clerk has recorded the placings.

COMPOSITE JUDGING SYSTEM

The Composite Judging System is used in some beef cattle shows. It is designed to reduce the inconsistencies inherent in traditional show-ring judging. It practically eliminates the personal prejudice, bias, and abrupt changes in type sometimes found in one-person judging; and it alleviates the dominant personality of a committee placing.

Three judges are used in the Composite Judging System. Ideally, each should represent a different segment of the industry, *i.e.*, breeder, commercial producer, and university personnel. At the beginning of each class, judges draw for their designated title in that class—*Judge A*, *Judge B*, and *Referee*. Then, regardless of title, each judge independently places each class, with no communication between them.

Show-ring procedure is much the same as in traditional one-person judging. The animals are lined up by age; then observed, walked, and judged. The referee instructs the ringman on when and where to walk the cattle and line up. The cattle are always lined up according to age until the final composite placing is announced; judges are not permitted to make pair comparisons, side by side.

Each judge selects enough top animals to fill the monied or ribbon placings. Thus, if there are 8 places, each judge selects 8 top animals for further consideration. These selections are written on the judge's card, which is turned in. Animals not picked by any 1 of the 3 judges may be either excused from the ring or left standing at one end. Of course, in small classes where all animals are in the money, the latter procedure is not necessary.

Next, each judge places, or ranks, the *top cut* animals on a scorecard, and turns the card into the ringman. Then the composite placing is calculated and posted by the ring secretary as the final placing.

In compiling the composite placing, only the highest placing animal on both Judge A's and Judge B's cards are considered. If these 2 animals are one and the same, that animal wins first place. If they are different animals, the referee's card is consulted to see which of these 2 animals ranked higher. The higher placing animal of the 2 on the referee's card then goes into the final placing. Once an animal is ranked in the final placing, the number is crossed off Judge A's and Judge B's cards.

This same procedure is repeated until all monied placings have been filed. Table 14-2, Composite Compilation Sheet, shows the placings on one class of cattle and illustrates how the Composite Judging System works.

TABLE 14-2
COMPOSITE COMPILATION SHEET

Placing	Judge A	Judge B	Referee	Final
1st	106	106	104	106
2nd	111	110	106	110
3rd	104	104	110	104
4th	110	111	111	111
5th	108	108	109	108
6th	105	105	108	105
7th	107	109	105	109
8th	109	107	107	107

As shown in Table 14-2, Judges A and B agree that entry 106 is first, so 106 wins the class. For second place, Judge A has 111 and Judge B has 110. They disagree, so the referee's card is consulted, with consideration given only to 111 and 110. The referee placed 110 over 111, so 110 places second in the class.

Using the highest placing animals on each card that have not been placed, consideration for third place is between 111 from Judge A's card and 104 from Judge B's card. Since they differ, the referee's card is again consulted; and because 104 placed higher, 104 goes in third place.

For fourth place, 111 is the highest unplaced animal on both Judge A's and Judge B's cards, so 111 wins fourth place. The fifth and sixth place animals are the same on both A's and B's cards, so they win fifth and sixth places, respectively, regardless of how the referee placed them. The two judges differ on seventh and eighth places, so the final placing is resolved by the same procedure previously followed—that of using the referee's placings.

In calculating the final placing, two basic principles are followed:

1. Only the top animal on both A's and B's card is considered.

2. The referee's card is considered only when disagreement exists between Judges A and B; then only those two animals are considered.

After the final composite placing is completed, the ringman lines up the cattle in the order of their final placing. Then the results are announced for exhibitor and spectator information.

The Composite Judging System does not require any longer than a committee or two judge system, once it is understood. There is no possibility of a deadlock. If the judges disagree, the referee's card can be used to make the final placing.

The Composite Judging System is a foolproof system of combining the judgment of three judges.

HOW TO DETERMINE THE AGE OF CATTLE BY THE TEETH

Many shows are mouthing cattle, as a means of assuring "honest" show classification according to age. Thus, it is important that the exhibitor know how to determine the age of cattle by the teeth. Also, because the life span of cattle is relatively short, the age of cattle is of practical importance to the breeder, the seller, and the buyer.

The approximate age of cattle can be determined by the teeth as described and illustrated herewith. There is nothing mysterious about this procedure. It is simply a matter of noting the time of appearance and the degree of wear of the temporary and the permanent teeth. The temporary or milk teeth are readily distinguished from the permanent ones by their smaller size and whiter color.

It should be realized, however, that theoretical knowledge is not sufficient and that anyone who would become proficient must also have practical experience. The best way to learn how to recognize the age of cattle is to examine the teeth of individuals of known ages.

Of course, age determination of cattle is an art, and not a science. "mouthing cattle" is subject to human judgment, much like a human medical diagnosis based on observation. Also, it should be noted that variation in an animal's dental development can result from several causes, but primarily from nutritional variation or stress from sickness and environmental conditions. It is affected little by genetics or breeds.

At maturity cattle have 32 teeth, of which 8 are incisors in the lower jaw. The 2 central incisors are known as pinchers; the next 2 are called first intermediates; the third pair is called second intermediates or laterals; and the outer pair is known as the corners. There are no upper incisor teeth; only the thick, hard dental pad.

Table 14-3 illustrates and describes how to determine the age of cattle by the teeth.

TABLE 14-3
GUIDE TO DETERMINING THE AGE OF CATTLE BY THE TEETH

Drawing of Teeth	Age of Animal	Description of Teeth
Fig. 14-30	At birth to 1 month	Two or more of the temporary incisor teeth present. Within first month, entire 8 temporary incisors appear.
Fig. 14-31	2 years	As a long-yearling, the central pair of temporary incisor teeth or pinchers is replaced by the permanent pinchers. At 2 years, the central permanent incisors attain full development.
Fig. 14-32	2½ years	Permanent first intermediates, one on each side of the pinchers, are cut. Usually these are fully developed at 3 years.
Fig. 14-33	3½ years	The second intermediates orlaterals are cut. They are on a level with the first intermediates and begin to wear at 4 years.
Fig. 14-34	4½ years	The corner teeth are replaced. At 5 years the animal usually has the full complement of incisors with the corners fully developed.
Fig. 14-35	5 or 6 years	The permanent pinchers are leveled, both pairs of intermediates are partially leveled, and the corner incisors show wear.
Fig. 14-36	7 to 10 years	At 7 or 8 years the pinchers show noticeable wear; at 8 or 9 years the middle pairs show noticeable wear; and at 10 years, the corner teeth show noticeable wear.
Fig. 14-37	12 years	After the animal passes the 6th year, the arch gradually loses its rounded contour and becomes nearly straight by the 12th year. In the meantime, the teeth gradually become triangular in shape, distinctly separated, and show progressive wearing to stubs. These conditions become more marked with increasing age.

AFTER THE SHOW IS OVER

Most shows have regulations requiring that all exhibits remain on the grounds until a specified time, after which signed releases must be secured from the superintendent of the show. Because most exhibitors are anxious to travel when the show is over and there is considerable confusion at this time, it is usually advisable to load all equipment, leftover feed, and other articles before the release of animals is secured. Then all that remains to be done is to load out the animals.

Upon returning to the farm or ranch, it is usually good policy to isolate the show herd for a period of 3 weeks. This procedure reduces the possibility of spreading diseases or parasites to the balance of the herd.

It is important that young stock to be developed for show purposes the following year continue to receive an adequate, though lighter, grain ration.

Where the herd is being exhibited on a circuit, the fitter must use great care in keeping the animals in show condition at all times. The peak condition should be reached at the strongest show. In order to be successful, showing on the circuit requires great skill on the part of the caretaker, especially from the standpoint of feeding and exercising the cattle.

QUESTIONS FOR STUDY AND DISCUSSION

1. Under what circumstances would you recommend that a purebred, a commercial, or a 4-H Club or FFA member (a) should show, and (b) should not show beef cattle?

2. Take and defend either the affirmative or the negative position of each of the following statements:

   a. Fitting and showing does not harm cattle.

   b. Livestock shows have been a powerful force in beef cattle improvement.

   c. Too much money is spent on livestock shows.

   d. Unless all cattle are fitted, groomed, and shown to the same degree of perfection, show-ring winnings are not indicative of the comparative quality of animals.

3. In recent years, most major shows have augmented visual appraisal judging with performance and carcass information. When conducted in this manner, will cattle shows make for genetic improvement?

4. What consideration should be given to each (a) type and (b) breeding and performance when selecting cattle for fitting and showing?

5. How would you improve upon the typical show classification for breeding beef cattle given in Table 14-1 of this chapter?

6. How would you improve upon the steer classification (both the on-foot and the carcass divisions) of the National Western Stock Show, Denver, as presented in this chapter?

7. In what ways do animals being fitted for show differ from animals not being fitted in each (a) management, (b) business records, and (c) shelter and care?

8. Wherein does feeding cattle for the show differ from feeding cattle in a commercial cattle feedlot?

9. How would you gentle, halter break, and teach a beef animal to pose?

10. How would you correct the following cattle leg problems: (a) bow-legged behind, (b) toe-out in front, (c) pigeon-toed in front, and (d) cow-hocked behind?

11. Why are show cattle washed so frequently when being fitted for show?

12. Why is hair growth important in show cattle?

13. Is clipping and blocking show cattle good or bad? Defend your position.

14. *Boning* the legs is a deceptive practice, designed for the purpose of (a) making the cannon bone appear larger, and/or (b) improving stance. Is this good or bad? Justify your answer.

15. Are livestock shows making for more double muscling? Justify your opinion.

16. Some livestock shows have an *expert* mouth cattle, in an attempt to keep ages *honest.* How accurate is such an age determination?

SELECTED REFERENCES

Title of Publication	Author(s)	Publisher
Beef Cattle, Seventh Edition	A. L. Neumann	John Wiley & Sons, Inc., New York, NY, 1977
Beef Cattle Production	J. F. Lasley	Prentice-Hall, Inc., Englewood Cliffs, NJ, 1981
Beef Production	R. V Diggins C. E. Bundy	Prentice-Hall Inc., Englewood Cliffs, NJ, 1958
Beef Production and Management, Second Edition	G. L. Minish D. G. Fox	Reston Publishing Company, Inc., Reston, VA, 1982
Livestock Judging, Selection and Evaluation, Third Edition	R. E. Hunsley W. M. Beeson	Interstate Publishers, Inc., Danville, IL, 1988
Stockman's Handbook, The, Seventh Edition	M. E. Ensminger	Interstate Publishers, Inc., Danville, IL, 1992

Aerial view of Excel's Dodge City, Kansas modern beef slaughter and fabrication facility, which has a capacity of 5,400 head per day (two shifts). (Courtesy, Don Fender, Field Sales Services, Excel, Wichita, KS)

MARKETING AND SLAUGHTERING CATTLE AND CALVES[1]

Contents *Page*

[1]Helpful review suggestions and/or pictures for this chapter were received from the following who have great expertise in the marketing and slaughtering of cattle and calves: Don Fender, Excel Corporation, Wichita, Kansas; H. Kenneth Johnson and Terry Dockerty, National Livestock and Meat Board, Chicago, Illinois; and Jim Wise, USDA, Livestock & Seed Division, Washington, DC.

Contents Page

Fig. 15-1. Two Egyptian cattlemen taking an ox to market; from Bas Relief found in Sakara in the tomb of King Ephto Stoptep. At first, meat animals were bartered for articles made by craftsmen. Eventually, bartering gave way to cash sales as coined money began to circulate. (Courtesy, The Bettmann Archive, Inc.)

Marketing is an integral part of modern cattle production. It is the end of the line; that part which gives point and purpose to all that has gone before. The importance of cattle marketing is further attested to by the following facts:

1. A total of 79.2 million head of cattle and calves were marketed in 1993, in comparison with 103.5 million hogs and 9.7 million sheep and lambs. (See Fig. 15-2, which also gives 1983, 1985, and 1990 figures.)

2. Cattle and calf transactions in 1993 had a total value of $40.1 billion, in comparison with $10.9 billion for hogs and $0.5 billion for sheep. (See Fig. 15-3, which also gives 1983, 1985, and 1990 figures.)

3. In 1993, U.S. farmers and ranchers received 23% of the total cash income which they derived from

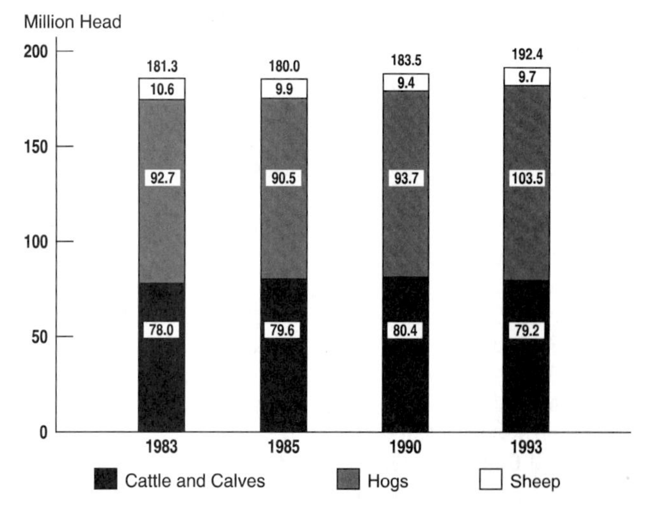

Fig. 15-2. Volume of livestock marketed by species, 1983-1993. (1993 from *Meat Animals*, April, 1994, USDA, pp. 4, 8, and 12; 1983–1990 from *Agricultural Statistics 1993*, USDA, pp. 236, 252, 264)

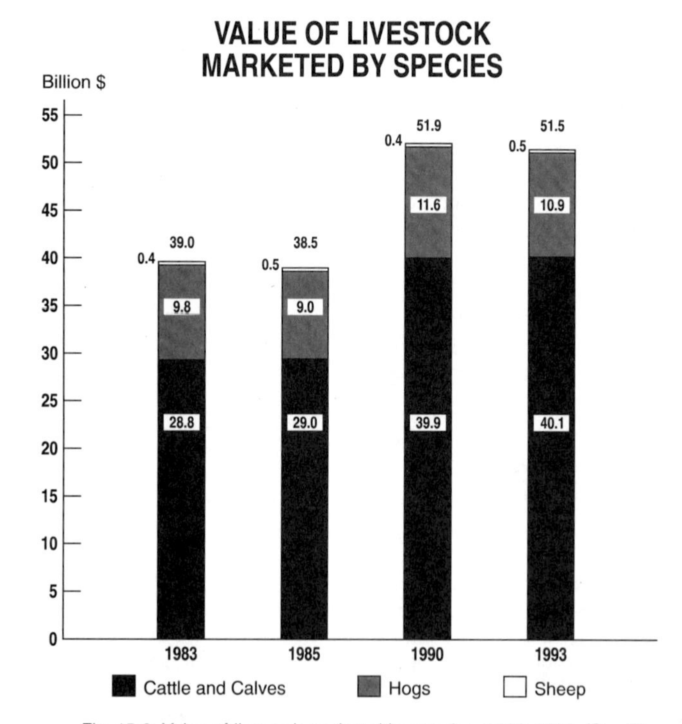

Fig. 15-3. Value of livestock marketed by species, 1983–1993. (See Fig. 15-2 for sources of data.)

livestock and livestock products *plus* crops from marketing cattle and calves.

4. In 1993, U.S. farmers and ranchers received 44% of the cash income which they derived from livestock and livestock products (exclusive of crops) from marketing cattle and calves.

5. In 1993, 33.3 million cattle were slaughtered in the United States.

6. Livestock markets establish values of all animals, including those on the farm or ranch. On January 1, 1993, there were 100,892,000 head of cattle in the United States, with an aggregate value of $65.5 billion, or $649 per head.[2]

[2]*Agricultural Statistics 1993*, USDA, p. 229, Table 370.

It is important, therefore, that the cattle producers know and follow good marketing practices.

MARKET PRICES

Market prices of cattle and beef are determined primarily by supply and demand, the structure of which is shown in Fig. 15-4. Also they are affected by marketing costs and government policy. Details follow.

1. **Supply of cattle.** This refers to the number of cattle that are available and which producers are willing to place on the market at the prices prevailing during a particular period of time.

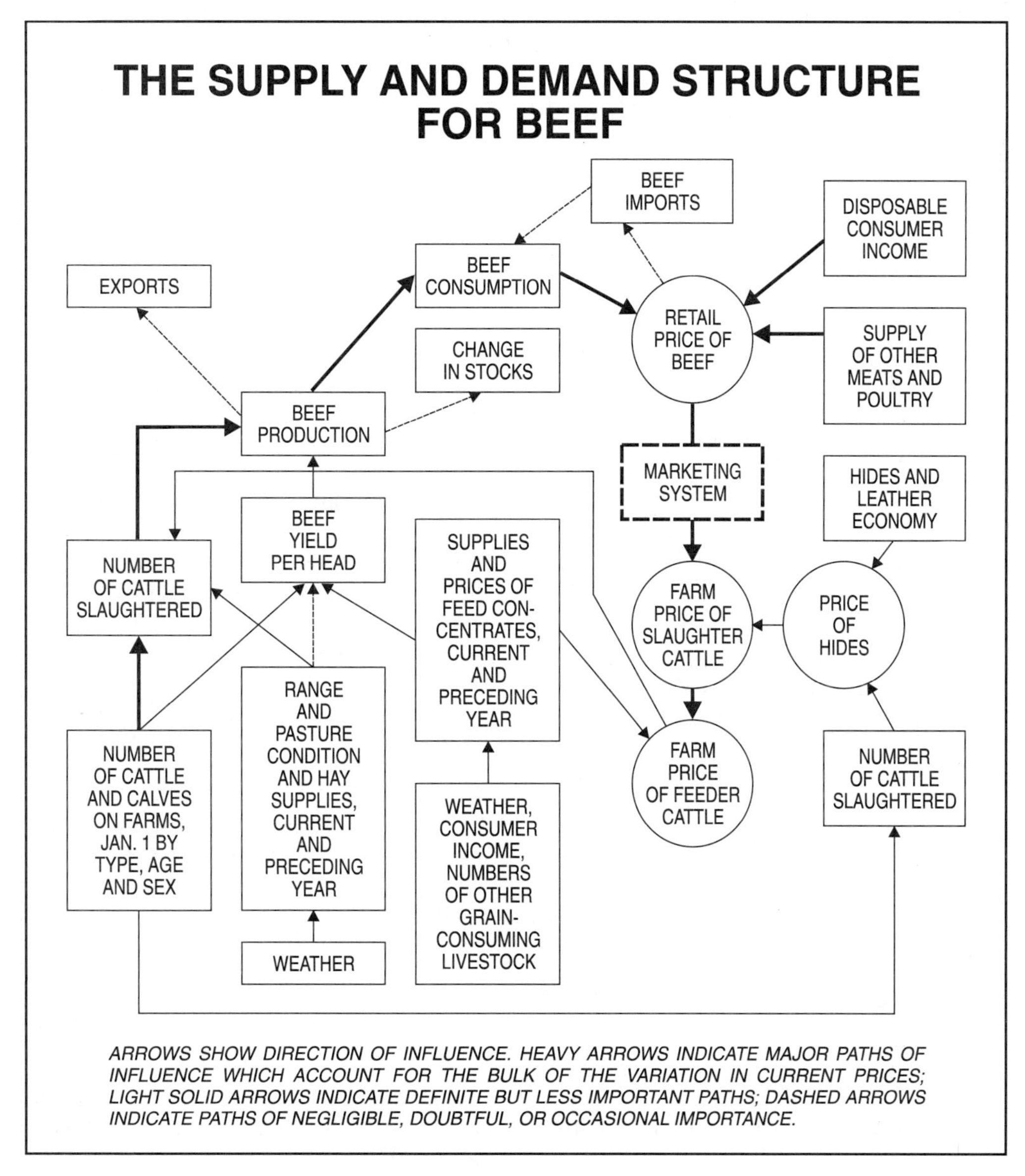

Fig. 15-4. Cattle prices are determined by two basic groups of factors—supply and demand. (USDA fig. redrawn)

2. **Consumer demand.** The demand for beef is influenced by six factors:

a. Human population.
b. Income of consumers.
c. Consistency and quality of beef.
d. Substitution of other products for beef. For example, if pork chops are selling at $3.24 per pound, and sirloin steaks are $3.91 per pound, the consumer has a tendency to buy pork chops rather than beef steak. Still other consumers may substitute poultry, fish, or cheese.
e. Tastes and preferences.
f. Habits, culture, and other environmental characteristics.

3. **Variation in marketing costs.** Marketing costs vary. They are affected by the market agency, transportation costs, etc.

4. **Government policy.** The government is deeply involved in policies favoring consumers. Thus, beef prices are affected by beef imports, government purchases of beef, etc.

In terms of individual sales, other factors that affect cattle prices are: size of lot, uniformity, quality, and record of desirability. The latter point refers to such things as feeder animals from a production tested herd, or finished animals with known high cutability.

In a private enterprise economy, the beef business competes with other enterprises for resources—for land, labor, capital, and management. Thus, if beef cattle are more profitable than alternate enterprises, resources will be allocated to cattle. However, if more money can be made raising vegetables, the resources may be switched to vegetable crops.

When cattle prices are depressed, the following psychological, and less tangible, factors come into play and tend to keep prices down.

1. **Pessimism.** Cattle producers tend to be either incurable optimists or unshakable pessimists. When the market price is up, bidding for cattle is lively. When the market goes down, buyers are as scarce as hen's teeth. This tends to exaggerate price fluctuations and extremes.

2. **Banker's mood.** Bankers are prone to wide fluctuations in mood—between mild pessimism (never optimism) and extreme pessimism. When prices are down, they get very nervous.

3. **Low equity levels.** Commercial cattle feeders normally operate with only 30 to 40% equity. Thus, a 35% drop in slaughter cattle prices may wipe out or leave equity levels perilously low (which bankers view dimly) and limit the feeder's ability to restock the feedlot.

4. **Commercial feeders lack the stability of the farmer-feeders.** Farmer-feeders, who produce most of their feed, are a tremendous stabilizing influence. They are able and willing to accept market risks. If the price is down, they may forego buying a new car and redecorating the house. But they continue to feed cattle. It's another story with their counterparts—the commercial feeders!

5. **Retail beef prices don't usually fall proportionally.** Packers and retailers may maintain or increase their margins during a depressed cattle market. As a result, there may be consumer resistance at the meat counter, further stymieing demand.

METHODS OF MARKETING CATTLE; MARKETING COSTS

Producers are confronted with the perplexing problem of determining where and how to market their animals. Usually there is a choice of market outlets, and the one selected often varies between classes and grades of cattle and among sections of the country. Thus, the method of marketing usually differs between slaughter and feeder cattle, and both of these differ from the marketing of purebreds.

Most market cattle are sold through the following channels: (1) direct, country dealers, etc.; (2) auctions; and (3) terminal markets.

Of course, it is generally recognized that these figures are continuing to shift, with terminal market sales decreasing and country and auction sales increasing.

Most U.S. livestock are marketed through any one of four channels—direct, auctions, terminals, and carcass grade and weight basis. But there are other methods, including country commission firms, local markets, concentration yards, order buyers, local plants and retailers, cooperative shipping associations, cooperative selling associations, telephone auctions, telephone direct selling, teletype auctions, electronic (video) marketing, and selling on consignment (custom method). It should be realized, however, that there is duplication in the listing. For example, order buyers operate at public stockyards and auctions, as well as conduct freelance country operations.

Producers need to be acquainted with livestock marketing costs, as well as with market channels. Although commission and yardage rates vary widely (1) according to size of consignment, and (2) between markets.

DIRECT METHOD (Country Dealers)

Direct selling, or country selling, refers to producers' sales of livestock directly to packers, local dealers, or farmers without the support of commission agents, selling agents, buying agents, or brokers. Direct selling

Fig. 15-5. Direct marketing of cattle between buyer *(left)* and seller *(right)*. (Courtesy, Smith Kline Animal Health Products, Philadelphia, PA)

does not involve a recognized market. The selling usually takes place at the farm, ranch, feedlot, or some other nonmarket buying station or collection yard.

Prior to the advent of public stockyards in 1865, country selling accounted for virtually all sales of livestock. Sales of livestock in the country declined with the growth of public stockyards until the latter method reached its peak of selling at the time of World War I. Country selling was accelerated by the large nationwide packers following World War I in order to meet the increased buying competition of the small interior packers.

Direct selling is similar to terminal market selling with respect to the price determination; both are by private treaty and negotiation. But it permits producers to observe and exercise some control over selling while it takes place; whereas consignment to distant terminal markets usually represents an irreversible commitment to sell. Larger and more specialized livestock farmers feel competent to sell their livestock direct.

Improved highways and trucking facilitated the growth of direct selling. Farmers were no longer tied to outlets located at important railroad terminals or river crossings. Livestock could move in any direction. Improved communications, such as the radio and telephone, and an expanded market information service, also aided in the development of country selling of livestock, especially in sales direct to packers.

Direct selling to meat packers is the most important outlet for slaughter cattle. Some packers buy cattle direct from producers at the plant; others send their buyers into the country, from farm to farm or feedlot to feedlot, where they make bids on the livestock that they inspect.

Packers also forward contract for cattle and have pricing arrangements in which cattle are priced according to a formula which usually includes futures prices and/or weekly top or average prices. These arrangements have been very controversial, as some people believe they are the wave of the future, while others believe they will lead to the downfall of the beef industry because of the loss of an active price discovery system for slaughter cattle.

- **Direct selling cost**—The out-of-pocket cost to the producer for direct selling is zero. The only expense is time, which, of course, is not a direct or out-of-pocket cost.

AUCTION MARKET METHOD

Livestock auctions (also referred to as sales barns, livestock auction agencies, community sales, and community auctions) are trading centers where animals are sold by public bidding to the buyer who offers the highest price per hundredweight or per head. Auctions may be owned by individuals, partnerships, corporations, or cooperative associations.

This method of selling livestock in this country is very old, apparently being copied from Great Britain where auction sales date back many centuries.

The auction method of selling was used in many of the colonies as a means of disposing of property, imported goods, secondhand household furnishings, farm utensils, and animals.

According to available records, the first public livestock auction sale was held in Ohio in 1836 by the Ohio Company, whose business was importing English cattle. This event also marked the first sale of purebred cattle ever held in America.

Although there are some records of occasional livestock auction sales during the 19th century, there is no indication of any auction market that continued operation throughout the period of the greatest development of public stockyards markets. It is within the

Fig. 15-6. Selling slaughter cattle at auction at the Omaha, Nebraska, stockyards. (Courtesy, *The Drovers Journal*, Kansas City, KS)

current century that present auction development had its beginnings. In fact, livestock auction markets had their greatest growth from 1930 to 1952.

Prior to the advent of livestock auctions, small livestock operators had two main market outlets for their animals: (1) shipping them to the nearest terminal market; or (2) selling them to buyers who came to their farm or ranch. Generally, the first method was too expensive because of the transportation distance involved and the greater expense in shipping small lots. The second method pretty much put producers at the mercy of the buyer, because they had no good alternative to taking the price offered, and often they did not know the value of their animals. By contrast, big operators are not particularly concerned about these things. Because of their large scale, usually they can take advantage of any of several terminal markets, and they know enough about values that they can deal satisfactorily with buyers who come to their feedlot, farm, or ranch. Thus, livestock auctions are really of greatest importance to small operators.

Rates charged for marketing livestock vary at different auctions. Services for which charges are levied may include selling, yardage, weighing, insurance, brand inspection, and health inspection. Many auctions do not provide all these services. A commission or selling fee, however, is charged at all markets and is the primary source of income to auction operators. At some auctions, the commission covers yardage and weighing in addition to the selling service. Some operators levy a separate charge for each service provided, while others charge a single rate to cover all services.

The auction market method of selling is similar to the terminal market in that both markets (1) are an assembly or collection point for livestock being offered for sale, (2) furnish or provide all necessary services associated with the selling activity, (3) are supervised by the federal government in accordance with the provisions of the National Packers and Stockyards Act, and (4) are characterized by buyers purchasing their animals on the basis of visual inspection.

But there are several important differences between terminals and auctions; among them, auction markets (1) are not always terminal with respect to livestock destination; (2) are generally smaller; (3) are usually single-firm operations; (4) sell by bid, rather than by offer and counter-offer; and (5) are completely open to the public with respect to bidding, and all buyers present have an equal opportunity to bid on all livestock offered for sale, whereas the terminal method is by private treaty (the negotiation is private).

■ **Auction charges**—Auction charges are typically on a percentage basis and vary, but average about as follows on a per head basis: feeder steers and heifers, $5.00; cows, $7.50; and cow-calf pairs, $12.50. Generally, additional charges are made for feed, brand inspection, insurance, health inspecting, and the beef checkoff program ($1.00).

FEEDER CALF AUCTIONS

Feeder calves are sold through the following market channels: auctions, direct, terminal markets, and dealers. In recent years, feeder calf sales, held auction style, have been especially popular in many states. Generally such sales are organized on a statewide or area basis, with the state agricultural extension service cooperating.

In advance of the auction, all entries are usually "sifted" by a committee, whose duty it is to assure the desired quality. The calves that pass the initial inspection are then delivered to the auction market a day ahead of the sale, at which time they are (1) tagged, (2) weighed, and (3) graded. Following this, each calf is penned with other calves of similar breed, sex, grade, and weight; thereby providing uniform lots for buyers. Usually, the pens are marked, giving the grade, sex, breed, and average weight of each lot of calves. The buyers are then given an opportunity to inspect the pens of calves prior to the sale. At the appointed time, each uniform lot of calves is offered for sale by the auction method.

For successful feeder calf auctions, it is necessary that there be both volume and quality. Large numbers attract more buyers and make it possible to sell calves in larger and more uniform lots.

The better managed feeder calf sales serve as an excellent education medium. Producers observe the grading demonstration, see how the weights of their calves compare with those of the neighbors, and realize the price spread between grades.

Two of the most important advantages to accrue from feeder calf sales are: (1) producers with a few head are provided market advantages comparable to large operators; and (2) cattle feeders are given an opportunity to purchase feeder calves of specified quality and in small or large lots, without the time and expense of shopping around.

VIDEO AUCTIONS

Pat Goggins, Billings, Montana, veteran auction owner and newspaper publisher, conducted the first video cattle auction at Billings, Montana, in September, 1976, when 21,000 head of video filmed cattle were sold from a television set—marking another first for America, another first for the world.

In 1992, the four major U.S. video auction companies sold a total of 1,136,131 head of cattle. Each year, more feeder cattle and breeding cattle are being marketed through video auctions, and this method is fast becoming the most popular method for marketing feeder cattle in certain areas of the United States for producers with larger groups of cattle.

Video auctions are conducted by showing buyers a 2 to 4 minute videotape taken at the farm or ranch of the cattle being offered for sale. The audio portion of the tape announces the weight, location, and background of the cattle; along with the weighing conditions, and delivery date(s).

Through a satellite hook-up, the videotape can be shown to buyers assembled at several locations. Also, anyone with a satellite "dish" can view consigned animals from any point in the continental United States and at other points throughout the world where special satellite televised network service is available. However, actual bidders must be pre-registered by most video auctions.

At the sale date and time, the auctioneer sells the cattle to the highest bidder. After the sale, seller and buyer arrange for the delivery of the animals.

Video marketing is an idea whose time has come. It's primary advantages in comparison with conventional auctions are:

1. **More cattle and more buyers.** A greater number of sale cattle are exposed to a greater number of buyers.
2. **Less stress and delivery expense.** Cattle offered for sale experience no handling stress or hauling expense prior to the day of delivery. The buyer is assured of farm-fresh cattle.

Since the dawn of the nation's cattle business, much of the annual multi-billion cattle transactions have been initiated by word of mouth, hip pocket banking, or matchbook accounting. Now, armchair video auctions are joining these time-honored and ingenious methods of selling cattle.

TERMINAL OR CENTRAL MARKET METHOD

Terminal or central markets (also referred to as terminal public markets, or central public markets) are livestock trading centers, which generally have several commission firms and an independent stockyards company.

The first terminal market was established at Chicago in 1865. Up through World War I, the majority of slaughter livestock in the United States was sold through terminal markets by farmers or by local buyers shipping to them. Since then, the importance of these markets has declined in relation to other outlets.

Formerly, terminal markets were synonymous with private treaty selling. Today, however, many terminal

Fig. 15-7. Terminal market. Note packer-buyer and commission representative negotiating. (Courtesy, Oklahoma State University)

markets operate their own sale ring and all, or almost all, of their livestock are sold by auction.

The terminal or central market method entails the following distinct steps:

1. The producer must decide when to sell animals.
2. The producer must deliver animals to the terminal.
3. The producer must consign animals to a commission firm.
4. The commission firm must accept the animals upon their arrival at the terminal.
5. The commission firm must pen, water, feed, sort, sell, and attend to all the necessary tasks from the time the animals arrive until they are sold.
6. The commission firm collects from the buyer and pays the seller.
7. The buyer takes title to the animals when they are weighed.

The stockyards company performs a number of useful services concerning the physical operation of the market. Generally, it takes no active part in either buying or selling animals.

LEADING PUBLIC MARKETS

Public markets are stockyards operating in commerce in the United States, including both auction and terminal markets, that are posted by the Packers and Stockyards Administration.

It is recognized that public markets have declined in importance in recent years, whereas direct marketing has increased. But figures on the number of cattle and calves marketed directly are difficult to come by. So, receipts of cattle and calves at the leading public markets are presented in this section because they are available and a good barometer of current location and activity.

Although public markets vary from year to year in total receipts, Table 15-1 shows the largest public markets in cattle receipts in 1993. It is noteworthy that Oklahoma City leads by quite a margin.

As would be expected, many market calves are of dairy breeding, especially the surplus bull calves that are not needed for breeding purposes. Of the remainder, a considerable number are culled out from beef herds because of undesirable type or breeding from the standpoint of future development. It can be expected, therefore, that the leading calf markets would not coincide with the leading cattle markets.

Table 15-2 shows the 10 leading public markets in calf receipts in the United States. Springfield, Missouri, with heavy receipts of dairy calves from the concentrated dairy area that it serves, leads.

In recent years, there has been an increasing tendency to market feeder cattle direct, without passing them through a public market. Also, an increasing number of producers are arranging to have their feeders custom fed.

TABLE 15-1
CATTLE RECEIPTS (Excluding Calves and Vealers) OF 10 LEADING PUBLIC MARKETS, BY RANK[1]

Market	1993
Oklahoma City, OK	498,803
Torrington, WY	413,880
LaJunta, CO	355,109
Sioux Falls, SD	320,666
Pratt, KS	310,782
Dodge City, KS	308,004
Lexington, KY	250,489
San Angelo, TX	243,377
Joplin, MO	231,524
Billings, MT	229,399
Total 28 markets	5,003,721

[1]Animal Products Branch, Economic Research Service, USDA.

TABLE 15-2
CALF RECEIPTS OF 5 LEADING PUBLIC MARKETS, BY RANK[1]

Market	1993
Springfield, MO	6,671
So. St. Paul, MN	4,800
Joplin, MO	4,551
Lancaster, PA	3,483
Louisville, KY	1,158
Total of 28 markets	19,563

[1]Animal Products Branch, Economic Research Service, USDA.

CARCASS GRADE AND WEIGHT BASIS

It is generally agreed that there is need for a system of marketing which favors payment for a high cutout value of primal cuts and a quality product. Selling on the basis of carcass grade and weight fulfills these needs.

The bargaining is in terms of the price to be paid per hundred pounds dressed weight for carcasses that meet certain grade specifications. It is the most accurate and unassailable evaluation of the value of a carcass. From the standpoint of the packer, this procedure is more time-consuming than the conventional basis of buying, and there is less flexibility in the operations.

In general, cattle producers who market superior animals benefit from selling on the basis of carcass grade and weight, whereas the producers of lower quality animals usually feel that this method unjustly discriminates against them.

The factors favorable to selling on the basis of carcass grade and weight may be summarized as follows:

1. It encourages the breeding and feeding of quality animals.
2. It provides the most unassailable evaluation of the product.
3. It eliminates wasteful filling on the market.
4. It makes it possible to trace losses from condemnations and price to the producer responsible for them.
5. It is the most effective approach to animal improvement.

The factors unfavorable to selling on the basis of carcass grade and weight are:

1. The procedure is more time-consuming than the conventional basis of buying.
2. There is less flexibility in the operations.
3. The physical difficulty of handling the vast U.S. volume of animals on this basis is great.
4. If all slaughter cattle were sold on this basis, there would be no active price discovery system to determine the value of slaughter cattle.

The first data for carcass purchasing of livestock was published by the U.S. Department of Agriculture for the year 1961. For the United States as a whole, the following percentages of total cattle and calf slaughter were purchased on a carcass grade and weight basis in 1990: All cattle, 38.2%; calves, 59.6%.[3]

[3]*Meat and Poultry Facts, 1994*, American Meat Institute, p. 11.

CHOICE OF MARKET CHANNEL

Marketing is dynamic; thus, changes are inevitable in types of market channels, market structures, and market services. Some outlets have gained in importance; others have declined.

The choice of a market channel represents the seller's evaluation of the most favorable market among the number of alternatives available. No simple and brief statement of criteria can be given as a guide to the choice of the most favorable market channel. Rather, an evaluation is required of the contributions made by alternative markets in terms of available services offered, selling costs, the competitive nature of the pricing process, and ultimately the producer's net return. Thus, an accurate appraisal is not simple.

From time to time, producers can be expected to shift from one type of market outlet to another. Because price changes at different market outlets do not take place simultaneously, nor in the same amount, nor even in the same direction, one market may be the most advantageous outlet for a particular class and grade of cattle at one time, but another may be more advantageous at some other time. The situation may differ for different classes and kinds of livestock and may vary from one area to another.

Regardless of the channel through which producers market their cattle, in one way or another, they pay or bear, either in the price received from the livestock or otherwise, the entire cost of marketing. Because of this, they should never choose a market because of convenience of habit, or because of personal acquaintance with the market and its operator. Rather, the choice should be determined strictly by the net returns from the sale of their livestock; effective selling and net returns are more important than selling costs. To arrive at net returns, the seller must deduct such indirect marketing costs as transportation and shrinkage. Likewise, it costs a packer money to keep a buyer at the market.

CATTLE MERCHANDISING

Because of the preponderance of finished cattle sold direct, the National Cattlemen's Beef Association prepared the following set of marketing guidelines to which it recommends that the industry adhere:

1. **Presentation of cattle.** Present only those cattle to each packer buyer which the feeder might reasonably expect to be bought.
2. **Delivery time.** All cattle should be sold for delivery within seven days of the sale date; and time extensions should not be allowed.
3. **Prompt payment.** Cattle should be paid for by check on the day of delivery to the buyer's account, or wire payment should be made to seller's bank by the close of the next business day after the transfer.
4. **Weighing conditions.** Cattle should be weighed early on the morning of delivery, without prior feeding that morning.
5. **Mud.** Excessive mud on cattle should not affect weighing conditions. Mud should be considered as a price factor and not a weight factor.
6. **Grade and yield selling.** Unless producers and feeders are paid a premium for superior meat-type animals, cattle should not be sold on a grade and yield basis, and all such transactions should have an agreement between packer and feeder on an established price. The yield should be on a hot weight basis and the packer should furnish a certified copy of the weight receipt.
7. **Rail killing.** Under no conditions should cattle be sold on carcass weight basis without a fixed price being established at the time of sale.
8. **Condemnation claims.** Any claims of packers should be supported by a previous agreement with the feeder.
9. **Point of change of ownership.** Ownership shall pass to buyer when cattle exit scale at time of weigh-up.
10. **Maintenance of current status at feedlot.** Feeders are encouraged to market cattle when they reach optimum weight and grade. They should never attempt to *bull* the market.
11. **Sales reporting.** Feeders should make a prompt (within the hour) report of all sales, to both *Cattle-Fax* and the nearest USDA market news office, in order to help keep the industry accurately informed on prices.

PREPARING AND SHIPPING CATTLE

Cattle are transported by truck, rail, and air. Regardless of the type of conveyance, proper preparing and shipping are much the same.

Improper handling of cattle immediately prior to and during shipment may result in excessive shrinkage; high death, bruise, and crippling losses; disappointing sales; and dissatisfied buyers. Unfortunately, many producers who do a superb job of producing cattle, dissipate all the good things that have gone before by doing a poor job of preparing and shipping. Generally speaking, such omissions are due to lack of know-how, rather than any deliberate attempt to take advantage of anyone. Even if the sale is consummated prior to delivery, negligence at shipping time will make for a dissatisfied customer. Buyers soon learn what to expect from various producers and place their bids accordingly.

The following general considerations should be

Fig. 15-8. Shipping cattle by truck. (Courtesy, Howard Barnes Livestock Trucking, Fayetteville, AR)

accorded in preparing cattle for shipment and in transporting them to market:

1. **Select the best suited method of transportation.** The producer should decide between truck, rail, and air transportation on the basis of which method best suits the particular situation.

The railroad system set the pattern for concentrated and centralized livestock marketing and meat packing locations. Then came the motor truck and highway system which were instrumental in reversal of the organizational structure—causing decentralization.

Plane shipments are a specialty, the details of which had best be left in the hands of an experienced person or agency, such as the livestock airlift specialist of an airline or an import-export company. At the present time, cattle shipments via air are largely confined to foreign movements, especially valuable breeding animals.

Fig. 15-9. Registered Hereford calves bred by Tequesquite Ranch, Albert, New Mexico, being loaded on a jet freighter at Amarillo, Texas, for air shipment to Parker Ranch, Hawaii. A total of one hundred and sixty 450-lb calves were loaded on this plane. (Courtesy, Tequesquite Ranch, Albert, NM)

2. **Feed properly prior to loading out.** Never ship cattle on an excess fill. Instead, withhold grain feeding 12 hours before loading (omit 1 feed), and do not allow access to water within 2 to 3 hours of shipment. Cattle may be allowed free access to dry, well-cured grass hay up to loading time, but more laxative-type hays, such as alfalfa or clover, should not be fed within 12 hours of shipment even if the animals were accustomed to them previously. Likewise, cattle on green or succulent feed should be conditioned to dry feeds prior to shipment.

Cattle that are too full of concentrated or succulent feeds or full of water at the time of loading will scour and urinate excessively. As a result, the floors become dirty and slippery and the animals befoul themselves. Such cattle undergo a heavy shrink and present an unattractive appearance when unloaded.

Abrupt ration changes of any kind prior to shipment should be avoided. Occasionally, a misinformed producer withholds water, but gives a liberal feeding of salt prior to shipment, to obtain maximum water consumption and fill on the market. This *sharp* practice cannot be condemned too strongly; it is cruel to animals, and experienced buyers are never deceived.

3. **Keep cattle quiet.** Prior to and during shipment, cattle should be handled carefully. Hot, excited animals are subject to more shrinkage, more disease and injury, and more dark cutting if slaughtered following shipment.

If the animals are trailed on-foot to the shipping point, they should be moved slowly and allowed to rest and to drink moderately prior to loading. Although loading may be exasperating at times, take it easy; never lose your temper. Avoid hurrying, loud hollering, and striking. Never beat an animal with such objects as pipes, sticks, canes, or forks; instead, use either (a) a flat, wide canvas slapper with a handle, or (b) an electric prod (the latter judiciously).

4. **Consider health certificates, permits, and brand inspection in interstate shipments.** When cattle are to be shipped into another state, the shipper should check into and comply with the state regulations relative to health certificates, permits, and brand inspection. Usually, the local veterinarian, railroad agent, or trucker will have this information. Should there be any question about the health regulations, however, the state livestock sanitary board (usually located at the state capital) of the state of destination should be consulted. Knowledge of and compliance with such regulations well in advance of shipment will avoid frustrations and costly delays.

5. **Comply with the 28-hour law in rail shipments.** *Note:* The following information is presented primarily for historic interest.

In the days of rail shipment, the shipper had no

alternative to complying with the 28-hour law. Details follow:

> By federal law, passed in 1873, livestock cannot be transported by rail for a longer period than 28 consecutive hours without unloading for the purpose of giving feed, water, and rest for a period of at least 5 consecutive hours before resuming transportation. The period may be extended to 36 hours upon written request from the owner of the animals; and most experienced cattle shippers routinely so request. With less than carload lots (LCL shipments) the owner may provide feed and water in the car with instructions that the animals be fed and watered en route.

Trucks were not included in the provisions of the original 28-hour law, simply because there were no trucks at the time. But most buyers of livestock insist on a rest stop in an extended truck shipment, just as is required by law with rail shipments.

6. **Use partitions when necessary.** When mixed loads (consisting of cattle, sheep, and/or hogs) are placed in the same truck or car, partition each class off separately. Also, partition calves from cattle, and separate out cripples and stags; tie bulls.

7. **Avoid shipping during extremes in weather.** Whenever possible, avoid shipping when the weather is either very hot or very cold. During such times, shrinkage and death losses are higher than normal. During warm weather, avoid transporting animals during the heat of the day; travel at night or in the evening or early morning.

Additional points pertinent to proper preparing and shipping cattle are covered in the four sections which follow.

HOW TO PREVENT BRUISES, CRIPPLING, AND DEATH LOSSES

Losses from bruising, crippling, and death that occur during the marketing process represent a part of the cost of marketing livestock; and, indirectly, the producer foots most of the bill.

The following precautions are suggested as a means of reducing cattle market losses from bruises, crippling, and death:

1. Dehorn cattle, preferably when young.
2. Remove projecting nails, splinters, and broken boards in feed racks and fences.
3. Keep feedlots free from old machinery, trash, and any obstacle that may bruise.
4. Do not feed grain heavily just prior to loading.
5. Use good loading chutes; not too steep.
6. Bed with sand free from stones, to prevent slipping.
7. For calves, cover sand with straw in cold weather.
8. Provide covers for trucks to protect from sun in summer and cold in winter.
9. Always partition mixed loads into separate classes, and partition calves from cattle.
10. Have upper deck of truck high enough to prevent back bruises on animals below.
11. Remove protruding nails, bolts, or any sharp objects in truck or car.
12. Load slowly to prevent crowding against sharp corners and to avoid excitement. Do not overload.
13. Use canvas slappers instead of clubs or canes.
14. Tie all bulls in truck or car, and partition stags and cripples.
15. Place bull board in position and secure before car door is closed on loaded cattle.
16. Drive trucks carefully; slow down on sharp turns and avoid sudden stops.
17. Inspect load en route to prevent trampling of animals that may be down.
18. Back truck slowly and squarely against unloading dock.
19. Unload slowly. Do not drop animals from upper to lower deck; use cleated inclines.

All these precautions are simple to apply; yet all are violated every day of the year.

NUMBER OF CATTLE IN A TRUCK

Overcrowding of market animals causes heavy losses. Sometimes a truck is overloaded in an attempt to effect a saving in hauling charges. More frequently, however, it is simply the result of not knowing space requirements.

The suggested number of animals in a truck (Table 15-3) should be tempered by such factors as distance of haul, class of cattle, weather, and road conditions.

KIND OF BEDDING TO USE FOR CATTLE IN TRANSIT

Among the several factors affecting livestock losses, perhaps none is more important than proper bedding and footing in transit. This applies to both truck and rail shipments, and to all classes of animals.

Footing, such as sand, is required at all times of the year, to prevent the car or truck floor from becoming wet and slick, thus predisposing animals to injury by slipping or falling. Bedding, such as straw, is recommended for warmth in the shipment of calves during extremely cold weather, and as cushioning for dairy

TABLE 15-3
NUMBER OF CATTLE FOR SAFE LOADING IN A TRUCK[1, 2]

Floor Length		Average Weight of Cattle, Lb					
		450	600	800	1,000	1,200	1,400
(ft)	*(m)*	*(no.)*	*(no.)*	*(no.)*	*(no.)*	*(no.)*	*(no.)*
8	*2.4*	8	7	5	4	4	3
10	*3.1*	10	8	7	6	5	4
12	*3.7*	13	10	8	7	6	5
15	*4.6*	16	13	10	9	8	7
18	*5.5*	20	16	13	11	9	8
20	*6.1*	22	18	14	12	10	9
24	*7.3*	27	22	17	15	13	11
28	*8.5*	31	25	20	17	15	13
30	*9.2*	34	27	22	19	16	14
32	*9.8*	36	[illegible]	23	20	17	15
36	*11.0*	41	33	26	22	19	17
42	*12.8*	48	39	31	28	22	20

[1]From the authoritative recommendations of Livestock Conservation, Inc.

[2]To make metric conversions, see the Appendix, "Section IV—Weights and Measures."

cows, breeding stock, or other animals loaded lightly enough to permit their lying down. Recommended kinds and amounts of bedding and footing materials are given in Table 15-4. (Because many loads are mixed, information relative to each class of animals is provided in Table 15-4.)

TABLE 15-4
GUIDE RELATIVE TO BEDDING AND FOOTING MATERIAL WHEN TRANSPORTING LIVESTOCK[1, 2, 3]

Class of Livestock	Kind of Bedding for Moderate or Warm Weather; Above 50°F *(10°C)*	Kind of Bedding for Cool or Cold Weather; Below 50°F *(10°C)*
Cattle	Sand, 2 in. *(5 cm)*	Sand; for calves use sand covered with straw
Sheep and goats	Sand	Sand covered with straw
Swine	Sand, ½ to 2 in. *(1 to 5 cm)*[4]	Sand covered with straw
Horses and mules	Sand	Sand

[1]Straw or other suitable bedding (covered with sand) should be used for protecting and for cushioning breeding animals that are loaded lightly enough to permit their lying down in the car or truck.

[2]Sand should be clean and medium-fine, and free from brick, stones, coarse gravel, dirt, or dust.

[3]*Fine* cinders may be used as footing for cattle, horses, and mules, but not for sheep or hogs. They are picked up by and damage the wool of sheep, and they damage hog casings.

[4]In hot weather, wet sand down before loading and while enroute. But never apply water to the backs of hogs—it may kill them.

SHRINKAGE IN MARKETING CATTLE

The shrinkage (or drift) refers to the weight loss encountered from the time animals leave the farm, ranch, or feedlot until they are weighed over the scales at their destination. Thus, if a steer weighed 1,000 lb at the feedlot and had a market weight of 970 lb, the shrinkage would be 30 lb or 3.0%. Shrink is usually expressed in terms of percentage. Most of this weight loss is due to excretion, in the form of feces and urine and the moisture in the expired air. On the other hand, there is some tissue shrinkage, which results from metabolic or breakdown changes.

The most important factors affecting shrinkage are:

1. **The fill.** Naturally, the larger the fill animals take upon their arrival at the market, the smaller the shrinkage.
2. **Time in transit.** The longer the animals are in transit and the greater the distance, the higher the total shrinkage.
3. **Season.** Extremes in temperature, either very hot or very cold weather, result in higher shrinkage.
4. **Age and weight.** Young animals of all species shrink proportionally more than older animals.
5. **Overloading.** Overloading always results in abnormally high shrinkage.
6. **Rough ride, abnormal feeding, and mixed loads.** Each of these factors will increase shrinkage.
7. **Auction sale or ranch origin.** Iowa State University reported an average shrink of 9.1% for cattle originating at auction sales compared with 7.2% shrink in cattle purchased directly from ranches. On the average, market cattle shrink from 3 to 10%, with younger animals shrinking more than older and fatter animals. Experienced cattle feeders report that it takes an average of 7 days after arrival at the feedyard to regain a shrink of 10% on feeder cattle.

SHRINKAGE TABLES

Cattle are sometimes bought or sold with a certain percentage shrink from actual weights. This is known as pencil shrinkage. Both cattle sellers and buyers must give consideration to such shrinkage. For example, if a buyer offers $60.00 per 100 with a 4% shrink allowance, the producer will want to know how much will actually be received. The answer can be quickly and easily obtained from Table 15-5 as follows: Look at $60.00 under column 1, headed "Offer." Go across to column headed "4%." As shown, the producer will receive $57.60.

If the producer has decided that $60.00 is the minimum asking price, the offer may be refused and the cattle shipped to market. Then, he/she will wish to know how much must be received in order to compensate for shrinkage. The answer can be obtained from

TABLE 15-5
SELLING CATTLE—NET PRICES AFTER ALLOWING FOR SHRINKAGE (Prices, per Cwt)

	Shrink				
Offer	2%	3%	4%	6%	8%
($)	($)	($)	($)	($)	($)
80.00	78.40	77.60	76.80	75.20	73.60
79.00	77.42	76.63	75.84	74.26	72.68
78.00	76.44	75.66	74.88	73.32	71.76
77.00	75.46	74.69	73.92	72.38	70.84
76.00	74.48	73.72	72.96	71.44	69.92
75.00	73.50	72.75	72.00	70.50	69.00
74.00	72.52	71.78	71.04	69.56	68.08
73.00	71.54	70.81	70.08	68.62	67.16
72.00	70.56	69.84	69.12	67.68	66.24
71.00	69.58	68.87	68.16	66.74	65.32
70.00	68.60	67.90	67.20	65.80	64.40
69.00	67.62	66.93	66.24	64.86	63.48
68.00	66.64	65.96	65.28	63.92	62.56
67.00	65.66	64.99	64.32	62.98	61.64
66.00	64.68	64.02	63.36	62.04	60.72
65.00	63.70	63.05	62.40	61.10	59.80
64.00	62.72	62.08	61.44	60.16	58.88
63.00	61.74	61.11	60.48	59.22	57.96
62.00	60.76	60.14	59.52	58.28	57.04
61.00	59.78	59.17	58.56	57.34	56.12
60.00	58.80	58.20	57.60	56.40	55.20
59.00	57.82	57.23	56.64	55.46	54.28
58.00	56.84	56.26	55.68	54.52	53.36
57.00	55.86	55.29	54.72	53.58	52.44
56.00	54.88	54.32	53.76	52.64	51.52
55.00	53.90	53.35	52.80	51.70	50.60
54.00	52.92	52.38	51.84	50.76	49.68
53.00	51.94	51.41	50.88	49.82	48.76
52.00	50.96	50.44	49.92	48.88	47.84
51.00	49.98	49.47	48.96	47.94	46.92
50.00	49.00	48.50	48.00	47.00	46.00
49.00	48.02	47.53	47.04	46.06	45.08
48.00	47.04	46.56	46.08	45.12	44.16
47.00	46.06	45.59	45.12	44.18	43.24
46.00	45.08	44.62	44.16	43.24	42.32
45.00	44.10	43.65	43.20	42.30	41.40
44.00	43.12	42.68	42.24	41.36	40.48
43.00	42.14	41.71	41.28	40.42	39.56
42.00	41.16	40.74	40.32	39.48	38.64
41.00	40.18	39.77	39.36	38.54	37.72
40.00	39.20	38.80	38.40	37.60	36.80
39.00	38.22	37.83	37.44	36.66	35.88
38.00	37.24	36.86	36.48	35.72	34.96
37.00	36.26	35.89	35.52	34.78	34.04
36.00	35.28	34.92	34.56	33.84	33.12
35.00	34.30	33.95	33.60	32.90	32.20
34.00	33.32	32.98	32.64	31.96	31.28
33.00	32.34	32.01	31.68	31.02	30.36
32.00	31.36	31.04	30.72	30.08	29.44
31.00	30.38	30.07	29.76	29.14	28.52
30.00	29.40	29.10	28.80	28.20	27.60
29.00	28.42	28.13	27.84	27.26	26.68
28.00	27.44	27.16	26.88	26.32	25.76
27.00	26.46	26.19	25.92	25.38	24.84
26.00	25.48	25.22	24.96	24.44	23.92

TABLE 15-6
BUYING CATTLE—CHANGE IN PRICE TO COMPENSATE FOR SHRINKAGE (Prices, per Cwt)

	Shrink				
Asking	2%	3%	4%	6%	8%
($)	($)	($)	($)	($)	($)
80.00	81.63	82.47	83.33	85.11	86.96
79.00	80.61	81.44	82.29	84.04	85.87
78.00	79.59	80.41	81.25	82.98	84.78
77.00	78.57	79.38	80.21	81.91	83.70
76.00	77.55	78.35	79.17	80.85	82.61
75.00	76.53	77.32	78.13	79.79	81.52
74.00	75.51	76.29	77.08	78.72	80.43
73.00	74.49	75.26	76.04	77.66	79.35
72.00	73.47	74.23	75.00	76.60	78.26
71.00	72.45	73.20	73.96	75.53	77.17
70.00	71.43	72.16	72.92	74.47	76.09
69.00	70.41	71.13	71.88	73.40	75.00
68.00	69.39	70.10	70.83	72.34	73.91
67.00	68.37	69.07	69.79	71.28	72.83
66.00	67.35	68.04	68.75	70.21	71.74
65.00	66.33	67.01	67.71	69.15	70.65
64.00	65.31	65.98	66.67	68.09	69.57
63.00	64.21	64.95	65.63	67.02	68.48
62.00	63.27	63.92	64.58	65.96	67.39
61.00	62.24	62.89	63.54	64.89	66.30
60.00	61.22	61.86	62.50	63.83	65.22
59.00	60.20	60.82	61.46	62.77	64.13
58.00	59.18	59.79	60.42	61.70	63.04
57.00	58.16	58.76	59.38	60.64	61.96
56.00	57.14	57.73	58.33	59.57	60.87
55.00	56.12	56.70	57.29	58.51	59.78
54.00	55.10	55.67	56.25	57.45	58.70
53.00	54.08	54.64	55.21	56.38	57.61
52.00	53.06	53.61	54.17	55.32	56.52
51.00	52.04	52.58	53.13	54.26	55.43
50.00	51.02	51.55	52.08	53.19	54.35
49.00	50.00	50.52	51.04	52.13	53.26
48.00	48.98	49.48	50.00	51.06	52.17
47.00	47.96	48.45	48.96	50.00	51.09
46.00	46.94	47.42	47.92	48.94	50.00
45.00	45.92	46.39	46.88	47.87	48.91
44.00	44.90	45.36	45.83	46.81	47.83
43.00	43.88	44.33	44.79	45.74	46.74
42.00	42.86	43.30	43.75	44.68	45.65
41.00	41.84	42.27	42.71	43.62	44.57
40.00	40.82	41.24	41.67	42.55	43.48
39.00	39.80	40.21	40.63	41.49	42.39
38.00	38.78	39.81	39.58	40.43	41.30
37.00	37.76	38.14	38.54	39.36	40.22
36.00	36.73	37.11	37.50	38.30	39.13
35.00	35.71	36.08	36.46	37.23	38.04
34.00	34.69	35.05	35.42	36.17	36.96
33.00	33.67	34.02	34.37	35.11	35.87
32.00	32.66	32.99	33.33	34.04	34.78
31.00	31.63	31.96	32.29	32.98	33.70
30.00	30.61	30.93	31.25	31.91	32.61

Table 15-6 as follows: Look at $60.00 under column 1, headed "Asking"; then read under the proper column to the right. Thus, if the animals shrink 4% during marketing, the price will have to be $62.50 in order to compensate for shrinkage.

MARKET CLASSES AND GRADES OF CATTLE

The generally accepted market classes and grades of live cattle are summarized in Table 15-7. The first five divisions and subdivisions include those factors that determine the class of the animal or the use to which it will be put. The grades, which are a combination of both their quality and yield grades (except that slaughter bulls are yield graded only), indicate how well the cattle fulfill the requirements to which they are put.

FACTORS DETERMINING MARKET CLASSES OF CATTLE

The market class of cattle is determined by (1) use selection, (2) sex, (3) age, and (4) weight (see Table 15-7).

CATTLE AND CALVES

All members of the bovine family are designated as calves until they are one year of age, after which they are known as cattle. Of the bovines marketed in 1994, 83% were cattle and 17% were calves.

BABY BEEF

In the 1920s and 1930s, the term *baby beef* was well known and widely used in cattle production and marketing circles. The ultimate in baby beef was the 4-H Club or FFA steer of the era. It referred to well-finished young animals, marketed at 12 to 16 months of age, weighing 700 to 1,000 lb, grading Good (Select) to Prime. The disrepute and demise of the term *baby beef* was caused by lack of enforceable high standards; to fill the demand, a wide assortment of cattle was marketed as *baby beef.*

In the 1970s, baby beef was again promoted, primarily as a means of reducing cattle numbers and lessening grain feeding. But it took on a new look! Today, baby beef connotes heavy calves that are fat enough for slaughter at weaning time, weighing 400 to 700 lb on foot, the meat of which is lean and tender.

USE SELECTION OF CATTLE AND CALVES

The cattle group is further divided into three use divisions, each indicating something of the purpose to which the animals will be put. These divisions are: (1) slaughter cattle; (2) feeder cattle; and (3) milkers and springers. Slaughter cattle include those which are considered suitable for immediate slaughter, feeders include those which are to be taken back to the country and grown for a time or fattened; and milkers and springers include those cows recently freshened or soon due to calve and which are sold for milk purposes.

The calf group is also subdivided into 3 classes: (1) vealers, including milk-fat animals under 3 months of age which are sold for immediate slaughter; (2) slaughter calves that are between the ages of 3 months and 1 year, which have usually received grain in addition to milk and which are fat enough for slaughter; and (3) feeder calves which are of weaning age and are sold to go back into the country for further growing or finishing.

In the selection of feeder cattle or calves, the sex, age, weight, and grade are of importance. In addition, consideration should be given to the following factors: (1) constitution and thrift; (2) natural fleshing; (3) breeding; (4) uniformity; (5) absence of horns; and (6) temperament and disposition.

As can be readily understood, the use to which animals are put is not always clear-cut and definite. Thus, when feed is abundant and factors are favorable for cattle finishing, feeders may outbid packer buyers for some of the animals that would normally go for slaughter purposes. On the other hand, slaughterers frequently outbid feeders for some of those animals that would normally go the feeder route.

THE SEX CLASSES

Cattle are divided into five sex classes: steers heifers, cows, bullocks, bulls. Each of these five groups has rather definite and easily distinguishable characteristics that are related to the commercial value of the carcass—especially in the cattle group—and which are important in determining the suitability of animals as stockers and feeders. In older cattle, sex is an important factor affecting carcass quality, finish, and conformation. The definition of each sex class follows:

1. **Steer.** A male bovine castrated when young and which has not begun to develop the secondary physical characteristics of a bull.
2. **Heifer.** An immature female bovine that has not developed the physical characteristics typical of cows.
3. **Cow.** A female bovine that has developed through reproduction or with age, the relatively promi-

TABLE 15-7
MARKET CLASSES AND QUALITY GRADES OF CATTLE[1]

Cattle or Calves	Use Selection	Sex Classes	Age	Weight Divisions: Wt. (Group)	Weight Divisions: (Lb)	Commonly Used Quality Grades[2]
Cattle	Slaughter cattle[2]	Steers	Yearlings	Light Medium Heavy	750 down 750–950 950 up	Prime, Choice, Select, Standard, Utility, Cutter, Canner
			2-year-olds and over	Light Medium Heavy	1,100 down 1,100–1,300 1,300 up	Prime, Choice, Select, Standard, Commercial, Utility, Cutter, Canner
		Heifers	Yearlings	Light Medium Heavy	750 down 750–900 900 up	Prime, Choice, Select, Standard, Utility, Cutter, Canner
			2-year-olds and over	Light Medium Heavy	900 down 900–1,050 1,050 up	Prime, Choice, Select, Standard, Commercial, Utility, Cutter, Canner
		Cows	All ages	All weights		Choice, Select, Standard, Commercial, Utility, Cutter, Canner
		Bullocks	24 mo. and under	All weights		Prime, Choice, Select, Standard, Utility
		Bulls		All weights		None (yield graded only)
	Feeder cattle	Steers	Yearlings	Light Medium Heavy Mixed		Prime, Choice, Select, Standard, Utility, Inferior
			2-year-olds and over	Light Medium Heavy Mixed		Prime, Choice, Select, Standard, Commercial, Utility, Inferior
		Heifers	Yearlings	Light Medium Heavy Mixed		Prime, Choice, Select, Standard, Utility, Inferior
			2-year-olds and over	Light Medium Heavy Mixed		Prime, Choice, Select, Standard, Commercial, Utility, Inferior
		Cows	All ages	All weights		Choice, Select, Standard, Commercial, Utility, Inferior
		Bullocks	24 mo. and under	All weights		Prime, Choice, Select, Standard, Utility, Inferior
		Bulls	24 mo. and over	All weights		None
	Milkers and springers	Cows (milkers and springers)	All ages	All weights		None
Calves	Vealers	No sex class (Sex characteristics of no importance at this age)	Under 3 mo.	Light Medium Heavy	110 down 110–180 180 up	Prime, Choice, Good, Standard, Utility
	Slaughter calves	Steers Heifers Bulls	3 mo. to 1 yr.	Light Medium Heavy	200 down 200–300 300 up	Prime, Choice, Good, Standard, Utility
	Feeder calves	Steers Heifers Bulls	Usually 6 mo. to 1 yr.	Light Medium Heavy Mixed		Prime, Choice, Good, Standard, Utility, Inferior

[1]To make metric conversions, see the Appendix, "Section IV—Weights and Measures."

[2]In addition to the quality grades, there are the following yield grades for all slaughter cattle, except bulls: Yield Grade 1, Yield Grade 2, Yield Grade 3, Yield Grade 4, and Yield Grade 5; with Yield Grade 1 representing the highest degree of cutability, and Yield Grade 5 the lowest. In 1989, the law was changed, separating quality and yield grades of beef, allowing packers to choose whether beef carcasses are graded for quality, yield, or for both quality and yield.

nent hips, large middle, and other physical characteristics typical of mature females.

4. **Bullock.** A young (under approximately 24 months of age) male bovine (castrated or uncastrated) that has developed or begun to develop the secondary physical characteristics of a bull.

5. **Bull.** A mature (approximately 24 months of age or older) uncastrated, male bovine. However, for the purpose of these standards, any mature, castrated, male bovine which has developed or begun to develop the secondary physical characteristics of an uncastrated male also will be considered a bull.

Calves are merely divided into three sex classes: steers, heifers, bulls. Because the secondary sex characteristics are not very pronounced in this group, the sex classes are of less importance for slaughter purposes than in older cattle. On the other hand, bull calves are not preferred as feeders because castration involves extra trouble and risk of loss.

AGE GROUPS

Because the age of cattle does affect certain carcass characteristics, it is logical that age groups should exist in market classifications. The terms used to indicate approximate age ranges for cattle are: vealers, calves, yearlings, 2-year-olds and older cattle. As previously indicated, vealers are under 3 months of age,[4] whereas calves are young cattle between the vealer and yearling stage. Yearlings range from 12 to 24 months in age, and 2-year-olds from 24 to 36 months. Older cattle are usually grouped along with the 2-year-olds as *2-year-olds and over.*

WEIGHT DIVISIONS

It is common to have three weight divisions: light, medium, and heavy. When several weight divisions are included together, they are referred to as *mixed weight.* The usual practice is to group animals by rather narrow weight divisions because purchasers are frequently rather "choosey" about weights, and market values often vary quite sharply with variations in weights.

[4]Vealers are generally over 21 days of age at time of slaughter, although federal and most state regulations governing meat inspection do not specify a minimum age. Rather, it is a matter of maturity. A big Holstein calf may, for example, make satisfactory veal at 2 weeks of age, whereas calves of some of the smaller dairy breeds might not pass inspection before 4 or 5 weeks of age. Animals that are too young or immature are generally considered unsuited for human consumption primarily for esthetic reasons rather than because of any harmful effect of the meat. On the market, underage veal calves are called *deacons* or *bob veal.*

MARKET GRADES OF SLAUGHTER CATTLE AND CALVES

While little official grading of live animals is done by the U.S. Department of Agriculture, market grades do form a basis for the uniform reporting of livestock marketed. The grade is the final step in classifying any kind of market livestock. It indicates the relative degree of excellence of an animal or group of animals. Slaughter cattle quality grades are based on an evaluation of factors related to the palatability of the lean, referred to as *quality.* The yield grades, which estimate the amount of salable meat a carcass will yield, are based on evaluations of muscling and fatness.

The current slaughter grades of cattle became effective April 9, 1989. USDA is scheduled to revise the quality grades, removing B-maturity carcasses with slight and small marbling scores from the Choice and Select grades, with the result that these cattle grade Standard. These revisions are scheduled to become effective January 31, 1997.

The U.S. quality grades of slaughter steers are shown in Fig. 15-10, and the U.S. yield grades of slaughter steers are shown in Fig. 15-11.

The use of both yield grading and quality grading is of much benefit. Formerly, most fed cattle were sold on the basis of *averages.* Now, cattle are sold more on the basis of their true worth. Among other things, the changes have resulted in shorter average feeding periods and in rewards to producers in all segments of the industry (breeder, producer, feeder) for producing cattle with more lean meat and less waste fat.

Table 15-7 lists the commonly used quality grades of cattle by classes. There also are 5 yield grades (see Table 15-7 footnote), designated by the numbers *1* through *5,* which are applicable to all classes of slaughter cattle, with Yield Grade 1 representing the highest cutability and Yield Grade 5 the lowest. As noted, the number of quality grades varies somewhat between classes chiefly because certain groups of animals present a wider range of variations in conformation, finish and quality than do other groups. Slaughter steers and heifers are divided into 8 grades: Prime, Choice, Select, Standard, Commercial, Utility, Cutter, Canner. As shown in Table 15-7, 7 grades apply to slaughter cows and 5 grades to slaughter bullocks. The Prime grade is not applied to cows, chiefly because of deficient conformation, finish, and quality in this class. Further, as in carcasses, bullocks on foot are always designated as *slaughter bullocks* since meat obtained from this class is never interchanged with meat carrying the same grade name from steers, heifers, and cows.

Cutter and *Canner* refer to the two lowest grades of slaughter cattle. Cutter cattle are so poor in form and lacking in muscle and fat covering that only such wholesale cuts as the loin and round are cut out and

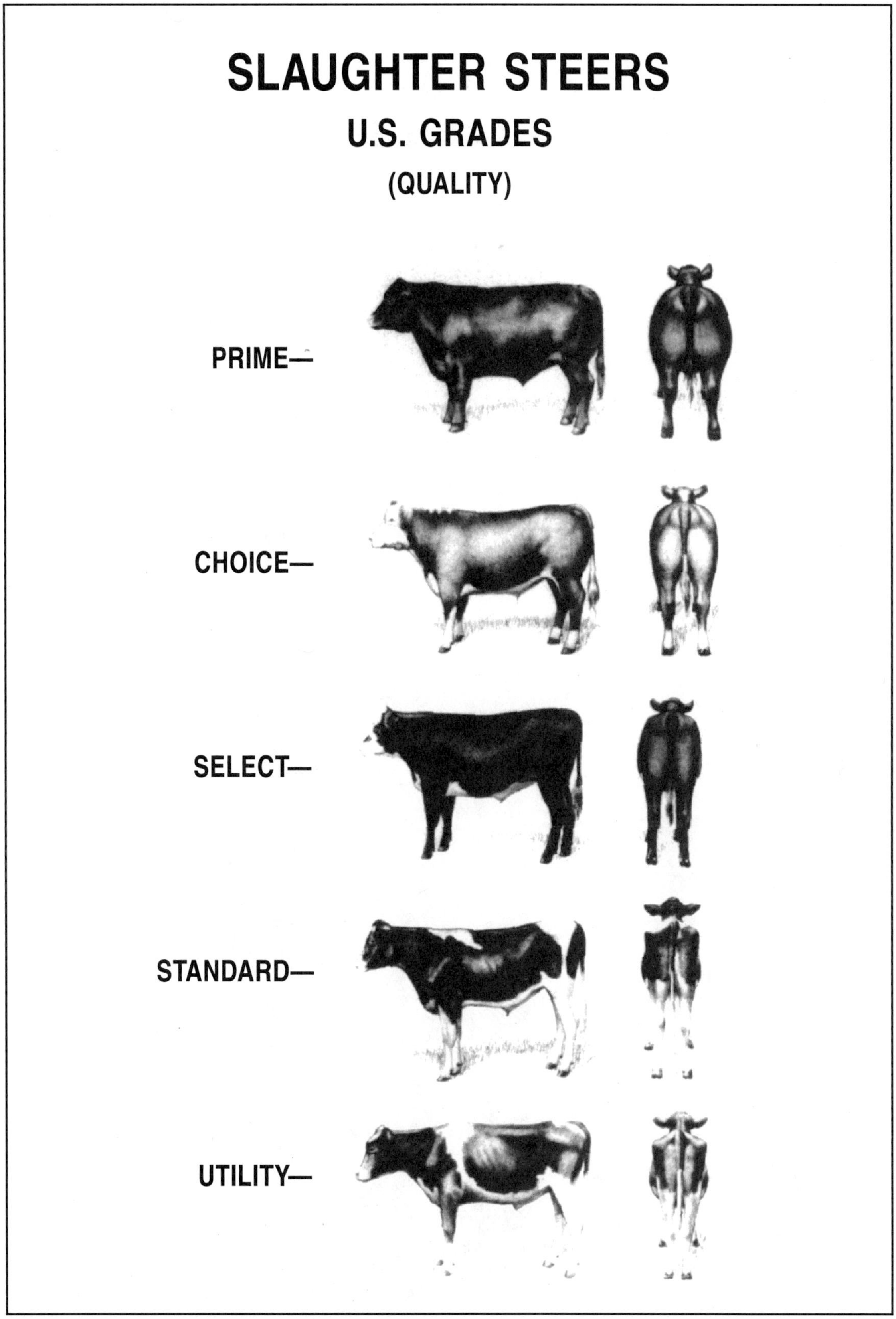

SLAUGHTER STEERS

U.S. GRADES

(QUALITY)

PRIME—

CHOICE—

SELECT—

STANDARD—

UTILITY—

Fig. 15-10. U.S. quality grades of slaughter steers. (Courtesy, USDA)

SLAUGHTER STEERS

U.S. GRADES

(YIELD)

YIELD GRADE 1—

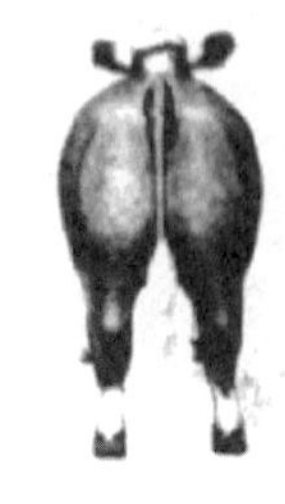

YIELD GRADE 2—

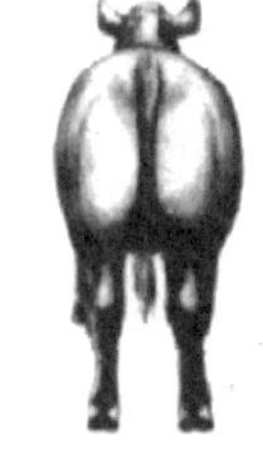

YIELD GRADE 3—

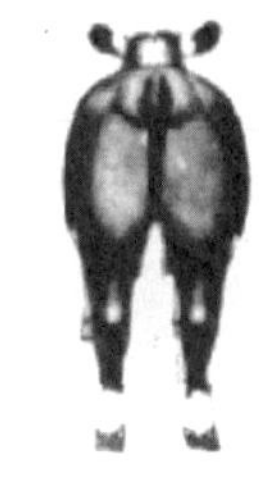

YIELD GRADE 4—

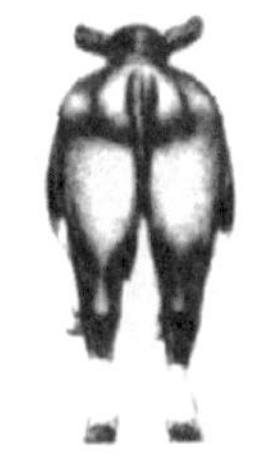

YIELD GRADE 5—

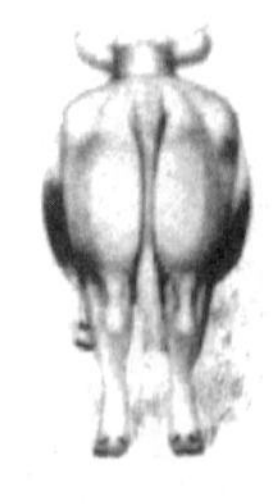

Fig. 15-11. U.S. yield grades of slaughter steers. (Courtesy, USDA)

sold over the block. The balance of the carcass is boned out and used in sausage and canned-meat products. Canners are almost entirely processed as ground and canned meats.

The grades of vealers and slaughter calves are: Prime, Choice, Good, Standard, and Utility. *Note*: The Good grade was retained in calves.

EVALUATING QUALITY AND YIELD GRADES OF CATTLE

Since carcass indices of quality and yield grades of slaughter cattle are not directly evident in live animals other factors in which differences can be noted must be used to evaluate quality and yield in on-foot cattle. These follow.

■ **Evaluating quality**—External finish and age are the primary determiners of quality.

The amount of external finish is a major quality grade factor, even though cattle with a specific degree of fatness may have widely varying degrees of quality. Differentiation between the quality of cattle with the same degree of fatness is based on distribution of finish and firmness of muscling.

Approximate maximum age limitations for the specific grades of steers, heifers, and cows are as follows: (1) Prime, Choice, Select, and Standard grades of steers, heifers, and cows—42 months; (2) Commercial grade for steers, heifers, and cows—only cattle over approximately 42 months; (3) no age limitations for the Utility, Cutter, and Canner grades; and (4) all grades of bullocks—approximately 24 months.

■ **Evaluating yield grade of live cattle**—The most practical method of appraising on-foot cattle for yield grade is to use only two factors—muscling and fatness. In this approach, evaluation of the thickness and fullness of muscling in relation to skeletal size largely accounts for the effects of two of the factors—area of ribeye and carcass weight. By the same token, an appraisal of the degree of external fatness largely accounts for the effects of thickness of fat over the ribeye and the percent of kidney, pelvic, and heart fat.

These fatness and muscling evaluations can best be made simultaneously. This is accomplished by considering the development of the various parts based on an understanding of how each part is affected by variations in muscling and fatness. While muscling of most cattle develops uniformly, fat is normally deposited at a considerably faster rate on some parts of the body than on others. Therefore, muscling can be appraised best by giving primary consideration to the parts least affected by fatness, such as the round and the forearm. Differences in thickness and fullness of these parts—with appropriate adjustments for the effects of variations in fatness—are the best indicators of the overall degree of muscling in live cattle.

The overall fatness of an animal can be determined best by observing those parts on which fat is deposited at a faster-than-average rate. These include the back, loin, rump, flank, cod or udder, twist, and brisket. As cattle increase in fatness, these parts appear progressively fuller, thicker, and more distended in relation to the thickness and fullness of the other parts, particularly the round. In thinly muscled cattle with a low degree of finish, the width of the back usually will be greater than the width through the center of the round. The back on either side of the backbone also will be flat or slightly sunken. Conversely, in thickly muscled cattle with a similar degree of finish, the thickness through the rounds will be greater than through the back and the back will appear full and rounded. At an intermediate degree of fatness, cattle which are thickly muscled will be about the same width through the round and back and the back will appear only slightly rounded. Thinly muscled cattle with an intermediate degree of finish will be considerably wider through the back than through the round and will be nearly flat across the back. Very fat cattle will be wider through the back than through the round, but this difference will be greater in thinly muscled cattle than in those that are thickly muscled. Such cattle with thin muscling also will have a distinct break from the back into the sides, while those with thick muscling will be nearly flat on top, but will have a less distinct break into the sides. As cattle increase in fatness, they also become deeper bodied because of large deposits of fat in the flanks and brisket and along the underline. Fullness of the twist and cod or udder and the bulge of the flanks, best observed when an animal walks, are other indications of fatness .

In determining yield grade, variations in fatness are much more important than variations in muscling.

PRICE SPREAD BETWEEN SLAUGHTER GRADES

Because the production of the better grades of cattle usually involves more expenditure in the breeding operations (due to the need for superior animals) and feeding to a higher degree of finish, there must be a spread in market grades in order to make the production of the top grades profitable. Fig. 15-12 shows the average wholesale price for 100-lb dressed, boxed, steer beef (carlots), Choice grade 550–700 lb vs Select grade, 1985 to 1993. Note that there is an average spread of $3.79 per 100 lb between Choice and Select grades of beef.

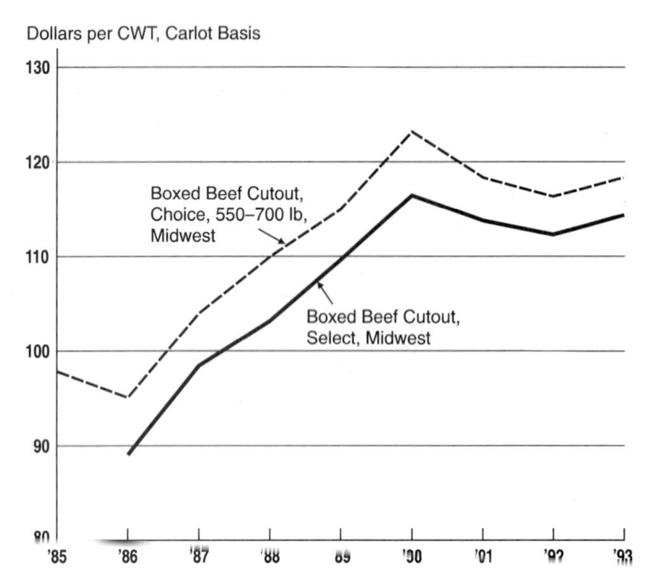

Fig. 15-12. Average wholesale price per 100-lb dressed (carlots) boxed steer beef. Choice grade, 550–700 lb vs Select grade, 1985 to 1993. (From: *Meat & Poultry Facts, 1994*, American Meat Institute)

Grades of slaughter (live) cattle are intended to be directly related to the grades of the carcasses that they will produce. To accomplish this, slaughter grade standards are based on factors which are related to the quality grade and the yield grade of beef carcasses.

MARKET GRADES OF FEEDER CATTLE

The current feeder cattle grades became effective September 2, 1979. The grades recognize 3 frame size grades and 3 thickness grades. In addition to 9 possible combinations (3 frame sizes, 3 muscle thicknesses) of feeder grades for thrifty animals, there is an inferior grade for unthrifty animals. Double muscled animals are included in the inferior grade.

The three frame size grades are: Large, Medium, and Small (see Fig. 15-13). The three thickness grades are: No. 1, No. 2, and No. 3 (see Fig. 15-14).

Younger and larger framed size feeders require a longer feeder period as is shown by the following figures:

	Avg. No. Days on Feed for Steers to Reach Choice Grade		
Age	**Large Frame**	**Medium Frame**	**Small Frame**
Calves (8 to 10 mo.)	210	180	150
Yearlings (14 to 16 mo.)	140	120	100

But that's not all! Larger framed, heavily muscled cattle are too heavy at market time; and small framed, light muscled cattle are too light. Packers prefer 1,100 lb Choice grade steers for slaughter.

SOME CATTLE MARKETING CONSIDERATIONS

Enlightened marketing practices generally characterize the successful cattle enterprise. Among the considerations of importance in marketing cattle are those which follow.

CYCLICAL TRENDS IN MARKET CATTLE

The price cycle as it applies to livestock may be defined as that period of time during which the price for a certain kind of livestock advances from a low point to a high point and then declines to a low point again. In reality, it is a change in animal numbers that represents the producer's response to prices. Although there is considerable variation in the length of the cycle within any given class of stock, in recent years it has been observed that the price cycle of cattle is about 10 years. (See Fig. 15-15.) Also, in recent years, it has been observed that the price cycle of cattle is shorter and less pronounced.

The specie cycles are a direct reflection of the rapidity with which the numbers of each class of farm animals can be shifted under practical conditions to meet consumer meat demands. Thus, litter-bearing and early-producing swine can be increased in numbers much more rapidly than either cattle or sheep.

When market cattle prices are favorable, established cattle enterprises are expanded, and new herds are founded, so that about every 10 years, on the average, the market is glutted and prices fall, only to rise again because too few cattle are being produced to take care of the demand for beef. Normal cycles are disturbed by droughts, wars, general periods of depression or inflation, and federal controls.

SEASONAL CHANGES IN MARKET CATTLE

Cattle prices vary by seasons, by classes, and by grades, as shown in Table 15-8, and Figs. 15-16, 15-17, 15-18, and 15-19. Consideration is given herein to slaughter steers, slaughter heifers, cows, and feeder steers.

1. **Slaughter steers.** Fig. 15-16 shows the seasonal variation in Choice slaughter steer prices. From

FEEDER CATTLE

OFFICIAL U.S. GRADES

FRAME SIZE

LARGE FRAME

- Tall and long-bodied
- 0.5 inch fat
- Steers over 1,200 lb
- Heifers over 1,000 lb

MEDIUM FRAME

- Slightly tall
- Slightly long-bodied
- 0.5 inch fat
- Steers 1,000–1,200 lb
- Heifers 850–1,000 lb

SMALL FRAME

- Short bodied
- Not as tall as medium
- 0.5 inch fat
- Steers—less than 1,000 lb
- Heifers—less than 850 lb

Fig. 15-13. U.S. frame size grades of feeder cattle. Large- and medium-frame pictures depict minimum grade requirements. The small-frame picture represents an animal typical of the grade. (Courtesy, USDA)

FEEDER CATTLE

OFFICIAL U.S. GRADES

THICKNESS GRADES

NUMBER 1

- Slightly thick
- Legs moderately wide apart
- High proportion beef breeding

NUMBER 2

- Narrow
- Legs set close together

NUMBER 3

- Less thickness than No. 2

Fig. 15-14. U.S. thickness grades of feeder cattle. The No. 1 and 2 thickness pictures depict minimum grade requirements. The No. 3 picture represents an animal typical of the grade.

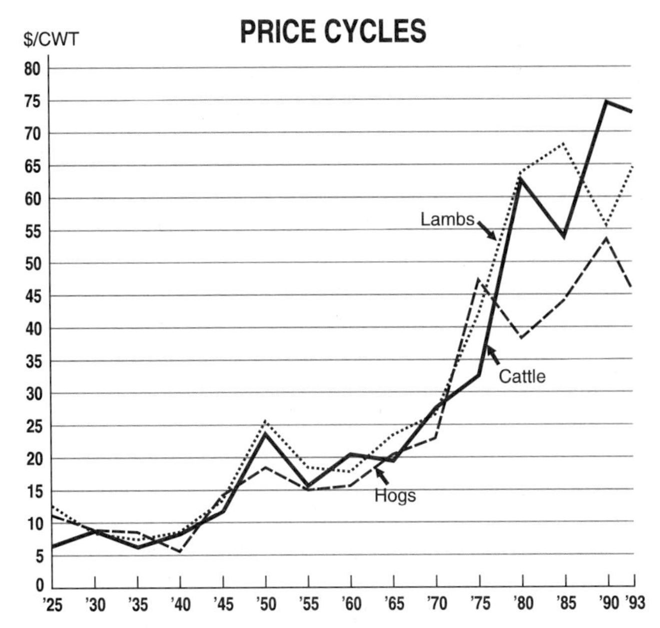

Fig. 15-15. Average price received by U.S. farmers for each class of livestock, 1925–93. In general, this shows that, currently, the price cycle of each class of animals is approximately as follows: hogs, 4 years; sheep, 9 to 10 years; and cattle, 10 years. (From: *Meat & Poultry Facts, 1994*, American Meat Institute, p. 8)

TABLE 15-8
WHEN TO BUY AND SELL CATTLE[1]

Kind of Class of Cattle	Lowest Prices (when to buy)	Highest Prices (when to sell)
Slaughter steers	July to Dec.	Jan. to June
Slaughter heifers	July to Dec.	Jan. to June
Cows (slaughter)	Oct. to March	April to Sept.
Feeder steers	Oct. to Feb.	March to Sept.

[1]Based on averages.

this and other market information, the following conclusions are drawn:

a. The cattle feeder will usually hit the highest market in January to June, and the lowest market from July to December.

b. Prices of steers of various grades parallel each other.

c. There is less spread in price between the Choice and Select grades than between the other grades.

2. **Slaughter heifers.** Fig. 15-17 shows the seasonal variation in slaughter heifer prices. From this and other market information, the following conclusions are drawn:

a. The price of slaughter heifers shows greatest strength in January to June, and greatest weakness in July to December.

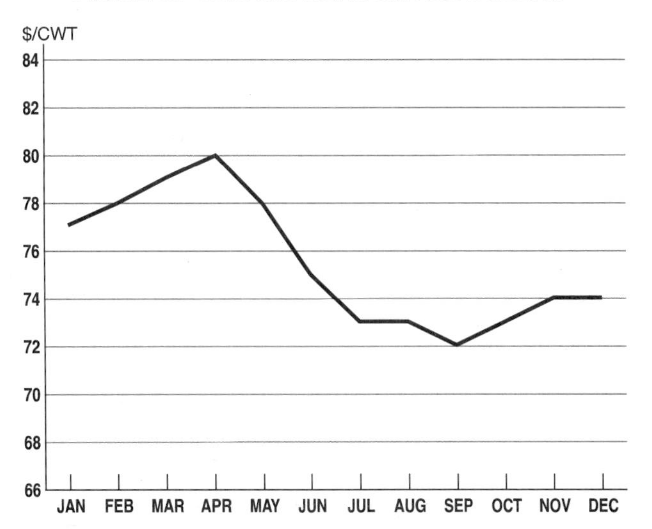

Fig. 15-16. Average price of Choice 2–4 slaughter steers 1,000–1,100 lb, Omaha, for the 4-year period 1990–93. (From: *Red Meats Yearbook*, USDA, ERS, Stat. Bull. No. 885, p. 63, Table 61)

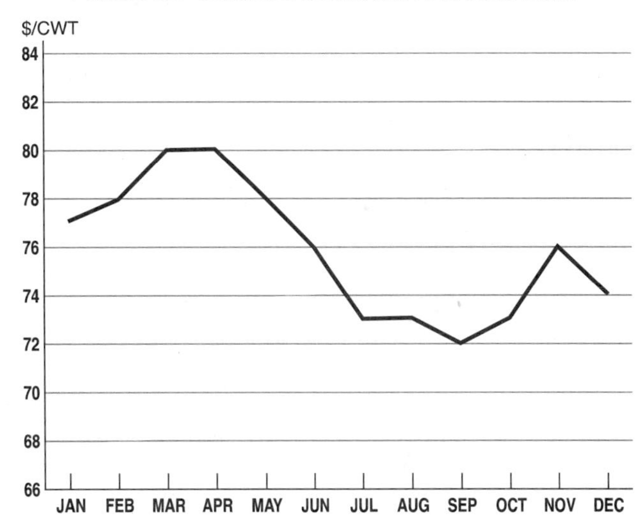

Fig. 15-17. Average price of Choice 2–4 slaughter heifers, 1,000–1,200 lb, Omaha, for the 4-year period 1990–1993. (From: *Red Meat Yearbook*, USDA, ERS, Stat. Bull. No. 885, p. 66, Table 64.

b. It is more difficult for the producer to take advantage of high market seasons with heifers than with steers, because the former are usually discounted in price after they pass the 750- to 900-lb mark and, therefore, must be sold.

3. **Cows.** Fig. 15-18 shows the seasonal variation in cow prices. From this and other market information, the following conclusions are drawn:

a. Usually, dry cows and cows that are to be

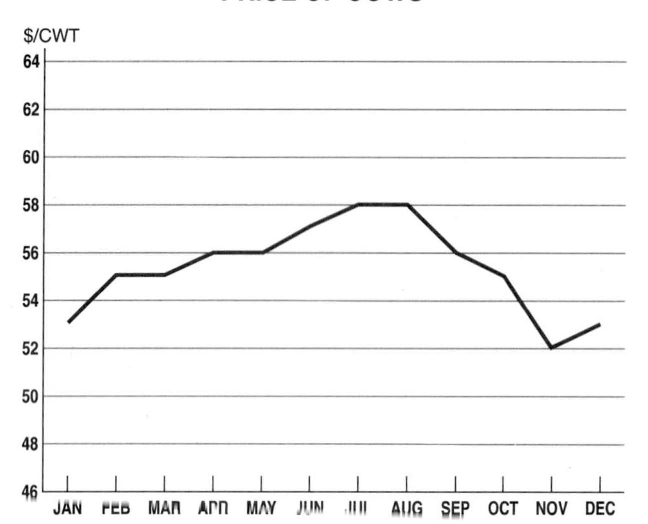

Fig. 15-18. Average price of cows. Commercial grade, Sioux Falls, 1990–1993. (From: *Red Meats Yearbook*, USDA, ERS, Stat. Bull. No. 885, p. 67, Table 65)

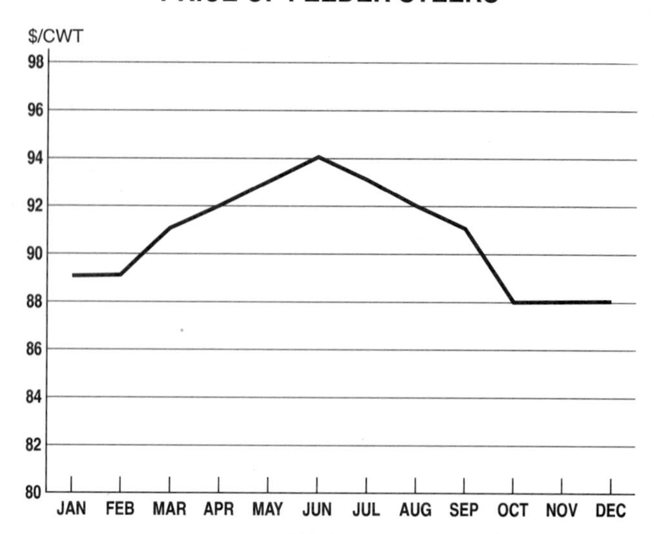

Fig. 15-19. Average price 1990–1993 of feeder steers, Medium No. 1, Oklahoma City, 600–650 lb. (From: *Red Meats Yearbook*, USDA, ERS, Stat. Bull. No. 885, p. 71, Table 69)

culled from the herd for one reason or another had best be marketed from April to September.

b. The lowest price for cows is from October to March.

c. The lower prices in the winter will likely nullify the value of any increased weight put on cows during the summer and fall. Thus, unless there is an overabundance of grass, or unless the cows can be raised a grade through increased weight gains, it is usually best to market them from April to September.

4. **Feeder cattle.** Fig. 15-19 shows the seasonal variation in feeder steer prices. To cattle feeders, this graph is a buying guide; to cattle producers, it is a selling guide. From this figure and other market information the following conclusions are drawn:

a. Feeder steer prices are usually lowest from October to February, and highest from March to September.

b. Feeder cattle receipts are greatest in the fall because this is the season when farmers and ranchers sell their surplus cattle off grass in preparation for winter, and when feeders want to fill their feedlots in order to utilize the new crop and available labor.

c. It is difficult for the producer of feeder cattle to take advantage of seasonal trends because pasture and range conditions often determine, or actually dictate, the time of sale from the range.

DOCKAGE

The value of some market animals is low because dressing losses are high, or because part of the product is of low quality. Cattle with lumpy jaw are usually bought subject to the amount of wastage.

LIVESTOCK MARKET NEWS SERVICES

Accurate market news is essential to the efficient marketing of livestock, both from the standpoint of the buyer and the seller. In the days of trailing, the meager market reports available were largely conveyed by word of mouth. Moreover, the time required to move livestock from the farm or ranch to market was so great that detailed market information would have been of little benefit even if it had been available. With the speed in transportation afforded by railroads and trucks, late information on market conditions became important.

■ **Federal Market News Service**—The Federal Market News Service was initiated by the U.S. Department of Agriculture beginning in 1916. This service was established for the purpose of providing unbiased and uniformly interpretable market information.

Today, livestock markets are covered by livestock, meat, and wool field offices maintained throughout the United States. This network provides coverage of terminal livestock markets, auctions, and major direct sales areas. Because of limitations in funds and personnel, market reporting in an era of increased auction sales and direct selling has been difficult, if not inadequate. The problem is accentuated because the Federal Market News Service depends on voluntary cooperation in gathering information. There is no legal compulsion for buyers and sellers to divulge purchase and sale information. Reports on direct sales are obtained largely by telephone, augmented by interviews made at feedlots, packing plants, and ranches.

The Federal Market News Service relies upon local and privately owned newspapers, radio stations, and T.V. stations, merely supplying them with the market reports. Because at least a part of the readers or listeners are interested in this type of information, the local papers and radio stations are usually glad to serve as media for disseminating these reports.

The chief contributions of the Federal Market News Service may be summarized as follows: (1) a common terminology has come to be established from market to market, thus a slaughter steer which will grade Select at Omaha would be designated the same grade at every other market in the country; and (2) facilities for the dissemination of market information have been provided.

■ **Market information provided by farm and trade magazines and market agencies**—Farm magazines are one of the most important sources of livestock market information.

Many market agencies—*i.e.*, commission firms, auction markets, and related organizations—prepare and distribute market information. By means of weekly market newsletters or cards, they commonly emphasize the price and market conditions of the particular market that they serve.

■ ***Cattle-Fax***—The National Cattlemen's Beef Association has been particularly effective in utilizing privately developed cattle market information. Under a separate corporation, known as Cattle Marketing Information Service, Inc. (CMIS), it operates its cattle marketing service, appropriately dubbed *Cattle-Fax*. CMIS has developed a system for determining numbers of cattle on feed, potential supplies at future dates, current prices, market conditions, and other market factors. It assembles, analyzes, and interprets this information centrally, then transmits it by phone and mail back to its members. Members utilize the information individually (or in local pooled arrangements) in negotiating direct sales with packers. *Cattle-Fax* subscribers pay a monthly fee, at a fixed minimum plus a charge per head up to a stipulated maximum total fee. The *Cattle-Fax* service is rather expensive for the small farmer-feeder, but pooled arrangements can be set up wherein several small operators share the cost of installation, which may be located in a bank or another business establishment convenient to the group.

FEDERAL MEAT INSPECTION

The federal government requires supervision of establishments which slaughter, pack, render, and prepare meats and meat products for interstate shipment and foreign export; it is the responsibility of the respective states to have and enforce legislation governing the slaughtering, packaging, and handling of meats shipped intrastate, but state standards cannot be lower than federal levels. The meat inspection laws do not apply to farm slaughtering for home consumption, although all states require inspection if the meat is sold.

The meat inspection service of the U.S. Department of Agriculture was inaugurated, and is maintained under, the Meat Inspection Act of June 30, 1906. This act was updated and strengthened by the Wholesome Meat Act of December 15, 1967. The latter statute (1) requires that state standards be at least to the levels applied to meat sent across state lines; and (2) assures consumers that all meat sold in the United States is inspected either by the federal government or by an equal state program. The Animal and Plant Health Inspection Service of the U.S. Department of Agriculture is charged with responsibility of meat inspection.

The purposes of meat inspection are (1) to safeguard the public by eliminating diseased or otherwise unwholesome meat from the food supply, (2) to enforce the sanitary preparation of meat and meat products, (3) to guard against the use of harmful ingredients, (4) to prevent the use of false or misleading names or statements on labels. Personnel for carrying out the provisions of the act are of two types: professionals or veterinary inspectors who are graduates of accredited veterinary colleges, and non-professional food inspectors who are required to pass a Civil Service examination. In brief, the inspections consist of the following two types:

1. **Antemortem (before death)** inspection is made in the pens or as the animals move from the scales after weighing. The inspection is performed to detect evidence of disease or any abnormal condition that would indicate a disease. Suspects are provided with a metal ear tag bearing the notation ""U.S. Suspect No. ...," and are given special postmortem scrutiny. If in the antemortem examination there is definite and conclusive evidence that the animal is not fit for human consumption, it is *condemned*, and no further postmortem examination is necessary.

2. **Postmortem (after death)** inspection is made at the time of slaughter and includes a careful examination of the carcass and the viscera (internal organs). All good carcasses are stamped "U.S. Inspected and Passed," whereas the inedible carcasses are stamped "U.S. Inspected and Condemned." The latter are sent

Fig. 15-20. Antemortem (before death) inspection of cattle being made by a federal veterinarian. Animals that are clearly diseased, emaciated, or otherwise unfit for human consumption are destroyed. Their carcasses may be used only in making inedible grease, fertilizer, or other nonfood products. Animals that appear slightly abnormal on foot are tagged "U.S. Suspect," and are given special postmortem scrutiny. (Courtesy, USDA)

to the rendering tanks, the products of which are not used for human food.

In addition to the antemortem and postmortem inspections referred to, the government meat inspectors have the power to refuse the application of the mark of inspection to meat products produced in a plant that is not sanitary. All parts of the plant and its equipment must be maintained in a sanitary condition at all times. In addition, plant employees must wear clean, washable garments, and suitable lavatory facilities must be provided for hand washing.

Meat inspection regulations require the condemnation of all or affected portions of carcasses of animals with various disease conditions, including pneumonia, peritonitis, abscesses and pyemia, uremia, tetanus, rabies, anthrax, tuberculosis, various neoplasms (cancer), arthritic, actinobacillosis, and many others.

The federal government developed a new meat inspection program in 1996. This program involved four main elements: (1) HACCP (hazard analysis and critical control points); (2) mandatory *E. coli* testing in slaughter plants; (3) pathogen reduction performance standards for salmonella; and (4) sanitation standard operating procedures. The new inspection program is going to be extremely costly and difficult for meat packers to implement. As part of the new regulations, meat plants were given from 6 to 18 months to implement the different components of the regulations.

The new regulations require every meat plant to develop and implement a HACCP plan. With HACCP, plants identify critical control points during their processes where hazards such as microbial contamination can occur. The plants must establish controls to prevent or reduce these hazards, and maintain records documenting the controls are working as intended.

The Food Safety and Inspection Service (FSIS) expects this combination HAACP, microbial testing, pathogen reduction performance standards, and sanitation standard operating procedures to significantly reduce contamination of meat and poultry with harmful bacteria and reduce the risk of foodborne illness.

Most of the larger meat packers are under federal inspection; hence, they are allowed to ship interstate.

(Also, see Chapter 16, section headed "Food Safety," sections on Irradiation, and Hazard Analysis Critical Control Point [HACCP].)

STATE MEAT INSPECTION

States have varying legislation governing the slaughtering and further processing of meats produced for intrastate commerce. However, the Wholesome Meat Act of 1967 requires that the state standards be equal to the federal standards. Inspection that was often formerly conducted under local ordinances is now conducted, with one exception, by state employees. The one exception is in the city of Chicago, where city employees conduct inspection under the overall supervision of the state.

The Wholesome Meat Act gave the states the option of either conducting their own inspection service, or turning the responsibility over to the federal government. In most of these states, the service is administered by the State Department of Agriculture. Quite frequently, they simply apply the federal regulations.

In 1996, USDA's Advisory Committee on Concentration in Agriculture recommended that the ban on interstate shipment of state-inspected meat and poultry should be repealed in an effort to increase competition in the meatpacking industry.

PACKER SLAUGHTERING AND DRESSING OF CATTLE

Table 15-9 shows the proportion of cattle and calves slaughtered commercially (meaning that they were slaughtered in federally inspected and other wholesale and retail establishments), and the proportion slaughtered on farms. The total figure refers to the number dressed in all establishments and on farms. As shown, farm slaughter is negligible, with a slightly higher percentage of calves farm slaughtered than of cattle.

Table 15-10 shows the states in which most of the commercial cattle are slaughtered. This table points

TABLE 15-9
PROPORTION OF CATTLE AND CALVES SLAUGHTERED COMMERCIALLY[1]

	Cattle			Calves		
Year	Total Number Slaughtered (commercially and farm)	Number Slaughtered Commercially	Percent Slaughtered Commercially	Total Number Slaughtered (commercially and farm)	Number Slaughtered Commercially	Percent Slaughtered Commercially
	(1,000 head)	*(1,000 head)*	*(%)*	*(1,000 head)*	*(1,000 head)*	*(%)*
1988	35,324	35,079	99.3	2,565	2,506	97.7
1989	34,106	33,917	99.4	2,223	2,172	97.7
1990	33,439	33,242	99.4	1,838	1,790	97.4
1991	32,885	32,690	99.4	1,483	1,436	96.9
1992	33,069	32,874	99.4	1,420	1,371	96.5

[1]*Agricultural Statistics 1993*, USDA, p. 237, Table 381.

TABLE 15-10
TOP 10 CATTLE SLAUGHTERING STATES, AND PERCENTAGE OF COMMERCIAL CATTLE SLAUGHTERED BY EACH IN 1993[1]

State	Commercial Cattle Slaughtered in 1993
	(%)
1. Nebraska	19.9
2. Kansas	18.7
3. Texas	18.2
4. Colorado	7.3
5. Iowa	5.0
6. Wisconsin	4.2
7. Minnesota	3.2
8. Pennsylvania	3.0
9. California	2.6
10. Washington	2.5

[1]*Meat and Poultry Facts, 1994*, p. 17.

up that by far the greatest number of cattle are slaughtered in Nebraska, Kansas, and Texas.

Cattle intended for slaughter purposes are bought primarily on the basis of projected quality and yield of carcasses. Upon reaching the packing house, they rapidly pass through the operations of killing and dressing. Unlike most manufacturing, meat packing is primarily a disassembly process wherein the manufacturing operation starts with a complete unit that is progressively broken down into its component parts. The various parts then are subjected to divergent processing operations. In most of the larger and newer slaughtering plants, cattle are processed with the endless chain method of dressing, similar to that used for dressing of calves, sheep, and hogs.

STEPS IN SLAUGHTERING AND DRESSING CATTLE

Although the procedure differs between plants, in general the endless chain method of slaughtering and dressing cattle involves the following steps, carried out in rapid succession:

1. **Rendering insensible.** The cattle are rendered insensible by a captive bolt stunner while cradled in a moving restrainer that allows a worker to shackle the leg of the animal from below. Also, carbon dioxide may be used for calves.

2. **Released from restrainer, sticking, and bleeding.** After stunning, the animal is released from the restrainer, and while shackled by a hind leg travels to a sticking area for bleeding.

3. **Skinning.** The hind shanks are skinned and removed at the hocks; beef hooks are inserted on the gam cord; the hide is opened along the median line of the belly and is removed from the belly and sides; then down pullers are used for removing the rest of the hide, including skinning the head. This process has greatly increased the efficiency of skinning and the yields of headmeat/cheekmeat. The head is not severed from the carcass until the hide is completely removed. The breast and aitch (rump) bones are split by sawing.

4. **Removing viscera.** All internal organs are removed except the kidneys. If the plant is under federal inspection, the carcass and viscera are examined at this stage in the slaughtering process.

5. **Splitting carcass and removing tail.** The carcass is then split through the center of the backbone and the tail is removed.

6. **Washing and drying.** The split carcasses or sides are washed with warm water under pressure.

7. **Sending to coolers.** Following slaughtering,

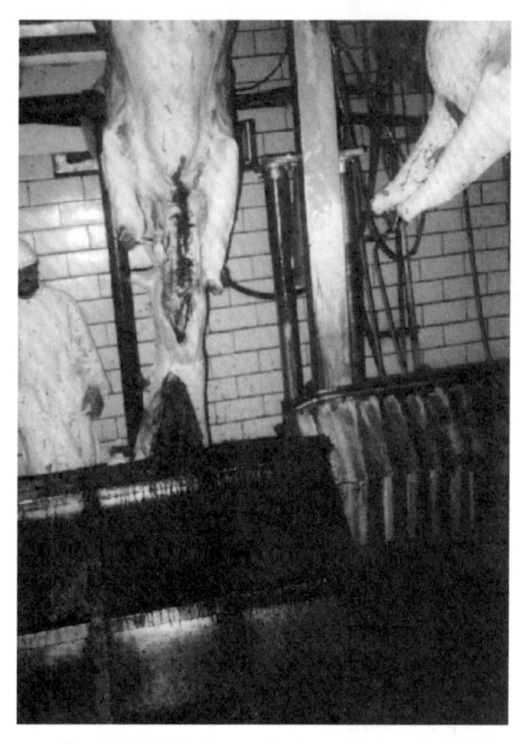

Fig. 15-21. A hide down-puller is used to remove the hide. Note the hide is pulled down off the carcass and head which keeps the edible carcass, head, etc. from coming in contact with the hair side of the hide. It also eliminates the need to remove the skin from the head prior to hide removal. (Courtesy, Excel Corp., Wichita, KS)

the sides are sent to the coolers where they are kept at a temperature of about 34°F for a minimum period of 24 hours before ribbing.

HOW SLAUGHTER OF VEAL CALVES DIFFERS

Because of their smaller size, calves are almost always dressed by the endless-chain method. A wheel hoist is used in lifting the shackled calves to the rail. They are then stuck, bled, dressed, and washed. Because of the high moisture content of veal, the hide is usually left on for the purpose of reducing evaporation. This also produces a more desirable carcass color. When the hide is left on, it is thoroughly washed before dressing.

KOSHER SLAUGHTER

Meat for the Jewish trade must come from animals slaughtered according to the rules of *Shehitah* (the ancient dietary rules). Although we usually think in terms of cattle when kosher slaughtering is mentioned, calves, sheep, lambs, goats, and poultry are slaughtered in a similar manner.

The killing is performed by a rabbi of the Jewish church or a specially trained representative; a person called the *shohet* or *shochet*, meaning slaughterer. In kosher slaughter, the animal is either hoisted without stunning or driven into a restrainer which holds it immobile, following which is cut across the throat with a special razor sharp knife, known as a *chalaf.* With one quick, clean stroke the throat is cut, through the jugular vein and other large vessels, together with the gullet and windpipe. Two reasons are given for using this method of killing instead of the more conventional method of stunning and sticking; namely, (1) it produces more instant death with less pain, and (2) it results in more rapid and complete bleeding, which Orthodox Hebrews consider essential for sanitary reasons.

The shohet also makes an inspection of the lungs, stomach and other organs while dressing. If the carcass is acceptable, it is marked on the brisket with a cross inside a circle. The mark also gives the date of slaughter and the name of the inspector.

Since neither packers nor meat retailers can hold kosher meat longer than 216 hours (and even then it must be washed at 72-hour intervals), rapid handling is imperative. This fact, plus the heavy concentration of Jewish folks in the eastern cities, results in large numbers of live cattle being shipped from the markets farther west to be slaughtered in or near the eastern consuming areas.

THE DRESSING PERCENTAGE

Dressing percentage may be defined as the percentage yield of hot carcass in relation to the weight of the animal on foot. For example, a steer which weighed 1,200 lb on foot and yielded a hot carcass weighing 720 lb may be said to have a dressing percentage of 60. The offal—so-called because formerly (with the exception of the hide, tallow, and tongue) the offal (waste) was thrown away—consists of the blood, head, shanks, tail, hide, viscera, and loose fat.

A high carcass yield is desirable provided it is due to muscle (lean meat), and not due to fat, because the carcass is much more valuable than the by-products. Although the packers have done a marvelous job in utilizing by-products, about 91% of the income from cattle and calves is derived from the sale of the carcass and only 9% from the by-products. Thus, the estimated dressing percentage of slaughter cattle is justifiably a major factor in determining the price or value of the live animal.

The chief factors determining the dressing percentage of cattle are: (1) the amount of fill, (2) the finish or degree of fatness; (3) the thickness of muscling; (4) the general quality and refinement (refinement of head, bone, hide, etc.); and (5) the size of udder. The better grades of steers have the highest dressing percentage, with thin Canner cows showing the lowest yield.

The highest dressing percentage on record was a yield of 76.75% made by a spayed Angus heifer at the Smithfield Fat Stock Show in England. The average liveweights of cattle and calves commercially slaughtered in the United States, and their percentage yield in meats, for the year 1993, are given in Table 15-11.

TABLE 15-11
AVERAGE LIVEWEIGHT, CARCASS YIELD, AND DRESSING PERCENTAGES OF ALL CATTLE AND CALVES COMMERCIALLY SLAUGHTERED IN THE UNITED STATES IN 1993[1]

	Average Liveweight		Dressed Weight		Dressing Percentage
	(lb)	*(kg)*	*(lb)*	*(kg)*	*(%)*
Cattle	1,161	*528*	688	*313*	59.3
Calves	388	*176*	224	*102*	57.8

[1]*Livestock Slaughter 1993 Summary*, March 1994, USDA.

FABRICATED, BOXED BEEF

Beef processing has gone modern as shown in Figs. 15-22, 15-23, and 15-24.

Formerly, beef was shipped in exposed halves, quarters, or wholesale cuts and divided into retail cuts in the backrooms of 200,000 supermarkets. This procedure left much to be desired from the standpoints of efficiency, sanitation, shrink, spoilage, and discoloration. To improve this situation, modern packers fabricate and box beef in their plants. Today, 75 to 80% of all beef is shipped to supermarkets and food service distributors fabricated into sub-primal boxed beef by a modern packing plant. Generally, fabricating modern meat packing plants consists of the following: After chilling, the carcass is subjected to a disassembly process, in which it is fabricated or broken into sub-primal cuts; vacuum-sealed; boxed; moved into storage by a fully automated, computer-controlled system; loaded by computer-automation into trailers; and shipped to retailers across the nation.

MOVEMENT OF FEEDER CATTLE

In the past, the traditional sources of feeder cattle for the Corn Belt and the West have been the cow-calf operations in the Southwest. However, in recent years, the number of cattle moving between these areas has decreased as the Southwest has kept back feeder

Fig. 15-22. A combination of power driven chains *(above)* and moving tables *(below)* create efficient high speed fabrication of beef carcasses through modern day large packing plants. (Courtesy, Excel Corp., Wichita, KS)

Fig. 15-23. Boxed beef. Product is identified by UCC 128 Random Weight Bar Code labels for efficient receiving and shipping throughout the distribution system. This new technology has added a considerable amount of cost savings to the handling, storage, and shipping of boxed beef at warehouses. (Courtesy, Excel Corp., Wichita, KS)

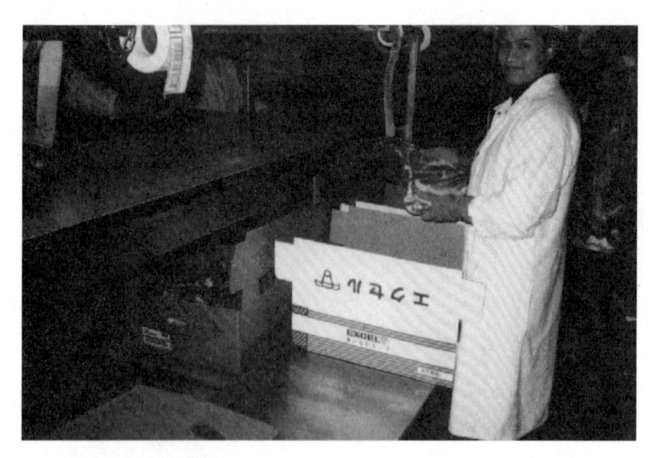

Fig. 15-24. As the export market provides sales for many beef products, countries such as Japan receive product in special cartons printed in their language. (Courtesy, Excel Corp., Wichita, KS)

cattle to supply its own ever-increasing and larger cattle feeding program. Texas and Kansas now feed more cattle than are available as feeders from their own yearly calf crop.

In the meantime, the southern and southeastern states have become more and more important as a source of feeder cattle. This transition is the result of (1) increases in the percent calf crop produced, and (2) a sizable reduction in veal calf slaughter.

In the future, the Corn Belt will produce more of its feeder cattle, almost out of necessity, and a number of feeder cattle deficit states will depend on feeder cattle from the southeastern states. Many of these may be grown on pastures in Kansas, Oklahoma, and Texas before reaching feedlots, as in the past. A predicted future direction of feeder cattle movement from their sources to feedlots for finishing is shown in Fig. 15-25.

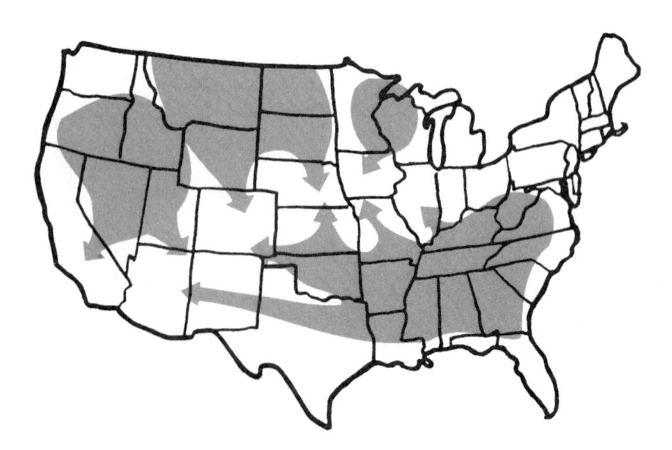

Fig. 15-25. Predicted directions of feeder cattle movements.

SELLING PUREBRED CATTLE

Selling purebred animals is a highly specialized and scientific business. Purebred animals are usually sold at private treaty directly to other purebred breeders or commercial producers or through auctions which may be sponsored by either one or a few breeders (joint sales or consignment sales).

In general, the vast majority of bulls saved for breeding purposes go into commercial herds. Only the elite sires are retained with the hope of affecting further breed improvement in purebred herds. On the other hand, the sale of purebred females is fairly well restricted to meeting the requirements for replacement purposes in existing purebred herds or for establishing new purebred herds.

Fig. 15-26. The success of a cattle auction depends upon many things in addition to the quality of the cattle; among them, (1) the number of buyers present, rather than just spectators; (2) the presence of a specialty consignment(s) which will attract buyers, who may bid on other consignments as well; and (3) the reputation of and homework done by the sales manager and consignors.

Altogether too often, a good or bad sale is a reflection of the above factors, rather than the quality of the cattle. (Courtesy, *Bull-O-Gram*, published by Litton Charolais Ranch, Chillicothe, MO)

Most consignment sales are sponsored by a breed association, either local, statewide, or national in character. Such auctions, therefore, are usually limited to one breed. Purebred auction sales are conducted by highly specialized auctioneers. In addition to being good salespeople, such auctioneers must have a keen knowledge of values and be familiar with the bloodlines of the breeding stock.

QUESTIONS FOR STUDY AND DISCUSSION

1. Many top authorities feel that marketing, not production, is the cattle producer's number one problem. Do you agree? Justify your answer.

2. Discuss the magnitude of cattle marketing.

3. U.S. farmers and ranchers receive a major portion of their cash income from cattle (a fantastic 23% of their total income from livestock and products plus crops; 44% of livestock and products—exclusive of crops). Is this good or bad from the standpoint of producers? Is it good or bad from the standpoint of consumers? Justify your answers.

4. Since World War I, terminal markets have declined in importance while country sales (direct marketing) have increased. Why has this happened?

5. Why are most calves sold through auctions, whereas most cattle are sold direct?

6. What market channel(s) would you recommend for the sale of cattle off your farm or ranch, or off a farm or ranch with which you are familiar? Justify your choice.

7. Does each market channel give adequate assurance of honesty, of sanitation, and of humane treatment of animals? Justify your answer.

8. Why is it important that a producer know the leading markets for each class of cattle?

9. How do you account for the fact that the leading public markets for finished cattle and calves do not coincide?

10. Which is the more important to the livestock seller: (a) low marketing costs; or (b) effective selling and net returns? Justify your answer.

11. Outline, step by step, how you would prepare and ship cattle.

12. Under what circumstances would you recommend the use of each of the following methods of transporting market cattle: (a) truck or (b) air?

13. Assume that you have two offers for a 1,000-lb steer: one buyer is offering $60.00/cwt with a 4% pencil shrink. The other buyer is offering $59.00/cwt with no pencil shrink. Which offer would you take? Show your computations.

14. Define on-foot market (a) classes, and (b) grades of cattle, and tell of their value.

15. Why was the *bullock* sex class added? Will we feed out more bulls in the future?

16. What's the difference between (a) quality grades and (b) yield grades?

17. Why is it important that a cattle producer know the market classes and grades of cattle and what each implies?

18. Why are grade specifications changed from time to time?

19. Since there is a rather uniform difference in the selling price of the different grades of cattle, with the top grades bringing the higher prices, why do not more cattle producers produce the top grades?

20. How may a buyer elevate the carcass quality and yield grades from looking at live cattle?

21. How do you explain the fact that there is a spread of $3.79 per 100 lb between Choice and Select grades of beef despite the fact that Select beef is leaner?

22. In your area, is there sufficient price spread between Choice and Select grade slaughter cattle to justify the added breeding and feeding cost of the Choice grade?

23. List and discuss the significance of (a) the three frame size grades and (b) the three thickness grades of feeder cattle.

24. How may a cattle producer take advantage of cyclical trends and seasonal changes?

25. Why are market news services important to cattle producers? What is *Cattle-Fax*; and of what value is it?

26. Give the steps, in order of application, involved in the commercial slaughtering of cattle.

27. How does kosher slaughter differ from nonkosher slaughter?

28. What is meant by fabricated, boxed beef?

29. Do packers control market cattle prices? Justify your answer.

30. Discuss the movement of feeder cattle, including (a) production, (b) the stocker stage, (c) feedlot, and (d) slaughter.

31. In what ways does the selling of purebred cattle differ from the selling of commercial cattle?

SELECTED REFERENCES

Title of Publication	Author(s)	Publisher
Animal Science, Ninth Edition	M. E. Ensminger	Interstate Publishers, Inc., Danville, IL, 1991
Essentials of Marketing Livestock	R. C. Ashby	The author, Sioux City, IA, 1953
Lessons on Meat, Second Edition	National Live Stock and Meat Board	National Live Stock and Meat Board, Chicago, IL, 1972
Livestock Marketing	A. A. Dowell K. Bjorka	McGraw-Hill Book Company, Inc., New York, NY, 1941
Livestock and Meat Marketing	J. H. McCoy	The Avi Publishing Company, Inc., Westport, CT, 1972
Marketing, The Yearbook of Agriculture 1954	Ed. by A. Stefferud	U.S. Department of Agriculture, Washington, DC, 1954
Meat Handbook, The, Fourth Edition	A. Levie	The Avi Publishing Company, Inc., Westport, CT, 1979
Meat We Eat, The	Romans, J. R., *et al.*	Interstate Publishers, Inc., Danville, IL, 1994
Problems and Practices of American Cattlemen, Wash. Ag. Exp. Sta. Bull. 562	M. E. Ensminger M. W. Galgan W. L. Slocum	Washington State University, Pullman, WA, 1955
Processed Meats	W. E. Kramlich A. M. Pearson F. W. Tauber	The Avi Publishing Company, Inc., Westport, CT, 1973
Stockman's Handbook, The, Seventh Edition	M. E. Ensminger	Interstate Publishers, Inc., Danville, IL, 1992
Uniform Retail Meat Identity Standards	Industrywide Cooperative Meat Identification Standards Committee	National Live Stock and Meat Board, Chicago, IL, 1973

Beef on the table. (Courtesy, National Live Stock and Meat Board, Chicago, IL)

BEEF AND VEAL, AND BYPRODUCTS FROM CATTLE SLAUGHTER[1]

Contents *Page*

[1]Helpful review suggestions and/or pictures for this chapter were received from the following who have great expertise in beef and veal, and byproducts from cattle slaughter: Don Fender, Excel Corporation, Wichita, Kansas; H. Kenneth Johnson and Terry Dockerty, National Livestock and Meat Board, Chicago, Illinois; and Jim Wise, USDA, Livestock & Seed Division, Washington, DC.

Contents *Page*

Fig. 16-1. Roast of beef. (Courtesy, National Live Stock and Meat Board, Chicago, IL)

Beef is the ultimate objective in producing cattle; it is the end product of all breeding, feeding, care and management, marketing, and processing. It is imperative, therefore, that the progressive producer, the student, and the beef cattle scientist have a working knowledge of beef and veal and of the byproducts from cattle slaughter. Such knowledge will be of value in selecting animals and in determining policies relative to their handling. To this end, this chapter is presented.

Of course, the type of animals best adapted to the production of meat over the block has changed in a changing world. Thus, in the early history of this country, the very survival of animals was often dependent upon their speed, hardiness, and ability to fight. Moreover, long legs and plenty of bone were important attributes when it came time for animals to trail hundreds of miles as drovers took them to market. The Texas Longhorn was adapted to these conditions.

With the advent of rail (and later truck) transportation and improved care and feeding methods, the ability of animals to travel and fight diminished in importance. It was then possible, through selection and breeding, to produce meat animals better suited to the needs of more critical consumers. With the development of large cities, artisans and craftsmen and their successors in industry required fewer calories than those who were engaged in the more arduous tasks of logging, building railroads, etc. Simultaneously, the American family decreased in size. The demand shifted, therefore, to smaller and less fatty cuts of meats; and, with greater prosperity, high-quality steaks and roasts were in demand. To meet the needs of the consumer, the producer gradually shifted to the breeding and marketing of younger animals with maximum cutout value, instead of marketing large, ponderous, fat 3- to 5-year-old steers.

Thus, through the years, consumer demand has exerted a powerful influence upon the type of cattle produced. To be sure, it is necessary that such production factors as prolificacy, economy of feed utilization, rapidity of gains, size, and longevity, receive due consideration along with consumer demands. But once these production factors have received due weight, cattle producers—whether they be purebred or commercial operators—must remember that meeting consumer demand at the retail case is the ultimate objective.

Fig. 16-2. A tiered meat case, with a customer selecting beef. (Courtesy, National Live Stock and Meat Board, Chicago, IL)

Now, and in the future, beef producers need to select and feed so as to obtain increased quality and cutout value, without excess fat.

QUALITIES IN BEEF DESIRED BY THE CONSUMER

Because consumer demands are such an important item in the production of beef, all members of the beef team—the farmer and rancher, the packer, and the meat retailer—should be familiar with these qualities, which are summarized as follows:

1. **Minimal outside fat.** For health and economic reasons, consumers insist on not more than ¼ to ⅛ inch outside fat, without concern as to whether the limited outside fat was obtained by breeding or close-trimming. The major packers estimate that more than 80% of the boxed beef was close-trimmed in 1996.
2. **Tenderness.** Consumers want fine-grained, tender meat, in contrast to coarse-grained, less tender meat.
3. **Flavor.** First and foremost, people eat meat because they like it. Palatability is influenced by the tenderness, juiciness, and flavor of the fat and lean.
4. **Attractiveness; eye appeal.** The general attractiveness is an important factor in selling meats to the consumer. The color of the lean, the degree of fatness, and the marbling are leading factors in determining buyer appeal. Most consumers prefer a white fat and a light or medium red color in the lean.
5. **Small cuts.** Most purchasers prefer to buy cuts of meat that are of a proper size to meet the needs of their respective families. Because the American family has decreased in size, this has meant smaller cuts. In turn, this has had a profound influence on the type of animals and on market age and weight.
6. **Ease of preparation.** In general, consumers prefer to select those cuts of beef that will give the greatest amount of leisure time. Steaks of beef can be prepared with greater ease and in less time than can roasts or stews. Hamburger and sausage are also easy to prepare.
7. **Consistency (repeatability).** The homemaker wants a cut of meat just like the one purchased last time, which calls for repeatability.

Fig. 16-3. Beef tip roast. (Courtesy, National Live Stock & Meat Board, Chicago, IL)

CATTLEMEN'S CARCASS DATA SERVICE

The Cattlemen's Carcass Data Service, which was initiated by the National Cattlemen's Association (NCA) March 1, 1992, is designed to increase the competitiveness of the beef industry by helping producers enhance the quality and consistency of beef carcasses. For a nominal charge of $3.00 to $5.00 per carcass, depending on how many carcasses are involved, cattle producers can obtain the quality and yield grades of their end product—beef. Also, by utilizing the Cattlemen's Carcass Data Service, producers can compare their cattle with the national average and see how well they are doing.

In addition to the NCA program, many of the state cattlemen's associations and breed registry associations sponsor carcass feedback programs. While each program is a little different, most of them concentrate on three things: (1) feedlot performance, (2) carcass characteristics, and (3) profitability.

In determining market value, the beef cattle industry, from producers to packers, needs to do away with dressing percentage and replace it with red meat yield and not more than ¼ inch outside fat.

THE FEDERAL GRADES OF BEEF

The grade of beef may be defined as a measure of its degree of excellence based on quality, or eating characteristics of the beef, and the yield, or total proportion, of primal cuts.

In 1926, a producer group, known as The Better Beef Association, petitioned the U.S. Department of Agriculture to set up a service for grading and stamping beef. Such a program was activated in 1927, and continues to this day. Since then, federal beef grades have been changed many times, in response to changes in consumer tastes and preferences, as well as changes in technology.

Unlike meat inspection, government meat grading is purely voluntary, on a charge basis. In 1995, a

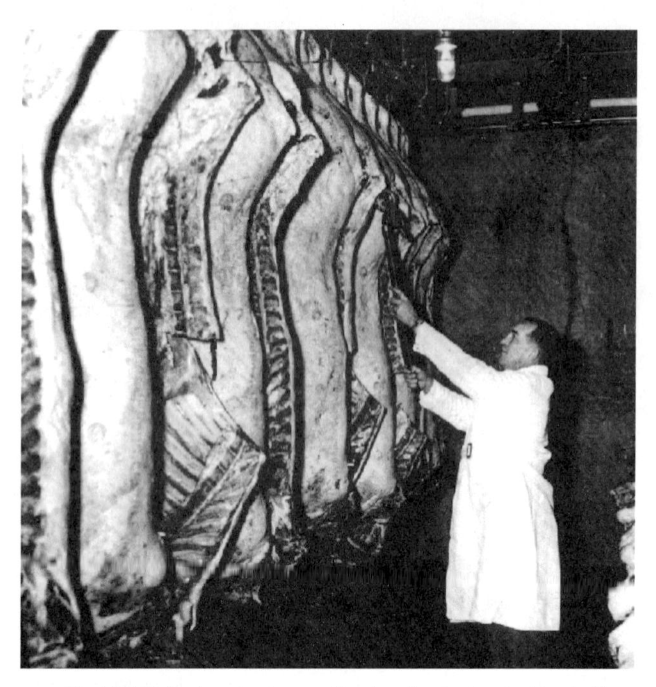

Fig. 16-4. Federal grader shown rolling (grading or stamping) beef with an edible vegetable dye. (Courtesy, *Livestock Breeders Journal*, Macon, GA)

government meat grader cost $36.60 per hour. The time includes both travel and the actual time required to do the grading.

The value of beef carcass depends chiefly upon two factors—the quality of the meat and the amount of salable meat the carcass will yield, particularly the yield of the high value, preferred retail cuts.

Fig. 16-5 shows the proportion of U.S. total meat beef, and veal-calf slaughter federally graded in 1993. It is noteworthy that more than 80% of the beef slaughter was federally graded in 1993.

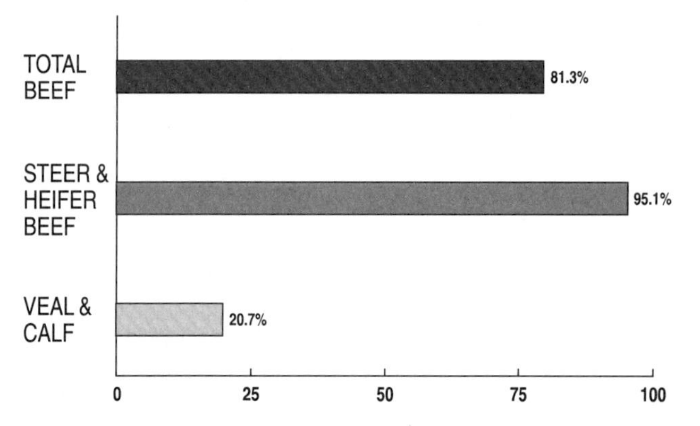

Fig. 16-5. Proportion of U.S. commercial beef and veal federally graded in 1993. (Data from Livestock and Seed Division, Agricultural Marketing Service, USDA)

Because the USDA beef grades have become so widely used in marketing, it is important that everyone from producer to consumer have an understanding of the federal standards for grades of beef. Producers need to understand grades in order to make wise production and marketing plans. Meat packers, wholesalers, and retailers also must be familiar with grades to facilitate their buying and selling operations. Consumers need to have an understanding of the grades in order to select the quality they desire. USDA yield grades also are important to consumers when they buy carcasses or wholesale cuts for the freezer.

Pertinent facts about both the quality grade and the yield grade follow:

1. **Quality and yield grades uncoupled.** In 1989, the law was changed, separating quality and yield grades of beef, allowing packers to choose whether beef carcasses shall be graded for quality, for yield, or for both quality and yield.

2. **Quality grade.** Quality refers to the palatability-indicating characteristics of the lean and is evaluated by considering the marbling (flecks of fat within the lean) and firmness of the lean as observed in a cut surface in relation to the apparent maturity of the animal from which the carcass was produced. The maturity of the carcass is determined by evaluating the size, shape, and ossification of the bones and cartilages—especially the split chine bones—and the color and texture of the lean flesh. Superior quality implies firm, well-marbled lean that is fine in texture and has a light red, youthful color.

USDA quality grades for beef—Prime, Choice, Select, Standard, Commercial, Utility, Cutter, and Canner—have for many years served as nationally reliable guides to the eating quality of beef—its tenderness, juiciness, and flavor.

Marbling and maturity are dependable and easily measured characteristics for determining the quality grade as a method of predicting differences in palatability.

In general, increases in marbling improve the palatability or eating quality of beef, but increases in maturity have the opposite effect. However, research has indicated that for younger cattle (under 30 months of age) maturity changes do not have an appreciable effect upon palatability. Therefore, within each of the Prime, Choice, Select, and Standard grades the minimum requirement for marbling is the same for all carcasses from animals under 30 months of age. This reduction in marbling requirement, which was promulgated by the USDA in 1976, resulted in slightly leaner beef with less excess fat, particularly in the Prime and Choice grades, and less grain required to produce cattle that would qualify for the top grades.

For cattle over 30 months of age, increased marbling is required with increasing maturity within each grade.

After the maturity and marbling are determined, these two factors are combined to determine the USDA quality grade. The relationship between marbling and maturity used to determine the quality grade of a carcass is presented in Fig. 16-6.

RELATIONSHIP BETWEEN MARBLING, MATURITY AND CARCASS QUALITY GRADE*

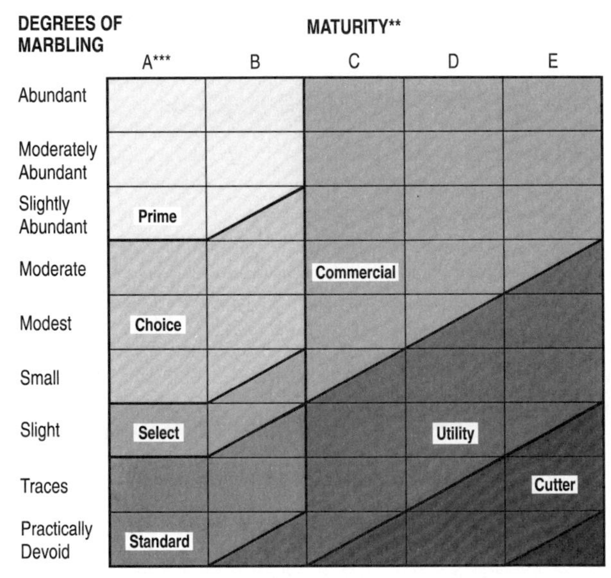

*Assumes that firmness of lean is comparably developed with the degree of marbling and that the carcass is not a "dark cutter."

**Maturity increases from left to right (A through E).

***The A maturity portion of the figure is the only portion applicable to bullock carcasses.

Fig. 16-6. USDA Beef Grading Chart. (Courtesy, USDA, Standardization Branch-AMS)

USDA has proposed revising the quality grading system by removing "B" maturity carcasses (from cattle approximately 30 to 42 months of age) with small or slight degrees of marbling from the Choice and Select grades and including them in the Standard grade. This revision is scheduled to become effective January 31, 1997.

3. **Yield grade.** It is recognized that variations in conformation (shape) due to differences in muscling do affect yields of lean—and carcass value. This is reflected by the yield grades. The significance of yield grades becomes evident when tests reveal that carcasses of the same quality grade—Choice for example—can vary in value by $100 or more due to differences in cutability.

The yield grade of a beef carcass is determined by considering four characteristics: (a) the amount of external fat; (b) the amount of kidney, pelvic, and heart fat; (c) the area of the ribeye muscle; and (d) the carcass weight.

Thus, in the same way that quality grades identify beef for differences in palatability, USDA yield grades for beef provide a nationally uniform method of identifying cutability differences among carcasses. Specifically, they are based on the percentage yields of boneless, closely trimmed retail cuts from the high value parts of the carcass—the round, loin, rib, and chuck—which account for more than 80% of its value. However, they also reflect differences in total yields of retail cuts.

There are five USDA yield grades numbered 1 through 5. Yield Grade 1 carcasses have the highest yield of retail cuts; Yield Grade 5 the lowest. A carcass which is typical of its yield grade would be expected to yield about 4.6% more in retail cuts than the next lower yield grade, when USDA cutting and trimming methods are followed. (Other differences in yield may result from different cutting procedures, but these should result in similar differences between yield grades.) When used in conjunction with quality grade, yield grades can be of benefit to all segments of the industry.

A discussion of each of the factors affecting yield grade follows:

a. **External fat.** The amount of fat over the outside of a carcass is the most important yield grade factor since it is a good indication of the amount of fat that is trimmed in making retail cuts. The less trimmable fat, the more desirable the yield grade and the more the carcass is worth. A single fat thickness measurement over the ribeye muscle has been found to be the most practical indicator of external fatness for use in a grading program. Four-tenths of an inch variation in thickness of fat over the ribeye makes a full yield grade change.

b. **Kidney, pelvic, and heart fat.** Fat deposits on the inside of the carcass around the kidney and in the pelvic and heart areas also affect yields. Since practically all of this fat is removed in trimming, increases in these fats decrease the yields of retail cuts. A change of 5% in these fats makes a full yield grade change.

c. **Area of ribeye.** The ribeye muscle lies on each side of the backbone and runs the full length of the back. It is the largest muscle in the carcass and one of the most palatable. When the side is separated into a hindquarter and a forequarter, a cross section of the ribeye muscle is exposed. Its area (in square inches) at this point is another factor used in determining the yield grade. Among carcasses of the same fatness and weight, an increase in the ribeye area indicates an increase in the yield of retail cuts. A change of 3 sq in. in ribeye area makes almost a full yield grade change.

d. **Carcass weight.** The warm carcass weight is the weight used in yield grading. When carcass weight is used in conjunction with the other three yield factors, an increase in weight indicates a decrease in yield of retail cuts. Weight is the least important yield grade factor. It takes a change of about 250 lb in carcass weight to make a full yield grade change.

Since yield grades essentially measure proportions of lean and fat, USDA meat graders can determine the correct yield grade of most carcasses after a simple, rapid, visual appraisal of the fatness and muscling of the carcass. But objective measurements can be made when necessary and the yield grade also can be calculated from an equation which gives each factor its proper weight. Very little extra time is required to grade carcasses for both quality and yield over that required to grade them only for quality.

TABLE 16-1
FEDERAL QUALITY AND YIELD GRADES OF BEEF AND VEAL[1]

Beef		Calf and Veal
Quality Grades	Yield Grades	(Quality Grades Only)
1. Prime[2]	1. Yield Grade 1	1. Prime
2. Choice	2. Yield Grade 2	2. Choice
3. Select	3. Yield Grade 3	3. Good
4. Standard	4. Yield Grade 4	4. Standard
5. Commercial[3]	5. Yield Grade 5	5. Utility
6. Utility		
7. Cutter		
8. Canner		

[1]In rolling meat, the letters *USDA* are included in a shield with each federal grade name. This is important as only government-graded meat can be so marked. For convenience, however, the letters *USDA* are not used in this table or in the discussion which follows.

[2]Cow beef is not eligible for the prime grade. The quality grade designations for bullock carcasses are Prime, Choice, Select, Standard, and Utility. Bull carcasses are eligible for yield grade only.

[3]No commercial grade of yearling steers and heifers.

Federally graded meats are so stamped (with an edible vegetable dye) that the grade will appear on the retail cuts as well as on the carcass and wholesale cuts. These are summarized in Table 16-1 for beef and veal.

As would be expected, in order to make the top grades, the carcass or cut must possess a very high degree of the attributes upon which grades are based. The lower grades of beef are deficient in one or more of these grade-determining factors. Because each grade is determined on the basis of a composite evaluation of all factors, a carcass or cut may possess some characteristics that are common to another grade. It must also be recognized that all of the wholesale cuts produced from a carcass are not necessarily of the same grade as the carcass from which they are secured.

Fig. 16-7 shows the percentage distribution of beef by grades of total beef federally graded in 1993.

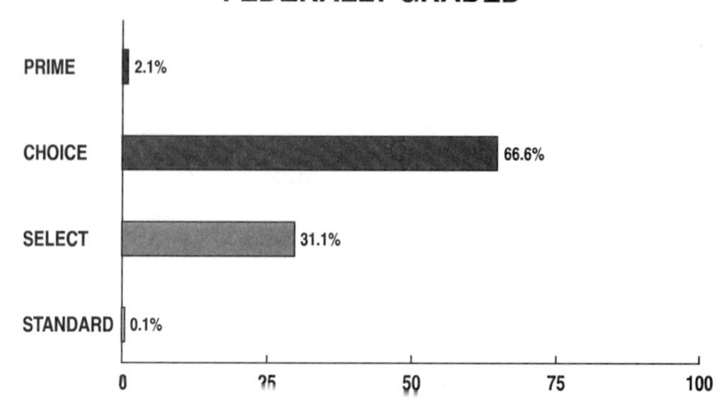

Fig. 16-7. Percentage distribution of beef by grades, in 1993. About 85% of steer and heifer beef was graded for quality in 1993. (Data provided by Livestock and Seed Division, Agricultural Marketing Service, USDA)

Some additional and pertinent facts relative to federal grades of beef are:

1. **There is no sex differentiation between steer, heifer, and cow beef.** Federal grades make no distinction between steer, heifer, and cow beef. It is not intended that this should be construed to imply that there is no carcass difference between these sex classes. Rather, there is no evidence that beef from steers, heifers, and cows with the same combination of quality-indicating characteristics is different in palatability. This step in simplification was taken so that it might be easier for the buyer or consumer to purchase meat on the basis of quality, without the added confusion of a more complicated system.

2. **Bull and bullock beef are identified.** Bull and bullock beef are identified by class as *bull* beef (from mature bulls) and *bullock* beef (from young bulls), respectively. Within the bullock class, there are the following five grades: (a) Prime; (b) Choice; (c) Select; (d) Standard; and (e) Utility. However, no designated grade of bullock beef is necessarily comparable in quality to a similarly designated grade of beef obtained from steers, heifers, or cows. Neither is the yield in a designated yield grade of bull or bullock beef comparable to a similarly designated yield grade of steer, heifer, or cow beef. There are no quality grades for bull beef.

3. **Lower grades seldom sold as retail cuts.** It is seldom that the lower grades (Standard, Commercial, Utility, Cutter, and Canner beef; and Standard and Utility calf and veal) are sold as retail cuts. The consumer, therefore, only needs to become familiar with the upper grades of each kind of meat.

4. **Yield grading.** Only carcass sides are eligible for grading. Quarters and wholesale/primal cuts are ineligible.

5. **Yield grading will reduce excess fat.** The real potential for reducing excess fat on beef is dependent upon the greater use of the yield grading system in reflecting appropriate price differentials.

6. **Use of quality and yield grades reflected from consumers to producers.** Grading beef carcasses for both quality and yield, rather than for one or the other, increases the effectiveness of the grades as a tool for reflecting consumer preferences back through marketing channels to producers. Also, if the market for beef and cattle reflects the full retail sales value differences associated with variations in both quality and cutability, producers respond by increasing the production of high-quality, high-cutability beef. This is advantageous to all segments of the industry and to consumers since it results in leaner beef with less waste in keeping with consumer tastes.

PACKER BRAND NAMES

Breed registry associations and breeders extol the virtues of their respective breeds. Likewise, some manufacturers and processors promote brand names through extensive advertising as a means of convincing consumers that their particular product is superior to all others. For example, orange growers have been successful in imaging *Sunkist* oranges; walnut growers in imprinting *Diamond* brand walnuts; and raisin growers in marketing *Sun-Maid* raisins.

However, branding and differentiating fresh meat is more difficult, with the result that only limited success has been achieved in this area. But a different situation exists in processed meats—hams, bacon, wieners, sausage, and luncheon meats. Here, processors have been able to develop brand-name products with individualized characteristics such as flavor, color, texture, and packaging. Then, by advertising and other promotional efforts, they have succeeded in differentiating their products and in influencing consumers. As a result, in certain sections of the country, some highly advertised brands of these products sell at higher prices than other brands or unadvertised brands.

Practically all packers identify their higher grades of meats with alluring private brands so that the consumer as well as the retailer can recognize the quality of a particular cut.

A meat packer's reputation depends upon consistent standards of quality for all meats that carry the company brand.

CERTIFIED BEEF OF BREED REGISTRIES

Beginning in the 1970s, consumers complained about the lack of quality and consistency of much of the beef. The time was ripe for breed registries to certify their own beef brands. The *Certified Angus Beef* (CAB) program began in January, 1978; and the *Certified Hereford Beef* (CHB) program followed in November, 1994. In addition, the Limousin, Red Angus, and Shorthorn associations have developed and implemented branded programs. No doubt additional breed programs will evolve.

- **Certified Angus Beef (CAB)**—The American Angus Association stipulates that Certified Angus Beef come from purebred Angus cattle, or from cattle with Angus-type characteristics provided they meet the evaluation of the USDA Grading Service based on marbling, maturity, and grade.

- **Certified Hereford Beef (CHB)**—The American Hereford Association launched two branded beef products: (1) Certified Hereford Beef Supreme (Choice grade), and (2) Certified Hereford Beef Special (upper Select grade). Only those animals meeting the criteria of "predominantly (51%) Hereford" are eligible for the CHB program. Cattle other than solid red or solid black with a predominantly white face are considered only if they exhibit markings of the Hereford breed. Upon initiating their program, the American Hereford Association stated their key reasons for marketing CHB as follows:

> To meet consumer demands for both consistency and quality.

The objectives of CAB, CHB, and the other breed programs are good. More than likely additional breeds will follow suit. The hazard is that during periods of high demand for beef and short supply someone will fill orders with beef that does not meet the standards. This is what happened with the *baby beef* program that was popular from about 1920 to 1950, with the result that *baby beef* ceased to exist.

AGING BEEF

Except for veal, fresh beef is not at optimum tenderness immediately after chilling. It must undergo an aging or ripening process before it really becomes tender.

The two most widely used methods of aging are:

1. **Traditional aging.** This refers to holding meat at temperatures of 34° to 38°F for 1 to 6 weeks. To be suitable for traditional aging, beef must have a fairly

thick covering of fat to prevent discoloration of the lean and keep evaporation to a minimum. There is a difference of opinion as to the best cooler humidity; some prefer 70 to 75% so that the exposed surface of meat remains dry, others use humidities of 85 to 90% in order to develop a mold growth on the outside of the meat and reduce evaporation losses.

2. **Vacuum packaging.** This refers to the use of a moisture-vapor proof film to protect the meat from the time it is fabricated until it reaches the consumer. Such packaging reduces weight loss and surface spoilage for 2 to 3 weeks. The majority of all commercially slaughtered and processed beef is vacuum packaged.

For maximum tenderness and flavor, beef should be aged in the vacuum bag and not used before 14 days following slaughter. Note the Beef Tenderness Chart (Fig. 16-8).

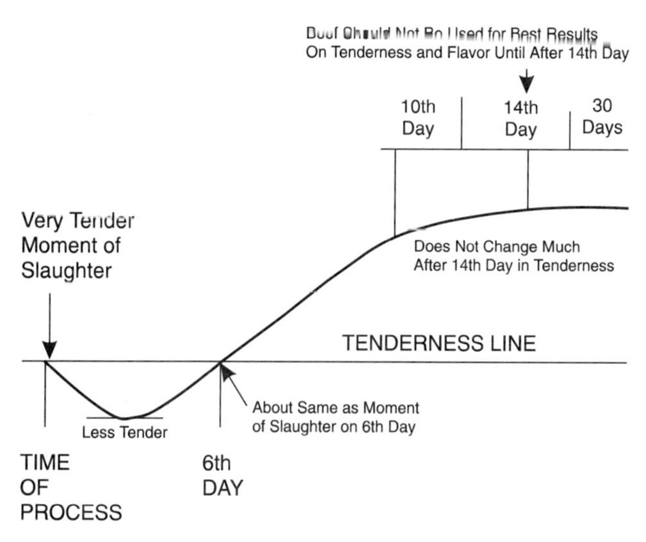

Fig. 16-8. Beef Tenderness Chart. (Courtesy, Excel Corp., Wichita, KS)

PROCESSING THE BEEF CARCASS

Beef carcasses are disposed of in one of four ways: (1) block beef; (2) boxed beef; (3) processed meats; or (4) case-ready beef.

Beef carcasses may be, and are being, processed in the following four ways: (1) block beef, (2) boxed beef), (3) processed beef, and (4) case-ready beef.

1. **Block beef.** Block beef refers to beef that is suitable for sale over the block. Such beef is purchased by the retailer in sides, quarters, or wholesale cuts. For the most part, block beef has been replaced by boxed beef.

2. **Boxed beef.** With the advent of the refrigerator car, meat was shipped in exposed halves, quarters, or wholesale cuts, and divided into retail cuts in the back rooms of meat markets. But this traditional procedure left much to be desired from the standpoints of efficiency, sanitation, shrink, spoilage, and discoloration. Today, modern packers fabricate and box meat in their plants, thereby freeing the back rooms of 200,000 supermarkets.

Fig. 16-9. Beef cuts are sealed in barrier bags by high speed rotary vacuum chamber machines before being placed in a box and stored at 28°F prior to shipment. (Courtesy, Excel Corp., Wichita, KS)

In 1990, on the average, the 20 largest U.S. meat packers boxed 95.8% of their beef. After chilling, the carcass is subjected to a disassembly process, in which it is fabricated or broken into counter-ready cuts; vacuum-sealed; moved into storage by an automated system; loaded into refrigerated trailers; and shipped to retailers across the nation.

The following benefits accrue from central fabricating, or cutting, of meats:

a. Twenty-five percent of the weight of the carcass is removed at the packing plant. This results in a reduction in shipping costs because bones and trim do not have to be transported.

b. Twenty to thirty percent of the carcass is bone and fat, which ends up as waste in the retail store. Where fabricating is done at the packing plant, trimmings, fats, and tallow are federally inspected and can be used as edible products; for example, fat and trim can be used in ground meats.

c. Processing equipment cost at the retail store level is usually high because of low volume and idle time. Hence, counter-ready cuts make for a saving in equipment cost at the retail store.

d. Central fabricating permits a regulated aging process.

e. Central fabricating aids in achieving uniformity of quality standards.

f. Central fabricating improves merchandising through more uniformity of cuts, new cuts, more variety of cuts, matching cuts to area preferences, creating a more tenable situation for retail meat sellers who lack expertise and interest in cutting,

Fig. 16-10. With the advent of close trim beef shipping to retail, a larger amount of trim *(left)* is retained at the packing plants and converted into ground beef *(right)* under strict HACCP (hazard analysis critical control points) guidelines. (Courtesy, Excel Corp., Wichita, KS)

and freeing meat managers from butchering—thereby permitting them to devote their time to personal selling and customer relations.

3. **Processed beef.** Beef that is not suitable for sale as block or boxed beef is (a) boned out and disposed of as boneless cuts, (b) canned, (c) made into sausage, or (d) cured by drying and smoking.

4. **Case-ready beef.** In the mid 1990s, almost all retail cuts were prepared at the store level. But the future direction is toward case-ready beef.

KOSHER BEEF[2]

Meat for the Jewish trade—known as kosher meat—is slaughtered, washed, and salted according to ancient Biblical laws, called *Kashruth,* dating back to the days of Moses, more than 3,000 years ago. The Hebrew religion holds that God issued these instructions directly to Moses, who, in turn, transmitted them to the Jewish people while they were wandering in the wilderness near Mount Sinai.

The Hebrew word *kosher* means *fit* or *proper,* and this is the guiding principle in the handling of meats for the Jewish trade. Also, only those classes of animals considered clean—those that both chew the cud and have cloven hooves—are used. Thus, cattle, sheep, and goats—but not hogs—are koshered (Deuteronomy 14:4 and 5 and Leviticus 11:1–8). Poultry is also koshered.

[2]Authoritative information relative to kosher meats was secured from: (1) Dr. Dayton I. Grunfeld, *The Jewish Dietary Laws,* The Soncino Press, London/Jerusalem/New York; (2) a personal letter to M. E. Ensminger from Rabbi Menachem Genack, Rabbinic Coordinator, Union of Orthodox Jewish Congregations of America, New York, NY; and (3) John J. Ensminger, Esq., New York, NY.

Both forequarters and hindquarters of kosher-slaughtered cattle, sheep, and goats may be used by Orthodox Jews. However, the Jewish trade usually confines itself to the forequarters. The hindquarters (that portion of beef carcass below the twelfth rib) are generally sold as nonkosher for the following reasons:

1. **The Sinew of Jacob (the *sinew that shrank,* now known as the *sciatic nerve*).** This nerve, which is found in the hindquarters only, must be removed by reason of the Biblical story of Jacob's struggle with the angel, in the course of which Jacob's thigh was injured and he was made to limp.

Actually, the sciatic nerve consists of two nerves; an inner long one located near the hip bone which spreads throughout the thigh, and an outer short one which lies near the flesh. Removal of the sinew (sciatic nerve) is very difficult.

The Biblical law of the sciatic nerve applies to cattle, sheep, and goats, but it does not apply to birds because they have no spoon-shaped hip (no hollow thigh).

2. **The very considerable quantity of forbidden fat (Heleb) found in the hindquarters, especially around the loins, flanks, and kidneys.** This must be removed—a process which is difficult and costly. Forbidden fat refers to fat (tallow) (a) that is not intermingled (marbled) with the flesh of the animal, but forms a separate solid layer; and (b) that is encrusted by a membrane which can be easily peeled off.

The Biblical law of the forbidden fat applies to cattle, sheep, and goats, but not birds and nondomesticated animals.

3. **The blood vessels must be removed.** The consumption of blood is forbidden, but the removal of blood vessels is especially difficult in the hindquarters.

Note well: Forbidden fat and blood (the blood vessels) must be removed from both fore and hindquarters, but such removal is more difficult in the hindquarters than in the forequarters. However, the Sinew of Jacob (the sciatic nerve), which must also be removed, is found in the hindquarters only.

Because of the difficulties in processing, meat from the hindquarters is not eaten by Orthodox Jews in many countries, including England. However, the consumption of the hindquarters is permitted by the Rabbinic authorities where there is a special hardship involved in obtaining alternative supplies of meat; thus, in Israel the sinews are removed and the hindquarters are eaten.

Because the forequarters do not contain such choice cuts as the hinds, the kosher trade attempts to secure the best possible fores; thus, this trade is for high grade slaughter animals.

Kosher meat must be sold by the packer or the retailer within 72 hours after slaughter, or it must be washed (a treatment known as *begiss*, meaning "to wash") and reinstated by a representative of the synagogue every subsequent 72 hours. At the expiration of 216 hours after the time of slaughter (and begissing 3 times), however, it is declared *trafeh*, meaning "forbidden food", and is automatically rejected for kosher trade. It is then sold in the regular meat channels. Because of these regulations, kosher meat is moved out very soon after slaughter.

Kosher sausage and prepared meats are made from kosher meats which are soaked in water ½ hour, sprinkled with salt, allowed to stand for an hour, and washed thoroughly. This makes them kosher indefinitely.

The Jewish law also provides that before kosher meat is cooked, it must be soaked in water for ½ hour. After soaking, the meat is placed on a perforated board in order to drain off the excess moisture. It is then sprinkled liberally with salt. One hour later, it is thoroughly washed. Such meat is then considered to remain kosher as long as it is fresh and wholesome.

Meats and fowl are the only food items which require ritual slaughter, washing, and salting before they are rendered kosher.

As would be expected, the volume of kosher meat is greatest in those Eastern Seaboard cities where the Jewish population is most concentrated. New York City alone uses about ¼ of all the beef koshered in the United States.

While only about half of the total of more than 6 million U.S. Jewish population is orthodox, most members of the faith are heavy users of kosher meats.

- **Porging**—This is making a carcass commercially clean by drawing out and removing the sinews, the forbidden fat, and the blood vessels.
- **Porger**—A person whose business is to do porging.

BEEF AND VEAL CARCASSES AND WHOLESALE CUTS

Fig. 16-11 shows the location of the wholesale cuts on a live animal.

There was a time when each area of the United States had its traditional cuts of beef and veal. However, increased central processing and boxed beef prompted the need for greater uniformity in cutting and labeling, among both packers and retailers. Out of this need arose a new nationwide Uniform Retail Meat Identity Standards, coordinated by the National Live Stock and Meat Board and adopted by the industry. The names for various cuts of beef, pork, and lamb sold in U.S. food stores were reduced from more than 1,000 to about 300. As a result of this system of uniform labeling, a ribeye steak is a ribeye steak—not a Delmonico steak at one place, a filet steak someplace else, or a spencer steak or beauty steak in still other stores, depending on the geographical area in the United States—or even where you shop in the same city.

Figs. 16-12 and 16-13 show for beef and veal, respectively, the primal area (or wholesale cuts) from which the retail cuts are derived.

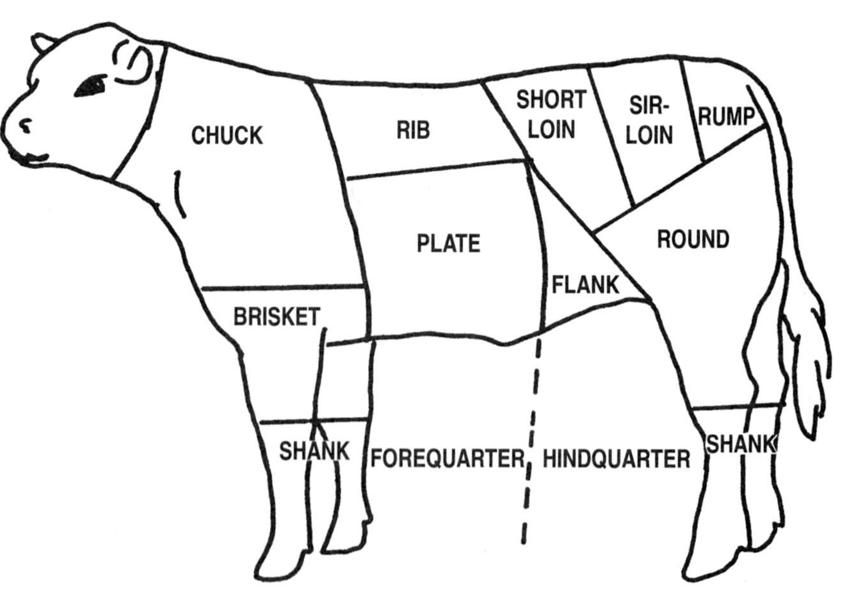

Fig. 16-11. Location of the wholesale cuts commonly derived from a beef animal.

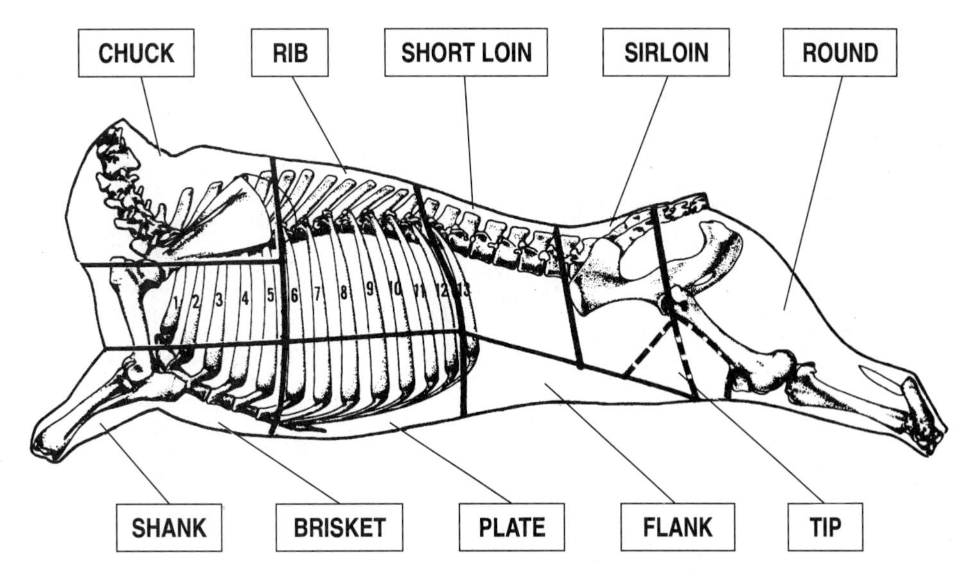

Fig. 16-12. Beef primal (wholesale) and subprimal cuts, and bone structure. (Courtesy, National Live Stock and Meat Board, Chicago, IL)

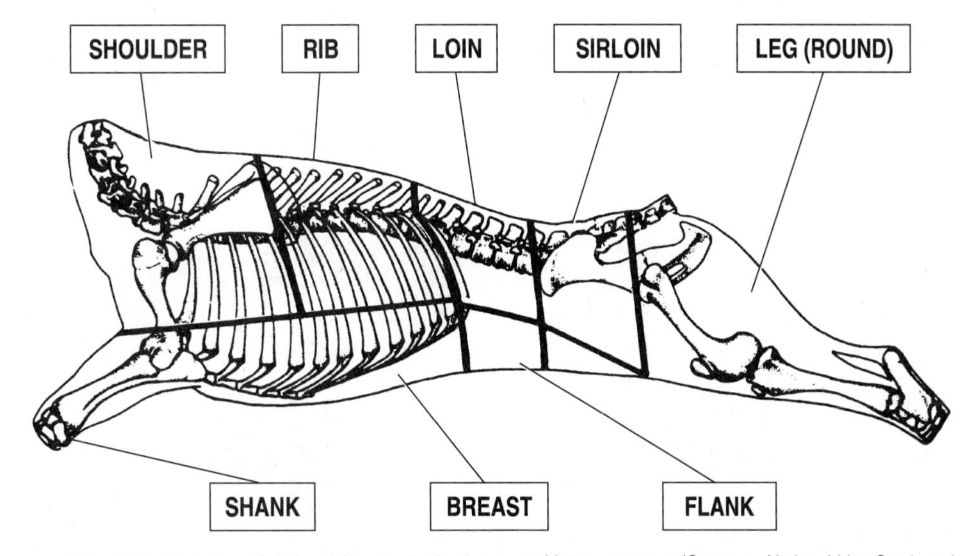

Fig. 16-13. Veal primal (wholesale) and subprimal cuts, and bone structure. (Courtesy, National Live Stock and Meat Board, Chicago, IL)

BEEF AND VEAL RETAIL CUTS AND HOW TO COOK THEM

Whether a beef carcass is cut up in the home or by an expert, it should always be cut across the grain of the muscle tissue and the thick cuts should be separated from the thin cuts and the tender cuts from the less tender cuts.

Fig. 16-14 shows the wholesale and retail cuts of beef and gives the recommended method(s) of cooking each. Fig. 16-15 presents similar information for veal.

In order to buy and/or process beef and veal wisely, and to make the best use of each part of the carcass, the consumer should be familiar with the types of cuts and how each should be processed.

Every grade and cut of meat can be made tender and palatable provided it is cooked by the proper method. Also, it is important that meat be cooked at low to moderate temperatures, usually between 300 and 325°F for roasting. At this temperature, it cooks slowly, and as a result is juicier, shrinks less, and has a better flavor than when cooked at high temperatures.

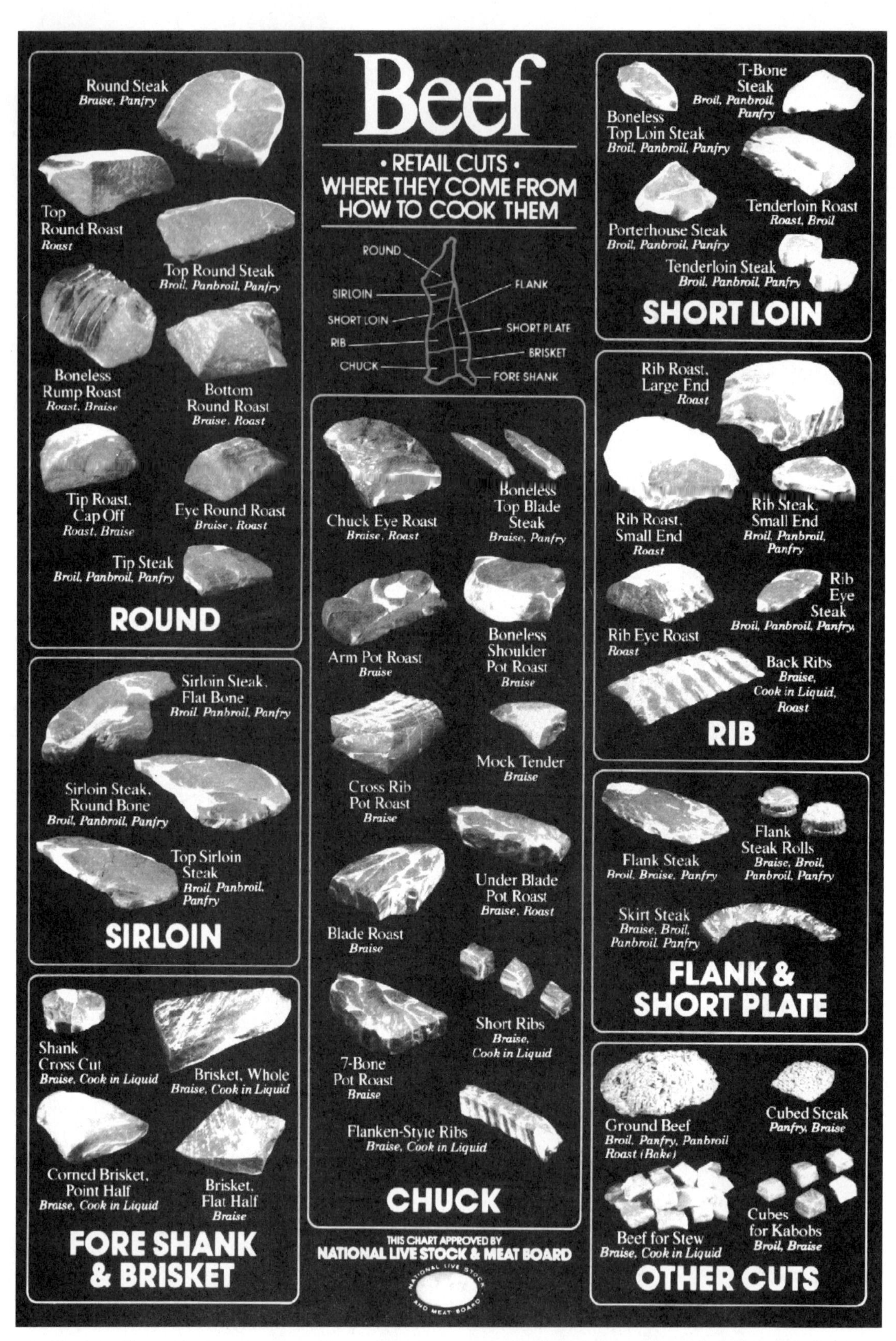

Fig. 16-14. The wholesale and retail cuts of beef, and the recommended method(s) of cooking each. (Courtesy, National Live Stock and Meat Board, Chicago, IL)

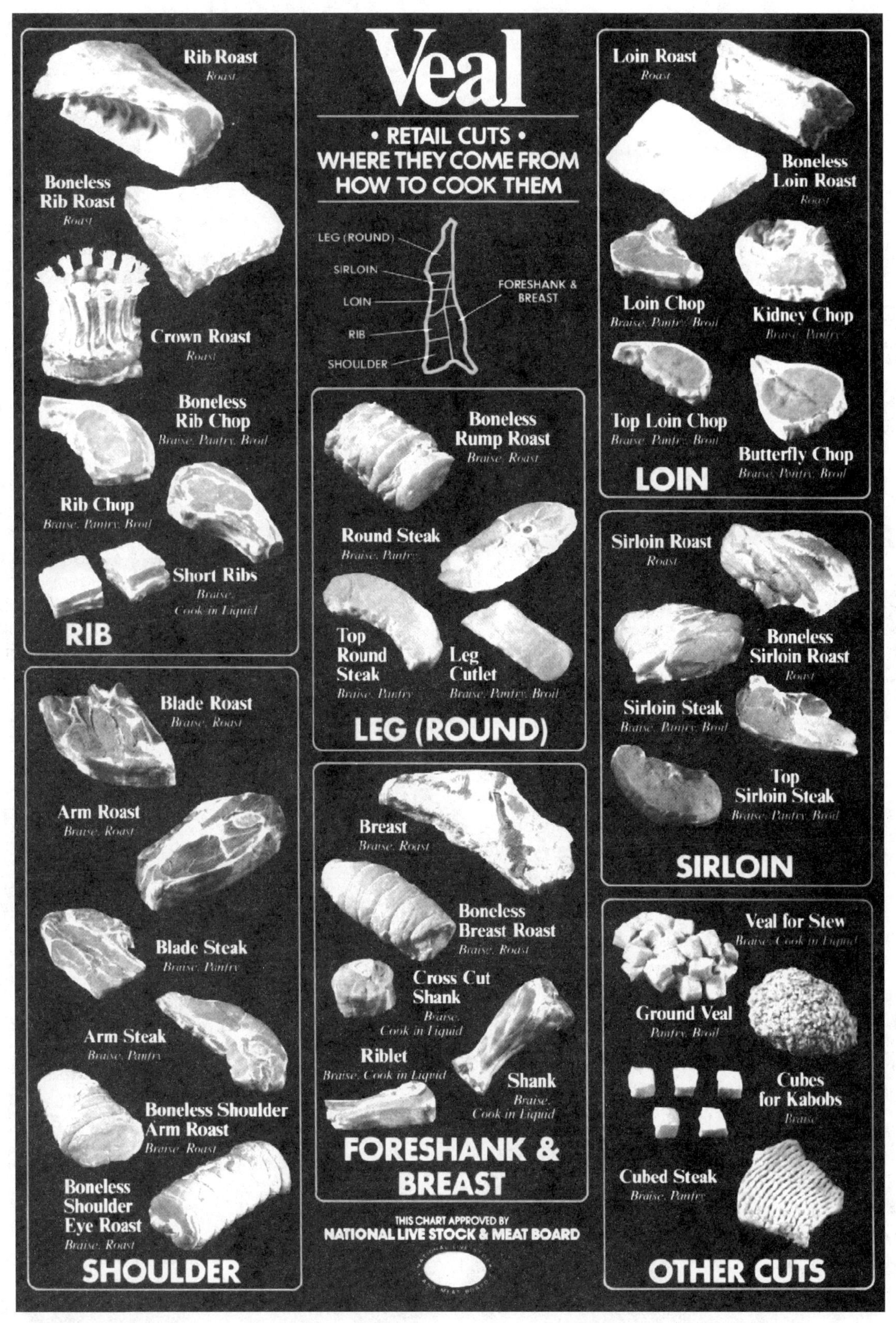

Fig. 16-15. The wholesale and retail cuts of veal, and the recommended method(s) of cooking each. (Courtesy, National Live Stock and Meat Board, Chicago, IL)

NUTRITIVE QUALITIES OF BEEF

Perhaps most people eat beef simply because they like it. They derive a rich enjoyment and satisfaction therefrom.

Fig. 16-16. Round roast of beef. (Courtesy, National Live Stock and Meat Board, Chicago, IL)

But beef is far more than just a very tempting and delicious food. From a nutritional standpoint, it contains certain essentials of an adequate diet; high-quality protein, energy, minerals, and vitamins. This is important, for how we live and how long we live are determined in large part by our diet.

Fig. 16-17 shows that beef is an excellent source of the essential nutrients, and that a 3-oz serving of cooked lean beef provides a very considerable percentage of the recommended daily allowance (RDA) of these nutrients.

Today's beef is lower in calories, but continuing to contribute a significant portion of many nutrients essential to people. Additional pertinent facts about the nutritive qualities of beef follow. (Where data are not available for beef alone, red meats [beef, pork, lamb, and veal] or animal proteins [red meat, poultry, milk, and eggs] are grouped in the presentation.)

1. **Protein.** The need for protein in the diet is actually a need for amino acids, of which 9 are essential for humans (10 are essential for animals; they require arginine, which is not required by humans). But all proteins are not created equal! Some proteins of certain foods are low or completely devoid of certain

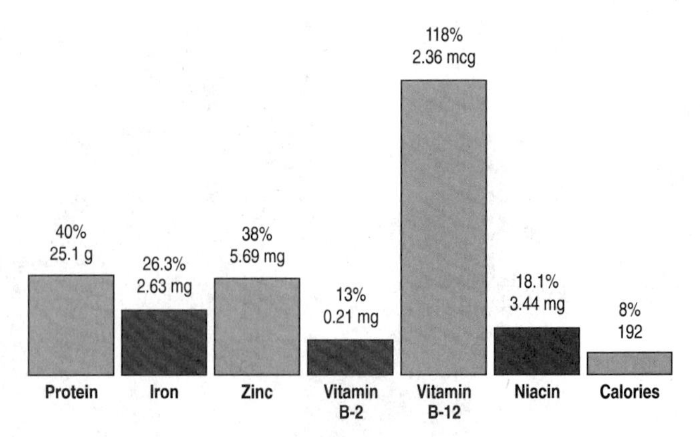

Fig. 16-17. The (1) nutrients, and (2) the percent of recommended daily allowance (RDA) provided by a 3-oz serving of cooked beef. (From: *Foods & Nutrition Encyclopedia*, 1994, CRC Press, Ensminger, *et al.*)

essential amino acids. Hence, meeting the protein requirement demands quality as well as quantity. It is noteworthy that all the essential amino acids are present in beef, and that they are in the proper ratio for human nutrition. However, there is considerable variation in the protein content of meat, depending on the degree of fatness and the water content; the total protein content of animal bodies ranges from about 10% in very fat, mature animals to 20% in thin, young animals.

The recommended daily allowance (RDA) for protein for a 25- to 50-year-old man is 63 g, of which 3 oz of cooked lean beef will supply 40%.

2. **Energy (calories).** The energy value of meat is largely dependent upon the amount of fat that it contains.

A 3-oz serving of cooked lean beef (the size serving often recommended by nutritionists) provides only 192 calories, or 8% of the average daily caloric intake of an adult male. Thus, meat is not a major contributor to excess energy intake and overweight.

Animal fats, which are highly digestible, also supply needed fatty acids, transport fat-soluble vitamins (A, D, E, K), provide protection and insulation to the human body, and add palatability to lean meat. Also, recent research on animals suggests that high amounts of polyunsaturates may suppress the body's natural immune responses, possibly increasing the risk of cancer.

There is no RDA for fat consumption since there is seldom a deficiency.

3. **Minerals.** Beef is a rich source of several minerals, but it is an especially rich source of phosphorus, copper, iron, and zinc.

- **Phosphorus**—Beef is a good source of phosphorus. Red meat (beef, pork, veal, lamb) (a) contributes 39.1% of the RDA of phosphorus, and (b) provides nearly 20% of the total phosphorus consumed in the United States.

■ **Iron**—Beef is a good source of dietary iron. Three ounces of cooked lean beef provides 26.3% of the adult male RDA for this essential mineral. Dietary iron is even more important for maturing and adult females, who require more iron than males; but a 3-oz serving of lean beef provides 18% of the females' RDA for iron. Also, and most important, 62% of the iron in beef is "heme" iron, which is far more usable in the body than the type of iron found in many other foods. Heme iron even makes iron from other sources more usable, as long as they are consumed at the same meal as the heme iron. Although beef muscle is a good source of iron, it is noteworthy that liver contains twice as much iron as muscle.

■ **Copper**—Organ meats (liver, kidney, brains) are rich sources of copper. Red meat supplies 10.5% of the RDA of copper.

■ **Zinc**—Beef is a good source of zinc. The animal proteins are generally much better sources of dietary zinc than plants because the phytate present in many plant sources complexes the zinc and makes it unavailable.

A 3-oz serving of cooked lean beef provides 38% of the adult male RDA for this nutrient.

4. **Vitamins.** Many phenomena of vitamin nutrition are related to solubility—vitamins are soluble in either fat or water. This is a convenient way in which to discuss vitamins in meats; hence, they are grouped and treated as fat-soluble vitamins or water-soluble vitamins.

The fat-soluble vitamins are stored in appreciable quantities in the animal body, whereas the water-soluble vitamins are not. Any of the fat-soluble vitamins can be stored wherever fat is deposited; and the greater the intake of the vitamin, the greater the storage. It follows that most meats are good sources of the fat-soluble vitamins.

■ **Vitamin A**—Liver is a rich food source of vitamin A, exceeding raw carrots.

Red meat (a) provides 45.5% of the RDA for an adult man, and (b) contributes 19% of the average daily per capita consumption of vitamin A (retinol).

The water-soluble vitamins are not stored in the animal body to any appreciable extent.

The following water-soluble vitamins are of importance in meats.

■ **Biotin**—Biotin is widely distributed in foods of animal origin, with liver and kidney being rich sources.

■ **Niacin (nicotinic acid; nicotinamide)**—Generally speaking, niacin is found in animal tissue as nicotinamide. Animal foods are excellent dietary sources, with the richest sources being liver, kidney, lean meats, poultry, fish, and rabbit.

A 3-oz serving of cooked lean beef provides 18.1% of the adult male RDA for niacin.

■ **Pantothenic acid (vitamin B-3)**—Pantothenic acid is widely distributed in animal foods, with organ meats (liver, kidney, heart) being particularly rich sources.

■ **Riboflavin**—The organ meats (liver, kidney, heart) are rich sources of riboflavin; lean meat (beef, pork, lamb), cheese, eggs, and bacon are good sources; while chicken and fish are only fair sources.

Red meat supplies (a) 26.9% of the RDA of riboflavin, and (b) 17.3% of the per capita intake of riboflavin per day.

■ **Thiamin (vitamin B-1)**—Some thiamin is found in a large variety of animal products, but it is abundant in few.

Red meat contributes (a) 35.7% of the RDA of thiamin, and (b) about 24% of the average per capita intake of thiamin per day.

■ **Vitamin B-6 (pyridoxine; pyridoxal; pyridoxamine)**—In animal tissues, vitamin B-6 occurs mainly as pyridoxal and pyridoxamine. Vitamin B-6 is widely distributed in foods of animal origin. Liver, kidney, lean meat, and poultry are good sources; eggs are a fair source; and fat, cheese, and milk are negligible sources.

Red meat (a) supplies 26.8% of the RDA of vitamin B-6, and (b) contributes about 25.5% of the total vitamin B-6 intake.

■ **Vitamin B-12 (cobalamins)**—Vitamin B-12 is found in all foods of animal origin. Plants cannot manufacture vitamin B-12; hence, except for trace amounts absorbed from the soil (because of soil bacteria, soil is a good source of B-12) by the growing plant, very little is found in plant foods.

Liver and other organ meats—kidney, heart—are rich sources of vitamin B-12; beef muscle a good source.

A 3-oz serving of cooked lean beef provides 118% of the adult male RDA for vitamin B-12. This means that most beef eaters have an abundance of dietary vitamin B-12.

5. **Cholesterol.** The National Cholesterol Education Program (NCEP) guidelines are:

> Intake of total fat less than 30% of the calories, saturated fat less than 10% of the calories, and cholesterol less than 300 mg/day.

A 3-oz serving of cooked lean beef contains only about 73 mg of cholesterol. How much cholesterol is 73 mg? The average individual ingests between 500 and 800 mg of cholesterol each day. One guideline accepted by some doctors is the American Heart Association's recommendation of 300 mg of dietary cholesterol per day. Thus, the cholesterol contained in the 3-oz serving of cooked lean beef amounts to only 24% of this recommendation.

An important fact, often ignored in the cholesterol controversy, is that the normal human body needs cholesterol—about 1,000 mg each day—and that the difference between the amount required and the amount consumed will be produced by the body itself.

6. **Digestibility.** Finally in considering the nutritive qualities of meats, it should be noted that it is highly digestible. About 97% of meat proteins and 95% of meat fats are digested.

That's the beef nutrition story! Although the full complement of protein, minerals, and vitamins can be obtained from other sources, unquestionably, beef is the favorite entreé when it comes to high-quality nutrients in a relatively low-calorie package that is tasty and satisfying.

MEAT MYTHS[3]

Much has been spoken and written linking the consumption of meat to certain health related problems in humans, including heart disease, cancer, high blood pressure (hypertension), and harmful residues. A summary of four incorrect statements, along with the correct facts, follows:

■ **Myth: Meat fats cause coronary heart disease**—Coronary heart disease (CHD) is the leading cause of death in the United States, accounting for one-third of all deaths. The major form of CHD results from atherosclerosis, a condition characterized by fatty deposits in the coronary arteries. These deposits are rich in cholesterol, a complex fat like substance. Also, in general, serum cholesterol levels are relatively high among individuals with atherosclerosis. This led some persons to hypothesize that high intakes of cholesterol and of saturated fats cause CHD. Because foods from animals are major dietary sources of cholesterol and saturated fats, it was deduced that they caused heart disease.

Fact: Both cholesterol and saturated fat are synthesized in the human body; and cholesterol is an essential constituent of all body cells.

Opinions of physicians are divided as to whether manipulating the dietary intake of cholesterol, saturated fats, and polyunsaturated fats has been or will be effective in combating heart disease.

Table 16-2 presents data showing the relationship between death rates from coronary heart disease and per capita meat consumption for selected countries. Although the data are epidemiological, there is no evident relationship between meat consumption and coronary heart disease. It is noteworthy, too, that deaths from heart disease in the United States have declined since 1950, although consumption of meat and poultry has greatly increased during this same period (from 169 lb in 1950 to 232 lb in 1993).

[3]In this section on "Meat Myths," all meats are grouped together because the available data are presented in this manner.

TABLE 16-2
COMPARISON BETWEEN THE RATE OF CORONARY HEART DISEASE AND PER CAPITA MEAT CONSUMPTION FOR SOME SELECTED COUNTRIES[1]

Country	CHD Death Rate[2]	Calories	Meat Consumption	
			(per capita/per year[3])	
		(kcal/day)	*(lb)*	*(kg)*
Finland	1,037	3,110	110	*50*
Australia	942	3,140	161	*73*
United States	359	3,700	163	[illegible]
New Zealand	889	3,490	120	*55*
Canada	832	3,020	150	*68*
United Kingdom	743	3,280	105	*48*
Denmark	586	3,370	92	*42*
Norway	583	2,930	79	*36*
Sweden	490	2,990	113	*52*
France	206	3,050	75	*34*

[1]Adapted by the senior author from Pearson, A. M., Department of Food Science and Human Nutrition, Michigan State University, Chapter 8 entitled "Meat and Health," *Developments in Meat Science 2*, published by Applied Science Publishers, Ltd., Ripple Road, Barking Essex, England; with permission of the author and the publisher.

[2]CHD—cornary heart disease per 100,000 between 55 and 64 years of age.

[3]Consumption of meats exclusive of poultry.

■ **Myth: Meat causes bowel cancer**—This question has been prompted by the following reports: (1) that the age-adjusted incidence of colon cancer has been found to increase with the per capita consumption of meat in countries; (2) that, in a study done in Hawaii, the incidence of colon cancer in persons of Japanese ancestry was found to be greater among those who ate Western-style meals, especially those who ate beef; and (3) that an examination of (a) international food consumption patterns, and (b) food consumption survey data from the United States showed that a higher incidence of colon cancer occurred in areas with greater beef consumption.

Fact: A direct cause-effect relationship between diet and cancer has not been established. Such studies as the three cited provide valuable leads for researchers who are trying to determine the cause of a certain disease such as colon cancer, but they do not establish the cause. The reason is that the factor measured and found associated with the incidence of colon cancer or other condition is not the only difference among the population groups studied, and the factor measured in the study may be only associated in some way with the real cause.

Rather than incriminating meat as the cause of cancer, the above studies raise the following questions:

1. In countries with high per capita meat consumption, the consumption of plant products and fiber tends to be low (for meat is devoid of fiber). It is conceivable that a low intake of fiber might cause colon cancer, but this hypothesis is without proof.

2. In the case of the Japanese, eating more beef wasn't the only change. They changed their life-styles, too; and they also ate less of some Japanese foods which could have been functioning as suppressants.

3. In the countries where meat consumption is high, some plausible associations of colon cancer with meat intake meriting further study are:

a. That the higher incidence of colon cancer may be due to higher fat consumption, part of which is derived from meat.

b. That the way in which the meat is prepared may be a causative factor.

■ **Myth: Meat causes high blood pressure (hypertension)**—Epidemiologic studies (studies of populations) have demonstrated a positive correlation between salt and hypertension. Moreover, salt restriction is used successfully to control, at least in part, high blood pressure. Also, some have implicated meat as a cause of high blood pressure.

Fact: There is no evidence that meat per se has any major effect on high blood pressure. However, consumption of cured meat containing large amounts of salt should be minimized as should the amount of salt used as a condiment on meat and other foods.

■ **Myth: Meat contains harmful residues**—Is there poison in Mrs. Murphy's stew? Do meats contain harmful toxic metals, pesticides, insecticides, animal drugs and additives?

Fact: If one pushed the argument of how safe is *safe* far enough, it would be necessary to forbid breast feeding as a source of food, because, from time to time, human milk has been found to contain DDT, antibiotics, thiobromine, caffeine, nicotine, and selenium.

Here are the facts relative to toxic metals, pesticides, insecticides, and animal drugs and additives in meats:

1. All metals are present in at least trace amounts in soil and water. It follows that they are also present in small amounts in plants and animals. Some of these metals, like copper and zinc, are essential to good health. Others, such as lead and mercury, do no observable harm in small amounts, nor do they have any known beneficial effects. But samplings of meat and poultry have not shown any toxic levels of these metals; hence, the normal intake of these metals in meats does not present any known hazard.

2. By properly applying pesticides and insecticides to control pests and insects, food and fiber losses can be, and are being, reduced substantially—perhaps by as much as 30 to 50%. When improperly used, pesticide and insecticide residues have been found in meat and poultry. But sometimes choices must be made; for example, between malaria-carrying mosquitoes and some fish, or between hordes of locusts and grasshoppers and the crops that they devour. This merely underscores the need for using pesticides and insecticides (1) in conformity with federal and state laws, and (2) in keeping with the instructions printed on the labels.

3. More than 1,000 drugs and additives are approved by the Food and Drug Administration (FDA) for use by livestock and poultry producers. They include products that fight disease, protect animals from infection, and make for higher and more efficient production of meat, milk, and eggs. Some of these drugs and additives can leave potentially harmful residues. For this reason, the FDA requires drug withdrawal times on some of them, in order to protect consumers from residues. Additionally, federal agencies (USDA and/or FDA), as well as certain state and local regulatory groups, conduct continuous surveillance, sampling programs, and analyses of meats and other food products on their content of drugs and additives.

MEAT BUYING[4]

Meat buying is important because (1) meat prices change, (2) one-fourth of the disposable personal income spent on food goes for red meats, and (3) buying food for a family may be a more intricate affair than its preparation at home. Hence, meat buying merits well informed buyers.

WHAT DETERMINES THE PRICE OF MEAT?

During those periods when meat is high in price, especially the choicest cuts, there is a tendency on the part of the consumer to blame any or all of the following: (1) the farmer or rancher, (2) the packer, (3) the meat retailer, and (4) the government. Such criticisms, which often have a way of becoming quite

[4]In this section on "Meat Buying," all meats/animal proteins are grouped together because they are interrelated from the standpoint of shoppers.

vicious, are not justified. Actually, meat prices are determined by the laws of supply and demand; that is, the price of meat is largely dependent upon what the consumers as a group are able and willing to pay for the available supply.

■ **The available supply of meat**—Because the vast majority of meats are marketed on a fresh basis rather than cured, and because meat is a perishable product, the supply of this food is dependent upon the number and weight of cattle, sheep, and hogs available for slaughter at a given time. In turn, the number of market animals is largely governed by the relative profitability of livestock enterprises in comparison with other agricultural pursuits. That is to say, farmers and ranchers—like any other good business people—generally do those things that are most profitable to them. Thus, a short supply of market animals at any given time usually reflects the unfavorable and unprofitable production factors that existed some months earlier and which caused curtailment of breeding and feeding operations.

Historically, when short meat supplies exist, meat prices rise, and the market price on slaughter animals usually advances, making livestock production profitable. But, unfortunately, livestock breeding and feeding operations cannot be turned on and off like a spigot. For example, a heifer cannot be bred until she is about 1½ years of age; the pregnancy period requires about 9 months; for various reasons only an average of 88 out of 100 cows bred in the United States conceive and give birth to young; and finally, the young are usually grown and fed until at least 1½ years of age before marketing. Thus, under the most favorable conditions, this production process, which is controlled by the laws of nature, requires about 4 years in which to produce a new generation of market cattle.

History also shows that if livestock prices remain high and feed abundant, the producer will step up the breeding and feeding operations as fast as possible within the limitations imposed by nature, only to discover when market time arrives that too many other producers have done likewise. Overproduction, disappointingly low prices, and curtailment in breeding and feeding operations are the result.

Nevertheless, the operations of livestock farmers and ranchers do respond to market prices, producing so-called cycles. Thus, the intervals of high production, or cycles, in cattle occur about every 10 years. In sheep, they occur about every 9 to 10 years, and in hogs—which are litter bearing, breed at an earlier age, have a shorter gestation period, and go to market at an earlier age—they occur every 4 years.

■ **The demand for meat**—The demand for meat is primarily determined by buying power and competition from other products. Stated in simple terms, demand is determined by the spending money available and the competitive bidding of millions of shoppers who are the chief home purchasers of meats. On a nationwide basis, a high buying power and great demand for meats exist when most people are employed and wages are high.

Also, it is generally recognized that in boom periods—periods of high personal income—meat purchases are affected in three ways: (1) More total meat is desired; (2) there is a greater demand for the choicest cuts; and (3) because of the increased money available and shorter working hours, there is a desire for more leisure time, which in turn increases the demand for those meat cuts that require a minimum of time in preparation (such as steaks, chops, and hamburger). Thus, during periods of high buying power, not only do people want more meats, but they compete for the choicer and easier prepared cuts of meats—porterhouse and T-bone steaks, lamb and pork chops, hams, and hamburger (chiefly because of the ease of preparation of the latter).

Because of the operation of the old law of supply and demand, when the choicer and easier prepared cuts of meat are in increased demand, they advance proportionately more in price than the cheaper cuts. This results in a great spread in prices, with some meat cuts very much higher than others. While porterhouse steaks, or pork or lamb chops, may be selling for 4 or 5 times the cost per pound of the live animal, less desirable cuts may be priced at less than the cost per pound of the animal on foot. This is so because a market must be secured for all the cuts.

But the novice may wonder why these choice cuts are so scarce, even though people are able and willing to pay a premium for them. The answer is simple. Nature does not make many choice cuts or top grades, regardless of price. Moreover, only two loins (a right and a left one) can be obtained from each carcass upon slaughter. In addition, not all weight on foot can be cut into meat. For example, the average steer weighing 1,150 lb on foot and grading Choice will only yield 567.8 lb of retail cuts (the balance consists of hide, internal organs, etc.). Secondly, this 567.8 lb will cut out about 29.4 lb of porterhouse and T-bone steaks. (See Fig. 16-18.) The balance of the cuts are equally wholesome and nutritious; and, although there are other steaks, many of the cuts are better adapted for use as roasts, stews, and soup bones. To make bad matters worse, not all cattle are of a quality suitable for the production of steaks. For example, the meat from most worn-out dairy animals and thin cattle of beef breeding is not sold over the block. Also, if the moneyed buyer insists on buying only the top grade of meat—namely, U.S. Prime or its equivalent—it must be remembered that only a small proportion of slaughter cattle produce carcasses of this top grade. To be sure, the lower grades are equally wholesome, but they are simply graded down because the carcass is

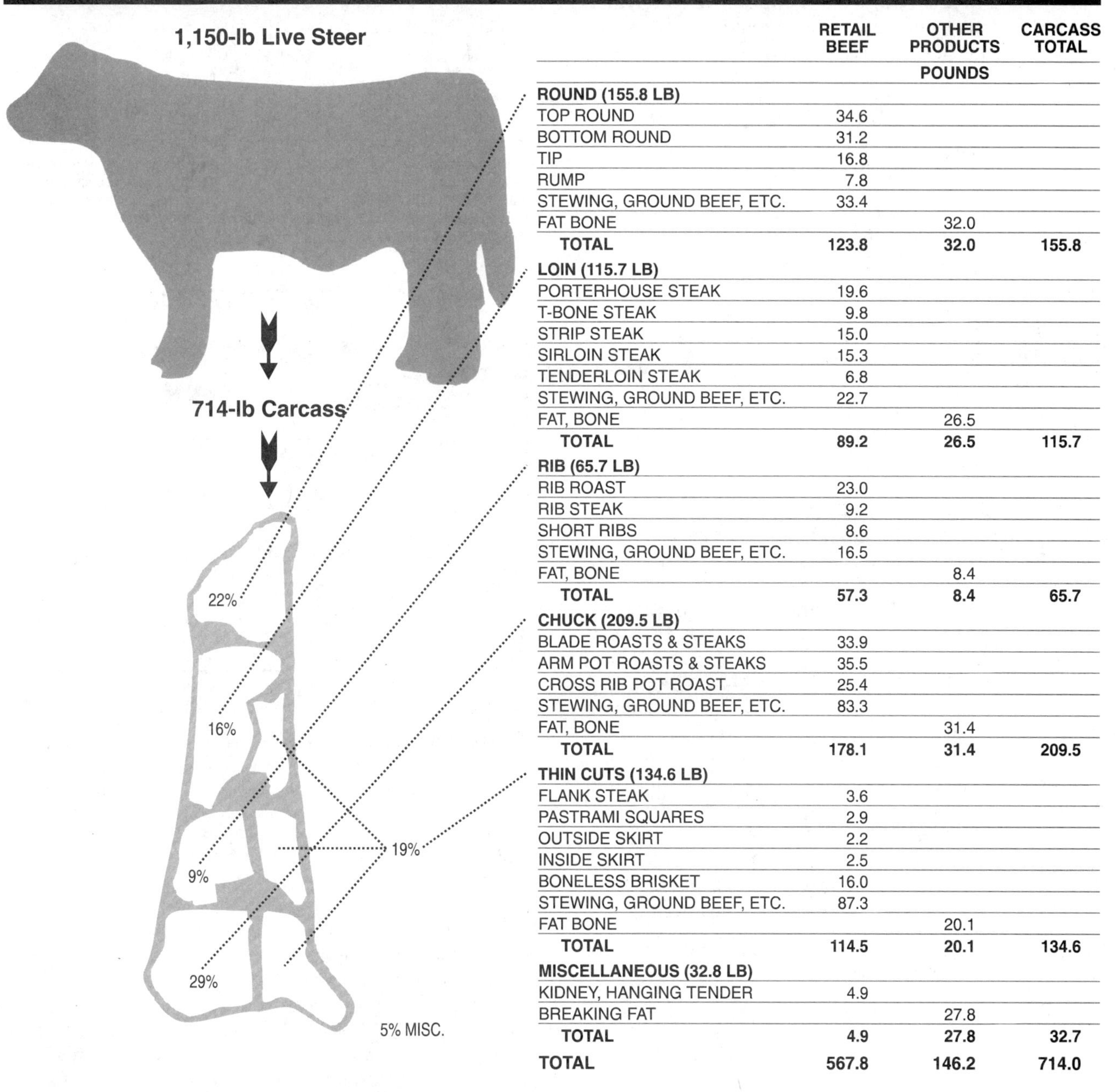

	RETAIL BEEF	OTHER PRODUCTS	CARCASS TOTAL
		POUNDS	
ROUND (155.8 LB)			
TOP ROUND	34.6		
BOTTOM ROUND	31.2		
TIP	16.8		
RUMP	7.8		
STEWING, GROUND BEEF, ETC.	33.4		
FAT BONE		32.0	
TOTAL	**123.8**	**32.0**	**155.8**
LOIN (115.7 LB)			
PORTERHOUSE STEAK	19.6		
T-BONE STEAK	9.8		
STRIP STEAK	15.0		
SIRLOIN STEAK	15.3		
TENDERLOIN STEAK	6.8		
STEWING, GROUND BEEF, ETC.	22.7		
FAT, BONE		26.5	
TOTAL	**89.2**	**26.5**	**115.7**
RIB (65.7 LB)			
RIB ROAST	23.0		
RIB STEAK	9.2		
SHORT RIBS	8.6		
STEWING, GROUND BEEF, ETC.	16.5		
FAT, BONE		8.4	
TOTAL	**57.3**	**8.4**	**65.7**
CHUCK (209.5 LB)			
BLADE ROASTS & STEAKS	33.9		
ARM POT ROASTS & STEAKS	35.5		
CROSS RIB POT ROAST	25.4		
STEWING, GROUND BEEF, ETC.	83.3		
FAT, BONE		31.4	
TOTAL	**178.1**	**31.4**	**209.5**
THIN CUTS (134.6 LB)			
FLANK STEAK	3.6		
PASTRAMI SQUARES	2.9		
OUTSIDE SKIRT	2.2		
INSIDE SKIRT	2.5		
BONELESS BRISKET	16.0		
STEWING, GROUND BEEF, ETC.	87.3		
FAT BONE		20.1	
TOTAL	**114.5**	**20.1**	**134.6**
MISCELLANEOUS (32.8 LB)			
KIDNEY, HANGING TENDER	4.9		
BREAKING FAT		27.8	
TOTAL	**4.9**	**27.8**	**32.7**
TOTAL	**567.8**	**146.2**	**714.0**

Fig. 16-18. Cattle are not all beef, and beef is not all steak! It is important therefore, that those who produce and slaughter animals and those who purchase wholesale and/or retail cuts know the approximate (1) percentage yield of chilled carcass in relation to the weight of the animal on foot, and (2) yield of different retail cuts. This figure illustrates these points. As noted, an average 1,150-lb steer grading Choice will yield about a 714-lb carcass, or 567.8 lb of retail cuts, only 29.4 lb of which will be porterhouse and T-bone steaks. (From: *Meat & Poultry Facts*, 1994, p. 22, American Meat Institute, Washington, DC)

somewhat deficient in conformation, finish, and/or quality.

Thus, when the national income is exceedingly high, there is a demand for the choicest but limited cuts of meat from the very top but limited grades. This is certain to make for high prices, for the supply of such cuts is limited, but the demand is great. Under these conditions, if prices did not move up to balance the supply with demand, there would be a market shortage of the desired cuts at the retail counter.

WHERE THE CONSUMER'S FOOD DOLLAR GOES

In recognition of the importance of food to the nation's economy and the welfare of its people, it is important to know (1) what percentage of the disposable income is spent for food, and (2) where the consumer's food dollar goes—the proportion of it that goes to the farmer, and the proportion that goes to the middleman.

In 1993, 14.5% or $684.5 billion, of the U.S. total disposable income of $4,706.7 billion was spent for food.[5] This vast sum included the bill for about 250,374 retail food stores, a large majority of the nation's 2,088,000 farms; thousands of wholesalers, brokers, eating establishments, and other food firms; and the transportation, equipment, and container industries. In 1995, food and kindred products employed 22.8 million people.

Table 16-3 reveals that of each food dollar in 1992, the farmer's share was only 26¢, the rest—74¢—went for processing and marketing. This means that about three-fourths of today's food dollar goes for preparing, packaging, and selling—and not for the food itself. Note, too, that the farmer's share of the food dollar has decreased from 34¢ to 26¢ since 1983; reflecting increased wages and costs of meat processing and retailing.

Fig. 16-19 reveals that of each food dollar in 1992, the farmer's share ranged from 7.4¢ for bakery products to 37.5¢ for poultry. For meat products, the farmer's share was 25.1¢ in 1992; the rest—74.9¢—went for processing and retailing. This means that three-fourths of today's meat dollar goes for meat packing and meat retailing.

The farmer's share of the retail price of poultry, eggs, dairy, and meat products is relatively high because processing is relatively simple, and transportation costs are low due to the concentrated nature of the products. On the other hand, the farmer's share of bakery and cereal products is low due to the high processing and container costs, and their bulky nature and costly transportation.

Among the reasons why farm prices of foods fail to go up, thereby giving the farmer a greater share of the consumer's food dollar, are (1) rapid technological advances on the farm, making it possible for the farmer to stay in business despite small margins; (2) overpro-

TABLE 16-3
WHERE THE CONSUMER'S DOLLAR GOES AS SHOWN BY THE MARKET BASKET[1]

Year	Retail Cost	Farm Value	Farm Retail Spread	Farm Value Share of Retail Cost
	- - - - - - *(Index, 1982–84 = 100)*	- - - - - -		*(%)*
1983	99	97	100	34
1985	104	96	108	32
1990	134	113	145	30
1992	138	103	157	26

[1]*Agricultural Statistics 1993*, USDA, p. 366, Table 550. The market basket contains the average quantities of U.S. farm food products purchased annually per household. The farm value is the return to the farmer for the farm product equivalent of foods in the market basket. The spread between the retail cost and farm value represents charges for processing and marketing the product.

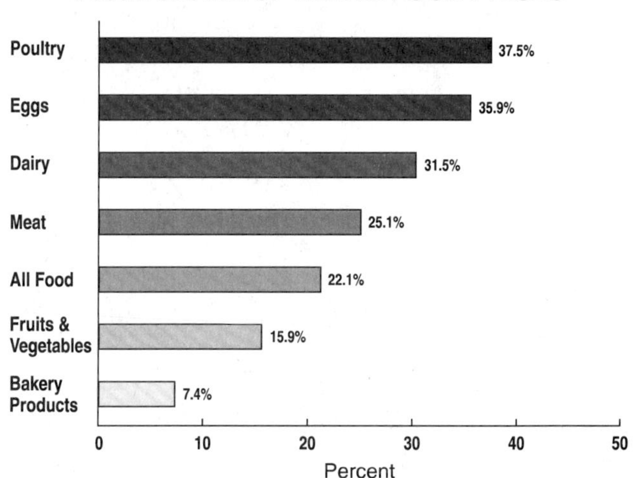

Fig. 16-19. Farm share of retail food prices, 1992. (From: *Farm & Food Facts*, Kiplinger Agricultural Letters, p. 34)

[5]It is noteworthy that food expenditures as a percent of income have decreased as follows:

Year	Percent of Income
	(%)
1970	16.2
1975	16.3
1980	15.7
1985	13.8
1990	13.9
1992	13.2
1993	13.2

(From: *Meat & Poultry Facts 1994*, American Meat Institute, p. 43.)

duction; and (3) the relative ease with which cost pressures within the marketing system can be passed backward rather than forward.

Over the years, processing and marketing costs have increased primarily because consumers have demanded, and gotten, more and more processing and packaging—more built-in services. For example, few consumers are interested in buying a live steer—or even a whole carcass. Instead, they want a roast—all trimmed, packaged, and ready for cooking. Likewise, few homemakers are interested in buying flour and baking bread. But consumers need to be reminded that, fine as these services are, they cost money—hence, they should be expected to pay for them. Without realizing it, American consumers have 1,675,600[6] people working for them in food processing and distributing alone. These mysterious persons do not do any work in the kitchen; they're the people who work on the food from the time it leaves the farms and ranches until it reaches the nation's kitchens. They're the people who make it possible for the homemaker to choose between quick-frozen, dry-frozen, quick-cooking, ready-to-heat, ready-to-eat, and many other conveniences. Hand in hand with this transition, and accentuating the demand for convenience foods, more women work outside the home. The proportion of the nation's labor force made up of women rose from 28% in 1947 to 45% in 1992. All this is fine, but it must be realized that the 1,675,600 individuals engaged in processing and distributing foods must be paid, for they want to eat, too.

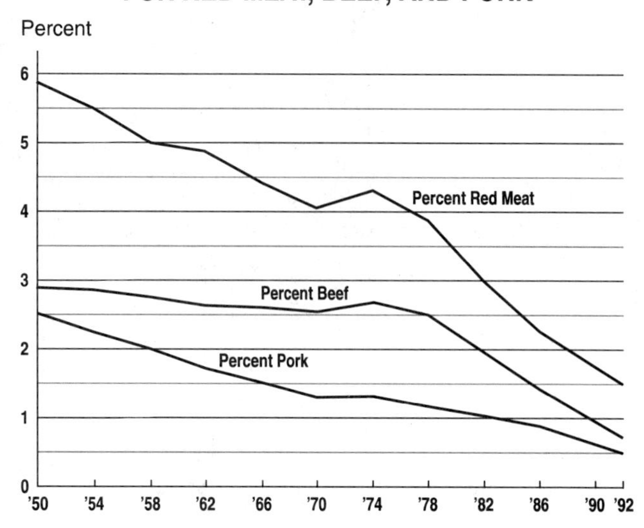

Fig. 16-20. Percent of U.S. disposable income spent for red meat, beef, and pork.

INCOME—PROPORTION SPENT FOR MEAT AND OTHER FOODS

U.S. consumers are among the most favored people in the world in terms of food costs and the variety of food products available. In 1992, only 13.2% of the U. S. disposable income was spent for food; 10.8% of this was spent for red meats, with 31% of the red meat share spent for beef and 23% of it spent for pork (see Fig. 16-20).

Food takes about one-fifth of the income in most other developed countries, and one-half or more of the income in most of the developing countries.

MEAT IS A GOOD BUY!

One of the best ways in which to evaluate whether a given product is a good or poor buy is in terms of the work hours required to purchase it. Today, it takes the average American wage earner only 18 minutes of work to buy 1 lb of beef. Fig. 16-21 presents

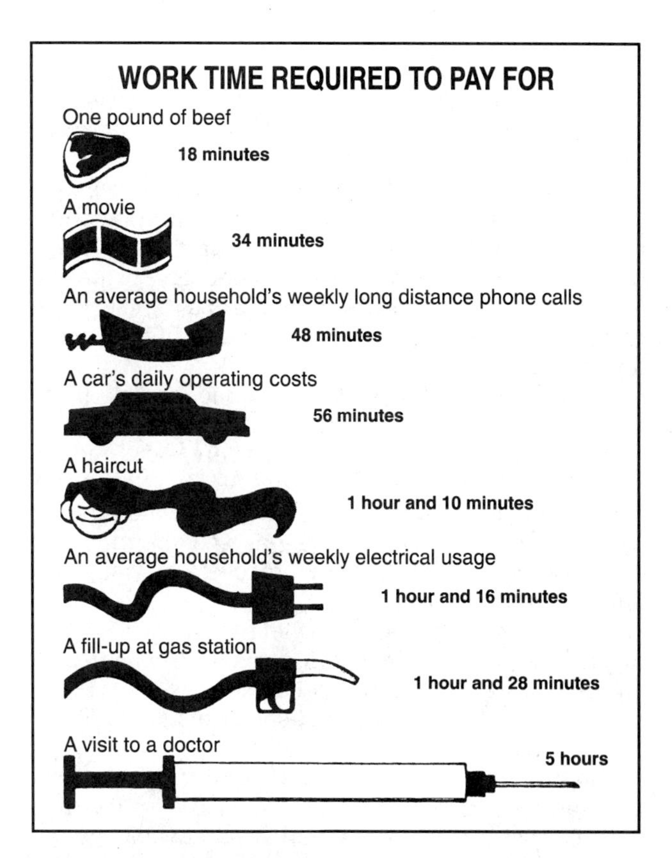

Fig. 16-21. Meat is a good buy! This shows the work time required to pay for 1 lb of beef in comparison with other costs. (Adapted by the authors from *Statistical Abstracts of the United States*, 1994)

[6]*Meat & Poultry Facts*, 1994, p. 32.

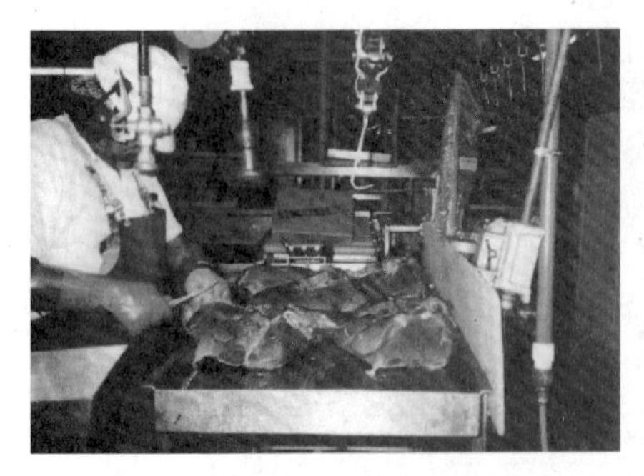

Fig. 16-22. Offals such as livers *(left)*, tongues *(center)*, and head meats *(right)*, are just some of the products saved for export and domestic markets. (Courtesy, Excel Corp., Wichita, KS)

convincing evidence that, in comparison with other costs, beef is a good buy.

Also, it is noteworthy that the 65 lb per capita annual beef consumption in the United States accounts for only 2.5% of the U.S. disposable income.

PACKINGHOUSE BYPRODUCTS FROM CATTLE SLAUGHTER

The meat or flesh of animals is the primary object of slaughtering. The numerous other products are obtained incidentally.

Byproducts include everything of value produced on the killing floor other than the dressed carcass; and they are classified as edible or inedible. The edible byproducts include blood, brains, casings, fats, gelatin, hearts, kidneys, livers, oxtails, sweetbreads, tongues, and tripe. The inedible byproducts include animal feeds, bone meal, bone products, cosmetics, fertilizer, glue, glycerin, hides and skins, lanolin, ligatures, lubricants, neat's-foot oil, pluck (lungs, etc.), and soap.

In addition to the edible (food) and inedible (nonfood) products, certain chemical substances useful as human drugs or pharmaceuticals are obtained as byproducts. Among such drugs are ACTH, cholesterol, estrogen, epinephrine, heparin, insulin, rennet, thrombin, TSH, and thyroid extracts—all valuable pharmaceuticals which are routinely recovered from meat animals in the United States.

So, upon slaughter, cattle yield an average of 40% of products other than carcass meat. When a meat packer buys a steer, more is bought than the cuts of meat that will eventually be obtained from the carcass; only about 60% is meat.

In the early days of the meat packing industry, the only salvaged cattle byproducts were hides, tallow, and tongue. The remainder of the offal was usually carted away and dumped into the river or burned or buried. In some instances, packers even paid for having the offal taken away. In due time, factories for the manufacture of glue, fertilizer, soap, buttons, and numerous other byproducts sprang up in the vicinity of the packing plants. Some factories were company-owned; others were independent industries. Soon much of the former waste material was being converted into materials of value.

Naturally, the relative value of carcass meat and byproducts varies both according to the class of livestock and from year to year. It is estimated that packers retrieve 11% of the live cost of slaughter cattle from the value of the byproducts (edible and inedible). Hides are the most important byproduct from cattle slaughter.

In contrast to the three early-day byproducts—hide, tallow, and tongue—modern cattle slaughter alone produces 134 pharmaceuticals plus numerous other byproducts which have a great variety of uses. Although many of the byproducts from cattle, sheep,

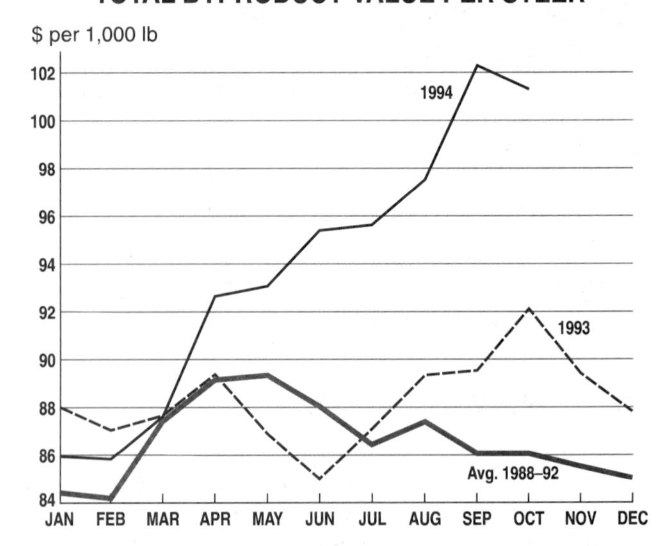

Fig. 16-23. In recent years, beef byproducts have represented about 11% of the value of slaughter steers, and wholesale beef has represented 89%. The hide is the most valuable byproduct, representing 59 to 68% of the total byproduct value. (Courtesy, USDA, Livestock Marketing Information Center, Lakewood, CO)

and hogs are utilized in a like manner, there are a few special products which are peculiar to each class of animal.

The complete utilization of byproducts is one of the chief reasons why large packers are able to compete so successfully with local butchers. Were it not for this conversion of waste material into salable form, the price of meat would be much higher than under existing conditions. In fact, under normal conditions, the wholesale value of the carcass is about the same as the cost of the animal on foot. The returns from the sale of byproducts cover all operating costs and return a reasonable profit.

It is not intended that this book should describe all of the byproducts obtained from cattle slaughter. Rather, only a few of the more important ones will be listed and discussed briefly (see Fig. 16-24).

1. **Hides.** Hides are particularly valuable as a byproduct of cattle slaughter. Thus, most of the discussion in this section will be especially applicable to cattle hides.

Cattle hides have been used by people since the dawn of time; and leather, particularly cowhide, has held an important place in commerce throughout recorded history. It was an important part of the clothing and armor of ancient and medieval times. Today, it has hundreds of industrial uses.

On the average, the hide represents 59 to 68% of the total byproduct value. There are two great classes of cattle hides, based on their place of origin: packer hides and country hides. Packer hides are the most valuable of the two because they are more uniform in shape, cure, and handling; much freer from cuts and gashes; and uniformly graded and available in larger lots.

The presence of needlessly large brands, or brands placed on the ribs, lowers the value of the hide. Cattle grubs (ox-warbles) also damage hides. It is estimated that one-third of all cattle hides produced in the United states are damaged by grubs. If there are five or more grub holes in the hide, it is classed as No. 2 and is discounted 1¢ per pound. Because of the larger throat cut, hides from kosher-killed cattle are less valuable.

The leather from animal hides is used for shoes, purses, wallets, coats, jackets, harnesses and saddles, belting, traveling bags, razor strops, footballs, baseballs, baseball mitts, "sheepskins" for diplomas, sweat bands for hats, gloves, and numerous other leather goods. On the average, one cowhide will make eight pairs of cowboy boots, or 18 pairs of shoes, or 144 baseballs.

2. **The fats.** Next to hides and pelts, the fats are the most valuable byproducts derived from slaughtering. Products rendered from them are used in the manufacture of oleomargarine, soaps, animal feeds,

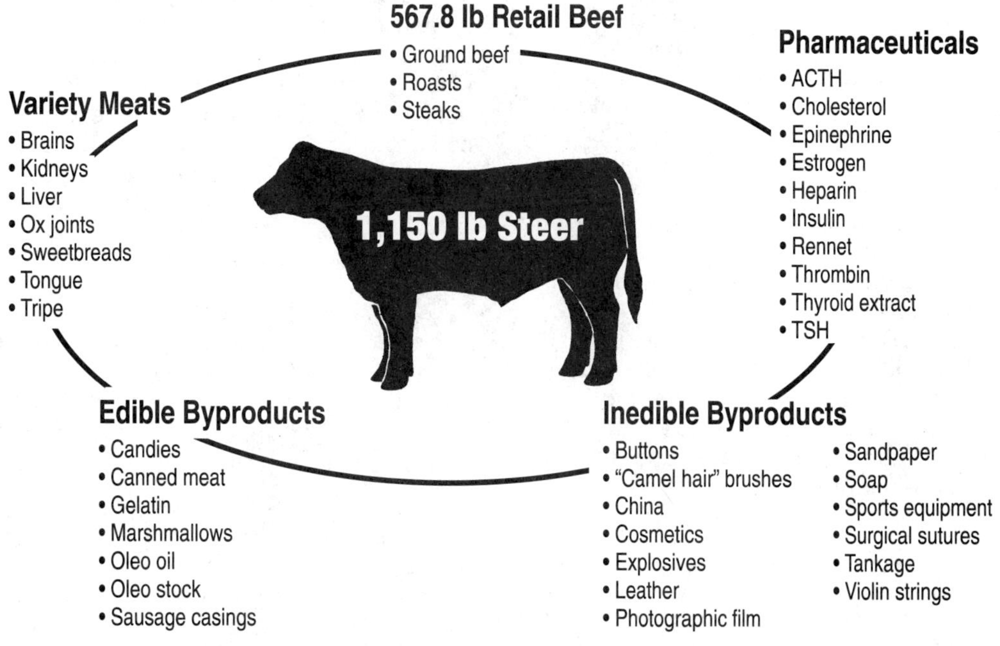

Fig. 16-24. Many good things come from cattle in addition to about 567.8 lb of steaks, roasts, and hamburger normally yielded by a 1,150-lb steer. Several of these products are shown in the above figure. (Courtesy, National Live Stock and Meat Board, Chicago, IL, adapted by the author)

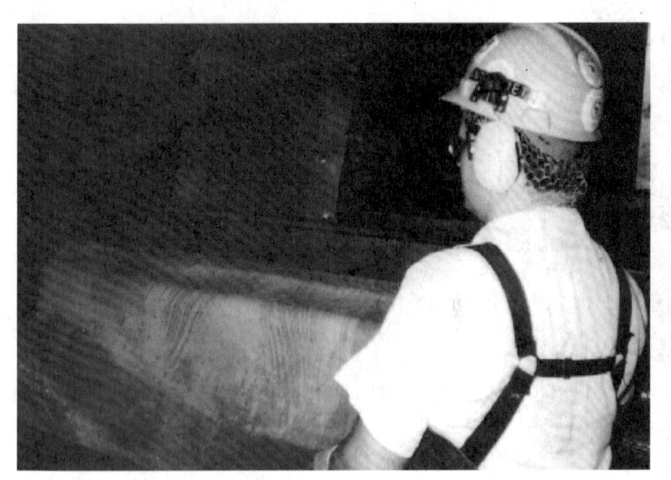

Fig. 16-25. Hides are fleshed *(left)* and brine-cured *(right)* prior to shipment to tanners. (Courtesy, Excel Corp., Wichita, KS)

industrial oils, lubricants, leather dressing, candles, fertilizer, and cosmetics (lipstick, face cream, and hand cream).

Oleomargarine, which is perhaps the best known of the products in which rendered animal fat is incorporated, is usually a mixture of vegetable oils and select animal fat.[7] Oleo oil, one of the chief animal fats of this product, is obtained from beef and mutton or lamb.

3. **Variety meats.** The heart, liver, brains, kidneys, tongue, cheek meat, tail, feet, sweetbreads (thymus and pancreatic glands), and tripe (pickled rumen of cattle and sheep) are sold over the counter as variety meats or fancy meats.

4. **Hair.** Artist and camel-hair brushes are made from the fine hair on the inside of the ears of cattle. Other hair from cattle is used in toothbrushes; paintbrushes, mattresses; upholstery for furniture, automobiles, and passenger planes; air filters; parachute seat pads; binders for plaster and asphalt paving; fan belts; felt pens; and home insulation.

5. **The horns and hoofs.** At one time considered a nuisance, horns and hoofs are now converted into napkin rings, goblets, tobacco boxes, knife and umbrella handles, combs, buttons, etc.

6. **Blood.** The blood is used in the refining of sugar, in making blood sausage and stock feeds, in making buttons, and in making shoe polish, etc.

7. **Meat scraps and muscle tissue.** After the grease is removed from meat scraps and muscle tissue, they are made into meat-meal or tankage.

[7]Oleomargarine is of two kinds: (a) a mixture of 50 to 80% animal fat and 20 to 50% vegetable oil, churned with pasteurized skimmed milk, or (b) 100% vegetable oil, churned with pasteurized skimmed milk. Oleomargarine was first perfected in 1869 by the Frenchman Mege, who won a prize offered by Napoleon III for a palatable table fat which would be cheaper than butter, keep better, and be less subject to rancidity.

8. **Bones.** The bones and cartilage are converted into bone china, stock feed, fertilizer, glue, crochet needles, dice, knife handles, buttons, teething rings, toothbrush handles, and numerous other articles.

9. **Intestines and bladders.** Intestines and bladders are used as sausage, lard, cheese, snuff, and putty containers. Lamb casings are used in making surgical sutures, strings for various musical instruments, and strings for tennis rackets.

10. **Glands.** Various glands of the body are used in the manufacture of numerous pharmaceutical preparations (see Fig. 16-26).

Proper preparation of glands requires quick chilling and skillful handling. Moreover, a very large number of glands must be collected in order to obtain any appreciable amount of most of these pharmaceutical products. For example, it takes pancreas glands from 1,500 cattle to produce 1 precious ounce of insulin. But, fortunately, only minute amounts of insulin are required.

Also, it is noteworthy that—

> It takes 40 beef thyroids to make 1 lb of thyroid extract.
>
> It takes 3,600 cattle to make 1 lb of parathyroid extract, a life-saving pharmaceutical for persons whose parathyroid gland has been removed.
>
> It takes the gall of 100 cattle to produce sufficient cortisone to treat the average patient for one week.

11. **Collagen.** The collagen of the connective tissues—sinews, lips, head, knuckles, feet, and bones—is made into glue and gelatin. The most important use for glue is in the woodworking industry. Gelatin is used in baking, ice cream making, capsules for medicine, coating for pills, photography, culture media for bacteria, etc. About 50% of the U.S. production of gelatin comes from veal.

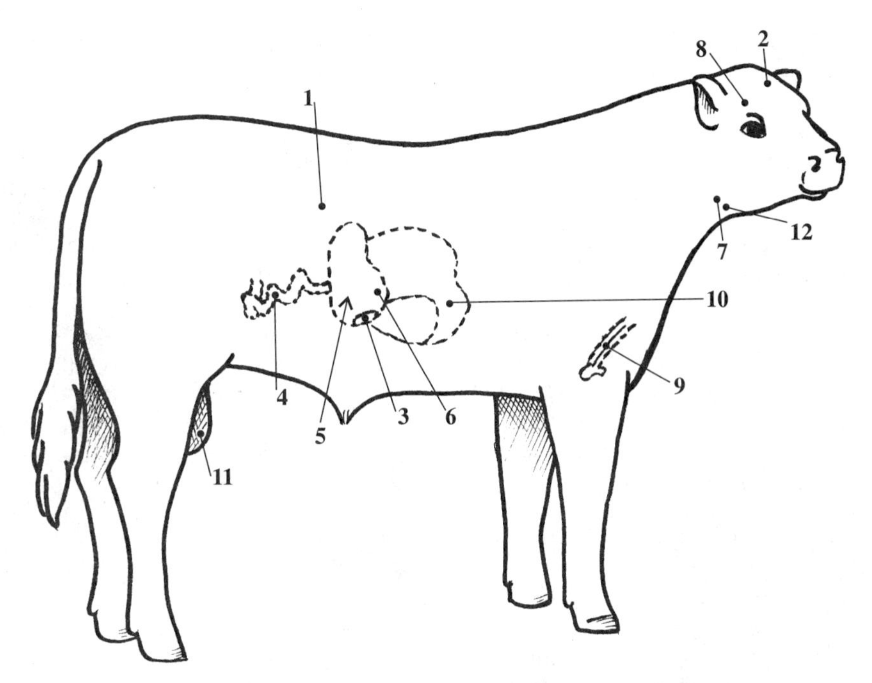

Fig. 16-26. Beef cattle are the source of 134 pharmaceuticals, which doctors and veterinarians administer daily to millions of people and animals to save lives, battle disease, relieve pain, and restore health. This figure shows the approximate location of a few of the glands and other tissues used in the manufacture of some of the pharmaceutical products of human and veterinary medicine.

1. **Adrenal (suprarenals).** Source of (a) epinephrine (used for asthma, hay fever, allergies, and shock), and (b) adrenal cortex extract (used for Addison's disease, and in postsurgical and burn shock).
2. **Brain.** Source of kephalin (or cephalin), used on oozing surfaces to check bleeding.
3. **Gallbladder.** Source of (a) bile salts and (b) dehydrocholic acid—used for gallbladder disturbances and abnormalities of fat digestion, and (c) cortisone (used for rheumatic fever, arthritis, various allergies, inflammatory eye diseases, etc.).
4. **Intestines.** Lamb intestines are used for surgical sutures.
5. **Liver.** Source of (a) liver extract (used for pernicious anemia), and (b) heparin (used to delay clotting of shed blood of ulcers and following surgery).
6. **Pancreas.** Source of (a) insulin (the only substance known to medical science which can control diabetes), (b) trypsin (the protein-digesting enzyme), (c) amalase (the starch-splitting enzyme), and (d) lipase (the fat-splitting enzyme). Each enzyme is used for digestion of these respective nutrients; and trypsin is also used to soften scar tissue or digest necrotic tissue in wounds and ulcers.
7. **Parathyroid.** Parathyroid extract is used for tetany, which follows removal of these glands.
8. **Pituitary.** Source of (a) posterior pituitary extract (used to increase blood pressure during shock, to promote uterine contraction during and after childbirth, and to control excessive urination of diabetes insipidus), and (b) ACTH (used for rheumatic fever, arthritis, acute inflammation of eyes and skin, acute alcoholism, severe asthma, and hay fever and other allergy conditions).
9. **Red bone marrow.** Bone marrow concentrates used in treatment of various blood disorders.
10. **Stomach.** Source of renin, used to aid milk digestion.
11. **Testes.** Source of the enzyme hyaluronidase.
12. **Thyroid.** Thyroid extract is used for malfunctions of the thyroid gland (some goiters, cretinism, and myxedema).
13. **Blood.** Source of thrombin, applied locally to wounds to stop bleeding.
14. **Bones and hides.** Source of gelatin, used as a plasma extender.

12. **Contents of the stomach.** Contents of the stomach are used in making fertilizer and feed.

Thus, in a modern packing plant, there is no waste; literally speaking, "everything but the squeal" is saved. These byproducts benefit the human race in many ways. Moreover, their utilization makes it possible to slaughter and process beef at a much lower cost. But this is not the end of accomplishment! Scientists are continually striving to find new and better uses for packinghouse byproducts in an effort to increase their value.

BEEF PROMOTION

Effective beef promotion—which should be conceived in a broad sense and embrace research, educational, and sales approaches—necessitates full knowledge of the nutritive qualities of the product. To this end, we need to recognize that (1) red meat contains 15 to 20% high-quality protein, on a fresh basis; (2) red meat is a good source of energy, the energy value being dependent largely upon the amount of fat it contains; (3) red meat is a rich source of several minerals, but it is especially good as a source of phosphorus and iron; (4) red meat is one of the richest sources of the important B group of vitamins, especially thiamin, riboflavin, niacin, and vitamin B-12; and (5) red meat is highly digestible, with about 97% of meat proteins and 95% of meat fats being digested. Thus, red meat is one of the best foods with which to alleviate human malnutrition, a most important consideration in light of the estimation that 35 to 40% of the U.S. population is now failing to receive an adequate diet.

Also, it is noteworthy that the per capita consumption of beef in two countries exceeds that of the United States: by rank, based on 1994 beef consumption, these are: (1) Uruguay, 189 lb and (2) Argentina, 144 lb. The United states ranks third in per capita beef consumption, with 98 lb (carcass weight basis).

Fig. 16-27. Beef is good, and good for you, two facts essential to effective beef promotion. (Courtesy, National Live Stock and Meat Board, Chicago, IL)

Thus, based on (1) its nutritive qualities and (2) per capita consumption in those two countries exceeding us, it would appear that there is a place, and a need, for increased beef promotion, thereby increasing beef consumption and price.

The 1985 Farm Bill contained the "Beef Promotion and Research Act"—enabling legislation which made it possible for the beef industry to establish a uniform national checkoff of $1.00 per head to fund promotion and research programs. So, beginning October 1, 1986, $1.00 per head was collected on all cattle sold or imported into the United States. From the beginning of the beef checkoff program to 1994, a total of $520 million had been collected. In the mid-1990s about $45 million per year was being collected.

In 1994, the University of Florida reported that their research showed that the beef industry received $5.40 return for every $1.00 of checkoff funds invested in promotion, consumer information, and industry information. A 113-member Cattlemen's Beef Promotion and Research Board—made up of representatives of the different states—has basic responsibility for the program, with 10 persons elected therefrom serving as the Beef Promotion and Operating Committee.

WHAT'S AHEAD FOR BEEF

In the 1990s, per capita beef consumption continued down, beef quality and consistency were variable, and beef was being underpriced by both broilers and pork. For sheer survival, changes were necessary. Among the changes in beef in progress, or being considered at that time, were those listed and detailed in an earlier section of this chapter headed, "Qualities in Beef Desired by the Consumers." Additionally, in the 1990s, the following programs/changes were formulated, researched, and/or tested:

- **The war on fat, 1990!**—This is the expressive title of a report dated August 1990, prepared by the "Value Based Task Force," assembled under the combined auspices of the National Cattlemen's Association

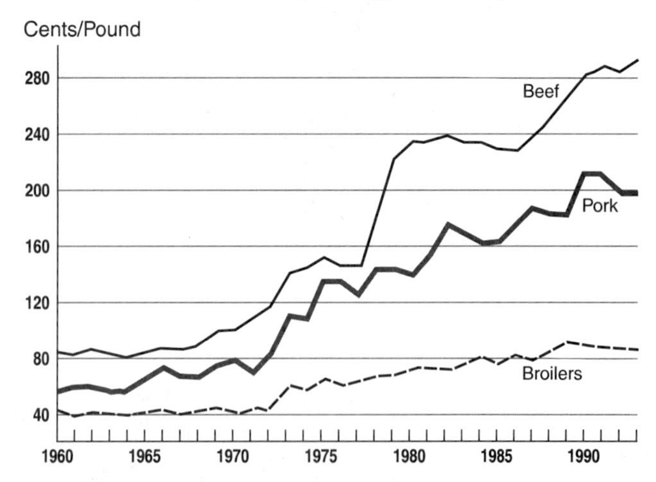

Fig. 16-28. Beef is underpriced by both broilers and pork; and the gap between beef and broilers is widening. (Courtesy, USDA, Livestock Marketing Information Center, Lakewood, CO)

(NCA) and the Beef Industry Council (BIC) of the National Live Stock and Meat Board, Chaired by Dr. H. Russell Cross.

The Task Force unanimously endorsed the belief that excess fat production is a detriment to the industry in terms of (1) production costs, and (2) meeting contemporary consumer demand. Further the Task Force listed recommendations and research needs to help accomplish the following primary objective:

> To improve production efficiency by reducing excess trimmable fat by 20% and increasing lean production by 6%, while maintaining the eating qualities of beef.

■ **National Beef Quality Audit, 1992**—This report was published by the National Cattlemen's Association in cooperation with Colorado State University and Texas A&M University.

Phase I of this study pinpointed quality defects and the relative costs associated with them. This information was obtained by face-to-face interviews of people involved with trade organizations, government agencies, laboratories and by-product users, purveyors, restaurateurs, retailers, and packers. According to those interviewed, beef's greatest challenges are excess trimmable fat, inconsistency, lack of palatability, and competitive pricing.

Phase II of the study consisted of a survey of federally inspected slaughter steers and heifers in 28 packing plants. The audit team assessed carcasses for Quality Grade and Yield Grade factors and summarized these data. The audit included 61.1% steers, 37.8% heifers, and 1.1% bullocks.

Phase III of the National Beef Quality Audit focused on the results gathered from Phases I and II, plus the input from 43 industry experts recognized as leaders in their respective fields. Their consensus follows:

> The overriding consensus of this group was that beef could be made more competitive in price with alternative protein-sources if it could be made more uniform and consistent.

The 43 participants also created a Top 12 List from the quality problems identified by purveyors, restaurateurs, retailers, and packers, and from the quality problems that emerged from the slaughter-floor and cooler audit (see Table 16-4).

Additionally, the participants outlined four key industry objectives, and assigned a dollar value to all quality problem areas. The four objectives: (1) attack waste, (2) enhance taste, (3) improve management, and (4) control weight.

The dollar figure attached to each of these four quality deficiencies was: $219.25 for waste, $28.81 for taste, $27.26 for management, and $4.50 for weight.

TABLE 16-4
BEEF QUALITY CONCERNS

1. Excessive external fat.
2. Excessive seam fat.
3. Low overall palatability.
4. Inadequate tenderness.
5. Low overall cutability.
6. Insufficient marbling.
7. Too frequent hide problems.
8. Too high incidence of injection-site blemishes.
9. Excessive weights/box.
10. Excessive live/carcass weights.
11. Inadequate understanding of the value of closer-trimmed beef.
12. Too large ribeyes/loineyes.

That's $279.82 of potential revenue gains if all cattle were perfect (see Table 16-5). However, perfection is not obtainable in a biological population—even Yield Grade 1 and Yield Grade 2 cattle have 82 lb of trimmable fat. The $279.82 figure provides a maximum target objective. Producers will obtain varying degrees

TABLE 16-5
QUALITY DEFICIENCIES AND THE MONETARY PENALTY OF EACH

Quality Deficiency	Loss Per Steer/Heifer
	($)
Waste — $219.25	
Excess external fat	111.99
Excess seam fat	62.94
Beef trim corrected to 20% fat	14.85
Muscling	29.47
Taste — $28.81	
Palatability	2.89
Marbling	21.68
Maturity	3.80
Gender	0.44
Management — $27.26	
Hide defects	16.88
Carcass pathology	1.35
Liver pathology	0.56
Tongue infection	0.35
Injection sites	1.74
Bruises	1.00
Dark cutters	5.00
Grubs, blood splash, calloused ribeyes, and yellow fat	0.38
Weight — $4.50	
Carcass weight (625–825 lb)	4.50
Total	279.82

of this value through genetic change, modification of management practices, and other innovative entrepreneurial initiatives.

■ **Strategic Alliances Field Study (SAFS), 1993**—Following publication of the National Beef Quality Audit (NBQA) in 1992, one question kept surfacing: How much of the $279.82 per head loss could the industry prevent or recover if proper precautions and strong remedies were taken to eliminate the non-conformities and quality defects within the beef industry? To answer this question, the Strategic Alliance Field Study followed in 1993. To cover the life stages of a slaughter animal, three industry segments—cow-calf producer, feedlot operator, and packer—were involved in the study. Also, because the partnership entailed risk, only cow-calf producers with more than 500 mother cows were invited to participate.

Fifteen cow-calf operators participated in the study, of which eight were straightbred British or British-crosses, and seven were Continental X British crosses. The Decatur County Feed Yard, Oberlin, Kansas, was selected for finishing the cattle; and Excel Corporation, Fort Morgan, Colorado, was selected as the packer. Safeway participated as the retailer. The 15 participating cow-calf operators shipped 1,253 calves to the Decatur County Feed Yard in November and December 1992.

The Strategic Alliance Field Study showed that, by working together, the beef team can (1) lower waste losses from $219.25 shown in the Strategic Alliance Field Study to $188.00 per steer, (2) lower management losses from $27.26 to $18.60, (3) lower weight losses from $4.50 to $0.84, and (4) lower total losses from $279.82 to $236.32. Thus, the SAFS cattle sustained $43.50 ($279.82 – $236.32) lower losses than the NBQA cattle. Taste remained the same in both studies. A statistical comparison of the NBQA versus SAFS loss per steer/heifer is presented in Table 16-6. *Note:* In an attempt to increase beef's competitiveness in the marketplace, the SAFS study included a side-study involving the feeding of 500 International Units (IU) of Vitamin E per head per day for the last 60 to 100 days of the feeding period.

TABLE 16-6
PREVENTABLE LOSSES WHICH COULD BE MADE BY THE BEEF TEAM

Quality Defect	NBQA Loss Per Steer/Heifer	SAFS Loss Per Steer/Heifer
	($)	*($)*
Waste		
Excess external fat	111.99	94.98
Excess seam fat	62.94	53.71
Beef trim corrected to 20% fat	14.85	12.92
Muscling	29.47	26.39
Subtotal	219.25	188.00
Taste		
Palatability	2.89	3.41
Marbling	21.68	25.43
Maturity	3.80	0
Gender	0.44	0.04
Subtotal	28.81	28.88
Management		
Hide defects	16.88	16.88
Carcass pathology	1.35	0
Liver pathology	0.56	0.46
Tongue infection	0.35	0.02
Injection sites	1.74	0
Bruises	1.00	0.85
Dark cutters	5.00	0.27
Grubs, bloodsplash, calloused ribeyes, and yellow fat	0.38	0.12
Subtotal	27.26	18.60
Weight		
Carcass weight (625–825 lb)	4.50	
Carcass weight (600–800 lb)		0.84
Subtotal	4.50	0.84
Total	279.82	236.32
Credit — Vitamin E		20.29
Grand Total	279.82	216.03

■ **National Beef Tenderness Conference Executive Summary, April 22–23, 1994, Denver, CO, National Cattlemen's Association**—In the introduction, this report presents the following startling facts:

1. One out of every four steaks is less than desirable in tenderness/palatability.
2. One tough beef carcass can affect as many as 542 consumers.
3. Tenderness within cuts of beef, as well as between cuts, varies greatly.

Sixty industry leaders, including animal and meat scientists, geneticists, feeders, professors, cow-calf producers, and representatives of breed associations, participated in this two-day National Beef Tenderness Conference.

The report contains the following pertinent background information:

1. Consumers consider three characteristics—flavor, juiciness, and tenderness—as they evaluate palatability and/or satisfaction from eating beef. Of the three characteristics, tenderness is most important.
2. Of the competing meats, veal, chicken, and turkey are almost never tough, and pork and lamb are seldom tough.
3. Most customers who have had a bad eating

experience with beef don't complain—they just don't come back.

4. Even if beef were offered to consumers at a more affordable price, two primary concerns—lack of consistency and lack of quality—would cause consumers to turn to other protein sources.

At the end of the two-day conference, participants devised 11 recommendations to combat the industry's tenderness issue. These are presented in Table 16-7.

TABLE 16-7
RECOMMENDATIONS TO COMBAT BEEF TENDERNESS

1. Establish and test a series of PACCP (Palatability Assurance Critical Control Points) models for beef tenderness variability.
2. Establish the value of improved tenderness and reduced variation in tenderness at retail.
3. Incorporate tenderness variation reduction values into the existing Computer Assisted Retail Decision Support (CARDS) program.
4. Establish and implement educational programs for all production segments in the use of existing technology and management in reducing tenderness and variation.
5. Encourage the commercialization of calcium salt and sodium salt solution injections as a means to reducing tenderness variation in selected cuts and quality grades.
6. Encourage the development of a rapid test for tenderness in carcasses.
7. Continue, and accelerate efforts, to find new predictors of genetic potential for tenderness in live cattle.
8. Encourage breed associations to collect progeny data relating to carcass traits and palatability.
9. Encourage producers to use tenderness technology to select against toughness, as it becomes available.
10. Standardize Warner-Bratzier shear force measurement protocols.
11. Establish minimum threshold standards for beef tenderness, once Warner-Bratzier shear force measurement protocols are standardized and the Customer Satisfaction Study is completed.

■ **National Non-Fed Beef Quality Audit (NNFBQA), December 1994**—The impact and importance of U.S. non-fed beef production is indicated by the following facts:

1. Sales of cows and bulls for slaughter account for 15 to 20% of producer revenues.
2. Non-fed beef represents 19 to 20% of U.S. beef production.
3. Approximately 33% of domestic beef production comes from dairy cows.
4. Ground beef, much of which comes from cows and bulls, accounts for 43 to 44% of all beef consumed in this country.
5. The annual ratio of bulls to cows for slaughter is a near-constant of 1 to 10.

The stated goal of the NNFBQA study: To conduct a national quality audit of U.S. non-fed slaughter cattle, their carcasses and their dress-off and offal items, for the U.S. beef industry, in order to establish baselines for present quality shortfalls and identify targets for desired quality levels.

The stated objectives: (1) To quantify, numerically and monetarily, the incidence of quality defects in U.S. non-fed slaughter cattle, their carcasses and their dress-off/offal items; (2) to characterize, through a substantive national audit in packing plants, as many as possible of the causes of these non-fed beef quality defects; (3) to identify strategies needed to reduce the incidence, or eliminate, specific quality defects; and (4) to determine which strategies to pursue and which tactics to employ in an effort to reduce and/or eliminate specific defects in quality of U.S. non-fed beef.

The methodology: **Phase I**—Face-to-face interviews with industry leaders to identify quality defects and to identify, numerically and monetarily, the incidence of quality defects in U.S. non-fed slaughter cattle, their carcasses and their dress-off/offal items.

Phase II—On site identified and quantified quality defects detected in non-fed cattle in packing plant holding pens, on kill floors, and in cooler rooms.

Phase III—A workshop to identify strategies needed to reduce and/or eliminate the incidence of specific defects, to determine which strategies to pursue to achieve the goals of correcting non-conformities, and to improve quality, consistency, and competitiveness of non-fed beef.

NNFBQA was launched in the summer of 1994. The study showed a total quality loss of $69.90 for each non-fed animal marketed in 1994, of which $14.60 per head could have been saved had the cattle been managed properly, $27.65 per head could have been saved had the cattle been monitored correctly, and $27.65 per head could have been saved had the cattle been marketed properly (see Table 16-8).

Participants made the following three recommendations to the industry:

1. **Manage** non-fed cattle to minimize defects and quality deficiencies.
2. **Monitor** the health and condition of non-fed cattle.
3. **Market** non-fed cattle in a timely manner.

Participants in the study also formulated the following 10 recommendations for improving the quality competitiveness and value of cull cows and bulls for beef:

1. Minimize condemnations by monitoring herd health and marketing non-fed cattle with physical disorders in a timely manner.
2. Effect end-product improvements by monitor-

TABLE 16-8
THE COSTS OF NON-CORFORMANCE IN NON-FED COWS AND BULLS

Defect	$/Head
	($)
Excess external fat	17.74
Inadequate muscling	14.43
Whole cattle and/or carcass condemnation	11.99
Brands	4.56
Bruises causing primal devaluation from bruise trim (attributing 33.3% of responsiblity to producers)	3.91
Light-weight carcasses (includes devaluation due to small ribeyes and tenderloins)	3.12
Latent defects and insect damage to hides	2.36
Yellow carcass fat	2.27
Carcasses passed with parts removed	2.13
Carcass weight lost to "zero tolerance" standards (attributing 36.4% of responsibility to producers)	1.87
Condemnation of edible offal	
Head	0.89
Tongue	0.46
Heart	0.17
Tripe	2.24
Liver	0.23
Handling of disabled cattle	0.78
Injection-site lesions in top butts and rounds	0.66
Dark-cutting beef	0.06
Carcasses passed for cooking	0.03
Total	69.90

ing and managing non-fed cattle and by marketing them before they become too fat or too lean, too light or too heavy, too thinly muscled or emaciated.

3. Decrease the hide damage by coordinating management and parasite-control practices and by developing new methods for permanent ownership identification of non-fed cattle.

4. Reduce bruises by dehorning, by correcting deficiencies in facilities, transportation and equipment, and by improving handling.

5. Encourage competitiveness by implementing non-fed cattle marketing practices that assure producer accountability.

6. Assure equity in salvage-value by requesting improved consistency of interpretation and application of federal meat inspection criteria among non-fed cattle slaughter establishments.

7. Improve beef safety by encouraging practices which reduce bacterial contamination of carcasses.

8. Prevent residues and injection-site lesions in non-fed cattle by ensuring responsible administration and withdrawal of all animal-health products.

9. Enhance price discovery by encouraging development of effective live and carcass grade standards for non-fed cattle.

10. Encourage on-farm euthanasia of disabled cattle and those with advanced bovine ocular neoplasia.

■ **Computer Assisted Retail Decision Support (CARDS) program**—This computer software package, developed by a research team from the National Live Stock and Meat Board and Texas A&M University, provides packers and retailers a tool to evaluate costs (including labor) associated with merchandising retail beef cuts from primals and subprimals.

■ **Food safety**—America has one of the safest food supplies in the world. Yet, FDA and the Centers for Disease Control and Prevention estimate that about 33 million people, or 14% of the population, become ill each year from microorganisms in food. In most of these cases, the health problem is a mere inconvenience. But it can be life threatening. A total of 9,000 deaths annually are attributed to food-borne diseases.

Beef safety and improved meat inspection became public issues in the wake of the tragic Seattle food-borne illness outbreak in January 1993, when four children died as a result of eating undercooked hamburgers contaminated with *E. coli.* Television's CBS and ABC, along with metropolitan newspapers across the nation, turned the heat on meat inspection. Several beef safety proposals followed; among them, the following:

1. **Irradiation.** Irradiation is referred to as "cold sterilization" because it does not appreciably raise the temperature or cook foods. Neither does it change the texture nor the flavor. The World Health Organization, the American Medical Association, and the International Atomic Energy Agency have approved the safety of the process. Also, irradiation of various foods has been approved in 37 countries. Yet, Americans are squeamish about eating something that they associate with a deadly force. So, much education must precede irradiation of beef.

2. **Hazard Analysis, Critical Control Point (HACCP).** The HACCP plan identifies hazards in food processing, followed by monitoring those crucial points. Problems are fixed as they occur. It is a way of processing safer food. In 1996, the federal government implemented a new inspection program for beef. One of the major components of the new regulations is that every processing plant must adopt and carry out its own HACCP plan. (Also see Chapter 15, section headed "Federal Meat Inspection.)

3. **Rapid microbial test.** This test, which adapts technology already being used in the pharmaceutical

and beer industries, takes five minutes and can be used in commercial meat plants. The rapid test provides a means of verifying that meat and poultry plants are operating under appropriate microbiological controls.

4. **Hot water/bactericidal rinse.** This method was developed by Colorado State University. It consists of high-pressure, hot water washing of beef carcasses followed by vacuuming to remove the water and bacteria. It has been proven to be more effective in removing physical contaminants from carcasses as compared to knife trimming. In 1996, USDA approved this process for use.

5. **Traceback.** This involves a livestock identification system which makes it possible to trace animals back to their origin in order to locate the source of contamination. Some countries already have mandatory identification requirements that allow them to trace animals from birth to slaughter.

■ **Food safe labeling**—Effective July 6, 1994, the food labeling rule became effective. This rule mandates safe cooking and handling labels for all uncooked meat and poultry products. The labels note that some food products may contain bacteria and can cause illness if mishandled or not cooked properly, and instruct consumers to keep raw meat and poultry refrigerated or frozen. Also, the labels warn that raw meat and poultry should be thawed only in a refrigerator or microwave, kept separate from other foods, cooked thoroughly, and refrigerated immediately or discarded.

■ **The USDA beef grades will change**—The USDA Quality Grades are used to predict the expected palatability of meat from a beef animal or carcass, using carcass physiological maturity and marbling to determine the USDA grade. USDA Yield Grades are used to estimate the expected lean meat, with a USDA YG1 being the leanest and a USDA YG5 being the fattest.

In 1994, the National Cattlemen's Association called for changes in both the quality and the yield grades. For the first time in history, the industry called for a narrowing, rather than a liberalization, of the beef quality grades. USDA has proposed changing the quality grades by removing "B" maturity carcasses with small or slight degrees of marbling from the Choice and Select grades and including them in the Standard grade. Although there is question as to whether the changes will be implemented, they are scheduled for implementation on January 31, 1997.

This is not the first time that beef grades have been changed, and they will be changed again in the future.

■ **The market for organically grown beef will increase**—Consumer demand for beef produced naturally, without hormones or antibiotics, is expected to increase moderately.

■ **Case-ready beef will increase**—First came carcass beef, then came boxed beef. In the future, case-ready beef will increase **provided** pre-cut, pre-wrapped beef can overcome some of its real technical hurdles to reach the retail shelf.

On the average, case-ready packaging can give ground beef 3 days on the shelf, and steaks 4 to 5 days. To date, no one has figured out how to extend the shelf life of roasts.

Researchers report that it's all in the packaging and atmosphere inside the package. New packaging and different gases inside the package will extend the shelf life of case-ready beef—at a cost. In the United States, most case-ready beef costs about 50¢ a pound more than similar products wrapped in the traditional manner.

As research overcomes the cost and shelf life of case-ready beef, and the volume of sales increases, case-ready beef will increase in volume. *Note:* Research at several U.S. universities has shown that feeding vitamin E before marketing imparts extended shelf life to beef.

SUMMARY OF WHAT'S AHEAD FOR BEEF

The entire beef industry is greatly concerned, and rightfully so, that a 10 percentage point decline in beef's market share occurred between 1982 and 1992, a downward spiral which is continuing unabated. So, the beef team is proposing several programs to stem the tide—all good, and all well conceived by able people in, and friends of, the industry. The briefs presented earlier in this section tell about these programs.

But the real challenge facing the beef industry is how to get 913,620 (data from *Statistical Abstracts 1993*, p. 244, Table 389) independently-minded operators to agree upon and follow any program on a voluntary basis. Additionally, there is the added problem created by the lack of consistency in the beef produced by many of the new breeds, crossbreds, and composites. Who is going to tell them to cease and desist in their breeding programs? *Note:* In the sixth edition of *Beef Cattle Science*, the senior author covered 52 beef breeds; in the current edition 83 breeds are covered.

The most effective approach would be to make sufficient price differential at the market place between those animals that meet the new standards and those that do not. Some of this will come to pass. But people being people, it will be spotty and partial. Nevertheless, such programs will be eminently worthwhile; they will slow the tide even though they will not stem it.

Note: The following proposal for lessening the beef dilemma merits the consideration of cow-calf producers:

> Evolve with a new federally graded *Baby Beef Program* consisting of marketing and slaughtering 500- to 700-lb calves at weaning which are never allowed to lose their baby fat. This calls for good pasture or range, cows that milk well, and/or creep feeding if necessary. The beef produced by these heavy calves would be consistently lean and tender, with small cuts.

(Also, see the senior author's preface to seventh edition of *Beef Cattle Science*.)

QUESTIONS FOR STUDY AND DISCUSSION

1. Why should the cattle producer have knowledge of the end products, beef and veal?

2. Studies indicate that consumers demand the following qualities in beef:

    a. Not more than 1/4 to 1/8 in. outside fat.
    b. Tenderness.
    c. Flavor.
    d. Consistency.

    Discuss the impact of each of these trends from the standpoint of the producer, the processor, and the consumer.

3. Are the above qualities reflected adequately in the top federal grades of beef?

4. Do you approve of federal grading of beef? Justify your answer.

5. What's the difference between quality grades and yield grades? Which is more important?

6. Why does Choice beef account for two-thirds of the total beef federally graded?

7. Why are separate (from steers, heifers, and cows) grade identifications used for bull and for bullock beef?

8. Why haven't packer brand names of beef become as widely known as Sunkist oranges, Diamond walnuts, and Sun-Maid raisins?

9. Why do most consumers prefer beef that has been properly aged?

10. Why has boxed beef replaced most of the shipping of halves, quarters, and wholesale cuts?

11. Explain why the Jewish trade prefers the forequarters of beef to the hindquarters.

12. Explain why uniform cutting and labeling of beef throughout the U.S. is important.

13. What facts relative to the nutritive qualities of beef are important in good eating and effective beef promotion?

14. Choose and debate either the affirmative or negative accusations about each of the following meat causes: (a) heart disease, (b) cancer, and (c) high blood pressure.

15. What determines the price of meat?

16. Choose and debate either the affirmative or negative of each of the following assertions:

    a. Beef prices are controlled by (1) the cattle producer, (2) the packer, or (3) the retailer.
    b. Excessive profits are made by (1) the cattle producer, (2) the packer, or (3) the retailer.

17. Fig. 16-16 shows that, on the average, an 1,150 lb steer yields 714 lb of carcass beef and 567.8 lb of retail beef. Compute percentages yielded of carcass beef and of retail beef.

18. For many years, the proportion of the consumer's dollar going to the farmer has decreased while the proportion going to the middleman has increased. What are the reasons for this shift? Is the shift justified?

19. Why has the percent of U.S. disposable income spent for red meat, beef, and pork gone down in recent years?

20. Packers retrieve about 11% of the live cost of slaughter cattle from the byproducts. Is this worth their effort?

21. Why do Uruguay and Argentina exceed the United States in per capita beef consumption?

22. Have beef promotion programs by cattle producer groups been effective in increasing beef consumption?

23. Why are consumers prone to complain about the price of beef more than the cost of a movie, average weekly long distance phone calls, the daily operation of a car, a haircut, weekly electrical usage, a fill-up at the gas station, or a visit to the doctor?

24. In the summary at the last of the section headed "What's Ahead For Beef," the authors state their concerns. Do you agree or disagree with the authors?

SELECTED REFERENCES

Title of Publication	Author(s)	Publisher
Adventures in Diet	V. Stefansson	Reprinted from *Harper's Magazine* by American Meat Institute, Washington, DC
Animal Science, Ninth Edition	M. E. Ensminger	Interstate Publishers, Inc., Danville, IL, 1991
Beef Production and Distribution	H. DeGraff	University of Oklahoma Press, Norman, OK, 1960
Byproducts in the Packing Industry	R. A. Clemen	University of Chicago Press Chicago, IL, 1927
Cattle and Beef Handbook	NCA Beef Promotion & Board	National Cattlemen's Assn., Englewood, CO, 1992
Food from Farmer to Consumer	National Commission on Food Marketing	U.S. Government Printing Office, Washington, DC, 1966
Hides and Skins	National Hide Association	Jacobsen Publishing Co., Chicago, IL, 1970
Lessons on Meat, Second Edition	National Live Stock and Meat Board	National Live Stock and Meat Board, Chicago, IL, 1972
Marketing of Livestock and Meat, The, Second Edition	S H. Fowler	The Interstate Printers & Publishers, Inc., Danville, IL, 1961 (out of print)
Meat, Poultry, and Seafood Technology	R. L. Henrickson	Prentice-Hall Inc., Englewood Cliffs , NJ, 1978
Meat Reference Book		American Meat Institute, Washington, DC
Meat for the Table	S. Bull	McGraw-Hill Book Co., New York, NY, 1951
Meat We Eat, The, Thirteenth Edition	J. R. Romans, *et al.*	Interstate Publishers, Inc., Danville, IL, 1994
Stockman's Handbook, The, Seventh Edition	M. E. Ensminger	Interstate Publishers, Inc., Danville, IL, 1992
Using Information in Cattle Marketing Decisions, A Handbook, WEMC Publication No. 5		Western Extension Marketing Committee Task Force on the Economics of Marketing Livestock, Fort Collins, CO, 1973

Also, literature on meats may be secured by writing to meat packers and processors and trade organizations; in particular, the following two trade organizations:

American Meat Institute
P. O. Box 3556
Washington, DC 20007

National Live Stock and Meat Board
444 N. Michigan Avenue
Chicago, IL 60611

Fig. 16-29. "Beef—It's what's for dinner." This shows a casserole of beef stew, and a pot roast cooked in a slow cooker. (Courtesy, California Beef Council, Burlingame, CA)

PART II

Cow-Calf System; Stockers

The cow-calf system refers to the breeding of cows and the raising of calves. In this system, the calves run with their dams, usually on pasture, until they are weaned, and the cows are not milked. It is the very foundation of beef production. Without the cow-calf system, there would be no stocker programs and no finishing operations, for there wouldn't be any raw material—calves and feeders. The importance of the cow-calf system in the agriculture of the nation rests chiefly upon the conversion of coarse forage and grass into palatable and nutritious food for human consumption. It is especially adapted to regions where pasture is plentiful and land is cheap; hence, it is the standard system of beef production on the western range. Reproduction—the production of calves—is the first and most important requisite of the cow-calf system, for if animals fail to reproduce, the breeder will soon be out of business. Nationally, cattle producers get an 88% calf crop, which is not good enough. It can and should be 95%.

Innovative cattle producers and new breeds of cattle have changed the cow-calf industry in America more in the last four decades than it has changed since the days of the Texas Longhorn and cattle trails. (Photo by Wyatt M. Casey, Jr.; Courtesy, The Lasater Ranch, Matheson, CO)

Stockers are young cattle that are fed and cared for so as to promote growth rather than finish. It includes steers and heifers that will be either (1) marketed as feeders, or (2) finished out by the owner who produced them. Strictly speaking, the term *stocker* does not include heifers intended for breeding purposes; the latter are more correctly designated as *replacement heifers*.

Since the 1960s, the cow-calf business of America has changed more than it has since the days of the cattle trails. One noteworthy evidence of this tremendous transition appears in the listing of breeds of beef cattle in college textbooks. A popular textbook published in 1920 listed 16 breeds of cattle, whereas the senior author's book, *Beef Cattle Science*, fourth edition, published in 1968, listed 21 breeds. Hence, only 5 more breeds were added in the 48-year period. The sixth edition of *Beef Cattle Science* listed and described 52 breeds—that's 31 new breeds, or more than double the number listed 18 years earlier in the fourth edition of this same textbook; and the current seventh edition lists 83 breeds.

Hand in hand with changes in the cow herd, stocker programs have changed. Many stockers no longer look like "peas in a pod"—and on the small side. Instead, more and more of them are large-framed multicolored crossbreds, with production tested ancestry, and weaned at heavy weights off milk and grass. Consequently, the trend is to shorten the finishing stage.

Indeed, humankind makes constant progress and nature undergoes constant change—they never remain the same. Therefore, the beef industry must go on researching, discovering, creating, and advancing. Chapters 17 through 25, which follow, will give a big assist in charting the course of the cow-calf system.

Cows and calves—the very foundation of beef production. (Courtesy, American-International Charolais Assn., Kansas City, MO)

CATTLE RAISING

Contents *Page*

Contents Page

Fig. 17-1. Partners in the cattle business.

Cattle raising must continue to improve and expand if consumers are to keep getting high-quality beef at reasonable prices. Changes within the beef industry, improved forage production and utilization, and favorable prices are major factors in encouraging cattle raising.

CONSIDERATIONS WHEN ESTABLISHING/EXPANDING THE BEEF ENTERPRISE

Whether or not cows and calves should be produced on a particular farm or ranch should be determined by a careful analysis of (1) the available resources (land, labor, capital, and managerial skills), and (2) the relative profitability of beef cattle in comparison with alternative enterprises.

Choices of alternate enterprises are limited on many ranches of the West and Southwest. Much of the land is suited only for grazing cattle or sheep. If beef cattle are selected rather than sheep, the only decision that remains is whether it should be strictly cow-calf, with all calves except replacements marketed at weaning time; a combination of cow-calf and stockers; or stockers only.

Farmers in the central and eastern United States have more options; hence, the choice of the enterprise becomes more difficult. They must first decide between grain and animal production, or some combination of the two. Additionally, if they decide to go the animal route, they must decide between beef cattle, dairy cattle, hogs, and sheep, or some combination of them.

Beef cattle do have certain advantages. Likewise, they have their disadvantages. These follow.

FACTORS FAVORABLE TO COW-CALF PRODUCTION

Some of the special advantages of cow-calf production as compared to other kinds of livestock on the farm or ranch are:

1. **It utilizes land not suited for grain production.** Beef cattle are well adapted to the use of the millions of acres of land unsuited for the production of grains or for any other type of farming, including humid areas for which the highest and best use is pasture, the arid and semiarid grazing lands of the West and Southwest, and the brush, forest, cut-over, and swamplands found in various sections of the United States. For the 50 states of the United States, 29.2% of the land area is used solely for grazing. An additional acreage is grazed part of the year (meadow aftermath

following a hay crop, cornstalks following harvest, cotton fields after harvest, winter wheat, etc.).

2. **It utilizes low-quality roughages.** Beef cattle efficiently utilize large quantities of coarse, relatively low-quality roughages produced on farms and ranches, including cornstalks, straw, and coarse or low-grade hays.

Fig. 17-2. Cows winter grazing cornstalks. Beef cows can utilize large quantities of low-quality roughages. (Courtesy, The University of Nebraska)

3. **It provides a profitable outlet for byproduct feeds.** Beef cattle provide a profitable outlet for many byproduct feeds, including corncobs, cottonseed hulls and gin trash, the oilseed meals, beet pulp, citrus pulp, molasses (cane, beet, citrus, and wood), wood byproducts (sawdust), rice bran and hulls, and fruit, nut, and vegetable refuse.

4. **It uses homegrown feeds.** Cattle can use the total homegrown production of grains and roughages, with or without the purchase of other feeds, more efficiently than any other class of livestock.

5. **It provides an elastic outlet for grain.** Beef cattle production provides an elastic outlet for grain. When plentiful, more grain can go into beef. When scarce, less grain and more grass and roughage will still produce beef.

6. **It maintains fertility.** Beef cattle provide an excellent way in which to maintain fertility on cultivated land. In addition to returning to the soil approximately 80% of the fertilizing constituents in the feeds, they offer a profitable way in which to utilize soil-building legumes that are usually a part of improved crop rotations .

7. **It requires a minimum of labor.** A beef cattle enterprise requires less labor than most other animal enterprises, with the result that it is relatively free from labor problems and adapted where labor is scarce and costly. Under average commercial range conditions, one person can care for approximately 300 cows.

8. **It distributes labor.** Beef cattle help to distribute the labor requirements throughout the year; they require but little attention except during the winter months.

9. **It requires small investment in buildings and equipment.** Beef cattle require a comparatively small investment in buildings and equipment.

10. **It entails little death risk.** Cattle past weaning age entail little death risk, as they are susceptible to comparatively few diseases and parasites. Death losses average about as follows: the mature beef herd, 2%; replacement heifers, 1%; and feedlot cattle, 1%.

11. **It is not normally a source of pollution.** Most cow-calf operations are not a source of pollution, because a minimum of confinement is involved and, for the most part, the animals defecate on the pasture or range—as wild animals do, and as nature intended.

12. **It produces the maximum amount of meat from milk and grass.** Beef cattle are adapted to the production of a maximum amount of meat from milk and grass, through (a) heavy weaning weights, or (b) finishing on pasture .

13. **It provides flexibility.** Cow-calf operations are flexible. Based primarily upon the availability and price of beef and the price of cattle, the options are: (a) sell all calves as weaners; (b) sell some calves as weaners and hold the balance over for sale as short or long yearlings; (c) buy additional calves and sell them as yearlings; or (d) feed home-raised calves to slaughter weight.

14. **It is suited to part-time farming.** Beef cattle are suited, as a supplementary enterprise, for part-time farming where the owner has off-farm employment or is semiretired, because of the relatively small and flexible labor needs.

15. **It is the preferred red meat.** Of the total per capita consumption of red meat (beef, veal, lamb, and pork) in 1995, 55% was beef.

16. **It results in 50% of the consumer's beef dollar going to the producer.** The farmer gets a larger share of the retail dollar spent for beef than for most products, simply because the cost of processing beef is much lower than the cost of processing products like a loaf of bread. In 1994, farmers got 22 cents of the American consumer's food dollar; in 1984, they got 26 cents; in 1974, they got 37 cents; in 1954, they got 42 cents.

17. **It need not compete with people for grains.** World food shortages favor the retention of beef cattle because they are adapted to the use of a maximum of such humanly inedible feeds as pasture and hay. Hence, cattle can adjust to the increased competition for grains for human consumption in the years ahead.

18. **It is in a favorable export situation.** U.S. beef cattle production and prices are in a favorable position due to world beef shortages and high beef prices abroad, factors which (a) lessen the consequences of beef imports through changing controls

(quotas, tariffs, and embargoes), and (b) create increased markets through exports.

19. **It imparts pride of ownership.** Ownership of beef cattle imparts pride of ownership and serves as a status symbol more than any other class of livestock, which is an important consideration for many people, farmers and nonfarm investors alike.

20. **It makes for a favorable balance of trade.** Through expanded exports, beef cattle can play a leading role in bringing about a favorable balance of trade, a more stable dollar, and a healthier world.

FACTORS UNFAVORABLE TO COW-CALF PRODUCTION

Factors which, under certain conditions, may be unfavorable to cow-calf production are:

1. **It requires considerable capital.** A cow-calf operation as normally conducted necessitates considerable capital. It requires an investment of $2,000 to $4,000 per cow for the animal, land, buildings, equipment, and machinery.

2. **It requires fencing and water.** Beef cattle require fencing and water—factors which have limited their expansion in many areas.

3. **It requires considerable knowledge and management ability.** A cow-calf operation requires know-how and managerial experience and ability commensurate with the size and sophistication of the operation.

4. **It isn't easy to comply with the grazing regulations on public lands.** The grazing regulations on the various federal- and state-controlled lands are often difficult with which to comply.

5. **It makes for high sire costs in boom periods.** During boom periods, high-quality purebred bulls are usually high in price and difficult to obtain.

6. **It is characterized by great spread between classes and grades.** The spread in price in market cattle is usually greater than is encountered in any other class of livestock. Shelly old cows that have outlived their usefulness in the breeding herd will bring comparatively less than an old brood sow. In other words, a greater spread usually exists between the price of Choice steers and Utility cows than between barrows and sows. In 1990, the comparative figures on the Omaha market were: the spread between Choice steers and Utility cows was $22.32 per 100 lb compared to a spread of only $4.67 per 100 lb between barrows and sows.

7. **It cannot be expanded or liquidated quickly.** Breeding herds cannot be expanded quickly, and the response to prices is slow. Ordinarily, a heifer does not calve until she is about 2 years of age, the pregnancy period requires another 9 months, and, finally, the young are usually grown 6 to 12 months before being sold to cattle feeders, who finish them for 4 months to a year; in total, involving about 4 years to produce a new generation.

Neither can breeding herds be liquidated quickly without great loss to the operator and much waste in terms of resources.

8. **It propagates slowly.** Cattle are neither as early breeders nor as prolific as hogs or sheep. Normally, a bovine female is not bred until she is 1½ years of age, gives birth only once per year, and produces only one young at a time. (See Fig. 17-3.)

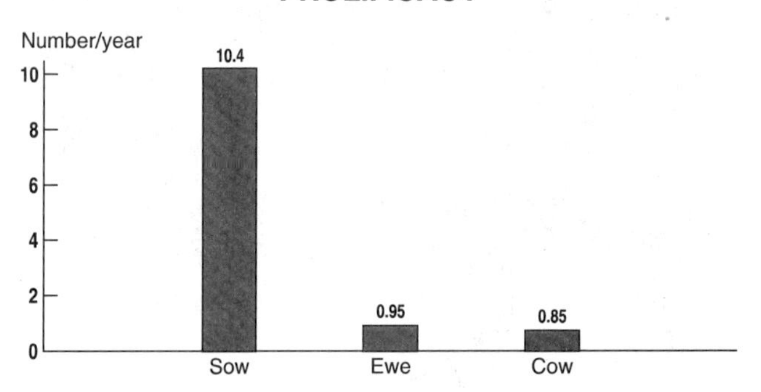

Fig. 17-3. Number of young raised to weaning age/year/breeding female. Generally, a cow produces only one calf per year, and only an 85% calf crop is raised to weaning age.

9. **It is subject to the hazard of such dreaded foreign diseases as foot-and-mouth disease.** The presence of foot-and-mouth disease, and of certain other diseases, in other countries always constitutes a hazard.

10. **It is inefficient in converting feed.** The beef cow is the least efficient of all feed converters, in pounds of feed required to produce 1 lb of product. (See Fig. 17-4.)

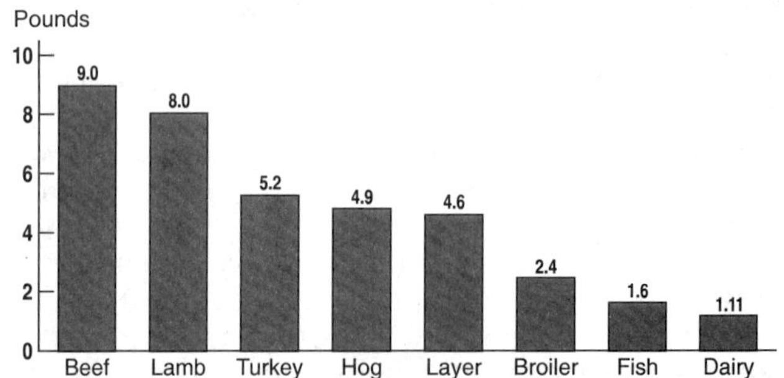

Fig. 17-4. Pounds of feed required to produce 1 lb of product—meat, milk, or eggs. This shows that it takes 9 lb of feed to produce 1 lb of on-foot beef, whereas it takes only 1.11 lb of feed to produce 1 lb of high-moisture milk. (Source: Chapter 8, Table 8-14, of this book)

11. **It is inefficient in converting protein.** Beef cattle are not efficient in converting protein in feed to ready-to-eat food. Only sheep are less efficient. (See Fig. 17-5.)

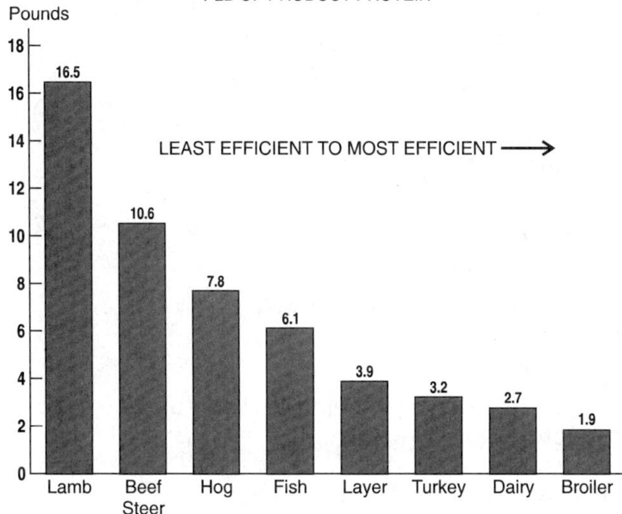

Fig. 17-5. Protein efficiency rating—protein in feed consumed converted to protein in ready-to-eat food, in percentage. This shows that it takes 10.6 lb of feed protein to produce 1 lb of beef protein, whereas only 1.9 lb of feed protein will produce 1 lb of broiler protein. (Source: Chapter 8, Table 8-14, of this book)

12. **It is costly to assemble feeder cattle.** Assembling uniform lots of feeder cattle from many small producers for movement to a few large feedlots presents structural problems.

13. **It is not very responsive to price and cost changes.** Because of the preponderance of small herds, the beef industry as a whole does not respond too well to price and cost changes, with the result that small herds tend to depress prices further and longer during times of overproduction.

14. **It is not very responsive to technological advances.** Because of the preponderance of small herds, the beef industry as a whole does not respond too well to technological advances, simply because most small operators cannot justify the cost of the technique. As a result, there is a tendency for the industry not to be as progressive as some.

15. **It overreacts to the economy.** Beef is the "Cadillac" of foods. Hence, like the "Cadillac" of cars, its sales slump faster and lower in times of a recession or depression than economy foods.

16. **It will constantly be threatened by soybeans and other meat substitutes.** Beef is somewhat vulnerable from substitutes by soybean and other proteins, provided the latter secure consumer acceptance and are low cost, especially if incomes fall.

17. **It will have increased competition from poultry.** More poultry, primarily broilers, will be consumed in the future. With it, beef consumption will trend downward.

BASIC RESOURCES FOR COW-CALF PRODUCTION

Basically, farmers and ranchers have four resources at their disposal with which to change cow numbers, that is, land, labor, capital, and managerial skills. In the years ahead, the most successful cattle producers will put these together in such manner as to maximize profits, followed by increased cow numbers. The cow-calf resource requirements, compared to alternate enterprises, are: (1) large acreages of land for which the highest and best use is pasture, along with considerable quantities of comparatively low-quality winter roughage; (2) available labor during the calving season (preferably with a liking for and knowledge of cattle); (3) adequate capital; and (4) able management commensurate with the size and sophistication of the operation.

KINDS OF COW-CALF OPERATIONS

Usually, cow-calf operators have several options. They may choose between (1) a farm or ranch herd, (2) running commercial or purebred cattle, (3) selling weaners or stockers, and (4) dual-purpose production.

FARM OR RANCH HERD

In general, beef cattle production in the farm states is merely part of a diversified type of farming.

Fig. 17-6. A farm herd of beef cattle. In general, farm herds are much smaller than range herds. Moreover, beef production in the farm states is usually part of a diversified type of farming.

Grain and pasture crops are produced; and, on the same farm, beef cattle may compete with dairy cattle, hogs, and sheep for the available feeds. This applies to practically all the farms located to the east of the 17 western range states. In general, farm beef cattle herds are much smaller than range herds of the West, and many of them lack the uniformity which prevails in range cattle.

More than half of all U.S. beef cattle are produced on the western range. Because a considerable portion of the range area is not suited to the production of grains, and because sheep, and in some areas big game, offer the only other major use of the grasses, it seems evident that range beef cattle production will continue to hold a place of prominence in American agriculture.

From the foregoing, it should be concluded that geographic location largely determines whether a herd of cattle shall be operated as a farm herd or range herd. Thus, the majority of the beef herds in the West and Southwest, except for relatively small herds in irrigated areas, are operated as range herds, whereas the vast majority of the herds in the central and eastern parts of the United States are operated as relatively small farm herds.

Fig. 17-7. Hereford cattle on a range in Oregon. South Sister Peak in the background. (Courtesy, U.S. Forest Service)

PUREBRED OR COMMERCIAL HERD

A purebred animal is a member of a breed, the animals of which possess a common ancestry and distinctive characteristics; and it is either registered or eligible for registry in the herd book of the breed.

Based on number of calves raised vs number of calves registered, the authors' computations show that only about 4% of U.S. cattle (beef and dairy) are registered purebreds. Hence, purebreds are a small, but mighty, minority.

Fig. 17-8. Purebred Angus herd. (Courtesy, American Angus Assn., St. Joseph, MO)

There is nothing sacred about purebreds, nor does the word itself imply any magic. It is generally agreed however, that purebreds have been the major factor in the beef improvement of the past, and that they will continue to exert a powerful influence in the future. Although limited in numbers, purebred herds are scattered throughout the United States and include both farm and ranch herds.

The purebred cattle business is a specialized type of production. Generally speaking, few producers should undertake the production of purebreds with the intention of furnishing foundation or replacement stock to other purebred breeders, or purebred bulls to the commercial producer. Although there have been many constructive purebred beef cattle breeders, and great progress has been made, it must be remembered that only a few master breeders such as Bakewell, Cruickshank, Gudgell and Simpson, and Congdon and Battles are among the immortals. Few breeders achieve the success that was theirs.

All nonpurebred cattle are known as commercial cattle. This includes the vast majority of the beef cattle of the United States—probably 96%. In general, however, because of the obvious merit of using well-bred bulls, most commercial calves are sired by purebreds. Beef over the block is the ultimate article of commerce of the commercial cow-calf producer; although in the process of getting from pasture to packer the calf may be subjected to 1 or 2 intermediate stages—as stockers, and in the feedlot.

Commercial producers are intensely practical. No cow meets with their favor unless she regularly produces the right kind of calf. Experience, industry, and good judgment are requisites to success in the commercial cow-calf business. Additionally, the commercial producer is a key in the nation's economy.

WEANERS OR STOCKERS

Commercial producers seldom adhere strictly to cow-calf operations as such. Rather, based primarily upon the price of feed and the price of cattle, they may option to (1) sell weaners, except for replacements, (2) carry all or part of their calf crop over to the yearling stage, (3) buy additional calves, and/or (4) finish out their home-produced calves, in their own feedlot or in a custom lot.

Fig. 17-9. Commercial beef cattle on a western Canadian ranch. They include Angus, Shorthorn, and Charolais crossbreds. (Courtesy, American Breeders Service, DeForest, WI)

DUAL-PURPOSE PRODUCTION

For the most part, dual-purpose production has been confined to the small farmer who lives upon the land and who makes a living therefrom. Cows of dual-purpose breeding are often referred to as the *farmer's cow*. In this type of production, an attempt is made to obtain, simultaneously, as much beef and milk as possible. That is to say, in its truest form, this type of management cannot be classified as either beef or dairy production.

One of the chief virtues of dual-purpose production is the flexibility which it affords. When labor is available and dairy products are high in price, the herd may be managed for market milk production. On the other hand, when labor is scarce and dairy products are low in price, calves may be left running with their dams, and emphasis may be placed on beef production.

Because of the very nature of dual-purpose production, it is not adapted to the extensive ranches of western and southwestern United States. Rather, it is practiced on a limited number of small farms scattered throughout the humid area of central and eastern United States. It is noteworthy, however, that many of

Fig. 17-10. Red Poll cow and bull calf. Dual-purpose cattle provide considerable flexibility. Thus, when labor is scarce and dairy products are low in price, the emphasis may be placed on beef production. (Courtesy, American Red Poll Assn., Louisville, KY)

the exotic breeds introduced into the United States and Canada, primarily for crossbreeding purposes in an effort to secure cows that will produce more milk for their calves, are known as dual-purpose breeds in the countries of their origin. Noteworthy, too, is the fact that these breeds are being used for crossbreeding in both farm and ranch herds.

BEEF NEEDS AND SUPPLIES

Dramatic changes have occurred in the beef industry since 1950. Pertinent information pertaining to past, present, and future cattle raising is given in Table 17-1 and in the discussion that follows.

- **Beef eaters**—In 1995, we had 260 million people

TABLE 17-1
THE U.S. PEOPLE AND THE BEEF AND VEAL SITUATION[1]

Year	No. of People	Per Capita Beef and Veal Consumption		No. of Cattle and Calves
	(million)	*(lb)*	*(kg)*	*(million)*
1950	151	71.4	*32.4*	77.96
1975	215	91.5	*41.6*	132.00
1985	237	81.0	*36.8*	110.00
1995	260	68.1	*40.0*	102.80
2005	285	65.0	*29.5*	94.00[2]

[1]Except where otherwise noted, the figures in this table were obtained from various USDA sources.

[2]Authors' projections. The year 2005 projection of 94 million cattle and calves takes into consideration both increased human population and decreased per capita beef and veal consumption.

to feed. By 2005, it is estimated that we shall have 285 million, an increase of 31 million people.

■ **Per capita beef and veal consumption**—Per capita beef and veal consumption rose to 97.7 lb in 1976 and since has declined. The authors project that per capita consumption will decrease from 68 lb in 1995 to 65 lb in the year 2005.

■ **Cattle and calf numbers**—Although human population has increased each year since 1950, per capita beef consumption has gone down since 1975 (see Table 17-1). The authors estimate that, in the year 2005, we shall have 285 million people, consuming an average of 65 lb of beef per year, and need 94 million cattle and calves—8 million fewer than in 1995. This is based on the assumptions that carcass weights will remain the same, and, that there will be no increase in beef imports.

■ **Veal slaughter; grain-fed slaughter**—In 1950, 10.5 million head of calves were slaughtered in the United States, representing 36% of all cattle and calves slaughtered that year. In 1980, 2.7 million head of calves were slaughtered, representing a mere 7.3% of all cattle and calves slaughtered that year. In 1994, only 1.27 million head of calves were slaughtered, representing a mere 4%. Hence, in terms of numbers, the relative importance of calf to total slaughter fell from 36% to 4% in the 44-year period from 1950 to 1994. As would be expected, per capita veal consumption declined sharply in this same period of time; it went from 8.0 lb in 1950 to 0.9 lb in 1994.

In 1947, only 3.6 million head of market cattle were grain fed, representing only 16% of the slaughter cattle that year. In 1980, 26.8 million head of grain-fed cattle were marketed, representing 79% of all cattle slaughtered that year. In 1995, 28.6 million head of feedlot cattle were slaughtered, representing 75% of all cattle slaughtered. Hence, in terms of numbers, the relative importance of fed cattle slaughter to total cattle slaughter rose from 16% to 79% in the 33-year period from 1947 to 1980; then, has stayed fairly constant from 1980 to 1995.

■ **Specialization in cattle feeding**—Cattle feeding has become increasingly specialized. It has changed from that of many small farmers feeding during the winter months to fewer and larger year-round commercial feeding operations. The percentage of cattle marketed from feedlots of greater than 1,000-head capacity increased from 36 to 87% between 1962 and 1993. Hand in hand with the development of large cattle feedlots, there has been a movement away from winter operations to a demand for feeder cattle more evenly distributed throughout the year.

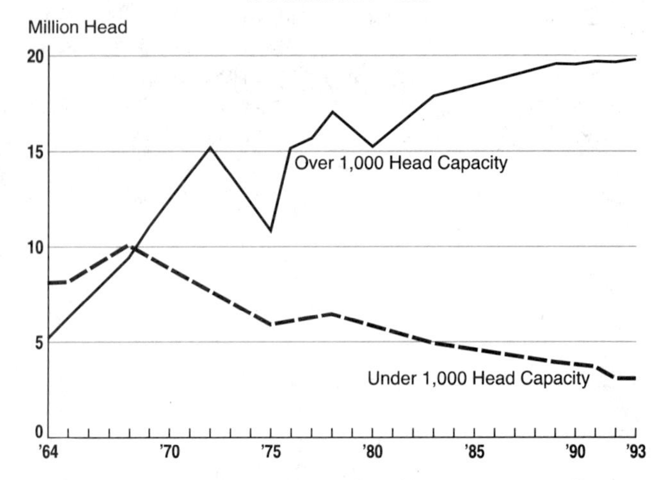

Fig. 17-11. Fed cattle marketings by size of feedlot, 1964–1993. (Sources: *Western Livestock Round Up*, Feb 1984, and *Cattle*, Final Estimates 1989–93, USDA, Sta. Bull. No. 905, pp. 65–74)

TRENDS IN BEEF AND MILK COW NUMBERS

Prior to 1950, there were more milk cows than beef cows in the United States. Beef cow numbers first exceeded milk cow numbers in 1954, and the gap widened until 1975. (See Fig. 17-12.)

Beef cow numbers grew from 16.7 million in 1950 to 35 million in 1995. (See Fig. 17-12.) Figure 17-12 also shows that in 1990 beef cattle numbers started upward. At the same time, beef cattle prices trended downward. *The forecast:* The market will be glutted and prices will fall.

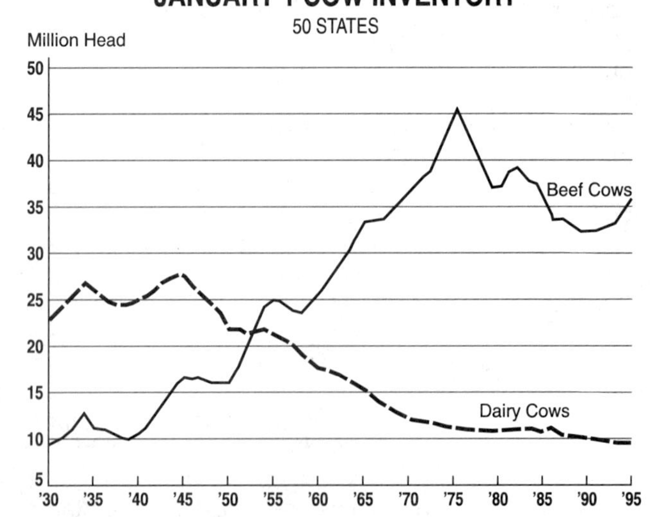

Fig. 17-12. Beef and milk cows on U.S. farms. (Courtesy, Livestock Marketing Information Center, Cooperative Extension Service, USDA)

GEOGRAPHIC TRENDS

As shown in Fig. 17-13 the trend in the beef calf crop was down during the 10-year period 1985-1994, reflecting the economic conditions that plagued the industry in this period. Further changes may be expected, due primarily to competing meat industries and the economy. However, differences among states and regions may be expected. Generally, beef cattle are expected to retain a strong position in the humid regions, where crop-livestock farms are dominant, and in those crop farming areas or regions, where ranching is the major type of agriculture. The Western area, with vast acreages of public land and open range, is well suited to beef raising and will continue as an important beef production center. The land's aridity and rather limited productivity, except where irrigated, tend to limit herd expansion, however. Region by region, here is the situation—past, present, and future—as the authors see it. (See Chapter 2, Fig. 2-20, for the six areas of U.S. beef production alluded to in the discussion that follows.)

NORTHEAST AREA

- **Present status**—The Northeast Area had only 3.9% of U.S. beef cattle in 1994. The beef herds of this area are small and often operated as a supplementary enterprise by part-time farmers. Hence, the area is of minor importance in beef production.

- **Future**—No significant change in beef cow numbers is expected in the Northeast. Dairy cattle will continue to be the dominant animal enterprise, to supply the highly concentrated human population with fresh milk.

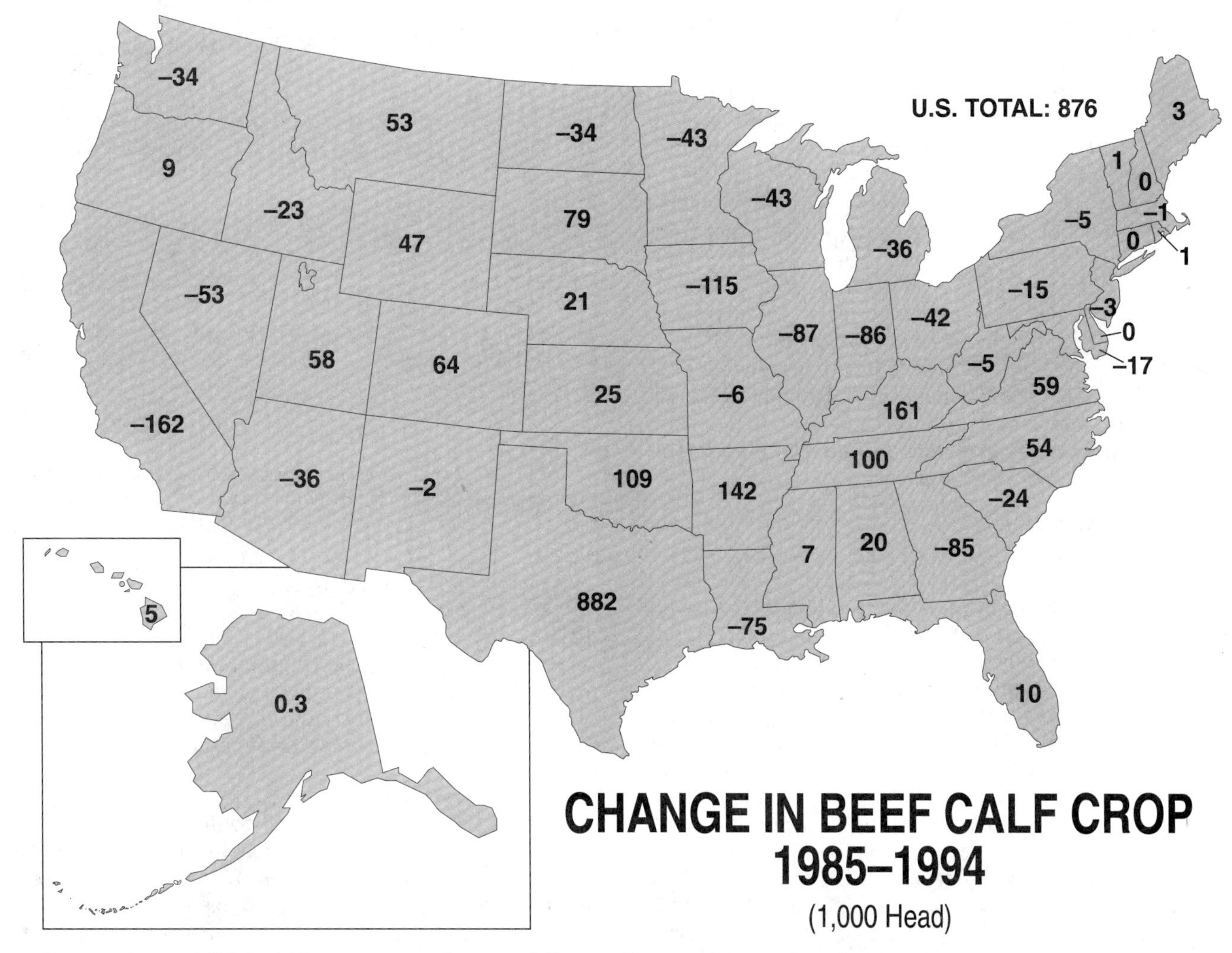

Fig. 17-13. Changes in U.S. beef calf crop, 1985–1994 (1,000 head). (Courtesy, Livestock Marketing Information Center, Cooperative Extension Service, USDA)

NORTH CENTRAL AREA

■ **Present status**—Even with its relative abundance of forage from crop residue, especially from corn, the North Central area has declined in beef cattle numbers in recent years. Among the explanations as to why beef cattle have declined in this area are: (1) decrease in pasture acreage; (2) the inability of beef cows to compete for land resources with crops which yield greater economic returns; (3) the beef cow herds are small on most farms, and small, fragmented landholdings do not lend themselves to enlarging the beef enterprise; and (4) lack of fencing, water, and managerial experience in beef production.

■ **Future**—In the future, cattle producers in the North Central area will likely (1) convert more marginal cropland to pasture, and (2) more effectively utilize relatively low-quality roughages, like cornstalks, for winter feed.

SOUTH AND EAST AREA

■ **Present status**—The South and East area had the most dramatic increase of any area during the period 1985-94. The factors which contributed most to the growth in beef cows in this area were: (1) the introduction and promotion of improved pastures, especially tall fescue; (2) the development of crossbred animals that can withstand hot, humid weather, diseases, and insects; (3) the increase in size of farms, with larger pasture acreage per farm; (4) the increase in part-time farmers (those who took off-farm employment while continuing to farm) who frequently choose beef cow enterprises because of the relatively small and flexible labor needs; and (5) the upward trend in calving percentages, reflecting better management practices.

■ **Future**—In the future, the South and East Area can weather adverse economic conditions in cattle production better than any other area of the U.S. Adequate rainfall and a long growing season contribute to good forage production, thus sustaining and encouraging cattle grazing. Generally speaking, the factors favorable to beef cattle growth until the middle 1990s will continue to influence the future of the cattle industry in the Southeast; namely, reduction in tobacco acreage, farm consolidation, scarce and costly farm labor, technological improvements in forage production and utilization, and a continuing trend toward part-time farming.

Factors having a restraining effect on beef cattle numbers in the Southeast are: rising land costs, high cotton and soybean prices, shifts of land to nonfarm uses, increase in calf grow-out (stockers), and grain finishing cattle on grass—which limits the pasture available for brood cows.

WEST NORTH CENTRAL AREA

■ **Present status**—Improved pasture and forage crops, decline in milk cow numbers, and increased farm size, along with larger acreages of pasture available per farm, have made the West North Central Area a great cattle country.

■ **Future**—In the North Central Area, the eastern part will continue to be most favorable to beef production as a result of improvement in the production and utilization of forage crops. Shifts in land use from grain to forage production will depend upon the relative profitability of grain crops vs beef cattle.

Fig. 17-14. Cows on winter fescue pasture, with round bales; on marginal cropland that was converted to pasture. (Courtesy, University of Illinois)

SOUTHWEST AREA

■ **Present status**—The highest and best use for much of the land in the Southwest, where combinations of private and public land often prevail, is for grazing. Hence, it has been a noted beef cattle area for a very long time.

■ **Future**—The Southwest will continue to be a well-stocked and highly specialized beef area. Modest increases in cattle numbers may come from improvement in forage production and expansion of cattle raising in the cropping areas. A shift from sheep and goats to beef will allow for some growth in beef cows.

Fig. 17-15. High-horned crossbreds on a Texas Ranch. (Courtesy, *Livestock Weekly*, San Angelo, TX)

WESTERN AREA

■ **Present status**—This area is characterized by three different kinds of cattle ranches: (1) mountain ranches, which, typically, combine irrigated meadows with range areas and private land with grazing permits to public lands; (2) desert ranches, involving public and private land, perhaps some irrigated hayland, and year-long grazing; (3) mixed crop–livestock farms; and (4) the Pacific states.

The Pacific states differ widely in climate and topography. Cow-calf operations are under mountain ranches, intermountain deserts, and mixed crop–livestock situations. In comparison with other regions, this region held more stable in milk cow numbers, which probably reflected increased human populations.

■ **Future**—Decline in beef cattle numbers may be expected in the Pacific states because of competing land uses and increased grazing fees and environmental controls of federal and state lands.

TRENDS IN NUMBERS OF HERDS AND HERD SIZE

About 52% of U.S. farmers own cattle (beef or milk cows).

Contrary to popular belief, the average U.S. cattle herd is relatively small. In 1995, 1,212,110 U.S. farms (including part-time and part-retirement farms) had 102,755,000 cattle (including milk cows), or an average of only 85 cattle and calves per herd.

Thus, the typical beef cow herd tends to be rather small and supplementary. Noteworthy, too, is the fact that further increases in part-time farming activities and continued farm consolidation may both be important in shaping future increases in beef supplies. Part-time farming favors a beef cattle enterprise which can be managed on a labor-extensive basis. As farms consolidate, small holdings of pasture are brought under one management, which makes it possible to support a profitable beef enterprise.

The preponderance of small beef cow herds is likely to affect the beef industry of the future in two ways: (1) They won't respond dramatically to price and cost changes—that is, part-time farmers are less responsive to prices than commercial farmers; and (2) they may limit technological change, simply because they cannot justify the cost of the technique.

Also, this means that (1) many producers will continue to lack sufficient volume to justify economically the use of many available technologies, and (2) assembling uniform lots of feeder cattle from many small producers for movement to a few large feedlots will continue to present structural problems for the entire beef industry. Nevertheless, the rising number of producers with larger herds can be expected to lead the way in the adoption of improved management and technology.

FACTORS INFLUENCING THE NUMBER OF BEEF COWS

Over a long period of time, economics determines the number of beef cows. In recent years, economic conditions have been unfavorable to beef cattle production. As a result, beef cow numbers have trended downward since 1975.

In 1995, there were 35,156,000 beef cows in the United States.

Beef cattle numbers are expected to decrease in the years ahead. A discussion of the influencing factors and the relative importance of each will follow.

SHIFTS IN CROP PRODUCTION

Traditionally, land for which the highest and best use is for pasture and hay has been used for beef cows. This practice will continue in the years ahead. However, forage crops produced with known technology will become more competitive with grain production. Forage for beef cows grown on land now occupied by wheat and cotton is uncertain, and largely dependent upon the world market for food and fiber. Also, should such acreage be released, the potential for soybeans and newly introduced cultural practices for raising corn will act as a counter force.

SHIFTS IN LIVESTOCK ENTERPRISES

Beef cow numbers already occupy much of the land formerly used by dairy cows, particularly in the humid regions.

Nationally, beef cows have been substituted for sheep through the years. Further decreases in sheep numbers in the West and the Southwest are anticipated, due primarily to difficulty in obtaining labor and losses from predators, followed by a conversion from sheep to beef.

Traditionally, few cattle enterprises are strictly cow-calf operations as such. Instead, the cattle producer exercises one or more of the options open, based primarily upon the availability and price of feed and the price of cattle. Some cow-calf outfits sell some calves and hold the remainder over for sale as short or long yearlings. Others buy additional calves and sell them as yearlings. Still others feed their calves to slaughter weight. Additionally, some beef cattle operators buy and sell stocker cattle only; they have no cows. Once weaned, a calf competes with beef cows for feed and forage and, consequently, acts as a constraint on production of additional calves.

Higher grain prices and small net profits per head in finishing operations will cause more Midwest farmers to discontinue feedlot operations, increase cow numbers, and push for heavier weaned calves. The resources thus freed from cattle finishing offer the greatest single potential for increased feeder calf production in this region.

Changes in the Southeast will curtail cow-calf expansion. Here the trend is toward more calf grow-out and grain feeding on pasture.

Southwestern cattle producers engage in three distinct cattle raising programs: (1) cow-calf; (2) cow-yearling; and (3) yearling-stocker. Except in Arizona, straight cow-calf programs are expected to decline somewhat throughout the Southwest. The slack will be taken up by cow-yearling programs in Texas and New Mexico and yearling-stocker programs in Oklahoma, where considerable wheat grazing will be used. This shift among cattle-raising systems will be prompted by the desire of cattle producers to capture returns from additional gains on the calves they produce, rather than raise more calves. The movement in this direction will be more noticeable in areas in close proximity to large feedlots. Of course, the price of weaner calves will determine the extent of this trend, for, given an option, operators will choose that system which is most profitable to them.

No significant changes in the existing cattle raising programs are expected in the Northeast and Western areas.

FORAGE PRODUCTION, HARVESTING, AND UTILIZATION

Most ranges are now fully stocked, some are over-grazed. Hence, increased beef cow numbers are dependent upon increased forage production. Available forage may be increased by—

1. Application of forage technology, including fertilization of pasture and hay crops, improvement of forage plant mixes, controlled grazing, range management renovation and reseeding of existing pastures and ranges, and use of herbicides to control undesirable plants.
2. Irrigation of forage crops in some of the western dryland areas.
3. Improvements in methods of salvaging crop residues, especially *husklage* and *stalklage* from corn.

Fig. 17-16. Harvesting corn stalklage with flail attachment. (Courtesy, Koehring Farm Division, Appleton, WI)

PUBLIC POLICIES

Public policies will change from time to time, but, for the most part, it is anticipated that they will not be adverse to the cattle industry. It is unlikely that there will be any longtime depressing effect on beef cattle numbers to accrue from crop control programs; from banning drugs; from livestock wastes; from imports and exports of beef; or from changes in federal grades of beef. However, cow numbers may be affected by the following:

■ **Charges and restrictions on the use of federal and state lands**—Increased grazing fees, environmental controls, and predator regulations are of concern to cattle producers who operate units by leases or use permits from federal or state agencies.

■ **Income tax regulations**—Current federal income tax regulations offer limited tax advantage to farmers to own beef cows. (See Chapter 13, Business Aspects of Beef Production, section headed "Income Taxes.") But it must be remembered that the federal income tax law is subject to frequent changes.

■ **Substitutes for beef**—Current public standards allow beef products to be extended by a specified amount with soybean and other proteins. If consumer acceptance is achieved and costs are lowered, extenders will likely be used in increasing amounts. However, except when incomes fall, neither meat extenders nor pork, nor other red meats are expected to have much bearing on beef consumption.

CATTLE PRICES

Anticipation of favorable feeder cattle prices is one of the strongest reasons for expanding beef cow numbers. People usually do those things which are most profitable to them—and cattle producers are people. World beef shortages, along with increased per capita beef consumption abroad, indicate that, generally speaking, beef prices and profits will be favorable for several years ahead.

POPULARITY OF CATTLE RAISING

Being a cattle producer serves as a status symbol for many people, farmers and nonfarm investors alike. Considerable romanticism has been attached to cattle raising through the years; and it has been enhanced by the recent publicity on crossbreeding, the exotic breeds, and Texas Longhorns. Ownership of cattle carries more prestige than ownership of any other kind of livestock.

SHIFTS IN LAND USE

There will be increased demands for grazing lands for industrial, residential, and recreational uses, all of which will mitigate against the expansion of cow-calf operations. In particular, pressures for recreational development and from the environmentalists will increase in the Western area.

ELDERLY AND PART-TIME FARMERS

The increasing age of farmers and the growth of part-time farming favor cow-calf operations. Farmers, like other people, are living longer. Many of them choose to pull up the reins gradually by shifting to less labor-intensive enterprises. A beef cow herd is often the enterprise of choice. Rising cost and scarcity of hired labor also favor beef cattle. More part-time cattle farming seems likely.

CHANGES IN PRODUCTION PER COW

A small addition to the supply of beef in the years ahead is expected to come from increased cow numbers, which will produce a larger number of calves to move into feedlots. In addition, the cow of tomorrow will produce more beef as an individual—she will be a product of research. In 1965, production per cow (carcass weight produced per cow) in the United States was less than 400 lb. By 1995, it increased to over 560 lb per cow. The major increase in beef to the year 2005 will come from more beef produced per cow.

Several potential changes in productivity per cow will directly affect the supply of beef obtainable from a given inventory of cows, among them those which follow.

INCREASED PERCENT CALF CROP

The authors estimate that there is a 90% calf crop in the U.S. This is based on calves born as a percentage of cows and heifers (beef and dairy cows combined) 2 years old and over January 1. No great improvement is seen in calf crop percentage ahead, primarily because gains become more difficult as calving percentages approach perfection—or 100%. It is estimated that the figure will move up only 2% by the year 2005, becoming about 92%. The largest gains are expected in the Southeast and Southwest, where rates have been the lowest.

Calf crop percentage on some cow-calf operations will be increased through selection; management practices, such as controlled breeding seasons (by use of hormones), fertility testing of bulls, and pregnancy testing of cows; improved nutrition, including improved pastures and ranges; and calving heifers at 2 years of age, instead of as 3-year-olds.

Superovulation (twinning and multiple births) could dramatically improve production per cow, if some of the obstacles could be overcome.

But the increased application of several practices that have an adverse effect on calving rate will partially offset the gains. This includes artificial insemination; ova transfer; the use of the larger breeds of cattle, which will make for more calving difficulty; shifts in the calving period; and more confinement cow operations.

REDUCED CALF LOSSES

United States calf death losses from birth to weaning average 8.0% of the total calf crop dropped.[1] Of course, there is room for improvement. Young calf losses will be reduced to some extent through the hybrid vigor of crossbreds and improved nutrition and management. However, these gains will be partially offset by increased confinement calving, which will make for more calf scours and other diseases. Also, any severe restriction in the use of antibiotics could have a negative effect. Although there are many unknowns, it would seem that a calf death loss of 6.0% from birth to weaning would be within the realm of possibility by the year 2005.

USE OF DAIRY CALVES

The average milk cow breaks down or is sold because of poor production after being in the milk string 4 years. To maintain a 100-cow dairy, therefore, 25 first-calf heifers must replace their elders on the milk line each year. But not all dairy heifer calves become tomorrow's cows! There are bull calves, calf losses, and calves that must be culled for one reason or another. To maintain status quo in a milking herd, therefore—with no provision whatsoever for expansion—each year a dairy operator must start with a minimum of 3 heifer calves for every 10 cows in the milking string. That's a 30% replacement. Of course, cull dairy cows and replacement heifers end up as beef, along with dairy steers and bulls.

In recent years, part of the demand for feeder calves has been met by increasing the proportion of the dairy calf crop fed out as dairy beef rather than vealed or kept for replacements. (Of course, fewer replacements, percentagewise, are kept when dairy cow numbers are being reduced instead of expanded.) Also, the need for dairy bulls has declined with increased artificial insemination. Although milk cow numbers halved in the 1950s and '60s, and some modest declines will occur in the years ahead, it is estimated that dairy beef, from cull dairy cows and dairy calves that are fed out, will provide about 15% of the nation's total annual beef supply to the year 2005.

In 1970, an estimated 40% of the nation's dairy calf crop was used for veal. This will change drastically, due to a continuing gain in demand for feeder calves relative to that for veal. By the year 2005, the authors predict that of all dairy calves born, 30% will be retained as replacements, 60% will be finished out as dairy beef and only 10% will be vealed. Each calf sent to the feedlot means about 800 lb more liveweight at slaughter than would have been obtained had the animal been vealed; hence, such a shift in disposition of dairy calves will increase production per cow.

[1]Authors' estimates.

WEIGHTS AND AGES OF CULL BREEDING STOCK

Cull cows contribute significantly to the total supply of beef, and account for nearly all nonfed beef produced in the United States. About 20% of beef cows and 25% of milk cows are culled each year and sent to slaughter. Beef cows are culled at an average of 7 to 9 years of age. No significant changes in age of culling are expected in the years ahead. Milk cows average only 4 years on the production line; they have a shorter life of usefulness than beef cows.

The trend is toward heavier weights of cull cows. Today, cull cows in most areas average around 1,100 lb in weight.

PRODUCTION TESTING AND CROSSBREEDING

Production testing and crossbreeding, combined, can increase beef yield per cow maintained over straightbreds by 15 to 25% depending on the choice of breeds and the breeding system. The 15 to 25% is achieved in two ways: (1) through selection, based on production testing, of the purebreds used in the crossbreeding program; and (2) through heterosis increase of the crossbreds.

Through production testing, it is possible to achieve (1) heavier weaning weights (20% heritability)—and young gains off milk and grass are very efficient; (2) higher daily gains from weaning to marketing, making for a shorter time in reaching market weight and condition, thereby effecting a saving in labor and making for a more rapid turnover in capital; and (3) greater efficiency of feed utilization, thereby making it possible to feed more cows and calves on a given quantity of feed.

A two-breed cross (in which only the calves are crossbred) gives an 8 to 10% increase in pounds of calf weaned per cow bred. Through a three-breed cross, it is possible to achieve even greater production per cow.

CALF WEIGHTS

Cow-calf producers will produce heavier calves in their efforts to increase efficiency and obtain a larger share of the total returns to the beef industry. Of course, heavier calves at weaning add to the total beef

Fig. 17-17. Heavier calves off milk and grass will increase the production per cow in the years ahead. (Courtesy, Beefmaster Breeders Universal, San Antonio, TX)

supply only if they are eventually carried to heavier slaughter weights. When calves are sold at around 400-lb weight, there is need for a stocker or backgrounding stage because most cattle feeders prefer cattle weighing 600 to 700 lb. Weaning heavier calves off milk and grass, especially if weights of 500 to 600 lb are achieved, will have a tendency to eliminate or shorten the stocker stage.

Larger cattle and heavier milking strains will result in heavier weaning weights in the years ahead. Much of this transition will come about as a result of exotic crosses and dairy crosses. Also, increased calf weights will result from selection of British breeding stock for more size and more milk, from improved nutrition of cows and suckling calves, and from greater use of production records to select breeding stock with superior growth rate potential.

The authors predict that the combination of breeding (improved British breeds, along with infusion of exotic and dairy breeding) plus improved nutrition, will result in an average of 500-lb weaning weights at 7 to 8 months of age by the year 2005, vs about 407 lb in 1970. This means that the better herds will be weaning off calves weighing 600 lb or better. The greatest increase will occur in the Southeast, but it should be added that they have the most room for improvement, because their weaning weights have always been lower than other sections of the country.

YEARLING WEIGHTS

Some cattle producers have long grown calves beyond weaning weight, with the prevalence of the practice varied by years according to available resources—particularly feed. It is expected that weaning calves at heavier weights will lessen the practice of holding them over as stockers except in the Southwest, where yearling cattle will remain important.

Short yearlings are weaner calves held over because they are too light to sell at weaning time and/or because surplus forage is available. Usually, they are sold at just under 1 year of age, although they may range up to 14 months in Arizona and New Mexico. Long yearlings average about 16 months of age.

Market weights of each group will gradually move upward, due primarily to improved forage supplies and selecting breeding stock for more size and more milk. The authors predict that, by the year 2005, without much change in age, short yearlings will weigh around 675 lb, 87 lb more than in 1970; and long yearlings will weigh around 750 lb, 60 lb more than in 1970.

QUESTIONS FOR STUDY AND DISCUSSION

1. Discuss the characteristics of, the relative importance of, and the relationship between (a) the cow-calf system, (b) the growing of stockers, and (c) drylot or pasture finishing.

2. Will the three phases of beef production—cow-calf, growing stockers, and drylot finishing—be more integrated or less integrated in the future?

3. What major changes have occurred in the cow-calf business since 1960? What forces have caused these changes?

4. Why have so many new breeds of beef cattle been developed or imported in recent years?

5. How and why have stocker programs changed in recent years?

6. What factors should determine whether cows-calves will be produced on a particular farm or ranch?

7. List and discuss what you consider to be the six *most favorable* factors to cow-calf production.

8. List and discuss what you consider to be the *most unfavorable* factors to cow-calf production.

9. Discuss the rating of beef cattle in comparison with other classes of animals from the standpoints of each (a) prolificacy, (b) feed efficiency, and (c) protein conversion.

10. What factors determine whether a beef operation shall be (a) strictly cow-calf, with all calves except replacements marketed at weaning time, (b) a combination cow-calf and stocker operation or (c) stockers only?

11. Why do farmers in central and eastern U.S. have more options in the choice of enterprises than western ranchers?

12. Basically, farmers and ranchers have four resources at their disposal with which to change cow numbers—land, labor, capital, and managerial skills. In the years ahead, how will the most successful cattle producers put these together to maximize profits?

13. How do the cow-calf resource requirements compare to alternate enterprises?

14. What are the pros, what are the cons, and what is your choice between (a) a farm or ranch herd, (b) purebred or commercial cattle, (c) selling weaners or stockers, and (d) dual-purpose production?

15. How do you account for the fact that probably fewer than 4.0% of U.S. cattle are registered purebreds?

16. In Table 17-1, the authors give their projections for the year 2005 in (a) number of people, (b) per capita beef and veal consumption, and (c) number of cattle and calves. You may challenge these figures, but be prepared to defend your own projections.

17. What changes do you foresee in cattle raising in the years ahead from the standpoints of (a) beef eaters and per capita consumption, (b) beef cow numbers, (c) veal slaughter, (d) percent of slaughter cattle that have been grain fed, and (e) specialized vs farmer-feeders?

18. Prior to 1950, there were more milk cows than beef cows in the U.S. But beef cows passed milk cows in 1954. By 1984, there were 3.4 times more beef cows than dairy cows in the United States. Why did such a shift from dairy cows to beef cows occur?

19. The North Central Area has an abundance of crop residue, especially from corn, most of which is left to rot in the field. Why, then, have beef cattle not increased in this area?

20. What factors have contributed to the dramatic growth of the beef industry in the South and East area?

21. What factors will have a restraining effect on future beef cattle growth in the South and East area?

22. Fig. 17-13 shows changes in the beef calf crop, 1985-1994. Why did the beef calf crop decline in most states during that period of time? How do you account for the fact that a few states showed increases?

23. If you were starting a beef cattle enterprise, and if you had the flexibility of choice, in what area would you locate, and why would you locate there?

24. Why are the vast majority of U.S. cattle herds small—averaging only 85 cattle and calves per herd in 1995?

25. How will the preponderance of small beef cow herds likely affect the beef cattle industry?

26. Do you feel that beef cattle will be able to compete with wheat, cotton, corn, and soybeans for the use of the land in the U.S.? Justify your answer.

27. What shifts in land use do you see in the years ahead: (a) from dairy to beef; (b) from sheep to beef; (c) from Corn Belt farmer-feeder to cow-calf operator; (d) from Southeast cow-calf expansion to growing stockers and pasture finishing; and (e) from Southwest cow-calf programs to cow-yearling programs?

28. Discuss the possibility of increasing beef production through increased pasture and range production by (a) fertilization, (b) improved forage plant mixes, (c) controlled grazing and range management, (d) renovation, (e) reseeding, (f) use of herbicides, and (g) irrigation.

29. Describe the modern methods that are being employed to salvage husklage and stalklage.

30. Discuss the impact on cow numbers of each of the following: (a) banning drugs; (b) environmental control; (c) imports and exports of beef; (d) changes in federal grades of beef; (e) income tax regulations; (f) substitutes for beef; (g) prices; (h) cattle serving as a status symbol; (i) increased demands for grazing lands for industrial, residential, and recreational use; and (1) part-time farmers.

31. How will the following changes in productivity per cow affect the supply of beef obtainable from a given inventory of cows: (a) percent calf crop; (b) reduced calf losses; (c) use of dairy calves; (d) weights and ages of cull breeding stock; (e) production testing and crossbreeding; (f) calf weights; and (g) yearling weights?

SELECTED REFERENCES

Title of Publication	Author(s)	Publisher
Animal Science, Ninth Edition	M. E. Ensminger	Interstate Publishers, Inc., Danville, IL, 1991
Beef Cattle, Seventh Edition	A. L. Neumann	John Wiley & Sons, Inc., New York, NY, 1977
Beef Cattle Production	J. F. Lasley	Prentice-Hall, Inc., Englewood Cliffs, NJ, 1981
Beef Cattle Production	K. A. Wagnon R. Albaugh G. H. Hart	The Macmillan Company, New York, NY, 1960
Beef Production and Management Decisions	R. E. Taylor	Macmillan Publishing Co., New York, NY, 1994
Beef Production and Management, Second Edition	G. L. Minish D. G. Fox	Reston Publishing Co., Reston, VA, 1982
Beef Cattle Science Handbook	Ed. by M. E. Ensminger	Agriservices Foundation, Clovis, CA, pub. annually 1964–81
Cattle Raising in the United States	R. N. Van Arsdall M. D. Skold	Economic Research Service, U.S. Department of Agriculture Washington, DC, 1973
Commercial Beef Cattle Production, Second Edition	Ed. by C. C. O'Mary I. A. Dyer	Lea & Febiger, Philadelphia, PA, 1978
Stockman's Handbook, The, Seventh Edition	M. E. Ensminger	Interstate Publishers, Inc., Danville, IL, 1992

A large and attractive cattle raising headquarters.

A beautiful environment, enhanced by farm windbreaks, on a ranch in North Dakota. (Courtesy, USDA Soil Conservation Service)

THE CATTLE FARM OR RANCH

Contents — *Page*

Contents	Page

Fig. 18-1. Attractive farmstead, enhanced by a fine herd of Polled Herefords and a good pasture. (Courtesy, *The Progressive Farmer*, Birmingham, AL)

"The past is prologue," according to a sign at the entrance to the National Archives Building in Washington, D.C. This gem of wisdom speaks loudly and clearly to cattle producers as they travel to the year 2000. The more successful operators will be those who recognize that the past is just the beginning (prologue)—those who focus their eyes on the future, then prepare for it. Indeed, the beef cattle industry is in the era of its greatest development as it explores and applies new genetic, nutrition, and marketing technology to meet worldwide demand for more beef. Many people are eager to get into the cattle business. But ownership of cows and a potential market are not enough. Land, feed, water, and management are necessary. Lest cattle producers forget the story of 1886, it should be told again and again. It was a severe winter of the type that is the bane of the cattle producer's existence. With the melting of the snow in the spring of 1887, thousands of cattle skeletons lay weathering on the western range, a grim reminder of overstocking and inadequate feed supplies. Many ranchers went broke, and the cattle industry of the West suffered a crippling blow that plagued it for the next two decades. Out of this disaster, however, ranchers learned the never-to-be-forgotten lessons of (1) avoiding overexpansion and too close grazing, and (2) the necessity of an adequate winter feed supply. This story is not retold for pessimistic reasons. Rather, it is repeated for purposes of emphasizing that in the future, as in the past, the farm or ranch is the most important basic resource to the success of a cow-calf operation.

Many people are interested in buying a cattle spread. Tenants are climbing the ladder to ownership. Present owners are mechanizing their holdings and borrowing more money, with the result that they want more acres. Farms and ranches are being bought for sons and daughters. City folks want to fulfill their dreams by semi-retiring on a cattle ranch. Investors, who have been disillusioned with the stock market, who are concerned about continued inflation, and who feel that manufacturing profits are on the wane, are looking for cattle farms and ranches. Some of these buyers will be happy with their purchases. Others will rue the day that they made the decision. How well the farm or ranch is selected, bought, and managed will make the difference. Indeed, no decision in the cattle business has greater consequences for the individual than selecting and purchasing the farm or ranch.

Although a cattle farm or ranch may be resold, most owners plan to operate the place that they acquire for a lifetime. This means that the fields and buildings are to be the purchaser's workshop, and that the alternatives open thereafter are greatly reduced. The owner can no longer consider operating in another area, even if the climate, feed, and market are more favorable. In many cases, there is a narrow range of alternatives, such as (1) cattle only, as on some holdings of the Southwest; (2) the particular combination of livestock, for example, hogs cannot be run on the western range; and (3) adapted crops and grasses. Moreover, except for an absentee landlord, the farm or ranch selected will be the family home. As such, the family will develop community ties that may last for years or even for generations, since many cattle farms and ranches are passed on from one generation to the next.

Indeed, the ultimate success of a cattle operation is determined by the careful selection of a farm or ranch; the proper combination of cattle, feed, and sometimes other livestock; the weather, disease, price fluctuations, and market demands; and the well-being and happiness of the family.

WHY OWN LAND?

In addition to providing self-employment, making a living, and obtaining a reasonable return on equity invested, ownership of land is attractive to investors for the following reasons:

1. **It is expected that land values will increase.** Although land values will level out, or even decrease slightly, from time to time, the long-time trend is up.

Three big economic forces are at work to raise land values: (a) the population is growing while land area remains about the same; (b) our rising standard of living requires land for many purposes; and (c) inflation raises the money value of land.

2. **It provides an opportunity to obtain relatively large, long-term capital gains.** In land, the major increase is in its value, rather than from operating income. In 1945, U.S. land (farm real estate, exclusive of livestock, bank deposits, etc.) was valued at $53.9 billion; by 1993 it was $684.6 billion. That's more than a 14-fold increase. It is expected that this trend will continue, but at a slower pace. There is a tremendous advantage in getting long-term capital gain from land, because it is not taxed as heavily as ordinary income.

3. **It provides an effective hedge against inflation.** If what you can purchase with your salary diminishes each year, and your savings account is shrinking, while, at the same time, the price of groceries is climbing, what do you do? The truth of the matter is that most folks don't do anything before it's too late. This is primarily because (a) they are too busy "grubbing" out a living, and (b) they do not recognize the erosive nature of inflation. However, those with perception have frequently read the signs and found the answer in land. Historically, land has increased in value faster than other prices. For example, from 1970 to 1993 the value of U.S. farmland increased from $196 to $700 per acre, an increase of 3.6 times, whereas, during this same period of time, consumer prices went up only 2.5 times.

Most economists agree that long-term inflation will not disappear. Thus, as inflation continues, people are likely to bid up land prices in an effort to protect themselves against the declining purchasing power of their dollars.

4. **It makes a good alternative to a jittery stock market.** The 1993 composite return on common stocks reported by Standard and Poor was 2.78%.

5. **It affords investment opportunity in America's biggest single industry.** On January 1, 1994, farm assets totaled $920 billion, which is equivalent to four-fifths the value of all stocks of all the corporations represented on the New York Stock Exchange.

6. **It's a business where the greatest need and action of the future lie.** The greatest need and action of the future will be that of providing food and fiber for the world's exploding human population—5.5 billion in 1994; and projected to be 8.5 billion by the year 2025.

7. **It provides a way through which to benefit from increased farm exports.** Expanded export opportunities for U.S. farm products assure a bright future for U.S. agriculture.

8. **It offers an escape from some of our environmental and social problems.** There is growing concern with environmental pollution, poverty, urban congestion, and social unrest. As a result, these forces are giving a boost to interest in rural living.

At the turn of the century, 75% of the nation's population lived in rural areas—mostly on farms. In 1990, only 24.8% were in rural areas, and only 1.9% lived on farms.

9. **It furnishes a recreational and vacation area of which the investor is part owner.** Such an arrangement (a) assures greater privacy than public recreational areas, and (b) makes it possible to combine business with pleasure.

10. **It provides good "leverage."** In most cases, people may borrow 65 to 90% of the cost of property, thereby tying up only a small amount of their own capital.

11. **It satisfies the psychological desire of people wanting to own some land.** Ownership of land—the good earth—imparts pride of ownership. Also, there is no security like the security of owning land.

TYPES OF CATTLE FARMS OR RANCHES

When selecting a cattle farm or ranch, the first major consideration should be the purpose of the ranch. For what use will it be put? Is it intended to be used as a full- or part-time cattle operation? Will it be used for cattle only; will it be used for cattle and one or more other classes of farm animals; or will it be a combination cattle and crop operation? Is it flexible enough to permit some choices should the economic conditions so indicate? Is it limited to a certain type of operation, like the Southwest where the choice, for the most part, is cattle or sheep? All these alternatives, and more, should be considered prior to purchasing a ranch. Of course, alternatives will also be considered, from time to time, in an established ranching operation. Nevertheless, the type of farm or ranch is important, for greatest success is usually achieved when the operator does those things which he/she likes best.

FULL-TIME CATTLE FARM OR RANCH

A full-time cattle farm or ranch is one in which the operator devotes full time to the enterprise and depends entirely upon cattle for income. As indicated later in this chapter (see section headed "Carrying Capacity; Size of Herd or Ranch"), (1) carrying capacity, not acreage, determines whether a unit is big enough to constitute a full-time operation, and (2) the authors estimate that a farm or ranch with a minimum carrying capacity of 300 head of brood cows (or equivalent; for example, 2 yearling stockers may be substituted for 1 cow) is necessary to be an economic unit for a full-time cattle operation. Without doubt, exceptions can be cited where operators with smaller herds are making a good living, with no other source of income. However, most operators with fewer than 300 brood cows are employed part-time off the farm or ranch or have another sizable farming enterprise in combination with the cattle.

In full-time cattle operations, it is important that the labor force be utilized efficiently throughout the year. This means that the size of the family, or the amount of hired help, should be considered. Generally speaking, an efficient full-time cattle operation should not have in excess of 1 person to 300 cows. When a cattle producer improves on that, the operation should be profitable; when the labor force is in excess of that, there may be trouble. Of course, the more specialized the cattle operation is, the more difficult it is to distribute labor properly throughout the year. For example, it is very difficult for a cattle producer who, during the

Fig. 18-2. Hilly and wooded farm not suited to crop production (Courtesy, American Angus Assn., St. Joseph, MO)

grazing season, pasture finishes steers in the Flint Hills of Kansas to use the labor throughout the year unless there is a cow-calf operation in addition. Likewise, it is easier to use labor effectively throughout the year where a good part or all of the winter feed is homegrown than where the cattle producer relies entirely on purchased winter feeds.

On some cattle ranches, none of the area is suitable for cultivation, or even for hay production, with the result that all supplemental feeds must be purchased. On still others, nearly year-round grazing is possible, with the result that little supplemental feed is needed. On the vast majority of the nation's cattle farms and ranches, however, the operator has the option of either (1) pasture only, with all supplemental feed purchased, or (2) pasture, with homegrown supplemental feed. The advantages and disadvantages of each of these systems follow:

PASTURE ONLY, WITH ALL SUPPLEMENTAL FEED PURCHASED

In comparison with an operation where supplemental feed is homegrown, the advantages and disadvantages of purchasing supplemental feed are:

- **Advantages**

1. More cattle can be kept.
2. Less machinery and equipment necessary.
3. More time can be spent caring for the cattle, especially when they need it—as when calving, or when there is a disease outbreak, etc.
4. Purchase of additional feed can give flexibility to the operation. When cattle are down in price and feeds are reasonable, cattle producers may buy feed and finish their own calves or even buy more calves.

■ **Disadvantages**

1. It is more difficult to distribute the labor throughout the year.

2. The peak pasture growth is usually not efficiently utilized—that is, pasture may go to waste during the lush growing season.

3. The feed supply is often uncertain and is subject to high prices during periods of drought or other times of scarcity.

4. It may be difficult to borrow money to buy feed when it is needed.

PASTURE, WITH HOMEGROWN SUPPLEMENTAL FEED

In comparison with a farm or ranch that purchases all supplemental feed, the advantages and disadvantages of home-growing supplemental feeds are:

■ **Advantages**

1. Labor can be used more effectively and efficiently throughout the year.

2. Land can be used for its highest and best use; that is, sometimes the highest and best use for land is to produce winter feed rather than to produce pasture.

3. It evens out costs, because it avoids the necessity of buying feed when it is scarce and high in price.

4. It makes it possible to rotate the use of land.

■ **Disadvantages**

1. It limits the number of animals, by the amount of land that must be used for supplemental feed.

2. It requires more able management to produce both cattle and supplemental feed than to produce cattle alone, simply because the knowledge of each type of operation is necessary.

3. It requires more machinery and equipment.

4. The labor force may be divided at a time when it is urgently needed for the cattle, as at calving time, when dehorning and castrating, or when there is a disease outbreak.

Fig. 18-3. Cattle and homegrown hay on a Corn Belt farm. (Courtesy, *The Corn Belt Farm Dailies*)

When selecting a cattle ranch, the above alternative management systems should be considered. What are the pros and cons for each system, then what is the decision?

More and more full-time cattle farms and ranches of the future will specialize in cattle only, without diversifying in cash crops or another class of farm animals. Also, cow herds will get larger.

PART-TIME CATTLE FARM OR RANCH

Most cattle operators with fewer than 200 brood cows are either employed part-time off the farm or ranch or semiretired and have another source of income.[1] In 1994, over 60% of the beef cattle producers had less than 50 head and these producers represented approximately 14% of total beef cattle numbers of the United States.

Many part-time cattle farmers or ranchers are individuals who have always wanted to own and operate a cattle ranch but who, because of limited funds, cannot acquire large enough spreads to make a living therefrom. The partial fulfillment of their dreams is realized by buying a small place on which they live, then they supplement their income from another job. They take care of their cattle before and after hours, and on weekends and holidays. Some of these part-time farmers hope eventually to farm full time. They plan to continue their nonfarm work until they accumulate enough capital for a full-time farm business, gaining valuable farming experience in the meantime on a part-time farm.

Many senior citizens and semiretired people find beef production a rewarding and remunerative experience. Folks are living longer. Many of them choose to pull up the reins gradually by shifting to less labor-intensive enterprises. Beef cattle fit their need. These retired or semiretired senior citizens are less dependent upon cattle prices and profits than full-time operators.

When it is planned that the operator seek off-farm

[1]On Jan. 1, 1995, 1,212,110 U.S. farms had 102,755,000 cattle and calves (beef and dairy), for an average of 85 cows and calves per herd. Obviously, therefore, many cattle producers secure a good deal of their income from sources other than cattle.

employment, it is important that the ranch be selected with this in mind. This means, that, in addition to all the other factors that should be weighed when selecting a cattle operation, the off-farm worker must consider the availability of additional work and the distance thereto. Likewise, they need to think how the cattle operation and the off-farm labor fit together. For example, it is very difficult to operate a part-time cattle farm or ranch when the operator must be away for 3 or 4 days each week, including overnight. Ideally, the part-time cattle producer should be home each night, so that the cows can be attended to morning and night. Likewise, weekends and holidays should be used for such things as mending fences, putting up hay, branding the calves, and so forth.

Part-time cattle enterprises will increase in the future. More and more folks will own a little cattle farm or ranch, as a source of some income and as a way of life. But they will derive most of their income from off-farm employment. An additional, and growing, number of part-time cattle farmers and ranchers will consist of senior citizens and semiretired folks. They are more interested in an enterprise that is relatively free from labor problems, and in the good life, than in monetary gain, for most of them have already made it.

KIND OF BEEF FARM OR RANCH

Traditionally, few beef cattle farms have been strictly cow-calf enterprises, limited to selling calves at weaning time. Some cow-calf producers sell part of the calf crop and hold the remainder over for sale as short or long yearlings, depending on the forage supply. Others buy additional calves and sell them as yearlings. Still others finish their home-raised calves out to slaughter weight. Additionally, there are cattle operations that own no cows; instead, they buy and sell stocker cattle.

In selecting a cattle farm or ranch, the above alternatives should be considered. Additionally, projected future changes in cattle raising in different regions should be considered. For example, because cattle feeding in the Corn Belt hasn't been too profitable in the past, it is expected that this area will go more to cow-calf production and push for heavier calves to increase returns. In the Southeast, more calves will be grown out to the yearling stage and grain finished on pasture. Cow-calf systems are expected to decrease somewhat in some of the southwestern states and to be replaced by more cow-yearling programs. In the Northern Plains the trend will be for more cow-calf programs and fewer cow-yearling and stocker programs.

CATTLE AND CASH CROP COMBINATIONS

Cattle and cash crop combination farm and ranches are preferred by some operators. It makes for desirable diversification, and, hopefully, cattle and the cash crop(s) will not be down in price at the same time.

When selecting a cattle-cash crop combination ranch, geographic location and crop adaptation must be considered. (See Chapter 2, Fig. 2-20, for geographical regions to which reference will be made in this chapter.) In the North Central states, for example, the cash crops might consist of corn and/or soybeans. In the West North Central area, small grain crops, like wheat, barley, and milo, might be considered. In the South and East, cotton, rice, milo, soybeans, and peanuts are possibilities.

Generally speaking, the most successful cattle and cash crop combination farms and ranches utilize those areas for which the highest and best use is pasture and hay for cattle. Then, the amount of cash crop acreage in the rest of the farm is determined primarily by prices. When cash crops are high in price, the maximum amount of area can be devoted to their production. On the other hand, when crop prices are down, more of the land can be converted to cattle feeds.

Still another type of flexibility exists on this type of operation. When crop prices are down and cattle prices are up, it may be advantageous to feed out the home raised calves or even buy more feeders. When grain prices are high, however, it may be advantageous to sell the cash crop and market the calves as weaners. Thus, the cash crop can be expanded or contracted as determined by price. Cattle-crop combinations also offer a fine opportunity to utilize land and crops to the highest level of efficiency. Small grain crops, like wheat and rye, can provide winter and spring grazing and after-harvest stubble grazing. Cornstalks can serve as a source of cheap roughage during the winter months.

Cattle-crop combinations provide an opportunity for specialization of labor. For example, in a family operation, in a partnership, or in a corporation, one or more persons may have expertise in cattle, whereas other members may have primary interest in machinery and crops.

Government crop programs may also play a part in cattle-crop combination farms and ranches. Like taxes however, such programs are subject to frequent changes; hence, a farm or ranch should not be purchased on the basis of the existing government program. Rather, like tax shelters, the farm or ranch owner should take advantage of whatever government crop programs exist at the time.

CATTLE AND OTHER CROP COMBINATIONS

Although they are not likely to be primary factors in determining the selection of a ranch, or how much will be paid for it, certain other cattle-crop combinations should not be overlooked. This includes both byproduct feeds and specialty crops.

Innumerable byproducts—both roughages and concentrates—from plant and animal processing, and from industrial manufacturing, are available and used as cattle feeds in different areas. Mention has already been made of small grain stubble fields and cornstalks. Cotton fields may also be pastured following harvest. Then, there are such additional byproduct feeds as cull potatoes, cottonseed hulls, corncobs, cull citrus, cannery refuse, beet tops, and a host of other similar products. Also, on many farms and ranches there are either low, wet areas, or rough, broken areas, which cannot be used profitably in crop production. Such areas can be fenced and made available to cattle.

Cotton and tobacco prices are subject to rather wide fluctuations. Also, an increasingly large proportion of these crops in the Southeast will be grown on fewer and larger farms. Forage for cows will be grown on some of the released cropland. But, of course, soybean and corn will be competing for the released acreage. Cattle and hay ranching make a good combination. Most always, some poor-quality hay is produced as a result of unfavorable haying conditions at harvest time, or during a wet season, silage may be made instead of hay. Rain-damaged hay and/or silage may be fed to beef cows. In some areas, cattle and timber make a good combination. This is true on much of the leased land under the supervision of the Bureau of Land Management and the U.S. Forest Service. Likewise, some small farms, particularly in the Southeast, are adapted to pine trees on the rough areas and cattle on the more level areas.

CATTLE AND OTHER LIVESTOCK COMBINATIONS

In many areas, and on many farms and ranches beef cattle and one or more other classes of livestock may be combined to advantage, thereby using the resources more efficiently and increasing income.

Farmers in the Corn Belt states long ago recognized the advantages of combining beef cattle and hogs. Regardless of the system of beef production—cow and calf proposition, the growing of stockers and feeders, finishing steers, dual-purpose production, or a combination of two or more of these systems—beef cattle and swine enterprises complement each other in balanced feeding. The cattle are able to utilize effectively great quantities of roughages, both dry forages (hay, cornstalks, etc.) and pastures; whereas pigs are fed primarily on concentrates. In brief, the beef cattle–hog combinations makes it possible to market efficiently all the forages and grains through livestock, with the manure being available for application back on the land. Such a combination makes for excellent distribution of labor. The largest labor requirements for both beef cattle and hogs come in the winter and early spring. During the growing and harvesting seasons, therefore, most of the labor is released for attention to the crops.

In the early history of the range livestock industry of both the United States and Canada, the cattle-sheep feuds frequently waxed hot. Each group warned the other away from its range. For the most part, however, the hatchet has long since been buried and only the legendary stories linger on. Today, many cattle producers would do well seriously to consider adding sheep to their enterprises. Limited experiments, along with observations, indicate that it is more effective to graze sheep and cattle together than to graze either species alone. Joint grazing results in (1) the production of more total pounds of meat, or greater carrying capacity, per acre, and (2) more complete and uniform grazing than pasturing by either species alone. This is attributed to the difference in the grazing habits of the two species. Cattle tend to leave patches of forage almost untouched, especially areas around urine spots and manure droppings. Also, cattle take larger bites and are less selective in their eating habits than sheep. Sheep tend to be selective of plant parts and will strip the leaves of plants. Also, sheep will graze many common weeds, even when good-quality grasses and legumes are abundant.

Beef cattle and dairy cattle compete for about the same feeds. Thus, a beef and market milk enterprise is seldom conducted in combination, with both as major enterprises on the same farm or ranch. In the East, cow-calf programs sometimes fit in where dairy farms have more feed than is needed in the dairy program. Also, some dairy operators, particularly small operators, have always used beef bulls as *cow fresheners* on lower producing cows, then marketed the calves as feeders. But a new type of beef-dairy combination has developed in recent years. It is known as *dairy beef.* Dairy beef is just what the term implies—beef derived from cattle of dairy breeding, or from dairy X beef crossbreds. Today, it is extolled with pride. The shift in consumer preference, along with rapidly expanding population, will result in the production of increased quantities of dairy beef.

Both commercial cattle feeders and dairy operators are interested in producing dairy beef, with the result that there is competition between them. Many dairy steers and cull heifers will continue to be finished out in commercial cattle feedlots. However, an increasing number of these animals will be finished out by dairy operators as a means of augmenting their income and diversifying.

Cattle and horses are found on many farms and ranches. In addition to being of use in working cattle, horses may contribute to the income of the operation. Also, it is noteworthy that cattle and horses are rotated on the pastures in the great horse breeding centers of the world (including the bluegrass area of Kentucky and the lush pastures of Ireland) for parasite control.

Combinations of beef cattle and poultry are occasionally seen, primarily as a means of disposing of poultry manure. Modern poultry operations require little land. However, manure disposal is a problem. Some poultry producers have solved their problem by having extensive pastures adjacent to the poultry operation on which the manure is spread. Such pastures produce an abundance of grass and are utilized by beef cattle.

IRRIGATED FARM OR RANCH

Rising land and labor costs favor more irrigation. Where the cost per cow-calf carrying capacity is cheaper under irrigation than a dryland operation, irrigation will increase. Likewise, intensive cow-calf operations under irrigation usually require less labor per cow-calf unit than more extensive dryland operations. Of course, once an area is irrigated, new crop alternatives are opened up, with the result that a determination will have to be made as to which will be the most profitable—cattle or crops.

ATTACHED GOVERNMENT AND PRIVATE LEASES

In the West, much of the grazing land that ranchers rely upon to maintain their cattle is built up into operating units by leasing or by obtaining use permits from several federal and state agencies. Overall, about half the range area in the 11 western states is federal or state land. This land is made available through permit or lease to nearby ranch operators, usually at a fixed annual fee per head of livestock. Although most of these leases are subject to ready cancellation, it is noteworthy that many of them have continued for years, even through 2 or 3 generations of the same family. Nevertheless, now and in the future, increasing pressure is being brought to use such public lands for recreation, wildlife, and environmental control. As a result, the future of such leases is less secure than in the past, and ranches made up principally of deeded land go for substantially more than those made up chiefly of government-leased land; and this gap will widen. Of course, the type of government lease, as well as the way the area lies in relationship to deeded land, can make a tremendous variation in land value.

The bulk of federal land is administered by the Bureau of Land Management and the U.S. Forest Service. Thus, where a cattle ranch is being acquired in the West, it is important that the cattle producer have knowledge of these particular agencies.

BUREAU OF LAND MANAGEMENT

The Bureau of Land Management of the U. S. Department of the Interior administers more than 40% of federally owned land. From the standpoint of the cattle producer, the most important function of the Bureau of Land Management is its administration of the grazing districts established under the Taylor Grazing Act of 1934 and of the unreserved public land situated outside of these districts which are subject to grazing lease under Section 15 of the act.

Grazing privileges are allocated to individual operators, associations, and corporations, and a fee is charged for grazing privileges. In 1996, the Bureau of Land Management charged $1.35 per animal unit month. (Also see Chapter 9, section entitled "Agencies Administering Public Lands.")

Fig. 18-4. Hereford breeding herd on the home ranch of a grazing permittee. These cattle summer in Colorado and winter in Utah. (Courtesy, Bureau of Land Management, U.S. Department of the Interior)

U.S. FOREST SERVICE

Approximately 51.5 million acres of the national forests are used for grazing under a system of permits issued to local farmers and ranchers by the Forest Service of the U.S. Department of Agriculture. These grazing allotments provide grazing for about 2.8 million livestock and about 3.4 million head of big game animals.

The Forest Service issues 10-year term permits to those who hold preferences and annual permits to those who hold temporary use. Among other things, the permit prescribes the boundaries of the range which they may use, the maximum number of animals allowed, and the season when grazing is permitted.

Forest Service grazing fees are based on a formula which takes into account livestock prices over the past 10 years, the quality of forage on the allotment, and the cost of ranch operation. In 1995, the charge came to $1.61 per animal unit month. (Also see Chapter 9, section entitled "Agencies Administering Public Lands.")

Fig. 18-5. For success and a heap of living, selecting and buying the cattle farm or ranch is most important. (Courtesy, Diamond Heart Farms, Irasburg, VT)

HOBBY FARM

In recent years, many people of wealth have established outstanding cattle herds, especially purebred herds. Most of these folks operate such cattle ranches as moneymaking enterprises. They are just as "money hungry" in the cattle business as they were in the industry from which they made their initial wealth. Also, conducting the cattle operation as a business is more of a challenge to them than if it were a hobby. Besides, the income tax regulations today are such as to make it impossible for many to afford not to operate a cattle farm or ranch as a business.

Internal Revenue Service agents are prone to attack the *farmer* status of absentee owners. Most challenges are raised where the taxpayer earns substantial off-farm or ranch income and is showing farm or ranch losses for a particular year or over a period of years. Though the concept is similar to the *hobby farm* challenge, the distinction exists in that the disallowance or challenge relates to capitalization of expenses. In hobby farm situations, the expenses are considered personal, and thus not deductible; nor can they be capitalized.

Of course, if a ranch is to be purchased and run as a hobby, the net return in terms of investment is likely to be of less concern to that individual.

SELECTING AND BUYING A BEEF CATTLE FARM OR RANCH

The first and most important requisite for success in the cattle business is proper selecting and buying of the beef cattle farm or ranch. The fundamental considerations will be discussed in the sections that follow.

CARRYING CAPACITY: SIZE OF HERD OR RANCH

Carrying capacity is defined as the number of animal units (1 cow, plus suckling calf—if there is a calf; or 1 heifer 2 years old or over) a property will carry on a year-round basis. This includes the land grazed plus the land necessary to produce the winter feed. Thus, if a 3,000-acre ranch provides all the pasture, along with winter feed, for a 300-cow herd plus their suckling calves, it has a carrying capacity of 10 acres per cow (3,000 ÷ 300 = 10). Two yearlings are considered equivalent to 1 cow-calf unit. The carrying capacity may vary anywhere from a productive irrigated farm with a carrying capacity of 1 acre per cow-calf to some ranges of the Southwest where grass and browse species are so sparse that it takes 60 acres or more to support one cow-calf. Most areas west of the 100th meridian require an average of about 30 acres of grazing land per animal unit.

Thus, when a realtor glibly refers to a certain farm or ranch as having a carrying capacity of 500 head, it is well to pin the realtor down by asking, "head of what?" Are their figures based on the definition given above, or are they counting calves, also; and is each yearling counted as ½ unit?

Admittedly, the carrying capacity of a ranch is difficult to determine. Past range management is a factor. If a range has been grazed too closely, carrying capacity is apt to be overrated. To continue stocking too heavily means lighter calf weaning weights, thinner cows, and perhaps a smaller calf crop. To rest and improve the range takes time and money.

From the above it should be deducted that carrying capacity, rather than number of acres, determines the number of cows that can be run on a given unit. No minimum or maximum figures can be given as to the

Fig. 18-6. An animal unit—one cow plus suckling calf. (Courtesy, International Braford Assn., Inc., Ft. Pierce, FL)

best size of herd. Rather, each case is one for individual consideration. It is noteworthy, however, that labor costs differ very little whether the herd numbers 80 or 300. Noteworthy, too, is the fact that the cost of purchasing and maintaining a herd bull comes rather high when too few females are kept. The extent and carrying capacity of the pasture, the amount of hay and other roughage produced, and the facilities for wintering stock are factors that should be considered in determining the size of a herd for a particular farm or ranch unit. The system of disposing of the young stock will also be an influencing factor. For example, if the calves are disposed of at weaning time and replacement heifers are bought, practically no animals other than the brood cows and herd bulls are kept. On the other hand, if the calves are carried over as stockers or finished at an older age, more feed, pasture, and shelter are required.

Then, too, whether the beef herd is to be a major or minor enterprise will have to be decided. Here again, each case is one for individual consideration. Big operations are getting bigger, with an increasing number of them involving multiple ownership. At the same time, there are an increasing number of small part-time herds, owned by individuals who have off-farm employment or are semiretired. Movement in both directions will continue.

The authors consider that a farm or ranch with a carrying capacity of 300 to 500 head of brood cows (or equivalent; for example, 2 yearling stockers may be substituted for 1 cow) is necessary to be an economic unit for a cattle rancher who devotes full time to the operation and depends entirely upon the cattle for income.

Since carrying capacity is the most important estimate in the entire appraisal, when buying a ranch it should be done by an expert who is completely familiar with the area. Generally, the seller is prone to overestimate carrying capacity; the seller's tendency is to estimate carrying capacity on the basis of the best season in 20 years, without including the other 19 years in the average.

COW-CALF PRODUCTION COSTS AND RETURNS

Those thinking of becoming cow-calf operators inevitably ask, "How much money will it take, and what can I make?" The operator needs to take stock of the financial situation. How much money is available now, how much credit, and how much debt is the operator willing to assume?

Cattle ranching is big business, requiring large amounts of capital. The capital requirements have risen as buyers have bid up the price of land and cattle, and as costs of machinery, equipment, and related items have advanced with the general price level.

Good cow-calf estimates of production costs and returns are essential for wise decision making and financial survival in the cattle business.

Fig. 18-7 shows the effect of herd size on cost per cow unit in the Western area. *Note well:* Larger herds can make better use of machinery, equipment, and management; hence, they generally have lower costs per cow unit.

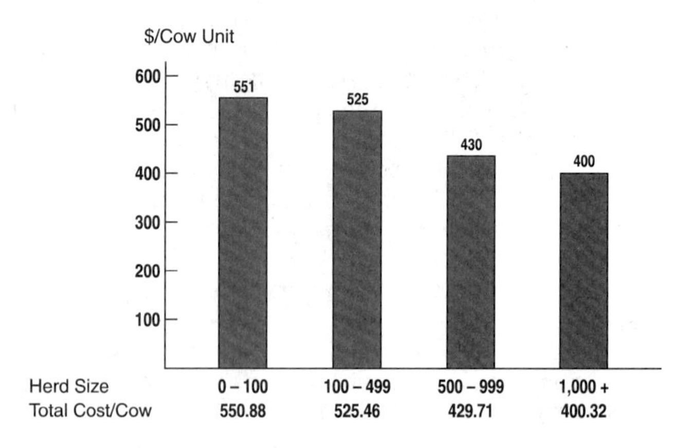

Fig. 18-7. Effect of herd size on cost per cow and profit or loss. (Source: *Cow/Calf Costs of Production, 1990–91*, USDA, Bull. 670, p. vi)

But factors other than cost of land and size of herd are involved in determining the cost of producing a calf. The three primary factors determining the cost of producing a calf are: (1) the annual cost per cow unit; (2) the percent of calves weaned per cow; and (3) the average weight of calves sold. (Also see Chapter 24, section on "Cow-Calf Costs and Returns.")

RISKS

The ownership of a farm or ranch does involve some risks, which are not pleasantly recalled by most current owners, and of which all prospective owners should be aware. Among them are the following:

1. **Drought.** A long dry period will shorten the grazing season; cut down on crops to be harvested for winter feed; result in loss in weight of cows; make for a smaller calf crop percentage; lower weaning weights; necessitate the purchase of feed; and/or even cause liquidation of all or part of the herd at unfavorable prices.
2. **Floods, storms, blizzards, fires.** Historically, some cattle are lost as a result of sudden rain storms and blizzards. Also, some are lost by fire.
3. **Disease outbreaks.** Traditionally, a cattle producer vaccinates against the diseases most common to the area, provided such a preventive exists; then, takes a calculated risk relative to a long list of other diseases. But costly disease outbreaks do occur, resulting in death losses and inefficiency among the living.

SELECTING THE LOCATION AND THE FARM OR RANCH

Choosing the right location and the right farm or ranch is a very important requisite for success. Nearness to friends and relatives is a major, but unmeasurable, factor in determining location, although it is less important than formerly with the development of more rapid transportation (improved highways and air travel) between areas. However, many nonsentimental factors should be considered in the selection of the particular location and the specific farm or ranch; among them, those which follow.

AREA AND CLIMATE

Area and climate (rain, snow, heat, cold, and wind) affect land use, cattle, and people; hence, they may determine the type of cattle farming or ranching of the particular area. In addition to soil, it takes moisture and reasonably warm weather to grow grass; and the longer the growing season, the more grass. Also, winter feeding is always more costly than summer grazing; hence, an area with mild winters has considerable advantage. So, the official weather bureau rainfall and temperature records—minimum, maximum, and average, by months and over a period of years—should be studied. Also, it is well to check the number of frost-free days, and even day lengths.

In the Corn Belt and Northern Plains, more than half of the land is cropland. In the southern states and mountain states, pasture and rangeland predominate. In the Northeast, forest is the major use of land, as it is in the Appalachian, Southeast, and Delta states, and on the Pacific Coast north of San Francisco.

Across the 48 contiguous states, there has been no appreciable overall change in land use since 1950. Cropland totals 17.4% of the land; forestland, 31.1%; grassland pasture, 29.3%; and various other uses, 22.2%. Urban land has doubled, mounting to nearly 2% of the total land area, and transportation land remains less than 1.5% of total acreage.

Area and climate also affect both cattle and people. For example, Brahman, or part Brahman cattle, are better adapted to hot, insect-infested areas than the European breeds. Climate affects people, too; hence, it must be considered by the cattle producers and their families. Some folks want to avoid hurricanes and tornadoes, no matter how lush the grass or how long the grazing season. Others want no part of shoveling snow and an area that may have to rely on emergency haylift by helicopter or airplane. Still others object to dust storms or to a hot, humid climate. Thus, in locating a farm or ranch, these factors are very important.

A fairly good idea of the prosperity of an area and of the kind of people who live there can be obtained by driving around and looking at the neighborhood. Painted and well-manicured buildings and surroundings indicate that the owners are doing all right. Poorly kept premises, with debris scattered about, give evidence of the *dry rot* of the occupants.

SOIL AND TOPOGRAPHY

Capacity to produce is the main thing for which to look when buying a cattle farm or ranch. Soil is the most basic thing affecting this ability. But soils differ, and their ability to produce is dependent on not only the soil itself but the use to be made of it. Thus, the prospective purchaser should learn about the main soil types in the area where they are planning to locate, so that they can recognize most of them. They should find out what kinds of yield of grass or crops can be expected on typical soils of each type. Do they have any special problems, such as alkali or poor drainage? If drainage is a problem, for example, they will want to find out if provisions have been made for tile or open ditch drainage on the farm or ranch, and if these are working satisfactorily.

Remember that soil is the *life blood* of the farm or ranch. To a very great extent, it sets the degree of profitability of a farm or ranch. It is normally cheaper to buy good soil than to purchase poor land and try to build it up.

There are several things a buyer can check on that will help determine the quality of soil. Study records of yields per acre, growth of crops, weeds, or trees on the land, soil tests, and production capacity maps. Note the depth of the soil, along road cuts or ditch banks or as determined by a soil auger. Land use capability maps are helpful to the buyer. Also, the seller or the Soil Conservation Service office may be able to provide complete plans showing soil types and classes, recommended crop use, and long-term development plans.

It has long been said that poor soils and poor people go together. Thus, one of the guidelines in estimating the worth of a farm or ranch is the prosperity of the entire community. If houses are modern, if buildings are painted, and if fences are in good repair, the area is likely productive and prosperous.

Pasture and hay crops—hence, cattle—are well adapted to hilly, rolling land. However, it must be realized that type of topography is a factor in erosion; and that erosion control measures are costly. Also, the cost of seedbed preparation and putting steep, barren slopes into grass is sometimes prohibitive, although initial costs might make the farm or ranch look attractive.

It is well to have an aerial photo of a farm or ranch at the time it is inspected in detail. This can usually be obtained from the ASC or SCS office in the county. The aerial photo will give a clue of the undesirable features of the property. It will show (1) how ditches cut through the place, (2) inaccessible or poorly watered pastures, (3) "point" rows due to curving roads or railroads, and (4) soils low in organic matter as indicated by light spots.

Grass is the cornerstone of successful cattle production. It is estimated that 85.7% of the total feed supply of U. S. beef cattle is derived from forage; in season, this means pasture. In purchasing a farm or ranch, therefore, it is important that there be abundant grass, preferably with a long grazing season.

Also, it is well to confer with the neighbors concerning the productivity of the land. They have the advantage of having seen it in all seasons, and over a period of years, whereas the prospective purchaser may be seeing it at a particularly favorable or unfavorable time. Belly deep grass makes quite a different impression than dry, parched pasture or a snow-covered range, yet neither represents average conditions.

Leasing a place with the option to purchase is the surest of all ways in which to try it out, or evaluate it. Thus, this possibility should not be overlooked, although it is recognized that it is seldom possible.

IMPROVEMENTS

The improvements—buildings, fences, corrals, scales, windmills, stock tanks, and irrigation ditches—are important on a cattle farm or ranch.

Fig. 18-8. An Iowa farmstead. (Courtesy, *The Corn Belt Farm Dailies*)

Hence, they should be carefully considered when buying a place. They should be adequate, without being elaborate, and they should meet the needs of the type of farm or ranch operation contemplated. Consider their suitability, flexibility, and state of repair. Old barns that cannot easily be adapted may be completely useless; worse yet, they may be a detriment. Most people find it better to purchase a farm or ranch with adequate buildings, fences, and corrals, rather than to attempt to build them. Construction costs are high. Moreover, if improvements have to be added immediately after purchase, they should be considered as part of the cost of the establishment; and it should be recognized that they will reduce working capital.

One should avoid over investment in buildings. The amount of barn rent for which a cow can pay is limited. There is an old saying in the industry that, "Fancy white houses and big red barns won't put fat on cattle. You just need grass and water." Yet, a cattle producer needs adequate facilities. In addition to their adequacy for shelter, cattle barns and shelters should be evaluated from the standpoint of storage of feed, saving of labor, and handling of manure.

Family happiness is important to the success of any cattle venture. Thus, the home should come in for its share of investigation, also. If it is not completely modern so far as plumbing, heating, and lighting are concerned, it will need to be made so as soon as possible. Of course, if these things have already been done, this can alleviate a sizable future expense.

Fences need to be considered. Repairing fencing or adding new fencing is expensive. For holding cattle, especially where calves and yearlings are involved, four barbed wires are minimum, and five would be better. It should be kept in mind that the life of the fence depends to great extent on the life of the posts, and the stability of the corner posts.

An essential part of a beef cow operation is a well-planned, workable corral system for handling cattle. Corrals should be located where they are accessible and, if practical, should contain a water supply.

A loading chute and a 2 ft wide working chute should be included in the facility. Also, it would be helpful if a platform type livestock scale (with panels), of 10,000-lb capacity, were a part of the corral. A workable set of corrals need not be elaborate, yet such things cost money. It is preferable, therefore, that a suitable corral be on the farm or ranch at the time of purchase. It will avoid time and money in planning and developing such handling facilities.

If the farm or ranch is in an area where drainage is a problem, get a tile map of the farm. Check the outlets and see if the system has been working. If there is a county drainage ditch involved, take a look to see if it will have to be cleaned out soon; this could mean a special assessment for the farm.

WIND DIRECTION, WINDBREAKS, AND NATURAL SHELTERS

While most farm families are conditioned to farm aromas, there is no use making for an unpleasant situation when it can be avoided. Thus, the house should be located on the windward side of the headquarters, with special consideration given to summer winds.

Unless hills form a natural windbreak, it is desirable to arrange a suitable tree planting for the farm or ranch headquarters. Usually, a tree windbreak is located 75 to 150 ft from the buildings to be protected, with 3 to 7 rows of trees 20 to 75 ft wide.

Natural shelters for beef cattle may consist of hills and valleys, timber, and other natural windbreaks. If natural windbreaks are adequate, it may avoid the necessity of constructing shelter.

Wind direction is also important from the standpoint of the location of open sheds. They should face away from the direction of the prevailing winds.

SERVICE FACILITIES, COMMUNITY, AND MARKETS

The service facilities, community, and markets are very important. It is important that the headquarters be near an all-weather road or highway that is well maintained. In some areas, dirt and gravel roads seem to be bottomless 3 to 4 months out of the year. In the northern states, snowplows are necessary at times. Nothing is more disturbing than trying to make delivery on contracted calves, only to find that your truck is hopelessly mired in a mudhole or stuck in a snowbank, with the animals shrinking all the while.

Normally, a location along an all-weather road has better access to electric and telephone lines, the school bus, the mail, religious and recreational facilities, and other services. Also, in irrigated areas, the irrigation turnout is usually near an all-weather road. Always make sure who is responsible for the maintenance of access roads, and how well the maintenance is done. The construction and maintenance of a road can be a considerable expense. When along dirt and gravel roads, the headquarters should be far enough away, taking into consideration the direction of the prevailing winds, to keep the dust from becoming a nuisance.

If possible, the headquarters should be near well-maintained telephone and electric lines. Farming or ranching is a business, and it is difficult to conduct any kind of business without access to a telephone. Likewise, electricity is essential for the operation of most modern facilities and automated equipment. Hence, power should be available at each area of main operation and not just at the headquarters.

The farm or ranch should have convenient access to an established mail route, a school bus, and delivery services (milk, laundry, bread, etc.). The availability of various social institutions is a good criterion of the community. Easy access to a good school, to adequate hospital facilities, to the church of choice, to recreational facilities of interest, and to farm, home, and youth organizations should not be overlooked when selecting a farm or ranch.

Cattle must be bought and sold; hence, distance from and kind(s) of market are important. Preferably, there should be a choice of markets for the kind and quality of animals that it is proposed to produce. Remember that raising cattle is a business, and that buying and selling cattle is a large part of the total operation; hence, it must be well done in order to be successful. Remember, too, that the greater the distance to market, the higher the trucking cost and shrinkage. Where supplemental feeds must be purchased, nearness to where they are produced cuts down on trucking expense.

WATER; WATER RIGHTS

Only seeing people without water is more disturbing than seeing cattle dying of thirst. Both people and cattle can survive longer without food than without water. Rainfall, wells, rivers, snow, streams, creeks, springs, lakes, ponds, or any other source of water should be considered. Because of the importance of water, the official weather rainfall records, by months and over a period of years, should be obtained.

In most states, domestic users have the first right to water. Agricultural uses usually rank second. In some instances, where industry is highly centralized and promotes the welfare of the public, manufacturing use is given priority to water over agriculture.

Plenty of good drinking water for human use is a high-priority item. Generally, this involves well water. Wells should be checked to make sure that they meet county health department standards, and that there is adequate water supply, for drilling is costly. The only accurate test of a well is to have it pumped dry and see how fast it fills up, or to see how much water can be pumped out of it and how fast, without substantially changing the water level. Of course, there is variation from season to season, and from year to year. In most western states, the general underlying water table is going down every year; hence, it is impossible to predict what it might be 10 years from date of purchase. Artesian wells are the most risky of all; nobody can predict when they're going to stop flowing.

From the standpoint of the water supply for cattle, consideration should be given to the distribution of rainfall, to snowfall in the northern areas and to river frontage wherever streams are found. Where running water is not available, artificial lakes, ponds, developed springs, and wells may be relied upon, but all these cost money. Hence, they should be considered at the time of purchase.

Where water is to be used for irrigation, both availability and cost must be considered. A cattle producer with free riparian water rights has considerable advantage over a neighbor who must pay an average of $20 to $40 per acre per season for the same amount of water.

Water rights have been of prime importance to the development of civilization. Nearly every society had its own system of regulating water. According to the Bedouin *code of the desert*, a traveler might drink of a well, but "should the well bear the *wasm* (camel brand) of a local tribe, and should the traveler, without permission, water his flocks and camels not bearing this brand, then should he be slain, and his body left to be devoured by the birds of the air and the beasts of the field."

Water rights have always been essential in arid countries where there is limited water supply. In western and southwestern United States, where the use of water is essential to the productivity of the land and to all living things on the land, the use of waterways is usually written into the deed of the farm or ranch. This water right becomes part and parcel of the land, meaning that it cannot be separated from the land. This is because much of the value of the property depends on accessibility to water. The rights are vested, and thus are considered private property.

Basically, there are two types of recognized doctrines, or water rights—the riparian right, based on English law, and the appropriative right.

■ **Riparian right**—A riparian right is the right of an owner who owns land adjacent to a body of water. Under riparian law, the property owner who has land lying next to a stream, or having a stream running through it, is entitled to water that is required for domestic consumption and for livestock. The English Common Law, from which the riparian doctrine stems, further states that the owner is entitled to have this stream flow undiminished in quantity or quality.

The Americanized version of the riparian right has a reasonable use clause which allows the riparian owner to make beneficial use of water so long as the quantity and quality of stream flow are not materially reduced. In other words, a downstream riparian landowner enjoys the same rights as an upstream owner. If, through unreasonable use, an upstream owner infringes upon the rights of a downstream owner and deprives the downstream owner of water, the latter could sue and possibly collect damages and halt the excessive use.

Groundwater may be a different story. Straight riparian rights apply to well-defined underground streams. But percolating ground water (water moving downward through the soil) is defined as real property in some states. As such, it is owned by the overlying landowner. Use of such water, even when it deprives a neighbor, cannot be contested.

There are many state variations and interpretations of riparian water rights. Beyond the right to use water for domestic consumption, riparian rules are usually vague. Ordinarily, this does not create a problem in the normally high rainfall areas of eastern United States. Yet, the irrigator does not really have adequate protection. Because of the irrigation boon, most riparian states have a special legislative committee studying water rights. They have established, or they will establish, priority among users.

In new irrigation areas, a land owner should consult an attorney on how to protect rights and investment. Records of the date irrigation started, the amount of water used, the acreage irrigated, and the return from the crops grown are information that can be valuable later if beneficial use must be proved.

When a person diverts water that other landowners have legal rights to, without being stopped, that person gains the rights to continued use. This is known as prescriptive rights. To gain prescriptive rights, the diversion must have been for a period of time designated by state statute, and the water must have been used openly.

■ **Appropriative rights**—An appropriative right is the right for a certain amount of water at a given place. It is not necessary for the water source to lie next to the land where the water is to be used. Rather, it is transported by such means as an irrigation ditch.

The appropriative rights for regulating water use are found mainly in the arid western states, although certain other states, including Minnesota and Mississippi, have adopted this type of law. Where states recognize both riparian and appropriative rights, the appropriative rights are dominant. Under the appropriative right, the individual may acquire the right to use water for beneficial purposes on a given tract of land by fulfilling certain requirements of written law. The appropriative right fully recognizes the public ownership of water. A landowner must apply to the controlling state agency to obtain the right to use water; and the right can be lost through nonuse after a stated period of time. In the appropriation states, rules vary on groundwater.

In appropriative rights, the right to use water depends upon prior claims made against the water sources. Anyone who filed to use water before you is entitled to their water needs before yours can be filled.

A vested right protects the rights of persons putting water to beneficial use before the appropriative law was passed. Usually, vested rights also apply to water applied beneficially within three years after passage of the law. Although a vested right has three years' priority, it can be lost through nonuse.

WHERE TO GO FOR WATER RIGHT HELP

It is very important that both prospective and present landowners obtain authoritative information relative to water rights.

In some riparian states, there are agencies that are studying water legislation and are in charge of regulating present water laws.

In states that use the appropriative doctrine, there is an authoritative agency, usually a chief engineer, to whom applications must be placed to obtain a right to use water beneficially. This official can give information concerning what must be done to abide by the statutes of the state.

You should request a copy of the water laws from either the state agency, your state senator or representative, or your state agricultural college. Where water rights may be an individual legal question, a competent attorney should be consulted.

MINERAL RIGHTS

Mineral rights are usually involved in the buying and selling of land, especially in oil and gas producing areas. Buyers of land should always have the title checked to see if the mineral rights have been severed. Generally, they are broken down into two broad classifications—surface and subsurface.

In the United States, the following two major theories prevail concerning the actual ownership of any subsurface wealth:

1. The ownership-in-place theory generally recognizes that any mineral deposit is actually a part of the land and is owned by the individual holding title to the land.
2. The nonownership theory in which the landowner does not have outright title to underlying mineral deposits but has the right to explore and retain any deposit developed.

Regardless of which theory is recognized in a particular state, the landowner has the following mineral rights:

1. Rights may be transferred to others.
2. Mineral deposits are part of the land.
3. Landowner has right to withdraw minerals, and, in the case of gas and oil, be free of liability for drainage (the pumping of oil or gas from under adjoining, nonowned property).

SEPARATING MINERAL RIGHTS

Surface and subsurface rights can be separated. In *ownership-in-place* states, you dispose of the minerals, while in *nonownership-in-place states*, it is simply the right to explore and retain any minerals recovered.

It should be noted that the term *minerals* refers to gas and oil, unless specifically stated otherwise. Moreover, any conveyance of a named mineral does not include other minerals unless so stated.

Surface and subsurface rights may be severed in any one of the following six basic ways:

1. By deed conveying all or part interest in the minerals.
2. By deed conveying land but retaining mineral rights.
3. By land contract excepting the minerals.
4. By mineral lease in an ownership-in-place jurisdiction. (Conveys present undivided ⅞ interest to the lessee on any mineral developed, except in Kansas.)
5. Court judgments setting aside mineral rights or part interest in a lawsuit.
6. Formation of a mining partnership.

Separating mineral interests can create problems, especially if interest is divided among many parties. For example, if the property has been through 3 or 4 different hands, possibly ⅛ of the minerals might be left for the new purchaser. Not only is such a small fraction of the mineral interest unattractive to the buyer, but widespread breakdown of the interest runs up the cost of bringing land abstracts up to date and often discourages leasing and well development. Separation of mineral interest reduces loan values and usually increases difficulty in obtaining credit.

As a general rule, it is not considered good business for a surface rights owner to dispose of over 50% of the subsurface rights.

OIL AND GAS LEASES

Oil and gas leases have become fairly standard. Nevertheless, one should check any proposed lease with an attorney before signing.

A general division on any developed oil and gas is ⅛ for the lessor (the mineral owner) and ⅞ for the lessee (the persons taking a lease on the land). Most leases are perpetual as long as certain qualifications are met. Leases are generally subject to termination by the lessee anytime within the base period by failure to begin a well or pay the delay rental by a stated date.

In some areas, mineral rights are more valuable than the surface.

TIMBER

Trees are pretty, but they may or may not have monetary value on a farm or ranch. Evaluation of timber is a job for an expert. The value of timber delivered to market may have little relationship to its quality as it stands. In some cases, the cost of cutting and transporting trees is exorbitant.

EASEMENTS; PROPERTY LINES

An easement is the right to go on and use the land of another in a particular manner. Two common types of easements are: (1) the grant of one landowner to another of the right to build or use a roadway across the land to provide access to another tract; and (2) where a power company purchases an easement to string an electric line across land. Before purchasing a farm or ranch, all easements should be checked because an easement (1) limits the landowner's use of the property, and (2) is valid against the purchaser. The senior author recalls one near-sale of a ranch that was being bought primarily because of a beautiful sight on which to build the new owner's dream house. Just as the deal was about to be closed, it was discovered that the county had a permanent easement for a 25-yd strip for a road right through the intended house location.

Three types of legal descriptions of farms and ranches are commonly used: (1) the rectangular survey, based on meridians and parallels; (2) metes and bounds (metes are measures of length—feet, inches, or perches—a perch equals 16½ ft; bounds are artificial boundaries such as roads, streams, adjoining farms), or (3) monuments (iron pin, blazed tree, lake, stream). Disagreements over property lines have led to feuds and lawsuits. Accordingly, before purchasing land, the property lines should be determined. Usually, it is wise to engage the services of a professional surveyor. The senior author knows of one case where failure to do this resulted in the cattle producer building a new ranch home only to discover, some years later, that the house was on an adjacent property—not his own.

NEARNESS TO FACTORIES OR CITY

Nearness to factories or city has both advantages and disadvantages from the standpoint of farm or ranch location. Some of these will be discussed.

1. **Off-farm employment.** Unless a cattle farm or ranch is large enough to support a 300 brood-cow herd, it is well to keep in mind that the operator will likely find it necessary to have part-time off-farm employment in order to make a go of it. This fact should be kept in mind at the time of purchase, even though the one making the purchase has sufficient income from other sources not to require part-time employment. This is so for resale reasons.

In a study conducted by the University of Arkansas, it was found that off-farm work was an important source of income for all of the upland areas of the state, with over 50% of the cattle producers working off-farm.[2] This means that a farm or ranch which will not support 300 brood cows should be located within a 50-mi commuting radius of a city or factory payroll.

2. **Big city attractions.** The new generation of cattle producers has gone modern. Although they may be perfectly content to live on an isolated farm or ranch and near a small town, most of them still like to be fairly near a large city. The big city is important to them from the standpoints of shopping, concerts, junior colleges, airports, etc.

3. **Air and noise pollution.** An increasingly important consideration in selecting a cattle farm or ranch is air and noise pollution. Thus, if the cattle operation being considered is located near a factory, an effort should be made to determine if the factory is likely to cause either water, air, or noise pollution. Although a national effort is being made to reduce such pollution, there continue to be claims, imagined or real, that the contamination from certain factories has polluted the air or water to the extent that the growth of grass and the well-being of cattle have been affected adversely. Also, there are reports where streams have been polluted to the extent of affecting certain species of fish, thereby reducing the value of the ranch from a recreational standpoint. Airports and highways are also sources of noise. Although animals adjust to usual noises, they may be excited and become nervous by unusual noises.

---

[2] *Production, Financing, and Off-Farm Employment Aspects of Beef Farming in Arkansas*, Ag. Exp. Sta. Bull 785. University of Arkansas, Fayetteville, p. 24.

EXPANSION POSSIBILITIES

Sooner or later, most cattle operators try to increase the size of their holdings. When that time comes, it is a tremendous advantage if they are not *hemmed in*—if there are expansion possibilities. Expansion possibilities usually exist where a farm or ranch is surrounded by other farms or ranches. Then, when a neighbor retires, first refusal may be obtained.

Where a farm or ranch is adjacent to an airport, factory, golf course, and the like, expansion possibilities are always limited.

In an Arkansas study, the typical beef cattle producer indicated that more forage land for the herd was desirable, but that lack of land was the most limiting factor.[3]

THINGS TO DO WHEN BUYING A FARM

After the prospective purchaser has found the farm or ranch desired and has determined that the price is right, the following things should be done:

1. **Have it appraised.** Buying a farm or ranch is a big financial transaction. Thus, it is good business to have an accredited rural appraiser make a detailed appraisal of the property. The appraiser's fee will vary depending upon the size of the ranch to be inspected and the time required to document income and expense items, search out and view comparable sales, and prepare a confidential written appraisal report.

There are three basic approaches to appraisal of land, or estimating its value. These are: (a) market value, or what similar land has sold for recently; (b) productive value, or net income the land will produce; and (c) present value of useful improvements.

The appraisal should show the fair market value of the property. This is frequently defined as *the price at which a willing seller would sell and a willing buyer would buy, neither being under abnormal pressure.* This definition assumes that both buyer and seller are fully informed as to the property and as to the state of the market for that type of property, and that the property has been exposed in the open market for a reasonable time.

The appraisal should include maps of the farm, showing the physical features and the various soils. There should be a summary sheet listing all the buildings and their size and description. The appraiser should also allocate a value to buildings, fences, tiling, wells, pipe lines, and other depreciable items for income tax purposes. This will give the buyer a reliable value from which to set up a depreciation schedule.

[3]*Ibid.*, p. 20.

The appraiser should evaluate improvements in terms of the owner's intended use. For example, a $40,000 turkey shed is worth only scrap lumber to an owner who is going to run cows. Worse yet, it may have a negative value because of taxes.

Preferably, the appraiser should be familiar with cow-calf operations. Such an appraiser can point out the highest and best use of the property for the intended purpose. Also, the appraiser may be able to indicate factors which would alter considerably the plans of the potential buyer.

2. **Check government programs.** Government programs change from time to time, but usually one or more government agencies are involved in most farms and ranches. Thus, it is well to check into the situation.

The county Agricultural Stabilization and Conservation Service (ASCS) offices advise on and administer commodity programs, including allotments and marketing quotas for the basic commodities. These offices can also supply information about the soil, water, timber, and wildlife conservation practices that the Agricultural Conservation Program helps carry out on individual farms. ASCS offices also are charged with the local administration of price support commodity loans made available through the Commodity Credit Corporation, certain emergency programs in designated areas affected by drought or floods, the feed grain program, and other farm programs.

The Soil Conservation Service (SCS) has offices in nearly every county. They provide technical assistance and information on soil and water conservation, land use alternatives, soil surveys, and resource use.

3. **Check courthouse records.** A courthouse check is standard procedure for most appraisers. However, the prospective owner should make sure that it is not overlooked.

The courthouse records should show what the property tax has been running on the farm or ranch in question, and if the property is subject to special levies for drainage or irrigation districts. The plat books should be checked to make certain of the boundaries of the property and how many acres are actually involved. What mortgages are on record against the farm? Check the recorder's office also for any special agreements in regard to property-line fences. The latter is especially important if there is a *water gap* where fences must frequently be rebuilt. Also, important water and mineral rights should be checked.

4. **Check mortgages.** The prospective purchaser should check the mortgage situation. Is there a mortgage on the property? If so, how much, at what interest rate; and can it be assumed without penalty? These questions are particularly important during times of scarce money and high interest rates.

THINGS TO AVOID WHEN BUYING A FARM

In the purchase of a farm or ranch, there are certain pitfalls which should be avoided; among them the following:

1. **Avoid legal problems.** Regardless of whether a farm is purchased direct from the seller on a first-hand negotiated basis or through a real estate broker, the buyer should have an attorney check the details, thereby lessening the hazard of legal problems.

2. **Beware of the glamour states.** Much land in California, Florida, and Arizona is priced so high that it is difficult to show a profit from the operation of a cattle farm or ranch. In these states, either a higher and better use must be considered, or the land must be purchased on the basis of speculation—its potential for recreational development, housing, etc. Beautiful mountains, trees, streams, and sunsets all make for a heap of living and enjoyment, but they don't feed a cow. Thus, they should be secondary in the selection of a farm or ranch.

3. **Avoid city suburbs and high taxes.** Where rapid-growth cities are involved, it is generally wise to stay at least 50 mi away if one wishes to develop a cattle farm or ranch. Of course, there are many small towns or cities that are not subject to rapid growth where it is possible to be closer in without the hazard of subdivisions or high taxes.

4. **Avoid overelaborate improvements.** Improvements are always expensive to maintain and they are subject to taxes. Hence, they should have utility value. No matter how attractive they may be, unless improvements contribute to the income of the farm or ranch, they have a negative value.

PURCHASE CONTRACT

After a prospective buyer has found the particular property desired and has agreed upon a price with the owner, an attorney should draft an agreement covering the terms of the purchase. Then the buyer should sign it and submit it as an offer to buy. It does not become a binding contract until the seller signs it, also. After both parties sign the contract, there is little bargaining power left; hence, the buyer should get all stipulations covered in the original contract.

The purchase contract should be relatively simple, amounting to a mere memorandum signed by buyer and seller, but it should clearly specify the following:

1. Amount and method of paying purchase price.
2. The amount of deposit or down payment, and the method of handling it. Is it to be applied to the total purchase price; when is it to be forfeited or returned; and is payment made to a responsible person?
3. Method of financing the purchase. Does purchaser assume and agree to pay existing mortgage? Will purchaser obtain a new mortgage loan?
4. Is seller to furnish an abstract of title brought up to date or a good and clear title that can be insured?
5. Date that possession can be taken.
6. A list of items that go with the property.
7. Who pays accrued and current taxes.
8. The legal rights of any tenant on the property.

It is important to remember that the buyer assumes risk of loss as soon as the contract is signed, even though the deed has not been delivered. If a barn burns and the seller has no insurance, the buyer could be forced to pay the full purchase price agreed upon, even though the barn has burned. So it is important that the buyer make certain that there is insurance on the buildings during the interim period.

It is customary to prorate annual taxes, with the buyer and seller assuming responsibility for the number of months that each has actual possession of the property.

The purchase contract should require the seller to deliver to the buyer an abstract of title for the property, certified to the date of sale. If the seller cannot produce a clear title or satisfactory title insurance by a certain date, the contract should call for a refund of the buyer's down payment.

Of course, the purchase contract should include the price being paid for the property and the date on which the seller guarantees possession. There should be a complete legal description of the holdings. If it is understood that certain portable buildings go with the farm, they should be itemized in the purchase contract.

The contract should detail how payment is to be made and the time of settlement. If the seller is to carry a mortgage on the farm or ranch for part of the purchase price, the buyer will need to work out the usual arrangements as to interest rates, prepayment privileges, etc.

If the buyer wishes to go on the place to do certain work in advance of taking actual possession—like repairing buildings, or reseeding a pasture—this should be spelled out in the contract.

Nothing should be left to oral agreements, because most such agreements are unenforceable. Also, the contract should be binding upon the heirs and assigns of the seller.

In summary, when completing the purchase of a farm or ranch, give attention to the following details:

1. Do you have satisfactory evidence that the seller has complete title to the property and can convey it to the purchaser by deed?
2. Examine the seller's deed. Does the wife re-

lease her dower; does seller warrant free and clear of all encumbrances; is the description of boundaries and acreage correct; are the easements or right-of-way for or against the farm; are mineral rights or water rights reserved; and has seller attached U. S. transfer stamps?

3. Examine mortgage and note before signing.

4. Immediately record deed with the Register of Deeds office in the county where the property is located.

5. Insure all uninsured buildings.

6. Make sure that expenses incurred in the last 60 days have been paid for materials or work done on buildings, for land clearing or leveling, and for wells or pipe lines.

WAYS OF ACQUIRING A FARM OR RANCH

Getting a suitable farm is a big problem, particularly for the beginner. One must compete for available farms with established farmers as well as with other beginners. Many established farmers need more land to enlarge their operations. Others move during the year, getting a better or more suitable farm. Some simply move to a new locality for personal reasons.

Farms or ranches may be acquired through gift or inheritance, by marriage, by renting or leasing, or by purchasing.

1. **Gift or inheritance.** Many farms and ranches are inherited and stay in the family for several generations. Certainly, inheriting a farm or ranch is a real advantage to anyone desiring to stay in the business and having the know-how and ability to operate the enterprise. Farmland and ranchland, and livestock (as well as corporate stocks, U.S. savings bonds, mutual funds, money, whole life and annuity life insurance, and commercial real estate) can be transferred to relatives and friends, with certain tax savings, provided certain well-established rules are observed. As evidence of the legitimate savings that may be affected through gift or inheritance, the following example is cited, in which the senior author served as consultant: Three owners, all in their 70s, with a cattle ranch valued at approximately $4 million, had done no estate planning. By having the property appraised and following a carefully planned gift program, $0.75 million in inheritance taxes were saved—all legally.

The basic rule in giving land or animals to another member of the family is that the donor (person giving the property) must give up control. Of course, if it actually or constructively passes through the hands of the donor, it is taxed to them.

A transfer of property either in trust, to a custodian, or outright, may incur federal gift tax. However, you can give away property valued at $10,000 each year ($20,000 if married and the spouse consents) to any number of donees without incurring a gift tax obligation.

You can give relatives or friends income-producing property and the income will be taxed to the person receiving the gift (donee). But the property must actually be transferred. Any strings attached to the ownership whereby the donor can have control over the property or get it back at some future time will not meet the requirements of the law. The basic elements are:

a. There must be intention to make a gift.

b. Transfer of legal title and control.

c. The donee accepts the gift.

d. No consideration (money or property) to be exchanged for the property.

If a person is expecting to inherit a farm or ranch, the recipient should become fully cognizant of the inheritance tax laws. Also, it is most important that the recipient have clear title to the land. If the title of ownership of a farm or ranch is left unsettled through 2 or 3 generations, the value of the land may almost be expended in settling the estate.

Where gift or inheritance money or property is involved, always seek the advice of a tax accountant and/or tax attorney.

(Also, see Chapter 13, Business Aspects of Beef Production, sections entitled "Income Taxes" and "Estate Planning.")

2. **Marriage.** Although young men wanting to enter farming or ranching are frequently admonished, facetiously, that they should "marry for love, but love a woman with plenty of money," there is more than a little bit of truth in the advice.

In some countries, the social orders call for the parents to arrange the marriages of sons and daughters, primarily to keep them within the same class strata, thereby not dividing property with those who "have not." However, this method is fast giving way to the new social order, which, like that in the United States, results in marriages between individuals of vastly different amounts of wealth.

Attitudes are important in any marriage, but they are doubly important in those marriages in which the wife contributes most of the wealth. The ability of the young man marrying the wealth, along with the attitude of the wife and the in-laws, can have a big part in contributing to the success or failure of both the marriage and the cattle farming or ranching.

3. **Rent or lease.** The main advantage in renting over buying is that less capital is required and less financial risk is involved. The main disadvantages are insecurity of tenure and that the farming enterprise may be limited in size or kind because the landowner is reluctant to make needed additional investments in

buildings and facilities. These disadvantages can be minimized and sometimes eliminated, by a suitable lease—an agreement between landlord and tenant under which the farm or ranch is rented and operated. Such a lease should always be in writing.

The most common types of cattle farm and ranch leases are:

a. **Cash lease.** This is a good type of lease for (1) the small farm or where the landlord lives at a distance, and (2) a tenant who has adequate livestock, equipment, and working capital. It encourages livestock farming because all of the crop can easily be fed on the farm. Also, it is simple, with little chance for controversy.

There are two types of cash leases: (1) that type in which a fixed rent per acre is agreed upon when the lease is drawn; and (2) that type in which the rent is adjusted to prices of farm products which prevail during the lease year. Under the second plan, the landlord bears part of the risk of price changes; however, it is difficult to keep cash rent in line with farm product prices. If product prices are used as a basis for rent changes, the products, markets, and dates should be specified.

Landlords may prefer a cash lease because (1) the amount paid is definite, and (2) it requires less supervision by the owner. On the other hand, it may not always be desirable from the standpoint of the landlord because (1) it generally makes for lower income, (2) it gives the landlord less control of the farm, and (3) it is difficult to collect rent if crops fail.

Tenants may prefer a cash lease because (1) it will make for more profit if they are successful managers, (2) it makes for more independence in the operation, and (3) it makes for more profit in the good years.

b. **Livestock share lease.** Cow-calf share leases vary considerably. But most of them provide for 50-50 ownership of the herd; 50-50 sharing of the costs of production—especially feed and veterinary expenses; and 50-50 division of the income from the sale of animals. Buildings are generally a cost borne by the landowner. Labor is the responsibility of the operator.

A livestock share lease fits the tenant who wants to raise livestock, but cannot finance a program. It is especially suited where tenant and landlord get along well and where the landlord can make a good contribution in management.

In order for this type of lease to work best, the landlord should live close to the farm, and either give it personal attention or arrange for adequate management help such as can be provided through a professional farm management service.

Landlords may prefer a livestock share lease because (1) it encourages more livestock and more manure, (2) low-quality crops can be utilized more easily, (3) they retain an active interest in management, and (4) it generally makes for more profit.

Tenants may prefer a livestock share lease because (1) the risk is less since rent is based on net income on the farm, (2) it requires less tenant capital, (3) the landlords are more willing to make improvements, and (4) they can gain experience from the guidance of a successful owner.

A careful determination of lease provisions and putting them in writing will result in a lease that is more equitable to both tenant and landowner, and will avert later misunderstandings and friction between the two parties. Standard lease forms are available, so the detailed provisions of a lease need not be spelled out in this book.

Renting or leasing might be a desirable way to start in the cattle business even if funds are available for purchase. This is particularly true where there is an option-to-buy clause. This gives renters an opportunity to study the ranch more carefully and gain additional management experience without committing their entire assets.

4. **Purchase.** Purchase of a cattle farm or ranch has the advantages of security of tenure and freedom to make management decisions. Earnings from the operator's equity capital may be added to labor and management earnings for living expenses, reinvestment in the business, or other uses. Also, the value of the land may rise over a period of time. On the other hand, ownership may involve substantial indebtedness. Also, risks of financial loss are greater than in renting.

Of course, ownership brings with it financial responsibility that is both greater and longer lasting than the financial responsibility which renting entails. Few persons buy more than one farm in a lifetime; moves are time-consuming and expensive.

Some part-time farmers, working at nonfarm jobs, use their off-farm income to move gradually into full-time farming. They use their initial savings to make a down payment on a small farm and to buy enough cattle and equipment to permit limited farming or ranching operations for the first few years. The off-farm income makes them better credit risks for lenders than if they were wholly dependent on farm earnings. They can continue to borrow to build up their farm business to a point where it will support their family to pay off previous loans. Such a gradual shift into full-time farming can usually be made with less sacrifice in family living standards and better chances of eventual success than an abrupt change to full-time farming.

Generally speaking, purchase of land is either by (a) land purchase contract, or (b) mortgage contract.

Purchase of land by use of land purchase contract has become much more important in recent years. These contracts allow the use of lower buyer down payments (usually from nothing to 29%) with the balance paid over a long period of years in annual payments. For the buyer, this offers a way in which to buy land without having to make a big down payment. For the seller, it has certain capital gain tax advantages and it usually attracts more prospective buyers and makes for a higher sale price. In order to qualify for special treatment on capital gains for federal income tax purposes the seller must not receive more than 30% of the purchase price in the year of sale.

Mortgage contracts differ from land purchase contracts in the following ways: (1) they are of longer duration—usually 20 to 30 years, or up to 40 years, whereas land purchase contracts are commonly for 10 years or less; (2) the law provides for specified grace periods after default in payments before the seller can foreclose the mortgage; and (3) larger down payments are normally required—frequently 40 or 50% of the purchase price.

WHAT'S A FARM OR RANCH WORTH?

The above question is best answered by still another question: How much will it make? The most logical answer to the latter question is that the farm or ranch should be expected to return to the prospective buyer as much on the investment as could be earned were the money invested in the best alternative enterprise of comparable risk. Thus, if an alternative enterprise of comparable risk will return 7% on investment, the farm or ranch should do likewise.

The above income-productivity approach does not take into consideration a possible tax shelter or increase in land values. However, neither of these should be counted upon. Although the prospective land buyer should have no objection to Uncle Sam being lenient on taxes, or to striking oil, or to having a fashionable summer resort go up on the adjacent property, none of these possibilities can be counted upon. The tax structure may change; and land values may not increase. Besides, the cattle producer living upon the land and depending upon earnings from the operation to pay interest on borrowed money and buy groceries for the family cannot rely on *paper profits* (unrealized profits).

The income-productivity approach, which calls for calculating receipts and expenses, should be based on projected longtime productivity levels, prices, and costs, without either undue optimism or pessimism.

In plain simple terms, the income-productivity approach is based on the capacity of the farm or ranch to make money—the more money it will make, the more it is worth. Step by step, it is determined as follows:

1. Record expected receipts and expenses from the operation of the cattle farm or ranch.
2. Determine the net returns to the land by subtracting all expenses from gross receipts, including a return to the operator for labor and management.
3. Divide the dollar returns to land by the rate of interest that you could get were your money invested in the best alternative enterprise of comparable risk to arrive at the *income-productivity value* of the farm or ranch. This value is the maximum that a buyer can pay for a farm or ranch based on its *productivity value*. If the price is below this figure, it's a good buy. If it's above, watch out.

The above procedure is given in Table 18-1, based on an actual ranch for which the senior author served as consultant.

Thus, based on income productivity, a buyer who wishes to allow $40,000 per year for labor and management, and who wishes to realize 10% on their

TABLE 18-1
INCOME PRODUCTIVITY VALUE OF BAR-NONE RANCH, A 500-COW OPERATION ON A 5,000-ACRE *(2,024-HA)* RANCH[1]

	(Amount)
Receipts:	
Cattle sales	$202,500
Expenses:	
Cash expenses:	
Feed purchases	24,230
Hired labor	10,000
Machinery	15,570
Property tax	8,050
Other	3,800
Noncash expenses:	
Depreciation	22,850
Interest on operating expenses	8,000
Operator labor and management	40,000
Total expenses	$132,500
Net return to land (ranch):	
Income productivity value at 10% rate of interest ($70,000 ÷ 0.10)	$700,000

[1]This is an abbreviated form. A competent appraiser will detail this. For example, under "cattle sales," there should be a breakdown as to number, weight, and price of steer calves; number, weight, and price of heifer calves; number, weight, and price of yearlings, by sex; number, weight, and price of cows; and number, weight, and price of bulls. Also, hay sales, crop sales, and pasture rents would be included—if income from these sources is expected.

Similarly, under expenses there should be a more detailed breakdown of costs.

investment, could consider the Bar-None Ranch—a 500-cow, 5,000-acre ranch—a good buy at up to $700,000, or $140 per acre. On the basis of a 500-cow carrying capacity, that's $1,400 per cow for the land and improvements.

Historically, the *income productivity* value of cattle farms and ranches has been below current market prices. This means that either land is too high or cattle are too cheap. Moreover, prospective purchasers must recognize that they will likely have to pay the going market price for a farm or ranch. That is, they will have to pay a price close to that for which comparable land is selling. Except when buying from relatives, there are precious few really good buys. However, there are two exceptions: (1) when an existing cattle producer enlarges his/her present holdings by acquiring nearby land, thereby lowering the cost per cow expenses—primarily through more efficient use of labor, management, and equipment; and (2) when unimproved land can be developed to where its income productivity exceeds its cost plus development.

Fig. 18-9 shows farmland value, 1960-1994. based on actual sales.

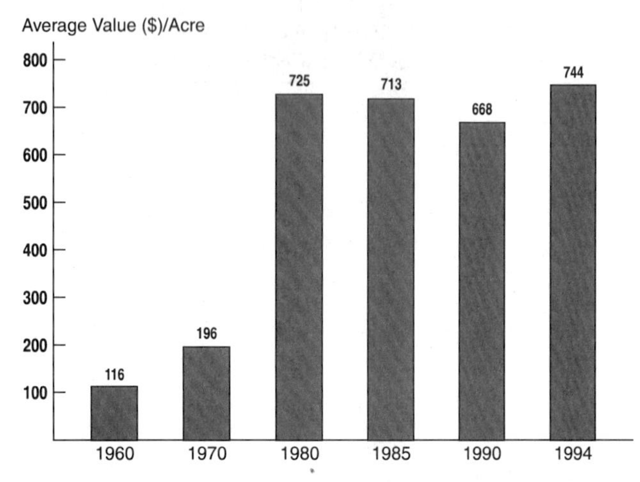

Fig 18-9. U.S. farmland value in dollars per acre, 1960-1993. (Sources: USDA & *Statistical Abstract of the United States 1994*, p. 668, Table 1086 and *Statistical Abstract of the United States 1995*, p. 675, Table 1107)

FINANCING THE FARM OR RANCH

Land cost per cow ranges from $2,000 to $4,000. As the authors have previously pointed out, a minimum of a 300-cow outfit is necessary for a profitable unit where the operator expects all of their income from the farm or ranch. This means that for a 300-cow unit, a minimum land cost of $600,000 is necessary. Of course, many cattle operations have more than 300 cows. Hence, a big cattle spread necessitates both money and knowledge of financing.

CREDIT

Credit may be defined as *belief in the truth of a statement, or in the sincerity of a person.* In farming and ranching, or in any other business transaction, credit means confidence that people will take care of their future obligations. Credit is the life blood of the cattle business. Without it, few large cattle operations would be possible, for not many people are able to provide all of the capital that they need.

Most commercial lenders have guides and standards that set upper limits on the amount they will lend. Usually, to get credit on a mortgage for buying a farm, the borrower is expected to make a down payment of 40 to 50% of the purchase price. Lenders usually will make loans on livestock and on new machinery for up to 80% of the purchase price.

In 1994, U.S. farm investments in land, improvements, machinery, equipment, animals, feed, and supplies totaled $920 billion. That same year, farm debt totaled $143 billion; and the proprietors' equities totaled $776 billion.

TYPES OF CREDIT OR LOANS

(See Chapter 13, Business Aspects of Beef Production, section entitled, "Types of Credit or Loans.")

SOURCES OF LOANS

Hand in hand with getting the right kind of loan, it is important that the best available source of the loan be secured. Table 18-2 shows the main sources of the three main kinds of loans. Fig. 18-10 shows where farmers borrow, with a breakdown into real estate loans and nonreal estate loans.

TABLE 18-2
PRINCIPAL SOURCES OF THREE MAIN KINDS OF FARM LOANS

Credit Source	Kind of Loan		
	Long-Term	Intermediate-Term	Short-Term
Commercial banks	x	x	x
Dealers and merchants		x	x
Farm mortgage companies	x		
Farmers Home Administration	x	x	x
Farm Credit Administration	x	x	
Individual lenders	x	x	x
Insurance companies	x		

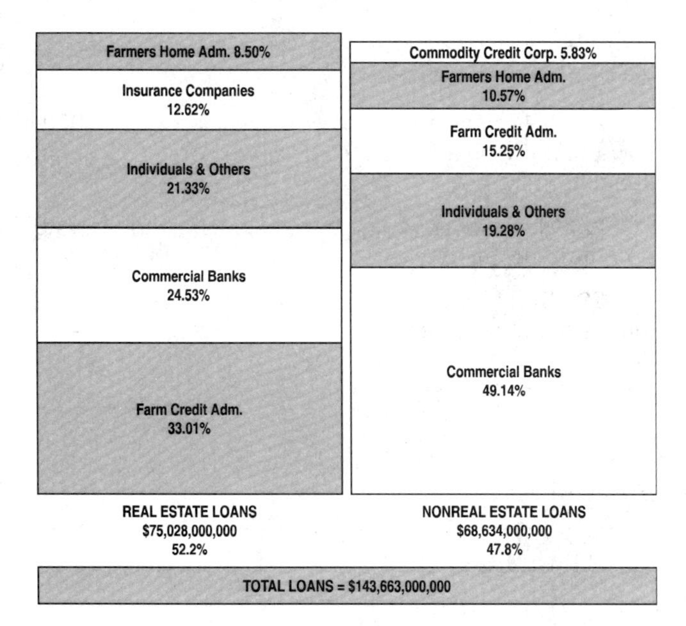

Fig. 18-10. Where farmers borrow (1992). (Source: Data provided by Farm Credit Administration, Washington, DC)

In seeking a loan, it usually pays to *shop around* in advance of actual need to see which source is best under the circumstances. Compare cost of credit, length of loan, loan fees, repayment privileges, and security required. Also, for long-term loans check the reputation of the lenders for staying with worthy borrowers in times of adversity.

In comparing costs of credit, look at the total dollar amounts, not just interest rates. Lenders figure charges in different ways. For example, if you buy feed or farm machinery on the time-purchase plan, you may find that you will pay more in interest than you would if you borrowed the money to pay for the purchase outright.

To obtain loans or information about loans from commercial banks, dealers and merchants, farm mortgage companies, individuals, and insurance companies, apply directly to these sources or to their local representatives. Local banks and farm real estate dealers often serve as loan correspondents for life insurance or farm mortgage companies; they can tell you about loan requirements, terms, conditions, and interest rates, and arrange for loans.

(Also see Chapter 13, Business Aspects of Beef Production, section entitled, "Sources of Capital.")

QUESTIONS FOR STUDY AND DISCUSSION

1. Why is the ultimate success of a cattle operation so dependent upon the selection of the proper farm or ranch?

2. Why is the ownership of land attractive to investors?

3. Explain how land provides a means of obtaining relatively large capital gains.

4. Explain how land provides an effective hedge against inflation.

5. Discuss the advantages and disadvantages of cattle farms and ranches with pasture only and with all supplemental feed purchased.

6. Discuss the advantages and disadvantages of cattle farms and ranches with pasture and home-grown supplemental feed.

7. Why are there so many part-time beef cattle ranches? From the standpoint of the beef cattle industry as a whole, is the preponderance of part-time farmers good or bad? Justify your answer.

8. Discuss the circumstances favoring combining cattle and (a) cash crops, (b) other crops, and (c) other livestock.

9. What factors are likely to favor an irrigated cattle farm or ranch over a dryland operation?

10. Why have ranches made up of deeded land increased in value more than those made up chiefly of government-leased land?

11. List and discuss the fundamental considerations when selecting (a) the geographical location for a farm or ranch, and (b) the specific farm or ranch.

12. Define *carrying capacity*. How important is a high carrying capacity?

13. Discuss the effect of herd size on cost per cow unit.

14. List the three primary factors which determine the cost of producing a calf, and discuss the importance of each of them.

15. Why is each of the following important when selecting the location of the farm or ranch: (a) area and climate; (b) soil and topography; (c) improvements; (d) wind direction, windbreaks, and natural shelters; and (e) service facilities, community, and markets?

16. How important are water and mineral rights to a land buyer?

17. List and discuss the important *things to do* when buying a farm.

18. When buying a farm, why should a purchaser avoid overelaborate improvements?

19. What are the advantages and disadvantages of acquiring a farm by (a) gift or inheritance, (b) marrying it, (c) rent or lease, and (d) purchase?

20. Show how the income-productivity approach may be used to determine what a farm or ranch is worth.

21. It is estimated that sufficient land for a 300-cow outfit will cost $600,000. How would you go about financing this?

22. How would you go about selecting and buying a cattle farm or ranch?

23. Can a young individual become a ranch owner today? If your answer is in the affirmative, explain how.

SELECTED REFERENCES

Title of Publication	Author(s)	Publisher
Cattle Raising in the United States	R. N. Van Arsdall M. D. Skold	Economic Research Service, U.S. Department of Agriculture, Washington, DC, 1973
Cowboy Arithmetic: "Cattle as an Investment," Fourth Edition	H. L. Oppenheimer	The Interstate Printers & Publishers, Inc., Danville, IL, 1985
Cowboy Economics: , "Rural Land as an Investment," Third Edition	H. L. Oppenheimer	The Interstate Printers & Publishers, Inc., Danville, IL, 1976
Cowboy Litigation: "Cattle and the Income Tax," Second Edition	H. L. Oppenheimer J. D. Keast	The Interstate Printers & Publishers, Inc., Danville, IL, 1972
Cowboy Securities	H. L. Oppenheimer S. K. Weber	The Interstate Printers & Publishers, Inc., Danville, IL, 1975
Do It Right the First Time	J. D. Keast, *et al.*	Doane Agricultural Service Inc., St. Louis, MO, 1973
Farm Management Economics	E. O. Heady H. R. Jensen	Prentice-Hall, Inc., Englewood Cliffs, NJ, 1954
Farm Records and Accounts	J. N. Efferson	John Wiley & Sons, Inc., New York, NY, 1949
Financing the Farm Business	I. W. Duggan R. U. Battles	John Wiley & Sons, Inc., New York, NY, 1950
Land Speculation: "An Evaluation and Analysis"	H. L. Oppenheimer	The Interstate Printers & Publishers, Inc., Danville, IL, 1972
Law on the Farm	H. W. Hannah	The Macmillan Company, New York, NY, 1950
Law and the Farmer	J. H. Beuscher	Springer Publishing Company, Inc. , New York NY, 1953

"Producing calves" is the name of the cow-calf game. Brood cows should be fed and managed so that they approach a 100% calf crop and wean off heavy calves. (Courtesy, University of Illinois)

FEEDING AND MANAGING BROOD COWS

Contents *Page*

Contents Page

Feed affects total profit and cow productivity. It accounts for 65 to 75% of the total cost of keeping cows, and it exerts a powerful influence on cow fertility and calf weaning weight—the two biggest success factors in the cattle business.

Fig. 19-1. Braunvieh herd on arid New Mexico range. (Courtesy, Spade Ranches, Lubbock, TX)

NUTRITIONAL REQUIREMENTS OF BROOD COWS (Also see Chapter 8 of this book)

Experiments and practical observations reveal that the period during which calf crop percentage is affected most by nutrition extends from 30 days before calving until 70 days after calving—until after rebreeding; a period of approximately 100 days. This, then, is the most critical period in the cow-calf business. It's when life begins—that period within which one calf is born and another is conceived. The needs for the cow during this most critical production period are approximately equal to her needs for the remainder of the year. This fact is pointed up in Fig. 19-2, showing the estimated energy requirements of a 1,000-lb beef cow during her 12-month reproductive cycle.

The average daily energy requirement is about 14.5 lb TDN (24 megacalories). However, as shown in Fig. 19-2, the requirements are above the average for nearly 6½ months of the year. This means that, for reasons of economy, the calving season should be timed so that much of the feed can be supplied by pasture and other economical sources of homegrown energy and protein.

A second important requisite of a sound beef cattle nutrition program is to feed animals according to their requirements. It is impossible to feed the herd properly where calving occurs the year around, or when dry pregnant cows, replacement heifers, and cows nursing calves are run together. This point becomes apparent in Figs. 19-3,[1] 19-4,[1] and 19-5,[1] which show certain nutritive requirements for the following classes of cattle: (1) dry mature cows, last 2 to 3 months of pregnancy, weighing 1,100 lb; (2) yearling heifers, last third of pregnancy, weighing 900 lb; and (3) cows nursing calves, superior milking ability, first 3 to 4 months after calving, weighing 1,100 lb.

Figs. 19-3, 19-4, and 19-5 show that the lactation requirements are the highest and most critical of all; in total feed consumed, in TDN and total protein of the ration, and in the major minerals—calcium and phosphorus. After a cow calves, her energy needs jump about 50%, her protein needs increase sharply, and

[1]*Note:* Before this book went to press, the new *Nutritive Requirements of Beef Cattle*, seventh edition, came off press. So, the new NRC requirements of beef cattle are presented in the Appendix, Section I, Table I-2, of this book; hence, the reader is referred thereto. Because allowances, rather than requirements, are used in ration formulations in order to provide margins of safety for animal and environmental differences, the principles illustrated in Figs. 19-3, 19-4, and 19-5 still apply.

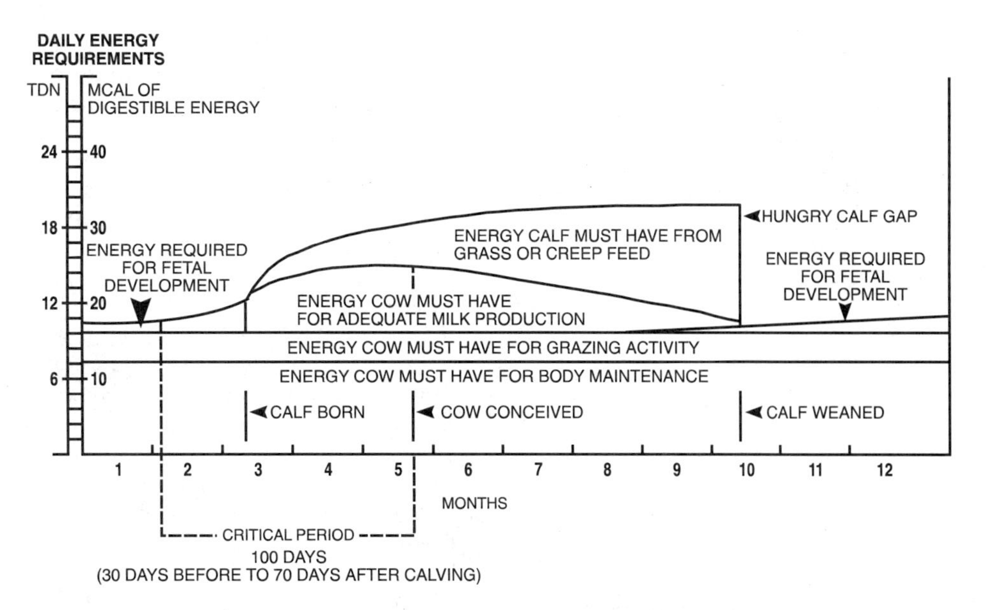

Fig 19-2. Estimated energy requirements of a mature 1,000-lb beef cow during her 12-month reproductive cycle; based on a 90-day calving season and 500-lb calf at 7 months of age. (Adapted by the authors from *Nutrient Requirements of the Cow and Calf*, Texas A&M University, B 1044, p. 7, Fig. 2)

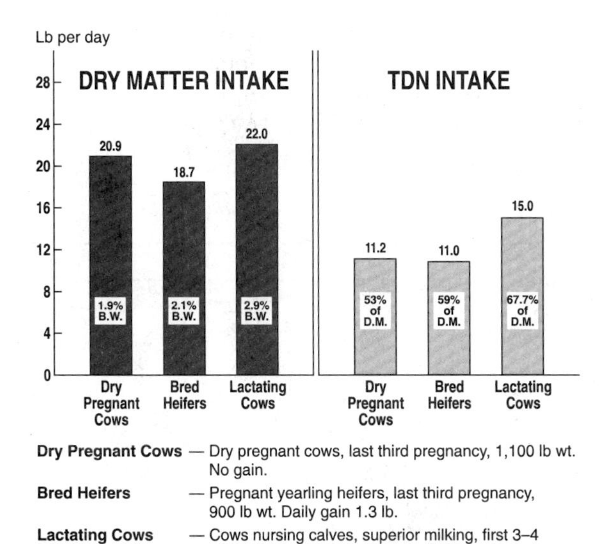

Fig. 19-3. Dry matter (in lb daily and in % of body weight) and TDN requirements (in lb daily and in % of the ration) of dry pregnant cows, bred heifers, and lactating cows. As shown, lactating cows require slightly more feed (22.0 vs 20.9 lb/day), along with a much higher energy ration (14.7% more), than dry pregnant cows. Replacement heifers, which must provide for the growth of their own bodies as well as for the growth of the fetus, consume more feed in relation to body weight (B.W.) than either dry or lactating cows. (Based on data from *Nutrient Requirements of Beef Cattle*, sixth revised edition, National Research Council-National Academy of Sciences, Washington, DC, 1984)

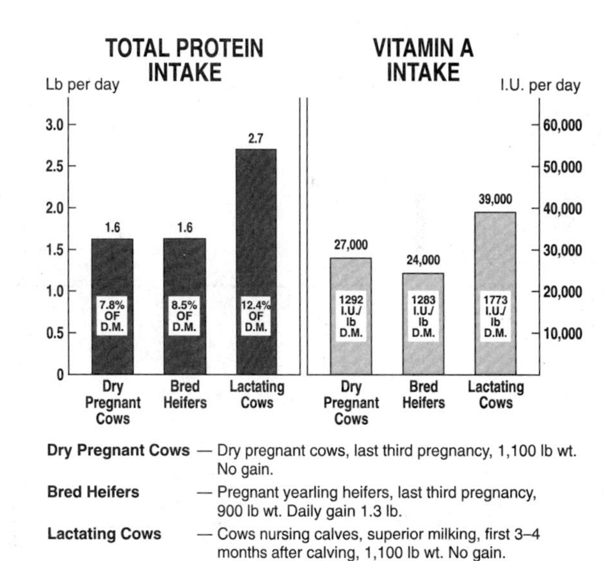

Fig. 19-4. Total protein and vitamin A requirements of dry pregnant cows, bred heifers, and lactating cows. Milk is high in protein; hence, it follows that for superior milk production the protein requirements of lactating cows are high, in terms of both crude protein intake and percent of the ration. Also, in order to meet the simultaneous protein requirements for body growth and fetal development, heifers require a higher protein ration than dry cows.

Lactating cows have the highest vitamin A requirement of the three groups, followed, in order, by dry pregnant cows and bred heifers. (Based on data from *Nutrient Requirements of Beef Cattle*, sixth revised edition, National Research Council-National Academy of Sciences, Washington, DC, 1984)

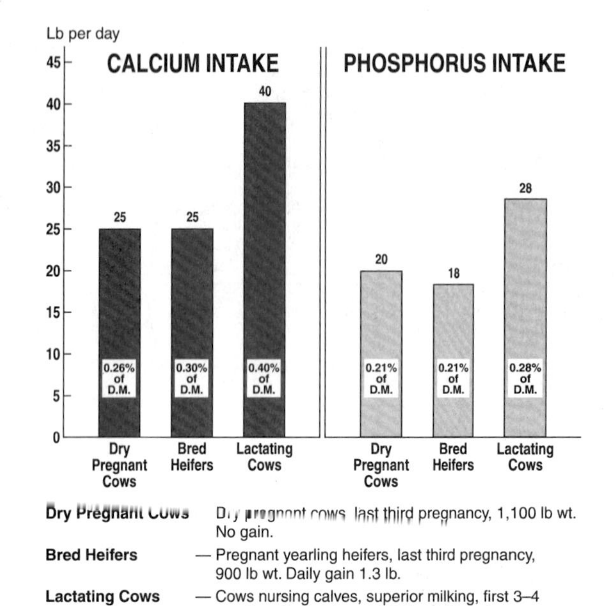

Fig. 19-5. Calcium and phosphorus requirements of dry pregnant cows, bred heifers, and lactating cows. Milk is a rich source of calcium and phosphorus; hence, it is no surprise to find that the requirements for each of these minerals are markedly higher for lactating cows than for either of the other groups. (Based on data from *Nutrient Requirements of Beef Cattle*, sixth revised edition, National Research Council-National Academy of Sciences, Washington, DC, 1984)

her calcium and phosphorus needs nearly double. These figures also show that the requirements of growing heifers are about the same as those of mature cows.

Weight also makes a difference, as shown in Table 19-1, which gives the daily nutrient requirements at various weights of (1) dry pregnant cows and (2) cows nursing calves.

Of course, these are minimum requirements. Hence, it would be well to add 1 percentage unit to the crude protein requirement and 3 percentage units to the TDN requirement, and to self-feed the minerals. This would take care of variations in feedstuffs and differences in requirements between individual cows within a herd.

Heavy grain feeding is uneconomical and unnecessary for the beef breeding herd. The nutrient requirements should be adequate merely to provide for maintenance, growth (if the animals are immature), and reproduction. Fortunately, these requirements can largely be met through feeding roughages.

NUTRITIONAL REPRODUCTIVE FAILURES IN COWS

Since cattle producers largely determine their own destiny when it comes to feeding, it is important that they know the causes of nutritional reproductive failures and how to rectify them.

TABLE 19-1
DAILY NUTRIENT REQUIREMENTS OF BEEF COWS[1, 2]

Body Weight		Total Protein		TDN		Calcium	Phosphorus
(lb)	*(kg)*	*(lb)*	*(kg)*	*(lb)*	*(kg)*	*(g)*	*(g)*
Dry pregnant mature cows (middle third of pregnancy)							
880	*400*	1.45	*0.66*	9.68	*4.4*	22	16
990	*450*	1.55	*0.70*	10.56	*4.8*	23	18
1,100	*500*	1.64	*0.74*	11.22	*5.1*	25	20
1,210	*550*	1.74	*0.79*	11.88	*5.4*	26	21
1,320	*600*	1.83	*0.83*	12.54	*5.7*	28	23
Cows nursing calves, first 3 to 4 months after calving (superior milking ability)							
880	*400*	2.42	*1.10*	13.0	*5.9*	37	25
990	*450*	2.61	*1.19*	14.1	*6.4*	39	26
1,100	*500*	2.74	*1.24*	15.0	*6.8*	40	28
1,210	*550*	2.86	*1.30*	15.6	*7.1*	42	30
1,320	*600*	2.97	*1.35*	16.5	*7.5*	43	31

[1]Adapted by the authors from *Nutrient Requirements of Beef Cattle*, sixth revised edition, National Research Council–National Academy of Sciences, Washington, DC, 1984, pp. 45 and 46; with U.S. Customary added by the authors.

[2]*Note:* Before this book went to press, the new *Nutrient Requirements of Beef Cattle*, seventh edition, came off press. So, the new NRC requirements of beef cattle are presented in the Appendix, Section I, Table I-2, of this book; hence, the reader is referred thereto. Because allowances, rather than requirements, are used in ration formulations in order to provide margins of safety for animal and environmental differences, the principles shown in Table 19-1 still apply.

A review of the literature clearly points to three important reproductive difficulties: (1) the small number of cows in heat and bred the first 21 days of the breeding season, (2) the low conception rate at first service, and (3) the excessive calf losses at birth or within the first 2 weeks of age. Also, it is noteworthy that each of the causes is more marked in young cows (first-calf heifers) than in mature cows.

Research throughout the country gives ample evidence that the real cause of most beef cow reproductive failures is a deficiency of one or more nutrients just before and immediately following calving—nutritive deficiencies during that critical 100-day period when life begins—a deficiency of energy, protein, minerals and/or vitamins. In the sections that follow, reproductive failures attributed to deficiencies of specific nutrients are summarized.

(Also see Chapter 6, section entitled, "Nutritional Factors Affecting Reproduction.")

ENERGY REPRODUCTIVE FAILURES

Energy is essential for the normal life processes of the cow, including body maintenance, reproduction, and lactation. Because energy is necessary for life itself, it is the most important nutrient.

■ **New Mexico workers showed** that inadequate energy intake, rather than a shortage of protein or vitamin A, was the major cause of lowered reproductive performance in cows grazing on semidesert grassland during drought years (Table 19-2). Supplementation started approximately 1 month (February 10) before calving and continued until an appreciable amount of green forage was available (June 1). Cows received approximately 1 lb of supplement before calving and approximately 2 lb after calving. The New Mexico Station was able to increase the percentage calf crop by 9% (from 78% to 87%) by providing 1 to 2 lb of supplemental energy feed (ground corn) to cows on semidesert grassland, starting about a month before calving and continuing until adequate green grass was available.

TABLE 19-2
RANGE CONDITIONS AND RESPONSE TO SUPPLEMENTAL FEEDING IN NEW MEXICO[1]

	Control	Ground Corn	Cottonseed Cake	Cottonseed Cake and Dehy. Alfalfa
Weight of calves:				
1. Years of average rainfall, lb *(kg)*	428 *(194)*	424 *(192)*	430 *(195)*	436 *(198)*
2. Drought years, lb *(kg)*	351 *(159)*	366 *(166)*	366 *(166)*	367 *(167)*
Percent of cows calving:				
1. After years of average rainfall, %	93	92	91	95
2. After drought years, %	78	87	84	84

[1]Knox, J. H., and W. E. Watkins, *Supplements for Range Cows*, New Mexico Ag. Exp. Sta. Bull. 425.

■ **In a Montana study,**[2] cows on heavily grazed (23.1 acres per cow yearly) ranges averaged 15% fewer calves born than those on moderately grazed (30.5 acres per cow yearly) or lightly grazed (38.8 acres per cow yearly) ranges. A shortage of forage appeared to be responsible for the results obtained.

■ **In an Oklahoma study,** the effect of winter plane of nutrition on the performance of spring-calving 3-year-old beef cows was studied (Table 19-3). The heifers were started on winter feed in early November of each year and fed each winter (approximately 160 days to mid-April) according to the following plan:

1. **Low plane.** No gain during the first winter as weaner calves, with a loss of at least 20% of fall weight during subsequent winters as bred females.
2. **Moderate plane.** Gains of 0.5 lb per head daily the first winter as weaner calves, with a loss of nearly 10% of fall weight during subsequent winters as bred females.

TABLE 19-3
EFFECT OF FOUR WIDELY DIFFERENT WINTER FEED LEVELS ON PERFORMANCE OF 3-YEAR-OLD BEEF COWS[1]

	Low	Medium	High	Very High
Winter feed/head, lb *(kg)*:				
Cottonseed cake	32 *(14)*	202 *(91)*	206 *(93)*	—
Ground milo	—	88 *(40)*	815 *(370)*	—
50% concentrate mix	—	—	—	4,660 *(2,118)*
Winter feed cost/cow, 3 years:	$74.62	$107.29	$147.87	$281.71
Calf crop weaned, %	71.4	85.7	92.9	84.6
Average birth weight, lb *(kg)*	69.5 *(31.5)*	73.8 *(33.5)*	75.3 *(34.2)*	77.0 *(35)*
Average weaning weight, lb *(kg)*	361 *(164)*	455 *(206)*	512 *(232)*	455 *(206)*

[1]Pinney, D., L. S. Pope, C. V. Cotthem, and K. Urban, "Effect of Winter Plane of Nutrition on the Performance of Three- and Four-Year-Old Beef Cows," *36th Annual Livestock Feeders' Day*, Oklahoma State University.

---

[2]Marsh, H., *et al.*, *Nutrition of Cattle on an Eastern Montana Range As Related to Weather, Soil, and Forage*, Mont. Ag. Exp. Sta. Bull. 549

3. **High plane.** Gains of approximately 1 lb per head daily during the first winter, with no loss in weight during subsequent winters.

4. **Very high plane.** Self-fed a 50% concentrate mixture to gain as rapidly as possible both as weaner calves and in subsequent winters.

The results obtained in this study point up the danger of underfeeding—low calf crop, smaller calves at birth and weaning, and depressed milk production. However, overfeeding resulted in more calving difficulty, depressed milk production, and a tremendous increase in feed cost. From the economic standpoint, a medium level of feeding appeared to be most desirable.

PROTEIN REPRODUCTIVE FAILURES

Protein is essential for maintenance and building of muscle tissue and bone, including growth of hair and hoofs; for the development of the fetus; for the growth of young stock; and for milk production.

■ **Workers at the Montana Station conducted** an experiment to compare the feeding of 1 lb per head per day of either a 10 or 20% protein supplement during the winter on the production of range cows. Hereford cows fed 1 lb of the 20% protein supplement daily weaned calves which, on a 205-day corrected weaning weight basis, averaged 22 lb heavier than those receiving 1 lb of the 10% supplement daily (Table 19-4).

TABLE 19-4
PERFORMANCE OF MATURE HEREFORD COWS FED 1 LB *(0.45 KG)* PER HEAD PER DAY OF EITHER A 10% OR A 20% PROTEIN SUPPLEMENT[1]

	Protein Content of Supplement	
	10%	20%
No. of head	24	24
Cows calving	24	24
Average birth wt., lb *(kg)*	78 *(35)*	82 *(37)*
Average weaning wt., lb *(kg)*	406 *(184)*	416 *(189)*
Average adjusted weaning wt., lb *(kg)* (205 days)	478 *(217)*	500 *(227)*

[1]Thomas, O. O., J. L. Van Horn, and F. S. Willson, "Effect of Level of Protein in Winter Rations and Added Feed at Calving upon Production of Range Cows," *8th Cattle Feeders' Day,* Montana State College.

■ **California workers found** that cows receiving supplemental feed during periods when forage was scarce had better reproductive performance than cows not receiving a supplement (Table 19-5). The supplementation consisted of 1 lb of 43% cottonseed cake beginning in August; 2 lb of cottonseed cake from calving time (October, November, and December) to first rains (usually January or February), and 2 lb of cottonseed cake plus 1 lb of barley from time of first rain until new forage growth was available. This study was conducted in the Sierra Nevada foothills where rainfall averaged 19.2 in. yearly for the 12 years of the experiment.

COMBINED ENERGY AND PROTEIN REPRODUCTIVE FAILURES

Lack of energy in cows results in poor reproduction: failure of some cows to show heat, more services per conception, lowered calf crops, and lightweight calves.

A severe shortage of protein causes depressed appetite, poor growth, loss of weight, reduced milk production, irregular estrus, and lowered calf crops.

■ **Bond and Wiltbank found** that there was a wide variation in reproductive performance in heifers fed different levels of energy and protein (Table 19-6). Angus heifers were placed on the experimental rations shortly after weaning and calved at approximately two years of age. Results observed in the heifers receiving low levels of protein were confounded because of inadequate intake. This low level intake resulted in an energy deficiency as well as a protein deficiency. It appeared that adequate energy was more important than adequate protein on reproductive performance.

■ **Nelson and coworkers in Oklahoma presented** the results of a study on wintering two groups of

TABLE 19-5
EFFECT OF SUPPLEMENT ON REPRODUCTION IN CALIFORNIA[1]

	Number of Cow Years	Cows Calving	Calf Crop	Weaning Weight		Weaning Age	Lb *(Kg)* of Calf Weaned per Cow Bred	
		(%)	*(%)*	*(lb)*	*(kg)*	*(days)*	*(lb)*	*(kg)*
Supplemented	478	90	83	464	*210*	240	385	*175*
Unsupplemented	469	75	66	406	*184*	230	270	*122*

[1]Wagnon, K. A., H. R. Guilbert, and G. H. Hart, *Beef Cattle Investigations on the San Joaquin Experimental Range,* California Ag. Exp. Sta. Bull. 765.

TABLE 19-6
EFFECT OF VARIOUS LEVELS OF PROTEIN AND ENERGY ON REPRODUCTIVE PERFORMANCE—MARYLAND AND LOUISIANA[1]

Energy Intake			Intake of DP			Showing Heat	Calves Lost at Birth	Interval from Calving to First Heat	Milk Production at 60 Days		Weight at First Heat	
	(lb)	*(kg)*		*(lb)*	*(kg)*	*(%)*	*(%)*	*(days)*	*(lb)*	*(kg)*	*(lb)*	*(kg)*
(H)	11.0	*5.0*	(H)	1.11	*0.50*	100	73	51	17.2	*7.8*	599	*272*
(H)	10.0	*4.5*	(M)	0.65	*0.29*	100	45	80	14.6	*6.6*	586	*266*
(H)	6.0	*2.7*	(L)	0.24	*0.11*	50	10	66	9.3	*4.2*	412	*187*
(M)	6.0	*2.7*	(H)	1.11	*0.50*	100	0	63	15.0	*6.8*	506	*230*
(M)	6.0	*2.7*	(M)	0.68	*0.31*	100	9	68	16.8	*7.6*	552	*250*
(M)	5.0	*2.2*	(L)	0.24	*0.10*	67	17	118	13.0	*5.9*	427	*194*
(L)	3.5	*1.59*	(H)	1.06	*0.48*	25	25	—	6.8	*3.1*	440	*200*
(L)	3.4	*1.54*	(M)	0.66	*0.30*	58	9	163	8.7	*4.0*	436	*198*
(L)	3.3	*1.5*	(L)	0.22	*0.10*	33	17	130	8.0	*3.8*	433	*196*

[1]*Beef Cattle Science Handbook*, Vol. 2, edited by M. E. Ensminger and published by Agriservices Foundation, based on work conducted by J. Bond and J. N. Wiltbank.

(H) = High; (M) = Medium; (L) = Low; DP = Digestible Protein.

3-year-old fall-calving beef cows (Table 19-7). During the summer, all cattle grazed the native grass pasture. Cows in the low group were fed 1.43 lb of pelleted cottonseed meal per head daily in addition to prairie hay starting October 27. The high group of cows was self-fed a cottonseed meal–milo-salt mixture (average consumption—1.80 lb cottonseed meal, 5.32 lb ground milo, 1.84 lb salt) as supplement to prairie hay. The level of winter feeding had a marked effect on weaning weight of calves. It is difficult to determine whether the additional protein or the energy accounted for the greater net return from the high group of cows.

TABLE 19-7
SUPPLEMENTAL WINTER FEEDING OF 3-YEAR-OLD BEEF COWS[1]

	Low	High
Winter loss in weight, lb *(kg)*	158 *(72)*	144 *(65)*
Birth weight of calves, lb *(kg)*	71 *(32)*	72 *(33)*
Spring weight of calves, lb *(kg)*	214 *(97)*	264 *(120)*
Weaning weight of calves, lb *(kg)*	382 *(178)*	473 *(215)*
Feed cost/cow	$ 57.73	$ 68.38
Selling value of calves	$100.42	$121.22
Selling value minus feed cost	$ 42.69	$ 52.84

[1]Nelson, A. B., R. D. Furr, and G. R. Waller, "Level of Wintering Fall-Calving Beef Cows," *36th Annual Livestock Feeders' Day*, Oklahoma State University.

PHOSPHORUS REPRODUCTIVE FAILURES

Cattle produced in phosphorus-deficient areas may have depraved appetites, may fail to breed regularly, and may produce markedly less milk. Growth and development are slow, and the animals become emaciated and fail to reach normal adult size. Death losses are abnormally high.

■ **Montana workers reported** the results of a study designed to determine the phosphorus requirements of breeding cows on winter range (Table 19-8). Heifers

TABLE 19-8
CALVING RECORDS OF 2-YEAR-OLD HEIFERS FED DIFFERENT LEVELS OF PHOSPHORUS IN SUPPLEMENTS[1]

	2 Lb *(0.9 Kg)* of a 20% Protein Supplement/Day			
	0.5% P	1.0% P	1.5% P	2.0% P
Number of head	14	14	13	14
Cows calving	11	11	13	13
Average birth weight, lb *(kg)*	71 *(32)*	70 *(31)*	64 *(29)*	64 *(29)*
Average weaning weight, lb *(kg)*	308 *(140)*	324 *(147)*	307 *(139)*	314 *(142)*
Average adjusted weaning weight, lb *(kg)* (205 days)	444 *(201)*	466 *(211)*	445 *(202)*	435 *(197)*

[1]*Beef Cattle Science Handbook*, Vol. 2, edited by M. E. Ensminger and published by Agriservices Foundation, based on work by O. O. Thomas, J. L. Van Horn, and F. S. Wilson.

fed the supplement containing 1.0% phosphorus weaned the heaviest calves.

VITAMIN A REPRODUCTIVE FAILURES

A severe deficiency of vitamin A may result in a low conception rate; a small calf crop; many calves weak or stillborn, with some born blind or without eyeballs; more cows with retained placentas; low gains; greater susceptibility to calf scours; and more respiratory troubles.

Lane summarized the results of work in Arizona with range cows.[3] Yearling heifers receiving a rumen injection of 1,000,000 IU of vitamin A plus 100,000 units of vitamin D_1 per cow and calving as 2-year-olds produced calves that averaged 20 lb heavier than those from untreated heifers. When treatment was prior to the breeding season, the percent calf crop increased 10 to 11%. The heifers were on ranches that were extremely droughty at the time of injection. Range conditions on each individual ranch should determine whether this injection should be given in the spring or fall.

[3]Lane, A., "Vitamin A Injections in Range Cows, *Arizona Cattle Feeders' Day*, University of Arizona.

■ **Verdugo, California State University–Fresno, summarized**[4] data collected for 26 years from a fall-calving Hereford herd in the foothills of the Sierra Nevada near Porterville, California (Fig. 19-6). Rainfall, which was highly variable and occurred in the winter, averaged slightly over 14 in. per year. Changes in a number of management and nutritional practices contributed to the increased calf crop over the years. The important change, particularly in the heifers, appeared to be the addition, beginning in 1948, of 10,000 IU of vitamin A per head daily to the supplement of alfalfa meal, cottonseed meal, and salt which was fed in the late summer or early fall until the dry grass season was over. Upon charting yearly calf crop and yearly rainfall, a striking parallel was noted between heifer calf crop and yearly rainfall up to 1948 when vitamin A supplementation was initiated (Fig. 19-6).

■ **Temple, Southern Regional Beef Cattle Research, reported**[5] that studies at Front Royal showed that (1) vitamin A injections during pregnancy lowered the incidence of stillborn calves by 2.2% and (2)

[4]Verdugo, W. R., "Recorded Research and Management Problems for 22 Years," AceHi Polled Hereford Ranch, Porterville, California, *Beef Cattle Science Handbook*, Vol. 2, edited by M. E. Ensminger and published by Agriservices Foundation.

[5]Summarized by the author in *The Stockman's Guide*, syndicated monthly livestock column.

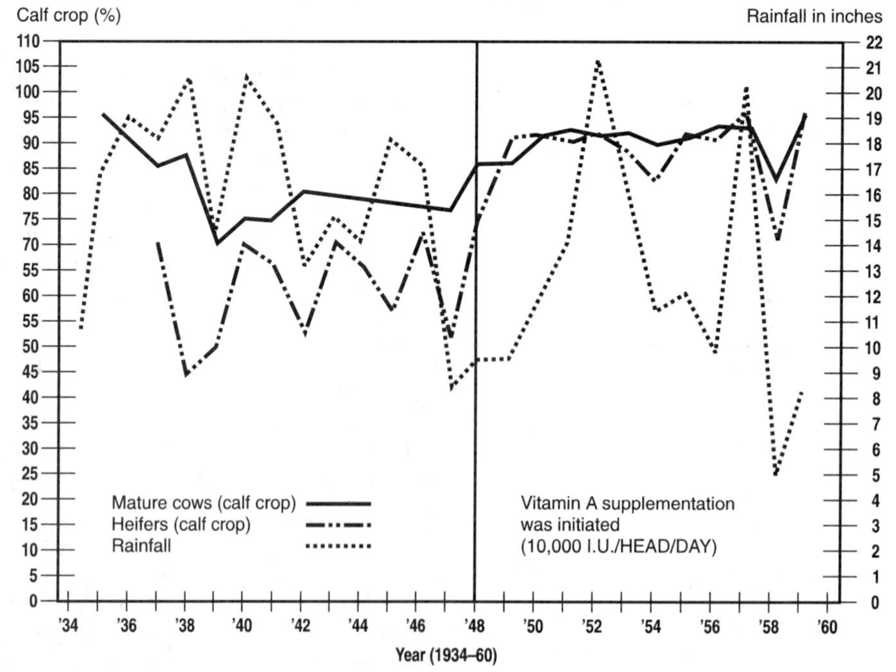

Fig. 19-6. Data on rainfall and calf crop. AceHi Hereford Ranch, Porterville, California.

vitamin A supplementation at birth lowered newborn calf mortality by 8.4%.

■ **Hentges, of the Florida Station in a review of the literature, reported**[6] that deficiencies of vitamin A in cows will cause delayed conception, retained placentas, and dead, weak, or malformed calves; in bulls, delayed puberty, testicular tissue degeneration, and pituitary cysts. Also, Hentges reported that a deficiency of vitamin D will cause infrequent heat periods and lowered reproduction.

Based on experiments plus experiences, it is recommended that vitamin A be supplied to cows and calves after periods of four months on dry, weathered forage—either winter pasture or harvested forage. This can be done either by injection of one million IU of vitamin A or by feeding a supplement.

SUMMARY OF NUTRITIONAL REPRODUCTIVE FAILURES IN COWS

From the above review of literature, the following conclusions seem justified:

1. Energy is more important than protein in reproduction.
2. Beef cows receiving inadequate energy reproduce at a low level.
3. Phosphorus supplementation of cows on range areas deficient in phosphorus increases the calf crop.
4. Administering additional vitamin A to heifers grazing dry forage increases the calf crop.
5. The level and kind of feed before and after calving will determine how many cows will show heat—and conceive. After calving, feed requirements increase tremendously because of milk production; hence, when a cow is suckling a calf, she needs approximately 50% greater feed allowance than during the pregnancy period (see Fig. 19-2). Otherwise, she will suffer a serious loss in weight and fail to come in heat and conceive.
6. Cows in average condition should gain a minimum of 100 lb during the pregnancy period, followed by a gain of ½ to ¾ lb daily after calving and extending through the breeding season. If they are on the thin side at calving time, they should gain 1½ to 2 lb daily after they drop calves. This calls for 7 to 12 lb of TDN daily before calving (which can be provided by feeding 14 to 22 lb of average-quality hay), and 10 to 17 lb of TDN after calving (which can be provided by feeding 14 to 28 lb of hay plus 4 lb of grain), with the lactating requirement dependent on both cow weight and milking ability. Additionally, there must be adequate protein, minerals, and vitamins.

---

[6]*Ibid.*

WINTER FEEDING BROOD COWS

In a country as large and diverse as the United States, wide variations exist in both the length of the winter season and the available feeds. But the same principles are applicable to all areas and enterprises, and the chief objective remains the same—economically to produce high percentage calf crops with heavy weaning weights.

Fig. 19-7. Winter feeding is often necessary; and is a must for cows with calves at side. (Courtesy, *Livestock Weekly*, San Angelo, TX)

Winter feeding is the most expensive time in cow-calf operations. From an economic standpoint, therefore, it is important that wintering practices be both knowledgeable and wise. The cheaper homegrown roughages should constitute the bulk of the winter ration for dry pregnant cows. Most of the grain and the higher class roughages may be used for other classes of livestock. A practical ration may consist of silage and/or dry roughages (legume or grass hays) combined with a small quantity of protein-rich concentrate (such as soybean meal or cottonseed meal). With the use of a leguminous roughage, the protein-rich concentrate may be omitted. Dusty or moldy feed and frozen silage should be avoided in feeding all cattle—especially in the case of the pregnant cow, for such feed may produce complications and possible abortion.

Except during the winter months, pastures constitute most of the feed of beef cattle. By fall, however, grass is usually in short supply and relatively poor as a source of protein, certain minerals (especially phosphorus), and carotene (provitamin A). To overcome these deficiencies, the cattle producer must resort to either (1) supplemental feeding on pasture, or (2) drylot feeding. At no other time in the operations is a

Fig. 19-8. Winter feed for brood cows. Laborsaving, tightly packed, loaf-like stacks of corn stover and hay. Such stacks may be deposited in the feeding area, then fenced, and cattle may be fed right from the stack. (Courtesy, Hesston Corporation, Hesston, KS)

possible profit so likely to be dissipated and replaced by a loss.

Fall feeding should not be delayed so long that animals begin to lose weight. The reason cattle often eat and get poor on dry, weathered grass is that it is low in energy, protein, carotene, and phosphorus and perhaps certain other minerals. These deficiencies become more acute and increase in severity as winter advances. Cattle simply cannot consume sufficient quantities of such bulky, low-quality roughage to meet their needs; and the younger the animal the more acute the problem. Under such circumstances, the maintenance needs are met by the breakdown of body tissues, accompanied by the observed loss in weight and condition. Young animals fail to grow; it makes for lightweight calves. Also, reproduction is affected adversely; serious underfeeding results in lowered calf crops. Supplementing fall grass with a concentrated type of supplement is the practical and ideal way in which to alleviate such nutritive deficiencies.

Likewise, spring feeding should be continued until grass has attained sufficient growth and sustenance. The calf crop will be adversely affected in pregnant cows and milk production will be lowered in lactating cows if they subsist for prolonged periods on insufficient amounts of forage or on frosted pastures that are low in protein, vitamin A, and minerals.

When roughage is scarce and high in price, feed less of it and more concentrate; conversely, feed more concentrate and less hay when grain is plentiful. Generally speaking, 1 lb of grain can replace 2 lb of dry roughage, providing they are of comparable quality.

The best calf crop is produced by cows that are kept in vigorous breeding condition—that are neither overfat nor thin and run-down. Generally speaking, this calls for winter feeding, with the maximum use of roughage. The kind and amount of concentrate needed will depend upon (1) the amount and kind of roughage given, (2) the age and condition of the cattle, and (3) whether the cows are dry or suckling calves. In total, it is important that the ration provides the kinds and amounts of nutrients needed, along with sufficient bulk to satisfy the appetite reasonably well. On a dry-feed basis, the daily requirement of dry pregnant cows is about as follows: thin cows, 2¼% of their liveweight; cows in average flesh, 2% of liveweight; and cows in good condition, 1¾% of liveweight. Cows suckling calves should receive approximately 50% more feed than dry cows of comparable weight and condition. From this it should be concluded that, unless the herd is so small as to make it impractical, dry cows should be wintered separately from those that are suckling calves. This makes it possible to limit the feed of dry cows and to effect certain other economies in their handling.

Dry pregnant cows in average condition should gain in weight sufficient to account for the growth of the fetus (60 to 90 lb) plus sufficient increase in weight and condition to carry them through the suckling period. In total, they should gain 100 to 150 lb during the pregnancy period, or at the rate of approximately ½ lb daily. Of course, the size and condition of the cow is the best gauge as to the feed allowance and desired gain. As previously noted, dry cows require less supplementation than cows suckling calves. Nevertheless, they should not be permitted to lose too much flesh, unless, of course they are overfat. Also, it is recognized that it requires less feed to keep cattle from losing flesh than it does to restore them to proper condition after they have become thin. Thus, it is good economy to start feeding before cows show any signs of malnutrition. Unless a good-quality legume roughage is fed, the concentrate should provide protein, energy, and needed minerals and vitamins.

Noteworthy, too, is the makeup of a calf at birth. An average 70-lb calf at birth is about 75% water, 20% protein, and 5% ash. The calf's 70-lb weight represents about 17.5 lb dry matter. From this, it is apparent that the dry gestation period does not create a heavy nutritional drain. Thus, this is a period when a cattle producer may economize by utilizing crop residues and winter pasture.

Cows with calves at side should be fed for the production of milk, for their feeding requirements are more rigorous than those during pregnancy. This is important because, until weaning time, the growth of the calf is determined chiefly by the nourishment available through the milk of its dam. The principal part of the calf's ration, therefore, may be cheaply and safely provided by giving its mother the proper feed for the production of milk. To stimulate milk flow, most beef cows need a concentrate during the winter months, and the poorer or the more limited the roughage the higher the supplement requirement. On the average, cows that calve in the fall should be fed a minimum of 4 to 6 lb concentrate daily throughout the winter.

Purebred breeders recognize that good condition and attractive appearance of breeding animals are important assets from the standpoint of favorably impressing prospective purchasers. Thus, a certain amount of the feed consumed by purebred cows—that which is above the amount needed for good health and vigor—can rightfully be charged to advertising costs. This is so because no more effective means of advertising exists than well-conditioned animals, regardless of the season of the year. Also, supplemental feeding of purebred cows that are suckling is favorably reflected in their calves—in greater growth, development, and bloom. In short, purebred breeders usually find it advantageous to feed a concentrate to cows during the winter months. Purebred breeders who are content with less than the maximum potential should not be in the purebred business in the first place; they should confine their efforts to commercial production.

RATIONS FOR DRY PREGNANT COWS

When winter grazing is not possible, the rations in Table 19-9 may be used to meet the daily needs for energy and protein of a 1,100-lb dry pregnant cow. A combination of legume roughage with lower quality roughage (such as stalklage, straw, corncobs, or cottonseed hulls) will meet both the energy and protein requirements without the use of a supplement.

TABLE 19-9
WINTERING RATIONS FOR A 1,100-LB *(500-KG)* DRY PREGNANT COW

	Rations				
	1	2	3	4	5
	- - - - - - - - - *(lb [kg]/day)* - - - - - - - - -				
Legume-grass hay	18 *(8)*				10 *(4.5)*
Legume-grass haylage[1]		30 *(12)*			
Corn or grain sorghum silage			35 *(15)*		
Stalklage or husklage				45 *(20)*	
Straw, cobs, or cottonseed hulls					10 *(4.5)*
Supplement[2]			0.5 *(0.2)*	1 *(0.45)*	

[1]Haylage figured at 55% dry matter, corn or grain sorghum silage at 35% dry matter, stalklage or husklage at 45% dry matter.

[2]Supplement figured at 48% crude protein. Quantity to be adjusted in keeping with the protein content of the supplement. For example, if a 24% crude protein supplement is fed, the quantity of supplement should be doubled.

RATIONS FOR COWS NURSING CALVES

The energy requirement of a cow nursing a calf is about 50% higher than that of a dry pregnant cow; and the protein, calcium, and phosphorus requirements are nearly double. Since the vast majority of the nation's cows with calves at side are on pasture most, if not all, of the lactation period, the only supplemental need is for salt and other minerals, unless the pasture is insufficient in quantity or quality of feed to support adequate milk production. The rations in Table 19-10 may be used for drylot feeding of beef cows nursing calves. Of course, the daily levels shown in Table 19-10 should be approached gradually so that nutritional scours will not develop in baby calves.

TABLE 19-10
WINTERING RATIONS FOR A 1,100-LB *(500-KG)* COW NURSING A CALF

	Rations				
	1	2	3	4	5
	- - - - - - - - - *(lb [kg]/day)* - - - - - - - - -				
Legume-grass hay	30 *(13)*			20 *(9)*	10 *(4.5)*
Legume-grass haylage[1]		50 *(22)*			
Corn or grain sorghum silage[1]			60 *(27)*		40 *(18)*
Grain				5 *(2)*	
Supplement[2]			1.5 *(0.6)*		

[1]Haylage figured at 55% dry matter, corn or grain sorghum silage at 35% dry matter.

[2]Supplement figured at 48% crude protein. Quantity to be adjusted in keeping with the protein content of the supplement. For example, if a 24% crude protein supplement is fed, the quantity of supplement should be doubled.

CROP RESIDUES AND WINTER PASTURES FOR BROOD COWS

Two requisites are important in wintering the cow herd: (1) bringing them through the winter in proper condition for calving, and (2) keeping feeding costs to the minimum consistent with nutritional demands. Meeting these requirements has prompted increased use of crop residues and winter pastures for brood cows. As the ever-increasing human population of the world consumes a higher proportion of grains and seeds, and their byproducts, directly, cattle will utilize a maximum of crop residues and pastures and a minimum of products suitable for human consumption. Thus, more and more farmers with crops will include a beef herd in their operations and realize a fair return from feeds which would otherwise be wasted.

CROP RESIDUES

Generally speaking, crop residues may be grazed, processed as dry feed, or made into silage. The important thing to remember is their relatively low value, in comparison with grains, necessitates low cost harvesting, storing, and feeding. Also, they must be fed to the right class of animals, and they must be properly supplemented. Remember, too, that there is a marked difference between economical wintering and deficient wintering.

In addition to the crop residues discussed in the sections that follow, good-quality roughages—such as alfalfa, bromegrass, and corn or grass silage—can be used to meet the nutritional requirements of the brood cow. But such high-quality feeds must be limited, otherwise animals will get too fat.

CORN RESIDUES

Of all crop residues, the residue of corn is produced in greatest abundance and offers the greatest potential for expansion in cow numbers. In 1993, 63 million acres of corn, yielding 100.7 bu per acre, were harvested in this country. For the most part, over and above the grain, 2½ to 3 tons of dry matter produced per acre (40 to 50% of the energy value of the total corn plant) were left to rot in the field. That's 250 million tons of potential cow feed wasted, enough to winter 192 million dry pregnant cows consuming an average of 22 lb of corn refuse per head per day during a 4-month period. Moreover, mature cows are physiologically well adapted to utilizing such roughage. And that's not all! When corn residue is used to the maximum as cow feed, acreage which would otherwise be used to pasture the herd is liberated to produce more corn and other crops. Also, there are many other crop residues, which, if properly utilized, could increase the 192 million figure given previously.

Fig. 19-9. Fox equipment with flail attachment harvesting corn stalklage. (Courtesy, Koehring Farm Division, Appleton, WI)

Although corn refuse offers tremendous potential as a cow feed, there are difficulties in harvesting and storing it. But science and technology have teamed up and are working ceaselessly away at solving these problems.

Broadly speaking, three alternate methods of salvaging corn refuse are being used: (1) grazing, (2) harvesting and dry feeding, and (3) ensiling; with different ways of accomplishing each. The choice of the method should be determined primarily by (1) cost, (2) the proportion of refuse utilized, and (3) how well it meshes in with other farm enterprises—for example, in some cases the need for fall plowing will necessitate removal of the material from the land and elimination of grazing as an alternative.

Grazing (Pasturing)

This refers to turning the animals directly into the stalk field—the traditional way of utilizing cornstalks. Letting the animals do their own harvesting is the simplest and least expensive method devised for utilizing a crop. However, there is considerable wastage, and it is not possible to prolong the winter feeding period. In an open fall and winter, 2 acres of cornstalks will carry a pregnant cow for 100 to 120 days. But the following problems are associated with this method of harvesting:

1. **Selective grazing.** Cows are selective grazers. They will consume the more palatable portions of corn refuse first, in the following order: corn ears, husks, leaves, and stalk.

2. **Waste.** Only an average of one-third of the stover is actually used, with the amount varying from 15 to 40%, depending primarily on weather conditions.

3. **Fencing.** Many cornfields are unfenced; hence, fence must be constructed in order to confine the animals. Also, strip grazing (grazing a part of the field at a time) will improve the utilization of stalks in large fields by making more uniform nutrition available throughout the grazing period. It prevents selective grazing over the entire field, with the result that the animals consume the more palatable portions of the plant first, and leave the bare stalks until last.

The fencing problem may be solved economically with electric fence.

4. **Snow cover; fall plowing; soil puddling; stock water.** In the northern part of the United States, snow cover prevents grazing for part of the winter and necessitates a reserve feed supply.

Another drawback is that grazing prevents fall plowing, a recommended practice on heavy soils. Also,

Fig. 19-10. Angus cows grazing cornstalks in the Corn Belt. (Courtesy, *Wallaces Farmer*, Des Moines, IA)

cattle may puddle and pack the soil, which lowers crop yields.

Frequently, supplying the herd with drinking water is costly. It may necessitate drilling a well or piping water from a distance.

Stalklage

Stalklage refers to all the residue remaining after harvesting corn with a combine or picker. It may be either stored as dry stalklage or ensiled.

■ **Dry stalklage**—Stalklage is more difficult to collect than husklage; and more expensive, since it involves more equipment and another trip across the field. A number of different machines for harvesting stalklage are being used; among them, forage harvesters, balers, stackers with flail pickups, and choppers and stackers. By operating the machine a few inches above the ground so as to prevent excess soil pickup, a yield of 1 to 3 tons of residue per acre may be obtained with the moisture content ranging from 20 to 55%, depending on the time of harvest. Stacked or baled cornstalks should be at the low end of this moisture range to reduce heating and spoilage.

Cows like dry stover. Self-feeders around a stack make feeding convenient. Leftover material may be used as bedding.

■ **Stalklage ensilage (stover silage)**—Stalklage may also be ensiled, producing a product known as corn stover silage or cornstalk silage. When this is done, the use of a forage harvester equipped with a screen or a recutter-blower at the silo is necessary in order to chop the material finely. Fine chopping will insure good packing and improve consumption by avoiding selectivity.

Where corn stover silage is to be made, the residue should be harvested as soon as possible after the grain is taken off, before the residue loses any moisture. At that time, the grain moisture will generally be under 30% and the refuse moisture will be above 48%. In an airtight silo, 40 to 45% moisture will suffice. In an unsealed or bunker silo, the moisture content should be 48 to 55% for proper lactic acid formation. Water may be added at the silo if necessary. As a precaution, some authorities recommend the addition of 56 lb of cornmeal (or other finely ground grain) per ton of corn stover silage, as a means of providing carbohydrates from which acids will form and act as a preservative. With husklage, the latter precaution is not necessary since there is sufficient grain remaining in the husk and cob.

The biggest deterrent to harvesting stalklage, in either dry or ensiled form, is the cost—primarily for the equipment. Rather than own such expensive equipment, which is used for a short period only, custom harvesting of stalklage is likely cheapest for most operators.

Husklage (Shucklage)

Husklage refers to the forage discharged from the rear of a combine when harvesting corn. It consists of the husks, cobs, and any grain carried over the combine, collected in a wagon or straw buncher pulled behind the combine. This operation minimizes labor and does not slow the grain harvest, because the husklage piles can be dumped at the end of the field for supplemental feeding or later pickup by a front-end loader and moved to another location for stacking or ensiling. The moisture content of this material will usually run between 30 and 40%, and the yields will be between 1 and 1.5 tons per acre.

The greatest difficulty encountered in feeding husklage dumps at the end of the field is waste. Depending on weather conditions, as much as 50% of the material may be wasted. But wastage of husklage dumps can be materially lessened by controlling access to them.

Stacking of husklage has been satisfactory for some producers.

Ensiling husklage, along with recutting and adding water, results in increased cow consumption and less rejection of cobs.

Since grazing stalk fields is widely practiced in the fall and winter, the feeding of baled, piled, or

Fig. 19-11. Cows on Circle S Ranch, Rockwood, Ontario, Canada, on a winter ration of corn husklage—the husks, cobs, and waste grain; supplemented with 2 lb of grain and 1 lb of a 32% protein concentrate daily, beginning 6 weeks before calving. (Courtesy, *Country Guide*, Winnipeg, Manitoba, Canada)

stacked corn residues in the field permits feeding cows on stalks most of the winter without supplemental feeding of hay or silage. Some molds may develop in the collected material, but usually they are not sufficient to affect the feed intake or health of mature cows.

Feeding Value of and Supplements for Corn Residues

Table 19-11 lists the daily nutritive requirements of a dry pregnant cow weighing 1,100 lb. Table 19-12 gives the nutritive composition of air-dry corn stover and husklage.

Not all corn refuse will be of the same composition as Table 19-12; some will be better, some will be poorer. The quality declines with the passing of time following grain harvest; and the more severe the weather, the greater the decline. Also, cultural practices during the growing season may alter corn residue quality.

Studies show that a 1,100-lb cow will eat approximately 22 to 24 lb per day of palatable air-dry stover, or 2 lb or more of air-dry stover per cwt per day. She will eat slightly larger amounts of husklage. This consumption, along with the information presented in Tables 19-11 and 19-12, suggests that stover and/or

TABLE 19-11
NUTRITIVE REQUIREMENTS OF A DRY PREGNANT COW
(Middle Third of Pregnancy)
WEIGHING 1,100 LB *(500 KG)*[1, 2]

Dry matter, daily	20.9 lb *(9.5 kg)*
TDN, daily	11.2 lb *(5.1 kg)*
Total protein, % of ration	7.8
Calcium, % of ration	0.26
Phosphorus, % of ration	0.21
Vitamin A, IU	27,000

[1]*Nutrient Requirements of Beef Cattle*, sixth revised edition, National Research Council–National Academy of Sciences, Washington, DC, 1984.

[2]*Note:* Before this book went to press, the new *Nutrient Requirements of Beef Cattle*, seventh edition, came off press. So, the new NRC requirements of beef cattle are presented in the Appendix, Section I, Table I-2, of this book; hence, the reader is referred thereto. Because allowances, rather than requirements, are used in ration formulations in order to provide margins of safety for animal and environmental differences, the principles shown in Table 19-11 still apply.

TABLE 19-12
ANALYSIS OF AIR-DRY CORN STOVER AND HUSKLAGE[1]

	Corn Stover	Husklage
	(%)	*(%)*
TDN	48	57
Crude protein	4.5	3.4
Calcium	0.4	0.02
Phosphorus	0.07	0.05
Vitamin A	—	—

[1]Cow-Calf Information Roundup, University of Illinois.

husklage rations will meet the daily energy (TDN) needs of dry pregnant cows, but such rations will be slightly deficient in protein, and low in phosphorus and vitamin A. Nevertheless, the highest and best use for corn residue is for dry pregnant cows for the period following conception to about 30 days before calving.

For corn refuse feeding, mature cows should be in medium to good condition at the start of the winter feeding period; and they should not be permitted to lose over 10 to 15% of their weight from fall through calving. Heifer weight losses should be under 5%. When weight loss approaches this limit, it's time to feed some grain or silage.

The following additional information is pertinent to the feeding value and supplementation of corn residues for cattle:

1. **Digestibility.** The components of corn residue rank as follows in digestibility, in descending order: remaining grain, husk, leaf, cob, and stalks.

2. **Energy.** Corn residues provide adequate energy to maintain dry pregnant cows, but they must be

supplemented with additional energy when fed to cows nursing calves or to young, growing animals.

3. **Protein.** The crude protein content of corn stover is on the low side for dry pregnant cows. It runs 4.5% (Table 19-12), whereas a dry pregnant cow requires 7.8% total protein (Table 19-11). Thus, it is recommended that ½ lb per head per day of a 30 to 40% crude protein equivalent (CPE) supplement be provided.

It follows that the protein content of corn refuse is much too low to support either productivity or growth. For example, a 1,100-lb lactating cow requires a daily allowance of 2.7 lb of total protein. However, a daily consumption of 24 lb of stover will provide only about 1.1 lb—less than half the need.

For nursing cows, the protein deficiency of stover and/or husklage may be corrected by supplementation with the following, on a per head per day basis: 2 lb of a 40% protein supplement, or 6 lb of a good legume hay. If desired, the protein supplement may be provided in the form of protein blocks, with one block provided for each 15 cows. Where hay is fed, it should be taken to the field, rather than fed in a feedlot, as this will encourage the cows to stay in the field and graze the cornstalks.

4. **Minerals.** Phosphorus should be provided to all cattle fed corn residue. Calcium may be deficient, especially for lactating cows. Also, some of the trace elements may be deficient. Hence, it is recommended that all cattle on high corn refuse have free access to a complete mineral. A mineral mixture with a Ca:P ratio of 1:2 is recommended for gestating cows. Lactating cows might perform better on a 1:1 ratio.

5. **Vitamin A.** Corn residue, along with other crop residues, is deficient in vitamin A. Hence, it must be supplemented. The precalving and postcalving (heavy milking) needs of approximately 27,000 and 39,000 IU per head per day, respectively, may be met by feeding vitamin A supplement, intramuscular injection of vitamin A solution, or by feeding adequate levels of green, leafy hay.

It is important that corn residue be tailored to match the cow's nutritional needs. This is relatively simple with dry pregnant cows, where supplementation with a high phosphorus mineral and vitamin A will usually suffice. Beginning 4 to 6 weeks before calving and continuing through the lactation period, much heavier supplementation is necessary; in addition to phosphorus and vitamin A, protein must be added, and preferably some energy and calcium for nursing cows.

OTHER CROP RESIDUES

A host of crop residues, other than corn residue, can be used for feeding cows. Some of these follow.

Sorghum (Milo)

Cows will make good use of sorghum as a winter feed. It can be grazed or harvested and stored as either dry feed or silage. The sorghum plant stays green late in the fall; hence, good sorghum stover silage can be made without additional water. In comparison with corn residue, sorghum residue (1) is higher in protein content (corn residue averages 4.5% crude protein, whereas sorghum residue averages 5.2%); (2) is less palatable (if given a choice, cows will select corn refuse in preference to sorghum refuse); (3) comprises a lower percentage of the total plant dry matter than corn (40% of the total plant dry matter of sorghum is residue compared with 40 to 50% for corn); and (4) is lower yielding.

After harvesting, sorghum will send up new shoots if moisture permits. The prussic acid content of these shoots may be harmful to grazing animals; hence, cattle producers should be aware of this possible poisoning. These shoots can be grazed safely 4 to 6 days after a hard killing frost.

Soybean Refuse

The stems and pods of soybean refuse available for feeding yield approximately ¼ ton per acre, with a ratio of stems to pods of about 2:1. The digestibility of stems is low—25 to 35%—due to their high lignin content (18 to 20% for the stalk portion). The digestibility of pods is much higher, ranging from 58 to 63%.

Small Grain Refuse

This refers to (1) straw and (2) tailings—the chaff and grain behind the combine. In the days of binders and threshing machines, straw stacks were commonplace; and they were extensively used for winter cattle feed. With the advent of combines, much of the straw was left to rot in the field. During periods of hay scarcity and high-priced hay, straw is frequently used as either a *hay-stretcher* or a *hay-replacer*.

Of the common cereal straws, oat straw is the most palatable and nutritious. Barley straw ranks second, and wheat straw is third.

Straw is a bulky feed; and it must be properly supplemented. It is low in protein (wheat straw averages about 3.6% crude protein), low in phosphorus, and low in vitamin A. Dry pregnant cows can be wintered on straw plus a daily allowance of either 5 to 6 lb of good-quality alfalfa hay or 1 to 2 lb of a 30 to 40% protein supplement, along with free access to a high-phosphorus mineral (one containing at least 12% phosphorus). If no legume hay is fed, vitamin A should be fed or injected. When oilseed meals are scarce and high in price, a slow-release non-protein nitrogen may be used.

The tailings—the chaff and grain behind the combine—are generally used by farmers and ranchers, either as dry feed or mixed with silage.

Legume and Grass Seed Straws

In addition to the cereal straws, other low-cost roughages available in certain sections of the United States are lentil straw, field pea straw, bean straw, clover straw, and bluegrass straw.

Cottonseed Hulls

Cottonseed hulls are one of the most important roughages in the South, especially for cattle. They supply 42% TDN, which is about as much as is furnished by late-cut grass hay or by oat straw. They are low in protein (4.1%)—and practically none of it is digestible—low in calcium (0.15%), very low in phosphorus (0.09%), and lacking in carotene. To correct these deficiencies when fed to dry pregnant cows, hulls should be supplemented with a daily allowance of either (1) 6 lb of a good-quality legume hay, or (2) 2 lb of a 30 to 40% protein supplement, along with free access to a complete mineral, high in phosphorus unless a phosphorus-rich supplement such as cottonseed meal is fed. If no legume is fed, vitamin A should be fed or injected.

When properly fed, cottonseed hulls are about equal in quality to fair-quality grass hay and are worth more per ton than corn or sorghum stover, straw, or poor hay. Also, they can be fed without further processing—there is no chopping; and they are well liked by cattle, even when fed as the only roughage.

Pelleted hulls are now on the market. In comparison with regular hulls, they are more digestible, require less transportation and storage space—because of their high density—and are easier to handle.

TREATING CROP RESIDUES TO INCREASE DIGESTIBILITY

Crop residues are inefficiently utilized by animals because of the high content and poor digestibility of the fibrous fraction. This poor digestibility is related to the extent of lignification of the cell wall component of these low-quality forages. Although crop residues provide a satisfactory ration for dry gestating cows, they do not provide sufficient energy for either young ruminants or lactating cows—they simply cannot hold enough of these low-quality roughages to provide adequate energy. This prompts interest in increasing the digestibility of these crop residues.

There are many approaches to delignifying and increasing the digestibility of crop residues; among them, treatment with sodium hydroxide, potassium hydroxide, ammonium hydroxide, anhydrous ammonia, and pressurized heating. The potential of such treatments becomes apparent when it is realized that straw, for example, is only 30 to 40% digestible before treatment. When pressure heated with water, it becomes 50 to 60% digestible; and digestibility increases to 70 to 80% when sodium hydroxide is added prior to cooking. By treating corn husklage and milo residue, the Nebraska Station was able to increase their energy value to 90% that of corn silage.

Lowering the cost of treating crop residues to increase digestibility is the primary area which must be researched before these procedures can be applied in more operations.

WINTER PASTURE

Where feasible, winter pasture offers cattle producers a means of reducing costs. By accumulating the feed in the field, rather than harvesting, storing, and handling the forage, the cost and labor of winter feeding can be substantially reduced. Also, costs of bedding and manure hauling can be eliminated.

Tall fescue is used as a winter pasture in the area to which it is adapted—Missouri, Illinois, Indiana, and Ohio. Usually, the new regrowth is baled in late June into round bales and left in the field. The round bales shed rain and snow and, together with the regrowth, make excellent late fall and winter grazing. Experience shows that field-stored forage has adequate quality to maintain beef cows in good condition.

Fescue is a cool-season grass; it actually grows some during the winter in the area to which it is

Fig. 19-12 Cows on winter fescue pasture, supplemented with round bales of fescue harvested the previous June and left in the field. (Courtesy, University of Illinois)

adapted; and it is more palatable during the fall and winter than any other season because of the high concentration of soluble sugars. Trampling during the fall, winter, and spring months does not injure the turf.

The Ohio Station reported that tall fescue winter pasture—including both standing growth and baled hay—carried 2 cows per acre for a 4-month period. The use of electric fence to strip graze the bales and regrowth increases carrying capacity by 50 to 60% over permitting the herd access to the entire field.

RANGE CATTLE SUPPLEMENTATION

Improved range should be the first goal of cattle producers, without using supplemental feeding as a substitute for good grass or as a crutch for poor range. Instead, the two—good range and proper supplemental feeding—go hand in hand.

RANGE NUTRIENT DEFICIENCIES

Growing grasses provide adequate nutrients for beef cattle in **unforced** production when (1) produced on fertile soils, (2) available in sufficient quantities, (3) not washy, and (4) not weathered, leached, or bleached. However, the simultaneous fulfillment of all these conditions is the exception, rather than the rule. Every cattle producer worthy of the name forces young stock for an early market; most soils are deficient in certain nutrients, which, in turn, affect the plants and the animals feeding thereon; during droughts and early and late in the season, feed may be in short supply (thereby limiting energy and other nutrients); early spring pastures are washy and lacking in energy; during droughts and late in the season, grasses become mature, leached, and bleached—they increase in fiber and decrease in protein, phosphorus, and carotene. To meet these conditions, a supplemental source of energy, protein, phosphorus, and vitamin A is necessary.

ENERGY

Hunger, due to lack of feed, is the most common deficiency on the western range. Thus, the first and most important range pasture need is that there be sufficient feed for the animal—to provide the necessary energy required to maintain the body. Over and above these needs, any surplus energy may be used for growth, reproduction, and conditioning.

With bulky, low-quality roughages—such as dry grass cured on the stalk, common to drought periods and late in the season—animals may not be able to consume sufficient quantities to meet their energy needs; and the younger the animal the more acute this problem. Also, very early spring grass is washy and lacking in energy. Under such energy-deficient circumstances, the maintenance needs of animals are met by the breakdown of body tissues, accompanied by the observed loss in weight and condition and the failure of young animals to grow. Also, reproduction is adversely affected; serious underfeeding results in the failure of some cows to show heat, more services per conception, lowered calf crops, and lightweight calves. Supplemental feeding is the practical way in which to alleviate such energy deficiencies.

PROTEIN

There is adequate evidence that a deficiency of protein results in depressed appetite, poor growth, loss of weight, reduced milk production, irregular estrus, and lowered calf crops.

The protein allowance of beef cattle, regardless of age, should be ample to replace the daily breakdown of tissues of the body, including the growth of hair, horns, and hoofs. In general, this need is most critical for the growth of the young calf and for the gestating-lactating cow.

Mature, weathered native range grass is almost always deficient in protein—being as low as 3%, or less. Protein leaching losses due to fall and winter rains may range from 37 to 73%.

Because protein supplements ordinarily cost more per ton than grains, beef cattle should not be fed larger quantities of them than are actually needed to balance the ration. But the temptation is to feed too little of them. When on mature, weathered grass, cows should receive about 2 lb of concentrate supplement daily—the exact amount depending upon the nutrient content of the supplement and other factors.

The protein and energy requirements are closely interdependent; hence, it follows that energy rather than feed intake should be the dietary component relative to which the nutrient needs are adjusted.

MINERALS

Growth, reproduction, and lactation require adequate minerals. Although the mineral requirements of dry pregnant cows and of lactating cows are much the same everywhere, it is recognized that age and individuality make a difference. Additionally, there are area and feed differences. Thus, the informed cattle producer will supply the specific minerals that are deficient in the ration and in the quantities necessary. Excesses and mineral imbalances should be avoided.

Salt should be available at all times, on the basis of about 25 lb per range cow annually.

Phosphorus deficiencies are rather common

among range beef cattle. A severe phosphorus deficiency will result in depraved appetite, emaciation, retarded growth and development, failure to reach normal adult size, failure to breed regularly, lowered calf crop, lowered milk production, and high death losses.

The New Mexico Station reported (1) phosphorus losses in grasses of 49 to 83% during the winter period, (2) increased average annual calf production per range cow of 53 lb through proper mineral supplementation, and (3) that the phosphorus supply should be continuous throughout the year, and not limited to the winter months.

Iodine, copper, cobalt, and selenium supplements should be added in those areas where deficiencies of these minerals are known to exist.

Other mineral elements are thought to be essential, but the picture is somewhat confused; and new findings clearly indicate that we have reason to be less certain of our mineral recommendations than heretofore. For the latter reason, the judicious use of certain trace minerals may be good insurance, even though not all the requirements are known.

VITAMINS

Under normal conditions, vitamin A is the vitamin most likely to be deficient in cattle rations, because dry, bleached range grass is very low in carotene (the precursor of vitamin A).

Inadequate amounts of vitamin A (carotene) during pregnancy may cause cows to abort or give birth to dead or weak calves. Extreme deficiencies may also impair the ability of cows to conceive. Bulls receiving insufficient vitamin A show a decline in sexual activity and semen quality.

In low sunshine areas, especially during the winter months, it is recommended that vitamin D be added to the ration.

INDIRECT DEFICIENCY LOSSES

In addition to the conditions given above, nutritional deficiencies on the range are accompanied by lowered resistance to parasites and diseases, and increased mortality of both breeding stock and calves. Also, where feed is scarce there are usually increased deaths from toxic plants.

SUPPLEMENTING EARLY SPRING RANGE

Turning out on the range when the first blades of green grass appear will usually make for a temporary deficiency of energy, due to (1) washy (high water content) grasses and (2) inadequate forage for the animal to consume. As a result, operators are often disappointed by the poor gains made during this period.

If there is good reason why grazing cannot be delayed until there is adequate spring growth, it is recommended that early pastures be supplemented with grass hay or straw (a legume hay will accentuate looseness, which usually exists under such circumstances), preferably placed in a rack; perhaps with an added high-energy concentrate.

SUPPLEMENTING DRY RANGE

Dry, mature, weathered, bleached grass characterizes (1) drought periods and (2) fall-winter range. Such cured-on-the-stalk grasses are low in energy, protein, carotene, and phosphorus, and perhaps certain other minerals. These deficiencies in range plants become more acute following frost and increase in severity as winter advances. This explains the often severe shrinkage encountered on the range following the first fall freeze.

In addition to the deficiencies which normally characterize whatever plants are available, dry ranges may be plagued by a short supply of feed.

Generally speaking, a concentrated type of supplement is best used during droughts or on fall-winter ranges. However, when there is an acute shortage of forage, hay or other roughage may be used with or without a concentrate.

SUPPLEMENTING THE PUREBRED HERD ON PASTURE

Purebred cattle breeders and commercial cattle producers, alike, recognize the following as important profit factors: (1) percent calf crop, (2) weaning weight, and (3) number of calves a cow produces in her lifetime. But, to be successful, the purebred breeder must go further. Attention must also be given to (1) saleability—sleek, bloomy, well-conditioned animals attract buyers and sell better; and (2) maximum development of genetic potential in characteristics of economic importance (rate of gain, feed efficiency, etc.), otherwise intelligent selections and breed progress cannot be made. Further, the acid test of the competence of the purebred breeder is that these objectives shall be achieved without jeopardizing the breeding performance of the animal—and that's not easy.

GOOD PASTURES AND GOOD PUREBREDS GO TOGETHER

Pastures are the foundation of successful purebred production. It has been well said that good purebred breeders, good pastures that are properly supplemented, and good cattle go hand in hand.

Fig. 19-13. Good purebred breeders, good cattle, and good pastures go hand in hand. This picture shows a purebred Brangus cow and calf in knee-deep grass. (Courtesy, International Brangus Breeders Assn., Inc., San Antonio, TX)

Under the old system of moderate growth, good pastures produced on well-fertilized soils met the nutritive requirements of most beef cattle. However, the nutritive requirements of animals—especially purebreds—have become more critical in recent years. Today, more and more purebreds are being production tested. Among other traits, they are being selected for more rapid and efficient gains—they are in forced production. Consequently, their nutritive requirements are more critical—especially from the standpoints of energy, protein, minerals, and vitamins.

EYE APPEAL IMPORTANT IN PUREBREDS

Successful purebred breeders recognize the importance of maintaining animals in good condition and attractive surroundings where they may be seen and admired by potential buyers. Certainly, a lush pasture is ideal from the standpoint of presenting beef cattle. Obtaining proper condition without overfatness and lowered reproduction is not easy. However, a combination of outdoor exercise, good grass, and proper supplemental feeding is the answer. Above all, replacement heifers should not be ruined by overfatness; it's expensive advertising and poor public relations.

Fig. 19-14. Eye-appealing purebred Beefmaster cows and calves. (Courtesy, Beefmaster Breeders Universal, San Antonio, TX)

BASIC CONSIDERATIONS IN SUPPLEMENTING PUREBREDS ON PASTURE

First and foremost, it must be recognized that, no matter how excellent pastures may be, they are roughages and not concentrates. Therefore, for the purebred herd, judicious supplemental feeding on grass may be warranted and profitable. In doing so, the following *to do* list is noteworthy:

1. **Practice economy, but avoid false economy.** Although the ration should be as economical as possible, condition and results are the primary objective, even at somewhat added expense.
2. **Supplement to achieve a balanced ration.** The supplement(s) should balance out the available grass, keeping in mind the varied requirements of (a) cattle of different ages, (b) lactating vs dry animals, and (c) other differences.
3. **Give an assist to early pastures.** If at all possible, allow 6 to 8 in. of growth before turning to pasture. If there are good reasons to turn out earlier, remember that young, tender grass is washy (high in moisture) and low in energy. Thus, it should be supplemented with carbohydrate feeds, preferably grass hay plus high-energy concentrates.
4. **Increase carrying capacity.** Supplemental feeding offers a practical way to step up the carrying capacity of pastures, especially prior to and following the period of peak pasture growth.
5. **Extend the grazing season.** Since there is no finer place for purebreds, the pasture season should be extended as much as possible; year-round grazing being the ultimate where the climate will permit it. Experiments show that it is possible in many areas to lengthen the grazing season by nearly two months by supplemental feeding on pasture.

6. **Prevent overgrazing or undergrazing.** Pastures should never be grazed more closely than 2 to 3 in. during the growing season, and 3 to 5 in. should be left for winter cover. Also, undergrazing should be avoided because (a) mature forage is unpalatable and of lower nutritive value, (b) tall-growing plants may drive out low-growing plants by shading, and (c) weeds, brush, and coarse grasses are apt to gain a foothold. Supplemental feeding is the answer to governing this situation.

SEPARATING PUREBREDS BY CLASSES AND AGES

It is important that different classes and ages of purebred cattle be kept in separate pastures and supplemented according to needs. The following groups should be sorted out:

1. **Heifers.** Because they are growing, heifers require more liberal feeding, especially from the standpoint of energy and proteins. However, this word of caution is in order: *Do not overfeed heifers* to the point that reproduction is adversely affected. A combination of grass, exercise, and proper supplementation offers the ideal way in which to condition and grow heifers.

2. **Pregnant cows.** Pregnant cows should be in healthy, vigorous condition. Mature cows that are overfat—such as show cows—should be let down through a combination of exercise and limited, but balanced, rations.

Most of the ration of pregnant cows should consist of pasture plus such supplements as required—with emphasis on proteins, minerals, and vitamin A; and the kind and level of supplementation should be varied according to the quality and quantity of grass available.

Fig. 19-15. Replacement heifers need to be separated from older cows. A combination of grass, exercise, and proper supplementation offers the ideal way in which to condition and grow heifers. (Courtesy, *Livestock Weekly*, San Angelo, TX)

3. **Lactating cows.** Immediately before and after calving, cows should be fed lightly and with laxative feeds. At this time, the amount of supplementation should be governed by the milk flow, the condition of the udder, the demands of the calf, and the appetite and condition of the cow.

Until weaning time, the growth of the calf is chiefly determined by the amount of milk available from its dam, plus whatever assist is given through creep feeding. With purebreds, where maximum early development of the calf is so important, it is generally good business to give the mother the proper feed for production of milk. A combination of pasture plus supplement is the practical way to bring this about.

END RESULT IN PUREBREDS IMPORTANT

In purebreds, the end result is all-important, even at somewhat added expense. In addition to supplying proper balance of nutrients, the supplement should provide variety and palatability. Also, it should not make for bloat, scours, or other digestive disturbances.

PASTURE AND RANGE SUPPLEMENTATION

Where dried grass cured on the stalk is grazed, or where insufficient pasture is available—perhaps due to drought or overstocking—supplemental feeding is necessary. Also, supplemental feeding is a way in which to extend the grazing season, both early and late.

SORTING PASTURE AND RANGE CATTLE

When supplemental feeding is planned, it is strongly recommended that cattle first be sorted by age and condition groups. Heifers should not be supplemented at the same levels as older cattle. They are growing; thus, they must be fed more liberally (see Fig. 19-3). Also, heifers have need for more protein (see Fig. 19-4), and they must be fed for a longer period. Likewise, thin cows should be placed where they can be given extra feed and special care. More energy, proteins, and minerals are required for a cow suckling a calf than for a pregnant or dry cow. The nutritional

Fig. 19-16. Feeding protein cubes or cakes on the ground. Also, this provides a method of checking the animals because they are attracted by the sight or sound of the vehicle when they know that there is something to eat. (Courtesy, Oklahoma State University)

requirement of cows nursing calves is approximately 50% higher than for pregnant cows.

HOW TO CHOOSE A PASTURE OR RANGE SUPPLEMENT

Every cattle producer faces the question of what supplement to use, when to feed, and how much to feed under the particular conditions.

In supplying a supplement to range cattle, the following requisites should be observed:

1. It should balance the diet of the animals to which it is fed, which means that it should supply all the nutrients needed by the animal which are missing in the forage.
2. It should be fed in such a way that each animal gets its proper portion.
3. It should be fed in a form that is convenient and practical from the standpoint of the feeder, and that will least disturb the animal.
4. It should be economical in terms of cost per unit of needed nutrients (see Chapter 8, section titled "How to Determine the Best Buy in Feeds").

The net profit resulting from the use of the supplement, rather than the cost per ton, should determine the choice of the supplement. This philosophy is the same as that which normally prevails in livestock marketing; where net receipts, rather than charges for marketing services, should be the determining factor.

TYPES AND SYSTEMS OF PASTURE AND RANGE SUPPLEMENTATION

There is no one best and most practical pasture or range supplement for any and all conditions. Many different feeds may be, and are, used; among them (1) ranch- or locally-produced hay, (2) alfalfa pellets or cubes, with or without fortification, and (3) supplements of various kinds.

Also, cattle producers can lessen the labor attendant to the daily feeding of a pasture or range supplement by (1) hand-feeding cubes at intervals, rather than daily, (2) using protein blocks, (3) using liquid protein supplements, or (4) self-feeding salt-feed mixtures. Where these feeding systems do not result in the neglect of the herd, there is no effect upon the health and weight of the cows, percent calf crop, or weaning weight of calves.

RANGE CUBES OR PELLETS

Traditionally, cattle have been supplemented either once or twice daily on pasture or range. Where this practice is followed, a urea-containing range cube or pellet, similar to the formulation shown in Table 19-13, may be used. Cubes may be scattered on the ground.

TABLE 19-13
RANGE CUBE OR PELLET, WITH UREA

Ingredient	Percent	Per Ton	
	(%)	*(lb)*	*(kg)*
Alfalfa meal	15.0	300	*136*
Soybean, cottonseed, linseed[1] and/or peanut meal	32.5	650	*296*
Urea, 45% grade	4.0	80	*36*
Corn, barley, wheat, oats, and/or milo	34.7	694	*315*
Molasses	10.0	200	*91*
Salt	1.0	20	*9*
Dical., or equivalent	2.0	40	*18*
Trace minerals	0.5	10	*5*
Vitamin A[2] (30,000 IU/gram potency)	0.3	6	*3*
Total	100.0	2,000	909

Calculated Analysis

	(%)
Crude protein[3]	32.2
Fiber	6.4
Fat	2.0
Calcium	0.9
Phosphorus	0.8
TDN	67.2

[1]If linseed is used, limit to 6% of the ration.

[2]In low sunshine areas, also add 6 million IU of vitamin D/ton of finished feed.

[3]This includes not more than 11.24% equivalent protein from nonprotein nitrogen; 34.9% of the total protein is furnished by urea.

Fig. 19-17. Range cubes fed on pasture or range. Many cattle producers prefer this method of supplementation, primarily for reasons of convenience and reducing losses from wind blowing. (Courtesy, Ralston Purina Company, St. Louis, MO)

Urea-containing supplements, particularly those containing high levels of urea, should not be fed at intervals on the range because (1) range forages are relatively low in energy, and (2) urea is extremely soluble and its nitrogen becomes available very quickly in the rumen. Where nonprotein nitrogen is used in a range cube or pellet, a slow-release product is safest.

HAND-FEEDING AT INTERVALS, RATHER THAN DAILY

The Texas Station compared (1) daily feeding, (2) twice weekly feeding, and (3) three times weekly feeding; with each group hand-fed. Cottonseed meal was used as the supplement at two levels: (1) 14 lb/head/week, and (2) 21 lb/head/week.

As a result of a four-year study, it was reported that—

1. The group fed twice weekly had a slight advantage over feeding daily and 3 times weekly in (a) weight change of cows, (b) percent calf crop weaned, and (c) weaning weight of calves.
2. Twice weekly feeding saved 60% in time and equipment over daily feeding.
3. Twice weekly feeding did not cause any digestive disturbances, even when fed 10½ lb at one time (one-half the allowance of the 21 lb/week level). It took these cows about 2 hours to consume their 10½ lb share of cottonseed cake, which gave the timid and the slow eaters an opportunity to get their share.
4. The cows fed twice weekly grazed more widely over their range than those fed daily.

Based on the above study, plus observations and experiences, the authors recommend feeding a non-urea range supplement twice weekly; allocating in each of the two feedings one-half as much supplement as would have been fed in a week on a daily feeding basis.

Protein cubes may be scattered on the ground—2 or 3 times a week. This offers a method of checking the animals because they are attracted by the sight or sound of the vehicle when they know that there is something to eat.

A suggested pasture-range supplement without urea is given in Table 19-14.

Twice weekly feeding has two distinct advantages over the use of salt-feed mixes: (1) It alleviates the cost of using excess salt, which has no nutritive value when so used; and (2) it forces inspection of the herd twice each week, which is as infrequent as is desirable.

TABLE 19-14
RANGE CUBE OR PELLET, WITHOUT UREA

Ingredient	Percent	Per Ton	
	(%)	(lb)	(kg)
Soybean or cottonseed meal (41%)	72.7	1,454	661
Alfalfa meal	15.0	300	136
Molasses	8.5	170	77
Salt	1.0	20	9
Dical., or equivalent	2.0	40	18
Trace minerals	0.5	10	5
Vitamin A[1] (30,000 IU/gram potency)	0.3	6	3
Total	100.0	2,000	909

Calculated Analysis

	(%)
Crude protein	32.6
Fiber	8.3
Fat	1.0
Calcium	0.9
Phosphorus	0.9
TDN	70.9

[1]In low sunshine areas, also add 6 million IU of vitamin D/ton of finished feed.

PROTEIN BLOCKS

Protein blocks are just what the designation implies. They are compressed protein blocks, generally weighing from 50 to 200 lb each.

Blocks may be placed in grazing areas where cattle have frequent access to them, with 1 block provided to 15 to 25 cows. Intake will vary with the feed supply and the type of block. Generally, it is planned to limit feed consumption to less than 2 lb per

Fig. 19-18. Protein block in use on pasture—a means of lessening the labor attendant to the daily feeding of a protein supplement on pasture or range. (Courtesy, Moorman Manufacturing Co., Quincy, IL)

head per day by hardness of block and salt and/or fat content.

LIQUID PROTEIN SUPPLEMENTS

Liquid supplements in a *lick* tank can be offered free-choice. This is a convenient and satisfactory way in which to supply protein, energy, and other nutrients, so long as the cattle do not consume more than they need.

In the United States, the vast majority of the liquid protein supplements are patented. Also, they are difficult to home mix. As a result, the universities have done little experimental work on them. However, Australia has long fed a great deal of molasses-urea on the range, due to its frequent droughts. Also, daily feeding on its vast stations (ranches), many of which stretch over many miles, isn't practical—it requires too much labor and travel. Hence, the Australians are more experienced than we are when it comes to using molasses-urea to supplement grass, and as a means of saving labor. Accordingly, the senior author has come to know their formulations and techniques, firsthand, in three visitations that he has made *down under.* Their feeding directions appear to be about as follows:

1. Do not feed the urea-containing supplement to cattle that have been without feed for 36 hours until they have had a chance to fill the rumen with grass.
2. Since some animals tend to overeat, it is desirable to restrict consumption to a desired level (1 to 2 lb/head/day). This may be accomplished by a free-turning plastic or wood wheel dipped in a tank or other similar equipment, or by adding an unpalatable ingredient.
3. It may take several days for cattle on dry grass to get accustomed to a liquid protein supplement.
4. Once you start feeding the liquid protein supplement, never let the animals run out of it until you discontinue feeding it entirely.
5. Self-feed regular minerals in addition.

SELF-FEEDING SALT-FEED MIXTURES

The practice of using salt as a governor to limit feed consumption on pasture or range has been around a very long time. It was ushered in as a laborsaving device for cattle and sheep in inaccessible and rough areas.

The proportion of salt to feed may vary anywhere from 5 to 40% (with 30 to 33⅓% salt content being most common), with the actual intake of feed supplement limited to 1 to 2½ lb daily. By varying the proportion of salt in the mixture, it is possible to hold the consumption of feed supplement to any level desired. In some range areas, a reduction of the salt level from 33⅓ to 24% will increase consumption by about 50%. When a liberal feeding of grain on pasture is desired, 5% salt may be sufficient.

Two suggested salt-meal supplements follow:

SALT-COTTONSEED OR SOYBEAN MEAL, 41% (Do Not Pellet)

- *1 part salt-2 parts meal (plus 4 lb vitamin A, 30,000 IU/g* potency, per ton of mix)—Guarantee: 27% crude protein, maximum of 35% salt, and 18,000 IU of vitamin A per pound. Cattle will generally consume about 1½ lb daily of such a mix.
- *1 part salt-3 parts meal (plus 3 lb vitamin A, 30,000 IU/g* potency, per ton of mix)—Guarantee: 30% crude protein, maximum of 27% salt, and 13,000 IU of vitamin A per pound. Cattle will generally consume about 2 lb daily of such a mix.

Based on experiments and experiences, the following points are pertinent to self-feeding salt-feed mixtures to range cattle:

1. The practice need not be limited to any specific protein supplement or feed.
2. It is best that salt mixes be in meal form, rather than pelleted. If pellets are small and soft, they will work satisfactorily. However, there is always the hazard that they will be hard enough to permit cows to swallow them without the salt being fully effective as an inhibitor, with the resulting overeating.
3. The proportion of salt and feed may vary anywhere from 5 to 40% (with 30 to 33⅓% salt content being most common), with the actual intake of feed supplement limited to 1 to 2½ lb daily. By varying the proportion of salt in the mixture, it is possible to hold the consumption of feed supplement to any level desired. In some range areas, a reduction of the salt

level from 33⅓ to 24% will increase consumption by about 50%. When a liberal feeding of grain on pasture is desired, 5% salt may be sufficient.[7]

4. The quantity of salt and the proportion of salt to supplement required to govern supplement consumption varies according to (a) the daily rate of feed consumption desired, (b) the age and weight of animals (higher quantities of salt are required in the case of older animals), (c) the fineness of the salt grind (fine grinding lowers the salt requirement), (d) the salinity of the water, (e) the severity of the weather, (f) the quality and quantity of forage, and (g) the length of the feeding period (as animals become accustomed to the mixture, it may be necessary to increase the proportion of salt).

5. It is common practice to prepare the starting feed by mixing 1 lb of salt to 4 lb of feed supplement, and to increase the proportion of salt in the mixture as the animals become accustomed to the feed.

6. It lessens the difficulty in starting cattle on a supplement, for sprinkling a little salt on the meal makes it more palatable.

7. It is recommended that animals be hand-fed a week or so before allowing free-choice to a salt-feed mixture; thus, getting them on feed gradually.

8. It is necessary to regulate or limit (by hand-feeding for a few days) the supply of salt-feed mixture when it is desired to shift animals from a straight feed supplement (such as cottonseed meal alone) to a salt-feed mixture. Otherwise, hungry animals may consume too much.

9. It is estimated that the practice increases the total salt consumption to 8 to 10 times that required in conventional salt feeding, and doubles or triples the water consumption.

10. If the salt-feed mixture is placed in close proximity to the water supply, it will make for restricted grazing distribution on the range, because of the greater intake of water on a high-salt diet. On the other hand, if the salt-feed mixture is shifted about on the range, it will make for desirable distribution, because of the animals following the feed supply.

11. It reduces the labor required in feeding, promotes more uniform feed consumption (among the greedy and the timid), and permits animals to eat at their leisure with less disturbance during blizzards or cold weather.

12. It lessens the space required for feed equipment (bunks or feeders) to 20% of that required in conventional hand-feeding, but makes it desirable that the feeder be constructed so as to protect the mixture from the weather (especially wind and rain).

[7]At the Irrigation Experiment Station, Prosser, Washington, it was found that 7 1/2% salt limited grain consumption by yearling steers on pasture to 10 to 12 lb daily; 5% salt limited grain consumption to 12 to 14 lb.

13. It is equally applicable to feeding during droughts, on dry summer range, and in the winter months.

14. It is commonly believed that under conditions of short feed supply (submaintenance) and relatively inaccessible water supply, animals may consume sufficient salt in this manner to produce toxic effects, especially during the winter months when low temperatures tend to lessen the water intake.

15. The practice of self-feeding salt-feed mixtures is well adapted to inaccessible and rougher areas, where daily feeding is difficult. In no case, however, should it be an excuse to neglect animals, for herds need to be checked often.

16. It reduces the consumption of minerals other than salt to practically nothing, with the result that mineral deficiencies must be considered.

FATTEN CULL COWS FOR MARKET

There is a new profit potential in the cow business—that of fattening cull cows for market. Where herd numbers are being held fairly constant, about 20% of the cows are culled each year, because of (1) poor calves, (2) being barren, (3) spoiled udders, (4) disease, (5) old age, and (6) miscellaneous reasons. Traditionally, these culls were sent to market—and the sooner the better. Today, good money can be made by holding and fattening these culls prior to slaughter.

Cutter and Canner cows are in demand; and the price spread between them and Choice steers has narrowed, both on foot and on the rail—especially the latter. The reasons: (1) the increased demand for hamburger and other ground and prepared meats, for which cow beef is admirably suited; (2) fewer cows being slaughtered, because decrease in dairy cows has slowed and beef cow numbers are expanding; (3) imports of manufacturing-type beef having held at rather stable levels; and (4) better quality cow carcasses available than formerly.

All this suggests the following to cow-calf producers:

1. More than ever, they cannot afford to keep marginal or barren cows. They should keep records, perform pregnancy tests, and remove loafers from the herd and fatten them for slaughter.

2. They should fatten cull cows on cheap, high-roughage rations, using a maximum of such feeds as silage, haylage, green chop, wheat pasture, irrigated pasture, etc. If they have good teeth and they are healthy, they will make remarkable gains.

3. They should compare selling on foot vs selling on the rail. The latter may be best.

Also, buying cull cows and finishing them commercially offers good opportunity, but numbers are limited, and buying such cows is a problem. They are scattered; and it is important that they have good teeth, and that they be healthy. Remember, too, that cows are usually culled because of feed shortages, breeding problems, disease, internal parasites, or age. In short, somebody had a problem with them.

FACILITIES

The investment in beef cattle buildings and equipment should be kept to a minimum. The farmstead should be neat and attractive, particularly where purebred cattle are sold for breeding purposes.

Buildings and yards should be located on a well-drained area; and plans should be made for the efficient feeding and management of the herd.

■ **Housing**—Housing for beef cows need not be elaborate or expensive. Allow the herd to be outside during the grazing season and most other times—even in winter. Beef cattle naturally grow long, thick hair coats in the fall. Except where it is extremely cold and windy, the most shelter they need is a wooded area or a hill for a windbreak. Generally speaking, the producers should give more attention to feed storage and to saving labor in feeding and manure handling than to the necessity of getting the cows inside. Also, it should be recognized that a combination of drafts, dampness, poor ventilation, and lack of sunlight creates hazards.

If a shed or barn is used, have it open to the south or east (away from the direction of prevailing winds), with an adjoining lot to permit them to stay indoors or run out at will. Mature brood cows require 50 to 60 sq ft of shelter, yearlings 35 to 40 sq ft, and weaned calves 25 to 30 sq ft. If the cattle are fed roughage under the shelter, more space will be needed—60 to 75 sq ft per cow, and a little more for young stock than the figures given.

Sheds more than 20 ft deep are preferred. The greater the depth, the warmer and drier the building.

Pole-type barns or sheds are excellent for beef cattle. They should be built on high ground, so that there will be good drainage away from them; they should have dirt floors; and they should be built high enough for convenience in removing manure—10 ft from the floor to the plate is sufficient.

Pole-type barns can be arranged so that roughage can be stored in the back of the shed, and so that, by use of movable racks, the cattle are permitted to eat their way back during the winter.

■ **Fencing**—All corral fences should be built of heavy board material, pole, or rail type construction.

For holding cattle, 3 or 4 barbed wires are sufficient. Four or five wires should be used along roads and boundary lines, and 3 wires are sufficient for other areas. Woven wire is satisfactory, but more expensive. The life of the fence depends to a great extent on the life of the posts and the stability of the corner posts.

Electric fences are satisfactory for temporary fencing or rotation grazing.

■ **Cattle-handling facilities**—Lack of adequate facilities for handling beef cattle prevents producers from carrying out practices which would otherwise be routine and would increase their returns from the beef cattle operation. Time and money spent in planning and developing handling facilities for cattle will return dividends in terms of added profits and greater efficiency. Cattle-handling facilities should include the following:

1. A permanent corral or holding pen located near the main livestock buildings, with a working alley and several attached smaller catch pens to help in sorting cattle.
2. A headgate, with a chute leading thereto.
3. A portable corral for use in pastures that are a considerable distance from headquarters, or a permanent corral constructed in such pastures.
4. A stationary or portable loading chute.
5. A scale, or ready access to a scale.
6. Cattle stocks where cattle are fitted for shows and sales, and for use in trimming feet.

■ **Other buildings and equipment**—Other types of buildings and equipment for the beef cattle establishment are covered in Chapter 11, Buildings and Equipment for Beef Cattle, of this book; hence, the reader is referred thereto.

QUESTIONS FOR STUDY AND DISCUSSION

1. How may feed costs, which account for 65 to 75% of the total cost of keeping cows, be lowered?

2. How may a practical cattle producer meet the added energy requirements of a brood cow during the critical 100 days, extending from 30 days before calving until 70 days after calving?

3. Compare and discuss the nutritive requirements of (a) dry pregnant cows, (b) bred heifers and

(c) lactating cows. From a practical standpoint does this mean that a cattle producer should separate different classes and ages of cattle, then feed them according to needs?

4. Cite experimental studies showing reproductive failure in cows due to a deficiency of each of the following: (a) energy, (b) protein, (c) phosphorus, and (d) vitamin A. How would you rectify each of these deficiencies?

5. Discuss the winter feeding of brood cows. How different are the rations of dry pregnant cows and cows nursing calves?

6. It is estimated that the U.S. annual production of corn residue totals more than 200 million tons, sufficient to winter 152 million pregnant cows. Yet, much of this potential cow feed is not utilized. Why is not more corn residue fed to cattle?

7. How would you supplement corn residue for each: (a) dry pregnant cows, or (b) cows suckling calves?

8. List and discuss the feeding value of crop residues, other than corn residue, that can be used for feeding cows.

9. Discuss nutrient deficiencies of cattle that are frequently encountered on U.S. ranges. How would you rectify each one?

10. Wherein does the supplementation of purebreds on pastures and ranges differ from the supplementation of commercial cattle?

11. What type and system of pasture and range supplementation would you recommend? Justify your answer.

12. Considering current feed and fat cow prices, present figures to show whether or not it will pay to fatten cull cows for market.

13. Select a particular area, then determine what facilities would be necessary and practical for the operation of a 300-cow herd.

SELECTED REFERENCES

Title of Publication	Author(s)	Publisher
Beef Cattle, Seventh Edition	A. L. Neumann	John Wiley & Sons, Inc., New York, NY, 1977
Beef Cattle Feeding and Nutrition, Second Edition	T. W. Perry M. J. Cecava	Academic Press, Inc., San Diego, CA, 1995
Beef Cattle Production	J. F. Lasley	Prentice-Hall, Inc., Englewood Cliffs, NJ, 1981
Beef Cattle Production	K. A. Wagnon R. Albaugh G. H. Hart	The Macmillan Company, New York, NY, 1960
Beef Cattle Science Handbook	Ed. by M. E. Ensminger	Agriservices Foundation, Clovis, CA, pub. annually 1964–1981
Beef Production and Management Decisions	R. E. Taylor	Macmillan Publishing Co., New York, NY, 1994
Beef Production and Management, Second Edition	G. L. Minish D. G. Fox	Reston Publishing Company, Inc., Reston, VA, 1982
Commercial Beef Cattle Production, Second Edition	Ed. by C. C. O'Mary I. A. Dyer	Lea & Febiger, Philadelphia, PA, 1978
Nutrient Requirements of Beef Cattle, Sixth Revised Edition	National Research Council	National Academy of Sciences, Washington, DC, 1984
Nutrient Requirements of Beef Cattle, Seventh Revised Edition	National Research Council	National Academy of Sciences, Washington, DC, 1996
Stockman's Handbook, The, Seventh Edition	M. E. Ensminger	Interstate Publishers, Inc., Danville, IL, 1992

Bulls, the most influential part of any beef cattle breeding program. (Courtesy, American Polled Hereford Assn., Kansas City, MO)

BULLS

Contents *Page*

Fig. 20-1. Beefmaster bull. (Courtesy, Beefmaster Breeders Universal, San Antonio, TX)

There are an estimated 2 million beef bulls of breeding age in the United States, and about 800,000 new bulls are selected each year. But numbers alone do not portray their importance. They are the most influential part of any beef cattle breeding program. Their selection can make or break any herd, either purebred or commercial. Since a much smaller proportion of males than of females are normally saved for replacements, it follows that selection among males can be more rigorous and that most of the genetic progress in a herd will be made from selection of males. Thus, if 2% of the males and 50% of the females in a given herd become parents, about 75% of the hereditary gain from selection will result from the selection of males and 25% from the selection of females, provided their generation lengths are equal. If the generation lengths of males are shorter than the generation lengths of females, as is true in most beef herds, the proportion of hereditary gain due to the selection of males will be even greater. Thus, it is estimated that between 80 and 90% of the progress realized in most beef herds comes about through the bulls used. There is little doubt, therefore, that proper bull selection is the most effective way in which to make genetic progress in a beef herd.

The importance of bull selection is sometimes confused by the statement that the bull is half the herd. This is true in the sense that half of the genetic material making up the calf crop comes from the sires used. However, each bull's contribution to the calf crop is 30 times greater than the average cow if we assume 30 cows per bull and a 100% calf crop from the cow herd.

Also, it is noteworthy that the bull's influence extends beyond the immediate calf crop in that his daughters, and perhaps sons, too, remain in production in the herd long after he is gone—they're constant *reminders*.

Another factor which causes the bull to play such an important role in herd improvement is the much greater selection differential that can be realized in selecting bulls compared to selecting females. This means that the bulls selected can be much better than the herd average compared to the heifer replacements selected. This is primarily because of the relative numbers of each that are required as replacements.

SOURCE OF BULLS

Generally speaking, cattle producers have three sources of herd sires: (1) purchase; (2) raising their own and (3) artificial insemination service. Additionally, there is a choice of sources when it comes to purchase or AI service.

PURCHASE OF BULLS

Bulls may be purchased from a breeder, at a central test station, or at a consignment sale.

■ **Purchase from a breeder**—Purchase directly from a breeder, either by private treaty or in a production sale, is the most common source of bulls. From the standpoint of the bull buyer, purchase from the breeder has the following advantages: (1) it provides an opportunity to buy a number of bulls of similar breeding, thereby leading to greater uniformity in the herd in which they are used; (2) it makes it possible to select from among animals that have been exposed to the same environment from birth, and that can be compared on the basis of their performance without environmental differences; (3) it is usually possible to inspect their sires, dams, and near relatives; and (4) there is adequate time for evaluation. The only real disadvantage in this method of buying is that it is more difficult to make comparisons with cattle bred by other breeders; hence, the bull buyer must be an astute cattle producer.

■ **Purchase at a central test station**—Over the years, there has been a trend toward evaluating bulls in central test stations, followed by an auction sale at the close of the test period. At these stations, bulls are entered for postweaning gain comparisons with bulls from other herds. The chief advantages of purchasing bulls from a test station are: (1) the chance for buyers to get acquainted with performance breeders who are willing to compete with each other on the basis of the performance of their breeding stock when placed in a comparable environment; (2) the opportunity to make gain-on-test comparisons between herds; (3) the likelihood that breeders have sent their best performance prospects to the test station for the promotional benefit, and that purchases can be made therefrom; and (4) the bulls are seldom heavily fitted. The disadvantage is that weight per day of age during the test period is affected to some degree by preweaning environment; hence, valid comparisons are possible only between bulls of the same herd. However, most test stations now require an adjustment period of 3 to 4 weeks prior to the actual test as a means of alleviating much of the difference due to compensatory gain.

- **Consignment sales**—Traditionally, consignment sales have been (1) a market place for establishing values back home, (2) a source of revenue of breed associations, (3) a meeting place for purebred breeders and commercial cattle producers, and (4) a stage for creating and promoting change in the kind of cattle produced. Normally, bulls in consignment sales are sifted, graded, and/or placed, and those with obvious defects or unsoundnesses are culled.

The chief advantages of a consignment sale from the standpoint of a bull buyer are (1) convenience, and (2) high-quality animals, due to (a) sifting, and (b) sellers using them to establish values back home. The disadvantages: (1) frequently bulls sold in consignment sales are overfitted; and (2) they provide little basis to compare growth rates of the consigned bulls, except within breeder groups.

RAISING YOUR OWN BULLS

Large purebred cattle establishments commonly raise a good part of their own replacement bulls. Some large commercial establishments also find it desirable to raise their own bulls. The main advantages from raising your own replacements are: (1) the bulls are raised and selected in the environment in which they are expected to perform; (2) management has complete control of the breeding program, with the result that they can provide specific lines and select for certain traits; and (3) the cost is usually less, because it alleviates transportation and selling charges. The disadvantages are that (1) a good commercial cattle producer may not necessarily be a good seed stock producer, and (2) it requires more skilled labor, more accurate records, and more facilities and fencing.

ARTIFICIAL INSEMINATION (AI) BULLS

Selection of artificial insemination sires involves essentially the same considerations as the selection of bulls for natural service. Additionally, they are usually progeny tested—their calves have been evaluated for performance. This means that for an AI bull it is usually possible to get more accurate information on birth weight and growth rate of his calves, ease of calving and mothering ability of his daughters, and carcass yield and grade of his progeny.

In addition, some of the major AI companies now score daughters of sires for many traits including udder characteristics, structural soundness, frame size, and femininity. This information is presented along with EPDs (expected progeny difference) for performance traits and the two combined provide a wealth of information on prospective sires.

BULL SELECTION—FACTORS TO CONSIDER

Cow-calf producers are striving for genetic improvement. They are hoping that the bulls selected are genetically superior to their cows—that they will transmit their apparent superiority to their offspring, thereby continuing to improve the herd. But how can he they sure?

Fig. 20-2. Brahman bulls, along with bulls of the newer breeds carrying Brahman blood, are selected by many cattle producers in the southern part of the U.S. This shows a 3-year-old Brahman bull, typical of the type used by many commercial producers in the South. (Courtesy, *The Florida Cattleman*)

Although there's no foolproof method through which herd-improving sires can be assured, application of the selection procedure and criteria which follow will come as near to doing so as is scientifically possible today.

Before starting out to buy a new herd sire, evaluate your cow herd and current calf crop. Where do they need the most improvement? Is it size, muscling, soundness, gaining ability, or some other trait? Then look for a bull to correct those weaknesses. But do not lose existing valuable traits to improve a weak one.

Decide which herds you should visit or which sales

you should attend. Buy from reputable breeders who are known to be doing a good job of production. Take time in making a selection. Start well ahead of the time you need a bull, so that you will have a better selection of bulls from which to choose.

Many bulls are selected on visual appraisal and age alone. But more and more cattle producers, both purebred and commercial, are selecting sires only after studying detailed performance records on not only the individual bulls but their near relatives. Since sire selection may be the most important factor in determining the success of a beef cattle breeding program, every breeder should use all the information available when selecting herd sires. At the outset, however, it should be recognized that a performance tested bull will seldom excel in all the economically important traits. The cattle producer must decide, therefore, how much importance will be accorded each trait—how much emphasis will be placed on yearling weight, rate of gain, and carcass evaluation, etc. This will vary from herd to herd, depending primarily on the level of performance in the trait already attained in the particular herd.

For maximum improvement in the herd, the factors which follow should be considered in bull selection. (Also see Chapter 7, section headed "Bull Breeding Evaluation.")

BREED

The commercial cattle producer has a wide germ plasm choice available, especially if artificial insemination is used. Over 300 breeds are available, along with all combinations of crosses. But the choice of the breed(s) or cross is not easy. Available experiment station, U.S. Department of Agriculture, and even field trial data must be interpreted and inferences made by the producer to management. Just how well will this new breed fit into beef production in the United States and on the particular farm or ranch? If replacement females are to be retained, care must be taken to evaluate the reproductive and maternal potentials of the breed. Will the breed have a terminal sire market only, or will it be a maternal breed?

Reynolds, W. L., *et al.*, USDA, Miles City, Montana, reported that cows mated to large sire breeds had longer gestation lengths, gave birth to heavier calves, and had more calving difficulty than cows mated to medium-sized sire breeds. Calves sired by the higher milk-producing breeds had heavier birth weights than calves sired by medium milk-producing breeds.[1]

Selection of the breed and planning the breeding program must go hand in hand. Remember that grading up takes a long time. Remember, too, that half the heterosis will be lost with each successive backcross—and this may result in disappointment.

AGE

The use of well-grown bulls, 14 to 15 months of age, is increasing. Performance tested bulls are generally large enough to use and sexually mature by the end of their test period. When you hold a bull out of service for 10 to 12 months longer than necessary, you've got nothing to show for it but a feed bill.

Where a breeder is making substantial progress in the herd, consideration may well be given to securing bulls out of young sires and young dams even though there is not as much data on which to make a decision.

An older bull that has been proven on the basis of his progeny to be a superior breeding sire, and that is free of reproductive diseases, is usually a sound investment. Such a bull may be available from a neighbor, on an exchange arrangement. Also, proven bulls should be available through artificial insemination.

HERD MERIT

The really difficult problem is the selection of the herd or source, once the breed has been decided upon. Whenever possible, bulls should be obtained

Fig. 20-3. WSU Cornerstone 704X, impressive production tested Polled Shorthorn bull, bred by Washington State University, shown in breeding condition. When fitted, this bull weighed 2,300 lb. (Courtesy, American Shorthorn Assn., Omaha, NE)

[1]Reynolds, W. L., *et al.*, "Biological Type Effects On Generation Length, Calving Traits, and Growth Rate," *Journal of Animal Science*, 1990, 68:630–639.

from herds having the highest average merit. But it is difficult to estimate the average merit of a herd in relation to other herds and the rest of the breed. However, by determining the production level of the herd and by observing how animals produced in the herd have performed for other breeders, an evaluation can be made.

Of course, not all animals produced in a herd will be breeding bull prospects. So, relative rank of a bull prospect among other bulls produced in the same calf crop and managed in the same way is important. The bull buyer should take the *top cut* of the performance tested bulls from a given, outstanding herd that is on production test. Weight ratios computed in most performance records are a convenient device to determine how an animal ranks among other animals raised at the same time in a herd and EPDs allow individual animals to be compared to all other animals in a breed.

Some important questions to ask of a seller of breeding stock as a means of determining the merit of the herd are:

1. How many years has this herd been selected for performance?
2. What is the average level of performance in your herd for the relevant traits?
3. What is your breeding program?
4. What is your management system?

WEANING WEIGHT (205-Day Weight)

A bull's own weaning weight (adjusted for age of dam) is both an indication of his genetic capability for growth and his dam's mothering (including milking) ability. A weaning weight ratio will indicate how a particular bull's weaning weight compared to the average of other bull calves raised at the same time under the same conditions. His weaning weight EPD will indicate how he compares to other animals in a breed.

POSTWEANING GROWTH RATE

A measure of gain from weaning time to one year of age is a good indication of a bull's genetic capability for gain. The gain period may be in a central test station in comparison with bulls from other herds or it may be on the farm or ranch in comparison with other bulls raised in the same herd. Either test is satisfactory. But the rate of gain itself must be evaluated in light of the feeding level and management conditions.

Post-weaning gain is about 30% heritable. (See Chapter 5, Table 5-5.) Yearling weight EPD is the best indicator or measure of postweaning growth rate.

CARCASS MERIT

The two main considerations in carcass merit are (1) quality, and (2) ratio of lean to fat and bone. At the present time, carcass quality can only be determined by slaughtering an animal and measuring the various things that contribute to quality in the carcass. This means that carcass quality information on prospective sires can only be obtained by slaughtering relatives, such as half-sibs from the same sires. Of course, once a bull is selected, he can be progeny tested by slaughtering some of his offspring and obtaining carcass information, but this takes time. Few breeders can supply meaningful records of carcass quality of their herds. Ultrasound is being used to estimate degree of marbling and quality grade of live animals.

Various methods can be used to estimate muscle content of prospective breeding animals directly. A trained judge can exhibit considerable accuracy by visual appraisal. Fat thickness can be obtained by mechanical means such as probing or by the use of ultrasonic devices. Knowing fat thickness helps in estimating muscle content of animals. Ultrasonic devices can be used to estimate the size of certain muscles (such as the loin eye muscle) which are related to total muscle content of the animal.

The K-40 whole body counter is being used to estimate the total muscle content of live beef cattle. By using this method, prospective breeding animals can be evaluated for muscle content and the better animals selected for breeding use or further testing. The K-40 count is highly related to total muscle content if the variables are fairly constant. Animals to be evaluated and compared should be of about the same age and weight and should come from the same management background. Also, there must be meaningful comparisons between animals. For example, it would do little good to obtain a K-40 count on a single sire in range condition.

The results of K-40 counter-evaluation are usually expressed as percent lean of liveweight, or as lean per day of age. To be good breeding prospects, bulls should grow rapidly to one year of age and still have a high percent lean of liveweight.

SIRE'S AND DAM'S PERFORMANCES

The records of the sire and dam of each prospective herd bull should be evaluated if possible. The dam's record may show her regularity of production, her mothering ability as reflected by the weaning weight of her calves, and the postweaning gain and carcass merit of her calves.

If sires have produced several calves, their EPDs for birth weight, weaning weight, yearling weight, milk production, and carcass merit should be evaluated, particularly relative to other sires in the same breed.

INDIVIDUALITY

Only bulls that are better than average should be used for breeding purposes.

The bull being considered for purchase should be well developed for his age, of good type, masculine, and structurally sound. Young bulls should be long, tall, and not excessively fat—indications that they will continue to grow. Bulls that show signs of early sexual maturity are not good prospects for continued rapid growth and large mature size.

The bull should have a strong, straight topline, well-sprung foreribs, and be well muscled. When viewed from the rear, the back and loin should slope gently downward, like a Quonset roof. Extreme flatness over the top indicates excess fatness, and is undesirable. The bull should be deeper in the fore flank than in the rear flank.

Look for a bulging arm, forearm, gaskin, and stifle; a rounded loin and round; and creases in the thighs. There should be a well-defined groove down the topline as a result of the loin eye bulging on each side of the backbone; and you should see the muscles when the animal walks.

If a bull is very smooth, he is likely either too fat or muscle deficient, or both. The bull should be masculine in front and have a sound pair of testicles behind; he should be alert—*on the look*, with head up and ears cocked; he should have a bold, masculine head, with reasonably prominent eyes, broad muzzle, and large nostrils; he should have a well-developed crest; he should have well-developed and clearly defined muscles, especially in the regions of the neck, loin, and thigh; and he should have well-developed external genitalia, with testicles of equal size and well defined, and a proper neck to the scrotum.

Fig. 20-4. An outstanding Angus bull, weighing 2,600 lb. (Courtesy, American Angus Assn., St. Joseph, MO)

Avoid heavy, pendulous sheaths, which tend to be a problem in certain breeds. Look for a sheath that is of medium size and neatly attached. It appears that bulls with pendulous sheaths transmit excessive development of sheath (and perhaps navel) to many of their offspring. Excessive sheath development often leads to injury to the sheath lining, and sometimes to the penis, especially when the bull is run on rough, brushy range.

The bull should be able to follow and breed cows on the pasture and range. This means that structural defects—particularly in the legs, feet, and joints—can shorten the productive life of any bull. This calls for structural soundness, especially in the underpinning. The legs should be straight, true, and squarely set. The foot should be large, deep at the heel, and have toes of equal size and shape that point straight ahead. The joints, particularly the hock and knee joints which are subject to great wear, should be correctly set, and without tendency toward puffiness or swelling. Sickle-hocked, post-legged, back at the knees (calf-kneed), or over at the knees (buck-kneed) are faults, and such animals should be scored down.

Where a purebred herd is involved, the bull should be true to breed type and color, factors of importance in merchandising cattle.

SEMEN AND FERTILITY EVALUATION

Semen quality is based on evaluating the ejaculate for (1) density of concentration, (2) motility, and (3) morphology. Also, the reproductive tract of the bull is palpated per rectum and evaluated.

It is advisable to have a semen or fertility evaluation made before you buy a bull, especially when purchasing for a single sire herd. The penalty for an infertile bull in a multi sire herd is not so great. Although a semen test will not provide absolute assurance that the bull will settle the females to which he is exposed, it is a strong indicator. Bulls that are not producing sperm cells, that are producing a high percentage of nonmotile or abnormal sperm cells, and those that have infections in the reproductive tract should be avoided. No matter how high the growth rate, and no matter how admirable all the other qualities, a bull that cannot sire a calf crop is useless.

PEDIGREE

In addition to recording lineage, pedigrees are

useful in selecting against undesirable recessive genes and such traits as vaginal or uterine prolapse.

The pedigree of a bull is not as important as it was once believed. When verbal descriptions were in vogue, names and numbers were sufficient. Today, selection pressure is being directed toward improving highly heritable traits, for which the bull's own performance is a good indicator of breeding value. Hence, the pedigree is less important.

In addition to identification of immediate ancestors, performance pedigrees, which list performance records, ratios, and EPDs of all immediate ancestors, are coming into use. Such pedigrees are an important aid to accurate sire selection and should be used when available.

The Beef Improvement Federation (BIF) recommends that the following basic performance information be included on each performance pedigree:

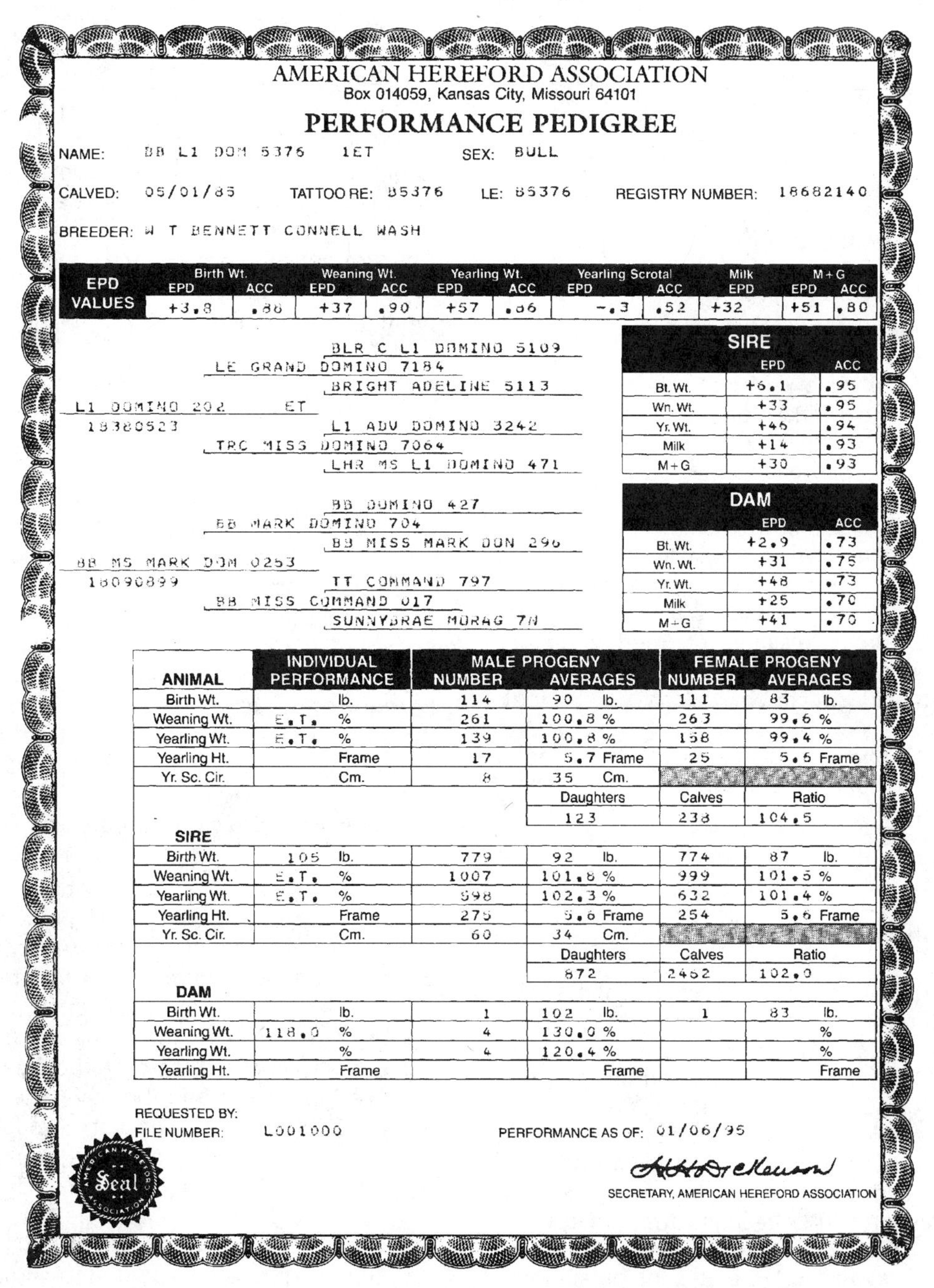

AMERICAN HEREFORD ASSOCIATION
Box 014059, Kansas City, Missouri 64101

PERFORMANCE PEDIGREE

NAME: BB L1 DOM 5376 1ET SEX: BULL

CALVED: 05/01/85 TATTOO RE: B5376 LE: B5376 REGISTRY NUMBER: 18682140

BREEDER: W T BENNETT CONNELL WASH

EPD VALUES	Birth Wt. EPD	Birth Wt. ACC	Weaning Wt. EPD	Weaning Wt. ACC	Yearling Wt. EPD	Yearling Wt. ACC	Yearling Scrotal EPD	Yearling Scrotal ACC	Milk EPD	M+G EPD	M+G ACC
	+3.8	.88	+37	.90	+57	.86	-.3	.52	+32	+51	.80

L1 DOMINO 202 ET
18380523
- LE GRAND DOMINO 7184
  - BLR C L1 DOMINO 5109
  - BRIGHT ADELINE 5113
- TRC MISS DOMINO 7064
  - L1 ADV DOMINO 3242
  - LHR MS L1 DOMINO 471

BB MS MARK DOM 0253
18090899
- BB MARK DOMINO 704
  - BB DOMINO 427
  - BB MISS MARK DON 296
- BB MISS COMMAND 017
  - TT COMMAND 797
  - SUNNYBRAE MORAG 7N

SIRE	EPD	ACC
Bt. Wt.	+6.1	.95
Wn. Wt.	+33	.95
Yr. Wt.	+46	.94
Milk	+14	.93
M+G	+30	.93

DAM	EPD	ACC
Bt. Wt.	+2.9	.73
Wn. Wt.	+31	.75
Yr. Wt.	+48	.73
Milk	+25	.70
M+G	+41	.70

ANIMAL	INDIVIDUAL PERFORMANCE	MALE PROGENY NUMBER	MALE PROGENY AVERAGES	FEMALE PROGENY NUMBER	FEMALE PROGENY AVERAGES
Birth Wt.	lb.	114	90 lb.	111	83 lb.
Weaning Wt.	E.T. %	261	100.8 %	263	99.6 %
Yearling Wt.	E.T. %	139	100.8 %	158	99.4 %
Yearling Ht.	Frame	17	5.7 Frame	25	5.6 Frame
Yr. Sc. Cir.	Cm.	8	35 Cm.		
			Daughters	Calves	Ratio
			123	238	104.5
SIRE					
Birth Wt.	105 lb.	779	92 lb.	774	87 lb.
Weaning Wt.	E.T. %	1007	101.8 %	999	101.5 %
Yearling Wt.	E.T. %	598	102.3 %	632	101.4 %
Yearling Ht.	Frame	275	5.6 Frame	254	5.6 Frame
Yr. Sc. Cir.	Cm.	60	34 Cm.		
			Daughters	Calves	Ratio
			872	2452	102.0
DAM					
Birth Wt.	lb.	1	102 lb.	1	83 lb.
Weaning Wt.	118.0 %	4	130.0 %		%
Yearling Wt.	%	4	120.4 %		%
Yearling Ht.	Frame		Frame		Frame

REQUESTED BY:
FILE NUMBER: L001000 PERFORMANCE AS OF: 01/06/95

SECRETARY, AMERICAN HEREFORD ASSOCIATION

Fig. 20-5. Performance pedigree of the Hereford bull, BB L1 Dom 5376, bred by W. T. Bennett, Connel, Washington. (Courtesy, American Hereford Assn., Kansas City, MO)

1. Birth weight, birth weight ratio, EPD, and accuracy.

2. Adjusted 205-day weight ratio, EPD, and accuracy.

3. 365-, 452-, or 550-day yearling weight ratio, EPD, and accuracy.

4. Number of contemporaries at weaning and as yearlings.

Individual weights should not be reported on performance pedigrees due to large environmental variations.

EPDs and progeny ratios may be added for additional traits that are considered important and may be included on the performance pedigree.

HEREDITARY DEFECTS (Undesirable Recessives)

Bull buyers should always be on the alert for hereditary defects. Even though the bull selected shows no evidence of defects, he may be a *carrier* if his relatives have a history of such defects. Among the more than 200 hereditary abnormalities in cattle to guard against are: dwarfism, double muscling, hernia, cryptorchidism, hydrocephalus, and arthrogryposis. Cattle producers commonly refer to such defects as *undesirable recessives* or *freaks*. Such traits range all the way from lethals, which cause the death of the affected animal, to those that cause only an economic loss.

In addition to the more simply inherited hereditary defects, there is now ample evidence that the predisposition to such undesirable characteristics as bloat, vaginal and uterine prolapse, and cancer eye is also heritable. Hence, one should not buy a bull whose relatives exhibit a high incidence of these problems.

PRICE

Once you have found him, what's a good bull worth?

One commercial cow-calf producer of the senior author's acquaintance pays an average of $3,000 for his bulls. He weans a 97% calf crop, with an average calf weight of 590 lb, and his reputation calves bring $2.00/cwt premium. Currently, he's netting $86 per cow. And that's not all! After this commercial cattleman is finished with his bulls, he sells them for an average of $1,500 per head. Many of them go into purebred herds, at a premium price because they have been progeny tested. This producer's success story can be attributed to two things: (1) the use of outstanding sires; and (2) the production of reputation calves.

Cowboy arithmetic shows that a cattle producer can afford to pay as much as $1,250 more for a good bull than for an average bull. Here is how most producers figure the situation: Calves sired by a superior bull will usually average 25 lb more at weaning time than calves by an ordinary bull. Let's assume that these calves bring 75¢ per pound, and that 25 calves are weaned annually over the 4 years the bull is in service. That's an extra $18.75 a calf, $468.75 a year, and $1,875 more for the 4 years. Usually, the superior bull will be more growthy than an ordinary bull by perhaps 300 lb. At $40/cwt, the bull will bring an extra $120 when he has passed his useful breeding age and is marketed for slaughter, bringing the total added value of the superior bull to $1,995. Thus, if a producer is willing to pay $1,500 for an ordinary bull, he can afford to pay up to $3,495 for a superior bull. Besides, if he is saving replacement heifers, a further bonus will accrue to the superior sire through his daughters.

A superior sire can improve immediate net profit through his calves, and make for capital improvement through his daughters. It follows, therefore, that the monetary value of these improvements represents the added worth of a superior bull over an ordinary bull. This may be accurately and scientifically computed as follows:

1. **Net profit from using a superior sire is the product of** (a) the number of calves produced per year, (b) the average pounds sold per calf, (c) the net value obtained per pound of calf, and (d) the number of years used, *less* this same product from an ordinary sire.

2. **Capital improvement through the daughters is the product of** (a) the number of daughters saved, (b) the average pounds sold per calf (maternal plus growth), (c) the net value obtained per pound of calf, and (d) the average number of calves produced per daughter, *less* the same product for daughters out of ordinary sires.

Based on the above, if a superior sire produces 100 calves that wean at 50 lb more at 10¢ net per pound, that's $500 more profit than from an ordinary sire whose calves are 50 lb lighter. Additionally, if 20 daughters are saved that produce 25 lb more calf because of improved maternal ability plus 25 lb more expected in growth and they average 5 calves, that's another $500 profit. Thus a cattle producer could afford to pay up to $1,000 more for a superior sire of this caliber than for an ordinary bull. Then, there is an added bonus. Bulls that increase net value per pound 1¢ on a 1,000-lb steer return $1,000 per 100 calves.

Of course, the purebred stock producer has even more to gain from using a top sire, since his worth is multiplied through both his sons and daughters.

Another formula that is sometimes used in pricing a bull is based on worth as determined by performance test. An example follows:

> Let's assume that as a result of a performance test, it is found that *bull A* has a weight per day of age advantage of 0.5 lb over *bull B*. Then calves from *bull A* would be expected to gain 0.15 lb per day faster up to 1 year of age than calves from *bull B* (0.5 × 0.6 heritability estimate for weight per day of age ÷ 2 for sire effect only). Let's further assume that 20 steer calves from each bull sell at 75 cents per pound at 1 year of age. The increased annual income from *bull A* over *bull B* would be $821.25. Let's further assume that these bulls are used 3 years. Thus, *bull A* is worth $2,463.75 more than *bull B*. The advantage from the heifer calves, many of which would likely be used as replacements, would be additional—and would be cumulative.

From the above, it may be concluded that the value of a bull is determined by his ability to produce calves that have enough quality to command the top dollar and enough pounds to make it pay.

GUARANTEE

Traditionally, beef bulls are *guaranteed breeders*, meaning that they are guaranteed to be fertile and to sire live calves. Liability is limited to an option to select a replacement of equal value. High priced bulls are sometimes sold under more detailed contracts, which may specify percent conception and name the committee that must certify to the fertility of the bull.

Where a bull is bought for an artificial insemination stud, the sale contract may stipulate that payment is contingent upon the bull producing semen in specified quantity and quality useful in AI.

Warranty for frozen semen is restricted to quality control characteristics, with liability limited to replacement in kind.

Of course, sellers of beef bulls cannot, and should not, be held responsible for infertility due to such things as infectious diseases, nutritional deficiencies, or mismanagement of the cow herd, or to injury to the bull after purchase. Likewise, those who sell frozen semen cannot be responsible for the cow herd's infertility for reasons such as those given above or for improper handling of semen or insemination procedure.

BULL SELECTION GUIDELINES

One of the most important management decisions a cattle producer makes is the selection of a herd sire. Careful thought and planning are required. After deciding on the breed, evaluate your cow herd and current calf crop. Where do they need improvement the most? Is it size, muscling, soundness, gaining ability, or some other trait?

Next, decide what herds you plan to visit or which sales you wish to attend. Buy from reputable breeders who are doing a good job of production and who will give a breeder's guarantee (and preferably a semen test) with the animals they sell. Patronize those who are cattle breeders in the truest sense—those who are making real progress in improving the quality and performance of their own cattle, and who are not just *multipliers*.

Take time in making a selection. Start well in advance of the time you need a bull. The earlier you start, the greater the number of bulls from which you may choose.

Select a bull that meets the following criteria:

- Performace tested, with a balanced set of EPDs for desired traits.
- Large framed, with plenty of size for his age.
- Well muscled.
- Structurally sound, including the feet and legs, and free from hereditary defects (and not a carrier of hereditary defects).
- Good disposition. Avoid flighty, nervous bulls; heifers sired by such bulls will likely exhibit the same traits.
- Sired by a bull that has been doing a good job of settling cows and of siring large-framed, fast-gaining calves.
- Out of a cow that consistently ranks in the top half of the herd in terms of production.
- Normal in testicular development—both testicles are present and they are fully descended, sound, and approximately equal in size.
- Has superior estimated or actual carcass EPDs for ribeye area, fat cover over the twelfth rib, cutability (yield grade), and quality grade.
- Free of reproductive diseases, as determined by blood tests and verified by health papers.

SIRE SELECTION IS A CONTINUOUS PROCESS

The sire selection process continues even after the bull has settled his cows. His progeny should be evaluated as soon as possible. Then, on the basis of progeny performance, the breeder should make the decision either to retain or reject the bull for future use. If the progeny performance is acceptable to the breeder and the sire is retained, the selection procedure dealing with semen and libido checks begin again during the next breeding season and should continue each breeding season that the sire is used. Thus, sire selection is truly a continuous process.

BULL GRADING

Bull grading programs have exerted a powerful influence in improving the commercial cattle on the western ranges of this country. Perhaps this movement received its greatest impetus in those areas where several owners run herds on unfenced public grazing lands. Formerly, those progressive ranchers who believed that only purebred beef bulls of high quality should be used could do nothing to prevent the presence of inferior bulls on these public ranges. The person who bought superior bulls got no more use from them than the neighbor who turned out scrubs because they could be purchased cheaply. This problem was finally solved when groups of cattle producers using common ranges decided to have their bulls classified and to use only bulls meeting certain grades. Today, grazing permits are sometimes refused or delayed because of ranchers refusing to use graded bulls.

In some cases in the West, all animals consigned to range bull sales are graded and individual ranchers grade their young bulls before turning them with the cow herd. In some consignment sales, bulls must be of a certain specified minimum grade in order to be sold. Grading of sale bulls is especially popular with most buyers, but some sellers object to it.

Many different systems of bull grading have been used over the years.

MANAGING AND FEEDING BULLS[2]

Frequently, little thought is given to the management and feeding of bulls except during the breeding season. Instead, the feeding program for herd bulls should be such as to keep them in a thrifty, vigorous condition at all times. They should be neither overfitted nor thin and run-down. Also, exercise is necessary for the normal well-being of the bull.

Periodic management practices for bulls should include:

1. External parasite control of lice and flies.
2. Fecal examination for internal parasites.

[2]The nutritive requirements of bulls for growth and maintenance, at different weights, are given in Chapter 8.

Fig. 20-6. Registered Brahman bulls on bur-clover pasture. (Courtesy, Soil Conservation Service, USDA)

3. Foot examination and care if needed—trimmed, corns removed, treatment of foot infections and other injuries. Neglected foot trimming has probably ruined more good bulls than deficiency in any other single management practice.
4. Eye examination, and treatment if necessary.
5. Reproductive examination, and semen evaluation if needed.

The feeding and management of bulls differ according to age and condition. For this reason, sale bulls, young bulls, and mature bulls are treated separately in the sections that follow:

SALE BULLS

Most bull sales are held in late winter and early spring, at which time mostly yearling and two-year-old bulls are sold. In order to attract buyers, they have usually been grain-fed since calfhood. Most bull buyers—especially commercial operators in rougher range areas—would rather have their new bulls in less than fitted sale condition. They find that such bulls are more fertile and more apt to range with the cows when turned to pasture during the breeding season.

Sale and show bulls should be acquired 2 to 3 months ahead of the breeding season, so that they may be conditioned, or let down. Also, bear in mind that it takes about 40 days from the time a sperm cell is formed until it is ready to be ejaculated. Since the stress of handling and hauling a bull can reduce his fertility for about 40 days, the rest period lets his body overcome these problems.

Handling highly conditioned sale bulls during the critical period—after the sale is over, and just ahead of the breeding season—is all-important. Experienced cattle producers *let them down* and yet retain strong, vigorous animals. They do this successfully by (1) providing plenty of exercise, (2) increasing the amount of bulky feeds, such as oats, in the ration, (3) cutting down gradually on the grain allowance, and (4) retaining the succulent feeds and increasing the pasture and hay.

Exercise is most important during this period, and the more excessive the finish the more vital the exercise. Heavily fitted bulls can best be exercised by leading as much as 2 mi daily. Moderately fitted bulls can be force-exercised by placing the feed and water on opposite sides of the field.

In summary, therefore, conditioning highly fitted sale bulls for breeding consists of the gradual elimination of high-energy rations along with forced exercise.

YOUNG BULLS

Lack of fertility in a bull may often be traced back to his early care and feeding. From weaning to 3 years of age, bulls should be kept separate by age groups. Young bulls should be fed more liberally than mature bulls because their growth requirements must be met before any improvement in condition can take place.

Following weaning, bulls should be fed and developed sufficiently to show their inherited characteristics, but without excessive finishing. Simultaneously, they should be given plenty of exercise. Overfeeding and lack of exercise are apt to result in infertility, low-quality sperm, and unsound feet and legs.

To achieve proper development, young bulls should gain at least 2½ lb daily from weaning to 12 to 15 months of age. This will necessitate a daily feed allowance equal to about 2½% of their body weight, with a ration comprised of 50% or more concentrate. From 15 months to 3 years old, they should make a daily gain of 2 to 2¼ lb and receive a feed allowance equal to 2 to 2¼% of their body weight, with the proportion of roughage increased after the first year.

Fig. 20-7. The least laborious and most desirable arrangement for handling young bulls consists of grain feeding on pasture a group of not to exceed 10 to 15 head of uniform size and age. (Courtesy, The American Hereford Assn., Kansas City, MO)

If desired, the roughage may be chopped and mixed with the concentrate; and the ration may be hand-fed or self-fed. If self-fed, the feed consumption may be held at the desired level by using salt as a regulator. Other cattle producers prefer to feed the roughage and concentrate separately; they usually free-choice the roughage and either self-feed a salt-concentrate mix or hand-feed the concentrate alone. When grain is fed separately from the roughage, about 1½ lb per 100 lb of body weight can be fed at the beginning, gradually decreasing the grain and increasing the roughage as the animal grows older.

Without doubt, the least laborious and most convenient management arrangement in handling young bulls consists of allowing a group of not to exceed 10 to 15 head of uniform size and age the run of a pasture or enclosure of ample size, thereby providing (1) exercise, and (2) pasture in season. Of course, wherever possible, bulls should be performance tested while being developed. Ideally, this calls for individual feed and body weight records, although group feeding plus individual weight records will suffice.

At intervals, young bulls should be thoroughly checked for feet and leg defects, such as sickle-hocked, post-legged, cow-hocked, bow-legged, toeing out, or toeing in; and corrective hoof trimming should be administered as needed.

Bulls handled as recommended above will generally attain half their mature weight by the time they are 14 to 15 months of age and may be used in limited service.

During the breeding season, young bulls should be fed a grain ration consistent with pasture quality and number of cows to be bred in order to promote proper growth and development. Drought, over pasturing, and poor-quality pastures are situations in which grain supplementation is particularly needed. Heavy service and poor pasture with no supplemental feeding may shorten the breeding career of a young bull.

After the breeding season, yearling bulls generally need 5 to 6 lb of grain along with good roughage.

MATURE BULLS

Winter is the proper time to condition bulls for the next breeding season. Bulls that have been running on pasture with the cows are likely to be thin; thus, they require sufficient concentrate to put them in proper flesh. Mature bulls will consume daily amounts

Fig. 20-8. Mature Hereford bulls being moved to their breeding ranges in southwestern U.S. (Courtesy, The American Hereford Assn., Kansas City, MO)

of feeds equal to 1½ to 3% of their liveweight, depending upon condition and individuality.

Outdoor exercise is also essential in keeping bulls virile and thrifty. The finest and easiest method of providing such exercise is to allow them the run of a well-fenced pasture. About two acres is a good size for one bull, with a larger enclosure where several bulls are run together.

The importance of having bulls in proper condition at the opening of the breeding season cannot be over-emphasized. Nothing is quite so disheartening or costly as a small calf crop, with many of the calves coming late. Lack of fertility in the bull may often be traced back to his care and feeding.

Feed mature bulls all the legume hay they will eat plus 3 to 5 lb of ground or rolled grain and 1 lb of a 32% protein supplement (or equivalent) per head per day. Also, provide free access to a suitable mineral mixture. About 60 days before the bulls are turned with the cows, increase the concentrate allowance by 25 to 50%, with the amount of the increase determined by the condition of the bulls.

The mature herd bull needs no additional feed when running with the cow herd on good summer pasture.

MANAGING BULLS DURING THE BREEDING SEASON

Even though a 100% calf crop is difficult to obtain, a herd owner should strive to approach this figure as closely as possible. Poor management of the herd at breeding time can greatly reduce the calving percentage.

The following management practices during the breeding season are recommended:

- **Purchase the bull early**—Bulls should be purchased early in the season when selection is best. Also, they should be on the farm or ranch where they are to be used 60 days prior to breeding. This will provide a period of isolation and give them a chance to adjust to the *bugs* on the new premises. Purchasing well in advance of the breeding season also give bulls a chance to overcome any fertility problems encountered from a sale or shipment.

- **Evaluate semen and reproductive soundness**—Perform a semen and reproductive soundness evaluation on all bulls two weeks ahead of the breeding season. This practice will detect sterile bulls and those with obvious low fertility prior to their use and will allow time to replace them.

  If foot trimming is required, do it at this time—at least two weeks before breeding.

- **Provide adequate bull power**—The proportion of bulls to cows is dependent on (1) the age of bulls—young or old bulls cannot carry their share of the breeding responsibility; (2) the topography and feed conditions—rough areas and sparse vegetation require more bulls; (3) condition of bulls—excessively fat or excessively thin bulls will handle fewer cows; and (4) the length of the breeding season—the shorter the breeding season, the greater the stress on the herd bull battery. Under average conditions, 1 yearling bull should be provided for each 15 to 20 cows and 1 mature bull for each 25 to 35 cows. With hand mating, more cows may be bred.

- **Shorten the breeding season**—Gradually shorten the breeding season each year until it is no longer than 60 to 75 days.

- **Check the breeding herd frequently**—Watch for and record cows in heat; and see that the bull is finding and breeding these cows. Bulls that have low libido and are reluctant to mate even though cows are showing visible signs of estrus should be culled immediately. Also, if a high percentage of the cows have not been settled after two heat periods, the bull should be replaced.

- **Modify hot weather**—Breeding during hot weather often reduces the quality of sperm and lowers conception rates. Because of this, it is important that cattle have access to plenty of fresh water, shelter from the sun, and protection from flies.

BULL COSTS; AI COSTS

With the increase in artificial insemination in recent years, a frequently asked question is: What's the cost of AI vs natural service?

The University of Nebraska's West Central Research and Education Center, North Platte, studied breeding costs by herd sizes of 30, 100, and 300 cows.

In 1994, they reported AI (70 days) vs natural service costs as shown in Table 20-1.[3]

TABLE 20-1
BREEDING SYSTEM COSTS, AI VS NATURAL SERVICE

Breeding System	Breeding Cost Per Pregnant Female in Herd Sizes of—		
	30 Head	100 Head	300 Head
	($)	*($)*	*($)*
AI, 70 days	46.20	40.00	38.20
Natural service	42.30	32.00	30.00

[3]*Drover's Journal*, February 1994, p. 31.

As shown in Table 20-1, on a per pregnant female basis, natural service costs less than AI.

But comparative cost figures alone do not tell the whole story. Artificial insemination eliminates the expense and problems with keeping bulls. Moreover, AI makes for improved production in the calves.

Based on a study of 37 commercial ranches in Wyoming, the Wyoming Experiment Station reported $1.87 greater cost per calf from AI than from natural service.[4] However, the AI-sired calves had a $7.05 per head greater value than the calves sired by natural service, leaving $5.18 net per calf in favor of AI This comparison included all charges related to breeding the cow herds.

[4]Stevens, D. M., and T. Mohr, *Artificial Insemination of Range Cattle in Wyoming: An Economic Analysis*, Wyoming Bull. 496.

QUESTIONS FOR STUDY AND DISCUSSION

1. Why does 80 to 90% of the hereditary gain in a given beef herd result from the selection of the bull?

2. What source of bulls do you favor—purchase from a breeder at private treaty or in a production sale, purchase at a central test station sale, raising your own, or artificial insemination; and why do you prefer it over other sources?

3. In a bull selection program, would you buy a young bull or a proved sire? Why?

4. How would you determine what herd has the highest merit—hence, the place for you to buy?

5. If you were limited to selection for just one of the following three production traits, which one would you choose, and why would you choose it: (1) weaning weight (205-day weight), (2) postweaning growth weight, and (3) carcass merit?

6. How important is *individuality* in selecting a bull?

7. Is a semen test 100% accurate in determining bull fertility?

8. Of what value is a regular pedigree in bull selection? Of what value is a performance pedigree?

9. Detail by example how you would arrive at the price that you would be willing to pay for a particular bull.

10. When buying a bull, what kind of a guarantee would you expect?

11. Briefly and concisely, list the criteria that you recommend using in selecting a bull.

12. How would you handle a heavily fitted sale bull from auction time to breeding season, which we shall assume to be a period of 60 days?

13. Wherein does the management of a young bull 13 to 14 months old differ from the management of a mature bull?

14. List and discuss the most important bull management practices during the breeding season.

15. Which will cost the most, and which will make for the highest net return—bulls used in natural service or artificial insemination? Justify your answer.

16. Select a particular farm or ranch, either your own or one with which you are familiar. Would you buy bulls or AI semen for this particular operation? Justify your answer.

SELECTED REFERENCES

Title of Publication	Author(s)	Publisher
Beef Cattle, Seventh Edition	A. L. Neumann	John Wiley & Sons, Inc., New York, NY, 1977
Beef Production and Management Decisions	R. E. Taylor	Macmillan Publishing Company, New York, NY, 1994
Commercial Beef Cattle Production, Second Edition	Ed. by C. C. O'Mary I. A. Dyer	Lea & Febiger, Philadelphia PA, 1978

Motherrrrrr!!!! (From a painting by the noted artist, Tom Phillips)

FEEDING AND HANDLING CALVES

Contents *Page*

Beef producers, as a whole, have lagged in applying much of what we know about feeding and handling calves. They're inclined to let mother cows and mother nature fend for the calves. As a result, there is an appalling calf loss of 7% between birth and weaning. Indeed, more proved good practices, based on both successful operations and research, need to be put to use in feeding and handling calves.

(The nutritive requirements of all cattle, including calves, are presented in Chapter 8 of this book; hence, the reader is referred thereto.)

MANAGEMENT AT BIRTH

Losing a calf means losing the profit on the cow for a whole year. Proper management at birth can make the difference.

Close observation of the cow herd at calving time is essential so that assistance can be provided when necessary. First-calf heifers will usually have more calving difficulty than older cows; hence, they should receive very close attention during the calving season.

Recommended calving-time management practices follow:

1. Have the cows in an area that can be checked easily.
2. Keep the cows in a clean place during calving—it's the best way to avoid scours and other diseases. In season, a clean pasture (one that has been idle for a time) is ideal. When calving in confinement in winter, early spring, or late fall, provide a clean, freshly bedded shed.

Fig. 21-1. Calving on clean ground makes for a good start in life and lessens calf scours. (Courtesy, Beefmaster Breeders Universal, San Antonio, TX)

3. If a cow is in true labor for more than 2 hours or in unusually severe labor, she should be examined. In normal presentation, the head is between and slightly above the front feet. The experienced caretaker may correct a minor problem, such as a front foot bent back or the calf's head turned back. For a difficult position problem, however, the veterinarian should be called. Difficult calving may be caused by any of the following conditions:
    a. The calf is very large, especially in the shoulders or hips—or both.
    b. The cow has a small pelvic area, or the pelvic area is filled with fat (as happens in excessively fat animals).
    c. The cow fails to dilate.
    d. The calf is presented backward (breech birth).
    e. One or both front legs are bent back.
    f. The head is bent back.
4. As soon as possible after birth, remove mucus from the calf's nose and mouth. If the calf does not start to breathe normally, check its airway and apply artificial respiration by alternate pressure and release on the rib cage.
5. Disinfect the navel with a 2% tincture of iodine solution; and inject the newborn calf with 250,000 to 1,000,000 IU of vitamin A (use the higher level in confinement production or where scours may be a problem).
6. See that the newborn calf nurses within 2 hours after birth. It is essential that it receive colostrum. The caretaker may have to assist a calf to nurse a dam that has very large teats or an udder that hangs very low. Also, weak calves should be helped to nurse.
7. If a cow does not claim her calf, put them in a separate pen and do not disturb them for a few hours.
8. If a cow fails to clean properly (has a retained placenta) within 12 hours after calving, call your veterinarian.
9. Keep the cows that have calved separate from those that are still to calve.
10. Keep a close watch for signs of mastitis or injury to udders. It may be necessary to milk out a few cows for the first 2 or 3 days after calving.
11. Be sure cows have access to plenty of clean, fresh water.
12. Ear tag or tattoo the calf, and record pertinent information on your calving record, including date of birth, ease of calving, abnormalities, availability of colostrum, strength of calf, birth weight of calf, and temperament of the cow.

Fig. 21-2. It is usually best to designate a small pasture adjoining headquarters as the calving area. The cows and calves should remain in this area for a few days, following which they may be turned with the main herd. Note that the cows and calves pictured above are separated from the rest of the herd. (Courtesy, Agricultural Research Service, USDA, Clay Center, NE)

(Also see Chapter 6, Physiology of Reproduction in Cattle.)

RAISING ORPHAN AND MULTIPLE BIRTH CALVES

Occasionally a cow dies during or immediately after parturition, leaving an orphan calf to be raised. Also, there are times when cows fail to give a sufficient quantity of milk for the newborn calf. Sometimes, there are multiple births.

If there are only a few orphans, usually they can be grafted onto (or adopted by) another cow—either one that has lost her calf or one that gives sufficient milk to raise two calves. Where such calves cannot be grafted, they must be raised by artificial methods—without a cow.

Regardless of whether orphans are grafted or raised artificially, the problem will be simplified if the calf receives colostrum, the first milk produced by a cow after giving birth to a calf, during the first 24 hours, and preferably for the first 3 days, of its life—from its mother, from another fresh cow, or from frozen-stored colostrum. Colostrum is higher than normal milk in dry matter, protein, vitamins, and minerals. Also, it contains antibodies (found in the gamma globulins) that give newborn calves a passive immunity against common calfhood diseases.

Because colostrum is so important for the newborn calf, cattle producers should store a surplus of it from time to time. It can be frozen and stored for a period of 1 year or longer, then, as needed, thawed and warmed to 100° to 105°F, and fed. Also, colostrum may be fermented and stored.

Fortunately, orphan calves can now be raised successfully on a milk replacer and calf starter ration, using them as directed. The milk replacer may be fed by using a bottle or pail equipped with a rubber nipple, or the calf may be taught to drink from a pail. It is important that all receptacles be kept absolutely clean and sanitary (clean and scald each time) and that feeding be at regular intervals. Dry feed should be started at the earliest possible time; not later than 1 week of age. With proper management, healthy calves may be switched entirely to a suitable dry feed at 4 to 5 weeks of age.

Basically, calves are fed according to 1 of 3 systems: (1) the whole milk system, (2) the combination whole milk-milk replacer system, or (3) the combination whole milk-calf starter system. Of course, various combinations of these 3 systems are used, also. A suggested schedule for each of these 3 systems is given in Table 21-1.

The whole milk method costs the most, but it produces the fastest gains, the best appearing calves, and requires the least skill of any system. Milk replacers can be fed as the only feed, immediately following the colostrum period; or, as shown in Table 21-1, they may replace whole milk beginning on about the seventh day.

There is hardly any limit to the number of calf starters on the market. Most of them are mixed commercially. Because of the difficulty in formulating a home-mixed calf starter ration, the purchase of a good commercial feed usually represents a wise investment.

Two suggested calf starter rations are given in Table 21-2.

Starter Ration A, of Table 21-2, is designed for feeding anytime after the first day following birth. Starter Ration B is designed for feeding beginning about 45 days of age. As is true in any ration change, the transition from Ration A to Ration B should be made gradually by blending the feeds over a period of 2 to 3 days.

Good-quality hay for young calves is essential; it provides an economical source of nutrients, helps maintain rate of gain, and speeds up the development of the rumen.

Many cattle producers make the mistake of placing calves on pasture at too early an age. Unless pastures are properly supplemented, young calves simply cannot hold enough grass, or other pasturage, to obtain

TABLE 21-1
SCHEDULE FOR FEEDING CALVES BY THREE DIFFERENT SYSTEMS

Age of Calf	Whole Milk System	Whole Milk–Milk Replacer System	Whole Milk–Calf Starter System
0 to 3 days	Calf should receive colostrum during first 3 days.	Calf should receive colostrum during first 3 days.	Calf should receive colostrum during first 3 days.
3 days	Start feeding whole milk at the rate of 1 lb *(0.45 kg)* milk to 10 lb *(4.5 kg)* body weight.[1]	Start feeding whole milk at the rate of 1 lb *(0.45 kg)* milk to 10 lb *(4.5 kg)* body weight.[1]	Start feeding whole milk at the rate of 1 lb *(0.45 kg)* milk to 10 lb *(4.5 kg)* body weight.[1]
7 days	Make grain available in box in pen (see Table 21-2).	Make calf starter available in box in pen.	Make calf starter available in box in pen.
7 to 10 days		Start replacing whole milk with fluid milk replacer. Replace 1 to 2 lb *(0.45 to 0.9 kg)* daily with fluid milk replacer until change is completed.	
14 days		Transition to milk replacer should be completed.	
21 days	Make good-quality hay available in rack in pen.	Make good-quality hay available in rack in pen.	Make good-quality hay available in rack in pen.
60 days		Discontinue milk replacer.	Discontinue feeding whole milk. Larger, more vigorous calves may have whole milk stopped as early as 42 days.
60 to 120 days	Permit calves to consume grain free-choice, up to 4 to 5 lb *(1.8 to 2.2 kg)* daily. Rest of nourishment should be obtained from hay.	Permit calves to consume calf starter free-choice, up to 4 to 5 lb *(1.8 to 2.2 kg)* daily. Rest of nourishment should be obtained from hay.	Permit calves to consume calf starter free-choice, up to 4 to 5 lb *(1.8 to 2.2 kg)* daily. Rest of nourishment should be obtained from hay.
90 days	Discontinue whole milk.		

[1]For economic reasons, it is never advisable to feed calves more than 12 lb (5.4 kg) whole milk daily during the entire milk feeding period.

sufficient nutrients for their growing bodies. Accordingly, growth will be retarded.

EARLY WEANING

Early weaning refers to the practice of weaning calves earlier than the usual weaning age of about 7 months, usually within the range of 45 days to 5 months of age. Although it is not common practice among U.S. beef producers, dairy operators have been weaning 3-day-old calves for years. Also, early weaning has long been an integral part of many of the beef programs of Europe.

Currently, there is much interest in early weaning because (1) it fits into a drylot cow-calf management system, and (2) it can give a big assist in getting females, especially two-year-old heifers, to rebreed in a short period of time.

The current interest in increasing the number of cows in the Corn Belt is largely predicated on more efficient use of crop residues, especially corn and sorghum residues. With heavy cropping, there is little or no pasture as such; hence, drylot management systems are evolving. When using crop residues as a basic feed source, the lactating cow is likely to need supplemental feed. Considering the low efficiency involved in converting supplemental energy to milk and in converting milk to meat, it is apparent that a more efficient use of feed could be achieved by giving the supplemental feed directly to the calf. A lactating cow requires about 50% more feed than a dry cow. So, rather than give her that additional feed, it is more efficient to give feed directly to the calf.

Weaning calves early from two-year-old first-calf heifers reduces the stress of milking and raising a calf. As a result, they recycle and rebreed earlier and grow out more rapidly. As heifers are bred for higher milk production, this reason for early weaning takes on greater importance, for the more milk they give, the slower they are to cycle. It appears doubtful that any level of nutrition will have the same effect on reproduction efficiency of the heifers as early weaning.

In addition to fitting into a drylot cow-calf system

TABLE 21-2
CALF STARTER RATIONS

Ingredients	Starter Ration A[1] (for feeding first 45 days, along with liquid skim milk)		Starter Ration B[1] (for feeding after first 45 days, with dry skim milk therein)	
	(lb)	*(kg)*	*(lb)*	*(kg)*
Dry skim milk	—	—	400	*182*
Soybean or cottonseed meal (41%)	560	*255*	450	*205*
Barley	1,000	*455*	750	*341*
Wheat bran	200	*91*	150	*68*
Dicalcium phosphate	20	*9*	20	*9*
Trace mineralized salt	20	*9*	20	*9*
Antibiotic (follow mfg's. directions)	10	*4*	10	*4*
Vitamin A	10,000 IU/lb[2]		2,000 IU/lb[2]	—
Vitamin D (not needed if calf is in sunlight)	2,000 IU/lb[2]		400 IU/lb[2]	—
Molasses	200	*91*	200	*91*
Total	2,100	*914*	2,000	*909*
Calculated Analysis:	*(%)*		*(%)*	
Crude protein	18.8		19.8	
Fiber	6.6		4.1	
Fat	2.3		1.4	
Calcium	0.43		0.69	
Phosphorus	0.77		0.74	
TDN	69.9		74.7	

[1]In ⅛- or 3⁄16-in. *(0.3 or 0.4 cm)* pellets.

[2]To convert to IU/kg, multiply vitamin quantities by 2.2.

and facilitating a program of calving two-year-old heifers, early weaning may be desirable for the following reasons:

- It may be the answer to getting one calf per cow every 12 months in intensive management systems.

- It may be the key to the most efficient feed utilization during times of droughts and other periods of feed shortages. Under such conditions, it might be advantageous to wean the calves early and to provide them with the highest quality feed available, while restricting the quality and quantity of feed fed to the cows.

- It fits in with fall calving where heavy winter feeding is required. As soon as the calves are weaned, cows may be turned to stalk or stubble fields to winter on cheap feed.

- It may make it possible to keep a particularly valuable old purebred cow in production longer.

- Young gains are cheap gains, due to (1) the higher water and lower fat content of young animals in comparison to older animals, and (2) the higher feed consumption per unit weight of young animals. Thus, the feed efficiency of early-weaned calves is excellent, ranging from 3 to 4 lb of TDN per pound of gain.

- Lactating cows decline in milk production after about 1 to 2 months following parturition.

- Parasite problems are minimized in an early weaning program.

Where early weaning is successful, the only responsibility of the beef cow is to produce a calf and give it a good start in life for a brief period, then go on a maintenance ration the rest of the year.

Like many good things in life, early weaning does have some disadvantages. To be successful, superior nutrition and management are essential; and the earlier the weaning age the more exacting these requirements.

HOW EARLY IS EARLY WEANING?

Experiments and experiences indicate that it is practical to wean calves as early as 45 days of age.

Fig. 21-3. A vigorous, growthy Angus calf at 45 days of age, old enough for early weaning. (Courtesy, Martin Jorgensen, Jr., Ideal, SD)

This is long enough to stress the cow a bit, but short enough to get her to recycling and rebred so that a calf will be produced each 12 months. Also, it allows the cow to function as a lactating animal. Weaning at 3 to 5 months of age doesn't make for early recycling; hence, it doesn't contribute to getting 1 calf per cow every 12 months.

WHAT TO FEED EARLY-WEANED CALVES

From 45 days of age on, early-weaned calves can be fed any good starter ration, most of which contain dry skim milk. One such ration is given in Table 21-2, Ration B. Most commercial feed companies manufacture a starter ration. Of course, the starter ration should be made available to the calves well ahead of weaning in order that they will be accustomed to it, thereby avoiding any setback.

MILK REPLACERS

Several reputable commercial companies now produce and sell milk replacers, which are composed of sizable amounts of milk byproducts, such as dry skim milk, buttermilk, and/or whey, along with additives.

Although scientists have not yet learned how to compound a synthetic product that will alleviate the necessity of colostrum, in certain other respects they have been able to improve upon nature's product milk. For example, it has long been known that milk is deficient in iron and copper, thus resulting in anemia in suckling young if proper precautions are not taken. In addition to correcting these deficiencies, synthetic milks are fortified with vitamins, minerals, and antibiotics. They can be fed as the only feed immediately following the colostrum period; or, as shown in Table 21-1, they may replace whole milk beginning about the seventh day.

From the standpoint of the beef cattle producer, a milk replacer is of primary interest for two uses; namely, (1) for raising orphaned and early-weaned calves, and (2) for replacing nurse cows. Also, it is a valuable adjunct in certain disease control programs, especially those diseases that may be transmitted from dam to offspring.

CREEP FEEDING

Creep feeding is the supplementation of calves while they are nursing their dams. It increases weaning weight. The basis for this response is related to the lactation curve of beef cows, the increasing nutrient requirements of the calf during the nursing period, and the decline in feed quality and quantity typical of most pastures or ranges which support the cows and calves during lactation. Studies reveal that milk production of dairy cows increases up to the fourth to sixth month following freshening, then declines gradually. By contrast, maximum milk production of beef cows occurs during the first two months after calving, then declines.

Fig. 21-4 shows why creep feeding is important. From birth to weaning, the protein and energy requirements of a growing calf increase well beyond the ability of most beef cows to meet those needs. For example, to meet the protein and energy requirements for growth, a 100-lb calf needs 10 lb of milk, whereas a 500-lb calf needs 50 lb of milk. Since the average beef cow gives only 13 lb of milk per day throughout a 7-month suckling period, a 500-lb calf lacks 40 lb of

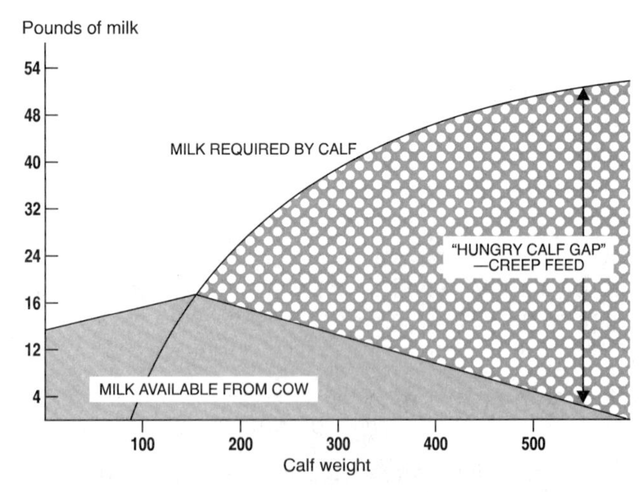

Fig. 21-4. Milk yield of a typical beef cow vs nutrient requirements of a nursing calf. This points up the need for creep feeding.

getting enough milk from its dam at this stage of lactation to meet its needs—that's the *hungry calf gap*.

To fill the *hungry calf gap*—the nutrient requirements over and above that provided by 13 lb of milk—would require the consumption of 50 lb of green grass daily. Of course, that's a physical impossibility, because a 500-lb calf simply cannot hold that much bulk. So, the best way to fill the *hungry calf gap* is to creep feed.

Creep feeding is no longer primarily an emergency program to supplement drought-stricken grasses and other conditions resulting in poor pastures. Rather, creep feeding of calves prior to weaning is on the increase because milk in quantity, and pasture in quality, are not normally available season-long to supply the necessary nutrition (1) to produce calves that meet today's market demands, and (2) to realize the maximum genetic potential from improved breeding.

Most calves will continue to be raised on their mother's milk plus whatever pasture or other feed they share with their dams. However, more and more of them will be creep fed in addition.

THE CREEP

A creep is an enclosure or a feeder for feeding purposes which is accessible to the calves but through which the cows cannot pass. It allows for the feeding of the calves but not their dams. For best results, the creep should be built at a spot where the herd is inclined to loiter; on high ground, in the shade, and near watering and salting grounds. The enclosing fence may be of board, pole, or metal construction, with an entrance 16 to 20 in. wide and 3 to 3½ ft high. Self-feeders, troughs, or racks may serve as feed containers; allowing 4 to 5 in. of space per calf for self-feeding and 8 to 12 in. for hand-feeding. Also, metal creep feeders are available consisting of a self-feeder to which the enclosing fence is firmly attached. These are especially suited for use on large range pastures, where frequent moving is necessary.

Fig. 21-5. A calf creep in the foreground, with purebred Hereford cows and calves in the background. (Courtesy, *Beef* magazine)

CREEP RATIONS; FEEDING DIRECTIONS

Creep-fed calves need special rations. They are bovine babies; and they are both in forced production and finishing. They are expected simultaneously to lay on fat and grow in protein tissues and skeleton. Consequently, their ration requirements are for feed high in protein, rich in readily available energy; fortified with vitamins, minerals, and unidentified factors; and with all the nutrients in proper balance. Also, the ration must be very palatable. This calls for an exacting ration. To meet these needs, more and more cattle producers are finding it practical to buy a commercially prepared complete creep feed, or a well fortified and highly concentrated supplement to add to locally available feeds, rather than purchase individual ingredients and mix from the ground up.

Tables 21-3 and 21-4 show two creep rations, formulated by the senior author, that have been widely and successfully used.

It takes considerable effort and patience to start calves on feed. Also, a little *calf psychology* helps;

TABLE 21-3
CALF CREEP RATION NO. 1[1]

Ingredient	Lb/Ton Mix[2]	Calculated Analysis	%
Oats	800	Crude protein	14.1
Corn	300	Fat	2.9
Barley	200	Fiber	8.4
Wheat bran	200	TDN	69.9
Dried molasses beet pulp	200	Calcium	0.37
Soybean meal	200	Phosphorus	0.50
Salt	10		
Dical	10		
Trace minerals[3]	1		
Vitamin A (30,000 IU/g)	1.5		
Molasses	100		
Total	2,022.5		

[1]Feed preparation: Preferable ⅛- or 3⁄16-in. *(0.32- or 0.48-cm)* pellets. Otherwise, steam roll grains, or grind grains coarsely.

[2]To convert the lb to kg, divide by 2.2.

[3]See Chapter 8, Tables 8-5 and 8-16, of this book for recommended trace mineral levels. Follow manufacturer's directions.

TABLE 21-4
CALF CREEP RATION NO. 2

Ingredient	Lb/Ton Mix[1]	Calculated Analysis	%
Alfalfa meal	450	Crude protein . . .	13.8
Soybean meal	124	Fat	3.2
Linseed meal	100	Fiber	12.8
Corncobs or grass hay .	200	TDN	60.5
Oats	400	Calcium	1.01
Corn	485	Phosphorus	0.71
Bran	100		
Dical	40		
Vitamin A (325,000 IU/g)	84 g[2]		
Trace minerals[3]	1		
Molasses .	100		
Total	2,000		

[1]To convert the lb to kg, divide by 2.2.

[2]When 4 lb/head/day of the calf creep ration is consumed, 54,600 IU of vitamin A will be obtained in the feed.

[3]See Chapter 8, Tables 8-5 and 8-16, of this book for recommended trace mineral levels. Follow manufacturer's directions.

remember that calves do not go for the privilege of eating, they need to be persuaded. One or more of the following techniques will usually prove helpful: Shut a gentle cow or a few calves in the creep, to serve as a decoy(s); scatter a little feed near the creep so that the cows will loiter nearby; and/or spread a little feed near and extending through the creep opening. It is also recognized that fall and early spring calves take to creep feeding better than late spring calves, simply because they have less grass and milk available. (Grass stimulates milk flow.)

When 3 to 4 weeks of age, calves should be started on feed very gradually. For the first 3 to 5 days, only about ¼ lb of feed per calf should be placed in the container(s) each day, and any leftover feed should be removed and given to the cows. In this manner, the feed will be kept clean and fresh. When calves are on lush pasture and their mothers are milking well, difficulty may be experienced in getting them to eat; but time and patience will pay off, and the results will become evident in 2 to 3 months.

After 5 to 7 days of hand-feeding, the creep ration can be left before the calves safely. Once they are on full feed, never let the feeder become empty; and avoid sudden changes. During the first 30 days, they will consume about 1 lb per head daily. Creep feeding has been proven to be more economical when consumption is less than 4 lb daily. In order to maintain consumption below 4 lb per head, salt can be added to the ration at a level of 3 to 10%. Once calves are started on feed, start adding salt gradually until the consumption reaches desired levels.

WHY CREEP FEED?

Unquestionably, the best yardstick for measuring performance in a beef breeding herd is pounds of calf weaned per cow bred. This fact, along with the demand for healthy, *gain-ready* feeder calves and the prices being paid, is causing cow-calf operators to take a new *pencil-pushing* look at the immediate and residual benefits of creep feeding.

Among the reasons for, or the benefits from, creep feeding are the following:

1. **It provides a way to fill the *hungry calf gap*.** Creep feeding provides a logical and practical way to compensate for insufficient milk which usually characterizes the following conditions: (a) the normal falloff in milk production of beef cows about two months after freshening; (b) periods of unfavorable feed conditions—droughts, overgrazing, early and late in the season, and fall-calving herds that are maintained on a low plane of winter nutrition—when the calf is shortchanged on both milk and pasture; (c) first-calf heifers, whose milk production is generally lower than mature cows; (d) shelly and poor-doing cows; and (e) poor milkers. Under such conditions, creep feeding makes for heavier weights.
2. **It makes for heavier weaning weights.** Creep feeding results in heavier weaning weight per calf, at no extra cost for the capital investment in land and cows.
3. **It facilitates fall calving.** Normally, fall-calving cows do not milk as well as spring-calving cows, because of the lack of lush grass to stimulate milk flow. As a result, fall calves neither gain as well nor have as heavy weaning weights as spring calves. It follows that creep feeding is more effective with fall calves than with spring calves. Generally, creep-fed fall calves make 20 lb more gain than creep-fed spring calves.
4. **The calves are more uniform.** Creep-fed calves are more uniform, because those that are getting less milk eat more feed in order to meet their nutritive requirements.
5. **Calves achieve full genetic growth potential.** Today, the emphasis in beef cattle is on size and growth, brought about through selection, introduction of the exotics, and crossbreeding. But most cattle producers have improved their cattle faster than they have improved their pastures. As a result, many calves are not getting sufficient nutrients from milk and grass to achieve the full genetic growth potential that is bred into them. Under these circumstances, creep feeding is the answer.

6. **Young gains are cheap gains.** Creep feeding makes for efficient and cheap gains, and profit.

The Virginia Station reported that drylot calves which only had access to cow's milk gained only 0.33 lb per day. Later, when creep fed (at 3 to 4 months of age), these same calves gained 2.2 lb per day. More remarkable yet, the creep-fed calves made 1 lb of gain for each 3 lb of creep feed.

7. **It is more efficient to feed calves directly than to feed cows too liberally.** This is so because when beef cows are fed above a certain level, they have a tendency to put tallow on their backs instead of milk in their bags. For this reason, creep feeding calves during prolonged periods when quantity or quality of forages is low is usually a good policy.

8. **It makes for attractive purebred calves.** Creep feeding is the ideal way in which to obtain the important development and bloom in purebred calves; for fat is a very pretty color. This is especially desirable if they are to be sold or shown at young ages.

9. **It makes it easy to reinforce and improve milk.** Creep feeding affords a convenient way in which to improve upon milk, chiefly by reinforcing it with certain vitamins and minerals; and in which to add such additives as desired.

10. **It controls parasites.** Creep feeding lessens parasites, simply because well-nourished animals have fewer parasites.

11. **It simplifies weaning.** Creep-fed calves rely less on the dam for nourishment and develop independence, with the result that there is less bawling, stress, and sickness when they are separated from the cows.

12. **It facilitates early weaning.** It makes it possible to wean calves at early ages—1½ to 5 months instead of the normal 7—a practice which is increasing.

13. **It makes for marketing flexibility.** Creep-fed calves may go the slaughter route—in addition to going into feedlots or as stockers; with the particular market avenue selected being determined by price.

14. **It narrows the price between heifers and steers.** It permits heifers to be marketed for slaughter at a weight when they will bring about as much as steers.

15. **It makes for better lifetime reproductive performance of heifers.** Creep-fed heifers are heavier at weaning, reach puberty at an earlier age, and tend to breed earlier in the season throughout life, than noncreep-fed heifers.

16. **They're *bunk-broke*.** It provides calves that are accustomed to feed, with the result that they will continue on feed in the feedlot with a minimum of stress, shrink, digestive disturbances, and death loss, and without the normal period of three weeks to get them on full feed. This, plus their higher initial condition, results in their being ready for market approximately 40 days sooner than noncreep-fed calves. In turn, the shorter feeding period makes for a saving in interest payment, and less exposure to disease, injuries, and weather.

17. **It makes for earlier cycling and conception.** Calves that are creep fed have a source of nutrients other than from their dams; hence, they nurse less and put less stress on their mothers than calves that are not creep fed. As a result, dams of creep-fed calves lose less weight and show heat and breed back earlier than the dams of noncreep-fed calves.

18. **Cows are in better condition.** It leaves the dams less suckled down and in better condition (25 to 50 lb heavier) at weaning time because the calves partially wean themselves, an important consideration relative to cows from the standpoint of either sale or the wintering period ahead.

19. **It gives first-calf heifers a needed assist.** Calves from first-calf heifers tend to be lighter at birth than calves from older cows; consequently, there is more potential for good conversion of creep feed to additional gains on them. Also, the milk production of two-year-old heifers is only about 70% of their mature production. Further, it is generally recognized that some of the stresses on first-calf heifers need to be lessened in order to get them rebred. It follows that calves from first-calf heifers generally give more response to creep feeding than calves from older cows, and that the heifers whose calves are creep fed show greater response in rebreeding than mature cows whose progeny are creep fed.

20. **It usually pays.** The potential profitability of creep feeding depends upon (a) the price of cattle and (b) the price of feed.

The following rule of thumb may be used to determine whether or not it will pay to creep feed: It pays to creep feed when the selling price per 100 pounds of calf is greater than the cost of ¼ ton (500 lb) of feed.

One cattle producer of the senior author's acquaintance summed up the economics of creep feeding in this way: "If your cost of gains runs 40¢ per pound, and you can sell the calves at 80¢ per pound, it's almost as good as finding a money tree." Assuming a $1/cwt selling advantage for noncreep-fed calves, because they're lighter and thinner, most cattle producers feel that from creep feeding, they'll net, $4 to $6 per calf more after feed and other costs are deducted.

LIMITATIONS OF CREEP FEEDING

Like many good things, creep feeding does have its limitations; among them, the following:

1. **It isn't always profitable.** Creep feeding may not be profitable because of the cost of the creep ration, low response, and/or price discrimination against fleshy weaner calves.
2. **It lowers feedlot gains and efficiency.** Fleshy creep-fed calves make slightly less rapid and efficient gains than calves not creep fed when they are (a) moved directly into the feedlot following weaning and (b) long-fed, because creep feeding alleviates compensatory gains. Also, it may cause small-type cattle to get too fleshy—to *stall* or stop growing before weaning. However, if the latter occurs, it's a sure sign that the wrong kind of cattle are being bred. However, these disadvantages may be compensated for, in part at least, by the shorter feedlot feeding period and more desirable market weights of creep-fed calves.
3. **It makes for less desirable stockers.** Creep-fed calves do not make as desirable stockers as calves that have not been creep fed, simply because the latter are normally placed on less nutritious growing rations consisting predominantly of roughages. This may be further explained in this way: One of the basic rules in feeding slaughter cattle is always to proceed to a higher plane of nutrition; never go down.
4. **It mitigates against selecting cows for milk production.** Creep feeding makes it difficult to put selection pressure on cows for good milk production since creep feeding cancels differences in weaning weight due to lactation differences in dams. This may be an important consideration in production testing programs.
5. **It is of limited benefit to calves on high-quality forage or on good milking cows.** Calves on high-quality forage or good milking cows do not benefit as much from creep feed as calves on poor-quality forage or poor milking cows.
6. **It is difficult in less remote areas.** Creep feeding is difficult on less accessible ranges, because the very nature of creep feeding calls for close attention.
7. **It cannot be done where there are hogs, sheep, or goats.** These animals can enter any creep opening that is big enough for a calf.

GROWTH STIMULANTS FOR CALVES

There are growth stimulants and implants which increase rate of gain and weaning weight of suckling calves. These are detailed in Chapter 30, Feeding Finishing (Fattening) Cattle; hence, the reader is referred thereto.

IDENTIFICATION

Fig. 21-6. Roping calves for tattooing and branding on Tequesquite Ranch, Albert, New Mexico. (Courtesy, Tequesquite Ranch, Albert, NM)

Calves should be identified as soon after birth as possible, and not later than three days of age. This subject is fully covered in Chapter 10, under the heading "Marking or Identifying"; hence, the reader is referred thereto.

DEHORNING

Dehorning is an economic necessity, because horned calves usually bring lower prices. in addition, dehorned and naturally polled animals do less damage to facilities and other animals than cattle that have horns.

All naturally horned animals should be dehorned preferably before they are two months old to minimize the *shock effects* of the operation. At that time, the blood vessels in the horn area are very small, which means less blood loss and minimum shock.

The subject of *Dehorning* is fully covered in Chapter 10; hence, the reader is referred thereto.

CASTRATING

Castration is recommended for all bull calves destined to be sold as feeders or finished in the feedlot. Bull calves and staggy-looking steer calves will not be accepted in many feeder calf sales. If sold as feeder animals, they usually bring a reduced price.

Castration time will vary according to method employed and management program, and it will be different for a commercial than for a purebred operation. Bull calves will weigh more at weaning than steer calves; however, younger calves are easier to restrain for castration and suffer less shock therefrom than older animals.

The commercial cow-calf operator should castrate

all bull calves before weaning, and preferably before they are four months old.

Most purebred breeders who raise bulls for sale may wish to wait until after weaning to castrate bull calves, so that they can evaluate weaning weights and use them in their bull selections. However, poor-quality purebred bull calves should be castrated earlier when it is evident that they are not bull prospects.

The subject of *Castrating* is fully covered in Chapter 10; hence, the reader is referred thereto.

WEANING

Weaning is a traumatic experience for a calf. It represents environmental, nutritional, psychological, and, altogether too often, vaccination-castration-dehorning changes—all of which make for great stress. Generally, at weaning time the calf is moved into a strange environment to which it must adjust—a new pasture, corral, or shelter; its food supply—milk—is suddenly removed; and its association with and protection by its mother is lost. Under such circumstances, it's small wonder that calves lose weight and become more susceptible to disease. The marvel is that they survive such mistreatment so well.

Calves should be weaned when they are 7 to 8 months old. Weaning at this age fits in well with the weight record-keeping requirements of most performance testing programs. Also, calves will be about the right age and weight for feeder calf sales.

Weaning earlier than 7 or 8 months may be necessary in years when pastures are short or when calves are from first-calf heifers.

The best way to wean is to remove the calves from their dams and keep them out of sight of each other. Cows and calves should never be turned together once the separation has been made. Such a practice will only prolong the weaning process, and it may also cause digestive disorders in the calf. Provide calves with plenty of water, free-choice hay, and 3 to 4 lb of grain per head per day. If calves were creep fed, continue their rations during the weaning period.

During the weaning process, calves should be confined to a small area to cut down on walking and shrinkage. In bad weather, protection should be provided from cold wind and rain; they should have access to a shed, wooded area, gorge, or other protection.

■ **How to minimize stress and weight loss**—Weaning calves is more a matter of preparation than of absolute separation from the dam. Minimizing stress and weight loss depends largely upon the thoroughness of the preparation. The following procedure is recommended:

1. Dehorn and castrate well ahead of weaning time.
2. Bring the cows and calves into a small pasture paddock 2 to 3 weeks before weaning. If weaning paddocks are not available, use small pastures. Avoid dusty pens in order to minimize respiratory diseases and pneumonia.
3. Creep feed the calves, so that they learn to eat. Use the same feed as the preconditioning ration (such as corn or sorghum silage plus a protein supplement; or a combination of a high-quality grass-legume hay plus a concentrate supplement; or a combination of these feeds); or if they have been on a creep ration all along, continue with the creep ration to which they are accustomed.

   Keep the feed clean and free from mold.
4. Give the first immunizations while the calves are still nursing, and follow with the booster shots as recommended by the manufacturer or the veterinarian.
5. Treat for parasites, internal and/or external, before weaning.
6. Wean the calves by moving the cows out of the weaning paddock(s) so that the calves remain in familiar surroundings.
7. Check calves 2 or 3 times daily. Remove sick ones and take them to the hospital area for treatment.
8. Administer TLC (tender, loving care) to calves being weaned. Always handle gently; never rope them.
9. Consider (confer with your veterinarian) the use of tranquilizers during the weaning process, as a means of calming calves and minimizing weight loss.

■ **Drying up the cow**—With higher milking strains of beef cattle, when drying up cows, cattle producers will have the same concerns as dairy operators—that of avoiding *spoiled udders*. To alleviate this problem, the following procedure is recommended:

1. Do not feed milk-stimulating feeds at weaning time. Either put the cows on poorer pastures or feed a nonlegume forage.
2. Let *back pressure* in the udder build up. Examine the udder at intervals, *but do not milk it out.* If the bag fills up and gets tight, rub spirits of camphor on it, but do not milk it out. At the end of 5 to 7 days, when the bag is soft and flabby, what little secretion remains (perhaps not more than half a cup) may be milked out if so desired.

CALF HEALTH-TREATMENT PROGRAM (Preconditioning)

This important program is fully covered in Chapter 12, under the heading "Calf Health-Treatment Program (Preconditioning)"; hence, the reader is referred thereto.

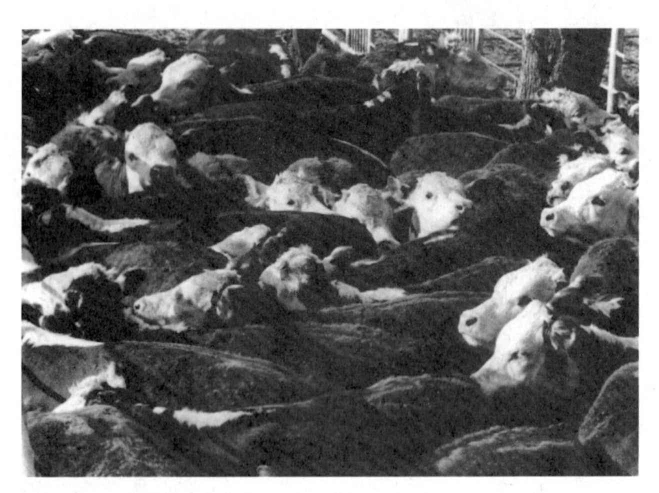

Fig. 21-7. Preconditioning, along with production testing, will be the trademark of the producer of reputation feeder calves of the future. (Courtesy, Ralston Purina Company, St. Louis, MO)

MARKETING ALTERNATIVES

Marketing the calf is the end of the line, it is that part of a cow-calf program which gives point and purpose to all that has gone before. After breeding, feeding, managing, and preconditioning the calf, it is still necessary to market it in such manner as to receive the greatest net income.

In recent years, the marketing of feeder calves has greatly benefited from numerous outlets, improved truck transportation, and improved market reporting. However, it requires more skill on the part of the cow-calf producer to benefit from today's modern marketing system than it did a generation ago. The producer must be aware of the numerous market channels, market news services, the value of the calves, and the economic and political factors affecting markets in order to sell the calf crop to the best advantage.

The following marketing alternatives are available to most cow-calf producers:

- **Selling feeder calves**—Spring-dropped calves may be weaned and sold as feeder calves in the fall when they are 7 to 8 months old. Steer calves normally weigh 400 to 600 lb. Usually heifer calves weigh about 5% less than steer calves.

Many feeder calf sales are held each fall; and a few are held in the spring. These provide a good outlet for locally produced calves fresh off the farm or ranch. Usually calves are sorted into uniform lots by weight and grade. Smaller producers benefit by being able to *pool* their calves with those of similar weight and quality, thus making a larger lot and bringing a higher price per cwt.

Cow-calf producers may also sell directly to cattle feeders, order buyers, or cattle dealers; or they may consign to terminal markets or decentralized buying stations. Also, if producers have adequate numbers of calves for "truck load" lots, they can market their calves through one of the satellite video auction companies. Many producers in the western United States prefer this method because of exposure to a large number of buyers.

- **Selling yearling feeders**—Some cattle producers have always grown calves beyond weaning weight, with the prevalence of the practice varied by years according to feed. This practice is most common in the southwestern United States.

Short yearlings are weaner calves held over because they are too light to sell at weaning time and/or because adequate forage is available. Usually, they are sold at just under 1 year of age, although they may range up to 14 months in Arizona and New Mexico. Long yearlings average about 16 months of age when sold to go into feedlots.

- **Finishing cattle**—This alternative is open to all cow-calf producers; either in their own facilities if they have them, or in custom feedlots. Calves may go directly into the feedlot at weaning to be marketed as finished cattle at 12 to 16 months old; or they may be grown out as stockers, then finished out at slightly older ages and heavier weights.

Producers who have the choice of selling their calves at weaning or as finished cattle can generally make more profit per cow by feeding their own calves, especially if they are sired by performance-tested bulls with genetic potential for rapid growth. They may as well reap the full benefits from their breeding program, rather than sell their calves and let someone else have a sizable share of the profits.

- **Replacement heifers**—The best source of replacement heifers is from the breeder's own herd, regardless of whether the operation is purebred or commercial. This is so because more is known about them—their age, immunization history, health status and performance—than replacements from any other source. Thus, at weaning time, the top-performing (heaviest) heifer calves should be retained—keeping 50% more than will actually be needed.

There is one exception where saving replacements would not be practical—that's a crossbreeding program where specialized F_1 females are mated to a terminal sire breed. In such a program, all calves are marketed at weaning or after finishing. Thus, it is necessary to purchase F_1 heifers from an outside source.

RETAINED OWNERSHIP

Unlike the vertically integrated poultry business, and the vertically integrated system that is now underway in the swine business, where ownership remains the same all through the production and processing phases to the retail outlet, the production of beef is segmented. Cattle operators *sell* weaned calves, *sell* stockers, *sell* feeders, *sell* fed cattle, and *sell* boxed beef. Each time that the animal/product changes hands and moves to another segment, the owner/handler expects a profit, and the inefficiency is increased.

Even though retained ownership has received increased publicity in recent years, less than 20% of the beef calves in the United States are retained from birth to slaughter, in a retained ownership program.

As future competition for the consumer's dollar increases, segmented cattle operators will, more and more, lose out to the unsegmented poultry and swine industries.

Retaining ownership through the cattle feedyard, and even through the packing plant, may be the answer. Besides, it will result in cow-calf producers (1) being paid for superior genetics, (2) seeing the product that they are producing for consumers, and (3) hopefully, reaping more profit in most years.

In a 12-year study, 1980–1991, Cattle Fax, the National Cattlemen's Beef Association market information service, found that cow-calf operators selling calves at weaning had profits in six years and losses in six years. By contrast, retained ownership programs were profitable in 11 out of the 12 years covered by the study.

In a study involving 53 separate tests, Kansas State University found that retained ownership of steers through the feedlot phase was profitable to producers in six out of nine years, and that the losses were large in only two years.[1]

The real challenge to the entire beef cattle industry is this: How can you get 907,630 (1992 figures from *Agricultural Statistics 1993*, p. 244, Table 389) independently-minded beef cow operators to agree upon and follow any kind of vertically integrated system?

Note: On a declining market, those who retain ownership may suffer big losses.

---

[1]Lambert, Chuck, *et al.*, *Kansas Steer Futurities, Summary of Nine Years of Retained Ownership*, Kansas State University, Manhattan, KS.

QUESTIONS FOR STUDY AND DISCUSSION

1. It is estimated that there is an appalling calf loss of 7% between birth and weaning. What can cattle producers do to lessen these losses?

2. Outline a management program for calving time.

3. How would you raise orphan calves or multiple birth calves? Give the feeds and the schedule of feeding for the first 120 days of a calf's life.

4. What are the advantages and disadvantages of early weaning? Outline a program for early weaning, giving the age of weaning, the feed, and the feeding schedule.

5. What are the advantages and disadvantages of creep feeding? Under what conditions would you recommend creep feeding; under what conditions would you recommend against creep feeding?

6. Should growth-promoting stimulants be used on calves? Justify your answer.

7. How would you go about weaning calves, from the standpoint of both the cows and calves?

8. Who benefits the most from a Calf Health-Treatment Program (preconditioning), the cow-calf producer or the cattle feeder?

9. What marketing alternative—selling feeder calves, selling yearling feeders, or finishing cattle—should a cow-calf producer select?

10. What factors favor a vertically integrated system in the beef cattle business?

11. How would you propose to get 907,630 independently-minded beef cow operators to agree upon and follow any kind of vertically integrated system?

SELECTED REFERENCES

Title of Publication	Author(s)	Publisher
Beef Cattle, Seventh Edition	A. L. Neumann	John Wiley & Sons, Inc., New York, NY, 1977
Beef Cattle Production	J. F. Lasley	Prentice-Hall, Inc., Englewood Cliffs, NJ, 1981
Beef Cattle Production	K. A. Wagnon R. Albaugh G. H. Hart	The Macmillan Company, New York, NY, 1960
Beef Cattle Science Handbook	Ed. by M. E. Ensminger	Agriservices Foundation, Clovis, CA, pub. annually 1964–1981
Beef Production and Management Decisions	R. E. Taylor	Macmillan Publishing Company, New York, NY, 1994
Beef Production and Management, Second Edition	G. L. Minish D. G. Fox	Reston Publishing Co. , Inc., Reston, VA, 1982
Commercial Beef Cattle Production, Second Edition	Ed. by C. C. O'Mary I. A. Dyer	Lea & Febiger, Philadelphia, PA, 1978
Stockman's Handbook, The, Seventh Edition	M. E. Ensminger	Interstate Publishers, Inc., Danville, IL, 1992

Limousin heifer calf. Cattle producers should match the productive potential of various breeds and breed crosses with the available resources. (Courtesy, *Limousin World*, Yukon, OK)

REPLACEMENT HEIFERS

Contents *Page*

Fig. 22-1. Replacement Santa Gertrudis heifers on King Ranch, Kingsville, Texas. (Courtesy, *Livestock Weekly*, San Angelo, TX)

There is no better or quicker way to improve the reproductive performance of a herd than through the selection and proper development of replacement heifers.

Where the beef cattle herd is neither being increased nor decreased in size, each year about 20% of the heifers, on the average, are retained as replacements, and about the same percentage of old cows is culled. That means that there is a complete turnover in the cow herd every five years.

But not all of today's heifers become tomorrow's cows! Statistics show that about 7% of all calves born die before reaching weaning age. Still others must be culled for one reason or another, either before or after weaning. Thus, to maintain *status quo* in a beef herd—with no provision for expansion whatsoever—each year the cattle producer should start with 50% more weaner heifers than are actually needed. This means that for every 100 cows in the herd, 20 replacement heifers are actually needed to maintain the same size herd. However, 30 weaner replacement heifer prospects (50% more than actually needed) should be held, simply because, based on averages, 10 of these will fall by the wayside—they will either die or have to be culled before they replace their elders.

CRITERIA FOR SELECTION OF REPLACEMENT HEIFERS

Selection of replacement heifers is an attempt to retain the best of those animals in the current generation as parents of the next generation. Obviously, the skill with which these selections are made is all-important in determining the future of the herd.

The producer who raises replacements is in a better position to evaluate the animal genetically than the operator who buys replacements, simply because their close relatives are available in the same herd under similar feed and management conditions.

The type and performance of each individual heifer and of her close relatives are the criteria to use in selecting replacement heifers. Since replacement heifers are not old enough to have progeny, the closest relatives are the parents and half-sibs (half-brothers and half-sisters).

Selection should be based on as accurate and complete records as possible, with consideration given to the criteria in the sections which follow.

INDIVIDUALITY

Each replacement heifer should be scored for conformation at the weanling stage, and again at the yearling stage. This score should relate the structure of the heifer to her function—that of producing calves, which, in turn, will produce quality beef efficiently. Thus, scoring should involve selecting animals for the following six traits:

1. Reproductive efficiency
2. Size
3. Muscling
4. Freedom from waste
5. Structural soundness
6. Breed type

PERFORMANCE TEST

Performance testing of heifers should include a record of (1) weaning weight, and (2) yearling weight, taken over scales. Heavy weaning weight is indicative of milking ability of the dam. Heavy yearling weight is indicative of size and growthiness and is correlated with feed efficiency. Basically, performance testing is for the purposes of improving growth rate, or rate of gain (which is 45% heritable), and feed efficiency (which is 40% heritable).

Traditionally, beef cattle performance testing is conducted as follows: Each animal is individually identified by means of an ear tattoo, ear tag, ear notches, brand, or neck strap. Soon after weaning, and following an adjustment period of 2 to 3 weeks, animals are individually weighed and individually full fed on weighed amounts of a high-energy feed for a period of 140 days, followed by individual weighing at the end of this period; thereby obtaining an individual record of both (1) rate of gain and (2) feed efficiency—the pounds of feed required to make 100 lb of gain.

Under practical conditions, performance testing of cattle is usually limited to bulls for the following reasons: (1) a bull produces in his lifetime many more offspring than a female, which means that his influence on the total genetic progress of the herd is greater; (2) many more replacement females than herd bulls are selected, with the result that the facilities, labor, and cost of individually testing them in the traditional manner would be overwhelming; and (3) if heifers are production tested in the traditional manner by full feeding on a high-energy ration, they may become so fat that it will impair their reproductive performance. Yet, where possible, heifers should be performance tested, with some modification of the above procedure, because greater genetic progress can be made thereby.

From the above, it may be concluded that the three major problems encountered in performance testing heifers in the traditional manner are:

1. The very considerable labor and expense involved in individually feeding large numbers of replacement heifers in order to measure feed efficiency.
2. The hazard of lowered reproductive performance as a result of heifers becoming too fat when full fed on a high-energy ration.
3. At the close of such a test, fat heifers must be placed on reducing rations, which means that much of the feed that went into making gains during the test period is lost.

To alleviate these three problems, it is recommended that heifers be performance tested somewhat differently than bulls; that their performance test be modified as follows:

1. **Measure rate of gain only.** Fortunately, efficiency of gain is seldom measured because there is a significant correlation between rate and efficiency of gain (see Chapter 5, Table 5-5); hence, selection for rapid gains should also improve efficiency of gains. Thus, it is recommended that heifers be tested only for rate of gain, which can be accomplished by individually weighing at the beginning and end of the test period.

By measuring rate of gain only and eliminating feed efficiency, heifers may be group fed; thereby alleviating the very considerable labor and facilities of individual feeding. Heifers gaining the fastest may then be selected for replacements.

2. **Full feed a high-roughage ration.** In order that heifers will not become too fat when full fed during the test period, they should be fed a high-roughage–low-energy ration. They may even be performance tested on pasture, as pasture gain has an estimated

Fig. 22-2. Yearling Angus heifers on summer range. Heifers may be performance tested on pasture, as yearling pasture gain has a heritability of 35% —in comparison with 40% for yearling feedlot gain. (Courtesy, Martin Jorgensen, Jr., Ideal, SD)

heritability of 35% (in comparison with 40% for feedlot gain).

By a high-roughage drylot ration is meant one in which the ratio of roughage to concentrate is somewhere between (a) 1 part of roughage to 1 part of concentrate, and (b) 2 parts of roughage to 1 part of concentrate.

In order to avoid selective eating, high-roughage rations should be fed as a complete mixed feed, pelleted or unpelleted. Also, the ration should be full fed, allowing each heifer to eat according to her appetite, because a good appetite is inherited and conducive to faster gains—a desirable characteristic, especially in feedlot cattle.

3. **Test for 140 days.** The performance test should cover a minimum of 140 days. A period of this length allows each heifer to express her potential growth rate and tends to average some of the environmental effects on growth rate during the feeding period—such as the condition (fatness or thinness) of each heifer when she is placed on test.

Usually, animals are performance tested soon after weaning, primarily for convenience reasons. It is noteworthy, however, that selection for faster growth rate during one period of life should improve growth rate in another period of life.

4. **Yearling weight (365 days) or long yearling weight (452 or 550 days) may be used.** Often it is not practical to performance test heifers by full feeding them in a drylot for 140 days or more, with weights taken at the beginning and the end of the period. It is generally more practical to feed them under normal conditions, then obtain a yearling weight (365 days) or long yearling weight (452 or 550 days). Heifers gaining the fastest may be selected for replacements.

The 365-day and 550-day weights may be calculated as follows:

$$\text{Adjusted 365-day weight} = \frac{\text{Actual final weight} - \text{Actual weaning weight}}{\text{Number of days between weights}} \times 160 + \text{205-day weaning weight adjusted for age of dam}$$

$$\text{Adjusted 550-day weight} = \frac{\text{Actual final weight} - \text{Actual weaning weight}}{\text{Number of days between weights}} \times 345 + \text{205-day weaning weight adjusted for age of dam}$$

To compute 452-day weight, substitute 247 for 345 in the above equation.

Selection of heifers on the basis of yearling weight (365 days) or long yearling weight (452 or 550 days) is generally much more practical than performance testing on a full feed.

(Also see Chapter 6, Some Principles of Cattle Genetics, section on "Production Testing Beef Cattle," and the subsections thereunder.)

RECORD OF THE SIRE AND DAM

Every animal receives one-half of its inheritance from each of its parents. Hence, the more superior the sire and dam are genetically, the more likely they are to transmit genetic superiority to their offspring. But even genetically superior parents will produce offspring which range from good to poor in type and performance. Nevertheless, the average of the progeny will be higher than the average of progeny from less superior parents.

The data presented in Table 22-1 show that the individual's own record is more important than the records of its parents or grandparents. These data also show that the individual's own records plus those of the parents increase the accuracy of selection very little over that attained when selection is based only on the individual's record.

RECORD OF THE HALF-SIBS

An average beef animal has very few full brothers and sisters during its lifetime, simply because only one offspring is normally produced at each birth. However, it may have many half-sibs born in the same herd and in the same season because a bull sires many calves in a given year. An animal is related to its half-sibs by about 25% if no inbreeding has been practiced.

The data presented in Table 22-2 show the relative accuracy of selection based on the records of the individual's half-sibs as compared to selection on its own records only.

Table 22-2 reveals the following: (1) the larger the

TABLE 22-1
ACCURACY OF SELECTION[1] BASED ON THE RECORDS OF THE PARENTS AND GRANDPARENTS COMPARED TO SELECTION BASED ON THE INDIVIDUAL'S OWN RECORDS[2]

Trait	Fertility	Pasture Gain	Yearling Pasture Weight
Heritability of each trait	0.10	0.30	0.40
Accuracy of selection based on the individual's records	0.32	0.55	0.63
Accuracy of selection based on records of the individual plus records of one parent	0.35	0.58	0.66
Accuracy of selection based on records of the sire, dam, and all four grandparents but not the individual's own records	0.27	0.43	0.48

[1]Perfect accuracy of selection would be 1.0.

[2]Lasley, John F., "Selection and Acquisition of Foundation Cows and Replacement Heifers," *Commercial Beef Cattle Production*, ed. by C. C. O'Mary and Irwin A. Dyer, pub. by Lea & Febiger, Philadelphia, PA.

TABLE 22-2
RELATIVE ACCURACY OF SELECTION BASED ON RECORDS OF THE INDIVIDUAL'S HALF-SIBS AS COMPARED TO SELECTION BASED ON THE INDIVIDUAL'S OWN RECORDS[1, 2]

Number of Half-Sibs	Fertility with Heritability of 0.10	Pasture Gain with Heritability of 0.30	Yearling Pasture Weight with Heritability of About 0.40
1	0.25	0.25	0.25
2	0.35	0.34	0.34
3	0.42	0.40	0.40
4	0.48	0.45	0.44
5	0.53	0.49	0.47
6	0.58	0.52	0.50
7	0.62	0.55	0.52
8	0.65	0.57	0.54
9	0.69	0.59	0.56
10	0.71	0.61	0.57
20	0.92	0.72	0.66
30	1.04	0.77	0.69
40	1.13	0.80	0.72
50	1.19	0.82	0.73
100	1.34	0.86	0.76

[1]Lasley, John F., "Selection and Acquisition of Foundation Cows and Replacement Heifers," *Commercial Beef Cattle Production*, ed. by C. C. O'Mary and Irwin A. Dyer, pub. by Lea & Febiger, Philadelphia, PA.

[2]Calculations consider many factors. A relative accuracy of 1.00 would mean that selection on the basis of half-sibs would be as accurate as selection based on the individual animal's own record. A relative accuracy of 0.25 would mean that selection on the basis of half-sib records would be only 25% as accurate as selection on the basis of the individual's own records.

number of half-sib records available, the more accurate the selection; and (2) selection on the basis of half-sib records is more accurate than selection only on the basis of the individual's own records (a) when the trait is lowly heritable, and (b) when records are available on 30 or more half-sibs.

Whenever possible, selection should be based on the individual animal's own records. Otherwise, selecting superior individuals from superior families is the method of choice.

SELECT AGAINST EXCESSIVELY FAT HEIFERS

A certain amount of condition improves the type and conformation of an animal, and makes for a pleasing appearance. But excessive feeding increases feed costs and produces too much fat, which is harmful from the standpoint of reproduction and milk production. Hence, in selecting replacement heifers, those which are excessively fat should be avoided. Among the reasons for selecting against excessive fatness are those which follow:

- **Excessive fatness has an adverse effect on reproduction**—A high plane of nutrition in growing heifers results in early sexual maturity, but it can be overdone. Excessive fatness has a detrimental effect on the reproductive process. It makes for calving problems, with heavy calf losses at and soon after birth as a result of excess fat decreasing the size of the birth canal and the calves being presented backward.

- **Excessive fatness has a detrimental effect on udder development and milk production**—Research with beef, dairy, and laboratory animals has demonstrated the adverse effect of very high planes of nutrition on mammary development and milk yield. Thus, in addition to increased calving difficulty and drastically increased feed costs, feeding replacement heifers excessively high amounts of feed results in decreased milk production. This was clearly demonstrated by the Oklahoma Station, in a study in which replacement Hereford heifers were allotted to four planes of winter feed: (1) low, (2) moderate, (3) high, and (4) very high. The *very high* group was self-fed a 50% concentrate mixture to gain as rapidly as possible, both as weanling calves, and in subsequent winters. The depression in milk yield is shown in Fig. 22-3.

The detrimental effects on milk production seem to increase from the first to subsequent lactations, according to the Cornell Station.[1] It would appear, therefore, that early overfatness has a detrimental residual effect on the development of secretory tissue which increases with time.

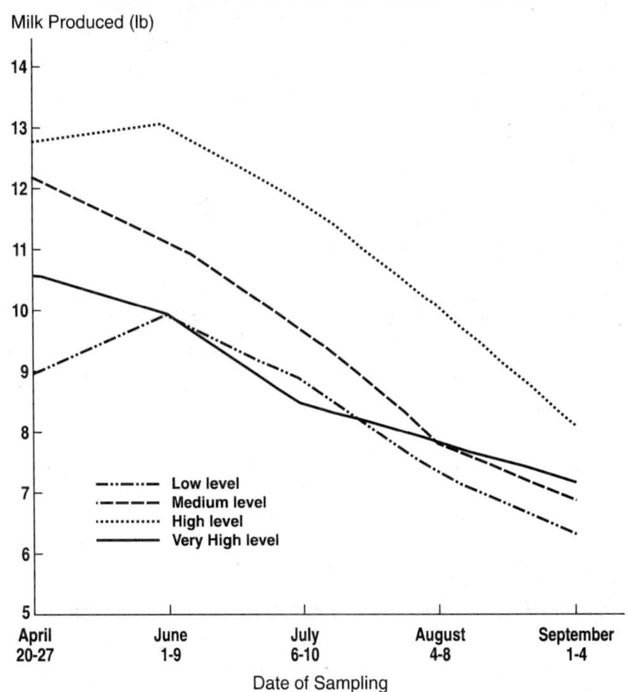

Fig. 22-3. Milk production of three-year-old Hereford cows wintered on four different levels of feed—low, medium, high, and very high. (Source: Oklahoma Ag. Exp. Sta., Oklahoma State University)

SELECT AGAINST UNDERFED HEIFERS

Restricted rations often occur during periods of drought, when pastures or ranges are overstocked, or when winter rations are skimpy. When such deprivations are extreme, there may be lowered reproductive efficiency on a temporary basis. It is noteworthy, however, that experimental results to date fail to show that the fertility of the germ cells is seriously impaired by uncommonly high or low feed and nutrient intakes.[2] In contrast to these findings, it is equally clear that the onset of heat and ovulation in young heifers is definitely and positively correlated with the level of nutrient intake; thus, if feed or nutrient intake is too low for normal rates of growth and development, the onset of reproductive function is delayed.[3]

---

[1]Reid, J. T., *et al., Causes and Prevention of Reproductive Failures in Dairy Cattle*, Cornell Univ. Ag. Exp. Sta. Bull. 987.

[2]Northeastern States Regional Bull. 32, Cornell Univ. Ag. Exp. Sta. Bull. 924, p. 30.

[3]*Ibid.*

Although healthy, undernourished heifers will usually recover when the level of feeding is returned to normal, it is not always possible to determine if poor condition is due to lack of feed, parasitism, or disease. Thus, there is reason to discriminate against poorly developed heifers.

CULLING REPLACEMENT HEIFERS

As previously indicated, when neither increasing nor decreasing the size of the herd, cattle producers should start with 50% more replacement heifers than they actually need. This means that for each 100 cows in the herd, 30 weaner heifer replacement prospects will be held; 20 head of which will actually be needed to replace their elders (20% replacement per year), and 10 of which will be culled.

Selection and culling is a continuous process. Some replacement heifers will be culled because they do not measure up from the standpoint of individuality. For example, when scoring (grading) them as weanlings and again as yearlings, it might be decided to cull all animals that are not in the upper 20% in each of the six traits—reproductive efficiency, size, muscling, freedom from waste, structural soundness, and breed type. Further, it might be decided to cull those in the lower 10% of the performance test—in rate of gain. Others may be culled because of sickness or injuries or because of such abnormalities as chronic bloat, founder, poor eyesight, or poor teeth. Still others may be culled because of such things as failing to conceive readily, calving difficulty, giving birth to a dead calf, or giving insufficient milk for the calf. A few heifers may even be carried until they wean their first calves, then culled because of the light weight of their calves, which is about 50% repeatable.

BREEDING TWO-YEAR-OLD HEIFERS

Surveys indicate that approximately 50% of two-year-old heifers require help at calving, and that 15% of the calves and 5% of the heifers are lost at parturition.[4] Thus, optimum reproductive performance of two-year-old heifers can be achieved only if there is application of the best in breeding, feeding, and management, along with knowledge of the factors affecting reproduction. (See Chapter 6 for "Factors Affecting Reproduction," many of which pertain to calving two-year-old heifers.)

Fig. 22-4. Well-grown Pinzgauer heifers calving as two-year-olds. (Courtesy, *Livestock Weekly*, San Angelo, TX)

At a symposium on "Management Considerations in Heifer Development and Puberty,"[5] Patterson, Perry, *et al.*, presented a landmark report based on a comprehensive literature review and evaluation of the current knowledge relating to the management aspects of heifer development and puberty. At the outset, these researchers stated that calving by 24 months of age is necessary to obtain maximum lifetime productivity; and that because the reproductive system is the last major organ system to mature, factors that influence puberty are critical. They found that the available information regarding puberty in heifers is voluminous, and that puberty in the female bovine is determined by an array of identifiable genetic and environmental variables. Further, they found that the management alternatives that ultimately affect lifetime productivity and reproductive performance of heifers begin at birth and include (1) growth-promoting implants, (2) creep-feeding, (3) breed type and/or species, (4) birth date and weaning weight, (5) social interaction, (6) sire selection, and (7) exogenous hormone treatment to synchronize and induce estrus. The researchers recommend that future basic and applied research efforts converge to match the production potential of the animal with the available resources.

AGE AND WEIGHT AT PUBERTY

Puberty may be defined as the age of sexual maturity, or the first ovulation and estrus. The normal age of puberty in cattle is 8 to 12 months, but it may range from 4 to 14 months. The age at which puberty

[4]Moore, D. G., *et. al.*, "Some Factors Affecting Difficulty at Parturition of Two-Year-Old Hereford Heifers," *Journal of Animal Science*, Vol. 15, No. 4.

[5]Patterson, D. J., R. C. Perry, *et al.*, "Management Considerations in Heifer Development and Puberty," *Journal of Animal Science*, December 1992, Vol. 70, No. 12, pp. 4018–4035.

is attained varies according to (1) breed, with the smaller, earlier maturing breeds having an earlier onset of puberty than the larger, slower maturing ones; and (2) nutritional and environmental factors.

The U.S. Meat Animal Research Center, Clay Center, Nebraska, where many breeds and crosses have been studied, found that Jersey crosses reached puberty at the youngest age (308 days) and lightest weight (518 pounds), while Zebu crosses were the slowest (414 and 429 days). Also, those types selected for high milk production reached puberty at an early age, regardless of type.

Donald D. Nelson, Washington State University, makes the following statement relative to the relationship between body weight and puberty in the heifer:[6]

> Within beef breeds, the onset of puberty occurs when heifers reach the approximate age and body weight, or critical minimum-breeding weight (CMBW), for that breed, or herd, within that environment. A heifer should reach 65% of her mature weight by breeding time and 85% of her mature weight at first calving time for optimum herd fertility. If all of the heifers reach this weight at the onset of the breeding season, you can expect about an 84% conception rate when using a 45-day breeding season.

AGE TO BREED HEIFERS

The age at which to breed heifers will vary with their growth and development. However, when heifers are reasonably well grown and weigh 600 to 700 lb, a safe rule is to breed at the first breeding season after they are 13 to 14 months old. Some breed registry associations will not register a calf born to the heifer under a certain stipulated age; thus, purebred breeders need to be informed relative to such rules.

The majority of commercial cattle producers breed heifers to calve as two-year-olds. The more successful early-breeding advocates feed such heifers rather liberally (1) by grazing them on the choicest range, and/or (2) by feeding added concentrates during the winter months. Also, more and more operators who calve out two-year-old heifers use a bull known to sire small calves at birth. Certainly, if the dam and the calf are not adversely affected, breeding at an early age is advantageous from the standpoint of cutting production costs.

[6]Nelson, Donald D., *Developing The Beef Replacement Heifer*, Washington State University, Cooperative Extension, Bulletin EB 1598, p. 3.

ADDED BEEF FROM TWO-YEAR-OLDS

On January 1, 1993, there were 6,183,000 beef replacement heifers in the United States.[7] If 80% of these heifers were bred to calve as two-year-olds rather than as three-year-olds; if they weaned an 80% calf crop, or 3,957,120 calves, averaging 400 lb; if all these calves were finished off to an average on-foot weight of 1,050 lb and an average carcass weight of 620 lb; that would make for a total of 2,453 million lb more beef produced than would have been obtained had all these heifers calved as three-year-olds. That's enough beef to feed 30.7 million people a whole year at the estimated per capita beef consumption level of 80 lb in the year 2005—more than sufficient beef to feed our increasing population.

Also, and most important, producing more beef by calving out two-year-olds instead of three-year-olds doesn't require more cow numbers. Hence, to produce the 3,957,120 head of 400-lb weaner calves would require few more acres and little more feed and equipment. This means that the capital investment is less than when cow numbers are increased. True enough, more feed will be required to finish the calves from weaning to slaughter. But the fact remains that greater efficiency can be achieved by putting a heifer to work early in life.

THE CHOICE—CALVE TWOS OR THREES

For practical reasons, most commercial cattle producers must either calve out as twos or threes. They have no in-between choice. This is so because of the common practice of one-season calving, spring or fall. But the choice makes a difference of 12 months—one whole year. In terms of the productive lifetime of an eight-year-old cow, a heifer calving at two will begin paying her keep after only 25% of her lifetime, whereas one calving at three will *freeload* 37.5% of her lifetime.

FACTORS FAVORABLE TO CALVING TWO-YEAR-OLDS

Based on experiments and experiences, the primary factors **favorable** to calving two-year-olds are:

1. On a lifetime basis, this will result in the production of about one more calf and an added calf weight of approximately 400 lb.
2. It is a way in which to increase beef production

[7]*Agricultural Statistics 1993*, p. 231, Table 373.

with no more cow numbers and little more feed and equipment.

3. Cow cost per 100 lb of weaned calf will be lower.

4. Increasing each animal's lifetime production efficiency provides a way in which to combat increased fixed costs of maintaining a beef animal.

FACTORS UNFAVORABLE TO CALVING TWO-YEAR-OLDS

The chief factors **unfavorable** to calving two-year-olds are:

1. The conception rate of young heifers bred when just reaching puberty is lower than in older ones. This may result in spreading the calving season over a longer period, with accompanying greater inconvenience and expense.
2. The percentage calf crop of heifers calving as twos will be about 10% lower than of older cows.
3. More calving troubles, and a higher death loss of both dams and calves will be encountered.
4. Early calving heifers will likely be somewhat undersized until they reach 4 to 5 years of age.
5. The calves will wean at 25 to 50 lb lighter weights than calves from older cows.
6. Heifers calving at two years of age will not breed back to have their second calves as three-year-olds if they do not receive enough energy during their first gestation and lactation periods, simply because they are developing and providing milk for a calf while still growing themselves.
7. It may not be desirable for the purebred breeder from the standpoints of (a) compliance with breed registry rules relative to minimum age at calving time (check them out; different breed registries have different rules), (b) selling of open heifers, (c) having well-grown young cows for visitors to see, and (d) distribution of birth dates so as to fill more show classifications; but, of course, these points do not affect commercial cattle producers.

FEEDING AND MANAGING REPLACEMENT HEIFERS

The feed and management program of replacement heifers will have a lifelong effect on their productivity. It will determine how young they may be bred, whether they calve early or late, whether they are good milkers or poor milkers, the weaning weight of their calves, and how long they remain in the herd. Also, feed accounts for 40 to 70% of the cost of raising replacement heifers; hence, it is important to know whether it is possible to effect savings on feed during the growing period without affecting reproduction adversely. It is even more important to know whether by suitable nutrition and management of heifers, we can enhance their performance as adult animals.

Fig. 22-5. Replacement heifers on pasture. Good pasture in season, water, shade, and minerals will make for proper growth and development of heifers. (Courtesy, The American Hereford Assn., Kansas City, MO)

The pregnancy requirements are really not too great. The body of an 80-lb newborn calf contains only about 12 lb of protein, 3.0 lb of fat, and 3.6 lb of mineral matter. But the lactation requirements are much more rigorous. If a two-year-old heifer gives her calf an average of 1¾ gal of milk per day over a seven-month suckling period, she will produce in that milk a total of 93 lb of protein, 107 lb of fat, 133 lb of sugar, and 20 lb of minerals.

Hence, the comparison: 12 lb of protein in the fetus vs 93 lb in the milk during the suckling period. This means that nearly 8 times more protein is required for 7 months of lactation than for 9 months of pregnancy.

Also, when breeding yearlings to calve as two-year-olds, producers should be aware that nature has ordained that the growth of the fetus, and the lactation which follows, shall take priority over the maternal requirements. Hence, when there is a nutritive deficiency, the young mother's body will be deprived, or even stunted, before the developing fetus or milk production will be materially affected.

NUTRIENT REQUIREMENTS OF REPLACEMENT HEIFERS

Meeting the nutrient requirements of heifers from weaning to first calving is of great importance. The requirements of heifers of different body weights and growth rates are given in Table 22-3.

TABLE 22-3
DAILY NUTRIENT REQUIREMENTS OF MEDIUM-FRAME GROWING HEIFERS[1, 2]

Body Weight		Daily Gain		Protein[3]	TDN[3]	Calcium	Phosphorus
(lb)	*(kg)*	*(lb)*	*(kg)*	*(lb)*	*(lb)*	*(g)*	*(g)*
400	*182*	2.00	*0.9*	1.29	7.7	26	13
500	*227*	2.00	*0.9*	1.34	9.1	24	13
600	*273*	2.00	*0.9*	1.40	10.4	23	14
700	*318*	2.00	*0.9*	1.45	11.7	22	15
800	*364*	1.40	*0.6*	1.60	10.4	25	16
900	*409*	1.40	*0.6*	1.60	11.3	26	18

[1]The above requirements for 800- and 900-lb *(364- and 409-kg)* weights are for pregnant yearling heifers last third of pregnancy. Adapted by the authors from *Nutrient Requirements of Beef Cattle*, sixth revised edition, National Research Council–National Academy of Sciences, Washington, DC, 1984.

[2]Before this book went to press, the new *Nutritive Requirements of Beef Cattle*, seventh edition, came off press. So, the new NRC requirements of beef cattle are presented in the Appendix, Section I, Table I-2, of this book; hence, the reader is referred thereto. Because allowances, rather than requirements, are used in ration formulations, in order to provide margins of safety for animal and environmental differences, the principles illustrated in Table 22-3 still apply.

[3]Pounds protein and TDN can be converted to kg by dividing by 2.2.

RATIONS FOR REPLACEMENT HEIFERS

In season, good pasture plus mineral supplements fed free-choice will meet the nutrient requirements for proper growth and development of heifers.

On the winter range, when dry forage is of low quality, and sometimes not too abundant, 1 to 2 lb of a protein supplement should be provided in the form of cubes, blocks, meal-salt, or liquid. When consumed at the intended level, the supplement should contain sufficient vitamin A to meet the requirements. Mineral supplements should also be provided, preferably free-choice.

Where winter grazing is not available, heifers must be drylotted and fed a complete ration. Sufficient nutrients should be provided to meet the requirements and to keep heifers in a thrifty condition, neither too fat nor too thin.

The wintering rations in Table 22-4 for 500-lb heifer calves should result in a rate of gain of 1 to 1.5 lb per day.

The wintering rations in Table 22-5 for 800- to 900-lb bred yearling heifers should allow a gain of 0.75 to 1 lb per day during the wintering period prior to calving.

Replacement heifers should be fed rather liberally—more so than stocker cattle which are being grown for the feedlot, to the end that they will acquire most of their growth and development before calving. With limited feeding, they will not have enough weight for age to breed when they are 15 months old; and it is best not to have them calve until they are 30 months

Fig. 22-6. A large group of well-grown, thrifty replacement heifers in a drylot. (Courtesy, Martin Jorgensen, Jr., Ideal, SD)

TABLE 22-4
DAILY RATIONS FOR HEIFER CALVES (500 lb)[1]

	Rations				
	1	2	3	4	5
	(lb/day)				
Legume-grass haylage	25				
Legume-grass hay		10	10		5
Corn or sorghum silage				30	20
Ground ear corn		4			
Corn, grain sorghum, or barley			3		
Supplement[2]				1	

[1]To convert lb to kg, divide by 2.2.

[2]Supplement contains 48% crude protein.

TABLE 22-5
DAILY RATIONS FOR BRED YEARLING HEIFERS
(800 to 900 lb)[1]

	Rations				
	1	2	3	4	5
	(lb/day)				
Corn or sorghum silage	45	25			
Legume-grass hay		10	20		15
Legume-grass haylage				35	
Corn, grain sorghum, or barley					3
Supplement[2]	1.5				

[1]To convert lb to kg, divide by 2.2.

[2]Supplement contains 48% crude protein.

of age. Cost of the ration is important, but too limited a ration may actually be costly. This is to say that it is usually cheaper in the long run to grow heifers out well rather than to delay their development and run into management problems, higher death losses, greater pasture costs, high interest on investment, and increased labor costs. Also, where breeding animals are sold, it is recognized that prospective purchasers are impressed by well-grown and well-conditioned young stock, whereas they are seldom interested in stunted animals at any price. The latter situation is accentuated in purebred heifers; hence, it is important that they be more liberally fed than replacement heifers for the commercial herd.

Occasionally, a replacement animal is injured by overfeeding or by fitting for the show, but such losses are insignificant compared with those resulting from the thousands of undersized, poorly developed animals that are grossly underfed.

During the winter months, the feeding of a ration containing adequate protein and energy, plus the required vitamins and minerals, is necessary to keep young stock growing and gaining in weight.

During their second winter, heifers bred to calve as two-year-olds should be fed more liberally than mature cows of comparable condition. The added feed is necessary because, in addition to maintenance and development of the fetus, provision must be made for body growth. Even then, these heifers should have close supervision at calving time and the calves should be weaned at an early age in order to alleviate the strain of lactation.

SEPARATE HEIFERS BY AGES

The nutritive requirements for heifers differ according to body weight and expected daily gain (Table 22-3). Consequently, the recommended ration for a 500-lb heifer calf (Table 22-4) differs from that of an 800- to 900-lb bred heifer (Table 22-5). It is important, therefore, that replacement heifers be separated by ages for wintering, with coming yearlings in one group and coming twos in another.

SUMMARY RELATIVE TO CALVING TWO-YEAR-OLDS

Fig. 22-7. Fast-growing F_1 heifers. Consideration should be given to the increased nutritional requirements of heifers with increased growth rates. (Courtesy, *Livestock Weekly*, San Angelo, TX)

From the above, it may be concluded that more breeders can, and should, breed yearling heifers to calve as two-year-olds. But, in doing so, the following practices should be observed:

1. Select the heaviest and highest scoring individual heifers at weaning. Weight at weaning is a means of evaluating the dam's milking ability.
2. Keep heifers separate from older cows.
3. Start with 50% more weaner replacement heifers than needed if it is the intent to maintain the same size herd—with no provision for expansion whatsoever. This means that for every 100 cows in the herd, 20 replacement heifers are actually needed each year in order to maintain the same size herd. (There is about a 20% replacement in each herd each year.) However, 30 weaner replacement prospects (50% more than actually needed) should be held simply because, based on averages, 10 of them will either die or have to be culled before they replace older cows.
4. Give consideration to the increased nutritional requirements of calves with increased growth rates.
5. Replacement heifers should be fed for gains of approximately 1.0 lb per head per day from weaning

Fig. 22-8. Replacement heifers. There is no better or quicker way in which to improve the reproductive performance of a herd than through the selection and proper development of replacement heifers. (Courtesy, Ralston Purina Company, St. Louis, MO)

to first breeding. Following the breeding season, heifers should be managed to assure continued growth and achieve 80 to 85% of expected mature weight at the time of first calving. From breeding until calving, 1¼ lb gain per day is about right.

6. Select yearlings and coming two-year-old heifers on the basis of individuality and rate of gain. Also, cull heifers with small pelvic openings; those with large pelvic openings (above 34 sq in.) have less calving difficulty. Avoid excessively fat heifers.

7. Breed only well-developed heifers, weighing 700 to 750 lb (depending on breed) at 13 to 14 months of age. Size at breeding is more important than age. Also, some breeds come in heat and mature a little earlier than others.

8. Breed heifers 20 days earlier than the cow herd and restrict the breeding season to 45 days. This gives a short concentrated calving period; therefore, proper attention and help can be given heifers at calving time.

9. *Flush* feed heifers to gain approximately 2.0 lb per head daily beginning 20 days before the start of and continuing through the breeding season.

10. Breed heifers to a bull known to sire small calves at birth.

11. Feed a well-balanced ration, and feed for continuous gain of 1.25 lb during the pregnancy period; but don't get them too fat.

12. Feed heifers to weigh at least 800 lb by 120 days before calving.

13. Feed heifers to gain 100 to 120 lb from 120 days prior to calving. Heifers should weigh at least 875 lb just before calving and approximately 775 lb shortly after calving.

14. Give heifers special care at calving time. This should include—

a. Providing adequate facilities, including (1) a pull stall, and (2) small pens each suitable for confining a heifer and her calf for approximately 24 hours of *mothering up*.

b. Move each heifer into the calving area approximately two weeks before the expected calving date.

c. Check heifers for calving at two-hour intervals.

d. Render assistance quickly and expertly when it is needed.

e. Remove heifer and calf from calving area within 24 hours after birth and put them into a clean, dry pasture or other similar area.

15. Provide superior nutrition—well balanced, and rather liberal—during the lactation period, because a heifer's nutritional requirements double after calving. This requires a good ration—one containing adequate energy and proteins, and fortified with the necessary vitamins and minerals. In season, usually this can be accomplished by keeping these heifers on good pastures, with or without supplemental feeding, both during pregnancy and lactation. When good grass is not available—in the winter, early and late, or during droughts—proper feeding must be relied upon.

16. If practical, wean early; at 4 to 6 months of age, rather than the normal 7 months.

17. Run heifers that calved as two-year-olds in a separate herd until after they have had their second calf.

18. Try it out on half of your replacement heifers to start with; make sure that you know what is involved before going all out.

Of course, the below-average breeder—the breeder who has lightweight, poorly developed heifers, and who wouldn't think of staying up nights and having cold, numb fingers while being nursemaid to a heifer and a newborn calf—should take another year and stick to calving out three-year-olds. But progressive, commercial producers should calve out more two-year-olds from the standpoint of cutting production costs and increasing profits.

QUESTIONS FOR STUDY AND DISCUSSION

1. If a cow herd is being built up through increasing numbers by 10% per year, how many replacement heifers at weaning time per 100 mature cows should be selected?

2. What consideration should be given to each of the following criteria for selection of replacement heifers: (a) individuality; (b) performance test; (c) record of the sire and dam; (d) record of half-sibs; (e) overly fat heifers; and (f) underfed heifers?

3. How do you account for the fact that, in a selection program, a heifer's own record is more important than the records of her parents or grandparents?

4. Discuss the economic and beef impact of breeding all U.S. heifers to calve as two-year-olds.

5. Do you agree or disagree with the authors' added beef projections to result from breeding two-year-old heifers? Justify your answer.

6. Why should producers match the production potential of various breeds and breed crosses with the available resources?

7. List and discuss each of the factors **favorable** to calving two-year-old heifers.

8. List and discuss each of the factors **unfavorable** to calving two-year-old heifers.

9. How do the nutrient requirements of heifers calving as two-year-olds differ from the nutrient requirements of heifers calving as three-year-olds?

10. Why should heifers be separated according to age, body weight, and expected daily gain?

11. Under what circumstances would you recommend buying replacement heifers, rather than raising them?

SELECTED REFERENCES

Title of Publication	Author(s)	Publisher
Beef Cattle, Seventh Edition	A. L. Neumann	John Wiley & Sons, Inc., New York, NY, 1977
Beef Cattle Production	J. F. Lasley	Prentice-Hall, Inc., Englewood Cliffs, NJ, 1981
Beef Cattle Production	K. A. Wagnon R. Albaugh G. H. Hart	The Macmillan Company, New York, NY, 1960
Beef Production and Management Decisions	R. E. Taylor	Macmillan Publishing Co., New York, NY, 1994
Beef Production and Management, Second Edition	G. L. Minish D. G. Fox	Reston Publishing Company, Reston, VA, 1982
Commercial Beef Cattle Production, Second Edition	Ed. by C. C. O'Mary I. A. Dyer	Lea & Febiger, Philadelphia, PA, 1978
Nutrient Requirements of Beef Cattle, Sixth Revised Edition	National Research Council	National Academy of Sciences, Washington, DC, 1984
Nutrient Requirements of Beef Cattle, Seventh Revised Edition	National Research Council	National Academy of Sciences, Washington, DC, 1996

Tom Brothers working facilities. From far left—scale covered by shed; combination sale barn and equipment storage barn; loading chute; two chutes leading into sale barn (one on left for mature animals; one on right for calves—also used to vaccinate calves). (Courtesy, Lytle Tom, Jr., Tom Brothers Ranch, Campbellton, TX)

CONFINEMENT (DRYLOT) BEEF COWS

Contents *Page*

Fig. 23-1. Beef cows in confinement, showing barns/sheds for cattle shelter and silos for feed storage. (Courtesy, *Livestock Breeders Journal*)

Confinement of cattle is not new. The pioneers planted trees and shrubs, such as osage orange, or built rail or stone fences, to hold their animals. Then, in 1873, Joseph F. Glidden of DeKalb, Illinois, invented barbed wire, or the *devil's rope*, as it was dubbed by those who considered the barbs inhumane or who opposed its use because it marked the beginning of the end of the open range and free grazing. But, for the most part, the wire that fenced the West confined cattle to large areas—pastures.

Today, experimentally and on a very limited commercial basis, beef cows are being confined to small quarters—to drylots, all or part of the year.

Historically, as countries become more densely populated, land values and tax rates rise, necessitating that the highest and best use of land be made in order to pay taxes and yield a satisfactory return on investment. It's the same principle that spawns multistory buildings—as urban land prices mount, skyscrapers appear.

The land area available for beef production is slowly decreasing due to urban development, highways, airports, and recreational areas; and precious little new land can be brought into production. Hand in hand with this transition, the price of grassland on a per animal unit basis has increased to the point where it no longer provides cheap animal feed for cattle. On most cow-calf operations, the largest single cost item is interest on the money invested in land. At $3,000 per cow unit and 12% interest, the yearly interest tab is $360 per cow. Under confinement, at least 60 cows can be handled on 1 acre. That's a land cost of $50 per cow ($3,000 ÷ 60), or an interest charge of $6 per year—a saving of $354 per cow on interest on land. This saving can be applied toward the increased feed and labor cost of a confinement operation. This situation has prompted interest in more intensive use of productive lands—in the production of more high yielding crops such as corn and sorghum. Although intensive production requires considerable investment in facilities, equipment, and labor, the point has been reached where these added costs are largely offset by the lower investment in land. With the demand for more beef and world beef shortages, this gap will widen in the years ahead. As a result, intensive drylot cow-calf production will increase on the more productive lands.

Although confinement production as a management technique in beef cow-calf production is new in the United States, the concept is very old. To gain efficiency and reduce cost of production, American dairy, poultry, and swine producers have, of necessity, invested in intensive production units. Also, confinement cattle production has long been traditional in China, Japan, and Europe—in the more densely populated areas of the world, where land is scarce and high in price.

WHAT DRYLOT COW-CALF EXPERIMENTS SHOW

Fortunately, several longtime research studies have been conducted on confinement cow-calf production. Thus, a considerable body of data and experience is available. A review of the experimental work involving drylotting cows the year-round follows:

▪ **Alabama Station, Lower Coastal Plain Station, Camden, Alabama[1]**—Two confinement systems and a pasture system were compared over a 5-year period by the Alabama Station. Cows under the pasture system were full-fed Coastal Bermudagrass hay plus 2 lb of cottonseed meal per head daily during the winter months—from November 1 until spring grazing. The balance of the year, they grazed Coastal Bermudagrass pasture at the rate of 1 cow-calf unit per acre. Cows in continuous confinement were fed either (1) sorghum silage plus 1½ lb of a 65% protein supplement per head daily for the first 180 days after calving, then 1 lb per day for the rest of the year; or (2) Coastal Bermudagrass hay plus 1 lb of a 65% protein supplement per head daily for the first 180 days postcalving and none for the rest of the year. Hereford cows calving in the spring were used the first 3 years and Angus-Hereford crossbreds calving in the fall were used the last 2 years. Hence, none of the cows were kept on the treatments for the duration of the study—5 years. Natural breeding was practiced.

Results of the 5-year trial are shown in Table 23-1.

[1]Harris, R. R., V. L. Brown, W. B. Anthony, and C. C. King, Jr., *Confined Feeding of Beef Brood Cows*, Alabama Ag. Exp. Sta. Bull. 411.

TABLE 23-1
COMPARATIVE PERFORMANCE OF BEEF COWS MANAGED UNDER TWO CONFINEMENT FEEDING SYSTEMS AND A CONVENTIONAL GRAZING SYSTEM FOR 5 YEARS, IN ALABAMA

Item	Management System		
	Summer Grazed, Drylot Wintered	Confined, Fed Silage	Confined, Fed Hay
Hay or silage . . . (tons/cow)	1.58	12.81	4.31
Hay or silage production . . . (tons/acre)	1.68	17.15	7.49
Land requirements; cow-calf units/acre (feed)	1.06	1.34	1.74
Calving rate . . . . (%)	95.0	92.0	82.0
Calf crop weaning . . . . (%)	92.0	88.0	84.0
Adj. calf weaning weight (lb)	453.0	551.0	511.0
(kg)	*206.0*	*250.0*	*232.0*

The Alabama experiment showed the following:

1. The confined cows performed satisfactorily in every respect except that their average calving percentages were lower than for the pasture group.

2. One acre of Coastal Bermudagrass fertilized with 200 lb of nitrogen provided grazing and hay for one conventionally managed cow-calf unit annually. One acre devoted to the production of sorghum silage or Coastal Bermudagrass for hay supported 1.34 and 1.74 cow units, respectively. Thus, only 0.58 acre was required to produce sufficient hay for 1 cow-calf unit kept in confinement.

3. A cow-calf unit in confinement the year around required 12.8 tons of sorghum silage (35% dry matter) or 4.3 tons of Coastal Bermudagrass annually.

4. Cows fed conventionally weaned a larger percent calf crop (92%) than did either of the confinement groups (88 and 84%, respectively).

5. The cows in confinement weaned calves that averaged 78 lb heavier than the pastured cows. However, their calves were creep fed and consumed an average of 1,422 lb of feed per calf; that's 18 lb of feed per pound of gain.

■ **Arizona Station**[2]—Total and partial confinement were studied at the Arizona Station. The partially confined cows were on irrigated pastures during a long grazing season, and oats and barley provided winter grazing and Sudangrass summer grazing. Alfalfa and fescue were added so as to extend the grazing season to nine months. The totally confined cows were fed a mixture of alfalfa hay and bermudagrass straw, with adjustments in proportions and quantities fed as well as in the use of supplements at different stages of the reproductive cycle. The calves in both groups were creep fed.

Annual feed costs were approximately $10 higher per cow and calf for the pairs totally confined than those partially confined. The Arizona investigators suggest that the cost of drylot cows could be lowered by 1¢ per cow per day by feeding on alternate days, based on their studies showing that feeding every other day is satisfactory.

■ **Illinois Station**[3]—In a 4-year trial, using spring-calving Shorthorn cows, the Illinois Station compared a year-round drylot cow-calf system with a summer grazing, winter drylot system. The pasture, which consisted of bromegrass, orchardgrass, and alfalfa, was grazed from May 15 to October 15. The cows in drylot were fed (1) cornstalk silage with 25 lb of dried molasses and 75 lb of cornmeal added per ton of forage at ensiling time, from November 1 to February 15; (2) corn silage from February 16 to June 30; and (3) haylage from July 1 until October 31. Urea, dicalcium phosphate, and trace mineralized salt were added to both corn forages at ensiling to provide a balanced ration of approximately 11% crude protein on an air-dry basis. Cows were bred by natural service. Calves in both groups were creep fed.

The Illinois experiment showed that (1) cows in drylot required fewer services per conception than those on pasture, (2) the drylot cows weaned slightly heavier calves than the pasture cows, and (3) the carrying capacity of the land was increased 47% by harvesting and feeding haylage rather than grazing the forage as pasture.

■ **Iowa Station**[4]—In a study involving cows nursing calves, the Iowa Station compared (1) drylot cows fed 114 lb of alfalfa-grass mixture as green chop, and (2) cows on pasture. The cows on green chop gained an average of 7 lb during the summer period and weaned calves weighing 405 lb, whereas the cows on pasture lost an average of 70 lb and weaned calves weighing 370 lb. Thus, the green-chop cows were 47 lb heavier and weaned off calves that were 35 lb heavier than the drylot cows. But the heavier calf weights were not sufficient to compensate for the added charges of harvesting and feeding the green chop; the cows on pasture returned $9.34 more per calf than the drylot cows fed green chop.

---

[2]McGintry, D. D., H. I. Essign, and E. Hussmann, "Production of Feeder Calves in Intensive Management Systems", *Cattle Feeders' Day*, Dept. of Animal Science, The University of Arizona.

[3]Albert, W. W., *Good Performance from Beef Cows Confined to Drylot the Year-Round*, Illinois Ag. Exp. Sta. IL Res.

[4]Woods, W., B. Taylor, and W. Burroughs, *Feeder Calf Production by Intensive Methods on Iowa Corn Land*, Animal Husbandry Leaflet 230, Iowa State University, Ames, Iowa.

■ **Minnesota Station**[5]—In a 5-year study, Angus and Hereford cows were maintained under two systems; one group was kept on legume-grass pasture during the grazing season and confined to drylot during the rest of the year, and the other group was confined to a drylot continuously for 5 years where they were fed legume-grass forage in season, similar to the forage grazed by the pastured herd. All cows were bred by natural service, and all calves were creep fed.

Table 23-2 shows the performance of the cows and calves.

The Minnesota investigators concluded that (1) about twice as many cows could be maintained per unit of land if they were confined instead of grazed in the conventional manner; (2) conception rates, calf crop percentages, and birth weights were similar; and (3) the drylot group had fewer losses, but lower weaning weights.

TABLE 23-2
COMPARATIVE PERFORMANCE OF BEEF COWS MAINTAINED IN PASTURE AND DRYLOT HERDS FOR 5 YEARS, IN MINNESOTA[1]

Treatment	Angus		Hereford	
	Pasture	Drylot	Pasture	Drylot
Conception rate . . . . . (%)	96	96	88	88
Calf birth rate . . . . . . (lb)	64	63	68	67
. . . . . . *(kg)*	*29*	*29*	*31*	*30*
Avg. adj. weaning weight 1963–1967 . . . . (lb)	470	450	490	426
. . . . *(kg)*	*214*	*205*	*223*	*194*

[1]Meiske, J. C., and R. D. Goodrich, *Drylot vs Conventional Cow-Calf Production*, Minnesota Res. Rep. Bull. 118, University of Minnesota.

■ **South Dakota Station**[6]—The South Dakota Station compared the performance of cows (1) maintained continuously in drylot, vs (2) cows summered on native pasture and wintered in drylot. They used Hereford-Angus crossbred cows, which were mated to Angus, Charolais, and Hereford bulls.

Results the first 5 years of the study showed that the pasture cows have a higher percentage of calves born (4% more) and weaned (5% more), and heavier weaning weights (18 lb more) than cows maintained continuously in drylot.

■ **Texas A&M Agricultural Research Station, Spur, Texas**[7]—Workers at the Rolling Plains Livestock Research Station at Spur, Texas, did a lifetime performance study of cows in total confinement compared with cows maintained continuously on native pasture. The experiment was started with 72 yearling Hereford heifers, and involved 2 systems—(1) pasture, and (2) drylot. Each system was further subdivided into 3 groups of 12 heifers each, which were fed rations of cottonseed meal, sorghum grain, and silage to provide 3 levels of energy designated as low, medium, or high during the winter period, November through April. Table 23-3 shows the digestible protein and metabolizable energy of the rations fed, and Table 23-4 shows the performance of the cows over an 11-year period. All animals, both drylot and pasture, had free access to plain block stock salt at all times; and during the period 4 to 6 weeks after calving, all cows had access to a 50/50 mixture of bone meal and salt. No vitamin supplements were used. Breeding was by natural service.

As shown in Table 23-4, the data was broken down to reflect the trouble years—the 3-year period, 1967-1969, when birth weights and calf survival were adversely affected by what subsequently appeared to be a protein deficiency.

Through the first 7 years of the Texas study, there was no significant difference in the average percent calf crop and weaning weights of the calves produced in drylot and on pasture. However, large differences between the 2 systems occurred in 1967 and continued through 1969. It was conjectured that the adverse effects of drylotting during this 3-year period—as evidenced in percentage calves weaned and lighter calves at birth and weaning—was caused by a cumulative nutritional deficiency (or deficiencies) over the period of years in confinement. So, a change was made in the rations during the winter of 1968-69, by increasing the protein and phosphorus intake. The change in protein was prompted so that the rations would meet National Research Council (NRC) requirements. The rations fed from 1959 to 1968 met the NRC digestible protein (DP) standards for dry pregnant cows, from November through March; but during lactation, from April through October, the cows received 0.5 lb less DP than the NRC recommendation. The investigators reasoned that this protein deficiency for the first 5 to 6 months of gestation while the cows were lactating, during the period of greatest stress on the cows, must have been a contributing factor in the birth of small, weak calves. The 1969-70 ration change consisted in increasing the DP in the ration by 0.21

[5]Meiske, J. C., and R. D. Goodrich, *Drylot vs Conventional Cow-Calf Production*, Minnesota Res. Rep. Bull. 118, University of Minnesota.

[6]Slyter, A. L., *Influence of Mating and Management Systems on the Performance of Beef Cows and Calves*, A. S. Series 72-5, South Dakota State University.

[7]Marion, P. T., J. K. Riggs, and J. L. Arnold, "Calving Performance of Drylot and Pasture Cows over an 11-Year Period," *Beef Cattle Research in Texas*, Consolidated PR 2963-2999.

TABLE 23-3
DIGESTIBLE PROTEIN (DP) AND METABOLIZABLE ENERGY (ME) FOR RATIONS FED 1959–68 AND 1969–70[1, 2]

Year	Winter Rations for Dry Cows, Level of Energy						Summer Ration, Lactating Cows	
	Low		Medium		High			
	DP	ME	DP	ME	DP	ME	DP	ME
	(lb)	*(Mcal)*	*(lb)*	*(Mcal)*	*(lb)*	*(Mcal)*	*(lb)*	*(Mcal)*
1959–68	0.57	8.70	0.56	11.88	0.53	15.90	0.70	15.80
1969–70	0.95	10.10	0.85	13.00	0.77	16.42	0.91	17.10

[1]*Beef Cattle Research in Texas*, Texas Ag. Exp. Sta. Consolidated PR-2963-2999, Texas A&M University.

[2]Lb can be converted to kg by dividing by 2.2.

TABLE 23-4
PERFORMANCE OF PASTURE VS DRYLOT COWS (1960–1970)[1, 2]

Period	Pasture				Drylot			
	Calf Crop				Calf Crop			
	Born	Weaned	Birth Wt.	Weaning Wt.	Born	Weaned	Birth Wt.	Weaning Wt.
	(%)	*(%)*	*(lb)*	*(lb)*	*(%)*	*(%)*	*(lb)*	*(lb)*
1960–1966—7-year avg. . . .	94.8	88	72.9	468	96.9	88	73.0	476
1967–1969—3-year avg. . . .	93.3	89	75.4	492	91.4	69	57.4	455
1970	100.0	90	87.4	618	90.4	86	68.7	527

[1]*Beef Cattle Research in Texas*, Texas Ag. Exp. Sta. Consolidated PR-2963-2999, Texas A&M University.

[2]Lb can be converted to kg by dividing by 2.2.

lb during lactation and 0.24 to 0.38 lb during the dry period (Table 23-3). Although this ration change did not affect the 1969 calf crop, it apparently greatly improved the birth weights and calf survival of the 1970 crop (Table 23-4).

The Texas experiment showed the following:

1. There were no significant differences in the average percent calf crop and weaning weight of calves produced by drylot and pasture cows through the seventh calf crop.
2. In drylot cow-calf production, it is important that the nutrient requirements be established and met, with increases during the lactating period when the requirements are most rigorous. Otherwise, a cumulative deficiency (or deficiencies) is likely to show up on cows confined for a number of years. In the Texas study, a protein deficiency became evident after seven years.
3. Overfeeding and excess fatness of drylot cows is costly and detrimental.
4. The teeth of the cows on pasture showed severe wear, whereas those of the drylot cows remained in good condition.
5. Eighteen cows—an equal number—in each group, drylot and pasture, had to be removed during the 12-year study. It is noteworthy, however, that five of the drylot cows were removed for being open two successive years, whereas no pasture cows were removed for this reason.
6. The average cow-calf consumption of the drylot cows was 4,403 lb of TDN.
7. Costs favored the pasture cows, primarily because summer feed for the drylot cows was expensive. The Texas workers suggested that if drylot cows could be grazed on more economical temporary pasture for 90 to 120 days during the summer, an additional $10 to $15 could be added to the net return per cow.
8. There is need for longtime studies (lifetime, and even generation after generation) where a cumulative nutritional deficiency, or deficiencies, may affect reproduction. In this study, it took 7 years for the protein deficiency to become evident.

ADVANTAGES AND DISADVANTAGES OF CONFINING BEEF COWS

The logical procedure to follow in arriving at any important business decision, including whether or not to switch to or start a confinement cow-calf business, involves three steps:

1. What are the pros?
2. What are the cons?
3. What is the decision?

Based on experiments and experiences, the following advantages and disadvantages are inherent in most confinement cow-calf operations in comparison with conventional pasture systems.

The **advantages**:

1. They require less investment in land per cow unit.
2. Cow numbers can be increased without obtaining more land.
3. They maximize feed production per acre of land, through utilizing harvested forages instead of grazing.
4. Hazards of drought and adverse weather conditions may be minimized through storing adequate feed reserves for future use.
5. Precise breeding programs can be designed and carried out more easily.
6. Control of estrus and use of artificial insemination programs become more practical.
7. Fewer services per conception are required in drylot than on pasture.
8. Individual performance records can be kept more easily.
9. Selection and culling are easier, because cattle can be observed more closely.
10. Plane of nutrition can be accurately known and controlled in accordance with the age and production needs of the cow; for example, nutrients can be increased during the critical 100 days, extending from 30 days before calving to 70 days after calving. After a calf is born, the cow's nutritional requirements rise approximately 50% over maintenance alone.
11. They make it easier to flush cows prior to the breeding season.
12. Byproducts and low-quality feeds may be utilized to advantage.
13. Partial environmental control may be achieved.
14. They permit close observation of cattle. Hence, illness and injuries are quickly detected and may be promptly treated.
15. Calving at various times of the year is more practical than in pasture handling.
16. They make for maximum flexibility in creep feeding. For example, steers and heifers can be fed separately.
17. Confinement-produced calves wean more easily and go on finishing rations more readily.

Fig. 23-2. Sorting confinement cows for artificial insemination on the Tom Brothers Ranch of Texas. Confinement cow-calf operations facilitate AI breeding because of the close observation accorded. (Courtesy, Lytle Tom, Jr., Tom Brothers Ranch, Campbellton, TX)

The **disadvantages**:

1. They usually require a higher investment in buildings and equipment.
2. They require more labor.
3. Labor must be provided seven days per week. Hence, it is more confining to the operator, unless additional help is available.
4. Disease problems, especially among calves, are generally more acute.
5. All feed must be harvested and moved to the feedlot, rather than being harvested directly by the cows.
6. An assured, adequate, and economic feed supply must be available. The capital tied up in stored feeds may be quite large.
7. More able management is required.
8. More knowledge of beef cow nutrition and ration formulation is needed.
9. Risks due to blizzards, hurricanes, tornadoes, floods, and lightning are increased due to the heavy concentration of animals.

KIND OF COW FOR CONFINEMENT (DRYLOT) PRODUCTION

What will the confinement cow of tomorrow look like? What kind of cow is best suited for confinement (drylot) production? How will she differ from her counterpart on the farm or range?

Under confinement, cows will be completely dependent upon the caretaker. They will no longer need to fight and fend for themselves, as did the Longhorn—or even the cow on the range. Will they, under such conditions, become timid and completely defenseless like sheep? Will they lose their independence of behavior and be unable ever to return to the wild state, or become feral? These traits in sheep are clearly the result of selection; and their dependence on a caretaker is a logical final result of domestication.

What size cow is best suited to confinement production, where the feed is harvested for her? Generally speaking, the European breeds of beef cattle which evolved under confined or semiconfined conditions were larger beasts than their counterparts that roamed the hills and pastures, gleaning the feed provided by nature.

Will the desirability of adaptation to environment be eliminated? Will heat and cold tolerance be necessary under confinement where partial environmental control is provided by shelters and shades, and where perishing in the winter and overheating in the summer are no longer hazards?

Answers to these questions, and more, will determine the ideal cow for confinement production. The authors' specifications of her follow:

1. She will be an F_1 (first cross) cow, which is the most productive of all cows; and she will be estrus-controlled and bred AI to a bull of a third breed to tailor her calves to command top price. Such a precise breeding program can be designed and carried out under confinement more easily than on pasture or range.
2. She will be a medium-sized cow, weighing around 1,000 lb rather than a huge beast, because:
   a. The maintenance requirement for energy is proportional to body weight to the ¾ power ($W^{0.75}$). This favors a small cow where feed costs are high.
   b. By breeding a medium-sized cow to a large bull (without accentuating calving problems), a rapid-gaining calf which will have large mature size can be secured without the added maintenance cost of a big cow.
3. Behavior-wise, she should (a) be docile and easily handled, (b) exhibit a minimum of independence and dominance, and (c) be able to tolerate crowding.
4. She should breed at an early age, and be highly fertile and prolific—twinning would be desirable if it becomes practical.
5. She may calve when desired, because calving at various times of the year is more practical under confinement than in pasture handling.
6. Heavy and persistent (over a long period of time) milking will not be essential, because early weaning at about two months of age will be practiced. It is more efficient to feed calves directly than it is to feed cows to produce milk to feed calves. Using Hereford and Charolais cows, Melton, *et al.*, showed that calves required an average of 2.1 lb TDN (including the TDN in both the dam's milk and the creep feed) per pound of gain from birth to 210 days of age, and 3.4 lb of TDN per pound of gain from birth to 365 days of age. When the 365-day TDN requirement of the dam was included, the TDN requirements per pound of calf produced increased to 8.5 and 7.2 lb at 210 and 365 days of age, respectively.[8]

   Also, it is easier to get cows to rebreed when they are not nursing a calf.
7. She will be of higher quality than pasture cows, because selection and culling are easier and cattle can be observed more closely under confinement.
8. She should be a very efficient converter of roughages and byproduct feeds.
9. She should be adapted to semiconfinement. In season, and if desired, she should be able to graze stalklage, or native or irrigated pasture.
10. She should not have horns—she should be either naturally polled or dehorned.
11. She should be more resistant to diseases than her less exposed counterpart on pastures and ranges.

FACILITIES FOR DRYLOTTING BEEF COWS

Facilities for drylotting beef cows must be adequate and functional, but they need not be elaborate. The requirements change somewhat according to the stage of the reproductive cycle and the area. For example, dry pregnant cows require less space than cows suckling calves. Dry and well-drained areas require less space than high rainfall, poorly drained areas. Cold areas require shelter, and hot areas require shade. However, the basic facility needs, subject to some adaptation, for a confinement cow-calf operation are as follows:

■ **Site**—Pollution is a most critical factor in site selection of a confinement cow-calf operation. Remoteness from urban development is recommended because of dust and odor. Also, before constructing facilities, the owner should be familiar with both state and federal regulations, then comply therewith.

■ **Corral space**—Under average conditions, and an unsurfaced lot, it is recommended that 300 sq ft of corral space be provided per cow-calf. More or less space may be provided, depending primarily on rainfall, drainage, dust control, pollution control, etc. Up to 400 sq ft may be required in a wet, muddy area. The Arizona Station provided 355 sq ft per cow and calf; the Texas Station provided 200 sq ft for a dry pregnant cow and 300 sq ft per cow-calf after calving.

---

[8]Melton, A. A., T. C. Cartwright, and W. E. Kruse, *Cow Size As Related to Efficiency of Beef Production*, Texas Ag. Exp. Sta. P.R. 2485.

In addition to corral space, provision must be made for feed storage, alleys, and working areas. So, under average conditions the total space requirements for confinement (drylot) beef cows is on the order of 1 acre for 60 to 65 cows.

■ **Fencing**—The specifications for a desirable fence are: 54 to 60 in. high; posts of 3- to 4-in. diameter pipe, or treated wood with 5-in. top diameter, set in concrete and spaced 8 to 12 ft apart; enclosed (except for the feed bunk area) by 4 strands of 3⁄8-in. steel cable spaced at intervals (from ground up) of 18 in., 12 in., 12 in., and 13 in.. Four strands of the cable should be placed above the feed bunk, with these spaced equal distances apart. The cable above the bunks should be so spaced as to keep a cow from butting a calf into the trough. Pipe or wooden rails may be used in place of steel cable if desired.

The above arrangement will confine the cows, but not the calves. Thirty-nine-inch woven wire, with number 9 top and bottom wires and 6-in. mesh, should be placed around the entire corral. Also, calves must be kept out of the feed bunks.

■ **Cows per pen**—Except at calving time, 50 to 100 cows may be run together in one group.

■ **Shelter and shade**—Where winters are severe and snowfall is heavy, a shelter should be provided. This is especially important for newborn calves. A high and dry, deep pole-type shed, opening away from the direction of prevailing winds is excellent. In areas with mild winters, a windbreak may be provided by hills, trees, or board fences.

In hot climates, a shade 12 ft or more high, oriented north-south (so that the sun will shine under the shade early in the morning and late in the evening), should be provided, with approximately 40 sq ft of area per cow-calf.

■ **Bunk and concrete apron**—Feed bunks, which form part of the pen fence, should provide 24 to 30 in. of space per cow. Bunks should be constructed of concrete with the outside (alley side) of the bunk 22 to 36 in. high, and the inside (the pen side) 22 to 24 in. high. The bottom of the bunk should be rounded and about 18 in. wide.

An 8-ft concrete apron (platform), 4 to 6 in. thick, should extend from the feed bunk into the pen to provide solid footing for the cattle. The apron should have a slope of 1 in./ft, which will make it nearly self-cleaning.

■ **Water troughs**—Confinement cows should have access to water at all times. One linear foot of trough space for each 8 cows is sufficient. There should be sufficient water supply to provide 20 gal per head daily. Shallow, low-capacity troughs are preferable, since frequent drainage is necessary to keep them clean.

Fig. 23-3. Confinement cow-calf feed bunk equipped with protective bar to keep cows from butting calves into the bunk. If such a bar or wire is not provided, some calves will be turned upside down in the bunk and die if the caretaker does not rescue them very soon. (Courtesy, Lytle Tom, Jr., Tom Brothers Ranch, Campbellton, TX)

Continuous-flow troughs are excellent and help to keep clean water. Also, in most areas, continuous-flow troughs will not freeze during the winter months; hence, heat is not required. Heated and insulated troughs are necessary in cold areas.

■ **Maternity stalls**—Where cows calve in confinement, maternity stalls 100 to 120 sq ft in size should be provided for occupancy during calving and continuing for 1 to 2 days thereafter. This is particularly true of first-calf heifers. When several heifers calve the same day in a large corral, they frequently claim the wrong calf, or no calf at all—with the result that they end up as dogies. Thus, it is best that a heifer and her calf be confined to a maternity stall until they pair up.

When calving during the winter months in cold areas, the maternity stalls should be in a barn or shed; and heat lamps should be provided. In warm areas, uncovered pens will suffice. All maternity stalls should be cleaned, disinfected, and freshly bedded for each birth.

■ **Calf creep**—Calves should be fed separately from the cows, in a creep.

■ **Hospital area**—A special hospital area should be provided, with individual pens in which to treat and isolate sick animals. It should be equipped with feed facilities for the sick animals and a chute for handling cows. In areas subject to severe weather, shelter should be provided for the hospitalized animals. Also, there should be a room equipped with refrigerator, running water, medicine, and equipment storage.

■ **Loading and unloading facilities, working chutes, squeeze, and scales**—An area should be provided for receiving and shipping cows, working animals, and weighing. The size and facilities of this area will be determined by the number of cows.

RATIONS FOR DRYLOT COWS

Rations for drylot cows generally consist of cheap roughages—such as crop refuse, straw, cottonseed hulls, and gin trash—supplemented with protein, grain, vitamins, and minerals as required. Where available, higher-quality roughages—such as silages, hays, and haylages—may be used, especially (1) during the critical 100 days, beginning 30 days before calving and extending 70 days after calving, and (2) for heifers calving as two-year-olds. Also, during the summer and fall, green chop is frequently fed. Cows in partial confinement may, in season, graze such forages as cornstalks, or irrigated or native pastures.

Phase-feeding according to stage of production and age of animals is recommended.

It is relatively easy to meet the nutritive requirements of a dry pregnant cow. Generally speaking, low-quality roughages or a combination of low-quality roughages and high-quality roughages, properly supplemented, will suffice. If only high-quality roughages are fed, they must be limited; otherwise, the cows will get too fat. However, such limited feeding does not meet the maximum fill requirements of the rumen, with the result that the cows nibble at the manure and pick up small feed particles scattered about the corral. This may make for disease and parasite problems. A simple solution to the weight and scavenger problems is to use a combination of low-quality and high-quality roughages.

The lactation requirements are much more rigorous than the dry pregnancy requirements; and the higher the milk production, the higher the nutritive requirements. (See Appendix, Section I, Tables I-1 and I-3, for the NRC requirements.)

Under a drylot system, heifers are commonly calved as two-year-olds. Such heifers are still growing. Thus, during lactation they have a nutritional requirement for both growth and lactation. The nutritional requirements for these animals will be large, particularly if they are crossbreds and bred for rapid growth, considerable size, and high milk production.

The mineral needs of confinement cows may be met either by incorporating the needed minerals in the supplement which is fed, or by feeding the required minerals free-choice.

Vitamin A supplementation is extremely important for drylot cows. The carotene content of the dry forage should be disregarded and the total vitamin A requirement met by supplementation. This can be done by feeding a supplement of 2 lb of mill waste containing I million IU of vitamin A per animal—feeding this vitamin A supplement once a month to heifers, and every other month to older cows. With older cows receiving high levels of dry forages containing normal amounts of carotene, it is probable that the vitamin A requirements are being met. However, it has been demonstrated under range conditions that (I) percent calf crop is markedly increased by supplementing with vitamin A during drought years, and (2) calves respond to vitamin A treatments given their dams 90 days prior to calving.

MANAGEMENT OF DRYLOT COWS

In order for a confinement cow-calf operation to be profitable, it is essential that it have a high percentage calf crop, heavy weaning weights, a low incidence of disease, and a low cost ration. To these ends, successful confinement operators pay close attention to the details of management. Among the good management techniques which are being used in intensified beef production are the following:

1. **Dehorning.** Cows kept in a drylot should be either naturally polled, dehorned, or have their horns tipped. (When the latter is done, leave only a 3- to 4-in. stub.)

2. **Flushing.** Females (both cows and heifers) should be flushed 2 to 3 weeks before breeding. This may be accomplished by replacing low-quality roughage with good alfalfa hay, or by adding about 4 lb of grain per head per day.

Semiconfined cows can be put on more lush pasture as a means of flushing.

3. **Breed yearlings; calve two-year-olds.** All heifers should be bred as yearlings to calve as two-year-olds. Because they are still growing, they require extra feed and attention, particularly if they are to be rebred to calve as three-year-olds. Also, they must be watched more closely, and assisted when necessary, at calving time.

4. **Pregnancy testing.** Pregnancy testing is essential following the breeding season so that barren cows will not be kept.

5. **Day of birth.** At birth, each calf should be weighed, tattooed, eartagged, injected with I million IU of vitamin A, and the navel should be treated with iodine.

6. **Excess milk.** With heavy milking cows, their feed should be limited until the calf is 10 days to 2 weeks old.

7. **Scours.** Calf scours is the bane of the confinement cow-calf operator. A drylot aggravates scours and favors a buildup of the scour problem over the years.

Wherever possible, it is strongly recommended that cows be removed from the drylot immediately before calving and placed on a sizable clean pasture (one that has been idle for a period of time) for calving out, thereby alleviating most, if not all, scouring. Also, a good pasture will stimulate milk flow and make for a good nutritional start in life for calves.

Where calving on clean pasture is not practical, the following precautions against scours should be taken:

a. Clean, disinfect, and bed the maternity stall after each birth.

b. Inject the newborn calf with 1 million IU of vitamin A.

c. Limit the feed of heavy producing cows until the calf is 10 days to 2 weeks old.

d. Where scours develop, see your local veterinarian for recommended treatment.

8. **Consider early weaning.** Weaning calves at two to four months of age will save feed and result in getting the cows rebred more quickly; hence, early weaning should be considered.

9. **Feed cost.** The key to a successful confinement cow-calf operation is to locate where a sure supply of inexpensive roughage (or byproduct feed) is available, with a minimum transportation cost.

10. **Amount of feed per cow.** An average cow will consume about 3¾ tons of air-dry feed per year. That's 20 lb per day. Many cows will stay in condition with only 14 lb of feed during the dry period, but they will need around 24 lb per day during the nursing period.

11. **Amount of creep feed.** Each calf will consume 200 to 500 lb of creep feed while nursing.

12. **Mineral-soil box.** A three-compartment mineral box should be provided for each pen. In it, place (a) salt, (b) a mixture of equal parts of salt and dicalcium phosphate (or bone meal), and (c) soil.

13. **Alternate day feeding.** Cows kept in confinement may be fed every other day without altering performance, thereby effecting a saving in labor.

14. **Parasites.** Both internal and external parasites should be controlled.

15. **Dust control.** In dry, windy areas, pens should be equipped with sprinklers for dust control.

SEMICONFINEMENT (OR PARTIAL CONFINEMENT) COW HERDS

A semiconfinement (or partial confinement) operation is one which takes advantage of grazing during part of the year, such as winter grazing of corn or sorghum stalks or seasonal grazing of pastures. In addition to providing low-cost feed and allowing the animals to do their own harvesting, breeding may be timed so that the calves will be dropped on clean pasture as a means of (l) preventing calf scours, and (2) stimulating milk flow.

Grazing crop residue and pastures is covered in Chapter 19 of this book; hence, the reader is referred thereto.

Fig 23-4 By going to semiconfinement, the Tom Brothers Ranch was able to increase cow numbers from 600 to 1,075 head, with no added acreage. (Courtesy, Lytle Tom, Jr., Tom Brothers Ranch, Campbellton, TX)

QUESTIONS FOR STUDY AND DISCUSSION

1. What factors favor increased confinement production in the future?

2. Why has confinement cow-calf production existed for so much longer in China, Japan, and Europe than in the United States?

3. Summarize what drylot cow-calf experiments show.

4. List the advantages of confinement cow-calf production.

5. List the disadvantages of confinement cow-calf production.

6. Describe the ideal cow for confinement (drylot) production.

7. List and describe the basic facility needs for a confinement cow-calf operation.

8. Discuss phase-feeding confinement cows according to stage of production and age of animals.

9. Outline a management program for a confinement cow-calf operation.

10. Discuss the advantages and disadvantages of (a) a semiconfinement vs (b) a year-round (total) confinement cow-calf operation.

11. Do confinement cow-calf operations lend themselves to (a) artificial insemination, and (b) early weaning? Justify your answer.

12. Which cow-calf system necessitates the most able management—(a) conventional pasture system, or (b) confinement?

SELECTED REFERENCES

Title of Publication	Author(s)	Publisher
Beef Cattle Science Handbook	Ed. by M. E. Ensminger	Agriservices Foundation, Clovis, CA, pub. annually 1964–1981
Commercial Beef Cattle Production, Second Edition	Ed. by C. C. O'Mary I. A. Dyer	Lea & Febiger, Philadelphia, PA, 1978

Beef cattle under an autumn sky. (Courtesy, J. C. Allen & Son, West Lafayette, IN)

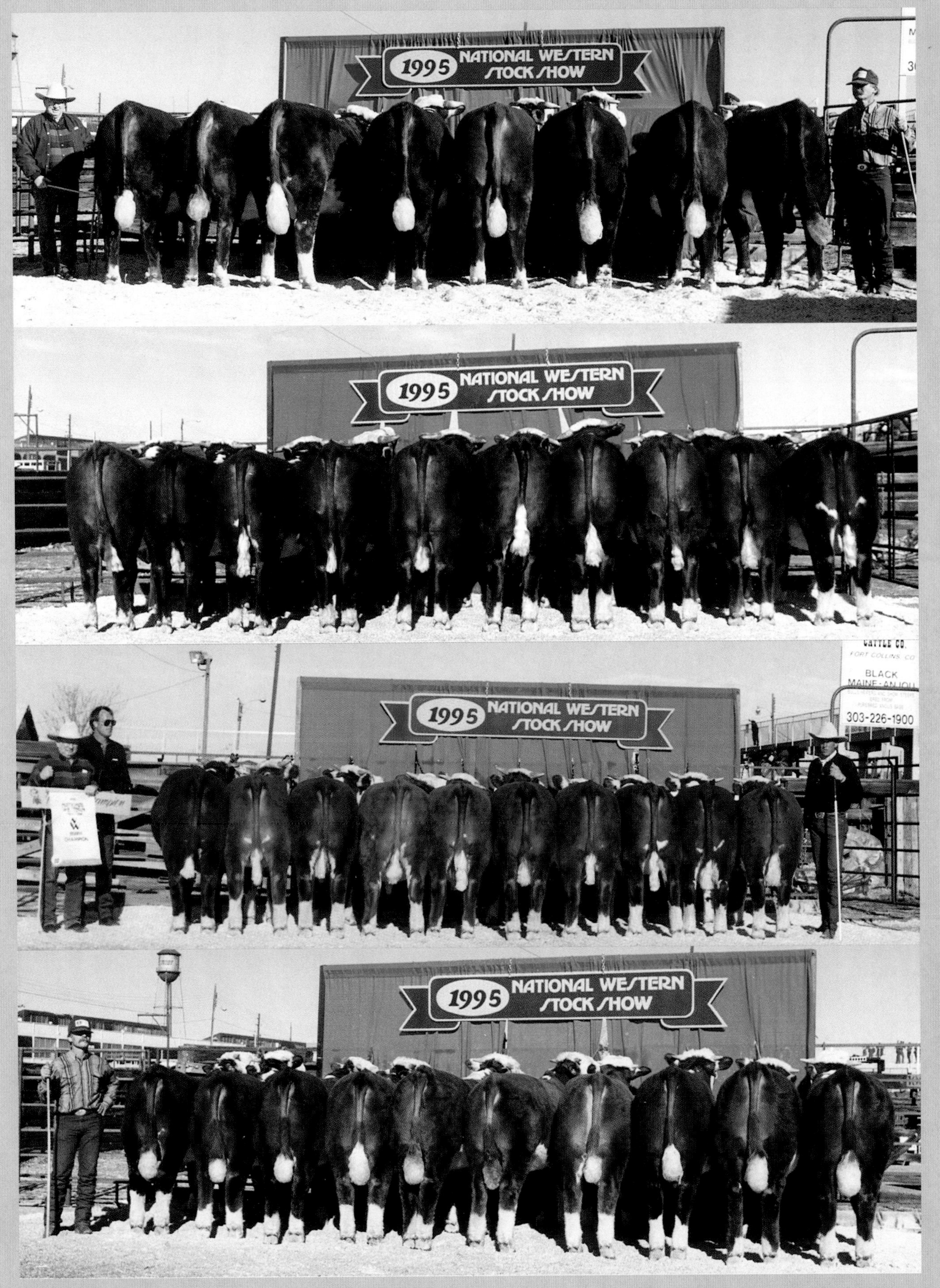

Plate 1. A lot of bulls! And good bulls, too. Four carloads of hereford bulls–40 head, bred by Bill Bennet, BB Cattle Co., Connell, WA; and shown at the National Western Stock Show, Denver, CO 1995. This was the first time in the history of the Denver show, that a single breeder won in all four classes. Top to bottom: 1. Champion yearling carload. 2. Reserve champion carload, senior calves. 3. Reserve grand champion carload, junior calves. 4. Champion carload of spring calves. (Note: Bill Bennett is one of my former students of whom I am very proud. Dr. E, author of *Beef Cattle Science*.)

Plate 2. Angus cows on pasture. (Courtesy, American Angus Assn., St. Joseph, MO)

Plate 3. Braford cows. (Courtesy, Adams Ranch, Fort Pierce, FL)

Plate 4. Beefmasters. (Courtesy, Beefmaster Breeders Universal, San Antonio, TX. Photo by B.E. Fichte, Rosebud Communications, Inc.)

Plate 5. Four-breed cross (Hereford X Braunvieh X Angus X Simmental) cows crossing the Canadian River. (Courtesy, Spade Ranches, Lubbock, TX)

Plate 6. Santa Cruz yearling replacement heifers. (Courtesy, King Ranch, Kingsville, TX)

Plate 7. Santa Gertrudis yearling replacement heifers. (Courtesy, King Ranch, Kingsville, TX)

Plate 8. Cowboy heeling a calf. (Courtesy, *American Hereford Journal*, Kansas City, Mo)

Plate 9. Tarentaise cows on Rogue River Ranch, Central Point, OR.
(Courtesy, Attache International, N. Kansas City, MO)

Plate 10. Polled Hereford cows drinking from a woodland pond.
(Courtesy, American Polled Hereford Assn., Kansas City, MO)

Plate 11. Beef in the counter. (Courtesy, National Live Stock and Meat Board, Chicago, IL)

Plate 13. Burgundy beef stew–bonless chuck.
(Courtesy, National Live Stock and Meat Board, Chicago, IL)

Plate 12. Beef rib eye roast with Madeira sauce.
(Courtesy, National Live Stock and Meat Board, Chicago, IL)

Plate 14. Rib eye steak with horseradish topped tomato.
(Courtesy, National Live stock and Meat Board, Chicago, IL)

Plate 15. Beef roast and fennel parmesan–cross rib roast.
(National Live Stock and Meat Board, Chicago, IL)

Plate 16. Classic meat loaf–ground beef.
(Courtesy, National Live Stock and Meat Board, Chicago, IL)

Plate 17. Beef, pepper, and mushroom kabobs.
(Courtesy, National Live Stock and Meat Board, Chicago, IL)

Plate 18. Hearty beef stew with dumplings for a winter meal by the fire.
(Courtesy, National Live Stock and Meat Board, Chicago, IL)

Land and cattle are the two biggest costs in the cattle business.

BUSINESS ASPECTS OF COW-CALF PRODUCTION

Contents *Page*

The cow business has changed from a way of life to a way of making a living. It's big and important business, and intricate too; and it is destined to get bigger and more complicated.

In 1935, there were 6,814,000 farms in the United States with an average size of 155 acres. By 1993, there were 2,068,000 farms, with an average size of 473 acres. Thus, within a span of 58 years, 4,746,080 farms—70% of our farms—disappeared from American agriculture; and the average size farm increased from 155 acres to 473 acres —more than double. With this transition, cow herds increased in size.

In the 42-year period 1952 to 1994, farm investment in land, buildings, livestock, and equipment increased more than 5.5-fold, rising from $167 billion to $920 billion. During this same period of time, farm debts increased more than 9.7-fold, going from $14.7 billion to $143 billion.

Fig. 24-1. Percent calf crop and weaning weight are important profit indicators on any cow-calf ranch. (Courtesy, *Livestock Weekly*, San Angelo, TX)

COW-CALF COSTS AND RETURNS

Those thinking of becoming cow-calf operators inevitably ask: How much money will it take, and what can I make?

It takes a lot of capital to own and operate a modern cattle farm or ranch. Thus, the owner must have all or part of the capital required, or be able to borrow all or part of it. If borrowed, the mortgage payments must be met.

Cow-calf costs are determined primarily by (1) size of operation, (2) location, and (3) commercial vs purebred cattle.

Cow-calf returns are largely determined by management; weather; calf crop dropped, calf crop weaned, and calf weaning weights; age and longevity of the cows; labor costs; and market price of cattle.

■ **Authors' estimated cow-calf costs and returns**—Table 24-1, cow-calf costs and returns, evolved out of an earlier consensus obtained by the senior author, subsequently updated for 1996. No claim is made relative to its scientific accuracy. Rather, it is presented because, to the knowledge of the senior author, it is the only information of its kind presently

TABLE 24-1
COW-CALF COSTS AND RETURNS[1]

	Average for U.S. Cow-Calf Operations			
	Commercial		Purebred	
	Average	Top 5%	Average	Top 5%
Investment/animal unit (one mature cow) in land and improvements (real estate) . . . . ($)	3,000	3,250	3,500	4,000
Percent calf crop dropped (based on no. cows bred) . . . . (%)	88	93	89	94
Percent of calf crop weaned (based on no. cows bred) . . . (%)	81	89	82	91
Weaning weight of calf at 7 mo. . . . . (lb) *(kg)*	450 *(205)*	550 *(250)*	475 *(216)*	575 *(261)*
Age and longevity of cows:				
Age when removed from herd . . . . (yr)	9	10	9.5	10.5
No. calves produced in lifetime of cow	6.0	8.0	7.0	8.5
Labor/cow/year . . . . (hr)	20.0	19.5	26.0	23.5
Net return per cow to management[2] . . . . ($)	27.25	41.50	38.00	95.00

[1]This is based on an earlier consensus, which the senior author subsequently updated to reflect 1996 cow-calf costs and returns. No claim is made relative to the scientific accuracy of the data; rather, it is presented (1) because it is the best information of its kind presently available on a nationwide basis, and (2) with the hope that it will stimulate needed research along these lines.

[2]Net return to management after deducting from gross receipts all costs, including depreciation on machinery, buildings and cattle, and interest on investment.

available on a nationwide basis, and with the hope that it will stimulate needed research along these lines. *Note:* Most cow-calf cost and return studies, including those presented in this chapter (see Tables 24-2 and 24-3) give costs on a per cow basis, without revealing the very considerable improvements (real estate) necessary for ownership of a cattle farm or ranch.

As shown in Table 24-1, the senior author estimates that, on a nationwide basis, a real estate investment of $3,000 to $4,000 per cow is required for the land and facilities of a cattle farm or ranch. By contrast, many small businesses can be started without any capital investment whatsoever—by the entrepreneurs merely hanging out their "shingles."

Note: In addition to the $3,000 to $4,000 for land and improvements shown in Table 24-1, the other major cost is for cattle. So, when launching a cow-calf business, provision should be made for the purchase of cattle at the price prevailing at the time.

It is noteworthy that Table 24-1 reveals that the top 5% operators have a higher investment/animal unit in land and improvements than their average counterparts. Obviously, better operators have better land and improvements; their savings are made in the handling of the herd.

The facts and figures presented in Table 24-1 point up the two main deterrents to entering the cow-calf business: (1) the inability to bring together the amount of capital required, and (2) the scarcity of top management.

- **USDA's cow-calf costs and returns**—The U.S. Department of Agriculture made a nationwide survey of the costs and returns of the nation's beef cattle operations in 1991. More than 98% of all U.S. beef cows were represented in the survey. Also, the results were reported by the following regions:

The West: CA, CO, ID, MT, NM, OR, UT, WA, and WY.

The Great Plains: KS, NE, ND, OK, SD, and TX.

The North Central: IL, IN, IA, MN, MO, and OH.

The South: AL, AR, FL, GA, KY, LA, MS, NC, TN, and VA.

Table 24-2 shows the cow-calf costs and returns per cow in each of the four regions in 1991.

The bottom line: Note that, on the average, producers in all four areas suffered losses in 1991, with the losses per cow ranging from $17.25 in the North Central Area to $123.28 in the South.

Herd size makes a difference. The USDA researchers also reported per-cow costs and returns for four size groups—0–100, 100–499, 500–999, 1,000+. The total fixed cash costs and capital expenditures decreased as the size of operations increased, with the result that the bottom line improved with increased

TABLE 24-2
COW-CALF COSTS AND RETURNS PER COW, 1991, IN FOUR U.S. REGIONS[1]

Item	West	Great Plains	North Central	South
Cash receipts:	- - - - - - *(dollars per cow)* - - - - - -			
Steer calves	51.55	64.20	56.18	106.21
Heifer calves	40.18	46.63	33.98	96.88
Yearling steers	147.51	142.92	134.56	78.95
Yearling heifers	68.91	92.62	87.07	34.32
Other cattle	136.11	143.39	126.33	80.27
Total	444.27	489.75	438.13	396.62
Cash expenses:				
Feeder cattle	23.87	22.07	2.38	21.95
Feed—				
Grain	3.89	7.70	14.24	6.52
Protein supplements	15.38	28.20	17.70	15.85
Byproducts	3.25	4.53	4.80	8.89
Harvested forages	58.24	54.35	50.62	40.68
Pasture	47.20	44.80	42.23	32.77
Total feed costs	127.96	139.57	129.59	104.72
Other				
Veterinary and medicine	12.99	12.44	14.99	10.34
Livestock hauling	3.11	1.49	0.61	2.02
Marketing	6.93	7.24	5.58	9.09
Custom feed mixing	0.34	0.74	0.23	0.23
Fuel and lube	17.49	14.06	13.01	17.73
Machinery and building repairs	25.95	28.36	28.24	34.69
Hired labor	32.58	27.00	14.65	26.63
Miscellaneous	4.94	6.62	5.26	7.56
Total, variable cash expenses	256.17	259.60	214.54	234.96
General farm overhead	43.81	35.35	34.65	35.27
Taxes and insurance	13.57	11.82	12.40	18.65
Interest (nonreal estate)	14.48	14.49	11.68	8.11
Interest (real estate)	44.51	23.81	28.18	25.13
Total, fixed cash expenses	116.36	85.47	86.91	87.16
Total, cash expenses	372.53	345.07	301.45	322.11
Cash receipts less cash expenses	74.71	144.68	136.68	74.51
Capital expenditures	105.04	117.29	102.79	124.64
Total cash expenses and capital expenditures	477.57	462.36	404.24	446.75
Net cash returns	–33.30	27.39	33.89	–50.13
Economic (full ownership) costs:				
Variable cash expenses	256.17	259.60	214.54	234.96
General farm overhead	43.81	35.35	34.65	35.27
Taxes and insurance	13.57	11.82	12.40	18.65
Capital expenditures	105.04	117.29	102.79	124.64
Operating capital	6.97	7.06	5.84	6.39
Other nonland capital	32.18	33.52	41.69	45.06
Land	88.52	92.31	43.47	54.94
Total cost/cow	546.25	556.95	455.38	519.90
Profit or loss	–101.98	–67.20	–17.25	–123.28

[1]Shapouri, H., K. H. Matthews, Jr., and P. Bailey, *Cow/Calf Costs of Production, 1990–91*, USDA, ERS, Agriculture Information Bulletin No. 670. Adapted by the authors.

herd size up to 1,000 head. (See Chapter 18, Fig. 18-7.)

■ **Washington State University's (WSU) cow-calf costs and returns**—In 1994, WSU estimated the costs and returns of a 100-head beef cow-calf enterprise (85 brood cows and 15 replacement heifers) in the Columbia Basin of Washington. The cattle enterprise which they studied was supplemental to an irrigated farm on which the cattle utilized stubble, corn stalks, and other available forages. Spring and summer feed was supplied by irrigated pastures.

Table 24-3 shows (1) the total costs and returns for the 100-head cow-calf enterprise, and (2) the costs and returns per cow.

TABLE 24-3
COW-CALF COSTS OF, AND RETURNS FROM, A 100-HEAD ENTERPRISE IN WASHINGTON IN 1994[1]

Returns	Number	Weight	Price/Lb	Total	Per Cow	Your Estimate
		(lb)	($)	($)	($)	
Steer calves	47	625	0.82	24,087	240.87	______
Heifer calves	47	575	0.77	20,809	208.09	______
Cull cows	14	1,100	0.47	7,238	72.38	______
Cull bulls	1	2,200	0.55	1,210	12.10	______
Total returns	109			53,344	533.44	

Operating Costs	Total	Per Cow	Your Estimate Total	Your Estimate Per Cow
	($)	($)		
Feed costs	23,343.80	233.44	______	______
Pregnancy check	250.00	2.50	______	______
Medical supplies	1,864.00	18.64	______	______
Implants (calves) 2x	188.00	1.88	______	______
Repair & maintenance	1,000.00	10.00	______	______
Fuel and oil	600.00	6.00	______	______
Utilities	150.00	1.50	______	______
Herd bull	2.500.00	25.00	______	______
Bred heifers 15 hd @ $850/hd	12,750.00	127.50	______	______
Brand inspection, beef promotion, and insurance	181.00	1.81	______	______
Sales commission (2.5%)	1,333.60	13.34	______	______
Hauling	250.00	2.50	______	______
Miscellaneous	275.00	2.75	______	______
Interest on operating capital ÷ 2 × 0.09	1,818.94	18.19	______	______
Total operating costs	46,504.34	465.04	______	______
Returns over operating costs	6,839.66	68.40	______	______
Less overhead costs				
Buildings and improvements	1,672.54	16.73	______	______
Machinery and equipment	1,938.75	19.39	______	______
Interest on livestock	5,308.97	53.08	______	______
Insurance	500.00	5.00	______	______
Total overhead costs	9,420.26	94.20	______	______
Total costs	55,924.60	559.24	______	______
Profit or loss	(2,580.60)	(25.80)	______	______

[1]Carkner, R. W., L. A. Mitchell, and the Cattlemen's Committee, *Estimated Costs and Returns for a 100-Head Beef Cow/Calf Enterprise, Grant-Adams Area, Washington, 1994.* Adapted by the authors.

Table 24-3 shows that in 1994 the 100-head cow-calf enterprise in Washington had—

1. Total returns of $53,344.00, or $533.44 for each cow in the breeding herd.
2. Total costs of $55,924.60, or $559.24 per cow.
3. Total losses of $2,580.60, or $25.80 per cow.

■ **Average cow-calf returns per cow**—Fig. 24-2 shows the estimated cow-calf returns per cow during the 20 year period 1974 to 1994. It is noteworthy that, during this 20 year period, producers made money 10 years and lost money 10 years.

ESTIMATED AVERAGE COW-CALF RETURNS

Returns Over Cash Costs

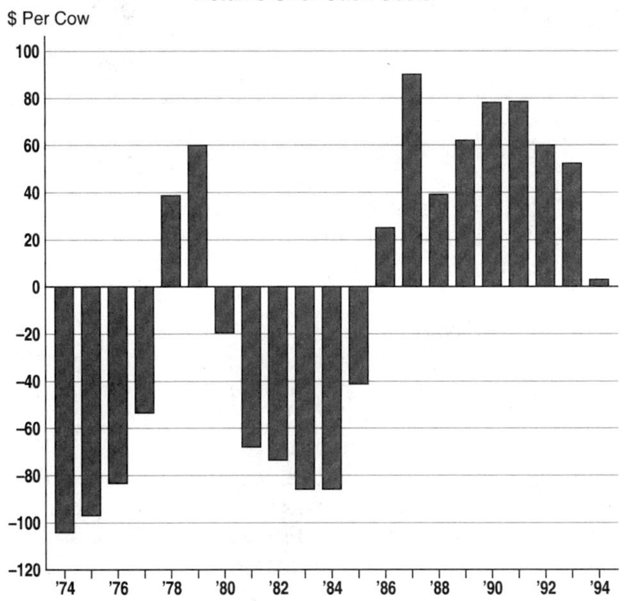

Fig. 24-2. Estimated average annual cow-calf returns per cow 1974 to 1994. (Courtesy, USDA, Livestock Marketing Information Center, Lakewood, CO)

For land and cattle to be profitable they should yield a return sufficient to the owner to (1) meet the interest payment on the investment, (2) retire a reasonable portion of the loan, and (3) provide satisfactory management return. But, there is no more reason why large land and cattle holdings should be debt free than there is for General Motors, or any other big corporation, to be debt free. Some land is overpriced, particularly in the more populous areas. Yet, there are compensating factors in many of the latter regions—their greater appreciation as potential building sites and recreational areas.

OTHER PLUS VALUES TO LAND AND CATTLE OWNERSHIP

Some buy land to balance operations and cut costs; some to keep the children down on the farm; others because it adjoins their present property; and still others because of pride of ownership. There are also other important plus values to land and cattle ownership. Among them:

1. Appreciation in land values. From 1960 to 1993, U.S. farmland values rose from $116 per acre to $700 per acre, an increase of 6 times.
2. As an effective hedge against inflation.
3. As a desirable alternative to a jittery stock market.
4. As a tax shelter.
5. Spreading roads and suburbs, and precious little new land that can be brought under production.

There is little doubt that, on a longtime basis, farm real estate values will continue to climb. The only question is: how soon and how much?

LAND AND CATTLE INVESTMENT COMPANIES WILL COME

Big land-cattle operations will get bigger, demanding more and more capital and top management. In the years ahead, many investors will become part owners in land and cattle, much as they now do through corporate stocks and bonds, and they will leave the management of the holdings to the professionals. Such an arrangement will also make it possible to (1) diversify in countries and types of invest-

Fig. 24-3. The major capital needs for a cow-calf enterprise are for land and cattle. Capital turnover is slow. (Courtesy, California Polytechnic State University, San Luis Obispo)

ments—in different areas of the United States, and in Australia, Canada, South America, and other areas where there are vast acreages of rangeland with great potential for improvement that can be secured at reasonable prices; (2) minimize risks of loss from droughts and local depressions; (3) obtain for investors, big and little, the benefits that accrue to bigness, such as lower investment per cow unit, and lower feed and labor costs; (4) furnish recreational and vacation areas on farms and ranches of which they are part owner; and (5) provide know-how, continuity, and able management.

Part ownership in land investment companies affords a modern way in which to spread investments and minimize risks—as is done with stocks and bonds, grain and livestock futures, and syndicated sires.

COW LEASING

Establishing a beef cattle herd is expensive. Getting out of it can also be costly if you haven't anticipated the tax bite. A cow lease arrangement can help at both ends. The challenge is to find someone you can work with, and a lease arrangement that suits both parties.

Leasing has its risks. Cows get sick and die, or lose value when the market drops. But if the arrangements come together with the right people, it is worth the effort for both parties.

Leasing suits a broad range of owner-operator combinations. Owners (lessors) may be farmers who wish to retire, but who do not want to give up the cattle business; or a nonfarmer who prefers investing in cattle rather than in stocks, bonds, or real estate.

Prime operators (lessees) are young farmers starting out, perhaps with some land and lots of energy, but with little capital; or perhaps they want to improve upon the genetics of their herd.

Although lease arrangements are as variable as the individuals involved, the following basic types of leases exist:

1. **Percentage share lease,** with the most common split being 60:40, with the operator covering all expenses, and taking 60% of the calves (or calf proceeds), and the owner getting 40%.
2. **Cash lease,** which makes for the lowest risk for the owner, but the greatest risk for the operator.
3. **Fixed-number lease,** in which the owner receives a pre-set number of calves or rent.
4. **Flexible share lease,** which provides for the operator to recover expenses, such as feed and veterinary costs first, even if the market return is low.

The most frequently asked question about cow leasing is: What's a fair return? Most owners expect a return of 8 to 12% on their investment.

RECORDS AND MONEY MANAGEMENT

In times of money scarcity and narrow profit margins, there will be greater stress on business and financial management skills. Also, cattle producers must realize that they are competing with many other users of credit, including retail merchants, manufacturers, home buyers, and professional people. Many of these borrowers can and do provide the lender with profit and loss statements and net worth statements, prepared by a CPA. Also, they usually submit annual budgets. As cattle producers increase the amount of borrowed capital in their operations, they, too, will be required to furnish adequate records and budgets if they are to compete successfully for the available capital.

TYPES OF RECORDS

Three general types of records are involved in a cow-calf operation. All are equally important to a well-run operation. They are:

- **Daily and weekly report**—A farm or ranch manager should keep a daily record, like the daily-weekly report shown in Fig. 24-4. It takes little time to keep it. Nevertheless, a certain time should be set aside daily for this purpose, thereby assuring that it will be kept—and that it will be kept accurately. For the manager, such a report provides an invaluable record of the day-by-day operations. For the owner, it's a quick and easy way to keep informed. This record should be filed, where it can be referred to as needed.

- **Production records**—Production records include such important profit indicators as percent calf crop dropped, percent calf crop weaned, weaning weight of calves at seven months, the pounds of beef per cow or per acre, the pounds of feed required to produce a pound of beef, death losses, etc.

- **Financial records**—The name of the game is *profit.* Thus, it is necessary that production be translated into dollars and weighed against costs involved in achieving that production.

Many different kinds of financial records are used by cattle producers; among the most common ones are the following: inventory, budgets, income, expenses, cash flow, depreciation schedule, annual net income, enterprise accounts, profit and loss statement, and net worth statement. (Also see Chapter 13, Business Aspects of Beef Production.)

BAR-NONE RANCH, FOR WEEK BEGINNING ______________________
(month, day, year)

Prepared by ______________________

	Mon.	Tues.	Wed.	Thur.	Fri.	Sat.	Sun.	Comments for Day or Week
WEATHER Rain or snow, in. . . . . . . Temp.—high . . . . low								
LABOR Accident or sickness . . . . Who, what, how . . . . Changes								
EQUIPMENT BREAKDOWN Kind and make . . . . Cause . . . . Cost to repair								
ANIMAL HEALTH PROGRAM (indicate cow-calf/cattle feedlot) Vaccination . . . . Treatment								
ANIMAL LOSSES (indicate cow-calf/cattle feedlot) No. . . . . Kind . . . . Cause								
ANIMALS RECEIVED (indicate cow-calf/cattle feedlot) No. . . . . Kind . . . . From . . . . Price or custom feed								
ANIMALS SOLD (indicate cow-calf/cattle feedlot) No. . . . . Kind . . . . To . . . . Price								
CROP HARVESTED Kind . . . . Field no. . . . . Acreage . . . . Yield/acre								
VISITORS Name and address								

Fig. 24-4. A good type of daily-weekly report.

WHY KEEP FINANCIAL RECORDS?

There are many reasons for keeping good financial records, the most important of which follow:

1. **To save on income taxes.** Good tax management and reporting consists in complying with the law, but in paying no more tax than is required. The cardinal principles of good tax management are: (a) maintenance of adequate records so as to assure payment of taxes and amounts no less or no more than required by law, and (b) conduct of business affairs to the end that the tax required by law is no greater than necessary.

Also, cattle producers need to recognize that good tax management and good farm management do not necessarily go hand in hand. In fact, they may be in conflict. When the latter condition prevails, the advantages of one must be balanced against the disadvantages of the other to the end that there shall be the greatest net return.

2. **To obtain needed credit.** Lenders need certain basic information in order to evaluate the soundness of a loan request. The financial record will show the borrower's current financial position, potential ahead, and liability to others.

3. **To guide changes in enterprises.** Farm and ranch records should provide information from which the farm business may be analyzed, with its strong and its weak points ascertained. From the facts thus determined, the manager may adjust current operations and develop a more profitable organization. The enterprise should be above average before the owner borrows to expand it. Is the ranch too small? Is it more profitable to sell weaners or yearlings? Should the farm or ranch produce or buy hay?

4. **To serve as a guide for current income and expenses.** Records of costs and returns of previous years are very valuable as guides, and as a means of spotting trouble. Items which deviate substantially from the historical record should be studied with care.

5. **To provide for continuity of the business.** *Barn door* and *memory* records are insufficient. They mitigate against continuity of the business. The sudden passing of a manager places a severe stress on a business even under the most favorable circumstances. However, a good set of financial records gives a big assist to those who must take over during such times.

KIND OF RECORD AND ACCOUNT BOOK

A farmer can make a record book by ruling off the pages of a bound notebook to fit specific needs, but the saving is negligible. Instead, it is recommended that a copy of a farm record book prepared for and adapted to the area be obtained. Such a book may usually be obtained at a nominal cost from the agricultural economics department of each state college of agriculture. Also, certain commercial companies distribute very acceptable farm record and account books at no cost.

KIND OF FINANCIAL RECORDS TO KEEP

At the outset it should be recognized that farm or ranch records should give the information desired to make a valuable analysis of the business. In general, the functions enumerated under the earlier section entitled, "Why Keep Financial Records?" can be met by the following kinds of records:

1. Annual inventory
2. Budget
3. Profit and loss (P&L) statement
4. Net worth statement
5. Enterprise accounts

Each of these records is discussed in Chapter 13; hence, the reader is referred thereto.

WHO SHALL KEEP THE RECORDS?

The records may be kept either by someone in the farm or ranch business, or by someone hired to perform this service—a professional.

■ **Farm or ranch help**—Very frequently, farm or ranch records are kept by the spouse. If there is adequate time, there is no reason why good records cannot be kept.

Most farmers are not good record keepers, primarily because the operation of an ordinary farm requires large amounts of physical labor. As a result, most farmers and ranchers are physically exhausted at the end of the day's work and have neither the time nor the ambition to record in at least four different places each transaction which occurred during the day, as would be necessary of the usual double-entry bookkeeping system.

■ **A professional**—There are, of course, individuals and firms who make a business of keeping farm or ranch records. Usually, they are accountants or farm management specialists.

HOW SHALL RECORDS BE KEPT?

Records may be kept either by hand or by computer. Accurate and up-to-the-minute records and controls have taken on increasing importance in all agriculture, including the cattle business, as the investment required to engage in farming and ranching has risen and profit margins have narrowed. Today's successful farmers and ranchers must have, and use, as complete records as any other business. Also, records must be kept current, it no longer suffices merely to know the bank balance at the end of the year.

■ **By hand**—The hand system can be used, but it is slow and tedious. This service is usually performed by a member of the family or by an accountant living in the community. Nevertheless, after learning what is wanted, a first-rate accounting system can be kept by hand.

■ **By computer**—Big and complex cattle enterprises have outgrown hand record keeping. It's too time consuming, with the result that it doesn't allow management enough time for planning and decision making. Additionally, it does not permit an all-at-once consideration of the complex interrelationships which affect the economic success of the business. This has prompted a computer technique known as linear programming.

In the past, the biggest deterrent to production testing on cow-calf operations, both purebred and commercial, has been the voluminous and time-consuming record keeping involved. Keeping records *per se* does not change what an animal will transmit, but records must be used to locate and propagate the genetically superior animals if genetic improvement is to be accomplished.

Production testing has been covered elsewhere in this book (see Chapter 5); hence, the reader is referred thereto. As has been pointed out, two factors contribute to optimum beef production; namely (1) heredity, and (2) environment. Records of traits must be adjusted for such well-known sources of variation as age of dam, age of calf, and sex. This is tedious and time consuming when records are kept and analyzed by hand. However, a computer can handle this assignment efficiently and with fewer risks of errors and omissions than the hand method.

In addition to their use in production testing, computerized records can be, and are, used for herd record purposes—as a means of keeping management up-to-date and serving as an alert on problems or work to be done. Each animal must be individually identified. Reports can be obtained at such intervals as desired, usually monthly or every two weeks. Also, the owner can keep as complete or as few records as desired, and the system may be adapted to either purebred or commercial herds. Of course, commercial operators do not need pedigrees. Here are several of the records that can be kept by computer:

1. Pedigrees.

2. Animals that need attention such as—
   a. Animals four months old that are unregistered.
   b. Females six months old that are not vaccinated.
   c. Bulls that are six months old, and which should either be marked to keep or be castrated.
   d. Heifers 18 months old that haven't been bred.
   e. Cows that have been bred two consecutive times.
   f. Cows not rebred three months after calving.
   g. Cows that have reached nine years of age, and that may be getting shelly and should be culled.
   h. Cows due to calve in 30 days.
   i. Calves seven months of age that haven't been weaned.
   j. Calves seven months of age that haven't been scored and weighed.

3. A running, or cumulative, inventory of the herd, by sex; including calves dropped, calves due, and purchases and sales—in number of animals and dollars.

4. The depreciation of purchased animals according to the accounting method of choice.

Whether a cattle producer should keep the records or hire a professional, and whether records should be kept by hand or by computer, each individual must decide. For the most part, the decision should be based on weighing the usefulness of the information each provides against the cost of obtaining it.

TAX AND ESTATE PLANNING

There are some things that cattle producers can do to lessen the tax bite. So, each producer is admonished to confer with a local tax advisor.

The subjects of Tax Management/Reporting and Estate Planning are fully covered in Chapter 13; hence, the reader is referred thereto.

COW-CALF PROFIT INDICATORS

Many factors determine the profitability of a cow-calf enterprise. Certainly, a favorable per animal unit capital investment in land and improvements is a first requisite. Additionally, percent calf crop weaned and weaning weight are exceedingly important, as shown in Table 24-4.

In Table 24-4, it is assumed that the yearly operating cost is $350 per cow. Then, the effect of size and weight of calf crop is computed. As shown, with a 90% calf crop and an average weaning weight per cow bred of 540 lb (600 × 90% = 540), a selling price of 64.8¢/lb will meet the break-even cost of $350. With a 75% calf crop, because of fewer pounds per cow bred (600 × 75% = 450 lb), the calves would have to bring 77.7¢/lb in order to break even.

AN INCENTIVE BASIS FOR THE HELP

On cow-calf enterprises, there is need for some system which will encourage caretakers to be good nursemaids to newborn calves, though it may mean the loss of sleep, and working with cold, numb fingers. Additionally, there is need to do all those things which make for the maximum percent calf crop weaned at a heavy weight.

From the standpoint of the owner of a cow-calf enterprise, production expenses remain practically unchanged regardless of the efficiency of the operation. Thus, the investment in land, buildings and equipment, cows, feed, and labor differs very little with a change (up or down) in the percent calf crop or the weaning weight of calves; and income above a certain break-even point is largely net profit. Yet, it must be remembered that owners take all the risks; hence, they should benefit the most from profits.

On a cow-calf operation, the authors recommend that profits beyond the break-even point (after deducting all expenses) be split on an 80:20 basis. This means that every dollar made above a certain level is split, with the owner taking 80¢ and the employees getting 20¢. Also, there is merit in an escalator arrangement with the split changed to 70:30, for example, when a certain plateau of efficiency is reached. Moreover, that which goes to the employees should be divided on the basis of their respective contributions, all the way down the line; for example, 25% of it might go to the manager, 25% might be divided among the foremen and 50% of it divided among the rest of the help.

A true profit-sharing system on a cow-calf outfit based on net profit has the disadvantages of (1) employees not benefiting when there are losses, as frequently happens in the cattle business, and (2) management opening up the books, which may lead to gossip, misinterpretation, and misunderstanding. An incentive system based on major profit factors alleviates these disadvantages.

Gross income in cow-calf operations is determined primarily by (1) percent calf crop weaned, (2) weaning weight of calves, and (3) price. The first two factors can easily be determined. Usually, enough calves are sold to establish price; otherwise, the going price can be used.

The incentive basis proposed in Table 24-5 for cow-calf operations is simple, direct, and easily applied. As noted, it is based on average pounds of calf weaned per cow which factor encompasses both percent calf crop and weaning weight.

TABLE 24-4
SIZE AND WEIGHT OF CALF CROP WEANED ARE IMPORTANT

	Yearly Operating Cost—$350 per Cow							
	Calf Weights							
Calf Crop	**600 lb *(273 kg)***	**Break-Even**	**500 lb *(227 kg)***	**Break-Even**	**450 lb *(204 kg)***	**Break-Even**	**400 lb *(181 kg)***	**Break-Even**
(%)		*($)*		*($)*		*($)*		*($)*
90	540 lb *(245 kg)*	0.648	450 lb *(205 kg)*	0.777	405 lb *(184 kg)*	0.864	360 lb *(163 kg)*	0.972
85	510 lb *(232 kg)*	0.686	425 lb *(193 kg)*	0.824	382 lb *(173 kg)*	0.916	340 lb *(154 kg)*	1.030
80	480 lb *(218 kg)*	0.729	400 lb *(182 kg)*	0.875	360 lb *(163 kg)*	0.972	320 lb *(145 kg)*	1.090
75	450 lb *(205 kg)*	0.777	375 lb *(170 kg)*	0.933	337 lb *(153 kg)*	1.040	300 lb *(136 kg)*	1.170
	Average break-even point @ $0.894/lb							

TABLE 24-5
A PROPOSED INCENTIVE BASIS FOR COW-CALF OPERATIONS

Average Lb[1] of Calf Weaned/ Cow Bred	Here's How It Works
(lb)	On this particular operation, the break-even point is assumed to be an average of 400 lb *(181 kg)* of calf weaned per cow bred; and, of course, this is arrived at after all costs of production factors have been included.
350	
375	
400 (break-even point)	Higher poundage of calf per cow bred can be achieved thorugh (1) increased calf crop percentage, (2) increased weaning weight of calf, and/or (3) a combination of the two.
425	
450	
475	Pounds *(kilograms)* of calf weaned per cow bred in excess of the break-even point (in this case 400 lb *[181 kg]*) are sold or evaluated at the going price.
500	
525	
550	If an average of 450 lb *(204 kg)* of calf per cow bred is weaned, and if mixed steers and heifers of this quality are worth 60¢ per pound, that's $30 more net profit per cow. In a 500-cow herd, that's $15,000. With an 80:20 division, $12,000 would go to the owner, and $3,000 would be distributed among the employees.
575	
600	Or, if desired, and if there is an escalator arrangement, there might be an 80:20 split at 425 lb *(193 kg)*, a 70:30 split at 450 lb *(204 kg)*, and a 65:35 split at 475 lb *(215 kg)*.

[1]To convert to kg, divide by 2.2.

QUESTIONS FOR STUDY AND DISCUSSION

1. Consider the following:

   a. In Table 24-1, the authors reported that it takes $3,000 to $4,000 per cow for land and improvements in order to be in the cow-calf business.

   b. In Table 24-2, the USDA reported losses in all four cow-calf areas of the U.S. in 1991.

   c. In Table 24-3, Washington State University reported that cow-calf operators in Washington lost money in 1994.

   Are these facts and figures conducive to entering the cattle business? Explain.

2. Take your own cattle farm or ranch, or one with which you are familiar, and see how you compare in cow-calf costs and returns with the figures given in Tables 24-1, 24-2, and 24-3.

3. What net return per calf weaned should a cattle producer reasonably expect?

4. Land values have risen rather sharply in recent years. What *plus* values do you see ahead to land and cattle ownership?

5. Would you recommend that a young couple launch a cow-calf enterprise by leasing cows?

6. How can business and financial management skills help a cattle producer in times of money scarcity, high interest rates, and narrow profit margins?

7. How would you improve on Fig. 24-4, daily and weekly report?

8. How will good financial records give an assist in (a) saving on income taxes, (b) obtaining needed credit, (c) guiding changes in enterprises, (d) guiding current income and expenses, and (e) providing for continuity of the business?

9. How would you go about setting up a budget for a cow-calf enterprise?

10. Who should keep the farm and ranch records—the farm and ranch help, or a professional? Justify your answer.

11. How should records be kept—by hand, or by computer?

12. What kind of cow-calf records can be kept by computer? What kind of questions can be answered by computer?

13. How do each of the following factors affect the profitability of a cow-calf enterprise: (a) percent calf crop weaned, and (b) weaning weight of calves?

14. Evaluate Table 24-5, A Proposed Incentive Basis for Cow-Calf Operations. How would you improve it?

SELECTED REFERENCES

Title of Publication	Author(s)	Publisher
Commercial Beef Cattle Production, Second Edition	Ed. by C. C. O'Mary I. A. Dyer	Lea & Febiger, Philadelphia, PA, 1978
Stockman's Handbook, The, Seventh Edition	M. E. Ensminger	Interstate Publishers, Inc., Danville, IL, 1992
Tax-sheltered Investments	W. J. Casey	Institute for Business Planning, Inc., New York, NY, 1973

Yearling steers grazing tall grass in the Sand Hills of Nebraska. (Courtesy, Panhandle Research Station, University of Nebraska, Scottsbluff)

STOCKER (FEEDER) CATTLE

Contents *Page*

Stocker cattle and stocker cattle programs have changed with the passage of time. Until the early 1900s, stockers involved growing purchased or home-grown calves or yearlings on grass and hay until they were 3 to 4 years of age. As calf weaning weights increased and finished slaughter cattle weights decreased, the amount of time and gain required to grow calves from weaning until the beginning of the finishing period was substantially shortened. The stocker cattle industry became a calf-yearling industry, usually starting with 350- to 550-lb calves and ending with a yearling sold to a feeder at 600 to 800 lb. Today, most stockers are calves under 550-lb weight.

The development of large feedlots and year-round feeding increased the demand for feeders ready to go on high-concentrate rations. The main reason that the larger feedlots like to purchase feeders ready to go on high-energy rations is that roughage is used more efficiently in growing cattle, whereas it usually is an expensive item to use in large feedlots.

Today, the stocker stage is changing again, as a result of forces working in opposite directions; with one force favoring lengthening of the stocker stage and the other favoring shortening it, as follows:

1. Scarce and high-priced grains and high interest rates favor more roughage feeding and less grain feeding, resulting in carrying stockers to older ages and heavier weights, followed by a shorter feedlot period.

2. Heavier milking cows and heavier weaning weights, coupled with high-priced land, favor shortening the stocker stage, or even eliminating it, as 550-lb, or heavier, weaning weights are achieved.

In the future, both types of stocker operations will prevail, with the choice determined primarily by (1) the price of grain and interest rates and (2) the weaning weight of the calves. Heavy weaned calves will likely go directly into the feedlot or for slaughter. Calves with

Fig. 25-1. Stocker calves on pasture.

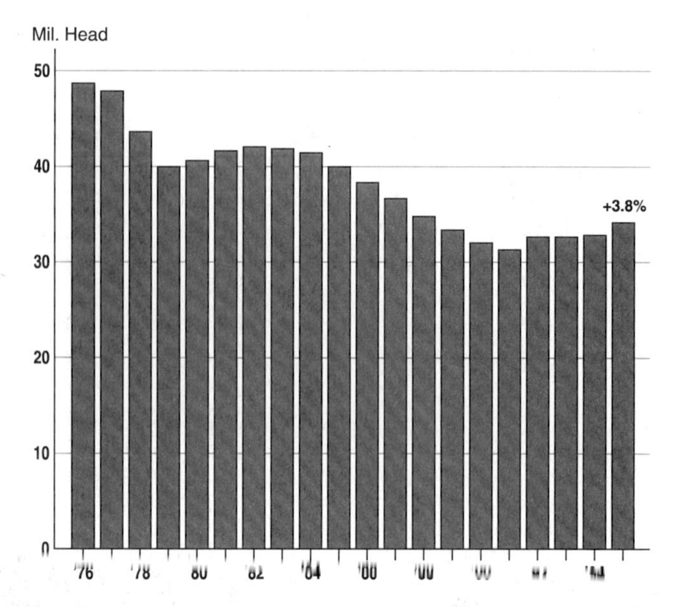

Fig. 25-2. January 1 feeder cattle supplies 1976 to 1994. (Courtesy, Livestock Marketing Information Center, Cooperative Extension Service, USDA)

light to average weaning weights will likely be carried as stockers to 700- to 800-lb weights, thereby shortening the feedlot period and lessening grain feeding.

Thus, the growing of calves from weaning until placing on finishing rations is not new. However, in recent years some new *wrinkles* have been added to the methods of conducting it. Today, most stockers are grown according to two systems: (1) calves or light yearlings are either roughed through the winter followed by grazing, or grazed only, then sold as feeders in late summer and fall; and (2) calves or yearlings are fed harvested roughage and grain in drylot, and then transferred to another location for finishing. Also, some new terms have evolved, definitions of which follow.

- ***Stockers*** *are calves and yearlings, both steers and heifers, that are intended for eventual finishing and slaughtering and which are being fed and cared for in such manner that growth rather than finishing will be realized. They are generally younger and thinner than feeder cattle.*

- ***Feeders*** *are calves and yearlings, both steers and heifers, carrying more weight and/or finish than stockers, which are ready to be placed on high-energy rations for finishing and slaughtering.*

- ***Replacement heifers*** *are the top end of the heifer calves selected to replace the older cows that are culled from the herd.*

- ***Preconditioning*** *refers to preparing the calf to withstand the stress and rigors of leaving its mother,* *learning to eat new kinds of feeds, and shipment from the farm or ranch to the feedlot or stocker grower.*

■ ***Backgrounding*** *is an old practice with a new emphasis and a new name. Actually backgrounding and the stocker stage are one and the same. Both refer to that period in the life of a calf from weaning to around 800-lb weight, when it is ready to go on a high-energy finishing ration.* However, the term *backgrounding*, which was ushered in with the development of large feedlots, indicates a shift in emphasis. The term *stocker stage* connotes emphasis on marketing roughages through thin cattle, whereas *backgrounding* connotes emphasis on growing out feeder calves ready to go on a high-energy finishing ration. Backgrounding may be done on pasture or in the drylot, or some combination of both. At its best, the animals should be in good health, bunk broke, and ready to go on full feed. (Also see section on "Backgrounding" in Chapter 31, Management of Feedlot Cattle.)

From the above, it may be concluded that in the variable period of a calf's life between weaning and finishing, it is usually classed as a stocker, a feeder, or a replacement heifer. Prior to weaning, calves may or may not be preconditioned.

The dividing line between stockers and feeders is not always as clear cut as the above definitions would indicate. That is, not all thin cattle are suitable for stockers. For example, very large yearlings and most heifers are usually sold as feeders, to be placed on high-energy feeds. Also, "Okie" type cattle are usually backgrounded for 50 to 60 days, then placed on a finishing ration.

TYPES OF STOCKER PROGRAMS

Sometimes the stocker operation is the only cattle enterprise on the farm or ranch. More frequently, however, it is conducted in conjunction with a cow-calf operation or it precedes the finishing program.

When the stocker enterprise is the only cattle enterprise on a farm or ranch, it is usually conducted in one of the following ways:

1. Calves or light yearlings are bought in the fall to be wintered on high-roughage rations in drylot and sold in the spring to buyers either (a) to go on grass for the summer, or (b) to go on a drylot finishing program.
2. Lightweight calves are bought in the fall to be wintered on roughage rations, then, under the same ownership, grazed throughout the following pasture season and sold in the fall. Under this plan, usually lighter weight calves are acquired and they are wintered at a lower rate of gain than in plan 1.
3. In Kansas, Oklahoma, and Texas, calves or light yearlings are bought in the fall and grazed on winter small grains, chiefly wheat. Good wheat pastures will produce very acceptable stocker gains. The main disadvantage to the program is that, due to weather conditions, winter wheat pasture cannot always be counted upon. When it fails, the stockers must either be sold or fed a higher cost roughage.
4. In the southeastern states, which is primarily a cow-calf area, winter oats and fescue are used extensively in stocker programs. This area is turning to stocker programs in order to utilize profitably winter pastures, and to satisfy the demand for 600- to 800-lb feeder steers as a result of the expansion of feedlots.

Fig. 25-3. Yearling stockers on pasture. (Courtesy, USDA)

There is a trend for more and more calves (not yearlings) to be handled according to plan 1—that is, bought in the fall, wintered on roughage, and sold directly into a finishing program. This trend will be accelerated because of heavier calves being weaned in the fall, and because it is more profitable either to use presently available pasture areas for brood cows to produce more calves or for crop production.

The most common type of operation is a combination stocker-feeder program, typical of the Corn Belt and the irrigated sections of the West, where high-yielding corn and sorghum crops are produced for silage. In these areas, cattle feeders usually purchase steer calves or light yearlings in the fall or late winter; fall-graze stalk fields and small-grain stubble where available; move into the drylot for the winter and feed corn or sorghum silage, supplemented with a legume hay or protein supplement; then finish on a high-energy ration either in the drylot or on pasture and sell for slaughter in the summer or fall.

An increasing number of feeders are grown on contract for and delivered to a feedlot for finishing. This trend has been prompted by the competition between feedlots. It is their way of assuring a continuous supply of feeders of the desired weights and quality. As a further inducement, many of the feedlots finance the grower (backgrounding) operation.

ADVANTAGES AND DISADVANTAGES OF A STOCKER PROGRAM

A stocker enterprise has both advantages and disadvantages in comparison with a cow-calf or a cattle finishing operation. These should always be weighed and balanced, especially where there is a choice.

A stocker program has the following **advantages** over other types of cattle programs:

1. **Flexibility.** A stocker operation is more flexible than a cow-calf operation from the standpoint of adjusting to feed supplies and cost, labor, and economic outlook. The number of stockers purchased each year may be altered accordingly.
2. **Efficient gains.** Stocker operators have the cattle when they make the most efficient gains.
3. **Low labor requirement.** Stocker cattle have a lower labor requirement than a cow-calf operation conducted on the same amount of land.
4. **Distribution of labor.** The peak labor requirement in wintering stockers in the drylot is completed ahead of spring and summer farm work.
5. **Quick returns.** Where a stocker program is limited either to wintering or pasturing only, returns come quickly, within 4 to 6 months. In some cases, this quick turnover permits handling 2 or 3 droves of stockers per year.
6. **Adapted to areas lacking accessibility to fat cattle markets or sources of grain.** A stocker operation is better adapted than a cattle finishing operation to areas lacking accessibility to slaughter cattle markets or sources of grain. Of course, such areas are also well suited to cow-calf operations; hence, a choice must be made.
7. **Utilize roughage and salvage feeds.** Stockers are adapted to the use of roughage and salvage feeds.
8. **Investment in buildings and equipment may be less.** If the stocker program is limited to grazing winter or summer pastures, a minimum of buildings and equipment is required.
9. **Contract basis requires little capital.** If a grower contract is arranged with a feedlot, as is sometimes possible, little capital is required.

There are also **disadvantages** to a stocker program, including the following:

1. **High risk.** It is a high risk venture. High risk results from seasonal and yearly price fluctuations in feeder cattle, and the fact that total gains are not large in proportion to the weight purchased. Also, stocker operators who rely on winter wheat or oat pastures or summer grazing may have to purchase cattle when the price is high and sell them when the market is flooded with similar cattle. Moreover, lack of rain may make for small gains and high costs.
2. **Cost of gain may have to offset negative margins.** Cost of gain must be kept down to offset negative margins that may prevail.
3. **High buying and selling skills are required.** Buying and selling skills are extremely important because (a) there is no established market as exists with slaughter cattle, and (b) the original weight purchased is a high percentage of the weight sold. It follows that any mistake made in buying and selling, such as mistakes in judging the quality and health of the stockers, has a greater influence on profits or losses than in a finishing program. Moreover, the entire livestock inventory is bought and sold at least once per year.
4. **Buying, selling, and shrink costs must be absorbed by a limited amount of gain.** Shrink on both ends and buying and selling costs can wipe out any economical, but relatively small, gains that can be made.
5. **High land, labor, and interest costs mitigate against stocker programs.** Because of the relatively small gains made by stockers, high land, labor, and interest costs mitigate against such programs. For this reason, growing calves at rates of less than 1 lb per head daily becomes increasingly difficult to justify.
6. **Disease can make for heavy losses.** The weanling calf is at the most susceptible stage in life to contagious diseases. Also, the limited weight gains in growing operations leave little opportunity to recover severe losses.
7. **High transportation costs.** Often the transportation cost is high because of long distance from the sources of supply and/or feedlots.

FACTS PERTINENT TO STOCKER PROGRAMS

The following points are pertinent to stocker programs:

- **Stocker programs are used by large cattle feedlots to insure continuous supply of feeder replacements**—Many large feedlots, which feed on a year-round basis, are effectively using stocker programs to insure a continuous supply of feeder replacements. They usually accomplish this by buying both calves and yearlings when they are available at favorable prices, then putting them on different stocker programs, often on a contract basis, designed to stagger their readiness to move to the feedlot at the weight and time desired.

■ **Stocker stage will be either lengthened or shortened, depending on the circumstances**—High-priced grains and cheap roughages, comparatively speaking, favor older and heavier stockers. So, yearling stockers will increase where and when pastures and dry roughages are relatively cheap. However, the following factors favor younger and lighter stockers—calves: (1) heavier weaning weight; (2) greater efficiency of feed utilization of younger cattle; (3) the need to place larger strains of cattle (the exotics) on finishing rations at younger ages—otherwise their carcasses become too heavy; (4) the demand for leaner beef; (5) higher remuneration from cow-calf operations; and (6) high land, labor, and interest costs mitigating against the small gains common to stocker operations.

■ **Diseases are a problem**—Diseases can take a tremendous toll in a stocker program. The weanling calf is at the most susceptible stage of life to contagious diseases. This coupled with the limited weight gains in growing operations leaves little opportunity to recover severe losses. Plans should be made for routine immunization, starting rations, and veterinary treatment. Also, the operator will be well paid for providing TLC (tender, loving care) and spotting sick animals early.

As every producer knows, losses due to diseases and accidents are higher in stocker calves than in yearlings. Studies show that death losses of calves usually run about 5%; yearlings about 1%; and two-year-old and older cattle about 0.5%. Most of these losses are due to stress and to viruses to which the animals are subjected between the farm or ranch and the final destination of the stockers. Shipping fever takes the heaviest toll. It and other diseases are fully covered in Chapter 12.

■ **Lightweight calves gain more efficiently**—The feed requirement for calf gain is directly related to the animal's weight. Because of low maintenance requirements, lightweight calves gain more efficiently than heavyweight calves. This is illustrated in Fig. 25-4 which shows the megacalories of digestible energy required for each 100-lb gain when calves are gaining at the rate of 1½ lb per day. All things being equal (health, genetics, etc.), calves will gain 100 lb on less feed at the lighter weights, which is an advantage for which most stocker buyers are looking.

■ **Lightweight calves require higher quality feed**—Although lightweight calves gain more efficiently than heavier calves, they require better quality feed for the same gain. For this reason, forage (pasture, hay, or silage) which will merely furnish a 300-lb calf sufficient feed to maintain body weight will allow a mature animal to gain weight. Fig. 25-5 illustrates this point. Note that 300-lb calves gaining 1½ lb per day

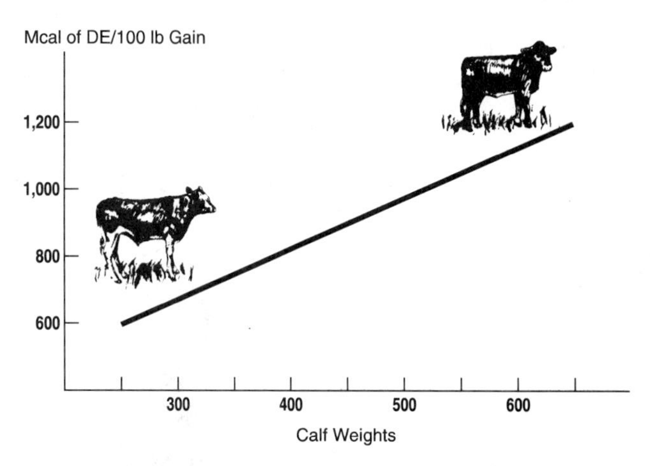

Fig. 25-4. Feed required per 100-lb gain for calves at different weights gaining 1½ lb per day. (Source: *Keys to Profitable Stocker Calf Operations*, MP-964, Texas A&M Univ. Ag. Ext. Serv.)

would need forage containing 1.3 megacalories of digestible energy per pound whereas 600-lb calves would gain at the same rate on forage with only 1.1 megacalories of energy per pound. This points up the fact that stocker operators should buy calves of the right age and weight to match their feed.

■ **Faster gains are cheaper**—Faster gains are cheaper because the maintenance requirement is the same regardless of the daily gain; hence, the maintenance requirement is a smaller part of the total feed consumed as gains increase. Fig. 25-6 illustrates this

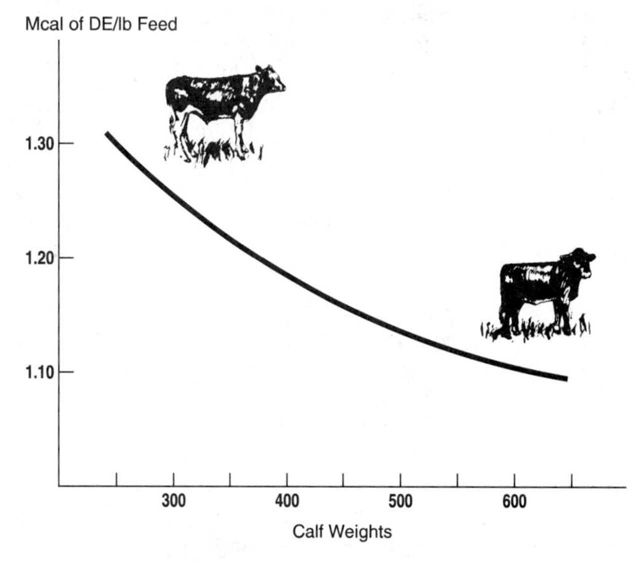

Fig. 25-5. Quality of feed required for calves of different weights to gain 1½ lb per day. (Source: *Keys to Profitable Stocker Calf Operations*, MP-964, Texas A&M Univ. Ag. Ext. Serv.)

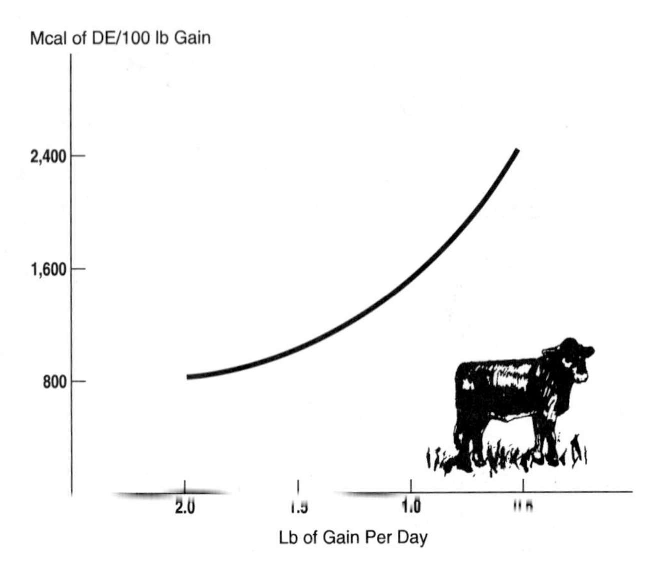

Fig. 25-6. Feed required per 100-lb gain for a 450-lb calf gaining at different rates per day. (Source: *Keys to Profitable Stocker Calf Operations*, MP-964, Texas A&M Univ. Ag. Ext. Serv.)

point. A calf gaining 2 lb per day requires only approximately 800 megacalories of digestible energy per 100-lb gain, whereas the energy requirements per pound of gain for a calf gaining 0.6 lb per day are doubled—to 1,600 megacalories. With interest rates, taxes, labor costs, etc., rising as they have, growing calves at rates of less than 1 lb per head daily becomes increasingly difficult to justify.

■ **Compensatory growth**—Compensatory growth is making up for a bad start in life. It is common practice for stocker cattle to be *roughed through* the winter as cheaply as possible, with limited daily gains. Then, in the spring, the animals are turned to lush spring pasture or put in a feedlot on a high-energy ration. Animals so managed exhibit the phenomena of *compensatory growth*; that is, on the high-energy diet they gain faster and more efficiently than similar cattle which were fed more liberally during the wintering period. Feedlot operators were quick to sense this situation, and to take advantage of it. This is the chief reason for the popularity of Okie-type cattle. They are animals whose growth has been held back to less than their genetic potential. When fed more liberally, they exhibit a surge in growth rate and feed efficiency. Large compensatory growth usually indicates that someone (the stocker operator) has lost money while someone else (the feeder) has made money. It is noteworthy that Holsteins and the larger exotics should never be handled so as to exhibit compensatory gains. If they're held back in the winter, they're too heavy when they finish.

■ **Avoid excess feeder condition**—If cattle get too fleshy as feeders, (1) they may reach market finish before they attain desirable market weight, and (2) they will tend to gain more slowly than desirable during the feedlot finishing period. The planned rate of gain should be determined by feeder finish, growth potential, sex, and the beginning weight of the feeder calves. It will also depend upon how much condition is acceptable to the buyer. British breed or crossbred steer calves gaining at the rate of 1¼ lb daily will about maintain their condition; at 1½ lb daily they will add some condition; at 2 lb daily they may be too fleshy by the time they reach 700- to 800-lb feeder weight. British breed or crossbred yearlings making daily gains of 0.9 lb will maintain growth without fattening. Larger cattle of some of the exotic breeds may make larger gains without fattening.

■ **Buy and sell carefully**—Because the original weight purchased in a stocker program is a high percentage of the weight sold by the grower, any mistake made in buying or selling has a greater impact on profits or losses than it would in a finishing program. If you pay too much, buy with too much fill, pay for quality that the feeder is not willing to pay for, or sell too low, profits will be rapidly eroded away.

■ **Steers are heavier**—Steer calves normally weigh about 5% more than heifer calves.

MARKET CLASSES AND GRADES OF STOCKERS

Stocker cattle are of many kinds, displaying a wide range of combinations of the various characteristics such as breeding, sex, age, weight, size, conformation, and condition. Fortunately, there is a market for each kind, with variation in prices reflecting both the supply of and the demand for each kind and the degree of suitability for the intended purpose. The use of uniform terms and descriptions of all classes and grades of cattle by all members of the beef team—cow-calf producers, feeders, packers, and buyers and sellers—throughout the country has contributed much to the orderly marketing of all cattle, including stockers. The market classes and grades of stocker cattle are fully covered in Chapter 15 of this book; hence, the reader is referred thereto.

SOURCES OF STOCKERS

Sources of stockers vary from area to area. Nationwide, most stockers are secured through auctions and direct purchase from cow-calf producers.

Stockers may be purchased to advantage in an

Fig. 25-7. Stocker cattle being sold at auction at Western Stockman's Market, Famoso Jct.., California. (Courtesy, *California Cattleman*, photo by Phil Raynard)

auction, but there are problems. They have already been stressed by being loaded, hauled, unloaded, and handled; in most cases, the buyer knows little or nothing of the origin of the cattle. and they may have come straight off native grass (which is good), or they may be making their third sale in the last 10 days (which is very bad).

The best doing stockers are usually those coming directly from cow-calf producers—preconditioned, and without passing through a market facility.

NATIVE (Homegrown or Local) OR WESTERN STOCKERS

In the midwest, native stockers are those coming from the farms of the Corn Belt, Great Lakes, and Southeast, whereas western stockers are the branded animals coming from the western ranges. For the most part, these cattle come from comparatively small farm herds. Calves produced in these herds are often fed out on the farms where they are produced. Others are offered for sale locally. There are both advantages and disadvantages to buying native stocker cattle.

Native stockers usually have the following **advantages**:

1. **Fewer disease problems.** Native stockers are subjected to less exposure and stress than western stockers, with the result that they have fewer diseases such as shipping fever. This is especially true if they are bought directly from the farm where produced.
2. **Lower freight and buying costs.** Usually freight cost is much less because of shorter distances; and buying costs may be less, especially where they are bought directly from the neighbors.
3. **Lower shrink.** Shrink, especially tissue shrink—the decrease in carcass weight, which occurs on long, extended hauls and during long periods of fast—is usually much lower.
4. **They're acclimated.** Native stockers do not have to adjust to the weather and altitude of the area.
5. **They're accustomed to the feed.** Native cattle are usually accustomed to the feeds of the area.

Native stockers usually have the following **disadvantages**:

1. **Lack uniformity.** They are usually less uniform than western stockers in breeding, quality, condition, and age.
2. **May be of lower quality.** On the average, there is little to choose between the quality of native and western cattle. However, quality is much more variable between native herds, with the result that they range from very high quality to very low quality.
3. **They're fleshier.** Native stockers are likely to be fleshier than western stockers, with the result that they are not apt to make as good gains when placed on high-roughage rations.
4. **Small lots must be combined.** Where a number of stockers is desired, it is usually necessary to buy and put together several small lots of cattle.
5. **Availability.** Native stockers may not be available in large enough numbers when desired.

METHOD OF BUYING STOCKERS

Stocker cattle may be purchased by stocker operators themselves, or by salaried buyers, order buyers, commission firms, or cattle dealers. Although owners or managers of stocker operations should be knowledgeable enough to buy stocker cattle, if they are large operators, they are generally too busy to do so if they are doing a proper job of running their outfits. Accordingly, the trend is to shift the buying responsibility to specialists—primarily order buyers.

TRANSPORTATION AND SHRINK OF STOCKER CATTLE

Improper handling of stocker cattle immediately prior to and during shipment may result in (1) excess shrinkage; (2) high death, bruise, and crippling losses; (3) disappointing sales; and (4) dissatisfied buyers. Unfortunately, many cow-calf operators who do a superb job of producing stockers dissipate all the good things that have gone before by doing a poor job of preparing and shipping. The subject of "Preparing and Shipping Cattle" is fully covered in Chapter 15 of this book; hence, the reader is referred thereto.

PRECONDITIONING; HANDLING NEWLY ARRIVED CATTLE

These important subjects are fully covered under similar headings in Chapters 12 and 33, respectively, of this book; hence, the reader is referred thereto.

FEEDING STOCKERS

For a stocker operation to be profitable, the grower must be ever aware of the following reasons back of it and feed stockers accordingly: (1) to provide a supply of the kind of cattle desired by finishing lots at the time needed, (2) to utilize roughages and other low cost feeds, and (3) to *cheapen down* the cattle.

Because of the very nature of the operation, the successful feeding of stockers requires the maximum of economy consistent with normal growth and development. This necessitates cheap feed—either pasture or range grazing or such cheap harvested roughage as hay, straw, fodder, and silage. In general, the winter feeds for stockers consist of the less desirable and less marketable roughages. It is important, therefore, that the high-roughage rations of young stockers be properly supplemented from the standpoints of proteins, minerals, and vitamins.

Fig. 25-8. Stocker cattle being fed a limited amount of concentrate. (Courtesy, *Beef* magazine)

The feed consumption of stockers will vary somewhat with the quality of the roughage available, the age of the cattle, and the rate of gain desired. As far as practical, the stocker ration should prepare the cattle for making maximum use of the feed which follows the stocker stage—either the finishing ration or grass. The rate and efficiency of gain in the feedlot or on pasture varies inversely with the amount of gain made during the stocker stage—that is, the smaller the stocker gains, the higher the finishing gains; and the higher the stocker gains, the lower the finishing gains. This phenomenon is known as *compensatory gain.*

Of course, too small gains may be unprofitable to the grower. Besides, young animals can be stunted. To make maximum growth without fattening—just to maintain condition—calves of the British breeds and crossbreds should gain 1.25 lb daily, and yearlings should gain 0.9 lb daily.

The amount of gain desired in a grower program depends largely on the way the cattle are to be handled in the next stage. For example, stockers that are to be grain fed on grass or go directly to the finishing lot can be wintered more liberally, and make a higher rate of gain, than cattle that are to be turned to pasture only. Also, winter and summer gains are not exactly inversely proportional. For example, if one lot of steers gains twice as much during the winter as a second lot, its summer gains won't be limited to half those of the second lot. Rather, they will likely be 70 to 90% as much. Thus, where the stocker grower retains ownership of the cattle through the finishing stage, the cattle that make the largest stocker gains may also make the most economical and largest total gain for the entire period—from weaning to slaughter.

Tables 25-1 and 25-2 contain some recommended rations for stocker cattle. Variations can and should be made in the rations used. The grower should give consideration to (1) the supply of homegrown feeds, (2) the availability and price of purchased feeds, (3) the class and age of cattle, (4) the health and condition of animals, and (5) the kind of feeder cattle in demand by feedlots.

In using Tables 25-1 and 25-2 as guides, it is to be recognized that feeds of similar nutritive properties can and should be interchanged as price relationships warrant. Thus, (1) the cereal grains may consist of corn, barley, wheat, oats, and/or sorghum; (2) the protein supplement may consist of soybean, cottonseed, peanut, linseed, safflower, and/or sunflower meal; (3) the roughage may include many varieties of hays and silages; and (4) a vast array of byproduct feeds may be utilized.

The following points are pertinent to the success of a stocker operation and should be kept in mind:

- **Recommended nutrient allowances**—Where grower rations are formulated on the basis of percentage of nutrients in the ration, the following allowances are recommended:

Protein

For up to 1.5 lb daily gain 10.0%
For 1.5 lb daily gain or more 10.5%

Calcium and phosphorus

For up to 500 lb liveweight . . . . . 0.3 to 0.5%
For over 500 lb liveweight 0.25%

TABLE 25-1
DAILY RATION FOR STOCKER CALVES (400–500 Lb *[182–227 Kg]*)

		Rations (Fed for Gains of 1.25 Lb/Head/Day *[0.57 Kg/Head/Day]*)									
		1	2	3	4	5	6	7	8	9	10
Legume hay or grass-legume mixed hay	(lb) *(kg)*	12–18 *5.4–8.2*		8–12 *3.6–5.4*		2–4 *0.9–1.8*		8–10 *3.6–4.5*			
Grass hay	(lb) *(kg)*		12–18 *5.4–8.2*	4–6 *1.8–2.7*			2–4 *0.9–1.8*			10–12 *4.5–5.4*	
Straw, corncobs, cornstalks, stalklage, cottonseed hulls	(lb) *(kg)*							2–4 *0.9–1.8*	2–3 *0.9–1.4*		2 *0.9*
Corn or sorghum silage	(lb) *(kg)*				25–40 *11.4–18.1*	20–30 *9.1–13.6*	20–30 *9.1–13.6*				
Legume-grass silage, or oat silage	(lb) *(kg)*								20–25 *9.1–11.4*		
Legume-grass haylage, or oat haylage	(lb) *(kg)*										20–25 *9.1–11.4*
Grain (corn, sorghum, barley, or oats)	(lb) *(kg)*							4–5 *1.8–2.3*		4–5 *1.8–2.3*	4–5 *1.8–2.3*
Protein supplement (41% or equivalent)	(lb) *(kg)*		1.25–1.5 *0.6–0.7*	0.25–1 *0.1–0.5*	1–1.25 *0.5–0.6*	0.75–1 *0.3–0.5*	1.25–1.5 *0.6–0.7*		1–1.5 *0.5–0.7*	1–1.5 *0.5–0.7*	
		With all rations, provide suitable minerals.									

TABLE 25-2
DAILY RATION FOR YEARLING STOCKERS (600–700 Lb *[273–318 Kg]*)

		Rations (Fed for Gains of 0.9 Lb/Head/Day *[0.4 Kg/Head/Day]*)									
		1	2	3	4	5	6	7	8	9	10
Legume hay or grass-legume mixed hay	(lb) *(kg)*	16–24 *7.3–10.9*		6–8 *2.7–3.6*		2–4 *0.9–1.8*		6–8 *2.7–3.6*			
Grass hay	(lb) *(kg)*		16–24 *7.3–10.9*	10–16 *4.5–7.3*			2–4 *0.9–1.8*			16–20 *7.3–9.1*	
Straw, corncobs, cornstalks, stalklage, cottonseed hulls	(lb) *(kg)*							12–15 *5.4–6.8*	10–12 *4.5–5.4*		2 *0.9*
Corn or sorghum silage	(lb) *(kg)*				45–55 *20.4–25.0*	40–50 *18.2–22.7*	40–50 *18.2–22.7*				
Legume-grass silage, or oat silage	(lb) *(kg)*								20 *9.1*		
Legume-grass haylage, or oat haylage	(lb) *(kg)*										35–40 *15.9–18.2*
Grain (corn, sorghum, barley, or oats)	(lb) *(kg)*							5–6 *2.3–2.7*		5–6 *2.3–2.7*	5–6 *2.3–2.7*
Protein supplement (41% or equivalent)	(lb) *(kg)*		1.5–1.75 *0.7–0.8*	1–1.5 *0.5–0.7*	1.25–1.5 *0.6–0.7*	0.75–1 *0.3–0.5*	1.25–1.5 *0.6–0.7*		1 *0.5*	1–1.5 *0.5–0.7*	
		With all rations, provide suitable minerals.									

Vitamin A

Air-dry feed (10% moisture) . . 800 to 1,000 IU per lb
. . 10,000 IU daily per head

Implant

Gains of more than 1.5 lb per head daily . . . . Include growth stimulant implant

■ **Protein**—Calves have a higher protein requirement per 100 lb of liveweight than older cattle and are more apt to be deficient on low-quality roughage. Extra energy will not be efficiently used unless protein intake is adequate.

Calves wintered on range will require 0.5 to 0.7 lb of crude protein from supplements to gain 0.5 lb daily. Calves wintered on good grass hay or meadow hay will require less supplemental protein and phosphorus than when wintered on range.

One pound of 41% supplement will supply about half the total protein requirement of a 500-lb calf. Three pounds of alfalfa hay or 2½ lb of dehydrated alfalfa will furnish about the same amount of protein but more energy than 1 lb of a 41% protein concentrate.

Urea is not well utilized as a protein supplement in high-roughage rations. Because of this, usually it is best to use a plant protein supplement or a slow release urea product in growing rations.

Work at Missouri and other stations indicates no advantage to feeding protein supplements on pasture which contains legumes or grasses in lush condition.

■ **Energy**—Satisfactory and efficient winter gains of weanling calves depend upon sufficient energy, along with a proper balance of protein and energy in the ration.

There is much information that indicates stocker cattle should not be winter-fed over 1 lb of grain per 100 lb of body weight if they are to make best use of pasture the following summer. Also, delayed grain feeding on pasture until after peak pasture growth is recommended.

■ **Vitamins and minerals**—All rations should provide adequate carotene or vitamin A, calcium, and phosphorus.

Most stockers are bought in the fall to be wintered on low-quality roughage or on winter wheat or other cool-season pasture. Some are bought in the spring to be placed on bluestem, Bermudagrass, or other native pastures. Such calves usually do not have a high store of vitamin A; hence, an intraruminal or intramuscular injection of 500,000 to 1,000,000 IU/head of vitamin A upon their arrival at the place where they will go on the stocker program could be helpful, especially if a nitrate problem is anticipated on wheat pastures.

■ **Roughage**—Feeding good-quality roughage in a drylot may be more desirable for weanling calves than wintering on the range. Winter range is most efficiently used by yearling steers and mature cattle. Meadow hay, a mixture of alfalfa and grass hay, good-quality upland hay, or silage provides a good basal diet for wintering calves.

Corn or sorghum silage does not need to be supplemented with additional grain or dry roughage unless it was put up too wet or had little grain as the result of drought. Sorghum silage or grass-legume silage that has more than 65 to 70% moisture will require more grain supplementation than drier silage of comparable quality.

Grass-legume silage cut at the proper stage of maturity and carefully ensiled is an excellent feed for stocker cattle. Such silage contains adequate protein and need not be supplemented with a protein concentrate. However, it is much lower in energy than corn or sorghum silage; hence, the gains will be smaller unless (1) 150 to 200 lb of grain per ton are added as a preservative, or (2) it is supplemented at feeding time with some grain or other energy source.

Grass-legume hay of good quality is excellent for stocker cattle. It may or may not be supplemented with grain.

Haylage, made from grass-legume or straight legume forage, wilted to about 50% moisture content, and commonly stored in an oxygen-free silo, is increasing as a stocker feed. Except for the difference in moisture content, haylage has a feeding value comparable to silage or hay made from a similar crop.

Grass hay, such as prairie hay, Bermuda, timothy, Sudan, Johnsongrass, etc., may make up most of the ration of stocker cattle. Energy supplementation is needed if improvement in the condition of stockers is desired. Also, such hays must be properly supplemented with protein, minerals, and vitamins.

Stubble and stalk fields furnish much feed for stocker cattle, especially yearlings, in the late fall and early winter. Unless there is access to a good winter pasture, cattle on stalk fields should be fed 4 to 6 lb of legume hay or 1½ lb of protein concentrate daily. In addition, minerals should be provided.

Occasionally, roughages are high in nitrates. Actually, calves can consume very high levels of nitrate provided (1) they are fed with regularity, (2) changes in feed are gradual, (3) feeds are mixed uniformly, and (4) ample water is provided at all times. Where nitrate is a problem, cattle may die of oxygen insufficiency. Others in the lot will show discoloration of nonpigmented areas of the epithelium and chocolate-colored blood.

■ **Keep roughage waste low**—Roughage need not

be processed. But it should be fed so that there will be a minimum of waste.

■ **Grain**—Calves are unable to consume enough dry roughage to gain more than 1 lb a day. Thus, grain should be added in the quantity necessary to achieve the desired gains. Bear in mind that with calves of the British breeds and crossbreds it takes a gain of about 1¼ lb daily to maintain condition; yearlings of the British breeds or crossbreds will maintain condition on a gain of about 0.9 lb daily.

Some grain should be included in the ration of stocker cattle when (1) they are to be finished immediately after the wintering period, (2) they weigh less than 350 lb when started on winter feeding, and (3) heifers are to be bred when they are 13 to 15 months old.

Calves that are full-fed corn or sorghum silage high in grain content, plus 1 lb of protein concentrate or 4 to 5 lb of legume hay, need not be fed grain.

■ **Level of wintering**—The level of wintering stockers affects the gains in the next stage. Thus, calves gaining the most during the winter make the least gains on pasture the following summer. This is clearly shown in Table 25-3.

TABLE 25-3
CALVES TO YEARLINGS—EFFECT OF WINTER GAINS ON SUBSEQUENT GAINS THE FOLLOWING SUMMER AS YEARLINGS[1]

Lot No.	Winter		Summer		Total	
	(lb)	*(kg)*	*(lb)*	*(kg)*	*(lb)*	*(kg)*
			Valentine, Nebraska			
1	115	*52*	205	*93*	320	*145*
2	120	*55*	181	*82*	301	*137*
3	127	*58*	183	*83*	310	*141*
4	129	*59*	183	*83*	312	*142*
5	150	*68*	170	*77*	321	*146*
6	157	*71*	143	*65*	300	*136*
7	164	*75*	170	*77*	304	*138*
8	179	*81*	160	*73*	339	*154*
9	184	*84*	150	*68*	324	*147*
10	186	*85*	152	*69*	338	*154*
11	186	*85*	162	*74*	348	*158*
			Fort Robinson, Nebraska			
1	67	*30*	226	*103*	309	*140*
2	83	*38*	218	*99*	308	*140*
3	87	*40*	231	*105*	298	*135*
4	90	*41*	221	*100*	308	*140*
5	104	*47*	202	*92*	306	*139*
6	136	*62*	183	*83*	319	*145*

[1]*Beef-Forage Notebook*, Cattle Management, 5-C-1, Univ. of Nebraska Ext. Svcs.

Calves wintered to gain 1 lb daily make satisfactory summer pasture gains. This level is recommended for calves to be grazed season-long the following summer, provided the same ownership is retained all the way through. One to two lb daily gain during the winter is usually desirable if calves (1) are to be sold in the spring, (2) will be on full feed 2 to 3 months after going to grass, (3) will be receiving a limited feed of grain on grass, or (4) are replacement heifers that are to be bred at 13 to 15 months of age.

Since yearlings are not growing as rapidly as calves, they may be fed for smaller gains than calves, and yet show comparable condition. Thus, for maximum growth without fattening (for just holding their condition) calves should gain approximately 1¼ lb daily, whereas yearlings need to gain only 0.9 lb daily.

■ **Pasture supplement**—Following wintering, many stockers are grazed throughout the pasture season. Supplementing grass with a protein supplement at the rate of 1 lb per head per day when summer pastures drop off in quantity and protein content (generally beginning about mid-July) will usually boost average daily gain by 0.4 lb.

■ **Additive**—Rumensin is approved by the FDA for increased growth in stocker cattle. In cattle that are gaining 0.75 to 2.00 lb per head per day, Rumensin will boost average daily gains by an additional 0.2 lb.

WINTER PASTURES

Wherever possible, stocker calf operations are planned around a winter pasture program. Weanling calves or lightweight, thin yearlings are purchased in the fall. In some cases, homegrown calves are retained and developed under this system for sale as yearlings. As would be expected, winter pasturing of stockers is largely limited to the southern part of the United States, with the kind of pasture varying from area to area.

■ **Winter wheat pastures**—Winter wheat pastures are widely used for stocker cattle in Kansas, Oklahoma, and Texas. When such pastures are good, cattle make very acceptable gains on them. However, wet weather or droughts make winter wheat pastures unreliable, with the result that it is important that there be flexibility in the stocker program, both in numbers and season of use.

■ **Other cool-season pastures**—In the southern states, extensive use is made of oats, rye, ryegrass, vetch, and fescue—a perennial grass that remains green throughout the winter. This area is turning more and more to winter grazing, as a means of making profitable year-round use of its land and labor and providing 600- to 800-lb feeder cattle in greatest demand by feedlots.

GRASS TETANY

Grass tetany (grass staggers or wheat pasture poisoning), a highly fatal nutritional ailment, is one of the hazards of winter pastures. Although it is more common among lactating cows, stocker cattle are affected. It generally occurs during the first two weeks of the pasture season, particularly in cattle grazing wheat or other cereal crops. The characteristic symptoms and signs are: nervousness, flickering of the third eyelid, twitching of the muscles (usually of the head and neck), head held high, accelerated respiration, high temperature, gnashing of the teeth, and abundant salivation. A slight stimulus may precipitate a crash to the ground and, finally, death. The condition follows a rapid course, with usually a lapse of only 2 to 6 hours between onset and death.

The exact cause of grass tetany is not completely understood. It is known that hypomagnesemia (low level of magnesium in the blood serum) is associated with grass tetany. But the causes of low levels of blood magnesium cannot be explained in all cases by rations being deficient in magnesium. Nevertheless, it has been established that supplemental magnesium will increase blood levels and alleviate at least in part the grass tetany problem. Approximately 6 g of magnesium per day (magnesium may be in the form of magnesium oxide, magnesium sulfate, or magnesium carbonate; since magnesium oxide is approximately 60% magnesium, it will require 10 g of it daily to provide 6 g of magnesium) added to the salt, mineral, or concentrate supplement will suffice. Commercial magnesium supplements are available in wheat pasture areas. These should be fed according to manufacturers' directions. Also, magnesium alloy "bullets" have been developed for cattle to give a slow release of magnesium in the rumen.

In addition to magnesium supplementation, access to some hay or dried mature grass during the first two weeks after stockers are first placed on winter pasture (wheat pasture or whatever kind) is also helpful in preventing grass tetany.

If grass tetany is suspected, a veterinarian should be called immediately for proper diagnosis and prescribed treatment. Normally, the prescribed treatment consists of an intravenous injection of at least 500 cc of a dextrose solution containing both magnesium and calcium. (Also see Chapter 8, Feeding Beef Cattle, section headed "Nutritional Diseases and Ailments," Table 8-15.)

STOCKER AND GROWER CONTRACTS

Hand in hand with the development of big feedlots and year-round feeding came the need for an assured supply of feeder cattle of the desired kind on a continuous basis. To meet this need, more and more feedlots have turned to contractual arrangements with stocker growers, with numerous kinds of contracts. Usually, the cattle are owned by the feedlot, most of which are large and in a stronger financial position than the majority of stocker growers. The two most common kinds of contracts are based on either (1) a fixed cost for the gain, or (2) an agreed feed cost plus an extra charge for labor and lot rental. Usually, there is provision for adjusting for death loss. Such contracts should always be in writing, with all provisions, including weighing conditions, spelled out.

Although the use of stocker and grower contracts has increased in recent years, the concept is not new. Many of the Kansas bluestem pasture owners have long grown out yearlings owned by Iowa and other Corn Belt feeders.

Today, many corn farmers in the fertile irrigated area around Greeley, Colorado, make corn silage and feed cattle on a contract basis to stockers owned by one of several large feedlots in the vicinity. Stocker cattle are also being grown under contract on the wheat pastures of Kansas, Oklahoma, and Texas; on hay and other roughages in the irrigated valleys of the West; and on sorghum silage and stalk fields throughout the Southwest.

STOCKER OPERATION COSTS AND RETURNS

Those thinking of launching a stocker operation inevitably ask: "How much money will it take, and what can I make?"

Answers to the above two-pronged question have been provided by *Cattle-Fax*, the cattle marketing arm of the National Cattlemen's Beef Association. As a result of a 12-year study, 1980–1991, Cattle-Fax reported that cow-calf operators selling calves at weaning had profits in six years and losses in six years. By contrast, retained ownership programs were profitable in 11 out of the 12 years covered by the study.

A summary of the Cattle-Fax study is presented in Table 25-4.

When sold as a weaner calf, returns per head ranged from a profit of $82 in 1990 to a loss of $77 in 1983. The annual average was a profit of $11.16.

The study showed that by retaining ownership, some producers had an average advantage of as much as $84 per head over sale at weaning.

As cattle numbers expand and per capita beef consumption continues downward, more and more cow-calf producers will engage in one or more of the programs open to them in retaining calves through one or more subsequent production steps beyond weaning and marketing at that time.

(Also see Chapter 21, section headed "Retained Ownership.")

TABLE 25-4
COMPARISON OF RETAINED OWNERSHIP OPTIONS TO SALE OF WEANED CALF IN $/HEAD 1980–1992

		If Sold As a Weaned Calf		Advantage of Programs Compared to Selling at Weaning										
Year	Average Cost of Produc-ing a Calf	Average Price Received	Profit or (Loss)	Dry Lot Winter Program	Dry Lot Winter Program and Summer Grass	Dry Lot Winter Program, Summer Grass, and Feedlot	Dry Lot Winter Program and Feedlot	Wheat Pasture	Wheat Pasture and Summer Grass	Wheat Pasture, Summer Grass, and Feedlot	Wheat Pasture and Feedlot	Back Ground	Back Ground and Feeding	Direct to Feedlot
	($)	($)	($)	($)	($)	($)	($)	($)	($)	($)	($)	($)	($)	($)
1980	345	376	31	—	—	—	—	—	—	—	—	—	—	—
1981	352	314	(38)	(106)	(78)	—	(95)	(21)	(25)	(38)	(12)	(98)	(84)	(5)
1982	351	315	(36)	(51)	22	(113)	32	18	81	42	73	(23)	27	126
1983	379	302	(77)	(13)	(21)	(5)	(8)	51	50	17	23	24	(1)	110
1984	369	326	(43)	(27)	2	55	(31)	65	78	94	48	(7)	(26)	28
1985	375	322	(53)	(31)	(60)	15	(117)	47	30	110	(68)	(4)	(134)	(32)
1986	354	328	(26)	(64)	19	(16)	1	(8)	4	118	42	(44)	(20)	(8)
1987	342	411	69	20	136	52	131	60	126	174	153	68	151	213
1988	368	443	75	24	82	139	61	71	116	101	93	72	47	120
1989	385	464	79	(48)	45	75	(11)	37	130	114	54	(2)	(31)	18
1990	390	472	82	(58)	73	79	35	13	123	147	84	(6)	12	69
1991	392	463	71	(18)	47	82	(79)	44	136	37	(39)	35	3	49
1992	—	—	—	(64)	56	(5)	(2)	2	83	91	49	(22)	(14)	38
Average advantage				(36)	27	32	(7)	32	78	84	42	(1)	(6)	60
Most profit				24	136	139	131	71	136	174	153	72	151	213
Least profit				(106)	(78)	(113)	(117)	(21)	(25)	(38)	(68)	(98)	(134)	(32)

QUESTIONS FOR STUDY AND DISCUSSION

1. How and why have stocker programs changed through the years?

2. Define the following terms: (a) *stockers*; (b) *feeders*; (c) *replacement heifers*; (d) *preconditioning*; and (e) *backgrounding*.

3. What are the common types of stocker programs, and what are the characteristics of each?

4. What are the advantages and disadvantages of a stocker program?

5. Do you feel that, for the most part, the stocker stage will be lengthened or shortened? Justify your answer.

6. Why do lightweight calves (a) gain more efficiently, (b) require higher quality feed, and (c) make faster and cheaper gains than heavyweight calves?

7. For owners who produce their own stocker calves, and who finish them out on a custom basis, is compensatory growth good or bad? Justify your answer.

8. What are the market classes and grades of stocker cattle? How do these relate to finished cattle grades?

9. How would you go about buying stocker cattle?

10. What are the advantages and disadvantages of (a) native stockers, and (b) western stockers?

11. Discuss rations and rate of gain of (a) stocker calves, and (b) stocker yearlings.

12. What are the primary differences between stocker and finishing rations?

13. Tell how the level of wintering stockers affects the next stage.

14. Discuss the favorable and unfavorable factors of winter wheat pastures.

15. Discuss the favorable and unfavorable factors of the common cool-season pastures of the Southeast.

16. How would you lessen the hazard of grass tetany?

17. What provisions should be incorporated in a stocker and grower contract?

18. What are the primary factors upon which profits from stocker operations depend?

SELECTED REFERENCES

Title of Publication	Author(s)	Publisher
Beef Cattle, Seventh Edition	A. L. Neumann	John Wiley & Sons, Inc., New York, NY, 1977
Beef Cattle Science Handbook	Ed. By M. E. Ensminger	Agriservices Foundation, Clovis, CA, pub. annually 1964–1981

PART III

Cattle Feedlots; Pasture Finishing

The finishing of cattle is what the name implies, the laying on of fat. The ultimate aim of the finishing process is to produce beef that will best answer the requirements and desires of the consumer. This is accomplished through an improvement in the flavor, tenderness, and quality of the lean beef.

In a general way, there are two methods of finishing cattle for market: (1) cattle feedlots, including confinement (sheltered) finishing; and (2) pasture finishing. Prior to 1900, the majority of fat cattle sent to market were 4- to 6-year-old steers that had been finished primarily on grass. Even today, the utilization of pastures continues to play an important role in all types of cattle finishing operations.

Cattle finishing of the 1950s and 1960s was characterized by a dramatic increase in grain feeding. In 1947, only about 6.9 million head of cattle were grain-fed, representing 30.1% of the slaughter cattle that year. In 1994, 27.6

Gelbvieh X Red Angus crossbred steers in a Colorado feedlot. (Courtesy, American Gelbvieh Assn., Westminster, CO)

million head of fed cattle were marketed, representing 80.7% of all the cattle slaughtered that year.

In the "good old days" of the horse-drawn wagon, scoop shovel, wicker basket, and processing ear corn by breaking the nubbins (ears) on the edge of the feed bunk, one caretaker fed 100 head of fattening cattle. Today, cattle feedlots with more than 1,000 capacity dominate cattle feeding; and in a modern cattle feedlot, one caretaker feeds 1,500 cattle, and in the more mechanized lots, one caretaker feeds as many as 2,500.

In 1973, beef prices spiraled to unprecedented heights, in response to increased costs of production brought on by scarce and high-priced grain. The reaction was loud and clear; neither beef consumption nor production is elastic at any price. Both are subject to the old law of supply and demand. Cattle feeders suffered staggering losses, and cattle investors fled. Out of this chaos, feedlot cattle increasingly became roughage burners. Cattle feeders came more and more to rely upon the ability of the ruminant to convert coarse forage, grass, and byproduct feeds, along with a minimum of grain, into palatable and nutritious food for human consumption; thereby not competing so much for humanly edible grains.

Chapters 26 to 35 will (1) give cattle feeders, and those who counsel with them, a big assist in the years ahead; and (2) provide students with scientific and technological information relative to modern cattle feeding.

A large, modern cattle feedyard—a beef factory. (Courtesy, Continental Grain Co., Ulysses, KS)

MODERN CATTLE FEEDING (FINISHING)[1]

Contents *Page*

[1]In the preparation of this chapter, the authors gratefully acknowledge the authoritative review and helpful suggestions, along with pictures, of the following person, who has great expertise in the operation of a modern cattle feedlot: Ron Baker, C & B Livestock Inc., Hermiston, Oregon.

Fig. 26-1. A Colorado feedyard. (Courtesy, Farr Feeders, Inc., Greeley, CO)

***Cattle feeding**, also termed **cattle finishing**, is that aspect of cattle production involving placing feeder cattle on high-energy rations to increase weight and market desirability.*

The ultimate aim of cattle finishing is to improve beef so that it will best meet the requirements and desires of the consumer. This is accomplished through increasing the marbling (the feathering of fat within the muscling), which imparts an improvement in flavor, tenderness, and quality of the lean.

In a general way, there are two methods of finishing cattle for market: (1) open lot and confinement (sheltered) finishing, and (2) pasture finishing. Prior to 1920, the majority of fat cattle sent to the market were 4- to 6-year-old steers that had been finished primarily on grass. But, today, pasture finishing plays a far less important role in most types of cattle finishing operations.

Cattle feeders are commonly classed as either (1) commercial feeders or (2) farmer-feeders, based largely on numbers. From the standpoint of statistical reporting, the U.S. Department of Agriculture commonly draws the line at 1,000 head. A commercial cattle feeding operation is defined as one having a capacity of 1,000 head or more, at any one time.

Traditional farmer-feeders evolved with Corn Belt farming, in the North Central area of the United States. Generally speaking, they market their crop, usually corn, through cattle (or hogs, or lambs), and spread the manure on the land. The purchase of feeder cattle for this type of enterprise is generally in the fall, with the actual feeding done during the winter months when labor is available due to limited field work. This traditional farmer-feeder type of operation has persisted to the present time, although it has been modernized through the years.

In addition to being larger, commercial cattle feeders generally differ from farmer-feeders in the following respects: (1) they usually feed cattle on a year-round basis, rather than during the winter months only; (2) they may grow little, or none, of their feed; (3) they are highly mechanized; and (4) they are knowledgeable of costs and returns, skillful buyers and sellers, and aware of market trends. Today, commercial feedlots with more than 1,000-head capacity dominate cattle feeding.

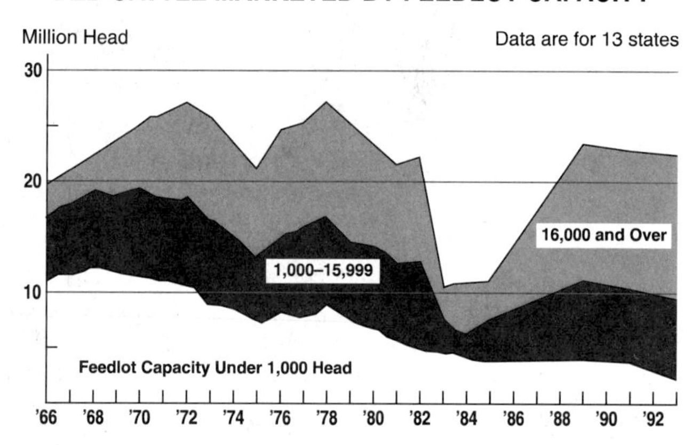

Fig. 26-2. Feedlots have gotten bigger! In 1962, feedlots with more than 1,000 head marketed 40% of the fed cattle. By 1993, 31 years later, feedlots with over 1,000 head marketed 87% of the nation's fed cattle. (From: *Meat & Poultry Facts 1994*, Am. Meat Inst., p. 10)

The trend is for larger but fewer feedlots. According to the U. S. Department of Agriculture, there were 43,632 cattle feedlots operating in the 13 major cattle feeding states in 1994. Of this number, only 1,820 lots, or 4%, had capacity of more than 1,000 head, but they accounted for 88% of the 22.9 million head of fed cattle marketed from the 13 states. This compared with 1,856 lots, or 1.5%, with capacity of more than 1,000 head in 1970, accounting for 58% of marketings.

In 1994, lots in the 16,000- to 31,999-head capacity range declined to 117 lots from 138 in 1992, yet these lots sold a total of 5,116,000 head, the same number as the 138 lots sold in 1992.

In 1994, the largest lots, those with 32,000-head capacity and over, increased in number by 9.6%, to 91 lots from 87. These lots sold a total of 8,387,000 head during 1994, compared to 7,873,000 the previous year. Texas had the most lots in the 32,000-head-and-over category—38 of them.

In 1994, the four leading states in total number of feedlots, by rank, were: Iowa, 15,800; Minnesota, 8,000; Illinois, 6,700; and Nebraska, 5,700. These states continued to lead the nation in number of lots, as they have over the past several decades.

In 1994, Texas led the nation in cattle feeding, marketing 5.6 million head. Nebraska was second, with 4.7 million head, followed by Kansas with 4.4 million, and Colorado with 2.3 million head.

It is predicted that the total number of cattle feedlots will continue to trend down, and that the proportion of fed cattle coming from the larger lots will continue to increase.

Feedlots exceeding 100,000 capacity are now in operation in Arizona, California, Colorado, and the northern Texas Panhandle. Even in the Corn Belt, feedlots are getting bigger.

It is important that all members of the beef cattle team, especially commercial cow-calf producers and purebred breeders, be knowledgeable relative to the role of cattle finishing in the marketing of the feeder cattle that are produced. More important still, again and again the story of efficiency and applied technology in cattle feeding needs to be told to consumers, in order that they will understand how the improved cutability and quality of the beef supplied to them was achieved.

THE GROWTH AND MATURITY OF CATTLE FEEDING

The growth years of beef and cattle feeding extended from 1947 to 1986. In 1947, only 6.9 million head of cattle were grain fed, representing 30.1% of the slaughter cattle that year. In 1947, the U.S. population was 144,034,000; and per capita beef consumption (carcass basis) was 69.1 lb.

In 1947, beef was king. It was not only good, but it was good for you. Consumers couldn't get enough of it; and they were able and willing to pay the price. Per capita beef consumption went up and up. The grain bins were bulging; and in most years cattle feeding was profitable. Cattle feeders were attuned to consumer demand for beef grading U.S. Choice, with repeatability—the same quality as they purchased last week. Large commercial cattle feeders fed cattle on a year-round basis, making grain-fed beef available to consumers throughout the year. Antibiotics, hormones, and other additives, along with new methods of processing grain, improved rate of gain and feed efficiency of cattle; marketing of finished cattle and beef packing became more efficient; and futures trading provided cattle feeders a means to lessen risks.

For 39 years, 1947 to 1986, beef cattle feeding went up and up. In 1986, 26.2 million head of cattle were grain fed, representing 70% of the slaughter cattle that year. In 1986, the U.S. population was 240,651,000; and per capita beef consumption (carcass basis) peaked at 107.6 lb.

But beginning about 1986 the American hunger for beef was blunted. The beef team—producers, processors, and retailers—was too little and too late in responding to consumer concerns about health and diet; and to their demand for less fat, higher quality, consistency (repeatability), and competitive price with other meats. Since 1986, U.S. per capita consumption of beef has gone down each year, from 107.6 lb in 1986 to 93 lb in 1993—a drop of 14.6 lb in seven years. Where it will end no one knows.

Now, and in the days ahead, many people will eat less meat; and many will substitute broilers, turkeys, and fish for steaks, roasts, and hamburgers. The future of beef is clouded by a multi-owned, segmented industry, out competed by well-organized and fully-integrated broiler and turkey industries that are controlled by few corporations. Thus, the high-flying days of beef are behind us.

Despite the decline in per capita beef consumption, cattle feeding has been sustained by increased human population—by more beef eaters. Thus, it is, and will remain, a formidable industry as evidenced by the following two charts:

Fig. 26-3 shows the growth and maturity of cattle feeding. Note the steep annual growth up to 1972, followed by leveling out.

Fig. 26-4 shows the composition of cattle slaughter. Since 1987, more than three-fourths of the slaughter steers and heifers have been grain fed.

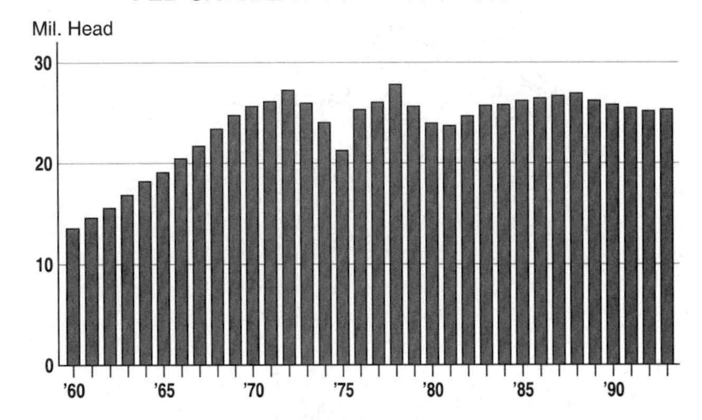

Fig. 26-3. This shows that the total U.S. fed cattle marketings increased sharply from 1960 to the peak year of 1972, following which they declined slightly. *Note:* This chart covers total U.S. cattle feeding, and is not limited to the 13 major cattle feeding states normally reported. (Data from: *Meat and Poultry Facts 1994*, American Meat Institute, Washington, DC)

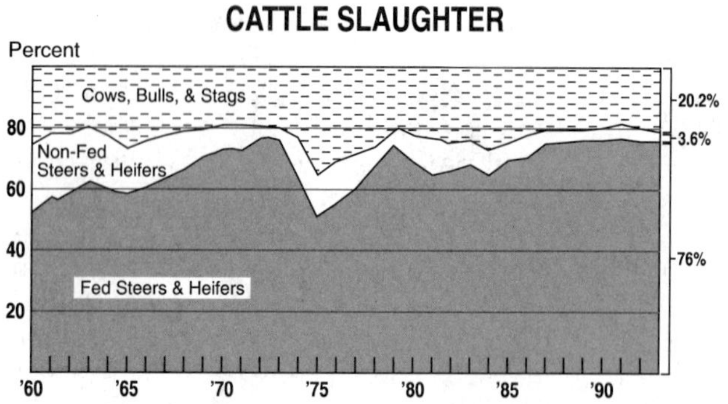

CATTLE SLAUGHTER

Percent

Cows, Bulls, & Stags — 20.2%

Non-Fed Steers & Heifers — 3.6%

Fed Steers & Heifers — 76%

'60 '65 '70 '75 '80 '85 '90

Fig. 26-4. Composition of cattle slaughter, 1960-1993. Few nonfed steers and heifers are now being marketed. (From: *Meat & Poultry Facts 1994*, American Meat Institute, p. 18)

Note: Also see Chapter 16, section headed "What's Ahead for Beef." As beef goes, cattle feeders and cow-calf producers go.

SHIFT IN GEOGRAPHY OF CATTLE FEEDING

Regional or geographical shift in cattle feeding is a matter of history. Such shifts in the location of cattle feeding are influenced by relative costs of production in the various feeding areas and states and by the alternatives for use of production resources.

Fig. 26-5 shows changes in fed cattle marketings, 1985–1994. As shown, Texas, Kansas, Colorado, Oklahoma, Idaho, and Nebraska showed big increases, while California, Illinois, and Iowa showed big decreases.

The primary reasons for the shifts in geography of cattle feeding are as follows:

- **Commercial cattle feeders are free to choose their locations**—Most farmer-feeders are *tied* to a particular farm; they cannot locate their feedyards

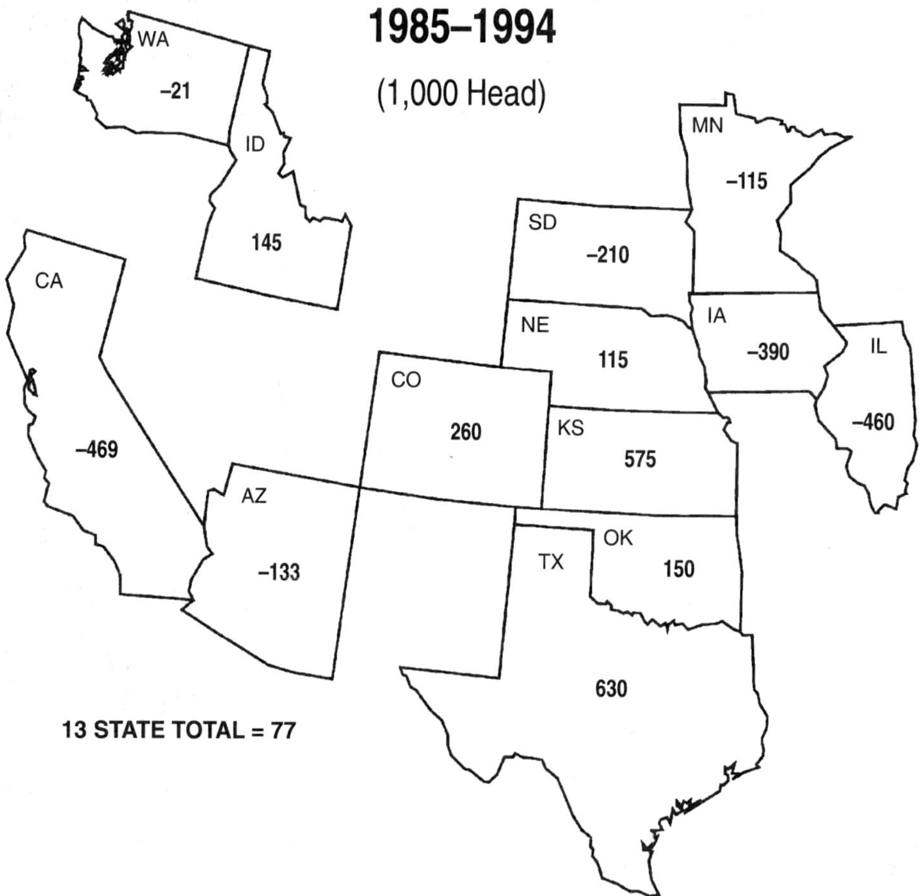

Fig. 26-5. Geographical shifts in cattle feeding, based on fed cattle marketings. Note that Texas, Kansas, Colorado, Oklahoma, Idaho, and Nebraska showed big increases in cattle feeding, while California, Illinois, and Iowa showed big decreases (Source: U.S. Department of Agriculture, Livestock Marketing Information Center, Cooperative Extension Service, Lakewood, CO)

elsewhere. However, most commercial cattle feeders are free to choose an area. As a result, they locate where they feel that they have the best combination of factors favorable to their success, with consideration given to availability and price of feed, supply of feeder cattle, weather, slaughter plants, etc. This makes for a shift in the geography of cattle feeding.

- **Large grain supplies, especially sorghum and corn in the Plains area**—Hand in hand with the increased production of sorghum and corn, more cattle feeding emerged. The economics of this close correlation between the area of sorghum production and cattle feeding becomes apparent when it is realized that it requires approximately 8 lb of feed to produce 1 lb of on-foot gain. Moreover, each $1 increase in freight cost per ton of feed increases feed cost per pound of gain by approximately 0.4¢. Thus, considerable saving in cost can be effected by moving the feeder cattle near the feed source, rather than moving the feed to the cattle.

- **Nearness of large feeder calf supplies in Texas and the Southeast**—For many years, the cow-calf population has been concentrated west of the Mississippi River. It was logical, therefore, that proximity to feeder cattle supplies (as well as feed) should be considered in locating cattle feedlots. This was a factor in the development of the Plains feeding area.

- **Decentralization of livestock marketing and meat packing**—Following World War II, terminal markets declined and meat packers decentralized. Both auction markets (especially feeder cattle auctions) and beef slaughterers located near concentrated cattle feeding areas.

IMPACT OF MODERN CATTLE FEEDING

Cattle feeding is a *beef factory.* More than 10 million tons of cattle, feed, beef, and other supplies must be transported annually to service this factory. In particular, feed grain growers and cow-calf operators benefit from cattle feedlots; and the nearer they are to one or more big feedlots, the greater the benefits.

The impact of one large commercial feedlot is enormous. For example, a 40,000-head capacity feedlot with a turnover of 2.5 times its capacity yearly, will turn out 100,000 finished animals, or 60 million lb of carcass beef each year. Large quantities of raw materials are required to keep this *beef factory* going. It will need the feeder calf production from 142,857 cows if 70% of the cows produce calves for the feedlot, or 1,428 herds each averaging 100 head of brood cows. To maintain these cows on the western range will require 1.5 million acres of average grazing land. To feed the cattle in this 40,000-head capacity lot (100,000 animals per year) will require 100,000 tons of corn and/or sorghum, or the production of nearly 100,000 acres averaging 60 bu per acre. And that's not all!

The cattle feeding industry purchases millions of dollars of pharmaceuticals, uses millions of dollars in electrical power, and pays millions of dollars for vehicles and related equipment. Additionally, the clothing merchant, the doctor, the lawyer, the banker, and the barber all benefit from the increased economic activity of the region. The local, state, and federal governments obtain added taxes. Indeed, the economic impact of the cattle feeding industry is enormous.

CHARACTERISTICS OF MODERN CATTLE FEEDING

The characteristics of the modern cattle feeding industry are:

- **It is unique to the United States**—Finishing (fattening) cattle by feeding grain, and other concentrates, is not practiced to any great extent elsewhere in the world, although interest is evolving in certain countries.

- **It is highly concentrated in a relatively few states**—Six states—Texas, Nebraska, Kansas, Colorado, Iowa, and Oklahoma—account for 85% of all the cattle fed in the United States.

- **It gave rise to a new *beef belt***—Beginning in the early 1950s, a massive *beef belt* formed, extending from the southern Great Plains through Nebraska. It was ushered in with increased irrigated feed crops, particularly sorghum, and corn of the area, along with

Fig 26-6. The modern shaded cattle feedlot of McElhaney Feeding Co., Yuma County, Arizona. (Courtesy, Sam McElhaney, Wellton, AZ)

new feedlot technology and mechanization. Today, this area has the largest concentration of feedlot cattle in the world.

■ **It is a high risk business**—In some years, the feeder may not even recover feed costs. Thus, cattle feeders should be in a relatively strong financial position.

■ **Facilities and worker-hours required for feeding vary**—Facilities vary from a dirt lot on a general farm, where a farmer may spend as much as 25 hours of labor to finish one steer, to elaborate, highly specialized facilities, costing almost as much per square foot as a modern home, where labor needs may drop to about one hour per head for the entire feeding period.

■ **Feed costs alone account for about 75% of the variable costs**—Variable costs include feed, interest on feed and feeder cattle, labor, death loss, and veterinary and medicine costs. Feed costs, which are the largest item, account for about 75% of these variable costs.

■ **Several methods of financing are being used**—Because of the large amounts of capital required, big feedlots are being financed by outside capital, much like any other big business.

■ **Type of legal ownership varies with size**—Although there is no set pattern, small commercial lots are generally single-proprietor owned; medium-sized lots are partnership owned; and large lots, with capacities of 10,000 head or more, are incorporated.

■ **Large amounts of investor money are involved**—Doctors, lawyers, merchants, industrialists are investing in cattle feeding; through consignment feeding (ownership) of cattle that are custom fed, limited partnerships, shareholders in a corporation, and other arrangements. Prior to the severe cattle feedlot losses of 1974, it was estimated that these "Wall Street cowboys" were financing up to one-half the cattle on feed.

But there has been a dramatic change in the type of cattle feedlot investors since the mid-1980s. Today's cattle feedlot investors are sophisticated and knowledgeable. Also, feedlot owners have been forced to own more of their cattle in order to keep their lots full; and feedlot managers have encouraged retained ownership. (See Chapter 21, section headed "Retained Ownership.")

■ **There is much vertical integration of lots, but little horizontal integration**—Vertical integration refers to control or ownership by other levels of the functional system; *e.g.*, cow-calf producers, stocker and yearling operators, cattle dealers, and packers. In 1993, the 15 major packers either fed or contracted for 17.4% of the fed steers and heifers which they slaughtered (known as captive supplies).

Horizontal integration is attained by ownership or control of similar functional levels; *e.g.*, a feedlot that merges with another feedlot. Some vertical integration of feedlots exists, but horizontal integration is limited.

■ **Larger lots do more custom feeding**—Most farmer-feeders and small commercial feeders own their cattle. Large commercial lots do a great deal of custom feeding; and, generally speaking, the proportion of cattle custom fed varies almost directly with size.

A Texas Station study showed the following types of ownership of custom-fed cattle on the southern plains; over 50% owned by farmers and ranchers; about one-third owned by cattle buyers, cattle dealers, and other types of investors; and 10 to 11% owned by packers.

■ **Most feeder cattle are bought through auctions or by direct purchase**—The two primary sources of feeders are: (1) auctions (including video auctions), and (2) direct purchase from producers. Feeder cattle are purchased by feedlot operators, salaried buyers of feedlots, order buyers, commission firms, and cattle dealers.

■ **Most finished cattle are sold directly to packing plants**—The majority of feedlot cattle are sold FOB the feedlot, on foot. In 1990, 38.2% of all cattle were purchased by packers on a carcass basis. In many areas, relatively small lots of finished cattle are sold through auctions and terminal markets.

■ **Most big lots buy every week; sell every week**—Few experienced commercial cattle feeders are *in-and-outers*. They assume that future price levels cannot be accurately predicted. Also, for efficiency reasons, it is necessary to maintain a constant flow of cattle in and out of the lot, thereby making for optimum operating capacity of the receiving chute crew, the doctoring crew, and the operation of the feed mill. So, most commercial cattle feeders buy feeder cattle and sell finished cattle on the market each week.

■ **Nutrition and animal health consultants are used**—The larger commercial feedlots usually rely on two different consultants: (1) a nutritionist for ration formulation and feed purchase advice, and (2) a veterinarian for animal health, disease, and sanitation decisions. The consequence of failing to keep up is too great to risk in the multimillion dollar business of today's *beef factory*.

CATTLE FEEDING COSTS AND RETURNS

Those thinking of feeding cattle inevitably ask: "How much money will it take, and what can I make?"

Answers to the above two-pronged question are provided in Table 26-1. In 1995, the average feedlot operation costs and returns were as shown in Table 26-1; under the assumptions shown in the same table.

As shown in Table 26-1 (under the assumptions made therein), in 1995 the typical U.S. cattle feedlot operation showed a total cost of gain of $48.60 per cwt; a break-even of $69.91 per cwt; and a net profit of $14.00 per head. Note that the 3 largest cost items, by rank, were: cost of the feeder steer, 70%; cost of feed, 24%; and operating interest, 3%.

Returns are determined by selling price and selling weight. In turn, selling price is determined by the supply of cattle and the consumer demand for beef.

Fig. 26-7 shows the average return per head to cattle feeders during the period 1985 to 1994. This points up that cattle feeding is a high risk business. In some years, the feeder may not even recover feed costs. Thus, cattle feeders should be in a strong financial position—so that they can absorb the losses of the bad years, and reap the profits of the good years.

TABLE 26-1
FEEDLOT OPERATION COSTS AND TYPICAL RETURNS, 1995[1]

Item	Total Cost	% of Total
	($/Head)	
Feeder steer	577	70
Feed	194	23.5
Operating interest	25	3
Death loss	7	1
Vet and medicine	10	1
Fixed overhead	12	1.5
Total costs/hd.	825	100

Total income/hd.	$839.00	Break-even ($/cst)	$0.6991
Net profit (loss)/hd.	$14.00	Total cost of gain ($/cwt)	$48.60

Average budget assumptions:

In weight (pay)	780 lb *(355 kg)*
Out weight (pay)	1,180 lb *(536 kg)*
Total gain	400 lb *(182 kg)*
Days on feed	133
Average daily gain	3.0 lb *(1.4 kg)*
Conversion (100% DM)	7.20 lb of feed/lb of gain *(3.3 kg of feed/ 0.45 kg of gain)*
In price	$74
Out price	$71
Ration cost (100% DM)	$130/ton
Operating loan interest rate	10%
Death loss	1%

[1]Data provided by Ron Baker, one of the senior author's former students of whom he is very proud, who has successfully owned and operated a 25,000 head capacity feedyard for many years.

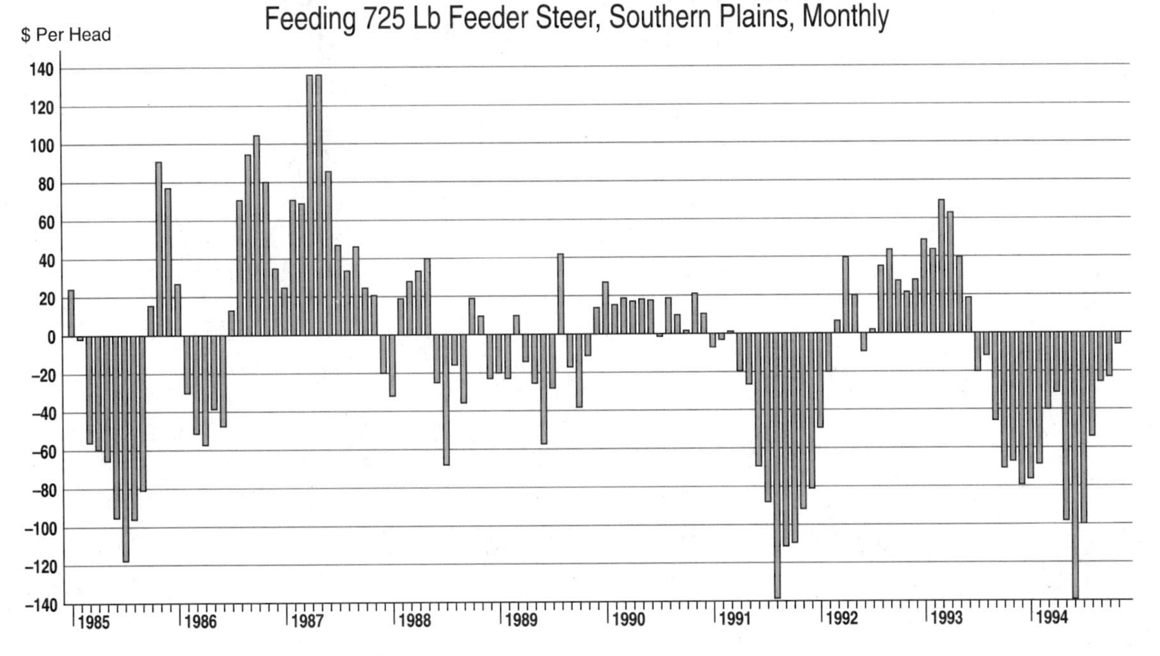

Fig. 26-7. Average returns per head to cattle feeders 1985 to 1994. Data for yearling steers, mostly in the Southern Plains—the leading cattle feeding area of the U.S. (Courtesy, USDA, Livestock Marketing Information Center, Lakewood, CO)

QUESTIONS FOR STUDY AND DISCUSSION

1. What is your prediction relative to the future of cattle finishing from the standpoints of—
    a. Open lot vs confinement (sheltered) finishing?
    b. Pasture finishing?
    c. Commercial feeders vs farmer-feeders?
    d. Larger feedlots?
    e. Shifts in geography?

2. During the period of a depressed fed cattle market and/or high feed prices, which is hurt most—a farmer-feeder or a commercial feedlot? Justify your answer.

3. When estimating profits, consumers are prone to compute from range (off grass) to range (stove), with no cattle finishing phase between. Why is this so?

4. Detail the reasons back of the phenomenal growth of cattle feeding—a growth which took it from 30.1% of all slaughter cattle grain-fed in 1947, to 70% of all slaughter cattle grain fed in 1986. Which force was most important in this growth?

5. What were the primary reasons for the shift in the center of the geography of cattle feeding from the Corn Belt to the Southwest? Which force was most important in this shift?

6. Cattle feeding is a *beef factory*. Hence, establishing a big cattle feedlot in a community makes for increased business. Will a realization of this fact make feedlot flies, dust, and odors less obnoxious to the nearby urban population? Justify your answer.

7. List and discuss the characteristics of modern cattle feeding.

8. List the six leading cattle feeding states of the nation. What factors are favorable to cattle feeding in each of them?

9. Are the "Wall Street cowboys" (the doctors, lawyers, merchants, and industrialists) who invest in cattle feeding good or bad for the cattle feeding industry? Justify your answer.

10. Would you recommend ownership of a commercial cattle feedlot as an investment? Justify your answer.

SELECTED REFERENCES

Title of Publication	Author(s)	Publisher
Beef Cattle Science Handbook	Ed. by M. E. Ensminger	Agriservices Foundation, Clovis, CA, pub. annually 1964–1981
Feeding Beef Cattle	J. K. Matsushima	Springer-Verlag, New York, NY, 1979
Feedlot, The, Third Edition	Ed. by G. B. Thompson C. C. O'Mary	Lea & Febiger, Philadelphia, PA, 1983

Futures trading on the Chicago Mercantile Exchange (CME) floor. (Courtesy, CME, Chicago, IL)

BUSINESS ASPECTS OF CATTLE FEEDING[1]

Contents *Page*

[1]In the preparation of this chapter, the authors gratefully acknowledge the authoritative review and helpful suggestions, along with pictures, of the following person, who has great expertise in the operation of a modern cattle feedlot: Ron Baker, C&B Livestock, Inc., Hermiston, Oregon.

Cattle feeding is big and important business, and intricate, too; and it is destined to get bigger and more complicated. Today, cattle feedlots with more than 16,000 capacity dominate cattle feeding.

Cattle feeders are in business and, like other business people, they hope to obtain a reasonably good return for the use of their capital, labor, and management. To this end, their business aspects must become more sophisticated and efficient; they must—

1. Compute break-even prices prior to buying feeder cattle, especially if they do not buy and sell each week.
2. Buy feeder cattle of the right size, quality, and price.
3. Sell the cattle to the best advantage.
4. Integrate when possible.
5. Feed cattle to weight and grade.
6. Evaluate performance.
7. Obtain economies with size.
8. Finance the feedlot and cattle properly and adequately.
9. Stay attuned to consumer demands.

Fig. 27-1. The business aspects of cattle feeding have become sophisticated and efficient, with computers and performance testing commonplace. This shows a personal computer. (Courtesy, California State University, Fresno)

Of course, the above nine points represent a great oversimplification of a complex business, but they do clearly set forth the main requisites for profitable cattle feeding. Anyone who wishes to make money feeding cattle must have expertise in these nine areas.

A new *breed* of cattle feeder is providing business acumen at a highly professional level, with the same degree of confidence that exists in any other big business. These people are attracting large amounts of new capital from sources outside of agriculture. Business aspects outweigh all other factors—feed additives, crossbreds, pollution control, etc.—producing change in cattle feeding. It is important, therefore, that feeders and those who counsel with them be thoroughly grounded in each of these areas.

COMPUTING BREAK-EVEN PRICES FOR CATTLE

Those who feed on a large scale and on a continuous basis try to build in some insurance against the consequences of price changes through their buying programs. When finished cattle are sold, they try to replace them with feeders bought at a price which would allow a suitable profit if they were sold at the same time as the finished cattle they replace. To the extent that prices of finished cattle and feeder cattle move together (both in direction and magnitude), this works reasonably well. But they don't always move together.

Except for big cattle feeders who buy and sell cattle each week as a means of hedging, cattle feeders should compute break-even prices before buying feeder cattle. Even the big operators who buy and sell weekly do not keep the same number of cattle on feed from year to year, or throughout the year. As a result, the demand for feeders tends to be buoyant following a period of good profits from cattle feeding and depressed following a period of low returns or losses.

A nomograph (a graph or chart) may be used in computing break-even prices for cattle. It can give a quick, preliminary idea of cost, price, and investment relationships. But a nomograph should not replace more detailed budgeting which should precede all major buying, selling, and investment decisions. Also, one should realize that a nomograph will give erroneous and misleading information unless based upon accurate and realistic cost and return data from the problem at hand.

Fig. 27-2 is a nomograph for use in making quick calculation of break-even buying and selling prices for cattle. The section that follows gives an example and step-by-step instructions on how to use it.

INSTRUCTIONS FOR USING NOMOGRAPH

This nomograph provides a quick method of computing the "break-even prices" a feeder can afford to pay for incoming cattle, based upon assumptions relative to costs, amount of gain, and the estimated selling price for finished cattle. Likewise, the break-even selling price for finished cattle can be determined, based upon assumptions concerning costs, amount of gain, and the prices of feeder cattle.

The following step-by-step procedure illustrates the use of the nomograph.

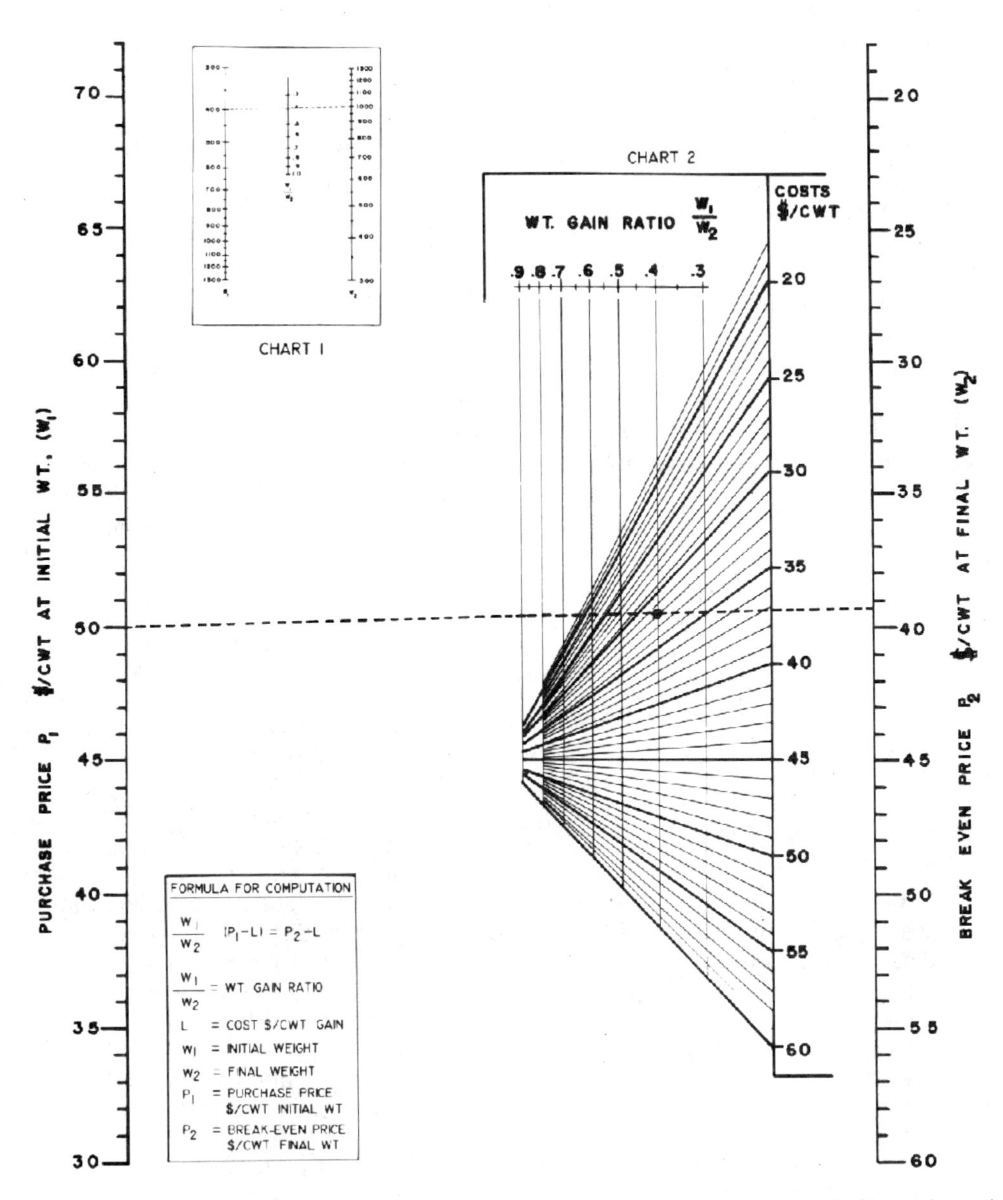

Fig. 27-2. Nomograph for computing the break-even prices for cattle. Beef cattle break-even price is determined by purchase price, feed cost ($/cwt), and weight-gain ratio. This nomograph, like any other figure of similar nature, can be updated by applying figures that are current at the time of use. (This ingenious nomograph, along with instructions on how to use it, was prepared by Robert M. George, University of Missouri.)

Data Needed		Example	Your Farm
1. Purchase weight of feeder cattle	(W_1)	400#	______
2. Selling weight of finished cattle	(W_2)	1,000#	______
3. Feed costs per cwt of gain	(L)	$32.00	______
4. Price per cwt of feeder cattle	(P_1)	$50.00	______
5. Price per cwt of finished cattle	(P_2)	$39.00	______

■ Procedure for determining the break-even price for feeder cattle—

Step 1. On Chart 1, locate the purchased weight of feeder cattle on scale W_1 and the final weight on scale W_2 and connect the two with a straight line.

Step 2. Read the W_1/W_2 weight-gain ratio on the center scale.

Step 3. On Chart 2, locate the same weight-gain ratio *vertical line*, W_1/W_2, on the scale to the immediate right.

Step 4. Follow this line straight down to the point where it intersects the *diagonal* line which represents the cost per cwt of gain—*mark this point.*

Step 5. On the vertical scale P_2 (break-even price), locate the expected selling price per cwt of finished cattle—*mark this point.*

Step 6. Draw a straight line to connect the two points located above. The point of intersection with the vertical scale P_1 on the left indicates the break-even price per unit for feeder cattle—the price which will allow recovery of costs.

Note well: In Step 4, a *diagonal cost line* may be selected to represent any cost per cwt to be recovered—feed costs only, total variable costs, total variable plus fixed costs, or total costs plus some desired profit per cwt of gain.

■ **Procedure for determining the break-even price for finished cattle**—To determine the break-even price one must receive for finished cattle, to recover the costs of feed, feeder cattle, etc., follow the same procedure down to Step 5, then locate the purchase price for feeder cattle on the vertical scale P_1, and, with a straight line, locate the break-even price for finished cattle on scale P_2.

Other factors (than the cost of feed, amount of gain, and price of slaughter cattle—the factors considered in the nomograph) affecting the price that a feeder can afford to pay for feeder cattle are:

1. Condition of the cattle. Thin cattle, if in good health, will make faster gains than fleshy cattle.
2. Growthy cattle—cattle that are big framed and on the rangy order—make better gains than the little, compact kind; and they may be carried to heavier weights. If the feeder cannot obtain cattle backed by production records, *eyeballing* will help.
3. Younger, lighter weight cattle tend to make more efficient gains.
4. Cattle of known, superior ancestry with gaining ability are worth more.
5. Higher grade cattle are worth more. This is so because better grades generally bring a higher selling price, and, therefore, a higher price is obtained on their gains made in the feedlot.
6. The higher the cost of feed, the greater the necessary margin between the cost of feeder cattle and the selling price of finished animals. This is so because of the high cost of gains as compared to their selling price.
7. Feeder steers are generally worth approximately $3 to $4 per cwt more than heifers. This is because they gain about 10% faster, require 5 to 10% less feed, and bring from $0.50 to $1.50 per cwt more than heifers when finished. Additionally, there is no pregnancy problem.
8. Good crossbreds will make 2 to 4% more rapid and efficient gains than the average of the parent breeds.

MARGIN

A positive margin exists when feeder cattle cost less than finished cattle. A negative margin exists when feeder cattle cost more than finished cattle. Cattle feeders will pay more for feeder cattle than they expect to receive for them as finished animals when there appears to be a favorable margin on the gain in weight; that is, when they can sell the gain in weight for considerably more than it cost to produce it.

Profits in cattle feeding come from two different kinds of margins—price margin and feeding margin.

Price margin is the difference between the cost per cwt of the feeder animal and the selling price per cwt of the same animal when finished.

For example, if a feeder pays $70 per cwt for a 600-lb steer and sells him for $65 per cwt, the price margin is a negative $5. This means that the cattle feeder would take a $30 loss on the original 600 lb.

Feeding margin is the difference between the cost of putting on 100 lb of gain and the selling price per cwt of the same animal when finished. Thus, if it costs $55 per cwt to put gain on yearling steers, and if finished cattle sell for $65 per cwt, the feeding margin would be $10 per cwt. Assuming a market weight of 1,000 lb, or of 400-lb gain, the feeder could expect to make about $40 on the feeding margin.

The amount a cattle feeder makes as a result of a good feeding margin can more than offset the losses accruing from a negative price margin, but it doesn't always work that way. It depends on many different things—the selling price, the cost of gain, the price paid for feeder animals, and other factors.

In the example just cited, the feeding margin amounted to $40 per animal. This is not to suggest, however, that cattle feeders should always put more gain on yearling steers. How much gain a cattle feeder should put on depends upon the kind of feeding program being followed, the kind and condition of the feeder cattle when they go into the lot, rate of gain, and several other factors. Research and experiences have clearly demonstrated that costs of gain go up pretty fast if cattle are fed much beyond Choice slaughter grade.

The principle of profits from price margin and feeding margin applies to feeder calves, also. But the relative importance of price margin vs feeding margin is not quite the same for calves as for yearlings. Let us analyze the situation further: The feeder who is feeding yearlings buys 600 lb of the 1,000 lb which will be marketed at the end of the feeding period. So, getting cattle bought right is pretty important. If $1 per cwt too much is paid for feeders, that takes $6 off the potential profit from price margin and from total profits. On the other hand, the feeder who buys calves is more interested in costs of gain and feeding margin than in price margin because about 60% of the weight which will be sold is from the gain put on in the feedlot. Thus, if the feeder pays $1 per cwt too much for a 400-lb feeder calf, it hurts, but not quite so much—$4 compared to $6 per head.

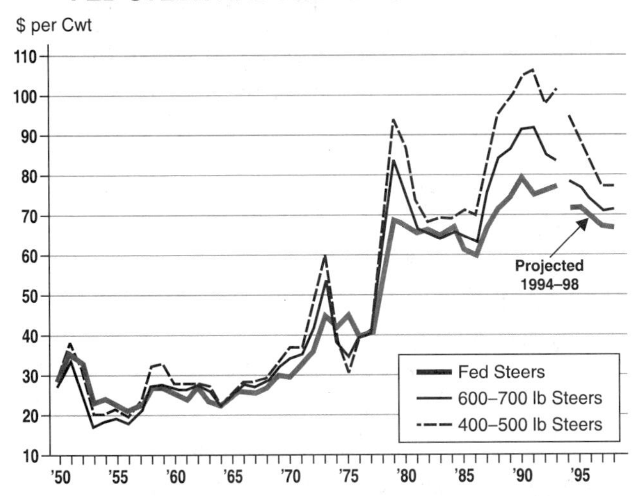

Fig. 27-3. Note negative margins, which refers to those times when feeder cattle cost more than finished cattle. (Courtesy, USDA, Livestock Marketing Information Center, Lakewood, CO)

If farmer-feeders just manage to balance gains from the feeding margin with their losses from the price margin, this does not necessarily mean that they should not feed cattle. Actually, they aren't in too bad shape. They're getting paid market price for their feed, a going wage for their labor, around 12% on their own capital, and they're getting enough to cover all fixed costs like depreciation, taxes, etc., on their facilities and equipment. With commercial feeders, it's another story. They usually buy most of their feed, and they operate on borrowed capital. Thus, to stay in business, they must turn a profit over and above these costs.

BUYING FEEDER CATTLE

When buying feeder cattle, it is well to keep in mind the old, but true, cliché: Cattle well bought are half sold. Indeed, wise buying will increase profits or minimize losses on any lot of cattle. The following factors are pertinent to buying feeder cattle:

- **Decide on kind of finished cattle**—Feeders should first decide on the kind of finished cattle that they wish to market. Then, they will be in a position to decide on the kind of feeder animals required. For example, certain feeders may decide that they wish to produce finished steers, weighing around 1,100 lb, grading 85% Choice, and yielding about 62% (cold) in the packing plant. Next, it may be decided that to meet this goal feeders will need to start with steers weighing 650 to 725 lb capable of reaching the desired weight and grade as quickly and as efficiently as possible. This weight of feeder steers may be dictated by the fact that it takes about 400 lb of gain in the feedlot to get such animals to grade Choice.

Fig. 27-4. Cattle well bought are half sold! These steers were bought as calves in Texas then finished in an Arizona feedlot. They are high performing, and they meet the market demand. (Courtesy, Red Rock Feeding Co., Red Rock, AZ)

- **Market sources of feeder cattle**—Most feeder cattle are secured through auctions and direct purchase, rather than through terminal markets.

The best cattle from the standpoint of feedlot performance are usually those feeders coming directly from cow-calf producers—preconditioned and without passing through a market facility.

Feeder cattle may be purchased to advantage in an auction, but the problems are greater; among them, (1) they have already been stressed by being loaded, hauled, unloaded, and handled; (2) the buyer knows little or nothing of the origin of the cattle in most cases; and (3) they may have come straight off native grass (which is good), or they may be making their third sale in the last 10 days (which is very bad).

- **Method of buying**—Feeder cattle may be purchased by feedlot operators themselves, or by salaried buyers, order buyers, commission firms, or cattle dealers. Although owners or managers of feedlots should be knowledgeable enough to buy feeder cattle, they are generally too busy to do so if they are doing a proper job of running their lots. Accordingly, the trend is to shift the buying responsibility to specialists—primarily order buyers. Specialized buyers purchase about three-fourths of all feeder cattle in California, Colorado, and the High Plains of Texas and Oklahoma.[2]

Specialized buyers offer several advantages. They are constantly on the market for feeder cattle; hence, they are knowledgeable relative to the availability of cattle, prices, and feeding and slaughter potential of various lots of cattle. Evaluating and pricing feeder cattle is not easy, for it involves a projection ahead relative to the finished product. Of course, much of the guess work is taken out of the projection where one can obtain and feed cattle with known performance records back of them.

■ **Shrinkage**—There are two types of weight loss or shrink: (1) excretory shrink, or loss of bellyfill; and (2) tissue shrink, or a decrease in the carcass weight of the animal. In the early part of shipment only excretory shrink occurs. Tissue shrink occurs on long, extended hauls or during long periods of fasting. It takes longer for animals to recover from tissue shrink than from excretory shrink. At an undefined stage in movement, both excretory and tissue shrinkage occur simultaneously. During the latter part in transit, tissue shrink is relatively more important.

In a study in Iowa involving 4,685 feeder cattle, it was found that an average shrink of 7.2% occurred with cattle purchased from a rancher as contrasted to 9.1% from those purchased from a sale yard. The cattle were shipped varying distances from 150 to 1,133 mi. It was found that there was a 0.61% shrink for each 100 mi in transit.

The cost of regaining weight loss during movement to the feedlot and getting feeder cattle back on a normal rate of gain is called *interruption cost.* Older feeders and preconditioned feeders have a lower interruption cost than younger feeders.

Factors affecting shrinkage, along with shrinkage tables designed to make adjustments to compensate for different shrinkage conditions, are presented in Chapter 15 of this book; hence, the reader is referred thereto.

■ **Preconditioning**—Preconditioning is not well defined in terms of standardized practices and procedures. In fact, it is often confused with backgrounding. The term *preconditioning* as used herein refers to dehorning and castrating well in advance of weaning; providing necessary vaccinations, treatment for control of parasites; and weaning at least three weeks before shipment. In addition, calves should be familiarized with drylot confinement, feed bunks, and water troughs.

The transition of calves from a cow herd in an open range environment to closely confined quarters, perhaps in a different climate hundreds of miles away, is a shock. This shock is a costly problem for cattle feeders.

Calves are under stress from a combination of factors such as weaning, medical treatment, rough handling, lack of feed and water during transit, strange environment, and change in feeds. Reduction or elimination of these factors lowers death loss, shrinkage, and time required to get cattle started on feed. Cattle feeders want cattle to reach their feedlots in the best possible condition. Thus, preconditioning is important to them.

(Also see Chapter 12, section headed "Calf Health-Treatment Record [Preconditioning].")

SELLING FINISHED (Fed) CATTLE

The feeding and marketing phases of cattle are closely related. So much so that selling should never end at the point of reaching an agreed price. Rather, feeders should follow through on the performance of the cattle. They should learn the final weight of the cattle at the plant, their grade and yield, their cutability, and how the retailer liked them from the standpoint of tenderness and salability. This type of total marketing knowledge is necessary for feeders to improve performance in the feedlot.

(See Chapter 16, section headed "Cattlemen's Carcass Data Service.")

■ **When to sell fed cattle**—Cattle should be sold when they are properly finished. This usually means that they should be sold as finished cattle when they reach the same grade they had as feeder cattle. In some cases, fed cattle may be carried to one grade higher. However, the hazard of overweight cattle should be avoided.

■ **Market channels for fed cattle**—Although there are area differences, fed cattle are marketed as follows: liveweight at the feedlot, carcass grade and weight basis, by auction, or through terminal markets.

During the 1950s, when cattle feeding was just entering its period of rapid growth, terminal public markets were the most widely used outlet for slaughter cattle in the United States. In 1955, over 38% of all fed cattle moved through terminal markets, and only

[2] *Cattle Feeding in the United States*, Economic Research Service USDA, Ag. Ec. Rpt. No. 186, p. 47.

12% were sold direct to packers. Today, direct selling (country; private treaty) is most common.

Carcass grade and weight selling of cattle is increasing somewhat in those areas where direct liveweight marketing is dominant, with 38.2% of all cattle sold on this basis in 1990.[3]

Direct liveweight, and grade and carcass weight sales increase as sales lot capacity increases. Commercial feedlots, with their large volume of fed cattle marketing, can attract buyers to the feedlot. Farmer-feeders, particularly the smaller ones, cannot attract buyers to the feedlot; hence, most of them sell through auctions and terminal markets.

Cattle feeders should study their market options carefully to determine which may be most desirable for them, with consideration given to (1) price, (2) marketing costs, and (3) convenience. Additionally, cattle feeders who sell their cattle at the lot will need to keep well informed about market conditions if they are to be on an equal bargaining basis with most cattle buyers.

■ **Factors influencing value of fed cattle**—Three major factors influence the value of slaughter cattle. They are:

1. Quality.
2. Outside fat, not more than ½ in.
3. Retail yield of saleable cuts.

In the past, packers bought cattle on a liveweight basis and sold beef on a carcass weight basis. As a result, only grade and dressing percent were considered in determining value. Additionally, variations in retail yield between carcasses should be considered to price live cattle accurately.

Research studies have shown wide differences in value due to retail yield (or cutout). At the present time, however, too small differentials, are being paid for the higher retail yielding cattle. It is important that the market system develop price differences that accurately reflect cutout value, because producing cattle that have a high percentage of retail cuts will contribute to a more efficient beef industry.

■ **Seasonality of prices**—The seasonality price pattern for fed cattle has not been consistent in recent years. This is because cattle feeding is now a year-round business, and adjustments can be made to increase or decrease marketings in any particular period. Consequently, it is expected that the seasonal pattern for fed slaughter cattle will continue unstable in the future.

■ **Shrinkage**—Shrinkage in market cattle is influenced by a number of factors, especially (1) time off feed and water, and (2) stresses of being hauled and handled.

Several studies have been made of the relationship of shrink to distance hauled. All of these studies show that shrinkage rises rapidly in the first few miles cattle are hauled and continues to increase, but at a decreasing rate as mileage accumulates. The results of one such study are given in Table 27-1.

TABLE 27-1
AVERAGE PERCENTAGE SHRINK OF SLAUGHTER CATTLE IN RELATION TO DISTANCE HAULED[1]

Distance Hauled		Shrink
(mi.)	*(km)*	*(%)*
0–34	*0–55*	1.36
35–70	*56–113*	1.43
71–104	*114–167*	2.07
105–144	*168–232*	2.47
145 & over	*233 & over*	2.83

[1]Henning, G. F., and P. R. Thomas, *Factors Influencing the Shrinkage of Livestock from the Farm to First Market*, AES Bull. 925, Ohio Ag. Exp. Sta., Columbus, Ohio.

The two shrinkage tables in Chapter 15 are designed to make price adjustments to compensate for different shrinkage conditions.

When cattle are sold private treaty, they are typically purchased or sold with a certain percentage shrink from actual weights. This is known as *pencil shrinkage.*

■ **Recommended handling of fed cattle prior to and during shipment**—Market cattle should be handled as follows prior to and during shipment:

1. Stop feeding all feed additives and drugs that require a certain time interval between use and slaughter. *Check directions and recommendations on containers and follow closely.*

2. If cattle are to be weighed at lot or off trucks, continue normal feeding schedule until they are weighed, with the exception of withdrawing additives or drugs as required.

3. If selling on a market where cattle will have access to feed and water before being weighed, as is the case at terminal markets and auctions—

   a. Withdraw protein from ration at least 48 hours prior to shipment.

   b. Reduce grain ration to half feed the day before shipment.

   c. Provide grass hay free-choice during the last 2 or 3 days prior to shipment.

   d. Don't allow animals to eat grain for 12 hours before shipment.

   e. Allow animals to have free access to water until shipment.

[3]*Meat and Poultry Facts 1994*, p. 11, American Meat Institute, Washington, DC.

4. If selling on a carcass grade and weight basis, all feed can be withdrawn 48 hours prior to slaughter. It should not be withdrawn more than 48 hours because there may be some tissue shrink. Animals sold on a carcass grade and weight basis can have free access to water until slaughter.

5. Move slowly and quietly when loading and handling cattle. Don't overcrowd trucks. Gate cattle into compartments on big trucks to take pressure off end animals when truck starts and stops.

INTEGRATE WHEN POSSIBLE

The integration of competitive meat-producing industries, particularly broiler production, has provided tough price competition to the beef industry and is forcing closer cooperation in the cattle industry so as to be able to compete.

There are two types of integration—horizontal and vertical. Horizontal integration is the combining of one or more firms engaged in the same enterprise into a larger firm; for example, when a cattle feeder in California acquires a feedlot in Texas. Vertical integration refers to combining two or more stages of production, processing, and distribution into a single entity; for example, a cow-calf producer entering the feedlot business, or a feedlot operator entering the packing business.

There has been, and will continue to be, some of each type of integration. But the greatest economic advantage can accrue from vertical integration; hence, it will be the dominant type of integration in the future, and it will be discussed herein.

- **Why integrate?**—

. . . The cow-calf producer sells to the stocker operator,
. . . and the stocker operator sells to the feedlot operator,
. . . and the feedlot operator sells to the packer,
. . . and commission and transportation costs multiply,
. . . and the calves shrink and get sick,
. . . and they're given shots on top of shots,
. . . and cattle die!

Indeed, the several ownerships characteristic of the traditional buy-and-sell routine make for a vicious cycle; and it's costly, too. There must be a better way—better for the cattle, and better for the pocketbooks of everybody along the production line. Vertical integration provides some answers.

Three Kansas State University economists estimated that the added costs due to lack of vertical integration—costs for commission, freight, pencil shrink, and extra veterinary and medicine costs due to stress and change in environment—may add nearly $40 to the overhead cost of a steer during his lifetime.

- **Who will do the integrating**—The integration operation could be set up in a number of ways. It could be done—

1. By cow-calf producers, who would retain ownership of their calves all the way through the feedlot.
2. By feedlots, who would buy the calves or get control of them by contract at weaning time.
3. By stocker operators, who would buy the calves and eventually have them finished on a custom basis in a commercial lot.

FEEDING CATTLE

Rations as such are covered later, in Chapter 30; thus, they will not be repeated herein. Rather, points pertaining to feeds and feeding not normally considered *nutritional*, but which affect profits and losses—the business aspects—will be considered here:

1. Consider the moisture level of feed when buying and selling, formulating and mixing the ration, and evaluating feed consumption.
2. Check the accuracy of scales used to weigh feed.
3. Where split loads are used in feeding two or more lots of cattle, make certain that each lot gets the intended amount of feed daily. This is easily accomplished by the use of electronic scales mounted on feed trucks or wagons which are in wide use today.
4. Feed cattle when they want to eat, rather than at the convenience of the feeder. For example, in a very hot climate, in the summertime, cattle may do 80% of their eating at night. In such areas, it is desirable to do feeding late in the evening and early in the morning.
5. The fixed nonfeed costs associated with the ownership of cattle—including interest, labor, equipment, drugs, veterinary expenses, death losses, and taxes—will add 15 to 20¢/lb to the cost of low gains of 1¼ lb/day in comparison with no more than 5 to 6¢/lb where gains of 3 lb/head/day are being obtained.

EVALUATING THE PERFORMANCE

Feeders cannot tell where they are going if they do not know where they have been. Thus, they must keep adequate records. Feedlot managers should have access to records that show them the cost of gain, rate of gain, and feed conversion of each pen of cattle that leaves the lot; the grade and yield of each lot of cattle slaughtered; inventory figures showing cattle and feed inventory on hand every day; death loss by pen or group, and the cause of death; and the air-dry feed consumption per head per day for each group or pen of cattle, as well as an idea of their average for the period.

The use of computers and a suitable software program can greatly facilitate record keeping.

Also, it is recommended that cattle feeders compute their profits or losses on a per head per day basis. This is important whether they own only the cattle, only the feedlot, or both. In any case, the important thing is to make the most profit per head of capacity per day. For example, it is better to make $10 per head on an animal that is fed 110 days ($0.0909/head/day) than to make $30 per head on an animal that is fed a whole year ($0.0822/head/day). The feedlot can also be evaluated on a profit per head per day basis to determine performance. Comparisons can be made in this way in the same lot with different numbers of cattle on feed, and between several lots of different sizes to determine relative performance.

OBTAIN ECONOMIES WITH SIZE

Big feedlots have gotten bigger. This trend will continue. The economies of size associated with cattle feedlots come about through the ability of the larger lots to obtain higher prices and buy at lower costs. Also, in some cases management simply wants a larger operation in order to increase total income.

Studies and practical observation show that larger feedlots have the following advantages in efficiency: they operate more nearly to capacity, with the feedlot fuller; there is greater specialization of labor; they are more highly mechanized; they can outdo smaller lots in buying and selling cattle, feeds, equipment, and supplies; and they have more opportunities to coordinate vertically with firms engaged in feed processing, byproduct feeds, transportation, meat packing, and retail outlets; and they attract more custom feeding.

Some inefficiencies arise with extremely large feedlots, chief among these are: a weakening of coordination among the many feedlot activities; a higher proportion of time needed to travel within the feedlot complex; a tendency to outgrow managerial capacities; and increased problems in manure disposal and pollution control.

(Also see Chapter 26, Fig. 26-2.)

FINANCING CATTLE FEEDLOTS

Big commercial cattle feedlots are now in the "era of money management," where success comes to those who obtain and use most effectively a quantity of borrowed capital.

Financing today's beef factories involves knowledge of, and finances for, fixed costs (feedlot and equipment), feed costs, and cattle costs. Also, it requires the preparation of budgets, and cash flow charts. It generally necessitates outside capital, which calls for knowledge of sources. The business aspects are much the same for all cattle feedlots regardless of size and location, but they must be tailored for each feedlot. Money matters are presented in Chapter 13, Business Aspects of Beef Production; hence, the reader is referred thereto.

Fig. 27-5. Feed mill and silos, major fixed costs of a cattle feedlot. (Courtesy, A. O. Smith Harvestore Products, Inc., Arlington Heights, IL)

CUSTOM (Contract) FEEDING

Custom cattle feeding is the feeding of cattle for a fee, usually without taking ownership of the animals.

Contract feeding is not new. It made rapid development after 1929, and there was much of it during the severe drought of 1934. From this time to World War II, contract feeding decreased in importance—a decline attributed to improved feed conditions on the western range, higher prices for feeder animals, and the availability of more credit through federal and private loan agencies.

Custom cattle feeding as we know it today paralleled the development of commercial feedlots. California pioneered it; thence, it spread to Arizona, the Northwest and other areas of the West, Nebraska, Texas, the Oklahoma Panhandle, and western Kansas. Even today, these are the principal custom feeding areas. Custom feeding provided a means of financing the rapid growth of cattle numbers needed to utilize the highly mechanized, large volume feeding operations. Individuals with limited capital could (1) build and operate a feedlot large enough to perform economically, and (2) acquire the necessary capital and volume by custom feeding cattle for others. In this way, part of the burden of providing capital for efficient operation of a large feedlot was shifted to outside interests.

Capital requirements, periods of severe economic conditions (like scarce money and high interest), times of depressed feeder cattle prices, and adverse pasture conditions, and the growing demand for fed beef caused custom feeding to grow following World War II. These same forces, along with the need for high occupancy (full feedlots) and increased integration, have resulted in further expansion of custom feeding.

Most custom feeders have developed large, highly mechanized, and very efficient plants. Usually, they have on their staffs highly trained consulting nutritionists who are charged with the responsibility of formulating rations and of obtaining maximum gains and feed efficiency at the lowest possible cost. Through custom feeding, they sell the use of their facilities, services, and know-how to cattle owners, usually with profit to each party.

The proportion of custom-fed cattle to cattle owned by the feedlot varies (1) *in period of time*—it increases in times of financial stress (when cattle feeding is not profitable, money is scarce, and interest is high); (2) *according to area*—for example, there is more custom feeding in Texas than in any other state; (3) *according to size of feedlot*—generally speaking, the larger the feedlot, the greater the percentage of custom feeding. Some feedlots do not do any custom feeding whatsoever; others are almost wholly on a custom basis; but most lots have part of each. Feedlots that do both—those in the dual role of custom feeding and owning cattle—vary in the proportion of cattle in each category, but most of them seem to prefer about two-thirds custom-fed cattle and one-third ownership. It's a good bread-and-butter division; in times when fed cattle lose money, such a feedlot has sufficient assured income to pay its bills.

The ownership of custom-fed cattle is diverse. It includes (1) cow-calf producers who wish to retain ownership of the cattle that they produce through the feedlot phase, (2) stocker operators, (3) packers, and (4) investors, including limited partnerships, corporations, cattle buyers, cattle dealers, and others.

PROVISIONS OF A CUSTOM FEEDING CONTRACT

Custom feeding contracts are usually detailed and in writing, for a good understanding is the best way to avoid a misunderstanding. Also, contracts should be fair to both parties—to both the feedlot owner and the cattle owner.

One large custom cattle feeder of the senior author's acquaintance never has a written contract; and he has never had a lawsuit. It is his opinion that written contracts encourage lawsuits.

The experience of feedlot owners and cattle owners, and the difficulties encountered, suggests that custom feeding contracts should include provision for the following:

■ **Ration**—Some basics about the rations—such as the different rations to be used when getting the cattle on full feed, proportion of concentration to roughage, and energy and protein content—should be spelled out in the contract. Yet, the feedlot operator should be permitted flexibility, so as to take advantage of price changes in ingredients, etc.

■ **Chute charge**—*The chute charge is the charge levied by the feedlot for processing cattle through the chute.*

All incoming cattle are processed through a chute at which time they are vaccinated, wormed, dehorned or horn-tipped, castrated, implanted, and treated for lice and grubs by a pour-on systemic, an injectable, or dipping. Unless they are branded or otherwise tagged, they run the risk of getting mixed up and difficult to identify.

Lighter cattle are often rerun through the chute at a later date for a second round of vaccinations and/or for reimplanting part way through the feeding period.

The chute charge varies according to cattle and products used. Generally, it runs from $5 to $8 per head for the initial processing and from $3 to $5 per head for reworking. The chute charge rate should be explained to the custom cattle owner.

■ **Veterinary and medication expenses**—This refers to the cost of caring for sick cattle, which are generally moved into a doctoring area where they receive special treatment and medication. These charges are generally handled in one or more of the following ways: (1) as a direct charge based on the type and amount of medication used; (2) as a prorated charge per head per month on all cattle, based on historical costs for such services and medications; (3) as part of the ration price mark-up.

Computers are now being used by many custom feedyards to keep sick pen records, and to provide the data for direct charge of medicines and a history of treatment and deaths of all cattle in the lot. Both managers and owners of custom fed cattle need this information. Several companies that sell cattle health products have developed software programs suitable for this purpose.

Usually veterinary and medication costs run from $5 to $8 per head during the feeding period. However, a high incidence of sickness can escalate such costs. The basis of charging for veterinary and medication expenses should be specified.

■ **Responsibility for death losses**—The contract should specify responsibility for death losses. The cattle owner normally assumes all losses prior to the arrival of the cattle at the feedlot.

In most of the larger custom feedyards, the cattle owner stands any death losses, unless they are the fault of the feedyard. Smaller lots and those specializing in calf growing often (1) guarantee the maximum death loss that the owner must stand, or (2) participate in the death loss by refunding the cost of feed and services on dead animals.

The death losses of yearling cattle in the feedlot generally do not average over 1%, but the death losses on calves will vary, depending on the backgrounding or preconditioning that the calves received prior to coming to the feedyard. Some feedyards have rigid requirements relative to calf backgrounding or preconditioning which must be met before they will accept the cattle.

Excessive doctoring of sick cattle can place a real strain on feedyard personnel and operational costs.

■ **Buying and selling services**—The buying and selling services provided by the feedlot are important to the success of the operation. In most cases, the feedlot is in the best position to sell the finished cattle. But this responsibility should be spelled out in the contract. If any charges are to be made for buying and selling, these should be made known in advance.

■ **Right to reject poor-doing cattle**—A feedlot using a payment-for-gain contract should reserve the right to reject poor-doing cattle. Of course, where the feedlot purchases the cattle for the owner, the feedlot accepts the responsibility to get good-doing cattle.

■ **Power of attorney for feedyard operator**—Under certain types of feeding contracts, the feedyard may need power of attorney for buying, selling, and borrowing for customers.

■ **Arbitration**—In cases of dispute and disagreement over a custom feeding contract that cannot be resolved by the owner of the cattle and the owner of the feedlot, the contract should provide for arbitration to be conducted by a committee of three—each party to the contract choosing a representative and these two then choosing a third party, to study the case and recommend settlement.

TYPES OF CUSTOM FEEDING CONTRACTS

The services rendered vary from feedlot to feedlot and according to the type of contract. In some instances, the services may be so complete that the customer never sees the cattle. The feedlot operator may buy the feeder cattle, feed them, market them, and send the customer (the client) a check for the balance, after deducting input costs, interest charges, and custom feeding charges. Less complete services are usually available to suit the customer.

Both the feedlot owner and the cattle owner should analyze different types of contracts and determine which best fits their respective circumstances. Some feedlots offer several types of contracts, thereby according the cattle owner a choice.

Competition may dictate the type of contract and the charges made. But by knowing the variables and managing them correctly, the feedlot owner can write and carry out a contract that will be fair to both the feedlot owner and the custom feeder.

Generally speaking, contracts with fixed charges are the most satisfactory and the most common, primarily because there is less room for misunderstanding.

Although there are many types of custom cattle feeding contracts, and many variations of each kind exist, most of them can be classified under one of the following types:

■ **Feed cost plus daily yardage fee per head**—This type of contract is based on the cost of feed plus an additional 20 to 30¢ per head per day to cover handling, yardage, feed grinding, and similar expenses. Generally, additional charges are made to cover chute handling and medication.

With this type of contract, the feedlot does not assume any risk whatsoever. It is merely its intent to sell feed, facilities, and services at an agreed price.

■ **Feed cost plus markup**—This type of contract calls for reimbursement on cost of feed plus a feed markup on either (1) a flat rate or (2) a percent of cost.

With a flat rate markup, a $20 to $30 charge per ton above feed cost is made to cover feed handling, grinding, and labor costs. An additional assessment is made to cover chute charges and medication; and the customer finances the purchase of cattle. Since actual feed milling costs (for labor, power, insurance, mill maintenance, etc.) run $9 to $15 per ton, profit to the feedlot accrues from having a higher markup than the milling cost. A flat markup per ton of feed favors the feedlot when prices fall, and the cattle owner when feed prices rise.

Also, markup on feed may be on a percentage of cost basis. With this arrangement, higher feed costs favor the feedlot, whereas lower feed costs reduce the actual return to the feedlot.

Any system of feed markup will be more profitable to the feedlot with heavy, high-performing cattle than with light, slow-gaining cattle, simply because the former eat more.

With the *feed cost plus markup contract*, the feedlot is essentially a feed manufacturer processing and delivering feed to its customers—the owners of the cattle.

■ **Feed cost plus (1) daily yardage fee per head and (2) markup per ton of feed**—This is a combination of the first two types. Those feedlots that charge the higher yardage rates per head daily add a smaller markup per ton of feed above actual ingredient prices; conversely, those that charge the lower yardage rates per head per day make a higher feed charge over and above actual ingredient cost. As a rule of thumb, feed markup is generally lowered by $1 per ton for each 1¢ per head per day of yardage charged. At the present time, it appears that the lower yardage cost and the higher feed markup is the favored basis; primarily because (1) the owner of the cattle is less inclined to object to such charges, and (2) increased competition has driven custom feeders to make their charges on the least conspicuous basis.

■ **Agreement to purchase contract**—In this plan the cattle feedlot operator buys the feeder cattle, usually with the client required to make a down payment of 20 to 30% of the purchase price. The client then executes an agreement to buy the cattle when they are ready for slaughter, including the original purchase price of the feeder cattle (less any down payment made) and all feeding, handling, and interest charges. (Interest charges are tax deductible.)

There are several variations of this type of contract. But all of them are much like buying commodities on the Chicago Mercantile Exchange.

■ **Payment for weight gained**—This plan is based on a charge per pound of gain. In this arrangement, the feeder is reimbursed on the basis of the gain in weight put on the cattle, at an agreed price of so many dollars per hundred. This type of contract has decreased in importance in recent years, because it frequently results in poor owner-feeder relations—due primarily to the following reasons: (1) it is impossible in advance of feeding to detect those lots of animals that will be *poor doers* because of such factors as nervousness, nutritional deficiencies, diseases, and/or parasites; (2) the longer the feeding period, the greater the cost of gains, and the length of the feeding period is seldom stipulated in such contracts; (3) because weather and disease, which are uncontrollable, affect rate of gain; and (4) the amount of fill or shrinkage when weighing animals in and out of the feedlot is of great importance to both the owner and the feeder, and a source of argument. Also, a major disadvantage of using a payment-for-gain system is that it does not adjust for feed costs; the feeder must absorb any increase in feed cost. However it does allow a feedlot to take advantage of opportunity feeds—such as down corn, or grass which cannot be delivered to the feedlot. Also, the small feedlot operator who does not have scales may use this type of custom contract.

Payment for weight gained may be used in growing and backgrounding operations, where it may be desirable to specify both the minimum and maximum rate of gain. Also, it may be used on cattle that are pastured for a time before being sent to the feedlot.

OPERATION PROCEDURES

The operation procedures followed by feedlots that custom feed vary. In particular there should be an understanding between the feedlot and the customer relative to billing and payment of feed costs, selling arrangement, and payment for slaughter cattle.

■ **Billing for feed or gain costs**—Most feedlots bill their clients for feed or gain costs either on the 1st and 15th of each month, at the end of the month, or at the end of the feeding period. The more frequent the billing, the smaller the amount of short-term capital required by the feedlot. Where feedlots carry feed costs longer than a month, they usually charge interest.

The vast majority of custom feedlots worthy of the name are equipped with platform or hopper scales and/or scale trucks with electric or mechanical scales mounted under the feeding box; hence, billings are based on actual weights. Occasionally, a farmer without scales will feed a few cattle on a custom basis. In the latter case, the feedlot owner and the cattle owner may agree to (1) compute feed costs on the basis of 3% of the incoming body weight of the cattle plus 7 lb per day (for example, the daily feed consumption of a steer weighing 700 lb at the time of delivery to the yard would be estimated at 28 lb; or (2) feed on a cost-of-gain basis, with payment delayed until the cattle are weighed and marketed at the end of the feeding period.

■ **Selling arrangement**—Feedlot managers generally handle the selling of custom-fed cattle in the same manner, and through the same channels, as cattle which the feedlot owns. Feedlot managers, or their representatives, are usually in a better position to estimate the weights and grades of cattle on feed than are their clients. Also, they are more familiar with the type and quality of fed cattle desired by various packer-buyers, and they are in a stronger position to bargain for the best possible price.

■ **Payment for custom-fed cattle**—Market payments for custom-fed cattle may be made directly to the owner of the cattle or to the feedlot, depending on prior arrangements between the feedlot and the client. However, commercial banks and other lending institutions generally retain a first lien on the client's cattle; and this must be satisfied. Likewise, finance agencies generally provide the necessary financing for feed and other custom feeding charges; hence, the feedlot operator ordinarily is assured of receiving full payment for feed bills and other services. In the event the client has outstanding bills with the custom feeder, feedlots handling payments for their clients are permitted to retain sufficient funds to satisfy these debts, after satisfying the first mortgage holder.

Some custom feedyards register with the Packers and Stockyards Administration and carry a registered account for the shipper's proceeds. This requires posting a bond and making an annual report to the P&S Administration. Although many feedyards have resisted this procedure, it is very professional because (1) of the large amounts of money handled, and (2) it establishes a clear separation of the client's proceeds from the operating funds of the feedyard. A good understanding between the client and the operator of the feedyard pertaining to the method of handling proceeds is very important.

WHAT CATTLE OWNERS EXPECT OF CUSTOM FEEDLOTS

The owner who entrusts valuable cattle to a feedlot for custom feeding rightfully expects certain things of the feedlot, chief of which are:

- **Profits**—Cattle owners assume considerable risk; hence, they expect profits commensurate therewith. They will contract with the feedlot which consistently returns the most on their investment.

- **Cattle feeding know-how, honesty, and integrity**—Cattle owners will not, knowingly, place cattle in a custom feedlot under inexperienced management. Also, because cattle owners are not involved in day-to-day management, the honesty and integrity of the custom feedlot management are very important to them.

- **Progress reports**—Cattle owners wish to keep informed of how well their investment is doing. To this end, they expect detailed monthly feed bills showing pounds of feed fed, average daily consumption, total cost per day, and death and sickness costs.

  At the end of the feeding period, a detailed feeding summary should be given to the owner of the cattle, giving average daily gain, cost per pound of gain, conversion ratios, and profit or loss summary.

- **Courteous customer treatment**—Like any other customer, a custom cattle owner likes to feel that the feedlot wants and appreciates the business.

- **Competitiveness**—The cattle owner expects the feedlot in which cattle are placed for custom feeding to be competitive in charges and performance with other feedlots of the area.

- **Satisfactory financial position**—The customer expects the feedlot to be in sufficiently strong financial position that it buys feeds when it is most advantageous (usually at harvest), and that it pays its bills regularly.

- **Knowledge of the financial position of packer-buyers**—The cattle owner expects that the feedlot owner know the financial position of the packer-buyer to whom cattle are sold, so that there can be no concern relative to payment. A few cattle owners who custom feed even require that the feedlot handling the sale of the finished cattle guarantee payment for them.

HOW TO ATTRACT CUSTOM FEEDERS

Success in attracting customers depends on reputation and performance. Reputation will attract new customers, but only performance will hold them.

If a feedlot relies on custom-fed cattle for a certain percentage of its capacity, it's important that there be enough clients with sufficient cattle to keep the lots filled, or nearly so, at all times.

- **Be successful**—The best way to recruit and retain customers is to be successful. After a customer has used the services of a given feedlot, its record will be known. Until then, the feedlot manager will have to convince the prospective customer of performance. Among the tools which the manager may use in proving ability are records and computers.

- **A complete set of records**—A good set of records will allow the feedlot manager to predict with confidence what can be done for the prospective custom feeding client in rate, efficiency, and cost of gain; in disease and death losses; and in buying and selling cattle.

- **A computerized system**—Prospective clients will be impressed with the sophistication of a computerized accounting system and a computerized closeout statement. It indicates the feedlot's capability of complete, accurate, and prompt records. In addition to being a good means of keeping customers well informed, a computerized system may be used to provide them with an accurate and complete set of records for accounting purposes. Also, records are useful to the feedlot, because they provide periodic analysis of progress and allow the manager to monitor the feeding program for needed changes.

- **Financing cattle and feed fed on a custom basis**—Financing helps to attract clients. Since feedlot operators are under pressure to keep their feedlots filled, financing of cattle and feed by custom operators will probably continue to be an important source of funds to those who place cattle in these lots.

  Commercial banks are the primary source of financing for cattle fed on a custom basis. They generally require a margin equivalent to 25 to 30% of the value of the feeder cattle. In addition, banks make loans to cover feeding charges. It is not uncommon for banks to finance 70% of the feeder cattle price plus all of the feeding charges. Depending on the reputation of the feedlot operator and the custom feeder, banks and other lending institutions may secure only the cattle as collateral for the loan. They may also specify that feeder cattle be hedged on a futures market before negotiating loans, although this has not been general practice to date. Some lending agencies will even advance funds for the margin on cattle futures contracts, as part of a feeder cattle loan.

Also, some custom feedlots finance cattle purchases in their own names, with the client executing an *agreement to purchase* the cattle when they are sold at a cost equal to the initial purchase price of the feeder plus all feeding, handling, and interest charges.

- **Bring prospective feeding customers together**—The feedlot manager may attract some customers by bringing them together, especially those who have insufficient funds to carry on a continuous feeding program by themselves. The feedlot manager can provide a real service by assisting them in securing funds.

COMPUTERS IN CATTLE FEEDLOTS

Ration formulation is only one use of computers in cattle feedlots. As every cattle producer knows, many management decisions are involved in obtaining the highest net returns from a given lot of cattle; and the bigger and the more complex the feedlot, the more important it is that the decisions be right.

Problems can be solved without the aid of a computer. But the machine has the distinct advantages of (1) speed, (2) coming up with answers to each of several problems simultaneously, and (3) offering the best single alternative, all factors considered. Among the possible uses of computers in cattle feedlots, particularly the largest operations, are:

1. Ration formulation (see Chapter 8).
2. How to determine the best ingredient buy.
3. The most profitable ration—the one that costs the least per pound of gain produced. In many cases, the cheapest ration will actually increase cost of gain. Also, rations that produce the most rapid gains may be too costly; for example, silage rations will not produce as high gains as an all-concentrate ration, yet the net profit from their use may be greater.
4. The most profitable kind of cattle—age, weight, sex, and grade—to feed in relation to available feeds and feed prices. (See Chapter 29, Kind of Cattle to Feed.)
5. Seasonal differences in performance of cattle in a given feedlot.
6. As a means of forecasting profits or losses; with all-at-once consideration given to feeder prices, slaughter cattle outlook, probable rate of gain, and interest and overhead costs.
7. As a means of keeping and updating the voluminous daily feed transfers from feed inventories, to mixed rations, to records for each lot of cattle. Accurate and current feed inventories are necessary for wise ingredient buying and for financing feed inventories; and accurate and current feed records by lots are important for both privately owned and custom-fed cattle.

AN INCENTIVE BASIS FOR THE HELP

An incentive basis for cattle feedlot help is needed for motivation purposes, just as it is in cow-calf operations. It is the most effective way in which to lessen absenteeism, poor processing and mixing of feeds, irregular and careless feeding, unsanitary troughs and water, sickness, shrinkage, and other profit-sapping factors.

Modern cattle feedlots are big business; cattle must be cared for seven days a week; and each assignment must be done well and on time. Generally, cattle feedlots are year-round operations; so, they can provide their employees with steady work and advancement. Additionally, dedication, a team approach, and pride are important. Management strives to obtain, and enhance, these traits in their help through the following approaches:

1. Selecting new employees with great care.
2. Providing a procedural manual, which includes an organization chart and a job description of each job.
3. Reviewing the performance of each employee annually, with a healthy exchange of work habits, performance, and future goals.
4. Providing the standard fringe benefits of paid vacations, health programs, and retirement and/or profit sharing programs.
5. Adopting and using the principles of total management. Although such a program must be tailored to each operation, the following provisions are generally common to all of them:
    a. Ask why, not who. Train all employees, then make them active, participating members of the team.
    b. Resolve problems by a team approach. Involve those who do the actual job.
    c. Provide methods of measuring the performance of each job. If it is worth doing, it is

worth measuring. Without measuring, management has no objective way in which to evaluate the progress made.

d. Make every employee a quality control person. Don't just dictate from the top down. The worker on the job often sees a solution that management may miss.

e. Make bonuses, salary increases, and profit sharing on an objective basis as determined by measuring.

Note: A Feedlot Quality Award Program is now being sponsored by the National Cattlemen's Beef Association. Information about this award may be obtained by writing to the National Cattlemen's Beef Association, 5420 S. Quebec St., Englewood, CO 80111.

Fig. 27-6. Chicago Mercantile Exchange (CME) trading floor. (Courtesy, CME, Chicago, IL)

FUTURES TRADING IN FINISHED CATTLE, FEEDER CATTLE, AND FEED[4]

The three big uncertainties in the cattle feeding business, any one of which can cause a cattle feeder to suffer heavy losses, are prices of (1) feeder cattle, (2) feed, and (3) finished cattle. Through futures contracts, the cattle feeder can now hedge all three. In advance of feeding, the price of feeder cattle, feed, and finished cattle can be locked in.

This discussion is devoted primarily to live (slaughter) beef cattle futures as they apply to cattle feedlot operators, because it is the highest risk phase of the cattle business, as well as the least flexible. Unless feeders contract ahead, they have no assurance of what their finished cattle will bring when they are ready to go. Moreover, there is little flexibility at market time, for the reason that excess finish is costly and unwanted by the consumer. As a result of this market price uncertainty, and in realization of the high risks involved, sleepless nights are rather commonplace among cattle feeders; they find it difficult to concentrate on the business at hand—the efficient feeding and management of cattle. Live (slaughter) beef cattle futures provide a means through which cattle feeders can fix their selling price at the outset of the feeding period.

The second major item of the triumvirate making for uncertainties in cattle feeding is the price of feeder cattle. Only by contracting ahead can cattle feeders be sure of the price that they will have to pay when they are ready to lay in feeder cattle. For many years, a fairly effective, albeit unorganized, cash contracting system has been operating relative to feeder cattle. Feeder cattle futures now offer, on an organized basis, a method for cattle feeders to lock in the price of feeder cattle well ahead of taking delivery, thereby alleviating possible heavy losses due to sharp price rises of feeder cattle. Without feeder cattle, a feedlot is not in business. Yet, much of the overhead cost for facilities and staff continues. Hence, a full feedlot is important. Cow-calf producers—the producers of feeder cattle—have more flexibility, and are less dependent on contracting ahead, than cattle feeders. If the feeder cattle market isn't good, they can hold their calf crop for a time; they may even carry them over for another year—to the yearling stage. Also, rather than accept what they consider to be an unfavorable price for their stockers, they can have them custom fed, or they can feed them out themselves. By retaining ownership for a longer period of time, they increase the probability of being able to price their cattle at a profit. Certainly, there are risks in the cow-calf business, but, in comparison with cattle feeding, there is more flexibility, and the timing is not so exacting.

Since feed represents such a large proportion of the cost of feeding cattle (amounting to approximately 70% of the costs exclusive of the purchase price of the feeder cattle), it is wise to set the price months in advance whenever possible. Usually, feed can be bought most advantageously at harvest time. Thus, cattle feedlot owners who have adequate storage and finances generally buy their main feed ingredients at that time. By so doing, they can project with reasonable accuracy what it will cost them to feed cattle. Corn and soybean meal futures permit the cattle feeder to accomplish the same thing without actually taking delivery on the feed and incurring storage costs and risks of physical deterioration. The cattle feeder can

[4]This entire section on beef futures was authoritatively reviewed by and helpful suggestions were received from J. Graham, Director Commodity Marketing and Education, Chicago Mercantile Exchange, Chicago, IL; and R. E. Sheldon, Manager- Agriculture Group, Economic Analysis and Planning Dept., Chicago Board of Trade, Chicago, IL.

use such futures to protect against increases in feed prices. The primary exchange dealing in grains and other feed-related commodities is the Chicago Board of Trade.

WHAT IS FUTURES TRADING?

Futures trading is not new. It is a well-accepted, century-old procedure used in many commodities; for protecting profits, stabilizing prices, and smoothing out the flow of merchandise. For example, it has long been an integral part of the grain industry; grain elevators, flour millers, feed manufacturers, and others have used it to protect themselves against losses due to price fluctuations. Also, a number of livestock products—hides, tallow, frozen pork bellies, and hams—were traded on the futures market before the advent of beef futures. Many livestock producers and processors prefer to forego the possibility of making a highly speculative profit in favor of earning a normal, but safe, margin through efficient operation of their business. They look to futures markets to provide (1) an insurance medium in the marketing field, and (2) the facilities and machinery for underwriting price risks.

A commodity exchange is a place where buyers and sellers meet on an organized market and transact business, without the physical presence of the commodity. The exchange neither buys nor sells; rather, it provides the facilities, establishes rules, serves as a clearing house, holds the margin money deposited by both buyers and sellers, and guarantees delivery on all contracts. Buyers and sellers either trade on their own account or are represented by brokerage firms.

The unique characteristic of futures markets is that trading is in terms of contracts to deliver or to take delivery, rather than on the immediate transfer of the physical commodity. In practice, however, very few contracts are held for delivery. The vast majority of them are canceled by offsetting transactions made before the delivery date.

Many cow-calf producers have long forward contracted their calves for future delivery without the medium of an exchange. They contract to sell and deliver to a buyer a certain number and kind of calves at an agreed upon price and place. Hence, the risk of loss from a decrease in price after the contract is signed is shifted to the buyer; by the same token, the seller foregoes the possibility of a price rise. In reality, such contracting is a form of futures trading. Unlike futures trading on an exchange, however, actual delivery of the cattle is a must. Also, such privately arranged contracts are not always available, the terms may not be acceptable, and the only recourse to default on the contract is a lawsuit. By contrast, futures contracts are readily available and easily disposed.

Livestock futures trading is relatively new. It was not until November 1964 that trading in live cattle futures opened on the Chicago Mercantile Exchange CME). Trading in live hogs began 15 months later. Since then, livestock futures trading has grown enormously.

WHAT IS A FUTURES CATTLE CONTRACT?

A futures contract is a standardized, legally binding transaction in which both parties promise to buy and sell a specified quantity and type of a commodity at a specified location(s) during a specified future month. The buying and selling are done through a third party (an exchange clearing member) so that the buyer and seller remain anonymous; the validity of the contract is guaranteed by all exchange clearing member firms, *i.e.*, the clearing house; and either buyer or seller can readily liquidate their positions by simply offsetting sale or purchase.

The Chicago Mercantile Exchange (CME) specifications of finished cattle and feeder cattle contracts follow:

- **Specifications for a live (slaughter) cattle contract are**—Stockyard or approved packing plant delivery of 40,000 lb of 55% Choice/45% Select USDA grade live steers (approximately 37 head) with 63% hot yield and weighing an average 1,050–1,250 lb live (stockyard delivery) or 600–900 lb carcasses (packing plant delivery). Units with variations in grade or hot yield from those above can be delivered at market-based premiums or discounts. Sellers may tender for delivery to Sioux City, IA; Omaha, Norfolk, North Platte, and Ogallala, NE; Dodge City and Pratt, KS; Amarillo, TX; Guymon, OK; and Clovis, NM; and buyers may take delivery at these stockyards, or at an approved packing plant within 150 miles of the stockyard, or the originating feedlot.

  Note: Under the current specifications, cattle futures may be settled (delivery against the contract made) on either a live cattle or carcass basis.

- **Specifications for a feeder cattle contract are**—50,000 lb of feeder steers averaging 700–800 lb (approximately 65 head), consisting of steers that will grade 60 to 80% Choice when fed to slaughter weight. The contract, on the last trading day, is cash-settled to the CME Weighted Average Price (CWAP), a volume weighted price from USDA reported steer sales within a 12-state central region.

COMMISSION FEES AND MARGIN REQUIREMENTS ON CATTLE CONTRACTS

The commission fee on all futures contracts covering both purchase and sale (called a round turn) is negotiable between the brokerage firm and customer. In 1995, the minimum hedge margin on live cattle was $450, and the speculative margin was $600. On feeders, the speculative margin was $600 and the hedge margin was $450. The margin deposit may be increased by the broker if the volatility of cattle prices should increase.

HEDGING/SPECULATING

Traditionally, futures contracts have been used for two purposes: (1) hedging, and (2) speculating.

The risks accepted by farmers take different forms; among them, loss of animals by disease, fire, lightning, and theft. Protection from most of these losses can be provided by livestock insurance. But insurance companies do not have policies that cover the loss in animal values that may occur due to price change. Fortunately, through the practice of hedging on commodities futures exchanges, it is possible to reduce risk and uncertainty due to falling prices.

Hedging is the purchase or sale of a futures contract as a temporary substitute for a merchandising transaction which will be made at a later date. Usually this involves opposite transactions in the futures market from those made in the cash market. Since the price movements in the two markets are related, it is anticipated that any loss in one market will be at least partially offset by a gain in the other, with the result that loss through price change will be reduced.

The purpose of the hedge is to protect the merchandising profit anticipated by a handler of a commodity. Merchandising profit is distinguished from speculative profit in that the merchandising profit results from producing and marketing the actual commodity (like finishing and marketing cattle), whereas speculative profit results solely from changes in price and is not the result of a producing or marketing function. Many cattle feeders prefer to leave the assumption of price risk to some other person who is interested in a product in the hope of reselling it at a profit without changing the nature of the product. This other person is referred to as a speculator. *A speculator is a person who is willing to accept the risks associated with price changes in the hope of profiting from increases or decreases in futures prices.* By assuming the risks of price change, the speculator provides many valuable economic functions, such as giving the market both liquidity and continuity.

BASIS

The essence of profit and loss in hedging of a commodity, like live cattle and feeder cattle, is the accurate calculation of the *basis* for a particular delivery month. *The difference between cash and futures prices is called the basis.* The difference between your own local cash price and the future price on sale day is *your basis.* The basis is the most important single factor in hedging regardless of the particular commodity involved.

With futures contracts such as those for live cattle and feeder cattle, which are continuously produced, nonstorable commodities, the relationship of the cash price to the futures price has relatively little meaning except during the contract month. Hence, *accurate estimation of the basis for a particular delivery month* is most important in effective hedging of live cattle and feeder cattle. If the estimated basis turns out to be the actual basis on sale day, a perfect price-protecting hedge is the result. However, if the estimated basis is incorrect, there will be a slight gain or loss from the expected results estimated earlier.

Basis variation usually represents an identifiable pattern which repeats itself from year to year and is mostly explainable by economic factors. Hence, accurate estimation of the basis for some future point in time, even if the basis varies during that time, is the key to successful hedging.

There are two methods of determining the basis for any local market: (1) historic price relationships, and (2) cost of delivery.

To calculate the basis with the first method, one must obtain past futures price data and compare those prices to one's local cash market prices. Hence, if one were calculating the basis for live cattle at Kansas City, Missouri, one might find that Kansas City cash prices have normally been 50¢ per cwt below the futures prices at the Chicago Mercantile Exchange in the delivery month. The basis would be—50¢.

To calculate the basis with the second method, one must obtain the actual cost of transporting the cattle from a local market to the delivery point designated by the futures contract; hence, to calculate the basis between Kansas City and Omaha, via the cost method, one should estimate the transportation cost (including shrink), interest charges, insurance charges, and the like.

There are many factors that cause the basis for any local market to vary over a period of time. These include such things as changes in local supply-demand factors, changes in local production costs, the size of a future crop, government programs, and local market receipts.

In established markets, *basis patterns between markets tend to repeat themselves from one year to the next.* Hence, experienced traders know that their local basis tends to be at a certain level during particular times of the year. The repetition of these patterns from one year to the next makes basis prediction more reliable than price prediction.

To operate successfully, hedgers must develop their own series of data. There is no shortcut to success. Only by keeping data will the hedger be familiar (day to day) with the basis situation, with the factors affecting basis, and with comparable basis situations at other times. Only in this way will the hedger have the hedging information pertinent to the hedging location.

The basis reflects the market. When the basis widens, it means that everybody is selling. When the basis narrows, the market is telling you to move your commodity to market.

The livestock producer (a cattle feeder) should be knowledgeable relative to the basis for that area before attempting to figure what price can be locked in by using futures.

EXAMPLES OF FUTURES HEDGING STRATEGIES

Examples of futures hedging by (1) a cattle feeder, (2) a cow-calf producer, (3) a packer, and (4) a cattle feeder using a long hedge to protect the price of feeder cattle replacements at the time of forward contracting, finished cattle follow. These illustrate hedging procedures, although it must be borne in mind that in actual application the hedges may not work out as perfectly as these.

■ **Example 1: A cattle feeder hedging to lock in price (see Table 27-2)**—It is now November, and the cattle feeder has just purchased feeder cattle to place in the feedlot. Based on past experience the feeder is quite confident that these cattle should be ready for market the following April. Through good record keeping, the feeder is also quite confident that production (including labor) and marketing costs should be about $67.50/cwt.

The cattle owner-feeder decides to hedge the cattle with the April futures contract which at the time is selling for $72.45/cwt. The feeder has also estimated the basis will be about $1.00/cwt in April. So, this figure is subtracted from the April futures price resulting in a localized futures price of $71.45/cwt or an estimated $3.95/cwt profit; hence, the feeder sells April futures.

The cattle feeder sold finished cattle on the cash market for $67.35/cwt which, after subtracting production costs of $67.50/cwt, gives a loss of $0.15/cwt. However, the feeder realized a profit of $3.60/cwt on the futures transactions, so that the total profit was $3.45/cwt.

This example illustrates what could happen on a declining market. The feeder still showed a profit, even though the cattle had to be sold in the cash market for a price lower than production costs because this loss was offset by a larger profit in the futures market. This is true because, as the cash price declined, the futures prices also declined.

If, however, the cash and futures prices had risen, the cattle feeder still could have made a profit, this time in the cash market. But because of a loss in the futures market, the total profit would have been less than had the feeder not hedged. Nevertheless, the feeder still received the price protection desired, which was the main purpose in hedging.

■ **Example 2: A cow-calf producer using a short hedge (see Table 27-3)**—During April, as the producer's calves are being born, a rancher decides to hedge these calves on a feeder cattle contract. Most of the calves will be sold as feeders during October. Through experience the cow-calf operator has estimated that it cost $78.25/cwt to produce these feeders. The futures market is showing October feeder cattle

TABLE 27-2
EXAMPLE OF A CATTLE FEEDER USING A SHORT HEDGE TO LOCK IN A PRICE

Cash Market	Futures Market	Basis
per cwt	per cwt	per cwt
Nov. 15:		
Expects to receive in April $71.45	Sells April futures at . . $72.45	–$1.00 Expected
April 10:		
Sells cattle on cash market at $67.35	Buys April futures at . . $68.85	–$1.50 Actual
Futures gain . . . $ 3.60		
Realized price . . . $70.95	Gain . . . . $ 3.60	Loss $0.50

TABLE 27-3
EXAMPLE OF A COW-CALF PRODUCER HEDGING TO LOCK IN A PRICE

Cash Market	Futures Market	Basis
per cwt	per cwt	per cwt
April 25:		
Expects to receive in Oct. $89.00	Sells Oct. futures at . . . $83.25	+$5.75 Expected
Oct. 10:		
Sells feeder cattle on cash market at $86.00	Buys Oct. futures at . . $80.00	+$6.00 Actual
Futures gain . . $ 3.25		
Realized price . . $89.25	Gain . . . . $ 3.25	Gain $0.25

at $83.25/cwt or at a localized price of $89/cwt. The rancher feels that this assures a reasonable profit; hence, the rancher sells October futures.

Even though the cash market was not as strong as the rancher had hoped, the desired profit was realized because of the hedge.

■ **Example 3: A meat-packer using a long hedge (see Table 27-4)**—The above examples are illustrations of short hedges. The following example will be of a long hedge, where the futures contract is bought.

The meatpacker has determined, from basis charts, that the February cattle futures are normally $1/cwt above the local cash price in February. This, then, assures the packer of the maximum amount necessary to pay for cattle in February—that is, if the basis does narrow to $1 during February, the cost of the slaughter cattle will be $1 below the futures price in February.

The meatpacker has also determined that the most that can be paid for the slaughter cattle and still make a profit is $70/cwt. Also, the packer is confident that the basis will be $1 in February.

The profit in the futures market of $7.20/cwt, when applied to the higher than expected cash prices, assured the meatpacker that slaughter cattle could be purchased at a price that allowed protecting the profit margin.

TABLE 27-4
EXAMPLE OF A PACKER USING A LONG HEDGE

Cash Market	Futures Market	Basis
per cwt	per cwt	per cwt
Sept. 27:		
Expects to pay in Feb. $69.80	Buys amount needed of Feb. futures at $70.80	−$1.00 Expected
Feb. 20:		
Buys slaughter cattle at $77.00	Sells Feb. futures at . . $78.00	−$1.00 Actual
Futures gain . . −$ 7.20	______	
Realized purchase price $69.80	Gain . . . . $ 7.20	

■ **Example 4: A cattle feeder hedging on feeder replacements (see Table 27-5)**—It is not uncommon for a feeder to contract slaughter cattle for future delivery at a set price to a packer before acquiring the necessary feeder cattle. If by the time of purchase the price of feeders increases beyond the feeder's expectations, the feeding margin may be substantially reduced or the feeder may even suffer a loss on the contract.

TABLE 27-5
A LONG HEDGE TO PROTECT FEEDING MARGIN ON FORWARD CONTRACTED FED CATTLE

Cash Market	Futures Market	Basis
per cwt	per cwt	per cwt
Jan. 15:		
Expects to pay in Feb. $85.60	Buys Feb. futures at . . $84.75	+$0.85 Expected
Feb. 15:		
Buys feeder cattle at $89.50	Sells Feb. futures at . . $89.15	+$0.35 Actual
Futures gain . . −$ 4.40	______	______
Realized purchase price $85.10	Gain . . . . $ 4.40	Gain $0.50

The futures contract in feeder cattle can be used in a long hedge to reduce or eliminate the risk involved in an adverse movement of feeder cattle prices. At the time that the feeder negotiated a forward contract with the packer, the feeder would buy feeder futures contracts, preferably for the month that it is actually planned to buy the feeders. If the price of the futures followed local market prices, the feeder would not care what happened to the level of feeder cattle prices because any loss or gain in the cash market would be offset by an opposite outcome in futures.

The arithmetic of this particular long hedge is illustrated in Table 27-5.

If the feeder had not hedged, it would have been necessary to pay $3.90/cwt more for feeder cattle than had been planned. By hedging, the feeder paid $85.10, plus hedging costs, or about what the contract price was based on in January.

It is advisable to place such a hedge in the futures contract month in which the cattle will be purchased because the cash-futures relationship is more predictable at this time.

A long hedge to fix the price of feeder cattle may be used in lieu of forward contracting to fix or cheapen the price of feeder cattle needed for replacements several weeks or months in advance of actual purchase. However, the same businesslike procedures are required to get a good buy in feeder futures as in forward contracting. The most likely time to buy futures is when the particular contract in which the feeder is interested is favorably priced relative to local feeder prices.

If futures and cash prices move together closely, the feeder can buy cattle locally when they are needed at about the price prevailing when the long hedge was placed. If futures advance relative to local cash, the feeder can buy cattle as needed and cheapen their cost by the gain in the futures transaction.

DELIVERY AGAINST THE CONTRACT

Although very few contracts, usually fewer than 3%, are consummated by actual delivery of the commodity, a hedger should consider delivery as one alternative, particularly when the cash and futures prices are out of line with each other. However, due consideration must be given to the costs of delivering or receiving delivery, since such costs may be of such magnitude as to offset the differences between the cash and futures prices.

It is not the function of the futures market to provide an alternative source of supply nor an alternate means of disposal of surplus commodities. The purpose of delivery is merely to serve as a safeguard to be used when all else fails.

FACTS ABOUT FUTURES CONTRACTS

A cardinal feature of any workable futures contract—whether it be steers, grain, or any other commodity—is that there shall be maintained a solid connection with the commodity; that is, cash and futures must be tied together.

Any contract held until maturity must be delivered or settled to an index of cash prices. This keeps the futures price in line with the cash price at the livestock market.

During the delivery month the cash and futures markets tend to come together at the point of delivery. If this were not so, traders would quickly take advantage of the situation. For example, if prior to the termination of trading on August live cattle futures, the price of U.S. slaughter steers on the terminal market was $5/cwt below August futures, traders could buy cattle and sell futures, then deliver on the contract for a profit of $5/cwt (less marketing and brokerage fees).

ADVANTAGES AND LIMITATIONS OF LIVE (*Slaughter*) CATTLE FUTURES

In this section, only live (slaughter) cattle futures will be discussed, simply because they constitute the greatest uncertainty, or risk, in the cattle business; hence, they will always dominate the futures market insofar as cattle feeding is concerned. Nevertheless, many of the same advantages and limitations apply to feeder cattle and feeds.

Live (slaughter) cattle futures are serving a useful purpose; and they are here to stay. Before using them, however, cattle producers should understand what they will and will not do.

Among the **advantages** of beef cattle futures are:

1. They serve as a price barometer for several months ahead, thereby increasing the range of information and judgments brought to bear on finished cattle prices and making it easier for feeders to choose a preferred course of action.

2. Through hedging, they can provide price protection or insurance against major breaks in the market.

3. They permit prices to be *locked in* anytime during the feeding period. Thus, they allow selectivity of the market time over the entire feeding period, rather than limit it to the one day that cattle are ready to go to market.

4. They make it possible for cattle producers to obtain credit more easily and to increase financial leverage. For example, let's assume that without hedging, a particular producer is able to borrow 70% of the cost of feeder cattle. Let's also assume that this producer has $90,000 in capital to invest. This will enable the producer to purchase $300,000 worth of feeder cattle. However, if the cattle producer hedges the cattle that are bought, the lender may be willing to lend up to 90% of their cost. The $90,000 of capital will then permit the purchase of $900,000 worth of feeder cattle. In this case, therefore, hedging tripled the number of cattle that could be purchased, and likewise increased the profit potential.

5. They make for a more stable market, with fewer peaks and valleys of price movements.

6. They make it possible for meatpackers to protect themselves when they contract with a feeder for delivery of finished cattle, (a) for a few months ahead, and (b) at the futures market price at the date of specified delivery. Thereupon, packers initiate hedges by selling futures to offset their purchase contracts.

Like many good things in life, live beef cattle futures are not perfect. They will not solve all the price problems, they will not raise longtime price levels, nor will they cause people to eat more beef. But these are not disadvantages, they are facts.

Among the **limitations** of live beef cattle futures are the following:

1. During an extended period of rising finished cattle prices, cattle feeders are disadvantaged when they fix a price for cattle in advance. One study revealed that over a 6-year-period the consistent hedger would have sacrificed 23% in profits to attain a 74% reduction in profit variability.[5]

---

[5]Curtis, C. E., Economic Research Service, USDA, "Beef Futures Trading Hit $16 Billion; No Let-Up in Sight," *Livestock Breeder Journal*, Aug. 1973, p. 12.

2. No provision for heifers is available; only steers may be delivered, however, heifers can be hedged if the basis, or differences, between steer prices and heifer prices can be determined.

3. Some delivery months may not move exactly as the cash market does.

4. A change in the basis (the spread between the cash price and the price of the futures) can mean a hedging loss as well as a hedging profit.

5. There is a relatively narrow range of time during which it is practical to hold slaughter cattle while waiting for a change in the basis.

6. The feeder must not forget to offset by purchase of another contract at the proper time; otherwise it may be necessary to deliver.

7. If the feeder sells futures for a greater or lesser amount than the finished weight of the cattle, the feeder is engaged in speculation for the amount of the excess.

8. There are some costs in futures which must be considered; namely, commission and interest on margin capital. These should be considered as costs of doing business; for the protection secured, the producer must pay a commission—much as is done for a life insurance policy.

9. Unless a feeder has maintained good and accurate records, and can project costs with reasonable accuracy, the feeder cannot intelligently determine if a futures price is favorable for placing a hedge.

HOW TO GO ABOUT HEDGING BEEF FUTURES

Here is the "how and where" that the livestock producer interested in hedging must follow:

1. Have good and accurate records of costs.

2. Contact a brokerage house that holds a membership in the commodity exchange.

3. Open up a trading account with the broker, by signing an agreement authorizing the broker to execute trades.

4. Deposit with the broker the necessary margin money for each contract desired. The broker will then maintain a separate account for the livestock producer. The commission fee is due when the contract is fulfilled by delivery, offsetting purchase, or sale of another contract.

5. Maintain basis charts, showing the relationship between (a) local prices of feeders and slaughter cattle, and (b) live beef futures.

LIVESTOCK OPTIONS

Livestock options are much newer than futures and put an interesting new twist on the concept of hedging. Live cattle options began trading on October 31, 1984. Feeder cattle options started on January 9, 1987.

Options are much more like insurance than futures. In exchange for a relatively small premium payment, the hedge buyer of an option receives protection against adverse price moves, but is able to take advantage of favorable price moves. Unlike futures, where a producer is locked in at a price and subject to margin calls, options have no margins and margin calls.

The specifications for options are the same as for futures. In fact these are options on the underlying futures. The owner of a put or call option has the *right* to take a position at any time in the futures contract. The only difference is that live cattle options expire on the first Friday of the delivery month. The feeder cattle options, since the contract is cash settled, expire on the same day as the futures. There are two types of options; *puts* and *calls*. Puts give the buyer the right to sell futures at a fixed price and increase in value if prices fall. Calls give the buyer the right to buy futures at a fixed price and increase in value if prices rise.

This fixed price is also known as the *strike price* or exercise price. The Chicago Mercantile Exchange sets the strike prices at $2, even-numbered intervals, *e.g.*, $70, $72, $74, etc. The buyer of an option gets to choose the strike price. The cost or *premium* of the option at each strike price will be different. Premium costs normally vary throughout the trading day.

A cattle feeder would most likely be a buyer of a live cattle put option, against cattle that will be sold in the cash market at a later date. A put hedge will give the cattle feeder a minimum (or floor) selling price, if price levels fall. Since the risk is limited to the premium paid, the cattle feeder would have no maximum price (less the premium lost), if price levels go higher.

A cattle feeder may also be a buyer of a *Feeder Cattle Call Option* to protect the cost of feeders that the feeder plans to buy. Buying a call could set a price ceiling, if prices rise, but no price floor, if prices fall. Once again, the rise is limited to the premium paid.

A rancher, or stocker operator, most likely would find *Feeder Cattle Put Options* useful in setting floor prices. There are many other option strategies, besides just buying puts and buying calls, which are beyond the scope of this section.

EXAMPLES OF OPTION PRICING

■ **Example 1: Buying a Live Cattle Put Option**—A producer purchases a February 74 put option at $2/cwt to price a group of cattle. At the time, February live cattle futures are at 74.75/cwt. Estimated basis for the end of January is $1/cwt. The producer's estimated minimum selling price would be the 74 strike, minus the premium of $2, and the estimated basis of $1, which would equal $71/cwt. Let's take a look at what happens in late January if the market goes up, stays about the same, or goes down.

At the end of January, the cattle are ready for market:

If Feb. Futures Are ($)	Value of Put ($)	A Put Net Gain or Loss ($)	B Local Cash Sale ($)	C Net Realized Price ($)
		(A + B = C)		
84	0	–2.00	83	81
74	0	2.00	73	71
64	10	8.00	63	71

As can be seen, when the futures price drops below the put strike price, the minimum selling price or insurance kicks in and protects the floor that was established when the 74 live cattle put was purchased. Should the market go higher, the increase less the cost of the premium will be realized while enjoying protection from a price drop.

A long hedger is one (such as a feedlot operator, a backgrounder, or a stocker operator) who needs a commodity at some point in the future and seeks to forward price the anticipated purchase. Again, choosing a particular hedging strategy depends upon the level of protection desired.

- **Example 2:** Buying a Feeder Cattle Call Option—A January 84 call option is purchased at $2.55/cwt to protect the purchase price of feeder cattle that will be needed in January. At the same time, January feeder cattle futures are at $85.50/cwt. Estimated basis for the end of January is +$3. The estimated maximum purchase price would be the 84 strike price, plus the premium of $2.55, plus the estimated basis of +$3 or a total of $89.55/cwt. Let's take a look at what happens in late January if the market goes up, stays the same, or goes down.

At the end of January when feeder cattle are purchased for feeding:

If Jan. Futures Are ($)	A Local Cash Purchase ($)	Value of Call ($)	B Call Gain or Loss ($)	C Net Realized Price ($)
		(A – B = C)		
94	97	10	7.45	89.55
84	87	0	–2.55	89.55
74	77	0	–2.55	79.55

Should the market rise between the time the 84 call was purchased and the time the animals were actually purchased, the feeder cattle cost would be limited to the ceiling price created. In this example an in-the-money call was purchased, thus benefiting from any price increase. Should the market fall after the 84 call was purchased, there is still, benefit from a lower feeder cattle purchase price (although the purchaser would be out the premium paid for the call).

GLOSSARY OF FUTURES MARKET TERMS

Futures markets have a jargon and language of their own. It is not necessary that livestock producers dealing in futures master all of them, but it will facilitate matters if they at least have a working knowledge of the following:

Basis—The difference or spread between the cash price at a particular market and the price of a futures contract. This spread differs from one market to another and changes with time.

Basis movement—The change which occurs in a particular cash-futures price relationship. It is the change in basis that determines the success or failure of a hedge, rather than changes in market prices. One should always hedge according to basis rather than price.

Bear market—A downward moving or lower market is considered *bearish*, because the bear strikes down its victim.

Bid—A bid subject to immediate acceptance made on the floor of an exchange to buy a definite quantity of a commodity future at a specified price.

Break—A more or less sharp price decline.

Broker—An agent who handles the execution of all trades. The broker may represent a clearinghouse member.

Bull market—An upward moving or higher market is considered *bullish*, because the bull tosses his victim upward on impaled horns.

Cash (spot)—The cash price refers to the price of live animals and not futures contract. Also known as spot commodity.

Cash market—Cattle bought and sold for immediate delivery. Also known as spot market.

Chicago Board of Trade—It was founded in 1848. The Chicago Board of Trade handles futures trading in such commodities as wheat, corn, oats, rye, soybeans, and soybean oil and meal.

Chicago Mercantile Exchange—It was founded

in 1919. The Chicago Mercantile Exchange handles futures trading in such commodities as live cattle, feeder cattle, live hogs, pork bellies, and random length lumber.

CFTC—The Commodity Futures Trading Commission, the independent federal agency created by Congress to regulate commodity futures trading. The CFTC Act of 1974 became effective April 21, 1975. Previously, futures trading had been regulated by the Commodity Exchange Authority of the USDA.

Commission—The charge made by a broker for buying or selling a futures contract.

Commission house—A firm which buys and sells actual commodities or futures contracts for the accounts of its customers.

Commitment—A trader is said to have a commitment, when the trader assumes the obligation to accept or make delivery on a futures contract.

Deferred futures—The futures, of those currently traded, that expire during the most distant months. (See Nearby.)

Delivery—The tender and receipt of the actual commodity, or warehouse receipts covering such commodity, in settlement of a futures contract.

Delivery points—Those points designated by futures exchanges at which the physical commodity covered by futures contract may be delivered in fulfillment of such a contract.

Discount to futures—When the cash price is under the futures price.

Forward contract—A forward contract calls for delivery at sometime in the future. In a forward contract, a livestock producer might make a deal with a buyer during the summer months that calls for delivery of cattle in the fall at the price agreed upon in the contract.

Futures—A term used to designate any and all contracts which are made or established subject to the rules for delivery at a later date.

Hedge—The purchase or sale of a futures contract as a temporary substitute for a merchandising transaction to be made at a later date. Usually it involves opposite positions in the cash market and the futures market at the same time.

Hedgers—Persons who desire to avoid risks, and who try to increase their normal profit margins through buying and selling futures contracts. They are feeders, packers, and others actually involved in production, processing, or marketing of beef. Their primary objective is to establish future prices and costs so that operational decisions can be made on the basis of known relationships.

Limit order—Placing price limitations on orders given the brokerage firm.

Long—The buying side of an open futures contract. A trader whose net position in the futures market shows an excess of open purchases over open sales is said to be *long.*

Long hedge—Buying on the futures market contracts against anticipated need in the future in order to protect against a rise in the market price. Thus, futures contracts in feeder cattle can be used in a long hedge to reduce or eliminate the risk involved in a rise of feeder cattle prices. At the time the feeder negotiates a forward contract with the packer, the feeder would buy feeder futures contracts, preferably for the month in which the feeder actually planned to buy the feeders.

Margin—Cash or equivalent posted as a guarantee of fulfillment of a futures contract (not a payment or purchase).

Margin call—If the market price of a futures contract changes after the feeder has sold or purchased a futures contract, the feeder will either make a profit or lose money. If the price moves in such a direction so that the feeder loses money, the broker will deduct the losses from the original *margin* and call for additional funds in order to bring the *margin* back up to the original amount.

For example, a feeder might have the broker sell a live cattle futures contract at $60/cwt. The feeder would deposit $700 margin with the broker. If the price of futures were to advance to $61, the feeder would have lost $1/cwt or $400. The broker would deduct the $400 from the original margin of $700, leaving $300. The broker would issue a *margin call* for an additional $400.

Nearby—The nearest active trading month of a futures market. (See Deferred futures.)

Offer—Indicates a willingness to sell a futures contract at a given price. (See Bid.)

Open interest—Number of open futures contracts. Refers to unliquidated purchases or sales but never to their combined total. (See Commitment.)

Pit—An octagonal platform on the trading floor of an exchange consisting of steps upon which traders and brokers stand while executing futures trades.

Premium—When the cash price is above the futures.

Rally—Quick advance in price following a decline.

Round turn—A purchase and its liquidating sale, or a sale and its liquidating purchase.

Security deposit (initial)—Synonymous with the term margin, a cash amount of funds which must be deposited with the broker for each contract as a guarantee of fulfillment of the futures contract. It is not considered as part payment of purchase.

Security deposit (maintenance)—A sum, usually smaller than, but part of, the original deposit or margin which must be maintained on deposit at all times. If a customer's equity in any futures positions drops to or under the maintenance level, the broker must issue a call for the amount of money required to restore the customer's equity in the account to the original margin level.

Settlement price—The daily price at which the clearinghouse clears all trades. The settlement price of each day's trading is based upon the closing range of that day's trading. Settlement prices are used to determine both margin calls and invoice prices for deliveries.

Short—The selling of an open futures contract. A trader whose net position in the futures market shows an excess of open sales over open purchases is said to be *short*.

Short hedge—When one owns an inventory of a commodity and hedges by selling an equivalent amount of futures contracts, one has sold short or is short futures and has what is called a short hedge. An example of a short hedge is selling on the futures markets contracts of live cattle which represent cattle that are on feed in the feedlot in order to protect the enterprise against a severe decline in the market.

Speculators—Persons who are willing to accept the risks associated with price changes in the hope of profiting from increases or decreases in futures prices.

Spot commodity—The actual physical commodity such as live cattle as distinguished from the futures. Also known as cash commodity.

Spread—A market position that is simultaneously long and short equivalent amounts of the same or related commodities. In some markets, the term *straddle* is used synonymously.

Ticker—A teletype machine which sends and receives futures market and cash information.

Trend—The direction prices are taking.

Volume—The number of purchases or sales of a commodity futures contract made during a specified period of time.

QUESTIONS FOR STUDY AND DISCUSSION

1. Use the nomograph (Fig. 27-2) to compute the break-even price that you can afford to pay for feeder cattle where the following circumstances prevail: purchase weight of feeder cattle, 500 lb; selling weight of fat cattle, 1,050 lb; feed cost per cwt gain, $55; price per cwt of fat cattle, $65.

2. Define (a) *price margin*, (b) *feeding margin*, (c) *positive margin*, and (d) *negative margin*. Why isn't the relative importance of price margin vs feeding margin the same for calves as for yearlings?

3. List and discuss each of the factors pertinent to buying feeder cattle.

4. Discuss each of the following points pertinent to selling finished (fed) cattle: (a) when to sell; (b) market channels; (c) factors influencing value; (d) seasonality of prices; (e) shrinkage; and (f) handling prior to and during shipment.

5. Define (a) *horizontal integration*, and (b) *vertical integration*. What are the advantages and disadvantages of each?

6. What efficiencies accrue to large cattle feedlots? What inefficiencies rise with extremely large feedlots?

7. Discuss each of the following points pertinent to custom feeding: (a) provisions of a contract; (b) types of contracts; (c) operation procedures; (d) what cattle owners expect of custom feedlots and (e) how to attract custom feeders.

8. What practical use can cattle feeders make of computers?

9. Detail a desirable and workable incentive basis for cattle feedlot help.

10. Discuss each of the following points pertinent to futures trading in finished cattle, feeder cattle, and feed: (a) definition of a futures cattle contract; (b) specifications for live (slaughter) cattle contracts, and for feeder cattle contracts; (c) commission fees and margin requirements; (d) how a futures contract works; (e) delivery against contract; (f) facts about futures contracts; (g) advantages and limitations of live (slaughter) cattle futures; and (h) how to go about hedging beef futures.

11. Definition of livestock options.

12. How do livestock options differ from livestock futures?

13. Define the following terms: (a) hedging, (b) speculating, (c) delivery, and (d) bull market.

14. Under what circumstances would you recommend that a cattle producer use beef futures? Under what circumstances would you recommend that a cattle producer not use beef futures?

SELECTED REFERENCES

Title of Publication	Author(s)	Publisher
Analysis of Beef Costs and Returns in California, An, AXT-258	A. D. Reed	Agricultural Extension Service, University of California, Davis, CA, 1971
Beef Cattle Futures: A Marketing Management Tool, Bull. 663	R. F. Bucher G. L. Cramer	Agricultural Experiment Station, Montana State University, Bozeman, MT, 1972
Budgeted Costs and Returns of Fifteen Cattle Feeding Systems in Four Areas of Texas, MP-1022	E. Williams D. E. Farris	Agricultural Experiment Station, Texas A&M University, College Station, TX, 1972
Business Management for Farmers	J. W. Looney	Doane Publishing, St. Louis, MO, 1983
Capital Structure and Financial Management Practices of the Texas Cattle Feeding Industry, The, B-1128	R. A. Dietrich J. R. Martin P. W. Ljungdahl	Agricultural Experiment Station, Texas A&M University, College Station, TX; Economic Research Service, U.S. Department of Agriculture, Washington, DC, 1972
Cattle Feeding in California	J. A. Hopkin R. C. Kramer	Bank of America, N.T. & S.A., San Francisco, CA, 1965
Cattle Feeding in the United States	R. A. Gustafson R. N. Van Arsdall	Economic Research Service, U.S. Department of Agriculture, Washington, DC, 1970
Cattle Futures Handbook	J. Sampier J. March	The authors, Chicago, IL, 1966
Commodity Futures Statistics, Stat. Bull. No 444		U.S. Department of Agriculture, Washington, DC, 1970
Commodity Markets and Futures Prices	R. M. Leuthold	Chicago Mercantile Exchange, Chicago, IL, 1979
Costs and Economies of Size in Texas-Oklahoma Cattle Feedlot Operations, B-1083	R. A. Dietrich	Agricultural Experiment Station, Texas A&M University, College Station, TX, 1969
Economic Comparison of Confinement, Conventional Drylot and Openlot Beef Feeding Systems, AE-4250	D. E. Erickson	Cooperative Extension Service, University of Illinois, Urbana, IL, 1970
Economics of Futures Trading for Commercial and Personal Profit	T. A. Hieronymus	Commodity Research Bureau, Inc., New York, NY, 1971
Evolution of Futures Trading	H. S. Irwin	Mimir Publishers, Inc., Madison, WI, 1954
Factors Affecting Cattle Prices	J. Ferris	Cooperative Extension Service, Michigan State University, East Lansing, MI, 1969

Feedlot, The, Third Edition	Ed. By G. B. Thompson C. C. O'Mary	Lea & Febiger, Philadelphia, PA, 1983
Futures Trading in Livestock—Origins and Concepts	Ed. by H. H. Bakken	Mimir Publishers, Inc., Madison, WI, 1970
Interregional Competition in the Cattle Feeding Economy with Special Emphasis on Economies of Size, B-1115	R. A. Dietrich	Agricultural Experiment Station, Texas A&M University, College Station, TX, 1971
Stockman's Handbook, The Seventh Edition	M. E. Ensminger	Interstate Publishers, Inc., Danville, IL, 1992
Texas-Oklahoma Cattle Feeding Industry, The, B-1079	R. A. Dietrich	Agricultural Experiment Station, Texas A&M University, College Station, TX, 1968
Trading in Live Beef Cattle Futures	R. P. Shiner D. G. Nash D. Schambach	Commodity Exchange Authority, U.S. Department of Agriculture, Washington, DC, 1970
Understanding the Commodity Futures Markets	Commodity Research Bureau, Inc.	Commodity Research Bureau, Inc., New York, NY, 19[illegible]3

Feed truck. Self-unloading truck, equipped with electronic scales, conveying feed through an outlet spout to the cattle feed bunk. (Courtesy, Butler Manufacturing Co., Oswalt Division, Garden City, KS)

FEEDLOT FACILITIES AND EQUIPMENT[1]

Contents *Page*

[1]In the preparation of this chapter, the authors gratefully acknowledge the authoritative review and helpful suggestions, along with pictures, of the following person, who has great expertise in the operation of a modern cattle feedlot: Ron Baker, C & B Livestock, Inc., Hermiston, Oregon.

Contents Page

Fig. 28-1. Modern cattle feedlot facilities. Note the neat, open pens, the shades, and the curved chute. (Courtesy, *Livestock Weekly*, San Angelo, TX)

Cattle feeding facilities and equipment are a manufacturing plant, wherein animate objects (cattle) convert feed into beef. Hence, they merit the same level of competence in planning and design as any other sophisticated manufacturing plant.

Because of variations in climatic conditions, number of cattle to be fed, and factors prevalent at the location where it is desired to construct the feedlot, no attempt will be made herein to present detailed facility and equipment plans and specifications. Rather, it is proposed to convey suggestions regarding the desirable features of cattle feeding facilities in various sections of the country. For detailed plans and specifications, the feeder should (1) study facilities and equipment in other feedlots, and (2) engage the services of a consultant(s) with expertise in cattle feeding facilities and equipment.

Some preliminary feedlot planning suggestions follow:

1. Decide on the number of cattle and the feed and storage requirements with provision for expansion.

2. Determine the justifiable investment in cattle feeding facilities. (See the section that follows.)

3. Select the facilities, equipment, and arrangement that best fit the management program you have chosen; for example, (a) fence-line bunks and a central feed processing plant, (b) upright storage with distributors and bunks, or (c) self-feeders.

4. Decide on the type of facilities: (a) feedlot (open pen); (b) cold confinement; or (c) warm confinement.

5. Design a system that is practical, laborsaving, environmentally suitable for economical gains of cattle, and attractive.

JUSTIFIABLE INVESTMENT IN CATTLE FEEDING FACILITIES

Cattle feeders need to know the size investment that they can justify in cattle feeding facilities. A nomograph may be used for this purpose. It can give a quick, preliminary idea of cost, gross profit, returns, and investment relationships. But a nomograph should not replace more detailed figuring which should precede all major investment decisions. Also, one should realize that a nomograph will give erroneous and misleading information unless based upon accurate and realistic cost and return data from the problem at hand.

Fig. 28-3 is a nomograph for use in making quick calculations of justifiable investments in beef cattle feeding systems. The section that follows gives an example and *step-by-step* instructions for using the nomograph. By working through an example, you will soon discover how quickly the nomograph can help in evaluating capital investments for cattle feeding enterprises.

Fig. 28-2. Cattle feeding has gone modern. *(Upper)* Yesterday. *(Lower)* Today. (Upper picture—courtesy, IBP, Inc., Dakota City, NE; lower picture—Pioneer Hi-Bred Corn Co., Des Moines, IA)

INSTRUCTIONS FOR USING NOMOGRAPH

This nomograph provides a quick method of computing the investment in cattle feeding facilities one can justify on the basis of three factors: (1) the percent of the total investment to *charge off* as annual costs each year; (2) the gross profit (GP) expected per head; and (3) the return per head (R/C & L) desired for labor, management, and interest on the capital required for cattle, feed, and miscellaneous equipment (*excluding* the investment in the feedlot facilities under consideration).

■ **Determining % annual costs**—Annual fixed costs include such items as interest, insurance, and taxes (usually from 4 to 6%) plus an allowance for depreciation or annual principal repayments required, if making a cash flow analysis (usually from 5 to 15%). These two percentages should be combined to get the proper *total annual costs, percent of investment for this analysis.*

Data Needed for Computations	Example	Your Farm
1. Percent (%) annual costs	15%	______
2. Gross profit (return over variable costs) per head	$20.00	______
3. Return to labor, mgt., and capital (R/C & L)	$14.00	______
4. Investment per head justified	$40.00	______

■ **Procedure for determining the investment justified in cattle feeding facilities—**

1. Locate the *vertical line* representing the total annual costs, % of investment (15% in the example).

2. Follow this line straight down to the point where it intersects with the diagonal line representing the expected gross profit (returns over variable costs) per head ($20.00 in the example)—**Mark this point.**

3. Locate the point on the left-hand vertical scale which represents the desired return ($/head) for labor, management, and capital invested in cattle, feed, and miscellaneous equipment ($14.00/head in example)—**Mark this point.**

4. Draw a straight line to connect the two points located above. The point of intersection with the right hand vertical scale indicates the justified dollar-per-head investment in facilities ($40.00/head in example).

■ **Procedure for determining the probable return ($/head) with a known, or contemplated, investment per head in facilities**—Follow the above procedure through point 2. Then, locate the dollar investment per head on the right-hand vertical scale ($40.00 per head in example). Connect a straight line from this point through the one located in point 2. The point of intersection with the left-hand vertical scale gives the probable $ return/head ($14.00 in example).

Other variations in the use of the nomograph are obvious. For example, if the investment per head in facilities is $50.00, a return of $15.00/head is desired, and the gross profit per head is $20.00—10% annual costs would be indicated.

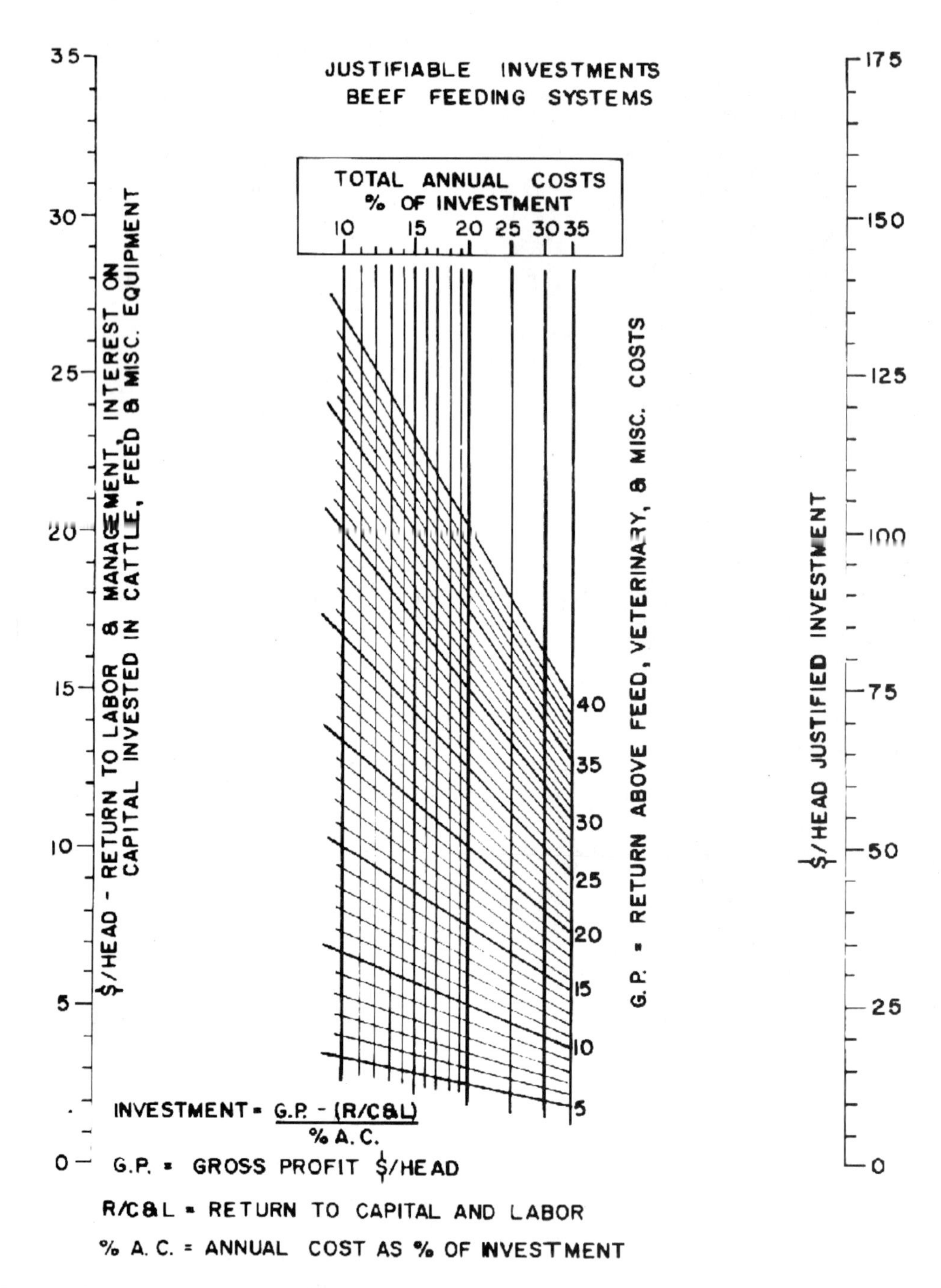

Fig. 28-3. Nomograph for computing justifiable investments in beef cattle feeding systems. (Nomograph, and instructions for using it, prepared by Robert M. George, University of Missouri)

FEEDLOT (Open Pen)

The vast majority of feedlot cattle are fed in open pens, without shelter. Some are provided wind protection (trees, hills, or wind fences). Still others are provided an open-front building, for shelter from wind, sun, rain, and snow.

The sections that follow present information pertinent to feedlot design and construction.

Fig. 28-4. Feedlot windbreak and cable fence. Upright fence boards are spiked to horizontal nail ties (20-in. lumber) for a windbreak in mounded feedlot. Windbreak, supported by treated poles, is vertical to the ground so it spans curved ground without leaning. Cable fence in foreground is braced with pipes. Pipes have "feet" welded to them that are spiked to the posts. Cables pass through holes drilled in posts. A cable brace with turnbuckle is anchored diagonally from gatepost to second post. (Photo by A. M. Wettach, Mount Pleasant, IA)

COST

Before constructing a feedlot, cost must be considered for two reasons: (1) capital must be secured; and (2) cost must be amortized. The usual basis of computing cost is on a *per animal unit capacity*. This will run about the same whether calves or yearlings are involved, because per unit capacity must consider carrying the animals to market time.

The area affects cost from the standpoint of shelter requirements and land values. Thus, because of the necessity for winter protection and shelters, feedlot costs are higher in the northern tier of states than in the South. Land values are higher in California than most areas of the United States, with the result that land costs become a factor.

Size of feedlot affects per animal cost. Most studies reveal that investment savings do accrue to the larger feedlots. Thus, the cost per animal usually decreases up to about 10,000-head capacity, then it increases slightly with larger lots. The slightly higher cost per head capacity of the larger lots appears to be due to duplication in equipment and the tendency to become more highly mechanized and elaborate.

An open lot without shelter is the cheapest type of feedlot construction of any. In the Southern Plains area, where the weather is mild and shelters are unnecessary, investment costs range from $100 to $125 per head of capacity.

Housing increases costs, and the more elaborate the housing the greater the cost. It is estimated that, on the average, the cost per head capacity where housing is involved is about as follows: open shed, $140; cold confinement, $185; warm confinement, $300.

LOCATION

In the present day and age, pollution control is the first and most important consideration in locating a cattle feedlot. The location should avoid (1) neighbors complaining about odors, flies, and dust; and (2) pollution of surface and underground water. Also, feedlots should be located on a well-drained site, with area available for expansion. Whenever possible, they should be built on a slope, preferably at the top of it. There should be a minimum amount of runoff from areas above lots (a diversion terrace can be used if necessary); and there should be ample space below feedlots for necessary water pollution control measures. Also, feedlots should be located where there is ample space for expansion, if and when desired. Of course, the space requirements will vary with the type of facility. Open lots require the most space; confinement housing the least. Minimum space requirements for an open feedlot—including lots, mill, office, etc.—are approximately $^{8}/_{10}$ acre per 100 head or 7 acres per 1,000 head. In order to allow for expansion to double this size, it is recommended that there be 1.4 acres per 100 head, or 12 acres per 1,000 head.

Fig. 28-5. In the present era, pollution control is most important. Besides, when a truck is bed deep in mud, it's difficult for a caretaker to remember to feed the cattle, and it's difficult for the cattle to know that they are supposed to eat. Few present-day cattle feeders permit such conditions as pictured above. (Courtesy, IBP, Inc., Dakota City, NE)

LAYOUT

Prior to starting construction, anyone contemplating a feedlot may avoid much subsequent difficulty and expense by doing some paper and pencil planning at the onset. First, decide on the size of the enterprise and the management system. Then, establish traffic routes for animals, feed, cleaning equipment, supply trucks, etc. Next, sketch out the facilities and equipment required to meet these needs in the most efficient and economical manner, including pens, mill, scales, office, etc. Where the area permits, the ideal feedlot should be U-shaped, with the following arrangement: the facilities for receiving and loading out cattle, scales, milling feed, office, and equipment barn should be located near the center of the U. Pen facilities should be located on the three closed sides of the U. The open end of the U should be connected to a public road. Also, trench silos should be located at the mouth of the U. The mouth of the U should be kept open to allow for the flow of livestock, feed, and visitors. The smallest pens should be located as near the feed mill as possible, in order to minimize travel time in feeding. Feed alleys should parallel the legs and closed end of the U, forming a semicircle around the feed mill area. The corners of the pens should be rounded to allow feed trucks to turn at all intersections.

But the layout of each feedlot is different, depending on such factors as topography, location of access roads and rail spurs, and traffic flow within the feedlot. If the movement of cattle, feed, cleaning equipment, supply trucks, and people can be accomplished efficiently, the design is good.

The above information, constituting the layout of operations, should first be put on paper in sketch form, by the cattle producer. From this, the engineer or consultant can design facilities and equipment which most effectively and economically meet the production requirements of the specific enterprise.

PENS

Pens are the working end of the business; they have much to do with the well-being and performance of the cattle. Hence, their design is most important; and the more severe the climate, the more important the design becomes. Consideration should be given to the following points when designing cattle pens:

1. **Mud.** Mud lowers feed intake, daily gain, and feed efficiency. Four to eight inches of mud in a feedlot will cause 8 to 15% decrease in feed intake, 14% decrease in daily gain, and 12 to 13% decrease in feed efficiency. Severe mud, 12 to 24 in. deep, can decrease feed intake up to 30%, decrease daily gain 25%, and decrease feed efficiency 20 to 25%. In a muddy lot, each steer will gain about ¾ lb less per day. This means that every 4 days spent in a muddy pen adds 1 day to the total time spent in the feedlot. So, 100 head of cattle in a muddy lot can cost an extra $1,000 per month.

Fig. 28-6. Well-fenced open pens with shades. (Courtesy, *Livestock Weekly*, San Angelo, TX)

2. **Drainage.** Muddy pens reduce cattle performance more than wind and rain; a muddy lot may reduce daily gains and feed efficiency by 25 to 35%. To supplement the natural drainage, surface grading should be done prior to starting construction. A grade of 4% should be established. Excessive slopes (above 10%) should be avoided, as they make for difficult footing and are subject to erosion.

3. **Mounds.** Mounds of dirt in each pen will provide a drier resting area for cattle. Preferably, mounds should be pushed up before the lot fences are installed. They should be 6 to 8 ft high, with a 4:1 or 5:1 (horizontal to vertical) slope on the sides. The top of the mound should be fairly narrow (about 10 ft wide) and crowned for good drainage. The size of the mound will vary with the size of the lot and the number of cattle for which it is intended that it shall provide a rest area. Each mound should be large enough so that the cattle can rest on the upper half of it; 10 to 15 sq ft of mound per animal will accomplish this, although not all animals will use mounds. The orientation of earth mounds is unimportant, so long as they do not block feedlot drainage. The mound should be built parallel with the general lot drainage to assure that liquids can readily drain from the mound area.

Mounds will require some maintenance. A logical time for this work is when manure is cleaned from the pen.

Fig. 28-7. This Caterpillar 613 self-loader works year-round in mounding pens at C & B Livestock, Inc. (Courtesy, C & B Livestock, Inc., Hermiston, OR)

4. **Surface.** When cattle are fed in outside pens during favorable weather, the type of pen surface (concrete vs dirt) is probably of little importance. Where good drainage cannot be provided, concrete surfaced pens should be considered. Cattle can be finished successfully in outside pens surfaced with concrete as evidenced by the feedlots in which such footing is provided.

5. **Size.** Climate and the amount of paving are the main factors determining pen size. It will vary anywhere from a minimum of 30 sq ft per animal with a surface lot and open housing to 400 sq ft per head for an unsurfaced lot in a wet, muddy area. With an open, dirt lot, 75 sq ft of pen space per head is adequate in a dry climate, whereas up to 400 sq ft per animal may be required in a wet climate. In a dry climate, excess pen space can make for dust problems; in a wet climate, too little space makes for mud problems. On the average, a pen space allowance of 125 to 200 sq ft per head is recommended if the lot is unpaved.

The number of animals per pen is variable. Investment may be reduced by using fewer but larger pens. However, visual inspection of cattle for sickness is easier in smaller pens, as are the separation and removal of individual animals. Custom feedlots require a variety of pen sizes in order to accommodate each customer's cattle in separate pens. In custom feedlots, the majority of pens with a capacity for 120 head appear to be desirable, because of customer convenience. From the standpoint of trucking requirements, pens should be sized to hold multiples of 60 heads. In most feedlots, where some custom feeding is done, it is recommended that feedlot sizes vary from 120-head capacity to 300-head.

Generally, most feedlots have too many large pens to permit them to feed the numerous small lots of cattle that they receive.

6. **Shape.** Pens should be rectangular in shape, preferably with rounded corners to allow the feed truck to turn at all intersections. The depth of the pen will depend upon the length of the bunk; there must be sufficient bunk space to accommodate the cattle in any given pen; then there should be sufficient depth to provide the number of square feet per animal intended.

7. **Fences.** A corral system is no more dependable than its fences. The specifications for the most desirable feedlot fence are: 54- to 60-in. high posts of 3- to 4-in. diameter pipe, or treated wood with 5-in. top diameter, set in concrete and spaced 8 to 12 ft apart; enclosed (except for the feed bunk area) by four strands of ⅜-in. steel cable spaced at intervals (from ground up) of 18 in., 12 in., 12 in., and 13 in. Three strands of the cable should be placed above the feed bunk, with these spaced equal distances apart. Steel or aluminum pipe, wood, or wire may be used in place of steel cable if desired.

An outside fence, enclosing the entire cattle feedlot complex, is also recommended. A 60-in. high chain link fence, with steel posts set in concrete, is very satisfactory for this purpose. Such a fence is effective in keeping prowlers, dogs, and predators out; and in holding cattle which, for one reason or another, have escaped from their pens or handling facilities.

Working corral fences should be higher and stronger than ordinary feedlot fences. They may be constructed with treated wood posts, 6 in. top diameter, spaced 8 ft apart and set in concrete, enclosed by 2-in. dimension lumber, or 4-in. diameter pipe and ⅞-in. sucker rod or equivalent may be used.

8. **Feed bunks and aprons.** Six to nine inches of bunk space per head are adequate when cattle have access to feed at all times. The bunk forms part of the fence for the pen. It should be constructed of concrete with the outside (alley side) of the bunk 22 to 36 in. high, and the inside (the pen side) 18 to 22 in. high.

Fig. 28-8. Concrete feed bunk designed for the way an animal eats feed—naturally. (Courtesy, Farr Feeders, Inc., Greeley, CO)

Fig. 28-9. A concrete apron adjacent to the bunk and a mound, connected by a gravel pad, avoids muddy conditions in the feedlot. (Courtesy, C & B Livestock, Inc., Hermiston, OR)

The bottom of the bunk should be rounded and about 18 in. wide.

A 6- to 8-ft concrete apron (platform), 4 to 6 in. thick should extend from the feed bunk into the pen to provide solid footing for the cattle. The apron should have a minimum slope of ½ to 1 in. per foot (a 1-in. per foot slope will be nearly self-cleaning). A lip at the low end of the concrete apron, to divert the water along the length of the slab and out the pen, rather than merely off the slab area and into the pen dirt area, will alleviate the low, muddy spot that usually develops at the juncture of the concrete and dirt.

9. **Water troughs.** Feedlot cattle should have access to clean, fresh water at all times. One linear foot of trough space for each 10 head of cattle is sufficient. Shallow, low capacity troughs are preferable, since frequent drainage is necessary to keep them clean. There should be sufficient water supply to provide 20 gal per head per day—the intake of a 1,000-lb animal on a hot day. Continuous-flow troughs are excellent and help to keep clean water. In most areas, continuous-flow troughs will not freeze during the winter months; hence, heat is not required. However, heated and insulated troughs are necessary in cold areas. Heat is usually provided by a 250-watt infrared reflector-type lamp, which is set to start when the temperature goes down to 33°F and to cut off when the temperature rises above 33°F.

Fig. 28-10. Durable concrete water trough in a finishing cattle lot, providing plenty of water at all times. (Courtesy, Feedlot Management, Minneapolis, MN)

10. **Windbreaks and shades.** In those sections of the country where snow and cold winds are a problem, windbreaks may be provided by hills, trees, or board fences. Where board fences are used, they are generally constructed of 1-in. lumber and are 7 ft high.

In hot climates, shades should be provided. (See section on "Shades" later in this chapter.)

11. **Gates.** Gates should be of the same height as the fences. The length of most gates should be coordinated with the width of alleyways and crowding areas. Thus, gates along a 12-ft alley should be 12 ft long and capable of swinging either way. All gates should have a substantial lock and catch to prevent breaking open under pressure. The easier operation, less sagging, and longer life of metal gates in comparison with wooden gates will usually more than compensate for their additional cost.

Most managers of big feedlots seem to agree that the most desirable arrangement consists in having two gates to each pen, with both gates offset to provide easy access for cattle and equipment. One of these gates should be located in the corner of the alley nearest the scales. The other should be located adjacent to the feed bunk, but in the side fence. The latter gate will facilitate the use of equipment in cleaning manure off aprons, and is useful for cowboys or equipment operators in moving from pen to pen without entering the cattle alley.

ALLEYS

Two types of alleys are common to most feedlots—feed alleys and drive alleys. In feedlots below approximately 5,000-head capacity, feed and drive alleys should be combined, for cost reasons. In larger feedlots, it is recommended that feed alleys be separated from drive alleys, so that feeding and cattle movement can be done without excessive interference. Feed alleys, and combined feed and drive alleys, should be at least 20 ft wide, so as to permit the passing of trucks. Working and drive alleys which do not handle truck traffic should be 12 ft wide. Alleyways should not occupy more than 20% of the total corral area unless their space can also be used effectively as pens.

CATTLE LOADING AND UNLOADING FACILITIES

The facilities for loading and unloading cattle are the connecting link between the corrals and the various types of vehicles used to transport cattle in and out. They should include a truck dock (with loading chute), scales, working alley, and holding pens. These facilities should be located centrally so that a smooth flow of cattle trucks in and out can be maintained. However, they should be somewhat removed from the feed-mill area, in order to separate the traffic flow of feed and cattle and alleviate congestion of traffic.

The truck dock generally consists of a platform, chute, and chute pen. Most loading platforms are 46 in. high, although they may range from 30 to 50 in. in order to accommodate various sizes of trucks. The ramp in a loading chute should provide a safe gradual slope for the cattle to climb. Ten feet is considered a minimum length, while many feedlot operators prefer loading chutes up to 16 ft long. Most chutes are 30 to 42 in. wide, with 48- to 54-in. high fences.

In order to have favorable weighing conditions for sale cattle, it is necessary that they be moved from their pens to the scale expeditiously and with a minimum of stress. Larger feedlots—those of over 30,000-head capacity—usually accomplish this by having two shipping areas. In such lots, a good design allows for the movement of cattle to and from the cattle scale at the same time other cattle are being loaded or unloaded through the loading chute.

Fig. 28-11. Cattle unloading facilities. Note the curved, solid steel fencing, which facilitates cattle movement quickly and quietly. (Courtesy, Farr Feeders, Inc., Greeley, CO)

Also, there is need for holding pens convenient to the loading and unloading facilities. The number of such pens will be determined by the size of the feedlot. In general, each pen should be of sufficient size to hold two loads of cattle. The shipping pens—those for outgoing cattle—need not contain feed and water facilities, since they are used only to hold cattle temporarily that are to be loaded onto trucks. However, receiving pens—those for incoming cattle—should have provision for fresh feed and water. This is especially true for cattle received during the night, which might otherwise be left in a pen or an alley without feed or water until the next morning.

SCALES

Smaller feedlots can use a combination cattle and commodity scale. With larger feedlots, however, it is recommended that there be separate scales for each of these uses.

The scale should have sufficient capacity to meet the largest anticipated volume demand, without the necessity of dividing a lot of cattle and weighing in two or more drafts. In the larger feedlots, this calls for a scale approximately 10 ft wide and 60 ft long, with a capacity of 100,000 lb. To avoid exceeding the weighing capacity of a scale, provide 1 sq ft or less of platform space for each 110 lb of rated capacity.

The scale should be equipped with a ticket printer, in order to verify weight records and avoid human error in market transactions.

Fig. 28-12. A load of finished cattle being weighed on the scales. (Courtesy, Benedict Feeding Co., Casa Grande, AZ)

CATTLE PROCESSING FACILITIES

A separate area should be provided where cattle can be branded, dehorned, castrated, vaccinated, and sprayed or dipped. This should include crowding areas and a curved chute 18 to 30 ft long to aid in the movement of the cattle to the squeeze chute. Generally cattle will work best in a chute constructed with a modest curvature; a sharp curve may spook them. A manually operated squeeze chute is satisfactory for a small feedlot, but larger feedlots should use a hydraulic squeeze chute for reasons of efficiency and ease of operation.

Where large numbers of cattle are to be processed daily, the processing pens should be equipped with feed and water facilities. Such provisions lessen stress and make for better performance of the cattle.

In areas subject to severe weather conditions—either very hot or very cold—the processing facilities should be covered. This makes for greater comfort of both the help and the cattle, thereby assuring better processing.

HOSPITAL AREAS

Hospital pens are for holding sick or injured animals. From 2 to 5% of the feedlot area should be allocated for this purpose in lots feeding mostly yearling cattle. Lots feeding calves will need more intensive care areas. Larger feedlots prefer to disperse the hospital areas into sections of the feedlot, with one such hospital area provided for about 6,000 head of cattle. This alleviates the necessity of moving sick cattle excessive distances for treatment. Each hospital area should be equipped with a squeeze chute, refrigerator, running water, medicine, equipment storage, and feed facilities for the sick animals. Two or three small pens at the hospital for cattle at various stages of recovery are recommended. In areas subject to severe weather, shelter should be provided for hospitalized animals.

FEED MILL

The feed mill should be located convenient to receiving feed, convenient to the feeding area, and away from cattle traffic patterns. Mill type will be determined by the number of cattle to be fed and the feeds to be used. Generally speaking, it is recommended that feedlots with fewer than 5,000 head of cattle use a self-mixing, self-unloading truck, especially if milling of the grain is not necessary immediately prior to feeding. For lots above 5,000-head capacity, a mixing mill and self-unloading feed truck are more economical. Mill type will be determined by the feeds that are to be used. If whole corn is fed, a minimum of mill facilities is needed. Where sorghum grain is the primary feed, steam-flaking equipment or other similar processing equipment must be considered.

Mill capacity is also important. For a 5,000-head feedlot, the mill should be able to process 20 tons of feed per hour; for a 10,000-head feedlot, it should be able to process 35 to 40 tons per hour. Modern mills automatically mix the feed, with operation from a central control panel.

Fig. 28-13. Cattle feed processing facilities at Farr Feeders Inc. Greeley, Colorado. (Courtesy, National Cattlemen's Beef Assn., Englewood, CO)

Fig. 28-14. A central computerized system housed in the feed mill controls all feed ingredient distribution in Farr Feeders' 80,000-head-per-year capacity feedlot near Greeley, Colorado. Here, the manager of operations is mixing and loading feed. Each truck load is assigned a specific pen in the feedlot, and that pen receives a precise ration to meet daily nutritional requirements. (Courtesy, Farr Feeders, Inc., Greeley, CO)

Plenty of extra space around the feed mill should be left for future expansion, keeping in mind that the trend is to larger and larger lots.

OFFICE AND PARKING

Since the office is the headquarters of the cattle feedlot business, its location is of importance. It should be located near the main access road in order to minimize nonessential traffic in the feedlot proper. In most feedlots, the truck scales are adjacent to the office, and weighings are made in the office. Where this arrangement exists, the location of the office should be such as to provide for a good traffic pattern for both cattle and feed trucks. The size of the office will vary. Generally speaking, custom feedlots need more office space than feedlots that own all of the cattle. Automobile parking space is generally provided near the office. As a rule of thumb, most commercial feedlots plan on having 2 or 3 parking spaces for every 1,000-head animal capacity.

EQUIPMENT STORAGE AND REPAIR BUILDING

Most commercial feedlots have a separate building in which they store trucks, tractors, silage loaders, spare parts, and miscellaneous supplies. Also, this building is used for repair work. The size of the building will vary with the size of the cattle feeding operation. Usually, such buildings are 30 ft deep and 30 to 100 ft long.

LIGHTS

Feedlot lights serve three purposes: (1) they have a calming and quieting effect on the cattle, although there is no conclusive experimental work to support the claim that lights will improve rate of gain or feed efficiency; (2) they prevent prowlers and pilfering; and (3) they make for convenience in working or loading cattle at night. Mercury vapor lights, equipped with a photo eye (which turns on automatically), are recommended.

POLLUTION CONTROL

Pollution control is a most critical factor in site selection and operation of a cattle feedlot. Remoteness from urban development is recommended because of dust and odor. Also, before constructing a cattle feedlot, the owner should be familiar with both federal and state regulations. Federal regulations apply to feedlots (1) with 1,000, or more, feeder cattle; or (2) in which animal wastes either empty directly into a stream that crosses the feedlot, or are conveyed directly into a nearby waterway by a pipe, ditch, or other artificial means. The state regulations can be secured from the state water board. They differ from state to state, but most states require a catch basin (detention pond) sufficient to contain the runoff from a storm of the magnitude of the largest rainfall during a 48-hour period of the most recent 10 years. A feedlot may minimize runoff by locating near the top of the slope and, if necessary, by using diversion embankments to divert runoff from other areas.

Various methods of handling manure are being studied, including recycling it for feed, the production of methane gas, firing boilers to generate electricity, the production of garden fertilizer, etc. Yet, today, and perhaps for sometime to come, the vast majority of manure will be spread on farmland for use as a fertilizer. Thus, it is imperative that adequate farmland be available, and that a suitable location be made for stockpiling waste until it is used. The amount of farmland necessary will vary. For example, less manure can be applied to farmland that is alkaline than to land that is acidic. A common rule of thumb, however, is that the feedlot should plan to have 1 acre of farmland for manure for each 3- to 5-head cattle capacity. In order to lessen hauling costs, this land should be as close as possible to the feedlot.

Dust from feedlots in arid areas is a problem. Likewise, dust from feed mills can cause damage to crops and discomfort to people. Also, dust can (1) aggravate respiratory diseases in cattle, and (2) create poor visibility and make for possible accidents on highways. Sprinkling and increased cattle density are the most effective methods of preventing feedlot dust in dry climates or during dry seasons.

Before building a new feedlot or expanding an old one, the owner should be fully aware of the following:

1. A feedlot may be required by its neighbors to

Fig. 28-15. Water wagon for dust control in alleys and pens. (Courtesy, Benedict Feeding Co., Casa Grande, AZ)

control odor. Both large units and periodic field distribution or large amounts of waste contribute to high odor levels.

2. Pollution of surface and underground water can occur, and should be avoided.

3. Runoff from lots must be held in detention ponds before disposal by hauling and spreading or by distribution in an irrigation system.

4. Runoff from fields may pollute surface water if wastes are spread on snow or frozen ground.

5. Leached nitrates from decomposing manure may pollute nearby shallow wells.

6. Wastes must be field-spread and absorbed, digested in an adequate lagoon, or otherwise handled to minimize pollution.

Fig. 28-17. Open feedlot with slotted floor and a deep pit in a Nebraska feedlot (Courtesy, *Beef* magazine)

CONFINEMENT FEEDING; SLOTTED (or Slatted) FLOORS

Currently, there is much interest in cattle confinement feeding and slotted floors. The main deterrent is cost; construction costs vary with type of structure and may range up to $300 per steer space.

Confinement cattle feeding refers to feeding in limited quarters, generally 20 to 25 sq ft per yearling animal, which is about one-eighth the space normally allotted to a yearling in an unsurfaced lot and one-third that of a paved lot. The confinement is usually on slotted floors.

Slotted floors are floors with slots through which the feces and urine pass to a storage area immediately below or nearby.

Fig. 28-16. Confinement cattle feeding facility in Nebraska. (Courtesy, University of Nebraska)

Interest in confinement feeding and slotted floors is prompted in an effort to (1) automate and save labor; (2) cut down on bedding and facilitate manure handling; (3) lessen mud, dust, odor, and fly problems; (4) increase gains and save feed; (5) require less land; and (6) lessen pollution.

Research has shown conclusively that cattle fed during the winter months in cold areas gain faster and more efficiently if they are sheltered. However, as pointed out earlier in this chapter under the section headed "Feedlot (Open Pen)—Cost," the per head cost is much higher for confined or sheltered cattle. Thus, the decision on whether or not cattle confinement can be justified, even in the northern part of the United States, should be determined by economics. Will the cattle in confinement quarters gain sufficiently more rapidly and efficiently to justify the added cost? Of course, manure disposal and pollution control should also be considered.

COLD CONFINEMENT[2]

Cold confinement refers to a more or less open shed for confining cattle; hence, winter temperatures therein are within a few degrees of outdoor temperatures. Open sheds should be faced away from the direction of the prevailing winds. Additionally, doors or other openings in the closed walls should be provided for summer ventilation.

---

[2]The terms *cold confinement* and *warm confinement* refer to winter conditions. Without mechanical cooling, both systems are *warm* during the summer months.

Fig. 28-18. A cold confinement cattle feeding facility in Iowa, open to the south. (Courtesy, Pioneer Hi-Bred Corn Co., Des Moines, IA)

WARM CONFINEMENT[3]

Warm confinement refers to a confinement building for cattle which is sufficiently insulated and ventilated to maintain inside winter conditions above 35°F in severe weather, and in the range of 50 to 60°F most of the time.

DESIGN REQUIREMENTS FOR CATTLE CONFINEMENT BUILDINGS AND SLOTTED FLOORS

The figures that follow, based on information and experiences presently available, may be used as guides. (Also see Chapter 11, Table 11-6, Space Requirements of Buildings and Equipment for Beef Cattle.)

- **Floor space**—Allot 15 to 30 sq ft per animal exclusive of the bunk and alley, with an average of 20 to 25 sq ft for a 1,000-lb animal.

- **Animals per pen**—25 to 100 head per pen, with 25 to 30 being most common.

- **Bunk**—Allow 6 to 18 in. of linear bunk space per animal, with the amount of feeding space determined by frequency of feeding and size of animal.

- **Waterers**—Locate one waterer per 25 head at the back (opposite feed bunk) of each pen, preferably in the center.

- **Slats**—Reinforced concrete or steel may be used. Most concrete slats are 5 to 6 in. wide across the top, 6 to 7 in. deep and tapered to 3 to 4 in. wide at the bottom; and placed so as to provide a slot width of 1½ to 1¾ in.

[3]*Ibid.*

Fig. 28-19. Cattle on concrete slats, over a 3-ft pit. The excreta solids, which run about 9%, are evacuated by pumps into irrigation water. The cattle are confined to 20 sq ft per animal. On either side of the slats is a 4-ft concrete apron with a 5% slope. The cattle move excreta into the pit by their natural movement. (Courtesy, E. S. Erwin and Associates, Tolleson, AZ)

- **Manure production and storage**—Manure production will vary with size of animal and kind of feed, but it will be approximately as follows:

Animal	Cu Ft/Day Solids & Liquid	% Water	Gal/Day
1,000-lb steer	1-½	80–90	7½–10¾

Here is how to determine how much manure will need to be stored:

Storage capacity = Number of animals × daily manure production × desired storage time (days) + extra water

A rule of thumb is that when the pit occupies the entire area beneath the cattle, it will fill at a rate of 8 to 10 in. per month.

A newer and less costly system than slotted floors, which seems to be gaining favor among some feedlot operators, consists of a concrete floor sloped so that manure drains into gutters spaced equidistant throughout the building. Periodically, water is flushed through the gutters, carrying the waste material to pits.

SHADES

A hot steer is a poor doer! Hence, providing adequate shade to protect cattle from the sun is among the more important and widely used devices for improving the environment of cattle in hot climates. Tests conducted by the University of California, at the Imperial Valley Field Station, showed that it required 200 to 300 lb more feed to make 100 lb gain during midsummer without shade than with shade. Based on work conducted at the Yuma, Arizona, Station, where summer temperatures of 110°F are not uncommon,

Fig. 28-20. Shades are necessary in a hot climate. This shows the shades in a pen at Red Rock Feeding Co., Red Rock, Arizona. (Courtesy, Red Rock Feeding Co., Red Rock, AZ)

University of Arizona workers found that good shades can increase feedlot gains by 20 to 25% and improve feed efficiency by 14 to 20%.

Shades should be 12 ft or more high; provide 20 to 25 sq ft per animal; and be oriented north and south, so that the sun will shine under the shade early in the morning and late in the evening. Shades may be run east-west in the hot deserts of the Southwest to take fullest advantage of the cooler north sky.

Fans, sprinklers, and other cooling devices increase the effectiveness of shades. At Yuma, Arizona, where August temperatures reach daily highs of 110°F, the Arizona Station reported that yearling steers in pens equipped with shades and evaporative coolers gained at the rate of 3.01 lb per head daily, compared with 2.66 lb daily for steers equipped with shades and sprinklers, and 2.62 lb per day in the control pens—shades only. The steers exposed to the evaporative coolers ate 17% more feed and gained 15% faster than the controls (shades only). Steers penned in the corrals with sprinklers ate 2% more feed and gained 29% *faster than the animals penned with shades only.*

■ **Temperature and humidity**—University of Missouri researchers found that Shorthorn calves gained about ½ lb more per head per day and required 0.49 fewer pound of TDN per pound of gain at 50°F than at 80°F, and that the heat-tolerant Brahmans were more efficient at 80°F than at 50°F.[4] At the University of California Station in the Imperial Valley, Herefords gained 0.36 lb more per head per day when they had access to shelter cooled 7°F below the high outside temperature.[5]

Somewhere between 50 and 75% relative humidity is considered optimum.

■ **Insulation**—In cold climates, provide for R = 14 in the walls and R = 23 in the ceiling. Protect the insulation with 4-ml polyethylene plastic or equivalent.

■ **Ventilation**—In most areas, summer and winter ventilation requirements are so different as to require two different systems. For a completely enclosed slotted floor cattle barn, the fans should provide—

Winter—75 to 80 cfm/ 1,000-lb animal
Summer—225 cfm/ 1,000-lb animal

But the summer requirements in a completely enclosed building can be lessened by providing doors and/or hinged walls (high up), which can be opened during warm weather. Besides, this is good insurance when, and if, the electricity fails. (Also see Table 11-5 Recommended Environmental Conditions for Animals.)

FEED DELIVERY EQUIPMENT

Most feedlots transport the feed from the mill to the feedlot by truck, using either (1) a self-unloading truck, or (2) a self-mixing, self-unloading truck.

■ **Self-unloading trucks**—These trucks do not mix the feed; they merely convey and unload it. The feed is mixed at the mill, loaded into the truck bed, driven to the lots, and augered into the bunks. Then, the truck returns to the feed mill or silage pit for another load. Self-unloading trucks are designed to convey the feed to one end of the bed and discharge it to one side.

Fig. 28-21. Self-unloading truck in operation, conveying feed through the outlet spout into the feed bunk. (Courtesy, Agricultural Research Service, USDA, Clay Center, NE)

[4]Brody, Samuel, *Climate Physiology of Cattle*, Jour. Series No. 1607, Mo. Ag. Exp. Sta.

[5]*Confinement Rearing of Cattle*, USDA, ARS 22-89.

Some of these trucks, used by the larger lots, are very large—up to 20-ton capacity.

- **Self-mixing, self-unloading trucks**—These trucks, which are usually limited to 6- to 8-ton capacity, are well suited for use in feedlots with fewer than 5,000 cattle. They are equipped with their own bed mixing mechanism, which makes it possible to alleviate mixing at the mill. The separate ingredients are loaded at the mill, delivered to the lots, mixed in the truck bed, and conveyed through the outlet spout into the feed bunk. Self-mixing, self-unloading trucks are also well suited to adding silage to the concentrate mix.

Scales may be installed on either type of truck. Generally speaking, pen size and ownership of the cattle determine whether or not trucks are equipped with scales. Where cattle are custom fed and/or small lots of cattle are involved, scales are important. For big lots of cattle, and where the cattle are owned by the feedlot, scales are not so necessary.

Two-way radios are standard equipment on most southwestern feedlots.

Some small- to medium-sized feedlots use conveyors (auger or belt-type) to move feed from the mill to the cattle. Unless the mill and the cattle are in close proximity (not more than 1,000 to 1,200 ft apart), conveyors are likely to be more costly than trucks. Also, when it comes to changing the feed formula of cattle, they are less flexible than trucks. Confinement systems, with a great concentration of cattle and short runs, make conveyors more practical.

Among other feed delivery systems that are being, or have been, used are the following: a slurry system, in which the feed and water are mixed together and moved to the cattle; and pneumatic delivery.

Fig. 28-22. A manure spreader in use at a cattle feedlot. (Courtesy, Sperry-New Holland, New Holland, PA)

Fig. 28-23. Renovating (loosening) manure prior to removal from pens. (Courtesy, Benedict Feeding Co., Casa Grande, AZ)

MANURE CLEANING EQUIPMENT

Pens are generally cleaned after the cattle have been removed and before a new lot is brought in. Sometimes, a tractor skiploader is used to push manure from the concrete slab adjacent to the bunks and from around waterers, even when cattle are in the pen. However, when conditions are bad enough to require machine work in lots occupied by cattle, the corral is usually so wet that the equipment cannot operate efficiently. Thus, it is best to avoid such a condition by reducing the concentration of cattle in wet weather.

The most common manure cleaning equipment consists of a front-end loader attached to a tractor, along with a dump truck. The front-end loader is a scoop 4 to 5 ft wide, and 15 in. deep. The truck is a 4-wheel dump. Typically, feedlots with 1,000-head capacity use 1 tractor-loader and 1 truck; 5,000-head capacity lots use 1 tractor and 2 trucks; and 10,000-head capacity lots use 2 tractor-loaders and 4 trucks. The number of trucks needed is dependent on the distance that the manure must be hauled. The average load hauled from the pens is 7,000 lb.

Feedlots with large pens sometimes use self-loading wheel scrapers.

SPRAYING PEN; DIPPING VAT

Periodic and regular treatment for the control of lice, grubs, summer flies, and other parasites is essential to good management. For this purpose, a spraying pen or a dipping vat will be needed.

A spray pen should not be over 15 ft wide, and it should have solid sides, good drainage, and a rough-paved or gravel surface. This type of pen will keep the animals close enough to the spray nozzle to give good spray penetration and reduce drift of the spray materials.

Fig. 28-24. Dipping vat facility with curved chute, round crowd pen, and wide curved alley, located at the Red River Feedyard, Stanfield, Arizona. Three people can move 300 cattle per hour through this vat. (Courtesy, Temple Grandin)

Dipping vats are increasing in popularity. A properly constructed system of chutes and pens with a dipping vat can provide fast, positive control. A metal vat can be purchased or a concrete vat can be built. The vat is usually 28 to 32 ft long, plus an entry chute at one end and a drip pen on the other.

BACK SCRATCHERS

Some feedlots provide back scratchers in their feeding pens. A back scratcher is a horizontal suspended arm with a burlap-type cloth attached at cattle back level. This device is so arranged that the cloth is always saturated with insect repellent. The repellent is placed on the animal's back through contact with the scratcher, thereby preventing flies and other insects from molesting the cattle.

QUESTIONS FOR STUDY AND DISCUSSION

1. Using the nomograph, make a quick calculation of the justifiable investment per head on the basis of the following three factors which you will stipulate: (a) the percent of total investment to charge off as annual costs each year, (b) the gross profit expected per head, and (c) the return per head desired for labor, management, and interest on the capital required.

2. Select a certain area for feeding cattle. Then, give for that particular area the pros and the cons for each: (a) an open feedlot, (b) cold confinement, and (c) warm confinement. Finally, give your recommendation.

3. What factors should be considered when locating a new feedlot?

4. Describe desirable features in layout and pens of a feedlot.

5. Describe a desirable feedlot mound.

6. Give the specifications for (a) a feedline bunk, and (b) an apron.

7. Why are so many cattle chutes curved?

8. What purposes do feedlot lights serve?

9. Before building a new feedlot or expanding an old one, of what pollution factors should the owner be apprised?

10. What are the advantages and disadvantages of slotted floors?

11. How much manure storage will be needed for a confinement building housing 200 head of 1,000-lb steers for 4 months?

12. Give the specifications for desirable cattle shades.

13. In confinement, how much ventilation should be provided (a) in the winter, and (b) in the summer?

14. Under what circumstances would you recommend each of the following feed delivery systems: (a) self-unloading truck, (b) self-mixing, self-unloading truck, and (c) conveyors (augers; belts)?

15. How, and how often, would you remove the manure from an open feedlot?

16. What are the advantages and what are the disadvantages of each: (a) dipping, and (b) spraying?

17. Of what value are back scratchers?

18. What major changes do you foresee in cattle feedlot facilities and equipment from now until 2005 A.D.?

SELECTED REFERENCES

Title of Publication	Author(s)	Publisher
Beef Housing and Equipment Handbook, Third Edition	Midwest Plan Service	Midwest Plan Service, Iowa State University, Ames, IA, 1987
Cattle Feeding in the United States, Ag. Econ. Rpt. No. 186	R. A. Gustafson R. N. Van Arsdall	Economic Research Service, U.S. Department of Agriculture, Washington, DC, 1970
Feedlot, The, Third Edition	Ed. by G. B. Thompson C. C. O'Mary	Lea & Febiger, Philadelphia, PA, 1983
Handbook of Building Plans, Third Revised Edition	Midwest Plan Service	Midwest Plan Service, Iowa State University, Ames, IA, 1984
Improved Methods and Facilities for Commercial Cattle Feedlots, Marketing Res. Rpt. No. 517		Agricultural Marketing Service, U.S . Department of Agriculture, Washington, DC, 1962
Livestock Waste Facilities Handbook	Midwest Plan Service	Midwest Plan Service, Iowa State University, Ames, IA, 1984
Stockman's Handbook, The, Seventh Edition	M. E. Ensminger	Interstate Publishers, Inc., Danville, IL, 1992
Structures and Environment Handbook, Eleventh Edition	Midwest Plan Service	Midwest Plan Service, Iowa State University, Ames, IA, 1983

An open-sided shed for confinement feeding. Note the drive-through alley for servicing the continuous feed bunk. (Courtesy, *Beef* Magazine)

The right kind to feed. (Courtesy, American Polled Hereford Assn., Kansas City, MO)

KIND OF CATTLE TO FEED[1]

Contents *Page*

Contents *Page*

[1]In the preparation of this chapter, the authors gratefully acknowledge the authoritative review and helpful suggestions, along with pictures, of the following person, who has great expertise in the operation of a modern cattle feedlot: Ron Baker, C & B Livestock, Inc., Hermiston, Oregon.

Fig. 29-1. Cattle high and dry in an open pen, well-fenced feedlot. (Courtesy, Continental Grain Co., Ulysses, KS)

All kinds of cattle may be, and are, fed. But, for maximum success, it is imperative that the right kind of cattle be selected for a particular feedlot. The cattle should match the operator's available feed, labor, shelter, and credit. Also, it is imperative that there be a suitable market outlet following finishing; for example, it would be unwise to feed lightweight heifers in an area where the strongest slaughter market is for heavy steers; nor should one finish out heavy Holstein steers where the primary interest of packers is for Choice beef. But, assuming that a satisfactory slaughter outlet exists for different kinds of cattle, the following general guides will be helpful in determining what kind of cattle to feed in a given lot.

Fig. 29-2. Straightbred Angus steers in the feedlot of an Iowa farmer-feeder. (Courtesy, American Angus Assn., St. Joseph, MO)

Fig. 29-3. Holstein steers on feed at the Davidson Feedyard, Pecos, Texas. (Courtesy, *Livestock Weekly*, San Angelo, TX)

AGE AND WEIGHT OF CATTLE

A generation ago the term *feeder steer* signified to both the rancher and the Corn Belt feeder a 2½- to 3-year-old animal weighing approximately 1,000 lb. Today, cattle are referred to by ages as calves yearlings, and 2-year-olds. This shift to younger cattle has been brought about primarily by consumer demand for smaller and lighter cuts of meat and improved feeding and management practices.

The age of cattle to feed is one of the most important questions to be decided upon by every practical cattle producer. The following factors should be considered in reaching an intelligent decision on this point:

- **Rate of gain**—When cattle are fed liberally from the time they are calves, the daily gains will reach their maximum the first year and decline with each succeeding year thereafter. On the other hand, when in comparable condition, thin but healthy 2-year-old steers will make more rapid gains in the feedlot than yearlings; likewise yearlings will make more rapid gains than calves. Table 29-1 illustrates this situation.

TABLE 29-1
EFFECT OF AGE OF CATTLE ON DAILY GAINS OF MEDIUM-FRAME CATTLE

Age	Daily Gains			
	Average of U.S. Feedlots		Top 5% of U.S. Feedlots	
	(lb)	*(kg)*	*(lb)*	*(kg)*
Calves	2.4	*1.09*	2.7	*1.23*
Yearlings	2.8	*1.27*	3.2	*1.45*
Two-year-olds	2.9	*1.32*	3.4	*1.54*

Fig. 29-4. Large-frame crossbred feeder calves of the kind that have grown steadily in popularity with cattle feeders as they choose between feeding calves, yearlings, and 2-year-olds. Their increased feed efficiency and high percentage of Yield Grade 2 carcasses is causing them to sell at a premium. (Courtesy, C & B Livestock, Inc., Hermiston, OR)

■ **Economy of gain**—Calves require less feed to produce 100 lb of beef than do older cattle. This may be explained as follows:

1. The increase in body weight of older cattle is largely due to the deposition of high-energy fat, whereas the increase in body weight of young animals is due mostly to the growth of muscles, bones, and organs. Thus, the body of a calf at birth usually consists of more than 70% water, whereas the body of a fat two-year-old steer will contain only 45% water. In the latter case, a considerable part of the water has been replaced by fat.

In passing, it is noteworthy that there is much room for improvement in cattle feedlot efficiency—even with calves, for the 8.0:1 feed efficiency of yearling cattle must compete with 2.1:1 feed efficiency of the modern broiler industry.

For feedlot efficiency figures to be very meaningful, it is necessary to know what kind of ration was fed; otherwise, on a poundage basis, it is not unlike comparing steaks and carrots in the human diet. The more concentrated and the higher the energy value of the ration, the fewer the pounds of feed required to produce 100 lb of gain. Yet, many times it is in the nature of good business to feed rations that necessitate more pounds of feed to produce 100 lb of gain simply because lower cost gains can be produced thereby. This applies to feeding corn silage, potato waste, and many by-product feeds.

2. Calves consume a larger proportion of feed in proportion to their body weight than do older cattle.

3. Calves masticate and digest their feed more thoroughly than older cattle. Despite the fact that calves require less feed per 100 lb of gain—because of the high-energy value of fat—older cattle store as much energy in their bodies for each 100 lb of total digestible nutrients consumed as do younger animals.

From the above, it is apparent that age of cattle affects the pounds of feed required to produce 100 lb of gain—that the younger the cattle the greater the feed efficiency.

Table 29-2 points up the effect of age of cattle on feedlot efficiency.

TABLE 29-2
EFFECT OF AGE ON FEED EFFICIENCY
OF MEDIUM-FRAME STEERS[1]

Age	Average of U.S. Feedlots			Top 5% of U.S. Feedlots		
	Feed/Lb Gain	TDN/Lb Gain	Mcal/Lb Gain[2]	Feed/Lb Gain	TDN/Lb Gain	Mcal/Lb Gain[2]
	(lb)	*(lb)*	*(lb)*	*(lb)*	*(lb)*	*(lb)*
Calves	7.5	6.0	9.86	7.0	5.6	9.20
Yearlings	8.0	6.4	10.52	7.5	6.0	9.86
Two-year-olds and over . . .	8.5	6.8	11.17	8.0	6.4	10.52

[1]Air-dry basis (approximately 90% dry matter).

[2]Mcal metabolizable energy (ME) was calculated by assuming 1.6434 Mcal ME = 1 lb TDN.

A more accurate measure of feed efficiency than pounds feed per pound gain can be obtained through the use of energy conversion; the TDN or Mcal required to produce 1 lb of beef. It alleviates much of the inevitable disadvantage to which a relatively low-energy, bulky ration (such as a high-silage ration) is put when it is compared on a poundage basis to a more highly concentrated feed (such as an all-concentrate ration). Thus, energy requirements (in both TDN and Mcal per pound of gain) are given in Table 29-2.

■ **Flexibility in marketing**—Calves will continue to make satisfactory gains at the end of the ordinary feeding period, whereas the efficiency of feed utilization decreases very sharply when mature steers are held past the time that they are finished. Therefore, under unfavorable market conditions, calves can be successfully held for a reasonable length of time, whereas prolonging the finishing period of older cattle is usually unprofitable.

■ **Length of feeding period**—Calves require a somewhat longer feeding period than older cattle to reach comparable finish. To reach Choice condition, steer calves are usually full-fed about 7 to 8 months; yearlings 4 to 5 months; and 2-year-olds only about 3 to 4 months. Table 29-3 points up this situation. The longer finishing period required for calves is due to the fact that they are growing as well as finishing.

TABLE 29-3
EFFECT OF AGE OF CATTLE ON LENGTH OF FEEDING PERIOD OF MEDIUM-FRAME CATTLE

Age	Average of U.S. Feedlots	Top 5% of U.S. Feedlots
	(days)	*(days)*
Steer calves	200	180
Yearlings	140	130
Two-year-olds	110	100

■ **Total gain required to finish**—In general, calves must put on more total gains in the feedlot than older animals to attain the same degree of finish. In terms of initial weight, calves practically double their weight in the feedlot. On the average, yearlings increase in weight about 400 lb, and 2-year-olds increase their initial feedlot weight about 320 lb. Table 29-4 illustrates this situation.

TABLE 29-4
EFFECT OF AGE OF CATTLE ON TOTAL GAIN REQUIRED TO FINISH MEDIUM-FRAME CATTLE

Age	Average of U.S. Feedlots	Top 5% of U.S. Feedlots
	(lb)	*(lb)*
Calves	550	525
Yearlings	400	380
Two-year-olds	320	300

■ **Total feed consumed**—Because of their smaller size, the daily feed consumption of calves is considerably less than for older cattle. However, as calves must be fed a longer feeding period, the total feed requirement for the entire finishing period is approximately the same for cattle of different ages.

■ **Experience of the feeder**—Young cattle are bovine *babies.* As such, they must be fed more expertly. Thus, the inexperienced feeder had best feed older cattle.

■ **Kind and quality of feed**—Because calves are growing, it is necessary that they have more protein in the ration. Since protein supplements are higher in price than carbonaceous feeds, the younger the cattle the more expensive the ration. Also, because of smaller digestive capacity, calves cannot utilize as much coarse roughage, pasture, or cheap by-product feeds as older cattle.

Calves also are more likely to develop peculiar eating habits than older cattle. They may reject coarse, stemmy roughages or moldy or damaged feeds that would be eaten readily by older cattle. Calves also require more elaborate preparation of the ration and attention to other small details designed to increase their appetite.

■ **Comparative costs**—Calves usually cost more per 100 lb as feeders than do older cattle.

■ **Dressing percentage and quality of beef**—Older cattle have a slightly higher dressing percentage than calves or baby beef. Moreover, many consumers have a decided preference for the greater flavor of beef obtained from older animals.

From the above discussion, it should be perfectly clear that there is no best age of cattle to feed under any and all conditions. Rather, each situation requires individual study and all factors must be weighed and balanced.

BABY BEEF

In the 1920s and 1930s, the term *baby beef* referred to 4-H Club or FFA steers, marketed at 12 to 16 months of age, weighing 700 to 1,000 lb, grading Good (Select) to Prime.

The disrepute and demise of the term *baby beef* was caused by lack of enforceable high standards; to fill the demand, a wide assortment of cattle was marketed as baby beef.

Today, *baby beef* connotes heavy calves that are fat enough for slaughter at weaning time, weighing 400 to 700 lb on foot, the meat from which is lean and tender.

In its truest form, the production of baby beeves involved the breeding, rearing, and finishing of calves on the same farm. The first requirement of baby beef production was superior breeding. Secondly, calves intended for the baby beef route were never allowed to lose their baby fat. This called for dams that milked well and/or creep feeding the calves.

Baby beef production was well suited to the production of the maximum amount of beef from milk and grass, with a minimum of grain.

SEX OF CATTLE

More steers than heifers are fed, simply because more of them are available. A portion of the heifers is held back for replacement purposes. In the future, more young bulls may be fed. Thus, the feedlot operator must give consideration to the sex of cattle fed.

First, and foremost, market outlets should be considered.

Table 29-5 shows the effect of sex on rate of gain and feed efficiency. It is noteworthy that bulls gain more rapidly on less feed than steers, and that steers gain more rapidly on less feed than heifers.

TABLE 29-5
EFFECT OF SEX ON GAIN OF MEDIUM-FRAME CATTLE[1]

Sex	Average of U.S. Feedlots		Top 5% of U.S. Feedlots	
	Daily Gain	Feed/Lb Gain[2]	Daily Gain	Feed/Lb Gain[2]
	(lb)	*(lb)*	*(lb)*	*(lb)*
Heifers	2.4	7.5	2.7	6.9
Steers	2.8	6.9	3.2	6.6
Bulls	3.0	6.7	3.4	6.4

[1]To convert lb to kg, divide by 2.2.

[2]Air-dry basis (approximately 90% dry matter).

STEERS VS HEIFERS

On the market, cattle are divided into five sex classes: steers, heifers, cows, bullocks, and bulls. The sex of feeder cattle is important to the producer from the standpoint of cost and selling price (or margin), the contemplated length of feeding period, quality of feeds available, and ease of handling. The consumer is conscious of sex differences in cattle and is of the impression that it affects the quality, finish, and conformation of the carcass.

Steers are by far the most important of any of the sex classes on the market, both from the standpoint of numbers and their availability throughout the year, whereas heifers are second.

The relative merits of steers vs heifers, both from the standpoint of feedlot performance and the quality of carcass produced, have long been a controversial issue. Based on experiments[2] and practical observations, the following conclusions and deductions seem to be warranted relative to this question:

- **Length of feeding periods**—Heifers mature earlier than steers and finish sooner, thus making for a shorter feeding period. In general, heifers may be ready for the market 20 to 30 days earlier than steers of the same age started on feed at the same time.

- **Market weight**—The most attractive heifer carcasses are obtained from animals weighing 1,050 to 1,200 lb on foot, showing good condition and finish but not patchy and wasty.

- **Rate and economy of gain**—Because of their slower daily gains and lower feed efficiency, the feedlot gains made by heifers are usually somewhat more costly than those made by steers of the same age.

- **Price**—Because of existing prejudices, feeder heifers can be purchased at a lower price per pound than steers, but they also bring a lower price when marketed. Thus, the net return per head may or may not be greater with heifers.

- **Carcass quality**—In England, there is no discrimination in price against well-finished heifers. In fact, the English argue that the grain of meat in heifer carcasses is finer and the quality superior. On the other hand, the hotels, clubs, and elite butcher shops in the United States hold a prejudice against heifer beef.

Carefully controlled experiments have now shown conclusively that when heifers are marketed at the proper weight and degree of finish, sex makes no appreciable difference in the dressing percentage, in the retail value of the carcasses, or in the color, tenderness, and palatability of the meat.

- **Ease of handling in the feedlot**—Because of disturbances at heat periods, many feeders do not like to handle heifers in the feedlot.

- **Flexibility in marketing**—If the market is unfavorable, it is usually less advisable to carry heifers on feed for a longer period than planned because (1) of possible pregnancies, and (2) they become too patchy and wasty.

- **Effect of pregnancy**—Packer-buyers have long insisted that they are justified in buying finished heifers at a lower price than steers of comparable quality and finish because: (1) some heifers are pregnant and have a lower dressing percentage; and (2) pregnant heifers yield less desirable carcasses. In realization that the packer will lower the price anyway, many feeders make it a regular practice to turn a bull with heifers about 3 to 4 months before the market period. Such feeders contend that the animals are then quieter and will make better feedlot gains.

The economic loss that accrues from pregnant feedlot heifers was pointed up by two different studies made and reported by Monfort of Greeley, Colorado.[3] As a result of a 1983 survey of a number of cattle

[2]Bull, Sleeter, F. C. Olson, and John H. Longwell, Illinois Ag. Exp. Sta. Bull. 355.

[3]Bennett, Bill, *Animal Nutrition & Health*, "The Pregnant Feedlot Heifer."

feeders and packers, Monfort reported the following relative to feedlot heifers: (1) a pregnancy rate of 16.5% on incoming feedlot heifers, (2) feeders placed a lower value of $30.32 per head on pregnant vs open heifers, (3) lower dressing percentage at slaughter of 3.3% on pregnant vs open heifers, and (4) a cost of $5.29 per head to pregnancy check and attempt an abortion on each incoming pregnant heifer. Additionally, in an actual study of 10,000 head of heifers and 1,000 heiferettes (heiferettes are large, heavy heifers possessing nearly the size and development of a mature cow) slaughtered at Monfort's Greeley, Colorado, plant in 1983–84, it was found that pregnancy lowered the dressing percentage on heifers by 5.6%, and on heiferettes 6.1%. Thus, the bottom line is that feeding pregnant heifers is costly.

In light of the preceding facts, the following management options may be considered when feeding heifers of unknown pregnancy status:

1. Feed heifers like steers, and meet the calving problems (difficult births, and caring for newborn calves) as they occur.

2. Buy only open or spayed heifers, the supply of which is limited.

3. Pregnancy examine all heifers and use an abortive agent on the pregnant ones, according to directions.

By pregnancy testing and the use of an abortifacient, the termination of early pregnancy can be brought about. However, the cost and setback in performance may not justify such action.

The trade in feeder cows and heifers assumes considerable volume only in the fall and early winter—at the close of the grazing season when the farmer or rancher is culling the herd and prior to the start of the wintering operations. When market conditions are favorable and an abundance of cheap roughage is available, cows may often be fed at a profit.

When there is considerable demand for cheap meats, the feeder may find it profitable to finish old bulls and stags. Usually it is difficult to purchase such animals in large numbers. Here, as with the finishing of old cows, the feeder should plan to utilize the maximum of cheap roughage.

SPAYED HEIFERS

Spaying prevents heifers from becoming pregnant and eliminates the necessity of separating heifers from bulls or steers. Also, some buyers pay a slight premium for spayed heifers. But, most early experiments showed that spayed heifers make less rapid gains and require more feed per 100-lb gain than open (control) heifers.[4] However, spaying by an experienced technician, using one of the new techniques, followed by a implanting the spays with a growth promotant(s), provides a sure way of preventing in-heat heifers and pregnancy, and assures top gains.

(Also, see Chapter 10, section on "Spaying.")

BULLS

The feeding of bulls (uncastrated males) instead of steers has been standard practice throughout Europe for many years. For example, since about 1954 Germany has, for the most part, fed out and slaughtered bulls as yearlings, instead of steers, because they obtain 10 to 15% greater rate of gain and feed efficiency thereby. The practice may increase in the United States now that carcasses from young bulls are federally graded as *bullock beef* rather than *bull beef*, thereby removing the connotation that the meat is inferior to or different from steer or heifer beef.

The carcasses from older bulls are still labeled *bull beef*, to differentiate them from the carcasses of younger bulls. Bullock beef from young bulls is graded according to the same quality standards as beef from steers and heifers.

Also, the economics of the situation favor the feeding of bulls instead of steers. The male hormones secreted by the testicles are excellent growth stimulants and will improve gain and feed efficiency by 10 to 15%. Also, bulls will produce more healthful lean meat than steers, and research has shown that bull meat is equal in value, quality, and palatability to steer meat.

Now that the carcasses from young bulls are differentiated from older bulls, only consumer acceptance remains.

The following guidelines are recommended in the feeding of bulls:

1. Start young bulls on full feed at weaning age (6 to 7 months) and feed out as rapidly as possible to a market weight of 1,100 lb.

2. Use high-energy rations for bull feeding, because they tend to grow rapidly and lay down less fat than steers.

3. Feed out bulls so that they are finished for market before 18 months of age.

4. Do not add new bulls to the pen after the weanling bulls are started on feed, because this tends to encourage fighting and riding, and results in reduced

---

[4]Dinussion, W. E., F. N. Andrews, and W. M. Beeson, *Journal of Animal Science*, Vol. 9, p. 321; Gramlich, H. J., and R. R. Thalman, Nebraska Ag. Exp. Sta. Bull. 252; Hart, G. H., H. R. Guilbert, and H. H. Cole, California Ag. Exp. Sta. Bull. 645; Langford L. H., R. J. Douglas, and M. L. Buchanan, North Dakota Ag. Exp. Sta. Bimonthly Bull., Vol. XVIII, No. 2.

gains. Even under the best of management, fighting and riding may be a problem.

5. Keep bulls separate from other cattle when marketing them. If possible, do not permit the bulls to stand in the pen overnight before slaughter.

6. Bulls of beef breeding are less nervous and ride less than bulls of dairy breeding.

GRADE OF CATTLE

The most profitable grade of cattle to feed will generally be that kind of cattle in which there is the greatest spread of margin between their purchase price as feeders and their selling price as fat cattle. As can be readily understood, one cannot arrive at this decision by merely comparing the existing price between the various grades at the time of purchase. Rather, it is necessary to project the differences that will probably exist, based on past records, when the animals are finished and ready for market.

As fewer grain-fed cattle are marketed in the summer and fall, the spread in price between Good- and Choice-fed cattle and those of the lower grades is usually the greatest during this season. On the other hand, the spread between these grades is likely to be least in late winter and early spring, when a large number of well-finished cattle are coming to market from the feedlots. However, such seasonal effects are minimal today, due to large feedlots and year-round feeding.

The length of the feeding period and the type of feed available should also receive consideration in determining the grade of cattle to feed. Thus, for a long feed and when a liberal allowance of grain is to be fed, only the better grades of feeders should be purchased. On the other hand, when a maximum quantity of coarse roughage is to be utilized and a short feed is planned, cattle of the medium or lower grades are most suitable. Thus, successful cattle feeders match the quality of the cattle selected with the quality of the available feed; the better the feed the higher the grade of cattle.

Cattle of the lower grades should be selected with very special care to make certain that only thrifty animals are bought. Ordinarily, death losses are much higher among low-grade feeder cattle, especially when the low-grade animals are calves. The death loss in handling average- or high-grade feeders seldom exceeds 1 to 2%; whereas with *cull* or *dogie* cattle, it frequently is 2 or 3 times this amount. Many low-grade cattle are horned, and dehorning further increases the death risk—in addition to the added labor and shrinkage resulting therefrom.

No given set of rules is applicable under any and all conditions in arriving at the particular grade of cattle to feed, but the following factors should receive consideration:

1. The feeding of high-grade cattle is favored when—
    a. The feeder is more experienced.
    b. A long feed with a maximum of grain in the ration is planned.
    c. Conditions point to a wide spread in price between grades at marketing time. Such conditions normally prevail in the late summer or early fall.
2. The feeding of average or low-grade cattle is favored when—
    a. The feeder is less experienced.[5]
    b. A short feed with a maximum of roughage or cheap by-products is planned.
    c. Conditions point to a narrow spread in price between grades at marketing time. Such conditions normally prevail in the spring.
3. In addition to the profit factors enumerated above, it should be pointed out that with well-bred cattle the following conditions prevail:
    a. Well-bred cattle possess greater capacity for consuming large quantities of feed than steers of a more common grade, especially during the latter part of the feeding period.
    b. The higher the grade of cattle, the higher the dressing percentage and the greater the proportionate development of the high-priced cuts.
    c. The higher the grade of the cattle, the greater the opportunities for both profit and loss.
    d. There is a great sense of pride and satisfaction in feeding well-bred cattle.

Certainly producers who raise their own feeder cattle should always strive to breed high-quality cattle, regardless of whether they finish them personally or sell them as feeders. On the other hand, purchasers of feeder steers can well afford to appraise the situation fully prior to purchasing any particular grade.

BREEDING AND TYPE OF CATTLE

Although the supporting data are rather limited, it is fully realized that there is considerable difference between individual animals insofar as rate and economy of gain is concerned.

The University of California has compared the feedlot performance and carcass quality of Okie and Hereford calves and yearlings. Contrary to the impli-

[5]In general the inexperienced feeder should stick to the middle kind and leave the extremes—the fancy and the plain cattle—to the feeder with experience.

cation of the name, Okie cattle do not necessarily originate in Oklahoma. Rather, they are cattle of nondescript breeding, including both beef and dairy background, that originate on small farms in the southern part of the United States.

The California studies showed that—

1. Herefords gained faster, utilized their feeds more efficiently, and graded higher than Okies.
2. Okies were upgraded from feeder to slaughter stage, whereas Herefords maintained their grade.

No claim can be made that the animals used in the California studies were representative of all cattle of the breeds and strains studied, for, when limited to small numbers, it is impossible accurately to select a representative sample of a widely scattered breed or strain.

CROSSBREDS

Good crossbreds will likely show 2 to 4% improvement over the average of the parent breeds for rate and efficiency of gains. Additionally, even larger advantages accrue to the cow-calf producer. Thus, it is inevitable that an increasing number of crossbred feedlot cattle will be seen. The primary characteristics desired in feedlot cattle are the same, whether they be crossbreds or straightbreds; namely, (1) high rate of gain, (2) efficient feed conversion, (3) high cutout percent, and (4) tender, palatable beef.

Fig. 29-5. Crossbred cattle showing Brahman breeding, in a Nebraska feedlot. (Courtesy, *Beef* magazine)

DAIRY BEEF

Dairy beef accounts for about 15% of the beef consumed in this country, with these animals marketed as veal calves, cull dairy cows and bulls, and finished dairy heifers and steers. Improvements in the science and technology of feeding and processing favor growing and finishing dairy beef, and minimum slaughter of veal calves.

Dairy beef is just what the term implies—beef derived from cattle of dairy breeding, or from dairy X beef crossbreds. Today, it's extolled with pride.

Fig. 29-6. Dairy beef in the making. Holstein steers on feed in a commercial feedlot near Corona, Calif. (Courtesy, Carnation Milling Co., Los Angeles, CA)

But dairy beef hasn't always enjoyed status. Prior to about 1960, few self-respecting cattle feeders would admit to finishing out dairy cattle. Given a choice between (1) topping the market with a uniform load of well-bred beef steers, even if they were fed at a loss, and (2) making money by feeding cattle of dairy breeding, most cattle feeders would have elected the first alternative—that is, they would have done so until recent years. They derived much satisfaction from topping the market, and they took pride in telling their neighbors about it. Likewise, meat packers were reluctant to have visitors see yellow-finished carcasses in their coolers, because of the yellow fat being indicative of dairy breeding or grass finish. Only the presence, suspicioned if not real, of goat carcasses was more humiliating to a packer. The near-contempt formerly evoked by cattle of dairy breeding was further evidenced on the nation's terminal markets by the names that were applied to them. Holsteins were known as *magpies*, and Jerseys were known as *yellow hammers*—terms which were neither endearing nor appetizing.

But time was! Today's cattle feeders are primarily concerned with rate and efficiency of gains, and net returns. As a result, most of them would just as soon feed steers of dairy breeding, either purebreds or crossbreds; some actually prefer them. Consumers demand beef that has a maximum of lean, with a minimum amount of waste fat, and which is tender and flavorful; and they couldn't care less whether it comes from a "critter" that was black, white-faced, roan, pink, yellow, or polka dot. As a result, more and more steers of dairy breeding are going the feedlot route, rather than as veal. As evidence of this transition, during the 50-year period 1942–1992, U.S. per capita veal consumption declined from 8.2 to 1.2 lb. Also, the shift in consumer demand to more lean and less fat has been reflected in the changed federal grades of beef. As a result, when properly fed, Holstein steers will make Choice, Select, or Standard grade.

■ **Britain's *barley beef***—The popularity of dairy beef traces to Europe. It was pioneered by Dr. T. R. Preston, well-known Scottish animal scientist, in Scotland and throughout the United Kingdom, where it became known as *barley beef.* Holstein-Friesian bull calves (not steers) were fed on all-concentrate rations consisting chiefly of barley. The British reported gains of 2½ to 3 lb per day, slaughter weights of 900 lb in less than 1 year of age, and lifetime feed conversions under 5 to 1.

■ **High growth thrust essential**—For a dairy beef program to be most successful, scientists and cattle producers in both Britain and the United States are agreed that the animals should have a high growth potential, as evidenced by heavy birth weight and heavy weight at maturity. Since Holsteins are heavy at birth and mature out at around 1,400 lb, in comparison with mature weights of 1,000 to 1,200 lb of the European beef breeds, it can be readily understood that Holsteins are ideal when it comes to producing dairy beef.

■ **Decide on feeding program**—There is, of course, no one best system of producing dairy beef for any and all conditions. As is true in any type of cattle feeding program, the operator should make the best use of those feeds that are readily available at the lowest possible cost. Then, these feeds must be combined into satisfactory rations, with consideration given to both economy and probable market price of finished cattle of various weights, grades, and degrees of finish.

■ **High-energy rations; light market weights**—If dairy steers are to be slaughtered at young ages and light weights, high-energy (low-roughage) rations are imperative. Under this system, usually young calves of either dairy or dairy X beef breeding are fed in confinement—in barns; and fed milk replacers from 1 to 4 days of age to 200 to 300 lb; and are full fed a high-concentrate ration from about 300 lb to market weight of 750 to 950 lb. Essentially, this is the *barley beef* program of Europe.

Crowding for market at an early age takes advantage of the fact that growth is generally most economical when most rapid, and that young gains are cheap gains. Also, experience shows that when Holstein calves are started on super energy rations at around 350 to 450 lb weight and marketed under 1,100 lb, (1) there's excellent marbling with very little bark (outside fat), and (2) many of these animals will grade Choice.

■ **High-roughage rations; heavy market weights**—If roughages are relatively more abundant and cheaper than concentrates, then it may be more remunerative to feed dairy beef more roughage and market at heavier weights—and with it to expect slower and less efficient gains. In any event, it's net returns that count, rather than rate of gain and pounds of feed required per pound of gain.

Under the high-roughage system, steers of dairy breeding are grown on maximum roughage to 600 to 750 lb weight, following which the ratio of concentrate to roughage is increased. Most dairy steers fed according to this system are marketed at weights of 1,150 to 1,400 lb grading Select or Commercial. (Most of them are too old to grade Standard, and lack the necessary marbling to grade Choice.)

■ **Dairy beef has good potential**—There is ample evidence that male calves of the larger dairy breeds (Holsteins and Brown Swiss), along with dairy beef crosses, have the potential for producing acceptable beef with good feed efficiency, under a system of either (1) full-feeding from an early age on a high-energy ration, or (2) growing and finishing on a maximum of roughage and marketing at older ages and heavier weights. In the final analysis, therefore, the system selected should be determined by net returns. Both methods necessitate the rearing of young I- to 4-day-old calves to weights of around 300 lb, with such early rearing done by either a calf-raising specialist or the cattle feeder who will do the ultimate finishing. Such calves must usually be obtained from over a wide area, and of variable ages and sizes; hence, they are difficult to come by. Also, death losses are frequently high and discouraging.

Both commercial cattle feeders and dairy operators are showing increased interest in producing dairy beef, with the result that there is competition between them. More dairy beef will be produced in the future.

QUESTIONS FOR STUDY AND DISCUSSION

1. Why is it important that the cattle should match the operator's available feed, labor, shelter, and credit?

2. Discuss how each of the following enters into the choice of the kind of cattle to feed in a given lot:
   a. Age and weight of cattle.
   b. Sex of cattle.
   c. Grade of cattle.
   d. Breeding and type of cattle.

3. Should cattle feeders achieve fewer pounds of feed per 100 lb gain by going to high-energy rations, like broiler rations?

4. Will more young bulls be fed in the U.S. in the future? Justify your answer.

5. Why do the English prefer heifer beef, whereas Americans prefer steer beef?

6. Would you recommend that a cattle feeder breed all heifers about 3 to 4 months before marketing? Justify your answer.

7. Discuss the advantages and the disadvantages of preventing heifers from coming in heat by spaying.

8. If the feeding of average or low-grade cattle is favored when the feeder is less experienced should a beginner who feeds out his/her own calves breed average or low-grade cattle? Justify your answer.

9. What are Okie steers? Under what circumstances would you buy and feed Okie steers instead of Hereford steers?

10. What advantages may accrue from feeding crossbreds rather than straightbreds?

11. Will more dairy beef be fed in the future? Who will feed the dairy beef of the future—the dairy operator, or the specialized feedlot feeder? Justify your answer.

12. What is *barley beef*?

13. What breeds of cattle are particularly suited to the production of dairy beef?

14. When feeding dairy beef, what will determine the choice between—(a) high-energy rations and light market weight; vs (b) high-roughage rations and heavy market weight?

SELECTED REFERENCES

Title of Publication	Author(s)	Publisher
Beef Cattle, Seventh Edition	A. L. Neumann	John Wiley & Sons, Inc., New York, NY, 1977
Beef Cattle Science Handbook	Ed. by M. E. Ensminger	Agriservices Foundation, Clovis, CA, pub. annually 1964–1981
Feedlot, The, Third Edition	Ed. by G. B. Thompson, C. C. O'Mary	Lea & Febiger, Philadelphia, PA, 1983

Finishing cattle on feed in a farmer-feeder lot. Note the board windbreak and the snow-covered bank in the background. (Courtesy, American-International Charolais Assn., Kansas City, MO)

FEEDING FINISHING (FATTENING) CATTLE[1]

Contents *Page*

[1]The principles of beef cattle nutrition are covered in Chapter 8; hence, they will not be repeated in this chapter. Instead, the application of nutrition to cattle finishing will be covered in this chapter.

Fig. 30-1. Feeding cattle the modern way. Feed truck, equipped with automatic augering system, is shown conveying feed through the outlet spout into the feed bunk. This is part of the facilities and equipment at Farr Feeders, Inc., Greeley, Colorado, a fully computerized commercial cattle feedlot capable of producing for market more than 80,000 head of cattle a year. (Courtesy, Farr Feeders, Inc., Greeley, CO)

An understanding of digestion is essential to intelligent feeding of finishing cattle. Fig. 30-2 shows the location of the parts of the ruminant stomach and the processes that occur in energy digestion and metabolism.

Digestion of feedstuffs in ruminants is primarily a fermentation process that occurs in the rumen. This allows ruminant animals to use both roughages and grains as sources of carbohydrates for energy. Part of the carbohydrates pass through the rumen and are digested in the abomasum and small intestine. Most carbohydrates in feeds are converted to either acetic, propionic, or butyric acid by rumen bacteria and protozoa. These short-chain fatty acids are then absorbed through the rumen wall into the bloodstream and are eventually used for energy in body tissue.

The major nutritional requirements of finishing cattle are: energy, protein, minerals, vitamins, and water.

About 75% of the cost of finishing cattle, exclusive of the purchase price of the feeders, is feedstuffs—grain, hay, silage, and miscellaneous wastes and

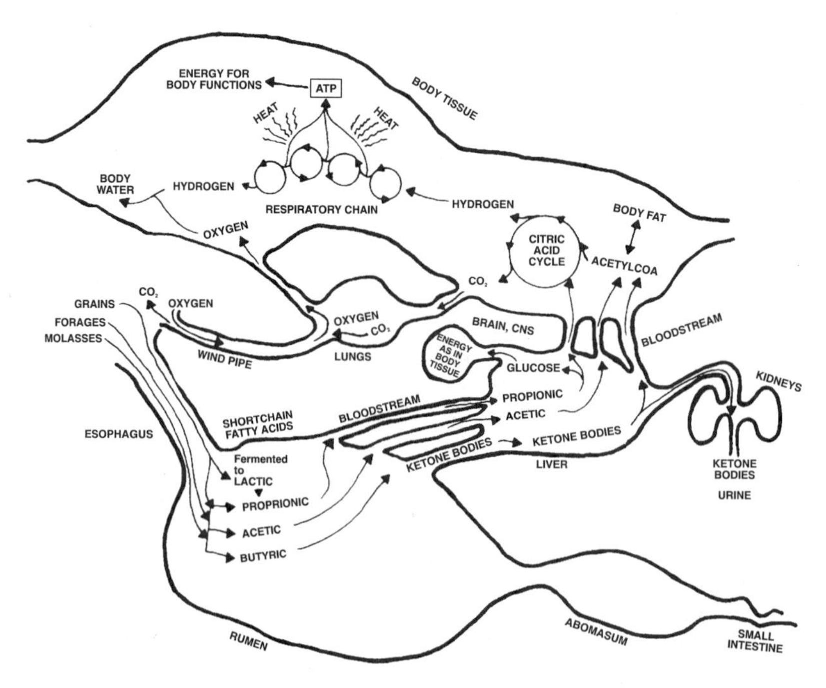

Fig. 30-2. Energy digestion and metabolism in ruminants. (Source: *Great Plains Beef Cattle Feeding Handbook*, GPE-1000)

byproducts. The greatest need is for energy. Of course, net profit depends on how much of that energy can be converted to pounds of gain—and how efficiently.

EXPRESSING ENERGY VALUES OF FEEDS

Energy is the largest cost item in the ration. Thus, the value of a feed ingredient is based primarily upon the energy which it will provide the animal, assuming it is not used specifically as a supplement to provide protein, minerals, vitamins, or additives. Cereal grains are higher than roughages in energy. Although grains usually cost more on a weight basis than roughages, they are often a cheaper source of energy for beef rations.

Broadly speaking, two methods of measuring energy are used in this country—the total digestible nutrient (TDN) system, and the calorie system. Both systems, along with energy losses and terms used to express the energy value of feeds, are discussed in Chapter 8 of this book, under the heading "Energy."

Calories are used to express the energy value of feedstuffs. One calorie is the amount of heat required to raise the temperature of 1 g of water 1°C. One kcal = 1,000 calories. One mcal or therm = 1,000,000 calories.

Through various digestive and metabolic processes, much of the energy in feed is dissipated as it passes through the animal's digestive system. About 60% of the total combustible energy in grain and about 80% of the total combustible energy in roughage is lost as feces, urine, gases, and heat.

Energy losses occur in the digestion and metabolism of feed. Measures that are used to express animal requirements and the energy content of feeds differ primarily in the digestive and metabolic losses that are included in their determination.

The net energy values of feeds are different for maintenance and production. Roughages compare more favorably with grain for maintenance than for production.

RATION FORMULATION

Feedlot cattle have access only to the rations provided by the caretaker. It is important, therefore, that cattle feedlot rations be balanced, and that they make for maximum net returns.

In addition to considering changes in availability of feeds and feed prices, ration formulation should be altered at stages to correspond to weight increases in the cattle.

Some suggested rations that may serve as useful guides are given in Chapter 8, Table 8-13. Also, methods of balancing rations are given in Chapter 8.

Ration formulation consists of combining feeds to make a ration that will be eaten in the amount needed to supply the daily nutrient requirements of the animal. The nutritive requirements of finishing cattle are given in the Appendix, Section I of this book. The nutrient compositions of feeds commonly fed to beef cattle are given in the Appendix, Section II. In these tables, the net energy of each feed is partitioned into energy used for body maintenance and for production; thus, the net energy values in megacalories (mcal) per pound are given for each feed for maintenance (NE_m) and for gain (NE_g).

Additional details relative to ration formulation are given in Chapter 8; hence, they will not be repeated in Chapter 30.

FEEDS

The growth of the cattle feeding industry of America has gone hand in hand with the production and feeding of more grains and byproduct feeds. Such feeds, which are high in energy and low in fiber, are usually the most economical part of a finishing ration. Hence, their availability influences the location and type of feeding program. Roughages, which require relatively more energy to digest and metabolize than grains, are used at low levels in most finishing rations. However, they are important in growing programs or in warm-up rations and they are even more important for maintenance of the breeding herd.

For convenience, the commonly used beef cattle finishing feeds are herein classified as (1) concentrates, (2) byproduct feeds, (3) protein supplements, (4) roughages, (5) minerals, (6) vitamins, and (7) additives.

CONCENTRATES

Concentrate feeds are those which are high in energy and low in fiber. Many different kinds of concentrate feeds can be, and are, used in beef cattle finishing. Availability and price are the two most important factors determining the choice of concentrates. Consideration of the latter factor—price—necessitates that the cattle feeder be a keen student of values, and must change the formulations of ration(s) in keeping with comparative feed prices.

Corn is the most common and the most desirable grain used in finishing cattle. In 1992, the United States produced 9.5 billion bu of corn, in comparison with 0.9 billion bu of grain sorghum that same year. Corn is palatable and rich in the energy-producing carbohydrates and fats, and low in fiber. Also, corn is easily stored, only moisture and carotene being lost over a period of time. However, corn has certain very definite limitations—it is low in protein and calcium.

The grain sorghums are assuming an increasingly important role in cattle feeding, particularly in the fringe areas of the Corn Belt, and in the South and Southwest where moisture conditions are less favorable. New and high-yielding varieties have been developed and have become popular. As a result, more and more grain sorghums are being fed to cattle. The chemical composition of sorghum (milo) is similar to corn except that the protein content is generally higher and more variable. Its feeding value is greatly enhanced by steam processing and other similar methods of preparation.

Although corn and sorghum are by far the most common grains used in finishing steers, such grains as barley, rye, oats, and wheat are used in many sections of the United States and Canada. The small grains are excellent for finishing cattle when properly used. In comparison with corn feeding: (1) barley-fed cattle are more susceptible to bloat (for this reason, it is best not to use a straight legume hay along with a grain ration high in barley; a mixture of barley and dried beet pulp is commonly used in the West); (2) barley- or wheat-fed animals are more apt to tire of their ration during a long feeding period; (3) rye should not constitute more than one-third of the grain ration because it is unpalatable; (4) more care is necessary to prevent wheat-fed cattle from going off feed; and (5) oats should not constitute more than one-half the ration, and preferably not more than one-third, because of its hulk. Fortunately, these limitations can be lessened considerably by mixing these feeds together, or by mixing the cereal grain with beet pulp, silage, or chopped hay. Also, it is important that the small grains be coarsely ground or properly rolled. It is recognized that wheat and oats are frequently too expensive to include in cattle finishing rations.

HIGH-MOISTURE (Early Harvested) GRAIN

High-moisture grain, especially corn and sorghum, is grain that is harvested at 24 to 30% moisture content and stored in an airtight silo. Experimental tests show that the feed efficiency of a high-moisture milo is improved by 8 to 15%, although there is little increase in daily gain. There is less improvement from high-moisture processing of shelled corn, and the results have been most variable. High-moisture milo and corn should be ground or rolled before it is fed. It is questionable, however, if it pays to process high-moisture corn in rations that have less than 15% roughage.

Fig. 30-3. Three silos in use at an Iowa cattle feedlot; two units for forage, and one unit for high-moisture corn. (Courtesy, Dave Brown and Associates, Oak Brook, IL)

■ **Moisture is important when buying feeds**—When buying grains, a feeder should never lose sight of how much water is being purchased. Table 30-1 illustrates the relative value (dry matter purchased) when paying for corn on a 15.5% moisture basis while actually

TABLE 30-1
RELATIVE VALUE OF U.S. NO. 2 CORN (15.5% Moisture) AS EFFECTED BY CHANGES IN MOISTURE[1]

Moisture	DM Basis Multiplier	Moisture	DM Basis Multiplier
(%)		(%)	
0	1.1834	19	0.9586
1	1.1716	20	0.9467
2	1.1598	21	0.9349
3	1.1479	22	0.9231
4	1.1361	23	0.9112
5	1.1243	24	0.8994
6	1.1124	25	0.8876
7	1.1006	26	0.8757
8	1.0888	27	0.8639
9	1.0769	28	0.8521
10	1.0651	29	0.8402
11	1.0533	30	0.8284
12	1.0414	31	0.8166
13	1.0296	32	0.8047
14	1.0178	33	0.7929
15	1.0059	34	0.7811
16	0.9941	35	0.7691
17	0.9822	36	0.7574
18	0.9704		

[1]If 15.5% moisture corn is the purchase basis, it will require 1.1834 units of purchase base corn to make 1 unit of 100% dry matter base corn. (From: *Great Plains Beef Cattle Feeding Handbook*, GPE-1602)

receiving corn of another moisture content. Thus, if the feeder were receiving 19% moisture corn and paying for 15.5% moisture, he/she would receive only 95.86% of the dry matter purchased. On the other hand, if corn is delivered with 7% moisture, while paying on a 15.5% moisture basis, the feeder would receive 110.06%.

■ **Moisture is important in formulating rations**—A careful feeder must constantly watch the moisture content of the feeds bought, and the effect of moisture on nutritional quality control. Most good feeders will readjust feeding formulas whenever moisture in a leading ingredient changes over 1%.

The way in which moisture changes cause imbalances is pointed up in the following example:

> Let's assume that a feeder is using a ration which has as one of its main ingredients corn silage with 68% moisture content, and that this ration requires 1.9% supplement on an *as fed* basis. Now assume that the moisture of the silage suddenly decreased to 55%, and with it the necessary supplement to balance the ration increased to 2.62%. Obviously, if the feeder did not adjust the feeding formula, a serious shortage of protein could result. In this case, the cattle would receive only 72.5% as much supplement as they should have since the mixing formula was not recalculated.

The multipliers in Table 30-2 may be used to determine the price per unit of dry matter simply by multiplying price times the appropriate factor for the indicated moisture.

■ **Acid-treated high-moisture grain**—In the past, high-moisture grain has been either (1) artificially dried, or (2) stored in an airtight silo, to prevent spoilage. Now there is a third alternative—the use of naturally occurring acetic and propionic acids for reducing mold growth and other deterioration of high-moisture grain. Tests with acid-treated (sprayed on at the time of storage) high-moisture grain suggest that rate of gain and dry matter conversion are at least equal, if not superior, to artificially dried or airtight stored grain.

TABLE 30-2
CORRECTION FACTORS TO USE WHEN CONVERTING FEEDS OF VARIOUS MOISTURE CONTENTS TO A 100% DRY MATTER BASIS (0% Moisture)[1]

Moisture	100% DM Basis Multiplier	Moisture	100% DM Basis Multiplier	Moisture	100% DM Basis Multiplier
0	1.0000	29	1.4084	58	2.3809
1	1.0101	30	1.4285	59	2.4390
2	1.0204	31	1.4492	60	2.5000
3	1.0309	32	1.4705	61	2.5641
4	1.0416	33	1.4925	62	2.6315
5	1.0526	34	1.5151	63	2.7020
6	1.0638	35	1.5384	64	2.7777
7	1.0752	36	1.5625	65	2.8571
8	1.0869	37	1.5873	66	2.9411
9	1.0989	38	1.6129	67	3.0303
10	1.1111	39	1.6393	68	3.1250
11	1.1235	40	1.6666	69	3.2258
12	1.1363	41	1.6949	70	3.3333
13	1.1494	42	1.7241	71	3.4482
14	1.1627	43	1.7543	72	3.5714
15	1.1765	44	1.7857	73	3.7037
16	1.1904	45	1.8181	74	3.8461
17	1.2048	46	1.8518	75	4.0000
18	1.2195	47	1.8867	76	4.1666
19	1.2345	48	1.9231	77	4.3478
20	1.2500	49	1.9607	78	4.5454
21	1.2658	50	2.0000	79	4.7619
22	1.2820	51	2.0408	80	5.0000
23	1.2987	52	2.0833	81	5.2631
24	1.3157	53	2.1276	82	5.5555
25	1.3333	54	2.1739	83	5.8824
26	1.3513	55	2.2222	84	6.2500
27	1.3698	56	2.2727	85	6.6666
28	1.3889	57	2.3255		

[1]From: *Great Plains Beef Cattle Feeding Handbook*, GPE-1602.

MOLASSES

Molasses—cane, beet, corn, and wood—is palatable (wood molasses is the least palatable) and a good source of energy; and cane molasses is also a good source of certain trace minerals. All molasses are low in protein. As a rule of thumb, molasses should not cost more than three-fourths as much as cereal grain, pound for pound, to be economical. Generally speaking, molasses is limited to 5 to 10% of the ration, although up to 15, or even 20%, can be added if the mixing facilities will handle it and the price is favorable.

FATS

Feeding of fats was prompted in an effort to find a profitable outlet for surplus packing house fats. For the most part fats were formerly used for soapmaking, but they are not used extensively in detergents. Thus, with the rise in the use of detergents in recent years, they became a *drug on the market*.

Animal and vegetable fats seem to be equally effective additions to feedlot rations; thus, selection should be determined solely by comparative price.

Ordinarily, animal fats are much cheaper than such vegetable fats as soybean or cottonseed oil.

Several different fat products are used as cattle feed. Each of them should be bought by specifications and guarantees. Pertinent specifications follow:

1. **Acidulated soap stock (foots).** This end product may be composed of acidulated soap stock from all vegetable and animal sources which have been washed free of mineral acids.
2. **Tallows.** This end product shall be composed of rendered, clean, disease-free, filtered animal fats.
3. **Greases (white and yellow).** This end product shall be composed of rendered, clean, disease-free, filtered animal fats and/or restaurant greases.
4. **Blended feeding fat.** This end product may be composed of rendered animal fat, animal grease, restaurant grease, vegetable oil, or acidulated soap stock in any combination.
5. **Other materials.** House grease, brown grease, sewer grease, modified yellow grease, and other low-grade materials are sometimes used.

Fat serves the following three practical functions when added to cattle rations:

1. **It increases the caloric density of the ration.** This appears to be very important with cattle on feed in hot climates and during the summer months. Since fat contains approximately 2¼ times as much energy as soluble carbohydrate, it is possible to increase the energy content with little increase of the bulk of the ration. Thus, with the same feed intake, energy intake is higher.
2. **It controls dust.** Thus, the addition of 1 to 2% fat materially lessens the dust involved in hay grinding. Also, it is well known that dusty rations are not consumed readily by cattle; hence, fat enhances consumption from this standpoint.
3. **It lessens the wear and tear on feed mixing equipment.** This is important, because both breakdowns and new equipment are costly.

It is recommended that 2 to 5% fat be added to high-concentrate rations in which milo, barley, and/or wheat are the chief grain sources. Higher levels of fat usually result in drastically lowered feed consumption. When fed at a 2 to 5% level, the energy value of fat is approximately 2¼ times that of the grains. When corn is the major source of grain, fat additions can be expected to be less useful than with the small grains. This is understandable when it is realized that corn contains approximately 4% fat as compared to 1 to 1½% for the other feed grains.

When fat is added, the calcium and phosphorus levels of the ration should be 0.55% and 0.33%, respectively.

A rule of thumb that feeders can use to determine how much they can afford to pay for supplemental fat is that it is worth 2.5 times the cost of grain. So, if corn costs 6.5¢ per pound, feeders can afford to pay 16.5¢ per pound for fat (6.5 × 2.5= 16.25).

ALL-CONCENTRATE AND HIGH-CONCENTRATE RATIONS

Based on experiments and experiences, the following conclusions relative to all-concentrate and high-concentrate rations appear to be justified:

1. Ruminants need some *roughness factor* or *scratch factor* to stimulate the rumen papillae for normal functioning. In high- or all-concentrate rations, this can be achieved partially by rolling or coarse grinding.
2. With the possible exception of whole shelled corn, and such high-fiber feeds as oats and barley, a high level of management is needed to make an all-concentrate system work under feedlot conditions. Problems associated with high-concentrate rations include acidosis, founder, and liver abscesses.
3. Some research indicates that continued high levels of performance can be better maintained by including 10 to 15% roughage in high-concentrate rations.
4. Rations having a concentrate content of 90% or more should be self-fed, and a liberal amount of feed should be available at all times.
5. Feed efficiency is improved in high- and all-concentrate rations, due to their high energy; but rate of gain is not materially affected.
6. The energy of a ration may be increased without eliminating much of the roughage by adding 4 to 5% fat.
7. Feed formulation and balance of nutrients become more critical on high- and all-concentrate rations; specifically—
   a. Higher levels of vitamin A must be added—50,000 to 100,000 IU/head/day.
   b. The ration must be fortified with the calcium, phosphorus, and trace elements that are normally provided through the roughage.
   c. The unidentified growth factors are usually reduced with a reduction of the roughage. To compensate therefore, 5% dehydrated alfalfa meal may be added to the ration.
8. Cattle on high- or all-concentrate rations stall and go off feed more frequently.
9. Pelleting high- or all-concentrate rations lowers daily gains.
10. In the final analysis, the comparative price of concentrates and roughages—the economics of the situation—along with management practices, will be the major determining factors.

BYPRODUCT FEEDS

Prior to 1900, byproduct feeds were unwanted, and practically unused. Wheat bran was dumped into the Mississippi River, because nobody wanted to buy it; cottonseed meal was used as a fertilizer; most of the linseed meal was shipped to Europe; and tankage had not been processed. Today, these are standard and valuable animal feeds. Innumerable other byproducts—both roughages and concentrates—from plant and animal processing, and from industrial manufacturing, are available and used as cattle feeds in different areas, including the following: potatoes and potato pulp, pea vines and corn refuse silage from the canning industry, and byproducts from numerous fruits and nuts.

As is true of any ration ingredient, the requisites to effective and profitable use of each byproduct feed in cattle feeding are: (1) that it be bought at a favorable price, nutritive composition considered; (2) that its proximate composition be known, and that it be incorporated in a balanced ration; (3) that it be palatable and consumed in adequate quantity; and (4) that it not adversely affect carcass quality, particularly from the standpoint of harmful chemical residues from pesticides applied to crops. Generally speaking, the use of byproduct feeds calls for ingenuity and experience in handling them, special knowledge relative to their nutritive qualities and use in balanced rations, and relatively high labor costs. As a result, many cattle feeders are not interested in using them, whereas others find it a lucrative business.

The feeding value and the maximum amount that can be fed to cattle of several byproduct feeds are given in Chapter 8, Table 8-10, Feed Substitution Table for Beef Cattle, As-Fed Basis.

ROUGHAGES

Roughages are used in feedlot rations to supply bulk, physical properties, energy, protein, minerals, and vitamins. They contain considerable fiber (cellulose, hemicellulose, and lignin); consequently, they have a lower available energy content than concentrates. For this reason, only limited amounts of roughages are incorporated in finishing rations, particularly toward the end of the feeding period. They are, however, used extensively in growing programs and in warm-up rations.

The amount of roughage in feedlot rations varies over a wide range—from roughage alone in some grower rations to all-concentrate rations in some finishing rations, and many roughage proportions between these two extremes. Each of these roughage to concentrate ratios may be highly successful and very practical under certain conditions. In the final analysis, therefore, the roughage-to-concentrate ratio for a given feedlot should be determined by (1) the available feeds and comparative prices, (2) feed processing facilities, (3) the feed handling charges, (4) the age and quality of cattle, (5) the stage in the feeding period (*e.g.*, starting vs finishing period), (6) temperature (decrease roughage and increase concentrate in hot summer months, because high-roughage rations produce more body heat), (7) the feeder, (8) the troubles encountered (off feed, founder, scours, bloat), and (9) the results obtained by previous experience.

Fig. 30-4. Drylot steers eating field-cured dry beet tops. (Courtesy, The Great Western Sugar Co., Denver, CO)

Also, step-wise reductions in the amount of roughage are usually made at least three times during the finishing period; for example—

Starter ration. 70% roughage, 30% concentrate
Intermediate ration. 30% roughage, 70% concentrate
Final ration. 10% roughage, 90% concentrate

Such step changes in roughage-to-concentrate ratios should be made gradually, by blending the two mixes for 2 to 4 days.

In drylot finishing, the kind of roughage fed varies from area to area. This is so because, normally, it is not practical to move roughages great distances. Thus, generally speaking, cattle feeders utilize those roughages that are most readily available and lowest in price.

■ **Hay**—Hay is the most important harvested roughage fed to feedlot cattle, although many other roughages can be, and are, utilized.

High-quality legume hays are superior from the standpoint of cattle performance—and, when they are fed, supplemental feed requirements are lower. However, lower-quality hays give satisfactory results when properly supplemented.

Alfalfa is the roughage of choice in commercial feedlots. It averages 17.1% protein on a dry (moisture-free) basis and contains 58% TDN. On a net energy basis for production (NE_p), it is estimated to contain 27 Mcal per cwt, in comparison with a value of 56 Mcal per cwt for milo.

■ **Silage**—Increasing quantities of corn and milo silage are being fed to finishing cattle. At the present time, it is estimated that 90% of the nation's silage is made from corn and sorghum and the other 10% from hay and pasture crops, small grains, byproducts, and other feeds. The following silage pointers, based on experiments and practical observation, are generally observed by successful cattle feeders:

1. Where cattle are given a liberal amount of grain, any silage that is fed is considered a part of the roughage ration; hence, the silage should be fed in accordance with the well-recognized rules for feeding roughages. Also, it can be assumed that about 2 lb of good corn silage may be substituted for 1 lb of hay in cattle rations.
2. The maximum use of silage is best obtained early in the finishing period or with more mature steers that possess a larger digestive capacity.
3. Low grade cattle may be fed either (a) entirely on silage or (b) liberally on silage throughout the finishing period, but cattle that grade Select or better may return more profit if they are full fed on grain during the last half of the feeding period.
4. Cattle can be placed on full feed of silage from the beginning of the feeding period without any detrimental effects.
5. When fed on silage alone, cattle on full feed will consume 6 to 7 lb per 100-lb weight.
6. Sorghum silage has 60 to 90% of the value of corn silage and should be supplemented in the same manner as corn silage.
7. Grass silage should be supplemented with additional energy feeds, such as cereal grain or molasses, to be of the same value as corn silage.

Fig. 30-5 Silage in a 6,000-ton modified trench-stack silo. (Courtesy, Agricultural Research Service, USDA, Clay Center, NE)

■ **Haylage**—Tests to date have shown that haylage, which generally contains about 40% moisture, is an excellent feed for finishing cattle. More research is needed, but indications are that costs of gains may be a little higher than for corn silage.

A fairly accurate rule of thumb is that it takes 1.5 lb of 40% moisture haylage to equal 1 lb of hay.

■ **Other roughages**—Among the other roughages (other than hays and silages) used for feedlot cattle are: cottonseed hulls, corncobs, cereal straws, milo stover, Bermuda straw, Sudanhay, sawdust and other wood products, oat hulls, beet tops, peanut hay, newspapers, and a host of others. When properly (1) combined with a high-quality legume roughage, and/or (2) supplemented with the necessary protein, minerals and vitamins, all of them are excellent feeds. Availability, costs, and results should be the determining factors in their use, just as the economics of the situation should determine the use of any other feed ingredient.

(Also see Chapter 8, Table 8-10, Feed Substitution Table for Beef Cattle, for the feeding value and the maximum amount that can be fed to cattle of each of several roughages.)

ROUGHAGE SUBSTITUTES

It appears that the rumen animal requires a minimum amount of roughage factor for normal rumen function. All-concentrate rations show a response to small amounts of roughage factor. As little as 1 lb per day of some low-quality roughage such as alfalfa stems, rice hulls, or sorghum straw has given improvements in efficiency ranging up to 15% of all-concentrate rations.

To meet the *roughage factor* need, different roughage substitutes have been developed and used with varying degrees of success.

PROTEIN SUPPLEMENTS

The daily protein requirements of growing-finishing cattle and the percentage of protein needed in the ration are given in the Appendix, Section I of this book. It should be noted, however, that a larger percentage of protein is needed in rations with higher energy density. This is because fewer pounds of the high-energy ration are needed daily to meet the animal's energy requirement but the protein needs stay the same. Other factors that must be considered in formulating feedlot rations are the variations in protein content of feedstuffs, the digestibility of the protein, metabolic efficiency of protein utilization, and the previous nutritional treatment of the cattle. Age, genetic background, and health may influence efficiency of protein utilization. Because of the many factors that affect protein requirements, it is advisable to include a safety factor when balancing for protein. Although excess protein in the ration can be partly utilized for energy, each 1% increase in protein above the required level may increase the cost of gain ¼ to ½¢ per pound. However, underfeeding protein can cost much more than overfeeding protein due to the slow gains and poor feed efficiency. There appears to be a trend among research workers and nutritionists to recommend higher protein levels than were believed necessary in recent years, especially on high-concentrate rations. Of course, price levels of protein supplements are a factor.

KIND OF PROTEIN

Quality of protein, or balance of essential amino acids, is not a critical factor in most beef cattle–finishing rations, because bacteria in the rumen *manufacture* proteins that are used by cattle. For this reason, it makes little difference to a steer whether his protein comes from one source or several. Yet, practical feeders recognize that protein mixtures may be more palatable than a single ingredient and cause animals to eat a little more feed, and that, on long-fed cattle, fed to high Choice or Prime grade, linseed meal produces extra *bloom* which makes it a little more valuable than its protein content would indicate.

The choice of a protein supplement should usually be determined by the comparative price of a pound of protein in the available supplements. (See Chapter 8, section on "How to Determine the Best Buy in Feeds.") The leading protein supplements for finishing cattle are soybean meal, cottonseed meal, linseed meal, urea, and slow released nonprotein nitrogen products.

AMOUNT OF PROTEIN SUPPLEMENT TO FEED

The percent protein supplement to add to the ration will depend upon the age of the cattle, the kind and amount of roughage, and the protein content of the grain(s), or other carbonaceous concentrate being fed. Also, more protein is needed in rations with higher energy density. Thus, the amount of supplement should be determined for each lot. On a percentage of total ration basis, it decreases as the cattle grow older. Here are the recommendations for crude protein in the total ration, on an air-dry basis (which is about 10% moisture):

	Stage in Feedlot	Crude Protein (%)
Calves	First 60–90 days	12–14
	Next 100–200 days	11–12
	200 days to market	10–11
Yearlings	First 60–90 days	11–12
	100 days to market	10–11

Based on experimental studies, the Ohio Station workers recommend that protein supplements be deleted from feedlot rations after cattle are past 750-lb weight. After removing the protein supplement, their test ration contained from 8.2 to 8.6% protein, a good 25% below the usual recommendations. They do warn that when lowering the protein the feeder must not decrease the levels of minerals and vitamins in the ration.

For cattle on relatively high-concentrate rations in the drylot, the following rules of thumb may be used:

1. Where no legume hay is fed, add 2 lb of oilseed protein supplement per head daily.
2. Where half the roughage consists of legume hay, add 1 lb of oilseed protein supplement per head daily

Because protein supplements are usually expensive, normally one should not add more than is required to balance the ration. Neither should they be shorted, for digestion of roughage is lowered if there is a lack of protein in the ration.

UREA

Urea is not a protein. It is a simple nitrogen compound, $NH_2-\overset{\overset{\displaystyle O}{\|}}{C}-NH_2$, from which the microorganisms can obtain nitrogen, synthesize amino acids, and finally bacterial protein—provided that all the nutrients essential for protein synthesis are present.

Each year, urea is replacing a larger percent of the supplementary protein in cattle feedlot rations. Because of the high price of oilseed proteins, due to their increasing use for human consumption and for monogastric animals, eventually urea and/or other nonprotein nitrogen compounds will be used as a major source of supplementary protein for feedlot cattle.

Urea or other nonprotein nitrogen compounds can furnish 33% of the total protein requirement of feedlot cattle; higher levels cause a depression in gain and feed efficiency. It is noteworthy, however, that, even in supplements in which 90% of the protein equivalent is from urea, only about one-third of the total protein in the ration is supplied from nonprotein nitrogen—the remainder is supplied from grain and roughages.

The following conditions are essential for proper urea utilization by feedlot cattle:

1. High level of bacteria in the rumen. Cattle off feed or sick cattle do not utilize urea very effectively.
2. Slow release of ammonia from urea.
3. Two- to four-week adjustment period.
4. Low level of natural protein in the diet. Urea utilization is depressed when used with increasing levels of natural protein. Microorganisms prefer the nitrogen from natural protein to nonprotein nitrogen because natural proteins, such as soybean meal, furnish other nutrients which are beneficial to bacterial and protozoan life.
5. High-quality ingredients are required in high-urea supplements. For this reason, high-fiber filler feeds (such as ground corncobs, oat hulls, rice hulls, cottonseed hulls, cellulose, paper, and sawdust) should not be used with urea.
6. Response to urea is greater on high-energy and limited roughage rations than on low-energy and high-roughage diets.
7. Urea supplements should be well mixed (homogeneous).
8. High-urea dry supplements should be protected from rain and kept as dry as possible, because urea is hygroscopic (it picks up moisture).
9. Essential nutritional factors including (a) readily available source of energy (such as grain or molasses); (b) adequate levels of calcium and phosphorus; (c) required level of trace mineral elements; (d) a nitrogen-sulfur ratio which is not wider than 15:1; (e) unidentified urea-protein synthesis factors (dehydrated alfalfa meal in dry urea mixes, and distillers' solubles in liquid supplements); (f) iodized salt, to improve the palatability and mask the taste of urea; (g) fortified with proper levels of synthetic vitamin A, to furnish a minimum of 20,000 IU of vitamin A daily for growing and finishing cattle; (h) fortified with vitamin D if cattle are confined, and (i) fortified with vitamin E if natural feedstuffs are low.
10. Do not feed urea or supplements containing urea to newly arrived or shipped-in cattle for a period of 21 to 28 days.
11. Do not feed urea to cattle that have been starved or off feed for 36 hours until they have had a chance to fill the rumen with feed.
12. For growing cattle, feed a maximum of 0.15 lb of urea daily.
13. For finishing steers or heifers on grain and roughage, do not feed more than 0.22 lb of urea per head daily.
14. Formulate complete cattle rations so that no more than 33% of the crude protein or nitrogen is derived from urea. Protein supplements may contain 85 to 90% of the protein from urea; but when blended with natural feedstuffs like grain and roughage, the total contribution of protein from urea is usually less than 33%.
15. Do not feed urea over and above the protein requirement; add only enough properly balanced urea supplement to meet the protein needs.
16. Urea should be either thoroughly mixed in a properly balanced supplement or incorporated in a complete ration.
17. If supplements containing urea are self-fed, the intake should be controlled by using a lick wheel for liquid supplements or incorporating high levels of salt and dry supplements.

(Also see Chapter 8, section entitled, "Nonprotein Nitrogen.")

PROTEIN POINTERS

The following points should be taken into consideration to assure proper protein utilization by finishing beef cattle:

1. The protein content of the major feed ingredient(s) should be known so that the ration formulation can be precise.
2. Consideration should be given to the protein content and digestibility of the grain, because the grain is the most economical source of protein and supplies the largest percent of the ration protein.
3. Higher levels of protein are needed in all-concentrate rations.
4. When possible, use a roughage high in protein so as to lessen the amount of supplemental protein necessary.
5. Urea can be successfully used at a level up to 1% of the total finishing ration to replace natural protein, provided the ration formulation permits optimum utilization of the urea by the rumen microorganisms.
6. When excessive protein levels are fed, the overage is wasted and performance may actually be reduced with high-performing animals.
7. Absolute protein requirements depend upon the age of the animal, and probably the energy of the ration.
8. When per head consumption of corn reaches a level of 18 lb per day, no supplemental protein is required, although other additives such as minerals and vitamins are necessary. Thus, protein supplements are not needed in the late finishing stages of yearling cattle.
9. The production of single-cell protein (SCP) from solid waste as a protein supplement for feedlot cattle will likely increase in importance.

MINERALS

Minerals play an important role in the nutrition of feedlot cattle. They help regulate the normal functions of the metabolic processes in the animal.

The amount of each mineral in the ration is important; both deficiencies and excesses are to be avoided. Therefore, an analysis is important, particularly where new feeds and feeds from new areas are involved. The following minerals have been established as dietary essentials for cattle:

- **Major minerals**—Calcium, phosphorus, magnesium, sodium, potassium, chlorine, and sulfur.

- **Trace minerals**—Cobalt, copper, iodine, iron, manganese, zinc, and selenium.

Where a complete mixed feed (roughage and concentrate combined) is fed to finishing cattle, it should contain 0.25 to 0.5% salt. Also, in the larger feedlots, the other needed minerals are usually incorporated in the ration as a special mineral supplement or in the protein supplement. For recommended kinds and allowances of minerals, see Appendix, Section I, NRC Nutrient Requirements of Beef Cattle. Special attention needs to be given to trace minerals in areas where there is a deficiency of one or more of them, when poor-quality roughage is fed, or when high- or all-concentrate rations are fed.

Even when minerals are added to the ration of finishing cattle, the senior author favors self-feeding them in addition. For this purpose, use a two-compartment mineral box, with (1) salt (iodized salt in iodine-deficient areas[2]) in one side, and (2) dicalcium phosphate, defluorinated phosphate, or a mixture of one-third salt (salt added for purpose of palatability) and two-thirds steamed bone meal, or a good commercial mineral mixture, in the other side. With this arrangement, if cattle need added minerals, they will consume them; if they don't need them, they'll pass them up. In particular, incoming feedlot cattle frequently crave minerals, due to deficiencies in their previous feeding. Such cattle, however, should be given only limited quantities of minerals until the danger of overeating has passed.

(Also see Chapter 8, section on "Minerals.")

[2]Unless cane molasses is included in the ration, a trace-mineralized salt may well be used. Cane molasses appears to be a good source of most of the needed trace minerals, and may supply sufficient trace minerals in certain areas and with certain feeds.

VITAMINS

Vitamins A, E, and in some cases vitamin D should be added to feedlot rations.

The rumen organisms synthesize adequate B vitamins and vitamin K, and nothing is gained by adding these to feedlot rations of healthy cattle. Likewise, no benefit has been reported from supplemental vitamin C. Vitamin D is produced in the skin of animals in direct sunlight, but during cloudy, winter weather, or when cattle are confined, it should be added to the ration.

Feeders should watch for the following symptoms of vitamin A deficiency in feedlot cattle: rough hair coat, watery eyes, loose and watery droppings, edema (stovepipe legs), and night blindness.

In Chapter 8, Tables 8-7 and 8-8 give the vitamin requirements of beef cattle, whereas Table 30-3 gives the recommended levels of vitamin A for feedlot cattle, with overage for safety.

TABLE 30-3
RECOMMENDED LEVELS OF VITAMIN A FOR FEEDLOT CATTLE

		Vitamin A/Ton of Supplement When It Is Fed at Level of—	
	Vitamin A/ Head/Day	1 Lb/Head/ Day	2 Lb/Head/ Day
	(IU)	*(IU)*	*(IU)*
Cattle on growing ration in winter	10,000	20,000,000	10,000,000
Cattle on full-fed finishing ration in winter	20,000	40,000,000	20,000,000
Cattle full-fed grain on pasture	20,000	40,000,000	20,000,000
Cattle fed in drylot in summer	30,000	60,000,000	30,000,000

A common guideline on the level of vitamin A for feedlot cattle is to use 3,000 IU per pound body weight, or 1,000 IU for each pound of total feed.

When vitamin D is needed and added to the ration, it is recommended that 4,000 to 6,000 IU of it be given per head per day. This is approximately 1/6 to 1/7 the recommended level of vitamin A.

Where grains are heat processed for feedlot cattle, some research shows that it may be advisable to provide supplemental vitamin E. The National Research Council requirement for vitamin E for growing and finishing cattle is given in the Appendix, Section I.

(Also see Chapter 8, section on "Vitamins.")

IMPLANTS AND GROWTH STIMULANTS

Feed additives first made headlines in 1952 when Iowa State University researchers announced the results of cattle feeding trials indicating a major breakthrough in lowering feed usage and increasing weight gains by feeding the compound diethylstilbestrol (DES).

For the next 20 years, cattle feeders and consumers greatly benefited from the increased rate and efficiency of gains accruing from the use of this product. Feeding and implanting DES increased growth rate 10 to 15%; improved feed efficiency 10%; put feedlot cattle on the market 20 to 30 days sooner; decreased feedlot costs by 3 to 3.5¢ per pound; and lowered the retail price of beef by 5 to 6¢ per pound. But the Food and Drug Administration banned the use of both oral (January 1, 1973) and implanted (April 25, 1973) DES, for the reason that stilbestrol is a carcinogen (in large enough quantities it can produce cancer in humans and animals); and the Delaney Amendment clearly states that foods can contain no residues (zero level) of carcinogens. Then, on January 24, 1974, the United States Court of Appeals invalidated the ban on DES and reinstated its use as an implant or feed additive. But the ruling of the court did nothing to change the Delaney Amendment. In the meantime, newer and more sophisticated assay techniques evolved which made it possible to detect the presence of DES residues in meat at lower levels than formerly. So, in 1979, the use of DES as feed or implants for cattle and sheep was banned. But several other additives were developed subsequently, and are now on the market.

Table 30-1 summarizes the growth stimulants that are presently available and can be used. All of these

TABLE
IMPLANTS AND GROWTH STIMULANTS

Class of Cattle	Additive	Method of Administering	Dosage	Cost	Increase in Daily Rate of Gain
Finishing steers	1. Antibiotic	Oral	10 mg/100 lb *(45 kg)* body wt. daily; or 70 to 75 mg/head daily	0.3¢; or 1¢/day	6%
	2. Bovatec (Lasalocid)	Oral	150–360 mg/day 250–360 mg/day	1.5¢/day for improved feed efficiency 1.5¢/day for improved feed efficiency and rate of gain	2%
	3. Cattlyst (Laidomycin propionate)	Oral	30–150 mg/head/day	3.5–4.0¢/day	5–6%
	4. Compudose	Implant	24 mg estradiol	$2.15/head	10–15%
	5. Impulse S	Implant	200 mg progesterone, 20 mg estradiol	$1.25–$1.55/head	10–15%
	6. Ralgro (Zeranol)	Implant	36 mg resorcyclic acid lactone	90¢/dose	10%
	7. Revalor-S	Implant	120 mg trenbolone acetate and 24 mg estradiol	$3.30/dose	15–25%
	8. Rumensin (Monensin)	Oral	50–360 mg/head/day, drylot; 50–200, pasture	1.5¢/day, 1.2¢–1.4¢/day	
	9. Synovex S (for steers)	Implant	200 mg progesterone, 20 mg estradiol benzoate	90¢/dose	10–15%

products have been shown to improve gain and feed efficiency of feedlot cattle significantly.

In considering the additives listed in Table 30-4, it should be noted that there is no evidence to indicate that the use of these products can or will alleviate the need for vigilant sanitation, improved nutrition, and superior management. Also, the benefits of each one must be weighed against its cost.

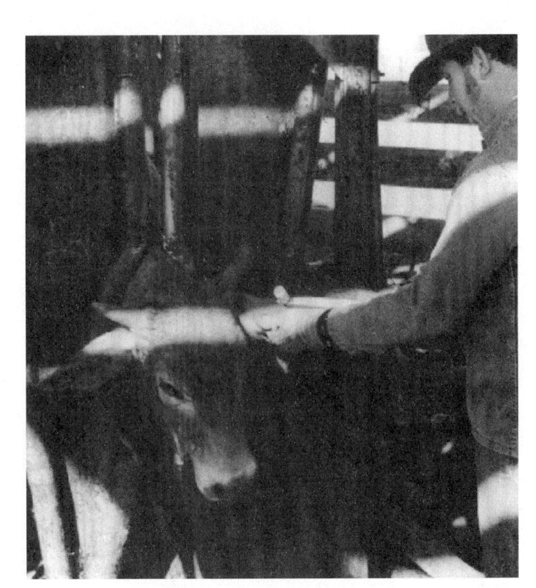

Fig. 30-6. Implanting a growth stimulant. (Courtesy, Benedict Feeding Co., Casa Grande, AZ)

30-4
FOR FINISHING CATTLE[1, 2]

Increase In Feed Efficiency	Effect on Carcass Quality	Other Comments	Withdrawal Period Prior to Slaughter
4%	Improves carcass quality slightly; more fat deposition and marbling. Decreases liver and rumen condemnations.	Antibiotics will also reduce the disease level. More effective on high-roughage rations than on high-concentrate rations.	No withdrawal required.
3–5%	No effect.	Alters rumen fermentation similiar to Monensin.	No withdrawal required.
6–10%	No effect.	Alters rumen fermentation. Reduces liver abscesses.	
5–10%	No effect.	Only one implant. Good for 200 days.	No withdrawal required.
5–10%	No effect.	Effective period of 140 days.	No withdrawal required.
5–10%	No effect.		No withdrawal required.
10–15%	No effect if properly implanted 100–110 days before slaughter.	Effective period of 120 days.	No withdrawal required.
3–6%	No effect.	Not a hormone. It results in more propionic acid and less butyric and acetic acids; hence, more energy.	No withdrawal required.
5–10%	No effect.		No withdrawal required.

(Continued)

TABLE 30-4

Class of Cattle	Additive	Method of Administering	Dosage	Cost	Increase in Daily Rate of Gain
Finishing heifers	1. Antibiotic	Oral	10 mg/100 lb *(45 kg)* body wt. daily; or 70 to 75 mg/head daily	0.3¢/day	6%
	2. Bovatec (Lasalocid)	Oral	150–360 mg/day	1.5¢/day	5%
	3. Cattlyst (Laidomycin propionate)	Oral	30–150 mg/head/day	3.5–4.0¢/day	5–6%
	4. Finaplix H	Implant	200 mg trenbolone acetate	$2.40/dose	15–20% when implanted with Synovex or implanted in heifers fed MGA
	5. Impulse H	Implant	Follow label directions.	$1.25–$1.55/head	10%
	6. MGA	Oral	0.25–0.50 mg daily melengestrol acetate	1¢/day	10%
	7. Ralgro (Zeranol)	Implant	36 mg resorcyclic acid lactone	85¢/dose	10%
	8. Rumensin (Monensin)	Oral	50–360 mg/head/day, drylot; 50–200, pasture	1.5¢/day	
	9. Synovex H (for heifers)	Implant	200 mg testosterone propionate, 20 mg estradiol benzoate	90¢/dose	10%
Suckling calves	1. Antibiotic	Oral (in creep feed)	15–20 mg/100 lb body wt. daily	0.1¢/day	6%
	2. Ralgro (Zeranol)	Implant	36 mg resorcyclic acid lactone	85¢/dose	5–6%
	3. Synovex C	Implant	10 mg estradiol benzoate and 100 mg progesterone	$1.00/head	5–6%

[1]FDA regulations are subject to change. Always follow the manufacturers' directions on the use of these products.

[2]Table 34-4 was authoritatively reviewed by and helpful suggestions for updating it were received from, Dr. Allen Trenkle, Iowa State University, Ames, IA.

(Continued)

Increase In Feed Efficiency	Effect on Carcass Quality	Other Comments	Withdrawal Period Prior to Slaughter
4%	Improves carcass quality slightly; more fat deposition and marbling.	Antibiotics will also reduce the disease level. More effective on high-roughage than on high-concentrate rations.	No withdrawal required.
8%	No effect.	Alters rumen fermentation similar to Monensin.	No withdrawal required.
5–10%	No effect.	Alters rumen fermentation.	
10–15%	No effect if properly implanted 100–110 days before slaughter.	Effective period of 120 days.	No withdrawal required.
5–10%	No effect.	Effective period of 140 days. For heifers over 400 lb.	No withdrawal required.
6%	MGA will lower the incidence of estrus in heifers and increase rate and efficiency of gain.	MGA is effective for heifers, but not for steers.	No withdrawal required. It is not effective with pregnant heifers.
5–10%	No effect.		No withdrawal required.
15%	No effect.	Not a hormone. It results in more propionic acid and less butyric and acetic acids; hence, more energy.	No withdrawal required.
5–10%	No effect.	Recommended for use in heifers during last 60 to 150 days of the finishing period.	No withdrawal required.
4%		Antibiotic will also reduce the disease level.	No withdrawal required. *Note well:* If fed at level of 350 mg or over/day, 48-hour withdrawal required.
5–10%	No effect.	It may be used in replacement heifers from 1 mo. of age to weaning.	No withdrawal required.
5–10%	No effect.	For calves weighing less than 100 lb.	No withdrawal required.

OTHER METHODS OF IMPROVING RATE AND EFFICIENCY OF GAIN

Several other methods, in addition to additives, can be used to increase the rate and efficiency of gain of feedlot cattle. Among them are the following:

1. Feed young bulls (uncastrated males) instead of steers. The male hormones secreted by the testicles are excellent growth stimulants and will improve gain and feed efficiency by 10 to 15%. Alternatives to bulls that merit consideration are short scrotum bulls (induced cryptorchidism), and Russian castrates. With these methods, testosterone is produced, yet, in comparison with bulls, the animals are easier to handle and may be carried to advanced ages without being labeled *bull beef*.

2. Take advantage of the genetic improvement of beef cattle by crossbreeding and the introduction of genes from the exotic breeds. This offers one of the most permanent ways of increasing the weaning weight of calves and improving performance in feedlot cattle. The selection of fast gaining and efficient cattle within the straightbreds, and in crossbreeding, may improve the efficiency of performance of feedlot cattle by 10% or more.

3. Reduce the cost of producing beef by improving the quality of cattle rations through grain processing and nutritionally balanced protein supplements.

4. Eliminate internal and external parasites, and protect cattle against the common diseases. This will save millions of dollars for both cattle feeders and consumers.

5. Keep abreast of new developments, including the discovery of new growth stimulants.

FEED PREPARATION

Prior to 1960, very little attention was given to feed processing for commercial cattle production, other than grinding or crushing grain and chopping forage. But in recent years great progress has been made and many new techniques have been developed.

The preparation of roughages has received less attention than the preparation of grains. However, with the increasing world food shortages, roughages will become more important. Under these circumstances, their processing will assume greater importance. Hence, roughage preparation will also be discussed herein.

Fig. 30-7. Cattle feed processing and storage facilities at Farr Feeders, Inc., Greeley, Colo. (Courtesy, Farr Feeders, Inc., Greeley, CO)

GRAIN PROCESSING METHODS

Modern day fattening rations usually contain from 75 to 95% concentrate. Moreover, grains supply up to 90% of the usable energy of the ration. Thus, any improvement in the efficiency of utilization of grain will be reflected in improved performance and feed requirement of fattening cattle.

The success of grain processing must result from physical and/or chemical changes. Physical changes include moisture level, heat, pressure, and particle size. Chemical factors may include structural changes in the starch, protein, and fat of grains resulting in changes in digestibility and metabolic end products. In some cases, so-called physiochemical changes occur in that both physical and chemical alterations are simultaneously apparent. Rate of ingesta passage and site of digestion within the GI tract are both likely end results of physiochemical changes in processed grains.

The primary reasons for processing grains for feedlot cattle are:

- To increase digestibility.
- To increase palatability.
- To increase surface area for greater microbial activity.
- To give rumen microorganisms and digestive enzymes easier access to the starches and readily utilizable nutrients.
- To affect the rate of passage of feed through the digestive tract, or to affect rumen mobility by increasing the bulk through certain processing methods.
- To increase feed efficiency through a combination of the above factors.

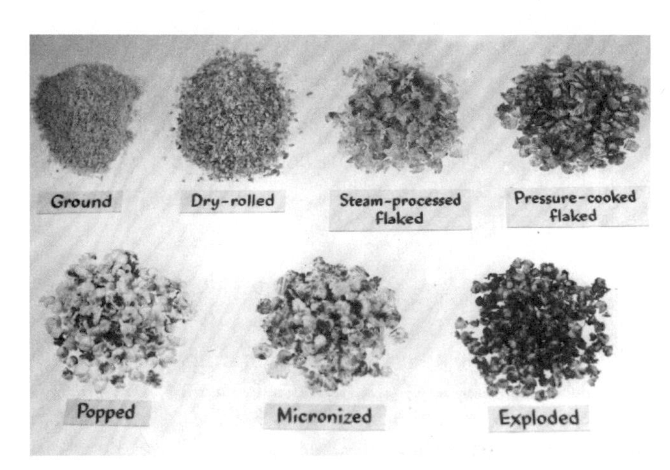

Fig. 30-8. Milo processed by several different methods. (Courtesy, Department of Animal Sciences, The University of Arizona)

Among the factors to consider when deciding on the grain processing method are the size of kernel, percentage of moisture, and percentage of concentrate in the ration.

When any of the dry processing methods are used, it is important that the kernel be broken, but that there be coarseness and relative freedom from fines.

Grain processing gives greater returns when feed intake of grains is high. Cattle on maintenance rations are not normally fed much grain; hence, the increase in feed efficiency may not return the added processing cost.

Several grain processing methods have evolved. A survey of the literature reveals that no process prior to the development and use of steam flaking for feedlot cattle improved performance so dramatically. Thus, most modern processing techniques have been developed in an attempt to obtain similar, and hopefully better, performance than steam flaking—with possibly a reduction in cost. Some are physical, others are chemical; some are dry processing, others are wet processing. It is recognized that any grouping of processing methods cannot be precise, for two or more processing treatments may be involved in a feed; for example, in making pellets, grinding is followed by adding moisture, thence pressure. Despite some overlapping, the authors have evolved with the following classification of grain processing methods for cattle:

Mechanical alterations

Dehulling
Extruding (gelatinization)
Grinding
Rolling
    Dry rolling (cracking, crushing)
    Steam rolling (crimping, steam crimping)

Heat treatments

Dry heat processing
    Micronizing
    Popping (jet-sploding)
    Roasting
Moist heat processing
    Cooking
    Exploding
Flaking
    Steam flaking
    Pressure flaking
Pelleting

Moisture alterations

Drying (dehydrating)
High-moisture (early harvested) grain
Reconstituted grain
Watered feeds

Blocks

Liquid supplements

Hydroponics (sprouted grain)

Unprocessed (whole) corn

MECHANICAL ALTERATIONS

The oldest and most widely used methods for processing grains are those which merely cause physical disruption of the cells by mechanical means. The fact that the more nutritious portions of the grain are surrounded by an outside coating or hull makes it easy to understand how the exposure of these nutrients to the action of digestion processes would increase the utilization of the nutrients. The mechanical methods by which the grain kernel is broken vary, but generally speaking, they involve either shearing, cutting, or mashing. In the milling of grains, there is also the abrasive action to scrub off the outer coats in processes referred to as burring, pearling, polishing, dehulling, and other similar terms.

DEHULLING

Dehulling is the process of removing the outer coat of grain, nuts, and some fruits. The hulls are high in fiber and low in digestibility.

The best known outer coverings of cereal grains are barley hulls, oat hulls, and rice hulls.

The best known of the oilseed hulls is cottonseed hulls. Today, cottonseed hulls are available in both loose and pelleted forms.

EXTRUDING (Gelatinization)

Extruding is a process by which feed is pressed, pushed, or protruded through constrictions under pressure.

Extruding usually involves grinding the grain, followed by heating with steam in order to soften it, then forcing the material through a steel tube by an auger. The softened material is then extruded through cone-shaped holes which are smaller where the feed enters and gradually enlarge where the feed is expelled. The expansion causes disruption, or granulation, of the starch granules. Various factors, including moisture of the grain, influence the character of the final product.

In Colorado work with finishing cattle, extruded milo produced results similar to corn processed in various ways. In Kansas trials, milo, processed by dry rolling, high-moisture storage, steam flaking and extruding, produced similar gains, but feed efficiency was 9 and 15% better for the flaked and extruding treatments, respectively, over dry rolled milo. In other Kansas trials, extruding improved feed efficiency by 11 and 15% over dry rolling. Results were similar to steam flaked and high-moisture grain.

GRINDING

Grinding is that process by which a feedstuff is reduced in particle size by impact, shearing, or attrition. It may change the digestibility of cellulose and protein.

Grinding is the most common, cheapest, and simplest method of feed preparation. It is usually accomplished by means of a hammer mill, which, by impact, reduces the particle size of the grain until it passes through a screen of a certain size. Medium-fine grinding, which can be distinguished by a gritty feeling as some of the feed is rubbed between the fingers, is best. Very fine grinding makes feeds dusty and lowers palatability. However, fine grinding may be desirable (1) where pelleting is to follow, or (2) where grains contain small weed seeds, the viability of which should be destroyed.

Factors influencing the nature of ground feed are screen size, hammer mill size, power and speed, type of grain, and moisture content of grain.

A major advantage of grinding compared with more sophisticated processing methods is the economic feasibility of having a hammer mill on the farm, or of having a custom grinder come to the farm or ranch periodically to process grain.

Differences in animal performance from ground grains as reported in the literature are partially due to variations in the fineness of grind used in the various experiments.

ROLLING

Rolling refers to the process by which grain is compressed into flat particles by passing it between rollers. The rolling may be accomplished without the addition of water (dry rolling) or after subjecting the grain to steam (steam rolling).

Dry Rolling (Cracking, Crushing)

Dry rolling, which is also called cracking or crushing, refers to passing grain, without steam, between a closely fitted set of steel rollers which are usually grooved on the surface. It breaks the hull and/or seed coat and results in an end product much like coarsely ground grain. Particle size varies from very small to very coarse and is influenced by pressure and spacing, moisture content of the grain, and rate of grain flow.

Steam Rolling (Crimping, Steam Crimping)

Steam rolling, which is also called crimping or steam crimping, refers to exposing grain to steam for a short period of time, usually 1 to 8 minutes, followed by rolling. The steam softens the kernel, producing a more intact, crimped-appearing product than that produced by dry rolling. Steam rolling offers little or no advantage in feed efficiency over grinding or dry rolling. However, the particle size and physical form of the steam rolled grain may improve palatability and animal acceptance in some instances.

The moisture content of steam rolled grain is increased slightly—perhaps by an average of 6%. Thus when whole grain is exchanged for processed grain on a ton for ton basis, as is frequently done by farmers who have grain processed at the local mill on a custom basis, there may be a loss of 6% of the grain in addition to the cost of processing. If grain is worth $6 per cwt, this would add 36¢ per cwt, or $7.20 per ton to the processing cost.

HEAT TREATMENTS

In recent years, several sophisticated heat treatments have evolved.

Excess heating damages some nutrients, such as the amino acids, and vitamins, whereas proper heating of protein sources (such as soybeans) and of carbohydrate sources (such as cereal grains, potatoes, and beans) results in better availability of nutrients. Heating soybeans destroys the trypsin inhibitor or a possible active protein fraction in raw soybeans, increases the amino acid availability, and results in better availability of the fat and increases metabolizable energy.

Proper heating of cereal grains, such as corn,

barley, and milo, will make for partial gelatinization and improve rate and efficiency of gains of cattle.

In general, heat treatments are the most successful of the newer feed processing techniques for ruminants.

DRY HEAT PROCESSING

Dry heat processing consists of surrounding the feed with dry air. It has the following advantages: (1) The temperature may be changed (turned on, or cut off) rapidly; (2) it does not add moisture to the feed, so it may be used on feeds that are to be stored following processing; and (3) it may remove some of the water from feeds that contain too much moisture for safe storage.

The common methods of processing by dry heat are micronizing, popping, and roasting.

Micronizing

Micronizing is a coined word used to describe a dry heat treatment of grain by microwaves emitted from infrared burners. In micronizing, grain is heated to 300°F by gas-fired infrared generators as it passes along an oscillating steel plate or skillet, following which it is dropped into knorling rolls. Micronized grain is not popped. It is reduced to about 7% moisture, then rolled to produce a uniform, stable, dry, free-flowing product. The product has an intact, flake like appearance, resembling some steam flaked grains. Densities of micronized grain normally range from 18 to 30 lb per bushel, with approximately 25 lb per bushel being recommended. Water is usually added just prior to feeding to adjust to a 10% moisture content.

Micronized grain sorghum compares favorably to steam flaked grain sorghum, from the standpoint of rate and efficiency of gain. However, cost of processing favors the micronizing technique over steam flaking because of a lower initial cost in equipment.

Popping (Jet-Sploding)

Popping is the exploding, or puffing out, of grain resulting from the rapid application of dry heat. Popping grain for cattle involves the same principle as processing popcorn for people, and the end results are similar.

The principle of popping is based on the use of super hot dry heat, with the grain quickly heated to a temperature of 300 to 310°F (preferably grain with 15 to 20% moisture content) and an exposure of 15 to 120 seconds (depending upon temperature). Rapid heating by dry heat volatizes the internal, natural moisture in the kernel until the pressure is great enough to explode it (to gelatinize and expand, or disrupt, the starch granules), causing the grain to puff out upon reaching atmospheric pressure. Usually not more than 45% of the grain pops, although the jet-sploding method approaches 100% popping. The percentage of grain popped depends primarily on moisture content, temperature, and rate of flow through the machine. The exploded product is then dry rolled.

All grains can be processed by this method, but it appears that it is especially effective in processing sorghum grain.

Several different popping methods have evolved. One of the most successful is known as jet-sploding, a dry heat processing method. High capacity can be built into this system without difficulty, and quality control is automatic. Also, the jet-sploder can change from one type of grain to another without adjustment. Studies show that jet-sploded milo compares favorably with steam processed and flaked milo.

Densities from various popping methods prior to rolling range from 6 to 15 lb per bushel. Remoisturizing, rolling, or regrinding increases the density.

Some conclusions relative to popping milo, based on experimental studies, are:

1. The resulting product very much resembles ordinary popcorn and has a moisture content of approximately 3%.
2. Popping increases digestibility and efficiency. However, there is disagreement among scientists as to whether the increased digestibility of starch is due to the expansion associated with popping *per se* or due to heat.
3. Popping causes disruption of the starch granule by using natural moisture in the kernel to steam, gelatinize, and expand the starch granules. Rolling and moisturization are usually essential.
4. Popped milo is palatable and very satisfactory for starting cattle on feed, but bulk densities are so light that they sometimes result in severely depressed feed intake and reduced daily gains.
5. Initial investment and operating costs per ton are lower than for steam flaking operations.
6. Popped milo requires more storage space than most other processing methods, due to its light density. Also, it may create handling problems in bin flow, hang up, bridging, and conveying.

Roasting

Roasting is a simple process of heating feed to the desired temperature in some form of oven for a period of time. It is another method of heat treatment.

The effects of roasting are not fully understood. But it appears to increase the availability of nutrients, possibly as a result of changes in the starch (perhaps gelatinization) due to the heat, along with some effect on the proteins.

In roasting corn the grain is heated to about 300°F. The roasted grain has a pleasant, *nutty* aroma and a puffed, caramelized appearance. Very few of the kernels are actually popped. However, there is some expansion during the roasting process; raw corn weighs 45 lb per cubic foot, whereas the roasted corn weighs only 39 lb per cubic foot. Also, the moisture content of the grain is decreased to 5 to 9%. Purdue University reports that, for fattening cattle, roasting improves feed efficiency by 10% and increases weight gains by 14% over ground corn.

MOIST HEAT PROCESSING

Moist heat processing consists of surrounding the feed by water or steam and (1) cooking either in a conventional vessel or under pressure, or (2) compressing.

The common moist heat processing methods are cooking, crumbling, exploding, flaking (steam flaking, pressure flaking), and pelleting.

Cooking

Cooking is processing by applying heat.

Professional caretakers have long cooked feed (especially barley) for show cattle, in order to increase palatability and feed consumption. However, cooking has the following limitations:

1. It is apt to decrease the digestibility of the proteins, even though it may slightly increase the digestibility of the starches.
2. It does not improve feeding value.

Exploding

Exploding is the swelling of grain, produced by steaming under pressure followed by releasing to the air. This technique involves delivering raw grain into high-tensile strength steel *bottles* which hold approximately 200 lb of grain. Live steam is injected into bottles until pressure reaches 250 psi. After about 20 seconds, a valve opens to let the grain escape as expanded balls with the hulls removed. Under the high pressure, moisture is forced into the kernels, which, when released into the air, swell to several times the original size. The product resembles puffed breakfast cereals. Excellent quality control and uniformity of product are possible with this process.

California workers compared exploded milo with steam flaking. The puffed material produced feed intake, gain, and feed efficiency comparable to the best performing flaked grain treatment.

FLAKING

Flaking is a modification of steam rolling in which the grain is subjected to steam either for a longer period of time or under pressure. Flaking is rolling into flat pieces following either (1) steaming at atmospheric pressure, or (2) steaming under pressure. The end product has a distinct and pleasant aroma, resembling cooked cereal. Proper flaking of grains renders the starch fraction more readily available to rumen microorganisms and enzyme degradation than conventional methods of steam or dry rolling.

The flaking process varies according to the grain. The grain that responds the most to flaking is milo. In comparison with dry rolling or grinding, cattle fed flaked milo will gain from 0.25 to 0.5 lb more, or about 10% more, per head per day and require 5 to 10% less feed. In studies with calves, the Arizona Station found that feed requirements were lowered by 10% by steam flaking grain sorghum as compared to steaming and rolling the grain in the older conventional manner, and that there was no difference between steam flaking and pressure flaking. In terms of improvement due to flaking, corn follows milo. Steam processing and flaking of barley and wheat appear to improve gain but not utilization of the grain. This is probably due to improved palatability and intake of the flaked product as compared to dry rolled or ground product.

Steam Flaking

This was the first modern technique which markedly increased feed efficiency and rate of gain in the case of milo. This process differs from steam rolling or crimping in that the grain is subjected to steam under atmospheric conditions for a longer period of time, usually 15 to 30 minutes, prior to rolling. Large, heavy roller mills set at near zero tolerance produce a very thin, flat flake which usually weighs from 22 to 28 lb per bushel and contains 16 to 20% moisture. The flaking process causes gelatinization of the starch granules (hydration or rupturing of the complex starch molecule), rendering them more digestible. The degree of flaking and level of gelatinization are influenced by such factors as steaming time, temperature, grain moisture, roller size and tolerance, processing rate, and type and variety of grain.

Pressure Flaking

In pressure flaking, the grain is subjected to steam under pressure for a short time, such as 50 psi for 1 to 2 minutes. A continuous flow cooker is operated by air lock valves to inject and eject grain. Steam is injected into the cooker at the desired pressure (somewhat like a pressure cooker used for food preparation). The grain in the chamber reaches a temperature approaching 300°F. When the grain is expelled from the cooker, it is generally cooled (by use of a cooling and drying tower) to below 200°F and 20% moisture before flaking. In comparison with steam flaking, flakes produced by pressure are less brittle and less subject to fragmenting during the mixing and feeding operation.

PELLETING

Pelleting is the agglomerating of feed by compacting and forcing it through die openings by a mechanical process. Pellets can be made into small chunks or cylinders of different diameters, lengths, and degrees of hardness. Large pellets—especially those large enough to be fed on pasture or range—are commonly called *range cubes.*

Grains and other concentrates are pelleted for the purposes of (1) facilitating mechanization in handling; (2) eliminating fines and dust, and increasing palatability; (3) alleviating separation of ingredients and sorting, (4) increasing feed density—thereby lessening transportation and labor costs; (5) reducing storage space; (6) making it possible to feed on the ground or in windy areas with little loss; and (7) improving the nutritional value of certain feedstuffs through the instantaneous heat and pressure.

Fig. 30-9. Range cubes fed to replacement heifers wintered on low-quality pasture. (Courtesy, Ralston Purina Company, St. Louis, MO)

On diets containing a low level of crude fiber, there is no advantage in pelleting feed for beef cattle.

Pelleting feeds may destroy vitamins A, E, and K, especially if the diet does not contain sufficient antioxidants to prevent the accelerated oxidation of these vitamins under conditions of moisture and high temperature.

The following concentrates may be pelleted: (1) the entire concentrate, (2) the fines only, (3) the protein supplement, and (4) the range supplements.

The pelleting of concentrates in cattle feedlot rations is generally limited to the protein supplement. If the protein supplement contains urea, trace minerals, vitamins, and/or antibiotics, it is usually pelleted.

MOISTURE ALTERATIONS

Water is important in feed preparation and processing. Sometimes the water content of a feed must be altered for proper feed storage, and sometimes it must be changed for feeding purposes.

Some feeds must be stored dry; others must be stored wet. Feeds carrying more than 15% moisture cannot be stored in bulk, for they will likely mold. For safe storage, therefore, grains with higher moisture content must be dried, ensiled, or acid treated.

The moisture content of forages that are to be preserved for ensiling is also of importance, since it affects the ease with which ensiling can be effected. Grass-legume forages must frequently be wilted to reduce moisture content to about 60 to 67%. On the other hand, mature forages often require the addition of water during the ensiling process.

Very dry feeds are often very dusty following grinding or dry rolling. Animals universally dislike dusty feeds; consequently, powdery rations are not eaten well. Dry, dusty rations may be improved by adding small quantities of water, by steaming, or by feeding the product in wet form.

DRYING (Dehydrating)

Drying is the removal of moisture by artificial or natural means. To avoid spoilage in storage, grains must be dry enough to prevent the growth of bacteria and molds.

Generally speaking, shelled or threshed grains stored in unventilated bins should not have more than about 15% moisture; preferably, it should not exceed 10 to 12%. Grain may be dried (1) by the use of fuel—artificially; (2) by natural air drying; or (3) by a combination of the two methods. Artificial drying is usually accomplished by running the grain through a heated chamber at a rate that will ensure its being adequately dried when it passes from the drier. The amount of heat and the drying time will vary with the amount of moisture to be removed. The process is expensive.

In addition to cost, prolonged drying at high temperature may adversely affect the feeding value of grain, especially the protein, carotene, and B vitamins.

Energy shortages and costs favor delaying harvest until grain is lower in moisture, along with maximum natural air drying. Also, the following alternatives to drying should be considered: (1) the immediate feeding of high-moisture grain, (2) storing it as high-moisture grain in an oxygen-limiting silo, or (3) treating it with an organic acid(s).

HIGH-MOISTURE (Early Harvested) GRAIN

High-moisture grain refers to grain that is harvested at a moisture level of 22 to 40% and stored without drying. Optimum conditions for ensiling high-moisture grain appear to be 25 to 32% moisture content. Correctly speaking, high-moisture grain does not involve moisture alteration.

The use of high-moisture grain for finishing cattle was prompted soon after 1900 when early frost terminated the natural maturity of corn in the Corn Belt. Kennedy, *et al.*, concluded that, on a dry matter basis, soft corn containing 35% moisture was equal to mature corn for finishing steers.[3]

But it remained for Indiana workers to rediscover and popularize high-moisture corn. In 1958, Beeson and Perry reported on a comparison of high-moisture and low-moisture corn for finishing cattle. They found no significant difference in rate of gain, but the cattle fed high-moisture corn utilized it 10 to 15% more efficiently on a dry matter basis than comparable cattle fed ground ear corn.[4]

Similar results have been obtained from high-moisture sorghum grain. The Texas Station harvested and stored sorghum grain with 23 to 32% moisture, then ground it as fed. Cattle fed ground moist grain required 11 to 26% less dry matter from grain than cattle fed ground dry grain.[5]

High-moisture grain may be successfully stored in either of two ways:

1. It may be ensiled (fermented) in an oxygen-limiting silo.
2. It may be preserved without ensiling and stored under atmospheric conditions by the addition of 1 to 1½% propionic acid (or a mixture of propionic acid with either acetic acid or formic acid) to inhibit mold or spoilage. Also, more recently, enzyme-like preparations have been developed which will preserve high-moisture grain without ensiling.

[3]Kennedy, W. J., *et al.*, *The Feeding Value of Soft Corn for Beef Production*, Iowa Ag. Exp. Sta. Bull. 75.

[4]Beeson, W. M., and T. W. Perry, "The Comparative Feeding Value of High-Moisture Corn and Low-Moisture Corn with Different Feed Additives for Fattening Cattle," *Journal of Animal Science*, Vol. 17, p. 368.

[5]Riggs, J. K., and D. D. McGintry, "Early Harvested and Reconstituted Sorghum Grain for Cattle," *Journal of Animal Science*, Vol. 31, No. 5, p. 991.

Fig. 30-10. A front-end loader scoops up a huge load of high-moisture corn (ground) in 1 of the 4 concrete-lined trench storage pits at the Farr Feeders, Inc., feedlot northeast of Greeley, Colorado. The high-moisture corn and corn silage are stored in the pits in large quantities (60,000 tons and 100,000 tons, respectively) to provide a year's continuous supply of feed. The two feeds make up the primary ingredients of the mixed ration, which, by computerized control in the feed mill, maintains a balanced nutritional diet for each animal. The storage pits are covered by polyethylene plastic and weighted down by tires. This protects the feed from the weather while providing excellent fermentation conditions. (Courtesy, Farr Feeders, Inc., Greeley, CO)

RECONSTITUTED GRAIN

Reconstituted grain is mature grain that is harvested at the normal moisture level (10 to 14% moisture), following which water is added to bring the moisture level to 25 to 30% and the wet product is stored in a suitable structure for 15 to 21 days prior to feeding. Thus, reconstituted grain involves processing that resembles soaking, and which results in an end product similar to high-moisture grain.

When stored in upright silos, the grain is stored whole, then rolled or ground at the time of removal. Reconstituted grain cannot be satisfactorily stored in horizontal silos as compaction cannot be obtained. Thus, an upright storage unit is necessary for reconstituting grain.

When reconstituting grain, the amount of moisture should be regulated so as to avoid getting too much or too little and reducing the benefits which may be derived. For example, a ton of dry grain normally contains 10 to 12% moisture, or 88 to 90% dry matter. Thus, the dry matter in a ton of dry grain usually totals 1,760 to 1,800 lb. If it is desired to increase the moisture content of this grain to 30%, it would require the addition of 500 to 570 lb of additional water per ton of grain. This means that the high-moisture grain would have the same nutrient value in 2,500 to 2,570 lb as the original grain would have in 2,000 lb. Feeding should be adjusted accordingly.

To make certain that reconstituted grain contains the desired amount of moisture, it is usually advisable to use a simple commercial moisture tester, into which is weighed a given amount of grain, followed by heating, then weighing again. The scales are calibrated in percentage of moisture.

Properly reconstituted milo and steam processed flaked milo give similar results with fattening cattle. Corn is also greatly improved by reconstituting, but there appears to be less advantage from reconstituting barley or wheat. It is noteworthy that, unlike most other methods of processing, no gelatinization of the starch occurs in reconstituted grain, yet the utilization of the starch is similar to that of other processing methods. Also, protein utilization of reconstituted grain is higher than that of other processing methods.

WATERED FEEDS

Water is frequently added to feed, with the amount varying from just enough for dust control to making a slop.

Ground and dry rolled grains, and finely ground alfalfa, tend to be dusty. The palatability of such feeds may be improved by adding a small amount of water at the time of feeding.

BLOCKS

Blocks are compressed packages, generally weighing from 30 to 50 lb each. Mineral blocks have been used for a very long time. These were followed by the development of protein blocks, primarily for supplementing cattle on the range and horses on pastures or in corrals. More recently, high-energy blocks evolved.

Fig. 30-11. Block in use on pasture—a means of lessening the labor attendant to the daily feeding of a supplement on pasture or range. (Courtesy, Moorman Manufacturing Co., Quincy, IL)

Blocks may be placed in grazing areas where cattle have frequent access to them, with one block provided to 15 cows. Intake will vary with the feed supply and the type of block. Generally, it is planned to limit feed consumption to about 2 lb per head per day by hardness of block and salt and/or fat content.

Range cattle operators use blocks as a means of (1) lessening the labor attendant to the daily feeding of a range supplement, and (2) alleviating the loss that accompanies feeding a meal.

LIQUID SUPPLEMENTS

Liquid supplements are supplements in liquid form. Many of them contain water, molasses, and urea, usually with added trace minerals and vitamins. This is a convenient way of feeding supplements to cattle on pasture or in a corral. Also, liquid supplements are sometimes added to complete ration mixes, either as part of the mix or as a top dressing.

The amount of molasses in most liquid supplements varies from 50 to 70% of the total weight. Most liquid supplements contain ½ to 2% phosphorus, often phosphoric acid. Other compounds that may be present in liquid feed supplements are fat, either animal or vegetable, to increase the amount of energy; alcohols—both ethyl alcohol and propylene glycol are used; and/or a product(s) to govern consumption.

Liquid supplements in a *lick* tank can be offered free-choice. This is a convenient and satisfactory way in which to supply protein, energy, and other nutrients, so long as the cattle do not consume more than they need.

UNPROCESSED (Whole) CORN

Unprocessed (whole) corn refers to shelled corn, the kernels of which have not been broken.

It is generally recognized that young cattle (animals under six months of age) masticate their feed well. Thus, although the digestibility of corn may be increased when it is processed for young bovines, the increased feeding value may not be sufficient to offset the added cost of processing. With the exception of young cattle, it has been assumed that corn should be ground, or otherwise processed, for cattle. Recent experiments at a number of experiment stations have indicated that there are exceptions—that the proportion of concentrate to forage is a factor in determining whether or not corn should be processed for cattle. Cattle on dry, whole shelled corn gain an average of

5% faster and require 7% less feed per pound of gain than cattle on ground or rolled corn *when high-concentrate rations are fed.* However, processing appears to have some value for dry shelled corn in rations with 20% or more roughage content or when corn is very dry—less than 12% moisture.

Eliminating processing costs is the main advantage from feeding whole corn.

ROUGHAGE PROCESSING METHODS

In recent years, researchers and feeders have been much interested in improved processing of grains. But little study has been made of forage preparation, except from the standpoint of mechanizing and ease in mixing. With the increased competition of grain for human consumption around the world, it is expected that roughage preparation will assume greater importance.

Before discussing each of the common methods of forage preparation, the following generalizations are pertinent to all of them:

1. In preparing forages, avoid processing those (a) with high moisture, which may heat and produce spontaneous combustion, and (b) in which there are foreign objects (wire and other hardware) which animals may not be able to select out, and which may generate sparks and ignite a fire during processing (grinding-chopping, conveying, or mixing).
2. Processing forages does increase cost from $2 to $10 a ton, depending on the method of processing. Therefore, cattle producers should apply their own cost figures, then determine which processing method would be most profitable. The important thing is that all costs be accounted for. For example, in computing the cost of baled hay, with which most processing methods are compared, such added "hidden" costs as losses in handling, shrinkage and wastage, grinding costs and losses, insurance, interest, and storage must be considered. Also, the age and grade of the animals, other available feeds and prices, and starter vs finishing rations must be considered.
3. Processed forages result in the forced feeding of the entire plant, including stems which may be of low nutritional value. With animals on high production, this may be a disadvantage.
4. The preparation of forages does not increase the value of the initial product.

The common methods of forage preparation are chopping, grinding, shredding, cubing (wafering), drying, ensiling, and pelleting.

CHOPPING, GRINDING, OR SHREDDING

Chopping, grinding, or shredding results in forages divided into smaller particles; but they differ from each other in how they section forage, and in the size of the particles. In comparison with a similar forage fed in long form, a forage subjected to any one of these three processes (1) is easier to handle and mechanize, (2) can be stored in a smaller area at less cost, (3) is fed with less feed refusal and waste, and (4) may make for slightly greater production.

Low-quality coarse forages usually benefit more from chopping than high-quality fine forages. This should not be construed as license to make poor-quality forage, then improve it by processing. Rather, processing makes for less waste, and perhaps some improvement in digestibility, but it does nothing to improve the nutrient content.

■ **Chopping**—This refers to cutting forage not less than 2 in. in length. (The 2 in. refers to the set of the choppers. Some of the material will be cut longer than this, and some shorter.)

Chopping has the disadvantage of being dusty. Also, there may be considerable leaf loss, or shattering, in field chopping because the hay must be drier than when it is baled or put up as long hay.

■ **Grinding**—This refers to processing forage less than 1 in. in length. Usually grinding is accomplished by means of a hammer mill, in which the forage is beaten by revolving metal hammers until it is small enough in size to pass through the screen placed in the grinder. Generally, screens with holes ¼ in. or larger are used so as to avoid pulverizing the hay. Chopping to a length of less than 1 in. is also referred to as grinding, even without the hammer mill treatment.

Fine grinding is more costly than coarse chopping; hence, it is less appealing from a practical standpoint. Yet, fine grinding is sometimes desirable when the material (either sun-cured or dehydrated) is to be incorporated in the rations of swine or poultry. Ground forages are less digestible for ruminants because they pass through the paunch more rapidly, with only limited bacterial action.

When it is advantageous to use ground hay in a ration, the addition of molasses, fat, or water will lessen the dustiness and reduce the air pollution. Some commercial mills spray a small amount of liquid fat on bales of hay just before they enter the grinder. Fat is easier to work with in a grinder or mixer than molasses, for the latter has a tendency to be sticky and "gum up" the equipment.

■ **Shredding**—This process is similar to chopping, except shredding tends to separate the stems longitudinally rather than cut them crosswise. Coarse forages, like fodder and stover, are better suited to shredding than to chopping and grinding. In some ways, it may be superior to chopping, because of exposing more of the inner part of the stem to fermenting bacteria in the rumen, thereby increasing the likelihood of better digestion. Shredding necessitates that hay be as dry as when it is chopped (10% or less); hence, it may result in as much leaf shattering as chopping. Shredding appears to have a more desirable image than chopping, with the result that hay chopping is sometimes referred to as shredding, when it is really nothing more than conventional chopping.

CUBING (WAFERING)[6]

When applied to forages, the term *cubing* (wafering) refers to the practice of compressing long or coarsely cut hay in cubes about 1¼ in. square and 2 in. long, with a bulk density of 30 to 32 lb per cubic foot. Cubes offer most of the advantages of pelleted forages, with few of the disadvantages. They alleviate fine grinding, and they facilitate automation in both haymaking and feeding. Cubing costs $5 to $8 per ton more than baling.

This method of haymaking is increasing, because it offers most of the advantages of pelleted forages, with few of the disadvantages. It alleviates fine grinding, and it lowers milk fat percentage only slightly, if at all.

DRYING

For safe storage, the moisture of hay must be lowered to the following levels: loose hay, 25%; baled hay, 20 to 22% (the lower figure for larger bales); field chopped hay, 18 to 20%; and cubes (wafers), 16 to 17%. These figures must be modified according to temperature; higher temperatures necessitate lower moisture.

Generally, hay moisture is lowered by field curing. However, artificial drying—artificial dehydrators, mow curing, and wagon dryers—may be used during times of inclement weather or when very high-quality forage is desired.

ENSILING

Ensiling refers to the changes which take place when forage or feed with sufficient moisture to cause fermentation is stored in a silo in the absence of air.

The entire ensiling process requires 2 to 3 weeks, during which time a small amount of oxygen is deleted with aerobic respiration, and aerobic fermentation occurs.

Ensiling is notable for its versatility. It can be conducted in facilities ranging from simple to sophisticated, and it can be applied to a wide variety of feedstuffs. Its greatest use is in preserving forages, with acetic, lactic, and other of the lower acids formed. The addition of grains (at the rate of about 150 lb per ton) as preservatives in ensiling forage crops also involves the principle of fermentation. Likewise, when high-moisture grain is stored in an oxygen-limiting silo, it undergoes a fermentation process.

A great variety of crops can be and are made into silage. A rule of thumb is that crops that are palatable and nutritious to animals as pasture, as green chop, or as dry forage also make palatable and nutritious silage. Likewise, crops that are unpalatable and unnutritious as pasture, as green chop, or as dry forage also make unpalatable and unnutritious silage.

Most silage in the United States is made from either corn or sorghum, with corn silage far in the lead—over 15 times as much corn silage as sorghum silage is made. Annually, about 110 million tons of corn silage and 7 million tons of sorghum silage are produced in the United States. At the present time, it is estimated that 70% of the nation's silage is made from corn and sorghum and 30% from grasses, legumes, and other feeds. In addition to the kinds of silage already mentioned, silage is made from sunflowers, the small grains, sugar beet tops, crop residues, aspen bark, wastes from food processing (sweet corn, green

Fig. 30-12. Silos at a South Dakota feedlot filled with forages and equipped with belt-type automated feedbunks. (Courtesy, *Beef* magazine)

[6]There is overlapping in the use of the word *cube*. The compressed long or coarsely cut hay packages about 1¼ in. square and 2 in. long are known as cubes. Also, pellets that are large enough to be fed on pasture or range are commonly called cubes.

beans, green peas), root crops, and various vegetable residues.

PELLETING[7]

When applied to forages, the term pelleting *refers to the process of forcing ground forage (usually with some added moisture) through a thick steel die and compressing it into a circular or rectangular mass which is cut at predetermined lengths.* They can be formed into shapes of varying thickness, length, and hardness. The larger shapes, such as are usually fed to cattle and sheep on the range, are referred to as cubes.

Binding agents are sometimes added to feedstuffs to regulate the hardness of pellets, especially forage pellets which bind less well than concentrates.

The two biggest deterrents to pelleting forages are (1) fine grinding, and (2) cost. From the standpoint of the animal, pelleted forages should be chopped coarsely in order to allow for optimum cellulose digestion in the rumen and to alleviate the incidence of bloat. As a rule of thumb, one would be on the safe side if the forage were not chopped more finely than silage. Also, there is a cost factor; processors charge $10 to $15 per ton for an all-forage pellet. Of course, the increased cost of pellets should be appraised against their increased value.

Because of the high cost of pellet mills and their lack of mobility, pelleting is largely confined to commercial feed companies and very large operations, who manufacture sufficient tonnage to make pelleting economically practical.

On the average, cattle on high-roughage (above 80% roughage) or all-roughage rations will eat about ⅓ more pellets than long or chopped hay, make about ½ to ¾ lb faster daily gains, and require 200 to 250 lb less feed per 100 lb of gain. Also, it is recognized that low-quality roughages are improved most by pelleting.

Cottonseed hulls, one of the most important roughages in the South—especially for cattle, are now on the market in pelleted form. In comparison with regular hulls, they are more digestible, require less transportation and storage space—because of their greater density, and are easier to handle.

The practice of pelleting forages will likely increase for the following reasons:

1. They are less bulky and easier to store and handle than any other form of forage, thus lessening storage and labor costs.
2. Pelleting forages prevents animals from selectivity, such as eating the leaves and leaving the stems.
3. Pelleting decreases wastage of relatively unpalatable forages, such as ground alfalfa.

Both cubing (wafering) and pelleting forages (1) simplify haymaking, (2) lessen transportation costs and storage space, (3) reduce labor, (4) make automatic hay feeding feasible, (5) decrease nutrient losses, and (6) eliminate dust.

With cubing or pelleting, the spread between high- and low-quality roughage is narrowed; that is, the poorer the quality of the roughage, the greater the improvement from cubing or pelleting. This is so because such preparation assures complete consumption of the roughage. Also, cubing or pelleting, especially the latter, tends to speed up the passage of roughage through the digestive system.

MIXED RATIONS VS FEEDING ROUGHAGE AND CONCENTRATE SEPARATELY

Most experiments and experiences have not shown any difference between mixed rations and the feeding of roughage and concentrates separately insofar as rate and efficiency of gain are concerned. However, a mixed ration has the following advantages:

1. It makes for greater efficiency in feeding and lessens the sorting at the feed bunk.
2. Where the roughage is relatively unpalatable, a mixed ration forces consumption.

Fig. 30-13. Finishing cattle at feed on a mixed ration. Feed accounts for approximately 75% of the cost of finishing cattle, exclusive of the purchase price of the feeders. (Courtesy, Benedict Feeding Co., Casa Grande, AZ)

[7]Pellets may refer to (a) the entire concentrate in pellet form, (b) the fines of the concentrate in pellet form, which are usually added back to the grain for feeding, (c) the forage in pellet form, (d) the protein supplement, or (e) the range supplements in pellet form.

3. Where it is desired to limit concentrate consumption, mixing with the roughage is desirable.

4. After cattle have become adjusted to the feedlot, a mixed ration makes it easier to get them on full feed.

Thus, feeders must make their own decision on the matter of mixed vs feeding roughage and concentrate separately, with relative costs and other factors considered. Most large feedlots use completely mixed rations.

QUESTIONS FOR STUDY AND DISCUSSION

1. Sketch the parts of a ruminant's stomach and describe the processes that occur in energy digestion and metabolism.

2. Explain why roughages compare more favorably with grain for maintenance than for production.

3. Why is it desirable to use net energy values to calculate rations for finishing cattle?

4. What cautions should be observed when feeding the following cereal grains to finishing cattle: barley, wheat, rye, oats?

5. Under what circumstances would you recommend feeding high-moisture grain to finishing cattle?

6. Explain why moisture is important when buying feeds, and when formulating rations.

7. What is acid-treated high-moisture grain? What are the virtues of this process?

8. Under what circumstances, and in what quantities, would you use (a) cane molasses, or (b) fat in the ration of finishing cattle?

9. Under what circumstances would you feed an all-concentrate ration to finishing cattle?

10. Discuss one byproduct feed that is being used by a cattle feeder of your acquaintance (or by a feeder with whom you will get acquainted). Among other things, determine the following relative to it: (a) price; (b) chemical composition; (c) quantity fed; and (d) replacement value.

11. Discuss the place and importance of each of the following roughages for finishing cattle: hay, silage, haylage.

12. Why is quality of protein, or balance of amino acids, not a critical factor in most beef cattle–finishing rations?

13. List the conditions that are essential for proper urea utilization by feedlot cattle.

14. What minerals would you provide feedlot cattle, in what quantities would you provide them, and how would you provide them?

15. What vitamins would you provide feedlot cattle, in what quantities would you provide them, and how would you provide them?

16. Based on net returns, what additive, if any, would you use for each: (a) finishing steers, (b) finishing heifers, and (c) suckling calves?

17. Discuss methods, other than additives, for improving rate and efficiency of gains.

18. List and discuss the advantages and disadvantages of each of the methods of feed preparation of (a) concentrates, and (b) roughages; then, indicate your preference, with justification for same.

19. List the advantages of mixed rations, with the roughage and concentrates combined for finishing cattle.

SELECTED REFERENCES

Title of Publication	Author(s)	Publisher
Beef Cattle Science Handbook	Ed. by M. E. Ensminger	Agriservices Foundation, Clovis, CA, pub. annually 1964–1981
Cattle Feeders Hand Book	R. M. Bonelli	Computer Publishing Co., Phoenix, AZ, 1968
Feeding Beef Cattle	J. K. Matsushima	Springer-Verlag, New York, NY, 1979
Feedlot, The, Third Edition	Ed. by G. B. Thompson C. C. O'Mary	Lea & Febiger, Philadelphia, PA, 1983
How to Make Money Feeding Cattle	L. H. Simerl B. Russell	United States Publishing Company, Indianapolis, IN, 1959
Stockman's Handbook, The, Seventh Edition	M. E. Ensminger	Interstate Publishers, Inc., Danville, IL, 1992
Third Dimension of Cattle Feeding, The	J. M. Hutchison	General Management Services, Inc., Phoenix, AZ, 1970

Stocker calves being winter-backgrounded on a high roughage ration. (Courtesy, *Livestock Breeder Journal*, Macon, GA)

MANAGEMENT OF FEEDLOT CATTLE

Contents *Page*

Contents Page

Fig. 31-1. Management gives point and purpose to everything else. This shows steers in the feedlot at Farr Feeders, Inc., Greeley, Colorado. (Courtesy, Farr Feeders, Inc., Greeley, CO)

Although it is not possible to arrive at any overall certain formula for success in operating a cattle feedlot, those operators who have made money have paid close attention to the details of management.

There are many facets of cattle management. Some are applicable to both cow-calf and cattle feedlot operations. These are covered in Chapter 10 of this book; hence, they will not be repeated. Other management practices are unique to cattle feedlots.

BACKGROUNDING

Backgrounding is the preparation of cattle from weaning until placing on finishing rations. It involves maximum roughage consumption and moderate gains.

The growing of calves from weaning until placing on finishing rations is not new. Only the term *backgrounding* is new. Likewise, some new *wrinkles* have been added to the method of conducting it.

Growing of calves to the yearling stage for placement in feedlots was, and still is, known as growing stockers. Farmers have, for many years, fed high-roughage rations to calves prior to marketing. Some ranch operators have, historically, retained calves for a second grazing season. Also, wintering cattle on small grain pasture is a well-established practice in the South and Lower Plains.

The term *backgrounding* came in with the development of large commercial cattle feedlots—outfits that usually had limited amounts of available roughage and other cheap feeds, and that had need for, on a year-round basis, growthy, but unfinished, cattle of a certain weight, usually within the range of 600 to 750 lb. Today, there is renewed interest in backgrounded cattle, due to high grain prices and the need to produce more beef from roughage. Also, it isn't particularly efficient for large feedlots to tie up capital for feeding cattle where limited gains are involved.

Fig. 31-2. Crossbred steer calves being backgrounded during October to December on turnip pasture prior to going into the feedyard. This is a popular program in the irrigated areas of the Pacific Northwest. (Courtesy, Ron Baker, C & B Livestock, Inc., Hermiston, OR)

KINDS OF BACKGROUNDING

Backgrounding of stockers and feeders can be divided into two systems: (1) backgrounding on pasture, in which calves or light yearlings are wintered and grazed, or grazed only, and sold as feeders in the late summer or fall; and (2) backgrounding in the drylot, in which the cattle are fed harvested roughage and grain and then transferred to another lot for finishing.

WHO DOES THE BACKGROUNDING?

Backgrounding is done by three different types of operators:

1. Cow-calf operators (farmers and ranchers) who either have a surplus of roughage or decide that it may be more profitable to market their calf crop at a later period.
2. Commercial finishing lots who do backgrounding as a means of assuring a supply of cattle to go into the finishing lot at the required times.
3. Specialized backgrounding lots. Such lots have evolved in recent years. Economics usually restrict these highly specialized operations to areas or large farming operations with high-roughage feed producing capacities, yet close enough to large feeder cattle production points. Investment costs in backgrounding feedyards can be considerably lower than for finishing yards. Lot space requirements are lower on a per unit basis, because cattle are never carried beyond yearling weights. Also, milling facilities and grain storage requirements are much less than for finishing operations.

KINDS OF CATTLE TO BACKGROUND

Generally speaking, the English beef breeds are best suited for backgrounding purposes. This is because they should be grown to approximately 600 to 750 lb before placing on finishing, or high-energy, rations. Holsteins and some of the larger, growthier exotics are not well suited to backgrounding, unless heavy finishing weights are planned. They need to be placed on high-energy rations at weaning time; otherwise, they will not finish out at desirable weights of 1,050 to 1,200 lb—instead, they will be too heavy at market time.

RATE OF GAIN OF BACKGROUNDED CATTLE

Properly backgrounded cattle should gain from 0.75 to 1.50 lb per head per day. Cattle finishers object to cattle that have made higher gains, because it lessens, or eliminates, compensatory growth. That is, when put on high-energy rations, animals that have been backgrounded so as to make minimal daily gains usually gain better than similar cattle that have been fed more liberally during the backgrounding period. For the latter reason, when contracting for backgrounding calves, feedlots commonly specify the kind of ration and the range in gains.

PROFIT POTENTIAL FROM BACKGROUNDING

Profit is the goal in any type of feeding operation, whether it be backgrounding or finishing. The profit potential in a background system is dependent upon (1) purchase cost of feeders, (2) total cost of gain, (3) amount of gain, and (4) selling price of the backgrounded feeders. The most profitable backgrounding operations are generally in areas where feed costs are low, where overhead costs are minimal, and where the health program is superior.

ADVANTAGES OF BACKGROUNDING

The **advantages** of the backgrounding system include the following: (1) owners have the cattle when they make the most efficient gains; (2) it is well adapted to the use of roughage and byproduct feeds; (3) it provides a way in which to make use of seasonal surplus labor or of buildings that are present on some farms or ranches; and (4) volume is more flexible than with cow-calf operations—that is, numbers can be easily changed to fit feed, labor, or economic outlook.

DISADVANTAGES OF BACKGROUNDING

The **disadvantages** of backgrounding include the following: (1) high buying and selling skills are required, because there is no established market as with slaughter cattle; (2) buying, selling, and shrink costs must be absorbed by a limited amount of gain; (3) cost of gain must be kept down to offset negative margins that usually prevail; (4) high risk results from seasonal and yearly price fluctuations in feeder cattle; and

(5) backgrounding operations cannot be located too far from finishing feedlots, because of high transportation costs.

It should be noted that backgrounding operations will have increasing competition for their finished products—calves ready to go into the finishing lot—from cow-calf operators who wean off heavier calves. The latter is being accomplished by crossbreeding, along with heavier milking cows.

PEOPLE MANAGEMENT

Big feedlots must rely on hired labor, all or in part. Good help is hard to come by; it's scarce, in strong demand, and difficult to keep. Moreover, the labor situation is going to become more difficult in the years ahead. It matters little if the feedlot facilities, the feed, and the cattle are the best if the labor fails to feed them properly or to treat the sick. It is important, therefore, in the operation of the feedlot that laborers be recruited with care, that there be an organization chart and job description, and that there be an incentive basis for the help. These points are covered in Chapter 13 of this book.

HOUSEKEEPING AND REPAIR

A well-kept and attractive feedlot makes for better employee attitude, which in turn makes for better cattle care. Fences, gates, equipment, and roads should be kept in constant repair. Unused materials should be stored or repaired. The entire premises should be tidied up at all times. Pride in the physical plant will be reflected in pride in the work.

RECORDS

Complete and well-kept records are a must in the operation of a cattle feedyard, even though they require a lot of time and expense. Deficient records and deficient managers generally go hand in hand.

RECORD FORMS

There is no limit to the number of different kinds of record forms that can be, and are, kept in a given feedlot. Also, there is little similarity in record forms between lots, due to differences between people, primarily managers and bookkeepers. The important things are that (1) record forms be so designed as to facilitate record keeping, with as much ease, efficiency, and accuracy as possible; and (2) records be kept.

Figs. 31-3 and 31-4 show two basic record forms; Fig. 31-3 is a daily record, whereas Fig. 31-4 is a monthly, cumulative, and final feed summary. Many variations of these can be made.

Among other necessary records are the following:

1. Feed costs, with this record kept by individual pens
2. Grain inventory
3. Roughage inventory
4. Feed projections ahead
5. Cattle receiving and movement records
6. Sick pen and movement records
7. Sick pen costs
8. Mortality slips and proof of death
9. Maintenance and repair costs
10. Routine office bookkeeping
11. Customer billing for feed
12. Closeout records

MAKE 28-DAY TEST WEIGHTS

Twenty-eight–day test weights will not adversely affect the performance of feedlot cattle provided the cattle are handled properly. Check weights should include a representative cross section of the cattle in the yard, including age, weight, type, background, and sex. Where it is not convenient, or it is not desired, to weigh an entire lot of cattle, *markers*—cattle of certain odd colors, animals with tail switches clipped, etc.—may be weighed. Also, it is important that weighing consistently be done at the same time of day, and that the lots be weighed in the same order, due to the effect of rumen fill.

MILL MANAGEMENT

Mill management includes many things; among them, the following: commodity scheduling and purchasing; inventory control; maintenance, repair, and new construction; mill cleanliness; commodity and ration quality control; and milling to meet daily feed needs. Quality control refers to being able to deliver to the cattle on a consistent basis the same quality or intended ration composition.

■ **Commodity quality control**—All feed ingredients should be bought by specification and grade. Then, each of them should be analyzed in order (1) to determine if the ingredient received meets these specifications, and (2) to know the composition of the ingredients used in ration formulation.

A chemical analysis is no better than the sampling. Thus, state feed control officials should be contacted

DAILY RECORD

Feedlot: ____________ **Pen No.** ________ **Date Started** ____________
Month Day Year

Day of Month	Head In						Death Losses	Head Out			Daily Feed		Sold			Comments
	No.	Origin	Pur. Price	Total Pur. Wt.	Total Wt. at Lot	Avg. Wt. at Lot	(cause)	No.	Total Wt.	Avg. Wt.	Total Lb	Lb/ Head/ Day	To	Price/ Cwt	Carcass Grade	
1																
2																
3																
4																
5																
6																
7																
8																
9																
10																
11																
12																
13																
14																
15																
16																
17																
18																
19																
20																
21																
22																
23																
24																
25																
26																
27																
28																
29																
30																
31																

Fig. 31-3. Form for daily record.

MONTHLY, CUMULATIVE, AND FINAL FEED SUMMARY

Feedlot: ____________ Period: ____________

Pen No. ________ No. Head ________ Date Started ________ Date Closed ________

RATION:

Ration No.	Total Pounds	Price/Ton	Total Cost
	(lb)	($)	($)
#1			
#2			
#3			
#4			
#5			
#6			

FEED ANALYSIS:

- Total feed fed ____________ ________ lb
- Total cost of feed to date ____________ ________ $
- Feed days (no. head × days) ____________ ________ no.
- Net weight out ____________ ________ lb
- Net weight in ____________ ________ lb
- Net gain ____________ ________ lb
- Avg. weight out ____________ ________ lb
- Avg. weight in ____________ ________ lb
- Feed per head per day ____________ ________ lb
- Cost per head per day ____________ ________ ¢
- Gain per head per day ____________ ________ lb
- Cost per lb gain ____________ ________ ¢
- Feed conversion (lb feed/lb gain) ____________ ________ lb

OTHER COSTS:

- Milling charges ____________ ________
- Mineral charges ____________ ________
- Medication ____________ ________
- Management ____________ ________
- Labor ____________ ________
- Physical plant (other than milling) ____________ ________

Fig. 31-4. Form for monthly, cumulative, and final feed summary.

Fig. 31-5. Cattle feed processing facilities at C & B Livestock, Inc., Hermiston, Oregon. (Courtesy, Ron Baker, C & B Livestock, Inc., Hermiston, OR)

for publications detailing recommended methods of sampling commodities and the equipment to use. All commodity samples should be properly labeled, including the yard name, name of the sampler, date sample was taken, commodity name, vendor of the commodity, invoice number, and car or truck number.

- **Ration quality control**—Rations should be analyzed according to an established schedule. Most of the very large cattle feedlots sample rations daily, then prepare a weekly composite of each ration, which is submitted to the laboratory for analysis. Additionally, a complete analysis (proximate, calcium, phosphorus, and nitrate) is made of each ration monthly. Always label each ration with care. This should include the identity of the sample or ration number, feedyard name, name of sampler, and date.

FEED BUNK MANAGEMENT

Feed bunk management is a combination of management factors involved with obtaining maximum performance, minimum digestive disorders, and keeping cattle on feed. Feed bunk management and quality control are directly involved with obtaining maximum and economical performance from cattle. It should be every feeder's goal to obtain maximum feed intake of a consistently high-quality ration, since both rate and efficiency of gain are directly related to nutrient intake.

SCHEDULE FOR GETTING CATTLE ON FEED

When new cattle arrive at the feedlot, the objective is to get them on full feed as rapidly as possible, without throwing them off feed. This is not easily accomplished because many factors influence the difficulties experienced in starting new cattle on feed, among them: (1) the length of time that the cattle have been without feed; (2) the kind of feed to which the cattle were accustomed prior to shipment; (3) the age of the cattle—young cattle adapt to a change in feed more easily than old cattle; (4) whether or not the cattle have been fed and watered out of troughs before; (5) the weather conditions; and (6) existing nutritional deficiencies.

Fig. 31-6. Calves that have been on feed 45 days, following shipment from south Texas to an Arizona feedlot. (Courtesy, Benedict Feeding Co., Casa Grande, AZ)

TRADITIONAL PROCEDURES OF GETTING OLDER CATTLE (Not Calves) ON FEED

When first brought into the feedlot, cattle that are not accustomed to grain may be started on feed by either of the following procedures:

1. Self-fed long grass hay (and/or corn or sorghum silage), and hand-fed concentrate according to the following schedule (with the cattle automatically lessening their self-fed hay consumption as the grain is increased):

First day—Feed 4 lb of concentrate/head/day, consisting of 2 lb of grain and 2 lb of protein supplement.

Daily increase—Step up the grain by 1 lb/head/day until cattle are receiving 1 lb/cwt body weight.

Increase every third day—After a level of 1 lb daily/cwt body weight is reached, make increases every third day as follows:

Yearlings—½ lb
2-year-olds—1 lb

2. Hand-fed a mixed ration of chopped grass hay (and/or corn or sorghum silage) and concentrate, with the proportion of roughage decreased and the grain increased according to the following schedule:

Day	Kind of Feed	Percent of Roughage
1	Grass hay and/or nonlegume silage	100
2–4	Grass hay plus starter	60–90
5–14	Starter ration	40–60
15–21	Transition ration	15–40
22 to market	Finisher ration	5–15

Although one of the above procedures may serve as a useful guide, it is recognized that no set of instructions can replace the cattle intuition and good judgment of an experienced feeder.

After cattle are on full feed, they may be either self-fed or hand-fed. Most large feedlots feed twice daily, barely letting the cattle clean up the previous feed before the next feeding.

SCHEDULE AND RATION FOR GETTING CALVES ON FEED

Most cattle feeders follow the procedure and type of ration given above for getting cattle on feed; they start them on a high-roughage ration, then work them over to a high-concentrate ration as they progress through the feeding program. However, based on University of California studies, it appears that for calves (not older cattle) a starting ration consisting of 28% hay (roughage) and 72% concentrate is best. This is similar to the procedure outlined in point 2 above for older cattle, except that the calves are immediately started on a lower roughage (28%), higher concentrate (72%) ration.

Among the problems encountered in new cattle are feed and water refusal, lacticacidosis, bloat, and diarrhea. Refusal of feed and water is generally due to the fact that the animals are not used to conventional troughs and/or the feed is so different.

Lacticacidosis generally results from feeding hungry cattle excessive levels of rapidly fermentable feeds. The condition is characterized by an accumulation of lactic acid in the rumen and a lowering of the pH in the blood and urine. The problem can be minimized by starting cattle on a high-roughage ration and shifting them gradually to a high-concentrate ration.

Bloat occasionally occurs in new cattle, although it is more frequent during the later stages of feeding. Bloat and diarrhea in new cattle can generally be prevented by feeding generous quantities of such roughages as straw, grass hay, cottonseed hulls, or corncobs.

FREQUENCY OF FEEDING

Experiments and experiences show that increased frequency of feeding (more than one time per day) improves performance above increased costs. For this reason, most commercial feedlot cattle are fed three times daily. Some experiments indicate that even more frequent feeding—more than three times daily—will produce slightly more rapid gains and result in greater feed efficiency. However, the improved performance is not enough to warrant the increased costs.

AMOUNT TO FEED; FULL VS LIMITED FEEDING

Feed intake is one of the key factors affecting feedlot performance. Perhaps no other factor has such overriding importance in determining rate and efficiency of gain, and, ultimately, the profit derived from feeding cattle. Of course, the reason for emphasis on high feed intake is that once a sufficient amount of the ration is consumed to meet the maintenance needs of a finishing animal, the remainder is converted to gain with remarkable efficiency. Thus, as shown in Fig. 31-7, by adding 4 lb to the daily feed intake of a 600-lb steer, rate of gain may be increased by $1\frac{1}{10}$ lb per

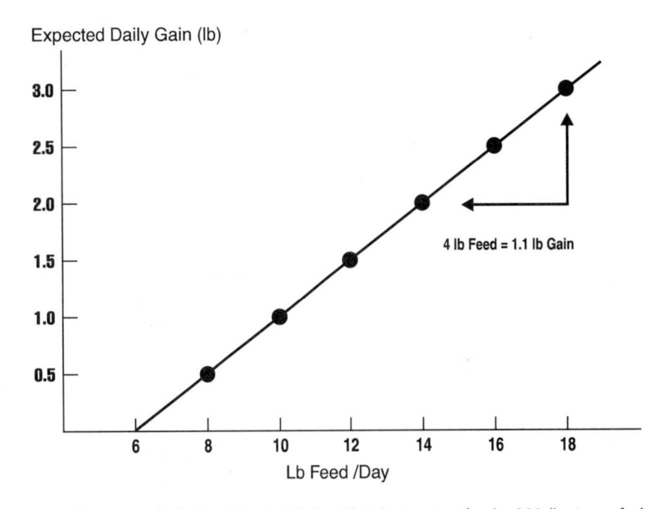

Fig. 31-7. Relationship of daily feed intake to rate of gain; 600-lb steers fed 85% concentrate ration.

day. Conversely, poor feed intake results in too high a percentage of the total nutrients being expended for maintenance.

Thus, finishing cattle should receive a maximum ration over and above the maintenance requirements. In general, they will consume daily an amount (on an air-dry basis) equal to 2.5 to 3.0% of their liveweight. Feed intake will vary according to the condition of the cattle, the palatability of the feeds, the energy of the ration (in general, animals eat to meet their energy needs), the weather conditions, and the management practices. For example, older and more fleshy cattle consume less feed per hundredweight than do younger animals carrying less condition; thus, mature, over-finished steers will consume feeds in amounts equal to about 1.5% of their liveweight, whereas thin steers under 2 years of age will consume fully twice as much feed per unit liveweight.

Overfeeding is also undesirable, being wasteful of feeds and creating a health hazard. When overfeeding exists, there is usually considerable leftover feed and wastage, and there is a high incidence of bloat, founder, scours, and even death. Animals that suffer from mild digestive disturbances are commonly referred to as *off feed.*

Limited feeding means just what the name indicates—not giving the animals all they want. Limited feeding generally decreases the rate of gain, adversely affects feed conversion, and increases cost of gains. Under most conditions, cattle should be full fed throughout the finishing period.

TABLE 31-1
AVERAGE EXPECTED DRY MATTER INTAKES PER HEAD DAILY FOR CATTLE[1, 2]

Body Weight		Expected Dry Matter Intake	
(lb)	*(kg)*	*(lb)*	*(kg)*
300	*136*	8.5	*4.0*
400	*182*	10.5	*4.8*
500	*227*	12.5	*5.7*
600	*273*	14.5	*6.6*
700	*318*	16.5	*7.5*
800	*364*	18.0	*8.2*
900	*409*	20.0	*9.1*
1,000	*455*	22.0	*10.0*
1,100	*500*	23.0	*10.5*
1,200	*545*	24.0	*10.9*

[1]From: *Great Plains Beef Cattle Feeding Handbook*, GPE-1100.3.

[2]To make metric conversions, see the Appendix, "Section IV—Weights and Measures."

FACTORS AFFECTING FEED INTAKE

A number of factors play major roles in governing feed intake; among them, the following:

■ **Age of cattle**—Calves consume a larger proportion of feed in proportion to their body weight than do older cattle. Table 31-1, which gives the average expected dry matter intakes per head daily for cattle, points up this situation.

■ **Propionic acid in the blood curbs appetite**—All mammals, including humans, have an *appetite center* at the base of the brain in the lateral region of the hypothalamus. Certain nerve cells actually regulate energy intake, causing the sensation of hunger, or preventing the animal from consuming too much. But there is a great species difference as to how much feed is enough—or too much. A growing boy *wolfs down* 8 to 9% of his body weight daily. He is outdone by a baby chick, which eats 10% of its body weight. A hog eats 5 to 6% of its body weight. But a fattening steer consumes only 2 to 3% of its body weight. Why the difference? It appears that propionic acid is the triggering mechanism which tells a steer when to stop eating. It acts on the nervous center of ruminants. Since increased grain levels tend to step up propionic acid, it is conjectured that grain rations, which increase propionic acid levels of the blood, are self-limiting when it comes to feed intake. For this reason, cattle eat fewer pounds per head daily of a high-concentrate ration than of a high-roughage ration, but both groups tend to take in about the same level of energy.

■ **Rumen *fill***—The *fill* in the rumen places a ceiling on feed intake. Since low-quality roughages pass through the rumen at a slower rate than high-quality roughages, they can limit the amount of total feed that the animal can consume in a 24-hour period. Of course, this is of little consequence with fattening-type rations, which normally contain limited amounts of roughages and roughages of good quality. The amount of concentrate in the ration has a marked effect on the total pounds of feed consumed daily. Thus, when on all-concentrate rations, the total energy intake of cattle over a 24-hour period is not much greater than with a bulky, low-concentrate ration. Apparently, this is due to the regulation of total calorie intake by the ruminant.

■ **Physical makeup of the ration**—Coarsely processed grains are more palatable to cattle than finely processed grains. Thus, they will eat more of such feed.

Energy density, or bulk of the ration—The weight per unit volume, or the bulk, of the ration affects total feed intake. That is, bulky rations which make for "fill" reduce dry matter consumption. Most feeders think of energy density in terms of roughage-concentrate ratio. When density is very low, as in a high-roughage ration, animals simply cannot hold enough to make gains. As energy density increases, gain increases.

■ **Heritability**—Experiments and experiences confirm that some cattle are better eaters than others—that they will consume more feed. This may be due either to a difference in rumen capacity and/or a difference in threshold for circulating metabolites in the bloodstream which affect the appetite center of the hypothalamus.

■ **Heat stress**—Cattle feeders have long known that finishing cattle consume less during hot weather. This was confirmed by the California Station in trials at the Imperial Valley Station. The lowering of feed consumption due to heat stress may be lessened by lowering the roughage in the diet, cooling the drinking water, providing shade, and adding higher levels of vitamin A.

■ **Protein or phosphorus deficiency**—A protein deficiency can markedly reduce feed intake by depressing the rumen bacterial count and the rate of breakdown of feeds. A phosphorus deficiency can cause a reduction in feed intake, and even a depraved appetite.

■ **Other factors affecting feed intake**—The feed intake of cattle is also affected by: (1) moisture level—very high-moisture feeds reduce total dry matter intake; (2) dustiness—dusty feeds lower total feed intake; (3) lack of water—depriving cattle of water will markedly reduce feed intake; (4) frequency of feeding—more frequent feeding results in higher feed consumption, and (5) freshness of feed—cattle will consume more clean, fresh feed than stale feed.

FEED REGULARLY

Feedlot cattle should be fed at regular times each day, by the clock. This means that in the larger lots cattle in each alley should be fed in the same order each day. Be prompt—remember that cattle are creatures of habit.

KEEP BUNKS CLEAN

All feed bunks should be cleaned thoroughly at least once a week during dry weather, and as needed during stormy weather. No feed should ever be allowed to spoil in the feed bunk. Manure should be cleaned from feed bunks daily.

WEATHER AFFECTS EATING AND DRINKING HABITS

During hot weather, feedlot cattle *peak* their eating during early morning and again during the evening hours—when it is cool. With heat, night drinking increases. In cool weather, they eat more during the midday than when it's hot. The feeder should sense these changes in cattle eating habits and program their feeding accordingly.

Cattle eat more following a bad storm or a hot spell. Thus, at such times the bunks may be *slick* for 2 to 3 hours and the cattle may line up waiting to be fed. When this happens, the ration should be increased. By going to a higher roughage ration at these times, the problems from acidosis and laminitis can be minimized.

WATER

Water is the cheapest feed! Thus, cattle should have access to plenty of clean, fresh water at all times. They will consume 7 to 12 gal per head per day. In cold climates, waterers should be equipped with heaters. Where the water supply is not limited by cost or volume, continuous-flow waterers are excellent. In order to keep the pathogen and algae content at a minimum, water tanks should be cleaned at least once a week in the winter and twice a week in the summer. In sick pens and pens of new cattle, the water tanks should be cleaned daily.

PROGRESSIVE CHANGES IN FEEDLOT CATTLE IN (1) RATE OF GAIN, (2) FEED CONSUMPTION, (3) FEED EFFICIENCY, AND (4) COST OF GAIN

It is important that cattle feeders be cognizant of the progressive changes that normally occur in feedlot cattle, from start to finish of the feeding period. Of course, it is recognized that many factors influence the degree of these changes.

The Arizona Station recorded the changes at 28-day intervals in 41 lots of feedlot cattle, with all lots taken from start to finish. Their findings are given in Table 31-2.

The Michigan Station used a slightly different approach to obtain changes at 28-day intervals. They fed 7 different lots of cattle, with each lot carrying a different length feeding period in increments of 28 days, ranging from 115 to 283 days on feed; and each group of cattle was closed out and slaughtered at the end of its feeding period. Their findings are given in Table 31-3.

Based on these two studies, the following deductions may be made relative to the progressive changes in feedlot cattle, from start to finish: (1) rate of gain decreases; (2) daily feed consumed per 100 lb of body weight decreases; (3) feed per 100 lb of gain increases; and (4) feed cost per 100 lb gain increases.

CULL OUT; TOP OUT

Obvious poor doers should be taken out early, and marketed at Standard grade. Where individual weighing can be made, consideration should be given to the practicality of individually tagging (with duplicate tags, one in each ear) and weighing incoming calves; weighing them again at the end of the grower-ration period

Fig. 31-8. Choice finished cattle being shipped to slaughter plants. Cattle should be sold when their genetic potential is reached. (Courtesy Farr Feeders, Inc., Greeley, CO)

TABLE 31-2
AVERAGE DAILY GAIN AND FEED PER POUND OF GAIN OF CATTLE AT 28-DAY INTERVALS[1]

28-Day Feeding Period	No. Days on Feed	Pay Weight of Cattle on Feed			Avg. Daily Gain	Avg. Daily Feed	Lb Air-Dry Feed/Lb Gain
		Begin.	End.	Avg.			
	(days)	(lb)			(lb)	(lb)	(lb)
1st	1–28	597	697	647	3.56	22.75	6.39
2nd	29–56	697	783	740	3.06	24.24	7.92
3rd	57–84	783	858	821	2.69	23.43	8.71
4th	85–112	858	923	890	2.30	23.83	10.36
5th	113–139	923	979	951	2.07	24.43	11.80

[1]Data collected by the Arizona Agricultural Experiment Station. It embraces nine experiments, 41 lots of yearling steers, and 640 animals. All cattle were started on relatively high-roughage rations, with the roughage decreased as the period progressed.

TABLE 31-3
EFFECT OF LENGTH OF FEEDING PERIOD ON (1) RATE OF GAIN, (2) FEED CONSUMPTION, AND (3) FEED EFFICIENCY[1, 2]

	Days on Feed						
	115	143	171	199	227	255	283
No. steers	8	8	8	8	8	8	8
Avg. initial weight (lb)	655	657	685	685	672	679	673
Avg. final weight (lb)	928	990	1,037	1,084	1,133	1,203	1,224
Avg. total gain (lb)	273	333	352	399	461	524	551
Avg. daily gain (lb)	2.73	2.33	2.06	2.01	2.03	2.05	1.95
Daily feed/100 lb body wt. (lb)	2.60	2.44	2.28	2.19	2.20	2.11	2.05
Total feed/100 lb gain (lb)	869	861	952	962	975	968	996
Concentrate-roughage ratio	77:23	77:23	77:23	77:23	77:23	77:23	77:23

[1]Merkel, R. A., H. E. Henderson, and H. W. Newland, *Effect of Length of Feeding Period on Rate, Composition and Cost of Gain*, Michigan State University.

[2]To make metric conversions, see the Appendix, "Section IV—Weights and Measures."

and prior to going on finishing rations; then culling out the bottom 10%.

Also, cattle should be sold when they make their grade; thereby avoiding loss in efficiency, excess finish, and too heavy weights. Usually, it is unwise to challenge a sagging market by holding and feeding for a higher market. There is no need to put feed and labor into heavy cattle at a cash discount when younger cattle will use these resources more efficiently.

OVERFINISHING

Excessive finishing is undesirable, both from the standpoint of the producer and the consumer. Experienced cattle feeders are fully aware of the fact that to carry finishing cattle to an unnecessarily high finish is usually prohibitive from a profit standpoint. This is true because the gains in weight then consist chiefly of fat but little water. In addition, a very fat animal eats less heartily, with the result that a small proportion of the nutrients, over and above the maintenance requirement, is available for making body tissue.

Fig. 31-9 shows that the heavier the cattle, the more expensive the gains. Also, this graph points up (1) the importance of topping out finished cattle, rather than waiting until the entire lot is ready; and (2) the reason why it is generally wise to sell cattle when they are ready to go, rather than to hold for a higher market.

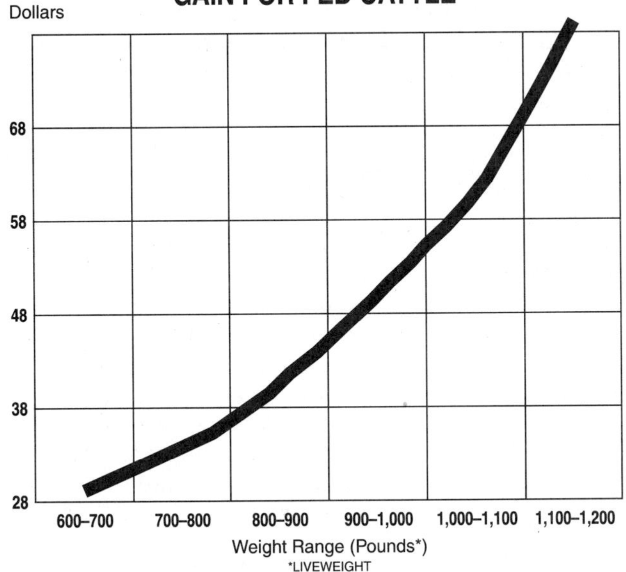

Fig. 31-9. This graph illustrates changes in feed conversion efficiency for cattle from normal feeder weights to slaughter weights. Note that feed costs per 100 lb of gain more than double from 600 to 700 lb to 1,000 to 1 ,100 lb, and that the conversion efficiency ratio changes even more sharply when cattle pass 1,100 lb.

ABORTING HEIFERS

Frequently feedlot operators have need to abort heifers. When such occasions arise, they want to know how to do it.

Caesareans on feedlot heifers must be performed by a veterinarian; hence, they are expensive. Moreover, such surgery lowers the rate and efficiency of gain very considerably.

Several drugs can be used. Prostaglandins and prostaglandin analogues are the abortifacients of choice during the first 150 days of pregnancy; beyond 150 days of pregnancy, additional products, such as dexamethasone or estradiol, may be needed.

The best nondrug method is to pregnancy check each heifer within 2 to 3 weeks after arrival in the feedlot. Those under 4 months pregnant can be aborted either by manually removing the corpus luteum on the ovary or using prostaglandin or a prostaglandin analog. Heifers over 4 months pregnant should be identified and sent to pasture for calving.

While termination of pregnancy is possible, the setback in performance may not justify the action. This is a management decision, and the ability to determine the stage of pregnancy may be the deciding factor.

BULLERS OR RIDERS

There are two types of bullers or riders: (1) those that persistently ride other cattle; and (2) those that other cattle pick out and ride. The first are often bulls or stags. A buller that is being ridden by a number of animals can receive serious injuries, and may even be killed.

There is a tendency for riding to occur (1) where too many animals are being confined in a small place; (2) where cattle are heavily infested with lice; and/or (3) where feeds high in estrogens are being fed.

The solution to the buller or rider problem is difficult—and often exasperating. One or more of the following control measures are generally applied by experienced cattle feeders:

1. Remove both the jumper animal(s) and the animal(s) being ridden, and put them in separate pens. After about 72 hours of isolation, return the animals to their original pen. This should be done during the morning feeding period, so that the animals being returned will be inconspicuous. If riding reoccurs, remove the animals again, before injury is inflicted.

2. Provide more space if the animals are crowded—spread them out.

3. Dip or spray the animals. This is an *odor control* approach—it makes them *smell* alike. Additionally, it will control lice if they are present.

MUD PROBLEM

University of California studies show that mud can reduce cattle gains by as much as 25 to 35%. Thus, it is important that the problem be minimized, especially in high rainfall areas. Good drainage is the first essential. This should be assured at the time the feedlot is located and constructed. Mounds, preferably perpendicular to the feed bunk, will provide cattle a dry place on which to lie down. Concrete aprons along the bunk will provide them with solid footing on which to stand and feed. Also, lessening of cattle density during the winter months—fewer animals per lot—is an effective method of controlling the mud problem. Thus, many feedlots plan to feed fewer cattle during the muddy season.

BEDDING

Open feedlots are usually mounded, but not bedded. Sometimes sawdust or shavings are placed on top of the mounds.

Shelters and confined houses are usually bedded, unless a liquid manure system is used. Whether or not bedding is used and the choice of kind of bedding depend primarily on the local cost and availability of bedding, and labor available for cleaning. For a full discussion of bedding cattle, see Chapter 10 of this book, under the heading entitled "Bedding Cattle."

BIRD CONTROL

Birds are gluttons and filthy; hence, they should be controlled. In a 5-month study of a 12,000-head cattle feedlot in California, University of California researchers found that the birds ate between 200 and 500 lb of feed each day, adding up to a total of 32,500 to 74,000 lb for the 5-month winter season. The bird population in the feedlot ranged from 10,000 to 20,000, of all species, with the most abundant being the house sparrow, which was estimated at from 6,000 to 8,000. Other feed-consuming bird species identified were starlings, brewer blackbirds, red-wing blackbirds, and cow-birds.

Nationwide, starlings constitute the major feedlot bird problem. Some large commercial feedlots estimate their starling population at 100,000 per lot. Iowa feeders figure that starlings add $3 to $4 to the cost of each steer marketed. Some western feedlot operators compute the cost for overwintering each 1,000 starlings at $100; others estimate that starling nuisance and feed costs add 2¢ to the cost of each pound of gain.

In addition to feed consumption, birds contaminate much feed and spread diseases—to both animals and humans. The starling has been incriminated in the spread of coccidiosis among animals, transmissible gastric enteritis (TGE) in swine, and histoplasmosis in humans.

Recordings of distressed bird calls, carbide cannons, and harassment or killing with guns achieve only partial control. Many chemicals and baits have been tested, and a few have been found to be effective. However, some states do not allow the use of chemicals in bird control. Therefore, before using any chemical, the cattle feeder should check with the appropriate federal, state, and local departments of health. Also, chemicals should always be handled with care; they should neither come in contact with the skin nor be inhaled. Gloves and a respirator should be worn when mixing or handling them.

FLY CONTROL

The housefly is the most common type of fly found around cattle feedlots. It is a scavenger and does not feed on animals, but it does cause irritation and annoyance. Stable flies, which are blood feeders, may also be present in certain areas and certain feedlots.

Effective housefly control requires proper animal waste management and good feedlot sanitation. The basic objective in fly control is to eliminate possible sources of fly development. This can be accomplished by the following steps: (1) provide proper drainage and avoid wet spots; (2) remove manure immediately after a pen is vacated; and (3) remove manure and spilled feed at important fly breeding areas such as fence lines, feed bunks, hospital pens, horse pens, truck washing stations, and receiving and shipping areas. Chemical control should be used in conjunction with the proper waste management techniques, and not as a sole means of control. Residual and space sprays aid in reduction of adult flies; and larvicides may be applied to areas of intense larval development such as manure stockpiles, hospital, and horse pens.

(Also, see Chapter 12, section entitled "Parasites of Beef Cattle.")

HOGS FOLLOWING CATTLE

There was a time when hogs following feedlot

Fig. 31-10. Hogs following cattle. One pig should follow every 1 to 3 steers, the ratio of pigs varying with the kind and preparation of the feed and the age of the cattle. Sometimes the only profit obtained is in the gains made by pigs following cattle. (Courtesy, American Feed Manufacturers Assn., Inc., Arlington, VA)

cattle was commonplace. But the practice declined with the advent of large, specialized commercial feedlots of 1,000-head, or larger, capacity. Today, only a few farmer-feeders of the Corn Belt have hogs following cattle, and the practice is almost nonexistent in the large commercial cattle feedlots of the nation. Because farmer-feeders account for relatively small numbers of cattle, in comparison with large commercial feeders, it follows that very few cattle are followed by hogs. The primary reasons given by cattle feeders for a decline in the practice are: (1) feeds are being processed in a more sophisticated manner than formerly, with the result that few grains pass through cattle whole; (2) hogs tend to get cattle up and to get into troughs; (3) an increase in fenceline feeders, which won't keep hogs in; and (4) few hogtight commercial feedlots.

Nevertheless, cattle feeders who have a convenient source of feeder pigs, who are not *allergic* to keeping hogs, and whose cattle lots are fenced hogtight, can add to their net income by having hogs follow cattle.

The following hog-cattle ratio is recommended, using 75- to 150-lb pigs:

	If Whole Shelled Corn Is Fed	If Ground or Rolled Grain Is Fed
	(Pig-Steer Ratio)	*(Pig-Steer Ratio)*
Calves	1:3	1:5
Yearlings	1:2	1:4
Two-year-olds	1:1½	1:3

For every 50 bushels of whole corn fed to yearling cattle, approximately 50 lb of pork will be produced. Allowing 55¢ for hogs, and subtracting $10 per pig for protein and other costs, that's $17.50 per pig.

Pigs sometimes inflict injury on heifers (injuring the vulva when they are lying down); therefore, their use is generally limited to steers.

Sows may be used, but because of their size, they may create problems from getting into the feed and water facilities.

CONDUCTING APPLIED FEEDLOT TESTS

When carefully conducted, and properly interpreted and used, feedlot trials can be a valuable adjunct in the operation of a large feedlot. Among their virtues, the feedlot operator can study area and feed differences. Among their limitations are usually less accuracy and fewer controls than most university-conducted experiments. For the latter reason, most of them should be looked upon as applied tests or demonstrations *per se*, rather than carefully controlled, basic experiments; terminology which doesn't detract from their value, but which does place them in proper perspective.

The number of pens which a feedlot should devote to test work will vary according to the size of the operation and the number of treatments planned at one time.

There should always be a minimum of two lots for controls, plus two lots for each treatment evaluated. Generally, the two control lots should be fed the standard feedlot ration, and two lots should be given each treatment evaluated.

The local county extension agent should be invited to participate in the test; usually the agent will welcome the opportunity.

The following procedure is recommended in conducting feedlot tests:

1. **Cattle.** The animals should be of uniform breeding, background, age, and weight, and of the same sex. Use cattle owned by the operator, rather than custom-fed animals.
2. **Number per lot.** Ten head if individually weighed; 20 to 40 head, or more, if group weighed.
3. **Randomization.** Gate or chute cut; one per treatment, or not more than five at a time.
4. **Identity.** Preferably (a) apply a different brand to each lot, and (b) individually identify each animal with duplicate numbers—one in each ear. For the latter, use plastic ear tags, the numbers on which can be easily read at a distance.
5. **Variables.** Have only one variable in each pair of treatment lots. Let us suppose, for example, that in

a given feedlot steers are now being implanted with Synovex S as standard procedure. However, the owner desires to determine if it would be practical to (a) switch to Ralgro implants, or (b) use a combination of both Ralgro implants and antibiotics. The design would be as follows:

	Lot	Treatment
Controls	1	Control (Standard ration with Synovex S implant)
	2	Control (Standard ration with Synovex S implant)
Treatment 1, Ralgro implants	3	Standard ration; Ralgro implant
	4	Standard ration; Ralgro implant
Treatment 2, Ralgro + antibiotics	5	Standard ration; Ralgro implant plus antibiotics
	6	Standard ration, Ralgro implant plus antibiotics

6. **Adjustment period.** After sorting cattle into test lots, allow a minimum adjustment period of 7 days; during which the cattle should be individually tagged and handled as necessary, and gradually accustomed to their new rations. In case of sickness, a longer adjustment period may be necessary—sometimes as much as 2 to 4 weeks.

7. **Weighing conditions.** Keep off feed and water overnight, then weigh the next morning. Weigh (preferably using a self-recording beam, so as to alleviate the human error) pens in the same order and at the same time each morning when (a) initiating the experiment, (b) at 28-day intervals, and (c) at the close of the test.

Also, weigh and record the amount of feed given to each lot of cattle; using a modified paired-feeding technique, in which the paired lots are limited in feed consumption to the lot consuming the least. Sometimes it is best to limit all lots (both controls and treatments) to the level of the lot consuming the least, although this will vary according to the treatment being evaluated.

8. **Carcass data.** Sell, or have custom slaughtered, with the stipulation that the slaughter plant provide individual (according to individual ear tags) (a) carcass weight and yield, and (b) federal grade. If slaughter data cannot be obtained on all cattle, get it for as many as possible and of the same number from each lot.

9. **Summarize results.** At the end of the trial summarize the results, using as criteria (a) rate of gain, (b) feed efficiency, and (c) carcass results.

10. **Determine the application.** If both lots of a given treatment are considerably better than the controls, decide (a) whether to repeat the test, or (b) adopt and use the new treatment throughout the feedlot. If the latter becomes the new standard, continue with it until a new and superior treatment evolves, based on new trials.

QUESTIONS FOR STUDY AND DISCUSSION

1. Define *backgrounding*. Then discuss each of the following aspects of backgrounding:
   a. Kinds of backgrounding.
   b. Who does the backgrounding.
   c. Kinds of cattle to background.
   d. Rate of gain of backgrounded cattle.
   e. Profit potential from backgrounding.
   f. Advantages of backgrounding.
   g. Disadvantages of backgrounding.

2. What is the difference between (a) preconditioning, (b) backgrounding, and (c) handling newly arrived feedlot cattle?

3. Detail the kind of feedlot records that should be kept.

4. How, and how frequently, should feedlot cattle be weighed?

5. Discuss the important methods of (a) commodity quality control, and (b) ration quality control.

6. Outline, step by step, a program for getting (a) yearling cattle and (b) calves on full feed.

7. Discuss full vs limited feeding of feedlot cattle.

8. Discuss the factors affecting feed intake of cattle.

9. How may weather affect the eating and drinking habits of cattle?

10. Outline the progressive changes that normally occur in feedlot cattle, from start to finish of the feeding period, including (a) rate of gain, (b) feed consumption, (c) feed efficiency, and (d) cost of gain.

11. How can you tell when cattle are ready to go to market?

12. Why is overfinishing undesirable?

13. Should feedlot heifers be aborted? If so, at what stage of pregnancy should it be done, and how should it be done?

14. Outline, step by step, a program for handling bullers or riders.

15. How may a feedlot mud problem be lessened?

16. How would you control (a) birds and (b) flies in a cattle feedlot?

17. Would you recommend that hogs follow feedlot cattle? Justify your answer.

18. How would you design and conduct a cattle feedlot test?

SELECTED REFERENCES

Title of Publication	Author(s)	Publisher
Beef Cattle Science Handbook	Ed. by M. E. Ensminger	Agriservices Foundation, Clovis, CA, pub. annually 1964–1981
Cattle Feeders Hand Book	R. M. Bonelli	Computer Publishing Co., Phoenix, AZ, 1968
Feeding Beef Cattle	J. K. Matsushima	Springer-Verlag, New York, NY, 1979
Feedlot, The, Third Edition	Ed. by G. B. Thompson C. C. O'Mary	Lea & Febiger, Philadelphia, PA, 1983
How to Make Money Feeding Cattle	L. H. Simerl B. Russell	United States Publishing Co., Indianapolis, IN, 1959
Stockman's Handbook, The, Seventh Edition	M. E. Ensminger	Interstate Publishers, Inc., Danville, IL, 1992
Third Dimension of Cattle Feeding, The	I. M. Hutchison	General Management Services, Inc., Phoenix, AZ, 1970

This shows a slotted floor in a beef confinement system placed beside the feed bunk to reduce manure build-up. A manure pit is located beneath the slats. (Courtesy, Portland Cement Assn.)

When properly handled, cattle feedlot manure can contribute to a sustainable agriculture. (Courtesy, Deere & Company, Moline, IL)

FEEDLOT POLLUTION CONTROL

Contents	*Page*

In recent years, there has been a worldwide awakening to the problem of pollution of the environment (air, water, and soil) and its effect on human health and on other forms of life. Much of this concern stems from the amount of manure produced by the sudden increase of animals in confinement. Certainly, there have been abuses of the environment (and they haven't been limited to agriculture). There is no argument that such neglect should be rectified in a sound, orderly manner, but it should be done with a minimum disruption of the economy and lowering of the standard of living.

In altogether too many cases, extreme environmentalists advocate policy changes and legislation that may in the end be detrimental to agriculture, to our food production potential, and to society in general. Frequently these new Messiahs have only used the data that support their theories about ecological doom. One of their favorite comparisons deals with the relative magnitude of the effect on the environment caused by animal manure, industrial waste, and municipal waste. Then, they add the "scare" to their story by citing the number of blue babies and suffocated fish caused by runoff from manure. In particular, they have incriminated cattle feedlots as major culprits. But many of their facts and figures have been in error. In order to set the records straight, and to assist cattle producers and others in controlling pollution to the maximum, this chapter is presented.

Invoking an old law (the Refuse Act of 1899, which gave the Corps of Engineers control over runoff or seepage into any stream which flows into navigable waters), the U.S. Environmental Protection Agency (EPA) launched the program to control water pollution by requiring that all cattle feedlots which had 1,000 head or more the previous year must apply for a permit by July 1, 1971. The states followed suit; although differing in their regulations, all of them increased legal pressures for clean water and air. Then followed the federal Water Pollution Control Act Amendments, enacted by Congress in 1971, charging the EPA with developing a broad national program to eliminate water pollution.

Increasingly, cattle feedlots will be subjected to more monitoring and to greater compliance with the regulations; and the closer the feedlots are to urban centers, the more demanding the requirements.

(*Note well:* Much of the concern pertaining to animal pollution can be materially lessened by giving proper attention to runoff control and pollution control, subjects which are covered in Chapter 28, Feedlot Facilities and Equipment, sections on "Location" and "Pollution Control.")

POLLUTION LAWS AND REGULATIONS

Both open lot and confinement cattle systems come under pollution regulations. Open lots present drainage and runoff problems. Confinement systems must be coordinated to the disposal area in order that pollution not be created when storage pits are emptied.

Registration of facilities and a permit to operate are the primary requirements that the federal government, and the various states, are, or will be, using to insure that livestock wastes are properly handled. Since state regulations vary somewhat, it is recommended that the cattle feeder check into the regulations of the state in which the feeder is operating or plans to operate. The federal guidelines, which are rather broad, are as follows:

1. **Who must apply.** The basic provisions of the federal regulations are (In some states 2 permits are needed—one state, the other federal.):

   a. Concentrated animal feeding operations with 1,000 or more animal units (1,000 cattle; 700 mature dairy cattle; 2,500 swine weighing over 55 lb; 500 horses; 10,000 sheep or lambs; 55,000 turkeys; 100,000 laying hens or broilers if the facility has continuous overflow watering, or 30,000 laying hens or broilers if the facility has a liquid manure system; or 5,000 ducks) must obtain a permit.

   b. Concentrated animal feeding operations with fewer than 1,000 animal units, but more than 300 animal units (more than 300 cattle; 200 mature dairy cattle; 750 swine weighing over 55 lb; 150 horses; 3,000 sheep or lambs; 16,500 turkeys; 30,000 laying hens or broilers if the facility has continuous overflow watering, or 9,000 laying hens or broilers if the facility has a liquid manure handling system; or 1,500 ducks), must obtain a permit if the facility discharges pollutants into waters of the U.S. either (1) through a conveyance constructed for the purpose, or (2) by waters that pass through the confined area.

Livestock confinement facilities include open feedlots, confined operations, stockyards, livestock auction barns, and buying stations. Also, regulations apply to any combination of species in the same feedlot.

The U.S. Environmental Protection Agency computes animal units as follows: It is assumed that 1 bovine slaughter or feeder animal represents 1 animal unit. Then, 1 bovine dairy cattle = 1.4 animal units; swine weighing over 55 lb = 0.4; sheep = 0.1; and horses = 2.0.

2. **How to apply.** Forms may be secured from the offices of the EPA and state environmental agencies, the county agent, or the Soil Conservation Service

(SCS) district offices. Fill out a *Short Form B* and send it, along with the $10 filing fee, to the EPA regional office. Then, either the federal EPA or the state agency will make an on-the-site inspection. It will draft a proposed permit, put it on public notice, and give the applicant and the public 30 days to comment on it. Then, if there are no protests, the federal discharge permit will be issued.

3. **Cost-sharing help.** In 1973–74, the Rural Environmental Assistance Program (REAP) provided cost-share payments on certain livestock waste storage and diversion facilities, with a limitation of 50% of the cost and a maximum of $2,500. The 1978 Revenue Act passed by Congress changed the treatment when applied to pollution control facilities; it increased to 100% the amount of investment that qualifies for the credit where the special 5-year amortization schedule is chosen. The cattle producer should check on the availability of cost-sharing funds and on tax treatment.

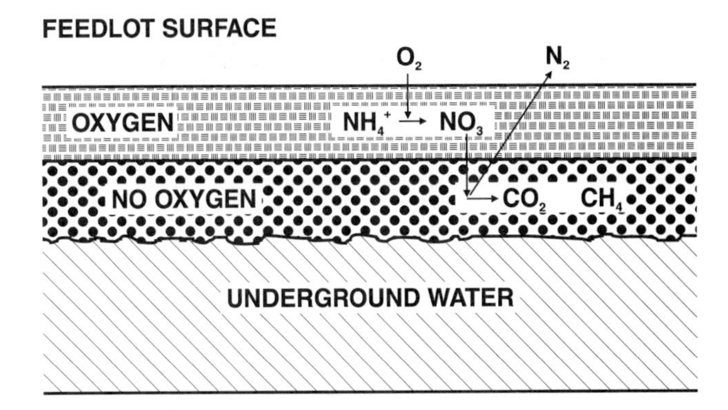

Fig. 32-1 Bacteria on feedlot surface utilize oxygen (O_2) and combine it with ammonium ions (NH_4^+) to produce nitrates (NO_3) However, nitrates do not go into groundwater because the bacteria in the lower layer of manure do not have free oxygen They must strip oxygen molecules from the nitrates. The result is free nitrogen (N_2). (Source: Paine, Myron D., "Confined Animals and Public Environment," *Great Plains Beef Cattle Feeding Handbook*, GPE-7000)

WHAT ARE THE CONCERNS?

Broadly speaking, there have been the following major concerns voiced by the environmentalists pertaining to animal pollution:

■ **Nitrates and blue babies**—Nitrates are a compound form of nitrogen, usually found in the soil. Plants utilize nitrates in the growth process to secure nitrogen. However, when the amount of nitrogen contained within nitrates exceeds 10 parts per million (ppm) in underground water, there is a chance that the excess nitrates may cause a disorder in human babies, commonly known as *blue babies*. The major concern, therefore, is that the subsurface water from a cattle feedlot, or other similar animal facility, might make for a nitrate buildup and cause blue babies. However, research evidence indicates that there is little nitrate buildup under active feedlots or runoff holding ponds. In fact, most studies indicate that there is actually a decrease in nitrates immediately beneath the surface of feedlots. There is a logical explanation of this phenomenon. The manure from the animals falls upon the feedlot surface. It is stirred by the animals' hoofs. As the manure continues to build up, the bottom layer becomes compacted while the surface layer remains loose. Because there is still energy in the manure, microorganisms continue their metabolic process. The microorganisms in the surface layer can use the oxygen that is mixed in the manure. Nitrates are one of the waste products of the metabolism of the organisms in the surface layer. (See Fig. 32-1.)

But the microorganisms in the compacted lower layer do not have oxygen available to complete their metabolism. They are desperate for oxygen molecules. Thus, they strip the oxygen from the nitrate compounds. Nitrogen gas, which constitutes 75.8% of the air, is created.

An exception to the above phenomenon does exist in feedlots that are lightly stocked, or that are abandoned for a portion of the year without cleaning. Under these circumstances, natural processes break up the compacted lower layer of manure and allow nitrates to move downward toward groundwater. However, an adequately stocked, active feedlot results in conditions that reduce nitrate movement to groundwater; hence, it does not materially affect nitrates in ground water, and it does not cause blue babies.

■ **Oxygen demand of runoff sludge suffocates fish**—In the 1960s, before feedlot control runoff measures were instituted, the runoff from the feedlots in the Great Plains area carried organic material that had a high oxygen demand. This runoff traveled as a sludge down the streams. The oxygen within the sludge was consumed. As the sludge passed along, fish were suffocated. The average pollutional loading on a stream was relatively minor, but the effect of a few hours without oxygen was spectacular. Cattle producers reacted swiftly. They constructed runoff control facilities in large commercial feedlots.

Fig. 32-2 shows fish kills in Kansas which were attributed to feedlots. Note that the largest number of fish killed occurred in the years 1964–67. Kansas regulations went into effect on January 1, 1968. The reduction of fish killed beginning in 1968 is partially credited to swift compliance with the regulations.

■ **Odors and dust**—Cattle feedlots located near centers of populations are having an increasing number of complaints lodged against them because of odors

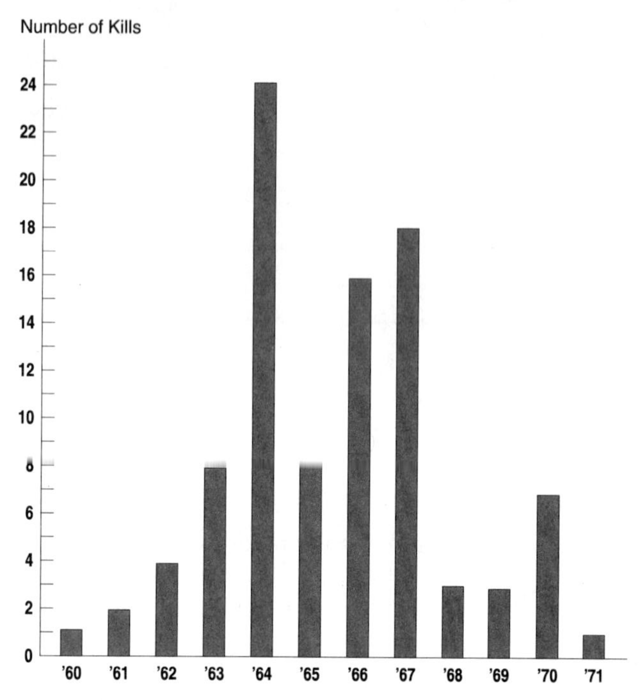

Fig. 32-2. Since 1968, fish kills caused by feedlots in Kansas have declined to previous low levels. (Source: Paine, Myron D., Confined Animals and Public Environment, *Great Plains Beef Cattle Feeding Handbook*, GPE-7000)

and dust. Lawsuits, based on the "nuisance law," are being filed against them.

MANURE

A general discussion of the subject of "Cattle Manure" appears in Chapter 10 of this book; hence, the reader is referred thereto. But feedlot manure as such will be covered in the sections that follow.

As shown in Table 10-5 of Chapter 10, approximately 1.3 billion tons of manure are produced annually in the United States. It should be noted, however, that this is all manure from all classes of animals, and without regard to confinement, moisture, or the ration. Much of this 1.3 billion tons of manure is distributed by the animals themselves onto pasture and grazing land, with the result that neither does it have to be hauled to the field nor is it a pollution problem.

MANURE PRODUCED VS MANURE HANDLED

Facts pertinent to calculating the amount of the 1.3 billion tons of manure produced that must be hauled to the field follow:

1. Only 25% of U.S. animals are in confinement.
2. Most of the animals kept in confinement are fed for meat production. These animals utilize a high-concentrate ration with a digestion between 75 and 90%. They eat a smaller amount of ration to secure their growth energy. The dry solid material produced by animals on a high-concentrate ration is about one-half (50%) of the dry solids coming from a dairy animal. Hence, a reduction factor for high-energy ration needs to be made.
3. The estimated 1.3 billion lb of manure produced annually is on a fresh manure basis, containing approximately 85% water. Following defecation, manure loses moisture by evaporation, Most manure that is hauled to the field has approximately 30 to 50% water. Thus, the manure that must be hauled may well be reduced by a factor from moisture evaporation of 50%.

Based on the above points, the calculations in Table 32-1 show the total U.S. manure production vs manure to be hauled:

TABLE 32-1
TOTAL U.S. MANURE PRODUCTION VS MANURE TO BE HAULED

Total manure produced	=	1.3 million tons
Reduction factor for confinement only		× 0.25
Manure produced in confinement	=	0.325 million tons
Reduction factor for high-energy ration		× 0.50
Manure from high-energy ration	=	162.5 million tons
Reduction factor for moisture evaporation		× 0.50
Total weight of manure hauled	=	81.25 million tons

The 81.25 million tons of manure that must be hauled include all classes of livestock, of which feedlot cattle are only a part.

Most feedlots estimate that one animal will produce about a ton of manure. Thus, the quantity is drastically reduced from the original quantity emitted by the animal (approximately 3.75 tons of fresh manure). The causes for this reduction are: (1) runoff, (2) evaporation, (3) leaching and percolation, and (4) decomposition. On the basis of l ton of manure produced per steer, at the time feedlots are cleaned, the 25.4 million feedlot cattle finished in 1993 produced that many tons of manure—25.4 million tons.

SOLID WASTE

As shown in the Table 32-1 calculations, approximately 81.25 million tons total weight of manure must be hauled to the fields, or disposed of otherwise. As shown in Fig. 32-3, this compares with 146 million tons of industry solid waste and 156 million tons of residential solid waste.

Thus, the manure that must be hauled to the field, or otherwise disposed of, is only 21% of the total solid waste produced by municipalities, industries, and animal systems.

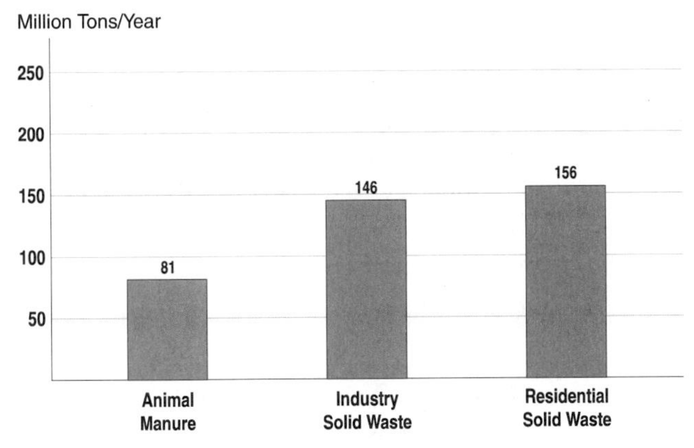

Fig. 32-3. The manure weight hauled from confined animals is only about one-fourth of the solid waste weight hauled from cities, industries, and animals.

MANURE DISPOSAL METHODS

The following methods are being used to dispose of manure:

1. Return to the soil, with or without composting.
2. Settling, flocculation, and dehydration, or other means of concentration with ultimate return to the soil.
3. Recycling, perhaps selectively, with or without processing as animal feed.
4. Incineration.
5. Generating energy.
6. Laissez-faire.

All of these methods have shortcomings. In many cases, the cost of waste disposal will exceed the value to the user. These costs may be reflected in price of product or in taxes.

It is noteworthy that the position of the Food and Drug Administration (FDA) relative to the feeding of animal waste (manure) is set forth by requiring submission of the following three basic categories of information: (1) establishing nutritive value (or efficacy); (2) determining safety to animals; and (3) determining that food from animals consuming such product is safe for humans. Research on the feeding value of manure is not discouraged by the FDA, but the

Fig. 32-4. Returning manure to the land. (Courtesy, Sperry-New Holland, New Holland, PA)

commercial marketing of manure is approved only within the framework given above.

HOW FEEDLOTS HANDLE THEIR MANURE

Most feedlot operators dispose of manure in two ways: (1) the solids are spread on land; and (2) detention ponds are used as a runoff control device.

■ **The solids are applied to the land**—Most feedlot operators keep the manure moved out as a matter of animal health. After they move out a lot of cattle they clean the pen. However, this soon results in a huge pile of manure, which must be disposed of. Most feedlot operators have solved this problem by developing close working relationships with local farmers, who take the manure for use as a fertilizer. Some of them are paying $4 to $5 per ton for it.

With the heavy concentration of cattle in one location, and with many lots ranging from 20,000 up to 100,000 head, the question is being asked: "How high rates of manure can be applied to the land without depressing crop yields, making for nitrate problems in feed, or contributing excess nitrate to groundwater or surface streams?" Based on earlier studies in the Midwest, before the rise of commercial fertilizers, it would appear that, on most soils and in most areas, one can apply up to 20 tons of manure per acre, year after year, with benefit. Heavier applications can be made, but probably should not be repeated every year. Excessive manure applications can (1) increase the potential for polluting surface of groundwater; (2) result in nitrogen concentrations in forage that pose a threat to animal health; and (3) cause salts such as sodium chloride to accumulate in concentrations that are toxic to plants and detrimental to soil structure. Of course,

Fig. 32-5. Farr Feeders, Inc., Greeley, Colorado, utilizes high-speed self-elevating scrapers, like the two pictured above, for pen and manure maintenance. This equipment is used to haul manure to a stockpile for later delivery to local farms. (Courtesy, Richard Farr)

salt accumulation can be controlled by lessening the salt content of the ration. Without doubt the maximum rate at which manure can be applied to the land will vary widely according to soil type, rainfall, and temperature.

State regulations differ in limiting the rate of manure application. Missouri limits it to 30 tons per acre on pasture, and 40 tons per acre on cropland. Indiana limits manure application according to the amount of nitrogen applied, with the maximum limit set at 225 lb per acre per year. Nebraska requires only ½ acre of land for liquid manure disposal per acre of feedlot, which appears to be the least acreage for manure disposal required by any state.

It is estimated that a 1,000-lb beef animal will excrete 120 lb of nitrogen per year.[1] Table 32-2 shows the average nitrogen losses expected on a year-round basis; whereas Table 32-3 shows the number of 1,000-lb beef animals necessary to provide 100 lb of nitrogen annually. [2]

Fig. 32-6. Loading solid feedlot manure by conveyor onto spreader trucks. (Courtesy, Benedict Feeding Co., Casa Grande, AZ)

TABLE 32-2
ESTIMATED NITROGEN LOSS DURING STORAGE, TREATMENT, AND HANDLING FOR VARIOUS WASTE MANAGEMENT SYSTEMS

System[1]	1	2	3	4	5	6
Total N loss, %	84	66	78	61	34	57

[1]System 1—Oxidation ditch, anaerobic lagoon, irrigation or liquid spreading.

System 2—Deep pit storage, liquid spreading.

System 3—Anaerobic lagoon, irrigation or liquid spreading.

System 4—Aerobic lagoon, irrigation or liquid spreading.

System 5—Bedded confinement, solid spreading.

System 6—Open lot (with or without shelter), solid spreading, runoff collected and irrigated or spread.

TABLE 32-3
NUMBER OF ANIMAL UNITS (AU)[1] PER ACRE OF DISPOSAL AREA TO PROVIDE 100 LB *(45 KG)* NITROGEN ANNUALLY

System[2]	1	2	3	4	5	6
Beef–AU	5.2	2.5	3.8	2.1	1.3	1.9

[1]Animal unit = 1,000-lb *(454-kg)* beef animal.

[2]Same as Table 32-2, footnote 1.

Thus, in Indiana, which allows for a maximum application of 225 lb of nitrogen per acre, the manure produced from 4.3 head of 1,000-lb steers ($1.9 \times 2.25 = 4.3$) could safely be applied to 1 acre of land. Since a feedlot steer averages about 750 lb throughout the feeding period only three-fourths of this amount of manure will be produced per animal.

Hence, 1 acre of Indiana land will take care of the production of 4.8 steers on full feed for a year. With a turnover of 2.25 steers per year, that means that the manure from 11 feedlot steers could be spread on 1 acre. An Indiana farmer feeding out 1,000 head of steers each year could spread all the manure on 90 acres of farmland.

As noted above, under the section entitled "Manure Produced vs Manure Handled," a total of 25.4 million tons of feedlot manure were produced in 1993, or 1 ton per animal. Assuming 2¼ turnover of cattle

[1]*Area Needed for Land Disposal of Beef and Swine Wastes*, Iowa State University, Pm-552.

[2]*Ibid.*

per feedlot per year, and the production of 1 ton of manure per animal, it would take 6 animals a whole year to produce enough manure to apply at the rate of 13.5 tons per acre. Thus, as shown in Fig. 32-7, it takes the production of about 18 acres of crops to produce enough feed for 6 feedlot cattle for a year. Yet, those 6 animals produce only sufficient manure for 1 acre, if manure is applied at the rate of 13.5 tons per acre.

Another noteworthy statistic is this: Since it takes 6 steers to produce enough manure for 1 acre of land, the nation's 25.4 million steers (1993 number) would produce enough manure for 4.23 million acres, an area slightly larger than the state of Connecticut.

■ **Detention ponds catch runoff**—In addition to disposing of the solids as indicated above, many feedlot operators are now using a detention pond as a runoff control device. Of the pollution control systems presently in operation, the detention pond, or some variation of this method, is probably the most common. As the name implies, this system detains the runoff until it can be disposed of. The detention pond is usually constructed from earth. In some cases, it will require soil sealer to prevent leaching. In a detention pond, the solids will settle out and usually very little decomposition or bacterial digestion will occur. Thus, solids will eventually accumulate and will have to be disposed of.

Disposal can generally be made in one of two ways: (1) the material can be pumped into a *honey wagon* and spread on the land; or (2) it can be pumped through an irrigation system. In either case, feeders who are using this method try to dispose of the suspended solids as well as the liquid through the disposal system. Most of the systems are agitated in some way before pumping to get as many solids into suspension as possible.

Sometimes a *settling basin* is used in combination with the detention pond. This is a small, shallow (2 to 4 ft deep) detention pond that will allow the solids to settle out before reaching the larger detention pond. The basin should be constructed of concrete or partly of concrete so that the solids can easily be cleaned out with a tractor and loader.

DUST CONTROL

In dry areas, feedlot dust can be a problem. But dust can be minimized by (1) control of cattle density—the more cattle per pen, the more "in pen" water (urine) application; (2) water application; and (3) manure removal. Increasing cattle density to 70 to 80 sq ft per head is effective in controlling dust; hence, this method should be used where shade area does not limit density. Also, water may be applied within lots by means of fence-line sprinklers or water trucks equipped to throw and spray into the pens. Dust from roadways, working alleys, and ground areas surrounding a feedlot may be controlled by water (water wagons and sprinklers), road oils, coarse gravel, or special chemicals designed for this purpose. Generally speaking, if the surface moisture of feedlots is maintained in excess of 20%, dust problems will be minimized. In addition to laying the dust, manure should be removed regularly, allowing less than 1 in. of a loose manure pad to remain.

ODOR CONTROL

Concentration of cattle into large feedlots tends to accentuate the odor problem. Also, odor may be-

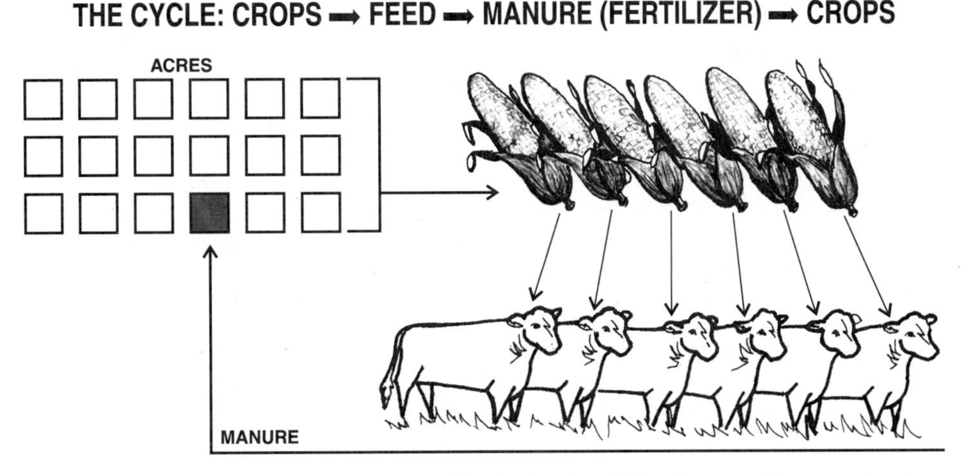

Fig. 32-7. Eighteen acres of Crops produce feed for 6 beef cattle on full feed for a year. The manure from these cattle will provide enough nutrients to refertilize only 1 acre, at the rate of 13.5 tons per acre.

come more intense on calm, humid, warm days just after a rain.

Odor is a subjective and emotional issue. Like most human traits, reaction to odors varies a great deal among individuals. Solving an odor problem can be complicated by the fact that it smells strong to some people and weak to others, and that it is not considered unpleasant by all people. Just the discussion of odor may encourage overreaction by an urban public.

Most cattle producers are concerned about odor control because they realize that odor does leave the confines of the feedlot, and, at present, there appears to be no economical method of positive control.

Lack of oxygen causes feedlot odor. Cattle manure contains energy for metabolism. Microorganisms in the manure accomplish this metabolism. The metabolic process converts complex carbohydrates, proteins, and fats to simpler compounds. When oxygen is present, the end products of metabolism are heat, CO_2, and H_2O. This process is called aerobic metabolism.

The oxygen transfer rate into manure must exceed the demand in order to prevent odor. When oxygen demand exceeds the transfer rate, anaerobic metabolism produces odor compounds.

Another odor produced in feedlots is ammonia (NH_3). Ammonia escapes as gas from urine. It is lighter than air and has an irritating smell.

The most common methods for controlling odor are sprinkling (water truck or sprinkler system), scraping, and chemical applications.

Sprinkling is perhaps the simplest way to control odor. Besides reducing dust, sprinkling provides moisture to enhance the biodegradation of manure in the presence of oxygen, thus limiting the nutrients available for anaerobic metabolism after compaction. In addition, ammonia is absorbed by the moist manure and less of it escapes into the air. A 40% moisture content of the surface manure is required for optimum odor control. If wet spots develop, a level of 25 to 30% may be more practical.

Scraping the surface of the pens exposes subsurface manure to air, with the result that oxygen is available to the microorganisms and desirable type degradation occurs at the exposed surface with little odor production. A regular scraping program will limit the amount of anaerobic metabolism that can occur as a result of compaction.

The most effective chemical compound to control odor is potassium permanganate ($KMNO_4$). A 1% solution applied by a water truck or sprinkler at the rate of 20 lb of $KMNO_4$ per acre will effectively control odors for several days.

The best management procedures for odor control are usually a combination of sprinkling, scraping, and chemical application. Pens should be scraped as often as practical and water should be applied at regular intervals. Chemical odor control measures should be used only when the odor becomes excessive.

TYPES OF ODOR CONTROL AGENTS

There are six main types of odor control agents:

1. **Digestive deodorants.** Digestive deodorants consist of a combination of digestive enzymes and aerobic/anaerobic bacteria. The logic appears to be that the enzymes or bacteria will create a digestive process that eliminates the odor.
2. **Masking agents.** Mixtures of aromatic oils which cover the odor of manure with a stronger, more *pleasant* odor.
3. **Deodorants.** Mixtures of chemicals designed to *kill* the odor of manure. The deodorant action may also kill the bacteria which produce the odor.
4. **Absorbents.** Products which have a large surface area, such as charcoal and clay, that absorb odorous chemicals before their release to the environment.
5. **Feed additives.** Certain products which, when added to the feed, reduce manure odors. Researchers at Colorado State University have discovered that feedlot odors are reduced or eliminated when small quantities of sagebrush are included in the ration of cattle. The volatile oils in sagebrush reduce the digestive tract's bacterial population that produces the disagreeable odors.
6. **Miscellaneous.** Products that do not belong to any of the other categories include a large number of chemicals such as chemical oxidants or germicides.

MEASURING ODORS

A portable machine that measures odors—a smell meter—was invented by Texas A&M engineers. The device works by comparing the odor intensity of the open air with that of a known concentration of butanol alcohol. The device may take some of the emotion and anger out of odor problems. A pollution agency could use this odor meter (1) to collect evidence, when there is a complaint, and (2) to develop standards of tolerance.

QUESTIONS FOR STUDY AND DISCUSSION

1. How far should a nation go in pollution control laws and regulations, bearing in mind (a) that pollution control measures cost money—hence, they will increase product prices to consumers; (b) that the ultimate in pollution control will lower food production potential in some cases; and (c) that many of the good things of life which contribute to our high standard of living, such as electricity, make for pollution?

2. How do the animal unit relationships between species used by the Environmental Protection Agency (EPA) compare with those given in the Appendix of this book?

3. Who must register and secure a permit to operate in compliance with federal and state livestock pollution laws and regulations? Give the step-by-step procedure involved in securing approval.

4. Debate either the affirmative or the negative of the following statements:

a. The runoff of cattle feedlots, and other similar facilities, causes blue babies.

b. The runoff of cattle feedlots, and other similar facilities, suffocates fish.

5. In arriving at the amount of manure that must be handled, the author assumed the following:

a. That only 25% of U. S. animals are in confinement.

b. That the dry solids of animals on high-energy rations kept in confinement are equivalent to only 50% of the dry solids coming from a dairy animal.

c. That 50% of the moisture of manure is lost before hauling.

Challenge each of the above assumptions.

6. Fig. 32-3 clearly shows that both residential solid waste and industrial solid waste exceed the total tonnage of manure that must be hauled to the field or otherwise disposed of. Then why are the environmentalists pointing such an accusing finger at livestock producers?

7. Discuss each of the six manure disposal methods listed in this chapter. Which two methods do you feel offer the most hope? Justify your choices.

8. What tonnage of manure per acre can be applied to the land on an annual basis?

9. The authors' computations show that it takes the production of about 18 acres of crops to produce enough feed for 6 feedlot cattle for a year. Yet, these 6 animals produce only sufficient manure for l acre, *if* manure is applied at the rate of 13.5 tons per acre. Challenge these computations.

10. What's a detention pond? How does it work?

11. How would you recommend that feedlot dust be controlled?

12. How would you recommend that feedlot odors be controlled?

SELECTED REFERENCES

Title of Publication	Author(s)	Publisher
Agriculture and the Environment	Alberta Institute of Agrologists	Alberta Institute of Agrologists, Edmonton, Alberta, Canada, 1971
Agriculture and the Quality of Our Environment	Ed. by N. C. Brady	American Association for the Advancement of Science, Washington, DC, 1967
Closing Circle, The, Nature, Man, and Technology	B. Commoner	Alfred A. Knopf, Inc., New York, NY, 1971
Distribution of Nitrates and Other Water Pollutants, Under Fields and Corrals in the Middle South Platte Valley of Colorado, ARS 41-134	B. A. Stewart, *et al.*	U.S. Department of Agriculture, Washington, DC, 1967
Environment, The, A National Mission for the Seventies	Editors of *Fortune*	Harper & Row, New York, NY, 1969

Environmental Science Challenge for the Seventies	National Science Board	National Science Board, National Science Foundation, Washington, DC, 1971
False Prophets of Pollution	R. M. Carleton	Trend Publication, Inc., Tampa, FL, 1973
Feedlot, The, Third Edition	G. B. Thompson C. C. O'Mary	Lea & Febiger, Philadelphia, PA, 1983
Guide to Environmental Research on Animals, A	National Research Council	National Academy of Sciences, Washington, DC, 1971
Health Hazards of the Human Environment	World Health Organization	World Health Organization, Geneva, Switzerland, 1972
NFC Directory of Environmental Information Sources, Second Edition	Ed. by C. E. Thibeau	The National Foundation for Environmental Control, Inc., Boston, MA, 1972
Our Living Land	Ed. by E. P. Essertier, *et al.*	U. S. Department of the Interior, Washington, DC, 1971
Population, Resources, Environment, Issues in Human Ecology, Second Edition	P. R. Ehrlich A. H. Ehrlich	W. H. Freeman and Company, San Francisco, CA, 1972
Principles of Animal Environment	M. L. Esmay	The Avi Publishing Company, Inc., Westport, CT, 1978
Proceedings of the International Livestock Environment Symposium	American Society of Agricultural Engineers	American Society of Agricultural Engineers, St. Joseph, MI, 1974
Water Encyclopedia, The	Ed. by D. K. Todd	Water Information Center, Inc., Port Washington, NY, 1970
Water Policies for the Future, The	National Water Commission	Water Information Center, Inc., Port Washington, NY, 1973

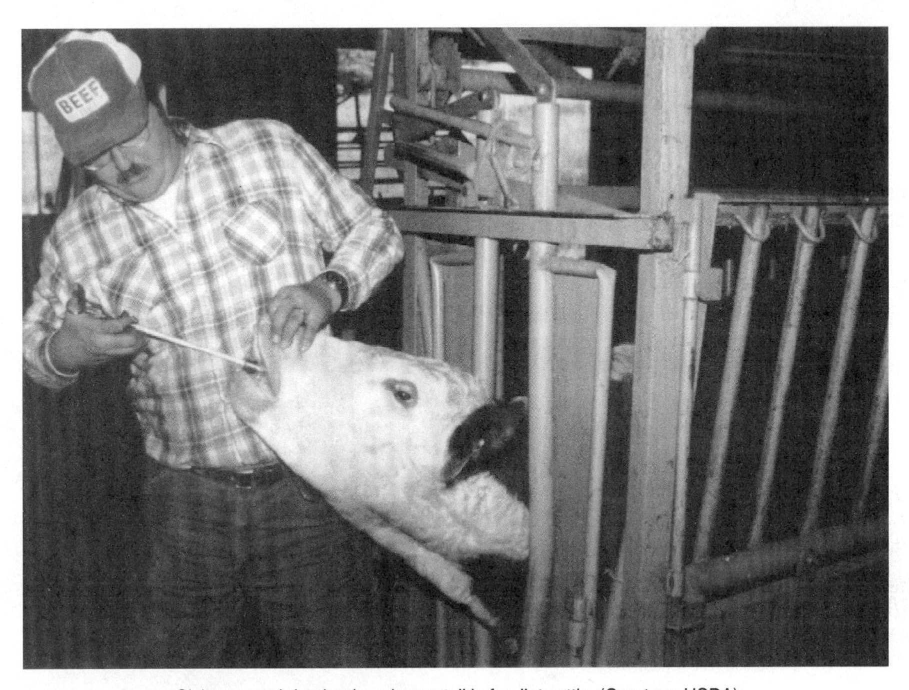

Sickness and death take a heavy toll in feedlot cattle. (Courtesy, USDA)

CATTLE FEEDLOT HEALTH

Contents *Page*

Fig. 33-1. Newly arrived, heavily stressed calves, trucked from south Texas to an Arizona feedlot. (Courtesy, Benedict Feeding Co., Casa Grande, AZ)

Loss from disease is greater in cattle feedlot operations than in any other type of cattle enterprise. The movement of cattle, stress conditions, methods of purchase, feeding of concentrated feeds, population density, sometimes unsanitary conditions, and the bigness and complexity of the operation all contribute to disease incidence; and disease incidence is directly proportional to population density.

On the average, 5% of the yearlings and 25% of the calves sicken, and 1% of the yearlings and 5% of the calves die. An estimated 762,000 cattle died in feedlots in 1993, at an estimated loss of $373,380,000.[1] But that's not all! The average sick animal shrinks 10 to 20%; and one time through the chute is equivalent to a 7-day feeding period. Considerable labor is required in treating sick animals, and medicines are expensive. Losses from sickness run 2 to 5 times as great as the actual death loss. It is estimated that the combined losses from death and sickness add $40 per head onto the cost of every finished animal. This situation can be greatly improved by (1) preconditioning, (2) moving cattle directly from the producer's farm or ranch to the feedlot (fewer than 20% of the cattle do so now), (3) reducing the time between ranch and feedlot, (4) lessening the amount of stress and exposure to infection during marketing and transportation periods, (5) providing the person receiving the cattle with more adequate medical and nutritional history of the cattle, (6) handling of incoming feedlot cattle properly, and (7) diagnosing and treating sick cattle early.

HANDLING NEWLY ARRIVED CATTLE

The most critical period for feeder cattle is the first 21 to 28 days in the feedlot. The following recommendations pertaining to incoming cattle will minimize death losses and maximize performance:

- **Provide clean, dry, comfortable quarters**—Whether it be an open lot or a building, incoming cattle should be provided with clean, dry, comfortable quarters. A dry and comfortable bed for resting is very essential because cattle are tired and have a low resistance to respiratory diseases.

- **Process upon arrival**—Processing usually consists of vaccinations, ear tagging, dipping or other pesticide treatment, putting growth promotant implants in the ear, and other treatments such as horn tipping and castration. The relative merits of processing calves (1) at point of origin, (2) upon arrival at destination, or (3) 2 to 3 weeks after arrival are often debated.

Fig. 33-2. Processing an incoming animal through a chute—in a squeeze. (Courtesy, Powder River, Provo, UT)

In a well-designed experiment, involving 358 calves, the University of California provided the answer to this question. The calves originated in Texas, and were in transit 32 to 38 hours with no rest stops. In all loads, one-third of the calves were processed at origin, one-third upon arrival at destination, and one-third were delayed 2 to 3 weeks after arrival. Processing consisted of branding, castration, ear tagging, use of a pour-on grubicide, vaccination (IBR, PI3, blackleg, and malignant edema), intramuscular injection of vitamins A, D, and E and 1 g of oxytetracycline.

Recognizing that the effect of time and place of processing on the entire feeding period from purchase to slaughter is the important thing, rather than the effect on the first month, the California workers very wisely carried part of these cattle from arrival through slaughter. A total of 120 of these calves. with an average purchase weight of 210 lb, were shipped from

[1]Authors' estimates: Of the 25.4 million fed, an estimated 3% died, at an average weight of 700 lb, valued at 70¢/lb.

Houston, Texas, and were in transit 38 hours to El Centro, California. The results of the entire 344-day feeding period, from purchase to slaughter, are summarized in Table 33-1.

The California study showed the following relative to time of processing:

1. For the entire feeding period—from purchase to slaughter—calves processed upon arrival gained at a slightly faster rate and had a higher feed efficiency than calves which were processed immediately prior to shipment, or calves which were delay-processed 13 days after arrival.

TABLE 33-1
PERFORMANCE FOR ENTIRE PERIOD[1, 2]

Items Compared	Time of Processing		
	Origin	Arrival	Delayed
Number of calves	40	39	37
Purchase weight per head (lb)	209	213	209
Net slaughter weight (lb)	970	1,019	989
Average days fed	344	341	344
Daily weight gain (lb)	2.21	2.36	2.27
Total feed consumed per head, lb:			
72% receiving ration	166	182	166
55% growing ration	1,012	1,036	1,053
90% finishing ration	3,336	3,314	3,404
Total	4,514	4,532	4,623
Average daily feed intake (lb)	13.12	13.29	13.44
Feed per pound of gain (lb)	5.93	5.63	5.93
Carcass data:			
Yield (%)	62.5	62.6	63.1
Quality grade score[3]	11.9	12.4	12.2
Cutability grade[4]	2.6	3.0	2.5
Receiving phase:			
Number of days	27	27	27
Gain from purchase weight (lb)	37	47	36
Cost per pound of gain[5] (¢)	27.49	22.89	28.72
Postreceiving period:			
Number of days	317	314	317
Postreceiving gain (lb)	724	759	744
Cost per pound of gain (¢)	31.51	30.03	31.41
Entire period, purchase to slaughter:			
Number of days	344	341	344
Total gain (lb)	761	806	780
Cost per pound of gain (¢)	31.32	29.62	31.28
Cost per pound of gain due to receiving period (¢)	1.34	1.33	1.33
Cost per pound of gain due to postreceiving period (¢)	29.98	28.29	29.95

[1]Lofgreen, G. P., University of California, El Centro, California, paper in *Beef Cattle Science Handbook*, Vol. 12, edited by M. E. Ensminger and published by Agriservices Foundation, Clovis, CA.

[2]To convert lb to kg, divide by 2.2.

[3]Quality grade score: Choice = 13; Low Choice = 12; High Select = 11.

[4]Cutability scored from 1 to 5, with 1 being the highest yield.

[5]Receiving costs include feed, processing, and medication. Feed costs used were $101, $85, and $111 per ton for the receiving, growing, and finishing rations, respectively.

2. At the end of the 27-day receiving period (the first 27 days after arrival), the calves processed on arrival had a weight advantage of 10 lb over those processed at origin and 11 lb over those delay-processed. At slaughter, these advantages had been increased to 45 and 26 lb, respectively. It appears, therefore, that if calves attain a weight gain advantage during the first month after arrival in the yard, they will retain that advantage throughout the entire feeding period.

3. Based on (a) rate of gain, (b) disease resistance, and (c) cost per pound of gain for feed, processing, and medication, these studies indicate that processing at arrival is best, and that processing at point of origin is preferable to delayed processing.

Note well: If the cattle are very tired and weak when they get off the truck, it may be advisable to delay processing for 24 hours, but processing should not be delayed more than 72 hours.

Fig. 33-3. Incoming cattle may be started on about 4 lb of concentrate/head/day, consisting of 2 lb of grain and 2 lb of protein supplement. (Courtesy, Ralston Purina Company, St. Louis, MO)

■ **Provide clean, fresh water**—Give the cattle easy access to clean, fresh water because they are usually dehydrated and thirsty upon arrival and will drink water before they eat feed. Open water tanks are preferable to automatic water bowls because most farm and ranch cattle are accustomed to drinking from tanks or ponds.

■ **Provide a palatable ration**—Feeding a palatable ration—one that cattle will start eating soon after they are unloaded in the feedlot—will reduce the incidence of shipping fever and make the cattle recover their weight loss more rapidly.

1. **Roughage.** The best roughage for newly arrived feedlot cattle is *long grass* hay, because it is very similar in composition and taste to the grass to which most feedlot cattle have been accustomed. Thus, cattle will usually eat long grass hay more quickly than any other roughage. In areas where grass hays are not available, or are too expensive to feed, any other nonlegume roughage can be fed, such as corn silage, sorghum silage, cottonseed hulls, corncobs, or grass-legume hay that contains more grass than legumes. Above all, do not feed high-quality alfalfa hay because it is too laxative and it will cause scouring which will trigger shipping fever. The same may be said relative to alfalfa haylage or alfalfa silage. Corn silage of approximately 65% moisture content is an excellent feed for new cattle. If cattle do not eat the corn silage too well at the outset, the feeder should sprinkle a little grass hay on the top of it to encourage them to start eating.

2. **Concentrate.** Incoming cattle may be fed approximately 4 lb of concentrate per head daily, with a breakdown between protein supplement and grains as follows:

a. Two pounds of a natural protein supplement, such as soybean oil meal, cottonseed meal, or a good commercial supplement, preferably with a little cane molasses added from the standpoint of palatability. The protein supplement should be fortified so as to provide 50,000 IU of vitamin A daily. For heavily stressed cattle, the protein supplement should also contain a high level of antibiotic, or a combination of antibiotic and a bactericidal agent such as sulfamethazine. The following level of antibiotic-sulfamethazine is recommended:

> Feed 350 mg of Aureomycin plus 350 mg of sulfamethazine per head daily to newly arrived cattle for a period of 28 days. With the antibiotic-sulfamethazine treatment, shipping fever is practically alleviated.

Do not feed urea for the first 28 days after the cattle arrive. Starvation destroys the ability of the rumen to utilize urea or other nonprotein nitrogen and makes cattle more sensitive to urea toxicity. Therefore, it is not wise to put extra stress on cattle by using urea during this adjustment period.

b. Two pounds of cereal grain per head daily, with the grain processed in the usual manner. The grain level can be raised at the rate of 1 lb per head daily if it seems desirable.

It has been, and still is, common practice to start cattle on a high-roughage ration, then gradually work them over to a high-concentrate ration as they progress through the feeding program. However, based on California and New Mexico studies, it appears that for calves, a starting ration consisting of 75% concentrate and 25% hay (roughage) is best.

■ **Satisfy mineral hunger**—Incoming cattle are usually hungry for minerals, especially if they have been on dry range forage. Thus, they should have access either to a mineral mixture consisting of 2 parts of dicalcium phosphate and 1 part of salt, or to a good commercial mineral.

■ **Observe, isolate, and treat sick animals**—Newly arrived cattle should be observed at least twice daily. Sick animals should be removed and treated. Treating sick animals promptly, rather than waiting until tomorrow, may mean the difference between life and death. Animals that show clinical signs of shipping fever—sunken eyes, runny nose, drooling at the mouth, labored breathing, and/or weaving (unsteady gait)—should be isolated in a separate *sick pen* or *hospital.*

Rest, fresh water, good feed, proper medication, and TLC (tender, loving care) are the cardinal essentials for preventing shipping fever and death losses.

CATTLE FEEDLOT DISEASE AND PARASITE CONTROL PROGRAM

A written-down cattle feedlot health program should be developed in cooperation with the feedlot veterinarian. The following outline will serve as a useful guide for this purpose.

1. Process (dehorn, castrate, and vaccinate) incoming cattle soon after they are unloaded from the truck rather than wait. While they are in the squeeze chute, first take their body temperature. Designate the animal as sick if the temperature is 104°F or greater. At that time, also implant with any approved growth stimulant intended to use.

   Newly arrived cattle are usually vaccinated against IBR, PI3, lepto, and blackleg; and wormed with a broad spectrum wormer. Also, they should be given a big injection of vitamin A if they are stressed (250,000 to 1,000,000 IU; depending on size of cattle and the degree of stressing).
2. For the first 28 days after arrival, fortify the natural protein supplement (do not feed incoming cattle urea) with the following per head per day: 350 mg of Aureomycin plus 350 mg of sulfamethazine (Aureo S-700, which is a combination of an antibiotic and the bactericidal agent sulfamethazine) *plus* 50,000 IU of vitamin A.
3. Administer plenty of TLC (tender, loving care) to the sick.
4. Treat sick animals three times in the first 24 hours. It will reduce repeats and chronic illness.
5. Keep hospital chutes and pens clean and well bedded.
6. Disinfect chutes, tanks and syringes, and balling guns to prevent spread of disease.
7. Seek professional advice if there is no response to medication within 48 hours.
8. Autopsy all dead animals for an accurate diagnosis. If one specific problem is causing a lot of deaths, seek prevention.
9. Don't change rations too fast.
10. Have all members of the health team—detection, treatment, and convalescence—performing at top level.
11. Control parasites:
    a. **Internal parasites.** Worm calves, if necessary, within two weeks after arrival in the feedlot, with one of the approved drugs, used according to manufacturer's directions.

       Treat only (1) if animals appear to be heavily parasitized or are from areas where previous experience has shown that they are heavily parasitized, or (2) if 300 or more eggs per gram (epg) of dry feces are found.

    b. **External parasites.** Treat for external parasites (commonly lice, grubs, flies, and ticks), if necessary, using a recommended insecticide and following the manufacturer's instructions on the label or container.

    c. **Control of flies around the feedlots and feedmill.** Prevent flies by starting the following program early in the season: sanitation, prevention of breeding areas, and use of residual sprays and mists before the fly population builds up. Before using any chemical in a control program directed toward either the adult fly or the larvae, read the label carefully to determine the usage and restrictions of the material.

CARE OF SICK CATTLE

Proper care of sick cattle necessitates two things: (1) suitable hospital facilities; and (2) prompt and correct diagnosis and treatment. Hospital facilities are covered in Chapter 28 of this book; hence, the discussion at this point will be limited to diagnosis and treatment.

Diagnosis and treatment are very important in the health program of any cattle feedlot. The pen checkers and hospital technician cannot completely operate this phase of the program alone. The yard foreman must supervise it, although he/she has too many responsibilities to be completely responsible for the program. The consulting veterinarian must establish general policies and treat unusual and difficult cases, although the veterinarian cannot examine every animal as it enters the hospital, simply because the health program costs would be too high. Thus, diagnosis and treatment is really a team approach. Every person involved should contribute to the program according to their responsibilities and abilities, thereby assuring the most effective and economical health program.

Diagnosis is the art or act of recognizing disease from its symptoms. Thus, correct diagnosis assumes that those responsible for the cattle feedlot health program recognize what is *normal*. Then, any deviation from the normal should be reason for concern and further study. Practical experience in both the normal and abnormal is essential in arriving at a correct diagnosis.

Treatment should be left in the hands of the veterinarian, who should give instructions for the proper use of drugs and biologicals—the administration, dosage levels, indications, and contraindications of a given product, and signs and lesions of specific diseases. It is the veterinarian's responsibility to keep management current on the latest findings regarding research, new products, and disease conditions.

NUTRITIONAL DISEASES

Several metabolic disorders, or diseases, in feedlot cattle are attributable wholly or in part to the feeding regimen. Among the more prevalent ones are: acidosis, bloat, liver abscesses, and urinary calculi (water belly; urolithiasis).

ACIDOSIS, OR LACTICACIDOSIS

Acidosis usually develops early during the fattening process when the ration is changed too rapidly from roughage to concentrate.

■ **Cause**—Aci dosis is caused by an increase in lactic acid–producing bacteria (both the D- and L-forms) and the rapid production of lactic acid. It commonly occurs when there is a sudden shift from a high-roughage to a high-concentrate ration. However, cattle maintained on high-energy rations are constantly in a marginal state of acidosis due to the formation of lactic acid in the rumen flora. Thus, ingredient changes, poor mixing of grain in the ration, or faulty feeding can promote acute acidosis.

■ **Symptoms and signs**—Marginal acidosis is characterized by poor performance and inconsistent feed ingestion. If ingredient changes or erratic feeding persist, acute acidosis may result, creating laminitis—and eventually *ski shoe* cattle. In severe cases, the rumen becomes immobilized, followed by increased pulse and respiration rate, variable rectal temperature, sunken eyes, loss of dermal elasticity, staggering, coma, and death.

■ **Prevention and treatment**—Prevention consists of starting cattle on a high-roughage ration and reducing the roughage in the ration over a 3- to 4-week period; and avoiding erratic feeding and abrupt ration changes. Different treatments have been used with varying degrees of success; among them: (1) removal of rumen contents and replacement by contents of an animal on a normal ration; (2) feeding a high level of penicillin (12 to 20 million units) to suppress lactic acid–producing bacteria; (3) drenching (or intravenous injection) with a solution of sodium bicarbonate to restore the acid-base balance; (4) daily intramuscular administration of antihistamines and cortical steroids for each of several days to help prevent intoxication and laminitis; or (5) backing the cattle down on both amount and kind of feed (lessening the ration, and returning to the mix that was being used before trouble was encountered).

BLOAT

Bloat in feedlot cattle can be costly, both in treatment and losses. A Kansas State University survey showed that 0.6% of the feedlot cattle bloated mildly to moderately, 0.2% bloated severely, and 0.1% died of bloat.

■ **Cause**—Bloat is usually considered to be a nutritional disease, because certain feeds ferment more readily and cause more bloat than others. The high-concentrate rations of feedlot cattle increase the number of slime-producing bacteria in the rumen. Slime traps fermentation gas and produces bloat. Both frothy and free gas bloat occur in feedlot cattle.

Feedlot bloat may occur during any month of the year. However, it is more common during hot, humid weather.

■ **Symptoms and signs**—Bloat is characterized by a greatly distended paunch, noticeable on the left side in front of the hip bone.

■ **Prevention and treatment**—When trouble strikes, alleviate all likely causes as quickly as possible. A high incidence of bloat frequently occurs in feedlot rations containing high levels of both barley and alfalfa; hence, replace some of the barley with dried beet pulp, oats, or ground corncobs, and replace part of the alfalfa with grass hay or straw. Also, less bloat is experienced on long or coarsely chopped hay than on finely ground hay; and less on well-flaked grain than on finely ground grain. In addition to rectifying these situations, allow access to a little coarse, long hay or straw.

When bloat is encountered, increase the proportion of roughage in the ration if feasible. Also, use poloxalene (Bloat Guard) or oxytetracycline (Terramycin/Neo-Terramycin) according to the manufacturer's directions.

Time permitting, bloated animals should be drenched with 1 to 2 oz of poloxalene. Then, 10 minutes after treatment, the gas should be relieved with a stomach tube.

Puncturing of the paunch should be a last resort.

LIVER ABSCESSES

Abscesses, as indicated by the name, are single or multiple abscesses on the liver, observed at slaughter. Usually the abscess consists of a central mass of necrotic liver surrounded by pus and a wall of connective tissue. At slaughter, most livers affected with abscesses are condemned for human food.

In some lots of cattle on high-concentrate rations, as high as 75% of the livers have been condemned. On the average, however, it will probably run between 5 and 10%. Since the liver of a 1,000-lb steer weighs approximately 11 lb, its condemnation represents a considerable monetary loss. The loss from reduced feed efficiency and gains may be even greater.

■ **Cause**—The direct cause of most bovine liver abscesses is *Spherophorus necrophorus*, the same bacteria which cause footrot. This organism, which is ever present in ruminal contents, penetrates the covering epithelium through points of injury, discontinuity, and necrosis. Factors back of rumenitis include: (1) rapid rate of change from a diet of roughage to one high in concentrate, (2) fattening with a diet containing more than 25% concentrate, (3) foreign body penetration of the wall, and (4) miscellaneous agents.

■ **Symptoms and signs**—Liver abscesses generally go undetected until cattle are slaughtered. However, reduced feed intake and gains near the end of the feeding period may be indicative.

■ **Prevention and treatment**—Liver abscesses can be greatly reduced, but not entirely eliminated, by continuous feeding during the fattening process of chlortetracycline at 70 mg/head/day. Also fewer ruminal lesions and liver abscesses develop when the ratio of concentrate to roughage is decreased, and when the transition period from a roughage to a finishing ration is increased.

URINARY CALCULI (Water Belly; Urolithiasis)

The term *urinary calculi* refers to mineral deposits in the urinary tract, which may block the flow of urine in the urethra, particularly in castrated male cattle. Prolonged blockage generally results in rupture of the urinary bladder or urethra, releasing urine into the surrounding tissue. This produces the condition commonly referred to as *water belly.*

The mineral deposits can be of variable sizes, shapes, and composition. In cattle, the phosphatic type predominates under feedlot conditions, and the siliceous type occurs most frequently in range cattle.

■ **Cause**—Experiments and experiences have shown

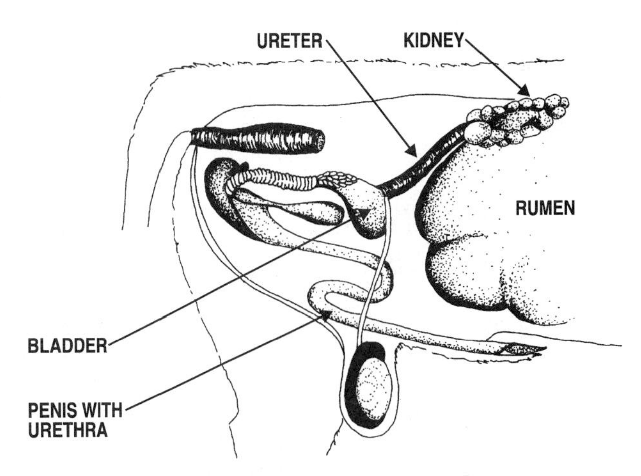

Fig. 33-4. Urinary tract of a bovine male. When the urethral tube becomes blocked, urine production continues causing extension and finally rupture of the bladder.

a higher incidence of urinary calculi when there is (1) a high potassium or phosphorus intake, (2) an incorrect Ca:P ratio, or (3) a high proportion of beet pulp or grain sorghum in the ration.

■ **Symptoms and signs**—At first, animals afflicted with urinary calculi appear restless and strain frequently in an unsuccessful attempt to urinate. They may repeatedly stamp their feet and kick their abdomen. If urinary blockage is not complete, urine may dribble slowly from the sheath. If the stone(s) fails to pass, there will likely be complete blockage of urine flow, followed by rupture of bladder and urethra and release of urine into the body cavity or surrounding tissues. Death follows.

■ **Prevention and treatment**—Most materials and practices offering some degree of protection against phosphatic urinary calculi appear to include at least one of the following:

1. **A lowering of the urinary phosphorus level.** A calcium-to-phosphorus ratio of approximately 2:1 is recommended.
2. **Acidification of the urine.** This may be achieved by the feeding of acid-forming salts. Ammonium chloride fed daily at the rate of 1¼ to 1½ oz to fattening cattle is effective.
3. **An increase in urine volume.** An increase in urine volume is dependent upon an increase in water consumption. This can be achieved by including salt in the diet at a level of 3 to 4%.

The only treatment that has been of demonstrable value is the use of urinary tract relaxants that aid in keeping the urethra open and allow passage of mineral deposits. Surgery represents the most effective treatment, with the stone(s) removed at the point of blockage. In steers, the urethra may be bisected and brought to the outside of the body to bypass the constricted portion of the tract. After a short period of time to eliminate any tissue residue of urine, such animals are marketable.

INFECTIOUS DISEASES AND PARASITES

In the above discussion, the most common noninfectious diseases of feedlot cattle have been covered. Additionally, feedlot cattle are subject to (1) infectious diseases—those caused by specific entities such as bacteria or viruses, most of which are contagious; and (2) parasitic diseases—those caused by parasites.

Respiratory diseases continue to be the major cause of disease loss in feedlot cattle. Viruses known or suspected to contribute to bovine respiratory diseases include: adenoviruses, bovine herpesvirus strain DN 599, bovine virus diarrhea (BVD), enteroviruses, infectious bovine rhinotracheitis (IBR) virus, parainfluenza 3 (PI3) virus, parvovirus, reoviruses, respiratory syncytial (RS) virus, and rhinoviruses.

The most important internal parasites affecting feedlot cattle are coccidia, flukes, and worms. The most important external parasites are flies, grubs, lice (biting and sucking), mites, and ticks. The infectious and parasitic diseases affecting feedlot cattle are the same as those that affect breeding cattle. Hence, they have been covered in Chapter 12 of this book.

QUESTIONS FOR STUDY AND DISCUSSION

1. It is estimated that the combined losses from death and sickness add $40 per head onto the cost of every feedlot finished animal. Outline how this cost could be lowered.

2. Detail a program for handling newly arrived cattle that will minimize death losses and maximize performance.

3. Traditionally, cattle are started on a high-roughage ration, then gradually worked over to a high-concentrate ration as they progress through the feeding program. However, experiments conducted by the California Station showed that for calves a starting ration consisting of 72% concentrate and 28% hay (roughage) is best. Which method do you favor—and why?

4. Outline a cattle feedlot disease and parasite control program.

5. Outline a fly control program for use around a cattle feedlot and feed mill.

6. Discuss the place of, and the relationship between: (a) the feedlot personnel, and (b) the veterinarian, in the diagnosis and treatment of feedlot diseases.

7. Discuss the (a) cause, (b) symptoms, and (c) prevention and treatment of each of the following disorders, or diseases, in feedlot cattle:

a. Acidosis.
b. Bloat.
c. Liver abscesses.
d. Urinary calculi.

SELECTED REFERENCES

Title of Publication	Author(s)	Publisher
Diseases of Cattle, Second Edition	Ed. by W. J. Gibbons	American Veterinary Publications, Inc., Santa Barbara, CA, 1963
Diseases of Feedlot Cattle, Third Edition	R. Jensen D. R. Mackey	Lea & Febiger, Philadelphia, PA, 1979
Feedlot, The, Third Edition	Ed. by G. B. Thompson C. C. O'Mary	Lea & Febiger, Philadelphia, PA, 1983
Merck Veterinary Manual, The, Seventh Edition	Ed. by C. M. Fraser	Merck & Co., Inc., Rahway, NJ, 1991

Stocker steer calves grazing irrigated alfalfa in the Texas Panhandle. Note sprinkler irrigation equipment in the distance. (Courtesy, *The Progressive Farmer*, Birmingham, AL)

PASTURE FINISHING (FATTENING)

Contents *Page*

Fig. 34-1. Most young cattle that are pastured are finished in a feedlot at the end of the grazing period. (Courtesy, Ron Baker, C & B Livestock, Inc., Hermiston, OR)

Pasture finishing of cattle has changed. Fewer and fewer steers are finished on grass alone, primarily because consumers no longer want heavy, grass-finished steers weighing 1,250 lb or more. Young cattle grow but do not reach market finish under usual pasture conditions. Thus, it is impossible to finish weaned calves and yearling cattle without supplemental feeding on pasture and/or lot finishing at the end of the grazing season.

Without doubt, world food shortages in the years ahead will result in a higher proportion of humanly edible grains and seeds being consumed directly by people, and increased pasture finishing of cattle.

Generally speaking, no cheaper method of harvesting forage crops has been devised than is afforded by harvesting directly by grazing animals. Moreover, even most seeded pastures last several years; thus, seeding costs may be distributed over the entire period. Naturally, the cash income to be derived from pastures will vary from year to year and from place to place, depending upon such factors as market price levels, class of animals, soil, season, and the use of adapted varieties.

ADVANTAGES AND DISADVANTAGES OF GRAIN FEEDING ON PASTURE

The **advantages** of grain feeding cattle on pasture, compared to strictly feedlot finishing, are:

1. Pasture gains are cheaper because: (a) less grain is required per 100 lb of gain; (b) grass is a cheaper roughage than hay or silage; and (c) less expensive protein supplement is required. Generally speaking, comparisons of self-feeding on pasture vs drylot indicate that pasture saves about 100 lb of dry feed per 100 lb of gain. Thus, if we assume that it requires 500 lb of gain to finish a steer, then each steer would require 500 lb less feed on pasture than in drylot. If feed costs 7¢ per pound, that's a saving of $35 per steer on feed cost.

2. Less labor is required because the cattle gather their own roughage and the labor required for feeding roughage is eliminated. In brief, grass-finished cattle do their own harvesting. Furthermore, it may be possible to get satisfactory results with but one grain feeding each day in finishing on pasture, or the animals may be self-fed, with the caretaker merely filling the feeder at intervals.

3. Handling of manure is eliminated, the maximum fertility value of the manure is conserved, and there is no pollution problem. When pastures are utilized by livestock, approximately 80% of the plant nutrients of the crop is returned to the soil.

4. Pasture finishing eliminates any requirement for buildings.

5. Finishing cattle on pasture is especially adapted to the small feeder and to areas where some of the land should be kept in permanent pasture.

The **disadvantages** of finishing cattle on pasture, compared to strictly feedlot finishing, are:

1. Most feeder cattle are marketed in the fall rather than in the spring. Therefore, feeder steers purchased in the spring and intended for pasture finishing are usually scarce and high in price.

2. Though less labor is required, less labor is available. The cropping season is a rush season.

3. During the midsummer, the combination of heat and flies may cause much discomfort to the animals and reduce the gains made.

4. Pastures may become dry and parched, reducing the gains made during dry seasons.

5. The manure is usually dropped on permanent pastures year after year, which may result in the neglect of the other fields.

6. In many pastures, availability of shade and water does not present a problem. However, some areas are less fortunate in this regard.

After both the advantages and disadvantages of pasture finishing are considered, the availability of cheap, rough pasture land and the price of concentrates will usually be the determining factors in deciding upon the system to follow.

SYSTEMS OF PASTURE FINISHING

When cattle are finished on pasture, any one of the following systems may be employed:

Fig. 34-2. These feeder cattle were raised by Oxbow Ranch, Prairie City, Oregon, and grazed in the Logan Valley. They went into the feedyard weighing 875 lb. (Courtesy, Ron Baker, C & B Livestock, Inc., Hermiston, OR)

1. Finishing on pastures alone—no concentrates being fed.
2. Limited grain allowance during the entire pasture period.
3. Full feeding during the entire pasture period.
4. Full or limited grain feeding on pasture following the period of peak pasture growth.
5. Short feeding (60 to 120 days) in the feedlot at the end of the pasture period.

The system of pasture finishing that will be decided upon will depend upon the age of the cattle, the quality of the pasture, the price of concentrates, the rapidity of gains desired, and the market conditions.

BASIC CONSIDERATIONS IN UTILIZING PASTURE FOR FINISHING CATTLE

The following points are basic in utilizing pastures for finishing cattle:

■ **Moderate winter feeding makes for most effective pasture utilization**—The more liberally beef cattle are fed during the winter, the less will be their effective utilization of pasture the following summer—the smaller the compensatory gains. Generally speaking, for maximum utilization of pasture, stocker calves should be fed for winter gains not in excess of 1.25 lb per head daily, and yearlings not in excess of 0.9 lb.

■ **Early pastures are *washy* but high in protein**—Cattle should not be turned to pasture too early. The first growth is extremely *washy*, possessing little energy. However, the crude protein content of the forages is high during the early stages of growth and rapidly decreases as the forages mature. This would indicate the importance of pasturing rather heavily during the period of maximum growth in the spring and early summer.

■ **Sudden changes are to be avoided**—Changes from drylot to pastures or from less succulent to more succulent pastures should be made with care; for grass is a laxative, and the cattle may shrink severely. Also, bloat may occur.

■ **Time of starting grain feeding on pastures is determined by condition of cattle and quality of pastures**—Cattle that have been fed grain rather liberally through the winter and are in good condition should usually be fed grain from the beginning of the grazing period. On the other hand, if they have been roughed through the winter, it may be just as well to feed the grain only during the last 80 to 120 days of the grazing season, after the season of peak pasture growth. The latter recommendation is made because it is sometimes difficult to get animals to consume grain when an abundance of palatable forage is available. At peak pasture growth, the animals should be started on feed and brought to full feed as rapidly as possible.

■ **Grain supplements on pastures usually make for larger daily gains and earlier marketing**—Young cattle (calves and yearlings) on summer pasture usually do not grow at their maximum potential due to energy and protein deficiencies in the feed at various times of the season. Thus, the addition of a grain

Fig. 34-3. Grain feeding finishing cattle on pasture. (Courtesy, Smith Kline Animal Health Products, Philadelphia, PA)

supplement for cattle on pasture makes for larger daily gains and earlier marketing—either directly off grass or with a shorter drylot finishing period. The owner thus avoids late fall competition and lower prices of strictly grass cattle. Also, because cattle that are grain fed on pasture can be marketed over a wider period of time, there is greater flexibility in the operations. However, cattle that are supplemented on summer pasture often sell for less to go into feedlots because feedlot operators fear that they may not gain as rapidly as cattle that are not supplemented on pasture. To answer this question, the Nevada workers studied the subsequent feedlot performance of yearling steers receiving a feed supplement while on summer pasture vs nonsupplemented cattle. The pasture-supplemented cattle were fed pellets on the ground 3 times daily, at an average daily rate of 2.18 lb per head per day, or slightly less than ½% of their body weight. The supplemented cattle out-gained the nonsupplemented cattle by 0.38 lb daily, or a total of 41 lb; and each pound of supplement produced 5.7 lb of gain. At the end of the pasture season, both groups of steers were placed in the feedlot. Subsequent feedlot gains of the two groups were essentially the same—407 and 406 lb for the supplemented and controls, respectively; and an average daily gain of 3.3 lb by each group. These results indicate that yearling steers may be profitably supplemented on summer pasture at the above rates without adversely affecting their subsequent feedlot performance.

- **Whole corn preferred to rolled corn**—When self-feeding steers on pasture, whole corn is preferred to rolled corn for the following reasons: (1) slightly less feed is required per 100 lb of gain; (2) it alleviates processing cost; and (3) it results in less incidence of founder and rumen parakeratosis because whole corn supplies some *roughness factor* in the ration to stimulate the rumen.

Fig. 34-4. Yearling steers grazing irrigated orchardgrass pasture, which is at the proper stage of growth for maximum animal gains. Orchardgrass is one of the top-yielding grasses for the northern humid states and the western irrigated areas. (Courtesy, Dr. Wilton W. Heinemann, Irrigated Agriculture Research and Extension Center, Washington State University)

- **Protein supplement not needed on good pasture**—As long as pasture is green and growing, no supplemental protein is required. During drought periods and late fall when the grass matures, extra protein is needed. At such times, it is good business to add 1 lb of protein supplement to each 8 to 12 lb of grain. Usually this will increase the rate and efficiency of gain.

- **Carrying capacity of pastures will vary**—The carrying capacity of pastures will vary with the amount of grain supplement, the quality of pasture, and the age and condition of the cattle. Because of these factors, the acreage per steer will vary all the way from 1 to 10.

- **Age is a factor**—Young cattle (yearlings) tend to grow as well as to fatten. Thus, older cattle (2 years or older) will reach a high degree of finish on pastures alone. As fine as pastures may be, it must be remembered that grass is still a roughage.

Fig. 34-5 Two-year-old Hereford steers on pasture. showing good finish. (Courtesy, Oklahoma State University)

- **Minerals are usually fed**—Salt is especially necessary when grass is being utilized. Finishing steers consume from ¾ to 1½ oz of salt per head daily. Also, cattle on pasture should have free access to a mineral mixture composed of 2 parts of dicalcium phosphate and 1 part of iodized salt or a commercial mineral mix.

- **Species of grasses or legumes will vary**—The best species of grasses or legumes or grass-legume mixtures to be seeded will vary according to the area, especially according to the soil and climatic conditions. Pasture yields vary greatly from area to area and season to season.

Temporary or supplemental pastures, such as Sudangrass or millet, are used for a short period and are usually more productive and palatable than permanent pastures. They are seeded for the purpose of providing supplemental grazing during the season

Fig. 34-6. Yearling Hereford steers on Sudangrass pasture near Weeping Water, Nebraska. (Courtesy. Soil Conservation Service, USDA)

when the regular permanent or rotation pastures are relatively unproductive.

■ **Grass vs grass-legume mixtures should be considered**—In general, where adapted legumes can be successfully grown—either alone or with grass mixtures—the results are superior to yields obtained from pure stands of the grasses. At Pullman, Washington, in a study of pure species of smooth brome and crested wheatgrass vs grass-alfalfa mixtures, it was found that: (1) the grasses produced an average of 87 lb of beef per acre, whereas the grass-alfalfa mixtures averaged 223 lb of beef per acre; (2) when based on forage yields at monthly intervals, the same mixtures produced 3 times as much oven-dry forage per acre as the pastures seeded to grasses alone; (3) the grass-legume mixtures provided a slightly longer grazing season; (4) the grass-legume mixtures provided a higher carrying capacity in terms of animals per acre; (5) the erosion-resisting characters of the soil were improved by the fibrous grass roots, both while the crop was growing and after the seeding had been plowed under; and (6) the addition of grasses to legumes tended to keep out cheatgrass and other undesirable plants.[1] The two latter points are based merely on careful observation, whereas the rest of the points were proved experimentally.

■ **Grain feeding will lengthen the grazing season**—At the Washington Agricultural Experiment Station, grain feeding cattle on pasture lengthened the grazing season by an average of 57 days.

■ **Self-feeding vs hand-feeding on pasture**—Self-feeding grain on pasture has generally proved superior to hand-feeding, as the animals consume more feed, make more rapid gains, and return more profit.

■ **Economy of grain feeding on pasture**—Whether or not it will be profitable to feed grain on pasture will depend primarily upon the price of grain, the premium paid for cattle of higher finish and grade, the season in which it is desired to market, and the area and quality of pasture.

The following practices will be helpful in reducing the bloat hazard:

1. Give a full feed of hay or other dry roughage before the animals are turned to legume pastures, to prevent the animals from filling too rapidly on the green material.
2. After the animals are once turned to pasture, they should be left there continuously. If they must be removed overnight or for longer periods, they should be filled with dry roughage before they are returned to pasture.
3. Mixtures that contain approximately half grasses and half legumes should be used.
4. Water and salt should be conveniently accessible at all times.
5. The animals should not be allowed to become empty when they congregate in a drylot for shade or insect protection and then be allowed to gorge themselves suddenly on the green forage.
6. Many practical cattle producers feel that the bloat hazard is reduced by mowing alternate strips through the pasture, thus allowing the animals to consume the dry forage along with the pasture. Others keep in the pasture a rack well filled with dry hay or straw.
7. Because of the many serrations on the leaves, Sudan hay appears especially effective in preventing bloat when fed to cattle on legume pastures.
8. Where legume bloat is encountered, use poloxalene (Bloat Guard), oxytetracycline (antibiotic), or polyoxyethylene (23) lauryl ether (Laureth-23/Enproal Bloat Blox) according to the respective manufacturer's directions.

Fig. 34-7. Yearling steers on alfalfa-orchardgrass, which they have grazed for about 120 days. Note stage of forage growth and condition of these 900-lb steers. (Courtesy, Dr. Wilton W. Heinemann, Irrigated Agriculture Research and Extension Center, Washington State University)

[1]Ensminger, M. E., *et al.*, Washington Ag. Exp. Sta. Bull. 444.

QUESTIONS FOR STUDY AND DISCUSSION

1. Why are so few cattle finished on pasture alone?

2. As grain becomes scarcer and higher in price, is it likely that more cattle will be grass finished, perhaps by supplemental grain feeding? Justify your answer.

3. The finish and marbling requirements of the top federal grades of beef have been lowered in recent years. Does this favor more pasture finishing? Justify your answer.

4. Using current feed prices, compute the value of grass on a per steer basis if it effects a saving of 100 lb of dry feed per 100 lb of gain. What additional advantages accrue from grain feeding cattle on pasture, compared to drylot finishing?

5. List and discuss the disadvantages of finishing cattle on pasture.

6. List and discuss each of the common systems of pasture finishing. What factors should determine the choice of the system?

7. Discuss each of the following basic points as it applies to pasture finishing:

   a. Moderate winter gains.

   b. Early, *washy* pasture.

   c. When to grain feed on pasture.

   d. Effect of supplementing young cattle on pasture on subsequent feedlot performance.

   e. Self-feeding whole corn vs rolled corn on pasture.

   f. Use of a protein supplement.

   g. Age of cattle.

   h. Grass vs grass-legume pastures.

   i. Self-feeding vs hand-feeding.

   j. Bloat control.

SELECTED REFERENCES

Title of Publication	Author(s)	Publisher
Beef Cattle, Seventh Edition	A. L. Neumann	John Wiley & Sons, Inc., New York, NY, 1977
Beef Cattle Production	K. A. Wagnon R. Albaugh G. H. Hart	The Macmillan Company, New York, NY, 1960
Beef Cattle Science Handbook	Ed. by M. E. Ensminger	Agriservices Foundation, Clovis, CA, pub. annually 1964–1981
Feeding Beef Cattle	J. K. Matsushima	Springer-Verlag, New York, NY, 1979
Stockman's Handbook, The, Seventh Edition	M. E. Ensminger	Interstate Publishers, Inc., Danville, IL, 1992

As beef goes, cattle finishing and the entire beef industry will go. (Courtesy, National Live Stock and Meat Board, Chicago, IL)

FUTURE OF CATTLE FINISHING (FATTENING)[1]

Contents	*Page*

[1]In the preparation of this chapter, the authors gratefully acknowledge the authoritative review and helpful suggestions, along with pictures, of the following person, who has great expertise in the operation of a modern cattle feedlot: Ron Baker, C & B Livestock, Inc., Hermiston, Oregon.

Fig. 35-1. Cattle feeding of the future will be more sophisticated. This shows Farr Feeders, feedlot, near Greeley, Colorado, capable of producing for market more than 80,000 head of cattle a year. The feedlot includes 160 pens, each engineered for drainage, encircled with steel fencing, and with a capacity of 250 head. There are $5\frac{1}{4}$ mi of concrete bunks, with 10-ft concrete platforms on which the cattle stand while they feed. The computerized feed mill, with mixing capacity of 2,100 lb per minute, is in the background. (Courtesy, Farr Feeders, Inc., Greeley, CO)

FUTURISTIC CATTLE FINISHING

The senior author's crystal ball shows that the cattle feeding industry in the future will be characterized by the following:

- **Lower profits per animal and greater difficulty in securing desirable financing**—Feedlot profits, on a per head basis, will likely decrease in the years ahead, due to (1) decreased demand for finished beef, (2) over-expansion of feedlots, (3) shortage of and higher priced feeder cattle, and (4) higher priced feeds as animals compete with humans for grains. Hand in hand with this, capital will be more difficult to secure.

- **Lots operating to capacity**—In the era of rapid expansion in the 1950s and 1960s, many lots were not operated to capacity, all or part of the year. Empty lots don't make money! Feedlots must stay 70 to 80% full to turn a profit. Hence, with smaller net profits per head in the future, feedlots will be forced to operate near capacity.

- **More integrated feedlots, both horizontally and vertically**—There will be more horizontal integration (one feedlot merging with another feedlot) motivated by (1) the desire of lots to get bigger, through acquisition, (2) the better financed lots buying the weaker financed ones, especially during periods of financial stress, and (3) spreading the area risks (of such things as droughts and feed shortages) by locating in different areas; for example, one lot might be in the Corn Belt and another in Texas.

  Also, greater efficiency will be achieved through increased vertical integration (the control or ownership at other levels of the functional system), with more feedlots owned by (1) large cow-calf operators and (2) packers. Increasingly, outside interests, with large amounts of outside capital, will engage in the cow-calf operations. Many of them will develop elaborate integrated systems which will permit them to retain ownership of the cattle through the feedlot—and even to the carcass hanging on the rail, and the retail meat market.

- **Lessened tax shelter advantages**—It seems reasonable to expect that the Internal Revenue Service will continue to lessen the tax shelter advantages formerly enjoyed by some *outside investors*—doctors, lawyers, merchants, chiefs.

- **Feedlot managers with superior business acumen**—The business aspects of the cattle feeding will become more important. More capital will be required, and there will be more competition for money available for lending; credit will be more important; production costs and labor problems will mount; inflation will be increasingly difficult to curb; computers will be used more extensively; beef futures will be more widely used; feeder calves will be in short supply; more grains will go for human consumption abroad, with the result that cattle feeders will increasingly find grains scarce and high in price. All these, and more, will call for top cattle feedlot managers with both superior cattle know how and business acumen.

- **Increased computerized management**—Feedlot managers will make increasing use of computers in decision making, beyond their use in balancing rations and inventory of cattle and feed. Among other things, they will be used to determine when to buy and sell cattle, when to buy feed, and how to guide the health treatments of both well and sick feedlot cattle.

- **Greater emphasis on public relations**—The cattle feeding industry will develop a more sophisticated and

Fig. 35-2. Crossbred cattle, products of performance testing, ready to go to market—the dual trademarks of finishing cattle of the future. (Courtesy, Ron Baker, C & B Livestock, Inc., Hermiston, OR)

costly public relations program on a continuing basis, rather than *put out fires*. The need for such a PR program is driven home by such things as (1) picketers protesting the high cost of beef, and (2) fewer and fewer legislators with farm background.

■ **Pollution control**—Environmentalists and neighbors will force pollution control. Feedlots will lessen harassments from these sources by better control of odors and dust and improved handling of manure.

■ **Recycling manure**—Recycling of feedlot manure will be perfected on a practical basis. Also, manure will be fed to fish, as is now being done in China.

■ **Fewer lots under 1,000-head capacity; more lots over 32,000-head capacity**—In recent years, there has been a decrease in small lots and an increase in big lots. In the 13 principal cattle feeding states, in 1993, 42,503 feedlots with capacity of less than 1,000 head marketed only 12.7% of the fed cattle while a mere 87 feedlots with 32,000 head or more marketed 35.1% of all fed cattle. This trend to bigness will continue.

■ **Increased mechanization**—Cattle feeding will become more mechanized, with the degree and expenditure for mechanization determined by the cost of labor replaced.

■ **More sheltered or confined feeding**—In the past, open feedlots tended to locate in areas with low rainfall and mild winters, to lessen mud and energy requirements. As sheltered, confined feeding becomes more practical (because, in comparison with open lots, of lower (1) land costs and (2) labor costs, due to better adaptation to mechanization), weather will be eliminated as a factor in determining location.

■ **More pasture finishing**—As grain for feeding livestock becomes scarcer and higher in price, due to increased human consumption of grains directly, grass finishing will increase; particularly in the Southeast and on smaller farms. With increased pasture finishing, pastures and ranges will be improved.

■ **Growth areas ahead**—Historically, it has been more efficient to take the cattle to areas of surplus grain for feeding rather than to transport the feed to the cattle. So, it is expected that the leading cattle feeding areas of the future will be (1) the Central Plains (Nebraska, Kansas, and Colorado), (2) the Southern Plains (Texas, Oklahoma, and New Mexico), and (3) the Corn Belt. However, impending water shortages may well affect the irrigated feed production of the Southern Plains.

■ **Further spreading over the country so as to be located near available feeds and cattle**—Certain highly concentrated feedlot areas appear to have approached the saturation point so far as proximity to available feed and cattle are concerned. When this point is reached, some other area will be favored for new lots.

■ **Not being limited to sparsely human populated areas for environmental reasons**—In the past, feedlots have commonly located in sparsely populated areas in order to alleviate the harassment of environmentalists, most of whom are more interested in pollution control than cattle feeding. As we learn to control feedlot odors and handle manure, pollution will be lessened as a factor in determining location.

■ **More feeders produced by cows kept for beef**—During the 1950s and 1960s, a very considerable proportion of increased feeder cattle numbers came from young animals of dairy breeding and animals that were formerly marketed as veal. These animals will continue to be fed, but no expansion from these sources can be expected. Thus, any increased feeders of the future must come from increased cows kept for beef.

■ **More crossbreds and composites**—Not only will there be more crossbreds and composites, but many of them will carry breeding of the newer breeds, recently imported. As a result, feedlots of the future will have a great array of colors and breeding.

■ **More efficient cattle being fed**—Feedlot cattle of the future will be bred for predictable rapid gains and high cutout value of high quality lean cuts.

■ **Stockers changed**—Stockers will be changed or eliminated. The stocker stage will be eliminated on calves out of good milking cows and with heavy weaning weights. Some will be slaughtered as heavy calves at weaning; others will go directly into feedlots.

But the great bulk of calves will continue to be handled as stocker calves or stocker yearlings before placement in feedlots.

■ **Most feeders being preconditioned**—With heavy, but young, calves being weaned and sent directly to feedlots, rather than spending time as stockers, preconditioning will be more important than ever. Thus, there will be more preconditioning in the future.

■ **Less total grain being fed**—This will be achieved in two ways: (1) by shortening the feeding period in the feedlot; and (2) by lessening the degree of finish. But grain will be fed to finishing cattle as long as grain-fed beef can be produced and marketed profitably.

■ **More nonprotein nitrogen (NPN)**—Worldwide protein shortages have resulted in increased quantities of the oilseed proteins being diverted for human consumption. As a result, greater quantities of nonprotein nitrogen in the form of liquid feed supplements are being fed to feedlot cattle.

■ **New additives**—New and effective noncarcinogenic growth stimulants will be developed and used.

■ **More bulls being fed**—With increasing acceptance of *bullock* beef—which may come, just as it has all over Europe—more bulls may be fed, to obtain faster gains, greater feed efficiency, and more red meat.

■ **More beef produced from grass and milk**—With rising feed costs due to animals competing with humans for the source of supply—around the world, a larger proportion of the product will be produced from grass and milk. This calls for better milking cows and heavier weaning weights.

■ **More byproduct feeds utilized in feedlots**—As human competition for grain increases around the world, more and more byproduct feeds will be fed to finishing cattle. Of course, byproduct feeds are being fed to cattle now. But many are not used. Some idea of the potential can be gained when it is realized that from 1 to 3 lb of byproducts are produced per pound of food produced.

■ **Feedlot cattle producing beef to more exacting specifications**—This will include weight, lean-to-bone ratio, amount of *bark* or outside fat, and palatability and tenderness. To meet this need, the cattle feeder will be called upon to satisfy a narrow range in market demand; in turn, the cow-calf producer will be asked to produce feeders of the right kind. This will be accomplished largely through production testing and selection of the purebreds back of the crossbreds, since carcass quality is little affected by heterosis.

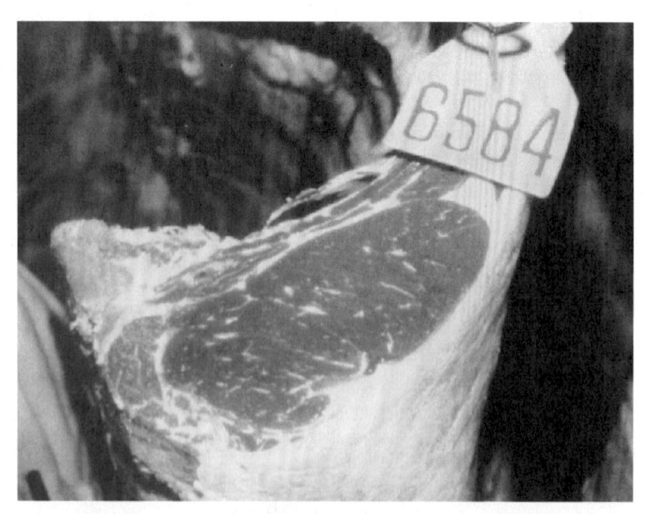

Fig. 35-3. This ribeye represents almost ideal beef specifications: yield grade 2.1 carcass, a small amount of marbling, 15.4 in. ribeye, 0.25 in. backfat, and Choice grade. (Courtesy, Ron Baker, C & B Livestock, Inc., Hermiston, OR)

■ **More beef being fabricated, boxed, and branded in packing plants**—In the future, more and more beef will be fabricated (cut and packaged in portion-ready cuts before shipping to retailers) in packing plants and warehouses, instead of the backrooms of supermarkets. This will result in (1) a saving in transportation (100 lb, or more, of fat and bone from each carcass unwanted by the supermarket will be left at the packing plant), (2) improved sanitation (beef packed in vacuum bags, rather than exposed in carcass or wholesale cuts), and (3) eliminating the present 200,000 inefficient fabricating units—the backrooms of the nation's supermarkets.

■ **Increased use of frozen beef**—The prejudice against frozen beef began disappearing with the coming of freezer lockers. In the future more beef will be frozen, thereby providing for a more orderly flow of the product.

■ **More beef tenderizers being used**—The use of beef tenderizers will become more widespread and effective.

■ **Inroads made by poultry and simulated meats (synthetic meats, or meat analogs)**—American consumers are eating more poultry each year, while the consumption of beef continues to decrease slightly.

The simulated meats, or meat analogs, especially those made from soybeans, will likely become more competitive with beef in the future, as their price becomes relatively more favorable and their taste and texture are improved. To meet this competition, the cattle industry of the future will place increasing emphasis on tenderness, leanness, and flavor.

■ **Improved market efficiency**—Currently, adequate feeder cattle and slaughter cattle market information is available. However, each feedlot is at liberty to buy and sell when it chooses, with the result that there are inevitable market gluts and scarcities, with prices varying accordingly. There is need for market controls which will allow feeders to retain most of their fierce independence, yet avoid wide swings in numbers and prices.

■ **Beef futures trading specifications that are more exacting**—As beef over the counter conforms to more exacting specifications, it would seem logical that such specifications should be reflected in beef futures trading. Thus, in addition to weight and grade, as now called for in beef futures trading, such things as guarantees of lean yield may be added.

■ **Beef imports causing constant apprehension among beef producers**—While the United States consumes about 25% of the world's beef and veal, it produces 21.1%. The difference is made up through importation. Despite some apprehension on the part of cattle producers, beef importations will continue, especially from Australia. But worldwide beef shortages and limited grain-feeding potential in other countries preclude any real foreign threat to U.S. cattle feeders, although we shall continue to import considerable quantities of lean, frozen beef.

- **Finished beef exported to Japan and perhaps to Europe**—Finished beef, produced in U.S. feedlots, and fabricated, packaged, and frozen in U.S. packing plants, will be transported via refrigerated jet freight and marketed in Japan and perhaps in Europe, at higher prices than can be secured at home.

- **Greater emphasis on tender, low fat, lean, flavorful beef**—Cattle producers should be ever mindful of what happened to pork, which was replaced by beef as the preferred red meat in 1953, primarily because the quality of pork failed to respond to consumer preference for a leaner product.

- **Meat myths will become more common**—More will be spoken and written linking the consumption of beef to certain health related problems in humans, including heart disease, cancer, high blood pressure (hypertension), and harmful residues. The effect of these myths on the future of cattle finishing will depend upon the effectiveness with which the cattle industry counters them with facts based on research.

- **More new technology**—New technology will have great impact on the future of cattle finishing; among such developments now being applied are hot boning to reduce energy costs, flaking and various formed steaks and roasts, and increased shelf life. Also, genetic engineering holds promise to improve and reduce the cost of drugs, feed additives, and growth hormones, and to improve animal immunity and lessen disease. New technology in future cattle feeding will be limited only by the research work that creates it.

QUESTIONS FOR STUDY AND DISCUSSION

1. Based on what you foresee as the future of the cattle feeding industry, would you recommend that a young person plan to become an owner-manager of a big cattle feedlot? Justify your answer.

2. Which type of integration will occur most frequently in cattle in the future—horizontal or vertical? Give the reasons for your answer.

3. Has the lessening of the tax shelter advantages of cattle feeding been good or bad? Justify your answer.

4. Outline a public relations program for the U.S. cattle industry designed to educate consumers and alleviate unjustified beef picket protesting.

5. How will feedlot pollution be controlled in the future?

6. How will cattle feedlot manure be used in the future?

7. Is the trend to bigness in cattle feedlots good or bad? Justify your answer.

8. As grain becomes more scarce and higher in price, will there be more pasture finishing, perhaps with grain supplementation? Justify your answer.

9. What growth areas in future cattle feeding do you see? Give the reasons for your prediction.

10. Will a higher proportion of feeders be produced by beef cows as such in the future, with a smaller proportion from dairy cows? Justify your answer.

11. Will feedlots of the future have more crossbreds? Justify your answer.

12. How will the stocker stage be changed in the future?

13. Will more feeders be preconditioned in the future? If so, why?

14. Will more bulls be fed in the future? Justify your answer. What problems go hand in hand with feeding bulls?

15. What suitable byproducts of which you have knowledge are not now being used for cattle feed?

16. What forces favor more beef being fabricated, boxed, and branded in packing plants in the future?

17. Is there any basis for the prejudice against frozen beef? Does it differ from frozen beef in a locker? Justify your answers.

18. What can, or should, the beef industry do to meet the likely increasing competition from synthetic meat analogs and poultry in the future?

19. What are the chances of the U.S. becoming a beef exporting nation in the future?

20. Is there hazard of neglecting carcass quality during an era of emphasis on lean beef? Justify your answer.

21. What is the best way in which to counter meat myths linking the consumption of beef to heart disease, cancer, high blood pressure, and harmful residues?

22. Discuss the relationship of new technology and research.

SELECTED REFERENCES

Title of Publication	Author(s)	Publisher
Beef Cattle Science Handbook	Ed. by M. E. Ensminger	Agriservices Foundation, Clovis, CA, pub. annually 1964–1981
Feedlot, The, Third Edition	Ed. by G. B. Thompson C. C. O'Mary	Lea & Febiger, Philadelphia, PA, 1983

PART IV

Behavior and Environment/Glossary

Today, there is great interest in animal behavior and environment. Those who grew up around animals and dealt with them in practical ways already have accumulated substantial workaday knowledge about their reaction to certain stimuli or their environment. But those who are less familiar with them may need to acquaint themselves with animal behavior in an unnatural environment, better to produce and care for them, and in order to recognize the signs when all is not well in the barnyard. Whether we come from farms or are city-bred, the principles and application of animal behavior and environment depend on understanding.

These cows and calves are grazing on a pasture which is irrigated with lagoon effluent from an on-site swine operation in North Carolina. (Courtesy, Dr. Garth W. Boyd and Ms. Rhonda Campbell, Murphy Farms, Inc., Rose Hill, NC)

A group of polled (hornless) Hereford cows drinking from a woodland pond. (Courtesy, American Polled Hereford Assn., Kansas City, MO)

Cattle exhibiting shelter-seeking behavior—in a ravine and facing away from the storm. (Courtesy, American Hereford Assn., Kansas City, MO)

BEEF CATTLE BEHAVIOR AND ENVIRONMENT

Contents *Page*

Contents	Page

Fig. 36-1. Contented cows! Shorthorn cows in the shade and ruminating—chewing their cuds.

This section is presented for the purpose of bridging the gap between something old and something new in animal behavior and environment.

ANIMAL BEHAVIOR

Animal behavior is the reaction of animals to certain stimuli, or the manner in which they react to their environment. Through the years, cattle behavior has received less attention than the quantity and quality of beef produced. But modern cattle production has brought renewed interest in behavior, especially as a factor in obtaining maximum production and efficiency. With the restriction of area over which they may roam, many abnormal behaviors have evolved to plague those who raise cattle. This has been due to a genetic time lag; producers have altered the environment faster than the genetic makeup of cattle.

This chapter is for the purpose of presenting some of the principles and applications of cattle behavior.

CAUSES OF ANIMAL BEHAVIOR

Animal behavior is caused by, or is the result of, three forces: (1) genetic, (2) simple learning (training and experience), and (3) complex learning (intelligence).

GENETIC

We need to breed beef cattle adapted to artificial environments. Confinement has not only limited space, but it has interfered with the habitat and social organization to which, through thousands of years of evolution, cattle became adapted and best suited. So, producers need to concern themselves more with the natural habitat of animals. Nature ordained that they do more than eat, sleep, and reproduce. We need to change cattle through heredity and selection as has been done in Israel, where they have selected intensively for docility for more than 35 years, with the result that the cows literally touch each other, with no antagonistic—or dominant—type response. (See subsequent section in this chapter headed "Breeding for Adaptation.")

SIMPLE LEARNING (Training and Experience)

In general, the behavior of animals depends upon the particular reaction patterns with which they were born. These are called instincts and reflexes. They are unlearned forms of behavior.

Cattle learn by experience. However, the training is only as effective as the inherited neural pathways will permit. Several types of learning processes are known; among them are those that follow:

HABITUATION

Habituation is getting used to, or ignoring, certain stimuli.

Bunk breaking calves is an example. If calves 6 to 7 months of age are weaned without prior bunk breaking, then suddenly transferred to a corral where there is no milk or grass, and where their feed must be obtained from a bunk, it is a traumatic experience for them. This is so because (1) there is a mother-young separation reaction, (2) they get homesick (and animals do get homesick), and (3) there is a change in feed and water. On the other hand, if they have been bunk broken prior to weaning, they take to the new feed bunk in the feedlot because they are used to it.

CONDITIONING (Operant Conditioning)

Conditioning is the type of learning in which the animal responds to a certain stimulus. For example, upon hearing the rolling of a barn door a cow may lick her tongue and moo, even though she can see no feed.

Artificial insemination techniques have been developed around the understanding of normal reproductive behaviors and the modifications of these behaviors. Semen collection routines are faced by behavioral responses that can change from impotence to optimum performance and high-quality semen. Proper stimulation of some bulls, for example, can increase sperm cell output by nearly 40%, compared to ejaculates after minimum stimulation.

Another example of conditioning is the use of an electric fence. When an electric fence is installed, the immediate instinct of cattle is to investigate—to touch it with their noses. Upon receiving a shock, they back off and let it alone. Thereafter, the electricity can be shut off for a considerable period of time before some animal again tests it.

Operant conditioning, or operant learning, refers to animal operation of some aspect of the environment to obtain access to feed or other animals. It is the learning of an act that has some consequence; *i.e.*, one that operates the environment—like pressing a bar that supplies some feed or turns off a light.

Broadly speaking, training is operant conditioning—it is an attempt to modify an animal's behavior. There are two types of training: (1) reinforced training, usually with positive rewards, and (2) forced training in which the animal is compelled to do certain things.

INSIGHT LEARNING (Reasoning)

Insight learning is the sudden adaptive reorganization of experience or sudden production of a new adaptive response not arrived at by overt trial-and-error behavior. It replaces trial-and-error. Of course, it is difficult to be certain in such cases that the animal did not have a similar type of problem before. Even so, the immediate application of past experience to a new situation is a noteworthy capacity.

The most important single factor to remember in training animals is that none of them (dogs included) can reason things out. An animal's mind functions by intuition, not logic. Moreover, it has no conscious sense of right and wrong. Thus, it is one of the trainer's tasks to teach an animal the difference between right and wrong—between good and bad. Although the animal cannot utilize pure reason, it can remember, and it has the ability to use the memory of one situation as it applies to another.

IMPRINTING (Socialization)

This is a form of early social learning which has been observed in some species. The pioneering work in this field was done by the Austrian zoologist Lorenz, with goslings. He found that if a baby gosling was exposed immediately after hatching to some moving object, especially if the object emitted sound, it would adopt that object as its parent-companion. Further studies revealed that goslings would adopt any other moving object in the same manner—dogs, cats, humans, and so forth. Also, it was found that the same principle applies to other fowl and to mammals.

Apparently, inheritance controls the time and the length of the critical period when an individual can be imprinted, the type of object to which it can be imprinted, the tendency to respond to the first object to which it is exposed, and the permanence of the attachment to the object following imprinting.

MEMORY

Memory is the ability to remember or keep in mind; the capacity to retain or recall that which is learned or experienced.

The existence of dominance in cattle is evidence that cattle do remember (recognize) each other; otherwise, bunting and hooking would be promiscuous and continue without end.

COMPLEX LEARNING (Intelligence)

Complex learning (intelligence) is the capacity to acquire and apply knowledge—the ability to learn from experience and to solve problems. It is the ability to solve complex problems by something more than simple trial-and-error, habit, or stimulus-response modifications. In humans, we recognize this capacity as the ability to develop concepts, to behave according to general principles, and to put together elements from past experience into a new organization.

Animals learn to do some things, whereas they inherit the ability to do others. The latter is often called *instinct.*

Generally speaking, behavioral scientists are agreed that each species has its own special abilities and capacities, and that it should only be tested on these. For example, the dog, pig, and rat are more adept at solving a maze test than cattle. Hence, solving a maze in order to find food favors the scavengers (and the dog, the pig, and the rat are all scavengers)—they have connived for their food since the beginning of time.

Thus, each species is uniquely adapted to only one ecological niche. Moreover, a niche is filled by the particular species that can solve food finding therein, and that is best adapted under the conditions that prevail. It follows that intelligence comparisons between species are not meaningful, and that it is absurd to say that one species is smarter than another.

HOW CATTLE BEHAVE—BEHAVIORAL SYSTEMS

Animals behave differently, according to species. Also, some behavioral systems or patterns are better developed in certain species than in others. Ingestive and sexual behavior systems have been most extensively studied because of their importance commercially. Nevertheless, most cattle exhibit the following nine general functions or behavioral systems:

1. Agonistic behavior (combat)
2. Allelomimetic behavior
3. Care-giving and care-seeking (mother-young) behavior
4. Eliminative behavior
5. Gregarious behavior
6. Ingestive behavior (eating and drinking)
7. Investigative behavior
8. Sexual behavior
9. Shelter-seeking behavior

AGONISTIC BEHAVIOR (Combat)

Fig. 36-2. Agonistic behavior—flight (distance between animals) exhibited by cows on the range. (Courtesy, American Hereford Assn., Kansas City, MO)

This type of behavior includes fighting, flight (distance between animals), and other related reactions associated with conflict. Among all species of farm mammals, males are more likely to fight than females. Nevertheless, females may exhibit fighting behavior under certain conditions. Castrated males are usually quite passive, which indicates that hormones (especially testosterone) are involved in this type of behavior. Thus, farmers have for centuries used castration as a means of producing docile males, particularly cattle, swine, and horses.

In combat, bulls paw the ground and bellow, followed by putting their heads together and butting.

Although young bulls raised together will seldom fight, a group of bulls may single out one individual and ride him to death, unless he is removed from the group.

Bringing together sexually mature strange bulls almost always results in a fight. Also, it is noteworthy that breeds of cattle differ in their agonistic behavior.

There is the hazard that bulls will be stifled as a result of fighting; hence, conditions that result in combat should be minimized.

Under range conditions, it is common for large numbers of bulls to be run together with a herd of cows. Even though many different bulls of different ages are included in the herd, fighting among them seldom occurs. Outside of the breeding season, as in the fall of the year, it is not uncommon to see bulls congregated together on the range, away from the cow herd.

It is noteworthy that breeds of cattle differ in their agonistic behavior.

ALLELOMIMETIC BEHAVIOR

Fig. 36-3. Allelomimetic behavior (steers competing with each other) usually results in higher feed consumption than when one steer is fed alone. (Courtesy, *Feedlot Management*, Minneapolis, MN)

Allelomimetic behavior is mutual mimicking behavior. Thus, when one member of a herd of cattle does something, another tends to do the same thing; and because others are doing it, the original individual continues.

In the wild state, this trait was advantageous in detecting the enemy, and in providing protection therefrom. Under domestication, animals are usually protected from predators. Nevertheless, the allelomimetic behavior still has important consequences.

Cows moving across a pasture toward a barn often display allelomimetic behavior. One cow starts toward the barn, and the others follow. Since the rest of the herd is following, the first cow proceeds on.

Because of stimulating and competing with each other, there is usually higher per steer feed consumption among a group of steers than by one steer alone. Thus one steer penned alone may eat "X" pounds of feed per day. However, when he is placed with other steers, his intake may be "X + Y" pounds. But, of course, the feed consumption advantage can be nullified when the animals are placed together too closely, with the result that the agonistic behavior comes into play.

CARE-GIVING AND CARE-SEEKING (Mother-Young) BEHAVIOR

The care-giving behavior is largely confined to females among domestic animals, where it is usually

Fig. 36-4. Care-giving and care-seeking behavior.

described as *maternal*; the care-seeking behavior is normal for young animals. This behavior begins shortly after birth and extends until the young are weaned. Care-giving and care-seeking vary widely among different species of farm animals.

Nature ordained that cows seek isolation at calving time. So, where possible, they will hide out.

Following birth, the care-giving behavior of the new mother becomes evident almost immediately. She gets up and begins to dry her newborn calf by licking it. Simultaneously, some cows "talk" to their newborn. They may become quite concerned and nervous as their "baby" first attempts to stand, takes a few footsteps—and falters. Aided by its mother's licking and encouraged by her "talking," eventually the calf makes it to its unsteady feet and commences to search for a teat.

A newborn calf cannot see too well, but it can smell, touch, and taste. It associates everything that is good and that cares for it with its mother. This is the beginning of herd instinct.

If on pasture, the new mother usually hides her calf. During the first day or two, the calf sleeps a great deal, while the mother grazes nearby. But a mother takes great pains not to disclose the hiding place of her calf. At intervals, she returns to feed it. If it is necessary for her to leave her calf in order to get water or supplemental feed, she does not tarry much along the way. Frequently, where there are a number of newborn calves, the cows "baby-sit" for each other. Part of the cows will leave for feed or water, but one or two will remain behind and guard all the calves. Then, when the first cows to leave have returned, the "baby-sitters" will take their turn and depart. In this

Fig. 36-5. Baby-sitter! This shows a cow baby-sitting for three other cows (note four calves; her own and three others) that have gone for feed or water. (Courtesy, Dickinson Studio, Calhan, CO)

manner, there are older cows with the calves at all times.

When a calf in hiding is approached by a human, it will usually lie as close to the ground as possible, without any movement except for its eyes. If picked up, and if scared, it may bawl (cry) for its mother. If the mother hears the call, she will come running—often ready to fight. Frequently, other cows in the vicinity, especially if they have calves of their own, may join in the response. If the disturbed calf runs away, it will return to the area after the danger has passed.

By the time the calf is 2 days old, the mother wanders more extensively, with the calf at her side. Soon, they rejoin the herd.

Recognition between mother and a calf is by smell (olfactory), sight (visual), and sound (auditory). Cows usually sniff their calves after being away for a time, and the calf recognizes its mother's call. The attachment of the mother to her calf is very strong. However, the calf accepts separation with less stress. Calves that are removed from their mothers during the critical period of an hour or so after birth, then resubmitted to their mothers, are frequently rejected.

Sometimes beef producers desire to change the normal mother-young relationship of a cow to multiple suckling. This can be done provided the calves to be adopted are properly "mothered up." Thus, if four calves are to be raised by a cow (her own and three others), stable bonding can be achieved by applying her birth fluids to the calves being "grafted" and putting them with her within the critical period after the birth of her own calf.

If a calf is stillborn, or dies soon after birth, some cows will leave the place where the fetus lies, never to return. Others may return to their dead calves at frequent intervals over a period of several days, smelling it and mooing gently.

Beef calves are normally weaned at about seven months of age. The bond between cows and calves is very considerable, with the result that the separation is a traumatic experience. Thus, both mothers and calves bawl, often in unison, for 2 to 3 days. In all cases, however, the weaning separation should be complete and final, preferably with no opportunity for the calf to see or hear its dam again. In no case should the cows and calves be turned together once the separation has been made, for it will only prolong the weaning process, and it may cause digestive disorders in the calf.

After weaning, the calf looks for care and shelter from the herd. Thus, if an animal is separated from the herd, it is stressed. It may even jump fences because of its strong instinct to rejoin the herd.

ELIMINATIVE BEHAVIOR

In recent years, elimination has become a most important phenomenon, and pollution has become a dirty word. Nevertheless, nature ordained that if animals eat, they must eliminate.

A full understanding of the eliminative behavior will make for improved animal building design and give a big assist in handling manure. Right off, it should be recognized that the eliminative behavior in farm animals tends to follow the general pattern of their wild ancestors; but it can be influenced by the method of management.

Cattle deposit their feces in a random fashion. Although cows can defecate while walking, with the result that their feces are scattered, generally they deposit their "chips" in neat piles. Most cows hump up to urinate, whereas bulls are inclined to stand squarely on all "fours."

GREGARIOUS BEHAVIOR

Gregarious behavior refers to the flocking, or herding, instinct of certain species. It is closely related to allelomimetic behavior. If animals imitate each other, they must stay together. If they stay together as a mobile group, they must use allelomimetic behavior to do so. All such behavior arises out of the process of social attachment.

Cattle tend to roam in groups of various sizes when a large herd is placed on a pasture. However, there is usually considerable space between the members of the herd. Moreover, on close observation it is evident that there are several small groups within a herd, each ranging from three to five head.

INGESTIVE BEHAVIOR (Eating and Drinking)

Fig. 36-6. Ingestive behavior—eating. Grazing occupies eight hours time each day. (Courtesy, American Hereford Assn., Kansas City, MO)

This type of behavior includes eating and drinking; hence, it is characteristic of animals of all species and all ages. It is very important because animals cannot live without feed and water. Moreover, for high production, animals must have aggressive eating habits; they must consume large quantities of feed.

The first ingestive behavior trait, common to all young mammals, is suckling.

Each species has its own particular method of ingesting feed. The natural feeding (grazing) position of cattle is heads down. In this position, they produce more saliva; and saliva aids digestion. When grazing, cattle wrap their tongues around grass, then jerk their heads forward so that the vegetation is cut off by the lower incisor teeth. (There are no upper incisor teeth, only the thick, hard dental pad.) When grazing, cattle also move their heads from side to side. This movement, aided by protuberant eyes and thin legs, gives them a continuous view of their entire surroundings, an essential for wild cattle in an environment containing dangerous predators. It is important that artificial feeding devices and arrangements not depart too far from this natural pattern, because cows have a built-in antipathy to being more or less blindfolded while eating.

Cattle are ruminants. Thus, they regurgitate their feed and chew it again in a process known as *rumination.*

Rumination is the act of chewing the cud, characteristic of herbivorous animals with split hoofs. It involves regurgitation of ingesta from the reticulo-rumen, swallowing of regurgitated liquids, remastication of the solids accompanied by resalivation, and reswallowing of the bolus. Rumination occupies about 8 hours of the cow's time each day. (In addition, the harvesting or grazing time may take another 8 hours. This means that cows may work a 16-hour day.)

Fig. 36-7. Ingestive behavior—drinking. Typical western range watering facility—well, windmill, and tank. (Courtesy, Lasater Ranch, Matheson, CO)

The Iowa Station reported that steers in lots on self-feeders spent 12 hours per day lying down, and this time was unaffected by shelter or season.

When the cow regurgitates, a soft mass of coarse feed particles, called a bolus, passes from the rumen through the esophagus in a fraction of a second. She chews each bolus for about one minute, then swallows the entire mass again. Originally, it was thought that the regrinding which occurred during rechewing helped the digestion by exposing a greater surface area to fiber-digesting microflora. But recent studies indicate that rechewing does not improve digestibility. Instead, rumination has an important effect on the amount of feed the animal can utilize. Feed particle size must be reduced to allow passage of the material from the rumen. It follows that high-quality forages require much less rechewing and pass out of the rumen at a faster rate; hence, they allow a cow to eat more.

INVESTIGATIVE BEHAVIOR

All animals are curious and have a tendency to explore their environment. Investigation takes place

Fig. 36-8. Investigative behavior (ears pointed forward and eyes focused) exhibited by British White heifers. (Courtesy, Walter Bohaty, Bellwood, NE)

Fig. 36-9. Sexual behavior displayed by a Texas Longhorn bull and cow. (From an original painting by artist Tom Phillips, 915 Fulton St., San Francisco, CA)

through seeing, hearing, smelling, tasting, and touching. Whenever an animal is introduced into a new area, its first reaction is to explore it. Experienced producers recognize that it is important to allow animals time for investigation before attempting to work them, either when they are placed in new quarters or when new animals are introduced into the herd.

If they are not afraid, cattle investigate a strange object at close range. They proceed toward it with their ears pointed forward and their eyes focused directly upon it. As they approach the object, they sniff and their nostrils quiver. When they reach the object, sniffing is replaced by licking; and if the object is small and pliable, they may chew it or even swallow it.

Cattle exhibit investigative behavior when placed in a new pasture or in a new barn. As a result, if there is an open gate in a pasture or a hole in the fence, they usually find it, then proceed to explore the new area.

Calves are generally more curious than older cattle. Perhaps this is due to the fact that older animals have seen more objects, with the result that fewer things are new or strange to them.

SEXUAL BEHAVIOR

Reproduction is the first and most important requisite of cattle breeding. Without young being born and born alive, the other economic traits are of academic interest only. Thus, it is important that all those who breed cattle should have a working knowledge of sexual behavior.

Sexual behavior involves courtship and mating. It is largely controlled by hormones, although males that are castrated after reaching sexual maturity (which are known as stags) usually retain considerable sex drive and exhibit sexual behavior. This suggests that psychological, or learned, as well as hormonal factors may be involved in sexual behavior.

Each animal species has a special pattern of sexual behavior. As a result, interspecies matings do not often occur.

Males in most species of farm animals detect females in heat by sight or smell. Also, it is noteworthy that courtship is more intense on pasture than under confinement, and that captivity has the effect of producing many distortions of sexual behavior compared to wild animals.

Today, livestock producers are attempting to control the sex life of cattle, by bringing about ovulation at the time of choice of the owner, rather than of the female.

Experienced producers can usually detect in-heat cows through one or more of the following characteristic symptoms: (1) restlessness; (2) mounting other cows, and standing to be mounted by another cow (standing heat appears to be the best single indicator of the proper time to breed); (3) a noticeable swelling of the labia of the vulva; (4) an inflamed appearance about the lips of the vulva; (5) frequent urination; (6) switching and raising the tail; and (7) a mucous discharge. A day or two following estrus, a bloody discharge is sometimes seen. Dry cows and heifers usually show a noticeable swelling or enlargement of the udder during estrus. When kept alone, some cows become restless, walk the fence, and bawl when they are in heat. Some may even jump the fence, or go through it, as they attempt to find a bull.

A bull can often detect a cow that is coming in heat 24 to 48 hours before she will mate, at which time he will remain in her company. Courtship of the bull consists of following the in-heat cow, licking and smelling the external genitalia, with the head extended horizontally and the lip upcurled, and chin-resting, with the chin and throat resting on the cow's rump.

Ovulation occurs late in the heat period, usually from about 28 to 30 hours after the onset of estrus.

Fig. 36-10. Shelter-seeking behavior—in the shade during the heat of the day. (Courtesy, *Livestock Weekly*, San Angelo, TX)

SHELTER-SEEKING BEHAVIOR

All species of animals seek shelter—protection from the sun, wind, rain and snow, insects, and predators.

Cattle are not as sensitive to extremes in temperature—heat and cold—as are swine. Nevertheless, they do seek shelter under natural conditions—this may consist of hills, valleys, timber, and other natural windbreaks; or they may even group closely together.

Cattle seem to be able to sense the coming of a storm, at which time they may race about and "act up." During a severe rain or snowstorm, they turn their rear ends to the storm and tend to drift away from the direction of the wind. By contrast, bison (buffalo) face a storm head on.

During the hot summer months, cattle seek either shade or a waterhole during the heat of the day. Then they graze in the cool of the evening or early morning. There are well-known breed differences in tolerance to heat. Brahman cattle can withstand more heat than the European breeds, whereas the heat tolerance of the Santa Gertrudis is intermediate.

SOCIAL RELATIONSHIPS OF CATTLE

Social behavior may be defined as any behavior caused by or affecting another animal, usually of the same species, but also in some cases, of another species.

Social organization may be defined as an aggregation of individuals into a fairly well integrated and self-consistent group in which the unity is based upon the interdependence of the separate organisms and upon their responses to one another.

The social structure and infrastructure in a herd of cattle is of great practical importance. Cattle producers should be knowledgeable relative to it. Then, if this social relationship is disturbed and/or modified under intensive, confined conditions, they will be better able to feed, care, and manage the animals with maximum consideration accorded to both economy of production and animal welfare.

Breed affects social stratification in cattle. For example, on pasture or range, Angus tend to dominate Shorthorns, while Herefords tend to submit to both breeds. Older cows generally dominate the younger ones, and the heavier animals (usually the older animals) tend to dominate the lighter; for this reason, two-year-old heifers should be segregated from older cows. However, among cows of similar age and breed, the smaller and more aggressive ones are most dominant. Also, cows with more seniority in the group and cows with horns tend to be of higher social rank. Aggression in cows appears to be ritualized, with most encounters taking place in the following sequence: approach, threat, and physical contact (or fighting).

Limited studies indicate a high relationship between social status of cattle and spacing, or social distance. The higher the social rank, the more likely cows are to be found near other members of the herd. Also, dominant cows tend to allow close approach by other cows more often than subordinates.

On the range, the following rank orders are evident in large heterosexual herds of cattle: (1) adult males, (2) adult females, and (3) juveniles. Adult males dominate adult females, which in turn, dominate juveniles. However, at about 1½ years of age, young males begin to fight with adult females, and by 2½ years of age, they dominate all the females, and join the adult male rank order.

When grouped together on the range, bulls are loners, and do not organize socially.

When moving on pasture or range, cows travel in a consistent leadership order. Mid-dominant cows tend to be in front of the group. However, the same individuals are seldom consistent leaders; instead, there is a pool of animals which tends to be in or near the lead. More consistency is found in the cows bringing up the rear of the moving herd. So, *rearship* is a more distinctive feature than *leadership*. The animals at the rear are usually the younger subordinate heifers. Also, in most herds there is a definite order in which cows enter a corral or pasture; the mid-age, mid-dominant cows tend to be first, followed by the older cows. Social dominance orders (called *bunt order* in polled cows and *hook order* in horned cows) become more complex as herd sizes increase.

Within a corral of feedlot (finishing) cattle, a linear-tending (a linear-tending hierarchy is a type of social hierarchy in which dominance-subordinance ranking includes a triangle or some more complex hierarchy loop) *peck order*, or dominance order, can be determined by observing agonistic encounters between pairs of animals within the group. The degree of linearity is greater with increases in heterogeneity of such factors as age, sex, weight, breed, and background. Linearity and stability of the dominance order tend to increase the longer the group is together, and linearity is greater among small groups.

SOCIAL ORDER (Dominance)

Fig. 36-11. Dominance. This shows a dominant cow attacking the neck of a subordinate. The latter submits and avoids a fight.

Within a herd or group of cattle, there is a well-organized social rank. When we restrict or confine them and force them into spaces that bring them within the natural, individual distance that has been established (the distance between each other when moving as a herd), we immediately create stress throughout the herd. Thereupon, the dominants have to pay more attention to maintaining their dominance. They have to be more aggressive in their reactions. The subservients become far more nervous, and their nervousness spreads throughout the herd.

In chickens, in which social order was first observed, the social rank order is called the *peck order*.

Thus, in a herd or group of cattle, the alpha animal in the herd will be dominant over all other individuals and the omega animal will be subordinate to all. In between, some animals will be subordinate in some relationships and dominant in others. Moreover, once these relationships are established, they seldom change. The social order is usually important only in females, because mature male animals are seldom run together in groups.

When several cows are brought together to form a herd, there is a substantial period during which there is much butting and threat posturing in order to establish a dominance hierarchy. This is disturbing to a herd and will result in reduced production. Usually the older and larger cows come out at the top of the hierarchy.

Once the social rank order is established, it results in a peaceful coexistence of the herd. Thereafter, when the dominant one merely threatens, the subordinate animal submits and avoids conflict. Of course, there are some pairs that fight every time they chance to meet. Also, if strange animals are introduced into such a group, social disorganization results in the outbreak of new fighting, as a new social rank order is established.

Among wild animals, social rank order is nature's way of giving mating priority to the top ranking males. Hence, they leave behind more of their progeny than do the less dominant males. Also, dominance establishes priority in feeding.

Social rank among cattle is of little consequence as long as they are on pasture, and if there is plenty of feed and water. But it becomes of very great importance when animals are placed in confinement. When cows are moved into limited quarters, social dominance decrees that replacement heifers be sorted out and fed separately, that young bulls be cared for in separate quarters, and that old cows with poor teeth be fed separately; otherwise, these animals will not get enough feed.

When self-feeders and central water tanks are used, care must be taken relative to providing both adequate space and proper placement; otherwise, submissive animals may find it difficult to get out of eye contact of dominant animals so as to eat and drink in peace.

Of course, social rank becomes doubly important if limited feeding is practiced. Under such circumstances, the dominant individuals crowd the subordinate ones away from the feed bunk, with the result that they may go hungry.

Several factors influence social rank; among them, (1) age—both young animals and those that are senile rank toward the bottom; (2) early experience—once a subordinate in a particular herd, usually always a subordinate; (3) weight and size; and (4) aggressiveness or timidity. Also, it is noteworthy that social rank is influenced by hormones.

In cattle confinement operations, social facilitation is of great practical importance. Dominants should be sorted out, and, if possible, grouped together. Of course, they will fight it out until a new social order is established. In the meantime, feed efficiency will suffer. But, as a result of removing the dominants, the feed intake of the rest of the animals will be improved, followed by greater feed efficiency, production, and profit. Among the more settled animals, social facilitation will become more evident. After the dominants have been removed, the rest of the animals will settle down into a new hierarchy, but within the limits of their dominance. Their interaction or social facilitation will be far more likely to have a calming effect on this group, to both the economic and practical advantages of the operator.

Dominance and subordination are not inherited as such, for these relations are developed by experience. Rather, the capacity to fight (agonistic behavior) is inherited, and, in turn, this determines dominance and subordination. Hence, when combat has been bred into the herd, such herds never have the same settled appearance and docility that is desired of high-production and intensive animals.

LEADER-FOLLOWER

Fig. 36-12. Leader-follower relationship of cattle evidenced by the long columns of cattle as they wind their way over the hills of the Canadian prairies to summer range. (Courtesy, Canadian National Film Board)

Leader-follower relationships are important in cattle. The young follow their mothers; hence, they continue to follow their elders.

The leader is the cow that is usually at the head of a moving column and often seems to initiate a new activity. The leader may be small, but she is always intelligent.

If the lead animal can be controlled, generally the remainder of the group can be moved easily.

Right off, it is important to distinguish leader-follower relationships from dominance; in the latter, the herd is driven, rather than led. After the dominants have been removed from the herd, the leader-follower phenomenon usually becomes more evident. It is well known that the dominant animal is not necessarily the leader; in fact, it is very rarely the leader. It pays too much attention to other matters of dominance in its relationship within the herd, with the result that it does not develop the qualities of leadership.

When a string of cows moves, the dominant animals are generally in the middle of the procession; with the leader in front, and the subordinate ones bringing up the rear.

INTERSPECIES RELATIONSHIPS

Social relationships are normally formed between members of the same species. However, they can be developed between two different species. In domestication this tendency is important (1) because it permits several species to be kept together in the same pasture or corral, and (2) because of the close relationship between caretakers and animals. Such interspecies relationships can be produced artificially, generally by taking advantage of the maternal instinct of females and using them as foster mothers.

All sorts of bizarre interspecies relationships have been arranged—including cows raising pigs.

PEOPLE-ANIMAL RELATIONSHIPS

Social relationships can also be transferred to human beings. As a result, a young calf associates everything good or bad with humans. Unfortunately, this is the period in life during which calves are dehorned, castrated, branded, and vaccinated; hence, it is no wonder that some cattle are hard to handle. In order to minimize the problem, calves should be worked as little as possible, with all such jobs done at one time.

Good cattle caretakers usually form a care-dependency relationship with the animals under their care, with the result that the cows readily come to them.

HOW CATTLE COMMUNICATE

Communication between cattle involves one individual giving some signal, which, on being received by another, influences its behavior.

SOUND

Sound communication is of special interest because it forms the fundamental basis of human language. The gift of language alone sets people apart from the rest of the animals and gives them enormous advantages in their adaptation to their environment and in their social organization.

Fig. 36-13. Communication by sound—a bellowing bull.

Sound is also an important means of communication among cattle. They use sounds in many ways; among them, (1) feeding, in sounds of hunger (bawling) by young; (2) distress calls like the bellowing of a bull; (3) sexual behavior and related fighting; (4) mother-young interrelations to establish contact and evoke care behavior; and (5) maintaining the group in its movements and assembly.

Cattle have a very acute sense of hearing, perceiving higher and fainter noises than the human ear.

SMELL

Cattle can smell at a greater distance than people. On a day with a 5-mile wind and a humidity of 75%, a cow can smell up to 6 miles away; as wind and humidity increase, she can smell even further.

In cattle, females in estrus secrete a substance that attracts males. Hence, bulls locate cows that are in heat by the sense of smell.

VISUAL DISPLAYS

When several strange cows are brought together, there is much threat posturing, as well as butting, in order to establish a dominance hierarchy. Also, bulls will strike a hostile stance prior to fighting.

Fig. 36-14. Agonistic behavior exhibited by a bull. Pawing the ground and bellowing are generally the first stage of combat in battle.

Birds are noted for their sexual behavior in the act of courtship. Visual displays during courtship are less evident among cattle, but they do occur to some extent.

HOMING AND ORIENTATION (Path-Finding, Navigation, Migration)

Through sound, scent, or some sense of which we do not know, when cattle are moved to distant places, they often find their way back home.

NORMAL CATTLE BEHAVIOR

Cattle producers need to be familiar with behavioral norms of cattle in order to detect and treat abnormal situations—especially illness. Many sicknesses are first suspected because of some change in behavior—loss of appetite (anorexia); listlessness; labored breathing; posture; reluctance or unusual movement; persistent rubbing or licking; and altered social behavior, such as one animal leaving the herd and going off by itself—these are among the useful diagnostic tools.

Also, it is important to know how cattle see and sleep.

A summary of normal cattle behavior follows.

HEALTH

Some of the signs of good health are:

1. Contentment
2. Alertness
3. Eating with relish and cudding
4. Sleek coat and pliable and elastic skin
5. Bright eyes and pink membranes
6. Normal feces and urine
7. Normal temperature, pulse rate, and breathing rate

Normal rectal temperature
Average, 101.5°F
Range, 100.4–102.8°F

Normal pulse rate
60–70/min.

Normal breathing rate
10–30/min.

SIGHT

The eyes of most animals are on the side of the head (the cat is an exception). This gives them an orbital, or panoramic, view—to the front, to the side, and to the back—virtually at the same time. Also, this is a rounded, or globular, type of vision. This leads to a different interpretation than that of the binocular type of vision of people.

The wide-set eyes of cattle enable them to have a large panoramic field of vision, even to the extent of seeing everything around them, with slight head movements. Only what is immediately behind their hindquarters is outside their field of view.

A cow does not see in color; she sees in shades of grays and blacks.

If a cow sees movement, her instinct is to escape; hence, movements around cattle should be made very quietly and slowly.

SLEEP

Normal behavior in sleep should be recognized, especially since it differs widely between species.

Cattle typically lie on the stomach or tilt to one side, with the forelimbs folded under the body; one hind limb extends forward, while the other protrudes toward the outside. Although cattle rest in this manner, they do not sleep in the sense that the term usually connotes. While lying down, they do shut their eyes for short periods of time.

Calves commonly spend up to ½ hour at a time with their heads turned back in the flank position.

ABNORMAL CATTLE BEHAVIOR

Abnormal behavior of domestic animals is not fully understood. As with human behavior disorders, more work is needed. However, we have learned from studies of captured wild animals that when the amount and quality, including variability, of the surroundings of an animal are reduced, there is increased probability that abnormal behavior will develop. Also, it is recognized that confinement of animals makes for lack of space; this often leads to unfavorable changes in habitat and social interactions.

Abnormal behavior in animals develops where there is a combination of confinement, excess stimulation, and forced production with a lack of opportunity to adapt to the situation.

Homosexual behavior is common among all species where adult mammals of one sex are confined together.

The *mean bull* complex is an example of abnormal behavior in cattle. Of course, there are inherited differences in the temperaments of cattle. Nevertheless, constant stress can change the temperament of an animal, just as it can in people. Thus, when a bull is kept for hand mating in a corral by which the cow herd passes each day, cows in heat, or coming in heat, stimulate his sexual behavior. Since he cannot respond naturally through coitus, he becomes a mean bull.

Pica in cattle (consumption of dirt, hair, bones, or feces) may develop, perhaps due to boredom, nutritional inadequacies, or physiological stress.

APPLIED CATTLE BEHAVIOR

In the beginning of this chapter, it was stated that this presentation is for the purpose of bridging the gap between the principles and the application of cattle behavior. So, let us next turn to some practical applications of cattle behavior.

BREEDING FOR ADAPTATION

The wide variety of livestock in different parts of the world reflects a continuous process of natural and artificial selection which has resulted in the survival of animals well adapted to climate and other environmental factors. Among the examples are *Bos indicus* (Zebu) types of cattle in tropical areas, and *Bos taurus* cattle in temperate zones. Such adaptations relate to survival of the animals, but they do not necessarily entail maximum productivity of food for people. European cattle usually have much higher yields of meat and milk than the breeds native to Africa or India. It is understandable, therefore, why there have been many attempts to introduce improved European cattle into countries in which the productivity of native stock is low. But there are many problems in breed replacement, with the result that a large number of experimental introductions of new breeds have not been successful. Tropical Africa provides an example. Because of disease problems, poor resistance to high temperatures, and limited feed supplies, many of the attempts made by former colonial powers to improve the output of native stock by replacing them with the European breeds failed. Breed replacement or a crossbreeding system might seem to be a simple panacea for low productivity. However, unless associated with special provisions for subsequent importation of breeding stock and simultaneous improvement of the nutritional, parasitological, disease, and husbandry environments of the crossbreds, it is not likely to succeed.

Selection should be from among animals kept in

an environment similar to that in which it is expected that their offspring shall perform—this requisite applies to animals brought in from another herd, either foundation or replacement animals. For example, animals that are going into a range herd should be selected from animals handled under range conditions, rather than from stall-fed animals.

We need to breed and select cattle that adapt quickly to an artificially-made environment—animals that not only survive, but thrive, under the conditions imposed upon them.

Also, early training and experience are extremely important. In general, as with humans, young animals learn more quickly and easily than adults; hence, advance preparation for adult life will pay handsome dividends. The optimum time for such training varies according to species.

Stress can be reduced or avoided entirely if animals proceed through a graduated sequence of events leading to an otherwise noxious experience. Preconditioning of cattle is an application of this principle to production practices. If calves are properly preconditioned (*e.g.*, started on feed, vaccinated, treated for parasites, etc.) prior to weaning, the stress of subsequent weaning and movement to a feedlot is minimized.

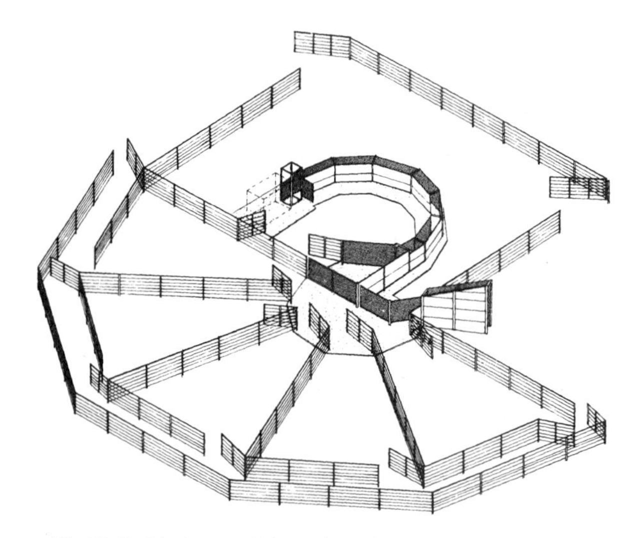

Fig. 26-15. Circular corral. A good corral is more than just a fence. It's a working area. It should have the following facilities: (1) crowding area, (2) working chute, (3) squeeze chute or head gate, and (4) loading chute. As shown above, it can be built initially with only part of the holding pens, then expanded as the herd gets bigger. Note the outside circular passageway and the long, curved working chute which will hold five or six cows at one time. (Adapted by the authors from *Plans and Building Information*, Extension Agricultural Engineering, Oklahoma State University, Stillwater)

QUICK ADAPTATION—EARLY TRAINING

We need to breed and select animals that adapt quickly to an artificial environment—animals that not only survive, but that thrive, under the conditions that people impose upon them.

Also, early training and experience are extremely important. In general, young animals learn more quickly and easily than adults; hence, advance preparation for adult life will pay handsome dividends. The optimum time for such training varies according to species.

Both good and poor patterns of behavior and productive traits come from early experience. For example, feeding group-reared calves from nipple pails (rather than buckets) reduces the vice of suckling each other.

Stress can be reduced or avoided entirely if animals proceed through a graduated sequence of events leading to an otherwise noxious experience. If calves are properly started on feed, vaccinated, treated for parasites prior to weaning, the stress of subsequent weaning and movement to another location is minimized.

LOADING CHUTE AND CORRAL DESIGN

With knowledge of basic behavior patterns and of social habits within the herd, we have at our disposal the necessary tools for designing facilities and housing which will enable us to make animals do what we want them to do, when we want them to do it—and with a saving in both labor and tempers. As an example, let us consider the matter of putting cattle through a chute on their own accord without interference or any extra driving from the handlers.

At the outset, it is recognized that cattle will follow the leader, and they will automatically try to escape through a gap, or opening. Hence, they will follow the leader through a curved chute more readily than through a straight one. As the lead animal approaches the chute and realizes that there is an opening either to the left or right, it goes forward with the idea of going through this gap—and escaping. Thus if one can get the leader-type cow into the chute first, it's a simple matter to get the rest of the herd to follow through a curved chute. So, the practical application of cattle behavior and social habits to handling facilities calls for a curved chute, with a curved entrance on the outer portion of the corral and the normal straight side on the funnel (inside) portion of the corral.

The corral should always match the work requirements, labor, and herd size. The chute should always be curved; and the corral should be designed so that there is always a gap to the left or right. Then, if the animals have been selected for their lack of dominance, they will automatically enter the funnel portion, thence the chute, thence proceed at their own pace throughout the whole of the facility.

■ **The circular corral**—By designing corral facilities in an entirely different fashion from the traditional—by abandoning the straight-sided corral—a facility can be designed that will lend itself to very rapid and adequate handling of stock, with a minimum of interference from the handlers. This has the advantages of (1) cutting down stress, (2) speeding up the work, and (3) distributing the cattle after an examination for a particular series of operations or cutouts. This takes the shape of a circular corral, with the diameter determined by the number of animals to be handled.

MANURE ELIMINATION

Body waste is a major concern; it is expensive, time-consuming, and a major pollution problem. But manure handling can be facilitated by an understanding and application of eliminative behavior.

Cattle are indiscriminate eliminators. Even so, this trait can be used effectively. For example, if cattle are fed at the same time each day, feed is released from the rumen into the true stomach regularly; and the moment the latter happens, there is a gastro-colic reflex. When this happens and cattle are put under slight stress, they defecate. Knowing this, cattle can be moved to the defecating area at the right time.

COMPANIONSHIP

Companionship in animals is of great practical importance. Except for the cat, all domestic animals are highly social and have constant need for companionship.

If not too crowded, placing cattle together sometimes accomplishes two things: (1) greater feed consumption, due to the competition between them (social facilitation); and (2) a quieting effect.

ANIMAL ENVIRONMENT

An animal is the result of two forces—heredity and environment. Heredity has already made its contribution at the time of fertilization, but environment works ceaselessly away until death. Since most animal traits are only 30 to 50% heritable, the expression of the rest (more than 50%) depends on the quality of all components of the environment. Thus, it is very important that the keeper of herds and flocks have enlightened knowledge of, and apply expert management to, animal environment.

Environment may be defined as all the conditions, circumstances, and influences surrounding and affecting the growth, development, and production of animals. The most important influences in the environment are the feed and quarters (space and shelter).

The branch of science concerned with the relation of living things to their environment and to one another is known as ecology.

Through the years, the domesticated animals best suited to a particular environment survived, and those that were poorly adapted either moved to a more favorable environment or perished. During the past two centuries, livestock producers have made great strides in the selection and propagation of animals suited to a particular environment, and during the past 50 years they have made progress in modifying the environment for the benefit of their animals and themselves.

It is becoming increasingly difficult to define environment, because scientists continue to discover important new environmental factors. Primitive people recognized that the sun and fire provided both heat and light, that body heat could be conserved by draping the body with animal skins, and that trees and caves provided protection from the weather. Today, it is recognized that these, along with a host of other environmental factors, affect animals and people.

The keepers of herds and flocks were little concerned with the effect of environment on animals so long as they grazed on pastures or ranges. But rising feed, land, and labor costs, along with the concentration of animals into smaller spaces, changed all this. Today, confinement production is increasing with all animals, including beef cattle.

Among animals, environmental control involves space requirements, light, air temperature, relative humidity, air velocity, wet bedding, ammonia buildup, dust, odors, and manure disposal, along with proper feed and water. Control or modification of these factors offers possibilities for improving animal performance. Although there is still much to be learned about environmental control, the gap between awareness and application is becoming smaller. Research on animal environment has lagged, primarily because it requires a melding of several disciplines—nutrition, physiology, genetics, engineering, and climatology. Those engaged in such studies are known as ecologists.

In the present era, pollution control is the first and most important requisite in locating a new livestock establishment, or in continuing an old one. The location should be such as to avoid (1) the neighbors complaining about odors, insects, and dust; and (2) pollution of surface and underground water. Without knowledge of animal behavior, or without pollution control, no amount of capital, native intelligence, and sweat will make for a successful livestock enterprise.

Over the long pull, selection provides a major answer to behavioral problems; we need to breed animals adapted to artificially made environments.

The following environmental factors are of special

importance in any discussion of animal behavior-environment:

1. Feed and nutrition
2. Water
3. Weather
4. Adaptation, acclimation, and habitation of species/breeds to the environment
5. Facilities
6. Health
7. Stress

FEED

The most important influence in the environment is the feed. Animals may be affected by (1) too little feed, (2) rations that are too low in one or more nutrients, (3) an imbalance between certain nutrients, or (4) objection to the physical form of the ration—for example, it may be ground too finely.

Forced production and the feeding of forages and grains which are often produced on leached and depleted soils have created many problems in nutrition. These conditions have been further aggravated through the increased confinement of animals, many animals being confined to stalls or lots all or a large part of the year. Under these unnatural conditions, nutritional diseases and ailments have become increasingly common.

Also, nutritional reproductive failures plague cattle operations. Generally speaking, energy is more important than protein in reproduction. The level and kind of feed before and after parturition will determine how many females will show heat—and conceive. After giving birth, feed requirements increase tremendously because of milk production; hence, a lactating female needs a much greater feed allowance than during the pregnancy period. Otherwise, she will suffer a serious loss in weight, and she may fail to come in heat and conceive. This basic fact, along with other pertinent findings, was confirmed by researchers at the Montana Agricultural Experiment Station. Based on 12 years research at Havre and Miles City Stations, they concluded that beef cattle size and milk production should be tailored to fit the environment. Big size and more milk are not better unless the range forage supply is better. The best size cow is one that fits the range conditions. Small cows do best on poor range because they can usually get 100% of their daily feed requirement for maintenance and milk production, whereas big cows on a poor range are borderline hungry all the time. Also, cows that give a lot of milk must have a good range; otherwise, they are stressed by lack of feed; and their fertility rate and calf crops drop. So, cow size and milk production should match their environment.

The next question is whether a breeding program can make maximum progress under conditions of suboptimal nutrition (such as is often found under some farm and range conditions). One school of thought is that selection for such factors as body form and growth rate in animals can be most effective only under nutritive conditions promoting the near maximum development of those characters of which the animal is capable. The other school of thought is that genetic differences affecting usefulness under suboptimal conditions will be expressed under such suboptimal conditions, and that differences observed under forced conditions may not be correlated with real utility under less favorable conditions. Those favoring the latter thinking argue, therefore, that the production and selection of breeding animals for the range should be under typical range conditions and that the animals should not be highly fitted in a box stall.

The results of a 10-year experiment conducted by the senior author and his colleagues at Washington State University, designed to study the effect of plane of nutrition on meat animal improvement, support the contention that selection of breeding animals should be carried on under the same environmental conditions as those under which commercial animals are produced.

The following additional feed-environmental factors are pertinent:

1. **Regularity of feeding.** Animals are creatures of habit; hence, they should be fed at regular times each day, by the clock.

2. **Underfeeding.** Too little feed results in slow and stunted growth of young stock; in loss of weight, poor condition, and excessive fatigue of mature animals; and in poor reproduction, failure of some females to show heat, more services per conception, lowered young crop, light birth weights, and lowered milk production.

3. **Overfeeding.** Too much feed is wasteful. Besides, it creates a health hazard; there is usually lowered reproduction in breeding animals, and a higher incidence of digestive disturbances (bloat, founder, and scours)—and even death. Animals that suffer from mild digestive disturbances are commonly referred to as *off feed.*

4. **Deficiency of nutrient(s).** A deficiency of any essential nutrient required by cattle will lower production and feed efficiency.

WATER

Water is one of the largest constituents in the animal body, ranging from 40% in very fat, mature cattle to 80% in newborn calves. Deficits or excesses of more than a few percent of the total body water are

Fig. 36-16. Cattle can survive for a longer period without feed than without water. (Courtesy, USDA)

incompatible with health, and large deficits of about 20% of the body weight lead to death.

The total water requirement of cattle varies primarily with the weather (temperature and humidity); feed (kind and amount); age, and weight of animal; and the physiological state. The need for water increases with increased intakes of protein and salt, and with increased milk production of lactating animals. Water quality is also important, especially with respect to the content of salts and toxic compounds.

It is generally recognized that animals consume more water in summer than in winter. Based on 5 summer and 4 winter trials, the Iowa State Agricultural Experiment Station reported that yearling cattle consumed an average of 8.5 gal per day in summer vs 5 gal per day in winter.

The water content of feeds ranges from about 10% in air-dry feeds to more than 80% in fresh, green forage. Feeds containing more than 20% water are known as *wet feeds*. The water content of feeds is especially important for animals which do not have ready access to drinking water.

Under practical conditions, the frequency of watering is best determined by the animals, by allowing them access to clean, fresh water at all times.

WEATHER

Webster defines weather as a state of the atmosphere with respect to heat or cold, wetness or dryness, calm or storm, clearness and cloudiness.

Extreme weather can cause wide fluctuations in animal performance. The difference in weather impact from one year to the next, and between areas of the country, causes difficulty in making a realistic analysis of buildings and management techniques used to reduce weather stress.

The research data clearly show that winter shelters and summer shades improve production and feed efficiency. The issue is clouded only because the additional costs incurred by shelters have frequently exceeded the benefits gained by the improved performance, particularly in those areas with less severe weather and climate.

During hot weather, feedlot cattle "peak" their eating during early morning and again during the evening hours—when it is cool. In cool weather, they eat more during midday than when it is hot. The feeder should sense these changes in cattle eating habits and program their feeding accordingly. Also, cattle eat more following a bad storm or a hot spell. At such times, the bunk may be "slick" for 2 to 3 hours and the cattle may line up waiting to be fed. When this happens, the ration should be increased. Also, by going to a higher roughage ration at these times, acidosis and laminitis can be minimized. The ability to recognize the "sign language of animals" and to change the feeding program accordingly is responsible for the oft-quoted statement that "the eye of the master fattens the cattle."

The maintenance requirements of animals increases as temperature, humidity, and air movement depart from the comfort zone. Likewise, the heat loss from animals is affected by these three factors. Animals adapt to weather as follows:

- **In cold weather,** the heating mechanisms are employed, including (1) increased insulation from growth of hair and more subcutaneous fat; (2) increase in thyroid activity; (3) seeking protective shelter and warming solar radiations (the animals sun themselves); (4) huddling together; (5) consumption of more feed, which increases the heat increment and warms animals; and (6) increasing activity. The most important cattle body heating mechanisms are amount of feed consumed and body activity, which are also evidenced in people. For example, after skiing in bitter cold weather, a skier feels comfortable after eating a beefsteak; and during a marathon race, a runner may feel quite warm when the temperature is near freezing (30°F).

- **In hot weather,** the cooling mechanisms are employed, including (1) moisture vaporization (from the skin and lungs), (2) avoidance of the heating solar radiation (the animals seek shade), (3) depression of thyroid activity, and (4) loafing (including lessening the production of meat).

THERMONEUTRAL ZONE (Comfort Zone)

Fig. 36-17 and the definitions that follow are pertinent to an understanding of thermal zones.

In Fig. 36-17 *heat production (metabolism)* is plotted against *ambient temperature* to depict the relationship between chemical and physical heat regulation.

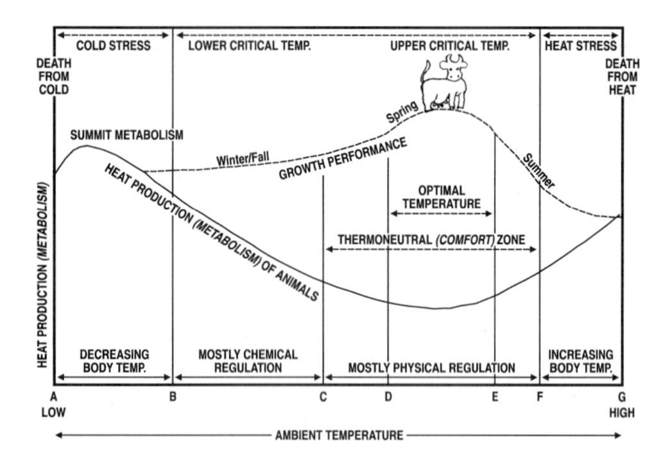

Fig. 36-17. Diagram showing (1) the influence of thermal zones and temperature on homeotherms (warm-blooded animals), and (2) the peak of meat and milk yields in the spring, followed by the summer slump due to high (hot) summer temperature and lignification of forage.

Note, too, the broad range of accommodation to low (cool) temperatures in contrast to the restricted range of accommodation to high (warm) temperatures. Definitions of terms pertaining to Fig. 36-17 follow.

- ***Thermoneutral (comfort) zone*** *(C to F) is the range in temperature within which the animal may perform with little discomfort, and in which physical temperature regulation is employed.*

- ***Optimum temperature*** *(D to E) is the temperature at which the animal responds most favorably, as determined by maximum production (gains, milk, wool, work, eggs) and feed efficiency.*

- ***Lower critical temperature*** *(B) is the low point of the cold temperature beyond which the animal cannot maintain normal body temperature.* The chemical temperature regulation is employed in the zone below C. When the environmental temperature reaches below point B, the chemical-regulation mechanism is no longer able to cope with cold, and the body temperature drops, followed by death. The French physiologist, Giaja, used the term *summit metabolism* (maximum sustained heat production) to indicate the point beyond which a decrease in ambient temperature causes the homeothermic mechanisms to break down, resulting in a decline in both heat production and body temperature and eventually death of the animal.

- ***Upper critical temperature*** *(F) is the high point on the range of the comfort zone, beyond which animals are heat stressed and physical regulation comes into play to cool them.*

The cow produces the maximum yield of milk during the spring when the temperature is optimum (D to E), and the minimum yield in the summer when it is hot (F to G).

The comfort zone, optimum temperature, and both upper and lower critical temperatures vary with different species, breeds, ages, body sizes, physiological and production status, acclimatizations, feed consumed (kind and amount), the activity of the animal, and the opportunity for evaporative cooling.

The temperature varies according to age, too. For example, the comfort zone of newborn lambs is 75 to 80°F, whereas the comfort zone of mature sheep is 45 to 75°F.

Animals that consume large quantities of roughage or high-protein feeds produce more heat during digestion; hence, they have a different critical temperature than the same animals fed a high-concentrate, moderate-protein ration. Because of this, experienced cattle feeders decrease the roughage and increase the concentrate of finishing cattle during the hot summer months.

Stresses of both high and low temperatures are increased with high humidity. The cooling effect of evaporating sweat is minimized and the respired air has less of a cooling effect. As humidity of the air increases, discomfort at any temperature, and nutrient utilization, decrease proportionately.

Air movement (wind) results in body heat being removed at a more rapid rate than when there is no wind. In warm weather, air movement may make the animal more comfortable, but in cold weather it adds to the stress temperature. At low temperatures, the nutrients required to maintain the body temperature are increased as the wind velocity increases. In addition to the wind, a drafty condition where the wind passes through small openings directly onto some portion or all of the animal body will usually be more detrimental to comfort and nutrient utilization than the wind itself.

ADAPTATION, ACCLIMATION, ACCLIMATIZATION, AND HABITUATION OF SPECIES/ BREEDS TO THE ENVIRONMENT

Every discipline has developed its own vocabulary. The study of adaptation/environment is no exception. So, the following definitions are pertinent to a discussion of this subject:

Adaptation refers to the adjustment of animals to changes in their environment.

Acclimation refers to the short-term (over days or weeks) response of animals to their immediate environment.

Acclimatization refers to evolutionary changes of

a species to a changed environment which may be passed on to succeeding generations.

Habituation is the act or process of making animals familiar with, or accustomed to, a new environment through use or experience.

Species differences in response to environmental factors result primarily from the kind of thermoregulatory mechanism provided by nature, such as type of coat (hair, wool, feathers), and sweat glands. Thus, hogs, which have a light coat of hair, are very sensitive to extremes of heat and cold. On the other hand, nature gave cattle an assist through growing more hair for winter and shedding hair for summer, with the result that they can withstand higher and lower temperatures than hogs. The long-haired, shaggy yak of Tibet and the wooly Scotch Highland cattle of Scotland are as cold tolerant as the arctic- dwelling caribou and reindeer.

From time to time, American buffalo *(bison bison)* and domestic beef cattle *(Bos taurus)* have been crossed to obtain a more hardy beast than cattle. The most publicized early work of this type was the development of the Cattalo (bison X domestic cattle), the initial cross for which was made at the Dominion Experiment Station, at Scott, Saskatchewan, in Canada, in 1915.

Fig. 36-18. Cattalo (¼ buffalo, ¾ domestic cattle) cow. The initial Cattalo breeding experiment was started by the Dominion Experimental Station, Scott, Saskatchewan, Canada, in 1915. The foundation herd consisted of 16 female and 4 male hybrids. (Courtesy, Research Station, Canada Department of Agriculture, Lethbridge, Alberta, Canada)

Male fertility and female reproductive rate have remained a problem in Cattalo. Although unquestionably hardy, bison X cattle crosses can be outperformed in nearly all environments by the currently available cattle breeds or crosses; and management procedures.

Also, there are breed differences, which make it possible to select animals well adapted to specific environments. Thus, the breeds of cattle that originated in the British Isles and Northern Europe are cold tolerant, whereas the Indian-evolved Zebu, or Brahman, cattle are heat tolerant.

Fig. 36-19. Brahman cattle are adapted to hot areas. (Courtesy, American Brahman Breeders Assn., Houston, TX)

In recent years, attempts have been made to combine the heat tolerance characteristics of tropical breeds with the high productive capacity of European stock. The best known of these planned beef breeds is Santa Gertrudis, developed on the famed King Ranch of Texas, in the early 1900s, which carry approximately ⅝ Shorthorn and ⅜ Brahman breeding.

FACILITIES

Optimum facility environments can only provide the means for animals to express their full genetic potential of production, but they do not compensate for poor management, health problems, or improper rations.

With the shift to confinement structures and high-density production operations, building design and environmental control became more critical. Limited basic research has shown that animals are more efficient—that they produce and perform better, and require less feed—if raised under ideal conditions of temperature, humidity, and ventilation. Properly designed barns and other shelters, shades, insulation, ventilation, and air-conditioning can be used to approach the environment that we wish.

However, the per head cost is much higher for environmentally controlled facilities. Thus, the decision on whether or not confinement and environmental control can be justified should be determined by economics.

There is still much to be learned about environmental control, but the gap between awareness and application is becoming smaller.

In hot climates, increased use is being made of shades, fans, sprinklers/sprayers/foggers, ventilation, and windbreaks.

Also, increasing attention needs to be given to other facility-related stress sources, such as space requirements, and the grouping of animals as affected by class, age, size, and sex.

HEALTH

Diseases and parasites (external and internal) are ever present animal environmental factors. Death takes a tremendous toll. Even greater economic losses in cattle result from lowered beef production, retarded growth of young stock, poor feed efficiency, and in labor and drug costs. The signs of good health are summarized earlier in this chapter in the section headed, "Normal Cattle Behavior"; hence, the reader is referred thereto.

Any departure from the signs of good health constitutes a warning of trouble. Most sicknesses are ushered in by one or more signs of poor health—by indicators that tell expert caretakers that all is not well—that tell them that their animals will go off feed tomorrow, and that prompt them to do something about it today.

Among the signs of cattle ill health are lack of appetite (the animal does not eat or graze normally); listlessness; droopy ears; sunken eyes; humped-up appearance; abnormal dung—either very hard or watery dung suggests an upset in the water balance or some intestinal disturbance following infection; abnormal urine (repeated attempts to urinate without success or off-colored urine should be cause for suspicion); abnormal discharges from the nose, mouth, and eyes, or a swelling under the jaw; unusual posture—such as standing with the head down or extreme nervousness; persistent rubbing or licking; hairless spots, dull hair coat, and dry, scurfy, hidebound skin; pale, red, or purple mucous membranes lining the eyes and gums; reluctance to move or unusual movements; higher than normal temperature; labored breathing—increased rate and depth; and altered social behavior such as leaving the herd and going off alone.

STRESS

Stress is defined as physical or psychological tension or strain. Stress of any kind affects animals. Among the external forces which may stress animals are previous nutrition, abrupt ration changes, change of water, space, level of production, number of animals together, changing quarters or mates, irregular care, transporting, excitement, presence of strangers, fatigue, previous training, illness, management, weaning, temperature, and abrupt weather changes.

Fig. 36-20. Heavily stressed yearling stocker cattle following a long shipment.

Separating a calf from its mother, followed in rapid succession by weaning, transporting many miles, going through an auction ring, and ending up in a corral without either milk or grass, where it is vaccinated, and sometimes even dehorned, is a traumatic experience and stress of the highest order. Such animals suffer from a lack of feed and water, digestive upsets, dehydration, and sometimes high fevers. The end result is lowered gains (usually shrinkage) and feed efficiency, illness, and sometimes even death. In order to get such environmentally stressed animals back on feed as soon as possible, it is important that the caretaker know what has happened to them and how to rectify the situation.

The weanling calf, being a ruminant, is dependent upon fermentation in the rumen for a major portion of its nutrients. Going without feed, whether voluntary or imposed, for 24 to 48 hours, will reduce rumen fermentation by as much as 70%. This is caused by a change in the ratio of rumen microorganisms which digest crude fiber, metabolize protein, and produce water-soluble vitamins. Thus, the ruminant loses its major source of energy, its ability to produce a balance of amino acids, and its primary source of vitamin C and the B vitamins. Since these vitamins are not stored in the body and are required on a daily basis, a vitamin deficiency develops very rapidly.

In addition to the rumen problem, the bacterial population of the lower digestive tract is altered, particularly if antibiotics have been administered to the stressed animal. This change paves the way for a rapid growth of *E. coli* and the development of scouring.

Thus, under stress, the animal suffers a severe malnutrition in terms of the major nutrients, along with an imbalance in blood electrolytes due to dehydration, acidosis, and scouring. This general breakdown in the normal state of the entire digestive system frequently results in the invasion of fungi, such as yeasts and molds.

Rations for animals under stress should be formulated to provide the best chance for the return of the digestive system to normalcy. To achieve this, the ration should provide a balance of nutrients ideally suited to the creation of normal rumen fermentation. This calls for the following:

1. **Restoration of fiber digestion.** The ideal feedstuffs for fiber-digesting organisms are those that contain highly digestible fiber, such as beet pulp, citrus pulp, soybean mill feed, and alfalfa meal. Also, *rumen stimulants*, such as those found in distillers' grains, molasses, and alfalfa meal, can be utilized to reduce the time required for normal rumen repopulation.
2. **High-quality proteins.** Since the ability to build amino acids in the usual way (by microorganisms) is reduced, nonprotein sources of nitrogen, such as urea, should be avoided. Thus, animals under stress should be fed oilseed or animal proteins.
3. **Restoration of flora in lower tract.** If left alone, the lower digestive tract will, in due time, restore its flora. But, since a speedy return to normal production is desired, some assist is usually desirable. The natural flora of the lower digestive tract include a large population of *Lactobacilli*, the presence of which creates a natural regulation of *E. coli* and a normal state of health in the lower digestive tract. The key to restoring the *Lactobacilli* is the use of organisms that will implant and grow in the tract.

Thus, the development of nutritional programs for animals under stress calls for the proper formulation of rations to correct the damage done by the stress. Since animals experience many stresses, the development of nutritional programs for periods of stress should receive as much attention as the medical treatment of ills. Also, and most important, stress should be kept to a minimum. For example, replacement heifers should be dehorned, vaccinated, treated for parasites, and on feed prior to movement to a new location.

In the life of an animal, some stresses are normal, and they may even be beneficial—they can stimulate favorable action on the part of an individual. Thus, we need to differentiate between stress and distress. Distress—not being able to adapt—is responsible for harmful effects. The trick is to manage stress so that it does not become distress and cause damage, and to recognize the warning signals of distress.

The principal criteria used to evaluate, or measure, the well-being or stress of people are: increased blood pressure, increased muscle tension, body temperature, rapid heart rate, rapid breathing, and altered endocrine gland function. In the whole scheme, the nervous system and the endocrine system are intimately involved in the response to stress and the effects of stress.

The principal criteria used to evaluate, or measure, the well-being or stress of cattle are: growth rate or production, efficiency of feed use, efficiency of reproduction, body temperature, pulse rate, breathing rate, mortality, and morbidity. Other signs of cattle well-being, any departure from which constitutes a warning signal are: contentment, alertness, eating with relish, cudding, sleek coat and pliable, elastic skin, bright eyes and pink eye membranes, and normal feces and urine.

Stress is unavoidable. Wild animals were often subjected to great stress; there were no caretakers to modify their weather, often their range was overgrazed, and sometimes malnutrition, predators, diseases, and parasites took a tremendous toll.

Domestic animals are subjected to different stresses than their wild ancestors, especially to more restricted areas and greater animal density. However, in order to be profitable, their stresses must be minimal.

CONTROL POLLUTION

Fig. 36-21. A well-managed range provides little pollution potential. (Courtesy, American Shorthorn Assn., Omaha, NE)

Pollution is the major issue of the decade. Anything that defiles, desecrates, or makes impure or unclean the surroundings pollutes the environment and can have a detrimental effect on animal health and performance. Thus, gases, odorous vapors, and dust particles from animal wastes (feces and urine) in buildings directly affect the quality of the environment. Muddy lots and stray electrical voltage may also pollute the environment. For healthy and productive animals, each of these pollutants must be maintained at an acceptable level. The most troublesome cattle pollutant is manure.

Before constructing a confinement beef cattle facility, the owner should become familiar with both state and federal regulations. The state regulations can be secured from the state water board. They differ from state to state, but most states require a catch basin (detention pond) sufficient to contain the runoff from a storm of the magnitude of the largest rainfall during a 48-hour period of the most recent ten years.

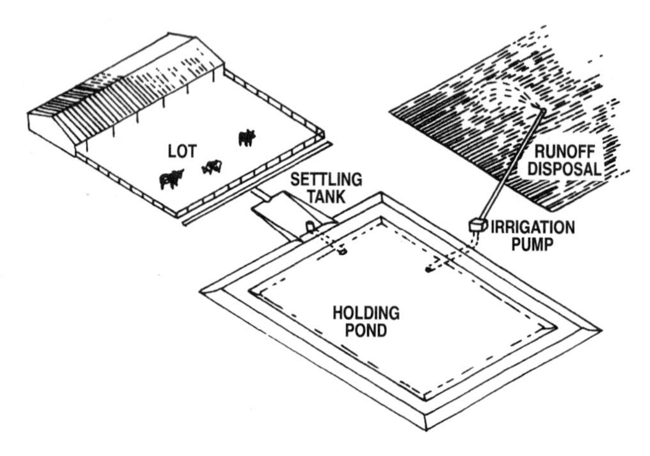

Fig. 36-22. A typical corral runoff pollution control system. Solids are collected in the settling tank, then hauled. Liquid runoff is stored in a holding pond, then returned to the land through the irrigation system.

Note: The subject of pollution control (including manure, dust, and odors), along with pertinent information about pollution laws and regulations, is fully covered in Chapter 32, Feedlot Pollution Control; hence, the reader is referred thereto.

ANIMAL WELFARE/ ANIMAL RIGHTS

In recent years, the behavior and environment of animals in confinement have come under increased scrutiny of animal welfare/animal rights groups all over the world. For example, in 1987 Sweden passed legislation designed (1) to phase out layer cages as soon as a viable alternative can be found; (2) to discontinue the use of sow stalls and farrowing crates; (3) to provide more space and straw bedding for slaughter hogs; and (4) to forbid the use of genetic engineering, growth hormones, and other drugs for farm animals except for veterinary therapy. Also, the law provides for fining and imprisoning violators.

Animal welfarists see many modern practices as unnatural, and not conducive to the welfare of animals.

Fig. 36-23. Animal welfare and animal profits are on the same side of the ledger. (Courtesy, American Milking Shorthorn Society, Beloit, WI)

In general, they construe animal welfare as the well-being, health, and happiness of animals; and they believe that certain intensive production systems are cruel and should be outlawed. The animal rightists go further; they maintain that humans are animals, too, and that all animals should be accorded the same moral protection. They contend that animals have essential physical and behavioral requirements, which, if denied, lead to privation, stress, and suffering; and they conclude that all animals have the right to live.

Cattle producers know that the abuse of animals in intensive/confinement systems leads to lowered production and income—a case in which decency and profits are on the same side of the ledger. They recognize that husbandry that reduces labor and housing costs often results in physical and social conditions that increase animal problems. Nevertheless, means of reducing behavioral and environmental stress are needed so that decreased labor and housing costs are not offset by losses in productivity. The welfarist/rightists counter with the claim that the evaluation of animal welfare must be based on more than productivity; they believe that there should be behavioral, physiological, and environmental evidence of well being, too. And so the arguments go!

To all animal caretakers, the principles and application of animal behavior and environment depend on understanding; and on recognizing that they should provide as comfortable an environment as feasible for their animals, for both humanitarian and economic reasons. This requires that attention be paid to environmental factors that influence the behavioral welfare of their animals as well as their physical comfort, with emphasis on the two most important influences of all in animal behavior and environment—feed and confinement.

Animal welfare issues tend to increase with urbanization. Moreover, fewer and fewer urbanites have farm backgrounds. As a result, the animal welfare gap between town and country widens. Also, both the news media and the legislators are increasingly from urban centers. It follows that the urban views that are propounded will have greater and greater impact in the years ahead.

FOOD SAFETY AND DIET/ HEALTH CONCERNS

Fig. 36-24. Pizza. America's food supply is the safest in the world. (Courtesy, National Live Stock and Meat Board, Chicago, IL)

Many food safety and diet/health concerns are unwarranted. American consumers are prone to overreact to rumors relative to their food. They care little about what they put on their backs, but they are greatly concerned about what goes into their stomachs.

America's food supply is the safest in the world! Nevertheless, there is need for constant vigilance and improvement, especially in animal products which are subject to all the hazards of other foods (spoilage, pesticides, toxicities), plus being capable of transmitting, or serving as passive carriers, of certain diseases to humans.

Because the welfare of the nation is dependent upon the health of its people, dairy (and other) products are carefully monitored by various government agencies to assure consumers that they are wholesome and safe; and because of recognizing the importance of consumers in the safety of their food, the private sector may do additional testing. The agencies most responsible for this important work are:

1. The U.S. Department of Health and Human Services, including the following agencies: The Center for Disease Control, the Food and Drug Administration (FDA), and the National Institute of Health.
2. The U.S. Department of Agriculture, including the following agencies: the Agricultural Research Service, the Animal and Plant Health Inspection Service, the Cooperative State Research Service, the Federal Extension Service, the Labeling and Registration Section, and the Veterinary Service Division.
3. State and local government agencies.
4. International organizations engaged in health and/or nutrition activities, including the World Health Organization (WHO), and Food and Agriculture Organization (FAO).
5. Private industry groups such as the National Dairy Council.
6. Professional organizations, including dentists, dietitians, doctors, health educators, nurses, and public health workers.
7. Food processors and retailers.

Note: The subject of beef safety is fully covered in Chapter 16, Beef and Veal, and Byproducts From Cattle Slaughter, in the section and subsections headed "What's Ahead For Beef," "Food Safety," and "Food Safe Labeling"; hence, the reader is referred thereto.

QUESTIONS FOR STUDY AND DISCUSSION

1. Define animal behavior.

2. Why has increased confinement of animals made for great interest in the subject of animal behavior?

3. Why has there been a genetic time lag; why have cattle producers altered the environment faster than the genetic makeup of animals?

4. Define each of the following causes of animal behavior: (a) genetic, (b) simple learning, (c) complex learning. How do each of these causes of animal behavior affect cattle?

5. List the nine general functions or behavior systems that cattle exhibit.

6. Discuss the following behavioral systems as they pertain to cattle:

   a. Care-giving and care-seeking behavior.
   b. Eliminative behavior.
   c. Ingestive behavior.
   d. Sexual behavior.

7. How is gregarious behavior related to allelomimetic behavior?

8. Describe the social organization in a cattle herd on pasture from the standpoints of (a) dominance and (b) spacing.

9. Describe the consistent order in which dairy cattle travel when moving from the corral to the pasture.

10. How is the dominance hierarchy established when several cows are brought together to form a herd? How is it maintained?

11. Explain the difference between dominance and leader-follower.

12. Discuss the importance of each of the following social relationships of cattle: (a) interspecies relationships and (b) people-animal relationships.

13. Discuss how cattle communicate with each other, and the importance of each method.

14 Those who care for cattle need to be familiar with behavioral norms in order to detect and treat abnormal situations—especially illness. Describe a normal cow. Describe (a) sight and (b) sleep of cattle.

15. Discuss each of the following abnormal behaviors of cattle: (a) homosexual, (b) the *mean bull* complex, and (c) pica.

16. List and discuss the significance of one example of the practical application of cattle behavior in each of the following areas:

   a. Breeding for adaptation.
   b. Quick adaptation-early training.
   c. Loading chute and corral design.
   d. Manure elimination.
   e. Companionship.

17. Discuss how each of the following environmental factors affects cattle:

   a. Feed.
   b. Water.
   c. Weather.
   d. Adaptation, acclimation, acclimatization, and habitation of species/breeds to the environment.
   e. Facilities.
   f. Health.
   g. Stress.

18. Define pollution. Why has pollution become such a great issue?

19. Discuss the animal welfare/animal rights issue.

20. Discuss food safety and diet/health concerns relative to beef.

SELECTED REFERENCES

Title of Publication	Author(s)	Publisher
Behavior of Domestic Animals, The, Third Edition	Ed. by E. S. E. Hafez	The Williams and Wilkens Company, Baltimore, MD, 1975
Bibliography of Livestock Waste Management	J. R. Miner D. Bundy G. Christenbury	Office of Research and Monitoring, U.S. Environmental Protection Agency, Washington, DC, 1972
Biology of Stress In Farm Animals: an integrative approach	P. R. Wiepkema P. W. M. van Adrichem	Kluwer Academic Publishers, Hingham, MA, 1987
Concise Survey of Animal Behavior, A	E. K. Honore P. H. Klopfer	Academic Press, Inc., Harcourt Brace Jovanovich, Publishers, San Diego, CA, 1990
Development and Evolution of Behavior	Ed. by L. R. Aronson, *et al.*	W. H. Freeman and Company, San Francisco, CA, 1970
Domestic Animal Behavior	J. V. Craig	Prentice-Hall, Inc., Englewood Cliffs, NJ, 1981
Effect of Environment on Nutrient Requirements of Animals	D. R. Ames, Chairman	NRC, National Academy Press, Washington, DC, 1981
Environmental and Functional Engineering of Agricultural Buildings	H. J. Barre L. L. Sammet G. L. Nelson	Van Nostrand Reinhold Co., New York, NY, 1988
Environmental Biology	P. L. Altman D. S. Dittmer	Federation of American Societies for Experimental Biology, Bethesda, MD, 1966
Environmental Control for Agricultural Buildings	M. L. Esmay J. E. Dixon	The AVI Publishing Company, Inc., Westport, CT, 1986
Environmental Management in Animal Agriculture	S. E. Curtis	Animal Environment Services, Mahomet, IL, 1981
Ethology, The Biology of Behavior, Second Edition	I. Eibl-Eibesfeldt	Holt, Rinehart and Winston, New York, NY, 1975
Farm Animal Manures: an overview of their role in the agricultural environment	J. Azevedo P. R. Stout	Agricultural Publications, University of California, Berkeley, 1974
Guide to Environmental Research on Animals, A	R. G. Yeck, Chairman	NRC, National Academy of Science, Washington, DC, 1971
Health Issues Related to Chemicals in the Environment: A Scientific Perspective	A. L. Craigmill, Chairman	Council for Agricultural Sciences and Technology, Ames, IA, 1987
Impact of Stress Proceedings, The,	Ed. by R. E. Moreng J. R. Herbertson	Colorado State University, Ft. Collins, 1986
Introduction to Animal Behavior, An: ethology's first century, Second Edition	P. H. Klopfer	Prentice-Hall, Inc., Englewood Cliffs, NJ, 1974
Livestock Behavior, a practical guide	R. Kilgour C. Dalton	Westview Press, Boulder, CO, 1984
Livestock Environment, Proceedings, Second International Livestock Environment Symposium	D. S. Bundy, Planning Chairman	American Society of Agricultural Engineers, St. Joseph, MI, 1982

Mechanisms of Animal Behavior	P. Marler W. J. Hamilton	John Wiley & Sons, New York, NY, 1966
Organic Farming: current technology and its role in a sustainable agriculture	Ed. by D. M. Kral	American Society of Agronomy, Madison, WI, 1984
Principles of Animal Behavior	W. N. Tavolga	Harper & Row, New York, NY, 1969
Principles of Animal Environment	M. L. Esmay	The AVI Publishing Company, Inc., Westport, CT, 1978
Readings in Animal Behavior	Ed. by T. E. McGill	Holt, Rinehart and Winston, New York, NY, 1973
Safe and Effective Use of Pesticides, The	P. J. Marer	University of California Publications, Oakland, 1988
Scientific Aspects of the Welfare of Food Animals	F. H. Baker, Chairman	Council for Agricultural Science and Technology, Ames, IA, 1981
Social Hierarchy and Dominance	Ed. by M. W. Schein	Dowden, Hutchinson & Ross, Inc., Stroudsburg, PA, 1975
Social Space for Domestic Animals	Ed. by R. Zayan	Kluwer Academic Publishers, Hingham, MA, 1985
Social Structure in Farm Animals	G. J. Syme L. A. Syme	Elsevier Scientific Publishing Co., Amsterdam, The Netherlands, 1979
Stress Physiology in Livestock	M. K. Yousef	CRC Press, Inc., Boca Raton, FL, 1985
Structures and Environment Handbook		Midwest Plan Service, Iowa State University, Ames, IA, 1972
Utilization, Treatment, and Disposal of Waste on Land, Proceedings	E. C. A. Runge, President of Society	Soil Science Society of America, Inc., Madison, WI, 1986

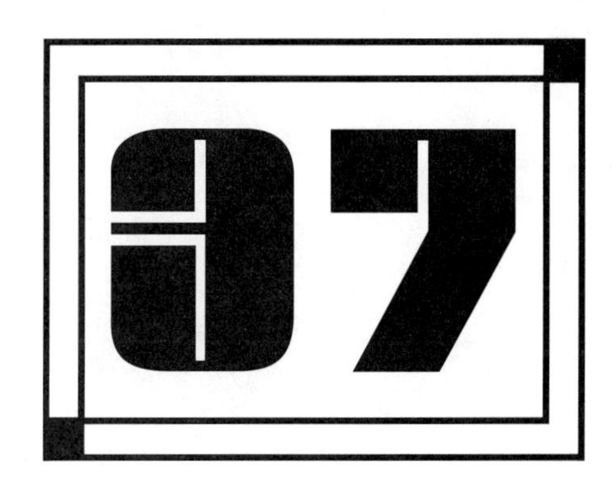

GLOSSARY OF CATTLE TERMS

A mark of distinction of good cattle producers is that they *speak the language*—they use the correct terms and know what they mean. Even though cattle terms are used glibly by people in the cattle business, often they are baffling to the newcomer.

Many terms that are defined or explained elsewhere in this book are not repeated in this chapter. Thus, if a particular term is not listed herein, the reader should look in the Index or in the particular chapter and section where it is discussed.

A

ABOMASUM. The fourth compartment of the ruminant stomach where enzymatic digestion occurs. It is often called the *true stomach.*

ABORTION. The expulsion or loss of the fetus before the completion of pregnancy, and before it is able to survive.

ABSORPTION. The process of being passed into or through, by osmosis or other processes. For example, nutrients are absorbed into the bloodstream through the wall of the gastrointestinal tract.

ACCLIMATIZATION. Complex of processes becoming accustomed to a new climate or other environmental conditions.

ACCURACY (OF SELECTION). Correlation between true breeding value and estimated breeding value.

ACIDOSIS. A high-acid condition in the rumen caused by rapid consumption or overconsumption of readily fermentable feed; it may cause digestive disturbance and/or death.

ACUTE. Having a rapid onset and pronounced signs of disease.

AD LIB FEEDING. Free-choice access to feed.

ADAPTATION. Adjustment of an organism to a new or changing environment.

ADDITIVE. An ingredient or a substance added to a basic feed mix, usually in small quantities, for the purpose of fortifying it with certain nutrients, stimulants, and/or medicines.

ADIPOSE TISSUE. Fatty tissue.

ADJUSTED WEANING WEIGHT. Weight adjusted to 205 days of age and for age of dam.

ADJUSTED YEARLING WEIGHT. Weight adjusted to 365, 452, or 550 days of age and for age of dam.

AEROBE. In the presence of air. The term usually applied to microorganisms that require oxygen to live and reproduce.

AEROBIC BACTERIA. Bacteria that require free elemental oxygen for growth. Oxygen in chemical combination will not support aerobic organisms.

AEROBIC DECOMPOSITION. Reduction of the net energy level of organic matter by aerobic microorganisms.

AFTERBIRTH. The placenta and fetal membranes expelled from the uterus at parturition. Cattle producers often refer to the expulsion of these membranes as *cleaning.*

AGONISTIC BEHAVIOR. Combat or fighting behavior.

AGRICULTURAL WASTES. Most such wastes are associated with the production of food and fiber on farms, ranges, and forests. These wastes normally include animal manure, crop residues, and dead animals. Agricultural chemicals, fertilizers, and pesticides which find their way into the soil and subsequently into the surface and subsurface water also are classified as agricultural wastes.

AI. Abbreviation for artificial insemination.

AIR-DRY ***(Approximately 90% dry matter).*** This refers to feed that is dried by means of natural air movement usually in the open. It may be either an actual or an assumed dry matter content; the latter is approximately 90%. Most feeds are fed in the air-dry state.

ALGAE. Primitive plants, one or many-celled, usually aquatic and capable of synthesizing their foodstuffs by photosynthesis.

ALKALI. It usually refers to a soluble salt or a mixture of soluble salts present in some soils of arid or semiarid regions in quantity detrimental to ordinary agriculture.

ALKALINITY. The capacity of water to neutralize acids, a property imparted by the water's content of carbonates, bicarbonates, hydroxides, and, occasionally, borates, silicates, and phosphates. It is expressed in milligrams per liter of equivalent calcium carbonate.

ALLELES. Alternative forms of a gene, which are located at the same point on each of a pair of chromosomes. One gene of each pair may have one effect and another gene of that same pair (allele) may have a different effect on the same trait.

ALLELOMIMETIC BEHAVIOR. Doing the same thing.

ALLERGY. A hypersensitivity to a particular substance which causes a body reaction such as hives, sneezing, labored breathing, etc., upon re-exposure to the substance.

AMBIENT TEMPERATURE. The prevailing or surrounding environmental temperature.

AMERICAN MEAT INSTITUTE (AMI). Association of meat-packing and processing companies. *The address:* P.O. Box 3556, Washington, DC 20007.

AMERICAN SOCIETY OF ANIMAL SCIENCE. A society of persons interested in animal science and livestock production. *The address:* 309 W. Clark Street, Champaign, IL 61820-4690.

AMERICAN VETERINARY MEDICAL ASSOCIATION. A professional organization of veterinarians and others who are interested in animal health. *The address:* 930 North Meacham Road, Schaumburg, IL 60196-0001.

AMINO ACIDS. Nitrogen-containing compounds that constitute the *building blocks* or units from which proteins are formed. They contain both an amino (NH_2) group and a carboxyl (COOH) group.

ANABOLISM. The conversion of simple substances into more complex substances by living cells (constructive metabolism).

ANAEROBE. A microorganism that normally does not require air or free oxygen to live and reproduce.

ANAEROBIC BACTERIA. Bacteria not requiring the presence of free or dissolved oxygen for metabolism. Strict anaerobes are hindered or completely blocked by the presence of dissolved oxygen and sometimes by the presence of highly oxidized substances, such as nitrates, nitrites, and, perhaps, sulfates. Facultative anaerobes can be active in the presence of dissolved oxygen, but do not require it.

ANAEROBIC DECOMPOSITION. Reduction of the net energy level and change in chemical composition of organic matter caused by microorganisms in an anaerobic environment.

ANATOMY. The science of the structure of the animal body and the relation of its parts.

ANEMIA. A deficiency in the quantity of blood or of one of its constituents—red blood cell count, hemoglobin concentration, or packed cell volume.

ANESTRUS. The nonbreeding season. Females not in heat (estrus).

ANIMAL PROTEIN. Protein derived from meat packing or rendering plants, surplus milk or milk products, or marine sources. It includes proteins from meat, milk, poultry, eggs, fish, and their products.

ANIMAL UNIT. Common animal denominator based on feed consumption. One mature cow, with or without an unweaned calf at side, represents one animal unit.

ANIMAL UNIT MONTH (AUM). Amount of feed or forage required to maintain one animal unit for one month.

ANOREXIA. A lack or loss of appetite for feed.

ANOXIA. A lack of oxygen in the blood or tissues. This condition may result from various types of anemia, reduction in the flow of blood to tissues, or lack of oxygen in the air at high altitudes.

ANTE MORTEM. Before death.

ANTHELMINTIC. A drug or chemical that kills or expels worms.

ANTIBIOTIC. A substance that destroys or inhibits the growth or action of microorganisms.

ANTIBODY. Protein produced by the body and carried in the blood that provides protection against a specific disease by interacting with the disease-causing agent and neutralizing it.

ANTIGEN. An enzyme, a toxin, or another *foreign* substance (usually protein) to which an animal body reacts by producing antibodies specific to the invading antigen.

ANTIHISTAMINES. A drug used to neutralize and treat allergic conditions in the body.

ANTIOXIDANT. A compound that prevents oxidative rancidity of polyunsaturated fats. Antioxidants are used to prevent rancidity in feeds and foods.

ANTISEPTIC. A chemical substance used on living tissue

that prevents the growth and development of microorganisms.

APPETITE. The immediate desire to eat when food is present. Loss of appetite in an animal is usually caused by illness or stress.

ARTIFICIAL INSEMINATION (AI). The deposition of spermatozoa in the female genitalia by artificial rather than by natural means.

ASEPTIC. Refers to something being free from pathogenic microorganisms.

AS-FED. This refers to feed as normally fed to animals. It may range from 0 to 100% dry matter.

ASH. The mineral matter of a feed. The residue that remains after complete incineration of the organic matter.

ASSAY. Determination of (1) the purity or potency of a substance, or (2) the amount of any particular constituent of a mixture.

ASSIMILATION. A physiological term referring to a group of processes by which the nutrients in feed are made available to and used by the body; includes digestion, absorption, distribution, and metabolism.

ASTRINGENT. A substance that causes shrinking or contraction in localized blood vessels, thereby reducing blood flow and discharge.

ATROPHY. A wasting away of a part of the body, usually muscular, induced by injury or disease.

AUTOPSY. Postmortem examination in which the body is dissected to determine cause of death.

AUTOSOMES. All chromosomes except the sex chromosomes.

AVERAGE DAILY GAIN (ADG). The average daily liveweight increase of an animal.

AVOIRDUPOIS WEIGHTS AND MEASURES. *Avoirdupois* is a French word, meaning *to weigh.* The old English system of weights and measures is referred to as the Avoirdupois System, or U.S. Customary Weights and Measures, to differentiate it from the Metric System.

B

BACKCROSS. The mating of a crossbred (F_1) animal to one of the parental breeds.

BACKFAT PROBE. Device for measuring the thickness of fat over the animal's back (usually at the twelfth to thirteenth rib).

BACKGROUNDING. An old practice with a new emphasis and a new name. Actually backgrounding and the stocker stage are one and the same. Both refer to that period in the life of a calf from weaning to around 800 lb, when it is ready to go on a high-energy finishing ration.

BACTERIA. Microscopic, single-cell plants, lacking chlorophyll, found in most environments, often referred to as microbes; some are beneficial, others are capable of causing disease.

BACTERICIDE. A product that destroys bacteria.

BACTERIN. A suspension of killed bacteria (vaccine) used to increase disease resistance.

BALANCED RATION. A ration which provides an animal the proper proportions and amounts of all the required nutrients for a period of 24 hours.

BARREN. A cow that is not in calf may be referred to as being barren.

BASAL METABOLIC RATE (BMR). The heat produced by an animal during complete rest (but not sleeping) following fasting, when using just enough energy to maintain vital cellular activity, respiration, and circulation, the measured value of which is called the basal metabolic rate (BMR).

BASIS. The difference or spread between the cash price of a commodity at a particular cash market and the price of a futures contract. This spread differs from one market to another and changes with time.

BASIS MOVEMENT. The change which occurs in particular cash-futures price relationship. It is the change in basis that determines the success or failure of a hedge, rather than changes in market price. One should always hedge according to basis rather than price.

BEAR MARKET. A downward moving or lower market is considered "bearish," because the bear strikes down its victim.

BEEF. Meat from cattle (bovine species) older than calves.

BEEF CHECKOFF PROGRAM. The Beef Promotion and Research Act passed in 1985 made it possible for the beef industry to establish a checkoff of $1.00/head to fund promotion and research programs. So, beginning October 1, 1986, $1.00/head has been collected on all cattle sold.

BEEF IMPROVEMENT FEDERATION (BIF). A federation or an organization of persons interested or involved in performance evaluation of beef cattle. The purposes of BIF are to bring about uniformity of procedures, development of programs, cooperation among interested entities, education of its members and its ultimate consumers, and confidence of the beef industry in the principles and potentials of performance testing.

BID. A bid subject to immediate acceptance made on the

floor of an exchange to buy a definite quantity of a commodity future at a specified price.

BIODEGRADATION *(Biodegradability).* The destruction or mineralization of either natural or synthetic organic materials by the microorganisms populating soils, natural bodies of water, or wastewater-treatment systems.

BIOLOGICAL VALUE OF A PROTEIN. The percentage of the protein of a feed or feed mixture which is usable as a protein by the animal. Thus, the biological value of a protein is a reflection of the kinds and amounts of amino acids available to the animal after digestion. A protein that has a high biological value is said to be of *good quality.*

BIOTECHNOLOGY. The application of biological science, particularly molecular biology, to economically and socially important problems.

BIRTH WEIGHT (BW). The weight of a calf taken within 24 hours after birth. Heavy birth weights tend to be correlated with calving problems, but the conformation of the calf and that of the cow are contributing factors.

BLOAT. A digestive disorder of ruminants usually characterized by an abnormal accumulation of gas in the rumen, usually seen on the animal's upper left side.

BLOCKY. A term which refers to an animal that is deep, wide, and low-set.

BLOOM.

- Said of an animal that has beauty and freshness. An animal in bloom has a glossy hair coat and presents an attractive appearance.
- The condition or time of flowering.

BOLUS.

- Regurgitated food that has been chewed and is ready to be swallowed.
- A large pill for animals.

BOMB CALORIMETER. An instrument used to measure the gross energy content of any material, in which the feed (or other substance) tested is placed and burned in the presence of oxygen.

BONING. A term used to describe the process by which hairs on the legs of a beef animal are pulled upward. *Boning* makes the animal appear larger-boned and is generally accomplished by using a bar of glycerine-base soap.

BOVINE. Refers to a general family grouping of cattle.

BOVINE VIRAL DIARRHEA (BVD). Viral disease in cattle that can cause diarrhea, lesions of the digestive tract, repeat breeding, abortion, mummification, and congenital defects.

BOXED BEEF. Prefabricated primal and sub-primal cuts that are delivered to retail stores in boxes.

BRAND. A mark used as a means of identification.

BRAND NAME. Any word, name, symbol, or device, or any combination of these, often registered as a trademark or name, which identifies a product and distinguishes it from others.

BREAK. A more or less sharp price decline.

BREAK-EVEN. A point in the accounting period when total accumulated revenues are equal to total variable expenses to that particular date, plus fixed expenses for the year.

BRED.

- Refers to an animal that is pregnant.
- Sometimes used synonymously with the term *mated.*

BREECH BIRTH OR DELIVERY. The birth of an animal with the buttocks or rear feet first rather than the front feet and head first.

BREED.

- Animals that are genetically pure enough to have similar external characteristics of color and conformation, and when mated together produce offspring with the same characteristics.
- The mating of animals.

BREED TYPE. The combination of characteristics that makes an animal better suited for a specific purpose.

BREEDER. Owner of the dam at the time of service who was responsible for the selection of the sire to which she was mated.

BRITISH BREEDS. Breeds of cattle such as Angus, Hereford, and Shorthorn native to Great Britain.

BRITISH THERMAL UNIT (Btu). The amount of energy required to raise 1 lb of water 1°F; equivalent to 252 calories.

BRIX. A term commonly used to indicate the sugar (sucrose) content of molasses. It is expressed in degrees and was originally used to indicate the percentage by weight of sugar in sucrose solutions, with each degree Brix being equal to 1% sucrose.

BROCKLE-FACED. White-faced with other colors splotched on the face and head.

BROKER. An agent who handles the execution of all trades.

BUCK-KNEED. Standing with the knees too far forward.

BUDGET. A projection of records and accounts and a plan for organizing and operating ahead for a specific period of time.

BUFFER. A substance in a solution that makes the degree of acidity (hydrogen-ion concentration) resistant to change when an acid or a base is added.

BULL. An uncastrated male bovine.

BULL MARKET. An upward moving or higher market is considered *bullish*, because the bull tosses his victim upward on impaled horns.

BULLOCK. Young bull, typically less than 20 months of age.

BULL TESTING. Method for evaluating the breeding soundness or performance of beef bulls.

BULLING HEIFER OR COW. A heifer or cow that is in estrus (heat).

BYPRODUCT. Products of considerably less value than the major product—beef. For example, the hide and offal are byproducts.

BYPRODUCT FEEDS. The innumerable roughages and concentrates obtained as secondary products from plant and animal processing, and from industrial manufacturing.

C

CAESAREAN SECTION. The surgical procedure of taking an unborn animal, at or near parturition, from the uterus by cutting through the abdominal and uterine wall. Caesarean sections are required when the pelvic opening is too small to accommodate the passage of the fetus or when the fetus is hopelessly malpositioned.

CAKE (*Presscake*). The mass resulting from the pressing of seeds, meat, or fish in order to remove oils, fats, or other liquids.

CALCIFICATION. The process by which organic tissue becomes hardened by a deposit of calcium salts.

CALF. The sexually immature young of cattle.

CALF CROP. The number or percentage of calves produced within a herd in a given year relative to the number of cows and heifers exposed to breeding.

CALF-KNEED. Standing with knees too far back; directly opposite to buck-kneed or knee-sprung. This condition causes more trouble than knee-sprung or buck-kneed.

CALORIC. Pertaining to heat or energy.

CALORIE. Amount of heat required to raise the temperature of 1 g of water 1°C (precisely from 16.5°C to 17.5°C).

CALORIMETER. An instrument for measuring the amount of energy.

CALVING. The act of giving birth.

CALVING DIFFICULTY *(Dystocia)*. Abnormal or difficult labor, causing difficulty in delivering the fetus and placenta.

CALVING SEASON. The season(s) of the year when the calves are born.

CARCASS. The dressed body of a meat animal, the usual items of offal having been removed.

CARCASS MERIT. The desirability of a carcass relative to quantity and quality of product.

CARCASS QUALITY GRADE. An estimate of palatability based primarily on marbling and maturity and generally to a lesser extent on color, texture, and firmness of lean.

CARCASS WEIGHT. The weight of the carcass of an animal following slaughter, as it hangs on the rail, expressed either as warm (hot) or chilled (cold) carcass weight.

CARCASS YIELD. The carcass weight as a percentage of the liveweight.

CARRIER.

- A disease-carrying animal.
- A heterozygous individual with one recessive gene and one dominant gene for a specific trait.
- An edible material to which ingredients are added to facilitate their uniform incorporation into feeds.

CARRYING CAPACITY. The number of animal units (one cow, plus suckling calf—if there is a calf; or one heifer two years old or over) a property will carry on a year-round basis.

CASE READY. Beef cuts received by the retailer that do not require further processing before they are put in the retail case for selling.

CASH *(Spot)*. The cash price refers to the price of live animals, not to a futures contract. Also, known as *spot commodity*.

CASH MARKET. Cattle bought and sold for immediate delivery. Also known as *spot market.*

CASTRATE.

- To remove the testicles.
- An animal that has had its testicles removed.

CASTRATING. The unsexing of a male animal.

CATTLE-FAX. Nonprofit marketing organization governed by cattle producers. Market analysis and information are provided to members by a staff of market analysts. *The address*: Cattle Fax, 5420 South Quebec Street, Englewood, CO 80111-1904.

CENTIGRADE (C). A means of expressing temperature. To convert to Fahrenheit, multiply by 9/5 and add 32.

CERTIFIED ANGUS BEEF. Branded beef product supplied by purebred cattle, or from cattle with Angus-type characteristics provided they meet the evaluation of the USDA Grading Service based on marbling, maturity, and grade.

CERTIFIED HEREFORD BEEF. This includes two branded beef products: (1) Certified Hereford Beef Supreme (Choice grade) and (2) Certified Hereford Beef Special (Upper Select grade). Certified Hereford Beef must come from predominantly (51%) Herefords, or from animals which exhibit markings of the Hereford breed.

CERVIX. The thick-walled structure of the reproductive tract located between the vagina and the uterus. The cervix is considered the opening to or the neck of the uterus.

CFTC. The Commodity Futures Trading Commission, the independent federal agency created by Congress to regulate commodity futures trading. The CFTC Act of 1974 became effective April 21, 1975. Previously, futures trading had been regulated by the Commodity Exchange Authority of the USDA.

CHEMICAL OXIDATION. Oxidation of organic substances without benefit of living organisms. Examples are by thermal combustion or by oxidizing agents such as chlorine.

CHICAGO BOARD OF TRADE. The Chicago Board of Trade, which was founded in 1948, handles futures trading in such commodities as wheat, corn, oats, soybeans, and soybean oil and meal.

CHICAGO MERCANTILE EXCHANGE. It was founded in 1919. The Chicago Mercantile Exchange handles trading in such commodities as fat cattle, feeder cattle, hogs, and pork bellies.

CHLORINATION. The application of chlorine to water, sewage, or industrial wastes, generally for the purpose of disinfection, but frequently for accomplishing other biological or chemical results.

CHOLESTEROL. A white, fat-soluble substance found in animal fats and oils, bile, blood, brain tissue, nervous tissue, the liver, kidneys, and adrenal glands. It is important in metabolism and is a precursor of certain hormones. Some have implicated cholesterol as a factor in arteriosclerosis.

CHROMOSOME. Chromosomes are long DNA molecules on which genes (the basic genetic codes) are located. Domestic cattle have 30 pairs of chromosomes.

CHRONIC. Refers to a condition that is continuous and long-lasting.

CLEANING. See **AFTERBIRTH**.

CLITORIS. A reproductive organ in the female that is the homologue of the penis of the male.

CLONE. A group of genetically identical organisms derived from a single individual by various types of asexual reproduction.

CLOSEBREEDING. Mating of closely related animals; such as sire to daughter, son to dam, and brother to sister.

CLOSED HERD. A herd in which no outside breeding stock (cattle) are introduced.

COARSE. Lacking in quality—evidenced in texture of hair and all-over lack of refinement.

COD. The small bag or pouch that remains in a steer after it is castrated; the remnants of the scrotum.

COEFFICIENT OF DIGESTIBILITY. The percentage value of a feed nutrient that is absorbed. For example, if a feed contains 10 g of nitrogen and it is found that 9.5 g are absorbed, the digestibility is 95%.

COLLATERAL RELATIVES. Relatives of an individual which are not its ancestors or descendants. Sibs are an example of collateral relatives.

COLOSTRUM. The milk secreted by mammalian females for the first few days before and following parturition, which is high in antibodies and laxative.

COMBUSTION. The combination of substances with oxygen accompanied by the liberation of heat.

COMMERCIAL FEEDS. Feeds mixed by manufacturers who specialize in the feed business.

COMMISSION. The charge made by a broker for buying or selling a cattle futures contract.

COMMISSION HOUSE OR BROKERAGE HOUSE. A firm which buys and sells actual commodities or futures contracts for the accounts of its customers.

COMMITMENT. A trader is said to have a commitment upon assumption of the obligation to accept or make delivery on a futures contract.

COMPENSATORY GAIN. Faster-than-normal rate of gain following a period of below-normal.

COMPLEMENTARY COMBINATIONS. Advantages of one cross over another cross or a purebred, resulting from the manner in which two or more characters combine or complement each other.

COMPLETE RATION. All feedstuffs (forages and grains) combined in one feed. A complete ration fits well into mechanized feeding and the use of computers to formulate least-cost rations.

COMPOSITE BREED. A breed formed by crossing two or more breeds.

COMPOSTING. Present-day composting is the aerobic, thermophilic decomposition of organic wastes to a relatively stable humus. The resulting humus may contain up to 25% dead or living organisms and is subject to further, slower decay, but should be sufficiently stable not to reheat or cause odor or fly problems. In composting, mixing and aeration are provided to maintain aerobic conditions and permit adequate heat development. The decomposition is done by aerobic organisms, primarily bacteria, actinomycetes, and fungi.

CONCENTRATE. A broad classification of feedstuffs which are high in energy and low in crude fiber (under 18%). For convenience, concentrates are often broken down into (1) carbonaceous feeds, and (2) nitrogenous feeds.

CONCEPTION. The fertilization of the ovum or egg. The process of conceiving or becoming pregnant.

CONDITION. The state of health, as evidenced by the coat, state of flesh, and general appearance.

CONFORMATION. Body shape or form.

CONGENITAL. Acquired during prenatal life. It exists at or dates from birth.

CONGENITAL DEFECTS. Defects that an animal shows evidence of having at birth.

CONJUNCTIVA. The mucous membrane that lines the inner surface of the eyelid.

CONTAGIOUS. Transmissible by contact.

CONTEMPORARY GROUP. Cattle that are of the same breed and sex and have been raised in the same management group (same location on the same feed and pasture).

CONTENTMENT. A stress-free condition exhibited by healthy animals.

CONTINENTAL BREED. Breeds originating in Europe other than England.

COOKED. Heated to alter chemical or physical characteristics or to sterilize.

CORPORATION. Device for carrying out a business enterprise as an entity entirely distinct from the persons who are interested in and control it.

CORPUS LUTEUM. The reddish-yellow mass which fills the cavity of the site where the ovum was released from the ovary. The corpus luteum secretes the hormone progesterone. It is sometimes called the *yellow body*.

CORRELATION. A measure of how two traits vary together. A correlation of +1.00 means that as one trait increases the other also increases—a perfect *positive* relationship. A correlation of –1.00 means that as one trait increases the other decreases—a perfect *negative*, or *inverse*, relationship. A correlation of 0.00 means that as one trait increases, the other may increase or decrease—no consistent relationship. Correlation coefficients may vary between +1.00 and –1.00.

COTYLEDONS. Button-like structures on the fetal placenta in ruminants which attach to the maternal caruncles in the uterus.

COW. A mature female of the bovine species, usually having had at least one calf.

COW-CALF SYSTEM. The breeding of cows and the raising of calves.

COW-HOCKED. Standing with the points of the hocks bent inward, with the toes pointing outward.

CREEP. An enclosure or a feeder for feeding purposes which is accessible to the calves but through which the cows cannot pass.

CREEP FEEDING. The supplementation of calves that are nursing their dams.

CREST. The top part of the neck. This may be well developed in bulls.

CROSSBRED. The offspring of a sire and dam of differing breeds.

CROSSBREEDING. The mating of animals of different breeds. Crossbreeding results in hybrid vigor or heterosis.

CRYPTORCHID. When one or both of the testicles of a male have not descended to the scrotum.

CUD. A bolus of regurgitated food, common only to ruminants.

CULL. An animal removed from the herd because it is below standards.

CULLING. The process of eliminating less productive or less desirable cattle from a herd.

CUSTOM CATTLE FEEDING. The feeding of cattle for a fee, usually without taking ownership of the animals.

CUTABILITY. An estimate of the percentage of salable meat from a carcass. Percentage of retail yield of carcass weight can be estimated by a USDA prediction equation that includes hot carcass weight, ribeye area, fat thickness, and estimated percent of kidney, pelvic, and heart fat.

CUTTING. The separation of one or more animals from a herd.

CYTOGENETICS. A branch of biology that deals with heredity at the cellular and genetic level.

CYTOPLASM. The semifluid substance of a cell exclusive of the nucleus.

D

DAM. The female parent of an animal.

DARK CUTTER. This refers to the dark appearance of the lean muscle of carcass beef, which is usually caused by stress (excitement) to the animal prior to slaughter.

DEFECATION. The evacuation of fecal material from the rectum.

DEFERRED FUTURES. The futures, of those currently traded, that expire during the most distant months.

DEFICIENCY DISEASE. A disease caused by a lack of one or more basic nutrients, such as a vitamin, a mineral, or an amino acid.

DEHYDRATION.

- The loss of water from the animal's body as a result of sickness or not drinking water.
- The removal of most or all moisture from a substance for the purpose of preservation, primarily through artificial drying.

DELAYED LETHALS. Gene changes which are expressed later in life.

DELIVERY. The tender and receipt of the actual commodity, or warehouse receipts covering such commodity, in settlement of a futures contract.

DELIVERY POINTS. Those points designated by futures exchange at which the physical commodity covered by a futures contract may be delivered in fulfillment of such contract.

DENTAL PAD. The very firm gums in the upper, front jaw of cattle, sheep, and goats.

DEPRECIATION. Costs represented by fixed assets, the original cost of which is partly recaptured, currently.

DESICCATE. To dry completely.

DETENTION POND. An earthen basin constructed to store runoff water until such time as the fluids may be recycled onto land.

DEVIATION. A difference between an individual record and the average for that trait for that contemporary group. These differences sum to zero when the correct average is used. A ratio deviation is the ratio less the average ratio or 100.

DIARRHEA. Frequent bowel movements of increased liquidity. Scours.

DIET. A feed ingredient or mixture of ingredients, including water which is consumed by animals.

DIGESTIBLE NUTRIENT. The part of each feed nutrient that is digested or absorbed by the animal.

DIGESTIBLE PROTEIN. That protein of the ingested food protein which is absorbed.

DIGESTION COEFFICIENT *(Coefficient of digestibility).* The difference between the nutrients consumed and the nutrients excreted expressed as a percentage.

DIHYBRID. An individual that is heterozygous with respect to two pairs of allelic genes.

DIRECT SELLING *(Country selling).* Producers' sales of livestock directly to packers, local dealers, or farmers without the support of commission firms, selling agents, buying agents, or brokers.

DISCOUNT TO FUTURES. When the cash price is lower than the futures price.

DISEASE. An illness, a sickness, or any deviation from a state of health.

DISINFECTANT. A chemical capable of destroying disease-causing microorganisms or which inactivates viruses.

DISINFECTION. The art of killing the larger portion of microorganisms in or on a substance with the probability that all pathogenic bacteria are killed by the agent used.

DISPOSITION. The temperament or spirit of an animal.

DIURESIS. An increased excretion of urine.

DIURETIC. An agent that increases the flow of urine.

DNA. Any of various nucleic acids that are the molecular basis of heredity; found chiefly in the nucleus of cells.

DOMINANCE *(Dominant).* The ability of one member of a pair of genes to *cover up* all outward evidence of the presence of the other member of the pair.

DRENCH. Liquid medicine given to an animal by mouth.

DRESSED BEEF. Carcasses from cattle.

DRESSING PERCENTAGE. The percentage yield of hot carcass in relation to the weight of the animal on foot.

DRIED. Materials from which water or other liquids have been removed.

DROVER. A word reminiscent of one of the most thrilling chapters in American history. Prior to the advent of railroads and improved highways, great herds of cattle, sheep, and hogs were driven on horseback over famous trails, often many hundreds of miles long. The crew of *drovers* usually consisted of the boss (often the owner of the herd), a rider along each side, and a fourth rider to lead. The *drovers*—those who did the driving—were rugged, and their lives were filled with adventure. The work was accompanied by an almost ceaseless battle with the elements, clashes with thieves, and no small amount of bloodshed.

DRUGS. Substances of mineral, vegetable, or animal

origin used in the relief of pain or for the treatment of a disease.

DRY. Nonlactating female. The dry period is the time between lactations (when a female is not secreting milk).

DRY MATTER BASIS. A method of expressing the level of a nutrient contained in a feed on the basis that the material contains no moisture—it is moisture free.

DRYING OFF (UP). The process of using certain management practices to stop milk production.

DRYLOT. A relatively small enclosure without vegetation, either (1) with shelter, or (2) on open yard, in which animals may be confined.

DUNG. The feces or excrement of animals.

DYSTOCIA *(Calving difficulty).* Abnormal or difficult labor, causing difficulty in delivering the fetus and placenta.

E

EAR NOTCHING. Making slits or perforations in an animal's ears for identification purposes.

EARLY MATURING. Completing sexual development at an early age.

EARLY WEANING. The practice of weaning young animals earlier than usual.

EASEMENT. The right to go on and use the land of another in a particular manner.

EASY KEEPER. An animal that grows or fattens rapidly on limited feed.

EC. European Economic Community.

E. COLI. *Escherichia coli*—a common intestinal bacterium, studied by generations of biologists, that has probably provided more knowledge of biochemistry and genetics than any other living thing.

ECONOMIC VALUE. The *net return* to an enterprise for making a unit change in a particular trait.

ECTOPARASITES. Parasites that live on the outside of the body of the host animal (*e.g.*, face flies, mites, ticks, and mosquitoes).

EDEMA. Swelling of a part or all of the body due to the accumulation of excess water.

EFFICIENCY OF FEED CONVERSION. In beef cattle, this usually refers to the units of feed per unit of body gain.

ELASTRATOR. A device used in bloodless castration and in the docking of tails of livestock, accomplished by applying a strong rubber band.

ELECTROLYTES. Body acids, bases, or salts; necessary for transmission of nerve impulses, oxygen and carbon dioxide transfer, digestive processes, and muscle contraction.

ELEMENT. One of the 103 known chemical substances that cannot be divided into simpler substances by chemical means.

EMACIATED. Excessive loss of flesh.

EMASCULATOR. An instrument used for castration of livestock. It contains both a pair of cutting blades used to remove the testicles or tail and crushing jaws to help prevent bleeding.

EMBRYO. A fertilized ovum (egg) in the earlier stages of prenatal development up to development of body parts.

EMBRYO SPLITTING. Dividing an embryo into two or more similar parts to produce several calves from a single embryo.

EMBRYO TRANSFER. Removing fertilized ova (embryos) from one cow (donor dam) and placing these embryos into other cows (host cows). More calves can be obtained from cows of superior breeding value by this technique. Only proven producers should become donor dams.

EMPHYSEMA. An abnormal lung condition resulting from the rupture of pulmonary alveoli, the symptoms of which are difficult breathing (dyspnea) and a forced, often double, expiratory effort.

ENDOCRINE. Pertaining to glands and their secretions that pass directly into the blood or lymph instead of into a duct (secreting internally). Hormones are secreted by endocrine glands. Examples of endocrine glands are the thyroid and the pituitary.

ENDOGENOUS. Originating within the body; *e.g.*, hormones and enzymes.

ENERGY FEEDS. Feeds that are high in energy and low in fiber (under 18%), and that generally contain less than 20% protein.

ENVIRONMENT. The sum total of all external conditions that affect the life and performance of a living creature.

ENVIRONMENTAL PROTECTION AGENCY (EPA). An independent agency of the federal government established to protect the nation's environment from pollution.

ENZYME. A protein that catalyzes one of life's chemical reactions.

EPD. Expected progeny difference.

EQUITY. Net worth.

ERGOSTEROL. A plant sterol which, when activated by ultraviolet rays, becomes vitamin D_1. It is also called provitamin D_2, and ergosterin.

ERGOT. A fungus disease of plants.

ERUCTION. The elimination of gas by belching.

ESCHERICHIA COLI (E. coli). One of the species of bacteria in the coliform group. Its presence is considered indicative of fresh fecal contamination.

ESSENTIAL AMINO ACIDS. Those amino acids which cannot be made in the body from other substances or which cannot be made in sufficient quantity to supply the animal's needs.

ESSENTIAL FATTY ACID. A fatty acid that cannot be synthesized in the body or that cannot be made in sufficient quantities for the body's needs.

ESTIMATED BREEDING VALUE (EBV). An estimate of an individual's true breeding value for a trait based on the performance of the individual and close relatives for the trait. EBV is a systematic way of combining available performance information on the individual, brothers and sisters of the individual, and the progeny of the individual, to improve the accuracy of selection compared with selection based on individual performance alone.

ESTRUS *(Heat).* The recurrent, restricted period of sexual receptivity in cows and heifers. Nonpregnant cows and heifers usually come into heat 18 to 21 days following their previous estrus.

ETHER EXTRACT (EE). Fatty substances of feeds and foods that are soluble in ether.

EUROPEAN ECONOMIC COMMUNITY (EC). The twelve countries (France, Germany, Italy, Belgium, Spain, the Netherlands, Luxembourg, Denmark, Portugal, Great Britain, Ireland, and Greece) whose major objective is to coordinate the development of economic activities. Also called EEC or Common Market.

EVISCERATE. To remove the viscera as in slaughtering or butchering an animal.

EXCRETA. The products of excretion—primarily feces and urine.

EXOGENOUS. Provided from outside the organism.

EXPECTED PROGENY DIFFERENCE (EPD). The difference in performance to be expected from future progeny of an individual, compared to that expected from future progeny of other individuals in the same breed. EPD is equal to ½ the estimate of breeding value obtainable from the progeny test records and individual performance.

EXPERIMENT. The word *experiment* is derived from the Latin *experimentum*, meaning "proved from experience." It is a procedure used to discover or to demonstrate a fact or general truth.

F

F_1. Offspring resulting from the mating of a purebred (straightbred) bull to females of another purebred (straightbred) breed.

FABRICATION. Breaking the carcass into primal, subprimal, or retail cuts. These cuts may be boned and trimmed of excess fat.

FAMILY. The lineage of an animal as traced through either the males or females, depending upon the breed.

FAT. The term *fat* is frequently used in a general sense to include both fats and oils, or a mixture of the two. Both fats and oils have the same general structure and chemical properties, but they have different physical characteristics. The melting points of most fats are such that they are solid at ordinary room temperatures, while oils have lower melting points and are liquids at these temperatures.

FAT THICKNESS. This commonly refers to the amount of fat that covers muscles; typically measured at the twelfth and thirteenth rib as inches of fat over the *longissimus dorsi* muscle (rib eye).

FATTENING. The deposition of energy in the form of fat within the body tissues.

FDA. The Food and Drug Administration (FDA), which is part of the U.S. Department of Health, Education, and Human Services, is charged with the responsibility of safeguarding American consumers against injury, unsanitary food, and fraud.

FECES. The excreta discharged from the digestive tract through the anus.

FECUNDITY. The ability to produce many offspring.

FED CATTLE. Steers and heifers that have been fed concentrates prior to slaughter.

FEED *(Feedstuff).* Any naturally occurring ingredient, or material, fed to animals for the purpose of sustaining them.

FEED ADDITIVE. An ingredient or a substance added to a feed to improve the rate and/or efficiency of gain of animals, prevent certain diseases, or preserve feeds.

FEED CONVERSION *(Feed efficiency).* Units of feed consumed per unit of weight increase. Also, the production (meat, milk) per unit of feed consumed.

FEED EFFICIENCY. The ratio expressing the number of units of feed required for one unit of production by an animal. This value is commonly expressed as pounds of feed eaten per pound of gain in body weight.

FEED GRAIN. Any of several grains most commonly used for livestock or poultry feed, such as corn, sorghum, oats, and barley.

FEED OUT. To feed an animal until it reaches market weight.

FEEDER CATTLE. Young animals that carry insufficient finish for slaughter purposes but will make good gains if placed on feed.

FEEDERS. Calves and yearlings, both steers and heifers, carrying more weight and/or finish than stockers, which are ready to be placed on high-energy rations for finishing and slaughtering.

FEEDER'S MARGIN. The difference between the cost per hundredweight of feeder animals and the selling price per hundredweight of the same animals when finished.

FEEDLOT. A lot or plot of land on which animals are fed or finished for market.

FEEDSTUFF. Any product of natural or artificial origin that has nutritional value in the diet when properly prepared.

FERAL. Domesticated animals which have reverted back to their original or untamed state.

FERMENTATION. Chemical changes brought about by enzymes produced by various microorganisms.

FERTILIZATION. The union of sperm with the ovum or egg to form a new life.

FETUS. A young organism in the uterus from the time the organ systems (body parts) develop until birth.

FIBER CONTENT OF A FEED. The amount of hard-to-digest carbohydrates. Most fiber is made up of cellulose and lignin.

FILL.

- The fullness of the digestive tract of an animal.
- With market animals, the amount of feed and water consumed upon their arrival at the market and prior to selling.

FINISH. To fatten an animal for slaughter. Also, the degree of fatness of such an animal.

FITTING. The conditioning of an animal for show or sale, which usually involves a combination of special feeding plus exercise and grooming.

FIXED EXPENSES. Expenses which do not change as business volume changes, such as real estate taxes, interest on mortgage, depreciation, insurance, manager's salary, and similar items.

FLORA. The plant life present. In nutrition it generally refers to the bacteria present in the digestive tract.

FLUSHING. The practice of feeding females more generously 2 to 3 weeks before breeding. The beneficial effects attributed to this practice are (1) more eggs (ova) are shed, and this results in more offspring; (2) the females come in heat more promptly; and (3) conception is more certain.

FOOD AND DRUG ADMINISTRATION (FDA). The federal agency in the Department of Health and Human Services that is charged with the responsibility of safeguarding American consumers against injury, unsanitary food, and fraud. It protects industry against unscrupulous competition, and it inspects and analyzes samples and conducts independent research on such things as toxicity (using laboratory animals), disappearance curves for pesticides, and long-range effects of drugs.

FORAGE. Vegetable material in a fresh (pasture), dried (hay), or ensiled (silage) state which is fed to livestock.

FORTIFY. Nutritionally, to add one or more feeds or feedstuffs.

FORWARD CONTRACT. A forward contract calls for delivery at some time in the future. In a forward contract, a cattle producer might make a deal with a buyer during the summer months that calls for delivery of cattle in the fall at the price agreed upon in the contract.

FRAME SCORE. A score based on subjective evaluation of height or measurement of hip height. This score is related to slaughter weights at which cattle will grade Choice or have ½ in. of fat cover over the loin eye at the twelfth to thirteenth rib.

FRATERNAL TWINS. Two individuals developing from the fertilization of two separate eggs, each by a different sperm cell. Carried during the same pregnancy. May be both males or females or one male and one female.

FREE-CHOICE. Free to eat a feed or feeds at will.

FREEMARTIN. Female born twin with a bull calf (approximately 9 out of 10 will not conceive).

FREEZE DRYING. The process of rapid freezing followed by drying under vacuum. The ice passes from the solid to the vapor state without melting.

FRESHEN. With reference to cattle, to give birth to or calve (parturition), and thereby begin a new lactation period. A fresh cow has recently calved and is actively lactating.

FULL BROTHERS (or sisters). Animals having the same sire and the same dam.

FULL-FEED. The term indicating that animals are being provided as much feed as they will consume safely without going off feed.

FUMIGANT. A liquid or solid substance that forms vapors that destroy pathogens, insects, and rodents.

FUNGI. Plants that contain no chlorophyll, flowers, or leaves; such as molds, mushrooms, toadstools, and yeasts. They may get their nourishment from either dead or living organic matter.

FUTURES. A term used to designate any and all contracts which are made or established subject to rules for delivery at a later date.

FUTURES CONTRACT. A standardized, legally binding transaction in which the seller promises to make delivery of a specified quantity and type of a commodity at a specified location(s) during a specified future month.

FUTURES TRADING. The futures market is a way in which to provide (1) an insurance medium in the marketing field, and (2) the facilities and machinery for under writing price risks.

G

GAMETE. A male or female reproductive cell before fertilization.

GASTROINTESTINAL. Pertaining to the stomach and intestines.

GENERATION INTERVAL. Average age of the parents when the offspring destined to replace them are born. A generation represents the average rate of turnover of a herd.

GENES. The basic units of heredity that occur in pairs and have their effect in pairs in the individual, but which are transmitted singly (one or the other gene at random of each pair) from each parent to offspring.

GENETIC CORRELATIONS. Correlations between 2 traits caused by the same genes influencing both traits or multiple gene action. Successful selection for 1 of 2 traits that are positively, highly correlated will result in an increase in the other trait. If the genes that influence 2 traits generally cause desirable performance in 1 trait and undesirable performance in the other trait, improvement by selection within 1 herd will be difficult.

GENETIC ENGINEERING. The manipulation of genetic material to modify organisms for specific uses and adaptations.

GENETICS. The science or study of inheritance.

GENOTYPE. Actual genetic makeup (constitution) of an individual determined by its genes or germplasm. For example, there are two genotypes for the polled phenotype: PP pure for polled (homozygous) and Pp polled (heterozygous).

GENOTYPE SELECTION. Selection of breeding stock not necessarily from the best appearing animals but from the best breeding animals, according to genetic makeup.

GERM PLASM. Germ cells and their precursors, bearers of hereditary characters.

GESTATION. Pregnancy. Refers to the carrying of the products of conception in the uterus from fertilization to parturition. Gestation period is the time between mating (conception) and parturition.

GET. The offspring of a male animal—his progeny.

GOITROGENIC. Producing or tending to produce goiter.

GONAD. The gland of a male or female that produces the reproductive cells. The testicle in the male and the ovary in the female.

GOSSYPOL. A toxic yellow pigment found in cottonseed.

GRADE.

- An animal having parents that cannot be registered by a breed association.
- A measure of how well an animal or product fulfills the requirements for the class; for example, the federal grades of cattle and their carcasses are a specific indication of the degree of excellence.

GRADING UP. The continued use of purebred sires of the same breed in a grade herd.

GRAIN. Seed from cereal plants.

GREGARIOUS. Fond of or liking to live in groups.

GRIND. To reduce to small segments by impact, shearing, or attrition (as in a mill).

GROW OUT. To feed animals so that they attain a certain desired amount of growth with little or no fattening.

GROWTH. May be defined as the increase in size of the muscles, bones, internal organs, and other parts of the body.

GROWTHY. Describes an animal that is large and well developed for its age.

H

HAIR. A slender outgrowth of the epidermis which performs a thermoregulatory function, to protect the animal from cold or heat. It becomes long and shaggy during the winter months in cold areas, especially if the animal is left outside. Then, during the warm season, or when the animal is blanketed, the animal sheds and the coat becomes short.

HALF-SIBS. Individuals having the same sire or dam. Half-brothers and half-sisters.

HAND-FEED. To provide a certain amount of a ration at regular intervals.

HAND MATING. A mating procedure in which the caretaker controls the pairings by allowing males to mate with females in heat, but keeps them separated before and after mating occurs.

HARD KEEPER. An animal that is unthrifty and grows or fattens slowly regardless of the quantity or quality of feed.

HAY. Dried forage.

HEAT *(Estrus)*. The period when the female will accept service by the male.

HEAT INCREMENT (HI). The increase in heat production following consumption of feed when the animal is in a thermoneutral environment.

HEAT SYNCHRONIZATION. Causing a group of cows or heifers to exhibit heat together at one time by artificial manipulation of the estrus cycle.

HEDGE. The purchase or sale of a futures contract as a temporary substitute for a merchandising transaction to be made at a later date. Usually it involves opposite positions in the cash market and the futures market at the same time.

HEDGERS. Persons who desire to avoid risks, and who try to increase their normal profit margins through buying and selling futures contracts. They are feeders, packers, and others actually involved in the production, processing, or marketing of beef or other products. Their primary objective is to establish future prices and costs so that operational decisions can be made on the basis of known relationships.

HEIFER. A female bovine that has not given birth to a calf. Sometimes used to denote females until their second calving.

HEMOGLOBIN. The pigment of the red blood cell which is involved in the transport of oxygen and carbon dioxide in the blood.

HEMORRHAGE. Bleeding. Loss of blood from the blood vessels.

HERD. A group of cattle (animals) collectively considered as a unit.

HEREDITY. The transmission of genetic or physical traits of parents to their offspring.

HERITABILITY. The proportion of the differences among cattle, measured or observed, that is transmitted to the offspring. Heritability varies from 0 to 1. The higher the heritability of a trait, the more accurately does individual performance predict breeding value and the more rapid should be the response due to selection for that trait.

HERITABILITY ESTIMATE. An estimate of the proportion of the total phenotypic variation between individuals for a certain trait that is due to heredity. More specifically, hereditary variation due to additive gene action.

HERMAPHRODITE. An individual whose genital organs have, in greater or lesser degree, the characters of both male and female.

HERNIA. The protrusion of some of the intestine through an opening in the body wall—commonly called a rupture.

HETEROSIS *(Hybrid vigor)*. Amount by which the crossbreds exceed the average of the two purebreds that are crossed to produce the crossbreds.

HETEROZYGOUS. Animals that have 1 dominant and 1 recessive factor in 1 or more pairs of genes.

HOMOGENIZED. Particles broken down into evenly distributed globules small enough to remain emulsified for long periods of time.

HOMOLOGOUS CHROMOSOMES. Chromosomes having the same size and shape, and containing the genes affecting the same characteristics.

HOMOZYGOTES. Individuals possessing identical genes. Refers particularly to any given pair of genes. Each pair of genes is made up of the same genes—they are pure for these characters.

HORMONE. A body-regulating chemical secreted by an endocrine gland into the bloodstream, thence transported to another region within the animal where it elicits a physiological response.

HOT FAT TRIMMING. Removal of excess surface fat while the carcass is still "hot," prior to chilling.

HULL. Outer covering of grain or other seed, especially when dry.

HUMUS. The dark or black carboniferous residue in the soil resulting from the decomposition of vegetable tissues of plants originally growing therein. Residues similar in appearance and behavior are found in composted manure and well-digested sludges.

HYBRID VIGOR OR HETEROSIS. Name given to the biological phenomenon which causes crossbreds to outproduce the average of their parents.

HYDROLYSIS. The splitting of a substance into the smaller units by chemically adding water to the material.

HYPERVITAMINOSIS. An abnormal condition resulting from the intake of an excess of one or more vitamins.

I

-ICIDE. Suffix denoting something that destroys or kills. A *bactericide* kills or destroys bacteria. A *fungicide* kills or destroys fungi. A *viricide* kills or destroys viruses.

IDENTICAL TWINS. Two individuals that developed as a result of the division of one fertilized egg; are of the same sex; and are extremely similar. Color markings of one may be mirror-image of the other. Often referred to as *monozygotic twins*.

IMMUNITY. The ability of an animal to resist or overcome an infection to which most members of its species are susceptible.

IMMUNIZATION. The process and procedures involved in creating immunity (resistance to disease) in an animal. Vaccination is a form of immunization.

IMMUNOGLOBULINS. A family of protein found in body fluids which have the property of combining with antigens; and, when the antigens are pathogenic, sometimes inactivating them and producing a state of immunity. Also called *antibodies*.

IMPACTION. A condition in which partially digested food material has been firmly wedged into a segment of the digestive system causing a stoppage of movement through the tract.

INBREEDING. The practice of mating closely related animals such as: full brother to sister, sire to daughter, son to dam. Inbreeding increases homozygosity.

INDEPENDENT CULLING LEVELS. Selection or culling based on cattle meeting specific levels of performance for each trait. For example, a breeder could cull all heifers with weaning weights below 400 lb (or those in the lower 20% on weaning weight) and yearling weights below 650 lb (or those in the lower 40%).

INFLAMMATION. Reaction of tissue to injury characterized by redness, swelling, pain, and heat.

INGEST. To eat or take in through the mouth.

INGESTA. Food or drink taken into the stomach.

INGESTION. The taking in of food and drink.

INGREDIENT. A constituent feed material.

INTERSEX. An individual showing both maleness and femaleness; in which the sex differences are not confined to clearly demarcated parts of the body, but blend more or less with one another.

INTRADERMAL. Into, or between, the layers of the skin.

INTRAMUSCULAR. Within the muscle.

INTRAMUSCULAR FAT. Fat within the muscle, or marbling.

INTRAPERITONEAL. Within the peritoneal cavity.

INTRAVENOUS. Within the vein or veins.

IONOPHORE. Feed additives that change the metabolism within the rumen by altering the rumen microflora to favor propionic acid production.

IRRADIATED. Having been treated with some form of radiant energy.

IRRADIATION. Exposure to ultraviolet light.

IU *(International Unit).* A standard unit of potency of a biologic agent (*e.g.*, a vitamin, a hormone, an antibiotic, an antitoxin) as defined by the International Conference for Unification of Formulae. Potency is based on bioassay that produces a particular effect agreed on internationally. Also called a *USP unit*.

J

JOULE. A proposed international unit (4,184 j = 1 calorie) for expressing mechanical, chemical, or electrical energy, as well as the concept of heat. In the future, energy requirements and feed values will likely be expressed by this unit.

K

KETOSIS. A metabolic disease characterized by hypoglycemia (low blood sugar). It is a disease of lactating ruminants which occurs within a few days after parturition. It is caused by an imbalance between nutrient intake and the nutrient requirements of the animal.

KNEE-SPRUNG. See **BUCK-KNEED**.

KOSHER MEAT. Meat from ruminant animals (with split hooves) that has been slaughtered according to Jewish law.

L

LACTATION. The secretion of milk. The period in which an animal is producing milk.

LAGOON. An all-inclusive term commonly given to a water impoundment in which organic wastes are stored or stabilized, or both. Lagoons may be described by the predominant biological characteristics (aerobic, anaerobic, or facultative), by location (indoor, outdoor), by position in a series (first stage, second stage, etc.), and by the organic material accepted (sewage, sludge, manure, or other).

LAMENESS. A defect that can be detected when the affected foot is favored when standing. In action, the load of the ailing foot is eased, and there is a characteristic bobbing of the head as the affected foot strikes the ground.

LESION. The change in the structure or form of an animal's body caused by a disease or an injury.

LETHALS. A gene which causes death of the calf carrying them prior to or shortly after birth.

LIBIDO. Sexual drive or sexual desire. Usually reserved to describe the male of the species.

LICE. Small, flattened, wingless insect parasites.

LIMESTONE. Rock formed primarily by the accumulation of organic remains (for example, shells or coral). Used as a calcium supplement.

LIMIT ORDER. Placing price limitations on orders given the brokerage firm.

LIMITED FEEDING. Feeding animals less than they would like to eat. Giving sufficient feed to maintain weight and growth, but not enough for their potential production or finishing.

LIMITED PARTNERSHIP. Arrangement in which two or more parties supply the capital, but only one partner is involved in the management.

LINEBREEDING. Mating of animals more distantly related than in closebreeding, and in which the matings are usually directed toward keeping the offspring closely related to some highly admired ancestor, such as half-brother and half-sister, female to grandsire, and cousins.

LINECROSS. A cross of two or more inbred lines.

LIQUID MANURE. A suspension of livestock manure in water, in which the concentration of manure solids is low enough so that the flow characteristics of the mixture are more like those of Newtonian fluids than of plastic fluids.

LIQUIDITY. Convertibility of assets to cash.

LIVESTOCK AUCTIONS. Trading centers where animals are sold by public bidding to the buyer who offers the highest price per hundredweight or per head.

LIVEWEIGHT. Weight of an animal on foot

LONG. The buying side of an open futures contract. A trader whose net position in the futures market shows an excess of open purchases over open sales is said to be *long*.

LONG HEDGE. Buying on the futures market contracts against anticipated need in the future in order to protect against a rise in the market price. Thus, futures contracts in feeder cattle can be used in a long hedge to reduce or eliminate the risk involved in a rise of feeder cattle prices. At the time the feeder negotiates a forward contract with the packer, the feeder would buy feeder cattle futures contracts, preferably for the month in which the feeder actually planned to buy the feeder cattle.

LONG YEARLING. A bovine animal between 18 months and 2 years of age.

LOW-SET. Designates a short-legged animal.

LYMPH. The slightly yellow, transparent fluid occupying the lymphatic channels of the body.

M

MACROMINERALS. The major minerals—calcium, phosphorus, sodium, chlorine, potassium, magnesium, and sulfur.

MAINTENANCE REQUIREMENT. A ration which is adequate to prevent any loss or gain of tissue in the body when there is no production.

MALNUTRITION. Any disorder of nutrition. Commonly used to indicate a state of inadequate nutrition.

MANAGEMENT. The act, or art, of managing, handling, controlling, or directing.

MANGE *(Scabies, scab, or itch).* A specific contagious disease caused by mites.

MANGY. Infected with a skin disease or parasite so that the skin is dry and scaly.

MANURE. Mixture of animal excrements (consisting of undigested feeds plus certain body wastes) and bedding.

MARBLING. The specks of fat (intramuscular fat) distributed in muscular tissue. Marbling is usually evaluated in the ribeye at the twelfth to thirteenth rib.

MARGIN. Cash or equivalent posted as guarantee of fulfillment of a futures contract.

MARGIN *(Spread).* The difference between the purchase price and the selling price.

MARGIN CALL. If the market price of a futures contract changes after the cattle producer has sold or purchased a futures contract, the producer will either make a profit or lose money. If the price moves in such a direction that the producer loses money, the broker will deduct the losses from the original *margin* and call for additional funds in order to bring the *margin* back up to the original amount.

MARKET CLASS. Animals grouped according to the use to which they will be put, such as slaughter or feeder.

MARKET GRADE. Animals grouped within a market class according to their value.

MASTICATION. The chewing of feed.

MASTITIS. Inflammation of the mammary gland.

MAVERICK. An unbranded animal, usually on the range.

MEAL.

- A feed ingredient having a particle size somewhat larger than flour.
- Mixtures of concentrate feeds, usually in which all of the ingredients are ground.

MEAT ANIMAL RESEARCH CENTER (MARC). USDA research center located in Clay Center, NE, which conducts beef cattle research.

MECONIUM. The dark-colored, semisolid fecal matter in the digestive tract of the fetus that forms the first bowel movement of the newborn.

MEDICATED FEED. Any feed which contains drug ingredients intended or represented for the cure, mitigation, treatment, or prevention of diseases of animals (other than humans).

METABOLIC BODY SIZE. The weight of the animal raised to the ¾ power ($W^{0.75}$); a figure to indicate level of metabolism to maintain a certain body weight.

METABOLISM. Refers to all the changes that take place in the nutrients after they are absorbed from the digestive tract, including (1) the building-up processes in which the absorbed nutrients are used in the formation or repair of body tissues, and (2) the breaking-down processes in which nutrients are oxidized for the production of heat and work.

METRITIS. Inflammation of the uterus.

MICROBE. Same as *microorganism.*

MICROFLORA. Microbial life characteristic of a region, such as the bacteria and protozoa populating the rumen.

MICROORGANISM. Any organism of microscopic size, applied especially to bacteria and protozoa.

MILL BY-PRODUCT. A secondary product obtained in addition to the principal product in milling practice.

MINERAL SUPPLEMENT. A rich source of one or more of the inorganic elements needed to perform certain essential body functions.

MINERALS *(Ash).* The inorganic elements of animals and plants, determined by burning off the organic matter and weighing the residue, which is called *ash.*

MITES. Very small parasites that cause mange (scabies, scab, or itch).

MOISTURE. A term used to indicate the water contained in feeds—expressed as a percentage.

MOISTURE-FREE *(M-F, oven-dry, 100% dry matter).* This refers to any substance that has been dried in an oven at 221°F until all the moisture has been removed.

MOLARS. The flattened back (jaw) teeth in animals that are used for grinding.

MOLDS *(Fungi).* Fungi which are distinguished by the formation of mycelium (a network of filaments or threads), or by spore masses.

MONOCLONAL ANTIBODIES. Identical antibodies cloned from a single source and targeted for a specific antigen.

MONOHYBRID. An individual that is heterozygous with respect to one pair of allelic genes.

MORBIDITY. A state of sickness or the rate of sickness.

MORTALITY. Death or death rate.

MOST PROBABLE PRODUCING ABILITY (MPPA). An estimate of a cow's future productivity for a trait (such as progeny weaning weight ratio) based on her past productivity. For example, a cow's MPPA for weaning ratio is calculated from the cow's average progeny weaning ratio, the number of her progeny with weaning records, and the repeatability of weaning weight.

MUCOUS MEMBRANES. The soft membranes that line the passages and cavities of the body such as the respiratory, alimentary, and genito-urinary systems.

MUTATION. A sudden variation which is later passed on through inheritance and which results from changes in a gene or genes.

MYCOTOXINS. Toxic metabolites produced by molds during growth. Sometimes present in feed materials.

N

NATIONAL ACADEMY OF SCIENCES. The academy was established in 1916 to promote the effective utilization of scientific and technical resources. *The address:* 2102 Constitution Avenue, N.W., Washington, DC 20418

NATIONAL CATTLEMEN'S BEEF ASSOCIATION (NCBA). National organization for cattle breeders, producers, feeders, and affiliated organizations. *The address:* 5420 S. Quebec Street, Englewood, CO 80111-1904.

NATIONAL LIVE STOCK AND MEAT BOARD (NLSMB). The board provides nutrition, research, education, and promotional information on beef, pork, and lamb. *The address:* 444 North Michigan Avenue, Chicago, IL 60611-3978.

NATIONAL RESEARCH COUNCIL (NRC). A division of the National Academy of Sciences established in 1916 to promote the effective utilization of scientific and technical resources. Periodically, this private nonprofit organization of scientists publishes bulletins giving nutrient requirements and allowances of domestic animals, copies of which are available on a charge basis through the National Academy of Sciences, National Research Council, 2101 Constitution Avenue, N.W., Washington, DC 20418.

NATIONAL SIRE EVALUATION. Programs of sire evaluation conducted by breed associations to compare sires on a progeny test basis. Carefully conducted national reference sire evaluation programs give unbiased estimates of expected progeny differences. Sire evaluations based on field data rely on large numbers of progeny per sire to compensate for possible favoritism or bias for sires within herds.

NAVEL CORD *(Umbilical cord).* The umbilical cord connects the fetus with the placenta and conveys food and removes waste products. It is severed or broken at birth.

NEARBYS. The nearest active trading month of a futures market. (See **DEFERRED FUTURES**.)

NET ENERGY. Gross energy minus fecal, urinary, gas, and heat losses; it is the energy available to the animal for maintenance and production.

NICKING. The way in which certain lines, strains, or breeds perform when mated together. If the offspring of certain matings are especially outstanding and better than their parents, breeders say that the animals nicked well.

NITROGEN. A chemical element essential to life. Animals get it from protein feeds; plants get it from the soil; and some bacteria get it directly from the air.

NITROGEN BALANCE. The nitrogen in the feed intake minus the nitrogen in the feces, minus the nitrogen in the urine.

NITROGEN FIXATION. Conversion of free nitrogen of the atmosphere to organic nitrogen compounds by symbiotic or nonsymbiotic microbial activity.

NITROGEN-FREE EXTRACT (NFE). It consists principally of sugars, starches, pentoses, and non-nitrogenous organic acids. The percentage is determined by subtracting the sum of the percentages of moisture, crude protein, crude fat, crude fiber, and ash from 100.

NONADDITIVE GENE EFFECTS. Favorable effects or actions produced by specific gene pairs or combinations. Nonadditive gene action is the primary cause of heterosis. Nonadditive gene action occurs when the heterozygous genotype is not intermediate in phenotypic value to the two homozygous genotypes.

NONPROTEIN NITROGEN (NPN). Nitrogen which comes from other than a protein source but may be used by a ruminant in the building of protein. NPN sources include compounds like urea and anhydrous ammonia, which are used in feed formulations for ruminants only.

NUCLEOTIDE. Building block of the DNA molecule composed of an organic base, a sugar, and a phosphate.

NUMBER OF CONTEMPORARIES. The number of animals of similar breed, sex, age, against which an animal was compared in performance tests. The greater the number of contemporaries, the greater the accuracy of comparisons.

NUTRIENT ALLOWANCES. Nutrient recommendations that allow for variations in feed composition; possible losses during storage and processing; day-to-day and period-to-period differences in needs of animals; age and size of animal; stage of gestation and lactation; the kind and degree of activity; the amount of stress; the system of management; the health, condition, and temperament of the animal; and the kind, quality, and amount of feed—all of which exert a powerful influence in determining nutritive needs.

NUTRIENT REQUIREMENTS. Meeting the animal's minimum needs, without margins of safety, for maintenance, growth, fitting, reproduction, lactation, and work. To meet these nutritive requirements, the different classes of animals must receive sufficient feed to furnish the necessary quantity of energy (carbohydrates and fats), protein, minerals, and vitamins.

NUTRIENTS. The chemical substances found in feed materials that can be used, and are necessary, for the maintenance, production, and health of animals. The chief classes of nutrients are carbohydrates, fats, proteins, minerals, vitamins, and water.

NUTRITION. The science encompassing the sum total of processes that have as a terminal objective the provision of nutrients to the component cells of an animal.

NUTRITIVE RATIO (NR). The ratio of digestible protein to other digestible nutrients in a feedstuff or ration. (The NR of shelled corn is about 1:10.)

O

OFFAL. All organs or tissues removed from the carcass in slaughtering.

OFFER. A willingness to sell a futures contract at a given price. (See **BID**.)

OFF FEED. Animals that refuse to eat, or that consume only small amounts of feed.

OMASUM. The third compartment of the ruminant stomach. It is often called the *manyplies* due to its structure.

ON FULL FEED. Animals that are receiving all the feed they will consume.

OPEN. A term commonly used to indicate a nonpregnant female.

OPEN INTEREST. Number of open futures contracts. Refers to unliquidated purchases or sales but never to their combined total. (See **COMMITMENT**.)

OPTIMUM LEVEL OF PERFORMANCE. The most profitable or favorable ranges in levels of performance for the economically important traits in a given environment and management system. For example, although many cows produce too little milk, in every management system there is a point beyond which higher levels of milk production will decrease profit.

ORGANIC MATTER. Chemical substances of animal or vegetable origin, or more correctly, of basically carbon structures, comprising compounds consisting of hydrocarbons and their derivatives.

OSSIFICATION. The process of bone formation; the calcification of bone with advancing maturity.

OSTEOMALACIA. A bone disease of adult animals caused by lack of vitamin D, inadequate intake of calcium or phosphorus, or an incorrect dietary ratio of calcium and phosphorus.

OSTEOPOROSIS. Abnormal porosity and fragility of bone as the result of (1) a calcium, phosphorus, and/or vitamin D deficiency, or (2) an incorrect ratio between the two minerals.

OUT OF. Refers to the female parent.

OUTCROSS. The introduction of genetic material from some outside and unrelated source, but of the same breed, into a herd which is more or less related.

OVARY. The female organ that produces eggs. There are two ovaries.

OVERBRAND. This term refers to leaving a freeze-branding iron on the hide of an animal long enough to produce a bald brand instead of white hair. This is usually done on white or cream-colored cattle.

OVERFEEDING. Feeding to excess.

OVERFINISHING. Excess finishing or fatness—a wasteful practice.

OVERSHOT JAW. The upper jaw protruding beyond the lower jaw. Same as *parrot mouth*.

OVULATION. The time when the follicle ruptures and the egg is released.

OVUM. Scientific name for egg, the female reproductive cell.

OWNER'S EQUITY *(Net Worth)*. Total investments made in the business plus profits earned through operations which have remained in the business (retained earnings).

OXIDATION. The combination with oxygen, or the loss of a hydrogen, or the loss of an electron, all of which render an ion more electropositive. The animal combines carbon from feedstuffs with inhaled oxygen to produce carbon dioxide, energy (as ATP), water, and heat.

OXIDATION DITCH. A modified form of the activated-sludge process. An aeration rotor supplies oxygen and circulates the liquid in an oval, racetrack-shaped, open-channel ditch.

OXIDATION POND. A basin used for retention of wastewater before final disposal, in which biological oxidation of organic material is effected by natural or artificially accelerated transfer of oxygen to the water from air.

P

PACKING PLANT. Facility in which animals are slaughtered and processed.

PALATABILITY. The result of the following factors sensed by the animal in locating and consuming feed: appearance, odor, taste, texture, and temperature. These factors are affected by the physical and chemical nature of the feed.

PALPATE. To investigate or examine by feeling or touching with the hands. An animal can be palpated to determine the state of pregnancy.

PARASITE. An organism that lives a part of its life cycle in or on another organism.

PARROT MOUTH. The upper jaw is longer than the lower jaw.

PARTNERSHIP. An association of two or more persons who, as co-owners, operate a business.

PARTS PER BILLION (PPB). It equals micrograms per kilogram or microliters per liter.

PARTS PER MILLION (PPM). It equals milligrams per kilogram or milliliters per liter.

PARTURITION. The act of giving birth.

PASTURE. An area with plants that may be harvested by grazing animals.

PASTURE ROTATION. The rotation of animals from one pasture to another so that some pasture areas have no livestock grazing on them during certain periods of time.

PATHOGEN. Any disease-producing organism.

PATHOGENIC. Disease causing.

PEDIGREE. A tabulation of names of ancestors, usually only those of the 3 to 5 closest generations.

PENCIL SHRINK. Deduction from an animal's weight to account for fill (usually 3% for off-pasture weights and 4% for fed-cattle weights).

PER CAPITA. Per person.

PER ORAL. Administration through the mouth.

PER OS. Oral administration (by the mouth).

PERFORMANCE DATA. The record of the individual animal for specific traits such as birth weight, weaning weight, post-weaning gain, yearling weight, etc.

PERFORMANCE PEDIGREE. A pedigree that includes performance records in addition to the usual pedigree information. Performance records for the individual and certain ancestors and progeny records of the individual and certain ancestors are included on the performance pedigree of some breed associations. Also, the performance information is systematically combined to list estimated breeding values on the pedigrees of some breed associations.

PERFORMANCE TEST. The evaluation of an animal by its own performance.

PERFORMANCE TESTING. Practice of evaluating and selecting animals on the basis of their individual merit.

pH. A measure of the acidity or alkalinity of a solution. Values range from 0 (most acid) to 14 (most alkaline), with neutrality at pH 7.

PHASE FEEDING. Refers to changes in the animal's diet (1) to adjust for age and stage of production, (2) to adjust for season of the year and for temperature and climatic changes, (3) to account for differences in body weight and nutrient requirements of different strains of animals, or (4) to adjust one or more nutrients as other nutrients are changed for economic or availability reasons.

PHENOTYPE. The characteristics of an animal that can be seen and/or measured.

PHENOTYPIC CORRELATIONS. Correlations between two traits caused by both genetic and environmental factors influencing both traits.

PHOTOSYNTHESIS. The process whereby green plants utilize the energy of the sun to build up complex organic molecules containing energy.

PIGEON-TOED. Pointing toes inward and heels outward.

PIT. An octagonal platform on the trading floor of an exchange consisting of steps upon which traders and brokers stand while executing futures trades.

PLACENTA. The membrane by which the fetus is attached to the uterus. Nutrients from the mother pass into the placenta and then through the navel cord to the fetus. When the animal is born, the placenta is expelled. It is commonly called the *afterbirth.*

PLASMID. A self-replicating circular DNA molecule found in bacteria and carrying two or more genes.

POLL. The top, crown, or back of the head of an animal. More specifically, it refers to that area directly between the ears of the animal.

POLLED. A naturally hornless animal.

POLLUTION. The presence in a body of water (or soil or air) of material in such quantities that it impairs the water's usefulness or renders it offensive to the senses of sight, taste, and/or smell. Contamination may accompany pollution. In general, a public health hazard is created, but in some instances, only economy or aesthetics is involved, as when waste salt brines contaminate surface waters or when foul odors pollute the air.

POLYUNSATURATED FATTY ACIDS. Fatty acids having more than one double bond.

POSTNATAL. Occurring after birth.

POSTPARTUM. Occurring after the birth of the offspring.

POUNDS OF RETAIL CUTS PER DAY OF AGE. A measure of cutability and growth combined, it is calculated as follows: (cutability × carcass weight) ÷ age in days. Also, it is reported as lean weight per day of age (LWDA) by some associations.

PRECONDITIONING. A way of preparing the calf to withstand the stress and rigors of leaving its mother, learning to eat new kinds of feed, and being shipped from the farm or ranch to the feedlot.

PRECURSOR. A compound that can be used by the body to form another compound; for example, carotene is a precursor of vitamin A.

PREHENSION. The seizing (grasping) and conveying of feed to the mouth.

PREMIUM. When the cash price is above the futures.

PRENATAL. Before birth.

PREPOTENCY. The ability of an individual to transmit its own qualities to its offspring, with the result that they resemble each other. Homozygous dominant individuals are prepotent. Also, inbred cattle tend to be more prepotent than outcross cattle.

PRESERVATIVES. A number of materials are available to incorporate into feeds, with claims made that they will improve the preservation of nutrients, nutritive value, and/or palatability of the feed.

PRICE CYCLE. That period of time during which the price for a certain kind of livestock advances from a low point to a high point and then declines to a low point again.

PRICE MARGIN. The difference between the cost per cwt of the feeder animal and the selling price per cwt of the same animal when finished.

PRIMAL CUTS. The wholesale cuts of beef—round, loin, flank, rib, chuck, brisket, plate, and shank.

PRODUCE. A female's offspring. The produce-of-dam commonly refers to two or more offspring of one dam.

PRODUCTION TESTING. A more inclusive term than performance or progeny testing. It includes performance testing and/or progeny testing.

PROFIT. Whenever revenues exceed expenses during a given accounting period, a *profit* has been made.

PROFIT AND LOSS (P&L) STATEMENT. A measure of the income generated, and the cost incurred, during a specific period of time—usually 1 year.

PROGENITOR. One that originates or precedes.

PROGENY. Offspring or descendants of one or both parents.

PROGENY RECORDS. The average comparative performance on the progeny of sires and dams.

PROGENY TESTING. An evaluation of an animal on the basis of the performance of its offspring.

PROGNOSIS. A prediction or an estimate of the progression and final outcome of a disease.

PROGRAM. A series of directives given to the computer which enables it to solve mathematical problems.

PROLAPSE. Abnormal protrusion of a part or an organ. For example, a prolapse of the uterus results when the uterus slips backward and is expelled from the reproductive tract through the vagina.

PROTEIN. From the Greek, meaning *of first rank, importance.* Complex organic compounds made up chiefly of amino acids present in characteristic proportions for each specific protein. At least 24 amino acids have been identified and may occur in combinations to form an almost limitless number of proteins. Protein always contains carbon, hydrogen, oxygen, and nitrogen; and, in addition, it usually contains sulfur and frequently phosphorus. Crude protein is determined by finding the nitrogen content and multiplying the result by 6.25. The nitrogen content of proteins averages about 16% ($100 \div 16 = 6.25$). Proteins are essential in all plant and animal life as components of the active protoplasm of each living cell.

PROTEIN SUPPLEMENTS. Products that contain more than 20% protein or protein equivalent.

PROUD FLESH. Excess flesh growing around a wound.

PROVITAMIN. The material from which an animal may produce vitamins; *e.g.*, carotene (provitamin A) in plants is converted to vitamin A in animals.

PROXIMATE ANALYSIS. A chemical scheme for evaluating feedstuffs, in which a feedstuff is partitioned into the six fractions: (1) moisture (water) or dry matter (DM); (2) total (crude) protein (CP or TP—N $\times$ 6.25); (3) ether extract (EE) or fat; (4) ash (mineral salts); (5) crude fiber (CF)—the incompletely digested carbohydrates; and (6) nitrogen-free extract (NFE)—the more readily digested carbohydrates (calculated rather than measured chemically).

PUBERTY. The age at which the reproductive organs become functionally operative—sexual maturity.

PUREBRED. Member of a breed, the animals of which possess a common ancestry and distinctive characteristics; and it is either registered or eligible for registry in the herd book of that breed.

PUTREFACTION. Biological decomposition of organic matter with the production of ill-smelling products associated with anaerobic conditions.

Q

QUALITATIVE TRAITS. Traits in which there is sharp distinction between phenotypes, usually involving only 1 or 2 pairs of genes. For example, black and white or horned and polled.

QUALITY.

- Refinement of an animal, as shown by a neat and well-chiseled head, fine texture of hair, and clean bone.
- A term used to denote the desirability and/or acceptance of an animal or feed product.

QUALITY GRADES. Beef quality grades (Prime, Choice, Select, etc.) are based on their palatability—indicating characteristics of the lean.

QUALITY OF PROTEIN. A term used to describe the amino acid balance of protein. A protein is said to be of good quality when it contains all the essential amino acids in proper proportions and amounts needed by a specific animal; and it is said to be poor quality when it is deficient in either content or balance of essential amino acids.

QUANTITATIVE TRAITS. Traits in which there is no sharp distinction between phenotypes, with a gradual variation from one phenotype to another. These include such economic traits as gestation length, birth weight, weaning weight, rate and efficiency of gain, and carcass quality.

QUARANTINE.

- Compulsory segregation of exposed susceptible animals for a period of time equal to the longest usual incubation period of the disease to which they have been exposed.
- An enforced regulation for the exclusion or isolation of an animal to prevent the spread of an infectious disease.

R

RADIOACTIVE. Giving off atomic energy in the form of alpha, beta, or gamma rays.

RALLY. Quick advance in prices following a decline.

RANCID. A term used to describe fats that have undergone partial decomposition.

RANDOM MATING. A system of mating where every female (cow and/or heifer) has an equal or random chance of being assigned to any bull used for breeding in a particular breeding season. Random mating is required for accurate progeny tests.

RATE OF GENETIC IMPROVEMENT. Rate of improvement per unit of time (year). The rate of improvement is dependent on: (1) heritability of traits considered; (2) selection differentials; (3) genetic correlations among traits considered, (4) generation interval in the herd; and (5) the number of traits for which selections are made.

RATE OF PASSAGE. The time taken by undigested residues from a given meal to reach the feces. (A stained undigestible material is commonly used to estimate rate of passage.)

RATIO. Used to indicate the performance of an individual in relation to the average of all animals of the same group. It is calculated as follows:

$$\frac{\text{Individual record}}{\text{Average of animals in group}} \times 100$$

It is a record or an index of individual deviation from the group average expressed in terms of percentage. A ratio of 100 is average for a particular group. Thus, ratios above 100 indicate animals above average, whereas ratios below 100 indicate animals below average.

RATION(S). The amount of feed supplied to an animal for a definite period, usually for a 24-hour period. However, in practical usage, the word *ration* implies the feed fed to an animal without limitation to the time in which it is consumed.

REACH. The difference between the average performance level of the cow herd and the performance level of the herd sire.

RECESSIVE CHARACTER. A characteristic which appears only when both members of a pair of genes are alike. Opposite of dominant.

RECESSIVE GENE. Recessive genes affect the phenotype only when present in a homozygous condition. Recessive genes must be received from both parents before the phenotype caused by the recessive genes can be observed.

RECOMBINANT DNA. DNA prepared in the laboratory by breaking up and slicing together DNA from different species of organisms.

RED MEAT. Meat that is red when raw. It includes beef, veal, pork, mutton, and lamb.

REFERENCE SIRE. A bull designated to be used as a benchmark in progeny testing other bulls (young sires). Progeny by reference sires in several herds enable comparisons to be made between bulls not producing progeny in the same herd(s).

REGISTERED. Designating purebred animals whose pedigrees are recorded in the breed registry.

REGRESSION *(Regressed).* A measure of the relationship between two variables. The response of value of one trait can be predicted by knowing the value of the other variable. For example, easily obtained carcass traits (hot carcass weight, fat thickness, ribeye area, and percent of internal fat) are used to predict percent cutability.

REGURGITATION. The casting up (backward flow) of undigested food from the stomach to the mouth, as by ruminants.

REPLACEMENT HEIFERS. The top end of the heifer calves selected to replace the older cows that are culled from the herd.

RETAIL CUTS. The cuts of meat that are purchased by the consumer.

RETAINED OWNERSHIP. This commonly refers to cow-calf producers maintaining ownership of their cattle through the feedlot.

RETAINED PLACENTA. A placenta not expelled at parturition or shortly thereafter.

RETICULUM. The second compartment of the ruminant stomach. It has a honeycomb-texture lining; so, it is often called the *honeycomb.*

RIB-EYE AREA (REA). Area of the *longissimus dorsi* muscle between the twelfth and thirteenth ribs, measured in square inches. It is also referred to as the loin-eye area.

ROTATIONAL CROSSBREEDING. System of crossing two or more breeds where the crossbred females are bred to bulls of the breed containing the least genes to that female's genotype. Rotation systems maintain relatively high levels of heterosis and produce replacement heifers from within the system. Opportunity to select replacement heifers is greater for rotation systems than for other crossbreeding systems.

ROUGHAGE. Feed consisting of bulky and coarse plants or plant parts, containing a high-fiber content and low total digestible nutrients, arbitrarily defined as feed with over 18% crude fiber. Roughage may be classed as either dry or green.

ROUND TURN. A purchase and its liquidating sale, or a sale and its liquidating purchase.

RUMEN. The first stomach compartment of a ruminant—the paunch. It is a large nutrient-producing vat.

RUMEN FLORA. The microorganisms of the rumen.

RUMENITIS. Inflammation of the wall of the rumen.

RUMINANTS. Even-toed, hoofed animals that ruminate (regurgitate and chew a cud) and have a complex four-compartment stomach. Cattle, sheep, and goats are ruminants.

RUN ON. To graze or pasture.

RUPTURE. See **HERNIA**.

S

SATURATED FAT. A completely hydrogenated fat—each carbon atom is associated with the maximum number of hydrogens; there are no double bonds.

SCABIES. See **MANGE**.

SCOURING. One of the major problems facing livestock producers is scouring (diarrhea) in young animals. It may be due to feeding practices, management practices, environment, or disease.

SCROTAL HERNIA. A weakness of the inguinal canal which allows part of the viscera to pass out into the scrotum.

SCROTUM. The saclike pouch that suspends the testicles outside the male animal.

SCURS. Small, rounded portions of horn tissue attached to the skin at the horn pits of polled animals. They may also be called *buttons.*

SECURITY DEPOSIT (Initial). Synonymous with the term *margin*, a cash amount of funds which must be deposited with the broker for each contract as a guarantee of fulfillment of the futures contract. It is not considered as part payment of purchase.

SECURITY DEPOSIT (Maintenance). A sum, usually smaller than, but part of, the original deposit or margin which must be maintained on deposit at all times. If a customer's equity in any futures position drops to or under the maintenance level, the broker must issue a call for the amount of money required to restore the customer's equity in the account to the original margin level.

SEEDSTOCK. Breeding animals.

SEEDSTOCK BREEDERS. Producers of breeding stock for purebred and commercial breeders. Progressive seedstock breeders have comprehensive programs designed to produce an optimum or desirable combination of economical traits (genetic package) that will ultimately increase the profitability of commercial beef production.

SELECTION. Determining which animals in a population will produce the next generation. Caretakers practice artificial selection while nature practices natural selection.

SELECTION DIFFERENTIAL *(Reach).* The difference between the average for a trait of the selected cattle and the average of the group from which they came. The expected response from selection for a trait is equal to selection differential times the heritability of the trait.

SELECTION INDEX. A formula which combines all important traits into one overall value or index.

SELENIUM. An element that functions with glutathione peroxidase, an enzyme which enables the tripeptide glutathione to perform its role as a biological antioxidant in the body. This explains why deficiencies of selenium and vitamin E result in similar signs—loss of appetite and slow growth.

SELF-FED. Provided with a part or all of the ration on a continuous basis, thereby permitting the animal to eat at will.

SELF-FEEDER. A feed container by means of which animals can eat at will. (See **AD LIB FEEDING**).

SEMEN. Sperm mixed with fluids from the accessory glands.

SEMILETHALS *(Sub-lethals).* Genetic factors which cause the death of the young after birth or sometime later in life if environmental situations aggravate the conditions.

SERUM. The colorless fluid portion of blood remaining after clotting and removal of corpuscles. It differs from plasma in that the fibrinogen has been removed.

SERVICE. The mating of a female by a male.

SETTLED. Used to indicate that the animal has become pregnant.

SETTLEMENT PRICE. The daily price at which the clearinghouse clears all trades. The settlement price of each day's trading is based upon the closing range of that day's trading. Settlement prices are used to determine both margin calls and invoice prices for deliveries.

SEX CELLS. The egg and the sperm, which unite to create life. They transmit genetic characteristics from the parents to the offspring.

SHORT. The selling of an open futures contract. A trader whose net position in the futures market shows an excess of open sales over open purchases is said to be short.

SHORT HEDGE. When the trader owns an inventory of a commodity and hedges by selling an equivalent amount of futures contracts, the trader has sold short and has what is called a short hedge. An example of a short hedge is selling on the futures markets contracts of live cattle which represent cattle that are on feed in the feedlot in order to protect the enterprise against a severe decline in the market.

SHORT YEARLING. A beef animal that is over 1 year of age but under 18 months of age.

SHOT. An injection.

SHOW BOX *(Tackbox).* A container in which to keep all show equipment and paraphernalia.

SHRINKAGE.

- The amount of loss in body weight when animals are exposed to adverse conditions, such as being transported, severe weather, or shortage of feed.
- The loss in carcass weight during the aging process.

SHY BREEDER. A reserved, or timid, breeder.

SIB. A brother or sister of an individual.

SIB TESTING. A method of selection in which an animal is selected on the basis of the performance of its brothers or sisters.

SICK PEN. Isolated pen in a feedlot where cattle are treated after they have been removed from a feedlot pen. Sometimes referred to as a hospital pen.

SIGNS. The word used when speaking of animal symptoms. Animals show *signs* of abnormality, whereas people can relate their symptoms of ill-health.

SIRE. The male parent of an animal. To father or beget.

SLOTTED FLOORS. Floors with slots through which the feces and urine pass to a storage area below or nearby.

SOFTWARE. Program instructions to make computer hardware function.

SORGHUM. The *Sorghum* genus can be broken down into four broad classifications according to use: (1) grain sorghum; (2) grass sorghums for hay, silage, and pasture; (3) syrup sorghums or sorgos; and (4) broomcorns for fiber production.

SPECIES. A group of animals possessing common characteristics that distinguish them from other animals.

SPECULATOR. A person who is willing to accept the risks associated with price changes in the hope of profiting from increases or decreases in futures prices.

SPERMATIC CORD. The *cord* that passes from the scrotum to the body and includes the vas deferens, muscle, artery, vein, nerves, and lymphatics leading to and from the testicles.

SPERMATOGENESIS. The formation and development of sperm in the testicles.

SPOOK. To scare or frighten animals.

SPORE. A bacterial cell that has an extra thick wall, making it resistant to most sterilization processes.

SPORT. An animal or a plant that deviates abruptly from the parent type—a mutation.

SPOT COMMODITY. The actual physical commodity such as live cattle as distinguished from futures. Also, known as *cash commodity*.

SPREAD. A market position that is simultaneously long and short equivalent amounts of the same or related commodities. In some markets, the term *straddle* is used synonymously.

STABILIZED. Made more resistant to chemical change by the addition of a particular substance.

STAG. A male that was castrated after the secondary sexual characteristics developed sufficiently to give the appearance of a mature male.

STANDING HEAT. Period in a female's heat (estrus) during which she will stand still when being mounted.

STEER. A male bovine castrated before the development of secondary sex characteristics.

STERILE. Incapable of reproducing.

STERILITY. Temporary or permanent reproductive failure, resulting from anestrus (lack of heat), failure to conceive, or abortion.

STERILIZATION. The process of sterilizing by boiling or by use of chemicals to make free of germs or microorganisms that could cause infections or disease. Primarily used on cutting or surgical tools used in livestock management.

STILLBORN. Born lifeless; dead at birth.

STOCKERS. Calves and yearlings, both steers and heifers, that are intended for eventual finishing and slaughtering and which are being fed and cared for in such manner that growth rather than finishing will be realized. They are generally younger and thinner than feeder cattle.

STOVER. Dried stalks and leaves, but not the grain portion, of corn or milo.

STRADDLE. The purchase in one market and the simultaneous sale of the same commodity in some other market, *e.g.*, the purchase of an October live cattle contract and the sale of a February contract. It may also refer to the purchase of one commodity against the sale of a different commodity, cattle vs hogs, *e.g.*, or cattle vs corn, both of which should normally be closely allied in price movements.

STRAGGLER. An animal that strays or wanders from a herd or flock.

STRAW. The plant residue remaining after separation of the seeds in threshing.

STRESS. Any physical or emotional factor to which an animal fails to make a satisfactory adaptation. Stress may be caused by excitement, temperament, fatigue, shipping, disease, heat or cold, nervous strain, number of animals together, previous nutrition, breed, age, or management. The greater the stress, the more exacting the nutritive requirements.

SUBCUTANEOUS. Situated or occurring beneath the skin.

SUBPRIMAL CUTS. Smaller-than-primal cuts, such as when the primal round is split into top round, bottom round, eye round, and sirloin tip. The term is used in boxed beef programs.

SUCKLE. To nurse at the mammary glands.

SUPPLEMENT. A feed or feed mixtures used to improve the nutritional value of basal feeds (*e.g.*, protein supplement—soybean meal). Supplements are usually rich in protein, minerals, vitamins, antibiotics, or a combination of part or all of these; and they are usually combined with basal feeds to produce a complete feed.

SYMMETRY. A balanced development of all parts.

SYNTHESIS. The bringing together of two or more substances to form a new material.

SYNTHETIC BREEDS. Also referred to as composite breeds. See **COMPOSITE BREED**.

SYNTHETICS. Artificially produced products that may be similar to natural products.

T

TALLOW. The extracted fat of cattle and sheep.

TANDEM SELECTION. Type of selection in which there is selection for only one trait at a time until the desired improvement in that particular trait is reached.

TATTOO. Permanent identification of animals produced by placing indelible ink under the skin; generally put in the ears of young animals.

TDN. See **TOTAL DIGESTIBLE NUTRIENTS.**

TERMINAL OR CENTRAL MARKETS. Livestock trading centers which generally have several commission firms and an independent stockyards company.

TERMINAL SIRES. Sires used in a crossbreeding system where all their progeny, both male and female, are marketed. For example, F_1 crossbred dams could be bred to sires of a third breed and all calves marketed. Although this system allows maximum heterosis and complementarity of breeds, replacement females must come from other herds.

TESTICLE. A male gland which produces sperm. There are two testicles.

TESTOSTERONE. Male sex hormone. Produced by the Leydig cells in the testicles.

THRIFTY. Healthy and vigorous in appearance.

TICKER. A teletype machine which sends and receives futures market and cash market information.

TOTAL DIGESTIBLE NUTRIENTS (TDN). A term which indicates the energy value of a feedstuff. It is computed by use of the following formula:

$$\% \text{ TDN} = \% \text{ DCP} + \% \text{ DCF} + \% \text{ DNFE} + (\% \text{ DEE} \times 2.25)$$

Where DCP = digestible crude protein: DCF = digestible crude fiber; DNFE = digestible nitrogen-free extract; and DEE = digestible ether extract. One pound of TDN = 2,000 kcal of digestible energy.

TOTAL SOLIDS. The sum of dissolved and undissolved constituents in water or wastewater, usually stated in milligrams per liter.

TOXEMIA. A general systemic toxification or poisoning caused by the absorption of toxic substances (toxins) produced by bacteria. The toxins are usually produced at and absorbed from a localized source of infection.

TOXIC. Of a poisonous nature.

TOXOID. A toxin that has been treated so that it has lost its toxicity but is still capable of causing the production of antitoxin when injected into an animal.

TRACE ELEMENT (*Mineral*). A chemical element used in minute amounts of organisms and held essential to their physiology. The essential trace elements are cobalt, copper, iodine, iron, manganese, selenium, and zinc.

TRAIT RATIO. An expression of an animal's performance for a particular trait relative to the herd or contemporary group average. It is usually calculated for most traits as:

$$\frac{\text{Individual record}}{\text{Average of animals in group}} \times 100$$

TRANQUILIZER. A drug used to reduce physical or nervous tension. It is used to quiet and calm nervous animals and thereby make them more receptive to handling attempts.

TRAY-READY BEEF. Retail cuts of beef that are cut and packaged at the packing plant for retail sales by retail stores.

TREND. The direction prices are taking.

TRIHYBRID. An individual that is heterozygous with respect to three pairs of allelic genes.

TRIPE. Edible meat from the walls of the stomach of ruminants.

TWENTY-EIGHT HOUR LAW IN RAIL SHIPMENTS. This law prohibits transporting livestock by rail for a longer period than 28 consecutive hours without unloading, feeding, watering, and resting 5 consecutive hours before resuming transportation. On request of the owner, the period can be extended to 36 hours.

TYPE.

- An ideal or a standard of perfection combining all the characters that contribute to the animal's usefulness for a specific purpose.
- Physical conformation of an animal.

U

UDDER. The encased group of mammary glands with each gland provided with a nipple or teat.

ULTRASOUND. This refers to high-frequency sound waves which show visual outlines of internal body structures. The machine sends sound waves into the animal and records these waves as they bounce off the tissues.

UNDERFEEDING. Usually, this refers to not providing sufficient energy. The degree of lowered production therefrom is related to the extent of underfeeding and the length of time it exists.

UNDERSHOT JAW. The lower jaw is longer than the upper jaw.

UNIDENTIFIED FACTORS. These are referred to as *unidentified* or *unknown* factors because they have not yet been isolated or synthesized in the laboratory. Nevertheless, rich sources of these factors and their effects have been well established. There is evidence that the growth factors exist in dried whey, marine and packing house byproducts, distillers' solubles, antibiotic fermentation residues, alfalfa meal, and certain green forages. Most of the unidentified factor sources are added to the diet at a level of 1 to 3%.

UNSATURATED FAT. A fat having one or more double bonds; not completely hydrogenated.

UNSATURATED FATTY ACID. Any one of several fatty acids containing one or more double bonds, such as oleic, linoleic, linolenic, and arachidonic acids.

UNTHRIFTINESS. Lack of vigor; poor growth or development; the quality or state of being unthrifty in animals.

URINARY CALCULI. *Stones* formed in any part of the urinary tract.

U.S. DEPARTMENT OF AGRICULTURE (USDA). An executive department of the U.S. government that serves all branches of the nation's agriculture. *The address:* 14th Street & Independence Ave. S.W., Washington, DC 20250.

USDA LIVESTOCK MARKETING INFORMATION CENTER. In cooperation with State Extension Services, the Marketing Center provides market information on cattle, hogs, lambs, beef, pork, and land. *The address:* 655 Parfet, Suite E310, Lakewood, CO 80215-5517.

USDA YIELD GRADE. Measurements of carcass cutability categorized into numerical categories with 1 being the leanest and 5 being the fattest. Yield grade and cutability are based on the same four carcass traits.

USP (*United States Pharmacopoeia*). A unit of measurement or potency of biologicals that usually coincides with an international unit. (See **IU**.)

UTERUS. Part of the reproductive tract in which the fertilized ovum attaches and develops into an embryo and a fetus.

V

VACCINATION (*Shot*). An injection of vaccine, bacterin, antiserum, or antitoxin to produce immunity or tolerance to disease.

VACCINE. A suspension of attenuated or killed microorganisms (bacteria, viruses, or rickettsiae) administered for the prevention, improvement, or treatment of infectious diseases.

VAGINA. The part of the female reproductive tract between the cervix and the external opening.

VARIANCE. Variance is a statistic which describes the variation in a trait. Without variation, no genetic progress is possible, since genetically superior animals would not be distinguishable from genetically inferior

VARIETY MEATS. Edible byproducts (*e.g.*, liver, heart, tongue, tripe).

VASECTOMY. The surgical removal of a section of the vas deferens thereby rendering a male sterile without affecting his libido. Sperm cannot be transported from the testicles at the time of ejaculation.

VEAL. Meat from a young calf, usually less than 3 months of age. In its strictest sense, veal production implies that the calves have been fed milk as their only source of nutrients.

VECTORS. Living organisms which carry pathogens.

VERMIFUGE (*Vermicide*). Any chemical substance given to animals to kill internal parasitic worms.

VETERINARIAN. One who treats diseases or afflictions of animals medically and surgically; a practitioner of veterinary medicine or surgery.

VIRULENT. Pathogenic or capable of causing disease.

VIRUS. One of a group of minute infectious agents. They lack independent metabolism and can only multiply within living host cells. Difficult to identify and control.

VITAMIN SUPPLEMENTS. Rich, synthetic or natural feed sources of one or more of the complex organic compounds, called vitamins, that are required in minute amounts by animals for normal growth, production, reproduction, and/or health.

VITAMINS. Complex organic compounds that function as parts of enzyme systems essential for the transformation of energy and the regulation of metabolism of the body, and required in minute amounts by one or more animal species for normal growth, production, reproduction, and/or health. All vitamins must be present in the ration for normal functioning, except for B vitamin in the ruminants (cattle and sheep) and vitamin C.

VOID. To evacuate feces and/or urine.

VOLATILE ACIDS. Fatty acids, containing six or less carbon atoms, that are soluble in water and that can be steam-distilled at atmospheric pressure. Volatile acids are commonly reported as acetic acid equivalent.

VOLUME. The number of purchases or sales of a commodity futures contract made during a specified period of time.

VOMITING. The forcible expulsion of the contents of the stomach through the mouth.

W

WASTY.

- A carcass with too much fat, requiring excessive trimming.
- Paunchy live animals.

WEANING. The stopping of young animals from suckling their mothers.

WEANLING. A weaned calf.

WEIGHT PER DAY OF AGE (WDA). Weight of an individual divided by days of age.

WILL. A set of instructions drawn up by or for an individual which details how he/she wishes his/her estate to be handled after his/her passing.

WITHDRAWAL TIME. The time required between the application or feeding of a drug or an additive and the slaughter of the animal to prevent any residue of the drug from remaining in the carcass. Withdrawal times are legally specified by the Food and Drug Administration (FDA).

WORKING CAPITAL. Excess of current assets over current liabilities.

WORMER. A commercial product containing an anthelmintic administered by stomach tube, in the feed or water, or by syringe to control internal parasites. Sometimes referred to as a dewormer.

Y

YEARLING. An animal that is 12 to 18 months of age.

YIELD GRADES. The USDA grades indicate the percent of trimmed, boneless major retail cuts to be derived from the beef carcass.

YOUR BASIS. The difference between your own local cash price and the future price on sale day.

Z

ZYGOTE. A fertilized egg.

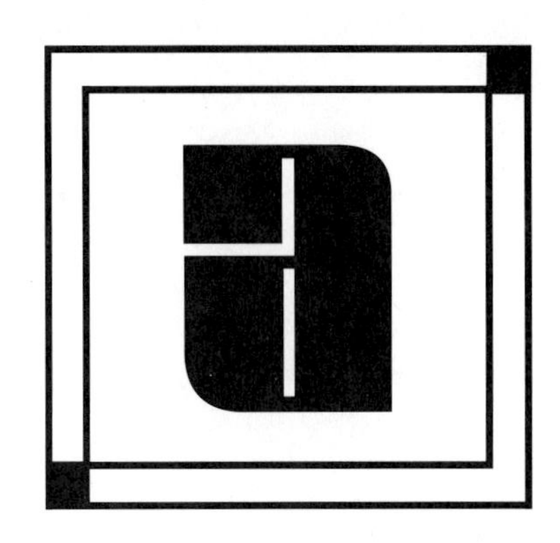

APPENDIX

SECTION I
NUTRIENT REQUIREMENTS OF BEEF CATTLE

TABLES I-1
1984 NATIONAL RESEARCH COUNCIL
NUTRIENT REQUIREMENTS OF BEEF CATTLE

TABLE I-1a
DAILY NUTRIENT REQUIREMENTS OF BREEDING CATTLE[1]

Weight[2]		Daily Gain[3]		Daily Consumption[4] As-fed		Daily Consumption[4] Moisture-free Dry Matter		Total Protein		Energy TDN		Energy ME	Energy NE$_m$	Energy NE$_g$	Calcium	Phos-phorus	Vitamin A[5]
(lb)	*(kg)*	*(lb)*	*(kg)*	*(lb)*	*(kg)*	*(lb)*	*(kg)*	*(lb)*	*(kg)*	*(lb)*	*(kg)*	*(Mcal)*	*(Mcal)*	*(Mcal)*	*(g)*	*(g)*	*(1,000s IU)*
Pregnant yearling heifers—Last third of pregnancy																	
700	*318*	0.9	*0.4*	**17.0**	***7.7***	15.3	*7.0*	1.3	*0.6*	8.5	*3.9*	13.9	7.95	NA[6]	19	14	19
700	*318*	1.4	*0.6*	**17.6**	***8.0***	15.8	*7.2*	1.4	*0.6*	9.6	*4.4*	15.7	7.95	0.87	24	15	20
700	*318*	1.9	*0.9*	**17.6**	***8.0***	15.8	*7.2*	1.5	*0.7*	10.6	*4.8*	17.4	7.95	1.89	27	16	20
750	*341*	0.9	*0.4*	**17.9**	***8.1***	16.1	*7.3*	1.3	*0.6*	8.9	*4.0*	14.6	8.25	NA	20	14	20
750	*341*	1.4	*0.6*	**18.4**	***8.4***	16.6	*7.5*	1.5	*0.7*	10.0	*4.5*	16.4	8.25	0.92	24	16	21
750	*341*	1.9	*0.9*	**18.4**	***8.4***	16.6	*7.5*	1.6	*0.7*	11.1	*5.0*	18.2	8.25	1.99	28	17	21
800	*364*	0.9	*0.4*	**18.7**	***8.5***	16.8	*7.6*	1.4	*0.6*	9.2	*4.2*	15.2	8.56	NA	21	15	21
800	*364*	1.4	*0.6*	**19.3**	***8.8***	17.4	*7.9*	1.5	*0.7*	10.4	*4.7*	17.1	8.56	0.96	25	16	22
800	*364*	1.9	*0.9*	**19.4**	***8.8***	17.5	*8.0*	1.6	*0.7*	11.6	*5.3*	19.0	8.56	2.09	28	17	22
850	*386*	0.9	*0.4*	**19.6**	***8.9***	17.6	*8.0*	1.4	*0.6*	9.6	*4.4*	15.7	8.85	NA	21	16	22
850	*386*	1.4	*0.6*	**20.2**	***9.2***	18.2	*8.3*	1.6	*0.7*	10.8	*4.9*	17.8	8.85	1.01	25	17	23
850	*386*	1.9	*0.9*	**20.3**	***9.2***	18.3	*8.3*	1.7	*0.8*	12.1	*5.5*	19.8	8.85	2.19	28	18	23
900	*409*	0.9	*0.4*	**20.3**	***9.2***	18.3	*8.3*	1.5	*0.7*	9.9	*4.5*	16.3	9.15	NA	22	17	23
900	*409*	1.4	*0.6*	**21.1**	***9.6***	19.0	*8.6*	1.6	*0.7*	11.3	*5.1*	18.5	9.15	1.05	26	18	24
900	*409*	1.9	*0.9*	**21.3**	***9.7***	19.2	*8.7*	1.7	*0.8*	12.5	*5.7*	20.6	9.15	2.28	28	19	24
950	*432*	0.9	*0.4*	**21.1**	***9.6***	19.0	*8.6*	1.5	*0.7*	10.3	*4.7*	16.9	9.44	NA	23	17	24
950	*432*	1.4	*0.6*	**22.0**	***10.0***	19.8	*9.0*	1.7	*0.8*	11.7	*5.3*	19.1	9.44	1.09	26	19	25
950	*432*	1.9	*0.9*	**22.2**	***10.1***	20.0	*9.0*	1.8	*0.8*	13.0	*5.9*	21.3	9.44	2.38	29	19	25
Dry pregnant mature cows—Middle third of pregnancy																	
800	*364*	0.0	*0.0*	**17.0**	***7.7***	15.3	*7.0*	1.1	*0.5*	7.5	*3.4*	12.3	6.41	NA	12	12	19
900	*409*	0.0	*0.0*	**18.6**	***8.5***	16.7	*7.6*	1.2	*0.5*	8.2	*3.7*	13.4	7.00	NA	14	14	21
1,000	*454*	0.0	*0.0*	**20.1**	***9.1***	18.1	*8.2*	1.3	*0.6*	8.8	*4.0*	14.5	7.57	NA	15	15	23
1,100	*500*	0.0	*0.0*	**21.7**	***9.9***	19.5	*8.9*	1.4	*0.6*	9.5	*4.3*	15.6	8.13	NA	17	17	25
1,200	*545*	0.0	*0.0*	**23.1**	***10.5***	20.8	*9.5*	1.4	*0.6*	10.1	*4.6*	16.6	8.68	NA	18	18	26
1,300	*591*	0.0	*0.0*	**24.4**	***11.1***	22.0	*10.0*	1.5	*0.7*	10.8	*4.9*	17.7	9.22	NA	20	20	28
1,400	*636*	0.0	*0.0*	**25.9**	***11.8***	23.3	*10.6*	1.6	*0.7*	11.4	*5.2*	18.7	9.75	NA	21	21	30
Dry pregnant mature cows—Last third of pregnancy																	
800	*364*	0.9	*0.4*	**18.7**	***8.5***	16.8	*7.6*	1.4	*0.6*	9.2	*4.2*	15.0	8.56	NA	20	15	21
900	*409*	0.9	*0.4*	**20.2**	***9.2***	18.2	*8.3*	1.5	*0.7*	9.8	*4.5*	16.2	9.15	NA	22	17	23
1,000	*454*	0.9	*0.4*	**21.8**	***9.9***	19.6	*8.9*	1.6	*0.7*	10.5	*4.8*	17.3	9.72	NA	23	18	25
1,100	*500*	0.9	*0.4*	**23.3**	***10.6***	21.0	*9.5*	1.6	*0.7*	11.2	*5.1*	18.3	10.28	NA	25	20	26
1,200	*545*	0.9	*0.4*	**24.8**	***11.3***	22.3	*10.1*	1.7	*0.8*	11.8	*5.4*	19.4	10.83	NA	26	21	28
1,300	*591*	0.9	*0.4*	**26.2**	***11.9***	23.6	*10.7*	1.8	*0.8*	12.5	*5.7*	20.4	11.37	NA	28	23	30
1,400	*636*	0.9	*0.4*	**27.7**	***12.6***	24.9	*11.3*	1.9	*0.9*	13.1	*6.0*	21.5	11.90	NA	29	24	32
Two-year-old heifers nursing calves—First 3–4 months postpartum—10 lb *(4.5 kg)* milk/day																	
700	*318*	0.5	*0.2*	**17.7**	***8.0***	15.9	*7.2*	1.8[7]	*0.8*	10.3	*4.7*	17.0	9.20[8]	0.87	26	17	28
750	*341*	0.5	*0.2*	**18.6**	***8.5***	16.7	*7.6*	1.8	*0.8*	10.8	*4.9*	17.7	9.51	0.92	26	18	30
800	*364*	0.5	*0.2*	**19.6**	***8.9***	17.6	*8.0*	1.9	*0.9*	11.2	*5.1*	18.4	9.81	0.96	27	19	31
850	*386*	0.5	*0.2*	**20.4**	***9.3***	18.4	*8.4*	1.9	*0.9*	11.6	*5.3*	19.1	10.11	1.01	27	19	33

(Continued)

TABLE I-1a (Continued)

Weight[2]		Daily Gain[3]		Daily Consumption[4] As-fed		Daily Consumption[4] Moisture-free Dry Matter		Total Protein		Energy TDN		Energy ME	Energy NE_m	Energy NE_g	Calcium	Phos-phorus	Vitamin A[5]
(lb)	*(kg)*	*(lb)*	*(kg)*	*(lb)*	*(kg)*	*(lb)*	*(kg)*	*(lb)*	*(kg)*	*(lb)*	*(kg)*	*(Mcal)*	*(Mcal)*	*(Mcal)*	*(g)*	*(g)*	*(1,000s IU)*
Two-year-old heifers nursing calves—First 3–4 months postpartum—10 lb *(4.5 kg)* milk/day (continued)																	
900	*409*	0.5	*0.2*	**21.3**	***9.7***	19.2	*8.7*	2.0	*0.9*	12.0	*5.5*	19.8	10.40	1.05	28	20	34
950	*432*	0.5	*0.2*	**22.2**	***10.1***	20.0	*9.0*	2.0	*0.9*	12.5	*5.7*	20.5	10.69	1.09	28	21	35
1,000	*454*	0.5	*0.2*	**23.1**	***10.5***	20.8	*9.5*	2.1	*1.0*	12.9	*5.9*	21.1	10.98	1.14	29	22	37
Cows nursing calves—Average milking ability—First 3–4 months postpartum—10 lb *(4.5 kg)* milk/day																	
800	*364*	0.0	*0.0*	**19.2**	***8.7***	17.3	*7.9*	1.8	*0.8*	10.1	*4.6*	16.6	9.81	NA	23	17	31
900	*409*	0.0	*0.0*	**20.1**	***9.1***	18.8	*8.5*	1.9	*0.9*	10.8	*4.9*	17.7	10.40	NA	24	19	33
1,000	*454*	0.0	*0.0*	**22.4**	***10.2***	20.2	*9.2*	2.0	*0.9*	11.5	*5.2*	18.8	10.98	NA	25	20	36
1,100	*500*	0.0	*0.0*	**24.0**	***10.9***	21.6	*9.8*	2.0	*0.9*	12.1	*5.5*	19.9	11.54	NA	27	22	38
1,200	*545*	0.0	*0.0*	**25.6**	***11.6***	23.0	*10.5*	2.1	*1.0*	12.8	*5.8*	21.0	12.09	NA	28	23	41
1,300	*591*	0.0	*0.0*	**27.0**	***12.3***	24.3	*11.0*	2.2	*1.0*	13.4	*6.1*	22.0	12.63	NA	30	25	43
1,400	*636*	0.0	*0.0*	**28.4**	***12.9***	25.6	*11.6*	2.3	*1.0*	14.0	*6.4*	23.0	13.15	NA	31	26	46
Cows nursing calves—Superior milking ability—First 3–4 months postpartum—20 lb *(9.1 kg)* milk/day																	
800	*364*	0.0	*0.0*	**17.4**	***7.9***	15.7	*7.1*	2.2	*1.0*	12.1	*5.5*	19.9	13.22	NA	34	22	28
900	*409*	0.0	*0.0*	**20.8**	***9.5***	18.7	*8.5*	2.4	*1.1*	13.1	*6.0*	21.5	13.81	NA	35	24	33
1,000	*454*	0.0	*0.0*	**22.9**	***10.4***	20.6	*9.4*	2.5	*1.1*	13.8	*6.3*	22.7	14.38	NA	36	25	37
1,100	*500*	0.0	*0.0*	**24.8**	***11.3***	22.3	*10.1*	2.6	*1.2*	14.5	*6.6*	23.8	14.94	NA	38	27	40
1,200	*545*	0.0	*0.0*	**26.4**	***12.0***	23.8	*10.8*	2.7	*1.2*	15.2	*6.9*	24.9	15.49	NA	39	28	42
1,300	*591*	0.0	*0.0*	**28.1**	***12.8***	25.3	*11.5*	2.8	*1.3*	15.9	*7.2*	26.0	16.03	NA	41	30	45
1,400	*636*	0.0	*0.0*	**29.7**	***13.5***	26.7	*12.1*	2.9	*1.3*	16.5	*7.5*	27.1	16.56	NA	42	31	47
Bulls, maintenance and slow rate of growth (regain body condition)																	
<1,300 For growth and development, use requirements for bulls in Table I-1c																	
1,300	*591*	1.0	*0.5*	**28.2**	***12.8***	25.4	*11.5*	1.9	*0.9*	14.2	*6.5*	23.3	9.22	2.20	25	22	45
1,300	*591*	1.5	*0.7*	**29.0**	***13.2***	26.1	*11.9*	2.0	*0.9*	15.6	*7.1*	25.5	9.22	3.43	28	23	46
1,300	*591*	2.0	*0.9*	**29.1**	***13.2***	26.2	*11.9*	2.2	*1.0*	16.8	*7.6*	27.6	9.22	4.71	31	24	46
1,400	*636*	1.0	*0.5*	**29.8**	***13.5***	26.8	*12.2*	2.0	*0.9*	15.0	*6.8*	24.6	9.75	2.33	26	23	48
1,400	*636*	1.5	*0.7*	**30.7**	***14.0***	27.6	*12.5*	2.1	*1.0*	16.5	*7.5*	27.0	9.75	3.63	29	24	49
1,400	*636*	2.0	*0.9*	**30.8**	***14.0***	27.7	*12.6*	2.2	*1.0*	17.8	*8.1*	29.1	9.75	4.98	31	25	49
1,500	*682*	0.0	*0.0*	**28.0**	***12.7***	25.2	*11.5*	1.7	*0.8*	12.2	*5.5*	20.0	10.26	NA	23	23	45
1,500	*682*	1.0	*0.5*	**31.4**	***14.3***	28.3	*12.9*	2.1	*1.0*	15.8	*7.2*	25.9	10.26	2.45	27	24	50
1,500	*682*	1.5	*0.7*	**32.2**	***14.6***	29.0	*13.2*	2.2	*1.0*	17.3	*7.9*	28.4	10.26	3.82	29	25	51
1,600	*727*	0.0	*0.0*	**29.4**	***13.4***	26.5	*12.0*	1.8	*0.8*	12.8	*5.8*	21.0	10.77	NA	23	24	47
1,600	*727*	1.0	*0.5*	**33.0**	***15.0***	29.7	*13.5*	2.2	*1.0*	16.6	*7.5*	27.2	10.77	2.57	29	26	53
1,600	*727*	1.5	*0.7*	**33.8**	***15.4***	30.4	*13.8*	2.3	*1.0*	18.2	*8.3*	29.8	10.77	4.01	31	27	54
1,700	*773*	0.0	*0.0*	**30.8**	***14.0***	27.7	*12.6*	1.9	*0.9*	13.4	*6.1*	22.0	11.28	NA	26	26	49
1,700	*773*	0.5	*0.2*	**32.9**	***15.0***	29.6	*13.5*	2.1	*1.0*	15.4	*7.0*	25.3	11.28	1.26	27	26	52
1,800	*818*	0.0	*0.0*	**32.1**	***14.6***	28.9	*13.1*	2.0	*0.9*	14.0	*6.4*	23.0	11.77	NA	27	27	51
1,800	*818*	0.5	*0.2*	**34.3**	***15.6***	30.9	*14.0*	2.2	*1.0*	16.1	*7.3*	26.4	11.77	1.31	28	28	55
1,900	*864*	0.0	*0.0*	**33.4**	***15.2***	30.1	*13.7*	2.0	*0.9*	14.6	*6.6*	23.9	12.26	NA	29	29	53
1,900	*864*	0.5	*0.2*	**35.8**	***16.3***	32.2	*14.6*	2.2	*1.0*	16.8	*7.6*	27.5	12.26	1.37	29	29	57
2,000	*909*	0.0	*0.0*	**34.8**	***15.8***	31.3	*14.2*	2.1	*1.0*	15.2	*6.9*	24.9	12.74	NA	30	30	55
2,100	*955*	0.0	*0.0*	**36.1**	***16.4***	32.5	*14.8*	2.2	*1.0*	15.7	*7.1*	25.8	13.21	NA	32	32	58
2,200	*1,000*	0.0	*0.0*	**37.3**	***17.0***	33.6	*15.3*	2.3	*1.0*	16.3	*7.4*	26.7	13.68	NA	33	33	60

[1]Adapted from *Nutrient Requirements of Beef Cattle*, sixth revised edition, National Research Council-National Academy of Sciences, 1984.

[2]Average weight for a feeding period.

[3]Approximately 0.9 ± 0.2lb of weight gain/day over the last third of pregnancy is accounted for by the products of conception. Daily 2.15 Mcal of NE_m and 0.1 lb of protein are provided for this requirement for a calf with a birth weight of 80 lb.

[4]Consumption should vary depending on the energy concentration of the diet and environmental conditions. These intakes are based on the energy concentration shown in the table and assuming a thermoneutral environment without snow or mud conditions. If the energy concentrations of the ration to be fed exceed the tabular value, limit feeding may be required.

[5]Vitamin A requirements per pound of diet are 1,273 IU for pregnant heifers and cows and 1,773 IU for lactating cows and breeding bulls.

[6]Not applicable.

[7]Includes 0.03 lb protein/lb of milk produced for all heifers and cows nursing calves.

[8]Includes 0.34 Mcal NE_m/lb of milk produced for all heifers and cows nursing calves.

TABLE I-1b
NUTRIENT REQUIREMENTS IN DIET FOR BREEDING CATTLE[1]

Weight[2]		Daily Gain[3]		Daily Consumption[4] As-fed		Daily Consumption[4] Moisture-free Dry Matter		Moisture Basis[5] A-F (As-fed) M-F (Moisture-free)	Total Protein	TDN	Energy ME (Mcal per)		Energy NE$_m$ (Mcal per)		Energy NE$_g$ (Mcal per)		Calcium	Phosphorus	Vitamin A (IU per)	
(lb)	(kg)	(lb)	(kg)	(lb)	(kg)	(lb)	(kg)		(%)	(%)	(lb)	(kg)	(lb)	(kg)	(lb)	(kg)	(%)	(%)	(lb)	(kg)
Pregnant yearling heifers—Last third of pregnancy																				
700	318	0.9	0.4	17.0	7.7	15.3	7.0	A-F	7.6	49.9	0.82	1.80	0.47	1.03	NA[6]		0.24	0.18	1,146	2,521
								M-F	8.4	55.4	0.91	2.00	0.52	1.14			0.27	0.20	1,273	2,801
700	318	1.4	0.6	17.6	8.0	15.8	7.2	A-F	8.1	54.3	0.89	1.96	0.54	1.19	0.31	0.68	0.30	0.19	1,146	2,521
								M-F	9.0	60.3	0.99	2.18	0.60	1.32	0.34	0.75	0.33	0.21	1,273	2,801
700	318	1.9	0.9	17.6	8.0	15.8	7.2	A-F	8.8	60.3	0.99	2.18	0.63	1.39	0.39	0.86	0.30	0.19	1,146	2,521
								M-F	9.8	67.0	1.10	2.42	0.70	1.54	0.43	0.95	0.33	0.21	1,273	2,801
750	341	0.9	0.4	17.9	8.1	16.1	7.3	A-F	7.5	49.6	0.84	1.78	0.47	1.03	NA		0.24	0.17	1,146	2,521
								M-F	8.3	55.1	0.90	1.98	0.52	1.14			0.27	0.19	1,273	2,801
750	341	1.4	0.6	18.4	8.4	16.6	7.5	A-F	8.0	53.9	0.88	1.94	0.54	1.19	0.30	0.66	0.29	0.19	1,146	2,521
								M-F	8.9	59.9	0.98	2.16	0.60	1.32	0.33	0.73	0.32	0.21	1,273	2,801
750	341	1.9	0.9	18.4	8.4	16.6	7.5	A-F	8.6	59.9	0.98	2.16	0.62	1.37	0.38	0.83	0.33	0.21	1,146	2,521
								M-F	9.5	66.5	1.09	2.40	0.69	1.52	0.42	0.92	0.37	0.23	1,273	2,801
800	364	0.9	0.4	18.7	8.5	16.8	7.6	A-F	7.4	49.3	0.81	1.78	0.46	1.01	NA		0.25	0.18	1,146	2,521
								M-F	8.2	54.8	0.90	1.98	0.51	1.12			0.28	0.20	1,273	2,801
800	364	1.4	0.6	19.3	8.8	17.4	7.9	A-F	7.9	53.6	0.88	1.94	0.53	1.17	0.30	0.66	0.30	0.19	1,146	2,521
								M-F	8.8	59.6	0.98	2.16	0.59	1.30	0.33	0.73	0.33	0.21	1,273	2,801
800	364	1.9	0.9	19.4	8.8	17.5	8.0	A-F	8.4	59.5	0.97	2.14	0.62	1.37	0.38	0.83	0.32	0.19	1,146	2,521
								M-F	9.3	66.1	1.08	2.38	0.69	1.52	0.42	0.92	0.35	0.21	1,273	2,801
850	386	0.9	0.4	19.6	8.9	17.6	8.0	A-F	7.4	49.1	0.80	1.76	0.46	1.01	NA		0.23	0.18	1,146	2,521
								M-F	8.2	54.5	0.89	1.96	0.51	1.12			0.26	0.20	1,273	2,801
850	386	1.4	0.6	20.2	9.2	18.2	8.3	A-F	7.7	53.4	0.87	1.92	0.53	1.17	0.29	0.63	0.27	0.19	1,146	2,521
								M-F	8.6	59.3	0.97	2.13	0.59	1.30	0.32	0.70	0.30	0.21	1,273	2,801
850	386	1.9	0.9	20.3	9.3	18.3	8.3	A-F	8.2	59.1	0.97	2.14	0.61	1.35	0.37	0.81	0.31	0.20	1,146	2,521
								M-F	9.1	65.7	1.08	2.38	0.68	1.50	0.41	0.90	0.34	0.22	1,273	2,801
900	409	0.9	0.4	20.3	9.3	18.3	8.3	A-F	7.3	48.9	0.80	1.76	0.46	1.01	NA		0.23	0.18	1,146	2,521
								M-F	8.1	54.3	0.89	1.96	0.51	1.12			0.26	0.20	1,273	2,801
900	409	1.4	0.6	21.1	9.6	19.0	8.6	A-F	7.7	53.2	0.87	1.92	0.52	1.15	0.29	0.63	0.27	0.19	1,146	2,521
								M-F	8.5	59.1	0.97	2.13	0.58	1.28	0.32	0.70	0.30	0.21	1,273	2,801
900	409	1.9	0.9	21.3	9.7	19.2	8.7	A-F	8.1	58.9	0.96	2.12	0.61	1.35	0.37	0.81	0.29	0.19	1,146	2,521
								M-F	9.0	65.4	1.07	2.35	0.68	1.50	0.41	0.90	0.32	0.21	1,273	2,801
950	432	0.9	0.4	21.1	9.6	19.0	8.6	A-F	7.2	48.7	0.80	1.76	0.45	0.99	NA		0.24	0.18	1,146	2,521
								M-F	8.0	54.1	0.89	1.96	0.50	1.10			0.27	0.20	1,273	2,801
950	432	1.4	0.6	22.0	10.0	19.8	9.0	A-F	7.6	53.0	0.87	1.92	0.52	1.15	0.29	0.63	0.26	0.19	1,146	2,521
								M-F	8.4	58.9	0.97	2.13	0.58	1.28	0.32	0.70	0.29	0.21	1,273	2,801
950	432	1.9	0.9	22.2	10.1	20.0	9.0	A-F	7.9	58.6	0.96	2.12	0.60	1.32	0.36	0.79	0.29	0.19	1,146	2,521
								M-F	8.8	65.1	1.07	2.35	0.67	1.47	0.40	0.88	0.32	0.21	1,273	2,801

(Continued)

TABLE I-1b (Continued)

Weight[2]		Daily Gain[3]		Daily Consumption[4] As-fed		Daily Consumption[4] Moisture-free Dry Matter		Moisture Basis[5] A-F (As-fed) M-F (Moisture-free)	Total Protein	Energy TDN	Energy ME (Mcal per)		Energy NE_m (Mcal per)		Energy NE_g (Mcal per)		Calcium	Phosphorus	Vitamin A (IU per)	
(lb)	(kg)	(lb)	(kg)	(lb)	(kg)	(lb)	(kg)		(%)	(%)	(lb)	(kg)	(lb)	(kg)	(lb)	(kg)	(%)	(%)	(lb)	(kg)
Dry pregnant mature cows—Middle third of pregnancy																				
800	364	0.0	0.0	17.0	7.7	15.3	7.0	A-F	6.4	43.9	0.72	1.58	0.38	0.83	NA		0.15	0.15	1,146	2,521
								M-F	7.1	48.8	0.80	1.76	0.42	0.92			0.17	0.17	1,273	2,801
900	409	0.0	0.0	18.6	8.5	16.7	7.6	A-F	6.3	43.9	0.72	1.58	0.38	0.83	NA		0.16	0.16	1,146	2,521
								M-F	7.0	48.8	0.80	1.76	0.42	0.92			0.18	0.18	1,273	2,801
1,000	454	0.0	0.0	20.1	9.1	18.1	8.2	A-F	6.3	43.9	0.72	1.58	0.38	0.83	NA		0.16	0.16	1,146	2,521
								M-F	7.0	48.8	0.80	1.76	0.42	0.92			0.18	0.18	1,273	2,801
1,100	500	0.0	0.0	21.7	9.9	19.5	8.9	A-F	6.3	43.9	0.72	1.58	0.38	0.83	NA		0.17	0.17	1,146	2,521
								M-F	7.0	48.8	0.80	1.76	0.42	0.92			0.19	0.19	1,273	2,801
1,200	545	0.0	0.0	23.1	10.5	20.8	9.5	A-F	6.2	43.9	0.72	1.58	0.38	0.83	NA		0.17	0.17	1,146	2,521
								M-F	6.9	48.8	0.80	1.76	0.42	0.92			0.19	0.19	1,273	2,801
1,300	591	0.0	0.0	24.4	11.1	22.0	10.0	A-F	6.2	43.9	0.72	1.58	0.38	0.83	NA		0.18	0.18	1,146	2,521
								M-F	6.9	48.8	0.80	1.76	0.42	0.92			0.20	0.20	1,273	2,801
1,400	636	0.0	0.0	25.9	11.8	23.3	10.6	A-F	6.2	43.9	0.72	1.58	0.38	0.83	NA		0.18	0.18	1,146	2,521
								M-F	6.9	48.8	0.80	1.76	0.42	0.92			0.20	0.20	1,273	2,801
Dry pregnant mature cows—Last third of pregnancy																				
800	364	0.9	0.4	18.7	8.5	16.8	7.6	A-F	7.4	49.1	0.80	1.76	0.46	1.01	NA[6]		0.23	0.18	1,146	2,521
								M-F	8.2	54.5	0.89	1.96	0.51	1.12			0.26	0.20	1,273	2,801
900	409	0.9	0.4	20.2	9.2	18.2	8.3	A-F	7.2	48.6	0.80	1.76	0.45	0.99	NA		0.24	0.19	1,146	2,521
								M-F	8.0	54.0	0.89	1.96	0.50	1.10			0.27	0.21	1,273	2,801
1,000	454	0.9	0.4	21.8	9.9	19.6	8.9	A-F	7.1	48.2	0.79	1.75	0.45	0.99	NA		0.23	0.18	1,146	2,521
								M-F	7.9	53.6	0.88	1.94	0.50	1.10			0.26	0.20	1,273	2,801
1,100	500	0.9	0.4	23.3	10.6	21.0	9.5	A-F	7.0	47.9	0.78	1.72	0.44	0.97	NA		0.23	0.19	1,146	2,521
								M-F	7.8	53.2	0.87	1.91	0.49	1.08			0.26	0.21	1,273	2,801
1,200	545	0.9	0.4	24.8	11.3	22.3	10.1	A-F	7.0	47.6	0.78	1.72	0.44	0.97	NA		0.23	0.19	1,146	2,521
								M-F	7.8	52.9	0.87	1.91	0.49	1.08			0.26	0.21	1,273	2,801
1,300	591	0.9	0.4	26.2	11.9	23.6	10.7	A-F	6.9	47.4	0.78	1.72	0.43	0.95	NA		0.23	0.19	1,146	2,521
								M-F	7.7	52.7	0.87	1.91	0.48	1.06			0.26	0.21	1,273	2,801
1,400	636	0.9	0.4	27.7	12.6	24.9	11.3	A-F	6.8	47.3	0.77	1.70	0.43	0.95	NA		0.23	0.19	1,146	2,521
								M-F	7.6	52.5	0.86	1.89	0.48	1.06			0.26	0.21	1,273	2,801
Two-year-old heifers nursing calves—First 3–4 months postpartum—10 lb *(4.5 kg)* milk/day																				
700	318	0.5	0.2	17.7	8.0	15.9	7.2	A-F	10.2	58.6	0.96	2.12	0.60	1.32	0.36	0.79	0.32	0.22	1,596	3,511
								M-F	11.3	65.1	1.07	2.35	0.67	1.47	0.40	0.88	0.36	0.24	1,773	3,901
750	341	0.5	0.2	18.6	8.5	16.7	7.6	A-F	9.9	58.0	0.95	2.10	0.59	1.31	0.36	0.79	0.31	0.22	1,596	3,511
								M-F	11.0	64.4	1.06	2.33	0.66	1.45	0.40	0.88	0.34	0.24	1,773	3,901
800	364	0.5	0.2	19.6	8.9	17.6	8.0	A-F	9.7	57.4	0.95	2.08	0.59	1.31	0.35	0.77	0.31	0.22	1,596	3,511
								M-F	10.8	63.8	1.05	2.31	0.66	1.45	0.39	0.86	0.34	0.24	1,773	3,901
850	386	0.5	0.2	20.4	9.3	18.4	8.4	A-F	9.5	56.9	0.94	2.06	0.59	1.29	0.34	0.76	0.30	0.21	1,596	3,511
								M-F	10.6	63.2	1.04	2.29	0.65	1.43	0.38	0.84	0.33	0.23	1,773	3,901

(Continued)

TABLE I-1b (Continued)

Weight[2]		Daily Gain[3]		Daily Consumption[4] As-fed		Daily Consumption[4] Moisture-free Dry Matter		Moisture Basis[5] A-F (As-fed) M-F (Moisture-free)	Total Protein	Energy TDN	Energy ME (Mcal per)		Energy NE_m (Mcal per)		Energy NE_g (Mcal per)		Calcium	Phosphorus	Vitamin A (IU per)	
(lb)	(kg)	(lb)	(kg)	(lb)	(kg)	(lb)	(kg)		(%)	(%)	(lb)	(kg)	(lb)	(kg)	(lb)	(kg)	(%)	(%)	(lb)	(kg)
Two-year-old heifers nursing calves—First 3–4 months postpartum—10 lb (4.5 kg) milk/day (continued)																				
900	409	0.5	0.2	21.3	9.7	19.2	8.7	A-F	9.4	56.4	0.93	2.04	0.58	1.27	0.33	0.73	0.29	0.21	1,596	3,511
								M-F	10.4	62.7	1.03	2.27	0.64	1.41	0.37	0.81	0.32	0.23	1,773	3,901
950	432	0.5	0.2	22.2	10.1	20.0	9.0	A-F	9.2	56.1	0.92	2.02	0.57	1.25	0.33	0.73	0.28	0.21	1,596	3,511
								M-F	10.2	62.3	1.02	2.24	0.63	1.39	0.37	0.81	0.31	0.23	1,773	3,901
1,000	454	0.5	0.2	23.1	10.5	20.8	9.5	A-F	9.0	55.7	0.92	2.02	0.56	1.22	0.32	0.71	0.28	0.21	1,596	3,511
								M-F	10.0	61.9	1.02	2.24	0.62	1.36	0.36	0.79	0.31	0.23	1,773	3,901
Cows nursing calves—Average milking ability—First 3–4 months postpartum—10 lb (4.5 kg) milk/day																				
800	364	0.0	0.0	19.2	8.7	17.3	7.9	A-F	9.2	52.4	0.86	1.90	0.51	1.13	NA		0.27	0.20	1,596	3,511
								M-F	10.2	58.2	0.96	2.11	0.57	1.25			0.30	0.22	1,773	3,901
900	409	0.0	0.0	20.1	9.1	18.8	8.5	A-F	8.9	51.6	0.85	1.86	0.50	1.09	NA		0.25	0.20	1,596	3,511
								M-F	9.9	57.3	0.94	2.07	0.55	1.21			0.28	0.22	1,773	3,901
1,000	454	0.0	0.0	22.4	10.2	20.2	9.2	A-F	8.6	50.9	0.84	1.85	0.50	1.09	NA		0.25	0.20	1,596	3,511
								M-F	9.6	56.6	0.93	2.05	0.55	1.21			0.28	0.22	1,773	3,901
1,100	500	0.0	0.0	24.0	10.9	21.6	9.8	A-F	8.5	50.4	0.83	1.82	0.49	1.07	NA		0.24	0.20	1,596	3,511
								M-F	9.4	56.0	0.92	2.02	0.54	1.19			0.27	0.22	1,773	3,901
1,200	545	0.0	0.0	25.6	11.6	23.0	10.5	A-F	8.4	50.0	0.82	1.80	0.48	1.05	NA		0.24	0.20	1,596	3,511
								M-F	9.3	55.5	0.91	2.00	0.53	1.17			0.27	0.22	1,773	3,901
1,300	591	0.0	0.0	27.0	12.3	24.3	11.0	A-F	8.2	49.6	0.81	1.78	0.47	1.03	NA		0.24	0.20	1,596	3,511
								M-F	9.1	55.1	0.90	1.98	0.52	1.14			0.27	0.22	1,773	3,901
1,400	636	0.0	0.0	28.4	12.9	25.6	11.6	A-F	8.1	49.2	0.81	1.78	0.46	1.01	NA		0.24	0.20	1,596	3,511
								M-F	9.0	54.7	0.90	1.98	0.51	1.12			0.27	0.22	1,773	3,901
Cows nursing calves—Superior milking ability—First 3–4 months postpartum—20 lb (9.1 kg) milk/day																				
800	364	0.0	0.0	17.4	7.9	15.7	7.1	A-F	12.8	70.0	1.14	2.51	0.77	1.68	NA		0.43	0.28	1,596	3,511
								M-F	14.2	77.3	1.27	2.79	0.85	1.87			0.48	0.31	1,773	3,901
900	409	0.0	0.0	20.8	9.5	18.7	8.5	A-F	11.6	62.8	1.04	2.28	0.67	1.47	NA		0.37	0.25	1,596	3,511
								M-F	12.9	69.8	1.15	2.53	0.74	1.63			0.41	0.28	1,773	3,901
1,000	454	0.0	0.0	22.9	10.4	20.6	9.4	A-F	11.1	60.3	0.99	2.18	0.63	1.39	NA		0.35	0.24	1,596	3,511
								M-F	12.3	67.0	1.10	2.42	0.70	1.54			0.39	0.27	1,773	3,901
1,100	500	0.0	0.0	24.8	11.3	22.3	10.1	A-F	10.7	58.7	0.96	2.12	0.60	1.32	NA		0.34	0.24	1,596	3,511
								M-F	11.9	65.2	1.07	2.35	0.67	1.47			0.38	0.27	1,773	3,901
1,200	545	0.0	0.0	26.4	12.0	23.8	10.8	A-F	10.4	57.3	0.95	2.08	0.59	1.29	NA		0.32	0.23	1,596	3,511
								M-F	11.5	63.7	1.05	2.31	0.65	1.43			0.36	0.26	1,773	3,901
1,300	591	0.0	0.0	28.1	12.8	25.3	11.5	A-F	10.1	56.3	0.93	2.04	0.58	1.27	NA		0.32	0.23	1,596	3,511
								M-F	11.2	62.6	1.03	2.27	0.64	1.41			0.36	0.26	1,773	3,901
1,400	636	0.0	0.0	29.7	13.5	26.7	12.1	A-F	9.9	55.5	0.91	2.00	0.56	1.22	NA		0.32	0.23	1,596	3,511
								M-F	11.0	61.7	1.01	2.22	0.62	1.36			0.35	0.26	1,773	3,901

(Continued)

TABLE I-1b (Continued)

Weight[2] (lb)	Weight[2] (kg)	Daily Gain[3] (lb)	Daily Gain[3] (kg)	Daily Consumption[4] As-fed (lb)	Daily Consumption[4] As-fed (kg)	Daily Consumption[4] Moisture-free Dry Matter (lb)	Daily Consumption[4] Moisture-free Dry Matter (kg)	Moisture Basis[5] A-F (As-fed) M-F (Moisture-free)	Total Protein (%)	Energy TDN (%)	Energy ME (Mcal per) (lb)	Energy ME (Mcal per) (kg)	Energy NE_m (Mcal per) (lb)	Energy NE_m (Mcal per) (kg)	Energy NE_g (Mcal per) (lb)	Energy NE_g (Mcal per) (kg)	Calcium (%)	Phosphorus (%)	Vitamin A (IU per) (lb)	Vitamin A (IU per) (kg)
Bulls, maintenance and slow rate of growth (regain body condition)																				
<1,300 For growth and development, use requirements for bulls in Tables I-1c and I-1d																				
1,300	591	1.0	0.5	28.2	12.8	25.4	11.5	A-F	6.8	50.2	0.83	1.82	0.48	1.05	0.25	0.56	0.20	0.17	1,596	3,511
								M-F	7.6	55.8	0.92	2.02	0.53	1.17	0.28	0.62	0.22	0.19	1,773	3,901
1,300	591	1.5	0.7	29.0	13.2	26.1	11.9	A-F	7.1	53.7	0.88	1.94	0.53	1.17	0.30	0.66	0.22	0.17	1,596	3,511
								M-F	7.9	59.7	0.98	2.16	0.59	1.30	0.33	0.73	0.24	0.19	1,773	3,901
1,300	591	2.0	0.9	29.1	13.2	26.2	11.9	A-F	7.4	57.6	0.95	2.08	0.59	1.29	0.35	0.77	0.23	0.18	1,596	3,511
								M-F	8.2	64.0	1.05	2.31	0.65	1.43	0.39	0.86	0.26	0.20	1,773	3,901
1,400	636	1.0	0.5	29.8	13.5	26.8	12.2	A-F	6.8	50.2	0.83	1.82	0.48	1.05	0.25	0.56	0.19	0.17	1,596	3,511
								M-F	7.5	55.8	0.92	2.02	0.53	1.17	0.28	0.62	0.21	0.19	1,773	3,901
1,400	636	1.5	0.7	30.7	14.0	27.6	12.5	A-F	6.9	53.7	0.88	1.94	0.53	1.17	0.30	0.66	0.21	0.17	1,596	3,511
								M-F	7.7	59.7	0.98	2.16	0.59	1.30	0.33	0.73	0.23	0.19	1,773	3,901
1,400	636	2.0	0.9	30.8	14.0	27.7	12.6	A-F	7.2	57.6	0.95	2.08	0.59	1.29	0.35	0.77	0.23	0.18	1,596	3,511
								M-F	8.0	64.0	1.05	2.31	0.65	1.43	0.39	0.86	0.25	0.20	1,773	3,901
1,500	682	0.0	0.0	28.0	12.7	25.2	11.5	A-F	6.2	43.6	0.71	1.57	0.37	0.81	NA		0.18	0.18	1,596	3,511
								M-F	6.9	48.4	0.79	1.74	0.41	0.90			0.20	0.20	1,773	3,901
1,500	682	1.0	0.5	31.4	14.3	28.3	12.9	A-F	6.7	50.2	0.83	1.82	0.48	1.05	0.25	0.56	0.19	0.17	1,596	3,511
								M-F	7.4	55.8	0.92	2.02	0.53	1.17	0.28	0.62	0.21	0.19	1,773	3,901
1,500	682	1.5	0.7	32.2	14.6	29.0	13.2	A-F	6.8	53.7	0.88	1.94	0.53	1.17	0.30	0.66	0.20	0.17	1,596	3,511
								M-F	7.6	59.7	0.98	2.16	0.59	1.30	0.33	0.73	0.22	0.19	1,773	3,901
1,600	727	0.0	0.0	29.4	13.4	26.5	12.0	A-F	6.2	43.6	0.71	1.57	0.37	0.81	NA		0.17	0.18	1,596	3,511
								M-F	6.9	48.4	0.79	1.74	0.41	0.90			0.19	0.20	1,773	3,901
1,600	727	1.0	0.5	33.0	15.0	29.7	13.5	A-F	6.6	50.2	0.83	1.82	0.48	1.05	0.25	0.56	0.20	0.17	1,596	3,511
								M-F	7.3	55.8	0.92	2.02	0.53	1.17	0.28	0.62	0.22	0.19	1,773	3,901
1,600	727	1.5	0.7	33.8	15.4	30.4	13.8	A-F	6.7	53.7	0.88	1.94	0.53	1.17	0.30	0.66	0.20	0.18	1,596	3,511
								M-F	7.4	59.7	0.98	2.16	0.59	1.30	0.33	0.73	0.22	0.20	1,773	3,901
1,700	773	0.0	0.0	30.8	14.0	27.7	12.6	A-F	6.1	43.6	0.71	1.57	0.37	0.81	NA		0.19	0.19	1,596	3,511
								M-F	6.8	48.4	0.79	1.74	0.41	0.90			0.21	0.21	1,773	3,901
1,700	773	0.5	0.2	32.9	15.0	29.6	13.5	A-F	6.3	46.8	0.77	1.63	0.42	0.93	0.20	0.43	0.18	0.17	1,596	3,511
								M-F	7.0	52.0	0.85	1.87	0.47	1.03	0.22	0.48	0.20	0.19	1,773	3,901
1,800	818	0.0	0.0	32.1	14.6	28.9	13.1	A-F	6.1	43.6	0.71	1.57	0.37	0.81	NA		0.19	0.19	1,596	3,511
								M-F	6.8	48.4	0.79	1.74	0.41	0.90			0.21	0.21	1,773	3,901
1,800	818	0.5	0.2	34.3	15.6	30.9	14.0	A-F	6.3	46.8	0.77	1.63	0.42	0.93	0.20	0.43	0.18	0.18	1,596	3,511
								M-F	7.0	52.0	0.85	1.87	0.47	1.03	0.22	0.48	0.20	0.20	1,773	3,901
1,900	864	0.0	0.0	33.4	15.2	30.1	13.7	A-F	6.1	43.6	0.71	1.57	0.37	0.81	NA		0.19	0.19	1,596	3,511
								M-F	6.8	48.4	0.79	1.74	0.41	0.90			0.21	0.21	1,773	3,901
1,900	864	0.5	0.2	35.8	16.3	32.2	14.6	A-F	6.2	46.8	0.77	1.63	0.42	0.93	0.20	0.43	0.18	0.18	1,596	3,511
								M-F	6.9	52.0	0.85	1.87	0.47	1.03	0.22	0.48	0.20	0.20	1,773	3,901
2,000	909	0.0	0.0	34.8	15.8	31.3	14.2	A-F	6.1	43.6	0.71	1.57	0.37	0.81	NA		0.19	0.19	1,596	3,511
								M-F	6.8	48.4	0.79	1.74	0.41	0.90			0.21	0.21	1,773	3,901
2,100	955	0.0	0.0	36.1	16.4	32.5	14.8	A-F	6.1	43.6	0.71	1.57	0.37	0.81	NA		0.20	0.20	1,596	3,511
								M-F	6.8	48.4	0.79	1.74	0.41	0.90			0.22	0.22	1,773	3,901
2,200	1,000	0.0	0.0	37.3	17.0	33.6	15.3	A-F	6.1	43.6	0.71	1.57	0.37	0.81	NA		0.20	0.20	1,596	3,511
								M-F	6.8	48.4	0.79	1.74	0.41	0.90			0.22	0.22	1,773	3,901

[1]Adapted from *Nutrient Requirements of Beef Cattle*, sixth revised edition, National Research Council-National Academy of Sciences, 1984.

[2]Average weight for a feeding period.

[3]Approximately 0.9 ± 0.2lb of weight gain/day over the last third of pregnancy is accounted for by the products of conception. Daily 2.15 Mcal of NE_m and 0.1 lb of protein are provided for this requirement for a calf with a birth weight of 80 lb.

[4]Consumption should vary depending on the energy concentration of the diet and environmental conditions. These intakes are based on the energy concentration shown in the table and assuming a thermoneutral environment without snow or mud conditions. If the energy concentrations of the diet to be fed exceed the tabular value, limit feeding may be required.

[5]As-fed was calculated using an average firgure of 90% dry matter. When using silages, roots, and other wet feeds, these feeds should be converted to a moisture-free basis and the ration calculated using the moisture-free data.

[6]Not applicable.

TABLE I-1c
DAILY NUTRIENT REQUIREMENTS OF GROWING AND FINISHING CATTLE[1, 2]

Weight		Daily Gain		Total Protein	Energy		Calcium	Phosphorus
					NE$_m$	NE$_g$		
(lb)	*(kg)*	*(lb)*	*(kg)*	*(g/day)*	*(Mcal/day)*	*(Mcal/day)*	*(g/day)*	*(g/day)*
Medium-frame steer calves								
330	*150*	0.4	*0.2*	343	3.30	0.41	11	7
		0.9	*0.4*	428	3.30	0.87	16	9
		1.3	*0.6*	503	3.30	1.36	21	11
		1.8	*0.8*	575	3.30	1.87	27	12
		2.2	*1.0*	642	3.30	2.39	32	14
		2.6	*1.2*	702	3.30	2.91	37	16
440	*200*	0.4	*0.2*	399	4.10	0.50	12	9
		0.9	*0.4*	482	4.10	1.08	17	10
		1.3	*0.6*	554	4.10	1.69	21	12
		1.8	*0.8*	621	4.10	2.32	26	13
		2.2	*1.0*	682	4.10	2.96	31	15
		2.6	*1.2*	735	4.10	3.62	35	16
550	*250*	0.4	*0.2*	450	4.84	0.60	13	10
		0.9	*0.4*	532	4.84	1.28	17	12
		1.3	*0.6*	601	4.84	2.00	21	13
		1.8	*0.8*	664	4.84	2.74	25	14
		2.2	*1.0*	720	4.84	3.50	29	16
		2.6	*1.2*	766	4.84	4.28	33	17
660	*300*	0.4	*0.2*	499	5.55	*0.69*	14	12
		0.9	*0.4*	580	5.55	1.47	18	13
		1.3	*0.6*	646	5.55	2.29	22	14
		1.8	*0.8*	704	5.55	3.14	25	15
		2.2	*1.0*	755	5.55	4.02	29	16
		2.6	*1.2*	794	5.55	4.90	32	17
770	*350*	0.4	*0.2*	545	6.24	*0.77*	15	13
		0.9	*0.4*	625	6.24	1.65	19	14
		1.3	*0.6*	688	6.24	2.57	22	15
		1.8	*0.8*	743	6.24	3.53	25	16
		2.2	*1.0*	789	6.24	4.51	28	17
		2.6	*1.2*	822	6.24	5.50	31	18
880	*400*	0.4	*0.2*	590	6.89	*0.85*	16	15
		0.9	*0.4*	668	6.89	1.82	19	16
		1.3	*0.6*	728	6.89	2.84	22	17
		1.8	*0.8*	780	6.89	3.90	25	17
		2.2	*1.0*	821	6.89	4.98	27	18
		2.6	*1.2*	848	6.89	6.69	29	19
990	*450*	0.4	*0.2*	633	7.52	*0.93*	17	16
		0.9	*0.4*	710	7.52	1.99	20	17
		1.3	*0.6*	767	7.52	3.11	22	18
		1.8	*0.8*	815	7.52	4.26	24	19
		2.2	*1.0*	852	7.52	5.44	26	19
		2.6	*1.2*	873	7.52	6.65	28	20
1,100	*500*	0.4	*0.2*	675	8.14	*1.01*	19	18
		0.9	*0.4*	751	8.14	2.16	21	18
		1.3	*0.6*	805	8.14	3.36	23	19
		1.8	*0.8*	849	8.14	4.61	24	20
		2.2	*1.0*	882	8.14	5.89	26	20
		2.6	*1.2*	897	8.14	7.19	27	21

(Continued)

TABLE I-1c (Continued)

Weight		Daily Gain		Total Protein	Energy		Calcium	Phosphorus
					NE_m	NE_g		
(lb)	*(kg)*	*(lb)*	*(kg)*	*(g/day)*	*(Mcal/day)*	*(Mcal/day)*	*(g/day)*	*(g/day)*
Medium-frame steer calves (continued)								
1,210	*550*	0.4	*0.2*	715	8.75	1.08	20	19
		0.9	*0.4*	790	8.75	2.32	22	20
		1.3	*0.6*	842	8.75	3.61	23	20
		1.8	*0.8*	883	8.75	4.95	24	21
		2.2	*1.0*	911	8.75	6.23	25	21
		2.6	*1.2*	921	8.75	7.73	26	21
Large-frame steer calves, compensating medium-frame yearling steers, and medium-frame bulls[3]								
330	*150*	0.4	*0.2*	361	3.30	0.36	11	7
		0.9	*0.4*	441	3.30	0.77	17	9
		1.3	*0.6*	522	3.30	1.21	22	11
		1.8	*0.8*	598	3.30	1.65	28	13
		2.2	*1.0*	671	3.30	2.11	33	14
		2.6	*1.2*	740	3.30	2.58	38	16
		3.1	*1.4*	806	3.30	3.06	44	18
		3.5	*1.6*	863	3.30	3.53	49	20
440	*200*	0.4	*0.2*	421	4.10	*0.45*	12	9
		0.9	*0.4*	499	4.10	0.96	17	10
		1.3	*0.6*	576	4.10	1.50	22	12
		1.8	*0.8*	650	4.10	2.06	27	14
		2.2	*1.0*	718	4.10	2.62	32	15
		2.6	*1.2*	782	4.10	3.20	37	17
		3.1	*1.4*	842	4.10	3.79	42	18
		3.5	*1.6*	892	4.10	4.39	47	20
550	*250*	0.4	*0.2*	476	4.84	*0.53*	13	10
		0.9	*0.4*	552	4.84	1.13	18	12
		1.3	*0.6*	628	4.84	1.77	23	13
		1.8	*0.8*	698	4.84	2.43	27	15
		2.2	*1.0*	762	4.84	3.10	31	16
		2.6	*1.2*	822	4.84	3.78	36	18
		3.1	*1.4*	877	4.84	4.48	40	19
		3.5	*1.6*	919	4.84	5.19	44	20
660	*300*	0.4	*0.2*	529	5.55	*0.61*	14	12
		0.9	*0.4*	603	5.55	1.30	19	13
		1.3	*0.6*	676	5.55	2.03	23	15
		1.8	*0.8*	743	5.55	2.78	27	16
		2.2	*1.0*	804	5.55	3.55	31	17
		2.6	*1.2*	859	5.55	4.34	35	18
		3.1	*1.4*	908	5.55	5.14	38	20
		3.5	*1.6*	943	5.55	5.95	42	21
770	*350*	0.4	*0.2*	579	6.24	*0.68*	16	13
		0.9	*0.4*	651	6.24	1.46	19	15
		1.3	*0.6*	722	6.24	2.28	23	16
		1.8	*0.8*	786	6.24	3.12	27	17
		2.2	*1.0*	843	6.24	3.99	30	18
		2.6	*1.2*	895	6.24	4.87	34	19
		3.1	*1.4*	938	6.24	5.77	37	20
		3.5	*1.6*	967	6.24	6.68	40	21

(Continued)

TABLE I-1c (Continued)

Weight		Daily Gain		Total Protein	Energy		Calcium	Phosphorus
					NE_m	NE_g		
(lb)	*(kg)*	*(lb)*	*(kg)*	*(g/day)*	*(Mcal/day)*	*(Mcal/day)*	*(g/day)*	*(g/day)*
Large-frame steer calves, compensating medium-frame yearling steers, and medium-frame bulls[3] (continued)								
880	*400*	0.4	*0.2*	627	6.89	0.75	17	15
		0.9	*0.4*	697	6.89	1.61	20	16
		1.3	*0.6*	766	6.89	2.52	24	17
		1.8	*0.8*	828	6.89	3.45	27	18
		2.2	*1.0*	881	6.89	4.41	30	19
		2.6	*1.2*	929	6.89	5.38	33	20
		3.1	*1.4*	967	6.89	6.38	36	21
		3.5	*1.6*	989	6.89	7.38	38	22
990	*450*	0.4	*0.2*	673	7.52	0.82	18	16
		0.9	*0.4*	742	7.52	1.76	21	17
		1.3	*0.6*	809	7.52	2.75	24	18
		1.8	*0.8*	867	7.52	3.77	27	19
		2.2	*1.0*	918	7.52	4.81	29	20
		2.6	*1.2*	961	7.52	5.88	32	21
		3.1	*1.4*	995	7.52	6.97	34	22
		3.5	*1.6*	1,011	7.52	8.07	37	22
1,100	*500*	0.4	*0.2*	719	8.14	0.89	19	18
		0.9	*0.4*	785	8.14	1.01	22	19
		1.3	*0.6*	850	8.14	2.98	24	20
		1.8	*0.8*	906	8.14	4.08	27	20
		2.2	*1.0*	953	8.14	5.21	29	21
		2.6	*1.2*	993	8.14	6.37	31	22
		3.1	*1.4*	1,022	8.14	7.54	33	22
		3.5	*1.6*	1,031	8.14	8.73	35	23
1,210	*550*	0.4	*0.2*	762	8.75	0.96	20	20
		0.9	*0.4*	827	8.75	2.05	23	20
		1.3	*0.6*	890	8.75	3.20	25	21
		1.8	*0.8*	944	8.75	4.38	27	22
		2.2	*1.0*	988	8.75	5.60	29	22
		2.6	*1.2*	1,023	8.75	6.84	30	23
		3.1	*1.4*	1,048	8.75	8.10	32	23
		3.5	*1.6*	1,052	8.75	9.38	34	24
1,320	*600*	0.4	*0.2*	805	9.33	1.02	22	21
		0.9	*0.4*	867	9.33	2.19	24	22
		1.3	*0.6*	930	9.33	3.41	25	22
		1.8	*0.8*	980	9.33	4.68	27	23
		2.2	*1.0*	1,021	9.33	5.98	28	23
		2.6	*1.2*	1,053	9.33	7.30	30	24
		3.1	*1.4*	1,073	9.33	8.64	31	24
		3.5	*1.6*	1,071	9.33	10.01	32	24
Large-frame bull calves and compensating large-frame yearling steers								
330	*150*	0.4	*0.2*	355	3.30	0.32	11	7
		0.9	*0.4*	438	3.30	0.69	17	9
		1.3	*0.6*	519	3.30	1.07	23	11
		1.8	*0.8*	597	3.30	1.47	28	13
		2.2	*1.0*	673	3.30	1.87	34	15
		2.6	*1.2*	745	3.30	2.29	40	17
		3.1	*1.4*	815	3.30	2.71	45	18
		3.5	*1.6*	880	3.30	3.14	51	20
		4.0	*1.8*	922	3.30	3.56	56	22

(Continued)

TABLE I-1c (Continued)

Weight		Daily Gain		Total Protein	Energy		Calcium	Phosphorus
					NE_m	NE_g		
(lb)	*(kg)*	*(lb)*	*(kg)*	*(g/day)*	*(Mcal/day)*	*(Mcal/day)*	*(g/day)*	*(g/day)*
Large-frame bull calves and compensating large-frame yearling steers (continued)								
440	*200*	0.4	*0.2*	414	4.10	0.40	12	9
		0.9	*0.4*	494	4.10	0.85	18	11
		1.3	*0.6*	574	4.10	1.33	23	12
		1.8	*0.8*	649	4.10	1.82	28	14
		2.2	*1.0*	721	4.10	2.32	34	16
		2.6	*1.2*	789	4.10	2.84	39	17
		3.1	*1.4*	854	4.10	3.36	44	19
		3.5	*1.6*	912	4.10	3.89	49	21
		4.0	*1.8*	942	4.10	4.43	54	22
550	*250*	0.4	*0.2*	468	4.84	0.47	13	10
		0.9	*0.4*	547	4.84	1.01	19	12
		1.3	*0.6*	624	4.84	1.57	23	14
		1.8	*0.8*	697	4.84	2.15	28	15
		2.2	*1.0*	765	4.84	2.75	33	17
		2.6	*1.2*	830	4.84	3.36	38	18
		3.1	*1.4*	890	4.84	3.97	42	20
		3.5	*1.6*	943	4.84	4.60	47	21
		4.0	*1.8*	962	4.84	5.23	51	22
660	*300*	0.4	*0.2*	519	5.55	0.54	15	12
		0.9	*0.4*	597	5.55	1.15	19	13
		1.3	*0.6*	672	5.55	1.80	24	15
		1.8	*0.8*	741	5.55	2.47	28	16
		2.2	*1.0*	807	5.55	3.15	33	18
		2.6	*1.2*	868	5.55	3.85	37	19
		3.1	*1.4*	924	5.55	4.56	41	20
		3.5	*1.6*	971	5.55	5.28	45	22
		4.0	*1.8*	980	5.55	6.00	49	23
770	*350*	0.4	*0.2*	568	6.24	0.60	16	13
		0.9	*0.4*	644	6.24	1.29	20	15
		1.3	*0.6*	718	6.24	2.02	24	16
		1.8	*0.8*	795	6.24	2.77	28	18
		2.2	*1.0*	847	6.24	3.54	32	19
		2.6	*1.2*	904	6.24	4.32	36	20
		3.1	*1.4*	956	6.24	5.11	40	21
		3.5	*1.6*	998	6.24	5.92	44	23
		4.0	*1.8*	997	6.24	6.74	47	23
880	*400*	0.4	*0.2*	615	6.89	0.67	17	15
		0.9	*0.4*	689	6.89	1.43	21	16
		1.3	*0.6*	761	6.89	2.23	25	18
		1.8	*0.8*	826	6.89	3.06	29	19
		2.2	*1.0*	885	6.89	3.91	32	20
		2.6	*1.2*	939	6.89	4.77	36	21
		3.1	*1.4*	986	6.89	5.65	39	22
		3.5	*1.6*	1,024	6.89	6.55	42	23
		4.0	*1.8*	1,013	6.89	7.45	45	24

(Continued)

TABLE I-1c (Continued)

Weight		Daily Gain		Total Protein	Energy		Calcium	Phosphorus
					NE_m	NE_g		
(lb)	*(kg)*	*(lb)*	*(kg)*	*(g/day)*	*(Mcal/day)*	*(Mcal/day)*	*(g/day)*	*(g/day)*
Large-frame bull calves and compensating large-frame yearling steers (continued)								
990	*450*	0.4	*0.2*	661	7.52	0.73	18	17
		0.9	*0.4*	733	7.52	1.56	22	18
		1.3	*0.6*	803	7.52	2.44	25	19
		1.8	*0.8*	866	7.52	3.34	29	20
		2.2	*1.0*	922	7.52	4.27	32	21
		2.6	*1.2*	973	7.52	5.21	35	22
		3.1	*1.4*	1,016	7.52	6.18	38	23
		3.5	*1.6*	1,048	7.52	7.15	41	24
		4.0	*1.8*	1,028	7.52	8.13	44	25
1,100	*500*	0.4	*0.2*	705	8.14	0.79	20	18
		0.9	*0.4*	770	8.14	1.69	23	19
		1.3	*0.6*	844	8.14	2.64	26	20
		1.8	*0.8*	905	8.14	3.62	29	21
		2.2	*1.0*	958	8.14	4.62	32	22
		2.6	*1.2*	1,005	8.14	5.64	35	23
		3.1	*1.4*	1,045	8.14	6.68	37	24
		3.5	*1.6*	1,072	8.14	7.74	40	25
		4.0	*1.8*	1,043	8.14	8.80	42	25
1,210	*550*	0.4	*0.2*	747	8.75	0.85	21	20
		0.9	*0.4*	817	8.75	1.82	24	21
		1.3	*0.6*	884	8.75	2.83	27	22
		1.8	*0.8*	942	8.75	3.88	29	22
		2.2	*1.0*	994	8.75	4.96	32	23
		2.6	*1.2*	1,037	8.75	6.06	34	24
		3.1	*1.4*	1,072	8.75	7.18	36	25
		3.5	*1.6*	1,095	8.75	8.31	39	25
		4.0	*1.8*	1,057	8.75	9.46	41	26
1,320	*600*	0.4	*0.2*	789	9.33	0.91	22	21
		0.9	*0.4*	857	9.33	1.94	25	22
		1.3	*0.6*	923	9.33	3.02	27	23
		1.8	*0.8*	979	9.33	4.15	30	24
		2.2	*1.0*	1,027	9.33	5.30	32	24
		2.6	*1.2*	1,067	9.33	6.47	34	25
		3.1	*1.4*	1,099	9.33	7.66	36	26
		3.5	*1.6*	1,117	9.33	8.87	38	26
		4.0	*1.8*	1,071	9.33	10.10	39	26
Medium-frame heifer calves								
330	*150*	0.4	*0.2*	323	3.30	0.49	10	7
		0.9	*0.4*	409	3.30	1.05	15	9
		1.3	*0.6*	477	3.30	1.66	20	10
		1.8	*0.8*	537	3.30	2.29	25	12
		2.2	*1.0*	562	3.30	2.94	29	13
440	*200*	0.4	*0.2*	374	4.10	0.60	11	9
		0.9	*0.4*	459	4.10	1.31	16	10
		1.3	*0.6*	522	4.10	2.06	20	11
		1.8	*0.8*	574	4.10	2.84	23	12
		2.2	*1.0*	583	4.10	3.65	27	14

(Continued)

TABLE I-1c (Continued)

Weight		Daily Gain		Total Protein	Energy		Calcium	Phosphorus
					NE_m	NE_g		
(lb)	*(kg)*	*(lb)*	*(kg)*	*(g/day)*	*(Mcal/day)*	*(Mcal/day)*	*(g/day)*	*(g/day)*
Medium-frame heifer calves (continued)								
550	*250*	0.4	*0.2*	421	4.84	0.71	12	10
		0.9	*0.4*	505	4.84	1.55	16	11
		1.3	*0.6*	563	4.84	2.44	19	12
		1.8	*0.8*	608	4.84	3.36	23	13
		2.2	*1.0*	603	4.84	4.31	26	14
660	*300*	0.4	*0.2*	465	5.55	0.82	13	11
		0.9	*0.4*	549	5.55	1.77	16	12
		1.3	*0.6*	602	5.55	2.79	19	13
		1.8	*0.8*	640	5.55	3.85	22	14
		2.2	*1.0*	621	5.55	4.94	24	15
770	*350*	0.4	*0.2*	508	6.24	0.92	14	13
		0.9	*0.4*	591	6.24	1.99	17	14
		1.2	*0.6*	638	6.24	3.13	19	14
		1.8	*0.8*	670	6.24	4.32	21	15
		2.2	*1.0*	638	6.24	5.55	23	16
880	*400*	0.4	*0.2*	549	6.89	1.01	16	14
		0.9	*0.4*	630	6.89	2.20	17	15
		1.3	*0.6*	674	6.89	3.46	19	16
		.8	*0.8*	700	6.89	4.78	20	16
		2.2	*1.0*	654	6.89	6.14	22	16
990	*450*	0.4	*0.2*	588	7.52	1.11	17	16
		0.9	*0.4*	669	7.52	2.40	18	16
		1.3	*0.6*	708	7.52	3.78	19	17
		1.8	*0.8*	728	7.52	5.22	20	17
		2.2	*1.0*	670	7.52	6.70	20	17
1,100	*500*	0.4	*0.2*	626	8.14	1.20	18	17
		0.9	*0.4*	706	8.14	2.60	19	18
		1.3	*0.6*	741	8.14	4.10	19	18
		1.8	*0.8*	755	8.14	5.65	19	18
		2.2	*1.0*	685	8.14	7.25	19	18
1,210	*550*	0.4	*0.2*	662	8.75	1.29	19	19
		0.9	*0.4*	742	8.75	2.79	19	19
		1.3	*0.6*	773	8.75	4.40	19	19
		1.8	*0.8*	781	8.75	6.07	19	19
		2.2	*1.0*	700	8.75	7.79	19	19
Large-frame heifer calves and compensating medium-frame yearling heifers								
330	*150*	0.4	*0.2*	342	3.30	0.43	11	7
		0.9	*0.4*	426	3.30	0.93	16	9
		1.3	*0.6*	500	3.30	1.47	21	10
		1.8	*0.8*	568	3.30	2.03	26	12
		2.2	*1.0*	630	3.30	2.61	31	14
		2.6	*1.2*	680	3.30	3.19	35	15
440	*200*	0.4	*0.2*	397	4.10	0.53	12	9
		0.9	*0.4*	480	4.10	1.16	16	10
		1.3	*0.6*	549	4.10	1.83	21	12
		1.8	*0.8*	613	4.10	2.62	25	13
		2.2	*1.0*	668	4.10	3.23	29	14
		2.6	*1.2*	708	4.10	3.97	33	16

(Continued)

TABLE I-1c (Continued)

Weight		Daily Gain		Total Protein	Energy		Calcium	Phosphorus
					NE_m	NE_g		
(lb)	*(kg)*	*(lb)*	*(kg)*	*(g/day)*	*(Mcal/day)*	*(Mcal/day)*	*(g/day)*	*(g/day)*
Large-frame heifer calves and compensating medium-frame yearling heifers (continued)								
550	*250*	0.4	*0.2*	449	4.84	0.63	13	10
		0.9	*0.4*	530	4.84	1.37	17	11
		1.3	*0.6*	596	4.84	2.16	21	13
		1.8	*0.8*	654	4.84	2.98	24	14
		2.2	*1.0*	703	4.84	3.82	28	15
		2.6	*1.2*	734	4.84	4.69	31	16
660	*300*	0.4	*0.2*	497	5.55	0.72	14	12
		0.9	*0.4*	577	5.55	1.57	17	13
		1.3	*0.6*	639	5.55	2.47	21	14
		1.8	*0.8*	693	5.55	3.41	24	15
		2.2	*1.0*	735	5.55	4.38	27	16
		2.6	*1.2*	758	5.55	5.37	30	17
770	*350*	0.4	*0.2*	543	6.24	0.81	15	13
		0.9	*0.4*	622	6.24	1.76	18	14
		1.3	*0.6*	681	6.24	2.78	21	15
		1.8	*0.8*	730	6.24	3.83	23	16
		2.2	*1.0*	767	6.24	4.92	26	17
		2.6	*1.2*	781	6.24	5.03	28	17
880	*400*	0.4	*0.2*	588	6.89	0.90	16	15
		0.9	*0.4*	665	6.89	1.95	19	15
		1.3	*0.6*	721	6.89	3.07	21	16
		1.8	*0.8*	765	6.89	4.24	23	17
		2.2	*1.0*	797	6.89	5.44	25	18
		2.6	*1.2*	803	6.89	6.67	27	18
990	*450*	0.4	*0.2*	631	7.52	0.98	17	16
		0.9	*0.4*	707	7.52	2.13	19	17
		1.3	*0.6*	759	7.52	3.35	21	17
		1.8	*0.8*	799	7.52	4.63	23	18
		2.2	*1.0*	826	7.52	5.94	24	18
		2.6	*1.2*	824	7.52	7.28	25	19
1,100	*500*	0.4	*0.2*	672	8.14	1.06	18	18
		0.9	*0.4*	747	8.14	2.31	20	18
		1.3	*0.6*	796	8.14	3.63	21	19
		1.8	*0.8*	833	8.14	5.01	22	19
		2.2	*1.0*	854	8.14	6.43	23	19
		2.6	*1.2*	844	8.14	7.88	24	20
1,210	*550*	0.4	*0.2*	712	8.75	1.14	20	19
		0.9	*0.4*	787	8.75	2.47	21	20
		1.3	*0.6*	832	8.75	3.90	22	20
		1.8	*0.8*	865	8.75	5.38	22	20
		2.2	*1.0*	881	8.75	6.91	23	20
		2.6	*1.2*	864	8.75	8.47	23	20
1,320	*600*	0.4	*0.2*	751	9.33	1.21	21	21
		0.9	*0.4*	825	9.33	2.64	22	21
		1.3	*0.6*	867	9.33	4.16	22	21
		1.8	*0.8*	896	9.33	5.74	22	21
		2.2	*1.0*	907	9.33	7.37	22	21
		2.6	*1.2*	883	9.33	9.03	22	21

[1]Adapted from *Nutrient Requirements of Beef Cattle*, sixth revised edition, National Research Council-National Academy of Sciences, 1984.
[2]Shrunk liveweight basis. This refers to the weight after an overnight feed and water shrink (generally equivalent to about 96% of the unshrunk weights taken in early morning).
[3]In *Nutrient Requirements of Beef Cattle*, 1984, the energy requirements for large-frame steers, compensating medium-frame yearling steers, and medium-frame bulls are listed together. However, the protein requirements for medium-frame bulls are listed separately from the protein requirements for large-frame steers and compensating medium-frame yearling steers, because the protein requirements for medium-frame bulls are somewhat lower than for large-frame steer calves and compensating medium-frame yearling steers. In Table I-1c, however, the protein requirements of all three groups are listed together in order to facilitate use and save space.

TABLE I-1d
NUTRIENT REQUIREMENTS IN DIET FOR GROWING AND FINISHING CATTLE[1, 2, 3, 4]

Weight		Daily Gain		Daily Consumption: As-fed		Daily Consumption: Moisture-free Dry Matter		Moisture Basis[5] A-F (As-fed) M-F (Moisture-free)	Total Protein	Energy: TDN	Energy: ME (Mcal per)		Energy: NE_m (Mcal per)		Energy: NE_g (Mcal per)		Calcium	Phosphorus
(lb)	*(kg)*	*(lb)*	*(kg)*	*(lb)*	*(kg)*	*(lb)*	*(kg)*		*(%)*	*(%)*	*(lb)*	*(kg)*	*(lb)*	*(kg)*	*(lb)*	*(kg)*	*(%)*	*(%)*
Medium-frame steer calves																		
300	*136*	0.5	*0.2*	**8.7**	***4.0***	7.8	*3.5*	**A-F**	**8.6**	**48.6**	**0.80**	***1.76***	**0.45**	***1.00***	0.23	0.50	**0.28**	**0.18**
								M-F	9.6	54.0	0.89	*1.96*	0.50	*1.10*	0.25	*0.55*	0.31	0.20
		1.0	*0.5*	**9.3**	***4.2***	8.4	*3.8*	**A-F**	**10.3**	**52.7**	**0.90**	***1.90***	**0.51**	***1.13***	**0.28**	***0.61***	**0.41**	**0.22**
								M-F	11.4	58.5	0.96	*2.11*	0.57	*1.25*	0.31	*0.68*	0.45	0.24
		1.5	*0.7*	**9.7**	***4.4***	8.7	*4.0*	**A-F**	**11.9**	**56.7**	**0.94**	***2.06***	**0.58**	***1.26***	**0.34**	***0.76***	**0.52**	**0.25**
								M-F	13.2	63.0	1.04	*2.29*	0.64	*1.40*	0.38	*0.84*	0.58	0.28
		2.0	*0.9*	**9.9**	***4.5***	8.9	*4.0*	**A-F**	**13.3**	**60.8**	**1.00**	***2.20***	**0.63**	***1.39***	**0.40**	***0.87***	**0.65**	**0.29**
								M-F	14.8	67.5	1.11	*2.44*	0.70	*1.54*	0.44	*0.97*	0.72	0.32
		2.5	*1.1*	**9.9**	***4.5***	8.9	*4.0*	**A-F**	**15.0**	**66.2**	**1.10**	***0.98***	**0.71**	***1.57***	**0.46**	***1.01***	**0.78**	**0.33**
								M-F	16.7	73.5	1.21	*1.09*	0.79	*1.74*	0.51	*1.12*	0.87	0.37
		3.0	*1.4*	**8.8**	***4.0***	8.0	*3.6*	**A-F**	**18.0**	**76.5**	**1.30**	***1.13***	**0.86**	***1.88***	**0.58**	***1.27***	**1.02**	**0.42**
								M-F	19.9	85.0	1.39	*1.25*	0.95	*2.09*	0.64	*1.41*	1.13	0.47
400	*182*	0.5	*0.2*	**10.8**	***4.9***	9.7	*4.4*	**A-F**	**8.0**	**48.6**	**0.80**	***1.76***	**0.45**	***1.00***	**0.23**	***0.50***	**0.24**	**0.16**
								M-F	8.9	54.0	0.89	*1.96*	0.50	*1.10*	0.25	*0.55*	0.27	0.18
		1.0	*0.5*	**11.6**	***5.3***	10.4	*4.7*	**A-F**	**9.3**	**52.7**	**0.90**	***1.90***	**0.51**	***1.13***	**0.28**	***0.61***	**0.34**	**0.19**
								M-F	10.3	58.5	0.96	*2.11*	0.57	*1.25*	0.31	*0.68*	0.38	0.21
		1.5	*0.7*	**9.7**	***4.4***	10.8	*4.9*	**A-F**	**10.4**	**56.7**	**0.94**	***2.06***	**0.58**	***1.26***	**0.34**	***0.76***	**0.42**	**0.22**
								M-F	11.5	63.0	1.04	*2.29*	0.64	*1.40*	0.38	*0.84*	0.47	0.25
		2.0	*0.9*	**12.2**	***5.5***	11.0	*5.0*	**A-F**	**11.4**	**60.8**	**1.00**	***2.20***	**0.63**	***1.39***	**0.40**	***0.87***	**0.50**	**0.23**
								M-F	12.7	67.5	1.11	*2.44*	0.70	*1.54*	0.44	*0.97*	0.56	0.26
		2.5	*1.1*	**12.2**	***5.5***	11.0	*5.0*	**A-F**	**12.8**	**66.2**	**1.10**	***0.98***	**0.71**	***1.57***	**0.46**	***1.01***	**0.61**	**0.27**
								M-F	14.2	73.5	1.21	*1.09*	0.79	*1.74*	0.51	*1.12*	0.68	0.30
		3.0	*1.4*	**11.1**	***5.0***	10.0	*4.5*	**A-F**	**14.9**	**76.5**	**1.30**	***1.13***	**0.86**	***1.88***	**0.58**	***1.27***	**0.77**	**0.33**
								M-F	16.6	85.0	1.39	*1.25*	0.95	*2.09*	0.64	*1.41*	0.86	0.37
500	*227*	0.5	*0.2*	**12.8**	***5.8***	11.5	*5.2*	**A-F**	**7.7**	**48.6**	**0.80**	***1.76***	**0.45**	***1.00***	**0.23**	***0.50***	**0.23**	**0.15**
								M-F	8.5	54.0	0.89	*1.96*	0.50	*1.10*	0.25	*0.55*	0.25	0.17
		1.0	*0.5*	**13.7**	***6.2***	12.3	*6.0*	**A-F**	**8.6**	**52.7**	**0.90**	***1.90***	**0.51**	***1.13***	**0.28**	***0.61***	**0.29**	**0.18**
								M-F	9.5	58.5	0.96	*2.11*	0.57	*1.25*	0.31	*0.68*	0.32	0.20
		1.5	*0.7*	**14.2**	***6.5***	12.8	*5.8*	**A-F**	**9.5**	**56.7**	**0.94**	***2.06***	**0.58**	***1.26***	**0.34**	***0.76***	**0.36**	**0.20**
								M-F	10.5	63.0	1.04	*2.29*	0.64	*1.40*	0.38	*0.84*	0.40	0.22
		2.0	*0.9*	**14.6**	***6.6***	13.1	*6.0*	**A-F**	**10.3**	**60.8**	**1.00**	***2.20***	**0.63**	***1.39***	**0.40**	***0.87***	**0.42**	**0.22**
								M-F	11.4	67.5	1.11	*2.44*	0.70	*1.54*	0.44	*0.97*	0.47	0.24
		2.5	*1.1*	**14.4**	***6.5***	13.0	*6.0*	**A-F**	**11.3**	**66.2**	**1.10**	***0.98***	**0.71**	***1.57***	**0.46**	***1.01***	**0.50**	**0.24**
								M-F	12.5	73.5	1.21	*1.09*	0.79	*1.74*	0.51	*1.12*	0.56	0.27
		3.0	*1.4*	**13.1**	***6.0***	11.8	*5.4*	**A-F**	**13.0**	**76.5**	**1.30**	***1.13***	**0.86**	***1.88***	**0.58**	***1.27***	**0.62**	**0.29**
								M-F	14.4	85.0	1.39	*1.25*	0.95	*2.09*	0.64	*1.41*	0.69	0.32

(Continued)

TABLE I-1d (Continued)

Weight		Daily Gain		Daily Consumption: As-fed		Daily Consumption: Moisture-free Dry Matter		Moisture Basis[5] A-F (As-fed) M-F (Moisture-free)	Total Protein	Energy: TDN	Energy: ME (Mcal per)		Energy: NE$_m$ (Mcal per)		Energy: NE$_g$ (Mcal per)		Calcium	Phos-phorus
(lb)	(kg)	(lb)	(kg)	(lb)	(kg)	(lb)	(kg)		(%)	(%)	(lb)	(kg)	(lb)	(kg)	(lb)	(kg)	(%)	(%)
Medium-frame steer calves (continued)																		
600	273	0.5	0.2	14.7	6.7	13.2	6.0	A-F	7.4	48.6	0.80	1.76	0.45	1.00	0.23	0.50	0.20	0.16
								M-F	8.2	54.0	0.89	1.96	0.50	1.10	0.25	0.55	0.23	0.18
		1.0	0.5	15.7	7.1	14.1	6.4	A-F	8.1	52.7	0.90	1.90	0.51	1.13	0.28	0.61	0.25	0.17
								M-F	9.0	58.5	0.96	2.11	0.57	1.25	0.31	0.68	0.28	0.19
		1.5	0.7	16.3	7.4	14.7	6.7	A-F	8.8	56.7	0.94	2.06	0.58	1.26	0.34	0.76	0.31	0.19
								M-F	9.8	63.0	1.04	2.29	0.64	1.40	0.38	0.84	0.35	0.21
		2.0	0.9	16.7	7.6	15.0	6.8	A-F	9.5	60.8	1.00	2.20	0.63	1.39	0.40	0.87	0.36	0.20
								M-F	10.5	67.5	1.11	2.44	0.70	1.54	0.44	0.97	0.40	0.22
		2.5	1.1	16.6	7.5	14.9	6.8	A-F	10.3	66.2	1.10	0.98	0.71	1.57	0.46	1.01	0.41	0.21
								M-F	11.4	73.5	1.21	1.09	0.79	1.74	0.51	1.12	0.46	0.24
		3.0	1.4	15.0	6.8	13.5	6.1	A-F	11.6	76.5	1.30	1.13	0.86	1.88	0.58	1.27	0.51	0.26
								M-F	12.9	85.0	1.39	1.25	0.95	2.09	0.64	1.41	0.57	0.29
700	318	0.5	0.2	16.4	7.5	14.8	6.7	A-F	7.1	48.6	0.80	1.76	0.45	1.00	0.23	0.50	0.20	0.16
								M-F	7.9	54.0	0.89	1.96	0.50	1.10	0.25	0.55	0.22	0.18
		1.0	0.5	17.6	8.0	15.8	7.2	A-F	7.7	52.7	0.90	1.90	0.51	1.13	0.28	0.61	0.24	0.16
								M-F	8.6	58.5	0.96	2.11	0.57	1.25	0.31	0.68	0.27	0.18
		1.5	0.7	18.3	8.3	16.5	7.5	A-F	8.3	56.7	0.94	2.06	0.58	1.26	0.34	0.76	0.28	0.18
								M-F	9.2	63.0	1.04	2.29	0.64	1.40	0.38	0.84	0.31	0.20
		2.0	0.9	18.7	8.5	16.8	7.6	A-F	8.8	60.8	1.00	2.20	0.63	1.39	0.40	0.87	0.31	0.19
								M-F	9.8	67.5	1.11	2.44	0.70	1.54	0.44	0.97	0.34	0.21
		2.5	1.1	18.6	8.5	16.7	7.6	A-F	9.5	66.2	1.10	0.98	0.71	1.57	0.46	1.01	0.36	0.20
								M-F	10.5	73.5	1.21	1.09	0.79	1.74	0.51	1.12	0.40	0.22
		3.0	1.4	16.9	7.7	15.2	6.9	A-F	10.5	76.5	1.30	1.13	0.86	1.88	0.58	1.27	0.44	0.23
								M-F	11.7	85.0	1.39	1.25	0.95	2.09	0.64	1.41	0.49	0.26
800	364	0.5	0.2	18.2	8.3	16.4	7.5	A-F	6.9	48.6	0.80	1.76	0.45	1.00	0.23	0.50	0.20	0.15
								M-F	7.7	54.0	0.89	1.96	0.50	1.10	0.25	0.55	0.22	0.17
		1.0	0.5	19.4	8.8	17.5	8.0	A-F	7.5	52.7	0.86	1.90	0.51	1.13	0.28	0.61	0.22	0.17
								M-F	8.3	58.5	0.96	2.11	0.57	1.25	0.31	0.68	0.24	0.19
		1.5	0.7	20.2	9.2	18.2	8.3	A-F	7.9	56.7	0.94	2.06	0.58	1.27	0.34	0.76	0.25	0.17
								M-F	8.8	63.0	1.04	2.29	0.64	1.41	0.38	0.84	0.28	0.19
		2.0	0.9	20.7	9.4	18.6	8.5	A-F	8.3	60.8	1.00	2.20	0.63	1.39	0.40	0.87	0.28	0.18
								M-F	9.2	67.5	1.11	2.44	0.70	1.54	0.44	0.97	0.31	0.20
		2.5	1.1	20.6	9.4	18.5	8.4	A-F	8.8	66.2	1.09	2.40	0.71	1.57	0.46	1.01	0.32	0.19
								M-F	9.8	73.5	1.21	2.66	0.79	1.74	0.51	1.12	0.35	0.21
		3.0	1.4	18.7	8.5	16.8	7.6	A-F	9.7	76.5	1.25	2.75	0.86	1.88	0.58	1.26	0.38	0.23
								M-F	10.8	85.0	1.39	3.06	0.95	2.09	0.64	1.40	0.42	0.25

(Continued)

TABLE I-1d (Continued)

Weight		Daily Gain		Daily Consumption: As-fed		Daily Consumption: Moisture-free Dry Matter		Moisture Basis[5] A-F (As-fed) M-F (Moisture-free)	Total Protein	Energy: TDN	Energy: ME (Mcal per)		Energy: NE$_m$ (Mcal per)		Energy: NE$_g$ (Mcal per)		Calcium	Phos-phorus
(lb)	*(kg)*	*(lb)*	*(kg)*	*(lb)*	*(kg)*	*(lb)*	*(kg)*		*(%)*	*(%)*	*(lb)*	*(kg)*	*(lb)*	*(kg)*	*(lb)*	*(kg)*	*(%)*	*(%)*
Medium-frame steer calves (continued)																		
900	*409*	0.5	*0.2*	**19.9**	***9.0***	17.9	*8.1*	**A-F**	**6.8**	**48.6**	**0.80**	***1.76***	**0.45**	***0.90***	**0.23**	***0.50***	**0.19**	**0.16**
								M-F	7.6	54.0	0.89	*1.96*	0.50	*1.10*	0.25	*0.55*	0.21	0.18
		1.0	*0.5*	**21.2**	***9.6***	19.1	*8.7*	**A-F**	**7.2**	**52.7**	**0.86**	***1.90***	**0.51**	***1.13***	**0.28**	***0.61***	**0.21**	**0.16**
								M-F	8.0	58.5	0.96	*2.11*	0.57	*1.25*	0.31	*0.68*	0.23	0.18
		1.5	*0.7*	**22.1**	***10.0***	19.9	*9.0*	**A-F**	**7.6**	**56.7**	**0.94**	***2.06***	**0.58**	***1.26***	**0.34**	***0.76***	**0.23**	**0.17**
								M-F	8.4	63.0	1.04	*2.29*	0.64	*1.40*	0.38	*0.84*	0.25	0.19
		2.0	*0.9*	**22.6**	***10.3***	20.3	*9.2*	**A-F**	**7.9**	**60.8**	**1.10**	***2.20***	**0.63**	***1.39***	**0.40**	***0.87***	**0.25**	**0.18**
								M-F	8.8	67.5	1.11	*2.44*	0.70	*1.54*	0.44	*0.97*	0.28	0.20
		2.5	*1.1*	**22.4**	***10.2***	20.2	*9.2*	**A-F**	**8.4**	**66.2**	**1.09**	***2.40***	**0.71**	***1.57***	**0.46**	***1.01***	**0.28**	**0.18**
								M-F	9.3	73.5	1.21	*2.66*	0.79	*1.74*	0.51	*1.12*	0.31	0.20
		3.0	*1.4*	**20.3**	***9.2***	18.3	*8.3*	**A-F**	**9.0**	**76.5**	**1.25**	***2.75***	**0.86**	***1.88***	**0.58**	***1.26***	**0.33**	**0.21**
								M-F	10.1	85.0	1.39	*3.06*	0.95	*2.09*	0.64	*1.40*	0.37	0.23
1,000	*454*	0.5	*0.2*	**21.4**	***9.7***	19.3	*8.8*	**A-F**	**6.8**	**48.6**	**0.80**	***1.76***	**0.45**	***0.90***	**0.23**	***0.50***	**0.19**	**0.16**
								M-F	7.5	54.0	0.89	*1.96*	0.50	*1.10*	0.25	*0.55*	0.21	0.18
		1.0	*0.5*	**23.0**	***10.5***	20.7	*9.4*	**A-F**	**7.0**	**32.7**	**0.86**	***1.90***	**0.51**	***1.13***	**0.28**	***0.61***	**0.19**	**0.16**
								M-F	7.8	58.5	0.96	*2.11*	0.57	*1.25*	0.31	*0.68*	0.21	0.18
		1.5	*0.7*	**23.9**	***10.9***	21.5	*9.8*	**A-F**	**7.3**	**56.7**	**0.94**	***2.06***	**0.58**	***1.27***	**0.34**	***0.76***	**0.22**	**0.16**
								M-F	8.1	63.0	1.04	*2.29*	0.64	*1.41*	0.38	*0.84*	0.24	0.18
		2.0	*0.9*	**24.4**	***11.1***	22.0	*10.0*	**A-F**	**7.6**	**60.8**	**1.10**	***2.20***	**0.63**	***1.39***	**0.40**	***0.87***	**0.23**	**0.17**
								M-F	8.4	67.5	1.11	*2.44*	0.70	*1.54*	0.44	*0.97*	0.25	0.19
		2.5	*1.1*	**24.3**	***11.0***	21.9	*10.0*	**A-F**	**7.9**	**66.2**	**1.09**	***2.40***	**0.71**	***1.57***	**0.46**	***1.01***	**0.24**	**0.17**
								M-F	8.8	73.5	1.21	*2.66*	0.79	*1.74*	0.51	*1.12*	0.27	0.19
		3.0	*1.4*	**22.0**	***10.0***	19.8	*9.0*	**A-F**	**8.6**	**76.5**	**1.25**	***2.75***	**0.86**	***1.88***	**0.58**	***1.26***	**0.29**	**0.20**
								M-F	9.5	85.0	1.39	*3.06*	0.95	*2.09*	0.64	*1.40*	0.32	0.22
Large-frame steer calves and compensating medium-frame yearling steers																		
300	*136*	0.5	*0.2*	**9.1**	***4.1***	8.2	*3.7*	**A-F**	**8.6**	**47.3**	**0.77**	***1.70***	**0.43**	***0.95***	**0.21**	***0.50***	**0.27**	**0.17**
								M-F	9.5	52.5	0.86	*1.89*	0.48	*1.06*	0.23	*0.51*	0.30	0.19
		1.0	*0.5*	**9.7**	***4.4***	8.7	*3.9*	**A-F**	**10.2**	**50.4**	**0.83**	***1.82***	**0.49**	***1.07***	**0.25**	***0.56***	**0.41**	**0.21**
								M-F	11.3	56.0	0.92	*2.02*	0.54	*1.19*	0.28	*0.62*	0.46	0.23
		1.5	*0.7*	**10.1**	***4.6***	9.1	*4.1*	**A-F**	**11.6**	**53.6**	**0.88**	***1.94***	**0.53**	***1.17***	**0.30**	***0.66***	**0.52**	**0.24**
								M-F	12.9	59.5	0.98	*2.16*	0.59	*1.30*	0.33	*0.73*	0.58	0.27
		2.0	*0.9*	**10.4**	***4.7***	9.4	*4.3*	**A-F**	**13.1**	**57.2**	**0.94**	***2.06***	**0.58**	***1.26***	**0.34**	***0.76***	**0.63**	**0.27**
								M-F	14.6	63.5	1.04	*2.29*	0.64	*1.40*	0.38	*0.84*	0.70	0.30
		2.5	*1.1*	**10.7**	***4.9***	9.6	*4.4*	**A-F**	**14.7**	**60.8**	**1.10**	***2.20***	**0.63**	***1.39***	**0.40**	***0.87***	**0.77**	**0.31**
								M-F	16.3	67.5	1.11	*2.44*	0.70	*1.54*	0.44	*0.97*	0.85	0.34
		3.0	*1.4*	**10.7**	***4.9***	9.6	*4.4*	**A-F**	**16.2**	**64.8**	**1.06**	***2.34***	**0.69**	***1.52***	**0.44**	***0.97***	**0.89**	**0.35**
								M-F	18.0	72.0	1.18	*2.60*	0.77	*1.69*	0.49	*1.08*	0.99	0.39
		3.5	*1.6*	**10.3**	***4.7***	9.3	*4.2*	**A-F**	**18.3**	**70.7**	**1.16**	***2.56***	**0.77**	***1.70***	**0.51**	***1.13***	**1.04**	**0.41**
								M-F	20.3	78.5	1.29	*2.84*	0.86	*1.89*	0.57	*1.25*	1.16	0.45

(Continued)

TABLE I-1d (Continued)

Weight		Daily Gain		Daily Consumption: As-fed		Daily Consumption: Moisture-free Dry Matter		Moisture Basis[5] A-F (As-fed) M-F (Moisture-free)	Total Protein	Energy: TDN	Energy: ME (Mcal per)		Energy: NE_m (Mcal per)		Energy: NE_g (Mcal per)		Calcium	Phos-phorus
(lb)	*(kg)*	*(lb)*	*(kg)*	*(lb)*	*(kg)*	*(lb)*	*(kg)*		*(%)*	*(%)*	*(lb)*	*(kg)*	*(lb)*	*(kg)*	*(lb)*	*(kg)*	*(%)*	*(%)*
Large-frame steer calves and compensating medium-frame yearling steers (continued)																		
400	*182*	0.5	*0.2*	**11.2**	***5.1***	10.1	*4.6*	**A-F**	**8.0**	**47.3**	**0.77**	*1.70*	**0.43**	*0.95*	**0.21**	*0.46*	**0.23**	**0.15**
								M-F	8.9	52.5	0.86	*1.89*	0.48	*1.06*	0.23	*0.51*	0.26	0.17
		1.0	*0.5*	**9.7**	***4.4***	10.8	*4.9*	**A-F**	**9.2**	**50.4**	**0.83**	***1.82***	**0.49**	***1.07***	**0.25**	***0.56***	**0.33**	**0.18**
								M-F	10.2	56.0	0.92	*2.02*	0.54	*1.19*	0.28	*0.62*	0.37	0.20
		1.5	*0.7*	**12.6**	***5.7***	11.3	*5.1*	**A-F**	**10.3**	**53.6**	**0.88**	***1.94***	**0.53**	***1.17***	**0.30**	***0.66***	**0.42**	**0.21**
								M-F	11.4	59.5	0.98	*2.16*	0.59	*1.30*	0.33	*0.73*	0.47	0.23
		2.0	*0.9*	**13.0**	***5.9***	11.7	*5.3*	**A-F**	**11.4**	**57.2**	**0.94**	***2.06***	**0.58**	***1.26***	**0.34**	***0.76***	**0.51**	[illegible]
								M-F	12.7	63.5	1.04	*2.29*	0.64	*1.40*	0.38	*0.84*	0.57	0.26
		2.5	*1.1*	**13.2**	***6.0***	11.9	*5.4*	**A-F**	**12.5**	**60.8**	**1.10**	***2.20***	**0.63**	***1.39***	**0.40**	***0.87***	**0.59**	**0.27**
								M-F	13.9	67.5	1.11	*2.44*	0.70	*1.54*	0.44	*0.97*	0.65	0.30
		3.0	*1.4*	**13.2**	***6.0***	11.9	*5.4*	**A-F**	**13.7**	**64.8**	**1.06**	***2.34***	**0.69**	***1.52***	**0.44**	***0.97***	**0.68**	**0.30**
								M-F	15.2	72.0	1.18	*2.60*	0.77	*1.69*	0.49	*1.08*	0.76	0.33
		3.5	*1.6*	**12.8**	***5.8***	11.5	*5.2*	**A-F**	**15.2**	**70.7**	**1.16**	***2.56***	**0.77**	***1.70***	**0.51**	***1.13***	**0.81**	**0.32**
								M-F	16.9	78.5	1.29	*2.84*	0.86	*1.89*	0.57	*1.25*	0.90	0.36
500	*227*	0.5	*0.2*	**13.3**	***6.0***	12.0	*5.5*	**A-F**	**7.7**	**47.3**	**0.77**	***1.70***	**0.43**	***0.95***	**0.21**	***0.46***	**0.22**	**0.15**
								M-F	8.5	52.5	0.86	*1.89*	0.48	*1.06*	0.23	*0.51*	0.24	0.17
		1.0	*0.5*	**14.2**	***6.5***	12.8	*5.8*	**A-F**	**8.6**	**50.4**	**0.83**	***1.82***	**0.49**	***1.07***	**0.25**	***0.56***	**0.30**	**0.17**
								M-F	9.5	56.0	0.92	*2.02*	0.54	*1.19*	0.28	*0.62*	0.33	0.19
		1.5	*0.7*	**14.9**	***6.8***	13.4	*6.1*	**A-F**	**9.4**	**53.6**	**0.88**	***1.94***	**0.53**	***1.17***	**0.30**	***0.66***	**0.35**	**0.19**
								M-F	10.4	59.5	0.98	*2.16*	0.59	*1.30*	0.33	*0.73*	0.39	0.21
		2.0	*0.9*	**15.3**	***7.0***	13.8	*6.3*	**A-F**	**10.3**	**57.2**	**0.94**	***2.06***	**0.58**	***1.27***	**0.34**	***0.76***	**0.41**	**0.22**
								M-F	11.4	63.5	1.04	*2.29*	0.64	*1.41*	0.38	*0.84*	0.46	0.24
		2.5	*1.1*	**15.6**	***7.1***	14.0	*6.4*	**A-F**	**11.2**	**60.8**	**1.00**	***2.20***	**0.63**	***1.39***	**0.40**	***0.87***	**0.50**	**0.23**
								M-F	12.4	67.5	1.11	*2.44*	0.70	*1.54*	0.44	*0.97*	0.55	0.25
		3.0	*1.4*	**15.6**	***7.1***	14.0	*6.4*	**A-F**	**12.1**	**64.8**	**1.06**	***2.34***	**0.69**	***1.52***	**0.44**	***0.97***	**0.57**	**0.25**
								M-F	13.4	72.0	1.18	*2.60*	0.77	*1.69*	0.49	*1.08*	0.63	0.28
		3.5	*1.6*	**15.1**	***6.9***	13.6	*6.2*	**A-F**	**13.2**	**70.7**	**1.16**	***2.56***	**0.77**	***1.70***	**0.51**	***1.13***	**0.66**	**0.29**
								M-F	14.7	78.5	1.29	*2.84*	0.86	*1.89*	0.57	*1.25*	0.73	0.32
600	*273*	0.5	*0.2*	**15.3**	***7.0***	13.8	*6.3*	**A-F**	**7.4**	**47.3**	**0.77**	***1.70***	**0.43**	***0.95***	**0.21**	***0.46***	**0.20**	**0.16**
								M-F	8.2	52.5	0.86	*1.89*	0.48	*1.06*	0.23	*0.51*	0.22	0.18
		1.0	*0.5*	**16.2**	***7.4***	14.6	*6.6*	**A-F**	**8.1**	**50.4**	**0.83**	***1.82***	**0.49**	***1.07***	**0.25**	***0.56***	**0.26**	**0.16**
								M-F	9.0	56.0	0.92	*2.02*	0.54	*1.19*	0.28	*0.62*	0.29	0.18
		1.5	*0.7*	**17.0**	***7.7***	15.3	*7.0*	**A-F**	**8.7**	**53.6**	**0.88**	***1.94***	**0.53**	***1.17***	**0.30**	***0.66***	**0.32**	**0.18**
								M-F	9.7	59.5	0.98	*2.16*	0.59	*1.30*	0.33	*0.73*	0.35	0.20
		2.0	*0.9*	**17.6**	***8.0***	15.8	*7.2*	**A-F**	**9.5**	**57.2**	**0.94**	***2.06***	**0.58**	***1.27***	**0.34**	***0.76***	**0.36**	**0.20**
								M-F	10.5	63.5	1.04	*2.29*	0.64	*1.41*	0.38	*0.84*	0.40	0.22
		2.5	*1.1*	**17.9**	***8.1***	16.1	*7.3*	**A-F**	**10.2**	**60.8**	**1.00**	***2.20***	**0.63**	***1.39***	**0.40**	***0.87***	**0.42**	**0.21**
								M-F	11.3	67.5	1.11	*2.44*	0.70	*1.54*	0.44	*0.97*	0.47	0.23

(Continued)

TABLE I-1d (Continued)

Weight		Daily Gain		Daily Consumption: As-fed		Daily Consumption: Moisture-free Dry Matter		Moisture Basis[5] A-F (As-fed) M-F (Moisture-free)	Total Protein	Energy: TDN	Energy: ME (Mcal per)		Energy: NE_m (Mcal per)		Energy: NE_g (Mcal per)		Calcium	Phos-phorus
(lb)	*(kg)*	*(lb)*	*(kg)*	*(lb)*	*(kg)*	*(lb)*	*(kg)*		*(%)*	*(%)*	*(lb)*	*(kg)*	*(lb)*	*(kg)*	*(lb)*	*(kg)*	*(%)*	*(%)*
Large-frame steer calves and compensating medium-frame yearling steers (continued)																		
600	*273*	3.0	*1.4*	**17.9**	***8.1***	16.1	*7.3*	**A-F**	**10.9**	**64.8**	**1.06**	***2.34***	**0.69**	***1.52***	**0.44**	***0.97***	**0.47**	**0.23**
								M-F	12.1	72.0	1.18	*2.60*	0.77	*1.69*	0.49	*1.08*	0.52	0.26
		3.5	*1.6*	**17.3**	***7.9***	15.6	*7.1*	**A-F**	**11.9**	**70.7**	**1.16**	***2.56***	**0.77**	***1.70***	**0.51**	***1.13***	**0.55**	**0.25**
								M-F	13.2	78.5	1.29	*2.84*	0.86	*1.89*	0.57	*1.25*	0.61	0.28
700	*318*	0.5	*0.2*	**17.1**	***7.8***	15.4	*7.0*	**A-F**	**7.1**	**47.3**	**0.77**	***1.70***	**0.43**	***0.95***	**0.21**	***0.46***	**0.19**	**0.15**
								M-F	7.9	52.5	0.86	*1.89*	0.48	*1.06*	0.23	*0.51*	0.21	0.17
		1.0	*0.5*	**18.2**	***8.3***	16.4	*7.5*	**A-F**	**7.7**	**50.4**	**0.83**	***1.82***	**0.49**	***1.07***	**0.25**	***0.56***	**0.24**	**0.17**
								M-F	8.6	56.0	0.92	*2.02*	0.54	*1.19*	0.28	*0.62*	0.27	0.19
		1.5	*0.7*	**19.1**	***8.7***	17.2	*7.8*	**A-F**	**8.3**	**53.6**	**0.88**	***1.94***	**0.53**	***1.17***	**0.30**	***0.66***	**0.28**	**0.17**
								M-F	9.2	59.5	0.98	*2.16*	0.59	*1.30*	0.33	*0.73*	0.31	0.19
		2.0	*0.9*	**19.8**	***9.0***	17.8	*8.1*	**A-F**	**8.8**	**57.2**	**0.94**	***2.06***	**0.58**	***1.27***	**0.34**	***0.76***	**0.32**	**0.19**
								M-F	9.8	63.5	1.04	*2.29*	0.64	*1.41*	0.38	*0.84*	0.36	0.21
		2.5	*1.1*	**20.0**	***9.1***	18.0	*8.2*	**A-F**	**9.5**	**60.8**	**1.00**	***2.20***	**0.63**	***1.39***	**0.40**	***0.87***	**0.36**	**0.20**
								M-F	10.5	67.5	1.11	*2.44*	0.70	*1.54*	0.44	*0.97*	0.40	0.22
		3.0	*1.4*	**20.0**	***9.1***	18.0	*8.2*	**A-F**	**10.0**	**64.8**	**1.06**	***2.34***	**0.69**	***1.52***	**0.44**	***0.97***	**0.41**	**0.21**
								M-F	11.1	72.0	1.18	*2.60*	0.77	*1.69*	0.49	*1.08*	0.45	0.23
		3.5	*1.6*	**19.4**	***8.8***	17.5	*8.0*	**A-F**	**10.8**	**70.7**	**1.16**	***2.56***	**0.77**	***1.70***	**0.51**	***1.13***	**0.47**	**0.23**
								M-F	12.0	78.5	1.29	*2.84*	0.86	*1.89*	0.57	*1.25*	0.52	0.26
800	*364*	0.5	*0.2*	**19.0**	***8.6***	17.1	*7.8*	**A-F**	**6.9**	**47.3**	**0.77**	***1.70***	**0.43**	***0.95***	**0.21**	***0.46***	**0.19**	**0.16**
								M-F	7.7	52.5	0.86	*1.89*	0.48	*1.06*	0.23	*0.51*	0.21	0.18
		1.0	*0.5*	**20.2**	***9.2***	18.2	*8.3*	**A-F**	**7.5**	**50.4**	**0.83**	***1.82***	**0.49**	***1.07***	**0.25**	***0.56***	**0.22**	**0.16**
								M-F	8.3	56.0	0.92	*2.02*	0.54	*1.19*	0.28	*0.62*	0.24	0.18
		1.5	*0.7*	**21.1**	***9.6***	19.0	*8.6*	**A-F**	**7.9**	**53.6**	**0.88**	***1.94***	**0.53**	***1.17***	**0.30**	***0.66***	**0.25**	**0.17**
								M-F	8.8	59.5	0.98	*2.16*	0.59	*1.30*	0.33	*0.73*	0.28	0.19
		2.0	*0.9*	**21.8**	***9.9***	19.6	*8.9*	**A-F**	**8.4**	**57.2**	**0.94**	***2.06***	**0.58**	***1.27***	**0.34**	***0.76***	**0.29**	**0.18**
								M-F	9.3	63.5	1.04	*2.29*	0.64	*1.41*	0.38	*0.84*	0.32	0.20
		2.5	*1.1*	**22.1**	***10.0***	19.9	*9.0*	**A-F**	**8.8**	**60.8**	**1.00**	***2.20***	**0.63**	***1.39***	**0.40**	***0.87***	**0.32**	**0.19**
								M-F	9.8	67.5	1.11	*2.44*	0.70	*1.54*	0.44	*0.97*	0.35	0.21
		3.0	*1.4*	**22.1**	***10.0***	19.9	*9.0*	**A-F**	**9.4**	**64.8**	**1.06**	***2.34***	**0.69**	***1.52***	**0.44**	***0.97***	**0.36**	**0.20**
								M-F	10.4	72.0	1.18	*2.60*	0.77	*1.69*	0.49	*1.08*	0.40	0.22
		3.5	*1.6*	**21.4**	***9.7***	19.3	*8.8*	**A-F**	**10.0**	**70.7**	**1.16**	***2.56***	**0.77**	***1.70***	**0.51**	***1.13***	**0.41**	**0.22**
								M-F	11.1	78.5	1.29	*2.84*	0.86	*1.89*	0.57	*1.25*	0.45	0.24
900	*409*	0.5	*0.2*	**20.7**	***9.4***	18.6	*8.5*	**A-F**	**6.8**	**47.3**	**0.77**	***1.70***	**0.43**	***0.86***	**0.21**	***0.46***	**0.18**	**0.16**
								M-F	7.6	52.5	0.86	*1.89*	0.48	*0.96*	0.23	*0.51*	0.20	0.18
		1.0	*0.5*	**22.0**	***10.0***	19.8	*9.0*	**A-F**	**7.2**	**50.4**	**0.83**	***1.82***	**0.49**	***1.07***	**0.25**	***0.56***	**0.20**	**0.16**
								M-F	8.0	56.0	0.92	*2.02*	0.54	*1.19*	0.28	*0.62*	0.23	0.18
		1.5	*0.7*	**23.1**	***10.5***	20.8	*9.5*	**A-F**	**7.7**	**53.6**	**0.88**	***1.94***	**0.53**	***1.17***	**0.30**	***0.66***	**0.24**	**0.16**
								M-F	8.5	59.5	0.98	*2.16*	0.59	*1.30*	0.33	*0.73*	0.27	0.18

(Continued)

TABLE I-1d (Continued)

Weight		Daily Gain		Daily Consumption: As-fed		Daily Consumption: Moisture-free Dry Matter		Moisture Basis[5] A-F (As-fed) M-F (Moisture-free)	Total Protein	Energy: TDN	Energy: ME (Mcal per)		Energy: NE_m (Mcal per)		Energy: NE_g (Mcal per)		Calcium	Phos-phorus
(lb)	*(kg)*	*(lb)*	*(kg)*	*(lb)*	*(kg)*	*(lb)*	*(kg)*		*(%)*	*(%)*	*(lb)*	*(kg)*	*(lb)*	*(kg)*	*(lb)*	*(kg)*	*(%)*	*(%)*
Large-frame steer calves and compensating medium-frame yearling steers (continued)																		
900	409	2.0	0.9	23.8	10.8	21.4	9.7	A-F	8.0	57.2	0.94	2.06	0.58	1.26	0.34	0.76	0.26	0.18
								M-F	8.9	63.5	1.04	2.29	0.64	1.40	0.38	0.84	0.29	0.20
		2.5	1.1	24.2	11.0	21.8	9.9	A-F	8.4	60.8	1.00	2.20	0.63	1.39	0.40	0.87	0.28	0.18
								M-F	9.3	67.5	1.11	2.44	0.70	1.54	0.44	0.97	0.31	0.20
		3.0	1.4	24.1	11.0	21.7	9.9	A-F	8.8	64.8	1.06	2.34	0.69	1.52	0.44	0.97	0.32	0.19
								M-F	9.8	72.0	1.18	2.60	0.77	1.69	0.49	1.08	0.36	0.21
		3.5	1.6	23.4	10.6	21.1	9.6	A-F	9.4	70.7	1.16	2.56	0.77	1.70	0.51	1.13	0.36	0.21
								M-F	10.4	78.5	1.29	2.84	0.86	1.89	0.57	1.25	0.40	0.23
1,000	454	0.5	0.2	22.4	10.2	20.2	9.2	A-F	6.8	47.3	0.77	1.70	0.43	0.86	0.21	0.46	0.18	0.15
								M-F	7.5	52.5	0.86	1.89	0.48	0.96	0.23	0.51	0.20	0.17
		1.0	0.5	23.9	10.9	21.5	9.8	A-F	7.0	50.4	0.83	1.82	0.49	1.07	0.25	0.56	0.20	0.15
								M-F	7.8	56.0	0.92	2.02	0.54	1.19	0.28	0.62	0.23	0.17
		1.5	0.7	25.0	11.4	22.5	10.2	A-F	7.4	53.6	0.88	1.94	0.53	1.17	0.30	0.66	0.23	0.16
								M-F	8.2	59.5	0.98	2.16	0.59	1.30	0.33	0.73	0.25	0.18
		2.0	0.9	25.8	11.7	23.2	10.5	A-F	7.7	57.2	0.94	2.06	0.58	1.26	0.34	0.76	0.24	0.16
								M-F	8.6	63.5	1.04	2.29	0.64	1.40	0.38	0.84	0.27	0.18
		2.5	1.1	26.2	11.9	23.6	10.7	A-F	8.0	60.8	1.00	2.20	0.63	1.39	0.40	0.87	0.26	0.17
								M-F	8.9	67.5	1.11	2.44	0.70	1.54	0.44	0.97	0.29	0.19
		3.0	1.4	26.2	11.9	23.6	10.7	A-F	8.4	64.8	1.06	2.34	0.69	1.52	0.44	0.97	0.29	0.18
								M-F	9.3	72.0	1.18	2.60	0.77	1.69	0.49	1.08	0.32	0.20
		3.5	1.6	25.3	11.5	22.8	10.4	A-F	8.8	70.7	1.16	2.56	0.77	1.70	0.51	1.13	0.32	0.19
								M-F	9.8	78.5	1.29	2.84	0.86	1.89	0.57	1.25	0.35	0.21
1,100	500	0.5	0.2	24.1	11.0	21.7	9.9	A-F	6.7	47.3	0.77	1.70	0.43	0.86	0.21	0.46	0.17	0.16
								M-F	7.4	52.5	0.86	1.89	0.48	0.96	0.23	0.51	0.19	0.18
		1.0	0.5	25.7	11.7	23.1	10.5	A-F	6.9	50.4	0.83	1.82	0.49	1.07	0.25	0.56	0.19	0.16
								M-F	7.7	56.0	0.92	2.02	0.54	1.19	0.28	0.62	0.21	0.18
		1.5	0.7	26.8	12.2	24.1	11.0	A-F	7.2	53.6	0.88	1.94	0.53	1.17	0.30	0.66	0.21	0.16
								M-F	8.0	59.5	0.98	2.16	0.59	1.30	0.33	0.73	0.23	0.18
		2.0	0.9	27.7	12.6	24.9	11.3	A-F	7.5	57.2	0.94	2.06	0.58	1.26	0.34	0.76	0.23	0.16
								M-F	8.3	63.5	1.04	2.29	0.64	1.40	0.38	0.84	0.25	0.18
		2.5	1.1	28.1	12.8	25.3	11.5	A-F	7.7	60.8	1.00	2.20	0.63	1.39	0.40	0.87	0.23	0.16
								M-F	8.5	67.5	1.11	2.44	0.70	1.54	0.44	0.97	0.26	0.18
		3.0	1.4	28.1	12.8	25.3	11.5	A-F	8.0	64.8	1.06	2.34	0.69	1.52	0.44	0.97	0.26	0.17
								M-F	8.9	72.0	1.18	2.60	0.77	1.69	0.49	1.08	0.29	0.19
		3.5	1.6	27.2	12.4	24.5	11.1	A-F	8.4	70.7	1.16	2.56	0.77	1.70	0.51	1.13	0.29	0.19
								M-F	9.3	78.5	1.29	2.84	0.86	1.89	0.57	1.25	0.32	0.21

(Continued)

TABLE I-1d (Continued)

Weight		Daily Gain		Daily Consumption: As-fed		Daily Consumption: Moisture-free Dry Matter		Moisture Basis[5] A-F (As-fed) M-F (Moisture-free)	Total Protein	Energy: TDN	Energy: ME (Mcal per)		Energy: NE_m (Mcal per)		Energy: NE_g (Mcal per)		Calcium	Phos-phorus
(lb)	*(kg)*	*(lb)*	*(kg)*	*(lb)*	*(kg)*	*(lb)*	*(kg)*		*(%)*	*(%)*	*(lb)*	*(kg)*	*(lb)*	*(kg)*	*(lb)*	*(kg)*	*(%)*	*(%)*
Medium-frame bulls																		
300	*136*	0.5	*0.2*	**8.7**	***4.0***	7.8	*3.5*	A-F	**8.7**	**48.2**	**0.79**	***1.75***	**0.44**	***0.97***	**0.22**	***0.48***	**0.28**	**0.18**
								M-F	9.7	53.5	0.88	*1.94*	0.49	*1.08*	0.24	*0.53*	0.31	0.20
		1.0	*0.5*	**9.2**	***4.2***	8.3	*3.8*	A-F	**10.4**	**51.8**	**0.85**	***1.86***	**0.50**	***1.11***	**0.27**	***0.59***	**0.43**	**0.22**
								M-F	11.6	57.5	0.94	*2.07*	0.56	*1.23*	0.30	*0.66*	0.48	0.24
		1.5	*0.7*	**9.6**	***4.4***	8.6	*3.9*	A-F	**12.1**	**55.4**	**0.91**	***2.00***	**0.56**	***1.22***	**0.32**	***0.69***	**0.56**	**0.25**
								M-F	13.4	61.5	1.01	*2.22*	0.62	*1.36*	0.35	*0.77*	0.62	0.28
		2.0	*0.9*	**9.8**	***4.5***	8.8	*4.0*	A-F	**13.7**	**59.0**	**0.97**	***2.14***	**0.61**	***1.35***	**0.40**	***0.81***	**0.68**	**0.30**
								M-F	15.2	65.5	1.08	*2.38*	0.68	*1.50*	0.41	*0.90*	0.75	0.33
		2.5	*1.1*	**9.9**	***4.5***	8.9	*4.0*	A-F	**15.3**	**63.0**	**1.04**	***2.28***	**0.67**	***1.47***	**0.42**	***0.93***	**0.83**	**0.33**
								M-F	17.0	70.0	1.15	*2.53*	0.74	*1.63*	0.47	*1.03*	0.92	0.37
		3.0	*1.4*	**9.7**	***4.4***	8.7	*4.0*	A-F	**17.4**	**68.9**	**1.13**	***2.49***	**0.76**	***1.67***	**0.49**	***1.07***	**0.98**	**0.39**
								M-F	19.3	76.5	1.26	*2.77*	0.84	*1.85*	0.54	*1.19*	1.09	0.43
400	*182*	0.5	*0.2*	**10.7**	***4.9***	9.6	*4.4*	A-F	**8.1**	**48.2**	**0.79**	***1.75***	**0.44**	***0.97***	**0.22**	***0.48***	**0.25**	**0.16**
								M-F	9.0	53.5	0.88	*1.94*	0.49	*1.08*	0.24	*0.53*	0.28	0.18
		1.0	*0.5*	**11.4**	***5.2***	10.3	*4.7*	A-F	**9.4**	**51.8**	**0.85**	***1.86***	**0.50**	***1.11***	**0.27**	***0.59***	**0.35**	**0.19**
								M-F	10.4	57.5	0.94	*2.07*	0.56	*1.23*	0.30	*0.66*	0.39	0.21
		1.5	*0.7*	**11.9**	***5.4***	10.7	*4.9*	A-F	**10.6**	**55.4**	**0.91**	***2.00***	**0.56**	***1.22***	**0.32**	***0.69***	**0.44**	**0.23**
								M-F	11.8	61.5	1.01	*2.22*	0.62	*1.36*	0.35	*0.77*	0.49	0.25
		2.0	*0.9*	**12.2**	***5.5***	11.0	*5.0*	A-F	**11.8**	**59.0**	**0.97**	***2.14***	**0.61**	***1.35***	**0.40**	***0.81***	**0.54**	**0.25**
								M-F	13.1	65.5	1.08	*2.38*	0.68	*1.50*	0.41	*0.90*	0.60	0.28
		2.5	*1.1*	**12.3**	***5.6***	11.1	*5.0*	A-F	**13.0**	**63.0**	**1.04**	***2.28***	**0.67**	***1.47***	**0.42**	***0.93***	**0.63**	**0.29**
								M-F	14.4	70.0	1.15	*2.53*	0.74	*1.63*	0.47	*1.03*	0.70	0.32
		3.0	*1.4*	**9.7**	***4.4***	10.8	*4.9*	A-F	**14.5**	**68.9**	**1.13**	***2.49***	**0.76**	***1.67***	**0.49**	***1.07***	**0.76**	**0.33**
								M-F	16.1	76.5	1.26	*2.77*	0.84	*1.85*	0.54	*1.19*	0.84	0.37
500	*227*	0.5	*0.2*	**12.7**	***5.8***	11.4	*5.2*	A-F	**7.7**	**48.2**	**0.79**	***1.75***	**0.44**	***0.97***	**0.22**	***0.48***	**0.23**	**0.15**
								M-F	8.6	53.5	0.88	*1.94*	0.49	*1.08*	0.24	*0.53*	0.25	0.17
		1.0	*0.5*	**13.4**	***6.1***	12.1	*5.5*	A-F	**8.7**	**51.8**	**0.85**	***1.86***	**0.50**	***1.11***	**0.27**	***0.59***	**0.32**	**0.18**
								M-F	9.7	57.5	0.94	*2.07*	0.56	*1.23*	0.30	*0.66*	0.35	0.20
		1.5	*0.7*	**14.1**	***6.4***	12.7	*5.8*	A-F	**9.6**	**55.4**	**0.91**	***2.00***	**0.56**	***1.22***	**0.32**	***0.69***	**0.38**	**0.21**
								M-F	10.7	61.5	1.01	*2.22*	0.62	*1.36*	0.35	*0.77*	0.42	0.23
		2.0	*0.9*	**14.4**	***6.5***	13.0	*5.9*	A-F	**10.5**	**59.0**	**0.97**	***2.14***	**0.61**	***1.35***	**0.40**	***0.81***	**0.44**	**0.23**
								M-F	11.7	65.5	1.08	*2.38*	0.68	*1.50*	0.41	*0.90*	0.49	0.25
		2.5	*1.1*	**14.6**	***6.6***	13.1	*6.0*	A-F	**11.5**	**63.0**	**1.04**	***2.28***	**0.67**	***1.47***	**0.42**	***0.93***	**0.53**	**0.24**
								M-F	12.8	70.0	1.15	*2.53*	0.74	*1.63*	0.47	*1.03*	0.59	0.27
		3.0	*1.4*	**14.2**	***6.5***	12.8	*5.8*	A-F	**12.7**	**68.9**	**1.13**	***2.49***	**0.76**	***1.67***	**0.49**	***1.07***	**0.62**	**0.28**
								M-F	14.1	76.5	1.26	*2.77*	0.84	*1.85*	0.54	*1.19*	0.69	0.31

(Continued)

TABLE I-1d (Continued)

Weight		Daily Gain		Daily Consumption: As-fed		Daily Consumption: Moisture-free Dry Matter		Moisture Basis[5] A-F (As-fed) M-F (Moisture-free)	Total Protein	Energy: TDN	Energy: ME (Mcal per)		Energy: NE_m (Mcal per)		Energy: NE_g (Mcal per)		Calcium	Phosphorus
(lb)	*(kg)*	*(lb)*	*(kg)*	*(lb)*	*(kg)*	*(lb)*	*(kg)*		*(%)*	*(%)*	*(lb)*	*(kg)*	*(lb)*	*(kg)*	*(lb)*	*(kg)*	*(%)*	*(%)*
Medium-frame bulls (continued)																		
600	*273*	0.5	*0.2*	**14.6**	***6.6***	13.1	*6.0*	**A-F**	**7.5**	**48.2**	**0.79**	***1.75***	**0.44**	***0.97***	**0.22**	***0.48***	**0.22**	**0.17**
								M-F	8.3	53.5	0.88	*1.94*	0.49	*1.08*	0.24	*0.53*	0.24	0.19
		1.0	*0.5*	**15.4**	***7.0***	13.9	*6.3*	**A-F**	**8.3**	**51.8**	**0.85**	***1.86***	**0.50**	***1.11***	**0.27**	***0.59***	**0.27**	**0.17**
								M-F	9.2	57.5	0.94	*2.07*	0.56	*1.23*	0.30	*0.66*	0.30	0.19
		1.5	*0.7*	**16.1**	***7.3***	14.5	*6.6*	**A-F**	**9.0**	**55.4**	**0.91**	***2.00***	**0.56**	***1.22***	**0.32**	***0.69***	**0.32**	**0.19**
								M-F	10.0	61.5	1.01	*2.22*	0.62	*1.36*	0.35	*0.77*	0.36	0.21
		2.0	*0.9*	**16.6**	***7.5***	14.9	*6.8*	**A-F**	**9.7**	**59.0**	**0.97**	***2.14***	**0.61**	***1.35***	**0.40**	***0.81***	**0.39**	**0.22**
								M-F	10.8	65.5	1.08	*2.38*	0.68	*1.50*	0.41	*0.90*	0.43	0.24
		2.5	*1.1*	**16.7**	***7.6***	15.0	*6.8*	**A-F**	**10.4**	**63.0**	**1.04**	***2.28***	**0.67**	***1.47***	**0.42**	***0.93***	**0.45**	**0.23**
								M-F	11.6	70.0	1.15	*2.53*	0.74	*1.63*	0.47	*1.03*	0.50	0.25
		3.0	*1.4*	**16.3**	***7.4***	14.7	*6.7*	**A-F**	**11.4**	**68.9**	**1.13**	***2.49***	**0.76**	***1.67***	**0.49**	***1.07***	**0.51**	**0.26**
								M-F	12.7	76.5	1.26	*2.77*	0.84	*1.85*	0.54	*1.19*	0.57	0.29
700	*318*	0.5	*0.2*	**16.3**	***7.4***	14.7	*6.7*	**A-F**	**7.2**	**48.2**	**0.79**	***1.75***	**0.44**	***0.97***	**0.22**	***0.48***	**0.21**	**0.16**
								M-F	8.0	53.5	0.88	*1.94*	0.49	*1.08*	0.24	*0.53*	0.23	0.18
		1.0	*0.5*	**17.3**	***7.9***	15.6	*7.0*	**A-F**	**7.9**	**51.8**	**0.85**	***1.86***	**0.50**	***1.11***	**0.27**	***0.59***	**0.25**	**0.18**
								M-F	8.8	57.5	0.94	*2.07*	0.56	*1.23*	0.30	*0.66*	0.28	0.20
		1.5	*0.7*	**18.1**	***8.2***	16.3	*7.4*	**A-F**	**8.5**	**55.4**	**0.91**	***2.00***	**0.56**	***1.22***	**0.32**	***0.69***	**0.29**	**0.18**
								M-F	9.4	61.5	1.01	*2.22*	0.62	*1.36*	0.35	*0.77*	0.32	0.20
		2.0	*0.9*	**18.6**	***8.5***	16.7	*7.6*	**A-F**	**10.0**	**59.0**	**0.97**	***2.14***	**0.61**	***1.35***	**0.40**	***0.81***	**0.34**	**0.20**
								M-F	10.1	65.5	1.08	*2.38*	0.68	*1.50*	0.41	*0.90*	0.38	0.22
		2.5	*1.1*	**18.7**	***8.5***	16.8	*7.6*	**A-F**	**9.7**	**63.0**	**1.04**	***2.28***	**0.67**	***1.47***	**0.42**	***0.93***	**0.39**	**0.22**
								M-F	10.8	70.0	1.15	*2.53*	0.74	*1.63*	0.47	*1.03*	0.43	0.24
		3.0	*1.4*	**18.3**	***8.3***	16.5	*7.5*	**A-F**	**10.5**	**68.9**	**1.13**	***2.49***	**0.76**	***1.67***	**0.49**	***1.07***	**0.44**	**0.23**
								M-F	11.7	76.5	1.26	*2.77*	0.84	*1.85*	0.54	*1.19*	0.49	0.25
800	*364*	0.5	*0.2*	**18.0**	***8.2***	16.2	*7.4*	**A-F**	**7.0**	**48.2**	**0.79**	***1.75***	**0.44**	***0.97***	**0.22**	***0.48***	**0.20**	**0.17**
								M-F	7.8	53.5	0.88	*1.94*	0.49	*1.08*	0.24	*0.53*	0.22	0.19
		1.0	*0.5*	**19.2**	***8.7***	17.3	*7.9*	**A-F**	**7.6**	**51.8**	**0.85**	***1.86***	**0.50**	***1.11***	**0.27**	***0.59***	**0.23**	**0.17**
								M-F	8.4	57.5	0.94	*2.07*	0.56	*1.23*	0.30	*0.66*	0.25	0.19
		1.5	*0.7*	**20.0**	***9.1***	18.0	*8.2*	**A-F**	**8.1**	**55.4**	**0.91**	***2.00***	**0.56**	***1.22***	**0.32**	***0.69***	**0.26**	**0.18**
								M-F	9.0	61.5	1.01	*2.22*	0.62	*1.36*	0.35	*0.77*	0.29	0.20
		2.0	*0.9*	**20.6**	***9.4***	18.5	*8.4*	**A-F**	**8.6**	**59.0**	**0.97**	***2.14***	**0.61**	***1.35***	**0.40**	***0.81***	**0.30**	**0.19**
								M-F	9.5	65.5	1.08	*2.38*	0.68	*1.50*	0.41	*0.90*	0.33	0.21
		2.5	*1.1*	**20.7**	***9.4***	18.6	*8.5*	**A-F**	**10.0**	**63.0**	**1.04**	***2.28***	**0.67**	***1.47***	**0.42**	***0.93***	**0.34**	**0.21**
								M-F	10.1	70.0	1.15	*2.53*	0.74	*1.63*	0.47	*1.03*	0.38	0.23
		3.0	*1.4*	**20.2**	***9.2***	18.2	*8.3*	**A-F**	**9.7**	**68.9**	**1.13**	***2.49***	**0.76**	***1.67***	**0.49**	***1.07***	**0.40**	**0.22**
								M-F	10.8	76.5	1.26	*2.77*	0.84	*1.85*	0.54	*1.19*	0.44	0.24

(Continued)

TABLE I-1d (Continued)

Weight		Daily Gain		Daily Consumption: As-fed		Daily Consumption: Moisture-free Dry Matter		Moisture Basis[5] A-F (As-fed) M-F (Moisture-free)	Total Protein	Energy: TDN	Energy: ME (Mcal per)		Energy: NE_m (Mcal per)		Energy: NE_g (Mcal per)		Calcium	Phos-phorus
(lb)	(kg)	(lb)	(kg)	(lb)	(kg)	(lb)	(kg)		(%)	(%)	(lb)	(kg)	(lb)	(kg)	(lb)	(kg)	(%)	(%)
Medium-frame bulls (continued)																		
900	*409*	0.5	*0.2*	**19.7**	***9.0***	17.7	*8.0*	**A-F**	**6.9**	**48.2**	**0.79**	***1.75***	**0.44**	***0.97***	**0.22**	***0.48***	**0.19**	**0.17**
								M-F	7.7	53.5	0.88	*1.94*	0.49	*1.08*	0.24	*0.53*	0.21	0.19
		1.0	*0.5*	**21.0**	***9.5***	18.9	*8.6*	**A-F**	**7.4**	**51.8**	**0.85**	***1.86***	**0.50**	***1.11***	**0.27**	***0.59***	**0.23**	**0.17**
								M-F	8.2	57.5	0.94	*2.07*	0.56	*1.23*	0.30	*0.66*	0.25	0.19
		1.5	*0.7*	**21.9**	***10.0***	19.7	*9.0*	**A-F**	**7.7**	**55.4**	**0.91**	***2.00***	**0.56**	***1.22***	**0.32**	***0.69***	**0.25**	**0.17**
								M-F	8.6	61.5	1.01	*2.22*	0.62	*1.36*	0.35	*0.77*	0.28	0.19
		2.0	*0.9*	**22.4**	***10.2***	20.2	*9.2*	**A-F**	**8.2**	**59.0**	**0.97**	***2.14***	**0.61**	***1.35***	**0.40**	***0.81***	**0.28**	**0.19**
								M-F	9.1	65.5	1.08	*2.38*	0.68	*1.50*	0.41	*0.90*	0.31	0.21
		2.5	*1.1*	**22.6**	***10.3***	20.3	*9.2*	**A-F**	**8.6**	**63.0**	**1.04**	***2.28***	**0.67**	***1.47***	**0.42**	***0.93***	**0.31**	**0.20**
								M-F	9.6	70.0	1.15	*2.53*	0.74	*1.63*	0.47	*1.03*	0.34	0.22
		3.0	*1.4*	**22.1**	***10.0***	19.9	*9.0*	**A-F**	**9.2**	**68.9**	**1.13**	***2.49***	**0.76**	***1.67***	**0.49**	***1.07***	**0.35**	**0.21**
								M-F	10.2	76.5	1.26	*2.77*	0.84	*1.85*	0.54	*1.19*	0.39	0.23
1,000	*454*	0.5	*0.2*	**21.3**	***9.7***	19.2	*8.7*	**A-F**	**6.8**	**48.2**	**0.79**	***1.75***	**0.44**	***0.97***	**0.22**	***0.48***	**0.19**	**0.16**
								M-F	7.5	53.5	0.88	*1.94*	0.49	*1.08*	0.24	*0.53*	0.21	0.18
		1.0	*0.5*	**22.7**	***10.3***	20.4	*9.3*	**A-F**	**7.2**	**51.8**	**0.85**	***1.86***	**0.50**	***1.11***	**0.27**	***0.59***	**0.22**	**0.16**
								M-F	8.0	57.5	0.94	*2.07*	0.56	*1.23*	0.30	*0.66*	0.24	0.18
		1.5	*0.7*	**23.7**	***10.8***	21.3	*9.7*	**A-F**	**7.6**	**55.4**	**0.91**	***2.00***	**0.56**	***1.22***	**0.32**	***0.69***	**0.23**	**0.17**
								M-F	8.4	61.5	1.01	*2.22*	0.62	*1.36*	0.35	*0.77*	0.26	0.19
		2.0	*0.9*	**24.2**	***11.0***	21.8	*9.9*	**A-F**	**7.8**	**59.0**	**0.97**	***2.14***	**0.61**	***1.35***	**0.40**	***0.81***	**0.25**	**0.17**
								M-F	8.7	65.5	1.08	*2.38*	0.68	*1.50*	0.41	*0.90*	0.28	0.19
		2.5	*1.1*	**24.4**	***11.1***	22.0	*10.0*	**A-F**	**8.2**	**63.0**	**1.04**	***2.28***	**0.67**	***1.47***	**0.42**	***0.93***	**0.28**	**0.18**
								M-F	9.1	70.0	1.15	*2.53*	0.74	*1.63*	0.47	*1.03*	0.31	0.20
		3.0	*1.4*	**23.9**	***10.9***	21.5	*9.8*	**A-F**	**8.6**	**68.9**	**1.13**	***2.49***	**0.76**	***1.67***	**0.49**	***1.07***	**0.32**	**0.20**
								M-F	9.6	76.5	1.26	*2.77*	0.84	*1.85*	0.54	*1.19*	0.35	0.22
1,100	*500*	0.5	*0.2*	**22.9**	***10.4***	20.6	*9.4*	**A-F**	**6.7**	**48.2**	**0.79**	***1.75***	**0.44**	***0.97***	**0.22**	***0.48***	**0.18**	**0.17**
								M-F	7.4	53.5	0.88	*1.94*	0.49	*1.08*	0.24	*0.53*	0.20	0.19
		1.0	*0.5*	**24.3**	***11.0***	21.9	*10.0*	**A-F**	**7.0**	**51.8**	**0.85**	***1.86***	**0.50**	***1.11***	**0.27**	***0.59***	**0.20**	**0.17**
								M-F	7.8	57.5	0.94	*2.07*	0.56	*1.23*	0.30	*0.66*	0.22	0.19
		1.5	*0.7*	**25.4**	***11.5***	22.9	*10.4*	**A-F**	**7.3**	**55.4**	**0.91**	***2.00***	**0.56**	***1.22***	**0.32**	***0.69***	**0.22**	**0.17**
								M-F	8.1	61.5	1.01	*2.22*	0.62	*1.36*	0.35	*0.77*	0.24	0.19
		2.0	*0.9*	**26.0**	***11.8***	23.4	*10.6*	**A-F**	**7.6**	**59.0**	**0.97**	***2.14***	**0.61**	***1.35***	**0.40**	***0.81***	**0.23**	**0.17**
								M-F	8.4	65.5	1.08	*2.38*	0.68	*1.50*	0.41	*0.90*	0.26	0.19
		2.5	*1.1*	**26.2**	***11.9***	23.6	*10.7*	**A-F**	**7.8**	**63.0**	**1.04**	***2.28***	**0.67**	***1.47***	**0.42**	***0.93***	**0.25**	**0.18**
								M-F	8.7	70.0	1.15	*2.53*	0.74	*1.63*	0.47	*1.03*	0.28	0.20
		3.0	*1.4*	**25.7**	***11.7***	23.1	*10.5*	**A-F**	**8.3**	**68.9**	**1.13**	***2.49***	**0.76**	***1.67***	**0.49**	***1.07***	**0.29**	**0.19**
								M-F	9.2	76.5	1.26	*2.77*	0.84	*1.85*	0.54	*1.19*	0.32	0.21

(Continued)

TABLE I-1d (Continued)

Weight		Daily Gain		Daily Consumption: As-fed		Daily Consumption: Moisture-free Dry Matter		Moisture Basis[5] A-F (As-fed) M-F (Moisture-free)	Total Protein	Energy: TDN	Energy: ME (Mcal per)		Energy: NE_m (Mcal per)		Energy: NE_g (Mcal per)		Calcium	Phos-phorus
(lb)	*(kg)*	*(lb)*	*(kg)*	*(lb)*	*(kg)*	*(lb)*	*(kg)*		*(%)*	*(%)*	*(lb)*	*(kg)*	*(lb)*	*(kg)*	*(lb)*	*(kg)*	*(%)*	*(%)*
Large-frame bull calves and compensating large-frame yearling steers																		
300	*136*	0.5	*0.2*	**8.8**	**4.0**	7.9	*3.6*	**A-F**	**8.7**	**47.3**	**0.77**	***1.70***	**0.43**	***0.95***	**0.21**	***0.46***	**0.28**	**0.18**
								M-F	9.7	52.5	0.86	*1.89*	0.48	*1.06*	0.23	*0.51*	0.31	0.20
		1.0	*0.5*	**9.3**	**4.2**	8.4	*3.8*	**A-F**	**10.5**	**50.4**	**0.83**	***1.82***	**0.49**	***1.07***	**0.25**	***0.56***	**0.42**	**0.22**
								M-F	11.7	56.0	0.92	*2.02*	0.54	*1.19*	0.28	*0.62*	0.47	0.24
		1.5	*0.7*	**9.8**	**4.5**	8.8	*4.0*	**A-F**	**12.2**	**53.6**	**0.88**	***1.95***	**0.53**	***1.17***	**0.30**	***0.66***	**0.57**	**0.25**
								M-F	13.5	59.5	0.98	*2.16*	0.59	*1.30*	0.33	*0.73*	0.63	0.28
		2.0	*0.9*	**10.0**	**4.5**	9.0	*4.1*	**A-F**	**13.6**	**56.3**	**0.93**	***2.04***	**0.57**	***1.25***	**0.33**	***0.73***	**0.68**	**0.29**
								M-F	15.1	62.5	1.03	*2.27*	0.63	*1.39*	0.37	*0.81*	0.76	0.32
		2.5	*1.1*	**10.2**	**4.6**	9.2	*4.2*	**A-F**	**15.3**	**59.9**	**0.98**	***2.16***	**0.62**	***1.37***	**0.38**	***0.83***	**0.82**	**0.32**
								M-F	17.0	66.5	1.09	*2.40*	0.69	*1.52*	0.42	*0.92*	0.91	0.36
		3.0	*1.4*	**10.2**	**4.6**	9.2	*4.2*	**A-F**	**16.9**	**63.5**	**1.04**	***2.30***	**0.68**	***1.49***	**0.42**	***0.93***	**0.97**	**0.39**
								M-F	18.8	70.5	1.16	*2.55*	0.75	*1.65*	0.47	*1.03*	1.08	0.43
		3.5	*1.6*	**10.1**	**4.6**	9.1	*4.1*	**A-F**	**18.8**	**68.0**	**1.12**	***2.46***	**0.74**	***1.62***	**0.48**	***1.05***	**1.12**	**0.43**
								M-F	20.9	75.5	1.24	*2.73*	0.82	*1.80*	0.53	*1.17*	1.24	0.48
		4.0	*1.8*	**9.1**	**4.1**	8.2	*3.7*	**A-F**	**22.2**	**77.4**	**1.27**	***2.79***	**0.86**	***1.90***	**0.59**	***1.31***	**1.38**	**0.53**
								M-F	24.7	86.0	1.41	*3.10*	0.96	*2.11*	0.66	*1.45*	1.53	0.59
400	*182*	0.5	*0.2*	**10.9**	**5.0**	9.8	*4.5*	**A-F**	**8.1**	**47.3**	**0.77**	***1.70***	**0.43**	***0.95***	**0.21**	***0.46***	**0.24**	**0.16**
								M-F	9.0	52.5	0.86	*1.89*	0.48	*1.06*	0.23	*0.51*	0.27	0.18
		1.0	*0.5*	**11.6**	**5.3**	10.4	*4.7*	**A-F**	**9.5**	**50.4**	**0.83**	***1.82***	**0.49**	***1.07***	**0.25**	***0.56***	**0.36**	**0.19**
								M-F	10.5	56.0	0.92	*2.02*	0.54	*1.19*	0.28	*0.62*	0.40	0.21
		1.5	*0.7*	**12.1**	**5.5**	10.9	*5.0*	**A-F**	**10.7**	**53.6**	**0.88**	***1.95***	**0.53**	***1.17***	**0.30**	***0.66***	**0.46**	**0.22**
								M-F	11.9	59.5	0.98	*2.16*	0.59	*1.30*	0.33	*0.73*	0.51	0.24
		2.0	*0.9*	**12.4**	**5.6**	11.2	*5.1*	**A-F**	**11.8**	**56.3**	**0.93**	***2.04***	**0.57**	***1.25***	**0.33**	***0.73***	**0.55**	**0.25**
								M-F	13.1	62.5	1.03	*2.27*	0.63	*1.39*	0.37	*0.81*	0.61	0.28
		2.5	*1.1*	**12.7**	**5.8**	11.4	*5.2*	**A-F**	**13.1**	**59.9**	**0.98**	***2.16***	**0.62**	***1.37***	**0.38**	***0.83***	**0.65**	**0.28**
								M-F	14.5	66.5	1.09	*2.40*	0.69	*1.52*	0.42	*0.92*	0.72	0.31
		3.0	*1.4*	**12.8**	**5.8**	11.5	*5.2*	**A-F**	**14.3**	**63.5**	**1.04**	***2.30***	**0.68**	***1.49***	**0.42**	***0.93***	**0.74**	**0.32**
								M-F	15.9	70.5	1.16	*2.55*	0.75	*1.65*	0.47	*1.03*	0.82	0.35
		3.5	*1.6*	**12.6**	**5.7**	11.3	*5.1*	**A-F**	**15.8**	**68.0**	**1.12**	***2.46***	**0.74**	***1.62***	**0.48**	***1.05***	**0.86**	**0.35**
								M-F	17.5	75.5	1.24	*2.73*	0.82	*1.80*	0.53	*1.17*	0.96	0.39
		4.0	*1.8*	**11.3**	**5.1**	10.2	*4.6*	**A-F**	**18.3**	**77.4**	**1.27**	***2.79***	**0.86**	***1.90***	**0.59**	***1.31***	**1.07**	**0.43**
								M-F	20.3	86.0	1.41	*3.10*	0.96	*2.11*	0.66	*1.45*	1.19	0.48
500	*227*	0.5	*0.2*	**12.9**	**5.9**	11.6	*5.3*	**A-F**	**7.7**	**47.3**	**0.77**	***1.70***	**0.43**	***0.95***	**0.21**	***0.46***	**0.23**	**0.17**
								M-F	8.6	52.5	0.86	*1.89*	0.48	*1.06*	0.23	*0.51*	0.25	0.19
		1.0	*0.5*	**13.7**	**6.2**	12.3	*5.6*	**A-F**	**8.8**	**50.4**	**0.83**	***1.82***	**0.49**	***1.07***	**0.25**	***0.56***	**0.32**	**0.19**
								M-F	9.8	56.0	0.92	*2.02*	0.54	*1.19*	0.28	*0.62*	0.36	0.21
		1.5	*0.7*	**14.3**	**6.5**	12.9	*5.9*	**A-F**	**9.8**	**53.6**	**0.88**	***1.95***	**0.53**	***1.17***	**0.30**	***0.66***	**0.39**	**0.20**
								M-F	10.9	59.5	0.98	*2.16*	0.59	*1.30*	0.33	*0.73*	0.43	0.22

(Continued)

TABLE I-1d (Continued)

Weight		Daily Gain		Daily Consumption: As-fed		Daily Consumption: Moisture-free Dry Matter		Moisture Basis[5] A-F (As-fed) M-F (Moisture-free)	Total Protein	Energy: TDN	Energy: ME (Mcal per)		Energy: NE$_m$ (Mcal per)		Energy: NE$_g$ (Mcal per)		Calcium	Phos-phorus
(lb)	*(kg)*	*(lb)*	*(kg)*	*(lb)*	*(kg)*	*(lb)*	*(kg)*		*(%)*	*(%)*	*(lb)*	*(kg)*	*(lb)*	*(kg)*	*(lb)*	*(kg)*	*(%)*	*(%)*
Large-frame bull calves and compensating large-frame yearling steers (continued)																		
500	*227*	2.0	*0.9*	**14.7**	***6.7***	13.2	*6.0*	**A-F**	**10.6**	**56.3**	**0.93**	***2.04***	**0.57**	***1.25***	**0.33**	***0.73***	**0.47**	**0.23**
								M-F	11.8	62.5	1.03	*2.27*	0.63	*1.39*	0.37	*0.81*	0.52	0.25
		2.5	*1.1*	**15.0**	***6.8***	13.5	*6.1*	**A-F**	**11.6**	**59.9**	**0.98**	***2.16***	**0.62**	***1.37***	**0.38**	***0.83***	**0.53**	**0.25**
								M-F	12.9	66.5	1.09	*2.40*	0.69	*1.52*	0.42	*0.92*	0.59	0.28
		3.0	*1.4*	**15.1**	***6.9***	13.6	*6.2*	**A-F**	**12.6**	**63.5**	**1.04**	***2.30***	**0.68**	***1.49***	**0.42**	***0.93***	**0.61**	**0.28**
								M-F	14.0	70.5	1.16	*2.55*	0.75	*1.65*	0.47	*1.03*	0.68	0.31
		3.5	*1.6*	**14.9**	***6.8***	13.4	*6.1*	**A-F**	**13.8**	**68.0**	**1.12**	***2.46***	**0.74**	***1.62***	**0.48**	***1.05***	**0.69**	**0.32**
								M-F	15.3	75.5	1.24	*2.73*	0.82	*1.80*	0.53	*1.17*	0.77	0.35
		4.0	*1.8*	**13.3**	***6.0***	12.0	*5.5*	**A-F**	**15.8**	**77.4**	**1.27**	***2.79***	**0.86**	***1.90***	**0.59**	***1.31***	**0.87**	**0.36**
								M-F	17.5	86.0	1.41	*3.10*	0.96	*2.11*	0.66	*1.45*	0.97	0.40
600	*273*	0.5	*0.2*	**14.8**	***6.7***	13.3	*6.0*	**A-F**	**7.5**	**47.3**	**0.77**	***1.70***	**0.43**	***0.95***	**0.21**	***0.46***	**0.21**	**0.16**
								M-F	8.3	52.5	0.86	*1.89*	0.48	*1.06*	0.23	*0.51*	0.23	0.18
		1.0	*0.5*	**15.7**	***7.1***	14.1	*6.4*	**A-F**	**8.3**	**50.4**	**0.83**	***1.82***	**0.49**	***1.07***	**0.25**	***0.56***	**0.28**	**0.18**
								M-F	9.2	56.0	0.92	*2.02*	0.54	*1.19*	0.28	*0.62*	0.31	0.20
		1.5	*0.7*	**16.4**	***7.5***	14.8	*6.7*	**A-F**	**9.1**	**53.6**	**0.88**	***1.95***	**0.53**	***1.17***	**0.30**	***0.66***	**0.33**	**0.19**
								M-F	10.1	59.5	0.98	*2.16*	0.59	*1.30*	0.33	*0.73*	0.37	0.21
		2.0	*0.9*	**16.9**	***7.7***	15.2	*6.9*	**A-F**	**9.8**	**56.3**	**0.93**	***2.04***	**0.57**	***1.25***	**0.33**	***0.73***	**0.40**	**0.21**
								M-F	10.9	62.5	1.03	*2.27*	0.63	*1.39*	0.37	*0.81*	0.44	0.23
		2.5	*1.1*	**17.2**	***7.8***	15.5	*7.0*	**A-F**	**10.6**	**59.9**	**0.98**	***2.16***	**0.62**	***1.37***	**0.38**	***0.83***	**0.46**	**0.23**
								M-F	11.8	66.5	1.09	*2.40*	0.69	*1.52*	0.42	*0.92*	0.51	0.26
		3.0	*1.4*	**17.2**	***7.8***	15.5	*7.0*	**A-F**	**11.4**	**63.5**	**1.04**	***2.30***	**0.68**	***1.49***	**0.42**	***0.93***	**0.52**	**0.24**
								M-F	12.7	70.5	1.16	*2.55*	0.75	*1.65*	0.47	*1.03*	0.58	0.27
		3.5	*1.6*	**17.0**	***7.7***	15.3	*7.0*	**A-F**	**12.3**	**68.0**	**1.12**	***2.46***	**0.74**	***1.62***	**0.48**	***1.05***	**0.59**	**0.27**
								M-F	13.7	75.5	1.24	*2.73*	0.82	*1.80*	0.53	*1.17*	0.66	0.30
		4.0	*1.8*	**15.3**	***7.0***	13.8	*6.3*	**A-F**	**14.0**	**77.4**	**1.27**	***2.79***	**0.86**	***1.90***	**0.59**	***1.31***	**0.73**	**0.33**
								M-F	15.6	86.0	1.41	*3.10*	0.96	*2.11*	0.66	*1.45*	0.81	0.37
700	*318*	0.5	*0.2*	**16.6**	***7.5***	14.9	*6.8*	**A-F**	**7.2**	**47.3**	**0.77**	***1.70***	**0.43**	***0.95***	**0.21**	***0.46***	**0.20**	**0.16**
								M-F	8.0	52.5	0.86	*1.89*	0.48	*1.06*	0.23	*0.51*	0.22	0.18
		1.0	*0.5*	**17.7**	***8.0***	15.9	*7.2*	**A-F**	**7.9**	**50.4**	**0.83**	***1.82***	**0.49**	***1.07***	**0.25**	***0.56***	**0.26**	**0.17**
								M-F	8.8	56.0	0.92	*2.02*	0.54	*1.19*	0.28	*0.62*	0.29	0.19
		1.5	*0.7*	**18.4**	***8.4***	16.6	*7.5*	**A-F**	**8.6**	**53.6**	**0.88**	***1.95***	**0.53**	***1.17***	**0.30**	***0.66***	**0.32**	**0.19**
								M-F	9.6	59.5	0.98	*2.16*	0.59	*1.30*	0.33	*0.73*	0.35	0.21
		2.0	*0.9*	**18.9**	***8.6***	17.0	*7.7*	**A-F**	**9.2**	**56.3**	**0.93**	***2.04***	**0.57**	***1.25***	**0.33**	***0.73***	**0.35**	**0.20**
								M-F	10.2	62.5	1.03	*2.27*	0.63	*1.39*	0.37	*0.81*	0.39	0.22
		2.5	*1.1*	**19.3**	***8.8***	17.4	*7.9*	**A-F**	**9.9**	**59.9**	**0.98**	***2.16***	**0.62**	***1.37***	**0.38**	***0.83***	**0.40**	**0.22**
								M-F	11.0	66.5	1.09	*2.40*	0.69	*1.52*	0.42	*0.92*	0.44	0.24
		3.0	*1.4*	**19.4**	***8.8***	17.5	*8.0*	**A-F**	**10.5**	**63.5**	**1.04**	***2.30***	**0.68**	***1.49***	**0.42**	***0.93***	**0.45**	**0.23**
								M-F	11.7	70.5	1.16	*2.55*	0.75	*1.65*	0.47	*1.03*	0.50	0.25

(Continued)

TABLE I-1d (Continued)

Weight		Daily Gain		Daily Consumption: As-fed		Daily Consumption: Moisture-free Dry Matter		Moisture Basis[5] A-F (As-fed) M-F (Moisture-free)	Total Protein	Energy: TDN	Energy: ME (Mcal per)		Energy: NE$_m$ (Mcal per)		Energy: NE$_g$ (Mcal per)		Calcium	Phos-phorus
(lb)	(kg)	(lb)	(kg)	(lb)	(kg)	(lb)	(kg)		(%)	(%)	(lb)	(kg)	(lb)	(kg)	(lb)	(kg)	(%)	(%)
Large-frame bull calves and compensating large-frame yearling steers (continued)																		
700	*318*	3.5	*1.6*	**19.1**	***8.7***	17.2	*7.8*	**A-F**	**11.3**	**68.0**	**1.12**	***2.46***	**0.74**	***1.62***	**0.48**	***1.05***	**0.50**	**0.25**
								M-F	12.5	75.5	1.24	*2.73*	0.82	*1.80*	0.53	*1.17*	0.56	0.28
		4.0	*1.8*	**17.2**	***7.8***	15.5	*7.0*	**A-F**	**12.7**	**77.4**	**1.27**	***2.79***	**0.86**	***1.90***	**0.59**	***1.31***	**0.63**	**0.30**
								M-F	14.1	86.0	1.41	*3.10*	0.96	*2.11*	0.66	*1.45*	0.70	0.33
800	*364*	0.5	*0.2*	**18.3**	***8.3***	16.5	*7.5*	**A-F**	**7.1**	**47.3**	**0.77**	***1.70***	**0.43**	***0.95***	**0.21**	***0.46***	**0.19**	**0.17**
								M-F	7.9	52.5	0.86	*1.89*	0.48	*1.06*	0.23	*0.51*	0.21	0.19
		1.0	*0.5*	**19.4**	***8.8***	17.5	*8.0*	**A-F**	**7.7**	**50.4**	**0.83**	***1.82***	**0.49**	***1.07***	**0.25**	***0.56***	**0.23**	**0.17**
								M-F	[illegible]	[illegible]	[illegible]	[illegible]	[illegible]	[illegible]	[illegible]	[illegible]	[illegible]	[illegible]
		1.5	*0.7*	**20.3**	***9.2***	18.3	*8.3*	**A-F**	**8.2**	**53.6**	**0.88**	***1.95***	**0.53**	***1.17***	**0.30**	***0.66***	**0.28**	**0.18**
								M-F	9.1	59.5	0.98	*2.16*	0.59	*1.30*	0.33	*0.73*	0.31	0.20
		2.0	*0.9*	**20.9**	***9.5***	18.8	*8.5*	**A-F**	**8.7**	**56.3**	**0.93**	***2.04***	**0.57**	***1.25***	**0.33**	***0.73***	**0.32**	**0.19**
								M-F	9.7	62.5	1.03	*2.27*	0.63	*1.39*	0.37	*0.81*	0.35	0.21
		2.5	*1.1*	**21.3**	***9.7***	19.2	*8.7*	**A-F**	**9.3**	**59.9**	**0.98**	***2.16***	**0.62**	***1.37***	**0.38**	***0.83***	**0.36**	**0.21**
								M-F	10.3	66.5	1.09	*2.40*	0.69	*1.52*	0.42	*0.92*	0.40	0.23
		3.0	*1.4*	**21.4**	***9.7***	19.3	*8.8*	**A-F**	**9.8**	**63.5**	**1.04**	***2.30***	**0.68**	***1.49***	**0.42**	***0.93***	**0.41**	**0.22**
								M-F	10.9	70.5	1.16	*2.55*	0.75	*1.65*	0.47	*1.03*	0.45	0.24
		3.5	*1.6*	**21.1**	***9.6***	19.0	*8.6*	**A-F**	**10.4**	**68.0**	**1.12**	***2.46***	**0.74**	***1.62***	**0.48**	***1.05***	**0.45**	**0.23**
								M-F	11.6	75.5	1.24	*2.73*	0.82	*1.80*	0.53	*1.17*	0.50	0.26
		4.0	*1.8*	**19.0**	***8.6***	17.1	*7.8*	**A-F**	**11.7**	**77.4**	**1.27**	***2.79***	**0.86**	***1.90***	**0.59**	***1.31***	**0.55**	**0.28**
								M-F	13.0	86.0	1.41	*3.10*	0.96	*2.11*	0.66	*1.45*	0.61	0.31
900	*409*	0.5	*0.2*	**20.0**	***9.1***	18.0	*8.2*	**A-F**	**6.9**	**47.3**	**0.77**	***1.70***	**0.43**	***0.95***	**0.21**	***0.46***	**0.20**	**0.16**
								M-F	7.7	52.5	0.86	*1.89*	0.48	*1.06*	0.23	*0.51*	0.22	0.18
		1.0	*0.5*	**21.3**	***9.7***	19.2	*8.7*	**A-F**	**7.5**	**50.4**	**0.83**	***1.82***	**0.49**	***1.07***	**0.25**	***0.56***	**0.23**	**0.16**
								M-F	8.3	56.0	0.92	*2.02*	0.54	*1.19*	0.28	*0.62*	0.25	0.18
		1.5	*0.7*	**22.2**	***10.1***	20.0	*9.1*	**A-F**	**7.9**	**53.6**	**0.88**	***1.95***	**0.53**	***1.17***	**0.30**	***0.66***	**0.26**	**0.18**
								M-F	8.8	59.5	0.98	*2.16*	0.59	*1.30*	0.33	*0.73*	0.29	0.20
		2.0	*0.9*	**22.9**	***10.4***	20.6	*9.4*	**A-F**	**8.9**	**56.3**	**0.93**	***2.04***	**0.57**	***1.25***	**0.33**	***0.73***	**0.30**	**0.18**
								M-F	9.2	62.5	1.03	*2.27*	0.63	*1.39*	0.37	*0.81*	0.32	0.20
		2.5	*1.1*	**23.3**	***10.6***	21.0	*9.5*	**A-F**	**8.8**	**59.9**	**0.98**	***2.16***	**0.62**	***1.37***	**0.38**	***0.83***	**0.32**	**0.19**
								M-F	9.8	66.5	1.09	*2.40*	0.69	*1.52*	0.42	*0.92*	0.36	0.21
		3.0	*1.4*	**23.4**	***10.6***	21.1	*9.6*	**A-F**	**9.3**	**63.5**	**1.04**	***2.30***	**0.68**	***1.49***	**0.42**	***0.93***	**0.36**	**0.21**
								M-F	10.3	70.5	1.16	*2.55*	0.75	*1.65*	0.47	*1.03*	0.40	0.23
		3.5	*1.6*	**23.1**	***10.5***	20.8	*9.5*	**A-F**	**9.8**	**68.0**	**1.12**	***2.46***	**0.74**	***1.62***	**0.48**	***1.05***	**0.41**	**0.22**
								M-F	10.9	75.5	1.24	*2.73*	0.82	*1.80*	0.53	*1.17*	0.45	0.24
		4.0	*1.8*	**20.8**	***9.5***	18.7	*8.5*	**A-F**	**10.9**	**77.4**	**1.27**	***2.79***	**0.86**	***1.90***	**0.59**	***1.31***	**0.48**	**0.25**
								M-F	12.1	86.0	1.41	*3.10*	0.96	*2.11*	0.66	*1.45*	0.53	0.28
1,000	*454*	0.5	*0.2*	**21.7**	***9.9***	19.5	*8.9*	**A-F**	**6.8**	**47.3**	**0.77**	***1.70***	**0.43**	***0.95***	**0.21**	***0.46***	**0.19**	**0.16**
								M-F	7.6	52.5	0.86	*1.89*	0.48	*1.06*	0.23	*0.51*	0.21	0.18

(Continued)

TABLE I-1d (Continued)

Weight		Daily Gain		Daily Consumption: As-fed		Daily Consumption: Moisture-free Dry Matter		Moisture Basis[5] A-F (As-fed) M-F (Moisture-free)	Total Protein	Energy: TDN	Energy: ME (Mcal per)		Energy: NE_m (Mcal per)		Energy: NE_g (Mcal per)		Calcium	Phosphorus
(lb)	(kg)	(lb)	(kg)	(lb)	(kg)	(lb)	(kg)		(%)	(%)	(lb)	(kg)	(lb)	(kg)	(lb)	(kg)	(%)	(%)
Large-frame bull calves and compensating large-frame yearling steers (continued)																		
1,000	454	1.0	0.5	23.0	10.5	20.7	9.4	A-F	7.3	50.4	0.83	1.82	0.49	1.07	0.25	0.56	0.23	0.17
								M-F	8.1	56.0	0.92	2.02	0.54	1.19	0.28	0.62	0.25	0.19
		1.5	0.7	24.1	11.0	21.7	9.9	A-F	7.7	53.6	0.88	1.95	0.53	1.17	0.30	0.66	0.24	0.17
								M-F	8.5	59.5	0.98	2.16	0.59	1.30	0.33	0.73	0.27	0.19
		2.0	0.9	24.8	11.3	22.3	10.1	A-F	7.9	56.3	0.93	2.04	0.57	1.25	0.33	0.73	0.27	0.18
								M-F	8.9	62.5	1.03	2.27	0.63	1.39	0.37	0.81	0.30	0.20
		2.5	1.1	25.2	11.5	22.7	10.3	A-F	8.4	59.9	0.98	2.16	0.62	1.37	0.38	0.83	0.30	0.18
								M-F	9.3	66.5	1.09	2.40	0.69	1.52	0.42	0.92	0.33	0.20
		3.0	1.4	25.3	11.5	22.8	10.4	A-F	8.7	63.5	1.04	2.30	0.68	1.49	0.42	0.93	0.32	0.19
								M-F	9.7	70.5	1.16	2.55	0.75	1.65	0.47	1.03	0.36	0.21
		3.5	1.6	25.0	11.4	22.5	10.2	A-F	9.3	68.0	1.12	2.46	0.74	1.62	0.48	1.05	0.36	0.22
								M-F	10.3	75.5	1.24	2.73	0.82	1.80	0.53	1.17	0.40	0.24
		4.0	1.8	22.4	10.2	20.2	9.2	A-F	10.2	77.4	1.27	2.79	0.86	1.90	0.59	1.31	0.43	0.24
								M-F	11.3	86.0	1.41	3.10	0.96	2.11	0.66	1.45	0.48	0.27
1,100	500	0.5	0.2	23.2	10.5	20.9	9.5	A-F	6.8	47.3	0.77	1.70	0.43	0.95	0.21	0.46	0.19	0.17
								M-F	7.5	52.5	0.86	1.89	0.48	1.06	0.23	0.51	0.21	0.19
		1.0	0.5	24.8	11.3	22.3	10.1	A-F	7.1	50.4	0.83	1.82	0.49	1.07	0.25	0.56	0.21	0.17
								M-F	7.9	56.0	0.92	2.02	0.54	1.19	0.28	0.62	0.23	0.19
		1.5	0.7	25.9	11.8	23.3	10.6	A-F	7.5	53.6	0.88	1.95	0.53	1.17	0.30	0.66	0.23	0.17
								M-F	8.3	59.5	0.98	2.16	0.59	1.30	0.33	0.73	0.26	0.19
		2.0	0.9	26.6	12.1	23.9	10.9	A-F	7.7	56.3	0.93	2.04	0.57	1.25	0.33	0.73	0.25	0.17
								M-F	8.6	62.5	1.03	2.27	0.63	1.39	0.37	0.81	0.28	0.19
		2.5	1.1	26.9	12.2	24.2	11.0	A-F	8.1	59.9	0.98	2.16	0.62	1.37	0.38	0.83	0.27	0.18
								M-F	9.0	66.5	1.09	2.40	0.69	1.52	0.42	0.92	0.30	0.20
		3.0	1.4	27.2	12.4	24.5	11.1	A-F	8.4	63.5	1.04	2.30	0.68	1.49	0.42	0.93	0.29	0.19
								M-F	9.3	70.5	1.16	2.55	0.75	1.65	0.47	1.03	0.32	0.21
		3.5	1.6	26.8	12.2	24.1	11.0	A-F	8.8	68.0	1.12	2.46	0.74	1.62	0.48	1.05	0.32	0.20
								M-F	9.8	75.5	1.24	2.73	0.82	1.80	0.53	1.17	0.36	0.22
		4.0	1.8	24.1	11.0	21.7	9.9	A-F	9.6	77.4	1.27	2.79	0.86	1.90	0.59	1.31	0.39	0.23
								M-F	10.7	86.0	1.41	3.10	0.96	2.11	0.66	1.45	0.43	0.25
Medium-frame heifer calves																		
300	136	0.5	0.2	8.3	3.8	7.5	3.4	A-F	8.6	50.4	0.83	1.82	0.49	1.07	0.25	0.56	0.26	0.19
								M-F	9.6	56.0	0.92	2.02	0.54	1.19	0.28	0.62	0.29	0.21
		1.0	0.5	8.9	4.0	8.0	3.6	A-F	10.3	55.8	0.92	2.03	0.57	1.25	0.32	0.71	0.40	0.20
								M-F	11.4	62.0	1.02	2.25	0.63	1.39	0.36	0.79	0.44	0.22
		1.5	0.7	9.1	4.1	8.2	3.7	A-F	11.8	61.7	1.02	2.24	0.65	1.42	0.40	0.87	0.53	0.24
								M-F	13.1	68.5	1.13	2.49	0.72	1.58	0.44	0.97	0.59	0.27

(Continued)

TABLE I-1d (Continued)

Weight		Daily Gain		Daily Consumption: As-fed		Daily Consumption: Moisture-free Dry Matter		Moisture Basis[5] A-F (As-fed) M-F (Moisture-free)	Total Protein	Energy: TDN	Energy: ME (Mcal per)		Energy: NE_m (Mcal per)		Energy: NE_g (Mcal per)		Calcium	Phosphorus
(lb)	*(kg)*	*(lb)*	*(kg)*	*(lb)*	*(kg)*	*(lb)*	*(kg)*		*(%)*	*(%)*	*(lb)*	*(kg)*	*(lb)*	*(kg)*	*(lb)*	*(kg)*	*(%)*	*(%)*
Medium-frame heifer calves (continued)																		
300	*136*	2.0	*0.9*	**8.9**	***4.0***	8.0	*3.6*	**A-F**	**13.6**	**69.3**	**1.13**	***2.50***	**0.76**	***1.67***	**0.50**	***1.09***	**0.67**	**0.30**
								M-F	15.1	77.0	1.26	*2.77*	0.84	*1.85*	0.55	*1.21*	0.74	0.33
400	*182*	0.5	*0.2*	**10.3**	***4.7***	9.3	*4.2*	**A-F**	**8.0**	**50.4**	**0.83**	***1.82***	**0.49**	***1.07***	**0.25**	***0.56***	**0.23**	**0.17**
								M-F	8.9	56.0	0.92	*2.02*	0.54	*1.19*	0.28	*0.62*	0.26	0.19
		1.0	*0.5*	**11.0**	***5.0***	9.9	*4.5*	**A-F**	**9.2**	**55.8**	**0.92**	***2.03***	**0.57**	***1.25***	**0.32**	***0.71***	**0.32**	**0.18**
								M-F	10.2	62.0	1.02	*2.25*	0.63	*1.39*	0.36	*0.79*	0.36	0.20
		1.5	*0.7*	**11.3**	***5.1***	10.2	*4.6*	**A-F**	**10.3**	**61.7**	**1.02**	***2.24***	**0.65**	***1.42***	**0.40**	***0.87***	**0.41**	**0.22**
								M-F	11.4	68.5	1.13	*2.49*	0.72	*1.58*	0.44	*0.97*	0.45	0.24
		2.0	*0.9*	**11.1**	***5.0***	10.0	*4.5*	**A-F**	**11.6**	**69.3**	**1.13**	***2.50***	**0.76**	***1.67***	**0.50**	***1.09***	**0.51**	**0.26**
								M-F	12.9	77.0	1.26	*2.77*	0.84	*1.85*	0.55	*1.21*	0.57	0.29
500	*227*	0.5	*0.2*	**12.2**	***5.5***	11.0	*5.0*	**A-F**	**7.7**	**50.4**	**0.83**	***1.82***	**0.49**	***1.07***	**0.25**	***0.56***	**0.22**	**0.16**
								M-F	8.5	56.0	0.92	*2.02*	0.54	*1.19*	0.28	*0.62*	0.24	0.18
		1.0	*0.5*	**13.1**	***6.0***	11.8	*5.4*	**A-F**	**8.5**	**55.8**	**0.92**	***2.03***	**0.57**	***1.25***	**0.32**	***0.71***	**0.27**	**0.19**
								M-F	9.4	62.0	1.02	*2.25*	0.63	*1.39*	0.36	*0.79*	0.30	0.21
		1.5	*0.7*	**13.4**	***6.1***	12.1	*5.5*	**A-F**	**9.3**	**61.7**	**1.02**	***2.24***	**0.65**	***1.42***	**0.40**	***0.87***	**0.34**	**0.20**
								M-F	10.3	68.5	1.13	*2.49*	0.72	*1.58*	0.44	*0.97*	0.38	0.22
		2.0	*0.9*	**13.1**	***6.0***	11.8	*5.4*	**A-F**	**10.3**	**69.3**	**1.13**	***2.50***	**0.76**	***1.67***	**0.50**	***1.09***	**0.41**	**0.22**
								M-F	11.4	77.0	1.26	*2.77*	0.84	*1.85*	0.55	*1.21*	0.45	0.24
600	*273*	0.5	*0.2*	**14.0**	***6.4***	12.6	*5.7*	**A-F**	**7.3**	**50.4**	**0.83**	***1.82***	**0.49**	***1.07***	**0.25**	***0.56***	**0.21**	**0.16**
								M-F	8.1	56.0	0.92	*2.02*	0.54	*1.19*	0.28	*0.62*	0.23	0.18
		1.0	*0.5*	**15.0**	***6.8***	13.5	*6.1*	**A-F**	**7.9**	**55.8**	**0.92**	***2.03***	**0.57**	***1.25***	**0.32**	***0.71***	**0.25**	**0.18**
								M-F	8.8	62.0	1.02	*2.25*	0.63	*1.39*	0.36	*0.79*	0.28	0.20
		1.5	*0.7*	**15.3**	***7.0***	13.8	*6.3*	**A-F**	**8.6**	**61.7**	**1.02**	***2.24***	**0.65**	***1.42***	**0.40**	***0.87***	**0.29**	**0.19**
								M-F	9.5	68.5	1.13	*2.49*	0.72	*1.58*	0.44	*0.97*	0.32	0.21
		2.0	*0.9*	**15.0**	***6.8***	13.5	*6.1*	**A-F**	**9.4**	**69.3**	**1.13**	***2.50***	**0.76**	***1.67***	**0.50**	***1.09***	**0.34**	**0.21**
								M-F	10.4	77.0	1.26	*2.77*	0.84	*1.85*	0.55	*1.21*	0.38	0.23
700	*318*	0.5	*0.2*	**15.7**	***7.1***	14.1	*6.4*	**A-F**	**7.1**	**50.4**	**0.83**	***1.82***	**0.49**	***1.07***	**0.25**	***0.56***	**0.20**	**0.17**
								M-F	7.9	56.0	0.92	*2.02*	0.54	*1.19*	0.28	*0.62*	0.22	0.19
		1.0	*0.5*	**16.8**	***7.6***	15.1	*6.9*	**A-F**	**7.6**	**55.8**	**0.92**	***2.03***	**0.57**	***1.25***	**0.32**	***0.71***	**0.23**	**0.17**
								M-F	8.4	62.0	1.02	*2.25*	0.63	*1.39*	0.36	*0.79*	0.25	0.19
		1.5	*0.7*	**16.9**	***7.7***	15.5	*7.0*	**A-F**	**8.1**	**61.7**	**1.02**	***2.24***	**0.65**	***1.42***	**0.40**	***0.87***	**0.25**	**0.18**
								M-F	9.0	68.5	1.13	*2.49*	0.72	*1.58*	0.44	*0.97*	0.28	0.20
		2.0	*0.9*	**16.9**	***7.7***	15.2	*7.0*	**A-F**	**8.6**	**69.3**	**1.13**	***2.50***	**0.76**	***1.67***	**0.50**	***1.09***	**0.29**	**0.20**
								M-F	9.6	77.0	1.26	*2.77*	0.84	*1.85*	0.55	*1.21*	0.32	0.22
800	*364*	0.5	*0.2*	**17.3**	***7.9***	15.6	*7.0*	**A-F**	**6.9**	**50.4**	**0.83**	***1.82***	**0.49**	***1.07***	**0.25**	***0.56***	**0.19**	**0.16**
								M-F	7.7	56.0	0.92	*2.02*	0.54	*1.19*	0.28	*0.62*	0.21	0.18
		1.0	*0.5*	**18.6**	***8.5***	16.7	*7.6*	**A-F**	**7.3**	**55.8**	**0.92**	***2.03***	**0.57**	***1.25***	**0.32**	***0.71***	**0.20**	**0.16**
								M-F	8.1	62.0	1.02	*2.25*	0.63	*1.39*	0.36	*0.79*	0.22	0.18

(Continued)

TABLE I-1d (Continued)

Weight		Daily Gain		Daily Consumption: As-fed		Daily Consumption: Moisture-free Dry Matter		Moisture Basis[5] A-F (As-fed) M-F (Moisture-free)	Total Protein	Energy: TDN	Energy: ME (Mcal per)		Energy: NE_m (Mcal per)		Energy: NE_g (Mcal per)		Calcium	Phosphorus
(lb)	*(kg)*	*(lb)*	*(kg)*	*(lb)*	*(kg)*	*(lb)*	*(kg)*		*(%)*	*(%)*	*(lb)*	*(kg)*	*(lb)*	*(kg)*	*(lb)*	*(kg)*	*(%)*	*(%)*
Medium-frame heifer calves (continued)																		
800	*364*	1.5	*0.7*	**19.1**	***8.7***	17.2	*7.8*	**A-F**	**7.7**	**61.7**	**1.02**	***2.24***	**0.65**	***1.42***	**0.40**	***0.87***	**0.22**	**0.17**
								M-F	8.5	68.5	1.13	*2.49*	0.72	*1.58*	0.44	*0.97*	0.24	0.19
		2.0	*0.9*	**18.7**	***8.5***	16.8	*7.6*	**A-F**	**8.1**	**69.3**	**1.13**	***2.50***	**0.76**	***1.67***	**0.50**	***1.09***	**0.25**	**0.18**
								M-F	9.0	77.0	1.26	*2.77*	0.84	*1.85*	0.55	*1.21*	0.28	0.20
900	*409*	0.5	*0.2*	**19.0**	***8.6***	17.1	*7.8*	**A-F**	**6.8**	**50.4**	**0.83**	***1.82***	**0.49**	***1.07***	**0.25**	***0.56***	**0.19**	**0.16**
								M-F	7.5	56.0	0.92	*2.02*	0.54	*1.19*	0.28	*0.62*	0.21	0.18
		1.0	*0.5*	**20.3**	***9.2***	18.3	*8.3*	**A-F**	**7.0**	**55.8**	**0.92**	***2.03***	**0.57**	***1.25***	**0.32**	***0.71***	**0.20**	**0.16**
								M-F	7.8	62.0	1.02	*2.25*	0.63	*1.39*	0.36	*0.79*	0.22	0.18
		1.5	*0.7*	**20.9**	***9.5***	18.8	*8.5*	**A-F**	**7.3**	**61.7**	**1.02**	***2.24***	**0.65**	***1.42***	**0.40**	***0.87***	**0.20**	**0.17**
								M-F	8.1	68.5	1.13	*2.49*	0.72	*1.58*	0.44	*0.97*	0.22	0.19
		2.0	*0.9*	**20.3**	***9.2***	18.3	*8.3*	**A-F**	**7.7**	**69.3**	**1.13**	***2.50***	**0.76**	***1.67***	**0.50**	***1.09***	**0.23**	**0.17**
								M-F	8.5	77.0	1.26	*2.77*	0.84	*1.85*	0.55	*1.21*	0.25	0.19
1,000	*454*	0.5	*0.2*	**20.6**	***9.4***	18.5	*8.4*	**A-F**	**6.7**	**50.4**	**0.83**	***1.82***	**0.49**	***1.07***	**0.25**	***0.56***	**0.18**	**0.17**
								M-F	7.4	56.0	0.92	*2.02*	0.54	*1.19*	0.28	*0.62*	0.20	0.19
		1.0	*0.5*	**22.0**	***10.0***	19.8	*9.0*	**A-F**	**6.8**	**55.8**	**0.92**	***2.03***	**0.57**	***1.25***	**0.32**	***0.71***	**0.18**	**0.16**
								M-F	7.6	62.0	1.02	*2.25*	0.63	*1.39*	0.36	*0.79*	0.20	0.18
		1.5	*0.7*	**22.6**	***10.3***	20.3	*9.2*	**A-F**	**7.0**	**61.7**	**1.02**	***2.24***	**0.65**	***1.42***	**0.40**	***0.87***	**0.19**	**0.16**
								M-F	7.8	68.5	1.13	*2.49*	0.72	*1.58*	0.44	*0.97*	0.21	0.18
		2.0	*0.9*	**22.0**	***10.0***	19.8	*9.0*	**A-F**	**7.3**	**69.3**	**1.13**	***2.50***	**0.76**	***1.67***	**0.50**	***1.09***	**0.20**	**0.17**
								M-F	8.1	77.0	1.26	*2.77*	0.84	*1.85*	0.55	*1.21*	0.22	0.19
Large-frame heifer calves and compensating medium-frame yearling heifers																		
300	*136*	0.5	*0.2*	**8.7**	***4.0***	7.8	*3.5*	**A-F**	**8.6**	**48.6**	**0.80**	***1.76***	**0.45**	***1.00***	**0.23**	***0.50***	**0.28**	**0.18**
								M-F	9.5	54.0	0.89	*1.96*	0.50	*1.10*	0.25	*0.55*	0.31	0.20
		1.0	*0.5*	**9.3**	***4.2***	8.4	*3.8*	**A-F**	**10.2**	**53.1**	**0.88**	***1.94***	**0.52**	***1.15***	**0.29**	***0.63***	**0.41**	**0.22**
								M-F	11.3	59.0	0.98	*2.16*	0.58	*1.28*	0.32	*0.70*	0.45	0.24
		1.5	*0.7*	**9.8**	***4.5***	8.8	*4.0*	**A-F**	**11.7**	**57.6**	**0.95**	***2.08***	**0.59**	***1.29***	**0.35**	***0.77***	**0.52**	**0.23**
								M-F	13.0	64.0	1.05	*2.31*	0.65	*1.43*	0.39	*0.86*	0.58	0.25
		2.0	*0.9*	**9.9**	***4.5***	8.9	*4.0*	**A-F**	**13.1**	**62.6**	**1.03**	***2.26***	**0.67**	***1.47***	**0.41**	***0.91***	**0.62**	**0.27**
								M-F	14.6	69.5	1.14	*2.51*	0.74	*1.63*	0.46	*1.01*	0.69	0.30
		2.5	*1.1*	**9.7**	***4.4***	8.7	*4.0*	**A-F**	**15.0**	**69.3**	**1.13**	***2.49***	**0.76**	***1.67***	**0.50**	***1.09***	**0.77**	**0.32**
								M-F	16.7	77.0	1.26	*2.77*	0.84	*1.85*	0.55	*1.21*	0.86	0.35
400	*182*	0.5	*0.2*	**10.8**	***4.9***	9.7	*4.4*	**A-F**	**8.0**	**48.6**	**0.80**	***1.76***	**0.45**	***1.00***	**0.23**	***0.50***	**0.24**	**0.16**
								M-F	8.9	54.0	0.89	*1.96*	0.50	*1.10*	0.25	*0.55*	0.27	0.18
		1.0	*0.5*	**11.7**	***5.3***	10.5	*4.8*	**A-F**	**10.0**	**53.1**	**0.88**	***1.94***	**0.52**	***1.15***	**0.29**	***0.63***	**0.32**	**0.19**
								M-F	10.1	59.0	0.98	*2.16*	0.58	*1.28*	0.32	*0.70*	0.36	0.21
		1.5	*0.7*	**12.1**	***5.5***	10.9	*5.0*	**A-F**	**10.2**	**57.6**	**0.95**	***2.08***	**0.59**	***1.29***	**0.35**	***0.77***	**0.41**	**0.20**
								M-F	11.3	64.0	1.05	*2.31*	0.65	*1.43*	0.39	*0.86*	0.45	0.22

(Continued)

TABLE I-1d (Continued)

Weight		Daily Gain		Daily Consumption: As-fed		Daily Consumption: Moisture-free Dry Matter		Moisture Basis[5] A-F (As-fed) M-F (Moisture-free)	Total Protein	Energy: TDN	Energy: ME (Mcal per)		Energy: NE_m (Mcal per)		Energy: NE_g (Mcal per)		Calcium	Phosphorus
(lb)	(kg)	(lb)	(kg)	(lb)	(kg)	(lb)	(kg)		(%)	(%)	(lb)	(kg)	(lb)	(kg)	(lb)	(kg)	(%)	(%)
Large-frame heifer calves and compensating medium-frame yearling heifers (continued)																		
400	182	2.0	0.9	12.3	5.6	11.1	5.0	A-F	11.3	62.6	1.03	2.26	0.67	1.47	0.41	0.91	0.49	0.23
								M-F	12.6	69.5	1.14	2.51	0.74	1.63	0.46	1.01	0.54	0.26
		2.5	1.1	9.7	4.4	10.8	4.9	A-F	12.7	69.3	1.13	2.49	0.76	1.67	0.50	1.09	0.59	0.28
								M-F	14.1	77.0	1.26	2.77	0.84	1.85	0.55	1.21	0.65	0.31
500	227	0.5	0.2	12.8	5.8	11.5	5.2	A-F	7.6	48.6	0.80	1.76	0.45	1.00	0.23	0.50	0.21	0.15
								M-F	8.4	54.0	0.89	1.96	0.50	1.10	0.25	0.55	0.23	0.17
		1.0	0.5	13.8	6.3	12.4	5.6	A-F	8.5	53.1	0.88	1.94	0.52	1.15	0.29	0.63	0.27	0.18
								M-F	9.4	59.0	0.98	2.16	0.58	1.28	0.32	0.70	0.30	0.20
		1.5	0.7	14.3	6.5	12.9	5.9	A-F	9.3	57.6	0.95	2.08	0.59	1.29	0.35	0.77	0.34	0.18
								M-F	10.3	64.0	1.05	2.31	0.65	1.43	0.39	0.86	0.38	0.20
		2.0	0.9	14.6	6.6	13.1	6.0	A-F	10.1	62.6	1.03	2.26	0.67	1.47	0.41	0.91	0.40	0.22
								M-F	11.2	69.5	1.14	2.51	0.74	1.63	0.46	1.01	0.44	0.24
		2.5	1.1	14.2	6.5	12.8	5.8	A-F	11.2	69.3	1.13	2.49	0.76	1.67	0.50	1.09	0.48	0.23
								M-F	12.4	77.0	1.26	2.77	0.84	1.85	0.55	1.21	0.53	0.26
600	273	0.5	0.2	14.7	6.7	13.2	6.0	A-F	7.3	48.6	0.80	1.76	0.45	1.00	0.23	0.50	0.20	0.16
								M-F	8.1	54.0	0.89	1.96	0.50	1.10	0.25	0.55	0.22	0.18
		1.0	0.5	15.7	7.1	14.1	6.4	A-F	8.0	53.1	0.88	1.94	0.52	1.15	0.29	0.63	0.25	0.17
								M-F	8.9	59.0	0.98	2.16	0.58	1.28	0.32	0.70	0.28	0.19
		1.5	0.7	16.4	7.5	14.8	6.7	A-F	8.6	57.6	0.95	2.08	0.59	1.29	0.35	0.77	0.30	0.17
								M-F	9.6	64.0	1.05	2.31	0.65	1.43	0.39	0.86	0.33	0.19
		2.0	0.9	16.7	7.6	15.0	6.8	A-F	9.3	62.6	1.03	2.26	0.67	1.47	0.41	0.91	0.34	0.20
								M-F	10.3	69.5	1.14	2.51	0.74	1.63	0.46	1.01	0.38	0.22
		2.5	1.1	16.2	7.4	14.6	6.6	A-F	10.1	69.3	1.13	2.49	0.76	1.67	0.50	1.09	0.40	0.22
								M-F	11.2	77.0	1.26	2.77	0.84	1.85	0.55	1.21	0.44	0.24
700	318	0.5	0.2	16.4	7.5	14.8	6.7	A-F	7.1	48.6	0.80	1.76	0.45	1.00	0.23	0.50	0.19	0.16
								M-F	7.9	54.0	0.89	1.96	0.50	1.10	0.25	0.55	0.21	0.18
		1.0	0.5	17.7	8.0	15.9	7.2	A-F	7.7	53.1	0.88	1.94	0.52	1.15	0.29	0.63	0.23	0.16
								M-F	8.5	59.0	0.98	2.16	0.58	1.28	0.32	0.70	0.25	0.18
		1.5	0.7	18.4	8.4	16.6	7.5	A-F	8.1	57.6	0.95	2.08	0.59	1.29	0.35	0.77	0.26	0.17
								M-F	9.0	64.0	1.05	2.31	0.65	1.43	0.39	0.86	0.29	0.19
		2.0	0.9	18.7	8.5	16.8	7.6	A-F	8.6	62.6	1.03	2.26	0.67	1.47	0.41	0.91	0.30	0.18
								M-F	9.6	69.5	1.14	2.51	0.74	1.63	0.46	1.01	0.33	0.20
		2.5	1.1	18.2	8.3	16.4	7.5	A-F	9.3	69.3	1.13	2.49	0.76	1.67	0.50	1.09	0.34	0.20
								M-F	10.3	77.0	1.26	2.77	0.84	1.85	0.55	1.21	0.38	0.22
800	364	0.5	0.2	18.2	8.3	16.4	7.5	A-F	6.9	48.6	0.80	1.76	0.45	1.00	0.23	0.50	0.18	0.15
								M-F	7.7	54.0	0.89	1.96	0.50	1.10	0.25	0.55	0.20	0.17
		1.0	0.5	20.0	9.1	17.6	8.0	A-F	7.4	53.1	0.88	1.94	0.52	1.15	0.29	0.63	0.22	0.16
								M-F	8.2	59.0	0.98	2.16	0.58	1.28	0.32	0.70	0.24	0.18

(Continued)

TABLE I-1d (Continued)

Weight		Daily Gain		Daily Consumption: As-fed		Daily Consumption: Moisture-free Dry Matter		Moisture Basis[5] A-F (As-fed) M-F (Moisture-free)	Total Protein	Energy: TDN	Energy: ME (Mcal per)		Energy: NE_m (Mcal per)		Energy: NE_g (Mcal per)		Calcium	Phosphorus
(lb)	(kg)	(lb)	(kg)	(lb)	(kg)	(lb)	(kg)		(%)	(%)	(lb)	(kg)	(lb)	(kg)	(lb)	(kg)	(%)	(%)
Large-frame heifer calves and compensating medium-frame yearling heifers (continued)																		
800	364	1.5	0.7	20.3	9.2	18.3	8.3	A-F	7.7	57.6	0.95	2.08	0.59	1.29	0.35	0.77	0.23	0.16
								M-F	8.6	64.0	1.05	2.31	0.65	1.43	0.39	0.86	0.25	0.18
		2.0	0.9	20.7	9.4	18.6	8.5	A-F	8.1	62.6	1.03	2.26	0.67	1.47	0.41	0.91	0.25	0.17
								M-F	9.0	69.5	1.14	2.51	0.74	1.63	0.46	1.01	0.28	0.19
		2.5	1.1	20.1	9.1	18.1	8.2	A-F	8.6	69.3	1.13	2.49	0.76	1.67	0.50	1.09	0.30	0.19
								M-F	9.6	77.0	1.26	2.77	0.84	1.85	0.55	1.21	0.33	0.21
900	409	0.5	0.2	19.8	9.0	17.8	8.1	A-F	6.8	48.6	0.80	1.76	0.45	1.00	0.23	0.50	0.18	0.16
								M-F	7.5	54.0	0.89	1.96	0.50	1.10	0.25	0.55	0.20	0.18
		1.0	0.5	21.3	9.7	19.2	8.7	A-F	7.1	53.1	0.88	1.94	0.52	1.15	0.29	0.63	0.20	0.16
								M-F	7.9	59.0	0.98	2.16	0.58	1.28	0.32	0.70	0.22	0.18
		1.5	0.7	22.2	10.1	20.0	9.1	A-F	7.4	57.6	0.95	2.08	0.59	1.29	0.35	0.77	0.21	0.16
								M-F	8.2	64.0	1.05	2.31	0.65	1.43	0.39	0.86	0.23	0.18
		2.0	0.9	22.6	10.3	20.3	9.2	A-F	7.7	62.6	1.03	2.26	0.67	1.47	0.41	0.91	0.23	0.16
								M-F	8.6	69.5	1.14	2.51	0.74	1.63	0.46	1.01	0.26	0.18
		2.5	1.1	22.0	10.0	19.8	9.0	A-F	8.1	69.3	1.13	2.49	0.76	1.67	0.50	1.09	0.26	0.18
								M-F	9.0	77.0	1.26	2.77	0.84	1.85	0.55	1.21	0.29	0.20
1,000	454	0.5	0.2	21.4	9.7	19.3	8.8	A-F	6.7	48.6	0.80	1.76	0.45	1.00	0.23	0.50	0.17	0.16
								M-F	7.4	54.0	0.89	1.96	0.50	1.10	0.25	0.55	0.19	0.18
		1.0	0.5	23.1	10.5	20.8	9.5	A-F	6.9	53.1	0.88	1.94	0.52	1.15	0.29	0.63	0.19	0.16
								M-F	7.7	59.0	0.98	2.16	0.58	1.28	0.32	0.70	0.21	0.18
		1.5	0.7	24.1	11.0	21.7	9.9	A-F	7.2	57.6	0.95	2.08	0.59	1.29	0.35	0.77	0.19	0.16
								M-F	8.0	64.0	1.05	2.31	0.65	1.43	0.39	0.86	0.21	0.18
		2.0	0.9	24.4	11.1	22.0	10.0	A-F	7.4	62.6	1.03	2.26	0.67	1.47	0.41	0.91	0.21	0.16
								M-F	8.2	69.5	1.14	2.51	0.74	1.63	0.46	1.01	0.23	0.18
		2.5	1.1	23.9	10.9	21.5	9.8	A-F	7.7	69.3	1.13	2.49	0.76	1.67	0.50	1.09	0.23	0.16
								M-F	8.6	77.0	1.26	2.77	0.84	1.85	0.55	1.21	0.25	0.18
1,100	500	0.5	0.2	23.1	10.5	20.8	9.5	A-F	6.6	48.6	0.80	1.76	0.45	1.00	0.23	0.50	0.17	0.16
								M-F	7.3	54.0	0.89	1.96	0.50	1.10	0.25	0.55	0.19	0.18
		1.0	0.5	24.8	11.3	22.3	10.1	A-F	6.8	53.1	0.88	1.94	0.52	1.15	0.29	0.63	0.18	0.16
								M-F	7.5	59.0	0.98	2.16	0.58	1.28	0.32	0.70	0.20	0.18
		1.5	0.7	25.9	11.8	23.3	10.6	A-F	6.9	57.6	0.95	2.08	0.59	1.29	0.35	0.77	0.18	0.16
								M-F	7.7	64.0	1.05	2.31	0.65	1.43	0.39	0.86	0.20	0.18
		2.0	0.9	26.2	11.9	23.6	10.7	A-F	7.1	62.6	1.03	2.26	0.67	1.47	0.41	0.91	0.19	0.16
								M-F	7.9	69.5	1.14	2.51	0.74	1.63	0.46	1.01	0.21	0.18
		2.5	1.1	25.7	11.7	23.1	10.5	A-F	7.4	69.3	1.13	2.49	0.76	1.67	0.50	1.09	0.20	0.16
								M-F	8.2	77.0	1.26	2.77	0.84	1.85	0.55	1.21	0.22	0.18

[1]Adapted from *Nutrient Requirements of Beef Cattle*, sixth revised edition, National Research Council-National Academy of Sciences, 1984.

[2]Shrunk liveweight basis. This refers to the weight after an overnight feed and water shrink (generally equivalent to about 96% of unshrunk weights taken in early morning).

[3]Vitamin A requirements are 1,000 IU per lb *(2,200 IU per kg)* of diet.

[4]This table gives reasonable examples of nutrient concentrations that should be suitable to formulate diets for specific management goals. It does not imply that diets with other nutrient concentrations when consumed in sufficient amounts would be inadequate to meet nutrient requirements.

[5]As-fed was calculated using an average figure of 90% of dry matter. When using silages, roots, and other wet feeds, these feeds should be converted to a moisture-free basis and the diet calculated using the moisture-free data.

TABLES I-2
1996 NATIONAL RESEARCH COUNCIL
NUTRIENT REQUIREMENTS OF BEEF CATTLE

TABLE I-2a
DAILY NUTRIENT REQUIREMENTS FOR GROWING AND FINISHING CATTLE[1, 2]

Wt. @ small marbling	**1,150 lb**
Weight range	**600–1,150 lb**
ADG range	**2.00–4.00 lb**
Breed code	**1 Angus**

Maintenance Requirements	**Body Weight (lb)**					
	600	**710**	**820**	**930**	**1040**	**1150**
NE_m . . . . (mcal/d)	5.16	5.85	6.52	7.17	7.80	8.41
MP . . . . (g/d)	255	289	322	354	385	415
Ca . . . . (g/d)	8.4	9.9	11.5	13.0	14.5	16.1
P . . . . (g/d)	6.4	7.6	8.8	9.9	11.1	12.3
Growth Requirements						
ADG	**NE_g Required for Gain (Mcal/d)**					
2.0 lb/d	3.13	3.55	3.95	4.34	4.72	5.09
2.5 lb/d	3.99	4.53	5.05	5.55	6.03	6.51
3.0 lb/d	4.88	5.54	6.17	6.78	7.37	7.95
3.5 lb/d	5.78	6.56	7.30	8.03	8.73	9.41
4.0 lb/d	6.69	7.59	8.45	9.29	10.10	10.90
	MP Required for Gain (g/d)					
2.0 lb/d	275	279	258	235	212	190
2.5 lb/d	339	343	316	286	257	229
3.0 lb/d	403	406	373	336	301	267
3.5 lb/d	465	468	429	385	343	303
4.0 lb/d	527	529	483	433	385	337
	Calcium Required for Gain (g/d)					
2.0 lb/d	21.5	19.7	18.0	16.4	14.8	13.3
2.5 lb/d	26.5	24.2	22.1	20.0	18.0	16.0
3.0 lb/d	31.4	28.7	26.1	23.5	21.0	18.6
3.5 lb/d	36.3	33.1	29.9	26.9	24.0	21.1
4.0 lb/d	41.1	37.4	33.8	30.3	26.9	23.6
	Phosphorus Required for Gain (g/d)					
2.0 lb/d	8.7	8.0	7.3	6.6	6.0	5.4
2.5 lb/d	10.7	9.8	8.9	8.1	7.3	6.5
3.0 lb/d	12.7	11.6	10.5	9.5	8.5	7.5
3.5 lb/d	14.7	13.4	12.1	10.9	9.7	8.5
4.0 lb/d	16.6	15.1	13.6	12.2	10.9	9.5

[1]Adapted from *Nutrient Requirements of Beef Cattle*, seventh revised edition, National Research Council-National Academy of Sciences, 1996.

[2]Concerning nutrient requirements for growing and finishing cattle, and growing bulls, requirements for net energy, metabolizable protein, calcium, and phosphorus are determined by adding maintenance requirements and requirements for the desired level of performance.

TABLE I-2b
DAILY NUTRIENT REQUIREMENTS FOR GROWING BULLS[1, 2]

Wt. @ maturity	**2,000 lb**
Weight range	**600–1,250 lb**
ADG range	**2.00–4.00 lb**
Breed code	**1 Angus**

Maintenance Requirements	**Body Weight (lb)**					
	600	**720**	**840**	**960**	**1080**	**1200**
NE_m . . . . (mcal/d)	5.93	6.80	7.64	8.44	9.22	9.98
MP . . . . (g/d)	255	292	328	362	396	428
Ca . . . . (g/d)	8.4	10.1	11.7	13.4	15.1	16.8
P . . . . (g/d)	6.4	7.7	9.0	10.2	11.5	12.8
Growth Requirements						
ADG	**NE_g Required for Gain (Mcal/d)**					
2.0 lb/d	3.03	3.47	3.90	4.31	4.71	5.09
2.5 lb/d	3.87	4.44	4.98	5.50	6.01	6.51
3.0 lb/d	4.73	5.42	6.08	6.72	7.34	7.95
3.5 lb/d	5.60	6.42	7.20	7.96	8.70	9.41
4.0 lb/d	6.48	7.43	8.34	9.22	10.07	10.90
	MP Required for Gain (g/d)					
2.0 lb/d	275	278	261	237	213	190
2.5 lb/d	339	342	320	289	259	229
3.0 lb/d	402	405	378	340	303	267
3.5 lb/d	465	467	435	389	345	303
4.0 lb/d	527	529	490	438	387	337
	Calcium Required for Gain (g/d)					
2.0 lb/d	21.9	20.0	18.3	16.5	14.9	13.3
2.5 lb/d	27.0	24.6	22.4	20.2	18.1	16.0
3.0 lb/d	32.1	29.2	26.4	23.7	21.1	18.6
3.5 lb/d	37.1	33.6	30.4	27.2	24.1	21.1
4.0 lb/d	42.0	38.1	34.3	30.6	27.0	23.6
	Phosphorus Required for Gain (g/d)					
2.0 lb/d	8.8	8.1	7.4	6.7	6.0	5.4
2.5 lb/d	10.9	10.0	9.0	8.2	7.3	6.5
3.0 lb/d	13.0	11.8	10.7	9.6	8.5	7.5
3.5 lb/d	15.0	13.6	12.3	11.0	9.7	8.5
4.0 lb/d	17.0	15.4	13.8	12.4	10.9	9.5

[1]Adapted from *Nutrient Requirements of Beef Cattle*, seventh revised edition, National Research Council-National Academy of Sciences, 1996.

[2]Concerning nutrient requirements for growing and finishing cattle, and growing bulls, requirements for net energy, metabolizable protein, calcium, and phosphorus are determined by adding maintenance requirements and requirements for the desired level of performance.

TABLE I-2c
DAILY NUTRIENT REQUIREMENTS OF PREGNANT REPLACEMENT HEIFERS[1]

Mature weight	**1,200 lb**
Calf birth weight	**80 lb**
Age @ breeding	**15 months**
Breed code	**1 Angus**

	Months Since Conception								
	1	2	3	4	5	6	7	8	9
NE_m required									
Maintenance . . . (Mcal/d)	6.08	6.24	6.40	6.56	6.72	6.88	7.03	7.19	7.34
Growth . . . . . . (Mcal/d)	2.35	2.41	2.47	2.53	2.60	2.66	2.72	2.78	2.84
Pregnancy . . . . (Mcal/d)	0.03	0.06	0.14	0.29	0.58	1.07	1.88	3.12	4.88
Total . . . . (Mcal/d)	8.46	8.72	9.02	9.39	9.89	10.61	11.63	13.09	15.05
MP required									
Maintenance . . . . . (g/d)	300	308	316	324	332	339	347	355	362
Growth . . . . (g/d)	121	121	121	122	121	119	117	115	113
Pregnancy . . . . . . (g/d)	2	3	7	13	24	45	80	137	227
Total . . . . (g/d)	422	432	444	458	478	504	544	607	702
Minerals									
Calcium required									
Maintenance . . . . . (g/d)	10	11	11	12	12	12	13	13	13
Growth . . . . (g/d)	9	9	9	9	8	8	8	8	8
Pregnancy . . . . . . (g/d)	0	0	0	0	0	0	11	11	11
Total . . . . (g/d)	20	20	20	20	20	21	32	32	32
Phosphorus required									
Maintenance . . . . . (g/d)	8	8	9	9	9	9	10	10	10
Growth . . . . (g/d)	4	4	4	3	3	3	3	3	3
Pregnancy . . . . . . (g/d)	0	0	0	0	0	0	5	5	5
Total . . . . (g/d)	12	12	12	12	13	13	17	18	18
ADG									
Growth . . . . (lb/d)	0.88	0.88	0.88	0.88	0.88	0.88	0.88	0.88	0.88
Pregnancy . . . . . (lb/d)	0.06	0.09	0.15	0.24	0.38	0.56	0.81	1.14	1.54
Total . . . . (lb/d)	0.93	0.97	1.03	1.12	1.25	1.44	1.69	2.01	2.42
Body weight									
Shrunk body . . . . . (lb)	747	773	800	827	853	880	907	933	960
Gravid uterus mass . (lb)	3	5	9	15	24	38	59	88	129
Total . . . . (lb)	750	779	809	842	878	918	966	1,022	1,089

[1]Adapted from *Nutrient Requirements of Beef Cattle*, seventh revised edition, National Research Council-National Academy of Sciences, 1996.

TABLE I-2d
DAILY NUTRIENT REQUIREMENTS OF BEEF COWS[1]

Mature weight	**1,200 lb**	**Milk fat**	**4.0%**
Calf birth weight	**80 lb**	**Milk protein**	**3.4%**
Age @ calving	**60 months**	**Calving interval**	**12 months**
Age @ weaning	**30 weeks**	**Time peak**	**8.5 weeks**
Peak milk	**20 lb**		
Breed code	**1 Angus**	**Milk SNF**	**8.3%**

	Months Since Calving											
	1	**2**	**3**	**4**	**5**	**6**	**7**	**8**	**9**	**10**	**11**	**12**
NE_m required												
Maintenance . . . (Mcal/d)	10.41	10.41	10.41	10.41	10.41	10.41	8.68	8.68	8.68	8.68	8.68	8.68
Growth . . . . . . (Mcal/d)	0.00	0.00	0.00	0.00	0.00	0.00	0.00	0.00	0.00	0.00	0.00	0.00
Lactation . . . . . (Mcal/d)	5.43	6.51	5.86	4.69	3.52	2.53	0.00	0.00	0.00	0.00	0.00	0.00
Pregnancy . . . . (Mcal/d)	0.00	0.00	0.01	0.03	0.06	0.14	0.29	0.58	1.07	1.88	3.12	4.88
Total . . . . (Mcal/d)	15.84	16.93	16.29	15.13	13.99	13.09	8.97	9.26	9.75	10.56	11.80	13.55
MP required												
Maintenance . . . . . (g/d)	428	428	428	428	428	428	428	428	428	428	428	428
Growth . . . . (g/d)	0	0	0	0	0	0	0	0	0	0	0	0
Lactation . . . . (g/d)	395	475	427	342	256	184	0	0	0	0	0	0
Pregnancy . . . . . . (g/d)	0	0	1	2	3	7	13	24	45	80	137	227
Total . . . . (g/d)	824	903	856	771	688	619	441	453	473	508	566	656
Calcium required												
Maintenance . . . . . (g/d)	17	17	17	17	17	17	17	17	17	17	17	17
Growth . . . . (g/d)	0	0	0	0	0	0	0	0	0	0	0	0
Lactation . . . . (g/d)	19	22	20	16	12	9	0	0	0	0	0	0
Pregnancy . . . . . . (g/d)	0	0	0	0	0	0	0	0	0	11	11	11
Total . . . . (g/d)	35	39	37	33	29	25	17	17	17	28	28	28
Phosphorus required												
Maintenance . . . . . (g/d)	13	13	13	13	13	13	13	13	13	13	13	13
Growth . . . . (g/d)	0	0	0	0	0	0	0	0	0	0	0	0
Lactation . . . . (g/d)	11	13	12	9	7	5	0	0	0	0	0	0
Pregnancy . . . . . . (g/d)	0	0	0	0	0	0	0	0	0	5	5	5
Total . . . . (g/d)	24	26	25	22	20	18	13	13	13	18	18	18
ADG												
Growth . . . . (lb/d)	0.00	0.00	0.00	0.00	0.00	0.00	0.00	0.00	0.00	0.00	0.00	0.00
Pregnancy . . . . . (lb/d)	0.00	0.00	0.03	0.06	0.09	0.15	0.24	0.38	0.56	0.81	1.14	1.54
Total . . . . (lb/d)	0.00	0.00	0.03	0.06	0.09	0.15	0.24	0.38	0.56	0.81	1.14	1.54
Milk . . . . (lb/d)	16.7	20.0	18.0	14.4	10.8	7.8						
Body weight												
Shrunk body . . . . . (lb)	1,200	1,200	1,200	1,200	1,200	1,200	1,200	1,200	1,200	1,200	1,200	1,200
Conceptus . . . . . . (lb)	0	0	2	3	5	9	15	24	38	59	88	129
Total . . . . (lb)	1,200	1,200	1,202	1,203	1,205	1,209	1,215	1,224	1,238	1,259	1,288	1,329

[1]Adapted from *Nutrient Requirements of Beef Cattle*, seventh revised edition, National Research Council-National Academy of Sciences, 1996.

TABLES I-3
1996 NATIONAL RESEARCH COUNCIL
DIET NUTRIENT DENSITY REQUIREMENTS

TABLE I-3a
DIET NUTRIENT DENSITIES FOR GROWING AND FINISHING CATTLE[1, 2, 3]

Body Weight	TDN	NE_m	NE_g	DMI	ADG	CP	Ca	P
(lb)	*(% DM)*	*(Mcal/lb)*	*(Mcal/lb)*	*(lb/day)*	*(lb/day)*	*(% DM)*	*(% DM)*	*(% DM)*
1,000 @ finishing (28% body fat—for feedlot steers and heifers) or maturity (replacement heifers)								
550	50	0.45	0.20	15.2	0.64	7.1	0.21	0.13
	60	0.61	0.35	16.1	1.77	9.8	0.36	0.19
	70	0.76	0.48	15.7	2.68	12.4	0.49	0.24
	80	0.90	0.61	14.8	3.34	14.9	0.61	0.29
	90	1.04	0.72	13.7	3.75	17.3	0.73	0.34
600	50	0.45	0.20	16.2	0.64	7.0	0.21	0.13
	60	0.61	0.35	17.2	1.77	9.5	0.34	0.18
	70	0.76	0.48	16.8	2.68	11.9	0.45	0.23
	80	0.90	0.61	15.8	3.34	14.3	0.56	0.27
	90	1.04	0.72	14.6	3.75	16.5	0.66	0.32
650	50	0.45	0.20	17.3	0.64	6.9	0.20	0.12
	60	0.61	0.35	18.2	1.77	9.2	0.32	0.17
	70	0.76	0.48	17.8	2.68	11.5	0.42	0.21
	80	0.90	0.61	16.8	3.34	13.7	0.52	0.26
	90	1.04	0.72	15.5	3.75	15.9	0.61	0.30
700	50	0.45	0.20	18.2	0.64	6.8	0.19	0.12
	60	0.61	0.35	19.3	1.77	8.8	0.30	0.16
	70	0.76	0.48	18.8	2.68	10.9	0.39	0.20
	80	0.90	0.61	17.8	3.34	13.0	0.48	0.24
	90	1.04	0.72	16.4	3.75	15.0	0.56	0.28
750	50	0.45	0.20	19.2	0.64	6.7	0.19	0.12
	60	0.61	0.35	20.3	1.77	8.5	0.28	0.16
	70	0.76	0.48	19.8	2.68	10.3	0.37	0.19
	80	0.90	0.61	18.7	3.34	12.2	0.45	0.23
	90	1.04	0.72	17.3	3.75	14.0	0.52	0.26
800	50	0.45	0.20	20.2	0.64	6.5	0.19	0.12
	60	0.61	0.35	21.3	1.77	8.1	0.27	0.15
	70	0.76	0.48	20.8	2.68	9.8	0.34	0.18
	80	0.90	0.61	19.6	3.34	11.5	0.42	0.22
	90	1.04	0.72	18.1	3.75	13.2	0.48	0.25
1,100 @ finishing (28% body fat—for feedlot steers and heifers) or maturity (replacement heifers)								
605	50	0.45	0.20	16.3	0.68	7.2	0.22	0.13
	60	0.61	0.35	17.3	1.88	10.0	0.36	0.19
	70	0.76	0.48	16.9	2.86	12.7	0.49	0.24
	80	0.90	0.61	15.9	3.56	15.3	0.61	0.29
	90	1.04	0.72	14.7	4.00	17.8	0.72	0.34

(Continued)

TABLE I-3a (Continued)

Body Weight	TDN	NE_m	NE_g	DMI	ADG	CP	Ca	P
(lb)	*(% DM)*	*(Mcal/lb)*	*(Mcal/lb)*	*(lb/day)*	*(lb/day)*	*(% DM)*	*(% DM)*	*(% DM)*
1,100 @ finishing (28% body fat—for feedlot steers and heifers) or maturity (replacement heifers) (continued)								
660	50	0.45	0.20	17.5	0.68	7.1	0.21	0.13
	60	0.61	0.35	18.4	1.88	9.7	0.34	0.18
	70	0.76	0.48	18.0	2.86	12.3	0.45	0.23
	80	0.90	0.61	17.0	3.56	14.7	0.56	0.27
	90	1.04	0.72	15.7	4.00	17.1	0.66	0.32
715	50	0.45	0.20	18.5	0.68	6.9	0.20	0.13
	60	0.61	0.35	19.6	1.88	9.2	0.32	0.17
	70	0.76	0.48	19.1	2.86	11.5	0.42	0.21
	80	0.90	0.61	18.1	3.56	13.7	0.52	0.26
	90	1.04	0.72	16.7	4.00	15.9	0.61	0.30
770	50	0.45	0.20	19.6	0.68	6.8	0.20	0.12
	60	0.61	0.35	20.7	1.88	8.8	0.30	0.16
	70	0.76	0.48	20.2	2.86	10.9	0.39	0.20
	80	0.90	0.61	19.1	3.56	12.9	0.48	0.24
	90	1.04	0.72	17.6	4.00	14.8	0.56	0.28
825	50	0.45	0.20	20.6	0.68	6.6	0.19	0.12
	60	0.61	0.35	21.8	1.88	8.4	0.28	0.16
	70	0.76	0.48	21.3	2.86	10.3	0.37	0.19
	80	0.90	0.61	20.1	3.56	12.1	0.44	0.23
	90	1.04	0.72	18.6	4.00	13.9	0.52	0.26
880	50	0.45	0.20	21.7	0.68	6.5	0.19	0.12
	60	0.61	0.35	22.9	1.88	8.1	0.27	0.15
	70	0.76	0.48	22.4	2.86	9.8	0.34	0.18
	80	0.90	0.61	21.1	3.56	11.4	0.42	0.22
	90	1.04	0.72	19.5	4.00	13.1	0.48	0.25
1,200 @ finishing (28% body fat—for feedlot steers and heifers) or maturity (replacement heifers)								
660	50	0.45	0.20	17.5	0.72	7.3	0.22	0.13
	60	0.61	0.35	18.4	2.00	10.2	0.36	0.19
	70	0.76	0.48	18.0	3.04	13.0	0.49	0.24
	80	0.90	0.61	17.0	3.78	15.8	0.61	0.29
	90	1.04	0.72	15.7	4.25	18.4	0.72	0.34
720	50	0.45	0.20	18.6	0.72	7.1	0.21	0.13
	60	0.61	0.35	19.7	2.00	9.7	0.34	0.18
	70	0.76	0.48	19.2	3.04	12.2	0.45	0.23
	80	0.90	0.61	18.2	3.78	14.6	0.56	0.27
	90	1.04	0.72	16.8	4.25	17.0	0.66	0.32
780	50	0.45	0.20	19.8	0.72	6.9	0.20	0.13
	60	0.61	0.35	20.9	2.00	9.2	0.32	0.17
	70	0.76	0.48	20.4	3.04	11.4	0.42	0.21
	80	0.90	0.61	19.3	3.78	13.6	0.52	0.26
	90	1.04	0.72	17.8	4.25	15.8	0.61	0.30

(Continued)

TABLE I-3a (Continued)

Body Weight	TDN	NE_m	NE_g	DMI	ADG	CP	Ca	P
(lb)	*(% DM)*	*(Mcal/lb)*	*(Mcal/lb)*	*(lb/day)*	*(lb/day)*	*(% DM)*	*(% DM)*	*(% DM)*
1,200 @ finishing (28% body fat—for feedlot steers and heifers) or maturity (replacement heifers) (continued)								
840	50	0.45	0.20	20.9	0.72	6.8	0.20	0.13
	60	0.61	0.35	22.1	2.00	8.8	0.30	0.16
	70	0.76	0.48	21.6	3.04	10.8	0.39	0.20
	80	0.90	0.61	20.4	3.78	12.8	0.48	0.24
	90	1.04	0.72	18.8	4.25	14.7	0.56	0.28
900	50	0.45	0.20	22.0	0.72	6.6	0.19	0.12
	60	0.61	0.35	23.3	2.00	8.4	0.28	0.16
	70	0.76	0.48	22.7	3.04	10.2	0.37	0.19
	80	0.90	0.61	21.5	3.78	12.0	0.44	0.23
	90	1.04	0.72	19.8	4.25	13.8	0.52	0.26
960	50	0.45	0.20	23.1	0.72	6.5	0.19	0.12
	60	0.61	0.35	24.4	2.00	8.1	0.27	0.15
	70	0.76	0.48	23.9	3.04	9.7	0.34	0.19
	80	0.90	0.61	22.5	3.78	11.3	0.41	0.22
	90	1.04	0.72	20.8	4.25	13.0	0.48	0.25
1,300 @ finishing (28% body fat—for feedlot steers and heifers) or maturity (replacement heifers)								
715	50	0.45	0.20	18.5	0.76	7.3	0.22	0.13
	60	0.61	0.35	19.6	2.11	10.2	0.36	0.19
	70	0.76	0.48	19.1	3.21	13.0	0.49	0.24
	80	0.90	0.61	18.1	3.99	15.7	0.61	0.29
	90	1.04	0.72	16.7	4.48	18.3	0.72	0.34
780	50	0.45	0.20	19.8	0.76	7.1	0.21	0.13
	60	0.61	0.35	20.9	2.11	9.6	0.34	0.18
	70	0.76	0.48	20.4	3.21	12.1	0.45	0.23
	80	0.90	0.61	19.3	3.99	14.5	0.56	0.27
	90	1.04	0.72	17.8	4.48	16.9	0.66	0.32
845	50	0.45	0.20	21.0	0.76	6.9	0.21	0.13
	60	0.61	0.35	22.2	2.11	9.1	0.32	0.17
	70	0.76	0.48	21.7	3.21	11.4	0.42	0.22
	80	0.90	0.61	20.5	3.99	13.6	0.51	0.26
	90	1.04	0.72	18.9	4.48	15.7	0.60	0.30
910	50	0.45	0.20	22.2	0.76	6.7	0.20	0.13
	60	0.61	0.35	23.5	2.11	8.7	0.30	0.17
	70	0.76	0.48	22.9	3.21	10.7	0.39	0.20
	80	0.90	0.61	21.6	3.99	12.7	0.48	0.24
	90	1.04	0.72	20.0	4.48	14.6	0.56	0.28
975	50	0.45	0.20	23.4	0.76	6.6	0.20	0.13
	60	0.61	0.35	24.7	2.11	8.3	0.28	0.16
	70	0.76	0.48	24.1	3.21	10.2	0.37	0.19
	80	0.90	0.61	22.8	3.99	11.9	0.44	0.23
	90	1.04	0.72	21.0	4.48	13.7	0.52	0.26

(Continued)

TABLE I-3a (Continued)

Body Weight	TDN	NE_m	NE_g	DMI	ADG	CP	Ca	P
(lb)	*(% DM)*	*(Mcal/lb)*	*(Mcal/lb)*	*(lb/day)*	*(lb/day)*	*(% DM)*	*(% DM)*	*(% DM)*
1,300 @ finishing (28% body fat—for feedlot steers and heifers) or maturity (replacement heifers) (continued)								
1,040	50	0.45	0.20	24.5	0.76	6.5	0.19	0.13
	60	0.61	0.35	25.9	2.11	8.0	0.27	0.15
	70	0.76	0.48	25.3	3.21	9.6	0.34	0.19
	80	0.90	0.61	23.9	3.99	11.3	0.41	0.22
	90	1.04	0.72	22.1	4.48	12.9	0.48	0.25
1,400 @ finishing (28% body fat—for feedlot steers and heifers) or maturity (replacement heifers)								
770	50	0.45	0.20	19.6	0.80	7.3	0.22	0.13
	60	0.61	0.35	20.7	2.22	10.1	0.36	0.19
	70	0.76	0.48	20.2	3.38	12.9	0.49	0.24
	80	0.90	0.61	19.1	4.20	15.6	0.61	0.29
	90	1.04	0.72	17.6	4.72	18.1	0.72	0.34
840	50	0.45	0.20	20.9	0.80	7.1	0.21	0.13
	60	0.61	0.35	22.1	2.22	9.6	0.34	0.18
	70	0.76	0.48	21.6	3.38	12.1	0.45	0.23
	80	0.90	0.61	20.4	4.20	14.5	0.56	0.27
	90	1.04	0.72	18.8	4.72	16.8	0.65	0.32
910	50	0.45	0.20	22.2	0.80	6.9	0.21	0.13
	60	0.61	0.35	23.5	2.22	9.1	0.32	0.17
	70	0.76	0.48	22.9	3.38	11.3	0.42	0.22
	80	0.90	0.61	21.6	4.20	13.5	0.51	0.26
	90	1.04	0.72	20.0	4.72	15.6	0.60	0.30
980	50	0.45	0.20	23.5	0.80	6.7	0.20	0.13
	60	0.61	0.35	24.8	2.22	8.7	0.30	0.17
	70	0.76	0.48	24.2	3.38	10.7	0.39	0.20
	80	0.90	0.61	22.9	4.20	12.6	0.47	0.24
	90	1.04	0.72	21.1	4.72	14.5	0.56	0.28
1,050	50	0.45	0.20	24.7	0.80	6.6	0.20	0.13
	60	0.61	0.35	26.1	2.22	8.3	0.28	0.16
	70	0.76	0.48	25.5	3.38	10.1	0.37	0.20
	80	0.90	0.61	24.1	4.20	11.9	0.44	0.23
	90	1.04	0.72	22.2	4.72	13.6	0.51	0.26
1,120	50	0.45	0.20	25.9	0.80	6.5	0.19	0.13
	60	0.61	0.35	27.4	2.22	8.0	0.27	0.16
	70	0.76	0.48	26.8	3.38	9.6	0.34	0.19
	80	0.90	0.61	25.3	4.20	11.2	0.41	0.22
	90	1.04	0.72	23.3	4.72	12.8	0.48	0.25

[1]Adapted from *Nutrient Requirements of Beef Cattle*, seventh revised edition, National Research Council-National Academy of Sciences, 1996.

[2]Shrunk liveweight basis. This refers to the weight after an overnight feed and water shrink (generally equivalent to about 96% of unshrunk weights taken in early morning).

[3]This table gives reasonable examples of nutrient densities or concentrations that should be suitable to formulate diets for specific management goals. It does not imply that diets with other nutrient concentrations when consumed in sufficient amounts would be inadequate to meet nutrient requirements.

TABLE I-3b
DIET NUTRIENT DENSITY REQUIREMENTS OF PREGNANT REPLACEMENT HEIFERS[1, 2, 3]

	Months Since Conception								
	1	2	3	4	5	6	7	8	9
1,000 lb mature weight									
TDN . . . . (% DM)	50.1	50.2	50.4	50.7	51.3	52.3	54.0	56.8	61.3
ME . . . . (Mcal/lb)	0.46	0.46	0.46	0.46	0.47	0.49	0.52	0.56	0.63
NE_m . . . . (Mcal/lb)	0.21	0.21	0.21	0.21	0.22	0.24	0.26	0.30	0.37
DMI . . . . (lb)	16.7	17.2	17.7	18.2	18.7	19.4	20.0	20.7	21.3
Target ADG	0.73	0.73	0.73	0.73	0.73	0.73	0.73	0.73	0.73
Shrunk body wt. . . . (lb)	622	644	667	689	711	733	756	778	800
CP . . . . (% DM)	7.18	7.16	7.16	7.21	7.32	7.56	7.99	8.74	10.02
Ca . . . . (% DM)	0.22	0.22	0.22	0.21	0.21	0.20	0.32	0.31	0.31
P . . . . (% DM)	0.17	0.17	0.17	0.17	0.17	0.16	0.23	0.23	0.22
1,100 lb mature weight									
TDN . . . . (% DM)	50.3	50.4	50.5	50.8	51.3	52.3	53.9	56.5	60.6
ME . . . . (Mcal/lb)	0.46	0.46	0.46	0.47	0.48	0.49	0.52	0.56	0.62
NE_m . . . . (Mcal/lb)	0.21	0.21	0.21	0.22	0.22	0.24	0.26	0.30	0.36
DMI . . . . (lb)	18.0	18.5	19.0	19.5	20.1	20.8	21.5	22.3	22.9
Target ADG	0.80	0.80	0.80	0.80	0.80	0.80	0.80	0.80	0.80
Shrunk body wt. . . . (lb)	684	709	733	758	782	807	831	856	880
CP . . . . (% DM)	7.20	7.17	7.17	7.21	7.32	7.54	7.93	8.63	9.80
Ca . . . . (% DM)	0.23	0.22	0.22	0.22	0.21	0.21	0.32	0.31	0.30
P . . . . (% DM)	0.18	0.17	0.17	0.17	0.17	0.17	0.23	0.22	0.22
1,200 lb mature weight									
TDN . . . . (% DM)	50.5	50.5	50.7	50.9	51.4	52.3	53.8	56.2	59.9
ME . . . . (Mcal/lb)	0.46	0.46	0.46	0.47	0.48	0.49	0.51	0.55	0.61
NE_m . . . . (Mcal/lb)	0.21	0.21	0.21	0.22	0.23	0.24	0.26	0.30	0.35
DMI . . . . (lb)	19.3	19.8	20.3	20.9	21.5	22.2	23.0	23.7	24.4
Target ADG	0.88	0.88	0.88	0.88	0.88	0.88	0.88	0.88	0.88
Shrunk body wt. . . . (lb)	747	773	800	827	853	880	907	933	960
CP . . . . (% DM)	7.21	7.19	7.18	7.22	7.31	7.52	7.89	8.53	9.62
Ca . . . . (% DM)	0.23	0.23	0.22	0.22	0.22	0.21	0.31	0.31	0.30
P . . . . (% DM)	0.18	0.18	0.18	0.17	0.17	0.17	0.23	0.22	0.22
1,300 lb mature weight									
TDN . . . . (% DM)	50.6	50.7	50.8	51.0	51.5	52.4	53.7	56.0	59.5
ME . . . . (Mcal/lb)	0.46	0.46	0.47	0.47	0.48	0.49	0.51	0.55	0.60
NE_m . . . . (Mcal/lb)	0.21	0.21	0.22	0.22	0.23	0.24	0.26	0.29	0.34
DMI . . . . (lb)	20.5	21.0	21.6	22.2	22.9	23.6	24.4	25.2	25.9
Target ADG	0.95	0.95	0.95	0.95	0.95	0.95	0.95	0.95	0.95
Shrunk body wt. . . . (lb)	809	838	867	896	924	953	982	1,011	1,040
CP . . . . (% DM)	7.23	7.20	7.20	7.22	7.31	7.50	7.85	8.45	9.46
Ca . . . . (% DM)	0.24	0.23	0.23	0.22	0.22	0.22	0.31	0.30	0.30
P . . . . (% DM)	0.18	0.18	0.18	0.18	0.18	0.17	0.23	0.22	0.22
1,400 lb mature weight									
TDN . . . . (% DM)	50.7	50.8	50.9	51.2	51.6	52.4	53.7	55.8	59.0
ME . . . . (Mcal/lb)	0.47	0.47	0.47	0.47	0.48	0.49	0.51	0.55	0.60
NE_m . . . . (Mcal/lb)	0.22	0.22	0.22	0.22	0.23	0.24	0.26	0.29	0.34
DMI . . . . (lb)	21.7	22.3	22.9	23.5	24.2	24.9	25.8	26.6	27.4
Target ADG	1.02	1.02	1.02	1.02	1.02	1.02	1.02	1.02	1.02
Shrunk body wt. . . . (lb)	871	902	933	964	996	1,027	1,058	1,089	1,120
CP . . . . (% DM)	7.25	7.22	7.21	7.23	7.31	7.48	7.81	8.38	9.33
Ca . . . . (% DM)	0.24	0.24	0.23	0.23	0.22	0.22	0.31	0.30	0.30
P . . . . (% DM)	0.18	0.18	0.18	0.18	0.18	0.18	0.23	0.22	0.22

[1]Adapted from *Nutrient Requirements of Beef Cattle*, seventh revised edition, National Research Council-National Academy of Sciences, 1996.

[2]Shrunk liveweight basis. This refers to the weight after an overnight feed and water shrink (generally equivalent to about 96% of unshrunk weights taken in early morning).

[3]This table gives reasonable examples of nutrient densities or concentrations that should be suitable to formulate diets for specific management goals. It does not imply that diets with other nutrient concentrations when consumed in sufficient amounts would be inadequate to meet nutrient requirements.

TABLE I-3c
DIET NUTRIENT DENSITY REQUIREMENTS OF BEEF COWS[1, 2, 3]

	Months Since Calving											
	1	2	3	4	5	6	7	8	9	10	11	12
1,000 lb mature weight, 10 lb peak milk												
TDN . . . . (% DM)	55.8	56.6	54.3	53.4	52.5	51.8	44.9	45.7	47.0	49.1	52.0	55.7
ME . . . . (Mcal/lb)	0.93	0.95	0.91	0.89	0.88	0.86	0.75	0.76	0.79	0.82	0.87	0.93
NE_m . . . . (Mcal/lb)	0.55	0.56	0.52	0.51	0.49	0.48	0.37	0.38	0.40	0.44	0.49	0.54
DM . . . . (lb)	21.6	22.1	23.0	22.5	22.1	21.7	21.1	21.0	20.9	20.8	21.0	21.4
Milk . . . . (lb/day)	8.3	10.0	9.0	7.2	5.4	3.9	0.0	0.0	0.0	0.0	0.0	0.0
CP . . . . (% DM)	8.70	9.10	8.41	7.97	7.51	7.14	5.98	6.16	6.47	6.95	7.66	8.67
Ca . . . . (% DM)	0.24	0.25	0.23	0.22	0.20	0.19	0.15	0.15	0.15	0.24	0.24	0.24
P . . . . (% DM)	0.17	0.17	0.16	0.15	0.14	0.14	0.11	0.11	0.11	0.15	0.15	0.15
1,000 lb mature weight, 20 lb peak milk												
TDN . . . . (% DM)	59.6	60.9	58.6	57.0	55.4	54.0	44.9	45.7	47.0	49.1	52.0	55.7
ME . . . . (Mcal/lb)	1.00	1.02	0.98	0.95	0.92	0.90	0.75	0.76	0.79	0.82	0.87	0.93
NE_m . . . . (Mcal/lb)	0.60	0.62	0.59	0.56	0.54	0.52	0.37	0.38	0.40	0.44	0.49	0.54
DM . . . . (lb)	24.0	25.0	25.4	24.4	23.5	22.7	21.1	21.0	20.9	20.8	21.0	21.4
Milk . . . . (lb/day)	16.7	20.0	18.0	14.4	10.8	7.8	0.0	0.0	0.0	0.0	0.0	0.0
CP . . . . (% DM)	10.54	11.18	10.38	9.65	8.86	8.17	5.98	6.16	6.47	6.95	7.66	8.67
Ca . . . . (% DM)	0.30	0.32	0.30	0.27	0.24	0.22	0.15	0.15	0.15	0.24	0.24	0.24
P . . . . (% DM)	0.20	0.21	0.19	0.18	0.17	0.15	0.11	0.11	0.11	0.15	0.15	0.15
1,000 lb mature weight, 30 lb peak milk												
TDN . . . . (% DM)	62.8	64.5	62.1	60.1	57.9	55.9	44.9	45.7	47.0	49.1	52.0	55.7
ME . . . . (Mcal/lb)	1.05	1.08	1.04	1.00	0.97	0.93	0.75	0.76	0.79	0.82	0.87	0.93
NE_m . . . . (Mcal/lb)	0.65	0.68	0.64	0.61	0.58	0.55	0.37	0.38	0.40	0.44	0.49	0.54
DM . . . . (lb)	26.4	27.8	27.8	26.4	24.9	23.7	21.1	21.0	20.9	20.8	21.0	21.4
Milk . . . . (lb/day)	25.0	30.0	27.0	21.6	16.2	11.7	0.0	0.0	0.0	0.0	0.0	0.0
CP . . . . (% DM)	12.06	12.86	12.00	11.07	10.04	9.09	5.98	6.16	6.47	6.95	7.66	8.67
Ca . . . . (% DM)	0.35	0.38	0.35	0.32	0.28	0.25	0.15	0.15	0.15	0.24	0.24	0.24
P . . . . (% DM)	0.22	0.24	0.22	0.21	0.19	0.17	0.11	0.11	0.11	0.15	0.15	0.15
1,200 lb mature weight, 10 lb peak milk												
TDN . . . . (% DM)	55.3	56.0	53.7	52.9	52.1	51.5	44.9	45.8	47.1	49.3	52.3	56.2
ME . . . . (Mcal/lb)	0.92	0.94	0.90	0.88	0.87	0.86	0.75	0.76	0.79	0.82	0.87	0.94
NE_m . . . . (Mcal/lb)	0.54	0.55	0.51	0.50	0.49	0.48	0.37	0.38	0.41	0.44	0.49	0.55
DM . . . . (lb)	24.4	24.9	26.0	25.6	25.1	24.8	24.2	24.1	24.0	23.9	24.1	24.6
Milk . . . . (lb/day)	8.3	10.0	9.0	7.2	5.4	3.9	0.0	0.0	0.0	0.0	0.0	0.0
CP . . . . (% DM)	8.43	8.79	8.13	7.73	7.33	7.00	5.99	6.18	6.50	7.00	7.73	8.78
Ca . . . . (% DM)	0.24	0.25	0.23	0.21	0.20	0.19	0.15	0.15	0.15	0.26	0.25	0.25
P . . . . (% DM)	0.17	0.17	0.16	0.15	0.14	0.14	0.12	0.12	0.12	0.16	0.16	0.16
1,200 lb mature weight, 20 lb peak milk												
TDN . . . . (% DM)	58.7	59.9	57.6	56.2	54.7	53.4	44.9	45.8	47.1	49.3	52.3	56.2
ME . . . . (Mcal/lb)	0.98	1.00	0.96	0.94	0.91	0.89	0.75	0.76	0.79	0.82	0.87	0.94
NE_m . . . . (Mcal/lb)	0.59	0.61	0.57	0.55	0.53	0.51	0.37	0.38	0.41	0.44	0.49	0.55
DM . . . . (lb)	26.8	27.8	28.4	27.4	26.5	25.7	24.2	24.1	24.0	23.9	24.1	24.6
Milk . . . . (lb/day)	16.7	20.0	18.0	14.4	10.8	7.8	0.0	0.0	0.0	0.0	0.0	0.0
CP . . . . (% DM)	10.10	10.69	9.92	9.25	8.54	7.92	5.99	6.18	6.50	7.00	7.73	8.78
Ca . . . . (% DM)	0.29	0.31	0.29	0.26	0.24	0.22	0.15	0.15	0.15	0.26	0.25	0.25
P . . . . (% DM)	0.19	0.21	0.19	0.18	0.17	0.15	0.12	0.12	0.12	0.16	0.16	0.16

(Continued)

TABLE I-3c (Continued)

	Months Since Calving											
	1	2	3	4	5	6	7	8	9	10	11	12
1,200 lb mature weight, 30 lb peak milk												
TDN . . . . (% DM)	61.6	63.2	60.8	59.0	57.0	55.2	44.9	45.8	47.1	49.3	52.3	56.2
ME . . . . (Mcal/lb)	1.03	1.06	1.02	0.99	0.95	0.92	0.75	0.76	0.79	0.82	0.87	0.94
NE_m . . . . (Mcal/lb)	0.64	0.66	0.62	0.59	0.56	0.54	0.37	0.38	0.41	0.44	0.49	0.55
DM . . . . (lb)	29.2	30.6	30.8	29.4	27.9	26.7	24.2	24.1	24.0	23.9	24.1	24.6
Milk . . . . (lb/day)	25.0	30.0	27.0	21.6	16.2	11.7	0.0	0.0	0.0	0.0	0.0	0.0
CP . . . . (% DM)	11.51	12.25	11.41	10.55	9.61	8.75	5.99	6.18	6.50	7.00	7.73	8.78
Ca . . . . (% DM)	0.34	0.36	0.34	0.31	0.27	0.25	0.15	0.15	0.15	0.26	0.25	0.25
P . . . . (% DM)	0.22	0.23	0.22	0.20	0.18	0.17	0.12	0.12	0.12	0.16	0.16	0.16
1,400 lb mature weight, 10 lb peak milk												
TDN . . . . (% DM)	54.9	55.5	53.3	52.5	51.8	51.2	45.0	45.8	47.3	49.5	52.6	56.6
ME . . . . (Mcal/lb)	0.92	0.93	0.89	0.88	0.86	0.86	0.75	0.77	0.79	0.83	0.88	0.95
NE_m . . . . (Mcal/lb)	0.53	0.54	0.51	0.49	0.48	0.47	0.37	0.39	0.41	0.44	0.49	0.56
DM . . . . (lb)	27.1	27.6	28.9	28.5	28.0	27.7	27.2	27.0	26.9	26.8	27.0	27.6
Milk . . . . (lb/day)	8.3	10.0	9.0	7.2	5.4	3.9	0.0	0.0	0.0	0.0	0.0	0.0
CP . . . . (% DM)	8.23	8.56	7.91	7.55	7.19	6.90	6.00	6.20	6.53	7.04	7.80	8.88
Ca . . . . (% DM)	0.23	0.25	0.23	0.21	0.20	0.19	0.16	0.16	0.16	0.27	0.26	0.26
P . . . . (% DM)	0.17	0.17	0.16	0.15	0.15	0.14	0.12	0.12	0.12	0.17	0.17	0.16
1,400 lb mature weight, 20 lb peak milk												
TDN . . . . (% DM)	58.0	59.1	56.8	55.5	54.1	53.0	45.0	45.8	47.3	49.5	52.6	56.6
ME . . . . (Mcal/lb)	0.97	0.99	0.95	0.93	0.90	0.89	0.75	0.77	0.79	0.83	0.88	0.95
NE_m . . . . (Mcal/lb)	0.58	0.60	0.56	0.54	0.52	0.50	0.37	0.39	0.41	0.44	0.49	0.56
DM . . . . (lb)	29.5	30.5	31.3	30.3	29.4	28.6	27.2	27.0	26.9	26.8	27.0	27.6
Milk . . . . (lb/day)	16.7	20.0	18.0	14.4	10.8	7.8	0.0	0.0	0.0	0.0	0.0	0.0
CP . . . . (% DM)	9.76	10.31	9.56	8.94	8.29	7.73	6.00	6.20	6.53	7.04	7.80	8.88
Ca . . . . (% DM)	0.28	0.30	0.28	0.26	0.24	0.22	0.16	0.16	0.16	0.27	0.26	0.26
P . . . . (% DM)	0.19	0.20	0.19	0.18	0.17	0.16	0.12	0.12	0.12	0.17	0.17	0.16
1,400 lb mature weight, 30 lb peak milk												
TDN . . . . (% DM)	60.7	62.2	59.8	58.1	56.2	54.7	45.0	45.8	47.3	49.5	52.6	56.6
ME . . . . (Mcal/lb)	1.01	1.04	1.00	0.97	0.94	0.91	0.75	0.77	0.79	0.83	0.88	0.95
NE_m . . . . (Mcal/lb)	0.62	0.64	0.61	0.58	0.55	0.53	0.37	0.39	0.41	0.44	0.49	0.56
DM . . . . (lb)	31.9	33.3	33.7	32.3	30.8	29.6	27.2	27.0	26.9	26.8	27.0	27.6
Milk . . . . (lb/day)	25.0	30.0	27.0	21.6	16.2	11.7	0.0	0.0	0.0	0.0	0.0	0.0
CP . . . . (% DM)	11.07	11.77	10.95	10.15	9.27	8.49	6.00	6.20	6.53	7.04	7.80	8.88
Ca . . . . (% DM)	0.33	0.35	0.32	0.30	0.27	0.24	0.16	0.16	0.16	0.27	0.26	0.26
P . . . . (% DM)	0.22	0.23	0.21	0.20	0.18	0.17	0.12	0.12	0.12	0.17	0.17	0.16

[1]Adapted from *Nutrient Requirements of Beef Cattle*, seventh revised edition, National Research Council-National Academy of Sciences, 1996.

[2]Shrunk liveweight basis. This refers to the weight after an overnight feed and water shrink (generally equivalent to about 96% of unshrunk weights taken in early morning).

[3]This table gives reasonable examples of nutrient densities or concentrations that should be suitable to formulate diets for specific management goals. It does not imply that diets with other nutrient concentrations when consumed in sufficient amounts would be inadequate to meet nutrient requirements.

SECTION II
COMPOSITION OF FEEDS[1]

Table II-1 gives the composition of feeds commonly used in beef cattle, and Table II-2 lists the minerals commonly used in diet formulations for beef cattle. Table II-3 lists bypass protein percentages in common feeds.

Additional feed compositions are given in *Feeds & Nutrition*, of which Dr. Ensminger is the senior author.

■ **Feed names**—Ideally, a feed name should conjure up the same meaning to all those who use it, and it should provide helpful information. This was the guiding philosophy of the authors when choosing the names given in the feed composition tables. Genus and species—Latin names—are also included. To facilitate worldwide usage, the International Feed Number of each feed is given. To the extent possible, consideration was also given to source (or parent material), variety or kind, stage of maturity, processing, part eaten, and grade. Where feeds are known by more than one name, cross-referencing is used.

■ **Moisture content of feeds**—It is necessary to know the moisture content of feeds in diet formulation and buying. Usually, the composition of feed is expressed according to one or more of three bases—as-fed, air-dry, and moisture free. In the feed composition and mineral composition tables of this book, Tables II-1 and II-2, respectively, values for as-fed (A-F) and moisture-free (M-F) are given. Definitions follow:

1. **As-fed: A-F (wet, fresh).** This refers to feed as normally fed to animals. It may range from 0 to 100% dry matter.
2. **Air-dry (approximately 90% dry matter).** This refers to a sample of feed that has been dried by means of natural air movement, usually in the open. It may be either an actual or an assumed dry matter content; the latter is approximately 90%. Most feeds are fed in an air-dry state.
3. **Moisture-free: M-F (oven-dry, 100% dry matter).** This refers to any substance that has been dried in an oven at 221°F until all the moisture has been removed.

In preparing Tables II-1 and II-2, all data were calculated on a 100% dry matter basis (moisture-free), then converted to an as-fed basis by multiplying the decimal equivalent of the DM content times the compositional value shown in the table.

■ **Vitamin A activity**—The vitamin A activity shown in the feed composition tables is based on the utilization of vitamin A and beta-carotene by the rat, with 1 mg of β-carotene equal to 1,667 IU of vitamin A, whereas it is estimated that 1 mg of β-carotene is equal to 400 IU of vitamin A for cattle.

---

[1]The feed composition and mineral composition tables (Table II-1 and Table II-2) presented in this section were secured from the International Feedstuffs Institute, College of Agriculture, Utah State University. Currently, the Feed Composition Data Bank is located at the following address: Feed Composition Data Bank, National Agricultural Library, 5th Floor, U.S. Department of Agriculture, Beltsville, MD 20705.

TABLE
COMPOSITION OF SOME

Entry Number	Feed Name Description	Inter-national Feed Number[2]	Moisture Basis: A-F (As-fed) or M-F (Moisture-free)	Chemical Analysis								
				Dry Matter	Ash	Crude Fiber	Cell Walls or NDF	Acid Detergent Fiber	Lignin	Ether Extract (Fat)	N-Free Extract	Crude Protein
				(%)	(%)	(%)	(%)	(%)	(%)	(%)	(%)	(%)
	ALFALFA (LUCERNE) *Medicago sativa*											
1	-PREBLOOM, FRESH	2-00-181	A-F	20	2.1	4.9	6.4	5.2	1.5	0.6	8.5	4.3
			M-F	100	10.2	24.2	31.7	25.5	7.5	2.9	41.6	21.1
2	-EARLY BLOOM, FRESH	2-00-184	A-F	24	3.1	6.6	9.0	7.2	2.0	0.7	8.2	5.4
			M-F	100	12.7	27.6	37.3	29.9	8.2	3.0	34.1	22.5
3	-MIDBLOOM, FRESH	2-00-185	A-F	24	2.2	7.2	9.1	8.5	2.5	0.6	8.9	4.8
			M-F	100	9.3	30.3	38.5	35.6	10.5	2.6	37.7	20.2
4	-FULL BLOOM, FRESH	2-00-188	A-F	24	2.4	7.2	9.6	9.0	3.2	0.6	9.0	4.6
			M-F	100	10.1	30.4	40.1	37.9	13.6	2.6	37.7	19.2
5	-HAY, EARLY BLOOM, SUN-CURED	1-00-059	A-F	91	8.4	25.8	36.8	29.0	5.8	2.6	35.8	17.9
			M-F	100	9.2	28.5	40.7	32.0	6.4	2.9	39.6	19.8
6	-HAY, MIDBLOOM, SUN-CURED	1-00-063	A-F	91	7.8	25.5	43.2	33.4	6.7	3.3	37.4	17.1
			M-F	100	8.6	28.0	47.4	36.7	7.4	3.6	41.1	18.8
7	-HAY, FULL BLOOM, SUN-CURED	1-00-068	A-F	91	7.1	27.3	45.0	35.2	6.9	3.1	37.9	15.5
			M-F	100	7.8	30.1	49.5	38.7	7.6	3.4	41.7	17.0
8	-HAY, MATURE, SUN-CURED	1-00-071	A-F	91	6.7	29.3	50.1	40.1	11.3	2.9	37.0	15.2
			M-F	100	7.4	32.1	55.0	44.0	12.4	3.2	40.6	16.7
9	-MEAL, DEHY, 17% PROTEIN	1-00-023	A-F	92	9.7	24.0	41.3	31.5	9.7	2.8	37.8	17.4
			M-F	100	10.6	26.2	45.0	34.3	10.6	3.0	41.2	18.9
10	-SILAGE, EARLY BLOOM, WILTED	3-00-216	A-F	35	2.8	11.0	—	13.7	—	1.1	13.9	5.8
			M-F	100	8.2	31.8	—	39.6	—	3.2	40.2	16.7
11	-SILAGE, MIDBLOOM, WILTED	3-00-217	A-F	38	3.0	11.9	—	—	—	1.2	15.6	6.4
			M-F	100	7.9	31.2	—	—	—	3.1	41.1	16.8
12	-SILAGE, FULL BLOOM, WILTED	3-00-218	A-F	36	2.8	12.1	—	—	—	1.0	15.1	5.4
			M-F	100	7.7	33.2	—	—	—	2.7	41.5	14.9
	ALMOND *Prunus amygdalus*											
13	-HULLS	4-00-359	A-F	90	6.1	13.5	—	29.8	—	2.9	63.8	4.1
			M-F	100	6.8	14.9	—	33.0	—	3.2	70.6	4.5
	APPLE *Malus spp*											
14	-POMACE, DEHY	4-00-423	A-F	89	3.0	16.2	—	—	—	4.3	61.2	4.4
			M-F	100	3.4	18.2	—	—	—	4.8	68.6	5.0
	BAHIAGRASS *Paspalum notatum*											
15	-FRESH	2-00-464	A-F	30	3.3	9.0	—	—	—	0.5	14.2	2.6
			M-F	100	11.1	30.4	—	—	—	1.6	48.0	8.9
16	-HAY, SUN-CURED	1-00-462	A-F	90	5.7	28.1	66.6	33.9	5.1	1.8	46.0	8.5
			M-F	100	6.3	31.2	73.9	37.7	5.7	2.0	51.1	9.5
	BAKERY											
17	-WASTE, DEHY (DRIED BAKERY PRODUCT)	4-00-466	A-F	91	3.7	1.2	—	1.6	—	10.9	65.3	10.1
			M-F	100	4.0	1.3	—	1.8	—	12.0	71.6	11.1
	BARLEY *Hordeum vulgare*											
18	-GRAIN, ALL ANALYSES	4-00-549	A-F	88	2.4	5.0	16.8	10.7	1.5	1.7	67.7	11.7
			M-F	100	2.7	5.7	19.0	12.1	1.7	1.9	76.5	13.2
19	-GRAIN, PACIFIC COAST	4-00-939	A-F	89	2.5	6.5	—	—	—	2.0	68.2	9.5
			M-F	100	2.8	7.3	—	—	—	2.2	76.9	10.8
20	-GRAIN SCREENINGS	4-00-542	A-F	89	3.1	7.5	—	9.8	—	2.3	64.5	11.5
			M-F	100	3.4	8.4	—	11.0	—	2.6	72.6	13.0
21	-HAY, SUN-CURED	1-00-495	A-F	88	6.6	23.6	—	—	—	1.9	48.5	7.8
			M-F	100	7.5	26.7	—	—	—	2.1	54.9	8.8
22	-SILAGE	3-00-512	A-F	31	3.2	9.4	—	—	—	1.2	14.3	3.2
			M-F	100	10.2	30.0	—	—	—	3.9	45.6	10.3
23	-STRAW	1-00-498	A-F	91	6.7	37.9	77.5	51.1	6.9	1.7	41.1	4.0
			M-F	100	7.3	41.5	84.8	55.9	7.6	1.9	44.9	4.4

II-1
BEEF CATTLE FEEDS[1]

Entry Number	TDN	Digestible Energy		Metabolizable Energy		Net Energy					
	Ruminant	Ruminant		Ruminant		Ruminant NE_m		Ruminant NE_g		Lactating Cows NE_{lc}	
	(%)	(Mcal)		(Mcal)		(Mcal)		(Mcal)		(Mcal)	
		(lb)	(kg)	(lb)	(kg)	(lb)	(kg)	(lb)	(kg)	(lb)	(kg)
1	12 61	0.24 1.20	0.54 2.65	0.20 0.99	0.44 2.17	0.12 0.60	0.27 1.31	0.07 0.34	0.15 0.74	0.13 0.63	0.28 1.38
2	15 63	0.30 1.25	0.66 2.76	0.25 1.06	0.56 2.34	0.15 0.63	0.34 1.40	0.09 0.37	0.20 0.82	0.15 0.64	0.34 1.42
3	14 61	0.29 1.20	0.63 2.65	0.24 1.01	0.53 2.23	0.14 0.60	0.31 1.32	0.08 0.34	0.18 0.74	0.14 0.59	0.31 1.29
4	13 55	0.28 1.17	0.62 2.58	0.23 0.98	0.51 2.16	0.15 0.63	0.33 1.39	0.09 0.37	0.19 0.81	0.15 0.64	0.34 1.41
5	52 58	1.15 1.27	2.54 2.80	0.96 1.06	2.12 2.35	0.60 0.67	1.33 1.47	0.36 0.40	0.80 0.88	0.55 0.61	1.21 1.33
6	52 57	1.12 1.23	2.46 2.70	0.94 1.03	2.07 2.28	0.52 0.57	1.14 1.25	0.28 0.31	0.62 0.68	0.51 0.56	1.13 1.24
7	51 56	1.09 1.19	2.39 2.63	0.88 0.97	1.95 2.14	0.49 0.54	1.09 1.20	0.26 0.29	0.57 0.63	0.52 0.58	1.15 1.27
8	55 61	1.22 1.33	2.68 2.94	0.93 1.03	2.06 2.26	0.47 0.52	1.04 1.14	0.24 0.26	0.53 0.58	0.51 0.56	1.12 1.23
9	55 60	1.12 1.22	2.47 2.69	0.96 1.05	2.12 2.32	0.60 0.65	1.32 1.44	0.36 0.39	0.79 0.86	0.57 0.62	1.25 1.37
10	18 52	0.39 1.13	0.86 2.49	0.33 0.94	0.72 2.07	0.17 0.48	0.36 1.05	0.08 0.22	0.17 0.50	0.18 0.53	0.40 1.16
11	23 61	0.47 1.22	1.03 2.70	0.39 1.03	0.87 2.28	0.23 0.61	0.51 1.34	0.13 0.35	0.29 0.76	0.24 0.62	0.52 1.37
12	21 59	0.43 1.17	0.94 2.59	0.36 0.98	0.79 2.16	0.20 0.56	0.45 1.23	0.11 0.30	0.24 0.66	0.21 0.59	0.47 1.29
13	66 73	1.34 1.48	2.95 3.27	1.15 1.28	2.54 2.82	0.77 0.85	1.70 1.88	0.51 0.56	1.12 1.24	0.66 0.73	1.46 1.61
14	60 68	1.21 1.35	2.66 2.98	1.04 1.16	2.28 2.56	0.66 0.74	1.46 1.63	0.42 0.47	0.91 1.03	0.64 0.72	1.42 1.59
15	16 54	0.32 1.08	0.70 2.37	0.26 0.88	0.57 1.95	0.15 0.51	0.33 1.13	0.08 0.26	0.17 0.57	0.16 0.55	0.36 1.22
16	46 51	0.96 1.07	2.12 2.36	0.79 0.88	1.74 1.93	0.41 0.45	0.90 1.00	0.18 0.20	0.40 0.45	0.46 0.51	1.01 1.13
17	82 89	1.62 1.78	3.57 3.91	1.43 1.57	3.16 3.46	1.00 1.09	2.20 2.41	0.70 0.77	1.54 1.69	0.85 0.94	1.88 2.06
18	75 85	1.55 1.75	3.42 3.86	1.17 1.32	2.57 2.91	0.78 0.89	1.73 1.95	0.52 0.59	1.16 1.31	0.82 0.93	1.81 2.05
19	75 85	1.51 1.70	3.32 3.75	1.34 1.51	2.96 3.34	0.81 0.91	1.77 2.00	0.54 0.61	1.20 1.35	0.76 0.86	1.67 1.89
20	71 80	1.43 1.60	3.14 3.53	1.26 1.42	2.77 3.12	0.78 0.88	1.73 1.94	0.52 0.59	1.15 1.30	0.74 0.84	1.64 1.84
21	50 57	0.99 1.11	2.17 2.46	0.79 0.89	1.73 1.96	0.45 0.51	0.99 1.11	0.22 0.25	0.49 0.56	0.50 0.57	1.10 1.25
22	16 51	0.32 1.01	0.70 2.22	0.25 0.80	0.55 1.76	0.13 0.42	0.29 0.93	0.05 0.17	0.12 0.38	0.16 0.51	0.35 1.13
23	43 47	0.84 0.92	1.86 2.03	0.64 0.70	1.40 1.54	0.29 0.32	0.64 0.70	0.07 0.08	0.15 0.16	0.42 0.45	0.91 1.00

(Continued)

TABLE II-1

Entry Number	Feed Name Description	Moisture Basis: A-F (As-fed) or M-F (Moisture-free)	Dry Matter	Macrominerals						
				Calcium (Ca)	Phos-phorus (P)	Sodium (Na)	Chlorine (Cl)	Mag-nesium (Mg)	Potassium (K)	Sulfur (S)
			(%)	(%)	(%)	(%)	(%)	(%)	(%)	(%)
	ALFALFA (LUCERNE) *Medicago sativa*									
1	-PREBLOOM, FRESH	**A-F**	**20**	**0.44**	**0.07**	**0.04**	**0.09**	**0.05**	**0.44**	**0.10**
		M-F	100	2.19	0.33	0.21	0.44	0.27	2.14	0.48
2	-EARLY BLOOM, FRESH	**A-F**	**24**	**0.39**	**0.07**	**0.04**	—	**0.12**	**0.88**	—
		M-F	100	1.61	0.28	0.18	—	0.49	3.64	—
3	-MIDBLOOM, FRESH	**A-F**	**24**	**0.40**	**0.07**	**0.03**	**0.11**	**0.10**	**0.79**	**0.07**
		M-F	100	1.69	0.31	0.14	0.45	0.41	3.31	0.29
4	-FULL BLOOM, FRESH	**A-F**	**24**	**0.28**	**0.06**	**0.04**	**0.10**	**0.10**	**0.86**	**0.07**
		M-F	100	1.19	0.26	0.16	0.43	0.40	3.63	0.31
5	-HAY, EARLY BLOOM, SUN-CURED	**A-F**	**91**	**1.48**	**0.20**	**0.14**	**0.34**	**0.31**	**2.32**	**0.27**
		M-F	100	1.63	0.22	0.15	0.38	0.34	2.56	0.30
6	-HAY, MIDBLOOM, SUN-CURED	**A-F**	**91**	**1.27**	**0.22**	**0.11**	**0.34**	**0.32**	**1.42**	**0.26**
		M-F	100	1.39	0.24	0.12	0.38	0.35	1.56	0.28
7	-HAY, FULL BLOOM, SUN-CURED	**A-F**	**91**	**1.08**	**0.22**	**0.06**	—	**0.25**	**1.42**	**0.27**
		M-F	100	1.19	0.24	0.07	—	0.27	1.56	0.30
8	-HAY, MATURE, SUN-CURED	**A-F**	**91**	**1.07**	**0.19**	**0.07**	—	**0.20**	**1.88**	**0.23**
		M-F	100	1.18	0.21	0.08	—	0.22	2.07	0.25
9	-MEAL, DEHY, 17% PROTEIN	**A-F**	**92**	**1.40**	**0.23**	**0.10**	**0.47**	**0.29**	**2.38**	**0.23**
		M-F	100	1.52	0.25	0.11	0.52	0.32	2.60	0.25
10	-SILAGE, EARLY BLOOM, WILTED	**A-F**	**35**	—	—	—	—	—	—	—
		M-F	100	—	—	—	—	—	—	—
11	-SILAGE, MIDBLOOM, WILTED	**A-F**	**38**	—	—	—	—	—	—	—
		M-F	100	—	—	—	—	—	—	—
12	-SILAGE, FULL BLOOM, WILTED	**A-F**	**36**	—	—	—	—	—	—	—
		M-F	100	—	—	—	—	—	—	—
	ALMOND *Prunus amygdalus*									
13	-HULLS	**A-F**	**90**	**0.19**	**0.09**	—	—	—	**0.48**	**0.10**
		M-F	100	0.21	0.10	—	—	—	0.53	0.11
	APPLE *Malus* spp									
14	-POMACE, DEHY	**A-F**	**89**	**0.11**	**0.10**	**0.12**	—	**0.06**	**0.43**	**0.02**
		M-F	100	0.13	0.12	0.14	—	0.07	0.49	0.02
	BAHIAGRASS *Paspalum notatum*									
15	-FRESH	**A-F**	**30**	**0.14**	**0.06**	—	—	**0.07**	**0.43**	—
		M-F	100	0.46	0.22	—	—	0.25	1.45	—
16	-HAY, SUN-CURED	**A-F**	**90**	**0.45**	**0.20**	—	—	**0.17**	—	—
		M-F	100	0.50	0.22	—	—	0.19	—	—
	BAKERY									
17	-WASTE, DEHY (DRIED BAKERY PRODUCT)	**A-F**	**91**	**0.14**	**0.22**	**1.02**	**1.47**	**0.16**	**0.40**	**0.02**
		M-F	100	0.15	0.24	1.12	1.61	0.18	0.43	0.02
	BARLEY *Hordeum vulgare*									
18	-GRAIN, ALL ANALYSES	**A-F**	**88**	**0.05**	**0.34**	**0.03**	**0.12**	**0.13**	**0.46**	**0.15**
		M-F	100	0.06	0.39	0.03	0.13	0.15	0.52	0.17
19	-GRAIN, PACIFIC COAST	**A-F**	**89**	**0.05**	**0.34**	**0.02**	**0.15**	**0.12**	**0.51**	**0.14**
		M-F	100	0.06	0.39	0.02	0.17	0.14	0.58	0.16
20	-GRAIN SCREENINGS	**A-F**	**89**	**0.32**	**0.29**	**0.02**	—	**0.12**	**0.80**	**0.13**
		M-F	100	0.36	0.33	0.02	—	0.14	0.90	0.15
21	-HAY, SUN-CURED	**A-F**	**88**	**0.21**	**0.25**	**0.12**	—	**0.14**	**1.30**	**0.15**
		M-F	100	0.24	0.28	0.14	—	0.16	1.47	0.17
22	-SILAGE	**A-F**	**31**	**0.11**	**0.09**	**0.00**	—	**0.04**	**0.63**	**0.04**
		M-F	100	0.34	0.28	0.01	—	0.13	2.01	0.11
23	-STRAW	**A-F**	**91**	**0.27**	**0.07**	**0.13**	**0.61**	**0.21**	**2.16**	**0.16**
		M-F	100	0.30	0.07	0.14	0.67	0.23	2.37	0.17

(Continued)

Entry Number	Microminerals							Fat-soluble Vitamins			
	Cobalt (Co)	Copper (Cu)	Iodine (I)	Iron (Fe)	Manganese (Mn)	Selenium (Se)	Zinc (Zn)	A (1 mg Carotene = 1,667 IU Vit. A)	Carotene (Provitamin A)	D	E (α-tocopherol)
	(ppm or mg/kg)	*(ppm or mg/kg)*	*(ppm or mg/kg)*	*(%)*	*(ppm or mg/kg)*	*(ppm or mg/kg)*	*(ppm or mg/kg)*	*(IU/g)*	*(ppm or mg/kg)*	*(IU/kg)*	*(ppm or mg/kg)*
1	**0.034**	**2.2**	**—**	**0.003**	**8.3**	**—**	**—**	**—**	**—**	**0**	**34.8**
	0.167	10.8	—	0.012	40.8	—	—	—	—	0	171.5
2	**0.107**	**4.4**	**—**	**0.008**	**33.2**	**—**	**9.6**	**69.9**	**41.9**	**—**	**—**
	0.443	18.3	—	0.032	138.3	—	40.0	291.0	174.6	—	—
3	**0.091**	**4.8**	**—**	**0.011**	**37.1**	**—**	**9.1**	**56.1**	**33.7**	**—**	**—**
	0.381	20.2	—	0.044	156.4	—	38.5	236.4	141.8	—	—
4	**0.117**	**3.6**	**—**	**0.007**	**26.4**	**—**	**8.5**	**—**	**—**	**—**	**—**
	0.489	14.9	—	0.030	110.9	—	35.9	—	—	—	—
5	**0.264**	**11.4**	**—**	**0.021**	**32.8**	**0.497**	**27.3**	**210.9**	**126.5**	**2**	**23.5**
	0.292	12.6	—	0.023	36.2	0.549	30.2	233.0	139.8	2	26.0
6	**0.359**	**16.1**	**—**	**0.021**	**55.1**	**—**	**28.1**	**50.5**	**30.3**	**1**	**—**
	0.394	17.7	—	0.023	60.5	—	30.9	55.5	33.3	2	—
7	**0.210**	**9.0**	**—**	**0.015**	**38.5**	**—**	**23.7**	**98.5**	**59.1**	**—**	**—**
	0.230	9.9	—	0.016	42.3	—	26.1	108.4	65.0	—	—
8	**0.370**	**12.5**	**—**	**0.015**	**35.1**	**—**	**20.1**	**17.6**	**10.6**	**1**	**—**
	0.406	13.7	—	0.017	38.5	—	22.1	19.3	11.6	1	—
9	**0.302**	**8.6**	**0.148**	**0.041**	**31.0**	**0.335**	**19.3**	**200.3**	**120.2**	**—**	**105.7**
	0.329	9.3	0.162	0.045	33.8	0.365	21.1	218.5	131.1	—	115.3
10	**—**	**—**	**—**	**—**	**—**	**—**	**—**	**—**	**—**	**—**	**—**
	—	—	—	—	—	—	—	—	—	—	—
11	**—**	**—**	**—**	**—**	**—**	**—**	**—**	**—**	**—**	**—**	**—**
	—	—	—	—	—	—	—	—	—	—	—
12	**—**	**—**	**—**	**—**	**—**	**—**	**—**	**—**	**—**	**—**	**—**
	—	—	—	—	—	—	—	—	—	—	—
13	**—**	**—**	**—**	**—**	**—**	**—**	**—**	**—**	**—**	**—**	**—**
	—	—	—	—	—	—	—	—	—	—	—
14	**—**	**—**	**—**	**0.027**	**7.2**	**—**	**—**	**—**	**—**	**—**	**—**
	—	—	—	0.030	8.1	—	—	—	—	—	—
15	**—**	**—**	**—**	**—**	**—**	**—**	**—**	**89.7**	**53.8**	**—**	**—**
	—	—	—	—	—	—	—	304.2	188.5	—	—
16	**—**	**—**	**—**	**0.006**	**—**	**—**	**—**	**—**	**—**	**—**	**—**
	—	—	—	0.006	—	—	—	—	—	—	—
17	**1.224**	**11.0**	**—**	**0.017**	**65.0**	**—**	**17.8**	**7.0**	**4.2**	**—**	**40.9**
	1.342	12.1	—	0.018	71.2	—	19.5	7.7	4.6	—	44.9
18	**0.171**	**7.6**	**0.044**	**0.008**	**16.0**	**0.158**	**39.3**	**3.4**	**2.0**	**—**	**23.2**
	0.193	8.6	0.050	0.009	18.1	0.179	44.4	3.8	2.3	—	26.2
19	**0.087**	**8.1**	**—**	**0.009**	**16.0**	**0.101**	**15.2**	**—**	**—**	**—**	**26.2**
	0.098	9.1	—	0.010	18.0	0.114	17.1	—	—	—	29.6
20	**—**	**—**	**—**	**0.006**	**—**	**—**	**—**	**—**	**—**	**—**	**—**
	—	—	—	0.006	—	—	—	—	—	—	—
21	**0.059**	**3.9**	**—**	**0.027**	**34.8**	**—**	**—**	**77.4**	**46.4**	**1**	**—**
	0.067	4.4	—	0.030	39.4	—	—	87.5	52.5	1	—
22	**0.212**	**1.5**	**—**	**0.009**	**24.8**	**0.029**	**7.0**	**13.4**	**8.0**	**—**	**—**
	0.674	4.9	—	0.028	78.9	0.092	22.4	42.6	25.6	—	—
23	**0.061**	**4.9**	**—**	**0.019**	**15.1**	**—**	**6.8**	**3.5**	**2.1**	**1**	**—**
	0.067	5.4	—	0.021	16.6	—	7.4	3.9	2.3	1	—

(Continued)

TABLE II-1

Entry Number	Feed Name Description	International Feed Number[2]	Moisture Basis: A-F (As-fed) or M-F (Moisture-free)	Chemical Analysis								
				Dry Matter	Ash	Crude Fiber	Cell Walls or NDF	Acid Detergent Fiber	Lignin	Ether Extract (Fat)	N-Free Extract	Crude Protein
				(%)	(%)	(%)	(%)	(%)	(%)	(%)	(%)	(%)
	BEAN *Phaseolus vulgaris*											
24	-SEEDS, NAVY	**5-00-623**	**A-F**	**89**	**4.0**	**4.4**	—	—	—	**1.4**	**56.7**	**22.9**
			M-F	100	4.5	5.0	—	—	—	1.5	63.4	25.6
	BEET, MANGEL *Beta vulgaris macrorrhiza*											
25	-ROOTS, FRESH	**4-00-637**	**A-F**	**11**	**1.1**	**0.8**	—	—	—	**0.1**	**7.7**	**1.3**
			M-F	100	9.6	7.4	—	—	—	0.7	70.5	11.8
	BEET, SUGAR *Beta vulgaris altissima*											
26	-TOPS WITH CROWNS, SILAGE	**3-00-660**	**A-F**	**25**	**8.6**	**3.2**	—	—	—	**0.7**	**9.3**	**3.4**
			M-F	100	34.0	12.8	—	—	—	2.9	37.0	13.4
	-MOLASSES (*SEE* MOLASSES AND SYRUP)											
27	-PULP, DEHY	**4-00-669**	**A-F**	**91**	**4.8**	**18.2**	**53.6**	**26.3**	**4.5**	**0.5**	**58.4**	**8.8**
			M-F	100	5.3	20.1	59.0	29.0	5.0	0.6	64.3	9.7
28	-PULP, WET	**4-00-671**	**A-F**	**11**	**0.5**	**2.3**	—	—	—	**0.2**	**6.7**	**1.2**
			M-F	100	4.7	21.3	—	—	—	2.1	60.8	11.2
29	-PULP WITH MOLASSES, DEHY	**4-00-672**	**A-F**	**92**	**5.7**	**15.2**	—	**24.5**	**2.4**	**0.6**	**61.1**	**9.3**
			M-F	100	6.2	16.6	—	26.6	2.6	0.6	66.5	10.1
	BERMUDAGRASS *Cynodon dactylon*											
30	-FRESH	**2-00-712**	**A-F**	**29**	**3.3**	**7.6**	—	—	—	**0.6**	**13.0**	**4.2**
			M-F	100	11.4	26.6	—	—	—	2.1	45.4	14.6
31	-HAY, SUN-CURED	**1-00-703**	**A-F**	**91**	**8.0**	**28.4**	—	—	—	**1.8**	**43.7**	**9.2**
			M-F	100	8.8	31.2	—	—	—	2.0	48.0	10.0
	BERMUDAGRASS, COASTAL *Cynodon dactylon*											
32	-FRESH	**2-00-719**	**A-F**	**29**	**1.8**	**8.3**	—	—	—	**1.1**	**13.6**	**4.4**
			M-F	100	6.3	28.4	—	—	—	3.8	46.6	15.0
33	-HAY, SUN-CURED	**1-00-716**	**A-F**	**91**	**6.3**	**27.0**	—	—	—	**2.0**	**43.9**	**11.7**
			M-F	100	6.9	29.7	—	—	—	2.2	48.3	12.8
	BLUEGRASS, CANADA *Poa compressa*											
34	-IMMATURE, FRESH	**2-00-763**	**A-F**	**26**	**2.4**	**6.6**	—	—	—	**1.0**	**11.2**	**4.9**
			M-F	100	9.1	25.5	—	—	—	3.7	43.0	18.7
35	-HAY, SUN-CURED	**1-00-762**	**A-F**	**92**	**6.5**	**27.6**	—	—	—	**2.4**	**45.8**	**9.5**
			M-F	100	7.0	30.1	—	—	—	2.6	49.9	10.3
	BLUEGRASS, KENTUCKY *Poa pratensis*											
36	-IMMATURE, FRESH	**2-00-777**	**A-F**	**31**	**2.9**	**7.8**	—	—	—	**1.1**	**13.7**	**5.4**
			M-F	100	9.4	25.2	—	—	—	3.6	44.4	17.4
37	-MILK STAGE, FRESH	**2-00-782**	**A-F**	**42**	**3.1**	**12.7**	—	—	—	**1.5**	**19.8**	**4.9**
			M-F	100	7.3	30.3	—	—	—	3.6	47.2	11.6
38	-HAY, SUN-CURED	**1-00-776**	**A-F**	**89**	**5.9**	**26.8**	—	—	—	**3.0**	**44.3**	**9.1**
			M-F	100	6.6	30.0	—	—	—	3.4	49.7	10.2
39	-HAY, MATURE, SUN-CURED	**1-00-774**	**A-F**	**89**	**6.5**	**29.7**	—	—	—	**2.6**	**44.9**	**5.6**
			M-F	100	7.3	33.3	—	—	—	2.9	50.3	6.3
	BLUESTEM *Andropogon* spp											
40	-IMMATURE, FRESH	**2-00-821**	**A-F**	**27**	**2.4**	**6.7**	—	—	—	**0.7**	**13.6**	**3.4**
			M-F	100	8.9	24.9	—	—	—	2.8	50.6	12.8
41	-MATURE, FRESH	**2-00-825**	**A-F**	**59**	**3.3**	**20.2**	—	—	—	**1.4**	**30.6**	**3.4**
			M-F	100	5.6	34.2	—	—	—	2.4	51.9	5.8
	BREWERS' GRAINS											
42	-DEHY	**5-02-141**	**A-F**	**92**	**3.6**	**13.0**	**38.7**	**23.9**	**4.6**	**6.6**	**41.6**	**27.3**
			M-F	100	4.0	14.1	42.0	26.0	5.0	7.1	45.2	29.6
43	-WET	**5-02-142**	**A-F**	**22**	**0.9**	**3.1**	**9.2**	**5.0**	**1.1**	**1.5**	**10.7**	**5.8**
			M-F	100	3.9	14.1	42.0	23.0	5.0	6.8	48.7	26.4

(Continued)

Entry Number	TDN	Digestible Energy		Metabolizable Energy		Net Energy					
	Ruminant	Ruminant		Ruminant		Ruminant NE_m		Ruminant NE_g		Lactating Cows NE_{lc}	
	(%)	(Mcal)		(Mcal)		(Mcal)		(Mcal)		(Mcal)	
		(lb)	(kg)	(lb)	(kg)	(lb)	(kg)	(lb)	(kg)	(lb)	(kg)
24	76	1.53	3.37	1.36	3.00	0.83	1.83	0.56	1.24	0.78	1.72
	85	1.71	3.77	1.52	3.36	0.93	2.05	0.63	1.39	0.87	1.93
25	9	0.18	0.39	0.15	0.34	0.09	0.21	0.06	0.14	0.09	0.20
	80	1.59	3.51	1.40	3.09	0.85	1.88	0.56	1.24	0.81	1.79
26	13	0.26	0.58	0.21	0.47	0.13	0.28	0.06	0.14	0.14	0.30
	52	1.04	2.29	0.84	1.86	0.50	1.11	0.25	0.55	0.55	1.20
27	67	1.31	2.89	1.10	2.43	0.73	1.60	0.47	1.03	0.70	1.55
	74	1.44	3.18	1.21	2.68	0.80	1.76	0.52	1.14	0.77	1.70
28	8	0.16	0.35	0.14	0.30	0.08	0.17	0.05	0.11	0.08	0.17
	72	1.44	3.18	1.25	2.76	0.71	1.57	0.44	0.97	0.70	1.55
29	69	1.39	3.05	1.21	2.67	0.73	1.61	0.47	1.04	0.71	1.55
	75	1.51	3.33	1.32	2.91	0.80	1.76	0.52	1.14	0.77	1.69
30	17	0.35	0.77	0.29	0.65	0.18	0.39	0.10	0.22	0.18	0.40
	61	1.22	2.68	1.02	2.26	0.62	1.36	0.36	0.78	0.63	1.39
31	43	0.88	1.93	0.64	1.41	0.29	0.65	0.07	0.16	0.40	0.88
	47	0.96	2.12	0.70	1.55	0.32	0.71	0.08	0.18	0.44	0.96
32	19	0.37	0.82	0.32	0.70	0.18	0.40	0.11	0.23	0.19	0.41
	64	1.28	2.83	1.09	2.41	0.63	1.39	0.37	0.81	0.64	1.41
33	49	1.04	2.30	0.66	1.44	0.31	0.68	0.09	0.20	0.53	1.16
	54	1.15	2.53	0.72	1.59	0.34	0.75	0.10	0.22	0.58	1.28
34	18	0.36	0.79	0.31	0.68	0.20	0.43	0.12	0.27	0.19	0.42
	71	1.37	3.03	1.18	2.61	0.76	1.67	0.48	1.06	0.74	1.62
35	57	1.14	2.52	0.97	2.13	0.61	1.35	0.37	0.81	0.61	1.35
	62	1.25	2.75	1.06	2.33	0.67	1.47	0.40	0.88	0.67	1.47
36	22	0.42	0.93	0.37	0.81	0.24	0.52	0.15	0.33	0.23	0.51
	72	1.38	3.03	1.19	2.61	0.77	1.70	0.49	1.08	0.75	1.64
37	26	0.52	1.15	0.44	0.97	0.27	0.59	0.15	0.34	0.27	0.59
	62	1.24	2.73	1.05	2.31	0.63	1.39	0.37	0.81	0.64	1.41
38	54	1.06	2.34	0.89	1.96	0.51	1.12	0.28	0.62	0.53	1.17
	61	1.19	2.63	1.00	2.20	0.57	1.26	0.31	0.69	0.60	1.31
39	41	0.89	1.97	0.72	1.59	0.45	1.00	0.23	0.50	0.49	1.08
	46	1.00	2.21	0.81	1.78	0.51	1.12	0.25	0.56	0.55	1.21
40	18	0.35	0.78	0.30	0.67	0.18	0.39	0.10	0.23	0.18	0.39
	68	1.32	2.90	1.13	2.48	0.65	1.44	0.39	0.86	0.66	1.45
41	31	0.67	1.47	0.55	1.22	0.35	0.76	0.19	0.43	0.36	0.79
	53	1.13	2.48	0.93	2.06	0.59	1.29	0.33	0.72	0.61	1.34
42	65	1.25	2.76	1.01	2.22	0.64	1.41	0.39	0.86	0.67	1.48
	71	1.36	2.99	1.09	2.41	0.69	1.53	0.42	0.93	0.73	1.61
43	15	0.32	0.70	0.27	0.60	0.18	0.40	0.12	0.26	0.17	0.38
	68	1.44	3.17	1.25	2.76	0.83	1.82	0.54	1.19	0.79	1.74

(Continued)

TABLE II-1

Entry Number	Feed Name Description	Moisture Basis: A-F (As-fed) or M-F (Moisture-free)	Dry Matter	Macrominerals						
				Calcium (Ca)	Phos-phorus (P)	Sodium (Na)	Chlorine (Cl)	Mag-nesium (Mg)	Potassium (K)	Sulfur (S)
			(%)	*(%)*	*(%)*	*(%)*	*(%)*	*(%)*	*(%)*	*(%)*
	BEAN *Phaseolus vulgaris*									
24	-SEEDS, NAVY	**A-F**	**89**	**0.17**	**0.54**	**0.04**	**0.06**	**0.13**	**1.31**	**0.23**
		M-F	100	0.19	0.61	0.05	0.06	0.15	1.47	0.26
	BEET, MANGEL *Beta vulgaris mcrorrhiza*									
25	-ROOTS, FRESH	**A-F**	**11**	**0.02**	**0.02**	**0.07**	**0.16**	**0.02**	**0.25**	**0.02**
		M-F	100	0.18	0.22	0.63	1.41	0.20	2.30	0.20
	BEET, SUGAR *Beta vulgaris altissima*									
26	-TOPS WITH CROWNS, SILAGE	**A-F**	**25**	**0.39**	**0.07**	**0.14**	**—**	**0.27**	**1.45**	**0.14**
		M-F	100	1.56	0.28	0.54	—	1.07	5.74	0.57
	-MOLASSES (*SEE* MOLASSES AND SYRUP)									
27	-PULP, DEHY	**A-F**	**91**	**0.63**	**0.09**	**0.19**	**0.04**	**0.26**	**0.18**	**0.20**
		M-F	100	0.70	0.10	0.21	0.04	0.28	0.20	0.22
28	-PULP, WET	**A-F**	**11**	**0.10**	**0.01**	**0.02**	**—**	**0.02**	**0.02**	**0.02**
		M-F	100	0.87	0.10	0.19	—	0.22	0.19	0.22
29	-PULP WITH MOLASSES, DEHY	**A-F**	**92**	**0.56**	**0.09**	**0.48**	**—**	**0.15**	**1.63**	**0.39**
		M-F	100	0.61	0.10	0.53	—	0.16	1.78	0.42
	BERMUDAGRASS *Cynodon dactylon*									
30	-FRESH	**A-F**	**29**	**0.16**	**0.06**	**0.13**	**—**	**0.07**	**0.55**	**—**
		M-F	100	0.55	0.21	0.44	—	0.24	1.92	—
31	-HAY, SUN-CURED	**A-F**	**91**	**0.43**	**0.16**	**0.07**	**—**	**0.16**	**1.40**	**0.19**
		M-F	100	0.47	0.17	0.08	—	0.17	1.53	0.21
	BERMUDAGRASS, COASTAL *Cynodon dactylon*									
32	-FRESH	**A-F**	**29**	**0.14**	**0.08**	**—**	**—**	**—**	**—**	**—**
		M-F	100	0.49	0.27	—	—	—	—	—
33	-HAY, SUN-CURED	**A-F**	**91**	**0.38**	**0.17**	**—**	**—**	**0.16**	**1.46**	**0.19**
		M-F	100	0.42	0.18	—	—	0.17	1.61	0.21
	BLUEGRASS, CANADA *Poa compressa*									
34	-IMMATURE, FRESH	**A-F**	**26**	**0.10**	**0.10**	**0.04**	**—**	**0.04**	**0.53**	**0.04**
		M-F	100	0.39	0.39	0.14	—	0.16	2.04	0.17
35	-HAY, SUN-CURED	**A-F**	**92**	**0.28**	**0.24**	**0.10**	**—**	**0.30**	**1.73**	**0.12**
		M-F	100	0.30	0.26	0.11	—	0.33	1.88	0.13
	BLUEGRASS, KENTUCKY *Poa pratensis*									
36	-IMMATURE, FRESH	**A-F**	**31**	**0.15**	**0.14**	**0.04**	**—**	**0.05**	**0.70**	**0.05**
		M-F	100	0.50	0.44	0.14	—	0.18	2.27	0.17
37	-MILK STAGE, FRESH	**A-F**	**42**	**—**	**—**	**—**	**—**	**—**	**—**	**—**
		M-F	100	—	—	—	—	—	—	—
38	-HAY, SUN-CURED	**A-F**	**89**	**0.40**	**0.27**	**0.10**	**0.55**	**0.19**	**1.66**	**0.12**
		M-F	100	0.45	0.30	0.11	0.62	0.21	1.87	0.13
39	-HAY, MATURE, SUN-CURED	**A-F**	**89**	**0.24**	**0.20**	**0.13**	**0.39**	**0.10**	**1.51**	**0.16**
		M-F	100	0.26	0.23	0.15	0.44	0.12	1.70	0.18
	BLUESTEM *Andropogon* spp									
40	-IMMATURE, FRESH	**A-F**	**27**	**0.17**	**0.05**	**—**	**—**	**—**	**0.46**	**—**
		M-F	100	0.63	0.20	—	—	—	1.72	—
41	-MATURE, FRESH	**A-F**	**59**	**0.23**	**0.07**	**—**	**—**	**0.04**	**0.30**	**—**
		M-F	100	0.40	0.12	—	—	0.06	0.51	—
	BREWERS' GRAINS									
42	-DEHY	**A-F**	**92**	**0.30**	**0.51**	**0.21**	**0.15**	**0.15**	**0.09**	**0.30**
		M-F	100	0.33	0.55	0.23	0.17	0.17	0.09	0.32
43	-WET	**A-F**	**22**	**0.06**	**0.12**	**0.06**	**0.03**	**0.03**	**0.02**	**0.07**
		M-F	100	0.29	0.54	0.28	0.13	0.15	0.09	0.34

(Continued)

Entry Number	Microminerals							Fat-soluble Vitamins			
	Cobalt (Co)	Copper (Cu)	Iodine (I)	Iron (Fe)	Manganese (Mn)	Selenium (Se)	Zinc (Zn)	A (1 mg Carotene = 1,667 IU Vit. A)	Carotene (Provitamin A)	D	E (α-tocopherol)
	(ppm or mg/kg)	*(ppm or mg/kg)*	*(ppm or mg/kg)*	*(%)*	*(ppm or mg/kg)*	*(ppm or mg/kg)*	*(ppm or mg/kg)*	*(IU/g)*	*(ppm or mg/kg)*	*(IU/kg)*	*(ppm or mg/kg)*
24	**—**	**9.9**	**—**	**0.010**	**21.1**	**—**	**—**	**—**	**—**	**—**	**1.0**
	—	11.0	—	0.012	23.6	—	—	—	—	—	1.1
25	**—**	**0.6**	**—**	**0.002**	**—**	**—**	**—**	**0.2**	**0.1**	**—**	**—**
	—	5.5	—	0.016	—	—	—	1.5	0.9	—	—
26	**—**	**—**	**—**	**0.006**	**—**	**—**	**—**	**—**	**—**	**—**	**—**
	—	—	—	0.020	—	—	—	—	—	—	—
27	**0.074**	**12.5**	**—**	**0.027**	**34.2**	**—**	**0.7**	**0.4**	**0.2**	**1**	**—**
	0.081	13.7	—	0.030	37.7	—	0.8	0.4	0.2	1	—
28	**—**	**—**	**—**	**0.004**	**—**	**—**	**1.1**	**—**	**—**	**—**	**—**
	—	—	—	0.033	—	—	10.0	—	—	—	—
29	**0.209**	**14.7**	**—**	**0.017**	**18.4**	**—**	**5.1**	**0.4**	**0.2**	**—**	**—**
	0.227	16.0	—	0.018	20.1	—	5.5	0.4	0.2	—	—
30	**0.022**	**1.6**	**—**	**0.033**	**28.6**	**—**	**—**	**147.8**	**88.7**	**—**	**—**
	0.075	5.7	—	0.112	100.1	—	—	517.3	310.3	—	—
31	**0.111**	**24.3**	**0.105**	**0.027**	**99.4**	**—**	**53.0**	**87.5**	**52.5**	**—**	**—**
	0.122	26.6	0.115	0.029	109.0	—	58.1	96.0	57.6	—	—
32	**—**	**—**	**—**	**—**	**—**	**—**	**—**	**160.3**	**96.1**	**—**	**—**
	—	—	—	—	—	—	—	550.9	330.5	—	—
33	**—**	**—**	**—**	**0.028**	**—**	**—**	**18.2**	**123.7**	**74.2**	**—**	**—**
	—	—	—	0.030	—	—	20.0	136.2	81.7	—	—
34	**—**	**—**	**—**	**0.008**	**—**	**—**	**—**	**172.6**	**103.5**	**—**	**—**
	—	—	—	0.030	—	—	—	665.1	399.0	—	—
35	**—**	**—**	**—**	**0.028**	**84.9**	**—**	**—**	**378.4**	**227.0**	**—**	**—**
	—	—	—	0.030	92.6	—	—	412.6	247.5	—	—
36	**—**	**—**	**—**	**0.010**	**—**	**—**	**—**	**247.6**	**148.5**	**—**	**47.8**
	—	—	—	0.030	—	—	—	803.4	481.9	—	155.0
37	**—**	**—**	**—**	**—**	**—**	**—**	**—**	**—**	**—**	**—**	**—**
	—	—	—	—	—	—	—	—	—	—	—
38	**—**	**8.8**	**—**	**0.025**	**76.2**	**—**	**—**	**—**	**—**	**—**	**—**
	—	9.9	—	0.028	85.6	—	—	—	—	—	—
39	**—**	**—**	**—**	**0.016**	**57.0**	**—**	**—**	**—**	**—**	**—**	**—**
	—	—	—	0.018	63.9	—	—	—	—	—	—
40	**—**	**12.6**	**—**	**0.024**	**28.5**	**—**	**—**	**97.9**	**58.7**	**—**	**—**
	—	47.0	—	0.090	106.3	—	—	365.3	219.1	—	—
41	**—**	**15.6**	**—**	**0.064**	**35.9**	**—**	**—**	**—**	**—**	**—**	**—**
	—	26.5	—	0.108	60.9	—	—	—	—	—	—
42	**0.076**	**21.7**	**0.066**	**0.024**	**37.2**	**—**	**27.3**	**0.8**	**0.5**	**—**	**26.7**
	0.083	23.6	0.072	0.026	40.4	—	29.6	0.8	0.5	—	29.0
43	**0.022**	**4.9**	**—**	**0.006**	**9.0**	**—**	**23.2**	**—**	**—**	**—**	**5.5**
	0.101	22.2	—	0.027	40.9	—	106.0	—	—	—	25.0

(Continued)

TABLE II-1

Entry Number	Feed Name Description	International Feed Number[2]	Moisture Basis: A-F (As-fed) or M-F (Moisture-free)	Dry Matter	Ash	Crude Fiber	Cell Walls or NDF	Acid Detergent Fiber	Lignin	Ether Extract (Fat)	N-Free Extract	Crude Protein
				(%)	(%)	(%)	(%)	(%)	(%)	(%)	(%)	(%)
	BROMEGRASS *Bromus* spp											
44	-IMMATURE, FRESH	**2-00-892**	**A-F**	**34**	**3.8**	**7.5**	**—**	**—**	**—**	**1.2**	**15.5**	**5.8**
			M-F	100	11.4	22.1	—	—	—	3.7	45.8	17.1
45	-HAY, PREBLOOM, SUN-CURED	**1-00-887**	**A-F**	**88**	**8.3**	**29.2**	**59.7**	**35.1**	**4.1**	**2.3**	**38.9**	**9.2**
			M-F	100	9.4	33.3	68.0	40.0	4.7	2.6	44.3	10.5
46	-HAY, MATURE, SUN-CURED	**1-00-889**	**A-F**	**92**	**6.0**	**33.0**	**—**	**—**	**—**	**1.7**	**45.2**	**6.1**
			M-F	100	6.6	35.9	—	—	—	1.8	49.1	6.6
	BROMEGRASS, SMOOTH *Bromus inermis*											
47	-FRESH	**2-00-963**	**A-F**	**27**	**—**	**7.7**	**—**	**—**	**—**	**0.8**	**13.1**	**3.1**
			M-F	100	—	28.4	—	—	—	3.0	48.6	11.4
48	-MATURE, FRESH	**2-08-364**	**A-F**	**55**	**3.8**	**19.1**	**—**	**—**	**—**	**1.3**	**27.4**	**3.3**
			M-F	100	6.9	34.8	—	—	—	2.4	49.9	6.0
49	-HAY, SUN-CURED	**1-00-947**	**A-F**	**90**	**7.1**	**29.0**	**70.5**	**33.4**	**5.1**	**2.6**	**38.4**	**12.4**
			M-F	100	8.0	32.4	78.7	37.3	5.7	2.9	42.9	13.9
	BUCKWHEAT, COMMON *Fagopyrum sagittatum*											
50	-GRAIN	**4-00-994**	**A-F**	**88**	**2.1**	**10.6**	**—**	**14.9**	**—**	**2.4**	**61.5**	**11.1**
			M-F	100	2.4	12.1	—	17.0	—	2.8	70.2	12.6
	BUFFALOGRASS *Buchloe dactyloides*											
51	-FRESH	**2-01-010**	**A-F**	**46**	**5.6**	**12.7**	**33.9**	**16.7**	**2.9**	**0.9**	**22.0**	**4.7**
			M-F	100	12.3	27.7	74.0	36.5	6.3	1.9	48.0	10.2
	CANARYGRASS, REED *Phalaris arundinacea*											
52	-FRESH	**2-01-113**	**A-F**	**23**	**2.3**	**5.6**	**10.6**	**6.5**	**1.0**	**0.9**	**10.1**	**3.9**
			M-F	100	10.2	24.4	46.4	28.3	4.3	4.1	44.4	17.0
53	-HAY, SUN-CURED	**1-01-104**	**A-F**	**89**	**7.3**	**30.2**	**62.9**	**32.7**	**—**	**2.7**	**40.0**	**9.1**
			M-F	100	8.1	33.9	70.5	36.6	—	3.0	44.8	10.2
	CARROT *Daucus* spp											
54	-ROOTS, FRESH	**4-01-145**	**A-F**	**11**	**1.0**	**1.1**	**—**	**—**	**—**	**0.2**	**8.1**	**1.2**
			M-F	100	8.4	9.5	—	—	—	1.3	70.7	10.0
	CASSAVA Manihot spp											
55	-TUBERS, FRESH	**4-01-150**	**A-F**	**32**	**1.3**	**1.5**	**—**	**—**	**—**	**0.3**	**28.2**	**1.2**
			M-F	100	3.9	4.6	—	—	—	1.0	86.9	3.6
56	-TUBERS, DEHY, MEAL	**4-01-152**	**A-F**	**91**	**2.1**	**4.1**	**—**	**—**	**—**	**0.6**	**81.5**	**2.2**
			M-F	100	2.3	4.5	—	—	—	0.7	90.1	2.4
	CATTLE *Bos taurus*											
57	-MANURE, WITHOUT BEDDING, DEHY	**1-01-190**	**A-F**	**93**	**17.9**	**30.7**	**—**	**43.8**	**—**	**2.5**	**30.1**	**12.0**
			M-F	100	19.1	33.0	—	47.0	—	2.7	32.3	12.9
	CEREAL											
58	-GRAIN SCREENINGS	**4-02-156**	**A-F**	**90**	**5.4**	**12.0**	**—**	**—**	**—**	**3.7**	**56.7**	**12.1**
			M-F	100	6.0	13.4	—	—	—	4.1	63.2	13.4
59	-GRAIN SCREENINGS, REFUSE	**4-02-151**	**A-F**	**91**	**9.0**	**18.6**	**—**	**36.3**	**—**	**4.4**	**46.2**	**12.6**
			M-F	100	9.9	20.5	—	40.0	—	4.9	50.9	13.8
60	-GRAIN SCREENINGS, UNCLEANED	**4-02-153**	**A-F**	**92**	**10.2**	**17.1**	**—**	**—**	**—**	**5.8**	**45.2**	**13.7**
			M-F	100	11.1	18.6	—	—	—	6.3	49.1	14.9
	CITRUS *Citrus* spp											
61	-PULP, SILAGE	**4-01-234**	**A-F**	**21**	**1.2**	**3.3**	**—**	**4.2**	**—**	**2.1**	**12.8**	**1.5**
			M-F	100	5.6	15.7	—	20.0	—	9.9	61.5	7.3
62	-PULP WITHOUT FINES, DEHY (DRIED CITRUS PULP)	**4-01-237**	**A-F**	**91**	**6.0**	**11.6**	**—**	**—**	**—**	**3.4**	**63.9**	**6.1**
			M-F	100	6.6	12.8	—	—	—	3.7	70.2	6.7

COMPOSITION OF FEEDS

(Continued)

Entry Number	TDN Ruminant	Digestible Energy Ruminant		Metabolizable Energy Ruminant		Net Energy Ruminant NE_m		Net Energy Ruminant NE_g		Net Energy Lactating Cows NE_{lc}	
	(%)	*(Mcal)*		*(Mcal)*		*(Mcal)*		*(Mcal)*		*(Mcal)*	
		(lb)	*(kg)*	*(lb)*	*(kg)*	*(lb)*	*(kg)*	*(lb)*	*(kg)*	*(lb)*	*(kg)*
44	25	0.50	1.11	0.44	0.97	0.24	0.53	0.15	0.33	0.24	0.52
	74	1.48	3.27	1.29	2.85	0.71	1.57	0.44	0.97	0.70	1.55
45	50	1.01	2.23	0.84	1.85	0.51	1.13	0.28	0.63	0.53	1.17
	58	1.15	2.54	0.96	2.11	0.58	1.28	0.32	0.71	0.60	1.33
46	47	0.93	2.06	0.76	1.67	0.43	0.94	0.20	0.44	0.48	1.05
	51	1.02	2.24	0.82	1.81	0.46	1.02	0.21	0.47	0.52	1.14
47	17	0.34	0.74	0.28	0.62	0.17	0.37	0.10	0.22	0.17	0.38
	62	1.24	2.73	1.05	2.31	0.62	1.38	0.36	0.80	0.63	1.40
48	29	0.61	1.34	0.50	1.10	0.31	0.68	0.17	0.37	0.32	0.71
	53	1.11	2.44	0.91	2.01	0.56	1.24	0.31	0.67	0.59	1.30
49	53	1.12	2.47	0.95	2.09	0.54	1.19	0.31	0.67	0.53	1.17
	60	1.25	2.76	1.06	2.33	0.60	1.33	0.34	0.75	0.59	1.30
50	63	1.26	2.77	1.09	2.40	0.65	1.43	0.41	0.90	0.63	1.40
	72	1.43	3.15	1.24	2.74	0.74	1.63	0.47	1.03	0.72	1.60
51	26	0.52	1.15	0.43	0.95	0.26	0.57	0.14	0.31	0.27	0.59
	56	1.14	2.51	0.94	2.08	0.56	1.24	0.30	0.67	0.59	1.30
52	14	0.29	0.64	0.24	0.54	0.15	0.34	0.09	0.20	0.15	0.34
	61	1.26	2.78	1.07	2.36	0.67	1.47	0.40	0.88	0.67	1.47
53	44	0.93	2.06	0.76	1.68	0.46	1.02	0.24	0.52	0.50	1.10
	49	1.05	2.31	0.85	1.88	0.52	1.15	0.27	0.59	0.56	1.23
54	10	0.19	0.43	0.17	0.38	0.10	0.23	0.07	0.16	0.10	0.22
	84	1.69	3.72	1.50	3.30	0.91	2.00	0.61	1.35	0.86	1.89
55	26	0.51	1.13	0.45	0.99	0.28	0.61	0.18	0.41	0.26	0.58
	80	1.58	3.48	1.39	3.07	0.86	1.89	0.57	1.25	0.82	1.80
56	72	1.43	3.16	1.26	2.79	0.79	1.73	0.52	1.15	0.75	1.64
	79	1.59	3.49	1.40	3.08	0.87	1.91	0.58	1.27	0.82	1.82
57	44	0.88	1.94	0.70	1.54	0.36	0.79	0.13	0.29	0.43	0.96
	47	0.94	2.08	0.75	1.65	0.39	0.85	0.14	0.31	0.47	1.03
58	62	1.23	2.71	1.06	2.34	0.63	1.39	0.39	0.85	0.62	1.37
	69	1.37	3.02	1.18	2.60	0.70	1.55	0.43	0.95	0.69	1.53
59	52	1.03	2.28	0.86	1.89	0.41	0.91	0.19	0.41	0.47	1.03
	57	1.14	2.51	0.95	2.08	0.46	1.00	0.21	0.45	0.51	1.13
60	60	1.19	2.62	1.01	2.23	0.62	1.36	0.37	0.81	0.62	1.36
	65	1.29	2.84	1.10	2.42	0.67	1.47	0.40	0.88	0.67	1.47
61	18	0.37	0.80	0.33	0.72	0.21	0.45	0.14	0.31	0.19	0.42
	88	1.76	3.87	1.57	3.46	0.99	2.18	0.68	1.50	0.92	2.04
62	75	1.41	3.10	1.12	2.46	0.74	1.63	0.48	1.06	0.77	1.71
	83	1.55	3.41	1.23	2.71	0.81	1.79	0.53	1.16	0.85	1.87

(Continued)

TABLE II-1

Entry Number	Feed Name Description	Moisture Basis: A-F (As-fed) or M-F (Moisture-free)	Dry Matter	Macrominerals						
				Calcium (Ca)	Phos-phorus (P)	Sodium (Na)	Chlorine (Cl)	Mag-nesium (Mg)	Potassium (K)	Sulfur (S)
			(%)	(%)	(%)	(%)	(%)	(%)	(%)	(%)
	BROMEGRASS *Bromus* spp									
44	-IMMATURE, FRESH	**A-F**	**34**	**0.20**	**0.13**	**0.01**	**—**	**0.06**	**1.46**	**0.07**
		M-F	100	0.59	0.37	0.02	—	0.18	4.30	0.20
45	-HAY, PREBLOOM, SUN-CURED	**A-F**	**88**	**0.28**	**0.33**	**0.02**	**—**	**0.08**	**2.04**	**0.18**
		M-F	100	0.32	0.37	0.02	—	0.09	2.32	0.20
46	-HAY, MATURE, SUN-CURED	**A-F**	**92**	**0.40**	**0.09**	**0.02**	**—**	**0.08**	**1.75**	**—**
		M-F	100	0.43	0.09	0.02	—	0.09	1.90	—
	BROMEGRASS, SMOOTH *Bromus inermis*									
47	-FRESH	**A-F**	**27**	**—**	**—**	**—**	**—**	**—**	**—**	**—**
		M-F	100	—	—	—	—	—	—	—
48	-MATURE, FRESH	**A-F**	**55**	**0.14**	**0.09**	**—**	**—**	**—**	**—**	**—**
		M-F	100	0.26	0.10		—	—	—	—
49	-HAY, SUN-CURED	**A-F**	**90**	**0.34**	**0.24**	**0.57**	**0.48**	**0.21**	**1.88**	**0.23**
		M-F	100	0.38	0.26	0.63	0.54	0.23	2.10	0.26
	BUCKWHEAT, COMMON *Fagopyrum sagittatum*									
50	-GRAIN	**A-F**	**88**	**0.10**	**0.33**	**0.05**	**0.04**	**0.10**	**0.45**	**0.14**
		M-F	100	0.11	0.37	0.06	0.05	0.12	0.51	0.16
	BUFFALOGRASS *Buchloe dactyloides*									
51	-FRESH	**A-F**	**46**	**0.26**	**0.09**	**—**	**—**	**0.06**	**0.33**	**—**
		M-F	100	0.57	0.21	—	—	0.14	0.71	—
	CANARYGRASS, REED *Phalaris arundinacea*									
52	-FRESH	**A-F**	**23**	**0.08**	**0.08**	**—**	**—**	**—**	**0.83**	**—**
		M-F	100	0.36	0.33	—	—	—	3.64	—
53	-HAY, SUN-CURED	**A-F**	**89**	**0.32**	**0.21**	**0.01**	**—**	**0.19**	**2.60**	**—**
		M-F	100	0.36	0.24	0.02	—	0.22	2.91	—
	CARROT *Daucus* spp									
54	-ROOTS, FRESH	**A-F**	**11**	**0.05**	**0.04**	**0.06**	**0.06**	**0.02**	**0.32**	**0.02**
		M-F	100	0.40	0.35	0.48	0.50	0.20	2.80	0.17
	CASSAVA *Manihot* spp									
55	-TUBERS, FRESH	**A-F**	**32**	**0.05**	**0.05**	**—**	**—**	**—**	**0.33**	**—**
		M-F	100	0.15	0.15	—	—	—	1.01	—
56	-TUBERS, DEHY, MEAL	**A-F**	**91**	**0.14**	**0.09**	**—**	**—**	**—**	**0.24**	**—**
		M-F	100	0.15	0.10	—	—	—	0.26	—
	CATTLE *Bos taurus*									
57	-MANURE, WITHOUT BEDDING, DEHY	**A-F**	**93**	**1.35**	**1.08**	**—**	**—**	**—**	**0.47**	**—**
		M-F	100	1.45	1.15	—	—	—	0.50	—
	CEREAL									
58	-GRAIN SCREENINGS	**A-F**	**90**	**0.33**	**0.35**	**0.40**	**—**	**0.12**	**0.30**	**—**
		M-F	100	0.37	0.39	0.45	—	0.14	0.34	—
59	-GRAIN SCREENINGS, REFUSE	**A-F**	**91**	**0.31**	**0.33**	**0.25**	**—**	**0.22**	**0.18**	**0.30**
		M-F	100	0.35	0.36	0.28	—	0.24	0.20	0.33
60	-GRAIN SCREENINGS, UNCLEANED	**A-F**	**92**	**0.37**	**0.41**	**0.26**	**—**	**0.22**	**0.18**	**0.30**
		M-F	100	0.40	0.45	0.28	—	0.24	0.20	0.33
	CITRUS *Citrus* spp									
61	-PULP, SILAGE	**A-F**	**21**	**0.42**	**0.03**	**0.02**	**—**	**0.03**	**0.13**	**0.00**
		M-F	100	2.04	0.15	0.09	—	0.16	0.62	0.02
62	-PULP WITHOUT FINES, DEHY (DRIED CITRUS PULP)	**A-F**	**91**	**1.69**	**0.12**	**0.07**	**—**	**0.15**	**0.71**	**0.17**
		M-F	100	1.86	0.13	0.08	—	0.17	0.78	0.19

(Continued)

Entry Number	Microminerals							Fat-soluble Vitamins			
	Cobalt (Co)	Copper (Cu)	Iodine (I)	Iron (Fe)	Manganese (Mn)	Selenium (Se)	Zinc (Zn)	A (1 mg Carotene = 1,667 IU Vit. A)	Carotene (Provitamin A)	D	E (α-tocopherol)
	(ppm or mg/kg)	*(ppm or mg/kg)*	*(ppm or mg/kg)*	*(%)*	*(ppm or mg/kg)*	*(ppm or mg/kg)*	*(ppm or mg/kg)*	*(IU/g)*	*(ppm or mg/kg)*	*(IU/kg)*	*(ppm or mg/kg)*
44	**—**	**—**	**—**	**0.007**	**—**	**—**	**—**	**259.6**	**155.7**	**—**	**—**
	—	—	—	0.020	—	—	—	765.9	459.4	—	—
45	**—**	**—**	**—**	**—**	**—**	**—**	**—**	**95.1**	**57.1**	**1**	**—**
	—	—	—	—	—	—	—	108.4	65.0	1	—
46	**—**	**—**	**—**	**—**	**—**	**—**	**—**	**40.0**	**24.0**	**1**	**—**
	—	—	—	—	—	—	—	43.5	26.1	1	—
47	**0.022**	**—**	**—**	**—**	**—**	**—**	**—**	**142.0**	**85.2**	**0**	**—**
	0.080	—	—	—	—	—	—	525.9	315.5	0	—
48	**—**	**1.2**	**—**	**—**	**—**	**—**	**—**	**—**	**—**	**—**	**—**
	—	2.2	—	—	—	—	—	—	—	—	—
49	**—**	**10.1**	**—**	**0.015**	**41.8**	**—**	**—**	**—**	**—**	**—**	**—**
	—	11.3	—	0.016	46.6	—	—	—	—	—	—
50	**0.049**	**9.5**	**—**	**0.005**	**33.7**	**—**	**8.8**	**—**	**—**	**—**	**—**
	0.056	10.8	—	0.005	38.4	—	10.0	—	—	—	—
51	**—**	**—**	**—**	**—**	**—**	**—**	**—**	**71.6**	**42.9**	**—**	**—**
	—	—	—	—	—	—	—	156.2	93.7	—	—
52	**—**	**—**	**—**	**—**	**—**	**—**	**—**	**—**	**—**	**—**	**—**
	—	—	—	—	—	—	—	—	—	—	—
53	**—**	**10.6**	**—**	**0.014**	**82.5**	**—**	**—**	**28.2**	**16.9**	**—**	**—**
	—	11.9	—	0.015	92.4	—	—	31.6	18.9	—	—
54	**—**	**1.2**	**—**	**0.002**	**3.6**	**—**	**—**	**129.9**	**77.9**	**—**	**6.9**
	—	10.4	—	0.013	31.5	—	—	1,129.4	677.5	—	60.2
55	**—**	**—**	**—**	**—**	**—**	**—**	**—**	**—**	**—**	**—**	**—**
	—	—	—	—	—	—	—	—	—	—	—
56	**—**	**—**	**—**	**—**	**—**	**—**	**—**	**—**	**—**	**—**	**—**
	—	—	—	—	—	—	—	—	—	—	—
57	**—**	**—**	**—**	**—**	**—**	**—**	**—**	**—**	**—**	**—**	**—**
	—	—	—	—	—	—	—	—	—	—	—
58	**—**	**—**	**—**	**—**	**44.4**	**—**	**—**	**—**	**—**	**—**	**—**
	—	—	—	—	49.4	—	—	—	—	—	—
59	**—**	**—**	**—**	**0.025**	**—**	**0.653**	**—**	**—**	**—**	**—**	**—**
	—	—	—	0.027	—	0.719	—	—	—	—	—
60	**—**	**—**	**—**	**0.025**	**—**	**—**	**—**	**—**	**—**	**—**	**—**
	—	—	—	0.027	—	—	—	—	—	—	—
61	**—**	**—**	**—**	**0.004**	**—**	**—**	**3.3**	**—**	**—**	**—**	**—**
	—	—	—	0.016	—	—	16.0	—	—	—	—
62	**0.169**	**5.0**	**—**	**0.033**	**6.6**	**—**	**13.7**	**0.4**	**0.2**	**—**	**—**
	0.185	5.4	—	0.036	7.3	—	15.1	0.4	0.2	—	—

(Continued)

TABLE II-1

Entry Number	Feed Name Description	International Feed Number[2]	Moisture Basis: A-F (As-fed) or M-F (Moisture-free)	Chemical Analysis								
				Dry Matter	Ash	Crude Fiber	Cell Walls or NDF	Acid Detergent Fiber	Lignin	Ether Extract (Fat)	N-Free Extract	Crude Protein
				(%)	(%)	(%)	(%)	(%)	(%)	(%)	(%)	(%)
	CLOVER, ALSIKE *Trifolium hybridum*											
63	-FRESH	**2-01-316**	**A-F**	**22**	**2.1**	**5.2**	**—**	**—**	**—**	**0.8**	**10.3**	**4.1**
			M-F	100	9.3	23.3	—	—	—	3.6	45.7	18.1
64	-HAY, SUN-CURED	**1-01-313**	**A-F**	**88**	**7.6**	**26.2**	**—**	**—**	**—**	**2.4**	**39.1**	**12.4**
			M-F	100	8.7	29.9	—	—	—	2.8	44.5	14.2
	CLOVER, CRIMSON *Trifolium incarnatum*											
65	-FRESH	**2-01-336**	**A-F**	**18**	**1.7**	**4.9**	**—**	**—**	**—**	**0.6**	**7.5**	**3.0**
			M-F	100	9.5	27.7	—	—	—	3.3	42.6	17.0
66	-HAY, SUN-CURED	**1-01-328**	**A-F**	**88**	**7.8**	**28.1**	**—**	**—**	**—**	**2.0**	**35.2**	**14.7**
			M-F	100	8.9	31.9	—	—	—	2.3	40.1	16.8
	CLOVER, LADINO *Trifolium repens*											
67	-FRESH	**2-01-383**	**A-F**	**18**	**1.9**	**2.5**	**—**			**0.9**	**8.1**	**4.4**
			M-F	100	10.5	14.2	—	—	—	4.8	45.7	24.7
68	HAY, SUN-CURED	**1-01-378**	**A-F**	**89**	**8.4**	**18.5**	**32.1**	**28.5**	**5.9**	**2.4**	**39.9**	**20.0**
			M-F	100	9.4	20.8	36.0	32.0	6.6	2.7	44.7	22.4
	CLOVER, RED *Trifolium pratense*											
69	-EARLY BLOOM, FRESH	**2-01-428**	**A-F**	**20**	**2.0**	**4.6**	**—**	**—**	**—**	**1.0**	**8.3**	**3.8**
			M-F	100	10.2	23.3	—	—	—	5.0	42.3	19.4
70	-FULL BLOOM, FRESH	**2-01-429**	**A-F**	**26**	**2.0**	**6.8**	**—**	**—**	**—**	**0.8**	**12.7**	**3.8**
			M-F	100	7.8	26.1	—	—	—	2.9	48.6	14.6
71	-HAY, SUN-CURED	**1-01-415**	**A-F**	**88**	**6.7**	**27.1**	**49.5**	**36.2**	**8.8**	**2.5**	**39.2**	**13.0**
			M-F	100	7.5	30.7	56.0	41.0	10.0	2.8	44.3	14.7
	COCONUT *Cocos nucifera*											
72	-KERNELS WITH COATS, MEAL MECH EXTD (COPRA MEAL)	**5-01-572**	**A-F**	**92**	**6.4**	**12.1**	**—**	**18.3**	**—**	**6.8**	**45.0**	**21.2**
			M-F	100	7.0	13.2	—	20.0	—	7.4	49.2	23.1
73	-KERNELS WITH COATS, MEAL SOLV EXTD (COPRA MEAL)	**5-01-573**	**A-F**	**91**	**6.0**	**14.4**	**—**	**21.9**	**—**	**2.1**	**47.3**	**21.3**
			M-F	100	6.6	15.8	—	24.0	—	2.3	52.0	23.4
	CORN *Zea mays*											
74	-FODDER WITH EARS, WITH HUSKS, SUN-CURED	**1-02-775**	**A-F**	**90**	**4.9**	**32.6**	**—**	**—**	**—**	**7.4**	**29.5**	**15.6**
			M-F	100	5.4	36.2	—	—	—	8.2	32.8	17.3
75	-FODDER WITH EARS, WITH HUSKS, MATURE, SUN-CURED	**1-02-772**	**A-F**	**82**	**4.4**	**18.6**	**—**	**—**	**—**	**1.9**	**50.6**	**6.6**
			M-F	100	5.4	22.6	—	—	—	2.3	61.7	8.0
76	-STOVER WITHOUT EARS, WITHOUT HUSKS, SUN-CURED	**1-02-776**	**A-F**	**85**	**6.1**	**29.3**	**—**	**—**	**—**	**1.1**	**43.2**	**5.4**
			M-F	100	7.2	34.4	—	—	—	1.3	50.8	6.4
77	-COBS, GROUND	**1-02-782**	**A-F**	**90**	**1.6**	**32.2**	**—**	**39.5**	**—**	**0.6**	**52.7**	**2.8**
			M-F	100	1.8	35.8	—	44.0	—	0.7	58.7	3.1
78	-DISTILLERS' GRAINS, DEHY	**5-02-842**	**A-F**	**93**	**2.2**	**11.5**	**—**	**—**	**—**	**8.9**	**43.1**	**27.8**
			M-F	100	2.4	12.3	—	—	—	9.5	46.2	29.7
79	-DISTILLERS' SOLUBLES, DEHY	**5-02-844**	**A-F**	**93**	**7.2**	**4.6**	**—**	**—**	**—**	**8.6**	**45.2**	**27.4**
			M-F	100	7.7	4.9	—	—	—	9.3	48.6	29.5
80	-GLUTEN MEAL	**5-02-900**	**A-F**	**91**	**3.1**	**4.5**	**—**	**—**	**—**	**2.2**	**38.4**	**43.2**
			M-F	100	3.4	4.9	—	—	—	2.4	42.0	47.3
81	-GLUTEN FEED	**5-02-903**	**A-F**	**90**	**6.6**	**8.7**	**—**	**—**	**—**	**2.1**	**49.4**	**23.0**
			M-F	100	7.4	9.7	—	—	—	2.4	55.0	25.6
	CORN, DENT YELLOW *Zea mays indentata*											
82	-EARS, GROUND (CORN-AND-COB MEAL)	**4-28-238**	**A-F**	**87**	**1.7**	**8.2**	**—**	**—**	**—**	**3.2**	**65.7**	**7.8**
			M-F	100	1.9	9.4	—	—	—	3.7	75.9	9.0
83	-EARS WITH HUSKS, SILAGE	**4-02-839**	**A-F**	**43**	**1.5**	**4.8**	**—**	**—**	**—**	**1.6**	**31.5**	**4.0**
			M-F	100	3.4	11.1	—	—	—	3.7	72.7	9.2
84	-GLUTEN MEAL, 60% PROTEIN	**5-28-242**	**A-F**	**90**	**1.7**	**1.8**	**—**	**—**	**—**	**2.1**	**23.7**	**60.8**
			M-F	100	1.9	2.0	—	—	—	2.3	26.3	67.5

(Continued)

Entry Number	TDN Ruminant (%)	Digestible Energy Ruminant (Mcal) (lb)	Digestible Energy Ruminant (Mcal) (kg)	Metabolizable Energy Ruminant (Mcal) (lb)	Metabolizable Energy Ruminant (Mcal) (kg)	Net Energy Ruminant NE_m (Mcal) (lb)	Net Energy Ruminant NE_m (Mcal) (kg)	Net Energy Ruminant NE_g (Mcal) (lb)	Net Energy Ruminant NE_g (Mcal) (kg)	Net Energy Lactating Cows NE_{lc} (Mcal) (lb)	Net Energy Lactating Cows NE_{lc} (Mcal) (kg)
63	16	0.32	0.71	0.28	0.61	0.17	0.37	0.11	0.24	0.17	0.36
	71	1.42	3.14	1.23	2.72	0.76	1.67	0.48	1.06	0.74	1.62
64	51	1.01	2.24	0.85	1.86	0.49	1.08	0.26	0.58	0.51	1.13
	58	1.16	2.55	0.96	2.12	0.56	1.23	0.30	0.66	0.59	1.29
65	11	0.23	0.50	0.19	0.42	0.11	0.25	0.07	0.15	0.12	0.25
	65	1.29	2.84	1.10	2.42	0.65	1.43	0.38	0.85	0.65	1.44
66	50	1.00	2.21	0.84	1.84	0.54	1.19	0.31	0.68	0.55	1.21
	57	1.14	2.52	0.95	2.10	0.61	1.35	0.35	0.77	0.63	1.38
67	13	0.27	0.60	0.24	0.52	0.15	0.33	0.10	0.22	0.14	0.32
	76	1.53	3.37	1.34	2.95	0.86	1.89	0.57	1.25	0.82	1.80
68	58	1.16	2.55	0.99	2.18	0.58	1.27	0.34	0.75	0.58	1.28
	65	1.30	2.87	1.11	2.44	0.65	1.43	0.38	0.84	0.65	1.44
69	14	0.27	0.58	0.23	0.50	0.14	0.32	0.09	0.20	0.14	0.31
	69	1.34	2.96	1.15	2.54	0.73	1.61	0.46	1.01	0.72	1.58
70	17	0.34	0.75	0.29	0.64	0.17	0.38	0.10	0.22	0.17	0.38
	64	1.30	2.86	1.11	2.44	0.65	1.44	0.39	0.86	0.66	1.45
71	52	1.21	2.67	0.95	2.10	0.50	1.11	0.28	0.61	0.53	1.16
	59	1.37	3.02	1.08	2.38	0.57	1.26	0.31	0.69	0.60	1.31
72	74	1.52	3.34	1.33	2.94	0.84	1.85	0.57	1.25	0.80	1.75
	81	1.66	3.65	1.46	3.21	0.92	2.03	0.62	1.37	0.87	1.91
73	69	1.37	3.01	1.20	2.64	0.74	1.63	0.48	1.06	0.70	1.55
	75	1.50	3.31	1.31	2.90	0.81	1.79	0.53	1.16	0.77	1.70
74	45	0.91	2.00	0.74	1.64	0.44	0.97	0.12	0.26	0.46	1.01
	50	1.01	2.22	0.83	1.82	0.49	1.08	0.13	0.29	0.51	1.12
75	56	1.04	2.30	0.89	1.95	0.59	1.30	0.37	0.81	0.58	1.28
	69	1.27	2.80	1.08	2.38	0.72	1.59	0.45	0.99	0.71	1.56
76	51	1.01	2.23	0.85	1.87	0.51	1.12	0.29	0.63	0.52	1.15
	59	1.19	2.62	1.00	2.19	0.60	1.32	0.34	0.74	0.61	1.35
77	44	0.91	2.00	0.74	1.62	0.39	0.87	0.17	0.37	0.44	0.96
	50	1.01	2.23	0.82	1.80	0.44	0.96	0.19	0.42	0.49	1.07
78	81	1.54	3.41	1.28	2.82	0.87	1.92	0.59	1.30	0.84	1.86
	87	1.65	3.64	1.37	3.02	0.93	2.05	0.63	1.39	0.90	1.99
79	80	1.55	3.42	1.34	2.96	0.92	2.03	0.63	1.40	0.86	1.89
	87	1.67	3.68	1.45	3.19	0.99	2.19	0.68	1.51	0.92	2.03
80	78	1.50	3.31	1.24	2.74	0.84	1.85	0.57	1.25	0.80	1.75
	85	1.64	3.62	1.36	2.99	0.92	2.03	0.62	1.37	0.87	1.92
81	75	1.44	3.17	1.21	2.67	0.82	1.80	0.55	1.21	0.78	1.72
	83	1.60	3.52	1.34	2.96	0.91	2.00	0.61	1.35	0.87	1.91
82	72	1.45	3.20	1.26	2.78	0.86	1.90	0.60	1.31	0.76	1.68
	83	1.68	3.70	1.46	3.21	1.00	2.20	0.69	1.52	0.88	1.95
83	31	0.66	1.45	0.57	1.27	0.33	0.74	0.21	0.47	0.32	0.71
	72	1.51	3.33	1.32	2.92	0.77	1.70	0.49	1.08	0.75	1.64
84	—	—	—	—	—	—	—	—	—	—	—
	—	—	—	—	—	—	—	—	—	—	—

(Continued)

TABLE II-1

Entry Number	Feed Name Description	Moisture Basis: A-F (As-fed) or M-F (Moisture-free)	Dry Matter	Macrominerals						
				Calcium (Ca)	Phos-phorus (P)	Sodium (Na)	Chlorine (Cl)	Mag-nesium (Mg)	Potassium (K)	Sulfur (S)
			(%)	(%)	(%)	(%)	(%)	(%)	(%)	(%)
	CLOVER, ALSIKE *Trifolium hybridum*									
63	-FRESH	**A-F**	**22**	**0.31**	**0.06**	**0.10**	**0.17**	**0.07**	**0.61**	**0.05**
		M-F	100	1.36	0.29	0.45	0.77	0.32	2.70	0.22
64	-HAY, SUN-CURED	**A-F**	**88**	**1.14**	**0.22**	**0.40**	**0.68**	**0.40**	**1.95**	**0.17**
		M-F	100	1.30	0.25	0.46	0.78	0.45	2.22	0.19
	CLOVER, CRIMSON *Trifolium incarnatum*									
65	-FRESH	**A-F**	**18**	**0.24**	**0.05**	**0.07**	**0.11**	**0.05**	**0.55**	**0.05**
		M-F	100	1.38	0.29	0.40	0.61	0.29	3.10	0.28
66	-HAY, SUN-CURED	**A-F**	**88**	**1.23**	**0.19**	**0.34**	**0.55**	**0.25**	**2.10**	**0.25**
		M-F	100	1.40	0.22	0.39	0.63	0.28	2.40	0.28
	CLOVER, LADINO *Trifolium repens*									
67	-FRESH	**A-F**	**18**	**0.22**	**0.07**	**0.02**	**—**	**0.09**	**0.33**	**0.02**
		M-F	100	1.27	0.42	0.12	—	0.40	1.87	0.12
68	-HAY, SUN-CURED	**A-F**	**89**	**1.30**	**0.30**	**0.12**	**0.27**	**0.42**	**2.17**	**0.19**
		M-F	100	1.45	0.34	0.13	0.30	0.47	2.44	0.21
	CLOVER, RED *Trifolium pratense*									
69	-EARLY BLOOM, FRESH	**A-F**	**20**	**0.45**	**0.08**	**0.04**	**—**	**0.10**	**0.49**	**0.03**
		M-F	100	2.26	0.38	0.20	—	0.51	2.49	0.17
70	-FULL BLOOM, FRESH	**A-F**	**26**	**0.27**	**0.07**	**0.05**	**—**	**0.13**	**0.51**	**0.05**
		M-F	100	1.01	0.27	0.20	—	0.51	1.96	0.17
71	-HAY, SUN-CURED	**A-F**	**88**	**1.22**	**0.22**	**0.16**	**0.28**	**0.34**	**1.60**	**0.15**
		M-F	100	1.38	0.25	0.18	0.32	0.38	1.81	0.16
	COCONUT *Cocos nucifera*									
72	-KERNELS WITH COATS, MEAL MECH EXTD (COPRA MEAL)	**A-F**	**92**	**0.19**	**0.60**	**0.04**	**—**	**0.30**	**1.65**	**0.34**
		M-F	100	0.21	0.65	0.04	—	0.33	1.80	0.37
73	-KERNELS WITH COATS, MEAL SOLV EXTD (COPRA MEAL)	**A-F**	**91**	**0.17**	**0.60**	**0.04**	**0.03**	**0.31**	**1.41**	**—**
		M-F	100	0.19	0.66	0.04	0.03	0.34	1.55	—
	CORN *Zea mays*									
74	-FODDER WITH EARS, WITH HUSKS, SUN-CURED	**A-F**	**90**	**0.45**	**0.23**	**0.03**	**0.17**	**0.26**	**0.84**	**0.13**
		M-F	100	0.50	0.25	0.03	0.19	0.29	0.93	0.14
75	-FODDER WITH EARS, WITH HUSKS, MATURE, A-F	**82**	**—**	**—**	**—**	**—**	**—**	**—**	**—**	
	SUN-CURED	M-F	100	—	—	—	—	—	—	—
76	-STOVER WITHOUT EARS, WITHOUT HUSKS, SUN-CURED	**A-F**	**85**	**0.49**	**0.08**	**0.06**	**—**	**0.34**	**1.24**	**0.15**
		M-F	100	0.57	0.10	0.07	—	0.40	1.45	0.17
77	-COBS, GROUND	**A-F**	**90**	**0.11**	**0.04**	**—**	**—**	**0.06**	**0.78**	**0.42**
		M-F	100	0.12	0.04	—	—	0.07	0.87	0.47
78	-DISTILLERS' GRAINS, DEHY	**A-F**	**93**	**0.09**	**0.39**	**0.09**	**0.07**	**0.07**	**0.16**	**0.43**
		M-F	100	0.10	0.42	0.09	0.08	0.07	0.18	0.46
79	-DISTILLERS' SOLUBLES, DEHY	**A-F**	**93**	**0.30**	**1.30**	**0.23**	**0.26**	**0.60**	**1.70**	**0.37**
		M-F	100	0.32	1.40	0.24	0.28	0.65	1.83	0.40
80	-GLUTEN MEAL	**A-F**	**91**	**0.15**	**0.46**	**0.09**	**0.06**	**0.06**	**0.03**	**0.20**
		M-F	100	0.16	0.51	0.10	0.07	0.06	0.03	0.22
81	-GLUTEN FEED	**A-F**	**90**	**0.32**	**0.74**	**0.12**	**0.22**	**0.33**	**0.57**	**0.21**
		M-F	100	0.36	0.82	0.14	0.25	0.36	0.64	0.23
	CORN, DENT YELLOW *Zea mays indentata*									
82	-EARS, GROUND (CORN-AND-COB MEAL)	**A-F**	**87**	**0.06**	**0.24**	**0.02**	**0.04**	**0.12**	**0.46**	**0.14**
		M-F	100	0.07	0.27	0.02	0.05	0.14	0.53	0.16
83	-EARS WITH HUSKS, SILAGE	**A-F**	**43**	**0.04**	**0.12**	**0.00**	**—**	**0.05**	**0.21**	**0.06**
		M-F	100	0.10	0.29	0.01	—	0.12	0.49	0.13
84	-GLUTEN MEAL, 60% PROTEIN	**A-F**	**90**	**0.07**	**0.45**	**0.05**	**0.09**	**0.08**	**0.18**	**0.65**
		M-F	100	0.08	0.50	0.05	0.10	0.09	0.20	0.72

(Continued)

Entry Number	Microminerals							Fat-soluble Vitamins			
	Cobalt (Co)	Copper (Cu)	Iodine (I)	Iron (Fe)	Manganese (Mn)	Selenium (Se)	Zinc (Zn)	A (1 mg Carotene = 1,667 IU Vit. A)	Carotene (Provitamin A)	D	E (α-tocopherol)
	(ppm or mg/kg)	*(ppm or mg/kg)*	*(ppm or mg/kg)*	*(%)*	*(ppm or mg/kg)*	*(ppm or mg/kg)*	*(ppm or mg/kg)*	*(IU/g)*	*(ppm or mg/kg)*	*(IU/kg)*	*(ppm or mg/kg)*
63	—	1.3	—	0.010	26.3	—	—	—	—	—	—
	—	6.0	—	0.044	117.1	—	—	—	—	—	—
64	—	5.3	—	0.023	60.5	—	—	272.1	163.2	—	—
	—	6.0	—	0.026	69.0	—	—	310.1	186.0	—	—
65	—	—	—	—	43.1	—	—	—	—	—	—
	—	—	—	—	245.8	—	—	—	—	—	—
66	—	—	0.059	0.062	183.3	—	—	32.9	19.8	—	—
	—	—	0.067	0.070	208.7	—	—	37.5	22.5	—	—
67	—	—	—	0.007	12.7	—	—	96.2	57.7	—	—
	—	—	—	0.037	71.7	—	—	545.2	327.0	—	—
68	0.144	8.4	0.268	0.042	109.7	—	15.2	239.5	143.7	—	—
	0.161	9.4	0.301	0.047	123.1	—	17.0	268.7	161.2	—	—
69	—	—	—	0.006	—	—	—	81.5	48.9	—	—
	—	—	—	0.030	—	—	—	412.6	247.5	—	—
70	—	—	—	0.008	—	—	—	90.6	54.4	—	—
	—	—	—	0.030	—	—	—	345.9	207.5	—	—
71	0.138	18.8	0.217	0.022	95.2	—	32.5	40.5	24.3	—	—
	0.156	21.2	0.245	0.024	107.7	—	36.7	45.9	27.5	—	—
72	0.127	16.7	—	0.068	70.1	—	48.5	—	—	—	—
	0.139	18.2	—	0.075	76.6	—	53.0	—	—	—	—
73	—	—	—	—	54.5	—	—	—	—	—	—
	—	—	—	—	59.8	—	—	—	—	—	—
74	—	6.9	—	0.009	61.4	—	—	6.6	4.0	1	—
	—	7.7	—	0.010	68.2	—	—	7.3	4.4	1	—
75	—	—	—	—	—	—	—	—	—	—	—
	—	—	—	—	—	—	—	—	—	—	—
76	—	4.3	—	0.018	115.9	—	—	6.3	3.8	1	—
	—	5.1	—	0.021	136.0	—	—	7.4	4.5	1	—
77	0.117	6.6	—	0.021	5.6	—	—	1.0	0.6	—	—
	0.130	7.3	—	0.023	6.2	—	—	1.2	0.7	—	—
78	0.076	38.9	0.048	0.020	19.3	0.352	41.7	5.2	3.1	—	—
	0.082	41.7	0.051	0.021	20.7	0.377	44.7	5.6	3.3	—	—
79	0.167	77.9	0.079	0.052	72.0	0.371	88.0	1.1	0.7	—	45.9
	0.180	83.9	0.085	0.056	77.6	0.400	94.8	1.2	0.7	—	49.4
80	0.077	27.7	—	0.039	7.7	1.015	173.7	27.3	16.3	—	29.3
	0.085	30.3	—	0.043	8.5	1.111	190.2	29.8	17.9	—	32.0
81	0.087	47.1	0.066	0.043	23.1	0.272	64.6	9.8	5.9	—	12.1
	0.097	52.3	0.074	0.048	25.7	0.302	71.8	10.9	6.5	—	13.5
82	0.273	6.8	0.023	0.008	19.9	0.074	12.1	5.3	3.2	—	17.5
	0.315	7.9	0.026	0.010	23.0	0.086	14.0	6.1	3.7	—	20.2
83	—	—	—	0.004	—	—	—	—	—	—	—
	—	—	—	0.008	—	—	—	—	—	—	—
84	0.045	26.1	0.018	0.023	6.3	0.829	30.6	—	—	—	14.6
	0.051	29.0	0.020	0.026	7.0	0.921	34.0	—	—	—	16.2

(Continued)

TABLE II-1

Entry Number	Feed Name Description	International Feed Number[2]	Moisture Basis: A-F (As-fed) or M-F (Moisture-free)	Chemical Analysis								
				Dry Matter	Ash	Crude Fiber	Cell Walls or NDF	Acid Detergent Fiber	Lignin	Ether Extract (Fat)	N-Free Extract	Crude Protein
				(%)	(%)	(%)	(%)	(%)	(%)	(%)	(%)	(%)
85	-GRAIN, GRADE 2, 54 LB/BU	4-02-931	A-F	87	1.2	2.1	—	—	—	4.0	71.3	8.9
			M-F	100	1.4	2.4	—	—	—	4.5	81.5	10.2
86	-GRAIN, HIGH-MOISTURE	4-20-770	A-F	77	1.2	2.1	17.5	3.8	1.3	3.3	61.9	8.1
			M-F	100	1.6	2.7	22.9	4.9	1.7	4.3	80.8	10.6
87	-GRAIN, FLAKED	4-28-244	A-F	89	0.9	0.6	—	—	—	2.0	75.4	9.9
			M-F	100	1.0	0.7	—	—	—	2.2	84.9	11.2
88	-HOMINY FEED	4-03-011	A-F	90	2.8	4.8	—	—	—	6.5	65.8	10.3
			M-F	100	3.1	5.3	—	—	—	7.2	72.9	11.4
89	-SILAGE, ALL ANALYSES	3-02-822	A-F	26	1.5	6.6	—	8.9	1.2	0.8	15.2	2.2
			M-F	100	5.6	25.1	—	34.0	4.4	3.2	57.8	8.3
90	-SILAGE, DOUGH STAGE	0 00 810	A-F	27	1.2	6.6	—	8.3	—	0.8	16.2	2.1
			M-F	100	4.6	24.5	—	30.9		2.8	60.3	7.7
91	-SILAGE, MATURE	3-02-820	A-F	30	1.4	6.5	—	—	—	1.3	18.7	2.5
			M-F	100	4.7	21.3	—	—	—	4.2	61.6	8.2
	CORN, SWEET *Zea mays saccharata*											
92	-CANNERY RESIDUE, FRESH	2-02-975	A-F	77	3.5	17.0	—	22.3	—	1.9	47.6	6.8
			M-F	100	4.6	22.2	—	29.0	—	2.5	62.0	8.8
93	-CANNERY RESIDUE, SILAGE	3-07-955	A-F	31	1.7	10.1	—	10.5	—	1.3	15.4	2.5
			M-F	100	5.4	32.7	—	34.0	—	4.3	49.6	8.0
	COTTON *Gossypium* spp											
94	-BOLLS, SUN-CURED	1-01-596	A-F	91	6.2	34.5	—	—	—	2.8	38.9	9.0
			M-F	100	6.8	37.8	—	—	—	3.0	42.6	9.8
95	-HULLS	1-01-599	A-F	90	2.6	43.2	79.8	58.7	21.0	1.5	39.3	3.8
			M-F	100	2.9	47.8	88.3	65.0	23.3	1.7	43.5	4.2
96	-SEEDS WITHOUT LINT	5-13-749	A-F	91	4.7	19.4	—	—	—	20.4	24.5	21.8
			M-F	100	5.2	21.4	—	—	—	22.5	27.0	24.0
97	-SEEDS, MEAL, MECH EXTD, 41% PROTEIN	5-01-617	A-F	93	6.1	11.9	25.9	18.5	5.6	4.7	28.9	41.0
			M-F	100	6.6	12.9	28.0	20.0	6.0	5.0	31.2	44.3
98	-SEEDS, MEAL, PREPRESSED, SOLV EXTD, 41% PROTEIN	5-07-872	A-F	90	6.4	12.9	—	—	—	1.0	28.8	41.3
			M-F	100	7.0	14.2	—	—	—	1.2	31.9	45.7
99	-SEEDS, MEAL, SOLV EXTD, 41% PROTEIN	5-01-621	A-F	91	6.5	12.1	23.6	18.4	5.5	1.5	29.6	41.2
			M-F	100	7.1	13.4	26.0	20.2	6.0	1.6	32.5	45.4
100	-SEEDS WITHOUT HULLS, MEAL, PREPRESSED, SOLV EXTD, 50% PROTEIN	5-07-874	A-F	93	6.6	8.2	—	—	—	1.3	26.8	50.3
			M-F	100	7.1	8.8	—	—	—	1.4	28.8	54.0
	COWPEA, COMMON *Vigna sinensis*											
101	-HAY, SUN-CURED	1-01-645	A-F	90	10.5	24.4	—	—	—	2.6	35.1	17.7
			M-F	100	11.7	27.1	—	—	—	2.9	38.8	19.6
	DESERT MOLLY (*SEE* SUMMER CYPRESS)											
	DISTILLERS' GRAINS (*SEE* CORN; *SEE* SORGHUM)											
	DROPSEED, SAND *Sporobolus cryptandrus*											
102	-STEM-CURED, FRESH	2-05-596	A-F	88	7.0	31.6	—	—	5.2	1.1	43.0	5.4
			M-F	100	7.9	35.9	—	—	5.9	1.2	48.8	6.1
	FATS AND OILS											
103	-FAT, ANIMAL, HYDROLYZED	4-00-376	A-F	99	—	—	—	—	—	98.4	—	—
			M-F	100	—	—	—	—	—	99.2	—	—
104	-FAT, ANIMAL—POULTRY	4-00-409	A-F	99	—	—	—	—	—	99.1	—	—
			M-F	100	—	—	—	—	—	100.0	—	—

COMPOSITION OF FEEDS

(Continued)

Entry Number	TDN	Digestible Energy		Metabolizable Energy		Net Energy					
	Ruminant	Ruminant		Ruminant		Ruminant NE_m		Ruminant NE_g		Lactating Cows NE_{lc}	
	(%)	*(Mcal)*		*(Mcal)*		*(Mcal)*		*(Mcal)*		*(Mcal)*	
		(lb)	*(kg)*	*(lb)*	*(kg)*	*(lb)*	*(kg)*	*(lb)*	*(kg)*	*(lb)*	*(kg)*
85	80	1.57	3.47	1.45	3.19	0.89	1.96	0.61	1.35	0.84	1.85
	92	1.80	3.96	1.66	3.65	1.01	2.24	0.70	1.55	0.96	2.11
86	71	1.41	3.11	1.27	2.80	0.79	1.75	0.55	1.22	0.74	1.63
	92	1.85	4.07	1.66	3.66	1.04	2.28	0.72	1.59	0.96	2.12
87	—	—	—	—	—	—	—	—	—	—	—
	—	—	—	—	—	—	—	—	—	—	—
88	84	1.69	3.73	1.29	2.85	0.88	1.95	0.61	1.34	0.91	2.00
	93	1.88	4.14	1.43	3.15	0.98	2.16	0.67	1.48	1.01	2.22
89	18	0.34	0.75	0.30	0.66	0.19	0.42	0.12	0.26	0.17	0.37
	68	1.30	2.86	1.14	2.50	0.73	1.61	0.46	1.01	0.65	1.42
90	19	0.36	0.79	0.30	0.67	0.19	0.43	0.12	0.27	0.18	0.40
	69	1.34	2.95	1.13	2.50	0.72	1.59	0.46	1.01	0.68	1.50
91	22	0.41	0.91	0.35	0.76	0.22	0.48	0.12	0.26	0.20	0.43
	74	1.36	3.00	1.14	2.51	0.72	1.58	0.38	0.84	0.65	1.42
92	54	1.05	2.31	0.90	1.99	0.57	1.26	0.36	0.79	0.56	1.23
	70	1.37	3.01	1.18	2.59	0.74	1.63	0.47	1.03	0.72	1.60
93	22	0.42	0.92	0.36	0.79	0.24	0.53	0.15	0.34	0.23	0.51
	72	1.35	2.97	1.16	2.55	0.77	1.70	0.49	1.08	0.75	1.64
94	40	0.80	1.77	0.62	1.37	0.28	0.62	0.06	0.14	0.38	0.83
	44	0.88	1.93	0.68	1.50	0.31	0.68	0.07	0.15	0.41	0.91
95	42	0.86	1.90	0.60	1.31	0.25	0.55	0.03	0.08	0.39	0.86
	47	0.95	2.10	0.66	1.45	0.28	0.61	0.04	0.08	0.43	0.95
96	87	1.57	3.46	1.39	3.07	0.97	2.13	0.67	1.49	0.91	2.01
	95	1.72	3.80	1.53	3.38	1.06	2.34	0.74	1.63	1.00	2.21
97	72	1.49	3.29	1.12	2.47	0.72	1.59	0.64	1.41	0.76	1.67
	77	1.61	3.56	1.21	2.67	0.78	1.72	0.69	1.52	0.82	1.81
98	72	1.51	3.33	1.14	2.51	0.70	1.55	0.64	1.41	0.76	1.67
	80	1.67	3.69	1.26	2.78	0.78	1.71	0.71	1.56	0.84	1.85
99	68	1.48	3.27	1.17	2.57	0.74	1.64	0.63	1.40	0.73	1.61
	75	1.63	3.60	1.28	2.83	0.82	1.80	0.70	1.54	0.80	1.77
100	70	1.43	3.15	1.25	2.76	0.76	1.67	0.49	1.08	0.73	1.60
	75	1.54	3.38	1.35	2.97	0.81	1.79	0.53	1.16	0.78	1.72
101	54	1.14	2.52	0.92	2.02	0.58	1.28	0.34	0.75	0.56	1.24
	60	1.26	2.78	1.02	2.24	0.64	1.41	0.38	0.83	0.62	1.38
102	52	0.96	2.12	0.83	1.82	0.47	1.04	0.25	0.54	0.50	1.11
	59	1.09	2.41	0.94	2.07	0.54	1.18	0.28	0.62	0.57	1.26
103	223	4.46	9.84	4.31	9.49	2.95	6.49	2.23	4.91	2.43	5.35
	225	4.50	9.92	4.34	9.57	2.97	6.55	2.25	4.95	2.45	5.39
104	188	3.57	7.87	3.40	7.51	2.65	5.85	1.99	4.38	2.03	4.48
	189	3.60	7.94	3.43	7.57	2.68	5.90	2.01	4.42	2.05	4.52

(Continued)

TABLE II-1

Entry Number	Feed Name Description	Moisture Basis: A-F (As-fed) or M-F (Moisture-free)	Dry Matter	Macrominerals						
				Calcium (Ca)	Phos-phorus (P)	Sodium (Na)	Chlorine (Cl)	Mag-nesium (Mg)	Potassium (K)	Sulfur (S)
			(%)	*(%)*	*(%)*	*(%)*	*(%)*	*(%)*	*(%)*	*(%)*
85	-GRAIN, GRADE 2, 54 LB/BU	**A-F**	**87**	**0.02**	**0.29**	**0.02**	**0.04**	**0.11**	**0.31**	**0.12**
		M-F	100	0.03	0.33	0.02	0.05	0.13	0.36	0.14
86	-GRAIN, HIGH-MOISTURE	**A-F**	**77**	**0.01**	**0.25**	**0.01**	**0.04**	**0.11**	**0.28**	**0.11**
		M-F	100	0.02	0.33	0.01	0.05	0.15	0.37	0.14
87	-GRAIN, FLAKED	**A-F**	**89**	—	—	—	—	—	—	—
		M-F	100	—	—	—	—	—	—	—
88	-HOMINY FEED	**A-F**	**90**	**0.05**	**0.51**	**0.08**	**0.05**	**0.24**	**0.59**	**0.03**
		M-F	100	0.05	0.57	0.09	0.06	0.26	0.65	0.03
89	-SILAGE, ALL ANALYSES	**A-F**	**26**	**0.08**	**0.07**	**0.01**	**0.05**	**0.06**	**0.32**	**0.03**
		M-F	100	0.31	0.27	0.03	0.18	0.22	1.22	0.12
90	-SILAGE, DOUGH STAGE	**A-F**	**27**	**0.07**	**0.05**	—	—	—	—	—
		M-F	100	0.27	0.19	—	—	—	—	—
91	-SILAGE, MATURE	**A-F**	**30**	**0.10**	**0.24**	—	—	**0.05**	—	—
		M-F	100	0.33	0.79	—	—	0.16	—	—
	CORN, SWEET *Zea mays saccharata*									
92	-CANNERY RESIDUE, FRESH	**A-F**	**77**	**0.25**	**0.54**	**0.02**	—	**0.18**	**0.88**	**0.10**
		M-F	100	0.32	0.70	0.03	—	0.24	1.15	0.13
93	-CANNERY RESIDUE, SILAGE	**A-F**	**31**	**0.10**	**0.24**	**0.01**	—	**0.07**	**0.36**	**0.03**
		M-F	100	0.32	0.77	0.03	—	0.24	1.15	0.11
	COTTON *Gossypium* spp									
94	-BOLLS, SUN-CURED	**A-F**	**91**	**0.72**	**0.13**	—	—	**0.24**	**2.42**	—
		M-F	100	0.78	0.14	—	—	0.27	2.64	—
95	-HULLS	**A-F**	**90**	**0.13**	**0.09**	**0.02**	**0.02**	**0.13**	**0.78**	**0.08**
		M-F	100	0.15	0.09	0.02	0.02	0.14	0.87	0.09
96	-SEEDS WITHOUT LINT	**A-F**	**91**	**0.14**	**0.69**	**0.03**	—	**0.32**	**1.11**	**0.24**
		M-F	100	0.16	0.76	0.03	—	0.35	1.22	0.26
97	-SEEDS, MEAL, MECH EXTD, 41% PROTEIN	**A-F**	**93**	**0.19**	**1.07**	**0.04**	**0.04**	**0.53**	**1.33**	**0.40**
		M-F	100	0.21	1.16	0.05	0.05	0.57	1.44	0.43
98	-SEEDS, MEAL, PREPRESSED, SOLV EXTD, 41% PROTEIN	**A-F**	**90**	**0.16**	**1.07**	**0.04**	**0.06**	**0.48**	**1.25**	**0.31**
		M-F	100	0.17	1.18	0.04	0.07	0.53	1.38	0.34
99	-SEEDS, MEAL, SOLV EXTD, 41% PROTEIN	**A-F**	**91**	**0.17**	**1.11**	**0.04**	**0.04**	**0.54**	**1.37**	**0.25**
		M-F	100	0.19	1.22	0.05	0.05	0.59	1.51	0.27
100	-SEEDS, WITHOUT HULLS, MEAL, PREPRESSED, SOLV EXTD, 50% PROTEIN	**A-F**	**93**	**0.18**	**1.16**	**0.05**	**0.05**	**0.46**	**1.45**	**0.52**
		M-F	100	0.19	1.24	0.06	0.05	0.50	1.56	0.56
	COWPEA, COMMON *Vigna sinensis*									
101	-HAY, SUN-CURED	**A-F**	**90**	**1.26**	**0.31**	**0.24**	**0.15**	**0.41**	**2.04**	**0.32**
		M-F	100	1.40	0.35	0.27	0.17	0.45	2.26	0.35
	DESERT MOLLY (*SEE* SUMMER CYPRES)									
	DISTILLERS' GRAINS (*SEE* CORN; *SEE* SORGHUM)									
	DROPSEED, SAND *Sporobolus cryptandrus*									
102	-STEM-CURED, FRESH	**A-F**	**88**	**0.40**	**0.07**	**0.01**	—	**0.06**	**0.28**	—
		M-F	100	0.45	0.08	0.01	—	0.06	0.32	—
	FATS AND OILS									
103	-FAT, ANIMAL, HYDROLYZED	**A-F**	**99**	—	—	—	—	—	—	—
		M-F	100	—	—	—	—	—	—	—
104	-FAT, ANIMAL—POULTRY	**A-F**	**99**	—	—	—	—	—	**0.23**	—
		M-F	100	—	—	—	—	—	0.23	—

(Continued)

Entry Number	Microminerals							Fat-soluble Vitamins			
	Cobalt (Co)	Copper (Cu)	Iodine (I)	Iron (Fe)	Manganese (Mn)	Selenium (Se)	Zinc (Zn)	A (1 mg Carotene = 1,667 IU Vit. A)	Carotene (Provitamin A)	D	E (α-tocopherol)
	(ppm or mg/kg)	*(ppm or mg/kg)*	*(ppm or mg/kg)*	*(%)*	*(ppm or mg/kg)*	*(ppm or mg/kg)*	*(ppm or mg/kg)*	*(IU/g)*	*(ppm or mg/kg)*	*(IU/kg)*	*(ppm or mg/kg)*
85	**0.029**	**3.8**	**—**	**0.003**	**5.3**	**—**	**13.7**	**2.9**	**1.7**	**—**	**21.6**
	0.033	4.3	—	0.003	6.1	—	15.6	3.3	2.0	—	24.7
86	**—**	**2.2**	**—**	**0.003**	**5.3**	**—**	**25.4**	**—**	**—**	**—**	**—**
	—	2.9	—	0.004	6.9	—	33.2	—	—	—	—
87	**—**	**—**	**—**	**—**	**—**	**—**	**—**	**—**	**—**	**—**	**0.8**
	—	—	—	—	—	—	—	—	—	—	0.9
88	**0.055**	**13.6**	**—**	**0.007**	**14.5**	**—**	**—**	**15.4**	**9.2**	**—**	**—**
	0.061	15.1	—	0.008	16.1	—	—	17.0	10.2	—	—
89	**0.026**	**2.4**	**—**	**0.005**	**10.8**	**—**	**5.5**	**15.2**	**9.1**	**0**	**—**
	0.097	9.2	—	0.018	41.1	—	21.2	58.1	34.9	0	—
90	**—**	**—**	**—**	**—**	**—**	**—**	**—**	**29.1**	**17.4**	**—**	**—**
	—	—	—	—	—	—	—	108.5	65.1	—	—
91	**0.019**	**—**	**—**	**—**	**—**	**—**	**—**	**7.9**	**4.8**	**—**	**—**
	0.060	—	—	—	—	—	—	26.1	15.7	—	—
92	**—**	**5.4**	**—**	**0.016**	**—**	**—**	**—**	**17.3**	**10.4**	**—**	**—**
	—	7.0	—	0.020	—	—	—	22.5	13.5	—	—
93	**—**	**—**	**—**	**0.007**	**—**	**—**	**—**	**6.9**	**4.2**	**—**	**—**
	—	—	—	0.020	—	—	—	22.3	13.4	—	—
94	**—**	**—**	**—**	**—**	**—**	**—**	**—**	**—**	**—**	**—**	**—**
	—	—	—	—	—	—	—	—	—	—	—
95	**0.018**	**12.0**	**—**	**0.012**	**107.8**	**—**	**19.8**	**—**	**—**	**—**	**—**
	0.020	13.3	—	0.014	119.2	—	21.9	—	—	—	—
96	**—**	**49.0**	**—**	**0.014**	**11.1**	**—**	**—**	**—**	**—**	**—**	**—**
	—	53.9	—	0.016	12.2	—	—	—	—	—	—
97	**0.626**	**18.5**	**—**	**0.018**	**22.3**	**—**	**61.8**	**0.4**	**0.2**	**—**	**32.3**
	0.676	20.0	—	0.019	24.1	—	66.8	0.4	0.2	—	34.9
98	**0.738**	**18.2**	**—**	**0.018**	**20.4**	**—**	**62.7**	**—**	**—**	**—**	**—**
	0.817	20.2	—	0.019	22.5	—	69.4	—	—	—	—
99	**0.483**	**19.5**	**—**	**0.019**	**20.6**	**—**	**60.7**	**—**	**—**	**—**	**14.6**
	0.531	21.4	—	0.021	22.7	—	66.7	—	—	—	16.1
100	**0.042**	**14.5**	**—**	**0.012**	**23.0**	**—**	**73.8**	**—**	**—**	**—**	**11.3**
	0.046	15.6	—	0.012	24.8	—	79.4	—	—	—	12.1
101	**0.064**	**—**	**—**	**0.055**	**438.4**	**—**	**—**	**52.7**	**31.6**	**—**	**—**
	0.070	—	—	0.060	485.1	—	—	58.3	35.0	—	—
102	**0.503**	**13.5**	**0.599**	**0.043**	**41.4**	**—**	**36.8**	**14.0**	**8.4**	**—**	**—**
	0.572	15.3	0.681	0.049	47.0	—	41.8	15.9	9.6	—	—
103	**—**	**—**	**—**	**—**	**—**	**—**	**—**	**—**	**—**	**—**	**—**
	—	—	—	—	—	—	—	—	—	—	—
104	**—**	**—**	**—**	**—**	**—**	**—**	**—**	**—**	**—**	**—**	**7.9**
	—	—	—	—	—	—	—	—	—	—	7.9

(Continued)

TABLE II-1

Entry Number	Feed Name Description	International Feed Number[2]	Moisture Basis: A-F (As-fed) or M-F (Moisture-free)	Chemical Analysis								
				Dry Matter	Ash	Crude Fiber	Cell Walls or NDF	Acid Detergent Fiber	Lignin	Ether Extract (Fat)	N-Free Extract	Crude Protein
				(%)	(%)	(%)	(%)	(%)	(%)	(%)	(%)	(%)
	FESCUE *Festuca* spp											
105	-HAY, SUN-CURED, ALTA	**1-05-684**	**A-F**	**89**	**5.9**	**32.6**	**61.7**	**35.6**	**—**	**2.0**	**41.4**	**7.2**
			M-F	100	6.6	36.6	69.3	40.0	—	2.2	46.5	8.1
106	-HAY, SUN-CURED, MEADOW	**1-01-912**	**A-F**	**88**	**7.9**	**28.0**	**74.0**	**43.8**	**—**	**2.4**	**41.0**	**8.2**
			M-F	100	9.0	32.0	84.6	50.0	—	2.7	46.9	9.4
	FLAX, COMMON *Linum usitatissimum*											
107	-SEED SCREENINGS	**4-02-056**	**A-F**	**91**	**6.2**	**12.1**	**—**	**—**	**—**	**9.3**	**47.1**	**16.6**
			M-F	100	6.8	13.2	—	—	—	10.2	51.6	18.2
108	-SEEDS, MEAL, SOLV EXTD, 35% PROTEIN (LINSEED MEAL)	**5-26-090**	**A-F**	**90**	**5.8**	**8.9**	**—**	**—**	**—**	**1.7**	**38.2**	**35.7**
			M-F	100	6.4	9.9	—	—	—	1.90	42.3	39.6
	GALLETA *Hilaria jamesii*											
109	-STEM-CURED, FRESH	**2-05-594**	**A-F**	**[illegible]**	**10.0**	**[illegible]**	**—**	**—**	**—**	**1.4**	**38.5**	**4.3**
			M-F	100	15.5	33.0	—	—	—	1.7	44.8	5.1
	GRAMA *Bouteloua* spp											
110	-IMMATURE, FRESH	**2-02-163**	**A-F**	**41**	**4.6**	**11.2**	**—**	**—**	**—**	**0.8**	**19.0**	**5.4**
			M-F	100	11.3	27.2	—	—	—	2.0	46.4	13.1
111	-MATURE, FRESH	**2-02-166**	**A-F**	**63**	**7.2**	**20.7**	**—**	**—**	**—**	**1.1**	**30.2**	**4.1**
			M-F	100	11.4	32.7	—	—	—	1.7	47.7	6.5
	GRAPE *Vitis* spp											
112	-POMACE, DEHY (MARC)	**1-02-208**	**A-F**	**90**	**7.5**	**27.9**	**48.1**	**49.2**	**31.8**	**7.6**	**35.3**	**12.1**
			M-F	100	8.3	30.9	53.2	54.4	35.2	8.4	39.0	13.4
	LESPEDEZA, COMMON—KOREAN *Lespedeza striata—stipulacea*											
113	-PREBLOOM, FRESH	**2-26-028**	**A-F**	**25**	**3.2**	**8.0**	**—**	**—**	**—**	**0.5**	**9.2**	**4.1**
			M-F	100	12.8	32.0	—	—	—	2.0	36.8	16.4
114	-EARLY BLOOM, FRESH	**2-20-880**	**A-F**	**28**	**—**	**6.2**	**—**	**—**	**—**	**—**	**—**	**4.6**
			M-F	100	—	22.0	—	—	—	—	—	16.4
115	-HAY, EARLY BLOOM, SUN-CURED	**1-26-025**	**A-F**	**93**	**6.3**	**28.5**	**—**	**—**	**—**	**4.1**	**39.6**	**14.5**
			M-F	100	6.8	30.6	—	—	—	4.4	42.6	15.6
116	-HAY, MIDBLOOM, SUN-CURED	**1-26-026**	**A-F**	**93**	**4.3**	**27.9**	**—**	**—**	**—**	**2.5**	**47.2**	**11.1**
			M-F	100	4.6	30.0	—	—	—	2.7	50.8	11.9
117	-HAY, FULL BLOOM, SUN-CURED	**1-26-027**	**A-F**	**93**	**5.3**	**30.2**	**—**	**—**	**—**	**3.0**	**41.1**	**13.4**
			M-F	100	5.7	32.5	—	—	—	3.2	44.2	14.4
	LINSEED (*SEE* FLAX, COMMON)											
	MEADOW PLANTS, INTERMOUNTAIN											
118	-HAY, SUN-CURED	**1-03-181**	**A-F**	**95**	**8.2**	**31.2**	**—**	**—**	**—**	**2.4**	**45.2**	**8.2**
			M-F	100	8.6	32.7	—	—	—	2.5	47.5	8.7
	MILLET, FOXTAIL *Setaria italica*											
119	-FRESH	**2-03-101**	**A-F**	**29**	**2.5**	**9.2**	**—**	**—**	**—**	**0.9**	**13.4**	**2.8**
			M-F	100	8.6	32.0	—	—	—	3.1	46.7	9.6
120	-GRAIN	**4-03-102**	**A-F**	**89**	**3.4**	**7.4**	**—**	**—**	**—**	**4.1**	**63.0**	**11.4**
			M-F	100	3.8	8.3	—	—	—	4.6	70.6	12.8
121	-HAY, SUN-CURED	**1-03-099**	**A-F**	**87**	**7.5**	**25.7**	**—**	**—**	**—**	**2.5**	**43.7**	**7.5**
			M-F	100	8.6	29.6	—	—	—	2.9	50.3	8.6
	MILLET, PROSO (BROOMCORN) *Panicum miliaceum*											
122	-GRAIN	**4-03-120**	**A-F**	**90**	**2.9**	**5.3**	**—**	**14.9**	**3.2**	**3.6**	**66.3**	**11.6**
			M-F	100	3.3	6.0	—	16.6	3.6	4.0	73.9	12.9
	MOLASSES AND SYRUP *Beta vulgaris altissima*											
123	-BEET, SUGAR, MOLASSES, MORE THAN 48% INVERT SUGARS, MORE THAN 79.5° BRIX	**4-00-668**	**A-F**	**78**	**8.9**	**—**	**—**	**—**	**—**	**0.2**	**62.2**	**6.6**
			M-F	100	11.4	—	—	—	—	0.2	79.9	8.5

COMPOSITION OF FEEDS

(Continued)

Entry Number	TDN Ruminant (%)	Digestible Energy Ruminant (Mcal) (lb)	Digestible Energy Ruminant (Mcal) (kg)	Metabolizable Energy Ruminant (Mcal) (lb)	Metabolizable Energy Ruminant (Mcal) (kg)	Net Energy Ruminant NE_m (Mcal) (lb)	Net Energy Ruminant NE_m (Mcal) (kg)	Net Energy Ruminant NE_g (Mcal) (lb)	Net Energy Ruminant NE_g (Mcal) (kg)	Net Energy Lactating Cows NE_{lc} (Mcal) (lb)	Net Energy Lactating Cows NE_{lc} (Mcal) (kg)
105	48	0.95	2.10	0.78	1.72	0.44	0.96	0.21	0.47	0.48	1.06
	54	1.07	2.36	0.88	1.93	0.49	1.08	0.24	0.53	0.54	1.19
106	53	1.06	2.35	0.90	1.98	0.55	1.20	0.32	0.70	0.56	1.22
	61	1.22	2.68	1.02	2.26	0.62	1.38	0.36	0.80	0.64	1.40
107	58	1.17	2.57	0.99	2.18	0.60	1.31	0.35	0.78	0.60	1.32
	64	1.28	2.82	1.09	2.39	0.65	1.44	0.39	0.86	0.66	1.45
108	70	1.41	3.10	1.24	2.73	0.75	1.65	0.49	1.08	0.72	1.58
	78	1.56	3.43	1.37	3.02	0.83	1.83	0.54	1.20	0.79	1.75
109	44	0.72	1.58	0.59	1.29	0.51	1.13	0.29	0.63	0.53	1.16
	51	0.83	1.83	0.68	1.50	0.59	1.31	0.33	0.74	0.61	1.35
110	25	0.51	1.12	0.43	0.94	0.25	0.54	0.14	0.31	0.25	0.56
	62	1.23	2.72	1.04	2.30	0.60	1.32	0.34	0.75	0.62	1.36
111	35	0.71	1.56	0.59	1.29	0.32	0.71	0.16	0.36	0.35	0.77
	56	1.12	2.46	0.92	2.04	0.51	1.13	0.26	0.57	0.55	1.22
112	24	0.48	1.06	0.30	0.66	—	—	—	—	—	—
	27	0.53	1.17	0.33	0.73	—	—	—	—	—	—
113	14	0.28	0.63	0.23	0.51	0.14	0.31	0.06	0.14	0.14	0.32
	57	1.13	2.50	0.93	2.05	0.56	1.23	0.25	0.54	0.58	1.27
114	16	0.33	0.73	0.27	0.59	—	—	—	—	—	—
	58	1.18	2.60	0.95	2.09	—	—	—	—	—	—
115	54	1.08	2.37	0.88	1.94	0.53	1.16	0.24	0.53	0.55	1.21
	58	1.16	2.55	0.95	2.09	0.57	1.25	0.26	0.57	0.59	1.30
116	55	1.09	2.41	0.89	1.97	0.54	1.19	0.26	0.57	0.56	1.23
	59	1.18	2.59	0.96	2.12	0.58	1.28	0.28	0.61	0.60	1.32
117	54	1.08	2.37	0.88	1.94	0.53	1.17	0.25	0.54	0.55	1.21
	58	1.16	2.55	0.95	2.09	0.57	1.26	0.26	0.58	0.59	1.30
118	55	1.06	2.34	0.88	1.93	0.47	1.04	0.23	0.51	0.52	1.14
	58	1.11	2.45	0.92	2.03	0.50	1.10	0.25	0.54	0.54	1.20
119	18	0.35	0.77	0.30	0.65	0.17	0.37	0.09	0.21	0.17	0.38
	63	1.22	2.69	1.03	2.27	0.58	1.29	0.33	0.72	0.61	1.33
120	76	1.45	3.19	1.28	2.82	0.75	1.65	0.49	1.09	0.71	1.58
	85	1.62	3.57	1.43	3.16	0.84	1.85	0.55	1.22	0.80	1.77
121	51	1.00	2.19	0.83	1.83	0.46	1.01	0.24	0.52	0.49	1.08
	59	1.15	2.52	0.95	2.10	0.53	1.16	0.27	0.60	0.56	1.24
122	74	1.47	3.25	1.31	2.88	0.75	1.66	0.50	1.09	0.72	1.59
	82	1.64	3.62	1.45	3.21	0.84	1.85	0.55	1.22	0.80	1.77
123	61	1.20	2.64	1.04	2.29	0.70	1.54	0.47	1.04	0.65	1.43
	78	1.54	3.38	1.33	2.94	0.90	1.98	0.60	1.33	0.83	1.83

(Continued)

TABLE II-1

Entry Number	Feed Name Description	Moisture Basis: A-F (As-fed) or M-F (Moisture-free)	Dry Matter	Macrominerals						
				Calcium (Ca)	Phos-phorus (P)	Sodium (Na)	Chlorine (Cl)	Mag-nesium (Mg)	Potassium (K)	Sulfur (S)
			(%)	(%)	(%)	(%)	(%)	(%)	(%)	(%)
	FESCUE *Festuca* spp									
105	-HAY, SUN-CURED, ALTA	**A-F**	**89**	**0.35**	**0.21**	**0.05**	**—**	**0.20**	**2.12**	**—**
		M-F	100	0.39	0.24	0.06	—	0.23	2.38	—
106	-HAY, SUN-CURED, MEADOW	**A-F**	**88**	**0.33**	**0.25**	**—**	**—**	**0.44**	**1.61**	**—**
		M-F	100	0.37	0.29	—	—	0.50	1.84	—
	FLAX, COMMON *Linum usitatissimum*									
107	-SEED SCREENINGS	**A-F**	**91**	**0.34**	**0.43**	**—**	**—**	**0.39**	**0.77**	**0.23**
		M-F	100	0.37	0.47	—	—	0.43	0.84	0.25
108	-SEEDS, MEAL, SOLV EXTD, 35% PROTEIN (LINSEED MEAL)	**A-F**	**90**	**0.40**	**0.82**	**0.14**	**—**	**0.60**	**1.37**	**0.39**
		M-F	100	0.44	0.91	0.15	—	0.66	1.52	0.43
	GALLETA *Hilaria jamesii*									
109	-STEM-CURED, FRESH	**A-F**	**86**	**0.60**	**0.06**	**0.01**	**—**	**0.07**	**0.41**	**0.09**
		M-F	100	0.70	0.07	0.01		0.08	0.48	0.10
	GRAMA *Bouteloua* spp									
110	-IMMATURE, FRESH	**A-F**	**41**	**0.22**	**0.08**	**—**	**—**	**—**	**—**	**—**
		M-F	100	0.53	0.19	—	—	—	—	—
111	-MATURE, FRESH	**A-F**	**63**	**0.22**	**0.08**	**—**	**—**	**—**	**0.22**	**—**
		M-F	100	0.34	0.12	—	—	—	0.35	—
	GRAPE *Vitis* spp									
112	-POMACE, DEHY (MARC)	**A-F**	**90**	**0.52**	**0.15**	**0.08**	**0.01**	**0.09**	**0.82**	**—**
		M-F	100	0.58	0.17	0.09	0.01	0.10	0.91	—
	LESPEDEZA, COMMON—KOREAN *Lespedeza striata—stipulacea*									
113	-PREBLOOM, FRESH	**A-F**	**25**	**0.28**	**0.07**	**—**	**—**	**—**	**0.32**	**—**
		M-F	100	1.12	0.28	—	—	—	1.28	—
114	-EARLY BLOOM, FRESH	**A-F**	**28**	**0.38**	**0.08**	**—**	**—**	**—**	**0.31**	**—**
		M-F	100	1.35	0.27	—	—	—	1.12	—
115	-HAY, EARLY BLOOM, SUN-CURED	**A-F**	**93**	**1.09**	**0.22**	**—**	**—**	**0.23**	**1.02**	**—**
		M-F	100	1.17	0.24	—	—	0.25	1.10	—
116	-HAY, MIDBLOOM, SUN-CURED	**A-F**	**93**	**1.11**	**0.24**	**—**	**—**	**0.25**	**0.93**	**—**
		M-F	100	1.19	0.26	—	—	0.27	1.00	—
117	-HAY, FULL BLOOM, SUN-CURED	**A-F**	**93**	**0.98**	**0.21**	**—**	**—**	**0.22**	**0.91**	**—**
		M-F	100	1.05	0.23	—	—	0.24	0.98	—
	LINSEED (*SEE* FLAX, COMMON)									
	MEADOW PLANTS, INTERMOUNTAIN									
118	-HAY, SUN-CURED	**A-F**	**95**	**0.58**	**0.17**	**0.11**	**—**	**0.16**	**1.50**	**—**
		M-F	100	0.61	0.18	0.12	—	0.17	1.58	—
	MILLET, FOXTAIL *Setaria italica*									
119	-FRESH	**A-F**	**29**	**0.09**	**0.05**	**—**	**—**	**—**	**0.56**	**—**
		M-F	100	0.32	0.19	—	—	—	1.94	—
120	-GRAIN	**A-F**	**89**	**—**	**0.41**	**—**	**—**	**—**	**0.31**	**—**
		M-F	100	—	0.46	—	—	—	0.35	—
121	-HAY, SUN-CURED	**A-F**	**87**	**0.29**	**0.16**	**0.09**	**0.11**	**0.20**	**1.69**	**0.14**
		M-F	100	0.33	0.18	0.10	0.13	0.23	1.94	0.16
	MILLET, PROSO (BROOMCORN) *Panicum miliaceum*									
122	-GRAIN	**A-F**	**90**	**0.03**	**0.30**	**—**	**—**	**0.16**	**0.43**	**—**
		M-F	100	0.03	0.34	—	—	0.18	0.48	—
	MOLASSES AND SYRUP *Beta vulgaris altissima*									
123	-BEET, SUGAR, MOLASSES, MORE THAN 48% INVERT SUGARS, MORE THAN 79.5° BRIX	**A-F**	**78**	**0.12**	**0.03**	**1.16**	**1.28**	**0.23**	**4.73**	**0.46**
		M-F	100	0.16	0.03	1.48	1.64	0.29	6.07	0.60

(Continued)

Entry Number	Microminerals							Fat-soluble Vitamins			
	Cobalt (Co)	Copper (Cu)	Iodine (I)	Iron (Fe)	Manganese (Mn)	Selenium (Se)	Zinc (Zn)	A (1 mg Carotene = 1,667 IU Vit. A)	Carotene (Provitamin A)	D	E (α-tocopherol)
	(ppm or mg/kg)	*(ppm or mg/kg)*	*(ppm or mg/kg)*	*(%)*	*(ppm or mg/kg)*	*(ppm or mg/kg)*	*(ppm or mg/kg)*	*(IU/g)*	*(ppm or mg/kg)*	*(IU/kg)*	*(ppm or mg/kg)*
105	— —	— —	— —	— —	— —	— —	— —	**30.8** 34.6	**18.5** 20.7	— —	— —
106	**0.119** 0.135	— —	— —	— —	**21.4** 24.5	— —	— —	**105.8** 120.9	**63.4** 72.5	— —	**118.6** 135.6
107	— —	— —	— —	**0.010** 0.010	— —	— —	— —	— —	— —	— —	— —
108	— —	— —	— —	— —	— —	— —	— —	— —	— —	— —	**5.9** 6.5
109	**0.591** 0.687	**16.3** 19.0	— —	**0.044** 0.052	**67.7** 78.7	— —	**19.5** 22.7	**0.3** 0.3	**0.2** 0.2	— —	— —
110	— —	**2.3** 5.5	— —	— —	**18.2** 44.3	— —	— —	— —	— —	— —	— —
111	**0.115** 0.181	**8.1** 12.8	— —	**0.083** 0.130	**30.0** 47.4	— —	— —	**32.2** 50.7	**19.3** 30.4	— —	— —
112	— —	— —	— —	— —	**36.8** 40.7	— —	**21.9** 24.2	— —	— —	— —	— —
113	— —	— —	— —	— —	— —	— —	— —	— —	— —	— —	— —
114	— —	— —	— —	**0.008** 0.025	**58.4** 208.6	— —	— —	— —	— —	— —	— —
115	**0.038** 0.040	**0.2** 0.2	— —	**0.038** 0.041	**237.6** 255.5	— —	— —	**213.2** 229.2	**127.9** 137.5	— —	— —
116	— —	— —	— —	**0.030** 0.033	**220.9** 237.5	— —	— —	**85.3** 91.7	**51.2** 55.0	— —	— —
117	— —	— —	— —	**0.030** 0.033	**140.9** 151.5	— —	— —	**19.4** 20.8	**11.6** 12.5	— —	— —
118	— —	— —	— —	— —	— —	— —	— —	**53.1** 55.8	**31.9** 33.5	— —	— —
119	— —	— —	— —	— —	— —	— —	— —	— —	— —	— —	— —
120	— —	— —	— —	**0.010** 0.011	— —	— —	— —	— —	— —	— —	— —
121	— —	— —	— —	— —	**120.1** 138.1	— —	— —	**86.9** 100.0	**52.1** 60.0	— —	— —
122	— —	— —	— —	**0.007** 0.008	— —	— —	— —	— —	— —	— —	— —
123	**0.362** 0.465	**16.8** 21.6	— —	**0.007** 0.009	**4.5** 5.8	— —	**14.0** 18.0	— —	— —	— —	**4.0** 5.1

(Continued)

TABLE II-1

Entry Number	Feed Name Description	International Feed Number[2]	Moisture Basis: A-F (As-fed) or M-F (Moisture-free)	Chemical Analysis								
				Dry Matter	Ash	Crude Fiber	Cell Walls or NDF	Acid Detergent Fiber	Lignin	Ether Extract (Fat)	N-Free Extract	Crude Protein
				(%)	(%)	(%)	(%)	(%)	(%)	(%)	(%)	(%)
124	-CITRUS, SYRUP (CITRUS MOLASSES)	4-01-241	A-F	67	5.1	—	—	—	—	0.2	55.7	5.8
			M-F	100	7.6	—	—	—	—	0.3	82.7	8.5
125	-SUGARCANE, MOLASSES, DEHY	4-04-695	A-F	94	12.5	6.3	—	—	—	0.9	65.0	9.7
			M-F	100	13.3	6.7	—	—	—	0.9	68.8	10.3
126	-SUGARCANE, MOLASSES, MORE THAN 46% INVERT SUGARS, MORE THAN 79.5° BRIX (BLACK STRAP)	4-04-696	A-F	74	9.8	0.4	—	0.3	0.2	0.2	59.7	4.3
			M-F	100	13.2	0.5	—	0.4	0.3	0.2	80.2	5.8
	NAPIERGRASS *Pennisetum purpureum*											
127	-PREBLOOM, FRESH	2-03-158	A-F	20	1.7	6.7	—	—	—	0.6	9.5	1.8
			M-F	100	8.6	33.0	—	—	—	3.0	46.7	8.7
128	-LATE BLOOM, FRESH	2-03-162	A-F	23	1.2	9.0	—	—	—	0.3	10.8	1.8
			M-F	100	5.3	39.0	—	—	—	1.1	46.8	7.8
	NEEDLE-AND-THREAD *Stipa comata*											
129	-STEM-CURED, FRESH	2-07-989	A-F	92	19.4	—	—	39.7	5.9	5.0	—	3.7
			M-F	100	21.1	—	—	43.2	6.4	5.4	—	4.1
	OATS *Avena sativa*											
130	-GRAIN, ALL ANALYSES	4-03-309	A-F	89	3.1	10.7	26.4	14.2	2.7	4.7	58.9	11.9
			M-F	100	3.4	11.9	29.6	15.9	3.0	5.2	66.1	13.3
131	-GRAIN, PACIFIC COAST	4-07-999	A-F	91	3.8	11.2	—	—	—	5.0	61.8	9.1
			M-F	100	4.2	12.3	—	—	—	5.5	68.0	10.0
132	-GROATS	4-03-331	A-F	90	2.1	2.5	—	—	—	6.2	63.0	15.8
			M-F	100	2.4	2.8	—	—	—	6.9	70.3	17.6
133	-HAY, SUN-CURED, ALL ANALYSES	1-03-280	A-F	91	7.2	29.1	—	34.8	—	2.2	43.6	8.6
			M-F	100	7.9	32.0	—	38.4	—	2.4	48.1	9.5
134	-HULLS	1-03-281	A-F	92	6.1	30.9	68.6	37.4	6.5	1.3	50.5	3.7
			M-F	100	6.6	33.4	74.3	40.5	7.0	1.4	54.7	4.0
135	-SILAGE, DOUGH STAGE	3-03-296	A-F	35	2.4	11.6	—	—	—	1.4	16.1	3.5
			M-F	100	6.9	33.0	—	—	—	4.1	46.0	10.0
136	-STRAW	1-03-283	A-F	92	7.2	37.2	65.7	43.1	7.0	2.0	41.6	4.1
			M-F	100	7.8	40.4	71.3	46.8	7.6	2.2	45.2	4.4
	ORCHARDGRASS *Dactylis glomerata*											
137	-IMMATURE, FRESH	2-03-439	A-F	23	2.6	5.7	9.8	5.2	0.7	1.1	7.6	5.5
			M-F	100	11.4	25.2	43.2	22.8	3.3	5.0	33.8	24.5
138	-MIDBLOOM, FRESH	2-03-443	A-F	27	2.1	9.2	15.8	9.8	2.1	1.0	12.4	2.8
			M-F	100	7.5	33.5	57.6	35.6	7.6	3.5	45.4	10.1
139	-HAY, SUN-CURED, ALL ANALYSES	1-03-438	A-F	89	6.5	31.0	64.1	36.0	—	2.8	39.7	9.4
			M-F	100	7.3	34.7	71.8	40.3	—	3.1	44.3	10.5
140	-HAY, FULL BLOOM, SUN-CURED	1-03-427	A-F	93	8.3	33.4	58.5	34.3	4.3	2.9	39.5	8.7
			M-F	100	8.9	36.0	63.1	37.0	4.6	3.1	42.6	9.4
	PANGOLAGRASS *Digitaria decumbens*											
141	-FRESH	2-03-493	A-F	20	1.5	6.6	—	7.5	1.0	0.5	9.8	1.8
			M-F	100	7.6	32.6	—	36.9	5.0	2.3	48.4	9.1
	PEA *Pisum* spp											
142	-SEEDS	5-03-600	A-F	89	2.9	5.5	—	—	—	1.1	56.7	23.2
			M-F	100	3.2	6.1	—	—	—	1.2	63.4	26.0
143	-STRAW	1-03-577	A-F	87	5.7	34.3	—	—	—	1.5	37.7	7.8
			M-F	100	6.5	39.5	—	—	—	1.8	43.4	8.9
144	-VINES (WITHOUT SEEDS, WITH PODS), SILAGE	3-03-596	A-F	25	2.2	7.3	—	—	—	0.8	11.0	3.2
			M-F	100	9.0	29.8	—	—	—	3.3	44.9	13.1
	PEANUT *Arachis hypogaea*											
145	-HAY, SUN-CURED	1-03-619	A-F	91	8.2	30.3	—	37.2	—	3.3	39.1	9.9
			M-F	100	9.0	33.4	—	41.0	—	3.6	43.1	10.9

(Continued)

Entry Number	TDN	Digestible Energy		Metabolizable Energy		Net Energy					
	Ruminant	Ruminant		Ruminant		Ruminant NE_m		Ruminant NE_g		Lactating Cows NE_{lc}	
	(%)	*(Mcal)*		*(Mcal)*		*(Mcal)*		*(Mcal)*		*(Mcal)*	
		(lb)	*(kg)*	*(lb)*	*(kg)*	*(lb)*	*(kg)*	*(lb)*	*(kg)*	*(lb)*	*(kg)*
124	51	1.01	2.22	0.86	1.89	0.57	1.26	0.38	0.83	0.52	1.15
	75	1.49	3.29	1.27	2.81	0.85	1.87	0.56	1.23	0.78	1.71
125	66	1.36	2.99	1.18	2.60	0.70	1.55	0.44	0.97	0.68	1.51
	70	1.44	3.17	1.25	2.75	0.74	1.64	0.47	1.03	0.73	1.60
126	60	1.22	2.68	1.12	2.46	0.77	1.70	0.53	1.18	0.64	1.41
	81	1.63	3.60	1.50	3.31	1.04	2.28	0.72	1.58	0.86	1.89
127	11	0.22	0.49	0.19	0.41	0.11	0.25	0.06	0.13	0.12	0.26
	55	1.11	2.44	0.91	2.01	0.55	1.21	0.29	0.64	0.58	1.28
128	12	0.24	0.54	0.20	0.44	0.13	0.28	0.07	0.15	0.13	0.29
	53	1.06	2.34	0.87	1.91	0.55	1.21	0.29	0.64	0.58	1.28
129	45	0.87	1.93	0.76	1.67	—	—	—	—	—	—
	49	0.95	2.09	0.82	1.81	—	—	—	—	—	—
130	69	1.36	3.00	1.19	2.62	0.80	1.77	0.54	1.19	0.70	1.55
	77	1.53	3.37	1.33	2.94	0.90	1.98	0.60	1.33	0.79	1.74
131	70	1.41	3.11	1.24	2.73	0.76	1.68	0.50	1.11	0.73	1.61
	78	1.55	3.42	1.36	3.00	0.84	1.85	0.55	1.22	0.80	1.77
132	87	1.72	3.80	1.46	3.21	1.02	2.24	0.72	1.58	0.88	1.94
	98	1.92	4.24	1.63	3.59	1.14	2.50	0.80	1.77	0.98	2.17
133	52	1.04	2.29	0.93	2.04	0.57	1.26	0.33	0.73	0.55	1.22
	57	1.15	2.53	1.02	2.25	0.63	1.39	0.37	0.81	0.61	1.35
134	32	0.64	1.41	0.46	1.01	0.16	0.36	–0.05	–0.12	0.30	0.67
	35	0.69	1.52	0.49	1.09	0.18	0.39	–0.06	–0.13	0.33	0.72
135	20	0.39	0.85	0.34	0.75	0.18	0.40	0.09	0.20	0.18	0.40
	57	1.10	2.43	0.97	2.13	0.52	1.14	0.26	0.58	0.52	1.15
136	46	1.15	2.53	0.78	1.72	0.43	0.94	0.20	0.43	0.48	1.06
	50	1.24	2.74	0.84	1.86	0.46	1.02	0.21	0.47	0.52	1.15
137	15	0.30	0.66	0.26	0.57	0.16	0.34	0.10	0.21	0.16	0.34
	67	1.33	2.93	1.14	2.51	0.69	1.52	0.42	0.93	0.68	1.51
138	16	0.33	0.73	0.28	0.61	0.16	0.36	0.09	0.20	0.17	0.37
	60	1.20	2.64	1.01	2.22	0.59	1.31	0.33	0.73	0.61	1.35
139	51	1.23	2.71	0.96	2.11	0.50	1.09	0.27	0.59	0.52	1.15
	57	1.38	3.03	1.07	2.36	0.56	1.22	0.30	0.66	0.58	1.29
140	50	1.02	2.25	0.83	1.82	0.46	1.01	0.22	0.49	0.50	1.10
	54	1.10	2.42	0.89	1.97	0.49	1.09	0.24	0.53	0.54	1.19
141	12	0.24	0.54	0.20	0.45	0.12	0.26	0.07	0.14	0.12	0.26
	60	1.20	2.65	1.01	2.23	0.58	1.28	0.32	0.71	0.57	1.26
142	77	1.56	3.44	1.39	3.07	0.87	1.91	0.59	1.31	0.81	1.79
	87	1.75	3.85	1.56	3.44	0.97	2.14	0.67	1.47	0.91	2.00
143	43	0.85	1.88	0.68	1.51	0.48	1.07	0.26	0.57	0.51	1.12
	49	0.98	2.16	0.79	1.73	0.56	1.23	0.30	0.66	0.59	1.29
144	14	0.28	0.61	0.23	0.51	0.13	0.29	0.07	0.15	0.14	0.31
	57	1.13	2.49	0.94	2.07	0.53	1.18	0.28	0.61	0.57	1.25
145	48	0.95	2.10	0.78	1.71	0.39	0.85	0.16	0.36	0.45	0.99
	53	1.05	2.32	0.86	1.89	0.43	0.94	0.18	0.40	0.49	1.09

(Continued)

TABLE II-1

Entry Number	Feed Name Description	Moisture Basis: A-F (As-fed) or M-F (Moisture-free)	Dry Matter	Macrominerals						
				Calcium (Ca)	Phos-phorus (P)	Sodium (Na)	Chlorine (Cl)	Mag-nesium (Mg)	Potassium (K)	Sulfur (S)
			(%)	(%)	(%)	(%)	(%)	(%)	(%)	(%)
124	-CITRUS, SYRUP (CITRUS MOLASSES)	**A-F** M-F	**67** 100	**1.18** 1.76	**0.09** 0.13	**0.28** 0.41	**0.07** 0.11	**0.14** 0.21	**0.09** 0.14	**0.16** 0.23
125	-SUGARCANE, MOLASSES, DEHY	**A-F** M-F	**94** 100	**1.04** 1.10	**0.42** 0.45	**0.19** 0.20	— —	**0.44** 0.47	**3.40** 3.60	**0.43** 0.46
126	-SUGARCANE, MOLASSES, MORE THAN 46% INVERT SUGARS, MORE THAN 79.5° BRIX (BLACK STRAP)	**A-F** M-F	**74** 100	**0.74** 1.00	**0.08** 0.11	**0.16** 0.22	**2.26** 3.04	**0.31** 0.42	**2.98** 4.01	**0.35** 0.47
	NAPIERGRASS *Pennisetum purpureum*									
127	-PREBLOOM, FRESH	**A-F** M-F	**20** 100	**0.12** 0.60	**0.08** 0.41	**0.00** 0.01	— —	**0.05** 0.26	**0.27** 1.31	**0.02** 0.10
128	-LATE BLOOM, FRESH	**A-F** M-F	**23** 100	**0.08** 0.35	**0.07** 0.30	**0.00** 0.01	— —	**0.06** 0.26	**0.30** 1.31	**0.02** 0.10
	NEEDLE-AND-THREAD *Stipa comata*									
129	-STEM-CURED, FRESH	**A-F** M-F	**92** 100	**0.99** 1.08	**0.06** 0.06	— —	— —	— —	— —	— —
	OATS *Avena sativa*									
130	-GRAIN, ALL ANALYSES	**A-F** M-F	**89** 100	**0.08** 0.09	**0.34** 0.38	**0.05** 0.06	**0.09** 0.10	**0.14** 0.16	**0.40** 0.45	**0.21** 0.23
131	-GRAIN, PACIFIC COAST	**A-F** M-F	**91** 100	**0.10** 0.11	**0.31** 0.34	**0.06** 0.07	**0.12** 0.13	**0.17** 0.19	**0.38** 0.42	**0.20** 0.22
132	-GROATS	**A-F** M-F	**90** 100	**0.08** 0.08	**0.43** 0.48	**0.05** 0.06	**0.08** 0.09	**0.11** 0.13	**0.35** 0.39	**0.20** 0.22
133	-HAY, SUN-CURED, ALL ANALYSES	**A-F** M-F	**91** 100	**0.29** 0.32	**0.23** 0.25	**0.17** 0.18	**0.47** 0.52	**0.26** 0.29	**1.35** 1.49	**0.21** 0.23
134	-HULLS	**A-F** M-F	**92** 100	**0.14** 0.15	**0.14** 0.15	**0.04** 0.04	**0.08** 0.08	**0.08** 0.09	**0.57** 0.62	**0.14** 0.15
135	-SILAGE, DOUGH STAGE	**A-F** M-F	**35** 100	**0.17** 0.47	**0.12** 0.33	— —	— —	— —	— —	— —
136	-STRAW	**A-F** M-F	**92** 100	**0.22** 0.24	**0.06** 0.07	**0.39** 0.42	**0.72** 0.78	**0.16** 0.17	**2.35** 2.55	**0.21** 0.23
	ORCHARDGRASS *Dactylis glomerata*									
137	-IMMATURE, FRESH	**A-F** M-F	**23** 100	**0.13** 0.57	**0.12** 0.54	— —	**0.02** 0.08	**0.06** 0.27	**0.73** 3.21	— —
138	-MIDBLOOM, FRESH	**A-F** M-F	**27** 100	**0.17** 0.60	**0.05** 0.17	**0.07** 0.26	— —	**0.09** 0.33	**0.57** 2.09	— —
139	-HAY, SUN-CURED, ALL ANALYSES	**A-F** M-F	**89** 100	**0.34** 0.38	**0.23** 0.26	**0.01** 0.02	**0.37** 0.41	**0.16** 0.18	**2.68** 3.00	**0.23** 0.26
140	-HAY, FULL BOOM, SUN-CURED	**A-F** M-F	**93** 100	— —	— —	— —	— —	— —	— —	— —
	PANGOLAGRASS *Digitaria decumbens*									
141	-FRESH	**A-F** M-F	**20** 100	**0.08** 0.38	**0.05** 0.22	— —	**0.04** 0.18	**0.29** 1.43	— —	— —
	PEA *Pisum* spp									
142	-SEEDS	**A-F** M-F	**89** 100	**0.12** 0.14	**0.41** 0.46	**0.04** 0.05	**0.05** 0.06	**0.12** 0.14	**0.95** 1.06	— —
143	-STRAW	**A-F** M-F	**87** 100	— —	— —	— —	— —	— —	— —	— —
144	-VINES (WITHOUT SEEDS, WITH PODS), SILAGE	**A-F** M-F	**25** 100	**0.32** 1.31	**0.06** 0.24	**0.00** 0.01	— —	**0.10** 0.39	**0.34** 1.40	**0.06** 0.25
	PEANUT *Arachis hypogaea*									
145	-HAY, SUN-CURED	**A-F** M-F	**91** 100	**1.12** 1.23	**0.14** 0.16	— —	— —	**0.44** 0.49	**1.25** 1.38	**0.21** 0.23

(Continued)

Entry Number	Microminerals							Fat-soluble Vitamins			
	Cobalt (Co)	Copper (Cu)	Iodine (I)	Iron (Fe)	Manganese (Mn)	Selenium (Se)	Zinc (Zn)	A (1 mg Carotene = 1,667 IU Vit. A)	Carotene (Provitamin A)	D	E (α-tocopherol)
	(ppm or mg/kg)	*(ppm or mg/kg)*	*(ppm or mg/kg)*	*(%)*	*(ppm or mg/kg)*	*(ppm or mg/kg)*	*(ppm or mg/kg)*	*(IU/g)*	*(ppm or mg/kg)*	*(IU/kg)*	*(ppm or mg/kg)*
124	**0.109**	**72.8**	**—**	**0.035**	**40.9**	**—**	**92.4**	**—**	**—**	**—**	**—**
	0.162	108.0	—	0.051	60.7	—	137.1	—	—	—	—
125	**1.145**	**74.9**	**—**	**0.024**	**54.1**	**—**	**31.2**	**—**	**—**	**—**	**5.2**
	1.213	79.4	—	0.025	57.3	—	33.0	—	—	—	5.5
126	**1.180**	**48.9**	**1.564**	**0.020**	**43.7**	**—**	**15.6**	**—**	**—**	**—**	**5.4**
	1.587	65.7	2.103	0.027	58.8	—	20.9	—	—	—	7.3
127	**—**	**—**	**—**	**—**	**—**	**—**	**—**	**—**	**—**	**—**	**—**
	—	—	—	—	—	—	—	—	—	—	—
128	**—**	**—**	**—**	**—**	**—**	**—**	**—**	**—**	**—**	**—**	**—**
	—	—	—	—	—	—	—	—	—	—	—
129	**—**	**—**	**—**	**—**	**—**	**—**	**—**	**—**	**—**	**—**	**—**
	—	—	—	—	—	—	—	—	—	—	—
130	**0.056**	**6.0**	**0.112**	**0.007**	**35.8**	**0.215**	**34.9**	**0.2**	**0.1**	**—**	**14.9**
	0.063	6.7	0.125	0.008	40.1	0.241	39.2	0.2	0.1	—	16.8
131	**—**	**—**	**—**	**0.008**	**38.0**	**0.076**	**—**	**—**	**—**	**—**	**20.2**
	—	—	—	0.008	41.8	0.084	—	—	—	—	22.2
132	**—**	**6.0**	**0.108**	**0.008**	**27.8**	**—**	**0.0**	**—**	**—**	**—**	**14.8**
	—	6.7	0.120	0.009	31.0	—	0.1	—	—	—	16.5
133	**0.067**	**4.4**	**—**	**0.037**	**89.6**	**—**	**40.8**	**45.0**	**27.0**	**1**	**—**
	0.073	4.8	—	0.041	98.7	—	45.0	49.6	29.7	2	—
134	**—**	**4.1**	**—**	**0.011**	**18.8**	**—**	**—**	**—**	**—**	**—**	**—**
	—	4.5	—	0.012	20.4	—	—	—	—	—	—
135	**—**	**—**	**—**	**—**	**—**	**—**	**—**	**35.1**	**21.1**	**—**	**—**
	—	—	—	—	—	—	—	100.0	60.0	—	—
136	**—**	**9.5**	**—**	**0.016**	**29.0**	**—**	**5.5**	**5.8**	**3.5**	**1**	**—**
	—	10.3	—	0.017	31.5	—	5.9	6.3	3.8	1	—
137	**—**	**1.6**	**—**	**—**	**7.1**	**—**	**—**	**236.1**	**141.6**	**—**	**—**
	—	7.0	—	—	31.4	—	—	1,044.6	626.6	—	—
138	**0.028**	**13.7**	**—**	**0.002**	**37.2**	**—**	**6.9**	**—**	**—**	**—**	**—**
	0.102	50.1	—	0.007	135.9	—	25.1	—	—	—	—
139	**0.339**	**12.9**	**—**	**0.014**	**162.7**	**—**	**32.0**	**28.9**	**17.3**	**—**	**170.7**
	0.379	14.5	—	0.015	182.3	—	35.8	32.4	19.4	—	191.1
140	**—**	**—**	**—**	**—**	**—**	**—**	**—**	**—**	**—**	**—**	**—**
	—	—	—	—	—	—	—	—	—	—	—
141	**—**	**—**	**—**	**—**	**—**	**—**	**—**	**—**	**—**	**—**	**—**
	—	—	—	—	—	—	—	—	—	—	—
142	**—**	**—**	**—**	**0.007**	**2.9**	**—**	**23.0**	**1.2**	**0.7**	**—**	**3.0**
	—	—	—	0.008	3.2	—	25.7	1.3	0.8	—	3.3
143	**—**	**—**	**—**	**—**	**—**	**—**	**—**	**—**	**—**	**—**	**—**
	—	—	—	—	—	—	—	—	—	—	—
144	**—**	**—**	**—**	**0.003**	**—**	**—**	**—**	**77.2**	**46.3**	**—**	**—**
	—	—	—	0.010	—	—	—	315.0	189.0	—	—
145	**0.072**	**—**	**—**	**—**	**—**	**—**	**—**	**52.6**	**31.5**	**—**	**—**
	0.080	—	—	—	—	—	—	58.0	34.8	—	—

(Continued)

TABLE II-1

Entry Number	Feed Name Description	International Feed Number[2]	Moisture Basis: A-F (As-fed) or M-F (Moisture-free)	Chemical Analysis								
				Dry Matter	Ash	Crude Fiber	Cell Walls or NDF	Acid Detergent Fiber	Lignin	Ether Extract (Fat)	N-Free Extract	Crude Protein
				(%)	(%)	(%)	(%)	(%)	(%)	(%)	(%)	(%)
146	-HULLS (PODS)	1-08-028	A-F	91	3.8	57.2	69.0	60.8	21.8	2.1	20.7	7.3
			M-F	100	4.2	62.9	75.8	66.8	23.9	2.3	22.8	8.0
147	-SEEDS WITHOUT HULLS, MEAL, MECH EXTD (PEANUT MEAL)	5-03-649	A-F	93	5.0	6.2	13.2	5.6	1.0	5.6	26.7	49.2
			M-F	100	5.4	6.7	14.2	6.1	1.1	6.0	28.8	53.1
148	-SEEDS WITHOUT HULLS, MEAL, SOLV EXTD (PEANUT MEAL)	5-03-650	A-F	93	5.8	7.7	—	—	—	2.2	27.9	49.0
			M-F	100	6.3	8.3	—	—	—	2.4	30.1	52.9
	PEARL MILLET *Pennisetum glaucum*											
149	-FRESH	2-03-115	A-F	21	1.9	6.5	—	—	—	0.6	9.7	2.1
			M-F	100	9.2	31.1	—	—	—	2.9	46.8	10.1
	PINEAPPLE *Ananas comosus*											
150	-CANNERY RESIDUE, DEHY (PINEAPPLE BRAN)	4-03-722	A-F	87	3.0	18.2	—	—	—	1.3	60.5	4.0
			M-F	100	3.5	20.9	—	—	—	1.5	69.5	4.6
	POTATO *Solanum tuberosum*											
151	-CANNERY RESIDUE, DEHY	4-03-775	A-F	89	3.0	6.5	—	—	—	0.3	71.5	7.4
			M-F	100	3.4	7.3	—	—	—	0.4	80.5	8.4
152	-TUBERS, FRESH	4-03-787	A-F	24	1.1	0.6	—	—	—	0.1	19.5	2.2
			M-F	100	4.8	2.4	—	—	—	0.4	83.2	9.3
153	-TUBERS, ALFALFA HAY ADDED, SILAGE	3-03-770	A-F	35	2.2	7.1	—	—	—	0.5	20.7	4.2
			M-F	100	6.3	20.5	—	—	—	1.4	59.7	12.2
	POULTRY											
154	-FEATHERS, MEAL, HYDROLYZED	5-03-795	A-F	93	3.2	1.4	—	6.1	—	5.1	—	83.8
			M-F	100	3.4	1.5	—	6.6	—	5.5	—	90.2
155	-MANURE WITHOUT LITTER, DEHY	5-14-015	A-F	90	29.2	12.0	36.3	14.4	2.1	2.0	21.6	25.4
			M-F	100	32.4	13.3	40.2	16.0	2.3	2.2	23.9	28.2
156	-MANURE WITH LITTER, DEHY	5-05-587	A-F	86	17.8	15.0	—	—	8.0	2.6	25.7	24.6
			M-F	100	20.8	17.5	—	—	9.4	3.0	30.0	28.7
	PRAIRIE GRASS, MIDWEST (PRAIRIE HAY)											
157	-HAY, SUN-CURED	1-03-191	A-F	91	7.2	30.7	—	—	—	2.1	45.2	5.8
			M-F	100	8.0	33.7	—	—	—	2.3	49.6	6.4
	RAPE *Brassica napus*											
158	-FRESH	2-03-867	A-F	17	2.1	2.4	—	—	—	0.6	8.5	2.9
			M-F	100	12.6	14.7	—	—	—	3.8	51.2	17.6
159	-SEEDS, MEAL, MECH EXTD	5-03-870	A-F	92	6.9	12.0	—	—	—	7.3	30.1	35.6
			M-F	100	7.5	13.1	—	—	—	7.9	32.7	38.7
160	-SEEDS, MEAL, SOLV EXTD, 34% PROTEIN	5-26-092	A-F	90	7.0	13.0	—	—	—	2.5	33.5	34.0
			M-F	100	7.8	14.4	—	—	—	2.8	37.2	37.8
	REDTOP *Agrostis alba*											
161	-FULL BLOOM, FRESH	2-03-891	A-F	26	1.8	6.6	—	—	—	0.9	14.8	2.1
			M-F	100	7.0	25.1	—	—	—	3.5	56.3	8.1
162	-HAY, SUN-CURED	1-03-885	A-F	92	6.0	28.4	—	—	—	2.8	47.4	7.4
			M-F	100	6.6	30.9	—	—	—	3.1	51.5	8.1
	RICE *Oryza sativa*											
163	-BRAN WITH GERMS (RICE, BRAN)	4-03-928	A-F	91	11.3	11.9	28.0	25.7	3.6	13.5	41.0	13.0
			M-F	100	12.5	13.1	30.9	28.4	4.0	14.9	45.2	14.3
164	-GRAIN, GROUND (GROUND ROUGH RICE, GROUND PADDY RICE)	4-03-938	A-F	89	5.3	8.6	—	—	—	1.6	65.9	7.5
			M-F	100	6.0	9.7	—	—	—	1.8	74.1	8.4
165	-HULLS	1-08-075	A-F	92	19.0	38.9	71.9	62.3	9.6	1.0	30.3	3.0
			M-F	100	20.6	42.2	78.0	67.6	10.4	1.1	32.9	3.2
166	-STRAW	1-03-925	A-F	91	15.4	31.9	64.4	50.1	4.4	1.3	38.2	3.9
			M-F	100	17.0	35.1	71.0	55.2	4.9	1.4	42.1	4.3

COMPOSITION OF FEEDS

(Continued)

Entry Number	TDN	Digestible Energy		Metabolizable Energy		Net Energy					
	Ruminant	Ruminant		Ruminant		Ruminant NE_m		Ruminant NE_g		Lactating Cows NE_{lc}	
	(%)	*(Mcal)*		*(Mcal)*		*(Mcal)*		*(Mcal)*		*(Mcal)*	
		(lb)	*(kg)*	*(lb)*	*(kg)*	*(lb)*	*(kg)*	*(lb)*	*(kg)*	*(lb)*	*(kg)*
146	16	0.33	0.72	0.14	0.32	–0.16	–0.36	–0.38	–0.83	0.11	0.24
	18	0.36	0.79	0.16	0.35	–0.18	–0.39	–0.41	–0.91	0.12	0.26
147	81	1.54	3.40	1.47	3.24	0.86	1.89	0.58	1.28	0.81	1.78
	87	1.66	3.67	1.59	3.50	0.92	2.04	0.63	1.38	0.87	1.92
148	73	1.43	3.15	1.29	2.84	0.78	1.72	0.51	1.13	0.74	1.63
	79	1.54	3.40	1.39	3.07	0.84	1.86	0.55	1.22	0.80	1.76
149	13	0.25	0.56	0.21	0.47	0.12	0.27	0.07	0.15	0.13	0.28
	62	1.22	2.68	1.02	2.26	0.59	1.29	0.33	0.72	0.61	1.34
150	64	1.27	2.81	1.11	2.45	0.70	1.55	0.46	1.00	0.67	1.49
	73	1.46	3.23	1.27	2.81	0.81	1.78	0.52	1.15	0.77	1.71
151	79	1.47	3.23	1.30	2.86	0.73	1.60	0.47	1.04	0.70	1.53
	90	1.65	3.64	1.46	3.22	0.82	1.80	0.53	1.18	0.78	1.73
152	19	0.38	0.84	0.34	0.74	0.20	0.45	0.14	0.30	0.19	0.43
	81	1.62	3.58	1.43	3.16	0.87	1.92	0.58	1.28	0.83	1.82
153	20	0.42	0.93	0.35	0.78	0.22	0.48	0.13	0.28	0.22	0.49
	59	1.21	2.68	1.02	2.25	0.63	1.40	0.37	0.81	0.64	1.41
154	67	1.21	2.66	0.84	1.84	0.48	1.05	0.24	0.53	0.65	1.43
	72	1.30	2.87	0.90	1.98	0.51	1.13	0.26	0.58	0.70	1.54
155	46	0.92	2.02	0.74	1.63	0.54	1.18	0.30	0.66	0.55	1.22
	51	1.01	2.24	0.82	1.81	0.59	1.31	0.33	0.73	0.61	1.35
156	47	0.99	2.19	0.83	1.83	0.52	1.16	0.30	0.66	0.54	1.18
	55	1.16	2.56	0.97	2.14	0.61	1.35	0.35	0.77	0.63	1.38
157	46	0.94	2.06	0.76	1.68	0.44	0.97	0.21	0.46	0.48	1.07
	50	1.03	2.27	0.84	1.84	0.48	1.06	0.23	0.51	0.53	1.17
158	13	0.25	0.54	0.21	0.47	0.12	0.26	0.07	0.16	0.12	0.26
	79	1.47	3.24	1.28	2.83	0.71	1.57	0.44	0.97	0.70	1.54
159	71	1.41	3.11	1.24	2.73	0.79	1.75	0.53	1.16	0.75	1.66
	77	1.53	3.38	1.35	2.97	0.86	1.90	0.57	1.26	0.82	1.81
160	—	—	—	—	—	—	—	—	—	—	—
	—	—	—	—	—	—	—	—	—	—	—
161	16	0.33	0.72	0.28	0.61	0.16	0.36	0.10	0.21	0.17	0.37
	62	1.24	2.73	1.04	2.30	0.62	1.38	0.36	0.80	0.64	1.40
162	50	1.02	2.25	0.84	1.86	0.50	1.11	0.27	0.59	0.53	1.17
	54	1.11	2.45	0.92	2.02	0.55	1.21	0.29	0.64	0.58	1.27
163	64	1.10	2.42	0.99	2.18	0.62	1.38	0.38	0.84	0.53	1.18
	71	1.21	2.67	1.09	2.40	0.69	1.52	0.42	0.93	0.59	1.30
164	68	1.35	2.98	1.18	2.61	0.71	1.57	0.46	1.01	0.68	1.51
	76	1.52	3.35	1.33	2.93	0.80	1.76	0.52	1.14	0.77	1.70
165	11	0.27	0.60	0.16	0.35	–0.26	–0.57	–0.47	–1.04	0.08	0.17
	12	0.30	0.65	0.17	0.38	–0.28	–0.62	–0.51	–1.13	0.08	0.19
166	40	0.80	1.76	0.62	1.37	0.34	0.75	0.12	0.26	0.41	0.91
	44	0.88	1.94	0.69	1.51	0.37	0.82	0.13	0.29	0.46	1.01

(Continued)

TABLE II-1

Entry Number	Feed Name Description	Moisture Basis: A-F (As-fed) or M-F (Moisture-free)	Dry Matter	Macrominerals: Calcium (Ca)	Phosphorus (P)	Sodium (Na)	Chlorine (Cl)	Magnesium (Mg)	Potassium (K)	Sulfur (S)
			(%)	(%)	(%)	(%)	(%)	(%)	(%)	(%)
146	-HULLS (PODS)	**A-F**	**91**	**0.24**	**0.06**	**0.12**	**—**	**0.15**	**0.87**	**0.09**
		M-F	100	0.26	0.07	0.13	—	0.17	0.95	0.10
147	-SEEDS WITHOUT HULLS, MEAL, MECH EXTD (PEANUT MEAL)	**A-F**	**93**	**0.20**	**0.56**	**0.12**	**0.03**	**0.26**	**1.16**	**0.22**
		M-F	100	0.22	0.61	0.13	0.03	0.28	1.25	0.24
148	-SEEDS WITHOUT HULLS, MEAL, SOLV EXTD (PEANUT MEAL)	**A-F**	**93**	**0.36**	**0.61**	**0.03**	**0.03**	**0.27**	**1.16**	**0.31**
		M-F	100	0.39	0.66	0.03	0.03	0.30	1.25	0.33
	PEARL MILLET *Pennisetum glaucum*									
149	-FRESH	**A-F**	**21**	**—**	**—**	**—**	**—**	**—**	**—**	**—**
		M-F	100	—	—	—	—	—	—	—
	PINEAPPLE *Ananas comosus*									
150	CANNERY RESIDUE, DEHY (PINEAPPLE BRAN)	**A-F**	**87**	**0.20**	**0.11**	**—**	**—**	**—**	**—**	**—**
		M-F	100	0.23	0.13	—	—	—	—	—
	POTATO *Solanum tuberosum*									
151	-CANNERY RESIDUE, DEHY	**A-F**	**89**	**0.14**	**0.23**	**—**	**—**	**—**	**—**	**—**
		M-F	100	0.16	0.25	—	—	—	—	—
152	-TUBERS, FRESH	**A-F**	**24**	**0.01**	**0.06**	**0.02**	**0.07**	**0.03**	**0.51**	**0.02**
		M-F	100	0.04	0.24	0.09	0.28	0.14	2.17	0.09
153	-TUBERS, ALFALFA HAY ADDED, SILAGE	**A-F**	**35**	**—**	**—**	**—**	**—**	**—**	**—**	**—**
		M-F	100	—	—	—	—	—	—	—
	POULTRY									
154	-FEATHERS, MEAL, HYDROLYZED	**A-F**	**93**	**0.30**	**0.62**	**0.63**	**0.28**	**0.18**	**0.27**	**1.50**
		M-F	100	0.33	0.67	0.68	0.30	0.19	0.29	1.61
155	-MANURE WITHOUT LITTER, DEHY	**A-F**	**90**	**8.07**	**2.22**	**0.61**	**0.86**	**0.56**	**1.99**	**0.16**
		M-F	100	8.95	2.46	0.68	0.96	0.62	2.20	0.18
156	-MANURE WITH LITTER, DEHY	**A-F**	**86**	**2.67**	**1.69**	**0.41**	**—**	**0.43**	**1.32**	**—**
		M-F	100	3.12	1.98	0.47	—	0.50	1.55	—
	PRAIRIE GRASS, MIDWEST (PRAIRIE HAY)									
157	-HAY, SUN-CURED	**A-F**	**91**	**0.32**	**0.13**	**—**	**—**	**0.24**	**0.98**	**—**
		M-F	100	0.35	0.14	—	—	0.26	1.08	—
	RAPE *Brassica napus*									
158	-FRESH	**A-F**	**17**	**0.25**	**0.07**	**—**	**—**	**0.01**	**0.56**	**0.11**
		M-F	100	1.47	0.43	—	—	0.06	3.37	0.68
159	-SEEDS, MEAL, MECH EXTD	**A-F**	**92**	**0.66**	**1.04**	**—**	**—**	**0.50**	**0.83**	**—**
		M-F	100	0.72	1.14	—	—	0.54	0.90	—
160	-SEEDS, MEAL, SOLV EXTD, 34% PROTEIN	**A-F**	**90**	**—**	**—**	**—**	**—**	**—**	**—**	**—**
		M-F	100	—	—	—	—	—	—	—
	REDTOP *Agrostis alba*									
161	-FULL BLOOM, FRESH	**A-F**	**26**	**0.16**	**0.10**	**0.01**	**—**	**0.07**	**0.62**	**0.04**
		M-F	100	0.62	0.37	0.05	—	0.25	2.35	0.16
162	-HAY, SUN-CURED	**A-F**	**92**	**0.39**	**0.20**	**0.06**	**0.06**	**0.20**	**1.74**	**0.23**
		M-F	100	0.43	0.22	0.07	0.07	0.22	1.89	0.25
	RICE *Oryza sativa*									
163	-BRAN WITH GERMS (RICE, BRAN)	**A-F**	**91**	**0.07**	**1.44**	**0.03**	**0.07**	**0.85**	**1.69**	**0.18**
		M-F	100	0.08	1.59	0.04	0.08	0.94	1.87	0.20
164	-GRAIN, GROUND (GROUND ROUGH RICE, GROUND PADDY RICE)	**A-F**	**89**	**0.07**	**0.32**	**0.06**	**0.07**	**0.13**	**0.47**	**0.05**
		M-F	100	0.07	0.36	0.07	0.08	0.14	0.53	0.05
165	-HULLS	**A-F**	**92**	**0.11**	**0.10**	**0.02**	**0.07**	**0.41**	**0.64**	**0.08**
		M-F	100	0.12	0.10	0.02	0.08	0.45	0.69	0.09
166	-STRAW	**A-F**	**91**	**0.19**	**0.07**	**0.28**	**—**	**0.10**	**1.20**	**—**
		M-F	100	0.21	0.08	0.31	—	0.11	1.32	—

(Continued)

Entry Number	Microminerals							Fat-soluble Vitamins			
	Cobalt (Co)	Copper (Cu)	Iodine (I)	Iron (Fe)	Manganese (Mn)	Selenium (Se)	Zinc (Zn)	A (1 mg Carotene = 1,667 IU Vit. A)	Carotene (Provitamin A)	D	E (α-tocopherol)
	(ppm or mg/kg)	*(ppm or mg/kg)*	*(ppm or mg/kg)*	*(%)*	*(ppm or mg/kg)*	*(ppm or mg/kg)*	*(ppm or mg/kg)*	*(IU/g)*	*(ppm or mg/kg)*	*(IU/kg)*	*(ppm or mg/kg)*
146	**0.109**	**16.2**	**—**	**0.029**	**62.5**	**—**	**21.9**	**1.3**	**0.8**	**—**	**—**
	0.119	17.8	—	0.032	68.7	—	24.1	1.5	0.9	—	—
147	**0.111**	**15.4**	**0.067**	**0.030**	**25.5**	**—**	**33.0**	**—**	**—**	**—**	**2.4**
	0.119	16.6	0.072	0.033	27.6	—	35.6	—	—	—	2.6
148	**—**	**—**	**—**	**—**	**—**	**—**	**—**	**—**	**—**	**—**	**2.9**
	—	—	—	—	—	—	—	—	—	—	3.2
149	**—**	**—**	**—**	**—**	**—**	**—**	**—**	**63.0**	**37.8**	**—**	**—**
	—	—	—	—	—	—	—	304.2	182.5	—	—
150	**—**	**—**	**—**	**0.049**	**—**	**—**	**—**	**78.4**	**47.0**	**—**	**—**
	—	—	—	0.057	—	—	—	90.0	54.0	—	—
151	**—**	**—**	**—**	**—**	**—**	**—**	**—**	**—**	**—**	**—**	**—**
	—	—	—	—	—	—	—	—	—	—	—
152	**—**	**6.7**	**—**	**0.002**	**9.8**	**—**	**—**	**—**	**—**	**—**	**—**
	—	28.4	—	0.008	41.7	—	—	—	—	—	—
153	**—**	**—**	**—**	**—**	**—**	**—**	**—**	**—**	**—**	**—**	**—**
	—	—	—	—	—	—	—	—	—	—	—
154	**0.116**	**7.3**	**0.044**	**0.023**	**11.9**	**0.913**	**71.9**	**—**	**—**	**—**	**—**
	0.125	7.9	0.047	0.025	12.9	0.983	77.3	—	—	—	—
155	**—**	**24.6**	**—**	**—**	**—**	**—**	**366.4**	**—**	**—**	**—**	**—**
	—	27.2	—	—	—	—	405.9	—	—	—	—
156	**—**	**283.3**	**—**	**0.046**	**281.1**	**0.559**	**359.4**	**—**	**—**	**—**	**—**
	—	331.0	—	0.054	328.4	0.653	419.9	—	—	—	—
157	**—**	**—**	**—**	**0.008**	**—**	**—**	**—**	**—**	**—**	**1**	**—**
	—	—	—	0.009	—	—	—	—	—	1	—
158	**—**	**1.4**	**—**	**0.004**	**7.7**	**—**	**—**	**—**	**—**	**—**	**—**
	—	8.1	—	0.019	46.0	—	—	—	—	—	—
159	**—**	**6.8**	**—**	**0.018**	**55.3**	**0.959**	**43.2**	**—**	**—**	**—**	**18.8**
	—	7.4	—	0.019	60.2	1.043	47.0	—	—	—	20.4
160	**—**	**—**	**—**	**—**	**—**	**—**	**—**	**—**	**—**	**—**	**—**
	—	—	—	—	—	—	—	—	—	—	—
161	**—**	**—**	**—**	**0.006**	**—**	**—**	**—**	**66.9**	**40.1**	**—**	**—**
	—	—	—	0.020	—	—	—	254.4	152.6	—	—
162	**0.134**	**3.6**	**0.092**	**0.015**	**207.7**	**—**	**—**	**6.1**	**3.7**	**—**	**—**
	0.146	3.9	0.100	0.016	225.5	—	—	6.6	4.0	—	—
163	**1.383**	**11.0**	**—**	**0.019**	**337.6**	**—**	**37.4**	**—**	**—**	**—**	**60.4**
	1.526	12.1	—	0.021	372.4	—	41.3	—	—	—	66.7
164	**—**	**—**	**—**	**—**	**18.0**	**—**	**15.0**	**—**	**—**	**—**	**14.0**
	—	—	—	—	20.2	—	16.9	—	—	—	15.7
165	**2.046**	**3.1**	**—**	**0.010**	**295.0**	**—**	**22.0**	**—**	**—**	**—**	**7.5**
	2.220	3.4	—	0.010	320.1	—	23.9	—	—	—	8.1
166	**—**	**—**	**—**	**—**	**313.9**	**—**	**—**	**—**	**—**	**—**	**—**
	—	—	—	—	345.8	—	—	—	—	—	—

(Continued)

TABLE II-1

Entry Number	Feed Name Description	International Feed Number[2]	Moisture Basis: A-F (As-fed) or M-F (Moisture-free)	Chemical Analysis: Dry Matter	Ash	Crude Fiber	Cell Walls or NDF	Acid Detergent Fiber	Lignin	Ether Extract (Fat)	N-Free Extract	Crude Protein
				(%)	(%)	(%)	(%)	(%)	(%)	(%)	(%)	(%)
	RYE *Secale cereale*											
167	-DISTILLERS' GRAINS, DEHY	5-04-023	A-F	92	2.3	12.3	—	—	—	6.0	48.3	23.0
			M-F	100	2.5	13.4	—	—	—	6.5	52.6	25.1
168	-FRESH	2-04-018	A-F	20	1.9	5.9	—	—	—	0.8	8.1	3.6
			M-F	100	9.3	29.1	—	—	—	3.9	40.0	17.6
169	-GRAIN, ALL ANALYSES	4-04-047	A-F	87	1.6	2.2	—	—	—	1.5	70.0	12.0
			M-F	100	1.9	2.5	—	—	—	1.7	80.1	13.8
170	-MILL RUN, LESS THAN 9.5% FIBER (RYE FEED)	4-04-034	A-F	88	4.7	4.9	—	—	—	3.1	59.9	15.6
			M-F	100	5.4	5.5	—	—	—	3.5	67.9	17.6
171	-STRAW	1-04-007	A-F	91	3.8	38.3	—	—	—	1.4	44.7	2.8
			M-F	100	4.2	42.1	—	—	—	1.5	49.2	3.0
	RYEGRASS, ITALIAN *Lolium multiflorum*											
172	-FRESH	2-04-073	A-F	23	3.9	4.7	—	—	—	0.9	9.0	4.0
			M-F	100	17.4	20.9	—	—	—	4.1	39.8	17.9
173	-HAY, SUN-CURED	1-04-057	A-F	88	7.1	25.3	—	—	—	1.8	46.2	7.5
			M-F	100	8.1	28.8	—	—	—	2.1	52.5	8.5
	SAFFLOWER *Carthamus tinctorius*											
174	-SEEDS	4-07-958	A-F	93	3.0	23.6	—	37.2	—	30.8	20.9	14.9
			M-F	100	3.2	25.3	—	40.0	—	33.1	22.4	16.0
175	-SEEDS, MEAL, SOLV EXTD, 20% PROTEIN	5-26-095	A-F	92	4.6	32.2	—	39.6	—	1.1	32.7	21.6
			M-F	100	5.0	34.9	—	43.0	—	1.2	35.5	23.4
176	-SEEDS WITHOUT HULLS, MEAL, SOLV EXTD, 42% PROTEIN	5-26-094	A-F	92	6.5	14.6	—	19.2	—	1.3	26.3	42.7
			M-F	100	7.2	16.0	—	21.0	—	1.5	28.8	46.7
177	-SEEDS WITHOUT HULLS, MEAL, MECH EXTD	5-08-499	A-F	91	6.5	12.8	—	—	—	6.0	23.9	42.0
			M-F	100	7.2	14.0	—	—	—	6.6	26.2	46.1
178	-SEEDS WITHOUT HULLS, MEAL, SOLV EXTD	5-07-959	A-F	91	7.7	13.1	—	—	—	1.2	26.2	42.8
			M-F	100	8.5	14.4	—	—	—	1.3	28.7	47.0
	SAGE, BLACK *Salvia mellifera*											
179	-BROWSE, STEM-CURED, FRESH	2-05-564	A-F	65	3.6	—	—	—	—	7.0	—	5.5
			M-F	100	5.5	—	—	—	—	10.7	—	8.5
	SAGEBRUSH, BIG *Artemisia tridentata*											
180	-BROWSE, STEM-CURED, FRESH	2-07-992	A-F	65	4.3	—	—	—	—	6.4	—	6.1
			M-F	100	6.6	—	—	—	—	9.8	—	9.3
	SAGEBRUSH, BUD *Artemisia spinescens*											
181	-BROWSE, IMMATURE, FRESH	2-07-991	A-F	23	4.9	—	—	—	—	1.1	—	4.0
			M-F	100	21.4	—	—	—	—	4.9	—	17.3
182	-BROWSE, FRESH	2-04-125	A-F	27	5.8	6.1	—	—	—	0.7	9.6	4.7
			M-F	100	21.6	22.7	—	—	—	2.5	35.7	17.5
	SAGEBRUSH, FRINGED *Artemisia frigida*											
183	-BROWSE, MIDBLOOM, FRESH	2-04-129	A-F	43	2.8	14.3	—	—	—	0.9	21.0	4.0
			M-F	100	6.5	33.2	—	—	—	2.0	48.9	9.4
184	-BROWSE, MATURE, FRESH	2-04-130	A-F	60	10.3	19.1	27.4	21.1	5.9	2.0	24.4	4.3
			M-F	100	17.1	31.8	45.6	35.1	9.8	3.4	40.6	7.1
	SALTBUSH, NUTTALL *Atriplex nuttallii*											
185	-BROWSE, STEM-CURED, FRESH	2-07-993	A-F	55	11.8	—	—	—	—	1.2	—	4.0
			M-F	100	21.5	—	—	—	—	2.2	—	7.2
	SALTGRASS *Distichlis* spp											
186	-OVERRIPE, FRESH	2-04-169	A-F	74	5.4	26.0	—	—	—	1.9	37.9	3.1
			M-F	100	7.3	34.9	—	—	—	2.6	51.0	4.2
187	-HAY, SUN-CURED	1-04-168	A-F	89	11.4	28.3	—	—	—	1.8	40.0	8.0
			M-F	100	12.7	31.6	—	—	—	2.1	44.7	8.9

(Continued)

Entry Number	TDN	Digestible Energy		Metabolizable Energy		Net Energy					
	Ruminant	Ruminant		Ruminant		Ruminant NE_m		Ruminant NE_g		Lactating Cows NE_{lc}	
	(%)	(Mcal)		(Mcal)		(Mcal)		(Mcal)		(Mcal)	
		(lb)	(kg)	(lb)	(kg)	(lb)	(kg)	(lb)	(kg)	(lb)	(kg)
167	54	1.08	2.38	0.90	1.99	0.45	1.00	0.22	0.49	0.50	1.09
	59	1.18	2.59	0.98	2.17	0.49	1.09	0.24	0.53	0.54	1.19
168	14	0.27	0.60	0.23	0.51	0.12	0.27	0.07	0.15	0.13	0.28
	67	1.34	2.96	1.15	2.54	0.61	1.34	0.35	0.77	0.62	1.37
169	73	1.42	3.12	1.18	2.60	0.80	1.75	0.54	1.18	0.74	1.63
	84	1.62	3.57	1.35	2.97	0.91	2.01	0.61	1.35	0.85	1.86
170	65	1.29	2.85	1.12	2.48	0.68	1.49	0.43	0.95	0.66	1.45
	73	1.46	3.23	1.27	2.81	0.77	1.69	0.49	1.08	0.74	1.64
171	39	0.78	1.72	0.60	1.32	0.25	0.55	0.03	0.07	0.35	0.78
	43	0.86	1.89	0.66	1.46	0.27	0.60	0.04	0.08	0.39	0.86
172	14	0.28	0.61	0.23	0.51	0.14	0.31	0.08	0.18	0.14	0.32
	61	1.23	2.70	1.03	2.28	0.62	1.38	0.36	0.80	0.64	1.40
173	53	1.06	2.34	0.89	1.97	0.54	1.19	0.31	0.68	0.55	1.22
	60	1.21	2.66	1.02	2.24	0.61	1.35	0.35	0.78	0.63	1.38
174	83	1.07	2.36	1.19	2.62	0.93	2.06	0.64	1.42	0.87	1.92
	89	1.15	2.53	1.27	2.81	1.00	2.21	0.69	1.52	0.94	2.06
175	46	0.87	1.92	0.86	1.90	0.51	1.12	0.27	0.59	0.40	0.89
	50	0.95	2.09	0.94	2.06	0.55	1.21	0.29	0.65	0.44	0.96
176	66	1.15	2.53	0.97	2.15	0.58	1.27	0.33	0.74	0.56	1.24
	72	1.26	2.77	1.06	2.35	0.63	1.39	0.37	0.81	0.61	1.35
177	70	1.40	3.08	1.22	2.70	0.76	1.67	0.50	1.10	0.72	1.60
	77	1.53	3.37	1.34	2.96	0.83	1.83	0.55	1.20	0.79	1.75
178	69	1.34	2.96	1.17	2.57	0.75	1.66	0.49	1.08	0.72	1.59
	76	1.47	3.25	1.28	2.83	0.83	1.82	0.54	1.19	0.79	1.74
179	32	0.62	1.38	0.31	0.68	—	—	—	—	—	—
	49	0.96	2.12	0.47	1.04	—	—	—	—	—	—
180	27	0.66	1.46	0.37	0.81	—	—	—	—	—	—
	42	1.02	2.25	0.56	1.24	—	—	—	—	—	—
181	12	0.27	0.59	0.21	0.46	—	—	—	—	—	—
	51	1.16	2.56	0.91	2.01	—	—	—	—	—	—
182	14	0.28	0.62	0.23	0.51	0.14	0.30	0.04	0.10	0.14	0.31
	52	1.04	2.30	0.86	1.89	0.51	1.12	0.16	0.36	0.53	1.16
183	26	0.53	1.16	0.45	0.98	0.26	0.57	0.15	0.32	0.27	0.59
	61	1.23	2.70	1.03	2.28	0.60	1.33	0.34	0.75	0.62	1.36
184	30	0.60	1.33	0.49	1.08	0.26	0.58	0.11	0.25	0.30	0.66
	50	1.01	2.22	0.81	1.79	0.44	0.96	0.19	0.41	0.50	1.10
185	20	0.37	0.82	0.33	0.73	—	—	—	—	—	—
	36	0.68	1.49	0.60	1.32	—	—	—	—	—	—
186	42	0.84	1.86	0.70	1.54	0.41	0.90	0.22	0.48	0.43	0.95
	57	1.13	2.50	0.94	2.08	0.55	1.20	0.29	0.64	0.58	1.27
187	45	0.90	1.99	0.73	1.60	0.39	0.87	0.17	0.38	0.45	0.99
	51	1.01	2.22	0.81	1.79	0.44	0.97	0.19	0.42	0.50	1.11

(Continued)

TABLE II-1

Entry Number	Feed Name Description	Moisture Basis: A-F (As-fed) or M-F (Moisture-free)	Dry Matter	Macrominerals: Calcium (Ca)	Phos-phorus (P)	Sodium (Na)	Chlorine (Cl)	Mag-nesium (Mg)	Potassium (K)	Sulfur (S)
			(%)	(%)	(%)	(%)	(%)	(%)	(%)	(%)
	RYE *Secale cereale*									
167	-DISTILLERS' GRAINS, DEHY	**A-F**	**92**	**0.15**	**0.48**	**0.17**	**0.05**	**0.17**	**0.07**	**0.44**
		M-F	100	0.16	0.52	0.18	0.05	0.18	0.08	0.48
168	-FRESH	**A-F**	**20**	**0.09**	**0.08**	**0.01**	—	**0.06**	**0.69**	—
		M-F	100	0.45	0.38	0.07	—	0.31	3.40	—
169	-GRAIN, ALL ANALYSES	**A-F**	**87**	**0.06**	**0.31**	**0.02**	**0.03**	**0.12**	**0.46**	**0.15**
		M-F	100	0.07	0.36	0.03	0.03	0.14	0.52	0.17
170	-MILL RUN, LESS THAN 9.5% FIBER (RYE FEED)	**A-F**	**88**	**0.07**	**0.63**	—	—	**0.23**	**0.81**	**0.04**
		M-F	100	0.08	0.71	—	—	0.26	0.92	0.04
171	-STRAW	**A-F**	**91**	**0.22**	**0.08**	**0.12**	**0.22**	**0.07**	**0.88**	**0.10**
		M-F	100	0.24	0.09	0.13	0.24	0.08	0.97	0.11
	RYEGRASS, ITALIAN *Lolium multiflorum*									
172	-FRESH	**A-F**	**23**	**0.15**	**0.09**	**0.00**	—	[illegible]	[illegible]	**0.02**
		M-F	100	0.65	0.41	0.01	—	0.35	2.00	0.10
173	-HAY, SUN-CURED	**A-F**	**88**	—	—	—	—	—	—	—
		M-F	100	—	—	—	—	—	—	—
	SAFFLOWER *Carthamus tinctorius*									
174	-SEEDS	**A-F**	**93**	**0.24**	**0.57**	**0.06**	—	**0.34**	**0.74**	**0.06**
		M-F	100	0.26	0.61	0.06	—	0.36	0.79	0.06
175	-SEEDS, MEAL, SOLV EXTD, 20% PROTEIN	**A-F**	**92**	**0.31**	**0.61**	—	—	**0.32**	**0.74**	**0.20**
		M-F	100	0.34	0.66	—	—	0.35	0.80	0.22
176	-SEEDS WITHOUT HULLS, MEAL, SOLV EXTD, 42% PROTEIN	**A-F**	**92**	**0.38**	**1.08**	—	—	**1.18**	**1.18**	**0.34**
		M-F	100	0.41	1.18	—	—	1.29	1.29	0.38
177	-SEEDS WITHOUT HULLS, MEAL, MECH EXTD	**A-F**	**91**	—	—	—	—	—	—	—
		M-F	100	—	—	—	—	—	—	—
178	-SEEDS WITHOUT HULLS, MEAL, SOLV EXTD	**A-F**	**91**	**0.35**	**1.42**	**0.04**	**0.16**	**0.92**	**1.05**	**0.06**
		M-F	100	0.38	1.56	0.05	0.18	1.01	1.15	0.06
	SAGE, BLACK *Salvia mellifera*									
179	-BROWSE, STEM-CURED, FRESH	**A-F**	**65**	**0.53**	**0.11**	—	—	—	—	—
		M-F	100	0.81	0.17	—	—	—	—	—
	SAGEBRUSH, BIG *Artemisia tridentata*									
180	-BROWSE, STEM-CURED, FRESH	**A-F**	**65**	**0.46**	**0.12**	—	—	—	—	—
		M-F	100	0.71	0.18	—	—	—	—	—
	SAGEBRUSH, BUD *Artemisia spinescens*									
181	-BROWSE, IMMATURE, FRESH	**A-F**	**23**	**0.22**	**0.08**	—	—	—	—	—
		M-F	100	0.97	0.33	—	—	—	—	—
182	-BROWSE, FRESH	**A-F**	**27**	**0.42**	**0.11**	—	—	**0.13**	—	**0.07**
		M-F	100	1.57	0.42	—	—	0.49	—	0.26
	SAGEBRUSH, FRINGED *Artemisia frigida*									
183	-BROWSE, MIDBLOOM, FRESH	**A-F**	**43**	—	—	—	—	—	—	—
		M-F	100	—	—	—	—	—	—	—
184	-BROWSE, MATURE, FRESH	**A-F**	**60**	—	—	—	—	—	—	—
		M-F	100	—	—	—	—	—	—	—
	SALTBUSH, NUTTALL *Atriplex nuttallii*									
185	-BROWSE, STEM-CURED, FRESH	**A-F**	**55**	**1.22**	**0.06**	—	—	—	—	—
		M-F	100	2.21	0.12	—	—	—	—	—
	SALTGRASS *Distichlis* spp									
186	-OVERRIPE, FRESH	**A-F**	**74**	**0.17**	**0.05**	—	—	**0.22**	—	—
		M-F	100	0.23	0.07	—	—	0.30	—	—
187	-HAY, SUN-CURED	**A-F**	**89**	—	—	—	—	—	—	—
		M-F	100	—	—	—	—	—	—	—

(Continued)

Entry Number	Microminerals							Fat-soluble Vitamins			
	Cobalt (Co)	Copper (Cu)	Iodine (I)	Iron (Fe)	Manganese (Mn)	Selenium (Se)	Zinc (Zn)	A (1 mg Carotene = 1,667 IU Vit. A)	Carotene (Provitamin A)	D	E (α-tocopherol)
	(ppm or mg/kg)	*(ppm or mg/kg)*	*(ppm or mg/kg)*	*(%)*	*(ppm or mg/kg)*	*(ppm or mg/kg)*	*(ppm or mg/kg)*	*(IU/g)*	*(ppm or mg/kg)*	*(IU/kg)*	*(ppm or mg/kg)*
167	**—**	**—**	**—**	**—**	**18.4**	**—**	**—**	**—**	**—**	**—**	**—**
	—	—	—	—	20.0	—	—	—	—	—	—
168	**—**	**—**	**—**	**—**	**—**	**—**	**—**	**115.0**	**69.0**	**—**	**—**
	—	—	—	—	—	—	—	571.1	342.6	—	—
169	**—**	**7.5**	**—**	**0.007**	**72.0**	**—**	**28.1**	**0.1**	**0.1**	**—**	**14.5**
	—	8.6	—	0.008	82.3	—	32.2	0.2	0.1	—	16.6
170	**—**	**—**	**—**	**—**	**—**	**—**	**—**	**—**	**—**	**—**	**—**
	—	—	—	—	—	—	—	—	—	—	—
171	**—**	**3.6**	**—**	**—**	**6.0**	**—**	**—**	**—**	**—**	**—**	**—**
	—	4.0	—	—	6.6	—	—	—	—	—	—
172	**—**	**—**	**—**	**0.023**	**—**	**—**	**—**	**—**	**—**	**—**	**—**
	—	—	—	0.101	—	—	—	—	—	—	—
173	**—**	**—**	**—**	**—**	**—**	**—**	**—**	**175.8**	**105.5**	**—**	**—**
	—	—	—	—	—	—	—	199.9	119.9	—	—
174	**—**	**10.0**	**—**	**0.032**	**1.1**	**—**	**30.0**	**—**	**—**	**—**	**—**
	—	10.7	—	0.035	1.2	—	32.2	—	—	—	—
175	**—**	**9.6**	**—**	**0.043**	**17.7**	**—**	**39.6**	**—**	**—**	**—**	**0.9**
	—	10.4	—	0.046	19.2	—	43.0	—	—	—	1.0
176	**1.832**	**80.6**	**—**	**0.091**	**36.6**	**—**	**168.5**	**—**	**—**	**—**	**0.6**
	2.000	88.0	—	0.100	40.0	—	184.0	—	—	—	0.7
177	**—**	**—**	**—**	**—**	**—**	**—**	**—**	**—**	**—**	**—**	**—**
	—	—	—	—	—	—	—	—	—	—	—
178	**2.022**	**88.6**	**—**	**0.082**	**40.3**	**—**	**186.3**	**—**	**—**	**—**	**0.7**
	2.221	97.3	—	0.090	44.2	—	204.6	—	—	—	0.8
179	**—**	**—**	**—**	**—**	**—**	**—**	**—**	**—**	**—**	**—**	**—**
	—	—	—	—	—	—	—	—	—	—	—
180	**—**	**—**	**—**	**—**	**—**	**—**	**—**	**17.3**	**10.4**	**—**	**—**
	—	—	—	—	—	—	—	26.6	15.9	—	—
181	**—**	**—**	**—**	**—**	**—**	**—**	**—**	**9.1**	**5.5**	**—**	**—**
	—	—	—	—	—	—	—	39.7	23.8	—	—
182	**—**	**—**	**—**	**—**	**—**	**—**	**—**	**10.7**	**6.4**	**—**	**—**
	—	—	—	—	—	—	—	39.7	23.8	—	—
183	**—**	**—**	**—**	**—**	**—**	**—**	**—**	**—**	**—**	**—**	**—**
	—	—	—	—	—	—	—	—	—	—	—
184	**—**	**—**	**—**	**—**	**—**	**—**	**—**	**—**	**—**	**—**	**—**
	—	—	—	—	—	—	—	—	—	—	—
185	**—**	**—**	**—**	**—**	**—**	**—**	**—**	**17.4**	**10.5**	**—**	**—**
	—	—	—	—	—	—	—	31.7	19.0	—	—
186	**—**	**—**	**—**	**—**	**—**	**—**	**—**	**—**	**—**	**—**	**—**
	—	—	—	—	—	—	—	—	—	—	—
187	**—**	**—**	**—**	**—**	**—**	**—**	**—**	**—**	**—**	**—**	**—**
	—	—	—	—	—	—	—	—	—	—	—

(Continued)

TABLE II-1

Entry Number	Feed Name Description	International Feed Number[2]	Moisture Basis: A-F (As-fed) or M-F (Moisture-free)	Chemical Analysis								
				Dry Matter	Ash	Crude Fiber	Cell Walls or NDF	Acid Detergent Fiber	Lignin	Ether Extract (Fat)	N-Free Extract	Crude Protein
				(%)	(%)	(%)	(%)	(%)	(%)	(%)	(%)	(%)
	SEAWEED (KELP) *Laminariales (order), Fucales (order)*											
188	-WHOLE, DEHY	1-08-073	A-F	91	35.0	6.5	—	—	—	0.5	42.4	6.5
			M-F	100	38.6	7.1	—	—	—	0.5	46.7	7.1
	SEDGE *Carex* spp											
189	-HAY, SUN-CURED	1-04-193	A-F	89	6.4	28.0	—	—	—	2.1	44.4	8.4
			M-F	100	7.2	31.3	—	—	—	2.4	49.7	9.4
	SESAME *Sesamum indicum*											
190	-SEEDS, MEAL MECH EXTD	5-04-220	A-F	93	10.3	5.6	—	—	—	8.7	23.0	45.0
			M-F	100	11.2	6.1	—	—	—	9.4	24.8	48.6
	SORGHUM *Sorghum bicolor*											
191	-FODDER WITH HEADS, SUN-CURED	[illegible]	[illegible]	[illegible]	[illegible]	25.6	—	—	—	2.0	47.4	6.2
			M-F	100	9.9	28.4	—			[illegible]	[illegible]	[illegible]
192	-STOVER WITHOUT HEADS, SUN-CURED	1-04-302	A-F	92	8.9	29.9	—	39.9	—	1.6	46.8	4.4
			M-F	100	9.7	32.6	—	43.6	—	1.8	51.1	4.9
193	-DISTILLERS' GRAINS, DEHY	5-04-374	A-F	94	4.3	12.1	—	—	—	8.3	38.3	30.8
			M-F	100	4.6	12.9	—	—	—	8.8	40.8	32.9
194	-GRAIN, LESS THAN 9% PROTEIN	4-08-138	A-F	89	2.1	2.2	—	—	—	2.9	72.4	8.9
			M-F	100	2.4	2.5	—	—	—	3.3	81.8	10.1
195	-GRAIN, 9–12% PROTEIN	4-08-139	A-F	89	1.9	2.4	—	—	—	2.7	72.2	9.8
			M-F	100	2.1	2.6	—	—	—	3.1	81.1	11.0
196	-GRAIN, MORE THAN 12% PROTEIN	4-08-140	A-F	89	2.3	1.8	—	—	—	1.5	71.8	11.6
			M-F	100	2.6	2.0	—	—	—	1.7	80.7	13.0
197	-SILAGE, MATURE	3-04-322	A-F	32	2.3	7.8	16.3	8.5	2.1	0.9	18.2	2.7
			M-F	100	7.1	24.3	51.0	26.6	6.5	2.8	57.1	8.6
	SORGHUM, JOHNSONGRASS *Sorghum halepense*											
198	-HAY, SUN-CURED	1-04-407	A-F	91	7.7	30.4	—	—	—	2.0	43.7	6.7
			M-F	100	8.6	33.6	—	—	—	2.2	48.3	7.5
	SORGHUM, SORGO *Sorghum bicolor saccharatum*											
199	-SILAGE	3-04-468	A-F	29	2.4	7.0	18.5	11.0	1.7	0.7	16.7	1.9
			M-F	100	8.3	24.4	64.0	38.0	6.0	2.5	58.1	6.8
	SORGHUM, SUDANGRASS *sorghum bicolor Sudanense*											
200	-IMMATURE, FRESH	2-04-484	A-F	18	1.5	6.7	—	—	—	0.6	6.8	2.6
			M-F	100	8.2	36.7	—	—	—	3.2	37.5	14.4
201	-MATURE, FRESH	2-04-487	A-F	30	2.4	10.6	—	—	—	0.5	14.5	1.6
			M-F	100	8.1	35.8	—	—	—	1.7	48.9	5.5
202	-HAY, SUN-CURED	1-04-480	A-F	91	10.7	26.2	60.2	20.4	35.5	1.6	41.7	10.9
			M-F	100	11.8	28.7	66.0	22.4	39.0	1.7	45.8	12.0
203	-SILAGE	3-04-499	A-F	23	2.1	7.9	16.4	1.2	9.7	0.7	9.9	2.6
			M-F	100	9.2	34.0	71.0	5.0	42.0	2.9	42.9	11.1
	SOYBEAN *Glycine max*											
204	-HAY, SUN-CURED	1-04-558	A-F	89	7.2	30.6	—	35.7	—	2.3	35.0	14.1
			M-F	100	8.0	34.3	—	40.0	—	2.5	39.3	15.8
205	-HULLS	1-04-560	A-F	91	4.6	36.2	59.4	42.4	1.8	2.0	37.0	10.8
			M-F	100	5.1	40.0	65.6	46.8	2.0	2.2	40.9	11.9
206	-SEEDS	5-04-610	A-F	92	5.1	5.4	—	—	—	17.2	25.9	38.4
			M-F	100	5.6	5.8	—	—	—	18.7	28.1	41.7
207	-SEEDS, MEAL, MECH EXTD, 41% PROTEIN	5-04-600	A-F	90	6.0	6.0	—	—	—	4.7	30.4	42.9
			M-F	100	6.7	6.7	—	—	—	5.2	33.8	47.7
208	-SEEDS, MEAL, SOLV EXTD, 44% PROTEIN	5-20-637	A-F	89	6.4	6.2	12.5	8.9	—	1.5	30.6	44.4
			M-F	100	7.2	7.0	14.0	10.0	—	1.7	34.3	49.8

(Continued)

Entry Number	TDN	Digestible Energy		Metabolizable Energy		Net Energy					
	Ruminant	Ruminant		Ruminant		Ruminant NE_m		Ruminant NE_g		Lactating Cows NE_{lc}	
	(%)	*(Mcal)*		*(Mcal)*		*(Mcal)*		*(Mcal)*		*(Mcal)*	
		(lb)	*(kg)*	*(lb)*	*(kg)*	*(lb)*	*(kg)*	*(lb)*	*(kg)*	*(lb)*	*(kg)*
188	29 32	0.58 0.63	*1.27* *1.40*	0.40 0.44	*0.87* *0.96*	— —	— —	— —	— —	— —	— —
189	47 52	0.96 1.08	*2.12* *2.38*	0.79 0.88	*1.74* *1.95*	0.47 0.53	*1.03* *1.16*	0.24 0.27	*0.53* *0.60*	0.50 0.56	*1.11* *1.24*
190	71 76	1.41 1.53	*3.12* *3.37*	1.24 1.34	*2.73* *2.95*	0.75 0.81	*1.66* *1.79*	0.49 0.53	*1.08* *1.16*	0.72 0.78	*1.59* *1.72*
191	51 56	1.02 1.13	*2.24* *2.49*	0.84 0.94	*1.86* *2.06*	0.51 0.56	*1.12* *1.24*	0.28 0.31	*0.61* *0.68*	0.53 0.59	*1.17* *1.30*
192	47 51	0.92 1.00	*2.02* *2.21*	0.74 0.81	*1.64* *1.79*	0.40 0.44	*0.89* *0.97*	0.18 0.19	*0.39* *0.43*	0.42 0.46	*0.93* *1.02*
193	78 83	1.56 1.67	*3.45* *3.68*	1.39 1.48	*3.06* *3.26*	0.85 0.91	*1.88* *2.00*	0.57 0.61	*1.27* *1.35*	0.80 0.86	*1.77* *1.89*
194	75 84	1.50 1.69	*3.30* *3.73*	1.33 1.50	*2.93* *3.31*	0.78 0.89	*1.73* *1.95*	0.53 0.59	*1.16* *1.31*	0.74 0.84	*1.64* *1.85*
195	73 82	1.42 1.60	*3.14* *3.53*	1.11 1.24	*2.44* *2.74*	0.73 0.82	*1.62* *1.82*	0.48 0.54	*1.06* *1.19*	0.76 0.85	*1.67* *1.88*
196	69 78	1.39 1.56	*3.06* *3.44*	1.22 1.37	*2.69* *3.02*	0.76 0.85	*1.67* *1.88*	0.50 0.56	*1.11* *1.24*	0.72 0.81	*1.59* *1.79*
197	17 53	0.33 1.03	*0.72* *2.26*	0.27 0.83	*0.58* *1.83*	0.16 0.50	*0.35* *1.10*	0.08 0.25	*0.17* *0.54*	0.16 0.51	*0.36* *1.12*
198	51 56	1.01 1.12	*2.23* *2.46*	0.84 0.93	*1.85* *2.04*	0.48 0.53	*1.06* *1.18*	0.25 0.28	*0.56* *0.61*	0.51 0.57	*1.13* *1.25*
199	17 58	0.34 1.17	*0.74* *2.58*	0.28 0.98	*0.62* *2.15*	0.16 0.55	*0.35* *1.22*	0.09 0.30	*0.19* *0.66*	0.17 0.58	*0.37* *1.29*
200	12 68	0.25 1.35	*0.54* *2.98*	0.21 1.16	*0.47* *2.56*	0.14 0.74	*0.30* *1.63*	0.09 0.47	*0.19* *1.03*	0.13 0.72	*0.29* *1.60*
201	19 65	0.36 1.21	*0.79* *2.67*	0.30 1.02	*0.67* *2.25*	0.16 0.53	*0.35* *1.17*	0.08 0.28	*0.18* *0.61*	0.17 0.57	*0.37* *1.25*
202	51 56	1.00 1.10	*2.20* *2.42*	0.82 0.90	*1.82* *1.99*	0.48 0.53	*1.07* *1.17*	0.25 0.28	*0.55* *0.61*	0.52 0.57	*1.14* *1.25*
203	13 57	0.26 1.14	*0.58* *2.51*	0.22 0.94	*0.48* *2.08*	0.13 0.55	*0.28* *1.21*	0.07 0.29	*0.15* *0.64*	0.13 0.58	*0.30* *1.28*
204	49 55	1.11 1.25	*2.45* *2.75*	0.86 0.97	*1.90* *2.13*	0.52 0.58	*1.14* *1.28*	0.29 0.32	*0.63* *0.70*	0.55 0.62	*1.21* *1.36*
205	69 77	1.20 1.33	*2.65* *2.92*	1.03 1.13	*2.26* *2.50*	0.75 0.83	*1.66* *1.84*	0.50 0.55	*1.09* *1.21*	0.72 0.80	*1.59* *1.75*
206	84 92	1.69 1.83	*3.72* *4.04*	1.52 1.65	*3.34* *3.63*	0.93 1.01	*2.04* *2.22*	0.64 0.70	*1.41* *1.53*	0.86 0.94	*1.90* *2.07*
207	77 85	1.53 1.70	*3.38* *3.76*	1.37 1.52	*3.01* *3.35*	0.86 0.95	*1.89* *2.10*	0.58 0.65	*1.29* *1.43*	0.81 0.90	*1.78* *1.98*
208	76 85	1.45 1.62	*3.19* *3.57*	1.17 1.32	*2.59* *2.90*	0.79 0.89	*1.74* *1.95*	0.53 0.59	*1.16* *1.30*	0.75 0.85	*1.66* *1.87*

(Continued)

TABLE II-1

Entry Number	Feed Name Description	Moisture Basis: A-F (As-fed) or M-F (Moisture-free)	Dry Matter	Macrominerals						
				Calcium (Ca)	Phos-phorus (P)	Sodium (Na)	Chlorine (Cl)	Mag-nesium (Mg)	Potassium (K)	Sulfur (S)
			(%)	*(%)*	*(%)*	*(%)*	*(%)*	*(%)*	*(%)*	*(%)*
	SEAWEED (KELP) *Laminariales (order), Fucales (order)*									
188	-WHOLE, DEHY	**A-F**	**91**	**2.47**	**0.28**	**—**	**—**	**0.85**	**—**	**—**
		M-F	100	2.72	0.31	—	—	0.93	—	—
	SEDGE *Carex* spp									
189	-HAY, SUN-CURED	**A-F**	**89**	**—**	**—**	**—**	**—**	**—**	**—**	**—**
		M-F	100	—	—	—	—	—	—	—
	SESAME *Sesamum indicum*									
190	-SEEDS, MEAL MECH EXTD	**A-F**	**93**	**2.01**	**1.36**	**0.05**	**0.07**	**0.46**	**1.25**	**0.33**
		M-F	100	2.17	1.46	0.05	0.07	0.50	1.35	0.35
	SORGHUM *Sorghum bicolor*									
191	-FODDER WITH HEADS, SUN-CURED	**A-F**	**90**	**0.56**	**0.17**	**0.02**	**—**	**0.27**	**1.12**	**—**
		M-F	100	[illegible]	0.19	0.02		0.30	1.24	—
192	-STOVER WITHOUT HEADS, SUN-CURED	**A-F**	**92**	**0.37**	**0.10**	**—**	**—**	**—**	**1.10**	**—**
		M-F	100	0.40	0.11	—	—	—	1.20	—
193	-DISTILLERS' GRAINS, DEHY	**A-F**	**94**	**0.15**	**0.69**	**0.05**	**—**	**0.18**	**0.36**	**0.17**
		M-F	100	0.16	0.74	0.05	—	0.19	0.38	0.18
194	-GRAIN, LESS THAN 9% PROTEIN	**A-F**	**89**	**0.03**	**0.27**	**0.04**	**—**	**—**	**0.35**	**—**
		M-F	100	0.03	0.31	0.05	—	—	0.40	—
195	-GRAIN, 9–12% PROTEIN	**A-F**	**89**	**0.03**	**0.27**	**0.02**	**—**	**0.15**	**0.34**	**0.14**
		M-F	100	0.04	0.30	0.02	—	0.17	0.38	0.16
196	-GRAIN, MORE THAN 12% PROTEIN	**A-F**	**89**	**0.03**	**0.29**	**0.04**	**—**	**0.17**	**0.34**	**0.16**
		M-F	100	0.03	0.32	0.05	—	0.19	0.38	0.18
197	-SILAGE, MATURE	**A-F**	**32**	**—**	**—**	**—**	**—**	**—**	**—**	**—**
		M-F	100	—	—	—	—	—	—	—
	SORGHUM, JOHNSONGRASS *Sorghum halepense*									
198	-HAY, SUN-CURED	**A-F**	**91**	**0.80**	**0.27**	**0.01**	**—**	**0.31**	**1.22**	**0.09**
		M-F	100	0.89	0.30	0.01	—	0.35	1.35	0.10
	SORGHUM, SORGO *Sorghum bicolor saccharatum*									
199	-SILAGE	**A-F**	**29**	**0.10**	**0.06**	**0.04**	**0.02**	**0.08**	**0.32**	**0.03**
		M-F	100	0.35	0.21	0.15	0.06	0.27	1.12	0.10
	SORGHUM, SUDANGRASS *Sorghum bicolor Sudanense*									
200	-IMMATURE, FRESH	**A-F**	**18**	**0.08**	**0.08**	**0.00**	**—**	**0.06**	**0.39**	**0.02**
		M-F	100	0.43	0.41	0.01	—	0.35	2.14	0.11
201	-MATURE, FRESH	**A-F**	**30**	**0.09**	**0.06**	**—**	**—**	**—**	**—**	**—**
		M-F	100	0.32	0.21	—	—	—	—	—
202	-HAY, SUN-CURED	**A-F**	**91**	**0.47**	**0.28**	**0.01**	**—**	**0.34**	**1.90**	**0.06**
		M-F	100	0.51	0.31	0.02	—	0.37	2.08	0.06
203	-SILAGE	**A-F**	**23**	**0.12**	**0.05**	**0.01**	**—**	**0.10**	**0.60**	**0.01**
		M-F	100	0.50	0.21	0.02	—	0.42	2.61	0.06
	SOYBEAN *Glycine max*									
204	-HAY, SUN-CURED	**A-F**	**89**	**1.13**	**0.22**	**0.10**	**0.13**	**0.72**	**0.92**	**0.25**
		M-F	100	1.26	0.24	0.11	0.15	0.81	1.04	0.28
205	-HULLS	**A-F**	**91**	**0.45**	**0.19**	**0.03**	**—**	**—**	**1.15**	**0.08**
		M-F	100	0.49	0.21	0.03	—	—	1.27	0.09
206	-SEEDS	**A-F**	**92**	**0.25**	**0.60**	**0.00**	**0.03**	**0.27**	**1.66**	**0.22**
		M-F	100	0.27	0.65	0.00	0.03	0.29	1.80	0.24
207	-SEEDS, MEAL, MECH EXTD, 41% PROTEIN	**A-F**	**90**	**0.26**	**0.61**	**0.18**	**0.07**	**0.26**	**1.79**	**0.33**
		M-F	100	0.29	0.68	0.20	0.08	0.29	1.98	0.37
208	-SEEDS, MEAL, SOLV EXTD, 44% PROTEIN	**A-F**	**89**	**0.35**	**0.64**	**0.03**	**—**	**0.27**	**1.98**	**0.41**
		M-F	100	0.40	0.71	0.04	—	0.31	2.22	0.47

COMPOSITION OF FEEDS

(Continued)

Entry Number	Microminerals							Fat-soluble Vitamins			
	Cobalt (Co)	Copper (Cu)	Iodine (I)	Iron (Fe)	Manganese (Mn)	Selenium (Se)	Zinc (Zn)	A (1 mg Carotene = 1,667 IU Vit. A)	Carotene (Provitamin A)	D	E (α-tocopherol)
	(ppm or mg/kg)	*(ppm or mg/kg)*	*(ppm or mg/kg)*	*(%)*	*(ppm or mg/kg)*	*(ppm or mg/kg)*	*(ppm or mg/kg)*	*(IU/g)*	*(ppm or mg/kg)*	*(IU/kg)*	*(ppm or mg/kg)*
188	**—**	**—**	**—**	**—**	**—**	**—**	**—**	**—**	**—**	**—**	**—**
	—	—	—	—	—	—	—	—	—	—	—
189	**—**	**2.9**	**—**	**—**	**—**	**—**	**—**	**—**	**—**	**—**	**—**
	—	3.3	—	—	—	—	—	—	—	—	—
190	**—**	**—**	**—**	**0.010**	**47.7**	**—**	**99.6**	**0.7**	**0.4**	**—**	**—**
	—	—	—	0.010	51.5	—	107.5	0.8	0.5	—	—
191	**—**	**—**	**—**	**—**	**—**	**—**	**—**	**—**	**—**	**—**	**—**
	—	—	—	—	—	—	—	—	—	—	—
192	**—**	**—**	**—**	**—**	**—**	**—**	**—**	**—**	**—**	**—**	**—**
	—	—	—	—	—	—	—	—	—	—	—
193	**—**	**—**	**—**	**0.005**	**—**	**—**	**—**	**—**	**—**	**—**	**—**
	—	—	—	0.005	—	—	—	—	—	—	—
194	**0.067**	**9.7**	**0.023**	**0.002**	**15.4**	**—**	**13.7**	**—**	**—**	**—**	**2.2**
	0.075	11.0	0.025	0.003	17.4	—	15.4	—	—	—	2.5
195	**0.067**	**9.7**	**0.023**	**0.004**	**15.4**	**—**	**13.7**	**—**	**—**	**—**	**1.3**
	0.075	10.9	0.025	0.004	17.3	—	15.4	—	—	—	1.5
196	**—**	**—**	**—**	**0.005**	**—**	**—**	**—**	**—**	**—**	**—**	**—**
	—	—	—	0.005	—	—	—	—	—	—	—
197	**—**	**—**	**—**	**—**	**—**	**—**	**—**	**—**	**—**	**—**	**—**
	—	—	—	—	—	—	—	—	—	—	—
198	**—**	**—**	**—**	**0.054**	**—**	**—**	**—**	**58.8**	**35.3**	**—**	**—**
	—	—	—	0.059	—	—	—	64.9	38.9	—	—
199	**—**	**9.0**	**—**	**0.006**	**17.6**	**—**	**—**	**20.4**	**12.2**	**—**	**—**
	—	31.1	—	0.020	61.0	—	—	70.7	42.4	—	—
200	**—**	**—**	**—**	**0.004**	**—**	**—**	**—**	**59.9**	**35.9**	**—**	**—**
	—	—	—	0.020	—	—	—	329.2	197.5	—	—
201	**—**	**—**	**—**	**—**	**—**	**—**	**—**	**—**	**—**	**—**	**—**
	—	—	—	—	—	—	—	—	—	—	—
202	**0.116**	**28.6**	**—**	**0.015**	**69.5**	**—**	**34.6**	**—**	**—**	**—**	**—**
	0.127	31.4	—	0.017	76.3	—	38.0	—	—	—	—
203	**0.063**	**8.5**	**—**	**0.003**	**22.8**	**—**	**—**	**40.6**	**24.3**	**—**	**—**
	0.270	36.6	—	0.012	98.8	—	—	175.4	105.2	—	—
204	**0.083**	**8.0**	**0.216**	**0.026**	**94.3**	**—**	**21.5**	**53.1**	**31.8**	**1**	**26.3**
	0.093	9.0	0.242	0.029	105.8	—	24.1	59.5	35.7	1	29.5
205	**0.109**	**16.1**	**—**	**0.030**	**9.9**	**—**	**21.8**	**—**	**—**	**—**	**6.6**
	0.121	17.8	—	0.033	11.0	—	24.1	—	—	—	7.3
206	**—**	**18.2**	**—**	**0.009**	**36.4**	**0.111**	**56.9**	**1.5**	**0.9**	**—**	**33.7**
	—	19.8	—	0.010	39.6	0.120	61.8	1.6	1.0	—	36.6
207	**0.178**	**21.7**	**—**	**0.017**	**31.3**	**0.102**	**57.2**	**0.4**	**0.2**	**—**	**6.5**
	0.198	24.1	—	0.019	34.8	0.113	63.6	0.4	0.2	—	7.3
208	**1.381**	**19.9**	**—**	**0.017**	**31.6**	**0.486**	**50.5**	**—**	**—**	**—**	**3.0**
	1.550	22.3	—	0.019	35.5	0.546	56.6	—	—	—	3.4

(Continued)

TABLE II-1

Entry Number	Feed Name Description	International Feed Number[2]	Moisture Basis: A-F (As-fed) or M-F (Moisture-free)	Chemical Analysis: Dry Matter	Ash	Crude Fiber	Cell Walls or NDF	Acid Detergent Fiber	Lignin	Ether Extract (Fat)	N-Free Extract	Crude Protein
				(%)	(%)	(%)	(%)	(%)	(%)	(%)	(%)	(%)
209	-SEEDS WITHOUT HULLS, MEAL, SOLV EXTD	5-04-612	A-F	90	5.8	3.4	6.9	5.5	—	0.9	30.3	49.7
			M-F	100	6.5	3.8	7.7	6.1	—	1.0	33.6	55.1
210	-SILAGE	3-04-581	A-F	30	3.0	9.0	—	—	—	0.8	12.2	5.2
			M-F	100	9.9	29.9	—	—	—	2.6	40.5	17.1
211	-STRAW	1-04-567	A-F	88	5.6	38.9	—	—	—	1.3	37.4	4.6
			M-F	100	6.4	44.3	—	—	—	1.5	42.7	5.2
	SPELT *Triticum spelta*											
212	-GRAIN	4-04-651	A-F	90	3.5	9.1	—	—	—	1.9	63.4	12.0
			M-F	100	3.9	10.2	—	—	—	2.1	70.5	13.3
	SQUIRREL TAIL *Sitanion* spp											
213	-STEM-CURED, FRESH	2-05-566	A-F	50	8.5	—	—	—	—	1.1	—	1.6
			M-F	100	17.0	—	—	—	—	2.2	—	3.1
	SUGARCANE *Saccharum officinarum*											
214	-BAGASSE, DEHY	1-04-686	A-F	91	2.9	42.3	78.8	54.5	12.8	0.7	43.8	1.4
			M-F	100	3.1	46.5	86.5	59.8	14.0	0.8	48.0	1.6
	-MOLASSES (*SEE* MOLASSES AND SYRUP)											
	SUMMER CYPRESS, GRAY *Kochia vestita*											
215	-STEM-CURED, FRESH	2-08-843	A-F	85	21.5	18.7	—	—	—	3.1	35.0	6.6
			M-F	100	25.3	22.0	—	—	—	3.7	41.2	7.8
	SUNFLOWER, COMMON *Helianthus annuus*											
216	-SEEDS WITHOUT HULLS, MEAL, MECH EXTD, 41% PROTEIN	5-26-097	A-F	92	6.7	13.2	—	—	—	7.5	23.9	40.7
			M-F	100	7.3	14.3	—	—	—	8.2	26.0	44.2
217	-SEEDS WITHOUT HULLS, MEAL, SOLV EXTD, 44% PROTEIN	5-26-098	A-F	93	7.7	11.0	—	—	—	2.9	24.6	46.8
			M-F	100	8.3	11.8	—	—	—	3.1	26.5	50.3
	SWEET CLOVER, YELLOW *Melilotus officinalis*											
218	-HAY, SUN-CURED	1-04-754	A-F	89	7.6	28.9	—	—	—	1.9	36.4	13.7
			M-F	100	8.6	32.6	—	—	—	2.2	41.1	15.4
	TIMOTHY *Phleum pratense*											
219	-PREBLOOM, FRESH	2-04-903	A-F	26	1.9	8.5	15.1	—	—	1.0	11.8	3.3
			M-F	100	7.0	32.1	57.0	—	—	3.8	44.5	12.5
220	-MIDBLOOM, FRESH	2-04-905	A-F	29	1.9	9.8	—	—	—	0.9	14.0	2.7
			M-F	100	6.6	33.5	—	—	—	3.0	47.9	9.1
221	-HAY, PREBLOOM, SUN-CURED	1-04-881	A-F	89	6.3	28.2	56.0	29.4	2.8	2.5	39.8	12.4
			M-F	100	7.1	31.6	62.7	32.9	3.1	2.8	44.6	13.9
222	-HAY, EARLY BLOOM, SUN-CURED	1-04-882	A-F	89	5.1	30.3	54.4	30.4	3.8	2.6	41.7	9.5
			M-F	100	5.7	33.9	61.0	34.1	4.3	2.9	46.8	10.7
223	-HAY, MIDBLOOM, SUN-CURED	1-04-883	A-F	89	5.6	30.3	58.0	33.6	4.4	2.3	42.1	8.6
			M-F	100	6.3	34.0	65.3	37.8	4.9	2.6	47.4	9.7
224	-HAY, FULL BLOOM, SUN-CURED	1-04-884	A-F	89	4.6	31.1	—	—	—	2.7	43.4	6.8
			M-F	100	5.2	35.1	—	—	—	3.0	48.9	7.7
225	-SILAGE, ALL ANALYSES	3-04-922	A-F	34	2.4	12.0	—	—	—	1.2	15.1	3.6
			M-F	100	7.1	35.0	—	—	—	3.4	44.0	10.5
	TOMATO *Lycopersicon esculentum*											
226	-POMACE, DEHY	5-05-041	A-F	92	6.8	25.0	50.4	46.5	10.5	9.8	29.3	21.0
			M-F	100	7.4	27.2	54.8	50.5	11.4	10.7	31.9	22.9
	TREFOIL, BIRDSFOOT *Lotus corniculatus*											
227	-FRESH	2-20-786	A-F	19	2.2	4.1	9.5	—	—	0.8	8.5	3.7
			M-F	100	11.2	21.2	49.4	—	—	4.0	44.3	19.3

(Continued)

Entry Number	TDN	Digestible Energy		Metabolizable Energy		Net Energy					
	Ruminant	Ruminant		Ruminant		Ruminant NE_m		Ruminant NE_g		Lactating Cows NE_{lc}	
	(%)	*(Mcal)*		*(Mcal)*		*(Mcal)*		*(Mcal)*		*(Mcal)*	
		(lb)	*(kg)*	*(lb)*	*(kg)*	*(lb)*	*(kg)*	*(lb)*	*(kg)*	*(lb)*	*(kg)*
209	78	1.54	3.39	1.37	3.02	0.84	1.85	0.57	1.26	0.79	1.74
	87	1.70	3.76	1.52	3.34	0.93	2.06	0.63	1.40	0.88	1.93
210	16	0.32	0.71	0.26	0.58	0.15	0.33	0.07	0.16	0.16	0.36
	53	1.07	2.35	0.87	1.92	0.50	1.10	0.25	0.54	0.54	1.20
211	36	0.71	1.57	0.54	1.19	0.21	0.46	0.00	0.00	0.32	0.71
	41	0.81	1.79	0.62	1.36	0.24	0.53	0.00	0.00	0.37	0.81
212	68	1.37	3.01	1.20	2.64	0.72	1.58	0.46	1.02	0.69	1.52
	76	1.52	3.35	1.33	2.94	0.80	1.75	0.51	1.13	0.77	1.69
213	25	0.46	1.01	0.39	0.85	—	—	—	—	—	—
	50	0.92	2.02	0.77	1.70	—	—	—	—	—	—
214	43	0.88	1.93	0.69	1.53	0.37	0.82	0.15	0.32	0.40	0.88
	48	0.96	2.12	0.76	1.68	0.41	0.90	0.16	0.36	0.44	0.96
215	43	0.77	1.71	0.68	1.51	0.33	0.73	0.12	0.27	0.40	0.88
	50	0.91	2.01	0.80	1.77	0.39	0.86	0.15	0.32	0.47	1.03
216	68	1.38	3.04	1.13	2.48	0.63	1.38	0.42	0.92	0.75	1.66
	74	1.50	3.30	1.23	2.70	0.68	1.50	0.45	1.00	0.82	1.80
217	65	1.22	2.70	1.01	2.23	0.59	1.30	0.34	0.74	0.63	1.40
	70	1.32	2.90	1.09	2.40	0.64	1.40	0.36	0.80	0.68	1.50
218	49	0.98	2.16	0.81	1.78	0.49	1.07	0.26	0.57	0.51	1.13
	55	1.11	2.44	0.91	2.01	0.55	1.21	0.29	0.64	0.58	1.28
219	17	0.34	0.75	0.29	0.64	0.18	0.41	0.11	0.25	0.18	0.40
	64	1.28	2.83	1.09	2.40	0.70	1.54	0.43	0.94	0.69	1.52
220	19	0.37	0.82	0.32	0.70	0.20	0.44	0.12	0.27	0.20	0.44
	64	1.28	2.81	1.08	2.39	0.68	1.51	0.41	0.91	0.68	1.50
221	57	1.15	2.53	0.97	2.15	0.57	1.25	0.33	0.73	0.58	1.27
	64	1.28	2.83	1.09	2.41	0.64	1.40	0.37	0.82	0.64	1.42
222	52	1.03	2.27	0.86	1.89	0.52	1.14	0.29	0.63	0.54	1.18
	58	1.16	2.55	0.96	2.12	0.58	1.28	0.32	0.71	0.60	1.33
223	53	1.06	2.33	0.88	1.95	0.51	1.13	0.28	0.63	0.53	1.18
	59	1.19	2.62	1.00	2.19	0.58	1.28	0.32	0.71	0.60	1.33
224	51	1.03	2.26	0.86	1.89	0.48	1.07	0.26	0.57	0.51	1.13
	58	1.16	2.55	0.97	2.13	0.55	1.20	0.29	0.64	0.58	1.27
225	20	0.41	0.90	0.34	0.75	0.20	0.44	0.11	0.25	0.21	0.46
	59	1.19	2.62	1.00	2.20	0.58	1.29	0.32	0.71	0.60	1.33
226	60	1.21	2.66	1.03	2.27	0.61	1.35	0.37	0.81	0.61	1.35
	66	1.31	2.89	1.12	2.47	0.67	1.47	0.40	0.88	0.67	1.47
227	13	0.26	0.58	0.23	0.50	0.16	0.35	0.10	0.22	0.15	0.33
	68	1.36	2.99	1.17	2.57	0.81	1.79	0.53	1.16	0.78	1.72

(Continued)

TABLE II-1

Entry Number	Feed Name Description	Moisture Basis: A-F (As-fed) or M-F (Moisture-free)	Dry Matter	Macrominerals						
				Calcium (Ca)	Phos-phorus (P)	Sodium (Na)	Chlorine (Cl)	Mag-nesium (Mg)	Potassium (K)	Sulfur (S)
			(%)	(%)	(%)	(%)	(%)	(%)	(%)	(%)
209	-SEEDS WITHOUT HULLS, MEAL, SOLV EXTD	**A-F**	**90**	**0.26**	**0.64**	**0.01**	**0.04**	**0.30**	**2.13**	**0.44**
		M-F	100	0.29	0.71	0.01	0.05	0.33	2.36	0.48
210	-SILAGE	**A-F**	**30**	**0.40**	**0.13**	**0.03**	—	**0.12**	**0.39**	**0.09**
		M-F	100	1.32	0.44	0.09	—	0.40	1.28	0.31
211	-STRAW	**A-F**	**88**	**1.40**	**0.05**	**0.11**	—	**0.81**	**0.49**	**0.23**
		M-F	100	1.59	0.06	0.12	—	0.92	0.56	0.26
	SPELT *Triticum spelta*									
212	-GRAIN	**A-F**	**90**	**0.12**	**0.38**	—	—	—	—	—
		M-F	100	0.13	0.42	—	—	—	—	—
	SQUIRREL TAIL *Sitanion* spp									
213	-STEM-CURED, FRESH	**A-F**	**50**	**0.19**	**0.03**	—	—	—	—	—
		M-F	100	0.37	0.06	—	—	—	—	—
	SUGARCANE *Saccharum officinarum*									
214	-BAGASSE, DEHY	**A-F**	**91**	**0.47**	**0.26**	**0.04**	—	**0.08**	**0.34**	**0.09**
		M-F	100	0.51	0.29	0.04	—	0.08	0.37	0.10
	-MOLASSES (SEE MOLASSES AND SYRUP)									
	SUMMER CYPRUS, GRAY *Kochia vestita*									
215	-STEM-CURED, FRESH	**A-F**	**85**	**2.09**	**0.12**	—	—	—	—	—
		M-F	100	2.46	0.14	—	—	—	—	—
	SUNFLOWER, COMMON *Helianthus annuus*									
216	-SEEDS WITHOUT HULLS, MEAL, MECH EXTD, 41% PROTEIN	**A-F**	**92**	—	—	—	—	—	—	—
		M-F	100	—	—	—	—	—	—	—
217	-SEEDS WITHOUT HULLS, MEAL, SOLV EXTD, 44% PROTEIN	**A-F**	**93**	—	—	—	—	—	—	—
		M-F	100	—	—	—	—	—	—	—
	SWEET CLOVER, YELLOW *Melilotus officinalis*									
218	-HAY, SUN-CURED	**A-F**	**89**	**1.44**	**0.24**	**0.08**	**0.33**	**0.39**	**1.35**	**0.42**
		M-F	100	1.63	0.27	0.09	0.37	0.44	1.53	0.47
	TIMOTHY *Phleum pratense*									
219	-PREBLOOM, FRESH	**A-F**	**26**	**0.11**	**0.07**	**0.03**	—	**0.04**	**0.72**	**0.03**
		M-F	100	0.40	0.26	0.11	—	0.16	2.73	0.13
220	-MIDBLOOM, FRESH	**A-F**	**29**	**0.11**	**0.09**	**0.06**	**0.19**	**0.04**	**0.60**	**0.04**
		M-F	100	0.38	0.30	0.19	0.64	0.14	2.06	0.13
221	-HAY, PREBLOOM, SUN-CURED	**A-F**	**89**	**0.41**	**0.36**	**0.06**	—	**0.10**	**2.72**	**0.12**
		M-F	100	0.45	0.40	0.07	—	0.11	3.05	0.13
222	-HAY, EARLY BLOOM, SUN-CURED	**A-F**	**89**	**0.46**	**0.25**	**0.09**	—	**0.11**	**2.14**	**0.12**
		M-F	100	0.51	0.29	0.10	—	0.13	2.41	0.13
223	-HAY, MIDBLOOM, SUN-CURED	**A-F**	**89**	**0.32**	**0.20**	**0.08**	—	**0.12**	**1.61**	**0.12**
		M-F	100	0.36	0.23	0.10	—	0.13	1.82	0.13
224	-HAY, FULL BLOOM, SUN-CURED	**A-F**	**89**	**0.36**	**0.21**	**0.11**	**0.55**	**0.10**	**1.77**	**0.12**
		M-F	100	0.41	0.24	0.12	0.62	0.11	2.00	0.13
225	-SILAGE, ALL ANALYSES	**A-F**	**34**	**0.19**	**0.10**	**0.04**	—	**0.05**	**0.50**	**0.05**
		M-F	100	0.57	0.29	0.11	—	0.15	1.69	0.13
	TOMATO *Lycopersicon esculentum*									
226	-POMACE, DEHY	**A-F**	**92**	**0.39**	**0.55**	—	—	**0.18**	**3.34**	—
		M-F	100	0.43	0.60	—	—	0.20	3.63	—
	TREFOIL, BIRDSFOOT *Lotus corniculatus*									
227	-FRESH	**A-F**	**19**	**0.34**	**0.05**	**0.02**	—	**0.08**	**0.63**	**0.05**
		M-F	100	1.74	0.26	0.11	—	0.40	3.26	0.25

(Continued)

Entry Number	Microminerals							Fat-soluble Vitamins			
	Cobalt (Co)	Copper (Cu)	Iodine (I)	Iron (Fe)	Manganese (Mn)	Selenium (Se)	Zinc (Zn)	A (1 mg Carotene = 1,667 IU Vit. A)	Carotene (Provitamin A)	D	E (α-tocopherol)
	(ppm or mg/kg)	*(ppm or mg/kg)*	*(ppm or mg/kg)*	*(%)*	*(ppm or mg/kg)*	*(ppm or mg/kg)*	*(ppm or mg/kg)*	*(IU/g)*	*(ppm or mg/kg)*	*(IU/kg)*	*(ppm or mg/kg)*
209	**0.066**	**20.3**	**0.109**	**0.014**	**37.2**	**0.101**	**57.2**	**—**	**—**	**—**	**3.3**
	0.073	22.5	0.121	0.015	41.3	0.112	63.5	—	—	—	3.7
210	**—**	**2.9**	**—**	**0.010**	**42.7**	**—**	**10.3**	**52.2**	**31.3**	**—**	**—**
	—	9.6	—	0.033	141.3	—	34.0	172.9	103.7	—	—
211	**—**	**—**	**—**	**0.027**	**44.9**	**—**	**—**	**—**	**—**	**—**	**—**
	—	—	—	0.030	51.1	—	—	—	—	—	—
212	**—**	**—**	**—**	**—**	**—**	**—**	**—**	**—**	**—**	**—**	**—**
	—	—	—	—	—	—	—	—	—	—	—
213	**—**	**—**	**—**	**—**	**—**	**—**	**—**	**—**	**—**	**—**	**—**
	—	—	—	—	—	—	—	—	—	—	—
214	**—**	**—**	**—**	**0.019**	**—**	**—**	**—**	**—**	**—**	**—**	**—**
	—	—	—	0.021	—	—	—	—	—	—	—
215	**—**	**—**	**—**	**—**	**—**	**—**	**—**	**18.1**	**10.9**	**—**	**—**
	—	—	—	—	—	—	—	21.3	12.8	—	—
216	**—**	**—**	**—**	**—**	**—**	**—**	**—**	**—**	**—**	**—**	**—**
	—	—	—	—	—	—	—	—	—	—	—
217	**—**	**—**	**—**	**—**	**—**	**—**	**—**	**—**	**—**	**—**	**—**
	—	—	—	—	—	—	—	—	—	—	—
218	**—**	**8.8**	**—**	**0.015**	**95.4**	**—**	**—**	**145.8**	**87.4**	**2**	**—**
	—	10.0	—	0.017	107.7	—	—	164.6	98.8	2	—
219	**0.040**	**2.4**	**—**	**0.004**	**33.5**	**—**	**9.4**	**103.1**	**61.8**	**—**	**—**
	0.150	8.9	—	0.014	126.8	—	35.7	390.1	234.0	—	—
220	**—**	**3.3**	**—**	**0.006**	**56.2**	**—**	**—**	**94.5**	**56.7**	**—**	**—**
	—	11.2	—	0.018	192.5	—	—	323.4	194.0	—	—
221	**—**	**23.0**	**—**	**0.021**	**79.5**	**—**	**59.8**	**186.1**	**111.6**	**—**	**—**
	—	25.8	—	0.024	89.0	—	67.0	208.4	125.0	—	—
222	**—**	**57.1**	**—**	**0.019**	**91.8**	**—**	**55.3**	**78.0**	**46.8**	**—**	**11.6**
	—	64.0	—	0.021	103.0	—	62.0	87.5	52.5	—	13.0
223	**—**	**14.3**	**—**	**0.014**	**49.9**	**—**	**38.2**	**79.0**	**47.4**	**2**	**—**
	—	16.0	—	0.015	56.1	—	43.0	88.9	53.3	2	—
224	**—**	**25.7**	**—**	**0.013**	**82.4**	**—**	**47.9**	**—**	**—**	**—**	**—**
	—	29.0	—	0.014	93.0	—	54.0	—	—	—	—
225	**—**	**1.9**	**—**	**0.004**	**30.9**	**—**	**—**	**51.3**	**30.8**	**—**	**—**
	—	5.5	—	0.011	90.2	—	—	149.6	89.8	—	—
226	**—**	**30.0**	**—**	**0.424**	**47.1**	**—**	**—**	**—**	**—**	**—**	**—**
	—	32.6	—	0.460	51.2	—	—	—	—	—	—
227	**0.094**	**2.5**	**—**	**0.006**	**16.0**	**—**	**6.0**	**—**	**—**	**—**	**—**
	0.487	12.8	—	0.031	82.9	—	31.1	—	—	—	—

(Continued)

TABLE II-1

Entry Number	Feed Name Description	International Feed Number[2]	Moisture Basis: A-F (As-fed) or M-F (Moisture-free)	Chemical Analysis: Dry Matter	Ash	Crude Fiber	Cell Walls or NDF	Acid Detergent Fiber	Lignin	Ether Extract (Fat)	N-Free Extract	Crude Protein
				(%)	(%)	(%)	(%)	(%)	(%)	(%)	(%)	(%)
228	-HAY, SUN-CURED	1-05-044	A-F	91	6.7	29.3	—	—	—	1.9	38.9	13.9
			M-F	100	7.4	32.3	—	—	—	2.1	42.9	15.3
	TRITICALE *Triticale hexaploide*											
229	-GRAIN	4-20-362	A-F	89	1.8	3.0	11.9	—	—	1.5	67.3	15.4
			M-F	100	2.0	3.3	13.3	—	—	1.7	75.7	17.3
	TURNIP *Brassica rapa rapa*											
230	-ROOTS, FRESH	4-05-067	A-F	9	0.8	1.1	—	1.4	—	0.2	5.9	1.2
			M-F	100	8.7	11.5	—	15.0	—	1.9	64.8	13.1
	UREA											
231	-45% NITROGEN, 281% PROTEIN EQUIVALENT	5-05-070	A-F	99	—	—	—	—	—	—	—	281.7
			M-F	100	—	—	—	—	—	—		[illegible]
	VETCH *Vicia* spp											
232	-HAY, SUN-CURED	1-05-106	A-F	89	7.8	24.8	—	—	—	2.7	35.1	18.4
			M-F	100	8.8	27.9	—	—	—	3.0	39.6	20.7
	WHEAT *Triticum aestivum*											
233	-BRAN	4-05-190	A-F	89	5.9	10.0	40.9	12.0	2.6	4.0	53.6	15.5
			M-F	100	6.7	11.2	45.9	13.5	3.0	4.5	60.2	17.5
234	-SHORTS, LESS THAN 7% FIBER	4-05-201	A-F	88	4.4	6.4	—	—	—	4.6	56.5	16.5
			M-F	100	5.0	7.2	—	—	—	5.2	63.9	18.7
235	-MIDDLINGS, LESS THAN 9.5% FIBER	4-05-205	A-F	89	4.7	7.7	—	9.8	—	4.3	55.7	16.4
			M-F	100	5.3	8.7	—	11.0	—	4.9	62.7	18.5
236	-IMMATURE, FRESH	2-05-176	A-F	22	3.0	3.9	10.2	6.3	1.0	1.0	8.3	6.1
			M-F	100	13.3	17.4	46.2	28.4	4.5	4.4	37.5	27.4
237	-GRAIN, ALL ANALYSES	4-05-211	A-F	89	1.8	2.6	—	—	—	1.8	69.7	13.1
			M-F	100	2.0	2.9	—	—	—	2.0	78.4	14.7
238	-GRAIN, HARD RED SPRING	4-05-258	A-F	88	1.7	2.6	37.9	11.0	—	1.8	67.4	14.2
			M-F	100	1.9	2.9	43.3	12.6	—	2.1	76.9	16.2
239	-GRAIN, HARD RED WINTER	4-05-268	A-F	89	1.7	2.6	24.8	3.9	0.9	1.6	69.8	12.8
			M-F	100	2.0	2.9	28.0	4.4	1.0	1.8	78.8	14.5
240	-GRAIN, SOFT RED WINTER	4-05-294	A-F	88	1.9	2.3	—	—	—	1.6	71.2	11.4
			M-F	100	2.1	2.6	—	—	—	1.8	80.5	12.9
241	-GRAIN, SOFT WHITE WINTER	4-05-337	A-F	90	1.5	2.3	—	—	—	1.5	75.0	10.2
			M-F	100	1.6	2.6	—	—	—	1.7	82.9	11.3
242	-GRAIN, SOFT WHITE WINTER, PACIFIC COAST	4-08-555	A-F	89	1.9	2.5	—	—	—	1.9	72.9	10.0
			M-F	100	2.1	2.8	—	—	—	2.2	81.7	11.2
243	-GRAIN SCREENINGS	4-05-216	A-F	89	3.2	5.3	—	—	—	2.8	64.1	13.3
			M-F	100	3.6	6.0	—	—	—	3.2	72.3	15.0
244	-HAY, SUN-CURED	1-05-172	A-F	89	7.0	25.7	—	—	—	2.0	46.4	7.7
			M-F	100	7.9	29.0	—	—	—	2.2	52.3	8.7
245	-MILL RUN, LESS THAN 9.5% FIBER	4-05-206	A-F	90	5.1	8.2	—	9.9	—	4.1	57.4	15.1
			M-F	100	5.7	9.1	—	11.0	—	4.6	63.9	16.7
246	-STRAW	1-05-175	A-F	90	6.9	37.4	70.3	47.7	8.4	1.8	40.4	3.2
			M-F	100	7.7	41.7	78.4	53.2	9.4	2.0	45.0	3.6
	WHEAT, DURUM *Triticum durum*											
247	-GRAIN	4-05-224	A-F	88	1.6	2.2	—	—	—	1.8	68.2	13.8
			M-F	100	1.8	2.6	—	—	—	2.0	77.8	15.7
	WHEATGRASS, CRESTED *Agropyron desertorum*											
248	-IMMATURE, FRESH	2-05-420	A-F	28	2.9	6.2	—	—	—	0.6	12.9	6.0
			M-F	100	10.0	21.6	—	—	—	2.2	45.2	21.0
249	-FULL BLOOM, FRESH	2-05-424	A-F	45	4.2	13.6	—	—	—	1.6	22.2	3.4
			M-F	100	9.3	30.3	—	—	—	3.6	49.3	7.5

(Continued)

Entry Number	TDN	Digestible Energy		Metabolizable Energy		Net Energy					
	Ruminant	Ruminant		Ruminant		Ruminant NE_m		Ruminant NE_g		Lactating Cows NE_{lc}	
	(%)	(Mcal)		(Mcal)		(Mcal)		(Mcal)		(Mcal)	
		(lb)	(kg)	(lb)	(kg)	(lb)	(kg)	(lb)	(kg)	(lb)	(kg)
228	54	0.91	2.01	0.83	1.84	0.55	1.22	0.32	0.69	0.57	1.25
	59	1.01	2.22	0.92	2.03	0.61	1.34	0.35	0.77	0.62	1.37
229	75	1.44	3.17	1.27	2.80	0.75	1.65	0.49	1.09	0.71	1.57
	84	1.62	3.56	1.43	3.15	0.84	1.86	0.56	1.22	0.80	1.77
230	8	0.16	0.34	0.14	0.31	0.09	0.19	0.06	0.13	0.08	0.18
	85	1.70	3.74	1.51	3.33	0.94	2.06	0.64	1.40	0.88	1.94
231	—	—	—	—	—	—	—	—	—	—	—
	—	—	—	—	—	—	—	—	—	—	—
232	55	1.09	2.41	0.92	2.04	0.55	1.22	0.32	0.71	0.56	1.24
	62	1.23	2.72	1.04	2.30	0.62	1.38	0.36	0.80	0.64	1.40
233	63	1.26	2.78	1.09	2.40	0.67	1.48	0.42	0.93	0.64	1.41
	70	1.42	3.12	1.22	2.70	0.75	1.66	0.48	1.05	0.72	1.58
234	76	1.47	3.24	1.25	2.75	0.85	1.88	0.58	1.28	0.79	1.75
	86	1.66	3.66	1.41	3.11	0.96	2.12	0.66	1.45	0.90	1.98
235	74	1.39	3.07	1.16	2.57	0.78	1.72	0.52	1.15	0.79	1.73
	83	1.57	3.45	1.31	2.89	0.88	1.94	0.59	1.30	0.88	1.95
236	17	0.33	0.73	0.29	0.64	0.19	0.42	0.13	0.28	0.18	0.40
	78	1.49	3.28	1.30	2.86	0.86	1.89	0.57	1.25	0.81	1.80
237	77	1.54	3.40	1.28	2.82	0.88	1.93	0.60	1.33	0.81	1.79
	87	1.73	3.82	1.44	3.17	0.98	2.17	0.68	1.49	0.92	2.02
238	78	1.56	3.43	1.39	3.07	0.87	1.91	0.60	1.31	0.81	1.78
	89	1.78	3.91	1.59	3.50	0.99	2.18	0.68	1.50	0.92	2.04
239	78	1.57	3.45	1.40	3.09	0.88	1.94	0.61	1.34	0.82	1.81
	88	1.77	3.90	1.58	3.49	1.00	2.19	0.69	1.51	0.93	2.05
240	78	1.57	3.45	1.40	3.09	0.89	1.95	0.61	1.35	0.83	1.82
	89	1.77	3.91	1.59	3.50	1.00	2.21	0.69	1.52	0.94	2.06
241	80	1.60	3.54	1.44	3.16	0.91	2.00	0.62	1.38	0.84	1.86
	89	1.77	3.91	1.59	3.50	1.00	2.21	0.69	1.52	0.93	2.06
242	79	1.58	3.48	1.41	3.11	0.88	1.95	0.61	1.34	0.82	1.82
	88	1.77	3.89	1.58	3.48	0.99	2.18	0.68	1.50	0.92	2.04
243	63	1.26	2.79	1.10	2.42	0.63	1.39	0.39	0.86	0.62	1.37
	71	1.43	3.14	1.24	2.73	0.71	1.57	0.44	0.97	0.70	1.54
244	49	0.94	2.07	0.79	1.74	0.45	0.99	0.22	0.49	0.54	1.19
	56	1.06	2.33	0.89	1.96	0.51	1.11	0.25	0.56	0.61	1.34
245	71	1.46	3.21	1.14	2.52	0.76	1.68	0.50	1.11	0.76	1.68
	79	1.62	3.58	1.27	2.81	0.85	1.87	0.56	1.23	0.85	1.87
246	40	0.86	1.90	0.70	1.54	0.36	0.79	0.14	0.30	0.39	0.86
	44	0.96	2.12	0.78	1.72	0.40	0.88	0.15	0.34	0.44	0.96
247	74	1.51	3.33	1.35	2.97	0.83	1.83	0.57	1.25	0.78	1.72
	85	1.72	3.80	1.54	3.39	0.95	2.09	0.65	1.43	0.89	1.96
248	21	0.41	0.90	0.35	0.78	0.20	0.45	0.13	0.28	0.20	0.44
	75	1.43	3.16	1.24	2.74	0.71	1.57	0.44	0.97	0.70	1.54
249	26	0.52	1.15	0.44	0.96	0.26	0.57	0.14	0.32	0.27	0.59
	58	1.16	2.56	0.97	2.14	0.58	1.27	0.32	0.70	0.60	1.32

(Continued)

TABLE II-1

Entry Number	Feed Name Description	Moisture Basis: A-F (As-fed) or M-F (Moisture-free)	Dry Matter	Macrominerals						
				Calcium (Ca)	Phos-phorus (P)	Sodium (Na)	Chlorine (Cl)	Mag-nesium (Mg)	Potassium (K)	Sulfur (S)
			(%)	*(%)*	*(%)*	*(%)*	*(%)*	*(%)*	*(%)*	*(%)*
228	-HAY, SUN-CURED	A-F	91	1.54	0.21	0.06	—	0.46	1.74	0.23
		M-F	100	1.70	0.23	0.07	—	0.51	1.92	0.25
	TRITICALE *Triticale hexaploide*									
229	-GRAIN	A-F	89	0.04	0.30	0.01	—	0.23	0.51	—
		M-F	100	0.04	0.34	0.01	—	0.26	0.57	—
	TURNIP *Brassica rapa rapa*									
230	-ROOTS, FRESH	A-F	9	0.06	0.03	0.01	0.06	0.02	0.26	0.04
		M-F	100	0.64	0.32	0.10	0.65	0.20	2.82	0.43
	UREA									
231	-45% NITROGEN, 281% PROTEIN EQUIVALENT	A-F	99	—	—	—	—	0.00	—	—
		M-F	100	—	—	—	—	0.00	—	—
	VETCH *Vicia spp*									
232	-HAY, SUN-CURED	A-F	89	1.21	0.30	0.46	—	0.24	1.88	0.13
		M-F	100	1.36	0.34	0.52	—	0.27	2.12	0.15
	WHEAT *Triticum aestivum*									
233	-BRAN	A-F	89	0.13	1.16	0.06	0.05	0.58	1.23	0.22
		M-F	100	0.14	1.30	0.06	0.06	0.65	1.38	0.25
234	-SHORTS, LESS THAN 7% FIBER	A-F	88	0.09	0.80	0.03	0.05	0.27	0.93	0.21
		M-F	100	0.10	0.91	0.03	0.06	0.31	1.05	0.23
235	-MIDDLINGS, LESS THAN 9.5% FIBER	A-F	89	0.13	0.89	0.01	0.04	0.34	0.98	0.17
		M-F	100	0.15	1.00	0.01	0.04	0.38	1.10	0.19
236	-IMMATURE, FRESH	A-F	22	0.09	0.09	0.04	—	0.05	0.78	0.05
		M-F	100	0.42	0.40	0.18	—	0.21	3.50	0.22
237	-GRAIN, ALL ANALYSES	A-F	89	0.05	0.35	0.06	0.08	0.14	0.41	0.18
		M-F	100	0.06	0.39	0.06	0.09	0.15	0.46	0.20
238	-GRAIN, HARD RED SPRING	A-F	88	0.04	0.37	0.02	0.08	0.14	0.36	0.15
		M-F	100	0.05	0.42	0.02	0.09	0.16	0.41	0.17
239	-GRAIN, HARD RED WINTER	A-F	89	0.04	0.37	0.02	0.05	0.12	0.43	0.14
		M-F	100	0.05	0.42	0.02	0.06	0.13	0.49	0.15
240	-GRAIN, SOFT RED WINTER	A-F	88	0.05	0.36	0.01	0.07	0.10	0.41	0.11
		M-F	100	0.06	0.40	0.01	0.08	0.11	0.46	0.12
241	-GRAIN, SOFT WHITE WINTER	A-F	90	—	0.40	0.02	—	—	—	0.12
		M-F	100	—	0.44	0.02	—	—	—	0.13
242	-GRAIN, SOFT WHITE WINTER, PACIFIC COAST	A-F	89	0.09	0.31	0.01	—	0.13	0.45	0.16
		M-F	100	0.10	0.35	0.01	—	0.15	0.51	0.18
243	-GRAIN SCREENINGS	A-F	89	0.13	0.34	0.05	—	0.21	0.81	0.20
		M-F	100	0.14	0.38	0.05	—	0.24	0.91	0.22
244	-HAY, SUN-CURED	A-F	89	0.13	0.18	0.19	—	0.11	0.88	0.19
		M-F	100	0.15	0.20	0.21	—	0.12	1.00	0.22
245	-MILL RUN, LESS THAN 9.5% FIBER	A-F	90	0.10	1.02	—	—	0.48	1.20	0.30
		M-F	100	0.11	1.13	—	—	0.53	1.34	0.34
246	-STRAW	A-F	90	0.16	0.05	0.13	0.29	0.11	1.27	0.17
		M-F	100	0.18	0.05	0.14	0.32	0.12	1.41	0.19
	WHEAT, DURUM *Triticum durum*									
247	-GRAIN	A-F	88	0.10	0.36	—	—	0.16	0.44	—
		M-F	100	0.11	0.41	—	—	0.18	0.50	—
	WHEATGRASS, CRESTED *Agropyron desertorum*									
248	-IMMATURE, FRESH	A-F	28	0.13	0.10	—	—	0.08	—	—
		M-F	100	0.44	0.33	—	—	0.28	—	—
249	-FULL BLOOM, FRESH	A-F	45	0.17	0.07	0.00	—	0.04	0.47	0.21
		M-F	100	0.37	0.15	0.01	—	0.09	1.04	0.47

(Continued)

Entry Number	Microminerals							Fat-soluble Vitamins			
	Cobalt (Co)	Copper (Cu)	Iodine (I)	Iron (Fe)	Manganese (Mn)	Selenium (Se)	Zinc (Zn)	A (1 mg Carotene = 1,667 IU Vit. A)	Carotene (Provitamin A)	D	E (α-tocopherol)
	(ppm or mg/kg)	*(ppm or mg/kg)*	*(ppm or mg/kg)*	*(%)*	*(ppm or mg/kg)*	*(ppm or mg/kg)*	*(ppm or mg/kg)*	*(IU/g)*	*(ppm or mg/kg)*	*(IU/kg)*	*(ppm or mg/kg)*
228	**0.100** 0.111	**8.4** 9.3	**—** —	**0.021** 0.023	**26.0** 28.7	**—** —	**69.9** 77.2	**217.8** 240.4	**130.6** 144.2	**1** 2	**—** —
229	**0.078** 0.087	**8.3** 9.3	**—** —	**0.005** 0.005	**42.5** 47.8	**—** —	**31.2** 35.1	**—** —	**—** —	**—** —	**—** —
230	**—** —	**2.0** 21.3	**—** —	**0.002** 0.012	**3.9** 42.7	**—** —	**2.7** 29.4	**—** —	**—** —	**—** —	**—** —
231	**—** —	**6.9** 7.0	**—** —	**0.018** 0.019	**—** —	**—** —	**6.9** 7.0	**—** —	**—** —	**—** —	**—** —
232	**0.315** 0.355	**8.8** 9.9	**0.437** 0.492	**0.044** 0.049	**53.9** 60.8	**—** —	**—** —	**—** —	**—** —	**—** —	**—** —
233	**0.075** 0.084	**11.0** 12.4	**0.066** 0.074	**0.015** 0.017	**114.9** 129.0	**0.641** 0.719	**94.6** 106.2	**4.4** 4.9	**2.6** 2.9	**—** —	**14.3** 16.0
234	**0.105** 0.119	**11.5** 13.0	**—** —	**0.008** 0.009	**114.1** 129.1	**0.476** 0.538	**102.4** 115.9	**5.1** 5.8	**3.1** 3.5	**—** —	**36.0** 40.7
235	**0.502** 0.565	**15.9** 17.9	**0.109** 0.123	**0.009** 0.011	**114.0** 128.3	**0.736** 0.828	**96.9** 109.1	**5.1** 5.8	**3.1** 3.5	**—** —	**23.8** 26.9
236	**—** —	**—** —	**—** —	**0.003** 0.010	**—** —	**—** —	**—** —	**192.5** 866.9	**115.4** 520.1	**—** —	**—** —
237	**0.442** 0.497	**5.8** 6.5	**0.090** 0.101	**0.006** 0.006	**41.5** 46.7	**0.256** 0.288	**31.4** 35.2	**—** —	**—** —	**—** —	**15.5** 17.4
238	**0.123** 0.140	**6.0** 6.8	**—** —	**0.006** 0.007	**37.0** 42.2	**0.263** 0.300	**37.9** 43.3	**—** —	**—** —	**—** —	**12.7** 14.4
239	**0.145** 0.163	**5.1** 5.7	**—** —	**0.004** 0.004	**30.4** 34.3	**0.289** 0.326	**35.2** 39.8	**—** —	**—** —	**—** —	**11.1** 12.5
240	**0.103** 0.117	**7.0** 8.0	**—** —	**0.003** 0.004	**33.4** 37.8	**0.042** 0.047	**42.1** 47.7	**—** —	**—** —	**—** —	**15.6** 17.7
241	**0.136** 0.150	**7.1** 7.8	**—** —	**0.004** 0.004	**36.2** 40.0	**0.046** 0.051	**27.1** 30.0	**—** —	**—** —	**—** —	**30.9** 34.2
242	**—** —	**—** —	**—** —	**0.006** 0.006	**—** —	**—** —	**—** —	**—** —	**—** —	**—** —	**—** —
243	**1.252** 1.412	**2.3** 2.6	**—** —	**0.012** 0.014	**28.9** 32.5	**0.603** 0.681	**38.9** 43.9	**—** —	**—** —	**—** —	**—** —
244	**—** —	**—** —	**—** —	**0.018** 0.020	**—** —	**—** —	**—** —	**126.3** 142.3	**75.8** 85.4	**1** 2	**—** —
245	**0.209** 0.232	**18.5** 20.6	**—** —	**0.010** 0.011	**104.1** 115.8	**—** —	**—** —	**—** —	**—** —	**—** —	**31.9** 35.5
246	**0.041** 0.046	**3.2** 3.6	**—** —	**0.015** 0.16	**36.7** 40.9	**—** —	**5.8** 6.5	**3.3** 3.7	**2.0** 2.2	**1** 1	**—** —
247	**0.079** 0.090	**6.8** 7.8	**—** —	**0.005** 0.005	**30.7** 35.0	**—** —	**19.3** 22.0	**—** —	**—** —	**—** —	**—** —
248	**—** —	**—** —	**—** —	**—** —	**—** —	**—** —	**—** —	**205.8** 722.9	**123.4** 433.6	**—** —	**—** —
249	**—** —	**2.9** 6.5	**—** —	**—** —	**19.5** 43.3	**—** —	**6.1** 13.5	**115.1** 255.8	**69.0** 153.4	**—** —	**—** —

(Continued)

TABLE II-1

Entry Number	Feed Name Description	International Feed Number[2]	Moisture Basis: A-F (As-fed) or M-F (Moisture-free)	Chemical Analysis								
				Dry Matter	Ash	Crude Fiber	Cell Walls or NDF	Acid Detergent Fiber	Lignin	Ether Extract (Fat)	N-Free Extract	Crude Protein
				(%)	(%)	(%)	(%)	(%)	(%)	(%)	(%)	(%)
250	-OVERRIPE, FRESH	2-05-428	A-F	**75**	**3.1**	**25.2**	—	—	—	**1.6**	**42.3**	**2.8**
			M-F	100	4.1	33.6	—	—	—	2.1	56.4	3.8
251	-HAY, SUN-CURED	1-05-418	A-F	**92**	**6.4**	**30.8**	—	**33.2**	**5.1**	**2.1**	**42.2**	**10.3**
			M-F	100	7.0	33.6	—	36.2	5.5	2.3	46.0	11.2
	WHEY, CATTLE *Bos taurus*											
252	-DEHY	4-01-182	A-F	**93**	**8.8**	**0.2**	**0.3**	**0.2**	—	**0.8**	**70.2**	**13.3**
			M-F	100	9.4	0.2	0.3	0.2	—	0.8	75.3	14.2
253	-FRESH	4-08-134	A-F	**7**	**0.7**	—	—	—	—	**0.3**	**5.1**	**0.9**
			M-F	100	9.4	—	—	—	—	4.3	73.9	13.2
254	-LOW LACTOSE, DEHY (DRIED WHEY PRODUCT)	4-01-186	A-F	**93**	**15.4**	**0.2**	—	—	—	**1.0**	**60.0**	**16.7**
			M-F	100	16.5	0.2	—	—	—	1.1	64.3	17.9
	WINTERFAT, COMMON *Eurotia lanata*											
255	-BROWSE, STEM-CURED, FRESH	2-26-142	A-F	**80**	**12.7**	—	—	—	—	**2.2**	—	**8.7**
			M-F	100	15.8	—	—	—	—	2.8	—	10.8
	YEAST, BREWERS' *Saccharomyces cerevisiae*											
256	-DEHY	7-05-527	A-F	**93**	**6.5**	**3.0**	—	**3.7**	—	**0.9**	**38.8**	**43.8**
			M-F	100	7.0	3.2	—	4.0	—	1.0	41.7	47.1
	YEAST, IRRADIATED *Saccharomyces cerevisiae*											
257	-DEHY	7-05-529	A-F	**94**	**6.2**	**6.2**	—	—	—	**1.1**	**32.4**	**48.1**
			M-F	100	6.6	6.5	—	—	—	1.2	34.5	51.2
	YEAST, TORULA *Torulopsis utilis*											
258	-DEHY	7-05-534	A-F	**93**	**8.0**	**2.5**	—	**3.7**	—	**1.6**	**31.5**	**49.6**
			M-F	100	8.6	2.7	—	4.0	—	1.7	33.8	53.3

[1]Adapted from *Nutrient Requirements of Beef Cattle*, sixth revised edition, National Research Council–National Academy of Sciences, 1984.

[2]The International Feed Number, consisting of six digits, is an identification system. The first digit is the feed class. Then, the remaining five digits code each feed class as follows: (1) dry forages and roughages; (2) pasture, range plants, and forages fed green; (3) silages; (4) energy feeds; and (5) protein supplements.

(Continued)

Entry Number	TDN Ruminant	Digestible Energy Ruminant		Metabolizable Energy Ruminant		Net Energy Ruminant NE_m		Net Energy Ruminant NE_g		Net Energy Lactating Cows NE_{lc}	
	(%)	(Mcal)		(Mcal)		(Mcal)		(Mcal)		(Mcal)	
		(lb)	(kg)	(lb)	(kg)	(lb)	(kg)	(lb)	(kg)	(lb)	(kg)
250	45	0.90	1.99	0.76	1.67	0.45	1.00	0.26	0.56	0.46	1.02
	60	1.21	2.66	1.01	2.23	0.60	1.33	0.34	0.75	0.62	1.36
251	49	0.99	2.19	0.82	1.80	0.48	1.06	0.25	0.55	0.52	1.14
	53	1.08	2.39	0.89	1.96	0.52	1.15	0.27	0.59	0.56	1.24
252	76	1.51	3.33	1.28	2.83	0.87	1.92	0.59	1.30	0.78	1.71
	82	1.62	3.57	1.38	3.03	0.93	2.06	0.63	1.40	0.83	1.83
253	7	0.13	0.29	0.12	0.26	—	—	—	—	—	—
	94	1.88	4.15	1.70	3.74	—	—	—	—	—	—
254	74	1.40	3.09	1.14	2.52	0.75	1.66	0.49	1.08	0.77	1.69
	79	1.50	3.31	1.22	2.70	0.81	1.78	0.52	1.15	0.82	1.81
255	28	0.60	1.33	0.48	1.05	—	—	—	—	—	—
	35	0.76	1.66	0.59	1.31	—	—	—	—	—	—
256	73	1.46	3.21	1.28	2.83	0.81	1.79	0.54	1.19	0.77	1.69
	78	1.57	3.45	1.38	3.04	0.87	1.92	0.58	1.28	0.83	1.82
257	72	1.43	3.16	1.26	2.77	—	—	—	—	—	—
	76	1.53	3.37	1.34	2.95	—	—	—	—	—	—
258	72	1.49	3.29	1.27	2.81	0.82	1.81	0.55	1.21	0.78	1.71
	78	1.60	3.53	1.37	3.02	0.88	1.95	0.59	1.30	0.84	1.84

TABLE II-1

Entry Number	Feed Name Description	Moisture Basis: A-F (As-fed) or M-F (Moisture-free)	Dry Matter	Macrominerals Calcium (Ca)	Phosphorus (P)	Sodium (Na)	Chlorine (Cl)	Magnesium (Mg)	Potassium (K)	Sulfur (S)
			(%)	(%)	(%)	(%)	(%)	(%)	(%)	(%)
250	-OVERRIPE, FRESH	A-F	75	0.20	0.11	—	—	—	—	—
		M-F	100	0.27	0.15	—	—	—	—	—
251	-HAY, SUN-CURED	A-F	92	0.24	0.14	—	—	—	—	—
		M-F	100	0.26	0.15	—	—	—	—	—
	WHEY, CATTLE *Bos taurus*									
252	-DEHY	A-F	93	0.86	0.76	0.62	0.07	0.13	1.11	1.04
		M-F	100	0.92	0.82	0.66	0.08	0.14	1.19	1.11
253	-FRESH	A-F	7	0.06	0.05	—	—	—	0.19	—
		M-F	100	0.81	0.71	—	—	—	2.75	—
254	-LOW LACTOSE, DEHY (DRIED WHEY PRODUCT)	A-F	93	1.49	1.11	1.44	1.03	0.21	2.95	1.07
		M-F	100	1.60	1.18	1.54	1.10	0.23	3.16	1.15
	WINTERFAT, COMMON *Eurotia lanata*									
255	-BROWSE, STEM-CURED, FRESH	A-F	80	1.58	0.09	—	—	—	—	—
		M-F	100	1.98	0.12	—	—	—	—	—
	YEAST, BREWERS' *Saccharomyces cerevisiae*									
256	-DEHY	A-F	93	0.14	1.36	0.07	0.07	0.24	1.69	0.43
		M-F	100	0.15	1.47	0.08	0.08	0.26	1.82	0.46
	YEAST, IRRADIATED *Saccharomyces cerevisiae*									
257	-DEHY	A-F	94	0.78	1.42	—	—	—	2.14	—
		M-F	100	0.83	1.51	—	—	—	2.28	—
	YEAST, TORULA *Torulopsis utilis*									
258	-DEHY	A-F	93	0.55	1.61	0.01	0.02	0.14	1.92	0.55
		M-F	100	0.59	1.73	0.01	0.02	0.15	2.06	0.59

TABLE
COMPOSITION OF MINERAL SUPPLEMENTS

Entry Number	Feed Name Description	International Feed Number	Moisture Basis: A-F (As-fed) or M-F (Moisture-free)	Chemical Analysis: Dry Matter	Ash	Crude Fiber	Ether Extract (Fat)	N-Free Extract	Crude Protein ($6.25 \times N$)	Digestible Protein: Ruminant
				(%)	(%)	(%)	(%)	(%)	(%)	(%)
1	AMMONIUM PHOSPHATE, MONOBASIC	6-09-338	A-F	98	53.0	—	—	—	69.4	—
			M-F	100	54.2	—	—	—	71.0	—
2	AMMONIUM PHOSPHATE, DIBASIC	6-00-370	A-F	98	35.5	—	—	—	112.9	—
			M-F	100	36.3	—	—	—	115.5	—
3	BONE, CHARCOAL	6-00-402	A-F	94	79.3	3.7	1.1	1.8	—	—
			M-F	100	84.6	3.9	1.1	1.9	—	—
4	BONE MEAL, STEAMED	6-00-400	A-F	95	67.3	1.9	3.6	3.8	18.6	—
			M-F	100	70.7	2.0	3.8	4.0	19.5	—
5	CALCIUM CARBONATE	6-01-069	A-F	100	97.1	—	—	—	—	—
			M-F	100	97.5	—	—	—	—	—

(Continued)

Entry Number	Microminerals							Fat-soluble Vitamins			
	Cobalt (Co)	Copper (Cu)	Iodine (I)	Iron (Fe)	Manganese (Mn)	Selenium (Se)	Zinc (Zn)	A (1 mg Carotene = 1,667 IU Vit. A)	Carotene (Provitamin A)	D	E (α-tocopherol)
	(ppm or mg/kg)	*(ppm or mg/kg)*	*(ppm or mg/kg)*	*(%)*	*(ppm or mg/kg)*	*(ppm or mg/kg)*	*(ppm or mg/kg)*	*(IU/g)*	*(ppm or mg/kg)*	*(IU/kg)*	*(ppm or mg/kg)*
250	**0.187** 0.250	**6.3** 8.4	— —	— —	**39.7** 52.9	— —	— —	**0.3** 0.4	**0.2** 0.2	— —	— —
251	**0.219** 0.239	— —	— —	— —	— —	— —	— —	**34.2** 37.2	**20.5** 22.3	— —	— —
252	**0.111** 0.119	**46.5** 49.9	— —	**0.017** 0.019	**5.9** 6.3	— —	**3.2** 3.4	— —	— —	— —	**0.2** 0.2
253	— —	— —	— —	**0.003** 0.029	**0.2** 3.2	— —	— —	— —	— —	— —	— —
254	— —	**7.0** 7.5	**9.854** 10.554	**0.025** 0.027	**8.0** 8.6	**0.052** 0.056	**7.9** 8.4	— —	— —	— —	— —
255	— —	— —	— —	— —	— —	— —	— —	**24.1** 30.2	**14.5** 18.1	— —	— —
256	**0.506** 0.544	**38.4** 41.3	**0.358** 0.384	**0.009** 0.009	**6.7** 7.2	**0.911** 0.979	**39.0** 41.9	— —	— —	— —	**2.1** 2.3
257	— —	— —	— —	— —	— —	— —	— —	— —	— —	— —	— —
258	**0.031** 0.033	**11.9** 12.8	**2.502** 2.689	**0.011** 0.012	**9.3** 10.0	— —	**99.5** 107.0	— —	— —	— —	— —

II-2
FOR BEEF CATTLE

Entry Number	Macrominerals							Microminerals							
	Calcium (Ca)	Phosphorus (P)	Sodium (Na)	Chlorine (Cl)	Magnesium (Mg)	Potassium (K)	Sulfur (S)	Cobalt (Co)	Copper (Cu)	Fluorine (Fl)	Iodine (I)	Iron (Fe)	Manganese (Mn)	Selenium (Se)	Zinc (Zn)
	(%)	*(%)*	*(%)*	*(%)*	*(%)*	*(%)*	*(%)*	*(ppm or mg/kg)*	*(ppm or mg/kg)*	*(ppm or mg/kg)*	*(ppm or mg/kg)*	*(%)*	*(ppm or mg/kg)*	*(ppm or mg/kg)*	*(ppm or mg/kg)*
1	**0.38** 0.39	**24.42** 24.99	**0.08** 0.08	— —	**0.46** 0.47	**0.14** 0.14	**0.82** 0.84	— —	**86** 88	**1,833** 1,876	— —	**0.991** 1.014	**462** 473	— —	**640** 655
2	**0.50** 0.52	**20.09** 20.54	**0.04** 0.04	— —	**0.45** 0.46	— —	**2.47** 2.53	— —	**81** 83	**1,548** 1,582	— —	**1.514** 1.548	**504** 516	— —	**303** 309
3	**31.92** 34.08	**14.84** 15.85	— —	— —	**0.55** 0.59	**0.15** 0.16	— —	— —	— —	— —	— —	— —	— —	— —	— —
4	**25.98** 27.31	**11.80** 12.40	**0.40** 0.42	**0.01** 0.01	**0.78** 0.82	**0.18** 0.19	**0.34** 0.36	**0** 0	**162** 170	**637** 669	**29** 31	**0.085** 0.089	**37** 39	— —	**362** 381
5	**37.97** 38.13	**0.04** 0.04	**0.07** 0.07	**0.04** 0.04	**0.41** 0.41	**0.04** 0.04	**0.08** 0.08	— —	**14** 14	**0** 0	— —	**0.059** 0.059	**159** 160	**0.07** 0.07	**17** 17

(Continued)

TABLE II-2

Entry Number	Feed Name Description	International Feed Number	Moisture Basis: A-F (As-fed) or M-F (Moisture-free)	Chemical Analysis: Dry Matter	Ash	Crude Fiber	Ether Extract (Fat)	N-Free Extract	Crude Protein (6.25 × N)	Digestible Protein: Ruminant
				(%)	(%)	(%)	(%)	(%)	(%)	(%)
6	CALCIUM PHOSPHATE, MONOBASIC, FROM DEFLUORINATED PHOSPHORIC ACID	6-01-082	A-F	99	87.1	—	—	—	—	—
			M-F	100	88.3	—	—	—	—	—
7	CALCIUM PHOSPHATE, DIBASIC, FROM DEFLUORINATED PHOSPHORIC ACID	6-01-080	A-F	97	89.7	—	—	—	—	—
			M-F	100	92.5	—	—	—	—	—
8	CALCIUM SULFATE DIHYDRATE (GYPSUM)	6-01-090	A-F	95	—	—	—	—	—	—
			M-F	100	—	—	—	—	—	—
9	COBALT CARBONATE	6-01-566	A-F	99	—	—	—	—	—	—
			M-F	100	—	—	—	—	—	—
10	COLLOIDAL CLAY (SOFT ROCK PHOSPHATE)	6-03-947	A-F	100	—	—	—	—	—	—
			M-F	100	—	—	—	—	—	—
11	COPPER (CUPRIC) SULFATE	6-01-719	A-F	99	—	—	—	—	—	—
			M-F	100	—	—	—	—	—	—
12	CURACAO PHOSPHATE	6-05-586	A-F	99	94.1	—	—	—	—	—
			M-F	100	95.0	—	—	—	—	—
13	ETHYLENEDIAMINE DIHYDROIODIDE	6-01-842	A-F	98	—	—	—	—	54.3	—
			M-F	100	—	—	—	—	55.4	—
14	IRON (FERROUS) SULFATE	6-20-734	A-F	99	—	—	—	—	—	—
			M-F	100	—	—	—	—	—	—
15	LIMESTONE, GROUND	6-02-632	A-F	100	93.8	—	—	—	—	—
			M-F	100	94.1	—	—	—	—	—
16	LIMESTONE, MAGNESIUM (DOLOMITE)	6-02-633	A-F	100	—	—	—	—	—	—
			M-F	100	—	—	—	—	—	—
17	MAGNESIUM CARBONATE	6-02-754	A-F	98	—	—	—	—	—	—
			M-F	100	—	—	—	—	—	—
18	MAGNESIUM OXIDE	6-02-756	A-F	98	98.3	—	—	—	—	—
			M-F	100	100.0	—	—	—	—	—
19	MANGANESE OXIDE	6-03-054	A-F	99	—	—	—	—	—	—
			M-F	100	—	—	—	—	—	—
20	OYSTERSHELL, GROUND (FLOUR)	6-03-481	A-F	99	79.0	1.8	0.3	17.0	0.7	—
			M-F	100	79.9	1.8	0.3	17.2	0.7	—
21	PHOSPHATE ROCK, DEFLUORINATED	6-01-780	A-F	100	99.3	—	—	—	—	—
			M-F	100	99.7	—	—	—	—	—
22	POTASSIUM BICARBONATE	6-09-337	A-F	99	—	—	—	—	—	—
			M-F	100	—	—	—	—	—	—
23	POTASSIUM CHLORIDE	6-03-755	A-F	100	98.9	—	—	—	—	—
			M-F	100	99.0	—	—	—	—	—
24	POTASSIUM IODIDE	6-03-759	A-F	—	—	—	—	—	—	—
			M-F	100	—	—	—	—	—	—
25	SODIUM BICARBONATE	6-04-272	A-F	100	—	—	—	—	—	—
			M-F	100	—	—	—	—	—	—
26	SODIUM CHLORIDE	6-04-152	A-F	97	93.0	—	—	—	—	—
			M-F	100	95.9	—	—	—	—	—
27	SODIUM PHOSPHATE, MONOBASIC, MONOHYDRATE	6-04-288	A-F	97	96.9	—	—	—	—	—
			M-F	100	99.8	—	—	—	—	—
28	SODIUM SELENITE	6-26-013	A-F	99	—	—	—	—	—	—
			M-F	100	—	—	—	—	—	—
29	SODIUM SULFATE DECAHYDRATE	6-04-291	A-F	97	—	—	—	—	—	—
			M-F	100	—	—	—	—	—	—
30	SODIUM TRIPOLYPHOSPHATE	6-08-076	A-F	97	89.7	—	—	—	—	—
			M-F	100	92.8	—	—	—	—	—
31	ZINC OXIDE	6-05-553	A-F	—	—	—	—	—	—	—
			M-F	100	—	—	—	—	—	—
32	ZINC SULFATE	6-05-555	A-F	99	—	—	—	—	—	—
			M-F	100	—	—	—	—	—	—

(Continued)

Entry Number	Macrominerals							Microminerals							
	Calcium (Ca)	Phosphorus (P)	Sodium (Na)	Chlorine (Cl)	Magnesium (Mg)	Potassium (K)	Sulfur (S)	Cobalt (Co)	Copper (Cu)	Fluorine (Fl)	Iodine (I)	Iron (Fe)	Manganese (Mn)	Selenium (Se)	Zinc (Zn)
	(%)	*(%)*	*(%)*	*(%)*	*(%)*	*(%)*	*(%)*	*(ppm or mg/kg)*	*(ppm or mg/kg)*	*(ppm or mg/kg)*	*(ppm or mg/kg)*	*(%)*	*(ppm or mg/kg)*	*(ppm or mg/kg)*	*(ppm or mg/kg)*
6	**18.55**	**20.98**	**0.06**	**—**	**0.81**	**0.40**	**0.81**	**5**	**5**	**1,410**	**—**	**1.007**	**201**	**—**	**419**
	18.80	21.27	0.06	—	0.82	0.41	0.82	5	5	1,429	—	1.021	204	—	424
7	**22.00**	**18.43**	**1.56**	**—**	**0.51**	**0.10**	**0.69**	**8**	**9**	**940**	**—**	**0.844**	**253**	**—**	**122**
	22.67	19.00	1.61	—	0.52	0.10	0.71	9	9	969	—	0.870	261	—	126
8	**21.86**	**—**	**—**	**—**	**0.46**	**—**	**16.20**	**—**	**—**	**27**	**—**	**—**	**—**	**—**	**—**
	23.01	—	—	—	0.48	—	17.05	—	—	29	—	—	—	—	—
9	**—**	**—**	**0.25**	**0.01**	**—**	**—**	**0.03**	**465,000**	**15**	**—**	**—**	**0.020**	**100**	**—**	**15**
	—	—	0.25	0.01	—	—	0.03	469,697	15	—	—	0.021	101	—	15
10	**16.01**	**9.00**	**0.10**	**—**	**0.38**	**—**	**—**	**—**	**—**	**12,061**	**—**	**1.911**	**995**	**—**	**—**
	16.09	9.05	0.10	—	0.38	—	—	—	—	12,121	—	1.920	1,000	—	—
11	**—**	**—**	**—**	**—**	**—**	**—**	**13.25**	**—**	**250,976**	**—**	**—**	**0.010**	**2**	**—**	**9**
	—	—	—	—	—	—	13.32	—	252,257	—	—	0.011	2	—	9
12	**35.10**	**14.24**	**0.20**	**—**	**0.80**	**—**	**—**	**—**	**—**	**5,445**	**—**	**0.347**	**—**	**—**	**—**
	35.45	14.38	0.20	—	0.81	—	—	—	—	5,500	—	0.350	—	—	—
13	**—**	**—**	**—**	**—**	**—**	**—**	**—**	**—**	**—**	**—**	**787,234**	**—**	**—**	**—**	**—**
	—	—	—	—	—	—	—	—	—	—	803,300	—	—	—	—
14	**—**	**—**	**—**	**—**	**0.21**	**—**	**11.00**	**—**	**100**	**—**	**—**	**20.899**	**0**	**—**	**100**
	—	—	—	—	0.21	—	11.06	—	101	—	—	21.006	0	—	101
15	**37.12**	**0.21**	**0.06**	**0.03**	**1.13**	**0.11**	**0.04**	**—**	**11**	**—**	**—**	**0.357**	**269**	**—**	**19**
	37.22	0.22	0.06	0.03	1.13	0.11	0.04	—	11	—	—	0.358	270	—	19
16	**20.61**	**0.02**	**0.38**	**0.12**	**10.37**	**0.27**	**0.01**	**—**	**20**	**—**	**—**	**0.053**	**—**	**—**	**—**
	20.65	0.02	0.38	0.12	10.39	0.27	0.01	—	20	—	—	0.053	—	—	—
17	**0.02**	**—**	**—**	**—**	**—**	**30.19**	**—**	**—**	**—**	**—**	**—**	**0.020**	**—**	**—**	**—**
	0.02	—	—	—	—	30.81	—	—	—	—	—	0.020	—	—	—
18	**1.66**	**—**	**—**	**—**	**55.19**	**—**	**0.10**	**501**	**5**	**251**	**—**	**1.048**	**80**	**0.35**	**9**
	1.69	—	—	—	56.15	—	0.10	510	5	255	—	1.066	82	0.35	9
19	**0.16**	**0.10**	**0.06**	**—**	**0.70**	**0.58**	**0.01**	**300**	**724**	**—**	**—**	**3.436**	**620,217**	**—**	**1,349**
	0.16	0.10	0.06	—	0.71	0.59	0.01	303	731	—	—	3.470	626,482	—	1,363
20	**35.85**	**0.10**	**0.21**	**0.01**	**0.24**	**0.10**	**—**	**—**	**15**	**—**	**—**	**0.254**	**178**	**—**	**7**
	36.27	0.10	0.21	0.01	0.24	0.10	—	—	15	—	—	0.257	180	—	7
21	**31.99**	**17.07**	**3.26**	**—**	**0.29**	**0.10**	**0.13**	**10**	**40**	**1,794**	**—**	**0.840**	**496**	**—**	**90**
	32.10	17.13	3.27	—	0.29	0.10	0.13	10	41	1,800	—	0.843	498	—	90
22	**—**	**—**	**—**	**—**	**—**	**38.67**	**—**	**—**	**—**	**—**	**—**	**—**	**—**	**—**	**—**
	—	—	—	—	—	39.06	—	—	—	—	—	—	—	—	—
23	**0.05**	**—**	**1.00**	**46.88**	**0.23**	**51.31**	**0.32**	**—**	**7**	**—**	**—**	**0.061**	**7**	**—**	**9**
	0.05	—	1.00	46.93	0.23	51.37	0.32	—	7	—	—	0.061	7	—	9
24	**—**	**—**	**—**	**—**	**—**	**—**	**—**	**—**	**—**	**—**	**—**	**—**	**—**	**—**	**—**
	—	—	0.01	—	—	21.00	—	—	—	—	681,700	—	—	—	—
25	**—**	**—**	**26.87**	**—**	**—**	**0.01**	**—**	**—**	**—**	**450,138**	**—**	**0.001**	**—**	**—**	**—**
	—	—	27.00	—	—	0.01	—	—	—	452,400	—	0.002	—	—	—
26	**—**	**—**	**38.17**	**58.46**	**—**	**—**	**—**	**—**	**—**	**—**	**—**	**—**	**—**	**—**	**—**
	—	—	39.34	60.26	—	—	—	—	—	—	—	—	—	—	—
27	**0.04**	**24.84**	**18.65**	**—**	**—**	**0.14**	**—**	**—**	**7**	**—**	**—**	**—**	**—**	**—**	**5**
	0.04	25.60	19.23	—	—	0.14	—	—	7	—	—	—	—	—	5
28	**—**	**—**	**26.40**	**—**	**0.01**	**—**	**—**	**—**	**10**	**—**	**—**	**0.031**	**—**	**452,927.78**	**—**
	—	—	26.60	—	0.01	—	—	—	10	—	—	0.031	—	456,386.34	—
29	**—**	**—**	**31.33**	**—**	**—**	**—**	**9.66**	**—**	**—**	**—**	**—**	**0.001**	**—**	**—**	**—**
	—	—	32.30	—	—	—	9.96	—	—	—	—	0.002	—	—	—
30	**—**	**24.53**	**30.18**	**—**	**—**	**—**	**—**	**—**	**—**	**247**	**—**	**0.004**	**—**	**—**	**—**
	—	25.38	31.23	—	—	—	—	—	—	256	—	0.004	—	—	—
31	**—**	**—**	**—**	**—**	**—**	**—**	**—**	**—**	**—**	**—**	**—**	**—**	**—**	**—**	**—**
	4.29	—	—	—	0.30	—	1.00	1,500	500	—	—	0.551	800	—	724,968
32	**0.05**	**—**	**—**	**0.20**	**—**	**17.62**	**—**	**—**	**55**	**—**	**—**	**0.053**	**169**	**99.24**	**359,073**
	0.05	—	—	0.20	—	17.76	—	—	56	—	—	0.053	171	100.00	361,815

TABLE II-3
BYPASS PROTEIN (PERCENT OF UNDEGRADED PROTEIN) IN COMMON FEEDS[1]

Feed[2, 3]	Bypass Protein	Feed[2, 3]	Bypass Protein	Feed[2, 3]	Bypass Protein
	(%)		*(%)*		*(%)*
Alfalfa, fresh	20	Corn, whole plant, pelleted	45	Sorghum grain (milo), flaked	50
Alfalfa meal, 17% protein[5]	59	Corn fodder	45	Sorghum silage	50
Alfalfa hay, early bloom	18	Cottonseed meal, screw-pressed, 41% protein[4]	50	Soybeans, whole[5]	25
Alfalfa hay, midbloom	22	Cottonseed meal, prepressed*	36	Soybean meal, solvent, 44% protein	26
Alfalfa hay, full bloom	28	Cottonseed meal, solvent, 41% protein[5]	43	Soybean meal, solvent, 49% protein	23
Alfalfa hay, mature	35	Cottonseed hulls	40	Soybean meal, dried 248°F *(120°C)*[4]	59
Alfalfa cubes	35	Distillers' grain, barley	60	Soybean meal, dried 266°F *(130°C)*[4]	71
Alfalfa stems	40	Distillers' grain, corn	57	Soybean meal, dried 284°F *(140°C)*[4]	82
Alfalfa silage[5]	23	Distillers' grain, with solubles	47	Soybean meal, HCHO (formaldehyde treatment)[4]	80
Alfalfa silage, wilted	22	Distillers' stillage, corn	65	Sunflower meal, solvent[5]	26
Bakery product, dried	20	Feathermeal, poultry[5]	76	Sunflower meal, with hulls	40
Barley[5]	27	Fescue, Kentucky 31, fresh	30	Timothy, fresh, pre bloom	20
Barley, flaked[4]	67	Fescue, Kentucky 31, hay, early bloom	30	Timothy hay, early bloom	25
Barley silage	25	Fescue, Kentucky 31, hay, mature	35	Timothy hay, full bloom	35
Barley silage, mature	35	Fish meal[5]	60	Timothy silage	25
Beet pulp, wet	30	Grass silage[4]	29	Triticale	25
Beet pulp, dried	35	Linseed meal, solvent[4]	35	Wheat, grain[5]	23
Beet pulp, wet with molasses	25	Meat meal, rendered[5]	56	Wheat, hard	35
Beet pulp, dried with molasses	25	Meat meal	63	Wheat, soft	35
Blood meal	82	Oats, grain[5]	17	Wheat, sprouted	20
Brewers' grains, wet	57	Oat groats	25	Wheat bran[5]	20
Brewers' grains, dried[5]	50	Oat middlings	20	Wheat middlings[4]	21
Coconut meal[4]	63	Oat silage	25	Wheat mill run	20
Corn, yellow dent[4]	52	Orchardgrass, fresh, immature	25	Wheat shorts	20
Corn, high moisture[4]	80	Orchardgrass hay	30	Wheat straw	80
Corn, steam flaked[4]	68	Peanut meal, solvent[5]	30	Wheat straw, ammoniated	25
Corn and cob meal	50	Peas[4]	22		
Corn cobs	50	Rapeseed meal, solvent (canola meal)[4]	28		
Corn gluten feed[5]	22	Rye[4]	19		
Corn gluten meal[5]	59	Ryegrass, fresh[5]	20		
Corn silage, milk stage	25	Sorghum grain (milo), ground	60		
Corn silage, mature, well eared	40				

[1]In addition to the bypass protein values given in this table, more complete compositions of these and other feeds are given in Appendix II, Table II-1.

[2]Feeds without an asterisk and without a superscript obtained from *Feedstuffs*, Oct. 12, 1987, pp. 18–26, by Dr. R. L. Preston, Animal Science Department, Texas Tech University, Lubbock; with the permission of Dr. Preston and *Feedstuffs*.

[3]Feeds followed by an asterisk obtained from *Ruminant Nitrogen Usage*, National Research Council, Washington, DC, 1985, p. 33, Table 6.

[4]Adapted by the authors from *Nutrient Requirements of Dairy Cattle*, 6th rev. ed., NRC, National Academy Press, 1988, pp. 113–115, Table 7-3.

[5]Adapted by the authors from *Nutrient Requirements of Beef Cattle*, 7th rev. ed., NRC, National Academy of Scienes, 1996, pp. 134–143, Table 11-1.

SECTION III
ANIMAL UNITS

An animal unit is a common animal denominator, based on feed consumption. It is assumed that one mature cow represents an animal unit. Then, the comparative (to a mature cow) feed consumption of other age groups or classes of animals determines the proportion of an animal unit which they represent. For example, it is generally estimated that the ration of 1 mature cow will feed 5 mature ewes, or that 5 mature ewes equal 1.0 animal unit.

The original concept of an animal unit included a weight stipulation—an animal unit referred to a 1,000-lb cow, with or without a calf at side. Unfortunately, in recent years, the 1,000-lb qualification has been dropped. Certainly, there is a wide difference in the daily feed requirements of a 900-lb range cow and of a 1,500-lb exotic cow. Both will consume dry matter on a daily basis at a level equivalent to about 2% of their body weight.

Hence, a 1,500-lb cow will consume 50% more feed than a 1,000-lb cow.

Also, the period of time to be grazed has an effect on the total carrying capacity. For example, if an animal is carried for one month only, it will take one-twelfth of the total feed required to carry the same animal for one year. For this reason, the term *animal unit months* is becoming increasingly important. So in addition to the weight factor, the time factor has a distinct bearing on the ultimate carrying capacity of a tract of land.

Table III-1 gives the animal units of different classes and ages of livestock.

TABLE III-1
ANIMAL UNITS

Type of Livestock	Animal Units
Cattle:	
Cow, with or without unweaned calf at side, or heifer 2 yrs. old or older	1.0
Bull, 2 yrs. old or older	1.3
Young cattle, 1 to 2 yrs. old	0.8
Weaned calves to yearlings	0.6
Horses:	
Horse, mature	1.3
Horse, yearling	1.0
Weanling colt or filly	0.75
Sheep:	
5 mature ewes, with or without unweaned lambs at side	1.0
5 rams, 2 yrs. old or older	1.3
5 yearlings	0.8
5 weaned lambs to yearlings	0.6
Swine:	
Sow	0.4
Boar	0.5
Pigs to 200 lb *(91 kg)*	0.2
Chickens:	
75 layers or breeders	1.0
325 replacement pullets to 6 mo. of age	1.0
650 8-week-old broilers	1.0
Turkeys:	
35 breeders	1.0
40 turkeys raised to maturity	1.0
75 turkeys to 6 mo. of age	1.0

SECTION IV
WEIGHTS AND MEASURES[2]

From time to time, cattle producers and those who counsel with time have need to refer to such weights and measures as follow:

TABLE IV-1
WEIGHTS AND MEASURES (METRIC AND U.S. CUSTOMARY)

LENGTH

Unit	Is Equal To	
Metric System		**(U.S. Customary)**
1 millimicron (mμ)	0.000000001 m	0.000000039 in.
1 micron (μ)	0.000001 m	0.000039 in.
1 millimeter (mm)	0.001 m	0.0394 in.
1 centimeter (cm)	0.01 m	0.3937 in.
1 decimeter (dm)	0.1 m	3.937 in.
1 meter (m)	1 m	39.37 in.; 3.281 ft; 1.094 yd
1 hectometer (hm)	100 m	328 ft, 1 in.; 19.8338 rd
1 kilometer (km)	1,000 m	3,280 ft, 10 in.; 0.621 mi
U.S. Customary		**(Metric)**
1 inch (in.)		25 mm; 2.54 cm
1 hand*	4 in.	
1 foot (ft)	12 in.	30.48 cm; 0.305 m
1 yard (yd)	3 ft	0.914 m
1 fathom** (fath)	6.08 ft	1.829 m
1 rod (rd), pole, or perch	16.5 ft; 5.5 yd	5.029 m
1 furlong (fur.)	220 yd; 40 rd	201.168 m
1 mile (mi)	5,280 ft; 1,760 yd; 320 rd; 8 fur.	1,609.35 m; 1.609 km
1 knot or nautical mile	6,080 ft; 1.15 land mi	
1 league (land)	3 mi (land)	
1 league (nautical)	3 mi (nautical)	

*Used in measuring height of horses.

**Used in measuring depth at sea.

CONVERSIONS

To Change	To	Multiply By
inches	centimeters	2.54
feet	meters	0.305
meters	inches	39.37
miles	kilometers	1.609
kilometers	miles	0.621

(To make opposite conversion, divide by the number given instead of multiplying)

(Continued)

[2]For additional conversion factors, or greater accuracy, see *Misc. Publ. 233*, National Bureau of Standards.

TABLE IV-1 (Continued)

SURFACE OR AREA

Unit	Is Equal To	
Metric System		**(U.S. Customary)**
1 square millimeter (mm^2)	0.000001 m^2	0.00155 $in.^2$
1 square centimeter (cm^2)	0.0001 m^2	0.155 $in.^2$
1 square decimeter (dm^2)	0.01 m^2	15.50 $in.^2$
1 square meter (m^2)	1 centare (ca)	1,550 $in.^2$; 10.76 ft^2; 1.196 yd^2
1 are (a)	100 m^2	119.6 yd^2
1 hectare (ha)	10,000 m^2	2.47 acres
1 square kilometer (km^2)	1,000,000 m^2	247.1 acres; 0.386 mi^2
U.S. Customary		**(Metric)**
1 square inch ($in.^2$)	1 in. × 1 in.	6.452 cm^2
1 square foot (ft^2)	144 $in.^2$	0.093 m^2
1 square yard (yd^2)	1,296 $in.^2$; 9 ft^2	0.836 m^2
1 square rod (rd^2)	272.25 ft^2; 30.25 yd^2	25.29 m^2
1 rood	40 rd^2	10.117 a
1 acre	43,560 ft^2; 4,840 yd^2; 160 rd^2; 4 roods	4,046.87 m^2; 0.405 ha
1 square mile (mi^2)	640 acres	2.59 km^2; 259 ha
1 township	36 sections; 6 mi^2	

CONVERSIONS

To Change	To	Multiply By
square inches	square centimeters	6.452
square centimeters	square inches	0.155
square yards	square meters	0.836
square meters	square yards	1.196

(To make opposite conversion, divide by the number given instead of multiplying.)

VOLUME

Unit	Is Equal To		
Metric System Liquid and Dry		**(U.S. Customary)**	
		(Liquid)	**(Dry)**
1 milliliter (ml)	0.001 liter	0.271 dram (fl)	0.061 $in.^3$
1 centiliter (cl)	0.01 liter	0.338 oz (fl)	0.610 $in.^3$
1 deciliter (dl)	0.1 liter	3.38 oz (fl)	
1 liter (l)	1,000 cc	1.057 qt; 0.2642 gal (fl)	0.908 qt
1 hectoliter (hl)	100 liter	26.418 gal	2.838 bu
1 kiloliter (kl)	1,000 liter	264.18 gal	1,308 yd^3

(Continued)

TABLE IV-1 (Continued)

VOLUME (Continued)

Unit	Is Equal To			
U.S. Customary *Liquid*		**(Ounces)**	**(Cubic Inches)**	**(Metric)**
1 teaspoon (t)	60 drops	0.1666		5 ml
1 dessert spoon	2 t			
1 tablespoon (T)	3 t	0.5		15 ml
1 fl oz		1	1.805	29.57 ml
1 gill (gi)	0.5 c	4	7.22	118.29 ml
1 cup (c)	16 T	8	14.44	236.58 ml; 0.24 l
1 pint (pt)	2 c	16	28.88	0.47 l
1 quart (qt)	2 pt	32	57.75	0.95 l
1 gallon (gal)	4 qt	8.34 lb	231	3.79 l
1 barrel (bbl)	31.5 gal			
1 hogshead (hhd)	2 bbl			
Dry		**(Ounces)**	**(Cubic Inches)**	**(Metric)**
1 pint (pt)	0.5 qt		33.6	0.55 l
1 quart (qt)	2 pt		67.20	1.10 l
1 peck (pk)	8 qt		537.61	8.81 l
1 bushel (bu)	4 pk		2,150.42	35.24 l
Solid **Metric System**		**(Metric)**	**(U.S. Customary)**	
1 cubic millimeter (mm^3)		0.001 cc		
1 cubic centimeter (cc)		1,000 mm^3	0.061 $in.^3$	
1 cubic decimeter (dm^3)		1,000 cc	61.023 $in.^3$	
1 cubic meter (m^3)		1,000 dm^3	35.315 ft^3; 1.308 yd^3	
U.S. Customary				**(Metric)**
1 cubic inch ($in.^3$)				16.387 cc
1 board foot (fbm)		144 $in.^3$		2,359.8 cc
1 cubic foot (ft^3)		1,728 $in.^3$		0.028 m^3
1 cubic yard (yd^3)		27 ft^3		0.765 m^3
1 cord		128 ft^3		3.625 m^3

CONVERSIONS

To Change	To	Multiply By
ounces (fluid)	cubic centimeters	29.57
cubic centimeters	ounces (fluid)	0.034
quarts	liters	0.946
liters	quarts	1.057
cubic inches	cubic centimeters	16.387
cubic centimeters	cubic inches	0.061
cubic yards	cubic meters	0.765
cubic meters	cubic yards	1.308

(To make opposite conversion, divide by the number given instead of multiplying.)

(Continued)

TABLE IV-1 (Continued)

WEIGHT

Unit	Is Equal To	
Metric System		**(U.S. Customary)**
1 microgram (mcg)	0.001 mg	
1 milligram (mg)	0.001 g	0.015432356 grain
1 centigram (cg)	0.01 g	0.15432356 grain
1 decigram (dg)	0.1 g	1.5432 grains
1 gram (g)	1,000 mg	0.03527396 oz
1 decagram (dkg)	10 g	5.643833 dr
1 hectogram (hg)	100 g	3.527396 oz
1 kilogram (kg)	1,000 g	35.274 oz; 2.2046223 lb
1 ton	1,000 kg	2,204.6 lb; 1.102 tons (short); 0.984 ton (long)
U.S. Customary		**(Metric)**
1 grain	0.037 dr	64.798918 mg; 0.064798918 g
1 dram (dr)	0.063 oz	1.771845 g
1 ounce (oz)	16 dr	28.349527 g
1 pound (lb)	16 oz	453.5924 g; 0.4536 kg
1 hundredweight (cwt)	100 lb	
1 ton (short)	2,000 lb	907.18486 kg; 0.907 (metric) ton
1 ton (long)	2,200 lb	1,016.05 kg; 1.016 (metric) ton
1 part per million (ppm)	1 microgram/gram; 1 mg/liter; 1 mg/kg; 0.0001%; 0.00013 oz/gal	0.4535924 mg/lb; 0.907 g/ton
1 percent (%) (1 part in 100 parts)	10,000 ppm; 10 g/liter; 1.28 oz/gal; 8.34 lb/100 gal	

CONVERSIONS

To Change	To	Multiply By
grains	milligrams	64.799
ounces (dry)	grams	28.35
pounds (dry)	kilograms	0.4535924
kilograms	pounds	2.2046223
milligrams/pound	parts/million	2.2046223
parts/million	grams/ton	0.90718486
grams/ton	parts/million	1.1
milligrams/pound	grams/ton	2
grams/ton	milligrams/pound	0.5
grams/pound	grams/ton	2,000
grams/ton	grams/pound	0.0005
grams/ton	pounds/ton	0.0022
pounds/ton	grams/ton	453.5924
grams/ton	percent	0.00011
percent	grams/ton	9,072
parts/million	percent	move decimal four places to left

(To make opposite conversion, divide by the number given instead of multiplying.)

WEIGHTS AND MEASURES PER UNIT

Unit	Is Equal To
Volume per Unit Area	
1 liter/hectare	0.107 gal/acre
1 gallon/acre	9.354 liter/ha
Weight per Unit Area	
1 kilogram/cm^2	14.22 lb/in.^{2}
1 kilogram/hectare	0.892 lb/acre
1 pound/square inch	0.0703 kg/cm^2
1 pound/acre	1.121 kg/ha
Area per Unit Weight	
1 square centimeter/kilogram	0.0703 in.^{2}/lb
1 square inch/pound	14.22 cm^2/kg

TEMPERATURE

One centigrade (C) degree is 1/100 the difference between the temperature of melting ice and that of water boiling at standard atmospheric pressure. *One centigrade degree equals 1.8°F.*

One Fahrenheit (F) degree is 1/180 the difference between the temperature of melting ice and that of water boiling at standard atmospheric pressure. *One Fahrenheit degree equals 0.556°C.*

To Change	To	Do This
Degrees centigrade	Degrees Fahrenheit	Multiply by 9/5 and add 32
Degrees Fahrenheit	Degrees centigrade	Subtract 32, then multiply by 5/9

WEIGHTS AND MEASURES OF COMMON FEEDS

In calculating rations and mixing concentrates, it is usually necessary to use weights rather than measures. However, in practical feeding operations it is often more convenient for the farmer or rancher to measure the concentrates. The following tabulation will serve as a guide in feeding by measure:

TABLE IV-2
WEIGHTS AND MEASURES OF COMMON FEEDS

Feed	Approximate Weight[1]		Feed	Approximate Weight[1]	
	Lb per Quart	Lb per Bushel		Lb per Quart	Lb per Bushel
Alfalfa meal	0.6	19	Meat scrap	1.3	42
Barley	1.5	48	Milo (grain sorghum)	1.7	56
Beet pulp (dried)	0.6	19	Molasses feed	0.8	26
Brewers' grain (dried)	0.6	19	Oats	1.0	32
Buckwheat	1.6	50	Oats, ground	0.7	22
Buckwheat bran	1.0	29	Oat middlings	1.5	48
Corn, husked ear	—	70	Peanut meal	1.0	32
Corn, cracked	1.6	50	Rice bran	0.8	26
Corn, shelled	1.8	56	Rye	1.7	56
Corn meal	1.6	50	Sorghum (grain)	1.7	56
Corn-and-cob meal	1.4	45	Soybeans	1.8	60
Cottonseed meal	1.5	48	Tankage	1.6	51
Cowpeas	1.9	60	Velvet beans, shelled	1.8	60
Distillers' grain (dried)	0.6	19	Wheat	1.9	60
Fish meal	1.0	35	Wheat bran	0.5	16
Gluten feed	1.3	42	Wheat middlings, standard	0.8	26
Linseed meal (old process)	1.1	35	Wheat screenings	1.0	32
Linseed meal (new process)	0.9	29			

[1]To convert to metric, refer to Table IV-1.

SECTION V ESTIMATING BEEF CATTLE WEIGHTS FROM HEART GIRTH MEASUREMENTS

Cattle feeders who finish large numbers of animals have scales in their feedyards for use in determining in-weights of feeder cattle, out-weights of finished cattle, and interim weight gains of cattle while they're on feed. Likewise, both purebred breeders and large commercial cow-calf operators usually have scales. However, those with only a few head of cattle—such as 4-H Club and FFA members, and part-time farmers—may not have scales. Under such circumstances, a simple but reasonably accurate method of estimating body weight is very useful. Fortunately, cattle weights may be determined with reasonable accuracy by taking two body measurements (length and circumference), then applying a certain formula. Here is how it works:

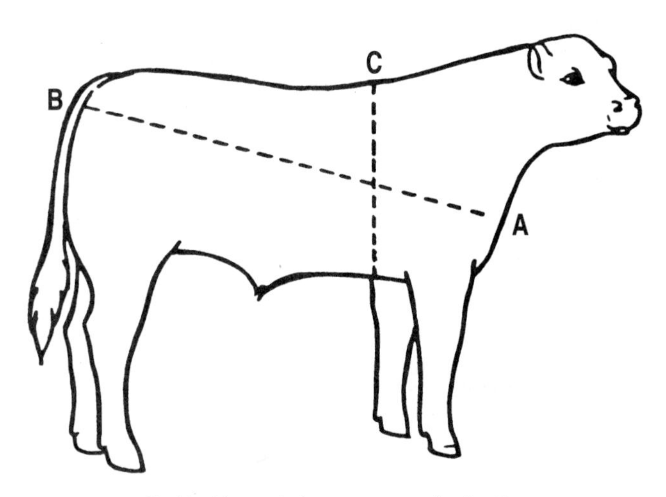

Fig. V-1. How and where to measure beef cattle.

1. *Step 1*—Measure the length of body, from the point of shoulder to the point of rump (pin bone), in inches (distance A-B of Fig. V-1).
2. *Step 2*—Measure the circumference (heart girth), from point slightly behind shoulder blade, thence down over fore-ribs and under body behind elbow (distance C of Fig. V-1).
3. *Step 3*—Take the values obtained in Steps 1 and 2 and apply the following formula to calculate body weight:

Heart girth × heart girth × body length ÷ 300 = weight in pounds.

4. *Example*—Assume that the heart girth measures 76 in. and the body length 66 in. How much does the animal weigh?

$76 \times 76 = 5{,}776$
$5{,}776 \times 66 = 381{,}216$
$381{,}216 \div 300 = 1{,}270$ lb

SECTION VI GESTATION TABLE

The cattle producer who has information relative to breeding dates can easily estimate calving dates from Table VI-1.

TABLE VI-1
GESTATION TABLE FOR COWS

Date Bred	Date Due, 283 Days	Date Bred	Date Due, 283 Days
Jan. 1	Oct. 11	July 5	Apr. 14
Jan. 6	Oct. 16	July 10	Apr. 19
Jan. 11	Oct. 21	July 15	Apr. 24
Jan. 16	Oct. 26	July 20	Apr. 29
Jan. 21	Oct. 31	July 25	May 4
Jan. 26	Nov. 5	July 30	May 9
Jan. 31	Nov. 10	Aug. 4	May 14
Feb. 5	Nov. 15	Aug. 9	May 19
Feb. 10	Nov. 20	Aug. 14	May 24
Feb. 15	Nov. 25	Aug. 19	May 29
Feb. 20	Nov. 30	Aug. 24	June 3
Feb. 25	Dec. 5	Aug. 29	June 8
Mar. 2	Dec. 10	Sept. 3	June 13
Mar. 7	Dec. 15	Sept. 8	June 18
Mar. 12	Dec. 20	Sept. 13	June 23
Mar. 17	Dec. 25	Sept. 18	June 28
Mar. 22	Dec. 30	Sept. 23	July 3
Mar. 27	Jan. 4	Sept. 28	July 8
Apr. 1	Jan. 9	Oct. 3	July 13
Apr. 6	Jan. 14	Oct. 8	July 18
Apr. 11	Jan. 19	Oct. 13	July 23
Apr. 16	Jan. 24	Oct. 18	July 28
Apr. 21	Jan. 29	Oct. 23	Aug. 2
Apr. 26	Feb. 3	Oct. 28	Aug. 7
May 1	Feb. 8	Nov. 2	Aug. 12
May 6	Feb. 13	Nov. 7	Aug. 17
May 11	Feb. 18	Nov. 12	Aug. 22
May 16	Feb. 23	Nov. 17	Aug. 27
May 21	Feb. 28	Nov. 22	Sept. 1
May 26	Mar. 5	Nov. 27	Sept. 6
May 31	Mar. 10	Dec. 2	Sept. 11
June 5	Mar. 15	Dec. 7	Sept. 16
June 10	Mar. 20	Dec. 12	Sept. 21
June 15	Mar. 25	Dec. 17	Sept. 26
June 20	Mar. 30	Dec. 22	Oct. 1
June 25	Apr. 4	Dec. 27	Oct. 6
June 30	Apr. 9		

SECTION VII
REGISTRY ASSOCIATIONS AND BREED MAGAZINES

A breed registry association consists of a group of breeders banded together for the purposes of: (1) recording the lineage of their animals; (2) protecting the purity of the breed; (3) encouraging further improvement of the breed; (4) promoting the interest of the breed.

The breed magazines publish news items and informative articles of special interest to cattle producers. Also, many of them employ field representatives whose chief duty is to assist in the buying and selling of animals.

In the compilation of the list herewith presented, no attempt was made to list the general livestock magazines of which there are numerous outstanding ones. Only those magazines which are devoted to a specific class or breed of beef cattle are included.

Both the breed registry associations and the breed magazines are listed in Table VII-1.

TABLE VII-1
BREED REGISTRY ASSOCIATIONS AND BREED MAGAZINES/NEWSLETTERS

Breeds & Associations	Magazines/Newsletters
Amerifax	
AMERIFAX CATTLE ASSN. P.O. Box 149 Hastings, NE 68901	*AMERIFAX, THE* P.O. Box 149 Hastings, NE 68901
Angus	
AMERICAN ANGUS ASSN. 3201 Frederick Blvd. St. Joseph, MO 64501	*ANGUS JOURNAL, THE* 3201 Frederick Blvd. St. Joseph, MO 64506
	ANGUS NEWS P.O. Box 20 Dowelltown, TN 37059
Ankole-Watusi	
ANKOLE-WATUSI INT'L. REGISTRY Route 1, Box 97 Spring Hill, KS 66083	
Barzona	
BARZONA BREEDERS ASSN. OF AMERICA P.O. Box 631 Prescott, AZ 86302	
Beefalo	
AMERICAN BEEFALO WORLD REGISTRY P.O. Box 12315 North Kansas City, MO 64116-0315	*BEEFALO NICKEL* P.O. Box 12315 North Kansas City, MO 64116
Beef Friesian	
BEEF FRIESIAN SOCIETY 25377 Weld County Road #17 Johnston, CO 80534	
Beefmaster	
BEEFMASTER BREEDERS UNITED 6800 Park Ten Blvd. Suite 290 West San Antonio, TX 78213-4284	*BEEFMASTER COWMAN, THE* 11201 Morning Court San Antonio, TX 78213
Belgian Blue	
AMERICAN BELGIAN BLUE ASSN. P.O. Box 307 Sulphur Springs, TX 75482-0307	
BELGIAN BLUE ASSN. OF THE AMERICAS P.O. Box 6111 Sarasota, FL 34278-6111	
Belted Galloway	
BELTED GALLOWAY SOCIETY, INC. 7118 Elliott Lane Leeds, AL 35094	*U.S. BELTED NEWS* 7118 Elliott Lane Leeds, AL 35094
Blonde D'Aquitaine	
AMERICAN BLONDE D'AQUITAINE ASSN. P.O. Box 12341 North Kansas City, MO 64116-0341	*BLONDE BULLETIN, THE* P.O. Box 12341 North Kansas City, MO 64116
Boran	
DEKKER NORTH AMERICA, INC. 8383 Greenway Blvd. P.O. Box 620676 Middleton, WI 53562-0676	

(Continued)

TABLE VI-1 (Continued)

Breeds & Associations	Magazines/Newsletters
Braford	
UNITED BRAFORD BREEDERS 422 East Main, Suite 218 Nacogdoches, TX 75961-5214	*BRAFORD NEWS* 422 East Main, Suite 218 Nacogdoches, TX 75961
Brahman	
AMERICAN BRAHMAN BREEDERS ASSN. 1313 LaConcha Lane Houston, TX 77054-1809	*BRAHMAN JOURNAL, THE* P.O. Box 220 Eddy, TX 76524
Brahmousin	
AMERICAN BRAHMOUSIN COUNCIL INC. P.O. Box 12363 North Kansas City, MO 64116	*THE BRAHMOUSIN CONNECTION* P.O. Box 12363 North Kansas City, MO 64116
AMERICAN BRAHMOUSIN COUNCIL Drawer V Dilley, TX 78017	
Bralers	
AMERICAN BRALERS ASSN. HC 61, Box 41 Ganado, TX 77962	
Brangus	
INT'L. BRANGUS BREEDERS ASSN., INC. P.O. Box 696020 San Antonio, TX 78269-6020	*BRANGUS JOURNAL* 5750 Epsilon San Antonio, TX 78249
Braunvieh	
BRAUNVIEH ASSN. OF AMERICA, THE P.O. Box 6396 Lincoln, NE 68506-0396	*BRAUNVIEH WORLD* 506 S.W. 10th Ave. Topeka, KS 66612-1606
British White	
BRITISH WHITE CATTLE ASSN. OF AMERICA P.O. Box 12702 North Kansas City, MO 64116	*BRITISH WHITE NEWS* P.O. Box 12702 North Kansas City, MO 64116
Buffalo (Bison)	
AMERICAN BUFFALO ASSN. Livestock Exchange Bldg. Room 324 P.O. Box 16660 Denver, CO 80216	
NATIONAL BUFFALO ASSN. 10 Main St. Fort Pierre, SD 57532	
Charbray	
AMERICAN-INT'L. CHAROLAIS ASSN. P.O. Box 20247 Kansas City, MO 64195-0247	
Charolais	
AMERICAN-INT'L. CHAROLAIS ASSN. P.O. Box 20247 Kansas City, MO 64195-0247	*CHAROLAIS JOURNAL* P.O. Box 20247 Kansas City, MO 64195
Char-Swiss	
CHAR-SWISS BREEDERS ASSN. 407 Chambers St. Marlin, TX 76661	
Chiangus	
COMPOSITE BREEDS REGISTRATION PROGRAM — CHIANGUS AMERICAN CHIANINA ASSN. P.O. Box 890 Platte City, MO 64079	
Chianina	
AMERICAN CHIANINA ASSN. P.O. Box 890 Platte City, MO 64079-0890	*AMERICAN CHIANINA JOURNAL* P.O. Box 890 Platte City, MO 64079
Chiford	
COMPOSITE BREEDS REGISTRATION PROGRAM — CHIFORD AMERICAN CHIANINA ASSN. P.O. Box 890 Platte City, MO 64079	
Chimaine	
COMPOSITE BREEDS REGISTRATION PROGRAM — CHIMAINE AMERICAN CHIANINA ASSN. P.O. Box 890 Platte City, MO 64079	
Cracker Cattle	
FLORIDA CRACKER CATTLE BREEDERS ASSN. Room 428, Mayo Bldg. Tallahassee, FL 32399	
Devon	
DEVON CATTLE ASSN. P.O. Box 61 The Plains, VA 22171	

(Continued)

TABLE VI-1 (Continued)

Breeds & Associations	Magazines/Newsletters	Breeds & Associations	Magazines/Newsletters
Dexter		**Herens**	
AMERICAN DEXTER CATTLE ASSN. Route 1, Box 378 Concordia, MO 64020	*ADCA BULLETIN* Route 1, Box 378 Concordia, MO 64020	AMERICAN HERENS ASSN. 122 N. Court St. Lewisburg, WV 24901	
El Monterey		**Indu-Brazil**	
ED MEARS 295 Corral De Tierra Salinas, CA 93908		INT'L. ZEBU BREEDERS ASSN. 2600 S. Loop W., Suite 310 Houston, TX 77054	
Galloway		**Limousin**	
AMERICAN GALLOWAY BREEDERS ASSN. 310 West Spruce Missoula, MT 59802	*GALLOWAY PRESS* 28289 Norris Road Bozeman, MT 59715 *MIDWEST GALLOWAY NEWS* North 4265 Country Road H Elkhorn, WI 53121	NORTH AMERICAN LIMOUSIN FOUNDATION P.O. Box 4467 Englewood, CO 80155-4467	*LIMOUSIN WORLD* P.O. Box 850870 Yukon, OK 73085
		Maine-Anjou	
		AMERICAN MAINE-ANJOU ASSN. 760 Livestock Exchange Bldg. Kansas City, MO 64102	*AMERICAN MAINE-ANJOU "VOICE"* 760 Livestock Exchange Bldg. Kansas City, MO 64102
Gelbray		**Mandalong Special**	
GELBRAY INT'L., INC. P.O. Box 2177 Ardmore, OK 73402		TRI-STATE BREEDERS E10890 Penny Lane Boraboo, WI 53913	
Gelbvieh		**Marchigiana**	
AMERICAN GELBVIEH ASSN. 10900 Dover Street Westminster, CO 80021-3993	*GELBVIEH WORLD* 10900 Dover Street Westminster, CO 80021	AMERICAN INT'L. MARCHIGIANA SOCIETY (Marky Cattle Assn.) P.O. Box 198 Walton, KS 67151-0198	*MARCHIGIANA NEWS* P.O. Box 198 Walton, KS 67151-0198
Geltex		**Murray Grey**	
GELTEX BREEDERS ASSN. Route 1, Box 114 Taylor, TX 76574		AMERICAN MURRAY GREY ASSN. P.O. Box 34590 North Kansas City, MO 64116-0990	*AMERICAN MURRAY GREY NEWS* P.O. Box 34590 North Kansas City, MO 64116-0990
Hays Converter		**Normande**	
THE CANADIAN HAYS CONVERTER ASSN. 509-6706 Elbow Dr. S.W. Calgary, Alberta CANADA T2V 0E5		AMERICAN NORMANDE ASSN. Rt. 1 Verdon, NE 68457	
Hereford		**Piedmontese**	
AMERICAN HEREFORD ASSN., THE P.O. Box 014059 Kansas City, MO 64101	*HEREFORD WORLD* 1501 Wyandotte P.O. Box 014059 Kansas City, MO 64101-4059 *TEXAS HEREFORD, THE* 4609 Airport Freeway Fort Worth, TX 76118	PIEDMONTESE ASSN. OF THE U.S. 4701 Marion St., Suite 108 Denver, CO 80216	*PIEDMONTESE PROFILE* 108 Livestock Exchange Bldg. Denver, CO 80216

(Continued)

TABLE VI-1 (Continued)

Breeds & Associations	Magazines/Newsletters
Pinzgauer	
AMERICAN PINZGAUER ASSN. 21555 State Hwy. 698 Jenera, OH 45841	*PINZGAUER JOURNAL* 216 Stevens Rd. Bolivar, TN 38008-1120
Polled Hereford	
AMERICAN HEREFORD ASSN., THE P.O. Box 014059 Kansas City, MO 64101	*HEREFORD WORLD* 1501 Wyandotte P.O. Box 014059 Kansas City, MO 64101-4059 *TEXAS POLLED HEREFORD* P.O. Box 70 Rio Vista, TX 76093
Ranger	
RANGER CATTLE COMPANY North Pecos Station P.O. Box 21300 Denver, CO 80221	
Red Angus	
RED ANGUS ASSN. OF AMERICA P.O. Box 776 4201 I-35 North Denton, TX 76207-3415	*AMERICAN RED ANGUS* 4201 I-35 North Denton, TX 76207-3415
Red Brangus	
AMERICAN RED BRANGUS ASSN. 3995 E. Hwy. 290 Dripping Springs, TX 78620-4205	
Red Poll	
AMERICAN RED POLL ASSN. P.O. Box 35519 Louisville, KY 40232	*RED POLL BEEF JOURNAL* *ATTACHE INT'L.* 1912 Clay Street North Kansas City, MO 64116
Romagnola	
AMERICAN ROMAGNOLA ASSN. P.O. Box 450 Navasota, TX 77868	
Salers	
AMERICAN SALERS ASSN. 5600 South Quebec Suite 220A Englewood, CO 80111-2208	*AMERICAN SALERS MAGAZINE* 5600 South Quebec Suite 220A Englewood, CO 80111-2207
Salorn	
INT'L. SALORN ASSN. P.O. Box 130B Riviera, TX 78379	
Santa Gertrudis	
SANTA GERTRUDIS BREEDERS INT'L. P.O. Box 1257 Kingsville, TX 78364-1257	*SANTA GERTRUDIS JOURNAL, THE* P.O. Box 938 Keller, TX 76248
Scotch Highland (or Highland)	
AMER. HIGHLAND CATTLE ASSN. 4701 Marion St., Suite 200 Denver, CO 80216	
Senopol	
SENEPOL CATTLE BREEDERS ASSN. P.O. Box 88 Louisa, VA 23093-0088	
Shorthorn	
AMERICAN SHORTHORN ASSN. 8288 Hascall St. Omaha, NE 68124-3293	*SHORTHORN COUNTRY* 8288 Hascall St. Omaha, NE 68124
Simbrah	
AMERICAN SIMMENTAL ASSN. 1 Simmental Way Bozeman, MT 59715-8599	*SIMBRAH WORLD* P.O. Box 21148 Fort Worth, TX 78121
Simmental	
AMERICAN SIMMENTAL ASSN. 1 Simmental Way Bozeman, MT 59715-8599	*THE REGISTER* 2 Simmental Way Bozeman, MT 59715
South Devon	
NORTH AMERICAN SOUTH DEVON ASSN. P.O. Box 68 Lynnville, IA 50153	
Sussex	
SUSSEX CATTLE ASSN. OF AMERICA Exec. Sec. P.O. Drawer AA Refugio, TX 78377	
Tarentaise	
AMERICAN TARENTAISE ASSN. P.O. Box 34705 North Kansas City, MO 64116	*TARENTAISE JOURNAL* P.O. Box 34705 North Kansas City, MO 64116

(Continued)

TABLE VI-1 (Continued)

Breeds & Associations	Magazines/Newsletters	Breeds & Associations	Magazines/Newsletters
Texas Longhorn		**Water Buffalo**	
TEXAS LONGHORN BREEDERS ASSN. OF AMERICA 2315 N. Main St., Suite 402 Fort Worth, TX 76106	*TEXAS LONGHORN TRAILS* 2315 N. Main, Suite 402 Fort Worth, TX 76106	AMERICAN WATER BUFFALO ASSN. 3028 McCarthy Hall University of Florida Gainesville, FL 32611	
CATTLEMEN'S TEXAS LONGHORN ASSN. Route 7, Box 260 Abilene, TX 79605	*TEXAS LONGHORN JOURNAL* Longhorn Pub. Inc. Box 311 Walsenburg, CO 81089	**Watusi**	
		WORLD WATUSI ASSN. P.O. Box 66 Crawford, NE 69339	
Texon		**Welsh Black**	
THE INT'L. TEXON CATTLE ASSN. Lakota Farm Flemington, VA 22734		WELSH BLACK CATTLE ASSN. Route 1, Box 76B Shelburn, IN 47879	*WELSH BLACK CATTLE WORLD* Route 1 Wahkon, MN 56386
Tuli		**White Park**	
Semen available from: DEKKER NORTH AMERICA, INC. P.O. Box 62076 Middleton, WI 53562-0676		WHITE PARK REGISTRY HC87, Box 2214 Big Timber, MO 59011	*PARK POST* 419 N. Water Madrid, IA 50156
Wagyu			
AMERICAN WAGYU ASSN. P.O. Box 4071 Bryan, TX 77805			

SECTION VIII U.S. STATE COLLEGES OF AGRICULTURE AND CANADIAN PROVINCIAL UNIVERSITIES

U.S. cattle producers can obtain a list of available bulletins and circulars, and other information regarding cattle by writing to (1) their state agricultural college (land-grant institution), and (2) the U.S. Superintendent of Documents, Washington, DC; or by going to the local county extension office (farm advisor) of the county in which they reside. Canadian producers may write to the Department of Agriculture of their province or to their provincial university. A list of U.S. land-grant institutions and Canadian provincial universities follows in Table VIII-1.

TABLE VIII-1
U.S. LAND-GRANT INSTITUTIONS AND CANADIAN PROVINCIAL UNIVERSITIES

State	Address
Alabama	School of Agriculture, Auburn University, Auburn, AL 36830
Alaska	Department of Agriculture, University of Alaska, Fairbanks, AK 99701
Arizona	College of Agriculture, The University of Arizona, Tucson, AZ 85721
Arkansas	Division of Agricutlure, University of Arkansas, Fayetteville, AR 72701
California	College of Agriculture and Environmental Sciences, University of California, Davis, CA 95616
Colorado	College of Agricultural Sciences, Colorado State University, Fort Collins, CO 80521
Connecticut	College of Agriculture and Natural Resources, University of Connecticut, Storrs, CT 06268
Delaware	College of Agricultural Sciences, University of Delaware, Newark, DE 19711
Florida	College of Agriculture, University of Florida, Gainesville, FL 32611
Georgia	College of Agriculture, University of Georgia, Athens, GA 30602
Hawaii	College of Tropical Agriculture, University of Hawaii, Honolulu, HI 96822
Idaho	College of Agriculture, University of Idaho, Moscow, ID 83843
Illinois	College of Agriculture, University of Illinois, Urbana–Champaign, IL 61801
Indiana	School of Agriculture, Purdue University, West Lafayette, IN 47907
Iowa	College of Agriculture, Iowa State University, Ames, IA 50010
Kansas	College of Agriculture, Kansas State University, Manhattan, KS 66506
Kentucky	College of Agriculture, University of Kentucky, Lexington, KY 40506
Louisiana	College of Agriculture, Louisiana State University and A&M College, University Station, Baton Rouge, LA 70803
Maine	College of Life Sciences and Agriculture, University of Maine, Orono, ME 04473
Maryland	College of Agriculture, University of Maryland, College Park, MD 20742
Massachusetts	College of Food and Natural Resources, University of Massachusetts, Amherst, MA 01002
Michigan	College of Agriculture and Natural Resources, Michigan State University, East Lansing, MI 48823
Minnesota	College of Agriculture, University of Minnesota, St. Paul, MN 55101
Mississippi	College of Agriculture, Mississippi State University, Mississippi State, MS 39762
Missouri	College of Agriculture, University of Missouri, Columbia, MO 65201
Montana	College of Agriculture, Montana State University, Bozeman, MT 59715

(Continued)

TABLE VIII-1 (Continued)

State	Address
Nebraska	College of Agriculture, University of Nebraska, Lincoln, NE 68503
Nevada	The Max C. Fleischmann College of Agriculture, University of Nevada, Reno, NV 89507
New Hampshire	College of Life Sciences and Agriculture, University of New Hampshire, Durham, NH 03824
New Jersey	College of Agriculture and Environmental Science, Rutgers University, New Brunswick, NJ 08903
New Mexico	College of Agriculture and Home Economics, New Mexico State University, Las Cruces, NM 88003
New York	New York State College of Agriculture, Cornell University, Ithaca, NY 14850
North Carolina	School of Agriculture, North Carolina State University, Raleigh, NC 27607
North Dakota	College of Agriculture, North Dakota State University, State University Station, Fargo, ND 58102
Ohio	College of Agriculture and Home Economics, The Ohio State University, Columbus, OH 43210
Oklahoma	College of Agriculture and Applied Science, Oklahoma State University, Stillwater, OK 74074
Oregon	School of Agriculture, Oregon State University, Corvallis, OR 97331
Pennsylvania	College of Agriculture, The Pennsylvania State University, University Park, PA 16802
Puerto Rico	College of Agricultural Sciences, University of Puerto Rico, Mayagüez, PR 00708
Rhode Island	College of Resource Development, University of Rhode Island, Kingston, RI 02881
South Carolina	College of Agricultural Sciences, Clemson University, Clemson, SC 29631
South Dakota	College of Agriculture and Biological Sciences, South Dakota State University, Brookings, SD 57006
Tennessee	College of Agriculture, University of Tennessee, P.O. Box 1071, Knoxville, TN 37901
Texas	College of Agriculture, Texas A&M University, College Station, TX 77843
Utah	College of Agriculture, Utah State University, Logan, UT 84321
Vermont	College of Agriculture, University of Vermont, Burlington, VT 05401
Virginia	College of Agriculture, Viriginia Polytechnic Institute and State University, Blacksburg, VA 24061
Washington	College of Agriculture, Washington State University, Pullman, WA 99163
West Virginia	College of Agriculture and Forestry, West Virginia University, Morgantown, WV 26506
Wisconsin	College of Agricultural and Life Sciences, University of Wisconsin, Madison, WI 53706
Wyoming	College of Agriculture, University of Wyoming, University Station, P.O. Box 3354, Laramie, WY 82070
Canada	**Address**
Alberta	University of Alberta, Edmonton, Alberta T6H 3K6
British Columbia	University of British Columbia, Vancouver, British Columbia V6T 1W5
Manitoba	University of Manitoba, Winnipeg, Manitoba R3T 2N2
New Brunswick	University of New Brunswick, Federicton, New Brunswick E3B 4N7
Ontario	University of Guelph, Guelph, Ontario N1G 2W1
Québéc	Faculty d'Agriculture, L'Université Laval, Québéc City, Québéc G1K 7D4; and Macdonald College of McGill University, Ste. Anne de Bellevue, Québéc H9X 1C0
Saskatchewan	University of Saskatchewan, Saskatoon, Saskatchewan S7N 0W0

INDEX

A

B

C

D

E

F

G

H

I

J

K

L

M

N

O

P

Q

R

S

T

U

V

W

X

Y

Z